De Vries · Kolb

Dictionary of Chemistry and Chemical Engineering

Wörterbuch der Chemie und der chemischen Verfahrenstechnik

VCH

L. De Vries · H. Kolb

Wörterbuch der Chemie und der chemischen Verfahrenstechnik

Zweite, überarbeitete und erweiterte Auflage

Band 2
Englisch/Deutsch

VCH

L. De Vries · H. Kolb

Dictionary of Chemistry and Chemical Engineering

Second, revised and enlarged edition

Volume 2
English/German

VCH

First edition: 1972
Second, revised and enlarged edition: 1979
First reprint of the second edition: 1984
Second reprint of the second edition: 1987

Library of Congress Card No. 87-10473

Deutsche Bibliothek Cataloguing-in-Publication Data:

De Vries, Louis:
Dictionary of chemistry and chemical engineering /
L. De Vries; H. Kolb. – Weinheim; New York: VCH
    Parallelsacht.: Wörterbuch der Chemie und der
    chemischen VBerfahrenstechnik. – Vol. 2, 1. Aufl.
    im Verl. Chemie, Weinheim u. bei Academic Press,
    New York, London. – Früher mit d. Verl.-Angabe
    Verl. Chemie, Weinheim, Deerfield Beach, Florida, Basel
    Bd. 1 u.d.T.: DeVries, Louis: Wörterbuch der
    Chemie und der chemischen Verfahrenstechnik
NE: Kolb, Helga:; HST
Vol. 2. English, German. – 2., rev. and enl. ed.,
2. reprint. – 1987.
    ISBN 3-527-25781-0 (Weinheim)
    ISBN 0-89573-025-1 (New York)

© VCH Verlagsgesellschaft mbH, D-6940 Weinheim, 1979, 1984, 1987

Composition: W. Girardet, Graphischer Betrieb, D-5600 Wuppertal
Printer: betz-druck gmbh, D-6100 Darmstadt 12
Bookbinder: Josef Spinner, D-7583 Ottersweier
Printed in the Federal Republic of Germany

Foreword to the Second, Revised Edition

The principles adopted in the first edition (compiled in collaboration with the late Professor Louis De Vries) have been adhered to and developed. The vocabulary has been expanded by inclusion of numerous expressions in the fields of pure and applied chemistry. Also included in this new edition of the dictionary are a number of useful terms from such diverse allied scientific disciplines as electronics, instrumentation, mathematics, data processing, medicine, etc. All the new terms and expressions have been carefully selected and compiled on the basis of a direct search of a wide range of the chemical literature, including textbooks, journals, patents, etc.

In the course of revising the first edition of this work special attention has been paid to the correction of printing errors, and an effort has been made to dispense with superfluous material. Naturally, the retention of some „obsolete words" is justifiable, since the older literature has to be consulted at times. Moreover, some obvious chemical terms, such as those which are similar or the same in the two languages have been retained, since some of the users of the dictionary may not be expert in chemistry.

The Appendix now includes an up-to-date periodic table of the elements, a conversion table of temperatures, rules for converting temperatures as well as a table for British and American measures and weights.

The first edition of this dictionary has already been widely acclaimed as a reliable aid in the translation of English/German chemical literature. Encouraged by this response every effort has been made to enhance the value of the contents and to make the dictionary a standard work of reference in chemistry and chemical engineering. We would be very grateful for comments and suggestions which might contribute to further improvement of future editions.

We express our sincere thanks to all those who have directly or indirectly contributed to the compilation of this work. Especial thanks are also due to Mr. Arthur Stimson and his colleagues of Verlag Chemie for their invaluable assistance.

Helga Kolb

Vorwort zur zweiten, überarbeiteten Auflage

Gegenüber der ersten Auflage des „Wörterbuches der Chemie und der chemischen Verfahrenstechnik", die zusammen mit dem inzwischen verstorbenen Herausgeber Prof. Louis De Vries erarbeitet wurde, ist der Wortschatz der vorliegenden Auflage durch Aufnahme zahlreicher Termini auf dem Gebiet der reinen und angewandten Chemie erweitert worden. Außerdem wurden wichtige Begriffe aus Fachgebieten, mit denen der Chemiker heute häufig zu tun hat, wie der Elektronik, Meßgeräteausrüstung, Mathematik, Datenverarbeitung, Medizin und anderen, aufgenommen. Die neuen Termini wurden aufgrund sorgfältiger Lektüre eines umfangreichen Schrifttums, einschließlich Lehrbüchern, Journalen, Patentschriften etc. gesammelt.

Bei der Überarbeitung sind Druckfehler und Ungenauigkeiten der ersten Auflage korrigiert worden; außerdem wurde versucht, Entbehrliches wegzulassen. In diesem Zusammenhang sei jedoch vermerkt, daß es durchaus berechtigt ist, einige der „obsoleten Ausdrücke" beizubehalten, da man auf diese in der älteren Literatur gelegentlich noch stößt. Dies trifft ebenfalls für chemische Begriffe zu, die in den beiden Sprachen ähnlich oder gleich sind, da nicht jeder Benutzer des Wörterbuches ein Fachmann auf chemischem Gebiet ist.

Der Anhang zum Wörterbuch umfaßt ein Periodensystem nach neuestem Stand, eine Temperaturumrechnungstabelle sowie entsprechende Umrechnungsregeln und eine Zusammenstellung britischer und amerikanischer Maße und Gewichte.

Benutzer des Wörterbuches haben in Zuschriften und Rezensionen bestätigt, daß das vorliegende Wörterbuch ein zuverlässiges Hilfsmittel bei englisch/deutschen Übersetzungen chemischer Texte darstellt. Dadurch ermutigt, wurde mit großer Mühe versucht, das Wörterbuch inhaltlich weiter zu bereichern, um es zu einem Standard-Referenzwerk für Chemie und chemische Verfahrenstechnik zu machen. Für Hinweise und Vorschläge, die zur weiteren Verbesserung zukünftiger Auflagen beitragen können, sind wir dankbar.

Dank gebührt all denen, die direkt oder indirekt zur Erarbeitung des Werkes beigetragen haben. Besonderen Dank verdienen Herr Arthur Stimson und seine Kollegen vom Verlag Chemie für ihre sehr wertvolle Mitarbeit.

Helga Kolb

Foreword to the First Edition

The present German/English technical dictionary with its comprehensive vocabulary of chemical and technical terms is equally suitable for students of science, practising scientists, particularly chemists, and for engineers and technical translators. The fields of organic, inorganic, physical, pharmaceutical and electro-chemistry have been dealt with thoroughly and many new terms have been listed. In addition great care has been taken in the compilation to ensure a detailed coverage of terms used in process technology, particularly in the metal, mineral oil, rubber, and textile industries; the most important words in use in the dye industry are also included.

For the chemist today, a rigid delineation of his scientific discipline is scarcely possible so great importance has been attached to the inclusion of words of common usage in other related branches of science such as physics, biology, medicine, mathematics, mineralogy, and crystallography.

In the final compilation of the vocabulary a large number of up-to-date newly-coined words have been carefully selected from recent literature on topics such as biochemistry, physiology, molecular biology, biophysics, atomic physics, spectroscopy, stereochemistry, reactor technology, and electronics thus enabling the user to cope with the more specialized problems of translation.

Numerous trivial names of chemical substances still in common use have been included, even in those cases where the English and German forms are similar or the same. When several English synonyms can be listed for a German entry they are separated by commas, whereas semicolons are employed for separation of alternative meanings. Further, when several meanings are given for a German word the particular subject field to which they belong is clearly indicated.

In view of the continual change in chemical terminology an attempt has been made to indicate those German words which are now considered obsolete (obs) and to give the presently valid English translation. American spelling has been used throughout; however, the variations in spelling (see under root word) as well as differences in American-English and British-English terminology are also included. Explanations are given in round brackets; parts of words or words which may be used or omitted as best suits the context are set off in square brackets, e.g. tyrosin[e]. Trade names have been taken into consideration; however, no guarantee can be given in this respect.

Expressions comprising several words e.g. "absoluter Alkohol" are also listed under the noun. All nouns are followed by an indication of gender: masculine (m), feminine (f), and neuter (n).

Irregular plurals are given in brackets, the modern spelling of Oxid (with i) has been used throughout, as has also that of „eszet", i.e. ß.

An alphabetical list of the chemical elements, a tabulation of German weights and measures, as well as a table and rules for temperature conversion and calculation, are given in the Appendix.

In their compilation of this dictionary the Authors have searched through a wide range of technical literature and consulted numerous German/English and English/German dictionaries. A detailed list of the sources from which information has been obtained is not given; however, the Authors wish to express their sincere thanks to all the editors of German and English dictionaries who have assisted in and contributed towards the materialization of this work.

Louis De Vries   Helga Kolb

Vorwort zur ersten Auflage

Das vorliegende deutsch-englische Fachwörterbuch ist mit seinem umfangreichen chemischen und technischen Wortschatz für Studenten der Naturwissenschaften, für Wissenschaftler, insbesondere für Chemiker, für Ingenieure und Fachübersetzer gleichermaßen geeignet. Es behandelt das Vokabular der organischen, anorganischen, physikalischen und pharmazeutischen Chemie sowie der Elektrochemie. Außerdem wurde sehr sorgfältig der Wortschatz der gesamten Verfahrenstechnik bearbeitet, unter besonderer Berücksichtigung der Metall-, Erdöl-, Gummi- und Textilindustrie. Hinzu kommen die wichtigsten Ausdrücke aus der Färberei.

Für den Chemiker ist eine strikte Begrenzung seines Fachgebietes heute kaum noch möglich. Aus diesem Grunde wurde besonderer Wert darauf gelegt, in diesem Wörterbuch neben dem chemischen auch das gängige Vokabular der Physik, Biologie, Medizin, Mathematik, Mineralogie und Kristallographie wiederzugeben.

Bei der endgültigen Bearbeitung dieses Wörterbuches wurden auch modernste, zum Teil erst in den letzten Jahren entstandene Ausdrücke einer Reihe von Fachgebieten erfaßt; diese sind der neueren Literatur der Biochemie, Physiologie, Molekularbiologie, Biophysik, Atomphysik, Spektroskopie, Stereochemie, Reaktortechnik und Elektronik entnommen.

Im Rahmen dieses Wörterbuches wurde eine Anzahl von chemischen Trivialnamen aufgenommen, und zwar – soweit es sich um gebräuchliche Begriffe handelt – auch dann, wenn der Ausdruck im Englischen die gleiche oder eine sehr ähnliche Form wie im Deutschen hat.

Werden für ein deutsches Stichwort mehrere englische Synonyme aufgeführt, so sind diese durch Kommata voneinander getrennt, während ein Semikolon immer dann gesetzt wurde, wenn ein prägnanter Bedeutungsunterschied besteht. Bei zahlreichen Fachausdrücken ist das Spezialgebiet, dem der Ausdruck angehört, angegeben.

Infolge der ständigen Wandlung chemischer Terminologie wurde der Versuch gemacht, veraltete deutsche Ausdrücke als solche zu kennzeichnen (obs) und in der englischen Übersetzung zusätzlich den derzeit gültigen Ausdruck aufzuführen. Allgemein wurde in diesem Wörterbuch der amerikanischen Schreibweise der Vorrang gegeben; jedoch wird auf Abweichungen in der Schreibweise (s. unter dem Stammwort) sowie auf terminologische Unterschiede im amerikanischen und britischen Englisch hingewiesen. Ein Ausdruck in runden Klammern dient der Erläuterung. Eckige Klammern schließen beliebige Weglassungen ein, wie z. B. tyrosin[e]. Markenschutz wurde berücksichtigt, jedoch erfolgt dies ohne Gewähr.

Begriffe, die sich aus mehreren Wörtern zusammensetzen, z. B. "absoluter Alkohol", sind unter dem Substantiv wiedergegeben. Allgemein sind Substantive durch maskulin (m), feminin (f) oder neutrum (n) gekennzeichnet. Unregelmäßige Plurale werden in Klammern angegeben.

Die neuere Schreibweise von Oxid (mit „i") wurde im Wörterbuch einheitlich durchgeführt, ebenso die Schreibung von „Eszet".

Im Anhang des Buches findet der Benutzer eine alphabetische Aufstellung der chemischen Elemente, eine Übersicht über deutsche Maße und Gewichte sowie eine Temperaturumrechnungstabelle und Umrechnungsregeln.

Die Verfasser haben bei ihren Arbeiten eine Vielzahl von Quellen benutzt, darunter zahlreiche deutsch-englische und englisch-deutsche Fachlexika sowie eine umfangreiche Fachliteratur. Aus diesem Grunde wurde von der Aufstellung eines detaillierten Quellenverzeichnisses abgesehen. Die Verfasser wollen jedoch an dieser Stelle allen Herausgebern von deutsch- und englisch-sprachigen Fachwörterbüchern, die ihnen Anregungen und Belehrungen vermittelten, ihren Dank aussprechen.

Louis De Vries   Helga Kolb

Table of Contents
Inhaltsverzeichnis

A

aa lava Aa-Lava *f* (Geol)
abaca Abaka *m,* Manilahanf *m*
abacus Abakus *m* (Math)
abandon *v* aufgeben
abasin Abasin *n*
abate *v* abflauen, beruhigen
abatement Abflauen, Abnehmen *n*
Abbe condenser Abbe-Kondensor *m* (Opt)
Abbe number Abbesche Zahl *f* (Opt)
Abbe sine condition Abbesche Sinusbedingung *f* (Opt)
Abbe theory of the resolution of a microscope Abbesche Theorie *f* für das Auflösungsvermögen eines Mikroskops (Opt)
abbreviate *v* abkürzen, verkürzen
abbreviated verkürzt
abbreviation Abkürzung *f,* Verkürzung, Abkürzung *f*
ABC weapons ABC-Waffen *pl* (atomare, biologische und chemische Waffen)
[Abderhalden] drying pistol Trockenpistole *f*
Abderhalden reaction Abderhaldensche Reaktion *f*
abdication Niederlegung *f* (Abdankung)
abdomen Abdomen *n* (Anat)
abduct *v* ableiten (Wärme)
Abel equation Abelsche Gleichung *f* (Mech)
Abelian field Abelsches Feld *n*
abelite Abelit *n* (Sprengstoff)
abelmosk fiber Abelmoschusfaser *f*
abelmosk seed Bisamkorn *n*
Abel's theorem Abelscher Satz *m*
Abel test Abeltest *m*
aberration Aberration *f* (Opt), Abweichung *f* (Opt), Fehler *m* (Opt), **chromatic ~** Farbabweichung *f,* **permissible ~** zulässige Abweichung, **spherical ~** sphärische Aberration
aberration function Fehlerfunktion *f* (Opt)
aberration of a lens Linsenabweichung *f*
abichite Abichit *m* (Min)
abies oil Abiesöl *n*
abietate Abietat *n* (Salz oder Ester der Abietinsäure), Resinat *n*
abietic acid Abietinsäure *f,* Sylvinsäure *f*
abietic anhydride Abietinsäureanhydrid *n*
abietin Abietin *n*
abietite Abietit *n*
ability Fähigkeit *f*
ability to extinguish Löschfähigkeit *f*
abiogenesis Abiogenese *f* (Biol), Urzeugung *f*
abiogenetic abiogenetisch (Biol)
abiosis Abiose *f* (Biol)
abiotic abiotisch (Biol)
ablation cooling Ablationskühlung *f* (Schmelzkühlung)

able fähig, tauglich
abnormal abnorm, anormal, normwidrig, regelwidrig
abnormality Abnormität *f*
abolish *v* abschaffen, abtun
aborine Aborin *n*
above oben, oberhalb
abrade *v* abschleifen, abreiben, abschaben, schmirgeln
abrasion Abnutzung *f* (durch Reibung), Abreibung *f,* Abrieb *m,* Abschleifung *f,* Abtragung *f,* Reibung *f,* Verschleiß *m,* **index of ~** Abriebzahl *f,* **resistance to ~** Reibfestigkeit *f,* **~ by rain** Abwaschen *n* [von Anstrichen] durch Regen
abrasion loss Abrieb *m*
abrasion of tire tread Abrieb *m* der Lauffläche (Reifen)
abrasion-proof abriebfest
abrasion property Scheuerfestigkeit *f*
abrasion resistance Abreibewiderstand *m,* Abreibfestigkeit *f,* Abriebfestigkeit *f,* Abriebwiderstand *m,* Abschleiffestigkeit *f*
abrasion test Abschleifversuch *m,* Verschleißprüfung *f*
abrasion wear Abriebfestigkeit *f*
abrasive Abreibungsmittel *n,* Scheuermittel *n,* Schleifmittel *n*
abrasive action Schleifwirkung *f*
abrasive cloth Schleifleinen *n,* Schmirgelleinen *n*
abrasive disc Schmirgelscheibe *f,* Schleifscheibe
abrasive dust Schleifstaub *m*
abrasive grain Schleifkorn *n*
abrasive hardness Schneidhärte *f* (Werkzeug)
abrasive paper Schleifpapier *n,* Schmirgelpapier *n*
abrasive paste Schleifpaste *f*
abrasive [powder] Schmirgel *m*
abrasive power Schleifwirkung *f*
abrasive sand Schleifsand *m*
abrasives industry Schleifmittelindustrie *f*
abrasive wear Abnutzung *f* durch Abrieb
abrasive wheel Schleifscheibe *f*
abrastol Abrastol *n,* Asaprol *n*
abraum salts Abraumsalze *pl*
abrazite Abrazit *m* (Min)
abridge *v* verkürzen
abrin Abrin *n*
abrupt plötzlich
abscess Geschwür *n* (Med)
abscissa Abszisse *f* (Geom), X-Achse *f*
absence Ausbleiben, Abwesenheit *n,* Fehlen *n*
absence of coarse particles Kornfreiheit *f*
absent abwesend, **to be ~** fehlen

absinth[e] Absinth *m*, Wermut *m*, **elixir of** ~
Absinthelixier *n*, **essence of** ~ Absinthessenz *f*
absinth[e] bitter Absinthwermut *m*
absinthial absinthartig
absinthiated wine Bitterwein *m*
absinthic acid Absinthsäure *f*
absinthin Wermutbitter *m*
absinthine absinthartig
absinthium Absinth *m*
absinthol Thujol *n*
absolute Absolute *f* (Math)
absolute *a* absolut, restlos, unbedingt
absolute alcohol wasserfreier Alkohol
absolute black Absolutschwarz *n*
absolute calculus Absolutrechnung *f*
absolute damping Absolutdämpfung *f*
absolute geometry Absolutgeometrie *f*
absolutely accurate haarscharf
absolute magnitude Absolutgröße *f*
absolute motion Absolutbewegung *f*
absolute pressure Absolutdruck *m*
absolute quantity Absolutgröße *f*
absolute space Absolutraum *m*
absolute unit Absoluteinheit *f*
absolute velocity Absolutgeschwindigkeit *f*
absorb *v* absorbieren, ansaugen, anziehen,
aufnehmen, aufsaugen, dämpfen (Stoß,
Geräusche), einsaugen, einziehen,
verschlucken
absorbability Absorbierbarkeit *f*,
Aufnahmefähigkeit *f*, Aufsaugbarkeit *f*,
Dämpfbarkeit *f*
absorbable absorbierbar, aufnahmefähig
absorbate Absorptiv *n*
absorbed energy verschluckte Energie
absorbency Extinktion *f* (Kolorimetrie),
Saugfähigkeit *f*
absorbent Absorbens *n*, Absorptionsmittel *n*
absorbent *a* absorbierend, absorptionsfähig,
aufsaugend, saugfähig
absorbent [agent] Absorbens *n* (pl. Absorbentia)
absorbent cotton Verbandwatte *f* (Pharm)
absorbent paper Saugpapier *n*
absorbent tissue Absorptionsgewebe *n*
absorber Absorber *m* (Kältemaschine);
Absorptionsmittel *n*, Aufnehmer *m*,
Aufsauger *m*
absorber control Absorberregelung *f*
(Kältemaschine)
absorber device Absorberelement *n* (Reaktor)
absorbing Absorbieren *n*, Dämpfen *n*,
Einsaugen *n*
absorbing *a* absorbierend, aufnehmend,
aufsaugend
absorbing medium Absorptionsmittel *n*
absorbing power for dyes Färbefähigkeit *f*
absorbing tissue Absorptionsgewebe *n*
absorptiometer Absorptiometer *n*,
Absorptionsmeßgerät *n*

absorptiometric absorptiometrisch
absorptiometry Absorptiometrie *f*
absorption Absorption *f*, Aufnahme *f*,
Aufsaugen *n*, Aufzehrung *f*, Dämpfung *f*,
Einsaugung *f*, Einziehung *f*, **degree of** ~
Absorptionsfähigkeit *f*, Aufzehrgrad *m*, **heat
of** ~ Absorptionswärme *f*, **law of** ~
Absorptionsgesetz *n*, **method of** ~
Absorptionsverfahren *n*,
Aufsaugeverfahren *n*, **process of** ~
Absorptionsverfahren *n*, **selective** ~
Selektivabsorption *f*
absorption apparatus Absorptionsapparat *m*
absorption axis Absorptionsachse *f*
absorption band Absorptionsband *n*,
Absorptionsstreifen *m*
absorption bottle Absorptionsflasche *f*
absorption capacity Aufnahmevermögen *n*,
Absorbierungsvermögen *n*,
Aufsaugevermögen *n*, Schluckvermögen *n*
absorption capacity for water
Wasseraufnahmefähigkeit *f*
absorption cell Absorptionsbehälter *m*,
Absorptionsküvette *f*
absorption chamber Absorptionsraum *m*
absorption coefficient Absorptionskoeffizient *m*
absorption coil Absorptionsschlange *f*
absorption color Absorptionsfarbe *f*
absorption column Absorptionskolonne *f*,
Absorptionsturm *m*, Waschturm *m*
absorption compound Absorptionsverbindung *f*
(Chem)
absorption cross section
Absorptionsquerschnitt *m*,
Auffangquerschnitt *m*
absorption current Absorptionsstrom *m* (Elektr)
absorption curve Extinktionskurve *f*
absorption discontinuity Absorptionskante *f*
(Spektr), Absorptionssprung *m*
absorption dynamometer
Absorptionsdynamometer *n*,
Bremsdynamometer *n*
absorption edge Absorptionskante *f* (Atom)
(Spektr)
absorption edge shift Verschiebung *f* der
Absorptionskante[n]
absorption equilibrium
Absorptionsgleichgewicht *n*
absorption equipment Absorptionsanlage *f*
absorption factor Absorptionsfaktor *m*,
Absorptionsvermögen *n*
absorption filter Absorptionsfilter *n*
absorption flask Absorptionsflasche *f*,
Aufnahmekolben *m*
absorption glass Filterglas *n*
absorption hygrometer Absorptionshygrometer *n*
absorption index Absorptionsindex *m* (Opt)
absorption installation Absorptionsanlage *f*

absorption limit Absorptionsgrenze *f,*
Absorptionskante *f* (Atom) (Spektr)
absorption line Absorptionslinie *f* (Spektr)
absorption liquid Absorptionsflüssigkeit *f*
absorption loss Absorptionsverlust *m*
absorption machine Absorptionsmaschine *f*
absorption measurement Absorptionsmessung *f*
absorption of carbon dioxide
Kohlensäureaufnahme *f*
absorption oil Waschöl *n*
absorption pipet[te] Absorptionspipette *f*
absorption plant Absorptionsanlage *f*
absorption power Aufnahmefähigkeit *f,*
Aufsaugevermögen *n*
absorption process Aufsaugverfahren *n*
absorption range Aufnahmebereich *m*
absorption rate Aufnahmegeschwindigkeit *f,*
Aufziehgeschwindigkeit *f* (Färb)
absorption refrigeration machine
Absorptionskältemaschine *f*
absorption region Absorptionsbereich *m*
absorption spectrum Absorptionsspektrum *n*
absorption spectrum analysis
Absorptionsspektralanalyse *f*
absorption tower Absorptionsturm *m*
absorption tube Absorptionsröhre *f,*
Eudiometer *n* (Phys)
absorption velocity Absorptionsgeschwindigkeit *f*
absorption vessel Absorptionsgefäß *n*
absorption wedge Absorptionskeil *m*
absorptive absorptionsfähig, aufnehmend,
aufsaugend, resorbierbar, saugfähig
absorptive capacity Absorptionsfähigkeit *f,*
Absorptionsleistung *f,*
Absorptionsvermögen *n,* Saugfähigkeit *f*
absorptive charcoal Absorptionskohle *f*
absorptive power Absorptionskraft *f,*
Absorptionsvermögen *n,*
Aufsaugungsvermögen *n,* Aufziehvermögen *n*
(Färb)
absorptivity Absorptionsvermögen *n,*
Aufnahmefähigkeit *f,* Aufsaugefähigkeit *f,*
Extinktionskoeffizient *m* (Kolorimetrie),
Saugfähigkeit *f*
abstract Abriß *m,* Kurzreferat *n*
abstract *a* abstrakt
abstract *v* absondern, abstrahieren, entziehen
abstracting Entziehen *n*
abstraction Abstrahieren *n,* Abstraktion *f,*
Entziehung *f*
abstricted abgesondert (Biol)
absurd widersinnig, sinnwidrig
abundance Ausgiebigkeit *f,* Fülle *f,* Häufigkeit *f,*
Überfluß *m*
abundance anomaly Häufigkeitsanomalie *f*
abundance curve Häufigkeitskurve *f* (Statist)
abundance ratio Häufigkeitsverhältnis *n* (Statist)
abundant ausgiebig, reichlich
abuse Mißbrauch *m*

abuse of drugs Drogenmißbrauch *m*
abutic acid Abutsäure *f*
abutment Widerlager *n*
acacatechin Acacatechin *n*
acacetine Acacetin *n*
acacia Akazie *f* (Bot)
acacia gum Akaziengummi *m*
acacia oil Akazienöl *n,* **white** ~
Weißakazienöl *n,* **yellow** ~ Gelbakazienöl *n*
acacia pod Akazienschote *f*
acaciine Acaciin *n*
academy Akademie *f*
acadialite Acadialith *m* (Min)
acajou Akajou *m* (Nierenbaum)
acajou gum Akajougummi *n,* Mahagongummi *n*
acajou resin Mahagonharz *n*
acanthite Akanthit *m* (Min)
acanthoidine Acanthoidin *n*
acanthoine Acanthoin *n*
acanthus oil Bärenklauöl *n*
acaricide Akaridentod *m* (Pharm),
Milbenbekämpfungsmittel *n* (Pharm)
acarid Milbe *f*
acaroid resin Akaroidharz *n,* Erdschellack *m*
accelerate *v* antreiben, beschleunigen,
hochfahren
accelerate catalytically *v* katalytisch
beschleunigen
accelerated test Kurzprüfung *f,* Kurzversuch *m*
accelerating chamber Beschleunigungskammer *f*
(Atom)
accelerating tube Beschleunigungsröhre *f*
(Atom)
acceleration Beschleunigung *f,* Akzeleration *f,*
angular ~ Winkelbeschleunigung *f,* **average**
~ mittlere Beschleunigung (Mech), **catalytic**
~ katalytische Beschleunigung, **law of** ~
Beschleunigungsgesetz *n,* **linear** ~ lineare
Beschleunigung, **moment of** ~
Beschleunigungsmoment *n,* **period of** ~
Beschleunigungszeit *f*
acceleration of fall Fallbeschleunigung *f* (Phys)
acceleration of gravity Schwerebeschleunigung *f*
accelerator Akzelerator *m,* Anreger *m* (Zement),
Beschleuniger *m,* Beschleunigungsanlage *f*
accelerator electrode Beschleunigungselektrode *f*
accelerator force Beschleunigungskraft *f*
accelerator globulin Akzelerin *n*
accelerator grid Beschleunigungsgitter *n*
accelerator nerve Beschleunigungsnerv *m* (Med)
accelerin Akzelerin *n*
accelerograph
Beschleunigungsregistrierapparat *m*
accelerometer Beschleunigungsmesser *m*
accentuate *v* akzentuieren, betonen
accept *v* annehmen (empfangen), abnehmen,
akzeptieren
acceptable quality level Annahmegrenze *f*
acceptance Aufnahme, Abnahme *f*

acceptance of electrons Aufnahme *f* von Elektronen *pl*
acceptance test Abnahmeversuch *m*
acceptor Akzeptor *m*
acceptor level Akzeptorniveau *n*
access Zugang *m*, Zutritt *m*
accessibility Lagerzugriff *m*, Zugänglichkeit *f*
accessible zugänglich
accessorial akzessorisch
accessories Zubehör *n*
accessory Beiwerk *n*, Neben-, Zubehörteil *n*
access time Zugriffszeit *f*, Zugriffzeit *f*
accident Unfall *m*, Panne *f*, Zufall *m*
accidental zufällig
accidental degeneracy zufällige Entartung *f*
accidental ground Masseschluß *m*
accident prevention Unfallschutz *m*, Unfallverhütung *f*
accident signalling system Unfallmeldeanlage *f*
acclimatization Akklimatisation *f*, Akklimatisierung *f*, Gewöhnung *f*
acclimatize *v* akklimatisieren, gewöhnen
accommodate *v* akkommodieren
accommodation, power of ~ Akkommodationsvermögen *n*
accommodation center Akkommodationszentrum *n*
accommodation coefficient Akkommodationskoeffizient *m* (Mech)
accommodation range Akkommodationsgebiet *n*
accommodative akkommodativ (Med)
accompany *v* begleiten
accompanying metal Begleitmetall *n*
accompanying substance Begleitstoff *m*
accomplish *v* bewerkstelligen, vollenden
account, on no ~ auf keinen Fall *m*
accountable haftbar
accracopalic acid Accracopalsäure *f*
accracopalinic acid Accracopalinsäure *f*
accretion Ansetzung *f*, Zuwachs *m;* (of crystals) Ansetzen *n* (von Kristallen)
accroides gum Acaroidharz *n*
accumulate *v* anhäufen, akkumulieren, anfallen, anlagern (anhäufen), ansammeln, aufhäufen, [auf]speichern, stapeln
accumulation Anhäufung *f*, Akkumulation *f*, Anlagerung *f*, Anreicherung *f*, Ansammlung *f*, Aufspeicherung *f*, Häufung *f*, Haufen *m*, Kumulation *f*, Speicherung *f*, Stauung *f*, **point of** ~ Häufungspunkt *m*
accumulation conveyor Stauförderer *m*
accumulation of cold Kältespeicherung *f*
accumulation point Häufungsstelle *f* (Mech)
accumulative anhäufend, kumulativ
accumulative process Anreicherungsprozeß *m*
accumulator Akkumulator, Akku *m*, Speicher *m*, Staukopf *m*, Stromsammler *m*, **to charge an** ~ einen Akkumulator *m* laden

accumulator acid Akkumulatorsäure *f*, Füllsäure *f*, Sammlersäure *f*
accumulator box Akkukasten *m*, Akkumulatorkasten *m*
accumulator jar Akkuflasche *f*
accumulator lead plate Akkumulatorbleiplatte *f*
accumulator pastes Akkumulatorstreichmassen *f pl*, Batteriemassen *f pl*
accumulator plate Akkumulatorplatte *f*
accumulator substation Akkumulatorunterstation *f*
accumulator tester Akkumulatorprüfer *m*
accumulator with pasted plates Masseakkumulator *m*
accuracy Genauigkeit *f*, Richtigkeit *f*, **attained** ~ erreichte Genauigkeit, **degree of** ~ Genauigkeitsgrad *m*, **five-figure** ~ fünfstellige Genauigkeit, **limit of** ~ Genauigkeitsgrenze *f*, **required** ~ verlangte Genauigkeit
accuracy in reading Ablesegenauigkeit *f*
accuracy of manufacture Herstellungsgenauigkeit *f*
accuracy of measurement Meßgenauigkeit *f*
accuracy requirements Maßgenauigkeit *f*
accurate genau
accuse *v* belangen, anklagen
accustom *v* gewöhnen
accustoming Gewöhnung *f*
aceanthrene Aceanthren *n*
aceanthrenequinone Aceanthrenchinon *n*
acecoline Acetylcholinchlorid *n*
aceconitic acid Aceconitsäure *f*
acenaphthalic acid Acenaphthalsäure *f*
acenaphthalide Acenaphthalid *n*
acenaphthazine Acenaphthazin *n*
acenaphthene Acenaphthen *n*, Azenaphthen *n*
acenaphthenedione Acenaphthenchinon *n*
acenaphthene picrate Acenaphthenpikrat *n*
acenaphthenequinone Acenaphthenchinon *n*
acenaphthenol Acenaphthenol *n*
acenaphthenone Acenaphthenon *n*
acenaphthindene Acenaphthinden *n*
acenaphthoquinoline Acenaphthochinolin *n*
acenaphthoylpropionic acid Acenaphthoylpropionsäure *f*
acenaphthylacetic acid Acenaphthylessigsäure *f*
acenaphthylene Acenaphthylen *n*, Azenaphthylen *n*
acenocoumarol Acenocumarol *n*
acentric azentrisch, nicht zentrisch
aceperimidine Aceperimidin *n*
acepleiadane Acepleiadan *n*
acepleiadiene Acepleiadien *n*
acepromazine Acepromazine *n*
acerbity Herbheit *f*
aceritannin Acer[i]tannin *n*
aceritol Acerit[ol]
acetal Acetal *n*
acetaldazine Acetaldazin *n*

acetaldehyde Acetaldehyd *m*, Äthanal *n*,
　Äthylaldehyd *m*
acetaldehyde ammonia Acetaldehydammoniak *n*
acetaldehyde cyanohydrin Milchsäurenitril *n*
acetaldehyde diethyl acetal
　Acetaldehyddiäthylacetal *n*
acetaldehyde resin Acetaldehydharz *n*
acetaldol Acetaldol *n*, Aldol *n*
acetaldoxime Acetaldoxim *n*
acetal resin Acetalharz *n*
acetamide Acetamid *n*
acetamidine Acetamidin *n*
acetamido- Acetamido-
acetaminomalonic acid Acetaminomalonsäure *f*
acetaminosalol Acetaminosalol *n*
acetanilide Acetanilid *n*, Antifebrin *n*
acetanisidine Acetanisidin *n*
acetarsol Acetarsol *n*
acetarsone Stovarsol *n*
acetate Acetat *n;* essigsaures Salz
acetate *a* essigsauer
acetate fiber Acetatfaser *f*
acetate film Acetatfolie *f*
acetate nitrate Acetatnitrat,
　Essigsalpetersäureester *m*
acetate rayon Acetatkunstseide *f,* Acetatreyon *n,*
　Acetatseide *f,* **spun ~** Acetatzellwolle *f*
acetate regenerating plant
　Acetatregenerierungsanlage *f*
acetate sheeting Acetatfolie *f*
acetate thiokinase Acetatthiokinase *f* (Biochem)
acetazolamide Acetazolamid *n*
acetbromanilide Antisepsin *n*
acetenyl- Acetenyl-
acethydrazide Acethydrazid *n*
acethydroxamic acid Acethydroxamsäure *f*
acethydroximic acid Acethydroximsäure *f*
acetic essigartig
acetic acid Essigsäure *f,* **salt or ester of ~**
　Acetat *n*
acetic [acid] amine Acetamid *n*
acetic [acid] anhydride Acetanhydrid *n*
acetic acid plant Essigsäureanlage *f*
acetic aldehyde Acetaldehyd *m*, Äthanal *n*
acetic anhydride Acetanhydrid *n,*
　Essigsäureanhydrid *n*
acetic ester Essigsäureäthylester *m,* essigsaures
　Äthyl, Äthylacetat *n*, Essigäther *m* (obs),
　Essigsäure[äthyl]äther *m* (obs)
acetic ether Essigsäureäthylester *m;* essigsaures
　Äthyl, Äthylacetat *n*, Essigäther *m* (obs),
　Essigsäure[äthyl]äther *m* (obs)
acetic fermentation Essig[säure]gärung *f*
acetic peroxide Acetoperoxid *n*
acetidin Acetidin *n*, Essigäther *m* (obs)
acetification Essigbildung *f,* Essiggärung *f,*
　Essigsäurebereitung *f*
acetify *v* in Essig verwandeln
acetimidic ester Acetiminoester *m*

acetin Acetin *n*
acetin blue Acetinblau *n*
acetmethylanilide Acetmethylanilid *n*
acetnaphthylamide Acetnaphthylamid *n*
acetoacetate Acetacetat *n*
acetoacetic acid Acetessigsäure *f,*
　Acetylessigsäure *f*
acetoacetic ester Acetessigester *m;*
　Acetessigsäureäthylester *m,*
　Äthylacetoacetat *n*
acetoacetic ester enol Acetessigesterenol *n*
acetoacetic ether Acetessigäther *m*
acetobacter aceti Essigsäurebazillus *m* (Bakt)
acetobacteria Essigbakterien *pl*
acetobromglucose Acetobromglucose *f*
acetogenins Acetogenine *pl* (Biochem)
acetoglyceral Acetoglyceral *n*
acetoguanamine Acetoguanamin *n*
acetoin Acetoin *n*
acetol Acetol *n*
acetolysis Acetolyse *f*
acetometer Acetometer *n*, Essig[säure]messer *m*
acetometric acetometrisch
acetometry Essigsäuremessung *f*
acetomycin Acetomycin *n*
aceton[a]emia Acetonämie *f* (Med)
acetonamine Acetonamin *n*
acetonaphthone Acetonaphthon *n*
acetone Aceton *n*, Dimethylketon *n*, **second**
　runnings from ~ Acetonnachlauf *m*
acetone acid Acetonsäure *f*
acetone alcohol Acetol *n*, Acetonalkohol *m*
acetone bisulfite Acetonbisulfit *n*
acetone body Acetonkörper *m* (Med)
acetone bromoform Acetonbromoform *n*
acetone carboxylic acid Acetoncarbonsäure *f*
acetone chloroform Acetonchloroform *n,*
　Chloreton *n*
acetone collodion Filmogen *n*
acetone cyanhydrin Acetoncyanhydrin *n*
acetone dehydration Acetonentwässerung *f*
acetone diacetic acid Acetondiessigsäure *f*
acetone dicarboxylic acid Acetondicarbonsäure *f*
acetone extraction Acetonextraktion *f*
acetone iodide Jodaceton *n*
acetone ketal Acetonketal *n*
acetone mannite Acetonmannit *m*
acetone oxime Acetoxim *n*
acetone powder Acetontrockenpulver *n*
acetone resin Acetonharz *n*
acetone sodium bisulfite Acetonnatriumbisulfit *n*
acetone-soluble acetonlöslich
acetone sulfite Acetonsulfit *n*
acetonic acid Acetonsäure *f*
acetonine Acetonin *n*
acetonitrile Acetonitril *n*, Cyanmethyl *n*,
　Methylcyanid *n*
acetonyl- Acetonyl-
acetonyl acetone Acetonylaceton *n*

acetonyl alcohol Acetol *n*
acetonyl salicylate Salacetol *n*
acetonyl urea Acetonylharnstoff *m*
acetophenetide Acetphenetid *n*, Phenacetin *n*
acetophenone Acetophenon *n*, Hypnon *n*,
  Phenylmethylketon *n*
acetophenone phenetidine Malarin *n*
acetopiperone Acetopiperon *n*
acetopropionic acid Acetylpropionsäure *f*
acetopseudocumene Acetopseudocumol *n*
acetoresorcinol Acetoresorcin *n*
acetotartrate *a* essigweinsauer
acetothienone Acetothienon *n*
acetous (acetose) essigartig, essigsauer
acetoxime Acetoxim *n*
acetoxy- Acetoxy-
acetoxyanthracene Acetoxyanthracen *n*
acetoxybenzpyrene Acetoxybenzpyren *n*
acetoxybutyraldehyde Acetoxybutyraldehyd *m*
acetoxymethylcholanthrene
  Acetoxymethylcholanthren *n*
acetoxynaphthoquinone
  Acetoxynaphthochinon *n*
acetozone Acetozon *n*
acetum Acetum *n*, Essig *m*
aceturic acid Acetursäure *f*
acetyl- Acetyl-
acetyl acetic acid Acetessigsäure *f*
acetylacetonate Acetylacetonat *n*
acetylacetone Acetylaceton *n*
acetylacetone ethylenediimine
  Acetylacetonäthylendiimin *n*
acetylamino- Acetylamin-, Acetamido-
acetylate *v* acetylieren
acetylated glucosyl bromide Acetobromglucose *f*
acetylating Acetylieren *n*
acetylating agent Acetylierungsmittel *n*
acetylating medium Acetylierungsmittel *n*
acetylating method Acetylierungsverfahren *n*
acetylating mixture Acetylierungsgemisch *n*
acetylation Acetylierung *f*
acetylation process Acetylierungsverfahren *n*
acetylatoxyl Arsacetin *n*
acetylbenzene Acetophenon *n*
acetylbenzoyl aconine Aconitin *n*
acetylbenzoyl japaconine Japaconitin *n*
acetylbenzoyl peroxide Acetozon *n*
acetylbromide Acetylbromid *n*
acetylcarbinol Acetol *n*
acetyl celluloid Acetylcelluloid *n*
acetylcellulose Celluloseacetat *n*
acetylcellulose nitrate Acetylnitrocellulose *f*
acetylchloride Acetylchlorid *n*
acetylcholine Acetylcholin *n*
acetylcholine chloride Acetylcholinchlorid *n*
acetyl CoA Acetyl-CoA *n* (Biochem)
acetyl coenzyme Acetylcoenzym *n*
acetyl coenzyme A Acetyl-CoA *n* (Biochem)
acetyl determination Acetylbestimmung *f*

acetyldigitoxin Acetyldigitoxin *n*
acetyl diisopropylamine Acetyldiisopropylamin *n*
acetylene Acetylen *n*, Äthin *n*, Carbidgas *n*,
  dissolved ~ Autogengas *n*
acetylene black Acetylenruß *m*
acetylene blowpipe lamp Acetylengebläselampe *f*
acetylene Bunsen burner
  Acetylenbunsenbrenner *m*
acetylene burner Acetylenbrenner *m*
acetylene carboxylic acid Propiolsäure *f*
acetylene dicarboxylic acid
  Acetylendicarbonsäure *f*
acetylene dichloride Acetylendichlorid *n*
acetylene dinitrile Kohlenstoffsubnitrid *n*
acetylene diurein Acetylendiurein *n*, Glykoluril *n*
acetylene gas Acetylengas *n*
acetylene generator Acetylenentwickler *m*,
  Acetylenerzeuger *m*
acetylene hydrocarbon
  Acetylenkohlenwasserstoff *m*
acetylene lamp Acetylenlampe *f*
acetylene linkage Acetylenbindung *f*
acetylene magnesium bromide
  Acetylenmagnesiumbromid *n*
acetylene oxide Acetylenoxid *n*
acetylene plant Acetylenanstalt *f*
acetylene purifier Acetylenreinigungsmasse *f*
acetylene purifying mass
  Acetylenreinigungsmasse *f*
acetylene series Acetylenreihe *f*
acetylene tetrabromide Acetylentetrabromid *n*
acetylene tetrachloride Acetylentetrachlorid *n*
acetylene urea Acetylenharnstoff *m*
acetylene welding Acetylenschweißung *f*
acetylene welding plant Acetylenschweißanlage *f*
acetylenyl- Acetylenyl-
acetylesterase Acetylesterase *f* (Biochem)
acetylglucosyl bromide Acetobromglucose *f*
acetylglycine Acetursäure *f*
acetylide Acetylid *n*
acetylidene Acetyliden *n*
acetylidene oxide Acetylidenoxid *n*
acetyliodide Acetyljodid *n*
acetylmethadol Acetylmethadol *n*
acetylmethionine Acetylmethionin *n*
acetylmethylcarbinol Acetoin *n*
acetyl number Acetylzahl *f*
acetyl oxide Acetanhydrid *n*
acetylphenylhydrazine Acetylphenylhydrazin,
  Pyrodin *n*
acetylphosphate Acetylphosphat *n*
acetylsalicylic acid Acetylsalicylsäure *f*, Acidum
  acetylsalicylicum *n* (Lat), Aspirin *n* (HN)
acetylsulfuric acid Acetylschwefelsäure *f*
acetyltannin Acetyltannin *n*, Diacetyltannin *n*,
  Tannigen *n*, Tannogen *n*
acetyl transacylase Acetyltransacylase *f*
  (Biochem)
acetylurea Acetylharnstoff *m*

acetyl value Acetylzahl *f*
ache Schmerz *m*
acherite Acherit *m* (Min)
achievement Errungenschaft *f*
achilleaic acid Achilleasäure *f*
achillea oil Achillenöl *n*
achilleine Achillein *n*
achmatite Achmatit *m* (Min)
achmite Achmatit *m* (Min)
achrematite Achrematit *m*
achrodextrin Achrodextrin *n*
achroite Achroit *m* (Min), farbloser Turmalin
   (Min)
achromacyte Achromacyt *m*, farbloser
   Erythrozyt
achromat Achromat *m* (Opt)
achromatic achromatisch, farblos
achromatic lens Achromat *m* (Opt)
achromatism Achromasie *f*, Achromatismus *m*,
   Farblosigkeit *f*, **D-G** ~ Achromasie *f* für die
   Linien D und G (Opt), **spherical** ~
   sphärischer Achromatismus
achromatization Achromatisierung *f*
achromatize *v* achromatisieren
achromatocyte Achromacyt *m*, farbloser
   Erythrozyt
achromatopsia Achromatopsie *f* (Med)
achroodextrin Achroodextrin *n*
achtaragite Achtaragit *m* (Min)
acicular nadelförmig
acicular lignite Nadelkohle *f*
acid Säure *f*, **anhydrous** ~ wasserfreie Säure,
   aqueous ~ wässrige Säure, **aromatic** ~
   aromatische Säure, **arsen[i]ous** ~ arsenige
   Säure, **attack by** ~ Säureangriff *m*,
   concentrated ~ konzentrierte Säure,
   containing ~ säurehaltig, **diluted** ~ verdünnte
   Säure, **fast to** ~ säurefest, **fast to boiling** ~
   säurekochecht, **free from** ~ säurefrei, **organic**
   ~ organische Säure, **oxygen-free** ~
   sauerstofffreie Säure, **persulfuric** ~ Carosche
   Säure, **polybasic** ~ mehrbasische Säure, **strong**
   ~ starke Säure, **tribasic** ~ dreibasische Säure,
   univalent ~ einwertige Säure, **unsaturated** ~
   ungesättigte Säure
acid *a* sauer
acid acceptor Säureakzeptor *m*
acid agitator Säurerührer *m*
acid albumen Acidalbumin *n*
acid albuminate Acidalbuminat *n*
acid amide Säureamid *n*
acid ammonium fluoride Ammoniumbifluorid *n*
acid ammonium oxalate Ammoniumbioxalat *n*
acid anhydride Säureanhydrid *n*
acid azide Säureazid *n*
acid-base catalysis Säure-Basen-Katalyse *f*
acid-base equilibrium or balance
   Säure-Basen-Gleichgewicht *n*
acid-base indicator Säure-Base-Indikator *m*

acid-base metabolism
   Säure-Basen-Stoffwechsel *m*
acid-base regulation Säure-Basen-Regulation *f*
acid bath Säurebad *n*, Einsäuerungsbad *n*
acid brittleness Beizbrüchigkeit *f*,
   Beizsprödigkeit *f*
acid capacity Acidität *f*
acid carbonate Bicarbonat *n*,
   Hydrogencarbonat *n*
acid carbonate *a* doppel[t]kohlensauer
acid carboy Säureballon *m*
acid catalysis Säurekatalyse *f*
acid catalyst Säurehärter *m*
acid centrifuge Säurezentrifuge *f*
acid chimney Säurekamin *m*
acid chloride Säurechlorid *n*
acid cistern Säureständer *m*
acid cleavage Säurespaltung *f*
acid concentration Säurekonzentration,
   Säuredichte *f*
acid concentration plant Säureeindampfanlage *f*
acid container Säurebehälter *m*
acid content Säuregehalt *m*
acid cooler Säurekühler *m*
acid corrosion Säureangriff *m*
acid-curing säurehärtend
acid density Säuredichte *f*
acid derivative Säurederivat *n*
acid determination Säurebestimmung *f*
acid dilution table Säureumrechnungstabelle *f*
acid dyestuff Säurefarbstoff *m*
acid egg Druckbirne *f*
acidemia Acidose *f* (Med)
acid equivalent Säureäquivalent *n*
acid ethyl sulfate Äthylschwefelsäure *f*
acid exchanger Säureaustauscher *m*
acid extractor Entsäuerungsapparat *m*
acid fermentation Säuregärung *f*
acid fluoride Doppelfluorid *n*
acid formation Säurebildung *f*
acid-forming säurebildend
acid-free säurefrei
acid fuchsine Säurefuchsin *n*
acid fume Säuredampf *m*
acid fume scrubber Säurenebelwascher *m*
acid fume venting plant Säureabluftanlage *f*
acid gate valve Säureschieber *m*
acid gutter Säurerinne *f*
acid halide Säurehalogenid *n*
acid heater Säureerhitzer *m*
acid hydrate Hydratsäure *f*, Säurehydrat *n*
acid hydrazide Säurehydrazid *n*
acid hydroextractor Säureschleuder *f*
acid hydrolysis Säurehydrolyse *f*
acidic sauer (Chem)
acidiferous säurehaltig
acidifiable säuerungsfähig, ansäuerbar
acidification Ansäuern *n*, Säuerung *f*,
   Säurebildung *f*

acidified angesäuert, abgesäuert
acidifier Säuerungsmittel *n*, Säurebildner *m*
acidify *v* ansäuern, absäuern, acidifizieren,
 aufsäuern, einsäuern, säuerlich machen,
 säuern, sauer machen
acidifying Ansäuern *n*, Säuern *n*
acidifying *a* säurebildend
acidifying agent Säuerungsmittel *n*
acidimeter Acidimeter *n*, Säuremesser *m*
acidimetric acidimetrisch
acidimetry Acidimetrie *f*, Azidimetrie *f*,
 Säuremessung *f*
acid imide Säureimid *n*
acidity Acidität *f*, Säure *f* (Schärfe),
 Säuregehalt *m*, **degree of** ~ Säuregrad *m*,
 determination of ~ Aciditätsbestimmung *f*
acid-like säureartig
acid liquor Sauerbrühe *f*
acid lithium carbonate Lithiumbicarbonat *n*
acid meter Säuremesser *m*
acid method Bessemerverfahren *n* (Metall)
acid mixture Säuregemisch *n*
acidness Säure *f*, Säuregehalt *m*
acid number (Am. E.) Säurezahl *f*
acidobutyrometer Acidobutyrometer *n*
acidobutyrometry Acidobutyrometrie *f*
acidol Acidol *n*
acidol chrome color Acidolchromfarbstoff *m*
acido ligand Acidoligand *m*
acidolysis Säurehydrolyse *f*
acidolytic azidolytisch
acidomycin Acidomycin *n*
acidosis Acidose *f* (Med)
acid outlet Säureabfluß *m*
acid plant Säureanlage *f*
acid poisoning Säurevergiftung *f*
acid precipitating bath Säurefällbad *n*
acid-proof säurebeständig, säurefest
acid-proof grease Säureschutzfett *n* (Akku)
acid-proof paint coating Säureschutzanstrich *m*
acid-proof structures Säureschutzbau *m*
acid proton Säureproton *n*
acid recovery plant
 Säure-Rückgewinnungsanlage *f*
acid refining (of oils) Säureraffination *f* (von
 Ölen)
acid residue Säurerest *m*, Säurerückstand *m*
acid resistance Säurefestigkeit *f*
acid-resistant säurefest
acid-resisting säurebeständig
acids, impervious to ~ säurefest, **resistance to** ~
 Säurebeständigkeit *f*, **stable toward** ~
 säurebeständig
acid separator Säureabscheider *m*
acid settling drum Säureabsatzbehälter *m*
acid silicate Doppelsilicat *n*
acid siphon Säureheber *m*
acid sludge Abfallsäure *f*, Säureabfall *m*
acid sludge pump Säureschlammpumpe *f*

acid sodium arsenate Dinatriumarsenat *n*
acid sodium citrate Dinatriumcitrat *n*
acid soluble säurelöslich
acid spinning process Säurespinnverfahren *n*
acid storage vessel Säurespeicher *m*
acid sulfate Bisulfat *n*, Hydrogensulfat *n*
acid sulfite Bisulfit *n*
acid sulfite *a* doppel[t]schwefligsauer
acid tank Säurekessel *m*
acid tank wagon Säurekesselwagen *m*
acid tartrate Bitartrat *n*
acid tartrate *a* doppeltweinsauer
acid tester Säureprüfer *m*
acid tower Säureturm *m*
acid treatment Säurebehandlung *f*
acidulate *v* ansäuern, [ab]säuern, säuerlich
 machen, sauer machen
acidulated angesäuert, abgesäuert
acidulated water Sauerwasser *n*
acidulating Säuern *n*
acidulation Ansäuern *n*, Säuerung *f*,
 Versäuerung *f*
acidulous säuerlich
acid value (Br. E.) Säurezahl *f*
acid vapor Säuredampf *m*
acid vat Säurekufe *f*, Säuretrog *m*
acid yellow Säuregelb *n*
acierate *v* verstählen
acierating Verstählen *n*
acieration Verstählung *f*
aci-form aci-Form *f* (Stereochem)
acinitrazole Acinitrazol *n*
acknowledge *v* anerkennen
acknowledge switch Löschschalter *m* (Comp)
aclinic aklinisch (Geogr)
aclinic line Nullisokline *f*
acme thread Trapezgewinde *n*
acme yellow Akmegelb *n*
acmite Akmit *m* (Min), Achmit *m* (Min),
 Ägirin *m* (Min), Ägirit *m* (Min)
acne Akne *f* (Med), Finne *f* (Med)
acocantherin Acocantherin *n*
acofriose Acofriose *f*
acoine Acoin *n*
acolycitine Acolycitin *n*
acolytine Acolytin *n*
aconelline Aconellin *n*
aconic acid Aconsäure *f*
aconine Aconin *n*, **extract of** ~ Aconinextrakt *m*
aconitase Aconitase *f* (Biochem)
aconitate Aconitat *n*
aconite Akonit *m* (Bot), Eisenhut *m* (Bot),
 Eisenhutkraut *n* (Bot), Sturmhut *m* (Bot),
 tincture of ~ Eisenhuttinktur *f*
aconite extract Eisenhutextrakt *m*
aconite root Akonitknolle *f* (Bot), Akonitwurzel *f*
 (Bot), Eisenhutknolle *f* (Bot)

aconitic acid Akonitsäure *f,* Achilleasäure *f,* Citridinsäure *f,* Equisetsäure *f,* **salt or ester of** ~ Aconitat *n*
aconitic tincture Aconittinktur *f*
aconitine Aconitin *n* (Pharm), Akonitin *n* (Pharm)
aconitine hydrochloride Akonitinchlorhydrat *n*
aconitinic acid Akonitinsäure *f*
acoranone Acoranon *n*
acorn Eichel *f,* **petrified** ~ Eichelstein *m*
acorn coffee Eichelkaffee *m*
acorn oil Eichenkernöl *n*
acorn sugar Quercit *m*
acorone Acoron *n*
acoustic[al] akustisch
acoustic frequency Schallfrequenz *f*
acoustic insulator Schallisoliermittel *n*
acoustic irradiation Beschallung *f*
acoustic measurement Schallmessung *f*
acoustic paint schalldämpfende Anstrichfarbe, Antidröhnlack *m,* Schallschlucklack *m*
acoustic panel Lärmschutzabdeckung *f* (für Flugtriebwerk), Schallisolierplatte *f,* Schallschutzplatte *f*
acoustic pressure Schalldruck *m*
acoustics Akustik *f* (Phys)
acoustic signal Hörzeichen *n*
acoustic wave Schallwelle *f*
acovenose Acovenose *f*
acquire *v* aneignen, erwerben
acrammerin Acrammerin *n*
acre Acre (= 4047 qm), Morgen; Acker *m*
acrichine Acrichin *n*
acrid scharf, beißend, scharf schmeckend
acridan Acridan *n*
acridic acid Acridinsäure *f,* Akridinsäure *f*
acridine Acridin *n,* Akridin *n*
acridine color Akridinfarbstoff *m*
acridine dye Akridinfarbstoff *m*
acridine orange Acridinorange *r*
acridine yellow Acridingelb *n,* Akridingelb *n*
acridity Schärfe *f* (Chem)
acridizine Acridizin *n*
acridol Acridol *n*
acridone Acridon *n,* Akridon *n*
acriflavine Acriflavin *n,* Akriflavin *n,* Trypaflavin *n*
acrifoline Acrifolin *n*
acrilan Acrilan *n*
acritol Acrit *m*
acrography Akrographie *f,* Hochätzung *f*
acrolactic acid Glucinsäure *f*
acrolein Acrolein *n,* Acrylaldehyd *m,* Allylaldehyd *m,* Propenal *n*
acromegaly Akromegalie *f* (Med)
acronycidinic acid Acronycidinsäure *f*
acronycinic acid Acronycinsäure *f*
acrose Acrose *f*
acrosome Acrosom *n*

acrylaldehyde Acrylaldehyd *m,* Acrolein *n,* Akrolein *n*
acrylamide Akrylamid *n*
acrylamide gel electrophoresis Akrylamid-Gel-Elektrophorese *f*
acrylate Acrylat *n,* Acrylester *m*
acrylate resin Acrylharz *n*
acryl glass Acrylglas *n*
acrylhydroxamic acid Acrylhydroxamsäure *f*
acrylic acid Acrylsäure *f*
acrylic aldehyde Propenal *n*
acrylic copolymers Acrylmischpolymere *n pl*
acrylic ester Acrylester *m*
acrylic glass Akrylglas *n* (Plexiglas, HN)
acrylic plastics Acrylkunststoffe *m pl*
acrylic resin Acrylharz *n*
acrylic rubber Acrylkautschuk *m*
acrylonitrile Acrylnitril *n*
act *v* eingreifen, einwirken, fungieren, sich verhalten, ~ **counter to** entgegenwirken, ~ **[on]** wirken [auf]
actamer Actamer *n*
ACTH Adrenocorticotropin *n,* Corticotropin *n*
actin Actin *n*
actinamine Actinamin *n*
acting, capable of ~ wirkungsfähig
actinic aktinisch
actinic screen Leuchtschirm *m*
actinides Actiniden, Aktiniden *pl*
actinide series Aktinidenreihe *f*
actinidine Actinidin *n*
actinism Aktinismus *m*
actinite Aktinit *m*
actinity Aktinität *f*
actinium Aktinium, Actinium *n* (Symb. Ac)
actinium emanation Aktiniumemanation *f* (Atom)
actinium radioactive series Aktinium-Zerfallsreihe *f* (Atom)
actinochemistry Aktinochemie *f*
actinocin Actinocin *n*
actinocinin Actinocinin *n*
actinoelectric aktinoelektrisch, lichtelektrisch
actinoelectricity Aktinoelektrizität *f*
actinogram Aktinogramm *n*
actinograph Aktinograph *m*
actinoid strahlenförmig
actinolite Aktinolith *m* (Min), Strahl[en]stein *m* (Min)
actinolitic schist Strahlsteinschiefer *m* (Min)
actinology Aktinologie *f,* Strahlenwirkungslehre *f*
actinometer Aktinometer *n,* Strahlenmesser *m,* **bimetal** ~ Bimetallaktinometer *n,* **shielded** ~ Panzeraktinometer *n*
actinometric aktinometrisch
actinometry Aktinometrie *f*
actinomyces Strahlenpilz *m* (Biol)

actinomycin Actinomycin *n*
actinon Actinon *n* (radioaktive Ausströmung von Actinium), Aktiniumemanation *f*
actinoscopy Aktinoskopie *f*
actinoslate Aktinolithschiefer *m* (Min)
actinospectinoic acid Actinospectinsäure *f*
actinouranium Aktinouran *n*
action Wirkung *f*, Aktion *f*, Einwirkung *f*, Vorgang *m* (Hergang), **antagonistic** ~ antagonistische Wirkung, **course of** ~ Aktionsverlauf *m*, **direction of** ~ Wirkungsrichtung *f*, **duration of** ~ Einwirkungsdauer *f*, **internal** ~ innere Einwirkung, **kind of** ~ Wirkungsart *f*, **radiobiological** ~ radiobiologische Wirkung, **range of** ~ Aktionsradius *m*, Wirkungsbreite *f*, **reducing** ~ reduzierende Wirkung, **sphere of** ~ Wirkungskreis *m*, **to set into** ~ in Betrieb setzen
action and reaction Druck und Gegendruck *m*
action constant Aktionskonstante *f*
action current Aktionsstrom *m*
action function Wirkungsfunktion *f*
action magnitude Wirkungsgröße *f*
action of acids Säurewirkung *f*
action of light Lichtwirkung *f*
action potential Aktionspotential *n*
action principle Wirkungsprinzip *n*
action quantity Wirkungsgröße *f*
action radius Ausladung *f* (Kran)
action spectrum Aktionsspektrum *n* (Festkörperphysik)
action variable with time zeitlich veränderlicher Vorgang
actiphenol Actiphenol *n*
actithiazic acid Acidomycin *n*, Actithiazsäure *f*
activate *v* aktivieren, anregen, beleben
activated charcoal Aktivkohle *f*, Adsorptionskohle *f*
activated charcoal in powder form Pulverkohle *f*
activated sludge (sewage disposal) Belebtschlamm *m*
activated sludge plant Belebungsanlage *f*
activated sludge process Belebungsverfahren *n*
activating Aktivieren *n*
activating agent Aktivierungsmittel *n*
activation Aktivierung *f*, **heat of** ~ Aktivierungswärme *f*
activation analysis Aktivierungsanalyse *f* (Atom)
activation detector Aktivierungsdetektor *m*
activation energy Aktivierungsenergie *f*
activation number Aktivierungszahl *f*
activation stress Aktivierungsspannung *f*
activator Aktivator *m*, Belebungsmittel *n*, Beschleuniger *m*
activator for curing agents Vulkanisationsaktivator *m*
activator for fillers Füllstoffaktivator *m*

active aktiv, wirksam, wirkungsfähig, **highly** ~ hochaktiv (Atom), starkwirkend
active carbon Aktivkohle *f*
active component Wirkkomponente *f*, Wirkstoff *m*
active current Wattstrom *m*
active ingredient Wirkstoff *m*
active lattice Reaktorgitter *n*
active pressure method Wirkdruckverfahren *n*
active storage area Schlackendepot *n* (Atom)
active zone Spaltzone *f* (Atom)
activin Aktivin *n*
activity Aktivität *f*, Wirken *n*, Wirksamkeit *f*, Wirkungsfähigkeit *f*, **distribution of** ~ Aktivitätsverteilung *f* (Atom), **measurement of** ~ Aktivitätsmessung *f*, **photochemical** ~ photochemische Wirkung
activity coefficient Aktivitätskoeffizient *m*
activity factor Aktivitätsfaktor *m*
activity gradient Aktivitätsgefälle *n*
actol Actol *n*
actomyosin Actomyosin *n*
actomyosin complex Actomyosinkomplex *m*
actual aktuell, faktisch, real, wirklich, tatsächlich
actual size or dimension Ist-Maß *n*
actuate *v* antreiben, betätigen (maschinell), bewegen
actuated intermittently absatzweise bewegt
actuating arm Angriffsschenkel *m*
actuation Betätigung *f*
actuator Sprühkopf *m* (Aerosoldose)
actuator cap Sprühkappe *f* (Aerosoldose)
actuator feeder Sprühkopfzuführung *f* (Aerosoldose)
acute akut, spitz (Winkel)
acute-angled spitzwinklig
acyclic azyklisch
acylate *v* acylieren
acyl-group Acylgruppe *f*
acyloin condensation Acyloin-Kondensation *f*
acyl phosphate Acylphosphat *n*
ad Reklameanzeige *f*, **classified** ~ Kleinanzeige *f* (Werbung), **commercial** ~ Geschäftsanzeige *f* (Werbung), **group** ~ gruppierte Anzeige *f* (Werbung), **solus** ~ alleinstehende Anzeige *f* (Werbung)
adaline Adalin *n*
adamantane Adamantan *n*
adamantine luster Diamantglanz *m*
adamantine spar Diamantspat *m* (Min)
adamantoblast Adamantoblast *m* (Schmelzzelle)
adamine Adamin *m* (Min)
Adams catalyst Adams-Katalysator *m*
adamsite Adamsit *m* (Kampfstoff)
Adant cube process Adantverfahren *n* (Zuck)
adapt *v* adaptieren, anpassen, einstellen, herrichten
adaptability Anpassungsfähigkeit *f*, Brauchbarkeit *f*, Verwendbarkeit *f*

adaptability range Verwendungsbereich *m*
adaptable anwendbar, verwendbar
adaptation Angleichung *f,* Anpassung *f*
adapted angepaßt
adapter Adapter *m* (Elektr), Angußbuchse *f*
(Techn), Anpaßstück *n,* Kassette *f* (Phot),
Retortenvorstoß *m* (Chem), Stutzen *m,*
Verlängerungsstück *n,* Vorstoß *m* (Chem),
Zwischenstecker *m* (Elektr), Zwischenstück *n*
adapter flange Übergangsflansch *m*
adapter pipe Übergangsrohr *n*
adapter plate Aufspannplatte *f*
adapter rim Aufsteckfassung *f*
adapter ring Einsatzring *m,* Haltering *m,*
Paßring *m*
adapter shaft Übergangskonus *m*
adapter spout Aufstecktülle *f*
adapter spring Angleichfeder *f*
adapting piece Paßrohr *n,* Formstück *n*
adaption Adaptierung *f*
adaptometer Adaptometer *n* (Opt)
adaptor see adapter
adatom Adatom *n*
add *v* addieren (Math), anfügen, beifügen,
beilegen, beimengen (Chem), beimischen
(Chem), eintragen, hinzurechnen (Math),
versetzen (Chem), zuführen, zugeben,
zusammenzählen (Math), zusetzen, ~ [to]
beigeben, hinzufügen, nachfüllen, ~ [up]
addieren, summieren, zusammenrechnen
addend Summand *m* (Math)
addendum Nachtrag *m,* Zusatz *m*
addendum angle of bevel gear Fußwinkel *m* des
Kegelrads
addendum circle (gears) Kopfkreis *m*
addicted süchtig
addiction Sucht *f*
adding Hinzufügen *n,* mistake in ~
Additionsfehler *m,* ~ up Addition *f*
adding device Addierwerk *n*
adding machine Addiermaschine *f*
Addison's disease Addisonsche Krankheit *f*
(Med), Bronzekrankheit *f* (Med)
addition Addition *f,* Addierung *f,* Anlagerung *f*
(Chem), Beigabe *f,* Beilage *f,* Beimengung *f*
(Chem), Beimischung *f* (Chem), Beisatz *m,*
Eintragung *f,* Hinzufügung *f,* Nachtrag *m,*
Versetzen *n* (Chem), Zufuhr *f,* Zusatz *m,*
Zuschlag *m* (Metall), capable of ~
additionsfähig, anlagerungsfähig, electrophilic
~ elektrophile Addition, ionic ~ ionische
Addition, liquid ~ flüssiger Zusatz,
stereospecific ~ stereospezifische Addition
addition agent Zusatzmittel *n*
additional Neben-
additional *a* zusätzlich
additional accelerator Zweitbeschleuniger *m*
additional air Nachluft *f*

additional charge Preisaufschlag *m,* Mehrpreis *m*
additional costs or effort Mehraufwand *m*
additional surface potential
Oberflächenzusatzpotential *n* (Phys)
additional treatment Nachbehandlung *f*
addition color Additionsfarbe *f*
addition complex Anlagerungskomplex *m*
addition compound Additionsverbindung *f,*
Anlagerungsverbindung *f*
addition constant Additionskonstante *f*
addition of acid Säurezusatz *m*
addition of carbon Anlagerung *f* von
Kohlenstoff
addition of electrolytes Elektrolytzusatz *m*
addition of impurities Fremdatomzusatz *m*
addition of liquid Zugießen *n*
addition of oil Ölzusatz *m*
addition of ores Zusetzen *n* von Erzen (pl)
addition of salt Salzzuschlag *m*
addition polymer Additionspolymerisat *n*
addition polymerization
Additionspolymerisation *f*
addition product Additionsprodukt *n,*
Anlagerungserzeugnis *n*
addition reaction Additionsreaktion *f,*
Anlagerungsreaktion *f*
addition reaction of an olefin
Olefin-Additionsreaktion *f*
addition set Zusatzaggregat *n*
addition sign Additionszeichen *n*
addition theorem Additionstheorem *n,*
Additionssatz *m*
additive Additiv *n* (Öl), Beimengung *f,*
Veredelungsstoff *m,* Zusatz *m* (Chem),
Zusatzmittel *n,* without ~ zusatzfrei
additive *a* addierend, additiv, zusätzlich
additive compound Anlagerungserzeugnis *n*
additive moment Zusatzmoment *n*
additive power Additionsfähigkeit *f*
additive reaction Additionsreaktion *f*
additivity Additivität *f,* principle of ~
Summierungsprinzip *n*
additivity relation Additivitätsbeziehung *f*
address Anschrift *f*
adduct Additionsprodukt *n,* Addukt *n*
adductor Adduktor *m* (Med)
adelite Adelit *m* (Min)
adelpholite Adelpholit *m* (Min)
adenase Adenase *f* (Biochem)
adenine Adenin *n,* Aminopurin *n*
adenine hexoside Adeninhexosid *n*
adenine riboside Adenosin *n,* Adeninribosid *n*
adenitis Adenitis, Drüsenentzündung *f* (Med)
adenocarpine Adenocarpin *n*
adenose Adenose *f*
adenosine Adenosin, Adeninribosid *n*
adenosine diphosphate Adenosindiphosphat *n*

adenosine monophosphate
Adenosinmonophosphat *n*, Adenylsäure *f*,
AMP
adenosine triphosphatase
Adenosintriphosphatase *f* (Biochem)
adenosyl homocysteine Adenosylhomocystein *n*
adenosyl methionine Adenosylmethionin *n*
adenovirus Adenovirus *m*
adenyl cyclase Adenylatcyclase *f* (Biochem)
adenylic acid Adenylsäure *f*,
Adenosinmonophosphat *n*
adeps Adeps *m*
adequate angemessen, ausreichend
adermin Adermin *n*, Pyridoxin, Vitamin B₆ *n*
adhere *v* anhaften, [an]kleben, haften
adherence Adhäsion *f*, Anhaftung *f*, Ankleben *n*,
Haften *n*, Haftenbleiben *n*, Haftfähigkeit *f*,
Haftvermögen *n*
adherent klebend
adhering anhaftend
adhesion Adhäsion *f*, Anhaftung *f*, Ankleben *n*,
Anziehung *f*, Haftbarkeit *f*, Haften *n*,
Haftfähigkeit *f*, Haftfestigkeit *f*, Haftkraft *f*,
Haftvermögen *n*, Klebfähigkeit *f*, **degree of** ~
Haftstärke *f*, **force of** ~ Adhäsionskraft *f*,
intensity of ~ Haftintensität *f*, **limit of** ~
Adhäsionsgrenze *f*, ~ **of the coating**
Verankerung *f* d. Anstrichs, ~ **of printing inks
[to]** Bedruckbarkeit *f* [von], ~
adhesion of the slag Schlackenansatz *m*
adhesion promoter Haftvermittler *m*
adhesion tension Adhäsionsspannung *f*
adhesion term Haftterm *m*
adhesive Adhäsionsmasse *f*, Haftmasse *f*,
Haftmittel *n*, Klebemittel *n*, Kleber *m*,
Klebstoff *m*, **cold setting** ~ kaltabbindender
Kleber *m*, **fast setting** ~ schnell abbindender
Klebstoff *m*, **high solids** ~ Klebstoff *m* mit
hohem Feststoffanteil, **moisture resistant** ~
Klebstoff *m* mit wasserabweisender
Eigenschaft, **synthetic resin** ~ Plastikkleber *m*,
thermoplastic ~ Schmelzkleber *m*, **hot-setting**
~ **heat-sealing adhesive** Heißkleber *m*
adhesive *a* [an]haftend, festhaltend, haftfähig,
klebend, klebrig
adhesive capacity Adhäsionsvermögen *n*,
Haftvermögen *n*
adhesive dispersion Dispersionskleber *m*,
Klebdispersion *f*
adhesive film Klebfolie *f*
adhesive friction Haftreibung *f*
adhesive grease or fat Adhäsionsfett *n*
adhesive label Aufklebeetikett *n*
adhesive lacquer Kleblack *m*
adhesive layer Klebschicht *f*
adhesiveness Klebrigkeit *f*, Adhäsionsfähigkeit *f*,
Adhäsionsvermögen *n*, Haftvermögen *n*,
Klebefähigkeit *f*

adhesive plaster Heftpflaster *n* (Med),
Klebpflaster *n*
adhesive power Adhäsionskraft *f*,
Adhäsionsvermögen *n*, Anziehungskraft *f*,
Haftvermögen *n*, Klebkraft *f*
adhesives industry Klebstoffindustrie *f*
adhesive strength Haftfestigkeit *f*,
Klebefestigkeit *f*
adhesive stress Haftspannung *f*
adhesive substance Klebemasse *f*, Klebstoff *m*
adhesive tape Heftpflaster *n* (Med),
Klebeband *n*, Klebfalz *m*, Klebstreifen *m*,
Leukoplast *n* (Med)
adhumulinone Adhumulinon *n*
adiabatic adiabatisch
adiabatic change adiabatische
Zustandsänderung *f*
adiabatic curve Adiabate *f*
adiabatic exponent Adiabatenexponent *m*
adiabatic hypothesis Adiabatenhypothese *f*
adiabatic theorem Adiabatensatz *m*
adiabetic adiabetisch (Med)
adiactinic adiaktinisch
adiathermal adiatherm
A die A-Düse *f*
adinole Feldstein *m* (Min)
adion Adion *n*
adipamide Adipinsäureamid *n*
adipate Adipinat *n*, Salz oder Ester der
Adipinsäure
adiphenine Adiphenin *n*
adipic acid Adipinsäure *f*, **salt or ester of** ~
Adipat *n*, Adipinat *n*
adipic aldehyde Adipinaldehyd *m*
adipic dinitrile Adipinsäuredinitril *n*
adipic ester Adipinsäureester *m*
adipiodone Adipiodon *n*
adipo-cellulose Adipocellulose *f*
adipocere Fettwachs *n*, Leichenwachs *n*
adipoceriform fettwachsartig
adipocerite Adipocerit *n*
adipocerous fettwachsartig
adipogenesis Fettbildung *f*
adipoin Adipoin *n*
adipomalic acid Adipomalsäure *f*
adipose adipös
adipose cell Fettzelle *f*
adipose tissue Fettgewebe *n*
adiposis Fettablagerung *f*
adipotartaric acid Adipoweinsäure *f*
adjacent aneinandergrenzend, angrenzend,
anliegend, anstoßend, benachbart,
naheliegend (räumlich), nebeneinander, **to be**
~ aneinandergrenzen
adjournment Aufschub *m*
adjunction Adjunktion *f* (Math)
adjust *v* justieren, abgleichen, akkommodieren
(Techn), angleichen, anpassen, ausgleichen,
ausrichten, berichtigen, einmitten, einpassen,

[ein]regeln, [ein]regulieren, einstellen, nachstellen, **~ an apparatus to a certain position** einen Apparat auf eine bestimmte Stelle einstellen
adjustability Regelbarkeit *f,* Regulierbarkeit *f,* Verstellbarkeit *f*
adjustable regulierbar, eichfähig, einstellbar, nachstellbar, regelbar, verstellbar
adjustable resistance Regelwiderstand *m,* Regulierwiderstand *m*
adjustable stop Einstellanschlag *m*
adjustable transformer Drehtransformator *m*
adjusting Adjustieren *n,* Einpassen *n*
adjusting balance Adjustierwaage *f* (Techn), Justierwaage *f*
adjusting collar Stellring *m*
adjusting device Regelvorrichtung *f,* Stellvorrichtung *f*
adjusting drive (for control pump) Stellantrieb *m*
adjusting error Abgleichfehler *m* (Elektr)
adjusting key Stellkeil *m*
adjusting knob Einstellknopf *m*
adjusting lever Einstellhebel *m,* Stellhebel *m,* Verstellhebel *m*
adjusting nut Stellmutter *f*
adjusting piece Paßstück *n,* Stellstück *n*
adjusting pin Paßstift *m,* Stellstift *m*
adjusting ring Justierring *m*
adjusting screw Justierschraube *f,* Berichtigungsschraube *f,* Einstellschraube *f,* Klemmschraube *f,* Richtschraube *f*
adjusting sleeve Einstellmuffe *f,* Stellbuchse *f*
adjusting stud Justierstift *m*
adjusting table Adjustiertisch *m*
adjusting tubing Aufbaugußring *m*
adjusting wedge Stellkeil *m*
adjusting wheel Lenkrad *n*
adjustment Justierung *f,* Adjustage *f,* Angleichung *f,* Anpassung *f,* Berichtigung *f,* Einstellung *f,* Regelung *f,* **method of ~** Einpaßverfahren *n,* **provision for ~** Verstellmöglichkeit *f,* **quick action ~** Schnellverstellung *f,* **range of ~** Regelbereich *m,* **time required for ~** Stellzeit *f*
adjustment of the bath Korrektur *f* der Badlösung
adjustment range Einstellbereich *m*
adjustment scale Einstellskala *f*
adjustment time Einstellzeit *f*
adjuvant Adjuvans *n,* Hilfsstoff *m*
adlumidine Adlumidin *n*
adlupulone Adlupulon *n*
administer *v* applizieren
administration building Verwaltungsgebäude *n*
admissibility Zulässigkeit *f*
admission Einlaß *m,* Zutritt *m,* **degree of ~** Füllungsgrad *m*
admission gear Einlaßsteuerung *f*
admission of air Luftzutritt *m*

admission of light Lichtzutritt *m*
admission pipe Eintrittsrohr *n*
admission pressure Admissionsdruck *m*
admission temperature Zulauftemperatur *f*
admission tension Admissionsspannung *f*
admission valve Einströmventil *n*
admit *v* anerkennen, zulassen, zugeben
admittance Admittanz *f* (Elektr), Richtleitwert *m* (Elektr), Scheinleitwert *m* (Elektr), Zutritt *m*
admittance matrix Wellenleitwert-Matrix *f*
admittance of light Lichtzutritt *m*
admix *v* beimengen, beimischen
admixed zugemischt
admixed air Beiluft *f*
admixing Beimengen *n*
admixture Beimengung *f,* Beimischung *f,* Beisatz *m,* Zusatz *m,* Zuschlag *m* (Metall)
adnamine Adnamin *n*
adonic acid Akonitsäure *f*
adonidin Adonidin *n*
adonin Adonin *n*
adonite Adonit *m*
adonitol Adonit *m*
adonitoxigenin Adonitoxigenin *n*
adopt *v* adoptieren
adrenal cortex Nebennierenrinde *f* (Anat)
adrenal cortex hormone Nebennierenrindenhormon *n*
adrenalectomy Adrenalektomie *f* (Med)
adrenal gland Nebenniere *f*
adrenal hormone Nebennierenhormon *n*
adrenalin[e] Adrenalin *n,* Epinephrin *n*
adrenalone Adrenalon *n*
adrenochrome Adrenochrom *n*
adrenocortical extract Nebennierenrindenextrakt *m*
adrenodoxin Adrenodoxin *n*
adrenoquinone Adrenochinon *n*
adrenosterone Adrenosteron *n*
adrenotropic or adrenocorticotropic hormone Adrenocorticotropin *n*
adronol acetate Adronolacetat *n*
adsorb *v* adsorbieren, aufnehmen
adsorbability Adsorbierbarkeit *f*
adsorbable adsorbierbar
adsorbate Adsorbat *n*
adsorbed film Adsorptionshaut *f*
adsorbed substance Adsorbat *n*
adsorbent Adsorbens *n,* Adsorptionsmittel *n*
adsorbent for gases Adsorptionsmittel *n* für Gase (pl)
adsorbent layer Adsorptionsschicht *f*
adsorber Adsorber *m*
adsorbing Adsorbieren *n*
adsorbing capacity Adsorptionsvermögen *n*
adsorbing substance Adsorbens *n*
adsorption Adsorption *f,* Haftung *f,* **activated ~** aktivierte Adsorption, **apolar ~** apolare

Adsorption (Phys), **degree of** ~
Adsorptionsgrad *m,* **heat of** ~
Adsorptionswärme *f,* **negative** ~ negative
Adsorption, **preferential** ~ bevorzugte
Adsorption, **selective** ~ selektive Adsorption
adsorption analysis Adsorptionsanalyse *f*
(Chromatographie)
adsorption capacity Beladefähigkeit *f*
adsorption catalysis Adsorptionskatalyse *f*
adsorption center Adsorptionszentrum *n*
adsorption column Adsorptionssäule *f*
adsorption compound Adsorptionsverbindung *f*
(Chem)
adsorption displacement
Adsorptionsverdrängung *f*
adsorption equilibrium
Adsorptionsgleichgewicht *n,* **to displace the** ~
das Adsorptionsgleichgewicht ändern
adsorption exponent Adsorptionsexponent *m*
adsorption indicator Adsorptionsindikator *m*
adsorption isotherm isotherme Adsorption *f,*
Adsorptionsisotherme *f*
adsorption layer Adsorptionsschicht *f*
adsorption phenomenon
Adsorptionserscheinung *f*
adsorption potential Adsorptionspotential *n*
adsorption process Adsorptionsvorgang *m*
adsorption pump Adsorptionspumpe *f*
adsorption shell Adsorptionshülle *f*
adsorptive capacity Adsorptionsfähigkeit *f*
adsorptive power Adsorptionskraft *f,*
Adsorptionsvermögen *n,* ~ **of the surface**
Oberflächenadsorption *f*
adsorptive property Adsorptionseigenschaft *f*
adstringent Adstringens *n* (pl Adstringentien)
adularia Adular *m* (Min), Edelspat *m* (Min)
adulterant Fälschungsmittel *n,*
Fälschungsstoff *m,* Streckungsmittel *n*
adulterate *v* [ver]fälschen, vergällen,
verschneiden
adulterating Verschneiden *n*
adulterating agent Verschnittmittel *n*
adulteration Fälschung *f,* Verfälschung *f,*
Vergällung *f,* ~ **of food**
Nahrungsmittelfälschung *f*
advance *v* voreilen, vorrücken
advantage Nutzen *m,* Vorteil *m*
advantageous vorteilhaft
adverse normwidrig
adversity Widerwärtigkeit *f*
advertise *v* Reklame machen, werben
advertisement Anzeige *f,* Reklame *f*
advertising Werbung *f,* Reklame *f,* **direct [mail]**
~ Direktwerbung *f*
advertising agency Werbefirma *f*
advertising article Werbeartikel *m*
advertising folder Prospekt *m*
advice Belehrung *f,* Beratung *f,* Rat *m*

advise *v* [an]raten, belehren, beraten, in
Kenntnis setzen
adviser Beirat *m*
advisory board Beirat *m*
advocate *v* befürworten
adynerin Adynerin *n*
adze Dechsel *f*
aegirine Ägirin *m* (Min), Ägirit *m* (Min)
aegirite Ägirin *m* (Min), Ägirit *m* (Min),
Akmit *m* (Min)
aeolipyle Dampfkugel *f*
aerate *v* durchlüften, lüften, mit Kohlensäure
sättigen
aerated kohlensauer, lufthaltig
aerated concrete Gasbeton *m*
aerated plastic Kunstschwamm *m,*
Schaum[kunst]stoff *m*
aerating agent Treibmittel *n* (Schaumstoff)
aeration Belüftung *f,* Durchlüftung *f,* Lüftung *f*
aeration apparatus Belüftungsapparat *m*
aeration device Belüftungsapparat *m*
aeration plant Belüftungsanlage *f*
aeration tube Belüftungsrohr *n*
aerial Antenne *f*
aerial cable Antennenkabel *n*
aerial line Oberleitung *f*
aerial nitrogen Luftstickstoff *m*
aerial oxygen Luftsauerstoff *m*
aerial photograph Luftaufnahme *f,* Luftbild *n*
aerial sulfur Luftschwefel *m*
aerial transmission Luftübertragung *f*
aeriferous lufthaltig
aeriform luftartig, luftförmig
aerobe Aerobe *f* (Biol), Arobier *m* (Biol)
aerobes Aerobien *pl* (Bakterien)
aerobian Aerobe *f* (Biol), Arobier *m* (Biol)
aerobic aerob, aerobiontisch, aerobisch
aerobic organism Aerobiont *m* (Biol)
aerobiology Aerobiologie *f*
aerobiosis Aerobiose *f*
aeroconcrete Schaumbeton *m*
aerodynamic aerodynamisch, windschnittig
aerodynamic drag Luftwiderstand *m*
aerodynamicist Aerodynamiker *m*
aerodynamics Aerodynamik *f,* Mechanik *f*
gasförmiger Körper, Strömungslehre *f* (Gase)
aero-elasticity Aeroelastizität *f*
aerofall mill Aerofallmühle *f*
aerofoil Tragfläche *f*
aerogel Aerogel *n* (Phys)
aerogene gas Aerogengas *n*
aerogenic aerogen
aerogenous aerogen
aerography Aerografie *f*
aerolite Aerolith *m* (Min), Meteorstein *m*
aeromechanics Aeromechanik *f*
aerometer Aerometer *n,* Luftdichtemesser *m,*
Tauchwaage *f*
aeronautic aeronautisch

aeronautics Aeronautik *f,* Flugwesen *n,*
  Luftfahrt *f*
aerophysical aerophysikalisch
aerosite dunkles Rotgilderz *n*
aerosol Aerosol *n*
aerosol can Aerosoldose *f,* Sprühdose *f*
aerosol packing Druckpackung *f*
aerosol propellant gas Aerosol-Treibgas *n*
aerostat Aerostat *m*
aerostatic aerostatisch
aerostatics Aerostatik *f*
aeruginous grünspanähnlich, grünrostig
aesculin Aesculin *n*
aethrioscope Differentialthermometer *n*
aetites Adlerstein *m* (Min), Erdstein *m* (Min)
affair Angelegenheit *f*
affect *v* beeinflussen, affizieren (Med),
  angreifen, beeinträchtigen, befallen, einwirken
affected behaftet, ~ by air luftempfindlich
affiliate *v* angliedern
affiliated company Filiale *f*
affiliated establishment Filialanstalt *f*
affiliation Angliederung *f*
affinated sugar Affinade *f* (Zuck)
affine affin
affinin Affinin *n*
affinity Affinität *f* (Chem), Aufziehvermögen *n,*
  Verbindungsfähigkeit *f* (Chem),
  Verwandtschaft *f* (Chem), direction of ~
  Affinitätsrichtung *f,* doctrine of ~
  Affinitätslehre *f,* unit of ~ Affinitätseinheit *f*
affinity constant Affinitätskonstante *f*
affinity curve Aufziehkurve *f*
affinity force Affinitätskraft *f*
affinity residue Affinitätsrest *m*
affirmation Behauptung *f,*
  Wahrheitsbekräftigung *f*
affix *v* aufsetzen
afflicted behaftet
affluence Zufluß *m,* Zuströmen *n*
afflux Zufluß *m,* Zustrom *m*
afflux velocity Zuflußgeschwindigkeit *f*
Afghan yellow Afghangelb *n*
AFL (Abk) Antifibrinolysin *n*
African green Afrikagrün *n*
afridol Afridol *n*
afromosin Afromosin *n*
AFT (Abk) Antifibrinolysintest *m*
after-annealing Nachtempern *n*
afterbake Nachhärtung *f* (Kunstharz)
after-burning Nachbrennen *n*
after-charge Nachbeschickung *f*
after-charging Nachbeschicken *n,* Nachsetzung *f*
afterchrome dyestuff Nachchromierfarbstoff *m*
afterchroming Nachchromieren *n*
after-copper *v* nachkupfern
after-crystallization Nachkristallisation *f*

afterdamp Nachdampf *m,* Nachschwaden *m*
  (Bergb)
aftereffect Nachwirkung *f,* magnetic ~
  magnetische Nachwirkung, to show an ~
  nachwirken
after-fermentation Nachgärung *f*
afterfix *v* nachfixieren (Phot)
after-flow Nachstrom *m* (Mech)
afterglow Nachglimmen *n,* Nachleuchten *n*
afterglow *v* nachleuchten
afterglow time Nachleuchtdauer *f*
afterheat Nachwärme *f* (Reaktor), Restwärme *f*
  (Reaktor)
after-mash Nachwürze *f,* Nachguß *m*
after-odor Nachgeruch *m*
afterproduct Nachprodukt *n*
after-purification Nachreinigung *f*
after-run Nachlauf *m*
after-smell Nachgeruch *m*
after-stretching Nachverstrecken *n*
aftertaste Beigeschmack *m,* Nachgeschmack *m*
aftertreat *v* nachbehandeln
aftertreatment Nachbehandlung *f*
afterworking, elastic ~ elastische Nachwirkung
afzelin Afzelin *n*
again wieder, erneut
agalite Agalit *m* (Min), Faserkalk *m* (Min)
agallol Agallol *n* (Pflanzenschutzmittel)
agalmatolite Agalmatolith *m* (Min), Bildstein *m*
  (Min)
agaphite Agaphit *m* (Min)
agar Nährboden *m,* solution of ~ Agarlösung *f*
agar-agar Agar-Agar *n,* japanische Gelatine
agaric acid Agaricinsäure *f*
agaricic acid Agaricinsäure *f*
agaricin Agaricin *n*
agaric mineral Bergmilch *f*
agaritine Agaritin *n*
agarobiose Agarobiose *f*
agarol Agarol *n*
agar plate Agarplatte *f*
agarythrine Agarythrin *n*
agate Achat *m* (Min), Siegelstein *m* (Min),
  botryoid ~ Traubenachat *m* (Min), brecciated
  ~ Trümmerachat *m* (Min), broken ~
  Brekzienachat *m* (Min), clouded ~
  Wolkenachat *m* (Min), smoky ~
  Rauchachat *m* (Min), translucent ~
  Kristallachat *m* (Min), uncolored ~
  Eisachat *m* (Min), variegated ~ Buntachat *m*
  (Min)
agate burnisher Vergoldestein *m*
agate cylinder Achatrolle *f*
agate dish Achatschale *f*
agate drill Achatbohrer *m*
agate edge Achatschneide *f*
agate flint Achatfeuerstein *m*
agate grinder Achatschleifer *m*
agate knife edge Achatschneide *f*

agate-like achatähnlich, achatartig
agate mortar Achatmörser *m*, Achatschale *f*
agate onyx Achatonyx *m* (Min)
agate paper Achatpapier *n* (Techn)
agate shellac Achatschellack *m*
agate stone, lenticular ~ Schwalbenstein (Min)
agate tissue paper Achatseidenpapier *n*
agateware Achatporzellan *n*
agathalene Agathalen *n*, Agathalin *n*
agathane Agathan *n*
agathic acid Agathsäure *f*
agatholic acid Agatholsäure *f*
agatiferous achathaltig
agatiform achatförmig
agatine achatähnlich, achatartig
agave fiber Agavenfaser *f*
agave hemp Agavenhanf *m*
agave sap Agavensaft *m*
agavose Agavose *f*
age *v* altern, zur Reife bringen, ablagern,
  auslagern (Duralum), ~ artificially tempern,
  warmaushärten
age coating Alterungsbelag *m*
aged gealtert
age determination Altersbestimmung *f*
ageharden *v* altern (Techn)
age-hardenable aushärtbar
age hardening Aushärten *n*,
  Ausscheidungshärtung *f*, Veredeln *n*,
  Vergüten *n*, Zeithärtung *f*
agency Agentur *f*
agent Agens *n* (Chem), Agent *m*,
  Behandlungsmittel *n* (Techn), Mittel *n*
  (Medium), Reaktionsmittel *n* (Chem),
  Vermittler *m*, Vertreter *m*, chemical ~
  chemisches Agens, extractive ~ selektives
  Lösungsmittel, oxidizing ~ oxydierendes
  Agens
age protector Alterungsschutzmittel *n*
ageratochromene Ageratochromen *n*
ageusia Ageusie *f* (Med, Geschmacksverlust)
aggerlit (HN) Aggerlit *m* (HN,
  Edelstahlformguß)
agglomerate Agglomerat *n*, Ballung *f*
agglomerate *v* agglomerieren, zusammenballen
agglomerating Agglomerieren *n*
agglomeration Agglomeratbildung *f*,
  Anhäufung *f*, Anlagerung *f* (Geol),
  Schwarmbildung *f* (Moleküle),
  Zusammenballung *f*
agglutinant Klebemittel *n*
agglutinate *v* zusammenkleben,
  zusammenklumpen, zusammenleimen
agglutination Agglutination *f*,
  Zusammenklebung *f*
agglutinin Agglutinin *n*
agglutinogen Agglutinogen *n*
agglutinoid Agglutinoid *n*

aggravate *v* erschweren, erschweren,
  verschlimmern
aggravation Erschwerung *f*
aggregate Aggregat *n*, Anhäufung *f*
aggregate *v* zusammenballen, aggregieren,
  anhäufen
aggregate fluidization Sprudelbett *n*
aggregate of molecules Molekülaggregat *n*
aggregate tension Spannungsanhäufung *f*
aggregation Aggregation *f* (Chem), Anhäufung *f*
aggressive aggressiv
aggressiveness Aggressivität *f*, Angriffslust *f*
aging Altern *n*, Vergüten *n* (Met), artificial ~
  Warmvergütung *f*, period of ~
  Alterungsdauer *f*, resistance to ~
  Alterungsbeständigkeit *f*, susceptibility to ~
  Alterungsempfindlichkeit *f*, ~ at room
  temperature Selbstalterung *f*
aging bombs Alterungsbomben *pl*
aging of coal Inkohlen *n*
aging plant Alterungsanlage *f*
aging process Alterungsverfahren *n*,
  Alterungsvorgang *m*
aging test Alterungsprüfung *f*,
  Alterungsversuch *m*, accelerated ~
  Schnellalterungsversuch *m*
aging time Nachhärtungsfrist *f*
agioneurosin Glycerinnitrat *n*
agitate *v* schütteln, quirlen, rühren
agitated autoclave Schüttelautoklav *m*
agitated dryer Drehtrockner *m*
agitating Schütteln *n*
agitating arm Rührarm *m*
agitating machine Rührmaschine *f*
agitating vane Rührflügel *m*
agitation Erregung *f*, Erschütterung *f*, thermal ~
  thermische Bewegung
agitation of the material Materialbewegung *f*
  (Färb)
agitator Rührapparat *m*, Rührer *m*,
  Rührvorrichtung *f*, Rührwerk *n*, double motion
  ~ gegenläufiges Doppelrührwerk
agitator arm Rührflügel *m*
agitator autoclave Rührautoklav *m*
agitator shaft Rührwerkswelle *f*
agitator vessel Rührkessel *m*
aglucone Aglucon *n*
aglycone Aglykon *n*
agmatine Agmatin *n*
agnolite Agnolith *m*
agnosterol Agnosterin *n*, Agnosterol *n*
agoniadin Agoniadin *n*
agree *v* übereinstimmen, beipflichten,
  beistimmen, kongruieren, ~ upon bedingen
agreement Abkommen *n*, Abmachung *f*,
  Einverständnis *n*, Genehmigung *f*,
  Übereinstimmung *f*, Zustimmung *f*
agricolite Agricolit *m* (Min)
agricultural chemicals Agrochemikalien *pl*

agricultural chemistry Ackerbauchemie *f*,
Agrikulturchemie *f*
agricultural[ly] landwirtschaftlich
agricultural physics Agrikulturphysik *f*
agricultural planning Landwirtschaftsplanung *f*
agricultural school Landwirtschaftsschule *f*
agricultural science Landwirtschaftslehre *f*
agriculture Ackerbau *m*, Agrikultur *f*,
Landwirtschaft *f*
agrimonol Agrimonol *n*
agroclavine Agroclavin *n*
agronomist Agronom *m*
agronomy Ackerbaukunde *f*
agropyrene Agropyren *n*
ague bark Rinde *f* des Fieberbaums
ague powder Fieberpulver *n*
aguilarite Aguilarit *m* (Min)
agurin Agurin *n*
aguttane Aguttan *n*
AHG (Abk) antihämatophiles Globulin
ahistan Ahistan *n*
aikinite Aikinit *m* (Min)
aileron Querruder, Quersteuer *n*
aim Ziel *n*
aim at *v* anvisieren
air Luft *f*, Preßluft *f*, **change of** ~
Luftveränderung *f*, **compressed** ~ Druckluft,
Preßluft *f*, **condition of the** ~
Luftbeschaffenheit *f*, **containing** ~ lufthaltig,
exclusion of ~ Luftausschluß *m*, **external** ~
Außenluft *f*, **free from** ~ luftfrei, **impermeable**
to ~ luftundurchlässig, **inflammable** ~
Brennluft *f*, **not affected by** ~ luftbeständig,
secondary ~ Zweitluft *f*, **separation of** ~ **into**
constituents Luftzerlegung *f*
air *v* lüften, auslüften, auswittern (Chem),
belüften, durchlüften
air accumulator Luftakkumulator *m*
air-activated gravity conveying
Fließrinnenförderung *f*
air-activated gravtiy conveyor Fließrinne *f*
air admission tube Belüftungsstutzen *m*
air agitator Luftrührer *m*
air bath Luftbad *n*
air blast Luftstoß *m*, Luftstrahl *m*
air blasting Verblasen *n* (Schmelze)
air-borne freischwebend
air-borne contamination Luftverseuchung *f*
air-borne dust Flugstaub *m*
air-borne infection Luftinfektion *f*
air-borne radioactivity Luftaktivität *f* (Atom),
Radioaktivität *f* der Luft
air-borne transmission Luftübertragung *f*
air brake Druckluftbremse *f*, Luftbremse *f*
air bridge Luftbrücke *f*
air-brush coating Luftbürstenauftrag *m*
air bubble Luftblase *f*, Gußblase *f* (Gieß),
entrapped ~ eingeschlossene Luftblase
air buffer Luftkissen *n*, Luftpuffer *m*

air cavity eingeschlossene Luftblase
air chamber Luftkammer *f*, Luftkessel *m*,
Windkammer *f*
air circulation Luftumwälzung *f*
air circulation system Umluftverfahren *n*
air classification Windsichten *n*, Windsichtung *f*
air classifier Windsichter *m*
air cleaning plant Luftreinigungsanlage *f*
air cock Entlüftungshahn *m*
air column Luftsäule *f*
air compressing engine
Luftkompressionsmaschine *f*
air compression Luftverdichtung *f*
air compressor Luftkompressor *m*,
Luftverdichtungsmaschine *f*
air compressor pump Luftdruckpumpe *f*
air condenser Luftkondensator *m*, Luftkühler *m*
(Chem), Luftverdichter *m*
air-condition *v* belüften (klimatisieren)
air conditioning Klimatisierung *f*, Bewetterung *f*,
Raumbewetterung *f*
air conditioning equipment Klimageräte *n pl*
air conditioning plant Klimaanlage *f*,
Luftreinigungsanlage *f*
air-conducting luftleitend
air contamination Luftverseuchung *f*,
Luftverunreinigung *f*
air content Luftgehalt *m*
air conveying passage Luftförderrinne *f*
air-cooled luftgekühlt
air cooler Luftkühler *m*
air cooling Luftkühlung *f*
air cooling plant Luftkühlanlage *f*
air-core coil eisenlose Spule *f*
air cure Luftvulkanisation *f*
air current Luftstrom *m*
air cushion Luftkissen *n*, Luftpuffer *m*
air-cushioned luftgefedert
air[-cushioned] bearing Luftlager *n*
air damping Luftdämpfung *f*
air damping machine Luftdurchfeuchter *m*
air densimeter Luftdichtigkeitsmesser *m*
air density Luftdichte *f*
air depression Luftunterdruck *m*
air diffuser Luftaustritt *m* (Gaschromat)
air discharge Luftentladung *f*
air-disinfecting luftreinigend
air-disinfecting apparatus Luftreinigungsgerät *n*
air disinfection Luftreinigung *f*
air displacement Luftverdrängung *f*
air distributing pipe Luftverteilungsrohr *n*
air distribution Windverteilung *f*
[air] draft Luftzug *m*
air-dried luftgetrocknet, lufttrocken
air-dried malt Luftmalz *n*
air-dry lufttrocken
air-dry *v* lufttrocknen
air dryer Lufttrockner *m*

air drying Lufttrocknen *n*, Lufttrocknung *f*,
   Trocknen *n* an der Luft
air-drying apparatus Lufttrockner *m*
air-drying plant Lufttrocknungsanlage *f*
air duct Fuchs *m* (Techn), Luftkanal *m*,
   Luftleitung *f*, Luftzuführungskanal *m*,
   Windkanal *m*
air electrometer Luftelektrometer *n*
air entrained concrete Luftporenbeton *m*
air escape valve Luftablaßventil *n*
air exhaustion Luftleere *f*
air extraction Entlüftung *f*
air fan Luftgebläse *n*
air film Luftschicht *f*
air filter Absolut-Filter *n*, Luftfilter *n*, *m*,
   Schwebstoff-Filter *n*, Schwebstoff-Luftfilter *n*
air filter oil Luftfilteröl *n*
air filter unit Luftreinigungsanlage *f*
air flap Luftklappe *f*
air force Luftwaffe *f*
air for combustion Brennluft *f*,
   Verbrennungsluft *f*
air freight Luftfracht *f*
air friction Luftreibung *f*
air furnace Zugofen *m*, Windofen *m*
air furnace slag Flammofenschlacke *f*
air gap Luftspalt *m*, Luftzwischenraum *m*
air gauge Luftdruckprüfer *m*, Luftmesser *m*
air hammer Lufthammer *m*
air-harden *v* an der Luft härten, lufthärten
air hardening Lufthärtung *f*, Luftstählung *f*
   (Met)
air-hardening steel Lufthärtestahl *m*
air heater Lufterhitzer *m*, Lufttrockner *m*,
   Winderhitzer *m*
air humidifier Luftanfeuchter *m*
air humidity regulating installation
   Luftfeuchteregelanlage *f*
air impermeability Luftundurchlässigkeit *f*
air improver Luftverbesserungsmittel *n*
air induction Luftinduktion *f*
airing Lüftung *f*, Belüftung *f*, Durchlüftung *f*
airing tube Bewitterungsleitung *f*
air injection Lufteinblasung *f*
air inlet Lufteinlaß *m*, Luftzuführungsöffnung *f*,
   Luftzufuhr *f*, Luftzutritt *m*
air inlet conduit Zuluftkanal *m*
air inlet hole Lufteinströmöffnung *f*
air inlet manifold Lufteinlaßventil *n*
air inlet screw Lufteinlaßschraube *f*
air inlet valve Lufteinlaßventil *n*,
   Lufteinströmklappe *f*, Windklappe *f*
air input Luftzuführung *f*
air insulation Luftisolation *f*
air intake Lufteinlaß *m*, Luftzutritt *m*
air jet Luftdüse *f*, Luftstrahl *m*
air jet lift Druckluftheber *m*
air knife Luftbürste *f*
air-knife coating Luftbürstenauftrag *m*

air leakage Austrittstelle *f* der Luft
air lift pneumatische Förderung
air-lift agitator Luftrührer *m*, Mischluftrührer *m*
air lift pump Druckluftheber *m*,
   Mammutpumpe *f*
air-like luftartig
air liquefaction plant Luftverflüssigungsanlage *f*
air liquefier Luftverflüssigungsmaschine *f*
air lock Wetterschleuse *f*
air main Hauptwindleitung *f*
air manometer Luftmanometer *n*
air meter Luftmengenmesser *m*
air mixture Luftgemisch *n*
air moistener Luftanfeuchter *m*,
   Luftbefeuchter *m*
air monitor Luftüberwachungsanlage *f*
air mortar Luftmörtel *m*
air nozzle Luftdüse *f*
airoform Airoform *n*, Airogen *n*
airogen Airoform, Airogen *n*
airol Airol *n*, Airoform *n*, Airogen *n*
air-operated preßluftbetätigt
air outlet Luftaustritt *m*
air outlet conduit Abluftkanal *m*, Abluftrohr *n*
air outlet pipe Abluftrohr *n*
air oxidation Luftoxidation *f*
air passage Luftzuführungskanal *m*
air permeability Luftdurchlässigkeit *f*
air pipe Luftleitung *f*, Windleitung *f*
air pocket Luftblase *f*, Lufteinschluß *m*;
   Luftloch *n*
air pollution Luftverpestung *f*,
   Luftverseuchung *f*, Luftverunreinigung *f*
airport storage tank Flugplatzlager *n* (für flüss.
   Brennstoffe)
air preheater Luftvorwärmer *m*
air pressure Luftdruck *m*, **positive ~**
   Luftüberdruck *m*
air pressure engine Luftdruckmaschine *f*
air pressure inlet Preßlufteintritt *m*
air pressure maintenance
   Luftdrucküberwachung *f*
air pressure manifold Preßluftanschluß *m*
air pressure recorder Winddruckmesser *m*
air-proof luftdicht
air-proof *v* luftdicht machen
air pump Luftpumpe *f*
air pump bell Luftpumpenglocke *f*
air pump cylinder Luftpumpenzylinder *m*
air pump disc Luftpumpenteller *m*
air pump liner Luftpumpenzylinderwand *f*
air pump piston Luftpumpenkolben *m*
air pump plate Luftpumpenteller *m*
air pump receiver Luftpumpenglocke *f*
air purification Luftreinigung *f*
air purifier Luftreiniger *m*
air purifying luftreinigend
air pyrometer Lufthitzemesser *m*,
   Luftpyrometer *n*

air quenching Luftstählung *f* (Met)
air receiver Luftaufnehmer *m*, Luftkessel *m*
air refractive power Luftbrechungsvermögen *n*
air release valve Entlüftungsventil *n*,
  Belüftungsventil *n*
air reservoir Windkammer *f*
air resistance Luftbeständigkeit *f*,
  Luftwiderstand *m*
air-resistant luftbeständig
air scoop Belüftungshaube *f*,
  Luftansaugestutzen *m*
air screw Luftschraube *f*
air search radar Radarsuchgerät *n*
air-season *v* lufttrocknen (Holz)
air-seasoned lufttrocken (Holz)
air separation Luftzerlegung *f*
air separation plant Luftrektifikationsanlage *f*,
  Lufttrennungsanlage *f*, Luftzerlegungsanlage *f*
air separator Windsichter *m*
air shaft Lüftungsschacht *m*, Luftschacht *m*
air space between the grate Rostspalt *m*
air spring Luftfeder *f*
airstat Airstat *m*, Warmluftthermostat *m*
air sterilization plant Luftentkeimungsanlage *f*
air stream Luftstrom *m*
air suction pipe Luftsaugeröhre *f*
air suction ventilator Luftabsaugvorrichtung *f*
air supply Luftzuführung *f*, Luftzufuhr *f*
air-suspended luftgefedert
air temperature Lufttemperatur *f*
air thermometer Gasthermometer *n*,
  Luftthermometer *n*
airtight luftdicht
airtightness Luftdichtheit *f*
airtight seal luftdichter Abschluß,
  Luftabschluß *m*
air valve Luftklappe *f*, Luftventil *n*
air vent Luftablaßventil *n*, Luftabzug *m*
air void Luftpore *f*
air-void *a* luftleer
air volume Luftmenge *f*
airwave Luftwelle *f*
airway Frequenzband *n* (Elektr),
  Umströmkanal *m*
aisle Gang *m*
ajacine Ajacin *n*
ajacol Guäthol *n*
ajakol Guäthol *n*
ajarmine Ajarmin *n*
ajava oil Adiowanöl *n*
ajmalicine Ajmalicin *n*
ajmalicinic acid Ajmalicinsäure *f*
ajmalidine Ajmalidin *n*
ajmyrine Ajmyrin *n*
ajour fabric Ajourware *f*
ajowan Ajowan *n*, Adiowansamen *m*
ajowan oil Adiowanöl *n*, Ajowanöl *n*
ajugose Ajugose *f*
akaustan Akaustan *n* (HN, Papierhilfsmittel)

akermanite Akermanit *m* (Min)
akrit Akrit *n* (HN, Hartlegierung)
akticit Akticit *n*
akundaric acid Akundarsäure *f*
alabamine Albamin *n*
alabandine Alabandin *m* (Min), Glanzblende *f*
  (Min), Manganblende *f* (Min)
alabandite Manganblende *f* (Min), Alabandit *m*
  (Min), Blumenbachit *m* (Min), Glanzblende *f*
  (Min), Manganglanz *m* (Min)
alabaster Alabaster *m* (Min), **calcareous** ~
  Kalkalabaster *m* (Min), **translucent** ~
  Eisalabaster *m* (Min)
alabasterglass Achatglas *n*, Alabaster-, Opalglas
alabaster white Alabasterweiß *n*
alabastrine process Alabasterverfahren *n*
alabastrite Alabastrit *m* (Min)
alacreatine Alakreatin *n*
alacreatinine Alakreatinin *n*
alaite Alait *m* (Min)
alalite Alalit *m* (Min)
alamosite Alamosit *m* (Min)
alangine Alangin *n*
alanine Alanin *n*, Aminopropionsäure *f*
alanine amide Alaninamid *n*
alanine chelate Alaninchelat *n*
alanine transaminase Alanintransaminase *f*
  (Biochem)
alaninol Alaninol *n*
alantic acid Alantsäure *f*
alantic anhydride Alantanhydrid *n*
alantin Alantin *n*, Alantcampher *m*,
  Alantstärkemehl *n*, Inulin *n*
alantolactone Alantolakton *n*
alantolic acid Alantolsäure *f*
alantotoxicum Fettgift *n*
alant root Inula *f* (Bot)
alant starch Alantstärkemehl *n*, Inulin *n*
alanyl- Alanyl-
alanylalanine Alanylalanin *n*
alanylglycine Alanylglycin *n*
alarm apparatus Warnvorrichtung *f*
alarm device Signalisiereinrichtung *f*
alarm switch Alarmschalter *m*
alarm thermometer Signalthermometer *n*
alaskaite Alaskait *m* (Min)
albafix Albafix *n* (Füllstoff)
albamine Albamin *n*
albane Alban *n*
albaspidin Albaspidin *n*
albedo limiting value Grenzalbedo *f*
alberene stone Alberenstein *m*
albertite Albertit *m* (Min)
albertol Albertol *n* (Lackindustrie)
alber[t]type Albertotypie *f*
  (Lichtdruckverfahren)
albidur Albidur *n*
albification Metallbleiche *f*
albigen Albigen *n* (Textilhilfsmittel)

albigenic acid Albigensäure *f*
albinism Albinismus *m* (Med)
albino rat Albino-Ratte *f*
albite Albit *m* (Min), Kieselspat *m* (Min),
Natronfeldspat *m* (Min), **finely granular** ~
Zuckerstein *m* (Min)
albitic albitartig
albitiferous albithaltig
albizziine Albizziin *n*
albocarbon Albocarbon *n*
albolene Alvolen *n*
albolite Albolith *m* (Min)
albomaculine Albomaculin *n*
albomycin Albomycin *n*
albumen Albumen *n*, Eiweißstoff *m*
albumiform eiweißartig
albumin Albumin *n*, Eiweißstoff *m*
albuminate Albuminat *n*
albumin content Albumingehalt *m*
albumin filament Albuminfaden *m*
albumin glue Albuminleim *m*, Eiweißleim *m*
albumin ichthyolate Ichthyoleiweiß *n*
albuminiferous albuminhaltig
albuminine Albuminin *n*
albuminization Albuminisierung *f*
albuminize *v* albuminisieren
albuminized paper Albuminpapier *n* (Phot),
Eiweißpapier *n*
albuminizing Albuminisieren *n*
albuminoid Albuminoid *n*
albuminoid *a* albuminartig, eiweißartig
albuminometer Albuminometer *n*
albuminone Albuminon *n*
albuminose Albuminose *f*
albuminous albuminhaltig, eiweißartig,
eiweißhaltig
albuminous compound Eiweißverbindung *f*
albuminous food Eiweißnahrung *f*
albuminous substance Eiweißstoff *m*,
Eiweißkörper *m*
albumin paper Albuminpapier *n* (Phot),
Eiweißpapier *n* (Phot)
albumin process Albuminprozeß *m* (Phot),
Albuminverfahren *n* (Phot)
albuminuria Albuminurie *f* (Med),
Eiweißharnen *n* (Med)
albumose Albumose *f*
alchemist Alchimist *m*
alchemistic alchimistisch
alchemy Alchimie *f*
alcogel Alkogel *n*
alcohol Alkohol *m*, Spiritus *m*, Sprit *m*, **addition
of** ~ Alkoholzusatz *m*, **containing much** ~
alkoholreich, **determination of the content of
the blood** ~ Blutalkoholbestimmung *f*,
insoluble in ~ alkoholunlöslich, **rectified** ~
Feinsprit *m*, **rich in** ~ alkoholreich, **soluble in**
~ alkohollöslich, **to extract** ~

entalkoholisieren, **percentage of** ~ **in the blood**
Blutalkohol *m*
alcohol acid Alkoholsäure *f*
alcoholate Alkoholat *n;*, Äthylat *n*
alcoholate *v* alkoholisieren
alcohol compound Alkoholverbindung *f*
alcohol content Alkoholspiegel *m*
alcohol content[s] Alkoholgehalt *m*
alcohol derivative Alkoholderivat *n*
alcohol deterrent Antialkoholikum *n* (Pharm)
alcohol evaporator Spiritusverdampfer *m*
alcohol ferment Branntweinhefe *f*
alcohol for engine operation Kraftsprit *m*
alcohol fractionation Alkoholfraktionierung *f*
alcohol-free alkoholfrei
alcohol group Alkoholgruppe *f*,
Hydroxylgruppe *f*
alcoholic alkoholisch, alkoholartig,
alkoholhaltig, spiritusartig
alcoholic content, to reduce the ~ entgeisten
alcoholic ether Alkoholäther *m*
alcoholic extract Alkoholauszug *m*
alcoholic fermentation Alkoholgärung *f*
alcoholic liquors Spirituosen *pl*
alcoholic poisoning Alkoholvergiftung *f*
alcoholic sodium hydroxide solution alkoholische
Natronlauge
alcoholism Alkoholismus *m* (Med),
Dipsomanie *f* (Med)
alcoholizable alkoholisierbar
alcoholization Alkoholbildung *f*,
Alkoholisierung *f*
alcoholize *v* alkoholisieren
alcoholizing Alkoholisieren *n*
alcohol lamp Spirituslampe *f*
alcohol-like alkoholartig
alcoholometer Alkoholmesser *m*,
Alkoholometer *n*, Branntweinprüfer *m*,
Branntweinwaage *f*, Spirituswaage *f*
alcoholometric alkoholometrisch
alcoholometric scale Alkoholometerskala *f*
alcoholometry Alkoholbestimmung *f*,
Alkoholometrie *f*
alcohol plant Alkoholanlage *f*
alcohol-proof alkoholfest
alcohol sulfonate Alkoholsulfonat *n*
alcohol test Alkoholprobe *f*
alcohol vapor Alkoholdampf *m*
alcohol vaporizer Spiritusverdampfer *m*
alcohol wet alkoholfeucht
alcoholysis Alkoholyse *f*
alcosol Alkosol *n*
aldazine Aldazin *n*
aldehyde Aldehyd *m;* Acetaldehyd *m*
aldehyde acid Aldehydsäure *f*
aldehyde ammonia Aldehydammoniak *n*
aldehyde compound Aldehydverbindung *f*
aldehyde condensation Aldehydkondensation *f*
aldehyde green Aldehydgrün *n*

aldehyde oxidase Aldehydoxidase *f*
aldehyde resin Aldehydharz *n*
aldehyde tannage Aldehydgerbung *f* (Gerb)
aldehydic aldehydhaltig, aldehydisch
aldehydic acid Aldehydsäure *f*
aldehydine Aldehydin *n*
alder charcoal Erlenkohle *f*
aldesulfone Aldesulfon *n*
aldime Aldim *n*
aldimine Aldimin *n*
aldimine chelate Aldiminchelat *n*
aldobiuronic acid Aldobiuronsäure *f*
aldohexose Aldohexose *f*
aldoketene Aldoketen *n*
aldol Aldol *n*, Acetaldol *n*
aldol alpha-naphthylamine
    Aldolalphanaphthylamin *n*
aldolase Aldolase *f* (Biochem)
aldol condensation Aldolkondensation *f*
aldomedone Aldomedon *n*
aldonic acid Aldonsäure *f*
aldopentose Aldopentose *f*
aldose Aldose *f*
aldosterone Aldosteron *n*
aldotripiperideine Aldotripiperidein *n*
aldoxime Aldoxim *n*, Acetaldoxim *n*
aldrin Aldrin *n* (Insektenmittel)
ale Malzbier *n*
alectoronic acid Alectoronsäure *f*
alembic Retorte *f*, Abziehblase *f*,
    Abziehkolben *m*, Blase *f*, Branntweinblase *f*,
    Destillationsgefäß *n*, Destillierapparat *m*,
    Destillierblase *f*
alepite Alepit *n*
alepopinic acid Alepopinsäure *f*
Aleppo combings Aleppokammwolle *f*
aletheine Alethein *n*
alethine Alethin *n*
aleudrine Aleudrin *n*
aleuritic acid Aleuritinsäure *f*
aleurometer Aleurometer *n*, Mehlprüfer *m*
aleuronate Aleuronat *n*
aleurone Aleuron *n*
aleuronic aleuronhaltig
alexandrite Alexandrit *m* (Min)
alexin Alexin *n*
alfa Alfa *f*, Alfagras *n*
alfa [grass] Alfagras *n*
alfalfa Luzerne *f* (Bot)
alfalfasaponin Alfalfasaponin *n*
alfalfol Alfalfol *n*
alfenide Alfenid *n*
alga (pl. algae) Alge *f*
algae, deposit of ~ Algenniederschlag *m*,
    formation of ~ Algenbildung *f*, **mucus of** ~
    Algenschleim *m*
algama green Algamagrün *n*
algaroth Antimonoxychlorid *n* (Min)
algaroth powder Algarotpulver *n*

algar[r]obilla Algarobilla *f* (Gerb)
algebra Algebra *f* (Math)
algebraic algebraisch
algebraic expression algebraischer Ausdruck
    (Math)
algerite Algerit *m*
algicide Algenbekämpfungsmittel *n*, Algizid *n*
algin Algin *n*
alginate Alginat *n*
algin fiber Alginatfaser *f*
alginic acid Alginsäure *f*, Algin *n*
algodonite Algodonit *m* (Min)
algol blue Algolblau *n*
algol-color Algolfarbe *f*
algoline Algolin *n*
alicant soda Alikantesoda *f*
alicyclic alizyklisch
alien [art]fremd
align *v* abgleichen, [ab]fluchten, ausrichten,
    gerade richten, orientieren, richten
aligned orientiert, justiert
alignment Ausrichtung *f*, Einstellung *f*,
    Gleichrichtung *f*, Richtung *f*, Zentrierung *f*,
    molecular ~ Gleichrichtung *f* der Moleküle
alignment nomogram Fluchtliniennomogramm *n*
alignment test Spurprüfung *f*
alimemazine Alimemazin *n*
alimentary deficiency Nahrungsmangel *m*
alimentary research Ernährungsforschung *f*
alimentary substance Ernährungsstoff *m*
aliphatic aliphatisch
aliphatic series Fettreihe *f* (Chem)
aliquant aliquant
aliquot Teilmenge *f* (Chem)
aliquot *a* aliquot
alisonite Alisonit *m* (Min)
alite Alit *m*
alite *v* alitieren
aliting process Alitierverfahren *n*
alitizing Alitieren *n*
alival Alival *n*
alizarin Alizarin *n*, Färberrot *n*,
    Krappfärbestoff *m*, Krapprot *n*
alizarin black Alizarinschwarz *n*,
    Naphthazarin *n*
alizarin blue Alizarinblau *n*, Anthracenblau *n*,
    acid ~ Säurealizarinblau *n*
alizarin bordeaux Alizarinbraun *n*
alizarin brown Alizarinbraun *n*
alizarin dye Alizarinfarbe *f*
alizarinic acid Alizarinsäure *f* (obs),
    Phthalsäure *f*
alizarin lake Alizarinfarblack *m*, Alizarinlack *m*
alizarin madder lake Alizarinkrapplack *m*
alizarin monosulfonic acid
    Alizarinmonosulfonsäure *f*
alizarin new red Alizarinneurot *n*
alizarin red Alizarinrot *n*
alizarin red *a* krapprot

alizarin sky blue Alizarinreinblau *n*
alizarin yellow Alizaringelb *n*
alizurol Alizurol *n* (Farbstoff, HN)
alkalescence Alkaleszenz *f*
alkali Alkali *n*, Lauge *f*, Laugensalz *n*, **amount of**
~ Alkalimenge *f*, **fast to** ~ alkaliecht, **loss in**
~ Alkaliverlust *m*, **poor in** ~ alkaliarm,
resistant to ~ alkalibeständig, **sensitive to** ~
alkaliempfindlich, **soluble in** ~ alkalilöslich,
swelling caused by ~ Alkaliquellung *f*, **to treat**
with ~ mit Alkali behandeln, alkalisieren
alkali albuminate Alkalialbuminat *n*
alkali amalgam Alkaliamalgam *n*
alkali atom Alkaliatom *n*
alkali-binding agent Alkalibindemittel *n*
alkali blue Alkaliblau *n*
alkali carbonate Alkalikarbonat *n*
alkali cell Alkalizelle *f*
alkali cellulose Alkalizellstoff *m*,
Alkalizellulose *f*
alkali chloride Alkalichlorid *n*, Chloralkali *n*
alkali chloride electrolyser
Alkalichloridelektrolyseur *m*
alkali circulation Laugenumlauf *m*
alkali compartment Laugenkammer *f*,
Laugenzelle *f*
alkali conduit Laugenleitung *f*
alkali content[s] Alkaligehalt *m*
alkali cyanide Alkalicyanid *n*, Cyanalkali *n*
alkali[e]s, resistant to ~ laugebeständig
alkali fast color Alkaliechtfarbe *f*
alkali fastness Alkaliechtheit *f*
alkali fast red Alkaliechtrot *n*
alkali feed Laugenzulauf *m*
alkali-free alkalifrei
alkali fusion Alkalischmelze *f*
alkaligenous alkalibildend
alkali halide Alkalihalogenid *n*
alkali halide crystal Alkalihalogenidkristall *m*
alkali halide film Alkalihalogenidschicht *f*
alkali halides, continuous spectra of ~
Alkalihalogenidkontinua *pl*
alkali humate Alkalihumat *n*
alkali hydroxide Alkalihydroxid *n*
alkali industry Alkaliindustrie *f*
alkali lye Alkalilauge *f*
alkali metal Alkalimetall *n*
alkali metal chelate Alkalimetallchelat *n*
alkali metal graphites
Graphit-Alkali-Verbindungen *pl*
alkali metal ion Alkalimetallion *n*
alkalimeter Alkalimesser *m*, Alkalimeter *n*,
Laugenmesser *m*
alkalimetric alkalimetrisch
alkalimetry Alkalimessung *f*, Alkalimetrie *f*
alkaline alkalihaltig, alkalisch, laugenartig,
slightly ~ schwach alkalisch, **to render** ~
alkalisch machen, alkalisieren
alkaline bath Laugenbad *n*

alkaline earth Erdalkali *n*
alkaline earth atom Erdalkaliatom *n*
alkaline earth chelate Erdalkalichelat *n*
alkaline earth halide Erdalkalihalogenid *n*
alkaline earth metal Erdalkalimetall *n*
alkaline earth oxide Erdalkalioxid *n*
alkaline earth salt Erdalkalisalz *n*
alkaline earth spectra, continuous ~
Erdalkalikontinua *pl*
alkaline liquor emptying apparatus
Laugenabfüllapparat *m*
alkaline salt Laugensalz *n*
alkaline solution Laugenflüssigkeit *f*
alkaline strenght Alkaligehalt *m*
alkalinity Alkaligehalt *m*, Alkalinität *f*,
Alkalität *f*, Alkalizität *f*, **total** ~
Gesamtalkalinität *f*
alkalinize *v* alkalisieren
alkali phenate Alkaliphenolat *n*
alkali phosphate Alkaliphosphat *n*
alkali photocell Alkaliphotozelle *f*
alkali production Alkaliherstellung *f*
alkali-proof alkalibeständig, alkalifest,
laugebeständig, laugenbeständig, laugenecht
alkali pump Laugenpumpe *f*
alkali-refined alkaliraffiniert
alkali reserve Alkalireserve *f*
alkali residue Alkalirückstand *m*
alkali resistance Laugenfestigkeit *f*
alkali-resistant alkalibeständig, alkalifest,
laugenfest
alkali-resistant material alkalibeständiger
Werkstoff
alkali-resisting alkalifest, laugenfest
alkali rock Alkaligestein *n*
alkalis Alkalien *pl*
alkali silicate Alkalisilikat *n*
alkali solution Alkalilösung *f*
alkali stannate Alkalistannat *n*
alkali sulfide Alkalisulfid *n*, Schwefelalkali *n*
alkali therapy Alkalitherapie *f*
alkali tolerance test Alkalitoleranzversuch *m*
alkali treatment plant Alkalisieranlage *f*
alkalizable alkalisierbar
alkalization Alkalisierung *f*
alkalize *v* alkalisch machen, alkalisieren
alkalizer Alkalisator *m*
alkalizing Alkalisieren *n*
alkaloid Alkaloid *n*
alkaloidal alkaloidartig
alkaloidal solution Alkaloidlösung *f*
alkaloid-like alkaloidartig
alkaloid salt Alkaloidsalz *n*
alkalosis Alkalose *f* (Med)
alkamine Alkamin *n*
alkane Alkan *n*, Grenzkohlenwasserstoff *m*,
Paraffinkohlenwasserstoff *m*
alkane sulfonic acid Alkansulfonsäure *f*

alkanet Alkannawurzel *f* (Bot), Färberalkanna *f* (Bot)
alkanization Alkanisierung *f*
alkanna extract Alkannaextrakt *m* (Färb)
alkanna red Alkannarot *n*
alkannin Alkannin *n*
alkaptonuria Alkaptonurie *f*
alkargen Alkargen *n*, Kakodylsäure *f*
alkarsine Kakodyloxid *n*
alkazid process Alkazidverfahren *n*
alkazid solution Alkazidlauge *f*
alkene Alken *n*
alkine Alkin *n*
alkoxy- Alkyloxy-
alkoxysilane Alkoxysilan *n*
alkydal Alkydal *n* (Lackrohstoff, HN)
alkyd resin Alkydharz *n*
alkyd resin varnish Alkydlack *m*
alkyl Alkyl *n*
alkylalkoxysilane Alkylalkoxysilan *n*
alkylamine Alkylamin *n*
alkylaromatic alkylaromatisch
alkyl aryl silicone Alkylarylsilicon *n*
alkylate *v* alkylieren
alkylated alkyliert
alkylating agent Alkylierungsmittel *n*
alkylation Alkylierung *f*
alkyl bromide Alkylbromid *n*
alkyl chloride Alkylchlorid *n*
alkyl compound Alkylverbindung *f*
alkyl cyanide Cyanalkyl *n*
alkyl derivative Alkylabkömmling *m*, Alkylderivat *n*
alkyl enamel Alkylharzfarbe *f*
alkylene Alkylen *n*
alkylene polysulfide Alkylenpolysulfid *n*
alkyl ester Alkylester *m*
alkyl fluoride Alkylfluorid *n*
alkyl group Alkylgruppe *f*, Alkylrest *m*
alkyl halide Alkylhalogenid *n*, Halogenalkyl *n*
alkyl hydrogen sulfide Alkylsulfhydrat *n*
alkylic ether Alkyläther *m*
alkylidene Alkyliden *n*
alkyl iodide Alkyljodid *n*, Jodalkyl *n*
alkyl magnesium halide Alkylmagnesiumhalogenid *n*
alkyl nitrate Alkylnitrat *n*
alkyl nitrite Alkylnitrit *n*
alkylphenolic resin Alkylphenolharz *n*
alkyl radical Alkylradikal *n*, Alkylrest *m*
alkyl residue Alkylrest *m*, Alkylradikal *n*
alkylsilane Alkylsilan *n*
alkylsilicone Alkylsilicon *n*
alkylsilicone resin Alkylsiliconharz *n*
alkyl sulfhydrate Alkylsulfhydrat *n*
alkyl sulfide Alkylsulfid *n*
alkylsulfuric acid Alkylschwefelsäure *f*
alkyl telluride Telluralkyl *n*
alkyltrichlorosilane Alkyltrichlorsilan *n*

alkyltriethoxysilane Alkyltriäthoxysilan *n*
alkyltrihalosilane Alkyltrihalogensilan *n*
allacite Allacit *m*
allactite Allaktit *m* (Min)
allanic acid Allansäure *f*
allanite Allanit *m* (Min), Orthit *m* (Min)
allantoic allantoisch
allantoic acid Allantoinsäure *f*
allantoin Allantoin *n*
allantoxaidine Allantoxaidin *n*
allantoxanic acid Allantoxansäure *f*
allanturic acid Allantursäure *f*
allel[e] Allel *n* (Biol), Erbfaktor *m* (Biol)
all-electric vollelektrisch
allelomorphic allelomorph
allelomorphism Allelomorphismus *m*
allelopathic Allelopathikum *n*
allelotropic allelotrop
allelotropy Allelotropie *f*
allemontite Allemontit *m* (Min), Antimonarsen *n* (Min), Arsenantimon *n* (Min)
allene Allen *n*
allene enantiomerism Allen-Enantiomerie *f* (Stereochem)
allenolic acid Allenolsäure *f*
allergy Allergie *f*
allethrin Allethrin *n*
allethrolone Allethrolon *n*
alleviate *v* erleichtern, lindern, mildern
allglass construction Allglasausführung *f*
allicin Allicin *n*
alligation Alligation *f* (Math), **calculation of ~** Alligationsrechnung *f*, **rule of ~** Alligationsregel *f*
Allihn filter tube Allihnsches Rohr *n*
alliin Alliin *n*
allithiamine Allithiamin *n*
allitol Allit *m*, Allodulcit *m*
allituric acid Allitursäure *f*
alliuminoside Alliuminosid *n*
all-mains receiver Allnetzgerät *n* (Elektr)
all-metal container Ganzmetallgehäuse *n*
alloaromadendrene Alloaromadendren *n*
allobarbital Allobarbital *n*
allocaffeine Allokaffein *n*
allocaine Allocain *n*
allochlorophyll Allochlorophyll *n*
allocholanic acid Allocholansäure *f*
allocholesterol Allocholesterin *n*
allocholic acid Allocholsäure *f*
allochroism Farbenwechsel *m*
allochroite Allochroit *m* (Min), brauner Eisengranat
allochromasy Farbenvertauschung *f*
allochromatic allochromatisch
allocinchonine Allocinchonin *n*
allocinnamic acid Allozimtsäure *f*
allocolchiceine Allocolchicein *n*

allocolchicine Allocolchicin *n*
allocryptopine Allokryptopin *n*
allocupreide Allocupreid *n*
allocyanine Allocyanin *n*
allodalbergin Allodalbergin *n*
allodulcitol Allodulcit *m*, Allit *m*
allodunnione Allodunnion *n*
alloechitamine Alloechitamin *n*
alloevodione Alloevodion *n*
allogeigeric acid Allogeigersäure *f*
allogibberic acid Allogibbersäure *f*
allogonite Allogonit *m*
allogyric birefringence allogyrische
    Doppelbrechung (Opt)
alloheptulose Alloheptulose *f*
alloibogaine Alloibogain *n*
alloimperatorin Alloimperatorin *n*
alloinositol Alloinosit *m*
alloisoleucine Alloisoleucin *n*
alloisomerism Alloisomerie *f*
alloite Alloit *m*
allokainic acid Allokainsäure *f*
allolactose Allolactose *f*
allomaltol Allomaltol *n*
allomatridine Allomatridin *n*
allomeric allomerisch
allomerism Allomerie *f*, Allomerismus *m*
allomethadione Allomethadion *n*
allomorphic allomorph
allomorphism Allomorphismus *m*
allomorphite Allomorphit *m* (Min)
allomorphous allomorph
allomucic acid Alloschleimsäure *f*
allomuscarine Allomuscarin *n*
allonic acid Allonsäure *f*
alloocimene Alloocimen *n*
allopalladium Allopalladium *n*
alloperiplocymarin Alloperiplocymarin *n*
allophanate Allophanat *n*
allophanate *a* allophansauer
allophane Allophan *m* (Min), Kollyrit *m* (Min)
allophanic acid Allophansäure *f*
allophite Allophit *m* (Min)
allopimarane Allopimaran *n*
allopregnane Allopregnan *n*
all-or-none law Alles-oder-Nichts-Gesetz *n*
all-or-none mechanism
    Alles-oder-Nichts-Mechanismus *m*
allosamine Allosamin *n*
allose Allose *f*
allosedamine Allosedamin *n*
allosedridine Allosedridin *n*
allosome Geschlechtschromosom *n* (Biol),
    Heterochromosom *n*
allosteric allosterisch (Biochem)
allot *v* aufteilen
alloteloidine Alloteloidin *n*
allothreonine Allothreonin *n*
allothreooxazoline Allothreooxazolin *n*

allotment Zuteilung *f*
allotriomorphic allotriomorph (Krist)
allotropic allotrop[isch]
allotropic change allotrope Modifikation
allotropism Allotropie *f*, Allotropismus *m*
allotropy Allotropie *f*
allow *v* erlauben, einräumen, zubilligen, **~ to
    settle** absetzen lassen, abstehen lassen
allowable limits Toleranzbereich *m*,
    Toleranzwerte *pl*
allowance Toleranz *f*, Zugabe *f*
allowed speed Normalgeschwindigkeit *f*
alloxan Alloxan *n*, Mesoxalylharnstoff *m*
alloxanate alloxansaures Salz *n*
alloxanate *a* alloxansauer
alloxan-diabetic alloxan-diabetisch
alloxanic acid Alloxansäure *f*
alloxantin Alloxantin *n*
alloxanylurea Alloxanylharnstoff *m*
alloxazin[e] Alloxazin *n*
alloxuric base Alloxurbase *f*
alloxuric body Alloxurkörper *m*
alloy Legierung *f*, Mischmetall *n*,
    Verschmelzung *f*, **aluminum-based ~**
    Legierung *f* auf Aluminiumgrundlage,
    antimony ~ antimonhaltige Legierung, **binary
    ~** binäre Legierung, Zweistofflegierung *f*,
    easily fusible ~ Leichtflußmetallegierung *f*,
    formation of ~ Legierungsbildung *f*, **fusible
    ~** schmelzbare Legierung,
    Schmelzlegierung *f*, **high-melting-point ~**
    hochschmelzende Legierung, **light [metal] ~**
    Leichtmetall-Legierung *f*, **low-melting-point ~**
    niedrigschmelzende Legierung, **non-magnetic
    ~** antimagnetische Legierung, **paramagnetic
    ~** paramagnetische Legierung
alloy *v* legieren, **~ gold or silver** karatieren, **~
    up** auflegieren
alloyability Legierbarkeit *f*
alloyable legierbar
alloyable metal legierbares Metall *n*
alloyage Legieren *n*, Metallbeschickung *f*,
    Münzbeschickung *f*
alloy balance Legierungswaage *f*,
    Metallmischungswaage *f*
alloy constituent Legierungskörper *m*
alloyed, highly ~ hochlegiert
alloyed state Legierungszustand *m*
alloying Legieren *n*, Metallverschmelzung *f*,
    Vermischen *n*, **~ of gold or silver**
    Karatieren *n*, Karatierung *f*
alloying addition Legierungszusatz *m*
alloying component Legierungsbestandteil *m*,
    Legierungselement *n*, **main ~**
    Hauptlegierungsbestandteil *m*
alloying constituent Legierungsbestandteil *m*
alloying content Legierungsgehalt *m*
alloying contribution Legierungsanteil *m*
alloying element Zusatzelement *n*

alloying metal Zusatzmetall *n*
alloying method Legierungsverfahren *n*
alloying property Legierbarkeit *f*
alloy ingredient, main ~
  Hauptlegierungsbestandteil *m*
alloy of copper and lead Kupferblei *n*
alloyohimbol Alloyohimbol *n*
alloy sheet Legierungsblech *n*
alloy steel Legierstahl *m*, Legierungsstahl *m*,
  Sonderstahl *m*, legierter Stahl
all-purpose adhesive Alleskleber *m*
all-purpose cleaner All[zweck]reiniger *m*
all-purpose tool Universalwerkzeug *n*
all rights reserved unter Vorbehalt aller Rechte
  pl
all-silk Reinseide *f*
allspice Allerleigewürz *n* (Bot),
  Jamaikapfeffer *m* (Bot), Nägeleinpfeffer *m*
  (Bot), Nelkenpfeffer *m* (Bot), Neugewürz *n*
  (Bot), Pimentpfeffer *m* (Bot)
alluaudite Alluaudit *m* (Min)
allulose Allulose *f*
alluranic acid Alluransäure *f*
alluvial alluvial, angeschwemmt
alluvial deposits angeschwemmter Boden (Geol)
alluvial gold Alluvialgold *n*
alluvial ore Seifenerz *n*
alluvial shell deposit Muschelgrus *m*
alluvial soil Alluvium *n* (Geol)
alluvium Alluvium *n* (Geol)
ally *v* vereinigen, alliieren
allyl- Allyl-
allylacetic acid Allylessigsäure *f*
allylacetone Allylaceton *n*
allyl alcohol Allylalkohol *m*
allyl aldehyde Allylaldehyd *m*, Propenal *n*
allylamine Allylamin *n*
allylbenzene Allylbenzol *n*
allyl bromide Allylbromid *n*
allyl chloride Allylchlorid *n*
allyl compound Allylverbindung *f*
allylcyanamide Sinamin *n*
allyl cyanide Allylcyanid *n*
allyl derivative Allylderivat *n*
allyldichlorosilane Allyldichlorsilan *n*
allyl disulfide Allyldisulfid *n*
allylene Allylen *n*
allylic rearrangement Allylumlagerung *f*
allyl iodide Allyljodid *n*, Jodallyl *n*
allyl isothiocyanate Allylsenföl *n*, Senföl *n*
allyl mustard oil Allylsenföl *n*
allyl phenyl methyl ether
  Allylphenylmethyläther *m*
allyl plastics Allylharzkunststoffe *pl*
allyl pyridine Allylpyridin *n*
allyl resin Allylharz *n*
allyl sulfide Allylsulfid *n*, Schwefelallyl *n*
allylsulfocarbamide Thiosinamin,
  Allylsulfocarbamid *n*

allyl thiocarbamide Allylsulfocarbamid *n*
allyl thiocyanate Allylrhodanid *n*
allyl thiourea Allylthioharnstoff *m*,
  Allylsulfocarbamid *n*, Thiosinamin *n*
allyltrichlorosilane Allyltrichlorsilan *n*
allyltriethoxysilane Allyltriäthoxysilan *n*
almagrerite Almagrerit *m* (Min)
almandine Almandin *m* (Min), roter
  Eisengranat (Min), Karfunkel[stein] *m* (Min)
almandine spinel Almandinspinell *m* (Min)
almandite Eisengranat *m* (Min),
  Eisentongranat *m* (Min), Karfunkelstein *m*
  (Min)
almasca Almaska *n*
almedine Almedin *n*
Almen-Nylander test Nylanders Reaktion *f*
almeriite Almeriit *m* (Min)
almond Mandel *f* (Bot), **sugared** ~
  Zuckermandel *f*
almond butter Mandelbutter *f*
almond furnace Scheideofen *m*
almond kernel Mandelkern *m*
almond meal Mandelkleie *f*
almond milk Mandelmilch *f*
almond oil Mandelöl *n*
almond [oil] soap Mandelseife *f*
almond ointment Mandelsalbe *f*
almond paste Mandelpaste *f*
almonds, emulsion of ~ Mandelmilch *f*, **syrup of**
  ~ Mandelsirup *m*
alnusene Alnusen *n*
alocemodin Rhabarberon *n*
aloe Aloe *f* (Bot)
aloe emodin Aloeemodin *n*
aloe extract Aloeauszug *m*, Aloeextrakt *m*
aloe fiber Aloefaser *f*, Aloehanf *m*
aloe hemp Aloehanf *m*, Agavenhanf *m*
aloe juice Aloesaft *m*
aloe red Aloerot *n*
aloes, extract of ~ Aloeextrakt *m*
aloe sol Aloesol *n*
aloetic aloehaltig, aloetisch
aloetic acid Aloesäure *f*, Aloetinsäure *f*
aloetic electuary Aloelatwerge *f* (Bot)
aloetic gum Aloeauszug *m*, Aloebitter *n*,
  Aloebitterstoff *m*
aloetic preparation Aloemittel *n*, Aloepräparat *n*
aloetic resin Aloeharz *n*
aloe-wood Aloeholz *n*
aloic acid Aloinsäure *f*
aloid aloeartig
aloin Aloin *n*, Aloealkaloid *n*, Aloeauszug *m*,
  Aloebitter *n*, Aloebitterstoff *m*
aloinose Aloinose *f*
'along' stretching device Längsreckvorrichtung *f*
aloxite Aloxit *n* (HN)
alpaca Alpaka *n*
alpaca hair Alpakawolle *f*

alpha-active alpha-radioaktiv (Atom),
alpha-aktiv (Atom)
alpha-activity Alpha-Aktivität *f* (Atom)
alphabetic interpreter Lochschriftübersetzer *m*
(Comp)
alpha bombardment Alphateilchenbeschuß *m*
alpha cellulose Alphacellulose *f*
alpha chloroacrylate Alphachloracrylat *n*
alpha decay Alphazerfall *m* (Atom)
alpha disintegration Alphazerfall *m* (Atom)
alpha disintegration energy
Alphazerfallsenergie *f*
alpha-emitting alphastrahlend (Atom)
alpha-helix Alpha-Helix *f*
alpha instability Alpha-Instabilität *f* (Atom)
alpha iron Alphaeisen *n*
alpha line Alphalinie *f*
alpha meter Alphastrahlenmesser *m*
alpha particle alpha-Teilchen *n* (Atom)
alpha particle bombardment Alphabeschuß *m*
(Atom)
alpha particle counter Alphateilchenzähler *m*,
Spitzenzähler *m*
alpha particle disintegration Alphazerfall *m*
(Atom)
alpha particle mass Alphateilchenmasse *f*
alpha particle model of nucleus
Alphateilchenkernmodell *n*
alpha particle source Alphateilchenquelle *f*
alpha particle spectrum Alphaspektrum *n*
alpha position Alphastellung *f*
alphaprodine Alphaprodin *n*
alpha projectile Alphageschoß *n* (Atom)
alpha radiator Alphastrahler *m*
alpha-radioactive alpha-[radio]aktiv (Atom)
alpha radioactivity Alpha-[Radio-]Aktivität *f*
(Atom)
alpha ray Alphastrahl *m*
alpha ray emitter Alphastrahler *m*
alpha ray source Alphastrahlenquelle *f* (Atom)
alpha ray spectrometer
Alphastrahlenspektrometer *n* (Atom)
alpha-reactive alpha-aktiv (Atom)
alpha state Alphazustand *m*
alpha survey meter Alphaüberwachungsgerät *n*
alphatron Alphatron *n*
alphenic Alphenicum *n*
alphol Alphol *n*
alphyl Alkylphenyl *n*, Alphyl *n*
alpinol Alpinol *n*
alquifou Glasurerz *n*
alsinit (sintered alumina) Alsinit *n* (gesintertes
Aluminiumoxid, HN)
alsol Alsol *n*
alstonidin Alstonidin *n*
alstonine Alstonin *n*, Chlorogenin *n*
alstonite Alstonit *m* (Min)
altaite Altait *m* (Min)

alter *v* ändern, abändern, abwandeln,
umändern, variieren, verändern
alterability Veränderlichkeit *f*
alterable veränderlich, abänderlich
alterant Alterans *n*
alteration Änderung *f*, Abänderung *f*,
Veränderung *f*
alternaric acid Alternarsäure *f*
alternariol Alternariol *n*
alternate *a* abwechselnd, alternierend
alternate *v* abwechseln, alternieren, wechseln
(Strom), wechselweise aufeinander folgen
alternate-immersion test Wechseltauchversuch *m*
alternately abwechselnd, wechselweise
alternate solution Ausweichlösung *f*
alternate stress test Wechselfestigkeitsprüfung *f*
alternating Alternieren *n*
alternating *a* alternierend, [ab]wechselnd
alternating bending test Wechselbiegeprüfung *f*
alternating component of voltage
Wechselspannungskomponente *f*
alternating current Wechselstrom *m*,
magnetization by ~
Wechselstrommagnetisierung *f*
alternating current accumulator
Wechselstromakkumulator *m*
alternating current ammeter
Wechselstromamperemeter *n*
alternating current arc Wechselstrombogen *m*
alternating current bridge Wechselstrombrücke *f*
alternating current circuit Wechselstromkreis *m*
alternating current commutator motor
Wechselstromkollektormotor
alternating current engineering
Wechselstromtechnik *f*
alternating current excitation
Wechselstromerregung *f*
alternating current furnace Drehstromofen *m*
alternating current generator
Wechselstromdynamo *m*
alternating current line Wechselstromleitung *f*
alternating current loss Wechselstromverlust *m*
alternating current motor Wechselstrommotor *m*
alternating current output
Wechselstromleistung *f*
alternating current period Wechselstromperiode *f*
alternating current plant Wechselstromanlage *f*
alternating current power Wechselstromleistung *f*
alternating current resistance
Wechselstromwiderstand *m*
alternating current shunt motor
Wechselstromnebenschlußmotor *m*
alternating current source Wechselstromquelle *f*
alternating current system Wechselstromsystem *n*
alternating current transformer
Wechselstromtransformator *m*
alternating current value Wechselstromgröße *f*
alternating current voltage
Wechselstromspannung *f*

alternating [cyclic] stress
  Dauerschwingbeanspruchung *f*
alternating effect Wechselwirkung *f*
alternating field Wechselfeld *n* (Elektr),
  Wechselstromfeld *n* (Elektr), **polyphase** ~
  mehrphasiges Wechsel[strom]feld, **single-phase**
  ~ einphasiges Wechsel[strom]feld
alternating flux Wechselkraftfluß *m*
alternating impact bending test
  Wechselschlagbiegeversuch *m*
alternating impact test Wechselschlagversuch *m*
alternating load deformation
  Wechselverformung *f*
alternating potential Wechselpotential *n*
alternating stress Wechselbeanspruchung *f*
  (Mech), Wechselspannung *f* (Mech)
alternating voltage Wechselspannung *f* (Elektr)
alternation Abwechslung *f,* Halbperiode *f*
  (Elektr), Stromwechsel *m* (Elektr), **law of** ~
  Alternationsgesetz *n* (Spektr)
alternations, number of ~ Wechselzahl *f*
alternative Alternative *f,* Wahl *f*
alternative stress Wechselbeanspruchung *f*
alternator Wechselstromdynamo *m,*
  Wechselstromgenerator *m*
alternidine Alternidin *n*
alth[a]ea Althee[wurzel] *f* (Bot)
alth[a]ea syrup Altheesirup *m,* Eibischsirup *m*
altheine Althein *n* (obs), Asparagin *n*
altimeter Altimeter *n,* Höhenmesser *m*
altimetry Höhenmessung *f*
altitude Höhe *f*
altitude meter Höhenmesser *m*
altitude of a triangle Höhe *f* eines Dreiecks
altometer Höhenmesser *m*
altoside Altosid *n*
altritol Altrit *m*
altroheptite Altroheptit *m*
altroheptulose Altroheptulose *f*
altronic acid Altronsäure *f*
altrose Altrose *f*
altruronic acid Altruronsäure *f*
aluchi Aluchi *n*
aluchi balsam Aluchibalsam *m*
aluchi resin Aluchiharz *n*
aludel Aludel *m*
aludel furnace Aludelofen *m*
aludrine Aludrin *n*
alum Alaun *m,* Aluminiumsulfat, **burnt** ~
  gebrannter Alaun, **calcined** ~ gebrannter
  Alaun, **common** ~ Kalialaun *m,* **exsiccated** ~
  gebrannter Alaun, **fibrous** ~ Faseralaun *m*
  (Min), **flowers of** ~ Alaunblumen *pl,*
  formation of ~ Alaunbildung *f,* **impure** ~
  Halbalaun *m,* **native** ~ Steinbutter *f* (Min),
  precipitated ~ Alaunmehl *n,* **production of** ~
  Alaunerzeugung *f*
alum *v* alaunen, alaunisieren

alum bath Alaunbad *n,* Alaunbeize *f,*
  Alaunbrühe *f*
alum boiler Alaunsieder *m*
alum boiling Alaunsieden *n*
alum cake Alaunkuchen *m* (Papier)
alum clay Alaunton *m*
alum content Alaungehalt *m*
alum crystal Alaunkristall *m*
alum earth Alaunerde *f,* Alaunton *m,* Tonerde *f*
alumed alaungar (Gerb)
alum flour Alaunmehl *n*
alum glue Alaunleim *m* (Papier)
alumina Tonerde *f,* Alaunerde *f,*
  Aluminiumoxid *n,* Felsenalaun *m* (Min),
  containing ~ tonerdehaltig, **emulsified** ~
  geschlämmte Tonerde, **hydrated** ~
  Tonerdehydrat *n,* **white** ~ Weißtonerde *f*
alumina content Tonerdegehalt *m*
alumina gel Alumogel *n*
alumina hydrate, native ~ Bauxit *m* (Min)
alumina plant Tonerdeanlage *f*
aluminate Aluminat *n*
aluminate *v* alaunen
aluminate liquor Aluminatlauge *f*
aluminate of potash Kalitonerde *f*
aluminate solution Aluminatlösung *f*
aluminiferous alaunhaltig, alaunig,
  aluminiumhaltig, tonerdehaltig
aluminite Aluminit *m* (Min), Steinalaun *m*
  (Min)
alumin[i]um (see also aluminium) Aluminium *n*
  (Symb. Al)
aluminize *v* mit Alaun behandeln; alitieren
aluminizing Aluminieren *n*
aluminon Aluminon *n*
aluminothermic aluminothermisch
aluminothermic process Thermitverfahren *n*
aluminothermics Aluminothermie *f*
aluminothermy Aluminothermie *f*
aluminous alaunartig, alaunhaltig, alaunig,
  alaunsauer, aluminiumhaltig
aluminous cement Schmelzzement *m*
aluminous flux Tonzuschlag *m*
aluminous limestone Alaunspat *m* (Min),
  Alaunstein *m* (Min)
aluminous mordant Alaunbeize *f*
aluminous pyrites Alaunkies *m* (Min)
aluminous soap Alaunseife *f*
aluminous water Alaunwasser *n*
aluminum (Symb. Al) Aluminium *n,* **crude** ~
  Rohaluminium *n,* **to plate with** ~ alumetieren
aluminum acetate Aluminiumacetat *n,* essigsaure
  Tonerde
aluminum acetotartrate
  Aluminiumacetotartrat *n,* Alsol *n,*
  essigweinsaure Tonerde
aluminum alcoholate Aluminiumalkoholat *n*
aluminum alkyl Aluminiumalkyl *n*

aluminum alloy Alulegierung *f,*
  Aluminiumlegierung *f*
aluminum and ferric oxide, mixture of ~
  Thermit *n*
aluminum block Aluminiumblock *m*
aluminum borate Aluminiumborat *n*
aluminum boride Aluminiumborid *n*
aluminum borohydride Aluminiumboranat *n*
aluminum borotartrate Boral *n*
aluminum brass
  Aluminiumkupferzinklegierung *f,*
  Aluminiummessing *n*
aluminum bromide Aluminiumbromid *n,*
  Bromaluminium *n*
aluminum bronze Aluminiumbronze *f*
aluminum cable Aluminiumkabel *n*
aluminum carbide Aluminiumcarbid *n*
aluminum carbonate Aluminiumkarbonat *n,*
  kohlensaure Tonerde *f*
aluminum casting Aluminiumguß *m*
aluminum chloride Aluminiumchlorid *n,*
  Chloraluminium *n*
aluminum coating Aluminiumüberzug *m*
aluminum contact Aluminiumkontakt *m*
aluminum content Aluminiumgehalt *m*
aluminum diffusion coating Alitieren *n*
aluminum enamel Aluminiumlackfarbe *f*
aluminum filing Aluminiumfeilspan *m*
aluminum fluoride Aluminiumfluorid *n,*
  Fluoraluminium *n*
aluminum fluosilicate Aluminiumfluorsilikat *n,*
  Kieselfluoraluminium *n*
aluminum foil Aluminiumfolie *f*
aluminum formate Aluminiumformiat *n*
aluminum foundry Aluminiumgießerei *f*
aluminum frame Aluminiumgestell *n*
aluminum furnace Aluminiumofen *m*
aluminum gallate Gallal *n*
aluminum garnet Spessartin *m* (Min)
aluminum hydride Aluminiumhydrid *n*
aluminum hydroxide Aluminiumhydroxid *n,*
  Aluminiumhydrat *n,* Aluminiumoxidhydrat *n,*
  Tonerdehydrat *n* (obs), **native** ~ Bauxit *m*
  (Min), Hydrargillit *m* (Min)
aluminum iodate Aluminiumjodat *n*
aluminum iodide Aluminiumjodid *n*
aluminum lactate Aluminiumlactat *n*
aluminum lead phosphate, native ~
  Gummi[blei]spat *m* (Min)
aluminum mordant Aluminiumbeize *f,*
  Tonerdebeize *f*
aluminum naphtholsulfonate
  Aluminiumnaphtholsulfonat *n,* Alumnol *n*
aluminum nitrate Aluminiumnitrat *n*
aluminum nitride Aluminiumnitrid *n*
aluminum ore Aluminiumerz *n*
aluminum oxide Aluminiumoxid *n,* Alaunerde *f,*
  Aloxit *n* (HN), Tonerde *f*
aluminum paint Aluminiumfarbe *f*

aluminum phenolate Phenolaluminium *n*
aluminum phenoxide Phenolaluminium *n*
aluminum phosphate, native ~ Lucinit *m* (Min)
aluminum plating Alumetieren *n*
aluminum potassium sulfate
  Aluminiumkaliumsulfat *n*
aluminum powder Aluminiumpulver *n*
aluminum pressure casting
  Aluminiumdruckguß *m*
aluminum rectifier
  Aluminiumzellengleichrichter *m*
aluminum rod Aluminiumstange *f*
aluminum rolling mill Aluminiumwalzwerk *n*
aluminum salicylate Salumin *n*
aluminum salt Aluminiumsalz *n,* Tonerdesalz *n*
  (obs)
aluminum section Aluminiumprofil *n*
aluminum sheet Aluminiumblech *n*
aluminum shot Aluminiumgrieß *m*
aluminum silicate Aluminiumsilicat *n,*
  kieselsaure Tonerde
aluminum silicofluoride Aluminiumfluorsilikat *n,*
  Kieselfluoraluminium *n*
aluminum silver alloy Drittelsilber *n*
aluminum sodium chloride
  Aluminiumnatriumchlorid *n*
aluminum sodium fluoride
  Natriumaluminiumfluorid *n*
aluminum sodium sulfate Natronalaun *m,*
  Natriumalaun *n*
aluminum stearate Aluminiumstearat *n*
aluminum sulfate Aluminiumsulfat *n,*
  schwefelsaure Tonerde (obs), **hydrated** ~
  Aluminiumsulfathydrat *n*
aluminum sulfide Aluminiumsulfid *n,*
  Schwefeltonerde *f*
aluminum sulfocyanate Aluminiumsulfocyanat *n*
aluminum tannate Tannal *n*
aluminum thallium(I) sulfate
  Thallium[aluminium]alaun *m*
aluminum thiocyanate Aluminiumrhodanid *n,*
  Rhodantonerde *f* (obs)
aluminum welding Aluminiumschweißung *f*
aluminum wire Aluminiumdraht *m*
aluminum zinc sulfate Zinkalaun *m*
alum liquor Alaunlauge *f*
alum making Alaunsieden *n*
alum mine Alaungrube *f*
alum mordant Alaunbrühe *f*
alumnol Alumnol *n,*
  Aluminiumnaphtholsulfonat *n*
alumocalcite Alumocalcit *m* (Min)
alumosulfate Alumosulfat *n*
alum pickle Alaunbrühe *f*
alum powder Alaunmehl *n*
alum root Alaunwurzel *f* (Bot)
alum schist Alaunschiefer *m* (Min)
alum shale Alaunschiefererz *n* (Min)

alum slate Alaunschiefer *m* (Min),
   Alaunschiefererz *n* (Min)
alum steep Alaunbrühe *f*
alum stone Alaunerz *n* (Min), Alaunit *m* (Min),
   Alaunspat *m* (Min), Alaunstein *m* (Min),
   Alunit *m* (Min), Bergalaun *m* (Min)
alum sugar Alaunzucker *m*, Zuckeralaun *m*
alum tanning Gerbung *f* (Weißgerbung),
   Glacégerbung *f*, Weißgerberei *f*
alum vat Alaunfaß *n*
alum water Alaunwasser *n*
alum works Alaunhütte *f*, Alaunsiederei *f*
alundum Alundum *n*
alunite Alunit *m* (Min), Alaunerz *n* (Min),
   Alaunschiefer *m* (Min), Alaunspat *m* (Min),
   Alaunstein *m* (Min), Bergalaun *m* (Min)
alunitization Alaunbildung *f*, Alunitisierung *f*
alunogen Alunogen *m* (Min), Federalaun *m*
   (Min), Halotrichit *m* (Min)
alveolar röhrenförmig, zellig, alveolar
alveolite Alveolit *m*
alvite Alvit *m* (Min)
alypine Alypin *n*
Amadori reaction or rearrangement
   Amadoriumlagerung *f*
amalgam Amalgam *n*, Quecksilberlegierung *f*,
   Quickbrei *m* (obs), **formation of an** ~
   Amalgambildung *f*
amalgamable amalgamierbar
amalgamate *v* amalgamieren, mit Quecksilber
   legieren, quicken (obs)
amalgamated amalgamiert
amalgamated metal Amalgam, Quickmetall *n*
amalgamating Amalgamieren *n*, Anquicken *n*
amalgamating barrel Amalgamierfaß *n*,
   Anquickfaß *n*, Quickfaß *n*
amalgamating bath Amalgambad *n*, Quickbeize *f*
amalgamating fluid Quickbeize *f*
amalgamating liquid
   Amalgamierungsflüssigkeit *f*
amalgamation Amalgamierung *f*,
   Amalgamation *f*, Anquickung *f*, Legieren *n*
   mit Quecksilber, Quickarbeit *f*
amalgamation process Amalgamverfahren *n*
amalgamator Amalgamiermaschine *f*
amalgam concentration Amalgamkonzentration *f*
amalgam decomposition Amalgamzersetzung *f*
amalgam electrode Amalgamelektrode *f*
amalgam filling Amalgamfüllung *f*
amalgam for silvering Brennsilber *n*
amalgamize *v* amalgamieren, quicken (obs)
amalgam of lead Bleiamalgam *n*
amalgam of tin Zinnamalgam *n*
amalgam solution Amalgambad *n*
amalic acid Amalinsäure *f*
amalinic acid Amalinsäure *f*
amandin Amandin *n* (Globulin)
amanitine Amanitin *n*
amanozine Amanozin *n*

amaranth Amarant *m, n*
amaranth bordeaux Amarantbordeaux *n*
amaranth red Amarantrot *n*
amarantite Amarantit *m* (Min)
amargoso bark Bitterrinde *f*
amaric acid Amarsäure *f*
amarine Amarin *n*
amaron Amaron *n*
amaryllidine Amaryllidin *n*
amasatine Isamid *n*
amasil Amasil *n* (Pflanzenschutzmittel)
amazonite Amazonenstein *m* (Min),
   Amazonit *m* (Min), Mikroklin *m* (Min),
   Smaragdspat *m* (Min)
amazon stone Amazonenstein *m* (Min),
   Amazonit *m* (Min), Smaragdspat *m* (Min)
ambazone Ambazon *n*
amber Bernstein *m*, Amber *m*, gelbes Erdharz,
   aluminous ~ Bernsteinalaun *m*, **common
   yellow** ~ Waschamber *m*, **containing** ~
   bernsteinhaltig
amber[-colored] bernsteinfarben
amber-containing earth Bernsteinerde *f*
amber glass Braunglas *n*
amber glass electrode Amberglaselektrode *f*
ambergris Amber *m* (Ambra)
ambergris fat Amberfett *n*
ambergris oil Ambraöl *n*
ambergris salt Ambrasalz *n*
amberite Amberit *n*
amberlike bernsteinartig
amberlite cation exchange resin
   Amberlitkationenaustauschharz *n* (HN)
amber mica Amberglimmer *m* (Min)
amber oil Amberöl *n*, Bernsteinöl *n*
amber [or musk] seed Ambrettekörner *pl* (Bot)
amber resin Bernsteinharz *n*
amber seed Abelmoschuskorn *n*
amber varnish Bernsteinfirnis *m*,
   Bernsteinlack *m*
ambient umgebend
ambient temperature Raumtemperatur *f*,
   umgebende Temperatur
ambiguity Doppeldeutigkeit *f*, Zweideutigkeit *f*
ambiguous zweideutig
amblygonite Amblygonit *m* (Min)
amblystegite Amblystegit *m* (Min)
amboside Ambosid *n*
ambrain Bernsteinfett, Amberfett *n*
ambrane Ambran *n*
ambreic acid Amberfettsäure *f*
ambrein Ambrein *n*, Amberfett *n*, Amberharz *n*,
   Amberstoff *m*
ambreinolide Ambreinolid *n*
ambrette musk Ambrettemoschus *m*
ambrette oil Ambretteöl *n*
ambrettolic acid Ambrettolsäure *f*
ambrettolide Ambrettolid *n*
ambrinol Ambrinol *n*

ambrite Ambrit *m* (Min)
ambroin Ambroin *n*
ambrosia Ambrosia *f*
ambrosine Ambrosin *n*
ambucaine Ambucain *n*
ambulance station Unfallstation *f*
ambulant ambulant, beweglich
ameba Amöbe *f*
amebicide Amöbizid *n*
ameloblast Adamantoblast *m* (Schmelzzelle)
amendment Abänderungsvorschlag *m* (Jur),
　Berichtigung *f*
americium (Symb. Am) Americium *n*
amesite Amesit *m* (Min)
amethyst Amethyst *m* (Min), **capillary** ~
　Haaramethyst *m* (Min)
amethyst color Amethystfarbe *f*
amethyst-colored amethystfarben,
　amethystfarbig
amethyst-like amethystartig
amianthine asbestartig, amiantartig,
　amiantförmig
amianthus Amiant *m* (Min), Chrysotilasbest *m*
　(Min), Serpentinasbest *m* (Min),
　Strahl[en]stein *m* (Min), **flexible** ~
　Federamiant *m* (Min), **ligneous** ~ holziger
　Bergflachs *m*
amicardine Amicardin *n*
amicetamine Amicetamin *n*
amicetaminol Amicetaminol *n*
amicetin Amicetin *n*
amicetose Amicetose *f*
amichin Amichin *n*
amicron Amikron *n*
amidase Amidase *f* (Biochem)
amidate *v* amidieren
amidated amidiert
amidating Amidieren *n*
amidation Amidierung *f*
amide Amid *n*
amide acid Amidsäure *f*
amide bond Amidbindung *f*
amide linkage Amidbindung *f*
amide nitrogen Amidstickstoff *m*
amidine Amidin *n*
amidinomycin Amidinomycin *n*
amidocarbonic acid Carbamidsäure *f,*
　Amidokohlensäure *f,* Aminokohlensäure *f*
amidogen Amidogen *n*
amido group Amidgruppe *f*
amidol Amidol *n*
amido-mercuric chloride weißes unschmelzbares
　Präzipitat (Chem)
amidon Amidon *n*
amidon bath Amidonbad *n*
amidopyrine Amidopyrin *n,* Pyramidon *n*
amidosulfonic acid Amidoschwefelsäure *f,*
　Amidosulfo[n]säure *f,* Aminosulfonsäure *f*

amidosulfuric acid Amidoschwefelsäure *f,*
　Amidosulfo[n]säure *f,* Aminosulfonsäure *f*
amidulin Amidulin *n*
aminase Aminase *f* (Biochem)
aminate *v* aminieren
aminating Aminieren *n*
amination Aminierung *f*
amine Amin *n,* **primary** ~ primäres Amin,
　quarternary ~ quartäres Amin, **secondary** ~
　sekundäres Amin, **tertiary** ~ tertiäres Amin
amine base Aminbase *f*
amine black Aminschwarz *n*
amine-like aminartig
aminitrozole Aminitrozol *n,* Acinitrazol *n*
amino- Amino-
aminoacetaldehyde Glycinaldehyd *m*
aminoacetanilide Aminoacetanilid *n*
aminoacetic acid Aminoessigsäure *f,* Glycin *n,*
　Glykokoll *n*
aminoaceto-phenetidine Phenokoll *n*
aminoacetophenone Aminoacetophenon *n*
amino acid Aminosäure *f,* **ketogenic** ~ ketogene
　Aminosäure *f,* **physiological** ~ physiologische
　Aminosäure *f,* **proteinogenic** ~ proteinogene
　Aminosäure *f,* **radiolabelled** ~ radioaktiv
　markierte Aminosäure *f*
amino acid amide Aminosäureamid *n*
amino acid composition
　Aminosäurezusammensetzung *f*
amino acid decarboxylase
　Aminosäuredecarboxylase *f* (Biochem)
amino acid oxidase Aminosäureoxydase *f*
　(Biochem)
amino acid radical Aminosäurerest *m*
amino acid replacement Aminosäureaustausch *m*
　(Mol. Biol)
amino acid residue Aminosäurerest *m*
amino acid sequence Aminosäuresequenz *f*
amino acid uptake Aminosäureaufnahme *f*
aminoacridine Aminoacridin *n*
amino alcohol Alkamin *n,* Aminoalkohol *m*
aminoalkylation Aminoalkylierung *f*
aminoanthracene Anthramin *n*
aminoazobenzene Aminoazobenzol *n*
aminoazobenzene base Aminoazobenzolbase *f*
aminoazobenzene disulfonic acid
　Aminoazobenzoldisulfonsäure *f*
aminoazobenzene hydrochloride
　Aminoazobenzolchlorhydrat *n*
aminoazotoluene Aminoazotoluol *n*
aminobarbituric acid Aminobarbitursäure *f,*
　Murexan *n,* Uramil *n*
amino-base primäres Amin
aminobenzene Aminobenzol *n,* Anilin *n*
aminobenzoic acid Aminobenzoesäure *f*
aminobutyric acid Aminobuttersäure *f*
aminobutyric acid chelate
　Aminobuttersäurechelat *n*
aminocaprolactam Aminocaprolactam *n*

aminocolchicide Aminocolchicid *n*
amino compound Aminoverbindung *f*
aminocresol Kresidin *n*
aminocymene Cymidin *n*
aminodurene Duridin *n*
aminoethanesulfonic acid Taurin *n*
aminoethanoic acid Glykokoll *n*
aminoethanol Aminoäthanol *n*
aminoethylbenzoic acid
   Aminoäthylbenzoesäure *f*
aminoethyl nitrate Aminoäthylnitrat *n*
aminoform Aminoform *n*
aminoformic acid Carbamidsäure *f*
aminoformyl Aminoformyl *n*
aminoglutaric acid Aminoglutarsäure *f*
amino group Aminogruppe *f*
aminohydroxytoluene Cresidin *n*
aminoisovaleric acid Valin *n*
amino ketone Aminoketon *n*
aminomethanamidine Guanidin *n*
aminomethane Methylamin *n*
aminomethylphosphonic acid
   Aminomethylphosphonsäure *f*
aminometradine Aminometradin *n*
aminonaphthalene Aminonaphthalin *n*
aminonaphthol Aminonaphthol *n*
aminonaphtholdisulfonic acid
   Aminonaphtholdisulfonsäure *f*
aminonaphtholsulfonic acid
   Aminonaphtholsulfonsäure *f*
aminonaphthoquinone Aminonaphthochinon *n*
aminooxine Aminooxin *n*
aminopeptidase Aminopeptidase *f* (Biochem)
aminophenetole Phenetidin *n*
aminophenol Aminophenol *n*
aminophenylglyoxylic acid Isatinsäure *f*
aminophylline Aminophyllin *n*
aminoplast Aminoplast *n*
aminoplastic aminoplastisch
aminoplastic resin Aminoplastharz *n*
aminoplastics Aminoplaste *pl*,
   Aminoplastkunststoffe *pl*
aminoplast molding compound
   Harnstoffpreßmasse *f*
aminoplast resin Harnstoffharz *n*
aminopolycarboxylic acid
   Aminopolycarbonsäure *f*
aminopromazine Aminopromazin *n*
aminopropane Propylamin *n*
aminopropanol Aminopropanol *n*
aminopropionic acid Alanin *n*,
   Aminopropionsäure *f*
aminopterin Aminopterin *n*
aminopurine Aminopurin *n*, **6-** ~ Adenin *n*
aminopyrazoline Aminopyrazolin *n*
aminopyridine Aminopyridin *n*
aminopyrine Pyramidon *n*
aminoquin Plasmochin *n*
aminosalicylic acid Aminosalicylsäure *f*

amino-succinamic acid Asparagin *n*
aminosuccinic acid Asparaginsäure *f*,
   Aminobernsteinsäure *f*
amino sugar Aminozucker *m*
amino terminal N-terminale Aminosäure
aminothiazole Aminothiazol *n*
aminothiophene Thiophenin *n*
aminotoluene Aminotoluol *n*
aminourazole Urazin *n*
aminourea Semicarbazid *n*
α-aminovaleric acid Norvalin *n*
amiphenazole Amiphenazol *n*
amitosis Amitose *f* (Biol), direkte Zellteilung
amizol Amizol *n*
ammelide Ammelid *n*, Melanurensäure *f*
ammeline Ammelin *n*
ammeter Ampèremesser *m*, Amperemeter *n*,
   Strommesser *m*
ammidin Ammidin *n*
ammine Ammin *n*, Ammoniakat *n*,
   ammoniakhaltiges Komplexsalz *n*
ammiol Ammiol *n*
ammiolite Ammiolith *m* (Min)
ammoidin Ammoidin *n*
ammoline Ammolin *n*
ammonia Ammoniak *n*, **addition of** ~
   Ammoniakzusatz *m*, **aqueous** ~
   Salmiakgeist *m*, **bicarbonate of** ~
   doppelkohlensaures Ammoniak, **combining
   with** ~ ammoniakbindend, **decomposition of**
   ~ Ammoniakzerfall *m*, **determination of** ~
   Ammoniakbestimmung *f*, **fixed** ~ gebundenes
   Ammoniak, **formation of** ~
   Ammoniakentwicklung *f*, **poor in** ~
   ammoniakarm, **rich in** ~ ammoniakreich
ammonia alum Ammoniakalaun *m*,
   Ammonalaun *m*
ammoniacal ammoniakalisch, ammoniakartig,
   ammoniakhaltig
ammoniacal copper oxide solution
   Kupferoxidammoniaklösung *f*, Blauwasser *n*
ammoniacal iron alum Ferriammonsulfat *n*,
   Eisenammonalaun *m*, Eisen(III)-ammonsulfat
ammoniacal liquor Ammoniakwasser *n*
ammoniacal nickelic oxide solution
   Nickeloxidammoniaklösung *f*
ammonia carboy Ammoniakflasche *f*
ammonia compound Ammoniakverbindung *f*
ammonia compression Ammoniakverdichtung *f*
ammonia compression system
   Ammoniakverdichtungsanlage *f*
ammonia compressor Ammoniakverdichter *m*
ammonia condenser Ammoniakverflüssiger *m*
ammonia distillation Ammoniakdestillation *f*
ammonia distillation plant
   Ammoniakdestillationsanlage *f*
ammonia distributor Ammoniakverteiler *m*
ammonia excretion Ammoniakausscheidung *f*

ammonia expansion pipe
  Ammoniakverdampfungsrohr *n*
ammonia leaching Ammoniaklaugung *f*
ammonia line Ammoniakleitung *f*
ammonia manuring Ammoniakdüngung *f*
ammonia meter Ammoniakmesser *m*
ammonia nitrogen Ammoniakstickstoff *m*
ammonia pipe Ammoniakrohr *n*
ammonia plant Ammoniakanlage *f*
ammonia process Ammoniakverfahren *n*
ammonia refrigerating coil
  Ammoniakkühlschlange *f*
ammonia refrigerating machine
  Ammoniakkältemaschine *f*
ammonia regulating valve
  Ammoniakregelventil *n*
ammonia residue Ammoniakrest *m*
ammonia scrubber Ammoniakwascher *m*
ammonia scrubbing Ammoniakwäsche *f*
ammonia separation Ammoniakscheidung *f*
ammonia separator Ammoniakabscheider *m*
ammonia soap Ammoniakseife *f*
ammonia soda Ammoniaksoda *f*, Solvaysoda *f*
ammonia soda process Ammoniaksodaprozeß *m*
ammonia superheater Ammoniaküberhitzer *m*
ammonia synthesis Ammoniaksynthese *f*
ammonia tank Ammoniakbehälter *m*
ammoniate Ammoniakat *n*
ammoniated iron Eisensalmiak *m*
ammoniated mercury ointment
  Quecksilberpräzipitatsalbe *f*
ammonia tester Ammoniakprüfer *m*
ammonia tube Ammoniakrohr *n*
ammonia vapors Ammoniakdämpfe *pl*
ammonia vat ammoniakalische Küpe
ammonia washer Ammoniakwascher *m*
ammonia water Salmiakgeist *m*, Ätzammoniak *n*,
  Ammoniakwasser *n*
ammonia water meter Ammoniakwassermesser *m*
ammonine Ammonin *n*
ammoniocupric chloride
  Kupferchloridammoniak *n*
ammonioferric sulfate Eisen(III)-ammonsulfat,
  Ferriammonsulfat *n*
ammoniometer Ammoniometer *n*
ammonite Ammonit *m* (Sprengstoff)
ammonium Ammonium *n*
ammonium acetate Ammoniumacetat *n*
ammonium alum Ammoniakalaun *m*,
  Ammoniumalaun *m*
ammonium aluminum sulfate
  Ammoniumalaun *m*
ammonium anacardate Anacardschwarz *n*
ammonium arsenate Ammoniumarsenat *n*
ammonium base Ammoniumbase *f*
ammonium benzoate Ammoniumbenzoat *n*
ammonium bicarbonate Ammoniumbicarbonat *n*
ammonium bichromate Ammoniumbichromat *n*
ammonium bifluoride Ammoniumbifluorid *n*

ammonium bi|n|oxalate Ammoniumbioxalat *n*
ammonium biphosphate
  Ammoniumbiphosphat *n*
ammonium bismuth citrate
  Ammoniumwismutcitrat *n*
ammonium bisulfite Ammoniumbisulfit *n*
ammonium borate Ammoniumborat *n*
ammonium bromide Ammoniumbromid *n*,
  Bromammonium *n*
ammonium bromocamphor sulfonate
  Ammoniumbromcamphersulfonat *n*
ammonium camphorate Ammoniumcamphorat *n*
ammonium carbamate Ammoniumcarbaminat *n*
ammonium carbonate Ammoniumcarbonat *n*,
  Ammoncarbonat *n*, **commercial ~**
  Hirschhornsalz *n*, **reaction with ~**
  Ammoncarbonisierung *f*
ammonium caseinate Caseinammoniak *n*,
  Eukasin *n*
ammonium chlorate Ammoniumchlorat *n*
ammonium chloride Ammoniumchlorid *n*,
  Chlorammonium *n*, Salmiak *m*
ammonium chloride solution
  Chlorammoniumlösung *f*, Salmiaklösung *f*
ammonium chloroplatinate
  Ammoniumchlorplatinat *n*,
  Ammoniumplatinchlorid *n*,
  Platinammoniumchlorid *n*
ammonium chlorostannate
  Ammoniumchlorostannat *n*,
  Ammoniumzinnchlorid *n*,
  Zinnammoniumchlorid *n*
ammonium chromate Ammoniumchromat *n*
ammonium citrate Ammoniumcitrat *n*
ammonium compound Ammoniakverbindung *f*
ammonium cyanide Cyanammonium *n*
ammonium cyanoferrate (III)
  Ammoniumferricyanid *n*
ammonium dichromate Ammoniumbichromat *n*,
  Ammoniumdichromat *n*
ammonium dihydrogen phosphate
  Ammoniumdihydrogenphosphat *n*
ammonium disulfide Ammoniumdisulfid *n*
ammonium ferric alum Eisen(III)-ammonsulfat *n*
ammonium ferric citrate Eisenammoncitrat *n*
ammonium ferric sulfate
  Ammoniumferrisulfat *n*, Eisenammonalaun *m*
ammonium ferricyanide Ammoniumferricyanid *n*
ammonium ferrosulfate Ammoniumferrosulfat *n*
ammonium ferrous sulfate Ammonferrosulfat *n*,
  Ferroammonsulfat *n*
ammonium fluoantimonate
  Antimonammoniumfluorid *n*
ammonium fluoride Ammoniumfluorid *n*,
  Fluorammonium *n*
ammonium fluosilicate
  Ammoniumsilicofluorid *n*
ammonium formate Ammoniumformiat *n*

ammonium glycerophosphate
Ammoniumglycerophosphat *n*
ammonium hexachlorostannate Pinksalz *n*
ammonium hippurate Ammoniumhippurat *n*
ammonium hydrochloride
Ammoniumhydrochlorid *n*
ammonium hydrogen fluoride
Ammoniumbifluorid *n*
ammonium hydrogen sulfite Ammoniumbisulfit *n*
ammonium hydrosulfide
Ammoniumhydrosulfid *n*,
Ammonsulfhydrat *n*
ammonium hydroxide Ammoniumhydroxid *n*,
Ätzammoniak *n*, Ammoniakwasser *n*,
Salmiakgeist *m*
ammonium hypophosphite
Ammoniumhypophosphit *n*
ammonium hyposulfite Ammoniumhyposulfit *n*
ammonium ichthyolsulfonate
Ammoniumsulfoichthyolat *n*, Bitumol *n*,
Ichthyol *n*
ammonium iodide Ammoniumjodid *n*,
Jodammonium *n*
ammonium iridium chloride
Ammoniumiridiumchlorid *n*
ammonium iron alum Ammoniumeisenalaun *m*
ammonium iron(III) citrate
Eisen(III)-ammoncitrat, Ferriammoncitrat *n*
ammonium iron(II) sulfate Ferroammonsulfat *n*
ammonium magnesium arsenate
Ammoniummagnesiumarsenat *n*
ammonium magnesium phosphate
Ammoniummagnesiumphosphat *n*
ammonium mercuric chloride
Ammonquecksilberchlorid *n*
ammonium meta-borate Ammoniummetaborat *n*
ammonium molybdate Ammoniummolybdat *n*
ammonium monosulfide
Ammoniummonosulfid *n*,
Einfachschwefelammonium *n*
ammonium monovanadate
Ammoniummonovanadat *n*
ammonium nickel sulfate
Nickelammoniumsulfat *n*
ammonium nitrate Ammoniumnitrat *n*,
Ammoniaksalpeter *m*, Ammonsalpeter *m*,
Gefriersalz *n*, Knallsalpeter *m*
ammonium nitrate explosive Ammonit *n*
(Sprengstoff)
ammonium nitrite Ammoniumnitrit *n*
ammonium palladium chloride
Ammoniumpalladiumchlorid *n*
ammonium permanganate
Ammoniumpermanganat *n*
ammonium persulfate
Ammoniumperoxydisulfat *n*,
Ammoniumpersulfat *n*
ammonium phosphate Ammoniumphosphat *n*,
Ammoniakphosphat *n*

ammonium phosphomolybdate
Ammoniumphosphormolybdat *n*
ammonium phosphotungstate
Ammoniumphosphorwolframat *n*
ammonium platinic chloride
Ammoniumplatinchlorid *n*
ammonium polysulfide Ammoniumpolysulfid *n*
ammonium potassium tartrate
Kaliumammoniumtartrat n,
Ammoniakweinstein *m*
ammonium purpurate Ammonpurpurat *n*,
Murexid *n*, Purpurcarmin *n*
ammonium radical Ammoniumradikal *n*,
Ammoniumrest *m*
ammonium residue Ammoniumrest *m*
ammonium rhodanate Ammoniumrhodanid *n*
ammonium salicylate Ammoniumsalicylat *n*
ammonium salt Ammoniumsalz *n*,
Ammoniaksalz *n*
ammonium silicofluoride
Ammoniumsilicofluorid *n*
ammonium sodium phosphate
Ammoniumnatriumphosphat *n*
ammonium stannic chloride
Ammoniumzinnchlorid *n*, Pinksalz *n*
ammonium succinate Ammonsuccinat *n*
ammonium sulfate Ammoniumsulfat *n*,
Ammonsulfat *n*, **mineral** ~ Mascagnin *m*
(Min)
ammonium sulfhydrate
Ammoniumhydrosulfid *n*,
Ammonsulfhydrat *n*
ammonium sulfide Ammoniumsulfid *n*,
Ammonsulfid *n*,
Einfachschwefelammonium *n*, **colorless** ~
farbloses Schwefelammon, **yellow** ~ gelbes
Schwefelammon
ammonium sulfite Ammoniumsulfit *n*
ammonium sulfoichthyolate
Ammoniumsulfoichthyolat *n*
ammonium thallium alum
Ammoniumthalliumalaun *m*
ammonium thallium sulfate
Ammoniumthalliumalaun *m*
ammonium thioarsenate
Ammoniumthioarsenat *n*
ammonium thioarsenite Ammoniumthioarsenit *n*
ammonium thiocyanate Ammoniumrhodanid *n*,
thiocyansaures Ammonium,
Rhodanammonium *n*
ammonium thiosulfate Ammoniumthiosulfat *n*
ammonium tungstate Ammoniumwolframat *n*
ammonium uranate Ammoniumuranat *n*
ammonium urate Ammonurat *n*
ammonium vanadate Ammoniumvanadat *n*
ammonium wolframate Ammoniumwolframat *n*
ammonolysis Ammonolyse *f*
ammonotelic ammonotelisch (Biol)
ammoresinol Ammoresinol *n*

ammunition Munition *f*
amnesia Amnesie *f* (Med)
amniotic acid Amnionsäure *f*
amniotic fluid Amnionflüssigkeit *f,*
Amnionwasser *n*
amobarbital Amobarbital *n*
amodiaquine Amodiachin *n*
amoeba (pl. amoebae) Amöbe *f*
amolanone Amolanon *n*
amopyroquine Amopyrochin *n*
amorphism Amorphismus *m,* Gestaltlosigkeit *f*
amorphous amorph, formlos, gestaltlos,
strukturlos
amorphousness Formlosigkeit *f*
amortization Amortisation *f,* Tilgung *f*
amosamine Amosamin *n*
amosaminol Amosaminol *n*
amount Menge *f,* Ausmaß *n,* Betrag *m,*
Quantität *f,* **lethal** ~ tödliche Dosis
amount formed Anfall *m*
amount of oversize Übermaß *n*
amount of ozone [in the air] Ozongehalt *m* [der
Luft]
amounts, by small ~ portionsweise
amount [to] *v* ergeben
amoxecaine Amoxecain *n*
AMP (Abk) Adenosinmonophosphat *n,*
Muskeladenylsäure *f*
ampeline Ampelin *n*
ampelite Ampelit *m* (Min)
ampelopsidine Ampelopsidin *n*
ampelopsine Ampelopsin *n*
ampeloptin Ampeloptin *n*
amperage Stromstärke *f,* Ampèrezahl *f,*
Stromintensität *f*
ampere Ampère *n* (Elektr)
ampere hour Ah, Ampèrestunde *f,*
Ampèrestunde *f*
ampere-hour capacity Ampèrestundenleistung *f*
amperemeter Ampèremesser *m,* Ampèremeter *w,*
Stromanzeiger *m*
ampere turn Ampèrewindung *f,*
Ampèrewindungszahl *f*
ampere winding Ampèrewindung *f*
ampere winding coefficient
Ampèrewindungsfaktor *m*
amperometer Ampèremesser *m*
amperometric amperometrisch
amperometry Ampèrometrie *f*
amphetamine Amphetamin *n*
amphetamine phosphate
Amphetaminphosphat *n,* Amphetaminum
phosphoricum *n*
amphetamine sulfate Amphetaminsulfat,
Amphetaminum sulfuricum *n*
amphibia Amphibien *pl*
amphibious animals Amphibien *f pl*
amphibole Amphibol *m* (Min)

amphibolic amphibolartig
amphibolic series Amphibolreihe *f*
amphiboliferous amphibolhaltig
amphibolite Amphibienstein *m* (Min),
Amphibol *m* (Min), Amphibolit *m* (Min),
Hornblendefels *m* (Min), **micaceous** ~
Glimmeramphibolit *m* (Min)
amphichroic amphichroitisch
amphichromatic amphichromatisch
amphide salt Amphidsalz *n*
amphidromic point Amphidromiepunkt *m*
amphifluoroquinone Amphifluorochinon *n*
amphigenite Amphigenit *m*
amphihexahedral doppelwürfelig
amphilogite Amphilogit *m*
amphioxus Lanzettfisch, Amphioxus *m* (Zool)
amphipathic amphipatisch
amphiphilic amphiphil
amphiprotic amphiprotisch
ampholyte Ampholyt *m*
amphoteric amphoter
amphoteric compound amphotere Verbindung *f*
amphotericin Amphotericin *n*
amphoteric ion Zwitterion *f*
amphotropine Amphotropin *n*
ample reichlich, umfangreich, umfassend
amplifiable verstärkbar (Elektr)
amplification Verstärkung *f* (Elektr), **degree of**
~ Verstärkungsgrad *m,* **linear** ~ lineare
Verstärkung, ~ **of the photocurrent**
Verstärkung *f* des Photostroms
amplification factor Verstärkungsfaktor *m*
amplification of gauge Spurerweiterung *f*
amplification range Verstellbereich *m* (Elektr)
amplification stage Verstärkerstufe *f*
amplified verstärkt
amplifier Meßverstärker *m,* Verstärker *m*
(Elektr), **feedback** ~ rückgekoppelter
Verstärker
amplifier action Verstärkerwirkung *f*
amplifier cascade Verstärkerkette *f* (Elektr)
amplifier circuit Verstärkerkreis *m* (Elektr)
amplifier effect Verstärkerwirkung *f*
amplifier gain Verstärkungsfaktor *m*
amplifier input Verstärkereingang *m* (Elektr)
amplify *v* verstärken (Elektr)
amplifying element Verstärkerelement *n* (Elektr)
amplifying valve Verstärkerröhre *f* (Elektr)
amplitude Amplitude *f,* Amplitudenweite *f,*
Ausschlag *m* (Pendel), Ausschlag[s]weite *f,*
Schwingungsweite *f,* **initial** ~
Anfangsausschlag *m,* **maximum** ~
Höchstausschlag *m,* Schwingungsbauch *m,*
variation of ~ Amplitudenänderung *f*
amplitude build-up Amplitudenaufschaukelung *f*
amplitude distortion Amplitudenverzerrung *f*
amplitude factor Scheitelfaktor *m*
amplitude fading Amplitudenschwund *m*
amplitude fluctuation Amplitudenschwankung *f*

amplitude limiter Amplitudenbegrenzer *m*
amplitude modulation Amplitudenmodulation *f*
amplitude regulation Amplitudenregelung *f*
amplitudes, correction of ~
   Amplitudenentzerrung *f*
amplitude variation Amplitudenschwankung *f*
ampoule Ampulle *f,* Ampullenflasche *f*
ampul[e] Ampulle *f*
ampullate ampullenartig
ampulliform ampullenförmig
ampul saw Ampullensäge *f*
amsonine Amsonin *n*
amurensin Amurensin *n*
amygdalate Amygdalat *n,* amygdalinsaures Salz
amygdalate *a* amygdalinsauer
amygdalic acid Mandelsäure *f,*
   Amygdalinsäure *f,* salt or ester of ~
   Amygdalat *n*
amygdalin Amygdalin *n,* Mandelstoff *m*
amygdaloid Mandelstein *m*
amygdaloid *a* mandelförmig, amygdaloid
amygdaloidal greenstone Diabasmandelstein *m*
   (Min)
amygdaloid storax Mandelstorax *m*
amygdalose Amygdalose *f*
amyl- Amyl-
amylaceous stärkeartig, stärkeführend,
   stärke[mehl]haltig
amylaceous substance Amyloid *n*
amyl acetate Amylacetat *n,*
   Essigsäureamylester *m*
amyl acetate lamp Amylacetatlampe *f*
amyl alcohol Amylalkohol *m,* Pentanol *n,*
   tertiary ~ tertiärer Amylalkohol
amyl aldehyde Amylaldehyd *m,* Pentanal *n*
amylamine Amylamin *n*
amylamine hydrochloride
   Amylaminchlorhydrat *n*
amylan Amylan *n*
amylaniline Amylanilin *n*
amylase Amylase *f* (Biochem), ~ in the blood
   serum Serumamylase *f*
amylate Amylat *n*
amyl benzoate Amylbenzoat *n*
amyl benzyl ether Amylbenzyläther *m*
amyl bromide Amylbromid *n*
amyl butyrate Amylbutyrat *n*
amyl chloride Amylchlorid *n,* Chloramyl *n*
amyl cinnamate Amylcinnamat *n*
amyl compound Amylverbindung *f*
amyl cyanide Capronitril *n*
amyldisulfocarbonic acid Xanthamylsäure *f*
amylene Amylen *n,* Valeren *n*
amylene chloral Amylenchloral *n,* Dormiol *n*
amylene hydrate Amylenhydrat *n*
amylene hydride Amylenhydrid *n*
amyl ester Amylester *m*
amyl ether Amyläther *m,* Amyloxid *n* (obs)
amyl ethyl ketone Amyläthylketon *n*

amylin Amylin *n,* Dextrin *n*
amyl iodide Amyljodid *n*
amyl mercaptan Pentanthiol *n*
amyl nitrite Amylnitrit *n*
amylobiose Amylobiose *f*
amyloclastic amyloklastisch, amylolytisch
amylodextrin Amylobiosedextrin *n*
amyloerythrine Amyloerythrin *n*
amylo-fermentation Stärkefermentation *f*
amyloform Amylobioseform *n*
amylogen Amylogen *n*
amyloid Amyloid *n*
amyloid *a* amyloid, stärkeartig
amyloid degeneration Amyloidentartung *f*
amyloid filament Amyloidfaden *m*
amyloin Amyloin *n,* Maltodextrin *n*
amylolysis Amylolyse *f,* Stärkespaltung *f*
amylolytic amylolytisch
amylolytic activity amylolytische Aktivität
amylometer Amylometer *n,* Stärkemesser *m*
amylopectin Amylopektin *n*
amyloplast Stärkebildner *m*
amylo process Amyloverfahren *n*
amylopsin Pankreasamylase *f* (Biochem)
amylose Amylose *f*
amylotriose Amylotriose *f*
amyl oxalate Amyloxalat *n*
amyl oxide Amyläther *m,* Amyloxid *n* (obs)
amylphenyl acetate Amylphenylacetat *n*
amyl rhodanate Amylrhodanid *n*
amyl salicylate Amylsalicylat *n,*
   Salicylamylester *m*
amyl silicone Amylsilicon *n*
amyl sulfide Amylsulfid *n*
amyl thiocyanate Amylrhodanid *n*
amylum Stärkemehl *n*
amyl valerate Amylvalerianat *n*
amyranone Amyranon *n*
amyrenol Amyrenol *n*
amyrin Amyrin *n*
amyrol Amyrol *n*
anabasine Anabasin *n*
anabatic anabatisch, konvektiv
anabiosis Anabiose *f*
anabiotic anabiotisch
anabolic anabolisch, aufbauend
anabolism Anabolismus *m* (Biol),
   Aufbaustoffwechsel *m* (Biol)
anacardate anacardinsauer, anacardsauer
anacardic acid Anacardinsäure *f,* Anacardsäure *f*
anacardium Anacardium *n*
anacardium oil Anacardienöl *n*
anacidity Anacidität *f*
anaclastic anaklastisch
anacyclin Anacyclin *n*
an[a]emia Blutmangel *m*
anaerobe Anaerobier *m,* Anaerobiont *m*
anaerobes Anaerobien *pl*
anaerobic anaerob, anaerobisch

anaerobiont Anaerobier *m*, Anaerobiont *m*
anaerobionts Anaerobien *pl*
anaerobiosis Anaerobiose *f*
anaerobism Anaerobiose *f*
anaerobium Anaerobier *m*, Anaerobiont *m*
anaerophyte Anaerophyt *m*
an[a]esthesia Narkose *f* (Med), **ether for** ~
Narkoseäther *m*
anaesthesin[e] Anaesthesin *n*
an[a]esthetize *v* narkotisieren (Med)
anaferine Anaferin *n*
anagyrine Anagyrin *n*
anagyrine hydrobromide
Anagyrinhydrobromid *n*
analcime Analzim *m* (Min), Analcim *m* (Min),
Analcit *m* (Min)
analcite Analcim *m* (Min), Analcit *m* (Min),
Schaumspat *m* (Min), Würfelzeolith *m* (Min)
analeptic Analeptikum *n* (Pharm), Weckmittel,
Anregungsmittel *n*
analeptic amine Weckamin *n*
analgen[e] Analgen *n*, Benzanalgen *n*
analgesia Schmerzbetäubung *f*,
Schmerzlosigkeit, Analgesie *f* (Med)
analgesic Analgeticum *n* (Pharm),
Schmerzbetäubungsmittel *n* (Pharm)
analgesic *a* schmerzstillend
analgesine Analgesin *n*
analgetic Analgetikum *n* (Pharm),
Schmerzmittel *n* (Pharm), schmerzstillend
analgin Creolin *n*, Kreolin *n*
analog Analogon *n*
analog computer Analog-Rechner *m* (Math)
analogon Analogon *n*
analogous ähnlich, vergleichbar, analog
analogue frequency changer
Frequenzanalogwandler *m*
analogy Analogie *f*
analyse *v* analysieren, bestimmen (Chem)
analysis (pl. analyses), Analyse *f*,
Gehaltsbestimmung *f*, Laborprüfung *f*,
Untersuchung *f*
biochemical ~ biochemische Analyse, **clinical**
~ klinische Analyse, **colorimetric** ~
kolorimetrische Analyse, **conductometric** ~
konduktometrische Analyse, **cost of** ~
Analysenkosten *pl*, **course of** ~
Analysengang *m*, **difference in** ~
Analysenunterschied *m*, **elementary** ~
Elementaranalyse *f*, **fluid** ~ Analyse auf
nassem Wege, **gravimetric** ~ gravimetrische
Analyse, **inorganic** ~ anorganische Analyse,
iodimetric ~ jodometrische Analyse, **method
of** ~ Analysenmethode *f*, **power of** ~
Auflösungsvermögen *n* (Opt), **process of** ~
Analysengang *m*, **quantitative** ~
Mengenbestimmung *f* (Chem), **ready for** ~
analysenfertig, **report of** ~
Analysenbericht *m*, **result of** ~
Analysenbefund *m*, **sample for** ~

Analysenprobe *f*, **scheme of** ~
Analysenschema *n*, **total** ~ Gesamtanalyse *f*,
ultimate ~ Elementaranalyse *f*, **weighed
portion for** ~ Analyseneinwaage *f*, ~ **by
boiling** Siedeanalyse *f*, ~ **by dry process**
Analyse *f* auf trockenem Wege, ~ **by
fractional distillation** Siedeanalyse *f*, ~ **by
refraction** refraktometrische Analyse, ~ **by
titration** Maßanalyse *f*, ~ **by wet process**
Analyse *f* auf nassem Wege
analysis formula Analysenformel *f*
analysis of path Bahnanalyse *f*
analysis of the atmosphere Luftanalyse *f*
analyst Analytiker *m*
analytical analytisch
analytical balance Analysenwaage *f*,
Laboratoriumswaage *f*
analytical error Analysenfehler *m*, **limit of** ~
Analysenfehlergrenze *f*
analytical filter Analysenfilter *n*
analytical finding Analysenbefund *m*
analytical funnel Analysentrichter *m*
analytical instruction Analysenvorschrift *f*
analytically pure analysenrein
analytical method Analysenmethode *f*
analytical procedure analytischer Trennungsgang
analytical quartz lamp Analysenquarzlampe *f*
analytical rapid-weighing balance
Analysenschnellwaage *f*
analytical result Analysenbefund *m*
analytical test Analysenprobe *f*
analytical weight Analysengewicht *n*
analytics Analytik *f*
analyzable analysierbar
analyzation Analysieren *n*
analyze *v* analysieren, auswerten, scheiden,
zerlegen
analyzer Analysator *m*, Meßzelle *f*, Prüfgerät *n*,
Untersuchungsapparat *m*
analyzing Analysieren *n*
anamesite Anamesit *m*
anamirtin Anamirtin *n*
anamorphism Anamorphismus *m*
anamorphosis Anamorphose *f*
anamorphous anamorph, verzerrt
anaphase Anaphase *f* (Biol)
anaphoresis Anaphorese *f*
anaphylaxis Anaphylaxie *f* (Med)
anaplerotic anaplerotisch
anastigmatic anastigmatisch (Opt)
anatabine Anatabin *n*
anatase Anatas *m* (Min), kristallines Titanoxid
(Min)
anatexis Anatexis *f* (Geol)
anatomical[ly] anatomisch
anatomy Anatomie *f*
anatoxin Anatoxin *n*
anatto Anatto *n* (Bot, Färb)
anauxite Anauxit *m* (Min)

anchoate anchoinsauer
anchor Anker *m*, Querriegel *m*
anchorage Verankerung *f*
anchor bolt Ankerbolzen *m*, Ankerrührer *m*,
　Fundamentanker *m*, Fundamentschraube *f*
anchoring cable Haltekabel *n*
anchor mixer Ankerrührer *m*
anchor screw Ankerrührer *m*
anchor steel Ankerstahl *m*
anchovy oil Anchovisöl *n*
anchusic acid Anchusasäure *f*
anchusin Anchusin *n*, Alkannarot *n*, Alkannin *n*
ancient uralt
ancillary processing Konfektionierung *f*
ancylite Ancylit *m* (Min)
andalusite Andalusit *m* (Min), Hartspat *m*
　(Min)
anda oil Andaöl *n*
Anderson bridge Anderson-Meßbrücke *f*
andesine Andesin *m* (Min),
　Natronkalkfeldspat *m* (Min)
andesite Andesit *m* (Min)
andirin Andirin *n*
andisine Andisin *n*
andorite Andorit *m* (Min)
andradite Andradit *m* (Min), Kalkgranat *m*
　(Min)
andreolite Andreolith *m* (Min)
andrewsite Andrewsit *m* (Min)
androgen Androgen *n*
andrographolide Andrographolid *n*
androkin Androsteron *n*
andromedotoxin Andromedotoxin *n*
androsine Androsin *n*
androstadiene Androstadien *n*
androstane Androstan *n*
androstanolone Androstanolon *n*, Androsteron *n*
androstenediol Androstendiol *n*
androsterone Androsteron *n*, Androstanolon *n*
androtin Androsteron *n*
anechoic room schalltoter Raum
anelastic anelastisch
anelasticity Anelastizität *f*
anelectric anelektrisch
anellate *v* anellieren
anellation Anellierung *f*
anemia Blutarmut *f*
anemic blutarm
anemometer Windgeschwindigkeitsmesser *m*,
　Windmesser *m*, Windstärkemesser *m*
anemometry Windmessung *f*
anemone camphor Anemoncampher *m*
anemonic acid Anemonsäure *f*
anemonin Anemonin *n*
anemoninic acid Anemoninsäure *f*
anemonolic acid Anemonolsäure *f*
anemosite Anemosit *m* (Min)

aneroid barometer Aneroidbarometer *n*,
　Dosenaneroid *n*, Dosenbarometer *n*,
　Federbarometer *n*
anesine Anesin *n*
aneson Chloreton *n*
anesthesia Anästhesie *f*
anesthesin Anaesthesin, Äthylaminobenzoat *n*
anesthetic Anästhetikum *n*, Betäubungsmittel *n*
anesthetize *v* anästhesieren, betäuben,
　ätherisieren
anesthol Anästhol *n*
anethine Anethin *n*
anethol Anethol *n*, Allylphenylmethyläther *m*,
　Aniscampher *m*
aneurin[e] Aneurin *n*, Thiamin *n*, Vitamin B$_1$ *n*
aneurine diphosphate Aneurindiphosphat *n*
aneurine disulfide Aneurindisulfid *n*
aneurine hydrochloride Aneurinhydrochlorid *n*
aneurine mononitrate Aneurinmononitrat *n*
aneurinethiol Aneurinthiol *n*
angelactic acid Angelactinsäure *f*
angelardite Angelardit *m* (Min)
angelica Angelika *f* (Bot), **extract of** ~
　Angelikaextrakt *m*
angelic acid Angelikasäure *f*
angelica lactone Angelikalacton *n*
angelic aldehyde Angelikaaldehyd *m*
angelica oil Engelwurzelöl *n*
angelica root Brustwurzel *f* (Bot), Engelwurzel *f*
　(Bot)
angelica spirit Angelikaspiritus *m*
angelicic acid Angelikasäure *f*
angelicin Angelicin *n*
angelicone Angelicon *n*
angelin Angelin *n*
angioma (pl. angiomata, angiomas) Angiom *n*
　(Med)
angiotensin Angiotensin *n*
angiotonin Angiotonin *n*
angle Winkel *m*, Kniestück *n*, Neigung *f*, **acute**
　~ spitzer Winkel, **complementary** ~
　Komplementärwinkel *m* (Math), **obtuse** ~
　stumpfer Winkel, **right** ~ rechter Winkel,
　supplementary ~ Supplementärwinkel *m*, ~
　with equal sides gleichschenkliges
　Winkeleisen *n*
angle bracket Winkelarm *m*
angle cock Eckhahn *m*, Winkelhahn *m*
angle cutter Winkelfräser *m*
angled eckig, gewinkelt
angle extrusion head Schrägspritzkopf *m*
angle iron Winkeleisen *n*, Eckeisen *n*
angle iron support Winkeleisengestell *n*
angle molding press Winkelpresse *f*
angle of contact Steuerwinkel *m*
angle of contingence Berührungswinkel *m*,
　Krümmungswinkel *m*
angle of curvature Krümmungswinkel *m*
angle of cut bias Schneidewinkel *m*

angle of direction Richtungswinkel *m*
angle of elevation Höhenwinkel *m*
angle of incline Neigungswinkel *m*
angle of reflection Reflexionswinkel *m*
angle of refraction Beugungswinkel *m*
angle ring Winkelring *m*
angles, having equal ~ gleichwinklig, having
  unequal ~ ungleichwinkelig
angle section V-Profil *n*
angle sheet iron Winkelblech *n*
anglesite Anglesit *m* (Min), Bleisulfat *n* (Min),
  Vitriolbleierz *n* (Min), Vitriolbleispat *m*
  (Min)
angle-type trolley Winkelkatze *f*
angle valve Eckventil *n*
angolensin Angolensin *n*
angophorol Angophorol *n*
angora wool Angorawolle *f*
angostura alkaloid Angosturaalkaloid *n*
angostura bark Angosturarinde *f*
angostura bitters Angosturabitter *m*
angosturine Angosturin *n*
Angström unit Angström-Einheit *f* (Maß)
angular eckig, gewinkelt, scharfkantig
angular acceleration Winkelbeschleunigung *f*
angular adjustment Winkeleinstellung *f*
angular deflection Winkelverdrehung *f*
angular deformation Winkeldeformation *f*
angular dependence Winkelabhängigkeit *f*
angular deviation Winkelabweichung *f*
angular displacement Winkelverschiebung *f*
angular distance Winkelabstand *m*
angular distribution Winkelverteilung *f*
angular drive Winkeltrieb *m*
angular frequency Kreisfrequenz *f*
angular height Höhenwinkel *m*
angularity Winkligkeit *f*
angular momentum Drall *m*, Drehimpuls *m*,
  Drehmoment *n*, total ~ Gesamtdrehimpuls *m*
angular momentum operator
  Drehimpulsoperator *m*
angular motion Winkelbewegung *f*
angular strain Winkelspannung *f*
angular sum Winkelsumme *f*
angular velocity Kreisfrequenz *f* (Elektr),
  Winkelgeschwindigkeit *f*
angular welding Winkelschweißen *n*
angustifoline Angustifolin *n*
angustione Angustion *n*
angustose Angustose *f*
anhalamine Anhalamin *n*
anhaline Anhalin *n*
anhalmine Anhalmin *n*
anhalonidine Anhalonidin *n*
anhalonine Anhalonin *n*
anharmonic nicht harmonisch
anhydremia Anhydrämie *f* (Med)
anhydride Anhydrid *n*
anhydride formation Anhydridbildung *f*

anhydrite Anhydrit *m* (Min), Muriazit *m* (Min),
  Würfelgips *m* (Min), Würfelspat *m* (Min)
anhydro- Anhydro-
anhydroacid Anhydrosäure *f*
anhydro alkannin Anhydroalkannin *n*
anhydro base Anhydrobase *f*
anhydrocarminic acid Anhydrocarminsäure *f*
anhydro cymarigenin Anhydrocymarigenin *n*
anhydrodigitic acid Anhydrodigitsäure *f*
anhydro digitoxigenin Anhydrodigitoxigenin *n*
anhydroformaldehyde aniline
  Anhydroformaldehydanilin *n*
anhydroformaldehyde paratoluidine
  Anhydroformaldehydparatoluidin *n*
anhydrogitalin Anhydrogitalin *n*
anhydrolycorine Anhydrolycorin *n*
anhydrous wasserfrei, absolut trocken,
  anhydrisch, entwässert, kristallwasserfrei
anhydrous boric acid Bortrioxid *n*
anhydrous crystal wasserfreier Kristall *m*
anhydrous salt wasserfreies Salz
anibine Anibin *n*
anil Anil *n*
anileridine Anileridin *n*
anilide Anilid *n*
aniline Anilin *n*, Aminobenzol *n*, Cyanol *n*,
  Phenylamin *n*
aniline black Anilinschwarz *n*, Nigranilin *n*,
  Nigrosin *n*
aniline blue Anilinblau *n*
aniline brown Anilinbraun *n*, Vesuvin *n*
  (Farbstoff)
aniline chlorate Anilinchlorat *n*
aniline color Anilinfarbe *f*
aniline dye Anilinfarbe *f*, Anilinfarbstoff *m*
aniline dye industry Anilinfarbenindustrie *f*
aniline dyes, residue of ~
  Anilinfarbenrückstand *m*
aniline hydrochloride Anilinchlorhydrat *n*,
  Anilinhydrochlorid *n*
aniline ink Anilintinte *f*
aniline nitrate Anilinnitrat *n*
aniline oil Anilinöl *n*
aniline oxalate Anilinoxalat *n*
aniline pink Anilinrosa *n*
aniline poisoning Anilinvergiftung *f*
aniline printing Anilindruck *m*
aniline printing ink Anilindruckfarbe *f*
aniline purple Mauvein *n*, Mauve *n*
aniline red Anilinrot *n*, Fuchsin *n*
aniline resin Anilinharz *n*
aniline salt Anilinsalz *n*
aniline sulfate Anilinsulfat *n*
anilinesulfonic acid Anilinsulfonsäure *f*
aniline violett Anilinviolett *n*
aniline works Anilinfabrik *f*
aniline yellow Anilingelb *n*, Echtgelb *n*
aniline zinc chloride Anilinchlorzink *n*
anilinism Anilinvergiftung *f*, Anilismus *n*

anilino- Anilino-
anilipyrine Anilipyrin *n*
animal Tier *n*
animal *a* animalisch, tierisch
animal ashes Tierkörperasche *f*
animal charcoal Tierkohle *f*, Beinschwarz *n*,
  Fleischkohle *f*, Knochenkohle *f*
animal charcoal dust Schaumschwärze *f*
animal chemistry Tierchemie *f*, Zoochemie *f*
animal experiment Tierversuch *m*
animal fat Adeps *m*, Tierfett *n*
animal feed Futtermittel *n*
animal fiber Tierfaser *f*
animal grease Adeps *m*
animalizing Animalisieren *n* (Wolle)
animal oil tierisches Öl *n*
animate *v* beleben, begeistern
animé Anime, Elemiharz *n*
animé gum Gummianime *n*
animé resin Animeharz *n*
animi gum (see also animé gum) Anime *n*,
  Animegummi *m*
animikite Animikit *m* (Min)
animi resin Anime *n*, Animeharz *n*,
  Gummianime *n*
aninsulin Aninsulin *n*
anion Anion *n*
anion-active anionaktiv
anion complex Anionkomplex *m*
anion conductivity Anionenleitfähigkeit *f*
anion exchanger Anionenaustauscher *m*
anionic anionisch, anionaktiv
anionic acid Anionsäure *f*
anionic dye Anionfarbstoff *m*
anionic ligand Acidoligand *m*
anionotropy Anionotropie *f*
anion vacancy Anionenfehlstelle *f*
anisal Anisal *n*, Anisyliden *n*
anisalcohol Anisalkohol *m*
anisaldehyde Anisaldehyd *m*
anisalhydantoin Anisalhydantoin *n*
anisate anisartig, anissauer
anise Anis *m* (Bot), **spirit of** ~ Anisgeist *m*
anise camphor Aniscampher *m*,
  Allylphenylmethyläther *m*, Anethol *n*
aniseed Anis *m* (Bot), Anissamen *m* (Bot)
aniseed oil Anisöl *n*
aniseed wood Anisholz *n*
anise oil Anisöl *n*
anisette Anisbranntwein *m*, Anisett *m*,
  Anisgeist *m*, Anislikör *m*, Aniswasser *n*
anishydramide Anishydramid *n*
anisidide Anisidid *n*
anisidin[e] Anisidin *n*
anisidino- Anisidino-
anisil Anisil *n*
anisilic acid Anisilsäure *f*
anisoelastic anisoelastisch
anisogamete Heterogamet *m*
anisoin Anisoin *n*

anisol[e] Anisol *n*
anisomeric anisomer (Phys)
anisometric anisometrisch
anisomycin Anisomycin *n*
anisotrope anisotrop
anisotropic anisotrop
anisotropy Anisotropie *f*
anisoyl- Anisoyl-
anisyl- Anisyl-
anisyl alcohol Anisalkohol *m*
anisylcyclone Anisylcyclon *n*
anisylidene Anisal *n*, Anisyliden *n*
ankerite Ankerit *m* (Min)
annabergite Annabergit *m* (Min), grüner
  Erdkobalt (Min), Nickelblüte *f* (Min)
annaline Annalin *n* (Pap)
annals Annalen *pl*
annatto Annatto *n*, Buttergelb *n*, Orlean *n*,
  dyeing with ~ Orleanfärberei *f*
annatto orange Orleanorange *n* (Färb)
anneal *v* ausglühen, abkühlen (Glas),
  adoucieren, anlassen (Met), ausbrennen,
  ausheizen, einbrennen, glühen (Met), tempern
  (Met), vergüten (Met), ~ **stress-free**
  spannungsfrei glühen
annealed angelassen, ausgeglüht, **bright** ~
  blankgeglüht
annealed wire Glühdraht *m*
annealer Ausglüher *m*
annealing Abglühen *n* (Met), Anlassen *n* (Met),
  Ausglühen *n*, Einbrennen *n*,
  Glühbehandlung *f*, Glühen *n* (Met),
  Tempern *n* (Met), Temperung *f* (Met),
  Vorglühen *n*, **black** ~ erste Glühung,
  intermediate ~ Zwischenglühung *f*, **white** ~
  zweite Glühung
annealing box Einsatzkasten *m*, Glühgefäß *n*,
  Glühkasten *m*
annealing color Anlaßfarbe *f*, Anlauffarbe *f*
  (Met), Einbrennfarbe *f*
annealing condition Temperbedingung *f* (Metall)
annealing effect Anlaßwirkung *f*
annealing furnace Glühofen *m*, Adoucierofen *m*,
  Ausglühflammofen *m*, Auswärmeofen *m*,
  Kühlofen *m* (Glas), Temperofen *m*,
  Vorglühofen *m*, **hood type** ~
  Haubenglühofen *m*, **non-oxidizing** ~
  Blankglühofen *m*
annealing hearth Glühherd *m*
annealing installation Glühanlage *f*
annealing lacquer Anlauflack *m*, Einbrennlack *m*
annealing method Glühverfahren *n*
annealing oven Anlaßofen *m*, Vorglühofen *m*
annealing pot Adouciergefäß *n*, Ausglühtopf *m*
annealing practice Glühbetrieb *m*
annealing process Temperverfahren *n*,
  Glühfrischverfahren *n*, Glühverfahren *n*
annealing temperature Anlauftemperatur *f*,
  Glühtemperatur *f*

annealing vat Abkühlfaß *n*
anneal treatment, strain-relief ~ spannungsfreies Glühen
annelid Annelid *m* (Zool), Ringelwurm *m*
annero[e]dite Annerödit *m* (Min)
annex Anhang *m,* Beilage *f*
annex *v* anfügen, annektieren, beifügen
annidalin Annidalin *n*
annihilate *v* vernichten, vertilgen
annihilation Annihilation *f* (Atom), Vernichtung *f*
annihilation radiation Vernichtungsstrahlung *f,* Annihilationsstrahlung *f,* **two photon** ~ Zweiquantenvernichtungsstrahlung *f*
annite Annit *m*
anniversary Jahrestag *m*
annivite Annivit *m* (Min)
annofoline Annofolin *n*
annotinine Annotinin *n*
announce *v* ankündigen, anmelden, anzeigen, bekanntgeben
announcement Bekanntmachung *f*
annoy *v* belästigen
annual general meeting Jahreshauptversammlung *f*
annul *v* annullieren, vernichten
annular ringförmig
annular cavity Kreisrinne *f*
annular clearance Ringspalt *m*
annular electrode Ringelektrode *f*
annular flow Ringströmung *f*
annular furnace Ringofen *m*
annular groove Ringnut *f,* Schmelzrinne *f*
annular magnet Ringmagnet *m*
annular opening kreisförmige Öffnung *f*
annular producer Ringgenerator *m*
annular solenoid Ringsolenoid *n*
annular tensile strength Ringzugfestigkeit *f*
annulment Nichtigerklärung *f*
annulus width Ringspaltweite *f*
anobial Anobial *n*
anode Anode *f,* positive Elektrode *f,* positiver Pol *m,* Sauerstoffpol *m,* **auxiliary** ~ Hilfsanode *f,* **hollow** ~ Hohlanode *f,* **movable** ~ Handanode *f,* Wanderanode *f,* **moving** ~ Rühranode *f,* **outer** ~ Außenanode *f,* **rolled** ~ Walzanode *f,* **soluble** ~ Lösungsanode *f,* ~ **with insulating surface** sperrende Anode
anode bar Anodenstange *f*
anode battery Anodenbatterie *f*
anode-brighten *v* elektrolytisch polieren
anode cage Anodenkorb *m*
anode clamp Anodenklemme *f*
anode dark space Anodendunkelraum *m*
anode density Anodendichte *f*
anode discharge Anodenentladung *f*
anode effect Anodeneffekt *m*
anode frame Anodenrahmen *m*
anode holder Anodenhalter *m*

anode loading Anodenbelastung *f*
anode mud Anodenschlamm *m* (Kupfergewinnung)
anode plate Anodenplatte *f*
anode [plate] circuit Anodenkreis *m*
anode [plate] current Anodenstrom *m*
anode plug Anodenstecker *m*
anode potential Anodenspannung *f*
anode ray Anodenstrahl *m*
anode rectification Anodengleichrichtung *f*
anode rectifier Anodengleichrichter *m* (Elektr)
anode region Anodenraum *m*
anode screen Anodenschutznetz *n*
anode slime Anodenschlamm *m* (Kupfergewinnung)
anode sludge Anodenschlamm *m* (Kupfergewinnung)
anode sponge Anodenschwamm *m* (Bleigewinnung)
anode terminal Anodenklemme *f*
anodic anodisch
anodic coating, insulating ~ sperrender Anodenüberzug *m*
anodic etching elektrolytisches Anätzen *n*
anodic treatment Eloxieren *n,* Eloxierung *f*
anodize *v* anodisch behandeln; eloxieren
anodizing Eloxalverfahren *n,* Eloxierung *f*
anodizing plant Eloxieranlage *f*
anodyne Anodyn *n*
anodynin Anodynin *n*
anogen Anogen *n*
anoint *v* einsalben
anol Anol *n*
anolobine Anolobin *n*
anolyte Anodenflüssigkeit *f,* Anolyt *m*
anomalous abweichend, anomal, anormal
anomaly Anomalie *f,* Abnormität *f,* Abweichung *f,* Unregelmäßigkeit *f*
anomer Anomer *n*
anomeric anomer
anomeric center Anomeriezentrum *n*
anomite Anomit *m* (Min)
anonaine Anonain *n*
anonol Anonol *n*
anophorite Anophorit *m* (Min)
anoretic, anorexigenic Appetitzügler *m*
anorthite Anorthit *m* (Min), Christianit *m* (Min), Kalkfeldspat *m* (Min), **decomposed** ~ Rosellan *m* (Min), Rosit *m* (Min)
anorthoclase Anorthoklas *m* (Min)
anosmia Anosmie *f* (Med)
anoxybiosis Anaerobiose *f*
ansa compound Ansa-Verbindung *f* (Stereochem)
anserine Anserin *n*
answer Antwort *f,* Bescheid *m,* Facit *n* (Math)
ant Ameise *f*
antabuse Antabus *n*
antacid säurewidrig, Säure neutralisierend

antagonism Antagonismus *m*
antagonist Antagonist *m*
antagonistic antagonistisch, entgegenwirkend
Antarctic Antarktis *f,* antarktisch
antarthritic Arthritismittel *n* (Pharm)
antecede *v* vorausgehen
antenna Antenne *f,* **corkscrew** ~
Wendelantenne *f,* **half-wave** ~
Halbwellenantenne *f,* **slotted** ~
Schlitzantenne *f*
anterior lobe of the pituitary gland
Hypophysenvorderlappen *m*
anthanthrone Anthanthron *n*
anthelmintic Anthelmintikum *n* (Pharm,
Wurmmittel), Wurmmittel *n* (Pharm)
anthelmintic *a* wurmvertilgend
anthelmintic herb Wurmkraut *n* (Bot)
anthemene Anthemen *n*
anthemol Anthemol *n,* Anthemolcampher *m*
antheraxanthin Antheraxanthin *n*
anthesterine Anthesterin *n*
anthochroite Anthochroit *m* (Min)
anthocyan Anthocyan *n*
anthocyanidin Anthocyanidin *n*
anthocyanin Anthocyanin *n*
anthophyllite Anthophyllit *m* (Min)
anthophyllite amphibol Anthophyllitamphibol *m*
anthosiderite Anthosiderit *m* (Min)
anthoxanthin Anthoxanthin *n,* Blumengelb *n*
anthracen Anthracen *n*
anthracene blue Alizarinblau *n,* Anthracenblau *n*
anthracene brown Alizarinbraun *n*
anthracene carboxylic acid Anthroesäure *f*
anthracenediol, 1,5- ~ Rufol *n*
anthracene dye Anthracenfarbstoff *m*
anthracene oil Anthracenöl *n*
anthracene pitch Anthracenpech *n*
anthracene tetrone Anthradichinon *n*
anthrachrysone Anthrachryson *n*
anthracine Anthracen *n*
anthracite Anthrazit *m,* Anthrazitkohle *f,*
Fettkohle *f,* Glanzkohle *f,* Kohlenblende *f,*
Steinkohle *f*
anthracite blast furnace Anthrazithochofen *m*
anthracite diamond Carbonado *m* (Min)
anthracite pig iron Anthrazitroheisen *n*
anthracitic anthrazitartig, anthrazithaltig,
anthrazitisch
anthracitization Anthrazitbildung *f*
anthracitoid anthrazitartig
anthracometer Kohlensäuremesser *m*
anthraconite Anthrakonit *m* (Min),
Kohlenkalkspat *m* (Min), Kohlenspat *m*
(Min), Stinkkalk *m* (Min), Stinkstein *m*
(Min)
anthracosis Anthrakose *f* (Med),
Kohlenstaublunge *f*
anthracyl- Anthracyl-, Anthrazyl-
anthradiol, 1,3- ~ Flavol *n,* **1,5-** ~ Rufol *n*

anthradiquinone Anthradichinon *n*
anthraflavic acid Anthraflavinsäure *f*
anthraflavine Anthraflavin *n* (Färb)
anthraflavone Anthraflavon *n*
anthrafuchsone Anthrafuchson *n*
anthragallol Anthragallol *n*
anthrahydroquinone Anthrahydrochinon *n,*
Oxanthranol *n*
anthraldehyde Anthraldehyd *m*
anthralur Anthralur *n* (HN, Aktivkohle)
anthramine Anthramin *n*
anthranil Anthranil *n,* Anthroxan *n*
anthranil aldehyde Anthranilaldehyd *m*
anthranilic acid Anthranilsäure *f,*
ortho-Aminobenzoesäure *f*
anthranilic acid lactam Anthranil *n*
anthranol Anthranol *n*
anthranone Anthranon *n,* Anthron *n*
anthranyl- Anthranyl-
anthraphenone Anthraphenon *n*
anthrapurpurin Anthrapurpurin *n*
anthrapyridine Anthrapyridin *n*
anthraquinoketene Anthrachinoketen *n*
anthraquinoline Anthrachinolin *n*
anthraquinone Anthrachinon *n*
anthraquinone acridine Anthrachinonacridin *n*
anthraquinone derivative
Anthrachinonabkömmling *m*
anthraquinone disulfonic acid
Anthrachinondisulfonsäure *f*
anthraquinone dye Anthrachinonfarbstoff *m*
anthraquinone fluresceine
Anthrachinonfluorescein *n*
anthraquinone sulfonic acid
Anthrachinonsulfonsäure *f*
anthrarobin Anthrarobin *n,* Cignolin *n* (HN),
Desoxyalizarin *n*
anthrarufin Anthrarufin *n*
anthrathiazine Anthrathiazin *n*
anthrazine Anthrazin *n*
anthricin Anthricin *n*
anthrodianthrene Anthrodianthren *n*
anthroic acid Anthroesäure *f*
anthrol Anthrol *n*
anthrone Anthron *n,* Anthranon *n*
anthropobiology Anthropobiologie *f*
anthropogenesis Anthropogenese *f,*
Anthropogenie *f* Entwicklungsgeschichte des
Menschen
anthropogenic anthropogen
anthropogeny Anthropogenese,
Entwicklungsgeschichte *f* des Menschen
anthropologist Anthropologe *m*
anthropology Anthropologie *f*
anthroxan Anthroxan *n*
anthroxan aldehyde Anthroxanaldehyd *m*
anthroxanic acid Anthroxansäure *f*
anthryl- Anthranyl-, Anthryl-
anthrylamine Anthramin *n*

anti-abrasive coating Verschleißschutzbelag *m*
antiacid säurewidrig
antiadipogenics Antiadiposita *pl* (Pharm)
anti-ager Alterungsschutzmittel *n*
antiagglutinin Antiagglutinin *n*
antiaggressin Antiaggressin *n*
antialbumose Antialbumose *f*
antialdoxime Antialdoxim *n*
antiallergic agent Antiallergikum *n* (Pharm)
antiar Antiar *n*
antiarigenin Antiarigenin *n*
antiarin Antiarin *n*
antiarol Antiarol *n*
antiarose Antiarose *f*
antibacteria Bakterienschutzmittel *n*
antibacterial antibakteriell, bakterienfeindlich,
  bakterienwachstumshemmend
anti-base Antibase *f*
antibiosis Antibiose *f* (Med)
antibiotic Antibiotikum *n* (pl Antibiotika),
  antibiotisch
antibody Abwehrstoff *m*, Antikörper *m*,
  Immunkörper *m*
antibody formation Antikörperbildung *f*
antibody level of the blood Immunkörperspiegel
  m des Blutes
antibonding state Antibindungszustand *m*
anti-bumping spiral Siedeverzugsspirale *f*
anticatalyst Katalytgift *n*
anticatalytic antikatalytisch, die Katalyse
  verhindernd
anticatalyzer Antikatalysator *m*
anticathode Antikathode *f*, Gegenkathode *f*
anticathode luminescence
  Antikathodenleuchten *n*
anticenter Antizentrum *n*
antichlor Antichlor *n*
anticlockwise entgegen dem Uhrzeigersinn,
  linksdrehend, to run ~ gegenlaufen
anti-clockwise direction Uhrzeigergegensinn *m*
anticlockwise rotation Linksdrehung *f*
anticoagulant Antikoagulierungsmittel *n*
anticoagulant *a* gerinnungshemmend
anti-coagulation agent Koagulierschutzmittel *n*
anti-coagulin Antikoagulierungsmittel *n*
anticodon Antikodon *m*
anticoincidence analyzer
  Antikoinzidenzanalysator *m*
anticoincidence arrangement
  Antikoinzidenzanordnung *f*
anticoincidence stage Antikoinzidenzstufe *f*
anticonfiguration Antistellung *f*
anticorrosion additive Korrosionsschutzmittel *n*
anti-corrosion paint Rostschutzfarbe *f*
anticorrosive Antikorrosionsmittel *n*,
  Korrosionsschutz *m*
anticorrosive *a* korrosionsverhindernd,
  rostverhütend

anticorrosive [agent] Korrosionsschutzmittel *n*,
  Rostschutzmittel *n*
anticorrosive coat of paint Schutzanstrich *m*
anticorrosive paint Korrosionsschutzfarbe *f*,
  Eisenschutzfarbe *f*
anticorrosive varnish Eisenschutzlack *m*
anticracking agent Antikrackmittel *n*
anticyclone Antizyklon *m*
anti-deuteron Antideuteron *n*
antidiabeticum Antidiabetikum *n* (Pharm)
antidiazotate Antidiazotat *n*
antidisintegrating abbauverhindernd
antidistortion device Entzerrungsanordnung *f*
antidiuretic Antidiuretikum *n* (Pharm)
antidotal giftabtreibend
antidote Antidot *n* (Pharm), Gegengift *n*
  (Pharm), Gegenmittel *n* (Pharm), Giftmittel *n*
  (Pharm), chemical ~ chemisches Gegengift
antidrumming compound Antidröhnmasse *f*
antiemetic Antiemetikum *n* (Pharm)
antiepileptic Antiepileptikum *n* (Pharm)
antifading control Schwundausgleich *m* (Radio)
antifebrin Antifebrin *n*, Acetanilid *n*
antifermentative gärungshemmend,
  gärungsverhindernd, gärungswidrig
antiferroelectric antiferroelektrisch
antiferroelectricity Antiferroelektrizität *f*
antiferroelectrics Antiferroelektrika *pl*
antiferromagnetic exchange integral
  Heisenbergsches Austauschintegral *n* (Magn)
antiferromagnetism Antiferromagnetismus *m*
antifibrinolysin Antifibrinolysin *n* (AFL)
antifibrinolysin test Antifibrinolysintest *m* (Med)
anti-flex cracking agent
  Ermüdungsschutzmittel *n*
anti-flex cracking properties Ermüdungsschutz *m*
antiflotation agent
  Ausschwimmverhütungsmittel *n*
antifluorite lattice Antiflußspatgitter *n* (Krist)
antifoam Antischaummittel *n*, Entschäumer *m*,
  schaumbekämpfend,
  Schaumverhütungsmittel *n*
antifoaming agent Antischaummittel *n*,
  Schaumverhütungsmittel *n*
antifoaming emulsion Antischaumemulsion *f*
anti-foaming splash head
  Schaumfängeraufsatz *m*
antifogging agent
  Beschlagverhinderungsmittel *n* (Glas)
antifouling fäulnisverhindernd,
  fäulnisverhütend, fäulniswidrig
antifouling coat Holzschutzanstrich *m*
antifouling composition Antifäulnisfarbe *f*,
  Bodenanstrich *m*
antifreeze Frostschutzmittel *n*,
  Gefrierschutzmittel *n*, Kälteschutzmittel *n*
antifreezing kältebeständig
antifreezing agent Gefrierschutzmittel *n*,
  Kälteschutzmittel *n*

antifreezing property Kältebeständigkeit *f*
antifriction alloy Lagerlegierung *f*
antifrictional qualities (of a bearing)
   Laufeigenschaften *pl* (eines Lagers)
antifriction bearing Antifriktionslager *n*,
   Walzlager *n*
antifriction metal Antifriktionsmetall *n*,
   Lagermetall *n*, Weißmetall *n*,
   Zapfenlagermetall *n*
anti-frostbite Frostsalbe *f* (Pharm)
antifroth schaumbekämpfend
antifrothing agent Antischaummittel *n*,
   Schaumverhütungsmittel *n*
antifroth oil [Anti-]Schaumöl *n*
antifungin Antifungin *n*
antigen Antigen *n*, Antigenkörper *m*
antigen-antibody reaction
   Antigen-Antiköperreaktion *f*,
   Antigen-Antikörper-Reaktion *f*
antigenic antigen
antigenicity Antigenität *f*, Antigenizität *f*
antigorite Antigorit *m* (Min),
   Blätterserpentin *m* (Min)
anti-gray [hair] factor Antigrau[haar]faktor *m*
anti-growth substance Antiwuchsstoff *m*
antih[a]emophilic globulin antihämatophiles
   Globulin *n* (AHG), antihämatophiles
   Globulin *n*
antihelix Antihelix *f*
antihemorrhagic antihämorrhagisch, blutstillend
antihistamine Antihistamin *n* (Pharm)
antihistamine preparation
   Antihistaminpräparat *n* (Pharm)
antihole Antiloch *n*
antihormone Antihormon *n*, Gegenhormon *n*
anti-icing fluid Vereisungsschutzflüssigkeit *f*
anti-immune immunitätshemmend
anti-incrustant Kesselstein[gegen]mittel *n*
anti-incrustant powder Kesselsteinpulver *n*
anti-incrustator Kesselstein[gegen]mittel *n*
antiisotypism Antiisotypie *f*
antikamine Antikamin *n*
antiknock klopffrei
antiknock [agent] Antiklopfmittel *n* (Mot),
   Klopfbremse *f* (Mot)
antiknock compound Antiklopfmittel *n* (Mot)
antiknock fuel Antiklopfbrennstoff *m*,
   klopffester Brennstoff *m*
antiknock[ing] *a* klopffest
antiknock property Antiklopfeigenschaft *f*
antiknock value Oktanwert *m*
antilog Numerus *m*
antilog *a* antilog, antimer, enantiomer
antilogarithm Antilogarithmus *m*, Numerus *m*
antilogarithmic antilogarithmisch (Math)
antilogs optische Antipoden *pl*
antiluetin Antiluetin *n*
antilysin Antilysin *n*
antilysis Antilysinwirkung *f*

antimagnetic antimagnetisch
antimatter Antimaterie *f* (Phys)
antimer antilog, antimer, enantiomer
antimere Antimer *n*, optische Antipode
antimetabolite Antimetabolit *m*
antimetric antimetrisch
antimicrobic bakterienwachstumshemmend
antimildew agent Antischimmelmittel *n*
antimonate Antimonat *n*, Antimoniat *n*, Stibiat *n*
antimonate *a* antimonsauer
antimonial antimonartig, antimonhaltig,
   antimonig, spießglanzartig (Min)
antimonial cinnabar Antimonzinnober *m* (Min)
antimonial copper glance
   Antimonkupferglanz *m* (Min)
antimonial gray copper Antimonfahlerz *n* (Min)
antimonial lead Antimon[al]blei *n*
antimonial lead ore Rädelerz *n* (Min)
antimonial nickel Antimonnickel *n* (Min),
   Antimonnickelkies *m* (Min)
antimonial saffron Antimonsafran *m*
antimonial silver Antimonsilber *n* (Min)
antimonial wine Brechwein *m*
antimonial zinc alloy Antimonzinklegierung *f*
antimonic acid Antimonsäure *f*
antimonic anhydride Antimonsäureanhydrid *n*
antimonic chloride Antimon(V)-chlorid *n*,
   Antimonpentachlorid *n*, Chlorspießglanz *m*
   (Min)
antimonic oxide Antimonpentoxid *n*
antimonic sulfide Antimonpentasulfid *n*
antimoniferous antimonhaltig
antimonine Antimonin *n*, Antimonlactat *n*
antimonious acid antimonige Säure *f*
antimonious sulfide Antimon(III)-sulfid,
   Antimonsulfür *n* (obs)
antimonite Antimonglanz *m* (Min),
   Antimonit *m* (Min), Grauspießglanz *m* (Min)
antimonite *a* antimonigsauer
antimonous antimonig, Antimon (III) -
antimonous acid antimonige Säure *f*
antimonous anhydride
   Antimonigsäureanhydrid *n*
antimonous chloride Antimontrichlorid *n*,
   Antimon(III)-chlorid
antimonous hydride Stibin *n*
antimonous iodide Antimontrijodid *n*
antimonous oxalate Antimonyloxalat *n*
antimonous oxide Antimontrioxid *n*
antimonous oxysulfide Antimonsafran *m*
antimonous sulfide Antimon(III)-sulfid,
   Antimonsulfür *n* (obs)
antimonous thiosulfate chelate
   Antimonthiosulfatchelat *n*
antimony Antimon *n* (Symb. Sb), Stibium *n*
   (Lat), **butter of** ~ Spießbutter *f* (obs),
   commercial ~ Handelsantimon *n*, **containing**
   ~ antimonhaltig, **crocus of** ~

Antimonsafran *m*, Spießglanzsafran *m*, **crude**
~ Rohantimon *n*, **flowers of** ~ Spießblumen *f*
pl (Min), **glass of** ~ Antimonglas *n*,
preparation containing ~ Antimonpräparat *n*,
regulus of ~ Spießglanzkönig *m*, **unrefined** ~
Rohantimon *n*, **white** ~ Valentinit *m* (Min),
Weißspießglanzerz *n* (Min)
antimony alloy Antimonlegierung *f*
antimony amalgam Antimonamalgam *n*
antimony ammonium fluoride
Antimonammoniumfluorid *n*
antimony arsenide Arsen[ik]antimon *n*
antimony ash Spießglanzasche *f*
antimony bath Antimonbad *n*
antimony blende Antimonblende *f* (Min),
Rotspießglanzerz *n* (Min), Spießblende *f*
(Min)
antimony bloom Antimonblüte *f* (Min)
antimony bromide Antimonbromid *n*,
Bromantimon *n*, Bronzespießglanz *m* (Min)
antimony chloride Antimonchlorid *n*,
Chlorantimon *n*
antimony electrode Antimonelektrode *f*
antimony fluoride Antimonfluorid *n*,
Fluorantimon *n*
antimony glance Antimonglanz *m* (Min),
Grauspießglanzerz *n* (Min), Stibnit *m* (Min)
antimony halide Antimonhalogenid *n*
antimony hydride Antimonwasserstoff *m*
antimony(III) chloride Antimonbutter *f*,
Antimonchlorür *n*, Antimon(III)-chlorid *n*,
Antimontrichlorid *n*
antimony(III) iodide Antimontrijodid *n*
antimony(III) oxide Antimontrioxid,
Antimonigsäureanhydrid *n*
antimony(III) sulfide Antimontrisulfid *n*,
Antimonsulfür *n* (obs)
antimony iodide Jodantimon *n*
antimonyl- Antimonyl-
antimony lactate Antimonin *n*, Antimonlactat *n*
antimonyl chloride Antimonylchlorid *n*
antimony lead sulfide Schwefelantimonblei *n*
(Min)
antimonyl oxalate Antimonyloxalat *n*
antimonyl potassium tartrate
Antimonylkaliumtartrat *n*
antimonyl radical Antimonylrest *m*
antimonyl salt Antimon(III)-oxidsalz *n*
antimony metal Antimonmetall *n*
antimony mirror Antimonspiegel *m* (Anal)
antimony nitrate Antimonnitrat *n*
antimony ocher Antimonocker *m*, Spießocker *m*
(Min), Stibiconit *m* (Min)
antimony ore Antimonerz *n* (Min), **gray** ~
Grauspießglanz *m* (Min)
antimony oxalate Antimonoxalat *n*
antimony oxide Antimonoxid *n*
antimony oxychloride Antimonoxychlorid *n*,
Algarotpulver *n*

antimony pentachloride Antimonpentachlorid *n*,
Antimonperchlorid *n*,
Fünffachchlorantimon *n*
antimony pentaiodide Antimonpentajodid *n*
antimony pentasulfide Antimonpentasulfid *n*,
Antimonsupersulfid *n*,
Fünffachschwefelantimon *n*; Goldschwefel *m*
antimony pentoxide Antimonpentoxid *n*,
Antimonsäureanhydrid *n*
antimony potassium salt Antimonkaliumsalz *n*
antimony potassium tartrate Brechweinstein *m*
antimony regulus Antimonregulus *m*
antimony salt Antimonsalz *n*
antimony sesquioxide Antimonsesquioxid *n*
antimony sodium tartrate
Natriumbrechweinstein *m*
antimony sulfide Antimonsulfid *n*,
Schwefelantimon *n*, **golden** ~
Goldschwefel *m*, **native red** ~ Kermesit *m*
(Min), Rotspießglanz[erz] (Min)
antimony tannate Antimontannat *n*
antimony tetroxide Antimontetroxid *n*
antimony trichloride Antimontrichlorid *n*,
Antimonchlorür *n* (obs), Bronziersalz *n*,
Dreifachchlorantimon *n*, Spießbutter *f* (obs)
antimony triiodide Antimontrijodid *n*,
Antimonjodür *n* (obs)
antimony trimethyl Trimethylstibin *n*
antimony trioxide Antimontrioxid *n*,
Antimonigsäureanhydrid *n*, Spießoxid *n*
(obs)
antimony triphenyl Triphenylstibin *n*
antimony trisulfide Antimontrisulfid *n*,
Antimonsulfür *n*
antimony(V) chloride Antimonpentachlorid *n*,
Antimonperchlorid *n*, Antimon(V)-chlorid *n*,
Fünffachchlorantimon *n*
antimony(V) oxide Antimonpentoxid *n*
antimony(V) sulfide Antimonpentasulfid *n*,
Antimonpersulfid *n*
antimony white Antimonweiß *n*, Spießweiß *n*
antimony yellow Antimongelb *n*, Neapelgelb *n*
antimutagen Antimutagen *n*
antimycin Antimycin *n*
antimycotic Antimykotikum *n* (Pharm)
antinarcotic antinarkotisch
antinervine Antinervin *n*
antineuralgic Antineuralgikum *n* (Pharm)
antineuralgic balsam Antineuralgiebalsam *m*
(Pharm)
antineutrino Antineutrino *n*
antineutron Antineutron *n* (Atom)
antinode Schwingungsbauch *m*
anti-noise paint Antidröhnlack *m*,
schalldämpfende Anstrichfarbe *f*,
Schallschlucklack *m*
antinonnin Antinonnin *n*, Viktoriagelb *n*
antinosin Antinosin *n*, Nosophennatrium *n*

antiodorin Antiodorin *n*
antiosmosis Antiosmose *f*
antioxidant Antioxidationsmittel *n*,
    Alterungsschutzmittel *n*, Antioxidans *n* (pl
    Antioxidantien)
antioxidizing agent Antioxidationsmittel *n*,
    Antioxidantium *n*
antioxygen Antioxygen *n*
antiparallel antiparallel
antiparasitic Antiparasitikum *n*,
    Parasitenmittel *n*
antiparticle Antiteilchen *n* (Atom),
    Gegenteilchen *n*
antipathy Abneigung *f*
antipellagra factor Antipellagrafaktor *m*
antiperthite Antiperthit *m* (Min)
antiphase Gegenphase *f*
antiphein Antiphein *n*
antiphlogistic antiphlogistisch
anti-pit agent Porenverhütungsmittel *n*
antipodal antipodisch; völlig entgegengesetzt
antipodal point Antipodenpunkt *m*
antipode [optischer] Antipode *m*
antipodes, optical ~ optische Antipoden *pl*
antipol Gegenpol *m*
antiprism Antiprisma *n*
antiproton Antiproton *n* (Atom)
antiputrefactive fäulnisverhindernd
antipyonin Antipyonin *n*
antipyretic Antipyretikum *n* (Med),
    Fieberarznei *f* (Med), Fiebermittel *n* (Med)
antipyretic *a* antipyretisch, fieberverhütend,
    fiebermildernd
antipyrine Antipyrin *n*, Analgesin *n*,
    Anodynin *n*, Metozin *n*
antirabic vaccination Tollwutimpfung *f*
antiradiation therapy Strahlenschutztherapie *f*
antiradiation tube Strahlenschutzröhre *f*
antirot fäulnisverhütend
antirust rostfrei
antirust agent Rost[schutz]mittel *n*
antirust paint Rostschutzanstrich *m*,
    Rostschutzfarbe *f*
antiscale-forming agent Kesselsteingegenmittel *n*
antiscorbutic antiskorbutisch
antisepsin Antisepsin *n*
antisepsis Antisepsis *f*
antiseptic Antiseptikum *n* (Pharm),
    antiseptisches Mittel *n*, Fäulnismittel *n*,
    keimtötendes Mittel *n*
antiseptic *a* antiseptisch, fäulnishemmend,
    fäulnisverhindernd, fäulnisverhütend,
    fäulniswidrig, gärungshemmend
antiseptin Antiseptin *n*
antiseptol Antiseptol *n*
antiserum Antiserum *n*, Heilserum *n*,
    Immunserum *n*
anti-set-off powder Druckbestäubungspuder *m*

antisettling agent Absetzverhinderungsmittel *n*,
    Absetzverhütungsmittel *n*, Antiabsetzmittel *n*,
    Schwebemittel *n*
anti-shrink process Antischrumpfbehandlung *f*
anti-sigma-minus-hyperon
    Anti-Sigma-Minus-Hyperon *n*
    (Elementarteilchen, 2300fache Masse e.
    Elektrons)
anti-skid rutschfest
anti-skid device Gleitschutz *m*
anti-skid property Rutschfestigkeit *f*
anti-skimming agent Antiemulgierungsmittel *n*
antiskinning agent Antihautmittel *n* (Lack),
    Hautverhinderungsmittel *n*,
    Hautverhütungsmittel *n* (Lack)
anti-slip surfacing compound
    Beschichtungsmittel *n* zum Verhindern des
    Gleitens
antisoftener Versteifungsmittel *n*
antispasmin Antispasmin *n*
antispasmodic Antispasmodikum *n* (Pharm),
    Krampfmittel *n* (Pharm), krampfstillend
antispastic Antispastikum *n*
antispastic *a* krampfstillend (Med)
antistatic antistatisch
antistatic agent Antistatikum *n*,
    Antistatischmittel *n*
antistatin Antistatin *n* (HN, Textilhilfsmittel)
antistokes line Antistokes'sche Linie *f*
antistreptolysin Antistreptolysin *n*
anti-swirl device (for petrol tank)
    Kraftstoffberuhigungstopf *m*
antisymmetric antisymmetrisch
antisymmetry Antisymmetrie *f*
anti-tarnishing bath Anlaufschutzbad *n*
antitartaric acid Antiweinsäure *f*
antitetanic injection Tetanusspritze *f* (Med)
antitetanic serum Antitetanusserum *n*,
    Tetanusantitoxin *n*
antithermin Antithermin *n*
antithesis Gegensatz *m*, Widerspruch *m*
anti-thixotropy Antithixotropie *f*
antithrombin Antithrombin *n*
antithrombinic factor Antithrombinfaktor *m*
antithromboplastin Antithromboplastin *n*
antithrombosin Antithrombosin *n*
antitoxic giftabtreibend, als Gegengift wirkend
antitoxic serum Heilserum *n*
antitoxic unit Immunitätseinheit,
    Antitoxineinheit *f*
antitoxin Antitoxin *n*, Gegengift *n*
antitoxin unit Immunitätseinheit *f*
antitumor agent Cytostatikum *n* (Pharm)
antityphoid vaccine Typhusimpfstoff *m*
antivenin Gegengift *n* (Schlangenserum)
anti-weathering agent
    Verwitterungsschutzmittel *n*
antizymotic gärungshemmend,
    gärungsverhindernd, gärungswidrig

antodine Antodin *n*
antogorite Antogorit *m* (Min)
antrimolite Antrimolith *m* (Min)
Antwerp blue Antwerpenerblau *n*
anvil Amboß *m*
anvil face Amboßbahn *f*
anvil plate Amboßbahn *f*
anysin Ichthyol *n*
anzic acid Anziasäure *f*
apart [from] abgesehen [von]
apatelite Apatelit *m* (Min)
apatite Apatit *m* (Min)
apatropine Apoatropin *n*
APC viruses APC-Viren *pl* (Adenoviren)
ape Affe *m*
aperiodic aperiodisch, nicht periodisch,
    unperiodisch
aperiodicity Aperiodizität *f*
apertural effect Unschärfering *m* (Opt),
    Zerstreuungskreis *m* (Opt)
aperture Apertur *f* (Opt), Öffnung *f* (Opt),
    Schlitz *m* (Opt), Spalt *m* (Opt)
apertured disk anode Lochscheibenanode *f*
apertured electrode disk of an electron microscope
    Lochelektrode *f* eines Elektronenmikroskops
aperture for filling Eingußloch *n*
aperture limitation Aperturbegrenzung *f* (Opt)
aperture ratio Öffnungsverhältnis *n*
aperture setting Blendeneinstellung *f* (Phot)
apex Spitze *f*, Apex *m*
aphanesite Strahlerz *n* (Min)
aphanin Aphanin *n*
aphanite Aphanit *m* (Min), basaltischer
    Grünstein *m* (Min), **calcareous ~**
    Kalkaphanit *m*
aphanitic aphanithaltig, aphanitisch
aphrite Aphrit *m* (Min)
aphrizite Graupenschörl *m* (Min)
aphrodisiac Aphrodisiakum *n* (Pharm)
aphrosiderite Aphrosiderit *m* (Min)
aphthalose Aphthalose *f*
aphthitalite Glaserit *m* (Min)
aphthonite Aphthonit *m* (Min)
aphtite Aphtit *n* (Leg)
aphyllidine Aphyllidin *n*
aphylline Aphyllin *n*
aphylline alcohol Aphyllinalkohol *m*
apigenidin Apigenidin *n*
apigenin Apigenin *n*
apiin Apiin *n*
apinol Apinol *n*
apiol aldehyde Apiolaldehyd *m*
apiol[e] Apiol *n*
apiolic acid Apiolsäure *f*
apionol Apionol *n*
apiose Apiose *f*
aplanatic aplanatisch (Astr)
aplanatic system Aplanat *m* (Astr)
aplite Aplit *m*

aplome Aplom *m* (Min)
aplotaxene Aplotaxen *n*
apoaromadendrone Apoaromadendron *n*
apoaspidospermine Apoaspidospermin *n*
apoatropine Apoatropin *n*, Atropamin *n*
apobornylene Apobornylen *n*
apocaffeine Apokaffein *n*
apocamphane Apocamphan *n*
apocamphenilone Apocamphenilon *n*
apocampholenic acid Apocampholensäure *f*
apocampholic acid Apocampholsäure *f*
apocamphor Apocampher *m*
apocardol Apocardol *n*
apocarnitine Apocarnitin *n*
apocarotenal Apocarotinal *n*
apocholic acid Apocholsäure *f*
apochromatic apochromatisch (Opt)
apochromat lens Apochromat *m* (Opt)
apocinchen Apocinchen *n*
apocinchonidine Apocinchonidin *n*
apocinchonine Apocinchonin *n*
apocitronellol Apocitronellol *n*
apocodeine Apocodein *n*, Apokodein *n*
apoconessine Apoconessin *n*
apocrenic acid Apokrensäure *f*, Krensäure *f*,
    Quellsäure *f*
apocyanine Apocyanin *n*
apocyclene Apocyclen *n*
apocynin Apocynin *n*
apocynol Apocynol *n*
apoenzyme Apoenzym *n*
apoephedrine Apoephedrin *n*
apoerythraline Apoerythralin *n*
apogalanthamine Apogalanthamin *n*
apogelsemine Apogelsemin *n*
apoglucic acid Apoglucinsäure *f*
apogossypol Apogossypol *n*
apoharmine Apoharmin *n*
apoharmyrine Apoharmyrin *n*
apoholarrhenine Apoholarrhenin *n*
apohydroquinine Apohydrochinin *n*
apohyoscine Apohyoscin *n*
apoisoborneol Apoisoborneol *n*
apolysin Apolysin *n*
apomethylbrucine Apomethylbrucin *n*
apomorphimethine Apomorphimethin *n*
apomorphine Apomorphin *n*
apomorphine hydrochloride
    Apomorphinhydrochlorid *n*
aponal Aponal *n*
aponarceine Aponarcein *n*
apophedrin Apophedrin *n*
apophyllenic acid Apophyllensäure *f*
apophyllite Apophyllit *m* (Min),
    Fischaugenstein *m* (Min),
    Ichthyophthalmit *m* (Min)
apopinene Apopinen *n*
apoplectic fit Apoplexie *f* (Med)

apoplexy Apoplexie *f* (Med), Hirnlähmung *f* (Med)
apopseudoionone Apopseudojonon *n*
apoquinine Apochinin *n*
aporeidine Aporeidin *n*
aporotaminic acid Aporotaminsäure *f*
aporphine Aporphin *n*
aposafranine Aposafranin *n*
aposafranone Aposafranon *n*
aposepedine Aposepedin *n*
aposorbic acid Aposorbinsäure *f*
apospermostrychnine Apospermostrychnin *n*
apoterramycin Apoterramycin *n*
apothecaries' weight Apothekergewicht *n*
apothecary's shop Apotheke *f*
apotheobromine Apotheobromin *n*
apothesine Apothesin *n*
apothiopyronine Apothiopyronin *n*
apotricyclol Apotricyclol *n*
apoyohimbic acid Apoyohimbinsäure *f*, Apoyohimboasäure *f*
apparatus Apparat *m*, Apparatur *f*, Gerät *n*, Vorrichtung *f*, **additional** ~ Zusatzgerät *n*, **construction of** ~ Apparatebau *m*, **manufacture of** ~ Apparatebau *m*, **pertaining to** ~ apparativ, **supplementary** ~ Zusatzgerät *n*
apparatus construction Apparatebau *m*
apparatus glass Geräteglas *n*
apparatus parts Apparateteile *pl*
apparent anscheinend, offensichtlich, scheinbar
apparent conductivity Scheinleitwert *m*
apparent density Aufschüttdichte *f*, Rüttelgewicht *n*, Schüttgewicht *n*
apparent modulus Kriechmodul *n*
apparent output Scheinleistung *f*
apparent resistance Impedanz *f* (Elektr), Scheinwiderstand *m* (Elektr)
apparent watts Scheinleistung *f* (Elektr)
appear *v* vorkommen, erscheinen
appearance Anschein *m*, Auftreten *n*, Aussehen *n*, Erscheinen *n*, Sichtbarwerden *n*, **superficial** ~ Oberflächenaussehen *n*
appendage Ansatz *m* (Med)
appendicitis Blinddarmentzündung *f* (Med)
appendix Anhang *m*, Beilage *f*, Blinddarm *m* (Med), Zusatz *m*
apple aphid Apfellaus *f* (Zool)
apple blossom weevil Apfelblütenstecher *m*
apple brandy Apfelbranntwein *m*
apple jack Apfelbranntwein *m*
apple oil Apfelöl *n*
apples, ferrated extract of ~ Apfeleisenextrakt *m*
apple sawfly Apfelsägewespe *f*
apple sucker Apfelblattsauger *m*
appliance Nutzanwendung *f;* Vorrichtung *f,* Werkzeug *n* (Hilfsgerät), Zubehör *n*

applicability Anwendbarkeit *f*, Anwendungsmöglichkeit *f*, Verwendungsfähigkeit *f*, **limit of** ~ Anwendungsgrenze *f*
applicable angebracht; anwendbar, benutzbar
applicant Bewerber *m*
applicate Applikate *f*
application Bewerbung *f;* Anwendung *f*, Applikation *f*, Auftragen *n* (z. B. Farbe), Auftragung *f* (z. B. Farbe), Nutzanwendung *f*, Verwendung *f*, **example of** ~ Anwendungsbeispiel *n*, **field of** ~ Anwendungsbereich *m*, Anwendungsgebiet *n*, **mode of** ~ Anwendungsweise *f*, **place of** ~ Angriffspunkt *m*, **range of** ~ Anwendungsbereich *m*, **rate of** ~ Aufwandmenge *f*, ~ **of load** Belastung *f*
applicator roll Auftragwalze *f*
applied angewandt
applied moment Angriffspunkt *m*
applied science angewandte Wissenschaft *f*
applied voltage Hilfsspannung *f*
apply *v* anwenden, applizieren, verwenden, ~ **for** beantragen, ~ **in one pass** in einem einzigen Arbeitsgang *m* auftragen
appoint *v* ernennen, bestellen
appointment Treffen *n*, Verabredung *f*
appraisable berechenbar, abschätzbar
appraise *v* [ab]schätzen
appraising Abschätzen *n*
appreciable bemerkenswert, nennenswert; berechenbar
appreciate *v* anerkennen, wertschätzen
appreciation Einschätzung *f*, Wertschätzung *f*, Würdigung *f*
apprentice Lehrling *m*, Volontär *m*
approach Annäherung *f* (e. g. to a large building), Anfahrbereich *m*, **velocity of** ~ Anflußgeschwindigkeit *f*
approach *v* annähern, anfahren
approach disposition Annäherungsaufstellung *f*
approach rate Annäherungsgeschwindigkeit *f*, Annäherungsverhalten *n*
approbation Billigung *f*
AP projection a-p-Projektion *f* (Atom)
appropriate passend, geeignet, angepaßt, einschlägig, sachgemäß, zutreffend, zweckentsprechend
appropriate *v* aneignen
appropriateness, to make with ~ sachgemäß herstellen
approval Begutachtung *f*, Billigung *f*, Genehmigung *f*, Zustimmung *f*
approve *v* billigen, beipflichten, gutheißen
approved bewährt
approximate *a* angenähert, approximativ
approximate *v* annähern

approximate calculation
Annäherungsrechnung *f*,
Näherungsrechnung *f* (Math)
approximate equation Näherungsgleichung *f*
(Math)
approximately annäherungsweise, ungefähr
approximate method Näherungsverfahren *n*
approximate quantity Näherungsgröße *f*
approximate value Näherungswert *m*,
Anhaltswert *m*
approximation Annäherung *f*,
Annäherungswert *m*, Fehlergrenze *f*,
Näherung *f* (Math), Näherungslösung *f*
(Math), Näherungswert *m*, **adiabatic** ~
adiabatische Näherung *f*, **degree of** ~
Annäherungsgrad *m*, **method of** ~
Annäherungsverfahren *n*, **stepwise** ~
schrittweise Näherung *f*, **strong-coupling** ~
mit fester Kopplung
approximation formula Näherungsformel *f*
approximation method Näherungsmethode *f*
approximations, theory of ~
Berichtigungsverfahren *n*
appurtenances Pertinenzien *pl* (Jur)
apricot essence Aprikosenessenz *f*
apricot kernel oil Aprikosenkernöl *n*,
Aprikosenäther *m*
apricot kernels, volatile oil of ~
Aprikosenäther *m*
apron conveyor Plattenband *n*,
Plattenbandförderer *m*
aptitude test Eignungsprüfung *f*
apyonin Apyonin *n*
apyrite Apyrit *m* (Min), Lithiumturmalin *m*
(Min)
apyron Apyron *n*, Grifa *n*
apyrous feuerfest, unschmelzbar, apyrisch
aqua fortis konzentrierte Salpetersäure *f*,
Ätzwasser *n*, Scheidewasser *n*, Silberwasser *n*
(obs)
aquamarine Aquamarin *m* (Min)
aqua mirabilis Wunderwasser *n* (Pharm)
aquamycin Aquamycin *n*
aqua regia Königswasser *n*, Goldscheidewasser *n*
aquastat Aquastat *m*, Tauchthermostat *m*
aqueous wäss[e]rig, wasserhaltig
aqueous alcoholic alkoholisch-wässerig
aqueous ammonia Ammoniakwasser *n*
aqueous condensate Niederschlagswasser *n*
aqueous solubility Wasserlöslichkeit *f*
aquinite Chlorpikrin *n*
aquocobalamin Aquocobalamin *n*
aquo compound Aquoverbindung *f*
aquo ion hydratisiertes Ion *n*
arabic acid Arabinsäure *f*, Gummisäure *f*
Arabic numeral arabische Ziffer *f*
arabin Arabin *n*
arabinal Arabinal *n*

arabinose Arabinose *f*, Gummizucker *m*,
Pektinose *f*
arabinosone Arabinoson *n*
arabite Arabit *m*
arabitic acid Gummisäure *f*
arabitol Arabit *m*
arable land Ackerland *n*
arable soil Ackerboden *m*, Ackererde *f*
araboflavine Araboflavin *n*
arabonic acid Arabonsäure *f*
arabonylglycine Arabonylglycin *n*
arabulose Arabulose *f*
araburonic acid Araburonsäure *f*
arachic acid Arachinsäure *f*, Erdnußsäure *f*
arachic alcohol Arachinalkohol *m*
arachidic acid Arachinsäure *f*, Erdnußsäure *f*
arachidic alcohol Eikosylalkohol *m*
arachidonic acid Arachidonsäure *f*
arachin Arachin *n*
arachis oil Erdnußöl *n*, Arachisöl *n*
arachis oil soap Erdnußölseife *f*
arachyl alcohol Arachylalkohol *m*
araeoxen Aräoxen *n* (Min)
aragonite Aragonit *m* (Min), Eisenblüte *f*
(Min), Erbsenstein *m* (Min), Faserkalk *m*
(Min), **fibrous** ~ Faseraragonit *m* (Min),
laminar and globular ~ Schalenkalk *m* (Min)
araroba Araroba *n*, Goapulver *n*
ararobinol Ararobinol *n*
arasan Arasan *n*
arbitrariness Willkürlichkeit *f*
arbitrary willkürlich
arbitration Begutachtung *f*
arbor Achse *f*, Auflage *f*, Baum *m* (Bot),
Dorn *m*, Kerneisen *n*, Spindel, Welle *f*,
Träger *m*
arbor dianae Dianenbaum *m* (Bot)
arborescent verzweigt, baumartig, baumförmig
arborescent agate Baumachat *m* (Min)
arborescin Arborescin *n*
arboricine Arboricin *n*
arboriform baumförmig
arborine Arborin *n*
arbor press Dornpresse *f*, Drehdornpresse *f*
arbor shaft Achswelle *f*
arbusterol Arbusterin *n*
arbutin Arbutin *n*
arc Bogen *m*, Flammenbogen *m*, Kreisbogen *m*,
Lichtbogen *m* (Elektr), **heating by the** ~
Lichtbogenheizung *f*, **materials exposed to
action of** ~ Lichtbogengut *n*, **temperature of
the** ~ Lichtbogenwärme *f*, **unsteady** ~
flackernder Lichtbogen *m*
arcaine Arcain *n*
arcanite Arkanit *m* (Min)
arc carbons, fragments of ~
Lichtbogenkohlebruchstücke *n pl*
arc cosine Arkuskosinus *m*
arc cotangent Arkuskotangens *m*

arc cutting Brennschneiden *n*
arc development Lichtbogenentfaltung *f*
arc discharge Bogenentladung *f,*
  Lichtbogenentladung *f*
arc discharge tube Lichtbogenrohr *n*
arc furnace Lichtbogenofen *m,*
  Lichtbogen-Schmelzofen *m*
arc gap Lichtbogenstrecke *f*
arc generator Lichtbogengenerator *m*
arch Bogen *m* (Arch), Gewölbe *n* (Arch)
archaeology Altertumskunde *f*
arc heating Lichtbogen[be]heizung *f,*
  Lichtbogenerhitzung *f*
archil Färberflechte *f* (Bot), Orseille *f*
  (Farbstoff)
archil carmine Orseillekarmin *n*
archil extract Orseilleextrakt *m*
archil paste Orseilleextrakt *m*
Archimedean screw Förderrohr *n,*
  Wasserschnecke *f*
Archimedes, helix of ∼ Archimedesschnecke *f*
Archimedes' principle Archimedisches Prinzip *n*
  (Phys)
archiplasm Urplasma *n*
architectural architektonisch
architectural filler paint Fassadenfüllfarbe *f*
archives Archiv *n*
arch lid Gewölbedeckel *m*
arch of the furnace Herdgewölbe *n*
arch plate Gewölbeplatte *f*
arc ignition Lichtbogenzündung *f*
arcing Lichtbogenbildung *f*
arc lamp Bogenlampe *f,* Lichtbogenlampe *f,*
  direct current ∼ Gleichstrombogenlampe *f*
arc lamp carbon Lichtbogenkohle *f*
arc light Bogenlicht *n*
arcose Arkose *f* (Geol)
arc physics Lichtbogenphysik *f*
arc rectifier Lichtbogengleichrichter *m*
arc reduction furnace
  Lichtbogen-Reduktionsofen *m*
arc resistance Lichtbogenfestigkeit *f,*
  Lichtbogenwiderstand *m*
arc resistance furnace
  Lichtbogenwiderstandsofen *m*
arc resitance, heating by ∼
  Lichtbogenwiderstandserhitzung *f*
arc secant Arkussekans *m* (Math)
arc sine Arkussinus *m* (Math)
arc source Lichtbogengenerator *m*
arc spectrum Bogenspektrum *n* (Spektr)
arc suppressor Funkenlöschkreis *m*
arcsutite Arksutit *m* (Min)
arc tangent Arkustangens *m* (Math)
arc tapping Lichtbogenabstich *m*
arctic arktisch
arctic zone Polarkreis *m*
arctigenin Arctigenin *n*
arctiin Arctiin *n*

arctisite Arktisit *m*
arc tracking Kriechwegbildung *f* durch
  Lichtbogen
arctuvine Arctuvin *n*
arcus, function of ∼ Arkusfunktion *f* (Math)
arc voltage Lichtbogenspannung *f*
arc welding elektrische Schweißung *f,*
  Abbrennschweißung *f,* Bogenschweißen *n,*
  Lichtbogenschweißung *f*
arc welding electrode
  Lichtbogenschweißelektrode *f*
arc welding method
  Lichtbogenschweißverfahren *n*
arc zone Lichtbogenbereich *m*
ardassine Ardassinestoff *m,* Perlseide *f*
ardennite Ardennit *m* (Min)
ardent brennend
ardometer Ardometer *n*
are Ar *n* (Maß)
area Fläche *f,* Areal *n,* Flächeninhalt *m,*
  Gebiet *n,* center of ∼ Flächenmittelpunkt *m,*
  effective ∼ wirksame Oberfläche *f,* scanned ∼
  abgetastetes Gebiet *n,* unit of ∼
  Flächeneinheit *f,* Oberflächeneinheit *f*
area function Areafunktion *f*
area of development Entwicklungsgebiet *n*
area of interface, specific ∼ spezifische
  Grenzfläche
area of pressure Druckfläche *f*
area preserving flächentreu
area ratio Flächenverhältnis *n*
areas, law of ∼ Flächensatz *m* (Mech)
arecaidine Arecaidin *n*
arecaine Arecain *n*
areca nut Arekanuß *f* (Gerb)
arecin Arecin *n*
arecolidine Arecolidin *n*
arecoline Arecolin *n*
arecoline hydrobromide Arecolinhydrobromid *n*
arecoline hydrochloride Arecolinhydrochlorid *n*
arecolone Arecolon *n*
arenaceous sandig
arenaceous quartz Quarzsand *m* (Min)
arendalite Arendalit *m* (Min), grüner Epidot *m*
  (Min)
arenobufagin Arenobufagin *n*
areolatin Areolatin *n*
areometer Aräometer *n,* Dichtemesser *m* (für
  Flüssigkeiten), Flüssigkeitswaage *f,*
  Senkwaage *f,* Spindel *f*
areometric aräometrisch
areometry Aräometrie *f*
areosaccharimeter Aräosaccharimeter *n*
areoxene Areoxen *n*
aretan Aretan *n* (HN, Naßbeizmittel)
arfvedsonite Arfvedsonit *m* (Min)
argand diagram Argand-Diagramm *n*
argand lamp Argandlampe *f,* Lampe *f* mit
  zylindrischem Docht

argentamine Argentamin *n*
argentan strap Argentanband *n*
argentic nitrate Höllenstein *m*
argentiferous silberhaltig, silberführend
argentiferous gold Goldsilber *n*
argentiferous lead silberhaltiges Blei *n*
argentiferous sand Silbersand *m*
argentiferous tetrahedrite Silberfahlerz *n* (Min),
  Weißgüldigerz *n* (Min)
argentine Argentin *n*, Neusilber *n*, versilbertes
  Weißmetall *n*
argentine *a* silberartig, silberig
argentite Silberglanz *m* (Min), Argentit *m*
  (Min), Argyrit *m* (Min), Glanzerz *n* (Min),
  Glanzsilber *n* (Min), Glaserz *n* (Min),
  Schwefelsilber *n* (Min), **earthy ~**
  Silberschwärze *f* (Min)
argentobismuthite Argentobismutit *m* (Min),
  Silberwismutglanz *m* (Min)
argentometer Argentometer *n*, Aräometer *n* für
  Silbernitrat (Phot), Silbermesser *m*
argentometric argentometrisch
argentometry Argentometrie *f*
argentopyrite Argentopyrit *m* (Min),
  Silberkies *m* (Min)
argentum Silber *n*, **~ cornu** Chlorargyrit *m*
  (Min), **~ fulminans** Knallsilber *n*
argillaceous lehmhaltig, lehmig, tonartig, tonig
argillaceous earth Tonerde *f*, **white ~**
  Weißtonerde *f*
argillaceous gypsum Tongips *m*
argillaceous marl Mergelton *m*
argillite Argillit *n* (Leg), Kieselton *m* (Min),
  Tonschiefer *m* (Min), **primitive ~**
  Urtonschiefer *m* (Min)
argilloarenaceous lehm- und sandhaltig
argillocalcareous ton- und kalkhaltig
argillocalcite Tonkalk *m* (Min)
argilloferruginous ton- und eisenhaltig
argillogypseous tongipshaltig
argillosiliceous tonkieselhaltig
arginase Arginase *f* (Biochem)
arginine Arginin *n*
arginine hydrochloride Argininhydrochlorid *n*
argininosuccinic acidemia
  Argininbernsteinsäurekrankheit *f*
arginosuccinic acid Arginobernsteinsäure *f*
arginylarginine Arginylarginin *n*
argochrome Argochrom *n*
argoflavine Argoflavin *n*
argon (Symb. A) Argon *n* (Symb. Ar)
argonal rectifier Argonalgleichrichter *m*
argon-arc welding Argonschweißung *f*
argonarc welding process Argonarc-Verfahren *n*
argon-clathrate Argon-Clathrat *n*
argon content Argongehalt *m*
argonin Argonin *n*
argument Argument *n*, Beweisartikel *m*

argumentation Argumentation *f*,
  Beweisführung *f*
argyrescine Argyrescin *n*
argyrite Argentit *m* (Min), Argyrit *m* (Min)
argyrodite Argyrodit *m* (Min)
argyrofelt Argyrofelt *n* (HN, Haarbeize)
argyrometric argyrometrisch
argyropyrite Argyropyrit *m* (Min)
aribine Aribin *n*, Harman *n*
aricine Aricin *n*
arid trocken, dürr
arine Arin *n*
arise *v* entstehen, hervorkommen
aristochin Aristochin *n*
aristochinine Aristochinin *n*
aristol Aristol *n*
aristolochic acid Aristolochiasäure *f*
aristolochine Aristolochin *n*
aristoquin Aristochin *n*
arite Arit *m* (Min)
arithmetic[ally] arithmetisch
arithmetical mean arithmetisches Mittel *n*
arithmetic mean arithmetischer Mittelwert *m*
arithmetic operation Rechenoperation *f*
arithmetics Arithmetik *f*
arithmetic shift Stellenverschiebung *f* (Comp)
arizonite Arizonit *m* (Min)
arjunolic acid Arjunolsäure *f*
arkansite Arkansit *m* (Min)
arm Arm *m* (Träger), Schenkel *m* (Winkel) (of a
  lever), Arm *m* (eines Hebels)
armalak Armalak *m*
armature Anker *m* (Magnet, Dynamo),
  Armatur *f* (Elektr), [Metall-]Beschlag *m*
armature body Ankerkörper *m*
armature bore Ankerbohrung *f*
armature coil Ankerwicklung *f*
armature core Ankerkern *m*
armature current Ankerstrom *m*
armature drop Ankerspannungsabfall *m*
armature end-plate Ankerisolierscheibe *f*
armature key Ankerkeil *m*
armature of a magnet Anker *m* eines Magneten
armature shaft Ankerwelle *f*
armature slot Ankernut *f*
armature spider Ankerbüchse *f* (Elektr)
armature turn Ankerwindung *f*
armature voltage Ankerspannung *f*
armature winding Ankerwicklung *f*
Armco-iron Armco-Eisen *n* (HN)
arm dowel pin Armprisonstift *m*
Armenian bole Färbererde *f*, Farberde *f*
armepavine Armepavin *n*
armored cable bewehrtes Kabel *n*
armored concrete Eisenbeton *m*
armoring Armierung *f*
armor plating Panzerung *f*
armoured glass Panzerglas *n*

arnica Arnica *f* (Bot), Arnicakraut *n*, **extract of**
  ~ Arnicaextrakt *m*
arnica flowers Arnicablüten *f pl*
arnica leaves Arnicakraut *n*
arnica oil Arnicaöl *n*
arnica rhizome Arnicawurzel *f*
arnica root Arnicawurzel *f*
arnica tincture Arnicatinktur *f*
arnicine Bitterharz *n*, Arnicin *n*
arnimite Arnimit *m* (Min)
Arnold's test Arnoldsche Probe *f* (Chem)
arnotta Arnotta *n*
arochlor Arochlor *n*
aroma Aroma *n*, Blume (Wein)
aromadendral Aromadendral *n*
aromadendrene Aromadendren *n*
aromadendrine Aromadendrin *n*
aromatic aromatische Substanz *f*, Gewürzstoff *m*
aromatic *a* aromatisch; gewürzartig,
  wohlriechend
aromatic essence Räucheressenz *f*
aromatic extract Gewürzextrakt *m*
aromatic principle Duftstoff *m*
aromatic substance Aromat *m* (Chem)
aromatic tincture Gewürztinktur *f*
aromatic vinegar Gewürzessig *m*
aromatic wine Gewürzwein *m*
aromatization Aromatisierung *f*
aromatize *v* aromatisieren, würzen
ar[o]moline Aromolin *n*
arouse *v* wecken, erregen, aufregen
aroyl- Aroyl-
aroylbenzoic acid Aroylbenzoesäure *f*
arquerite Arquerit *m* (Min)
arrack Arrak *m*, Reisbranntwein *m*
arrange *v* abmachen, anordnen, ansetzen
  (Math), einrichten, herrichten, ordnen, reihen,
  richten, vermitteln, ~ **in layers** einschichten,
  in Schichten ordnen, ~ **in lines** staffeln, ~ **in**
  series anreihen
arranged side by side nebeneinander angeordnet
arrangement Abkommen *n*, Abmachung *f*,
  Anordnung *f*, Aufstellung *f*, Disposition *f*,
  Einrichtung *f*, Ordnung *f*, Vereinbarung *f*,
  disordered ~ ungeordneter Zustand *m*
arrangement of particles Packungsart *f* (Krist)
arrangement of points Punktanordnung *f*
arrest Haft *f*, Hemmung *f*
arrest *v* zurückhalten, hindern, arretieren
arresting device Arretiervorrichtung *f*
arrestment Anhalten *n*, Arretierung *f*
arrhenal Arrhenal *n*
Arrhenius equation Arrhenius-Gleichung *f*
Arrhenius law Arrhenius-Gesetz *n*
arrive *v* ankommen, einlaufen
arrow poison Pfeilgift *n*
arsacetin Arsacetin *n*
arsafluorinic acid Arsafluorinsäure *f*
arsanilate arsanilsauer

arsanilic acid Arsanilsäure *f*, Atoxylsäure *f*
arsanthracene Arsanthracen *n*
arsanthrene Arsanthren *n*
arsanthrenic acid Arsanthrensäure *f*
arsanthridine Arsanthridin *n*
arsedine Arsedin *n*
arsenate Arsenat *n*
arsenate *a* arsensauer, arseniksauer
arsenate of lime Pharmakolith *m* (Min)
arsenazo Arsenazo *n*
arsenic (Symb. As) Arsen *n*, **containing** ~
  arsenhaltig, **flowers of** ~ Giftmehl *n*,
  Hüttenmehl *n*, **free from** ~ arsenfrei, **native** ~
  Fliegenstein *m* (Min), **removal of** ~
  Entarsenierung *f*, **ruby** ~ Rauschrot *n* (Min),
  Realgar *m* (Min), **test for** ~ Arsenprobe *f*
  (Anal), **to remove** ~ entarsenizieren, **tower to**
  catch the ~ Giftfang *m*, **white** ~
  Arsenikglas *n*, Giftstein *m*, Hüttenmehl *n*, **red**
  ~ **[sulfide]** Rauschrot *n* (Min), Realgar *m*
  (Min)
arsenic acid Arsensäure *f*, **salt of** ~
  Arseniksalz *n*, **salt or ester of** ~ Arsenat *n*
arsenic [acid] anhydride Arsensäureanhydrid *n*,
  Arsenpentoxid *n*, Arsen(V)-oxid *n*
arsenical Arsenmittel *n*
arsenical *a* arsenhaltig, arsenikalisch,
  arsenikhaltig
arsenical cadmia Giftstein *m*
arsenical cobalt ore Speiskobalt *m* (Min)
arsenical iron Arseneisen *n* (Min)
arsenic alloy Arsenlegierung *f*
arsenical nickel Rotarsennickel *n* (Min)
arsenical preparation Arsenpräparat *n*,
  Arsenikpräparat *n*
arsenical pyrite, white ~ Silberkies *m* (Min)
arsenical pyrites Arsen[ik]kies *m* (Min),
  Giftkies *m* (Min)
arsenicals Arsenikalien *pl*
arsenical silver Arseniksilber *n*
arsenical silver glance Arsensilberblende *f* (Min)
arsenic antidote Arsen[ik]gegengift *n*
arsenic bath Arsenbad *n*
arsenic black Arsenikschwarz *n*
arsenic blende Arsenblende *f* (Min), **red** ~
  Rubinschwefel *m* (Min), **yellow** ~
  Rauschgelb *n*, Reißgelb *n*
arsenic bloom Arsenblüte *f* (Min)
arsenic bromide Arsenbromid *n*, Bromarsen *n*
arsenic butter Arsenikbutter *f*, Arsentrichlorid *n*
arsenic chloride Arsenchlorid *n*, Chlorarsen *n*
arsenic compound Arsenikverbindung *f*
arsenic-containing arsenhaltig
arsenic content Arsengehalt *m*
arsenic-cured arseniziert
arsenic determination Arsenbestimmung *f*, **tube**
  for ~ Arsenbestimmungsrohr *n*
arsenic diiodide Arsen(II)-jodid *n*, Arsenjodür *n*
  (obs)

arsenic dimethyl Arsendimethyl *n*
arsenic disulfide Arsendisulfid *n*, Rauschrot *n*
(Min), Realgar *m* (Min)
arsenic fluoride Arsenfluorid *n*, Fluorarsen *n*
arsenic-free arsenfrei
arsenic glass Arsen[ik]glas *n*, **red** ~ Rauschrot *n*
(Min), Realgar *m* (Min)
arsenic halide Arsenhalogenid *n*
arsenic hydride Arsenwasserstoff *m*
arsenic(III) bromide Arsen(III)-bromid *n*,
Arsenikbromid *n*
arsenic(III) chloride Arsen(III)-chlorid *n*,
Arsentrichlorid *n*
arsenic(III) iodide Arsen(III)-jodid *n*
arsenic(II) iodide Arsen(II)-jodid *n*,
Arsen[ik]jodür *n*
arsenic(III) oxide Arsentrioxid *n*,
Arsenigsäureanhydrid *n*, Arsen(III)-oxid *n*,
Arsenik *n*
arsenic(III) sulfide Arsentrisulfid *n*
arsenic iodide Jodarsen *n*
arsenic lattice Arsengitter *n* (Krist)
arsenic lime Arsenikäscher *m*
arsenic metal metallisches Arsen *n*
arsenic mineral Arsenikmineral *n*
arsenic mirror Arsen[ik]spiegel *m* (Anal)
arsenic ore Arsen[ik]erz *n* (Min), Gifterz *n*
(Min)
arsenic pentachloride Arsenpentachlorid *n*
arsenic pentafluoride Arsenpentafluorid *n*
arsenic pentasulfide Arsenpentasulfid *n*
arsenic pentoxide Arsenpentoxid *n*,
Arsensäureanhydrid *n*, Arsen(V)-oxid *n*,
hydrated ~ Arsensäurehydrat *n*
arsenic powder, white ~ Giftmehl *n*
arsenic ruby Arsendisulfid *n* (Min)
arsenic salt Arsensalz *n*
arsenic silver blende Arsensilberblende *f* (Min)
arsenic skimmings Arsenabstrich *m*
arsenic stain Arsenfleck *m*
arsenic sulfide Arsensulfid *n*, Schwefelarsen *n*,
red ~ rote Arsenblende *f*
arsenic tribromide Arsen(III)-bromid *n*,
Arsentribromid *n*
arsenic trichloride Arsen(III)-chlorid *n*,
Arsentrichlorid *n*, Chlorarsenik *n*
arsenic triiodide Arsen(III)-jodid *n*,
Arsentrijodid *n*
arsenic trioxide Arsen(III)-oxid *n*,
Arsenigsäureanhydrid *n*, Arsenik *n*,
Arsentrioxid *n*, Giftstein *m*, Hüttenrauch *m*,
powdered ~ Giftmehl *n*, **sublimed** ~
Giftrauch *m*, **vitreous** ~ Arsen[ik]glas *n*
arsenic trisulfide Arsentrisulfid *n*, **native** ~
Operment *n* (Min)
arsenic tube Arsenrohr *n*
arsenic vapor Arsendampf *m*
arsenic(V) fluoride Arsenpentafluorid *n*

arsenic(V) oxide Arsenpentoxid *n*,
Arsensäureanhydrid *n*, Arsen(V)-oxid *n*,
hydrated ~ Arsensäurehydrat *n*
arsenic works Arsenikhütte *f*
arsenide Arsenid *n*, Arsenür *n* (obs)
arseniferous arsen[ik]haltig
arseniopleite Arseniopleit *m* (Min)
arseniosiderite Arseniosiderit *m* (Min)
arsenious arsenig
arsenite arsenigsaures Salz *n*, Arsenit *n*
arsenite *a* arsenigsauer
arsenobenzene Arsenobenzol *n*
arsenobenzoic acid Arsenobenzoesäure *f*
arsenoferrite Arsenoferrit *m* (Min)
arsenohippuric acid Arsenohippursäure *f*
arsenolamprite Arsenolamprit *m* (Min)
arsenolite Arsenolith *m* (Min), Arsenblüte *f*
(Min), Arsenikkalk *m* (Min)
arsenomethanol Arsenomethylalkohol *m*
arsenophenol Arsenophenol *n*
arsenopyrite Arsenopyrit *m* (Min),
Arsenikkies *m* (Min), Arsenkies *m* (Min),
Giftkies *m*, Weißerz *n* (Min), Weißkies *m*
(Min)
arsenostibinobenzene Arsenostibinobenzol *n*
arsenostovaine Arsenostovain *n*
arsenous arsenig
arsenous acid arsenige Säure *f*, **salt of** ~
Arsenigsäuresalz *n*, Arsenit *n*
arsenous [acid] anhydride
Arsenigsäureanhydrid *n*, Arsentrioxid *n*
arsenous bromide Arsentribromid *n*,
Arsen[ik]bromid *n*
arsenous chloride Arsentrichlorid *n*,
Arsenikbutter *f*, Chlorarsenik *n*
arsenous hydride Arsenwasserstoff *m*, Arsin *n*
arsenous iodide Arsen(III)-jodid *n*
arsenous oxide Arsen(III)-oxid *n*,
Arsenigsäureanhydrid *n*, Arsenik *n*,
Arsentrioxid *n*
arsenous sulfide Arsentrisulfid *n*, Auripigment *n*
(Min)
arsindoline Arsindolin *n*
arsine Arsenwasserstoff *m*
arsphenamine Arsphenamin *n*
arsthinol Arsthinol *n*
arsuline Arsulin *n*
arsulolidine Arsulolidin *n*
arsuloline Arsulolin *n*
artabotrine Artabotrin *n*
artabsin Artabsin *n*
artamine Artamin *n*
artarine Artarin *n*
art bronze Kunstbronze *f*
art-dye *v* kunstfärben
art-dyeing Kunstfärben *n*
artebufogenin Artebufogenin *n*
artefact Artefakt *n*
artemazulene Artemazulen *n*

artemetin Artemetin *n*, Artemisetin *n*
artemisetin Artemetin *n*, Artemisetin *n*
artemisia Beifuß *m* (Bot)
artemisia ketone Artemisiaketon *n*
artemisia oil Alpenbeifußöl *n*
artemisic acid Artemissäure *f*
artemisin Artemisin *n*
arterenol Arterenol *n*
arterenone Arterenon *n*
arterial arteriell
arterial blood arterielles Blut *n*, Arterienblut *n*
arterial tissue Arteriengewebe *n* (Med)
arterial wall Arterienwand *f*
arteriole Arteriole *f* (Med), kleine Arterie *f* (Med)
arteriosclerosis Arterienverkalkung *f* (Med), Arteriosklerose *f* (Med)
artery Arterie *f* (Med), Schlagader *f* (Med)
arthranitin Arthranitin *n*, Cyclamin *n*
arthritis Arthritis *f* (Med), Gelenkentzündung *f* (Med)
arthroplasty Gelenkechirurgie *f*
arthropods Gliederfüßer *pl*
article Abhandlung *f*, Artikel *m*, Fabrikat *n*, Gegenstand *m*, **finished** ~ Fertigfabrikat *n*, **molded** ~ Formartikel *m*, **semimanufactured** ~ Halbfabrikat *n*
article of daily use Gebrauchsartikel *m*
articulate gegliedert
articulated conveyor belt Scharnierband *n*
articulated front-end loader knickgelenkter Frontlader *m*
articulated jib Knickausleger *m*
articulated joint Gelenkverbindung *f* (Techn)
articulated pipe Gelenkrohr *n*
articulated steering Knicklenkung *f*
articulated trolley jib Laufkatzenknick-Ausleger *m*
articulation Fugengelenk *n*, Gliederfuge *f*, Gliederung *f*
artifact Artefakt *n*
artificial künstlich, synthetisch, unecht
artificial aging Ausscheidungshärtung *f* (Leichtmetall)
artificial atmosphere furnace Schutzgasofen *m*
artificial gem Glaspaste *f*
artificial gum Dextrin *n*
artificial marble Marmorzement *m*, Alaungips *m*
artificial silk fabric Kunstseidengewebe *n* (Text)
artist's color Malerfarbe *f*
artocarpetin Artocarpetin *n*
art of etching Ätzkunst *f*
artosine Artosin *n*
art paper Kreidepapier *n*, Kunstdruckpapier *n*
art silk Glanzstoff *m* (HN)
arubren Arubren *n* (HN, Gummiindustrie)
arum Aronstab *m* (Bot), Arum *n*
arum root Aronsknolle *f*, Aronswurzel *f* (Bot)
arum starch Aronsstärke *f*

aryl- Aryl-
arylamine Arylamin *n*
arylate *v* arylieren
arylation Arylierung *f*
arylbutyric acid Arylbuttersäure *f*
aryl halide Arylhalogenid *n*, Halogenaryl *n*
arylmethylbenzoic acid Arylmethylbenzoesäure *f*
arylpropionic acid Arylpropionsäure *f*
aryl silicone Arylsilicon *n*
aryl sulfonate Arylsulfonat *n*
arzrunite Arzrunit *m* (Min)
asafetida Asa Foetida *f*, Stinkasant *m*, Teufelsdreck *m*
asafetida oil Asafötidaöl *n*, Asantöl *n*
asafoetida Asa Foetida *f*
asafoetida oil Asafötidaöl *n*
asaprol Asaprol *n*, Abrastol *n*
asarabacca Haselwurz *f* (Bot)
asarin Asarin *n*, Asaron *n*
asarinin Asarinin *n*
asarone Asarin *n*, Asaron *n*, Asarumcampher *m*
asaronic acid Asaronsäure *f*
asarum Haselwurz *f* (Bot)
asarum camphor Asarumcampher *m*
asarum oil Asarumöl *n*
asaryl- Asaryl-
asarylic acid Asaronsäure *f*, Asarylsäure *f*
asbestiform asbestartig
asbestin Asbestin *n*
asbestoid asbestartig
asbestos Asbest *m* (Min), Bergflachs *m* (Min), Berghaar *n* (Min), Bergkork *m* (Min), Bergwolle *f* (Min), Bergzunder *m* (Min), Flachsstein *m* (Min), **dressing of** ~ Asbestaufbereitung *f*, **flaked** ~ Flockenasbest *m* (Min), **flexible** ~ Federasbest *m* (Min), **ligneous** ~ Holzasbest *m*, **palladized** ~ Palladiumasbest *m*, **raw** ~ Rohasbest *m*, **vein of** ~ Asbestader *f*
asbestos bed Asbestlager *n*
asbestos board Asbestpappe *f*
asbestos cement Asbestzement *m*, Eternit *n* (HN, Asbestzement)
asbestos cloth Asbestgewebe *n*, Asbestnetz *n*
asbestos-coated wire gauze Asbestdrahtnetz *n*
asbestos cord Asbestschnur *f*
asbestos covering Asbestbekleidung *f*
asbestos fabric Asbestgewebe *n*, Asbestnetz *n*
asbestos fiber Asbestfaser *f*
asbestos fiber bundle Asbestfaserbündel *n*
asbestos fiber sheet Asbestfaserplatte *f*
asbestos filler Asbestfüllstoff *m*
asbestos filter Asbestfilter *n*
asbestos glove Asbesthandschuh *m*
asbestos goods Asbestwaren *pl*
asbestos insulating plate Asbestisolierplatte *f*
asbestos joint Asbestdichtung *f*
asbestos [joint] ring Asbestring *m*

asbestos layer Asbesteinlage *f,* Asbestschicht *f*
asbestos milk Asbestaufschlämmung *f*
asbestos packing Asbestdichtung *f*
asbestos paper Asbestpapier *n*
asbestos quarry Asbestgrube *f*
asbestos-rubber high-pressure seals It-Platten *pl*
asbestos screen Asbestschirm *m*
asbestos sheet Asbestplatte *f*
asbestos slate Asbestschiefer *m* (Min)
asbestos sponge Asbestschwamm *m*
asbestos stove Asbestofen *m*
asbestos stratum Asbestschicht *f*
asbestos suit Asbestanzug *m*
asbestos suspension Asbestaufschlämmung *f*
asbestos washer Asbestscheibe *f*
asbestos wire gauze Asbestdrahtnetz *n,*
  Drahtasbestgewebe *n*
asbestos wire net Asbestdrahtnetz *n*
asbestos wool Asbestflocken *f pl*
asbestos wrapping Asbestumwicklung *f*
asbolane Asbolan *m* (Min), kobalthaltiger
  Braunstein *m* (Min), Kobaltmanganerz *n*
  (Min), Kobaltschwärze *f* (Min), Rußkobalt *m*
  (Min)
asbolin Asbolin *n*
asbolite Asbolan *m* (Min), Erdkobalt *m* (Min),
  Hornkobalt *m* (Min), kobalthaltiger
  Braunstein *m* (Min), Kobaltmanganerz *n*
  (Min), Kobaltschwärze *f* (Min), Rußkobalt *m*
  (Min)
ascaridic acid Ascaridinsäure *f*
ascaridol Ascaridol *n*
ascaridolic acid Ascaridinsäure *f*
ascaroside Ascarosid *n*
ascarylitol Ascarylit *m*
ascarylose Ascarylose *f*
ascend *v* steigen, ansteigen
ascending [auf]steigend
ascending [air] current Steigwind *m*
ascending force Auftrieb *m*
ascending pipe Steigleitung *f*
ascending pipe line Steigrohrleitung *f*
ascending power Auftrieb *m*
ascent Aufstieg *m,* Gefälle *n,* Steigung *f,* method
  of steepest ~ Methode *f* des steilsten Anstiegs
ascertain *v* bestimmen, ermitteln, feststellen,
  konstatieren
ascertainable feststellbar
Aschan's dichloride Aschan-Dichlorid *n*
ascharite Ascharit *m* (Min)
ascinin Ascinin *n* (HN, Lackhilfsstoff)
asclepiol Asclepiol *n*
ascomycetes Askomyzeten *pl,* Schlauchpilze *pl*
ascorbic acid Ascorbinsäure *f,* Vitamin C *n*
ascorbic acid dehydrogenase
  Ascorbinsäuredehydrogenase *f* (Biochem)
ascorbic acid oxidase Ascorbinsäureoxidase *f*
  (Biochem)
ascorbigen Ascorbigen *n*

ascorbylpalmitate Ascorbylpalmitat *n*
ascosterol Ascosterin *n*
asduana Asduana *f*
asebogenin Asebogenin *n*
asebotin Asebotin *n*
asellic acid Asellinsäure *f*
asepsis Asepsis *f* (Med)
aseptic aseptisch, fäulnishemmend, keimfrei,
  steril
aseptic filling aseptische Abfüllung *f*
asepticism Aseptik *f*
aseptic technique Aseptik *f*
aseptine Aseptin *n* (Med)
aseptol Aseptol *n,* Sozolsäure *f*
ash Asche *f* (Chem), determination of ~
  Aschenermittlung *f,* percentage of ~
  Aschegehalt *m,* rich in ~ aschereich
ash *v* veraschen, einäschern
ash bark Eschenrinde *f*
ash bath Aschenbad *n*
ash blower Flugascheausblaser *m*
ash-colored aschfarben
ash constituent Aschenbestandteil *m*
ash content Aschegehalt *m,* poor in ~
  aschenarm, rich in ~ aschenreich
ash determination Aschegehaltsbestimmung *f,*
  Aschenbestimmung *f*
ash discharge Entaschung *f* (Treibstoff)
ash ejector Aschenauswerfer *m*
ashes Asche *f,* Verbrennungsrückstand *m,* free
  from ~ aschenfrei, free from volatile ~
  flugaschenfrei, fusing of the ~ (clinkering)
  Aschenverflüssigung *f,* hot ~ Glühasche *f,*
  lixiviated ~ Kalkäscher *m,* lye from ~
  Aschenlauge *f,* removal of ~ Aschenabfuhr *f,*
  to burn to ~ einäschern, to reduce to ~
  einäschern
ash furnace Aschenofen *m* (Glasherstellung),
  Frittofen *m,* Glasschmelzofen *m*
  (Glasherstellung)
ash heap Aschenhalde *f*
ashing crucible Gekrätzprobentiegel *m*
ash pan Aschenkasten *m*
ash pit Aschengrube *f,* Aschenraum *m,*
  Feuergrube *f*
ash pocket Aschensack *m*
ash removal Entaschung *f* (Treibstoff)
ash removal plant Entaschungsanlage *f*
ash tree oil Eschenholzöl *n*
ashy aschfarben, aschig
asiatic acid Asiatsäure *f*
asiaticoside Asiaticosid *n*
aside seitwärts, abseits
asiderite Asiderit *m* (Min)
asimina Asimina *f* (Bot)
asmanite Asmanit *m* (Min)
asordin Asordin *n* (HN, Lösungsmittel)
asparacemic acid Asparacemsäure *f*
asparaginase Asparaginase *f* (Biochem)

asparagine Asparagin *n*, Althein *n* (obs),
   Asparamid *n* (obs)
asparaginic acid Asparaginsäure *f*
asparagus stone Spargelstein *m* (Min)
asparamide Asparagin *n*, Asparamid *n* (obs)
aspartal Aspartal *n*
aspartamic acid Asparagin *n*
aspartase Aspartase *f* (Biochem)
aspartate asparaginsaures Salz *n*, Aspartat *n*
aspartate *a* asparaginsauer
aspartic acid Asparaginsäure *f*,
   Aminobernsteinsäure *f*, Aspartinsäure *f*
aspasiolite Aspasiolith *m* (Min)
aspect Aspekt *m*
aspergillic acid Aspergillsäure *f*
aspergillus (pl. aspergilli) Aspergillus *m* (Pilz)
asperolite Asperolith *m*
asperthecin Asperthecin *n*
asphalin Asphalin *n*
asphalite Asphalit *m*
asphalt Asphalt *m*, Bergpech *n*, Bitumen *n*,
   Erdpech *n*, artificial ~ Teerpech *n*, hard ~
   Steinpech *n*, native ~ Asphaltstein *m*, soft ~
   Goudron *m*, Weichasphalt *m*, to coat with ~
   asphaltieren
asphalt *v* asphaltieren, mit Asphalt bestreichen
asphalt-bearing asphalthaltig
asphalt concrete Asphaltbeton *m*
asphalted cardboard Dachpappe *f*
asphalted paper Asphaltpappe *f*
asphalted wire Asphaltdraht *m*
asphaltic asphaltartig, asphalthaltig, asphaltisch,
   erdharzartig, erdpechhaltig
asphaltic cement Asphaltzement *m*,
   Asphaltmastix *m*
asphalt-impregnated felt Asphaltfilz *m*
asphalting Asphaltieren *n*
asphalt-laminated kraft paper Asphaltpapier *n*
asphalt mastic Asphaltmastix *m*, Asphaltkitt *m*
asphaltotype Asphaltnegativ *n*
asphalt paper Bitumenpapier *n*
asphalt paving Asphaltpflaster *n* (Straßenbau)
asphalt pulp Asphaltbrei *m*
asphalt rock Pechgang *m*
asphalt testing apparatus Asphaltprüfgerät *n*
asphaltum Asphalt *m*
asphalt varnish Asphaltfirnis *m*, Asphaltlack *m*
aspherical asphärisch, nicht kugelförmig
asphodel Asphodill *m* (Bot)
asphyxia Asphyxie *f* (Med)
asphyxiate *v* ersticken (durch Gas)
asphyxiation Erstickung *f*
aspidin Aspidin *n*
aspidinol Aspidinol *n*
aspidocarpine Aspidocarpin *n*
aspidolite Aspidolith *m*
aspidosamine Aspidosamin *n*
aspidosine Aspidosin *n*
aspidospermine Aspidospermin *n*

aspirate *v* absaugen (Gas), ansaugen (Gas)
aspirated volume Ansaug[e]volumen *n*
aspirating unit Luftsaugeraggregat *n*
aspiration Ansaugen *n*, Atmen *n*, Aufsaugen *n*;
   Einsaugen *n*
aspirator Abklärflasche *f*, Aspirator *m* (Med),
   Luftsaugepumpe *f*, Luftsauger *m*,
   Saugapparat *m*, Sauger *m*,
   Wasserstrahlpumpe *f*
aspirator [bottle] Saugflasche *f*
aspire *v* streben, aspirieren
aspirin Acetylsalicylsäure *f*, Aspirin *n* (HN)
assamar Assamar *n*
assay Analyse *f*, Gehaltsbestimmung *f*, Probe *f*,
   Prüfung *f*, Untersuchung *f*, Versuch *m*
assay *v* erproben, untersuchen, probieren,
   kapellieren (Met), prüfen
assay balance Probewaage *f*
assay crucible Probiertiegel *m*
assayed by cupelling abgetrieben
assayer Prüfer *m*, Anrichter *m*
assay furnace Kapellenofen *m*, Muffelofen *m*
assay lead Kornblei *n*, Probierblei *n*
assay of tapped metal Stichprobe *f* (Metall)
assay sensitivity Nachweisempfindlichkeit *f*
   (Anal)
assay weight Probiergewicht *n* (Metall)
assemblage Montage *f*, Montierung *f*
assemble *v* montieren, aufstellen,
   zusammenfügen, zusammensetzen
assembler Monteur *m*
assembling workshop Montagehalle *f*
assembly Versammlung *f*; Montage *f*,
   Zusammensetzung *f*, Zusammenstellung *f*,
   final ~ Fertigmontage *f*
assembly adhesive Montagekleber *m*,
   Montageklebstoff *m*
assembly line Fließband *n*, Montageband *n*
assembly line work Fließbandarbeit *f*
assembly point Sammelpunkt *m*
assembly time Verleimzeit *f*
assent *v* beipflichten, beistimmen
assert *v* behaupten, bekräftigen
assertion Behauptung *f*, Erklärung *f*
assign *v* zuordnen
assigned zugeordnet
assignment Abtretung *f* (Jur), Anweisung *f*,
   Bestimmung *f*, method of ~
   Zuordnungsverfahren *n*
assimilability Assimilierbarkeit *f*
assimilable assimilierbar
assimilate *v* angleichen, assimilieren
assimilated, capable of being ~ assimilierbar
assimilating process Assimilationsvorgang *m*
   (Biol)
assimilation Angleichung *f*, Assimilation *f*
   (Biol), Aufnahme *f* (Stoffwechsel)
assimilation [process] Assimilationsprozeß *m*
assimilative process Assimilationsprozeß *m*

assimilatory assimilatorisch, assimilierbar
assist v beisteuern, beitragen, helfen,
  unterstützen, assistieren, beistehen
assistance Beistand m, Mithilfe f,
  Zuhilfenahme f
assistant Assistent m, Beistand m, Gehilfe m
assisting plug Hilfsstempel m, Vorziehstempel m
associate v assoziieren, vereinigen, hinzufügen
associate professor Extraordinarius m
associate professorship Extraordinariat n
association Assoziation f, Verbindung f,
  Vereinigung f, **degree of** ~
  Assoziationsgrad m
assort v [as]sortieren, aussuchen
assortment Auswahl f; Sortieren n, Sortiment n
assuasive ointment Linderungssalbe f (Pharm)
assume v annehmen
assuming angenommen
assumption Annahme f, Voraussetzung f
assurance Gewährleistung f
assure v gewährleisten
astacein Astacein n
astacin Astacin n
A-state A-Zustand m
astatic astatisch (Magnet), unstet, unstabil
astatin[e] Astat n, Astatin n (Symb. At)
astatization Astatisierung f
asterane Asteran n
asterine Asterin n
asterism Asterismus m (Spektr)
asterric acid Asterrsäure f
asterubin Asterubin n
asthenosphere Asthenosphäre f
astigmatic astigmatisch
astigmatism Astigmatismus m (Phys)
astilbin Astilbin n
astragalin Astragalin n
astrakanite Astrakanit m (Min), Blödit m (Min)
astralite Astralit m
astralon (vinyl copolymer) Astralon n (HN,
  Vinylmischpolymerisat)
astraphloxine Astraphloxin n
astra yellow Astragelb n
astringent Adstringens n (pl Adstringentien),
  blutstillendes Mittel n
astringent a adstringierend, zusammenziehend
astrochemistry Astrochemie f
astrocyte Astrozyt m (Histol), Sternzelle f
  (Histol)
astrol Astrol n
astronautics Astronautik f, Kosmonautik f,
  Raumfahrt f, Weltraumschiffahrt f
astronomer Astronom m, Sternkundiger m
astronomic[al] astronomisch
astronomy Astronomie f, Sternkunde f
astrophyllite Astrophyllit m (Min)
astrophysics Astrophysik f
astropyrine Astropyrin n
astroscope Astroskop n

astrospectroscope Sternspektroskop n
asulgan (fur industry) Asulgan n (HN,
  Pelzindustrie)
asurol Asurol n
asymmetric[al] asymmetrisch, nicht symmetrisch,
  ungleichförmig, ungleichmäßig,
  unsymmetrisch
asymmetry Asymmetrie f, Unsymmetrie f, **atomic**
  ~ atomare Asymmetrie (Stereochem),
  conformational ~ konformative Asymmetrie
  (Stereochem), **degree of** ~
  Unsymmetriegrad m, **formation of** ~
  Ausbildung f der Asymmetrie, **molecular** ~
  molekulare Asymmetrie (Stereochem), ~ **due**
  to restricted rotation Asymmetrie f infolge
  Rotationsbehinderung (Stereochem)
asymptote Asymptote f (Math)
asymptotic asymptotisch
asymptotic behavior Asymptotik f
asymptotic expression asymptotischer Ausdruck
asymptotic focal length
  Asymptoten-Brennweite f
asymptotic image Asymptoten-Abbildung f
asynchronous asynchron
asynchronous motor Asynchronmotor m,
  Induktionsmotor m
atabrine Atebrin n (HN, Pharm)
atacamite Atakamit m (Min), Salzkupfererz n
atactic ataktisch
AT-cut crystal AT-Kristallschnitt m (Krist)
atebrin Atebrin n (HN, Pharm)
atelestite Atelestit m (Min)
atelite Atelit m (Min)
atephen Atephen n (Kunstharz)
athamantin Athamantin n
athebenol Äthebenol n
atheriastite Atheriastit m
athermal athermisch
athermanous atherman, wärmeundurchlässig
atherosclerosis Arterienverkalkung f (Med),
  Aderverkalkung f (Med), Arteriosklerose f
  (Med)
atherospermine Ätherospermin n
atidine Atidin n
atisine Atisin n
atlantone Atlanton n
atlas blue Atlasblau n
atlasite Atlasit m (Min)
atmolysis Atmolyse f
atmometer Verdunstungsmesser m,
  Atmometer n, Ausdunstungsmesser m,
  Dunstmesser m
atmosphere Atmosphäre f, Außenluft f, Luft f,
  Lufthülle f, **residual** ~
  Atmosphärenrückstand m
atmospheric atmosphärisch
atmospheric action Lufteinwirkung f
atmospheric condition Luftzustand m
atmospheric conditions Witterungsverhältnisse pl

atmospheric cracking Lichtrisse *m pl*
(Prüftechnik)
atmospheric density gauge
Luftverdichtungsmesser *m*
atmospheric disturbance atmosphärische
Störung *f*
atmospheric electricity Luftelektrizität *f*
atmospheric engine Luftmaschine *f*
atmospheric exposure Witterungseinwirkung *f*
atmospheric moisture Luftfeuchtigkeit *f*
atmospheric nitrogen Luftstickstoff *m*, **fixation of**
~ Luftstickstoff-Fixierung *f*
atmospheric oxygen Luftsauerstoff *m*
atmospheric phenomenon Lufterscheinung *f*
atmospheric pressure Atmosphärendruck *m*,
Außendruck *m*, Luftdruck *m*
atmospheric reflection Luftspiegelung *f* (Opt)
atmospheric resistance Luftwiderstand *m*
atmospheric stone Meteorstein *m*
atmospheric substances Atmosphärilien *pl*
atmospheric sulfur Luftschwefel *m*
atmospheric water Luftwasser *n*
atom Atom *n*, **foreign** ~ Fremdatom *n*, **impurity**
~ Fremdatom *n*, **interstitial** ~
Gitterhohlraumatom *n* (Krist), **peripheric** ~
Außenatom *n*, **stripped** ~ hochionisiertes
Atom
atom conversion Atomumwandlung *f*
atom excitation Atomerregung *f*
atom fission Atomspaltung *f*
atom gun Atomkanone *f*
atomic atomar, atomisch
atomic absorption analysis
Atomabsorptionsanalyse *f*
atomic absorption spectrometry
Atom-Absorptionsspektrometrie *f*
atomic aeroengine Atomflugmotor *m*
atomic affinity Atomaffinität *f*
atomic age Atomzeitalter *n*
atomic arrangement Gitteraufbau *m*
atomic bomb Atombombe *f*
atomic bombardment Atombeschießung *f*,
Atombeschuß *m*
atomic bombardment particle
Atomgeschoßteilchen *n*
atomic bond Atombindung *f*
atomic burst Atomexplosion *f*
atomic charge Atomladung *f*
atomic chart Atomgewichtstabelle *f*,
Atomgewichtstafel *f*
atomic chemistry Atomchemie *f*
atomic clock Atomuhr *f*
atomic combining power Atombindungskraft *f*
atomic conductance Atomleitfähigkeit *f*
atomic core Atomkern *m*
atomic decay Atomzerfall *m*
atomic decomposition Atomzerlegung *f*
atomic density Atomdichte *f*
atomic diameter Atomdurchmesser *m*

atomic disintegration Atomzerfall *m*
atomic dispersion Atomdispersion *f*
atomic displacement Atomverschiebung *f*,
Verschiebung *f pl* von Atomen
atomic dust Atomstaub *m*
atomic electric station Atomkraftwerk *n*
atomic energy Atomenergie *f*
atomic energy generation
Atomenergieerzeugung *f*
atomic energy machine Atomkraftmaschine *f*
atomic envelope Atomhülle *f*
atomic explosion Atomexplosion *f*
atomic explosive Atomsprengstoff *m*
atomic frequency Atomfrequenz *f*
atomic fuel Atombrennstoff *m*
atomic fusion Atomverschmelzung *f*
atomic heat Atomwärme *f*
atomic heat of fusion Atomschmelzwärme *f*
atomic hydrogen [arc] welding
Arcatomschweißung *f*
atomic hydrogen welding Arkatomschweißen *n*
atomic hypothesis Atomhypothese *f*
atomic index Atomzahl *f*, Atomziffer *f*,
Kernladungszahl *f*
atomicity Atomigkeit *f*, Atomismus *m*,
Atomistik *f*, Atomizität *f*, Valenz *f*,
Wertigkeit *f*
atomic level Atomhüllenniveau *n*
atomic linkage Atombindung *f*,
Atomverkettung *f*
atomic magnetism Atommagnetismus *m*
atomic mass Atommasse *f*
atomic mass constant Atommassenkonstante *f*
atomic mass unit Atommasseeinheit *f*
atomic medicine Atommedizin *f*
atomic model Atommodell *n*
atomic motion Atombewegung *f*
atomic nucleus Atomkern *m*
atomic nucleus explosion Atomkernsprengung *f*
atomic number Kernladungszahl *f* (Atom),
Ordnungszahl *f*
atomic orbital Atomorbital *n*
atomic oscillation Atomschwingung *f*
atomic pile Atomreaktor *m*, Kernreaktor *m*,
Reaktor *m*
atomic plane Netzebene *f* (Krist)
atomic position Atomlage *f*
atomic power Atomkraft *f*
atomic-powered atombetrieben, mit Atomkraft
betrieben
atomic-powered submarine Atom-U-Boot *n*
atomic power plant Atomkraftanlage *f*,
Atomkraftwerk *n*
atomic power station Atomkraftanlage *f*,
Atomkraftwerk *n*
atomic projectile Atomgeschoß *n*
atomic propulsion Atomantrieb *m*

atomic radiation atomare Strahlung *f,*
  Atomstrahlung *f,* **injury caused by** ~
  Atomstrahlungsschäden *pl*
atomic radius Atomhalbmesser *m,* Atomradius *m*
atomic ratio Atomverhältnis *n*
atomic ray Atomstrahl *m*
atomic refraction Atomrefraktion *f*
atomic relation Atomverhältnis *n*
atomic research Atomforschung *f*
atomic residue Atomrest *m,* Atomrumpf *m*
atomic resistivity Atomwiderstand *m*
atomic rotation Atomrotation *f*
atomics Atomphysik *f*
atomic science Atomwissenschaft *f*
atomic scientist Atomforscher *m*
atomic shell Atomschale *f*
atomic sign Atomzeichen *n*
atomic space lattice Atomgitter *n*
atomic spectrum Atomspektrum *n*
atomic state Atomzustand *m*
atomic structure Atom[auf]bau *m,* Atomismus *m,*
  Atomstruktur *f,* Raumgitter *n* (Krist)
atomic symbol Atomzeichen *n*
atomic synthesis Atomsynthese *f*
atomic test Atomversuch *m*
atomic theory Atomtheorie *f*
atomic timing device Atomuhr *f*
atomic transformation Atomumwandlung *f*
atomic union Atomverband *m*
atomic valence Atomwertigkeit *f,* Atomaffinität *f,*
  Atomvalenz *f*
atomic volume Atomvolumen *n*
atomic warhead Atomsprengkopf *m*
atomic waste, removal of ~
  Atommüllbeseitigung *f*
atomic weapon Atomwaffe *f,* Kernwaffe *f*
atomic weight Atomgewicht *n,* **determination of**
  ~ Atomgewichtsbestimmung *f*
atomic weights, table of ~ Atomgewichtstafel *f*
atomism Atomismus *m*
atomistic atomistisch
atomistics Atomistik *f,* Atomwissenschaft *f*
atomization Atomisierung *f,* Zerstäubung *f*
atomize *v* atomisieren; verstäuben, zerstäuben
atomized fuel Ölnebel *m*
atomized metal Metallnebel *m*
atomizer Feinstzerstäuber *m,*
  Flüssigkeitszerstäuber *m,* Sprühapparat *m,*
  Sprühpistole *f,* Verstäuber *m,*
  Verstäubungsapparat *m,* Zerstäuber *m,*
  centrifugal disc ~ Rotations-Zerstäuber *m,*
  return-flow ~ Rücklaufzerstäuber *m,*
  rotary-cup ~ Drehzerstäuber *m,* **sonic** ~
  Schallzerstäuber *m,* **steam** ~
  Öldruck-Dampfdruck-Zerstäuber *m,* **tubular**
  ~ Rohrzerstäuber *m,* **steam** ~ , **Y-atomizer**
  Y-Zerstäuber
atomizer cone Zerstäuberkegel *m*
atomizing agent Zerstäubungsmittel *n*

atomizing crystallizer
  Zerstäubungs-Kristallisator *m*
atomizing purifier Zerstäubungsreiniger *m*
atomizing valve Zerstäuberventil *n*
atom model Atomkalotte *f*
atom nucleus Atomkern *m*
atom number Atomzahl *f*
atoms, cluster of ~ Atomhaufen *m,* **hypothesis of**
  ~ Atomhypothese *f,* **linking of** ~
  Atomverkettung *f,* **movement of** ~
  Atombewegung *f,* **ring of** ~ Atomring *m,*
  union of ~ Atomverband *m*
atom smasher Atomzertrümmerungsmaschine *f,*
  Teilchenbeschleuniger *m*
atom smashing Atomzertrümmerung *f*
atom smashing experiment
  Atomzertrümmerungsversuch *m*
atom smashing machine
  Atomzertrümmerungsmaschine *f*
atom splitting Atomaufspaltung *f,*
  Atomkernspaltung *f,* Atomzertrümmerung *f*
atom theory Atomtheorie *f*
atophan Atophan *n,* Cinchophen *n*
atopite Atopit *m* (Min)
atoxic atoxisch, ungiftig
atoxicocaine Atoxicocain *n*
atoxyl Atoxyl *n* (HN)
atoxylic acid Arsanilsäure *f,* Atoxylsäure *f*
atractylene Atractylen *n,* Machilen *n*
atractylol Atractylol *n,* Machilol *n*
atrament process Atramentverfahren *n*
atranol Atranol *n*
atranoric acid Atranorsäure *f*
atranorin Atranorin *n*
atranorinic acid Atranorinsäure *f*
atrolactic acid Atrolactinsäure *f*
atromentic acid Atromentinsäure *f*
atromentin Atromentin *n*
atropamine Atropamin *n,* Apoatropin *n*
atrophy Atrophie *f* (Med), Ausfallerscheinung *f*
  (Muskel), Entartung *f,* Verkümmerung *f*
atropic acid Atropasäure *f*
atropin[e] Atropin *n*
atropine hydrobromide Atropinhydrobromid *n*
atropine methylbromide Atropinmethylbromid *n,*
  Atropinbrommethylat *n*
atropine methyl nitrate Atropinmethylnitrat *n*
atropine poisoning Atropinvergiftung *f* (Med)
atropine salicylate Atropinsalicylat *n*
atropine sulfate Atropinsulfat *n*
atropism Atropinsucht *f* (Med),
  Atropinvergiftung *f* (Med)
atropisomerism Atropisomerie *f* (Stereochem)
atropurol Atropurol *n*
atroscine Atroscin *n*
atrovenetin Atrovenetin *n*
attach *v* anfügen, angliedern, anheften,
  ankleben, ankuppeln, aufstecken, befestigen,
  beifügen, knüpfen, verbinden

attachable ansetzbar, aufsetzbar, aufsteckbar
attached, to be ~ anhaften
attached piece Ansatzstück *n*
attached tube Ansatzrohr *n*
attachment Anhaftung *f,* Anhang *m,* Aufsatz *m,*
  Befestigung *f,* Beiwerk *n,* Gestellteil *n,* **point of**
  ~ Ansatzpunkt *m,* Haftstelle *f*
attachment plug Zwischenstecker *m*
attack Anfall *m;* (Med), Angriff *m;* Befall *m,*
  Inangriffnahme *f,* **chemical** ~ chemischer
  Angriff, **line of** ~ Angriffslinie *f,* **means of** ~
  Bekämpfungsmittel *n*
attack *v* angreifen, anfallen; anfechten;
  anfressen; befallen
attackable angreifbar
attacking agent Angriffsmittel *n*
attainable erreichbar
attemperator Bottichkühler *m,*
  Temperaturregulator *m*
attempt to interpret Deutungsversuch *m*
attend *v* bedienen, begleiten, warten (Techn)
attendance Wartung *f,* Bedienung *f*
attendant Bedienungsmann *m*
attention! Achtung!, **to pay** ~ achtgeben
attenuate *v* abschwächen, verkleinern, dämpfen
  (Schwingungen)
attenuation Abdämpfung *f* (Akust),
  Abschwächung *f,* Dämpfung *f*
  (Schwingungen), Schwächung *f,* ~ **per unit**
  length Dämpfung *f* je Längeneinheit
attenuation coefficient
  Schwächungskoeffizient *m,*
  Abschwächungskoeffizient *m* (Atom),
  Dämpfungskonstante *f*
attenuation coil Abflachungsdrossel *f*
attenuation constant Dämpfung *f* je
  Längeneinheit
attenuation distortion Dämpfungsverzerrung *f*
attenuation equalization Dämpfungsausgleich *m,*
  Dämpfungsentzerrung *f*
attenuation factor Abschwächungsfaktor *m,*
  Abschwächungskoeffizient *m* (Atom)
attenuation of sound Abfall *m* der
  Schallintensität
attenuation region Dämpfungsbereich *m* (Akust)
attenuator Empfindlichkeitsregler *m*
  (Gaschromat)
attest *v* beglaubigen, bescheinigen, beurkunden
attitude Haltung *f,* Stellung *f*
attract *v* anziehen
attractant Lockstoff *m*
attraction Anziehung *f,* Anziehungskraft *f,*
  Reiz *m,* **center of** ~ Anziehungszentrum *n,*
  effect of ~ Anziehungseffekt *m,* **law of** ~
  Anziehungsgesetz *n*
attraction constant Anziehungszahl *f*
attraction of gravity Gravitationskraft *f*
attraction potential Anziehungspotential *n*
attractive capacity Anziehungskraft *f*

attractive force Anziehung *f,* Anziehungskraft *f*
attribute *v* beilegen, beimessen
attrition Abreibung *f,* Abrieb *m,*
  Schleifwirkung *f,* Zerreibung *f*
attrition mill, toothed ~ Zahnscheibenmühle *f,*
  two-roll ~ Zweirollen-Scheibenmühle *f*
AU (antitoxin unit) Antitoxineinheit *f* (AE)
auburn kastanienbraun, Kastanienbraun *n*
auburn *a* kastanienbraun, goldbraun
aucubin Aucubin *n*
audibility Hörbarkeit *f,* **limit of** ~
  Hörbarkeitsgrenze *f,* **range of** ~
  Hörbereich *m*
audible hörbar, vernehmbar
audience Zuhörerschaft *f*
audio amplifier Niederfrequenzverstärker *m*
audio frequency Niederfrequenz *f,*
  Schallfrequenz *f*
audio frequency spectrometer
  Tonfrequenzspektrometer *n*
audiogram Audiogramm *n*
audiometry NF-Messung *f* (Elektr)
audion receiver Audionempfänger *m*
audio transformer
  Niederfrequenztransformator *m*
auditorium Hörsaal *m,* Zuschauerraum *m*
auerbachite Auerbachit *m* (Min)
auerlite Auerlith *m* (Min)
auer metal Auermetall *n*
auger [großer] Bohrer *m,* Hohlbohrer *m,*
  Holzbohrer *m*
auger bit Bohrspitze *f*
Auger effect Augereffekt *m* (Phys)
Auger electron Auger-Elektron *n* (Atom)
Auger electron spectroscopy
  Auger-Spektroskopie *f*
Auger transition Augerübergang *m*
augite Augit *m* (Min), Malakolith *m* (Min),
  Omphazit *m* (Min), **foliated** ~ Blätteraugit *m*
  (Min)
augite porphyry Augitporphyr *m* (Min),
  Melaphyr *m* (Min)
augite series Augitreihe *f*
augitic augitartig (Min), augithaltig (Min)
augitite Augitit *m* (Min)
aulamine Aulamin *n*
aural akustisch
auramine Auramin *n,* Apyonin *n*
auramine base Auraminbase *f*
auramine hydrochloride Auraminhydrochlorid *n*
auranetin Auranetin *n*
aurantia Aurantia *n*
aurantiacin Aurantiacin *n*
aurantine Aurantin *n*
aurantiogliocladin Aurantiogliocladin *n*
aurantium Naringin *n*
aurate Aurat *n,* Goldoxidsalz *n*
aureate goldgelb
aureine Aurein *n*

aureolin Aureolin *n*
aureomycin Aureomycin *n*, Chlortetracyclin *n*
aureomycinic acid Aureomycinsäure *f*
aureone amide Aureonamid *n*
aureothinic acid Aureothinsäure *f*
aureothricin Aureothricin *n*
aureusidin Aureusidin *n*
aureusin Aureusin *n*
auri-argentiferous gold- und silberhaltig
auric acid Goldsäure *f*
auric bromide Gold(III)-bromid *n*
auric chloride Aurichlorid *n*, Gold(III)-chlorid *n*
auric compound Auriverbindung *f*,
  Gold(III)-Verbindung *f*,
  Goldoxidverbindung *f*
auric cyanide Auricyanid *n*, Gold(III)-cyanid *n*
aurichalcite Aurichalcit *m* (Min),
  Messingblüte *f* (Min)
aurichloride Gold(III)-chlorid *n*
aurichlorohydric acid Aurichlorwasserstoffsäure,
  Gold(III)-chlorwasserstoffsäure *f*
auric hydroxide Aurihydroxid *n*,
  Gold(III)-hydroxid *n*, Goldsäure *f*
auric iodide Gold(III)-jodid *n*
auric oxide Aurioxid, Gold(III)-oxid *n*
auric salt Gold(III)-Salz *n*, Goldoxidsalz *n*
auric selenide Goldselenid *n*
auric sulfide Aurisulfid *n*, Gold(III)-sulfid *n*
auricyanic acid Gold(III)-cyanwasserstoffsäure *f*
auriferous goldführend, goldhaltig, goldreich
auriferous earth Golderde *f*
auriferous iron pyrites Goldschwefelkies *m*
  (Min)
auriferous pyrites Goldkies *m*
auriferous sinter Goldsinter *m*
auriferous stone Goldstein *m*
aurin Aurin *n*, Pararosolsäure *f*
aurine Corallin *n*
auripigment Auripigment *n* (Min),
  Rauschgelb *n* (Min), Reißgelb *n* (Min)
aurithiocyanic acid
  Aurirhodanwasserstoffsäure *f*,
  Gold(III)-rhodanwasserstoffsäure *f*
aurocantane Aurocantan *n*
aurochin Aurochin *n*
aurocyanic acid Aurocyanwasserstoffsäure *f*,
  Gold(I)-cyanwasserstoffsäure *f*
aurofelt Aurofelt *n* (Haarbeize)
auronal dye Auronalfarbe *f*
aurone Auron *n*
aurophenine Aurophenin *n*
aurora australis Südlicht *n* (Astr)
aurora borealis Nordlicht *n* (Astr)
auroral line Aurora-Linie *f* (Spektr), **green ~**
  Nordlichtlinie *f* (Astr)
aurothiocyanic acid
  Aurorhodanwasserstoffsäure *f*,
  Gold(I)-rhodanwasserstoffsäure *f*
aurothiosulfuric acid Goldthioschwefelsäure *f*

aurotine Aurotin *n*
aurous bromide Aurobromid, Gold(I)-bromid *n*
aurous chloride Aurochlorid, Gold(I)-chlorid *n*,
  Goldchlorür *n* (obs)
aurous compound Auroverbindung *f*,
  Gold(I)-Verbindung *f*,
  Goldoxydulverbindung *f* (obs)
aurous cyanide Aurocyanid, Gold(I)-cyanid *n*,
  Goldcyanür *n* (obs)
aurous iodide Gold(I)-jodid, Goldjodür *n*
aurous oxide Aurooxid *n*, Gold(I)-oxid *n*
aurous salt Gold(I)-Salz *n*
aurous sulfide Gold(I)-sulfid *n*
auroxanthin Auroxanthin *n*
aurum (Lat) Gold *n*
austenite Austenit *m*
austenitic austenitisch (Met)
austenitic manganese steel Manganhartstahl *m*
austenitic steel Austenitstahl *m*
australene Australen *n*
Australian tannage Australgerbung *f*
australite Australit *m* (Min)
australol Australol *n*
autan Autan *n*
authenticate *v* beurkunden
authenticity, certificate of ~ Echtheitszeugnis *n*,
  proof of ~ Echtheitsbeweis *m*
author Autor *m*, Schriftsteller *m*, Urheber *m*,
  Verfasser *m*
author index Autorenregister *n*
authority Behörde *f*, Instanz *f*,
  Sachverständiger *m*
authorization Bevollmächtigung *f*, Befugnis *f*,
  Berechtigung *f*
authorize *v* bevollmächtigen, beauftragen,
  befugen, berechtigen
authorized agent Bevollmächtigter *m*
auto-agglutination Autoagglutination *f*
autobarotropic autobarotrop
autobarotropy Autobarotropie *f*
autobody sheet Karosserieblech *n* (Kfz)
autocatalysis Autokatalyse *f*
autocatalyst Autokatalysator *m*
autocatalytic autokatalytisch
autochrome Autochrom *n*
autochrome process Autochromverfahren *n*
  (Phot)
autoclave Autoklav *m*, Dampfkochkessel *m*,
  Druckkessel *m*, Drucktopf *m*,
  Schnellkochtopf *m*, **~ with stirrer**
  Rührautoklav *m*
autoclave *v* sterilisieren
autoclave method Autoklavenverfahren *n*
autoclave single heater Autoklaveneinzelheizer *m*
autocollimator Autokollimator *m* (Opt)
autocorrelation function
  Autokorrelationsfunktion *f*
autodiffusion Selbstdiffusion *f*
autodigestion Autodigestion *f*, Autolyse *f*

autodyne Autodyn *n*
autoelectric effect Schottky-Effekt *m*
autoelectronic emission
　Autoelektronenemission *f*
autofining process Autofining-Verfahren *n*
autogenous autogen
autogenous cutting autogenes Schneiden *n*
autogenous welding Autogenschweißung *f,*
　Gasschweißen *n,*
　Wasserstoff-Sauerstoff-Schweißung *f*
autogenous welding process
　Autogenschweißverfahren *n*
autographic autographisch
autography Autographie *f*
autoh[a]emolysin Autohämolysin *n*
autoh[a]emolysis Autohämolyse *f*
autoh[a]emolytic autohämolytisch
autointoxication Autointoxikation *f,*
　Selbstvergiftung *f*
autoionization Autoionisation *f,*
　Selbstionisation *f*
autolysis Autolyse *f*
automate *v* automatisieren
automated [voll]automatisiert
automatic automatisch, selbsttätig, **fully** ~
　vollautomatisch
automatic air filter Drehfilter *n,*
　Rollband-Luftfilter *n*
automatically controlled automatisch betätigt
automatic arc welding machine
　Lichtbogenschweißautomat *m*
automatic batch centrifugal pusher
　Schubzentrifuge *f*
automatic circuit breaker Selbstunterbrecher *m*
　(Elektr)
automatic control Automatisierung *f*
automatic control technology Regelungstechnik *f*
automatic defroster Abtauautomatik *f*
　(Kühlschrank)
automatic die cutter vollautomatischer
　Stanzautomat *m*
automatic discharge Selbstentladung *f,*
　Selbstentleerung *f*
automatic exposure timer Belichtungsautomat *m*
automatic grab Drehschaufelbagger *m,*
　Greifbagger *m*
automatic gripping action Selbstgreiferbetrieb *m*
automatic lathe Drehautomat *m*
automatic lighter Selbstzünder *m*
automatic machine Automat *m*
automatic pressure apparatus Druckautomat *m*
automatic regulation Selbstregulierung *f,*
　Selbststeuerung *f*
automatic release Selbstauslöser *m* (Phot)
automatic resilience Eigenfederung *f*
automation Automation *f,* Automatisierung *f,*
　complete ~ Vollautomatisierung *f*
automobile Kraftfahrzeug *n,* Kraftwagen *m*
automobile body Karosserie *f* (Kfz)

automobile exhaust gas Autoabgas *n*
automobile finish Autolack *m*
automolite Automolith *m* (Min)
automorphic automorph, idiomorph (Krist)
automorphism Automorphie *f*
autonomic autonom
autopolymerization Autopolymerisation *f*
autopsy Leichenöffnung *f*
autoracemization Autoracemisierung *f*
autoradiogram Autoradiogramm *n*
auto-refrigerated cascade ARC-Verfahren *n*
autostereo-regulation Autostereoregulierung *f*
autotrophy Autotrophie *f* (Biol)
autotype Autotypie *f* (Buchdr), Rasterätzung *f*
　(Buchdr)
autotypy Autotypie *f* (Buchdr), Rasterätzung *f*
　(Buchdr)
autoxidation Autoxidation *f,* Selbstoxidation *f*
autoxidizability Aut[o]oxidationsfähigkeit *f*
autoxidizable aut[o]oxidationsfähig, autoxidabel
autoxidizer Autoxidator *m*
autunite Autunit *m* (Min), Kalkuranglimmer *m*
　(Min), Kalkuranit *m* (Min)
auxiliary Hilfsmittel *n*
auxiliary *a* zusätzlich
auxiliary appliances Nebenapparate *pl,*
　Zusatzapparate *pl*
auxiliary attachment Hilfsvorrichtung *f*
auxiliary condenser Hilfskondensator *m*
auxiliary container Nebenkasten *m*
auxiliary current Hilfsstrom *m*
auxiliary engine Hilfsmaschine *f,* Hilfsmotor *m*
auxiliary equation Hilfsgleichung *f* (Math)
auxiliary equipment Hilfseinrichtung *f*
auxiliary feeding stuff Beifuttermittel *n*
auxiliary force Hilfskraft *f*
auxiliary line Hilfslinie *f*
auxiliary main Hilfsleitung *f*
auxiliary mordant Hilfsbeize *f*
auxiliary phase Hilfsphase *f*
auxiliary pole Hilfspol *m*
auxiliary product Hilfsmittel *n,* Hilfsprodukt *n*
auxiliary quantity Hilfsgröße *f* (Math)
auxiliary series Nebenreihe *f*
auxiliary transformer Hilfstransformator *m*
auxiliary voltage Hilfsspannung *f*
auxiliary controlled variable Hilfsstellgröße *f*
　(Regeltechn)
auxin Auxin *n*
auxochrome Auxochrom *n*
auxochromic auxochrom
availability Zugänglichkeit *f*
available verfügbar, erhältlich, zugänglich
available load Nutzlast *f*
avalanche Lawine *f,* **hot** ~ Glutwolke *f*
avalanche breakdown Lawinendurchschlag *m*
avalanche fluctuation Lawinenschwankung *f*
avalite Avalit *m*

avenin Avenin *n*, Legumin *n*
avens root Nelkenwurzel *f* (Bot)
aventurine Aventurin *m* (Min),
  Glimmerquarz *m* (Min), Goldsiegellack *m*
aventurine feldspar Aventurinfeldspat *m* (Min),
  Aventurinstein *m* (Min)
aventurine [glass] Aventuringlas *n*
aventurine quartz Aventurinquarz *m* (Min)
aventurize *v* aventurisieren
average Durchschnitt *m*, Durchschnittszahl *f*,
  Mittel *n*, Mittelwert *m*, **general** ~
  Gesamthäufigkeit *f*, **to form the** ~ ausmitteln
average *a* durchschnittlich, gemittelt
average output Durchschnittsausbringung *f*
average performance Durchschnittsleistung *f*
average price Durchschnittspreis *m*
average quality Durchschnittsqualität *f*
average roughness number Glättungstiefe *f*
average sample Durchschnittsprobe *f*
average speed Durchschnittsgeschwindigkeit *f*
average tension Mittelspannung *f*
average value Durchschnittswert *m*
average voltage Mittelspannung *f*
average weight Durchschnittsgewicht *n*
averaging Mittelung *f*
avert *v* abwenden, verhüten
avertin Avertin *n* (HN), Tribromoäthanol *n*
aviation Flugwesen *n*, Luftfahrt *f*, **interplanetary**
  ~ Raumfahrt *f*
aviation fuel Flugkraftstoff *m*
aviation gasoline Flugbenzin *n*
aviation industry Flugzeugindustrie *f*
avicine Avicin *n*
avicularin Avicularin *n*
avidin Avidin *n*
avidity Avidität *f* (obs), starke Affinität *f*
avitaminosis Avitaminose *f* (Med)
Avogadro constant Avogadrosche Konstante *f*
Avogadro number Avogadrosche Konstante *f*
Avogadro's constant Avogadrosche Zahl *f*
Avogadro's law Avogadrosches Gesetz *n*
avoid *v* vermeiden, entgehen, verhindern,
  verhüten
avoidance Vermeidung *f*
avoirdupois [weight] Handelsgewicht *n*
await *v* abwarten
award, to accord an ~ prämieren
awaruite Awaruit *m* (Min)
away entfernt
awl Ahle *f*, Handbohrer *m*, Pfrieme *f*
awning Markisenstoff *m*, Zeltplane *f*, Zeltstoff *m*
axerophthene Axerophthen *n*
axerophthol Axerophthol *n*,
  Epithelschutzvitamin *n*
axes, distance between ~ Achsabstand *m*
  (Math), **intersection of** ~ Achsenkreuz *n*
axial achsenförmig, axial, **boat** ~ axial in der
  Wannenform (Stereochem)
axial angle Achsenwinkel *m*

axial compression Knickbeanspruchung *f*
axial direction Achsenrichtung *f*
axial-fed die axial gespeiste Düse *f*
axial-flow compressor Axialverdichter *m*
axial-flow pump Axialpumpe *f*
axial force Axialkraft *f*
axial head Längsspritzkopf *m*
axially symmetrical achsensymmetrisch
axial plane Achsenebene *f* (Krist)
axial pressure Achsdruck *m*
axial pump Schraubenpumpe *f*
axial ratio Achsenverhältnis *n*
axial sponginess sekundärer Lunker *m*
axial stress Axialbeanspruchung *f*,
  Axialspannung *f*
axial symmetry Achsensymmetrie *f* (Krist),
  Spiegelsymmetrie *f*
axial thrust Axialschub *m*
axial tube Achsenrohr *n*
axin Axin *n*
axinic acid Axinsäure *f*
axinite Afterschörl *m* (Min), Axinit *m* (Min),
  Glasschörl *m* (Min)
axiom Axiom *n*, Grundregel *f*, Grundsatz *m*
axiometer Axiometer *n*
axis Achse *f*, Mittellinie *f*, **crystallographic** ~
  kristallographische Achse, **free** ~ freie Achse,
  horizontal ~ horizontale Achse,
  Horizontalachse *f*, **inclination of an** ~
  Achsenneigung *f*, **magnetic** ~ magnetische
  Achse, **major** ~ große Achse (Ellipse), **minor**
  ~ kleine Achse (Ellipse), **optic** ~ optische
  Achse, **parallactic** ~ parallaktische Achse,
  principal ~ Hauptachse *f*, **section through the**
  ~ Achsenschnitt *m*, **to rotate around one's own**
  ~ sich um seine eigene Achse drehen, **turning**
  ~ Drehungsachse *f*, **vertical** ~ vertikale
  Achse, ~ **parallel to the path** Bahnachse *f*
axis intercept Achsenabschnitt *m*
axis of ordinate Ordinatenachse *f*
axis of rotation Rotationsachse *f*,
  Bewegungsachse *f*, Impulsachse *f*, **fixed** ~
  feste Drehungsachse (Mech)
axis of symmetry Symmetrieachse *f*, **principal** ~
  Hauptsymmetrieachse *f*
axis of tilt Hauptwaagerechte *f*
axis of vibration or oscillation
  Schwingungsachse *f*
axle Achse *f* (z. B. Rad), Drehachse *f*, Welle *f*
axle base Achsabstand *m*
axle bearing Achs[en]lager *n*, Achsbüchse *f*
axle box Achsbüchse *f*
axle box guide Achsbüchsenführung *f*
axle box liner Achsgleitplatte *f*
axle casing Achsengehäuse *n*
axle end pivot Achszapfen *m*
axle friction Achsenreibung *f*
axle grease Achsenschmiere *f*, Achsfett *n*,
  Wagenschmiere *f*

axle journal Achszapfen *m*
axle lathe Achsendrehbank *f*
axle load Achsbelastung *f,* Achsdruck *m*
axle oil Achsenöl *n*
axle pressure Achsendruck *m*
axle ring Achsenring *m,* Buchsring *m*
axle seat Achs[en]lager *n*
axle sleeve Achsengehäuse *n*
axle spring Achsfeder *f*
axle suspension Achsaufhängung *f,*
Achsfederung *f*
axstone Beilstein *m* (Min), Nephrit *m* (Min)
ayanin Ayanin *n*
ayapanin Ayapanin *n*
ayapin Ayapin *n*
azacyclonol Azacyclonol *n*
azafrin Azafrin *n*
azaleatin Azaleatin *n*
azaleine Azalein *n*
azamethonium Azamethonium *n*
azarsine Azarsin *n*
azaserine Azaserin *n*
azaurolic acid Azaurolsäure *f*
azedarach oil Margosöl *n*
azelaic acid Azelainsäure *f*
azelaic semialdehyde Azelainhalbaldehyd *m*
azelain Azelain *n*
azelaoin Azelaoin *n*
azelaone Azelaon *n*
azelate anchoinsauer
azeotrope azeotropes Gemisch *n*
azeotropic azeotrop
azeotropic point Azeotroppunkt *m*
az▮▮py Azeotropie *f*
az▮▮e Azepin *n*
az▮▮ine Azetidin *n*
az▮▮zil Azibenzil *n*
a▮▮ Azid *n,* Triazo-
a▮▮blue Azidinblau *n*
▮▮e color Azidinfarbe *f*
▮▮ne fast yellow Azidinechtgelb *n*
▮o- Azido-
▮oacetic acid Azidoessigsäure *f*
▮thane Aziäthan *n*
▮methane Diazomethan *n,* Azimethan *n*
▮methylene Diazomethan *n*
▮mino- Azimino-
▮iminobenzene Aziminobenzol *n*
▮imuth Azimut *m, n,* Scheitelbogen *m,*
Scheitelwinkel *m*
zimuthal mode Azimutalschwingung *f*
zimuth [angle] Azimut *n,* Höhenwinkel *m*
azimuth compass Azimutkompaß *m*
azimuth of course Kurswinkel *m*
azine Azin *n*
azine color Azinfarbstoff *m*
azine dye Azinfarbstoff *m*
azine dyestuff Azinfarbstoff *m*

azine green Azingrün *n*
azipyrazole Azipyrazol *n*
azipyridil Azipyridil *n*
aziridine Aziridin *n*
azlactone Azlacton *n*
azlactone synthesis Azlactonsynthese *f*
azlon fiber Azlonfaser *f*
azo acid blue Azosäureblau *n*
azo acid yellow Azosäuregelb *n*
azoanisole Azoanisol *n*
azobenzene Azobenzol *n*
azobenzol Azobenzol *n*
azoblue Azoblau *n*
azo body Azokörper *m*
azo carmine Azokarmin *n*
azocinnamic acid Azozimtsäure *f*
azo color Azofarbe *f,* acid ~
Azosäuregerbstoff *m,* fast ~ Azoechtfarbe *f*
azo compound Azoverbindung *f,* Azokörper *m*
azocyclic azozyklisch
azo derivative Azoverbindung *f,* Azokörper *m*
azodicarbonamide Azodicarbonamid *n,*
Azoformamid *n*
azodicarbonic acid Azoameisensäure *f*
azodicarboxylic acid Azodicarbonsäure *f*
azodimethylaniline Azodimethylanilin *n*
azodulcin Azodulcin *n*
azo dye Azofarbe *f,* Azofarbstoff *m,* acid ~
Azosäuregerbstoff *m,* fast ~ Azoechtfarbe *f*
azo dyestuff Azofarbe *f*
azoflavin Azoflavin *n,* Azogelb *n,*
Azosäuregelb *n*
azoformamide Azodicarbonamid *n,*
Azoformamid *n*
azoformic acid Azoameisensäure *f*
azo group Azogruppe *f*
azoic developing bath Azo-Entwicklungsbad *n*
azoic print Azodruck *m*
azoimide Azoimid *n*
azole Azol *n,* Pyrrol *n*
azolitmin Azolitmin *n*
azolitmin paper Azolitminpapier *n*
azometer Azometer *n*
azomethane Azomethan *n*
azomethine Methylenimin *n*
azomethine chelate Azomethinchelat *n*
azomycin Azomycin *n*
azonaphthalene Azonaphthalin *n*
azophenetole Azophenetol *n*
azophenol Azophenol *n*
azophosphin Azophosphin *n*
azophthalic acid Azophthalsäure *f*
azopiperonal Azopiperonal *n*
azo piperonylic acid Azopiperonylsäure *f*
azorite Azorit *m* (Min)
azosalicylic acid Azosalicylsäure *f*
azotide Azotid *n*
azotine Azotin *n*

azotoluene Azotoluol *n*
azotometer Azotometer *n*
azoturia Nitrokörperausscheidung *f* (Med)
azoxazole Furazan *n*
azoxyanisole Azoxyanisol *n*
azoxybenzene Azoxybenzol *n*
azoxybenzoic acid Azoxybenzoesäure *f*
azoxycinnamic acid Azoxyzimtsäure *f*
azoxy compound Azoxyverbindung *f*
azoxy derivative Azoxyabkömmling *m*
azoxynaphthalene Azoxynaphthalin *n*
azoxyphenetole Azoxyphenetol *n*
azoxyphenol Azoxyphenol *n*
aztequine Aztechin *n*
azulene Azulen *n*
azulin Azulin *n*
azulmic acid Azulminsäure *f*
azulmin Azulminsäure *f*
azure Azurblau *n,* Himmelblau *n*
azure spar Blauspat *m* (Min), Lasurspat *m*
(Min), Lazulith *m* (Min)
azurestone Lapislazuli *m* (Min), Lasurstein *m*
(Min), Ultramarin *n* (Min)
azurine Azurin *n*
azurite Azurit *m* (Min), Azurstein *m* (Min),
Bergasche *f* (Min), Chessylit[h] *m* (Min),
Kupfercarbonat *n* (Min), Lasurit *m* (Min)
azurol blue Azurolblau *n*

B

bababudanite Bababudanit *m* (Min)
babassu oil Babassuöl *n*
babbitt metal Babbitmetall *n*, Lagermetall *n*,
    Weißmetall *n*
babel-quartz Babelquarz *m* (Min),
    Babylonquarz *m* (Min)
babingtonite Babingtonit *m* (Min)
baby food Säuglingsnahrungsmittel *n*
bacillary spore Bazillenspore *f*
bacillary strain Bazillenstamm *m*
bacilli, free from ~ bazillenfrei
bacillicidal bazillentötend, bazillenvernichtend
bacilliculture Bazillenkultur, Bakterienkultur *f*,
    Bazillenzüchtung *f*
bacilli focus Bazillenherd *m*
bacillus (pl. bazilli) Stäbchenbakterie *f*, Bacillus
    m, Bazillus *m* (pl. Bazillen), Bazille *f*
bacillus carrier Bazillenträger *m*
bacillus of timothy grass Grasbakterium *n*
bacillus tetani Starrkrampfbazillus *m*
bacitracin Bacitracin *n*
back *v* kaschieren (Pap)
back borehole Firstenbohrloch *n*
back coupling Rückkopplung *f*
back diffusion Rückdiffusion *f*
back diffusion loss Rückdiffusionsverlust *m*
back discharge Rückentladung *f*
back draft Gegenzug *m*
backed kaschiert
backer Unterlage *f* (Verstärkung)
back-filling Auffüllung *f*
back finish Rückseitenappretur *f* (Text)
backfire Frühzündung *f*, Rückschlag *m*
backfiring Flammenrückschlag *m*,
    Rückschlagen *n*
backflow Rückfluß *m*, Rückstrom *m*
background Hintergrund *m*, Untergrund *m*
background count Nulleffektimpuls *m*
background current Grundstrom *m*
background memory Hintergrundspeicher *m*
background monitoring
    Untergrundüberwachung *f*
background noise Störhintergrund *m*,
    Nebengeräusch *n*
background radiation Nulleffekt *m*
    (Radioaktivität), Untergrundstrahlung *f*
    (Radioaktivität)
background scattering Untergrundstreuung *f*
backing Schutzfolie *f*, Steifleinen *n*, Unterlage *f*,
    Verstärkung *f*
backing material Mitläufer *m*
backing metal Hintergießmetall *n*
backing plate Aufspannplatte *f*, Stützplatte *f*
backing roll Stützwalze *f*, Gegendruckwalze *f*,
    Leitwalze *f*

backlash Rücksog *m*, Einbauflankenspiel *n*;
    toter Gang *m*, Getriebespiel *n*, Rückprall *m*,
    Spiel *n*; Zahnflankenspiel *n* (of gear),
    Flankenspiel *n*, free from ~ spielfrei
back-lighting Gegenlicht *n* (Phot)
backmixing Rückvermischen *n*,
    Rückvermischung *f*
back of fabric Rückseite *f* des Gewebes (Text),
    Stoffrückseite *f* (Text)
back plate Aschenzacken *m*, Hinterzacken *m*
back pressure Gegendruck *m*, Staudruck *m*
back pressure engine Gegendruckmaschine *f*
back pressure valve Rückschlagventil *n*
back-purge system Rückspülsystem *n*
    (Gaschrom)
back reaction Gegenreaktion, Rückreaktion *f*
back-scattered rückwärts gestreut
back scattering Rückstreuung *f*
backscattering saturation Rückstreusättigung *f*
backstroke Rückgang *m* (Kolben)
backstromite Backströmit *m* (Min)
back taper Hinterschneidung *f*
back-titrate *v* zurücktitrieren (Anal)
back-titrating Zurücktitrieren *n* (Anal)
back-titration Rücktitration *f*, Zurücktitrierung *f*
    (Anal)
backward rückwärts, rückläufig
back[ward] reaction Rückreaktion *f*
backwash Rückstrom *m*
backwater Rückwasser *n*
bacteria Bakterien *pl*, aerobic ~ aerobe
    Bakterien, anaerobic ~ anaerobe Bakterien,
    carrier of ~ Bakterienträger *m*, demonstration
    of ~ Bakteriennachweis *m*, denitrifying ~ pl
    denitrifizierende Bakterien, disturbed growth
    of ~ Dysbakterie *f* (Med), free of ~
    bakterienfrei, putrefactive ~
    Fäulnisbakterien *pl*, rod-shaped ~
    Stäbchenbakterien *pl* (Bakt), staining of ~
    Bakterienfärbung *f*
bacterial bakteriell, bakterienartig
bacterial bate Bakterienbeize *f*
bacterial cell Bakterienzelle *f*
bacterial count Bakterienzählung *f*,
    Keimgehalt *m*, Keimzahl *f*
bacterial counting Keimzählung *f*
bacterial cultures, solution for ~
    Bakteriennährlösung *f*
bacterial growth Bakterienwachstum *n*
bacterial plaque Zahnbelag *m* (Zahnmed)
bacterial strain Bakterienstamm *m*
bacterial suspension Bakterienaufschwemmung *f*
bacterial toxin Bakteriengift *n*, Bakterientoxin *n*
bacterial warfare Bakterienkrieg *m*
bacteria resistance Bakterienresistenz *f*
bacteria-resistant bakterienfest

bactericidal bakterizid, antibakteriell, bakterienfeindlich, bakterientötend, bakterienvernichtend
bactericide Bakterizid *n*, Bakteriengift *n*
bacteriemia Bakteriämie *f* (Med)
bacteriform bakterienförmig
bacteriochlorine Bakteriochlorin *n*
bacteriogenic bacteriogen, bakteriogen
bacteriogenous bacteriogen, bakteriogen
bacteriohemolysin Bakteriohämolysin *n*
bacterioid bakterienartig
bacteriological bakteriologisch
bacteriologist Bakteriologe *m*
bacteriology Bakteriologie *f*, Bakterienkunde *f*, Bakterienlehre *f*
bacteriolysin Bakteriolysin *n*
bacteriolysis Bakteriolyse *f*
bacteriolytic bakteriolytisch
bacteriophage Bakteriophage *m* (Biol)
bacteriophage treatment Phagenbehandlung *f* (Med)
bacteriopheophytin Bakteriophäophytin *n*
bacterioscopic bakterioskopisch
bacterioscopy Bakterioskopie *f*
bacteriosis Bakteriose *f* (Bot)
bacteriostasis Bakterienwachstumshemmung, Bakteriostase *f*
bacteriostatic *a* bakterienwachstumshemmend, bakteriostatisch, wachstumshindernd (Bakt)
bacteriostatic Bakteriostatikum *n*
bacteriotherapeutic bakteriotherapeutisch
bacteriotoxin Bakteriengift *n*
bacterium Bakterium *n* (pl Bakterien) (pl. bacteria), Bakterie *f*, **chromogenic** ~ Pigmentbakterie *f*, **nitrifying** ~ Nitrosebakterium *n*, **threadlike** ~ Fadenbakterium *n*
bacteriuria Bakteriurie *f* (Med)
bacteroid bakterienartig, bakteroid
bacteroidal bakterienartig, bakteroid
bad schlecht
baddeleyite Baddeleyit *m* (Min)
baden acid Badische Säure *f*
badger fat Dachsfett *n*
bad liquor Fusel *m*
bad spirits Fusel *m*
baeckeol Baeckeol *n*
Baeyer strain theory Baeyersche Spannungstheorie *f* (Chem)
Baeyer test Baeyersche Probe *f*
baffle Trennblech *n*, Leitblech *n*, Leitzunge *f*, Lenkblech *n*, Prallblech *n*, Prallhaube *f*, Resonanzwand *f*, Schallschluckplatte *f*, Schikane *f*, Strombrecher *m*, Umlenkblech *n*
baffle *v* drosseln
baffle member Staublech *n*
baffle plate Scheidewand *f*, Baffle *n*, Fangblech *n*, Gichtzacken *m*, Prallschirm *m*,

Schutzplatte *f*, Staublech *n*, Stauscheibe *f*, Wallplatte *f*, Windzacken *m*
baffle-ring centrifuge Prallringzentrifuge *f*
baffle sheet Auffangblech *n*
baffle stone Wallstein *m*
bag Sack *m*, Beutel *m*, Handtasche *f*, Tasche *f*, ~ **with punched hand grip and gussetted bottom** Griffloch-Tasche *f* mit Bodenfalt
bag *v* einsacken, bombieren (auswölben)
bagasse Bagasse *f*, Zuckerrohrrückstände *m pl*
bagasse cellulose Bagassezellstoff *m*
bagasse furnace Bagassefeuerung *f*
bagatelle Bagatelle, Kleinigkeit *f*
bag conveyor Sackförderer *m*
bag filter Beutelfilter *n* (Zuck), Reihenfilter *n*, Sackfilter *n*
bagging Einsacken *n*, Rupfen *m*; Sackleinwand *f*
bagging [and weighing] machine Absackwaage *f*
bagging plant Absackanlage *f*
bagging scales Absackwaage, Einsackwaage *f*
bag molding Presse *f* mit Gummisack
bagrationite Bagrationit *m* (Min)
bag sealer Beutelverschließgerät *n*
bag sealing Beutelschließen *n*, Beutelschweißen *n*
bag with mo[u]lded handle Bügelgriff-Tasche *f*
bahia rubber Bahiagummi *n*
baicalein Baicalein *n*
baicalin Baicalin *n*
baikalite Baikalit *m* (Min)
baikiaine Baikiain *n*
bait Köder *m*
bake *v* backen, anbacken, brennen (Keram), dörren, einbrennen, sintern, zusammenbacken
baked varnish coat Einbrennfarbe *f*
bakelite Bakelit *n* (HN)
bakelite varnish Bakelitlack *m*
bakerite Bakerit *m* (Min)
baker's yeast Backhefe *f*
bakery pyrometer Backofenpyrometer *n*
baking Backen *n*, Brand *m* (Keram), Sintern *n*, Sinterung *f*, **capable of caking or** ~ backfähig
baking characteristics Backfähigkeit *f*
baking enamel Einbrennemail *n*, Einbrennlack *m*
baking finish Einbrennfarbe *f*
[baking] oven Backofen *m*, Härteofen *m*
baking plant Brennhaus *n*
baking powder Backpulver *n*
baking soda Natriumbicarbonat *n*, Natriumhydrogencarbonat *n*
baking temperature Einbrenntemperatur *f*
baking varnish Einbrennlack *m*
baking varnishing Einbrennlackierung *f*
bakuin Bakuin *n*
bakuol Bakuol *n*
balance Gleichgewicht *n*, Abgleich *m*, Balance *f*, Bilanz *f*; Waage *f*, Wiegevorrichtung *f*,

analytical ~ analytische Waage, **automatic** ~
automatische Waage, **hydrostatic** ~
hydrostatische Waage, **magnetic** ~
magnetische Waage, **metallometric** ~
metallometrische Waage, **precision** ~
analytische Waage, **quick** ~ Schnellwaage *f*,
rapid ~ Schnellwaage *f*, **standing** ~ ruhendes
Gleichgewicht, **static** ~ Gleichgewicht *n* der
Ruhelage
balance *v* abgleichen, abstimmen, abwägen,
aufwiegen, ausbalancieren, ausgleichen,
auswichten, auswuchten, balancieren, im
Gleichgewicht halten, wiegen
balance arm Waagebalken *m*
balance beam Waagebalken *m*
balance blade Waageschneide *f*
balance case Waagekasten *m*, Waagengehäuse *n*
balanced abgeglichen, ausgeglichen, ausgewogen
(im Gleichgewicht befindlich), halbberuhigt
(Metall)
balance dough Auswuchtpaste *f*, Unwuchtpaste *f*
balance error Abgleichfehler *m*, Unwucht *f*
balance frame Balancierspant *n*
balance galvanometer Vertikalgalvanometer *n*
(Elektr)
balance indicator Ausgleichsanzeiger *m*
balance mechanism Ausgleichsvorgang *m*
balance method Abgleichverfahren *n*
balance pan Waagschale *f*
balancer Ausgleichsregler *m*,
Auswuchtmaschine *f*, Stabilisator *m*; Wippe *f*
balance range Abgleichsbereich *m*
balance recorder Ausgleichsanzeiger *m*
balance room Wägezimmer *n*
balance sensitivity Empfindlichkeit *f* der Waage
balance stock Unwuchtpaste *f*
balancing Gewichtsausgleichung *f*, Abgleich *m*,
Ausbalancierung *f*, Ausgleich *m*,
Auswuchtung *f*, Entlastung *f*
balancing current Ausgleichsstrom *m*
balancing device Auswuchtvorrichtung *f*
balancing fixture Auswuchtvorrichtung *f*
balancing frequency Abgleichfrequenz *f*
balancing mandrel Auswuchtdorn *m*
balancing tester Abgleichprüfer *m*
balancing transformer
Ausgleichtransformator *m*,
Dreileitertransformator *m*,
Spannungsteilertransformator *m*
balancing voltage Kompensationsspannung *f*
(Elektr)
balanophorin Balanophorin *n*
balas ruby Balasrubin *m* (Min)
balata Balata *f*
balata belt Balatariemen *m*
balata gum Balatagummi *n*
bale Ballen *m*, Emballage *f*
bale *v* ballen, emballieren
bale cutting machine Ballenschneidemaschine *f*

baleen Fischbein *n*
bale-shaped ballenförmig
balfourodine Balfourodin *n*
baling press Ballenpresse *f*
ball Kugel *f*, Ball *m*, Knäuel *m*, *n*
ball *v* **together** zusammenballen, ~ **up** sich
knäueln
ball and socket joint Gelenklager *n*,
Kugelgelenk *n*, Kugelscharnier *n*
ball and stick model Kugel-Stab-Modell *n*
(Stereochem)
ballast Ballast *m*
ballast material Ballaststoff *m*
ballast port Ballastpforte *f*
ballast tank Ballastbehälter *m*
ball bearing Kugellager *n*, **oblique** ~
Schrägkugellager *n*, **self-aligning** ~
Pendelkugellager *n*, **single row** ~ einreihiges
Kugellager
ball bearing grease Kugellagerfett *n*
ball bearing housing Kugellagergehäuse *n*
ball-bearing test Kugeldruckprobe *f*
ball check nozzle Kugelverschlußdüse *f*
ball condenser Kugelkühler *m*
ball-gauge Kugellehre *f*
ball hardness test Kugeldruckprobe *f*,
Kugelhärteprobe *f*
ball hardness testing machine
Kugelhärteprüfer *m*
ball inclinometer Kugelneigungsmesser *m*
ball-indentation test Kugeldruckprobe *f*
balling Ballung *f*, Zusammenballen *n*,
Zusammenballung *f*
balling pan Pelletierteller *m*
balling property Ballungsfähigkeit *f*
ball iron Luppeneisen *n*
ballistic ballistisch
ballistic galvanometer Stoßgalvanometer *n*
ballistic power Perkussionskraft *f*
ballistics Ballistik *f*, Wurflehre *f*
ballistite Ballistit *n*
ball joint Kugelgelenk *n* (Med, Techn)
ball lightning Kugelblitz *m*
ball mill Kugelmühle *f*, **inclined** ~
Schrägkugelmühle *f*, **stirred** ~
Rührwerksmühle *f*, ~ **with air drier**
Luftstrommühle *f*, ~ **with closed circuit**
mechanical classification Siebkugelmühle *f*
ball molecule Kugelmolekül *n*
balloon Ballon *m* (Chem)
balloon control device Ballonfänger *m*
balloon fabric Ballonstoff *m*
balloon flask Ballon *m* (Chem), Kugelflasche *f*
balloon tire Ballonreifen *m*
ball piercing apparatus Kugeldurchschlaggerät *n*
ball pin Kugelbolzen *m*
ball point pen[cil] Kugelschreiber *m*
ball press Luppenhammer *m*
ball-pressure hardness Kugeldruckhärte *f*

ball-pressure test Kugeldruckprobe *f*
ball race Führungsring, Kugellagerring *m*
balls, to form ~ sich ballen
ball shape Kugelform *f*
ball-shaped kugelförmig, ballfömig
ball stop-cock Kugelhahn *m*
ball thrust bearing Kugeldrucklager, Druckkugellager, Kugelspurlager *n*, Längskugellager *n*
ball-thrust test Kugeldruckprobe *f*
ball valve Kugelventil *n*
balm Balsam *m*, Linderungsbalsam *m* (Pharm)
balm apple Balsamapfel *m*
Balmer series Balmerserie *f* (Spektr)
balm leaves Melissenblätter *pl*
balm [mint] Melisse *f* (Bot)
balm mint oil Melissenöl *n*
balm of Gilead Mekkabalsam *m*
balm of Mecca Mekkabalsam *m*
balm of roses Rosenbalsam *m*
balm wood Balsamholz *n*
balmy balsamisch
balsam Balsam *m*
balsam apple Balsamapfel *m*
balsam fir Balsamtanne *f*
balsamic balsamisch
balsamic resin Balsamharz *n*
balsamiferous balsamerzeugend
balsam of fir Kanadabalsam *m*
balsam of sulfur Schwefelleinöl *n*
balsa wood Balsaholz *n*
baltimorite Baltimorit *m* (Min)
Baly cell Balysche Zelle *f* (Spektr)
Baly tube Balyrohr *n* (Spektr), Balysche Zelle *f* (Spektr)
bamboo Bambus *m*, Bambus[rohr] *n*
bamboo cane Bambusrohr *n*
bamboo grass Bambusgras *n*
bamboo sugar Bambuszucker *m*
ban *v* bannen
banana plug Bananenstecker *m* (Elektr)
banca tin Bancazinn *n* (Leg), Bankazinn *n* (Leg)
Bancora process Bancora-Verfahren *n* (Wolle)
band Band *n*, Bande *f*, Binde *f*, **endless** ~ Band *n* ohne Ende
bandage Binde *f* (Med), Verband *m* (Med)
bandage *v* einbinden (Med)
bandage sterilizer Verbandstoffsterilisator *m*
bandaging material Verbandstoff *m* (Med), Verbandzeug *n* (Med)
band analysis Bandenanalyse *f*
band brake Bandbremse *f*
band center Nullinie *f* (Spektr)
banded agate Bandachat *m* (Min)
band gap Bandabstand *m* (Festkörperphysik)
band head Bandenkopf *m* (Spektr)
band iron Bandeisen *n*
band line Bandenlinie *f*
band magnet Bandmagnet *m*

bandoline Bandolin *n*
band-pass filter Siebkette *f*
band pulley Riemenscheibe *f*
band saw Blattsäge *f*, Bandsäge *f*
band selector switch Bandbreitenschalter *m*
band spectrum Bandenspektrum *n*
band steel Bandstahl *m*
band width Bandbreite *f*
band width control Bandbreiteneinstellung *f*, Bandbreitenregelung *f*
baneberry root Schwarzwurzel *f* (Bot)
Bang reagent Bangreagenz *n*
Bang's solution Bangsche Lösung *f*
banish *v* [ver]bannen
banisterine Banisterin *n*
bank Bank *f* (Geol), Rollwulst *m* (Walzwerk)
bank guard Fangblech *n*
banking of blood Blutkonservierung *f*
baobab oil Affenbrotöl *n*, Baobaböl *n*
baobab tree Affenbrotbaum *m*
baphiin Baphiin *n*
baphinitone Baphiniton *n*
baptifoline Baptifolin *n*
baptisin Baptisin *n*
bar Barre *f*, Barren *m*, Nocken *m*, Schiene *f*, Stab *m*, Stange *f* (international unit of pressure), Bar *n*, **center of** ~ Stabkern *m*, **cross section of a** ~ Stabquerschnitt *m*, **unwrought** ~ unbearbeiteter Stab
bar *v* abriegeln, absperren, versperren
barbaloin Barbaloin *n*
barbatic acid Rhizonsäure *f*
barbatolic acid Barbatolsäure *f*
barbed wire Stacheldraht *m*
barberry Berberin *n*, Berberitze *f* (Bot), Sauerdorn *m* (Bot)
barberry bark Berberisrinde *f* (Bot)
barberry juice Berberinsaft *m*
barbierite Barbierit *m* (Min)
barbital Barbital *n*, Veronal *n* (HN)
barbituric acid Barbitursäure *f*, Malonylharnstoff *m*
barcenite Barcenit *m* (Min)
baregin Baregin *n*
bare wire electrode Blankdrahtelektrode *f*
bar expansion Stabdehnung *f*
bar feed Stangenvorschub *m*
bar head Knebelkopf *m*
bar heating Knüppelerwärmung *f*
barilla [soda] Barilla *n*
bar iron Stabeisen *n*, Stangeneisen *n*, **figured** ~ Formeisen *n*, **flat** ~ Flacheisen *n*
barite Baryt, Schwerspat *m* (Min), Barytstein *m* (Min), Blanc fixe *n*, **columnar** ~ Stangenspat *m* (Min)
barite[s] Schwerspat *m* (Min)
barite white Barytweiß *n*, Bariumsulfat *n*, Blanc fixe *n*
barium Barium *n* (Symb. Ba)

barium acetate Bariumacetat *n*
barium acetylide Bariumcarbid *n*
barium alloy Bariumlegierung *f*
barium aluminate Bariumaluminat *n*
barium binoxide Bariumdioxid *n*,
Bariumperoxid *n*, Bariumsuperoxid *n*
barium borate Bariumborat *n*
barium borate glass Bariumboratglas *n*
barium bromate Bariumbromat *n*
barium bromide Brombarium *n*
barium carbide Bariumcarbid *n*
barium carbonate Bariumcarbonat *n*, **native** ~
Witherit *m* (Min)
barium chloride Bariumchlorid *n*
barium chromate Bariumchromat *n*,
Steinbühlergelb *n*
barium cyanide Bariumcyanid *n*, Cyanbarium *n*
barium cyanoplatinate(II) Platincyanbarium *n*,
Bariumplatincyanür *n* (obs)
barium dioxide Bariumdioxid *n*,
Bariumhyperoxid *n*, Bariumperoxid *n*,
Bariumsuperoxid *n*
barium feldspar Bariumfeldspat *m* (Min),
Hyalophan *m* (Min)
barium fluoride Bariumfluorid, Fluorbarium *n*
barium fluosilicate Bariumsilicofluorid *n*
barium hydride Bariumhydrid *n*
barium hydroxide Bariumhydroxid *n*, Ätzbaryt *m*
barium hydroxide solution Barytlauge *f*,
Barytlösung *f*, Barytwasser *n*
barium hypophosphate Bariumhypophosphat *n*
barium hypophosphite Bariumhypophosphit *n*
barium hyposulfate Bariumhyposulfat *n*
barium hyposulfite Bariumhyposulfit *n*
barium iodide Bariumjodid, Jodbarium *n*
barium lake Bariumlack *m*
barium manganate Bariummanganat *n*, Kasseler
Grün *n*
barium meal Bariumbrei *m*
barium monosulfide Bariummonosulfid *n*
barium monoxide Bariumoxid *n*, Baryt *m;*
Baryterde *f*
barium nitrate Bariumnitrat *n*, Barytsalpeter *m*
barium nitrite Bariumnitrit *n*
barium oxide Bariumoxid *n*, Baryterde *f* (Min),
Schwererde *f* (Min)
barium oxide hydrate Bariumoxidhydrat *n*
barium permanganate Bariumpermanganat *n*
barium peroxide Bariumdioxid *n*,
Bariumhyperoxid *n*, Bariumperoxid *n*,
Bariumsuperoxid *n*
barium peroxide hydrate Bariumperoxidhydrat *n*,
Bariumsuperoxidhydrat *n*
barium plaster Bariumzement *m*
barium platinocyanide Platincyanbarium *n*,
Bariumplatincyanür *n* (obs)
barium protoxide Bariumoxid *n*, Baryt *m;*
Baryterde *f*
barium pyrophosphate Bariumpyrophosphat *n*

barium selenate Bariumselenat *n*
barium silicate Bariumsilikat *n*
barium silicofluoride Bariumsilikofluorid *n*
barium sulfate Bariumsulfat *n*, schwefelsaures
Barium, Blanc fixe *n;* Schwerspat *m* (Min);,
native ~ Baryt *m* (Min), Barytstein *m* (Min)
barium sulfide Bariumsulfid *n*,
Bariummonosulfid *n*, Schwefelbarium *n*
barium sulfide furnace Schwefelbariumofen *m*
barium superoxide Bariumsuperoxid *n*,
Bariumdioxid *n*, Bariumperoxid *n*
barium thiocyanate Bariumrhodanid,
Rhodanbarium *n*
barium thiosulfate Bariumthiosulfat *n*
barium tungstate Bariumwolframat *n*,
Wolframweiß *n*
barium uranite Bariumuranit *n*
barium yellow Barytgelb *n*, Steinbühlergelb *n*
bark Baumrinde *f*, Rinde *f* (Baum)
bark *v* ablohen, abrinden, abschwarten
bark liquor Lohbrühe *f*
barkometer Barkometer *n*, Brühmesser *m*,
Gerbsäuremesser *m*, Lohmesser *m*
bark shredding machine
Rindenzerfaserungsmaschine *f*
bark-tanned lohgar
bark tannery Rotgerberei *f*
barley Gerste *f*, **to soak the** ~ Gerste *f* keimen
lassen
barley dressing Gerstenschrotbeize *f*
barley malt Gerstenmalz *n*
barley seed oil Gerstensamenöl *n*
barley sugar Gerstenzucker *m*, Benitzucker *m*
barley water Gerstenwasser *n*
barm Hefe *f*, Bierhefe *f* (Brau), Faßbärme *f*
(Brau), Faßhefe *f* (Brau)
bar magnet Magnetstab *m*, Stabmagnet *m*
bar mo[u]ld Backenwerkzeug *n*,
Schieberwerkzeug *n*
barmy hefig
barn Barn *n* (Flächeneinheit des
Kernquerschnitts)
barnhardtite Barnhardtit *m* (Min)
baroclinicity vector Baroklinievektor *m*
barograph Barograph *m*, Luftdruckschreiber *m*
barolite Barokalzit *m* (Min)
barometer Luftdruckmesser *m*, Barometer *n*,
Baroskop *n*, **aneroid** ~ luftleeres Barometer
barometer box Barometerdose *f*
barometer flower Barometerblume *f*
barometer reading Barometerablesung *f*
barometric barometrisch, baroskopisch
barometrical variation Barometerschwankung *f*
barometric column Barometersäule *f*,
Quecksilbersäule *f*
barometric height Barometerhöhe *f*,
Barometerstand *m*
barometric pressure Barometerdruck *m*,
Barometerstand *m*, Luftdruck *m*

barometric reading Luftdruckstand *m*
barometric variation Luftdruckschwankung *f*
barometry Barometrie *f* (Meteor)
baroscope Baroskop *n*, Auftriebswaage *f*
baroscopic baroskopisch
baroselenite Baroselenit *m* (Min)
barostat Barostat *m*
barotropic barotrop
barotropy Barotropie *f*
barrandite Barrandit *m* (Min)
barras Barras *m* (Harz)
barrel Faß *n*, Barrel *n*, Tonne *f*
barrel chime and bead machine
　Faßzargenrichtmaschine *f*
barrel converter Trommelkonverter *m*
barrelene Barrelen *n*
barrel heating mantle Fässerheizmantel *m*
barrel mill Trommelmühle *f*
barrel mixer Trommelmischer *m*
barrel plating Trommelgalvanisierung *f* (Galv)
barrel pump Fässerpumpe *f*
barrels Fustage *f*
barrel shaking machine Fässerrüttelmaschine *f*
barrel sleeve Zylinderbüchse *f*
barrel stacking machine Fässerstapler *m*
barrel straightening machine
　Faßausbeulmaschine *f*
barren trocken (öde)
barrenness Unfruchtbarkeit *f*
barricade *v* abriegeln, versperren
barrier Barriere *f*, Schutzwand *f*
barrier height Höhe *f* des Potentialwalles,
　Schwellhöhe *f*
barrier layer photoelectric effect
　Sperrschichtphotoeffekt *m*
barrier layer theory Sperrschichttheorie *f*
barrier material Sperrschichtmaterial *n*
barrier penetration factor Gamowfaktor *m*
barrier-plane photocell Halbleiterphotozelle *f*
barrier properties Abdichtungseigenschaften *pl*
barrier sheet Abschirmschicht *f*,
　Zwischenschicht *f*
barrier transistor Sperrschichttransistor *m*
barrier-type cell Sperrschichtzelle *f*, Sperrzelle *f*
barringtogenic acid Barringtogensäure *f*
bar rolling mill Stabwalzwerk *n*
barrow charging Karrenbegichtung *f*
barthite Barthit *m* (Min)
bartholomite Bartholomit *m* (Min)
bar tin Stangenzinn *n*
barutin Barutin *n*
barwood afrikanisches Sandelholz
barycentric baryzentrisch
barylite Barylit[h] *m* (Min)
baryon Baryon *n*
barysilite Barysilit *m* (Min)
barystrontianite Barystrontianit *m* (Min)
baryta Bariumoxid *n*, Baryt *m* (Min),
　Schwererde *f* (Min)

baryta feldspar Barytfeldspat *m* (Min)
baryta paper Barytpapier *n*,
　Chromoaristopapier *n*, Kreidepapier *n*
baryta solution Barytlauge *f*, Barytlösung *f*,
　Barytwasser *n*
baryta water Barytwasser *n*, Barytlauge *f*,
　Barytlösung *f*
baryta yellow Barytgelb *n*
barytes Bariumsulfat *n*, Baryt *m* (Min),
　Barytstein *m* (Min), Schwerspat *m* (Min)
barytharmotome Barytharmotom *m* (Min)
barytic barytartig, barytführend,
　schwerspathaltig
barytic fluorspar Bariumflußspat *m* (Min)
barytiferous barytführend
barytine Barytin *m* (Min)
barytocalcite Barytocalcit *m* (Min)
barytofluorite Bariumflußspat *m* (Min)
barytophyllite Barytphyllit *m* (Min)
barytron schweres Elektron
basal area Basisfläche *f*
basal cell Basalzelle *f* (Biol)
basal metabolic rate Basalstoffwechsel *m* (Med),
　Grundumsatz *m* (Med), **determination of the**
　~ Grundumsatzbestimmung *f* (Med)
basal metabolism Grundstoffwechsel *m*
basal plane Basisfläche *f* (Krist)
basal surface Basisfläche *f*, Grundfläche *f*
basalt Basalt *m* (Min), **containing** ~
　basalthaltig, **tabular** ~ Tafelbasalt *m* (Min),
　to convert slag into a material resembling ~
　basaltieren
basal temperature Basaltemperatur *f* (Med)
basalt glass Basaltglas *n* (Min)
basaltic basaltähnlich, basalthaltig, basaltisch
basaltic iron ore Basalteisenerz *n*
basaltic jasper Basaltjaspis *m* (Min)
basaltic rock Basaltfelsen *m*, Basaltgestein *n*
basaltiform basaltförmig
basaltine Basaltin *m* (Min)
basaltlike basaltartig
basaltoid basaltähnlich
basalt sand Basaltsand *m*
basalt ware Basaltsteingut *n*
basalt wool Basaltwolle *f*
basanite Basanit *m* (Geol)
base Ausgangspunkt *m*, Base *f* (Chem); Basis *f*
　(Math), Grundfläche *f*, Grundlage *f*, Lauge *f*
　(Chem), Sockel *m* (Arch), Unterbau *m*,
　Untergrund *m*, **setting of the** ~
　Basiseinstellung *f*
base *a* unedel (Met)
base *v* basieren
base adjustment Basiseinstellung *f*
base alloy Grundlegierung *f*
base analog Basenanalogon *n*
base anhydride Basenanhydrid *n*
base block Bodenstein *m*

base board Grundbrett *n*, Grundplatte *f*,
	Scheuerleiste *f*
basebox Basiskiste *f* (112 Tafeln Blech 20x14″)
base box with flange Fußkasten *m* mit Muffe
base broom Färberkraut *n*
base burner Moderierofen *m*
base catalysis Basenkatalyse *f*
base-centered basiszentriert (Krist)
base charge Grundladung *f*
base coat erster Anstrich, Grundstrich *m*
base coating Grundlackierung *f*
base drift Grundliniendrift *f*
base exchange Basenaustausch *m*
base exchanger Basenaustauscher *m*
base exchanging compound Basenaustauscher *m*
base former Basenbildner *m*
base-forming basenbildend
baseless grundlos
base line Grundlinie *f*, Basislinie *f*
base material Ausgangsmaterial *n*,
	Ausgangsstoff *m*, Grundwerkstoff *m*
basement floor Kellerboden *m*
base mix Grundmischung *f*
base number Grundzahl *f*
base oil Grundöl *n*
base pair Basenpaar *n* (Mol. Biol)
base pairing Basenpaarung *f* (Mol. Biol)
base paper Rohpapier *n*
base plate Unterlagsplatte *f*, Bodenplatte *f*,
	Fußgestell *n*, Grundplatte *f*, Säulenfuß *m*,
	Sohlplatte *f*, Unterlage *f*
base point Stützpunkt *m*, Trägerpunkt *m*
base pressure Sohldruck *m*
base product Ausgangsprodukt *n*
base range finder Basisentfernungsmesser *m*
base rim Felgenkörper *m*
base ring Fußring *m*
base value Basenwert *m*
basic basisch (Chem), grundlegend
basic antimony chloride Algarotpulver *n*
basic capacity Basizität *f*
basic catalysis Basekatalyse *f*
basic chemicals Grundchemikalien *pl*
basic converter steel Thomasflußstahl *m*
basic course Grundkurs *m*
basic design Grundentwurf *m*
basic equation Grundgleichung *f*
basic form Grundform *f*
basic idea Grundgedanke *m*
basic instrument Grundgerät *n*
basic invention Grunderfindung *f*
basicity Basizität *f*
basicity number Basizitätszahl *f*
basicity value Basizitätszahl *f*
basic load Einheitsbelastung *f*
basic material Ausgangsstoff *m*, Grundstoff *m*
	(Rohstoff)
basic materials industry Grundstoffindustrie *f*
basic particle Grundpartikel *n*

basic pattern Grundprinzip *n*
basic press time Preßgrundzeit *f*
basic process Elementarvorgang *m*
basic range Anfangsskala *f*
basic region Grundgebiet *n*
basic state Grundzustand *m*
basic substance Grundsubstanz *f*
basic workhardening Grundverfestigung *f*
basidiomycetes Basidienpilze *pl* (Bot),
	Basidiomyceten *pl*
basidiospore Basidiospore *f*
basidium Basidie *f* (Bot), Basidium *n*
basification Basischmachen *n*
basifier Basenbildner *m*
basifying basenbildend
basigenous basenbildend
basil Basilie *f*, Basilikumkraut *n*
basilicon Basilikumsalbe, Königssalbe *f*
basilicon ointment Königssalbe *f*
basil oil Basilienöl *n*, Basilikumöl *n*
basin Becken *n*, Bassin *n*, Behälter *m*, Kessel *m*,
	Mulde *f*, ~ **with handle** Kasserolle *f*
basis Basis *f* (Math), Grund *m* (Grundlage),
	Grundfläche *f* (Math), Grundlinie *f*
basis rate system Mengenregelung *f*
basis weight Flächengewicht *n* (Papier)
basket Korb *m*
basket centrifuge Korbzentrifuge *f*
basket extractor Korbbandextrakteur *m*
basket weave Panamabindung *f*,
	Panamagewebe *n*
Basle blue Baseler Blau *n*
Basle green Baseler Grün *n*
basocyte Basozyt *m* (basophiler Leukozyt)
basophil basophil (Färb)
basophilous basophil (Färb)
bassanite Bassanit *m* (Min)
bass bow Baßbogen *m*
bassetite Bassetit *m* (Min)
bassia Bassia *f*
bassia oil Bassiafett *n*, Bassiaöl *n*
bassic acid Bassiasäure *f*
bassora gum Bassoragummi *n*
bassora rubber Torgummi *n*
bassoric acid Bassorinsäure *f*
bassorin Bassorin *n*, Tragantstoff *m*
basswood oil Lindenöl *n*
bast Bast *m*
bastard Bastard *m*
bastard file Bastardfeile *f*
bastard saffron Färberdistel *f* (Bot),
	Färbersaflor *m*
bastard scarlet Halbscharlachfarbe *f*
bastard sugar Basternzucker *m*
bast black Bastschwarz *n*
bast cell Bastzelle *f*
bast fiber Bastfaser *f*
bastite Bastit *m* (Min)
bastnasite Bastnäsit *m* (Min), Hamartit *m* (Min)

bast paper Bastpapier *n*
batatic acid Batat[in]säure *f*
batch Beschickungsansatz *m*, Charge *f*, Flotte *f*
(Färb); Glassatz *m*, Haufen *m*; Partie *f*
(Reaktionsführung)
batch annealed haubengeglüht
batch distillation Blasendestillation *f*
batchelorite Batchelorit *m* (Min)
batches, in ~ schubweise, stoßweise
batching Batschen *n*
batch mixer Chargenmischer *m*
batch off mill Ausschneidewerk *n*
batch operation Satzbetrieb *m*
batch process Chargenprozeß *m*
batch production Chargenbetrieb *m*, partieweise
Herstellung
batch retort Chargenautoklav *m*
batch sampling Musternahme *f*
batchwise satzweise
bate Beizbrühe *f* (Gerb), Beizflüssigkeit *f*
(Gerb), Gerberbeize *f*
bate stone Beizstein *m* (Gerb)
bath Bad *n*, agitation of the ~ Badbewegung *f*,
constant-temperature ~ Bad *n* mit konstanter
Temperatur, electroplating ~ galvanisches
Bad, impoverishment of the ~
Badverarmung *f*, preparation of the ~
Badbeschickung *f*, to decopper the ~ das
Bad *n* entkupfern, to prepare a ~ ein Bad
ansetzen
bath addition agent Badzusatz *m*
bath compartment Badunterabteilung *f*
bath current regulator Badstromregler *m* (Galv)
bath liquid Badflüssigkeit *f*
Bath metal Bathmetall *n*
bathochrome Bathochrom *n* (Opt)
bathochromic bathochrom
bath potential Badspannung *f* (Galv)
bath reaction Badreaktion *f*
bath resistance Badwiderstand *m* (Galv)
bath salt Badesalz *n*, Mutterlaugensalz *n*
bath sample Badprobe *f*
bath solution Badflüssigkeit *f*
bathymetric chart Tiefenkarte *f*
bating Beizen *n* (Gerb), Beizung *f* (Gerb),
determination of the strength of ~
Beizwertbestimmung *f*
bating bath Beizbad *n* (Gerb)
bating process Beizprozeß *m* (Gerb)
batiste Batist *m* (Text)
batrachite Batrachit *m* (Min), Froschstein *m*
(Min)
battery Batterie *f* (Elektr), Element *n* (Elektr),
Zelle *f* (Elektr), galvanic ~ galvanische
Batterie, reversible ~ Kippbatterie *f*,
secondary ~ Ladungssäule *f*, voltaic ~
galvanische Batterie
battery acid Akkumulatorsäure *f*
battery boiler Doppelkessel *m*

battery booster Spannungserhöher *m*
battery box Batteriegehäuse *n*
battery capacity Batteriekapazität *f*
battery case Akkumulatorkasten *m*
battery cell Batterieelement *n*, Batteriezelle *f*
battery charge Akkumulatorladung *f*
battery charger Batterieladegerät *n*,
Batterieladesatz *m*, Ladegerät *n*
battery charging Akkumulatorladung *f*,
Batterieaufladung *f*, Laden *n* eines
Akkumulators
battery connection Batterieanschluß *m* (Elektr)
battery discharge Akkumulatorentladung *f*,
Batterieentladung *f*
battery gauge Batteriegalvanometer *n*
battery ignition Batteriezündung *f*
battery jar Batterieglas *n*
battery limits Anlagengrenzen *pl*
battery-operated batteriebetrieben
battery resistance Batteriewiderstand *m*
battery scrap Akkumulatorschrott *m*
battery terminal Batterieanschlußklemme *f*,
Batterieklemme *f*
battery voltage Batteriespannung *f*
batwing burner Schlitzbrenner, Fächerbrenner *m*,
Schmetterlingsbrenner *m*, Schnittbrenner *m*
batyl alcohol Batylalkohol *m*
baudisserite Baudisserit *m* (Min)
bauerenol Bauerenol *n*
baulite Baulit *m* (Min)
Baumé degree Baumégrad *m*
Baumé spindle Bauméspindel *f*
Baumé standard (American standard scale for
density) Bauméstandard *m*
baumhauerite Baumhauerit *m* (Min)
baumlerite Bäumlerit *m* (Min)
bauxite Bauxit *m* (Min)
bauxite brick Bauxitziegel *m*
bauxite kiln Bauxitofen *m*
bauxite lixiviation Bauxitlaugerei *f*
bauxite mill Bauxitmühle *f*
Bavarian blue Bayrischblau *n*
bay Lorbeer *m* (Bot), oil of ~ Lorbeeröl *n*
bayberry Lorbeere *f* (Bot)
bayberry wax Myrtenwachs *n*
Bayer 205 Bayer 205 *n* (HN, Pharm)
Bayer process Bayerprozeß *m*
bayer's acid Bayer-Säure (HN), Bayersche
Säure *f*
bayin Bayin *n*
bay leaf Lorbeerblatt *n*
bay oil Bayöl *n*, Lorbeeröl *n*
bayonet fitting Bajonettanschluß *m*
bayonet lock Bajonettverschluß *m*
bayonet socket Bajonetthülse *f*
bayonet tube Bajonettrohr *n*
bay rum Bayrum *m*, Lorbeerspiritus *m*,
Pimentrum *m*
bays Hallenbauten *pl*

bay salt Baysalz, Seesalz *n*
bazzite Bazzit *m* (Min)
bdellium Bdellium *n* (Chem)
bead Perle *f,* Falzverbindung *f* (Bördelkante),
  Hohlkugel *f,* Kügelchen *n,* Sicke *f* (Dose),
  Wulst *m* (Reifen)
bead *v* ausfräsen (Masch), bördeln (sicken)
bead area Wulstpartie *f*
bead base dimensions Wulstanschlußmaße *pl*
bead [core] Wulstkern *m*
bead cover fabric Drahtkernummantelung *f,*
  Kernumwicklung *f* (Reifen)
beaded rim bottle Rollrandflasche *f*
bead filler Profilstreifen *m,* Kernreiter *m*
  (Reifen)
bead heel Wulstferse *f*
beading Bördeln *n*
beading electrode Umbügelelektrode *f*
beading machine Bördelmaschine *f,*
  Sickenmaschine *f* (f. Dosen)
beading ring Bördelring *m*
beading roll Bördelrolle *f*
bead [of tire] Laufdeckenwulst *m*
bead polymers Perlpolymerisate *pl*
bead region Wulstpartie *f*
bead reinforcement Kernreiter *m* (Reifen),
  Profilstreifen *m*
bead seat Felgenschulter *f*
bead seat band Schrägschulterring *m*
bead seat taper Schulterschräge *f* (Reifen)
bead setter Kernanlegering *m*
bead stitcher Wulstanrollvorrichtung *f*
bead test Bördelversuch *m* (Techn),
  Perlenprobe *f* (Anal)
bead toe Wulstspitze *f,* Wulstzehe *f*
bead tube Perlrohr *n*
bead wire insulation Drahtkernumspritzung *f*
bead wrap Kernummantelung *f* (Reifen)
bead wrapper Wulstkernbelag *m*
beaker Becher *m,* Becherglas *n,* Kochbecher *m*
beam Balken *m,* Balancier *m,* Baum *m* (Mech),
  Bündel *n* (Licht), Schiene *f,* Strahl *m*
  (Lichtstrahl), Träger *m,* **focused** ~
  gebündelter Strahl, **heterogeneous** ~
  heterogenes Strahlenbündel (n), **wide-flanged**
  ~ Breitflanschträger *m*
beam arresting device Balkenarretierung *f*
beam balance Hebelwaage *f*
beam deflection Strahlablenkung *f*
beam end Balkenende *n,* Balkenknie *n*
beam hole Strahlenöffnung *f*
beaming Bündelung *f* (Phys)
beam loading Balkenbelastung *f*
beam microbalance Balkenmikrowaage *f*
beam of light Lichtstrahlenbündel *n*
beam producing system
  Bündelerzeugungssystem *n*
beams Gebälk *n*
beam scale Balkenwaage *f,* Hebelwaage *f*

beam splitter Strahlenteiler *m*
beam spring indicator Stabfederindikator *m*
beam tetrode Bündeltetrode *f*
bean curd Bohnengallerte *f*
bean flour Bohnenmehl *n*
bean meal Bohnenmehl *n*
bean ore Bohnenerz *n* (Min)
bear Eisensau *f* (Metall)
bearberry, common ~ Bärentraube *f* (Gerb)
bearing Lager *n* (Techn), Richtung *f*
  (Orientierung), **big end** ~
  Kurbelwellenlager *n*, **cooling of** ~
  Lagerkühlung *f,* **laminated** ~ Lager *n* aus
  Schichtstoff, **lubrication of the** ~
  Lagerschmierung *f,* **self-lubricating** ~
  selbstschmierendes Lager, **spring** ~ federndes
  Lager, **to adjust the** ~ das Lager nachstellen,
  to bush the ~ das Lager ausbuchsen, **to grease
  the** ~ das Lager schmieren, **to line the** ~ das
  Lager ausfüttern, **to lubricate the** ~ das Lager
  schmieren, **to oil the** ~ das Lager schmieren,
  to scrape the ~ das Lager einschaben, **wear of**
  ~ Lagerabnutzung *f,* Lagerverschleiß *m,* ~
  for reciprocating crankshaft
  Schüttlerwellenlager *n*
bearing alloy Lagerlegierung *f*
bearing ball Lagerkugel *f*
bearing beam Lagerbalken *m*
bearing block Stehbock, Lagerbock *m*
bearing bracket Lagerarm, Lagerbock *m*
bearing bush Lagerbüchse *f,* Lagerfutter *n,*
  Lagerschale *f*
bearing cage Lagergehäuse *n,* Lagerkäfig *m*
bearing capacity Tragfähigkeit *f*
bearing carrier Lagerbalken *m*
bearing casing Lagergehäuse *n*
bearing clearance Lagerspiel *n*
bearing frame Lagerrahmen *m*
bearing friction Lagerreibung *f*
bearing grease Lagerschmierfett *n*
bearing journal Tragzapfen *m*
bearing metal Babbitmetall *n,* Büchsenmetall *n,*
  Lagermetall *n,* **zinc base** ~ zinkreiches
  Lagermetall
bearing needle Lagernadel *f*
bearing race Laufring *m*
bearing retainer Lagerbefestigung *f*
bearing ring Gleitring *m,* Tragring *m*
bearing rod Tragstange *f*
bearing roller Lagerrolle *f*
bearing spring Tragfeder *f*
bearing strength Tragfähigkeit *f*
bearing stress Belastung *f* durch Eigengewicht
bearing surface Auflagefläche *f,* Lagerfläche *f,*
  Lageroberfläche *f,* Stützpunkt *m*
bearing testing machine Lagerprüfmaschine *f*
bear's breech oil Bärenklauöl *n*
bear's grease Bärenfett *n*
bear's weed Bergbalsam *m*

bear's wort oil Bärwurzöl *n*
beat Schlag *m*, Anschlag *m*, Takt *m*
beat *v* schlagen, abklopfen; übertreffen
beaten aluminum Blattaluminium *n*
beaten lead Bleifolie *f*
beaten silver Blattsilber *n*
beater Holländer *m* (Pap), Klöppel *m*,
  Mahlholländer *m* (Pap), Stoffmühle *f* (Pap)
beater bar Schlägerstange *f*, Schlägerwalze *f*
beater mill Schlagmühle *f*
beater roll Messerwalze *f*
beating Mahlarbeit *f* (Pap), Mahlung *f* (Pap)
beating engine Holländer *m* (Pap)
beating iron Pocheisen *n*, Pochschuh *m*
beating machine Ausklopfmaschine *f*,
  Feinzeugholländer *m*
beating mill Stoßkalander *m* (Pap)
beating plant Mahlanlage *f* (Pap)
beating vat Schlagküpe *f*
beaumontite Beaumontit *m* (Min)
beaverite Beaverit *m* (Min)
bebeerine Bebeerin *n*, Bebirin *n*
bebeeru bark Bebeerurinde *f*
bechilite Bechilit *m* (Min)
beck flacher Bottich *m*, Kufe *f*
beckelite Beckelith *m* (Min)
Beckmann rearrangement Beckmannsche
  Umlagerung *f*
Beckmann thermometer
  Beckmannthermometer *n*
beclamide Beclamid *n*
becquerelite Becquerelit *m* (Min)
Becquerel rays Becquerelstrahlen *pl*
bed Lage *f* (Bergb), Lager *n* (Geol),
  Lagerstätte *f* (Bergb)
bed bug Wanze *f* (Zool)
bed die Lochring *m*
bedeguar Rosenschwamm *m* (Bot)
bed plate Auflageplatte *f*, Grundplatte *f*,
  Sohlplatte *f*, Unterlagsplatte *f*
bedrock Grundgestein *n* (Geol)
beech gall Buchengallapfel *m*
beechnut Buchecker *f* (Bot)
beechnut oil Bucheckernöl *n*
beech pitch Buchenholzpech *n*
beech tar creosote Buchenholzteerkreosot *n*
beech tar oil Buchenholzteeröl *n*
beech tree Buche *f*
beechwood ashes Buchenasche *f*
beef marrow fat Rindermarkfett *n*
beef suet Rinderfett *n*
beef tallow Rindertalg *m*
beegerite Beegerit *m* (Min)
bee glue Bienenharz *n*, Kittwachs *n*,
  Klebwachs *n*, Pichwachs *n*
bee-keeper Imker *m*
beekeeping Bienenzucht *f*
beer Bier *n*, **bottom fermentation** ~ untergäriges
  Bier, **dark** ~ dunkles Bier, **light** ~ helles Bier,

low fermentation ~ untergäriges Bier, **new** ~
  Jungbier *n*, **top fermentation** ~ obergäriges
  Bier
beer barrel Bierfaß *n*
beer brewing Bierbrauen *n*
beer haze Biertrübung *f*
beer scale destroying agent Biersteinentferner *m*
beer vinegar Bieressig *m*
beer wort cooler Bierwürzekühler *m*
beeswax Bienenwachs *n*
beet juice Rübensaft *m*
beetle Käfer *m* (Zool)
beetroot (Am. E.) Runkelrübe *f*, Zuckerrübe *f*,
  molasses of ~ Runkelrübenmelasse *f*
beetroot molasses Rübenmelasse *f*
beetroot sugar Rübenzucker *m*,
  Runkelrübenzucker *m*
beet slice Schnitzel *m n* (Zucker)
beet syrup Rübensirup *m*
bee venom Bienengift *n*
befall *v* befallen, sich ereignen
beforehand zuvor, pränumerando
begin *v* aufnehmen, einsetzen (beginnen), ~ **to**
  boil ankochen, zu kochen beginnen, ~ **to burn**
  anbrennen, ~ **to corrode** anätzen, ~ **to dry**
  antrocknen, ~ **to heat** anwärmen, ~ **to rot**
  anfaulen
beginning Beginn, Anfang *m*, ~ **of setting**
  Erhärtungsbeginn *m*
behave *v* sich verhalten, fungieren
behavior Verhalten *n* (Br. E. behaviour)
behavior towards dyes Anfärbbarkeit *f* (Färb)
behenic acid Behensäure *f*, Dokosansäure *f*
behen nut Behennuß *f*
behen oil Behenöl *n*, Moringaöl *n*
behenolic acid Behenolsäure *f*
behenone Behenon *n*
beige beige
Beilby layer Beilby-Schicht *f*
Beilstein test Beilsteinprobe *f*
bel Bela *f*
belemnite Belemnit *m* (Min), Donnerstein *m*
  (Min), Fingerstein *m* (Min), Teufelsstein *m*
  (Min), Wetterstein *m* (Min)
belite Belit *m* (Min)
bell Glocke *f*
belladine Belladin *n*
belladonna Tollkirsche *f* (Bot), Belladonna *f*
  (Med), **extract of** ~ Tollkirschenextrakt *m*
belladonna alkaloid Belladonnaalkaloid *n*
belladonna leaves Belladonnablätter *pl* (Bot)
belladonna root Tollkrautwurzel *f* (Bot)
belladonnine Belladonnin *n*
bellamarine Bellamarin *n*
bell counter Glockenzählrohr *n*
bellcrank lever Kniehebel *m*
bell crusher Glockenmühle *f*
bell insulator Isolierglocke *f*
bellite Bellit *m* (Min)

bell jar Glasglocke *f,* Rezipient *m*
bell-jar plate Rezipiententeller *m*
bell lifting rod Hubstange *f*
bell metal Glockenmetall *n,* Glockenbronze *f,*
    Glockengut *n,* Glockenspeise *f*
bell method Glockenverfahren *n*
bellmouth intake Einlauftrompete *f*
bellow-ring packing Balglinse *f*
bellows Balg *m,* Blasebalg *m,* Faltenbalg *m,*
    Gebläse *n*
bellows extension Balgauszug *m*
bellows vacuum gauge Federmanometer *n* für
    Vakuumtechnik
bell pepper Paprika *m*
bell shape Glockenform *f*
bell-shaped glockenförmig
bell-shaped insulator Isolierglocke *f*
bell valve Glockenventil *n*
bell winch Glockenwinde *f*
bell wire Klingelleitung *f*
belmontit Belmontit *m* (Min)
belonging [to] zugehörig, angehörig
belonite Belonit *m* (Min)
belt Riemen *m,* Band *n,* Gürtel *m,* **contact side of
    the ~** Laufseite *f* des Riemens, **endless ~**
    Band *n* ohne Ende, **width of ~**
    Riemenbreite *f*
belt assembly Fließbandmontage *f*
belt carrier Bandförderer *m*
belt center line Fluchtlinie *f*
belt conveyance Bandförderung *f*
belt conveyer Bandtransporteur *m,*
    Förderband *n,* Fördergurt *m,* Gurtförderer *m,*
    Gurttransporteur *m,* **horizontal ~**
    Horizontalförderband *n,* **~ for coal**
    Kohlenförderband *n*
belt coupling Bandkupplung *f*
belt crystallizer Kristallisierband *n*
belt diameter Laufdurchmesser *m* (Keilriemen)
belt drier Bandtrockner *m*
belt drive Riemenantrieb *m*
belt-driven grinding machine
    Riemenschleifmaschine *f*
belt-driven hammer Transmissionshammer *m*
belt-driven pump Riemenpumpe *f*
belt extractor Bandextrakteur *m*
belt filter Bandfilter *n*
belt grinding machine Bandschleifmaschine *f*
belt guide Riemenführer *m,* Riemengabel *f,*
    Riemenleiter *m*
belting Riemenleder *n*
belt length Riemenlänge *f*
belt line Fluchtlinie *f*
belt pulley Riemenscheibe *f*
belt separator Bandscheider *m,* Bandseparator *m*
belt shifting Riemenverschiebung *f*
belt thickness Riemendicke *f*
belt transmission Riemenübertragung *f*

belt weigher Förderbandwaage *f*
bemegride Bemegrid *n*
bementite Bementit *m* (Min)
bemidone Bemidon *n*
Bence-Jones protein Bence-Jonesscher
    Eiweißkörper *m*
bench Bank *f*
bench clamp Tischklemme *f*
bench lathe Tischdrehbank *f*
bench mark Fixpunkt *m,* Nivellierungszeichen *n*
bench test Laborversuch *m*
bench vice Bankschraubstock *m*
bend Biegung *f,* Bogen *m,* Bogenstück *n,*
    Krümmling *m,* Krümmung *f;* Kurve *f,* **flanged
    ~** Krümmer *m* mit Flanschen (pl), **~ with
    multiple connection** Destillierspinne *f,* **~ with
    vent** Bogen *m* mit Rohransatz
bend *v* biegen, anspannen, beugen, krümmen,
    neigen, **~ cold** kaltbiegen, **~ in** einbiegen,
    einbeulen, **~ off** abbiegen, **~ over** umbiegen,
    ~ sharply abknicken, **~ through** durchbiegen
bend allowance Biegungseinräumung *f*
bending Biegung *f,* Krümmung *f,* **resistant to ~**
    biegesteif, **~ off** Abbiegung *f*
bending and folding machine Abkant- und
    Falzmaschine *f*
bending at angles Abkröpfung *f*
bending coefficient Biegegröße *f*
bending device Biegevorrichtung *f*
bending fatigue Biegungsermüdung *f*
bending fixture Biegevorrichtung *f*
bending force Biegekraft *f*
bending force constant Kraftkonstante *f*
bending impact test Biegeschlagversuch *m*
bending limit of spring plate Federbiegegrenze *f*
bending machine Abkantmaschine *f* (Met),
    Biegemaschine *f*
bending mandrel Biegedorn *m*
bending moment Biegemoment *n,*
    Biegungsmoment *n*
bending point Biegestelle *f,*
    Wärmestandfestigkeit *f* (Kerzen)
bending press Biegepresse *f*
bending pressure Biegedruck *m*
bending property Biegefähigkeit *f*
bending radius Biegehalbmesser *m,*
    Biegungshalbmesser *m*
bending resistance Biegesteifigkeit *f*
bending roll Biegewalze *f*
bending strain Beanspruchung *f* auf Biegung,
    Biegebelastung *f,* Biegedruck *m,*
    Biegungsbeanspruchung *f*
bending strength Bieg[e]festigkeit *f,*
    Biegungsfestigkeit *f,* **transverse ~**
    Durchbiegungsfestigkeit *f*
bending stress Biegebeanspruchung *f,*
    Biegespannung *f,* Biegungsspannung *f,*
    repetition of ~ Dauerbiegebeanspruchung *f*

bending stress by wind pressure Beanspruchung *f*
auf Biegung durch den Winddruck
bending stress tester Biegespannungsmaschine *f*
bending support Abstandsklappe *f*
bending test Biegeprobe *f*, Biegeprüfung *f*, ~ in
tempered state Abschreckbiegeprobe *f*,
Härtungsbiegeprobe *f*
bending-test shackle Biegevorrichtung *f*
bending value Biegezahl *f*
bend test Faltversuch *m*, notched bar ~
Kerbbiegeprobe *f*
bend test apparatus Biegeprüfer *m*
beneficial förderlich
benefit Nutzen *m*
bengal blue Bengalblau *n*
bengal cachou Bengalcachou *n*
bengal gelatin Agar-Agar *n*
Bengal light or fire Bengalisches Feuer *n*
bengal rose Bengalrosa *n*
benihiol Benihiol *n*
benincopal acid Benincopalsäure *f*
benincopalinic acid Benincopalinsäure *f*
benitoite Benitoit *m* (Min)
ben nut Behennuß *f*
ben oil Behenöl *n*
bent Abkantung *f*
bent *a* gekrümmt (gebogen), krumm, ~ at a
right angle rechtwinklig gebogen, ~ in a circle
kreisförmig gebogen
bent lever Winkelhebel, Kniehebel *m*
bentonite binder Bentonitbinder *m*
bent pipe Knierohr *n*
bent syphon tube Winkelheber *m*
bent thermometer Winkelthermometer *n*
bent tube Winkelrohr *n*
benzaconine Napellin *n*
benzacridine Benzacridin *n*,
Phenonaphthacridin *n*
benzal Benzal-, Benzyliden-
benzalacetaldehyde Cinnamylaldehyd *m*
benzal acetic acid Cinnamylsäure *f*
benzalacetone Benzalaceton *n*
benzalacetophenone Benzalacetophenon *n*
benzalaniline Benzalanilin *n*
benzalazine Benzalazin *n*, Benzaldazin *n*
benzalcamphor Benzalcampher *n*
benzalchloride Benzalchlorid *n*, Chlorobenzal *n*
benzalcyanohydrin Mandelsäurenitril *n*,
Benzaldehydcyanhydrin *n*
benzaldazine Benzalazin *n*, Benzaldazin *n*
benzaldehyde Benzaldehyd *m*,
Benzoylwasserstoff *m*
benzaldehyde cyanohydrin
Benzaldehydcyanhydrin *n*,
Mandelsäurenitril *n*
benzaldehyde oxime Benzaldoxim *n*
benzaldoxime Benzaldoxim *n*
benzal green Benzalgrün *n*, Malachitgrün *n*

benzalizarin Benzalizarin *n*
benzalphenylhydrazone Benzalphenylhydrazon *n*
benzamarone Benzamaron *n*
benzamide Benzamid *n*
benzamidine Benzamidin *n*
benzamine blue Benzaminblau *n*
benzaminic acid Benzaminsäure *f*
benzaminoacetic acid Hippursäure *f*
benzanalgen Benzanalgen *n*
benzanilide Benzanilid *n*
benzaniside Benzanisid *n*
benzanthracene Benzanthracen *n*
benzanthraquinone Benzanthrachinon *n*
benzanthrene Benzanthren *n*,
Naphthanthracen *n*
benzanthrone Benzanthron *n*
benzatropine Benzatropin *n*
benzaurine Benzaurin *n*
benzazimide Benzazimid *n*
benzedrine Benzedrin *n*, Aktedon *n*
benzene Benzol *n*, Cyclohexatrien *n*, commercial
~ Handelsbenzol *n*, rectified ~ Reinbenzol *n*
benzenearsonic acid Benzolarsonsäure *f*
benzeneazonaphthol Benzolazonaphthol *n*
benzeneazophenol Benzolazophenol *n*
benzenecarbonyl chloride Benzoesäurechlorid *n*
benzene derivative Benzolabkömmling *m*,
Benzolderivat *n*
benzene diamine Phenylendiamin *n*
benzenediazoic acid Benzoldiazosäure *f*
benzenediazonium chloride
Benzoldiazoniumchlorid *n*,
Diazobenzolchlorid *n*
benzenediazonium hydroxide
Benzoldiazoniumhydroxid *n*
benzene diazotate Benzoldiazotat *n*
benzene dicarboxylic acid, 1,2- ~ Phthalsäure *f*,
1,3- ~ Metaphthalsäure *f*
benzenediol, 1,2- ~ Pyrocatechin *n*, 1,3- ~
Resorcin *n*
benzene disulfonic acid Benzoldisulfonsäure *f*
benzene hexabromide Benzolhexabromid *n*
benzenehexacarboxylic acid Mellitsäure *f*
benzene hexachloride Benzolhexachlorid *n*
benzene linkage Benzolbindung *f*
benzene nucleus Benzolkern *m*
benzene pentacarboxylic acid
Benzolpentacarbonsäure *f*
benzene plant Benzolanlage *f*
benzene poisoning Benzolvergiftung *f*
benzene residue Benzolrest *m*
benzene ring Benzolring *m*
benzene series Benzolreihe *f*
benzenestibonic acid Phenylantimonsäure *f*
benzenesulfinic acid Benzolsulfinsäure *f*
benzene sulfonamide Benzolsulfonamid *n*
benzenesulfonic benzolsulfonsauer (Salz)
benzenesulfonic acid Benzolsulfonsäure *f*
benzenesulfonyl chloride Benzolsulfonylchlorid *n*

benzene tetracarboxylic acid
  Benzoltetracarbonsäure *f*
benzene tricarboxylic acid
  Benzoltricarbonsäure *f*
benzenetriol Trioxybenzol *n*, **1,2,3-** ~
  Pyrogallol *n*, Pyrogallussäure *f*
benzenyl Benzenyl-
benzenyl chloride Benzotrichlorid *n*
benzerythrene Benzerythren *n*
benzhydrazide Benzhydrazid *n*
benzhydrol Benzhydrol *n*
benzhydroxyanthrone Benzhydroxyanthron *n*
benzhydryl Benzhydryl-
benzhydrylamine Benzhydrylamin *n*
benzhydryl ether Benzhydryläther *m*
benzidine Benzidin *n*
benzidine brown Benzidinbraun *n*
benzidine disulfonic acid Benzidindisulfonsäure *f*
benzidine hydrochloride Benzidinhydrochlorid *n*
benzidine rearrangement Benzidin- Umlagerung
benzidine sulfonic acid Benzidinsulfonsäure *f*
benzidine transformation Benzidinumlagerung *f*
benzil Benzil *n*, Dibenzoyl *n*
benzildianil Benzildianil *n*
benzildioxime Benzildioxim *n*
benzilic acid Benzilsäure *f*
benzilic acid rearrangement
  Benzilsäureumlagerung *f*
benzilide Benzilid *n*
benzilimine oxime Benziliminoxim *n*
benzilmethylimine oxime
  Benzilmethyliminoxim *n*
benzimidazole Benzimidazol *n*
benzimido Benzimido-
benzin[e] Benzin *n*, **medium** ~ Mittelbenzin *n*
benzine blowtorch Benzinlötlampe *f*
benzine burner Benzinbrenner *m*
benzine plant Benzinanlage *f*
benzine pump Benzinpumpe *f*
benzine soap Benzinseife *f*
benzine substitute Benzinersatz *m*
benzine vapor Benzindampf *m*
benzoate Benzoat *n*, Benzoesäureester *m*,
  Benzoesäuresalz *n*
benzoate *a* benzoesauer
benzoated verbunden mit Benzoesäure
benzoated lard Benzoeschmalz *n*
benzoazurine Benzoazurin *n*
benzo black blue Benzoschwarzblau *n*
benzocaine Benzocain *n*, Äthylaminobenzoat *n*,
  Anaesthesin *n*
benzo copper dye Benzokupferfarbstoff *m*
benzocoumarane Naphthofuran *n*
benzodiazine, 1,8- ~ Naphthyridin *n*
benzodioxan Benzodioxan *n*
benzodioxole Benzodioxol *n*
benzo fast color Benzoechtfarbe *f*
benzoflavine Benzoflavin *n*
benzofuran Benzofuran *n*, Cumaron *n*

benzofurazan Benz[o]furazan *n*
benzofuroxan Benz[o]furoxan *n*
benzoguanamine Benzoguanamin *n*
benzohydroxamic acid Benzhydroxamsäure *f*
benzoic benzoehaltig
benzoic acid Benzoesäure *f*, Acidum
  benzoicum *n*, **ester of** ~ Benzoesäureester *m*,
  salt of ~ Benzoesäuresalz *n*, **soap of** ~
  Benzoesäureseife *f*, **solution of** ~
  Benzoesäurelösung *f*
benzoic anhydride Benzoesäureanhydrid *n*
benzoic ester Benzoesäureester *m*
benzoic ether Benzoesäureäthylester *m*
benzoin Benzoin *n*, **flowers of** ~ Benzoeblumen *f*
  pl
benzoin condensation Benzoinkondensation *f*
benzoin gum Benzoeharz *n*, Benzoe *f*,
  Benzoegummi *n*
benzoin imide Amaron *n*
benzoinoxime Benzoinoxim *n*
benzoin resin Benzoeharz *n*, Benzoe *f*,
  Benzoegummi *n*
benzoin tree Benzoebaum *m*
benzol Benzol *n*, **commercial** ~
  Handelsbenzol *n*, **crude** ~
  Benzolvorprodukt *n*
benzol recovery by washing Benzolwäsche *f*
benzol scrubber Benzolwäscher *m*
benzol still Benzolabtreiber *m*
benzol tincture Benzoltinktur *f*
benzol varnish Benzollack *m*
benzol washer Benzolwäscher *m*
benzol wash oil Benzolwaschöl *n*
benzonaphthol Benzonaphthol *n*
benzonatate Benzonatat *n*
benzonitrile Benzonitril *n*
benzo-nitro-benzoic anhydride
  Nitrobenzoesäureanhydrid *n*
benzooxine Benzooxin *n*
benzophenanthrene Chrysen *n*
benzophenazine Phenonaphthazin *n*
benzophenol Benzophenol *n*
benzophenone Benzophenon *n*, Diphenylketon *n*
benzophenone sulfide Thioxanthon *n*
benzophenone sulfone Benzophenonsulfon *n*
benzopinacol Benzpinakol *n*, Benzpinakon *n*
benzo pure yellow Benzoreingelb *n*
benzopurpurin Benzopurpurin *n*
benzopyran Benzopyran *n*
benzopyrazole Indazol *n*
benzopyrene Benz[o]pyren *n*
benzopyrone Benzopyron *n*, **1,2-** ~ Cumarin *n*,
  1,4- ~ Chromon *n*
benzopyrrole Indol *n*
benzoquinhydrone Benzochinhydron *n*
benzoquinol Benzochinol *n*
benzoquinoline Benzochinolin *n*
benzoquinone Benzochinon *n*, Chinon *n*
benzoquinoxaline Naphthopyrazin *n*

benzoresorcinol Benzoresorcin *n*
benzosalin Benzosalin *n*
benzoselenazole Benzselenazol *n*
benzoselenodiazole Benzselendiazol *n*
benzosol Benzosol *n*, Guajacolbenzoat *n*
benzosulfimide Benzoesäuresulfimid *n*,
　Saccharin *n*
benzotetronic acid Benzotetronsäure *f*
benzothialene Benzothialen *n*
benzothiazine Benz[o]thiazin *n*
benzothiazole Benz[o]thiazol *n*
benzothiazone Benzthiazon *n*
benzothiophene Benzothiophen *n*
benzothiopyran Thiochromen *n*
benzothiopyrone Thiochromon *n*
benzotriazine Benztriazin *n*
benzotriazole Benz[o]triazol *n*, Aziminobenzol *n*
benzotrichloride Benzotrichlorid *n*,
　Phenylchloroform *n*
benzoxazine Benzoxazin *n*
benzoxdiazine Benzoxdiazin *n*
benzoxdiazole Benzoxdiazol *n*
benzoxthiole Benzoxthiol *n*
benzoyl- Benzoyl-
benzoylacetone Benzoylaceton *n*
benzoylacetone amine Benzoylacetonamin *n*
benzoylaminomalonic ester
　Benzoylaminomalonester *m*
benzoylanisylcarbinol Benzoylanisylcarbinol *n*
benzoylate *v* benzoylieren
benzoylation Benzoylierung *f*
benzoylbenzamide Dibenzamid *n*
benzoylbenzoic acid Benzoylbenzoesäure *f*
benzoylcamphor Benzoylcampher *m*
benzoyl chloride Benzoesäurechlorid *n*,
　Benzoylchlorid *n*, Chlorbenzoyl *n*
benzoylconine Napellin *n*
benzoylecgonine Benzoylecgonin *n*
benzoylene Benzoylen-
benzoylene urea Benzoylenharnstoff *m*
benzoylglycine Benzoylglykokoll *n*,
　Hippursäure *f*
benzoylglycocoll Benzoylglykokoll *n*,
　Hippursäure *f*
benzoyl hydride Benzoylwasserstoff *m*
benzoylleucine Benzoylleucin *n*
benzoylnaphthol Naphtholbenzoat *n*
benzoyl peroxide Benzoylperoxid *n*,
　Benzoylsuperoxid *n*, Lucidol *n*
benzoylpseudotropeine Benzoylpseudotropein *n*,
　Tropacocain *n*
benzoylpyruvic acid Benzoylbrenztraubensäure *f*
benzoyl radical Benzoyl-Radikal *n*
benzoylsalicin Benzoylsalicin *n*
benzoylsulfonic imide Saccharin *n*
benzoyltrifluoroacetone Benzoyltrifluoraceton *n*
benzozone Acetozon *n*
benzpyrene Benzpyren *n*
benzthiophenide Benzthiophenid *n*

benzyl- Benzyl-
benzyl acetate Benzylacetat *n*
benzyl alcohol Benzylalkohol *m*
benzylamine Benzylamin *n*
benzylaniline Benzylanilin *n*
benzylation Benzylierung *f*
benzylbenzene Benzylbenzol *n*,
　Diphenylmethan *n*
benzyl benzoate Benzoesäurebenzylester *m*,
　Benzylbenzoat *n*
benzyl bromide Benzylbromid *n*
benzyl butyl adipate Benzylbutyladipinat *n*
benzyl butyl phthalate Benzylbutylphthalat *n*
benzylcarbinol Benzylcarbinol *n*
benzyl cellulose Benzylcellulose *f*
benzyl chloride Benzylchlorid *n*, Benzylchlorid,
　Chlorbenzyl *n*
benzyl cinnamate Zimtsäurebenzylester *m*,
　Benzylcinnamat *n*, Cinnamein *n*
benzyl cyanide Benzylcyanid *n*
benzylene Benzylen-
benzyl ester Benzylester *m*
benzyl ether Benzyläther *m*
benzyl ethylaniline Benzyläthylanilin *n*
benzyleugenol Benzyleugenol *n*
benzylidene Benzal-, Benzyliden-
benzylidene acetone Benzalaceton *n*
benzylidene aniline Benzalanilin *n*
benzylidene azine Benzalazin *n*, Benzaldazin *n*
benzylidene chloride Benzalchlorid *n*
benzylidenemalonic acid Benzylidenmalonsäure *f*
benzylidenemalonic ester
　Benzylidenmalonsäureester *m*
benzylisoeugenol Benzylisoeugenol *n*
benzyl isothiocyanate Benzylsenföl *n*
benzylmalonic acid Benzylmalonsäure *f*
benzylmalonic ester Benzylmalonsäureester *m*
benzyl mercaptan Benzylmercaptan *n*
benzylmethylaniline Benzylmethylanilin *n*
benzylmethylglyoxime Benzylmethylglyoxim *n*
benzylmorphine hydrochloride Peronin *n*
benzyl octyl adipate Benzyloctyladipat *n*
benzyl orange Benzylorange *n*
benzyloxyacetophenone Benzyloxyacetophenon *n*
benzylphenol Benzylphenol *n*
benzylresorcinol Benzylresorcin *n*
benzylsilicochloroform
　Benzylsiliciumtrichlorid *n*
benzyl silicone Benzylsilicon *n*
benzyl silicon trichloride
　Benzylsiliciumtrichlorid *n*
benzyl succinate Benzylsuccinat *n*
benzyl sulfide Benzylsulfid *n*
benzyl trichlorosilane Benzylsiliciumtrichlorid *n*,
　Benzyltrichlorsilan *n*
benzyne Benzyn *n*
beomycesic acid Baeomycessäure *f*
beraunite Beraunit *m* (Min)
berbamine Berbamin *n*

berban Berban *n*
berberal Berberal *n*
berberidene Berberiden *n*
berberidic acid Berberidinsäure *f*
berberilic acid Berberilsäure *f*
berberinal Berberinal *n*
berberine Berberin *n*, Sauerdornbitter *m*
berberine base Berberinbase *f*
berberine hydrate Berberinhydrat *n*
berberine hydrochloride Berberinhydrochlorid *n*
berberine sulfate Berberinsulfat *n*
berberine yellow Berberisgelb *n*
berberinium hydroxide Berberiniumhydroxid *n*
berberinol Berberinol *n*
berberis Berberitze *f* (Bot)
berberonic acid Berberonsäure *f*
berberrubic acid Berberrubinsäure *f*
berberrubine Berberrubin *n*
berberrubinol Berberrubinol *n*
berberrubinone Berberrubinon *n*
berberry Sauerdorn *m* (Bot)
berbine Berbin *n*
berengelite Berengelit *m* (Min)
beresite Beresit *m* (Min)
beresovite Beresowit *m* (Min)
bergamot, oil of ~ Bergamottöl *n*
bergamot camphor Heraclin *n*
bergamot essence Bergamottessenz *f*
bergamot oil Bergamottessenz *f*
bergamottin Bergamottin *n*, Bergaptin *n*
bergaptenquinone Bergaptenchinon *n*
bergaptin Bergaptin *n*
bergenine Bergenin *n*
Bergius process Bergiusverfahren *n*
bergmannite Bergmannit *m* (Min)
berilic acid Berilsäure *f*
berkelium (Symb. Bk) Berkelium *n*
berlambine Berlambin *n*
Berlin blue Berlinerblau *n*, Preußischblau *n*
berlinite Berlinit *m* (Min)
Berlin red Berliner Rot *n*
Berl saddle Berlsattel *m* (Dest)
bernotar Bernotar *n*
Bernoulli distribution Binomialverteilung *f*
Bernoulli's equation Bernoullische Gleichung *f*
berry pigment Beerenfarbstoff *m*
berthierite Berthierit *m* (Min),
   Eisenantimonglanz *m* (Min),
   Eisenspießglanzerz *n* (Min)
bertrandite Bertrandit *m* (Min)
beryl Beryll *m* (Min)
beryllate Beryllat *n*
beryllia Beryllerde *f*, Berylliumoxid *n*,
   Glucinerde *f*, Glycinerde *f*, Süßerde *f*
beryllide Beryllid *n*
berylline beryllartig
beryllium (Symb. Be) Beryllium *n*
beryllium bromide Berylliumbromid *n*
beryllium carbide Berylliumcarbid *n*

beryllium chloride Berylliumchlorid *n*
beryllium fluoride Berylliumfluorid *n*
beryllium methylsalicylate
   Berylliummethylsalicylat *n*
beryllium nitrate Berylliumnitrat *n*
beryllium nitroacetate chelate
   Berylliumnitroacetatchelat *n*
beryllium orthosilicate Berylliumorthosilikat *n*
beryllium oxide Berylliumoxid *n*, Beryllerde *f*,
   Glucinerde *f*, Glycinerde *f*
beryllium reactor Berylliumreaktor *m*
beryllium reflector Berylliumreflektor *m*
beryllium salt Berylliumsalz *n*
beryllium sulfate Berylliumsulfat *n*
beryllon Beryllon *n*
beryllonite Beryllonit *m* (Min)
berzelianite Berzelianit *m* (Min), Selenkupfer *n*
   (Min)
berzeliite Berzeliit *m* (Min)
Berzelius spirit lamp Berzelius'sche
   Weingeistlampe *f*
Bessel functions Bessel-Funktionen *pl*
Bessel's interpolation formula
   Bessel-Interpolationsformel *f*
bessemer converter Bessemer Birne *f*, small ~
   Kleinbessemerbirne *f*
bessemer foundry Bessemerei *f*
bessemer heat Bessemer Schmelze *f*
bessemer ingot iron Bessemerflußeisen *n*
bessemer iron *v* bessemern
bessemerization Bessemerverfahren *n*,
   Bessemern *n*
bessemerize *v* bessemern
bessemerizing Verblasen *n* in flüssigem Zustand
bessemer pig iron Bessemerroheisen *n*, basic ~
   Thomasroheisen *n*
bessemer plant Bessemerei *f*
bessemer process Bessemerprozeß *m* (Metall),
   Bessemerverfahren *n*, Windfrischverfahren *n*
   (Metall), basic ~ Thomas-Verfahren *n*
bessemer slag Bessemerschlacke *f*
bessemer steel Bessemerstahl *m*,
   Bessemereisen *n*, saures Eisen *n*
bessemer steel works Bessemerstahlwerk *n*
bessisterol Bessisterin *n*
best-equipped bestausgerüstet
best performance Bestleistung *f*
best shape Höchstform *f*
beta acid Betasäure *f*
beta-active beta-aktiv (Atom), beta-radioaktiv
   (Atom)
beta background Betauntergrund *m*
beta-biotin Vitamin H *n*
beta counter Betazählrohr *n* (Atom)
beta counting Betazählung *f*
beta decay Betazerfall *m*
beta disintegration Betaumwandlung *f* (Atom),
   Betazerfall *m*

beta disintegration energy Beta-Zerfallsenergie *f* (Atom)
beta emission Betastrahlung *f*
beta emitter Betastrahler *m*
beta-emitting betastrahlend
betafite Betafit *m* (Min)
beta gauge Betadickenmesser *m* (Atom)
beta globulin Betaglobulin *n*
betain[e] Betain *n*
betaine hydrochloride Betainhydrochlorid *n*, Acidol *n*, Betainchlorhydrat *n*, Betainchlorid *n*
betain formula Betainformel *f*
beta instability Beta-Instabilität *f* (Atom)
beta iron Betaeisen *n*
beta-oxydation of fatty acids beta-Oxidation der Fettsäuren pl
beta particle Betateilchen *n*
beta-particle spectrum Betaspektrum *n* (Atom)
betaprodine Betaprodin *n*
beta-radioactive beta-radioaktiv (Atom), beta-aktiv (Atom)
beta ray gauge Betastrahlenmeßgerät *n*
beta-ray source Betastrahlenquelle *f*
beta-ray spectrometer Betastrahlspektrometer *n* (Atom), Betaspektrometer *n* (Atom)
beta-ray spectroscopy Betaspektroskopie *f* (Atom)
beta-ray spectrum Betaspektrum *n* (Atom)
beta-ray therapy Betastrahlentherapie *f*
beta-sensitive betaempfindlich (Atom)
beta sensitivity Betaempfindlichkeit *f* (Atom)
beta thickness gauge Betadickenmeßgerät *n*
beta transformation Betaumwandlung *f* (Atom), Betazerfall *m* (Atom)
beta transition Beta-Übergang *m* (Atom)
betatron Betatron *n*, Elektronenbeschleuniger *m*, Elektronenschleuder *f*
betatron radiation Betatronstrahlung *f*
beta uranium Beta-Uran *n*
betazole Betazol *n*
betel [nut] Arekanuß *f* (Gerb)
betel [pepper] Betelpfeffer *m*, Kaupfeffer *m*
bethogenin Bethogenin *n*
betol Betol *n*, Naphthalol *n*, Salinaphthol *n*
betonicine Betonicin *n*
betony Betonienkraut *n* (Bot)
betorcinol Betorcinol *n*
betruxic acid Betruxinsäure *f*
Betti's base Bettibase *f*
betulafolientriol Betulafolientriol *n*
betula oil Birkenöl *n*, Birkenrindenöl *n*
betulene Betulen *n*
betuligenol Betuligenol *n*
betulin Birkencampher *m*
betulinic acid Betulinsäure *f*
betulinol Betulin *n*
betulol Betulol *n*
betuloside Betulosid *n*

beudantite Beudantit *m* (Min)
bevatron Bevatron *n* (Atom)
bevel Abschrägung *f*, Fase *f*; Gehre *f*; Kegel *m* (abgeschrägter Teil), Schrägmaß *n*
bevel *a* kegelig, schief (schräg); schrägwinklig
bevel *v* abflachen, abkanten, ausschärfen (Techn), ~ **off** abschrägen
bevel angle Öffnungswinkel *m*
bevel-fit valve Schrägsitzventil *n*
bevel friction wheel Kegelreibrad *n*
bevel gear Ritzel *n*
bevel[l]ed abgeflacht, abgeschrägt
bevel[l]ing Abschrägen *n*
bevel[l]ing machine Abkant- und Falzmaschine *f*, Kegelradmaschine *f*
bevel protractor Schrägmaß *n*, Schmiege *f*, Stellwinkel *m*
bevel [spur] gear Kegelrad *n*
bevel wheel Kegelrad *n*
beverage Getränk, Gebräu *n*, **alcoholic** ~ alkoholisches Getränk *n*
beyond repair nicht mehr instandsetzungsfähig, irreparabel
beyrichite Beyrichit *m* (Min)
bezoar Bezoar *m*, Ziegenstein *m*
bezoar powder Bezoarpulver *n*
bezoar stone Bezoarstein *m*
biacene Biacen *n*
biacenone Biacenon *n*
biacetyl Diacetyl *n*
biacetylene Diacetylen *n*
bialamicol Bialamicol *n*
biallyl Diallyl *n*, Biallyl *n*
biallylene Diallylen *n*
bianisoyl Anisil *n*
bianthracyl Dianthracyl *n*
bianthranol Dianthranol *n*, Dianthron *n*
bianthranyl Dianthranyl *n*
bianthrone Bianthron *n*
bianthryl Dianthryl *n*
bias Vorspannung *f* (Elektr)
bias *a* diagonal
bias angle Schnittwinkel *m* (Reifen), Schrägschnittwinkel *m* (Reifen)
bias band Reiterband *n*
bias cutter Gewebeschneidmaschine *f*, Schrägschneider *m*
bias cutting machine Diagonalschneider *m*, Schrägschneider *m*
biasing resistor Gitterwiderstand *m*
bias voltage Gittervorspannung *f*
biatomic doppelatomig
biatomic molecule zweiatomiges Molekül *n*
biaxial biaxial, zweiachsig
biaxiality Zweiachsigkeit *f*
bibasic doppelbasisch
bibenzal Stilben *n*
bibenzaldehyde Diphenaldehyd *m*
bibenzenone Diphenonchinon *n*

bibenzoic acid Diphensäure *f*
bibenzoyl Benzil *n*
bibenzyl Dibenzyl *n*, Diphenyläthan *n*
biberine chloroaurate Bibirinchloraurat *n*
biberine chloromercurate
  Bibirinchloromercurat *n*
biberine chloroplatinate Bibirinchloroplatinat *n*
biberine gold chloride Bibirinchloraurat *n*
biberine hydrochloride Bibirinchlorhydrat *n*
biberine mercury chloride
  Bibirinchloromercurat *n*
biberine platinum chloride
  Bibirinchloroplatinat *n*
bibirine Bebeerin *n*
bibliography Bibliographie, Literaturangabe *f*,
  Literaturzusammenstellung *f*
bibliolite Bibliolit *m* (Min), Blätterschiefer *m*
  (Min)
bibornyl Dibornyl *n*
bibrocathol Bibrocathol *n*
bibulous schwammig (Papier)
bicamphene Dicamphen *n*
bicamphoryl Dicamphoryl *n*
bicarbonate Bikarbonat *n*, Hydrogenkarbonat *n*
bicarbonate of potash Kaliumbicarbonat *n*
bicathode tube Bikathodenröhre *f*
bicentric orbital Zweizentrenorbital *n* (Atom)
bicetyl Dicetyl *n*
bichloride Doppelchlorid *n*
bichromate Dichromat *n*, Bichromat *n*, saures
  Chromat
bichromate cell Chromsäureelement *n*,
  Bichromatgefäß *n*
bichromate of potash Kaliumbichromat *n*
bichromate plant Bichromatanlage *f*
bichromate titration Bichromattitration *f*
bicinnamic acid Dizimtsäure *f*
bicinnamylidene Dicinnamyliden *n*
biconcave bikonkav
biconical doppeltkonisch
biconvex bikonvex, runderhaben (auf beiden
  Seiten)
bicresol Dikresol *n*
bicucine Bicucin *n*
bicuculline Bicucullin *n*
bicumenyl Dicymol *n*
bicuminal Cuminil *n*
bicycle gear case Kettenschutz *m* (Fahrrad)
bicyclic bicyclisch
bicyclobutane Bicyclobutan *n*
bicycloheptanol Bicycloheptanol *n*
bicycloheptanone Bicycloheptanon *n*
bicyclohexane Bicyclohexan *n*
bicyclohexanone Bicyclohexanon *n*
bicyclononatriene Bicyclononatrien *n*
bicyclooctadiene Bicyclooctadien *n*
bicymol Dicymol *n*
bidesyl Bidesyl *n*, Didesyl *n*
bid evaluation Angebotsvergleich *m*

bidistillate Bidestillat *n*
bidistiller Bidestillator *m*
bieberite Bieberit *m* (Min)
Biebrich scarlet Biebricher Scharlach *m*
bienanthic acid Diönanthsäure *f*
bietamiverine Bietamiverin *n*
bifidus factor Bifidus-Faktor *m*
bifilar bifilar, zweifädig, doppelgängig
bifilar galvanometer Bifilargalvanometer *n*
biflorine Biflorin *n*
bifluorenylidene Difluorenyliden *n*
bifluoride Doppelfluorid *n*
bifocal bifokal (Opt)
biformin Biformin *n*
biformyl Diformyl *n*, Glyoxal *n*
bifunctional bifunktionell
bifurcate gabelförmig, doppelgängig
bifurcate *v* abzweigen
bifurcated gabelförmig, gegabelt, gespalten
bifurcated pipe Gabelrohr *n*
big industry Großindustrie *f*
biglandulinic acid Biglandulinsäure *f*
big mill Grobeisenstraße *f*, Grobstrecke *f*
bigrid Doppelgitter *n*
biguanide Diguanid *n*, Guanylguanidin *n*
biindone Biindon *n*
bikhaconitine Bikhaconitin *n*
bilateral beidseitig, bilateral, doppelseitig,
  zweiseitig
bilberry Schwarzbeere *f* (Bot)
bile Galle *f* (Med)
bile acid Gallensäure *f*
bile pigment Gallenfarbstoff *m*
bile pigments, excretion of ~
  Gallenfarbstoffausscheidung *f*
bile substance Gallenstoff *m*
bilge pipe Bilgenrohr *n*
bilge pump Bilgenpumpe, Sodpumpe *f*,
  Leckwasserpumpe *f*, Lenzpumpe *f*
bilge valve Lenzschieber *m*
bilge water Bilgenwasser *n*
bilharzia Bilharzia *f* (Zool, Med)
bilharziasis Bilharziose *f* (Med)
bilharziosis Bilharziose *f* (Med)
bilianic acid Biliansäure *f*
biliary gallig
biliary calculus Gallenstein *m* (Med)
biliary excretion Gallenexkretion *f*
biliary pigment Gallenfarbstoff *m*
bilicyanin Bilicyanin *n*
biliflavin Biliflavin *n*
bilifuscin Bilifuscin *n*
biligenesis Biligenese *f*
bilihumin Bilihumin *n*
bilin Bilin *n*, Gallenstoff *m*
bilineurine Bilineurin *n*, Chilin *n*
bilinigrin Bilinigrin *n*
biliobanic acid Biliobansäure *f*
bilious gallig

biliprasin Biliprasin *n*
bilipurpurin[e] Bilipurpurin *n*
bilirubin Bilirubin *n*, Gallenfarbstoff *m*,
  Hämatoidin *n*
bilirubin content Bilirubingehalt *m*
bilirubinic acid Bilirubinsäure *f*
bilirubin level Bilirubinspiegel *m*
bilisoidanic acid Bilisoidansäure *f*
biliverdic acid Biliverdinsäure *f*
biliverdin Biliverdin *n*
bilixanthine Bilixanthin *n*
billet for extruding Puppe *f* (Techn)
billet roll Vorwalze *f*
billion (Am. E.) Milliarde *f* (Br. E.), Billion *f*
billitonite Billitonit *m* (Min)
billon silver Scheidmünzsilber *n*
billow cloud Wogenwolke *f*
bilobol Bilobol *n*
biloidanic acid Biloidansäure *f*,
  Norsolanellsäure *f*
bimalonic ester Bimalonester *m*
bimenthene Dimenthen *n*
bimesityl Dimesityl *n*
bimetal Bimetall *n*
bimetallic bimetallisch, zweimetallisch
bimetallic bar Doppelmetallstange *f*
bimetallic strip Bimetallstreifen *m*
bimetallic wire Bimetalldraht *m*
bimetal switch Bimetallschalter *m*
bimetal thermometer Bimetallthermometer *n*
bimethacrylic acid Dimethacrylsäure *f*
bimolecular bimolekular
bimorph cell bimorpher Piezokristall *m* (Elektr)
bin Behälter *m*, Bunker *m*; Verschlag *m*
binaphthoquinone Dinaphthochinon *n*
binaphthyl Dinaphthyl *n*
binaries Doppelsterne *m pl* (Astr)
binarite Markasit *m* (Min)
binary zweizählig, binär, zweigliedrig (System)
binary code Binärcode *m* (Comp)
binary component Doppelsternkomponente *f*
binary compound binäre Verbindung,
  Zweifachverbindung *f*
binary mixture Zweistoffgemisch *n*
  (Verfahrenstechn)
binary number Dualzahl *f*
binary scaler Zweifachuntersetzer *m*
binary system Zweistoffgemisch *n*,
  Zweistoffsystem *n* (Verfahrenstechn);
  Dualsystem *n*
bind *v* binden, anbinden, anlagern (Chem),
  einbinden, verknüpfen, ~ **[together]**
  verbinden, ~ **up** einschnüren
binder Bindemittel *n*, Binder *m*, Lackkörper *m*
bindheimite Bindheimit *m* (Min),
  Antimonbleispat *m* (Min), Bleiniere *f* (Min)
binding Binden *n*, Bindung *f* (Chem);
  Einband *m* (Buch), Verbindung *f* (chem.
  Vereinigung)

binding agent Bindemittel *n*
binding energy Bindungsenergie *f*, **nuclear** ~
  Kernbindungsenergie *f* (Atom)
binding force Bindungskraft *f*
binding material Bindematerial *n*, Bindemittel *n*
binding of water Wasserbindung *f*
binding power Bindekraft *f*, Bindevermögen *n*
binding property Bindefähigkeit *f*,
  Bindevermögen *n*, Klebfähigkeit *f*
binding rivet Heftniet *m*
binding screw Klemmschraube *f*
binding wire Bindedraht *m*
Bindschedler's green Bindschedlers Grün *n*
bin filter (in pneumatic conveyor systems)
  Bunkeraufsatzfilter *n*
bin for product Sinterbunker *m*
bin for return fines Rückgutbunker *m*
Bingham body Binghamsches Medium *n*
Bingham flow Binghamsche Strömung
biniodide of mercury Doppeljodquecksilber *n*
binitrotoluene Binitrotoluol *n*
binnite Binnit *m* (Min)
binocular binokular
binodal curve Binodalkurve *f*
binodal surface Binodalfläche *f*
binode Binode *f*
binomial binomial, binomisch (Math),
  zweigliedrig (Math)
binomial distribution Binomialverteilung *f*
binomial equation Binomialgleichung *f* (Math)
binomial formula Binomialformel *f* (Math)
binomial series Binomialreihe *f* (Math)
binomial theorem Binomialsatz *m* (Math)
binominal coefficient Binomialkoeffizient *m*
  (Math)
binoxide Doppeloxid *n*
bin scale Bunkerwaage *f*
binuclear zweikernig
bio-aeration Belebungsverfahren *n*
bio-assay Bio-Analyse *f*
biocatalyst Biokatalysator *m*
biochanin Biochanin *n*
biochemical biochemisch
biochemical oxygen demand (B.O.D)
  biochemischer Sauerstoffbedarf *m* (B.S.B.)
biochemical product biochemisches Produkt
biochemicals Biochemikalien *pl*
biochemist Biochemiker *m*
biochemistry Biochemie *f*
biocide Biozid *n*
bioclimatology Bioklimatik *f*, Bioklimatologie *f*
biocycle Biozyklus *m*
biocytin Biocytin *n*
biocytoculture Zellkultur *f*
biodynamic biodynamisch
biodynamics Biodynamik *f*
bio-electric bioelektrisch
bioenergetics Bioenergetik *f*
bioengineering Biotechnik *f*

biofilter Biofilter *n*
biogenesis Biogenese *f* (Biol),
  Entwicklungsgeschichte *f* (Biol)
biogenetic[al] biogenetisch
biological biologisch
biological activity biologische Wirksamkeit
biological apparatus biologischer Apparat
biological effectiveness biologische Wirksamkeit
biological sludge Belebtschlamm *m*
biologist Biologe *m*
biology Biologie *f*
bioluminescence Biolumineszenz *f*
biolysis Biolyse *f*
biomechanics Biomechanik *f*
biometric biometrisch
biometry Biometrie *f*
biomicroscope Mikroskop *n* zur Untersuchung
  lebender Organismen
biomicroscopy Biomikroskopie *f*
biomorphosis Biomorphose *f*
bionics Bionik *f*
bionomics Ökologie *f*
biopathology Biopathologie *f*
biophene Biophen *n*
biophore Biophor *m*
biophysical biophysikalisch
biophysics Biophysik *f*
bioplasm Bioplasma *n* (Biochem)
bioplast Biophor *m*
biopolymer Biopolymer *n*
biopsy Biopsie *f* (Med)
biopterin Biopterin *n*
bioptic bioptisch
biorhythm Biorhythmus *m*
bioscopic[al] bioskopisch
bioscopy Bioskopie *f*
biose Biose *f*
biosphere Biosphäre *f*
biostatics Biostatik *f*
biostimulation Biostimulation *f*
biosynthesis Biogenese *f*, Biosynthese *f*
  (Biochem)
biotic[al] biotisch
biotin Biotin *n*
biotinol Biotinol *n*
biotite Biotit *m* (Min), schwarzer Glimmer
  (Min), Rhombenglimmer *m* (Min)
biotite series Biotitreihe *f*
biotropism Biotropismus *m*
biotype Biotypus *m*
bipartite zweigeteilt, zweiteilig
biphenanthryl Diphenanthryl *n*
biphenetidine Diphenetidin *n*
biphenyl Biphenyl *n*, Diphenyl *n*, Xenyl-
biphenylyl Diphenylyl *n*
biphthalic acid Diphthalsäure *f*
biphthalyl Diphthalyl *n*
bipindaloside Bipindalosid *n*
bipindogenin Bipindogenin *n*

bipiperidine Bipiperidin *n*
bipolar bipolar, doppelpolig, zweipolig
bipolarity Bipolarität *f*
bipolar particulate bed electrode bipolar
  arbeitende Festbettelektrode *f*
bipotential equation Bipotentialgleichung *f*
biprism Doppelprisma *n*
biprism method Biprismaversuch *m*
bipropargyl Dipropargyl *n*, Hexadiin *n*
bipyramid Bipyramide *f* (Krist),
  Doppelpyramide *f* (Krist)
bipyridine Dipyridyl *n*
bipyridyl Bipyridyl, Dipyridyl *n*
biquinaldine Dichinaldin *n*
biquinolyl Dichinolyl *n*
biradical Biradikal *n*
birch camphor Birkencampher *m*, Betulin *n*
birch charcoal Birkenkohle *f*
birch juice Birkensaft *m*
birch oil Birkenöl *n*, Birkenrindenöl *n*
birch tar Birkenteer *m*
birch water Birkenwasser *n*
bird glue Vogelleim *m*
bird lime Vogelleim *m*
bird manure Guano *m*
bird shot Tarierschrot *m*
birefraction Doppelbrechung *f* (Opt)
birefractive doppelbrechend (Opt), zweifach
  brechend (Opt)
birefringence Doppelbrechung *f* (Opt)
birefringent doppelbrechend (Opt)
biresorcinol Diresorcin *n*
birotation Birotation *f*
bisabolene Bisabolen *n*
bisanthene Bisanthen *n*
bisanthranil Bisanthranil *n*
bisazobenzene Bisazobenzol *n*
bisbeeite Bisbeeit *m* (Min)
bischofite Bischofit *m* (Min)
biscuit Metallschwamm *m* (Atom)
bisdiazoacetic acid Bisdiazoessigsäure *f*
bisect *v* halbieren
bisecting line of an angle Winkelhalbierende *f*
bisection, point of ~ Halbierungspunkt *m*
bisector Halbierungslinie *f* (Math), ~ **of an**
  angle Winkelhalbierende *f*
bisectric line Halbierungslinie *f* (Math)
bisectrix Halbierende *f* (Math),
  Halbierungslinie *f* (Math)
bisglyoxaline Glykosin *n*
bisilicate Doppelsilicat *n*
Bismarck brown Bismarckbraun *n*, Vesuvin *n*
  (Farbstoff)
bismite Bismit *m* (Min), Wismutblüte *f* (Min),
  Wismutocker *m* (Min)
bismuth (Symb. Bi) Wismut *n*, **butter of** ~
  Wismutbutter *f*, **needle-shaped sulfuretted** ~
  Federwismut *n* (Min), **spongy** ~

Wismutschwamm *m*, **to solder with** ~
wismuten
bismuthate Wismutat *n*
bismuthate *a* wismutsauer (Salz)
bismuth benzoate Wismutbenzoat *n*
bismuth blende Wismutblende *f* (Min),
Kieselwismut *n*
bismuth bromide Bromwismut *n*
bismuth bronze Wismutbronze *f*
bismuth chloride Wismutchlorid *n*,
Chlorwismut *n*
bismuth cinnamate Wismutcinnamat *n*
bismuth citrate Wismutcitrat *n*
bismuth content Wismutgehalt *m*
bismuth dithiosalicylate, basic ~ Thioform *n*
bismuth fission Wismutspaltung *f*
bismuth gallate Wismutgallat *n*
bismuth glance Wismutglanz *m* (Min),
Bismutit *m* (Min)
bismuth hydride Wismutwasserstoff *m*
bismuth hydroxide Wismuthydroxid *n*
bismuth hydroxyiodotannate Ibit *n*
bismuthic acid Wismutsäure *f*
bismuthic anhydride Wismutoxid *n*
(Wismut(V)-oxid), Wismutpentoxid *n*,
Wismutsäureanhydrid *n*
bismuthic compound Wismut(V)-Verbindung *f*
bismuthic oxide Wismutoxid *n*
(Wismut(V)-oxid), Wismutpentoxid *n*,
Wismutsäureanhydrid *n*
bismuthiferous wismuthaltig
bismuthine Bismutin *m* (Min)
bismuthinite Bismutin *m* (Min), Wismutglanz *m*
(Min)
bismuth iodide Jodwismut *n*
bismuthite Wismutit *m* (Min), Wismutspat *m*
(Min)
bismuth lead ore Wismutbleierz *n* (Min)
bismuth litharge Wismutglätte *f*
bismuth loaded wismutgetränkt
bismuth metal Wismutmetall *n*
bismuth naphtholate Naphtholwismut *n*
bismuth nickel Saynit *m* (Min)
bismuth nitrate Wismutnitrat *n*
bismuth ocher Wismutocker *m* (Min), Bismit *m*
(Min), Wismutblüte *f* (Min)
bismuth ore Wismuterz *n*
bismuthous benzoate Wismutbenzoat *n*
bismuthous chloride Wismutchlorid *n*
(Wismut(III)-chlorid), Wismutbutter *f*
bismuthous compound Wismut(III)-Verbindung *f*
bismuthous hydroxide Wismuthydroxid *n*
(Wismut(III)-hydroxid)
bismuthous nitrate Wismutnitrat *n*
(Wismut(III)-nitrat)
bismuthous valerate Wismutvalerianat *n*
bismuth oxide Wismutoxid *n*, Wismutglätte *f*
(Min)
bismuth oxychloride Wismutoxychlorid *n*

bismuth oxyiodide Wismutoxyjodid *n*
bismuth oxyiodide subgallate Airol *n* (HN),
Airoform *n* (HN), Airogen *n* (HN)
bismuth pentoxide Wismutpentoxid *n*,
Wismutsäureanhydrid *n*
bismuth phenolate Phenolwismut *n*
bismuth phenoxide Phenolwismut *n*
bismuth precipitate Wismutniederschlag *m*
bismuth pyrogallate Pyrogallolwismut *n*,
Helcosol *n*, Wismutpyrogallat *n*
bismuth salicylate Wismutsalicylat *n*
bismuth selenide Wismutselenid *n*
bismuth silicate Wismutsilikat, Kieselwismut *n*
bismuth solder Wismutlot *n*
bismuth subgallate Wismutsubgallat *n*,
Dermatol *n*
bismuth subnitrate Wismutweiß *n*
bismuth sulfide Wismutsulfid, Schwefelwismut *n*,
Bismutin *m* (Min)
bismuth telluride Tellurwismut *n*,
Wismuttellur *n*, **native** ~ Markasitglanz *m*
(Min)
bismuth tellurite Joseit *m* (Min)
bismuth tetraiodophenolphthalein Eudoxin *n*
bismuth tetroxide Wismuttetroxid *n*
bismuth tribromophenolate Xeroform *n*
bismuth trichloride Wismut(III)-chlorid *n*,
Wismutbutter *f*
bismuth trinitrate Wismut(III)-nitrat *n*
bismuth valerate Wismutvalerianat *n*
bismuth(V) oxide Wismutpentoxid *n*,
Wismutsäureanhydrid *n*, Wismut(V)-oxid *n*
bismuth white Wismutweiß *n*
bismuthyl chloride Wismutoxychlorid *n*
bismuthyl iodide Wismutoxyjodid *n*
bismutite Bismutit *m* (Min)
bismutoferrite Bismutoferrit *m* (Min)
bispidine Bispidin *n*
bister Bister *m n*, Bisterbraun *n*
bistort Krebswurzel *f* (Bot), Wiesenknöterich *m*
(Bot)
bistyrene Distyrol *n*
bisulfate Bisulfat *n*, Disulfat *n*, Doppelsulfat *n*,
Hydrogensulfat *n*
bisulfate *a* doppel[t]schwefelsauer
bisulfite Bisulfit *n*, Doppelsulfit *n*
bisulfite *a* doppel[t]schwefligsauer
bisyringyl Bisyringyl *n*
bit Bohreisen *n*, Schneide *f*, Spitzbohrer *m*
bitartrate Bitartrat *n*
bitartrate *a* doppeltweinsauer
bite *v* ätzen, kaustizieren
bitetralyl Ditetralyl *n*
bit gauge Meißelschablone *f*
bithiophene Thiophthen *n*
bithymol Dithymol *n*
biting beißend, scharf
bitolyl Ditolyl *n*
bits Lackteilchen *pl*, **freedom from** ~
Kornfreiheit *f*

bitter bitter, herb, gallig, **slightly** ~ schwach bitter
bitter almond, essence of ~ Bittermandelessenz *f,* **oil of** ~ Bittermandelöl *n,* **spirit of** ~ Bittermandelgeist *m*
bitter almond oil, artificial ~ künstliches Bittermandelöl
bitter almond soap Bittermandelseife *f*
bitter ash Bitteresche *f*
bitter bark Bitterrinde *f*
bitterdock root Grindwurzel *f* (Bot)
bitter earth Bittererde *f* (Min), Magnesia *f,* Magnesiumoxid *n*
bittering Bitterstoff *m* (Brau)
bitter mineral water Bitterling *m,* Bitterwasser *n*
bittern Bitterling *m* (Brau), Salzmutterlauge *f*
bitterness Herbheit *f,* **to deprive of** ~ entbittern
bitter principle Bitterstoff *m*
bitter resin Bitterharz *n*
bitter salt Bittersalz *n*
bitter salt water Bittersalzwasser *n*
bitter spar Bitterspat *m* (Min), Dolomit *m* (Min), **ferruginous** ~ Eisenbraunkalk *m* (Min)
bittersweet Bittersüß *n* (Bot), **extract of** ~ Bittersüßextrakt *m*
bitter water Bitterwasser *n*
bitter wood Bitterholz *n* (Bot), Fliegenholz *n* (Bot), Jamaicabitterholz *n* (Bot), Quassiaholz *n* (Bot)
bitter wort Enzian *m* (Bot)
bittiness Pünktchenbildung *f* (Farben)
bitts Beting *f* (Schiff)
bitumen Asphalt *m,* Bergharz *n,* Bitumen *n,* Erdharz *n,* Erdpech *n,* **color of** ~ Erdharzfarbe *f,* **soft tarlike** ~ Bergteer *m,* **viscous** ~ Bergteer *m*
bitumen-laminated bitumenkaschiert
bitumen pavement Bitumendecke *f* (Straßenbau)
bitumen plant Asphaltanlage *f*
bitumen process Asphaltverfahren *n*
bituminiferous asphalthaltig, erdharzhaltig
bituminization Bituminisierung *f,* Imprägnierung *f* mit Erdpech, Verwandlung *f* in Erdpech
bituminize *v* in Erdharz verwandeln
bituminized fiber pipe Pechfaserrohr *n*
bituminous bituminös, asphaltartig, bergharzig, erdharzartig, erdharzhaltig, erdpechartig, erdpechhaltig, pechartig
bituminous clay Brauseton *m,* Kohlenblume *f*
bituminous coal Harzkohle *f,* Pechkohle *f,* Pechsteinkohle *f,* Steinkohle, Fettkohle *f*
bituminous coal deposit Steinkohlenvorkommen *n*
bituminous coal tar Steinkohlenteer *m*
bituminous earth Pecherde *f*
bituminous emulsion Kaltasphalt *m*

bituminous lignite ölreiche Braunkohle
bituminous marl Stinkmergel *m*
bituminous mastic Asphaltmastix *m*
bituminous paint Bitumenfarbe *f*
bituminous paper Bitumenpapier *n*
bituminous peat Specktorf *m*
bituminous pitch Asphaltpech *n*
bituminous red clay Brauseerde *f*
bituminous roofing paper Bitumenisolierpappe *f*
bituminous shale Brandschiefer *m,* Kohlenbrandschiefer *m,* Kohlenschiefer *m*
bituminous surface Bitumendecke *f*
bituminous varnish Asphaltlack *m,* Bitumenlack *m*
bityite Bityit *m* (Min)
biurate Biurat *n*
biurea Diharnstoff *m,* Hydrazodicarbonamid *n,* Hydrazoformamid *n*
biuret Biuret *n*
biuret reaction Biuretreaktion *f*
bivalence Zweiwertigkeit *f* (Chem)
bivalent bivalent, doppelwertig, zweiwertig
bivector Bivektor *m*
bivinyl Butadien *n,* Divinyl *n,* Erythren *n*
bixbyite Bixbyit *m* (Min)
bixin Bixin *n,* Orleanrot *n*
bixylenol Dixylenol *n*
bixylyl Dixylyl *n*
bjelkite Bjelkit *m* (Min)
black schwarz, dunkel
black aldertree bark Faulbaumrinde *f*
black-and-white film Schwarzweißfilm *m*
black-and-white photograph Schwarzweißphoto *n*
blackband Kohleneisenstein *m* (Min)
black body radiation Hohlraumstrahlung *f*
black body radiator Schwarzstrahler *m*
black-boy gum Acaroidharz *n*
black brown schwarzbraun
black [char]coal Schwarzkohle *f*
black cobalt ocher Kobaltmulm *m* (Min)
black-colored schwarzfarbig
black-colored cerussite Bleimulm *m*
black copper Kupferschwärze *f*
black copper ore Schwarzkupfererz *n*
black dyeing Schwarzfärberei *f*
black earth Dammerde *f*
blacken *v* schwärzen, abschwärzen, anblaken, anschwärzen, einschwärzen
black enamel Schwarzschmelz *m*
blackened with smoke angeblakt, angeschmaucht, rauchig
blackening Schwärzung *f,* ~ **with soot** Anblaken *n*
blackening limit Schwärzungsgrenze *f* (Phot)
black glass filter Schwarzglasfilter *n*
black haw bark Viburnumrinde *f* (Bot)
black heart malleable iron Schwarzkernguß *m*
black hornblende Kohlenhornblende *f*
blacking Schuhwichse *f,* Schwärze *f*

black iron plate Schwarzblech *n*
black lake Lackschwarz *n*
black lead Aschblei *n* (Min), Bleischwärze *f*
(Min), Graphit *m* (Min), Graphitstift *m*,
Reißblei *n* (Min), **powdered** ~
Ofenschwärze *f*
blacklead *v* graphitieren
black lead crucible Reißbleitiegel *m*
black lead ore Bleimulm *m*
black lead powder Eisenschwärze *f*
black line paper Heliographiepapier *n*
black liquor Schwarzlauge *f*
black manganese ore Hartmanganerz *n* (Min)
blackness Schwärze *f*, **degree of** ~
Schwärzegrad *m*
black oak Färbereiche *f*
black oil Schwarzöl *n*
blackout Verdunk[e]lung *f*, Abdunkelung *f*,
Stromausfall *m*
black out *v* verdunkeln
blackout curtain Verdunkelungsvorhang *m*
blackout paint Verdunkelungsanstrich *m*
black pitch Schwarzpech *n*
blackplate Schwarzblech *n*
black precipitate Quecksilber(I)-oxid *n*
black radiation Hohlraumstrahlung *f*
black rot Schwarzfäule *f*
black sheet [iron] Schwarzblech *n*
black-short schwarzbrüchig
[black]smith Schmied *m*
black spinel Eisenspinell *m* (Min), Kandit *m*
(Min)
blackthorn Schlehdorn *m* (Bot)
black ware Basaltsteingut *n*
black wash Gießereischwärze *f*
blackwash *v* schwärzen (Gieß), schlichten (Gieß)
bladder Blase *f* (Anat), Harnblase *f*
bladder wrack Blasentang *m*
blade Blatt *n* (Messer), Flügel *m* (Rührwerk),
Flügelblatt *n*, Klinge *f*, Messer *n* (Klinge),
Schaufel *f* (Turbine), **scraper** ~ **or doctor
blade** Abschabemesser *n*
blade angle Blattwinkel *m*, Schaufelwinkel *m*
blade dryer Schaufeltrockner *m*
bladed wheel Schaufelrad *n*
blade mixer Blattrührer *m*
blade of agitator Rührwerksflügel *m*
bladhianin Bladhianin *n*
blading Beschaufelung *f*, Schaufelung *f*
blanc fixe Bariumsulfat *n*, Barytweiß *n*, Blanc
fixe *n*
blanch *v* ansieden, aussieden, blanchieren,
bleichen, verzinnen (Techn), weißsieden
(Silber)
blanched weißfarbig (gebleicht)
blanchimeter Blanchimeter *n*
blanching Beizen *n*, Beizung *f*
blanching bath Beizbad *n*
blanching liquor Weißsiedlauge *f*

blanching solution Weißsud *m*
blank Rohling *m*
blank *v* ausstanzen, löschen (Comp)
blank cartridge Platzpatrone *f*
blank condition, in ~ unbearbeitet
blanket Aufzug *m* (Buchdr), Brutmantel *m*
(Reaktor), Decke *f*, Mantel *m* (Atom)
blanket material exposure Bestrahlung *f* des
Brutstoffes
blank experiment Leerversuch *m*
blank flange Blindflansch *m*, Blindscheibe *f*
blank form, in ~ unbearbeitet
blank hardening Blindhärten *n*
blanking Stanzen *n*, **freehand** ~ Ausstanzen von
Hand
blanking die Stanzform *f*
blanking tool Stanzwerkzeug *n*
blanks Lehrenkörper *m*
blank test Blindprobe *f* (Analyse),
Blindversuch *m*, Leerversuch *m*,
Vergleichsversuch *m*, **numerical result of** ~
Blindwert *m*
blank titration Blindtitration *f*
blank trial Blindversuch *m*
blast Gebläse *n*, Gebläsewind *m*, Schuß *m*
(Bergb), Wind *m*, **action of** ~ Blaswirkung *f*,
calculation of ~ Windberechnung *f*, **cold** ~
kalter Wind, **continuous** ~ gleichmäßiger
Windstrom, **cutting** ~ Wind *m* mit starker
Pressung, **hot** ~ heißer Wind, **soft** ~ Wind *m*
mit geringer Pressung, **to put on the** ~
anblasen (Schachtofen), **uniform** ~
gleichmäßiger Windstrom
blast *v* sprengen, abstrahlen, aussprengen,
schießen (Bergb), ~ **off** absprengen,
lossprengen
blast air Gebläseluft *f*
blast apparatus Blasapparat *m*, Gebläse *n*,
Gebläsevorrichtung *f*, Gebläsewerk *n*
blast box Windkasten *m*
blast connection Düsenstock *m*
blast consumption Windverbrauch *m*
blast cylinder Gebläsezylinder *m*,
Windzylinder *m*
blast distribution Windverteilung *f*
blast drawing Düsenblasverfahren *n*
blast drying Windtrocknung *f*
blast engine Gebläsemaschine *f*
blast forge Glutesse *f*
blast furnace Hochofen *m*, Gebläseofen *m*,
Gebläseschachtofen *m*, Schachtofen *m*, **to tap
the** ~ den Hochofen abstechen, ~ **of medium
height** Halbhochofen *m* (Metall), ~ **with
bucket hoist** Hochofen *m* mit
Kübelbegichtung, ~ **without appendices**
Hochofenrumpf *m*, ~ **with skip hoist**
Hochofen *m* mit Kippgefäßbegichtung
blast furnace blowing engine Hochofengebläse *n*
blast furnace bricks Hochofensteine *pl*

blast furnace burden Möller *m*
blast furnace casting Hochofenabstich *m*
blast furnace cement Hochofenzement *m*
blast furnace charging Hochofenbegichtung *f,*
  Hochofenbeschickung *f*
blast furnace coke Hochofenkoks *m*
blast furnace crucible Hochofengestell *n*
blast furnace drawing Hochofenzeichnung *f*
blast furnace dust Gichtstaub *m*
blast furnace [equipment] Hochofenanlage *f*
blast furnace fittings Hochofenarmaturen *f pl*
blast furnace gas Gichtgas *n,* Hochofengas *n,*
  Schwachgas *n*
blast furnace gas cleaning Gichtgasreinigung *f*
blast furnace gas engine Gichtgasmaschine *f*
[blast] furnace gas main Gichtgasleitung *f*
blast furnace gas purifying plant
  Hochofengasreinigungsanlage *f*
blast furnace gas valve Hochofengasschieber *m*
blast furnace gun Stichlochstopfmaschine *f*
blast furnace hearth or well Hochofengestell *n*
blast furnace hoist Hochofenaufzug *m*
blast furnace jacket Hochofenpanzer *m*
blast furnace lime Hüttenkalk *m*
blast furnace lines Hochofenprofil *n*
blast furnace lining Hochofenfutter *n*
blast furnace lintel plate Hochofentragring *m*
blast furnace mixer Roheisenmischer *m*
blast furnace operation Hochofengang *m*
blast furnace plant Hütte *f*
blast furnace process Hochofenverfahren *n*
blast furnace ring or mantle Hochofentragring *m*
blast furnace settler Hochofenvorherd *m*
blast furnace slag Hochofenschlacke *f*
blast furnace slag sand
  Hochofenschlackensand *m*
blast furnace smelting Hochofenschmelze *f,*
  Schachtofenarbeit *f,* Schachtofenschmelzen *n*
blast furnace throat or mouth Hochofengicht *f*
blast gauge Windmesser *m*
blast heating Winderhitzung *f*
blasthole Sprengloch *n*
blasting Sprengung *f*
blasting agent Sprengmittel *n,* Strahlmittel *n*
blasting cap Zündhütchen *n*
blasting gelatin Nitrogelatine *f*
blasting oil Glycerinnitrat *n*
blasting powder Sprengpulver *n*
blasting process Sprengvorgang *m*
blast inlet Windzuführung *f,* Windloch *n*
blast lamp Gebläselampe *f,* Lötlampe *f*
blast lamp table Blasetisch *m*
blast main Hauptwindleitung *f*
blast meter Gebläsemesser *m*
blast nozzle Blasdüse *f*
blastogenesis Blastogenese *f* (Biol)
blastolysis Blastolyse *f*
blastomere Blastomere *f* (Biol)
blastomycic acid Blastmycinsäure *f*

blastomycin Blastmycin *n*
blast pipe Blasrohr *n,* Düse *f,* Windleitung *f*
  (Bergb)
blast pressure Detonations-, *m,*
  Detonationsdruck *m,* Windspannung *f*
blast roasting Verblaserösten *n*
blast superheater Luftüberhitzer *m*
blast tank Windkessel *m*
blast tube Blas[e]rohr *n,* Strahlrohr *n* (Rakete)
blast valve Windventil *n*
blaze *v* flammen, aufglühen, fackeln, ~ off *v*
  abbrennen
blazed off abgebrannt (Chem)
blazing flammend
blazing fire Flammfeuer *n*
blazing glow Feuerglut *f*
blazing off Abbrennen *n*
bleach, resistance to ~ Bleichechtheit *f*
bleach *v* bleichen, entfärben, weißen, ~ by
  decoloring ausbleichen
bleachability Bleichfähigkeit *f*
bleacher Bleicher *m,* Klärkübel *m*
bleachery Bleichanlage *f,* Bleichanstalt *f,*
  Bleiche *f,* Bleicherei *f*
bleaching Ausbleichen *n,* Bleiche *f,* Bleichen *n,*
  Entfärbung *f,* chemical ~ chemische Bleiche,
  quick chemical ~ Schnellbleiche *f,* subsequent
  ~ Nachbleiche *f,* ~ with bleaching powder
  Chlorkalkbleiche *f,* ~ of colored goods
  Buntbleiche *f,* ~ of cotton Weißbleiche *f* (Text)
bleaching *a* entfärbend
bleaching acid Bleichsäure *f*
bleaching action Bleichwirkung *f*
bleaching agent Bleichmittel *n,* Weißtöner *m*
  (Text)
bleaching apparatus Bleichapparat *m*
bleaching assistant Bleichhilfsmittel *n*
bleaching auxiliary Bleichhilfsmittel *n*
bleaching bath Bleichbad *n*
bleaching capacity Bleichvermögen *n*
bleaching clay Bleicherde *f*
bleaching earth Bleicherde *f*
bleaching effect Bleichwirkung *f*
bleaching frame Bleichrahmen *m*
bleaching liquor Bleichbad *n,* Bleichflotte *f,*
  Bleichflüssigkeit *f,* Bleichlauge *f*
bleaching lye Bleichlauge *f*
bleaching machine Bleichmaschine *f,* open width
  ~ Breitbleichmaschine *f*
bleaching mordant Bleichbeize *f*
bleaching of cellulose Bleichen *n* des Zellstoffes
bleaching-out process Ausbleichverfahren *n*
bleaching plant Bleichanlage *f,* Bleicherei *f*
bleaching powder Bleichkalk *m,* Bleichpulver *n,*
  Chlorkalk *m,* Entfärbungspulver *n*
bleaching powder solution Bleichkalklösung *f,*
  Chlorkalklösung *f*

bleaching power Bleichkraft *f*, Bleichvermögen *n*,
  Entfärbungskraft *f*
bleaching process Bleichprozeß *m*,
  Bleichverfahren *n*
bleaching product Bleichprodukt *n*
bleaching salt Bleichsalz *n*
bleaching soap Bleichseife *f*
bleaching soda Bleichsoda *f*
bleaching solution Bleichlauge *f*
bleaching stain Bleichfleck *m*
bleaching tower Bleichturm *m*
bleaching vat Bleichfaß *n*, Bleichkasten *m*
bleaching vessel Bleichgefäß *n*
bleaching works Bleichanstalt *f*
bleach spot Bleichfleck *m*
bleed *v* bluten, ~ color verwaschen (Färb),
  verlaufen (Färb)
bleeder current Anzapfstrom *m*
bleeder electrode Ableitelektrode *f* (Elektr)
bleeder resistance Ableitwiderstand *m* (Telev)
bleeder screw Ablaßschraube *f*
bleeder valve Anzapfventil *n*
bleeding Anzapfung *f*, Bluten *n* (Anstrich),
  Blutung *f*, Farbbluten *n*, ~ (of colors)
  Ausbluten *n*, Ausblutung *f*
bleeding point Anzapfstelle *f*
bleeding the column Abdampfen *n* der
  Trennflüssigkeit (Gaschromatographie)
bleeding through (e. g. of an adhesive)
  Durchschlagen *n* (Klebstoff)
bleed screw Entlüftungsschraube *f*
blemish Fehler *m* (Holz)
blend Mischung *f*, Vermischung *f* (Biol),
  Verschnitt *m* (Spirituosen)
blend *v* mischen, anmengen, durchmischen,
  melieren, vermengen, vermischen,
  verschmelzen (Färb), verschneiden
  (Spirituosen)
blended mixture Verschnitt *m*
blender Mischer, Mixer *m*, Menger *m*
blende roasting furnace Blenderöstofen *m*
blending Vermischen *n*, Mengen *n*, Mischen *n*,
  Mischung *f*, Vermengung *f*, Verschneiden *n*
  (Spirituosen), Verschnitt *m*
blending agent Zusatzmittel *n*
blending machine Egalisiermaschine *f*
blessed thistle, extract of ~
  Kardobenediktenextrakt *m*
bliabergite Bliabergit *m* (Min)
blight Getreidebrand *m*, Kornbrand *m*,
  Kornfäule *f*
blind *v* blenden
blind alley Sackgasse *f*
blind coal Taubkohle *f* (Bergb)
blind current Blindstrom *m*
blind flange Blindflansch *m*
blind hole Sackloch *n*
blind pass Blindkaliber *n*
blind shell Blindgänger *m*

blind spot Schattenstelle *f*
blister Bläschen *n*, Blase *f* (Pathol) (Schweiß),
  Gasblase *f*, Preßfehler *m*
blister *v* Blasen bilden
blister box Blasenkammer *f* (Farben)
blister copper Blasenkupfer *n*, Rohkupfer *n*
blistered blasig
blistered condition Blasigkeit *f*
blistered state Blasigkeit *f*
blister formation Blasenbildung *f*
blistering Ausblühung *f* (Schweiß),
  Blasenbildung *f*
blistering *a* bläschenziehend, blasenziehend
blister-like bläschenartig
blister plaster Blasenpflaster *n*
blister resistance Beulsteifigkeit *f*
blisters, to get ~ blasig werden, to raise ~
  Blasen ziehen
blister steel Blasenstahl *m*, Brennstahl *m*,
  blasiger Stahl, Zementstahl *m*
blistery blasig
blob density Klecksdichte *f*
Bloch wall Bloch-Wand *f*
block Klotz *m*, Block *m*, Hindernis *n;* Klischee *n*
block *v* absperren, blocken, blockieren,
  hemmen, hindern, sperren (blockieren),
  versperren, ~ off *v* abschirmen
blockade Absperrung *f*
block and tackle Flaschenzug *m*, Rollenzug *m*
block chain Blockkette *f*
block condenser Blockkondensator *m*
block conductor pattern Blockleitersystem *n*
block faulting Blockverwerfung *f*
block formation Blockbildung *f*
blocking Blockierung *f*, Hemmen *n*, Kleben *n*
blocking *a* absperrend, blockierend
blocking layer Sperrschicht *f*
blocking oscillator Sperrschwinger *m*
blocking voltage Sperrspannung *f*
block letter Druckbuchstabe *m*
block method Blockmethode *f*
block mill Blockmühle *f*
block of iron Eisenblock *m*
block of stone Steinquader *m*
block pig Gans *f* (Metall)
block polymer Blockpolymer *n*,
  Blockpolymerisat *n*
block polymerization Blockpolymerisation *f*,
  Polymerisation *f* in Masse
block press Blockpresse *f*, Kochpresse *f*,
  Kofferpresse *f*
block printing Handdruck *m*
block process Blockbetrieb *m*
block relaxation Blockrelaxation *f*
block sheave Blockscheibe *f*
block slicing machine Blockschneidmaschine *f*,
  Blockteilmaschine *f*
block tackle Handflaschenzug *m*

Name

Street

City/Post Code

Country

Profession

Comments

This information will be computerized.

Stamp

Reply Card

VCH Verlagsgesellschaft
Product Information
P.O. Box 1260/1280

D-6940 Weinheim
Federal Republic of Germany

Name

Street

City/Post Code

Country

Profession

Comments

This information will be computerized.

Stamp

Reply Card

VCH Verlagsgesellschaft
Product Information
P.O. Box 1260/1280

D-6940 Weinheim
Federal Republic of Germany

We would like to inform you regularly about our books and journals. Please check the subjects that interest you.

01 00 Chemistry

- 01 01 General Chemistry
- 01 05 Inorganic Chemistry
- 01 10 Analytical Chemistry
- 01 15 Physical Chemistry
- 01 20 Theoretical Chemistry
- 01 25 Nuclear Chemistry
- 01 30 Organic Chemistry
- 01 31 Organometallic Chemistry
- 01 35 Macromolecular Chemistry
- 01 40 Biochemistry/Chemical Physiology
- 01 45 Foodstuff Chemistry
- 01 50 Neurochemistry
- 01 52 Medicinal Chemistry
- 01 54 Clinical Chemistry
- 01 56 Toxicology
- 01 58 Environmental Chemistry
- 01 60 Water Chemistry
- 01 61 Detergents/Cosmetics
- 01 62 Corrosion
- 01 64 Materials Chemistry
- 01 65 Paint Chemistry/Surface Treatment
- 01 66 Construction Chemistry
- 01 68 Technical Chemistry
- 01 70 Agricultural Chemistry
- 01 72 Soil Chemistry/Geochemistry
- 01 74 Biotechnology
- 01 76 Legislation/Regulations/ Specifications
- 01 78 Information/Documentation
- 01 80 Didactics of Chemistry
- 01 97 History of Chemistry
- 01 99 All Fields

02 00 Physics

- 02 01 General Physics
- 02 05 Mechanics
- 02 10 Thermodynamics
- 02 15 Electromagnetism
- 02 20 Optics
- 02 25 Atomic Physics
- 02 30 Nuclear Physics
- 02 35 Fundamental Particles/ High Energy Physics
- 02 40 Solid State Physics
- 02 50 Molecular Physics
- 02 52 Plasma Physics
- 02 54 Astronomy/Astrophysics
- 02 56 Biophysics/Medicinal Physics
- 02 58 Geophysics
- 02 60 Applied Physics
- 02 62 Measuring Instruments/ Measuring Methods
- 02 64 Didactics of Physics
- 02 99 All Fields

03 00 Biology

- 03 01 General Biology
- 03 05 Botany
- 03 10 Zoology
- 03 15 Systematical Zoology
- 03 20 Evolution
- 03 25 General Morphology/Physiolo
- 03 30 Cytology
- 03 35 Histology
- 03 40 Microbiology
- 03 45 Genetics
- 03 50 Molecular Biology
- 03 52 Biophysics
- 03 54 Ecology
- 03 56 Ethology
- 03 58 Applied Biology
- 03 60 Palaeobiology
- 03 62 Legislation/Regulations/ Specifications
- 03 64 Didactics of Biology
- 03 97 History of Biology
- 03 99 All Fields

04 00 Pharmacy

- 04 01 Pharmacy
- 04 05 Pharmacology
- 04 10 Therapeutics
- 04 15 Pharmaceutical Chemistry
- 04 20 Pharmaceutical Analysis
- 04 25 Pharmaceutical Technology
- 04 30 Pharmaceutical Biology
- 04 35 Instruments/Apparatus/ Equipment
- 04 40 Legislation/Regulations/ Specifications
- 04 97 History of Pharmacy
- 04 99 All Fields

✂

We would like to inform you regularly about our books and journals. Please check the subjects that interest you.

07 00 Mathematics

- 07 01 Mathematical Logics
- 07 05 Theory of Numbers
- 07 10 Geometry
- 07 15 Algebra, Analytical Geometry
- 07 18 Topology
- 07 20 Analysis
- 07 21 Sequences and Series
- 07 25 Calculus
- 07 26 Differential Equations
- 07 27 Theory of Functions
- 07 28 Analysis of Functions
- 07 29 Differential Geometry
- 07 30 Permutations, Combinations
- 07 35 Numerical Optimization Methods
- 07 40 Special Functions
- 07 45 Vector Analysis
- 07 50 Applied Mathematics, Numerical Methods
- 07 55 Numerical Analysis
- 07 60 Kybernetics
- 07 62 Artifical Intelligence
- 07 65 Laplace-Transforms
- 07 70 Theory of Probability
- 07 75 Statistics
- 07 80 Applied Mathematics
- 07 85 Mathematics for Physicians, Biologists, Chemists
- 07 97 History of Mathematics
- 07 99 All Fields

08 00 Chemical Engineering

- 08 01 Process Engineering
- 08 05 Plant Construction
- 08 10 Apparatus Construction
- 08 15 Measuring and Automatic Control Technology
- 08 20 Materials Technology
- 08 25 Corrosion Protection
- 08 30 Energy Technology
- 08 35 Biotechnology
- 08 40 Operations Technology
- 08 97 History of Technology
- 08 99 All Fields

09 00 Other Sciences

- 09 01 Minerology
- 09 05 Geology
- 09 10 Geography
- 09 15 Oceanography
- 09 20 Meteorology/Climatology
- 09 22 Planetology
- 09 25 Soil Science
- 09 30 Plant Nutrition

10 00 Agricultural Scienc

- 10 99 All Fields

11 00 Forestry

- 11 99 All Fields

12 00 Nutrition Sciences

- 12 99 All Fields

15 00 Other Engineering Sciences

- 15 01 Metallurgy
- 15 02 Mining/Ores/Minerals
- 15 03 Coal/Coal Industry
- 15 04 Petroleum/Mineral Oil Industry
- 15 05 Energy Technology
- 15 06 Mechanical Engineering
- 15 07 Civil Engineering
- 15 08 Automobile Construction/ Transportation
- 15 09 Aeronautics and Astronautics

20 00 Jurisprudence

- 20 01 Patent and Copyright Law
- 20 05 Industrial Property Law

block tin Blockzinn *n*
block valve Abteilventil *n*
blodite Blödit *m* (Min)
bloedite Blödit *m* (Min)
blomstrandite Blomstrandin *m* (Min)
blood, arterial ~ sauerstoffreiches Blut, **clotted**
~ geronnenes Blut, **coagulated** ~ geronnenes
Blut, **decomposition of the** ~ Blutzersetzung *f,*
defibrinated ~ defibriniertes Blut, **peripheral**
~ peripheres Blut, **stored** ~ konserviertes
Blut, **to supply with** ~ durchbluten, **venous** ~
sauerstoffarmes Blut
blood albumin Blutalbumin *n*
blood alcohol Blutalkohol *m*
blood alcohol concentration Blutalkoholgehalt *m*
blood alcohol determination
Blutalkoholbestimmung *f*
blood analysis Blutanalyse *f*
blood bank Blutbank *f*
blood bank refrigerator Blutbankkühlschrank *m*
blood buffer Blutpuffer *m*
blood calcium level Blutcalciumspiegel *m*
blood cell Blutkörperchen *n*
blood cells, formation of ~ Blutbildung *f*
blood charcoal Blutkohle *f*
blood circulation Durchblutung *f*
blood clot Blutgerinnsel *n* (Med),
Blutklumpen *m*, Blutkuchen *m* (Med),
Blutpfropf *m*
blood clotting Blutgerinnung *f*
blood coagulation Blutgerinnung *f*
blood-colored agate Blutachat *m* (Min)
blood composition Blutzusammensetzung *f*
blood conservation Blutkonservierung *f*
blood constituent Blutbestandteil *m*
blood corpusc[u]le Blutkörperchen *n*
blood count Blutbild *n*, Blutbildkontrolle *f,*
Blutkörperchenzählung *f*
blood crystal Blutkristall *m*
blood deficiency Blutmangel *m*
blood disease Blutkrankheit *f*
blood donor Blutspender *m*
blood fibrin Blutfibrin *n*
blood-forming blutbildend
blood group Blutgruppe *f*
blood group classification
Blutgruppeneinteilung *f*
blood group compatibility
Blutgruppenverträglichkeit *f*
blood group incompatibility
Blutgruppenunverträglichkeit *f*
blood-grouping Blutgruppenbestimmung *f,*
Blutgruppenfeststellung *f*
blood groups, incompatibility of the ~
Blutgruppendissonanz *f*
bloodless blutarm
blood level Blutspiegelwert *m*
blood matching Blutgruppenabstimmung *f*
blood meal Blutmehl *n*

blood molasses Blutmelasse *f*
blood picture Blutbild *n*
blood pigment Blutfarbstoff *m*
blood plaque Blutplättchen *n*
blood platelet Blutplättchen *n*, Thrombozyt *m*
blood poisoning Blutvergiftung, Sepsis *f* (Med),
Toxämie *f* (Med)
blood powder Blutmehl *n*
blood processing plant Blutverarbeitungsanlage *f*
blood sample Blutprobe *f*
blood sedimentation Blutsenkung *f*
blood serum Blutserum *n*
blood serum bottle Blutserumflasche *f*
blood specimen Blutprobe *f*
blood stain Blutfleck *m*
blood-stanching blutstillend
bloodstone Blutjaspis *m* (Min), Blutstein *m*
(Min), Brünierstein *m*, Eisennuß *f,* Hämatit *m*
(Min)
blood stream Blutbahn *f*
blood substitute Blutersatz *m*,
Plasmaersatzstoff *m*
blood sugar Blutzucker *m*, **increase of the** ~
Blutzuckeranstieg *m*
blood-sugar determination
Blutzuckerbestimmung *f*
blood-sugar level Blutzuckergehalt *m*,
Blutzuckerspiegel *m*
blood supply Blutzufuhr *f*
blood test Blutprobe *f,* Blutuntersuchung *f*
(Med)
blood toxin Blutgift *n*
blood transfusion Bluttransfusion *f* (Med),
Blutübertragung *f* (Med)
blood-typing Blutgruppenbestimmung *f,*
Blutgruppenfeststellung *f*
blood vessel Blutader *f* (Med), Blutgefäß *n*
bloom Barren *m*, Eisenblock *m*, Fettreif *m*
(Schokolade), Flor *m* (Bot), Luppe *f,*
Massel *f,* Stahlblock *m*, Walzblock *m*
bloom *v* anlaufen (Met), ausblühen,
ausschwefeln; blühen; effloreszieren
bloom end Blockende *n*
bloomer Blockwalze *f*
bloomery fire Luppenfrischfeuer *n*,
Frischfeuer *n*, Rennfeuer *n*
bloomery forge Rennfeuer *n*
bloomery iron Herdfrischeisen *n*
blooming Anlaufen *n* (Met), Ausblühen *n*,
Ausschwefeln *n*; Blauanlaufen *n* (Met),
Blühen *n* (Bot)
blooming mill Vorwalzwerk *n*, Blockstraße *f,*
Blockwalzwerk *n*, Luppenwalzwerk *n*,
Vorstraße *f,* Vorstrecke *f*
blooming mill stand Blockgerüst *n*
blooming pass Blockkaliber *n*,
Blockwalzkaliber *n*
blooming rolls Luppenwalzwerk *n*
bloom iron Luppeneisen, Frischeisen *n* (Metall)

bloom (on varnish) Hauch *m*
bloom pass Blockwalzkaliber *n*
bloom roll Vorwalze *f*
blooms Grobeisen *n*
bloom steel Luppenstahl *m*
blossom Blüte *f* (Bot)
blossom oil Blütenöl *n*
blot *v* fließen (von Tinte), beflecken, beklecksern
blotch printing Zweiphasendruckverfahren *n*
blotting pad Schreibunterlage *f*
blotting paper Fließpapier *n*, Löschpapier *n*
blow Stoß *m*, **strength of** ~ Schlagstärke *f*,
Schlagwirkung *f*
blow *v* wehen, anfachen, blasen, bombieren;
durchbrennen (Sicherung); verblasen, ~ **in**
einblasen, anblasen, ~ **off** ausblasen (Dampf),
ablassen (Dampf), absprengen, ~ **off air** *v*
Luft abblasen, ~ **out** abschmelzen (Sicherung),
ausblasen, durchschlagen (Sicherung),
durchschmelzen (Sicherung), ~ **up** *v*
aufblasen, aufsprengen
blow bending test Schlagbiegeprobe *f*
blow case Druckbirne *f*
blow-down piping Abblaseleitung *f*
blow-down tank Abblasetank *m*
blower Gebläse *n*, Gebläsemaschine *f*,
Gebläsewerk *n*, Lader *m*, Luftgebläse *n*
blower agitator Rührgebläse *n*
blower casing Ladergehäuse *n*
blower ratio Ladeübersetzung *f*
blow extrusion Blasspritzen *n* von Filmen
blow forming Blasen *n*, Blasformung *f*,
Blasverformung *f*
blow gun Druckluft-Pistole *f*
blowhead Blaskopf *m*
blowhole Blase *f* (Schweiß, Gieß),
Blasenraum *m*, Gasblase *f*, Gußblase *f*,
Hohlraum *m*, Innenlunker *m*
blowhole formation Lunkerung *f*
blowholes, formation of ~ Blasenbildung (Gieß)
blowing Blasen *n*, Aufblähung *f*,
Durchschmelzen *n* (Sicherung), Leerblasen *n*
(Dest), Verblasen *n*, **rate of** ~
Blasgeschwindigkeit *f*, ~ **off** Abblasen *n*, ~
out Ausblasen *n*, Niederblasen *n*, ~ **up**
Auftreibung *f*, Sprengung *f*
blowing agent Blähmittel *n*, Treibmittel *n*
(Schaumstoff)
blowing effect Blaswirkung *f*
blowing furnace Blasofen *m*
blowing iron Pfeife *f* (Glashütte)
blowing machine Gebläsemaschine *f*
blowing mill Gebläsemühle *f*
blowing pressure Blasdruck *m*
blowing process Blasverfahren *n*, Blasvorgang *m*
blow lamp Abbrennlampe *f*
blow mold Blasform *f*

blow molding unit with plunger-type extruder
Blasanlage mit Staukopf *m*
blow mold parting line Trennfuge *f* der Blasform
blow mo[u]lding Blasverfahren *n*
blown film Blasfolie *f*
blow-off cock Ablaßhahn *m*, Abblasehahn *m*,
Abblaseventil *n*
blow-off loss Blasverlust *m*
blow-off pipe Abblaserohr *n*, Ablaßleitung *f*,
Dampfabblaserohr *n*, Dampfableitungsrohr *n*
blow-off pressure Abblasedruck *m*,
Ablaßdruck *m*
blow-off valve Ablaßventil *n*, Abblaseventil *n*,
Ausblaseventil *n*, Entlastungsventil *n*,
Überdruckventil *n*
blow-out Funkenlöschung *f*, Platzen *n* (Reifen)
blow-out coil Funkenlöschspule *f*
blowpipe Lötrohr *n*, Blasrohr *n*, Brenner *m*
(Schweiß), Glasbläserpfeife *f*,
Schneidbrenner *m*
blowpipe analysis Lötanalyse *f*, Lötrohranalyse *f*,
Lötrohrprobe *f*, Lötrohruntersuchung *f*
blowpipe apparatus Lötrohrapparat *m*
blowpipe burner Gebläsebrenner *m*
blowpipe experiment Lötrohrversuch *m*
blowpipe flame Lötflamme *f*, Lötrohrflamme *f*,
Stichflamme *f*
blowpipe flux Lötrohrfluß *m*
blowpipe for glass Glasblaseröhre *f*
blowpipe lamp Gebläselampe *f*, Lötrohrlampe *f*
blowpipe nipple Lötrohrspitze *f*
blowpipe proof Lötprobe *f*
blowpipe reagent Lötrohrreagens *n*
blowpipe table Blasetisch *m*
blowpipe test Lötrohrprobe *f*, Lötversuch *m*
blowpipe testing outfit Lötrohrprüfgerätschaft *f*
blowpipe with bellows Lötrohrgebläse *n*
blow stress Beanspruchung *f* auf
Schlagfestigkeit, Schlagbeanspruchung *f*
blowtorch Gebläselampe *f*, Lötlampe *f*
blow valve Durchblaseventil *n* (Techn),
Schnarrventil, Durchblaseventil *n*
blubber Brauntran *m*, Walspeck *m*
blue, deep ~ sattblau, **light** ~ hellblau, **to color**
~ anbläuen (Färb)
blue *v* anbläuen (Färb)
blue annealing Blauglühen *n* (Metall)
blue ashes Kalkblau *n*
blueberry Schwarzbeere *f* (Bot)
blue black schwarzblau
blue brittleness Blaubruch *m* (Metall),
Blaubrüchigkeit *f* (Metall)
blue carbon paper Blaupapier *n*
blue color Blaufärbung *f*
blue coloration Blaufärbung *f*
blue coloring Blaufärbung *f*
blue disease Blaufäule *f* (Holz)

blue earth Bernsteinerde *f*
blue filter Blaufilter *n*
blue fracture test Blaubruchversuch *m*
blue gas Blaugas *n*
blue-green algae Blaualgen *pl*
blue hot blauglühend
blu[e]ing Bläuen *n,* Blauanlaufen *n*
blue iron earth Blaueisenerde *f*
blue iron ore Blaueisenerz *n* (Min)
blue lead ore Blaubleierz *n*
blue light mirror Blauspiegel *m*
blueness Bläue *f*
blue ointment Quecksilbersalbe *f* (Med),
    Mercurialsalbe *f* (Med)
blue ore Blauerz *n* (Min)
blue pill Quecksilberpille *f* (Med),
    Mercurialpille *f* (Med)
blueprint Lichtpause *f,* Blaudruck *m,*
    Blaupause *f,* Cyanotyp *n*
blueprint apparatus Lichtpausapparat *m,*
    Lichtpausgerät *n*
blueprinting Blaudruck *m,* Cyanotypie *f,*
    Cyanotypverfahren *n*
blueprinting frame Lichtpausapparat *m*
blueprinting machine Lichtpauseinrichtung *f*
blueprinting paper Lichtpauspapier *n*
blueprinting process Lichtpausverfahren *n*
blueprint lamp Pauslampe *f*
blueprint paper Pauspapier *n*
blueprint process Blaudruck *m*
blueprint tracing Blaupause *f*
blue short blaubrüchig
blue shortness Blaubruch *m*
blue spar Blauspat *m* (Min), Lazulith *m* (Min)
blue stain Blaufäule *f* (Holz)
blue staining of wood Bläuebefall *m* (Holz)
bluestone Blaustein *m* (Min), Kupfervitriol *n*
blue test Blauprobe *f*
blue turquoise Agaphit *m* (Min)
blue vat Blauküpe *f* (Färb), Vitriolküpe *f* (Färb)
blue verdigris Kupferblau *n*
blue verditer basisches Kupferkarbonat,
    Kalkblau *n,* Neuwiederblau *n*
blue violet blauviolett
blue vitriol Blaustein *m* (Min), Blauvitriol *n,*
    Kupfersulfat *n,* Kupfervitriol *n,* **ammoniacal
    solution of** ~ Blauwasser *n*
bluish (also blueish) bläulich
bluish black blauschwarz
bluish cast Blaustich *m*
bluish gray blaugrau
bluish green blaugrün
bluish red blaurot
bluish tinge Blaustich *m*
bluish tint Blaustich *m,* **to give a** ~ überblauen
blunder Fehlgriff *m*
blunt stumpf (Schneide)
blunt *v* abstumpfen (Spitze)
blunt-edged stumpfkantig

bluntness Abstumpfung *f* (Schneide)
blur *v* bekleckern, verwackeln (Phot);
    verwaschen
blurred unscharf, trüb[e], verwischt
blurring Verschmierung *f* (Röntgenstrahlen),
    Verwaschung *f* (Phot)
blushing Trübung *f* (Lackfilm), Weißanlaufen *n*
board Brett *n,* Karton *m* (Papier), Platte *f,*
    Tafel *f* (Holz), **air-dried** ~ luftgetrockneter
    Karton, **coated** ~ beschichteter Karton *m,*
    combined ~ kaschierter Karton, **grained** ~
    gemusterter Karton, **lacquered** ~ gelackter
    Karton, **laminated** ~ kaschierter Karton, **lined**
    ~ gegautschter Karton, kaschierter Karton,
    machine-glazed ~ maschinengeglätteter
    Karton, **mildew-proof** ~ schimmelfester
    Karton, **mold-resistant** ~ schimmelfester
    Karton, **pasted** ~ geklebter Karton,
    patent-coated ~ gegautschter Karton, **pressed**
    ~ gepreßter Karton, **reinforced** ~ verstärkter
    Karton, **single-ply** ~ einlagiger Karton,
    varnished ~ gelackter Karton, **block** ~ **or core
    board** Tischlerplatte *f*
board *v* verschalen
board drop hammer Brettfallhammer *m*
board glazing Decksatinage *f* (Papier)
boarding Verschalung *f*
board of directors Vorstand *m*
boat Schiffchen *n* (Labor)
boat form Wannenform *f* (Stereochem)
boat varnish Bootslack *m*
bobbierite Bobbierit *m* (Min)
bobbin Bobine *f,* Gurttrommel *f,* Spule *f* (Kern
    zum Aufwickeln), Spulenkern *m,* Wicklung *f*
    (Elektr)
bobbin height Spulenlänge *f*
bobbin overall diameter mittlere Spulenweite
bobbin spinning Spulenspinnverfahren *n*
bodenite Bodenit *m* (Min)
bodied oil Lackkörper *m,* Standöl *n*
bodily körperlich
bodkin Ahle, Schnürnadel *f*
body Körper *m,* Deckkraft *f,* Füllkraft *f,*
    Karosserie *f* (Kfz), Rumpf *m,* **black** ~
    schwarzer Körper; **full** ~ gute Füllkraft *f,*
    intermediate ~ Zwischenkörper *m,* **part of the**
    ~ Körperteil *m,* **rigid** ~ starrer Körper
    (Mech)
body *v* eindicken (von Ölen)
body-centered innenzentriert (Krist),
    raumzentriert (Krist)
body collisions Körperstöße *m pl*
body color Deckfarbe *f*
body deodorant Körperdesodorant *n* (Kosm)
body displacement Körperverschiebung *f*
body fluid Körperflüssigkeit *f*
body force Massenkraft *f*
body function Körperfunktion *f*
body heat Körperwärme *f*

body joint Fabrikkante *f* (Karton)
body surface Körperoberfläche *f*
body temperature Körpertemperatur *f*
Boettger test Böttgersche Probe *f*
bog bean Fieberklee *m*
bog earth Moorerde *f*
boghead coal Bogheadkohle *f*
boghead naphtha Braunkohlenbenzin *n*
bogie truck Drehgestell *n*
bog iron ore Eisensumpferz *n* (Min),
    Modererz *n* (Min), Rasen[eisen]erz *n* (Min),
    Sumpfeisenstein *m* (Min), Sumpferz *n* (Min),
    Torfeisenerz *n* (Min), Wiesenerz *n* (Min)
bog iron ore purifier Rasenerzreiniger *m*
bog iron schorl Eisenschörl *m* (Min)
bog manganese Manganschaum *m*,
    Braunsteinrahm *m*, Braunsteinschaum *m*,
    Schaumerz *n*
bog manganite Wad *n* (Min)
bog ore Morasterz *n* (Min), Limonit *m* (Min)
Bohemian brown topas Aftertopas *m* (Min)
Bohr orbit Bohrscher Kreis *m* (Atom)
Bohr's atom model Bohrsches Atommodell *n*
Bohr's hydrogen radius Bohrscher
    Wasserstoffradius *m*
Bohr's magneton Bohrsches Magneton *n*
Bohr theory Bohrsche Theorie *f* (Atom)
boil, at the ~ kochend
boil *v* kochen, abbrühen, aufkochen, brodeln,
    sieden, wallen, **to bring to** ~ zum Sieden
    bringen, **to half** ~ souplieren (Seide), ~ **off**
    entbasten, ~ **off a second time** repassieren
    (Seide), ~ **away** *v* verkochen (konzentrieren),
    wegkochen, ~ **down** *v* abdampfen, eindicken,
    einkochen (konzentrieren), verkochen
    (konzentrieren), ~ **off** *v* wegkochen, ~ **out** *v*
    auskochen, aussieden, ~ **over** *v* überkochen,
    ~ **partially** *v* assouplieren, ~ **up** *v* aufkochen
    [lassen], aufbrodeln, aufwallen
boiled-off silk Cuiteseide *f*
boiled oil Firnis *m*
boiled out ausgekocht
boiler Kessel *m*, Aufkochgefäß *n*, Kocher *m*,
    Kochkessel *m*, Kochtopf *m*, Küpe *f*,
    Siedegefäß *n*, Siedekessel *m*,
    Warmwasserspeicher *m*, Wasserkessel *m*,
    automatically controlled ~ selbstregelnder
    Kessel, **auxiliary** ~ Hilfskessel *m*,
    high-pressure ~ Hochdruckkessel *m*, **jacketed**
    ~ ummantelter Kessel, **revolving** ~
    rotierender Kocher, **sectional** ~
    Gliederkessel *m*, **vertical** ~ stehender Kessel
boiler anti-scaling composition
    Kesselsteinverhütungsmittel *n*
boiler blow-off cock Kesselablaßhahn *m*
boiler casing Kesselbekleidung *f*
boiler compound Kesselstein[gegen]mittel *n*
boiler draft Kesselzug *m*
boiler end Kesselboden *m*

boiler explosion Kesselexplosion *f*
boiler feed pump Kesselspeisepumpe *f*
boiler feed water Kesselspeisewasser *n*
boiler firing [Dampf]Kesselfeuerung *f*
boiler flue gas Kesselgas *n*
boiler furnace Kesselfeuerung *f*
boiler house Kesselhaus *n*
boiler insulation Kesselisolierung *f*
boiler jacket Kesselbekleidung *f*
boiler output Kesselleistung *f*
boiler pipe Siederohr *n*
boiler plant Kesselanlage *f*
boiler plate Kesselblech *n*, Grobblech *n*
boiler pressure Kesseldruck *m*
boiler scale Kesselstein *m*, Kesselniederschlag *m*,
    Pfannenstein *m*
boiler scaling tool or hammer
    Kesselsteinabklopfer *m*
boiler slag Kesselasche *f*
boiler steam Kesseldampf *m*
boil eruption period Kocheruptionsperiode *f*
boiling Kochen *n*, Absieden *n*, Ansud *m*,
    Aufkochen *n*, Aufwallen *n*, Brodeln *n*,
    Gebrodel *n*, Sieden *n*, Wallen *n*, **delay in** ~
    Siedeverzug *m*, **fast to** ~ abkochecht,
    kochbeständig, **forced convection** ~ Sieden *n*
    bei Zwangskonvektion, **free convection** ~
    Sieden *n* bei freier Konvektion, **local** ~
    örtliches Sieden, **loss by** ~ Kochverlust *m*, **to
    separate by** ~ auskochen
boiling *a* kochend
boiling agent Kocherlauge *f*
boiling apparatus Kocher *m*
boiling bath Abkochbad *n*
boiling capillary Siedekapillare *f*
boiling chip Siedestein *m*
boiling curve, shape of the ~ Verlauf *m* der
    Siedekurve
boiling down Abkochen *n*, Einkochen *n*
boiling flask Kochflasche *f*
boiling heat Siedehitze *f*
boiling lye Siedelauge *f*
boiling of the bath Aufkochen *n* des Bades
boiling period Kochzeit *f*
boiling plate Siedeblech *n*
boiling point Siedepunkt *m*, Kochpunkt *m*,
    having a high ~ hochsiedend, **of low** ~
    leichtsiedend, **raising of the** ~
    Siedepunktserhöhung *f*
boiling point apparatus
    Siedepunktbestimmungsapparat *m*
boiling [point] curve Siedekurve *f*
boiling product Siedeprodukt *n*
boiling range Siedebereich *m*, Siedegrenze *f*
boiling sediment Siedeabfall *m*
boiling sheet Siedeblech *n*
boiling stage [of iron] Rohfrischperiode (des
    Eisens) *f*
boiling stone Siedestein *m*

boiling temperature Siedetemperatur *f*
boiling test Kochprobe *f*
boiling tub Brühfaß *n*
boiling vessel Aufkocher *m*, Aufkochgefäß *n*,
  Kochkessel *m*, Siedegefäß *n*
boiling water reactor Verdampfungsreaktor *m*
boiling with steam generation Sieden *n* mit
  Dampferzeugung
boil-proof kochfest
boil strength Kochfestigkeit *f*
boil string-proof *v* verkochen (Zuck)
boil until clear *v* klarkochen
boivinose Boivinose *f*
bolar bolusartig
bold halbfett (Buchdr)
boldawite Bouteillenstein *m* (Min)
boldine Boldin *n*
bole Boluserde *f*
boleite Boleit *m* (Min)
boletic acid Boletsäure *f*
boletol Boletol *n*
boll weevil Baumwollkapselkäfer *m* (Zool)
Bologna flask Bologneser Flasche *f*
Bologna spar Bologneser Spat *m*
Bologna stone Bologneser Spat, Strahlbaryt *m*
  (Min)
bolometer Bolometer *n*
bolster Kissen *n*, Polster *n*, Pressenunterteil *n*
bolster *v* polstern, unterstützen
bolt Bolzen *m*, Riegel *m* (Schloß), Schraube (mit
  Mutter) *f*, **nut for a ~** Bolzenmutter *f*, **washer
  for a ~** Bolzenscheibe *f*
bolt *v* abriegeln, verriegeln *n*, **~ on** *v*
  anschrauben
bolt base Schloßfuß *m*
bolt circle Lochkreis *m*
bolted joint Schraubenverbindung *f*,
  Überschieber *m*
bolter Beutelsieb *n*
bolt-hole Bohrung *f*
bolt-hole circle Schraubenlochkreis *m*
bolt hook Hakenschraube *f*
bolting Beuteln *n*
bolting apparatus Beutelzeug *n*
bolting cloth Beuteltuch *n*, Beutelzeug *n*
bolting sieve Beutelsieb *n*
bolt key Bolzenkeil *m*
bolt of flange Flanschenschraube *f*
boltonite Boltonit *m* (Min)
bolt [pitch] circle Schraubenlochkreis *m*
bolt with head Kopfschraube *f*
Boltzmann's constant Boltzmannkonstante *f*
Boltzmann's function Boltzmannsche Funktion *f*
bolus Arzneikugel *f*, Bolus *m* (Min)
bolus alba Kaolin *n* (Min)
bomb Bombe *f*, **shell of ~** Bombenkörper *m*

bombard *v* beschießen (Atom), bombardieren
bombarding particle Beschußteilchen *n*,
  Geschoßteilchen *n* (Atom)
bombardment Beschuß *m*, Bombardement *n*
bombardment damage Schaden *m* durch
  Teilchenbeschuß (Atom)
bombardment particle Geschoßteilchen *n*
  (Atom)
bombay cachou Bombaycachou *n*
bomb calorimeter Bombenkalorimeter *n*,
  kalorimetrische Bombe,
  Explosionskalorimeter *n*
bomb fusion process Bombenaufschluß *m*
bomb oven Schießofen *m*
bomb tube Bombenrohr *n*
bombycol Bombykol *n*
bombyx Seidenraupe *f*
bond Bindung *f* (Chem), Zusammenschluß *m*,
  atomic ~ Atombindung, **axial ~** axiale
  Bindung (Stereochem), **conjugated ~**
  konjugierte Bindung, **coordinate ~**
  koordinative Bindung, **equatorial ~**
  äquatoriale Bindung (Stereochem),
  heteropolar ~ heteropolare Bindung,
  homopolar ~ homöopolare Bindung,
  molecular ~ Molekülbindung, **semipolar ~**
  semipolare Bindung, **spatially undirected ~**
  nicht räumlich gerichtete Bindung *f*
bond *v* kaschieren (Text), verkleben
bondable bindungsfähig
bond angle Bindungswinkel *m*
bond breaking Bindungsbruch *m*
bonded fiber fabrics Fließware *f*
bonded rubber mounting Schwingmetallagerung *f*
bond energy Bindungsenergie *f*
bonding Verleimung *f*
bonding coat Haftbrücke *f*
bonding material Imprägnierungsmittel *n*
bonding properties Verklebbarkeit *f*
bonding strength Haftfestigkeit *f*,
  Klebefestigkeit *f*, Leimfestigkeit *f*
bonding test Klebfestigkeitsprüfung *f*
bond length Bindungsabstand *m*,
  Bindungslänge *f*
bond moment Bindungsmoment *n*
bond order Bindungsgrad *m*
bond refraction Bindungsrefraktion *f*
bond stress Haftspannung *f*
bond structure Bindungsstruktur *f*
bond type Bindungsart *f* (Chem),
  Bindungstyp *m* (Chem)
bond value Haftwert *m*
bone Knochen *m*, Bein *n*
bone ash Klärstaub *m*, Knochenasche *f*,
  Knochenerde *f*
bone atrophy Knochenatrophie *f* (Med)

bone black Knochenkohle *f,* Beinschwarz *n,*
Elfenbeinschwarz *n,* Knochenschwarz *n,*
Spodium *n*
bone black filter Kohlenfilter *n*
bone black furnace Knochenkohleglühofen *m*
bone carbonizing oven
Knochenverkohlungsofen *m*
bone cell Knochenzelle *f*
bone charcoal Knochenkohle *f,* Beinschwarz *n,*
Spodium *n*
bonedry staubtrocken
bone dust Knochenmehl *n*
bone earth Knochenerde *f*
bone fat Knochenfett *n*
bone formation Knochenbildung *f* (Med)
bone gelatin Knochengallerte *f*
bone glass Milchglas, Opalglas *n*
bone glue Knochenleim *m*
bone injury Knochenschaden *m* (Med)
bone manure Knochendünger *m*
bone [manure] meal Knochendüngemehl *n*
bone marrow Knochenmark, Knochenmark *n*
(Med)
bone meal Knochenmehl *n,* Beinmehl *n*
bone meal ammonium nitrate
Knochenmehlammonsalpeter *m*
bone meal manuring Knochendüngung *f*
bone oil Knochenöl *n*
bone steamer Knochendämpfapparat *m*
bone steaming apparatus
Knochendämpfapparat *m*
bone tallow Knochentalg *m*
bone turquoise Beintürkis *m* (Min)
boninic acid Boninsäure *f*
bonnet Haube *f,* Kappe *f,* Kuppel *f*
bonus Prämie *f*
bony knochig
bookcase (Bücher)Regal *n*
booklet Broschüre, Heft *n*
book stone Bibliolit *m* (Min), Blätterschiefer *m*
(Min)
boom Hochkonjunktur *f*
boost *v* aufladen, verstärken
boost control Ladedruckregler *m*
booster Auflader *m,* Kompressor *m,*
Spannungsverstärker *m* (Elektr),
Verstärker *m* (Elektr), Zusatzgerät *n*
booster aggregate Zusatzaggregat *n*
booster amplifier Zusatzverstärker *m* (Elektr)
booster battery Zusatzbatterie *f* (Elektr)
booster channel Verstärkerkanal *m* (Elektr)
booster circuit Zusatzstromkreis *m*
booster dryer Zusatztrockner *m* (Tauchanlage)
booster for long-distance gas mains
Ferngasverdichter *m*
booster generator Generator *m* zur Verstärkung
booster motor Hilfsmotor *m*
booster nozzle Treibdüse *f*

booster pump Förderpumpe, Hilfspumpe *f,*
Verstärkerpumpe *f,* Zusatzpumpe *f*
booster transformer Zusatztransformator *m*
(Elektr)
booster type diffusion pump Treibdampfpumpe *f*
boot and shoe industry Schuhindustrie *f*
boothite Boothit *m* (Min)
boracic acid Borsäure *f*
boracite Boracit *m* (Min), Boraxspat *m* (Min),
Würfelspat, Borazit *m* (Min)
borage Boretsch *m* (Bot)
boral Boral *n*
borane Boran *n,* Borwasserstoff *m*
borate Borat *n,* borsaures Salz
borate *a* borsauer
borate of lime, hydrous ~ Hydroborocalcit *m*
(Min)
borax Borax *m,* **boiled** ~ gebrannter Borax,
native ~ Tinkal *m* (Min)
borax bead Boraxperle *f* (Anal)
borax glass Boraxglas *n*
borax honey Boraxhonig *m*
borax lake Boraxsee *m*
borax plant Boraxanlage *f*
borax soap Boraxseife *f*
borazarene Borazaren *n*
Bordeaux red Bordeauxrot
border Grenze *f,* Abgrenzung *f,* Borte *f*
(Umrandung), Kante *f,* Kimme *f,* Rand *m*
border *v* bördeln, grenzen, ~ **on** *v* angrenzen
borderline case Grenzfall *m*
border region Grenzgebiet *n*
border stone Kantenstein *m*
bore Bohrung *f,* Kernmaß *n;* **center of** ~
Bohrlochachse *f*
bore *v* bohren, anbohren, aufbohren, drillen,
teufen (Bergb), ~ **by percussion** stoßbohren
(Bergb), ~ **out** *v* ausbohren, ~ **through** *v*
durchbohren
bore chip Bohrspan
bore frame Bohrgestell *n*
borehole Bohrloch *n,* **center of** ~ Achsmitte *f*
des Bohrlochs
borehole gate valve Bohrlochschieber *m*
borehole logging Bohrlochuntersuchung *f*
borehole pump Bohrlochpumpe *f*
borer Bohrer *m*
borethane Boräthan *n*
borethyl Boräthyl *n*
boric borartig
boric acid Borsäure *f,* Acidum boricum *n* (Lat),
Boraxsäure *f* (obs), **anhydrous** ~
Borsäureanhydrid *n,* **native** ~ Sassolin *m*
(Min), **soaked in** ~ borsäuregetränkt
boric acid chelate Borsäurechelat *n*
boric acid methyl ester Borsäuremethylester *m*
boric acid ointment Borsalbe *f*

boric acid soap Borsäureseife *f*
boric acid solution Borsäurelösung *f,*
  Borwasser *n*
boric anhydride Borsäureanhydrid *n,*
  Bortrioxid *n*
boric cotton [wool] Borwatte *f*
borickite Borickit *m* (Min)
boriding Borieren *n*
borine Borin *n*
boring Bohren *n,* Bohrloch *n;* Bohrspan *m;*
  Bohrung *f*
boring apparatus Bohrvorrichtung *f*
boring bench Bohrbank *f*
boring frame Bohrgestell *n*
boring machine, automatic ~ Bohrautomat *m*
boring mill Bohrbank *f*
boring rod Bohrgestänge *n*
borings Bohrspäne *pl,* Feilspäne *pl*
boring socket Bohrfutter *n*
boring spindle Bohrwelle *f*
boring stock Bohrstock *m*
boring tolerance Bohrungstoleranz *f*
boring tool Bohrstahl *m*
boring tower Bohrturm *m*
boring tube Bohrröhre *f*
borinic acid Borinsäure *f*
bormethyl Bormethyl *n*
bornane Bornan *n*
borneo-camphor Borneocampher *m,* Borneol *n*
borneol Borneol *n*
Borneo tallow Illipebutter *f*
bornesitol Bornesit *m*
Born-Haber cycle Born-Haber-Kreisprozeß *m*
bornite Bornit *m* (Min), Buntkupfererz *n* (Min)
bornyl- Bornyl-
bornyl alcohol Borneol *n*
bornylamine Bornylamin *n*
bornyl chloride Bornylchlorid *n*
bornylene Bornylen *n*
bornylone Bornylon *n*
bornyl salicylate Salit *n*
bornyl valerate Bornylvalerianat *n*
bornyval Bornyval *n*
borobutane Borbutan *n*
borocalcite Borocalcit *m* (Min)
borocitrate Borocitrat *n*
boroethane Boräthan *n*
borofluohydric acid Borfluorwasserstoffsäure *f*
borofluoric acid Borfluorwasserstoffsäure *f*
boroformic acid Borameisensäure *f*
boroglyceride Boroglycerid *n*
borol Borol *n*
boron (Symb. B) Bor *n,* crystallized ~
  Borkristalle *pl,* graphitic ~ graphitartiges Bor
boronatrocalcite Boronatrocalcit *m* (Min)
boron bromide Bor[tri]bromid *n*
boron carbide Borkarbid *n*
boron carbide rod Borkarbidstab *m*
boron chamber Bor-Ionisations-Kammer *f*

boron chloride Borchlorid *n,* Bor[tri]chlorid *n*
boron compound Borverbindung *f*
boron content Borgehalt *m*
boron crystals Borkristalle *pl*
boron equivalent Boräquivalent *n*
boron ethyl Boräthyl *n*
boron-filled borgefüllt
boron fluoride Bor[tri]fluorid *n*
boron hydride Borwasserstoff *m,* Boran *n,*
  Borhydrid *n*
boronic acid Boronsäure *f*
boron iodide Borjodid *n,* Bortrijodid *n*
boron methyl Bormethyl *n*
boron nitride Bornitrid *n,* Borstickstoff *m,*
  Stickstoffbor *n*
boron oxide Bortrioxid, Borsäureanhydrid *n*
boronsalicylic acid chelate
  Borsalicylsäurechelat *n*
boron shield Borschirm *m*
boron silicide Borsilicid *n*
boron steel Borstahl *m* (Atom)
boron target Borschirm *m*
boron tribromide Bortribromid *n*
boron trichloride Bortrichlorid *n*
boron triethyl Boräthyl, Bortriäthyl *n*
boron trifluoride Bortrifluorid *n,* Borfluorid *n*
boron triiodide Bortrijodid *n,* Borjodid *n*
boron trimethyl Bormethyl *n*
boron trioxide Bortrioxid, Borsäureanhydrid *n*
borosalicylic acid Borsalicylsäure *f*
borotannate Borotannat *n*
borotartrate Borotartrat *n*
borotungstic acid Borwolframsäure *f*
borovertine Borovertin *n*
bort Bort *m* (Industriediamant), Carbonado *m*
  (Min), Diamantabfall *m*
boryl- Boryl; Äthylborsalizylat *n*
boryltartaric acid Borylweinsäure *f*
bosh Rast *f* (Hochofen), angle of ~
  Rastwinkel *m,* height of ~ Rasthöhe *f,*
  pressure on the ~ Rastdruck *m*
bosh brickwork Rastmauerung *f*
bosh stones Raststeine *m pl* (Hochofen)
boson Boson *n*
boss Vorsprung *m* (Bosse)
bosshead Muffe *f*
boswellic acid Boswellinsäure *f*
boswellonic acid Boswellonsäure *f*
botanical botanisch
botanist Botaniker *m*
botany Botanik *f,* Pflanzenkunde *f*
bother *v* belästigen
botogenin Botogenin *n*
botryolite Botryolith *m* (Min), Traubenstein *m*
  (Min)
bottle Flasche *f,* small ~ Flakon *m n,* stoppered
  ~ Stöpselflasche *f,* wide-mouth ~
  Weithalsflasche *f* (Chem), wide-necked ~
  Weithalsflasche *f* (Chem)

bottle v abfüllen (in Flaschen)
bottle battery Flaschenelement n
bottle brush Flaschenbürste f
bottle cap Flaschenkappe f, Flaschenkapsel f, Flaschenverschluß m
bottle capsule lacquer Flaschenkapsellack m
bottle cell Flaschenelement n
bottle cleaner Flaschenreinigungsmaschine f
bottle filling apparatus Flaschenfüllapparat m
bottle glass Flaschenglas n
bottlegreen Glasgrün n
bottle in a wicker case Korbflasche f
bottleneck Flaschenhals m
bottle seal Flaschenverschluß m
bottlestone Tektit m (Min)
bottle with cap Kappenflasche f
bottle with rolled flange Rollrandflasche f
bottling Abfüllung f, Flaschenfüllung f, **date of ~ (filling or racking)** Abfülldatum n
bottling line Abfüllanlage f (Flaschen)
bottling machine Abfüllmaschine f, Flaschenfüller m, Flaschenfüllmaschine f
bottling plant Abfüllanlage f (f. Flaschen)
bottling room Flaschenabfüllraum m
bottom Boden m, Grund m, Kolonnensumpf m (Dest), Sumpf m (Dest)
bottom blow valve Bodenventil n
bottom box Unterkasten m
bottom-cast v steigend gießen (Gieß)
bottom casting steigender Guß
bottom color Grundfarbe f
bottom disc Bodenscheibe f
bottom discharge conduit Grundablaß m
bottom ejection Ausdrücken n von unten (aus dem Gesenk)
bottom electrode Bodenelektrode f
bottom fermentation Untergärung f (Brau)
bottom fermented untergärig (Brau)
bottom flue Sohlenkanal m
bottom heating Bodenheizung f
bottom layer Grundierung f, Unterschicht f
bottom liner of board Kartonrückseite f
bottom mold Patrize f
bottom of melting pan Düsenboden m
bottom of shell Gehäuseboden m
bottom part Unterstück n, Unterteil n
bottom plate Bodenplatte f, Grundplatte f, Sohlplatte f
bottom plug Unterstempel, Matrizeneinsatz m
bottom-pour v steigend gießen (Gieß)
bottom pouring steigender Guß
bottom pressure Sohldruck m
bottom product (Dest) Ablauf m, Schlempe f, Sumpfprodukt n
bottom ram press Unterdruckpresse f, Unterkolbenpresse f
bottom roll Unterwalze f, Matrizenwalze f
bottoms Bodensatz m
bottom sealing Sohlendichtung f

bottom sediments Bodensatz m
bottom stamp Unterstempel m
bottom stone Bodenstein m
bottom view Ansicht f von unten
bottom yeast Unterhefe f (Brau)
Bottone's scale of hardness Bottone's Härteskala f
botton steering Knopfsteuerung f
botulism Botulismus m
Boudouard equilibrium Boudouard-Gleichgewicht n
bougie Bougie f (Med)
boulangerite Antimonbleiblende f (Min), Boulangerit m (Min), Schwefelantimonblei n (Min)
boulder Flußstein m
boulder clay Blocklehm m, Geschiebelehm m
bounce Abprall m, Aufprallen n, Aufschlagen n, Sprung m
bounce v abprallen, springen (schnellen)
bouncing pin (antiknock test) Springstift m
bound angebunden, gebunden
bound v aufprallen, begrenzen
boundary Grenze f, Abgrenzung f, Begrenzung f, Begrenzungsfläche f, Umgrenzung f, Zone f
boundary angle (of surface tension) Randwinkel m (Oberflächenspannung)
boundary break-away separation Grenzschichtablösung f
boundary condition Grenz[flächen]bedingung f, Randbedingung f
boundary density Randdichte f
boundary diffusion Grenzflächendiffusion f
boundary effect Randeffekt m
boundary energy Grenzenergie f
boundary flow Grenzströmung f
boundary layer Grenzschicht f, Grenzfläche f
boundary layer friction Grenzflächenreibung f
boundary layer influence Grenzschichtbeeinflussung f
boundary layer theory Grenzschichttheorie f
boundary line Grenzlinie f, Begrenzungslinie f, Scheidelinie f
boundary problem Randproblem n
boundary radius Grenzradius m
boundary resistance Grenzwiderstand m
boundary scattering Streuung f an Grenzflächen (pl)
boundary surface Begrenzungsoberfläche f, Grenzfläche f
boundary treatment Randbehandlung f
boundary value Randwert m
boundary value problem Randwertaufgabe f, Randwertproblem n
boundary wave Grenzflächenwelle f, Grenzwelle f
boundless unbegrenzt
boundlessness Unbegrenztheit f

bournonite Bournonit *m* (Min),
Antimonbleikupferblende *f* (Min),
Antimonkupferglanz *m* (Min), Bleifahlerz *n*
(Min), Rädelerz *n* (Min),
Schwarzspießglanzerz *n* (Min),
Spießbleierz *n* (Min)
bour[r]ette silk Bourrette[seide] *f*
bovista Bovist *m*
bovocryptoside Bovokryptosid *n*
bovogenin Bovogenin *n*
bovoside Bovosid *n*
bovosidol Bovosidol *n*
bovovaccine Bovovakzine *f*
bow Bügel *m*
bow *v* beugen, biegen
bow collector Kontaktbügel *m* (Elektr)
bow compasses Krummzirkel *m*
bowden rope Bowdenseil *n*
bowden stranded wire Bowdenlitze *f*
Bowditch's law Alles-oder-Nichts-Gesetz *n*
(Med)
bowenite Bowenit *m* (Min)
bowk *v* beuchen
bowking Beuchen *n*
bowl Schale *f*, Schüssel *f*
bowl centrifugal decanter
Dekantierkorbzentrifuge *f*
bowl for overflowing oil Ölauffang *m*
bowl mill Pendelmühle *f*, Schüsselmühle *f*
bowl-shaped schalenförmig
bowmanite Bowmanit *m* (Min)
box Schachtel *f*, Büchse *f*, Dose *f*, Emballage *f*,
Etui *n*, Kasten *m*, Kiste *f*, Trog *m* (Metall)
box *v* in Schachteln packen, emballieren
box annealing Kastenglühung *f*
box blower Kastengebläse *n*
boxboard Hartpappe *f*, Karton *m*
box casting Kastenguß *m*, Ladenguß *m*
box filling machine Dosenfüllmaschine *f*
box furnace Kammerofen *m*
box jib [crane] Kastenausleger *m*
box manufacturing and paperworking industry
Kartonagenindustrie *f*
box pass Flachkaliber *n*
box potential Kastenpotential *n*
box press Packpresse *f*
box thread Innengewinde *n*
box with movable part Abziehformkasten *m*
box with perforated cap Streubüchse *f*
boxwood Franzosenholz *n*, Guajacholz *n*
Boyle-Mariotte's law Boyle-Mariottesches
Gesetz *n*
brace Klammer *f*, Band *n*, Brasse *f*, Bügel *m*,
Stützbalken *m*, Verstrebung *f*
brace *v* abspreizen, abstützen, aussteifen,
verklammern, verspannen, versteifen
brace back *v* backbrassen
brace block Brassenblock *m*
braces [geschweifte] Klammern *f*

brachiose Brachiose *f*
brachistochrone Brachistochrone *f* (Math)
brachyaxis Brachyachse *f* (Krist)
brachyprism Brachyprisma *n*
bracing Abstützung *f* (Techn)
bracing of furnace Ofenverspannung *f*, lateral ~
Seitenverspannung *f* des Ofens
brackebuschite Brackebuschit *m* (Min)
bracket Auflage *f*, Gestell *n*, Konsole *f*,
Krücke *f* (Rohrstütze), Stehlager *n*, Träger *m*
bracket crane Konsolkran *m*
bracket drop hanger Armhängelager *n*
bracket plate Stützplatte *f*
bracket relation Klammerrelation *f*
bracket rim Tragkranz *m*
Brackett series Brackettlinien *pl* (Spektr),
Brackett-Serie *f* (Spektr)
brackish brackig, leicht salzig
brackish water Brackwasser *n*
bradykinin Bradykinin *n*
Bragg angle Braggscher Winkel *m* (Opt)
Bragg method Braggmethode *f*
Bragg plane Braggebene *f*
Bragg's law Braggsches Gesetz *n*
Bragg spectrometer Kristallspektrometer *n*
bragite Bragit *m* (Min)
braid Borte *f*, Litze *f*
braid *v* flechten, [um]flechten
braided metal packing Gewebepackung *f* (Dest)
braided wire Litzendraht *m* (Elektr)
braiding Umklöppelung *f*
braiding machine Flechtmaschine *f*,
Klöppelmaschine *f*
braid lacquer Litzenlack *m*
brain Gehirn *n*, Verstand *m*
brain cell Gehirnzelle *f*
brain center Gehirnzentrum *n*
brain substance Gehirnsubstanz *f*
brake Bremse *f*
brake *v* [ab]bremsen
brake adjusting bearing Bremsregulierlager *n*
brake band Bremsband *n*
brake band impregnation plant
Bremsbandimprägnieranlage *f*
brake band support Bremsbandstütze *f*
brake band tension Bremsbandspannung *f*
brake block Bremsklotz *m*
brake block holder Bremsklotzsohle *f*
brake clutch Bremsbacke *f*
brake control Bremsantrieb *m*
brake cylinder Bremsbehälter *m*,
Bremszylinder *m*
brake disc Bremsscheibe *f*
brake distance Bremsbahn *f*
brake effect Bremswirkung *f*
brake equipment Bremsausrüstung *f*
brake fluid Bremsflüssigkeit *f*
brake horsepower Bremskraft *f*, Bremsleistung *f*
brake light Bremsleuchte *f*

brake lining Bremsbelag *m*, Bremsfutter *n*
brake load Bremsbelastung *f*
brake pressure Bremsdruck *m*, Bremskraft *f*
brake pulley Bremsscheibe *f*
brake rod Bremsstange *f*
brake shaft Bremswelle *f*
brake shoe Bremsbacke *f*, Bremsklotz *m*,
  Bremsschuh *m*
brake wheel Bremsrad *n*
braking Bremsung *f*, Abbremsung *f*, **energy of** ~
  Bremsenergie *f*
braking area Bremsfläche *f*
braking articulation Bremsgelenk *n*
braking distance Bremsbahn *f*
braking effect Bremsarbeit *f*
braking friction Bremsreibung *f*
braking moment Bremsmoment *n*
braking radiation Bremsstrahlung *f*
bran Kleie *f*, **bread made of** ~ Kleiebrot *n*
branch Zweig *m*, Ableitung *f*; Abzweigstelle *f*,
  Abzweigung *f*, Bein *n* (Gestell); Fach *n*
  (Fachgebiet), Filiale *f*, Schenkel *m* (Zirkel),
  Zweigniederlassung *f*
branch *v* verzweigen, ~ **off** *v* abbiegen, ableiten,
  abzweigen
branch box Abzweigdose *f*, Abzweigkasten *m*
branch circuit Zweigstromkreis *m*, Ableitung *f*,
  Stromzweig *m*
branch cock Verteilungshahn *m*
branch conductor Leitungszweig *m*
branch current Teilstrom *m*, Zweigstrom *m*
branched verzweigt
branch establishment Filialanstalt *f*
branching Abzweigung *f*, Aufzweigung *f* (Atom)
branching *a* abzweigend, doppelgängig
branching factor Verzweigungsfaktor *m*
branching model Verzweigungsmodell *n*
branching of a conductor Leitungsabzweigung *f*
  (Elektr)
branching [off] Verzweigung *f*
branching-off point Abzweigstelle *f*
branching point Verzweigungspunkt *m*
branching probability
  Verzweigungswahrscheinlichkeit *f*
branching ratio Abzweigungsverhältnis *n*,
  Verzweigungsverhältnis *n*
branchless unverzweigt
branch [line] Abzweigleitung *f*,
  Verzweigungslinie *f*, Zweigleitung *f*
branch of manufacture Betriebszweig *m*
branch piece Abzweigstück *n*, Abzweigstutzen *m*
branch pipe Abzweigrohr *n*, Nebenrohr *n*,
  Zweigrohr *n*
branch point Verzweigungsstelle *f*
branch sleeve Abzweigmuffe *f*
branch socket Abzweigmuffe *f*
brand Marke *f*, Fabrikmarke *f*, Markenname *m*,
  Warenzeichen *n*
brand *v* einbrennen

brandisite Brandisit *m* (Min)
brand-new fabrikneu, nagelneu
brandtite Brandtit *m* (Min)
brandy Branntwein *m*, **to distill** ~
  Branntwein *m* brennen, ~ **made from wine**
  Weinbrand *m*
bran dye bath Kleienbad *n*
brandy vinegar Branntweinessig *m*,
  Weingeistessig *m*
bran liquid Kleienbeize *f*
bran molasses Kleiemelasse *f*
brannerite Brannerit *m* (Min)
branny kleienartig, kleiig
bran of almonds Mandelkleie *f*
bran vinegar Kleienessig *m*
bran water Kleienwasser *n*
brasilic acid Brasilsäure *f*
Brasil wax Karnaubawachs *n*
brass Messing *n*, **beaten** ~ Schlagmessing *n*;
  crude ~ Rohmessing *n*, **dipping liquid for** ~
  Gelbbrenne *f*, **foliated** ~ Flittergold *n*,
  hard-drawn ~ Hartmessing *n*, **polished** ~
  Glanzmessing *n*, **to pickle** ~ Messing
  abbrennen, **to refine** ~ Messing brennen,
  white ~ Weißmessing *n*, **yellow** ~
  Neumessing *n*
brass anode Messinganode *f*
brass bath Messingbad *n*
brass clippings Krätzmessing *n*
brass-colored messingfarben
brass column Messingsäule *f*
brass dipping plant Gelbbrennanlage *f*
  (Keramchemie)
brass foil Messingblatt *n*
brass founder Gelbgießer *m*
brass foundry Gelbgießerei *f*
brass gauze cathode Messingdrahtnetzkathode *f*
brassic acid Brassidinsäure *f*, Brassinsäure *f*
brassicasterol Brassicasterin *n*
brassidic acid Brassidinsäure *f*, Brassinsäure *f*,
  Isoerucasäure *f*
brassidone Brassidon *n*
brassiness messingartige Beschaffenheit,
  Messingartigkeit *f*
brass in lumps Brockenmessing *n*
brassmaking Messingbrennen *n*
brass scrap Bruchmessing *n*
brass scum, yellow ~ Messingabstrich *m*
brass sheet Messingblech *n*
brass solder Messingschlaglot *n*
brass wire Messingdraht *m*
brass wire brush Messingkratzbürste *f*
brassy messingartig
brassylic acid Brassylsäure *f*
braunite Braunit *m* (Min), Hartmangan *n* (Min)
braunite cast Hartmanganstahlguß *m*
Braun tube Braunsche Röhre *f*,
  Kathodenstrahloszillograph *m*

bravaisite Bravaisit *m* (Min), Glaukonit *m* (Min)
brayera powder Kussopulver *n*
brazability Hartlötbarkeit *f*
brazan Brasan *n*
brazanquinone Brasanchinon *n*
braze *v* hartlöten, löten mit Hartlot, löten
brazed hartgelötet
brazed-plate heat exchanger Plattenwärmeaustauscher *m*
Brazil brilliant Brasildiamant *m*
brazilcopalic acid Brasilkopalsäure *f*
brazilcopalinic acid Brasilkopalinsäure *f*
brazilein Brasilein *n*
Brazilian cocoa Guarana *f* (Bot)
brazilin Brasilin *n* (Färb)
brazilinic acid Brasilinsäure *f*
brazilite Brazilit *m* (Min)
Brazil-nut oil Paranußöl *n*
Brazil tea Matéblätter *n pl*
Brazil wood Brasilholz *n*, Fernambukholz *n*, **extract of** ~ Fernambukextrakt *m*, **red dyestuff of** ~ Fernambukrot *n*
Brazil wood lacquer Fernambuklack *m*
brazing Hartlöten *n*, Hartlötung *f*, Löten *n* (Hartlöten), Lötstelle *f*
brazing solder Hartlot *n* (Metall)
breach Einbruchstelle *f*
bread, aerated ~ Kohlensäurebrot *n*
[bread] dough Brotteig *m*
breadth Breite *f*
break Knick *m*, Bruch *m*, Haltepunkt *m*; Unterbrechung *f*; **elongation at** ~ Bruchdehnung *f*, **extension at** ~ Bruchdehnung *f*, **point of** ~ Brechpunkt *m*
break *v* brechen, durchreißen, knicken, springen (bersten), zerschlagen, ~ **down** *v* zusammenbrechen, abbauen, aufspalten, durchschlagen (Elektr), spalten, zerfallen, ~ **loose** *v* ablösen, ~ **off** *v* abbrechen, ablösen, absprengen, ~ **open** *v* aufbrechen, ~ **out** *v* ausbrechen, losbrechen
break through *v* durchbrechen
breakthrough curve Durchbruchkurve *f* (Adsorption)
break up *v* aufbrechen, aufschließen, ausmahlen
break up into small pieces *v* zerkleinern
break up roughly *v* halbmahlen
breakability Brechbarkeit *f*
breakable zerbrechlich
break a bond *v* eine chem. Bindg. spalten
breakage Zerbrechen *n*
breakage-proof bruchfest
break-away Ausbrechen '*n*
breakaway torque Losbrechkraft *f* (Bagger)
breakdown Zusammenbruch *m*, Ausfall *m*, Betriebsstörung *f*, Durchschlag *m* (Elektr), Panne *f*, Störung *f*, Verfall *m*, **initial** ~ Anfangsdurchschlag *m*

breakdown channel Entladungskanal *m*
breakdown field Durchbruchsfeldstärke *f*
breakdown filament Durchschlagskanal *m*
breakdown mill Mastizierwalzwerk *n*
breakdown moment Abfallmoment *n*
breakdown potential Durchschlagspotential *n*
breakdown resistance Durchschlagsfestigkeit *f*
breakdown strength Durchschlagsfestigkeit *f*
breakdown test Durchschlagsprobe *f* (Elektr)
breakdown torque Abfallmoment *n*
breakdown voltage Durchschlagsspannung *f*, Zündspannung *f*
breakdown voltage in air Luftdurchschlagspannung *f*
breaker Brecher *m*, Brechkapsel *f*; Brechtopf *m*, Halbstoffholländer *m*, Mahlholländer *m*, Mahlwerk *n*, Unterbrecher *m*
breaker beater Auflöseholländer *m*, Brechholländer *m*, Schlagholländer *m*
breaker fabric Zwischenbaugewebe *n* (Reifen), Zwischenlagegewebe *n* (Reifen)
breaker mouth Brechmaul *n*
breaker plate Lochscheibe *f*, Stauscheibe *f*
breaker type Brechertyp *m*
break impulse Unterbrechungsimpuls *m*
breaking Brechung *f*, Knickung *f*, Springen *n*, **preliminary** ~ Vorzerkleinern *n*, ~ **into small pieces** Zerstückelung *f*, ~ **of the cords** Reißen *n* der Fäden (Reifen), ~ **of the core** Zertrümmerung *f* des Kernes
breaking-down mill Grobeisenstraße *f*, Grobstrecke *f*
breaking-down point Bruchfestigkeitsgrenze *f*
breaking-down temperature Abbautemperatur *f*
breaking limit Bruchgrenze *f*, Zerreißgrenze *f*
breaking load Zerreißbelastung *f*, Bruchbelastung *f*, Bruchlast *f*, Knickbeanspruchung *f*
breaking load test Belastungsprobe *f* bis zum Bruch
breaking modulus Festigkeitsmodul *m*
breaking moment Bruchmoment *n*
breaking off Abbruch *m*
breaking out Durchbruch *m*
breaking point Zerreißgrenze *f*, Knickpunkt *m*
breaking resistance Knickfestigkeit *f*
breaking shaft Brechachse *f*
breaking strain Zugfestigkeit *f*
breaking strength Bruchfestigkeit *f*, Bruchwiderstand *m*
breaking stress Bruchbeanspruchung *f*, Bruchspannung *f*, Zerreißbelastung *f*
breaking tension Bruchdehnung *f*
breaking test Zerreißprobe *f*, Brechprobe *f*, Bruchprobe *f*, Reißprobe *f*
breaking tester Bruchfestigkeitsprüfer *m*
breaking through Durchbruch *m*
breaking up Aufbrechen *n*, Auflockerung *f*, Aufschließung *f*, Aufschluß *m*, Spaltung *f*

break [into pieces] v zerbrechen
break point Durchbruch m
break point instruction, conditional ~ bedingter
Stoppbefehl (Comp)
break resistance Bruchfestigkeit f, ~ at low
temperature Kältebruchfestigkeit f
break-resistant bruchfest, bruchsicher
break spark Unterbrechungsfunke m (Elektr)
breakthrough Durchbruch m
breath Atemzug m, Hauch m, shortness of ~
Atemnot f
breathe v atmen, ~ in v einatmen, ~ out v
ausatmen
breather Entlüfter f, Entlüftungsstutzen m (Pkw
Kraftstoffbehälter)
breathing Atmung f, Lüften n (des Werkzeugs)
breccia Breccie f, Brekzie f (Geol),
Brockengestein n, calcareous ~
Kalkkonglomerat n
breccia structure Brecciengefüge n
brecciated breccienartig
brecciated agate Breccienachat m (Min)
brecciated marble Brecciemarmor m
Bredt rule Bredtsche Regel f (Stereochem)
breech mechanism Ladevorrichtung f
Breecht's double salt Breechtsches Doppelsalz n
breed Rasse f (Tier)
breed v züchten
breeder Brüter m (Atom), Brutreaktor m
(Atom), Züchter m
breeder material Brutstoff m
breeding Brüten n (Kerntechnik),
Brutvorgang m, Haltung f; Züchtung f
breeding apparatus Brutapparat m
breeding cycle Brutzyklus m (Biol)
breeding gain Brutgewinn m (Atom)
breeding of animals Tierzucht f
breeding ratio Brutverhältnis n
breeding reactor Brutreaktor m (Atom)
breeze Grus m, Wind m
brein Brein n
breislakite Breislakit m (Min)
breithauptite Breithauptit m (Min),
Antimonnickel n (Min)
Bremen blue Bremerblau n
Bremen green Bremergrün n
bremsstrahlung Bremsstrahlung f (Atom)
brenzcain Brenzcain n
brescian steel Münzstahl m
breunnerite Breunnerit m (Min), Mesitin m
(Min), Mesitinspat m (Min)
brevifolin Brevifolin n
brevium Brevium n
brew v brauen
brewer's grains Malztreber pl
brewers mash flask Maischekolben m
brewer's pitch Brauerpech n (Brau)
brewer's yeast Bierhefe f (Brau)
brewery Brauerei, Bierbrauerei f

brewing, period of ~ Sudzeit f (Brau)
brewing house Sudhaus n (Brau)
brewing machine Brauereimaschine f
brewing malt Braumalz n
brewing method Braumethode f
brewing vat Braubottich m (Brau)
Brewster angle Brewsterscher Winkel,
Polarisationswinkel m
Brewster fringes Brewstersche Streifen pl (Opt)
brewsterite Brewsterit m (Min)
brick Ziegel m, Brandstein m, refractory ~
feuerfester Stein, solid ~ Vollziegel m,
sun-dried ~ Lehmstein m, unburnt ~
Lehmstein m, Rohziegel m, unburnt sun-dried
~ Lehmbaustein m
brick v einmauern
brick clay Ziegelerde f, Ziegelton m
brick color Ziegelfarbe f
brick-colored ziegelfarbig
brick dust Ziegelmehl n
brick earth Ziegelerde f
brick hardening plant Steinhärteanlage f
brick kiln Backsteinofen m
bricklayer Maurer m
brick lining Ausmauerung f
brick products Ziegelware f
brick red Ziegelfarbe f
brick-red a hochrot, ziegelrot
brick rim Mauerkranz m
brick's shape Ziegelform f
brick trough Backsteintrog m
brick wall Mauersockel m, Steinsockel m
brick-walled ausgemauert
brickworks Ziegelei f
brickyard Ziegelei f
bridge Brücke f, Überführung f
bridge atom Brückenatom n (Stereochem)
bridge circuit Brückenschaltung f
bridge crane Brückenkran m
bridged system Brückensystem n (Stereochem)
bridge head Brückenkopf m
bridgehead atom Brückenkopfatom n
(Stereochem)
bridgehead carbanion Brückenkopf-Carbanion n
(Stereochem)
bridgehead radical Brückenkopf-Radikal n
(Stereochem)
bridge method Brückenschaltung f
bridge rectifier Gleichrichterelement n
bridge relay Brückenrelais n
bridge-ring structure Brückenringstruktur f
(Stereochem)
bridge-ring system Brückenringsystem n
(Stereochem)
bridge-shaped stillhead Destillierbrücke f
bridging atom Brückenatom n (Stereochem)
bridging jack Parallelklinke f

bridging oxygen Brückensauerstoff *m*
bridging plug Brückenstecker *m*
brief kurz, knapp
Briehl radiator Hellstrahler *m*
brier wood Bruyèreholz *n*
Brigg's logarithm Briggscher Logarithmus *m*
bright blank, grell, hell, leuchtend, licht,
   lichtstark
bright adaptation Hellanpassung *f* (Opt)
bright annealing Blankglühen *n*
brighten [up] *v* aufhellen, avivieren, beleben
brightener Glanzzusatz *m*
brightening Aufhellung *f*, Avivage *f*
brightening agent Aufheller *m*,
   Aufhellungsmittel *n* (Mikroskopie),
   Aviviermittel *n*
brightening dyestuff Schönungsfarbstoff *m*
brightening fastness Avivierechtheit *f* (Text)
brightening fluid Schönungsflüssigkeit *f*
bright field illumination Hellfeldabbildung *f*
bright field image Hellfeldbild *n*
bright gilding Glanzvergoldung *f*
bright ground illumination Hellfeldbeleuchtung *f*
bright luster Hochglanz *m*
brightness Helligkeit *f*, Leuchtstärke *f*,
   Lichtglanz *m*, Schein *m*, degree of ~
   Beleuchtungsstärke *f*
brightness change Helligkeitsänderung *f*
brightness compensation Helligkeitsangleichung *f*
brightness distribution Helligkeitsverteilung *f*
brightness reference value Hellbezugswert *m*
brightness sensation Helligkeitsempfindung *f*
brightness value Helligkeitswert *m*
bright red hochrot
bright section Blankprofil *n*
bright side Glanzseite *f*, Lichtseite *f*
Brigl's anhydride Brigls Anhydrid *n*
brilliance Glanz *m*, Glanzeffekt *m*, Helligkeit *f*,
   Leuchtkraft *f*
brilliancy Helligkeit *f*, [high] ~ Hochglanz *m*
brilliancy of colors Farbenschiller *m*
brilliant Brillant *m*
brilliant *a* farbenprächtig, glanzvoll
brilliant acid blue Brillantsäureblau *n*
brilliant acid green Brillantsäuregrün *n*
brilliant alizarin blue Brillantalizarinblau *n*
brilliant azurine Brillantazurin *n*
brilliant black Brillantschwarz *n*
brilliant carmine Brillantkarmin *n*
brilliant carmoisine Brillantcarmoisin *n*
brilliant color Glanzfarbe *f*
brilliant cresyl blue Brillantkresylblau *n*
brilliant crocein Brillantcrocein *n*
brilliant dye Brillantfarbstoff *m*
brilliantine Brillantine *f*
brilliant luster Lichtglanz *m*
brilliant oil Glanzöl *n*
brilliant pink Brillantrosa *n*

brilliant varnish Glanzlack *m*
brilliant white Glanzweiß *n*
brilliant yellow Brillantgelb *n*
Brillouin zones Brillouin-Zonen *pl* (Krist)
brim Rand *m*, to fit with a ~ kimmen
brimful bis zum Rand voll
brimstone Schwefel *m*
brimstone impression Schwefelabdruck *m*
brimstone pit Schwefelgrube *f*
brine Lauge *f*, Kochsalzlösung *f*, Lake *f*,
   Salzbrühe *f*, Salzlauge *f*, Salzlösung *f*,
   Salzwasser *n*, Siedesole *f*, Sole *f*, Solwasser *n*,
   raw ~ Rohsole *f*
brine bath Salzbad *n*
brine cistern Solbehälter *m*
brine concentrator Soleeindämpfer *m*
brine cooler Salzwasserkühler *m*, Solekühler *m*
brine cooling Salzwasserkühlung *f*, Solekühlung *f*
brine gauge Salzwaage *f*, Gradierwaage *f*,
   Laugenwaage *f*, Solwaage *f*
brine inlet Solezulauf *m*, Solezuleitung *f*
brine leaching Salzlaugung *f*
Brinell ball-hardness test
   Brinell-Kugeldruckprobe *f*
Brinell hardness Brinellhärte *f*
Brinell hardness number Brinellsche Härtezahl *f*,
   Brinellzahl *f*
Brinell hardness test Brinell-Messerverfahren *n*
Brinell unit Brinell-Einheit *f*
brine mixer Solebereiter *m*
brine outlet Soleableitung *f*, Soleaustritt *m*
brine pit Solbehälter *m*
brine pond Salzteich *m*
brine salt Solsalz *n*, Brunnensalz *n*
brine spring Solquelle *f*
brine strainer Solefilter *m*
brine temperature Laugentemperatur *f*
brine tub Solfaß *n*
bring forth *v* hervorbringen
briny salzig, salinisch
briquet[te] Brikett *n*, Brennstoffziegel *m*,
   Patentkohle, Preßkohle *f*, wet-pressed ~
   Nußbrikett *n*
briquet[te] *v* brikettieren, pressen (Kohle)
briquet[te] cement Brikettbindemittel *n*
briquetting Brikettierung *f*, Formung *f* (Kohle),
   Ziegelung *f*
briquetting method Formpreßverfahren *n*
briquetting plant Brikettierungsanlage *f*
briquetting process Brikettierverfahren *n*
briquetting property Brikettierfähigkeit *f*
briquetting rolls Walzenpresse *f*
brisance Brisanz *f*, Sprengkraft *f*
brisk heftig, lebhaft
bristle Borste *f*, Haar *n*, Pinselhaar *n*
bristly borstig, struppig
Britannia metal Metallsilber, Britanniametall *n*
British antilewisite British Antilewisit *n*
British gum Britischgummi *n*, Dextrin *n*

brittle brüchig, beizbrüchig, bröckelig, faulbrüchig, glashart, kurzbrüchig, spröde, zerbrechlich (spröde), ~ **at blue heat** blaubrüchig, ~ **from cold** kaltbrüchig, ~ **in the hot state** heißbrüchig
brittle fracture Sprödbruch *m*
brittleness Brüchigkeit *f*, Faulbruch *m*, Sprödigkeit *f*, Zerbrechlichkeit *f*, **cold** ~ Kaltsprödigkeit *f*, **development of** ~ Versprödung *f*, **limit of** ~ Alterungsgrenze *f*
brittleness temperature Kältebiegeschlagwert *m*
brittle point Sprödigkeitspunkt *m*
broach Glättahle *f*, Reibahle *f*
broaching Anstich *m* (Brau)
broaching device Räumvorrichtung *f*
broaching machine Räummaschine *f*
broad breit, weit
broad axe Breitbeil *n*
broad beam absorption Großfeldabsorption *f* (Atom)
broadbeam measurement Großfeldmessung *f*
broadbeam shielding Großfeldabschirmung *f*
broadcast *v* funken (Radio), übertragen (Rad)
broadcasting Radio *m* (Funk), Rundfunk *m*
broaden *v* weiten
broadening Verbreiterung *f*
broad-faced steel gear Kammwalze *f*
broadside Querformat *n* (Buchdr)
broad tool Breiteisen *n*
brocade Brokat *m*
brocade color Brokatfarbe *f*
brocade dye Brokatfarbe *f*
brocade finish Runzellack *m*
brocade leather Brokatleder *n*
brocatelle Brokatell *m* (Text)
brochantite Brochantit *m* (Min)
bröggerite Bröggerit *m* (Min)
broil *v* braten, grillen
broken gebrochen, zerbrochen, gestrichelt (Linie), kaputt
broken brass Bruchmessing *n*
broken chips Bröckelspan *m*
broken coke Bruchkoks *m*
broken copper Bruchkupfer *n*
broken glass Bruchglas *n*, **fragment of** ~ Glassplitter *m*, **piece of** ~ Glassplitter *m*
broken gold Bruchgold *n*
broken iron Brucheisen *n*
broken lead Bruchblei *n*
broken piece Bruchstück *n*
broken silver Bruchsilber *n*
broken stone Bruchstein *m*
bromacetic acid Bromessigsäure *f*
bromacetol Bromacetol *n*
bromal Bromal *n*
bromal hydrate Bromalhydrat *n*
bromalin Bromalin *n*, Bromäthylformin *n*
bromamide Bromamid *n*
bromanil Bromanil *n*

bromanilic acid Bromanilsäure *f*
bromaniline Bromanilin *n*
bromargyrite Bromargyrit *m* (Min)
bromate Bromat *n*
bromate *a* bromsauer
bromate *v* bromieren
bromate discharge Bromatätze *f*
bromation Bromieren *n*, Bromierung *f*
bromatography Bromatographie *f*
bromatology Bromatologie *f*
bromatometric bromatometrisch
bromcarmine Bromcarmin *n*
brome grass Trespe *f* (Bot)
bromeine Bromein *n*
bromelin Bromelin *n*
bromethylformin Bromäthylformin *n*, Bromalin *n*
brometone Brometon *n*
bromic acid Bromsäure *f*, **salt or ester of** ~ Bromat *n*
bromide Bromid *n*, Bromsalz *n*, Bromür *n* (obs)
bromide paper Bromidpapier *n*
bromide process Bromsilberverfahren *n*
brominate *v* bromieren
bromination Bromierung *f*, Bromieren *n*
bromination reaction Bromierungsreaktion *f*
bromine (Symb. Br) Brom *n*, **additive compound of** ~ Bromadditionsprodukt *n*, **containing** ~ bromhaltig, **raw** ~ Rohbrom *n*
bromine atom Bromatom *n*
bromine bottle Bromflasche *f*
bromine chloride Chlorbrom *n*
bromine compound Bromverbindung *f*
bromine content Bromgehalt *m*
bromine derivative Bromderivat *n*
bromine fluoride Bromfluorid *n*, Fluorbrom *n*
bromine number Bromzahl *f*
bromine poisoning Bromvergiftung *f*, Bromismus *m* (Med)
bromine preparation Brompräparat *n*
bromine solution Bromlösung *f*
bromine value Bromzahl *f*
bromine vapor Bromdampf *m*
bromine washing bottle Bromwaschflasche *f*
bromine water Bromwasser *n*
brominism Bromvergiftung, Bromismus *m* (Med)
brominize *v* bromieren
brominizing Bromieren *n*
bromiodobenzene Bromjodbenzol *n*
bromipin Bromipin *n*
bromipin capsule Bromipinkapsel *f*
bromism Bromismus *m* (Med), Bromvergiftung *f* (Med)
bromisoval Bromisoval *n*
bromite Bromit *m* (Min)
bromlite Bromlit *m* (Min), Alstonit *m* (Min)
bromoacetic acid Bromessigsäure *f*
bromoacetone Bromaceton *n*

bromoacetylene Bromacetylen *n*
bromobenzene Brombenzol *n*
bromobenzoic acid Brombenzoesäure *f*
bromocamphor Bromcampher *m*,
  Monobromcampher *m*
bromocholine bromide Bromcholinbromid *n*
bromocodide Bromocodid *n*
bromocoll Bromocoll *n*
bromocoll ointment Bromocollsalbe *f*
bromocoll soap Bromocollseife *f*
bromocoll solution Bromocollösung *f*
bromocresol green Bromkresolgrün *n*
bromocresol purple Bromkresolpurpur *n*
bromocyanogen Bromcyan *n*
bromodiethyl acetamide Neuronal *n*
bromodifluoromethane Bromdifluormethan *n*
bromodimethylarsine Kakodylbromid *n*
bromoethane Äthylbromid *n*, Bromäthyl *n*
bromoethene Bromäthylen *n*
bromoethoxybenzene Bromphenetol *n*
bromoethylene Bromäthylen *n*
bromoethyne Bromacetylen *n*
bromoform Bromoform *n*
bromoguanide Bromguanid *n*
bromohydrin Bromhydrin *n*
bromohydroquinol Bromhydrochinon *n*
bromoiodonaphthalene Bromjodnaphthalin *n*
bromoisocaproic acid Bromisocapronsäure *f*
bromoisocaproyl chloride
  Bromisocaproylchlorid *n*
bromol Bromol *n*
bromometric bromometrisch
bromometry Bromometrie *f*, bromometrische
  Titration *f*
bromomorphide Bromomorphid *n*
bromonaphthalene Bromnaphthalin *n*
bromonaphthoquinone Bromnaphthochinon *n*
bromonitrobenzene Bromnitrobenzol *n*
bromonium ion Bromoniumion *n*
bromophenanthroline Bromphenanthrolin *n*
bromophenetol Bromphenetol *n*
bromophenol Bromphenol *n*
bromophenol blue Bromphenolblau *n*
bromophenylenediamine Bromphenylendiamin *n*
bromophenylhydrazine Bromphenylhydrazin *n*
bromophosgene Bromphosgen *n*
bromopicrin Brom[o]pikrin *n*
bromoplatinate Bromoplatinat *n*
bromopyrine Bromopyrin *n*
bromoquinal Bromochinal *n*
bromostannic acid Zinnbromwasserstoffsäure *f*
bromostyrene Bromstyrol *n*
bromosuccinic acid Brombernsteinsäure *f*
bromotannin Bromtannin *n*
bromothymol blue Bromthymolblau *n*
bromotoluene Bromtoluol *n*
bromoxylenol blue Bromxylenolblau *n*
bromthymol blue Bromthymolblau *n*, ~ **broth
  base** Bromthymolblau-Grundnährboden *m*

bromural Bromural *n*
bromyrite Bromyrit *m* (Min), Bromit *m* (Min),
  Bromspat *m* (Min)
broncholytic preparation Broncholytikum *n*
  (Pharm)
brongniardite Brongniardit *m* (Min)
Bronsted-Lowry definition of acids and bases
  Definition *f* der Säuren und Basen nach
  Brönsted und Lowry
bronze Bronze *f*, Bronzemetall *n*, Gießerz,
  Glockenmetall *n*, **red** ~ rote Bronze, **white** ~
  weiße Bronze, **yellow** ~ gelbe Bronze
bronze *v* bronzieren, brünieren
bronze bush Bronzefutter *n*
bronze coating Bronzebezug *m*
bronze color Bronzefarbe *f*, Brokatfarbe *f*,
  Erzfarbe *f*
bronze-colored bronzefarbig
bronze cyanide bath Bronzebad *n*
bronze diabetes Bronze-Diabetes *m* (Med)
bronze for bells Glockenspeise *f*
bronze founder Bronzegießer *m*
bronze foundry Bronzegießerei *f*
bronze lacquer Bronzelack *m*
bronze-like bronzeartig
bronze liquor Bronzetinktur *f*
bronze luster Bronzeglanz *m*
bronze metal Bronzemetall *n*
bronze packing Bronzefutter *n*
bronze paper Bronzepapier *n*
bronze plating Bronzebezug *m*
bronze powder Bronzepulver *n*, Bronzierpulver *n*
bronze printing Bronzedruck *m* (Buchdr)
bronze sleeve Bronzebezug *m*
bronze varnish Bronzelack *m*
bronze wire Bronzedraht *m*
bronzine Bronzin *n*
bronzing Bronzeglanz *m*, Bronzieren *n;*
  Bronzierung *f*
bronzing lacquer Bronzelack *m*
bronzing liquid Bronzelack *m*, Bronzetinktur *f*,
  Bronzierflüssigkeit *f*
bronzing pickle Brünierbeize *f*
bronzing powder Bronzierpulver *n*
bronzing salt Bronziersalz *n*
bronzite Bronzit *m* (Min)
bronzy bronzeartig
brookite Brookit *m* (Min), Arkansit *m* (Min),
  Titankiesel *m* (Min)
broom Besen *m*, Pfriemenkraut *n* (Bot)
broom flowers Ginsterblüten *f pl* (Bot)
broparoestrol Broparöstrol *n*
broth [culture] Bouillon
brown *a* braun, **dark** ~ dunkelbraun,
  schwarzbraun, tiefbraun
brown *v* bräunen (Färb), anbräunen, brünieren
brown algae Braunalgen *pl* (Bot)
brown black braunschwarz

brown coal Braunkohle *f,* Erdkohle *f,* ~ **for**
low-temperature retort process Schwelkohle *f*
brown coal bitumen Braunkohlenbitumen *n*
brown coal briquet[te] Braunkohlenbrikett *n*
brown coal coke Braunkohlenkoks *m*
brown coal grit Braunkohlensandstein *m*
brown-coal mining Braunkohlenförderung *f*
brown coal plant Braunkohlenanlage *f*
brown coal tar Braunkohlenteer *m*
brown coal wax Braunkohlenwachs *n*
brown coloring Bräunung *f,* Braunfärbung *f*
brown green Braungrün *n*
brown hematite Brauneisenstein *m* (Min),
Limonit *m* (Min)
Brownian motion Molekularbewegung *f*
Brownian movement Brownsche
Molekularbewegung *f* (Phys)
Brownian particle Brownsches Teilchen *n*
browning Bräunen *n,* Anbräunen *n,* Bräunung *f,*
Brünieren *n*
brown iron ocher Brauneisenocker *m* (Min)
brown iron ore Brauneisen *n* (Min),
Brauneisenstein *m* (Min), Braunerz *n* (Min)
brown iron ore nodules
Brauneisensteinknollen *pl* (Min)
brownish bräunlich
brownish red braunrot
brownish yellow braungelb
brown lake Braunlack *m*
brown lead oxide Bleiperoxid *n,* Bleihyperoxid *n,*
Bleisuperoxid *n*
brown lead oxide plate Bleisuperoxidplatte *f*
brown ocher Braunocker *m*
brown ore Braunerz *n* (Min)
brown paper Packpapier *n*
brown pearlspar Braunspat *m* (Min)
brown rot Braunfäule *f* (Bot)
brown spar Braunspat *m*
brown sugar Sandzucker *m*
brucic acid Brucinsäure *f*
brucidine Brucidin *n*
brucine Brucin *n,* Bruzin *n*
brucine hydrochloride Brucinchlorhydrat *n*
brucine nitrate Brucinnitrat *n*
brucine quinone Brucinchinon *n*
brucine solution Brucinlösung *f*
brucine sulfate Brucinsulfat *n*
brucinic acid Brucinsäure *f*
brucinolic acid Brucinolsäure *f*
brucinolone Brucinolon *n*
brucinolonic acid Brucinolonsäure *f*
brucinonic acid Brucinonsäure *f*
brucite Brucit *m* (Min), Nemalith *m* (Min),
ferruginous ~ Eisenbrucit *m* (Min)
brugnatellite Brugnatellit *m* (Min)
bruise damage Stoßverletzung *f*
brunsvigite Brunswigit *m*

brush Bürste *f,* Pinsel *m,* Stromabnehmer *m*
(Elektr), ~ **for producing a mat surface**
Mattierbürste *f*
brush *v* überstreichen, ~ **on** *v* aufstreichen
brushability Streichbarkeit *f,* Streichfähigkeit *f,*
Verstreichbarkeit *f*
brushable streichfähig
brushable plaster for outdoors Füllfarbe *f*
brush contact Bürstenkontakt *m* (Elektr)
brush coppering Pinselverkupferung *f*
brush discharge Büschelentladung *f,*
Glimmentladung *f,* Sprühentladung *f*
brush gilding Pinselvergoldung *f*
brush hair Pinselhaar *n*
brushing Streichen *n*
brushing liquid Kratzwasser *n*
brushing on Aufstrich *m*
brushing property Streichfähigkeit *f*
brushite Brushit *m* (Min)
brush proof Bürstenabzug *m*
brush rod Kontaktschlitten *m*
brush spreader Bürstenstreichmaschine *f*
brush spreading Bürstenstreichverfahren *n*
brush viscosity Streichviskosität *f*
bryogenine Bryogenin *n*
bryonin Bryonin *n*
bryonol Bryonol *n*
bryoresin Bryoresin *n*
BT-cut crystal BT-Kristallschnitt *m* (Krist)
bubble Blase *f* (Luft, Gas), Folienblase *f,*
Lunker *m* (Kunstharz), Preßfehler *m* (Blase),
small ~ Bläschen *n*
bubble *v* aufwallen, brodeln, schäumen,
sprudeln (schäumen), wallen, ~ **through**
durchperlen
bubble over *v* übersprudeln, aufgischen
bubble up *v* aufbrodeln, aufsprudeln, aufsteigen
(Gasblasen)
bubble and skin packaging Konturen- und
Hautpackungen *pl*
bubble by bubble blasenweise
bubble cap Haube *f* (Dest)
bubble-cap column Glockenbodenkolonne *f*
bubble-cap columns Blasensäule *f*
bubble-cap plate Glockenboden *m* (Dest)
bubble-cap tray Glockenboden *m* (Dest)
bubble chamber Blasenkammer *f*
bubble control Blasenregelung *f*
bubble counter Blasenzähler *m*
bubble flow laminare Blasenströmung *f*
bubble formation Blasenbildung *f*
bubble gauge Blasenzähler *m,*
Gasblasenströmungsmesser *m*
bubble growth Blasenwachstum *n*
bubble level Wasserwaage *f*
bubble model Blasenmodell *n* (v. Fließbetten)
bubble plate Glockenboden *m* (Dest)
bubbler Gluckertopf *m*
bubble rising Blasenaufstieg *m*

bubble rising velocity
Blasenaufstiegsgeschwindigkeit *f*
bubble test Blasprobe *f* (Zuck)
bubble tray Glockenboden *m* (Dest)
bubble vaporization Blasenverdampfung *f* (Öl)
bubble visco[si]meter Luftblasen-Viskosimeter *n*
bubbling Blasenbildung *f,* Brodeln *n,* Gärung *f,*
Gebrodel *n,* Sprudeln *n,* Wallen *n*
bubbling column Gasblasenwäscher *m*
bubbling promoter Schaumverstärker *m*
(Siebbodenkolonne)
bubbling throttle Einperldrossel *f*
bubbling-type electrode
Durchperlungselektrode *f*
bubbly blasig
buccocamphor Buccocampher *m*
bucco leaves Buccoblätter *n pl*
Buchner funnel Büchnertrichter *m* (Chem),
Filterkolben *m* (Chem)
buchu Buccoblätter *n pl*
buchu camphor Buccocampher *m,* Diosphenol *n*
buck Beuche *f,* Beuchwasser *n*
buck *v* beuchen, laugen
buck ashes Laugenasche *f*
buck bean Fieberklee *m* (Bot)
bucket Eimer *m,* Becher *m,* Kübel *m,*
Ladeschaufel *f,* Schaufel *f* (Turbine),
Schöpfzelle *f*
bucket conveyor Becheraufzug *m,* Becherwerk *n*
bucket-conveyor extractor
Becherwerksextrakteur *m*
bucket elevator Becherbandförderer *m,*
Becherwerk *n,* Eimerwerk *n,* Kübelaufzug *m*
bucket grab Greiferbügel *m*
buckets Schaufelung *f*
bucket trough Schaufelmulde *f*
bucket wheel Becherrad *n,* Schöpfrad *n,*
Zellenrad *n*
bucket wheel excavator Schaufelradbagger *m*
bucket-wheel extractor Zellenradextrakteur *m*
bucking Auslaugen *n,* Beuchen *n*
bucking cloth Laugentuch *n*
bucking iron Erzpocheisen *n*
bucking kier Beuchkessel *m,* Auskocher *m*
bucking lye Beuche *f,* Beuchlauge *f*
bucking tub Beuchfaß *n*
bucklandite Bucklandit *m* (Min)
buckle *v* krümmen, sich verziehen, sich werfen
buckling Knickerscheinung *f,* Knickung *f,*
Verbiegung *f,* ~ **of the ring** Verziehen *n* des
Ringes
buckling load Knicklast *f*
buckling or crippling test Knickversuch *m*
buckling strength Knickfestigkeit *f*
buckling stress Knickbeanspruchung *f,*
Knickspannung *f*
buckram Glanzleinwand *f*

buckskin Wildleder *n*
buck-tallow Bockstalg *m*
buckthorn Wegedorn *m* (Bot), **syrup of** ~
Kreuzbeerensirup *m*
buckthorn berry Kreuzdornbeere *f* (Bot)
buckthorn oil Kreuzdornöl *n*
buckwheat Buchweizen *m* (Bot)
buclizine Buclizin *n*
bud Keim *m* (Bot), Knospe *f*
bud *v* keimen (Bot)
buddle schlämmen (Bergb)
buddling Schlämmen *f* (Bergb)
buds, formation of ~ Knospenbildung *f*
Büchner funnel Büchner-Trichter *m*
bufagin Bufagin *n*
bufalin Bufalin *n*
buff Ledergelb *n,* Isabellfarbe *f*
buff *a* lederfarben
buff *v* mit Leder polieren; aufrauhen
buff-colored gelbbraun, lederfarben
buffer Glättmaschine *f,* Puffer *m* (Chem)
buffer *v* puffern, abpuffern, abstumpfen
buffer acid Puffersäure *f*
buffer action Pufferwirkung *f*
buffer battery Pufferbatterie *f*
buffer capacity Pufferkapazität *f*
buffer contact Pufferkontakt *m*
buffered gepuffert
buffer effect Pufferwirkung *f*
buffer gas Schutzgas *n*
buffering Pufferung *f*
buffering action Pufferwirkung *f*
buffer reagent Puffergemisch *n*
buffer solution Pufferlösung *f*
buffer spring Pufferfeder *f,* Spiralfeder *f*
buffer system Puffersystem *n*
buffer voltage Pufferspannung *f*
buffing Polieren, Aufrauhen *n,* Rauhen *n,*
Schwabbeln *n*
buffing agent Schwabbelmittel *n*
buffing brush Schleifbürste *f,* Borstenscheibe *f*
buffing lathe Reifenabschleifmaschine *f*
buffing machine Aufrauhmaschine *f*
buffing rib Schleifkante *f*
buffing wheel Schwabbelscheibe *f*
buff liquor Gelbbeize *f*
bufogenin Bufogenin *n*
bufotalic acid Bufotalsäure *f*
bufotalin Bufotalin *n*
bufotenidine Bufotenidin *n*
bufotenine Bufotenin *n*
bufothionine Bufothionin *n*
bufotoxine Bufotoxin *n*
bug Käfer *m* (Zool)
bugloss Ochsenzungenkraut *n* (Bot)
bug[s] Ungeziefer *n*
build *v* bauen, mauern ~ **up** *v* aufbauen
builder Aufbaustoff *m* (Waschmittel)
builders Gerüstsubstanzen *f pl*

building Bauwerk *n*
building board Baupappe *f*, Bauplatte *f*
building brick Mauerziegel *m*
building industry Bauindustrie *f*
building machinery Baumaschinen *pl*
building material Baumaterial *n*, Baustoff *m*
building materials testing Baustoffprüfung *f*
building preservative agent Bautenschutzmittel *n*
building proofing material Bautenschutzmittel *n*
building purposes Bauzwecke *m pl*
building regulations Bauvorschriften *pl*
building trade Baugewerbe *n*
building under construction Neubau *m*
building-up form Reifenwickeltrommel *f*
building-up former Aufbautrommel *f* (Reifen)
building-up process Aufschaukelvorgang *m*
build-up factor Aufbaufaktor *m*,
  Zuwachsfaktor *m*
built-in stirrer Einbaurührer *m*
built-in unit Einbauaggregat *n*
built-up welding Auftragsschweißen *n*
bulb Ballon *m*, Birne *f* (Elektr), Knolle *f* (Bot),
  Küvette *f*, Kugel *f* (Thermometer-Kugel),
  Zwiebel *f* (Knolle)
bulb *v* bombieren
bulb angle Winkelwulsteisen *n*
bulb barometer Gefäßbarometer *n*
bulb flats Wulsteisen *n*
bulb iron Wulsteisen *n*, **flat** ~ Flachwulsteisen *n*
bulb level Kugelniveau *n*
bulb-like zwiebelartig
bulbocapnine Bulbocapnin *n*
bulbous bauchig, zwiebelartig
bulb pipet[te] Vollpipette *f* (Anal)
bulb rail Flanschwulsteisen *n*
bulb shape Kölbchenform *f*
bulb stopper Kugelstopfen *m*
bulb tube Kugelröhre *f*, Kugelrohr *n*
bulge Bauchung *f*, Auftreibung *f*, Ausbuchtung *f*
bulge *v* ausbauchen, aufweiten
bulged ausgebaucht, angeschwollen, ballig,
  bauchig
bulge testing apparatus Einbeulapparat *m*
bulging Wölbung *f*, Aufweiten *n*
bulging test Anschwellprobe *f*,
  Aufweitversuch *m* (an Rohren),
  Stauchprobe *f*, Stauchversuch *m*
bulk Masse, Menge *f*, Volumen *n*
bulk articles Massengüter *pl*
bulk density Schüttdichte *f*, Schüttgewicht *n*
bulk factor Füllfaktor *m*, Füllkonstante *f*,
  Verdichtungsgrad *m*
bulk liquid Flüssigkeitsmasse *f*
bulk modulus Kompressionsmodul *m*
bulk packing lose Verpackung *f*
bulk polymerization Blockpolymerisation *f*,
  Polymerisation *f* in Masse,
  Trockenpolymerisation *f*

bulk resilience Bauschelastizität *f*
bulk viscosity, volume or ~ Volumenviskosität *f*
bulk volume Schüttvolumen *n*
bulk weight Schüttgewicht *n*
bulky sperrig, voluminös
bulla Blase, Hautblase *f* (Pathol)
bullatenone Bullatenon *n*
bulldog metal Puddelschlacke *f*
bulldozer Planierraupe *f*
bullet Kugel *f*
bulletin Bulletin n, kurzer Bericht m, Heft *n*
bullet-proof glass Panzerglas *n*
bump Schlag *m*, Stoß *m*
bumper Stoßdämpfer *m*
bumping stoßweises Sieden, Siedeverzug *m*
bumping mechanism Prellvorrichtung *f*
buna Buna *n* (HN)
buna rubber Bunagummi *n*
bunching Bündelung *f*
bundle Bündel *n*
bundled gebündelt
bung Spund *m*
bunker Bunker *m*
bunker belt system Bunkerbandsystem *n*
bunker coal Bunkerkohle *f*
Bunsen burner Bunsenbrenner *m*
Bunsen cell Bunsenelement *n* (Elektr)
Bunsen flame Bunsenflamme *f*
bunsenine Bunsenin *n*
bunsenite Bunsenin *n*, Bunsenit *m* (Min)
Bunsen screen Bunsenschirm *m*
Bunsen valve Bunsenventil *n*
Bunte gas burette Buntebürette *f*
Bunte salts Buntesalze *pl*
buoyancy Auftrieb *m*, Schwimmfähigkeit *f*,
  Schwimmvermögen *n*, **center of** ~
  Auftriebsmittelpunkt *m*, **plane of** ~
  Schwimmebene *f*
buoyancy axis Schwimmachse *f*
buoyancy bulb Schwimmkugel *f*
buoyancy coefficient Auftriebkoeffizient *m*
buoyancy in water Wasserauftrieb *m*
buoyant schwimmfähig
buoyant density Schwebedichte *f*
buphanidrine Buphanidrin *n*
buphanisine Buphanisin *n*
buphanitine Buphanitin *n*
bur Klette *f* (Bot)
Burawoy ketone Burawoy-Keton *n*
burden Last *f*, Beschickung *f*, Beschwerung *f*
burden *v* begichten (Metall), beladen, belasten,
  bepacken
burdock Klette *f* (Bot)
bureau of standards Eichamt *n*
buret[te] Bürette *f*, Maßröhre *f*, Meßröhre *f*
buret[te] clip Bürettenquetschhahn *m*
buret[te] cock Bürettenhahn *m*
buret[te] head Bürettenkappe *f*
buret[te] holder Bürettenhalter *m*

buret[te]stand Bürettengestell *n*,
  Bürettenständer *m*
buret[te] tip Bürettenausflußspitze *f*
buret[te] valve Bürettenhahn *m*
burin Stichel, Meißel *m*
burn Brandwunde *f*, first degree ~
  Verbrennung *f* ersten Grades
burn *v* brennen, anbrennen, ausbrennen,
  verbrennen, verschmoren (Elektr), ~ out
  durchglühen
burn again *v* nachbrennen
burn in *v* einbrennen
burn off *v* abbrennen, wegbrennen
burn off paint *v* Farbe abbrennen
burn out *v* durchbrennen (Lampen)
burn slowly *v* schwelen
burn through *v* durchbrennen
burn to ashes *v* einäschern
burn together *v* zusammenbrennen
burnable [ver]brennbar
burned fireclay Schamotte *f*
burned-off vollständig gar (Metall)
burner Brenner *m*, multiple tube ~ Vielfach-
  Brenner, turbulent ~ Turbobrenner *m*
burner equipment Brennerausrüstung *f*
burner for oil residues Rückstandsbrenner *m*
  (Öl)
burner nozzle Brennerdüse *f*,
  Brennrohrmundstück *n*
burner tip Brennerkopf *m*
burnettize burnettisieren
burning Brand *m*, Brennarbeit *f*, Brennen *n*,
  Verbrennen *n*, slow ~ Schwelung *f*
burning *a* brennend, scharf (beißend)
burning glass Brennglas *n*
burning hour Brennstunde *f*
burning of glue line Brennen *n* der Leimfuge
burning oil Brennöl *n*
burning oven Brennofen *m*
burning period Brennzeit *f*
burning rate Brenngeschwindigkeit *f*
burning test Brennbarkeitsprobe *f*,
  Brennversuch *m*
burning the filter Einäschern *n* des Filters
burning through Durchbrennen *n*
burning wood Brennholz *n*
burnish *v* polieren, bräunen (Gieß), brünieren,
  glätten, glanzschleifen, hochglanzpolieren
burnisher Brüniereisen *n* (Techn), Brünierer *m*,
  Gerbstahl *m*, Glänzer *m*, Glätter *m*,
  Glattschleifer *m*, Poliereisen *n*, Polierer *m*,
  Polierstahl *m*
burnishing Brünieren *n*, Hochglanzerzeugung *f*
burnishing bath Glanzbrenne *f*
burnishing gold Poliergold *n*
burnishing iron Brüniereisen *n* (Techn)
burnishing platinum Polierplatin *n*
burnishing roll Polierwalze *f*

burnishing silver Poliersilber *n*
burnishing stone Brünierstein *m*, Polierstein *m*
burn lime *v* Kalk brennen
burnout Heizflächenzerstörung *f*
burnout feature Ausbrennverhalten *n*
burns ointment Brandsalbe *f* (Pharm)
burnt verbrannt, brenzlig
burnt gas Abzugsgas *n*, Auspuffgas *n*,
  Feuergas *n*, Rauchgas *n*
burnt-in color Muffelfarbe *f*
burnt iron Brandeisen *n*
burnt island red Aluminiummennige *f*
burnt spot Verbrennungsmarkierung *f*
burnt sugar, tincture of ~ Zuckercouleur *f*
burn-up Abrand *m* (Atom),
  Spaltstoffausnutzung *f*
burr Grat *m* (Techn), Walzenbart *m*
burring machine Abgratmaschine *f*
burseracine Burseracin *n*
burst *v* bersten, aufbrechen, aufplatzen,
  aufspalten, explodieren, platzen, reißen
  (zerreißen), springen (bersten), zerplatzen,
  zerspringen, ~ forth *v* herausbrechen
burst off *v* absprengen
burst open *v* aufplatzen, sprengen (aufsprengen)
bursting Bersten *n*, Bruch *m*, Platzen *n* (Reifen),
  Sprengung *f*, Springen *n*
bursting charge Sprengladung *f*
bursting disc Berstscheibe *f*, Brechscheibe *f*,
  Reißscheibe *f*, Zerreißplättchen *n*
bursting plate Brechplatte *f*
bursting pressure Berstdruck *m*
bursting strength Berstdruckfestigkeit *f*,
  Berstfestigkeit *f*, Berstwiderstand *m*
bursting tester Berstdruckprüfer *m*
burstwort Bruchkraut *n*
bus bar Sammelschiene *f*, ~ of the bath
  Schiene *f* des Bades
bush Busch *m*
bushing Buchsring *m*, Hülse *f* (Techn),
  Lagerbüchse *f*, Radbüchse *f*
bushing chain Stiftkette *f*
bush metal Büchsenmetall, Hartguß *n*
bushy stunt virus Bushy-Stunt-Virus *m* (Biol)
business Beschäftigung *f*, Geschäft *n*
business administration Betriebswissenschaft *f*
business knowledge Fachkenntnis *f*
bustamite Bustamit *m* (Min)
busulfan Busulfan *n*
busy arbeitsam, fleißig
busycon Busycon *n*
butacaine Butacain *n*
butadiene Butadien *n*, Divinyl *n*
butadiene caoutchouc Butadienkautschuk *m*
butadiene quinone Butadienchinon *n*
butadiene sulfone Butadiensulfon *n*
butadiine Diacetylen *n*
butalbital Butalbital *n*

butamin Tutocain *n*
butanal Butylaldehyd *m*, Butyraldehyd *m*
butanamide Butyramid *n*
butane Butan *n*
butanedial Succindialdehyd *m*
butanediamine, 1,4- ~ Putrescin *n*
butanedicarboxylic acid, 1,4- ~ Adipinsäure *f*
butanedioic acid Bernsteinsäure *f*
butanedioyl Succinyl-
butanol Butanol *n*, Butylalkohol *m*
butanol acetone fermentation
    Butanolacetonvergärung *f*
butanol production plant
    Butanolgewinnungsanlage *f*
Butenandt acid Butenandt-Säure *f*
Butenandt ketone Butenandt-Keton *n*
butene Buten *n*, Butylen *n*
butenol Crotylalkohol *m*
butenolide Butenolid *n*
butenyl Butenyl-
butesin Butesin *n*
buthalital Buthalital *n*
butine Butin *n*
butine diol Butindiol *n*
butolane Butolan *n*
butrin Butrin *n*
butropine Butropin *n*
butter, sweet ~ ungesalzene Butter
butter fat Butterfett *n*
butterfly Schmetterling *m* (Zool)
butterfly nut Flügelmutter *f*
butterfly twin Schwalbenschwanzzwilling *m*
    (Krist)
butterfly valve Drehklappe *f*, Drosselventil *n*,
    Flügelventil *n*, Wechselklappe *f*
butter-milk Buttermilch *f*
butter of antimony Antimonbutter *f*,
    Antimon(III)-chlorid *n*
butter of tin Zinnbutter *f*
buttery butterähnlich, butterartig
butter yellow Buttergelb *n*,
    Dimethylaminoazobenzol *n*
butt joined stumpf angefugt
butt joint Stumpfnaht *f*, Stumpfstoß *m*
butt-joint rivetting Laschennietung *f*
buttmuff coupling Muffenkupplung *f*
button Knopf *m*
buttonhead Halbrundkopf *m*
buttress thread Sägengewinde *n*,
    Sägeprofilgewinde *n*
butt-solder *v* stumpf löten
butt-splice *v* stumpf zusammensetzen
buttstrap Lasche *f*, Stoßblech *n*
butt weld Stumpfnaht *f*, Stumpfstoß *m*
butt-weld *v* stumpf schweißen
butt welding Stumpfschweißen *n*,
    Stumpfstoßschweißung *f*
butyl Butyl-
butyl acetate Butylacetat *n*

butyl acetyl acetone Butylacetylaceton *n*
butyl alcohol Butanol *n*, Butylalkohol *m*, **tertiary**
    ~ tertiärer Butylalkohol,
    Pseudobutylalkohol *m*
butyl aldehyde Butylaldehyd *m*
butyl amine Butylamin *n*
butyl benzoate Butylbenzoat *n*
butyl butyrate Butylbutyrat *n*
butyl cellosolve Butylcellosolve *f*
butylchloral hydrate Butylchloralhydrat *n*
butyl chloride, tertiary ~ Pseudobutylchlorid *n*
butyl cyanide Butylcyanid *n*, Valeronitril *n*
butylene Buten *n*, Butylen *n*
butylene diamine Putrescin *n*
butylene glycol Butylenglykol *n*
butylene sulfide Butylensulfid *n*
butyl group Butylgruppe *f*
butyl halide Butylhalogenid *n*
butyl hypnal Butylhypnal *n*
butyl iodide Butyljodid *n*
butyl lactate Butyllactat *n*
butyl malonate Butylmalonat *n*
butylmalonic acid Butylmalonsäure *f*
butylmalonic ester Butylmalon[säure]ester *m*
butyl phenyl Butylphenyl *n*
butyl phthalate Butylphthalat *n*
butyl resorcinol Butylresorcin *n*
butyl rubber Butylkautschuk *m*
butyl salicylate Butylsalicylat *n*
butyl silicone Butylsilicon *n*
butyl tartrate Butyltartrat *n*
butyl trichlorosilane Butyltrichlorsilan *n*
butyne Butin *n*
butyne diol Butindiol *n*
butynoic acid Tetrolsäure *f*
butyraceous butterähnlich, butterartig,
    butterhaltig
butyral Butylaldehyd *m*, Butyraldehyd *m*
butyraldehyde Butylaldehyd *m*, Butyraldehyd *m*
butyraldol Butylaldol *n*
butyramide Butyramid *n*
butyramidine Butyramidin *n*
butyrate Butyrat *n*
butyrate *a* buttersauer
butyric acid Buttersäure *f*, **radical of** ~
    Buttersäureradikal *n*
butyric anhydride Buttersäureanhydrid *n*
butyric fermentation Buttersäuregärung *f*
butyric oil Butteröl *n* (obs)
butyrin Butyrin *n*
butyrobetaine Butyrobetain *n*
butyroin Butyroin *n*
butyrolactone Butyrolacton *n*
butyrometer Buttermesser *m*, Butyrometer *n*
butyrone Butyron *n*, Dipropylketon *n*
butyronitrile Butyronitril *n*
butyrophenone Butyrophenon *n*
butyrospermadiene Butyrospermadien *n*
butyryl Butyryl-, Buttersäurerest *m*

buxine Buxin *n*
buyer's guide Bezugsquellenverzeichnis *n*
buyer's inspection Abnahmeprüfung *f*
buyer's specification Gütevorschrift *f*
buzzer Summer *m*
buzzer excitation Summererregung *f*
byakangelicin Byakangelicin *n*
byakangelicol Byakangelicol *n*
by-mordant Hilfsbeize *f*
by-pass Nebenleitung *f*, Ableitung *f*,
    Kurzschlußleitung *f*, Nebenauslaß *m*,
    Nebenrohr *n*, Nebenschlußleitung *f* (Elektr),
    Umlaufkanal *m*, Umwegleitung *f*, Weiche *f*
    (pipe), Umführungsleitung *f* (tubing),
    Verzweigung *f*, **two position** ∼
    Außenwegschaltung *f*
by-pass *v* umgehen
by-pass [duct] Umgehungsleitung *f*
by-passing conveyor Überholförderer *m*
by-pass line Umgehungsleitung *f*
by-pass regulator Umlaufregler *m*
by-pass valve Umgehungsventil *n*,
    Abschaltventil *n*, Umlaufventil *n*
by-product Nebenerzeugnis *n*, Abfallerzeugnis *n*,
    Abfallprodukt *n*, Ausscheidungsprodukt *n*,
    Nebenprodukt *n*
by-product[s] Abfall *m* (Met, Holz)
byssolite Byssolith *m*, Muschelseidenstein *m*
    (Min)
bytownite Bytownit *m* (Min)

C

cabane Verspannung *f* (Spannturm)
cabbage butterfly Kohlweißling *m*
cabinet Schrank *m*
cabinet dryer Trockenschrank *m*
cable Kabel *n*, Kabelleitung *f*, Seil *n* (Kabel),
   laying of a ~ Kabellegung *f*, plastic insulated
   ~ Kunststoffkabel *n*, round ~ Rundseil *n*,
   solid ~ Vollseil *n*, submarine ~
   unterseeisches Kabel *n*, underground ~
   unterirdisches Kabel *n*
cable *v* telegraphieren, kabeln
cable board Kabelbrett *n*
cable branching Kabelverzweigung *f*
cable clamp Kabelklemme *f*
cable compound Kabelmasse *f*
cable cord Kabelcord *m*
cable cord V-belt Kabelcordkeilriemen *m*
cable covering Kabelmantel *m*
cable crane Kabelkran *m*
cable drum Seiltrommel *f*
cable drying plant Kabeltrockenanlage *f*
cable extruder head Umlenkkopf *m*
cable impregnation plant
   Kabelimprägnieranlage *f*
cable joint Seilschloß *n*
cable junction Kabelanschluß *m*
cable oil Kabelöl *n*
cable protection Kabelschutz *m*
cable railway Drahtseilbahn *f*
cable section Kabelformstück *n*
cable sheathing Kabelmantel *m*,
   Kabelummantelung *f*
cable size Cordnummer *f*
cable sleeve Kabelschelle *f*
cable terminal Kabelendverschluß *m*
cable testing equipment Kabelprüfgerät *n*
cable twist Drehung *f* des Cordfadens
cable wax Kabelwachs *n*
cable winch Kabelwinde *f*
cabrerite Cabrerit *m* (Min)
cabreuva oil Cabreuvaöl *n*
caceres phosphate Caceresphosphat *n*
cacholong Kascholong *m* (Min)
cacholong opal Kascholongopal *m* (Min)
cacholong quartz Kascholongquarz *m* (Min)
cachou Cachou *n* (Pharm)
c-acid C-Säure *f*
cacodyl Kakodyl *n*, Tetramethyldiarsin *n*
cacodylate Kakodylat *n*
cacodylate *a* kakodylsauer
cacodyl bromide Kakodylbromid *n*
cacodyl carbide Kakodylcarbid *n*
cacodyl chloride Kakodylchlorid *n*
cacodyl hydride Kakodylwasserstoff *m*
cacodylic acid Kakodylsäure *f*, Alkargen *n*,
   Dimethylarsinsäure *f*

cacodyl oxide Kakodyloxid *n*
cacodyl preparation Kakodylpräparat *n*
cacogenin Cacogenin *n*
cacotheline Kakothelin *n*
cacothelinium hydroxide
   Kakotheliniumhydroxid *n*
cacoxene Kakoxen *m* (Min)
cacoxenite Kakoxen *m* (Min)
cactus Kaktus *m* (Bot)
cadalene Cadalin *n*
cadaver Leiche *f*
cadaverine Cadaverin *n*
cade oil Cadeöl *n*, Wacholderteer *m*
Cadet's fuming liquid Cadetsche Flüssigkeit *f*
cadinane Cadinan *n*
cadinene Cadinen *n*
cadinol Cadinol *n*
cadmate Cadmat *n*
cadmia Zinkofenbruch *m*, Zinkschwamm *m*
cadmiferous cadmiumhaltig
cadmine Cadmin *n*
cadmium (Symb. Cd), Kadmium *n*, Cadmium *n*,
   containing ~ cadmiumhaltig, metallic ~
   Cadmiummetall *n*
cadmium acetate Cadmiumacetat *n*
cadmium alloy Cadmiumlegierung *f*
cadmium amalgam Cadmiumamalgam *n*
cadmium arsenate Cadmiumarsenat *n*
cadmium arsenide Cadmiumarsenid *n*
cadmium aurichloride
   Cadmiumgold(III)-chlorid *n*
cadmium bromide Bromcadmium *n*,
   Cadmiumbromid *n*
cadmium carbonate, native ~ Otavit *m* (Min)
cadmium cement Cadmiumzement *m*
cadmium chloride Chlorcadmium *n*
cadmium chloroaurate(III)
   Cadmiumgold(III)-chlorid *n*
cadmium color Cadmiumfarbe *f*
cadmium compound Cadmiumverbindung *f*
cadmium content Cadmiumgehalt *m*
cadmium control rod Cadmiumkontrollstab *m*
   (Atom)
cadmium electrode Cadmiumelektrode *f*
cadmium ethylenediamine chelate
   Cadmiumäthylendiaminchelat *n*
cadmium fluoride Cadmiumfluorid *n*
cadmium hydroxide Cadmiumhydroxid *n*,
   Cadmiumoxidhydrat *n*
cadmium iodide Cadmiumjodid *n*,
   Jodcadmium *n*
cadmium metal Cadmiummetall *n*
cadmium metasilicate Cadmiummetasilikat *n*
cadmium-plate *v* kadmieren
cadmium plating Kadmieren *n*
cadmium salicylate Cadmiumsalicylat *n*

cadmium salt Cadmiumsalz *n*
cadmium selenide Cadmiumselenid *n*
cadmium solution Cadmiumlösung *f*
cadmium suboxide Cadmiumsuboxid *n*
cadmium sulfate Cadmiumsulfat *n*
cadmium sulfide Cadmiumsulfid *n*,
   Cadmiumgelb *n*, Schwefelcadmium *n*, native
   ~ Greenockit *m* (Min)
cadmium telluride Cadmiumtellurid *n*
cadmium tungstate Cadmiumwolframat *n*
cadmium yellow Cadmiumgelb *n*, Schwefelgelb *n*
cadogel Cadogel *n*
caesium also cesium (Symb. Cs), Cäsium,
   Caesium *n*
cafestadiene Cafestadien *n*
caffalic acid Kaffalsäure *f*
caffeic acid Kaffeesäure *f*
caffeidine Kaffeidin *n*
caffeine Koffein *n*, Caffein *n*, Coffein *n*,
   Trimethylxanthin *n*
caffeine benzoate Coffeinbenzoat *n*
caffeine citrate Coffeincitrat *n*
caffeine hydrobromide Coffeinbromhydrat *n*
caffeine hydrochloride Coffeinchlorhydrat *n*
caffeine oxalate Coffeinoxalat *n*
caffeine poisoning Coffeinvergiftung *f*
caffeine salt Coffeinsalz *n*
caffeine sodium benzoate
   Coffeinnatriumbenzoat *n*
caffeine sodium salicylate
   Coffeinnatriumsalicylat *n*
caffeine valerate Coffeinvalerianat *n*
caffeinism Coffeinvergiftung *f* (Med),
   Kaffeevergiftung *f* (Med)
caffetannic acid Kaffeegerbsäure *f*
caffolide Kaffolid *n*
caffoline Kaffolin *n*
caffuric acid Kaffursäure *f*
cage Käfig *m*
cage compound Käfigverbindung *f* (Chem)
cage effect Käfigeffekt *m* (Stereochem),
   Käfig-Wirkung *f* (Stereochem)
cage press Seiherpresse *f*
cahinca root Caincawurzel *f* (Bot),
   Kahinkawurzel *f*
cahincic acid Caincasäure *f*, Caincin *n*
cainca root Caincawurzel *f* (Bot)
caincic acid Caincasäure *f*, Caincin *n*
caincin Caincasäure *f*, Caincin *n*
cairngorm [stone] Cairngorm *m* (Min)
caisson disease Caissonkrankheit *f* (Med)
caje nut Anacardium *n*
cajeput Kajeput *n*, spirit of ~ Cajeputgeist *m*
cajeput oil Cajeputöl *n*
cajeputol Eukalyptol *n*
cajuput Kajeput *n*
cajuput oil Cajeputöl *n*
cajuputol Cajeputöl *n*
cake Kuchen *m*, Stück *n* (Seife)

cake *v* sintern, zusammenbacken
   ~ together *v* festbacken
cake discharge Kuchenabwurf *m*
cake filtration Kuchenfiltration *f*
caking Backen *n*
caking coal Backkohle *f*
caking property Backfähigkeit *f*
calabar bean Calabarbohne *f*, extract of ~
   Calabarbohnenextrakt *m*
calabarine Calabarin *n*
calabarol Calabarol *n*
calafatite Calafatit *m*
calaic acid Calainsäure *f*
calambac Calambakholz *n*
calamene Calamen *n*
calamenediol Calamendiol *n*
calamenene Calamenen *n*
calameone Calameon *n*
calamine [edler] Galmei *m* (Min), Kalamin *m*
   (Min), Kieselzinkerz *n* (Min),
   Kohlengalmei *m* (Min), Zinkspat *m* (Min),
   cellular ~ Zellengalmei *m* (Min), containing
   ~ galmeihaltig, green ~ Aurichalcit *m* (Min),
   siliceous ~ gewöhnlicher Galmei,
   Kieselgalmei *m* (Min), Zinkglas *n* (Min),
   Zinkkiesel *m* (Min)
calamine blende Galmeiblende *f* (Min)
calamine of copper Kupfergalmei *m* (Min)
calamint Bergmelisse *f*
calamite Calamit *m*
calamus Kalmus *m* (Bot)
calamus camphor Kalmuscampher *m*
calamus oil Kalmusöl *n*
calamus root Kalmuswurzel *f* (Bot)
calaverite Calaverit *m* (Min), Tellurgold *n*
calcar Ofen *m* (Glasherstellung)
calcar arch Frittofen *m*
calcareous kalkartig, kalkerdig, kalkhaltig,
   kalkig, kalkreich
calcareous barite Kalkbaryt *m* (Min)
calcareous cement Kalkkitt *m*
calcareous deposit Kalklager *n*
calcareous earth Kalkerde *f*
calcareous flux Kalk[stein]zuschlag *m*,
   Zuschlagkalkstein *m*
calcareous gravel Kalkkies *m*
calcareous mountain-chain Kalkgebirge *n*
calcareousness Kalkartigkeit *f*
calcareous niter Mauersalpeter *m*
calcareous oolite Kalkoolith *m* (Min)
calcareous phyllite Kalkphyllit *m* (Min)
calcareous sand Kalksand *m*
calcareous sandstone Kalksandstein *m* (Min)
calcareous sediment Kalkablagerung *f*
calcareous silex Kalkkiesel *m*
calcareous sinter Kalksinter *m*, Kalktropfstein *m*
calcareous slate clay Kalkschieferton *m* (Min)
calcareous soil Kalkboden *m*
calcareous spar Doppelstein *m* (Min)

calcareous talc Kalktalk *m* (Min)
calcareous tartar Kalkweinstein *m*
calcareous tufa Kalksinter *m*
calcareous uranite Kalkuranglimmer *m* (Min)
calcareous whiteware Kalksteingut *n*
calcibram Calcibram *n*
calciferol Calciferol *n*, Vitamin D$_2$ *n*
calciferous kalkhaltig
calcification Kalkablagerung *f* (Physiol),
　Kalkbildung *f*, Verkalkung *f*
calciform kalkförmig
calcify *v* verkalken, kalzifizieren, calcifizieren
calciglycine Calciglycin *n*
calcihyde Calcihyd *n*
calcimeter Kalzimeter *n*, Kalkmesser *m*
calcimurite Calcimurit *m* (Min)
calcinable calcinierbar
calcinating furnace Calciniertrommel *f*
calcination Kalzinierung *f*, Abbrand *m*,
　Calcinieren *n*, Einäscherung *f*, Glühen *n*,
　Röstung *f*, Verkalkung *f* (Chem), **method of**
　~ Röstverfahren *n*, ~ **of iron ores** Rösten *n*
　der Eisenerze, ~ **of limestone** Kalkbrennen *n*
calcination assay Röstprobe *f*
calcination gas Röstgas *n*
calcination plant Kalzinieranlage *f*
calcination test Röstprobe *f*
calcine *v* kalzinieren, abschwefeln, abschwelen,
　ansintern, ausglühen, brennen (z. B. Kalk),
　einäschern, glühen, rösten (Erze), verkalken
calcined abgeschwelt, geglüht, abgebrannt
calcined product Röstprodukt *n*, kalziniertes
　Produkt *n*, Rösterzeugnis *n*
calcined wine lees Weinhefenasche *f*
calciner Kalzinierofen *m*, Röster *m*, Röstofen *m*
calciner gas cleaning plant
　Röstgasreinigungsanlage *f*
calcining Kalzinieren *n*, Abbrennen *n*,
　Brennen *n*, Einäschern *n*, Rösten *n*, ~ **at**
　white heat Weißbrennen *n*
calcining apparatus Röstapparat *m*
calcining crucible Kalziniertopf *m*
calcining furnace Brennofen *m*, Kalzinierofen *m*,
　Röstofen *m*
calcining hearth Kalzinierherd *m*
calcining heat Brennwärme *f*
calcining kiln Brennofen *m*, Kalzinierofen *m*
calcining oven Äscherofen *m* (Keram)
calcining plant Röstanlage *f*
calcining process Brennprozeß *m*, Röstprozeß *m*
calcining temperature Rösttemperatur *f*
calcining test Brennprobe *f*
calcioferrite Calcioferrit *m* (Min)
calciostrontianite Calciostrontianit *m* (Min)
calcite Kalkspat *m* (Min), Atlasspat *m* (Min),
　Calcit *m* (Min), Kalzit *m* (Min), **fibrous** ~
　Faserkalk *m*
calcite interferometer Kalkspatinterferometer *n*

calcium (Symb. Ca) Calcium *n*, **metallic** ~
　Calciummetall *n*
calcium-12-hydroxy grease
　Kalk-12-Hydroxyfett *n*
calcium acetate Calciumacetat *n*, Holzkalk *m*
　(obs)
calcium acetylide Calciumcarbid *n*
calcium acetylsalicylate Calciumacetylsalicylat *n*
calcium alloy Calciumlegierung *f*
calcium aluminate Calciumaluminat *n*
calcium aluminum silicate, native ~ Margarit *m*
　(Min)
calcium aminoethylphosphonic acid
　Calciumaminoäthylphosphonsäure *f*
calcium aminoethylsulfonic acid
　Calciumaminoäthylsulfonsäure *f*
calcium antimonate, native ~ Romeit *m* (Min)
calcium arsenate Arsenikkalk *m* (Min),
　Calciumarsenat *n*
calcium arsenide Calciumarsenid *n*
calcium arsenite Calciumarsenit *n*
calcium ascorbate Calciumascorbat *n*
calcium balance Kalkhaushalt *m* (Physiol)
calcium benzoate Calciumbenzoat *n*
calcium bicarbonate
　Calciumhydrogencarbonat *n*
calcium bichromate Calciumbichromat *n*
calcium bisulfite Calciumbisulfit *n*
calcium borate Calciumborat *n*, Boraxkalk *m*,
　Borkalk *m*, **native** ~ Meyerhofferit *m* (Min),
　Rhodizit *m* (Min)
calcium bromide Calciumbromid *n*,
　Bromcalcium *n*
calcium carbide Calciumcarbid *n*
calcium carbonate Calciumcarbonat *n*,
　kohlensaures Calcium, Kalkspat *m* (Min),
　Kreide *f*, **acid** ~ Calciumhydrogencarbonat *n*,
　earthy ~ Kreidegur *f*, **precipitated** ~
　Kalkniederschlag *m*
calcium chelate Calciumchelat *n*
calcium chelating capacity
　Kalkbindevermögen *n* (Waschmittel)
calcium chelating power
　Calciumchelatbildungsvermögen *n*
calcium chlorate Calciumchlorat *n*
calcium chloride Calciumchlorid *n*,
　Chlorcalcium *n*, **containing** ~
　chlorcalciumhaltig
calcium chloride cylinder
　Calciumchlorid-Trockenturm *m*,
　Chlorcalciumzylinder *m*
calcium chloride solution Calciumchloridlösung *f*
calcium chloride tube Calciumchloridrohr *n*
calcium chromate Calciumchromat *n*
calcium-chromium garnet Chromgranat *m* (Min)
calcium citrate Calciumcitrat *n*
calcium complex grease Kalkkomplexfett *n*
calcium compound Calciumverbindung *f*,
　Kalkverbindung *f*

calcium content Calciumgehalt *m*
calcium cyanamide Calciumcyanamid *n*,
  Cyanamidcalcium *n*, Kalkstickstoff *m*
calcium cyanamide fertilizer
  Kalkstickstoffdünger *m*
calcium cyanide Calciumcyanid *n*
calcium cyclamate Calciumcyclamat *n*
calcium deficiency Kalkmangel *m* (Biol, Med)
calcium dibromobehenate Sabromin *n*
calcium dichromate Calciumbichromat *n*
calcium difluoride Calciumfluorid *n*
calcium ferricyanide Calciumferricyanid *n*
calcium ferrocyanide Calciumferrocyanid *n*
calcium fluoride Calciumfluorid *n*,
  Fluorcalcium *n*, **native** ~ Flußspat *m* (Min)
calcium fluosilicate Calciumsilicofluorid *n*,
  Kieselfluorcalcium *n*
calcium formate Calciumformiat *n*
calcium glycerophosphate
  Calciumglycerophosphat *n*
calcium grease Kalkfett *n*
calcium hardness Kalkhärte *f*
calcium hippurate Calciumhippurat *n*
calcium hydride Calciumhydrid *n*, Hydrolith *m*
calcium hydrogen carbonate
  Calciumhydrogencarbonat *n*
calcium hydrogen orthophosphate
  Dicalciumphosphat *n*
calcium hydrogen sulfite Calciumbisulfit *n*
calcium hydrosulfide Calciumhydrosulfid *n*,
  Calciumsulfhydrat *n*
calcium hydrosulfite Calciumhydrosulfit *n*
calcium hydroxide Calciumhydroxid *n*,
  gelöschter Kalk *m*
calcium hydroxide solution Kalklauge *f*,
  Kalklösung *f*
calcium hypobromite Calciumhypobromit *n*
calcium hypochlorite Calciumhypochlorit *n*,
  Bleichkalk *m*
calcium hypophosphite Calciumhypophosphit *n*
calcium iodide Calciumjodid *n*, Jodcalcium *n*
calcium iodobehenate Sajodin *n*
calcium lactate Calciumlactat *n*
calcium level Calciumspiegel *m*
calcium light Kalklicht *n*
calcium lignosulfate Calciumlignosulfat *n*
calcium malonate Calciummalonat *n*
calcium malonate chelate
  Calciummalonsäurechelat *n*
calcium metal Calciummetall *n*
calcium metaphosphate Calciummetaphosphat *n*
calcium metasilicate Calciummetasilikat *n*
calcium methylamine diacetate chelate
  Calciummethylamindiessigsäurechelat *n*
calcium monosulfide Calciummonosulfid *n*,
  Einfachschwefelcalcium *n*
calcium naphtholsulfonate
  Calciumnaphtholsulfonat *n*

calcium nitrate Calciumnitrat *n*, Kalksalpeter *m*,
  Norgesalpeter *m*
calcium nitrate efflorescence Salpeterschaum *m*
calcium nitride Calciumnitrid *n*,
  Stickstoffcalcium *n*
calcium nitrite Calciumnitrit *n*
calcium nitroacetate Calciumnitroessigsäure *f*
calcium orthoplumbate Calciumorthoplumbat *n*
calcium orthosilicate Calciumorthosilikat *n*
calcium oxalate Calciumoxalat *n*
calcium oxide Calciumoxid *n*, gebrannter
  Ätzkalk *m*, Kalkerde *f*
calcium pantothenate Calciumpantothenat *n*
calcium perborate Calciumperborat *n*
calcium permanganate Calciumpermanganat *n*
calcium peroxide Calciumperoxid *n*,
  Calciumsuperoxid *n*
calcium phenate Carbolkalk *m*
calcium phenolsulfonate
  Calciumphenolsulfonat *n*
calcium phenoxide Phenolcalcium *n*
calcium phosphate Calciumphosphat *n*, **impure**
  ~ Beinasche *f*, Knochenasche *f*, **neutral** ~
  Tricalciumphosphat *n*
calcium phosphide Calciumphosphid *n*,
  Phosphorkalk *m*
calcium phospholactate Calciumphospholactat *n*
calcium plumbate Calciumplumbat *n*
calcium preparation Calciumpräparat *n*
calcium pyrophosphate Calciumpyrophosphat *n*
calcium rhodanide Calciumrhodanid *n*
calcium saccharate Calciumsaccharat *n*
calcium saccharose phosphate
  Hesperonalcalcium *n*
calcium salicylate Calciumsalicylat *n*
calcium salt Calciumsalz *n*
calcium selenate Calciumselenat *n*
calcium silicate Calciumsilikat *n*
calcium silicide Calciumsilicid *n*
calcium silicofluoride Calciumsilicofluorid *n*,
  Kieselfluorcalcium *n*
calcium soap Kalkseife *f*
calcium sodium sulfate, native ~ Glauberit *m*
  (Min)
calcium stearate Calciumstearat *n*
calcium succinate Calciumsuccinat *n*
calcium succinate chelate
  Calciumbernsteinsäurechelat *n*
calcium sulfate Calciumsulfat *n*, **containing** ~
  gipshaltig
calcium sulfate dihydrate Gips *m*
calcium sulfate solution Gipslösung *f*
calcium sulfhydrate Calciumsulfhydrat *n*
calcium sulfide Calciumsulfid *n*,
  Einfachschwefelcalcium *n*,
  Kalkschwefelleber *f* (obs), Schwefelcalcium *n*
calcium sulfite Calciumsulfit *n*, **acid** ~
  Calciumbisulfit *n*

calcium thiocyanate Calciumrhodanid *n*,
Rhodancalcium *n*
calcium thiosulfate Calciumthiosulfat *n*
calcium titriplex Calciumtitriplex *n*
calcium tungstate Calciumwolframat *n*
calcoferrite Calcoferrit *m* (Min)
calcouranite Autunit *m* (Min)
calc-sinter Kalksinter *m* (Min)
calcspar Kalkspat *m* (Min)
calc-tuff Kalktuff *m* (Min)
calculability Berechenbarkeit *f*
calculable berechenbar; verläßlich
calculate *v* kalkulieren, ausrechnen, [be]rechnen,
einschätzen, ~ **a charge** gattieren, ~ **again**
nachrechnen, ~ **in advance** vorausberechnen
calculated errechnet
calculating machine Rechenmaschine *f*,
automatic ~ Rechenautomat *m*
calculating rule Rechenschieber *m* (Math),
Rechenstab *m*
calculation Kalkulation *f*, Ausrechnung *f*,
Berechnung *f*, Rechnung *f*, Schätzung *f*,
according to ~ rechnungsmäßig, **basis for** ~
Berechnungsgrundlage *f*, **error in** ~
Rechenfehler *m*, **method of** ~
Berechnungsweise *f*, **mode of** ~
Berechnungsweise *f*, **preliminary** ~
Vorkalkulation *f*, ~ **of most probable values**
Ausgleichsrechnung *f*
calculator Rechenmaschine *f*, Rechner *m*
calculus Kalkül *n* (Math)
calculus of variations Variationsrechnung *f*
caldariomycin Caldariomycin *n*
Caledonian brown Caledonischbraun *n*
caledonite Caledonit *m* (Min), Kaledonit *m*
(Min)
calefaction Erhitzung *f*, Erwärmung *f*
calefactory erwärmend
calender Kalander *m*, Glättmaschine *f*,
Glattpresse *f*, Kalandrierwalze *f*, Walzwerk *n*,
Waschsieb *n*, **glazing** ~ Glanzkalander *m*
calender *v* glätten, kalandern, satinieren
calender coater Auftragkalander *m*
calender coating Aufkalandrieren *n*,
Kalanderauftragsverfahren *n*
calender dyeing Kalanderfärbung *f*
calendering Kalandrieren *n*, Walzen *n* (Kunstst)
calendering machine Kalander *m*
calender paper Kalanderpapier *n*
calender roller Satinierwalze *f*
calender stack Glättwerk *n*
calender staining Kalanderfärbung *f*
calendulin Calendulin *n*
calescence Kaleszenz *f*, **point of** ~
Kaleszenzpunkt *m*
caliatour wood Kaliaturholz *n*
caliber Kaliber *n*, Kernmaß *n*
caliber gauge Kaliberlehre *f*

calibrate *v* kalibrieren, austarieren, auswägen,
eichen, normen, normieren
calibrated geeicht, graduiert
calibrater Eicher *m*, Eichmeister *m*
calibrating Eichen *n*, Kalibrieren *n*
calibrating pipette Kalibrierpipette *f*
calibrating plot Eichkurve *f*
calibrating ring Kalibrierring *m*
calibration Eichung *f*, Kalibrierung *f*,
Normung *f*, **certificate of** ~ Eichschein *m*,
error of ~ Eichfehler *m*, Nennwertfehler *m*,
permanent ~ Eichtreue *f*
calibration constant Eichfaktor *m*
calibration error Eichfehler *m*
calibration mark Eichstrich *m*
calibration office Eichamt *n*
calibration record Eichprotokoll *n*
calibration value Eichwert *m*
caliche Caliche *m*, roher Chilesalpeter *m*
calico Kaliko *m* (Text), Baumwolle *f* (Text),
Baumwollstoff *m* (Text), Kattun *m* (Text),
printed ~ Druckkattun *m* (Text)
calico printing Kalikodruck *m* (Text),
Kattundruck *m* (Text)
californite Californit *m* (Min)
californium (Symb. Cf) Californium *n*,
Kalifornium *n*
calin alloy Calinlegierung *f*
caliper gauge Tasterlehre *f*
calipers Dickenmesser *m*, Greifzirkel *m*,
Kaliberzirkel *m*, Rachenlehre *f*, Taster *m*
caliper slide Schieblehre *f*, Schublehre *f*
caliper square Schublehre *f*
calisaya bark Calisayarinde *f*, Chinarinde *f*
calk *v* kalfatern, abdichten
calked joint Stemmfuge *f*
calking chisel Fülleisen *n*
calking tool Stemmsetze *f*
call *v* rufen, benennen
callaite Kallait *m* (Min)
callicrein Kallikrein *n* (Biochem)
callilite Kallilith *m* (Min)
cal[l]iper *v* abtasten (Techn)
callistephin[e] Callistephin *n*
callitrol Callitrol *n*
callochrome Kallochrom *m*
callopisminic acid Callopisminsäure *f*
callus Knochensubstanz *f*
calm still, ruhig
calmeghine Kalmeghin *n*
calming agent Beruhigungsmittel *n* (Metall)
calmonal Calmonal *n*
calmopyrine Kalmopyrin *n*
calnitro Kalkammonsalpeter *m* (Kunstdünger)
calomel Kalomel *n*, Calomel *n*,
Hornquecksilber *n* (Min), Mercurochlorid *n*,
Quecksilber(I)-chlorid *n*, **colloidal** ~
Calomelol *n*, **native** ~ Quecksilberhornerz *n*

calomel electrode Kalomelelektrode *f*, fiber type
~ Kalomelfaserelektrode *f*, normal ~
Standardkalomelelektrode *f*
calophyllic acid Calophyllsäure *f*
calophyllolide Calophyllolid *n*
calorescence Calorescenz *f*
caloric kalorisch
caloric content Kaloriengehalt *m*
caloric receptivity Wärmeaufnahmefähigkeit *f*
caloric unit Kalorie *f*
calorie Calorie *f*, Kalorie *f*, gram ~ kleine
Kalorie *f*, kilogram ~ große Kalorie *f*, large
~ Kilogrammkalorie, große Kalorie *f*, small
~ Grammkalorie, kleine Kalorie *f*
calorie requirement Kalorienbedarf *m*
calories, rich in ~ kalorienreich
calorific kalorisch, wärmeerzeugend
calorification Wärmeerzeugung *f* (Biol)
calorific balance Wärmebilanz *f*
calorific effect Heizwert *m*
calorific efficiency Wärmewirkungsgrad *m*
calorific power Heizkraft *f*, Heizwert *m*
calorific value Kalorienwert *m*, Brennwert *m*,
Heizkraft *f*, Heizwert *m*, Kalorienwert *m*,
Wärmewert *m*, determination of ~
Brennwertbestimmung *f*, gross ~ oberer
Heizwert (Therm), lower ~ unterer Heizwert,
net ~ unterer Heizwert, upper ~ oberer
Heizwert
calorific value determination
Heizwertbestimmung *f*
calorific value meter Heizwertmesser *m*
calorific value regulator Heizwertregler *m*
calorimeter Kalorimeter *n*, Heizwertmesser *m*,
Wärmemesser *m*
calorimeter accessories Kalorimeterzubehör *n*
calorimeter equipment Kalorimeterzubehör *n*
calorimeter tube Kalorimeterrohr *n*
calorimeter vessel Kalorimetergefäß *n*
calorimetric kalorimetrisch
calorimetric bomb Wärmemeßbombe *f*
calorimetric test Heizwertuntersuchung *f*,
kalorimetrische Bestimmung *f*
calorimetry Kalorimetrie *f*,
Brennwertbestimmung *f*, Kalorimetrie *f*,
Wärmemengenmessung *f*
calorize *v* kalorisieren
calorizing Kalorisieren *n*, Kalorisierung *f*
calory Kalorie *f*
calotropagenin Calotropagenin *n*
calotypy Kalotypie *f*
calsil Calsil *n*
calsolene oil Calsolenöl *n*
calspirine Calspirin *n*
calumba Colombowurzel *f*
calumbic acid Colombosäure *f*
caluszite Kaluszit *m* (Min)
calutron Calutron *n*, Isotopentrennungsanlage *f*
calvacin Calvacin *n*

calycanine Calycanin *n*
calycanthine Calycanthin *n*
calycanthoside Calycanthosid *n*
calycine Calycin *n*
calycopterin Calycopterin *n*
calythrone Calythron *n*
calyx Kelch *m* (Bot)
cam Daumen *m*, Daumennocke *f*, Exzenter *m*,
Hebedaumen *m*, Hebekopf *m*, Hubkurve *f*,
Knagge *f*, Nocke *f*, Nocken *m*, Steuernocke *f*
camber Ausbauchung *f*, Bombierung *f*,
Krümmung *f*, Sturz *m* (Felge)
cambering Bombieren *n* (Techn)
cambium Cambium *n* (Bot)
cambopinenic acid Cambopinensäure *f*
cambopinonic acid Cambopinonsäure *f*
cam control Nockensteuerung *f*
cam disc Daumenscheibe *f*
cam drive Nockenantrieb *m*
cam drum Daumentrommel *f*
camel hair Kamelhaar *n*
camellin Camellin *n*
camera Kamera *f* (Phot), Photoapparat *m*
cam gear Daumenantrieb *m*,
Knaggensteuerung *f*
camomile Kamille *f* (Bot), German ~ echte
Kamille (Bot), scentless ~ falsche Kamille
(Bot)
camomile oil Kamillenöl *n*
camouflage Tarnanstrich *m*, Tarnung *f*
camouflage *v* tarnen
camouflaged verschleiert
camouflage paint Tarnfarbe *f*
Campbell method Campbellverfahren *n*
campeachy tree Blauholzbaum *m*
campeachy wood Blauholz *n*, Blankholz *n*,
Blutholz *n*, Campecheholz *n*
campestadienol Campestadienol *n*
campestanol Campestanol *n*
campesterol Campesterin *n*
camphaline Camphalin *n*
camphane Camphan *n*
camphanic acid Camphansäure *f*
camphanol Camphanol *n*
camphene Camphen *n*, Kamphen *n*
camphene chlorohydrin Camphenchlorhydrin *n*
camphene glycol Camphenglykol *n*
camphene hydride Camphenhydrid *n*
camphenelauronolic acid
Camphenlauronolsäure *f*
camphenic acid Camphensäure *f*
camphenilaic acid Camphenilansäure *f*
camphenilaic aldehyde Camphenilanaldehyd *m*
camphenilane Camphenilan *n*
camphenilanic acid Camphenilansäure *f*
camphenilic acid Camphenilsäure *f*
camphenilol Camphenilol *n*
camphenilolic acid Camphenilolsäure *f*
camphenilone Camphenilon *n*

camphenilonic acid Camphenilonsäure *f*
camphenol Camphenol *n*
camphenolic acid Camphenolsäure *f*
camphenone Camphenon *n*
camphenonic acid Camphenonsäure *f*
campherol Campherol *n*
camphidine Camphidin *n*
camphidone Camphidon *n*
camphine Camphen *n*
camphocamphoric acid Camphocamphersäure *f*
camphocarboxylic acid Camphocarbonsäure *f*
camphoceenic acid Camphoceensäure *f*
camphogen Camphogen *n*
camphoglycuronic acid Camphoglykuronsäure *f*
camphoid Camphoid *n*
camphoindole Camphoindol *n*
camphoketene Camphoketen *n*
campholactone Campholacton *n*
campholamine Campholamin *n*
campholene Campholen *n*
campholene aldehyde Campholenaldehyd *m*,
　Camphon *n*
campholenic acid Campholensäure *f*
campholic acid Camphol[an]säure *f*
campholide Campholid *n*
campholonic acid Campholonsäure *f*
campholytic acid Campholytsäure *f*
camphonanic acid Camphonansäure *f*
camphone Camphon *n*
camphonene Camphonen *n*
camphonenic acid Camphonensäure *f*
camphonitrile Camphonitril *n*
camphonolic acid Camphonolsäure *f*
camphononic acid Camphononsäure *f*
camphor Campher *m*, Kampfer *m*, **containing** ~
　kampferhaltig, **dibromated** ~
　Campherdibromid *n*, **dichlorated** ~
　Campherdichlorid *n*, **flowers of** ~
　Campherblumen *pl*, **monobromated** ~
　Monobromcampher *m*
camphoraceous kampferartig
camphoraldehydic acid Campheraldehydsäure *f*
camphoramic acid Campheramidsäure *f*
camphorate Camphorat *n*, kampfersaures Salz *n*
camphorate *a* kampfersauer
camphorate *v* mit Kampfer behandeln
camphorated kampferhaltig
camphorated spirit Camphergeist *m*
camphor dibromide Campherdibromid *n*
camphor dichloride Campherdichlorid *n*
camphor distillation plant
　Campherdestillationsanlage *f*
camphorene Camphoren *n*
camphoric acid Camphersäure *f*, Kampfersäure *f*
camphoric anhydride Camphersäureanhydrid *n*
camphor-like kampferartig
camphor liniment Campherliniment *n*,
　Campheröl *n*
camphornitrilo-acid Camphernitrilsäure *f*

camphor oil Kampferöl *n*
camphor ointment Kampfersalbe *f* (Pharm)
camphorone Camphoron *n*
camphoronic acid Camphoronsäure *f*
camphor phorone Campherphoron *n*
camphor pinacone Campherpinakon *n*
camphor quinone Campherchinon *n*
camphor soap Kampferseife *f*
camphor substitute Campherersatz *m*,
　Kampferersatz *m*
camphorsulfonic acid Camphersulfonsäure *f*
camphor water Campherwasser *n*
camphor wood Campherholz *n*, Kampferholz *n*
camphor wood oil Campherholzöl *n*
camphotamide Camphotamid *n*
camphovinic campherweinsauer
camphovinic acid Campherweinsäure *f*
camphrene Camphren *n*
camphrenic acid Camphrensäure *f*
camphyl Camphyl-
camphylamine Camphylamin *n*
camphylglycol Camphylglykol *n*
camphylic acid Camphylsäure *f*
campnospermonol Campnospermonol *n*
cam press Nockenpresse *f*
camptonite Camptonit *m*
campylite Kampylit *m* (Min)
camshaft Nockenwelle *f*, Daumenwelle *f*,
　Exzenterwelle *f*, Steuerwelle *f*
cam shoe Frosch *m* (Metall)
cam wheel Nockenrad *n*, Daumenrad *n*
camwood Kamholz *n*, Camwood *n*, Gabanholz *n*,
　Kambalholz *n*, afrikanisches Sandelholz
can Blechdose *f*, Dose *f*, Kanister *m*, **beaded** ~
　gesickte Dose *f*, **bimetallic** ~ Dose *f* aus
　Stahl- und Aluminiumblech, **compartment** ~
　Dose *f* mit Einsatz, **drawn** ~ gezogene Dose *f*,
　plain ~ unlackierte Dose *f*, **seamless** ~
　nahtlose Dose *f*, **squat** ~ Dose *f* im
　Flachformat
can *v* eindosen, konservieren (in Dosen)
Canada balsam kanadischer Balsam,
　Kanadabalsam *m*
canadic acid Canadinsäure *f*
canadine Canadin *n*
canadinium hydroxide Canadiniumhydroxid *n*
canadinol Canadinol *n*
canadinolic acid Canadinolsäure *f*
canadol Canadol *n*, Kanadol *n*
canadolic acid Canadolsäure *f*
canal Graben *m*, [künstlicher] Kanal *m*
canaline Canalin *n*
canalization Kanalisation *f*, Kanalisierung *f*
canalize *v* kanalisieren
canalizing Kanalisieren *n*
canal ray discharge Kanalstrahlentladung *f*
canal rays, analysis of ~ Kanalstrahlenanalyse *f*
canal ray tube Kanalstrahlröhre *f*
cananga oil Canangaöl *n*

canarin Canarin *n*, Kanarin *n*
canary color Kanarienfarbe *f*
canary yellow kanariengelb
canavaline Canavalin *n*
canavanine Canavanin *n*
cancel *v* annullieren, aufheben (Math), kürzen (Math), löschen (Math)
cancellation Annullierung *f*, Kürzung *f* (Math)
cancelling Streichung *f*, Tilgung *f*
cancer Krebs *m* (Med), Karzinom *n* (Med), cause of ~ Krebsursache *f* (Med), combat against ~ Krebsbekämpfung *f*, growth of a ~ Krebswachstum *n*, indicative of ~ krebsverdächtig Symptom, pitch-worker's ~ teerbedingter Krebs, soot ~ rußbedingter Krebs, suspicion of ~ Krebsverdacht *m* (Med), ~ of the respiratory system Krebs *m pl* der Luftwege
canceration Krebsbildung *f* (Med)
cancer cell Krebszelle *f*
cancer control Krebsbekämpfung *f*
cancer diagnosis Krebserkennung *f* (Med)
cancer growth Krebswachstum *n*
cancerogenic karzinogen (Med), krebsbildend, krebserregend
cancerologist Krebsforscher *m*
cancerous kanzerös, krebsartig
cancerous tissue Krebsgewebe *n*
cancerous tissue fragment Krebsgewebebruchstück *n*
cancerous ulcer Krebsgeschwür *n*
cancer prevention Krebsverhütung *f* (Med)
cancer prophylaxis Krebsprophylaxe *f*, Krebsvorbeugung *f*
cancer research Krebsforschung *f*
cancer research center Krebsforschungszentrum *n*
cancer research institute Krebsforschungsinstitut *n*
cancer tissue Krebsgewebe *n*
cancriform krebsförmig, krebsähnlich
cancrinite Kankrinit *m* (Min)
cancroid krebsartig, krebsähnlich
candelilla plant Candelillapflanze *f*
candelilla wax Candelillawachs *n*
candicine Candicin *n*
candidate Bewerber *m*, Kandidat *m*
candidate for a doctor's degree Doktorand *m*
candiolin Kandiolin *n*
candite Kandit *m* (Min)
candle Kerze *f*
candle filter Filterkerze *f*
candle power Lichtstärke *f*, Kerzenstärke *f*, Lichteinheit *f*
candle stick Kerzenhalter *m*
candogenin Candogenin *n*
can dryer Zylindertrockner *m*
candy *v* kandieren
candying Kandieren *n*

cane Rohr *n* (Bot), Rohrstock *m*
cane juice Zuckerrohrsaft *m*
canella Kaneel *m*
canella bark Canellarinde *f*, Kaneelrinde *f*
canella oil Canellenöl *n*
cane mill Rohrmühle *f* (Zucker)
canescic acid Canescinsäure *f*
canescine Canescin *n*
cane sugar Rohrzucker *m*
cane sugar mill Rohrzuckerfabrik *f*
cane trash Bagasse *f*, Zuckerrohrrückstände *m pl*
canfieldite Canfieldit *m* (Min)
canister Kanister *m*
canister filter Büchsenfilter *m*
can life Topfzeit
can making industry Dosenindustrie *f*
cannabidiol Cannabidiol *n*
cannabin Cannabin *n*
cannabinol Cannabinol *n*
cannabinone Cannabinon *n*
cannabiscetin Cannabiscetin *n*
canna starch Cannastärke *f*
canned beer Dosenbier *n*
canned food Dosenkonserve *f*
canned fruit Obstkonserven *f pl*
canned milk Dosenmilch *f*
canned motor Spaltrohrmotor *m*
cannel coal Kännelkohle *f*, Cannelkohle *f*, Fackelkohle *f*, Fettkohle *f*
canner Konservenhersteller *m*
cannery Konservenfabrik *f*
canning Eindosen *n*, Einhülsen *n* (Atom)
canning factory Konservenfabrik *f*
canning industry Konservenindustrie *f*
Cannizzaro reaction Cannizzarosche Reaktion *f*
cannogenin Cannogenin *n*
can opener Büchsenöffner *m*, Dosenöffner *m*
can stability Lagerbeständigkeit *f* (Konserven), Standzeit *f* (Konserven)
cantharene Cantharen *n*
cantharenol Cantharenol *n*
cantharides tincture Cantharidentinktur *f*
cantharidin[e] Cantharidin *n*
cantharolic acid Cantharolsäure *f*
canthaxanthin Canthaxanthin *n*
canthine Canthin *n*
cantilever Auslegearm *m*, Tragarm *m*
cantilever *a* freitragend
cantilever [beam] ausladender Balken (oder freitragender)
canvas Zeltleinwand *f*, Blachenstoff *m*, Markisenstoff *m*, Stramin *m* (Text), Zeltstoff *m*, coarse ~ Sackleinwand *f*
canvas air conduit Tuchlutte *f*
canvas shoe Leinenschuh *m*
caoutchene Kautschin *n*
caoutchouc Kautschuk *m*, Gummi *n*, hardened ~ Hartkautschuk *m*
caoutchouc cement Kautschukkitt *m*

caoutchouc hose Kautschukschlauch *m*
caoutchoucin Kautschucin *n*, Kautschuköl *n*
caoutchouc milk Kautschukmilch *f*
caoutchouc paste Kautschukmasse *f*
caoutchouc solution Kautschuklösung *f*
caoutchouc substitute Kautschukersatzstoff *m*
caoutchouc tetrabromide
   Kautschuktetrabromid *n*
caoutchouc tree Kautschukbaum *m*, common ~
   Federharzbaum *m*
caoutchouc tube Kautschukschlauch *m*
caoutchouc ware Kautschukware *f*
caoutchouc waste Kautschukabfall *m*
cap Deckel *m*, Haube *f*, Kappe *f*
capability Fähigkeit *f*, Leistungsfähigkeit *f*,
   Vermögen *n*
capable fähig, ~ of absorbing absorptionsfähig,
   aufnahmefähig, ~ of being hardened or
   cemented härtbar, ~ of opening aufklappbar,
   ~ of transmitting heat diatherman
capacitance kapazitive Impedanz (Elektr),
   Kapazität *f* (Elektr), Kapazitätsreaktanz *f*
   (Elektr), Kapazitanz *f* (Elektr), kapazitive
   Reaktanz *f* (Elektr), kapazitiver
   Widerstand *m* (Elektr), Kondensanz *f*
   (Elektr), geometric ~ geometrische Kapazität,
   natural ~ Eigenkapazität *f* (Elektr)
capacitance range Kapazitätsbereich *m*
capacitive kapazitiv (Elektr)
capacitive reactance Kapazitanz *f* (Elektr),
   kapazitiver Widerstand *m* (Elektr)
capacitor Kondensator *m* (Elektr), discharge of
   a ~ Kondensatorentladung *f*
capacitor rating Kondensatorleistung *f*
capacitron Kapazitron *n*
capacity Leistungsfähigkeit *f*, Belastbarkeit *f*
   (Dest), Fähigkeit *f*, Fassungsraum *m;*
   Fassungsvermögen *n*, Inhalt *m*, Kapazität *f*
   (Elektr), Raumgröße *f*, Rauminhalt *m*,
   coefficient of ~ Kapazitätskoeffizient *m*,
   decrease of ~ Kapazitätsabnahme *f*,
   diminution of ~ Kapazitätsabnahme *f*,
   Kapazitätsschwund *m*, initial ~
   Anfangskapazität *f*, limit of ~
   Leistungsgrenze *f*, loss of ~
   Kapazitätseinbuße *f*, Kapazitätsschwund *m*,
   magnetic ~ magnetische Kapazität, maximum
   ~ Spitzenleistung *f*, nominal or rated ~
   Nennleistung *f*, of too low ~
   unterdimensioniert, range of ~
   Leistungsbereich *m*, rated ~ Nennkapazität *f*,
   total ~ Gesamterzeugung *f*, unit of ~
   Kapazitätseinheit *f*
capacity balance Kapazitätsausgleich *m*
capacity for resistance Widerstandsfähigkeit *f*
capacity index Kapazitätsindex *m*
capacity reactance Kapazitätsreaktanz *f*,
   kapazitive Reaktanz *f*
capacity test Kapazitätsprobe *f*

capacity value Kapazitätsgröße *f*
capacity voltage Kapazitätsspannung *f* (Elektr)
capauridine Capauridin *n*
capaurine Capaurin *n*
cap [closure] Verschlußkappe *f*
caperatic acid Caperatsäure *f*
cape weed Färberflechte *f* (Bot)
capillaceous kapillarförmig
capillarity Kapillarität *f*, Kapillarwirkung *f*,
   coefficient of ~ Kapillaritätskoeffizient *m*
capillary Kapillare *f*, Saugröhrchen *n*
capillary *a* kapillar, haarförmig, kapillarförmig
capillary action Kapillarwirkung *f*
capillary active kapillaraktiv, oberflächenaktiv
capillary activity Kapillaraktivität *f*
capillary affinity Kapillaraffinität *f*
capillary alum Haaralaun *m*
capillary analysis Kapillaranalyse *f*
capillary attraction Kapillaranziehung *f*,
   Kapillarität *f*
capillary bottle Kapillarfläschchen *n*
capillary column Kapillarsäule *f*
capillary combustion tube
   Verbrennungskapillare *f*
capillary condensation Kapillarkondensation *f*
capillary constant Kapillaritätskonstante *f*
capillary depression Kapillardepression *f*
capillary duct Haarkanal *m*
capillary effect Kapillarwirkung *f*
capillary electrode Kapillarelektrode *f*
capillary electrometer Kapillarelektrometer *n*
capillary fall Kapillarsenkung *f*
capillary filament Haarfaser *f* (Biol)
capillary fluid Kapillarflüssigkeit *f*
capillary force Kapillarkraft *f*
capillary gap cell Kapillarspaltzelle *f*
   (Elektrolyse)
capillary gold Haargold *n* (Min)
capillary nickel Haarnickel *n*
capillary ore Haarerz *n* (Min)
capillary permeability, reduction of ~
   Gefäßabdichtung *f*
capillary potential Kapillarpotential *n*
capillary pressure Kapillardruck *m*
capillary stress Kapillarspannung *f*
capillary tension Kapillarspannung *f*
capillary theory Kapillartheorie *f*
capillary tube Kapillare *f*, Haarröhrchen *n*,
   Kapillarrohr *n*, small ~ Kapillarröhrchen *n*,
   ~ with several bores Mehrfachkapillare *f*
capillary tube melting-point apparatus
   Schmelzpunktbestimmungsapparat *m* mit
   Kapillarröhrchen
capillary viscosimeter Kapillarviskosimeter *n*
capillene Capillen *n*
capilliform haarförmig (Bot)
capital stock Grundkapitel *n*
capnoidine Capnoidin *n*
capnoscope Kapnoskop *n*

cap nut Hutmutter *f,* Überwurfmutter *f*
caporcianite Caporcianit *m*
cappelenite Cappelinit *m* (Min)
capping Deckeln *n* (Tabletten)
cappite Kappit *m*
capraric acid Caprarsäure *f*
caprate Caprinat *n*
caprate *a* caprinsauer
capreomycin Capreomycin *n* (Pharm)
Capri blue Capriblau *n*
capric acid Caprinsäure *f,* Decylsäure *f*
caprin Caprin *n*
caprinone Caprinon *n*
caproaldehyde Capronaldehyd *m*
caproate capronsauer
caproic acid Capronsäure *f*
caproin Caproin *n,* Capronfett *n*
caprolactam Caprolactam *n*
caprolactone Caprolacton *n*
capronamide Capronamid *n*
caprone Capron *n*
capronic acid Capronsäure *f,* Hexansäure *f*
capronitrile Capronitril *n*
capronoine Capronoin *n*
capronophenone Capronophenon *n*
caproyl Caproyl-
capryl Capryl-, Hexyl-
capryl alcohol Caprylalkohol *m*
caprylaldehyde Caprylaldehyd *m*
caprylamine Caprylamin *n*
caprylate Caprylat *n*
caprylene Caprylen *n*
caprylic acid Caprylsäure *f,* Octylsäure *f*
caprylic aldehyde Octanal *n*
caprylidene Capryliden *n*
caprylone Caprylon *n*
capryloyl Capryloyl-
capsacutine Capsacutin *n*
capsaicin Capsaicin *n*
capsanthin Capsanthin *n*
capsanthone Capsanthon *n*
cap screw Kopfschraube *f,* Überwurfschraube *f,*
    square head ~ Kopfschraube *f* mit
    Zylinderkopf
capsicine Capsicin *n*
capsicol Capsicol *n*
capsicum Cayennepfeffer *m,* Schotenpfeffer *m*
capsids Capside *pl*
capsomere Capsomer *n*
capsorubin Capsorubin *n*
capsorubone Capsorubon *n*
capstan Winde *f,* Drehhaspel *f*
capstan lathe Revolverdrehbank *f*
capstan spike Stellstift *m* (Mech)
cap stopper Flaschenkappe *f*
capsular kapselartig, kapselförmig
capsularin Capsularin *n*
capsule Kapsel *f*
capsule lacquer Kapsellack *m*

captan Captan *n*
caption Titel *m,* Überschrift *f*
captol Captol *n*
capture Einfang *m,* Einfangreaktion *f*
capture *v* abfangen (Atom), [ein]fangen
capture cross section Einfangquerschnitt *m*
capture gamma rays Einfanggammastrahlen *m pl*
capture probability Einfangwahrscheinlichkeit *f*
capturing process Einfangprozeß *m*
caracurine Caracurin *n*
carajurin Carajurin *n*
caramel Karamel *m,* gebrannter Zucker *m,*
    Zuckercouleur *f,* to form ~ karamelisieren
caramelan Caramelan *n*
caramelene Caramelen *n*
caramelin Caramelin *n*
caramelization Karamelisierung *f*
caramelize *v* karamelisieren
caramel smell Karamelgeruch *m*
caramine Caramin *n*
caramiphen Caramiphen *n*
carane Caran *n*
caranine Caranin *n*
caran[n]a balsam Carannabalsam *m*
caran[n]a resin Carannagummi *n,* Carannaharz *n*
carap[a] oil Carapafett *n,* Carapaöl *n*
carat Diamantgewicht *n,* Edelsteingewicht *n,*
    Karat *n*
carat weight Karatgewicht *n*
caraway oil Kümmelöl *n*
caraway [seeds] Kümmel *m*
carbachol Carbachol[in] *n*
carballylic acid Carballylsäure *f*
carbamate Carbamat *n,* Carbaminat *n*
carbamate *a* carbaminsauer
carbamic acid Carbamidsäure *f,*
    Amidokohlensäure *f*
carbamide Carbamid *n,* Harnstoff *m*
carbamide chloride Harnstoffchlorid *n*
carbamidine Guanidin *n*
carbamido acetic acid Glykolursäure *f,*
    Hydantoinsäure *f*
carbamidobarbituric acid, 5- ~
    Pseudoharnsäure *f*
carbaminate Carbaminat *n*
carbaminate *a* carbaminsauer
carbamyl Carbamyl-, Aminoformyl *n*
carbamyl carbamic acid Allophansäure *f*
carbamyl chloride Harnstoffchlorid *n*
carbamylglycine Glykolursäure *f,*
    Hydantoinsäure *f*
carbamylguanidine Guanylharnstoff *m*
carbamyloxamic acid Oxalursäure *f*
carbamyltoluene Toluamid *n*
carbamylurea Biuret *n*
carbanil Carbanil *n*
carbanilic acid Carbanilsäure *f*
carbanilide Carbanilid *n*

carbanion Carbanion *n*, **intermediary** ~
intermediäres Carbanion, ~ **at bridgehead**
Carbanion *n* an einem Brückenkopf
(Stereochem)
carbanionic cleavage of a C-C-bond Carbanion *n*
durch Sprengung einer C-C-Bindung
(Stereochem)
carbarsone Carbarson *n*
carbazide Carbazid *n*
carbazine Carbazin *n*
carbazinic acid Carbazinsäure *f*
carbazochrome Carbazochrom *n*
carbazolate, potassium ~ Carbazolkalium *n*
carbazole Carbazol *n*, Diphenylenimid *n*
carbazole blue Carbazolblau *n*
carbazole yellow Carbazolgelb *n*
carbazotate Pikrat *n*
carbazylic acid Carbazylsäure *f*
carbene Carben *n*
carbeniate ion Carbeniat-Ion *n*
carbeniate structure Carbeniat-Struktur *f*
carbenium structure Carbenium-Struktur *f*
carbethoxy Carbäthoxy-
carbethoxycyclopentanone
Carbäthoxycyclopentanon *n*
carbethoxyl Carbäthoxyl-
carbide Karbid *n*
carbide carbon Karbidkohle *f*
carbide furnace Karbidofen *m*
carbide fusion Karbidschmelze *f*
carbide lamp Karbidlampe *f*
carbide of iron Zementit *m* (Metall)
carbide pocket Karbidnest *n*
carbide process Karbidverfahren *n*
carbide slag Karbidschlacke *f*
carbide-tipped tool Hartmetallwerkzeug *n*,
Werkzeug *n* mit Hartmetallschneide
carbide tipping Hartmetallbestückung *f*
carbimazole Carbimazol *n*
carbimide Carbimid *n*, Isocyanat *n*
carbindigo Carbindigo *n*
carbinol Methylalkohol *m*, Carbinol *n*,
Holzgeist *m*, Holzspiritus *m*, Methanol *n*
carbinoxamine Carbinoxamin *n*
carbitol Carbitol *n*
carboanhydrase Carboanhydrase *f* (Biochem)
carboazotine Carboazotin *n*
carbobenzoxy chloride Carbobenzoxychlorid *n*
carbobenzoxy derivative Carbobenzoxyderivat *n*
carbobenzoxy glutamic acid
Carbobenzoxyglutaminsäure *f*
carbobenzoxy synthesis Carbobenzoxysynthese *f*
carbocyanine Carbocyanin *n*
carbocyclic carbocyclisch, carbozyklisch
carbodiimide Carbodiimid *n*
carbodiphenylimide Carbodiphenylimid *n*
car body Karosserie *f* (Kfz)
carbodynamite Carbodynamit *n*
carboferrite Carboferrit *m* (Min)

carbohydrase Carbohydrase *f* (Biochem)
carbohydrate Kohlenhydrat *n*
carbohydrate catabolism Kohlenhydratabbau *m*
carbohydrate depot Kohlenhydratdepot *n*
carbohydrate metabolism
Kohlenhydratstoffwechsel *m*
carbohydrates, rich in ~ kohlenhydratreich
carbohydrazide Carbohydrazid *n*
carbol Carbol *n*
carbolate Phenolat *n*
carbolate *a* carbolsauer
carbolate *v* mit Carbolsäure tränken
carbol fuchsin Carbolfuchsin *n*
carbolic acid Phenol *n*, Acidum carbolicum *n*
(Lat), Carbol *n*, Carbolsäure *f*, **solution of** ~
Carbolsäurelösung *f*
carbolic acid bath Carbolsäurebad *n*
carbolic acid carboy Carbolsäureflasche *f*
carbolic acid oil Carbolöl *n*
carbolic acid soap Carbolsäureseife *f*
carbolic acid solution Phenollösung *f*
carbolic glycerol soap Carbolglyzerinseife *f*
carbolic oil Phenolöl *n*
carbolic powder Carbolpulver *n*
carbolic soap Carbolseife *f*, Karbolseife *f*
carbolignum Carbolignum *n*
carboline Carbolin *n*
carbolineum Carbolineum *n*, Karbolineum *n*,
coat of ~ Carbolineumanstrich *m*
carbolize *v* mit Carbolsäure tränken
carbolxylene Carbolxylol *n*
carbomethoxy Carbomethoxy-
carbomycin Carbomycin *n*
carbon (Symb. C) Kohlenstoff *m;* Kohle *f*
amorphous ~ amorphe Kohle, **amphoteric** ~
amphoterer Kohlenstoff; **artificial** ~
Kunstkohle *f*, **consumption of** ~
Kohlenstoffverbrauch *m*, Kohlenabbrand *m*,
containing ~ kohlenstoffhaltig; **determination
of** ~ Kohlenstoffbestimmung *f*, **fixed** ~
gebundener Kohlenstoff, **radioactive** ~
radioaktiver Kohlenstoff, **separation of** ~
Kohlenstoffausscheidung *f*, **variety of** ~
Kohlenstoffart *f*, ~ **for the Bunsen cell**
Bunsenkohle *f*
carbonaceous kohlenstoffhaltig, kohleartig,
kohlenhaltig
carbonaceous lining Kohlenfutter *n*
carbonado Carbonado *m* (Min)
carbon anode Kohleanode *f*
carbon arc Kohlebogen *m*, Kohlelichtbogen *m*
(Techn)
carbon arc lamp Kohlenbogenlampe *f*
carbon arc welding Kohlelichtbogenschweißung *f*
carbonate Carbonat *n*, Karbonat *n*, kohlensaures
Salz, **containing** ~ carbonathaltig
carbonate *a* kohlensauer (Chem)
carbonate *v* carbonisieren
carbonated kohlensäurehaltig

carbonated water Kohlensäurewasser *n*
carbonate hardness Carbonathärte *f*
carbonate ion Carbonat-Ion *n*
carbonate of copper, green ~ Schiefergrün *n*
carbonate of iron Eisenkarbonat *n*
carbonate of lime Calciumcarbonat *n*, Kreide *f*
carbonate water Carbonatwasser *n*
carbonation Carbonisierung *f,* Carbonisation *f,*
   Karbonisation *f* (Versetzen mit Kohlensäure)
carbonato chelate Carbonatochelat *n*
carbon atom Kohlenstoffatom *n*, **asymmetric** ~
   asymmetrisches Kohlenstoffatom, **terminal** ~
   endständiges Kohlenstoffatom
carbon atoms, skeleton of ~ Kohlenstoffskelett *n*
carbonato tetramine Carbonatotetramin *n*
carbon black Lampenruß *m*, Gasruß *m*,
   Lampenschwarz *n*, Ruß *m*
carbon block Kohleblock *m*
carbon boat Kohleschiffchen *n*
carbon bond Kohlenstoffbindung *f*
carbon borer Kohlenbohrer *m*
carbon brush Kohlebürste *f* (Elektr)
carbon cell Kohleelement *n*
carbon chain Kohlenstoffkette *f*
carbon clock Kohlenstoffuhr *f*
carbon compound Kohlenstoffverbindung *f*
carbon contact Kohlekontakt *m*
carbon content Kohlenstoffgehalt *m*, **total** ~
   Gesamtkohlenstoffgehalt *m*
carbon [copy] Kopie *f*
carbon crucible Kohletiegel *m*
carbon dioxide Kohlendioxid *n*,
   Kohlensäureanhydrid *n*, **assimilation of** ~
   Kohlensäureassimilation *f,* **containing** ~
   kohlensäurehaltig, **solid** ~ Trockeneis *n*
carbon dioxide assimilation
   Kohlendioxidassimilation *f*
carbon dioxide content Kohlensäuregehalt *m*
carbon dioxide fire extinguisher
   Kohlensäurelöscher *m*
carbon dioxide fittings Kohlensäurearmaturen *f*
   pl
carbon dioxide gas Kohlensäuregas *n*
carbon dioxide indicator Kohlensäureanzeiger *m*
carbon dioxide recorder or monitor
   Kohlensäureschreiber *m*
carbon dioxide separator
   Kohlensäureabscheider *m*
carbon dioxide snow Kohlendioxidschnee *m*,
   Kohlensäureschnee *m*
carbon disulfide Schwefelkohlenstoff *m*,
   Kohlenstoffdisulfid *n*, **production of** ~
   Schwefelkohlenstoffgewinnung *f*
carbon disulfide plant
   Schwefelkohlenstoffanlage *f*
carbon double bond Kohlenstoffdoppelbindung *f*
carbon electrode Kohleelektrode *f*
carboneum Carboneum *n*
carbon filament Kohlefaden *m*

carbon filament lamp Kohlefadenlampe *f,*
   Kohlenfadenlampe *f*
carbon-free kohlenstofffrei, kohlefrei
carbon halide Halogenkohlenstoff *m*
carbon holder Kohle[n]halter *m*
carbonic acid Kohlensäure *f,* **containing** ~
   kohlensäurehaltig, **salt or ester of** ~
   Carbonat *n*
carbonic acid compressor
   Kohlensäurekompressor *m*
carbonic acid current Kohlensäurestrom *m*
carbonic acid gas Kohlensäuregas *n*,
   Kohlensäuredampf *m*
carbonic acid gas drive Kohlensäureantrieb *m*
carbonic acid hardening Kohlensäurehärtung *f*
carbonic anhydrase Carboanhydrase *f*
   (Biochem)
carbonic anhydride Kohlendioxid *n*,
   Kohlensäureanhydrid *n*
carbonic ester Kohlensäureester *m*
carbonide Carbid *n*
carboniferous kohlehaltig, kohlenstoffhaltig
carboniferous limestone Kohlenkalk *m*,
   Kohlenkalkstein *m*
carboniferous sandstone Kohlensandstein *m*
   (Bergb)
carbonium ion Carbonium-Ion *n*
carbonization Inkohlung *f* (Geol),
   Karbonisation *f* (Chem), Karbonisierung *f*
   (Chem), Kohlung *f,* Trockendestillation *f*
   (Techn), Verkohlung *f,* Verkokung *f* (Techn),
   Verrußung *f* (Katalysator), **degree of** ~
   Kohlungsgrad *m*, **low-temperature** ~
   Schwelen *n*, Schwelung *f,*
   Tieftemperaturvergasung *f,* Verschwelung *f,*
   partial ~ (of coal) Halbkokung *f*
carbonization gas Schwelgas *n*
carbonization of bituminous coal
   Steinkohlendestillation *f*
carbonization [of coal] Verkokung *f*
carbonization of wood Holzverkohlung *f*
carbonization plant Kokerei[anlage] *f,*
   [low-temperature] ~ Schwelanlage *f*
carbonization process Schwelvorgang *m*
carbonization retort Verkokungsretorte *f*
carbonization zone Schwelzone *f*
carbonize *v* verkohlen, auskohlen, einsatzhärten
   (aufkohlen), inkohlen (Geol), in Kohlenstoff
   verwandeln, karbonisieren (Chem), verkoken
carbonize at low temperature *v* schwelen
carbonized ausgekohlt
carbonize under vacuum *v* abschwelen
carbonizing Verkohlen *n*, Auskohlen *n*,
   Carbonisieren *n* (Wolle), Garung *f*
carbonizing action Karbonisationswirkung *f*
carbonizing apparatus Auskohlvorrichtung *f,*
   Karbonisierapparat *m*
carbonizing bath Auskohlbad *n*,
   Karbonisierbad *n*

carbonizing chamber Schwelraum *m*
carbonizing liquor Karbonisierflüssigkeit *f*
carbonizing of rags Lumpenauskohlung *f*
carbonizing period Garungsdauer *f,*
　Garungszeit *f* (Metall)
carbonizing plant Auskohlungsanlage *f,*
　Karbonisieranlage *f*
carbonizing process Auskohlungsverfahren *n*
carbonizing space Schwelraum *m*
carbonizing stove Auskohlungsofen *m,*
　Karbonisationsofen *m,* Karbonisierofen *m*
carbonizing temperature Karbonisiertemperatur *f*
carbonizing works Karbonisieranstalt *f*
carbonizing zone Kohlungszone *f*
carbon-like kohleartig; kohlenstoffartig
carbon linkage Kohlenstoffbindung *f*
carbon metabolism Kohlenstoffkreislauf *m*
carbon monoxide Kohlen[mon]oxid *n,*
　Kohlengas *n,* formation of ~
　Kohlenmonoxidentwicklung *f*
carbon monoxide apparatus
　Kohlenmonoxidapparat *m*
carbon monoxide recorder or monitor
　Kohlenmonoxidschreiber *m*
carbon nitride Kohlenstoffnitrid *n*
carbon-nitrogen cycle
　Stickstoff-Kohlenstoffkreislauf *m*
carbon nucleus Kohlenstoffkern *m*
carbon of an arc lamp Beleuchtungskohle *f*
carbonometer Kohlensäuremesser *m* (Med)
carbon oxide Kohlenoxid *n*
carbon oxide detector Kohlenoxidanzeiger *m*
carbon oxide gas Kohlenoxidgas *n*
carbon oxide indicator Kohlenoxidanzeiger *m*
carbon oxide poisoning Kohlenoxidvergiftung *f*
carbon oxybromide Kohlenstoffoxybromid *n*
carbon oxychloride Phosgen *n,*
　Kohlenstoffoxychlorid *n*
carbon oxysulfide Kohlenoxysulfid *n*
carbon paper Kohlepapier *n*
carbon pencil Kohlestift *m*
carbon pile graphitmoderierter Reaktor *m*
carbon removal Kohlenstoffentziehung *f*
carbon replica Kohlehüllen *pl*
　(Elektronenmikroskopie)
carbon rod Kohlestab *m* (Elektr)
carbon sand mixture Kohle-Sand-Gemischsand *n*
carbon silicide Kohlenstoffsilicium *n*
carbon skeleton Kohlenstoffskelett *n*
carbon sliding piece Kohleschleifstück *n*
carbon slip bow Kohleschleifbügel *m*
carbon source Kohlenstoffquelle *f*
carbon starter Kohlenanlasser *m*
carbon steel Flußstahl *m,* Kohlenstoffstahl *m,*
　unlegierter Stahl
carbon subnitride Kohlenstoffsubnitrid *n*
carbon suboxide Kohlen[stoff]suboxid,
　Karbodicarbonyl *n*
carbon telluride Tellurkohlenstoff *m*

carbon tetrabromide Bromkohlenstoff *m,*
　Kohlenstofftetrabromid *n*
carbon tetrachloride Tetrachlorkohlenstoff *m*
carbon tetrafluoride Tetrafluorkohlenstoff *m,*
　Fluorkohlenstoff *m*
carbon tetraiodide Tetrajodkohlenstoff *m,*
　Jodkohlenstoff *m*
carbon tool steel Kohlenstoffwerkzeugstahl *m*
carbon trichloride Hexachloräthan *n,* Hexoran *n*
carbon uranium pile Urangraphitreaktor *m*
carbon welding plate Formkohleplatte *f*
carbonyl Carbonyl-
carbonylation Carbonylierung *f*
carbonyl bromide Bromphosgen *n,*
　Kohlenstoffoxybromid *n*
carbonyl chloride Phosgen *n,* Carbonylchlorid *n,*
　Kohlensäurechlorid *n,*
　Kohlenstoffoxychlorid *n*
carbonyl compound Carbonylverbindung *f*
carbonyl group Carbonylgruppe *f*
carbonyl hemoglobin Kohlenoxidhämoglobin *n*
carbonyl iodide Jodphosgen *n*
carbonyl oxygen Carbonylsauerstoff *m*
carbonyl sulfide Carbonylsulfid *n,*
　Kohlenoxysulfid *n*
carbon-zinc cell Kohlezinkelement *n*
carbopyrride Carbopyrrid *n*
carborane Carboran *n*
carborundum Karborund[um] *n,*
　Siliciumcarbid *n*
carborundum brick Karborundumstein *m*
carborundum furnace Karborundumofen *m*
carborundum manufacture
　Karborundumdarstellung *f*
carborundum mill Karborundmühle *f*
carborundum paper Schmirgelpapier *n*
carbosilane Carbosilan *n*
carbostyril Carbostyril *n*
carbothialdine Carbothialdin *n*
carbothiocyanin Carbothiocyanin *n*
carboxonium salt Carboxoniumsalz *n*
carboxy Carboxy[l]-
carboxyacetylene Propiolsäure *f*
carboxyl Carboxy[l]-
carboxylase Carboxylase *f* (Biochem)
carboxylation Carboxylierung *f*
carboxylbenzoylacetic acid
　Carboxybenzoylessigsäure *f*
carboxyl group Carboxylgruppe *f*
carboxylic acid Carbonsäure *f*
carboxyllignin Carboxyllignin *n*
carboxy methyl cellulose
　Carboxymethylcellulose *f*
carboxy methyl chloride Chloressigsäure *f*
carboxypeptidase Carboxypeptidase *f* (Biochem)
carboxyphenylacetonitrile
　Carboxyphenylacetonitril *n*
carboxy terminal amino acid C-terminale
　Aminosäure

carboy Ballon *m* (Chem), Korbflasche *f,*
  Wasserballon *m*
carboy emptier Ballonentleerer *m*
carboy filling apparatus Ballonabfüller *m*
carboy filter Ballonfilter *n*
carboy hamper Ballonkorb *m*
carboy holder Ballongestell *n*
carboy inclinator Ballonkipper *m*
carboy protective basket Ballonschutzkorb *m*
carboy pump Ballonheber *m*
carboy tipper Ballonkipper *m*
carbromal Carbromal *n*
carbuncle Karfunkel *m* (Med)
carburate *v* carburieren, karburieren
carburating Carburieren *n*
carburet *v* carburieren, karburieren
carburetion Karburierung *f,*
  Vergaseranordnung *f,* Vergasung *f*
carburetor (Br. E. carburettor) Karburator *m*
  (Chem), Vergaser *m*
carburetor engine Vergasermotor *m*
carburetor float Vergaserschwimmer *m*
carburetor float spindle Vergasernadel *f*
carburetor nozzle Vergaserdüse *f*
carburetor primer Tipper *m* des Vergasers
carburetter (Br. E.) Carburator *m*
carburet[t]ing Karburierung *f*
carburet[t]or Vergaser *m* (Auto)
carburet[t]or fuel Vergaserkraftstoff *m*
carburization Einsatzhärtung *f* durch
  Aufkohlung, Karburierung *f,* Kohlung *f*
carburization material Kohlungsstoff *m*
carburization zone Kohlungszone *f*
carburize *v* aufkohlen (Metall), karbonisieren
  (Chem), karburieren (Chem), kohlen
carburizer Aufkohlungsmittel *n*
carburizing Einsatzhärten *n,* Karburieren *n*
carburizing agent or material Einsatzpulver *n*
carburizing granulate Kohlungsgranulat *n*
carburizing salt Härtesalz *n,* Kohlungssalz *n*
carburometer Carburometer *n*
carbylamine Carbylamin *n,* Isocyanid *n,*
  Isonitril *n*
carbyl sulfate Carbylsulfat *n*
carcass Gewebeunterbau *m* (Reifen); Kadaver *m;*
  Karkasse *f* (Kfz), durability of ~
  Karkassfestigkeit *f*
carcass break Gewebebruch *m* (Reifen),
  Karkassenbruch *m* (Kfz)
carcinogen Carcinogen *n,* Krebserreger *m,*
  Krebserzeuger *m*
carcinogenesis Cancerogenese *f* (Med),
  Karzinogenese *f* (Med), Krebsentstehung *f*
  (Med)
carcinogenic karzinogen, krebsauslösend,
  krebsbildend, krebserregend, krebserzeugend
carcinogenic agent Krebserreger *m*
carcinolysis Karzinolyse *f*

carcinolytic karzinolytisch,
  krebszellenzerstörend
carcinoma Karzinom *n* (Med), Krebs *m* (Med),
  Krebsgeschwür *n* (Med), Krebsgeschwulst *n*
  (Med), formation of a ~ Krebsbildung *f*
  (Med), initial ~ Krebsherd *m* (Med)
carcinomatous cell Krebszelle *f*
carcinosis Karzinose *f,* Krebserkrankung *f*
carcinostatic krebshemmend
card Karte *f,* Kratze *f* (Wolle)
card *v* kardieren (Wolle), krempeln (Wolle),
  rauhen (Wolle)
cardamom Kardamom *m*
cardamom oil Kardamomöl *n*
cardamon Kardamom *m*
Cardan joint Kardangelenk *n*
cardanolide Cardanolid *n*
cardan shaft Gelenkwelle *f* (Techn),
  Kardanwelle *f*
cardboard Hartpappe *f,* Karton[papier] *n,*
  Pappe *f,* glazed ~ Glanzkarton *m,* paraffined
  ~ paraffinierte Pappe
cardboard articles Kartonagen *pl*
cardboard box Pappschachtel *f*
cardbord case Papphülse *f*
card catalog[ue] Kartei *f*
carded yarn Streichgarn *n* (Text)
card face Lochkartenvorderseite *f* (Comp)
card file Kartei *f,* Kartothek *f*
cardiac glycoside Herzglykosid *n*
cardiac poison Herzgift *n*
cardiac stimulant Herzanregungsmittel *n*
cardiac stimulants Cardiaka *pl* (Pharm)
cardiac valve Herzklappe *f*
cardiac vein Herzvene *f*
cardiazol Cardiazol *n*
cardinal Kardinalrot *n*
cardinal number Grundzahl *f,* Kardinalzahl *f*
cardinal point Grundpunkt *m* (Opt),
  Kardinalpunkt *m* (Opt)
card index Kartei *f,* Kartothek *f*
carding Karden *n,* Kardieren *n,* Krempeln *n*
carding cloth Kratzentuch *n*
cardiogram Kardiogramm *n* (Med)
cardioid Herzkurve *f,* Kardioide *f*
cardiolipin Cardiolipin *n*
cardiotonine Cardiotonin *n*
cardol Cardol *n*
card punch Locher *m* (Comp)
card record system Karteisystem *n*
card setting Krempeleinstellung *f*
card waste Kardierabfall *m*
card wire stitcher Kartonheftmaschine *f*
care Achtsamkeit *f,* Pflege *f,* Schonung *f,*
  Sorgfalt *f,* Vorsicht *f*
careful vorsichtig, achtsam, behutsam, mäßig,
  schonend
careless achtlos, fahrlässig, unsorgfältig
carelessness Fahrlässigkeit *f*

carene Caren n
carex root Queckenwurzel f (Bot), Sandsegge f
  (Bot)
car fittings Wagenbeschlag m
caries Karies f (Zahnmed), Zahnfäule f
carissone Carisson n
Carius tube Bombenrohr n, Einschmelzrohr n
carlic acid Carlinsäure f, Carlsäure f
carlina oil Eberwurzelöl n
carlina oxide Carlinaoxid n
carline root Eberwurzel f
car load Wagenladung f
carlosic acid Carlossäure f
Carlsbad salt Karlsbader Salz n, Sprudelsalz n
Carlsbad twin Karlsbadzwilling m (Krist)
carmalum solution Karmalaunlösung f
carmelite water Karmeliterwasser n,
  Melissengeist m
carminative Carminativum n (Pharm)
carmine Karmin n, Carmin n, Carminfarbe f, to
  dye in ~ carminrot färben
carmine a carminrot
carmine azarine Carminazarin n
carmine azarine quinone Carminazarinchinon n
carmine lake Karminlack m, Carminlack m,
  Florentiner Lack m, Pariser Lack m
carmine naphtha Carminnaphtha n
carmine paper Carminpapier n
carmine red Karminrot n
carmine red a carminrot
carmine spar Karminspat m (Min)
carminic acid Carminsäure f, Karminsäure f
carminite Karminspat m
carminone Carminon n
carminoquinone Carminochinon n
carmoisine Carmoisin n
carnallite Carnallit m (Min), Karnallit m (Min)
carnauba substitute Carnaubaersatz m
carnauba [wax] Carnaubawachs n,
  Karnaubawachs n
carnaubic acid Carnaubasäure f,
  Karnaubasäure f
carnaubyl alcohol Carnaubylalkohol m
carnegieite Carnegieit m (Min)
carnegine Carnegin n
carnelian Carneol m (Min), Karneol m (Min)
carneol Carneol m (Min)
carnine Carnin n
carnitine Carnitin n
carnitine ethyl ester Oblitin n
carnitine fatty acyl transferase
  Carnitin-Fettsäure-Acyltransferase f
  (Biochem)
carnivorous fleischfressend
carnomuscarine Carnomuscarin n
carnosidase Carnosidase f
carnosinase Carnosinase f
carnosine Carnosin n, Ignotin n

Carnot cycle Carnotscher Kreisprozeß m
  (Therm)
Carnot efficiency Carnotscher Wirkungsgrad m
carnotite Carnotit m (Min), Karnotit m (Min)
carob bean Johannisbrot n, Karobe f
carolathine Carolathin m
carolic acid Carolsäure f
carolinic acid Carolinsäure f
carone Caron n
caronic acid Caronsäure f
Caro's acid Carosche Säure f
carotene Carotin n, Mohrrübenfarbstoff m
carotenoid Carotinoid n
carotin Carotin n
carotinoid Carotinoid n
carotin yellow Carotingelb n
carotol Carotol n
carpaine Carpain n
car paint shop Autolackiererei f
carpamic acid Carpamsäure f
carpamol Carpamol n
carpet backing Grundgewebe n für Teppiche,
  Teppichgrundgewebe n
carpholite Karpholith m (Min), Strohstein m
  (Min)
carphosiderite Karphosiderit m (Min)
carpilic acid Carpilinsäure f
carpiline Carpilin n
carpogenin Carpogenin n
car polish Autopolitur f, Wagenpflegemittel n
carpyrinic acid Carpyrinsäure f
carquejol Carquéjol n
carrabiose Carrabiose f
carrag[h]een Karrag[h]eenmoos n (Bot)
carrag[h]een moss Karrag[h]eenmoos, Perlmoos n
carrara marble Carraramarmor m
carriage Wagen m, Ausschubwagen m,
  Schlitten m (Masch)
carriage frame Wagengestell n
carrier Bazillenausscheider m, Schlitten m
  (Masch), Träger m (Chem), Transporteur m,
  Überträger m (Med)
carrier beam Trägerbalken m, Trägerbündel n
carrier bolt Mitnehmerbolzen m
carrier density Trägerdichte f
carrier diffusion Trägerdiffusion f
carrier distillation Trägerdampfdestillation f
carrier electrode Trägerelektrode f
carrier foil Trägerfolie f
carrier frequency Trägerfrequenz f
carrier gas Trägergas n
carrier injection Trägerinjektion f
carrier lifetime Lebensdauer des Trägers f
carrier [material] Träger m
carrier mobility Trägerbeweglichkeit f
carrier of reaction Reaktionsträger m
carrier pin Haltestift m
carrier precipitation Trägerfällung f
carrier protein Trägerprotein n (Biochem)

carrier replenishment Trägerauffüllung *f*
carrier roller Tragrolle *f*, Transportband *n*
carrier solution Trägerlösung *f*
carrier [substance] Trägersubstanz *f*
carrier trapping Trägereinfang *m*
carrier voltage Trägerspannung *f*
car[r]otene Carotin *n*, Möhrenfarbstoff *m*
carrotin Carotin *n*
carroting Beizen *n* (Pelz), Beizung *f* (Pelz)
carroting bath Beizbad *n* (Pelz)
carrot oil Möhrenöl *n*
carrots, juice of ~ Möhrensaft *m*
carry Übertrag *m* (Comp), **complete** ~
    vollständiger Übertrag (Comp),
    self-instructed ~ autonomer Übertrag
    (Comp), **separately instructed** ~ gesteuerter
    Übertrag (Comp)
carry *v* tragen, verschleppen (Krankheit), ~
    along *v* mitführen, mitreißen, ~ **away** *v*
    abschleppen, abtragen, ~ **down** *v* mitreißen
    (Niederschlag), ~ **on** *v* weiterführen, ausüben,
    betreiben, ~ **out** *v* ausführen, bewerkstelligen,
    durchführen, ~ **over** *v* hinübertragen,
    mitreißen (Dest)
carrying arm Tragarm *m*
carrying cable Tragseil *n*
carrying capacity Belastungsfähigkeit *f*,
    Fassungsvermögen *n* (z. B. Lastwagen),
    Tragfähigkeit *f*
carrying over Übertragung *f*
carrying power Tragkraft *f*
carrying rope Tragseil *n*
carrying side Tragseite *f* (Transportbänder)
carrying strap Tragriemen *m*
carry over Endübertrag *m*
car spring Wagenfeder *f*
carthamic acid Carthamin *n*, Carthaminsäure *f*
carthamin Carthamin *n*, Carthaminsäure *f*,
    Saflorrot *n*, **extract of** ~ Carthaminextrakt *m*
carthamin tincture Carthamintinktur *f*
carthamone Carthamon *n*
carthamus paint Safranlack *m*
cartilage Knorpel *m*, Knorpelband *n*
cartilage cell Knorpelzelle *f*
cartilaginification Verknorpelung *f*
cartilaginous knorpelartig, verknorpelt, **to**
    become ~ verknorpeln
cartilaginous tissue Knorpelband *n*
carton Karton *m*
cartridge Spaltstoffhülse *f*, Spaltstoffstab *m*
cartridge filter Kerzenfilter *n*
cartridge heater Heizpatrone *f*,
    Patronenheizkörper *m*
cartridge paper Karduspapier *n*
cartridge wire Zünddraht *m* (Bergb)
carubin coffee Carubinkaffee *m*
carvacrol Carvacrol *n*
carvacrol phthalein Carvacrolphthalein *n*

carvacrol quinone Carvacrolchinon *n*
carvacrol sulfophthalein
    Carvacrolsulfophthalein *n*
carvacryl Carvacryl-
carvacrylamine Carvacrylamin *n*
carvacrylarabinoside Carvacrylarabinosid *n*
carvacrylxyloside Carvacrylxylosid *n*
carve *v* schnitzen
carvelone Carvelon *n*
carvene Carven *n*, Hesperiden *n*
carvenene Carvenen *n*
carvenol Carvenol *n*
carvenolic acid Carvenolsäure *f*
carvenolide Carvenolid *n*
carvenone Carvenon *n*
carveol Carveol *n*
carvestrene Carvestren *n*
carvine Carvin *n*
carving knife Schnitzmesser *n*,
    Tranchiermesser *n*
carviolin Carviolin *n*
carvol Carvol *n*, Carvon *n*
carvomenthane Carvomenthan *n*
carvomenthol Carvomenthol *n*
carvomenthone Carvomenthon *n*
carvone Carvol *n*, Carvon *n*
carvone borneol Carvonborneol *n*
carvone camphor Carvoncampher *m*
carvopinone Carvopinon *n*
carvotanacetone Carvotanaceton *n*
carvyl Carvyl-
carvylamine Carvylamin *n*
caryin Caryin *n*
caryinite Caryinit *m* (Min)
caryolane Caryolan *n*
caryolysine Caryolysin *n*
caryophyllane Caryophyllan *n*
caryophyllene Caryophyllen *n*
caryophyllin Caryophyllin *n*, Oleanolsäure *f*
caryophyllinic acid Caryophyllinsäure *f*
caryophyllol Caryophyllol *n*
caryoterpine Caryoterpin *n*
carzenide Carzenid *n*
cascade Kaskade *f*
cascade amplification Kaskadenverstärkung *f*
cascade connection Kaskede[nschaltung] *f*,
    Hintereinanderschaltung *f*, Reihenschaltung *f*,
    Stufenschaltung *f*
cascade control Folgeregelung *f* (Regeltechn),
    Kaskadenregelung *f*
cascade decay Kaskadenzerfall *m*
cascade demagnetization
    Kaskadenentmagnetisierung *f*
cascade detection Kaskadennachweis *m*
cascade dryer Rieseltrockner *m*
cascade generator Stufengenerator *m*
cascade method of isotope separation
    Kaskadenmethode *f* der Isotopentrennung
cascade mill Kaskadenmühle *f*

cascade of reactors Kaskade *f* von Reaktoren, Reaktorkaskade *f*
cascade preheater Kaskadenvorwärmer *m*
cascade process Kaskadenverfahren *n*, Stufenverfahren *n*
cascade rectifier Kaskadengleichrichter *m*
cascade scrubber Kaskadenwäscher *m*
cascade shower Kaskadenschauer *m*
cascade transition Kaskadenübergang *m*
cascade tube Kaskadenröhre *f*
cascading Abrollen *n* (in einer Kugelmühle)
cascara bark Cascararinde *f*
cascara sagrada Cascarasagrada *f*, Sagradarinde *f*, **extract of** ~ Sagradaextrakt *m*
cascarilla bark Cascarillarinde *f*
cascarilla extract Cascarillaextrakt *m*
cascarilla oil Cascarillaöl *n*
cascarilline Cascarillin *n*
cascarol Cascarol *n*
case Behälter *m*, Fall *m* (Umstand), Grund *m* (Ursache), Hülse *f*, Kapsel *f*, Kasten *m*, Mantel *m* (Gehäuse), **in** ~ falls, **in any** ~ auf jeden Fall *m*
case *v* einschalen, ummanteln, füttern (auskleiden), verkleiden, verrohren
casealutine Casealutin *n*
caseanine Caseanin *n*
caseation Käsebildung *f*
case-casting Hartguß *m*, Kapselguß *m*
cased butt coupling Muffenkupplung *f*
cased-in blower Kapselwerk *n*
caseharden *v* einsatzhärten, anstählen, zementieren
casehardened im Einsatz gehärtet (Met)
casehardened casting Schalenguß *m*
casehardened steel Schalengußstahl *m*
casehardening Einsatzhärten *n*, Einsatzhärtung *f*, Kastenhärtung *f*, Oberflächenhärtung *f*, Oberflächenzementierung *f*
casehardening agent Einsatzhärtemittel *n*
casehardening bath Härtebad *n*
casehardening crucible Hartgußtiegel *m*
casehardening furnace Härteofen *m*
casehardening powder Einsatzaufstreupulver *n*, Einsatz[härte]pulver *n*
casehardening steel Einsatzstahl *m*
casein Casein *n*, Kasein *n*
casein borax Kaseinborax *m*
casein coating color Caseindeckfarbe *f*
casein fiber Kaseinfaser *f*
casein formaldehyde resin Kasein-Kunsthorn *n*
casein glue Kaseinleim *m*
caseinogen Caseinogen *n*
casein ointment Kaseinsalbe *f*
casein paint Kaseinfarbe *f*
casein plastic Caseinkunststoff *m*
casein rayon Kaseinseide *f*
casein silk Kaseinseide *f* (Text)

casein wool Kaseinfaser *f*
caseous käseartig, käsig
case-shaped kapselförmig
cashew Akajoubaum *m*, Nierenbaum *m*
cashew nut Akajounuß *f*
cashew nut oil Akajoubalsam *m*, Akajouöl *n*
cashew oil Akajoubalsam *m*
cashew resin Akajouharz *n*
cashmere Kaschmir *m* (Text)
cashmere hair Kaschmirhaar *n* (Text)
cashmere wool Kaschmirwolle *f* (Text), Kaschmirhaar *n* (Text)
cashoo Catechu *n*
casimiroedine Casimiroedin *n*
casimiroic acid Casimirosäure *f*
casimiroin Casimiroin *n*
casimiroinol Casimiroinol *n*
casimiroitine Casimiroitin *n*
casing Bohrrohr *n* (Bergb), Einschalung *f*, Futter *n*, Futterrohr *n*, Gehäuse *n*, Hülse *f*, Kapsel *f*, Karkasse *f* (Kfz), Mantel *m* (Gehäuse), Umkleidungsplatte *f*, Verkleidung *f*, Verrohrung *f*
casing flange Futterrohrflansch *m*
casing iron Einschaleisen *n*
casing pipe Bohrrohr *n*
Casing process Casing-Verfahren *n* (Kunststoffe)
casing shoe Rohrschuh *m*
casing strength Karkassfestigkeit *f*
casing tube Hüllrohr *n*
cask Faß *n*, Tonne *f*
cask deposits, impure ~ Faßgelager *n*
cask fermentation Faßgärung *f*
cask hoist Faßheber *m*
casks Fustage *f*
cassaic acid Cassainsäure *f*
cassaidic acid Cassaidinsäure *f*
cassaine Cassain *n*
cassamic acid Cassaminsäure *f*
cassamine Cassamin *n*
cassanic acid Cassansäure *f*
cassava Kassawamehl *n*, Mandioka, Maniok[stärke f] *m*
cassava starch Maniokstärke *f*
Cassel blue Kasseler Blau *n*
Cassel brown Kasseler Braun *n*
Cassel green Kasseler Grün *n*, Mangangrün *n*
Cassel yellow Kasseler Gelb *n*, Chemischgelb *n*
casserole Kasserolle *f*
cassette Kassette *f* (Phot)
cassia Cassia *f*
cassia bark Cassiarinde *f*, Holzzimt *m*
cassia bud Zimtblüte *f*
cassia flask Kassiakolben *m*
cassia oil Kassiaöl *n*, Kassienblütenöl *n*, Zimtblütenöl *n*, chinesisches Zimtöl
cassia pulp Kassienmark *n*, Röhrenkassie *f*
cassic acid Cassinsäure *f*

cassinite Cassinit *m*

cassiopeium (Symb. Lu) Cassiopeium *n*,
  Lutetium *n*

cassiterite Kassiterit *m* (Min), Bergzinn[erz] *n*
  (Min), Zinnerz *n* (Min), Zinnstein *m* (Min),
  acicular ~ Nadelzinnerz *n* (Min), **fibrous ~**
  Holzzinnerz *n* (Min)

cast Guß *m*, Gußform *f*, Gußstück *n*, Gußwerk *n*,
  Wurf *m*, **ready to ~** vergießbar, **structure of ~**
  Gußgefüge *n*, **to hollow ~** hohlgießen, **capable
  of being ~** gießbar

cast *v* gießen (Gieß), werfen, **~ in brass** *v*
  gelbgießen, **~ off** *v* abketten (Strickerei),
  abstoßen, abwerfen, **~ on** *v* angießen, **~ out** *v*
  ausstoßen, ausschießen, **~ upon a core** *v*
  hohlgießen

castability Gießbarkeit *f* (Metall),
  Gießfähigkeit *f* (Metall), Vergießbarkeit *f*

castable gußfähig

cast aluminum Aluminiumguß *m*,
  Gußaluminium *n*

cast angle (of manifold die) Spritzwinkel *m*

castanite Castanit *m* (Min)

cast anode Gußanode *f*

cast basalt Schmelzbasalt *m*

cast brass Gußmessing *n*

cast bronze Gußbronze *f*

cast cold *v* kaltvergießen

cast concrete Gußbeton *m*

castelagenine Castelagenin *n*

castelamarin Castelamarin *n*

casteline Castelin *n*

castelnaudite Castelnaudit *m*

caster Gießer *m*, Laufrolle *f* (an Möbeln)

cast film Gießfolie *f* (Plast)

cast housing Gußgehäuse *n*

Castile soap Marseiller Seife *f*, Ölsodaseife *f*,
  Olivenölseife *f*

castillite Castillit *m*

casting Bewurf, Kalkverputz *m*, Formteil *n*,
  Gießen *n*; Guß *m*, Gußarbeit *f*, Gußstück *n*,
  Gußteil *n*, Gußwerk *n*, Spritzling *m*,
  Verziehen *n*, Werfen *n*; **case-hardened ~**
  Kokillenguß *m*, **cold ~** Kaltguß *m*, **continuous
  ~** Strangguß *m*, **depth of ~** Abgußtiefe *f*,
  finished ~ Fertigguß *m*, **first quality ~**
  Feinguß *m*, **kind of ~** Gußart *f*, **malleable ~**
  Temperguß *m*, **open sand ~** Herdguß *m*,
  rough ~ Rohguß *m*, **small ~** Kleinguß *m*,
  solid ~ Vollguß *m*, **spoiled ~** Fehlguß *m*,
  Kaltguß *m*, Wrackguß *m*, **tempered ~**
  schmiedbarer Guß, **~ of the pig** Gießen *n* des
  Roheisens

casting alloy Gußlegierung *f*

casting base Gießunterlage *f*

casting bed Gießbett *n*, Masselbett *n*

casting box Formkasten *m*

casting burr Gußnaht *f*

casting cement Gußmörtel *m*

casting composition Ausgußmasse *f*

casting compound Abgußmasse *f*

casting concrete Gußmörtel *m*

casting cone Gießbuckel *m*, Gußkegel *m*

casting core Gußkern *m*

casting crane Gießkran *m*

casting defect Gußfehler *m*

casting from the cupola Kupolofenguß *m*

casting funnel Gußtrichter *m*

casting gutter Gußrinne *f*

casting hole Gußloch *n*, **to fill up the ~** das
  Gußloch verstopfen

casting house Gießhalle *f*

casting in chills Kapselguß *m*

casting in crucibles Tiegelguß *m*

casting in flasks Ladenguß *m*

casting ladle Gießkelle *f*, Gießlöffel *m*,
  Gießpfanne *f*

casting machine Gießmaschine *f*

casting metal Gießmetall *n*

casting mold Gießform *f*, Gießkasten *m*

casting off Abstoßung *f*

casting pattern Gießmodell *n*, Gußmodell *n*

casting pit Gießgrube *f*

casting practice Gießtechnik *f*

casting process Gießverfahren *n*

casting resin Gießharz *n*, Schmelzharz *n*,
  Vergußharz *n*

casting roll Auftragwalze *f*

castings Gußwaren *f pl*, **set of ~** Gußsatz *m*

casting sand plant Formsandanlage *f*

casting sand testing equipment
  Formsandprüfgeräte *n pl*

casting scrap Gußschrott *m*

casting speed Gießgeschwindigkeit *f*

casting stress Gußspannung *f*

casting temperature Gießtemperatur *f*,
  Vergießtemperatur *f*

casting wax Gußwachs *n*

casting wheel Gießrad *n*, Gießtrommel *f*

cast iron Gußeisen *n*, Gießereiroheisen *n*,
  Roheisen *n*, **acidproof ~** säurebeständiger
  Guß, **charcoal hearth ~**
  Frischfeuerroheisen *n*, **heatproof ~**
  feuerbeständiger Guß, **high-strength ~**
  Festigkeitsguß *m*, **high-test or high-quality ~**
  hochwertiges Gußeisen, **malleable ~**
  Tempereisen *n*, Glühstahl *m*, schmiedbarer
  Guß, Tempergußeisen *n*, **malleable hard ~**
  weißes Gußeisen, **porous ~** undichter Guß,
  special ~ Spezialgußeisen *n*, **spongy ~**
  schwammiger Guß, **white ~** Weißguß *m*

cast-iron *a* gußeisern

cast iron brazing Graugußlöten *n*

cast iron casting Grauguß *m*, Roheisenguß *m*

cast iron chip Gußeisensplitter *m*

cast iron column Gußeisensäule *f*

cast iron crown Gußeisenkranz *m*

cast iron disc Gußeisenscheibe *f*

cat iron flange Gußeisenkranz *m*
cast iron grid Gußeisengitter *n*
cast iron housing Gußgehäuse *n*
cast iron mold Gußkokille *f*, Gußschale *f*,
  Kokille *f*
cast iron rim Gußeisenkranz *m*
cast iron shell Gußeisengehäuse *n*
cast iron splinter Gußeisensplitter *m*
cast iron turnings Guß[eisen]späne *pl*
cast iron ware Eisengußware *f*
cast lead Gußblei *n*
cast metal Gußmetall *n*, Metallguß *m*
Castner's process Castnerprozeß *m*
castor bean Castornuß *f*
castor beans Rizinussamen *m*
castoreum Bibergeil *n*, Castoreum *n*
castoric acid Castorinsäure *f*
castorin Castorin *n*, Bibergeilcampher *m*
castor oil Rizinusöl *n*, Castoröl *n*,
  Christpalmöl *n*, **dehydrated** ~ Ricinenöl *n*,
  sulfonated ~ geschwefeltes Rizinusöl
castor oil plant Rizinus *m* (Bot)
castor oil soap Rizinusölseife *f*
cast [piece] Gußstück *n*
cast resin Edelkunstharz *n*
cast sheet gegossene Folie
cast solid *v* massiv gießen
cast steel Gußstahl *m*, Stahlguß *m*, **soft** ~
  schweißbarer Gußstahl
cast steel casing Stahlgußgehäuse *n*
cast steel crucible Gußstahltiegel *m*
cast steel mountings Stahlgußarmaturen *f pl*
cast steel plate Gußstahlblech *n*
cast steel wire Gußdraht *m*
cast steel wire brush Gußstahldrahtbürste *f*
cast zinc Gußzink *n*
cat Katze *f* (Transport)
catabolic product Abbauprodukt *n* (Biol)
catabolism Abbau *m* (Physiol), abbauender
  Stoffwechsel *m* (Physiol), Katabolismus *m*
  (Physiol)
catabolite Katabolit *m* (Physiol),
  Stoffwechselendprodukt *n*
catalase Katalase *f* (Biochem)
catalase action Katalasewirkung *f*
catalase activity Katalaseaktivität *f* (Biochem)
catalog (Am. E.) Katalog *m*
catalog number Verkaufsnummer *f*
catalogue (Br. E.) Katalog *m*
catalysis Katalyse *f*, Reaktionsbeschleunigung *f*,
  contact or surface ~ Kontaktkatalyse *f*,
  heterogeneous ~ heterogene Katalyse,
  homogeneous ~ homogene Katalyse, **negative**
  ~ negative Katalyse, Reaktionshemmung *f*,
  photochemical ~ photochemische Katalyse
catalyst Katalysator *m*, Kontakt *m*,
  Kontaktmittel *n*, Reaktionsbeschleuniger *m*,

fixed ~ Feststoffkatalysator *m*, **fluid** ~
  Fließkatalysator *m*, **grid type** ~
  Hordenkontakt *m*, **mixed** ~
  Mischkatalysator *m*, **poisoning of the** ~
  Katalysatorvergiftung *f*
catalyst gauze Netzkatalysator *m*
catalyst poison Katalysatorgift *n*, Kontaktgift *n*
catalyst poisoning Kontaktvergiftung *f*
catalyst severity Kontaktbelastung *f*
catalyst space Kontaktraum *m*
catalyst surface area Katalysatoroberfläche *f*
catalytic katalytisch, kontaktwirksam
catalytic agent Katalysator *m*
catalytic furnace Kontaktofen *m*
catalytic poison Katalytgift *n*, Kontaktgift *n*
catalytic process Kontaktverfahren *n*
catalyze *v* katalysieren, katalytisch beeinflussen
catalyzer Katalysator *m*
cataphoresis Kataphorese *f*
cataplasm[a] Breiumschlag *m* (Med),
  Kataplasma *n* (Med)
cataplasm of herbs Kräuterpflaster *n*
cataracting Kugelfall *m* (in einer Kugelmühle)
catch Haken *m*, Knagge *f*, Knaggen *m*, Öse *f*
catch *v* [ab]fangen, auffangen, einspringen (z. B.
  Schloß), ~ **up** *v* aufholen
catcher Greifvorrichtung *f*
catch fire *v* sich entzünden
catch pan Auffanggefäß *n*
catchpot Auffangtasse *f*
catch spring Einschnappfeder *f*
catchword Stichwort *n*
cat cracker Schwebekrackanlage *f* (Öl)
catechin Catechin *n*, Catechinsäure *f*,
  Catechusäure *f*
catechol Brenzcatechin *n*, Catechin *n*,
  Catechusäure *f*
catecholamine Catecholamin *n*
catechol tan Katechingerbstoff *m*
catechol tannin Brenzkatechingerbstoff *m*
catechu Catechu *n*, Katechin *n* (Färb), **tannin of**
  ~ Katechugerbsäure *f*, **to dye with** ~
  cachoutieren
catechu brown Cachoubraun *n*
catechudiamine Catechudiamin *n*
catechuic acid Catechinsäure *f*, Catechusäure *f*
catechutannic acid Catechugerbsäure *f*
catechu tannin Catechugerbsäure *f*
catechu tincture Catechutinktur *f*
catechu-ulmine Catechuulmin *n*
category Kategorie *f*, Art *f*, Begriffsfach *n*,
  Gattung *f*
catena-polysulfur catena- Polyschwefel
catenarin Catenarin *n*
caterpillar Raupe *f* (Zool), Raupenfahrzeug *n*
  (Techn), Raupenkette *f* (Techn)
caterpillar dozer Raupendozer *m*

cat gold Katzenglas *n* (Min), Katzenglimmer *m* (Min), Katzengold *n* (Min)
catgut string Saite *f* (Darm)
cathartic Abführmittel *n* (Pharm), Laxativ *n* (Pharm)
cathartin Kathartin *n*
cathartomannitol Cathartomannit *m*
cathepsin Kathepsin *n*
cathine Cathin *n*
cathodal kathodisch, negativ elektrisch
cathode Kathode *f,* negative Elektrode *f,* negativer Pol, Zinkpol *m* (obs), **cold** ~ kalte Kathode, **concave** ~ Hohlkathode *f,* **directly heated** ~ direkt geheizte Kathode, **incandescent** ~ glühende Kathode, **indirectly heated** ~ indirekt geheizte Kathode, **isopotential** ~ indirekt geheizte Kathode, **plate-shaped** ~ Flächenkathode *f,* **powder** ~ Sinterkathode *f*
cathode amplifier Kathodenverstärker *m*
cathode bar Kathodenstange *f*
cathode beam Elektronenbündel *n,* Elektronenstrahl *m*
cathode bias Kathodenvorspannung *f*
cathode coating Kathodenbelag *m*
cathode copper Kathodenkupfer *n*
cathode core sheet Mutterblech *n*
cathode current Emissionsstrom *m,* Kathodenstrom *m*
cathode dark space Kathodendunkelraum *m*
cathode density Kathodendichte *f*
cathode deposit Kathodenniederschlag *m*
cathode disintegration Kathodenabnutzung *f*
cathode evaluation Kathodenberechnung *f*
cathode evaporation Kathodenzerstäubung *f*
cathode filament Kathodenfaden *m*
cathode fluorescence Kathodenfluoreszenz *f*
cathode glow Kathodenglimmlicht *n,* Kathodenleuchten *n,* Kathodenlicht *n*
cathode grid Kathodenrost *m*
cathode heating time Kathodenanheizzeit *f*
cathode luminescence Kathodenleuchten *n*
cathode mechanism Kathodenmechanismus *m*
cathode of a phototube Fotokathode *f*
cathode ray Kathodenstrahl *m*
cathode ray oscillograph Elektronenstrahloszillograph *m,* Kathoden[strahl]oszillograph *m*
cathode ray tube Bildröhre *f* (Telev), Braunsche Röhre *f,* Kathodenstrahlröhre *f*
cathode region Kathodenraum *m*
cathode resistance Kathodenwiderstand *m*
cathode space Kathodenraum *m*
cathode sputtering Kathodenzerstäubung *f*
cathode surface Kathoden[ober]fläche *f*
cathode tube Kathodenröhre *f,* ~ **with oxide-coated filament** Oxidkathodenröhre *f*
cathodic kathodisch

cathodic protection Fernschutzwirkung *f* (Korr), kathodischer Schutz *m*
cathodic sputtering Kathodenzerstäubung *f*
cathodoluminescence Kathodolumineszenz *f*
catholyte Katholyt *m,* Kathodenflüssigkeit *f*
catine Catin *n*
cation Kation *n*
cation acid Kationsäure *f*
cation-active kation[en]aktiv
cation base Kationbase *f*
cation exchange Kationenaustausch *m*
cation exchange resin Kationenaustauscher[harz] *m*
cation exchange separation Kationenaustauschtrennung *f*
cation hole Kationenlücke *f*
cationic kationaktiv
cationic detergent Invertseife *f*
cationic exchanger Kationenaustauscher *m*
cation migration Kationenwanderung *f*
cativic acid Cativinsäure *f*
catlinite Pfeifenstein *m* (Min)
catophorite Katophorit *m* (Min)
catovit Katovit *n*
Cat-Ox-process Cat-Ox-Verfahren *n*
cat's eye Augenachat *m* (Min), Katzenauge *n* (Min), Schillerquarz *m* (Min)
cat silver Katzensilber *n* (Min)
cat thyme Amberkraut *n* (Bot), Katzenkraut *n* (Bot), Moschuskraut *n* (Bot)
cattle Rinder *n,* Rindvieh *n*
cattle breeding Rinderzucht *f*
Cauchy multiplication (of convergent series) Cauchy-Multiplikation *f*
caulk *v* abdichten, kalfatern, stemmen
caulking Abdichtung *f*
caulking hammer Dichthammer *m,* Stemmhammer *m*
caulking machine Dichtmaschine *f*
caulking material Dichtungsmaterial *n,* Dichtungsmittel *n*
caulking seam Dichtungsnaht *f*
caulk welding Dichtungsschweißung *f*
caulophylline Caulophyllin *n*
caulophyllosapogenin Caulophyllosapogenin *n*
caulophyllosaponin Caulophyllosaponin *n*
caulosapogenin Caulosapogenin *n*
caulosaponin Caulosaponin *n,* Leontin *n*
cauls Preßbleche *pl*
cauprene Kaupren *n*
causality Kausalität *f*
causality principle Kausalitätsprinzip *n*
causality requirement Kausalitätsforderung *f*
cause Anlaß *m,* Anstoß *m,* Grund *m,* Ursache *f*
cause *v* anrichten, bewirken, verursachen
causeless grundlos
caustic Ätzmittel *n,* Ätzstoff *m,* Beize *f,* Beizmittel *n,* Kaustikum *n,* **resisting** ~ ätzbeständig, **to make** ~ ätzend machen

caustic *a* ätzend, beizend, kaustisch
caustic alkali Ätzalkali *n*
caustic alkaline ätzalkalisch
caustic alkaline solution Ätzalkalilösung *f*
caustic ammonia Ätzammoniak *n*
caustic baryta Bariumhydroxid *n*
caustic barytes Ätzbaryt *m*
caustic ink Ätztinte *f*
causticity Ätzkraft *f,* Beizkraft *f,* Kaustizität *f*
causticize *v* ätzend machen, kaustifizieren, kaustizieren
causticizer Kaustifizieranlage *f*
caustic lime Ätzkalk *m*
caustic liquid Ätzflüssigkeit *f*
caustic liquor Ätzlauge *f,* Beizbrühe *f,* Beizflüssigkeit *f,* Beizwasser *n,* **weak ~** Abrichtelauge *f*
caustic lye Ätzlauge *f,* Seifensiederlauge *f*
caustic magnesia Ätzmagnesia *f*
caustic poison Ätzgift *n*
caustic pot Laugenkessel *m*
caustic potash Ätzkali *n,* Ätzstein *m,* Kaliumhydroxid *n*
caustic potash plant Ätzkalianlage *f*
caustic potash solution Ätzlauge *f,* Kalilauge *f*
caustic powder Ätzpulver *n*
caustic process Ätzverfahren *n*
caustic[s] Kaustik *f,* **to destroy by ~** ausätzen
caustic salt Ätzsalz *n*
caustic soda Ätznatron *n,* Laugenstein *m,* Natriumhydroxid *n,* kaustisches Natron, **electrolysis of ~** Ätznatronelektrolyse *f*
caustic soda dust Natronstaub *m*
caustic soda lye Ätznatronlösung *f*
caustic soda melt Ätznatronschmelze *f*
caustic soda plant Ätznatronanlage *f*
caustic soda solution Ätznatronlösung *f,* Natronlauge *f*
caustic solution Ätzlauge *f,* Ätzlösung *f*
caustic stick Ätzstift *m*
caustic stone Ätzstein *m*
caustic surface Brennfläche *f,* Kaustikfläche *f*
caustic tip Kaustikspitze *f*
causticum Kaustikum *n*
caustic washing Laugenwäsche *f*
caustic water Ätzwasser *n,* Ätze *f*
caustification Kausti[fi]zierung *f*
caustify *v* kaustifizieren
caustobiolith Kaustobiolith *m*
cauterization Abbeizung *f* (Med), Ätzung *f,* Anätzung *f,* Ausätzung *f,* Kaustizierung *f,* Kauterisation *f*
cauterize *v* abätzen (Med), ätzen, anätzen, ausätzen, [ein]brennen, kauterisieren, zerfressen
cauterizing medium Ablaugmittel *n*
caution Vorsicht *f*
cautious vorsichtig, behutsam, schonend
caviar Kaviar *m*

cavitation Kavitation *f,* Hohlraumbildung *f,* Hohlsog *m* (in metals), Lunkerung *f,* **resistant to ~** kavitationsbeständig
cavity Hohlraum *m,* Aussparung *f,* Höhlung *f,* Loch *n,* Lunker *m* (Metall), Mulde *f* (Bergb), Vertiefung *f*
cavity growth Hohlraumwachstum *n*
cavity plate Matrizenplatte *f*
cavity plug Backe *f*
cavity radiation Hohlraumstrahlung *f*
cavity resonator Hohlraumresonator *m*
cavity retainer plate Gesenkplatte *f*
cayenne pepper Cayennepfeffer *m*
CDP (Abk) Cytidindiphosphat *n*
ceara rubber Cearakautschuk *m*
cease *v* nachlassen, aufhören
cease dripping *v* auströpfeln
cedar Zeder *f* (Bot)
cedar-camphor Zederncampher *m*
cedar oil Zedern[holz]öl *n*
cedar resin Zedernharz *n*
cedar wood oil Zedernholzöl *n*
cede *v* abtreten
cedrene Cedren *n*
cedrene camphor Cedrol *n,* Zederncampher *m*
cedrene guriunene Cedrengurjunen *n*
cedrenic acid Cedrensäure *f*
cedrenol Cedrenol *n*
cedrenolic acid Cedrenolsäure *f*
cedric acid Cedrinsäure *f*
cedrin Cedrin *n*
cedriret Cedriret *n*
cedrium Cedrium *n,* Zedernharz *n*
cedrol Cedrol *n,* Zederncampher *m*
cedrolic acid Cedrolsäure *f*
cedron Cedron *n*
ceiling Decke *f,* Plafond *m,* **false, drop or counter ~** abgehängte Decke, Lüftungsrasterdecke *f,* **illuminated ~** Deckeneinbauleuchten *pl*
ceiling duct Deckendurchführung *f*
ceiling fan Deckenventilator *m*
celadonite Grünerde *f* (Min), Seladonit *m* (Min)
celandine Schöllkraut *n* (Bot)
celandine oil Schöllkrautöl *n*
celastin Menyanthin *n*
celaxanthin Celaxanthin *n*
celestial mechanics Himmelsmechanik *f*
celestine Cölestin *n,* Schützit *m* (Min), Zölestin *m* (Min), **fibrous ~** Faserzölestin *m* (Min)
celestine blue Cölestinblau *n*
celestite Zölestin *m* (Min)
cell Zelle *f,* Hohlraum *m,* Küvette *f;* **aerobic ~** aerobe Zelle *f;* **anaerobic ~** anaerobe Zelle *f,* **basophilic ~** basophile Zelle, **ciliated ~** Geißelzelle *f* (Histol), **connective-tissue ~** Fibroblast *m* (Histol), **cylindrical ~** Zylinderzelle *f,* **diploid ~** diploide Zelle *f,*

flagellate ~ Geißelzelle f (Histol), **galvanic** ~
galvanische Kette, **gram-positive** ~
grampositive Zelle f, **haploid** ~ haploide
Zelle f, **light-reactive** ~ lichtempfindliche
Zelle, **malignant** ~ bösartige Zelle f, **mature**
~ reife Zelle, **photo[electric]** ~ lichtelektrische
Zelle, lichtempfindliche Zelle, **round** ~
Kugelzelle f
cell activity Zellenaktivität f
cellar Keller m
cellar floor Kellerfußboden m
cellar laboratory Kellerlaboratorium n
cellase Cellase f (Biochem), Zellase f (Biochem)
cell cavity Porenraum m
cell cover Küvettenverschluß m
cell culture Zellkultur f
cell differentiation Zelldifferenzierung f (Biol)
cell division Zellteilung f (Biol), **direct** ~
Amitose f (Biol), direkte Zellteilung, **indirect**
~ indirekte Zellteilung
cell enzyme Zellenzym n
cell formation Zellenbildung f
cell-forming zellenbildend
cell fractionation Zellfraktionierung f
cell fusion Zellverschmelzung f
cell inclusion Zelleneinschluß m
cell insulator Zellenisolator m
cell-like zellenartig
cell membrane Plasmahaut f
cellobial Cellobial n
cellobiitol Cellobiit m
cellobionic acid Cellobionsäure f
cellobiose Cellobiose f, Cellose f
cellobiuronic acid Cellobiuronsäure f
celloidin Celloidin n, Zelloidin n
celloidin paper Zelloidinpapier n (Phot)
cellon Cellon n, Tetrachloräthan n (HN)
cellon varnish Cellonlack m
cellophane Cellophan n, Cellulosefolie f,
Zellglas n
cellose Cellobiose f, Cellose f
cellosolve Cellosolve f
cellotetrose Cellotetrose f
cellotriose Cellotriose f
cellotropin Cellotropin n
cell parasite Zellparasit m
cell pigment Zellpigment n
cell plasm Zellplasma n, Zytoplasma n
cell reaction Zellreaktion f
cell sap Zellsaft m
cell space Porenraum m
cell structure Zellenaufbau m, Zellstruktur f
cell surface Zelloberfläche f
cell switch Zellenschalter m
cell test Zellenprüfung f
cellucotton (HN) Papierwatte f
cellular porig, zellenartig, zell[en]förmig, zellig
cellular body Zellkörper m
cellular chalk Zellenkalk m

cellular concrete Porenbeton m, Zellenbeton m
cellular cooler Zellenkühler m
cellularity Zellenförmigkeit f
cellular membrane Zellmembran f
cellular method Zellenmethode f
cellular structure Zellenbau m
cellulase Cellulase f, Zellulase f
cellule kleine Zelle
cellulith Cellulith m
celluloid Celluloid n (HN), Zelluloid n (HN)
celluloid dish Celluloidschale f
celluloidine Celluloidin n
celluloid paper Celluloidpapier n
celluloid varnish Celluloidlack m, Kristalline f
cellulose Cellulose f, Holzfaserstoff m,
Zellstoff m, Zellulose f, **hydrated** ~
Hydratcellulose f, **nitrated** ~ Nitrocellulose f,
regenerated ~ Regeneratcellulose f, **without** ~
holzfrei (Pap)
cellulose acetate Acetylcellulose f,
Celluloseacetat n
cellulose acetate butyrate
Celluloseacetobutyrat n
cellulose acetate flakes Celluloseacetatflocken f
pl
cellulose acetate lacquer Cellonlack m,
Celluloseacetatlack m
cellulose acetate sheeting Celluloseacetatfolie f
cellulose constituent Cellulosebestandteil m
cellulose derivative Celluloseabkömmling m,
Cellulosederivat n
cellulose digester Zellstoffkocher m
cellulose drying Zellstofftrocknung f
cellulose ester Celluloseester m
cellulose finish Zelluloselack m
cellulose formate Formylcellulose f
cellulose hydrate Cellulosehydrat n,
Hydratcellulose f
cellulose ion exchanger
Cellulose-Ionenaustauscher m
cellulose kier Zellstoffkocher m
cellulose lacquer Zelluloselack m
cellulose nitrate Cellulosenitrat n,
Nitrozellulose f, **acetylated** ~
Acetylnitrocellulose f
cellulose nitrate filament Pyroxylinfaden m
cellulose nitrate rayon Nitroseide f, Silikin n
cellulose paper Zellstoffpapier n
cellulose powder Cellulosepulver n
cellulose pulp Zellulosebrei m
cellulose silk Zellstoffseide f (Text)
cellulose solution Zellstofflösung f
cellulose tetracetate Cellulosetetracetat n
cellulose thiocarbonate Cellulosethiocarbonat n
cellulose triacetate Cellulosetriacetat n
cellulose trinitrate Cellulosetrinitrat,
Schießbaumwolle f
cellulose wadding Zellstoffwatte f
cellulose xanthate Cellulosexanthogenat n

cellulose xanthogenate Cellulosexanthogenat *n*
cellulose yarn Zellstoffgarn *n* (Text)
cell voltage Akkumulatorspannung *f*,
Badspannung *f* (Galv)
cell wall Zellwand *f*
celosia oil Celosiaöl *n*
celsian Celsian *m* (Min)
Celsius degree Grad Celsius
celtium Celtium *n*, Hafnium *n*
celtrobiose Celtrobiose *f*
cembrene Cembren *n*
cement Zement *m*, Bindemittel *n*, Kitt *m*,
Klebstoffkitt *m*, Steinmörtel *m*, **acid-proof** ~
Säurekitt *m*, **coat of** ~ Einstrich *m*, **fireproof**
~ Feuerkitt *m*, **mixed** ~ Mischzement *m*,
natural ~ Romanzement *m*, **production of** ~
Zementerzeugung *f*, **quick-setting** ~
Schnellbinder *m*, schnell bindender Zement,
slow-setting ~ langsam bindender Zement, **to**
fasten with ~ ankitten, **to fill with** ~
auskitten, ~ **consisting of Portland cement and**
finely ground blast furnace slag
Hochofenzement *m*, **to allow** ~ **to set**
abbinden lassen
cement *v* zementieren, [an]kitten, aufkohlen
(Metall), backen (Stahl), festkitten, verkitten,
mit Zement verstreichen, ~ **on** *v* aufkitten
cementation Zementierung *f*, Aufkohlen *n*
(Metall), Brennstahlbereitung *f* (Metall),
Kohlung *f* (Metall), Oberflächenkohlung *f*
(Metall), Verkittung *f*, Zementeinspritzung *f*
cementation furnace Härteofen *m*,
Zementierofen *m*
cementation powder Härtepulver *n*,
Zementierpulver *n*
cementation process Zementieren *n*,
Einsatzhärten *n*, Glühfrischen *n*
cementation steel Zementstahl *m*
cementation vat Zementationsgefäß *n*
cement clay Kitterde *f*
cement clinker Zementklinker *m*
cement concrete Zementbeton *m*
cement copper Kupferzement *m*, Zementkupfer *n*
cement dust Zementstaub *m*
cemented zusammengekittet
cemented carbide Aufschweiß-Hartlegierung *f*,
Sinterhartmetall *n*
cemented steel Brennstahl *m*
cement factory Zementfabrik *f*
cement felt Zementfilz *m*
cement for stone Steinkitt *m*, Steinleim *m*
cement foundation Zementsockel *m*
cement grout dünnflüssiger Zement,
Zementmilch *f*
cement gun Zementspritzapparat *m*
cement industry Zementindustrie *f*
cementing Zementieren *n*, Aufgipsen *n*,
Verkitten *n*, Zusammenkitten *n*, **multiple-stage**

~ stufenweise Zementierung, ~ **by jet**
Zementieren *n* im Spritzverfahren
cementing agent Bindemittel *n*,
Zementiermittel *n*
cementing box Zementierkiste *f*, Glühtopf *m*
cementing powder Einsatzpulver *n*,
Härtepulver *n*, Zementierpulver *n*
cementing power Bindevermögen *n*
cementing property Backfähigkeit *f*
cementing water Zementwasser *n*
cementite Zementit *m* (Metall), **granular** ~
körniger Zementit, **nodular** ~ kugeliger
Zementit, **spheroidal** ~ kugeliger Zementit
cementite disintegration Zementitzerfall *m*
cement kiln Zement[brenn]ofen *m*
cement-like zementartig, kittartig
cement-lime mortar Zementkalkmörtel *m*
cement metal Zementmetall *n*
cement mill Zementmühle *f*
cement mortar Zementmörtel *m*
cement silver Zementsilber *n*
cement slab Betonplatte *f*
cement slurry Zementbrei *m*
cement stone Zementmergel *m*, Zementstein *m*
cement tank Zementtrog *m*
cement-temper *v* zementhärten
cement works Zementfabrik *f*
censure *v* kritisieren, bekritteln
centaureidin Centaureidin *n*
centaureine Centaurein *n*
centaurin Centaurin *n*
centennial hundertjährig
center Mittelpunkt *m*, Knotenpunkt *m*, Mitte *f*,
Zentrum *n*, **active** ~ aktives Zentrum
(Biochem), **dead** ~ toter Punkt
center *a*, **off** ~ ausmittig, außenmittig
center *v* zentrieren, anbohren (zum Zentrieren),
auswichten, einmitten
center axis Mittelachse *f*
center bit Anbohrer *m*
centered mittenrichtig
center groove Mittelrille *f*
centering Einmitten *n*, Zentrieren *n*
centering cone Klemmkegel *m*
centering disc Zentrierscheibe *f*
centering frame Zentrierrahmen *m*,
Ausrichtrahmen *m*, Durchzugring *m*
centering lathe Spitzendrehbank *f*,
Zentrierdrehbank *f*
centering molding machine Zentriermaschine *f*
(Gieß)
centering ring Durchzugring *m*, Zentrierring *m*
center lathe Spitzendrehbank *f*
center line Mittelachse *f*, Fluchtlinie *f*,
Mittellinie *f*, Zentrallinie *f*
center-loaded zentralbelastet
center mark Ankörnung *f*, Körner *m*
center neck Mittelhals *m*

center of curvature Krümmungsmittelpunkt *m*
center of diffraction Beugungszentrum *n*
center of gravity Schwerpunkt *m*, **determination of the** ~ Schwerpunktbestimmung *f,*
eccentricity of ~ Schwerpunktsverlagerung *f,*
location of the ~ Schwerpunktslage *f,* **radius of** ~ Schwerpunktsradius *m*
center of gravity system Schwerpunktssystem *n*
center of mass Trägheitszentrum *n*
center of mass coordinate
  Schwerpunktskoordinate *f*
center of mass motion Schwerpunktsbewegung *f*
center of percussion Perkussionszentrum *n*
center of symmetry Symmetriezentrum *n*
center piece Mittelteil *m*, Zentralstück *n*
center position Mittelstellung *f*
center punch Ankörner *m*, Locheisen *n*
center-punch *v* ankörnen
center rib Mittelrippe *f*
center tap Mittelabgriff *m*
centesimal hundertteilig
centigrade Grad *m* (Celsius)
centigrade *a* hundertgradig, hundertteilig
centigrade thermometer Celsiusthermometer *n*
centigram Zentigramm *n* (Maß)
centimeter Zentimeter *n m* (Maß)
centimeter-gram-second system
  Zentimeter-Gramm-Sekunde-System *n*
  (CGS-System)
centipoise Centipoise *n* (Maß), Zentipoise *n*
  (Maß)
centistoke Centistoke *n* (Maß)
centner Zentner *m* (Maß)
central zentral, mittelständig, zentrisch
central axis Zentralachse *f*
central body Zentralkörperchen *n*, Centriole *f*
central field Zentralfeld *n*
central field approximation
  Zentralfeldnäherung *f*
central force Zentralkraft *f*
central force motion Zentralbewegung *f*
central gate Eingußrohr *n*
central heating Zentralheizung *f*
central heating system Zentralheizung *f*
central intensity Zentralintensität *f*
centralite Zentralit *n*
centralization Zentralisierung *f*
centralize *v* zentralisieren
central lubrication Zentralschmierung *f*
central nervous system Zentralnervensystem *n*
central piece Mittelstück *n*
central point Mittelpunkt *m*, Zentralpunkt *m*
central position Mittellage *f*
central processing unit Zentraleinheit *f* (Comp)
central register Zentralregister *n*
central-symmetric zentralsymmetrisch
central symmetry Zentralsymmetrie *f*
central tube Zentralrohr *n*
centre (Br. E.) Mittelpunkt *m*, Zentrum *n*

centric zentral, zentrisch, mittig
centricity zentrale Lage *f*, Zentralität *f*
centrifiner Zentrifugalsortierer *m*
centrifugal zentrifugal
centrifugal acceleration Fliehbeschleunigung *f*
  (Phys), Zentrifugalbeschleunigung *f* (Phys)
centrifugal acid pump Säurekreiselpumpe *f*
centrifugal apparatus Schleuder *f*
centrifugal applicator
  Lackschichtschleudergerät *n*
centrifugal atomizer Zentrifugalzerstäuber *m*
centrifugal ball mill Fliehkraft-Kugelmühle *f*
centrifugal belt conveyor
  Schleuderbandförderer *m*
centrifugal blower Schleudergebläse *n*,
  Zentrifugalgebläse *n*
centrifugal casting Zentrifugalguß *m*,
  Schleuderguß *m*, Schleudern *n*
centrifugal classifier Klärzentrifuge *f,*
  Kreiselsichter *m*
centrifugal compressor Kreiselkompressor *m*
centrifugal decanter Dekantierzentrifuge *f*
centrifugal disc atomizer Rotationszerstäuber *m*
centrifugal dryer Trockenzentrifuge *f*
centrifugal extractor Zentrifugalextraktor *m*
centrifugal field Zentrifugalfeld *n*
centrifugal filter Siebschleuder *f,*
  Zentrifugalfilter *m*
centrifugal force Fliehkraft *f* (Phys),
  Zentrifugalkraft *f* (Phys), **to expel by** ~
  ausschleudern
centrifugal hydroextractor Schleudertrockner *m*
centrifugal lubrication Zentrifugalschmierung *f*
centrifugal machine Schleudermaschine *f,*
  Schwungmaschine *f*
centrifugal method Zentrifugalprozeß *m*
centrifugal mill Schleudermühle *f*, Schlagmühle *f*
centrifugal mixer Schleudermischer *m*
centrifugal motion Zentrifugalbewegung *f*
centrifugal pendulum Fliehpendel *n*,
  Zentrifugalpendel *n*
centrifugal pump Kreiselpumpe *f,*
  Schleuderpumpe *f*, Zentrifugalpumpe *f*
centrifugal rag engine Scheibenholländer *m*
  (Pap)
centrifugal roll mill Pendelmühle *f,*
  Rollenmühle *f*, Walzringmühle *f*
centrifugal separator Separatorzentrifuge *f,*
  Zentrifugalabscheider *m*, Zentrifugalsichter *m*
centrifugal sieve Schwingsiebschleuder *f*
centrifugal sprayer Zentrifugalzerstäuber *m*
centrifugal strainer Zentrifugalsortierer *m*
centrifugal supercharger Kreiselgebläse *n*,
  Kreiselvorverdichter *m*
centrifugation Zentrifugation *f*
centrifuge Zentrifuge *f,* Ausschleudermaschine *f,*
  Schleuder[maschine] *f*, **air-knife**
  or suction discharge ~
  Absaug-Filterzentrifuge *f,* **baffle-ring** ~

Prallringzentrifuge *f*, **basket** ~
Korbzentrifuge *f*, **dewatering** ~
Entwässerungszentrifuge *f*, **differential volute**
~ Leitkanal-Zentrifuge *f*, **disc** ~
Tellerzentrifuge *f*, **discontinuous** ~
diskontinuierliche Zentrifuge, **double-cone** ~
Doppelkegelzentrifuge *f*, **drum of the** ~
Zentrifugentrommel *f*, **filter belt** ~
Filterbandzentrifuge *f*, **fluted screen** ~
Faltensiebzentrifuge *f*, **impulse** ~
Freistrahlzentrifuge *f*, **multi-chamber** ~
[Ring]Kammerzentrifuge *f*, **nozzle discharge** ~
Düsen-Teller-Zentrifuge *f*, **oscillating screen**
~ Schwingsiebzentrifuge *f*, **overflow** ~
Überlaufzentrifuge *f*, **raker blade** ~
Raumschaufelzentrifuge *f*, **solid bowl** ~
Vollwandzentrifuge *f*, **suspended** ~
Hänge[korb]-Zentrifuge *f*, **torsional** ~
Torsionschwing- Zentrifuge *f*, **tubular bowl** ~
Röhrenzentrifuge *f*, **decanting** ~ **or centrifugal**
decanter Dekantierzentrifuge *f*, **sedimentation**
~ **or centrifugal settler** Absetzzentrifuge *f*
centrifuge *v* zentrifugieren, [aus]schleudern
centrifuge decanter Klassierdekanter *m*
centrifuge drier Schleudertrockner *m*
centrifuge separator Trennschleuder *f*
centrifuge tray Zentrifugalboden *m* (Dest)
centrifuge tube Zentrifugenglas *n*
centriole Zentralkörperchen *n*, Centriole *f*
centripetal zentripetal
centripetal classifier Kanalradsichter *m*,
Kanalsichtrad *n*
centripetal force Zentripetalkraft *f*,
Anstrebekraft *f*
centro-dissymmetry Zentroassymmetrie *f*
centrosome Zentrosom *n* (Biol), Centriole *f*
(Biol), Zentralkörperchen *n* (Biol)
centrosymmetric zentralsymmetrisch
cephaeline Cephaelin *n*
cephalanthin Cephalanthin *n*
cephalic acid Cephalinsäure *f*
cephalin Cephalin *n*, Kephalin *n*
cephalosporin Cephalosporin *n*
ceracyanin Keracyanin *n*
ceraic acid Cerainsäure *f*
ceraine Cerain *n*
ceramet Keramikmetallgemisch *n*
ceramic keramisch
ceramic art Töpferei *f*, Töpferkunst *f*
ceramic cooling coil Tonkühlschlange *f*
ceramic goods Keramikwaren *pl*
ceramics Keramik *f*, Keramikgegenstände *pl*,
Töpferkunst *f*, **structural** ~ Baukeramik *f*
ceramic tile Kachel *f*
ceramic varnish Einbrennlack *m*
ceramide Ceramid *n*
ceramidonine Cöramidonin *n*
cerane Ceran *n*

cerargyrite Cerargyrit *m* (Min), Hornsilber *n*
(Min), Kerargyrit *m* (Min), Silberhornerz *n*
(Min)
cerasein Cerasin *n*
cerasin Cerasin *n*
cerasite Cerasit *m* (Min), Kerasit *m* (Min)
cerate Wachssalbe *f*, Cerat *n*
cerate paper Ceratpapier *n*
ceratin Ceratin *n*
ceratite Ceratit *m*
cerberin Cerberin *n*
cereal pest Getreideschädling *m*
cereals Getreide *n*, Korn *n*
cereal seed oil Getreidekeimöl *n*
cerebral blood supply Hirndurchblutung *f* (Med)
cerebric acid Cerebrinsäure *f*
cerebrin Cerebrin *n*
cerebron Cerebron *n*
cerebronic acid Cerebronsäure *f*
cerebrose Cerebrose *f*, Galaktose *f*
cerebroside Cerebrosid *n*, Galaktosid *n*
cerebrospinal fluid Markflüssigkeit *f* (Med)
cerebrosterol Cerebrosterin *n*
cerenox Cerenox *n*
ceresin Ceresin *n*
ceria Ceroxid *n*
ceric ammonium nitrate Cerammoniumnitrat *n*
ceric ammonium sulfate Cerammoniumsulfat *n*
ceric chloride Cer(IV)-chlorid *n*
ceric compound Ceriverbindung *f*,
Cer(IV)-Verbindung *f*
ceric nitrate Cer(IV)-nitrat *n*
ceric oxalate Cer(IV)-oxalat *n*
ceric oxide Cerdioxid *n*, Cerioxid *n*, Ceroxid *n*
ceric salt Cer(IV)-Salz *n*
ceric sulfate Cerisulfat *n*
cerimetric cerimetrisch
cerin Cerin *n*
cerise kirschrot
cerite Cererz *n* (Min), Cerit *m* (Min),
Kieselcerit *m* (Min)
cerium (Symb. Ce) Cer *n*, **metallic** ~ Cermetall *n*
cerium acetylide Cercarbid *n*
cerium carbide Cercarbid *n*
cerium chloride Cerchlorid *n*
cerium compound Cerverbindung *f*
cerium dioxide Cerdioxid *n*, Cerioxid *n*,
Ceroxid *n*
cerium fluoride Cerfluorid *n*
cerium(III) compound Cer(III)-Verbindung *f*,
Ceroverbindung *f*
cerium(IV) compound Ceriverbindung *f*,
Cer(IV)-Verbindung *f*
cerium(IV) oxide Cerdioxid *n*, Cerioxid *n*,
Ceroxid *n*
cerium(IV) sulfate Cerisulfat *n*
cerium metal Cermetall *n*
cerium ore Cererz *n* (Min)
cerium oxalate Ceroxalat *n*

cerium silicide Cersilicid *n*
cerium sulfate Cersulfat *n*
cerium sulfide Cersulfid *n*
cermet Keramikmetallgemisch *n*
cernuin Cernuin *n*
cernuoside Cernuosid *n*
cerolein Cerolein *n*
cerolite Kerolith *m* (Min)
ceroptene Ceropten *n*
cerosin Cerosin *n*
cerosinyl cerosate Cerosin *n*
cerotate Cerotat *n*
cerotene Ceroten *n*
cerothian Coerthian *n*
cerothiene Coerthien *n*
cerothione Coerthion *n*
cerotic acid Cerotinsäure *f,* Zerotinsäure *f*
cerotin Cerotin *n*
cerotinic acid Cerotinsäure *f*
cerotol Cerylalkohol *m*
cerous compound Cer(III)-Verbindung *f,*
    Ceroverbindung *f*
cerous salt Cer(III)-Salz *n*
ceroxonone Cöroxonon *n*
certainty Gewißheit *f,* Sicherheit *f*
certifiable anzeigepflichtig (Med)
certificate Attest *n,* Bescheinigung *f,* Zeugnis *n*
certification Beurkundung *f*
certified chemist Diplomchemiker *m*
certified milk Vorzugsmilch *f*
certify *v* beglaubigen, bescheinigen
cerulean [blue] Himmelblau *n*
cerulein Cörulein *n*
ceruleolactine Cöruleolactin *n*
ceruleum Coeruleum *n* (Färb)
cerulignol Cörulignol *n*
cerulignone Cörulignon *n*
cerulin Cerulin *n*
ceruloplasmin Caeruloplasmin *n*
cerussite Cerussit *m* (Min), Bleispat *m* (Min),
    Weißbleierz *n* (Min)
cervantite Cervantit *m* (Min), Antimonocker *m*
    (Min), Gelbantimonerz *n* (Min)
ceryl Ceryl-
ceryl alcohol Cerylalkohol *m*
ceryl cerotate Cerotin *n*
cesium (Symb. Cs) Cäsium *n*
cesium alum Cäsiumalaun *m*
cesium aluminum sulfate Cäsiumalaun *m*
cesium arc Cäsiumlichtbogen *m*
cesium azide Cäsiumazid *n*
cesium carbonate Cäsiumcarbonat *n*
cesium cell Cäsiumphotozelle *f*
cesium chloride Cäsiumchlorid *n*
cesium chloride lattice Cäsiumchloridgitter *n*
cesium compound Cäsiumverbindung *f*
cesium fluosilicate Cäsiumsilicofluorid *n*
cesium hydride Cäsiumhydrid *n*
cesium iodate Cäsiumjodat *n*

cesium iodide Cäsiumjodid *n*
cesium nitrate Cäsiumnitrat *n*
cesium peroxide Cäsiumsuperoxid *n*
cesium salicylaldehyde Cäsiumsalicylaldehyd *m*
cesium silicofluoride Cäsiumsilicofluorid *n*
cesium sulfate Cäsiumsulfat *n*
cesium sulfide Cäsiumsulfid *n*
cesol Cesol *n*
cessation of work Arbeitseinstellung *f*
cession Abtretung *f*
cetane Cetan *n,* Hexadecan *n*
cetane number Cetanzahl *f*
cetene Ceten *n,* Cetylen *n,* Hexadecylen *n*
cetin Cetin *n*
cetobemidone Cetobemidon *n*
cetoleic acid Cetoleinsäure *f*
cetraria Moosgallerte *f*
cetraric acid Cetrarsäure *f*
cetrarin Cetrarin *n,* Flechtenbitter *n*
cetrarinic acid Cetrarin *n,* Flechtenbitter *n*
cetyl Cetyl-
cetyl alcohol Cetylalkohol *m,* Äthal *n,*
    Hexadecanol *n,* Hexadecylalkohol *m*
cetylate Palmitat *n,* Palmitinsäuresalz *n*
cetylcetylate Cetin *n*
cetylene Ceten *n,* Cetylen *n*
cetylic acid Cetylsäure *f,* Palmitinsäure *f*
cetyl iodide Cetyljodid *n*
cetyl palmitate Palmitinsäurecetylester *m*
cevadic acid Sabadillsäure *f*
cevadilla Sabadillsamen *m*
cevadine Cevadin *n,* Veratrin *n*
cevagenine Cevagenin *n*
cevane Cevan *n*
cevanthridine Cevanthridin *n*
cevine Cevin *n*
cevinilic acid Cevinilsäure *f*
ceylanite Ceylanit *m* (Min)
Ceylon cinnamon Ceylonrinde *f,* Ceylonzimt *m*
Ceylon gelatin Agar-Agar *n*
ceylonite Ceylonit *m,* Eisenspinell *m* (Min)
Ceylon moss Ceylonmoos *n*
Ceylon oil Ceylonöl *n*
Ceylon ruby Ceylonrubin *m* (Min)
CFR-engine Klopfmotor *m*
c. g. s. system CGS-System *n,*
    Zentimeter-Gramm-Sekunde-System *n*
chabazite Chabasit *m* (Min)
chaconine Chaconin *n*
chacotriose Chacotriose *f*
chafe *v* scheuern, reiben
chaff Häcksel *n,* Spreu *f* [¹
chain Kette *f,* **branched** ~ verzweigte Kette *f*
    (Chem), **endless** ~ Band *n* ohne Ende, **length
    of** ~ Kettenlänge *f,* **straight** ~ gerade Kette,
    twist link ~ Kette *f pl* mit gedrehten Gliedern
chain belt endlose Kette *f*
chain branching Kettenverzweigung *f,*
    Verzweigung *f pl* von Ketten

chain breaking Kettenabbruch *m*
chain bucket elevator Kettenbecherwerk *n*
chain conveyor Kettenförderer *m*
chain cutter molding machine Kettenfräse *f*
chain drive Ketten[rad]antrieb *m*
chain drive spanner Kettenspanner *m*
chain entanglement Kettenverhakung *f*
chain formation Verkettung *f*
chain-forming kettenbildend (Bakt)
chain grate Wanderrost *m*
chain-grate stoker Schubrost *m*
chain growth Kettenwachstum *n*
chain hoist Kettenhebevorrichtung *f*,
　Kettenhebewerk *n*
chain isomerism Kettenisomerie *f*
chain lattice Fadengitter *n*
chain length Kettenlänge *f*
chain-like kettenförmig
chain link Schake *f*
chain molecule Kettenmolekül *n*
chain of carbon atoms Kohlenstoffkette *f*
chain pulley Haspelrad *n*
chain pulley block Kettenflaschenzug *m*
chain pump Kettenpumpe *f*
chain-reacting plant Kettenreaktionsanlage *f*
chain reaction Kettenreaktion *f*, **controlled ~**
　gelenkte Kettenreaktion
chain rule Kettenregel *f*
chain structure Kettenstruktur *f*
chain termination Kettenabbruch *m*
chain transfer Kettenübertragung *f*
chain transfer agent
　Kettenübertragungsreagenz *n*
chain transfer constant (in polymerization)
　Übertragungskonstante *f*
chain weighing column Kettenwiegeapparat *m*
chain wheel Kettenrad *n*
chain winch Kettenwinde *f*
chair-chair conformation
　Doppelsessel-Konformation *f* (Stereochem)
chair-form Sesselform *f* (Stereochem)
chakranine Chakranin *n*
chaksine Chaksin *n*
chaksinic acid Chaksinsäure *f*
chalcanthite Chalkanthit *m* (Min)
chalcedonic chalcedonartig
chalcedonic chalcite Chalcedonchalcit *m* (Min)
chalcedoniferous chalcedonhaltig
chalcedony Chalcedon *m* (Min), **bright red ~**
　Karneol *m* (Min)
chalcedony quartz Onyx *m* (Min)
chalcocite Chalkozit *m* (Min), Graufahlerz *n*
　(Min), Graukupfererz *n* (Min),
　Kupferglanz *m* (Min)
chalcodite Melanglimmer *m* (Min)
chalcographic chalkographisch
chalcography Chalkographie *f*
chalcolamprite Chalkolamprit *m* (Min)

chalcolite Chalkolith *m* (Min),
　Kupferuranglimmer *m* (Min),
　Kupferuranit *m* (Min)
chalcome Chalkom *n*
chalcomenite Chalkomenit *m* (Min)
chalcone Chalkon *n*
chalcophanite Chalkophanit *m* (Min)
chalcophyllite Chalkophyllit *m* (Min),
　Kupferglimmer *m* (Min)
chalcopyrite Chalkopyrit *m* (Min),
　Gelbkupfererz *n* (Min), Halbkupfererz *n*
　(Min), Kupferkies *m* (Min)
chalcose Chalcose *f*
chalcosiderite Chalkosiderit *m* (Min)
chalcosine Chalkosin *m* (Min)
chalcosphere Chalkosphäre *f*
chalcostibite Chalkostibit *m* (Min),
　Antimonkupferglanz *m* (Min),
　Kupferantimonglanz *m* (Min)
chalcotrichite Chalkotrichit *m* (Min),
　Rotkupfer *n* (Min)
chalilite Chalilith *m*
chalinasterol Chalinasterin *n*
chalk Kalk *m*, Kalkstein *m* (Min), Kreide *f*;
　Kreidestift *m*, **black ~** Rußkreide *f*, **common**
　~ Schreibkreide *f*, **containing ~** kreidehaltig,
　gray ~ Graukalk *m*, **like ~** kreideartig,
　powdered ~ Kreidemehl *n*, Kreidepulver *n*,
　siliceous ~ Kieselkreide *f*
chalk *v* ankreiden
chalk ground Kreidegrund *m*
chalkiness Kalkartigkeit *f*
chalking Kreiden *n*, Ausschwitzen *n*, kalkiger
　Beschlag
chalking resistance Abkreidefestigkeit *f*
chalking the wire Kalken *n* des Drahtes
chalk liming Äschern *n*, Äscherung *f*
chalk marl Kreidemergel *m*
chalk mill Kreidemühle *f*
chalk overlay paper Kreidepapier *n*
chalk powder Kreidepulver *n*
chalk slate Kreideschiefer *m*
chalk whiting Marmorweiß *n*
chalky kalkig, kreideartig, kreidehaltig
chalky soil Kreideboden *m* (Geol)
chalky white kreideweiß
challenge *v* anfechten, herausfordern
chalmersite Chalmersit *m* (Min)
chalybeate tartar Stahlstein *m* (Min)
chalybeate water Eisensäuerling *m*,
　Eisenwasser *n*, Stahlwasser *n*
chalybite Chalybit *m* (Min)
chamazulene Chamazulen *n*
chamber Kammer *f*, Raum *m*
chamber acid Kammersäure *f*
chamber crystal [Blei]Kammerkristall *m* (Chem)
chamber drier Kammertrockner *m*
chamber filter press Kammerfilterpresse *f*
chamber furnace Kammerofen *m*

chamber reactor Kammerreaktor *m*
chameleon Chamäleon *n* (Zool)
chameleon mineral Chamäleon *n* (Min)
chamenol Chamenol *n*
chamfer Abkantung *f,* Abschrägung *f*
chamfer *v* abfasern, abkanten, abschrägen, auskehlen, kannelieren, verjüngen
chamfered abgeschrägt, ausgekehlt
chamfered edge Abkantung *f*
chamfering Abschrägung *f,* Auskehlung *f,* Kannelierung *f*
chamfering machine Abschrägmaschine *f*
chamic acid Chamsäure *f*
chaminic acid Chaminsäure *f*
chamois Chamois *n,* Sämischleder *n*
chamois *v* sämisch gerben
chamois-dressing Sämischgerben *n*
chamoising Sämischgerben *n*
chamo[i]sit Chamosit *m* (Min)
chamois [leather] Sämischleder *n,* Chamoisleder *n,* Fensterleder *n,* Ölleder *n;* Putzleder *n,* Waschleder
c[h]amomile, large ~ Edelkamille *f* (Bot)
chamomile camphor Anthemol *n*
chamomilla alcohol Chamomillaalkohol *m*
chamomilla ester Chamomillaester *m*
chamomillol Chamomillol *n*
chamotte Schamotte *f*
chamotte crucible Schamottetiegel *m*
champacol Champacol *n*
champagne Champagner *m,* Schaumwein *m,* Sekt *m*
chance Zufall *m,* **law of** ~ Zufallsgesetz *n*
change Änderung *f,* Abwechslung *f,* Umformung *f,* Umschlag *m* (Chem), Umwandlung *f,* Veränderung *f,* Wechsel *m,* **capacity for** ~ Umwandlungsfähigkeit *f,* **dimensional** ~ Formänderung *f,* **intramolecular** ~ intramolekulare Umlagerung, **isotropic** ~ isotrope Zustandsänderung (Phys), **rate of** ~ Änderungsgeschwindigkeit *f,* **to undergo a** ~ eine Veränderung durchmachen
change *v* [ab]ändern, abwandeln, abwechseln, umändern, umwandeln, verändern, verwandeln, wechseln
changeability Veränderlichkeit *f*
changeable auswechselbar, wandelbar, unbeständig
changeable color Schillerfarbe *f*
changeable luster Schillerglanz *m*
change colors *v* schillern
change gear Wechselgetriebe *n*
change of air Luftwechsel *m*
change of matter Aggregatzustandsänderung *f*
change of state Zustandsänderung *f,* Aggregatzustandsänderung *f*
change of voltage Spannungsänderung *f* (Elektr)
change point Umschlag[s]punkt *m*

changing Veränderung *f,* Verwandeln *n,* Wechsel *m*
channel Kanal *m* (natürlicher), Abflußrinne *f,* Auskehlung *f;* Eingußkanal *m,* Frequenzband *n,* Furche *f;* Hohlkehle *f* (Masch), Kanal *m* (TV), Rinne *f*
channel *v* auskehlen, kannelieren, nuten
channel black Gasruß *m,* Kanalruß *m,* Lampenruß *m*
channel coal Ampelit *m* (Min)
channel cracking Profilrisse *pl,* Rißbildung *f* [im Profil]
channel depth Gangtiefe *f,* Gewindetiefe *f*
channeled gerieft, geriffelt, rinnenförmig
channel flow Kanalströmung *f*
channeling Bachbildung *f* (Füllkörperkolonne), Gassenbildung *f* (Füllkörperkolonne), Kanalbildung *f,* Kannelierung *f,* Strähnenbildung *f* (Dest)
channel iron U-Eisen *n*
channel mold Kanalform *f*
channel section U-Profil *n*
channel steel U-Stahl *m*
chanoclavine Chanoclavin *n*
char *v* verkohlen, versengen, **partially** ~ ankohlen
char through *v* durchschmoren (Kabel)
characteristic Charakteristik *f,* Eigenart *f,* Eigenfunktion *f,* Kennlinie *f,* Kennzeichen *n*
characteristic *a* ausgeprägt, charakteristisch, bezeichnend, arteigen
characteristically bezeichnenderweise
characteristic atom radiation Atomeigenstrahlung *f*
characteristic curve Kennlinie *f*
characteristic data or values Kenndaten *pl*
characteristic equation Kenngrößengleichung *f,* Kennliniengleichung *f*
characteristic feature Charakteristikum *n,* Unterscheidungsmerkmal *n*
characteristic [film] curve Schwärzungskurve *f* (Phot)
characteristic frequency Eigenfrequenz *f*
characteristic impedance Kennwiderstand *m,* Wellenwiderstand *m*
characteristic number Kenngröße *f,* Kennzahl *f,* Kennziffer *f*
characteristic period Eigenperiode *f*
characteristics, family of ~ Kennlinienfeld *n*
characteristics method Charakteristikenverfahren *n*
characteristic speed Eigengeschwindigkeit *f*
characteristic state Eigenzustand *m*
characteristic value Kenngröße *f,* Kennwert *m*
characterization Kenntlichmachung *f,* Kennzeichnung *f*
characterization factor Charakterisierungsfaktor *m* (Therm)
characterize *v* charakterisieren

charcoal Holzkohle *f*, Meilerkohle *f*, **ground** ~ Kohlenpulver *n*, **heating with** ~ Holzkohlenfeuerung *f*
charcoal ashes Holzkohlenasche *f*
charcoal black Holzkohlenschwärze *f*, Kohlenschwarz *n*
charcoal breeze Holzkohlenlösche *f*
charcoal burning Holzverkohlung *f*, Grubenverkohlung *f*, Kohlenbrennen *n*, ~ **in piles** Haufenverkohlung *f*
charcoal burning plant Brennerei *f* für Holzkohle
charcoal crayon Reißkohle *f*
charcoal dust Holzkohlenstaub *m*, Holzkohlenlösche *f*, Lösche *f* (obs)
charcoal filter Kohlefilter *n*
charcoal finery Feinfeuer *n*
charcoal fining process Löscharbeit *f*
charcoal fire Holzkohlenfeuer *n*
charcoal for refining sugar Zuckerkohle *f*
charcoal gas Gasogen *n*
charcoal heap Holzkohlenmeiler *m*
charcoal hearth cast iron Herdfrischroheisen *n*, Holzkohlenroheisen *n*
charcoal hearth steel Herdfrischstahl *m*, Holzkohlenfrischstahl *m*
charcoal iron Frischfeuereisen *n*, Holzkohleneisen *n*
charcoal pencil Kohlestift *m*
charcoal pig iron Holzkohlenroheisen *n*
charcoal pile Kohlenmeiler *m*
charcoal powder Holzkohlenmehl *n*, Holzkohlenpulver *n*, Holzkohlenstaub *m*, Kohlenpulver *n*
charcoal stick Kohlenstift *m*
charcoal tablets Kohletabletten *pl* (Pharm)
charcoal wood Meilerholz *n*
Chardonnet silk Chardonnetseide *f*
charge Ladegewicht *n*, Besatz *m* (Ofen), Beschickung *f* (Ofen), Beschickungsgut *n* (Ofen), Beschickungsmaterial *n* (Ofen), Beschwerung *f*, Charge *f* (Ofen), Einsatz *m* (Ofen); Einsatzmaterial *n* (Ofen), Füllgut *n* (Ofen), Ladung *f* (Elektr), Last *f*, Möller *m* (Ofen); Preis *m*, Gebühr *f*; Schuß *m* (Bergb)
average density of ~ mittlere Belastungsdichte *f* (Elektr), **change of** ~ Ladungsänderung *f*, **coefficient of** ~ Ladungskoeffizient *m*, **dead** ~ leere Gicht, **degree of** ~ Füllungsgrad *m*, **double measurement of** ~ Doppelkraftmessung *f*, **dropping of the** ~ Niedergehen *n* der Gicht, **electric** ~ elektrische Ladung, **electrostatic** ~ elektrostatische Aufladung *f*, **excess of** ~ Ladungsüberschuß *m*, **fictitious** ~ fiktive Ladung, **formal** ~ formale Ladung, **free** ~ freie Ladung, **induced** ~ induzierte Ladung, **kind of** ~ Ladungsart *f*, **loss of** ~ Ladeverlust *m*, **multiplication of** ~

Ladungsvervielfachung *f*, **negative** ~ negative Ladung, **opposite** ~ entgegengesetzte Ladung, **positive** ~ positive Ladung, **specific** ~ spezifische Ladung, **static** ~ statische Ladung, **[unit of]** ~ Ladungseinheit *f*
charge *v* belasten, anrechnen (berechnen), aufgichten (Metall), aufladen (Batterie), begichten (Metall), beladen, bepacken, beschicken (Metall), beschweren, laden (Elektr)
chargeable belastbar
charge book Schmelzbuch *n*
charge bridge Gichtbrücke *f*
charge capacity Ladekapazität *f*, Ladungsvermögen *n*
charge carrier Ladungsträger *m*
charge carrier density Ladungsträgerdichte *f*
charge carrier generation Ladungsträgerproduktion *f*
charge cavity Füllraum *m*
charge cloud Ladungswolke *f*
charge coke Satzkoks *m*
charged geladen, stromführend, **positively** ~ positiv geladen
charged atom geladenes Atom
charge density Ladedichte *f*
charge distortion Ladungsverformung *f*
charge distribution Ladungsverteilung *f*
charge door Beschickungstür *f*, Beschickungsklappe *f*
charge equalization Ladungsausgleich *m*
charge exchange Umladung *f*
charge exchange cross section Umladungsquerschnitt *m*
charge independence Ladungsunabhängigkeit *f*, Ladungsinvarianz *f*
charge independent ladungsunabhängig
charge level Beschickungshöhe *f*, Setzboden *m*
charge material Ausgangsmaterial *n*
charge multiplet Ladungsmultiplett *n*
charge number Ladungszahl *f*
charge of a limekiln Brand *m* eines Kalkofens
charge of an electron Ladung *f* eines Elektrons
charge of ore Erzgicht *f*, Erzsatz *m*
charge pig iron Einsatzroheisen *n* (Metall)
charge quantity Chargengröße *f*
charger Füller *m* (Hochofen), Ladevorrichtung *f* (Elektr), **mechanical** ~ Beschickungsanlage *f*
charge radius ratio Ladung-Radius-Verhältnis *n*
charge separation Ladungstrennung *f* (Phys), ~ **of the interface** Grenzflächenaufladung *f* (Phys)
charge shovel Einsatzschaufel *f*
charge space Füllraum *m*, Ladungsraum *m*
charge spin Ladungsspin *m*
charge stock Einsatzprodukt *n*
charge storage Ladungsspeicherung *f* (Festkörperphys)

charge supply Ladungsbereitstellung *f*
charge symmetry Ladungssymmetrie *f*
charge tank Zugabegefäß *n*
charge to mass ratio Ladungsmasseverhältnis *n*
charge-to-tap time Schmelzzeit *f*
charge transfer Ladungsaustausch *m*,
  Ladungsübergang *m* (Phys),
  Ladungsübertragung *f*
charge-transfer band Charge-Transfer-Bande *f*
charge-transfer complex
  Charge-Transfer-Komplex *m*
charge transfer factor
  Ladungsdurchtrittsfaktor *m*
charge-transfer spectrum
  Ladungsaustauschspektrum *n* (Spektr)
charge transport Ladungstransport *m*
charge wagon Beschick[ungs]wagen *m*
charging Belasten *n*, Beschicken *n* (Metall),
  Beschweren *n*, Bestürzen *n* (Metall),
  Gichten *n* (Metall), Rostbeschickung *f*
  (Metall), **method of** ~ Beschickungsweise *f*,
  sequence of ~ Beschickungsfolge *f*, **time
  required for** ~ Aufladezeit *f*, ~ **of a condenser**
  Laden *n* des Kondensators, ~ **of ores**
  Einfüllen *n* der Erze (pl), ~ **of slag**
  Einwerfen *n* von Schlacke
charging apparatus Gichtaufzug *m*
charging bin Füllzylinder *m*
charging box Beschickungsmulde *f*,
  Chargiermulde *f*, Lademulde *f*
charging bucket Chargierkübel *m*
charging capacitor Ladekondensator *m*
charging capacity Tragkraft *f*
charging cone Abrutschkegel *m*
charging connection Ladeanschluß *m*
charging crane Beschickungskran *m*,
  Chargierkran *m*, Einsetzkran *m*
charging current Aufladestrom *m*, Ladestrom *m*,
  strength of ~ Ladestromstärke *f*
charging delay Beschickungsverzögerung *f*,
  Einsatzunterbrechung *f*
charging density Ladedichte *f*
charging device Aufgabevorrichtung *f*,
  Begichtungsvorrichtung *f*,
  Beschickmaschine *f*,
  Beschickungsvorrichtung *f*,
  Chargiervorrichtung *f*
charging door Beschickungstür *f*, Einsatztür *f*,
  Füllklappe *f*
charging equipment Beschickungsanlage *f*,
  Ladevorrichtung *f* (Elektr)
charging floor Begichtungsbühne *f*,
  Beschickungsbühne *f*, Chargierbühne *f*,
  Gichtbühne *f*
charging funnel Einfülltrichter *m*
charging gallery foot plates Gichtbelag *m*
charging hole Einschüttöffnung *f* (Metall),
  Füllöffnung *f*

charging hopper Einfülltrichter *m*,
  Füllvorrichtung *f*, Füllzylinder *m*
charging machine Beschickungsmaschine *f*,
  Chargiermaschine *f*, Chargiervorrichtung *f*
charging mechanism Aufgabevorrichtung *f*
charging opening Einsatzöffnung *f*, Füllöffnung *f*
charging peel Beschickungslöffel *m*,
  Chargierlöffel *m*
charging period Beschickungsdauer *f*,
  Beschickungszeit *f*, Chargierzeit *f* (battery),
  Ladedauer *f* (der Batterie)
charging plant Begichtungsanlage *f*
charging platform Beschickungsbühne *f*,
  Chargierbühne *f*, Ladebühne *f*, Setzboden *m*
  (of a blast furnace), Gichtbühne *f*
charging potential Ladepotential *n*
charging rate Chargiergeschwindigkeit *f*,
  Chargierleistung *f*
charging rectifier Ladegleichrichter *m* (Elektr)
charging resistance Aufladewiderstand *m*
  (Kondensator), Ladewiderstand *m*
charging sample Beschickungsprobe *f*
charging set Ladeaggregat *n*
charging side Vorderseite *f* (Metall)
charging spoon Beschickungslöffel *m*,
  Chargierlöffel *m*
charging station Ladestation *f* (Batterie)
charging the anode with oxygen
  Sauerstoffbeladung *f* der Anode
charging the furnace Begichtung *f* des Ofens
charging tray Fülltablett *n*
charging voltage Ladespannung *f*
charlton white Charltonweiß *n*
charred verkohlt
charred horn Hornkohle *f*
charring Verkohlen *n*, Verkohlung *f*
charring mound Holzkohlenmeiler *m*
charring of wood Holzverkohlung *f*
chart Tabelle *f*, graphische Darstellung *f*,
  Seekarte; Übersichtstabelle *f*
chart paper Registrierpapier *n*
chart speed Diagrammvorschub *m*,
  Papiervorschub *m*, **basic** ~ Grundvorschub *m*
chase Formrahmen *m*, Matrize *f*
chase in walls (for pipes) Durchbruch *m* (Rohr)
chaser Gewindestrehler *m*, Strehler *m*
chase ring Prägerahmen *m*, Prägering *m*
chaser tooth Kammzahn *m*
chasing tool Gewindestrehler *m*, Strehler *m*
chassis Chassis *n*, Fahrgestell *n*
chassis dynamometer Rollenprüfstand *m*
chatter marks Rattermarken *pl*
chaulmoogra butter Chaulmoograbutter *f*
chaulmoogra oil Chaulmoograöl *n*
chaulmoogric acid Chaulmoograsäure *f*
chaulmosulfone Chaulmosulfon *n*
chavibetol Chavibetol *n*
chavicic acid Chavicinsäure *f*

chavicine Chavicin *n*
chavicinic acid Chavicinsäure *f*
chavicol Chavicol *n*
chavosot Chavosot *n*
chebulic acid Chebulsäure *f*
check Kontrolle *f*, Probe *f*; Scheck *m*; Viereck *n*
check *v* kontrollieren, nacheichen;
    [nach]prüfen; nachrechnen, überprüfen
check analysis Kontrollanalyse *f*
check column method Zeilensummenprobe *f*
    (Math)
checked kariert
checker *v* karieren
checkerberry oil Bergteeöl *n* (Bot)
checkered kariert, scheckig
checkered sheet Riffelblech *n*
checker valve Umsteuerventil *n*
check flask Kontrollkolben *m*
check gauge Prüfmaß *n*
checking Kontrolle *f*, Prüfung *f*, Revision *f*,
    Rißbildung *f* (Lack), Überwachung *f*
checking equipment Kontrollanlage *f*
checking plane Hemmungsebene *f*
check nut Gegenmutter *f*
check or control determination
    Kontrollbestimmung *f*
check screw Begrenzungsschraube *f*,
    Halteschraube *f*, Stellschraube *f*
check test Kontrollversuch *m*
check valve Absperrventil *n*, Rückschlagventil *n*,
    Sperrventil *n*, Steuerventil *n*
check weight Kontrollgewicht *n*
cheddite Cheddit *m* (Sprengstoff)
cheek Wange *f*, Backe *f*, Einspannbacke *f*
cheekbone Backenknochen *m*
cheese Käse *m*, Kreuzspule *f*
cheese curds Käsebutter *f*
cheese factory Käserei *f*
cheese glue Käseleim *m*
cheese spread Streichkäse *m*
cheesy käseartig, käsig
cheilanthifoline Cheilanthifolin *n*
cheiranthic acid Cheiranthussäure *f*
cheirantin Cheiranthin *n*
cheiroline Cheirolin *n*
chelalbine Chelalbin *n*
chelate Chelat *n*, bivalent metal ~ Chelat *n pl*
    zweiwertiger Metalle
chelate complex Chelatkomplex *m*
chelate ring Chelatring *m*
chelate stability Chelatstabilität *f*
chelate stability constant
    Chelatstabilitätskonstante *f*
chelate-stabilizing chelatstabilisierend
chelating agent Chelatbildner *m*,
    Komplexbildner *m*
chelation Chelatbildung *f*
chelatometry Chelatometrie *f*
chelerythrine Chelerythrin *n*

cheleutite Cheleutit *m* (Min),
    Wismutkobaltkies *m* (Min)
chelidamic acid Chelidamsäure *f*
chelidonic acid Chelidonsäure *f*
chelidonine Chelidonin *n*
chelidonium oil Schöllkrautöl *n*
chellolglucosid Chellolglucosid *n*
chemical Chemikalie *f*
chemical *a* chemisch
chemical activity chemische Aktivität *f*
chemical affinity chemische Affinität *f*
chemical [and process] engineering
    Verfahrenstechnik *f*, chemische Technik
chemical change chemische Umwandlung
chemical discharge Ätzmittel *n*
chemical displacement chemische Verschiebung *f*
chemical documentation chemische
    Literaturerfassung *f*
chemical dressing agent Appreturzusatzmittel *n*
chemical engineer Verfahrenstechniker *m*
chemical kinetics Reaktionskinetik *f*
chemical lead Feinblei *n*, Kupferfeinblei *n*
chemically combined chemisch gebunden
chemically pure chemisch rein
chemically resistant chemisch widerstandsfähig
chemical mechanism Chemismus *m*
chemical peeling Laugenschälen *n*
chemical reaction chemische Umsetzung
chemical technician Chemotechniker *m*
chemical works chemische Fabrik *f*
chemic blue Chemischblau *n*
chemic green Chemischgrün *n*
chemicking Chlorkalkbleiche *f*
chemicoceramic keramchemisch
chemico-physics Chemiephysik *f*
chemic vat Chlorkalkbad *n*
chemigraphy Chemigraphie *f*
chemiluminescence Chemilumineszenz *f*,
    chemisches Leuchten *n*
chemiluminescence continuum
    Chemilumineszenzkontinuum *n*
chemism Chemismus *m*
chemisorption Chemisorption *f*
chemist Chemiker *m*
chemistry Chemie *f*, analytical ~ analytische
    Chemie, applied ~ angewandte Chemie,
    biological ~ biologische Chemie,
    experimental ~ experimentelle Chemie,
    inorganic ~ anorganische Chemie,
    low-temperature ~ Kryochemie *f*, nuclear ~
    Kernchemie *f* (Atom), organic ~ organische
    Chemie, pathological ~ Chemopathologie *f*
    (Med), pharmaceutical ~ pharmazeutische
    Chemie, physical ~ physikalische Chemie,
    physiological ~ physiologische Chemie,
    preparative ~ präparative Chemie, structural
    ~ Strukturchemie *f*, theoretical ~
    theoretische Chemie, toxicological ~
    toxikologische Chemie

chemistry of explosives Explosivstoffchemie *f*
chemistry of salts Halochemie *f,* Salzchemie *f*
chemistry of the tissues Histochemie *f*
chemist's mortar Pulvermörser *m*
chemist's shop Apotheke *f*
chemitype Chemitypie *f*
chemocoagulation Chemokoagulation *f*
chemolithotrophy Chemilithotrophie *f*
chemoprophylaxis Chemoprophylaxe *f* (Med)
chemoreception Chemorezeption *f* (Med)
chemoreceptor Chemorezeptor *m* (Med)
chemoreflex Chemoreflex *m*
chemoresistance Chemoresistenz *f*
chemoresistant chemoresistent
chemosensitive chemosensibel
chemosensitivity Chemosensibilität *f*
chemosorption Chemisorption *f*
chemosynthesis Chemosynthese *f*
chemotaxonomy Chemotaxonomie *f*
chemotherapy Chemotherapie *f*
chemotropism Chemotropismus *m* (Med)
chemurgy Chemurgie *f*
chenevixite Chenevixit *m* (Min)
chenocholenic acid Chenocholensäure *f*
chenocholic acid Chenocholsäure *f*
chenodeoxycholic acid Chenodesoxycholsäure *f*
chequer (Br. E.) Karo *n*
chequer *v* (Br. E.) karieren
chequered (Br. E.) kariert
chequered plate (Br. E.) Riffelblech *n*
cherry Kirsche *f*
cherry brandy Kirschbranntwein *m,*
   Kirschlikör *m*
cherry coal Sinterkohle *f,* weiche Kohle *f*
cherry-colored kirschfarben
cherry gum Kirschgummi *n*
cherry juice Kirschsaft *m*
cherry laurel Kirschlorbeer *m*
cherry red Kirschrot *n*
cherry red *a* kirschrot
cherry rum Kirschrum *m*
cherry spirit Kirschbranntwein *m*
cherry syrup Kirschensirup *m*
chert Bergkiesel *m* (Min), Hornstein *m* (Min)
chert gravel Hornsteinkies *m*
cherty hornsteinhaltig
chervil oil Kerbelöl *n*
chessylite Chessylit[h] *m* (Min), Azurit *m* (Min),
   Azurstein *m* (Min), Bergasche *f* (Min),
   Kupferlasur *f*
chest Kiste *f*
chestnut Kastanie *f,* liquid extract of ~
   Kastanienfluidextrakt *m*
chestnut [brown] Kastanienbraun *n*
chestnut [brown] *a* kastanienbraun
chevilling Chevillieren *n*
chewing gum Kaugummi *m*
chewing tobacco Kautabak *m*

chiastolite Chiastolith *m* (Min), Andreolith *m*
   (Min), Hohlspat *m* (Min)
chicago acid Chicagosäure *f*
chichipegenin Chichipegenin *n*
chicken-pox Windpocken *pl* (Med)
chickenwire glass Zelldrahtglas *n*
chiclafluavil Chiclafluavil *n*
chiclalbane Chiclalban *n*
chicoric acid Chicorsäure *f*
chicory Zichorie *f* (Bot), Zichorienkraut *n* (Bot)
chicory brown Zichorienbraun *n*
chicory coffee Zichorienkaffee *m*
chicory root Wegwartwurzel *f* (Bot),
   Zichorienwurzel *f* (Bot)
chief constituent Hauptbestandteil *m*
chief laboratory Hauptlaboratorium *n*
chief portion Hauptanteil *m*
chigadmarene Chigadmaren *n*
childrenite Childrenit *m* (Min)
chile copper bars Chilekupfer *n* in Barren (pl)
chileite Chileit *m* (Min)
chilenite Chilenit *m* (Min)
chile niter Chilesalpeter *m*
chile saltpeter Chilesalpeter *m,* Caliche *m,*
   Natronsalpeter *m*
chill Kapsel *f* (Gieß)
chill *v* [ab]kühlen, abschrecken, in Kokillen
   gießen (Gieß)
chillagite Chillagit *m* (Min)
chill-cast *v* hart gießen
chill casting Hartguß *m,* Kapselguß *m,*
   Kokillenguß, Schalenguß *m*
chill casting mold Hartgußform *f*
chilled iron roll Hartgußwalze *f*
chilled work Kapselguß *m* (Gieß)
chill foundry pig iron Hartgußroheisen *n*
chill haze Kältetrübung *f* (Brauw),
   Kühltrübung *f* (Brauw)
chilliness Kälte *f*
chilling Abkühlen *n,* Abschrecken *n*
chilling point Kristallisationsbeginn *m*
chilling time Abschreckzeit *f*
chill mold Kokille *f,* Abschreckform *f,*
   Brammenkokille *f,* Gußschale *f*
chill molds, group of ~ Kokillengespann *n*
chillproofing of beer Klärung *f* von Bier
chill roll Auflaufwalze *f* (Extrusion)
chilly kalt
chimaphilin Chimaphyllin *n*
chime Glockenspiel *n*
chimney Schornstein *m,* Esse *f,* Kamin *m,*
   Qualmabzugsrohr *n,* Schlot *m*
chimney cooler Kaminkühler *m*
chimney draft or draught Schornsteinzug *m*
chimney flue Essenkanal *m,* Rauchkanal *m,*
   Zugkanal *m*
chimney formation Kaminverband *m* (Säcke)
chimney gas Abzugsgas *n*
chimney hole Rauchabzugsöffnung *f*

chimney hood Funkenesse *f*, Rauchfang *m*
chimney platform Kaminbühne *f*
chimney soot Kaminruß *m*, Büttenruß *m*
chimney stack Schornstein *m*
chimney ventilator Schornsteinventilator *m*
chimyl alcohol Chimylalkohol *m*
china Porzellan[geschirr] *n*
China clay Porzellanerde *f*, Chinaclay *m* (Min),
　Kaolin *n* (Min), Porzellanton *m* (Min),
　Weißtonerde *f* (Min)
china evaporating basin
　Porzellanabdampfschale *f*
China grass Ramie *f* (Bot), Ramiefaser *f*
china painting Porzellanmalerei *f*
China root Chinawurzel *f* (Bot)
China silk chinesische Seide
China silver Chinasilber *n*
chinaware Porzellanwaren *pl*
chiné chiniert (Text)
chineonal Chineonal *n*
Chinese blue Chinablau *n*
Chinese green Chinesischgrün *n*
Chinese red Chinesischrot *n*
Chinese silk chinesische Seide
Chinese wax Chinawachs *n*
Chinese wood oil Tungöl *n*
Chinese yellow Königsgelb *n*
chinic acid Chinasäure *f*, Chininsäure *f*
chinide Chinid *n*
chinidine Cinchonidin *n*
chiniofon Chiniofon *n*
chink Riß *m*
chinona Calisayarinde *f*
chinosol Chinosol *n*
chinotoxine Chinotoxin *n*
chinovin Chinovin *n*
chintz Chintz *m* (Text), Druckkattun *m* (Text)
chintzing calender Chintzkalander *m*
chintz paper Kattunpapier *n*
chiolite Chiolith *m* (Min)
chip Schnitzel *m n*, Span *m*, Splitter *m*
chip *v* abraspeln, abspänen, ~ **off** *v* abblättern,
　abbröckeln, abspringen
chip board Holzspanplatte *f*, Preßspanplatten *pl*,
　Spanholzplatte *f*
chip breaker Spanbrechernute *f*
chip flow Spanabfluß *m*
chip formation Spanbildung *f*
chip-oil extraction Späneentölung *f*
chip paper Schrenzpapier *n*
chipper Hackmaschine *f* (Holz)
chipper knife Hackmesser *n*
chipping Abbröckeln *n*, Abspringen *n*
chipping off Abschälen *n*
chippings Abfall *m* (Stein, Holz etc.), Späne *pl*
chips Späne *m pl*, Abfall *m* (Stein, Holz etc.),
　Abschlag *m*, Granulat *n*
chip wood Spanholz *n*
chirality Chiralität *f*

chiralkol Chiralkol *n*
chisel Drehstahl *m*, Gradiereisen *n*,
　Haumeißel *m*, Meißel *m*
chisel bit Gesteinsbohrer *m*
chisel steel Diamantstahl *m*, Meißelstahl *m*
chitaric acid Chitarsäure *f*
chitenine Chitenin *n*
chitin Chitin *n*
chitinase Chitinase *f*
chitobiose Chitobiose *f*
chitonic acid Chitonsäure *f*
chitopyrrol Chitopyrrol *n*
chitosamine Chitosamin *n*
chitosan Chitosan *n*
chitosazone Chitosazon *n*
chitose Chitose *f*
chiviatite Chiviatit *m* (Min)
chladnite Chladnit *m* (Min)
chlinochlorite Klinochlor *m* (Min)
chloanthite Chloanthit *m* (Min),
　Arseniknickelkies *m* (Min)
chloracetic acid Chloressigsäure *f*
chloracetyl Chloracetyl-
chloracetylacetone Chloracetylaceton *n*
chloracetyl chloride Chloracetylchlorid *n*
chloral Chloral *n*
chloralamide Chloralformamid *n*
chloralantipyrine Hypnal *n*
chloral caffeine Chloralcoffein *n*
chloral formamide Chloralformamid *n*,
　Chloramid *n*
chloral hydrate Chloralhydrat *n*
chloralide Chloralid *n*
chloralimide Chloralimid *n*
chloralism Vergiftung *f* durch Chloral
chlorallyl Chlorallyl-
chloralose Chloralose *f*
chloral poisoning Chloralvergiftung *f*
chloralum Chloralaun *n*
chloraluminite Chloraluminit *m* (Min)
chloral urethane Chloralurethan *n*,
　Urethanchloral *n*
chlorambucil Chlorambucil *n*
chloramide Chloramid *n*
chloramine Chloramin *n*
chloramine brown Chloraminbraun *n*
chloramine dye Chloraminfarbstoff *m*
chloramine-T Chloramin-T *n*, Chlorsulfamin *n*
chloramine yellow Chloramingelb *n*
chloramphenicol Chloramphenicol *n*
chloranil Chloranil *n*, Tetrachlorchinon *n*
chloranilic acid Chloranilsäure *f*
chloranol Chloranol *n*
chlorapatite Chlorapatit *m* (Min)
chlorargyrite Chlorargyrit *m* (Min)
chlorastrolite Chlorastrolith *m* (Min)
chlorate Chlorat *n*
chlorate *a* chlorsauer

chlorate aniline black base
Chloratanilinschwarzbase *f*
chlorate discharge Chloratätze *f*
chlorate explosive Chloratsprengstoff *m*
chloratranorin Chloratranorin *n*
chlorazanil Chlorazanil *n*
chlorazodin Chlorazodin *n*
chlorazol Chlorazol *n*
chlordane Chlordan *n*
chlordimorine Chlordimorin *n*
chlore *v* chloren
chlorella Chlorella *f*
chlorendic acid Het-Säure *f*
chlorethene Chloräthylen *n*
chloretone Chloreton *n*
chloretyl Chloretyl *n*
chlorhemine Chlorhämin *n*
chlorhexidine Chlorhexidin *n*
chlorhydrate Hydrochlorid *n*, Chlorhydrat *n*
chlorhydrin Chlorhydrin *n*
chloric acid Chlorsäure *f*, **salt of** ~ Chlorat *n*
chloride Chlorid *n*, Muriat *n* (obs)
chloride accumulator Chloridakkumulator *m*
chloride ion Chloridion *n*
chloride of lime bleaching Chlorkalkbleiche *f*,
Fixbleiche *f*
chloride of lime plant Chlorkalkanlage *f*
chloride of potash Chlorkali *n*
chloridizable chlorierungsfähig
chloridization Chlorierung *f*
chloridize *v* chlorieren
chloridized chloriert
chloridizing Chloren *n*, Chlorieren *n*
chloridometer Chlormesser *m*
chlorimetry Chlorbestimmung *f*
chlorinate *v* chlorieren, chloren
chlorinated choleriert, ~ **in the nucleus**
kernchloriert (Chem)
chlorinated hydrocarbon
Chlorkohlenwasserstoff *m*
chlorinated lime, solution of ~
Chlorkalklösung *f*
chlorinated paraffin Chlorparaffin *n*
chlorinating Chlorieren *n*, Chloren *n*
chlorinating agent Chlorierungsmittel *n*
chlorinating plant Chlorungsanlage *f*
chlorinating pyrolysis Chlorolyse *f*
chlorinating tower Chlorierturm *m*
chlorination Chlorierung *f*, Chloren *n*,
Chlorieren *n*, Chlorung *f*, **vapor-phase** ~
Gasphasechlorierung *f*
chlorinator Chlorierer *m*
chlorine Chlor *n* (Symb. Cl), **active** ~
Bleichchlor *n*, aktives Chlor, **aqueous solution**
of ~ Chlorwasser *n*, **containing** ~ chlorhaltig,
current of ~ Chlorstrom *m*, **fast to** ~
chlorecht, **free from** ~ chlorfrei, **generation of**
~ Chlorentwicklung *f*, **like** ~ chlorähnlich,
chlorartig, **to treat with** ~ chloren, chlorieren

chlorine-alkali electrolysis
Chloralkalielektrolyse *f*
chlorine bleaching Chlorbleiche *f*
chlorine carrier Chlorüberträger *m*
chlorine cell Chlorzelle *f*
chlorine compartment Chlorkammer *f*
chlorine compound Chlorverbindung *f*
chlorine content Chlorgehalt *m*
chlorine cyanide Chlorcyan *n*, Cyanchlorid *n*
chlorine detonating gas Chlorknallgas *n*
chlorine dioxide Chlordioxid *n*
chlorine dioxide plant Chlordioxidanlage *f*
chlorine fumigation Chlorräucherung *f*
chlorine gas Chlorgas *n*
chlorine gas cell Chlorgaselement *n*
chlorine generating flask
Chlorentwicklungsflasche *f*,
Chlorentwicklungskolben *m*
chlorine generator Chlorentwicklungsapparat *m*
chlorine heptoxide Chlorheptoxid *n*,
Perchlorsäureanhydrid *n*
chlorine hydrate Chlorhydrat *n*
chlorine hydrogen explosion
Chlorknallgasexplosion *f*
chlorine iodine compound Chlorjodverbindung *f*
chlorine liquefaction Chlorverflüssigung *f*
chlorine liquefying plant
Chlorverflüssigungsanlage *f*
chlorine monoxide Chlormonoxid *n*,
Chloroxydul *n* (obs),
Unterchlorigsäureanhydrid *n*
chlorine odor Chlorgeruch *m*
chlorine outlet Chlorableitung *f*
chlorine oxide Chloroxid *n*
chlorine peroxide Chlordioxid *n*
chlorine peroxide bleach Chlorsuperoxidbleiche *f*
chlorine plant Chloranlage *f*
chlorine soap Chlorseife *f*
chlorine water Chlorwasser *n*, Bleichwasser *n*
chlorine water pump Chlorwasserpumpe *f*
chlorinity Chlorgehalt *m*
chlorinous chlorähnlich, chlorartig
chlorisatic acid Chlorisatinsäure *f*
chlorisatin Chlorisatin *n*
chlorisondamine Chlorisondamin *n*
chlorite Chlorit (Min), Chlorit *n* (Chem),
containing ~ chlorithaltig
chlorite *a* chlorigsauer
chlorite bleaching Chloritbleiche *f*
chlorite slate Chloritschiefer *m* (Min)
chloritoid Chloritoid *m* (Min), Phyllit *m* (Min)
chloritous chloritartig, chloritführend,
chlorithaltig
chlormerodrine Chlormerodrin *n*
chlormethine Chlormethin *n*
chlormezanone Chlormezanon *n*
chlornaphazine Chlornaphazin *n*
chloroacetamide Chloracetamid *n*
chloroacetate Chloracetat *n*

chloroacetate *a* chloressigsauer
chloroacetic acid Chloressigsäure *f*
chloroacetic ethyl ester
 Chloressigsäureäthylester *m*
chloroacetol Chloracetol *n*
chloroacetone Chloraceton *n*
chloroacetyl Chloracetyl-
chloroacetyl-1,2-benzenediol
 Chloracetobrenzcatechin *n*
chloroacetyl chloride Chloracetylchlorid *n*
chloroacetylglycylglycine
 Chloracetylglycylglycin *n*
chloroacetylpyrocatechol
 Chloracetobrenzcatechin *n*
chlor[o]amine Monochloramin *n*
chloroaminobenzene Chloranilin *n*
chloroaminotoluene Chloraminotoluol *n*
chloroaniline Chloranilin *n*
chloroaniline hydrochloride
 Chloranilinhydrochlorid *n*
chloroanthraquinone Chloranthrachinon *n*
chloroauric(I) acid Aurochlorwasserstoffsäure *f*
chloroauric(III) acid
 Gold(III)-chlorwasserstoffsäure *f*
chloroaurous acid Aurochlorwasserstoffsäure *f*
chlorobenzal Chlorobenzal *n*
chlorobenzanthrone Chlorbenzanthron *n*
chlorobenzene Chlorbenzol *n*, Phenylchlorid *n*
chlorobenzoic acid Chlorbenzoesäure *f*
chlorobenzoyl Chlorbenzoyl *n*
chlorobenzyl Chlorbenzyl-
chlorobenzylidene Chlorobenzal *n*
chlorobromacetic acid Chlorbromessigsäure *f*
chlorobromobenzene Chlorbrombenzol *n*
chlorobutadien Chloropren *n*
chlorobutanol Chlorobutanol *n*,
 Acetonchloroform *n*, Chloreton *n*
chlorocalcite Chlorocalcit *m* (Min)
chlorocarbonate Chlorkohlensäureester *m*
chlorocarbonate *a* chlorkohlensauer
chlorocarbonic acid Chlorkohlensäure *f*
chlorocarbonic ester Chlorkohlensäureester *m*
chlorochromate chlorchromsauer
chlorochromic acid Chlorchromsäure *f*
chlorocodide Chlor[o]codid *n*
chlorocresol Chlorkresol *n*
chlorocrotonic acid Chlorcrotonsäure *f*
chlorocruorin Chlorocruorin *n*
chlorocyclizine Chlorcyclizin *n*
chlorodimethylarsine Kakodylchlorid *n*
chloroethanamide Chloracetamid *n*
chloroethane Äthylchlorid *n*
chloroethyl chloride Chloroäthylchlorid *n*
chloroethylene Chloräthylen *n*, Vinylchlorid *n*
chloroethylene chloride Chloräthylenchlorid *n*
chloroethylidene Chloräthyliden *n*
chloroform Chloroform *n*
chloroform *v* chloroformieren
chloroformate Chlorkohlensäureester *m*

chloroformate *a* chlorkohlensauer
chloroformic acid Chlorkohlensäure *f*
chloroformic ester Chlorkohlensäureester *m*
chloroforming Chloroformieren *n*
chlorogallol Chlorogallol *n*
chlorogenic acid Chlorogensäure *f*
chlorogenine Chlorogenin *n*, Alstonin *n*
chlorohydrin Chlorhydrin *n*
chlorohydroquinol Chlorhydrochinon *n*
chlorohydroquinone Chlorhydrochinon *n*
chloroisocyanuric acid Chlorisocyanursäure *f*
chloromalic acid Chloräpfelsäure *f*
chloromelanite Chloromelanit *m* (Min)
chloromercuriphenol Chlormercuriphenol *n*
chlorometer Chlormesser *m*, Chlorometer *n*
chloromethane Methylchlorid *n*
chloromethyl Chlormethyl-
chloromethylation Chlormethylierung *f*
chloromethyl menthyl ether
 Chlormethylmenthyläther *m*
chloromethyl naphthalene
 Chlormethylnaphthalin *n*
chlorometric chlorometrisch
chlorometry Chlorometrie *f*
chloromorphide Chloromorphid *n*
chloromycetin Chloromycetin *n*
chloronaphthalene Chlornaphthalin *n*
chloronitrobenzene Chlornitrobenzol *n*
chloronitrobenzenesulfonic acid
 Chlornitrobenzolsulfonsäure *f*
chloronitroparaffin Chlornitroparaffin *n*
chlorooxine Chloroxin *n*
chloropal Chloropal *m* (Min)
chloropalladic acid
 Palladium(IV)-chlorwasserstoff *m*
chloroparaffin Chlorparaffin *n*
chlorophaite Chlorophait *m*
chlorophane Chlorophan *m* (Min),
 Pyrosmaragd *m* (Min)
chlorophenol Chlorphenol *n*
chlorophenylamine Chloranilin *n*
chlorophenylenediamine Chlorphenylendiamin *n*
chlorophenyl salicylate Chlorsalol *n*
chlorophorin Chlorophorin *n*
chlorophthalic acid Chlorphthalsäure *f*
chlorophyll Chlorophyll *n*, Blattgrün *n*, rich in ~
 chlorophyllreich
chlorophyllaceous chlorophyllhaltig
chlorophyllane Chlorophyllan *n*
chlorophyll-bearing chlorophyllhaltig
chlorophyll coloring matter
 Chlorophyllfarbstoff *m*
chlorophyllide Chlorophyllid *n*
chlorophyllin Chlorophyllin *n*
chlorophyllin ester Chlorophyllid *n*
chlorophyllin paste Chlorophyllinpaste *f*
chlorophyllite Chlorophyllit *m* (Min)
chloropicrin Chlorpikrin *n*
chloroplast Chlorophyllkorn *n*, Chloroplast *m*

chloroplatinate Chloroplatinat *n*
chloroplatinic acid Chlorplatinsäure *f,*
    Platinchlorwasserstoffsäure *f*
chloroplatinic(IV) acid
    Platinichlorwasserstoffsäure *f*
chloroplatinous acid
    Platinochlorwasserstoffsäure *f*
chloropolypropylene Chlorpolypropylen *n*
chloroporphyrin Chloroporphyrin *n*
chloroprene Chloropren *n*
chloroprocaine Chlorprocain *n*
chloropropanolone Chloracetol *n*
chloropyramine Chloropyramin *n*
chloropyrene Chlorpyren *n*
chloropyridine Chloropyridin *n,* Chlorpyridin *n*
chloropyrilene Chlorpyrilen *n*
chloroquine Chlorochin *n*
chlororicinic acid Chlorricin[in]säure *f*
chlorosalicylaldehyde Chlorsalicylaldehyd *m*
chlorosalol Chlorsalol *n*
chlorosilane Chlorsilan *n*
chlorosis Bleichsucht *f* (Med), Chlorose *f* (Med)
chlorospinel Chlorospinell *m* (Min)
chlorostannic acid Zinnchlorwasserstoffsäure *f,*
    Stannichlorwasserstoffsäure *f*
chlorostannous acid
    Stannochlorwasserstoffsäure *f*
chlorosuccinic acid Chlorbernsteinsäure *f*
chlorosulfonation Chlorsulfonierung *f*
    (Sulfochlorierung)
chlorosulfonic acid Chlorsulfonsäure *f*
chlorothiazide Chlorothiazid *n*
chlorothymol Chlorthymol *n*
chlorotoluene Chlortoluol *n*
chlorotrianisene Chlorotrianisen *n*
chlorotriphenylsilane Triphenylsiliciumchlorid *n*
chlorourethane Chlorurethan *n*
chlorous chlorig
chlorous acid chlorige Säure *f,* salt or ester of ~
    Chlorit *n*
chloroxalethyline Chloroxaläthylin *n,*
    Chloroxalsäure *f*
chloroxalic acid Chloroxalsäure *f,*
    Chloroxaläthylin *n*
chlorozone Chlorozon *n*
chlorphenamine Chlorphenamin *n*
chlorphenesin Chlorphenesin *n*
chlorphenoctium Chlorphenoctium *n*
chlorpromazine Chlorpromazin *n*
chlorpropamide Chlorpropamid *n*
chlorquinaldol Chlorchinaldol *n*
chlorspar Mendipit *m* (Min)
chlorsuccinate Chlorsuccinat *n*
chlortetracycline Chlortetracyclin *n*
chlorthion Chlorthion *n*
chloryl Chloryl-
chlorylene Chlorylen *n*
chock Keil *m* (Hemmvorrichtung)
chocolate Schokolade *f*

choice Auswahl *f,* Wahl *f,* by ~ wahlweise
choice *a* vorzüglich
choke Drossel[spule] *f,* Luftklappe *f* (des
    Vergasers)
choke *v* [ab]drosseln, ersticken
choke damp Grubengas *n,* Nachdampf *m,*
    Nachschwaden *m,* schlagendes Wetter *n,*
    Schwaden *m*
choker bar (for regulating flow of material in die
    of extruder for sheets) Staubalken *m*
choke tube Luftdüse *f*
choking Abdrosseln *n,* Abdrosselung *f,*
    Drosselung *f,* Ersticken *n,* Verstopfen *n*
choking *a* erstickend
choking down Abblenden *n* (Düsen in
    Extrusion)
choladiene Choladien *n*
choladienic acid Choladiensäure *f*
cholalic acid Cholsäure *f*
cholanamide Cholanamid *n*
cholane Cholan *n*
cholanic acid Cholansäure *f*
cholanthrene Cholanthren *n*
cholate *a* cholalsauer, cholsauer
cholatriene Cholatrien *n*
cholatrienic acid Cholatriensäure *f*
cholecyanin Bilicyanin *n*
choleh[a]ematin[e] Bilipurpurin *n*
choleine Cholein *n*
cholera Cholera *f* (Med)
cholera inoculation Choleraschutzimpfung *f*
cholera virus Choleravirus *m n*
cholestane Cholestan *n*
cholestanediol Cholestandiol *n*
cholestanol Cholestanol *n*
cholestanone Cholestanon *n*
cholestanonol Cholestanonol *n*
cholestene Cholesten *n*
cholestenone Cholestenon *n*
cholesterate Cholesterat *n*
cholesterate *a* cholesterinsauer
cholesterin Cholesterin *n,* Cholesterol *n*
cholesterinate *a* cholesterinsauer
cholesterol Cholesterin *n,* Cholesterol *n*
cholesterol ester Cholesterinester *m*
cholesterol level Cholesterinspiegel *m*
cholesterol-like cholesterinähnlich
cholesterol metabolism
    Cholesterinstoffwechsel *m*
cholesterol wax Cholesterinwachs *n*
cholesterylene Cholesterylen *n*
cholestrophan Cholestrophan *n*
cholic acid Cholsäure *f*
choline Cholin *n*
choline base Cholinbase *f*
choline bitartrate Cholinbitartrat *n*
choline chloride Cholinchlorid *n*
choline dehydrogenase Cholindehydrase *f*
    (Biochem)

choline esterase Cholinesterase *f* (Biochem)
choline ether Cholinäther *m*
choline nitrate Nitratocholin *n*
choline salt Cholinsalz *n*
choloidanic acid Choloidansäure *f*
chondocurarine Chondocurarin *n*
chondocurine Chondocurin *n*
chondodendrine Chondodendrin *n*
chondridine Chondridin *n*
chondrification Knorpelbildung *f*(Biol),
   Verknorpelung *f*
chondrillasterol Chondrillasterin *n*
chondrin Chondrin *n*, Knorpelleim *m*
chondrite Chondrit *m* (Min)
chondroarsenite Chondroarsenit *m* (Min)
chondroblast Knorpelzelle *f*
chondrocyte Knorpelzelle *f*
chondrodine Chondrodin *n*
chondrodite Chondrodit *m* (Min)
chondroitin Chondroitin *n*
chondroitin sulfate Chondroitinsulfat *n*
chondroma Chondrom *n*, Knorpeltumor *m*
chondronic acid Chondronsäure *f*
chondroninic acid Chondroninsäure *f*
chondrosamic acid Chondrosaminsäure *f*
chondrosamine Chondrosamin *n*
chondrosin Chondrosin *n*
chondrosinic acid Chondrosinsäure *f*
chonemorphine Chonemorphin *n*
choose *v* [aus]wählen
chop *v* hacken, hauen, ~ **off** *v* abhacken,
   abschroten
chop-out die Stanzform *f*
chopped meat Hackfleisch *n*
chopped straw Häcksel *n*
chopper Hacker *m*, Hackmaschine *f*,
   Schopper *m*, Unterbrecher *m* (Elektr)
chopper bar controller Fallbügelregler *m*
chopper disk Fizeausches Rad *n*
chopping block Hackblock *m*
chopping board Hackbrett *n*
chopping machine Hackmaschine *f*
chord Akkord (Akust), Saite *f*, Sehne *f* (Kreis)
chor[i]oid membrane (of the eye) Aderhaut *f* [des
   Auges]
chovismic acid Chovisminsäure *f*
christianite Christianit *m* (Min), Anorthit *m*
   (Min), Kalkharmotom *m* (Min)
chroman Chroman *n*
chromanol Chromanol *n*
chromanone Chromanon *n*
chromate Chromat *n*, Chromsäuresalz *n*
chromate *a* chromsauer
chromate black Chromatschwarz *n*
chromatic chromatisch, farbig
chromatic aberration chromatische Aberration *f*,
   Farbfehler *m*
chromatic defect Farbenfehler *m*

chromaticity Chromatizität *f* (Opt),
   complementary ~ komplementäre
   Chromatizität (Opt)
chromaticity coordinate Normfarbwertanteil *m*
chromaticity coordinates Farbwertanteile *pl*
chromaticity diagram Farbdreieck *n* (Opt),
   Farbtafel *f*
chromaticness Chromatizität *f* (Opt)
chromatic phenomenon Farbenerscheinung *f*
chromatic printing Chromatdruck *m* (Buchdr),
   Farbendruck *m*
chromatics Farbenlehre *f*
chromatid Halbchromosom *n* (Biol)
chromatin Chromatin *n* (Biol, Med)
chromatogenic farbenerzeugend
chromatogram Chromatogramm *n*
chromatograph *v* chromatographieren
chromatographic chromatographisch
chromatographic analysis Chromatographie *f*
chromatographic column chromatographische
   Säule
chromatographic separation chromatographische
   Trennung
chromatography Chromatographie *f*, **ascending**
   ~ aufsteigende Chromatographie, **descending**
   ~ absteigende Chromatographie,
   high-pressure liquid ~
   Hochdruck-Flüssigkeitschromatographie *f*,
   liquid ~ Flüssigkeits-Chromatographie *f*
chromatography tank Trennkammer *f*
chromatology Farbenlehre *f*
chromatometer Farbenmesser *m*
chromatophore Chromatophor *n* (Bot),
   Farbenzelle *f*, Pigmentzelle *f*
chromatoscope Chromatoskop *n*
chromatrope Chromatrop *n*, Farbenschiller *m*
chromazurol Chromazurol *n*
chrome (Symb. Cr) Chrom *n*, **fast to** ~
   chromecht
chrome *v* chromieren
chrome alum Chromalaun *m*, Gerbsalz *n*
chrome alum solution Chromalaunlauge *f*
chrome ammine Chromiak *n*
chrome black Chromschwarz *n*, **fast** ~
   Chromechtschwarz *n*
chrome blue Chromblau *n*
chrome brown Chrombraun *n*
chrome color Chromfarbe *f*
chrome dyestuff Chromierfarbstoff *m*
chrome green Chromgrün *n*, Chromoxidgrün *n*,
   Deckgrün *n*, Neapelgrün *n*
chrome iron ore Chromeisen[erz] *n* (Min),
   Chromeisenstein *m* (Min), Chromit *m* (Min)
chrome leather Chromleder *n*
chrome leather black Chromlederschwarz *n*
chrome leather color Chromlederfarbe *f*
chrome lye Chromlauge *f*
chrome mica Chromglimmer *m* (Min)
chrome mordant Chrombeize *f*

chrome-mordant v chrombeizen
chromene Chromen n
chrome ocher Chromocker m (Min)
chrome orange Chromorange n
chrome oxide green Chrom[oxid]grün n
chrome-plate v verchromen
chrome-plated verchromt
chrome-plating Verchromen n
chrome red Chromrot n
chrome resinate Chromresinat n
chrome spinel Picotit m (Min)
chrome steel Chromstahl m
chrome subsulfate Chromsubsulfat n
chrome tannage Chromgerbung f
chrome-tanned chromgar
chrome-tanned leather Chromleder n
chrome tannery Chromgerberei f
chrome tanning Chromgerbung f, **dry** ~
   Chromtrockengerbung f
chrome tanning extract Chromgerbeextrakt m
chrome-tungsten steel Chromwolframstahl m
chrome-vanadium steel Chromvanadiumstahl m
chrome violet Chromviolett n
chrome yellow Chromgelb n, Königsgelb n,
   Leipzigergelb n
chromic acetate Chromiacetat n,
   Chrom(III)-acetat n
chromic acid Chromsäure f, **salt of** ~
   Chromsäuresalz n
chromic acid cell Chrom[säure]element n,
   Flaschenelement n
chromic anhydride Chromsäureanhydrid n,
   Chromtrioxid n
chromic chloride Chromichlorid n,
   Chrom(III)-chlorid n, Chromtrichlorid n
chromic compound Chrom(III)-Verbindung f,
   Chromiverbindung f
chromic fluoride Chrom[tri]fluorid n
chromic hydroxide Chromihydroxid n,
   Chrom(III)-hydroxid n
chromic iron Eisenchrom n
chromic oxide Chrom(III)-oxid n,
   Chrom[oxid]grün n, Chromsesquioxid n
chromic potassium alum Kaliumchromalaun m
chromic salt Chrom(III)-Salz n, Chromisalz n
chromic sulfate Chromisulfat n
chromicyanic acid Chromicyanwasserstoffsäure f
chromiferous chromhaltig
chrominance Chrominanz f (Opt)
chromite Chromit m (Min), Chromeisen[erz] n
   (Min), Chromeisenstein m (Min)
chromite brick Chromitstein m
chromithiocyanic acid
   Chromirhodanwasserstoffsäure f
chromium Chrom n, **containing** ~ chromhaltig
chromium acetate Chromacetat n
chromium carbide Chromcarbid n
chromium chlorate Chromchlorat n

chromium chloride Chlorchrom n,
   Chromchlorid n, **basic** ~ Chromsubchlorid n
chromium compound Chromverbindung f
chromium content Chromgehalt m
chromium dichloride Chromdichlorid n
chromium difluoride Chromdifluorid n
chromium fluoride Chromfluorid n,
   Fluorchrom n
chromium fluosilicate Chromsilicofluorid n
chromium glass Chromglas n
chromium hydroxide Chromhydroxid n,
   Chromoxidhydrat n
chromium(II) acetate Chrom(II)-acetat n,
   Chromoacetat n
chromium(II) chloride Chromchlorür n (obs),
   Chromdichlorid n, Chrom(II)-chlorid n,
   Chromochlorid n
chromium(II) compound
   Chrom(II)-Verbindung f,
   Chromoverbindung f,
   Chromoxydulverbindung f
chromium(II) fluoride Chromdifluorid n
chromium(II) hydroxide Chromhydroxydul n,
   Chrom(II)-hydroxid n, Chromohydroxid n
chromium(III) acetate Chromiacetat n,
   Chrom(III)-acetat n
chromium(III) chloride Chromichlorid n,
   Chrom(III)-chlorid n, Chromtrichlorid n
chromium(III) compound
   Chrom(III)-Verbindung f,
   Chromiverbindung f
chromium(III) fluoride Chrom(III)-fluorid n,
   Chromtrifluorid n
chromium(III) hydroxide Chromihydroxid n,
   Chrom(III)-hydroxid n, Chromoxidhydrat n
chromium(II) ion Chromoion n
chromium(III) oxide Chrom(III)-oxid n,
   Chromsesquioxid n
chromium(III) salt Chrom(III)-Salz n,
   Chromisalz n
chromium(III) sulfate Chromisulfat n
chromium(II) oxide Chrom(II)-oxid n,
   Chromoxydul n
chromium(II) salt Chrom(II)-Salz n
chromium(II) sulfide Chrom(II)-sulfid n,
   Chromsulfür n (obs)
chromium metal Chrommetall n
chromium metaphosphate Chrommetaphosphat n
chromium-nickel steel Chromnickelstahl m
chromium nitrate Chromnitrat n
chromium ore Chromerz n (Min)
chromium oxide Chromoxid n,
   Chrom[oxid]grün n, **native** ~ Chromocker m
   (Min)
chromium oxychloride Chromylchlorid n,
   Chromoxychlorid n, Chromsubchlorid n
chromium oxyfluoride Chromoxyfluorid n
chromium oxysulfate Chromsubsulfat n
chromium peroxide Chromperoxid n

chromium phosphide Chromphosphid *n*
chromium-plate *v* verchromen
chromium-plated verchromt
chromium-plating Verchromung *f*, **hard ~**
　Hartverchromung *f* (Elektrochem)
chromium potassium sulfate Chromalaun *m*
chromium resinate Chromresinat *n*
chromium salt Chromsalz *n*
chromium sesquioxide Chrom(III)-oxid *n*,
　Chromsesquioxid *n*, **hydrated ~** Afrikagrün *n*
chromium silicide Chromsilicid *n*
chromium silicofluoride Chromsilicofluorid *n*
chromium steel Chromstahl *m*
chromium subchloride Chromsubchlorid *n*
chromium sulfate Chromsulfat *n*, **basic ~**
　Chromsubsulfat *n*
chromium sulfide Chromsulfid *n*
chromium sulfite Chromsulfit *n*
chromium trichloride Chromtrichlorid *n*
chromium trifluoride Chromtrifluorid *n*
chromium trioxide Chromtrioxid *n*,
　Chromsäureanhydrid *n*
chromium tungstate Chromwolframat *n*
chromium(VI) oxide Chromsäureanhydrid *n*,
　Chromtrioxid *n*
chromizing process Inchromverfahren *n*
chromoacetic acid Chromoessigsäure *f*
chromocyclite Chromocyclit *m* (Min)
chromodiacetic acid Chromodiessigsäure *f*
chromoform Chromoform *n*
chromogen Chromogen *n*, Farbenerzeuger *m*
chromogenic chromogen, Farbe erzeugend
chromograph Chromograph *m*
chromoisomer Chromoisomer *n*
chromoisomerism Chromoisomerie *f*
chromolithograph Chromolithographie *f*,
　Farbensteindruck *m*
chromolithographer Chromolithograph *m*
chromolithography Chromolithographie *f*,
　Farbenlithographie *f*, Farbensteindruck *m*
chromolysis Chromolyse *f*
chromomere Chromomer *n*
chromomonoacetic acid
　Chromomonoessigsäure *f*
chromone Chromon *n*
chromonitric acid Chromsalpetersäure *f*
chromonol Chromonol *n*
chromopaper Chromopapier *n*
chromophore Chromophor *m*, Farbträger *m*
chromophoric chromophor
chromophorous chromophor
chromophotography Chromophotographie *f*,
　Farbphotographie *f*
chromophotolithography
　Chromphotolithographie *f*
chromophototypy Chromophototypie *f*
chromophotoxylography
　Chromophotoxylographie *f*

chromophotozincography
　Chromophotozinkographie *f*
chromopicotite Chrompicotit *m*
chromoplast Chromoplast *m* (Biol)
chromoprotein Chromoproteid *n*
chromopyrometer Chromopyrometer *n*
chromosantonin Chromosantonin *n*
chromosomal chromosomal
chromosomal abnormality
　Chromosomenabnormität *f*
chromosomal make-up
　Chromosomenzusammensetzung *f*
chromosome Chromosom *n* (Gen), **heterotypical**
　~ Heterochromosom *n*
chromosome count Chromosomenzahl *f*
chromosome loop Chromosomenschleife *f*
chromosome map Chromosomenbild *n*
chromosomes, arrangement of ~
　Chromosomenanordnung *f*, **clumping of ~**
　Chromosomenzusammenballung *f*, **exchange**
　of ~ Chromosomenaustausch *m*, **fusion of ~**
　Chromosomenverschmelzung *f*, **number of ~**
　Chromosomenzahl *f*, **pairing of ~**
　Chromosomenpaarung *f*, **pair of ~**
　Chromosomenpaar *n*, **reduplication of ~**
　Chromosomenverdopplung *f*, **set of ~**
　Chromosomensatz *m*
chromosphere Chromosphäre *f* (Astr)
chromospheric chromosphärisch
chromosulfuric acid Chromschwefelsäure *f*,
　Schwefelchromsäure *f*
chromotrop[e] chromotrop
chromotropic acid Chromotropsäure *f*
chromotypography Chromotypographie *f*,
　Farbenbuchdruck *m*
chromotypy Chromotypie *f*
chromous acetate Chrom(II)-acetat *n*,
　Chromoacetat *n*
chromous chloride Chromdichlorid *n*,
　Chromchlorür *n* (obs), Chromochlorid *n*
chromous compound Chrom(II)-Verbindung *f*,
　Chromoverbindung *f*,
　Chromoxydulverbindung *f* (obs)
chromous fluoride Chromdifluorid *n*
chromous hydroxide Chrom(II)-hydroxid *n*,
　Chromhydroxydul *n* (obs),
　Chromohydroxid *n*
chromous ion Chromoion *n*
chromous oxide Chrom(II)-oxid *n*,
　Chromoxydul *n* (obs)
chromous salt Chrom(II)-Salz *n*
chromous sulfide Chrom(II)-sulfid *n*,
　Chromsulfür *n* (obs)
chromoxylograph Chromoxylograph *m*
chromoxylography Chromoxylographie *f*,
　Farbenholzschnitt *m*
chromozincotypy Färben *n* von Zinkographien
　(pl)
chromyl Chromyl-

chromyl chloride Chromylchlorid *n*, Chromoxychlorid *n*
chromyl fluoride Chromoxyfluorid *n*
chromyl sulfate Chromsubsulfat *n*
chronic chronisch
chronological chronologisch, zeitlich aufeinanderfolgend
chronological sequence zeitliche Reihenfolge *f*
chronometer Chronometer *n*, Zeitmeßgerät *n*
chronometry Zeitmessung *f*
chronopotentiometry Chronopotentiometrie *f*
chronoscope Chronoskop *n*
chronotron Chronotron *n*
chrysalis (pl. chrysalises o. chrysalides), Insektenpuppe *f*, Puppe *f*, Schmetterlingspuppe *f*, Larve *f*, **to change into a ~** verpuppen (Biol)
chrysalis oil Chrysalidenöl *n*
chrysamine Chrysamin *n*
chrysammic acid Chrysamminsäure *f*, Tetranitrochrysazin *n*
chrysanilic acid Chrysanilsäure *f*
chrysaniline Chrysanilin *n*, Ledergelb *n*
chrysanisic acid Chrysanissäure *f*
chrysanthemine Chrysanthemin *n*
chrysanthem[um]ic acid Chrysanthem[um]säure *f*
chrysanthenone Chrysanthenon *n*
chrysarine Chrysarin *n*
chrysarobin Chrysarobin *n*
chrysarobin soap Chrysarobinseife *f*
chrysarone Chrysaron *n*
chrysatropic acid Chrysatropasäure *f*, Gelseminsäure *f*
chrysazin Chrysazin *n*
chrysean Chrysean *n*
chrysene Chrysen *n*
chrysene quinone Chrysenchinon *n*
chrysenic acid Chrysensäure *f*
chryseoline Chryseolin *n*
chrysidan Chrysidan *n*
chrysidine Chrysidin *n*
chrysin Chrysin *n*
chrysoberyl Chrysoberyll *m* (Min), Goldberyll *m* (Min)
chrysocale Chrysocal *n*
chrysocetraric acid Chrysocetrarsäure *f*
chrysocolla Chrysokoll *m* (Min), Hepatinerz *n* (Min), Kieselkupfer *n* (Min), Kieselmalachit *m* (Min), Pechkupfer *n* (Min)
chrysoeriol Chrysoeriol *n*
chrysofluorene Chrysofluoren *n*
chrysogen Chrysogen *n*
chrysography Chrysographie *f*, Goldschrift *f*
chrysoid Chrysoid *n*
chrysoidine Chrysoidin *n*, Akmegelb *n*
chrysoine Chrysoin *n*
chrysoine brown Chrysoinbraun *n*
chrysoketone Chrysoketon *n*
chrysolepic acid Pikrinsäure *f*

chrysoline Chrysolin *n*
chrysolite Chrysolith *m* (Min), Olivin *m* (Min), Peridot *m* (Min), Tektit *m* (Min)
chrysone Chryson *n*
chrysonetine Chrysonetin *n*
chrysophane Holmit *m* (Min)
chrysophanic acid Chrysophanol *n*, Chrysophansäure *f*
chrysophanol Chrysophanol *n*, Chrysophansäure *f*
chrysophenic acid Chrysopheninsäure *f*
chrysophenine Chrysophenin *n*
chrysophyll Chrysophyll *n*
chrysophyscin Chrysophyscin *n*, Physciasäure *f*, Physcion *n*
chrysopicrin Chrysopikrin *n*, Vulpinsäure *f*
chrysopontine Chrysopontin *n*
chrysoprase Chrysopras *m* (Min)
chrysopterin Chrysopterin *n*
chrysoquinone Chrysochinon *n*
chrysorhaminine Chrysorhaminin *n*
chrysotile Chrysotil[asbest] *m* (Min)
chrysotoxin Chrysotoxin *n*
chuck Aufspannfutter *n*, Futter *n*, Klemmfutter *n*
chucking device Spannvorrichtung *f*
chucking mechanism Spanngetriebe *n*
chuck [jaws] Futterbacken *pl*, Spannfutter *n*
churchite Churchit *m* (Min)
churn Rollfaß *n*
churn *v* Schaum schlagen
churning machine Kirnvorrichtung *f*
chute Fallschacht *m*, Gleitbahn *f*, Gleitfläche *f*, Rutsche *f*, Stürzrinne *f*
chute grate Schüttrost *m*
chydenanthine Chydenanthin *n*
chyle Chylus *m* (Med), Milchsaft *m* (Bot)
chylomicrons Chylomikronen *pl*
chymase Labferment *n* (Biochem)
chyme Chymus *m* (Med)
chymosin Chymosin *n*
chymotrypsin Chymotrypsin *n* (Biochem)
chymotrypsinogen Chymotrypsinogen *n* (Biochem)
chyraline Chyralin *n*
C.I. (Abk.) Färbeindex *m*
cichorigenin Cichorigenin *n*
cichoriin Cichoriin *n*
cicutine Cicutin *n*, Koniin *n*
cicutol Cicutol *n*
cicutoxine Cicutoxin *n*
cider Apfelwein *m*
cider press Apfelpresse *f*
C.I.F. Hamburg Lieferung *f* frei Hamburg
cigaret[te] paper Zigarettenpapier *n*
cignolin Cignolin *n*
cilianic acid Ciliansäure *f*
ciloidaic acid Ciloidansäure *f*
cimicic acid Cimicinsäure *f*
cimicifugin Cimicifugin *n*

cimicine Cimicin *n*
cimolite Cimolit *m* (Min)
cinchaine Cinchain *n*
cinchamidine Cinchamidin *n*
cinchene Cinchen *n*
cinchocaine Cinchocain *n*
cinchoic acid Cinchonsäure *f*
cincholic acid Cincholsäure *f*
cincholidine Cincholidin *n*
cincholoipone Cincholoipon *n*
cincholoiponic acid Cincholoiponsäure *f*
cinchomeronic acid Cinchomeronsäure *f*
cinchona Chinarindenbaum *m*, Cinchona *f*,
    Perurinde *f*, essence of ~ Chinaessenz *f*,
    tincture of ~ Chinatinktur *f*
cinchona alkaloid Chinaalkaloid *n*
cinchona bark Chinarinde *f*, Jesuitenrinde *f*
cinchona base Chinabase *f*, Cinchonabase *f*
cinchonamine Cinchonamin *n*
cinchona red China[rinden]rot *n*
cinchona toxin Chinatoxin *n*
cinchona tree Chinabaum *m*
cinchonhydrin Cinchonhydrin *n*
cinchonic acid Cinchonsäure *f*
cinchonicine Cinchonicin *n*
cinchonidine Cinchonidin *n*
cinchonidine bisulfate Cinchonidinbisulfat *n*
cinchonidine hydrochloride
    Cinchonidinhydrochlorid *n*
cinchonidine hydrogen sulfate
    Cinchonidinbisulfat *n*
cinchonidine sulfate Cinchonidinsulfat *n*
cinchonifine Cinchonifin *n*
cinchonigine Cinchonigin *n*
cinchoniline Cinchonilin *n*
cinchoninal Cinchoninal *n*
cinchonine Cinchonin *n*
cinchonine bisulfate Cinchoninbisulfat *n*
cinchonine hydrochloride
    Cinchoninhydrochlorid *n*
cinchonine hydrogen sulfate Cinchoninbisulfat *n*
cinchonine nitrate Cinchoninnitrat *n*
cinchonine sulfate Cinchoninsulfat *n*
cinchoninic acid Chinchoninsäure *f*
cinchoninone Cinchoninon *n*
cinchoniretine Cinchoniretin *n*
cinchonirine Cinchonirin *n*
cinchonism Cinchoninvergiftung *f*
cinchophen Cinchophen *n*
cinchotenine Cinchotenin *n*
cinchoticine Cinchoticin *n*
cinchotidine Cinchotidin *n*
cinchotine Cinchotin *n*
cinchotinone Cinchotinon *n*
cinchotoxine Cinchotoxin *n*, Cinchonicin *n*
cinchotoxol Cinchotoxol *n*
cinder Abbrand *m*, Kohlenschlacke *f*, Lösche *f*,
    Zinder *m*
cinder bed Schlackenbett *n*

cinder cement Schlackenzement *m*
cinder charging Schlackenzusatz *m*
cinder handling Abbrandtransport *m*
cinder handling plant Abbrandförderanlage *f*
cinder iron Schlackeneisen *n*
cinder notch Abschlacköffnung *f*,
    Schlackenstichloch *n*
cinders Asche *f*
cinder stone Schlackenstein *m*
cinder tap Abschlacköffnung *f*, Schlackenauge *n*,
    Schlackenstichloch *n*
cindery schlackig
cinematography Kinematographie *f* (obs)
cinene Cinen *n*
cinenic acid Cinensäure *f*
cineol[e] Cineol *n*, Eukalyptol *n*
cineolic acid Cineolsäure *f*
cineraceous aschig, aschenartig
cinerin Cinerin *n*
cinerolone Cinerolon *n*
cinerone Cineron *n*
cinnabar Zinnober *m* (Min), Merkurblende *f*
    (Min), crystallized ~ Zinnoberspat *m* (Min),
    hepatic ~ Quecksilberlebererz *n*, inflammable
    ~ Quecksilberbranderz *n*, native ~
    Bergzinnober *m*
cinnabar green grüner Zinnober
cinnabarine Cinnabarin *n*
cinnabarite Cinnabarit *m* (Min)
cinnabar ore Zinnobererz *n*
cinnabar scarlet Zinnoberscharlach *n*
cinnamaldehyde Zimtaldehyd *m*,
    Cinnamylaldehyd *m*
cinnamate Cinnamat *n*
cinnamate *a* zimtsauer (Salz)
cinnamein Cinnamein *n*
cinnamene Cinnamen *n*, Styrol *n*
cinnamenyl Cinnamenyl-
cinnamic zimtsauer
cinnamic acid Zimtsäure *f*, Cinnamylsäure *f*
    (obs), Phenylakrylsäure *f*, salt or ester of ~
    Cinnamat
cinnamic alcohol Zimtalkohol *m*, Peruvin *n*
cinnamic aldehyde Zimtaldehyd *m*,
    Cinnamylaldehyd *m* (obs)
cinnamic benzyl ester Zimtsäurebenzylester *m*
cinnamic ether Zimtäther *m*
cinnamic ethyl ester Zimtsäureäthylester *m*
cinnamoin Cinnamoin *n*
cinnamon Zimt *m*, broken ~ Kaneelbruch *m*,
    spirit of ~ Zimtbranntwein *m*, syrup of ~
    Zimtsirup *m*
cinnamon bark Kaneel *m*, Zimtrinde *f*
cinnamon brown Zimtbraun *n*
cinnamon-colored zimtfarben
cinnamon flower Zimtblüte *f*
cinnamon oil Zimtöl *n*

cinnamon stone Hessonit *m* (Min),
   Hyacinthgranat *m* (Min), Kaneelgranat *m*
   (Min)
cinnamon water Zimtwasser *n*
cinnamon wax Kaneelwachs *n*
cinnamoyl Cinnamoyl-
cinnamyl Cinnamyl-
cinnamyl alcohol Zimtalkohol *m*,
   Cinnamylalkohol *m*, Styron *n*
cinnamyl chloride Cinnamylchlorid *n*
cinnamyl cinnamate Styracin *n*
cinnamyl cocaine Cinnamylcocain *n*
cinnamyl ecgonine Cinnamylecgonin *n*
cinnamylidene Cinnamyliden-
cinnamylidene acetone Cinnamylidenaceton *n*
cinnaquinone Cinnachinon *n*
cinnarizine Cinnarizin *n*
cinnolic acid Cinnolinsäure *f*
cinnoline Cinnolin *n*
cinobufagin Cinobufagin *n*
cipher Chiffre *f*, Code *m*, Zahl *f*
cipolin Cipollin *m* (Min),
   Kalkglimmerschiefer *m* (Min),
   Zwiebelmarmor *m* (Min)
circle Kreis *m*, **area of a** ~ Kreisinhalt *m*, **center
of a** ~ Kreismittelpunkt *m*, **complete** ~
Vollkreis *m*, **full** ~ Vollkreis *m*, **to move in a**
~ kreisen, **whole depth** ~ Fußkreis *m*
circle *v* kreisen, ~ **around** umkreisen
circle of curvature Krümmungskreis *m*
circling kreisend
circuit Kreis *m* (Elektr), Kreislauf *m*,
   Stromkreis *m* (Elektr), Umkreis *m*,
   Umlauf *m*, **basic or fundamental** ~
   Grundschaltung *f* (Elektr), **closed** ~
   geschlossener Stromkreis, **element of the** ~
   Stromkreiselement *n*, **external** ~ äußerer
   Stromkreis, **integrating** ~ integrierender
   Kreis (Elektr), **internal** ~ innerer Stromkreis,
   open ~ offener Stromkreis, **out of** ~ stromlos,
   to break the ~ den Stromkreis öffnen, **to close
the** ~ den Stromkreis schließen, **to put into the**
~ in den Stromkreis einschalten
circuit breaker Stromunterbrecher *m*,
   Abschalter *m*, Ausschalter *m*, **automatic** ~
   Selbstausschalter *m*
circuit-breaker chamber for high-voltage
   switchgear Hochspannungsschalterrohr *n*
circuit closer Einschalter *m*
circuit cut-out switch Stromunterbrecher *m*
circuit diagram Schaltanordnung *f*, Schaltbild *n*,
   Schaltskizze *f*
circuit technique Schaltungstechnik *f*
circular kreisförmig, [kreis]rund, ringförmig;
   regelmäßig
circular aperture Kreisellochblende *f*
circular area Kreisfläche *f*, Kreisinhalt *m*
circular burner Mischbrenner *m*
circular chart Diagrammscheibe *f*, Kreisblatt *n*

circular chart recorder Kreis[blatt]schreiber *m*
circular cross section Kreisquerschnitt *m*
circular current Kreisstrom *m*
circular dichroism Circulardichroismus *m*
   (Spektr), Zirkulardichroismus *m*
circular disc Kreisscheibe *f*
circular disc with a hole Kreislochplatte *f*
circular feeder Rundbeschicker *m*
circular form Ringform *f*
circular function Kreisfunktion *f*,
   trigonometrische Funktion *f*
circular furnace Rundofen *m*
circular galvanometer Dosengalvanometer *n*
circular gate runner Ringeinguß *m*,
   Grateinguß *m*
circular graduation Kreis[ein]teilung *f*
circular-grate stoker runder Wanderrost *m*
circular knife Kreismesser *n*
circular layer Kreisschicht *f*
circular line Kreislinie *f*
circular-magnet belt separator
   Ringbandscheider *m*
circular magnet separator Magnetringscheider *m*
circular motion Kreisbewegung *f*,
   Achsendrehung *f*
circular path Kreisbahn *f*
circular pitch Schaufelabstand *m* (Turbine),
   Zahnkreisteilung *f*
circular polariscope Zirkularpolariskop *n*
circular polarization Zirkularpolarisation *f*
circular saw Kreissäge *f*
circular scale Kreisteilung *f*, Tellerskala *f*
circular scale thermometer Kreisthermometer *n*
circular shape Kreisform *f*
circular shelf Drehteller *m*
circular slide rule Rechenscheibe *f*
circular slide valve Rundschieber *m*
circular slitting knife Tellermesser *n*
circular slot burner Rundlochbrenner *m*
circular surface Kreisfläche *f*
circular tank Ringkessel *m*
circular trough Kreisrinne *f*
circular twist Rundschlag *m*
circulate *v* kreisen, zirkulieren
circulated, widely ~ weitverbreitet
circulated gas Wälzgas *n*
circulating air Umluft *f*, **heating by** ~
   Umluftheizung *f*
circulating air conveyor Umluftförderanlage *f*
circulating amount Umlaufmenge *f*
circulating dryer Umlufttrockner *m*
circulating evaporator Umlaufverdampfer *m*
circulating fuel reactor Kreislaufreaktor *m*,
   Zirkulationsreaktor *m*
circulating liquid Zirkulationsflüssigkeit *f*
circulating pump Umlaufpumpe *f*,
   Umwälzpumpe *f*
circulating storage Umlaufspeicher *m* (Comp)
circulating system Zirkulationssystem *n*

circulating system lubrication
  Umlaufschmierung *f*
circulating tank Kreislaufbehälter *m*
circulation Blutkreislauf *m* (Med), Kreislauf *m*,
  Kreisströmung *f*, Umlauf *m*, Umwälzung *f*,
  Zirkulation *f*, **disturbed** ~
  Durchblutungsstörung *f*, **forced** ~
  Zwangsumlauf *m*, **rate of** ~
  Zirkulationsgeschwindigkeit *f*
circulation apparatus Zirkulationsapparat *m*
circulation boiler Umlaufkessel *m*
circulation by pumping Umpump *m*
circulation channel Umströmkanal *m*
circulation evaporator Umlaufverdampfer *m*
circulation flow Zirkulationsströmung *f*
circulation heating Umlaufheizung *f*
circulation loop Umlaufleitung *f* (Atom)
circulation of air Luftumlauf *m*, Luftzirkulation *f*
circulation of blood, stimulating the ~
  durchblutungsfördernd
circulation of oil Ölumlauf *m*, Ölzirkulation *f*
circulation of the lye Laugenkreislauf *m*
circulation passage Umströmkanal *m*
circulation pump Zirkulationspumpe *f*
circulation regulator Umlaufregler *m*
circulation system Zirkulationssystem *n*
circulatory collapse Kreislaufkollaps *m* (Med),
  Kreislaufversagen *n* (Med)
circulatory cyclone evaporator
  Umlaufverdampfer *m*
circulatory disease Kreislaufkrankheit *f* (Med)
circulatory disturbance Kreislaufstörung *f* (Med)
circulatory function Kreislauffunktion *f* (Med)
circulatory insufficiency Kreislaufschwäche *f*
  (Med)
circulatory preparation Kreislaufmittel *n*
  (Pharm)
circulatory system Blutsystem *n* (Med)
circumanthracene Circumanthracen *n*
circumference Umfang *m*, Kreislinie *f*,
  Umkreis *m*
circumference [of a circle] Kreisumfang *m*
circumferential crack Umfangsriß *m* (;)
circumferential stress Umfangsspannung *f*
circumnuclear kernumgebend
circumpolar zirkumpolar
circumscribe *v* begrenzen, umschreiben
circumscription Umschreibung *f*
circumstance Umstand *m*, Verhältnis *n*
circumstances Sachverhalt *m*
circumstantial durch die Umstände bedingt,
  umständlich
circumstantial evidence Indizienbeweis *m*
cis-effect cis-Effekt *m* (Stereochem)
cis-form cis-Form
cismollan Cismollan *n*
cis-position cis-Stellung *f*
cis-rule cis-Regel *f*
cissoid Zissoide *f* (Math)

cistern Wasserbehälter *m*, Zisterne *f*
cistern barometer Gefäßbarometer *n*
cis-trans isomerism Cis-Trans-Isomerie *f*
  (Stereochem)
cis-trans isomerization
  Cis-Trans-Isomerisierung *f* (Stereochem)
cistron Cistron *n*
citarin Citarin *n*
cite *v* anführen
citracone anil Citraconanil *n*
citraconic acid Citraconsäure *f*
citraconic anhydride Citraconsäureanhydrid *n*
citraconic ester Citraconsäureester *m*
citraconyl Citraconyl-
citral Citral *n*, Geranial *n*, Neral *n*
citramalic acid Citramalsäure *f*
citranilide Citranilid *n*
citrate Citrat *n*, Zitronensäureester *m*,
  zitronensaures Salz *n*, **sodium anhydro
  methylene** ~ Citarin *n*
citrate *a* zitronensauer (Salz, Ester)
citrate cycle Citratcyclus *m*
citrate of lime Calciumcitrat *n*
citraurin Citraurin *n*
citrazinic acid Zitrazinsäure *f*
citrene Citren *n*, Hesperiden *n*
citreorosein Citreorosein *n*
citric acid Zitronensäure *f*, **salt or ester of** ~
  Citrat *n*
citric [acid] cycle Tricarbonsäurezyklus *m*,
  Zitronensäurezyklus *m*
citric ester Zitronensäureester *m*
citric ether Zitronenäther *m*
citridic acid Akonitsäure *f*, Citridinsäure *f*
citrifoliol Citrifoliol *n*
citrin Citrin *n*, Hesperidin *n*
citrine Zitrin *m* (Min)
citrine *a* zitronenfarben, zitronengelb
citrinin Citrinin *n*
citromycetin Citromycetin *n*
citronellal Citronellal *n*
citronella oil Citronyl *n*, Grasöl *n*,
  Zitronell[a]öl *n*
citronellic acid Citronell[a]säure *f*
citronellol Citronellol *n*
citronetin Citronetin *n*
citronyl Citronyl *n*, Zitronell[a]öl *n*
citrophen Citrophen *n*
citroptene Citropten *n*, Limettin *n*
citrostadienol Citrostadienol *n*
citrostanol Citrostanol *n*
citrovorum factor Citrovorumfaktor *m*
citroxanthin Citroxanthin *n*
citrulline Citrullin *n*
citrullol Citrullol *n*
citrus fruit Zitrusfrucht *f*
citryl Cityl-
city gas Leuchtgas *n*
civet Zibet *m*

civetane Zibetan *n*
civetone Zibeton *n*
civil law Zivilrecht *n*
clack valve Pumpenventil *n*, Rückschlagventil *n*
cladding Plattieren *n*, Plattierung *f,*
  Walzplattieren *n*
cladding materials by welding
  Schweißplattieren *n*
cladestic acid Cladestinsäure *f*
cladestine Cladestin *n*
cladinose Cladinose *f*
cladonine Cladonin *n*
claim Anrecht *n*, Anspruch *m* (Patent)
claim *v* anfordern, beanspruchen
Claisen flask Claisenkolben *m*
Claisen rearrangement Claisensche
  Umlagerung *f* (Chem)
Claisen-Schmidt condensation
  Claisen-Schmidt-Kondensation *f*
Claisen stillhead Claisenaufsatz *m*
clamp Klammer *f,* Einspannbacke *f,*
  Haltevorrichtung *f,* Klemme *f,* Klemmmuffe *f,*
  jaw of ~ Einspannklaue *f,* Einspannklemme *f*
clamp *v* einklemmen, einspannen,
  festklammern, verspannen, ~ **firmly**
  festklemmen, ~ **tightly** *v* festspannen
clamp bolt Spannschraube *f*
clamp coupling Klemmkupplung *f*
clamp for stand Ständerklemme *f* (Chem)
clamping Einspannen *n*, Einspannung *f*
clamping arrangement Spannvorrichtung *f*
clamping bolt Anziehbolzen *m,*
  Befestigungsschraube *f*
clamping capacity Schließleistung *f*
clamping device Aufspannvorrichtung *f,*
  Einspannvorrichtung *f,* Klemmvorrichtung *f,*
  Spannvorrichtung *f*
clamping force Einspanndruck *m*
clamping fork Klammergabel *f*
clamping frame Spannrahmen *m,*
  Halterahmen *m,* Streckrahmen *m*
clamping jaw Spannbacke *f*
clamping lever Spannhebel *m*
clamping plate Aufspannplatte *f,* Klemmplatte *f*
clamping plunger Druckkolben *m*
clamping power Spannkraft *f*
clamping pressure Spanndruck *m,*
  Schließdruck *m*
clamping ring Spannring *m*, Ziehring *m*
clamping screw Klemmschraube *f*
clamping sleeve Klemmuffe *f*
clamping strap Spannriemen *m*
clamping time (cold bonding) Preßzeit *f*
clamping tool Spannwerkzeug *n*
clamp [iron] Schraubzwinge *f,* Zwingeisen *n*
clamp-on stirrer Anklemmrührer *m*
clamp plate Klemmplatte *f*
clamp roll Spannwalze *f*
clamp screw Spannschraube *f,* Bündelschraube *f*

clamp welding machine
  Einspannschweißvorrichtung *f*
clam-shell dredge Greifbagger *m*
clandestinine Clandestinin *n*
clapper Klöppel *m*
claret red Bordeauxrot, Rotweinfarbe *f*
clarification Klärung *f,* Abschlämmung *f*
  (Zucker), Abschleimung *f,* Abseihung *f,*
  Defäkation *f,* Läuterung *f,* Scheidung *f,*
  method of ~ Abklärungsmethode *f,* **process of**
  ~ Klärungsprozeß *m*
clarification plant Kläranlage *f*
clarification tank Klärgrube *f*
clarified geklärt, abgeschleimt, geläutert
clarified liquid Kläre *f*
clarifier Klärmittel *n,* Kläranlage *f,* Klärgefäß *n;*
  Klärkessel *m,* Klärtopf *m,* Läuterpfanne *f,*
  Scheidepfanne *f*
clarify *v* aufklären, [ab]klären, [ab]läutern,
  abschleimen, aufhellen
clarifying Klären *n,* Abklären *n;* Scheiden *n*
clarifying *a* klärend
clarifying agent Abklärungsmittel *n,* Klärmittel *n*
clarifying apparatus Klärapparat *m,* Klärgefäß *n*
clarifying basin Kläranlage *f*
clarifying centrifuge Klärzentrifuge *f*
clarifying sump Klärsumpf *m*
clarifying tub Läuterbottich *m*
Clark cell Clarkelement *n* (Elektr)
clarone Claron *n*
clash *v* kollidieren
clasp Klammer *f,* Bandklammer *f,* Haken *m,*
  Krampe *f*
clasp *v* haken, krampen
clasping Einhaken *n*
class Klasse *f,* Art *f* (Gattung), Rang *m,* Sorte *f*
class frequency Klassenhäufigkeit *f*
classification Klassifikation *f,* Bezeichnung *f,*
  Einteilung *f,* Fachordnung *f,*
  Klasseneinteilung *f,* Klassierung *f* (Trennung
  nach Größe), Ordnung *f*
classified ad Kleinanzeige *f* (Werbung)
classifier Klassierapparat *m,* Klassierer *m,*
  Klassifikator *m,* Sichter *m,* **fluted** ~
  Faltenschlämmer *m,* **hindered-settling** ~
  Horizontalschlämmer *m,* **performance of a** ~
  Sichtleistung *f,* ~ **with closed air circuit**
  Streusichter *m*
classifier with circumferential screen
  Korbsichter *m*
classify *v* klassifizieren, [auf]gliedern,
  einordnen, einteilen, klassieren (Techn)
classifying plant, grading or ~ Klassieranlage *f*
classifying screen Klassiersieb *n*
class limit Klassengrenze *f*
class range Klassenbreite *f*
clastic klastisch (Geol)
clathrate Clathrat *n,* Einschlußverbindung *f*

clathrate compound Klathrat *n* (Chem)
clatter *v* klappern
claudetite Claudetit *m* (Min), Rhombarsenit *m* (Min)
Claus catalyst Clauskatalysator *m*
Clausius-Clapeyron equation Clausius-Clapeyronsche Gleichung *f* (Therm)
clausthalite Clausthalit *m* (Min), Selenblei *n* (Min)
clavatin Clavatin *n*
clavatol Clavatol *n*
clavicepsin Clavicepsin *n*
claviformin Claviformin *n*
clavolonine Clavolonin *n*
claw Klaue *f*, Pranke *f*
claw belt fastener Riemenkralle *f*
claw magnet Klauenmagnet *m*
clay Ton[erde] *f*, Lehm *m*, Mergel *m*, **ball ~** Bindeton *m*, **bonding ~** Bindeton *m*, **calcareous ~** kalkhaltiger Ton, **colored ~** Färberde *f*, Farberde *f*, **foliated ~** Blätterton *m*, **meager ~** magerer Ton, **of ~** tönern, **plastic ~** bildsamer Ton, **refractory ~** feuerfester Ton, **soapy ~** fetter Ton, **tempered ~** Tonspeise *f*, **to wash the ~** den Ton schlämmen, **white plastic ~** Bindeton *m*
clay *a* irden
clay *v* verstopfen (Bohrloch)
clay brick Lehmziegel *m*
clay cell Tonzelle *f*
clay cement Tonbindemittel *n*
clay-colored tonfarbig
clay containing iron eisenschüssiger Ton
clay cutter Tonschneider *m*
clay dish Tonteller *m*
clay disk Tonplatte *f*, Tonscheibe *f*
clayey lehmhaltig, lehmig, tönern, tonartig, tonig
clay filter Tonfilter *n*
clay for refining sugar Zuckererde *f*
clay furnace Tonofen *m*
clay gun Stichlochstopfmaschine *f*
clay industry Tonindustrie *f*
clay ironstone Toneisenstein *m* (Min), Eisenton *m* (Min), Sphärosiderit *m* (Min), **yellow ~** Gelbeisenerz *n* (Min)
clay layer Tonschicht *f*
clay marl Lehmmergel *m*, Tonmergel *m*
clay mixer Tonmischer *m*
clay mortar Lehmmörtel *m*, Tonspeise *f*
clay pit Lehmgrube *f*
clay plate Tonteller *m*, Lehmplatte *f*
clay processing Tonverarbeitung *f*
clay retort Tonretorte *f*, Tonmuffel *f*
clay slate Tonschiefer *m* (Min), Kieselton *m* (Min), Klebschiefer *m* (Min)
clay slip Tonbrei *m* (Keram), Tonspeise *f* (Keram)
clay soil Tonboden *m*
clay stone Tonstein *m*

clay treatment Bleicherdebehandlung *f*
clay vessel Tongefäß *n*
clay wash Tonschlämme *f*
clean sauber, blank, klar, rein
clean *v* reinigen, [ab]putzen, abwischen, säubern, spülen, waschen, **~ with soap** abseifen, **~ iron** *v* Eisen ausschweißen
clean area aktivitätsfreier Raum *m* (Atom)
cleaner Reiniger *m*, Fleckenreiniger *m*
cleaning Reinigen *n*, Reinigung *f*, Spülung *f*, **~ with soap** Abseifung *f*
cleaning basin Reinigungsbassin *n*, Reinigungsbecken *n*
cleaning bath Reinigungsbad *n*
cleaning door Reinigungstür *f*
cleaning effect Waschwirkung *f*
cleaning equipment Reinigungsgeräte *n pl*
cleaning field Reinigungsfeld *n*
cleaning hole Putzöffnung *f*, Reinigungsöffnung *f*
cleaning liquid Fleckenwasser *n*
cleaning powder Putzmittel *n*
cleaning process Waschvorgang *m*
cleaning rag Putzlappen *m*
cleaning trench Reinigungsgraben *m*
cleaning wool Putzwolle *f*
cleanliness Reinlichkeit *f*, Sauberkeit *f*
clean-out Reinigungsöffnung *f*
cleanse *v* reinigen, [ab]putzen, abspritzen, abspülen, abwaschen, säubern, **~ by scrubbing** abscheuern, **~ from soap** entseifen
cleanser Reinigermasse *f*, Reinigungsmittel *n*
cleansing Reinigung *f*, Abspülen *n*, **~ from soap** Entseifen *n*
cleansing agent Abwaschmittel *n*
cleansing apparatus Schlämmapparat *m*
cleansing cream Reinigungscreme *f* (Kosm)
cleansing device Schlämmvorrichtung *f*
cleansing drum Rollfaß *n*
cleansing gas Spülgas *n*
cleansing machine Putzmaschine *f*
cleansing material Putzmittel *n*
cleansing process Abspritzverfahren *n*, Reinigungsvorgang *m*
cleansing system Ausstoßsystem *n*, Reinigungsmethode *f*
cleansing tool Putzwerkzeug *n*
cleansing vat Nachgärungsbottich *m*; Nachgärungsfaß *n*
clear klar, blank, durchsichtig, hell, kenntlich, rein
clear *v* [ab]klären, [ab]läutern, [ab]räumen, leeren, roden, **~ away** aufräumen, wegräumen, **~ from mud** entschlammen, **~ of fumes** entnebeln, **~ of trees** abholzen, **~ up** *v* aufhellen, aufklären, aufräumen
clearance Abstand *m*, Aussparung *f*, Spielraum *m*, Zwischenraum *m*
clearance of mud Abschlämmung *f*

cleared from dregs abgeklärt, **~ from dross** abgeschlackt
cleared woodland Rodung *f*
clearer Klärbecken *n*
clearer board Putzbrettchen *n*, Reinigungsbrettchen *n*
clear face varnish Silberlack *m*
clearing Klärung *f*, Abklären *n*, Ausräumen *n*, Defäkation *f*, Klären *n*, Läuterung *f*, Leerung *f*, Scheiden *n*, Scheidung *f*
clearing agent Abschwächer *m* (Phot), Klär[ungs]mittel *n*
clearing bath Klärbad *n*
clearing cask Klärungsfaß *n*
clearing field Reinigungsfeld *n*
clearing liquor Kläre *f*
clearing pan Klärpfanne *f*, Läuterpfanne *f*
clearing the bath Reinigung *f* des Bades
clearing up Aufräumen *n*
clearing vat Klärbottich *m*
clear lacquer Klarlack *m*
clear liquor sugar Klärsel *n*
clearness Klarheit *f*, Helligkeit *f*, Reinheit *f*, Schärfe *f* (Phot)
clear rinsing agent Klarspülmittel *n*
cleat Leiste *f;* Holzleiste *f*, Klotz *m* des Reifenprofils
cleat *v* mit Leisten versehen
cleat insulator Klemmisolator *m*
cleat work Leistenarbeit *f*
cleavability Spaltbarkeit *f*
cleavable [ab]spaltbar
cleavage Abspaltung *f*, Aufblättern *n* (v. Schichtstoffen), Aufspaltung *f*, Spalt *m* (Bergb), **line of ~** Abbruchlinie *f* (Geol)
cleavage brittleness Spaltbrüchigkeit *f*
cleavage crystal Spaltungskristall *m*
cleavage face Spaltfläche *f*
cleavage plane Ablösungsrichtung *f* (Krist), Spaltfläche *f*, Spalt[ungs]ebene *f*
cleavage product Spaltstück *n*, Spalt[ungs]produkt *n*
cleavage surface Spaltfläche *f*, Ablösungsfläche *f*
cleave, to begin to ~ anspalten
cleave *v* [ab]spalten, schlitzen, spleißen
cleavelandite Cleavelandit *m* (Min)
cleaving Abtrennung *f*, Spaltung *f*
cleaving *a* [ab]spaltend
clemizole Clemizol *n*
clenching Einhaken *n*
clever geistreich
Cleve's acid Clevesäure *f*
clevis plate Festhalteplatte *f*
cliché Klischee *n*
clicker press Ausstanzer *m*
clicking press Stanzpresse *f*
clidinium Clidinium *n*

cliftonite Cliftonit *m* (Min)
climate Klima *n*
climatic klimatisch
climatic chamber Klimaprüfschrank *m*
climatology Klimatologie *f*, Klimalehre *f*
climax Höhepunkt *m*, Gipfel *m*, **to reach a ~** gipfeln, einen Höhepunkt erreichen
climbing crane Kletterkran *m*
climbing film evaporator Aufsteigschichtverdampfer *m*, Kletterfilmverdampfer *m*
clincher Haspe *f*, Klammer *f*, Klampe *f*, Niet *m*
clincher tire Wulstreifen *m*
cling *v* anhaften
clingmannite Clingmannit *m* (Min)
clinic Klinik *f* (Med), Klinikum *n*, Universitätskrankenhaus *n*
clinical klinisch
clinker hartgebrannter Ziegel *m;* Hartziegel *m*, Klinker *m*, Schlacke *f*, Sinterschlacke *f;*, **formation of ~** Schlackenbildung *f*, **forming tough ~** zähschlackig, **to clear of ~** ausschlacken
clinker *v* sintern, [ver]schlacken
clinker cake Schlackenkuchen *m*
clinker concrete Klinkerbeton *m*
clinker content[s] Schlackengehalt *m*
clinker crust Belag *m*
clinker grate Schlackenrost *m*
clinkering Backen *n*, Sintern *n*, Sinterung *f*, **~ of coal** Backen *n* der Kohle
clinkering *a* schlackend
clinker utilization machine Klinkerverwertungsmaschine *f*
clinkery schlackig
clink stone Klingstein *m* (Min)
clino axis Klinoachse *f*
clinochlore Klinochlor *m* (Min)
clinoclase Klinoklas *m* (Min)
clinoclasite Klinoklas *m* (Min), Strahlenkupfer *n* (Min), Strahlerz *n* (Min)
clinodiagonal Klinodiagonale *f*
clinoedrite Klinoedrit *m* (Min)
clinoenstatite Klinoenstatit *m* (Min)
clinohedrite Klinoedrit *m*
clinohumite Klinohumit *m* (Min)
clinometer Gefällemesser *m*, Klinometer *n*, Libelle *f* (Wasserwaage), Neigungsmesser *m*
clinopyramid Klinopyramide *f* (Krist)
clinozoisite Klinozoisit *m* (Min)
clinquant Flittergold *n*
clintonite Clintonit *m* (Min)
clionasterol Clionasterin *n*
clip Klammer *f*, Klemme *f*, Querhahn *m*, Quetschhahn *m*
clip *v* [ab]schneiden, befestigen (anstecken), stutzen
clip bolt Hakenschraube *f*
clip-on cap Kapselverschluß *m*

clip-on stirrer Anklemmrührer *m*
clippers Schneidezange *f*
clipper stage Impulsbegrenzerstufe *f*
clippings Schneidabfall *m*, Blechabfälle *m pl*,
  Schnitzel *pl*
clip spring switch Federschalter *m*
clivonine Clivonin *n*
clockwise im Uhrzeigersinn
clockwise rotation Rechtsdrehung *f*
clockwork Zeigerwerk *n*
clod Klumpen *m* (Erde), Klunker *m*, Scholle *f*
  (Landw)
clod breaker Schollenbrecher *m* (Landw)
clog Knebel *m*
clog *v* Klumpen *pl* bilden, sich verstopfen,
  blocken, ~ **[up]** *v* sich verstopfen
cloisonné enamel Cloisonnéemail *n*
close nah, eng, dicht, klamm
close *v* [ab]schließen, verschließen, zudrehen,
  zumachen, ~ **by melting** *v* zuschmelzen
closed circuit geschlossener Kreislauf *m*
closed-circuit conveyor Kreisförderer *m*
closed-circuit grinding Kreislaufmahlung *f*
closed-in pressure (at well head)
  Kopfschließdruck *m*
closed loop Regelkreis *m*
close fit Edelpassung *f*
close fitting knapp
close-grained feinkörnig, kleinluckig
close-meshed engmaschig
close-off rating Schließdruck *m*
close pig feinkörniges Roheisen
close-range action Nahwirkung *f*
closest packed hexagonal lattice hexagonal
  dichteste Kugelpackung *f*
closet Schrank *m*, Wandschrank *m*
close-up Nahaufnahme *f* (Phot)
closing Schließen *n*, Schließung *f*
closing cover Verschlußdeckel *m*
closing device Abschlußvorrichtung *f*,
  Absperrvorrichtung *f*, Verschluß *m*
closing disk Abschlußscheibe *f*
closing head Schließkopf *m*
closing joint Werkzeugschluß *m*
closing machine Verschließmaschine *f* (Dosen)
closing motion Schließbewegung *f*
closing valve Abschlußventil *n*
clostridium Klostridium *n* (Bakt)
closure Schließen *n*, Verschluß *m*, **tear-off** ~ **or**
  pull-off closure Abreißverschluß *m*
closure chain Verschlußkette *f*
clot geronnenes Blut (Med), Brocken *m*,
  Klunker *m*, Koagulat *n*
clot *v* stocken, gerinnen
cloth Tuch *n* (Text), Leinwand *f* (Text), Stoff *m*
  (Text)
cloth air conduit Tuchlutte *f*

clothe *v* bekleiden, umhüllen
cloth filter Stoffilter *n*, Tuchfilter *n*
clothing, protective ~ Schutz[be]kleidung *f*
clothing industry Bekleidungsindustrie *f*
cloth printer Tuchdrucker *m* (Text)
cloth scarlet Tuchscharlach *n*
clotted geronnen, klumpig
clotting Festwerden *n* (Blut), ~ **of the red cells**
  Hämagglutination *f*
clotting delay Gerinnungsverzögerung *f*
clotting factor Gerinnungsfaktor *m*
clotting process Gerinnungsvorgang *m*
clotty klumpig
cloud Wolke *f*, **radioactive** ~ radioaktive Wolke,
  ~ **of electrons** Elektronenwolke *f*
cloud chamber Nebelkammer *f*, Wilsonkammer *f*
cloud chamber expansion
  Nebelkammerexpansion *f*
cloud chamber photograph
  Nebelkammeraufnahme *f*,
  Wilsonkammeraufnahme *f*
cloud discharge Wolkenentladung *f*
cloud formation Wolkenbildung *f*
cloudiness Schleierbildung *f*, Trübe *f*, Trübung *f*,
  Unreinheit *f* (Kristall)
clouding of the glass Glastrübung *f*
cloud limit Nebelbildungsgrenzwert *m*
cloud point Trübungspunkt *m*,
  Kristallisationsbeginn *m*
cloud track Nebelspur[bahn] *f*
cloudy wolkig, getrübt, trüb[e]
clout Blattnagel *m*, Hartnagel *m*
clovane Clovan *n*
clovanediol Clovandiol *n*
clove Gewürznelke *f*, Nelke *f* (Bot)
cloven [auf]gespalten
clovene Cloven *n*
clovenic acid Clovensäure *f*
clover, white ~ Steinklee *m* (Bot)
cloves, oil of ~ Gewürznelkenöl *n*, Nelkenöl *n*
club Keule *f*
club moss Bärlapp *m* (Bot), Drudenfuß *m* (Bot),
  Erdschwefel *m* (Bot), Lycopodium *n* (Bot)
clue Anhaltspunkt *m*, Stichwort *n*
clumsy ungeschickt, unhandlich
clupanodenic acid Clupanodensäure *f*
clupein Clupein *n*
Clusius column Clusiussches Trennrohr *n*
cluster Haufen *m*, Schwarm *m*, ~ **of crystals**
  Druse *f* (Min)
cluster gear Stufenzahnrad *n*
clustering Schwarmbildung *f* (Moleküle)
cluster ion Cluster-Ion *n*
cluster of wells Bohrungsgruppe *f*
cluster-roll mill Mehrwalzenstuhl *m*
clutch Klemmvorrichtung *f*, Kupplung *f*,
  lösbare, **hydraulic** ~ Flüssigkeitskupplung *f*,
  hydraulische Kupplung
clutch box Kupplungsgehäuse *n*

clutch [coupling] Schaltkupplung *f*
clutch disk Kupplungsscheibe *f*
clutch facing Kupplungsverkleidung *f*
clutch housing Kupplungsgehäuse *n*
clutch lever Schalthebel *m*, Ausrückhebel *m*,
  Kupplungspedal *n*
clutch lining Kupplungsbelag *m*
clutch pedal Kupplungsfußhebel *m*,
  Kupplungspedal *n*
clutch shaft Kupplungswelle *f*
cluthalite Cluthalit *m* (Min)
cluytianol Cluytianol *n*
cluytic acid Cluytinsäure *f*
cluytyl alcohol Cluytylalkohol *m*
CMP (Abk) Cytidinmonophosphat *n*
cnicin Cnicin *n*
cnidionic acid Cnidiumsäure *f*
cnidium lactone Cnidiumlacton *n*
CoA (Abk) Coenzym A *n*
coacervation Koazervation *f*
coagulability Gerinnbarkeit *f*,
  Gerinnungsfähigkeit *f*
coagulable gerinnbar, gerinnungsfähig,
  koagulierbar
coagulant Gerinnungsmittel *n*, Koagulans *n*,
  Koagulationsmittel *n*, Koagulierungsmittel *n*
coagulate Koagulat *n*
coagulate *v* gerinnen, festwerden, gelieren,
  koagulieren
coagulated geronnen, koaguliert
coagulated mass Gerinnsel *n*
coagulating Koagulieren *n*
coagulating *a* erstarrend
coagulating agent Gerinnungsmittel *n*,
  Koagulierungsmittel *n*
coagulating bath Erstarrungsbad *n*
coagulating liquid Erstarrungsflüssigkeit *f*,
  Koagulierungsflüssigkeit *f*
coagulating property Koagulierungsfähigkeit *f*,
  Erhärtungsfähigkeit *f*
coagulation Gerinnung *f*, Ausflockung *f*,
  Festwerden *n*, Flockenbildung *f*, Käsen *n*,
  Koagulation *f*, Zusammenballung *f*, **defective**
  ~ Gerinnungsdefekt *m*, **heat of** ~
  Koagulationswärme *f*, **inducing** ~
  gerinnungsbegünstigend, **mechanical** ~
  mechanische Koagulation
coagulation analysis Gerinnungsanalyse *f* (Med)
coagulation factor Gerinnungsfaktor *m*
coagulator Gerinnstoff *m*, Gerinnungsmittel *n*
coagulum Blutklumpen *m* (Med), Blutkuchen *m*
  (Med), Gerinnsel *n*, Gerinnungsmasse *f*,
  Koagulat *n*
coal Kohle *f* (Min), **bituminous** ~ Fettkohle *f*,
  backende Kohle, **containing** ~ kohlenhaltig,
  dressing of the ~ Aufbereitung *f* der Kohle,
  dry burning ~ magere Kohle *f*, Magerkohle *f*,
  fat ~ Fettkohle *f*, fette Kohle, **fibrous** ~
  Faserkohle *f*, **free-ash** ~ magere Kohle *f*, **hard**

~ Steinkohle *f*, **high-ash** ~ aschereiche
Kohle, **high-volatile** ~ Gaskohle *f*, **large** ~
Stückkohle *f*, **lean** ~ Magerkohle *f*, **lump** ~
grobstückige Kohle, **mother of** ~ Fusain *n*,
nonbaking ~ Magerkohle *f*, **nonbituminous** ~
Magerkohle *f*, **non caking** ~ magere Kohle *f*,
Magerkohle *f*, **non clinkering** ~
schlackenreine Kohle *f*, **open burning** ~
Flammkohle *f*, **poor** ~ geringwertige Kohle,
powdered ~ Staubkohle *f*, **pulverized** ~
feingemahlene Kohle, **rough** ~ Förderkohle *f*,
Rohkohle *f*, **semibituminous** ~ halbfette
Kohle *f*, **short flaming** ~ kurzflammige
Kohle, **size of** ~ Kohlengröße *f*, **small** ~
Grießkohle *f*, Grus *m*, Kohlegrus *m*,
Kohlengestübe *n*, Kohlenklein *n*,
Kohlenmulm *m*, **smoke of** ~ Kohlendunst *m*,
to convert into ~ verkohlen, **working of** ~
Kohlenabbau *m* (Bergb), ~ **for decoloring**
E-Kohle *f*, ~ **for gas and vapor adsorption**
A-Kohle *f*, ~ **for gasmasks** G-Kohle *f*, ~ **for**
water purification WR-Kohle *f*, ~ **in powder**
form Pulver-Kohle *f*, **to pulverize** ~ Kohle *f*
zermahlen
coal analysis Kohlenanalyse *f*
coal ashes Kohlenasche *f*, Bockasche *f*
coal auger Kohlenbohrmaschine *f*
coal bearing shale Steinkohlenschiefer *m*
coal black Kohlenschwarz *n*
coal brass Markasit *m* (Min)
coal breaker Kohlenbrecher *m*
coal breaking Kohlenzerkleinerung *f*
coal breeze Kohlengestübe *n*, Kohlenmulm *m*
coal briquet[te] Preßkohle *f*
coal burning [furnace] Kohlefeuerung *f*
coal burning with flame Flammkohle *f*
coal cake Patentkohle *f*
coal carbonization Kohleentgasung *f* (Metall),
  Steinkohlenschwelung *f*
coal carbonizing plant Kokerei *f*
coal carbonizing practice Kokereitechnik *f*
coal charge Kohlengicht *f*
coal chute Kohlensturzbahn *f*
coal consumption Kohlenverbrauch *m*
coal crusher Kohlenbrecher *m*
coal crushing Kohlenzerkleinerung *f*
coal distillation Kohlenvergasung *f*
coal dressing Kohlenaufbereitung *f*
coal dust Kohlenstaub *m*, Staubkohle *f*,
  containing ~ grushaltig, **fine** ~
  Steinkohlenmehl *n*
coal dust explosion Kohlenstaubexplosion *f*
coal dust furnace Kohlenstaubfeuerung *f*
coalesce *v* ineinanderfließen, verschmelzen,
  koalisieren
coalescence Koaleszenz *f*, Vereinigung *f*,
  Verschmelzung *f*, Zusammenfließen *n*
coal firing Kohle[n]feuerung *f*, **pulverized** ~
  [Kohlen]Staubfeuerung *f*

coal gas Kohlengas *n*, Leuchtgas *n*,
    Steinkohlengas *n*
coal gas manufacture Steinkohlengaserzeugung *f*
coal gasoline Steinkohlenbenzin *n*
coal heap Kohlenhalde *f*
coal hopper Kohlentrichter *m*
coal hydrogenation Kohlehydrierung *f*
coal hydrogenizing plant Kohlehydrieranlage *f*
coal igniter Kohlenanzünder *m*
coalite Halbkoks *m* (HN)
coal-like kohlenartig
coal loader Räummaschine *f*
coal mine Kohlengrube *f*, Kohlenzeche *f*
    (Bergb), Steinkohlenbergwerk *n*
coal mining Kohlenbergbau *m*
coal mud Kohlenschlamm *m*, Schlammkohle *f*
coal oil Fließkohle *f*
coal pile Kohlenhalde *f*
coal pit Zeche *f* (Bergb)
coal plant Bekohlungsanlage *f*
coal pulverizer Kohlemahlanlage *f*
coal pulverizing plant Kohlenstaubanlage *f*
coal separating plant Kohlenscheider *m*
coal slate Kohlenschiefer *m*, Brandschiefer *m*
coal slimes Kohlenschlamm *m*
coal stamping machine Kohlenstampfmaschine *f*
coal tar Kohlenteer *m*
coal tar asphalt Teerpech *n*
coal tar dye Teerfarbstoff *m*
coal tar naphtha Steinkohlenteeröl *n*
coal tar oil Kohlenteeröl *n*, Steinkohlenöl *n*
coal tar pitch Steinkohlenteerpech *n*,
    Teerasphalt *m*
coal tipper Kohlenkipper *m*
coal vein Kohlenader *f* (Bergb)
coal washer Kohlenwaschmaschine *f*
coal washings Kohlenschlamm *m*
coarse grob, rauh
coarse adjustment Grobeinstellung *f*
coarse control Grobregelung *f*
coarse crush *v* grobzerkleinern
coarse-crystalline grobkristallin
coarse-fibered grobfaserig
coarse flour Aftermehl *n*
coarse-grained grobkörnig
coarse granulation Grobbruch *m*
coarse grind *v* grobzerkleinern
coarse lump [großer] Klumpen *m*
coarsely granular grobkörnig
coarse-meshed weitmaschig
coarseness Derbheit *f*, Grobheit *f*, Rauhigkeit *f*
coarse spiegeleisen Grobspiegeleisen *n*
coarse-threaded grobfädig, grobdrahtig
coarse tuning Grobabstimmung *f* (Radio)
coat Aufstrich *m*, Auftrag *m*, Beschichtung *f*,
    Hülle *f*, Mantelform *f* (Gieß), Überzug *m*,
    application of a ~ Auftragen *n* eines Striches
    (Pap), first ~ Grundanstrich *m*,

Grundieranstrich *m*, Grundierung *f*, to give
    the first ~ grundieren
coat *v* auftragen, aufstreichen, ausfüttern,
    auskleiden, bekleiden, beschichten,
    bestreichen, dragieren (Pharm), überziehen,
    umkleiden, ~ with tin *v* anzinnen
coated überzogen; angelaufen; imprägniert
    (Text), to become ~ sich beschlagen, sich
    überziehen, ~ on both sides beiderseitig
    aufgetragen, ~ on one side einseitig
    aufgetragen
coated fabric Streichstoff *m*
coated paper Lackpapier *n*
coater Auftragmaschine *f*,
    Beschichtungsmaschine *f*,
    Lackierungsmaschine *f*, Streichanlage *f*
coating Bekleidung *f*, Angußfarbe *f* (Keram),
    Aufdampfen *n*; Auflage *f* (Belag), Auftrag *m*
    (Belag), Bedampfung *f*, Beschichten *n*,
    Beschichtung *f* (Belag), Beschlag *m* (Belag),
    Schicht *f* (Belag), Streichen *n*, Überzug *m*
    (Belag), fireproof ~ Feuerschutzanstrich *m*,
    recessed ~ ausgesparte Lackierung *f*, ~ of ice
    Eisbelag *m*
coating activation Überzugaktivierung *f*
coating apparatus for paper
    Papierstreichvorrichtung *f*
coating composition Anstrichfarbe *f*, Lack *m*
coating compound Anstrichfarbe *f*,
    Anstrichmittel *n*, Beschichtungsmasse *f*,
    Beschichtungsmaterial *n*, Streichlack *m*,
    Überzugsmasse *f*
coating conductivity Überzugsleitfähigkeit *f*
coating drum Kandiertrommel *f*
coating machine Streichmaschine *f*,
    Auftragmaschine *f*, Beschichtungsmaschine *f*,
    Lackiermaschine *f*
coating mass Streichmasse *f*
coating material Anstrichmittel *n*,
    Beschichtungsstoff *m*
coating mixture Streichmischung *f*
coating of paint, first ~ Grundierschicht *f*
coating press Streichpresse *f*
coating process Überzugsverfahren *n*,
    Metallisierungsverfahren *n*
coating resin Überzugsharz *n*, Lackharz *n*
coating substance Streichmasse *f*
coating thickness Auftragsdicke *f* (z. B. Lack)
coating unit Beschichtungsanlage *f*
coating varnish Imprägnierlack *m*,
    Überzugslack *m*
coating weight Auflagegewicht *n*,
    Trockenauftragsmenge *f* (Lack)
coat of oil, first ~ Ölgrund *m*
coat protein Hüllprotein *n*
coat weight Auftragsgewicht *n*, Streichgewicht *n*
coaxial koaxial, konzentrisch, achsengleich
cobalamine (vitamin B$_{12}$) Cobalamin *n*

cobalt (Symb. Co) Kobalt *n*, **amorphous gray** ~
  Kobaltgraupen *pl*, **earthy** ~ Hornkobalt *m*
  (Min), **red** ~ Kobaltblüte *f* (Min), **specular** ~
  Kobaltspiegel *m* (Min), **tin-white** ~
  Glanzkobalt *m* (Min), **yellow earthy** ~
  Lederkobalt *m* (Min)
cobalt acetate Kobaltacetat *n*
cobalt alloy Kobaltlegierung *f*
cobalt aluminate Kobaltaluminat *n*
cobaltammine Kobaltammin *n*
cobalt ammonium sulfate
  Kobaltammoniumsulfat *n*
cobalt arsenate Kobaltarsenat *n*
cobalt bath Kobaltbad *n*
cobalt black Kobaltschwärze *f* (Min)
cobalt bloom Kobaltblüte *f* (Min)
cobalt blue Kobaltblau *n*, Königsblau *n*,
  Leydenerblau *n*, Schmaltblau *n*
cobalt bomb Kobaltbombe *f*
cobalt boride Kobaltborid *n*
cobalt bronze Kobaltbronze *f*
cobalt carbonate Kobaltcarbonat *n*
cobalt chloride Kobaltchlorid *n*
cobalt-chromium steel Kobaltchromstahl *m*
cobalt coloring matter Kobaltfarbstoff *m*
cobalt cyanide Cyankobalt *n*
cobalt diarsenide, native ~ Graupenkobalt *n*
  (Min)
cobalt dibromide Kobalt(II)-bromid *n*,
  Kobaltbromür *n* (obs)
cobalt dichloride Kobalt(II)-chlorid *n*,
  Kobaltchlorür *n* (obs)
cobalt diiodide Kobalt(II)-jodid *n*,
  Kobaltjodür *n* (obs)
cobalt fluoride Kobaltfluorid *n*
cobalt glance Kobaltglanz *m* (Min), Kobaltin *m*
  (Min), Kobaltit *m* (Min)
cobalt glass Kobaltglas *n*
cobalt green Kobaltgrün *n*,
  Kobaltoxydulzinkoxid *n* (obs)
cobalt hydrate Kobalthydroxid *n*
cobalt hydrogen carbonyl
  Kobaltcarbonylwasserstoff *m*
cobalt hydroxide Kobalthydroxid *n*
cobaltic chloride Kobalt(III)-chlorid *n*,
  Kobaltichlorid *n*
cobaltic compound Kobalt(III)-Verbindung *f*,
  Kobaltiverbindung *f*
cobaltic cyanide Kobalticyanid *n*,
  Kobalt(III)-cyanid *n*
cobaltic fluoride Kobalt(III)-fluorid *n*
cobaltichloride Kobaltichlorid *n*,
  Kobalt(III)-chlorid *n*
cobaltic oxide Kobalt(III)-oxid, Kobaltioxid *n*,
  Kobaltsesquioxid *n* (obs)
cobaltic potassium nitrite Kaliumkobaltnitrit *n*,
  Kobaltikaliumnitrit *n*
cobaltic salt Kobalt(III)-Salz *n*
cobalticyanic acid Kobalticyanwasserstoff[säure]

cobalticyanide Kobalticyanid *n*,
  Kobalt(III)-cyanid *n*
cobaltiferous kobalthaltig
cobalt(II) bromide Kobalt(II)-bromid *n*,
  Kobaltbromür *n* (obs)
cobalt(II) chloride Kobalt(II)-chlorid *n*,
  Kobaltchlorür *n* (obs), Kobaltochlorid *n*
cobalt(II) compound Kobalt(II)-Verbindung *f*,
  Kobaltoverbindung *f*,
  Kobaltoxydulverbindung *f* (obs)
cobalt(III) chloride Kobaltichlorid *n*,
  Kobalt(III)-chlorid *n*
cobalt(III) compound Kobalt(III)-Verbindung *f*,
  Kobaltiverbindung *f*
cobalt(III) cyanide Kobalticyanid *n*,
  Kobalt(III)-cyanid *n*
cobalt(III) fluoride Kobalt(III)-fluorid *n*
cobalt(II) iodide Kobalt(II)-jodid *n*,
  Kobaltjodür *n* (obs)
cobalt(III) oxide Kobalt(III)-oxid *n*,
  Kobaltioxid *n*, Kobaltsesquioxid *n* (obs)
cobalt(III) salt Kobalt(III)-Salz *n*
cobalt(II) nitrate Kobalt(II)-nitrat *n*,
  Kobaltonitrat *n*
cobalt(II) oxide Kobalt(II)-oxid *n*,
  Kobaltooxid *n*, Kobaltoxydul *n* (obs)
cobalt(II) salt Kobalt(II)-Salz *n*, Kobaltosalz *n*,
  Kobaltoxydulsalz *n* (obs)
cobalt(II) sulfate Kobalt(II)-sulfat *n*,
  Kobaltosulfat *n*
cobaltine Kobaltin *m* (Min), Kobaltit *m* (Min)
cobaltite Kobaltit *m* (Min), Glanzkobalt *m*
  (Min), Kobaltglanz *m* (Min), Kobaltin *m*
  (Min)
cobalt lattice Kobaltgitter *n*
cobalt-like kobaltartig
cobalt monoboride Kobaltborid *n*
cobalt monoxide Kobalt(II)-oxid *n*,
  Kobaltoxydul *n* (obs)
cobalt mordant Kobaltbeize *f*
cobalt nickel pyrite Nickelkobaltkies *m* (Min),
  Linneit *m* (Min)
cobalt nitrate Kobaltnitrat *n*
cobalt nitrite Kobaltnitrit *n*
cobaltocalcite Kobaltocalcit *m* (Min)
cobaltocobaltic oxide Kobalt(II)(III)-oxid *n*,
  Kobaltoxyduloxid *n*
cobaltocyanic acid
  Kobaltocyanwasserstoffsäure *f*
cobaltomenite Kobaltomenit *m* (Min)
cobalt ore Kobalterz *n* (Min), **black** ~
  Kobaltschwärze *f* (Min), **black earthy** ~
  Rußkobalt *m* (Min), **gray** ~ Graukobalterz *n*
  (Min), Speiskobalt *m* (Min), **specular** ~
  Spiegelkobalt *n* (Min), **transparent** ~
  Kobaltspiegel *m* (Min)
cobaltothiocyanic acid
  Kobaltorhodanwasserstoffsäure *f*

cobaltous bromide Kobalt(II)-bromid *n*,
 Kobaltbromür *n* (obs)
cobaltous chloride Kobalt(II)-chlorid *n*,
 Kobaltchlorür *n* (obs), Kobaltochlorid *n*
cobaltous compound Kobalt(II)-Verbindung *f*,
 Kobaltoverbindung *f*,
 Kobaltoxydulverbindung *f* (obs)
cobaltous iodide Kobalt(II)-jodid *n*,
 Kobaltjodür *n* (obs)
cobaltous nitrate Kobalt(II)-nitrat *n*,
 Kobaltonitrat *n*
cobaltous oxide Kobalt(II)-oxid *n*,
 Kobaltooxid *n*, Kobaltoxydul *n* (obs)
cobaltous salt Kobalt(II)-Salz *n*, Kobaltosalz *n*,
 Kobaltoxydulsalz *n* (obs)
cobaltous sulfate Kobalt(II)-sulfat *n*,
 Kobaltosulfat *n*
cobalt oxalate Kobaltoxalat *n*
cobalt phosphide Kobaltphosphid *n*
cobalt pigment Kobaltfarbe *f*, Kobaltfarbstoff *m*
cobalt plating Kobaltplattierung *f*
cobalt potassium nitrite Kobaltkaliumnitrit *n*
cobalt protoxide Kobalt(II)-oxid *n*,
 Kobaltooxid *n*, Kobaltoxydul *n* (obs)
cobalt pyrites Graupenkobalt *n* (Min),
 Kobaltarsenkies *m* (Min), Kobaltkies *m*
 (Min)
cobalt sesquioxide Kobaltioxid *n*,
 Kobaltsesquioxid *n*
cobalt siccative Kobaltsikkativ *n*
cobalt silicate Kobaltsilikat *n*
cobalt silicide Kobaltsilicid *n*
cobalt speiss Kobaltspeise *f*
cobalt sulfate Kobaltsulfat *n*
cobalt sulfide Kobaltsulfid *n*, Schwefelkobalt *m*
 (Min)
cobalt tetracarbonyl Kobalttetracarbonyl *n*
cobalt trichloride Kobalt(III)-chlorid *n*,
 Kobalttrichlorid *n*
cobalt trifluoride Kobalt(III)-fluorid *n*
cobalt ultramarine Königsblau *n*
cobalt vitriol Kobaltvitriol *n*
cobalt yellow Kobaltgelb *n*, Indischgelb *n*
cobalt zincate Kobaltgrün *n*, Kobaltzinkat *n*
cobamic acid Cobamsäure *f*
cobamide Cobamid *n*
cobamide coenzyme Cobamid-Coenzym *n*
cobbing Voranreicherung *f*
cobbler's wax Schuhpech *n*, Schusterpech *n*
cob coal Würfelkohle *f*
cobinic acid Cobinsäure *f*
cobwebbing Fadenziehen *f*, Fadenziehen *n*
 (Lack)
coca Coca *f*
cocaic acid Cocasäure *f*
cocain[e] Cocain *n*, Kokain *n*
cocaine hydrochloride Cocainhydrochlorid *n*
cocaine poisoning Cocainvergiftung *f*
cocaine substitute Cocainersatz *m*

cocainism Cocainvergiftung *f*
cocainize *v* mit Cocain behandeln
coca leaves Cocablätter *pl*
cocamine Cocamin *n*
cocarboxylase Cocarboxylase *f* (Biochem)
coccelic acid Coccelsäure *f*
cocceryl alcohol Coccerylalkohol *m*
coccic acid Coccinsäure *f*
coccine Coccin *n*
coccinine Coccinin *n*
coccinite Coccinit *m* (Min)
coccinone Coccinon *n*
coccolith Kokkolith *m* (Min)
cocculin Cocculin *n*
cochalic acid Cochalsäure *f*
cochenillic acid Cochenillesäure *f*
cochenilline Cochenillin *n*
cochineal Cochenille *f*, Cochenillenfarbstoff *m*,
 Koschenille *f*, Scharlachrot *n*
cochineal red Cochenillerot *n*, Koschenillerot *n*
cochineal scarlet Cochenillenscharlach *m*
cochineal tincture Cochenillentinktur *f*,
 Koschenilletinktur *f*
cochinilin Karminsäure *f*
cochin oil Cochinöl *n*
cocine Cocin *n*
cocinic cocinsauer
cocinine Cocinin *n*
cock Hahn *m* (Chem), Hahnventil *n* (Chem),
 self-acting ~ selbsttätiger Hahn, two-way ~
 Zweiwegehahn *m*
coclanoline Coclanolin *n*
coclaurine Coclaurin *n*
cocoa Kakao *m*
cocoa bean Kakaobohne *f* (Bot)
cocoa butter Kakaobutter *f*, Kakaoöl *n*
cocoa flavor Kakaoaroma *n*
cocoa husks Gruskakao *m*
cocoa oil Kakaoöl *n*
co-condensation Mischkondensation *f*
coco[nut] Kokosnuß *f*
coconut fat Kokosfett *n*
coconut meal Kokosschrot *n*
coconut milk Kokosmilch *f*
coconut oil Cocos[nuß]öl *n*, Palmnußöl *n*
coconut tallow Cocostalg *m*
cocoon Kokon *m*
cocoonase Cocoonase *f*
cod Kabeljau *m*
codamine Kodamin *n*
code Kennziffer *f*, Kodex *m*, Schlüssel *m*
codecarboxylase Codecarboxylase *f* (Biochem)
codehydrase I Codehydrase I *f* (Biochem),
 Cozymase I *f* (Biochem),
 Diphosphopyridinnucleotid *n* (Biochem)
codeine Codein *n*, Kodein *n*
codeine bromomethylate Kodeinbrommethylat *n*
codeine hydrobromide Codeinhydrobromid *n*

codeine hydrochloride Codeinchlorhydrat *n*,
 Codeinhydrochlorid *n*
codeine phosphate Codeinphosphat *n*,
 Kodeinphosphat *n*
codeinone Codeinon *n*, Kodeinon *n*
code name Deckname *m*
code of practice Richtlinien *f pl*
codeonal Codeonal *n*
codetermination Mitbestimmung *f*
codethyline Kodäthylin *n*
code word Deckname *m*
codimer Codimer *n*
codimerisation Codimerisation *f*
codiran Codiran *n*
codling moth Apfelwickler *m* (Zool)
cod-liver oil Lebertran *m*, **ferrated** ~
 Eisenlebertran *m*, **iodated** ~ Jodlebertran *m*
cod oil Codöl *n*, **thick** ~ Brauntran *m*
codol Retinol *n*
codon Codon *m*
coefficient Koeffizient *m*
coefficient equation Kennziffergleichung *f*
 (Math)
coefficient of expansion
 Ausdehnungskoeffizient *m*
coefficient of friction Reibungskoeffizient *m*
coefficient of hardness Brinellsche Härtezahl *f*
 (Brinell-Zahl)
coefficient of harmonic distortion Klirrfaktor *m*
 (Radio)
coefficient of performance Nutzeffekt *m*
coefficient of reduction Abnahmezahl *f*
coefficient of resistance Festigkeitskoeffizient *m*
coefficient of superficial thermal expansion
 Flächenausdehnungszahl *f*
coefficient of thermal expansion
 Wärmeausdehnungszahl *f*
coenzyme Coenzym *n* (Biochem)
coenzyme A Coenzym A *n*
coenzyme F Coenzym F *n*, Citrovorumfaktor *m*
coenzyme I Coenzym I *n*, Codehydrase I *f*,
 Cozymase I *f*, Diphosphopyridinnucleotid *n*
coenzyme Q Coenzym Q *n*
coercimeter Koerzitivkraftmesser *m*
coercive field Koerzitivfeld *n*
coercive force Koerzitivkraft *f*, **intrinsic** ~
 Induktionskoerzitivkraft *f* (Magn)
coercive force, coercivity Koerzitivität *f*
coercivity Koerzitivkraft *f*
coerulein Coerulein *n*
coerulinsulfuric acid Indigocarmin *n*
coexist *v* koexistieren
coexistence Koexistenz *f*
coexistence equation Koexistenzgleichung *f* (f.
 zweiphas. Zweistoffsystem)
cofactor Cofaktor *m*
coferment Coferment *n*
coffalic acid Kaffalsäure *f*
coffamine Coffamin *n*

coffearine Coffearin *n*
coffee Kaffee *m*, **cereal substitute for** ~
 Kaffeesurrogat *n*, **ingredient added to** ~
 Kaffeezusatz *m*
coffee bean Kaffeebohne *f*
coffee-colored kaffeebraun
coffee flavor Kaffeearoma *n*
coffee grounds Kaffeesatz *m*
coffee substitute Kaffee-Ersatz *m*
coffee surrogate Kaffee-Ersatz *m*
coffeine Koffein *n*, Coffein *n*, Kaffeebitter *n*
coffeinism Koffeinvergiftung *f*
coffer dam Fangdamm *m*
coffered ceiling Kassettendecke *f*
cog Zahn *m* eines Rades
cog *v* (Br. E.) blocken (Metall)
cogged gezahnt
cogged belt Zahnriemen *m* (Keilriemen)
cogged cylinder Kammwalze *f*
cogging mill Vorstraße *f*, Blockstraße *f*,
 Blockwalzwerk *n*, Vorwalzwerk *n*
cogging pass Blockwalzkaliber *n*
cogging roll Blockwalze *f*
cognac Kognak *m*
cogwheel Zahnrad *n*
cogwheel ore Bournonit *m* (Min)
cohenite Cohenit *m* (Min)
cohere *v* fritten (Radio), kohärieren
coherence Frittung *f* (Radio), Kohärenz *f*,
 Kohäsion *f*, **concept of** ~ Kohärenzbegriff *m*
coherency Kohärenz *f*, Kohäsion *f*,
 Zusammenhalt *m*
coherent kohärent, zusammenhängend
coherent radiation Kohärenzstrahlung *f*
coherer Fritter *m*, Kohärer *m* (Elektr)
coherer resistance Fritterwiderstand *m*
coherer terminal Fritterklemme *f*
coherer tube Fritterröhre *f*, Kohärerröhre *f*
cohesion Kohäsion *f*, Bindekraft *f*,
 Zusammenhalt *m*
cohesional force Kohäsionskraft *f* (Phys)
cohesion energy Kohäsionskraft *f* (Phys)
cohesion pressure Kohäsionsdruck *m* (Phys)
cohesion strength Kohäsionsfestigkeit *f*
cohesive kohäsiv, zäh, zusammenhängend
cohesive friction Haftreibung *f*
cohesiveness Kohäsionsvermögen *n*,
 Klebfähigkeit *f*, Kohäsionskraft *f* (Phys)
cohesive property Kohäsionseigenschaft *f*
cohesive resistance Trennwiderstand *m*
cohumulone Cohumulon *n*
Co I Codehydrase I *f* (Biochem), Coenzym I *n*
 (Biochem), Cozymase I *f* (Biochem)
coil Spirale *f*, Drahtring *m*, Knäuel *m*, Rolle *f*
 (Garn etc.), Schlange *f*, Schlangenrohr *n*,
 Spule *f* (Wicklung), Windung *f* (Draht), **axis
 of** ~ Spulenachse *f*, **solid** ~ Vollspule *f*
coil *v* aufspulen, wickeln
coil carrier Spulenhalter *m*

coil coating Bandlackierverfahren *n*
coil condenser Kühlschlange *f,*
　Schlangenkühler *m,* Spiralkühler *m*
coil cooler Schlangenkühler *m*
coil current supply Spulenstromversorgung *f*
coiled aufgerollt, aufgewickelt, gewunden,
　spiralförmig
coiled column Ringsäule *f* (Gaschromat)
coiled filament Spirale *f*
coiled pipe Rohrschlange *f*
coil evaporator Wendelrohrverdampfer *m*
coil field Spulenfeld *n*
coil galvanometer Spulengalvanometer *n*
coil of piping Röhrenstrang *m,* Rohrstrang *m*
coil spring Spiralfeder *f*
coil-type absorber Schlangenabsorber *m*
coin *v* prägen (massivprägen)
coinage metal Münzmetall *n*
coincide *v* kongruieren (Geom),
　zusammenfallen, zusammentreffen, ~ with
　sich decken mit (Geom)
coincidence Zufall *m,* Zusammentreffen *n*
coincidence analyzer Koinzidenzanalysator *m*
coincidence circuit Koinzidenzschaltung *f*
coincident gleichzeitig, [zeitlich]
　zusammenfallend
coining Prägen *n*
coir Kokosfaser *f*
coixol Coixol *n*
coke Koks *m,* **crushed** ~ Brechkoks *m,* **hard
　lumpy** ~ stückiger Koks, **heating with** ~
　Koksheizung *f,* **layer of** ~ Koksschicht *f,* **lean**
　~ Magerkoks *m,* **production of** ~
　Kokserzeugung *f,* **sifted** ~ Siebkoks *m,* **yield
　of** ~ Koksausbeute *f,* **granular** ~ **from lignite**
　Grudekoks *m,* **to wet** ~ **with water** Koks *m*
　ablöschen
coke *v* verkoken, backen, carbonisieren
coke basket Kokskorb *m*
coke [blast] furnace Kokshochofen *m*
coke breeze Grudekoks *m,* Koksgrus *m*
coke cake Kokskuchen *m*
coke dross Koksklein *n*
coke dust Koksstaub *m*
coke filter Koksfilter *m*
coke fork Koksgabel *f*
coke furnace Koks[brenn]ofen *m*
coke gas Koksgas *n*
coke oven Koksofen *m,* Kokerei *f,* **battery of** ~
　Koksofenbatterie *f,* **door of** ~ Koksofentür *f*
coke oven gas Koksofengas *n,* Kokereigas *n*
coke pig iron Koksroheisen *n*
coke plant Kokerei *f*
coke powder Kokspulver *n*
coke pusher Koksausdruckmaschine *f*
coke pushing machine Koksausdruckmaschine *f*
coke scrubber Kokswäscher *m*
coke sizing plant Koksklassieranlage *f*
coke storing place Kokslagerplatz *m*

coke washer Kokswäscher *m*
coke waste Koksabfall *m*
coking Verkokung *f,* Backen *n,*
　Kohlenvergasung *f,* Koksbereitung *f,* **rate of**
　~ Verkokungsgeschwindigkeit *f,* **temperature
　of** ~ Verkokungstemperatur *f,* ~ **in ovens**
　Ofenverkokung *f,* ~ **of pit coal** Verkokung *f*
　von Steinkohle
coking arch Feuergewölbe *n*
coking capability Verkokungsfähigkeit *f*
coking capacity Verkokungsfähigkeit *f*
coking chamber Schwelraum *m,*
　Verkokungskammer *f*
coking cylinder Schwelzylinder *m*
coking duff Koksgrus *m*
coking kiln Koksbrennofen *m*
coking period Garungszeit *f* (Metall)
coking plant Kokerei *f,* Verkokungsanlage *f*
coking practice Kokereiwesen *n*
coking process Verkokungsvorgang *m*
coking test Verkokungsprobe *f*
cola, liquid extract of ~ Colafluidextrakt *m*
colamine Colamin *n,* Äthanolamin *n*
colander Durchschlag *m,* Filtriertrichter *m*
cola nut Cola[nuß] *f,* Kola[nuß] *f*
colatannin Colatin *n*
colatin Colatin *n*
colchamine Colchamin *n*
colchic acid Colchinsäure *f*
colchicamide Colchicamid *n*
colchiceine Colchicein *n*
colchicide Colchicid *n*
colchicine Colchicin *n,* Kolchizin *n*
colchicinic acid Colchicinsäure *f*
colchicoside Colchicosid *n*
colchicum extract Colchicumextrakt *m*
colchicum preparation Colchicumpräparat *n*
colchicum seeds Herbstzeitlosensamen *m,*
　tincture of ~ Colchicumtinktur *f*
colchicum wine Colchicumwein *m*
colchide Colchid *n*
colchinol Colchinol *n*
colcothar Colcothar *m,* Eisenrot *n,*
　Englischrot *n,* Glanzrot *n,* Kaiserrot *n,*
　Totenkopf *m* (Min)
cold Kälte *f,* **getting** ~ Kaltwerden *n,* **susceptible
　to** ~ kälteempfindlich, **to accumulate** ~ Kälte
　aufspeichern
cold *a* kalt
cold age-hardening Kaltaushärtung *f*
cold aging Kaltaushärtung *f*
cold area aktivitätsfreier Raum *m* (Atom)
cold bath Kaltbad *n*
cold bending Kaltbiegen *n,* Kaltverformen *n*
cold bending test Kaltbiegeprobe *f*
cold-bend test Biegeprüfung *f* in der Kälte
coldblast *v* kalt blasen
cold blast cupola Kaltwindkupolofen *m*
cold blast practice Kaltwindbetrieb *m*

cold blast slide Kaltwindschieber *m*
cold boiler Vakuumkocher *m*
cold-brittleness Kaltbruch *m*, Kaltbrüchigkeit *f*
cold charge kalter Einsatz
cold charging Kalteinsatz *m*
cold chisel Hartmeißel *m*, Schrotmeißel *m*
cold cream Frostsalbe *f* (Pharm)
cold cure Kaltvulkanisation *f*
cold curing Kaltvulkanisation *f*
cold-cut varnish Kaltansatzlack *m*, Kaltlack *m*
cold-draw *v* kaltziehen
cold drawability Kaltverstreckbarkeit *f*
cold drawing Kaltziehen *n*
cold drawn kalt gezogen, kaltverstreckt
cold-dress *v* kaltpressen (Text)
cold dyeing Kaltfärben *n*
cold dyeing process Kältefärbeverfahren *n*
cold end Kaltlötstelle *f*, kalte Lötstelle
cold extractor Kaltextraktor *m*
cold finger Einhängekühler *m*
cold-form *v* kaltverformen, kaltprofilieren
cold forming Kaltverformen *n*, Kaltverformung *f*
cold gilding Kältevergoldung *f*
cold glaze Kaltglasur *f*
cold-hammer *v* hartschlagen, kalt hämmern
cold-harden *v* kalthärten
cold-hardening Kalthärten *n*
cold insulation Kälteisolierung *f*
cold insulator Kälteisoliermittel *n*,
    Kälteschutzmittel *n*
cold junction Kaltlötstelle *f*
cold-mold *v* kalt pressen (Kunstharz)
cold molding compound Kaltpreßmasse *f*
coldness Kälte *f*
cold patch Reparaturplättchen *n* zum
    Kaltauflegen
cold [plastic] paint Kaltanstrich *m*
cold-press *v* kalt pressen (Text)
cold-press welding Kaltpreßschweißung *f*
cold quenching Kalthärten *n*
cold resistance Kältebeständigkeit *f*,
    Kältefestigkeit *f*, Kaltwiderstandsfähigkeit *f*
cold-resistant kältebeständig
cold resisting property Frostbeständigkeit *f*
cold-roll *v* kalt walzen
cold rolled kalt gewalzt
cold rolled section Kaltprofil *n*
cold rolling mill Kaltwalzwerk *n*
cold rolling practice Kaltwalzerei *f*
cold room Kühlraum *m*
cold saw Kaltkreissäge *f*, Kaltsägemaschine *f*
cold sealable kaltsiegelfähig
cold sealing Kaltsiegelung *f*, Selbstklebung *f*
cold-setting kaltabbindend, kalthärtend
cold-setting adhesive kalthärtender Leim *m*
cold shaping Kaltverformung *f*
cold-short kaltbrüchig
cold-shortness Kaltbruch *m*, Kaltbrüchigkeit *f*,
    Kaltrissigkeit *f*

cold shrinking Kaltschrumpfen *n*
cold soldering Lötinseln *pl*
cold-soluble kaltlöslich
cold [spark] plug Kerze *f* hohen Wärmewertes
cold spots Lötinseln *pl*
cold-spray *v* kaltspritzen
cold storage Kühllagerung *f*
cold storage house Kühlhaus *n*
cold straining Kaltbeanspruchung *f*,
    Kaltreckung *f*
cold-stretch *v* kaltstrecken
cold tempering Kaltaushärtung *f*
cold test Kaltprobe *f*
cold trap Kühlfalle *f*
cold vulcanizing Kaltvulkanisation *f*
cold-water bath Kaltwasserbad *n*
cold-water paint Kaltwasserfarbe *f*
cold-work hardening Kaltverfestigung *f*
cold working Kaltbearbeitung *f*, Kaltformen *n*
    (Met), Kaltformgebung *f*, Kaltverarbeitung *f*,
    Kaltverfestigung *f*, Kaltverformung *f*,
    Rohgang *m*
cold working property Kaltverformbarkeit *f*
colemanite Colemanit *m* (Min)
coleol Coleol *n*
coli-bacillus Kolibazillus *m*, Escherichia coli,
    Kolibakterie *f*
colititol Colitit *m*
colitose Colitose *f*
collaborate *v* zusammenwirken
collaborator Mitarbeiter *m*
collagen Collagen *n*, Kollagen *n*, Leimgewebe *n*
    (Bindegewebssubstanz), Ossein *n*
collagenase Kollagenase *f* (Biochem)
collapse Zusammenbruch *m*, Durchbruch *m*,
    Schrumpfanfälligkeit *f*
collapse *v* durchbrechen, zusammenbrechen,
    zusammenfallen
collapse load Knicklast *f*
collapsible zusammenklappbar, zusammenlegbar
collapsible tube Quetschtube *f*
collar Hals *m*, Hubbegrenzer *m* (Masch),
    Kragen *m*, Manschette *f*, Reif *m*, Reifen *m*, ~
    of the roll Rand *m* der Walze, Ring *m* der
    Walze
collar beam Querbalken *m*, Sattel *m*
collar bearing Halslager *n*
collargol Kollargol *n*, Collargol *n*, Credesches
    Silber *n*
collargol ointment Collargolsalbe *f*
collar head screw Vierkantschraube *f*
collar journal Kammzapfen *m*
collar socket Halsdose *f*
collar thrust bearing Kammlager *n*
collar vortex Vortexring *m*
collatolic acid Collatolsäure *f*
collatolone Collatolon *n*
collaurin Collaurin *n*
colleague Mitarbeiter *m*

collect *v* [an]sammeln, auffangen, aufsammeln, vereinigen
collecting bar Sammelschiene *f*
collecting basin Auffangschale *f*
collecting brush Abnehmerbürste *f,* Überträgerbürste *f*
collecting chamber Sammelraum *m*
collecting cylinder Sammelzylinder *m*
collecting electrode Sammelelektrode *f*
collecting flask Auffangkolben *m*
collecting flue Sammelfuchs *m* (Metall)
collecting funnel Auffangtrichter *m*
collecting pipe Sammelleitung *f,* Sammelrohr *n*
collecting point Saugspitze *f*
collecting ring Sammelring *m,* Sammelrinne *f,* Saugring *m*
collecting side Saugseite *f*
collecting tank Sammelbehälter *m*
collecting tray Tropfschale *f*
collecting vessel Auffangbehälter *m,* Auffanggefäß *n,* Sammelbecken *n,* Sammelgefäß *n*
collection Sammlung *f*
collective advertising Gemeinschaftswerbung *f*
collective model Kollektivmodell *n*
collector Sammelbehälter *m,* Sammler *m,* Stromabnehmer *m* (Elektr)
collector anode Sammelanode *f*
collector comb Saugkamm *m*
collector electrode Abnahmeelektrode *f,* Auffangelektrode *f*
collector ring Kollektorring *m* (Elektr), Schleifring *m* (Elektr)
collector tank Leckbehälter *m*
collector vessel Sammelvorlage *f*
college Akademie *f*
collemplastrum Collemplastrum *n*
collet Fassung *f,* Metallring *m*
collide *v* kollidieren, zusammenprallen, zusammenstoßen
collidine Collidin *n,* Kollidin *n*
colliding Zusammenstoßen *n*
collimate *v* ausblenden (Opt), zusammenfallen lassen
collimating cone Kegelblende *f*
collimating slit Spaltblende *f* (Opt)
collimation Kollimation *f*
collimator Kollimator *m*
collineation Kollineation *f* (Math)
collinic acid Gelatinesäure *f*
collinol Collinol *n*
collision Kollision *f,* Aufeinanderprall *m,* Aufprall *m,* Stoß *m,* Zusammenprall *m,* Zusammenstoß *m,* duration of ~ Stoßdauer *f,* inelastic ~ unelastischer Stoß, probability of ~ Stoßwahrscheinlichkeit *f* (Atom), ~ of the first kind Stoß *m* erster Art (Atom)
collision complex Stoßkomplex *m*
collision cross section Stoßquerschnitt *m*

collision damping Stoßdämpfung *f*
collision deactivation Stoßentaktivierung *f*
collision density Kollisionsdichte *f* (Atom), Stoßdichte *f* (Atom)
collision energy Stoßenergie *f*
collision excitation Stoßanregung *f*
collision factor Stoßfaktor *m*
collision frequency Stoßhäufigkeit *f*
collision frequency per unit Kollisionszahl *f* (Atom)
collision impulse Stoßanregung *f*
collision mechanism Stoßvorgang *m*
collision momentum Stoßmoment *n*
collision multiplication Stoßvervielfachung *f*
collision particle Stoßteilchen *n*
collision partner Stoßpartner *m*
collision period Stoßdauer *f*
collision problem Stoßproblem *n*
collisions, effect of ~ Stoßeinfluß *m,* number of ~ Stoßzahl *f* (Atom)
collision strength Stoßstärke *f*
collision transition Stoßübergang *m*
collodion Kollodium *n,* Colloxylin *n,* Kollodion *n,* flexible ~ Kollodiumlösung *f*
collodion containing silver bromide Bromsilberkollodium *n*
collodion cotton Nitrozellulose *f,* Pyroxylin *n* Schießbaumwolle
collodion film Kollodiumschicht *f,* Kollodiumüberzug *m*
collodion-like kollodiumähnlich
collodion plate Kollodiumplatte *f*
collodion process Kollodiumverfahren *n*
collodion silk Kollodiumseide *f*
collodion solution Kollodiumlösung *f*
collodion substitute Kollodiumersatz *m*
collodion wool Kollodiumwolle *f,* Celloidin *n,* Collodiumwolle *f*
collodiotype Kollodiumbild *n*
collodium Collodium *n,* Kollodium *n*
collodium paper Kollodiumpapier *n*
collodium wool Kollodiumwolle *f*
colloid Kolloid *n,* addition of a ~ Kolloidzusatz *m,* hydrophilic ~ hydrophiles Kolloid, hydrophobic ~ hydrophobes Kolloid, lyophilic ~ lyophiles Kolloid, lyophobic ~ lyophobes Kolloid, natural ~ natürliches Kolloid
colloid *a* kolloidal
colloidal kolloidal, gallertähnlich, gallertartig
colloidal coal Kolloidkohle *f*
colloidal color Kolloidfarbe *f*
colloidal emulsion Emulsionskolloid *n,* Emulsoid *n*
colloidal gel Kolloidgel *n*
colloidality Kolloidcharakter *m*
colloidally disperse kolloiddispers
colloidal particle Kolloidteilchen *n*
colloidal silver Kollargol *n*

colloidal state Kolloidzustand *m*
colloidal substance Kolloid *n*
colloidal sulfur Kolloidschwefel *m*, Sulfidal *n*
colloid chemistry Kolloidchemie *f*
colloid diffusion Kolloiddiffusion *f*
colloid filter Gallertfilter *m*
colloid mill Kolloidmühle *f*
colloidochemical kolloidchemisch
colloid rectifier Kolloidgleichrichter *m*
colloid stability Kolloidbeständigkeit *f*
colloturine Colloturin *n*
collotype Kollotypie *f*, Lichtdruckverfahren
collotypy Kollotypie *f*, Lichtdruckverfahren *n*
colloxylin Colloxylin *n*
collyrite Kollyrit *m* (Min)
Cologne brown Kölnischbraun *n*
Cologne earth Braunkohlenpulver *n*
Cologne water Kölnisches Wasser *n*
colophene Colophen *n*, Kolophen *n*
colopholic acid Kolopholsäure *f*,
  Kolophonsäure *f*
colophonic harzsauer
colophonic acid Harzsäure *f*, Kolopholsäure *f*,
  Kolophonsäure *f*
colophonite Kolophonit *m* (Min), Pechgranat *m*
  (Min)
colophonium Colophonium *n*, Kolophonium *n*
colophonone Kolophonon *n*
colophony Kolophonium *n*, Geigenharz *n*,
  Terpentinharz *n*
color Farbe *f*, Färbung *f*, absence of ~
  Farblosigkeit *f*, addition of ~ Farbzusatz *m*,
  adjective ~ adjektive Farbe, alteration of ~
  Farbenänderung *f*, analysis of ~
  Farbenzerlegung *f*, change in ~
  Farbumschlag *m* (Chem), characteristic ~
  Eigenfarbe *f*, complementary ~
  Komplementärfarbe *f*, depth of ~
  Farbintensität *f*, Farbtiefe *f*, difference in ~
  Farbunterschied *m*, intensity of ~
  Farb[en]intensität *f*, intermediate ~
  Zwischenfarbe *f*, of the same ~ gleichfarbig,
  of uniform ~ homochrom, richness in ~
  Farbenreichtum *m*, substantive ~ substantive
  Farbe, thick ~ Dickfarbe *f*, to change ~ sich
  verfärben, to fix a ~ eine Farbe haltbar
  machen, to lose ~ ausbleichen, verblassen, ~
  of red wine Rotweinfarbe *f*
color *v* [ab]färben, anfärben, anstreichen,
  buntfärben, kolorieren, ~ red rot färben
colorability Anfärbevermögen *n*, Färbbarkeit *f*
colorable [an]färbbar
Colorado beetle Kartoffelkäfer *m* (Zool)
coloradoite Coloradoit *m* (Min)
color and dye industry Farbenindustrie *f*
colorant in paste form Abtönpaste *f*
colorant acceptance Pigmentverträglichkeit *f*
coloration Farbengebung *f*, Kolorieren *n*, ~ of
  smoke Rauchfärbung *f*

color batch Farbcharge *f*
color binder Farbenbindemittel *n*
color black Farbruß *m*
color-blind farbenblind
color blindness Farbenblindheit *f*
color boiler Farbkochapparat *m*
color boiling apparatus Farbkochapparat *m*
color brilliancy Farbenglanz *m*
color carrier Farbträger *m*, Chromophor *m*
color cast Farbstich *m*
color center Farbzentrum *n*
color centrifuge Farbenzentrifuge *f*
color change Farbänderung *f*,
  Farb[en]umschlag *m* (Chem)
color chart Farbtafel *f*, Farbenskala *f*,
  Farbentafel *f*, Farbstoffkarte *f*
color chemistry Farbenchemie *f*
color coat Farbenschicht *f*
color combination Farbzusammensetzung *f*
color constituent Farbenbestandteil *m*
color cycle system Farbringsystem *n*
color defect Farb[en]fehler *m*, Farbfehler *m*
color depth Farbenstärke *f*
color deviation Farbabweichung *f*
color disc Farbenscheibe *f*
color dispersion Farbenzerlegung *f*
color dryer Farbentrockner *m*
colored farbig, gefärbt, deeply ~ tiefgefärbt
colored clay Angußfarbe *f* (Keram)
colored earth Farbenerde *f*
colored glass Farbglas *n*
colored paper Buntpapier *n*
colored pencil lead Farbstiftmine *f*
color effect Farbeffekt *m*
color fastness Farbbeständigkeit *f*, Farbechtheit *f*
color fastness test Farbechtheitsprüfung *f*
color film Farbfilm *m*
color filter Farbfilter *n* (Opt)
color for porcelain painting
  [Porzellan]Einbrennfarbe *f*
color grad[u]ation Farbabstufung *f*
color grating Farbgitter *n*
color grid Farbgitter *n*
color grinder Farbenreiber *m*
color grinding Farbenreiben *n*
color grinding mill Farb[reib]mühle *f*
color grinding stone Farbenreibstein *m*
color homogenizer
  Farbenhomogenisiermaschine *f*
color homogenizing machine
  Farbenhomogenisiermaschine *f*
colorimeter Kolorimeter *n*, Farb[en]messer *m*
colorimetric kolorimetrisch
colorimetric analysis Kolorimetrie *f*
colorimetry Kolorimetrie *f*, Farbenmessung *f*,
  Farbmetrik *f*
colorin Colorin *n*
color in artificial light Abendfarbe *f*
color index Färbeindex *m*, Farbindex *m*

coloring Abtönen *n*, Färbung *f*, Farbgebung *f*,
Kolorit *n*, **art of** ~ Farbgebung *f*
coloring *a* färbend
coloring agent Färbehilfsmittel *n*, Farbmittel *n*
coloring liquor Farbenbrühe *f*
coloring material Farbstoff *m*
coloring matter Färbemittel *n*, Farbkörper *m*,
Farbsubstanz *f*, Pigment *n*, **indigoid** ~
indigoider Farbstoff
coloring power Farbkraft *f*, Anfärbevermögen *n*,
Färbevermögen *n*
coloring solution Farblösung *f*
coloring strength Farbstärke *f*
coloring substance Farbkörper *m*
color internegative Farbzwischennegativ *n*
(Phot)
colorizing glaze Farbglasur *f*
color lake Farblack *m*, Verschnittfarbe *f*
color layoff paper Farbfreilegungspapier *n*
colorless farblos, ungefärbt
colorlessness Farblosigkeit *f*
colorless oil coating Firnis *m*
color marking Kennfarbe *f*
color matching Farbangleichung *f*,
Farbanpassung *f*
color matching experiment
Abmusterungsversuch *m*
color mill Farbenmühle *f*
color milling machine Farbenreibmaschine *f*
color mixture curve Farbmischungskurve *f* (Opt)
color paste Farbpaste *f*
color pencil Farbstift *m*
color perception Farbempfinden *n*,
Farbenemfindung *f*, Farbensehen *n*, **true** ~
Farbentüchtigkeit *f*
color photography Farbphotographie *f*
color phototypy Lichtfarbendruck *m*
color picture Farbbild *n* (Phot)
color pit Farbengang *m*
color print Buntdruck *m*, Farbabzug *m* (Phot),
Farbenphotographie *f*
color printing Buntdruck *m*, Farbendruck *m*
color producing Farbe erzeugend
colorproof farbecht
color reaction Farbreaktion *f* (Chem)
color-react paper
Reaktionsdurchschreibepapier *n*
color refraction Farbenbrechung *f* (Opt)
color remover Farbenabbeizmittel *n*
color rendering Farbwiedergabe *f*
color reproduction Farbwiedergabe *f*
color reversal film Farbumkehrfilm *m* (Phot)
colors, change of ~ Farbenwechsel *m*, **gradation
of** ~ Farbenabstufung *f*, **laying-on of** ~
Farbenauftrag *m*, **mixture of** ~
Farb[en]mischung *f*, **of different** ~
ungleichfarbig, **production of** ~
Farbenerzeugung *f*, **theory of** ~ Farbenlehre *f*,
variety of ~ Farbenreichtum *m*

color saturation Vollfarbigkeit *f*
color scale Farbskala *f*
color scale of temperature Thermofarbenskala *f*
color schlieren method Farbschlierenverfahren *n*
color screen Farbfilter *n* (Opt), Farbraster *m*
(Opt)
color sensation Farbenempfindung *f*
color-sensitive farbempfindlich,
farbenempfindlich
color sensitivity Farbempfindlichkeit *f*
color separation Farbtrennung *f*
color shade Farbschattierung *f*, Farbstufe *f*
color sieve Farbensieb *n*
color slide Farbendiapositiv *n* (Phot)
color solution Farblösung *f*
color stability Farbbeständigkeit *f*,
Farbechtheit *f*
color television Farbfernsehen *n*
color temperature Farbtemperatur *f* (Opt)
color threshold Farbschwelle *f* (Opt)
color treatment Farbenbehandlung *f*
color triangle Farbendreieck *n*
color value Farbzahl *f*
color varnish Farbenlack *m*
colostrum Kolostrum *n* (Med)
colour (Br. E.) Farbe *f* (see color)
colter Pflugmesser *n*
colubridine Colubridin *n*
colubrine Colubrin *n*
columbamine Columbamin *n*
columbamine base Columbaminbase *f*
columbate Niobat *n*
Columbia black Kolumbiaschwarz *n*
columbia copalic acid Columbiakopalsäure *f*
Columbia green Kolumbiagrün *n*
Columbia yellow Columbiagelb *n*
columbic acid Niobsäure *f*, Colombosäure *f*
columbic anhydride Niobpentoxid *n*
columbic compound Niob(V)-Verbindung *f*
columbic fluoride Niobpentafluorid *n*
columbic oxide Niobpentoxid *n*
columbin Columbin *n*, Columbobitter *n*
columbite Kolumbit *m* (Min), Columbeisen *n*
(Min), Niobit *m* (Min)
columbium (Symb. Cb) Columbium *n*,
Niob[ium] *n*
columbium hydride Niobwasserstoff *m*
columbium oxide Nioboxid *n*
columbium pentafluoride Niobpentafluorid *n*
columbium pentoxide Niobpentoxid *n*
columbium(V) fluoride Niobpentafluorid *n*
columbium(V) oxide Niobpentoxid *n*
columbous compound Niob(III)-Verbindung *f*
column Turm *m*, Kolonne *f*, Pressenholm *m*,
Reihe *f*, Säule *f* (Chem); Spalte *f* (Buchdr),
chromatographic ~ chromatographische
Säule, **head of** ~ Kolonnenkopf *m*
column apparatus Kolonnenapparat *m*
columnar säulenförmig

columnar coal Stangenkohle *f*
columnar recombination Säulenrekombination *f*
columnar resistance Säulenwiderstand *m*
column balance Säulenwaage *f*
column chromatography
    Säulenchromatographie *f*
column efficiency Trennleistung *f*
column index Spaltenindex *m* (Buchdr)
column lamp Mastleuchte *f*
column of a press Preßsäule *f*
column packing Kolonnenpackung *f* (Dest)
column press Säulenpresse *f*
column shell Kolonnenmantel *m*
column steam apparatus Dampfzylinder *m*
column volume Kolonnenvolumen *n*
colupulone Colupulon *n*
colza Raps *m*
colza oil Colzaöl *n*, Rapsöl *n*, [refined] ~
    Rüböl *n*
comanic acid Komansäure *f*
comb Kamm *m*, Riet *n* (Web), Rietblatt *n*
    (Web), Scherblatt *n* (Web)
combat *v* bekämpfen
combat measure Bekämpfungsmaßnahme *f*
combinal Combinal *n*
combination Kombination *f*, Assoziation *f*,
    Verbindung *f*, Vereinigung *f*, Verknüfung *f*,
    Zusammenstellung *f*, heat of ~
    Bindungswärme *f*, mode of ~ Bindungsweise *f*
combination bevel Doppelschmiege *f*
combination boiler Verbundkessel *m*
combination color Mischfarbe *f*
combination fuse Doppelzünder *m*
combination lock Alphabetschloß *n*,
    Kombinationsschloß *n*
combination mill Verbundmühle *f*
combination seed disinfecting agent
    Kombibeizmittel *n*
combination stage Verbindungsstufe *f*
combinatorial analysis Kombinatorik *f*
combine *v* verbinden, kombinieren, vereinigen,
    zusammensetzen
combined gemischt, gemeinsam, verbunden,
    vereinigt, zusammengesetzt
combined action Zusammenwirkung *f*
combined effect Zusammenwirkung *f*
combined material Verbundstoff *m*
combined plastic Verbundpreßstoff *m*
combining Verbinden *n*, Verbindung *f* (chem.
    Vereinigung), Zusammenfügen *n*
combining ability Verbindungsfähigkeit *f*
    (Chem)
combining nozzle Mischdüse *f*
combining power Bindekraft *f*, Bindevermögen *n*
combining proportion Verbindungsverhältnis *n*
    (Chem)
combining volume Verbindungsvolumen *n*
combining weight Verbindungsgewicht *n*
combitron Combitron *n*

combustibility Brennbarkeit *f*, Verbrennbarkeit *f*
combustible Brennstoff *m*
combustible *a* [ver]brennbar, abbrennbar,
    feuergefährlich
combustible gas Brenngas *n*
combustible material Brennmaterial *n*
combustible mixture Brenngemisch *n*
combustion Verbrennung *f*, Entzündung *f*,
    accelerated ~ beschleunigte Verbrennung, air
    for ~ Feuerluft *f*, analysis by ~
    Verbrennungsanalyse *f*, complete ~
    vollständige Verbrennung, course of ~
    Verbrennungsablauf *m*, duration of ~
    Brenndauer *f*, Brennzeit *f*, heat of ~
    Verbrennungswärme *f*, imperfect or incomplete
    ~ unvollkommene Verbrennung, intensity of
    ~ Brennkraft *f*, Verbrennungsintensität *f*,
    phenomenon of ~ Verbrennungserscheinung *f*,
    product of ~ Verbrennungserzeugnis *n*, rapid
    ~ lebhafte Verbrennung, rate of ~
    Verbrennungsgeschwindigkeit *f*, result of ~
    Verbrennungsergebnis *n*, retarded ~
    verzögerte Verbrennung, slow ~ schleichende
    Verbrennung, träge Verbrennung, state of ~
    Verbrennungszustand *m*, temperature of ~
    Brenntemperatur *f*, zone of ~ Brennzone *f*, ~
    producing little smoke rauchschwache
    Verbrennung *f*
combustion analysis Verbrennungsanalyse *f*
combustion boat Glühschiffchen *n*,
    Verbrennungsschälchen *n*
combustion bomb Explosionskalorimeter *n*
combustion chamber Feuerraum *m*,
    Verbrennungskammer *f*,
    Verbrennungsraum *m*
combustion drag Brennwiderstand *m*
combustion energy Verbrennungsenergie *f*
combustion engine Explosionsmotor *m*,
    Verbrennungs[kraft]maschine *f*
combustion enthalpy Verbrennungsenthalpie *f*
combustion furnace Verbrennungsofen *m*
combustion gas Feuergas *n*, Heizgas *n*
combustion glass Einschmelzglas *n*
combustion method Verbrennungsmethode *f*
combustion motor Verbrennungsmotor *m*
combustion nozzle Verbrennungsdüse *f*
combustion period Brennperiode *f*
combustion pipet[te] Verbrennungspipette *f*
combustion plant Verbrennungsanlage *f*
combustion pressure Verbrennungsdruck *m*
combustion process Verbrennungsprozeß *m*
combustion product Verbrennungsprodukt *n*
combustion rate Verbrennungsgeschwindigkeit *f*
combustion residue Verbrennungsrückstand *m*
combustion shaft Verbrennungsschacht *m*
combustion space Verbrennungsraum *m*
combustion tube Verbrennungsröhre *f*,
    Verbrennungsrohr *n*
combustion tuyère Verbrennungsdüse *f*

combustion velocity
Verbrennungsgeschwindigkeit *f*
combustion zone Verbrennungszone *f*
comenamic acid Komenaminsäure *f*
comenic acid Komensäure *f*
come off *v* abgehen, ablösen
cometary orbit Kometenbahn *f* (Astr)
comfortable komfortabel, bequem
command Befehl *m*
command *v* beherrschen
command reference input Führungsgröße *f*
(Regeltechn)
command voltage Steuerspannung *f*
comment Stellungnahme *f*
commerce Handel *m*
commercial handelsüblich, käuflich
commercial ad Geschäftsanzeige *f*
commercial article Handelsartikel *m*
commercial chemist Handelschemiker *m*
commercial customs, pertaining to ~
handelsüblich
commercial grade Handelssorte *f*
commercial iron Handelseisen *n*
commercially available käuflich
commercial product Handelsware *f*
commercial quality Handelsqualität *f*
commercial rubber Handelskautschuk *m*
commercial standard Industrienorm *f*
commission Auftrag *m*, Vermittlungsgebühr *f*
commission *v* beauftragen
commission dyer Lohnfärber *m*
committee Ausschuß *m*, ~ **of inquiry**
Untersuchungsausschuß *m*
committee of experts Fachausschuß *m*
commodities Bedarfsartikel *m*, Waren *f pl*
commodity Gebrauchsgegenstand *m*,
Gebrauchsgut *n*, Handelsartikel *m*
common gewöhnlich, allgemein, gemein,
gemeinsam
common camomile Feldkamille *f* (Bot)
common cold Erkältung *f* (Med),
Erkältungskrankheit *f* (Med)
common denominator Hauptnenner *m* (Math),
Generalnenner *m* (Math)
common fraction echter Bruch (Math)
common logarithm Dezimallogarithmus *m*
common main Sammelleitung *f*
common nettle Brennesselkraut *n* (Bot)
common pitch Pichpech *n*
commonplace alltäglich, abgedroschen, banal,
gewöhnlich
common privet Liguster *m* (Bot)
common rubber tree Federharzbaum *m*
common salt Kochsalz *n*, Natriumchlorid *n*,
containing ~ kochsalzhaltig
common salt solution Kochsalzlösung *f*,
Natriumchloridlösung *f*
common tin Probezinn *n*
commotion Erregung *f*, Erschütterung *f*

communicable übertragbar
communicating kommunizierend
communicating gallery Verbindungskanal *m*
communication Mitteilung *f*, Nachricht *f*, **system**
of ~ Nachrichtensystem *n*
communication cable Schwachstromkabel *n*
commutable vertauschbar, austauschbar
commutable matrix vertauschbare Matrize *f*
commutating kommutierend
commutation Kommutation *f*, Stromumkehr *f*,
Stromwendung *f* (Elektr), Vertauschung *f*
commutation relation Vertauschungsrelation *f*
commutation rule Vertauschungsregel *f*
commutative auswechselbar, kommutativ
commutative law Kommutativität *f* (Math)
commutator Kollektor *m* (Elektr),
Kommutator *m* (Math), Stromwender *m*
(Elektr), Umschalter *m* (Motor),
Zündverteiler *m*, **cooling of the** ~
Kollektorkühlung *f*, **diameter of** ~
Kollektordurchmesser *m*, **length of** ~
Kollektorlänge *f*, **size of** ~ Kollektorgröße *f*,
~ **with two directions** Doppelunterbrecher *m*
commutator bars, number of ~
Kollektorlamellenzahl *f*
commutator dimension Kollektorabmessung *f*
commutator grinding device
Kollektorschleifvorrichtung *f*
commutator insulation Kollektorisolierung *f*,
Stromwenderisolation *f*
commutator lubricant Kollektorschmiere *f*
commutator motor Kollektormotor *m*
commutator nut Kollektormutter *f*
commutator polishing paper
Kollektorschleifpapier *n*
commutator rectifier Kommutatorgleichrichter *m*
commutator ring Kollektorring *m*
commutator switch Kreuzschalter *m*
commutator turning device
Kollektorabdrehvorrichtung *f*
commutator voltage Kollektorspannung *f*
commute *v* kommutieren (Elektr), vertauschen,
auswechseln
compact dicht, fest, kompakt, raumsparend
compact *v* verdichten, zusammendrücken
compacter Müllverdichter *m*
compacting rolls, paired ~
Doppelwalzenpresse *f*
compactness Festigkeit *f*, Kompaktheit *f*
comparability Vergleichbarkeit *f*
comparable vergleichbar, ähnlich
comparative relativ, verhältnismäßig
comparative dyeing Vergleichsfärbung *f* (Färb)
comparative experiment Vergleichsversuch *m*
comparative figure Vergleichszahl *f*
comparatively verhältnismäßig
comparative method Vergleichsverfahren *n*
comparative value Vergleichswert *m*

comparator Komparator *m*,
Komparatorschaltung *f* (Elektr)
comparator circuit Komparatorschaltung *f*
(Elektr), Vergleichsschaltung *f* (Elektr)
compare *v* vergleichen, gegenüberstellen
comparison Vergleich *m*, basis of ~
Vergleichsunterlage *f*, numerical ~
zahlenmäßiger Vergleich, standard of ~
Vergleichsmaßstab *m*
comparison dyeing Vergleichsfärbung *f* (Färb)
comparison electrode Vergleichselektrode *f*
comparison function Vergleichsfunktion *f*
comparison lamp Vergleichslichtquelle *f* (Opt)
comparison line Vergleichslinie *f*
comparison measuring Vergleichsmessung *f*
comparison method Vergleichsmethode *f*
comparison section Vergleichsstrecke *f*
comparison solution Vergleichslösung *f* (Math)
comparison spectrum Vergleichsspektrum *n*
(Spektr)
comparison test Vergleichsversuch *m*
compartment Abteilung *f*, Fach *n*, Zelle *f*
compartment mill Mehrkammermühle *f*
compass Kompaß *m*
compass deviation Kompaßabweichung *f*
compasses, pair of ~ Zirkel *m*
compass needle Magnetnadel *f*
compass set Zirkelkasten *m*
compatibility Kompatibilität *f*, Vereinbarkeit *f*,
Verträglichkeit *f* (Pharm)
compatibility condition
Kompatibilitätsbedingung *f*
compatibility problem
Materialauswahlproblem *n* (Atom)
compatibility relation Kompatibilitätsrelation *f*
compatible verträglich, vereinbar, partly ~
teilverträglich
compendium Handbuch *n*
compensate *v* kompensieren, angleichen,
aufheben (ausgleichen), ausbalanzieren,
ausgleichen, auswuchten, vergüten
compensated hangers Kompensationsgehänge *n*
compensated state Ausgeglichenheit *f*
compensating allowance Ausgleichszulage *f*
compensating current Ausgleichsstrom *m*
compensating curve Ausgleichskurve *f*
compensating gear Ausgleichsrad *n*
compensating hanging
Kompensationsaufhängung *f*
compensating heat Ausgleichswärme *f*
compensating jet Ausgleichdüse *f*
compensating magnet Ausgleichmagnet *m*
compensating network preamplifier Entzerrer *m*
compensating pipe Ausgleichrohr *n*
compensating pole Kompensationspol *m*
compensating resistance
Kompensationswiderstand *m* (Elektr),
Regulierwiderstand *m* (Elektr),
Vorwiderstand *m* (Elektr)

compensating suspension Ausgleichsaufhängung *f*
compensating voltage Ausgleichspannung *f*
compensation Ausgleich *m*, Entschädigung *f*,
Ersatz *m*, Kompensation *f*
compensation claim Schadenersatzanspruch *m*
compensation developer Ausgleichentwickler *m*
(Phot)
compensation line Ausgleichsleitung *f*
compensation measurement Ausgleichsmessung *f*
compensation method Kompensationsmethode *f*,
Kompensationsverfahren *n* (Elektr)
compensation set Ausgleichaggregat *n*
compensation theorem Kompensationstheorem *n*
(Elektr)
compensation value Ausgleichswert *m*
compensation water Zuschußwasser *n*
compensator Ausgleicher *m*,
Ausgleichsvorrichtung *f*, Kompensator *m*
(Elektr)
compete, able to ~ konkurrenzfähig
compete *v* konkurrieren
competent einschlägig, fachkundig
competing konkurrierend
competition Wettbewerb *m*, Konkurrenz *f*
competitive konkurrenzfähig, konkurrierend,
wettbewerbsfähig
competitive inhibitor kompetitiver Inhibitor *m*
competitor Konkurrent *m*, Wettbewerber *m*
compilation Zusammenstellung *f*
compile *v* zusammenstellen (z. B. Material)
complain *v* reklamieren, sich beschweren
complaint Beschwerde *f*, Reklamation *f*
complementarity Komplementarität *f*
complementary komplementär, ergänzend
complementary chromaticity
Komplementärfarbigkeit *f* (Opt)
complementary color Komplementärfarbe *f*
complementary surface Ergänzungsfläche *f*
complete vollständig, komplett, restlos, total,
vollkommen
complete *v* vervollständigen, abschließen,
ergänzen, fertigstellen, vollenden
complete blowpipe Lötbesteck *n*
completed vollendet
completeness Vollständigkeit *f*
completeness relation Vollständigkeitsrelation *f*
completion Vollendung *f*, Abschluß *m*,
Fertigstellung *f*, Vervollständigung *f*
complex Komplex *m*, activated ~ aktivierter
Komplex, inert ~ inerter Komplex, ionogenic
~ ionogener Komplex, stable ~ beständiger
Komplex, unstable ~ unbeständiger Komplex
complex *a* komplex, mehrteilig, schwierig,
zusammengesetzt
complex *v* komplexieren, in einem Komplex
binden
complex chemistry Komplexchemie *f*
complex compound Komplexverbindung *f*
complex formation Komplexbildung *f*

complexing Komplexbildung *f*
complexing agent Komplexbildner *m*,
  Maskierungsmittel *n*
complex ion Komplexion *n*
complex metal salt Metallkomplexsalz *n*
complex number Komplexzahl *f* (Math)
complexometric komplexometrisch
complexon Komplexon *n*
complex salt Komplexsalz *n*
complicate *v* erschweren, komplizieren
complicated kompliziert
complication Erschwerung *f*, Komplikation *f*,
  Verwick[e]lung *f*
comply [with] *v* befolgen
component Komponente *f*, Bestandteil *m*,
  Ingrediens *n*, **basic** ~ Grundbestandteil *m*,
  primary ~ Grundbestandteil *m*, **volatile** ~
  flüchtige Komponente
component of current sequence
  Stromsequenzkomponente *f*
component resistance
  Komponentenwiderstand *m*
compose *v* zusammensetzen
composed zusammengesetzt, **to be** ~ **of** bestehen
  aus
composite construction Mischbauweise *f*
composite container Kombidose *f*
composite feed Mischfutter *n*
composite film Verbundfolie *f*
composite fuel Gemischkraftstoff *m*
composite material Verbund[werk]stoff *m*
composite mold Mehrfachform *f*
composite structure Verbundkonstruktion *f*
composition Zusammensetzung *f* (Chem),
  chemical ~ chemische Zusammensetzung,
  critical ~ kritische Zusammensetzung,
  percentage ~ prozentuale Zusammensetzung,
  variable ~ variable Zusammensetzung, ~ **of**
  material Stoffzusammensetzung *f*
composition board Holzfaserplatte *f*
composition metal Legierung *f*,
  Kompositionsmetall *n*
composition plane Verwachsungsebene *f* (Krist)
composition surface Verwachsungsfläche *f*
  (Krist)
compost Kompost *m* (Agr), Düngeerde *f* (Agr),
  Mischdünger *m* (Agr)
composting Düngung *f*
compound Verbindung *f* (Chem), **alicyclic** ~
  alicyclische Verbindung, **aliphatic** ~
  aliphatische Verbindung, **asymmetric** ~
  asymmetrische Verbindung, **binary** ~ binäre
  Verbindung, **conjugated** ~ konjugierte
  Verbindung, **cyclic** ~ cyclische Verbindung,
  exothermic ~ exotherme Verbindung,
  heterocyclic ~ heterocyclische Verbindung,
  insoluble ~ unlösliche Verbindung (Chem),
  intermetallic ~ intermetallische Verbindung,
  isomeric ~ isomere Verbindung, **metastable**

~ metastabile Verbindung, **naturally**
  occurring ~ in der Natur vorkommende
  Verbindung, **organic** ~ organische
  Verbindung, **physical** ~ physikalische
  Verbindung, **polar** ~ polare Verbindung,
  polycyclic ~ mehrkernige Verbindung,
  resonance ~ mesomere Verbindung,
  unsaturated ~ ungesättigte Verbindung
compound *v* [ver]mischen, legieren,
  zusammensetzen
compound casting Verbundguß *m* (Gieß)
compound cast rail Verbundgußschiene *f*
compound color zusammengesetzte Farbe
compound dynamo Verbundmaschine *f*
compounded oil Compoundöl *n*
compound engine Compoundmaschine *f*,
  Verbundmaschine *f*
compound excitation Doppelschlußerregung *f*,
  Verbunderregung *f*
compound eye Facettenauge *n* (Zool)
compound fertilizer Mischdünger *m*,
  Volldünger *m*
compound fraction Doppelbruch *m*,
  zusammengesetzter Bruch *m*
compound generator Compoundmaschine *f*
  (Elektr)
compound glass Verbundglas *n*
compound impregnated paper
  Compoundpapier *n*
compounding Vermischen *n*, Vermischung *f*
compounding procedure Mischungsherstellung *f*
compounding room Mischraum *m*
compound lens Verbundlinse *f* (Opt)
compound magnet Blättermagnet *m*,
  Lamellenmagnet *m*, Magnetbündel *n*
compound matrix Kombinationsmatrize *f*
  (Math)
compound nuclear model
  Compoundkernmodell *n* (Atom)
compound nucleus Compoundkern *m* (Atom)
compound pipe Verbundrohr *n*
compound steam engine
  Verbunddampfmaschine *f*
compound steam pump Verbunddampfpumpe *f*
compound steel Compoundstahl *m*
compound system Gesamtsystem *n*
compound[-wound] dynamo
  Doppelschlußgenerator *m* (Elektr)
compound [-wound] motor Kompoundmotor *m*
  (Elektr)
compregnate *v* komprimieren (Holz)
compregnated wood Preßholz *n*
comprehend *v* begreifen, erfassen, umfassen
comprehensible verständlich
comprehension Fassungsvermögen *n*
comprehensive umfassend, reichhaltig
compress Kompresse *f* (Med), Umschlag *m*
  (Med)

compress *v* komprimieren, stauchen, verdichten, zusammendrücken, zusammenpressen
compressable zusammenpreßbar
compressed, highly ~ hochkomprimiert
compressed air Druckluft *f,* Preßluft *f*
compressed air bottle Druckluftflasche *f*
compressed air conditioning Zusatzbelüftung *f*
compressed-air cylinder Preßluftflasche *f*
compressed air deoiling Druckluftentölung *f*
compressed air equipment Druckluftausrüstung *f*
compressed air fittings Druckluftarmaturen *pl*
compressed air hammer Luftdruckhammer *m*
compressed air lamp Druckluftleuchte *f*
compressed air motor Preßluftmotor *m*
compressed air mounting Preßluftarmatur *f*
compressed air plant Druckluftanlage *f*
compressed air receiver Druckluftbehälter *m*
compressed air spraying apparatus
Druckluftspritzapparat *m*
compressed air stirrer Preßluftrührwerk *n*
compressed air tubing Druckluftschlauch *m*
compressed air valve Druckluftstutzen *m*
compressed carbon mass Kohlenstampfmasse *f*
compressed cone connection
Klemmkegelverbindung *f*
compressed gas Druckgas *n*
compressed gas cylinder Druckgasflasche *f*
compressed oxygen Hochdrucksauerstoff *m*
compressed steel Preßstahl *m*
compressed vapor refrigerator
Dampfkompressionskältemaschine *f*
compressibility Zusammendrückbarkeit *f,*
Kompressibilität *f,* Komprimierbarkeit *f,*
Verdichtbarkeit *f,* **effect of** ~
Kompressibilitätseinfluß *m,* **isentropic** ~
isentropische Kompressibilität *f,* **isothermal** ~
isotherme Kompressibilität *f*
compressibility effect Kompressibilitätseinfluß *m*
compressibility factor Kompressibilitätsfaktor *m*
(Mech), Realgasfaktor *m,*
Verdichtungsfaktor *m*
compressible zusammendrückbar, kompressibel,
komprimierbar, verdichtbar
compressing Komprimieren *n,* Verdichten *n*
compressing *a* komprimierend, **high** ~
hochverdichtend
compression *v* Kompression *f,* Druck *m,*
Druckbeanspruchung *f,* Komprimierung *f,*
Stauchen *n,* Stauchung *f,* Verdichtung *f,*
adiabatic ~ adiabatische Kompression
(Therm), adiabatische Verdichtung, **degree of**
~ Kompressionsgrad *m,* **end-to-end** ~
Stauchen *n* in Längsrichtung, **heat of** ~
Kompressionswärme *f,* Verdichtungswärme *f,*
isothermal ~ isothermische Verdichtung, **loss
of** ~ Kompressionsverlust *m,* **side-to-side** ~
Stauchen *n* in Querrichtung, **top-to-bottom** ~
vertikales Stauchen *n*
compressional vibration Druckschwingung *f*

compression coupling Druckkupplung *f,*
Klemmkupplung *f*
compression cylinder Preßzylinder *m*
compression direction Stauchrichtung *f*
compression elasticity Druckfestigkeit *f*
compression heat Kompressionswärme *f*
compression ignition Verdichtungszündung *f*
compression ignition engine Dieselmotor *m*
compression joint Klemmverbindung *f*
compression load Druckbelastung *f*
compression mold Preßform[maschine] *f*
compression molding Formpressen *n*
compression molding material Formpreßstoff *m*
compression pressure Verdichtungsdruck *m*
compression ratio Kompressionsverhältnis *n,*
Verdichtungsgrad *m,*
Verdichtungsverhältnis *n*
compression refrigerating machine
Kompressionskältemaschine *f*
compression refrigerator
Kompressionskühlschrank *m*
compression resistance
Kompressionswiderstand *m*
compression ring Kolbendichtungsring *m*
compression section Druckzone *f*
compression space Kompressionsraum *m*
compression spring Druckfeder *f*
compression strain Beanspruchung *f* auf Druck,
Druckspannung *f*
compression stress Druckbeanspruchung *f,*
Druckspannung *f*
compression stroke Verdichtungshub *m* (Mot)
compression surge drum Windkessel *m*
compression test Druckversuch *m,*
Stauchversuch *m*
compression vacuum gauge
Kompressionsvakuummeter *n*
compression wave Verdichtungswelle *f*
compression work Verdichtungsarbeit *f*
compression yield point Quetschgrenze *f*
compression zone Verdichtungszone *f*
compressive cleaving Druckspaltung *f*
compressive force Druckkraft *f*
compressive impact stress
Schlagdruckbeanspruchung *f*
compressive strength Druckfestigkeit *f*
compressive stress Druckbeanspruchung *f,*
Druckspannung *f*
compressive twin formation
Druckzwillingsbildung *f* (Krist)
compressor Kompressor *m,* Verdichter *m*
compressor casing Verdichtergehäuse *n*
compressor oil Kompressorenöl *n*
Compton effect Comptoneffekt *m*
Compton electron Comptonelektron *n,*
Rückstoßelektron *n*
comptonite Comptonit *m* (Min)
Compton radiation Compton-Strahlung *f*
Compton scattering Compton-Streuung *f*

Compton shift Comptonverschiebung *f*
Compton-Simon experiment
  Compton-Simonscher Versuch *m* (Atom)
Compton wave-length Compton-Wellenlänge *f*
compulsion Zwang *m*
compulsory obligatorisch
compulsory license regulation
  Zwangslizenzbestimmung *f*
computability Berechenbarkeit *f* (Math)
computable berechenbar
computation Berechnung *f,* Ausrechnung *f,*
  Kalkulation *f,* Rechnen *n,* **formula of** ~
  Berechnungsformel *f*
computational specifications Berechnungsnorm *f*
computation of fields Feldberechnung *f*
compute *v* berechnen, [aus]rechnen, kalkulieren
computer Rechner *m,* Computer *m,*
  Impulsrechner *m,* Rechenautomat *m,*
  Rechenmaschine *f,* **electronic** ~
  Elektronengehirn *n*
computer color matching Farbrezeptberechnung *f*
computer science Informatik *f*
computing machine Rechenmaschine *f*
computor Rechenanlage *f*
conamine Conamin *n*
conanine Conanin *n*
conarrhimine Conarrhimin *n*
concave konkav, hohlgeschliffen, hohl[rund]
concave fillet weld Hohlkehlschweißung *f*
concave gitter spectrograph
  Konkavgitterspektrograph *m*
concave glass Hohlglas *n*
concave grating konkaves Gitter (Opt),
  Konkavgitter *n*
concave ground hohlgeschliffen
concave lens Hohllinse *f* (Opt), Konkavlinse *f*
  (Opt), Zerstreuungslinse *f*
concave mirror Vergrößerungsspiegel *m,*
  Brennspiegel *m,* Hohlspiegel *m,*
  Konkavspiegel *m*
concave mirror cathode Hohlspiegelkathode *f*
concavity Einbuchtung *f,* Hohlrundung *f*
concavo-concave bikonkav
concavo-convex hohlerhaben, konkav-konvex
  (Opt)
conceal *v* verbergen, verstecken
concealed verdeckt
conceive *v* begreifen, auffassen
concentrate Konzentrat *n,*
  Aufbereitungsprodukt *n*
concentrate *v* konzentrieren, anreichern,
  aufbereiten (Techn), eindicken, einengen
  (Chem), entwässern, konzentrieren,
  verdichten
concentrate by evaporation *v* eindampfen
concentrated konzentriert, angereichert,
  eingedickt, verstärkt, **highly** ~
  hochkonzentriert
concentrated compound Stammschmelze *f*

concentrated metal Dublierstein *m*
concentrates Schlamm *m,* ~ **of ore** Erzschlich *m*
concentrating Konzentrieren *n,* Bündelung *f*
  (Strahlen), Eindampfen *n,* Eindicken *n,*
  Einengen *n* (Chem), Verdichten *n*
concentrating column (in distillation plant)
  Verstärkersäule *f*
concentrating plant Eindickanlage *f,*
  Konzentrationsanlage *f*
concentration Konzentrierung *f,* Anreicherung *f,*
  Eindampfung *f,* Eindickung *f,*
  Entwässerung *f,* Gehalt *m,* Konzentration *f;*
  Lösungsdichte *f,* **alteration of** ~
  Konzentrationsänderung *f,* **at the maximum of**
  ~ höchstprozentig, **degree of** ~
  Sättigungsgrad *m,* **dependence of** ~
  Konzentrationsabhängigkeit *f,* **initial** ~
  Anfangskonzentration *f,* **molar** ~ molare
  Konzentration, **range of** ~
  Konzentrationsbereich *m,* **ratio of** ~
  Konzentrationsverhältnis *n,* ~ **by boiling**
  Einkochung *f,* ~ **of solution** Konzentration *f*
  der Lösung
concentration cell Konzentrationselement *n,*
  Konzentrationskette *f* (Elektrochem)
concentration change Konzentrationsänderung *f*
concentration current Konzentrationsstrom *m*
concentration distribution
  Konzentrationsverteilung *f*
concentration excess
  Konzentrationsüberschuß *m*
concentration gradient Konzentrationsgefälle *n*
concentration melting
  Konzentrationsschmelzen *n*
concentration of a saturated solution
  Sättigungskonzentration *f*
concentration of ions Ionenverdichtung *f*
concentration plant Eindampfanlage *f*
concentration polarization
  Konzentrationspolarisation *f,*
  Verdichtungspolarisation *f*
concentration slag Spurschlacke *f*
concentration unit Konzentrationseinheit *f*
concentrative effect Packungseffekt *m*
concentrator Eindickzylinder *m,*
  Konzentrationsapparat *m*
concentric gasket Manschettendichtung *f*
concept Begriff *m,* **basic** ~ Grundbegriff *m*
conception Begriff *m,* Anschauung *f*
  (Vorstellung), Auffassung *f*
conceptual begrifflich
concern Angelegenheit *f*
concern *v* betreffen, anbelangen
concerning betreffend
concession Bewilligung *f,* Konzession *f,*
  Zugeständnis *n*
conchairamidine Conchairamidin *n*
conche Konche *f*
conche *v* konchieren (Schokolade)

conchinine Chinidin *n*, Conchinin *n*
conchite Conchit *m*
conchoidal muschelig
conchoidal fracture muscheliger Bruch
conchoidal iron ore Muschelerz *n* (Min)
concise knapp, kurz gefaßt
conclude ableiten, folgern (Folgerung)
conclude *v* beenden, beschließen
conclusion Abschluß *m*, Folgerung *f,*
    Rückschluß *m*, Schlußfolgerung *f*
conclusive abschließend, beweiskräftig
concrete Beton *m*, Steinzement *m*, **dense** ~
    Schwerbeton *m*, **fine asphaltic** ~
    Asphaltfeinbeton *m*, **heaped** ~ Schüttbeton *m*,
    layer of ~ Betonschicht *f,* **light-weight** ~
    Porenbeton *m*, **poured** ~ Schüttbeton *m*,
    prestressed ~ Spannbeton *m*, **ready-mixed** ~
    Fertigbeton *m*, **stamped** ~ Stampfbeton *m*,
    tamped ~ Stampfbeton *m*, **vibrated** ~
    Rüttelbeton *m*, ~ **sufficiently wet to flow**
    Gußbeton *m*
concrete *v* betonieren
concrete base Betonunterlage *f*
concrete building brick Betonstein *m*
concrete construction Betonbau *m*
concrete evidence Beweisstück *n*, Sachbeweis *m*
concrete for facing Vorsatzbeton *m*
concrete foundation Betonsockel *m*
concrete hardening material Betonhartstoff *m*
concrete mixer Betonmischmaschine *f*
concrete paint Betonfarbe *f,* Anstrichfarbe *f* für
    Betonflächen
concrete sealing agent Betondichtungsmittel *n*
concrete shield Betonabschirmung *f* (Reaktor)
concrete shuttering Betonverschalung *f*
concrete steel Stahlbeton *m*, Betonstahl *m*
concrete stone Betonstein *m*
concrete subfloor Betonunterlage *f*
concrete thinner Betonverflüssiger *m*
concreting Betonieren *n*
concretion Ablagerung *f* (Med)
concur *v* zusammentreffen; beistimmen
concurrent gleichlaufend, durch denselben Punkt
    gehend
concurrent flow Parallellauf *m*
concussion Erschütterung *f,* Stoß *m*
concussion burst Stoßbruch *m*
concussion fracture Stoßbruch *m*
concussion-free erschütterungsfrei
concussion spring Federdämpfer *m*,
    Stoßdämpfer *m*
condensability Kondensierbarkeit *f,*
    Verdichtbarkeit *f*
condensable kondensierbar, eindickbar,
    verdichtbar
condensate Kondensat *n*, Dampfwasser *n*,
    Kondenswasser *n*, Niederschlag *m* (durch
    Kondensation)

condensate collector
    Kondensatsammelbehälter *m*
condensate deoiling plant
    Kondenswasserentöler *m*
condensate divider Kondensatteiler *m*
condensate pump Kondensatpumpe *f*
condensate separator Kondensatabscheider *m*
condensate tank Vorlage *f* (Techn)
condensate treatment Kondensatbehandlung *f*
condensation Kondensation *f,* Kondensierung *f,*
    Niederschlag *m* (Kondensat), Verdichtung *f,*
    Verflüssigung *f,* **heat of** ~
    Kondensationswärme *f,*
    Niederschlagswärme *f,* **intermolecular** ~
    zwischenmolekulare Kondensation, **loss due**
    to ~ Kondensationsverlust *m*, **partial** ~
    Teilkondensation *f,* **water of** ~
    Niederschlagswasser *n*, ~ **by contact**
    Oberflächenkondensation *f*
condensation accumulator vessel
    Kondensatsammelgefäß *n*
condensation chamber Abscheidekammer *f,*
    Kondensationskammer *f*
condensation core Kondensationskern *m*
condensation loss Kondensationsverlust *m*,
    Abkühlverlust *m*
condensation point Kondensationspunkt *m*
condensation polymer Kondensationsprodukt *n*
    (Kunststoff), Polykondensat *n*
condensation polymerization
    Kondensations-Polymerisation *f*
condensation product Kondensat *n*,
    Kondensationsprodukt *n*
condensation pump Kondensationspumpe *f*
condensation resin Kondensationsharz *n*
condensation trail Kondensstreifen *m*
condensation trap Kondensatsammelgefäß *n*
condensation tube Kondensationsrohr *f*
condensation water Kondenswasser *n*,
    Niederschlagswasser *n*
condense *v* kondensieren, niederschlagen,
    verdichten, [ver]kürzen
condensed kondensiert
condensed milk Kondensmilch *f*
condensed steam pump Kondensatpumpe *f*
condensed water Kondensationswasser *n*,
    Kondenswasser *n*, Niederschlagswasser *n*
condenser Kondensationsapparat *m*,
    Kondensator *m* (Elektr), Kondensor *m*
    (Opt), Kühlapparat *m*, Kühler *m*, Kühlrohr *n*,
    Kühlzylinder *m*, Retortenvorstoß *m* (Chem),
    Verdichter *m*, Verflüssiger *m*, **counter-current**
    ~ Gegenstromkondensator *m* (Elektr),
    pear-shaped ~ Birnenkühler *m* (Chem),
    variable ~ Drehkondensator *m*
condenser battery Kondensatorelement *n*
condenser coil Kühlschlange *f*
condenser effect Kondensatorwirkung *f*

condenser electroscope
Kondensatorelektroskop *n*
condenser gauge Dampfdichtigkeitsmesser *m*
condenser jacket Kühlermantel *m*
condenser load kapazitive Belastung
condenser paper Kondensatorpapier *n*
condenser pipe Kondensatorröhre *f*
condenser plate Kondensatorplatte *f*
condenser retort Kühlerretorte *f*
condenser temperature Kondensatortemperatur *f*
condenser tube Kondensatorröhre *f*
condenser unit Kondensationsanlage *f*
condenser water Brüdenwasser *n*
condensing Kondensieren *n*, Verdichten *n*
condensing *a* kondensierend
condensing air-pump Luftverdichtungspumpe *f*
condensing apparatus Kondensationsapparat *m*
condensing chamber Kondensationskammer *f*,
Flugstaubkammer *f*
condensing engine
Kondensationsdampfmaschine *f*
condensing funnel Ablauftrichter *m*
condensing jet Einspritzstrahl *m*
condensing kettle Kondensationsblase *f*
condensing lens Kondensationslinse *f* (Opt),
Sammellinse *f* (Opt)
condensing plant Kondensationsanlage *f*
condensing tower Kondensationsturm *m*
condensing tube Kondensationsrohr *n*
condensing vessel Kondensationsgefäß *n*
condensing water Kondensationswasser *n*,
Kondenswasser *n*
condiment Gewürz *n*, Würze *f*
condition Zustand *m*, Bedingung *f*, Verfassung *f*,
Voraussetzung *f*, **anti-bonding** ~
nichtbindender Zustand (Atom), **bonding** ~
bindender Zustand, **repulsing** ~ abstoßender
Zustand (Atom), **steady** ~ gleichbleibender
Zustand
condition *v* bedingen, klimatisieren,
konditionieren
conditional bedingt
conditionally correct bedingt richtig (Math)
conditioned klimatisiert, normalfeucht (Text)
conditioned tenacity Normalfestigkeit *f* (Text)
conditioner Ätzlösung (Galvanik),
Konditioniervorrichtung *f*
conditioning Konditionieren *n*
conditioning balance Konditionierwaage *f*
conditioning cabinet Klimaschrank *m*
conditioning period Nachhärtungsfrist *f*
conditioning room Klimaraum *m*
conditioning time Beizzeit *f*
conditions Bestimmungen *f pl*
conditions of operation Betriebsbedingungen *f pl*
conducive förderlich, zweckdienlich
conduct *v* abführen (Wärme), betreiben
(Geschäft), leiten (Elektr)

conductance Konduktanz *f*, Leitfähigkeit *f*,
Leitvermögen *n*, Wirkleitwert *m*, **diagram of**
~ Dreieck *n* der Leitwerte, **dielectric** ~
dielektrische Leitfähigkeit, **magnetic** ~
magnetische Leitfähigkeit, **of ideal** ~ ideal
leitend, **of perfect** ~ ideal leitend
conductance cell, high pressure ~ Autoklav *m*
für Leitfähigkeitsmessungen
conductance of heat Wärmeleitung *f*
conductance parameter Leitwertsparameter *m*
conductance ratio Leitwertsverhältnis *n* (Elektr)
conductibility Leitfähigkeit *f*,
Leit[ungs]vermögen *n*
conductimeter Konduktometer *n*
conductimetric konduktometrisch
conducting leitend, leitfähig
conducting capacity Leitfähigkeit *f*
conducting coat leitender Belag
conducting coil Induktionsspule *f*
conducting paint Leitlack *m*
conducting salt leitendes Salz *n*
conducting surface Leitfläche *f*
conducting wire Leitungsdraht *m*
conduction Leitung *f* (Wärme, Elektr), **heat of** ~
Leitungswärme *f*, **metallic** ~ metallische
Leitung
conduction band Leitungsband *n* (Elektr)
conduction current Leitungsstrom *m*
conduction current density Leitungsstromdichte *f*
conduction electron Leitelektron *n*
conduction process Leitungsprozeß *m*
conduction property Leit[ungs]eigenschaft *f*
conductive leitend, leitfähig, **to render** ~ leitend
machen
conductive hole, energy of ~ **state** energetische
Lage *f* des Defektelektronen-leitenden
Zustandes
conductivity Leitfähigkeit *f*, Leitvermögen *n*,
Leitwert *m*, **acoustic** ~ akustisches
Leitvermögen, **asymmetrical** ~
richtungsabhängige Leitfähigkeit, **coefficient
of** ~ Leitungskoeffizient *m*, **combined** ~
Kombinationsleitwert *m*, **determination of** ~
Leitfähigkeitsbestimmung *f*, **molar** ~ molare
Leitfähigkeit, **thermal** ~
Temperaturleitfähigkeit *f*, **vessel for measuring**
~ Leitfähigkeitsgefäß *n*
conductivity band Leitfähigkeitsband *n*
conductivity cell Leitfähigkeitszelle *f*
conductivity counter Leitfähigkeitsmesser *m*
conductivity decay Leitfähigkeitsabfall *m*
conductivity electron Leitfähigkeitselektron *n*
conductivity indicator Leitfähigkeitsmeßbrücke *f*
conductivity measurement
Leitfähigkeitsbestimmung *f*
conductivity meter Leitfähigkeitsmesser *m*
conductivity of transfer Übergangsleitfähigkeit *f*
conductivity water Leitfähigkeitswasser *n*

conductometer Konduktometer *n*,
   Leitfähigkeitsmesser *m*
conductometric konduktometrisch
conductometric analysis Konduktometrie *f*
conductor Blitzableiter *m*, Konduktor *m*
   (Elektr), Leiter *m* (Elektr), **poor ~** schlechter
   Leiter
conductor circuit Leiterkreis *m*
conductor of heat Wärmeleiter *m*
conductor rail Leitschiene *f* (Elektr)
conductor resistance Leiterwiderstand *m*
conduit Leitung *f*, Röhre *f*, Röhrengang *m*,
   Rohrleitung *f*
conduit clip Rohrschelle *f*
conduit pipe Ableitungsröhre *f*
conduit wire Rohrdraht *m*
condurangin Condurangin *n*
condurango, liquid extract of ~
   Kondurangofluidextrakt *m*
condurango bark Kondurangorinde *f*
condurango liquid extract
   Kondurangofluidextrakt *m*
condurite Condurit *m*
conduritol Condurit *m*
cone Kegel *m*, Konus *m*, **axis of ~** Kegelachse *f*,
   blunt ~ stumpfer Kegel *m*, **tapered ~**
   Klemmkonus *m*, **truncated ~** abgestumpfter
   Kegel, Kegelstumpf *m*
cone and plate viscometer
   Kegel-Platte-Viskosimeter *n*
cone belt drive Keilriemenantrieb *m*
cone brake Kegelbremse *f*
cone clamp Klemmkegel *m*
cone crusher Kegelbrecher *m*
cone fusion test Kegelfallpunkt-Prüfung *f*
   (Email)
cone impeller Kegelkreiselmischer *m*,
   Kegelschnellrührer *m*
conelike kegelähnlich
cone membrane Konusmembran *f*
cone mill Glockenmühle *f*, Trichtermühle *f*
cone of light Lichtkegel *m*
cone pelletizer Pelletierkonus *m*
conephrin Conephrin *n*
cone pulley Kegelscheibe *f*
cone-shaped kegelförmig, kegelig
conessi bark Conessirinde *f*
conessidine Conessidin *n*
conessine Conessin *n*
cone wheel Kegelrad *n*, Stufenscheibe *f*
confectioner's sugar Puderzucker *m*
confectionery Konfekt *n*
confer *v* verleihen
conference Beratung *f*, Konferenz *f*, Tagung *f*
confidence level Konfidenzzahl *f* (Statist)
confidence limits Konfidenzintervall *m* (Statist)
configuration Anordnung *f* im Raum,
   Konfiguration *f*, Strukturschema *n*, **absolute**
   ~ absolute Konfiguration, **determination of**

~ Konfigurationsbestimmung *f*, **inversion of**
   ~ Konfigurationsumkehrung *f*, **puckered ~**
   geknickte Konfiguration, **relative ~** relative
   Konfiguration
configurational base unit
   Konfigurationsbaustein *m*
configurational entropy Konfigurationsentropie *f*
configurational relationship
   Konfigurationsverwandtschaft *f*
configuration in space räumliche Konfiguration
configuration interaction
   Konfigurationswechselwirkung *f*
configuration in the plane ebene Konfiguration
configuration of saddle point
   Sattelpunktskonfiguration *f*
configuration space Konfigurationsraum *m*
confine *v* begrenzen, einengen, einschränken
confining liquid Absperrflüssigkeit *f*
confirm *v* bestätigen, beglaubigen, begründen,
   bekräftigen, bestärken
confirmation Bekräftigung *f*, Bestätigung *f*
confiscate *v* beschlagnahmen
conformable [to] angemessen
conformation Konformation *f* (Stereochem),
   doubly skewed ~ doppelt windschiefe
   Konformation, **eclipsed ~** Stellung *f* auf
   Deckung, ekliptische Konformation,
   staggered ~ gestaffelte Konformation,
   Stellung *f* auf Lücke
conformational analysis Konformationsanalyse *f*
conformational change
   Konformationsänderung *f* (Biochem)
conformational interconversion
   Konformationswechsel *m*
conformity Formgleichheit *f*, Maßgenauigkeit *f*,
   Übereinstimmung *f*
conform to *v* entsprechen
confront *v* gegenüberstellen
confused verwirrt, fassungslos
confusion Durcheinander *n*, Verwirrung *f*
congeal *v* erstarren, gefrieren, gerinnen
congealability Gefrierbarkeit *f*
congealable gefrierbar, gerinnbar,
   gerinnungsfähig
congealed gefroren, geronnen
congealer Gefrierer *m*, Gefriervorrichtung *f*
congealing erstarrend
congealing point Eispunkt *m*
congelation Erstarrung *f*, Gefrieren *n*,
   Gerinnen *n*
congenital kongenital (Med)
conglobate *a* zusammengeballt
conglobate *v* sich zusammenklumpen
conglomerate Konglomerat *n*, Gemenge *n*
   (Geol), Menggestein *n* (Geol),
   Trümmergestein *n* (Geol)
conglomerate *v* konglomerieren, sich
   zusammenballen, sich zusammenklumpen

conglomerate integration kombinierte
Integration *f*
conglomerate structure Konglomeratgefüge *n*
conglomeration Anhäufung *f,*
Zusammenballung *f*
conglomerone Conglomeron *n*
conglutinin Konglutinin *n*
Congo blue Kongoblau *n*
congocidine Congocidin *n*
Congo fast blue Kongoechtblau *n*
Congo paper Kongopapier *n*
Congo red Kongorot *n*
Congo rubber Kongogummi *n*
Congo yellow Kongogelb *n*
congruence Kongruenz, Kongruenz *f* (Math),
Übereinstimmung *f,* **assumption of** ~
Kongruenzannahme *f* (Math)
congruent deckungsgleich (Math), kongruent
congruent melting point
Kongruenzschmelzpunkt *m* (Phys)
congruent transformation
Kongruenztransformation *f* (Math)
congruous, to be ~ kongruieren
conhydrine Conhydrin *n,* Conydrin *n*
conhydrinone Conhydrinon *n*
conic konisch
conic acid Coniumsäure *f*
conical konisch, kegelähnlich, kegelförmig,
kegelig
conical conductor Kegelkonduktor *m*
conical dense-medium vessel
Konus[sink]scheider *m*
(Schwertrübetrennung)
conical drum Kegeltrommel *f*
conical flask Erlenmeyerkolben *m*
conical funnel Hüttentrichter *m*
conical gate Keileinguß *m*
conical grinder Kegelmühle *f,* Glockenmühle *f,*
Kegelbrecher *m*
conical mill Konusmühle *f*
conical pivot Spitzzapfen *m*
conical point Kegelpunkt *m* (einer Fläche)
conical runner Keileinguß *m*
conical screen centrifuge with differential
conveyor Dünnschichtzentrifuge *f*
coniceine Conicein *n*
conic form Kegelgestalt *f*
conichalcite Konichalcit *m* (Min)
conic[ic] acid Coniumsäure *f,* Schierlingssäure *f*
conicine Conicin *n*
conicine balsam Conicinbalsam *m*
conicity Kegelform *f,* Konizität *f*
conic section Kegelschnitt *m*
conidendrin Conidendrin *n*
conidine Conidin *n*
conifer Konifere *f,* Nadelbaum *m*
coniferaldehyde Coniferylaldehyd *m*
coniferin Coniferin *n*
coniferol Coniferylalkohol *m,* Lubanol *n*

coniferoside Coniferosid *n*
coniferous forest Nadelwald *m*
coniferyl Coniferyl-
coniferyl alcohol Coniferylalkohol *m*
coniform kegelförmig, kegelig, konisch
coniic acid Coniumsäure *f,* Schierlingssäure *f*
coniine Coniin *n*
coniine extract Coniinextrakt *m*
coniine hydrobromide Coniinhydrobromid *n*
coniine ointment Coniinsalbe *f*
conima Conima[harz] *n*
conimeter Konimeter *n*
conine Coniin *n*
conium Schierlingsgift *n*
conjugate *v* konjugieren, paaren
conjugate complex konjugiert komplex
conjugated konjugiert (Chem)
conjugated double bond konjugierte
Doppelbindung *f*
conjugation Konjugation *f* (Chem), Paarung *f*
conjugative effect Konjugationseffekt *m*
conjunctiva (pl. conjunctivas, conjunctivae)
Bindehaut *f* (Med)
conkurchine Conkurchin *n*
connate water Haftwasser *n*
connect *v* verbinden, aneinanderschließen,
anknüpfen, ankuppeln, koppeln, verknüpfen,
zusammenfügen
connected verbunden, angeschlossen,
zusammenhängend, **to be** ~
zusammenhängen
connecting Verbinden *n,* Koppeln *n,*
Zusammenfügen *n*
connecting box Anschlußdose *f*
connecting branch Einsatzstutzen *m*
connecting bridge Verbindungsbrücke *f*
connecting busbar Verbindungsschiene *f* (Elektr)
connecting cable Anschlußkabel *n* (Elektr),
Anschlußleitung *f* (Elektr),
Verbindungskabel *n* (Elektr)
connecting cock Verbindungshahn *m*
connecting cord Verbindungsschnur *f* (Elektr)
connecting flange Anschlußflansch *m*
connecting gear Kuppelrad *n*
connecting lead Experimentierschnur *f*
connecting line Verbindungsgerade *f,*
Verbindungslinie *f*
connecting link Bindeglied *n,* Zwischenglied *n*
connecting main Verbindungsleitung *f* (Elektr)
connecting member Zwischenglied *n,*
Bindeglied *n*
connecting membrane Verbindungshaut *f* (Biol)
connecting oxygen Brückensauerstoff *m*
connecting piece Verbindungsstück *n,*
Ansatzstück *n,* Anschlußstutzen *m,*
Zwischenstück *n*
connecting plate Verbindungsblech *n*
connecting plug Anschlußstecker *m* (Elektr),
Stecker *m* (Elektr)

connecting rack Verbindungsgestell *n*
connecting rod Kurbelstange *f,* Pleuelstange *f,*
Verbindungsstange *f*
connecting rod bearing Pleuellager *n*
connecting screw Anschlußschraube *f*
connecting sleeve Verbindungsmuffe *f* (Elektr)
connecting strip Verbindungsblech *n*
connecting tap Verbindungshahn *m*
connecting terminal Verbindungsklemme *f*
connecting tube Ansatzrohr *n,*
Verbindungsröhre *f,* Winkelstück *n*
connection Verbindung *f,* Anschluß *m* (Elektr),
Anschlußstutzen *m;* Bindeglied *n,*
Kopplung *f,* Verbindungsstück *n,* **total** ~
Gesamtanschluß *m*
connection branch Anschlußstutzen *m*
connection for exhaust Absaug[e]stutzen *m*
connection in parallel Nebeneinanderschaltung *f*
connection in series Reihenschaltung *f*
connection screw Anschlußverschraubung *f*
connective tissue Bindegewebe *n*
connector Verbindungsklemme *f*
connellite Connellit *m* (Min)
connesine Connesin *n*
conoid kegelähnlich, kegelig
conoidal kegelähnlich
conopharyngine Conopharyngin *n*
conopharynginic acid Conopharynginsäure *f*
conormal Konormale *f* (Math)
conquinamine Conchinamin *n*
consecutive aufeinanderfolgend, hintereinander
consecutive reaction Folgereaktion *f* (Chem),
zusammengesetzte Reaktion
consequence Folge[erscheinung] *f,* Konsequenz *f,*
Resultat *n*
consequential condition Folgebedingung *f*
consequent[ly] folgerichtig
consequent reaction Folgereaktion *f* (Chem)
conservation Aufbewahrung *f,* Erhaltung *f,*
Haltbarmachung *f,* Konservierung *f,* ~ **of**
energy Erhaltung *f* der Energie, ~ **of mass**
Masseerhaltung *f,* ~ **of momentum**
Erhaltung *f* des Impulses, ~ **of wood**
Holzaufbereitung *f*
conservation metabolism
Erhaltungsstoffwechsel *m*
conservation of food
Lebensmittelkonservierung *f*
conserve *v* haltbar machen, konservieren,
konservieren, erhalten, bewahren
conserving Haltbarmachen *n*
consider *v* abwägen, bedenken, berücksichtigen
considerable ansehnlich, beachtlich, beträchtlich,
einschneidend
considerate rücksichtsvoll, schonend
consideration Bedenken *n,* Berücksichtigung *f,*
Erwägung *f,* Rücksicht *f,* **auxiliary** ~
Hilfsbetrachtung *f,* **in** ~ **[of]** in Anbetracht *m*
consignee Adressat *m,* Empfänger *m*

consignor Übersender *m*
consist [of] *v* bestehen aus
consistency Konsistenz *f,* Beständigkeit *f,*
Bestand *m,* Festigkeit *f,* Zusammenhalt *m,*
measure of ~ Konsistenzmaß *n*
consistency factor Festigkeitszahl *f*
consistency variable Konsistenzparameter *m*
consistent konsistent
consistometer Konsistenzmesser *m*
console Konsole *f,* Stütze *f*
consolidate *v* festigen, fest werden (Tumor),
konsolidieren, sich verfestigen
consolidation Festwerden *n,* Verdichtung *f,*
Verfestigung *f,* **degree of** ~
Verfestigungsgrad *m*
consonance Einklang *m,* Harmonie *f,*
Konsonanz *f* (Akust)
conspicuous auffallend, kenntlich
constancy Konstanz *f,* Beständigkeit *f,*
Gleichmäßigkeit *f,* Unveränderlichkeit *f*
constant Konstante *f,* Festwert *m,* Kennzahl *f*
constant *a* konstant, beständig, gleichbleibend,
gleichmäßig, ständig, stetig
constant *v,* **to keep** ~ konstant halten, **to**
maintain ~ konstant halten, **to remain** ~
gleich bleiben
constantan Konstantan *n*
constant boiling mixture Mischung *f* mit
konstantem Siedepunkt (Therm)
constant boiling point Fixpunkt *m*
constant direction error Halbkreisfehler *m*
constant excitation Konstanterregung *f*
constant pitch gleichbleibende Steigung
constant pressure change Volumenarbeit *f*
constant pressure line Linie *f* gleichen Drucks
constant-temperature pressure welding
Abbrennschweißung *f*
constipation Verstopfung *f* (Med)
constituent Bestandteil *m* (Chem),
Komponente *f* (Chem), Konstituent *m,*
incombustible ~ unverbrennbarer Bestandteil,
minor ~ Nebenbestandteil *m,* **volatile** ~
flüchtiger Bestandteil
constitution Bau *m,* Konstitution *f,*
Zusammensetzung *f,* **water of** ~
Konstitutionswasser *n*
constitutional formula Strukturformel *f*
constitutional influence Konstitutionseinfluß *m*
constitutional isomerism Konstitutionsisomerie *f*
constrained zwangsläufig, gezwungen
constraint Zwang *m,* **principle of least** ~ Prinzip
des geringsten Zwanges
constrict *v* zusammenziehen, einschnüren
constricting head Staukopf *m*
constriction Einschnürung *f,* Drosselung *f,*
Verengung *f*
constringe *v* einengen, zusammenziehen,
konstringieren
constringent einengend, konstringierend

construct *v* bauen, herstellen, konstruieren
constructable konstruierbar
construction Konstruktion *f,* Bau *m,* Bauart *f,*
  Bauweise *f,* **cost of** ~ Anlagekosten *pl,* **error
  of** ~ Konstruktionsfehler *m,* **solid** ~
  Vollbauweise *f* Massivbauweise, **unit or
  element of** ~ Konstruktionsglied *n*
constructional baulich
constructional material
  Konstruktionswerkstoff *m*
constructional proofing Bautenschutz *m*
constructional unit Bauelement *n*
construction gluing Montageleimung *f*
construction material, processing of ~
  Werkstoffverfeinerung *f*
construction principle Aufbauprinzip *n*
construction work Bauarbeiten *pl* (Bauw)
constructive aufbauend, konstruktiv
constructive metabolism Aufbaustoffwechsel *m*
  (Biol)
consult *v* beraten, beratschlagen
consultation Beratung *f,* Konsultation *f*
consume *v* verbrauchen, aufbrauchen,
  vernichten, verzehren
consumer Abnehmer *m,* Konsument *m,*
  Verbraucher *m*
consuming heat endotherm
consumption Verbrauch *m,* Aufzehrung *f,*
  Ausnutzung *f,* Bedarf *m,* Konsum *m,* **area of**
  ~ Konsumgebiet *n,* Verbrauchsgebiet *n,*
  center of ~ Konsumschwerpunkt *m,*
  Konsumzentrum *n,* **density of** ~
  Konsumdichte *f,* **internal** ~
  Eigenverbrauch *m,* **place of** ~
  Verbrauchsstelle *f,* **region of** ~
  Verbrauchsgebiet *n,* **total** ~
  Gesamtverbrauch *m,* ~ **of fuel**
  Brennstoffaufwand *m,* ~ **of water**
  Wasserverbrauch *m,* ~ **under idling condition**
  Leerlaufverbrauch *m*
consumption density Verbrauchsdichte *f*
contact Kontakt *m,* Anschluß *m* (Elektr),
  Berührung *f,* **area of** ~ Kontaktfläche *f,*
  defective ~ Wackelkontakt *m* (Elektr),
  flexible ~ federnder Kontakt, **place of** ~
  Berührungsstelle *f,* **point of** ~
  Berührungspunkt *m,* Anknüpfungspunkt *m,*
  Anstoß *m,* Berührungsstelle *f,* **to be in** ~
  aneinandergrenzen, **to close the** ~ den
  Kontakt schließen, **to come in** ~
  aneinandergrenzen, **to make** ~ abtasten, ~
  with air Liegen *n* an der Luft
contact acid Kontaktsäure *f*
contact action Kontaktwirkung *f*
contact adhesive Haftkleber *m,*
  Kontaktklebstoff *m*
contact area Berührungsfläche *f*
contact blade Kontaktfinger *m*
contact block Kontaktklemme *f* (Elektr)

contact box Übergangsdose *f*
contact breaker Unterbrecherkontakt *m,*
  automatic ~ selbsttätiger Unterbrecher
  (Elektr)
contact breaking brush Abbrennbürste *f* (Elektr)
contact button Kontaktknopf *m*
contact carriage Kontaktschlitten *m*
contact catalysis Kontaktkatalyse *f,* heterogene
  Katalyse *f*
contact cathode Berührungskathode *f*
contact clearance Kontaktweite *f*
contact copper-plating Kontaktverkupferung *f*
contact corrosion Berührungskorrosion *f*
  (corrosion at a contact with a second metal),
  Kontaktkorrosion *f*
contact current Kontaktstrom *m*
contact curve Walzkurve *f*
contact detector Kontaktgleichrichter *m*
contact discontinuity Kontaktunstetigkeit *f*
contact dryer Kontakttrockner *m*
contact effect Kontaktwirkung *f*
contact electricity Berührungselektrizität *f*
contact electrode Kontaktelektrode *f*
contact finger Kontaktfinger *m*
contact fire Berührungszündung *f*
contact goniometer Anlegegoniometer *n*
contact heating Berührungsheizung *f*
contacting seal Berührungsdichtung *f*
contact inhibition Kontakthemmung *f*
contact insecticide Berührungsinsektizid *n*
contact interface Kontaktfläche *f*
contact jaw Einspannbacke *f,* Spannbacke *f*
contact lens Haftlinse *f* (Opt), Haftschale *f,*
  Kontaktlinse *f*
contact maker Kontaktgeber *m,*
  Stromschließer *m*
contact murmur Kontaktrauschen *n* (Elektr)
contactor Kontaktgeber *m* (Elektr), **auxiliary** ~
  Haltekontakt *m* (am Wendeschütz)
contact piece, sparking ~
  Abbrennkontaktstück *n*
contact plant Kontaktanlage *f*
contact plate Kontaktboden *m,* Kontaktplatte *f*
contact plating Kontaktgalvanisierung *f*
contact plug Kontaktstöpsel *m*
contact point Kontaktstelle *f*
contact pointer Kontaktzeiger *m*
contact poison Kontaktgift *n*
  (Schädlingsbekämpfung)
contact potential Kontaktpotential *n*
contact powder Kontaktpulver *n*
contact print method Kontaktabdruckmethode *f*
contact process Kontaktverfahren *n*
contact resistance Kontaktwiderstand *m,*
  Übergangswiderstand *m*
contact roller Führungsrolle *f,* Kontaktrolle *f*
contact spacing Kontaktabstand *m*
contact spring Kontaktfeder *f*
contact substance Kontaktmittel *n*

contact surface Kontaktfläche *f,* Auflagefläche *f,*
Berührungsfläche *f,* Grenzfläche *f*
contact theory Kontakttheorie *f*
contact thermometer Kontaktthermometer *n*
contact transformation
Berührungstransformation *f*
contact twin Berührungszwilling *m* (Krist),
Juxtapositionszwilling *m* (Krist)
contact-type regulator Kontaktregler *m*
contact value Anschlußwert *m*
contact voltage Berührungsspannung *f,*
Kontaktspannung *f*
contact wire Kontaktdraht *m*
contagious ansteckend (Med), übertragbar (Med)
contain *v* enthalten, fassen
container Behälter *m,* Bottich *m,* Gefäß *n,*
Kanister *m,* Tank *m,* **reusable** ~
wiederverwendbare Verpackung *f,*
non-returnable ~ **or disposable container**
Einwegverpackung *f*
containing alcohol alkoholhaltig, ~ **alkali**
alkalihaltig, ~ **ammonia** ammoniakhaltig, ~
iron eisenhaltig, ~ **lead** bleihaltig
contaminant Verseuchungsstoff *m*
contaminate *v* verunreinigen, kontaminieren,
vergiften, verseuchen
contaminated verunreinigt, unrein (Chem),
verschmutzt
contamination Verunreinigung *f,*
Kontamination *f,* Unreinheit *f* (Schmutz),
Vergiftung *f,* Verschmutzung *f,* Verseuchung *f,*
danger of ~ Verschmutzungsgefahr *f,*
radioactive ~ radioaktive Verseuchung, **source**
of ~ Verunreinigungsherd *m*
conteben Conteben *n*
contemplate *v* betrachten, beabsichtigen
contemporary Zeitgenosse *m*
contemporary *a* zeitgenössisch
content Fassungsvermögen *n,* Gehalt *m,*
Inhalt *m,* Raumgröße *f,* **having a high** ~ **[of]**
stark durchsetzt
content[s] Gehalt *m,* Inhalt *m,* ~ **of a solution**
Gehalt *m* einer Flüssigkeit
contergan Contergan *n* (Pharm)
context Zusammenhang *m*
contiguous angrenzend
continental block Kontinentalscholle *f* (Geol)
continental climate Binnenklima *n* (Meteor)
continental drift Kontinentalverschiebung *f*
(Geol)
continental margin Kontinentalrand *m*
continental solid point Erstarrungspunkt *m* am
rotierenden Thermometer
continuance Bestand *m,* Fortbestehen *n,*
Fortdauer *f*
continuation Fortbestand *m,* Fortsetzung *f*
continue *v* andauern, beharren, verbleiben,
weiterführen
continued fraction Kettenbruch *m* (Math)

continuity Kontinuität *f,* Stetigkeit *f,* **condition**
for ~ Stetigkeitsbedingung *f,* ~ **of a function**
Stetigkeit *f* einer Funktion (Math)
continuity equation Kontinuitätsgleichung *f*
continuity limit Stetigkeitsgrenze *f*
continuous andauernd, anhaltend, durchgehend,
endlos, kontinuierlich, ständig, stetig,
unaufhaltsam
continuous assembly belt Arbeitsband *n*
continuous charging grate Schüttrost *m*
continuous chip Bandspan *m*
continuous-coil evaporator
Einspritzverdampfer *m*
continuous current arc Gleichstromlichtbogen *m*
continuous current arc lamp
Gleichstrombogenlampe *f*
continuous current circuit Gleichstromkreis *m*
continuous current converter
Gleichstromumformer *m*
continuous current excitation
Gleichstromerregung *f*
continuous current generator
Gleichstromgenerator *m*
continuous current line Gleichstromleitung *f*
continuous current plant Gleichstromanlage *f*
continuous current system Gleichstromsystem *n*
continuous discharge Dauerentladung *f*
continuous evaporator Durchlaufverdampfer *m*
continuous filament process
Düsenziehverfahren *n*
continuous finishing mill for pipe
[Rohr]kontistraße *f*
continuous-flow conveyor Stetigförderer *m*
continuous flow heater Durchflußerhitzer *m*
continuous line recorder Linienschreiber *m*
continuous load Dauerbeanspruchung *f*
continuous[ly] durchgehend, fortlaufend,
ununterbrochen
continuous paper Rollenpapier *n*
continuous process Kreisprozeß *m*
continuous rare gas spectra Edelgaskontinua *n pl*
continuous rectification stetige Rektifikation *f*
continuous scales for dosing Dosierbandwaage *f*
continuous spectrum kontinuierliches
Spektrum *n,* Kontinuum *n*
continuous test Dauerprüfung *f*
continuum kontinuierliches Spektrum *n,*
Kontinuum *n*
contorted geknäuelt (Molekül)
contortion Krümmung *f,* Verzerrung *f*
contour Kontur *f,* Profil *n,* Umriß *m*
contour *v* fassonieren
contour diagram Umrißdiagramm *n*
contour line Isohypse *f*
contour map Höhenlinienkarte *f*
contour vibration Formschwingung *f*
contraceptive Antikonzeptionsmittel *n* (Pharm),
Empfängnisverhütungsmittel *n,*
Verhütungsmittel *n* (Pharm)

contract Vertrag *m,* Abkommen *n,* Abmachung *f,* Akkord *m*

contract *v* sich zusammenziehen, kontrahieren, schrumpfen, schwinden, verengen, verkürzen, zusammenschrumpfen

contracted verkürzt

contractible zusammenziehbar

contractile zusammenziehbar

contracting Kontrahieren *n,* Zusammenziehen *n*

contraction Kontraktion *f,* Schrumpfung *f,* Schwinden *n,* Schwund *m* (durch Zusammenziehen), Verkürzung *f,* Zusammenziehung *f,* **amount of** ~ Schwindmaß *n,* ~ **of the steel** Verziehen *n* des Stahls

contraction allowance Schrumpfübermaß *n*

contraction crack Schwundriß *m*

contraction theory Kontraktionstheorie *f*

contractor Lieferant *m,* Unternehmer *m*

contract packager Abpackbetrieb *m*

contract packaging plant Abpackbetrieb *m*

contract research Auftragsforschung *f*

contradiction Widerspruch *m*

contradictory widerspruchsvoll

contradistinction Gegensatz *m*

contraindication Gegenanzeige *f,* Kontraindikation *f*

contramine Intramin *n*

contrarotating gegenläufig

contrarotation ball-race-type pulverizing mill Federkraft-Kugelmühle *f*

contrary *a* entgegengesetzt

contrast Gegensatz *m,* Kontrast *m*

contrast *v* abstechen, entgegensetzen, Kontrast geben, ~ **[with]** sich abheben

contrast coloring Gegenfärbung *f*

contrast filter Kontrastfilter *n*

contrasting *a* abstechend

contrast medium Kontrastmittel *n*

contrast medium method Kontrastmethode *f* (Röntgenologie)

contrast photometer Kontrastphotometer *n*

contrast radiograph Kontrastdarstellung *f* (Med)

contrast ratio Kontrastverhältnis *n*

contrast rendition Kontrastwiedergabe *f*

contrast roentgenography Kontrastverfahren *n* (Röntgenologie)

contrast sensibility Kontrastempfindlichkeit *f*

contrast solution Kontrastlösung *f* (Röntgenologie)

contrast staining Kontrastfärbung *f*

contravariant kontravariant (Math)

contrayerva Bezoarwurzel *f*

contribute *v* beisteuern, beitragen

contributing form mesomere Grenzform (Chem)

contribution Beitrag *m*

contributor Mitarbeiter *m*

contrivance Vorrichtung *f*

contrive *v* ersinnen, anstiften

control Überwachung *f,* Aufsicht *f,* Beherrschung *f,* Kontrolle *f,* Regulierung *f,* Steuerorgan *n,* **automatic** ~ automatische Regelung, **centralized** ~ Zentralsteuerung *f;,* **floating** ~ schwebende Regelung, **local** ~ örtliche Steuerung (Comp), **method of** ~ Bekämpfungsmethode *f,* **modulating** ~ stetige Regelung, **open-loop** ~ einfache offene Steuerung (Comp), rückführungslose Steuerung (Comp), **photoelectric** ~ photoelektrische Steuerung, **range of** ~ Regelbereich *m,* ~ **of operations** Betriebskontrolle *f*

control *v* kontrollieren, beaufsichtigen, beherrschen, bekämpfen, betätigen (maschinell), nachprüfen, regeln, regulieren, steuern, ~ **a measurement** nachmessen

control accuracy Regelgenauigkeit *f*

control agent Bekämpfungsmittel *n*

control amplifier Regelverstärker *m*

control analysis Kontrollanalyse *f*

control animal Kontrolltier *n*

control arm Lenkarm *m*

control board Steuerpult *n*

control cabinet Schaltschrank *m*

control cable lacquer Schaltdrahtlack *m*

control circuit Regelschaltung *f,* Steuerstromkreis *m* (Elektr)

control cylinder Steuerzylinder *m*

control diagram Regelschema *n*

control element Regelglied *n,* Regelvorrichtung *f*

control engineering Regelungstechnik *f*

control equipment Regelvorrichtung *f,* Regler *m*

control experiment Gegenversuch *m,* Kontrollversuch *m*

control finger Schaltfinger *m*

control form Regelweise *f*

control gear Kontrollvorrichtung *f,* Steuerung *f*

control generator Steuergenerator *m*

control grid Kontrollrost *m,* Steuergitter *n*

control grid tube Steuergitterröhre *f*

control impulse transmitter Regelsignalgeber *m*

control knob Schaltknopf *m,* Bedienungsgriff *m,* Bedienungsknopf *m,* Drehknopf *m,* Einstellknopf *m*

controllability Regelbarkeit *f,* Regulierbarkeit *f*

controllable einstellbar, kontrollierbar, lenkbar (Chem), regelbar, regulierbar

controllable hub Verstellnabe *f*

control lag Anlaufzeit *f*

control lamp Kontrollampe *f*

controlled gesteuert, **separately** ~ fremdgesteuert

controlled atmosphere storage CA-Lagerung *f*

controlled condition Regelgröße *f*

controlled gap seal Spaltdichtung *f*

controlled variable (Am. E.) Regelgröße *f*

controller Kontrolleur *m,* Fahrschalter *m,* Regler *m* (Elektr), Schalter *m;*

Überwachungsgerät *n*, **integral (or floating) action** ~ integral wirkender Regler *m*
controller output Stellgröße *f*
control lever Bedienungshebel *m*, Einrückhebel *m*
control limit Toleranzgrenze *f*
controlling the press Steuern *n* der Presse
control[ling] device Kontrollvorrichtung *f*
controlling magnet Richtmagnet *m*
controlling means Regler *m*
controlling mechanism Steuerapparat *m*, Schaltwerk *n*
controlling pressure gauge Kontrollmanometer *n*
controlling strip Schablonenstreifen *m*
control loop Regelkreis *m*
control loop elements Regelkreisglieder *pl*
control measure Bekämpfungsmaßnahme *f*
control mechanism Steuergerät *n*
control operation Steuerungsvorgang *m*
control panel Bedienungspult *n*, Instrumentenbrett *n*, Schalttafel *f*
control plant Überwachungsanlage *f*
control plate Leitplatte *f*
control platform Steuerbühne *f*
control point Istwert *m* (Regelgröße), Meßstelle *f*, Regelpunkt *m*
control process Regelvorgang *m*
control pyrometer Kontrollpyrometer *n*
control register Steuerbefehlspeicher *m* (Comp)
control reset button Kontrollauslöseknopf *m*
control rod Steuerstange *f* (Atomtechnik)
control roll Leitwalze *f*
control room Meßwarte *f*, Meßzentrale *f*, Regelraum *m*
controls Regelvorrichtung *f*, Steuerung *f*, Steuerwerk *n*
control sample or test Vergleichsprobe *f*
control series Kontrollreihe *f*
control strip Regulatorstreifen *m*
control switch Bedienungsschalter *m*, Betätigungsschalter *m*
control system Überwachungsanlage *f*(Am. E.), Regelkreis *m*
control system of a process (Am. E.) Regelstrecke *f*
control test Gegenversuch *m*, Kontrollversuch *m*
control unit Steuerwerk *n*
control valve Regelventil *n*, Regulierventil *n*, Schaltventil *n*
control voltage Regelspannung *f*
contusion Anprall *m*, Quetschung *f* (Med)
conurbation Ballungszentrum *n*
convallamarin Convallamarin *n*
convallamarogenin Convallamarogenin *n*
convallaretin Convallaretin *n*
convallarin Convallarin *n*
convallatoxin Convallatoxin *n*
convalloside Convallosid *n*
convect *v* mitbewegen

convection Konvektion *f*, Mitbewegung *f*, **forced** ~ erzwungene Konvektion, **free** ~ freie Konvektion
convection current Konvektionsstrom *m*
convection dryer Heißluftstromtrockner *m*
convection heat Konvektionswärme *f*
convection heater Strahlungsofen *m*
convection heating Konvektionsheizung *f*, Strahlungsheizung *f*
convection-type boiler Rauchrohrkessel *m*
convective konvektiv
convention Tagung *f*
conventional design Standardausführung *f*
conventional[ly] konventionell
converge *v* zusammenlaufen (Linien), zusammentreffen, konvergieren
convergence Konvergenz *f* (Math), **circle of** ~ Konvergenzkreis *m*, **region of** ~ Konvergenzgebiet *n*, ~ **of a series** Konvergenz *f* einer Reihe (Math)
convergence angle Konvergenzwinkel *m*
convergence pressure Konvergenzdruck *m*
convergency Konvergenz *f* (Math)
convergent konvergent
converging lens Konvexlinse *f* (Opt), Sammellinse *f* (Opt)
converse umgekehrt
conversion Konvertierung *f*, Überführung *f* (Chem), Umformung *f* (Strom), Umkehrung *f*, Umrechnung *f* (Math), Umsatz *m* (Chem), Umsetzung *f* (Chem), Umwandlung *f*, Verwandlung *f*, **process of** ~ Umwandlungsprozeß *m*, **velocity of** ~ Umsetzungsgeschwindigkeit *f*, ~ **into carbon** Carbonisieren *n*, ~ **into rosettes** Scheibenreißen *n*, **finishing** ~ **of pigments to powders, pastes, etc.** Formieren *n* (v. Pigmenten)
conversion constant Umrechnungskonstante *f*
conversion factor Konversionsfaktor *m* (Atom), Umrechnungsfaktor *m*, Umrechnungswert *m*, Umwandlungsfaktor *m*
conversion process Umwandlungsvorgang *m*
conversion ratio Umwandlungsverhältnis *n*, Konversionskoeffizient *m*, Umsetzungszahl *f* (Atom), Umwandlungskoeffizient *m*
conversion saltpeter Konversionssalpeter *m*
conversion spectrum Umwandlungsspektrum *n*
conversion table Umrechnungstabelle *f*, Umrechnungstafel *f*
conversion variable Umsatzvariable *f*
convert *v* umwandeln, konvertieren, umändern, umrechnen (Math), verwandeln, weiterverarbeiten (Text), ~ **into an amide** amidieren (Chem), ~ **into an amine** aminieren
converted umgesetzt (Chem)
converted gas Konvertgas *n*
converted steel Blasenstahl *m*, Brennstahl *m*

converter Konverter *m* (Metall), Birne *f*
(Metall), Drehofen *m* (Metall),
Drehtransformator *m* (Elektr), Umformer *m*
(Elektr), Windfrischapparat *m* (Metall),
barrel or horizontal ~ liegender Konverter,
continuous to threephase ~
Gleichstromdrehstromumformer *m* (Elektr),
discharge from the ~ Birnenauswurf *m*
(Metall), **fixed or stationary** ~ feststehender
Konverter, **tilting or tipping** ~ kippbarer
Konverter, **tipping the** ~ Drehen *n* der Birne,
to raise the ~ die Birne aufrichten (Metall)
converter bottom Konverterboden *m*
converter charge Konvertereinsatz *m*
converter charging platform or gallery
Konverterbühne *f*
converter copper Konverterkupfer *n*
converter film Konverterfilm *m*
converter iron Flußeisen *n*
converter lining Konverterauskleidung *f*,
Konverterfutter *n*
converter plant Konverteranlage *f*
converter practice Konverterbetrieb *m* (Metall)
converter process Thomasverfahren *n* (Metall),
Birnenverfahren *n* (Metall),
Windfrischverfahren *n* (Metall)
converter reactor Konverterreaktor *m*
converter shed Konverterhalle *f*
converter shop Konverterhalle *f*
converter stand Birnenständer *m*
converter steel, basic ~ Thomasstahl *m*
converter trunnion ring Konverterring *m*
converter waste Birnenauswurf *m* (Metall),
Konverterauswürfe *m pl*
convertible konvertierbar, umwandelbar
converting Umsetzen *n*, Verwandeln *n*,
Windfrischen *n* (Metall)
converting furnace Brennstahlofen *m*
converting process, oxidizing or ~
Frischungsprozeß *m* (Metall)
convex konvex, erhaben, gewölbt, runderhaben
convexity konvexe Form *f*, Wölbung *f*
convex lens Konvexlinse *f* (Opt), Sammellinse *f*
(Opt)
convex mirror Konvexspiegel *m* (Opt),
Vollspiegel *m* (Opt)
convexo-concave konvex-konkav (Opt)
convexo-convex bikonvex (Opt)
convex reflector Konvexspiegel *m* (Opt)
convey *v* befördern, weiterleiten
conveyance Förderung *f*, Transport *m*,
underground ~ unterirdische Förderung
conveyer (also see conveyor) Förderanlage *f*,
Förderapparat *m*, Fördergerät *n*, **small** ~
Kleinförderanlage *f*
conveyer belt Fließband *n*, Förderband *n*,
Fördergurt *m*, Transportband *n*
conveyer bucket Fördergefäß *n*, Förderkübel *m*
conveyer chain Förderkette *f*

conveyer chains Transportketten *pl*
conveyer chute Förderrinne *f*
conveyer felt Transportfilz *m* (Papier)
conveyer pelletizer Pelletierband *n*
conveyer pipe Förderrohr *n*
conveyer pipeline Förderleitung *f*
conveyer plant Förderanlage *f*
conveyer system Förderanlage *f*
conveyer trough Förderrinne *f*
conveying Förderung *f*, Verfrachtung *f*, ~ **of**
charge Gichtbeförderung *f*
conveying agent Förderorgan *n*
conveying belt Förderriemen *m*
conveying capacity Förderleistung *f*
conveying chute Förderrinne *f*, Förderrutsche *f*
conveying means Fördermittel *n*
conveying plant Förderanlage *f*
conveying pump Förderpumpe *f*
conveying stock Fördergut *n*
conveying table Fließarbeitstisch *m*
conveying trough Förderrinne *f*,
Trogförderband *n*
conveyor cf. conveyer, Förderanlage *f*,
articulated or bucket-belt ~
Faltenband-Förderer *m*, **branch** ~
Stichbahn *f*, **fixed-screw** ~
Schneckenrohrförderer *m*, **worm [or screw]** ~
Schneckenförderer *m*, **apron** ~ **or**
articulated-belt conveyor
Gliederbahnförderer *m*
convicin Convicin *n*
convincing überzeugend, beweiskräftig
convolamine Convolamin *n*
convolution Schraubengang *m*, Windung *f*
convolution integral Faltungsintegral *n* (Math)
convolvidine Convolvidin *n*
convolvine Convolvin *n*
convolvulic acid Convolvulinsäure *f*
convolvulin Convolvulin *n*
convulsion Krampf *m* (Med), Zuckung *f* (Med)
Conway micro diffusion dish Conwayschale *f*
conydrine Conhydrin *n*, Conydrin *n*
conylene Conylen *n*
conyrine Conyrin *n*
cook *v* kochen
cookeite Cookeit *m* (Min)
cooker Kocher *m*, Kochkessel *m*
cooking apparatus Kochapparat *m*
cooking fat Backfett *n*
cooking lye Kochlauge *f*
cooking utensils Kochgerät *n*, Kochgeschirr *n*
cool *a* kalt
cool *v* [ab]kühlen, erkalten, ~ **completely**
auskühlen, ~ **down** abkühlen, ~ **suddenly**
abschrecken, ~ **thoroughly** auskalten,
auskühlen
coolant Kühlmittel *n*, Kühlflüssigkeit *f*,
Kühlöl *n*, Kühlwasser *n*
coolant circuit Kühlkreislauf *m*
coolant jacket Kühlmantel *m*

coolant passage Kühlmitteldurchlauf *m*
cooled [down] abgekühlt
cooler Kühler *m*, Kühlkasten *m*,
Kühlvorrichtung *f*, **intermediate** ~
Zwischenkühler *m*
cooler area Kühlerfläche *f*
cooler surface Kühlerfläche *f*
coolgardite Coolgardit *m*
cool hammer *v* kalt hämmern, kalt schmieden,
hartschlagen
cooling Erkalten *n*, Kaltwerden *n*, Kühlen *n*,
Kühlung *f*, Wärmeentzug *m*, **method of** ~
Kühlmethode *f*, **slow** ~ Tempern *n*, ~ **by**
expansion Entspannungsabkühlung *f*, ~ **by**
means of circulating water
Wasserumlaufkühlung *f*, ~ **due to expansion**
Entspannungsabkühlung *f*, ~ **of furnace floor**
Bodenkühlung *f*, ~ **with ice** Eiskühlung *f*
cooling *a* [ab]kühlend, erkaltend
cooling agent Kühlmittel *n*, Abschreckmittel *n*,
Kühlsubstanz *f*
cooling air Kühlluft *f*
cooling apparatus Kühlapparat *m*
cooling arrangement Kühlvorrichtung *f*
cooling bath Kältebad *n*
cooling bed Kühlbett *n* (Metall)
cooling box Kühlkasten *m*
cooling centrifuge Kühlzentrifuge *f*
cooling chamber Kühlkammer *f*, Kühlraum *m*
cooling channel Kühlkanal *m*
cooling coefficient Kühlungskoeffizient *m*
cooling coil Kühlschlange *f*, Kühlspirale *f*
cooling conveyer Kühlband *n*
cooling copper Kühlschiff *n*
cooling curve Abkühlungskurve *f* (Therm)
cooling cycle Kühlkreislauf *m*
cooling cylinder Kühlzylinder *m*
cooling device Kühlvorrichtung *f*
cooling disc Kühlplatte *f*, Kühlscheibe *f*
cooling [down] Abkühlen *n*, Abkühlung *f*
cooling drum Kühltrommel *f*
cooling effect Kühlwirkung *f*
cooling fin Kühlrippe *f*
cooling fluid Kühlflüssigkeit *f*
cooling jacket Kühlmantel *m*, **crack in the** ~
Kühlmantelriß *m*
cooling liquid Kühlflüssigkeit *f*
cooling medium Kälteträger *m*, Kühlmittel *n*
cooling method Kühlverfahren *n*
cooling oil Kühlöl *n*
cooling pan Kühlpfanne *f*, Kühlschiff *n*
cooling performance Kühlleistung *f*
cooling period Abkühlungszeit *f*
cooling plant Kälteanlage *f*, Kühlanlage *f*,
Kühlwerk *n*
cooling pond Kühlteich *m*
cooling roll[er] Kühlwalze *f*
cooling speed Abkühlgeschwindigkeit *f*
cooling strain Abkühlungsspannung *f*

cooling stress Abkühlungsspannung *f*
cooling surface Kühlfläche *f*,
Abkühlungs[ober]fläche *f*
cooling tank Kühlbehälter *m*, Kühltank *m*
cooling test Abkühlungsversuch *m*
cooling time Abkühlungszeit *f*
cooling tower Kondensationsturm *m*,
Kühlturm *m*, Rieselwerk *n*
cooling trap Kühlfalle *f*, Ausfrierfalle *f*
cooling tray Kühlschiff *n*
cooling trough Kühltrog *m*
cooling tube Kühlrohr *n*
cooling vat Abkühlfaß *n*, Kühlfaß *n*
cooling vessel Abkühlfaß *n*, Abkühlkessel *m*
cooling water Kühlwasser *n*, **consumption of** ~
Kühlwasserverbrauch *m*, **quantity of** ~
Kühlwassermenge *f*, **temperature of** ~
Kühlwassertemperatur *f*
cooling water tank Kühlwasserbehälter *m*
cooling worm Kühlschnecke *f*
cooper Böttcher *m*
cooperate *v* kooperieren, zusammenarbeiten,
zusammenwirken
cooperation Mithilfe *f*, Zusammenarbeit *f*,
Zusammenwirkung *f*
cooperative kooperativ, zusammenwirkend
cooperative effort Gemeinschaftsarbeit *f*
cooperative research Gemeinschaftsforschung *f*
cooper's pitch Faßpech *n*
coordinate Achse *f* (Math), Koordinate *f*
(Math), **angular** ~ Winkelkoordinate *f*
coordinate *v* abstimmen (z. B. Arbeiten),
ausrichten, gleichschalten, gleichstellen,
koordinieren
coordinate axis Koordinatenachse *f*
coordinate direction Koordinationsrichtung *f*
coordinate induction koordinative Induktion *f*
coordinate plane Koordinatenebene *f*
coordinate recorder Koordinatenschreiber *m*
coordinates, origin of ~
Koordinatennullpunkt *m*, **system of** ~
Koordinatensystem *n*, Achsenkreuz *n*,
Achsensystem *n*, **transformation of** ~
Koordinatentransformation *f*
coordinate system Koordinatensystem *n*, **origin**
of the ~ Ursprung *m* des
Koordinatensystems
coordination Koordinierung *f*, Zusammenspiel *n*,
index of ~ Koordinationszahl *f* (Krist)
coordination compound
Koordinationsverbindung *f* (Chem)
coordination formula Koordinationsformel *f*
coordination isomerism Koordinationsisomerie *f*
coordination lattice Koordinationsgitter *n*
coordination number Koordinationszahl *f* (Krist)
coordination polyhedron
Koordinationspolyeder *n* (Krist)
coordination theory Koordinationslehre *f*,
Koordinationstheorie *f*

coordinative koordinativ
coorongit Coorongit *m* (Min)
copaene Copaen *n*
copahuvic acid Copaivasäure *f*
copaiba Copaiva *f*
copaiva Copaiva *f*
copaiva balsam Copaivabalsam *m*
copaiva oil Copaiva[balsam]öl *n*
copaivic acid Copaivasäure *f*
copal Copalharz *n*, Flußharz *n*, Kopal *m*,
  Lackharz *n*
copal esterification plant
  Kopalveresterungsanlage *f*
copalic acid Kopalsäure *f*
copalin Kopalin *n* (Min), Kopalinharz *n*
copalin balsam Kopalinbalsam *m*
copalite Kopalin *m* (Min)
copal lacquer Kopallack *m*
copal oil Kopalöl *n*
copal resin melting plant Kopalschmelzanlage *f*
copal substitute Kopalersatz *m*
copal varnish Kopalfirnis *m*, Kopallack *m*
cope Oberkasten *m* (Gieß)
copellidine Kopellidin *n*
cophasal gleichphasig, in Phase [mit]
copiapite Copiapit *m* (Min), Gelbeisenstein *m*
  (Min)
copolycondensation Copolykondensation *f*
copolymer Mischpolymer *n*, Copolymer *n*,
  Mischpolymerisat *n*, **alternative** ~
  alternierendes Copolymer
copolymerization Mischpolymerisation *f*,
  Copolymerisation *f*, Heteropolymerisation *f*
copper (Symb. Cu) Kupfer *n*, **black** ~
  Schwarzkupfer *n*, **capillary** ~ Haarkupfer *n*,
  carbonate of ~ Kupfercarbonat *n*, **containing**
  ~ kupferhaltig, **crude** ~ Rohkupfer *n*,
  Gelbkupfer *n*, Kupferstein *m*, **dry or**
  underpoled ~ übergares Kupfer, **electrolytic**
  ~ galvanisch gefälltes Kupfer, **flat** ~
  Flachkupfer *n*, **granulated** ~
  Kupfergranalien *pl*, Kornkupfer *n*,
  Kupferkörner *n pl*, Kupferschrot *m*, **hard**
  drawn ~ Hartkupfer *n*, **ocher of** ~
  Kupferocker *m* (Min), **red oxide of** ~
  Kuprit *m* (Min), **refined** ~ Raffinatkupfer *n*,
  Rosenkupfer *n*, **separation of** ~ Kupfergare *f*,
  shine of ~ Kupferblick *m*, **soft** ~
  Weichkupfer *n*, **test for** ~ Kupferprobe *f*,
  tough pitch ~ Zähkupfer *n*, **white** ~
  Weißkupfer *n*, **yellow** ~ Messing *n*, ~ **in bars**
  Stangenkupfer *n*, ~ **in disks**
  Rosettenkupfer *n*, ~ **in grains** Perlkupfer *n*, ~
  in rolls Rollenkupfer *n*, ~ **in sheets**
  Kupferblech *n*, **vitreous** ~ **ore**
  Glaskupfererz *n* (Min), ~ **from waste**
  Krätzkupfer *n*, ~ **obtained from dross**
  Abzugskupfer *n*
copper *a* kupfern

copper *v* verkupfern, **to free from** ~ entkupfern,
  to plate with ~ verkupfern, **to treat with** ~
  verkupfern
copper accumulator Kupferakkumulator *m*
copper acetate Kupferacetat *n*
copper acetoarsenite Kupferarsenacetat *n*,
  Pariser Grün *n*, Schöngrün *n*
copper acetylide Acetylenkupfer *n*,
  Kupferacetylen *n*
copper alanine Kupferalanin *n*
copper albuminate Albuminkupfer *n*,
  Kupferalbuminat *n*
copper alloy Kupferlegierung *f*
copper alum Kupferalaun *m*
copper aluminate Kupferalaun *m*
copper amalgam Kupferamalgam *n*
copper aminoacetate Kupferglykokoll *n*
copper aminopropionate Kupferalanin *n*
copper ammonium sulfate
  Kupferammoniumsulfat *n*
copper antimony carbonate, native basic ~
  Rivotit *m* (Min)
copper arch Kupferbügel *m*
copper arsenate, hydrated ~ Chlorotil *m* (Min),
  native ~ Olivenit *m* (Min), Holzkupfer *n*,
  Olivenerz *n*
copper arsenide Kupferarsenid *n*
copper arsenite Kupferarsenit *n*, **acid** ~
  Scheelesches Grün *n*
copperas Eisenvitriol *n*, Melanterit *m* (Min),
  grüner Vitriol *n*, **blue** ~ Kupfersulfat *n*, **white**
  ~ Augenstein *m* (Min), Zinksulfat *n*
copper-asbestos gasket Kupferasbestdichtung *f*
copper ashes Kupferschlag *m*
copper assay Kupferprobe *f*
copperas vat Vitriolküpe *f*
copper autotype process Kupferautotypie *f*
copper band Kupferband *n*
copper bar Kupferbarren *m*
copper bath Kupferbad *n*
copper-bearing kupferführend
copper bismuth sulfide, native ~ Klaprotholit *m*
  (Min)
copper black Kupferschwarz *n*
copper blast furnace Kupferschachtofen *m*
copper bloom Kupferblüte *f* (Min),
  Kupferfedererz *n* (Min)
copper blue Kupferblau *n*, Kupferlasur *f*
copper bolt Kupferbolzen *m*
copper bottom Bodenkupfer *n*
copper bromate Kupferbromat *n*
copper bromide Kupferbromid *n*
copper brown Kupferbraun *n*, Kupferocker *m*
  (Min)
copper calciner Kupferröstofen *m*
copper calx (obs) Kupferkalk *m*
copper cap Zündhütchen *n*

copper carbonate Kupferkarbonat *n*, **blue ~**
Bremerblau *n*, **hydrated basic ~** Bergasche *f*
(Min), Chessylit[h] *m* (Min)
copper cellulose Kupfercellulose *f*
copper chelate Kupferchelat *n*
copper chlorate Kupferchlorat *n*
copper chloride Kupferchlorid *n*
copper chromate Kupferchromat *n*
copper citrate Kupfercitrat *n*
copper coating Kupferüberzug *m*
copper coil Kupferschlange *f*
copper coin Kupfermünze *f*
copper color Kupferfarbe *f*
copper-colored kupferfarben, kupfern, kupferrot
copper-constantan couple
Kupferkonstantanelement *n*
copper content Kupfergehalt *m*
copper cyanide Kupfercyanid *n*
copper cyanoferrate(II) Kupferferrocyanid *n*
copper cylinder Kupferwalze *f*
copper damper Kupferdämpfer *m*
copper damping Kupferdämpfung *f*
copper deposit Kupferniederschlag *m*,
Kupferüberzug *m*
copper dichloride Kupfer[di]chlorid *n*
copper dish test Kupferschalentest *m*
copper disk Kupferscheibe *f*
copper dross Schmelzkupfer *n*
copper electrode Kupferelektrode *f*
copper engraving Kupferstechen *n*,
Kupferstich *m*
copper enzyme Kupferenzym *n* (Biochem)
copper etching Kupferradierung *f*
copper-faced mit Kupfer überzogen
copper ferrocyanide Ferrokupfercyanid *n*,
Kupferferrocyanid *n*
copper filings Kupferfeilicht *n*,
Kupferfeilspäne *n pl*
copper finery Kupferfrischofen *m*
copper fluosilicate Kupfersilicofluorid *n*
copper flux Kupferzuschlag *m* (Gieß)
copper foil Blattkupfer *n*, Kupferblech *n*,
Kupferfolie *f*
copper forged work Kupferschmiedearbeit *f*
copper forging Kupferschmiedestück *n*
copper foundry Kupfergießerei *f*
copper froth Tyrolit *m* (Min)
copper fulminate Kupferfulminat *n*
copper fumes Kupferrauch *m*
copper furnace Kupfergarherd *m*,
Kupfersteinröstofen *m*
copper galvanoplastic process
Kupfergalvanoplastikverfahren *n*
copper glance Kupferglanz *m* (Min)
copper glycerol Kupferglycerin *n*
copper glycine Kupferglykokoll *n*
copper glycocoll Kupferglykokoll *n*
copper granules Kupfergranalien *pl*

copper green Kupfergrün *n*
copper hydride Kupferwasserstoff *m*
copper hydroxide Kupferhydroxid *n*
copper(I) bromide Kupfer(I)-bromid *n*,
Kupferbromür *n* (obs)
copper(I) chloride Cuprochlorid *n*,
Kupferchlorür *n* (obs), Kupfer(I)-chlorid *n*
copper(I) compound Kupfer(I)-Verbindung *f*,
Cuproverbindung *f*,
Kupferoxydulverbindung *f* (obs)
copper(I) cyanide Kupfer(I)-cyanid *n*,
Kupfercyanür *n* (obs)
copper(I) hydroxide Kupfer(I)-hydroxid *n*,
Kupferhydroxydul *n* (obs),
Kupferoxydulhydrat *n* (obs)
copper(II) bromide Kupfer(II)-bromid *n*
copper(II) carbonate Cupricarbonat *n*
copper(II) chloride Cuprichlorid *n*,
Kupferbichlorid *n*
copper(II) chromate Cuprichromat *n*
copper(II) compound Kupfer(II)-Verbindung *f*,
Cupriverbindung *f*, Kupferoxidverbindung *f*
(obs)
copper(II) iodide Kupfer(II)-jodid *n*
copper(I) iodide Kupfer(I)-jodid *n*,
Kupferjodür *n* (obs)
copper(II) salt Kupfer(II)-Salz *n*,
Kupferoxidsalz *n* (obs)
copper(II) sulfate Kupfer(II)-sulfat *n*,
Kupfervitriol *n*
coppering Verkupfern *n*, Verkupferung *f*
copper iodide Kupferjodid *n*
copper ion Kupferion *n*
copper(I) oxide Cuprooxid *n*, Kupfer(I)-oxid *n*,
Kupferoxydul *n* (obs)
copper(I) salt Kupfer(I)-Salz *n*, Cuprosalz *n*
copper(I) sulfide Kupfer(I)-sulfid *n*,
Halbschwefelkupfer *n*, Kupfersulfür *n* (obs)
copper(I) sulfite Cuprosulfit *n*
copper(I) thiocyanate Kupfer(I)-rhodanid *n*,
Kupferrhodanür *n* (obs)
copper jacket Kupfermantel *m*
copper lead selenide Selenkupferblei *n* (Min)
copper-like kupferartig, kupferig
copper lining Kupferverkleidung *f*
copper matte Kupferstein *m*
copper mordant Kupferbeize *f*
copper nickel alloy Kupfernickellegierung *f*
copper nitrate Kupfernitrat *n*, **native basic ~**
Gerhardtit *m* (Min)
copper number Kupferzahl *f*
copper oleate Kupferoleat *n*
copperon Cupferron *n*
copper ore Kupfererz *n* (Min), **capillary red ~**
Kupferblüte *f* (Min), **gray ~** Kupferfahlerz *n*
(Min), Graufahlerz *n* (Min), Graugültigerz *n*
(Min), **green ~** Erdgrün *n* (Min), **hepatic ~**
Kupferlebererz *n* (Min), **purple ~**

Buntkupfererz *n* (Min), **red** ~ Kupferrot *n*
(Min), **variegated** ~ Buntkupfererz *n* (Min),
Buntkupferkies *m* (Min), **white** ~
Weißkupfererz *n* (Min), Kieskupfererz *n*,
yellow ~ Gelbkupfererz *n* (Min), Brüherz *n*
(Min)

copper-ore smelting Kupferverhüttung *f*

copper oxalate Kupferoxalat *n*

copper oxide Kupferoxid *n*, **ammoniacal** ~
Cupramminbase *f*, **black** ~ Kupferschwärze *f*,
native ~ Melakonit *m* (Min), **native black** ~
Tenorit *m* (Min), **red** ~ Kupfer(I)-oxid *n*,
Kupferoxydul *n* (obs)

copper oxide bridge Kupferoxidbrücke *f* (Elektr)

copper oxide cell Kupronelement *n*

copper oxychloride Atakamit *m* (Min),
Kupferoxychlorid *n* (Chem)

copper packing Kupferdichtung *f*

copper phenolsulfonate Kupferphenolsulfonat *n*

copper phosphide Kupferphosphid *n*,
Phosphorkupfer *n*

copper pigment Kupferfarbe *f*

copper pipe Kupferrohr *n*

copperplate Kupferblech *n*

copper-plate *v* verkupfern

copperplate black Kupferdruckerschwärze *f*

copper-plated steel wire Kupferstahldraht *m*

copperplate print Kupferstich *m*

copperplate printing Kupferdruck *m*

copper-plating Verkupfern *n*, Verkupferung *f*, ~
by immersion Tauchverkupferung *f*

copper poisoning Kupfervergiftung *f*

copper pole Kupferpol *m*

copper powder rote Bronze

copper precipitate Kupferniederschlag *m*

copper press Kupferpresse *f*

copper protoxide Kupfer(I)-oxid *n*,
Kupferoxydul *n* (obs)

copper pyrites Gelbkupfererz *n* (Min), Gelferz *n*
(Min), Kupferkies *m* (Min)

copper rain Kupferregen *m*, Spratzkupfer *n*

copper red kupferrot

copper refinery Kupferraffinerie *f*

copper refining Kupferreinigung *f*,
Garmachen *n* des Kupfers,
Kupferraffination *f*

copper refining hearth Rosettenherd *m*

copper resinate Kupferresinat *n*

copper rust Grünspan *m*

copper saccharate Kupfersaccharat *n*

copper salt Kupfersalz *n*

copper salt solution Kupfersalzlösung *f*

copper sample Kupferprobe *f*

copper sand Kupfersand *m*

copper scale Kupferhammerschlag *m*,
Kupferasche *f*, Kupfersinter *m*

copper schist Kupferschiefer *m* (Min)

copper scrap Kupferabfälle *m pl*

copper scum Kupferschaum *m*

copper selenide Selenkupfer *n* (Min)

copper sheeting Kupferbeschlag *m*

copper silicide Kupfersilicid *n*

copper silicofluoride Kupfersilicofluorid *n*

copper silver alloy Kupfersilber *n*

copper slag Kupferschlacke *f*,
Garkupferschlacke *f*, **coarse** ~
Schwarzkupferschlacke *f*

copper slick Kupferschlich *m*

copper smelting plant Kupferhütte *f*

coppersmith's cement Kupferkitt *m*

copper sodium cellulose Kupfernatroncellulose *f*

copper solder Kupferlot *n*

copper solution Kupferlösung *f*

copper sponge Kupferschwamm *m*,
Schwammkupfer *n*

copper spray Kupferspritzmittel *n*

copper stannate Kupferstannat *n*

copper storage battery Kupferakkumulator *m*

copper strand Kupferlitze *f*

copper strip test Kupferstreifenprobe *f*

copper sulfantimonide, native ~
Kupferantimonglanz *m* (Min)

copper sulfate blauer Galitzenstein *m* (Min),
Kupfersulfat *n*, blauer Vitriol *n*, **hydrated** ~
Blauvitriol *n*, **native basic** ~ Kamarezit *m*
(Min)

copper sulfate crystal Kupfervitriolkristall *m*

copper sulfate plant Kupfersulfatanlage *f*

copper sulfide Kupfersulfid *n*, Schwefelkupfer *n*

copper sulfocarbolate Kupfersulfophenylat *n*

copper sulfophenate Kupfersulfophenylat *n*

copper thiocyanate Kupferrhodanid *n*,
Rhodankupfer *n*

copper tube Kupferrohr *n*

copper turnings Kupferdrehspäne *pl*

copper uranite Chalkolith *m* (Min)

copper value Kupferzahl *f*

copper vitriol Kupfervitriol *n*, Blauvitriol *n*

copper voltameter Kupfervoltameter *n*

copper waste Kupferabfälle *m pl*

copper welding Kupferschweißung *f*

copper wire Kupferdraht *m*, **hard-drawn** ~
Hartkupferdraht *m*, **soft** ~
Weichkupferdraht *m*

copper wire coil Kupferdrahtspule *f*

copper [wire] gauze Kupferdrahtnetz *n*, **roll of** ~
Kupferdrahtnetzrolle *f*

coppery kupferartig, kupferig, kupfern, kupferrot

coppery luster Kupferglanz *m*

copper zinc accumulator
Kupferzinkakkumulator *m*

copper zinc storage battery
Kupferzinkakkumulator *m*

coppite Koppit *m* (Min)

coppying attachment Kopiervorrichtung *f*

copra Kopra *f*

coprecipitate *v* mitfällen

coprecipitation Mitfällung *f*

coprolite Koprolith *m*
coprophilic koprophil (Bakt)
coprophilous koprophil (Bakt)
coproporphyrin[e] Koproporphyrin *n*
coprostane Koprostan *n*
coprostanic acid Koprostansäure *f*
coprostanol Koprostanol *n*
coprosterol Koprostanol *n*
coptisine Coptisin *n*
copy Kopie *f,* Abbild *n,* Abdruck *m,* Abzug *m*
    (Buchdr, Phot), Durchschlag *m*
    (Durchschrift), Imitation *f,* Photokopie *f*
copy *v* kopieren, abbilden, abdrucken,
    abklatschen (Buchdr), abzeichnen,
    nachahmen, nacharbeiten, nachbilden,
    nachformen
copying, method of ~ Abdruckverfahren *n*
    (Buchdr)
copying apparatus Kopierapparat *m,*
    Kopiergerät *n,* Vervielfältigungsapparat *m*
copying ink Kopiertinte *f*
copying machine Vervielfältigungsmaschine *f*
copying mill Kopierfräsmaschine *f*
copying milling machine Kopierfräsmaschine *f*
copying paper Kopierpapier *n*
copying process Kopierverfahren *n*
copy paper Durchschlagpapier *n*
copyright Urheberrecht *n*
copy-righted gesetzlich geschützt
copyrine Copyrin *n*
coquimbite Coquimbit *m* (Min)
coracite Coracit *m* (Min)
coral Koralle *f*
coral agate Korallenachat *m* (Min)
coral color Korallenfarbe *f*
coral-colored korallenfarben
coral lac Korallenlack *m*
corallin Corallin *n,* Rosolsäure *f*
corallite Korallenversteinerung *f*
coral ore Korallenerz *n* (Min)
coral powder Korallenpulver *n*
coral red Korallenrot *n*
coralydine Coralydin *n*
coralydinium hydroxide Coralydiniumhydroxid *n*
coralyne Coralyn *n*
coralyne base Coralynbase *f*
coralyne hydrate Coralynhydrat *n*
coramine Coramin *n*
coranene Coranen *n*
corazole Corazol *n*
corbadrine Corbadrin *n*
corbisterol Corbisterin *n*
corchorolic acid Corchorolsäure *f*
cord Schnur *f,* Bindfaden *m,* Klafter *f m n*
    (Maß), Kordel *f,* Leitungsschnur *f,* Litze *f*
    (Elektr), Zwirn *m*
cord angle Fadenwinkel *m* (Reifen),
    Zenitwinkel *m* (Reifen)
cord count Fadendichte *f* (Reifen)

cord fabric Cordgewebe *n*
cordierite Cordierit *m* (Min), Iolith *m* (Min)
cordite Cordit *n*
cordofan gum Kordofangummi *n*
cord pulley Schnurrolle *f,* Schnurwirtel *m*
cordrastine Cordrastin *n*
cord rupture Gewebebruch *m*
cord weight Schnurgewicht *n*
cordycepic acid Cordycepinsäure *f*
cordycepin Cordycepin *n*
cordycepose Cordycepose *f*
cordylite Cordylit *m* (Min)
core Kern *m,* Dorn *m,* Gußkern *m* (Gieß),
    Innere[s] *n,* Kernhaus *n* (Frucht), Kernzone *f,*
    Mark *n,* Zentralstück *n,* air in the ~
    Kernluft *f*
core *v* durchbohren
coreactant Reaktionspartner *m*
core binder Kernbindemittel *n*
core box Formkasten *m* (Gieß), Kernkasten *m*
core cement Trommelklebgummi *m* (Reifen)
core changer Dornwechsler *m*
core circuit Kernkreislauf *m*
core compound Kernmasse *f*
CO-recorder Kohlenoxidschreiber *m,*
    Kohlensäureschreiber *m*
cored carbon Dochtkohle *f* (Elektr)
cored casting Kernguß *m*
core diameter Kerndurchmesser *m*
core drying stove Kerntrockenofen *m*
cored work Kernguß *m*
core fork Kerngabel *f*
core frame Kerngerippe *n* (Gieß),
    Kernrahmen *m* (Gieß)
coregonin Coregonin *n*
core groove Kerneinschnürung *f*
core iron Eiseneinlage *f,* Kerneisen *n*
core isomerism Rumpfisomerie *f*
core lathe Kerndrehbank *f*
core maker Kernmacher *m* (Gieß)
core material Spaltmaterial *n* (Atom)
core memory Kernspeicher *m*
core model Rumpfmodell *n*
core molding Kernformerei *f* (Gieß)
coreopsin Coreopsin *n*
core part Kernstück *n* (Gieß)
core pin Gewindestift *m,* Kernlochstift *m,*
    Lochstift *m,* plain ~ Lochstift *m*
core pin plate Lochstiftplatte *f*
corepressor Corepressor *m* (Mol. Biol)
core radius Kernhalbmesser *m*
core recess Kerneinschnürung *f*
core roasting Kernröstung *f*
core sand Kernsand *m,* condition of ~
    Kernsandbeschaffenheit *f*
core spider Dornhalter *m*
core spindle Kernspindel *f*
core strickle Kernschablone *f*
core template Kernschablone *f*

core transformer Kerntransformator *m*
core winding machine Hülsenwickelmaschine *f*
coreximine Coreximin *n*
coriaceous lederähnlich, lederartig
coriamyrtin Coriamyrtin *n*
coriander oil Korianderöl *n*
coriander seed Koriander *m*
coriandrol Coriandrol *n*, Linalool *n*
coridine Coridin *n*
coring Entmischung *f* (Krist)
coriolis force Corioliskraft *f* (Phys)
cork Kork *m*, Korken *m*, Stöpsel *m* aus Kork,
   agglomerated ~ gepreßter Kork, **burnt** ~
   Korkkohle *f*
cork *v* verkorken, zukorken
cork black Korkschwarz *n*
cork board Korkstein *m*
cork borer Korkbohrer *m*
cork boring machine Korkbohrmaschine *f*
cork brick Korkstein *m*
cork charcoal Korkkohle *f*
cork disc Korkscheibe *f*
cork fat Suberin *n*
corking machine Flaschenverkorkmaschine *f*
cork insulation Korkisolation *f*
corkite Corkit *m* (Min)
cork knife Korkmesser *n*
cork-like korkähnlich, korkartig, korkig
cork plate Korkplatte *f* (Bauw)
cork pliers Korkzange *f*
cork polishing wheel Korkschleifscheibe *f*
cork powder Korkmehl *n*
cork press Korkpresse *f*
cork processing Korkverarbeitung *f*
cork pulley Korkscheibe *f*
corkscrew Korkenzieher *m*
cork slab Korkplatte *f* (Bauw)
cork [stopper] Korkstopfen *m*
cork tree Korkbaum *m*
corky korkähnlich, korkartig, korkig
corlumine Corlumin *n*
corn Getreide *n*, Korn *n* (Am. E.), Mais *m*
corn *v* einpökeln; körnen
corn brandy Kornbranntwein *m*,
   Kornschnaps *m*, Whisky *m*
corn cob Maiskolben *m*
cornelian Karneol *m* (Min)
corneous hornig
corneous silver Hornchlorsilber *n* (Min),
   Hornsilber *n* (Min)
corner Ecke *f*
corner block Eckblock *m*
corner chisel Winkelmeißel *m*
corner column Ecksäule *f*
cornered eckig, winkelig
cornering thrust Kurvenschiebung *f* (Kfz)
corner pillar Eckpfeiler *m*
cornerstone Grundstein *m* (Arch)
corner valve Winkelventil *n*

corner weld Ecknaht *f* (Schweiß),
   Kehlschweißung *f* (Schweiß)
corn flour Maismehl *n*
cornflower blue Kornblumenblau *n*
cornic acid Cornin *n*
cornice Gesims *n* (Arch)
cornicularine Cornicularin *n*
cornin Cornin *n*
Cornish boiler Einflammrohrkessel *m*
cornite Kornit *n*
corn oil Maiskeimöl, Maisöl *n*
corn starch Maisstärke *f*, Amylum *n*
corn sugar Maisdextrose *f*, Maiszucker *m*
cornubianite Cornubianit *m*
cornutine Kornutin *n*
cornwallite Cornwallit *m* (Min)
coroglaucigenin Coroglaucigenin *n*
corollary Folgesatz *m* (Math)
corona Hof *m* (Opt), Korona *f*,
   Koronaentladung *f*, stille Entladung (pl.
   coronas, coronae), Glimmentladung *f*
corona arc Koronabogen *m*
corona [discharge] Koronaentladung *f*,
   Sprühentladung *f*
corona discharge electrode Sprühelektrode *f*
coronadite Coronadit *m* (Min)
corona effect Koronaeffekt *m*
coronal line Koronalinie *f*
corona losses Glimmverluste *pl*
coronaridine Coronaridin *n*
coronillin Coronillin *n*
coronium Coronium *n* (Leg)
coronium line Coroniumlinie *f*
coronopic acid Coronopsäure *f*
corotoxigenin Corotoxigenin *n*
corpaverine Corpaverin *n*
corphin Corphin *n*
corpse Leiche *f*, Kadaver *m*
corpulent dick
corpuscle Blutkörperchen *n* (Med),
   Korpuskel *n* (Chem, Phys)
corpuscular korpuskular
corpuscular radiation Korpuskularstrahlung *f*
corpuscular rays Materiestrahlen *m pl*
corpuscular theory of light Korpuskulartheorie *f*
   des Lichtes (Opt)
corpuscule Korpuskel *n* (Chem, Phys)
corpus luteum Gelber Körper *m* (Physiol)
corpus luteum hormone
   Corpus-luteum-Hormon *n*, Luteohormon *n*
correct korrekt, genau, richtig, wahr, **to be** ~
   stimmen, richtig sein
correct *v* berichtigen, umändern, korrigieren,
   verbessern
corrected verbessert
correction Berichtigung *f*, Korrektur *f*,
   Richtigstellung *f*, Verbesserung *f*
correction [of mistakes] Fehlerberichtigung *f*
correction factor Korrekturfaktor *m*

correction for profile shift
Profilverschiebungsfaktor *m*
correction method Berichtigungsverfahren *n*
corrective Besserungsmittel *n*, Korrektivmittel *n*
(Med)
corrective error Ausgleichsfehler *m*
corrective factor Berichtigungsfaktor *m*
correctness Genauigkeit *f*, Richtigkeit *f*
correct product (screening) Normalkorn *n*
correlate *v* korrelieren
correlation Korrelation *f*, Wechselbeziehung *f*,
wechselseitige Beziehung *f*, Wechselwirkung *f*
correlation effect Korrelationseinfluß *m*
correlation energy Korrelationsenergie *f*
correlation factor Korrelationskoeffizient *m*
correllogenin Correllogenin *n*
correspond *v* korrespondieren
correspondence Korrespondenz *f*
correspondence principle
Korrespondenzprinzip *n* (Quant)
corresponding korrespondierend, entsprechend
corrin Corrin *n*
corroborate *v* bestärken, bestätigen
corrode *v* korrodieren, [ab]ätzen, [an]fressen,
angreifen, [an]rosten, beizen, verrosten,
zerfressen
corroded korrodiert, abgefressen, eingeätzt,
verrostet
corrodibility Korrodierbarkeit *f*, Ätzbarkeit *f*,
Korrosionsempfindlichkeit *f*,
Rostempfindlichkeit *f*
corrodible ätzbar, angreifbar (Met),
korrodierbar, rostempfindlich
corroding ätzend, beizend
corroding agent Rostbildner *m*
corroding bath Beizbad *n*
corroding brittleness Beizsprödigkeit *f*
corroding proof Ätzprobe *f*
corrodkote process Corrodkote-Verfahren *n*
corrogate *v* wellen
corrole Corrol *n*
corrosion Korrosion *f*, Ätzung *f* (Chem),
Anfressung *f*, chemischer Angriff (Chem),
Beizen *n* (Chem), Rostbildung *f*, Rostfraß *m*,
Verrostung *f*, Zerfressung *f*, **aqueous** ~
Feuchtigkeitskorrosion *f*, **atmospheric** ~
atmosphärische Korrosion *f*, **attack by** ~
Korrosionsangriff *m*, **channeling** ~ langadrige
Rostanfressung, **danger of** ~ Rostgefahr *f*,
electrochemical ~ elektrochemische
Korrosion *f*, **external** ~ Außenkorrosion *f*,
flat etching type of ~ Oberflächenkorrosion *f*,
form of ~ Korrosionserscheinung *f*, **fretting**
~ Lochfraßkorrosion *f*, **high-temperature** ~
Hochtemperaturkorrosion *f*, **honeycomb** ~
Rostanfressung *f* bis zu tiefen Grübchen,
inhibiting ~ korrosionshemmend, **internal** ~
Innenkorrosion *f*, **localized** ~ örtliche (lokale)

Korrosion, **low-temperature** ~
Tieftemperaturkorrosion *f*, **pitting** ~
Lochfraßkorrosion *f*, **resistance to atmospheric**
~ Witterungsbeständigkeit *f*, **resistant to** ~
korrosionsbeständig, **selective** ~ selektive
Korrosion, **sensitiveness to** ~
Korrosionsempfindlichkeit *f*,
Rostempfindlichkeit *f*, **sensitive to** ~
rostempfindlich, **stimulating** ~
korrosionsfördernd, **susceptibility to** ~
Korrosionsneigung *f*, **susceptible to** ~
korrosionsempfindlich, **tendency to** ~
Korrosionsneigung *f*, **to eat away by** ~
abätzen, **tubercular** ~ grübchenartige
Rostanfressung, ~ **by condensed moisture**
Schwitzwasserkorrosion *f*, ~ **by niter**
Salpeterfraß *m*, **intergranular** ~ **or**
intercrystalline corrosion
Korngrenzenkorrosion *f*, **transcrystalline** ~ **or**
intracrystalline corrosion transkristalline
Korrosion *f*, ~ **under mechanical stress**
Korrosion *f* unter gleichzeitiger mechanischer
Beanspruchung
corrosion brittleness Beizbrüchigkeit *f*
corrosion cell Korrosionselement *n*
corrosion fatigue Korrosionsermüdung *f*,
Schwingungsrißkorrosion *f*
corrosion figure Ätzfigur *f*
corrosion inhibition Korrosionsverhinderung *f*
corrosion inhibitor Korrosionsinhibitor *m*,
Korrosionsschutzmittel *n*
corrosion phenomenon Korrosionserscheinung *f*
corrosion pit Korrosionsnarbe *f*, Rostnarbe *f*
corrosion preventive Korrosionsschutzmittel *n*
corrosion preventive oil Korrosionsschutzöl *n*
corrosion problem Korrosionsproblem *n*
corrosion-prone korrosionsempfindlich
corrosion-proof korrosionsbeständig,
rostbeständig
corrosion protection Korrosionsschutz *m*
corrosion resistance Rostbeständigkeit *f*
corrosion-resistant korrosionsbeständig,
rostsicher
corrosion resisting quality
Korrosionsbeständigkeit *f*
corrosion test, accelerated ~
Schnellkorrosionsversuch *m*
corrosion testing Korrosionsprüfung *f*
corrosive Korrosionsmittel *n*, Ätzmittel *n*,
Angriffsmittel *n*, Beize *f*, Beizmittel *n*, **to**
remove with ~ ausbeizen
corrosive *a* korrodierend, beizend, kaustisch
corrosive action Korrosionsvorgang *m*,
Rostangriff *m*
corrosive agent Korrosionsmittel *n*
corrosive attack Anfressung *f*, Rostangriff *m*
corrosive effect Korrosionseinfluß *m*
corrosive liquid Beizflüssigkeit *f*, Beizwasser *n*
corrosiveness Ätzkraft *f*, Schärfe *f* (Chem)

corrosive paste Ätzpaste *f*
corrosive power Beizkraft *f*
corrosive salt Ätzsalz *n*
corrosive sublimate Ätzsublimat *n*,
 Mercurichlorid *n*
corrosive [substance] Kaustikum *n* (Chem)
corrugate *v* riffeln
corrugated geriffelt, gewellt
corrugated board Falzbaupappe
corrugated box Faltkiste *f* aus Wellpappe
corrugated iron Wellblech *n*
corrugated plate Wellblech *n*
corrugated roll Riffelwalze *f*
corrugated rubber tubing Faltenschlauch *m*
corrugated sheet Wellblech *n*
corrugated sheets, press for ~ Wellblechpresse *f*
corrugating roll Wellblechwalze *f*
corrugating rolling mill Wellblechwalzwerk *n*
corrugation Riffelung *f*, Wellenbildung *f*
 (Techn)
corrugation profile Wellenprofil *n*
Corsican moss Wurmmoos *n* (Bot)
corsite Corsit *m* (Min), Napoleonit *m* (Min)
cortex Baumrinde *f*, Rinde *f*
cortical hormone Rindenhormon *n*
cortication Rindenbildung *f* (Bot)
corticin Corticin *n*, Rindenstoff *m*
corticinic acid Corticinsäure *f*
corticoid Corticoid *n*
corticosterone Corticosteron *n*
corticotropin Corticotropin *n*
corticrocin Corticrocin *n*
cortisalin Cortisalin *n*
cortisol Cortisol *n*
cortisone Cortison *n*
cortol Cortol *n*
cortolone Cortolon *n*
corundellite Korundellit *m*
corundophilite Korundophilit *m* (Min)
corundum Korund *m* (Min), Diamantspat *m*
 (Min)
corundum disc Korundscheibe *f*
coruscate *v* koruszieren, funkeln
coruscating koruszierend
corybulbine Corybulbin *n*
corycavidine Corycavidin *n*
corycavine Corycavin *n*
corycavinemethine Corycavinmethin *n*
corydaline Corydalin *n*
corydalis alkaloid Corydalisalkaloid *n*
corydine Corydin *n*
corydinic acid Corydinsäure *f*
coryline Corylin *n*
corylopsin Corylopsin *n*
corynantheal Corynantheal *n*
corynantheane Corynanthean *n*
corynantheic acid Corynantheinsäure *f*
corynantheine Corynanthein *n*
corynanthic acid Corynanthinsäure *f*

corynanthidine Corynanthidin *n*
corynanthine Corynanthin *n*
corynanthyrine Corynanthyrin *n*
coryneine Corynein *n*
corynine Yohimbin *n*
corynite Arsenantimonnickelglanz *m* (Min)
corynomycolenic acid Corynomycolensäure *f*
corynomycolic acid Corynomycolsäure *f*
corynoxeine Corynoxein *n*
coryocavidine Coryocavidin *n*
coryocavine Coryocavin *n*
corypalline Corypallin *n*
corypalmine Corypalmin *n*
corytuberine Corytuberin *n*
corytuberinemethine Corytuberinmethin *n*
cosalite Cosalit[h] *m* (Min)
cosecant Kosekante *f* (Math)
coseparation Mitabtrennung *f*
cosine Kosinus *m* (Math), **hyperbolical** ~
 hyperbolischer Kosinus
cosine emission law Kosinusgesetz *n* (Opt)
cosine law Kosinussatz *m* (Math)
cosmene Cosmen *n*
cosmetic Hautpflegemittel *n*,
 Körperpflegemittel *n*, Kosmetikum *n*,
 Schönheitsmittel *n*
cosmetic *a* kosmetisch
cosmetics Kosmetik *f*, Kosmetika *pl*,
 Schönheitspflege *f*
cosmic kosmisch
cosmic chemistry Kosmochemie *f*
cosmic radiation Höhenstrahlung *f*,
 Ultrastrahlung *f*
cosmic ray Ultrastrahl *m*
cosmic ray background kosmische
 Grundstrahlung
cosmic ray investigation
 Höhenstrahluntersuchung *f*
cosmic ray meson Höhenstrahlmeson *n* (Atom)
cosmic ray neutron Höhenstrahlneutron *n*,
 Neutron *n* der kosmischen Strahlung
cosmic ray particle Höhenstrahlteilchen *n*
cosmic rays Höhenstrahlen *pl*,
 Weltraumstrahlung *f*, **primary** ~ primäre
 Höhenstrahlen, **secondary** ~ sekundäre
 Höhenstrahlen
cosmic ray track Nebelspur *f pl* kosmischer
 Strahlen
cosmic showers kosmische Schauer *pl*
cosmic space Weltraum *m*
cosmonaut Kosmonaut *m*, Weltraumfahrer *m*
cosmonautics Kosmonautik *f*, Raumfahrt *f*
cosmos Kosmos *m*, Universum *n*, Weltall *n*
cosmosphere Weltkugel *f*
cosmotron Kosmotron *n* (Phys)
cossaite Cossait *n*
cossa salt Cossasalz *n*
cossyrite Cossyrit *m* (Min)
costaclavine Costaclavin *n*

costene Costen *n*
cost estimate Kostenvoranschlag *m*
costic acid Costussäure *f*
costly kostspielig
cost of materials Materialpreis *m*
cost of production Gestehungspreis *m*,
  Herstellungskosten *pl*
costol Costol *n*
cost price Einkaufspreis *m*, Fabrikpreis *m*,
  Gestehungspreis *m*, Selbstkosten *pl*
cost reduction [Kosten]Einsparung *f*
costunolide Costunolid *n*
costusic acid Costussäure *f*
costus lactone Costuslacton *n*
costus oil Costusöl *n*
costyl chloride Costylchlorid *n*
cotangent Kotangens *m* (Math), Kotangente *f*
  (Math)
cotangent law Kotangentensatz *m* (Math)
cotarnic acid Kotarnsäure *f*
cotarnine Kotarnin *n*
cotarnine hydrochloride Kotarninhydrochlorid *n*,
  Stypticin *n*
cotarnine phthalate Styptol *n*
cotarnone Kotarnon *n*
cotinine Cotinin *n*
coto bark Cotorinde *f*
cotoin Cotoin *n*
cottage cheese Quark *m*
cottage steamer Dämpfkasten *m*,
  Runddämpfer *m*
cotter Keil *m* (Querkeil), Querriegel *m*,
  Vorsteckkeil *m*
cotter [pin] Splint *m*, Vorstecker *m*
cotton Baumwolle *f*, baumwollen, Watte *f*,
  artificial ~ Zellstoffwatte *f*
cotton azodye Baumwollazofarbstoff *m*
cotton beaver Baumwollflanell *m* (Text)
cotton boll Baumwollkapsel *f* (Bot)
cotton cellulose Gossypin *n*
cotton cloth Baumwollgewebe *n*
cotton dressing Baumwollaufbereitung *f*
cotton dyeing Kattunfärberei *f*
cotton effect Cotton-Effekt *m* (Spektr), multiple
  ~ mehrfacher Cotton-Effekt
cotton fiber Baumwollfaser *f*
cotton fiber paper Baumwollfaserpapier *n*
cotton filter Wattefilter *m*
cotton flocks Baumwollflocken *f pl*
cottonize *v* cottonisieren (Text)
cottonizing Kotonisieren *n*
cottonlike baumwollartig
cotton mill Baumwollspinnerei *f*
cotton mordant Baumwollbeize *f*
cotton oil Baumwollöl *n*, Cottonöl *n*
cotton padding Baumwollwatte *f*
cotton plant Baumwollpflanze *f* (Bot)
cotton plug Wattepfropfen *m*
cotton printing Kattundruckerei *f*

cottonseed Baumwollsamen *m*
cottonseed oil Baumwoll[kern]öl *n*,
  Baumwollsamenöl *n*
cotton thread Baumwollfaden *m*
cotton twill Baumwollköper *m* (Text)
cotton wadding Baumwollwatte *f*
cotton waste Baumwollabfall *m*,
  Baumwollputzwolle *f*, Putzbaumwolle *f*
cotton [wool] Watte *f*
cotton yarn Baumwollgarn *n*
Cottrell [electrical precipitation] process
  Cottrell-Verfahren *n*
Cottrell hardening Verfestigung *f* durch
  Cottrell-Effekt (Krist)
Cottrell locking Cottrell-Blockierung *f* (Krist)
cotunnite Cotunnit *m* (Min)
couch *v* gautschen (Papier)
coucher Gautsche *f*, Saugwalze *f*
coucher jacket Gautschmantel *m*
couch-grass Quecke *f* (Bot)
couchgrass root Queckenwurzel *f* (Bot)
Couette flow Couette-Strömung *f*
cough Husten *m* (Med)
cough drop Hustenbonbon *m*
cough lozenge Hustenbonbon *m*, Malzbonbon *m*
cough remedy Hustenmittel *n* (Pharm)
coulomb Coulomb *n* (Maß)
coulomb energy Coulombsche Energie *f* (Elektr)
coulomb forces Coulombsche Kräfte *f pl*
  (Elektr)
coulombian repulsion Coulombsche Abstoßung *f*
coulombmeter Coulombmeßgerät *n* (Elektr)
Coulomb's balance Coulombsche Waage *f*
Coulomb's law Coulombsches Gesetz *n* (Elektr)
coulometer Coulombmeßgerät *n* (Elektr),
  Coulombmeter *n* (Elektr)
coulometric coulometrisch
coumalic acid Cumalinsäure *f*
coumalin Cumalin *n*
coumaraldehyde Cumaraldehyd *m*
coumaran Cumaran *n*
coumaranone Cumaranon *n*
coumarazone Cumarazon *n*
coumaric acid Cumarinsäure *f*, Cumarsäure *f*
coumarilic acid Cumarilsäure *f*
coumarin Cumarin *n*
coumarinic acid Cumarinsäure *f*, Cumarsäure *f*
coumarinoline Cumarinolin *n*
coumarin quinone Cumarinchinon *n*
coumarone Cumaron *n*
coumarone picrate Cumaronpikrat *n*
coumarone resin Cumaronharz *n*
coumestan Cumöstan *n*
coumestrol Cumöstrol *n*
coumon oil Coumonöl *n*
count *v* [zusammen]zählen, anrechnen, ~ again
  nachzählen
countable [ab]zählbar
count capacity Zählkapazität *f*

counter Ladentisch *m*, Zähler *m*, Zählrohr *n*,
  Zählwerk *n*, **boron-filled** ~ Zähler *m* mit
  Borfüllung;, **gamma sensitive** ~
  Gammazähler *m*
counteract *v* bekämpfen, entgegenwirken
counteracting force Gegenkraft *f*
counteraction Gegenmaßnahme *f*,
  Gegenwirkung *f*
counteractive entgegenwirkend
counterbalance Gegengewicht *n*,
  Massenausgleich *m*
counterbalance *v* aufwiegen, ausbalancieren,
  auswuchten, kompensieren
counterbalance weight Ausgleichgewicht *n*
counterbalancing Ausbalancierung *f*
counterbore Senker *m*
counterbrace Gegendiagonale *f*
counter calibration Zählrohreichung *f*
counter cast Gegenguß *m*
countercell gegengeschaltete Zelle *f*
counter characteristic Zählrohrcharakteristik *f*
counterclockwise entgegen dem Uhrzeigersinn,
  linksdrehend, linksgängig
counterclockwise motion Linksdrehung *f*
counter connection Gegenschaltung *f*
counter control Zählrohrsteuerung *f*
countercurrent Gegenlauf *m*, Gegenstrom *m*
countercurrent *a* gegenläufig
countercurrent apparatus Gegenstromapparat *m*
countercurrent boiler Gegenstromkessel *m*
countercurrent braking Gegenstrombremsung *f*
  (Mech)
countercurrent column Gegenstromkolonne *f*
countercurrent condenser Gegenstromkühler *m*
countercurrent cooler Gegenstromkühler *m*
countercurrent distillation
  Gegenstromdestillation *f*
countercurrent distribution
  Gegenstromverteilung *f*
countercurrent electrode Gegenstrom-Elektrode *f*
countercurrent electrolysis
  Gegenstromelektrolyse *f*
countercurrent feed heater Gegenstromwärmer *m*
countercurrent ionophoresis
  Gegenstromionophorese *f*
countercurrent pan mixer
  Gegenstromtellermischer *m*
countercurrent principle Gegenstromprinzip *n*
countercurrent process Gegenstromverfahren *n*
counter dead time Zählrohrtotzeit *f*
counter diagonal Gegendiagonale *f*
counter efficiency Ansprechvermögen *n* (Atom),
  Zählerausbeute *f*, Zählrohrausbeute *f*
counter-electrode Gegenelektrode *f*
counterenamel Gegenemail *n*, Gegenschmelz *m*
counterfeit Fälschung *f*
counterfeit *a* gefälscht
counterfeit *v* fälschen
counterflange Gegenflansch *m*, Widerlager *n*

counterflow principle Gegenstromprinzip *n*
counterforce Gegenkraft *f*
counterion Gegenion *n*
counter life Zählrohrlebensdauer *f*
countermeasure Gegenmaßnahme *f*
countermovement Gegenbewegung *f*
counterpart Gegenstück *n*
counterpiston Gegenkolben *m*
counterpoise Gegengewicht *n*
counterpoise *v* abgleichen (Gewichte),
  ausbalancieren
counterpoison Gegengift *n*
counterpressure Gegendruck *m*, **area of** ~
  Gegendruckfläche *f*
counter quench circuit Zählrohrlöschkreis *m*
counter range Anlaufgebiet *n* (Atom)
counterrotating gegenläufig
counterrotation Gegendrehung *f*, Gegenlauf *m*
countershaft Vorgelegewelle *f*
countershaft gear Vorgelege *n*
countersink Versenkbohrer *m* (Techn)
countersink *v* ausfräsen
countersinking bit Ankörnbohrer *m*
counterstaining Kontrastfärbung *f*
countersteam Gegendampf *m*
counterstream principle Gegenstromprinzip *n*
counterstress Gegenkraft *f*
countersunk screw Flachkopfschraube *f*,
  Senkschraube *f*
counterthrust Gegenstoß *m*
counter [tube] Zählrohr *n*
counterweight Gegengewicht *n*,
  Ausgleichgewicht *n*
counting Zählen *n*, Zählung *f*
counting device Zählwerk *n*
counting diamond Zähldiamant *m*
counting efficiency Zählausbeute *f* (Atom)
counting ionization chamber Zählkammer *f*
  (Atom)
counting machine Rechenmaschine *f*
counting mechanism Zählwerk *n*
counting response Zählrohrcharakteristik *f*
counting technique Meßtechnik *f* (Atom)
countless unzählbar
couple Paar *n*, Koppel *f*, Riemen *m*
couple *v* verbinden, anflanschen; [an]kuppeln,
  einkuppeln, koppeln, paaren, zusammenfügen
coupled gekoppelt, gepaart
coupled jib Koppelausleger *m*
coupling Verbindung *f*, Anschluß *m* (Masch),
  Koppeln *n* (Elektr), Kopplung *f* (Elektr),
  Kupplung *f*, Paarung *f*, Verbinden *n*, **elastic**
  ~ biegsame (elastische) Kupplung, **flexible** ~
  Gelenkkupplung *f*, biegsame (elastische)
  Kupplung, ~ **for rubber tubing**
  Schlauchkupplung *f*
coupling box Kupplungsmuffe *f*, Muffenhülse *f*
coupling chain Kupplungskette *f*
coupling coil Kopplungsspule *f* (Elektr)

coupling component Kupplungskomponente *f*
coupling constant Kopplungskonstante *f*, **nuclear**
  ~ Kernkopplungskonstante *f*
coupling factor Kopplungsfaktor *m*
coupling lever Auslösehebel *m*
coupling nut Überwurfmutter *f*
coupling pin Kupplungsbolzen *m*
coupling rod Kuppelstange *f*
coupling spindle Kupplungsspindel *f*
coupling through-pin Mitnehmerstift *m*
coupling value Kupplungswert *m*
coupon Gutschein *m*, Kontrollabschnitt *m*
course Kurs *m*, Lehrgang *m*, Verlauf *m*, **~ of**
  manufacture Verlauf *m* der Herstellung, **~ of**
  reaction Reaktionsverlauf *m*
covalence Kovalenz *f* (Chem)
covalency Kovalenz *f* (Chem)
covalent kovalent (Chem)
covariance (of random variables) Kovarianz *f*
covariant Kovariante *f* (Math)
covariant *a* kovariant
covelline Covellin *m* (Min), Covellit *m* (Min)
covellite Covellin *m* (Min)
cover Auflage *f* (Belag), Decke *f*, Deckel *m*,
  Einband *m* (Buch), Klappe *f*,
  Ummantelung *f*, **~ impregnated with salts**
  Salzdecke *f*
cover *v* abdecken, abschirmen, bedecken,
  bekleiden, einhüllen, überdecken, überziehen,
  umhüllen, verhüllen, verkleiden, **~ with a**
  layer überschichten, **~ with concrete**
  betonieren, **~ with soot** einrußen
cover bolt Deckelschraube *f*, Deckschraube *f*
  (Techn)
cover cell Deckzelle *f*
cover disc Abdeckscheibe *f*
covered bedeckt, umhüllt, **~ with a layer**
  überschichtet
cover glass Deckglas *n*
cover glass gauge Deckglastaster *m*
covering Auskleidung *f*, Bedeckung *f*,
  Bekleidung *f*, Belag *m*, Decke *f*, Mantel *m*,
  Überzug *m*, Umhüllung *f*, Verkleidung *f*
covering color Deckfarbe *f*
covering device Abdeckvorrichtung *f*
covering flange Deckflansch *m*
covering liquor Deckkläre *f*
covering machine Überspritzmaschine *f*
covering material Bespannstoff *m*
covering medium Deckmittel *n*
covering of cables, protective ~ Kabelschutz *m*
covering panel Verkleidungsblech *n*
cover[ing] plate Abdeckplatte *f*
covering power Deckkraft *f*, Deckfähigkeit *f*,
  Deckvermögen *n*, **of good ~** deckfähig
covering slab Deckplatte *f*
covering varnish Deckfirnis *m*, Decklack *m*
cover joint Deckeldichtung *f*,
  Deckelverschraubung *f*

cover nut Hutmutter *f*
cover plate Abdeckplatte *f*, Lasche *f*
cover sheet Deckbogen *m*
cover strip Abdeckleiste *f*, Befestigungsleiste *f*,
  Blendleiste *f*
cover tile Deckziegel *m*
cover tube Verkleidungsrohr *n*
cover valve Deckelschieber *m*
coveyer chain Förderkette *f* (Techn)
Cowper stove Cowper *m*
cowpox Kuhpocken *pl* (Med)
cowslip flower Primel *f* (Bot)
cow vaccine Bovovaccin *n*
cozymase I Codehydrase I *f* (Biochem),
  Coenzym I *n* (Biochem), Cozymase I *f*,
  Diphosphopyridinnucleotid *n*
cozymase II Cocarboxylase *f* (Biochem),
  Cozymase II *f*, Thiaminpyrophosphat *n*
crab *v* einbrennen (Text)
crabbing Einbrennen *n* (Text)
crabbing jack Brennblock *m*
crab cider Holzapfelwein *m*
crack Riß *m*, Fehlstelle *f* (Schweiß), Härteriß *m*,
  Spalt *m*, Sprung *m*, **~ in pipes** Rohrbruch *m*
crack *v* [ab]spalten, einen Sprung bekommen,
  krachen, kracken (Erdöl), platzen,
  zersplittern, [zer]springen, **~ [open]** *v*
  aufknacken
crack arrest temperature (CAT)
  Rißauffangtemperatur *f*
crack detector Anrißsucher *m*
cracked rissig
cracked gas Krackgas *n*, Spaltgas *n*
cracked gasoline Crackbenzin *n*
cracker Brecher *m*, Brechwalze *f*,
  Knallbonbon *m*
crack extension force Rißausweitungskraft *f*
crack formation Rißbildung *f*
crack growth Rißwachstum *n* (Prüftechnik)
cracking Springen *n*, Kracken *n* (Erdöl),
  Rißbildung *f*, Spaltdestillation *f* (Erdöl),
  Spalten *n* (Erdöl), **catalytic ~** katalytisches
  Kracken, **freedom from ~** Rißfreiheit *f*,
  thermal ~ thermisches Kracken, **vapor-phase**
  ~ Gasphase-Krackverfahren *n*, **~ of**
  petroleum Spaltung *f* von Erdöl
cracking *a* [ab]spaltend
cracking coal Sprengkohle *f*
cracking distillation Krackdestillation *f*
cracking plant Krackanlage *f*, Spaltanlage *f*
cracking process Krackprozeß *m*,
  Krackverfahren *n*
cracking temperature Abbautemperatur,
  Spalttemperatur *f*
cracking tower Fraktionierkolonne *f* (Dest)
crackle *v* knarren, knistern, verpuffen
crackle finish Reißlack *m*
crackle varnish Krakelierlack *m*

crackling Knattern *n*, Knistern *n*, ~ of tin
Zinngeschrei *n* (Chem)
crack-proof bruchfest
crack propagation Rißausbreitung *f*
cracks, free from ~ rißfrei
crack strength Rißfestigkeit *f*
cracky brüchig
craftsman Handwerker *m*
crag Crag *m*
crag formation Cragformation *f* (Geol)
cramp Klammer *f*, Krampe *f*, Krampf *m* (Med)
cramp *v* anklammern, ankrampen, befestigen
(Klammern)
cramp iron Haspe *f*, Krampe *f*
cranberry Preiselbeere *f*
crandallite Crandallit *m* (Min)
crane Kran *m*, Aufzug *m*, Hebelade *f*, ~ for the
reheating furnace Blockziehkran *m*, ~ with
grab gear Greifkran *m*
crane beam Kranbalken *m*
crane cable Krankabel *n*, Kranseil *n*
crane carriage Laufkatze *f*
crane chain Krankette *f*
crane hook Kranhaken *m*
crane ladle Kranpfanne *f*
crane magnet Kranmagnet *m*
crane pneumatic rammer
Kranpreßluftstampfer *m*
crane rope Kranseil *n*
crane spreader Krantraverse *f*
crane travel Kranfahren *n*
crane trolley Laufkatze *f*
cranium Hirnschädel *m*
crank Drehkurbel *f*, Kurbel *f*, Kurbelarm *m*,
inside ~ beiderseitig gelagerte Kurbel, outside
~ einseitig gelagerte Kurbel, overhang ~
einseitig gelagerte Kurbel
crank *v* kurbeln
crank axle Kurbelachse *f*, Kurbelwelle *f*
crank brace Bohrkurbel *f*
crankcase Kurbelwanne *f*
crankcase ventilation Kurbelwannenentlüftung *f*,
positive ~ positive (geschlossene)
Kurbelwannenentlüftung
crank distance Kurbelabstand *m*
crank drive Kurbelantrieb *m*
cranked gekröpft (Techn)
crank gear Kurbelwerk *n*
crank guide Kurbelschleife *f*
crank handle Kurbelgriff *m*
crank mechanism Kurbelmechanismus *m*
crank pin Kurbelgriff *m*, Kurbelzapfen *m*,
Treibzapfen *m*
crank press Kurbelpresse *f*
crank pressure Kurbeldruck *m*
crankshaft Kurbelwelle *f*, body of ~
Kurbelwellenschaft *m*
crankshaft bearing Kurbelwellenlager *n*

crankshaft grinder
Kurbelwellenschleifmaschine *f*
crankshaft lathe Kurbelwellendrehbank *f*
crank web Kurbelblatt *n*
crape Krepp *m*
crash Sturz *m*, Zusammensturz *m*
crash *v* krachen
crash [against] *v* kollidieren (Auto)
crash barriers Leitplanken *pl*
crataegin Crataegin *n*
crate Lattenkiste *f*, Verschlag *m*
crater Krater *m* (Geol)
crawl *v* kriechen
crawler Laderaupe *f*
crawler crane Raupenkran *m*
crawling Kriechen *n*, Runzelbildung *f*,
Zusammenziehen *n*
crayon Kreidestift *m*, Pastellstift *m*
craze Riß *m* (Steingut)
crazed haarrissig
crazing Elefantenhautbildung *f* (Gummi),
Haarrißbildung *f*
cream Creme *f*, Rahm *m*, extraction of ~
Rahmgewinnung *f*, formation of ~
Rahmbildung *f*
cream *v* aufrahmen (Latex), ~ up aufrahmen
(Latex)
cream color Cremefarbe *f*, Isabellfarbe *f*
cream-colored cremefarben, gelbweiß
cream cooler Rahmkühler *m*
creamery Butterei *f*, Molkerei *f*
cream freezer Rahmgefriermaschine *f*
creaming [up] Aufrahmung *f*
creasability Rillbarkeit *f*
crease Falte *f*, Kniff *m*, Knitter *m*, Rillung *f*,
depth of ~ Rilltiefe *f*
crease *v* knittern, rillen
creased faltenreich, knitterig
crease-proof knitterfest, knitterfrei
crease proofing Knitterfestmachen *n*
creaseproofing finish Knitterfestausrüstung *f*
crease resistance Knitterfestigkeit *f*
crease-resistant knitterfest, knitterfrei
creasing Faltenwerfen *n*, Fettung *f*
creasing angle Abkantwinkel *m*
creasing bar Abkantstab *m*
creasing support Abkantklappe *f*
create *v* erschaffen, hervorbringen
creatine Kreatin *n*
creatine phosphate Kreatinphosphat *n*
creatine phosphate kinase
Kreatinphosphokinase *f* (Biochem)
creatine synthesis Kreatinsynthese *f*
creatinine Kreatinin *n*
creation Erschaffung *f*, Erzeugung *f*
creative schöpferisch
creatone Kreaton *n*
creator Urheber *m*
crebanine Crebanin *n*

credentials Beglaubigungsschreiben *n*
credit Guthaben *n*, Kredit *m*
crednerite Crednerit *m* (Min),
   Mangankupfererz *n* (Min)
creel Spulengatter *n* (Spinnerei), Weidenkorb *m*
creel fabric Fadenstoff *m* (Reifen),
   Gattergewebe *n*
creep Kriechen *n*, kalter Fluß (Prüftechnik),
   Kriechdehnung *f*, **elongation at break in** ~
   Zeitbruchdehnung *f*,
   Zeitstandbruchdehnung *f*, **reduction in area in**
   ~ Zeitbrucheinschnürung *f*,
   Zeitstandbrucheinschnürung *f*
creep *v* kriechen
creep behavior Kriechverhalten *n*, **compressive** ~
   Druckfließverhalten *n*
creep curve Dehnverlauf *m* (Prüftechnik),
   Zeitdehnlinie *f*
creep distance Kriechstrecke *f*
creep hoist speed Feinhubgeschwindigkeit *f*
creeping Kriechen *n*, Zusammenziehen *n*
creeping current Kriechstrom *m* (Elektr)
creeping property Kriechverhalten *n*
creeping wave Kriechwelle *f*
creep limit Kriechgrenze *f*
creep rate Kriechgeschwindigkeit *f*
creep resistance Dauerstandfestigkeit *f*,
   Kriechfestigkeit *f*
creep strength Kriechgrenze *f*,
   Dauerstandfestigkeit *f*, Zeitstandfestigkeit *f*
creep stress, limiting ~
   Kriechgrenzenspannung *f*
creep test Kriechprobe *f* (Anal),
   Kriechversuch *m*, Zeitstandversuch *m*
crenelate *v* krenelieren
crenic acid Krensäure *f*, Quellsäure *f*
creolin Kreolin *n*
creosol Kreosol *n*
creosotal Kreosotal *n*, Kreosotcarbonat *n*
creosote Kreosot *n*, **containing** ~ kreosothaltig
creosote carbonate Kreosotal *n*,
   Kreosotcarbonat *n*
creosote oil Kreosotöl *n*
creosote oleate Oleokreosot *n*
creosote water Kreosotwasser *n*
creosotic kreosothaltig
creosoting Kreosottränkung *f*
crepe Krepp *m*, **bark-like** ~ Borkenkrepp *m*
crepe de Chine Kreppseide *f* (Text)
crepe paper Kreppapier *n*
creping calander Kreppkalander *m* (Text)
crepitant knisternd
crepitate *v* knistern (Chem)
crepitating Knistern *n*
crepitation Knistern *n*
cresalol Kresalol *n*
cresatin Kresatin *n*
cresaurin Kresaurin *n*
cresidine Kresidin *n*

cresineol Cresineol *n*
cresol Hydroxytoluol *n*, Kresol *n*,
   Methylphenol *n*
cresol indophenol Kresolindophenol *n*
cresol powder Kresolpuder *m*
cresol purple Kresolpurpur *n*
cresol red Kresolrot *n*
cresol resin Kresolharz *n*
cresol soap Kresolseife *f*, **liquid** ~
   Kresolseifenlösung *f*
cresol-sulfuric acid mixture
   Kresolschwefelsäure *f*
cresorcinol Kresorcin *n*
cresosteril Kresosteril *n*
cresotate *a* kresotinsauer
cresotic acid Kresotinsäure *f*
cresotine yellow Kresotingelb *n*
cresotinic acid Kresotinsäure *f*
cress Kresse *f* (Bot)
cress oil Kressenöl *n*
crest Höhepunkt *m*, Kamm *m*, Spitze *f*
crest angle Scheitelwinkel *m*
crest clearance Spitzenspiel *n*
crest length Kammlänge *f*
crest power Höchstleistung *f*
crest template Spitzenlehre *f*
crest voltage Spitzenspannung *f* (Elektr)
cresyl Cresyl-, Kresyl-
cresyl acetate Kresylacetat *n*
cresyl carbonate Kresylcarbonat *n*
cresylic acid Kresolsäure *f*, Kresol *n*,
   Kresylsäure *f*
cresylic alcohol Kresylalkohol *m*
cresylic resin Kresolharz *n*
cresylite Cresylit *n* (Sprengstoff)
cresyl methyl ether Kresylmethyläther *m*,
   Kresylmethylester *m*
cresyl phosphate Kresylphosphat *n*
cresylsulfuric acid Kresylschwefelsäure *f*
cresyl toluenesulfonate Kresyltoluolsulfonat *n*
cresyl violet Kresylviolett *n*
cretaceous kreideartig, kreidehaltig
cretaceous structure Kreidegebilde *n*
cretaceous system Kreideformation *f* (Geol)
crevasse Gletscherspalte *f*, Riß *m*, Spalte *f*
crevassed rissig
crevice Spalte *f*
crevice corrosion Spaltkorrosion *f*, ~ **at a contact**
   with non-metallic material
   Berührungskorrosion *f*
creviced rissig
crew Belegschaft *f*, Mannschaft *f*
crichtonite Crichtonit *m* (Min)
crimp Kräuselung *f*, Sicke *f* (Dose)
crimp *v* kräuseln
crimping Faltenwerfen *n*, Kräuseln *n*
crimping machine Kniffmaschine *f*
crimp seal Kräuselverschluß *m*
crimson Karmesin *n*, Karmin *n*

crimson *a* karmesinrot
crimson *v* karmesinrot färben
crimson-colored karmesinfarbig
crimson lake Karmesinlack *m*
crinamine Crinamin *n*
crinane Crinan *n*
crinidine Crinidin *n*
crinine Crinin *n*
crinkle *v* runzeln, kräuseln
crinkle leather Knautschlack[leder n] *m*
cripple *v* lähmen
crisp knusperig, knusprig
crispatic acid Crispatsäure *f*
cristobalite Cristobalit *m* (Min)
cristophite Cristophit *m* (Min)
criterion ([pl. criteria, criterions]), Kriterium *n*,
  Kennzeichen *n*, Anhaltspunkt *m*
crithmene Crithmen *n*
crithminic acid Crithminsäure *f*
critical kritisch, bedenklich, **below the ~**
  unterkritisch
critical angle Grenzwinkel *m*
critical density kritische Dichte
critical heat Umwandlungswärme *f*
critical limit Abfallgrenze *f*
critical path analysis Netzplantechnik *f*
critical point kritischer Punkt, Gefahrpunkt *m*,
  Haltepunkt *m*, Umwandlungspunkt *m*
critical pressure kritischer Druck
critical resistance kritischer Widerstand
criticism Kritik *f*, Rezension *f*, Tadel *m*
criticize *v* kritisieren, bekritteln, rezensieren
criwelline Criwellin *n*
crocalite Krokalith *m*
croceic acid Croceinsäure *f*
crocein Crocein *n*
crocein orange Croceinorange *n*
crocein scarlet Croceinscharlach *m*
crocetane Crocetan *n*
crocetin Crocetin *n*
crocetin dialdehyde Crocetindialdehyd *m*
crochet *v* häkeln
crocic acid Krokonsäure *f*
crocidolite Krokydolith *m* (Min), faseriger
  Eisenblauspat (Min)
crocin Crocin *n*
crockery Steingut *n*, Tonware *f*
crocodile clip Schnabelklemme *f*
crocoisite Krokoit *m* (Min), Rotbleierz *n* (Min)
crocoite Krokoit *m* (Min), Chromblei *n* (Min),
  Rotbleierz *n* (Min)
crocolite Krokoit *m* (Min)
croconic acid Krokonsäure *f*
crocose Crocose *f*
crocus of iron Eisensafran *m*
cromfordite Hornblei *n* (Min)
crook *v* krümmen
crooked krumm, ungerade, verbogen
crookesite Crookesit *m* (Min)

crop *v* ernten, scheren
crop of tin Zinnerzsand *m*
crop protection Pflanzenschutz *m* (Agr)
crop protection product Pflanzenschutzmittel *n*
cross Kreuz *n;* Kreuzstück *n*
cross *v* [durch]kreuzen, durchschneiden, sich
  schneiden
cross-axing (of calender rolls) Schrägverstellung
crossbar Querriegel *m*, Querstab *m*,
  Querträger *m*, Traverse *f*
crossbeam Querträger *m*
cross beater mill Schlagkreuzmühle *f*
crossbelt Querförderband *n*
cross-belt separator Kreuzbandscheider *m*
cross-blade mixer Kreuzbalkenrührer *m*
cross blending Verschneiden *n* (Gummichemie)
cross bonds, formation of ~ Vernetzung *f*
cross boring Querbohrung *f*
cross brace Querverstrebung *f*, Strebe *f*
cross-breaking test Querbiegeversuch *m*
crossbreed gekreuzte Rasse *f* (Zool)
cross-breeding Artkreuzung *f* (Biol)
cross chisel Breiteisen *n*, Kreuzmeißel *m*
cross conjugation Kreuzkonjugation *f*
cross connection Kreuzverbindung *f*
cross-country geländegängig (Fahrzeug)
cross coupling wilde Kopplung (Elektr)
cross crack Querriß *m*, Kantenriß *m*
cross current Querströmung *f*, Gegenstrom *m*
cross-current classifier Querstromsichter *m*
crosscut *v* abschroten
crosscut saw Fuchsschwanz *m*
crosscutter Querschneider *m*
cross drift Querdrift *f*
cross-dyeing Überfärben *n*, **fast to ~**
  überfarbecht (Färb)
crossed gekreuzt, verschränkt
crossed coil Kreuzspule *f* (Elektr)
crossed position gekreuzte Anordnung *f* (Opt)
cross [extruder] head Querspritzkopf *m*
cross field Querfeld *n* (Elektr)
cross flow Kreuzstrom *m* (Hydrodynamik),
  Querstrom *m* (Hydrodynamik)
cross flow heat exchanger
  Kreuzstromwärmeaustauscher *m*
cross flow tray Kreuzstromboden *m*
cross flue Querzug *m*
cross flux Streufluß *m* (Elektr), **magnetic ~**
  magnetische Streuung
cross girder Querträger *m*
cross-grain quer zur Laufrichtung *f*, quer zur
  Faserrichtung
cross-hairs Fadenkreuz *n*
crosshatch *v* [kreuz]schraffieren
crosshatched schraffiert
crosshead Kopfplatte *f*, Kreuzkopf *m*,
  Querhaupt *n*
crosshead die Querkopfziehform *f*
cross induction Querinduktion *f*

crossing Kreuzung f, Übergang m
crossing a schneidend
crossing over Genaustausch m (Gen)
crossing sleeve Kreuzmuffe f
crossing valve Kreuzventil n
cross iron Kreuzeisen n
crossite Crossit m (Min)
cross joint Kreuzverbindung f
cross laminated kreuzweise geschichtet
crossline Querline f, Querstrich m
crosslines Fadenkreuz n
crosslink Querverbindung f
cross-link v vernetzen
cross linkage Querverbindung f, Vernetzung f
cross linkage position Vernetzungsstelle f
cross-linked vernetzt
cross-linker Vernetzer m
cross-linking Brückenbindung f (Chem)
cross-linking agent Vernetzer m
cross-linking site Vernetzungsstelle f
cross-link universal coupling
    Kreuzgelenkkupplung f
cross magnetization Quermagnetisierung f
cross-magnetize v quermagnetisieren
cross-magnetizing effect
    Quermagnetisierungseffekt m
cross member Querstück n
crossover experiment Kreuzungsversuch m
cross-over point Kreuzungspunkt m
crosspiece Kreuzstück n, Quersteg m
crosspoint Kreuzungspunkt m
cross polarization Kreuzpolarisation f (Elektr)
cross ratio Doppelverhältnis n
cross-reference Querverweis m
cross relation Wechselbeziehung f
cross resistance Querwiderstand m (Elektr)
cross section Querschnitt m, Querprofil n,
    differential ~ differentieller
    Wirkungsquerschnitt m, effective ~
    Wirkungsquerschnitt m, reduction of ~
    Querschnittsverringerung f, total ~
    Gesamtquerschnitt m, Vollprofil n, variation
    of ~ Querschnittsveränderung f, ~ of a fiber
    Faserquerschnitt m, ~ of arms
    Schenkelquerschnitt m, ~ of passage
    Durchflußquerschnitt m
cross sectional area Querschnittsfläche f,
    Durchgangsquerschnitt m, change of ~
    Querschnittsveränderung f, reduction of ~
    Querschnittsverminderung f
cross slide Kreuzschlitten m
cross slip Quergleiten n, Quergleitung f
cross slip line Quergleitlinie f
cross stone Kreuzstein m, Harmotom m (Min)
cross supports Querverankerung f
cross-table Kreuztisch m
cross tie Querschwelle f
cross twill Kreuzköper m (Text)
cross weave Schußfaden m (Web)

cross-wires Fadenkreuz n
cross[wise] quer
crosswise direction Querrichtung f
cross wound spool Kreuzspule f
crotalotoxin Crotalotoxin n
crotamiton Crotamiton n
crotonaldehyde Crotonaldehyd m
crotonase Crotonase f (Biochem)
crotonate Crotonat n
crotonate a crotonsauer
croton betaine Crotonbetain n
crotonic acid Crotonsäure f, salt or ester of ~
    Crotonat n
crotonic ester Crotonsäureester m
crotonin Crotonin n
croton lactone Crotonlacton n
croton oil Crotonöl n
crotonolic acid Crotonolsäure f
croton resin Crotonharz n
croton seeds Crotonkörner n pl
crotonyl Crotonyl-
crotonylene Crotonylen n
crotyl alcohol Crotylalkohol m
crotyl chloride Crotylchlorid n
crowbar Brecheisen n, Brechstange f,
    Hebeeisen n, Hebestange f, Schürstange f,
    Stemmeisen n
crowding, effect of ~ Abdrängeffekt m
    (Stereochem)
croweacic acid Croweacinsäure f
croweacin Croweacin n
crow foot Kreuzköper m (Text)
crown angle Fadenwinkel m (Reifen)
crown block rating (drilling) Kronenregellast f
crown burner Kronenbrenner m
crown drum Hochschultertrommel f
crown gear Zahnscheibe f
crown glass Kronglas n (Opt)
crown glass prism Kronglasprisma n
crown of calender roll Bombierung f der
    Kalanderwalze
crown of the tire Zenitpartie f des Reifens
crown pin Kronenstift m
crown plate Gewölbeplatte f
crown region Lauffflächenzone f
crown wheel Zahnscheibe f
croze Kröse f
crucible Glühtiegel m, Gußtiegel m,
    Schmelztiegel m, ~ for distillation
    Destillationstiegel m
crucible belly Tiegelbauch m
crucible cast steel Tiegel[guß]stahl m,
    Tiegelflußstahl m
crucible charge Tiegelinhalt m
crucible coking test Tiegelverkokung f
crucible contents Tiegelinhalt m
crucible drying apparatus Tiegeltrockner m
crucible furnace Tiegelofen m, Blockofen m

crucible hearth of the blast furnace Kasten *m* des Hochofens
crucible lid Tiegeldeckel *m*
crucible lining Tiegelfutter *n*
crucible material Tiegelmasse *f,* old ~ Tiegelscherben *pl*
crucible melting furnace Tiegelschmelzofen *m*
crucible melting plant Tiegelschmelzerei *f*
crucible melting process Tiegelschmelzverfahren *n*
crucible mold Tiegelhohlform *f*
crucible oven Tiegelbrennofen *m*
crucible press Tiegelpresse *f*
crucible ring Tiegelring *m*
crucibles, manufacture of ~ Tiegelherstellung *f*
crucible steel process Tiegelstahlherstellung *f*
crucible steel works Tiegelstahlhütte *f*
crucible test Tiegelprobe *f*
crucible tongs Schmelz[tiegel]zange *f,* Tiegelzange *f*
crucible triangle Tiegeldreieck *n,* Glühring *m*
crucible works Tiegelfabrik *f*
cruciform twin group Durchkreuzungszwilling *m* (Krist)
crucilite Crucilith *m*
crude roh (Öl, Metalle, Leder), unbearbeitet
crude asphalt Asphaltgestein *n*
crude benzene Rohbenzol *n*
crude bottom Bodensatz (nach der Destillation von Rohöl) *m*
crude carbide mixture Rohkarbidmischung *f*
crude cresol Rohkresol *n*
crude fiber Rohfaser *f*
crude gas Rohgas *n*
crude homogenate Rohhomogenat *n* (Biochem)
crude litharge Abstrichblei *n* (Metall)
crude matte Rohstein *m*
crude oil Rohöl *n,* Petroleum *n,* mixed base ~ gemischtbasisches Rohöl, naphthenic base ~ naphthenbasisches Rohöl, paraffin base ~ paraffinbasisches Rohöl
crude oil emulsion Rohölemulsion *f*
crude ore Bergerz *n,* Roherz *n*
crude potash Pottaschefluß *m*
crude state Rohzustand *m*
crude tartar Argal *m*
crumb Krümel *m*
crumble *v* zerbröckeln, grießeln, krümeln, zerdrücken, zerfallen, ~ off abbröckeln
crumbling of the ore Zerfallen *n* des Erzes
crumbly abbrüchig, bröcklig, krümelig
crumple *v* knittern
crumpled knitterig
crunch *v* knirschen
crunching Geknister *n*
crurin Crurin *n*
crush *v* [grob]zerkleinern, pochen (Erz), quetschen, zerdrücken, zermalmen, zerquetschen, zerreiben, zerschlagen, zerstoßen
crushed, coarsely ~ grobgepulvert
crushed coke Knabbelkoks *m*
crushed rock Schotter *m*
crushed sand Brechsand *m*
crusher Brecher *m,* Brechwerk *n,* Quetschmühle *f,* Zerkleinerungsmaschine *f,* gyratory ~ Walzenbrecher *m,* jaw ~ Backenbrecher *m*
crusher index Stauchwert *m*
crusher wheels Brecherräder *n pl*
crushing Zerkleinerung *f,* Zerquetschen *n,* coarse to medium ~ Vorzerkleinern *n,* primary ~ Vorzerkleinern *n,* ~ of hard materials Hartzerkleinerung *f*
crushing effect Stauchwirkung *f*
crushing load Druckspannung *f*
crushing machine Quetschmaschine *f,* Zerkleinerungsmaschine *f*
crushing mill Brechwalzwerk *n,* Quetschmühle *f,* Zerkleinerungsmühle *f*
crushing plant Zerkleinerungsanlage *f,* Brechanlage *f,* Mahlanlage *f*
crushing plate Zerkleinerungsplatte *f*
crushing stress Druckkraft *f*
crushing test Stauchversuch *m*
crushing tool Quetsche *f*
crushing unit Zerkleinerungsanlage *f*
crush-proof knitterfrei, überrollbar
crusocreatinine Crusocreatinin *n*
crust Kruste *f,* Haut *f,* Rinde *f* (Erde etc.), Überzug *m,* to cover with a ~ bekrusten
crust *v* verkrusten
crustacean Krustentier *n*
crustaceans Krebstiere *pl*
crustaceous überkrustet, krustig
crustal rocks Krustengestein *n*
crusted verkrustet, krustig
crusty krustig
cryobiology Kryobiologie *f*
cryochemistry Kältechemie *f,* Kryochemie *f*
cryoconite Kryokonit *m*
cryofluorane Cryofluoran *n*
cryogen Kühlmittel *n*
cryogenic expander Entspannungsmaschine *f* für Tieftemperaturtechnik
cryogenics Tieftemperaturforschung *f,* Tieftemperaturtechnik *f*
cryogenin Kryogenin *n*
cryogrinding Tieftemperaturmahlen *n*
cryohydrate Kryohydrat *n*
cryohydric kryohydratisch
cryolite Kryolith *m* (Min), Aluminiumnatriumfluorid *n* (Min)
cryolite soda Kryolithsoda *f* (Min)
cryomagnetic kryomagnetisch
cryometer Gefrierpunktsmesser *m*
cryophorus Kryophor *m*

cryophyllite Kryophyllit *m* (Min)
cryopump Kryopumpe *f*
cryopumping Kryopumpen *n*
cryoscope Gefrierpunktsmesser *m*, Kryoskop *n*
cryoscopic kryoskopisch
cryoscopy Kryoskopie *f*
cryosel Kryohydrat *n*
cryostat Kryostat *m*, Kälteregler *m*,
    Thermostat *m* für tiefe Temperaturen
cryosurgery Kryochirurgie *f*
cryotrapping Kryopumpen *n*
cryotrimming Tieftemperatur-Entgraten *n*
cryotumbling Tieftemperatur-Schleudern *n*
cryptaustoline Cryptaustolin *n*
cryptic verborgen
cryptobrucinolone Kryptobrucinolon *n*
cryptocrystalline kryptokristallin
cryptogenin Kryptogenin *n*
cryptograndoside Kryptograndosid *n*
cryptolepine Kryptolepin *n*
cryptolite Kryptolith *m*
cryptomerene Cryptomeren *n*
cryptomeriol Cryptomeriol *n*, Kryptomeriol *n*
cryptomorphite Kryptomorphit *m*
cryptopidene Kryptopiden *n*
cryptopidic acid Kryptopidinsäure *f*
cryptopidiol Kryptopidiol *n*
cryptopimaric acid Kryptopimarsäure *f*
cryptopine Kryptopin *n*
cryptopinone Cryptopinon *n*
cryptopleurine Kryptopleurin *n*
cryptopyrrole Kryptopyrrol *n*
cryptoscope Kryptoskop *n*
cryptostrobin Kryptostrobin *n*
cryptowolline Cryptowollin *n*
cryptoxanthin Kryptoxanthin *n*
crysanthemaxanthin Chrysanthemaxanthin *n*
crysoidine orange Chrysoidinorange *n*
crysolgan Krysolgan *n*
crystal (also see crystals) Kristall *m*, edge of a ~
    Kristallkante *f*, hemihedral ~ hemiedrischer
    Kristall, idiochromatic ~ idiochromatischer
    Kristall, imperfectly formed ~ Kristallit *m*,
    liquid ~ flüssiger Kristall, mixed ~
    Mischkristall *m* (Krist), needle-shaped ~
    Nadelkristall *m* (Krist), perfect ~ Idealkristall
    m, piezoelectric ~ piezoelektrischer Kristall,
    powdered ~ Kristallmehl *n*, Kristallpulver *n*,
    uniaxial ~ einachsiger Kristall
crystal analysis Kristallanalyse *f*
crystal angle Kristallwinkel *m*
crystal axis Kristallachse *f*
crystal boundary Korngrenze *f* (Krist, Metall)
crystal center Kristallkern *m*
crystal chemistry Kristallchemie *f*
crystal collimator Kristallkollimator *m*
crystal conduction counter Kristallzähler *m*
crystal core Kristallkern *m*

crystal cut Kristallschnitt *m*
crystal defect Kristallbaufehler *m*
crystal detector Kristalldetektor *m* (Atom)
crystal druse Kristalldruse *f*
crystal filter Kristallfilter *n*
crystal finish Eisblumenlack *m*
crystal flexural vibration
    Kristallbiegungsschwingung *f*
crystal form Kristallform *f*
crystal germ Kristallkeim *m*
crystal glass Kristallglas *n*
crystal glaze, frosted ~ Kristallglasur *f*
crystal goniometry Kristallwinkelmessung *f*
crystal grain Kristallkorn *n*
crystal grating Kristallgitter *n* (Spektr)
crystal growth Kristallwachstum *n*
crystal growth inhibition
    Kristallisationsverzögerung *f*
crystal hyperfine structure
    Kristallhyperfeinstruktur *f*
crystal lacquer Eisblumenlack *m*
crystal lattice Kristallgitter *n* (Krist),
    Raumgitter *n* (Krist), disordered ~ gestörtes
    Kristallgitter
crystal lattice model Kristallgittermodell *n*
crystal lattice scale Kristallgitterskala *f*
crystal lattice spacing Kristallgitterabstand *m*
crystal lens Kristallinse *f*
crystal-like kristallartig
crystallin Kristallin *n*
crystalline kristallin[isch], kristallähnlich,
    kristallartig, kristallen
crystalline acid Eisessig *m*
crystalline crust Kristallhaut *f*
crystalline deposit Kristallansatz *m*
crystalline feldspar Eisspat *m* (Min)
crystalline field interaction
    Kristallwechselwirkung *f*
crystalline field splitting
    Kristallfeldaufspaltung *f*
crystalline film Kristallhäutchen *n*
crystalline glaze Kristallglasur *f*
crystalline grain Kristallkorn *n*, formation of a ~
    Kristallkornbildung *f*
crystalline growth Kristallwachstum *n*,
    Kornwachstum *n*, Kristall[aus]bildung *f*, rate
    of ~ Kristallisationsgeschwindigkeit *f*
crystalline humidity Kristallfeuchtigkeit *f*
crystalline layer Kristallschicht *f*
crystalline lens Kristallinse *f*
crystalline refinement Kornverfeinerung *f*
crystalline sodium sulfate Duisburger
    Natriumsulfat *n*
crystalline state Kristallform *f*, kristalliner
    Zustand *m*
crystalline varnish Kristallisierlack *m*
crystallite Kristallit *m*
crystallizability Kristallisationsfähigkeit *f*,
    Kristallisationsvermögen *n*

crystallizable kristallisationsfähig,
kristallisierbar
crystallization Kristallisierung *f,*
Auskristallisierung *f,* Kristallanschuß *m,*
Kristallisation *f,* Kristall[kern]bildung *f,* **center
of** ~ Kristallisationskern *m,*
Kristallisationszentrum *n,* **fractional** ~
fraktionierte Kristallisation, **heat of** ~
Kristallisationswärme *f,* **nucleus of** ~
Kristallisationskern *m,* **primary** ~ primäre
Kristallisation, **secondary** ~ sekundäre
Kristallisation, **water of** ~
Kristallisationswasser *n,* ~ **with spiral**
Kristallisierschnecke *f*
crystallization column Kristallisationskolonne *f*
crystallization interval Erstarrungsintervall *m*
crystallization plant Kristallisieranlage *f*
crystallization point Kristallisationsbeginn *m*
crystallization vessel Kristallisationsgefäß *n*
crystallize *v* kristallisieren, ~ **[out]**
auskristallisieren
crystallized kristallisiert
crystallizer Kristallisationsapparat *m,*
Kristallisationsgefäß *n*
crystallizing Kristallisieren *n*
crystallizing *a* kristallisierend
crystallizing dish Kristallisierschale *f*
crystallizing fuming sulfuric acid Kristallsäure *f*
crystallizing pan Kristallisationspfanne *f,*
Kristallisierschale *f,* Soggenpfanne *f*
crystallizing plant Kristallisationsanlage *f*
crystallizing process Kristallisationsvorgang *m*
crystallizing rolls Kristallisierwalze *f*
crystallizing vessel Anschießgefäß *n*
crystallogeny Kristallbildungslehre *f*
crystallogram Kristallbild *n,* Kristallogramm *n*
crystallographer Kristallforscher *m,*
Kristallograph *m*
crystallographic kristallographisch
crystallographic axial ratio kristallographisches
Achsenverhältnis *n*
crystallographic axis Kristallachse *f*
crystallographic axis orientation
Kristallachsenrichtung *f*
crystallographic class Kristallklasse *f*
crystallographic plane Kristallebene *f*
crystallographic system Kristallsystem *n*
crystallography Kristallographie *f,* Kristallehre *f,*
Kristallkunde *f*
crystalloid Kristalloid *n*
crystalloid *a* kristallähnlich, kristallartig,
kristalloid
crystalloidal *a* kristalloid
crystalloluminescence Kristallolumineszenz *f*
crystallometric kristallometrisch
crystallometry Kristallmessung *f,*
Kristallometrie *f*
crystallose Kristallose *f*

crystal monochromator
Kristallmonochromator *m*
crystal morphology Kristallmorphologie *f*
crystal nucleus Kristallkeim *m*
crystal optics Kristalloptik *f*
crystal orientation Kristallrichtung *f,*
Kristallorientierung *f*
crystal oscillation Kristallschwingung *f*
crystal oscillator Schwingkristall *m*
crystal pattern Kristallbild *n*
crystal pebble Kristallkiesel *m*
crystal periodicity Kristallperiodizität *f*
crystal photoelectric effect Kristallphotoeffekt *m*
crystal physics Kristallphysik *f*
crystal plane Kristallebene *f*
crystal pulling Kristallziehen *n,* ~ **from melt**
Kristallziehen *n* aus der Schmelze
crystal pulling machine Kristallziehofen *m*
crystal pulling method Kristallziehverfahren *n*
crystal rectifier Kristallgleichrichter *m,*
Quarzgleichrichter *m*
crystals, cluster of ~ Kristalldruse *f,* **formation
of** ~ Kristall[aus]bildung *f,*
Auskristallisation *f,* **to form** ~
auskristallisieren
crystal scintillator Kristallszintillator *m*
crystal slab Kristallblock *m*
crystal spectrometer Kristallspektrometer *n*
crystal spectrum Kristallspektrum *n*
crystal structure Kristallbau *m,* Kristallstruktur *f*
crystal [structure] analysis
Kristallstrukturuntersuchung *f*
crystal structure determination
Kristallstrukturanalyse *f*
crystal surface Kristalloberfläche *f*
crystal symmetry Kristallsymmetrie *f*
crystal system Kristallsystem *n*
crystal triode Kristalltriode *f*
crystal vibration Kristallschwingung *f*
crystal violet Kristallviolett *n*
C-terminal [amino acid] C-terminale Aminosäure
CTP (Abk) Cytidintriphosphat *n*
cubane Cuban *n*
cubanite Cubanit *m* (Min), Weißkupfererz *n*
(Min)
cubature Kubikberechnung *f*
cuba wood Kubaholz *n*
cube Würfel *m,* dritte Potenz *f,* Kubus *m*
cube *v* kubieren, zur dritten Potenz erheben
cubeb, extract of ~ Cubebenextrakt *m*
cubebene Cubeben *n*
cubebic acid Cubebinsäure *f*
cubebin Cubebin *n*
cubebin ether Cubebinäther *m*
cubebinol Cubebinol *n*
cubebinolide Cubebinolid *n*
cubeb oil Cubebenöl *n*
cubebs Cubebenpfeffer *m*
cube edge Würfelkante *f*

cube face Würfelfläche *f*
cube ore Würfelerz *n* (Min)
cube shape Würfelform *f*
cube spar Würfelgips *m* (Min), Würfelspat *m*
  (Min)
cube sugar Würfelzucker *m*
cubic kubisch, würfelförmig
cubical contents Rauminhalt *m*
cubic alum Würfelalaun *m*
cubic body-centered kubisch raumzentriert
  (Krist)
cubic capacity Kubikinhalt *m*
cubic centimeter Kubikzentimeter *n*
cubic closest packed kubisch dichtest gepackt
  (Krist)
cubic dilatation Volumdilatation *f*
cubic face-centered kubisch flächenzentriert
  (Krist)
cubic head Würfelkopf *m*
cubicite Cubicit *m* (Min), Cuboit *m* (Min),
  Würfelzeolith *m* (Min)
cubic measure Raummaß *n,* Körpermaß *n*
cubic meter Kubikmeter *n,* Raummeter *n*
cubic millimeter Kubikmillimeter *n*
cubic root Kubikwurzel *f* (Math)
cubit Elle *f* (Maß)
cuboid würfelförmig
cuboite Cuboit *m* (Min)
cubo octahedron Kubooktaeder *n*
cubosilicite Cubosilicit *m* (Min)
cucoline Cucolin *n*
cucurbitaceac oil Cucurbitaceenöl *n*
cucurbitacin Cucurbitacin *n*
cucurbitol Cucurbitol *n*
cue Übertragungsbefehl *m* (Comp)
cuite-silk Cuiteseide *f* (Text)
cularimine Cularimin *n*
cularine Cularin *n*
culilaban oil Culilawanöl *n*
cullet Bruchglas *n,* Glasabfall *m,*
  Glasbrocken *m,* Glasscherbe *f*
culm Kohlenstaub *m,* Lösche *f,* Staubkohle *f*
culminate *v* gipfeln, kulminieren
culminating edge Scheitelkante *f*
culmination Kulmination *f*
culmination point Gipfelpunkt *m*
culsageeite Culsageeit *m* (Min)
cultivate *v* kultivieren, bebauen, züchten
cultivation Kultivierung *f,* Pflege *f,*
  Urbarmachung *f,* Zucht *f,* ~ of land
  Ackerbau *m,* ~ of vine Weinbau *m*
cultivation experiment Kulturversuch *m,*
  Züchtungsversuch *m*
cultivation of plants Pflanzenzüchtung *f*
culture Kultur *f* (Bakt), pure ~
  Bazillenreinkultur *f* (Bakt), Reinkultur *f*
  (Bakt)
culture bottle Kulturflasche *f*
culture disk Kulturschale *f*

culture extract Kulturextrakt *m*
culture filtrate Kulturfiltrat *n*
culture flask Kulturflasche *f,* Kulturkolben *m,*
  Pasteurkolben *m*
culture medium Kulturboden *m* (Bakt),
  Nährboden *m* (Bakt)
culture tube Kulturröhrchen *n*
culture tubes, filling attachement for ~
  Füllschutzglocke *f*
culture yeast Kulturhefe *f*
cumaldehyde Cuminaldehyd *m*
cumalin Cumalin *n,* Kumalin *n*
cumalinic acid Kumalinsäure *f*
cumaric acid Kumarsäure *f*
cumarin Cumarin *n,* Kumarin *n*
cumbersome umständlich, beschwerlich
cumene Cumol *n,* Kumol *n*
cumengeite Cumengeit *m* (Min)
cumenol Cumenol *n*
cumenyl Cumyl-
cumic acid Cuminsäure *f,* Kuminsäure *f*
cumic alcohol Cuminalkohol *m*
cumic aldehyde Cuminaldehyd *m*
cumidic acid Cumidinsäure *f*
cumidine Cumidin *n,* Kumidin *n*
cumin Cumin *n,* Kümmel *m* (Bot)
cuminalcohol Cuminol *n*
cuminaldazine Cuminaldazin *n*
cuminaldehyde Cuminaldehyd *m*
cuminic acid Cuminsäure *f,* Kuminsäure *f*
cuminil Cuminil *n*
cumin oil Kümmelöl *n*
cuminoin Cuminoin *n*
cuminol Cuminol *n,* Cumenol *n*
cuminone Cuminon *n*
cuminyl Cuminyl *n*
cummin Kronkümmel *m*
cummingtonite Cummingtonit *m* (Min)
cumolene Cumolen *n*
cumoquinol Cumochinol *n*
cumoquinone Cumochinon *n*
cumulating kumulierend
cumulation Anhäufung *f,* Kumulation *f*
cumulative kumulativ, anhäufend, kumulierend,
  zunehmend
cumulative frequency curve
  Verteilungssummenkurve *f*
cumulenes Kumulene *pl*
cumuliform kumulusartig
cumyl Cumyl-
cumylic acid Cumylsäure *f,* Durylsäure *f*
cuneate keilähnlich, keilförmig
cuneiform keilähnlich, keilförmig
cuorin Cuorin *n*
cup Becher *m,* Kelch *m* (Bot), Manschette *f*
  (Dichtung), Schale *f* (Gefäß), little ~
  Schälchen *n,* one-use ~ or disposable hot drink
  cup Einwegtasse *f*
cup and cone bearing konisches Kugellager *n*

cup anemometer Schalenanemometer *n*,
Schalenkreuzanemometer *n*,
Schalenkreuzwindmesser *m*
cuparene Cuparen *n*
cuparenic acid Cuparensäure *f*
cupboard Schrank *m*, Wandschrank *m*
cupel Glühschale *f* (Metall), Kapelle *f* (Metall),
Scheidekapelle *f* (Metall)
cupel *v* kupellieren, abtreiben, läutern
cupel furnace Kapelle *f* (Metall), **small** ~
Abtreibescherbe *f*
cupellation Kupellieren *n* (Metall),
Abtreibung *f* (Metall), Treibarbeit *f* (Metall),
Treibprozeß *m* (Metall), Treibverfahren *n*
(Metall), **refining by** ~ Abtreibearbeit *f*
(Metall)
cupelling furnace Kapellenofen *m*, Treibofen *m*
cupel mold Kapellenform *f*
cupel pyrometer Legierungspyrometer *n*
cupel stand Kapellenstativ *n*
cupel test Kapellenprobe *f*
cupferron Cupferron *n*
cup flow figure Becherfließzahl *f*
cup grease Staufferfett *n*
cup leather Ledermanschette *f*
cup leather packing Manschettendichtung *f*
cupola Kuppel *f*, Haube *f*, Kugelgewölbe *n*,
Kupolofen *m* (Metall), **basic-lined** ~
basischer Kupolofen, **compound** ~
Kupolofen *m* mit erweitertem Herd,
hand-charged ~ Kupolofen *m* mit
Handbeschickung, **reservoir** ~ Kupolofen *m*
mit erweitertem Herd, **small** ~
Kleinkupolofen *m*
cupola blast furnace Kupolhochofen *m*
cupola brick Kupolstein *m*
cupola casting Kupolguß *m*
cupola furnace Kupolofen *m*, Kuppelofen *m*,
Schachtofen *m*, **melting in a** ~
Kupolofenschmelzen *n*, **small** ~
Kleinkupolofen *m*
cupola kiln Kupolofen *m*
cupola lining Kupolofenauskleidung *f*
cupola mixture Kupolofengattierung *f*
cupola supporting frame Abfangplatte *f* (Metall)
cupping test Tiefungsversuch *m*, Tiefziehprobe *f*
cupral Cupral *n*
cuprammine base Cupramminbase *f*
cuprammonia Cupramminbase *f*,
Kupferammoniak *n*
cuprammonia solution Kupferammoniaklösung *f*
cuprammonium Kupferammonium *n*
cuprammonium chloride
Kupferchloridammoniak *n*
cuprammonium compound
Kupferammoniumverbindung *f*
cuprammonium hydroxide Cupramminbase *f*
cuprammonium hydroxide solution
Kupferammoniumlösung *f*, Blauwasser *n*

cuprammonium rayon
Kupfer[ammonium]kunstseide *f* (Text),
Glanzstoff *m* (Text), Kupferseide *f* (Text)
cuprammonium staple Kupferspinnfaser *f*
cuprammonium sulfate
Kupferammoniumsulfat *n*
cuprated silk Glanzstoff *m* (Text)
cupreane Cuprean *n*
cupreidane Cupreidan *n*
cupreidine Cupreidin *n*
cupreine Cuprein *n*
cuprene Cupren *n*
cupreous kupfern, kupferartig, kupferhaltig,
kupferig
cupreous calcium polysulfide
Kupferschwefelkalk *m*
cupreous manganese ore Kupfermanganerz *n*
(Min)
cupretenine Cupretenin *n*
cupric acetate Kupferacetat *n*, Grünspan *m*
cupric acetate arsenite Kupferacetatarsenit *n*,
Kupferarsenacetat *n*
cupric acetoarsenite (see also cupric acetate
arsenite) Baseler Grün *n*
cupric acetylacetonate Kupferacetylacetonat *n*
cupric ammonium chloride
Kupferammoniumchlorid *n*
cupric ammonium sulfate
Kupferammoniumsulfat *n*
cupric arsenite Kupferarsenit *n*
cupric bromate Kupferbromat *n*
cupric bromide Kupfer(II)-bromid *n*
cupric carbonate Cupricarbonat *n*, Grünspan *m*,
basic ~ Kupferlasur *f*
cupric chelate Kupferchelat *n*
cupric chlorate Kupferchlorat *n*
cupric chloride Cuprichlorid *n*,
Kupferbichlorid *n*, Kupfer(II)-chlorid *n*, **basic**
~ Kupferoxychlorid *n*
cupric chromate Kupferchromat *n*,
Cuprichromat *n*
cupric citrate Kupfercitrat *n*
cupric compound Kupfer(II)-Verbindung *f*,
Cupriverbindung *f*
cupric cyanoferrate(II) Ferrokupfercyanid *n*
cupric ferrocyanide Ferrokupfercyanid *n*
cupric hydroxide Kupfer(II)-hydroxid *n*
cupric iodide Kupfer(II)-jodid *n*
cupric nitrate Kupfer(II)-nitrat *n*
cupric oxalate Cuprioxalat *n*,
Kupfer(II)-oxalat *n*
cupric oxide Kupfer(II)-oxid *n*, Cupriöxid *n*
cupric oxide plate Kupferoxidplatte *f*
cupric oxychloride Kupferoxychlorid *n*
cupric salt Kupfer(II)-Salz *n*
cupric sulfate Kupfer(II)-sulfat *n*,
Kupfervitriol *n*
cupric sulfide Kupfer(II)-sulfid *n*, Cuprisulfid *n*

cupric thiocyanate Kupfer(II)-rhodanid *n*,
　Rhodankupfer *n*
cupricyanic acid Cupricyanwasserstoffsäure *f*
cupriferous kupferführend, kupferhaltig
cupriferous slate Kupferschiefer *m* (Min)
cupritartaric acid Cupriweinsäure *f*
cuprite Cuprit *m* (Min), Rotkupfererz *n* (Min),
　capillary ～ Kupferblüte *f* (Min),
　Kupferfedererz *n* (Min), **earthy ferruginous** ～
　Kupferbraun *n* (Min), Kupferocker *m* (Min)
cuproadamite Cuproadamit *m* (Min)
cuprocupric sulfite Cuprocuprisulfit *n*
cuprocyanic acid Cuprocyanwasserstoffsäure *f*
cuprodescloizite Cuprodescloizit *m* (Min)
cuprofulminate Cuprofulminat *n*
cuprohemol Cuprohämol *n*
cupromagnesite Cupromagnesit *m* (Min)
cupromanganese Mangankupfer *n*,
　Cupromangan *n*, Kupfermangan *n*
cupron Cupron *n*
cupron cell Cupronelement *n*
cupronine Cupronin *n*
cuproplumbite Cuproplumbit *m* (Min),
　Kupferbleiglanz *m* (Min)
cupropyrite Cupropyrit *m* (Min)
cuproscheelite Cuproscheelit *m* (Min)
cuprotungstite Cuprotungstit *m* (Min)
cuprotypy Cuprotypie *f*
cuprous acetylide Kupfercarbid *n*
cuprous bromide Kupferbromür *n* (obs),
　Kupfer(I)-bromid *n*
cuprous carbide Kupfercarbid *n*
cuprous chloride Cuprochlorid *n*,
　Kupferchlorür *n* (obs), Kupfer(I)-chlorid *n*
cuprous coal Kupferbranderz *n*
cuprous compound Kupfer(I)-Verbindung *f*,
　Cuproverbindung *f*,
　Kupferoxydulverbindung *f* (obs)
cuprous cyanide Kupfer(I)-cyanid *n*,
　Kupfercyanür *n* (obs)
cuprous hydroxide Kupfer(I)-hydroxid *n*,
　Kupferhydroxydul *n* (obs)
cuprous iodide Kupfer(I)-jodid *n*, Kupferjodür *n*
　(obs)
cuprous iodomercurate(II) Cupromercurijodid *n*
cuprous ion Cuproion *n*
cuprous mercuric iodide Cupromercurijodid *n*
cuprous oxide Cuprooxid *n*, Kupfer(I)-oxid *n*,
　Kupferoxydul *n* (obs)
cuprous phosphide Kupferphosphid *n*
cuprous salt Kupfer(I)-Salz *n*, Cuprosalz *n*
cuprous sulfide Kupfer(I)-sulfid *n*,
　Halbschwefelkupfer *n*, Kupfersulfür *n* (obs)
cuprous sulfite Cuprosulfit *n*
cuprous thiocyanate Kupfer(I)-rhodanid *n*,
　Kupferrhodanür *n* (obs)
cup shape Schalenform *f*
cup-shaped becherförmig, kelchförmig,
　schalenförmig

cup-shaped hammer Schellhammer *m*
cup vane Schalenflügel *m*
curable heilbar
curacose Curacose *f*
curara Curare *n*
curare Curare *n*
curari Curare *n*
curarine Curarin *n*
curaroid Curaroid *n*
curative heilkräftig, Heilstoff *m*
curative power Heilkraft *f*
curative water Heilwasser *n*
curb *v* dämmen
curbine Curbin *n*
curb press Kelterpresse *f*, Korbpresse *f*
curbstone Pflasterstein *m*
curcas oil Curcasöl *n*, Höllenöl *n*
curcuma Gelbwurzel *f* (Bot), Kurkume *f* (Bot)
curcuma paper Curcumapapier *n*
curcumene Curcumen *n*
curcumic acid Curcumasäure *f*
curcumin Curcumin *n*, Curcumagelb *n*
curcumone Curcumon *n*
curd Gerinnsel *n*, geronnene Milch *f*
curd *v* gerinnen; käsen
curdle *v* gerinnen, erstarren, fest werden,
　gelieren, sauer werden (Milch), stocken
curdler Gerinnungsmittel *n*
curdling Gerinnung *f*, Festwerden *n*, Käsen *n*
curds käsiger Niederschlag *m*, Quark *m*
curdy käsig, klümperig, quarkartig
cure, continuance of ～ Nachvulkanisation *f*,
　degree of ～ Aushärtungsgrad *m* (Kunststoff),
　level of ～ Vulkanisationsniveau *n*
cure *v* heilen, [aus]härten (Gummi, Kunststoff),
　einpökeln, vulkanisieren (Gummi)
cure-all Allheilmittel *n* (Pharm);
　Universalarznei *f* (Pharm), Universalmittel *n*
　(Pharm)
cured good Vulkanisat *n*
cured malt Darrmalz *n*
curie Curie *n* (Phys)
Curie constant Curiekonstante *f*
curiegraph Curiegraph *m*
Curie point Curiepunkt *m*
curiescopy Curieskopie *f*
Curie's law Curiesches Gesetz *n*
Curie strength Curiegehalt *m*
curie temperature Curie-Temperatur *f*
curie unit Curie *n* (Phys)
curine Curin *n*
curing Aushärten *n*, Härten *n*, Vulkanisation *f*,
　Vulkanisieren *n*
curing agent Aushärtungskatalysator *m*,
　Härter *m*, Vulkanisiermittel *n*
curing bag Heizschlauch *m* (Gummi)
curing band Heizband *n*
curing oven Härteofen *m*
curing period Abbindezeit *f*

curing rim Heizfelge *f*
curing time Härtezeit *f,* Pökeldauer *f*
curing tube, full circle ~ Kreisheizschlauch *m*
curite Kurit *m* (Min)
curium (Symb. Cm) Curium *n*
curium isotope Curiumisotop *n*
curl *v* kräuseln, rollen (Papier)
curral Curral *n*
currant Johannisbeere *f* (Bot), Korinthe *f,*
Rosine *f*
currant juice Johannisbeersaft *m*
currant wine Johannisbeerwein *m*
current Lauf *m,* Strömung *f,* Strom *m* (Elektr),
absence of ~ Stromausfall *m,*
Stromlosigkeit *f,* **active** ~ Wirkstrom *m,*
alternating ~ Wechselstrom *m,* **backward or**
inverse ~ Sperrstrom *m* (Elektr), **branching of**
~ Stromverzweigung *f,* **commutation of** ~
Stromumkehrung *f,* **compensation of** ~
Stromausgleich *m,* **constant** ~ gleichbleibende
Stromstärke, **consumption of** ~
Stromverbrauch *m,* Stromentnahme *f,*
continuous ~ Gleichstrom *m,* **damping of** ~
Stromdämpfung *f,* **decaying** ~ abklingender
Strom, **decay of** ~ Abfall *m* des Stromes
(Elektr), **decrease of** ~ Stromabnahme *f,*
dephased ~ phasenverschobener Strom, **direct**
~ Gleichstrom *m,* **direction of** ~
Stromrichtung *f* (Elektr), **displacement of** ~
Stromverdrängung *f,* **fluctuating** ~
unregelmäßiger Strom, **fluctuation of** ~
Stromschwankung *f,* **generation of** ~
Stromerzeugung *f,* **increase of** ~
Stromzunahme *f,* **inducing** ~ induzierender
Strom, **interrupted** ~ unterbrochener Strom,
interruption of the ~ Stromunterbrechung *f,*
leakage of ~ Stromverlust *m,* **loss of** ~
Stromverlust *m,* **out-of-phase** ~
phasenverschobener Strom, **path of** ~
Stromweg *m,* **phase-displaced** ~
phasenverschobener Strom, **photoelectric** ~
lichtelektrischer Strom, **pulsating** ~
pulsierender Strom, **rectified** ~
gleichgerichteter Strom, **retardation of** ~
Stromverzögerung *f,* **reversal of** ~
Stromumkehrung *f,* Stromwendung *f,* **ripple**
~ pulsierender Strom, **strength of** ~
Stromstärke *f,* **supply of** ~ Stromversorgung *f,*
to cut off the ~ den Strom sperren, **total** ~
Gesamtstrom *m,* **transport of the** ~
Stromtransport *m,* **traversed by a** ~
stromdurchflossen, **type of** ~ Stromart *f,*
unidirectional ~ Strom *m* gleicher Richtung,
weak ~ Schwachstrom *m,* **weakening of the** ~
Stromschwächung *f,* ~ **from external source[s]**
vagabundierender Strom *m,* ~ **in solid**
conductor Körperstrom *m,* ~ **near the edge of**
the field Randstrom *m,* ~ **of air** Luftzug *m*
current amplification Stromverstärkung *f*

current attenuation Stromdämpfung *f*
current breaker Stromunterbrecher *m*
current capacity Stromkapazität *f*
current carrier mobility Beweglichkeit *f* der
Stromträger
current carrying stromdurchflossen,
stromführend
current change Stromänderung *f*
current circuit Stromkreis *m*
current collector Stromabnehmer *m* (Elektr)
current component Stromkomponente *f*
current concentration Stromkonzentration *f*
current conducting clamp
Stromzuführungsschelle *f*
current conduction Stromleitung *f*
current damper Stromdämpfer *m*
current density Stromdichte *f,* **cathodic** ~
kathodische Stromdichte, **distribution of** ~
Stromdichteverteilung *f,* **limiting** ~
Grenzstromdichte *f,* ~ **at the electrode**
Elektrodenstromdichte *f*
current diagram Stromdiagramm *n*
current discharge capacity
Stromabgabefähigkeit *f*
current distribution Stromverteilung *f*
current efficiency Stromausbeute *f*
current field Stromfeld *n* (Elektr)
current generating plant
Stromerzeugungsanlage *f*
current generating set
Stromerzeugungsaggregat *n*
[current] generator Stromerzeuger *m*
current impulse Stromstoß *m*
current indicator Stromanzeiger *m*
current intensity Stromintensität *f,* Stromstärke *f*
current lead-in Stromdurchführung *f*
currentless stromlos
current loop Stromschleife *f*
current meter Stromzähler *m,* Stromuhr *f*
current multiplication Stromvervielfachung *f*
current network Stromnetz *n*
current passage Stromdurchgang *m*
current path Strombahn *f*
current paths, dispersion of the ~
Stromlinienstreuung *f*
current phase Stromphase *f* (Elektr)
current pulsation Stromstoß *m*
current rectifier Stromgleichrichter *m*
current regulator Stromregler *m*
current reverser Stromwender *m*
current rise Stromanstieg *m*
current sensitivity Stromempfindlichkeit *f*
current stabilization Stromstabilisierung *f*
current strength Stromintensität *f,* Stromstärke *f*
current transformer Stromtransformator *m,*
Stromwandler *m*
current transmission Stromübertragung *f*
current triangle Stromdreieck *n*
current unit Stromeinheit *f*

current vector Stromvektor *m*
current velocity Stromgeschwindigkeit *f*
current-voltage characteristic
  Stromspannungskennlinie *f*
current yield Stromausbeute *f*
curriculum Lehrgang *m*, Studienplan *m*
curriculum vitae (Lat) Lebenslauf *m*
currier's ink Eisenschwärze *f*
curry Curry *m*
curry *v* gerben, ~ **again** nachgerben
currying Gerben *n*, Gerbung *f*, Lohgare *f*
curtail *v* abkürzen, beschneiden
curtain Vorhang *m*
curtain machine Gießmaschine *f*
Curtius reaction Curtiusscher Abbau *m* (Chem)
Curtius rearrangement Curtius-Umlagerung *f*
curvature Krümmung *f*, Biegung *f*, Rundung *f*,
  Wölbung *f*, **axis of** ~ Krümmungsachse *f*,
  total ~ Gesamtkrümmung *f*, ~ **of field**
  Bildfeldwölbung *f* (Opt)
curvature radius Krümmungsradius *m*
curve Kurve *f*, Biegung *f*, Krümmung *f*,
  Linienzug *m*, **branch of a** ~ Kurvenast *m*
  (Math), **dotted** ~ punktierte Kurve,
  Punktkurve *f*, **logarithmic** ~ logarithmische
  Kurve, **solid** ~ ausgezogene Kurve
curve *v* abrunden, biegen, krümmen
curved gebogen, gekrümmt, krumm
curve fitting Kurvenanpassung *f* (Math)
curve of solubility Löslichkeitskurve *f*
curvilinear krummlinig
curvilinear integral Kurvenintegral *n*
curving, amount of ~ Krümmung *f*
cuscamidine Cuscamidin *n*
cusco bark Cuscorinde *f*
cuscohygrine Cuscohygrin *n*, Cuskhygrin *n*
cusconidine Cusconidin *n*
cushion Kissen *n*, Polster *n*
cushion *v* dämpfen (Stoß)
cushioning Dämpfung *f*, Prellvorrichtung *f*
cusparcine Cusparcin *n*
cuspareine Cusparein *n*
cusparidine Cusparidin *n*
cusparine Cusparin *n*
cuspidine Cuspidin *m* (Min)
cusso extract Kosoextrakt *m*
custerite Custerit *m* (Min)
custody Bewachung *f*, Haft *f*
customary gebräuchlich, gewöhnlich
customer pick-up Selbstabholung *f*
customs office Zollamt *n*
cusylol Cusylol *n*
cut Schnitt *m*, Einschnitt *m*, Fasson *f*, Schnitt *m*
  (Dest), **width of** ~ Schnittweite *f*
cut *v* schneiden, durchstechen, hauen, stutzen,
  zerschneiden, ~ **across** durchschneiden, ~
  away wegschneiden, ~ **in[to]** einhauen,
  einschneiden, ~ **off** abschneiden, abhauen,
  abholzen, abkappen, abschalten (Elektr),

absperren, ~ **open** aufschneiden, ~ **out**
ausschneiden, ausstechen, ~ **through**
durchschneiden, ~ **to size** zuschneiden
cutaneous fungus Dermatophyt *m* (Bakt),
  Hautpilz *m*
cutaneous respiration Hautatmung *f*
cutaneous tissue Hautgewebe *n*
cutch Cachou *n*, Catechu *n*
cut edge Schnittrand *m*
cut growth Schnittwachstum *n* (Reifen)
cutic acid Cutinsäure *f*
cuticle Nagelhaut *f*
cut-in Zwischenschaltung *f*
cutinase Cutinase *f* (Biochem)
cutinic acid Cutininsäure *f*
cutlery Besteck *n*
cut off *a* abgeschaltet
cut-off device Abstellvorrichtung *f*
cut-off frequency Grenzfrequenz *f*,
  Abschneidefrequenz *f*
cut-off relay Abschaltrelais *n*
cut-off valve Absperrventil *n*
cut-off voltage Unterdrückungsspannung *f*
  (Elektr)
cutoff wall Herdmauer *f*
cut-off wavelength Grenzwellenlänge *f*
cutose Cutose *f*
cutout Ausschnitt *m*
cutout valve Absperrventil *n*
cut peat Stechtorf *m*
cut resistance Schnittfestigkeit *f*
cut size Teilungsdichte *f* (screening),
  Kornscheide *f*, Trenngrenze *f*
cutter Schneidmaschine *f*, Fräser *m*, Messer *n*,
  Schneideapparat *m*, **face** ~ **or plain milling**
  cutter Walzenstirnfräser *m*
cutter bar Bohrwelle *f*
cutter blade Schneidmesser *n*
cutter clearance gauge Fräserschleiflehre *f*
cutter for milling Halbkreisformfräser *m*
  (Techn)
cutter-loader Abbau-Vortriebsmaschine *f*
  (Bergb), Fräslader *m* (Bergb), Schrämlader *m*
  (Bergb)
cutting Schneiden *n*, Abtrennung *f*,
  Einschnitt *m*, Span *m*, spanabhebende
  Bearbeitung *f*, ~ **of lines of force** Schneiden *n*
  pl von Kraftlinien, ~ **with a die** Ausstanzen *n*
cutting *a* schneidend, spanabhebend
cutting alloy Hartmetall *n*, Schneidmetall *n*
cutting angle Schneidwinkel *m*, Schnittwinkel *m*
cutting apparatus Schneidewerkzeug *n*
cutting blade Klinge *f*
cutting die Stanzmesser *n*
cutting edge Schneide *f*, Schneidkante *f*
cutting efficiency Schnittleistung *f*
cutting face Anschnitt *m* (Fläche)
cutting flame Schneidflamme *f*
cutting force (lathe cutting) Schnittkraft *f*

cutting hardness Schneidhärte *f* (Werkzeug)
cutting machine Fräsmaschine *f,*
  Planschneider *m,* Schneid[e]maschine *f*
cutting mill Schneidmühle *f*
cutting-off machine, automatic ~
  Abstechautomat *m* (Techn)
cutting oil Bohröl *n,* Kühlöl *n*
cutting operation Zerspanung *f*
cutting pliers Schneidezange *f*
cutting press Stanzpresse *f*
cutting process Schneidvorgang *m*
cutting property Zerspanungseigenschaft *f*
cutting rule Schneidlineal *n*
cuttings Abfall *m* (Met, Holz), Verschnitt *m*
cutting through Durchschneiden *n*
cutting tool Drehstahl *m,* Schneidwerkzeug *n*
cutting torch Schneidbrenner *m*
cutting velocity Schnittgeschwindigkeit *f*
cut work Einschnittarbeit *f*
cuvette Küvette *f*
cyamelide Cyamelid *n*
cyanacetamide Cyanacetamid *n*
cyanalcohol Cyanhydrin *n*
cyanamide Cyanamid *n*
cyananil Cyananil *n*
cyananilic acid Cyananilsäure *f*
cyananilide Cyananilid *n*
cyananthrene Cyananthren *n*
cyanate Cyanat *n*
cyanate *a* cyansauer
cyanation Cyanierung *f*
cyanatobenzene Phenylcyanat *n*
cyan blue Cyanblau *n*
cyanchloride Cyanchlorid *n*
cyancyanide Cyanogen *n*
cyanhydrate Hydrocyanid *n*
cyanic cyanblau
cyanic acid Cyansäure *f,* **salt or ester of** ~
  Cyanat *n*
cyanidan Cyanidan *n*
cyanidation Cyanidlaugung *f*
cyanide Cyanid *n,* **free from** ~ cyanidfrei
cyanide *a* blausauer, cyanwasserstoffsauer
cyanide bath Cyanbad *n*
cyanide hardening Cyanhärtung *f*
cyanide hematoporphyrin
  Cyanidhämatoporphyrin *n*
cyanide liquor Cyanidlauge *f*
cyanide mesoporphyrin Cyanidmesoporphyrin *n*
cyanidenolone Cyanidenolon *n*
cyanide preparation Cyanpräparat *n*
cyanide process Cyanidlaugung *f,*
  Cyanidverfahren *n*
cyanidin[e] Cyanidin *n*
cyaniding Cyanhärtung *f*
cyanidylin Cyanidylin *n*
cyanine Cyanin *n,* Cyaninfarbstoff *m*
cyanite Cyanit *m* (Min), blättriger Beryll (Min),
  Blauschörl *m* (Min), blauer Disthen (Min)

cyanization Cyanisierung *f*
cyanize *v* cyanisieren
cyanmethine Cyanmethin *n*
cyano-2-nitroethanamide, 2- ~ Fulminursäure *f*
cyanoacetamide Cyanacetamid *n*
cyanoacetate *a* cyanessigsauer
cyanoacetic acid Cyanessigsäure *f*
cyanoacetic ester Cyanessigester *m*
cyanoacetyl chloride Cyanessigsäurechlorid *n*
cyanoalkyl Cyanalkyl *n*
cyanoanilide Cyananilid *n*
cyanoanthrene Cyananthren *n*
cyanobenzene Cyanbenzol *n*
cyanocadmic acid Cadmiumcyanwasserstoff *m*
cyanocarbamic acid Cyanamidokohlensäure *f*
cyanocarbonic acid Cyankohlensäure *f*
cyanocobalamin Cyanocobalamin *n,* Vitamin
  B$_{12}$ *n*
cyanoethylation Cyanäthylierung *f*
cyanoferrate(II) Ferrocyanid *n*
cyanoferrate(III) Ferricyanid *n*
cyanoferric cyaneisenhaltig
cyanoform Cyanoform *n*
cyanoformic acid Cyankohlensäure *f*
cyanogen Cyan *n,* Cyanogen *n,* Dicyan *n,*
  Oxalnitril *n,* **containing** ~ cyanhaltig; cyanig
cyanogen azide Cyanazid *n*
cyanogen bromide Bromcyan *n,* Cyanbromid *n*
cyanogen chloride Chlorcyan *n,* Cyanchlorid *n*
cyanogen compound Cyanverbindung *f*
cyanogen discrepancy Cyandiskrepanz *f*
cyanogen gas Cyangas *n*
cyanogen iodide Cyanjodid *n,* Jodcyan *n*
cyanogen soap Cyanseife *f*
cyanogen sulfide Schwefelcyan *n*
cyano group Cyangruppe *f*
cyanoguanidine Cyanoguanidin *n,*
  Dicyandiamid *n*
cyanohydrin Cyanhydrin *n*
cyanohydrin synthesis Cyanhydrinsynthese *f*
cyanol Anilin *n,* Cyanol *n*
cyanol green Cyanolgrün *n*
cyanomaclurin Cyanomaclurin *n*
cyanometer Cyanometer *n*
cyanomethane Methylcyanid *n*
cyanomustard oil Cyansenföl *n*
cyanonitride Cyanstickstoff *m*
cyanonitroacetamide Fulminursäure *f*
cyanopathy Cyanose *f* (Med)
cyanophilous cyanophil (Bakt)
cyanophylline Cyanophyllin *n*
cyanoplatinic acid Platincyanwasserstoff *m*
cyanoporphyrine Cyanoporphyrin *n*
cyanopropionic acid Cyanpropionsäure *f*
cyanopyronine Cyanpyronin *n*
cyanosilane Cyanosilan *n*
cyanosine Cyanosin *n*
cyanosis Blausucht *f* (Med), Cyanose *f* (Med)
cyanothiamine Cyanthiamin *n*

cyanotic cyanotisch (Med)
cyanotoluene Cyantoluol *n*, Tolunitril *n*
cyanotrichite Sammeterz *n* (Min)
cyanotype Blaupause *f,* Cyanotypie *f*
cyanotype paper Cyanotyppapier *n*
cyanotype process Cyanotypie *f*
cyanphenanthrene Cyanphenanthren *n*
cyanur Cyanur-
cyanurate Cyanurat *n*, Tricyansäureverbindung *f*
cyanuric acid Cyanursäure *f,* Tricyansäure *f,*
    compound of ~ Tricyansäureverbindung *f*
cyanuric chloride Cyanurchlorid *n*,
    Cyanurtrichlorid *n*, Trichlorcyan *n*,
    Tricyanchlorid *n*
cyanuric cyanide Cyanurtricyanid *n*
cyanuric hydrazide Cyanurtrihydrazid *n*
cyanurotriamide Cyanurtriamid *n*, Melamin *n*,
    Tricyansäuretriamid *n*
cyanurotricarboxylic acid
    Cyanurtricarbonsäure *f*
cyanurotrichloride Cyanurtrichlorid *n*
cyanurotricyanide Cyanurtricyanid *n*
cyanurotrihydrazide Cyanurtrihydrazid *n*
cyarsal Cyarsal *n*
cybernetics Kybernetik *f*
cybertron Cybertron *n*
cycasin Cycasin *n*
cyclamate Cyclamat *n*
cyclamenaldehyde Cyclamenaldehyd *m*
cyclamic acid Cyclaminsäure *f*
cyclamin Arthranitin *n*, Cyclamin *n*
cyclamose Cyclamose *f*
cyclandelate Cyclandelat *n*
cyclanoline Cyclanolin *n*
cyclazin Cyclazin *n*
cycle Kreislauf *m*, Kreisprozeß *m*, Periode *f*
    (Elektr), Turnus *m*, Umlauf *m*, Zyklus *m*,
    complete ~ Vollperiode *f* (Elektr), **full** ~
    Vollschwingung *f*
cycleanine Cycleanin *n*
cyclene Cyclen *n*
cycle per second Hertz *n* (Elektr)
cyclic zyklisch, ringförmig (Chem)
cyclic adenosine monophosphate zyklisches
    Adenosinmonophosphat *n*
cyclic adenylic acid zyklisches
    Adenosinmonophosphat *n*
cyclic AMP zyklisches
    Adenosinmonophosphat *n*
cyclic compound Ringverbindung *f* (Chem)
cyclic homolog[ue] Ringhomologe[s] *n*
cyclic ketone Ringketon *n*
cyclic loading Wechselbiegebeanspruchung *f*
cyclic process Kreisprozeß *m*, Kreislauf *m*,
    Kreisvorgang *m*
cyclic sulfur Cycloschwefel *m*
cyclic system Ringsystem *n*
cycling time Laufzeit *f* (einer Maschine)

cyclization Cyclisierung *f,* Ringbildung *f,*
    Ringschließung *f,* Ringschluß *m*,
    Zyklisierung *f*
cyclizine Cyclizin *n*
cycloaddition Cycloaddition *f*
cycloaliphatic cycloaliphatisch
cycloalkane Cycloalkan *n*
cycloalkannine Cycloalkannin *n*
cycloalliin Cycloalliin *n*
cycloartane Cycloartan *n*
cycloartenol Cycloartenol *n*
cycloartenone Cycloartenon *n*
cyclobarbital Cyclobarbital *n*, Phanodorm *n*
cyclobutadiene Cyclobutadien *n*
cyclobutane Cyclobutan *n*
cyclobutanone Cyclobutanon *n*
cyclobutene Cyclobuten *n*
cyclobutyl Cyclobutyl-
cyclobutylene Cyclobutadien *n*
cyclocamphidine Cyclocamphidin *n*
cyclocamphor Cyclocampher *m*
cyclocholestan Cyclocholestan *n*
cyclocitral Cyclocitral *n*
cyclocoloranol Cyclocoloranol *n*
cyclocolorenone Cyclocolorenon *n*
cyclodecandione Cyclodecandion *n*
cyclodecane Cyclodecan *n*
cyclodiastereomerism Cyclodiastereomerie *f*
    (Stereochem)
cyclodimerisation Cyclodimerisation *f*
cycloenantiomer Cycloenantiomer *n*
    (Stereochem)
cycloenantiomerism Cycloenantiomerie *f*
    (Stereochem)
cycloeucalane Cycloeucalan *n*
cyclofenchene Cyclofenchen *n*
cycloform Cycloform *n*
cyclogallipharaol Cyclogallipharaol *n*
cyclogeranic acid Cyclogeraniumsäure *f*
cyclogeraniol Cyclogeraniol *n*
cyclogeraniolene Cyclogeraniolen *n*
cycloheptadecanone Cycloheptadecanon *n*
cycloheptadiene Cycloheptadien *n*
cycloheptane Cycloheptan *n*, Heptamethylen *n*,
    Suberan *n*
cycloheptanol Cycloheptanol *n*, Suberol *n*
cycloheptanone Cycloheptanon *n*, Suberon *n*
cycloheptatriene Cycloheptatrien *n*, **1,3,5-** ~
    Tropiliden *n*
cycloheptene Cyclohepten *n*
cycloheptine Cycloheptin *n*
cyclohexadiene Cyclohexadien *n*
cyclohexane Cyclohexan *n*, Hexahydrobenzol *n*,
    Hexamethylen *n*
cyclohexanediol Cyclohexandiol *n*
cyclohexanedionedioxime
    Cyclohexandiondioxim *n*
cyclohexanepentol Quercit *m*
cyclohexanol Cyclohexanol *n*, Hexalin *n*

cyclohexanone Cyclohexanon *n*, Pimelinketon *n*
cyclohexanone oxime Cyclohexanonoxim *n*
cyclohexatriene Cyclohexatrien *n*
cyclohexene Cyclohexen *n*
cyclohexenol Cyclohexenol *n*
cyclohexenone Cyclohexenon *n*
cyclo hexenyl Cyclohexenyl-
cycloheximide Cycloheximid *n*
cyclohexine Cyclohexin *n*
cyclohexyl Cyclohexyl-
cyclohexyl acetate Cyclohexylacetat *n*
cyclohexylamine Cyclohexylamin *n*
cyclohexylethylamine Cyclohexyläthylamin *n*
cyclohexylidene Cyclohexyliden *n*
cyclohexyl methacrylate
   Cyclohexylmethacrylat *n*
cyclohexyl silicone Cyclohexylsilicon *n*
cyclohexyl trichlorosilane
   Cyclohexyltrichlorsilan *n*
cyclohomocitral Cyclohomocitral *n*
cyclohomogeranic acid
   Cyclohomogeraniumsäure *f*
cyclohomogeraniol Cyclohomogeraniol *n*
cycloid Radlinie *f*, Zykloide *f* (Math)
cyclolavandulol Cyclolavandulol *n*
cyclomethycaine Cyclomethycain *n*
cyclometry Kreismessung *f* (Math)
cyclone Zyklon *m*
cyclone dust collector
   Fliehkraftstaubabscheider *m*
cyclone firing Zyklonfeuerung *f*
cyclone heat exchanger
   Zyklonwärmeaustauscher *m*
cyclone impeller Zyklonrührer *m*
cyclone scrubber Zyklonwäscher *m*
cyclone separator Zyklonabscheider *m*
cyclone smelting furnace Schmelzzyklon *m*
cyclonite Cyclonit *n* (Sprengstoff)
cyclononane Cyclononan *n*
cyclonovobiocic acid Cyclonovobiocinsäure *f*
cyclo-octadiene Cyclooctadien *n*
cyclo-octane Cyclooctan *n*
cyclo-octanone Cyclooctanon *n*
cyclo-octatetraene Cyclooctatetraen *n*
cyclo-octene Cycloocten *n*
cyclo-olefine Cycloolefin *n*
cyclooligomerisation Cyclooligomerisation *f*
cyclopaldic acid Cyclopaldsäure *f*
cycloparaffin Cycloparaffin *n*
cyclopentadecanone Cyclopentadecanon *n*
cyclopentadiene Cyclopentadien *n*
cyclopentamethylene Cyclopentamethylen *n*,
   Cyclopentan *n*
cyclopentamine Cyclopentamin *n*
cyclopentane Cyclopentamethylen *n*,
   Cyclopentan *n*
cyclopentanecarboxylic acid
   Cyclopentancarbonsäure *f*
cyclopentanol Cyclopentanol *n*

cyclopentanone Cyclopentanon *n*,
   Ketopentamethylen *n*
cyclopentene Cyclopenten *n*
cyclopentenone Cyclopentenon *n*
cyclopentyl Cyclopentyl-
cyclophorase Cyclophorase *f*
cyclophosphamide Cyclophosphamid *n*
cyclopin Cyclopin *n*
cyclopite Cyclopit *m*
cyclopolic acid Cyclopolsäure *f*
cyclopolymerization Cyclopolymerisation *f*
cyclopregnol Cyclopregnol *n*
cyclopropane Cyclopropan *n*, Trimethylen *n*
cyclopropene Cyclopropen *n*
cyclopropenone Cyclopropenon *n*
cyclopropyl Cyclopropyl-
cyclopterin Cyclopterin *n*
cyclopyrethrosin Cyclopyrethrosin *n*
cycloserine Cycloserin *n*
cyclosilane Cyclosilan *n*
cyclosiloxane Cyclosiloxan *n*
cyclosis Plasmaströmung *f* (Bot)
cyclosolanidane Cyclosolanidan *n*
cyclostereoisomerism Cyclostereoisomerie *f*
   (Stereochem)
cyclosteroid Cyclosteroid *n*
cyclothreonine Cyclothreonin *n*
cyclotron Zyklotron *n*, Teilchenbeschleuniger *m*,
   frequency-modulated ~ frequenzmoduliertes
   Zyklotron
cyclotron frequency Zyklotronumlauffrequenz *f*
cyclotron resonance experiment
   Zyklotronresonanzexperiment *n*
cycloundecane Cycloundecan *n*
cycrimine Cycrimin *n*
cylinder Zylinder *m*, Rolle *f*, Trommel *f*, Walze *f*,
   air-cooled ~ luftgekühlter Zylinder, **cubic
   capacity of the** ~ Zylindervolumen *n*,
   graduated ~ graduierter Zylinder (Chem),
   inverted ~ hängender Zylinder, **revolving** ~
   umlaufender Zylinder, **single acting** ~
   einfachwirkender Zylinder, **upright** ~
   stehender Zylinder, **water-cooled** ~
   wassergekühlter Zylinder, **~ with stirrer**
   Rührzylinder *m*
cylinder block thermometer Motorthermometer *n*
cylinder boring machine Zylinderbohrmaschine *f*
cylinder clothing Zylinderummantelung *f*
cylinder coking Verkokung *f* in Retorten
cylinder cover Zylinderdeckel *m*
cylinder dryer Trockenzylinder *m*,
   Schachttrockner *m*
cylinder function Zylinderfunktion *f* (Math)
cylinder furnace Zylinderofen *m*
cylinder gas Flaschengas *n*
cylinder head Zylinderkopf *m* (Mot)
cylinder head joint Zylinderkopfdichtung *f*
cylinder journal Walzenzapfen *m* (Gummi)

cylinder liner Einsatzzylinder *m*,
  Führungsbüchse *f*, Zylinderauskleidung *f*
cylinder lubrication Zylinderschmierung *f*
cylinder mixer Trommelmischer *m*
cylinder oil Zylinderöl *n*
cylinder press Zylinder[schnell]presse *f*,
  Zuckermühle *f*
cylinder printing machine
  Rouleauxdruckmaschine *f*
cylinder support Walzenstuhl *m* (Buchdr)
cylinder valve Flaschenventil *n*
cylinder volume Zylinderinhalt *m*,
  Zylindervolumen *n*
cylindrical rollenförmig, walzenförmig,
  zylinderförmig
cylindrical blower Zylindergebläse *n*
cylindrical cathode Hohlkathode *f*
cylindrical cell Zylinderzelle *f*
cylindrical condenser Zylinderkondensator *m*
cylindrical conductor Zylinderkonduktor *m*
cylindrical gasket Lippendichtung *f*
cylindrical grinder Rundschleifmaschine *f*
cylindrical lens Zylinderlinse *f* (Opt)
cylindrical magnet Zylindermagnet *m*
cylindrical shape Walzenform *f*
cymaric acid Cymarinsäure *f*
cymarigenin Cymarigenin *n*
cymarin Cymarin *n*
cymarol Cymarol *n*
cymarose Cymarose *f*
cymatolite Cymatolith *m*
cymene Cymol *n*, Camphogen *n*
cymidine Cymidin *n*, Carvacrylamin *n*
cymogene Cymogen *n*
cymometer Cymometer *n*, Wellenmesser *m*
cymophane Cymophan *m* (Min)
cymopyrocatechol Cymobrenzcatechin *n*
cymoscope Cymoscop *n* (Elektr), Zymoskop *n*
  (Elektr)
cymylamine Cymylamin *n*
cynanchotoxine Cynanchotoxin *n*
cynapine Cynapin *n*
cynocannoside Cynocannosid *n*
cynoctonine Cynoctonin *n*
cynodontin Cynodontin *n*
cynoglossophine Cynoglossofin *n*
cyperene Cyperen *n*
cyperone Cyperon *n*
cypress oil Cypressenöl *n*, Zypressenöl *n*
cyprinine Cyprinin *n*
Cyprus blue Cypernblau *n*
cyrtolite Cyrtolith *m* (Min)
cyrtominetin Cyrtominetin *n*
cyrtopterinetin Cyrtopterinetin *n*
cyst Zyste *f* (Biol)
cystamine Cystamin *n*, Cystinamin *n*
cystathionase Cystathionase *f* (Biochem)
cystathione Cystathion *n*
cystathionine Cystathionin *n*

cystazol Cystazol *n*
cysteamine Cysteamin *n*
cysteic acid Cysteinsäure *f*
cysteine Cystein *n*
cysteine desulfhydrase Cysteindesulfhydrase *f*
  (Biochem)
cysteine hydrochloride Cysteinhydrochlorid *n*
cysteine reductase Cystein-Reduktase *f*
  (Biochem)
cystinal Cystinal *n*
cystine Cystin *n*
cystine amine Cystinamin *n*
cystine hydantoin Cystinhydantoin *n*
cystinol Cystinol *n*
cystinosis Cystinose *f*
cystinuria Cystinurie *f* (Med)
cystolite Cystolith *m*
cystopurine Cystopurin *n*
cystosine Cystosin *n*
cytase Zytase *f* (Biochem), Alexin *n*
cytidine Cytidin *n*
cytidine diphosphate Cytidindiphosphat *n*
cytidine monophosphate Cytidinmonophosphat *n*
cytidine triphosphate Cytidintriphosphat *n*
cytidylic acid Cytidylsäure *f*
cytisine Cytisin *n*
cytisolidine Cytisolidin *n*
cytisoline Cytisolin *n*
cytobiology Zellenbiologie *f*
cytochemistry Cytochemie *f*, Zytochemie *f*
cytochrome Cytochrom *n*
cytochrome oxidase Cytochromoxydase *f*
  (Biochem)
cytochrome reductase Cytochromreduktase *f*
  (Biochem)
cytocidal zell[en]tötend, zellvernichtend
cytoclastic zellvernichtend
cytode Cytode *f* (Biol)
cytodeuteroporphyrin Cytodeuteroporphyrin *n*
cytogenesis Cytogenese *f*, Zellenbildung *f*,
  Zellproduktion *f*
cytogenic zellenbildend
cytogenitics Cytogenetik *f*
cytoglobin Cytoglobin *n*
cytohormone Zytohormon *n*
cytokinase Biochem Zytokinase *f*
cytokinesis Cytokinese *f* (Biol)
cytology Cytologie *f*, Zellforschung *f*
cytolysin Cytolysin *n*
cytolysis Zytolyse *f*, Zelltod *m*, Zellzerfall *m*
cytolytic zytolytisch
cytometric zytometrisch
cytometry Zytometrie *f*
cytomorphology Zellmorphologie *f*
cytomorphosis Zellveränderung *f*
cytophysiological zellphysiologisch
cytophysiology Zellphysiologie *f*
cytoplasm Cytoplasma *n* (Biol), Zellplasma *n*
  (Biol)

cytoplast Cytoplasma *n* (Biol)
cytopyrrolic acid Cytopyrrolsäure *f*
cytosamine Cytosamin *n*
cytosine Cytosin *n*
cytosol Cytosol *n*
cytosylic acid Cytosylsäure *f*
cytozoon (pl cytozoa) Zellparasit *m*

D

dab v abtupfen, betupfen
dacite Dacit m
dacrene Dacren n
dacron Dacron n, Terrylen n (HN)
dacrydene Dacryden n
dactylin Dactylin n
dagingolic acid Dagingolsäure f
dagingoresene Dagingoresen n
dahlia Dahlie f (Bot)
dahlia violet Dahliaviolett n
dahlin Dahlin n
dahllite Dahllit m (Min)
Dahl's acid Dahlsche Säure f
dahmenite Dahmenit n
daidzein Daidzein n
daidzin Daidzin n
daily output Tagesleistung f, Tagesproduktion f
daily use Handgebrauch m
dairy Molkerei f
Dakin oxidation Dakin-Oxidation f
Dakin reaction Dakin-Reaktion f
Dakin's solution Dakinsche Lösung f
dalbergin Dalbergin n
dalton Dalton n (Maß)
Dalton's law of partial pressures
    Partialdruckgesetz n von Dalton
dalzin Dalzin n
dam Damm m, Staumauer f
dam v dämmen, stauen (Wasser)
damage Beschädigung f, Schaden m,
    Schädigung f, Verletzung f
damage v beschädigen, schädigen, verletzen
damage compensation Schadenersatz m
damaged beschädigt
damaged by air luftverdorben
damage rate in fatigue pl
damascene v damaszieren (Met)
damascened steel Damaststahl m
damascenine Damascenin n
damascening Damaszierung f (Met)
damasceninic acid Damasceninsäure f
damask v damaszieren (Met)
damask steel Damaststahl m,
    Damaszener Stahl m
dambonitol Dambonit m
dambose Dambose f
damiana Damiana f (Pharm)
damiana oil Damianaöl n (Pharm)
dammar Dammar[harz] n
dammarane Dammaran n
dammaranol Dammaranol n
dammaric acid Dammarsäure f
dammaroresene Dammaroresen n
dammar resin Dammarharz n,
    Katzenaugenharz n
dammar varnish Dammarfirnis m

dammaryl Dammaryl n
damming Stauung f, **height of** ~ Stauhöhe f
damnacanthal Damnacanthal n
damnacanthol Damnacanthol n
damourite Damourit m (Min)
damp feucht, naß
damp v bedampfen, befeuchten, benetzen,
    dämpfen (Licht), feucht machen
damp dispersion plant Entdunstungsanlage f
dampen v anfeuchten, befeuchten, benetzen,
    dämpfen
dampening Befeuchtung f
dampening agent Befeuchtungsmittel n
dampening air Befeuchtungsluft f
damper Dämpfer m; Dämpfungsvorrichtung f,
    Drossel f, Drosselklappe f
damper wing Dämpferflügel m
damping Dämpfung f (Elektr) ~ **down of the coke**
    Löschen n des Kokses
damping capacity Dämpfungsfähigkeit f
damping chamber Dämpferkammer f
damping circuit Dämpfungs[strom]kreis m
damping constant Abklingkonstante f
damping device Dämpfer m,
    Dämpfungsvorrichtung f
damping-down tower Löschturm m
damping magnet Dämpfungsmagnet m
damping parameter Dämpfungsparameter m
damping power Dämpfungsvermögen n
damping regulator Dämpfungsregler m
damping roller Anfeuchtwalze f, Feuchtwalze f
damping rolls Anfeuchtmaschine f (Pap)
damping vane, arm of ~ Dämpferflügelarm m
dam plate Wallplatte f
dampness Feuchtheit f, Feuchtigkeit f, Nässe f;
    ~ **of the atmosphere** Luftfeuchtigkeit f
damp-proof sheeting Dichtungsbahn f
damsin Damsin n
dam stone Wallstein m
danain Danain n
danaite Danait m (Min), Kobaltarsenkies m
    (Min)
danalite Danalith m (Min)
danburite Danburit m (Min)
dancer roll Losrolle f, Tänzerrolle f (for tension
    control), Tänzerwalze f
dandruff Kopfschuppen pl
danger! Achtung! Gefahr!, Gefährdung f,
    Gefahr f
danger coefficient Gefährdungskoeffizient m
danger of ignition Entzündungsgefahr f
danger point Gefahrpunkt m
danger signal Warnungssignal n, Gefahrsignal n,
    Notsignal n
danger zone Gefahrenbereich m
Daniell cell Daniellelement n

Danish white Dänisch Weiß *n*
dannemorite Dannemorit *m* (Min)
dansylamino acid Dansyl-Aminosäure *f*
dansyl chloride Dansylchlorid *n*
danthron Danthron *n*
Danzig blue Danziger Blau *n*
Danzig brandy Danziger Goldwasser *n*
Danzig water Danziger Goldwasser *n*
daphnandrine Daphnandrin *n*
daphnetin Daphnetin *n*
daphnin Daphnin *n*
daphnite Daphnit *m* (Min)
daphnoline Daphnolin *n*
dappled gefleckt, scheckig
darapskite Darapskit *m* (Min)
dark blind (Metall), dunkel, tief (Farbe)
dark adaptation Dunkeladaption *f* (Opt),
  Dunkelanpassung *f* (Opt)
dark-adapted dunkeladaptiert
dark blue dunkelblau
dark body radiator Dunkelstrahler *m*
dark brown dunkelbraun, braunschwarz
dark-colored dunkelfarbig
dark conduction Dunkelleitung *f* (Elektr)
dark conductivity Dunkelleitfähigkeit *f*
dark contrast method Dunkelfeldverfahren *n*
  (Opt)
dark current Dunkelstrom *m* (Photozelle)
dark discharge Dunkelentladung *f*, Entladung *f*
  ohne Funkenbildung
darken *v* [ab]dunkeln, nachdunkeln, schwärzen,
  verdunkeln
darkening Dunkelfärbung *f*, Nachdunkeln *n*,
  Verdunkeln *n*
darkening dye Abtrübungsfarbe *f*
darker, to become ~ nachdunkeln
dark field illumination Dunkelfeldabbildung *f*
  (Opt)
dark green dunkelgrün
dark-line spectrum Spektrum *n pl* mit dunklen
  Linien
darkness adaptability
  Dunkelanpassungsfähigkeit *f*
dark nickeling Dunkelvernicklung *f*
dark oil Dunkelöl *n*
dark pulse Dunkelstromimpuls *m*
dark reaction Dunkelreaktion *f*
dark red dunkelrot
dark resistance Dunkelwiderstand *m*
  (Photozelle)
darkroom Dunkelkammer *f* (Phot)
darkroom equipment Dunkelkammergerät *n*
dark yellow dunkelgelb
darnel Taumelkorn *n* (Bot)
Darwinian theory Darwinsche Theorie *f* (Biol)
darwinism Darwinismus *m*
darwinite Darwinit *m* (Min)
dash Schlag *m*, Strich *m*
dash *v* schlagen, schmettern, prallen

dash-dot method Strich-Punkt-Verfahren *n*
  (Elektron)
dashed gestrichelt
dash-lined strichliert
dash pot Dämpfungszylinder *m*
dasymeter Dasymeter *n*, Dichtemesser *m*
data Daten *pl*, Angaben *pl*, Information *f*,
  Unterlagen *pl*, **initial** ~ Anfangsdaten *pl*,
  observational ~ Beobachtungsdaten *pl*
data gathering Datenerfassung *f*
data input equipment Datenerfassungsgerät *n*
data logger Datenspeicher *m*,
  Meßwertsammler *m*, Meßwertspeicher *m*
data-processing datenverarbeitend,
  Datenverarbeitung *f*
data-processing plant
  Datenverarbeitungsanlage *f*
data sheet Liste *f pl* technischer Angaben,
  Merkblatt *n*
data transfer rate Datenübertragungsrate *f*
date Dattel *f* (Bot); Datum *n*, Datumsangabe *f*,
  Zeitangabe *f*
date mark Datumsaufdruck *m*
date of maturity Fälligkeitstermin *m*
datestone Datolith *m* (Min)
date sugar Dattelzucker *m*
date wine Dattelwein *m*
dating Altersbestimmung *f*, Datierung *f*
datiscetin Datiscetin *n*
datiscin Datiscin *n*, Datiscagelb *n*
datolite Datolith *m* (Min)
daturic acid Daturinsäure *f*, Margarinesäure *f*
daturine Daturin *n*, Atropin *n*, Hyoscyamin *n*
daub *v* **with ashes and lime** einschwöden
daubreeite Daubreeit *m* (Min)
daubreelite Daubreelith *m* (Min)
daucol Daucol *n*
daucosterol Daucosterin *n*
daughter activity Tochteraktivität *f*
daughter atom Folgeelement *n* (Atom)
daughter cell Tochterzelle *f* (Biol)
daughter colony Tochterkolonie *f* (Bakt)
daughter nucleus Folgekern *m*, Tochterkern *m*
daunomycin Daunomycin *n*
dauphinite Dauphinit *m* (Min)
dauricine Dauricin *n*
Davy lamp Davysche Sicherheitslampe *f*
  (Bergb)
davyn Davyn *m* (Min)
Davy's safety lamp
  Davysche Sicherheitslampe *f* (Bergb)
dawsonite Dawsonit *m* (Min)
day and night service Tag- und Nachtbetrieb *m*
dayfly Eintagsfliege *f* (Zool)
daylight lichte Einbauhöhe *f* (Presse),
  Tageslicht *n*
daylight fluorescent colors
  Tageslichtleuchtfarben *pl*
day shift Tagschicht *f*

dazzle Blendung *f*
dazzle *v* blenden
dazzlement Blendung *f*
dazzle paint Tarnanstrich *m*
dazzling blendend, grell
d. c. potentiometer
  Gleichspannungskompensator *m* (Elektr)
DDTC Abk Diäthyldithiocarbaminsäure *f*
deacidification Entsäuerung *f*
deacidified entsäuert
deacidify *v* entsäuern
deacidifying Entsäuern *n*
Deacon process Deacon-Prozeß *m*
deactivate *v* abklingen (Katalysator),
  desaktivieren, entaktivieren
deactivation Desaktivierung *f,* Entaktivierung *f,*
  Inaktivierung *f*
deacylation Deacylierung *f*
dead tot, abgestanden (Gips), [ab]gestorben,
  glanzlos, matt, spannungslos (Elektr),
  stromlos (Elektr); taub (Bergb)
dead angle toter Winkel *m*
dead band (Am. E.) Regelunempfindlichkeit *f*
  (Regeltechnik)
dead beat aperiodische Dämpfung
dead-burn *v* totbrennen
dead center toter Punkt *m,* Körnerspitze *f*
dead center position Totpunktlage *f*
dead dip Mattbrenne *f*
dead-dipping of brass Mattieren *n* des Messings
deaden *v* [ab]dämpfen, abstumpfen, schwächen,
  mattieren (Met)
deadening schalldämpfend
deadening color Mattfarbe *f*
dead gilding Mattvergoldung *f*
dead-gilt mattvergoldet
dead gold Mattgold *n*
dead green mattgrün
deadhead Gießkopf *m* (Gieß)
dead line Stichtag *m,* Fälligkeitstermin *m,*
  Indifferenzlinie *f,* Terminplan *m*
dead load Eigenbelastung *f,* Eigenlast *f;*
  Leerlast *f,* Ruhebelastung *f*
deadly tödlich, lebensgefährlich, letal
deadly nightshade Belladonna *f* (Bot)
dead melt *v* abstehen lassen (Metall)
dead point toter Punkt
dead roasting Totbrennen *n*
deads Totliegendes *n* (Bergb)
dead-smooth spiegelglatt
dead space toter Raum
dead steam Abdampf *m,* Abgangsdampf *m*
dead stop method Nullpunktmethode *f*
dead-sure totsicher
dead time Sperrzeit *f,* Totzeit *f* (Regeltechnik)
  (Am. E.), Totzeit *f* (Regeltechn)
dead weight Eigengewicht *n,* Leergewicht *n,*
  Tara *f*

dead zone (Br. E.) Regelunempfindlichkeit *f*
  (Regeltechnik)
de-aerate *v* entlüften
de-aerating plant Entlüftungsvorrichtung *f*
de-aeration Entlüftung *f*
de-aerator Entlüfter *m*
deaf taub
deal Bohle *f,* Brett *n*
deal *v* handeln
dealcoholization Entziehung von Alkohol
dealcoholize *v* entalkoholisieren, entgeisten
dealer Händler *m,* Lieferant *m*
dealkylate *v* entalkylieren
deamidate *v* desamidieren
deamidize *v* desamidieren
deaminate *v* desaminieren
deamination Desaminierung *f*
deaminize *v* desaminieren
dearomatization Desaromatisierung *f*
dearsenification Entarsenierung *f*
dearsenification plant Entarsenieranlage *f*
deasphaltization Entasphaltierung *f*
deathly pale kreideweiß, leichenblaß
debark *v* abrinden, abschälen, entrinden
debenzolize *v* Benzol abtreiben
debilic acid Debilinsäure *f*
debris Abbruch *m,* Aufschüttung *f* (Geol),
  Trümmer *pl*
de Broglie wave Materiewelle *f*
debrominate *v* entbromen
debugging Störbeseitigung *f* (Comp)
deburr *v* abgraten
Debye-Hückel theory Debye-Hückel-Theorie *f*
Debye-Scherrer ring or circle
  Debye-Scherrer-Ring *m*
decachlorobenzophenone Perchlorbenzophenon *n*
decacyclene Dekacyclen *n*
decade counter Dekadenzähler *m*
decade scaler Dekadenuntersetzer *m*
decadic dekadisch
decagon Zehneck *n*
decagram Dekagramm *n* (10 Gramm)
decahedral dekaedrisch
decahedron Dekaeder *n*
decahydrate Dekahydrat *n*
decahydronaphthalene Dekalin *n,* Naphthan *n,*
  Perhydronaphthalin *n*
decalcified entkalkt
decalcify *v* entkalken
decalene Dekalen *n*
decalescence Dekaleszenz *f* (Phys), sprunghafte
  Temperaturabnahme
decalin Dekalin *n,* Naphthan *n*
decalol Dekalol *n*
decameter Dekameter *n*
decamethonium Decamethonium *n*
decamethylene glycol Decamethylenglykol *n*
decanal Decylaldehyd *m*
decanaphthene Dekanaphthen *n*

decanaphthenic acid Dekanaphthensäure *f*
decane Decan *n*
decane diol Decandiol *n*
decanol Decanol *n,* Decylalkohol *m*
decant *v* [ab]dekantieren, abgießen, abklären, abschlämmen, umgießen
decantation Dekantierung *f,* Abgießen *n,* Abklärung *f,* Schlämmung *f*
decantation apparatus Dekantierapparat *m*
decanted abgeschlämmt
decanter Dekantiergefäß *n,* Abklärgefäß *n,* Abtreibapparat *m*
decanting Dekantieren *n,* Abgießen *n,* Abklären *n,* Abschlämmen *n,* Abseihen *n*
decanting bottle Klärflasche *f*
decanting centrifuge Dekantierzentrifuge *f,* Klärzentrifuge *f*
decanting flask Klärflasche *f*
decanting machine Schlämmaschine *f*
decanting vessel Dekantiergefäß *n,* Abklärgefäß *n*
decapitate *v* enthaupten, abkappen
decarbonization Dekarbonisierung *f,* Entkohlung *f,* Kohlenstoffentziehung *f*
decarbonize *v* dekarbonisieren, entkohlen, von Kohlenstoff befreien
decarbonizing Dekarbonisieren *n,* Entkohlen *n*
decarboxylase Decarboxylase *f* (Biochem)
decarboxylate *v* entkarboxylieren, Carboxyl entfernen
decarboxylation Decarboxylierung *f*
decarburization Kohlenstoffentziehung *f,* Dekarbonisierung *f,* Entkohlung *f,* Frischen *n* (Metall), Garen *n* (Metall)
decarburize *v* dekarbonisieren, entkohlen, frischen (Metall), garen (Metall)
decarburizing Dekarbonisieren *n*
decascaler Dekadenuntersetzer *m*
decatize *v* dekatieren
decatizing Dekatieren *n,* Dekatur *f*
decatizing apparatus Dekatierapparat *m*
decatizing establishment Dekatieranstalt *f*
decatizing machine Dekatiermaschine *f*
decatoic acid Decylsäure *f*
decatylene Decatylen *n*
decausticize *v* dekaustizieren
decay Anbrüchigkeit *f,* Fäulnis *f,* Moder *m,* Verderb *m,* Verwesung *f,* Zerfall *m* (Atom), Zersetzung *f,* **liable to** ~ verweslich, **radioactive** ~ radioaktiver Zerfall *m,* **rate of** ~ Zerfallsgeschwindigkeit *f,* **smell of** ~ Moderduft *m,* **spontaneous** ~ Spontanzerfall *m* (Atom)
decay *v* ausschwingen (im Schwung nachlassen); faul werden, sich zersetzen, [ver]faulen, [ver]modern, verwesen, verwittern
decay chain, radioactive ~ radioaktive Zerfallsreihe

decay characteristic Abklingcharakteristik *f* (Akust)
decay constant Abklingkonstante *f* (Akust), Zerfallskonstante *f* (Atom)
decay curve Abklingkurve *f* (Radioaktivität), Zerfallskurve *f*
decayed anbrüchig, moderig, morsch, schwammig (Holz), verfault, zersetzt, **thoroughly** ~ rotfaul
decay electron (resulting from disintegration) Zerfallselektron *n* (Atom)
decaying Verwittern *n*
decaying *a* morsch
decay law Zerfallsgesetz *n*
decay modulus Zerfallsmodul *m*
decay path Zerfallsweg *m*
decay period Abklingzeit *f* (Akust), Zerfallszeit *f*
decay probability Zerfallswahrscheinlichkeit *f*
decay product Zerfallsprodukt *n* (Atom)
decay scheme Zerfallsschema *n*
decay sequence Zerfallsfolge *f* (Atom)
decay series Zerfallsreihe *f*
decay time Abklingzeit *f* (Akust), Zerfallszeit *f*
decelerate *v* abbremsen, abflauen, verlangsamen, verzögern
deceleration negative Beschleunigung, Bremsung *f,* Geschwindigkeitsabnahme *f,* Moderierung *f* (Atom), Verlangsamung *f,* Verzögerung *f,* ~ **of electrons** Elektronenbremsung *f*
decene Decylen *n*
decenedicarboxylic acid Decendicarbonsäure *f*
decentralization Dezentralisierung *f*
decevinic acid Decevinsäure *f*
dechenite Dechenit *m* (Min)
dechlorinate *v* entchloren, entchloren, Chlor entfernen
dechlorination Entchlorung *f,* Entfernung von Chlor, ~ **of potable water** Entchloren *n* von Trinkwasser
decibel Dezibel *n* (Phys, Maßeinheit f. Dämpfung)
decide *v* beschließen, entscheiden
decidecameter Dezidekameter *n*
deciduous tree Laubbaum *m*
deciduous wood Laubwald *m*
decigram Dezigramm *n*
deciliter Zehntelliter *n m*
decimal dezimal
decimal balance Dezimalwaage *f*
decimal classification Dezimalklassifikation *f*
decimal fraction Dezimalbruch *m*
decimal number Dezimalzahl *f*
decimal place Dezimalstelle *f*
decimal point Dezimalstrich *m,* Komma *n* (Math)
decimal power Zehnerpotenz *f*
decimal resistance Dekadenwiderstand *m* (Elektr)

decimal rheostat Dekadenrheostat *m*
decimal system Dezimalsystem *n* (Math),
Zehnersystem *n* (Math)
decimeter Dezimeter *m n*
decine Decin *n*
decinene Decin *n*
decinormal zehntelnormal (Chem)
decipher *v* entziffern
decision Entscheidung *f,* Entschluß *m,*
confirmatory ~ Bestätigungsurteil *n*
decisive ausschlaggebend
declaration Behauptung *f,* Erklärung *f*
declare *v* erklären, angeben; verzollen
declination Ablenkung *f,* Deklination *f*
declination axis Deklinationsachse *f* (Astr)
declination chart Deklinationskarte *f*
declination compass Abweichungskompaß *m*
declination map Deklinationskarte *f*
declination tides Deklinationstiden *f pl*
decline Abnahme *f,* Abschwächung *f*
decline *v* abfallen (Straße etc.), verfallen,
zurückgehen, abnehmen
declining needle Deklinationsnadel *f*
declinograph Deklinograph *m*
declivity Abhang *m,* Abschüssigkeit *f,* Neigung *f*
decoct *v* abkochen, absieden
decoctible abkochbar
decocting Absieden *n,* Auskochen *n*
decoction Abkochung *f,* Absud *m* (Brau),
Auskochung *f,* Dekokt *n,* ~ **of oak bark**
Eichenrindeabkochung *f*
decoction apparatus Auskochapparat *m*
decoction medium Abkochmittel *n*
decoction press Dekoktpresse *f*
decode *v* entziffern
decogenin Decogenin *n*
decohere *v* entfritten (Elektr)
decoherence Entfrittung *f* (Elektr)
decoherer Entfritter *m* (Elektr)
decoking Abkohlung *f*
decolor *v* entfärben
decolorant Bleichmittel *n,* Entfärber *m,*
Entfärbungsmittel *n*
decolorimeter Entfärbungsmesser *m,*
Decolorimeter *n*
decoloring Entfärben *n*
decolorization Entfärbung *f*
decolorize *v* entfärben, verfärben
decolorizing Bleichen *n,* Entfärbung *f*
decolorizing *a* entfärbend, bleichend
decolorizing agent Entfärber *m,*
Entfärbungsmittel *n*
decolorizing carbon Entfärbungskohle *f*
decolorizing filter Entfärbungsfilter *n*
decolorizing power Entfärbungsvermögen *n*
decomposability Zerlegbarkeit *f,* Zersetzbarkeit *f*
decomposable zerlegbar, zersetzbar, zersetzlich,
easily ~ leicht zersetzlich

decompose *v* zersetzen, abbauen (Chem),
aufschließen (Chem), aufspalten (Chem),
verwittern, zerfallen, zerlegen, ~ **slowly**
verwesen
decomposing Zersetzen *n,* Faulen *n,* Zerfallen *n,*
Zerlegen *n*
decomposing agent Zersetzungsmittel *n,*
Abbaumittel *n,* Aufschlußmittel *n*
decomposition Zersetzung *f,* Abbau *m* (Chem),
Auflösung *f* (Chem), Aufschließen *n* (Anal),
Aufschluß *m* (Chem), Fäulnis *f,*
Vermoderung *f,* Zerfall *m* (Chem),
Zerlegung *f,* **acid** ~ saurer Aufschluß *m*
(Chem), **alkaline** ~ alkalischer Aufschluß *m*
(Chem), **curve of** ~ Zersetzungskurve *f,* **degree
of** ~ Zerfallsgrad *m,* **double** ~
Wechselzersetzung *f,* **dry** ~ trockener
Aufschluß *m* (Chem), **heat of** ~
Zerfallswärme *f,* Zersetzungswärme *f,* **point of**
~ Zersetzungspunkt *m,* **potential of** ~
Zersetzungspotential *n,* **process of** ~
Zersetzungsvorgang *m,* Fäulnisprozeß *m,* **rate
of** ~ Zersetzungsgeschwindigkeit *f,* **thermal** ~
thermische Zersetzung, ~ **by acid**
Säureaufschluß *m,* ~ **of a force**
Kraftzerlegung *f,* ~ **of water**
Wasserzersetzung *f,* ~ **by ignition**
Glühaufschluß *m*
decomposition catalyst Zersetzungskatalysator *m*
decomposition curve Zerfallskurve *f,*
Zersetzungskurve *f*
decomposition flask Zersetzungskolben *m*
decomposition limit Zerfallsgrenze *f*
decomposition potential Zersetzungspotential *n*
decomposition process Aufschlußverfahren *n*
(Chem), Zerfallsprozeß *m*
decomposition product Zerfallsprodukt *n,*
Abbauprodukt *n* (Biol), Fäulnisprodukt *n,*
Verwitterungsprodukt *n,*
Zersetzungsprodukt *n*
decomposition reaction Zerfallsreaktion *f*
decomposition residue Zersetzungsrückstand *m*
decomposition stage Zerfallsstufe *f*
decomposition tank Zersetzungsbottich *m*
decomposition temperature Abbautemperatur *f*
decomposition value Zersetzungswert *m*
decomposition voltage Zersetzungsspannung *f*
(Galv)
decompress *v* dekomprimieren, den Druck
wegnehmen
decompression Druckverminderung *f,*
Kompressionsentspannung *f*
decompression lever Entkompressionshebel *m*
decontaminate *v* entgiften, entgasen, entseuchen
decontaminating agent Entgiftungsmittel *n*
decontamination Dekontamination *f* (Radioakt),
Entgiftung *f*
decontamination factor (or index)
Entseuchungsfaktor *m* (Atom)

decontamination plant Entgiftungsanlage *f*
decopper *v* entkupfern
decopperization Entkupferung *f* ·
decoration, overglaze ~ Verzierung *f* über der
   Glasur (Keram)
decoration film Dekorfilm *m* (Pap)
decorative coating Schmucküberzug *m*
decorative laminate Dekorationsplatte *f*
decorative paint Dekorationslack *m*
decorative sheet Dekorationsplatte *f*
decorator Dekorateur *m,* Maler *m*
decorticate *v* abrinden, entrinden, abschwarten
decorticated abgeschält (ohne Rinde oder Hülse)
decorticating Abschälen *n,* Entrinden *n*
decortication Entrindung *f*
decoside Decosid *n*
decoupling Entkoppelung *f*
decrapitating Dekrepitieren *n*
decrease Abfall *m,* Abnahme *f,* Erniedrigung *f,*
   Rückgang *m,* Schwinden *n,* Sinken *n,*
   Verkleinerung *f,* Verminderung *f,*
   Verringerung *f,* ~ **in volume**
   Volumenabnahme *f,* ~ **of solubility**
   Löslichkeitsverminderung *f,* ~ **of the**
   conductivity Rückgang *m* der Leitfähigkeit
decrease *v* abfallen, abnehmen, erniedrigen,
   herabsetzen, schwinden, sinken, [ver]mindern,
   verringern
decreasing Kleinerwerden *n*
decree *v* verordnen, beschließen
decrement Abnahme *f,* Dekrement *n* (Math),
   Verringerung *f*
decremeter Dämpfungsmesser *m* (Elektr)
decrepit altersschwach, verbraucht
decrepitate *v* verknistern (Chem)
decrepitating salt Knistersalz *n*
decrepitation Verknisterung *f*
decrepitation water Verknisterungswasser *n*
   (Chem)
decrescence allmähliche Abnahme *f,*
   Dekreszenz *f*
decrolin Dekrolin *n*
decussatin Decussatin *n*
decyl alcohol Decanol *n,* Decylalkohol *m*
decylaldehyde Decylaldehyd *m*
decylene Decylen *n*
decylenic acid Decylensäure *f*
decylic acid Decylsäure *f*
decyne Decin *n*
dedicate *v* widmen
dedication Widmung *f*
deduce *v* ableiten, folgern
deduct *v* abrechnen, abziehen
deduction Abschlag *m,* Abzug *m,* Rabatt *m*
dedulcify *v* entzuckern
deeckeite Deeckit *m* (Min)
de-emulsification Demulgierung *f*
de-enamelling Entemaillieren *n*
de-energize *v* entmagnetisieren

deep dunkel (Färb), tief
deep bed filter Bettfilter *n,* Tiefenfilter *n*
deep bed filtration Tiefenfiltration *f*
deep cherry glow Dunkelrotglut *f*
deep cooling Tiefkühlung *f*
deep-draw *v* tiefziehen
deep-drawing Tiefziehen *n,* Tiefziehverfahren *n*
deep-drawing property Tiefziehfähigkeit *f*
deep-drawing test Tiefungsversuch *m*
deep-drawn sheet metal Tiefziehblech *n*
deep-draw press Tiefziehpresse *f*
deepen *v* vertiefen, ausschachten, abteufen
   (Bergb)
deepening Vertiefung *f,* ~ **in color**
   Dunkelfärbung *f*
deep-etch test Tiefätzprobe *f*
deep-freeze grinding Gefriermahlung *f*
deep-freeze package Gefrierpackung *f*
deep freezer Tiefgefrierschrank *m,*
   Tiefkühlanlage *f,* Tiefkühlapparat *m,*
   Tiefkühltruhe *f*
deep freezing Tiefkühlen *n,* Tiefkühlung *f*
deep freezing apparatus Tiefkühlanlage *f,*
   Tiefkühlapparat *m*
deep freezing method Tiefkühlverfahren *n*
deep frozen food Tiefkühlkost *f*
deep-groove ball bearing Rillenkugellager *n*
deep mine workings Tiefbau *m* (Bergb)
deerskin polishing disk Wildlederpolierscheibe *f*
deexcitation Abregung *f*
deexcitation photon Abregungsphoton *n*
defat *v* entfetten
defatting Fettentziehung *f*
default Nichterscheinung *f,* Unterlassung *f*
defecating pan Scheidepfanne *f*
defecation Läuterung *f,* Defäkation *f,*
   Reinigung *f,* Scheidung *f* (Zucker)
defecation scum Scheideschlamm *m* (Zucker)
defecation slime Scheideschlamm *m* (Zucker)
defecation tank Scheidepfanne *f*
defecator Dekanteur *m* (Zucker),
   Läuterkessel *m,* Läuterpfanne *f,* Reiniger *m,*
   Scheidepfanne *f* (Zucker), Scheidungsfilter *m*
defect Defekt *m,* Fehler *m,* Fehlordnung *f,*
   Fehlstelle *f,* Makel *m,* Störstelle *f*
defect conductor Defektleiter *m,* Mangelleiter *m*
defect electron Defektelektron *n*
defect generation Fehlstellenerzeugung *f*
defective baufällig, beschädigt, fehlerhaft,
   lückenhaft, mangelhaft, unvollkommen
defective casting Fehlguß *m,* Gußschaden *m*
defectiveness Fehlerhaftigkeit *f*
defective portion Fehlerstelle *f*
defect mobility Störstellenbeweglichkeit *f*
defence mechanism Abwehrmechanismus *m*
defence reaction Abwehrreaktion *f* (Biol)
defend *v* verteidigen, beschützen
defensive patent Abwehrpatent *n*
defer *v* verschieben, aufschieben

deferrization Enteisenung *f*
defibrator Entwässerungspresse *f* (Pap)
defibrinated defibriniert
deficiency Unzulänglichkeit *f,* Defekt *m,*
Fehlen *n,* Fehler *m,* Lücke *f,*
Mangelhaftigkeit *f,* ~ **in weight**
Mindergewicht *n*
deficiency disease Mangelkrankheit *f* (Med)
deficiency of air Luftmangel *m*
deficiency symptom Ausfallerscheinung *f*
deficient mangelhaft, unzulänglich
deficit Ausfall *m,* Defizit *n*
definable definierbar, abgrenzbar
define *v* definieren, abgrenzen, begrenzen,
präzisieren
defined, as ~ definitionsgemäß
defining equation Bestimmungsgleichung *f*
(Math)
definite bestimmt, endgültig, definitiv; begrenzt
definition Begriff *m,* Begriffsbestimmung *f,*
Definition *f,* **according to** ~
definitionsgemäß
definition range Definitionsbereich *m*
deflagrability Abbrennbarkeit *f,* Brennbarkeit *f,*
Deflagrierbarkeit *f*
deflagrate *v* deflagrieren, aufflackern,
explosionsartig verbrennen, verpuffen
deflagrated abgebrannt (Chem)
deflagrating Abbrennen *n*
deflagrating jar Abbrennglocke *f*
deflagrating spoon Abbrennlöffel *m,*
Phosphorlöffel *m*
deflagration Verpuffung *f,* Abbrennung *f,*
Aufflackern *n,* Deflagration *f*
deflashing machine Entgratungsmaschine *f*
deflate *v* entleeren (Gas)
deflated luftleer
deflect *v* ablenken, abweichen, ausschlagen
(Zeiger), durchbiegen, ~ **inwards** einbeulen
deflectability Ablenkbarkeit *f*
deflectable durchbiegungsfähig
deflected abgelenkt (Opt)
deflecting coil Ablenkspule *f* (Elektr)
deflecting electrode Ablenkelektrode *f*
deflecting force Ablenkungskraft *f*
deflecting magnet Ablenkungsmagnet *m*
deflecting magnetic field
Ablenkungsmagnetfeld *n*
deflecting system Ablenksystem *n*
deflection Abbiegung *f,* Ablenkung *f,*
Abweichung *f,* Beugung *f* (Lichtstrahlen),
Biegung *f,* Durchbiegung *f,* Durchfederung *f,*
Ausschlag *m* (der Nadel), **angle of** ~
Ablenkungswinkel *m,* Abprallwinkel *m,*
Ausschlagwinkel *m,* **rate of** ~
Durchbiegungsgeschwindigkeit *f,* **vertical** ~
Vertikalablenkung *f,* ~ **from the vertical**
Lotabweichung *f,* ~ **of a beam** Balkenbiegung
f, ~ **of the needle**

Nadelabweichung *f,* Nadelausschlag *m,* ~ **to
the right** Rechtsablenkung *f*
deflection aberration Ablenkfehler *m*
deflection astigmatism Ablenkastigmatismus *m*
deflection distortion Ablenkverzeichnung *f*
deflection hardness Auslenkhärte *f*
deflection indicator Biegungsmesser *m*
deflection multiplier Ablenkungsvervielfacher *m*
deflection registration Ablenkregistrierung *f*
deflection test Biegeversuch *m,*
Durchbiegeversuch *m*
deflectometer Biegungsmesser *m*
deflector Ablenkelektrode *f,* Ablenker *m,*
Prallschirm *m*
deflector field Ablenkfeld *n*
deflocculate *v* entflocken
defoamer Antischaummittel *n,*
Entschäumungsmittel *n*
defoaming Schaumzerstörung *f*
defoaming agent Entschäumungsmittel *n,*
Demulgator *m,* Entschäumer *m*
defocus *v* defokussieren, entbündeln
Defo hardness Defohärte *f*
defoliant Entlaubungsmittel *n*
defoliate abblättern, entlauben (Bot)
deforest *v* abholzen
deform *v* deformieren, umformen, verformen,
verstümmeln, verzerren, verziehen
deformability Deformierbarkeit *f,* Formbarkeit *f,*
Formveränderungsvermögen *n,*
Verformbarkeit *f*
deformable [ver]formbar, deformierbar,
deformierbar, verbiegbar
deformation Deformation *f,* Deformierung *f,*
Form[ver]änderung *f,* Gestaltsveränderung *f,*
Verbiegung *f,* Verformung *f,* deformierende
Wirkung, **capacity for** ~
Formänderungsvermögen *n,* **cold** ~
Kaltverformung *f,* **elastic** ~ elastische
Formänderung, **measure of** ~
Deformationsmaß *n,* **mechanism of** ~
Deformationsmechanismus *m,*
Verformungsmechanismus *m,* **plastic** ~
plastische Verformung, **range of** ~ gesamte
Formänderung, **rate of** ~
Deformationsgeschwindigkeit *f,* **region of** ~
Formänderungsbereich *m,* **resistance to** ~
Deformationswiderstand *m,*
Verformungswiderstand *m,* **work of** ~
Formänderungsarbeit *f,* **resistant to** ~
formbeständig
deformation resistance
Formänderungswiderstand *m*
deformation [strain] Verzerrung *f* (Mech)
deformation tensor Verzerrungstensor *m*
deformation value Defowert *m*
defray *v* bestreiten (Kosten)
defrost *v* abtauen, auftauen, enteisen
defroster Enteisungsanlage *f,* Entfroster *m*

degas *v* entgasen
degasification Entgasung *f,* heat of ~
Entgasungswärme *f*
degasification agent Entgasungsmittel *n*
degasify *v* entgasen
degassing Entgasen *n,* Entgasung *f,* ~ of the
furnace Entgasung *f* des Ofens
degassing furnace Entgasungsofen *m*
degassing plant Entgasungsanlage *f*
degate *v* den Anguß *m* entfernen
degauss *v* entmagnetisieren
degaussing Entmagnetisierung *f*
degeneracy Degeneration *f,* Entartung *f,* ~ of
ground state Entartung *f* des Grundzustandes,
~ of spin states Entartung *f* von
Spinzuständen
degenerate ausgeartet, degeneriert
degenerate *v* degenerieren, entarten,
verkümmern
degenerate electron gas entartetes
Elektronengas *n*
degenerate electron state entarteter
Elektronenzustand *m*
degenerate species Abart *f* (Biol)
degeneration Degeneration *f,* Entartung *f,*
Verkümmerung *f,* phenomenon of ~
Degenerationserscheinung *f*
degerminate *v* entkeimen
degermination Entkeimung *f* (Getreide)
degeroite Degeroit *m*
degradation Abbau *m* (Chem), Degeneration *f*
(Biol), Herabsetzung *f*
degradation energy Dissipationsenergie *f* (Mech)
degradation of energy Energieverminderung *f,*
law of ~ Entropiesatz *m* (Therm)
degradation product Abbauprodukt *n* (Biol)
degradation with acid Säureabbau *m*
degrade *v* abbauen (Chem)
degraded abgebaut
degras Degras *n,* Lederschmiere *f*
degrease *v* entfetten, abfetten, ~ cold *v* kalt
entfetten, ~ hot *v* heiß entfetten
degreased entfettet
degreasing Entfetten *n,* Fettentziehung *f,* ~ of
wool Wollentfettung *f*
degreasing *a* entfettend
degreasing agent Entfettungsmittel *n*
degreasing apparatus Entfettungsapparat *m*
degreasing bath Entfettungsbad *n*
degreasing plant Entfettungsanlage *f*
degreasing process Entfettungsprozeß *m,*
Entfettungsvorgang *m*
degree Abstufung *f,* Grad *m* (Temperatur,
Winkel), Rang *m,* Stufe *f,* tenth of a ~
Zehntelgrad *m*
degree below zero Kältegrad *m*
degree centigrade Celsiusgrad *m*

degree Kelvin Kelvingrad *m*
degree Oechsle Oechsle-Grad *n*
degree of compaction Lagerungsdichte *f*
degree of fineness by sieving Sichtfeinheit *f*
degree of formation Bildungsgrad *m*
degree of freeness Mahlgrad *m* (Pap)
degree of heat Hitzegrad *m*
degree of purity Reinheitsgrad *m*
degree of softness Weichheitsgrad *m*
degrees, by ~ stufenweise
deguelidiol Deguelidiol *n*
deguelin Deguelin *n*
degum *v* degummieren, entbasten,
entgummieren, entleimen
degumming Entbasten *n* (Text), Kotonisieren *n*
(Text)
degumming bath Entbastungsbad *n* (Text),
Entbastungsflotte *f* (Text)
degumming yield Entbastungsausbeute *f* (Text)
dehalogenate *v* enthalogenieren
dehalogenation Dehalogenierung *f*
dehumidification Feuchtigkeitsentzug *m,*
Trocknen *n*
dehumidifier Trockenmittel *n*
dehydracetic acid Dehydracetsäure *f*
dehydrase Dehydrase *f* (Dehydrogenase)
dehydratase Dehydratase *f*
dehydrate *v* dehydratisieren, dörren, entwässern,
Wasser *n* entziehen
dehydrated entwässert, wasserfrei
dehydrated alum entwässerter Alaun
dehydrating wasserentziehend
dehydrating agent Dehydratisierungsmittel *n,*
Trockenmittel *n,* Trocknungsmittel *n,*
Wasserentziehungsmittel *n*
dehydration Absolutierung *f,* Dehydratation *f,*
Dehydratisierung *f,* Entwässerung *f,*
Wasserentziehung *f,* ~ of oil Öltrocknung *f*
dehydration plant Entwässerungsanlage *f*
dehydration unit Trockenpatrone *f*
dehydroabietic acid Dehydroabietinsäure *f*
dehydroamaric acid Dehydroamarsäure *f*
dehydroascorbic acid Dehydroascorbinsäure *f*
dehydrobenzene Dehydrobenzol *n*
dehydroberberine Dehydroberberin *n*
dehydrobilic acid Dehydrobilinsäure *f*
dehydrobrucinolone Dehydrobrucinolon *n*
dehydrocamphenic acid Dehydrocamphensäure *f*
dehydrocamphenilanic acid
Dehydrocamphenilansäure *f*
dehydrocamphenylic acid
Dehydrocamphenylsäure *f*
dehydrocamphoric acid Dehydrocamphersäure *f*
dehydrocauprene Dehydrokaupren *n*
dehydrocholeic acid Dehydrocholeinsäure *f*
dehydrocholesterol Dehydrocholesterin *n*
dehydrocholic acid Dehydrocholsäure *f*
dehydrocorticosterone Dehydrocorticosteron *n*
dehydrocortisone Dehydrocortison *n*

dehydrocorycavamine Dehydrocorycavamin *n*
dehydrocorydaline base Dehydrocorydalinbase *f*
dehydrocryptopidene Dehydrokryptopiden *n*
dehydrocyclization Dehydrocyclisierung *f*
dehydrodesoxycholic acid
    Dehydrodesoxycholsäure *f*
dehydrodimerization Dehydrodimerisation *f*
dehydroemetine Dehydroemetin *n*
dehydroemodineanthranol
    Dehydroemodinanthranol *n*
dehydroepiandrosterone
    Dehydroepiandrosteron *n*
dehydrofenchocamphoric acid
    Dehydrofenchocamphersäure *f*
dehydrofenchoic acid Dehydrofenchosäure *f*
dehydrofluorindine Dehydrofluorindin *n*
dehydrogenase Dehydrogenase *f* (Biochem)
dehydrogenate *v* dehydrieren, von Wasserstoff
    befreien
dehydrogenating dehydrogenierend,
    wasserstoffentziehend
dehydrogenation Dehydrierung *f*,
    Wasserstoffentziehung *f*
dehydrogenation plant Dehydrierungsanlage *f*,
    DHD-Anlage *f*
dehydrogenization Dehydrierung *f*, Entzug *m*
    von Wasserstoff
dehydrogenize *v* dehydrieren, Wasserstoff
    entziehen
dehydroglaucine Dehydroglaucin *n*
dehydrohalogenation Dehydrohalogenierung *f*,
    Halogenwasserstoffabspaltung *f*
dehydrohydantoic acid Dehydrohydantoinsäure *f*
dehydroisodypnopinacol
    Dehydroisodypnopinakol *n*
dehydroisolapachone Dehydroisolapachon *n*
dehydrolaurolenic acid Dehydrolaurolensäure *f*
dehydrolithocholic acid Dehydrolithocholsäure *f*
dehydromovraic acid Dehydromovrasäure *f*
dehydromucic acid Dehydroschleimsäure *f*
dehydronaphthol Dehydronaphthol *n*
dehydronerolidol Dehydronerolidol *n*
dehydrophellandrene Dehydrophellandren *n*
dehydroprogesterone Dehydroprogesteron *n*
dehydroquinacridone Dehydrochinacridon *n*
dehydroquinine Dehydrochinin *n*
dehydrothiotoluidine Dehydrothiotoluidin *n*
dehydrothymol Dehydrothymol *n*
dehydroxycodeine Dehydroxykodein *n*
deice *v* enteisen
de-icer Entfroster *m*
de-inking solution Entfärbelösung *f*, Entfärber *m*
deionization Deionisation *f*, Entionisierung *f*
de-ionize *v* entionisieren
de-ionizing property Deionisiervermögen *n*,
    Entionisiervermögen *n*
de-ironing Enteisenung *f*
dekalin Dekalin *n*, Naphthan *n*
delafossite Delafossit *m* (Min)

delaminate *v* in Schichten abblättern
delamination Abblätterung *f*, Entleimung *f*,
    Schichtentrennung *f*
delatynite Delatynit *m* (Min)
delay Aufschub *m*, **without** ~ unverzüglich
delay *v* aufschieben, vertagen, verzögern,
    zeitlich verlegen
delay demodulator Verzögerungsdemodulator *m*
    (Elektr)
delayed action Verlangsamung *f*, Verzögerung *f*
delead entbleien
delessite Delessit *m* (Min)
delf[t] Fayence *f*
deliberate *v* beratschlagen
delicacy Feinheit *f*, Zartheit *f*
delicate zart, delikat, schmackhaft
delime *v* abkalken, entkalken
deliming Entkalken *n*, Entkalkung *f*
deliming agent Entkalkungsmittel *n*
delimit *v* abgrenzen
delimitation Abgrenzung *f*
delineate *v* skizzieren; abgrenzen
deliquesce *v* zerfließen
deliquescence Zerfließbarkeit *f*, Zerfließen *n*
deliquescent zerfließend
deliquescent property Zerfließbarkeit *f*
deliver *v* abgeben (Phys), entbinden (Med),
    liefern
delivery Entbindung *f* (Med), Lieferung *f*, **scope**
    of ~ Lieferumfang *m*, **time of** ~ Lieferfrist *f*
delivery air vessel Druckwindkessel *m*
delivery bill Lieferschein *m*
delivery cock Ablaßhahn *m*
delivery head Förderhöhe *f*, Zuführungskopf *m*
delivery pipe Ableitungsröhre *f*, Ausgußrohr *n*,
    Ausströmungsrohr *n*; Druckleitung *f*,
    Druckrohr *n*, Steigrohr *n*
delivery pipet[te] Abfüllpipette *f*
delivery pressure Förderdruck *m*
delivery socket-pipe Muffendruckrohr *n*
delivery temperature Abgabetemperatur *f*
delivery tube Abgaberohr *n*, Abzugsrohr *n*,
    Einleitungsrohr *n*
delivery valve Ablaßventil *n*
delocalisation energy Delokalisationsenergie *f*
delocalization Delokalisierung *f*
delocalization energy Mesomerieenergie *f*
    (Chem)
delocalize *v* delokalisieren
delocanic acid Delokansäure *f*
delorenzite Delorenzit *m* (Min)
delphidan Delphidan *n*
delphinate Delphinat *n*
delphin blue Delphinblau *n*
delphinic acid Delphinsäure *f*, **salt of** ~
    Delphinsalz *n*, **salt or ester of** ~ Delphinat *n*
delphinidine Delphinidin *n*
delphinin Delphinin *n*
delphinite Delphinit *m*

delphinoidine Delphinoidin *n*
delphisine Delphisin *n*
delphonine Delphonin *n*
delsine Delsin *n*
delta Dreieck *n* (Elektr)
delta connection Dreieckschaltung *f* (Elektr)
delta current Dreieckstrom *m*
delta function Deltafunktion *f*
delta high-tension insulator Deltaisolator *m* (Elektr)
deltaline Deltalin *n*
delta metal Deltametall *n*
deltamine Deltamin *n*
delta purpurine Deltapurpurin *n*
delta rays Deltastrahlen *pl* (Phys)
delta ring seal Deltaring *m*
delta voltage Dreiecksspannung *f*
delta voltage sweep method Dreieckspannungsmethode *f*
deltohedron Deltoeder *n* (Krist)
deltoiddodecahedron Deltoiddodekaeder *n* (Krist)
deltruxic acid Deltruxinsäure *f*
delusion, optical ~ optische Täuschung
deluster *v* mattieren
delusterant Mattierungsmittel *n*
delustered matt[iert]
delustering Mattieren *n*, Mattierung *f*
delustering agent Mattierungsmittel *n*
delvauxite Delvauxit *m* (Min)
demagnetization Entmagnetisierung *f*
demagnetize *v* entmagnetisieren
demagnetizing factor Demagnetisierungsfaktor *m*, Entmagnetisierungsfaktor *m*
demand Anforderung *f*, Anspruch *m*, Bedarf *m*, **to satisfy the** ~ den Bedarf decken
demand *v* anfordern, beantragen
demand for energy Energiebedarf *m*
demand setter Leistungsvorwahlschalter *m*
demanganization Entmangan[ier]ung *f*
demanganizing of water Wasserentmangan[ier]ung *f*
demantoid Demantoid *m* (Min)
demarcate *v* abgrenzen
demarcation Abgrenzung *f*
demarcation line Demarkationslinie *f*
demargarinating process Demargarinierungsprozeß *m*
dematerialization Entmaterialisierung *f* (Atom)
demecolcine Demecolcin *n*
demerol Demerol *n*
demethylate *v* demethylieren, entmethylieren
demethylation Demethylierung *f*
demidowite Demidowit *m*
demigranite Halbgranit *m* (Min)
demijohn Korbflasche *f* (Chem), Ballon *m* (Chem)

demineralization Entmineralisierung *f*, Entsalzung *f*
demineralization of water Wasserentsalzung *f*
demineralize *v* entmineralisieren, entsalzen (Quellwasser)
demineralizer Entsalzungsvorrichtung *f*
demissidine Demissidin *n*
demissine Demissin *n*
demisting plant Entnebelungsanlage *f*
demodulator Demodulator *m* (Elektr)
demolish *v* [ab]sprengen, einreißen, niederreißen, zerstören, zertrümmern
demolition Abbruch *m*, Zerstörung *f*, Zertrümmerung *f*, **due for** ~ abbruchreif
demolition work Abbrucharbeit *f*
demonstrate *v* beweisen, demonstrieren, vorführen, zeigen
demonstration Beweisführung *f*, Demonstration *f*, Vorführung *f*
demonstration model Lehrmodell *n*
demonstration transformer Experimentiertransformator *m*
demulcent Milderungsmittel *n* (Med)
demulsification Entmischung *f*
demulsifier Demulgator *m*
demulsify *v* entmischen
demulsifying Entmischen *n*
denaturant Denaturant *m* (Atom), Denaturierungsmittel *n*, Vergällungsmittel *n*
denaturation Denaturierung *f*, Vergällung *f*
denature *v* denaturieren, vergällen
denatured denaturiert
denaturing Denaturieren *n*, Vergällen *n*, **of alcohol** ~ Alkoholvergällung *f*
denaturization Denaturierung *f*
denaturize *v* denaturieren
denaturizing Denaturieren *n*
denaturizing agent Denaturierungsmittel *n*
dendrite Dendrit *m* (Min)
dendritic baumähnlich, verzweigt
dendritic agate Baumachat *m* (Min), Dendrachat *m*, Dendritenachat *m* (Min)
dendritic formation Dendritenbildung *f*
dendritic structure Dendritenstruktur *f*
dendroketose Dendroketose *f*
dendrology Dendrologie *f*, Holzkunde *f*
denial Verweigerung *f*
denicotinize *v* entnikotinisieren
denier Denier *n* (Maß)
denier variation Titerabweichung *f* (Text)
denitrate *v* denitrieren
denitrating Denitrieren *n*
denitrating agent Denitriermittel *n*
denitrating bath Denitrierbad *n*
denitrating plant Denitrieranlage *f*
denitration Denitrierung *f*
denitrification Denitrifikation *f*, Denitrifizierung *f*
denitrify *v* denitrifizieren

denomination Benennung f (Math)
denominator Nenner m (Math), **lowest common**
~ kleinster gemeinsamer Nenner
denote v bedeuten
dense dicht, undurchdringlich
dense-media cyclone Waschzyklon m
denseness Dichte f
densified verdichtet
densified laminated wood Preßschichtholz n
densified wood Preßholz n
densimeter Densimeter n, Aräometer n,
Dichtemesser m, Hydrometer n
densimetric densimetrisch
densipimaric acid Densipimarsäure f
densitometer Densitometer n,
Schwärzungsmesser m
density Dichte f, Dichtigkeit f, Schwärzung f
(Phot), **alteration of** ~ Dichteänderung f,
determination of ~ Dichtebestimmung f
density anisotropy Dichteanisotropie f
density balance Dichtewaage f
density by volume Raumdichte f
density determination Dichtebestimmung f
density distribution Dichteverteilung f
density field Dichtefeld n
density gradient Dichtegradient m
density-gradient centrifugation
Dichtegradient-Zentrifugation f
density graduation Schwärzungsabstufung f
(Opt)
density number Dichtezahl f
density of charge Ladungsdichte f
density of the brine Grädigkeit f der Sole
density range Schwärzungsbereich m
density value Dichtewert m
denso band Densobinde f (Densoschutzbinde)
densograph Densograph m
dent Beule f, Delle f, Einbeulung f, Kerbe f
dental zahnärztlich
dental calculus Zahnstein m
dental caries Zahnkaries f
dental cement Zahnkitt m
dental gas Lachgas n (Chem, Med)
dentalite Dentalith m
dental material Dentalmaterial n
dental mold Zahngipsabdruck m
dental plaque Zahnbelag m
dental plaster Dentalgips m
dentate gezahnt, zähnig
dented ausgekerbt
dentine Dentin n, Zahnsubstanz f
denting stress Beulspannung f
dentinification Dentinbildung f (Zahnmed)
dentinogenic dentinbildend (Zahnmed)
dentinogenous dentinbildend (Zahnmed)
dentist Zahnarzt m
dentistry Zahnheilkunde f
denudation Abtragung f (Geol)
deny v verweigern

deodorant Desodorisationsmittel n
deodorants De[s]odorantien pl
deodorization Desodor[is]ierung f,
Geruchfreimachen n, Geruchlosmachung f,
Geruchsbekämpfung f, Geruchsverbesserung f
deodorize v desodorieren
deodorizer Desodorisationsmittel n
deodorizing geruchsbeseitigend
deodorizing plant Desodorieranlage f
deoil v entölen, degrassieren
deoiler Fettabscheider m, Fettausscheider m
deoiling Entölung f
deoiling plant Entölungsanlage f,
Ölextraktionsanlage f
deoxidant Desoxidationsmittel n
deoxidate v desoxidieren, desoxydieren
deoxidating Desoxidation f, Desoxidieren n
deoxidating agent Desoxidationsmittel n
deoxidation Desoxidation f,
Sauerstoffentziehung f
deoxidize v desoxidieren, dekapieren (Met),
entsäuern, Sauerstoff m entziehen
deoxidizer Desoxidationsmittel n (Chem)
deoxidizing Desoxidieren n
deoxidizing agent Desoxidationsmittel n
deoxidizing slag Reduktionsschlacke f
deoxy Desoxy-
deoxyfulminuric acid Desoxyfulminursäure f
deoxygenate v Sauerstoff m entziehen
deoxygenation Entziehung f von Sauerstoff,
Sauerstoffentzug m
deoxyhemoglobin Desoxyhämoglobin n
deoxyinosine triphosphate
Desoxyinosintriphosphat n
deoxymyoglobin Desoxymyoglobin n
deoxyribonuclease Desoxyribonuclease f
deoxyribose Desoxyribose f
deoxy sugar Desoxyzucker m
deoxyxanthine Desoxyxanthin n
department Abteilung f
departmental manager Abteilungsleiter m
Department of Agriculture (Am. E.)
Landwirtschaftsministerium n
department of development
Entwicklungsabteilung f
department-of-health regulations
gesundheitspolizeiliche Vorschriften pl
departure Abflug m (Luftf)
depend v abhängen
dependability Verläßlichkeit f
dependable verläßlich
dependence Abhängigkeit f
dependent abhängig
dephanthanic acid Dephanthansäure f
dephanthic acid Dephanthsäure f
dephased außer Phase, verschiedenphasig
dephased condition Phasenverschiebung f
dephenolation plant Entphenolungsanlage f
dephenolization Entphenolung f

dephlegmate *v* rektifizieren, dephlegmieren
dephlegmation Dephlegmation *f*
dephlegmator Dephlegmator *m*
dephosphorate *v* entphosphoren
dephosphoring period Entphosphorungsperiode *f*
dephosphorization Entphosphorung *f*
dephosphorize *v* entphosphoren, von Phosphor
  befreien
dephosphorizing period
  Entphosphorungsperiode *f*
depilate *v* enthaaren
depilating Enthaaren *n*
depilation Depilation *f*, Enthaarung *f*, **enzymatic**
  ~ Enzymenthaarung *f* (Gerb)
depilatory Depilatorium *n*, Enthaarungsmittel *n*,
  Haarbeize *f*
depilatory *a* enthaarend
depletion Abreicherung *f* (Atom), Verarmung *f*
depletion of ions Ionenverarmung *f*
depolarization Depolarisation *f*,
  Entpolarisierung *f*
depolarization factor Depolarisationsgrad *m*
depolarize *v* depolarisieren, entpolarisieren
depolarizer Depolarisator *m* (Elektr)
depolarizing ability Depolarisationsfähigkeit *f*
depolymerization Depolymerisation *f*,
  Depolymerisierung *f*, Entpolymerisierung *f*
depolymerize *v* depolymerisieren
deposit Ablagerung *f*, Absatz *m* (Niederschlag),
  Abscheidungsprodukt *n*, Ansatz *m* (Belag),
  Auflage *f* (Galv), Aufschüttung *f* (Geol),
  Ausfällung *f* (Niederschlag), Beschlag *m*,
  Bodensatz *m* (Niederschlag),
  Fällungsprodukt *n* (Niederschlag),
  Kesselschlamm *m*, Lagerstätte *f* (Bergb),
  Satz *m* (Niederschlag), Sediment *n*
  (Niederschlag), Vorkommen *n* (Min), **active**
  ~ aktiver Niederschlag *m*, **alluvial** ~
  angeschwemmte Lagerstätte *f*, **amount of** ~
  Ablagerungsmenge *f*, **keeping back the** ~
  Aussparen *n* des Niederschlags, **muddy** ~
  schlammiger Niederschlag, **weight of the** ~
  Niederschlagsgewicht *n*, ~ **from hot springs**
  Sprudelstein *m*
deposit *v* abscheiden (Galv), absetzen, absitzen,
  anlegen, ansetzen (Schmutz), ausscheiden
  (Chem), aussedimentieren, Bodensatz *m*
  ablagern, niederschlagen
deposit attack Belagkorrosion *f*
deposited abgelagert (Niederschlag),
  ausgeschieden (Niederschlag)
depositing Absetzen *n*
depositing of mud Anschlämmen *n*
depositing tank Reinigungsbassin *n*
deposition Ablagerung *f*, Absetzung *f*,
  Ausfällung *f*, Sedimentbildung *f*
depositional environment Ablagerungsbereich *m*
deposition potential Abscheidungspotential *n*
deposition tube Absitzrohr *n*

deposit mud *v* anschlämmen
deposit of ice Eisansatz *m*
deposit of oxide films Ablagerung *f* von
  Oxidkrusten
deposit of scale Wassersteinansatz *m*
deposit tracking Kriechwegbildung *f* durch
  leitende Ablagerung
deposit welding Auftragschweißung *f*
depot Ablage *f*, Depot *n*, Lagerplatz *m*
depot drug Depotpräparat *n* (Pharm)
depot fat Depotfett *n*, Fettreservemittel *n*
depot insulin Depotinsulin *n*
depot iron Depoteisen *n*
depot protein Depoteiweiß *n*
depreciate *v* entwerten
depreciation Herabsetzung *f* des Werts
depress *v* herunterdrücken, senken
depressant Bremsmittel *n* (Flotation),
  Sinkmittel *n* (Flotation)
depression Depression *f*, Druckerniedrigung *f*,
  Tiefdruck *m* (Meteor), Unterdruck *m*,
  Vertiefung *f*
depression box Unterdruckkammer *f*
depression chamber Unterdruckkammer *f*
depressor test Depressortest *m*
deprivation Entziehung *f*
deprive *v* berauben, entziehen, entblößen, ~ **of**
  tar *v* entteeren
deprived [of] entblößt
depriving Entziehen *n*
depsane Depsan *n*
depsene Depsen *n*
depsidan Depsidan *n*
depside Depsid *n*
depsipeptide Depsipeptid *n*
depth Tiefe *f*
depth adjustment Tiefeneinstellung *f* (Opt)
depth gauge Tiefenlehre *f*, Ahming *f* (Schiff),
  Pegel *m*
depth indicator Tiefenmesser *m*, Pegel *m*
depth of cavity Lochtiefe *f*
depth of cut Spantiefe *f*
depth of focus Tiefenschärfe *f* (Phot), **lack of** ~
  Tiefenunschärfe *f* (Opt)
depth of the color Tiefe *f* der Färbung
depulp *v* entpulpen
depurative blutreinigend
dequalinium Dequalinium *n*
derbylite Derbylith *m* (Min)
Derbyshire spar, blue ~ Saphirfluß *m* (Min)
de-repression Derepression *f*
deresinate *v* entharzen
deresination Entharzung *f*
dericin oil Dericinöl *n*
derivable ableitbar
derivate abgeleitete Funktion *f* (Math),
  Derivat *n*
derivation Ableitung *f* (Math), Abstammung *f*,
  Abzweigung *f*

derivation law Ableitungsgesetz *n*
derivative Abkömmling *m* (Chem), Derivat *n*
(Chem), **higher** ~ höhere Ableitung *f,*
logarithmic ~ logarithmische Ableitung *f,*
partial ~ partiale Ableitung *f*
derivative action differenzierend wirkender
Einfluß *m* (Regeltechn), Vorhalt *m*
(Regeltechn)
derivative action time (Brit. E.) Vorhaltzeit *f*
derivative of a function Ableitung *f* einer
Funktion (Math)
derive [from] *v* ableiten (Math)
derive advantage [from] *v* Nutzen *m* ziehen [aus]
derived, to be ~ **from** abstammen, herstammen
derived circuit Abzweigstromkreis *m*
derived current Abzweigstrom *m*
dermatin Dermatin *n*
dermatol Dermatol *n*, Wismutsubgallat *n*
dermatologist Dermatologe *m* (Med)
dermatology Dermatologie, Hautlehre *f* (Med)
dermatolysis Dermatolyse *f* (Med)
dermatomyces Dermatophyt, Hautpilz *m* (Bakt)
dermatopathia Hautkrankheit *f* (Med)
dermatophyte Dermatophyt *m* (Bakt),
Hautpilz *m* (Bakt)
dermatoplastic hautplastisch
dermatoplasty Hautplastik *f*
dermatosis Dermatose *f,* Hautkrankheit *f*
dermatozoon Hautparasit *m*
dermocybine Dermocybin *n*
derric acid Derrsäure *f*
derrick Bohrturm *m*
derrick [crane] Derrickkran *m*
derricking jib crane Auslegerdrehkran *m*
derrid Derrid *n*
de-rust *v* entrosten
de-rusting Entrosten *n*
desaccharification Entzuckerung *f*
desalination Entsalzung *f*
desalination of seawater Meerwasserentsalzung *f*
desalination of water Wasserentsalzung *f*
desalt *v* entsalzen
desalting, crude oil ~ Entsalzen *n* von Rohöl
desalting of water Wasserentsalzung *f*
desaminochondrosamine
Desaminochondrosamin *n*
de-saponify *v* entseifen
desarogenin Desarogenin *n*
desaspidin Desaspidin *n*
desaspidinol Desaspidinol *n*
desaurine Desaurin *n*
descale *v* dekapieren, entkrusten, entzundern
descaling Entzundern *n*
descend *v* abstammen, abwärts bewegen,
abwärts gleiten, [her]absteigen
descendant Abkömmling *m*
descent Abkunft *f,* Abstieg *m,* Gefälle *n* (einer
Straße), Gefällestrecke *f,* **method of steepest** ~
Methode *f* des steilsten Abstiegs

descloizite Descloizit *m* (Min)
description Beschreibung *f,* Bezeichnung *f,*
Erläuterung *f*
descriptive anschaulich
desensitization Desensibilisierung *f* (Phot)
desensitize *v* desensibilisieren (Phot)
desensitizer Desensibilisator *m*
deserpidinol Deserpidinol *n*
desiccant Exsikkatorfüllung *f,* Trockenmittel *n,*
Trocknungsmittel *n*
desiccate *v* trocknen, abdarren (Früchte),
ausdörren, austrocknen, eintrocknen,
entwässern (im Exsikkator), Wasser *n*
entziehen
desiccated getrocknet, eingedörrt, wasserfrei
(Chem)
desiccated medium Trockenmittel *n,*
Trockennährboden *m*
desiccating Trocknen *n,* Abdörren *n*
desiccating *a* wasserentziehend
desiccation Austrocknung *f,* Eintrocknung *f,*
Entwässern *n,* Trocknung *f,* Vertrocknung *f,*
Wasserentziehung *f*
desiccator Exsikkator *m* (Chem), Entfeuchter *m,*
Lufttrockner *m,* Trockner *m*
desiccator cabinet Trockenschrank *m*
desiccator cover Exsikkatoraufsatz *m*
desiccator plate Exsikkatoreinsatz *m*
design Entwurf *m,* Dessin *n,* Gestaltung *f,*
Plan *m,* Skizze *f,* Zeichnung *f*
design *v* entwerfen, gestalten, zeichnen
designate *v* bezeichnen, benennen, beschriften
designation Benennung *f,* Beschriftung *f,*
Bezeichnung *f*
design elements Entwurfsbestandteile *m pl*
designer Gebrauchsgraphiker *m,*
Konstrukteur *m,* Musterausnehmer *m*
designing engineer Konstrukteur *m*
designing machine Schablonenmaschine *f*
design of experiments Versuchsplanung *f*
design strength Gestaltfestigkeit *f*
desilicate *v* entkieseln
desilicating Entkieseln *n*
desilicification Entkieselung *f*
desilicification process Entkieselungsverfahren *n*
desilicify *v* entkieseln
desiliconization Entsilizierung *f*
desiliconize *v* entsilizieren
desilver *v* entsilbern
desinter *v* entsintern
desirable wünschenswert
desirous begierig
desist *v* abstehen
desk Pult *n*
deslag *v* entschlacken
deslagging Entschlacken *n*
deslime *v* entschleimen
desludger Schlammseparator *m*
desludging Schlammabscheidung *f*

desmin Desmin *m* (Min)
desmo-enzyme Desmoenzym *n* (Biochem)
desmolase Desmolase *f* (Biochem)
desmosine Desmosin *n*
desmosterol Desmosterin *n*
desmotropic desmotrop
desmotropism Desmotropie *f*
desmotropoartemisine Desmotropoartemisin *n*
desmotroposantonine Desmotroposantonin *n*
desmotropy Desmotropie *f* (Chem)
desomorphine Desomorphin *n*
desorb *v* desorbieren
desorption Desorption *f*
desorption column or tower Abblasekolonne *f*
desosamine Desosamin *n*
desoxindigo Desoxindigo *n*
desoxine Desoxin *n*
de[s]oxy Desoxy-
de[s]oxyalizarin Desoxyalizarin *n*
de[s]oxyallocaffuric acid Desoxyallokaffursäure *f*
de[s]oxyandrographolide
 Desoxyandrographolid *n*
de[s]oxyanisoine Desoxyanisoin *n*
de[s]oxybiliaic acid Desoxybiliansäure *f*
de[s]oxybiliobanic acid Desoxybiliobansäure *f*
de[s]oxycantharidic acid
 Desoxycantharidinsäure *f*
de[s]oxycantharidine Desoxycantharidin *n*
de[s]oxycarminic acid Desoxycarminsäure *f*
de[s]oxycholic acid Desoxycholsäure *f*
de[s]oxycodomethine Desoxykodomethin *n*
de[s]oxycorticosterone Desoxycorticosteron *n*
de[s]oxyflavopurpurine Desoxyflavopurpurin *n*
de[s]oxyhematoporphyrin
 Desoxyhämatoporphyrin *n*
de[s]oxymesityl oxide Desoxymesityloxid *n*
de[s]oxy-nor-ephedrine Desoxy-nor-ephedrin *n*
desoxyquinine Chinan *n*
de[s]oxyribonucleic acid
 Desoxyribonukleinsäure *f* (DNS)
desoxythebacodine Desoxythebacodin *n*
despumate *v* entschäumen
destain *v* abfärben
destimulator Destimulator *m* (Korr)
destination Bestimmungsort *m*
destination positioning Zielkennzeichnung *f*
destrictic acid Destrictasäure *f*
destrictinic acid Destrictinsäure *f*
destroy *v* zerstören, abtöten, verderben,
 vernichten, zertrümmern
destructibility Zersetzbarkeit *f*, Zerstörbarkeit *f*
destructible zerstörbar, zersetzlich
destruction Vernichtung *f*, Zerstörung *f*,
 Zertrümmerung *f*, ~ **of red blood cells**
 Erythrozytenzerfall *m* (Med)
destruction limit Zerreißgrenze *f*
destructive zerstörend, vernichtend, schädlich
destructive distillation Entgasung *f*
destructive effect Zerstörungseffekt *m*

desulfinase Desulfinase *f* (Biochem)
desulfonation Desulfonierung *f*
desulfurating Abschwefeln *n*
desulfuration Abschwefelung *f*
desulfurization Entschwefelung *f,*
 Abschwefelung *f,* Desulfurierung *f*
 (Entschwefelung), Entfernung *f* des
 Schwefels, **process of** ~
 Entschwefelungsprozeß *m*
desulfurize *v* entschwefeln, abschwefeln
desulfurizing Entschwefelung *f,* Abschwefeln *n,*
 Schwefelentfernung *f*
desulfurizing agent Entschwefelungsmittel *n*
desulfurizing furnace Entschwefelungsofen *m*
desulfurizing slag Entschwefelungsschlacke *f*
desyl Desyl-
desyl chloride Desylchlorid *n*
detach *v* ablösen, abmachen, abreißen,
 absondern, abtrennen, loslösen
detachable abnehmbar, abspaltbar, abtrennbar,
 auswechselbar, lösbar
detached abgelöst, lose
detachment Ablösung *f,* Absonderung *f*
detachment of electrons Elektronenablösung *f*
detail Detail *n,* Einzelheit *f,* Feinheit *f*
detain *v* abhalten, hindern, zurückhalten
de-tan *v* entgerben
detar *v* entteeren
detarring Entteerung *f*
detect *v* auffinden, nachweisen
detectable nachweisbar
detection Beobachtung *f,* Erkennung *f,*
 Nachweis *m*
detection limit Nachweisgrenze *f*
detection method Nachweismethode *f*
detection of leakages Lecksuche *f*
detection sensitivity Nachweisempfindlichkeit *f*
detector Anzeiger *m,* Detektor *m* (Elektr),
 Fühlgerät *n,* Galvanoskop *n* (Elektr),
 Nachweisinstrument *n,* Spürgerät *n*
detector box Störungskontrollgerät *n*
detent gear of an indicator
 Indikatoranhaltevorrichtung *f*
detention Einbehaltung *f,* Vorenthaltung *f*
detent pin Anschlagstift *m*
detergent Reinigungsmittel *n,* Reinigermasse *f,*
 Waschmittel *n,* Waschpulver *n*
detergent base material Waschrohstoff *m*
deteriorate *v* verschlechtern, altern, entarten
deterioration Verschlechterung *f,* Verderb *m,*
 Verfall *m,* Zerstörung *f*
determinable bestimmbar, feststellbar
determinant Bestimmungsgröße *f* (Math),
 Determinante *f* (Math)
determinants, expansion of ~ Entwicklung v.
 Determinanten *pl* (Math)
determination Bestimmung *f,* Entscheidung *f,*
 calorimetric ~ kalorimetrische Bestimmung,

method of ~ Bestimmungsmethode *f,*
Ermittlungsverfahren *n*
determination by separate operation
Einzelbestimmung *f*
determination of the content[s]
Gehaltsbestimmung *f*
determination tube Bestimmungsrohr *n*
determinative ausschlaggebend, entscheidend
determine *v* beschließen, bestimmen (Chem),
ermitteln, festlegen, feststellen, ~ **empirically**
empirisch bestimmen, ~ **in advance**
vorausbestimmen
determinental equation Bestimmungsgleichung *f*
(Math)
determining ausschlaggebend
detin *v* entzinnen
detinning Entzinnung *f,* **degree of** ~
Entzinnungsgrad *m*
detinning bath Entzinnungsbad *n*
detinning plant Entzinnungsanlage *f*
detonate *v* detonieren, explodieren, knallen,
verpuffen
detonating ball Knallerbse *f*
detonating cap Zündhütchen *n,* Sprengkapsel *f*
detonating charge Zündladung *f*
detonating composition Kompositionssatz *m*
detonating cord Knallzündschnur *f*
detonating gas Knallgas *n*
detonating powder Knallpulver *n*
detonating priming Knallzündmittel *n*
detonation Detonation *f,* Explosion *f,* Knall *m,*
Verpuffung *f*
detonation velocity Detonationsgeschwindigkeit *f*
detonation wave Detonationswelle *f*
detonator Sprengkapsel *f,* Detonator *m,*
Knallkörper *m,* Knallzünder *m,* Zünder *m,*
Zündhütchen *n,* Zündkapsel *f*
detoxicant Entgiftungsmittel *n*
detoxicate *v* entgiften
detoxicating agent Entgiftungsmittel *n*
detoxication Entgiftung *f*
detoxication plant Entgiftungsanlage *f*
detoxification Entgiftung *f*
detreader Schälmaschine *f*
detreading maschine Schälmaschine *f*
detriment Benachteiligung *f,* Nachteil *m,*
Schaden *m*
detrimental schädlich, nachteilig
detritus Detritus *m* (Geol), Geröll *n* (Geol)
deuterate *v* deuterieren
deuteration Deuterieren *n*
deuterium Deuterium *n,* schwerer Wasserstoff *m*
(Deuterium)
deuterium exchange Deuterium-Austausch *m*
deuterium-moderated reactor
deuteriummoderierter Reaktor *m*
deuterium oxide Deuteriumoxid *n,* schweres
Wasser *n*
deuterohemin Deuterohämin *n*

deuterolysis Deuterolyse *f*
deuteron Deuteron *n* (Atom)
deuteron beam Deuteronenstrahl *m* (Atom)
deuteron formation Deuteronenbildung *f*
deuteron-induced activation Aktivierung *f* durch
Deuteronen
deuteron photodisintegration
Deuteronphotoeffekt *m*
deuteroporphyrin Deuteroporphyrin *n*
deuton Deuteron *n* (Atom), Deuton *n*
deutoplasma Deutoplasma *n*
devaluate *v* abwerten
devaluation Abwertung *f*
devalue *v* abwerten
devaporation Entnebelung *f*
devaporize *v* entnebeln
devaporizing Entnebeln *n*
devaporizing plant Entdunstungsanlage *f,*
Entnebelungsanlage *f*
Devarda's alloy Devardasche Legierung *f*
develop *v* entwickeln, ausbilden, hervorbringen
developable entwicklungsfähig, entwickelbar
developed, highly ~ hochentwickelt
developer Entwickler *m* (Phot),
Entwicklungsbad *n* (Phot)
developing Entwickeln *n* (Phot)
developing bath Entwicklungsbad *n* (Phot)
developing dyestuff Entwicklungsfarbstoff *m*
development Entwicklung *f,* Ausbildung *f,*
Bildung *f,* **continuous** ~ Entwicklung *f* im
Durchlauf (Phot), **full** ~ Ausentwicklung *f,*
line of ~ Entwicklungslinie *f,* **rate of** ~
Entwicklungstempo *n*
development drilling Erweiterungsbohrtätigkeit *f*
development of piping Lunkerbildung *f*
development product Versuchsprodukt *n*
deviate *v* abweichen, abarten, abbiegen,
ablenken, abwandern
deviating abweichend
deviating prism Umlenkprisma *n*
deviation Abweichung *f,* Ablenkung *f,* Abtrift *f,*
Ausweichen *n,* Streuung *f,* Ungenauigkeit,
angle of ~ Ausschlagwinkel *m,* **magnetic** ~
magnetische Abweichung, **offset or sustained**
~ bleibende Regelabweichung (Regeltechn),
~ **from set value** Regelbandbreite *f*
device Vorrichtung *f,* Apparat *m,* Apparatur *f,*
Gerät *n,* Hilfseinrichtung *f,* Hilfsgerät *n*
devitrify *v* entglasen
devote *v* widmen
devotion Hingabe *f,* Widmung *f*
devulcanization Regeneration *f* (Gummi)
devulcanize *v* regenerieren (Gummi)
devulcanized rubber Regeneratkautschuk *m*
dew Tau *m*
Dewar [vessel] Dewargefäß *n*
dewater *v* entwässern
dewatering Entwässerung *f*
dewatering centrifuge Entwässerungszentrifuge *f*

dewaxer Entparaffinierungsvorrichtung *f*
dewaxing Entparaffinierung *f*
dewaxing plant Entparaffinieranlage *f*
deweylite Deweylit *m* (Min), Eisengymnit *m*
dew point curve Taukurve *f*
dewy tauig
dexamethasone Dexamethason *n*
dexamphetamine Dexamphetamin[um] *n*
dextran[e] Dextran *n*
dextrantriose Dextrantriose *f*
dextrin Dextrin *n*, Britischgummi *n*,
　　Stärkegummi *n*
dextrinase Dextrinase *f*
dextrinized starch Dextrinstärke *f*
dextrin syrup Dextrinsirup *m*
dextro-acid Rechtssäure *f*
dextroamphetamine sulfate
　　Dextroamphetaminsulfat *n*
dextrocamphor Rechtscampher *m*
dextro-compound rechtsdrehende Verbindung *f*
dextromethorphan Dextromethorphan *n*
dextromoramide Dextromoramid *n*
dextronic acid Dextronsäure *f*, Gluconsäure *f*
dextropimarene Dextropimaren *n*
dextropimaric acid Dextropimarsäure *f*
dextropimarine Dextropimarin *n*
dextropimarol Dextropimarol *n*
dextropropoxyphene Dextropropoxyphen *n*
dextrorotation Rechtsdrehung *f*
dextrorotatory rechtsdrehend
dextrorphan Dextrorphan *n*
dextrose Dextrose *f*, Glucose *f*
dextrotartaric acid Rechtsweinsäure *f*
dezinc[k]ing Entzinkung *f*
dhurrin Dhurrin *n*
diabantite Diabantit *m* (Min)
diabase Diabas *m* (Min), Grünstein[schiefer] *m*
　　(Min)
diabetes Diabetes *m* (Med), Zuckerkrankheit *f*,
　　neurogenous ~ neurogener Diabetes,
　　temporary ~ wechselnder Diabetes, true ~
　　echter Diabetes
diabetic Diabetiker *m*, zuckerkrank
diabetic bread Diabetikerbrot *n*
diabetic coma Diabeteskoma *n*
diabetic diet Diabetesdiät *f*, Diabetikerdiät *f*
diabetic milk Diabetikermilch *f*
diabetin Diabetin *n*
diabetogenic diabetogen (Med)
diabetogenic factor Diabetesfaktor *m*
diabetometer Diabetometer *n*
diaboline Diabolin *n*
diacetamide Diacetamid *n*
diacetate Diacetat *n*
diacetic acid Diessigsäure *f*
diacetimide Diacetimid *n*
diacetin Diacetin *n*
diacetonamine Diacetonamin *n*
diacetone alcohol Diacetonalkohol *m*

diacetone sorbose Diacetonsorbose *f*
diacetonide Diacetonid *n*
diacetyl Diacetyl *n*
diacetyldiphenolisatin Diacetyldiphenolisatin *n*
diacetylene Diacetylen *n*
diacetylmorphine Diacetylmorphin *n*, Heroin *n*
diacetylmutase Diacetylmutase *f* (Biochem)
diacetyl peroxide Acetoperoxid *n*
diacetylphenylenediamine
　　Diacetylphenylendiamin *n*
diacetylresorcinol, 4,6- ~ Resodiacetophenon *n*
diacetyltannin Diacetyltannin *n*
diachylon Diachylonpflaster *n* (Med), gummed
　　~ Gummipflaster *n* (Med)
diachylon ointment Diachylonsalbe *f* (Med),
　　Bleipflastersalbe *f* (Med)
diachylum Diachylonpflaster *n* (Med)
diacid zweisäurig
diaclasite Diaklasit *m* (Min)
diactinic die aktinischen Strahlen durchlassend,
　　diaktinisch (Phys)
diadelphite Diadelphit *m* (Min)
diadoch diadoch
diadochite Diadochit *m* (Min)
diagnose *v* diagnostizieren (Med)
diagnosis, early ~ Früherkennung *f* (Med)
diagnostic diagnostisch
diagnostic[s] Diagnostik *f* (Med)
diagonal Diagonale *f*
diagonal *a* diagonal, quer, schräg
diagonal cutter Diagonalschneider *m*
diagonalising Diagonalisierung *f* (Math)
diagonalizing (of matrices) Diagonalisierung *f*
　　(v. Matrizen) (Math)
diagonal matrix Diagonalmatrize *f* (Math)
diagonal relationship Schräganalogie *f*
diagonal rider Diagonalschiene *f*
diagonal tie Diagonalband *n*
diagram Diagramm *n*, Abbildung *f*,
　　Aufzeichnung *f*, graphische Darstellung *f*,
　　Schaubild *n*
diagrammatic[ally] schematisch
diagrammatic section Prinzipschnitt *m*,
　　Schemabild *n*
diagram of connections Stromlaufschema *n*
　　(Elektr)
diagram of forces Kräftediagramm *n*
diagram of magnetic field Kraftlinienbild *n*
diagram of solidification Erstarrungsdiagramm *n*
diakinesis Diakenese *f* (Biol)
dial Zifferblatt *n*, Skala *f*, Wählerscheibe *f*,
　　Zahlenscheibe *f*
dial *v* wählen
dialdehyde Dialdehyd *m*
dial face Zifferblatt *n*
dialin Dialin *n*
dialkyl Dialkyl *n*
dialkylamine Dialkylamin *n*
dialkylene Dialkylen *n*

dialkyl sulfide Alkylsulfid *n*
dialkyl telluride Telluralkyl *n*
diallage Diallag *m* (Min)
diallage rock Gabbro *m*
diallydene Diallyden *n*
diallyl Diallyl *n*
diallylbarbituric acid Diallylbarbitursäure *f*
diallyldichlorosilane Diallyldichlorsilan *n*
diallylene Diallylen *n*
diallylmorphimethine Diallylmorphimethin *n*
diallyl sulfide Allylsulfid *n*, Schwefelallyl *n*
diallylurea Sinapolin *n*
dialogite Dialogit *m* (Min)
dial plate Zifferblatt *n*
dial switch Fingerscheibe *f*
dial thermometer Zeigerthermometer *n* ˙
dialurate Dialurat *n*
dialuric acid Dialursäure *f*
dialysis Dialyse *f*
dialytic dialytisch
dialytical dialytisch
dialyzable dialysierbar
dialyzation Dialyse *f*
dialyzator Dialysator *m*
dialyze *v* dialysieren
dialyzed iron Eisendialysat *n*
dialyzer Dialysator *m*
dialyzing Dialysieren *n*
dialyzing tube Dialysierschlauch *m*
diamagnetic diamagnetisch
diamagnetism Diamagnetismus *m*
diamantiferous diamantführend, diamanthaltig
diameter Dicke *f*, Durchmesser *m*, Stärke *f*,
  inside ~ Innendurchmesser *m*, lichter
  Durchmesser *m*, lichte Weite *f*, **internal** ~
  lichter Durchmesser, **nominal** ~
  Solldurchmesser *m*, **nuclear** ~
  Kerndurchmesser *m* (Atom), **overall** ~
  Gesamtdurchmesser *m*
diameter of a circle Kreisdurchmesser *m*
diameter of bore Kaliber *n*
diameter of hearth Gestellweite *f* des Ofens
diameter of nucleus Kerndurchmesser *m* (Atom)
diametrical diametral, diametrisch, genau
  entgegengesetzt
diametrically opposed grundsätzlich verschieden
diamide Diamid *n*
diamidogen Hydrazin *n*
diamine Diamin[o]verbindung *f*, Hydrazin *n*
diamine blue Diaminblau *n*, Trypanblau *n*
diamine fast red Diaminechtrot *n*
diamine pure blue Diaminreinblau *n*
diaminoacridine Diaminoacridin *n*
diaminobenzene Diaminobenzol *n*,
  Phenylendiamin *n*
diamino-biphenyl, 4,4- ~ Benzidin *n*
diaminocyclopentane Diaminocyclopentan *n*
diaminodecane Diaminodecan *n*
diaminodiethylamine Diaminodiäthylamin *n*

diaminodiethylsulfide Diaminodiäthylsulfid *n*
diaminodiphenylmethane
  Diaminodiphenylmethan *n*
diaminodiphenylsulfone
  Diaminodiphenylsulfon *n*
diaminodiphenylurea
  Diaminodiphenylharnstoff *m*
diaminofluoran, 3,6- ~ Rhodamin *n*
diaminofluorene Diaminofluoren *n*
diaminogene blue Diaminogenblau *n*
diaminomethoxybenzene
  Diaminomethoxybenzol *n*
diaminonaphthalene Naphthylendiamin *n*
diaminopentane, 1,5- ~ Cadacerin *n*
diaminophenol Diaminophenol *n*
diaminophenol hydrochloride, 3,4- ~ Amidol *n*
diamino-phenylphenazine chloride, 3,6- ~
  Phenosafranin *n*
diaminoresorcinol Diaminoresorcin *n*
diaminostilbene Diaminostilben *n*
diaminothiodiphenylamine Leukothionin *n*
diaminotoluene Diaminotoluol *n*
diammine mercuric chloride weißes schmelzbares
  Präzipitat (Chem)
diammonium phosphate
  Diammoniumphosphat *n*
diamol Diamol *n*
diamond Diamant *m* (Min), Raute *f* (Math),
  Rhombus *m* (Math), **imitation** ~
  Glasdiamant *m*, **rough** ~ Rohdiamant *m*,
  Käsestein *m*
diamond black Diamantschwarz *n*
diamond blue Diamantblau *n*
diamond cement Diamantzement *m*
diamond cutter Diamantschleifer *m*,
  Diamantschneider *m* (Techn)
diamond design Raupenmuster *n*
diamond drill Diamantbohrer *m* (Techn)
diamond dust Diamantpulver *n*
diamond fuchsine Diamantfuchsin *n*
diamondiferous diamantführend
diamond ink Diamanttinte *f*
diamond lattice Diamantgitter *n* (Krist)
diamond mortar Diamantmörser *m*
diamond pass Vierkantkaliber *n*
diamond scale Diamantwaage *f*
diamond-shaped rautenförmig
diamond structure Diamantstruktur *f* (Krist)
diamond yellow Diamantgelb *n*
diamyl amine Diamylamin *n*
diamyl ether Diamyläther *m*, Amyläther *m*
diamyl ketone Diamylketon *n*
diamylose Diamylose *f*
diamyl phthalate Diamylphthalat *n*
diamyl sulfide Amylsulfid *n*
dianil blue Dianilblau *n*
dianisidine Dianisidin *n*
dianisidine blue Dianisidinblau *n*
dianol Dianol *n*

dianthine Dianthin *n*
dianthracyl Dianthracyl *n*
dianthranilide Dianthranilid *n*
dianthranyl Dianthranyl *n*
dianthraquinone Dianthrachinon *n*
dianthraquinonyl Dianthrachinonyl *n*
dianthrene blue Dianthrenblau *n*
dianthrimide Dianthrimid *n*
dianthrone Dianthron *n*
dianthryl Dianthryl *n*, Dianthranyl *n*
diaphanometer Diaphanometer *n*,
Lichtdurchlässigkeitsmesser *m*,
Transparenzmesser *m*
diaphanoscope Diaphanoskop *n*
diaphanous diaphan, durchsichtig, transparent
diaphenylsulfone Diaphenylsulfon *n*
diaphorase Diaphorase *f* (Biochem)
diaphoretic schweißtreibendes Mittel *n* (Pharm)
diaphoretic *a* schweißtreibend (Pharm)
diaphoretic powder Schweißpulver *n* (Pharm)
diaphorite Diaphorit *m* (Min)
diaphragm Diaphragma *n*, Aperturblende *f*
(Opt), Blende *f* (Opt), Membran *f*,
Membranfilter *n*, Scheidewand *f*,
Zwerchfell *n* (Med), **semi-permeable** ~
halbdurchlässige Membran, ~ **with several
grades** Stufenblende *f* (Phot)
diaphragm-actuated jig Membransetzmaschine *f*
diaphragm bush Leitradbuchse *f*
diaphragm compressor Membranverdichter *m*
diaphragm current Diaphragmenstrom *m*
diaphragm liner Leitradbuchse *f*
diaphragm oscillation Membranschwingung *f*
diaphragm packing Membrandichtung *f*
diaphragm plane Blendenebene *f* (Opt)
diaphragm position Blendenstellung *f* (Opt)
diaphragm process Diaphragmaverfahren *n*,
Diaphragmenverfahren *n*
diaphragm pump Diaphragmapumpe *f*,
Membranpumpe *f*
diaphragm scale Blendenskala *f*
diaphragm seal Folienabdeckung *f*,
Membrandichtung *f*
diaphragm setting Blenden[ein]stellung *f*
diaphragm vaccum gauge Membranmanometer *n*
für Vakuumtechnik
diaphragm valve Membranventil *n*
diapositive Diapositiv *n* (Phot)
diapurine Diapurin *n*
diarrhoea Diarrhoe *f* (Med)
diarsenobenzene Diarsenobenzol *n*
diaspirin Diaspirin *n*
diaspore Diaspor *m* (Min)
diastase Diastase *f* (Biochem)
diastatic diastatisch
diastereoisomer Diastereoisomer *n*
(Stereochem), Diastereomer *n* (Stereochem)
diastereo[iso]meric diastereomer
diastereomer Diastereomer *n* (Stereochem)

diastereomerism Diastereomerie *f* (Stereochem)
diastrophism Diastrophismus *m* (Geol)
diaterebinic acid Diaterebinsäure *f*
diathermal diatherman
diathermancy Durchlässigkeit *f pl* für infrarote
Strahlen, Wärmedurchlässigkeit *f*
diathermanous diatherm[an], wärmedurchlässig,
infrarotdurchlässig
diathermic diatherm, diatherman,
infrarotdurchlässig, wärmedurchlässig
diatomacae Kieselalgen *pl*
diatomaceous diatomeenartig
diatomaceous earth Diatomeenerde *f*,
Infusorienerde *f*, Kieselgur *f*
diatomic doppelatomig, zweiatomig
diatomite Diatomit *m*, Berggur *f*
diatomite layer Diatomitschicht *f*
diatophane Diatophan *n*
diatretyne Diatretin *n*
diazene Diazen *n*
diazepine Diazepin *n*
diazete Diazen *n*
diazide Diazid *n*
diazin[e] Diazin *n*, **1,2-** ~ Pyridazin *n*, **1,4-** ~
Pyrazin *n*
diazine blue Diazinblau *n*
diaziridine Diaziridin *n*
diazoacetic acid Diazoessigsäure *f*
diazoacetic ester Diazoessigester *m*
diazo acetone Diazoaceton *n*
diazoaminobenzene Diazoaminobenzol *n*
diazoamino compound Diazoaminoverbindung *f*
diazobenzene Diazobenzol *n*
diazobenzene chloride Diazobenzolchlorid *n*
diazobenzene sulfonic acid
Diazobenzosulfonsäure *f*
diazobenzoic acid Diazobenzoesäure *f*
diazo black Diazoschwarz *n*
diazocamphor Diazocampher *m*
diazo compound Diazoverbindung *n*
diazo coupling Diazokupplung *f*
diazocystine Diazocystin *n*
diazodinitrophenol Diazodinitrophenol *n*
diazo dye Diazofarbstoff *m*
diazoethane Diazoäthan *n*, Aziäthan *n*
diazoformic acid Diazoameisensäure *f*
diazoimide Diazoimid *n*, Azoimid *n*
diazoisatin Diazoisatin *n*
diazole Diazol *n*, **1,3-** ~ Glyoxalin *n*, Imidazol *n*
diazomalonic acid Diazomalonsäure *f*
diazo metal compound Diazometallverbindung *f*
diazomethane Diazomethan *n*, Azimethan *n*
diazonaphtholsulfonic acid
Diazonaphtholsulfo[n]säure *f*
diazonium compound Diazoniumverbindung *f*
diazonium hydroxide Diazoniumhydroxid *n*
diazonium salt Diazoniumsalz *n* (Färb)
diazooxalacetic acid Diazooxalessigsäure *f*
diazooxide Diazooxid *n*

diazoparaffin Diazoparaffin *n*
diazophenol Diazophenol *n*
diazo solution Diazolösung *f*
diazosuccinic acid Diazobernsteinsäure *f*
diazotate Diazotat *n*
diazotetrazole Diazotetrazol *n*
diazotizable diazotierbar
diazotization Diazotierung *f*
diazotize *v* diazotieren
diazotized metal compound
  Diazometallverbindung *f*
diazotizing Diazotieren *n*
diazotype Diazotypie *f*
diazotype paper Diazopapier *n*
diazotype printing Diazodruck *m*
diazthine Thiodiazin *n*
dibasic zweibasisch, doppelbasisch
dibasic calcium phosphate Dicalciumphosphat *n*
dibasic potassium phosphate
  Dikaliumphosphat *n*
dibasic sodium phosphate
  Dinatrium[hydrogen]phosphat *n*
dibemethine Dibemethin *n*
dibenzacridine Dibenzacridin *n*,
  Naphthacridin *n*
dibenzalhydrazine Benzalazin *n*
dibenzamide Dibenzamid *n*
dibenzanthracene Dibenzanthracen *n*
dibenzanthraquinone Dibenzanthrachinon *n*
dibenzanthrone Dibenzanthron *n*
dibenzanthronyl Dibenzanthronyl *n*
dibenzcarbazole Dibenzcarbazol *n*
dibenzcoronene Dibenzcoronen *n*
dibenzfluorene Dibenzfluoren *n*
dibenzhydryl ether Benzhydryläther *m*
dibenzo-1,4-dithiin Thianthren *n*
dibenzo-1,4-pyran Xanthen *n*
dibenzodioxin Diphenylendioxid *n*,
  Phendioxin *n*
dibenzofuran Diphenylenoxid *n*
dibenzofurfuran Diphenylenoxid *n*
dibenzophenazine Phenophenanthrazin *n*
dibenzopyrone Dibenzopyron *n*
dibenzo thiazyl disulfide
  Dibenzothiazyldisulfid *n*
dibenzoyl Dibenzoyl-
dibenzoylate *v* dibenzoylieren
dibenzoylbibenzyl Didesyl *n*
dibenzoyldinaphthyl Dibenzoyldinaphthyl *n*
dibenzoylethane Diphenazyl *n*
dibenzoylheptane Dibenzoylheptan *n*
dibenzoyl peroxide Dibenzoylperoxid *n*,
  Benzoylperoxid *n*, Lucidol *n*
dibenzoylperylene Dibenzoylperylen *n*
dibenzphenanthrene Dibenzphenanthren *n*
dibenzpyrenequinone Dibenzpyrenchinon *n*
dibenzyl Dibenzyl-
dibenzyl dichlorosilane Dibenzyldichlorsilan *n*
dibenzyl ether Dibenzyläther *m*, Benzyläther *m*

dibenzyl ketone Dibenzylketon *n*
dibenzyl silanediol Dibenzylsilandiol *n*
diborane Diboran *n*
dibornyl Dibornyl *n*
diboron trioxide Dibortrioxid *n*
dibrazanquinone Dibrasanchinon *n*
dibromanthracene Dibromanthracen *n*
dibrom- (dibromo-) Dibrom-
dibromide Dibromid *n*
dibromindigo Dibromindigo *n*
dibromoacetophenone Dibromacetophenon *n*
dibromoanthracene Dibromanthracen *n*
dibrom[o]anthraquinone Dibromanthrachinon *n*
dibromobenzene Dibrombenzol *n*
dibromobenzoic acid Dibrombenzoesäure *f*
dibromobutene Dibrombuten *n*
dibromocamphor Campherdibromid *n*
dibromo-dihydroxyanthracene
  Dibromdihydroxyanthracen *n*
dibromodiphenyl Dibromdiphenyl *n*
dibromoethane Dibromäthan *n*
dibromofumaric acid Dibromfumarsäure *f*
dibromohydrin Dibromhydrin *n*
dibromo-hydroxyquinoline Dibromoxychinolin *n*
dibromo-ketodihydronaphthalene
  Dibromketodihydronaphthalin *n*
dibromomaleic acid Dibrommaleinsäure *f*
dibromomenthone Dibrommenthon *n*
dibromomethane Dibrommethan *n*
dibromonaphthalene Dibromnaphthalin *n*
dibromonaphthol Dibromnaphthol *n*
dibromonaphthylamine Dibromnaphthylamin *n*
dibromopropane Dibrompropan *n*
dibromoquinone chlorimide
  Dibromchinonchlorimid *n*
dibromosalicylic acid Dibromsalicylsäure *f*
dibromosuccinic acid Dibrombernsteinsäure *f*
dibromotoluidine Dibromtoluidin *n*
dibromotyrosine Dibromtyrosin *n*
dibrompropamidine Dibrompropamidin *n*
dibutyl Octan *n*
dibutyl adipate Dibutyladipinat *n*
dibutyl ether Dibutyläther *m*
dibutyl ketone Dibutylketon *n*
dibutyl phthalate Dibutylphthalat *n*,
  Butylphthalat *n*
dibutyrin Dibutyrin *n*
dibutyroolein Oleodibutyrin *n*
dicacodyl oxide Kakodyloxid *n*
dicadic acid Dicadisäure *f*
dicamphene Dicamphen *n*
dicamphoketone Dicamphoketon *n*
dicamphoryl Dicamphoryl *n*
dicamproin Dicamproin *n*
dicarbonate Dikarbonat *n*
dicarboxylate Dicarboxylat *n*
dicarboxylic acid Dicarbonsäure *f*
dice Würfel *m*
dicentrin[e] Dicentrin *n*

dice-shaped würfelförmig
dicetyl Dicetyl *n*
dicetyl sulfone Dicetylsulfon *n*
dichloramidobenzosulfonic acid
   Sulfondichloramidobenzoesäure *f*
dichloramine Dichloramin *n*
dichlorcarbene Dichlorcarben *n*
dichloride Dichlorid *n*, Doppelchlorid *n*
dichlorisatin Dichlorisatin *n*
dichloroacetic acid Dichloressigsäure *f*
dichloroacetone Dichloraceton *n*
dichloroaniline Dichloranilin *n*
dichlorobenzal chloride Dichlorbenzalchlorid *n*
dichlorobenzaldehyde Dichlorbenzaldehyd *m*
dichlorobenzene Dichlorbenzol *n*,
   Bichlorbenzol *n*
dichlorobenzene sulfonic acid
   Dichlorbenzolsulfonsäure *f*
dichlorobenzidine Dichlorbenzidin *n*
dichlorobenzoic acid Dichlorbenzoesäure *f*
dichlorocamphor Campherdichlorid *n*
dichlorodiethyl ether Dichlordiäthyläther *m*
dichlorodiethyl sulfide, 2,2- ~ Yperit *n*
dichloro-diphenylacetic acid
   Dichlor[o]diphenylessigsäure *f*
dichlorodiphenyl trichloroethane (insecticide)
   Dichlorodiphenyltrichloroäthan *n*
   (Insektenvertilgungsmittel)
dichloroethane Dichloräthan *n*
dichloroethanoic acid Dichloressigsäure *f*
dichloroether Dichloräther *m*
dichloroethylene Dichloräthylen *n*
dichloroethyl ether Dichloräthyläther *m*
dichloroethyl sulfide Dichloräthylsulfid *n*
dichlorohydrin Dichlorhydrin *n*
dichlorohydrin palmitate Palmitodichlorhydrin *n*
dichloromalealdehydic acid Mucochlorsäure *f*
dichloromethane Dichlormethan *n*
dichloropentane Dichlorpentan *n*
dichlorophenarsine Dichlorophenarsin *n*
dichlorophene Dichlorophen *n*
dichloro-phenoxyacetic acid
   Dichlorphenoxyessigsäure *f*
dichloro-phenylmercaptan
   Dichlorphenylmercaptan *n*
dichlorophthalic acid Dichlorphthalsäure *f*
dichloropropane Dichlorpropan *n*
dichlororesorcinol Dichlorresorcin *n*
dichlorosilane Dichlorsilan *n*
dichlorosuccinic acid Dichlorbernsteinsäure *f*
dichloro-thiophenol Dichlorphenylmercaptan *n*
dichlorotoluene Dichlortoluol *n*
dichlorvos Dichlorvos *n* (Insektizid)
dichroic dichroitisch (Krist), doppelfarbig,
   zweifarbig
dichroine Dichroin *n*
dichroism Dichroismus *m* (Opt),
   Doppelfarbigkeit *f* (Opt), Zweifarbigkeit *f*
   (Opt)

dichroite Dichroit *m* (Min), Cordierit *m* (Min),
   Iolith *m* (Min)
dichromate Dichromat *n*, Bichromat *n*
dichromate *a* dichromsauer
dichromate cell Chromsäureelement *n*
dichromatic dichromatisch, doppelfarbig,
   zweifarbig
dichromic acid Dichromsäure *f*
dichroscope Dichroskop *n* (Opt)
dicinnamic alcohol Dizimtalkohol *m*
dicinnamylidene Dicinnamyliden *n*
dick Dick *n* (Kampfgas)
dicobalt trioxide Kobaltsesquioxid *n*
dicodeine Dicodein *n*
dicodide Dicodid *n*
diconchinine Dichinidin *n*
dicoumarol Dicumarol *n*
dicresol Dikresol *n*
dicrotaline Dicrotalin *n*
dictamnic acid Dictamninsäure *f*
dictamnine Dictamnin *n*
dictyosome Dictyosom *n* (Biol)
dicyan Dicyan *n*
dicyandiamide Dicyandiamid *n*,
   Cyanoguanidin *n*
dicyandiamidine Dicyandiamidin *n*,
   Guanylharnstoff *m*
dicyanimide Dicyanimid *n*
dicyanin[e] Dicyanin *n*
dicyanogen Dicyan *n*
dicyclohexylbenzene Dicyclohexylbenzol *n*
dicyclohexyl phthalate Dicyclohexylphthalat *n*
dicyclopentadiene Dicyclopentadien *n*
dicycloverine Dicycloverin *n*
dicymylamine Dicymylamin *n*
didecyl phthalate Didecylphthalat *n*
didehydrocorycavamine
   Didehydrocorycavamin *n*
didehydrocorycavine Didehydrocorycavin *n*
didesyl Didesyl *n*
didodecahedral didodekaedrisch (Krist)
didymolite Didymolit *m* (Min)
die Gesenk *n*, Gußform *f*, Lochstempel *m*,
   Matrize *f*, Stempel *m*, Zieheisen *n* (Draht)
die *v*, ~ **away** abklingen (Akust), ~ **down**
   ausschwingen (im Schwung nachlassen), ~
   off absterben, ~ **out** aussterben
die adapter system Düsenanschlußstück *n*
die block Werkzeughalter *m*
die body Düsengehäuse *n*
die bottom Düsenunterteil *n*
die-cast aluminum Aluminiumspritzguß *m*
die casting Formguß *m*, Formgußteil *n*,
   Gesenkgußstück *n*, Schalenguß *m*,
   Spritzguß *m*
die casting alloy Spritzgußlegierung *f*
die cast steel Schalengußstahl *m*
die-cast threads *v* Gewinde spritzen
die channel Düsenkanal *m*

Dieckmann ester Dieckmann-Ester *m*
die-cut *v* ausstanzen
die-cutter Stanze *f*
die cutting Stanzen *n*
die forging Gesenkschmieden *n*
die grid Isolierrost *m*
die head Schneidkopf *m*, Spritzkopf *m*
dielectric Dielektrikum *n*
dielectric *a* dielektrisch
dielectric absorption dielektrische Absorption *f* (Elektr)
dielectric breakdown dielektrischer Durchschlag *m* (Elektr)
dielectric coefficient Dielektrizitätskonstante *f*
dielectric constant Dielektrizitätskonstante *f*
dielectric dryer Hochfrequenztrockner *m*
dielectric fatigue dielektrische Nachwirkung *f*
dielectric material Isolierstoff *m*
dielectric strength dielektrische Widerstandsfähigkeit *f*, Durchschlagfeldstärke *f*, Durchschlagsfestigkeit *f*
Diels' acid Dielssäure *f*
Diels-Alder-reaction Diels-Alder-Reaktion *f*
Diels hydrocarbon Diels-Kohlenwasserstoff *m*
dienanthic acid Diönanthsäure *f*
diene Dien *n*
dienestrol Dienöstrol *n*
diene synthesis Diensynthese *f*
dienol Dienol *n*
die orifice Düsenaustritt *m*, Düsenmund *m*, Düsenöffnung *f*, Mundstück *n*
die plate Stempelplatte *f*
die pressing Formstanzen *n*
die restriction Stauscheibe *f*
die ring Düsenprofil *n*, Düsenring *m*
diesel drive Dieselantrieb *m*
Diesel engine Dieselmotor *m*
Diesel fuel Dieselkraftstoff *m*
Diesel generating set Dieselaggregat *n*
Diesel oil Dieselöl *n*
die sinking mill Nachformfräsmaschine *f*
die stock Kluppe *f*
diet Diät *f*, Ernährung *f*, **high-calorie** ~ kalorienreiche Kost *f*, **vegetarian** ~ vegetarische Kost *f*
dietetic diätetisch
dietetics Diätetik *f*, Diätkunde *f*, Diätlehre *f*, Ernährungskunde *f*
diethanolamine Diäthanolamin *n*
diethazine Diethazin *n*
dietherate Diätherat *n*
diethyl acetic acid Diäthylessigsäure *f*
diethylamine Diäthylamin *n*
diethylamine hydrochloride Diäthylaminchlorhydrat *n*
diethylaminoethanol Diäthylaminoäthanol *n*
diethylaniline Diäthylanilin *n*

diethylbarbituric acid Barbital *n*, Diäthylbarbitursäure *f*, Veronal *n* (HN)
diethylbenzene Diäthylbenzol *n*
diethylbromacetamide Diäthylbromacetamid *n*, Neuronal *n*
diethylcarbamazine Diäthylcarbamazin *n*
diethylcarbocyanine iodide Diäthylcarbocyaninjodid *n*
diethyl carbonate Kohlensäurediäthylester *m*
diethylcyanine iodide Diäthylcyanjodid *n*
diethyldichlorosilane Diäthyldichlorsilan *n*
diethyldiethoxysilane Diäthyldiäthoxysilan *n*
diethyldiphenylurea Diäthyldiphenylharnstoff *m*
diethyl dithiocarbamic acid Diäthyldithiocarbaminsäure *f*
diethylenediamine Diäthylendiamin *n*, Piperazin *n*
diethylene disulfide Diäthylendisulfid *n*, Dithian *n*
diethylene glycol Diäthylenglykol *n*
diethylenetriamine Diäthylentriamin *n*
diethyl ether Diäthyläther *m*, Äthyläther *m*, Äthyloxid *n* (obs)
diethyl glutarate Glutarsäurediäthylester *m*
diethyl itaconate Itaconsäurediäthylester *m*
diethylketone Metaceton *n*
diethyl malate Äpfelsäurediäthylester *m*
diethyl maleate Maleinsäurediäthylester *m*
diethyl malonate Diäthylmalonat *n*
diethyl malonic acid Diäthylmalonsäure *f*
diethyl mesotartrate Diäthylmesotartrat *n*
diethyl metanilic acid Diäthylmetanilsäure *f*
diethyl nitrophenyl thiophosphate Diäthylnitrophenylthiophosphat *n* (E 605)
diethyl oxalacetate Oxalessigsäureester *m*
diethyl oxalate Oxalsäurediäthylester *m*
diethyl oxide Äthyläther, Äthyloxid *n*
diethyl peroxide Äthylperoxid *n*
diethyl phthalate Diäthylphthalat *n*, Äthylphthalat *n*
diethyl succinate Bernsteinsäurediäthylester *m*, Äthylsuccinat *n*, Diäthylsuccinat *n*
diethyl sulfate Diäthylsulfat *n*, Äthylsulfat *n*
diethyl sulfide Diäthylsulfid *n*, Äthylsulfid *n*
diethyl tartrate Diäthyltartrat *n*, Äthyltartrat *n*
[di]ethyl telluride Tellurdiäthyl *n*
diethyl toluidine Diäthyltoluidin *n*
diethylvaleramide Valyl *n*
diethyl zinc Zinkäthyl *n*
dietrichite Dietrichit *m* (Min)
diferulaic acid Diferulasäure *f*
differ *v* sich unterscheiden, sich nicht decken
difference Unterschied *m*, Abweichung *f*, Differenz *f*, Verschiedenheit *f*, ~ **in length** Längenunterschied *m*, ~ **in solubility** Löslichkeitsunterschied *m*, ~ **of atmospheric pressure** Luftdruckunterschied *m*
difference scheme method Differenzenschemaverfahren *n*

different verschieden, andersartig, unterschiedlich, verschiedenartig

differentiability Differenzierbarkeit *f* (Math)

differentiable differenzierbar (Math)

differential Differential *n* (Math)

differential aeration unterschiedliche Belüftung *f*

differential aeration cell Belüftungselement *n* (Korr)

differential barometer Differentialbarometer *n*

differential blood picture Differentialblutbild *n*

differential calculus Differentialrechnung *f*

differential centrifugation Differentialzentrifugation *f,* differentielle Zentrifugation *f*

differential chain block Differentialflaschenzug *m*

differential coefficient or quotient Differentialquotient *m* (Math)

differential coil Differentialspule *f*

differential condenser Differential[dreh]kondensator *m*

differential coupling Differentialkupplung *f*

differential current Differenzstrom *m*

differential draft gauge Differenzzugmesser *m*

differential equation Differentialgleichung *f* (Math), **elliptical** ~ elliptische Differentialgleichung *f,* **functional** ~ Funktionaldifferentialgleichung *f,* **higher order** ~ Differentialgleichung *f* höherer Ordnung, **homogeneous** ~ homogene Differentialgleichung *f,* **hyperbolic** ~ hyperbolische Differentialgleichung *f,* **linear** ~ lineare Differentialgleichung *f,* **non-linear** ~ nichtlineare Differentialgleichung *f,* **order of a** ~ Ordnung einer Differentialgleichung *f,* **ordinary** ~ gewöhnliche Differentialgleichung *f,* **parabolic** ~ parabolische Differentialgleichung *f,* **second order** ~ Differentialgleichung *f* zweiter Ordnung, ~ **for length of arc** Bodenlängendifferential *n* (Math), ~ **of the first order** Differentialgleichung *f* erster Ordnung

differential formula Differentialformel *f* (Math)

differential gauge Differenzmanometer *n*

differential gear Differentialgetriebe *n*

differential high-speed pump Differentialeilpumpe *f*

differential manometer Differentialdruckmesser *m,* Differentialmanometer *n*

differential measurement Differenzmessung *f* (Elektr)

differential piston Differentialkolben *m*

differential pressure Differenzdruck *m*

differential pump Stufenkolbenpumpe *f*

differential refraction Differentialbrechung *f*

differential relay Differentialrelais *n*

differential screw Differentialschraube *f*

differential shaft Ausgleichswelle *f*

differential system Differentialschaltung *f*

differential thermal analysis Differentialthermoanalyse *f*

differential thermometer Differentialthermometer *n*

differential thermostat Differenzthermostat *m*

differential windlass Differentialwinde *f*

differentiate *v* differenzieren, unterscheiden

differentiating circuit Differenzierschaltung *f* (Rdr)

differentiation Differentiation *f* (Math), Differenzierung *f* (Math)

difficult schwer, schwierig

difficult to volatilize schwerflüchtig

difficulty Erschwerung *f,* Schwierigkeit *f*

diffract *v* beugen (Opt)

diffractaic acid Diffractasäure *f*

diffraction Beugung *f* (Opt), Ablenkung *f* (Opt), Diffraktion *f*

diffraction angle Diffraktionswinkel *m*

diffraction camera Interferenzapparatur *f*

diffraction fringe Beugungsstreifen *m* (Opt)

diffraction grating Beugungsgitter *n* (Opt), Diffraktionsgitter *n* (Opt), ~ **spectroscope** Gitterspektroskop *n*

diffraction of light Lichtzerlegung *f*

diffraction of rays Strahlenbrechung *f*

diffraction pattern Beugungsbild *n* (Opt), Beugungsdiagramm *n* (Opt), Beugungsfigur *f* (Opt)

diffraction spectroscope Beugungsspektroskop *n*

diffraction spectrum Gitterspektrum *n* (Spektr), Beugungsspektrum *n* (Spektr), Normalspektrum *n* (Spektr)

diffraction theory Beugungstheorie *f*

diffraction zone Beugungszone *f*

diffractometer Beugungsmesser *m,* Diffraktometer *n*

diffusate Diffusat *n* (Atom)

diffuse diffus, zerstreut, unscharf, **to** ~ **back** rückdiffundieren, **having** ~ **edges** randverschmiert

diffuse *v* diffundieren, verbreiten, wandern, zerstreuen, ~ **into** hineindiffundieren, ~ **out** herausdiffundieren

diffuser Diffuseur *m,* Diffusionsapparat *m,* Zerstäuber *m*

diffusibility Diffusionsfähigkeit *f,* Diffusionsvermögen *n,* Zerstreuungsvermögen *n*

diffusible diffusionsfähig

diffusing Diffundieren *n,* Verteilen *n*

diffusion Diffusion *f,* Ausbreitung *f,* **velocity of** ~ Diffusionsgeschwindigkeit *f,* **working fluid for** ~ Diffusionsbetriebswasser *n,* ~ **accompanied by reaction** Diffusion *f* bei gleichzeitiger Reaktion, ~ **in solids** Diffusion *f pl* in Festkörpern

diffusional jog Diffusionssprung *m*
diffusion analysis Diffusionsanalyse *f* (Phys)
diffusion apparatus Diffuseur *m*
diffusion barrier Diffusionswand *f* (Phys)
diffusion cell Diffusionszelle *f*
diffusion cloud chamber
   Diffusionsnebelkammer *f*
diffusion coefficient Diffusionskoeffizient *m*
diffusion column Diffusionskolonne *f*
diffusion cross section Diffusionsquerschnitt *m*
diffusion current Grenzstrom *m* (Elektr)
diffusion electrophoresis
   Diffusionselektrophorese *f*
diffusion equation Diffusionsgleichung *f*
diffusion flame reactor
   Diffusionsflammenreaktor *m*
diffusion heat Diffusionswärme *f*
diffusion path Diffusionsweg *m*
diffusion potential Diffusionspotential *n*
diffusion process Diffusionsverfahren *n*
diffusion pump Diffusionspumpe *f*, **high-vacuum**
   ~ Hochvakuum-Diffusionspumpe *f*
diffusion rate Diffusionsgeschwindigkeit *f*
diffusion resistance Ausbreitungswiderstand *m*
diffusion separation method
   Diffusionstrennverfahren *n*
diffusion tensor Diffusionstensor *m*
diffusion transport Diffusionstransport *m*
diffusiveness Ausbreitungsfähigkeit *f*
diffusivity Diffusionsfähigkeit *f*,
   Diffusionsvermögen *n*
diffusor Auslaugeturm *m*
difluan Difluan *n*
difluorenyl Difluorenyl *n*
difluorenylidene Difluorenyliden *n*
difluoride Doppelfluorid *n*
difluorobiphenyl Difluordiphenyl *n*
difluorodiphenyl Difluordiphenyl *n*
difluoroethylene Difluoräthylen *n*
difluoroxenene Difluordiphenyl *n*
diformamide Diformamid *n*
diformyl Diformyl *n*, Glyoxal *n*
difurylglyoxal Furil *n*
dig *v* graben, ~ **in** eingraben, ~ **up** aufgraben
digalene Digalen *n*
digallic acid Digallussäure *f*
digenite Digenit *m* (Min)
digentisic acid Digentisinsäure *f*
digest *v* verdauen, digerieren
digester Aufgußgefäß *n*, Autoklav *m*,
   Dampfkochtopf *m*, Druckflasche *f*, Kocher *m*,
   Regeneratkessel *m*, **revolving** ~ rotierender
   Kocher
digester gas Faulgas *n*
digestibility Verdaulichkeit *f*
digestible verdaulich
digesting Digerieren *n*, Verdauen *n*
digesting flask Digestionskolben *m*

digestion Verdauung *f* (Med), Aufschließung *f*
   (Atom), Aufschluß *m* (Chem), Digestion *f*
   (Med), Verdauungsprozeß *m* (Med)
digestion bottle Digerierflasche *f*
digestion flask Digerierflasche *f*,
   Digestionskolben *m*
digestion process Verdauungsvorgang *m* (Med)
digestive apparatus Digerierofen *m*
digestive organ Verdauungsorgan *n* (Med)
digestive process Verdauungsprozeß *m* (Med)
digestive salt verdauungsförderndes Salz *n*
digestive system Verdauungssystem *n*
digestor Dampffaß *n* (Brau), Digestor *m* (Brau)
digger Bagger *m*
diginatigenin Diginatigenin *n*
diginatin Diginatin *n*
diginin Diginin *n*
diginose Diginose *f*
digipan Digipan *n*
digital digital
digital computer Digitalrechenmaschine *f*,
   Digitalrechner *m*
digital display Digitalanzeige *f*
digitalein Digitalein *n*
digitalic salt Digitalsalz *n*
digitaligenine Digitaligenin *n*
digitalin Digitalin *n*
digitalis extract Fingerhutextrakt *m* (Bot)
digitalis glucoside Digitalisglucosid *n*
digitalis leaves Fingerhutblätter *n pl* (Bot),
   Digitalisblätter *pl* (Bot)
digitalism Digitalisvergiftung *f*
digitalis tincture Fingerhuttinktur *f*
digitalonic acid Digitalonsäure *f*
digitalose Digitalose *f*
digital printer Meßwertdrucker *m*
digital recorder Digitalschreiber *m*
digitane Digitan *n*
digitic acid Digitsäure *f*
digitogenic acid Digitogensäure *f*
digitogenin Digitogenin *n*
digitoic acid Digitosäure *f*
digitoleine Digitolein *n*
digitolutein Digitolutein *n*
digitonide Digitonid *n*
digitonin Digitonin *n*
digitophylline Digitophyllin *n*
digitoxigenin Digitoxigenin *n*
digitoxin Digitoxin *n*
digitoxose Digitoxose *f*
diglycerin[e] Diglycerin *n*
diglycerol Diglycerin *n*
diglycerophosphoric acid
   Diglycerinphosphorsäure *f*
diglycol Diglykol *n*
diglycollic acid Diglykolsäure *f*
diglycylcystine Diglycylcystin *n*
diglycylglycine Diglycylglycin *n*
digoxigenin Digoxigenin *n*

diguanylamine Diguanylamin *n*
diguanyldisulfide Diguanyldisulfid *n*
dihalide Dihalogenid *n*
dihedral Dieder *m*, Zweiflächner *m*
diheptylamine Diheptylamin *n*
diheteroatomic diheteroatomig
dihexadecyl ether Margaron *n*
dihexagonal dihexagonal (Krist)
dihexagonalpyramidal dihexagonalpyramidal
(Krist)
dihexahedral dihexaedrisch (Krist),
doppelsechsflächig (Krist)
dihexahedron Dihexaeder *n* (Krist),
Doppelsechsflächner *m* (Krist)
dihexyl Dihexyl-
dihexyl ether Dihexyläther *m*
dihexylketone Önanthon *n*
dihexyverine Dihexyverin *n*
dihydracrylic acid Dihydracrylsäure *f*
dihydralazine Dihydralazin *n*
dihydrate Dihydrat *n*
dihydrite Dihydrit *m*
dihydroanthracene Dihydroanthracen *n*
dihydroanthraquinoneazine
Dihydroanthrachinonazin *n*
dihydrobenzacridine carboxylic acid Tetrophan *n*
(HN)
dihydrobenzene Dihydrobenzol *n*
dihydrobenzothiopyran Thiochroman *n*
dihydroberberine Dihydroberberin *n*
dihydrocholesterol Koprostanol *n*
dihydrocodeinone Dicodid *n*
dihydrocoumarin, 3,4- ~ Melilotol *n*
dihydrodianthrone Dianthranol *n*, Dianthron *n*
dihydrodiketonaphthalene Naphthochinon *n*
dihydrodimerization Dihydrodimerisation *f*
dihydroemetine Rubremetin *n*
dihydroergocristine Dihydroergocristin *n*
dihydroergotamine Dihydroergotamin *n*
dihydroergotamine tartrate
Dihydroergotamintartrat *n*
dihydroergotoxine Dihydroergotoxin *n*
dihydrofolic acid Dihydrofolsäure *f*
dihydrofolic acid reductase
Dihydrofolsäurereduktase *f* (Biochem)
dihydrofuran Dihydrofuran *n*
dihydrogen phosphate primäres Phosphat,
Dihydrogenphosphat *n*
dihydroharmine, 3,4- ~ Harmalin *n*
dihydrohydroxy codeinone Eukodal *n*
dihydroketoanthracene Anthranon *n*
dihydrolipoic acid Dihydroliponsäure *f*
dihydronaphthalene Dialin *n*,
Dihydronaphthalin *n*
dihydroorotase Dihydroorotase *f* (Biochem)
dihydroorotate dehydrogenase
Dihydroorotsäurereduktase *f* (Biochem)
dihydroorotic acid Dihydroorotsäure *f*
dihydrophenanthrene Dihydrophenanthren *n*

dihydrophytyl bromide Dihydrophytylbromid *n*
dihydropyrazole Pyrazolin *n*
dihydropyrrol[e] Pyrrolin *n*
dihydroquinine Dihydrochinin *n*
dihydrostreptitol Dihydrostreptit *m*
dihydrostreptomycin Dihydrostreptomycin *n*
dihydrotachysterol Dihydrotachysterin *n*,
Dihydrotachysterol *n*
dihydrothiamine Dihydrothiamin *n*
dihydrotoxiferine Dihydrotoxiferin *n*
dihydrourete, 1,2- ~ Uretin *n*
dihydroxy-1,3,5-trinitrobenzene, 2,4- ~
Styphninsäure *f*
dihydroxy-1-methylanthraquinone, 2,4- ~
Rubiadin *n*
dihydroxy-6-amino-1,3,5-triazine, 2,4- ~
Melanurensäure *f*
dihydroxyacetone Dihydroxyaceton *n*,
Dioxyaceton *n*
dihydroxyacetone phosphate
Dihydroxyacetonphosphat *n*
dihydroxyacetophenone, 2,4- ~
Resacetophenon *n*
dihydroxyadipic acid Adipoweinsäure *f*
dihydroxyanthracene Dihydroxyanthracen *n*, 1,5-
~ Rufol *n*
dihydroxyanthraquinone Dioxyanthrachinon *n*,
1,2- ~ Alizarin *n*, Krappfärbestoff *m*, **1,3-** ~
Xanthopurpurin *n*, **1,5-** ~ Anthrarufin *n*, **2,6-**
~ Anthraflavinsäure *f*
dihydroxybenzaldehyde, 3,4- ~
Protocatechualdehyd *m*
dihydroxybenzene Dihydroxybenzol *n*,
Dioxybenzol *n*, **1,2-** ~ Brenzcatechin *n*,
Pyrocatechin *n*, **1,3-** ~ Resorcin *n*
dihydroxybenzoic acid Dihydroxybenzoesäure *f*,
Dioxybenzoesäure *f*, **3,4-** ~
Protocatechusäure *f*
dihydroxycoumarin, 7,8- ~ Daphnetin *n*
dihydroxydibenzanthracene
Dihydroxydibenzanthracen *n*
dihydroxydiethylstilbene
Dihydroxydiäthylstilben *n*
dihydroxydinaphthyl Dihydroxydinaphthyl *n*
dihydroxydiphenylmethane
Dihydroxydiphenylmethan *n*
dihydroxydiquinoyl Rhodizonsäure *f*
dihydroxyethane Glykol *n*
dihydroxyethylene sulfide Thiodiglykol *n*
dihydroxy fluoran, 3,6- ~ Resorcinphthalein *n*
dihydroxylene Cantharen *n*
dihydroxymenthane Terpin *n*
dihydroxynaphthalene Dihydroxynaphthalin *n*,
1,2- ~ Naphthobrenzcatechin *n*, **1,3-** ~
Naphthoresorcin *n*
dihydroxyperylene Dihydroxyperylen *n*
dihydroxyphenylacetic acid
Dihydroxyphenylessigsäure *f*

dihydroxyphenylalanine
  Dihydroxyphenylalanin *n*, 3,4- ~ Dopa *n*
dihydroxypropane Propandiol *n*
dihydroxypurine, 2,6- ~ Xanthin *n*
dihydroxypyrene Dihydroxypyren *n*
dihydroxypyrimidine, 2,6- ~ Uracil *n*
dihydroxyquinoline Dioxychinolin *n*
dihydroxyquinone Dioxychinon *n*
dihydroxystearic acid Dihydroxystearinsäure *f*
dihydroxysuccinic acid Dioxybernsteinsäure *f*
dihydroxytoluene, 2,4- ~ Kresorcin *n*, 3,5- ~
  Orcin *n*
dihydroxyviolanthrone Dihydroxyviolanthron *n*
dihydroxyxanthone, 1,7- ~ Euxanthon *n*
dihypnal Dihypnal *n*
diimide Diimid *n*
diindene Diinden *n*
diindole Diindol *n*
diindole indigo Diindolindigo *n*
diindolyl Diindolyl *n*
diindone Diindon *n*
diindyl Diindyl *n*
diiodide Dijodid *n*
diiodoacetylene Dijodacetylen *n*
diiodocarbazole Dijodcarbazol *n*
diiododithymol Dijododithymol *n*
diiodoform Dijodoform *n*
diiodo-hexamethylenetetramine Dijodurotropin *n*
diiodohydrin Dijodhydrin *n*
diiodohydroxypropane Jothion *n*
diiodomethane Methylenjodid *n*
diiodo-p-phenolsulfonic acid Sozojodol *n*,
  Sozojodolsäure *f*
diiodopropyl alcohol Jothion *n*
diiodoresorcin potassium monosulfonate Pikrol *n*
diiodosalicylic acid Dijodsalicylsäure *f*
diiodothymol Annidalin *n*
diiodotyrosine Dijodtyrosin *n*, 3,5- ~
  Jodgorgosäure *f*
diiodourotropine Dijodurotropin *n*
diisatogen Diisatogen *n*
diisoamyl ether Diisoamyläther *m*
diisobutylamine Diisobutylamin *n*
diisobutyl ketone Diisobutylketon *n*, Valeron *n*
diisochavibetol Diisochavibetol *n*
diisocyanate Diisocyanat *n*
diisodecyl adipate Diisodecyladipat *n*
diisodecyl phthalate Diisodecylphthalat *n*
diisoeugenol Diisoeugenol *n*
diisooctyl adipate Diisooctyladipat *n*
diisooctyl azelate Diisooctylazelat *n*
diisooctyl phthalate Diisooctylphthalat *n*
diisopropylamine Diisopropylamin *n*
di-iso-propylbarbituric acid Proponal *n*
diisopropyl benzene Diisopropylbenzol *n*
diisopropyl benzoin Cuminoin *n*
diisopropyl carbinol Diisopropylcarbinol *n*
diisopropyl ether Diisopropyläther *m*

diisopropyl fluorophosphate
  Diisopropylfluorphosphat *n*
diisopropylidene acetone Phoron *n*
diisopropyl ketone Diisopropylketon *n*
dika butter Dikabutter *f*
dika fat Dikabutter *f*
diketene Diketen *n*
diketoapocamphoric acid
  Diketoapocamphersäure *f*
diketocamphoric acid Diketocamphersäure *f*
diketocholanic acid Diketocholansäure *f*
diketone Diketon *n*
diketopiperazine Diketopiperazin *n*
diketotetrahydroquinazoline
  Benzoylenharnstoff *m*
diketotriazolidine Urazol *n*
diketoxime chelate Diketoximchelat *n*
dilactamic acid Dilactamidsäure *f*
dilactic acid Dilactylsäure *f*, Dimilchsäure *f*
dilapidated baufällig
dilatability Dehnbarkeit *f*,
  Ausdehnungsfähigkeit *f*
dilatable [aus]dehnbar, ausweitbar
dilatancy Dilatanz *f* (Phys)
dilatant dilatant (Phys)
dilatation Ausdehnung *f* (Phys), Dilatation *f*,
  Expansion *f*
dilate *v* [aus]dehnen, erweitern (Pupille)
dilation Dehnung *f*, Erweiterung *f* (Pupille)
dilatometer Dilatometer *n*,
  Ausdehnungsmesser *m*,
  Dehnbarkeitsmesser *m*
dilatometry Dilatometrie *f* (Phys)
dilaurin Dilaurin *n*
dilaurylamine Dilaurylamin *n*
dilimonene Dilimonen *n*
dilituric acid Dilitursäure *f*
dill Aneth *m* (Bot), Dill *m* (Bot)
dill apiole Dillapiol *n*
dillnite Dillnit *m* (Min)
dill oil Dillöl *n* (Min)
dill seed oil Dillöl *n*
diluent Verdünnungsmittel *n*,
  Abschwächungsmittel *n*, Streckmittel *n*,
  Verschnittmittel *n*
dilutability Verdünnbarkeit *f*
dilutable verdünnbar
dilute verdünnt
dilute *v* verdünnen, strecken, wässern
diluted verdünnt
diluting Verdünnen *n*
diluting agent Verdünnungsmittel *n*,
  Verschnittmittel *n*
dilution Verdünnung *f*, **degree of** ~
  Verdünnungsgrad *m*, **heat of** ~
  Verdünnungswärme *f*
dilution analysis Verdünnungsanalyse *f*
dilution capacity Verschnittverhalten *n*
dilution clause Verwässerungsschutzklausel *f*

dilution heat Verdünnungswärme *f*
dilution law Verdünnungsgesetz *n*
diluvial ore Wascherz *n*
diluvium Diluvium *n* (Geol)
dim glanzlos, matt, trübe, blaß, ~ **red** mattrot
dim *v* abblenden, abdunkeln, dämpfen (Licht)
dimagnesium phosphate
  Magnesiumhydrogenphosphat *n*
dimazole Dimazol *n*
dimedone Dimedon *n*
dimenoxadole Dimenoxadol *n*
dimension Dimension *f,* Abmessung *f,*
  Ausmaß *n,* Größe *f,* Größenmaß *n,* Maß *n,*
  external ~ Außenabmessung *f,* **outer** ~
  Außenmaß *n*
dimension *v* dimensionieren
dimensional dimensional
dimensional analysis
  Ähnlichkeitskennzahlmethode *f*
dimensional change Größenänderung *f*
dimensional deviation Maßabweichung *f*
dimensional factor Maßgröße *f*
dimensionally stable formbeständig, formstabil
dimensional stability Maßbeständigkeit *f*
dimensional tolerance Maßtoleranz *f*
dimension analysis Dimensionsanalyse *f* (Phys)
dimensioned, fully ~ volldimensioniert
dimensioned drawing Maßzeichnung *f*
dimension in depth Tiefendimension *f*
dimensioning Dimensionierung *f*
dimensionless dimensionslos
dimension number Dimensionszahl *f*
dimensions, proportion of ~ Größenverhältnis *n*
dimenthene Dimenthen *n*
dimenthyl Dimenthyl *n*
dimepheptanol Dimepheptanol *n*
dimer Dimer *n*
dimercaprol Dimercaprol *n*
dimercaptopropanol Dimercaptopropanol *n*
dimercaptothiodiazol Dimercaptothiodiazol *n*
dimercaptotoluene Dimercaptotoluol *n*
dimeric dimer, zweiteilig
dimerization Dimerisation *f,* Dimerisierung *f*
dimerize *v* dimerisieren
dimesityl Dimesityl *n*
dimetanilic acid Dimetanilsäure *f*
dimethacrylic acid Dimethacrylsäure *f*
dimethiodal Dimethiodal *n*
dimethoxy acetophenone
  Dimethoxyacetophenon *n*
dimethoxy anthraquinone
  Dimethoxyanthrachinon *n*
dimethoxy benzaldehyde
  Dimethoxybenzaldehyd *m,* **3,4-** ~
  Veratraldehyd *m*
dimethoxybenzene, 1,2- ~ Veratrol *n*
dimethoxy benzidine Dimethoxybenzidin *n*
dimethoxybenzoic acid, 3,4- ~ Veratrinsäure *f*
dimethoxybenzoin Anisoin *n*

dimethoxy benzoquinone
  Dimethoxybenzochinon *n*
dimethoxylepidine Dimethoxylepidin *n*
dimethoxymethane Formal *n,* Methylal *n*
dimethoxyphthalic acid, 3,4- ~ Hemipinsäure *f*
dimethoxy succinic acid
  Dimethoxybernsteinsäure *f*
dimethyl-2,6-dihydroxypurine, 1,3- ~
  Theophyllin *n*
dimethyl-6-pyridone, 2,4- ~ Lutidon *n*
dimethyl acetal Dimethylacetal *n*
dimethyl acetophenone Dimethylacetophenon *n*
dimethyl acetylene Crotonylen *n,*
  Dimethylacetylen *n*
dimethyl acridine Dimethylacridin *n*
dimethyl acrylic acid Dimethylacrylsäure *f*
dimethylamine Dimethylamin *n*
dimethylamine hydrochloride
  Dimethylaminhydrochlorid *n*
dimethylaminoazobenzene
  Dimethylaminoazobenzol *n*
dimethylaminobenzaldehyde
  Dimethylaminobenzaldehyd *m,* Ehrlichs
  Aldehydreagens *n*
dimethylaminobenzalrhodanine
  Dimethylaminobenzalrhodanin *n*
dimethylaminobenzene Dimethylaminobenzol *n*
dimethylaminobenzophenone
  Dimethylaminobenzophenon *n*
dimethylaminophenol Dimethylaminophenol *n*
dimethylaniline Dimethylanilin *n*
dimethylanthracene Dimethylanthracen *n*
dimethylanthraquinone Dimethylanthrachinon *n*
dimethylarsenic monobromide Kakodylbromid *n*
dimethylarsenic monochloride Kakodylchlorid *n*
dimethylarsine Kakodylwasserstoff *m*
dimethylarsinic acid Kakodylsäure *f*
dimethylated dimethyliert
dimethylbenzacridine Dimethylbenzacridin *n*
dimethylbenzanthracene
  Dimethylbenzanthracen *n*
dimethylbenzene Dimethylbenzol *n,* Xylol *n*
dimethylbenzidine Tolidin *n*
dimethylbenzil Tolil *n*
dimethylbenzimidazole Dimethylbenzimidazol *n*
dimethylbenzoic acid Xylylsäure *f,* **2,3-** ~
  Hemellitylsäure *f,* **3,5-** ~ Mesitylensäure *f*
dimethylbenzthiophanthrene
  Dimethylbenzthiophanthren *n*
dimethylbutadiene Dimethylbutadien *n*
dimethylbutadiene caoutchouc
  Dimethylbutadienkautschuk *m*
dimethylbutane Dimethylbutan *n,* **2,2-** ~
  Neohexan *n*
dimethylbutylquinoline Dimethylbutylchinolin *n*
dimethylbutylsulfonium iodide
  Dimethylbutylsulfoniumjodid *n*
dimethylbutyne Dimethylbutin *n*
dimethylcarbinol Dimethylcarbinol *n*

dimethylchrysene Dimethylchrysen *n*
dimethylcyclohexanedione Dimedon *n*,
Dimethylcyclohexandion *n*
dimethylcyclopentane Dimethylcyclopentan *n*
dimethyldichlorosilane Dimethyldichlorsilan *n*
dimethyldiethoxysilane Dimethyldiäthoxysilan *n*
dimethyldihydroxyanthracene
Dimethyldihydroxyanthracen *n*
dimethyldiphenyl Dimethyldiphenyl *n*
dimethyldiphenylurea
Dimethyldiphenylharnstoff *m*
dimethyldipyrrylmethene
Dimethyldipyrrylmethen *n*
dimethyldisilane Dimethyldisilan *n*
dimethylenediamine Dimethylendiamin *n*
dimethylene oxide Äthylenoxid *n*
dimethyl ether Dimethyläther *m*, Holzäther *m*
(obs), Methyläther *m*
dimethylethylene Dimethyläthylen *n*
dimethylethylenediamine
Dimethyläthylendiamin *n*
dimethylethylmethane Dimethyläthylmethan *n*
dimethylethylpyrrole Dimethyläthylpyrrol *n*
dimethylformamide Dimethylformamid *n*
dimethylfuranecarboxylic acid Uvinsäure *f*
dimethylglucose Dimethylglucose *f*
dimethylglyoxal Dimethylglyoxal *n*
dimethylglyoxime Dimethylglyoxim *n*,
Tschugajeffs Reagens *n*
dimethyl hydantoin formaldehyde
Dimethylhydantoinformaldehyd *m*
dimethylhydroquinol Dimethylhydrochinon *n*,
Xylohydrochinon *n*
dimethylisobutenylcyclopropane
Dimethylisobutenylcyclopropan *n*
dimethylketone Aceton *n*, Dimethylketon *n*
dimethyllevulinic acid, 2,2- ~ Mesitonsäure *f*
dimethylmaleic anhydride
Dimethylmaleinsäureanhydrid *n*
dimethylmalonate Methylmalonat *n*
dimethylmalonic acid Dimethylmalonsäure *f*
dimethylmandelic acid Dimethylmandelsäure *f*
dimethylmorphine Thebain *n*
dimethylnaphthalene Dimethylnaphthalin *n*
dimethylnaphthalene picrate
Dimethylnaphthalinpikrat *n*
dimethylnaphthidine Dimethylnaphthidin *n*
dimethylnaphthoquinone
Dimethylnaphthochinon *n*
dimethyloxalate Methyloxalat *n*
dimethyloxaloacetic acid
Dimethyloxalessigsäure *f*
dimethyloxamide Dimethyloxamid *n*
dimethyloxazole Dimethyloxazol *n*
dimethyl parabanic acid Cholestrophan *n*
dimethylparaconic acid, 2,2- ~ Terpentinsäure *f*
dimethylpentane Dimethylpentan *n*
dimethylphenol Dimethylphenol *n*, Xylenol *n*

dimethylphenylenediamin
Dimethylphenylendiamin *n*
dimethylphenylpyrazolone
Dimethylphenylpyrazolon *n*
dimethylphosphine Dimethylphosphin *n*
dimethyl phthalate Dimethylphthalat *n*,
Phthalsäuredimethylester *m*
dimethylpiperazine Dimethylpiperazin *n*
dimethylpiperazine tartrate, 2,5- ~ Lycetol *n*
dimethylpiperidine, 2,6- ~ Lupetidin *n*
dimethylpropanal, 2,2- ~ Pivalinaldehyd *m*
dimethylpropane, 2,2- ~ Neopentan *n*
dimethylpropanoic acid, 2,2- ~ Pivalinsäure *f*
dimethylpyridine Dimethylpyridin *n*, Lutidin *n*
dimethylpyrone Dimethylpyron *n*
dimethylpyrrole Dimethylpyrrol *n*
dimethylquinoline Dimethylchinolin *n*
dimethylresorcinol Dimethylresorcin *n*,
Xylorcin *n*
dimethylsalicylaldehyde
Dimethylsalicylaldehyd *m*
dimethyl sebacate Dimethylsebacat *n*
dimethylsilane Dimethylsilan *n*
dimethylsilanediol Dimethylsilandiol *n*
dimethylsilicone Dimethylsilicon *n*
dimethylsiloxane Dimethylsiloxan *n*
dimethyl sulfate Dimethylsulfat *n*,
Schwefelsäuredimethylester *m*
dimethyl sulfide Methylsulfid *n*
dimethyl sulfone Methylsulfon *n*
dimethyl sulfoxide Dimethylsulfoxid *n*,
Methylsulfoxid *n*
dimethyltetrahydrobenzaldehyde
Dimethyltetrahydrobenzaldehyd *m*
dimethylthallium Dimethylthallium *n*
dimethylthetin Dimethylthetin *n*
dimethylthiambutene Dimethylthiambuten *n*
dimethylthiophene Thioxen *n*
dimethylthreonamide Dimethylthreonamid *n*
dimethyltoluidine Dimethyltoluidin *n*
dimethylurea Dimethylharnstoff *m*
dimethylxanthine Dimethylxanthin *n*, 1,3- ~
Theophyllin *n*, 1,7- ~ Paraxanthin *n*, 3,7- ~
Theobromin *n*
dimethyl yellow Dimethylgelb *n*
dimethyl zinc Zinkmethyl *n*
diminish *v* [ab]schwächen, abflauen, abnehmen,
geringer werden, verkleinern, vermindern,
verringern
diminished verkürzt, vermindert
diminishing Abnehmen *n*
diminishing glass Verkleinerungsglas *n*
diminution Verringerung *f*, Abnahme *f*,
Kleinerwerden *n*, Minderung *f*, Schwächung *f*,
Verkleinerung *f*, Verminderung *f*, ~ of the red
cells Erythrozytenverminderung *f* (Med)
dimmer Abblendvorrichtung *f*
dimming Abdunkelung *f*
dimming device Abblendvorrichtung *f*

dimming switch Abblendschalter *m*
dimness Mattheit *f*
dimorphecolic acid Dimorphecolsäure *f*
dimorphic dimorph (Krist), zweigestaltig
dimorphism Dimorphie *f* (Krist),
Dimorphismus *m* (Krist)
dimorphous dimorph (Krist), zweigestaltig
dimoxyline Dimoxylin *n*
dimyristin Dimyristin *n*
din Getöse *n*
dinaphthacridine Dinaphthacridin *n*
dinaphthanthradiquinone
Dinaphthanthradichinon *n*
dinaphthanthrone Dinaphthanthron *n*
dinaphthazine Dinaphthazin *n*,
Phenophenanthrazin *n*
dinaphthazinium hydroxide
Dinaphthaziniumhydroxid *n*
dinaphthofluorindine Dinaphthofluorindin *n*
dinaphthol Dinaphthol *n*
dinaphthoquinone Dinaphthochinon *n*
dinaphthoxanthene Dinaphthoxanthen *n*
dinaphthyl Dinaphthyl *n*
dinaphthylene Dinaphthylen *n*
dinaphthyl ether Dinaphthyläther *m*
dinaphthyline Dinaphthylin *n*
dinaphthylphenylenediamine
Dinaphthylphenylendiamin *n*
dinas brick Dinasstein *m*, Dinasziegel *m*
dineutron Doppelneutron *n*
dinicotinic acid Dinicotinsäure *f*
dinite Dinit *m*
dinitro-1-naphthol-7-sulfonic acid, 2,4- ~
Flaviansäure *f*
dinitroaminophenol Dinitroaminophenol *n*
dinitroaniline Dinitroanilin *n*
dinitroanisol Dinitroanisol *n*
dinitroanthraquinone Dinitroanthrachinon *n*
dinitrobenzene Dinitrobenzol *n*
dinitrochlorobenzene Dinitrochlorbenzol *n*
dinitrodihydroxy-benzoquinone Nitranilsäure *f*
dinitrodiphenic acid Dinitrodiphensäure *f*
dinitrodiphenyl Dinitrodiphenyl *n*
dinitrodiphenylamine Dinitrodiphenylamin *n*,
Citronin *n*
dinitrofluorobenzene Dinitrofluorbenzol *n*
dinitrogen monoxide Distickstoffoxid *n*,
Stickstoffoxydul *n* (obs)
dinitrogen pentoxide Stickstoffpentoxid *n*
dinitrogen tetroxide Distickstofftetroxid *n*,
Stickstofftetroxid *n*
dinitrogen trioxide Distickstofftrioxid *n*,
Salpetrigsäureanhydrid *n*,
Stickstoffsesquioxid *n*, Stickstofftrioxid *n*
dinitroglycerol Dinitroglyzerin *n*
dinitroglycerol blasting charge
Dinitroglyzerinsprengstoff *m*
dinitromethylaniline Dinitromethylanilin *n*
dinitronaphthalene Dinitronaphthalin *n*

dinitronaphthol Dinitronaphthol *n*
dinitrophenol Dinitrophenol *n*
dinitrophenylacridine Dinitrophenylacridin *n*
dinitrophenyl amino acid
Dinitrophenylaminosäure *f*
dinitrophenylhydrazine Dinitrophenylhydrazin *n*
dinitrophenyl osazone Dinitrophenylosazon *n*
dinitrotoluene Dinitrotoluol *n*
dioctahedral dioktaedrisch (Krist),
doppelachtflächig (Krist)
dioctahedron Dioktaeder *n* (Krist),
Doppelachtflächner *m* (Krist)
dioctyl Hexadecan *n*
dioctyl phthalate Dioctylphthalat *n*
diode Diode *f* (Elektr), Doppelelektrode *f*
(Elektr), Zweipolröhre *f* (Elektr), **point
contact** ~ Spitzendiode *f*
diode path Diodenstrecke *f* (Elektr)
diode rectification Zweipolgleichrichtung *f*
diode rectifier Diodengleichrichter *m*
diode tube Zweipolröhre *f*
diodone Diodon *n*
diogenal Diogenal *n*
diol Diol *n*
diolefine Diolefin *n*
diolein Diolein *n*
dioleopalmitine Dioleopalmitin *n*
dioleostearin Dioleostearin *n*
dionine Dionin *n*, Äthylmorphinhydrochlorid *n*
diopside Diopsid *m* (Min), Mussit *m* (Min)
diopside variety Grünspat *m* (Min)
dioptase Dioptas *m* (Min), Kupfersmaragd *m*
(Min)
diopter Dioptrie *f* (Phys)
dioptrics Brechungslehre *f*, Dioptrik *f*
diorite Diorit *m* (Min), Aphanit *m* (Min)
dioritic diorithaltig
diorsellinic acid Diorsellinsäure *f*,
Lecanorsäure *f*
diorthotolyl guanidine Diorthotolylguanidin *n*
dioscin Dioscin *n*
dioscorine Dioscorin *n*
dioscorinol Dioscorinol *n*
diosgenin Diosgenin *n*
diosmetine Diosmetin *n*
diosmin Diosmin *n*
diosmine Buccobitter *n*
diosphenol Diosphenol *n*
diospyrol Diospyrol *n*
dioxaborole Dioxaborol *n*
dioxan[e] Dioxan *n*
dioxaphospholane Dioxaphospholan *n*
dioxazine Dioxazin *n*
dioxazole Dioxazol *n*
dioxene Dioxen *n*
dioxide Dioxid *n*, Doppeloxid *n*
dioxime Dioxim *n*
dioxin Dioxin *n*
dioxindole Dioxindol *n*

dioxolane Dioxolan *n*
dioxole Dioxol *n*
dioxopiperazine Dioxopiperazin *n*
dioxybenzene Dioxybenzol *n*
dioxydiphenyl Dioxydiphenyl *n*
dioxydiphenyl urethane Dioxydiphenylurethan *n*
dioxyquinoline Dioxychinolin *n*
dioxyquinone Dioxychinon *n*
dip Bad *n* (Färb), Brenne *f*
dip *v* [ein]tauchen, abbeizen (Metall), abbrennen
    (Metall), einsenken, versenken, ~ **dead** matt
    brennen, ~ **[into]** hineintauchen
dipalmitin Dipalmitin *n*
dipalmitoolein Oleodipalmitin *n*
dipalmitostearin Dipalmitostearin *n*
dip and drain equipment Tauchlackierapparat *m*
dip brazing Tauchhartlöten *n*
dip coat Überzug *m*
dip coating wax Tauchwachs *n*
dipentene Dipenten *n*
dipeptidase Dipeptidase *f* (Biochem)
dipeptide Dipeptid *n*
diperodon Diperodon *n*
dip gilding Eintauchvergoldung *f*
dip-hardening Tauchhärtung *f*
diphase zweiphasig
diphemanil Diphemanil *n*
diphenaldehyde Diphenaldehyd *m*
diphenanthryl Diphenanthryl *n*
diphenazyl Diphenazyl *n*
diphenetidine Diphenetidin *n*
diphenhydramine Diphenhydramin *n*
diphenic acid Diphensäure *f*
diphenine Diphenin *n*
diphenol Diphenol *n*
diphenolic acid Diphenolsäure *f*
diphenoquinone Diphenonchinon *n*
diphenyl Diphenyl *n*
diphenylacetylene Tolan *n*
diphenylamine Diphenylamin *n*
diphenylamine-arsine chloride Adamsit *m*
    (Kampfstoff)
diphenylamine blue Diphenylaminblau *n*
diphenylarsenous chloride
    Diphenylarsenchlorid *n*
diphenyl benzoquinone Diphenylbenzochinon *n*
diphenylboron chloride Diphenylborchlorid *n*
diphenylbutadiene Diphenylbutadien *n*
diphenylcarbamide Diphenylcarbamid *n*
diphenylcarbinol Benzhydrol *n*
diphenylcarbodiimide Carbodiphenylimid *n*
diphenyl carbonate Diphenylcarbonat *n*
diphenyldecapentaene Diphenyldecapentaen *n*
diphenyldecatetraene Diphenyldecatetraen *n*
diphenyldiazomethane Diphenyldiazomethan *n*
diphenyldichlorosilane Diphenyldichlorsilan *n*
diphenyl diketone Benzil *n*
diphenylene Diphenylen *n*

diphenylene dioxide Diphenylendioxid *n*,
    Phendioxin *n*
diphenylene disulfide Thianthren *n*
diphenylene imide Diphenylenimid *n*
diphenylene ketone Fluorenon *n*
diphenylene oxide Diphenylenoxid *n*
diphenylene sulfide Diphenylensulfid *n*
diphenylene sulfone Diphenylensulfon *n*
diphenylethane Diphenyläthan *n*, **1,2-** ~
    Dibenzyl *n*
diphenyl ether Diphenyläther *m*
diphenylethylamine Diphenyläthylamin *n*
diphenylethylene Diphenyläthylen *n*, Stilben *n*
diphenylglyoxal Benzil *n*
diphenylglyoxime Diphenylglyoxim *n*
diphenylguanidine Diphenylguanidin *n*
diphenylhexadiene Diphenylhexadien *n*
diphenylhexatriene Diphenylhexatrien *n*
diphenylhydrazine Diphenylhydrazin *n*,
    Hydrazobenzol *n*
diphenyline Diphenylin *n*
diphenyliodonium hydroxide
    Diphenyljodoniumhydroxid *n*
diphenyliodonium iodide
    Diphenyljodoniumjodid *n*
diphenyl isomerism Diphenylisomerie *f*
diphenyl ketone Benzophenon *n*,
    Diphenylketon *n*
diphenylmethane Diphenylmethan *n*
diphenylmethane dyestuff
    Diphenylmethanfarbstoff *m*
diphenylnaphthylmethyl
    Diphenylnaphthylmethyl *n*
diphenyloctatetraene Diphenyloctatetraen *n*
diphenyl oxalate Phenostal *n*
diphenyl phenylene diamine
    Diphenylphenylendiamin *n*
diphenylpolyene Diphenylpolyen *n*
diphenylquinaldine blue
    Diphenylchinaldinblau *n*
diphenylquinoxaline Diphenylchinoxalin *n*
diphenylsilanediol Diphenylsilandiol *n*
diphenylstannic chloride Zinndiphenylchlorid *n*
diphenyl sulfone Diphenylsulfon *n*,
    Sulfobenzid *n*
diphenylthiocarbazone Diphenylthiocarbazon *n*,
    Dithizon *n*
diphenylthiourea Diphenylthioharnstoff *m*,
    Sulfocarbanilid *n*
diphenyl tin chloride Zinndiphenylchlorid *n*
diphenylurea Carbanilid *n*
diphenylyl Diphenylyl *n*
diphonia, dynamic ~ dynamische Diphonie *f*
    (Akust)
diphosgene Diphosgen *n*, Surpalit *n*
diphosphate Pyrophosphat *n*
diphosphoglyceraldehyde
    Diphosphoglycerinaldehyd *m*
diphosphoglyceric acid Diphosphoglycerinsäure *f*

diphosphoinositide Diphosphoinositid *n*
diphosphopyridine nucleotide
    Diphosphopyridinnucleotid *n*, **reduced** ~
    reduziertes Diphosphopyridinnucleotid *n*
diphosphoric acid Diphosphorsäure *f*
diphosphothiamine Diphosphothiamin *n*
diphthalic acid Diphthalsäure *f*
diphthalyl Diphthalyl *n*
diphtheria Diphtherie *f* (Med)
diphtheria bacillus Diphtheriebazillus *m*
diphyscion Diphyscion *n*
dipicolinate Dipicolinat *n*
dipicolinic acid Dipicolinsäure *f*
dipicric acid Dipikrinsäure *f*
dipicrylamine Dipikrylamin *n*
dipipanone Dipipanon *n*
dipipecolinic acid Dipipecolinsäure *f*
diplogen Deuterium *n*
diploid Diploeder *n* (Krist)
diploid *a* diploid, zweifach (Biol)
diplosal Diplosal *n*, Disalicylsäure *f*,
    Salicylosalicylsäure *f*
diploschistesic acid Diploschistessäure *f*
diplosome doppeltes Zentralkörperchen *n*
diplospartyrine Diplospartyrin *n*
dipolar dipolar
dipolar ion Dipolion *n*, Zwitterion *n*
dipole Dipol *m*, **permanent** ~ permanenter
    Dipol
dipole *a* zweipolig
dipole approximation Dipolnäherung *f*
dipole array Dipolanordnung *f*
dipole axis Dipolachse *f*
dipole dipole broadening
    Dipoldipolverbreiterung *f*
dipole dipole interaction
    Dipoldipolwechselwirkung *f*
dipole molecule Dipolmolekül *n*
dipole moment Dipolmoment *n*
dipole rule Dipol-Regel *f* (Stereochem)
dipole surface density Dipolflächendichte *f*
dipole transition Dipolübergang *m*
dipotassium phosphate Dikaliumphosphat *n*
Dippel's oil Dippelsöl *n*, Knochenöl *n*
dipper Eintaucher *m*
dipping Eintauchen *n*, Abbeizen *n*,
    Dekapieren *n* (Met), Gelbbrennen *n* (Met)
dipping battery Tauchbatterie *f* (Elektr)
dipping counter Tauchzähler *m*
dipping electrode Tauchelektrode *f*
dipping frame Tauchrahmen *m*
dipping lacquer Tauchlack *m*
dipping machine Tauchmaschine *f*
dipping mold Tauchform *f*
dipping needle Inklinationsnadel *f*
dipping paint Tauchfarbe *f*
dipping pan Abbrennkessel *m*
dipping pot Gelbbrenntopf *m* (Met)
dipping process Tauchverfahren *n*

dipping pyrometer Eintauchpyrometer *n*
dipping rod Wünschelrute *f*
dipping sieve Gelbbrennsieb *n* (Met)
dipping varnish Tauchlack *m*
dip pipe Tauchrohr *n*, Verschlußrohr *n*
diprene Dipren *n*
diprismatic doppeltprismatisch
dipropargyl Dipropargyl *n*
dipropionyl methane Dipropionylmethan *n*
dipropylamine Dipropylamin *n*
dipropylbarbituric acid Dipropylbarbitursäure *f*
dipropylenetriamine Dipropylentriamin *n*
dipropyl ether Dipropyläther *m*
dipropyl ketone Dipropylketon *n*, Butyron *n*
dipropylmalonylurea
    Dipropylmalonylharnstoff *m*
dipropylmethane Heptan *n*
dip soldering Eintauchlöten *n*, Tauchlöten *n*
dipsomania Dipsomanie *f* (Med), Trunksucht *f*
dipstick Meßstab *m* (Öl)
dip tank Eintauchtrog *m*, Tauchbehälter *m*,
    Tauchtank *m*
dipterine Dipterin *n*
dipterocarpol Dipterocarpol *n*
dipterocarpone Dipterocarpon *n*
dip-tinning Tauchverzinnung *f*
dip tube Steigrohr *n* (Aerosoldose)
dip welding Eintauchschweißen *n*
dipyrazolanthronyl Dipyrazolanthronyl *n*
dipyre Dipyr *m* (Min), Schmelzstein *m* (Min)
dipyridine coproporphyrin
    Dipyridinkoproporphyrin *n*
dipyridine hematoporphyrin
    Dipyridinhämatoporphyrin *n*
dipyridine mesoporphyrin
    Dipyridinmesoporphyrin *n*
dipyridine protoporphyrin
    Dipyridinprotoporphyrin *n*
dipyridyl Dipyridyl *n*
dipyrrylmethene Dipyrrylmethen *n*
diquinaldine Dichinaldin *n*
diquinidine Dichinidin *n*
diquinoline Dichinolin *n*
diquinolyl Dichinolyl *n*
direct direkt, unmittelbar
direct *v* hinweisen, lenken, richten, orientieren
direct-arc furnace
    Lichtbogen-Widerstandsofen *m*
direct black Direktschwarz *n*
direct connection Direktschaltung *f*
direct contact condensation Mischkondensation *f*
direct current Gleichstrom *m*, **capacity of** ~
    Gleichstrombelastbarkeit *f*
direct current armature Gleichstromanker *m*
direct current converter Gleichstromumformer *m*
direct current [electric] energy
    Gleichstromenergie *f*
direct current motor Gleichstrommotor *m*
direct current plant Gleichstromanlage *f*

direct current system Gleichstromkreis *m*,
　Gleichstromsystem *n*
direct current terminal Gleichstromklemme *f*
direct current value Gleichstromgröße *f*
direct dyestuff Direktfarbstoff *m*
directed orientiert
directing magnet Richtmagnet *m*
direct-injection molding angußloses Spritzen *n*
direction Richtung *f*, Anweisung *f*, Leitung *f*,
　Richtlinie *f*, Vorschrift *f*, **axial** ~ achsiale
　Richtung *f*, **change of** ~ Richtungsänderung *f*,
　Richtungswechsel *m*, **effective** ~
　Wirkungsrichtung *f*, **independent of** ~
　richtungsunabhängig, **indication of** ~
　Richtungsangabe *f*, **in opposite** ~ in
　entgegengesetzter Richtung *f*, widersinnig,
　opposite ~ Gegensinn *m*, **preferred** ~
　Vorzugsrichtung *f*, **privileged** ~ bevorzugte
　Richtung *f* (Opt), **vertical** ~ senkrechte
　Richtung *f*
directional gerichtet (Valenz, Kristall etc.)
directional breakdown Richtungsdurchschlag *m*
directional coincidence Richtungskoinzidenz *f*
directional counter Richtzähler *m* (Atom)
directional dependence Richtungsabhängigkeit *f*
directional diagram Richtcharakteristik *f*
directional distribution Richtungsverteilung *f*
directional focus[s]ing Richtungsfokussierung *f*
directional property Richtfähigkeit *f*
directional quantity Richtgröße *f*
directional quantization Richtungsquantelung *f*
direction finding Ortung *f*
direction focus[s]ing richtungsfokussierend
direction indicator Richtungsanzeiger *m*,
　Folgezeiger *m*, Winker *m* (Auto)
direction in space Raumrichtung *f*
direction of arrival Einfallsrichtung *f*
direction of rotation Drehungssinn *m*
directions for testing Prüfungsvorschriften *pl*
direction[s] for use Gebrauchsanweisung *f*,
　Gebrauchsvorschrift *f*
direction uncertainty Richtungsunschärfe *f*
directory Verzeichnis *n*
direct printing Direktdruck *m*, Direktkopieren *n*
direct process Erzfrischverfahren *n* (Metall)
direct-process malleable iron Renneisen *n*
direct radiation Direktstrahlung *f*
direct-reading instrument Ablesegerät *n*
directrix Direktrix *f* (Geom), Leitkurve *f*
　(Geom)
direct steam Freidampf *m*
direct voltage Gleichspannung *f* (Elektr)
diresorcinol Diresorcin *n*
diresorcylic acid Diresorcylsäure *f*
dirhein Dirhein *n*
dirhodan Dirhodan *n*
diricinolein Diricinolein *n*
diricinolic acid Diricinolsäure *f*
dirt Schmutz *m*

dirt dissolving schmutzlösend
dirt trap Schmutzfänger *m*
dirty schmutzig
dirty water Schmutzwasser *n*
disable *v* verstümmeln, außerstand setzen
disaccharide Disaccharid *n*, Biose *f*
disaccustom *v* entwöhnen
disadvantage Nachteil *m*
disadvantage *v* benachteiligen
disadvantage factor Absenkungsfaktor *m* (Atom)
disadvantageous nachteilig, ungünstig
disaggregation Desaggregation *f*
disalicylic acid Disalicylsäure *f*, Diplosal *n*
disalicylide Disalicylid *n*, Salosalicylid *n*
disalignment Lageänderung *f*
disappear *v* verschwinden, ausbleiben
disappearance Verschwinden *n*, Ausbleiben *n*
disarranged ungeordnet
disarrangement Fehlordnung *f*, Störung *f*,
　Unordnung *f*
disassemble *v* auseinandernehmen,
　abmontieren, ausbauen, demontieren,
　zerlegen
disassembly Demontage *f*, Zerlegung *f*
disc Scheibe *f* (see also disk)
disc arc Lichtbogenscheibe *f*
discard Abfall *m*
discard *v* abwerfen, ablegen
discatol Diskatol *n*
disc atomizer Scheibenzerstäuber *m*
disc attrition mill Scheibenmühle *f*
disc centrifuge Tellerzentrifuge *f*
disc cooler Scheibenkühler *m*
disc drier Tellertrockner *m*
disc filter Scheibenfilter *n*
discharge Ausfluß *m*, Abfluß *m*, Abführung *f*,
　Ablaß *m*, Ausladung *f* (Fracht), Auslauf *m*,
　Ausströmen *n*, Auswurf *m*, Entladung *f*
　(Elektr), **aperiodic** ~ aperiodische
　Entladung *f*, **assisted** ~ unselbständige
　Entladung *f*, **cathodic** ~ kathodische
　Entladung *f*, **coefficient of** ~
　Ausflußkoeffizient *m*, **colored** ~ Buntätze *f*,
　dead-beat ~ aperiodische Entladung, **duration**
　of ~ Entladedauer *f*, **impulsive** ~
　aperiodische Entladung *f*, **non-oscillatory** ~
　aperiodische Entladung *f*, **non-self-sustained**
　~ unselbständige Entladung *f*, **process of** ~
　Abflußvorgang *m* (Techn), **rate of** ~
　Ausströmgeschwindigkeit *f* (Techn),
　Entladegeschwindigkeit *f* (Elektr),
　self-sustained ~ selbständige Entladung *f*,
　silent ~ stille Entladung *f*, **spontaneous** ~
　selbständige Entladung *f*, **strength of** ~
　Entladestärke *f* (Elektr), **time of** ~
　Ausflußzeit *f* (Techn), Entladezeit *f* (Elektr),
　unassisted ~ selbständige Entladung *f*,
　unidirected ~ gleichgerichtete Entladung *f*,

volumetric rate of ~ Ausstoßvolumen *n* (Techn)
discharge *v* abfließen, abladen (Fracht), ablassen, absondern, ausfließen, ausladen (Fracht), ausscheiden (Med), ausschießen, ausströmen, entladen (Elektr), entlasten, entleeren, münden
dischargeability Ätzbarkeit *f*
dischargeable ätzbar; entladbar; ausladbar
discharge action Ätzwirkung *f*
discharge afterglow Entladungsnachglimmen *n* (Elektr)
discharge aperture Abstichloch *n* (Metall)
discharge apparatus Abzugsvorrichtung *f*
discharge belt Abzugband *n*, Bandabsetzer *m*
discharge branch Druckstutzen *m*
discharge capacity Entladekapazität *f* (Elektr)
discharge cock Ablaßhahn *m*, Ausflußhahn *m*, Entweichungshahn *m*
discharge coefficient Ausflußzahl *f*
discharge color Ätzfarbe *f*
discharge connection Ausgußstutzen *m*
discharge current Entladestrom *m* (Elektr), **strength of** ~ Entladestromstärke *f* (Elektr)
discharge device Entladungsapparat *m*
discharge door Entleerungstür *f*, Falltür *f*
discharge door top Sattel *m* der Falltür
discharged water Ablaufwasser *n*
discharge end Ablaufende *n*
discharge flow Ausflußströmung *f*
discharge gas Abgas *n*
discharge hopper Ablauftrichter *m*, Abwurftrichter *m*
discharge jet Ausflußstrahl *m*
discharge lake Ätzlack *m*
discharge liquid Ablaufflüssigkeit *f*
discharge loss Austrittverlust *m*
discharge mordant Ätzbeize *f* (Färb)
discharge nozzle Blasdüse *f*, Zapfhahn *m*
discharge opening Ausflußloch *n*, Ausflußöffnung *f*, Durchflußöffnung *f*
discharge pipe Abflußrohr *n*, Abfallrohr *n*, Abflußleitung *f*, Ablaufrohr *n*, Ableitungsrohr *n*, Absaug[e]schlot *m*, Abzugskanal *m*, Abzugsrohr *n*, Ausströmungsrohr *n*, Druckrohr *n*
discharge piping Ableitung *f*, Druckleitung *f*
discharge potential Entladungspotential *n*, Entladungsspannung *f* (Elektr)
discharge pressure Abspritzdruck *m*, Enddruck *m*
discharge printing Ätz[beiz]druck *m*
discharge process, cathodic ~ kathodischer Entladungsvorgang
discharge pulse Entladungsstoß *m* (Elektr)
discharge quantity Abflußmenge *f*, Ausflußmenge *f*
discharger Ablader *m*, Elektrizitätsentlader *m*, Entlader *m* (Elektr)

discharge rate Entladegeschwindigkeit *f*, Fördergeschwindigkeit *f* (b. Rühren)
discharge space Entladungsraum *m*
discharge spark Entladungsfunken *m*
discharge stroke Entladungsschlag *m*
discharge test Entladeprobe *f*
discharge tip Auslaufspitze *f*
discharge tube Ausführungsrohr *n*, Entladungsröhre *f*
discharge valve Abblaseventil *n*, Abflußventil *n*, Ausflußventil *n*
discharge velocity Abflußgeschwindigkeit *f*, Ausflußgeschwindigkeit *f*
discharge voltage Abgabespannung *f*, Entladespannung *f* (Elektr)
discharge water Abflußwasser *n*
discharge yellow Enlevagegelb *n*
discharging Abblasen *n* (Dampf), Ausladen *n* (Fracht), Ausströmen *n*, Entladen *n* (Elektr), Entleeren *n*
discharging agent Ätzmittel *n*
discharging apparatus Entladevorrichtung *f*
discharging auxiliary Ätzhilfsmittel *n*
discharging chamber Entleerungskammer *f*
discharging current Entladestrom *m*
discharging process Entladevorgang *m* (Elektr)
disc harrow Egge *f*
disclose *v* aufdecken, enthüllen, offenbaren (Patent)
disclosure Offenbarung *f* (Patent)
disc machine Scheibenmaschine *f*
disc mill Räderwalzwerk *n*
disc mixer Scheibenrührer *m*
discolith Scheibenstein *m* (Geol)
discolor *v* entfärben, ausbleichen
discolorable abfärbig
discoloration Entfärbung *f*, Farbenveränderung *f*, Verfärbung *f*, **degree of** ~ Verschießungsgrad *m* (Text)
discoloration factor (of oil) Entfärbungszahl *f* (von Öl)
discolored entfärbt, ausgebleicht, mißfarbig
discoloring Abziehen *n* (Text), verfärbend
discomycetes Scheibenpilze *pl*
disconcerted fassungslos
disconnect *v* unterbrechen, abhängen, aushängen, [ab]lösen, abschalten, abstellen, auskuppeln, ausschalten, trennen
disconnectable ausrückbar
disconnected abgetrennt; zusammenhangslos; abgeschaltet
disconnecting Auskuppeln *n*, Trennen *n* (Elektr)
disconnecting box Trennmuffe *f*
disconnecting contact Ausschaltkontakt *m*
disconnecting insulator Trennisolator *m*
disconnecting lever Ausschalthebel *m*
disconnecting magnet Abschaltmagnet *m*
disconnecting switch Ausschalter *m*
disconnection Abschaltung *f*, Entkoppelung *f*

discontinuance Trennung *f*
discontinue *v* abbrechen, unterbrechen
discontinuity Diskontinuität *f* (Math), Sprung *m*
(Math), Unstetigkeit *f* (Math), **point of** ~
Unstetigkeitsstelle *f* (Math), **surface of** ~
Unstetigkeitsfläche *f*
discontinuity condition Unstetigkeitsbedingung *f*
discontinuity interaction
Unstetigkeitswechselwirkung *f*
discontinuity line Unstetigkeitslinie *f*
discontinuity surface Unstetigkeitsfläche *f*
discontinuous unterbrochen, diskontinuierlich,
ruckweise, sprunghaft, ungleichmäßig,
unstetig (Math)
discontinuous adsorption isotherm unstetige
Adsorptionsisotherme *f*
discontinuous curled chip Lamellenspan *m*
discount Rabatt *m*
discover *v* entdecken, auffinden, erforschen,
erschließen, nachweisen
discoverer Entdecker *m*, Erforscher *m*
discovery Entdeckung *f*, **place of** ~ Fundort *m*
disc pelletizer Pelletierteller *m*
disc piston Scheibenkolben *m*
discrasite Silberspießglanz *m* (Min)
discrepancy Abweichung *f*, Diskrepanz *f*,
Widerspruch *m*
discretisation error Diskretisierungsfehler *m*
discriminant Diskriminante *f* (Math)
discriminate *v* unterscheiden
discrimination Diskriminierung *f*,
Trennschärfe *f*, Unterscheidung *f*,
Unterschied *m*
discrimination index Unterscheidungsfaktor *m*
(Opt)
disc roller Profilrolle *f*
disc separator Scheibenrost *m*,
Tellerseparator *m*, Tellerwäscher *m*
disc-shaped scheibenförmig
disc-shaped conductor Scheibenkonduktor *m*
discuss *v* diskutieren, abhandeln, absprechen
discussion Diskussion *f*, **model for further** ~
Denkmodell *n*
disc valve Scheibenventil *n*
disease Erkrankung *f*, Krankheit *f* (Med),
Sucht *f* (Med), **cause of a** ~
Krankheitsursache *f*, **causing** ~
krankheitserregend, **chronic** ~ chronische
Krankheit, **communicable** ~ übertragbare
Krankheit, **contagious** ~ ansteckende
Krankheit, **course of a** ~
Krankheitsverlauf *m* (Med), **focus of a** ~
Krankheitsherd *m* (Med), **infectious** ~
Infektionskrankheit *f*, ansteckende Krankheit,
professional ~ Berufskrankheit *f*
disease carrier Krankheitsträger *m*
disease fungus pathogener Pilz *m*
disembitter *v* entbittern
disembittered entbittert

disemulsify *v* entemulgieren, entemulsionieren
disengage *v* [ab]lösen, ausklinken, ausschalten,
befreien, entbinden, in Freiheit setzen,
loskuppeln
disengaged frei, entbunden
disengaging Entbinden *n*
disengaging clutch Ausrückkupplung *f*,
Ausrückmuffe *f*
disengaging fork Ausrückergabel *f*
disengaging gear Auslösevorrichtung *f*,
Ausrückvorrichtung *f*
disengaging lever Ausrückhebel *m*
disengaging shaft Ausrückwelle *f*
disequilibrium Ungleichgewicht *n*, Labilität *f*
disfavor *v* benachteiligen
disfigure *v* entstellen
disgusting ekelerregend
dish Schale *f*, Schüssel *f*, Teller *m*
disharmonious unharmonisch
dish drainer Abtropfgestell *n*
dished (e. g. surface) einwärts gekrümmt
dished electrode Schalenelektrode *f*
dish-washer Spülmaschine *f*
dish washing agent Geschirrspülmittel *n*
disilane Disilan *n*, Siliciumäthan *n*, Silicoäthan *n*
disiloxane Disiloxan *n*
disincrustant Kesselsteinbeseitigungsmittel *n*,
Kesselsteingegenmittel *n*,
Kesselsteinlösemittel *n*
disinfect *v* desinfizieren, entgiften, entkeimen,
entpesten, entseuchen, reinigen
disinfectant Desinfektionsmittel *n*,
Desinfiziens *n*, Entgiftungsmittel *n*,
Entkeimungsmittel *n*
disinfectant soap Desinfektionsseife *f* (Pharm)
disinfecting Beizen *n* (Techn)
disinfecting apparatus Desinfektionsapparat *m*
disinfecting bath Beizbad *n* (Techn)
disinfecting liquid Beizflüssigkeit *f* (Techn)
disinfecting liquor Desinfektionswasser *n*
disinfecting paper Desinfektionspapier *n*
disinfecting power Beizkraft *f* (Techn),
Desinfektionskraft *f*
disinfection Desinfektion *f*, Desinfizierung *f*,
Entgiftung *f*, Entkeimung *f*, Entpestung *f*,
Entseuchung *f*
disinfection treatment, short ~ Kurzbeize *f*
(Techn)
disinfector Desinfektor *m*
disinsectization Insektenausrottung *f*
disintegrant Abbaumittel *n*
disintegrate *v* zerfallen, abbauen (Chem),
auflockern, aufschließen (Chem),
auseinanderfallen, desintegrieren, entmischen,
sich in seine Bestandteile auflösen, verwittern,
zerkleinern, zersetzen
disintegrated verwittert
disintegrated atom gespaltenes Atom *n*

disintegrating Zersetzen *n*, Verwittern *n*, Zerfallen *n*
disintegrating agent Aufschlußmittel *n* (Chem)
disintegrating machine Zerkleinerungsmaschine *f*
disintegration Zersetzung *f*, Abbau *m* (Chem), Auflockerung *f*, Auflösung *f*, Aufschluß *m* (Chem), Entmischung *f*, Trennung *f*, Verfall *m*, Verwitterung *f*, Zerfall *m*, Zerkleinerung *f*, Zerlegung *f*, **electrical** ~ elektrische Zerstäubung *f*, **process of** ~ Auflösungsprozeß *m*, **total** ~ Gesamtzerfall *m*
disintegration constant Zerfallskonstante *f* (Atom), Umwandlungskonstante *f* (Atom)
disintegration energy Zerfallsenergie *f*
disintegration law Zerfall[s]gesetz *n*
disintegration of floating material Schwimmstoffzerkleinerung *f*
disintegration plant Zerkleinerungsanlage *f*
disintegration probability Zerfallswahrscheinlichkeit *f*
disintegration process Aufschlußverfahren *n* (Chem)
disintegration rate Zerfallsgeschwindigkeit *f*
disintegration time Zerfall[s]zeit *f*
disintegration velocity, radioactive ~ radioaktive Zerfallsgeschwindigkeit *f*
disintegration voltage kritischer Spannungsabfall *m*
disintegrative zersetzend
disintegrator Desintegrator *m*, Knotenbrecher *m* (Zuck), Schleudermühle *f*, Zerkleinerer *m*
disjoin *v* absondern
disk Scheibe *f* (see also disc)
disk brake Scheibenbremse *f*
disk cell Scheibenzelle *f*
disk clutch Lamellenkupplung *f*, Scheibenkupplung *f*
disk cover Scheibenschützer *m*
disk dryer Tellertrockner *m*
disk filter Scheibenfilter *n*
disk holder Scheibenhalter *m*
disk impeller [mixer] Scheibenkreiselmischer *m*
disk mill, toothed ~ Zahnscheibenmühle *f*
disk of refined copper Garscheibe *f*
disk valve Tellerventil *n*
disk winch Scheibenhaspel *f*
dislocate *v* verlagern
dislocation Dislokation *f*, Fehlordnung *f*, Verlagerung *f*, Verschiebung *f* (Geol), Versetzung *f*, **energy of** ~ Versetzungsenergie *f* (Krist), **partial** ~ Teilversetzung *f*
dislocation density Versetzungsdichte *f*
dislocation distribution Versetzungsverteilung *f*
dislocation jog Versetzungssprung *m*
dislocation kernel Versetzungskern *m*
dislocation line Versetzungslinie *f*
dislocation motion Versetzungsbewegung *f*
dislocation network Versetzungsnetzwerk *n*

dismantle *v* demontieren, abmontieren, abtakeln (Schiff), abwracken (Wrack), auseinandernehmen
dismantling Abmontierung *f*
dismemberment Zerstückelung *f*
dismiss *v* entlassen
dismissing Entlassen *n*
dismount *v* demontieren, abmontieren, ausbauen (Teile), auseinandernehmen
dismountable abnehmbar, zerlegbar
dismounting Abmontierung *f*, Demontage *f*, ~ **of a machine** Abbau *m* einer Maschine
dismutation Dismutation *f*
disodium citrate Dinatriumcitrat *n*
disodium [hydrogen] arsenate Dinatriumarsenat *n*
disodium malonate [Di]Natriummalonat *n*
disodium salt Dinatriumsalz *n*
disodium tetraborate Dinatriumtetraborat *n*
disorder Unordnung *f*, Fehlordnung *f* (Krist), Störung *f*
disordered ungeordnet
disorderly unordentlich
disorder order transition Unordnungsordnungsumwandlung *f* (Chem)
disorder scattering Fehlordnungsstreuung *f*
dispatch Abfertigung *f*, Entnahme *f* (aus einem Lager)
dispatch *v* abfertigen, versenden
dispatch aisle Entnahmegang *m*
dispatch pail Versandeimer *m*
dispensable entbehrlich
dispensatory Apothekerordnung *f*, Arzneibuch *n*, Pharmakopöe *f*
dispense *v* austeilen; dispensieren
dispenser Abfüllvorrichtung *f*, Verteiler *m*
dispenser, dispensing machine Abrollapparat *m*
dispensing Arzneizubereitung *f*
dispersal effect Entleerungseffekt *m* (Atom)
dispersant Dispersionsmittel *n*
disperse, coarsely ~ grobdispers
disperse *a* dispers
disperse *v* [zer]streuen, dispergieren, verteilen, zerlegen (Opt), zerstäuben
dispersed dispers, **finely** ~ fein verteilt
dispersed dyestuff Dispersionsfarbstoff *m*
disperse system Dispersoid *n*
dispersing Verteilen *n*
dispersing action Dispersionswirkung *f*
dispersing agent Dispersionsmittel *n*, Auflockerungsmittel *n*, Dispergens *n*, Verteilungsmittel *n*
dispersing tool Dispergiergerät *n*
dispersion Dispersion *f*, Streuung *f*, Zerstäubung *f*, Zerstreuung *f*, **acoustic** ~ Dispersion *f* des Schalles, **coefficient of** ~ Zerstreuungskoeffizient *m*, **degree of** ~ Dispersionsgrad *m*, **linear** ~ lineare Streuung *f* (Spektr), **magnetic** ~ magnetische

Streuung *f*, **rotary ~** Drehungsdispersion *f*,
spectral ~ spektrale Auflösung *f* (Opt)
dispersional frequency Grenzfrequenz *f*
dispersion angle Streuungswinkel *m*
dispersion binder Bindemitteldispersion *f*
dispersion coefficient Streufaktor *m*
dispersion colloid Dispersionskolloid *n*
dispersion effect Dispersionseffekt *m*
dispersion force Dispersionskraft *f*
dispersion kneader Dispersionskneter *m*
dispersion lens Streulinse *f*
dispersion measurement Dispersionsmessung *f*
dispersion medium Dispersionsmittel *n*
dispersion method Dispersionsmethode *f*
dispersion of a light source Streuung *f* einer
    Lichtquelle
dispersion of light Lichtstreuung *f*
dispersion of rotation Rotationsdispersion *f*
    (Opt)
dispersion phase Streuphase *f*
dispersion range Streubereich *m*
dispersity Dispersionsgrad *m*
dispersive dispergierend
dispersive power Dispersionsvermögen *n*,
    Zerstreuungsvermögen *n*
dispersoid Dispersoid *n* (Chem)
dispiro compound Dispiroverbindung *f*
displace *v* verdrängen, deplacieren, ersetzen
    (Chem), verlagern, verlegen, verschieben
displaceability Austauschbarkeit *f* (Chem),
    Verschiebbarkeit *f*
displaceable verschiebbar; ersetzbar
displacement Verdrängung *f*, Deplacierung *f*,
    Ersatz *m* (Chem); Verlagerung *f*,
    Verrückung *f*, Verschiebung *f*, **computation of**
    ~ Deplacementsberechnung *f*, **horizontal ~**
    Horizontalverschiebung *f*, **scale of ~**
    Deplacementsskala *f*
displacement chromatography
    Verdrängungschromatographie *f*
displacement current Verschiebungsstrom *m*
displacement field Verschiebungsfeld *n*
displacement in potential Potentialverschiebung *f*
displacement law Verschiebungsgesetz *n*
displacement law of Soddy
    Soddy[-Fajans]scher *m* Verschiebungssatz *m*
displacement of the absorption band
    Verschiebung *f* der Absorptionsbande
displacement process Verschiebungsvorgang *m*
displacement pump, positive ~
    Verdrängerpumpe *f*
displacement-type motor
    Kolbenflüssigkeitszähler *m*
displacer piston Verdrängungskolben *m*
display *v* ausbreiten, ausstellen, entfalten
display package Leerpackung *f*
display packing Schaupackung *f*
disposable package Einwegverpackung *f*

disposal Beseitigung *f*, **radioactive waste ~**
    Beseitigung *f* radioaktiver Abfälle (pl)
dispose *v* disponieren, entscheiden, verteilen, **~**
    [of] verwerfen (Niederschlag)
disposition Disposition *f*, Veranlagung *f*
dispossess *v* enteignen
disproportion Mißverhältnis *n*
disproportionate unproportioniert,
    ungleichmäßig, unverhältnismäßig
disproportionate *v* disproportionieren
disproportionation Disproportionierung *f*
dispute *v* diskutieren, bestreiten
disqualification Ausschließung *f*,
    Disqualifikation *f*
disregard *v* vernachlässigen
disregarded unberücksichtigt
disrotatory disrotatorisch
disrupt *v* zerbrechen, zerreißen
disruption Bruch *m*, Zerreißung *f*
disruptive disruptiv (Elektr)
disruptive strength Durchschlagsfestigkeit *f*
disruptive voltage Durchschlagsspannung *f*
dissect *v* sezieren (Med), präparieren (Med),
    zerschneiden
dissectible zerlegbar
dissecting lens Präparierlupe *f*
dissecting microscope Präpariermikroskop *n*
dissection Sezieren *n*, Zerlegung *f*,
    Zerschneidung *f*
disseminate *v* ausstreuen, einstreuen, verbreiten
dissemination Einsprengung *f* (Geol)
dissertation Abhandlung *f*, Dissertation *f*,
    Doktorarbeit *f*
dissimilar ungleich, verschieden
dissimilarity Ungleichartigkeit *f*,
    Ungleichförmigkeit *f*, Ungleichheit *f*
dissimilation Dissimilation *f* (Biol)
dissipate *v* abgeben (Phys), zerstreuen
dissipation Dissipation *f*, Verteilung *f*,
    Zerstreuung *f*, **heat of ~** Dissipationswärme *f*
dissipation factor Verlustfaktor *m*
dissipationless line verlustlose Leitung *f* (Elektr)
dissipation of heat Wärmeableitung *f* (Phys)
dissociable dissoziierbar (Chem)
dissociate *v* abtrennen, dissoziieren (Chem),
    zerfallen
dissociating Zerfallen *n*
dissociation Abtrennung *f*, Aufspaltung *f*
    (Chem), Dissoziation *f* (Chem),
    Dissoziierung *f* (Chem), Zerfall *m* (Chem),
    coefficient of ~ Dissoziationskonstante *f*,
    electrolytic ~ elektrolytische Dissoziation,
    equilibrium of ~
    Dissoziationsgleichgewicht *n*, **heat of ~**
    Dissoziationswärme *f*, **retrogression of ~**
    Dissoziationsrückgang *m*
dissociation condition Dissoziationszustand *m*
dissociation constant Dissoziationskonstante *f*
dissociation degree Dissoziationsgrad *m*

dissociation energy Dissoziationsenergie *f*
dissociation equation Dissoziationsgleichung *f*
dissociation equilibrium
  Dissoziationsgleichgewicht *n*
dissociation formula Dissoziationsformel *f*
dissociation isotherm Dissoziationsisotherme *f*
dissociation limit Dissoziationsgrenze *f,*
  Zerfallsgrenze *f*
dissociation power Dissoziationsvermögen *n*
dissociation process Dissoziationsvorgang *m*
dissociation product Zerfallsprodukt *n* (Chem)
dissociation temperature
  Dissoziationstemperatur *f*
dissociation theory Dissoziationstheorie *f*
dissociative zersetzend
dissolution Auflösung *f,* Lösung *f* (Chem),
  Trennung *f,* **velocity of** ~
  Auflösungsgeschwindigkeit *f*
dissolvable auflösbar; löslich
dissolve, difficult to ~ schwerlöslich
dissolve *v* [auf]lösen (Chem), in Lösung gehen
  (Chem), solvieren (Chem), ~ **and reprecipitate**
  umfällen, ~ **by heat** aufschmelzen, ~ **out**
  herauslösen
dissolved acetylene [gas] Dissousgas *n*
dissolved substance Gelöste *n*
dissolvent Lösungsmittel *n,* Auflösungsmittel *n*
dissolver Auflösebehälter *m,* Auflöser *m*
dissolving Auflösen *n,* **rate of** ~ ,
  Lösegeschwindigkeit *f*
dissolving *a* [auf]lösend, solvierend
dissolving apparatus Löseapparat *m*
dissolving capacity Auflösungsvermögen *n*
  (Chem)
dissolving drum Lösekessel *m*
dissolving intermediary Lösungsvermittler *m*
dissolving liquid Auflösungsflüssigkeit *f*
dissolving machine Aufschließmaschine *f*
dissolving pan Klärpfanne *f,* Läuterpfanne *f*
dissolving power Auflösungskraft *f,*
  Lösungsvermögen *n* (Chem)
dissolving tank Lösekessel *m*
dissolving test Lösungsversuch *m*
dissolving vessel Auflösungsgefäß *n*
dissymmetry Asymmetrie *f,* Dissymmetrie *f,* **axial**
  ~ Axialdissymmetrie *f* (Stereochem),
  molecular ~ Molekeldissymmetrie *f,* ~ **due to**
  molecular overcrowding Dissymmetrie *f* durch
  räumliche Überlappung (Stereochem)
dissymmetry grouping dissymmetrische
  Gruppierung *f* (Stereochem)
distance Abstand *m,* Entfernung *f,* Strecke *f,*
  Weglänge *f,* **action at a** ~ Fernwirkung *f*
distance between grips Einspannlänge *f*
distance between points of support
  Knotenabstand *m*
distance between rolls Walzenabstand *m*
distance bolt Distanzschraube *f*
distance circle Entfernungskreis *m*

distance control Fernbedienung *f*
distance in air Luftstrecke *f*
distance meter Entfernungsmesser *m*
distance piece Abstandbuchse *f,*
  Abstandsstück *n,* Einsatzstück *n,*
  Zwischenstück *n*
distance switch Fernschalter *m*
distance thermometer Fernthermometer *n*
distance-velocity log (Br. E.) Totzeit *f*
  (Regeltechn)
distant entfernt
distearin Distearin *n*
distearopalmitin Distearopalmitin *n*
distemonanthin Distemonanthin *n*
distemper Temperafarbe *f,* Wasserfarbe *f,*
  non-washable ~ Leimfarbe *f,* **washable** ~
  Kaseinfarbe *f*
distend *v* ausweiten, dehnen
disthene Disthen *m* (Min), blättriger Beryll *m*
  (Min)
distichine Distichin *n*
distilbene Distilben *n*
distil[l] *v* destillieren, brennen (Alkohol), ~ **at**
  low temperature schwelen, ~ **off**
  abdestillieren, abtreiben, ~ **over**
  überdestillieren, ~ **singlings** luttern, ~ **weak**
  brandy luttern
distillability Destillierbarkeit *f*
distillable destillierbar, ~ **with [water] steam** mit
  Wasserdampf flüchtig
distillate Destillat *n,* ~ **of mineral oil**
  Öldestillat *n*
distillating flask Destillationskolben *m*
distillation Destillation *f,* Abziehen *n,*
  Brennen *n* (Alkohol), **azeotropic** ~ azeotrope
  Destillation, **cut in** ~ Fraktionsschnitt *m,*
  destructive ~ Crackdestillation *f,* trockene
  Destillation, Trockendestillation *f,*
  Vergasung *f,* Zersetzungsdestillation *f,* **dry** ~
  Zersetzungsdestillation *f,* **extractive** ~
  extraktive Destillation, **fractional** ~
  fraktionierte Destillation, **non-destructive** ~
  schonende Destillation *f*
  ~ **with steam** Wasserdampfdestillation *f*
distillation apparatus Destillationsapparatur *f,*
  Destillierapparat *m*
distillation characteristics Siedeverhalten *n*
distillation cut Destillationsschnitt *m*
distillation flask Siedegefäß *n,* Siedekolben *m*
distillation head Destillationsaufsatz *m,*
  Destillieraufsatz *m*
distillation loss Verlust *m* bei der Destillation
distillation of wood Holzvergasung *f*
distillation plant Destillationsanlage *f*
distillation pot Destilliertopf *m*
distillation process Destillationsverfahren *n*
distillation product Destillat *n,*
  Destillationsprodukt *n,*
  Entgasungserzeugnis *n*
distillation receiver Destillationsvorlage *f*

distillation residue Destillationsrückstand *m*
distillation thermometer
  Destillationsthermometer *n*
distillation vessel Destilliertopf *m*
distilled destilliert
distilled gas Destillationsgas *n*
distilled water destilliertes Wasser *n*, Aqua
  destillata (Lat)
distiller Branntweinbrenner *m*
distiller's solubles, dried ~ [Mais]Schlempe *f*
distillery Destillierhaus *n*, Brennerei *f* für
  Branntwein, Destillieranlage *f*, Garanlage *f*,
  Schnapsbrennerei *f*
distillery charcoal Destillierholzkohle *f*
distillery mash Destillatmaische *f*
distillery waste water Brennereiabwässer *pl*
distilling Destillieren *n*, ~ **over**
  Überdestillieren *n*
distilling apparatus Brennapparat *m*,
  Destillationsapparatur *f*, Destillierapparat *m*
distilling barley Brenngerste *f*
distilling column Destillierkolonne *f*
distilling filter Destillierfilter *m*
distilling flask Destilliergefäß *n*,
  Destillierkolben *m*
distilling head Destillierhelm *m*
distilling pipe Destillationsrohr *n*
distilling plant Destillationsanlage *f*
distilling stove Destillationsofen *m*
distilling tower Destillierkolonne *f*
distilling trap Fraktionsaufsatz *m*
distilling vessel Destillationsgefäß *n*
distinct ausgeprägt, kenntlich, klar, verschieden
distinction Auszeichnung *f*, Unterscheidung *f*
distinctive auffallend, charakteristisch
distinctive feature Unterscheidungsmerkmal *n*
distinctness Klarheit *f*
distinguish *v* unterscheiden, auszeichnen
distinguishable unterscheidbar, kenntlich
distinguished ausgezeichnet
distort *v* verzerren, deformieren, entstellen,
  verziehen
distorted verzerrt
distorted wave method Störwellenmethode *f*
distortion Verdrehung *f*, Deformation *f*,
  Deformierung *f*, Verformung *f*, Verzerrung *f*,
  amplitude ~ nichtlineare Verzerrung (Elektr),
  freedom from ~ Verzerrungsfreiheit *f*,
  nonlinear ~ nichtlineare Verzerrung (Elektr),
  permanent ~ bleibende Formänderung *f*,
  stationary ~ stationäre Verzerrung, **tolerance
  of** ~ Verzerrungstoleranz *f*, ~ **due to
  hardening** Härteverzug *m*, ~ **of amplitude**
  Dämpfungsverzerrung *f*
distortion broadening Verzerrungsverbreiterung *f*
distortion factor Klirrfaktor *m* (Elektr)
distortion-free verzerrungsfrei
distortionless verzerrungsfrei

distortion phase Störphase *f*
distribute *v* verteilen, aufteilen, austeilen,
  verbreiten
distributed, evenly or continuously ~
  gleichmäßig verteilt, **randomly** ~ statistisch
  verteilt, ~ **in lumps** punktförmig verteilt
distributed capacity Kapazitätsbelag *m*
distributing Verteilen *n*
distributing box Abzweigkasten *m* (Elektr)
distributing cock Steuerungshahn *m*
distributing equipment Zuteilvorrichtung *f*
distributing main Verteilungsleitung *f*
distributing mixer Verteilungsmischer *m* (Zuck)
distributing pan (cooling tower) Spritzteller *m*
distributing panel Verteilertafel *f*
distributing pipes Rohrnetz *n*
distributing station Verteilungszentrale *f*
distributing tube Verteilungsröhre *f*
distribution Verteilung *f*, Ausbreitung *f*,
  Verbreitung *f*, **law of** ~ Verteilungsgesetz *n*,
  spatial ~ räumliche Verteilung *f*, ~ **of
  electricity** Elektrizitätsverteilung *f*, ~ **of
  energy** Kraftverteilung *f*, ~ **of light**
  Lichtverteilung *f*, ~ **of power**
  Kraftverteilung *f*, ~ **of pressure**
  Druckverteilung *f*
distribution board Verteilertafel *f*
distribution chromatography
  Verteilungschromatographie *f*
distribution cock Verteilungshahn *m*
distribution coefficient Verteilungskoeffizient *m*
distribution constant Verteilungsquotient *m*
distribution curve Verteilungskurve *f*
distribution diagram Steuerungsdiagramm *n*
distribution function Verteilungsfunktion *f*
  (Math)
distribution law of Nernst Nernstsches
  Verteilungsgesetz *n*
distribution piece Verteilungsstück *n*
distribution pipe Verteilungsrohr *n*,
  Windkranz *m*
distribution plug Abzweigstecker *m*
distribution ratio Trennkoeffizient *m* (Dest)
distribution velocity Verteilungsgeschwindigkeit *f*
distributive law Distributivgesetz *n*
distributivity Distributivität *f* (Math)
distributor Verteiler *m*
distributor shaft Verteilerschacht *m*
distributor tube Entnahmestutzen *m*
district heating Fernheizung *f*
disturb *v* stören
disturbance Störung *f*, **center of** ~
  Störungszentrum *n*, **local** ~ stationäre
  Störung, **magnetic** ~ magnetische Störung,
  point of ~ Störstelle *f*
disturbance calculation Störungsrechnung *f*
disturbance feed-forward control
  Störgrößenaufschaltung *f* (Regeltechn)

disturbance value Störgröße *f*
disturbed gestört
disturbed function Ausfallerscheinung *f* (Med)
distylin Distylin *n*
distyrinic acid Distyrinsäure *f*
disubstituent Disubstituent *m*
disulfamic acid Disulfaminsäure *f*
disulfate Disulfat *n*
disulfide Disulfid *n*
disulfide bridge Disulfidbrücke *f*
disulfide cleavage Disulfidspaltung *f*
disulfide formation Disulfidbildung *f*
disulfiram Disulfiram *n*
disulfole Dithiol *n*
disulfonic acid Disulfosäure *f*
disulfuric acid Dischwefelsäure *f,*
Pyroschwefelsäure *f*
dita bark Ditarinde *f*
ditaine Ditain *n,* Echitamin *n*
ditamine Ditamin *n*
ditantalum tetroxide Tantaltetroxid *n*
ditartaric acid Diweinsäure *f*
ditch Graben *m*
diterpenes Diterpene *pl*
ditetragonalbipyramidal ditetragonalbipyramidal
(Krist)
ditetragonalpyramidal ditetragonalpyramidal
(Krist)
ditetrahedral ditetraedrisch (Krist)
ditetralyl Ditetralyl *n*
dithiane Dithian *n,* Diäthylendisulfid *n*
dithiazole Dithiazol *n*
dithietane Dithietan *n*
dithiin Dithiin *n*
dithiobenzoate Dithiobenzoat *n*
dithiobiuret Dithiobiuret *n*
dithiocarbamate Dithiocarbamat *n*
dithiocarbamic acid Dithiocarbaminsäure *f*
dithiocarbamylurea Dithiobiuret *n*
dithiocarbazic acid Dithiocarbazidsäure *f*
dithiocarbonic acid Dithiokohlensäure *f*
dithiocoumarin Dithiocumarin *n*
dithiodiglycolic acid Dithiodiglykolsäure *f*
dithioerythritol Dithioerythrit *m*
dithiofluorescein Dithiofluorescein *n*
dithioformic acid Dithioameisensäure *f*
dithiolane Dithiolan *n*
dithiole Dithiol *n*
dithion Dithion *n*
dithionaphthene indigo Dithionaphthenindigo *n*
dithionate Dithionat *n*
dithionic acid Dithionsäure *f,* salt of ~
Dithionat *n*
dithionite Dithionit *n*
dithionous acid Dithionigsäure *f*
dithiooxalate chelate Dithiooxalatchelat *n*
dithiophosphoric acid Dithiophosphorsäure *f*
dithiosalicylic acid Dithiosalicylsäure *f*
dithiothreitol Dithiothreitol *n*

dithiourethane Dithiourethan *n*
dithizonate Dithizonat *n*
dithizone Dithizon *n*
dithizone chloroform Dithizonchloroform *n*
dithranol Dithranol *n*
dithymol Dithymol *n*
ditolyl Ditolyl *n*
ditolyldichlorosilane Ditolyldichlorsilan *n*
ditrigonal ditrigonal (Krist)
ditrigonalbipyramidal ditrigonalbipyramidal
(Krist)
ditrigonalpyramidal ditrigonalpyramidal (Krist)
dittany root Eschenwurzel *f* (Bot)
dittmarite Dittmarit *m* (Min)
ditungsten carbide Diwolframcarbid *n*
ditungstic acid Diwolframsäure *f*
diuranate Diuranat *n*
diurea Diharnstoff *m*
diureide Diureid *n*
diuretic Diureticum *n* (Pharm), Harnmittel *n*
(Pharm), harntreibend, harntreibendes Mittel
(Pharm)
diuretic[al] harntreibend (Pharm)
diuretine Diuretin *n*
divalent bivalent (Chem), zweiwertig (Chem)
divalonic acid Divalonsäure *f*
divaric acid Divarsäure *f*
divaricatic acid Divaricatsäure *f*
divarinol Divarin *n*
diverge *v* abweichen, sich nicht decken,
divergieren
divergence Abweichung *f,* Auseinanderlaufen *n,*
Divergenz *f,* ~ of a series Divergenz *f* einer
Reihe (Math), ~ of fluid Strömungsdivergenz *f*
divergence loss Divergenzverlust *m* (Akust)
divergent abweichend, auseinanderstrebend,
divergierend
divergent lens, image formed by a ~
Zerstreuungsbild *n*
diverging divergierend
diverging lens Hohllinse *f,* Konkavlinse *f* (Opt)
diverse verschieden
diversification Differenzierung *f*
diversified verschiedenartig
diversine Diversin *n*
diversity Verschiedenheit *f,* Mannigfaltigkeit,
Vielseitigkeit *f*
divert *v* ableiten, ablenken, umlenken
diverting relay Umlenkrelais *n*
divest *v* abstreifen
divicine Divicin *n*
dividable [ein]teilbar
divide *v* abteilen, dividieren (Math), [ein]teilen
divided sheet Einteilungsbogen *m*
dividend Dividend *m* (Math)
dividers Stechzirkel *m,* Spitzzirkel *m*
dividing Trennen *n,* Verteilen *n*

divi-divi Dividivi *pl*
diving bell Taucherglocke *f*
diving body Senkkörper *m*
divinyl Divinyl *n*, Butadien *n*
divinylacetylene Divinylacetylen *n*
divinyl caoutchouc Divinylkautschuk *m*
divinyldichlorosilane Divinyldichlorsilan *n*
divisibility Teilbarkeit *f*
divisible [zer]teilbar
division Trennung *f*, Division *f* (Math),
   Einteilung *f*, Fach *n*, Spaltung *f* (Biol),
   Teilung *f* (Math), Verteilung *f*, degree of ~
   Zerteilungsgrad *m*, error of ~
   Teilungsfehler *m* (Math)
division line Teilstrich *m* (Math)
division mark Teilstrich *m* (Math)
division plane Trennungsebene *f*
division sign Teilungszeichen *n* (Math)
division wall Zwischenwand *f*
divisor Divisor *m* (Math), Teiler *m* (Math),
   common ~ gemeinsamer Teiler
divostroside Divostrosid *n*
dixanthogen Dixanthogen *n*
dixanthylene Dixanthylen *n*
dixylyl Dixylyl *n*
dizziness Gleichgewichtsstörung *f* (Med)
djenkolic acid Djenkolsäure *f*
dl-aspartic acid Asparacemsäure *f*
dl-hyoscyamine Atropin *n*
d-limonene Carven *n*, Hesperiden *n*
D-line D-Linie *f* (Spektr)
dl-lobelin Lobelidin *n*
dl-narcotine Gnoscopin *n*
d-lyxomethylose Rhodeotetrose *f*
D. M. D. (Abk) Doktor *m* der Zahnheilkunde
   (Dr. med. dent.)
docosane Docosan *n*
docosanoic acid Dokosansäure *f*
docosanol Dokosylalkohol *m*
docosyl alcohol Dokosylalkohol *m*
doctor Abschab[e]eisen *n* (Techn), Arzt *m*,
   Kratzeisen *n* (Techn), Rakel *f* (Techn), ~ of
   dental surgery Doktor *m* der Zahnheilkunde
   (Dr. med. dent.), ~ of medicine Doktor *m* der
   Medizin (Dr. med.), ~ of pharmacy Doktor *m*
   der Pharmazie, ~ of science (D. Sc) Doktor *m*
   der Naturwissenschaften *pl* (Dr. rer. nat.)
doctor *v*, ~ off abstreifen
doctorate, to attain a ~ promovieren
doctor bar Abstreichmesser *n*
doctor blade Abstreifmesser *n*, Rakel *f*,
   Schaber *m*
doctor finish Rakelappretur *f* (Text)
doctor knife Rakel *f*
doctor roll Abstreifwalze *f*, Auftragwalze *f*,
   Dosierwalze *f*, Walzenauftragmaschine *f*
doctor roll[er] Abquetschwalze *f*
doctor table Abstreiftisch *m*

doctrine Lehre *f*
document Dokument *n*, Beleg *m*, Urkunde *f*
document *v* dokumentieren
documentary analysis Beleganalyse *f*
documentation Dokumentation *f*
document for optical character reader
   Klarschriftbeleg *m* (Informatik)
dodecahedral dodecaedrisch (Krist), zwölfflächig
   (Krist)
dodecahedral slip Dodekaedergleitung *f*
dodecahedrane Dodecahedran *n*
dodecahedron Dodekaeder *n* (Krist),
   Zwölfflächner *m* (Krist), rhombic ~
   Granatdodekaeder *n* (Krist), Granatoeder *n*
   (Krist)
dodecahydrotriphenylene
   Dodecahydrotriphenylen *n*
dodecanal Dodecylaldehyd *m*, Laurinaldehyd *m*
dodecane Dodecan *n*, Dihexyl *n*
dodecanic acid Dodecansäure *f*, Laurinsäure *f*
dodecanoic acid Dodecansäure *f*, Dodecylsäure *f*,
   Laurinsäure *f*
dodecine Dodecin *n*
dodecyl Dodecyl-
dodecyl alcohol Dodecylalkohol *m*
dodecylene Dodecylen *n*
dodecylenic acid Dodecylensäure *f*
dodecyne Dodecin *n*
Döbereiner's lamp Platinfeuerzeug *n*,
   Wasserstofflampe *f*
Döbereiner's matchbox Platinfeuerzeug *n*
Doebner's violet Döbners Violett *n*
doeskin Rehleder *n*
dogleg Krümmung *f*
dohexacontane Dohexacontan *n*
doisynolic acid Doisynolsäure *f*
dolantin Dolantin *n*
dolerite Dolerit *m*, Flözgrünstein *m*
doleritic dolerithaltig
dolerophanite Dolerophanit *m* (Min)
dolomite Dolomit *m* (Min), Bitterkalk *m* (Min),
   Bitterspat *m* (Min), Magnesiakalk *m* (Min),
   crude ~ Rohdolomit *m* (Min), deadburned ~
   totgebrannter Dolomit, ferruginous ~
   Eisenbraunkalk *m* (Min), powdered ~
   Bergkreide *f*, raw ~ Rohdolomit *m* (Min),
   secondary ~ Flözdolomit *m* (Min)
dolomite brick Dolomitstein *m*
dolomite calcining kiln Dolomitbrennofen *m*
dolomite mill Dolomitmühle *f*
dolomite mixer Dolomitmischer *m*
dolomite powder Dolomitmehl *n*
dolomite sand Dolomitsand *m*
dolomitic dolomithaltig
dolomitization Dolomitbildung *f* (Geol)
dolphin oil Delphinöl *n*, Delphintran *m*
domain Bereich *m*, Gebiet *n*, Sachgebiet *n*
domain collocation Gebietskollokation *f*

domain memory, magnetic ~ magnetischer
Domänenspeicher *m*
domain of integrity Integritätsbereich *m*
domain orientation Hauptrichtung *f*
dome Dom *m*, Gewölbe *n*, Kuppel *f*
domesticine Domesticin *n*
domestic preservation Haushaltkonservierung *f*
domeykite Domeykit *m* (Min), Arsenkupfer *n*
(Min), Weißkupfer *n* (Min)
doming Wölben *n*
doming *a* gewölbt
domite Domit *m*
donarite Donarit *n* (Sprengstoff)
donation of electrons Abgabe *f pl* von
Elektronen
donaxine Donaxin *n*
Donnan effect Donnan-Effekt *m*
Donnan equilibrium Donnan-Gleichgewicht *n*,
membrane or ~ Membrangleichgewicht *n*
donor Donator *m*, Spender *m*
donor atom Donatoratom *n*
donor enzyme Donatorenzym *n* (Biochem)
donor level Donatorenniveau *n*
donor migration Donatorenwanderung *f*
donor's blood Spenderblut *n*
door latch Türverschluß *m*
door panel Türfüllung *f*
dopa Dopa *n*
dopamine Dopamin *n*
dope Klebelack *m*, Rauschgift *n*
dope-dyed spinngefärbt
dope-fiend (slang) Rauschgiftsüchtiger *m*
doping Dotieren *n* (Elektr)
Dopol furnace Dopol-Ofen *m*
Doppler broadening Doppler-Verbreiterung *f*
(Spektr)
Doppler displacement Doppler-Verschiebung *f*
(Spektr)
dopplerite Dopplerit *m* (Min)
Doppler shift Doppler-Verschiebung *f* (Spektr)
Doppler's principle Dopplereffekt *m*
doremol Doremol *n*
doremone Doremon *n*
dormancy Wachstumsruhe *f* (Biol)
dormiol Dormiol *n*, Amylenchloral *n*
dormonal Dormonal *n*
dosage Dosierung *f*, Dosis *f*
dose Dosis *f*, **curative** ~ kleinste therapeutische
Dosis *f*, **lethal** ~ tödliche Dosis *f*, **maximum
permissible** ~ höchstzulässige Dosis, **minimum
effective** ~ kleinste wirkungsvolle Dosis,
sublethal ~ nichttödliche Dosis
dose *v* dosieren, abwiegen
dose pump Dosierungspumpe *f*
dosimeter Dosimeter *n*, Tropfenzähler *m*
dosimetry Dosierungskunde *f*, Dosimetrie *f*
dosing Dosieren *n*, Dosierung *f*
dosing apparatus Dosieranlage *f*

dosing machine Dosiermaschine *f*,
Abwiegemaschine *f*
dosing plant Dosieranlage *f*
dot Punkt *m*
dot-dash line, indicated by a ~ strichpunktiert
dotriacontane Dotriacontan *n*, Dicetyl *n*
dotriacontanic acid Laccersäure *f*
dotted gepunktet, punktiert
dotted line Strichlinie *f*, **to mark with a** ~
stricheln
double doppelt, Doublé *n*, zweifach
double *v* verdoppeln
double acting doppelwirkend
double-arm kneader Doppelarmkneter *m*
double bandage Doppelverband *m*
double bath Doppelbad *n*
double-beam instrument Zweistrahlgerät *n*
double-bed sintering
Zweischichtensinterverfahren *n*
double bifilar gravimeter
Doppelbifilargravimeter *n*
double blade doppelschauflig
double boiler Doppelkessel *m*
double bond Doppelbindung *f* (Chem),
Zweifachbindung *f* (Chem)
double boss Kreuzmuffe *f*
double bottom Doppelboden *m*
double branch gate Gabeleinguß *m*
double buffing machine Doppelpoliermaschine *f*
double carbonate Doppelkarbonat *n*
double chloride Doppelchlorid *n*
double clamp Doppelklemme *f*
double clamp connection
Doppelschellenverbindung *f*
double collision Zweierstoß *m*
double-column rectification
Zweisäulen-Rektifikation *f*
doublecone blender Doppelkonusmischer *m*
double cone centrifuge Doppelkegelschleuder *f*
doublecone filter Doppelkegelfilter *m*
doublecone impeller [mixer]
Doppelkegelkreiselmischer *m*
doublecone mixer
Doppelkegeltrommelmischer *m*,
Doppelkonusmischer *m*
double connector Doppelklemme *f*
double-convex bikonvex (Opt)
double current Doppelstrom *m*
double current dynamo Doppelstromerzeuger *m*
double cyanide Cyandoppelsalz *n*
double decomposition Doppelzersetzung *f*
double diode Binode *f*
double-drum dryer Zweiwalzentrockner *m*
double dye *v* zweimal färben
double-edged zweischneidig
double-ended doppelendig
double-ended boiler Doppelkessel *m*
double flame furnace Doppelflammenofen *m*
double-flanged bobbin Scheibenspule *f*

double flue boiler Zweiflammenrohrkessel *m*
double fluoride Doppelfluorid *n*
double force mold Doppelstempelform *f*
double furnace Doppelofen *m*
double gate (in pneumatic conveyor systems) Doppelpendelklappe *f*
double hanger frame Hängebock *m*
double helical gear Pfeilrad *n*, Pfeilverzahnung *f*
double helix Doppelhelix *f*
double-holed capstan head Kreuzlochkopf *m*
double ignition Doppelzündung *f*
double image Doppelbild *n*
double inlet doppelseitiger Einlaß
double integral Doppelintegral *n* (Math)
double jet carburet[t]or Doppeldüsenvergaser *m*
double layer Doppelschicht *f*
double-layer belt Doppelriemen *m*
double linkage Doppelbindung *f*
double molecule Doppelmolekül *n*
double nut Doppelmutter *f*
double ore separator Doppelerzscheider *m*
double oxide Doppeloxid *n*
double pipe Doppelrohr *n*
double-pipe condensor
    Doppelrohrkondensator *m*
double-pipe heat exchanger
    Doppelrohrwärmeaustauscher *m*
double-piston compressor
    Boxer-Kolbenverdichter *m*
double piston pump Doppelkolbenpumpe *f*
double plug Doppelstecker *m*
double producer Doppelgenerator *m*
doubler (for plastics sealing) Umbugschieber *m*
double ram press Doppelkolbenpresse *f*
double-ray betatron Zweistrahlbetatron *n*
double recorder Doppelbandschreiber *m*
double-refracting doppelbrechend (Opt)
double refraction Doppelbrechung *f* (Opt)
double-roll crusher Walzen- Brecher *m*
double-roll mill Walzenbrecher *m*
double salt Doppelsalz *n*, Mischsalz *n*, Zwillingssalz *n*
double scattering experiment
    Doppelstreuversuch *m*
double screw Doppelschnecke *f*
double seam Doppelfalz *m* (Dose)
double seat valve Doppelsitzventil *n*
[double] shear steel Zementgerbstahl *m*
double-sided grinder Doppelschleifmaschine *f*
double silicate Doppelsilicat *n*
double skeleton electrode
    Doppelschicht-Elektrode *f*
double slit Doppelspalt *m* (Opt)
double slit method Doppelblendenmethode *f* (Opt)
double spiral condenser Doppelwendelkühler *m*
double spiral mixer Doppelschneckenmischer *m*
double-stranded doppelstrangig

double-strand platform conveyor
    Doppelstrang-Kettenförderer *m* mit Tragrahmen
double sulfate Doppelsulfat *n*
double sulfite Doppelsulfit *n*
double superphosphate Doppelsuperphosphat *n*
double switch Doppelweiche *f*
doublet Doppellinie *f* (Opt), Dublette *f*
double-tapered muff coupling
    Doppelkegelkupplung *f*
double terminal Doppelklemme *f*
double T-girder Doppel-T-Träger *m*
double thermoscope Doppelthermoskop *n*
double thread Doppelgewinde *n*
double throw Hebelumschalter *m*
double T-iron Doppel-T-Eisen *n* (Techn), H-Eisen *n* (Techn)
doublet line Dublettlinie *f*
double tray Doppelboden *m* (Dest)
doublet ring Zweierschale *f* (Atom)
doublet spacing Dublettabstand *m*
doublet spectrum Dublettspektrum *n*
doublet splitting Dublettaufspaltung *f*
doublet structure Dublettstruktur *f*
double union Doppelbindung *f*
double void Doppelleerstelle *f*
double voltage Doppelspannung *f*
double wall box Herdeisen *n*
double-walled doppelwandig
double weighing Doppelwägung *f*
double X-ray flash tube
    Doppelröntgenblitzröhre *f*
doubling Verdoppelung *f*
doubt, case of ~ Zweifelsfall *m*
doubtful bedenklich, ungewiß
dough Teig *m*
dough kneading machine Teigknetmaschine *f*
dough mixing machine Kneter *m*
dough molding Teigpressen *n*
dough molding process Teigpreßverfahren *n*
doughnut Flußverstärker *m* (Atom)
dough scraper Teigschaber *m*
doughy teigig
douglasite Douglasit *m* (Min)
Dove prism Dove-Prisma *n* (Opt), Wendeprisma *n* (Opt)
dovetail *v* einzahnen
dove-tailed schwalbenschwanzförmig
dowel Dübel *m*, Führungsstift *m*, Paßstift *m*
dowel bushel kalibermäßige Bohrung *f*
dowel hole Führungsbohrung *f*, Führungsloch *n*
dowel pin Dübel *m*
dowel screw Führungsschraube *f*
dowex cation exchange resin
    Dowexkationenaustauschharz *n* (HN)
downcomer Ablaufstutzen *m* (Kolonne), Fallrohr *n*, Standrohr *n*, Zulaufstutzen *m* (Bodenkolonne)
downcomer performance Schluckvermögen *n*

downcomer [tube] Gichtgasabzugsrohr *n*
downdraft Fallstrom *m*
downdraft gasification Fallstromschwelung *f*
downdraft-type furnace Ofen *m* mit
niedergehender Flamme
downgate Eingußtrichter *m*
downmain Fallrohr *n*
down milling Gleichlauffräsen *n*
downpipe Fallrohr *n*
down sand Dünensand *m*
down spout Traufe *f*
downstroke Hauptkolbenhub *m*
downstroke press Formungsanlage *f* mit
Oberstempel, Oberdruckpresse *f*,
Oberkolbenpresse *f*
downtime Stillstandszeit *f*
downward abwärts
downward motion Abwärtsbewegung *f*
downward path Fallweg *m*
downward pressure Druck nach unten
downy weich
dowson gas Halbwassergas *n*
doxylamine Doxylamin *n*
dozen Dutzend *n*
dozent Dozent *m*
DPA (Abk) Diphenolsäure *f*
DPN (Abk) Diphosphopyridinnucleotid *n*,
Codehydrase I *f*, Cozymase I *f*
DPNH (Abk) reduziertes
Diphosphopyridinnucleotid *n*
DPPD (Abk) Diphenylphenylendiamin *n*
dracoic acid Dracosäure *f*
dracorhodin Dracorhodin *n*
dracorubin Dracorubin *n*
draft Zugluft *f*, Abriß *m* (Zeichnung),
Durchzug *m*, Gebläsewind *m*, Luftzufuhr *f*,
Riß *m* (Zeichnung), Skizze *f* (Zeichnung),
forced ~ Gebläseluft *f*, **intensity of** ~
Zugstärke *f*
draft *v* zeichnen
draft angle Neigungswinkel *m*
draft beer Bier *n* vom Faß, Faßbier *n*
draft gauge Zugmesser *m*
drafting force Zugkraft *f*
draft meter Gebläsemesser *m*
draft regulation Zugregelung *f*
draft regulator Zugregler *m*
drag Unterkasten *m* (Gieß), Widerstand *m*
(Mech), **induced** ~ induzierter Widerstand,
total ~ Gesamtwiderstand *m* (Mech)
drag *v* schleppen, zerren, ziehen
drag bar Kuppelstange *f*
drag chain Hemmkette *f*, Redler-Filter *n*
drag-chain conveyor Schleppkettenförderer *m*
drag chain filter Redlerfilter *n*
drag coefficient Widerstandszahl *f*
drag conveyor Trogkettenförderer *m*
dragee Dragee *n*
dragging Mitbewegung *f*

dragon's blood Drachenblut *n*
dragon's blood gum Drachenblutgummi *n*
drag strut Versteifungsstrebe *f*
drag turf Modertorf *m*
drain Abfluß[kanal] *m*, Abflußrinne *f*, Ablaß *m*,
Ableitungskanal *m*, Abwasserleitung *f*,
Abzugskanal *m*, Entwässerung *f*, Kloake *f*
drain *v* ablaufen lassen, ausleeren, aussaugen,
austrocknen, dränieren, entleeren, entwässern,
~ **by pumping** abpumpen, ~ **off** abführen
(Dampf), ablaufen, abnutschen, absaugen
(Flüssigkeit), abtröpfeln, abtropfen, ~
through sewers kanalisieren
drainable entwässerbar
drainage Wasserabfluß *m*, Abflußrinne *f*,
Entwässern *n*, Trockenlegung *f*,
Wasserableitung *f*, **system of** ~
Entwässerungssystem *n*
drainage channel Ablaufrinne *f*, Ableitung *f*
drainage tap Ablaßhahn *m*
drain board Ablaufbrett *n*, Trockengestell *n*
drain chest Abtropfkasten *m* (Pap)
drain cock Auslaufhahn *m*, Ablaßhahn *m*,
Entwässerungshahn *m*, Zapfhahn *m*
drained off abgeführt
drainer Abtropfgefäß *n*, Trockner *m*
drainer plate Filtrierplatte *f*
drain field Zugfeld *n* (Festkörperphys)
drain hole Ausgußloch *n*
draining Austrocknen *n*, Entleeren *n*,
Entwässern *n*, ~ **by means of sewers**
Kanalisieren *n*
draining board Abtropfblech *n*, Tropfbrett *n*
draining dish Abtropfschale *f*
draining flask Absaug[e]kolben *m*
draining funnel Ablauftrichter *m*
draining machine Ausschleudermaschine *f*
draining medium Ableitungsmittel *n*
draining stand Abtropfgestell *n*,
Abtropfständer *m*
draining tank Absetzbehälter *m*
drain pan Abtropfpfanne *f*
drain pipe Abflußrohr *n*, Abfallrohr *n*,
Ablaufrohr *n*, Ableitungsrinne *f*,
Abwasserrohr *n*, Abzugsrohr *n*, Ausgußrohr *n*,
Entwässerungsrohr *n*
drain plug Abflußstopfen *m*, Ablaßschraube *f*
drain screw Ablaßschraube *f*
drain shaft Ablaufschacht *m*
drain sluice valve Ablaßschieber *m*
drain table Abtropfkasten *m* (Met)
drain tap Ablaßhahn *m*
drain tube Entschlammungsrohr *n*
drain valve Ablaßventil *n*, Abzapfhahn *m*
drain water Kanalisationsabwasser *n*
dralon Dralon *n* (HN, Text)
drapery Dekorationsstoff *m*
drastic drastisch, starkwirkend

draught (Br. E.) Luftzug *m*, Luftzufuhr *f*, Zugluft *f*
draught beer Faßbier *n*
dravite Dravit *m* (Min)
draw, extent of ~ Ziehtiefe *f*
draw *v* [an]ziehen, auftragen (Kurven etc.), [auf]zeichnen (skizzieren), strecken, verstrecken (Kunststoffäden), **to hot** ~ warm ziehen, ~ **back** zurückziehen, ~ **forth** hervorziehen, ~ **forward** herausziehen, ~ **in** einziehen, ~ **off** abdekantieren, abführen, ablassen, ableiten, abschleppen, abstechen (Metall), ~ **out** ausdehnen, herausziehen, ~ **up** hinaufziehen
drawability Tiefziehbarkeit *f*
draw-back ram Rückdruckkolben *m*
draw bar Kuppelstange *f*
drawer Schublade *f*
draw former Ziehbiegemaschine *f*
draw gear Zugvorrichtung *f*
draw grid Ziehgitter *n*
draw hole Ziehloch *n*
drawing Ziehen *n*, Streckziehen *n*, Verstrecken *n*, Zeichnung *f* (Skizze), Ziehverfahren *n*, **engineering** ~ technisches Zeichnen *n*
drawing bars Stangenzug *m*
drawing board Reißbrett *n*, Zeichenbrett *n*
drawing chalk Zeichenkreide *f*
drawing die Ziehform *f*, Ziehwerkzeug *n*
drawing frame Streckmaschine *f*
drawing-in air Ansaugen *n* von Luft
drawing ink Ausziehtusche *f*, Zeichentinte *f*
drawing instrument Zeichengerät *n*
drawing instruments Reißzeug *n*
drawing machine with pliers Schleppzangenziehbank *f*
drawing machine with spool Leierziehbank *f*, Scheibenziehbank *f*
drawing off Ablassen *n*, Abstich *m*
drawing-off of steam Dampfentnahme *f*
drawing off the sodium metal Ablassen *n* des Natriummetalls
drawing-out worm Ausziehschnecke *f*
drawing paper Zeichenpapier *n*
drawing pen Reißfeder *f*
drawing pencil Zeichenstift *m*
drawing pin Heftzwecke *f*, Reißnagel *m*
drawing process Streckziehverfahren *n*, Ziehvorgang *m*
drawing property Ziehfähigkeit *f*
drawing set Reißzeug *n*
drawing temperature Ablaßtemperatur *f*
drawn, bright ~ blankgezogen
drawn off abgeführt
drawn out ausgezogen
drawn part Ziehteil *n*
draw-off cock Abzapfhahn *m*
draw-off roll Abzugswalze *f*
draw plate Zieheisen *n*

draw ratio Längsblasverhältnis *n*
draw tongs Froschklemme *f*, Schleppzange *f*
drawworks (drilling) Hebewerk *n*
dredged peat Baggertorf *m*
dredger Bagger *m*, Streubüchse *f*
dredge rope Baggerseil *n*
dredging Baggerbetrieb *m*
dregs Bodenhefe *f*, Bodensatz *m*, Drusen *pl* (obs), Rückstand *m*, **oil made from** ~ **of wine** Drusenöl *n*
drench Beizbrühe *f*, Beizflüssigkeit *f*
drench *v* durchwässern, nachbeizen
drenching Beizen *n*
drenching bath Beizbad *n*
drepanocyte Sichelzelle *f*
dress *v* bekleiden; aufbereiten, abbeizen (Gerb), appretieren (Text), garen (Gerb), putzen, schlichten (Weberei), zubereiten, zurechtschneiden, zurichten (Gerb)
dressed gar (Häute), ~ **with alum** alaungar (Gerb), ~ **with lime** eingekalkt, ~ **with oil** fettgar
dressing Appretieren *n* (Text), Appretur *f* (Text), Aufbereitung *f*, Nachbearbeitung *f*, Zubereitung *f*, **loss due to** ~ Aufbereitungsverlust *m*, **product of** ~ Aufbereitungsgut *n*
dressing agent Appreturmittel *n*
dressing hammer Abrichthammer *m*
dressing liquor Garbrühe *f*
dressing machine Appreturmaschine *f* (Text)
dressing oil Appreturöl *n*
dressing plant Aufbereitungsanlage *f*
dressing size Appreturleim *m*
dressing tool Richtwerkzeug *n*
dressing tub Garfaß *n*
dressing vat Garfaß *n*
dried getrocknet
dried blood Blutmehl *n*
dried distiller' solubles Schlempe *f*
dried vegetables Trockengemüse *n*
drier Trockner *m*
drift Strömung *f*, Abtrift *f*, Diluvium *n* (Geol)
drift current Triftstrom *m*
drift gauge Abtriftmesser *m*
drifting test Aufdornprobe *f*
drift meter Abtriftmesser *m*
drift of electrons Elektronenfluß *m*
drift velocity Wanderungsgeschwindigkeit *f*
drill Bohreisen *n*, Bohrer *m*, Drillbohrer *m*, **pointed** ~ Spitzbohrer *m*
drill *v* [an]bohren
drill bit Bohrerspitze *f*
drill chuck Bohrfutter *n*
drill core Bohrkern *m*
drilling Bohren *n*
drilling bench Bohrbank *f*
drilling head Bohrkopf *m*
drilling machine Bohrmaschine *f*

drilling mud Bohrschlamm *m*
drilling rig Bohranlage *f*, Bohrkran *m*
drilling slime Bohrschlamm *m*
drill pipe Gestängerohr *n*
drill poles Bohrgestänge *n*
drill press Bohrmaschine *f*
drill test Bohrprobe *f*
drimanic acid Drimansäure *f*
drimanol Drimanol *n*
drimenol Drimenol *n*
drimic acid Drimsäure *f*
drimone Drimon *n*
drink alkoholisches Getränk *n*, Gebräu *n*
drinkable trinkbar
drinking water Trinkwasser *n*
drinking water supply Trinkwasserversorgung *f*
drip *v* [ab]triefen, tröpfeln, ∼ off abtropfen,
   abtröpfeln
drip board Abtropfer *m*
drip cock Entwässerungshahn *m*, Tropfhahn *m*
drip cup Tropfbecher *m*, Auffangglocke *f*
drip funnel Tropftrichter *m*
drip nozzle Tropfdüse *f*
drip pan Tropfschale *f*, Auffangglas *n*,
   Auffangschale *f*, Ölfänger *m*
dripping Abtröpfeln *n*, Tröpfeln *n*
dripping board Tropfbrett *n*
drip[ping] cup Tropfbehälter *m*
drip[ping] pan Tropfbehälter *m*, Tropfpfanne *f*
drip pipe Ablaufrohr *n*
drip ring Tropfring *m*
dripstone Tropfstein *m* (Min)
drip water Tropfwasser *n*
dristributor Verteilerboden *m* (Dest)
drive Antrieb *m*, Triebwerk *n*, diagram of ∼
   Getriebeschema *n*, intermediate ∼
   Zwischengetriebe *n*, separate ∼
   Fremdantrieb *m*
drive *v* treiben, ∼ away vertreiben, ∼ over *v*
   überdestillieren, ∼ out verdrängen, verjagen,
   ∼ through durchtreiben
drive belt Treibriemen *m*
drive pulley Antriebsscheibe *f*,
   Antriebstrommel *f*
drive pulse Steuerimpuls *m*
driver's stand Führerstand *m*
drive screw Schlagschraube *f*
drive system Antriebssystem *n*
driving, mode of ∼ Antriebsart *f*
driving axle Antriebsachse *f*, Treibachse *f*
driving belt Treibriemen *m*
driving engine Antriebsmaschine *f*
driving force Antriebskraft *f*
driving gear Antrieb *m*, Getriebe *n*, differential
   ∼ Ausgleichgetriebe *n*
driving jet Treibstrahl *m*
driving machine Antriebmaschine *f*
driving mechanism Antriebswerk *n*
driving motor Antriebsmotor *m*

driving pin Mitnehmerstift *m*
driving plunger Antriebkolben *m*
driving power Antriebskraft *f*, Triebkraft *f*
driving pressure Antriebsdruck *m*
driving pulley Treibscheibe *f*
driving shaft Antriebswelle *f*, Hauptwelle *f*
driving spring Triebfeder *f* (Uhr)
drizzle Nieselregen *m*, Sprühregen *m*
drizzle *v* nieseln, sprühen
drizzling fog Sprühnebel *m*
dromoran Dromoran *n*
drop Tropfen *m*, Abnahme *f*, Fall *m*, Sinken *n*,
   Verminderung *f*; (of a river) Gefälle *n* (z. B.
   eines Flusses), form of a ∼ Tropfenform *f*,
   height or length of ∼ Fallhöhe *f*, small ∼
   Tröpfchen *n*, sudden ∼ Sturz *m*, ∼ of
   pressure Druckverlust *m*
drop *v* tropfen, [ab]fallen, abwärts gleiten,
   abwerfen, ∼ in einträufeln; eintröpfeln
drop analysis Tüpfelanalyse *f*
drop by drop tropfenweise
drop capillary Tropfkapillare *f*
drop center rim Tiefbettfelge *f*
drop collector Tropfenfänger *m*
drop condensation Tropfenkondensation *f*
drop counter Tropfenzähler *m*
drop dispenser Tropfeinsatz *m*
drop-feed oiling Tropfschmierung *f*
drop forge die Schmiedegesenk *n*
drop forging Gesenkschmieden *n*
drop forging article Hammerschmiedestück *n*
drop hammer Fallhammer *m*, Parallelhammer *m*,
   Rahmenhammer *m*
drop hammer die Fallhammerform *f*, Stanzform *f*
drop hammer tester Fallhammergerät *n*
droplet Tröpfchen *n*
droplet growth Tropfenwachstum *n*
droplets, formation of ∼ Tröpfchenbildung *f*
dropout Ausfall *m* von Zeichen (Comp),
   Aussetzer *m* (Comp)
droppable abwerfbar
dropper closure Tropfverschluß *m*
dropping ball method Fallkugelmethode *f*
drop[ping] bottle Tropfflasche *f*
dropping dart test Fallbolzentest *m*
dropping electrode Rieselelektrode *f*,
   Rieselfilmelektrode *f*, Tropfelektrode *f*
dropping funnel Tropftrichter *m*
dropping lubricator Tropföler *m*,
   Tropfschmiergefäß *n*
drop[ping] point Tropfpunkt *m*
droppings Abfallwolle *f*
drops, formation of ∼ Tropfenbildung *f*, in form
   of ∼ in Tropfenform
drop-shaped tropfenförmig
dropstone Tropfstein *m*
drop sulfur Tropfschwefel *m*
drop tin Tropfzinn *n*
drop weight Fallkugel *f*

dropwise tropfenweise
drop zinc Tropfzink *n*
droserone Droseron *n*
drosophila Taufliege *f* (Zool)
drosophilin Drosophilin *n*
dross Abschaum *m*, Abstrich *m*, Glasschlacke *f*, Grus *m*, Hüttenafter *n*, Metallgekrätz *n*, Schlacke *f*, **fusible** ~ Nasenschlacke *f*, **rough** ~ Fördergrus *m* (Bergb), **to clear from** ~ abschlacken
dross hood Abzug *m* (für Abgase)
drossing Entschlickerung *f*
dross of copper Darrgekrätz *n*, Darrsohle *f*
dross of pig iron Roheisenschlacke *f*
drossy schlackenartig, schlackig
Drude equation Drudesche Gleichung *f* (Opt)
drug Arznei *f*, Arzneimittel *n*, Droge *f*, Heilmittel *n*, Medikament *n*, Medizin *f*, Rauschgift *n*
drug abuse Arzneimittelmißbrauch *m*
drug addict Rauschgiftsüchtiger *m*
drug addiction Arzneimittelsucht *f;* Rauschgiftsucht *f*
druggist Drogist *m*, Apotheker *m*
druggist's scale Apothekerwaage *f*
drug intoxication Arzneimittelvergiftung *f*
drug poisoning Arzneimittelvergiftung *f*
drug resistance Arzneimittelresistenz *f*
drug retardant Depot-Tabletten *pl*
drugs, standardization of ~ Arzneimittelstandardisierung *f*
drugstore Apotheke *f*, Drogerie *f*
drug synthesis Arzneimittelsynthese *f*
drug tolerance Arzneimittelgewöhnung *f*
drug traffic Rauschgifthandel *m*
drum Trommel *f*, Walze *f*, **flat** ~ Flachschultertrommel *f*, **[full aperture]** ~ Hobbock *m*
drum apparatus Trommelapparat *m* (Galv)
drum camera Trommelkamera *f*
drum cell Trommelzelle *f*
drum drier Trommeltrockner *m*, Walzentrockner *m*, ~ **with heating tubes** Röhrentrommeltrockner *m*
drum dumper Trommelkippanlage *f*
drum filter Trommelfilter *n*, **single-compartment** ~ Trommelsaugfilter *n*
drum mixer Trommelmischer *m*, Zylindermischer *m*, **end-over-end type** ~ aufrechtstehender Trommelmischer
Drummond light Drummondsches Licht *n*, Hydrogenlicht *n*, Kalklicht *n*
drum pelletizer Pelletiertrommel *f*
drum regenerator Trommelregenerator *m*
drum rheostat Walzenwiderstand *m*
drum rotor Trommelrotor *m*
drum sander Zylinderschleifmaschine *f*
drum separator Trommelscheider *m*

drum set Trommeleinstellung *f*
drum sextant Trommelsextant *m*
drum spider Trommelstern *m*
drum washer Trommelwascher *m*, Trommelwaschmaschine *f*
drum wheel Trommelrad *n*
drum winch Trommelhaspel *f*
druse Druse *f* (Min)
drusy drusenförmig
drusy cobalt Drusenkobalt *n* (Min)
dry trocken, übergar (Metall), **absolutely** ~ absolut trocken, **bone** ~ absolut trocken, **half** ~ halbtrocken
dry *v* [ab]trocknen, austrocknen, darren, dörren, eintrocknen, ~ **through** durchtrocknen, ~ **up** vertrocknen, austrocknen, dörren, versiegen
dry assay Trockenprobe *f*
dry bleaching Trockenbleiche *f*
dry blend Krümmelmasse *f*
dry-bright emulsion Selbstglanzemulsion *f*
dry cell Trockenelement *n*, Trockenzelle *f*
dry cleaning chemische Reinigung *f*, Chemisch-Reinigen *n*, Trockenreinigung *f*
dry cleaning oil Putzöl *n*
dry cleanser Trockenreiniger *m*
dry color Trockenfarbe *f*
dry coloring Trockeneinfärbung *f*
dry compression machine Trockenkompressionsmaschine *f*
dry content[s] Trockengehalt *m*
dry crushing Trockenvermahlung *f* (Bergb)
dry cycle time Leerlaufspritzzeit *f*
dry-dial water meter Trockenläufer *m*
dry distil *v* entgasen
dry distillation Trockendestillation *f*
dry dressing Trockenbeize *f*
dry egg Trockenei *n*
dryer (also drier) Trockner *m*, Trockenapparat *m*, Trockenkasten *m*, **continuous** ~ Durchlauftrockner *m*, **rotary** ~ Trommeltrockner *m*, ~ **with agitator** Rührwerkstrockner *m*
dryer section Trockenpartie *f*
dryer tray Trockenblech *n*
dry extract Trockenauszug *m*
dry felt-sheeting Rohfilzbahnen *pl*
dry filling Trockenfüllung *f*
dry filter Trockenfilter *n*
dry finishing Trockenappretur *f*
dry gas Trockengas *n*
dry grinding Trockenvermahlung *f*, Trockenmahlen *n*, Trockenschleifen *n*
dry ice Trockeneis *n*, Kohlendioxidschnee *m*, Kohlensäureschnee *m*
dry ice pan Trockeneistrog *m*
drying Trocknen *n*, Austrocknen *n*, Darren *n*, Eintrocknen *n*, Feuchtigkeitsentzug *m*, **fast** ~ schnelltrocknend, **forced** ~ Beschleunigungstrocknung *f*, **quick** ~

schnelltrocknend, **rapid** ~
Schnelltrocknung *f*, ~ **in the air** Trocknen *n*
an der Luft, ~ **of the electrode** Trocknen *n*
der Elektrode, ~ **of the molds** Trocknen *n* der
Formen (pl)
drying *a* trocknend
drying accelerator Trocknungsbeschleuniger *m*
drying agent Trockenmittel *n*
drying and roasting apparatus Trocken- und
Röstapparat *m*
drying apparatus Trockenapparat *m*,
Trockenpistole *f*, Trocknungsanlage *f*
drying auxiliary Trockenhilfsmittel *n*
drying board Darrbrett *n*
drying bottle Trockenflasche *f*
drying cabinet Trockenschrank *m*,
Schranktrockner *m*
drying cell Trockenpatrone *f*
drying centrifuge Trockenschleuder *f*
drying chamber Trockenkammer *f*, Darraum *m*,
Darrkammer *f*, Kammertrockner *m*,
Trockenkasten *m*, Trockenraum *m*
drying closet Trockenschrank *m*,
Trockenkasten *m*
drying cup Eindampfschale *f*
drying cupboard Darrschrank *m*,
Labortrockenschrank *m*
drying cylinder Trockentrommel *f*,
Trockenzylinder *m*
drying dish Trockenschale *f*
drying drum Trockentrommel *f*, Trockenwalze *f*,
Trockenzylinder *m*
drying equipment Trocknungsanlage *f*
drying felt Trockenfilz *m*
drying flask Trockenflasche *f*
drying floor Darrboden *m*
drying frame Trockengestell *n*,
Trockenrahmen *m*
drying house Darrkammer *f*
drying kiln Trockenofen *m*
drying oil Trockenöl *n*, schnell trocknendes Öl *n*,
Trockenfirnis *m*
drying oven Trockenofen *m*, Darrofen *m*,
Darrschrank *m*, Wärmeschrank *m*
drying plant Trockenanlage *f*
drying process Trockenverfahren *n*,
Eintrocknungsprozeß *m*, **final or second** ~
Nachtrocknen *n*
drying rack Trockengestell *n*
drying roller Trockenwalze *f*
drying room Trockenraum *m*, Darre *f*,
Trockenkammer *f*
drying section Trockenpartie *f*
drying stand Trockengestell *n*
drying sterilizer Trockensterilisator *m*
drying stove Trockenofen *m*, Darrkammer *f*
drying surface Darrfläche *f*
drying time Trockenzeit *f*
drying tower Trockenturm *m*

drying tube Trockenröhre *f*
drying tunnel Trockenkanal *m*
drying zone Trockenzone *f*
dry joint schlechte Lötstelle *f*
dry lacquer compound Trockenlackmasse *f*
dry measure Trockenhohlmaß *n*
dryness Dürre *f*, Trockenheit *f*, Trockne *f*
(Chem), **degree of** ~ Trockenheitsgrad *m*
dry operation Trockenlauf *m*
dry preparation Trockenpräparat *n*
dry-pressed briquet[te] Trockenbrikett *n*
dry puddling Trockenpuddeln *n*
dry purification Trockenreinigung *f*
dry residue Trockenrückstand *m*
dry rot Hausschwamm *m*, Holzfäule *f*,
Trockenfäule *f*
dry rubber content Trockenkautschukgehalt *m*
dry sand Formmasse *f* (Gieß), Streusand *m*
dry sand casting Masseguß *m*
dry sifting Trockensiebung *f*
dry-sinter process Trockenaufschluß *m*
dry spinning Trockenspinnverfahren *n*
dry spot Luftstelle *f*
dry strength Trockenfestigkeit *f*
dry substance getrockneter Stoff *m*,
Trockengut *n*
dry tack Trockenklebrigkeit *f*
dry test trockenchemische Untersuchung *f*
dry weight Trockenauftragsmenge *f* (Lack),
Trockengewicht *n*, **absolute** ~
Absoluttrockengewicht *n*
dry well Sickergrube *f*
dry wine Sekt *m*
dry yeast Trockenhefe *f*
d-tartaric acid Rechtsweinsäure *f*
dual bead tire Zweikernreifen *m*
dual cycle Doppelkreislauf *m*
dual grid Doppelgitter *n*
dualin Dualin *n*
dualin dynamite Dualindynamit *n*
dualism Dualismus *m*, ~ **of wave and particle**
Dualismus *m* von Welle und Teilchen
dualistic dualistisch
duality Dualität *f*
duality principle Dualitätsprinzip *n*
dual magneto Doppelzündmagnet *m*
dual purpose zwei Zwecken dienend
dual recorder Doppelschreiber *m*
dual-spraying equipment
Zweikomponentenspritzvorrichtung *f*
dual tire Zwillingsbereifung *f*
dual worm Doppelschnecke *f*
dubamine Dubamin *n*
dubatol Dubatol *n*
dubbing Lederschmiere *f*, Lederöl *n*,
Schuhschmiere *f*
dubinidine Dubinidin *n*
duboisine Duboisin *n*
duct Kanal *m*, Leitung *f*, Röhre *f*

ductile [aus]dehnbar, [aus]ziehbar, biegsam, duktil, geschmeidig, hämmerbar, schmiedbar, streckbar
ductile fracture Verformungsbruch *m*
ductility Duktilität *f* (Met), Ausdehnungsvermögen *n* (Met), Ausziehbarkeit *f* (Met), Biegsamkeit *f* (Met), Dehnbarkeit *f* (Met), Formänderungsfähigkeit *f* (Met), Geschmeidigkeit *f* (Met), Streckbarkeit *f* (Met), Verformbarkeit *f* (Met), Zähigkeit *f* (Met), **hot ~** Warmbildsamkeit *f* (Met)
dud Blindgänger *m*
dudleyite Dudleyit *m* (Min)
due fällig
Dufour effect (inverse thermodiffusion) Diffusions-Thermoeffekt *m* (Dufour-Effekt)
dufrenite Dufrenit *m* (Min), Grüneisenerde *f* (Min)
dufrenoysite Dufrenoysit *m* (Min)
dugong oil Dugongöl *n*
dug peat Stichtorf *m*
dulcamarin[e] Dulcamarin *n*
dulcin Dulcin *n*, Sucrol *n*
dulcite Dulcit *m*, Melampyrin *n*
dulcitol Dulcit *m*, Melampyrin *n*
dull blind (Metall), dumpf (hohlklingend), glanzlos, matt, stumpf (Farbe)
dull *v* abdunkeln, abstumpfen (schwächen), matt brennen, mattieren
dull blue mattblau
dull clear varnish Mattlack *m*
dull coal Mattkohle *f*
dull fabric Mattgewebe *n*
dull finish Mattanstrich *m*, Mattglanz *m*
dull finish lacquer Schleiflack *m*
dull-grind *v* mattschleifen
dulling Blindwerden *n*, Glanzverlust *m*, Mattierung *f*
dulling agent Mattierungsmittel *n*, Trübungsmittel *n* (Email)
dulling dye Abtrübungsfarbe *f*
dull luster Mattglanz *m*
dullness Abstumpfung *f* (Schneide), Glanzlosigkeit *f*, Mattheit *f*, Trübe *f*
dull pickle Mattbrenne *f*
dull pickling Mattbrennen *n*
dull polish Mattierung *f*
dull red dunkle Rotglut *f*
dull silvering Mattversilberung *f*
dull silvery luster Edelglanz *m*
dull white mattweiß
Dulong and Petit's law Dulong-Petitsches Gesetz *n*
duly ordnungsgemäß, vorschriftsmäßig
duly dressed fellgar
dumasin Dumasin *n*
dumbbell model Hantelmodell *n*

dummy Attrappe *f*, Leerpackung *f*, Schaupackung *f*
dummy hole Blindloch *n*
dummy piston Gegendruckkolben *m*
dummy rivet Heftniet *m*
dumortierite Dumortierit *m* (Min)
dump *v* abladen, abwerfen, **~ out** auskippen, ausschütten
dumper Muldenkipper *m*
dumping grate Schlackenrost *m*
dumping ground Abladeplatz *m*, Absturzhalde *f*, Müllabladeplatz *m*
dump slag Haldenschlacke *f*
dump valve Ablaßventil *n*
dump wagon Kippwagen *m*
dun fahl
dun-colored graubraun
dundasite Dundasit *m* (Min)
dung Dünger *m*, Dung *m*, Kot *m*, Mist *m*
dung-salt Düngesalz *n*
dunite Dunit *m*
dunnione Dunnion *n*
duodecimal system Duodezimalsystem *n*
duodenum Duodenum *n* (Med)
duotal Duotal *n*
duplex doppelt
duplex board Duplexkarton *m*
duplex connection Duplexschaltung *f*
duplex metal Bimetall *n*
duplex molding Duplexpressen *n*
duplex paper Doppelpapier *n*
duplex pump Doppelzylinderpumpe *f*
duplex recorder Zwillingsschreiber *m*
duplex slip Doppelgleitung *f*
duplex star connection Doppelsternschaltung *f* (Elektr)
duplex system Duplexsystem *n*
duplicate Duplikat *n*, Kopie *f*, Zweitschrift *f*
duplicate *v* verdoppeln, nachahmen
duplicate determination Kontrollbestimmung *f*
duplicate main Doppelleitung *f*
duplicate pipe line Doppelleitung *f*
duplicate plate Auswechselplatte *f*
duplicate wire Doppelleitung *f*
duplicating apparatus Vervielfältigungsapparat *m*
duplicating machine Vervielfältigungsmaschine *f*
duplicating mill Kopierfräsmaschine *f*
duplicating milling machine Kopierfräsmaschine *f*
duplicating paper Abzugspapier *n*
duplicating stencil Dauerschablone *f*
duplicator Vervielfältigungsapparat *m*
duplicity theory (of the eye) Duplizitätstheorie *f* (des Auges)
duponol Duponol *n*
duprene Neopren *n*
durability Beständigkeit *f*, Dauerhaftigkeit *f*, Festigkeit *f*, Haltbarkeit *f*, Lebensdauer *f*

durable dauerhaft, haltbar, stabil
durain Mattkohle *f*
duralium Duralium *n*
duralumin Duralumin *n*, Dural[uminium] *n*
duramycin Duramycin *n*
durana metal Duranametall *n*
durangite Durangit *m* (Min)
duran glass Duranglas *n*
duration Bestand *m* (Fortbestehen), Fortdauer *f*,
    Länge *f* (Zeit), Zeitdauer *f*, ~ of load
    application Belastungsdauer *f*, ~ of
    short-circuit Kurzschlußperiode *f*, ~ of
    tempering Anlaßdauer *f* (Met)
duratol Duratol *n*
durax glass Duraxglas *n*
durene Durol *n*
durenol Durenol *n*
durex carbon black Durexruß *m*
durferrit pot furnace Durferrittiegelofen *m*
duridine Duridin *n*
duriron Duriron *n*
durohydroquinol Durohydrochinon *n*
durol Durol *n*
durometer Härtemesser *m*, Härteprüfer *m*,
    Härteprüfgerät *n*
duroplast Duroplast *n*
duroplastic duroplastisch
duroplastics härtbare Plaste *pl*
duroquinone Durochinon *n*
duryl Duryl-
durylic acid Durylsäure *f*, Cumylsäure *f*
dust Staub *m*, **dry as** ~ staubtrocken, **formation**
    of ~ Staubentwicklung *f*, ~ **from grinding**
    Schleifstaub *m*, ~ **from throat of furnace**
    Gichtstaub *m*, ~ **of picked ore** Scheidemehl *n*
    (Metall), ~ **of roasted ore** Röststaub *m*
    (Metall)
dust *v* bestäuben, einstäuben, stäuben, ~ **[off]**
    abstauben
dust bin Mülleimer *m*, Staubsack *m*
dust box Staubfang *m*, Staubkammer *f*
dust catcher Staubabscheider *m*, Exhaustor *m*,
    Flugstaubkammer *f*
dust chamber Staubkammer *f*
dust coal Grießkohle *f*, Staubkohle *f*
dust collection Staubabscheidung *f*
dust collector Entstäuber *m*, Staubabscheider *m*,
    Staubfänger *m*, Staubfang *m*,
    Staubsammler *m*
dust content of the air Staubgehalt *m* der Luft
dust conveyor Staubförderanlage *f*
dust core Staubkern *m*
dust deposit Staubablagerung *f*
dust dry staubtrocken
dust emission Staubauswurf *m*
duster Haderndrescher *m* (Pap)
dust explosion Staubexplosion *f*
dust extraction plant Entstaubungsanlage *f*
dust figure, electrical ~ elektrisches Staubbild *n*

dust filter Staubfilter *n*
dust formation Staubbildung *f*
dust-free staubfrei
dusting Abstauben *n*, Bestäubung *f*,
    Einstäubung *f*, Staubbildung *f*
dusting gold Pudergold *n*
dusting platinum Puderplatin *n*
dusting silver Pudersilber *n*
dust-like staubartig, pulverig
dust loss Verstäubungsverlust *m*
dust measuring instrument Staubmeßgerät *n*
dust meter Staubmeßgerät *n*
dust mixture Staubgemisch *n*
dust nuisance Staubbelästigung *f*
dust particle Staubteilchen *n*, **coarse** ~
    Staubkorn *n*
dust pocket Staubsack *m*
dust prevention Staubverhütung *f*
dustproof staubdicht, staubsicher
dustproof packing Staubdichtung *f*
dust removal plant Entstaubungsanlage *f*
dust remover Staubentferner *m*
dust roaster Staubröstofen *m*
dust-separating method
    Staubabscheideverfahren *n*
dust separation Staubabscheidung *f*
dust separator Staubabscheider *m*,
    Entstaubungsanlage *f*, Rauchgasentstäuber *m*
dust settling chamber Staubkammer *f*
dust stop Staubabdichtung *f*
dust-tight staubdicht
dusty staubig, staubartig, staubförmig
Dutch balance Holländerwaage *f*
Dutch foil Rauschgold *n*
Dutch gold Rauschgold *n*
Dutch metal Flittergold *n*, Schaumgold *n*
Dutch tile Fliese *f*, Kachel *f*
Dutch white Holländerweiß *n*
dutiable zollpflichtig
duty Nutzeffekt *m* (Techn), Pflicht,
    Wirkleistung *f* (Techn), Zoll *m*, **high** ~
    Hochleistung *f*
dwarfism Zwergwuchs *m*
dwarf runner bean Buschbohne *f*
dwell Entlüftungspause *f*
dwell time (injection molding) Verweilzeit *f*
dyad Dyade *f* (Chem)
dyakisdodecahedral dyakisdodekaedrisch (Krist)
dyakisdodecahedron Dyakisdodekaeder *n* (Krist)
dyclonine Dyclonin *n*
dye Farbstoff *m*, Farbe *f*, **direct** ~ substantiver
    Farbstoff *m*, **substantive** ~ substantiver
    Farbstoff *m*
dye *v* färben, anfärben, einfärben, ~ **again**
    auffärben, nachfärben, ~ **inside out**
    linksfärben, ~ **on the wrong side** linksfärben,
    ~ **to pattern** bemustern (Färb)
dyeability Färbbarkeit *f*
dyeable [an]färbbar

dye absorption Farbstoffaufnahme *f*
dye affinity Farbstoffaffinität *f,* Anfärbbarkeit *f,*
   Anfärbevermögen *n*
dye bath Färbebad *n,* Farbflotte *f,* Flotte *f*
dye beaker Färbebecher *m*
dye beck Färberküpe *f*
dyed gefärbt
dyed-in-the-grain waschecht (Text)
dyed-in-the-wool waschecht (Text)
dye fluid Färbebad *n,* Färbeflüssigkeit *f,*
   Färberflotte *f*
dye house Färberei *f*
dyeing Färben *n,* Färberei *f* (Gewerbe),
   Färbung *f,* **goods for** ~ Farbgut *n,* **rate of** ~
   Färbegeschwindigkeit *f,* **to finish** ~ ausfärben
dyeing *a* färbend
dyeing auxiliary Färbehilfsmittel *n,*
   Färbereihilfsmittel *n* (Färbehilfsmittel)
dyeing capacity Färbekraft *f*
dyeing carthamus Färberdistel *f* (Bot)
dyeing machine Färbemaschine *f*
dyeing vat Färbekufe *f* (Färb)
dye jig Färbejigger *m* (Färb)
dye liquor Farbbrühe *f,* Flotte *f*
dye nozzle Farbdüse *f*
dyer Färber *m*
dyer's bath Färberflotte *f*
dyer's broom Färberginster *m* (Bot),
   Färberkraut *n* (Bot), Färberpfrieme *f* (Bot)
dyer's buckthorn Färberkreuzdorn *m* (Bot)
dyer's bugloss Färberalkanna *f* (Bot)
dyer's lichen Färbermoos *n* (Bot)
dyer's madder Färberröte *f* (Bot),
   Färberwurzel *f* (Bot)
dyer's moss Färberflechte *f* (Bot), Färbermoos *n*
   (Bot)
dyer's mulberry Färbermaulbeerbaum *m* (Bot)
dyer's oak Färbereiche *f* (Bot)
dyer's oil Färberöl *n*
dyer's rocket Harnkraut *n* (Bot)
dyer's safflower Färberdistel *f* (Bot),
   Färbersaflor *m* (Bot)
dyer's woad Färberwaid *m* (Bot), **common** ~
   Indigkraut *n* (Bot)
dyer's woodruff Färberwaldmeister *m* (Bot)
dye sensitivity Farbenempfindlichkeit *f*
dye solution Farb[stoff]lösung *f*
dye stone Farbstein *m*
dyestuff Farbstoff *m*
dyestuff factory Farbstoffabrik *f*
dyestuff for coloring lacquers Lackfarbstoff *m*
dyestuffs industry Farbenindustrie *f*
dyestuff solvent Farbstofflösungsmittel *n*
dyestuff yield Farbausbeute *f*
dye tanning Färbegerberei *f*
dye test Farbenprobe *f*
dye tub Färberküpe *f*
dyewood Färbeholz *n,* Farbholz *n*
dyewood extract Farbholzextrakt *m*

dye works Farbenfabrik *f,* Zeugfärberei *f*
dying-out of the oscillation Abklingen *n* der
   Schwingung
dymal Dymal *n*
dynamdon brick Dynamdonstein *m*
dynamic dynamisch
dynamic compression stress
   Schlagdruckbeanspruchung *f*
dynamic compression test Schlagdruckversuch *m*
dynamic cooling Expansionskühlung *f* (Therm)
dynamic creep test
   Dauerstandfestigkeitsprüfung *f*
dynamic hardness testing machine
   Schlaghärteprüfer *m*
dynamic pressure Staudruck *m*
dynamic range Betriebsdynamik *f,* Dynamik *f*
   (Magnetband)
dynamics Dynamik *f*
dynamic strength dynamische Festigkeit *f*
dynamic tensile test Schlagzerreißversuch *m*
dynamite Dynamit *n,* Sprengstoff *m,* **explosion of**
   ~ Dynamitexplosion *f,* **stick of** ~
   Dynamitstange *f*
dynamite charge Dynamitladung *f*
dynamo Dynamo *m* (Elektr), **balance** ~
   Ausgleichdynamo *m,* **method of driving the** ~
   Dynamoantrieb *m,* **size of** ~ Dynamogröße *f,*
   supplementary ~ Zusatzdynamo *m*
dynamo brush Dynamobürste *f*
dynamo electric machine Dynamomaschine *f*
dynamo mains Dynamoleitung *f*
dynamometer Dynamometer *n,* Kraftmesser *m,*
   Pendelmotor *m* (Transmission)
dynamo oil Dynamoöl *n*
dynamo rule Dynamoregel *f*
dyne Dyn *n* (Maß)
dyneine Dynein *n* (Biochem)
dynode Dynode *f,* Zwischenelektrode *f*
dynode structure Dynodenstruktur *f*
dypnone Dypnon *n*
dyrophanthine Dyrophanthin *n*
dysacousia Lärmempfindlichkeit *f* (Med)
dyscrasite Dyskrasit *m* (Min), Antimonsilber *n*
   (Min), Drittelsilber *n* (Min)
dysentery Ruhr *f* (Med)
dysentery bacillus Ruhrbazillus *m*
dysluite Dysluit *m* (Min)
dysodil Dysodil *m*
dysprosium (Symb. Dy) Dysprosium *n*
dysprosium chloride Dysprosiumchlorid *n*
dystectics Dystektikum *n*

E

eager eifrig, begierig
eaglestone Adlerstein *m* (Min), Ätit *m* (Min),
  Eisenniere *f* (Min), Klapperstein *m* (Min)
ear Ähre *f* (Getreide), Ohr *n* (Med)
earliest section Anfangsquerschnitt *m*
early cropping Ernteverfrühen *n*
earn *v* verdienen, erwerben
earnings yield Kapitalrendite *f*
ear-stone Otolith *m* (Min)
earth Erde *f* (Chem, Elektr) (Planet), Erdung *f*
  (Elektr), **containing** ~ erdehaltig, **crust of the**
  ~ Erdrinde *f,* **ferriferous** ~ eisenhaltiger
  Boden *m,* **heat of the** ~ Erdwärme *f,* **interior**
  of the ~ Erdinnere *n,* **rotation of the** ~
  Erdumdrehung *f,* **rotten** ~ Mododererde *f,*
  surface of the ~ Erdoberfläche *f*
earth *v* erden (Elektr)
earth alkaline erdalkalisch
earth bank Erddamm *m*
earth boring Erdbohrung *f*
earth cobalt Kobaltmanganerz *n* (Min)
earth-colored erdfahl, erdfarben, erdfarbig
earth conductor Nulleitung *f*
earth conductor cable Nulleiterkabel *n*
earth connection Erdleitung *f* (Elektr), Erdung *f*
  (Elektr)
earthed geerdet
earthed coating geerdeter Belag *m*
earthen irden
earthen vessel Tongefäß *n*
earthenware irdenes Geschirr *n,* Steingut *n,*
  Tonwaren *f pl*
earthenware cooling coil Tonkühlschlange *f*
earthenware flanged tube Tonflanschenrohr *n*
earthenware pipe Steinzeugrohr *n*
earthenware tank Steingutwanne *f*
earth flattening Erdabplattung *f*
earth for molds Gießerde *f*
earth[ing] Erdschluß *m* (Elektr), Erdung *f*
  (Elektr)
earthing conductor Erdleiter *m* (Elektr)
earth magnetic erdmagnetisch
earth magnetism Erdmagnetismus *m*
earthmannite Erdmannit *m* (Min)
earth metal Erdmetall *n*
earth orbit Erdumlaufbahn *f*
earth pitch Bergpech *n,* Petroleumasphalt *m*
earth potential Erdpotential *n*
earthquake Erdbeben *n*
earthquake focus Erdbebenherd *m*
earths, rare ~ seltene Erden *pl*
earth switch Erdungsschalter *m* (Elektr)
earth wax Erdwachs *n,* Bergtalg *m,* Bergwachs,
  Ceresin *n*
earth wire Erdungsdraht *m* (Elektr),
  Erdungsleitung *f*

earthy erdig
earthy calcite Mehlkreide *f*
earthy cerussite Bleierde *f*
earthy coal Erdkohle *f*
earthy cobalt Asbolan *m* (Min)
earthy ferruginous cuprite Kupferziegelerz *n*
  (Min)
earthy fluorite Flußspaterde *f*
earthy gypsum Mehlgips *m*
earthy hornsilver Buttermilcherz *n* (Min),
  Buttersilber *n* (Min)
earthy lead carbonate Bleierde *f*
earthy magnetite Eisenschwärze *f*
earthy marl Mergelerde *f*
earthy pit coal Rußkohle *f*
earthy slag Erdschlacke *f*
earthy smell Erdgeruch *m*
earthy talc Erdtalk *m*
earthy vivianite Eisenblauerde *f*
earthy white lead ore Bleierde *f*
ease *v* erleichtern
easing valve Entlastungsventil *n*
easy leicht, **very** ~ spielend leicht
easy care finishing Pflegeleicht-Ausrüstung *f*
eat away *v* abfressen, verzehren
eau de Cologne Kölnisches Wasser *n*
eau de Javelle Eau de Javelle *n,* Bleichwasser *n,*
  Javellesche Lauge *f,*
  Kaliumhypochloritlösung *f*
ebb [tide] Ebbe *f*
eblanin Eblanin *n*
ebonite Ebonit *n,* Hartgummi *n,*
  Hartkautschuk *m*
ebonite goods Hartgummiartikel *m pl*
ebonite insulator Hartgummiisolator *m*
ebonite plate Hartgummiplatte *f*
ebonite powder Hartgummimehl *n*
ebonite rod Ebonitstange *f*
ebonite support Untersatz *m* aus Hartgummi
ebonite tube Ebonitrohr *n*
ebonize *v* schwarz beizen
ebon-wood Ebenholz *n*
ebony Ebenholz *n,* **black** ~ echtes Ebenholz *n*
ebony black Elfenbeinschwarz *n*
ebulliometer Ebulliometer *n*
ebullioscope Ebullioskop *n*
ebullioscopic ebullioskopisch
ebullioscopy Ebullioskopie *f*
ebullition Aufwallen *n,* Brodeln *n,* Sieden *n,*
  retardation of ~ Siedeverzug *m,* ~ **of lime**
  Aufgehen *n* des Kalkes
ebullition regulating coil Siedeverzugsspirale *f*
eburicane Eburican *n*
eburicoic acid Eburicosäure *f*
eburine Eburin *n*
eburnean elfenbeinartig, elfenbeinfarbig

ecboline Ekbolin *n*
eccaine Ekkain *n*
eccentric Exzenter *m*
eccentric *a* exzentrisch, nicht durch den
  Mittelpunkt gehend, unrund
eccentric attachment Exzenter *m*
eccentric bolt Exzenterbolzen *m*
eccentric breaker Exzenterpresse *f*
eccentric disk Exzenterscheibe *f,* Kropfscheibe *f*
eccentric drive Exzenterantrieb *m*
eccentric fork Exzentergabel *f*
eccentric governor Exzenterregulator *m*
eccentric grinder Exzenterschleifmaschine *f,*
  Unrundschleifmaschine *f*
eccentric hoop Exzenterbügel *m*
eccentricity Exzentrizität *f,* Außermittigkeit *f*
eccentric lathe Exzenterdrehbank *f,*
  Unrunddrehbank *f*
eccentric press Exzenterpresse *f*
eccentric punch Exzenterstanze *f*
eccentric ring Exzenterring *m*
eccentric rod Exzenterstange *f*
eccentric sheave Exzenterscheibe *f*
eccentric single-rotor screw pump
  Exzenterschneckenpumpe *f*
eccentric strap Exzenterbügel *m*
ecdemite Ekdemit *m* (Min)
E.C.G. (Abk) Elektrokardiogramm *n* (Med)
ecgonidine Ekgonidin *n*
ecgonine Ekgonin *n*
ecgoninic acid Ekgoninsäure *f*
ecgoninol Ekgoninol *n*
echelon grating Stufengitter *n*
echelon lens Stufenlinse *f* (Opt)
echelon prism Stufenprisma *n* (Opt)
echicerin Echicerin *n*
echinacein Echinacein *n*
echinatine Echinatin *n*
echinenone Echinenon *n*
echinite Echinit *m,* versteinerter Seeigel *m*
echinochrome Echinochrom *n*
echinomycin Echinomycin *n*
echinopsine Echinopsin *n*
echinulin Echinulin *n*
echitamine Echitamin *n,* Ditain *n*
echiumine Echiumin *n*
echo Echo *n* (Phys), Widerhall *m*
echujin Echujin *n*
eclipse Eklipse *f* (Astr), Verdunk[e]lung *f* (Astr),
  Verfinsterung *f* (Astr)
eclipsed ekliptisch (Konformation)
eclogite Eklogit *m* (Min)
eclogite shell Eklogitschale *f*
ecology Ökologie *f*
economic[al] wirtschaftlich, rationell, sparsam
economic crisis Wirtschaftskrise *f*
economic situation Wirtschaftslage *f*

economizer Abgasvorwärmer *m,* Anwärmer *m,*
  Ekonomiser *m,* Heizgasvorwärmer *m,*
  Rauchgasvorwärmer *m*
economizing transformer Spartransformator *m*
economy Wirtschaft[lichkeit] *f,* Sparsamkeit *f*
economy in use Ergiebigkeit *f*
economy size pack Familienpackung *f*
ecospecies Ökospecies *f* (Biol)
ecrasite Ekrasit *n*
ecru silk Ekrüseide *f*
ecsantalal Eksantalal *n*
ecsantalic acid Eksantalsäure *f*
ecsantalol Eksantalol *n*
ectoparasite Außenschmarotzer *m* (Zool),
  Ektoparasit *m* (Zool)
ectophyte Hautpilz *m*
ectoplasm Ektoplasma *n*
ectoplasmic ektoplasmisch
ectosite Ektosit *m* (Darmschmarotzer)
ectozoon Ektozoon *n* (tierischer Hautparasit)
ectylcarbamide Ektylcarbamid *n*
eczema Ekzem *n* (Med)
eddy Wirbel *m*
eddy conductivity Scheinleitfähigkeit *f*
eddy current Wirbelstrom *m* (Elektr),
  Induktionsstrom *m* (Elektr), Streustrom *m*
  (Elektr)
eddy current circuit Wirbelstromkreis *m* (Elektr)
eddy current constant Wirbelstromkonstante *f*
  (Elektr)
eddy diffusion Streudiffusion *f,* Wirbeldiffusion *f*
eddy diffusion coefficient Mischkoeffizient *m*
eddy energy Turbulenzenergie *f*
eddy flow Wirbelströmung *f*
eddy flux Turbulenzfluß *m,*
  Turbulenztransport *m*
eddy viscosity coefficient Turbulenzkoeffizient *m*
  (Viskosität)
edestin Edestin *n*
edetates Edetate *pl* (Kurzbez. f. Salze der
  Äthylendiamintetraessigsäure)
edetic acid Edetinsäure *f*
  (Äthylendiamintetraessigsäure)
edge Ecke *f,* Kante *f,* Rand *m,* Saum *m,* **blunt** ~
  stumpfe Schneide *f,* **on** ~ hochkant[ig], **sharp**
  ~ scharfe Schneide *f,* **to set on** ~ kanten,
  upper ~ Oberkante *f*
edge *v* ausfräsen, bördeln (Blech)
edge angle Kantenwinkel *m* (Krist)
edge bond Eckverband *m*
edge crack Kantenriß *m*
edge dislocation Stufenversetzung *f*
edge effect Kanteneffekt *m*
edge emission Kantenemission *f*
edge gumming Randgummierung *f*
edge joint Eckverband *m*
edge length Kantenlänge *f*
edge mill Kollergang *m,* Kollermühle *f*
edge milling machine Kantenfräsmaschine *f*

edge-punched card Lochstreifenkarte *f* (Comp)
edge runner, grinding on the ~ Kollerung *f,*
   mixture ground on the ~ Kollerung *f*
edge runner [mill] Kollergang *m*
edge shift Kantenverschiebung *f*
edge tools Schneidwerkzeuge *pl*
edgeways, to place ~ hochkantig stellen
edgewise hochkant[ig]
edgewise scale Profilskala *f*
edging Borte *f*
edible eßbar, genießbar
edible oil Speiseöl *n*
edifice Bauwerk *n,* Gebäude *n*
edingtonite Edingtonit *m* (Min)
edinol Edinol *n*
Edison-battery Edison-Akkumulator *m* (Elektr)
edition Auflage *f* (Buch), complete ~
   Gesamtausgabe *f*
editor Herausgeber *m*
Edman degradation Edman-Abbau *m*
edrophonium Edrophonium *n*
education Ausbildung *f,* Bildung *f*
edulcorate *v* entsäuern, absüßen, adoucieren,
   auswaschen
edulcorating basin Absüßschale *f* (Zuck),
   Absüßwanne *f* (Zuck)
edulcorating tank Absüßkessel *m* (Zuck)
edulcorating tub Absüßbottich *m* (Zuck)
edulcorating vessel Adouciergefäß *n*
edulcoration Absüßung *f*
edulein Edulein *n*
E.E.G. (or e.e.g.) (Abk)
   Elektroenzephalogramm *n* (Med)
effect Wirkung *f,* Auswirkung *f,* Effekt *m,*
   Einfluß *m,* Einwirkung *f,* Ergebnis *n,*
   bactericidal ~ bakterizide Wirkung *f,*
   bathochromic ~ bathochromer Effekt *m,*
   biological ~ biologische Wirkung *f,* bleaching
   ~ Bleicheffekt *m,* demagnetizing ~
   entmagnetisierende Wirkung *f,* duration of ~
   Wirkungsdauer *f,* elastic ~ elastischer
   Effekt *m,* inductive ~ induktiver Effekt *m,*
   photoelectric ~ lichtelektrischer Effekt *m,*
   photoelektrischer Effekt *m,* photoemissive ~
   äußerer lichtelektrischer Effekt *m,*
   physiological ~ physiologische Wirkung *f,*
   radioactive ~ radioaktive Wirkung *f,*
   second-order ~ Effekt *m* zweiter Ordnung,
   steric ~ sterische Wirkung *f,* thermoelectric ~
   thermoelektrische Wirkung *f*
effect *v* bewerkstelligen, bewirken
effect enamel Effektlack *m*
effective wirksam, effektiv, nutzbar,
   wirkungsfähig, wirkungsvoll, highly ~
   hochwirksam
effective attenuation Betriebsdämpfung *f*
effective charge Effektivladung *f*
effective coverage Deckvermögen *n*

effective current Effektivstrom *m* (Elektr),
   Leistungsstrom *m*
effective force Wirkungskraft *f*
effective heat Nutzwärme *f*
effective height Effektivhöhe *f*
effectiveness Wirksamkeit *f,* Wirkungsfähigkeit *f,*
   Wirkungsgrad *m,* range of ~
   Wirkungsbereich *m*
effectiveness factor Effektivitätsfaktor *m*
   (heterogene Katalyse)
effective output Nutzleistung *f*
effective power Nutzeffekt *m,* Wirkleistung *f*
   (Elektr)
effective range Wirkungssphäre *f*
effective resistance Wirkwiderstand *m* (Elektr)
effective value Effektivwert *m*
effective work Nutzarbeit *f,* Kraftleistung *f*
effectors Effektoren *pl*
effervesce *v* [auf]brausen, [auf]schäumen,
   aufwallen, efferveszieren, moussieren,
   sprudeln
effervescence Aufbrausen *n,* Aufwallung *f,*
   Moussieren *n,* Schäumen *n*
effervescent [auf]brausend, moussierend,
   schäumend
effervescent lithium citrate Brauselithiumcitrat *n*
effervescent mixture Brausemischung *f*
effervescent powder Brausepulver *n,*
   Sodapulver *n*
effervescent salt Brausesalz *n*
efficacious wirksam, wirkungsvoll
efficacy Wirksamkeit *f*
efficiency Wirksamkeit *f,* Arbeitsleistung *f,*
   Austauschwirkung *f* (Wärmeübergang),
   Leistung *f,* Leistungsfähigkeit *f,* Nutzeffekt *m,*
   Trennwirkung *f* (Dest), Wirkungsfähigkeit *f,*
   Wirkungsgrad *m,* adiabatic ~ adiabatischer
   Wirkungsgrad, gross ~ gesamter
   Wirkungsgrad, high ~ Hochleistung *f,* loss of
   ~ Wirkungseinbuße *f* (Droge), manometric
   ~ manometrischer Wirkungsgrad, mechanical
   ~ mechanischer Wirkungsgrad, optical ~
   optischer Wirkungsgrad, overall ~
   Gesamtwirkungsgrad *m,* total ~
   Gesamtnutzeffekt *m,* Gesamtwirkungsgrad *m*
efficiency data Gütedaten *pl*
efficiency factor Güteziffer *f* (Dest)
efficiency of a luminous source Lichtausbeute *f*
efficiency of exchanger
   Austauscherwirkungsgrad *m*
efficiency per unit Einzelwirkungsgrad *m*
efficiency study Nutzwertanalyse *f*
efficiency tensor Ausbeutetensor *m*
efficiency test Wirkungsgradbestimmung *f*
efficiency value Gütewert *m*
efficient leistungsfähig, wirksam, wirkungsvoll,
   wirkungsfähig
effloresce *v* auswittern (Chem), effloreszieren

efflorescence Ausblühung *f*, Auswitterung *f*, Effloreszenz *f*
efflorescing Ausblühen *n*
effluence Ausfluß *m*
effluent Abfluß *m*, Abwasser *n*
effluent *a* ausfließend, ausströmend
effluent water testing Abwasseruntersuchung *f*
effluent [water] treatment Abwasserbehandlung *f*
effluvium Effluvium *n*, Miasma *n*
efflux Ausfluß *m*, Ausströmen *n*, Ausströmung *f*, Austritt *m* (Austreten v. Flüssigkeiten)
efflux coefficient Ausflußkoeffizient *m*
efflux condenser Abflußkühler *m*
effort Anstrengung *f*, Mühe *f*
effusiometer Ausströmungsmesser *m*, Effusiometer *n*
effusion Effusion *f*
effusive rock Effusivgestein *n* (Geol), Ergußgestein *n* (Geol)
egeline Aegelin *n*
E.G.G. (Abk) Elektrogastrogramm *n* (Med)
egg albumin Ovalbumin *n*
egg cell Eizelle *f*
egg preservation Eierkonservierung *f*
egg preservative Eierkonserve *f*
egg shape Eiform *f*
egg size Eigröße *f*
egg white das Weiße des Eies
egg yolk Eidotter *m*, Eigelb *n*
egonol Egonol *n*
Egyptian blue Ägyptisch Blau *n*
Egyptian cotton Mako *m*
Egyptian pebble Kugeljaspis *m* (Min)
ehlite Ehlit *m* (Min)
Ehrlich's reagent Ehrlichs Aldehydreagens *n*
eiconal equation Eikonalgleichung *f* (Opt)
eicosane Eikosan *n*
eicosanic acid Eikosansäure *f*
eicosanol Eikosylalkohol *m*
eicosantrienonic acid Eicosantrienonsäure *f*
eicosene Eikosen *n*
eicosyl Eikosyl-
eicosyl alcohol Eikosylalkohol *m*, Arachinalkohol *m*
eicosylene Eikosylen *n*
eicosylic acid Eikosylsäure *f*
eigenfunction Eigenfunktion *f*
eigenstate Eigenzustand *m* (Chem)
eigenvalue Eigenwert *m* (Math)
eigenvalue problem Eigenwertproblem *n*
eight, figure of ~ Achter *m*
eightangled achteckig
eightfold achtfach, achtfältig
eighth [part] Achtel *n*
eight-membered ring Achtring *m* (Chem)
eight-sided achtseitig
eikonogen Eikonogen *n*
eikosane Eikosan *n*
einstein Einstein *n* (Maß)

Einstein equation for heat capacity Einsteinsche Gleichung *f* für spezifische Wärme (Therm), kalorische Zustandsgleichung *f* (Therm)
Einstein equation of mass-energy equivalence Einsteinsche Masse-Energie-Äquivalentgleichung *f*
einsteinium Einsteinium *n* (Symb. Ei)
Einstein shift Einsteinsche Verschiebung *f*
eject *v* ausschleudern, ausspritzen, ausstoßen, auswerfen
ejecting device Auswurfvorrichtung *f*
ejection Ausdrücken *n*, Ausstoß *m* (Phys), Ausstoßung *f*, Auswerfung *f*, Auswurf *m* (Vulkan)
ejection plate Ausdrückplatte *f*
ejection ram Ausdrückkolben *m*, Ausheberbolzen *m*
ejection tie-bar Ausdrücker-Verbindungsstange *f*
ejector Ausdrücker *m*, Ausheber *m*, Auswerfer *m*, Ejektor *m*, Saugstrahlpumpe *f*, Strahlsauger *m*
ejector connecting bar Ausdruckbolzen *m*, Druckbalken *m*
ejector device Ausstoßvorrichtung *f*
ejector frame Ausdrückrahmen *m*, Ausheberrahmen *m*
ejector frame guide Ausdrücker-Führung *f*
ejector mixer Hohlrührer *m*
ejector pin Ausdrückstift *m*, Ausheberstift *m*, Auswerferstift *m*, Druckstift *m*
ejector pin mark Markierung *f* des Ausdrückstifts
ejector plate Ausdrückplatte *f*, Ausheberplatte *f*
[ejector plate] return pin Rückdruckstift *m*
ejector pump Dampfstrahlpumpe *f*
ejector ram Auswerferkolben *m*
ejector rod Ausdruckbolzen *m*, Ausheberbolzen *m*, Auswerferbolzen *m*
ejector tool Auswerfer *m*
eka-aluminum Ekaaluminium *n*, Gallium *n*
eka-boron Ekabor *n*, Skandium *n*
eka-silicon Ekasilizium *n*, Germanium *n*
ektogan Ektogan *n*
elaeagnine Elaeagnin *n*
elaeolite Eläolith *m* (Min)
elaeo-margaric acid Eläomargarinsäure *f*
elaeoptene Eläopten *n*
elaidic acid Elaidinsäure *f*
elaidic soap Elaidinseife *f*
elaidin Elaidin *n*
elaidinization Elaidinisierung *f*
elaidone Elaidon *n*
elaidyl alcohol Elaidylalkohol *m*
elain (obs) Äthylen *n*
elaiomycin Elaiomycin *n*
elapse *v* ablaufen (zeitlich)
elarsone Elarson *n*
elarsonic acid Elarsonsäure *f*
elastase Elastase *f* (Biochem)

elastic Gummiband *n*, Gummielastikum *n*,
Gummifaden *m*
elastic *a* elastisch, dehnbar, federnd
elasticator Elastifizierungsmittel *n*, Elastikator *m*
elastic band Gummiband *n*
elastic bitumen Elaterit *m*
elastic curve Bieglinie *f*
elastic force Federkraft *f*
elastic gum Federharz *n*
elasticity Elastizität *f*, Dehnbarkeit *f*,
Federkraft *f*, Federung *f*, Schnellkraft *f*,
Spannkraft *f*, **adiabatic** ~ adiabatische
Elastizität (Therm), **coefficient of** ~
Dehnungskoeffizient *m*, Elastizitätsbeiwert *m*,
degree of ~ Elastizitätsgrad *m*, **dynamic** ~
Elastodynamik *f*, **fundamental law of** ~
Elastizitätsgrundgesetz *n*, **imperfect** ~
unvollkommene Elastizität, **law of** ~
Elastizitätsgesetz *n*, **limit of** ~
Elastizitätsgrenze *f*, **modulus of** ~
Dehnungsmodul *m* (Techn), Dehnungszahl *f*,
perfect ~ vollkommene Elastizität, **theory of**
~ Elastizitätstheorie *f*, **transverse** ~ elastische
Durchbiegung *f*, **modulus of** ~ **in tension**
Zugelastizitätsmodul *m*
elasticity of compression Druckelastizität *f*
elasticity of extension Dehnungselastizität *f*
elasticity of flexure Biegefaktor *m*
elastic limit Elastizitätsgrenze *f*,
Dehnungsgrenze *f*, Streckgrenze *f*
elastic ratio Elastizitätsgrad *m*
elastic spring Sprungfeder *f*
elastic thread Gummifaden *m*
elastic tube pump Schlauchpumpe *f*
elastic web manufacture Gummibandweberei *f*
elastin Elastin *n*
elastomer Elastomer *n*, elastische Masse *f*
elastomeric gummiartig, elastomer
elastomers, synthetic ~ synthetische
Elastomere *pl*
elastometer Elastizitätsmesser *m*
elateridine Elateridin *n*
elateridoquinone Elateridochinon *n*
elaterin Elaterin *n*
elaterin glycoside Elateringlykosid *n*
elaterite Elaterit *m*, Erdharz *n*, Federharz *n*
elaterium Elaterium *n* (Pharm)
elaterone Elateron *n*
elatic acid Elatsäure *f*
elatinic acid Elatinsäure *f*
elatinolic acid Elatinolsäure *f*
elbon Elbon *n*
elbow Knie[stück] *n*, Rohrknie *n*
elbow box thermometer
Winkelkastenthermometer *n*
elbow joint Knieverbindung *f* (Techn)
elbow pipe Knierohr *n*
elbow thermometer Winkelthermometer *n*
elbow tube knieförmig gebogenes Rohr *n*

Elbs reaction Elbs-Reaktion *f*
elder Holunder *m* (Bot)
elderberry Holunderbeere *f* (Bot)
elderberry oil Holunderbeerenöl *n*
elder tea Holunderblütentee *m*
elder wine Holunderwein *m*
elecampane Alant *m* (Bot), Inula *f* (Bot)
elecampane camphor Alantcampher *m*
elecampane oil Alantöl *n*
elecampane root Alantwurzel *f* (Bot)
electric[al] elektrisch
electrical conductivity of ions Ionenleitfähigkeit *f*
[electrical control] panel Schaltbrett *n*
electrical energy Elektroenergie *f*
electrical engineer Elektroingenieur *m*,
Elektrotechniker *m*
electrical engineering Elektrotechnik *f*
electrical friction machine
Reibungselektrisiermaschine *f*
electrical gas purification plant elektrische
Gasreinigungsanlage *f*
electrical heat Elektrowärme *f*
electrical industry Elektroindustrie *f*
electrical insulation value
Durchschlagsfestigkeit *f*
electrically heated elektrisch beheizt
electrically neutral elektroneutral
electric appliance Elektrogerät *n*
electric arc Flammenbogen *m*, Lichtbogen *m*
electric arc furnace Elektrolichtbogenofen *m*,
Lichtbogenofen *m*
electric barrel pump Elektrofaßpumpe *f*
electric bill Elektrizitätsrechnung *f*
electric breakdown elektrischer Durchschlag *m*
electric bulb lacquer Glühlampenlack *m*
electric calamine Hemimorphit *m* (Min)
electric charge elektrische Ladung *f*
electric crucible furnace Elektrotiegelofen *m*
electric double layer elektrochemische
Doppelschicht *f*
electric drive Elektroantrieb *m*
electric eye Lichtschranke *f*, Photozelle *f*
electric furnace Elektroofen *m*
electric furnace steel Elektrostahl *m*
electric fuse Abschmelzsicherung *f*
electric hammer drill Schlagbohrmaschine *f*
electric highspeed conveying plant
Elektroschnellförderer *m*
electrician Elektriker *m*, Elektromechaniker *m*
electricity Elektrizität *f*, **dynamic** ~ dynamische
Elektrizität, **galvanic** ~ galvanische
Elektrizität, **generation of** ~
Elektrizitätserzeugung *f*, **hightened** ~
hochgespannte Elektrizität, **quantity of** ~
Elektrizitätsmenge *f*, **source of** ~
Elektrizitätsquelle *f*, **static inductional** ~
Influenzelektrizität
electricity produced by pressure
Druckelektrizität *f*
electricity generating plant
Stromerzeugungsanlage *f*

electricity meter Elektrizitätszähler *m*,
Elektrozähler *m*
[electricity] meter box Zählergehäuse *n* (Elektr)
electricity of opposite sign ungleichnamige
Elektrizität
electricity of same sign gleichnamige Elektrizität
electric line of force elektrische Kraftlinie *f*
electric match or fuse elektrischer Zünder *m*
electric melting furnace Elektrostahlofen *m*
electric osmosis Elektroosmose *f* (Biochem)
electric piler Elektrostapler *m*
electric potential Potentialfunktion *f* der
Elektrizität
electric potential gradient Spannungsgefälle *n*
(Elektr)
electric power Kraftstrom *m*
electric power plant Stromanlage *f*
electric pulley block Demagzug *m*
electric separator Elektrofilter *n*
electric shaft furnace Elektrohochofen *m*,
Elektroschachtofen *m*
electric shock elektrischer Schlag *m*
electric sign advertising Leuchtreklame *f*
electric smelting furnace Elektrohochofen *m*
electric stacker Elektrostapler *m*
electric steel Elektrostahl *m*
electric steel casting Elektrostahlformguß *m*
electric steel manufacture
Elektrostahlerzeugung *f*
electric supply lines Stromnetz *n*
electric susceptibility Elektrisierbarkeit *f*
electric traversing gear Elektrofahrwerk *n*
electric vehicle Elektrofahrzeug *n*,
Elektromobil *n*
electric welding Elektroschweißung *f*
electric whirl Flugrad *n*
electric wire covering Kabelumspinnung *f*
electrifiable elektrisierbar
electrification Elektrisierung *f*,
Elektrizitätserregung *f*
electrify *v* elektrisieren
electro- Elektro-, Galvano-
electroaffinity Elektroaffinität *f*
electroanalysis Elektroanalyse *f*
electro-analytical separation elektroanalytische
Trennung *f*
electrobiology Elektrobiologie *f*
electrobrighten *v* elektrolytisch polieren
electrobronze Galvanobronze *f*
electrocaloric elektrokalorisch
electrocardiogram Elektrokardiogramm *n* (Med)
electrocardiograph EKG-Apparat *m*,
Elektrokardiograph *m*
electrocardiography Elektrokardiographie *f*
electrocatalysis Elektrokatalyse *f*
electrochemical elektrochemisch
electrochemical series Spannungsreihe *f*

electrochemical transport elektrochemische
Überführung *f*
electrochemical valence elektrochemische
Wertigkeit *f*
electrochemist Elektrochemiker *m*
electrochemistry Elektrochemie *f*
electrocollargol (Am. E.) kolloidales Silber *n*
electrocopying process Elektrokopierverfahren *n*
electrocorundum Elektrokorund *m*
electrode Elektrode *f*, auxiliary ~
Hilfselektrode *f*, bipolar particulate bed ~
bipolar arbeitende Festbettelektrode *f*, coated
~ ummantelte Elektrode *f*, continuous ~
Dauerelektrode *f*, covered ~ ummantelte
Elektrode, Mantelelektrode *f*, destruction of
the ~ Aufzehrung *f* der Elektrode, double
skeleton ~ Doppelschichtelektrode *f*,
fluidized ~ fluidisierte Elektrode, hollow-cone
~ kegelige Mantelelektrode *f*, negative ~
Kathode *f*, Minuselektrode *f*, operational ~
Arbeitselektrode *f*, polarizable ~
polarisierbare Elektrode, positive ~ Anode *f*,
primary ~ Elektrode *f* erster Art, secondary
~ Elektrode *f* zweiter Art, (radio)
Hilfselektrode *f*, sheathed ~
Mantelelektrode *f*, vibrating ~
Zitterelektrode *f*, ~ of the first type
Elektrode *f* erster Art, ~ of the second type
Elektrode *f* zweiter Art
electrode box Elektrodenkasten *m*
electrode bundles, withdrawal of the ~
Ausfahren *n* der Elektrodenbündel (pl)
electrodecantation Elektrodekantation *f*
electrode carbon Elektrodenkohle *f*
electrode chain Elektrodenkette *f*
electrode contact Elektrodenkontakt *m*
electrode gap Elektrodenabstand *m*
electrode holder Elektrodenhalter *m*
electrode jacket Elektrodenmantel *m*
electrode kinetics Elektrodenkinetik *f*
electrode of coherer Fritterelektrode *f*
electrode passage Elektrodendurchführung *f*
electrode pick-up Auflegieren *n* der Elektroden
electrodeposit galvanischer Überzug *m*
electrodeposit *v* galvanisch niederschlagen
electrodeposition elektrolytische Fällung *f*,
Galvanisieren *n*, Galvanisierprozeß *m*,
Galvanisierung *f*, Galvanoplattierung *f*,
Galvanostegie *f*
electrodeposition equivalent Stromausbeute *f*
electrodeposition paint Elektrotauchlack *m*
electrode potential Elektrodenpotential *n*,
Elektrodenspannung *f*, single ~
Galvani-Potential *n*
electrode press Elektrodenpresse *f*
electrode reaction Elektrodenreaktion *f*
electrode roller Elektrodenwalze *f*
electrodes, adjustment of the ~
Elektrodenregelung *f*, arrangement of ~

Elektrodenanordnung *f,* **changing the ~**
Elektrodenwechsel *m,* **system of ~**
Elektrodensystem *n*
electrode shield Elektrodenschutzglocke *f*
electrode socket Kohlenhalter *m*
electrode surface Elektrodenoberfläche *f*
electrode tube Elektrodenröhre *f*
electrodialysis Elektrodialyse *f*
electrodispersion Elektrodispersion *f*
electrodynamic[al] elektrodynamisch
electrodynamics Elektrodynamik *f*
electrodynamics in vacuum
Vakuumelektrodynamik *f*
electro-encephalogram Elektroenzephalogramm
(EEG)
electroendosmosis Elektroendosmose *f*
electro-engraving galvanische Ätzung *f*
electroenzephalogram
Elektroenzephalogramm *n* (Med)
electroerosion Elektroerosion *f*
electro-etching galvanische Ätzung *f*
electrogalvanic elektrogalvanisch
electrogastrogram Elektrogastrogramm *n* (Med)
electrographite Elektrographit *m*
electrography Elektrographie *f,*
Galvanographie *f*
electrogravimetry Elektrogravimetrie *f*
electrograving apparatus
Elektrograviervorrichtung *f*
electro-inductive elektroinduktiv
electrokinetic elektrokinetisch
electrokinetic potential Zetapotential *n*
electrokinetics Elektrokinetik *f*
electroluminescence Elektrolumineszenz *f*
electrolysis Elektrolyse *f,* Zersetzung *f* durch
Elektrolyse
electrolysis beaker Elektrolysierbecher *m*
electrolysis of fused salts Elektrolyse *f* im
Schmelzfluß
electrolysis of sodium chloride
Kochsalzelektrolyse *f*
electrolysis of water Wasserelektrolyse *f*
electrolyte Elektrolyt *m,* **colloidal ~** kolloidaler
Elektrolyt (Elektr), **fused ~** geschmolzener
Elektrolyt, **molten ~** geschmolzener
Elektrolyt, **strong ~** starker Elektrolyt,
supplementary ~ Zusatzelektrolyt *m,* **triionic**
~ dreiioniger Elektrolyt, **univalent ~**
einwertiger Elektrolyt, **weak ~** schwacher
Elektrolyt
electrolyte acid Füllsäure *f*
electrolyte bridge Elektrolytbrücke *f*
electrolyte metabolism Elektrolythaushalt *m*
electrolyte resistance Elektrolytwiderstand *m*
electrolyte tank Elektrolytbehälter *m*
electrolyte volume Elektrolytvolumen *n*
electrolytic elektrolytisch
electrolytic analysis Elektroanalyse *f*
electrolytic bath Elektrolysenbad *n*

electrolytic cell elektrolytische Zelle *f,*
Badkasten *m*
electrolytic cell rectifier
Aluminiumzellengleichrichter *m*
electrolytic coagulation Elektrolytkoagulation *f*
electrolytic copper Elektrolytkupfer *n*
(Elektrochem)
electrolytic dissociation elektrolytische
Dissoziation *f,* Ionisierung *f* (Chem)
electrolytic gold Elektrolytgold *n*
electrolytic interrupter Elektrolytunterbrecher *m*
electrolytic iron Elektrolyteisen *n*
electrolytic lead sheet Elektrolytbleiblech *n*
electrolytic oxidation Eloxierung *f*
electrolytic oxidation process Eloxalverfahren *n*
(Elektrochem)
electrolytic pickling Elektrolytbeize *f*
electrolytic rectifier Elektrolytgleichrichter *m*
electrolytic refining of lead Bleielektrolyse *f*
electrolytic refining process Elektrolyseprozeß *m*
electrolytic resistance Zersetzungswiderstand *m*
(Elektr)
electrolytic separation elektrolytische Trennung *f*
electrolytic sheet copper Elektrolytkupferblech *n*
electrolytic silver Elektrolytsilber *n*
electrolytic slime Elektrolysenschlamm *m*
electrolytic tinplate elektrolytisch verzinntes
Blech *n,* Elektrolytblech *n*
electrolytic zinc Elektrolytzink *n*
electrolyze *v* elektrolysieren, zersetzen durch
Elektrolyse
electrolyzing apparatus
Elektrolysiervorrichtung *f*
electromagnet Elektromagnet *m,* **fremderregter**
Magnet *m,* **semiannular ~**
Halbringelektromagnet *m,* **semicircular ~**
Halbringelektromagnet *m*
electromagnetic elektromagnetisch
electromagnetic furnace Elektromagnetofen *m*
electromagnetic isotope separation
elektromagnetische Isotopentrennung *f*
[electro]magnetic leak magnetischer
Nebenschluß *m*
electromagnetism Elektromagnetismus *m,*
Galvanomagnetismus *m*
electromechanic Elektromechaniker *m*
electromechanical elektromechanisch
electromechanics Elektromechanik *f*
electromercurol (Am. E.) kolloidales
Quecksilber *n*
electrometallurgy Elektrometallurgie *f,*
Galvanoplastik *f*
electrometer Elektrometer *n,* **gold leaf**
electroscope or ~ Goldblattelektrometer *n*
electrometric elektrometrisch
electromigration studies
Elektromigrationsuntersuchungen *f*
electromobile Elektromobil *n*
electromotion Elektrizitätserregung *f*

electromotive elektromotorisch
electromotive force elektromotorische Kraft *f*
(EMK), back ~ elektromotorische
Gegenkraft *f* (Elektr), counter ~
elektromotorische Gegenkraft *f* (Elektr),
induced ~ Induktionsspannung *f*
electromotive series Spannungsreihe *f*
electromotor Elektromotor *m*
electron Elektron *n*, disintegration ~
ausgestoßenes Elektron, inner ~
Rumpfelektron *n*, lone ~ Einzelelektron *n*
(Atom), orbital ~ Bahnelektron *n*, kernfernes
Elektron, outer ~ Außenelektron *n*, outer
level ~ kernfernes Elektron, peripheral ~
Bahnelektron *n*, kernfernes Elektron, positive
~ Positron *n*, positives Elektron, secondary ~
Sekundärelektron *n*, triggering ~
Auslöseelektron *n*
electron absorption Elektronenabsorption *f*
electron accelerator Elektronenbeschleuniger *m*
electron acceptance Elektronenaufnahme *f*
electron acceptor Elektronenakzeptor *m*
electron accumulation Elektronenstauung *f*
electron affinity Elektronenaffinität *f*
electron angular momentum Elektronenspin *m*
electron arrangement Elektronenanordnung *f*
electron-attachment mass spectrography Elektro-
nenanlagerungs-Massenspektrographie *f*
electron avalanche Elektronenlawine *f*
electron beam Elektronenbündel *n*,
Elektronenstrahl *m*
electron beam welding
Elektronenstrahlschweißverfahren *n*
electron bombardment Elektronenbeschuß *m*
electron capture Elektronenauffang *m*,
Elektroneneinfang *m*
electron capture detector
Elektroneneinfang-Detektor *m* (Gaschromat)
electron carrier Elektronenüberträger *m*
electron charge density Elektronenladedichte *f*
electron cloud Elektronenwolke *f*
electron collision Elektronenstoß *m*
electron decay Elektronenzerfall *m*
electron-defect conductivity
Elektronenmangelleitung *f*
electron deficiency Elektronenmangel *m*
electron deficient binding
Elektronenmangelbindung *f* (Chem)
electron deficient compound
Elektronenmangelverbindung *n*
electron deficit position Elektronenleerstelle *f*
electron density Elektronendichte *f*, distribution
of ~ Elektronendichteverteilung *f*
electron diffraction Elektronenbeugung *f*
electron discharge Elektronenentladung *f*
electron displacement Elektronenverschiebung *f*
electron distribution Elektronenverteilung *f*
electron donor Elektronendonator *m*

electronegative elektronegativ, negativ elektrisch,
unedel (Met)
electronegativity Elektronegativität *f*
electron-electron collision Stoß *m* zwischen
Elektronen (pl)
electron emission Elektronenemission *f*,
Elektronenstrahlung *f*, secondary ~
Sekundärelektronenemission *f*
electron emission area Elektronenabgabefläche *f*
electron-emitting elektronenaussendend
electron energy Elektronenenergie *f*
electroneutrality principle
Elektroneutralitätsprinzip *n*
electron evaporation glühelektrische
Elektronenemission *f*
electron excess conduction
Elektronenüberschußleitung *f*
electron excess conductor
Elektronenüberschußleiter *m*, n-Leiter *m*
electron excess semiconducting
elektronenüberschußleitend
electron focus[s]ing Elektronenbündelung *f*
electron friction Elektronenreibung *f*
electron gap Elektronenlücke *f*
electron gas Elektronengas *n*
electron generator Elektronengenerator *m*
electron gun Elektronenkanone *f*,
Kathodenstrahlerzeuger *m*
electron hole mobility
Defektelektronenbeweglichkeit *f*
electron-hole pair Elektron-Loch-Paar *n*
electronic elektronisch
electronic band spectrum
Elektronenbandenspektrum *n* (Spektr)
electronic brain Elektronengehirn *n*
electronic charge Elektronenladung *f*
[electronic] computer elektronische
Rechenmaschine *f*
electronic conductivity Elektronenleitung *f*
electronic configuration Elektronenanordnung *f*
electronic control elektronische Steuerung *f*
electronic current Emissionsstrom *m*
electronic flash Elektronenblitzgerät *n*
electronic formula Elektronenformel *f*
electronic heat sealing Hochfrequenzsiegeln *n*
electronic isomerism Elektronenisomerie *f*
electronic lens Elektronenlinse *f*
electronic mode (of the electron) Freiheitsgrad *m*
des Elektrons
electronic peak-reading voltmeter elektronisches
Voltmeter *n* mit Spitzenablesung
electronic polarizability
Elektronenpolarisierbarkeit *f*
electronics Elektronik *f*, Elektronenlehre *f*,
Elektronentechnik *f*
electron image Elektronenabbildung *f*
electron image tube Elektronenbildröhre *f*
electron impact Elektronenstoß *m*
electron injector Elektroneninjektor *m* (Atom)

electron ion recombination
Elektronionrekombination *f*
electron ion wall recombination
Elektronionwandrekombination *f*
electron jump Elektronensprung *m*
electron jump spectrum
Elektronensprungspektrum *n*
electron level Elektronenniveau *n*
electron microscope Elektronenmikroskop *n*,
field emission ~ Feldelektronenmikroskop *n*
electron microscopic elektronenmikroskopisch
electron microscopy Elektronenmikroskopie *f*
electron multiplier Elektronenvervielfältiger *m*,
secondary ~ Sekundärelektronenverstärker *m*
electron multiplier valve
Elektronenverstärkerröhre *f*
electron neutron interaction
Elektronneutronwechselwirkung *f*
electron optical elektronenoptisch
electron optics Elektronenoptik *f*
electron oscillation Elektronenschwingung *f*
electron pair Elektronenduett *n*,
Elektronenpaar *n*, **lone** ~ freies
Elektronenpaar *n*
electron particle density
Elektronenteilchendichte *f*
electron path Elektronenbahn *f*
electron-positron pair Elektron-Positron-Paar *n*
electron pressure Elektronendruck *m*
electron probe microanalysis
Elektronenstrahl-Mikroanalyse *f*
electron promotion Elektronenübergang *m*,
Quantenzahlenvergrößerung *f* bei Elektronen
(pl)
electron relay Elektronenrelais *n*
electron resonance Elektronenresonanz *f*,
secondary ~ Sekundärelektronenresonanz *f*
electrons, chromophoric ~
Chromophorelektronen *pl*, **unpaired** ~
ungepaarte Elektronen *pl*
electron scattering Elektronenstreuung *f*
electron sextet[te] Elektronensextett *n*
electron shell Elektronenhülle *f*, **complete** ~
abgeschlossene Elektronenschale *f* (Atom),
outer ~ Außenelektronenschale *f*
electron source Elektronenquelle *f*
electron spectroscopy Elektronenspektroskopie *f*
electron spin Elektronenspin *m*,
Elektroneneigendrehimpuls *m*
electron spin quantum number
Elektronenspinquantenzahl *f*
electron spin resonance Elektronenspinresonanz *f*
electron state Elektronenzustand *m*
electron stream Elektronenstrahl *m*
electron swarm Elektronenschar *f*
electron synchrotron Elektronensynchrotron *n*
electron term Elektronenterm *m*
electron transfer Elektronenüberführung *f*,
Elektronenübertragung *f*

electron transition Elektronenübergang *m*
electron transition probability
Elektronenübergangswahrscheinlichkeit *f*
electron transit time Elektronenlaufzeit *f*
electron-transmitting elektronendurchlässig
electron transport Elektronentransport *m*
electron trap Elektronenfänger *m*
electron tube, high-vacuum ~ (Br. E. valve)
Hochvakuumelektronenröhre *f*
electron vacancy Elektronenlücke *f*
electron volt Elektronenvolt *n*
electroosmosis Elektroosmose *f*
electropalladiol (Am. E.) kolloides Palladium *n*
electropathy Elektrotherapie *f*
electropherogram Elektropherogramm *n*
electropherography Elektropherographie *f*
electrophilic elektrophil
electrophone Elektrophon *n*
electrophoresis Elektrophorese *f*, **carrierless** ~
trägerfreie Elektrophorese *f*, **high-voltage** ~
Hochspannungselektrophorese *f*
electrophoresis equipment
Elektrophoreseapparatur *f*
electrophoretic painting Elektrotauchlackierung *f*
electrophorus Elektrophor *m*
electrophorus bottom plate Elektrophorteller *m*
electrophorus disc Elektrophordeckel *m*,
Elektrophorkuchen *m*
electrophotography Elektrophotographie *f*
electrophysics Elektrophysik *f*
electrophysiological elektrophysiologisch (Med)
electrophysiology Elektrophysiologie *f* (Med)
electroplate *v* elektroplattieren, galvanisieren,
~ **with silver** galvanisch versilbern
electroplater Galvanotechniker *m*
electroplating Elektroplattierung *f*,
Galvanisation *f*, galvanischer Überzug *m*,
Galvanoplattierung *f*
electroplating copper bath, rapid ~
Schnellkupfergalvanoplastikbad *n*
electroplating of plastics
Kunststoffgalvanisierung *f*
electroplating plant Galvanisieranlage *f*
electroplating process Elektroplattierverfahren *n*
electroplating with steel galvanische
Verstählung *f*
electropneumatic elektropneumatisch
electropositive elektropositiv, positiv elektrisch
electroprecipitation Elektroabscheidung *f*
electro-refining Elektroraffination *f*
electro-regulator Elektroregler *m*
electroretinogram Elektroretinogramm *n* (Med)
(Abk), ERG
electrorhodial (Am. E.) kolloidales Rhodium *n*
electroscope Elektroskop *n*,
Elektrizitätsanzeiger *m*, **gold leaf** ~
Goldblattelektroskop *n*
electroscopic elektroskopisch
electroscopy Elektroskopie *f*

electrosilver *v* galvanisch versilbern
electroslag remelting
   Elektro-Schlacke-Umschmelzverfahren *n*,
   ESU-Verfahren *n*
electroslag welding
   Schlackenschweißverfahren *n*
electrostatic elektrostatisch
electrostatic attraction elektrostatische
   Anziehung *f*
electrostatic discharge Elektrizitätsentladung *f*
electrostatic filter or precipitator
   Elektroabscheider m, Elektrofilter *n*
electrostatic flux Induktionsfluß *m*
electrostatic generator Elektrisiermaschine *f*
electrostatic induction Influenz *f*
electrostatics Elektrostatik *f*
electrostatic separation elektrische Sortierung *f*
electrostatic unit (esu.) elektrostatische Einheit *f*
   (ESE)
electrostriction Elektrostriktion *f* (Elektr)
electrotechnics Elektrotechnik *f*
electrotherapeutics Elektropathie *f*,
   Elektrotherapie *f*
electrothermal elektrothermisch
electrothermic elektrothermisch
electrothermostat Elektrothermostat *m*
electro-tinning galvanische Verzinnung *f*
electrotype Elektrotype *f*, Galvano *n*
electrotype *a* galvanographisch
electrotype *v* klischieren
electrotypic galvanoplastisch
electrotyping process Galvanoplastik *f*
electrotypy Galvanotypie *f*
electrovalence Elektrovalenz *f*
electrovalency Elektrovalenz *f*
electroweld *v* elektrisch schweißen
electrowelding elektrische Schweißung *f*,
   Elektrodenschweißung *f*
electrum Elektrum *n*, Goldsilber *n*, silberhaltiges
   Gold *n*
electuary Latwerge *f* (Bot, Pharm)
eledonine Eledonin *n*
elegance Eleganz *f*, Feinheit *f*
elemadienolic acid Elemadienolsäure *f*
elemane Eleman *n*
elemene Elemen *n*
element Element *n* (Chem),
   Grundbestandteil *m*, Grundstoff *m*,
   Urstoff *m* (Naturphilosophie), Zelle *f*
   (Elektr), **artificially radioactive** ~ künstliches
   radioaktives Element *n*, **electronegative** ~
   elektronegatives Element, **electropositive** ~
   elektropositives Element, **gaseous** ~
   gasförmiges Element, **radioactive** ~
   radioaktives Element, ~ **with a large number**
   of spectral lines linienreiches Element
   ~ **of resistance** Widerstandselement *n*
elemental elementar
elementary einfach, ursprünglich, elementar,
   grundlegend, fundamental

elementary analysis Elementaranalyse *f*
elementary building block Elementarbaustein *m*
elementary cell Elementarzelle *f* (Chem),
   Urzelle *f* (Biol)
elementary charge Elementarladung *f*
elementary color Primärfarbe *f*
elementary constituent Urbestandteil *m*
elementary dipole Elementardipol *m*
elementary event Elementarereignis *n* (Statist)
elementary microanalysis
   Mikroelementaranalyse *f*
elementary photochemical reaction
   photochemische Elementarreaktion *f*
elementary physics Elementarphysik *f*
elementary process Elementarprozeß *m*,
   Elementarvorgang *m*
elementary quantum Elementarquantum *n*
elementary reaction Elementarreaktion *f*
elementary sulfur Elementarschwefel *m*
elements, mixed ~ Mischelemente *n pl*
elemi Elemi *n*, Ölbaumharz *n*
elemic acid Elemisäure *f*
elemicin Elemicin *n*
elemi oil Elemiöl *n*
elemol Elemol *n*
elemolic acid Elemolsäure *f*
elemonic acid Elemonsäure *f*
eleolite Eleolith *m* (Min)
eleonorite Eleonorit *m* (Min)
eleostearic acid Eläostearinsäure *f*
eleutherin Eleutherin *n*
eleutherinol Eleutherinol *n*
eleutherolic acid Eleutherolsäure *f*
elevate *v* erhöhen, heben, hochheben
elevated erhöht
elevating platform truck Hubwagen *m*
elevating tube Steigrohr *n*
elevation Anstieg *m*, Erhöhung *f*,
   Hochförderung *f*, Steighöhe *f*, **angle of** ~
   Elevationswinkel *m*, Steigwinkel *m*
elevator Fahrstuhl *m*, Aufzug *m*,
   Becherhebewerk *n*, Elevator *m*, Lift *m*
elevator bucket Elevatoreimer *m*, Förderschale *f*
elevator cable Aufzugsseil *n*
elevator cage Förderschale *f*
elevator frame Fördergerüst *n*, Förderturm *m*
elfwort Inula *f* (Bot)
elgadienol Elgadienol *n*
elgenol Elgenol *n*
eliasite Eliasit *m* (Min)
eliminate *v* beseitigen, ausmerzen, ausscheiden
   (Med, Chem), aussondern, ausstoßen,
   eliminieren, entfernen, tilgen, weglassen
eliminated beseitigt
elimination Abspaltung *f*, Ausmerzung *f*,
   Ausscheidung *f* (Med), Ausstoßung *f*,
   Beseitigung *f*, Elimination *f* (Math),
   Eliminierung *f* (Chem), Entfernung *f*,

Wegnahme *f*, **bimolecular** ~ bimolekulare
Eliminierung, **intramolecular** ~
intramolekulare Eliminierung
elimination of acid Entsäuerung *f*
elimination of water Wasserabgabe *f*
eliminator Sieb *n* (Elektr)
elixir Elixier *n*
elk fat Elchfett *n*
ell Elle *f* (Maß)
ellagate Ellagat *n*
ellagic acid Ellagsäure *f*, Ellagengerbsäure *f*
ellagorubin Ellagorubin *n*
ellipse Ellipse *f*
ellipsoid Ellipsoid *n*, **oblate** ~ abgeplattetes
Ellipsoid *n*, **prolate** ~ gestrecktes Ellipsoid *n*
ellipsoidal property Ellipsoideigenschaft *f*
ellipsoid of revolution Rotationsellipsoid *n*
ellipsoid of rotation Rotationsellipsoid *n*
elliptic elliptisch
elliptical orbit or path Ellipsenbahn *f*
elliptic compass Ellipsenzirkel *m*
ellipticine Ellipticin *n*
ellipticity Abplattung *f*, Elliptizität *f*
elliptone Ellipton *n*
elm Rüster *f*, Ulme *f*
elm bark Rüsterrinde *f*
elon Elon *n*
elongate *v* verlängern, ausrecken (Metall),
dehnen, strecken
elongated verlängert
elongation Verlängerung *f*, Ausdehnung *f*,
Dehnung *f*, Streckung *f*
elongation at break Reißdehnung *f*
elongation in cross direction Querdehnung *f*
elongation resistance Streckfestigkeit *f*
elpidite Elpidit *m* (Min)
elscholtzic acid Elscholtziasäure *f*
elsholtzione Elsholtzion *n*
eluate Eluat *n* (Chromatogr)
elucidate *v* aufklären, beleuchten, erklären
elucidation Erklärung *f*, Klarstellung *f*
elusive ausweichend
elute *v* eluieren
elution Elution *f*
elutriate *v* abklären, [aus]schlämmen, reinigen
elutriated abgeschlämmt
elutriating Schlämmarbeit *f*
elutriating apparatus Schlämmapparat *m*
elutriating process Schlämmverfahren *n*
elutriation Schlämmung *f*, Abschlämmung *f*,
Abseihung *f*, Anschlämmung *f* (Chem),
Auslaugung *f*, Ausspülung *f*, Reinigung *f*,
Schlämmanalyse *f*
elutriator Aufstromklassierer *m*,
Schlämmapparat *m*, Schlämmer *m*,
Schlämmgerät *n*
elymoclavine Elymoclavin *n*
elytal process Elytal-Verfahren *n*
emaciate *v* abzehren, ausmergeln

emaciation Abmagerung *f*, Auszehrung *f*
emanate *v* ausfließen, ausschleudern (Atom),
ausströmen
emanation Emanation *f* (Chem)
emanium Emanium *n*
emanometer Emanometer *n*
embalm *v* [ein]balsamieren
embalming Einbalsamieren *n*
embalming fluid Einbalsamierungsflüssigkeit *f*
embalmment Einbalsamierung *f*
embarine Embarin *n*
embarrassed befangen
embed *v* einbetten, eingraben, einhüllen,
einschichten
embedding Einbetten *n*, Einbettung *f*,
Einmauerung *f*
embedding medium Einbettungsmittel *n*
embelic acid Embeliasäure *f*
embelin Embelin *n*
embellic acid Embeliasäure *f*
embers Glühasche *f*, Glutasche *f*
embitter *v* bitter machen, vergällen
embody *v* verkörpern
embolism Embolie *f* (Med)
embolite Embolit *m* (Min), Chlorbromsilber *n*
(Min)
embonic acid Embonsäure *f*
emboss *v* ausbauchen, erhaben ausarbeiten,
gaufrieren, hämmern, prägen (hohlprägen)
embossed printing Prägedruck *m*
embossed sheet geprägte Folie *f*, geprägte Folie *f*,
Prägefolie *f*
embossing Prägen *n*, Gaufrage *f* (Text, Papier),
Gaufrieren *n*, Gravierung *f*, Narben *n*,
Prägung *f*
embossing calender Gaufrierkalander *m*,
Grainierkalander *m*, Prägekalander *m*
embossing machine Gaufriermaschine *f*
embossing plate Prägeplatte *f*
embossing roll Prägewalze *f*
embossment Relief[arbeit] *f*
embrittle *v* verspröden
embrittlement Brüchigwerden *n*, Versprödung *f*
embrocation Einreibemittel *n* (Pharm),
Einreibung *f*
embryo (Biol) Embryo *m*, Frucht *f*, Keim *m*,
embryology Embryologie *f*
embryonal keimförmig, embryonal
embryonal cell Elementarzelle *f* (Biol)
embryonated angebrütet (Ei)
embryonic embryonal
embryonic cell Bildungszelle *f* (Biol),
Elementarzelle *f*
emerald Smaragd *m* (Min), **false** ~ Atlaserz *n*
(Min), **pseudo** ~ grüner Flußspat *m*
emerald *a* smaragdfarben, smaragdgrün
emerald copper Dioptas *m* (Min),
Kupfersmaragd *m* (Min)

emerald green Smaragdgrün *n*, Neapelgrün *n*
emeraldine Emeraldin *n*
emerald nickel Nickelsmaragd *m* (Min)
emerge *v* auftauchen, herauskommen
emergence Auflauf *m* (Herbicid)
emergency Notstand *m*
emergency device Notbehelf *m*
emergency door Notausgang *m*
emergency exit Notausgang *m*
emergency lighting Notbeleuchtung *f*
emergency release Schnellablaß *m*
emergency shut-down Notabschaltung *f*,
  Sicherheitsabschaltung *f*
emergency signal Notsignal *n*
emergency ward Unfallstation *f*
emery kleinkörniger Korund *m*, Schmirgel *m*,
  apparatus for decanting ~
  Schmirgelschlämmapparat *m*, to grind with ~
  schmirgeln, to rub with ~ schmirgeln
emery cloth Schmirgelleinwand *f*,
  Schmirgeltuch *n*
emery disc Schmirgelscheibe *f*
emery dust Schmirgelstaub *m*
emerylite Emerylith *m* (Min)
emery paper Schmirgelpapier *n*, Schleifpapier *n*
emery powder Schmirgelpulver *n*
emery wheel Schmirgelscheibe *f*
emetallyline Emetallylin *n*
emetamine Emetamin *n*
emetethyline Emetäthylin *n*
emetic Brechmittel *n* (Pharm), Emetikum *n*
  (Pharm)
emetic *a* emetisch
emetic wine Brechwein *m*
emetine Emetin *n*
emetine hydrochloride Emetinhydrochlorid *n*
emetinemethine Emetinmethin *n*
emetinium hydroxide Emetiniumhydroxid *n*
emetoline Emetolin *n*
emetpropyline Emetpropylin *n*
E.M.F. (Abk) elektromotorische Kraft
emission Ausstrahlung *f*, Aussendung *f*,
  Ausströmung *f*, Emission *f*, secondary ~
  Sekundärausstrahlung *f*, velocity of ~
  Emissionsgeschwindigkeit *f*
emission band Emissionsbande *f*
emission capability Emissionsfähigkeit *f*
emission cell Photozelle *f* mit äußerem
  lichtelektrischen Effekt
emission coefficient Emissionskoeffizient *m*
emission control Emissionssteuerung *f*
emission current Emissionsstrom *m*
emission law Entladungsgesetz *n*
emission of electrons Elektronenabgabe *f*,
  Elektronenausstrahlung *f*, thermionic ~
  glühelektrische Elektronenabgabe
emission spectrum Emissionsspektrum *n*,
  continuous ~ Emissionskontinuum *n*,

continuous molecular ~
  Molekülemissionskontinuum *n* (Spektr)
emission spectrum of crystals
  Emissionskristallspektrum *n*
emissive ausstrahlend
emissive power Ausstrahlungsvermögen *n*,
  Emissionsfähigkeit *f*, Emissionsvermögen *n*
emissivity Emissionsvermögen *n*,
  Emissionsfähigkeit *f*, Emissionskoeffizient *m*,
  photoelectric ~ photoelektrische Ausbeute *f*
emit *v* abstrahlen, aussenden, ausstrahlen,
  emittieren
emittance Emissionsfähigkeit *f*,
  Emissionsvermögen *n*
emitter Strahler *m*, Aussender *m*
emitting area Strahlungsfläche *f*
emitting electron Leuchtelektron *n*
emitting surface Strahlungsfläche *f*
emodic acid Emodinsäure *f*
emodin Emodin *n*
emodinanthranol Emodinanthranol *n*
emodinol Emodinol *n*
emollient Aufweichmittel *n*,
  Erweichungsmittel *n*, Weichmacher *m*
emollient *a* aufweichend
emotion Affekt *m*, Gefühlsbewegung *f*
emphasis Bekräftigung *f*
emphasize *v* bekräftigen, betonen
empiric[al] empirisch, erfahrungsmäßig
empirical facts Erfahrungstatsache *f*
empirical formula Bruttoformel *f*, empirische
  Formel *f*
empirical knowledge Empirie *f*
empirical law Erfahrungssatz *m*
empirical result Erfahrungsergebnis *n*
empirical science Erfahrungswissenschaft *f*
empirical value Erfahrungswert *m*
emplectite Emplektit *m* (Min)
employ *v* anstellen, beschäftigen, anwenden,
  benutzen, verwenden
employability Verwendungsfähigkeit *f*
employee Angestellter *m*, Arbeitnehmer *m*
employees Belegschaft *f*
employer Arbeitgeber *m*, Unternehmer *m*
employment Beschäftigung *f*, Anwendung *f*
empower *v* bevollmächtigen
empressite Empressit *m* (Min)
emptiness Leere *f*
empty leer, hohl
empty *v* [ent]leeren, ausgießen, ausschütten,
  räumen
empty chamber Hohlraum *m*,
  Hohlraumkammer *f*
emptying Entleeren *n*, Entleerung *f*, ~ the
  furnace Entleerung *f* des Ofens
emptying funnel Ablaßtrichter *m*
empyroform Empyroform *n*
empyroform paste Empyroformpaste *f*
emulate *v* rivalisieren

emulgent Reinigungsmittel *n* (Med)
emulsifiability Emulgierfähigkeit *f*
emulsifiable emulgierbar, emulsionsfähig
emulsification Emulgierung *f,*
  Emulsionsbildung *f*
emulsification equipment Emulgiergeräte *n pl*
emulsification stirrer Emulsionsrührer *m*
emulsified binder Emulsionsbinder *m*
emulsified cream eye shadow Lidstrichemulsion *f*
emulsifier Emulgator *m,* Emulgiermaschine *f*
emulsify *v* emulgieren, emulsionieren
emulsifying agent Emulgier[ungs]mittel *n*
emulsifying centrifuge Emulgierzentrifuge *f*
emulsifying machine Emulgiermaschine *f*
emulsin Emulsin *n* (Biochem)
emulsion Emulsion *f,* **breaking of an ~**
  Desemulgierung *f,* Emulsionszerstörung *f,*
  camphorated ~ Camphermilch *f,*
  deuterium-loaded ~ deuterierte Emulsion
emulsion binder Bindemittelemulsion *f*
emulsion breaking Emulsionszerstörung *f*
emulsion coating Emulsionsguß *m* (Phot)
emulsion copolymerization
  Emulsionsmischpolymerisation *f*
emulsionize *v* emulgieren
emulsion matrix Trägersubstanz *f*
emulsion mortar Emulsionsmörser *m*
emulsion ointment Emulsionssalbe *f*
emulsion paint Emulsionsfarbe *f,* Binderfarbe *f,*
  Dispersionsfarbe *f*
emulsion polymerization
  Emulsionspolymerisation *f*
emulsion process Emulsionsverfahren *n*
emulsion stability Emulsionsstabilität *f*
emulsion stack Emulsionspaket *n*
emulsion varnish Lackemulsion *f*
enable *v* befähigen
enamel Emaille *f,* Email *n,* Emailglas *n,*
  glasartiger Überzug *m,* Glasierung *f,*
  Glasschmelz *m,* Glasur *f,* Glasurmasse *f,*
  Lack *m,* Schmalt *m* (Email), Schmelzglas *n,*
  Schmelzglasur *f,* Zahnschmelz *m* (Zahnmed),
  baking ~ Ofenemaillelack *m,* **base ~**
  Grundemail *n,* **coat of ~** Emailschicht *f,*
  colorless ~ Bleifluß *m* (Keram), **embossed ~**
  Hochemail *n,* **finishing ~** Überzugsemaille *f,*
  white ~ Weißlack *m* (Keram)
enamel *v* emaillieren, glasieren, lackieren, mit
  Email überziehen
enamel builder Adamantoblast *m* (Schmelzzelle)
enamel color Emailfarbe *f,* Schmelzfarbe *f*
enamel[l]ed emailliert, glasiert
enamel[l]ing Emaillieren *n,* Emaillierung *f,*
  Glasieren *n,* Lackierung *f,* Überschmelzung *f*
enamelling furnace Emaillierofen *m,*
  Farbenschmelzofen *m*
enamelling soda Emailliersoda *f*
enamel mill Emailmühle *f*
enamel mixing machine Emailmischmaschine *f*

enamel stripping Entemaillieren *n*
enamel stripping plant Entemaillierungsanlage *f*
enamel varnish Emaillack *m*
enamel vitrifying color Emailschmelzfarbe *f*
enamel-ware Emailwaren *pl*
enamel work Emaillierung *f*
enamines Enamine *pl*
enantate Enantat *n*
enanthal Önanthaldehyd *m*
enanthaldehyde Önanthaldehyd *m*
enanthetol Önanthetol *n*
enanthic acid Normalheptylsäure *f,*
  Önanthsäure *f,* Önanthylsäure *f*
enanthic aldehyde Önanthaldehyd *m*
enanthic ether Önanthäther *m*
enanthin Önanthin *n*
enanthoin Önanthoin *n*
enanthol Önanthalkohol *m*
enanthole Önanthol *n*
enanthone Önanthon *n*
enanthotoxin Önanthotoxin *n*
enanthyl Önanthyl-
enanthylic acid Önanthsäure *f,* Önanthylsäure *f*
enantiomer enantiomer
enantiomerism Enantiomerie *f,* optische
  Isomerie *f*
enantiomorphic enantiomorph, spiegelbildisomer
enantiomorphism Spiegelbildisomerie *f*
enantiotropic enantiotrop
enantiotropism Enantiotropie *f*
enantiotropy Enantiotropie *f*
enargite Enargit *m* (Min)
encapsulate *v* einbetten (Elektr), einkapseln,
  einschließen, verkapseln
encapsulated, to become ~ eingekapselt werden
encapsulation Einkapselung *f*
encase *v* umhüllen
encaustic Enkaustik *f*
encaustic *a* enkaustisch
encaustic painting Wachsmalerei *f*
encephalitis Gehirnentzündung *f* (Med)
encephalograph Enzepholagramm *n* (Med)
encephaloma Hirntumor *m* (Med)
enclose *v* beifügen, beilegen, einschließen,
  umgeben
enclosed beiliegend
enclosed thermometer Einschlußthermometer *n*
enclosure Anlage *f* (Brief etc.), Beilage *f* (Brief
  etc), Einschalung *f,* Einschluß *m*
encounter Zusammenstoß *m,* Zusammentreffen *n*
encounter *v* zusammenstoßen, treffen [auf]
encrinitic rock Enkrinitenkalk *m*
encroach upon *v* beeinträchtigen
encrusting inkrustierend
encyclopedia Nachschlagewerk *n*
encyst *v* einkapseln
encystation Abkapselung *f* (Med, Biol),
  Verkapselung *f*
end Ende *n,* Schluß *m*

endanger *v* gefährden
end box Kabelendverschluß *m* (Elektr)
end burner Kopfbrenner *m*
end-centered basisflächenzentriert (Krist)
end cutting pliers Kneifzange *f*
endeavor *v* bemühen, befleißigen
endeiolite Endeiolith *m* (Min)
endergonic endergon
end face Endfläche *f* (Krist)
end group Endgruppe *f* (Chem)
end group analysis Endgruppenbestimmung *f*
endiols Endiole *pl*
endless endlos, unendlich
endless wire Langsieb *n* (Pap)
endlichite Endlichit *m* (Min)
end mill Langlochfräser *m*, Stirnfräser *m*
end milling cutter Stirnfräser *m*, Fingerfräser *m*
endocrine endokrin
endocrine glands endokrine Drüsen *pl*
endocrinological endokrinologisch
endocrinologist Endokrinologe *m*
endocrinology Endokrinologie *f*
endo deoxyribonuclease
  Endodesoxyribonuklease *f* (Biochem)
endoenzyme Endoenzym *n*, Zellenzym *n*
endo-exo-isomerism endo-exo-Isomerie *f*
endogenic endogen
endogenous endogen, von innen herauswachsend
  (Biol)
endoiminotriazole Endoiminotriazol *n*
endomitosis Endomitose *f* (Biol)
endomorphic endomorph
endomorphism Endomorphismus *m*
endonuclease Endonuklease *f* (Biochem)
endoparasite Entoparasit *m*,
  Innenschmarotzer *m* (Biol), im Innern des
  Wirtes lebender Parasit
endopeptidase Endopeptidase *f* (Biochem)
endoperoxide Endoperoxid *n*
endoplasm Endoplasma *n*, innere
  Plasmaschicht *f*
endoplasmic endoplasmatisch
endosmosis Endosmose *f*
endosmotic endosmotisch
endosperm Endosperm *n* (Bot), Nährgewebe *n*
  (Bot)
endothelial cell Endothelzelle *f* (Med)
endotheliocyte Endothelwanderzelle *f*
endotheliosis Endothelzellenvermehrung *f*
  (Path)
endothermal endotherm
endothermic endotherm
endothianin Endothianin *n*
endothiotriazole Endothiotriazol *n*
endotoxin Endotoxin *n* (Bakt)
endoxytriazole Endoxytriazol *n*
end phase Endphase *f*
end piece Endstück *n*
end plane Endfläche *f* (Krist)

end plate Endplatte *f*
end point Äquivalenzpunkt *m* (Chem),
  Endpunkt *m*
end position Endstellung *f*, Endlage *f*
end product Endprodukt *n*
end-product repression Endproduktrepression *f*
  (Biochem)
endrin Endrin *n*
endurance Ausdauer *f*, Dauerhaftigkeit *f*,
  Geduld *f*
endurance bending test Dauerbiegeversuch *m*
endurance limit Dauerfestigkeitsgrenze *f*,
  Ermüdungsgrenze *f*, Haltbarkeitsgrenze *f*
endurance limit of stress Dauerfestigkeit *f*
endurance strength Widerstandsfestigkeit *f*
endurance tension test Zugversuch *m* mit
  Dauerbeanspruchung
endurance test Ermüdungsversuch *m*,
  Belastungsprobe *f*, Dauerprüfung *f*
endurance-testing machine
  Dauerfestigkeitsprüfmaschine *f*
endure *v* ausdauern, fortbestehen
enduring ausdauernd, dauerhaft
end value Grenzwert *m*
energetic energisch, energetisch, kräftig
energetics Energetik *f*, Energielehre *f*
energize *v* erregen
energizing circuit Erregerkreis *m* (Elektr)
energizing current Erregerstrom *m*
energizing voltage Erregerspannung *f*
energy Energie *f*, Arbeitsvermögen *n* (Phys),
  Kraft *f*, **accumulation of** ~
  Energieaufspeicherung *f*, **addition of** ~
  Energiezufuhr *f*, **amount of** ~
  Energiebetrag *m*, **characteristic** ~
  Eigenenergie *f*, **chemical** ~ chemische
  Energie, **density of** ~ Energiedichte *f*,
  development of ~ Energieentwicklung *f*,
  emission of ~ Energieemission *f*, **form of** ~
  Energieform *f*, **free** ~ freie Energie *f*, **gain of**
  ~ Energiezuwachs *m*, **initial** ~
  Anfangsenergie *f*, **internal** ~ innere Energie,
  intrinsic ~ wahre Energie, **kinetic** ~
  kinetische Energie, **liberation of** ~
  Energiefreisetzung *f*, **loss of** ~
  Energieverlust *m*, **measuring of** ~
  Energiemessung *f*, **potential** ~ potentielle
  Energie, **radiation of** ~ Energieausstrahlung *f*,
  reactive ~ Blindenergie *f*, **regenerated or**
  restored ~ zurückgewonnene Energie, **saving**
  in ~ Energieersparnis *f*, **source of** ~
  Energiequelle *f*, **threshold of** ~
  Energieschwelle *f*, **total** ~ Gesamtenergie *f*,
  utilisation of ~ Energieausnutzung *f*, **waste of**
  ~ Energieverschwendung *f*, ~ **required**
  Energieaufwand *m*, ~ **of discharge**
  Entladungsenergie *f*, ~ **of formation**
  Bildungsenergie *f*, ~ **of generation**
  Bildungsenergie *f*
energy balance Energiebilanz *f*,
  Energiehaushalt *m*

energy band Energieband *n*, **allowed** ~ erlaubtes Energieband, **forbidden** ~ verbotenes Energieband
energy barrier Energieschwelle *f*
energy conservation Energieerhaltung *f*
energy consumption Energieaufwand *m*
energy-containing energiehaltig
energy content Energieinhalt *m*
energy conversion Energieumwandlung *f*, **direct** ~ Energie-Direktumwandlung *f*
energy converter Energieumformer *m*, Energieumwandler *m*
energy cycle Energiekreislauf *m*
energy decrease Energieabnahme *f*
energy decrement, logarithmic ~ logarithmisches Energiedekrement *n*
energy depression Energiemulde *f*
energy dissipation Energiezerstreuung *f*
energy dissipation density Energiedissipationsdichte *f*
energy distribution Energieverteilung *f*, **curve of** ~ Energieverteilungskurve *f*
energy drop Energieabfall *m*
energy dynamo Energiedynamo *m*
energy equation Energiegleichung *f*
energy flow Energiefluß *m*, Energiestrom *m*
energy gain Energiegewinn *m*
energy hill or barrier Energieberg *m*
energy level Energieniveau *n*, Energiestufe *f*
energy level diagram Energieniveaudiagramm *n*
energy-mass relation Energiemassebeziehung *f*
energy momentum tensor Energieimpulstensor *m*
energy output Energieabgabe *f*
energy-poor energiearm
energy position energetische Lage *f*
energy principle Energieprinzip *n*
energy quantization Energiequantelung *f*
energy quantum Energiequant *n*
energy range Energiebereich *m*, Reichweite *f* (Atom)
energy region Reichweite *f* (Atom)
energy regulator Impulsregler *m* (Elektr), Wattregler *m* (Elektr)
energy release Energiefreigabe *f*
energy-rich energiereich
energy spacing Energieabstand *m*
energy spectrum Energiespektrum *n*
energy state Energiezustand *m*
energy [state] term Energieterm *m* (Atom)
energy storage Energie[auf]speicherung *f*
energy storage electrode Speicherelektrode *f*
energy term Energiestufe *f*
energy theorem Energiesatz *m*
energy transfer Energieübertragung *f*
energy unit Energieeinheit *f*
energy uptake Energieaufnahme *f*
energy yield Energieausbeute *f*
enesol Enesol *n*

engage *v* ankuppeln (Techn), anstellen, beschäftigen, [ein]kuppeln (Techn), verpflichten
engaging coupling Einrückkuppelung *f*
engaging gear Einrückvorrichtung *f*
engaging lever Einrückhebel *m*
engelitin Engelitin *n*
Engel process Engel-Verfahren n, Sinterverfahren *n* (Polyäthylen)
engine Maschine *f*, Lokomotive *f*, Motor *m*, **multicycle** ~ Mehrphasenmotor *m*, **to start the** ~ den Motor anlassen, **to stop the** ~ den Motor abstellen
engineer Ingenieur *m*, Techniker *m*, **consulting** ~ beratender Ingenieur, ~ **in charge** Betriebsingenieur *m*
engineering Apparatewesen *n*, Ingenieurwesen *n*, Maschinenbau *m*, Technik *f*, **chemical** ~ Verfahrenstechnik *f*, **chemical and process** ~ Chemie-Ingenieur-Technik *f*, **heavy** ~ Großmaschinenbau *m*, **high tension** ~ Hochspannungstechnik *f*
engineering chemistry Verfahrenstechnik *f*
engineering costs Planungskosten *pl*
engineering table Konstruktionsmerkblatt *n*
engine failure Motorpanne *f*
engine gas Maschinengas *n*
engine lathe Leitspindeldrehbank *f*, Maschinendrehbank *f*
engine noise Motorengeräusch *n*
engine operator Maschinist *m*
engine output Maschinenleistung *f*
engine primer Einspritzanlasser *m*
engine room Maschinenraum *m*
engine shaft Motorwelle *f*
engine speed Motordrehzahl *f*
engine trouble Motorpanne *f*
English blue Englischblau *n*
English red Englischrot *n*, Colcothar *n*, Glanzrot *n*, Kaiserrot *n*
engobe Begußmasse *f*, Engobe *f*, Überzug *m*, **to coat with** ~ engobieren
engobe color Engobefarbe *f*
engrave *v* gravieren, eingraben, einschleifen
engraver Bildstecher *m*, Graveur *m*
engraving Kupferstich *m*, Gravierkunst *f*, Gravur *f*
engraving machine, electric ~ Elektrograviermaschine *f*
engraving mill Gravierfräsmaschine *f*
enhanced erhöht
enhancement Verstärkung *f*, Anreicherung *f*
enhydrite Hydrochalcedon *m*
enhydros Enhydros *m* (Min)
enidin Önidin *n*
enin Önin *n*
enlarge *v* erweitern, vergrößern

enlargement Vergrößerung *f,* **scale of** ~ Vergrößerungsmaßstab *m*
enlarger Vergrößerungsapparat *m* (Phot), **autofocus** ~ Vergrößerungsapparat *m* mit selbsttätiger Einstellung
enneacontahedron Neunzigflächner *m* (Krist)
enneacosane Enneakosan *n*
enneagon Neuneck *n*
enneahedral neunflächig
enneahedron Neunflächner *m* (Krist)
enol Enol *n*
enolase Enolase *f* (Biochem)
enol form Enolform *f*
enolization Enolisierung *f*
enomorphone Enomorphon *n*
enosmite Campherharz *n*
enoylhydratase Enoylhydratase *f* (Biochem)
enquiry specification Anfragespezifikation *f*
enrage *v* aufregen
enrich *v* anreichern
enriched angereichert
enriched pile Sekundärreaktor *m* (Atom)
enriching zone Verstärkungsteil *m*
enrichment Anreicherung *f,* Anreicherungsprozeß *m,* Bodenverstärkung *f* (Rektifikation), Konzentrierung *f,* Vered[e]lung *f*
enrichment factor Anreicherungsfaktor *m*
enrichment plant Anreicherungsanlage *f*
enrichment ratio Verstärkungsverhältnis *n*
enrol[l] *v* immatrikulieren
ensilage *v* einsäuern (Futter)
ensnare *v* verstricken
enstatite Enstatit *m* (Min), Amblystegit *m*
entangle *v* verstricken
entanglement Gewirr *n,* Verwick[e]lung *f*
enter *v* eingehen, einlaufen, eintragen, ~ **[into]** eindringen, ~ **into combination** eine Verbindung eingehen (Chem)
entering Eintragen *n*
entering angle Eintrittswinkel *m*
enteritis Dünndarmentzündung *f* (Med)
enterokinase Enterokinase *f* (Biochem)
enterosane Enterosan *n*
enteroseptyl Enteroseptyl *n*
enterozoon (pl enterozoa) Enterozoon *n* ((pl. Enterozoen))
enterprise Unternehmen *n*
enthalpy Enthalpie *f,* **excess** ~ Überschußenthalpie *f,* **free** ~ freie Enthalpie, ~ **at absolute zero** Nullpunktsenthalpie *f,* ~ **of bonding** Bindungsenthalpie *f,* ~ **of formation** Bildungsenthalpie *f,* ~ **of reaction** Reaktionsenthalpie *f,* ~ **of activation** Aktivierungsenthalpie *f,* ~ **of mixing** Mischungsenthalpie *f,* ~ **of reaction** Reaktionsenthalpie *f,* **standard** ~ Standardreaktionsenthalpie *f*

enthalpy change Enthalpieänderung *f*
enthalpy effect Enthalpie-Effekt *m*
entire ganz, komplett, vollständig
entitle *v* befugen, berechtigen
entitled berechtigt
entomologist Entomologe *m*
entomology Entomologie *f,* Insektenkunde *f*
entoparasite Entoparasit *m*
entoplasm Entoplasma *n,* innere Plasmaschicht *f*
entozoon (pl. entozoa) Entozoon, im Inneren des Wirtes lebender Parasit *m*
entrain *v* mitreißen
entrainer Schleppmittel *n* (Rektifikation)
entrainer disk Mitnehmerscheibe *f*
entrainment Mitführung *f* (Rektifikation), Mitreißen *n* (Rektifikation)
entrainment point Mitreißgrenze *f*
entrance Eingang *m,* Eintritt *m*
entrance examination Aufnahmeprüfung *f*
entrance loss Eintrittsverlust *m*
entrance phase Eintrittsphase *f*
entrance slit Eingangsspalt *m* (Opt)
entrapped air Lufteinschluß *m*
entropy Entropie *f,* **calorimetric** ~ kalorimetrische Entropie *f,* **constant** ~ unveränderliche Entropie, **excess** ~ Überschußentropie *f,* **normal** ~ Normalentropie *f,* **spectroscopic** ~ spektroskopische Entropie *f,* **standard** ~ Standardentropie *f,* **virtual** ~ virtuelle Entropie (Atom), ~ **at absolute zero** Nullpunktsentropie *f,* ~ **of activation** Aktivierungsentropie *f,* ~ **of disorder** Entropie *f* bei ungeordneter Verteilung (Phys), ~ **of fusion** Schmelzentropie *f* (Phys), ~ **of mixing** Mischungsentropie *f* (Phys), ~ **of solution** Lösungsentropie *f,* ~ **principle** Entropieprinzip *n*
entropy change Entropieänderung *f*
entropy density Entropiedichte *f*
entropy effect Entropie-Effekt *m*
entropy equation Entropiegleichung *f*
entrust *v* beauftragen
entry Eintrag *m*
entry line Eingangsbahn *f*
entry side Eintrittsseite *f*
enumerate *v* aufzählen
envelop *v* umhüllen, einschlagen, einwickeln
envelope Hülle *f,* Mantel *m,* Umhüllung *f,* Umschlag *m* (Brief)
envelope curve Einhüllende *f* (einer Kurvenschar)
envelope form Briefumschlagform *f* (Stereochem)
enveloping, condition of ~ Einhüllbedingung *f*
environment Umwelt *f*
environmental umweltbedingt
enzymatic enzymatisch

enzymatic action Enzymwirkung *f*
enzymatic activity enzymatische Aktivität *f*
enzymatic degradation enzymatischer Abbau *m*
enzymatic filter aid Filtrationsenzym *n*
enzyme Enzym *n* (Biochem), Ferment *n*
  (Biochem), **adaptive** ~ adaptives Enzym,
  allosteric ~ allosterisches Enzym, **auxiliary** ~
  Hilfsenzym *n*, **constitutive** ~ konstitutives
  Enzym, **digestive** ~ Verdauungsenzym *n*,
  extracellular ~ extrazelluläres Enzym,
  hydrolytic ~ hydrolytisches Enzym, **inducible**
  ~ induzierbares Enzym, **intracellular** ~
  intrazelluläres Enzym, **protective** ~
  Abwehrenzym *n* (Biochem)
enzyme activation Enzymaktivierung *f*
enzyme activity Enzymaktivität *f*
enzyme assay Enzymbestimmung *f*
enzyme diagnostic Fermentdiagnostikum *n*
enzyme induction Enzyminduktion *f*
enzyme inhibition Enzymhemmung *f*
enzyme-inhibitor complex
  Enzym-Inhibitor-Komplex *m*
enzyme purification Enzymreinigung *f*
enzyme repression Enzymrepression *f*
enzyme specificity Enzymspezifität *f*
enzyme-substrate complex
  Enzym-Substrat-Komplex *m*
enzyme turnover number Enzymumsatzzahl *f*
enzyme unit Enzymeinheit *f*
enzymology Enzymology *f*
eosin[e] Eosin *n*
eosine acid Eosinsäure *f*
eosphorite Eosphorit *m* (Min)
eperuic acid Eperusäure *f*
ephedrine Ephedrin *n*
ephedrine hydrochloride Ephedrinhydrochlorid *n*
ephedrine sulfate Ephedrinsulfat *n*
ephemera Eintagsfliege *f* (Zool)
epiallomuscarine Epiallomuscarin *n*
epiandrosterone Epiandrosteron *n*
epiarabinose Epiarabinose *f*
epiarabite Epiarabit *m*
epiarabonic acid Epiarabonsäure *f*
epiasarinin Epiasarinin *n*
epiblast Epiblast *n* (Biol)
epiborneol Epiborneol *n*
epiboulangerite Epiboulangerit *m* (Min)
epibromohydrin Epibromhydrin *n*
epicamphor Epicampher *m*
epicarissone Epicarisson *n*
epicatechol Epicatechin *n*
epicenter Mittelpunkt *m* (of an earthquake),
  Epizentrum *n* (eines Erdbebens)
epichitosamine Epichitosamin *n*
epichitose Epichitose *f*
epichlorohydrin Epichlorhydrin *n*
epicholestanol Epicholestanol *n*
epicoprostanol Epikoprostanol *n*
epicoprosterol Epikoprosterin *n*

epicyclic epizyklisch
epicycloid Epizykloide *f*, Radlinie *f*
epidemic Epidemie *f* (Med),
  Massenerkrankung *f* (Med), Seuche *f* (Med),
  Sucht *f* (Med)
epidemic *a* epidemisch
epidermic tissue Oberhautgewebe *n* (Med)
epidermis Epidermis *f*, Oberhaut *f*
epidesmin Epidesmin *n*
epidiascope Bildwerfer *m*, Epidiaskop *n*
epidichlorohydrin Epidichlorhydrin *n*
epididymite Epididymit *m* (Min)
epidosite Epidosit *m* (Min)
epidote Epidot *m* (Min), Pistazit *m* (Min)
epidotic epidotähnlich
epidotiferous epidotführend
epierythrose Epierythrose *f*
epiestriol Epiöstriol *n*
epifucose Epifucose *f*
epigalactose Epigalactose *f*
epigallocatechol Epigallocatechin *n*
epigenite Epigenit *m* (Min)
epiglycose Epiglykose *f*
epiguanine Epiguanin *n*
epigulite Epigulit *n*
epigulose Epigulose *f*
epihaemanthidine Epihämanthidin *n*
epihydrin alcohol Epihydrinalkohol *m*
epihydrinic acid Glycidsäure *f*
epiinositol Epiinosit *m*
epimer Epimer *n*
epimerization Epimerisierung *f*
epinephrin[e] Adrenalin *n*, Epinephrin *n*
epinine Epinin *n*
epiphyte Aerophyt *m* (Biol)
epiquercitol Epiquercit *m*
epiquinamine Epichinamin *n*
epiquinidine Epichinidin *n*
epirockogenin Epirockogenin *n*
episaccharic acid Epizuckersäure *f*
episamogenin Episamogenin *n*
episcope Episkop *n*
episesamin Episesamin *n*
episterol Episterin *n*
epistilbite Epistilbit *m* (Min)
epistolite Epistolit *m* (Min)
epitaxy Epitaxie *f* (Krist)
epitestosterone Epitestosteron *n*
epithermal epithermisch
epitruxillic acid Epitruxillsäure *f*
epixylose Epixylose *f*
epizeorin Epizeorin *n*
epoch-making bahnbrechend
epoxypropionic acid Glycidsäure *f*
epoxy resin Epoxyharz *n*
epsomite Epsomit *m* (Min)
epsom salt Bittersalz *n*
equal gleich[wertig]
equal-area projection flächentreue Abbildung *f*

equal-armed gleicharmig
equality Gleichheit *f,* **~ of surfaces** Gleichheit *f*
der Flächen (pl)
equalization Ausgleich[ung] *f,* Gleichsetzung *f,*
Gleichstellung *f,* **~ of charges** Ausgleich *m* der
Ladungen (pl)
equalize *v* abgleichen, ausgleichen,
kompensieren, egalisieren, gleichsetzen,
gleichstellen
equalized abgeglichen
equalizer Ausgleicher *m,* Ausgleichleitung *f,*
Stabilisator *m* (Techn)
equalizing connection Äquipotentialverbindung *f*
equalizing current Ausgleichsstrom *m*
equalizing evaporator Ausgleichsverdampfer *m*
equalizing mains Ausgleichleitung *f*
equalizing power Egalisierungsvermögen *n*
equalizing pressure Ausgleichdruck *m,*
Ausgleichspannung *f*
equalizing reservoir Ausgleichsgefäß *n*
equalizing tank Ausgleichbehälter *m*
equalizing temperature Ausgleichtemperatur *f*
equate *v* angleichen, ausgleichen, gleichsetzen
(Math), gleichstellen
equation Gleichung *f,* **chemical ~** chemische
Gleichung, **collinear ~** kollineare Gleichung
(Opt), **derived ~** abgeleitete Gleichung
(Math), **determinate ~** bestimmte Gleichung
(Math), **first order or linear ~** lineare
Gleichung erster Ordnung, **general ~**
allgemeine Gleichung, **quadratic ~**
quadratische Gleichung (Math), **setup of an
~** Gleichungsansatz *m,* **to fit an ~** einer
Gleichung entsprechen, **to form an ~** eine
Gleichung ansetzen, **to solve an ~ with respect
to x** eine Gleichung nach x auflösen
equation of formation Bildungsgleichung *f*
equation of ions Ionengleichung *f*
equation of moments Hebelgesetz *n* (Phys)
equation of the third order Gleichung *f* dritten
Grades
equations, system of ~ Gleichungssystem *n*
equator Äquator *m*
equatorial äquatorial
equiangular gleichwinklig, isogonal, winkeltreu
equiangularity Gleichwinkligkeit *f*
equiareal gleichflächig
equiatomic gleichatomig
equiaxial gleichachsig
equidirectional gleichgerichtet
equidistant abstandsgleich, äquidistant
equiformity Gleichförmigkeit *f*
equilenane Equilenan *n*
equilenin Equilenin *n*
equilibrate *v* abgleichen (Elektr),
ausbalancieren, im Gleichgewicht halten, ins
Gleichgewicht bringen

equilibration Äquilibrierung *f,*
Ausbalancierung *f,* Ausgleichung *f,*
Herstellung *f* des Gleichgewichts
equilibrium Gleichgewicht *n,* **apparent ~**
scheinbares Gleichgewicht, **asymmetric
transformation of ~** asymmetrische
Gleichgewichtsumlagerung, **biphase or diphase
~** zweiphasiges Gleichgewicht, **condition of ~**
Gleichgewichtsbedingung *f,* **displacement of
~** Gleichgewichtsstörung *f,*
Gleichgewichtsverschiebung *f,* **dynamic ~**
dynamisches Gleichgewicht, **establishment of
~** Gleichgewichtseinstellung *f,* **false ~**
scheinbares Gleichgewicht, **kinetic ~**
kinetisches Gleichgewicht, **metastable ~**
metastabiles Gleichgewicht, **monophase ~**
einphasiges Gleichgewicht, **neutral ~**
indifferentes Gleichgewicht, **nonvariant ~**
unfreies Gleichgewicht, **preceding ~**
vorgelagertes Gleichgewicht, **radioactive ~**
radioaktives Gleichgewicht, **stable ~** stabiles
Gleichgewicht, **state of ~**
Gleichgewichtszustand *m,* **stationary ~**
stationäres Gleichgewicht, **theory of ~**
Gleichgewichtslehre *f,* **unstable ~** labiles
Gleichgewicht, **~ of forces**
Kräftegleichgewicht *n,* **~ of reaction**
Reaktionsgleichgewicht *n*
equilibrium apparatus Gleichgewichtsapparat *m*
equilibrium concentration
Gleichgewichtskonzentration *f*
equilibrium constant Gleichgewichtskonstante *f*
equilibrium curve Gleichgewichtskurve *f*
equilibrium diagram Gleichgewichtsdiagramm *n*
equilibrium equation Gleichgewichtsgleichung *f*
equilibrium position Gleichgewichtslage *f*
equilibrium potential Gleichgewichtspotential *n,*
Ruhepotential *n*
equilibrium ratio Gleichgewichtsverhältnis *n*
equilibrium reaction Gleichgewichtsreaktion *f*
equilibrium state Ausgleichszustand *m,*
Gleichgewichtszustand *m*
equilibrium temperature Beharrungstemperatur *f*
equilibrium voltage Gleichgewichtsspannung *f*
equimolecular äquimolekular
equinox Tagundnachtgleiche *f*
equinuclear gleichkernig
equip *v* ausrüsten, ausstaffieren, ausstatten,
einrichten, installieren
equipartition Gleichteilung *f,* Gleichverteilung *f,*
law of ~ Gleichverteilungssatz *m*
equipment Apparatewesen *n,* Ausrüstung *f,*
Ausstattung *f,* Einrichtung *f,* **basic ~**
Grundausrüstung *f*
equipoise Gleichgewicht *n*
equiponderous von gleichem Gewicht
equipotential Äquipotential[linie *f*] *n*
equipotential *a* äquipotential

equipotential line Äquipotentiallinie *f,* Linie *f*
gleichen Potentials
equipotential surface Äquipotentialfläche *f,*
Niveaufläche *f*
equisetic acid Equisetsäure *f*
equisetrin Equisetrin *n*
equivalence Äquivalenz *f,* Gleichwertigkeit *f*
equivalence law Äquivalenzgesetz *n*
equivalence point Äquivalenzpunkt *m*
equivalence principle Äquivalenzprinzip *n*
equivalence unit Äquivalenzeinheit *f*
equivalent Äquivalent *n*
equivalent *a* äquivalent (Chem),
gleichbedeutend, gleichwertig (Chem)
equivalent charge Äquivalenzladung *f*
equivalent circuit Ersatzschaltbild *n,*
Ersatzschaltung *f*
equivalent concentration
Äquivalentkonzentration *f*
equivalent conductance Äquivalentleitfähigkeit *f*
equivalent conductivity Äquivalentleitfähigkeit *f,*
Äquivalentleitvermögen *n*
equivalent mixture äquivalente Mischung *f*
equivalent of heat Wärmeäquivalent *n,*
mechanical ~ Wärmearbeitswert *m*
equivalent weight Äquivalenzgewicht *n*
era Zeitalter *n*
eradicate *v* ausmerzen, ausrotten, vertilgen
eradicator Vertilgungsmittel *n*
erase *v* ausradieren, löschen
erase field Löschfeld *n* (Comp)
erase head Löschkopf *m* (Comp)
eraser Radiergummi *m*
erasing Löschung *f*
erasing pattern Radierschablone *f*
erbia Erbinerde *f,* Erbiumoxid *n*
erbium (Symb. Er) Erbium *n*
erbium compound Erbiumverbindung *f*
erbium content Erbiumgehalt *m*
erbium oxide Erbiumoxid *n*
erbium preparation Erbiumpräparat *n*
erbium sulfate Erbiumsulfat *n*
erdin Erdin *n*
erect *v* aufrichten; aufschlagen, aufstellen,
errichten, montieren
erection Aufstellung *f,* Errichtung *f,* Montage *f,*
Montierung *f*
erector Monteur *m*
eremacausis Eremakausie *f,* langsame
Verbrennung *f*
eremophilone Eremophilon *n*
erepsin Erepsin *n*
ergine Ergin *n*
ergobasine Ergobasin *n*
ergocalciferol Ergocalciferol *n*
ergocristine Ergocristin *n*
ergocryptine Ergokryptin *n*
ergocryptinine Ergokryptinin *n*
ergodic hypothesis Ergodensatz *m*

ergoline Ergolin *n*
ergometer Dynamometer *n,* Energiemesser *m*
ergometrine Ergometrin *n*
ergometrine maleate Ergometrinmaleat *n*
ergometrinine Ergometrinin *n*
ergonovine Ergonovin *n*
ergosine Ergosin *n*
ergosomes Ergosomen *pl* (Biol)
ergostane Ergostan *n*
ergostanol Ergostanol *n*
ergosterin Ergosterin *n*
ergosterol Ergosterin *n*
ergosterone Ergosteron *n*
ergostetrine Ergostetrin *n*
ergot Mutterkorn *n* (Bot)
ergotachysterol Ergotachysterin *n*
ergot alkaloid Ergot-Alkaloid *n,*
Mutterkornalkaloid *n*
ergotamine Ergotamin *n*
ergotamine bitartrate Ergotaminbitartrat *n*
ergot extract Mutterkornextrakt *m,* Ergotin *n*
ergothioneine Ergothionein *n*
ergotic acid Mutterkornsäure *f,* Sklerotinsäure *f*
ergotine Ergotin *n*
ergotinine Ergotinin *n*
ergotism Ergotismus *m,* Mutterkornbrand *m*
ergotocin Ergometrin *n*
ergotoxine Ergotoxin *n*
erianthin Erianthin *n*
ericin Mesotan *n*
ericite Heidenstein *m* (Min)
Ericsson cupping or ductility test
Ericssontiefziehprobe *f*
erigeron oil Erigeronöl *n*
erinite Erinit *m* (Min)
eriochrome black Eriochromschwarz *n*
eriochrome verdone Eriochromverdon *n*
eriocitrin Eriocitrin *n*
eriodictyol Eriodictyol *n*
erionite Erionit *m* (Min)
Erlanger blue Erlanger Blau *n*
Erlenmeyer flask Erlenmeyerkolben *m,*
narrow-mouth ~ enghalsiger
Erlenmeyerkolben, **wide-mouth** ~ weithalsiger
Erlenmeyerkolben
ermine Hermelin *m* (Pelz)
erode *v* erodieren, anfressen, auswaschen
erosion Abnützung *f,* Abtragung *f* (Geol),
Auswaschung *f* (Geol), Erosion *f* (Geol),
Verschleiß *m*
erosion by water Wassererosion *f*
erosnin Erosnin *n*
erroneous[ly] irrtümlich
error Fehler *m,* Irrtum *m,* **casual** ~ zufälliger
Fehler, **curve of** ~ Fehlerlinie *f,* **determination
of** ~ Fehlerbestimmung *f,* **ellipsoid of** ~
Fehlerellipsoid *n,* **estimation of** ~
Fehlerabschätzung *f,* **limit of** ~
Fehlergrenze *f,* **mean** ~ mittlerer Fehler,

personal ~ Beobachtungsfehler *m*, **possibility of** ~ Fehlermöglichkeit *f*, **probable** ~ wahrscheinlicher Fehler, **source of** ~ Fehlerquelle *f*, **systematic** ~ systematischer Fehler, **total** ~ Gesamtfehler *m*, ~ **in reading** Ablesefehler *m*, Ablesungsfehler *m*
error calculation Fehlerrechnung *f*
error distribution Fehlerverteilung *f*
error equation Fehlergleichung *f*
error function Fehlerfunktion *f*
error integral Fehlerintegral *n* (Math)
error law of Gauss Gaußsches Fehlergesetz *n*
error limit Fehlergrenze *f*, Meßfehlergrenze *f*
error matrix Fehlermatrix *f* (Math)
error of analysis, limit of ~ Analysenfehlergrenze *f*
error probability Irrtumswahrscheinlichkeit *f*
errors, compensation of ~ Fehlerausgleichung *f*, **law of** ~ Fehlergesetz *n*
ersbyite Ersbyit *m* (Min)
erubescite Bornit *m* (Min)
erucic acid Erucasäure *f*
erucine Erucin *n*
erucone Erucon *n*
erucyl alcohol Erucylalkohol *m*
eruginous grünspanähnlich
eruption Ausbruch *m*, Durchbruch *m*, Eruption *f*
eruptive eruptiv (Geol)
eruptive rocks Eruptivgestein *n* (Geol), Auswurfgestein *n* (Geol)
ervasine Ervasin *n*
erysodine Erysodin *n*
erysothiopine Erysothiopin *n*
erythraline Erythralin *n*
erythramine Erythramin *n*
erythrene Erythren *n*
erythrene caoutchouc Erythrenkautschuk *m*
erythrene glycol Erythrenglykol *n*
erythricine Erythricin *n*
erythrite Erythrit *m* (Min), Kobaltblüte *f* (Min), Kobaltblume *f* (Min)
erythritol Erythrit *m*
erythroaphin Erythroaphin *n*
erythro arrangement erythro-Anordnung *f* (Stereochem)
erythrocruorin Erythrocruorin *n*
erythrocyte Erythrozyt *m*, rotes Blutkörperchen *n*
erythrocyte resistance Erythrozytenresistenz *f*
erythrodextrin Erythrodextrin *n*
erythrogenic acid Erythrogensäure *f*
erythroglaucin Erythroglaucin *n*
erythroidine Erythroidin *n*
erythro isomers pl
erythrol Erythrit *m*
erythronic acid Erythronsäure *f*
erythrophloeine Erythrophloein *n*
erythrose Erythrose *f*

erythrose phosphate Erythrosephosphat *n*
erythrosiderite Erythrosiderit *m* (Min)
erythrosin Erythrosin *n*, Jodeosin *n*
erythrosone Erythroson *n*
erythrozincite Erythrozinkit *m* (Min)
erythrulose Erythrulose *f*
escape Austritt *m* (Dampf), Entrinnen *n*, Entweichung *f*, Flucht *f*, Weichen *n*
escape *v* [ent]fliehen, abziehen (Rauch), ausströmen (Dampf), austreten (Dampf), ausweichen, entweichen
escape channel Abzugskanal *m*
escape device Abzugsvorrichtung *f*
escape orifice Ausströmungsöffnung *f*
escape pipe Ausflußrohr *n*
escape reaction Ausweichreaktion *f*
escape steam Auspuffdampf *m*
escape tube Abzugsrohr *n*
escape valve Ablaßventil *n*, Abflußventil *n*, Auslaßventil *n*
escaping Entweichen *n*
escaping *a* entweichend
Escherichia coli Kolibazillus *m*
escholerine Escholerin *n*
eschynite Äschynit *m* (Min)
esco mordant Eskobeize *f* (HN)
esculetin Äsculetin *n*
esculetinic acid Äsculussäure *f*
esculin Äsculin *n*
escutcheon Schlüsselrosette *f*
eserethole Eseräthol *n*
eseridine Eseridin *n*
eserine Eserin *n*, Physostigmin *n*
eserine oil Eserinöl *n*
eserine salicylate Eserinsalicylat *n*
eserine sulfate Eserinsulfat *n*
eseroline Eserolin *n*
esmarkite Esmarkit *m* (Min)
esparto Alfa *f*
esparto paper Alfapapier *n* (Buchdr)
essence Auszug *m* (Chem), Essenz *f* (Chem), Hauptinhalt *m*
essential wesentlich, ätherisch (Chem), essentiell
essonite Hyacinthgranat *m* (Min), Kaneelgranat *m* (Min)
establish *v* aufstellen, begründen, einrichten, festsetzen, herstellen, nachweisen
establishment Einrichtung *f*, Anlage *f*, Begründung *f*, Errichtung *f*
esteem *v* achten
ester Ester *m* (Chem)
esterase Esterase *f* (Biochem)
ester condensation Esterkondensation *f*
ester formation Esterbildung *f*
ester gum Estergummi *n*, Harzester *m*
esterifiable veresterbar
esterification Esterbildung *f*, Veresterung *f*
esterification plant Veresterungsanlage *f*
esterify *v* verestern

ester interchange Umesterung *f*
ester number Esterzahl *f*
ester pyrolysis Esterpyrolyse *f*
ester recovery plant
  Esterrückgewinnungsanlage *f*
estimate Schätzung *f,* Kalkulation *f,*
  Kostenschätzung *f,* Kostenvoranschlag *m*
estimate *v* [ab]schätzen, beurteilen, bewerten,
  einschätzen, vorausbestimmen
estimated value Schätzwert *m*
estimating Bewerten *n*
estimation Schätzung *f,* Bestimmung *f,*
  Bewertung *f,* Vorausberechnung *f,*
  Wertschätzung *f*
estomycin Estomycin *n*
estradiol Östradiol *n*
estradiol benzoate Östradiolbenzoat *n*
estragole Estragol *n*
estrane Östran *n*
estriol Östriol *n*
estrodienol Östrodienol *n*
estrogen Östrogen *n*
estrone (Am. E.) Östron *n*
etamiphyllin Etamiphyllin *n*
etamycin acid Etamycinsäure *f*
Etard's reaction Etardreaktion *f* (Chem)
etch *v* [an]ätzen, einätzen, kupferstechen,
  radieren, ~ **upon** aufbeizen
etched, capable of being ~ ätzbar
etched pit Ätzgrübchen *n*
etcher Ätzer *m,* Kupferstecher *m*
etching Ätzen *n,* Ätzdruck *m,* Ätzung *f,* Beizen *n*
  (Met), Einätzen *n,* Radierung *f,* **coarse** ~
  Grobätzung *f,* **deep** ~ Tiefätzung *f,* **ground**
  section for ~ Ätzschliff *m,* **to remove [metal] by**
  ~ abätmen (Metall)
etching acid Ätzflüssigkeit *f*
etching bath Ätzbad *n*
etching board Ätzbrett *n*
etching figure Ätzfigur *f*
etching ground Ätzgrund *m*
etching ink Ätztinte *f*
etching lye Ätzlauge *f*
etching needle Ätznadel *f*
etching paste Ätzpaste *f*
etching polishing Ätzpolieren *n*
etching powder Ätzpulver *n*
etching primer Haftgrund[ier]mittel *n*
etching process Ätzverfahren *n*
etching sample Ätzprobe *f*
etching surface Ätzgrund *m*
etching test Ätzprobe *f*
etching time Beizzeit *f*
etching tool Reißnadel *f*
etching varnish Radierfirnis *m*
etch pattern Ätzbild *n*
etch-polish Ätzpolitur *f*
etch-polish *v* ätzpolieren
ethacridine Aethacridin *n*

ethal Cetylalkohol *m,* Äthal *n*
ethanal Acetaldehyd *m,* Äthanal *n*
ethanal acid Glyoxylsäure *f*
ethanamide Acetamid *n*
ethane Äthan *n*
ethanedial Glyoxal *n*
ethanediamide Oxamid *n*
ethane dinitrile Cyanogen *n,* Dicyan *n*
ethanedioic acid Oxalsäure *f*
ethane nitrile Methylcyanid *n*
ethanesulfonic acid Äthylsulfonsäure *f*
ethanethiol Äthylmercaptan *n*
ethanoic acid (obs) Essigsäure *f*
ethanol Äthanol *n,* Äthylalkohol *m*
ethanolamine Äthanolamin *n,* Aminoäthanol *n*
ethanolic äthylalkoholisch
ethanoyl (obs) Acetyl-
ethanoyl bromide Acetylbromid *n*
ethanoyl chloride Acetylchlorid *n*
ethaverine Ethaverin *n*
ethebenine Äthebenin *n*
ethene Äthen, Äthylen *n*
ethenol Vinylalkohol *m*
ethenyl Vinyl-
ether Äther *m,* **to convert into an** ~ in einen
  Äther verwandeln, veräthern, **to extract with**
  ~ ausäthern, **to shake out with** ~ mit Äther
  ausschütteln, ausäthern
etherate Ätherat *n*
ethereal ätherisch, ätherähnlich, ätherartig
ethereal extract Ätherauszug *m*
ether formation Ätherbildung *f*
etherifiable ätherisierbar
etherification Ätherbildung *f*
etherify *v* veräthern, in einen Äther verwandeln
etherizable ätherisierbar
etherize *v* ätherisieren, narkotisieren (mit Äther)
ether-like ätherähnlich
ether machine Äthermaschine *f*
ether narcosis Äthernarkose *f* (Med)
ether plant Ätheranlage *f*
ether resin Ätherharz *n,* Äthoxylinharz *n*
ether spray Ätherzerstäuber *m*
ether vapor Ätherdampf *m*
ether washing process Ätherwaschverfahren *n*
ethine (obs) Acetylen *n*
ethinyl- Acetenyl-
ethionic acid Äthionsäure *f*
ethiops mineral Mineralmohr *m*
ethoxalyl Äthoxalyl-
ethoxide Äthylat *n*
ethoxy- Äthoxy-, Äthoxyl-
ethoxybenzene Äthylphenyläther *m,* Phenetol *n,*
  Phenyläthyläther *m*
ethoxycaffeine Äthoxycoffein *n*
ethoxychrysoidine hydrochloride
  Äthoxychrysoidinhydrochlorid *n*
ethoxyethane Äthyläther *m*
ethoxyl Äthoxy-, Äthoxyl-

ethoxylation Äthoxylierung *f*
ethoxysalicylic aldehyde Äthoxysalicylaldehyd *m*
ethusanol Aethusanol *n*
ethusin Aethusin *n*
ethyl- Äthyl-
ethyl abietate Abietinsäureäthylester *m*
ethyl acetanilide Äthylacetanilid *n*
ethyl acetate Essigsäureäthylester *m*, Acetidin *n*,
    Äthylacetat *n*, Essigester *m*, essigsaures
    Äthyl *n*, **solution in** ~ Essigesterlösung *f*
ethyl acetoacetate Acetessigester *m*,
    Acetessigsäureäthylester *m*
ethyl acrylate Acrylsäureäthylester *m*
ethylal Acetaldehyd *m*
ethyl alcohol Äthanol *n*, Äthylalkohol *m*, Sprit *m*
ethylalcoholic äthylalkoholisch
ethyl aldehyde Acetaldehyd *m*
ethylamine Äthylamin *n*
ethyl aminobenzoate Äthylaminobenzoat *n*,
    Benzocain *n*
ethyl aminobenzoic acid
    Äthylaminobenzoesäure *f*
ethylaniline Äthylanilin *n*
ethyl arsine Äthylarsin *n*
ethylate Äthylat *n*
ethylate *v* äthylieren
ethylation Äthylierung *f*
ethyl benzene Äthylbenzol *n*
ethyl benzene sulfonate Äthylbenzolsulfonat *n*
ethyl benzoate Äthylbenzoat *n*,
    Benzoesäureäthylester *m*
ethyl benzyl aniline Äthylbenzylanilin *n*
ethyl benzyl ketone Äthylbenzylketon *n*
ethyl bromide Äthylbromid *n*, Bromäthyl *n*
ethyl butyl malonate Äthylbutylmalonat *n*
ethyl butyrate Äthylbutyrat *n*, Ananasäther *m*,
    Ananasessenz *f*
ethyl caprate Caprinsäureäthylester *m*
ethyl caproate Capronsäureäthylester *m*
ethyl caprylate Caprylsäureäthylester *m*
ethyl carbamate Äthylcarbamat *n*,
    Äthylurethan *n*, Carbaminsäureäthylester *m*,
    Urethan *n*
ethyl carbinol Äthylcarbinol *n*
ethyl carbylamine Äthylcarbylamin *n*
ethyl cellosolve Äthylcellosolve *f*
ethyl cellulose Äthylcellulose *f*
ethyl chloride Äthylchlorid *n*, Kelen *n* (HN)
ethyl chloroacetate Äthylchloracetat *n*,
    Chloressigsäureäthylester *m*, chloressigsaures
    Äthyl *n*
ethyl chloroformate Äthylchlorformiat *n*,
    Chlorkohlensäureäthylester *m*
ethyl chlorosilane Äthylchlorsilan *n*
ethyl chlorostannic acid Äthylchlorzinnsäure *f*
ethyl cinnamate Zimtsäureäthylester *m*,
    Äthylcinnamat *n*
ethyl compound Äthylverbindung *f*

ethyl cresylate Kresyläthyläther *m*,
    Kresyläthylester *m*
ethyl cyanide Äthylcyanid *n*, Propionitril *n*
ethyl dichloroarsine Dick *n* (Kampfgas)
ethyldimethylsulfonium salt
    Äthyldimethylsulfoniumsalz *n*
ethyl disulfide Äthyldisulfid *n*
ethyl dithiocarbamate Dithiourethan *n*
ethylene Äthylen *n*, Äthen *n*
ethylene aldehyde Propenal *n*
ethylene benzene Äthylenbenzol *n*
ethylene bromide Äthylen[di]bromid *n*
ethylene carbonate Äthylencarbonat *n*
ethylene chloride Äthylenchlorid *n*
ethylene chlorohydrin Äthylenchlorhydrin *n*
ethylene cyanide Äthylencyanid *n*,
    Succinonitril *n*
ethylene cyanohydrin Äthylencyanhydrin *n*
ethylenediamine Äthylendiamin *n*,
    Dimethylendiamin *n*
ethylene diamine tetraacetic acid
    Äthylendiamintetraessigsäure *f* (EDTA)
ethylene dibromide Äthylendibromid *n*
ethylenedicarboxylic acid Maleinsäure *f*
ethylene dichloride Äthylen[di]chlorid *n*
ethylene dicyanide Äthylencyanid *n*
ethylene diiodide Äthylenjodid *n*
ethylene glycol Glykol *n*, Äthylenglykol *n*
ethylenehydrinsulfonic acid Isäthionsäure *f*
ethylene imine Äthylenimin *n*
ethylene iodide Äthylenjodid *n*
ethylene ketal Äthylenketal *n*
ethylene linkage Äthylenbindung *f*
ethylene naphthene Acenaphthen *n*
ethylene oxide Äthylenoxid *n*, Oxiran *n*
ethylene plastics Äthylenkunststoffe *pl*,
    Kunststoffe *pl* aus Äthylenharzen (pl)
ethylene radical Äthylenradikal *n*, Äthylenrest *m*
ethylene recovery plant
    Äthylen[rück]gewinnungsanlage *f*
ethylene series Äthylenreihe *f*
ethylene sulfonic acid Äthionsäure *f*
ethylene thioketal Äthylenthioketal *n*
ethylenimine Piperazin *n*
ethyl ether Äthyläther *m*, Äthoxyäthan *n*
ethyl ethoxysilane Äthyläthoxysilan *n*
ethyl fluoride Äthylfluorid *n*
ethyl formate Äthylformiat *n*,
    Ameisensäureäthylester *m*
ethyl glucoside Äthylglucosid *n*
ethyl glutarate Glutarsäurediäthylester *m*
ethyl glycol acetate Äthylglykolacetat *n*
ethyl hexanol Äthylhexanol *n*
ethyl hydrogen sulfate Äthylschwefelsäure *f*
ethyl hydrosulfide Äthylmercaptan *n*
ethylidene bromide Äthylidenbromid *n*
ethylidene chloride Äthylidenchlorid *n*,
    Chloräthyliden *n*
ethylidene dibromide Äthylidenbromid *n*

ethylidene dichloride Äthylidenchlorid *n*
ethylidene iodide Äthylidenjodid *n*
ethylidenelactic acid Äthylidenmilchsäure *f*
ethylidene urea Äthylidenharnstoff *m*
ethyl iodide Äthyljodid *n*, Jodäthyl *n*
ethyl isobutyrate Äthylisobutyrat *n*
ethyl isosuccinate Äthylisosuccinat *n*
ethyl isothiocyanate Äthylsenföl *n*
ethyl lactate Milchsäureäthylester *m*
ethyl malonate Äthylmalonat *n*,
  Malonsäurediäthylester *m*
ethylmalonic acid Äthylmalonsäure *f*
ethyl mercaptan Äthylmercaptan *n*,
  Äthylsulfhydrat *n*
ethyl methylacetic acid Äthylmethylacetsäure *f*
ethyl morphine Äthylmorphin *n*
ethyl morphine hydrochloride
  Äthylmorphinhydrochlorid *n*
ethyl mustard oil Äthylsenföl *n*
ethyl nitrate Äthylnitrat *n*
ethyl nitrite Äthylnitrit *n*
ethyl nitrobenzoate Äthylnitrobenzoat *n*
ethyl nitrocinnamate Nitrozimtsäureäthylester *m*
ethyl orthoformate Äthylorthoformiat *n*
ethyl oxalate Äthyloxalat *n*, Oxalester *m*
ethyl oxamate Oxamethan *n*
ethyl oxide Äthyläther *m*, Äthoxyäthan *n*,
  Äthyloxid *n*
ethylpentane Äthylpentan *n*
ethyl peroxide Äthylperoxid *n*
ethyl phenylacetate Äthylphenylacetat *n*
ethyl phenyl carbonate Äthylphenylcarbonat *n*
ethyl phenyl dibromopropionate Zebromal *n*
ethyl phenyl ether Äthylphenyläther *m*
ethyl phenyl ketone Äthylphenylketon *n*
ethyl phthalate Äthylphthalat *n*
ethyl propionate Äthylpropionat *n*
ethyl propyl carbinol Äthylpropylcarbinol *n*
ethyl propyl ether Äthylpropyläther *m*
ethyl propyl ketone Äthylpropylketon *n*
ethyl pyruvate Äthylpyruvat *n*
ethyl racemate Äthylracemat *n*
ethyl red Äthylrot *n*
ethyl resorcinol Äthylresorcin *n*
ethyl salicylate Äthylsalicylat *n*,
  Salicylsäureäthylester *m*
ethyl selenide Selenäthyl *n*
ethyl stannic acid Äthylzinnsäure *f*
ethyl succinate Äthylsuccinat *n*
ethyl sulfate Äthylsulfat *n*
ethyl sulfide Äthylsulfid *n*
ethyl sulfonic acid Äthylsulfonsäure *f*
ethyl sulfuric acid Äthylschwefelsäure *f*
ethyl tartaric acid Äthylweinsäure *f*
ethyl tartrate Äthyltartrat *n*
ethyl thiocyanate Äthylrhodanid *n*
ethyl toluene sulfonamide
  Äthyltoluolsulfonamid *n*
ethyl toluene sulfonate Äthyltoluolsulfonat *n*

ethylurethane Äthylurethan *n*, Urethan *n*
ethyl valerate Valeriansäureäthylester *m*
ethyl vinyl ketone Äthylvinylketon *n*
ethyl violet Äthylviolett *n*
ethylxanthate Äthylxanthogenat *n*
ethylxanthogenate Äthylxanthogenat *n*
ethyne Acetylen *n*, Äthin *n*, Carbidgas *n*
ethynyl- Acetylenyl-
ethynylation Äthinylierung *f*
ethypicone Ethypicon *n*
etioergosterol Ätioergosterin *n*
etiology Ätiologie *f* (Med)
etiomesoporphyrin Ätiomesoporphyrin *n*
etiophyllin Ätiophyllin *n*
etioporphyrin Ätioporphyrin *n*
etoxeridine Etoxeridin *n*
ettringite Ettringit *m* (Min)
eucaine Eucain *n*, Eukain *n*
eucaine hydrochloride Eucainhydrochlorid *n*
eucairite Eukairit *m* (Min)
eucalyptol[e] Eukalyptol *n*, Cajeputöl *n*, Cineol *n*
eucalyptus leaves Eukalyptusblätter *n pl*
eucalyptus oil Eukalyptusöl *n*
eucamptite Eukamptit *m* (Min)
eucarv[e]ol Eucarv[e]ol *n*
eucarvone Eucarvon *n*
eucasin Eukasin *n*, Kaseinammoniak *n*
eucatropine Eucatropin *n*
euchinine Euchinin *n*
euchlorine Chloroxydul *n*, Euchlorin *n*
euchroic acid Euchronsäure *f*
euchroite Euchroit *m* (Min),
  Smaragdmalachit *m* (Min)
euchrysine Euchrysin *n*
euclase Euklas *m* (Min)
eucodal Eukodal *n*
eucolite Eukolit *m* (Min)
eucolloid Eukolloid *n* (Phys)
eucrasite Eukrasit *m* (Min)
eucryptite Eukryptit *m* (Min)
eucupine Eucupin *n*
eucupinic acid Eucupinsäure *f*
eudalene Eudalin *n*
eudesmane Eudesman *n*
eudesmene Eudesmen *n*, Machilen *n*
eudesmic acid Eudesminsäure *f*
eudesmine Eudesmin *n*
eudesmol Eudesmol *n*, Machilol *n*
eudialite Eudialyt *m* (Min)
eudidymite Eudidymit *m* (Min)
eudiometer Eudiometer *n*, Gasmeßrohr *n*,
  Gasprüfer *m*
eudiometric eudiometrisch
eudnophite Eudnophit *m* (Min)
eudoxine Eudoxin *n*
eugallol Eugallol *n*
eugene glance Eugenglanz *m* (Min)
eugenic acid Eugensäure *f*
eugenin Eugenin *n*

eugenitin Eugenitin *n*
eugenol Eugenol *n*
eugentiogenin Eugentiogenin *n*
eugenyl acetate Eugenylacetat *n*
eugenyl benzoate Eugenylbenzoat *n*
eugenyl benzyl ether Eugenylbenzyläther *m*
euglobulin Euglobulin *n*
eugoform Eugoform *n*
eukanol binder Eukanolbinder *m* (HN)
eulatin Eulatin *n*
eulaxane Eulaxan *n*
Euler-Lagrange equations Eulersche
  Gleichungen *pl* (Mech)
eulysite Eulysit *m*
eulytine Demantblende *f*, Eulytin *m* (Min)
eulytite Wismutblende *f* (Min), Demantblende *f*
  (Min), Kieselwismut *n* (Min)
eumenol Eumenol *n*
eumydrine Eumydrin *n*
eunatrol Eunatrol *n*
euonymus Spindelbaum *m* (Bot)
eupatorin Eupatorin *n*
euphorbin Euphorbin *n*
euphorbium Euphorbiengummi *m*,
  Euphorbium *n*
euphorine Euphorin *n*, Phenylurethan *n*
euphosterol Euphosterol *n*
euphthalmine Euphthalmin *n*
euphylline Euphyllin *n*
euphyllite Euphyllit *m*
eupittonic acid Pittacol *n*
euporphine Euporphin *n*
eupyrchroite Eupyrchroit *m* (Min)
eupyrine Eupyrin *n*
euralite Euralith *m* (Min)
euresol Euresol *n*
eurhodine Eurhodin *n*
eurite Eurit *m*
European Atomic Energy Community Euratom,
  Europäische Gemeinschaft für Atomenergie
europhen Europhen *n*
europium (Symb. Eu) Europium *n*
eusantonane Eusantonan *n*
eusapyl Eusapyl *n*
eustenin Eustenin *n*
eusynchite Eusynchit *m* (Min)
eutannin Eutannin *n*
eutectane Eutectan *n*
eutectic Eutektikum *n*
eutectic *a* eutektisch
eutectic alloy Eutektikum *n*
eutectic mixture Eutektikum *n*, eutektische
  Mischung *f*
eutectoid Eutektoid *n*, eutektoide Legierung *f*
euthiochronic acid Euthiochronsäure *f*
eutrophic nährstoffreich
euxanthane Euxanthan *n*
euxanthic acid Euxanthinsäure *f*
euxanthin Euxanthin *n*

euxanthinic acid Euxanthinsäure *f*
euxanthogene Euxanthogen *n*
euxanthone Euxanthon *n*
euxenite Euxenit *m* (Min)
evacuate *v* ausleeren, entleeren, evakuieren,
  luftleer machen
evacuated evakuiert, luftleer, luftverdünnt
evacuation Evakuierung *f*, Abführung *f*;
  Auspumpen *n*, Leerung *f*
evaluable auswertbar
evaluate *v* bewerten, auswerten, durchrechnen,
  [ein]schätzen
evaluating Bewerten *n*
evaluation Bewertung *f*, Auswertung *f*,
  Beurteilung *f*
evaluation method Auswerteverfahren *n*
evansite Evansit *m* (Min)
evaporable verdunstbar
evaporableness Verdunstbarkeit *f*
evaporate *v* abdampfen, abrauchen, einkochen
  (z. B. Milch), evaporieren (z. B. Milch),
  verdampfen, verdunsten, sich verflüchtigen,
  ~ **to dryness** bis zur Trockne eindampfen
  (Chem)
evaporated verdampft; evaporiert, ~ **to dryness**
  abgeraucht (Chem)
evaporated acid Pfannensäure *f*
evaporated contact Aufdampfkontakt *m*
evaporated film Aufdampfschicht *f*
evaporated milk Kondensmilch *f*
evaporating Verdunsten *n*, Abrauchen *n*,
  Eindampfen *n*, Verdampfen *n*, ~ **the sud**
  Eindampfen *n* der Lauge
evaporating apparatus Eindampfapparat *m*,
  Verdunstungsapparat *m*
evaporating boiler Abdampfkessel *m*,
  Siedepfanne *f*
evaporating capacity Verdampfungsvermögen *n*
evaporating chamber Abrauchraum *m*
evaporating dish Abdampfschale *f*,
  Abrauchschale *f*
evaporating equipment Abdampfvorrichtung *f*
evaporating flask Abdampfkolben *m*
evaporating funnel Abdampftrichter *m*
evaporating liquor Siedelauge *f*
evaporating machine Verdampfungsmaschine *f*
evaporating pan Abdampfkasserolle *f*,
  Abdampfpfanne *f*, Abdampfschale *f*,
  Eindampfschale *f*
evaporating plant Verdampfanlage *f*
evaporating vessel Abdampfgefäß *n*
evaporation Verdampfung *f*, Abdampfen *n*,
  Abrauchen *n*, Eindampfen *n*, Einkochen *n*,
  Verdunstung *f*, Verflüchtigung *f*, **coefficient of**
  ~ Verdampfungskoeffizient *m*, **cooling by** ~
  Verdampfungskühlung *f*, **equilibrium of** ~
  Verdampfungsgleichgewicht *n*, **flash** ~
  Verdampfung *f* durch Entspannung, **heat of**
  ~ Verdampfungswärme *f*, **loss by** ~

Verdampfungsverlust *m*,
Verdunstungsverlust *m*, **rate of** ~
Verdampfungsgeschwindigkeit *f*, **time of** ~
Verdampfungszeit *f*, ~ **of water**
Wasserverdunstung *f*
evaporation condenser
Verdunstungskondensator *m*
evaporation cooling Kühlung *f* durch
Verdampfung, Verdampfungskühlung *f*
evaporation curve Verdampfungskurve *f*
evaporation enthalpy Verdampfungsenthalpie *f*
evaporation number Verdunstungszahl *f*
evaporation pan Verdampfpfanne *f*
evaporation plant Eindampfanlage *f*
evaporation point Verdampfungspunkt *m*
evaporation regulator Verdampfungsregler *m*
evaporation residue Abdampf[ungs]rückstand *m*,
Siederückstand *m*, Verdampfungsrückstand *m*
evaporation surface Verdampfungsoberfläche *f*,
Verdunstungsfläche *f*
evaporation temperature Abdampftemperatur *f*,
Verdampfungstemperatur *f*
evaporation time Verdunstungszeit *f*
evaporative capacity Verdampfungsfähigkeit *f*
evaporative condenser
Verdunstungsverflüssiger *m*
evaporative cooling Siedekühlen *n* (v.
Reaktoren), Verdampfungskühlung *f*,
Verdunstungskühlung *f*
evaporative value Verdampfungswert *m*
evaporator Verdampfer *m*, Abdampfapparat *m*,
Abdampfvorrichtung *f*, Eindampfkessel *m*,
Eindampfschale *f*, Evaporator *m*, **film-type** ~
Filmverdampfer *m*, **flooded** ~ überfluteter
Verdampfer *m*, **forced circulation** ~
Umlaufverdampfer *m*, **rapid action** ~
Schnellumlaufverdampfer *m*
evaporator assembly Verdampfersystem *n*
evaporator coil Verdampferschlange *f*
evaporator condenser Brüdenkondensator *m*
evaporator feed pump Verdampferspeisepumpe *f*
evaporator furnace Verdampfungsofen *m*
evaporator residue Verdampferrückstand *m*
evaporator tube Verdampferrohr *n*
evaporimeter Verdampfungsmesser,
Verdunstungsmesser *m*
evaporometer Evaporometer *n*,
Verdunstungsmeßgerät *n*,
Verdunstungsmesser *m*
evasion Ausweichen *n*, Entkommen *n*
evasive ausweichend
evatromonoside Evatromonosid *n*
evection Evektion *f* (Astr)
evectional tides Evektionstiden *f pl*
even eben, flach, gleichmäßig (einheitlich)
even *v* ebnen, egalisieren, ~ **up** abflachen
even and friction motion Gleichlauf *m* und
Friktion *f*

even-grained gleichgekörnt
evening shade Abendfarbe *f*
evenness Glätte *f*, Gleichmäßigkeit *f*
even-numbered geradzahlig, gradzahlig
event Begebenheit *f*; Ereignis *n*, Vorgang *m*, **in
the** ~ **that** falls
event field Ereignisfeld *n* (Statist)
events timing Zeitintervallmessung *f*
evernic acid Evernsäure *f*
everninic acid Everninsäure *f*
evernuric acid Evernursäure *f*
evidence Anzeichen *n*, Beleg *m*, Beweis *m*,
Beweismaterial *n*
evident einleuchtend, offensichtlich
evigtokite Evigtokit *m* (Min)
evipan Evipan *n* (HN)
evodene Evoden *n*
evodiamine Evodiamin *n*
evodine Evodin *n*
evodionol Evodionol *n*
evolatine Evolatin *n*
evolitrine Evolitrin *n*
evolution Entwicklung *f*, Evolution *f*,
Freisetzung *f* (Gas), **theory of** ~
Evolutionstheorie *f* (Biol), ~ **of man**
Anthropogenese *f*
evolve *v* entwickeln, entfalten
evomonoside Evomonosid *n*
evoxanthine Evoxanthin *n*
exact exakt, genau
exact measurement genaues Maß *n*
exactness Genauigkeit *f*
exaggerate *v* übertreiben
exaggeration Übertreibung *f*, Aufbauschung *f*
exalgin Exalgin *n*, Methylacetanilid *n*
exaltone Exalton *n*
examination Prüfung *f*, Probe *f* (Kontrolle),
Test *m*, Überprüfung *f*, Untersuchung *f*
(Med), **method of** ~ Untersuchungsmethode *f*
examination of material Stoffprüfung *f*
examination procedure Prüfungsverfahren *n*
examine *v* untersuchen, besichtigen,
[über]prüfen, ~ **microscopically**
mikroskopisch untersuchen, mikroskopieren
examinee Prüfling *m*
examiner Prüfer *m*
example Beispiel *n*
excavate *v* ausgraben
excavated material Abtrag *m*
excavation Ausgrabung *f*, Aushöhlung *f*
excavator Bagger *m*, **high-capacity bucket-wheel**
~ Großschaufelradbagger *m*
exceed *v* überschreiten, übersteigen, übertreffen
excel *v* übertreffen
excellence Vorzüglichkeit *f*
excellent vorzüglich
excelsior Holzwolle *f*, kleine weiche
Holzspäne *pl*
except ausschließlich

exception Ausnahme *f*, Regelabweichung *f*
exceptional ungewöhnlich
exceptionally ausnahmsweise
excerpt Auszug *m*, Exzerpt *n*
excess Überschuß *m*, Übermaß *n*, in ~
  überschüssig, ~ of acid Säureüberschuß *m*, ~
  of air Luftüberschuß *m*
excess air Falschluft *f*
excess air coefficient Luftüberschußzahl *f*,
  Luftzahl *f*
excess base Basenüberschuß *m*
excess charge Überschußladung *f*
excess content Mehrgehalt *m*
excess factors (for non-ideality of mixtures)
  Exzeßgrößen *pl*
excessive übermäßig, übertrieben
excess pressure Überdruck *m*
excess pressure container Überdruckbehälter *m*
excess salt überschüssiges Salz *n*
excess semiconductor n-Leiter *m*
excess vapor pressure Dampfüberdruck *m*
excess weight Mehrgewicht *n*
exchange Austausch *m*, Auswechslung *f*,
  Umtausch *m*, Wechsel *m*, ~ of bases
  Basenaustausch *m*, ~ of ideas
  Gedankenaustausch *m*, ~ of knowhow
  Erfahrungsaustausch *m*
exchange *v* austauschen, auswechseln,
  permutieren (Math), umspeichern (Comp),
  vertauschen
exchangeability Austauschbarkeit *f*,
  Auswechselbarkeit *f*
exchangeable austauschbar, auswechselbar
exchange acidity Austauschazidität *f*
exchange action Austauschwirkung *f*
exchange algorithm Austauschalgorithmus *m*
exchange chromatography
  Austauschchromatographie *f*
exchange coefficient Austauschgröße *f*
exchange correction Austauschkorrektur *f*
exchange half-life Austauschhalbwertzeit *f*
exchange limitation Austauschbeschränkung *f*
exchange mass correction Massenkorrektur *f* für
  Austausch
exchange potential Austauschpotential *n*
exchanger Austauscher *m*
exchange reaction Austauschreaktion *f*
exchange term Austauschterm *m*
exchanging Auswechslung *f*, Umspeicherung *f*
  (Comp)
excipient Arzneimittelträger *m*
excise *v* ausschneiden
excision Ausschneidung *f*, ~ of an adrenal gland
  Adrenalektomie *f* (Med)
excitable erregbar
excitant Erregermasse *f*, Stimulans *n*
excitation Anregung *f* (Chem, Elektr),
  Erregung *f*, Reizung *f*, foreign ~
  Fremderregung *f*, indirect ~

Umweganregung *f*, method of ~
Erregungsart *f*, separate ~ Fremderregung *f*,
work of ~ Erregerarbeit *f*, ~ of electrons
  Elektronenanregung *f*, Anhebung *f* von
  Elektronen
excitation collision Anregungsstoß *m*
excitation condition Anregungsbedingung *f*
excitation current Anregungsstrom *m*
excitation defect Erregungsfehler *m*
excitation energy Anregungsenergie *f*
excitation frequency Erregerfrequenz *f*
excitation intensity Anregungsintensität *f*
excitation level Anregungsniveau *n*
excitation potential Anregungsspannung *f*
excitation process Anregungsvorgang *m*
excitation stage Anregungsstufe *f*
excitation voltage Anregungsspannung *f*,
  Erregerspannung *f*
excite *v* anregen, aufregen, erregen
excited angeregt, erregt, separately ~
  fremderregt
excited atom angeregtes Atom *n*
excited state Anregungszustand *m*
exciter Erreger *m*
exciter coil Erregerspule *f* (Elektr)
exciter voltage Erregerspannung *f*
exciting anregend, aufregend
exciting circuit Erregerkreis *m*,
  Erregerstromkreis *m*
exciting current Erregerstrom *m*
exciting dynamo Dynamoerreger *m*
exciting field Erregerfeld *n* (Elektr)
exciting rectifier Erregergleichrichter *m* (Elektr)
exclude *v* ausschalten, ausschließen
exclusion Ausschließung *f*, ~ of air
  Luftabschluß *m*
exclusion of single observations Ausscheiden
  einzelner Meßpunkte
exclusion principle Ausschlußprinzip *n* (Atom),
  Besetzungsverbot *n* (Atom)
exclusion rule Alternativ-Verbot *n*,
  Ausschließungsregel *f*
exclusion sphere Ausschlußbereich *m*
exclusive exklusiv
excorticated entrindet
excrement Ausscheidung *f*, Auswurf *m*,
  Exkrement *n*, Kot *m*
excrescence Auswuchs *m*, Wuchs *m*
excreta Ausscheidungsstoffe *pl*, Kot *m*
excrete *v* abscheiden, absondern, ausscheiden
  (Med), aussondern, sezernieren (Med)
excretion Ausscheidung, Auswurf *m*, Exkretion *f*
excuse *v* entschuldigen; dispensieren
execute *v* ausführen, durchführen
executive committee Arbeitsausschuß *m*
exemplary musterhaft, exemplarisch
exempt *v* befreien, dispensieren
exercise Übung *f*

exercise *v* [aus]üben, praktizieren
exergonic exergon
exert *v* anstrengen, ausüben
exfoliate *v* abblättern, abrinden, abschälen, schiefern
exfoliation Abblätterung *f*
exfoliative Abblätterungsmittel *n*
exhalation Ausatmung *f*, Ausdünstung *f*, Brodem *m*, Dunst *m*
exhale *v* ausatmen; ausdünsten, aushauchen
exhaust Ableitung *f*, Abzug *m* für Abgase, Auspuff *m*
exhaust *v* abführen (Dampf), ausblasen (Dampf), ausmergeln (Med), auspuffen, auspumpen, aussaugen, ausströmen (Dampf), austreten (Dampf), entkräftigen (Med), erschöpfen (Med), herausziehen (Chem), luftleer machen (Chem), verarmen, versiegen, ~ **by suction** absaugen
exhaust air Abluft *f*, Abwetter *n* (Bergb)
exhaust air duct Abluftlutte *f*
exhaust balance line Abgasausgleichleitung *f*
exhaust collector Auspuffsammler *m*
exhaust duct Entlüftungsleitung *f*
exhausted erschöpft; luftleer, evakuiert
exhauster Absaug[e]anlage *f*, Absauger *m*, Exhaustergebläse *n*, Exhaustor *m*
exhaust fan Absaugventilator *m*
exhaust filter Auspuffilter *m*
exhaust flame damper Auspuffflammendämpfer *m*
exhaust gas Abgas *n*, Abluft *f*, Auspuffgas *n*, Rauchgas *n*
exhaust gas analyzer Abgasprüfgerät *n*
exhaust gas jet thrust Abgasstrahlschub *m*
exhaust gas plant Abgasanlage *f*
exhaust gas stack Abluftkamin *m*
exhaust gas thermometer Rauchgasthermometer *n*
exhaust head Auspuffhaube *f*
exhaustible erschöpfbar
exhausting by suction Absaugen *n*
exhausting device Absaugvorrichtung *f*
exhaustion Absaugung *f*, Erschöpfung *f*, Verarmung *f*
exhaustion hypothesis Erschöpfungshypothese *f*
exhaustive erschöpfend; schwächend
exhaustive methylation erschöpfende Methylierung *f*
exhaust main Auspuffleitung *f*
exhaust manifold Auspuffrohr *n*
exhaust passage Dampfausgang *m*
exhaust pipe Ausströmungsrohr *n*, Abdampfleitung *f*, Auspuffleitung *f*, Auspuffrohr *n*
exhaust port Auspufföffnung *f*, Auspuffstutzen *m*; Ausströmkanal *m*, Druckstutzen *m*
exhaust process Ausziehverfahren *n* (Färberei)

exhaust pump Absaugpumpe *f*
exhaust side Auspuffseite *f*
exhaust steam Abdampf *m*, Auspuffdampf *m*, Rückdampf *m* (Zucker)
exhaust steam collector Abdampfsammelstück *n*, Abdampfverflüssiger *m*
exhaust steam main Abdampfleitung *f*
exhaust-steam oil separator Abdampfentöler *m*
exhaust steam pipe Abdampfrohr *n*, Dampfauslaßrohr *n*
exhaust steam piping Abdampfrohrnetz *n*
exhaust-steam pressure regulator Abdampfdruckregler *m*
exhaust steam turbine Abdampfturbine *f*
exhaust stroke Auspuffhub *m*
exhaust tail pipe Abgasabführungsrohr *n*
exhaust tube Gasableitungsrohr *n*
exhaust valve Ablaßventil *n*, Abblaseventil *n*, Auslaßventil *n*, Auspuffventil *n*
exhibit *v* aufweisen, ausstellen
exhibition Ausstellung *f*
exhibition room Ausstellungsraum *m*
eximine Eximin *n*
exist, beginning to ~ naszierend
exist *v* bestehen, existieren
existence Bestehen *n*, Existenz *f*, **incapable of** ~ nicht existenzfähig
existing befindlich, **capable of** ~ existenzfähig
exit Austritt *m*
exit gas Abzugsgas *n*
exit heat Abzugswärme *f*
exit side Austrittseite *f*
exit slit Ausgangsspalt *m* (Opt), Austrittsblende *f* (Opt)
exit temperature Abzugstemperatur *f*
exit velocity Austrittgeschwindigkeit *f*
exocellular extrazellulär
exocyclic exocyclisch
exoelectrons Exoelektronen *pl*
exo-enzyme Exoenzym *n*, extrazelluläres Enzym *n*
exogenic exogen
exograph Röntgenstrahlaufnahme *f*
exonuclease Exonuklease *f* (Biochem)
exopeptidase Exopeptidase *f* (Biochem)
exoplasm äußere Plasmaschicht *f*, Ektoplasma *n*, Exoplasma *n*
exosmosis Exosmose *f* (Phys)
exosmotic exosmotisch
exothermal exotherm[isch], wärmeabgebend, wärmeliefernd
exothermic exotherm[isch], wärmeerzeugend
exotoxin Exotoxin *n*
expand, tendency to ~ Ausdehnungsdrang *m*
expand *v* ausbreiten, ausdehnen, entspannen, erweitern, expandieren, schäumen (Plast), weiten
expandability Ausdehnbarkeit *f*
expanded flange Aufwalzflansch *m*

expanded metal Streckmetall *n*
expanded rubber Schwammgummi *m*,
  Zellgummi *m*
expander Entspannungsmaschine *f*
expanding action Spreizwirkung *f*
expanding agent Schaummittel *n*, Treibmittel *n*
  (Schaumstoff)
expanding machine Aufspannmaschine *f*
expanding mandrel Spreizdorn *m*
expanding or drift test Aufdornversuch *m*
expanding ring Spreizring *m*
expanding screw thread Spreizgewinde *n*
expanding spring Spreizfeder *f*
expanding wedge Spreizkeil *m*
expansibility Ausdehnungsfähigkeit *f*,
  Ausdehnungsvermögen *n*, Dehnkraft *f*,
  Dehnvermögen *n*, Spannkraft *f*
expansible [aus]dehnbar
expansin Expansin *n*
expansion Erweiterung *f*, Ausdehnung *f*,
  Dehnung *f*, Entspannung *f*, Entwicklung *f*
  (Math), Expansion *f*, **adiabatic** ~ adiabatische
  Ausdehnung *f* (Therm), **adiabatic curve of** ~
  Expansionsadiabate *f*, **asymptotic** ~
  asymptotische Entwicklung *f*, **coefficient of** ~
  Ausdehnungskoeffizient *m*, **coefficient of**
  linear ~ linearer Ausdehnungskoeffizient,
  force of ~ Ausdehnungskraft *f*, **linear** ~
  Längenausdehnung *f*, **power of** ~ Dehnkraft *f*,
  thermal ~ thermische
  Ausdehnung *f*, **total** ~ Gesamtdehnung *f*,
  work done on ~ Ausdehnungsarbeit *f*
expansion arch Dehnungsbogen *m*
expansion bolt Bolzenschraube *f*
expansion chamber Wilsonkammer *f*,
  Ausdehnungsraum *m*, Expandierraum *m*,
  Expansionskammer *f* (Atom)
expansion coefficient Ausdehnungskoeffizient *m*,
  Ausdehnungszahl *f*
expansion compensator Dehnungsausgleicher *m*
expansion cooler Entspannungskühler *m*
expansion cylinder Verdampfungszylinder *m*
expansion engine Expansionsmaschine *f*, ~ **for**
  air Luftexpansionsmaschine *f*
expansion fitting Ausdehnungsarmatur *f*
expansion joint dehnbare Verbindung *f*,
  Ausdehnungskupplung *f*,
  Dehnungsausgleicher *m*
expansion loop Dehnungsausgleicher *m*
expansion piece Ausdehnungsstück *n*
expansion reservoir Überlaufbehälter *m*
expansion ring Spannring *m*
expansion screw Spreizschraube *f*
expansion space Ausdehnungsraum *m*
expansion tank Ausgleichsbehälter *m*
expansion theory Entwicklungssatz *m* (Math)
expansion turbine Expansionsturbine *f*
expansion valve Entspannungsventil *n*,
  Expansionsventil *n*, Überdruckventil *n*

expansion vessel Ausgleichsbehälter *m*,
  Ausdehnungsgefäß *n*
expansive ausdehnbar
expansive capacity Expansionsvermögen *n*
expansive force Expansionskraft *f*
expansivity Expansionsvermögen *n*
expect *v* entgegensehen, erwarten
expectation value Erwartungswert *m*
expected, as ~ erwartungsgemäß
expectorant Expektorans *n* (Pharm),
  schleimlösend (Med), schleimlösendes
  Mittel *n*
expedient Behelf *m*, Behelfslösung *f*,
  Behelfsmittel *n*, Hilfsmittel *n*, Notbehelf *m*
expedient *a* praktisch, nützlich; ratsam
expediting Terminkontrolle *f*
expel *v* abtreiben (Chem), ausstoßen,
  herausschleudern, verdrängen, verjagen,
  vertreiben
expelled gas Abtreibgas *n*
expeller cake Preßkuchen *m*
expelling Austreiben *n*
expenditure Aufwand *m*, Ausgabe *f*,
  Kostenaufwand *m*
expenditure of force Kraftaufwand *m*
expenses Kosten *pl*, pl, **fixed** ~ laufende
  Ausgaben *pl*, **living** ~
  Lebenshaltungskosten *pl*, **working** ~
  Betriebskosten *pl*
expenses covered kostenfrei
expensive kostspielig, teuer
experience Erfahrung *f*, Sachkenntnis *f*, **from** ~
  erfahrungsgemäß
experienced erfahren, sachkundig, erprobt,
  sachverständig
experienced personnel Fachpersonal *n*
experiment Versuch *m*, Experiment *n*, Probe *f*,
  result of ~ Versuchsergebnis *n*
experiment *v* experimentieren, versuchen
experimental experimentell
experimental animal Versuchstier *n*
experimental arrangement Versuchsanordnung *f*
experimental boiler Versuchskocher *m*
experimental chemistry Experimentalchemie *f*
experimental condition Versuchsbedingung *f*
experimental direction Versuchsvorschrift *f*
experimental error Versuchsfehler *m*
experimentalist Experimentator *m*
experimentally versuchsmäßig, auf
  experimentellem Wege, versuchsweise
experimental material Versuchsmaterial *n*
experimental method Versuchsmethode *f*
experimental physics Experimentalphysik *f*
experimental plant Versuchsanlage *f*,
  Versuchspflanze *f* (Bot)
experimental point Meßpunkt *m*
experimental procedure Versuchsdurchführung *f*,
  Versuchsvorgang *m*
experimental reactor Versuchsreaktor *m*

experimental result Meßergebnis *n,*
Versuchsergebnis *n*
experimental scale Versuchsmaßstab *m*
experimental stage Versuchsphase *f,*
Versuchsstadium *n*
experimental station Versuchsanstalt *f*
experimental time Versuchsdauer *f*
experimental value Meßwert *m*
experimental voltage Versuchsspannung *f*
experimental works Versuchsbetrieb *m*
experiment field Versuchsfeld *n*
experimenting table Experimentiertisch *m*
experimentor Experimentator *m*
expert . . . fach-, Fachmann *m,* Sachkundiger *m,*
Sachverständiger *m,* Spezialist *m*
expert *a* fachkundig, fachmännisch, sachkundig,
sachverständig
expert knowledge Fachkenntnis *f,* Sachkenntnis *f*
expert opinion Gutachten *n*
expiration Ablauf *m,* Ausatmung *f*
expire *v* ablaufen (zeitlich), ausatmen
explain *v* erklären; auseinandersetzen, erläutern
explanation Auslegung *f,* Erklärung *f,*
Erläuterung *f,* Klarstellung *f*
explicit explizit (Math)
explode *v* explodieren, aufplatzen, bersten,
detonieren, verpuffen, [zer]platzen,
zerspringen
exploded wire Metalldrahtentladung *f*
exploded wire continuum
Metalldrahtentladungskontinuum *n*
exploding explodierend
exploit *v* abbauen (Bergb)
exploitation Abbau *m* (Bergb), Ausbeutung *f,*
Ausnutzung *f*
exploration Erforschung *f*
exploration drilling Aufschlußbohrtätigkeit *f*
exploration well Aufschlußbohrung *f,*
Suchbohrung *f*
explore *v* erforschen
exploring electrode Abtastelektrode *f*
explosion Explosion *f,* Bersten *n,*
Erschütterung *f,* Knall *m,* Sprengung *f,*
Verpuffung *f,* **cause of** ~ Explosionsursache *f,*
danger of ~ Explosionsgefahr *f*
explosion bulb Explosionskugel *f*
explosion hazard Explosionsgefahr *f*
explosion impulse Explosionsstoß *m*
explosion limit Explosionsgrenze *f,* **lower** ~
untere Explosionsgrenze, **upper** ~ obere
Explosionsgrenze
explosion method Verpuffungsverfahren *n*
explosion pressure Explosionsdruck *m,*
Verpuffungsspannung *f*
explosion-proof explosionssicher
explosion pyrometer Knallpyrometer *n*
explosion reaction Explosionsreaktion *f*
explosion spectrum Explosionsspektrum *n*
explosion wave Explosionswelle *f*

explosive Sprengstoff *m,*
Explosivstoff *m,* Sprengmittel *n,* **initial** ~
Initialzündmittel *n*
explosive *a* explosiv, explosionsfähig,
sprengfähig, **highly** ~ hochexplosiv,
hochbrisant
explosive action Sprengwirkung *f*
explosive cartridge Sprengpatrone *f*
explosive charge Sprengladung *f,*
Sprengfüllung *f,* Sprengkörper *m*
explosive cotton Nitrozellulose *f,*
Schießbaumwolle *f*
explosive effect Explosionswirkung *f,*
Sprengwirkung *f*
explosive force Explosionskraft *f,* Sprengkraft *f*
explosive gelatin Sprenggelatine *f*
explosive impact Explosionsstoß *m*
explosive liquid Sprengflüssigkeit *f*
explosive mixture Explosionsgemisch *n*
explosiveness Explodierbarkeit *f,*
Explosionsfähigkeit *f*
explosive powder Sprengpulver *n*
explosive power Sprengkraft *f,* Brisanz *f,*
Explosivkraft *f*
explosive-powered tool Bolzensetzer *m,*
Bolzensetzwerkzeug *n*
explosive process Explosionsvorgang *m*
explosive property Explosionsfähigkeit *f*
exponent Exponent *m* (Math), Hochzahl *f*
(Math), Potenz *f* (Math), **fractional** ~
gebrochener Exponent (Math)
exponential Exponentialgröße *f,* Exponential-
exponential *a* exponentiell
exponential curve Exponentialkurve *f*
exponential distribution Potenzverteilung *f*
exponential equation Exponentialgleichung *f*
(Math)
exponential function Exponentialfunktion *f*
exponential integral Exponentialintegral *n*
(Math)
exponential law Exponentialgesetz *n*
exponential series Exponentialreihe *f*
export Ausfuhr *f,* Export *m*
export *v* ausführen, exportieren
exportable ausführbar, exportfähig
exportation Ausfuhr *f,* Export *m*
expose *v* aussetzen, belichten (Phot), bloßlegen,
entblößen, exponieren, ~ **to rays** bestrahlen
exposed, to be ~ **to** ausgesetzt sein
exposed concrete Sichtbeton *m*
exposure, duration of ~ Bestrahlungsdauer *f*
(Med), **stage of** ~ Belichtungsstufe *f,* **time of**
~ Belichtungsdauer *f* (Phot),
Belichtungszeit *f* (Phot), Bestrahlungszeit *f,* ~
[to light] Belichtung *f* (Phot), ~ **to rays**
Bestrahlung *f*
exposure chart Belichtungskarte *f* (Phot)
exposure-density relationship Gradationskurve *f*
(Phys)

exposure difference Belichtungsdifferenz *f*
exposure factor Expositionsfaktor *m*
exposure hole Bestrahlungskanal *m*
exposure lid Belichtungsschieber *m* (Phot)
exposure meter Belichtungsmesser *m* (Phot)
exposure shutter Belichtungsschieber *m* (Phot)
exposure site Bewitterungsstelle *f*
exposure table Belichtungstabelle *f*
exposure trials Bewitterungsversuche *pl*
exposure value Belichtungswert *m*
express *v* äußern, ausdrücken
expressed oil Ablauföl *n*
expression Äußerung *f,* Ausdruck *m* (Math), ~
    containing four terms Quadrinom *n* (Math)
expression in parentheses (or brackets or braces)
    Klammerausdruck *m* (Math)
expropriate *v* enteignen
expulsion Ausstoßung *f,* Austreibung *f,* **heat of**
    ~ Austreibungswärme *f*
expulsion tube Ausstoßrohr *n*
exsiccate *v* austrocknen, eindörren
exsiccated getrocknet, eingedörrt
exsiccating Abdörren *n,* Austrocknen *n*
exsiccator Exsikkator *m*
exsiccator grease Exsikkatorfett *n*
extemporize *v* extemporieren
extend *v* [aus]dehnen, ausbreiten, ausrecken
    (Metall), [aus]strecken, [aus]weiten, ausziehen,
    erweitern, recken, sich erstrecken, verlängern
extended breit (Buchdr), verlängert, erweitert
extended metal Streckmetall *n*
extended standard breit mager (Buchdr)
extended surface Rippenoberfläche *f*
extended-surface heat exchanger
    Rippenrohrwärmeaustauscher *m*
extender Streckmittel *n,* Verdünnungsmittel *n,*
    Verschnittmittel *n*
extendibility Ausdehnbarkeit *f,*
    Ausdehnungsfähigkeit *f*
extending Ausziehen *n*
extending slider Verlängerungsreiter *m*
extensibility Ausdehnungsvermögen *n,*
    Dehnbarkeit *f,* Dehnfähigkeit *f,* Spannkraft *f,*
    Streckbarkeit *f*
extensible ausdehnbar, ausweitbar, ausziehbar,
    spannbar
extension Ausdehnung *f,* Dehnung *f,*
    Erweiterung *f,* Verlängerung *f,* **elastic** ~
    elastische Ausdehnung, **rate of** ~
    Dehnungsgeschwindigkeit *f,* **total** ~
    Gesamtdehnung *f*
extension cable Verlängerungsschnur *f* (Elektr)
extension piece Ansatzstück *n,*
    Verlängerungsstück *n*
extension pipe Ansatzrohr *n*
extension table Ausziehtisch *m*
extension tongs Fernbedienungszange *f*
extensive ausgedehnt, umfangreich

extensometer Dehnbarkeitsmesser *m,*
    Dehnungsmesser *m*
extent Ausdehnung *f,* Ausmaß *n,* Größe *f*
exterior äußerlich
exterior angle (of a triangle) Außenwinkel *m*
    (eines Dreiecks)
exterior coating Außenanstrich *m,*
    Schutzüberzug *m*
exterior paint Außen[anstrich]farbe *f*
exterior varnish Außenlack *m*
exterminate *v* ausrotten, vertilgen
extermination Ausrottung *f,* Vertilgung *f*
external äußerlich, außenseitig
external air Außenluft *f*
external autoignition Fremdzündung *f*
external furnace Außenfeuerung *f*
external gauge Rachenlehre *f*
external grinder Außenrundschleifmaschine *f*
external pole Außenpol *m*
external pole generator Außenpolgenerator *m*
external resistance Außenwiderstand *m* (Elektr)
external shoe brake Außenbackenbremse *f*
external thread Außengewinde *n*
extinct ausgelöscht, erloschen, **to become** ~
    aussterben
extinction Auslöschung *f,* Extinktion *f*
extinction coefficient Extinktionskoeffizient *m,*
    molar ~ molarer Extinktionskoeffizent
extinction curve Extinktionskurve *f*
extinguish *v* [aus]löschen, vernichten
extinguishable spark gap Löschfunkenstrecke *f*
extinguisher Löschgerät *n*
extinguishing effect Löschwirkung *f*
extinguishing substance Löschmittel *n*
extirpate *v* ausrotten, vertilgen
extirpation Ausrottung *f*
extirpator Vertilgungsmittel *n*
extra zusätzlich
extraaxial außerachsial
extrabold fett (Buchdr)
extracellular extrazellulär
extract Extrakt *m,* Absud *m,* Auszug *m,*
    Exzerpt *n,* **concentrated** ~ Quintessenz *f,*
    content of ~ Auszuggehalt *m*
extract *v* extrahieren, auskochen, auslaugen,
    ausschütteln, herauslösen, ziehen (Wurzel), ~
    dust [from] entstauben, ~ **fat** entfetten, Fett
    extrahieren, ~ **lead [from]** entbleien, ~ **sugar**
    entzuckern, ~ **water [from]** entwässern
extractable ausziehbar, extrahierbar
extract content Extraktgehalt *m*
extracted ausgelaugt, ausgezogen, extrahiert
extracted liquor Extraktbrühe *f*
extractibility Extrahierbarkeit *f*
extractible extrahierbar
extracting Extrahieren *n,* Ausziehen *n,* ~ **with**
    ether Ausäthern *n*
extracting agent Extraktionsmittel *n*

extraction Extraktion *f*, Auslaugung *f*,
  Ausschütteln *n*, Ausziehung *f*, Auszug *m*,
  Entziehung *f*, Extrakt *m*, **continuous** ~
  kontinuierliche Extraktion, **counter-current** ~
  Gegenstromextraktion *f*, **counterflow** ~
  Gegenstromextraktion *f*, **liquid-liquid** ~
  flüssig-flüssig Extrahieren *n*, **time required for**
  ~ Extraktionsdauer *f*, ~ **of acid**
  Entsäuerung *f*, ~ **of benzene**
  Entbenzolierung *f*, ~ **of cube root**
  Kubikwurzelziehen *n* (Math), ~ **of metals**
  Metalldarstellung *f*, Metallgewinnung *f*, ~ **of**
  water Entwässerung *f*, ~ **under pressure**
  Druckextraktion *f*
extraction analysis Extraktionsanalyse *f*
extraction apparatus Extraktionsapparat *m*,
  Auslaugeapparat *m*, Auslauger *m*
extraction battery Extraktionsbatterie *f*
extraction centrifuge Extraktionszentrifuge *f*
extraction column Extraktionskolonne *f*
extraction crucible Einsatztiegel *m*
extraction filter Einsatzfilter *m*
extraction funnel Auslaugtrichter *m*
extraction naphtha Extraktionsbenzin *n*
extraction plant Extraktionsanlage *f*
extraction thimble Extraktionshülse *f*
extraction vessel Extraktionsgefäß *n*
extractive extraktiv
extractive distillation extraktive Rektifikation *f*
extractive matter Extraktivstoff *m*
extractive principle Extraktivstoff *m*
extractive substance Extraktstoff *m*
extractor Auskocher *m*, Auslauger *m*,
  Extraktor *m*, Hilfsvorrichtung zum
  Ausdrücken, Trockenschleuder *f*,
  counter-current ~ Gegenstromextraktor *m*,
  direct-flow ~ Durchflußextraktor *m*
extractor attachment Extraktoraufsatz *m*
extra expenses Mehraufwand *m*
extra hard steel Diamantstahl *m*
extramolecular extramolekular
extraneous unwesentlich, nicht zugehörig
extraordinary außerordentlich, ungewöhnlich
extrapolate *v* extrapolieren
extrapolation Extrapolation *f*
extrapolation chamber Extrapolationskammer *f*
extrapolation formula Extrapolationsformel *f*
extra resonance energy Extraresonanzenergie *f*
extras Sonderausrüstung *f*
extraterrestrial extraterrestrisch (Geol)
extra yield Extraausbeute *f*
extreme Extrem *n*
extreme *a* extrem, äußerst
extreme case Extremfall *m*
extremely außerordentlich, ungemein
extremely thin hauchdünn
extreme values Extremwerte *pl* (v. Funktionen)
extremity Extrem *n*, Extremität *f* (Anat)
extricate *v* entwinden

extrudability Spritzbarkeit *f*
extrude *v* auspressen, spritzen, strangpressen
extruded stranggepreßt
extruded bar Profilstab *m*
extruded shape gespritztes Profil *n*
extruder Extruder *m*, Spritzmaschine *f*,
  Strangpresse *f*
extruder head Spritzkopf *m*
extruder pelletizer Krümelspritzmaschine *f*
extruding Extrudieren *n*
extruding head Spritzkopf *m*
extruding press Strangpresse *f* (Met)
extruding screw Schnecke *f* (der Strangpresse)
extrusion Strangpressen *n*, Ziehprofil *n*, **rate of**
  ~ Spinngeschwindigkeit *f*
extrusion aid Spritzbarmacher *m*
extrusion chamber Schneckengehäuse *n*
extrusion die Spritzform *f*, Spritzwerkzeug *n*,
  Strangpreßform *f*, ~ **for tubing**
  Schlauchspritzform *f*
extrusion molding Strangpressen *n*
extrusion press Spritzpresse *f*, Strangpresse *f*, ~
  for tubes Rohrpresse *f*
extrusion temperature Spinntemperatur *f*
  (Kunststoff)
extrusive rocks Extrusivgestein *n*
exuberance Wucherung *f* (Bot)
exudate Exsudat *n*
exudate *v* ausscheiden
exudation Ausscheidung *f*, Ausschwitzen *n*,
  Ausschwitzung *f*
exude *v* ausschwitzen
eye Öhr *n*, Öse *f*
eyebolt Augbolzen *m*, Ringschraube *f*
eyeglass Okular *n* (Opt)
eye-holing als Porenbildung sichtbare
  Lackverlaufsstörung *f*
eye lens Okularlinse *f* (Opt)
eyelet Öhr *n*, Öse *f*
eye nut Ringmutter *f*
eyepiece Okular *n*, **focussing of the** ~
  Okulareinstellung *f*
eyepiece cell Okularfotozelle *f*
eyepiece diaphragm Okularblende *f*
eye shadow Lidschatten *m*
eyeshield Augenmuschel *f* (Atom)
eye spot fungus Halmbruchpilz *m*
eye-witness Augenzeuge *m*

F

fabiatrin Fabiatrin *n*
fabric Gewebe *n* (Text), Stoff *m* (Text), Tuch *n* (Text), **impregnated** ~ getränktes Gewebe (Text), **man-made** ~ Chemiefaser *f,* **open** ~ weitgestelltes Gewebe, **pick weave** ~ beschossenes Gewebe, **raised** ~ Rauhfaser *f* (Text), **synthetic** ~ Chemiefaser *f,* synthetisches Gewebe *n,* **water-repellent** ~ Trockengewebe *n* (Text), wasserabstoßendes Gewebe *n,* **weftless** ~ schußloses Gewebe
fabricate *v* [an]fertigen, fabrizieren
fabricating Verarbeitung *f*
fabricating technique Bearbeitungsverfahren *n,* Herstellungsverfahren *n*
fabrication Fabrikation *f,* Fertigung *f*
fabricator Hersteller *m,* Verarbeiter *m*
fabric clippings Gewebeschnitzel *pl* (Web)
fabric coating Kunstharzüberzug *m*
fabric collector Tuchfilter *n*
fabric covering Stoffbespannung *f*
fabric-filled molding compound Preßmasse *f* mit Gewebe-Füllstoff
fabric insert Gewebeeinlage *f*
fabric lining Stoffauskleidung *f*
fabrics, blended ~ Mischgewebe *n* (Text), **knitted** ~ Strickwaren *f pl* (Text)
fabric type Gewebetyp *m* (Web)
facade Fassade *f,* Stirnseite *f*
face Vorderseite *f,* ~ **of a crystal** Kristallfläche *f*
face *v* gegenüberstehen, fräsen (Techn), plandrehen (Techn)
face-centered flächenzentriert (Krist)
face-centered lattice flächenzentriertes Raumgitter *n*
face flange Planflansch *m*
face gear Zahnscheibe *f*
face guard Schutzmaske *f*
face hammer Bahnhammer *m*
face lathe Plandrehbank *f*
face lotion Gesichtswasser *n* (Kosm)
face mill Planfräser *m*
face milling Planfräsen *n*
face-milling cutter Flächenfräser *m* (Techn)
face milling machine Planfräsmaschine *f*
face mix Vorsatzbeton *m*
face-out Funkschwund *m*
face plate Planscheibe *f* (Drehbank)
face pressure Flankendruck *m* (Keilriemen)
faces, number of ~ Flächenzahl *f* (Krist)
face stitcher Anroller *m,* Anrollvorrichtung *f,* Planroller *m*
facet Facette *f,* Kristallfläche *f*
faceted facettiert
facets, number of ~ Flächenzahl *f*
face value Nennwert *m,* Nominalwert *m,* Sollwert *m*

face wall Stirnwand *f* (Bauw)
face wheel Planrad *n*
facilitate *v* erleichtern
facility Einrichtung *f,* Vorrichtung *f*
facing sand fein gesiebter Sand (Gieß)
facsimile printing Faksimiledruck *m* (Buchdr)
facsimile recorder Faksimileschreiber *m*
factice Faktis *m,* Gummi-Ersatz *m,* Ölkautschuk *m,* **mass of** ~ Faktismässe *f*
factor Faktor *m,* Beiwert *m,* ~ **of capacity** Quantitätsfaktor *m*
factor analysis Faktorenanalyse *f* (Math)
factor group Faktorgruppe *f*
factorial Fakultät *f* (Math)
factorial design faktorielle Versuchplanung *f*
factorization Faktorenzerlegung *f*
factorization method Faktorisierungsmethode *f* (Math)
factory Fabrik *f,* Betrieb *m,* **by** ~ **methods** fabrikmäßig
factory defect Fabrikationsfehler *m*
factory doctor Betriebsarzt *m*
factory hand Fabrikarbeiter *m*
factory hygiene Betriebshygiene *f*
factory molasses Rohzuckermelasse *f*
factory operation Fabrikbetrieb *m*
factory plant Fabrikanlage *f*
factory practice Fabrikpraxis *f*
factory price Fabrikpreis *m*
factory scale, on a ~ fabrikmäßig
factory worker Fabrikarbeiter *m*
faculty Fähigkeit *f,* Fakultät *f*
FAD (Abk) Flavinadenindinucleotid *n*
fade *v* abklingen (Akust); [aus]bleichen (Farbe), gelb färben (Farbe), verblassen (Farbe), welken (Bot), ~ **out** ausschwingen, im Schwung nachlassen
faded verblaßt
faden thermometer Fadenthermometer *n*
fadeometer Fadeometer *n,* Farbechtheitsprüfer *m*
fade-proof lichtecht
fading Bleichung *f* (Farbe), Fading *n* (Radio), Farbenschwund *m,* Nachlassen *n* (Farbe), Schwund *m* (Abklingen), Verfärben *n* (Farbe), Verschießen *n* (Farbe), **degree of** ~ Verschießungsgrad *m* (Text), **gas fume** ~ Ausbleichen *n* durch Abgase (pl)
fading control Fadingregulation *f* (Radio)
faecal fäkal
fagaramide Fagaramid *n*
fagarasterol Fagarasterol *n*
fagaric acid Fagarsäure *f*
fagarine Fagarin *n*
fagarol Fagarol *n*
Fageren flotation cell Fageren-Zelle *f*

faggoting (Br. E.) Paketieren *n*
fagine Fagin *n*
fagot Garbe *f,* Schweißpaket *n* (Techn)
fagot filling Faschinenausfüllung *f*
fagoting (Am. E.) Paketieren *n*
fahlband Fahlband *n*
fahl copper ore Fahlkupfererz *n*
fahlerz Fahlerz *n* (Min)
fahlore Fahlerz *n* (Min), tetraedrischer
Dystomglanz *m*
Fahrenheit temperature scale Temperaturskala *f*
nach Fahrenheit
Fahrenheit thermometer
Fahrenheit-Thermometer *n*
Fahrenwald flotation cell Fahrenwald-Zelle *f*
faience Fayence *f* (Tonware)
fail *v* fehlschlagen, scheitern, versagen
failing-film evaporator Fallfilmverdampfer *m*
failure Versagen *n*, Defekt *m*, Fehlschlag *m*,
Mißerfolg *m*, Panne *f,* Störung *f,* ~ **of**
machinery Maschinendefekt *m*
failure sequence Fehlsequenz *f* (Biochem)
faint matt, schwach, undeutlich
faintness Schwäche; Mattheit *f*
fairfieldite Fairfieldit *m* (Min)
fairly ziemlich, halbwegs
falcatine Falcatin *n*
fall Fall *m*, Absturz *m*, Gefälle *n* (eines Flusses),
Sturz *m*, **free** ~ freier Fall, **height of** ~
Fallhöhe *f,* **rate of** ~ Fallgeschwindigkeit *f*
(Phys)
fall *v* [ab]sinken, fallen, ~ **apart**
auseinanderfallen, ~ **down** herunterfallen,
abstürzen, ~ **in** einfallen (Strahlen), ~ **into**
pieces zusammenbrechen, ~ **off** abfallen, ~
out ausfallen, ~ **over** umfallen, ~ **through**
durchfallen
falling ball test Kugelfallprobe *f*
falling body tube Fallröhre *f*
falling film Fallfilm *m*, Rieselfilm *m*
falling film column Rieselfilmkolonne *f*
falling film cooler Fallfilmkühler *m*
falling film evaporator Fallfilmverdampfer *m*,
Fallstromverdampfer *m*
falling film scraped surface evaporator
Fallfilmkratzverdampfer *m*
falling-film scrubber Filmwäscher *m*
falling film vaporizer Fallfilmverdampfer *m*,
Fallstromverdampfer *m*
falling out Ausfall *m* (Haar)
falling tide Ebbe *f*
falling time Fallzeit *f*
fall line Fallinie *f*
fall-out Atomstaub *m* (Atom), Ausfallen *n*,
radioaktiver Niederschlag *m* (Atom)
fallow fahl, falb
fallow colored fahlfarben
fallowness Fahlheit *f*
fall test Fallprobe *f*

false falsch, unecht, künstlich
false bottom Zwischenboden *m*,
Doppelboden *m*, Läuterboden *m*
false diet Ernährungsfehler *m*
false light Falschlicht *n* (Opt)
falsely fälschlich
false pass Blindkaliber *n*
falsification Fälschung *f*
falsify *v* fälschen
falunite Falunit *m*
famatinite Famatinit *m* (Min)
familiarize *v* gewöhnen
family Schar *f* (Math)
family of curves Kurvenschar *f* (Math)
family of lines Geradenbüschel *n*
fan Fächer *m*, Flügel *m*, Gebläse *n*, Ventilator *m*
fan *v* anblasen; anfachen; [an]fächeln
fan blade Lüfterflügel *m*, Lüfterschaufel *f*
fan blower Flügelgebläse *n*, Schleudergebläse *n*
fan coupling Lüfterkupplung *f*
fan cupola Saugkupolofen *m*, Saugkuppelofen *m*
fancy modisch
fancy article Modeartikel *m*
fancy soap Feinseife *f*
fancy woven buntgewebt
fan drive Ventilatorantrieb *m*
fan engine Gebläsemaschine *f*
fan gate Angußverteiler *m*, Schirmanguß *m*
fangchinoline Fangchinolin *n*
fanlight Oberlicht *n*
fan shaft Flügelwelle *f*
fan-tailed burner Fächerbrenner *m*,
Rundstrahlbrenner *m*
fantastic phantastisch
far entfernt
farad Farad *n* (Elektr)
faraday Faraday *n* (Elektr)
Faraday cage Faradayscher Käfig *m*
Faraday constant Faradaykonstante *f* (Elektr)
Faraday equivalent Faraday-Äquivalent *n*
(elektrochem. Äquivalent)
Faraday's law of electromagnetic induction
Induktionsgesetz *n* von Faraday,
Faraday-Gesetz *n* der elektromagnetischen
Induktion (Elektr, Magn)
faradiol Faradiol *n*
farina Mehl *n*
farinaceous mehlig, kleienartig, mehlhaltig
farine Mehl *n*
farinose Farinose *f*
farinose *a* mehlhaltig
farming Ackerbau *m*
farm land Ackerland *n*
farnesal Farnesal *n*
farnesene Farnesen *n*
farnesenic acid Farnesensäure *f*
farnesiferol Farnesiferol *n*
farnesinal Farnesinal *n*

farnesol Farnesol *n*
farnesyl bromide Farnesylbromid *n*
farnesylic acid Farnesylsäure *f*
farnoquinone Farnochinon *n*, Vitamin K₂ *n*
faroelite Faroelith *m* (Min)
farrerol Farrerol *n*
fascine Faschine *f*
fascine filling Faschinenausfüllung *f*
fashion Fasson *f*, Mode *f*
fashion *v* bilden, fassonieren; (Text)
fashionable modisch
fashioning Fassonierung *f*, Formgebung *f*
fashion shade Saisonfarbe *f*
fast echt (Farbe), fest, haltbar, rasch, schnell,
    waschecht (Text), ~ to light farbtonbeständig,
    not ~ to light lichtunecht, ~ to rubbing
    abrußecht, ~ to staining abfleckecht, ~ to
    washing waschecht
fast acid fuchsin Echtsäurefuchsin *n*
fast acid ponceau Echtsäureponceau *n*
fast base Echtbase *f*
fast color Echtfarbe *f*
fast [color] base Echtbase *f*
fast cotton blue Echtbaumwollblau *n*
fast diazo color Diazoechtfarbe *f*
fast dyeing Echtfärben *n*
fasten *v* befestigen, anheften, anklammern,
    anknüpfen, festmachen, krampen, verfestigen,
    ~ the stitches abketten (Text), ~ with cement
    einkitten
fastening Befestigung *f*, Festmachen *n*
fastening clip Befestigungsschelle *f*
faster moving schnellerbewegt
fast fission effect Schnellspaltungseffekt *m*
    (Atom)
fast fission factor Schnellspaltungsfaktor *m*
fast flowing schnellfließend
fast green Echtgrün *n*
fast mordant color Echtbeizenfarbe *f*
fastness Echtheit *f*, Haltbarkeit *f* (von Farben
    (pl)), degree of ~ Echtheitsgrad *m*, ~ to
    acid[s] Säureechtheit *f*, ~ to scraping
    Scheuerfestigkeit *f*
fastness test Echtheitsprüfung *f* (Färb)
fast neutral violet Echtneutralviolett *n*
fast orange Echtorange *n*
fast ponceau Echtponceau *n*
fast pulley Festscheibe *f*
fast-scanning schnellabtastend
fast scarlet Echtscharlach *m*
fast steam color Echtdampffarbe *f*
fast yellow Echtgelb *n*
fat Fett *n*, containing ~ fetthaltig, determination
    of ~ Fettbestimmung *f*, hydrolysis of ~
    Fettspaltung *f*, oily ~ öliges Fett, oxidation of
    ~ Fettoxydation *f*, poor in ~ fettarm, removal
    of ~ Entfettung *f*, resembling ~ fettähnlich,
    solid ~ festes Fett, soluble in ~ fettlöslich,

storage of ~ Fettanlagerung *f*, Fettreservoir *n*,
    taste of ~ Fettgeschmack *m*
fat *a* fett
fat *v*, to cleave the ~ Fett spalten, to extract ~
    ausfetten, to saponify ~ Fett verseifen, to split
    the ~ Fett spalten
fatal tödlich, letal
fat alcohol Fettalkohol *m*
fata morgana Luftspiegelung *f* (Opt)
fat catching vessel Fettauffanggefäß *n*
fat cell Fettzelle *f*, Fettzelle *f*, Lipozyt *m*
fat cleavage Fettspaltung *f*
fat cleaving fettspaltend
fat cleaving agent Fettspalter *m*
fat content Fettgehalt *m*, determination of the ~
    Fettgehaltsbestimmung *f*, ~ of milk
    Milchfettgehalt *m*
fat deficiency Fettmangel *m*
fat deterioration Fettverderb *m*
fat dissolver Fettlösungsmittel *n*
fat dissolving fettlösend
fat dyestuff Fettfarbstoff *m*
fat-edge formation Tropfnasenbildung *f* (Lack)
fat elimination Entfetten *n*
fat extracting agent Fettextraktionsmittel *n*
fat extraction Fettentziehung *f*, Fettextraktion *f*,
    method of ~ Fettextraktionsverfahren *n*
fat formation Fettbildung *f*
fat hardening Fetthärtung *f*
fathom Klafter *f* (Maß)
fathom *v* loten
fatigue Ermüdung *f*, magnetic ~ magnetische
    Nachwirkung *f*, ~ of material
    Materialermüdung *f*, ~ under scrubbing
    Walkermüdung *f*
fatigue bending Dauerbiegung *f*
fatigue bending test Dauerbiegeversuch *m*
fatigue crack Daueranriß *m*, Ermüdungsriß *m*
fatigue durability Dauerhaltbarkeit *f*
fatigue failure Ermüdungsbruch *m*
fatigue fracture Ermüdungsbruch *m*
fatigue impact strength Dauerschlagfestigkeit *f*
fatigue impact test Dauerschlagversuch *m*
fatigue limit Dauerfestigkeitsgrenze *f*,
    Ermüdungsgrenze *f*
fatigue loading Dauerbeanspruchung *f*,
    Dauerschwingbeanspruchung *f*
fatigue phenomenon Ermüdungserscheinung *f*
fatigue-proof ermüdungsfrei
fatigue ratio Dauerfestigkeits-Verhältnis *n*
fatigue sign Ermüdungserscheinung *f*
fatigue strength Biegefestigkeit *f*,
    Dauerschwingungsfestigkeit *f*,
    Ermüdungsbeständigkeit *f*,
    Ermüdungsfestigkeit *f* (of large structures),
    Dauerhaltbarkeit *f*, flexural ~
    Biegungs-Dauerschwingfestigkeit *f*, ~ in
    tension Zugschwellfestigkeit *f*, ~ under
    alternating tensile stress Dauerfestigkeit *f* im

Zug-Wechselbereich, ~ **under pulsating [oscillating, fluctuating] compressive stress** Dauerfestigkeit *f* im Druck-Schwellbereich, ~ **under repeated (pulsating, fluctuating) bending stresses** Biegedauerfestigkeit *f* im Schwellbereich, ~ **under reversed bending stresses** Biegewechselfestigkeit *f*

fatigue strength diagram Dauerfestigkeitsschaubild *n*

fatigue strength for finite life Zeitfestigkeit *f*

fatigue tension test Zugversuch *m* mit Dauerbeanspruchung

fatigue test Ermüdungsversuch *m,* Dauerprüfung *f,* Dauerschwingversuch *m,* Wechselfestigkeitsprüfung *f,* ~ **at elevated temperature** Dauerschwingversuch *m* in der Wärme, ~ **at low temperature** Dauerschwingversuch *m* in der Kälte, ~ **under corrosion** Dauerschwingversuch *m* unter Korrosion, ~ **under fretting corrosion** Dauerschwingversuch *m* unter Reibkorrosion, ~ **under service condition** (at mixed amplitudes, at operational stresses or at working stress levels) Betriebs-Schwingversuch *m,* ~ **with rotating beam** Umlaufbiegeprüfung *f*

fatigue testing machine Dauerversuchsmaschine *f*

fatigue tube test Schlauchbiegeermüdungsprüfung *f*

fatigue [vibration] failure Dauerbruch *m,* Dauerschwingbruch *m*

fatiguing Altern *n*

fat-like fettähnlich, fettartig

fat lime Fettkalk *m,* Speckkalk *m*

fat liquor Fettlicker *m* (Gerbung), Fettschmiere *f*

fat metabolism Fettstoffwechsel *m* (Biochem)

fatness Fettigkeit *f*

fat pot Fettkessel *m*

fat removal Fettentziehung *f*

fat removing plant Entfettungsanlage *f*

fat scrapings Abstoßfett *n*

fat-soluble fettlöslich

fat solvent Fettlösungsmittel *n*

fat splitting Fettspaltung *f*

fat splitting *a* fettspaltend

fat splitting agent Fettspalter *m*

fat splitting plant Fettspaltanlage *f*

fat synthesis Fettsynthese *f*

fat-tight fettdicht

fat trap Fettauffanggefäß *n*

fatty fettartig, fettig

fatty acid Fettsäure *f,* **crude** ~ Rohfettsäure *f,* **emulsion of a** ~ Fettsäureemulsion *f,* **ester of a** ~ Fettsäureester *m,* **nitrated** ~ Nitrofettsäure *f,* **nonesterified** ~ unveresterte Fettsäure *f,* **of or combined with a** ~ fettsauer,

phenylated ~ Phenylfettsäure *f,* **polyunsaturated** ~ mehrfach ungesättigte Fettsäure *f,* **saturated** ~ gesättigte Fettsäure *f,* **unsaturated** ~ ungesättigte Fettsäure *f*

fatty acid distillation plant Fettsäuredestillationsanlage *f*

fatty acid ester Fettsäureester *m*

fatty acid pitch Fettpech *n*

fatty acid synthetase Fettsäuresynthetase *f* (Biochem)

fatty acid thiokinase Fettsäurethiokinase *f* (Biochem)

fatty substance Fettkörper *m,* Fettsubstanz *f*

fatty sweat Fettschweiß *m*

fatty yeast Fetthefe *f*

faucet Hahn *m,* Wasserhahn *m*

faujasite Faujasit *m* (Min)

fault Bruch *m* (Geol), Fehler *m,* Gußnarbe *f* (Gieß), Gußschaden *m* (Gieß), Makel *m,* Spalte *f* (Geol), ~ **in material** Materialfehler *m*

fault detection Fehlerauffindung *f*

faultiness Fehlerhaftigkeit *f*

faultless einwandfrei, fehlerfrei

faultlessness Fehlerlosigkeit *f*

fault localization Fehlerbegrenzung *f*

faulty fehlerhaft, mangelhaft, schadhaft

faulty alignment Fehlabgleichung *f*

fauna Fauna *f*

fauserite Fauserit *m* (Min)

favor Vergünstigung *f*

favor *v* begünstigen, bevorzugen

fawcettiine Fawcettiin *n*

fawn colored rehfarben

fayalite Fayalit *m* (Min), Eisenchrysolith *m* (Min), Eisenglas *n* (Min), Eisenolivin *m* (Min), Eisenperidot *m* (Min)

feasibility Ausführbarkeit *f,* Brauchbarkeit *f,* Durchführbarkeit *f*

feasible ausführbar, durchführbar

feather Feder *f*

feather agate Federachat *m* (Min)

feather alum Federalaun *m* (Min), Schieferalaun *m* (Min)

feather-like federähnlich, federartig

feather ore Federerz *n* (Min), Jamesonit *m* (Min)

feather-weigth, feather-light federleicht

feathery federartig

feature Eigenschaft *f,* Kennzeichen *n,* Merkmal *n*

feaze *v* ausfasern

febrifugal fiebermildernd (Med)

febrifuge Fieberarznei *f* (Med), Fiebermittel *n* (Med)

febrifugin Febrifugin *n*

fecal fäkal

fecal substances Fäkalien *pl,* Fäkalstoffe *m pl*

fecal water Fäkalwasser *n*

feces Fäkalien *pl*, Fäkalstoffe *m pl*, Kot *m*
Fecht acid Fechtsäure *f*
fecosterol Faecosterin *n*
fecula Bodenmehl *n*, Satzmehl *n*
fecundate *v* befruchten
Federal Health Office Bundesgesundheitsamt *n*
Federal Research Institute
    Bundesforschungsanstalt *f*
fedorwite Fedorwit *m*
fee Gebühr *f*, Vermittlungsgebühr *f*
feed Futter *n*, Ausgangsstoff *m*, Beschickung *f*,
    Vorschub *m*, Zufuhr *f*; Zulauf *m* (Dest),
    Zuleitung *f*, change of ~ Vorschubänderung *f*,
    range of ~ Vorschubbereich *m*
feed *v* füttern, beschicken, [ein]füllen, ernähren,
    speisen; zuführen
feedback Rückwirkung *f*, Rückführung *f*,
    Rückkopplung *f* (Elektr),
    Rückkopplungsleitung *f* (Elektr), elastic ~
    nachgebende Rückführung (Regeltechn)
feedback circuit Rückkopplungsschaltung *f*
feedback control system (Am. E.) Regelkreis *m*
feedback inhibition Endprodukthemmung *f*
    (Biochem)
feedback regulation Selbstregulierung *f*
feedback signal Rückkoppelungssignal *n*
feed box Eingußkasten *m*
feed bush Angußbuchse *f*
feed cable Speisekabel *n*
feed capillary Zuführungskapillare *f*
feed chute Fülltrichter *m*
feed coil Speisedrossel *f*
feed control Schaltsteuerung *f*
feed cylinder Füllzylinder *m*
feeder Aufgabevorrichtung *f*,
    Beschickungsanlage *f*, Einfüller *m*,
    Ladegerät *n* (Elektr), Speiseleitung *f* (Elektr),
    Zubringerband *n*, Zuleitung *f* (Elektr)
feeder head Vorschubkopf *m*
feeder line Zubringerleitung *f*
feed forward control Steuerung *f* (Regeltechn)
feed funnel Aufgabetrichter *m*
feed gas Betriebsgas *n*, Speisegas *n*
feed-gear mechanism Vorschubgetriebe *n*
feedhead Anguß *m* (Gieß)
feed heater Anwärmer *m*, Vorwärmer *m*
feed hole Führungsloch *n*
feed hopper Aufgabetrichter *m*,
    Beschickungstrichter *m*, Einfülltrichter *m*
feed hopper door Beschickungsklappe *f*,
    Beschickungstür *f*
feeding Beschickung *f*, Einleiten *n* (Techn),
    Speisung *f*, Zuführung *f*, Zuleitung *f*
feeding belt Aufgabeband *n*
feeding conveyor Zubringerband *n*
feeding device Aufgabevorrichtung *f*
feeding door Chargiertür *f*
feeding hopper Aufgabetrichter *m*,
    Aufschüttrichter *m*

feeding installation Beschickungsanlage *f*
feeding ladle Aufgießlöffel *m*
feeding plant Beschickungsanlage *f*
feeding plate Aufgabeblech *n*
feeding point Speisepunkt *m*, main ~
    Hauptspeisepunkt *m*
feed inlet Einlaßöffnung *f*, Materialzuführung *f*
feed line Speiseleitung *f*
feed load Zulaufbelastung *f*
feed magnet Nachschubmagnet *m*
feed material Ausgangsmaterial *n*
feed opening Einfüllöffnung *f*
feed orifice Anschnitt *m*
feed pipe Zuleitungsrohr *n*
feed piping Speiseleitung *f*
feed plate Aufgabeboden *m*, Einlaufboden *m*,
    Verteilboden *m*, Zulaufboden *m*
feed pump Förderpumpe *f*, Zubringerpumpe *f*
feed roll Vorschubrolle *f*
feed roller Vorschubwalze *f*, Zuführwalze *f*
feed scoop Schöpfbecher *m*
feed section Beschickungszone *f*
feed-stock Einsatzmaterial *n*, Einsatzprodukt *n*
feed strip Fütterungsstreifen *m*
feed supplement Futterzusatz *m*
feed temperature Speisetemperatur *f*
feed trough Zulauf *m* (Bergb)
feed valve Einlaßventil *n*
feed voltage Vorschubspannung *f*
feed water Speisewasser *n*
feed water connection Speisewasseranschluß *m*
feed water heater Vorwärmer *m* für Speisewasser
feed [water] pump Speisepumpe *f*
feed water regulator Wasserspeiser *m*
feed zone Beschickungszone *f*
feel *v* abtasten, [be]fühlen, tasten
feeler Abtaster *m*, Fühler *m*
feeler gauge Fühlerlehre *f*, Spion *m*
Fehling reagent Fehlingsches Reagenz *n*
Fehling's solution Fehlingsche Lösung *f*
feldspar Feldspat *m* (Min), changeable ~
    Schillerquarz *m* (Min), common ~ gemeiner
    Feldspat, containing ~ feldspathaltig,
    flesh-colored ~ edler Feldspat, glossy ~
    Glanzspat *m* (Min), green ~ Smaragdspat *m*
    (Min)
feldspathic feldspatähnlich, feldspathaltig
feldspathic rock Feldspatgestein *n* (Min)
feldspathic ware Hartsteingut *n*
felicur Felicur *n*
felinine Felinin *n*
felite Felith *m* (Min)
fell *v* [um]fällen
Fellgett advantage Multiplex-Vorteil *m*
fellitine Fellitin *n*
felloe Felge *f*, Radkranz *m*
fellowship Forschungsstipendium *n*
felly Felge *f*
felsite Felsit *m* (Min), Bergkiesel *m* (Min)

felsobanyite Felsobanyit *m* (Min)
felt Filz *m,* **asphalt ~ or roofing felt**
Dachpappe *f*
felt *v* verfilzen
felt blanket Filzdecke *f,* Filzunterlage *f*
felt buffing wheel Filzpolierscheibe *f*
felt cover Filzüberzug *m*
felted filzig, verfilzt
felt guide Filzführung *f*
felting Verfilzen *n*
felt insulation plate Filzisolierplatte *f*
felt jacket Filzschlauch *m,* Filzüberzug *m*
felt-like filzig
felt polishing wheel Filzpolierscheibe *f,*
Polierfilzscheibe *f*
felt washer Filzdichtung *f*
female die Matrize *f,* Gesenk *n,* Hohlform *f*
female ground joint Hülsenschliff *m*
female mold Matrize *f*
female mold method Negativverfahren *n*
female screw Hohlschraube *f,* Mutterschraube *f*
female thread Hohlgewinde *n,* Innengewinde *n,*
Muttergewinde *n*
feminell Feminell *n*
femur Oberschenkelknochen *m* (Anat)
fenchaic acid Fenchansäure *f*
fenchane Fenchan *n*
fenchene Fenchen *n*
fenchene hydrate Fenchenhydrat *n*
fenchenic acid Fenchensäure *f*
fenchenonic acid Fenchenonsäure *f*
fenchenylanic acid Fenchenylansäure *f*
fenchobornylene Fenchobornylen *n*
fenchocamphoric acid Fenchocamphersäure *f*
fenchocamphorol Fenchocamphorol *n*
fenchol Fenchol *n,* Fenchylalkohol *m*
fencholane Fencholan *n*
fencholauronolic acid Fencholauronolsäure *f*
fencholene amine Fencholenamin *n*
fencholenic acid Fencholensäure *f*
fencholic acid Fencholsäure *f*
fenchone Fenchon *n*
fenchosantenone Fenchosantenon *n*
fenchyl Fenchyl- *n*
fenchyl alcohol Fenchol *n,* Fenchylalkohol *m*
fenchylene Fenchylen *n*
fenethazine Fenethazin *n*
fennel Fenchel *m* (Bot)
fennel apple Fenchelapfel *m*
fennel brandy Fenchelbranntwein *m*
fennel odor Fenchelgeruch *m*
fennel oil Fenchelöl *n,* **common ~**
Bitterfenchelöl *n*
fenugreek Bockshornsamen *m* (Bot)
ferberite Ferberit *m* (Min)
ferghanite Ferghanit *n* (Min)
fergusonite Fergusonit *m* (Min)
ferment Enzym *n* (Biochem), Ferment *n*
(Biochem), Gärungserreger *m* (Biochem),

Gärungsmittel *n* (Biochem), Gärungsstoff *m*
(Biochem), **digestive ~** abbauendes Ferment
ferment *v* gären, aufbrausen, efferveszieren,
faulen, fermentieren, treiben, **~ again**
nachgären, **~ sufficiently** durchgären, **to ~**
completely endvergären
fermentability Gärungsfähigkeit *f,*
Gärvermögen *n,* Vergärbarkeit *f*
fermentable [ver]gärbar, gärungsfähig
fermentation Fermentation *f,* Gärung *f,*
Gärungsprozeß *m,* Vergärung *f,* **accelerated ~**
Schnellgärung *f,* **amylolytic ~**
Stärkefermentation *f,* **causing ~**
gärungserregend, **degree of ~**
Vergärungsgrad *m,* **delayed ~** Spätgärung *f,*
duration of ~ Gärdauer *f,* Gärungszeit *f,* **heat
of ~** Gärungswärme *f,* **high ~** Obergärung *f*
(Brau), **method of ~** Gärungsverfahren *n,*
Gärführung *f,* **putrefactive ~** faulende
Gärung, **quick ~** Schnellgärung *f,* **stopping ~**
gärungshemmend, **strong smell during ~**
Faulgeruch *m,* **time of ~** Gärdauer *f,* **to enter
into ~** in Gärung übergehen, **~ in the bottle**
Flaschengärung *f,* **~ of grapes**
Traubengärung *f,* **~ of sugar** Zuckergärung *f,*
~ under pressure Druckgärung *f*
fermentation amyl alcohol
Gärungsamylalkohol *m*
fermentation butyric acid Gärungsbuttersäure *f*
fermentation cellar Gärkeller *m*
fermentation chemistry Gär[ungs]chemie *f*
fermentation ethyl alcohol Gärungsalkohol *m*
fermentation fungus Gärungspilz *m*
fermentation gas Biogas *n*
fermentation lactic acid Gärungsmilchsäure *f*
fermentation method Faulverfahren *n* (Seide)
fermentation period Ausgärzeit *f*
fermentation process Gär[ungs]vorgang *m,*
Fermentierungsprozeß *m,* Gärverfahren *n*
fermentation product Gärerzeugnis *n,*
Gärprodukt *n,* Gärungsprodukt *n*
fermentation salt Gärsalz *n*
fermentation sample Gärprobe *f*
fermentation test Gärprobe *f*
fermentation vat Gärkasten *m,* Gärungsküpe *f*
fermentation yeast juice Hefepreßsaft *m*
fermentative gärungserregend, fermentativ
fermentative activity Gärtätigkeit *f*
fermentative power Gärungskraft *f,*
Gärvermögen *n*
ferment diagnostic Fermentdiagnostikum *n*
fermented vergoren (Brau), **~ from top** obergärig
(Brau)
fermenter Gärbottich *m,* Gärbütte *f*
fermenting Fermentieren *n,* Vergären *n*
fermenting *a* gärend, **highly ~** hochvergärend
fermenting acid Gärsäure *f*
fermenting agent Gärungsmittel *n,*
Gärungsstoff *m*

fermenting capacity Gärungsfähigkeit *f,*
Gärvermögen *n*
fermenting cellar Gärkeller *m*
fermenting house Gärraum *m*
fermenting pit Faulgrube *f*
fermenting power Gär[ungs]kraft *f*
fermenting room Gärkammer *f,* Gärkeller *m*
fermenting tank Gärtank *m*
fermenting temperature Gärtemperatur *f*
fermenting trough Faulbütte *f,* Gärbütte *f*
fermenting tub Gärbottich *m,* Gärbütte *f*
fermenting vat Gärbottich *m,* Gärbütte *f*
fermenting vessel Gärgefäß *n*
Fermi age model Fermizeittafel *f*
Fermi constant Fermikonstante *f*
Fermi contact term Fermi-Kontakt-Term *m*
fermi level Fermi-Niveau *n* (Phys)
Fermi limit Fermigrenze *f*
fermipile Fermireaktor *m* (Phys)
fermium Fermium *n* (Symb. Fm)
fermium isotope Fermiumisotop *n*
fermorite Fermorit *m* (Min)
fern Farnkraut *n* (Bot)
fernambuco wood Brasilholz *n*
fernandinite Fernandinit *m* (Min)
Fernholz acid Fernholz-Säure *f*
fern oil Farnkrautöl *n,* **sweet** ~ Comptoniaöl *n*
fern root Farnwurzel *f* (Bot)
feronialactone Feronialacton *n*
ferralum Ferralum *n*
Ferrari cement Ferrarizement *m*
ferrate Ferrat *n*
ferrate *a* eisensauer
ferratin Ferratin *n*
ferredoxin Ferredoxin *n*
ferreirin Ferreirin *n*
ferreous eisenhaltig
ferrescasane Ferrescasan *n*
ferriammonium citrate Eisen(III)-ammoncitrat *n,*
Ferriammoncitrat *n*
ferriammonium sulfate Ferriammonsulfat *n,*
Ammoniumferrisulfat *n,*
Eisen(III)-ammonsulfat *n*
ferriarsenite Eisen(III)-arsenit *n,* Ferriarsenit *n*
ferric Eisen(III)-, Ferri-
ferric acetate Eisen(III)-acetat *n,* Ferriacetat *n*
ferric alum Eisenalaun *m*
ferric ammonium citrate
Eisen(III)-ammoncitrat *n,* Ferriammoncitrat *n*
ferric ammonium sulfate
Eisenammoniakalaun *m,*
Eisen(III)-ammonsulfat *n*
ferric arsenite Eisen(III)-arsenit *n,* Ferriarsenit *n*
ferric bromide Ferribromid *n*
ferric chloride Eisen(III)-chlorid *n,*
Eisensesquichlorid *n,* Eisentrichlorid *n,*
Ferrichlorid *n,* **ethereal** ~ **tincture**
Goldtropfen *m pl*
ferric chloride wool Eisenchloridwatte *f* (Pharm)

ferric chromate Eisen(III)-chromat *n,*
Ferrichromat *n*
ferric citrate Eisen(III)-citrat *n,* Ferricitrat *n*
ferric compound Eisen(III)-Verbindung *f,*
Ferriverbindung *f*
ferric cyanide Eisen(III)-cyanid *n,* Ferricyanid *n*
ferric cyanoferrate(II) Ferriferrocyanid *n,*
Ferrocyaneisen *n*
ferric ferrocyanide Ferriferrocyanid *n,*
Berlinerblau *n,* Ferrocyaneisen *n,*
Preußischblau *n*
ferric fluoride Eisen(III)-fluorid *n,*
Eisentrifluorid *n*
ferric glycerophosphate
Eisen(III)-glycerophosphat *n*
ferrichloric acid Ferrichlorwasserstoff *m*
ferric hydroxide Eisen(III)-hydroxid *n,*
Eisenoxidhydrat *n* (obs), Ferrihydroxid *n*
ferric iodate Eisen(III)-jodat *n,* Ferrijodat *n*
ferric iodide Eisen(III)-jodid *n*
ferric ion Eisen(III)-ion *n,* Ferriion *n*
ferric lactate Ferrilactat *n*
ferric metaphosphate Eisen(III)-metaphosphat *n*
ferric nitrate Ferrinitrat *n*
ferric oxalate Eisen(III)-oxalat *n,* Ferrioxalat *n*
ferric oxide Eisen(III)-oxid *n,* Eisensesquioxid *n,*
Ferrioxid *n,* Polierrot *n,* **containing** ~
eisenoxidhaltig, **crystalline hydrated** ~
Hydrohämatit *m* (Min), **native** ~ Martit *m*
(Min), **poor in** ~ eisenoxidarm, **red** ~
Berlinerrot *n,* Caput mortuum *n*
ferric phosphate Ferriphosphat *n*
ferric potassium cyanide
Kaliumeisen(III)-cyanid *n,* rotes
Blutlaugensalz *n*
ferric potassium sulfate Ferrikaliumsulfat *n,*
Eisenkaliumalaun *m*
ferric rhodanate Ferrirhodanid *n*
ferric saccharate Eisenzucker *m*
ferric salt Eisen(III)-Salz *n,* Ferrisalz *n*
ferric sodium saccharate Natriumferrisaccharat *n*
ferric sulfate Ferrisulfat *n*
ferric sulfide Eisen(III)-sulfid *n,*
Eisensesquisulfid *n,* Ferrisulfid *n*
ferric tannate Eisen(III)-tannat *n,* Ferritannat *n*
ferric tartrate Eisen(III)-tartrat *n,* Ferritartrat *n*
ferric thiocyanate Eisen(III)-rhodanid *n,*
Ferrirhodanid *n,* Rhodaneisen *n*
ferric valerate Eisen(III)-valerianat *n*
ferricyanide Ferricyanid *n*
ferricyanogen Ferricyan *n*
ferriferous eisenführend, eisenhaltig,
eisenschüssig (Min)
ferriferous cassiterite Eisenzinnerz *n* (Min)
ferriferous quartz Eisenquarz *m* (Min)
ferriferous smithsonite Zinkeisenspat *m* (Min)
ferriferrocyanide Ferriferrocyanid *n,*
Ferrocyaneisen *n*
ferriferrous oxide Eisen(II)(III)-oxid *n*

ferriheme Eisenhäm *n*
ferrihemochromogen Eisenhämochromogen *n*
ferrihemoglobin Eisenhämoglobin *n*
ferrioxamine Ferrioxamin *n*
ferripotassium oxalate Kaliumferrioxalat *n*
ferripotassium sulfate Kaliumferrisulfat *n*
ferripyrine Ferripyrin *n*
ferrite Ferrit *m* (Min)
ferrite rod Ferritstab *m*
ferrite slab Ferritplatte *f*
ferrite sphere Ferritkugel *f*
ferrite yellow Eisenoxidgelb *n*
ferritic ferritisch (Stahl)
ferritin Ferritin *n*
ferrivine Ferrivin *n*
ferroalloy Ferrolegierung *f*
ferroaluminum Aluminiumeisen *n*
ferroammonium sulfate Ammoniumferrosulfat *n*,
  Ferroammonsulfat *n*
ferroboron Ferrobor *n*
ferrobronze Eisenbronze *f*
ferrocene Ferrocen *n*
ferrochrome Eisenchrom *n*, Ferrochrom *n*
ferrochromium Eisenchrom *n*, Ferrochrom *n*
ferroconcrete Eisenbeton *m*, Stahlbeton *m*
ferrocyanic acid Ferrocyanwasserstoff *m*,
  Eisenblausäure *f*
ferrocyanide Cyaneisenverbindung *f*,
  Ferrocyanid *n*, **potassium** ~
  Cyaneisenkalium *n*
ferrocyanogen Ferrocyan *n*
ferroelectric ferroelektrisch
ferroelectricity Ferroelektrizität *f*
ferroelectrics Ferroelektrika *n pl*
ferroelectric substances Ferroelektrika *n pl*
ferroferricyanide Ferroferricyanid *n*
ferrohemol Ferrohämol *n*
ferroin solution Ferroinlösung *f*
ferromagnesian limestone Eisenbitterkalk *m*
  (Min)
ferromagnesium Ferromagnesium *n*
ferromagnetic ferromagnetisch
ferromagnetics Ferromagnetika *n pl*
ferromagnetic substances Ferromagnetika *n pl*
ferromagnetism Ferromagnetismus *m*
ferromanganese Ferromangan *n*, Manganeisen *n*
ferromanganese peptonate
  Eisenmanganpeptonat *n*
ferromanganese saccharate
  Eisenmangansaccharat *n*
ferromanganese silicon Ferromangansilicium *n*
ferromanganese titanium Ferromangantitan *n*
ferromanganese tungstate
  Eisenmanganwolframat *n*
ferromolybdenum Molybdäneisen *n*,
  Eisenmolybdän *n*, Ferromolybdän *n*
ferronickel Eisennickel *n*, Ferronickel *n*,
  Nickeleisen *n*

ferrophosphorus Phosphoreisen *n*,
  Eisenphosphid *n*, Eisenphosphor *m*
ferrophosphorus nickel Phosphoreisennickel *n*
ferroprussiate process Eisenblaudruck *m*
ferropyrine Ferropyrin *n*
ferrosilico manganese Ferromangansilicium *n*,
  Siliciumspiegel *m*
ferrosilicon Ferrosilicium *n*, Siliciumeisen *n*
ferrosilicon furnace Ferrosiliciumofen *m*
ferrosilicon production Ferrosiliciumerzeugung *f*
ferrosilicon tapping Ferrosiliciumabstich *m*
ferrosoferric chloride Eisen(II)(III)-chlorid *n*
ferrosoferric iodide Ferriferrojodid *n*,
  Eisenjodürjodid *n* (obs)
ferrosoferric oxide Eisen(II)(III)-oxid *n*,
  Eisenoxidoxydul *n* (obs), Ferriferrooxid *n*
ferrotartrate Eisenweinstein *m*
ferrotartrate *a* eisenweinsteinsauer
ferrotitanite Ferrotitanit *m*
ferrotitanium Eisentitan *n*, Ferrotitan *n*
ferrotungsten Eisenwolfram *n*, Ferrowolfram *n*
ferrotype process Ferrotypie *f*
ferrous Ferro-, Eisen(II)-
ferrous *a* eisenhaltig, eisenschüssig (Min)
ferrous acetate Eisen(II)-acetat *n*, Ferroacetat *n*
ferrous ammonium sulfate Mohrsches Salz *n*,
  Eisenoxydulammonsulfat *n* (obs)
ferrous arsenate Ferroarsenat *n*
ferrous bromide Eisendibromid *n*,
  Eisenbromür *n* (obs), Ferrobromid *n*
ferrous carbonate Eisen(II)-carbonat *n*,
  Ferrocarbonat *n*
ferrous chloride Eisendichlorid *n*,
  Eisenchlorür *n* (obs), Eisen(II)-chlorid *n*,
  Ferrochlorid *n*
ferrous compound Eisen(II)-Verbindung *f*,
  Eisenoxydulverbindung *f* (obs),
  Ferroverbindung *f*
ferrous cyanide Ferrocyanid *n*, Eisencyanür *n*
  (obs), Eisen(II)-cyanid *n*
ferrous cyanoferrate(III) Ferroferricyanid *n*,
  Turnbulls Blau *n*
ferrous ferricyanide Ferroferricyanid *n*,
  Turnbulls Blau *n*
ferrous fluoride Eisendifluorid *n*,
  Eisen(II)-fluorid *n*
ferrous hydroxide Ferrohydroxid *n*,
  Eisenhydroxydul *n* (obs),
  Eisen(II)-hydroxid *n*
ferrous iodide Eisendijodid *n*, Eisen(II)-jodid *n*,
  Eisenjodür *n* (obs), Ferrojodid *n*
ferrous ion Eisen(II)-Ion *n*, Ferroion *n*
ferrous lactate Ferrolactat *n*, Ferrum lacticum
  (Lat)
ferrous manganite Bixbyit *m* (Min)
ferrous metallurgy Eisenindustrie *f*
ferrous nitrate Ferronitrat *n*
ferrous oxalate Ferrooxalat *n*

ferrous oxide Eisen(II)-oxid *n*, Eisenmonoxid *n*, Eisenoxydul *n* (obs), Ferrooxid *n*
ferrous phosphate Ferrophosphat *n*
ferrous potassium cyanide gelbes Blutlaugensalz *n*, Kalium-eisen(II)-cyanid *n*, Kaliumferrocyanid *n*
ferrous potassium sulfate Ferrokaliumsulfat *n*
ferrous potassium tartrate Ferrokaliumtartrat *n*
ferrous salt Eisen(II)-Salz *n*, Eisenoxydulsalz *n* (obs), Ferrosalz *n*
ferrous selenide Seleneisen *n*, Eisenselenür *n* (obs)
ferrous sulfate Eisen(II)-sulfat *n*, Ferrosulfat *n*, grüner Vitriol *n*
ferrous sulfide Eisen(II)-sulfid *n*, Eisenmonosulfid *n*, Eisensulfür *n* (obs), Ferrosulfid *n*, Schwefelkies *m* (Min)
ferrous thiocyanate Eisen(II)-rhodanid *n*, Eisenrhodanür *n* (obs)
ferrovanadium Eisenvanadium *n*, Ferrovanadium *n*, Vanadiumeisen *n*
ferroverdin Ferroverdin *n*
ferroxyl indicator Ferroxyl-Indikator *m*
ferrozirconium Ferrozirkonium *n*
ferrugin[e]ous limestone Kalkeisenstein *m* (Min)
ferruginol Ferruginol *n*
ferruginous eisenhaltig, eisenähnlich, eisenartig, eisenschüssig (Min)
ferruginous antimony Eisenantimon *n*
ferruginous calamine Eisenzinkspat *m* (Min)
ferruginous color Eisenrostfarbe *f*
ferruginous earth Eisenerde *f*
ferruginous jasper Eisenjaspis *m* (Min)
ferruginous magnesia limestone Eisenbitterkalk *m* (Min)
ferruginous mud Eisenschlamm *m*
ferruginous opal Eisenopal *m* (Min)
ferruginous quartz Eisenkiesel *m*
ferruginous quinine citrate Eisenchininicitrat *n*
ferruginous remedy Eisenarznei *f* (Pharm)
ferruginous sand Eisensand *m*
ferruginous tincture Eisentinktur *f*
ferrule Ringbeschlag *m*, Zwinge *f*
fertile fruchtbar
fertile material Brutstoff *m*, Spaltrohstoff *m* (Atom)
fertility Fruchtbarkeit *f*
fertilization Düngung *f*
fertilize *v* befruchten (Biol), düngen (Agr)
fertilizer Düngemittel *n;* Dünger *m*, Kunstdünger *m*, **artificial** ~ künstlicher Dünger, **complete** ~ Volldüngemittel *n*, **growth stimulating** ~ Reizdünger *m*, **mixed** ~ Mischdünger *m*
fertilizer factory Düngemittelfabrik *f*
fertilizer salt Düngesalz *n*
fertilizer urea Düngeharnstoff *m*
fertilizing Düngen *n*
fertilizing substance Dünger *m*

ferulaldehyde Coniferylaldehyd *m*, Ferulaaldehyd *m*
ferulene Ferulen *n*
ferulic acid Ferulasäure *f*
fervenuline Fervenulin *n*
fesogenin Fesogenin *n*
festoon dryer Hängetrockner *m*
festuclavine Festuclavin *n*
fetal fötal (Med)
fetch *v* abholen
fetid übelriechend
fetid cinnabar Stinkzinnober *m* (Min)
fetid marl Stinkmergel *m* (Min)
fetid stone Stinkstein *m* (Min)
fetterine Fetterin *n*
fetus Fötus *m* (Med)
Feulgen method Feulgentest *m*
feverfew Mutterkraut *n* (Bot)
feverfew oil Mutterkrautöl *n*
fiber (Am. E.) Faser *f*, Faden *m*, Fiber *f*, **artificial** ~ Kunstfaser *f*, Chemiefaser *f*, Kunstfaden *m*, **chemical** ~ Chemiefaser *f*, Chemiefaden *m*, **color of the** ~ Faserfärbung *f*, **man-made** ~ Chemiespinnfaser *f*, synthetische Faser *f*, **natural** ~ natürliche Faser *f*, **synthetic** ~ Kunstfaser *f*, synthetische Faser *f*, **vulcanized** ~ Vulkanitfaser *f*
fiber ash Faserasche *f*
fiber board [Holz]Faserplatte *f*, **coarse** ~ Heraklitplatte *f* (HN)
fiber content Fasergehalt *m*
fiber direction Faserrichtung *f*
fiber dust Faserstaub *m*
fiber end Faserende *n*
fiber fleece Faservlies *n*
fiber formation Faserbildung *f*
fiberglas (HN) Glasfaser *f*
fiberglas reinforcement Glasfaserverstärkung *f*
fiber insulation Faserisolation *f*
fiber-like faserartig, faserförmig
fiber material Fasermaterial *n*
fiber metallurgy Fasermetallurgie *f*
fiber molecule Fadenmolekül *n*
fiber optic Faseroptik *f*
fiber pattern Faserstruktur *f*, Fasermuster *n*
fiber pinion Fiberrad *n*, Fiberritzel *n*
fiber reinforcement Faserverstärkung *f*
fibers, to separate into ~ auffasern
fiber slurry Faseraufschlämmung *f*, Faserbrei *m*
fiber structure Faseraufbau *m*, Faserstruktur *f*
fiber suspension Fadenaufhängung *f*
fiber tip Faserende *n*
fibre (Br. E.) (see also fiber) Faser *f*, Faden *m*
fibril Fibrille *f*, kleine Faser *f*
fibrillation Faserbildung *f* (Metall), Faserung *f* (Biol), Fibrillierung *f* (Papier)
fibrin Fibrin *n*, Blutfaserstoff *m*
fibrin clot Fibringerinnsel *n*

fibrinogen Fibrinogen *n*, **lack or deficiency of** ~ Fibrinogenmangel *m*
fibrinogenesis Fibrinbildung *f*, Fibrinogenese *f*
fibrinogenous fibrinbildend
fibrinolysis Fibrinolyse *f*
fibrinoplastic fibrinoplastisch
fibrin plug Fibrinpfropf *m*
fibroblast Fibroblast *m* (Histol)
fibroblast culture Fibroblastenkultur *f*
fibrocellule Faserzelle *f* (Bot)
fibro-crystalline faserkristallin
fibrocyte Fibroblast *m* (Histol)
fibroferrite Fibroferrit *m* (Min)
fibroin Fibroin *n*
fibrolite Fibrolith *m* (Min), Buchholzit *m* (Min), Faserkiesel *m* (Min), Faserstein *m* (Min)
fibrolysin Fibrolysin *n*
fibrous faserig, fadenförmig, faserähnlich, faserartig, fibrös, gefasert, sehnig
fibrous asbestos Faserasbest *m*
fibrous material Faserstoff *m*
fibrous serpentine Metaxit *m* (Min)
fibrous substance Faserstoff *m*
fibrous tissue Fasergewebe *n*
ficin Ficin *n*
Fick's law Ficksches Gesetz *n*
ficoceryl alcohol Ficocerylalkohol *m*
fictitious unwirklich, fiktiv
ficusin Ficusin *n*
fidelity, high ~ hohe Wiedergabetreue *f* (Akust)
fiedlerite Fiedlerit *m* (Min)
field Acker *m* (Agr), Fachgebiet *n*, Feld *n* (Agr), Gebiet *n*, Sachgebiet *n*, **electromagnetic** ~ elektromagnetisches Feld, **geomagnetic** ~ erdmagnetisches Feld, **intensity of** ~ Feldstärke *f*, **magnetic** ~ magnetisches Feld, **non-uniform** ~ inhomogenes Feld, **strengthening of** ~ Feldverstärkung *f* (Elektr), **strength of** ~ Feldstärke *f*, ~ **of action** Wirkungsfeld *n*, ~ **of application** Verwendungsbereich *m*, ~ **of operation** Betriebsfeld *n*, ~ **of same or opposite polarity** gleichnamiges oder ungleichnamiges Feld *n*
field change Feldsprung *m*
field cypress Feldzypresse *f* (Bot)
field dependence Feldabhängigkeit *f*
field desorption Felddesorption *f*
field distortion Feldverzerrung *f*
field electron Feldelektron *n*
field emission Feldemission *f*
field emission microscope Feldemissionsmikroskop *n*
field emitter source Feldemissionsquelle *f*
field energy Erregerenergie *f*
field equation Feldgleichung *f*
field excitation Felderregung *f*
field experiment Feldversuch *m*
field fluctuation Feldschwankung *f*

field intensity Feldintensität *f*, Feldstärke *f*, **free-space** ~ Feldstärke *f* im freien Raum (Elektr)
field ion mass spectrometer Feldionen-Massenspektrometer *n*
field ion mass spectroscopy Feldionenmassenspektroskopie *f*
field magnet Feldmagnet *m*
field operation Geländearbeit *f*
field particle Feldteilchen *n* (Atom)
field pest Ackerschädling *m*
field quantization Feldquantelung *f*
field quantum Feldquant *n* (Atom)
field resistance Magnetwicklungswiderstand *m*
field reversal Feldumkehr *f* (Elektr)
field rheostat Feldwiderstand *m* (Elektr)
field strength Feldstärke *f*, **electrostatic** ~ elektrostatische Feldstärke, **magnetic** ~ magnetische Feldstärke
field system Feldsystem *n*, **multipolar** ~ mehrpoliges Feldsystem
field trial Anbauversuch *m* (Agr)
field voltage Feldspannung *f*
fiery feurig
fiery red knallrot
fig Feige *f* (Bot)
fig coffee Feigenkaffee *m*
figurative bildlich
figure Figur *f*, Abbildung *f*, Bild *n*, Nummer *f* (Math), Zahl *f* (Math); Ziffer *f* (Math)
figure *v* berechnen, fassonieren
figured agate Bilderachat *m* (Min)
fig wine Feigenwein *m*
filament Faden *m;* Faser *f*, Haar *n*, **synthetic** ~ Kunstfaden *m*
filamentary fadenförmig
filamentary transistor Fadentransistor *m* (Elektr)
filament battery Heizbatterie *f* (Elektr)
filament bridge Faserbrücke *f*
filament current Heizstrom *m*
filament electrode Strichelektrode *f*
filament growth method Aufwachsverfahren *n* (Met)
filament lamp Glühlampe *f*
filamentous faserig, faserartig, faserförmig
filament rheostat Heizwiderstand *m* (Elektr)
filament seal Einschmelzstelle *f* des Glühfadens
filament temperature Glühfadentemperatur *f*
filament winding Heizwicklung *f*, Wickelverfahren *n*
filament yarn Endlosgarn *n*
filaria (pl. filariae) Fadenwurm *m*
file Ablage *f* (Akten), Akte *f*, Feile *f* (Techn), **coarse** ~ Grobfeile *f*, **stroke of a** ~ Feilstrich *m*
file *v* abheften, einreichen, feilen (Techn), ~ **off** abfeilen (Techn)
file cover Aktendeckel *m*

file cut Feilenhieb *m*
file cutter Feilenhauer *m* (Techn)
filed, capable of being ~ feilbar
file dust Feilstaub *m*
file handle Feilenheft *n*
file hardening furnace Feilenhärteofen *m*
file number Aktennummer *f,* Aktenzeichen *n*
files, consumption of ~ Feilenverbrauch *m*
file tooth Feilenzahn *m*
filicic acid anhydride Filicin *n*
filicin Filicin *n*
filicinic acid Filicinsäure *f*
filiform fadenförmig
filiform corrosion Fadenkorrosion *f*
filigree [work] Filigran *n*
filing Feilspan *m*
filing cabinet Aktenschrank *m*
filing card Karteikarte *f*
filing machine Feilmaschine *f*
filings Feilspäne *m pl,* Feilicht *n,* Feilstaub *m*
filing system Registratursystem *n*
filipin Filipin *n*
filixic acid Filixsäure *f*
fill *v* [ab]füllen, ausfüllen, einfüllen, einnehmen,
  ~ up auffüllen, nachfüllen, tanken (Benzin),
  verstreichen (Fugen), vollfüllen
fill box Einfüllkasten *m*
filled [ab]gefüllt
filled band vollbesetztes Energieband *n*
filled lattice energy bands besetzte
  Gitterenergiebänder *pl*
filler Füller *m,* Füllmaterial *n,* Füllmittel *n,*
  Füllstoff *m,* Grundierfirnis *m* (Malerei),
  Kitt *m* (Email), Spachtelmasse *f,*
  Streckmittel *n,* Vergußmasse *f,* Zusatzmittel *n*
filler cap Füllstutzen *m*
filler gas Füllgas *n*
filler plate Füllplatte *f*
filler plug Füllschraube *f*
filler rod Schweißdraht *m,* Schweißstab *m*
filler speck Füllstoffnest *n*
filler strip Füllstreifen *m*
fillet Abrundung *f* von Übergängen,
  Ausrundung *f*
fillet radius Abrundungshalbmesser *m*
fillet weld Kehlnaht *f*
fillet welding Kehlschweißung *f*
filling Füllen *n,* Ausfüllmasse *f,* Einlage *f,*
  Füllmasse *f,* Füllmittel *n;* Füllung *f,* Plombe *f*
  (Zahnmed)
filling and counting machinery Abfüll- und
  Zählmaschinen *pl*
filling and dosing machine Abfüll- und
  Dosiermaschine *f*
filling and sealing machinery Abfüll- und
  Verschließmaschinen *pl*
filling apparatus Füllapparat *m*
filling compound Füllmasse *f*
filling cylinder Füllflasche *f*

filling device Abfüllvorrichtung *f,*
  Abfüllapparat *m*
filling factor Füllfaktor *m*
filling frame Aufsatzrahmen *m*
filling ladle Füllöffel *m*
filling level Füllstrich *m*
filling machine Abfüllmaschine *f* (f. Pulver etc.)
filling material Füllmaterial *n,* Füllkörper *pl,*
  Füllmittel *n,* Füllstoff *m*
filling piece Füllstück *n*
filling pipe Füllrohr *n*
filling plant Abfüllanlage *f* (f. Pulver etc.),
  Füllanlage *f*
filling plug Füllschraube *f*
filling pressure Fülldruck *m*
filling ratio Füllungsverhältnis *n*
filling screw Füllschraube *f*
filling station Tankstelle *f*
filling stock Füllmischung *f*
filling timber Füllspant *m*
filling-up color Spachtelfarbe *f*
filling vent Einfüllstutzen *m*
fillowite Fillowit *m* (Min)
fill-weigher Abfüllwaage *f*
film Film *m,* Folie *f,* Häutchen *n,* Kahm *m,*
  Überzug *m,* gaseous ~ molekulare
  Gasschicht *f,* monomolecular ~
  monomolekulare Schicht *f,* to wash the ~ den
  Film wässern, unimolecular ~
  monomolekulare Schicht *f,* welded ~
  geschweißte Folie *f*
film applicator Filmaufziehgerät *n,* Lackhantel *f*
filmaron Filmaron *n*
film base Filmträger *m* (Phot), Filmunterlage *f*
  (Phot)
film boiling Filmsieden *n,* Schichtsieden *n*
film casting Filmgießen *n*
film coating Befilmen *n*
film column Filmkolonne *f*
film condensation Filmkondensation *f*
film creep Kriechen *n* dünner
  Flüssigkeitsschichten (pl)
film dosimetry Filmdosimetrie *f*
film dryer Dünnschichttrockner *m*
film evaporation Dünnschichtverdampfung *f,*
  Filmverdampfung *f*
film evaporator Dünnschichtverdampfer *m,*
  Fallfilmverdampfer *m*
film flow Filmströmung *f,* Ringströmung *f*
film formation Hautbildung *f*
film former Filmbildner *m*
film forming filmbildend
film-forming agent Filmbildner *m*
film growth Schichtwachstum *n*
film illuminator Negativschaukasten *m*
filmogen Filmogen *n*
film printing Filmdruck *m*
film regeneration Filmregenerierung *f* (Phot)
film thickness Foliendicke *f*

film thickness measurement
Schichtdickenmessung *f*
filter Filter *n m*, Filtrierapparat *m*, Seiher *m*,
balancing ~ Ausgleichfilter *m*, **equalizing** ~
Ausgleichfilter *m*, **folded** ~ Faltenfilter *m*
(Chem), **glass** ~ Glasfilter *m*, **gravity** ~ mit
Doppelboden, **monochromatic** ~
monochromatisches Filter *n* (Opt), **residue on
the** ~ Filterrückstand *m*, **spherical** ~
Kugelfilter *m*, **ultra** ~ Ultrafilter *n*, **vacuum** ~
Vakuumfilter *n*, **yellow** ~ Gelbfilter *m* (Phot),
~ for condensation water
Kondenswasserfilter *n*, **~ for insertion**
Einsatzfilter *m*, **pressure leaf** ~ **with
centrifugal sluicing**
Zentrifugalreinigungs-Filter *n*
filter *v* filtrieren, abklären, abseihen,
durchgießen, filtern, kolieren, **~ by suction**
abnutschen, **~ by vacuum** absaugen, **~ off**
abfiltrieren, abfiltern, **~ through**
durchfiltrieren, **~ with suction** absaugen
filterable filtrierbar, filtrationsfähig, filtrierbar
filter aid Filterhilfsmittel *n*, Filtermittel *n*
filter area Filterfläche *f*
filter attachment Filtrieraufsatz *m*
filter bag Filterbeutel *m*, Filtersack *m*,
Filtriersack *m*
filter beaker Filtrierbecher *m*
filter bed Filterschicht *f*, Kläranlage *f*
filter bed scrubber Filterschichtenwäscher *m*
filter belt centrifuge Filterbandzentrifuge *f*
filter bottom Filterboden *m*
filter box Filterkasten *m*
filter cake Filterkuchen *m*
filter candle Filterkerze *f*
filter cartridge Filterkerze *f*
filter casing Filtergehäuse *n*
filter cell Filterelement *n*
filter chamber Filterkammer *f*
filter charcoal Filterkohle *f*
filter circuit Siebkreis *m* (Elektr)
filter cleaning machine
Filterreinigungsmaschine *f*
filter cloth Filtertuch *n*, Filtriertuch *n*,
Koliertuch *n*, Stoffilter *m*
filter cloth diaphragm Filtertuchdiaphragma *n*
filter cone Filterkegel *m*, Trichtereinlage *f*,
Trichtereinsatz *m*
filter crucible Filtertiegel *m*, Filtriertiegel *m*
filter disc Filterplatte *f*
filtered filtriert, abgeklärt, abgeseiht, gefiltert, ~
by suction abgenutscht
filter fabrics Filtergewebe *n*
filter flakes Filterflocken *pl*
filter flask Filterkolben *m*
filter frame Filterrahmen *m*, Kolierrahmen *m*
filter funnel Filtertrichter *m*
filter gauze Filtergaze *f*
filter glass Filterglas *n*

filter gravel Filterkies *m*
filter holder Filterhalter *m*, Filterträger *m*
filter hose Filterschlauch *m*
filtering Filtrieren *n*, Abklären *n*, Durchseihen *n*,
Filtern *n*, Kolieren *n*, **rate of** ~
Filtriergeschwindigkeit *f*, **time of** ~
Filtrierdauer *f*, **~ by suction** Abnutschen *n*, **~
off** Abfiltrierung *f*
filtering apparatus Filtrierapparat *m*,
Filtrationseinrichtung *f*
filtering area Filtrierfläche *f*
filtering auxiliary Filterhilfsmittel *n*
filtering bag Filtrierbeutel *m*
filtering basin Klärbecken *n*,
Reinigungsbehälter *m*
filtering basket Filterkorb *m*
filtering cloth Filtriertuch *n*
filtering earth Filtriererde *f*
filtering flask Absaug[e]flasche *f*, Nutsche *f*,
Saugflasche *f*
filtering layer Filterschicht *f*
filtering material Filtermaterial *n*, Filterstoff *m*
filtering pipette Filterstechheber *m*
filtering property Filtrierbarkeit *f*
filter[ing] sand Filtriersand *m*
filtering screen Siebfilter *m*
filtering sieve Filtersieb *n*
filtering stone Filtrier[kalk]stein *m*
filtering tub Filterbottich *m*
filtering tube Filterröhre *f*
filter mass Filtermasse *f*
filter media Filtermittel *n*
filter medium Filtermaterial *n*
filter-medium filtration Klärfiltration *f*
filter mud Filterschlamm *m*
filter pad Filterkissen *n*
filter paper Filterpapier *n*, Filtrierpapier *n*,
incineration of ~ Filterveraschung *f*
filter plate Filterplatte *f*, Filterscheibe *f*,
Siebplatte *f*
filter pocket Filtertasche *f*
filter press Filterpresse *f*, Filtrierpresse *f*,
Plattenfilter *n*, **electroosmotic** ~
elektroosmotische Filterpresse
filter-press-cake removing device
Filterkuchenabnahmevorrichtung *f*
filter press frame Filterpressenrahmen *m*
filter press plate Filterpressenplatte *f*
filter press ring Filterpreßring *m*
filter pump Filterpumpe *f*, Filtrierpumpe *f*
filter ring Filterhalter *m*, Filtrierring *m*
filter sand Filtriersand *m*
filter shifter Filterschieber *m*
filter sieve Filtriersieb *n*
filter slide Filterschieber *m*
filter stand Filtriergestell *n*, Filtrierstativ *n*
filter stick Eintauchnutsche *f*, Filterstäbchen *n*,
Saugstäbchen *n*
filter tank Reinigungsbehälter *m*

filter tower Filtrierturm *m*
filter tube Filterröhre *f,* Filtrierrohr *n*
filth Schmutz *m*
filthy schmutzig
filtrate Filtrat *n,* Durchgeseihtes *n*
filtrate *v* filtern, filtrieren
filtration Filtrierung *f,* Durchseihung *f,*
 Filterung *f,* Filtration *f,* **rapid** ~
 Schnellfiltrieren *n,* **rate of** ~
 Filtrationsgeschwindigkeit *f*
filtration chamber Filterraum *m*
filtration leaf Handfilterplatte *f*
filtration plant Filteranlage *f*
filtration residue Filterrückstand *m*
filtron Filtron *n*
fin Finne *f* (Zool), Flosse *f* (Zool),
 Walzenbart *m*
final endgültig, abschließend
final cleaning Nachreinigung *f*
final condenser Endkühler *m*
final control element Stellorgan *n* (Regeltechnik)
final filtering Nachfiltern *n*
final image luminescent screen
 Endbildleuchtschirm *m*
final liquor Endlauge *f*
final nucleus Endkern *m*
final pickle Glanzbrenne *f*
final position Endlage *f*
final product Enderzeugnis *n,* Endprodukt *n*
final purification Fertigfrischen *n* (Met)
final purifier Nachreiniger *m*
final remark Schlußwort *n*
final report Abschlußbericht *m,*
 Abschlußmeldung *f*
final result Endergebnis *n,* Endresultat *n*
final rolling Fertigwalzen *n*
final shape Endform *f*
final sintering operation Hochsintern *f*
final temperature Endtemperatur *f*
final test Endabnahme *f*
find *v* finden, antreffen
findings Befund *m*
fine fein, zart, dünn
fine adjustment Feineinstellung *f,*
 Feinregulierung *f*
fine-adjustment screw Feinstellschraube *f*
fine beam tube Fadenstrahlrohr *n*
fine casting Edelguß *m*
fine chemicals Feinchemikalien *pl*
fine coal Feinkohle *f*
fine control Feinregelung *f*
fine copper Feinkupfer *n*
fine crusher Feinbrecher *m*
fine crushing Feinzerkleinerung *f,* Feinmahlen *n*
fine-crystalline feinkristallin
fine-fibered feinfaserig
fine-fibrous feinfaserig
fine filter Feinfilter *n*
fine gold Feingold *n,* Kapellengold *n*

fine grain Feinkorn *n* (Phot)
fine grain developer Feinkornentwickler *m*
 (Phot)
fine-grained feinkörnig, feinfaserig
fine grain film Feinkornfilm *m* (Phot)
fine grain mixture Kleinkornmischung *f* (Beton)
fine granulation Feinbruch *m*
fine-grind *v* feinschleifen
fine grinder Feinmahlmaschine *f*
fine grinding Feinmahlen *n,* Feinschliff *m,*
 Feinzerkleinerung *f*
fine-ground finish Feinschliff *m*
finely crushed feingepulvert
finely crystalline kleinkristallin
finely dispersed feinverteilt
finely distributed feinverteilt
finely divided feinzerteilt
finely ground feingemahlen, feinzerrieben
finely ground charcoal Holzkohlenmehl *n*
finely pored feinporig
finely powdered feinpulverisiert, feinpulvrig,
 feinverteilt
finely pulverized feinmehlig, feinpulverisiert
fine-mechanical feinmechanisch
fine-meshed feinmaschig, kleinluckig
fine metal contact Edelmetallkontakt *m*
fineness Feingehalt *m* (z. B. Gold), Feinheit *f,*
 Feinheitsgrad *m,* **degree of** ~
 Feinheitsgrad *m,* Reinheitsgrad *m*
fine ore Feinkies *m,* **contents of** ~
 Feinerzgehalt *m*
fine polishing Hochglanzerzeugung *f*
fine range diffraction Feinbereichsbeugung *f*
fine reading Feinablesung *f*
fine regulation valve Feinregulierventil *n*
finer's metal Feineisen *n*
finery Frischerei *f* (Metall), Raffinerie *f,* **first**
 process of ~ Rohfrischen *n* (Metall)
finery cinders Eisenschlacke *f*
finery fire Frischfeuer *n*
finery iron Frischeisen *n* (Metall)
finery process Frischfeuerbetrieb *m* (Metall),
 Frischprozeß *m* (Metall)
fines Abrieb *m,* Erzstaub *m,* Feinkies *m,*
 Feinkohle *f,* Feinstkohle *f,* Grubenklein *n,*
 Verschnitt *m* (Mahlen)
fine sieve Feinsieb *n*
fine silver Feinsilber *n,* Kapellensilber *n*
fine smelting Feinarbeit *f* (Metall)
fine steel Edelstahl *m,* Herdstahl *m*
fine structure Feinstruktur *f,* Feingefüge *n*
fine structure constant Feinstrukturkonstante *f*
 (Spektr)
fine structure splitting Feinaufspaltung *f*
 (Spektr)
fine thread Feingewinde *n*
fine-threaded feinfädig
fine tin Feinzinn *n,* Klangzinn *n*
fine wood Edelholz *n*

fine working Feinarbeit *f*
finger Finger *m*
finger disk Wählscheibe *f*
finger nail test Daumennagelprobe *f*
finger paddle agitator Fingerrührer *m*
finger paddle mixer Fingerrührer *m*
finger plate Fingerplatte *f*
finger print Fingerabdruck *m*
fingerstone Belemnite *m* (Min)
fingertip dispenser Druckzerstäuber *m*
finger-type contact Fingerkontakt *m* (Elektr)
fining Frischen *n* (Metall), Klärung *f,* ~ **in two operations** Kaltfrischen *n*
fining process Frischarbeit *f* (Metall), Frischmethode *f* (Metall), Frischverfahren *n* (Metall)
fining slag Rohfrischschlacke *f*
fining work Gararbeit *f*
finish Politur *f,* Appretur *f* (Text), Avivage *f* (Färb), Deckanstrich *m,* Oberschicht *f,* **brushable textured** ~ Streichputz *m,* **dry** ~ rauhe Oberfläche *f* (Papier), **gel coat** ~ gelartiger Überzug *m,* **gloss** ~ glänzender Überzug *m,* **high** ~ Hochglanz *m,* **semigloss** ~ halbglänzender Überzug *m,* ~ **of setting** Erhärtungsende *n*
finish *v* beenden, appretieren (Text), fertigmachen, fertigstellen, garen (Chem), vollenden, ~ **extra bright** hochglanzpolieren, ~ **mashing** abmaischen, ~ **off** Schluß machen, abrunden
finish appearance Außenbeschaffenheit *f*
finished product Fertigteil *n,* Ganzfabrikat *n*
finisher Glänzer *m*
finishes, reflow or flow-back ~ Schwitzlacke *pl*
finishing Appretieren *n* (Text), Appretur *f* (Text), Avivage *f* (Färb), Fertigmachen *n,* Fertigstellung *f,* Veredeln *n* (Text), Vered[e]lung *f* (Text), Vollenden *n,* Zubereitung *f,* Zurichtung *f,* **unaffected by** ~ appreturecht
finishing agent Appreturmittel *n*
finishing bath Appreturbad *n*
finishing coat Deckanstrich *m,* Deckschicht *f,* Glattstrich *m* (Verputz), letzter Anstrich *m*
finishing drum Fertigwaschtrommel *f*
finishing industry Vered[e]lungsindustrie *f* (Text)
finishing lacquer Decklack *m*
finishing liquor Garbrühe *f*
finishing machine Abdrehmaschine *f* (Techn), Appreturmaschine *f* (Text), Feinmühle *f* (Pap), Schlichtmaschine *f* (Techn)
finishing-off room Fertigstellraum *m*
finishing paint Deckfarbe *f*
finishing process Appreturverfahren *n* (Text), Endbearbeitung *f*
finishing roasting Garrösten *n*
finishing roll Fertigwalze *f*
finishing rolling mills Feinwalze *f*

finishing rolls Fertigstraße *f,* Fertigstrecke *f,* Fertigwalzwerk *n*
finishing slag Endschlacke *f*
finishing temperature Abdarrtemperatur *f* (Brau)
finned pipe Rippenrohr *n*
finned tube Rippenrohr *n*
fiolaxglass Fiolaxglas *n*
fiorite Fiorit *m*
fir Tanne *f* (Bot)
fir cone oil Tannenzapfenöl *n*
fire Feuer *n,* Brand *m,* Feuerung *f,* **cascade of** ~ Feuerregen *m,* **danger of** ~ Feuergefahr *f,* **early stages of a** ~ Entstehungsbrand *m,* **liability to catch** ~ Feuergefährlichkeit *f,* **resistance against** ~ Feuerbeständigkeit *f,* **risk of** ~ Feuergefahr *f,* **roasting** ~ Garfeuer *n,* ~ **of reverberatory furnace** Streichfeuer *n*
fire *v* abfeuern (Schuß), anfeuern, backen (Keram), [be]feuern, heizen; kündigen; schießen, ~ **off** abschießen, ~ **to maturity** garbrennen
fire alarm Feueralarm *m* ;, Feuermelder *m*
fire-alarm system Feuermeldeanlage *f*
fire arch Frittofen *m*
fire arm Schußwaffe *f*
fire assay Brandprobe *f*
fire bar Roststab *m*
fire bar bearer Rostschwelle *f,* Rostträger *m*
fire bomb Brandbombe *f*
firebox Feuerkammer *f,* Feuerungsraum *m,* Heizkammer *f*
firebox plate Feuerblech *n*
fire brick feuerfester Ziegel *m,* Brennziegel *m,* Schamotteziegel *m*
fire bridge Feuerbrücke *f,* Herdbrücke *f*
fire chamber Feuerraum *m,* Heizraum *m,* Verbrennungsraum *m*
fireclay feuerfester Ton *m,* Feuerton *m,* Schamotte *f*
fireclay box Tonkiste *f*
fire[clay] brick Schamottestein *m*
fireclay crucible Schamottetiegel *m*
fireclay mortar Tonmörtel *m*
fireclay retort Schamotteretorte *f*
fire-colored feuerfarbig
fire damage Brandschaden *m*
firedamp Grubengas *n* (Bergb), Schlagwetter *n* (Bergb)
firedamp indicator Schlagwetteranzeiger *m*
firedamp tester Grubengasprüfer *m*
fire door Heiztür *f,* Heizloch *n*
fire drying apparatus Feuertrockner *m*
fire escape Feuerleiter *f,* Notausgang *m*
fire extinguisher Feuerlöscher *m,* Feuerlöschapparat *m*
fire extinguishing Feuerlöschen *n*
fire-extinguishing foam Löschschaum *m*
fire extinguishing substance Feuerlöschmittel *n*
firefighting installation Feuerschutzanlage *f*

fire flue boiler Flammrohrkessel *m*
fire gilding Feuervergoldung *f*
fire-gilt feuervergoldet
fire insurance Brandversicherung *f*
fire-lighter Feueranzünder *m*,
   Kohlenanzünder *m*
fire link Rostglied *n*
fireman Heizer *m*
fire marble Muschelmarmor *m* (Min)
fire opal Feueropal *m* (Min)
fireplace Feuerraum *m*, Heizraum *m*
fire-polish *v* feuerpolieren
fire-polished feuerpoliert
fire polishing Feuerpolitur *f* (Glas)
fire prevention Feuerverhütung *f*
fireproof feuersicher, brandfest, feuerbeständig,
   feuerfest, **highly** ~ hochfeuerfest
fireproof cement Brandkitt *m*
fireproof clay Schamotte *f*
fireproof color Scharffeuerfarbe *f*
fireproofing Feuerfestmachen *n*
fireproof lute Brandkitt *m*
fireproofness Feuerbeständigkeit *f*,
   Feuerfestigkeit *f*
fireproof paint Brandschutzanstrichfarbe *f*
fire protecting agent Feuerschutzmittel *n*
fire protection Brandschutz *m*, Feuerschutz *m*
fire rake Krücke *f*
fire-resistant quality Feuerbeständigkeit *f*,
   Feuerfestigkeit *f*
fire-resisting feuerbeständig, feuerhemmend
fire-resisting paint Brandschutzanstrichfarbe *f*
fire-resisting property
   Feuerwiderstandsfähigkeit *f*
fire-retardant feuerhemmend
fire-retardant paint Flammschutzmittel *n*
fire-retarding feuerhemmend
fire sheaf Feuergarbe *f*
fire silvering Feuerversilberung *f*
fire stain Feuerflecken *m*
firestone Feuerstein *m*
fire stop Feuerbrücke *f*
fire test Brandprobe *f*, Feuerprobe *f*
fire tube Feuerkanal *m*, Feuerrohr *n*,
   Flammrohr *n*, Heizrohr *n*
fire-tube boiler Heizröhrenkessel *m*,
   Rauchrohrkessel *m*
fire wall Brandmauer *f*
firewood Brennholz *n*, Holzscheit *n*
firework Feuerwerk *n*
fireworks Feuerwerkskörper *m pl*
firing Anfeuern *n*, Befeuerung *f*, Feuerung *f*,
   Zündung *f*, **type of** ~ Feuerungsart *f*, ~ **the**
   enamel Einbrennen *n* des Emails
firing door Einsatztür *f*
firing floor Heizerstand *m*
firing order Zündfolge *f*
firing plant Feuerungsanlage *f*
firing range Brennbereich *m*

firing tape Zündschnur *f*
firm fest, baufest, beständig, hart, konsistent,
   stabil, stark
firmness Beständigkeit *f*, Festigkeit *f*, **degree of**
   ~ Festigkeitsgrad *m*
firpene Firpen *n*
first absorption Vorwäsche *f* (durch Absorption)
first-aid box Verbandkasten *m*
first-aid post Unfallstation *f*
first breaking-up Rohaufbrechen *n*
first coat Voranstrich *m*, Grundierung *f*,
   Unterschicht *f*
first cost Anschaffungskosten *pl*
first evaporator Erstverdampfer *m*
first filter Vorfilter *m*
first harmonic current Grundstrom *m*
first harmonic voltage Grundspannung *f*
first light oil Vorlauf *m* (Dest)
first member Anfangsglied *n*
first pass Anfangskaliber *n*
first pickling Vorbeizen *n*
first principles Grundlehre *f*
first product Anfangsprodukt *n*, Vorprodukt *n*
first proof erster Abdruck *m* (Buchdr)
first running[s] Vorlauf *m* (Dest)
first shape Vorform *f*
first step Vorstufe *f*
first stuff Lumpenbrei *m* (Papier)
fir wood oil Fichtenholzöl *n*
fischerite Fischerit *m* (Min)
Fischer-Tropsch gasoline synthesis
   Fischer-Tropsch-Benzinsynthese *f*
fisetic acid Fisetin *n*
fisetin Fisetin *n*
fisetinidin Fisetinidin *n*
fisetinidol Fisetinidol *n*
fisetol Fisetol *n*
fish-bone Gräte *f*
fish eyes Fischaugen *pl* (Metall)
fisheye stone Fischaugenstein *m* (Min)
fish glue Fischleim *m*, Hausenblase *f*,
   Ichthyocoll *n*
fish grid Fischrechen *m*
fish-liver oil Lebertran *m*
fish manure Fischdünger *m*
fish meal Fischmehl *n*
fish oil Tran *m*, Fischöl *n*, **clear light-yellow** ~
   Helltran *m*
fish oil fatty acid Tranfettsäure *f*
fish oil soap Transeife *f*
fish-plate Lasche *f*, Schienenlasche *f*
fish-plate bolt Laschenbolzen *m*
fish-plate pass Laschenkaliber *n*
fish poison Fischgift *n*
fish scale Fischschuppe *f*
fishtail burner Flachbrenner *m*
fishtail twin Schwalbenschwanzzwilling *m*
   (Krist)
fisidan Fisidan *n*

fisidin Fisidin *n*
fisidylin Fisidylin *n*
fissile spaltbar
fissile material Spaltmaterial *n* (Atom)
fissility Spaltbarkeit *f*
fission Spaltung *f,* Aufspaltung *f* (Atom),
Teilung *f,* **fast** ~ Schnellspaltung *f* (Atom),
spontaneous ~ Spontanspaltung *f,* ~ **of cells**
Zellspaltung *f* (Biol), ~ **of molecules**
Molekularzertrümmerung *f*
fissionability Spaltbarkeit *f*
fissionable spaltbar, zerlegbar
fission chain reaction
Kernspaltungskettenreaktion *f*
fission chamber Spaltkammer *f* (Atom)
fission counting Spaltzählung *f*
fission cross section Spaltungsquerschnitt *m*
fission decay chain Spaltzerfallskette *f*
fission detection Spaltungsnachweis *m*
fission energy Spalt[ungs]energie *f* (Atom)
fission fragment Kernspaltungsfragment *n*
(Atom), Spaltbruchstück *n*
fission fragments Splattrümmer *pl* (Atom)
fission fungus Spaltpilz *m*
fission investigation Spaltungsuntersuchung *f*
fission material Spaltmaterial *n* (Atom)
fission monitor Spaltüberwachungsgerät *n*
fission neutron Spaltneutron *n* (Atom)
fission poison Spaltgift *n* (Atom)
fission process Spaltprozeß *m*
fission product Spalt[ungs]produkt *n* (Atom),
Spaltstück *n* (Atom), Zerfallsprodukt *n*
(Atom)
fission-product poisoning
Spaltproduktvergiftung *f* (Atom)
fission reaction Aufspaltreaktion *f*
fission reactor Kernspaltungsreaktor *m* (Atom),
Spaltreaktor *m* (Atom)
fission recoil particle Rückstoßteilchen *n* bei
Spaltung
fission threshold Spaltschwelle *f*
fission yield curve Spaltausbeutekurve *f*
fissure Spalt *m,* Einriß *m,* Riß *m,* Ritze *f,*
Spaltung *f,* Sprung *m,* **capillary** ~ Haarriß *m*
fissure *v* sich spalten
fissured rissig
fissures, full of ~ zerklüftet
fist size Faustgröße *f*
fistular röhrenartig
fit Passung *f,* Sitz *m,*
fit *a* passend, geeignet, tauglich
fit *v* passen, einpassen, montieren, sitzen,
zurichten, ~ **in** einfügen, ~ **into each other**
ineinanderpassen
fitchet fat Iltisfett *n*
fitness Brauchbarkeit *f,* Tauglichkeit *f*
fitted ausgerüstet
fitted bolt Paßschraube *f*
fitter Monteur *m,* Schlosser *m*

fitter cup Extraktionshülse *f*
fitting Einpassen *n,* Einrichten *n,* Montierung *f,*
Passung *f*
fitting *a* passend, geeignet, angebracht
fitting piece Kupplungsstück *n;* Paßstück *n;*
Rohrverbindung *f*
fitting pin Paßstift *m*
fitting ring Einsatzring *m*
fittings Armaturen *f pl,* Beschläge *m pl,*
Beschlagteile *pl*
fitting tolerance Paßtoleranz *f*
five component system Fünfstoffsystem *n*
five-electrode tube Fünfpolröhre *f* (Elektr),
Pentode *f*
five-figure fünfstellig
fivefold fünffach
five-membered fünfgliedrig
five-membered ring Fünfring *m* (Chem)
five-phase system Fünfphasensystem *n*
five-place fünfstellig
five-prong base Fünfsteckersockel *m*
five-sided fünfseitig
fix *v* befestigen, festlegen, festmachen, fixieren,
~ **completely** ausfixieren (Phot), ~ **in plaster**
eingipsen, ~ **together** zusammenfügen
fixable fixierbar
fixation Befestigen *n,* Bindung *f* (Chem),
Festmachen *n,* Fixieren *n* (Phot), Fixierung *f*
(Phot), ~ **of nitrogen** Stickstoffgewinnung *f*
fixation process Fixierprozeß *m* (Histol)
fixative Fixateur *m* (Parfümerie), Fixativ *n*
(Phot), Fixier[ungs]mittel *n* (Phot)
fixed fest angeordnet, feststehend, gebunden
(Chem), stationär, unbeweglich
fixed bed Schüttschicht *f*
fixed-bed adsorption unit
Festbettadsorptionsanlage *f*
fixed-bed catalyst fest angeordneter
Katalysator *m*
fixed head storage unit
Festkopfplattenspeicher *m* (Comp)
fixed ladle Standpfanne *f*
fixed plate Stammgesenk *n,* Stammplatte *f*
fixed point Festpunkt *m,* Fixpunkt *m*
fixed roll Festrolle *f*
fixed value Fixwert *m*
fixer Fixierungsmittel *n* (Phot), Fixativ *n*
(Phot), Fixierlösung *f* (Phot)
fixing Befestigen *n,* Fixieren *n* (Phot),
Fixierung *f* (Phot)
fixing agent Befestigungsmittel *n,*
Fixier[ungs]mittel *n* (Phot)
fixing apparatus Aufspannapparat *m*
fixing bath Fixierbad *n* (Phot),
Fixierflüssigkeit *f* (Phot)
fixing bolt Befestigungsschraube *f*
fixing clay Fixierton *m*
fixing device Befestigungsvorrichtung *f*
fixing liquid Fixierflüssigkeit *f*

fixing liquor Fixierflotte *f,* Fixierflüssigkeit *f*
fixing plate Befestigungsplatte *f*
fixing salt Fixiersalz *n* (Phot), Fixiernatron *n*
fixing solution Fixierlösung *f* (Phot), Fixativ *n*
(Phot)
fixing temperature Fixiertemperatur *f*
fixture Armatur *f,* Beschlag *m*
fizz Zischen *n*
fizz *v* aufschäumen, [auf]zischen, moussieren
fizzing zischend
flabby schlaff
flacon Fläschchen *n,* Flakon *m*
flagellar membrane Flagellenmembran *f*
flagellate Flagellat *m* (Zool) (pl flagellata),
Geißeltierchen *n*
flagellin Flagellin *n*
flagpole antenna Stabantenne *f*
flake Flocke *f,* Blättchen *n,* Plättchen *n,*
Schuppe *f*
flake *v* [ab]blättern, abschiefern, abschilfern,
[ab]schuppen, absplittern
flake graphite Flockengraphit *m* (Min)
flake ice Scherbeneis *n*
flake-like flockenartig
flake powder Blättchenpulver *n*
flaking Abblättern *n,* Abschälen *n,* Beflocken *n*
flaky flockenartig, flockig
flame Flamme *f,* **carbonizing** ~ reduzierende
Flamme, **explosive** ~ Stichflamme *f,* **fine
pointed** ~ Stichflamme *f,* **luminous** ~
aufleuchtende Flamme, **oxidizing** ~
oxidierende Flamme, **reducing** ~
reduzierende Flamme, Reduktionsflamme *f,*
roaring of the ~ Brausen *n* der Flamme,
sootless ~ nicht rußende Flamme
flame *v* flammen, ~ **up** aufflammen, aufglühen,
auflodern
flame arc Flammenbogen *m*
flame arrestor Flammenrückschlagsicherung *f*
flame carbon Effektkohle *f* (Elektr)
flame cleaning Flammentrosten *n,*
Flammstrahlen *n*
flame coloration Flammenfärbung *f*
flame cone Flammenkegel *m*
flame control device Flammenüberwachung *f*
flame cutting Brennschneiden *n*
flame descaling Flammentrosten *n,*
Flammstrahlen *n*
flame duct, upturned ~ Flammenstulpe *f*
flame emission continuum
Flammenemissionskontinuum *n*
flame emission spectrum Flammenspektrum *n*
flame front Flammenfront *f*
flame furnace Flammenofen *m*
flame hardening Flammenhärtung *f* (Metall),
Oberflächenhärtung *f* (Metall)
flame hole Flammenloch *n*
flame ionization detector
Flammenionisationsdetektor *m*

flameless flammenlos
flame line Flammenlinie *f*
flame photometer Flammenphotometer *n*
flame-polished feuerpoliert
flame polishing Flammpolieren *n*
flameproof flammensicher, nicht entflammbar
flameproof lamp Sicherheitslampe *f*
flameproofness Unentflammbarkeit *f*
flame resistance Flammwidrigkeit *f*
flame-resistant nicht entflammbar
flame-resistant impregnation
Flammfestausrüstung *f*
flame-retardant feuerhemmend,
Flammschutzmittel *n*
flame-retardant paint Flammschutzfarbe *f*
flame spectrography Flammenspektrographie *f*
flame spectrometry Flammenspektrometrie *f*
flame sprayed flammgespritzt
flame spraying Flammenspritzen *n,*
Flammspritzen *n,* Flammspritzverfahren *n*
flame test (for tin) Leuchtprobe *f* ((für Zinn)
Anal)
flame tester Flammensonde *f*
flame thrower Flammenwerfer *m*
flame tip Düse *f* (Gaschromat)
flame trap Rückschlagsicherung *f*
flame treatment process
Flammbehandlungsverfahren *n*
flame welding Flammenschweißen *n*
flaming flammend
flaming arc Flammenbogen *m*
flammability Brennbarkeit *f,* Entflammbarkeit *f,*
Entzündbarkeit *f*
flammable entflammbar, entzündbar
flange Flansch *m,* Bördel *n,* Randscheibe *f,*
brazed ~ Auflötflansch *m,* **small** ~
Kleinflansch *m,* **soldered** ~ Auflötflansch *m,*
threaded ~ Gewindeflansch *m,* ~ **of the rim**
Felgenhorn *n*
flange *v* [ein]flanschen, bördeln, ~ **on**
anflanschen
flange bolt Flanschenbolzen *m,*
Flanschenschraube *f*
flange closure Flanschverschluß *m*
flange coupling Flanschbefestigung *f,*
Scheibenkupplung *f*
flanged ball tee Kugel-T-Stück *n*
flanged bearing Flanschlager *n*
flanged connection Flanschverbindung *f*
flanged coupling Flanschverbindung *f*
flanged fitting Flanschenrohrformstück *n*
flanged globe tee Kugel-T-Stück *n*
flanged joint Flanschverbindung *f*
flanged pipe Flanschenrohr *n*
flanged plate Bördelblech *n,* Krempblech *n*
flanged reducing tee Verengerungs-T-Stück *n*
flanged ring Flanschenring *m*
flanged sheetwork Bördelarbeit *f*

flanged socket Flanschenmuffenstück *n*
(E-Stück)
flanged stopper Deckstopfen *m*
flanged tee T-Stück *n*
flange height Felgenhornhöhe *f*
flange joint Flanschverbindung *f*
flange lathe Flanschendrehbank *f*
flange ring Flanschring *m*
flange shaft Flanschwelle *f*
flange-type pipe Anflanschrohr *n*
flanging Bördeln *n*, Einflanschen *n*,
Umbördelung *f*
flanging machine Bördelmaschine *f*,
Flanschenaufwalzmaschine *f*, **automatic** ~
Bördelautomat *m*
flanging press Bördelpresse *f*
flanging roll Bördelwalze *f*,
Flanschenaufwalzmaschine *f*
flank of thread Gewindeflanke *f*
flank gear Zahnflanke *f*
flannel polishing wheel Flanellpolierscheibe *f*
flap Klappe *f*, Falltür *f*, Felgenband *n* (Reifen),
Leitzunge *f*
flapper Strahlabschneiderfahne *f*
flap valve Klapp[en]ventil *n*, Klappe *f*,
Klappenverschluß *m*
flare Fackel *f*, Flackern *n*
flare *v* flackern, ~ **up** aufflackern, aufflammen
flareback Flammenrückschlag *m*
flare cartridge Leuchtpatrone *f*
flash Abquetschgrat *m*, Aufblitzen *n*,
Entflammen *n*, Grat *m* (Schmiedegrat),
Preßgrat *m*
flash *v* auflodern, blitzen, entflammen, ~ **over**
überspringen (Elektr)
flashback Flammendurchschlag *m*,
Flammenrückschlag *m*
flash discharge Lichtblitzentladung *f*
flash distillation Kurzwegdestillation *f*,
Spontanverdampfung *f*
flash dryer Rohrtrockner *m*
flash evaporation Entspannungsverdampfung *f*,
Stoßverdampfung *f*, stoßartige Verdampfung
flash groove Abquetschnute *f*, Stoffabflußnute *f*
flash gun Blitzleuchte *f*
flash illumination Blitzbeleuchtung *f*
flashing Aufblitzen *n*, Überlaufnähte *pl*
(Spritzguß), ~ **over** Überspringen *n*
flashlight Blitzlicht *n* (Phot), Taschenlampe *f*
flashlight mixture Blitzlichtmischung *f*
flash line Gratlinie *f* (z.B. Preßgratlinie)
flash mold Abquetschwerkzeug *n*
flash-off Entlüfung *f*
flash-off period (time between spraying and
stoving) Abluftzeit *f*
flashover Überschlag *m* (Elektr)
flash [overflow] mold Abquetschform *f*
flashover voltage Überschlagsspannung *f*
flash photolysis Blitzlicht-Photolyse *f*

flash point Entzündungspunkt *m*,
Entflammungstemperatur *f*, Flammpunkt *m*,
Leuchtpunkt *m*
flash point apparatus Flammpunktprüfer *m*
flash point tester Flammpunktprüfer *m*
flash ridge Abquetschrand *m*
flash ring Abquetschrand *m;* Gratring *m*
flash roasting Sinterröstung *f*
flash test Flammpunktbestimmung *f*,
Entflammungsprobe *f*
flash welding Abschmelzschweißung *f*
flask Kolben *m* (Chem), Flasche *f* (Chem),
Glasballon *m* (Chem), **flat-bottom[ed]** ~
Stehkolben *m* (Chem), **half-liter** ~
Halbliterkolben *m* (Chem), **neck of a** ~
Kolbenhals *m* (Chem), **overflow** ~
Auslaufflasche *f*, **round[-bottom]** ~
Rundkolben *m* (Chem), **stoppered** ~
Stöpselflasche *f*, **tapered** ~ Spitzkolben *m*,
volumetric ~ Meßkolben *m*, **wide-necked** ~
Weithalskolben *m* (Chem), ~ **with flat bottom**
Standflasche *f*, Stehkolben *m*
flask casting Kastenguß *m*
flask stand Kolbenträger *m* (Chem)
flat flach, eben, abgestanden, fade, schal
flat bag Flachbeutel *m*
flat bar steel Flachstahl *m*
flatbed printing Flachbettdruck *m*
flat bending test Flachbiegeversuch *m*
flat billet Breiteisen *n*
flat bloom Bramme *f*
flat bottom rail Vignolschiene *f*
flat cable Bandseil *n*
flat coil Flachteilspule *f*
flat cross Flachkreuz *n*
flat face of a nut Schlüsselfläche *f* einer Mutter
flat fishplate Flachlasche *f*
flat flange joint Planschliffverbindung *f*
flat furnace mixer Flachherdmischer *m*
flat grate Planrost *m*
flathead screw Flachkopfschraube *f* (Am. E.),
Senkschraube *f*
flat lacquer Mattlack *m*
flatness Flachheit *f*
flat paint Mattanstrichfarbe *f*
flat pliers Flachzange *f*
flat polish Mattpolitur *f*
flat rate Pauschalgebühr *f*
flat rope Bandseil *n*
flat sheet Flachtafel *f*
flat sieve Plansieb *n*
flat spray nozzle Flachstrahldüse *f*
flat spring Bandfeder *f*, Blattfeder *f*
flat steel Flacheisen *n*
flat stopper Deckelstopfen *m*
flatten *v* abflachen, abplatten, flachdrücken,
flachpressen, glätten, plätten
flattened abgeflacht, abgeplattet

flattening Abflachung *f,* Abplattung *f,* ~ **out** Ausbreitung *f*
flattening test Ausbreiteprobe *f*
flatting agent Mattierungsmittel *n*
flatting varnish Schleiflack *m*
flat varnish Mattlack *m*
flat wedge Flachkeil *m*
flavacidin Flavacidin *n*
flavacol Flavacol *n*
flavan Flavan *n*
flavaniline Flavanilin *n*
flavanol Flavanol *n*
flavanomarein Flavanomarein *n*
flavanone Flavanon *n*
flavanookanin Flavanookanin *n*
flavanthrene Flavanthren *n*
flavanthrine Flavanthrin *n*
flavanthrone Flavanthron *n*
flavaspidic acid Flavaspidsäure *f*
flavazine Flavazin *n*
flavazole Flavazol *n*
flavellagic acid Flavellagsäure *f*
flavene Flaven *n*
flavianic acid Flaviansäure *f*
flavicin Flavicin *n*
flavin dehydrogenase Flavindehydrogenase *f* (Biochem)
flavinduline Flavindulin *n*
flavin[e] Flavin *n*
flavin[e] adenine dinucleotide Flavinadenindinucleotid *n*
flavin[e] mononucleotide Flavinmononucleotid *n*
flavin reduction Flavinreduktion *f*
flavin synthesis Flavinsynthese *f*
flaviolin Flaviolin *n*
flavipin Flavipin *n*
flavochrome Flavochrom *n*
flavocoryline Flavocorylin *n*
flavogallol Flavogallol *n*
flavoglaucin Flavoglaucin *n*
flavognost Flavognost *n*
flavokinase Flavokinase *f* (Biochem)
flavol Flavol *n*
flavomycoin Flavomycoin *n*
flavone Flavon *n*
flavonol Flavonol *n*
flavopereirine Flavopereirin *n*
flavophene Flavophen *n*
flavophenine Chrysamin *n*
flavoprotein Flavoprotein *n*
flavopurpurin Flavopurpurin *n*
flavor (Am. E.) Beigeschmack *m,* Geschmack *m,* **aromatic** ~ Würzgeschmack *m,* **development of** ~ Aromaentwicklung *f,* **spicy** ~ Würzgeschmack *m*
flavor *v* würzen
flavoring Geschmackstoff *m,* Würze *f*
flavoring essence Aromakonzentrat *n*
flavorless geschmackfrei, geschmacklos

flavotine Flavotin *n*
flavour (Br. E.) Geschmack *m* (see also flavor)
flavoxanthin Flavoxanthin *n*
flaw Fehler *m,* Bruch *m,* Einriß *m,* Fehlstelle *f* (Schweiß), Gußblase *f,* Gußnarbe *f,* Gußschaden *m,* Härteriß *m,* Makel *m,* Riß *m,* Sprung *m,* ~ **in material** Materialfehler *m,* ~ **in the casting** Gußfehler *m*
flawlessness Fehlerlosigkeit *f*
flaws, free from ~ rißfrei
flax Flachs *m,* Lein *m*
flaxen flachsfarben, flachsgelb
flaxseed oil Flachssamenöl *n*
flax wax Flachswachs *n*
flaxweed salve Leinsalbe *f*
flaxy flachsfarben
fleabane Flohkraut *n* (Bot)
fleece Schurwolle *f,* Vlies *n*
fleecy wollähnlich, wollig
flesh Fleisch *n*
flesh-colored fleischfarben, fleischrot
flesh meal Fleischmehl *n*
flesh side Fleischseite *f*
flex *v* walken (Reifen)
flex cracking Biegerisse *pl,* Ermüdungsrisse *pl* (Reifen)
flex fatigue Biegeermüdung *f*
flexibility Biegbarkeit *f,* Biegefähigkeit *f,* Biegsamkeit *f,* Dehnbarkeit *f,* Dehnfähigkeit *f,* Elastizität *f,* Geschmeidigkeit *f,* Schmiegsamkeit *f*
flexible biegsam, beweglich, biegungsfähig, dehnbar, elastisch, geschmeidig, schmiegsam
flexible connection Federungskörper *m,* flexible Verbindung *f*
flexible coupling Ausdehnungskupplung *f,* Gelenkkupplung *f*
flexible metal tube Metallschlauch *m*
flexible package Folienpackung *f,* Weichpackung *f*
flexible sheet Weichfolie *f*
flexible tubing Schlauch *m*
flexing, resistance to ~ Walkwiderstand *m*
flexing area Walkzone *f* (Reifen)
flexinine Flexinin *n*
flexion Beugung *f,* Biegung *f*
flexographic printing Gummidruck *m*
flexographic printing inks Gummidruckfarben *pl*
flex resisting ermüdungstüchtig
flexural elasticity Biegungselastizität *f*
flexural endurance properties Dauerbiegefestigkeit *f*
flexural strain Biegebeanspruchung *f,* Biegespannung *f*
flexural strength Biegefestigkeit *f,* Dauerbiegefestigkeit *f,* ~ **at maximum deflection** Grenzbiegespannung *f*
flexural stress Biegebeanspruchung *f,* Biegungsbeanspruchung *f*

flexure Beugung *f,* **elasticity of** ~
  Biegungselastizität *f,* **moment of** ~
  Biegungsmoment *n,* **plane of** ~
  Biegungsebene *f,* **strength of** ~
  Biegungsfestigkeit *f*
flexure mode Biegungsschwingung *f*
flicker *v* flackern
flickering Geflacker *n*
flicker photometer Flimmerphotometer *n*
flight, line of ~ Flugbahn *f,* Fluglinie *f*
flighted dryer Rieseltrockner *m*
flight path Flugbahn *f,* Wurfbahn *f*
flindersiamine Flindersiamin *n*
flindersine Flindersin *n*
flindersinic acid Flindersinsäure *f*
fling *v* schleudern, werfen, ~ **back**
  zurückschnellen, ~ **out** herausschleudern
flinkite Flinkit *m* (Min)
flint Feuerstein *m,* Flint *m,* Kiesel *m,* **crushed** ~
  Flintmehl *n,* **hard as** ~ kieselhart
flint glass Flintglas *n,* Kristallglas *n*
flint glass insulator Flintglasisolator *m*
flint glass prism Flintglasprisma *n*
flint hardness Kieselhärte *f*
flint meal Flintmehl *n*
flintrock Kieselfels *m*
flintstone Kiesel *m*
flinty kieselartig, felsenhart, feuersteinartig
flinty earth Kieselerde *f* (Min)
flinty slate Kieselschiefer *m* (Min)
flip-flop circuit Kippschaltung *f*
flip-over process Umklappprozeß *m* (Elektr)
flipper Kernfahne *f*
flipper stitcher Kernfahnenanrollvorrichtung *f*
flipping machine Maschine *f* zum Umlegen der
  Ummantelung
flitch plates Universaleisen *n*
float Schwimmkugel *f*
float *v* schwimmen, flotieren, schweben
floatable schwimmfähig, flotationsfähig
floating Ausschwimmen *n* (Pigmente),
  Schwimmen *n*
floating *a* schwimmend, treibend, schwebend
floating control Integralregelung *f*
floating crane Schwimmerkran *m,*
  Schwimmkran *m*
floating knife Luftrakel *f*
floating knife coater Luftrakelstreichmaschine *f*
floating knife coating
  Luftrakelstreichverfahren *n*
floating magnet Schwebemagnet *m*
floating plate Zwischenplatte *f*
floating roll Tänzerrolle *f*
floating speed Stellgeschwindigkeit *f*
floating thermometer Schwimmthermometer *n*
float needle valve Schwimmernadelventil *n*
floatstone Schwimmstein *m* (Min)
float valve Schwimmerventil *n,* Vergaserventil *n*
flocculant Flockungsmittel *n*

flocculate *v* ausflocken
flocculating agent Flockungsmittel *n*
flocculation Ausfällung *f,* Ausflockung *f,*
  Flockenbildung *f*
flocculation point Ausflockungspunkt *m*
flocculation tendency Ausflockbarkeit *f*
flocculent flockenartig, flockig
flocculent structure Flockenstruktur *f*
flock Flocke *f* (Wolle)
flock-coating Beflocken *n* (ganze Oberfl.)
flock paper Flockenpapier *n,* Samttapete *f*
flock-printing Beflocken *n* (Muster),
  Flockdruck *m*
flocky flockig
flokite Flokit *m* (Min)
flong Mater *f*
flood Flut *f,* Überschwemmung *f*
flood *v* überfließen, ausschwemmen,
  anschwöden (Gerb)
flooding Ausschwimmen *n* (Pigmente),
  Überflutung *f,* Überschwemmung *f,*
  [commencement of] ~ Flutbeginn *m* (Dest)
flooding point Flutbelastung *f* (Dest),
  Flutgrenze *f* (Dest), Flutpunkt *m,*
  Staupunkt *m*
flood light Flutlicht *n,* Scheinwerferlicht *n*
flood point Flutpunkt *m*
floor Boden *m,* Fußboden *m,* Stockwerk *n*
floorborne vehicles Flurfördermittel *n*
  (Flurförderer)
floor contact Tretkontakt *m*
floor covering Bodenbelag *m,*
  Fußbodenbelag *m,* **carrier material for**
  needled felt ~ Grundgewebe *n* für
  Nadelvliesbodenbelag
floor finish Fußbodenanstrich *m,*
  Fußbodenlack *m*
floor frame Stehbock *m*
flooring Bodenbelag *m,* Oberboden *m*
flooring cement Fußbodenkitt *m*
flooring oil Glanzöl *n*
flooring plaster Estrichgips *m*
floor level Bodenhöhe *f*
floor paint Fußboden[grundier]farbe *f*
floor painting Bodenanstrich *m*
floor panel Fußbodenplatte *f*
floor polish Fußbodenpflegemittel *n*
floor space Grundfläche *f*
floor stand Stehbock *m*
floor tile Fußbodenfliese *f,* Fußbodenplatte *f*
floor wax Bohnerwachs *n*
flora Flora *f* (Bot), **abnormal bacterial** ~
  Dysbakterie *f* (Med), **intestinal** ~
  Darmflora *f*
florantyrone Florantyron *n*
florence stone Landschaftsstein *m* (Min)
florencite Florencit *m* (Min)
Florentine lake Florentiner Lack *m*
Florentine marble Landschaftsstein *m* (Min)

Florentine receiver Florentiner Flasche *f,*
Ölvorlage *f* (Chem)
Florida bleaching earth Floridableicherde *f*
Florida phosphate Floridaphosphat *n*
floridin Floridableicherde *f*
floridoside Floridosid *n*
florigen Florigen *n*
floropryl Floropryl *n*
flos ferri Faseraragonit *m* (Min)
flotation Flotation *f,* Naßaufbereitung *f,*
Schwemmverfahren *n,* **axis of** ~
Schwimmachse *f*
flotation agent Flotationsmittel *n,*
Schwebemittel *n*
flotation machine, agitation-type ~
Rührwerksapparat *m*
flotation method Flotationsverfahren *n*
flotation oil Trägeröl *n*
flotation plant Flotationsanlage *f*
flotation process Aufschlämmverfahren *n,*
Flotationsverfahren *n*
flotation save-all Flotationsstoffänger *m*
(Papier)
flour Mehl *n,* **superfine** ~ Auszugmehl *n*
flour dust Mehlstaub *m*
flour mill by-products Mühlennachprodukte *pl*
flour tester Mehlprüfer *m*
floury mehlig
flow Fließen *n,* Ausfluß *m,* Durchfluß *m,*
Durchflußmenge *f,* Fluß *m,* Strömung *f,*
Strom *m,* **cold** ~ kalter Fluß *m* (Prüftechnik),
compressible ~ kompressible Strömung *f,*
direction of ~ Durchflußrichtung *f,*
Strömungsrichtung *f,* **laminar** ~ laminare
Strömung *f,* **mechanism of** ~
Strömungsvorgang *m,* **plastic** ~ plastisches
Fließen *n,* plastische Strömung *f,* **process of** ~
Strömungsvorgang *m,* **rate of** ~
Durchflußgeschwindigkeit *f,*
Durchströmungsgeschwindigkeit *f,*
Fließgeschwindigkeit *f,* **stationary** ~
stationäre Strömung *f,* **steady** ~ stationäre
Strömung *f,* **telescopic** ~ laminares Fließen *n,*
tendency to ~ Fließfähigkeit *f,* **time of** ~
Durchflußzeit *f,* **turbulent** ~ turbulente
Strömung *f,* **uniform** ~ gleichförmige
Strömung *f* (Mech), **velocity of** ~
Strömungsgeschwindigkeit *f,* ~ **of electrons**
Elektronenfluß *m*
flow *v* fließen, rinnen, strömen, ~ **back**
zurückfließen, zurückströmen, ~ **in**
einfließen, einströmen, ~ **into** einmünden, ~
off abfließen, ~ **out** ablaufen, ausfließen,
auslaufen, austreten, ~ **over** überströmen, ~
through durchfließen, durchströmen
flowability Fließfähigkeit *f*
flowable fließbar
flow apparatus Ausflußapparat *m*
flow area Durchflußquerschnitt *m*

flow back valve Rückschlagventil *n*
flow behavior Fließverhalten *n*
flow-brightening Glanzschmelzen *n*
(Wiederaufschmelzen, z.*VB. der Verzinnung
von Blech)
flow cell Durchflußzelle *f*
flow chain Durchflußkette *f*
flow chart Fließdiagramm *n,* Schaubild *n*
flow coating Flutlackieren *n,* Gießen *n* (Lack)
flow controller Durchflußregler *m*
flow diagram Strömungsdiagramm *n*
flow equation Strömungsgleichung *f*
flower Blüte *f* (Bot), Blume *f* (Bot)
flower-cup Kelch *m* (Bot)
flower pigment Blütenfarbstoff *m*
flowers, essence of ~ Blütenöl *n*
flowers of salt Salzblumen *f pl*
flow glaze Laufglasur *f*
flow guard Strömungswächter *m*
flow head Gießkopf *m*
flow heater, continuous ~ Durchlauferhitzer *m*
flow improver Verlaufmittel *n* (Lack)
flowing Strömen *n,* Fließen *n,* Fluten *n,*
Flutlackieren *n,* ~ **through** Durchfluß *m*
flowing furnace Blauofen *m,* Flußofen *m*
flowing water Fließwasser *n*
flow-in pressure Einströmdruck *m*
flow limit Fließgrenze *f*
flow line Fließlinie *f,* Schweißnaht *f*
flow mark Fließlinie *f,* Fließmarkierung *f*
flow measurement Durchflußmessung *f,* ~ **by
pressure drop** Durchflußmessung *f* aus
Druckverlust [an Einbauten], ~ **with
rotameters** Durchflußmessung *f* mit
Schwimmkörpern
flow medium Strömungsmittel *n*
flowmeter Durchflußmengenmeßgerät *n,*
Durchflußmesser *m,* Fließprüfer *m,*
Flüssigkeitszähler *m,* Strömungsmesser *m*
flow method Durchströmungsmethode *f*
flow mixer Durchflußmischer *m,*
Flüssigkeitsmischer *m*
flow model Strömungsmodell *n*
flow molding Preßspritzen *n*
flow-off valve Abzugsklappe *f*
flow pattern Fließdiagramm *n,* Strömungsbild *n*
flow point Fließpunkt *m*
flow pressure Fließdruck *m*
flow process Fließvorgang *m*
flow properties Fließvermögen *n,*
Verlaufseigenschaften *pl*
flow property Fließeigenschaft *f*
flow rack Durchlaufregal *n*
flow rate Durchflußrate *f,* Durchsatzmenge *f,*
Strömungsgeschwindigkeit *f*
flow reactor Durchflußreaktor *m*
flow recorder Mengenschreiber *m*
flow setting Zuflußeinsteller *m*

flow sheet Fließbild *n*, Arbeitsschema *n*,
Betriebsschema *n*, Fließschema *n*,
Schemazeichnung *f*
flow stress Fließwiderstand *m*, **change in ~**
Fließsprung *m*
flow system, return ~ Gegenflußsystem *n*
flow temperature Fließtemperatur *f*
flow test Ausbreiteprobe *f*, **continuous ~**
Durchflußversuch *m*
flow tube Strömungsrohr *n*
flow type Strömungstyp *m*
flow-type electrode Durchflußelektrode *f*
flow vector Flußvektor *m*
flow velocity Durchflußgeschwindigkeit *f*
fluate Fluat n, Fluorosilicat
fluavil Fluavil *n*
fluctuate *v* schwanken, abwechseln, fluktuieren
fluctuating schwankend, fluktuierend,
unregelmäßig
fluctuation Schwankung *f*, Fluktuation *f*, **violent**
~ Sprunghaftigkeit *f*, **~ of pressure**
Druckschwankung *f*
fludrocortisone acetate Fludrocortisonacetat *n*
flue Rauchfang *m*, Abzug *m* für Abgase,
Abzugsschacht *m*, Esse *f*, Feuerrohr *n*,
Kamin *m*, Rauchabzugskanal *m*, Schlot *m*,
Schornstein *m*
flue ashes Flugasche *f*
flue boiler Flammrohrkessel *m*
flue bridge Fuchsbrücke *f*
flue dust Flugasche *f*, Flugstaub *m*,
Gichtgasstaub *m*, Gichtstaub *m*,
Hüttenrauch *m*, **deposit of ~**
Flugascheablagerung *f*
flue dust briquetting plant
Gichtstaubbrikettierungsanlage *f*
flue dust separation Gichtstaubabscheidung *f*
flue gas Rauchgas *n*, Abgas *n*, Abzugsgas *n*,
Fuchsgas *n*, Röstgas *n*, Verbrennungsgas *n*,
cleaning of ~ Rauchgasreinigung *f*
flue gas analysis Rauchgasanalyse *f*
flue gas analyzer Abgasprüfgerät *n*,
Rauchgasanalysator *m*
flue gases, composition of ~
Rauchgaszusammensetzung *f*
flue gas explosion Rauchgasexplosion *f*
flue gas flap Rauchgasklappe *f*
flue gas heating Rauchgasvorwärmung *f*
flue gas preheater Rauchgasvorwärmer *m*
flue gas tester Rauchgasprüfer *m*
flue gas testing apparatus
Rauchuntersuchungsapparat *m*
flue gas utilization Rauchgasverwertung *f*
fluellite Fluellit *m* (Min)
flue roof Fuchsdecke *f*
fluffy flaumig, flockig
fluid Flüssigkeit *f*, **Newtonian ~**
Newtonsche Flüssigkeit *f*, reinviskose
Flüssigkeit *f*, **refractive ~** lichtbrechende

Flüssigkeit, **thixotropic** ~ thixotrope
Flüssigkeit, **weight of a** ~
Flüssigkeitsgewicht *n*
fluid *a* flüssig, fließend
fluid balance Flüssigkeitshaushalt *m* (Physiol)
fluid bed Fließbett *n*, Wirbelschicht *f*
fluid bed catalysis Fließbettkatalyse *f*
fluid bed cooler Schwebekühler *m*
fluid bed process Fließbettverfahren *n*
fluid coking Fließkoksverfahren *n*
fluid content Flüssigkeitsgehalt *m*
fluid displacement Flüssigkeitsverdrängung *f*
fluid energy mill Strahlmühle *f*
fluid extract Fluidextrakt *m*
fluid flow in open channels Kanalströmung *f*
fluid flow stagnation point Staupunkt *m* der
Flüssigkeitsströmung
fluidification Fluidifikation *f*
fluid inlet Einlaßstutzen *m*
fluid intake Flüssigkeitsaufnahme *f*
fluidity Fließfähigkeit *f*, flüssiger Zustand *m*,
Fluidität *f*, Zähigkeitskehrwert *m*
fluidization, bubbling or aggregative ~
inhomogene Fluidisation *f*,
Gemischwirbelung *f*
fluidization process Staubfließverfahren *n*
fluidize *v* aufwirbeln (Wirbelschichtverfahren),
fluidifizieren, verflüssigen
fluidized bed Fließbett *n*, Wirbelschicht *f*
fluid[ized] bed apparatus Wirbelschichtanlage *f*
fluid[ized] bed dryer Fließbetttrockner *m*,
Wirbelschichttrockner *m*
fluid[ized] bed furnace Wirbelschichtofen *m*
fluid[ized] bed mixer Fließbettmischer *m*
fluid[ized] bed plant Wirbelschichtanlage *f*
fluid[ized] bed process Wirbelschichtverfahren *n*,
Wirbelsinterverfahren *n*
fluidized bed reactor Wirbelschichtreaktor *m*
fluidized catalyst Fließbettkatalysator *m*,
Fließkontakt *m*
fluidized dust Flugstaub *m*
fluidized electrode fluidisierte Elektrode *f*
fluidized flow process Wirbelfließverfahren *n*
fluidizing classifier Wirbelbettstromklassierer *m*
fluid lubrication Schwimmreibung *f*
fluid metal Flußmetall *n*
fluid meter Flüssigkeitsmesser *m*,
Flüssigkeitszähler *m*
fluid output Flüssigkeitsausscheidung *f*
fluid packing Dichtung durch Sperrflüssigkeit
fluid pressure Flüssigkeitsdruck *m*
fluid replacement Flüssigkeitsersatz *m*
fluid requirement Flüssigkeitsbedarf *m* (Physiol)
fluid reservoir Ausgleichsbehälter *m*,
Flüssigkeitsreservoir *n*
fluids, intake and output of ~
Flüssigkeitshaushalt *m* (Physiol)

fluid-type reforming plant
Fließbettreformieranlage *f*
fluo-aluminic acid
Aluminiumfluorwasserstoffsäure *f*
fluobenzene Fluorbenzol *n*
fluoborate fluorborsaures Salz *n*
fluoborate *a* borflußsauer, fluorborsauer
fluoboric acid Borflußsäure *f*
fluocerite Fluocerit *m* (Min)
fluoflavine Fluoflavin *n*
fluoform Fluoroform *n*
fluogermanic acid
Germaniumfluorwasserstoffsäure *f*
fluolite Fluolith *m* (Min)
fluophosphate Fluorphosphat *n*
fluoradene Fluoraden *n*
fluoran[e] Fluoran *n*
fluoranil Fluoranil *n*
fluoranthene Fluoranthen *n*
fluoranthene quinone Fluoranthenchinon *n*
fluorapatite Fluorapatit *m* (Min)
fluorazene Fluorazen *n*
fluorazol Fluorazol *n*
fluor earth Flußerde *f*
fluorenacene Fluorenacen *n*
fluorenaphene Fluorenaphen *n*
fluorene Fluoren *n*
fluorenone Fluorenon *n*
fluorenyl Fluorenyl-
fluoresce *v* fluoreszieren, schillern
fluorescein Fluorescein *n*, Resorcinphthalein *n*
fluorescence Fluoreszenz *f*, Fluoreszieren *n*,
  quenching of ~ Fluoreszenzlöschung *f*,
  secondary ~ Sekundärfluoreszenz *f*, **sensitized**
  ~ Sekundärfluoreszenz *f*
fluorescence analysis Fluoreszenzanalyse *f*
fluorescence equipment Fluoreszenzgeräte *n pl*
fluorescence spectroscopy
Fluoreszenzspektroskopie *f*
fluorescence spectrum Fluoreszenzspektrum *n*
fluorescence yield Fluoreszenzausbeute *f*,
Fluoreszenzstrom *m*
fluorescent fluoreszierend
fluorescent glow Fluoreszenzleuchten *n*
fluorescent hand lamp Leuchtstoffhandlampe *f*
fluorescent ink Leuchtfarbe *f*
fluorescent lamp Fluoreszenzlampe *f*,
Leuchtstofflampe *f*
fluorescent material Leuchtstoff *m*
fluorescent paint Fluoreszenzfarbe *f*
fluorescent response
Fluoreszenzansprechvermögen *n*
fluorescent screen Fluoreszenzschirm *m*,
Leuchtschirm *m*, Röntgenschirm *m*
fluorescent tube Leucht[stoff]röhre *f*
fluorescin Fluorescin *n*
fluorescyanine Fluorescyanin *n*
fluorhydrate Hydrofluorid *n*
fluorhydric fluorwasserstoffsauer

fluorhydrocortisone Fluorhydrocortison *n*
fluoridation Fluoridierung *f*
fluoride Fluorid *n*, flußsaures Salz *n*
fluoride complex Fluoridkomplex *m*
fluoride of lime Calciumfluorid *n*
fluorimeter Fluoreszenzmesser *m*, Fluorimeter *n*,
Fluorometer *n*
fluorimetric fluorimetrisch
fluorimetry Fluorimetrie *f*
fluorinate *v* fluorieren
fluorination Fluorierung *f*
fluorindene Fluorinden *n*
fluorindine Fluorindin *n*
fluorine Fluor *n*, **containing** ~ fluorhaltig
fluorine compound Fluorverbindung *f*
fluorine content Fluorgehalt *m*
fluorite Flußspat *m* (Min), Fluorit *m* (Min),
  bituminous ~ Stinkflußspat *m* (Min)
fluorite lattice Flußspatgitter *n* (Krist)
fluoritic flußspathaltig
fluoroacetic acid Fluoressigsäure *f*
fluorobenzene Fluorbenzol *n*
fluorocarbon Fluorkohlenstoff *m*
fluorocitrate Fluorcitrat *n*
fluorocyclene Fluorocyclen *n*
fluoroethylene Fluoräthylen *n*
fluoroform Fluoroform *n*
fluorography Fluoreszenzaufnahme *f*
fluoroid Fluoroid *n*
fluorometer Fluorometer *n*,
Fluoreszenzmesser *m*, Fluorimeter *n*
fluorometric fluorimetrisch
fluorometry Fluorimetrie *f*
fluorone Fluoron *n*
fluorophore Fluorophor *n*
fluoroprene Fluoropren *n*
fluoroprene rubber Fluoroprenkautschuk *m*
fluorosalicylaldehyde Fluorsalicylaldehyd *m*
fluoroscope Fluoroskop *n*
fluoroscopy Fluoroskopie *f*
fluorotoluene Fluortoluol *n*
fluorspar Fluorit *m* (Min), Flußspat *m* (Min),
  bituminous ~ Stinkflußspat *m* (Min), **compact**
  ~ Flußstein *m* (Min), **fetid** ~
  Stinkflußspat *m* (Min)
fluorspar powder Flußspatpulver *n*
fluorubine Fluorubin *n*
fluorylidene Fluoryliden *n*
fluosilicate Fluat n, Fluorsilikat,
Fluorsiliciumverbindung *f*,
Siliciumfluorverbindung *f*, Silicofluorid *n*
fluosilicate *a* flußkieselsauer,
kieselfluorwasserstoffsauer
fluosilicic acid Fluorkieselsäure *f*,
Kieselflußsäure *f*,
Siliciumfluorwasserstoffsäure *f*, **salt of** ~
Fluorsilicat *n*, Silicofluorid *n*
fluotantalate *a* fluortantalsauer
fluotantalic acid Fluortantalsäure *f*

fluotitanic acid Titanfluorwasserstoffsäure *f,*
Titanflußsäure *f*
fluotoluene Fluortoluol *n*
flush *v* ausschwemmen, [aus]spülen
flush box Spülkasten *m*
flushing Spülen *n,* Ausschwemmen *n,*
Flushingverfahren *n* (Web), Spülung *f*
flushing gate Spülventil *n*
flushing hole Schlackenstichloch *n*
flushing water Spülwasser *n*
flush ring Einbauring *m*
flush weld Flachnaht *f*
flute Hohlkehle *f,* Rille *f,* Rinne *f*
flute *v* auskehlen, kannelieren, riefeln
fluted ausgekehlt, gerillt, rinnenförmig
fluted bar iron Hohlkanteisen *n*
fluted screen centrifuge Faltensiebzentrifuge *f*
flute length Spirallänge *f* (Bohrer)
fluting Hohlkehle *f,* Kannelierung *f,* Kehlung *f*
fluting plane Kehlhobel *m*
fluttering contact Flackerkontakt *m*
flux Fluß *m,* Flußmittel *n,* Lötsalz *n,*
Schmelzmittel *n,* Zuschlag *m* (Metall), **to need
no ~** kahl gehen, **~ for making slag**
Schlackenbildner *m*
flux *v* schmelzen, verschlacken
flux converter Flußumwandler *m*
flux density Flußdichte *f,* Kraftflußdichte *f*
flux depression Flußabsenkung *f*
flux distortion Flußverzerrung *f*
fluxing Schmelzen *n,* Weichmachen *n*
fluxing agent Flußmittel *n*
fluxing medium Flußmittel *n*
fluxing ore Zuschlagerz *n* (Metall)
fluxing power Verschlackungsfähigkeit *f*
flux linkage Flußverkettung *f*
fluxmeter Durchflußmeßgerät *n,*
Strommesser *m* (Elektr)
flux oil Restöl *n,* Stellöl *n*
flux powder Flußpulver *n*
flux refraction Brechung *f* der Flußlinien (pl)
(Magn)
flux variation, total ~
Gesamtkraftlinienänderung *f*
fly agaric Fliegenpilz *m*
fly amanita Fliegenpilz *m*
fly around *v* herumfliegen,
durcheinanderfliegen
fly ash Flugasche *f*
flying brands Flugfeuer *n,* **fires caused by ~**
Flugfeuer *n*
flying jib Spitzenausleger *m*
fly paper Fliegenfänger *m*
fly poison Fliegengift *n*
fly powder Fliegenpulver *n*
fly stone Fliegenstein *m* (Min)
fly trap Fliegenfalle *f*
flywheel Schwungrad *n* (Mech)
flywheel drive Schwungradantrieb *m*

flywheel rim Schwungradkranz *m*
flywheel steam pump Schwungraddampfpumpe *f*
foam Bierschaum *m* (Brau), Schaum *m,* **height
of ~** Schaumhöhe *f,* **prevention of ~**
Schaumverhütung *f,* **rigid ~** fester Schaum, **to
cover with ~** beschäumen
foam *v* schäumen, gischen, **~ over**
überschäumen, **~ up** aufschäumen,
aufgischen
foam breaker Schaumbrecher *m*
foam concrete Schaumbeton *m*
foamed plastic Kunstschwamm *m,*
Schwammkunststoff *m*
foam extinguisher Schaumfeuerlöscher *m*
foam flow Schaumströmung *f*
foam formation Schaumbildung *f*
foam glue Schaumkleber *m*
foam holding schaumbeständig
foaminess Schäumigkeit *f*
foaming Schäumen *n,* Gebrodel *n,*
Schaumbildung *f,* Schaumentwicklung *f,*
prevention of ~ Entschäumung *f,* **to cease ~**
aufhören zu schäumen, ausgischen
foaming *a* schäumend
foaming agent Blähmittel *n,* Schaummittel *n,*
Treibmittel *n* (Schaumstoff)
foaming property Schaumbildungsvermögen *n,*
Schaumfähigkeit *f*
foaming quality Schaumbildungsvermögen *n*
foam inhibition Schaumzerstörung *f*
foamless schaumlos
foam-like schaumähnlich
foam mixing chamber Schaummischkammer *f*
foam plastics Schaumstoff *m*
foam rubber Schaumgummi *m,* Latexschaum *m,*
Zellgummi *m*
foam skimmer Schaumabstreifer *m*
foam suppressor Antischaummittel *n*
foam tube Schaumröhre *f*
foamy schaumig, schaumähnlich
foam[y] plastic Schaum[kunst]stoff *m*
focal axis Brennachse *f* (Opt)
focal collimator Meßkollimator *m* (Opt)
focal distance Brennpunktabstand *m,*
Brennweite *f*
focal length Brennpunktabstand *m,*
Brennweite *f,* Fokuslänge *f,* **front ~**
dingseitige Brennweite (Opt), **setting of the ~**
Brennweiteneinstellung *f,* **~ of the eyepiece**
Brennweite *f* des Okulars
focal line Brennlinie *f* (Opt)
focal plane Bildebene *f,* Brenn[punkt]ebene *f*
focal point Brennpunkt *m* (Opt), **principal ~**
Hauptbrennpunkt *m*
focal spot characteristic Brennfleckausdehnung *f*
focimeter Brennweitenmesser *m,* Fokometer *n*
(Phys), Fokusmesser *m*
focometer (see also focimeter) Fokometer *n*
(Phys)

focus Brennpunkt *m* (Opt), Fokus *m* (Opt),
　　Herd *m* (Mittelpunkt), Sammelpunkt *m*
　　(Opt), **astigmatic** ~ astigmatischer
　　Brennpunkt (Opt), **depth of** ~ Fokustiefe *f*
　　(Phot), **in** ~ scharf eingestellt, **out of** ~ nicht
　　im Brennpunkt, **principal** ~
　　Hauptbrennpunkt *m*, **to bring into** ~ in den
　　Brennpunkt bringen
focus *v* akkommodieren (Opt), fokussieren,
　　sammeln, scharf einstellen
focus lens Kondensorlinse *f*
focus[s]ed scharf eingestellt
focus[s]ing Fokussierung *f*, Bündelung *f* (Opt),
　　Einstellung *f*, **exact** ~ Scharfeinstellung *f*
　　(Opt), **strong** ~ Courantfokussierung *f*
focus[s]ing electrode Bündelungselektrode *f*
focus[s]ing lens Bündelungslinse *f*,
　　Sucherokular *n*
focus[s]ing magnifier Einstellupe *f* (Opt)
focus[s]ing screen Mattscheibe *f* (Opt)
focus[s]ing screw Einstellschraube *f*
fodder Futter *n;* Viehfutter *n*
foetus Frucht *f* (Biol)
fog Nebel *m*, Schleier *m* (Phot)
fog disc Nebelblendscheibe *f*
foggy neblig, nebelartig
fog veil process Nebelschleierverfahren *n*
foil Folie *f*, **to cover with** ~ foliieren
foil activation Folienaktivierung *f*
foil brush Blätterbürste *f*
foil detector Aktivierungsfolie *f*
foil lacquer Folienlack *m*
fold Abkantung *f*(Met), Falte *f*, Falz *m*,
　　Knitter *m*
fold *v* falten, abkanten, falzen, knicken, ~ [up]
　　zusammenfalten
foldback Raupenbildung *f* (Kunstharz)
folded *a* gefaltet, faltenreich
folder Aktendeckel *m*, Faltprospekt *m*,
　　Falzmaschine *f* (Papier)
folding Abkanten *n*, Falten *n*, ~
　　Zusammenlegen *n*, **point of** ~
　　Faltungspunkt *m*
folding blade Spreizflügel *m*
folding carton Faltschachtel *f* (aus Karton)
folding endurance Falzfestigkeit *f*,
　　Falzwiderstand *m*
folding machine Abkantmaschine *f*,
　　Bördelmaschine *f*, Falzmaschine *f* (Papier)
folding press Abkantmaschine *f*
folding resistance Falzwiderstand *m*
folding strength Knickfestigkeit *f*
folding test Faltversuch *m*
folgerite Folgerit *m* (Min)
foliaceous spar Blätterspat *m* (Min)
foliage Laub *n*, Laubwerk *n*
foliated blätterig, blattförmig, dünntafelig,
　　lamellar, schiefrig (Min)
foliated coal Blätterkohle *f*

foliated gypsum Schaumgips *m*
foliated spar Mengspat *m* (Min)
foliated tellurium Blättertellur *n* (Min)
foliated zeolite Blätterzeolith *m* (Min)
folic acid Folsäure *f*, Vitamin Bc
folic acid deficiency Folsäuremangel *m*
folinic acid Folinsäure *f*
folin reagent Folinreagenz *n*
folio Folio[blatt] *n*
folliculin Follikulin *n*
follow *v* nachfolgen, nacheilen
follower Nachlaufwerk *n*, Nebenrad *n*
follower controller Folgeregler *m* (Regeltechn)
follower pin Mitnehmerstift *m*
follow-up system Folgezeigersystem *n*
foment *v* bähen, erwärmen (Med)
food Lebensmittel *n pl*, Nährmittel *n*,
　　Nahrungsmittel *n*, Nahrungsstoff *m*, **tinned or**
　　canned ~ Konserve *f*
food additive Lebensmittelzusatz *m*
food adulteration Lebensmittelfälschung *f*,
　　Nahrungsmittelfälschung *f*
food analysis Lebensmitteluntersuchung *f*
food chemist Lebensmittelchemiker *m*
food chemistry Lebensmittelchemie *f*,
　　Nahrungsmittelchemie *f*
food color Lebensmittelfarbstoff *m*
food compartments of refrigerators
　　Kühlschrankinnenbehälter *pl*
food control Nahrungsmittelkontrolle *f*
food flavoring Geschmacksstoff *m* für
　　Nahrungsmittel (pl)
food for thought Denkansatz *m*
food hygiene Nahrungsmittelhygiene *f*
food industry Lebensmittelindustrie *f*
food inspection Nahrungsmittelkontrolle *f*
food intake Nahrungszufuhr *f*
food investigation Lebensmitteluntersuchung *f*
food law Lebensmittelgesetz *n*
food poisoning Lebensmittelvergiftung *f*,
　　Nahrungsmittelvergiftung *f*
food [processing] industry
　　Nahrungsmittelindustrie *f*
foodstuff Nahrungsmittel *n*, Nahrungsstoff *m*
foodstuff chemist Lebensmittelchemiker *m*
food substance Nährstoff *m*
food technology Lebensmitteltechnologie *f*
food value Nährwert *m*
food wrapping Lebensmittelverpackung *f*
foolproof absolut sicher, narrensicher
fool's gold Gelbeisenkies *m* (Min),
　　Narrengold *m*
foot Fuß *m*, Tragfuß *m*, ~ **of a perpendicular**
　　Fußpunkt *m* des Lotes
foot bellows Tretgebläse *n*
foot brake pedal Bremsfußhebel *m*
footnote Fußnote *f*
foot roller Fußrolle *f*
foots Endlauge *f*

foot valve Fußventil *n*, Grundventil *n*
forage Futter *n*, Viehfutter *n*
forage plant Futterpflanze *f*
force Kraft *f*, Druck *m*, Stärke *f*, acting ~
angreifende Kraft, coming into ~
Inkrafttreten *n*, direction of ~ Kraftrichtung *f*,
electromotive ~ elektromotorische Kraft,
external ~ äußere Kraft, intermolecular ~
zwischenmolekulare Kraft, line of ~
Kraftlinie *f*, magnetomotive ~
magnetomotorische Kraft, photoelectromotive
~ photoelektromotorische Kraft, repellent ~
abstoßende Kraft, resultant ~ resultierende
Kraft, to exert a ~ eine Kraft ausüben, unit of
~ Krafteinheit *f*, ~ acting in any direction
beliebig gerichtete Kraft, ~ of gravity
Schwerkraft *f*
force *v* zwingen, einen Druck ausüben, ~ apart
spreizen, ~ on aufzwängen, aufzwingen, ~
open aufsprengen, ~ upon aufdrängen
force arm of a lever Kraftarm *m* eines Hebels
forced angestrengt, zwangsläufig
forced circulation boiler
Zwangsdurchlaufkessel *m*
forced circulation cooling
Zwangsumlaufkühlung *f*
forced circulation cooling tower
Ventilatorkühlturm *m*
forced circulation evaporator
Zwangsumlaufverdampfer *m*
forced circulation heating
Zwangsumlaufheizung *f*
forced circulation mixer Gegenstrom-Zwangs-
Mischer *m*
forced draft cooling Druckkühlung *f*,
Druckluftkühlung *f*
forced draft furnace Unterwindfeuerung *f*
forced feed Druckschmierung *f*
forced-flow once-through boiler
Zwangsumlaufkessel *m*
force equilibrium Kräfteausgleich *m*
force field Kraftfeld *n*
force impact point Angriffspunkt *m* der Kraft
force parallelogram Kräfteparallelogramm *n*,
Krafteck *n*
force plate Stemmplatte *f*
force plug Preßstempel *m*, Stempel *m*
force point Kraftpunkt *m*
force polygon Kräftepolygon *n*
forceps Pinzette *f* (Med), Zange *f* (Med)
force pump Druckpumpe *f*
force radius Kraftradius *m*
forces, chemical ~ chemische Kräfte *pl*,
concurrent ~ Kräfte, die in einem Punkt
angreifen (Mech), Coulomb ~ Coulombsche
Kräfte, electrostatic ~ elektrostatische Kräfte,
resolution of ~ Kräftezerlegung *f*, short-range
~ Kräfte *pl* geringer Reichweite, ~ having
opposite directions entgegengesetzte Kräfte *pl*,

~ having the same direction gleichgerichtete
Kräfte *pl*
forcherite Forcherit *m* (Min)
forcible heftig, kräftig, gewaltsam
forcing test Forcierungsversuch *m*
forcite Forcit *n*
Ford cup Fordbecher *m*
Ford viscosimeter Fordbecher *m*
forecast Voraussage *f*, Vorhersage *f*
forecast *v* voraussagen, vorhersagen
forecooler Vorkühler *m*
forecooling Vorkühlung *f*
foreground Vordergrund *m*
forehearth Vor[glüh]herd *m*, Vorwärmeherd *m*
foreign [art]fremd
foreign body Fremdkörper *m*
foreign gas recorder Fremdgasschreiber *m*
foreign matter Fremdbestandteil *m*,
Fremdkörper *m*, Fremdstoff *m*
foreign metal Begleitmetall *n*
foreign nucleus Fremdkeim *m*
foreign substance Fremdkörper *m*
foreign trade Außenhandel *m*
foreman Aufseher *m*, Meister *m*, Vorarbeiter *m*
forensic forensisch, gerichtsmedizinisch
forensic chemistry Gerichtschemie *f*
forerunner Vorläufer *m*
fore runnings Kopfprodukt *n*, Vorlauf *m* (Dest)
forescreen Vorsieb *n*
foresee *v* vorhersehen
foresite Foresit *m* (Min)
forest, mixed ~ Mischwald *m*
forestry Forstwirtschaft *f*, Waldwirtschaft *f*
forevacuum Vorvakuum *n*, stability of ~
Vorvakuumbeständigkeit *f*
for example beispielsweise
forge Schmiede *f*, Eisenfrischherd *m*,
Eisenhütte *f*, Eisenschmiede *f*, Esse *f*,
Hammer *m*
forge *v* schmieden, hämmern, pinken, ~ on
anschmieden
forgeability Schmiedbarkeit *f*, Hämmerbarkeit *f*,
Warmbildsamkeit *f*
forgeable schmiedbar, hämmerbar (heiß)
forged geschmiedet, cold ~ kalt geschmiedet
forged steel Schmiedestahl *m*
forge fire Schmiedefeuer *n*
forge hammer Eisenhammer *m*
forge hearth Frischherd *m*
forge pig iron Puddelroheisen *n*
forge pigs Schmiederoheisen *n*
forger Schmied *m*
forge scale Hammerschlag *m*
forge smith Hammerschmied *m*
forge welding Feuerschweißen *n*
forging Fälschung *f*, Schmieden *n*,
Schmiedstück *n*, final ~
Weiterverschmieden *n*, hot ~

Warmschmieden *n*, **rough** ~ geschmiedeter
Rohling *m*
forging furnace Schmiedeofen *m*
forging press Schmiedepresse *f*
forging property Schmiedbarkeit *f*
forging quality Schmiedbarkeit *f*
forging steel Schmiedeeisen *n*
forging temperature Schmiedetemperatur *f*
forging test Schmiedeprobe *f*
fork Gabel *f*
forked gegabelt, doppelgängig, gabelförmig
forked prop Gabelstütze *f*
forked tube Gabelrohr *n*
fork guide Gabelführung *f*
fork joint Gabelgelenk *n*
fork lift Gabelstapler *m*
fork lift truck Gabelhubstapler *m*
fork truck Gabelstapler *m*
form Gestalt *f*, Form *f*, Formblatt *n*, **change of** ~
Deformation *f*, Formänderung *f*, **region of**
change of ~ Formänderungsbereich *m*, **solid**
~ massive Form; ~ **of construction**
Ausführungsform *f*, ~ **of matter**
Erscheinungsform *f*
form *v* formen, bilden, fassonieren, ~ **a**
complex einen Komplex bilden,
komplexieren, anlagern (Chem), ~ **a**
precipitate einen Niederschlag bilden (Chem),
~ **back** rückbilden, ~ **clinkers** festbacken
formal Formal *n*, Methylal *n*
formaldazine Formaldazin *n*
formaldehyde Formaldehyd *m*, **fixing with** ~
Formaldehydhärtung *f* (Gerb)
formaldehyde acetamide Formicin *n*
formaldehyde bisulfite Formaldehydbisulfit *n*
formaldehyde hydrogen sulfite
Formaldehydbisulfit *n*
formaldehyde oxime Formoxim *n*
formaldehyde resin Formaldehydharz *n*
formaldehyde sodium bisulfite
Formaldehydnatriumbisulfit *n*
formaldehyde sodium hydrogen sulfite
Formaldehydnatriumbisulfit *n*
formaldehyde solution Formaldehydlösung *f*,
aqueous ~ Formalin *n*
formaldehyde sulfoxylate
Formaldehydsulfoxylat *n*
formaldehyde sulfoxylic acid
Formaldehydsulfoxylsäure *f*
formaldoxime Formaldoxim *n*, Formoxim *n*
formalin Formalin *n*
formalin tanning Formalingerbung *f* (Gerb)
formamide Formamid *n*
formamidine Formamidin *n*
formanilide Formanilid *n*
formate ameisensaures Salz *n*, Formiat *n*
formation Bildung *f*, Ausbildung *f*, Entstehung *f*,
Entwicklung *f* (Chem), Formung *f*, Gebilde *n*,
Gestaltung *f*, **time of** ~ Aufbauzeit *f*;

Entstehungszeit *f*, ~ **of a shoulder**
Gratbildung *f*, ~ **of carbonic acid**
Kohlensäureentwicklung *f*, ~ **of crystals**
Kristallbildung *f*, ~ **of ions** Ionenbildung *f*,
Ionenbildungsgeschwindigkeit *f*, ~ **of mucus**
Schleimbildung *f*, ~ **of nuclei** Keimbildung *f*
formation condition Entstehungsbedingung *f*
formation constant Bildungskonstante *f*
formation enthalpy Bildungsenthalpie *f*
formative cell Bildungszelle *f* (Biol)
formative time Bildungsdauer *f*
formazan Formazan *n*
formazyl Formazyl *n*
formed piece Formteil *n*
former Former *m*, Gießer *m*, Schablone *f*
former block Formblock *m*
formhydrazidine Formhydrazidin *n*
formhydroxamic acid Formhydroxamsäure *f*
formic acid Ameisensäure *f*, **salt or ester of** ~
Formiat *n*
formic acid plant Ameisensäureanlage *f*
formic aldehyde Formaldehyd *m*
formic anhydride Ameisensäureanhydrid *n*
formicin Formicin *n*
formic tincture Ameisensäuretinktur *f*
formimidyl chloride Formimidchlorid *n*
formin Formin *n*
formine Hexamethylenamin *n*
forming die Biegegesenk *n*, Formwerkzeug *n*,
Ziehstempel *m*
forming lathe Fassondrehbank *f*
forming pad Gummikissen *n*, Preßkissen *n*
forming process Verformungsvorgang *m*
forming tool Verformungswerkzeug *n*
forminitrazole Forminitrazol *n*
formless gestaltlos
formlessness Formlosigkeit *f*
formocholine Formocholin *n*
formoguanamine Formoguanamin *n*
formoguanine Formoguanin *n*
formol Formol *n*
formonitrile Ameisensäurenitril *n*, Blausäure *f*
formononetin Formononetin *n*
formose Formose *f*
formosulfite Formosulfit *n*
formotannin Formotannin *n*
formoxime Formoxim *n*
form quotient Formzahl *f*
form-seal equipment Form-, Füll- und
Schließsystem *n*
formula Rezept *n*, Formel *f* (Chem),
Herstellungsvorschrift *f*, **determination of** ~
Formelbestimmung *f*, **empirical** ~
Summenformel *f*, empirische Formel *f*, **flying**
wedge ~ Keilstrich-Formel *f* (Stereochem),
structural ~ Strukturformel *f*, Formelbild *n*
(Chem)
formula conversion Formelumsatz *m*
formula sign Formelzeichen *n*

formulate *v* formulieren, ~ **more precisely** präzisieren
formulation Ansatz *m* (Math), Formulierung *f*, ~ **of the problem** Problemstellung *f*
formula weight Formelgewicht *n*
formyl Formyl-
formylacetic acid Malonaldehydsäure *f*
formyl amine Formamid *n*
formylaniline Formanilid *n*
formylate *v* formylieren
formylation Formylierung *f*
formylcamphor Formylcampher *m*
formylcellulose Formylcellulose *f*
formyl fluoride Formylfluorid *n*
formylic acid Ameisensäure *f*, Formylsäure *f*
forsterite Forsterit *m* (Min)
fortify *v* [ver]stärken, kräftigen
fortoin Fortoin *n*
forty-fold vierzigfach
forward *v* [be]fördern, nachsenden
forward characteristic Durchlaßkennlinie *f*
forward direction (of a rectifier) Durchlaßrichtung *f* (eines Gleichrichters)
forwarder Aufgeber *m*
forwarding Beförderung *f*
forwarding office Expedition *f*
forward reaction Hinreaktion *f*
forward voltage Durchlaßspannung *f*
fossil Fossil *n*, Versteinerung *f*
fossiliferous fossilführend
fossilification Fossilienbildung *f*
fossilization Fossilienbildung *f*, Versteinerung *f*
fossilized versteinert, fossil
fossil meal Erdmehl *n*
fossil resin Erdharz *n*
fossil wax Ozokerit *m*
foul air Abluft *f*
foulard Foulard *m* (Text)
fouling Verschmutzung *f*
found *v* [be]gründen, fundieren
foundation Stiftung *f*, Fundament *n*, Grundlage *f*, Grundmauer *f*, Unterbau *m*, Unterlage *f*
foundation wall Grundmauer *f*
founder Gießer *m*
founder's dust Formstaub *m*
founding Gießen *n*
founding furnace Schmelzofen *m*
founding metal Gußmetall *n*
foundry Gießerei *f*, Hüttenwerk *n*, Schmelzanlage *f*
foundry coke Gießereikoks *m*, Schmelzkoks *m*
foundry cupola Gießereikupolofen *m*
foundry ladle Gießpfanne *f*
foundry pig iron Gießereieisen *n*
foundry pit Dammgrube *f*
foundry practice Gießereitechnik *f*
foundry product Gießereierzeugnis *n*

foundry reverberatory furnace Gießereiflammofen *m*
foundry sand Formsand *m*
foundry worker Hüttenarbeiter *m*
fountain künstlicher Brunnen *m*, Quelle *f*
fountain pen Füller *m*, Füllfederhalter *m*
fountain roll Farbauftragswalze *f*, Lackauftragswalze *f*, Tauchwalze *f*
four-axle vierachsig
four-center addition Vierzentrenaddition *f* (Stereochem)
four-color printing Vierfarbendruck *m*
four-component balance Vierkomponentenwaage *f*
four-component system Vierstoffsystem *n*
four-cornered viereckig
four-cycle Viertakt *m*
four-cycle engine Viertaktmotor *m*
four-digit vierstellig
four-dimensional vierdimensional
four-dimensionality Vierdimensionalität *f*
fourdrinier Langsiebmaschine *f*
four-faced vierflächig
four-figure vierstellig
four flow paths [at 90°] Angußkreuz
fourfold vierfach, quartär, vierzählig
Fourier analysis Fourier-Analyse *f* (Math)
Fourier series Fouriersche Reihe *f* (Math)
fourmarierite Fourmarierit *m* (Min)
four-membered viergliedrig
four-membered ring Vierring *m* (Chem)
four-phase vierphasig (Elektr)
four-phase star connection Vierphasenkreuzschaltung *f* (Elektr)
four-phase system Vierphasensystem *n*
four-polar vierpolig
four-roll calender Vierwalzenkalander *m* (Papier)
four-sided vierseitig, quadrilateral
four-speed gear Vierganggetriebe *n*
four-stroke cycle Viertakt *m*
fourth Viertel *n*
four-way cock Kreuzhahn *m*, Vierwegehahn *m*
four-way cross Kreuzstück *n*
four-way fork lift truck Vierwege-Stapler *m*
four-wire system Vierleitersystem *n* (Elektr)
fowlerite Fowlerit *m* (Min)
Fowler's solution Fowlersche Lösung *f* (Pharm)
fox fat Fuchsfett *n*
foxglove Fingerhut *m* (Bot)
foxglove leaves Fingerhutblätter *n pl* (Bot)
fraction Anteil *m*, Bruch *m* (Math), Bruchteil *m*, Fraktion *f* (Chem), **high boiling** ~ hochsiedende Fraktion, **low-boiling** ~ niedrigsiedende Fraktion
fractional fraktioniert, absatzweise
fractional crystallization Kristallisationsdifferentiation *f*

fractional distillation flask Fraktionskolben *m*
(Chem)
fractional line Bruchstrich *m* (Math)
fractional mathematics Bruchrechnen *n* (Math)
fractional weight Aufsetzgewicht *n*,
Bruchgewicht *n*
fractionate *v* fraktionieren, rektifizieren (Dest)
fractionating Fraktionieren *n*
fractionating apparatus Fraktionierapparat *m*
fractionating attachment Fraktionieraufsatz *m*
(Dest), Fraktionsaufsatz *m* (Dest)
fractionating brush Fraktionierbürste *f*
fractionating column Fraktionierkolonne *f*
(Dest), Rektifikationskolonne *f* (Dest)
fractionating device Fraktioniereinrichtung *f*
fractionating flask Fraktionierkolben *m*
fractionating pump Dosierpumpe *f*
fractionating tower Fraktionierkolonne *f* (Dest),
Fraktionierturm *m*
fractionating tube Fraktionierrohr *n*
fractionation Fraktionierung *f*, Rektifikation *f*
(Dest), **simple** ∼ Kurzwegfraktionierung *f*
fraction collector Fraktionssammler *m*
fraction receiver Destilliervorlage *f*
fracture Bruch *m*, Anbruch *m*, Riß *m*,
Zerbrechen *n*, **character of** ∼ Bruchgefüge *n*,
conchoidal ∼ muschelige Bruchfläche *f*,
energy of ∼ Brucharbeit *f*, **fibrous** ∼ faserige
Bruchfläche *f*, **initiation of** ∼
Bruchbeginn *m*, **irregular or uneven** ∼
irreguläre Bruchfläche *f*, **location of** ∼
Bruchstelle *f*, **method of** ∼ Bruchart *f*,
propagation of ∼ Bruchausbreitung *f*, **risk of**
∼ Bruchgefahr *f*, **spot of** ∼ Bruchstelle *f*,
surface of ∼ Bruchfläche *f*, **tendency to** ∼
Bruchneigung *f*, ∼ **of chain** Kettenbruch *m*,
∼ **of roll** Walzenbruch *m*
fracture *v* [zer]brechen
fractured surface Bruchfläche *f*
fracture formation Bruchbildung *f*
fracture-like bruchartig
fracture mechanics Bruchmechanik *f*
fracture origin Bruchentstehung *f*
fracture pattern Bruchgefüge *n*
fracture velocity Bruchgeschwindigkeit *f*
fragarin Fragarin *n*
fragile zerbrechlich, brüchig
fragility Zerbrechlichkeit *f*, Brüchigkeit *f*
fragment Bruchstück *n*, Fragment *n*,
Spaltprodukt *n* (Phys), Splitter *m*, Teil *m*
fragmentary fragmentarisch, bruchstückartig
fragmentation Fragmentierung *f*
fragments Abschlag *m*
fragrance Aroma *n*, Duft *m*, Wohlgeruch *m*
fragrant wohlriechend, duftend, duftig,
geruchverbreitend
fragrant essence Räucheressenz *f*
fragrant oil Duftöl *n*

frame Rahmen *m*, Bock *m* (Techn), Fassung *f*
(Techn), Gestell *n* (Maschine), Gußform *f*
(Gieß), Spant *n*
frame *v* [ein]fassen, einrahmen
frame bolster Matrizenrahmen *m*
frame filter Rahmenfilter *m*
frame filter press Rahmen[filter]presse *f*
frame guide Bildschieber *m*
frame mounting machine
Spantenaufsetzmaschine *f*
frame roller Verteilerwalze *f*
frames [for dishes in dishwashing machines]
Geschirrspülkörbe *pl*
frame saw Gattersäge *f*, Klobensäge *f*
framework Fachwerk *n*
framing iron Einschaleisen *n*
francium (Symb. Fr) Francium *n*
franckeite Franckeit *m* (Min)
francolite Francolith *m* (Min)
frangula emodin Frangula-Emodin *n*
frangulic acid Frangulinsäure *f*
frangulin Frangulin *n*, Faulbaumbitter *n*
frangulinic acid Frangulinsäure *f*
Frankfort black Frankfurter Schwarz *n*,
Druckerschwärze *f*, Rebenschwarz *n*
frankincense Weihrauch *m*, Weihrauchharz *n*
frankincense oil Weihrauchöl *n*
franklandite Franklandit *m* (Min)
franklinite Franklinit *m* (Min), Zinkeisenerz *n*
(Min), Zinkeisenstein *m* (Min)
Fraunhofer lines Fraunhoferlinien *pl*
fraxin Fraxin *n*
fray *v* abstoßen, abnützen, ausfransen, fasern
freckles Legierungsflecke *pl*
free frei, ungebunden
free *v* [from] befreien, ∼ **from acid** entsäuern, ∼
from alcohol entalkoholisieren, ∼ **from dust**
entstauben, ∼ **from mildew** entschimmeln, ∼
from rust entrosten, ∼ **from silver** entsilbern,
∼ **from slime** entschleimen, ∼ **from soap**
entseifen, ∼ **from sulfur** entschwefeln,
abschwefeln
free at works frei Hütte
free burning coal Sandkohle *f*
free-cutting steel Automatenstahl *m*
freed from fat abgefettet
freedom, degree of ∼ Freiheitsgrad *m* (Atom)
free-enterprise system soziale Marktwirtschaft *f*
free falling mixer Freifallmischer *m*
free-falling velocity Schwebegeschwindigkeit *f*
free fit Bewegungssitz *m*, Gewindepassung *f*
free-flowing rieselfähig (Pulver)
freeing from dust Entstauben *n*
free lance freier Mitarbeiter *m*
free molecule diffusion Diffusion *f* freier
Moleküle (pl)
freeness, low ∼ Schmierigkeit *f* (Papier)
free nuclear induction, method of ∼
Spin-Echo-Verfahren *n* (Atom)

free on board bordfrei, frei an Bord,
[Lieferung] *f* frei Schiff
free on trucks frei Waggon
free roller Losrolle *f*
free-settling hydraulic classifier
Freifallstromklassierer *m*
freestone Quader *m*
freewheeling freilaufend
freezable gefrierbar
freeze *v* [ge]frieren, einfrieren, erstarren,
tiefkühlen, ~ **hard** festfrieren, ~ **in** einfrieren,
einwintern, ~ **on** anfrieren, ~ **out** ausfrieren,
~ **together** zusammenfrieren, ~ **up** zufrieren
freeze-dried gefriergetrocknet, gefriergetrocknet,
lyophilisiert
freeze-dry *v* gefriertrocknen
freeze drying Gefriertrocknung *f*
freeze drying apparatus Gefriertrockenapparat *m*
freeze drying plant Gefriertrocknungsanlage *f*
freeze plug Erstarrungsstopfen *m*
freezer Gefrierfach *n* (Kühlschrank),
Tiefkühltruhe *f*, **sharp** ~
Schnellgefrierraum *m*
freezer burn Gefrierbrand *m*
freezing Gefrieren *n*, Einfrieren *n*, Erfrieren *n*,
Erstarren *n*, Erstarrung *f*, **to concentrate by** ~
ausfrieren, ~ **out** Ausfrieren *n*, ~ **out the ore**
Ausfrieren *n* der Erze (pl), ~ **up of the tap**
hole Zufrieren *n* des Abstichloches
freezing *a* eisig
freezing apparatus Gefrierapparat *m*
freezing assembly Ausfriervorrichtung *f*
freezing chamber Gefrierkammer *f*
freezing compartment Gefrierfach *n*
(Kühlschrank)
freezing instrument, [quick] ~
[Schnell]-Gefriervorrichtung *f*
freezing-in temperature Einfriertemperatur *f*
freezing method Ausfrierverfahren *n*
freezing mixture Gefriermischung *f*,
Kältemischung *f*, Kühlmittel *n*
freezing nucleus Gefrierkern *m*
freezing-out temperature Ausfriertemperatur *f*
freezing pipe Gefrierrohr *n*
freezing plant Gefrieranlage *f*
freezing point Gefrierpunkt *m*, Eispunkt *m*,
Frostpunkt *m*, **lowering of the** ~
Gefrierpunktserniedrigung *f*
freezing point apparatus
Gefrierpunktsbestimmungsapparat *m*
freezing point curve Erstarrungskurve *f*
freezing point thermometer
Gefrierthermometer *n*
freezing process Ausfrierverfahren *n*
freezing salt Gefriersalz *n*
freezing section Gefrierschnitt *m*
freezing temperature Gefriertemperatur *f*
freezing test Gefrierversuch *m*
freezing trap Kühlfalle *f*, Ausfrierfalle *f*

freibergite Freibergit *m* (Min)
freieslebenite Freieslebenit *m* (Min)
freight Fracht *f*
freight-charging Ladung *f*
freight elevator Lastenaufzug *m*
fremontite Fremontit *m* (Min)
French brandy Franzbranntwein *m*
French chalk Federweiß *n*
French curve Kurvenlineal *n*
French polish Schellackpolitur *f*
French scarlet Kermesscharlach *m*
frenzelite Frenzelit *m* (Min)
freon ((HN)) Freon *n*, Frigen *n*
frequency Frequenz *f* (Elektr), Häufigkeit *f*,
Periodenzahl *f*, **angular** ~ Winkelfrequenz *f*,
assigned ~ Kennfrequenz *f*, **of high** ~
hochfrequent, **variation of** ~
Frequenzschwankung *f*, ~ **of flutter**
Schwankungsfrequenz *f*
frequency band Frequenzband *n* (Elektr)
frequency control Frequenzstabilisierung *f*
(Elektr)
frequency controller Frequenzregler *m*
frequency curve Häufigkeitskurve *f* (Statist)
frequency data Frequenzangaben *f pl*
frequency dependence Frequenzverhalten *n*
frequency-dependent frequenzabhängig
frequency distribution Häufigkeitsverteilung *f*
frequency distribution curve
Häufigkeitsverteilungskurve *f*
frequency divider Frequenzuntersetzer *m*,
Impulsfrequenzuntersetzer *m*
frequency equation Frequenzgleichung *f*
frequency factor Häufigkeitsfaktor *m* (Statist),
Stoßfaktor *m* (Therm)
frequency function Häufigkeitsfunktion *f* (Math)
frequency inversion Frequenzumkehrung *f*
frequency meter Frequenzmesser *m* (Elektr)
frequency-modulated frequenzmoduliert
frequency number Schwingungszahl *f*, Stoßzahl *f*
(Atom), Wechselzahl *f* (Atom)
frequency polygon Häufigkeitspolygon *n*
(Statist)
frequency radiation Frequenzstrahlung *f*
frequency range Frequenzbereich *m*
frequency-range control Bandbreitenregelung *f*
(Elektr)
frequency response (Am. E.) Frequenzgang *m*
(Regeltechn)
frequency response [characteristic]
Frequenzwiedergabe *f*
frequency response [function] Frequenzgang *m*
(b. Transformationen)
frequency stability Beständigkeit *f* der Frequenz
(Elektr)
frequency standard Frequenznormal *n*
frequency sweep method
Frequenzwobbelverfahren *n*
frequency unit Frequenzeinheit *f*

frequency value Häufigkeitswert *m* (Statist)
frequent häufig
frequent *v* frequentieren
fresco painting Gipsmalerei *f*
fresh air Frischluft *f*
fresh cell therapy, embryonal ~
  Frischzellentherapie *f*
freshen *v* erfrischen, ~ **up** auffrischen
freshener Belebungsmittel *n*
fresh food packaging Frischhalteverpackung *f*
freshwater Süßwasser *n,* Frischwasser *n,*
  Trinkwasser *n*
freshwater fish Süßwasserfisch *m*
Fresnel diffraction Fresnelsche Beugung *f* (Opt)
Fresnel's mirror Fresnelspiegel *m* (Opt)
fretting Abriebkorrosion *f*
fretting corrosion Lochfraßkorrosion *f*
  (destruction of metal surfaces by the
  combined action of corrosion and very slight
  frictional movement of the surfaces in
  contact), Reibkorrosion *f*
Freundlich adsorption isotherm Freundlichsche
  Adsorptionsisotherme *f*
freyalite Freyalith *m* (Min)
friability Brüchigkeit *f,* Sprödigkeit *f*
friable [ab]brüchig, bröcklig, krümelig, leicht
  zerbröckelnd, mürbe, spröde
friableness Bröcklichkeit *f,* Sprödigkeit *f*
friction Reibung *f,* Friktion *f,* **angle of** ~
  Reibungswinkel *m,* **coefficient of** ~
  Reibungsfaktor *m,* Reibungskoeffizient *m,*
  Reibungszahl *f,* **heat due to** ~
  Reibungswärme *f,* **internal** ~ innere Reibung,
  rolling ~ gleitende Reibung, rollende
  Reibung, **sliding** ~ gleitende Reibung, **without**
  ~ reibungsfrei, ~ **at maximum load**
  Höchstlastreibung *f,* ~ **of a liquid**
  Flüssigkeitsreibung *f*
friction accumulating conveyor
  Reibstauförderer *m*
frictional area Reibfläche *f*
friction[al] coefficient Reibungskoeffizient *m,*
  Reibungszahl *f*
frictional diameter Reibungsdurchmesser *m*
frictional electricity Reibungselektrizität *f*
frictional force Reibungskraft *f*
frictional heat Reibungswärme *f*
friction[al] loss Reibungsverlust *m*
frictional property Reibeigenschaft *f*
frictional resistance Reibungswiderstand *m*
frictional surface Reibfläche *f*
frictional wear Reibungsabnutzung *f*
frictional work Reibungsarbeit *f*
friction bearing Friktionslagermetall *n*
friction brake Reibungsbremse *f*
friction calender Friktionskalander *m* (Gummi)
friction clip Hülsenkupplung *f*
friction clutch Friktionskupplung *f,*
  Reibungskupplung *f*

friction coating Friktionieren *n*
friction compound Friktionsmischung *f*
  (Gummi)
friction cone Reibungskegel *m*
friction coupling Reibungskupplung *f*
friction disk Reibscheibe *f,* Friktionsscheibe *f,*
  Reibrolle *f,* Reibteller *m*
friction drive Reibradantrieb *m,*
  Reibungsantrieb *m*
friction drum Reibtrommel *f*
friction electrostatic machine
  Reibungselektrisiermaschine *f*
friction factor Reibungsfaktor *m*
friction gear Friktionsgetriebe *n,*
  Reibradgetriebe *n*
friction-glazed kalandriert
friction hammer Reibungshammer *m*
friction heating Reibungserwärmung *f*
friction layer Reibungsschicht *f*
frictionless reibungsfrei, reibungslos
friction let-off gebremster Abwickelbock *m,*
  Reibungsabzug *m*
friction machine Reibungsmaschine *f*
friction meter Friktionsmesser *m*
friction press Friktionspresse *f*
friction pressure drop Reibungsdruckabfall *m*
friction ring Reibring *m*
friction roll Friktionswalze, Reibwalzwerk *n,*
  Schleppwalze *f*
friction roller Reibtrommel *f,* Reibwalze *f*
friction starting clutch
  Antriebsreibungskupplung *f*
friction stress Reibungsspannung *f*
friction surface Lauffläche *f,* Reibungsfläche *f*
friction test Reibungsprobe *f*
friction welding Reibungsschweißen *n*
friction wheel Reibrad *n,* Reibscheibe *f*
Friedel-Crafts reaction Friedel-Crafts-Reaktion *f*
Friedel-Crafts rearrangement Umlagerung *f* bei
  Friedel-Crafts-Reaktionen (pl)
friedelin Friedelin *n*
friedelite Friedelit *m* (Min)
friedonic acid Friedonsäure *f*
frieseite Frieseit *m* (Min)
Fries rearrangement Friessche Umlagerung *f*
  (Chem)
frigen Frigen *n*
frigid kalt
frigorific Kälte *f* erzeugend, kälteerzeugend
frigorific mixture Kältemischung *f*
frigorimeter Kältemesser *m*
fringe Einfassung *f,* Rand[zone *f*] *m*
fringe pattern Streifenbild *n* (Opt)
fringing Ausfasern *n*
frit Fritte *f,* Glassatz *m*
frit *v* fritten, sintern, zusammenbacken
frit fly Fritfliege *f* (Zool)
frit kiln Frittofen *m*
frit porcelain Fritt[en]porzellan *n*

fritting Sintern *n*, Ausbrennen *n*, Frittung *f*, Sinterung *f*
fritzscheite Fritzscheit *m* (Min)
frog Frosch *m* (Zool)
frog's leg Froschschenkel *m* (Zool)
front Fassade *f*, Vorderseite *f*
frontal attack Frontalangriff *m*
frontal view Stirnansicht *f*
front axle Vorderachse *f*
front diaphragm Vorderblende *f* (Opt)
front-end forklift [truck] Frontgabelstapler *m*
front face Stirnfläche *f*
frontier Grenze *f*
frontier orbital Grenzorbital *n*
frontier point Randpunkt *m*
front lens Vorsatzlinse *f* (Opt)
front milling tool Stirnfräser *m*
front page Titelblatt *n*
front plate Vorderplatte *f*
front roll Vorderwalze *f*
front shoe Düsenflansch *m*, Düsenplatte *f*
front surface Stirnfläche *f*
front view Vorderansicht *f*
front wall Fassade *f*
front-wheel Vorderrad *n*
front-wheel alignment Vorderradausrichtung *f*
front-wheel brake Vorderradbremse *f*
front-wheel drive Vorderantrieb *m*
frost Frost *m*, Kälte *f*
frost *v* mattieren (Glas)
frosted mattiert (Glas)
frosting Eisbildung *f*, Eisblumenbildung *f*
frosting lacquer Eisblumenlack *m*
frost period Frostzeit *f*
frost-proof frostbeständig
frost-resistant frostbeständig
frost-resisting kältefest
frost-sensitive frostempfindlich
frost shield Frostschutzscheibe *f*
frost test Gefrierprobe *f*
froth Bierschaum *m* (Brau), Schaum *m*
froth *v* [auf]schäumen, gischen, moussieren, ~ over überschäumen
frother Schäumer *m*
froth flow turbulente Blasenströmung *f*
frothing Schaumbildung *f*, Schaumentwicklung *f*, **to cease** ~ ausschäumen, ~ **of the melt** Schäumen *n* der Schmelze
frothing agent Schäumer *m*, Sammlerschäumer *m*
frothing quality Schaumbildungsvermögen *n*
frothing reagent Schaumerzeuger *m* (Metall)
frothless schaumlos
froth stabilizing agent Schaumstabilisator *m*
frothy schaumig, schaumähnlich
frozen [ein]gefroren, festgefroren
frozen meat Gefrierfleisch *n*
frozen section Gefrierschnitt *m*

fructokinase Fructokinase *f* (Biochem)
fructosamine Fructosamin *n*
fructosan Lävulosan *n*
fructosazine Fructosazin *n*
fructosazone Fructosazon *n*
fructose Fructose *f*, Fruchtzucker *m*, Lävulose *f*
fructose biphosphatase Fructosebisphosphatase *f* (Biochem)
fructose biphosphate Fructosebisphosphat *n*
fructose diphosphoric acid Fructosediphosphorsäure *f*
fructosone Fructoson *n*
fruit Frucht *f*, Obst *n*, ~ **of the soil** Ackerfrucht *f*
fruit acid Fruchtsäure *f*
fruit brandy Obstbranntwein *m*
fruit culture Obstbau *m*
fruit essence Fruchtäther *m*, Fruchtessenz *f*
fruit growing Obstbau *m*
fruit juice Fruchtsaft *m*
fruit-like fruchtartig
fruit processing Früchteverarbeitung *f*
fruit pulp Fruchtfleisch *n*
fruit spirit Obstbranntwein *m*
fruit sugar Fruchtzucker *m*, Fructose *f*, Lävulose *f*
fruit tannin Fruchtgerbstoff *m*
fruit vinegar Fruchtessig *m*
fruit wine Obstwein *m*
fruity fruchtartig
fry *v* braten
fuchsin Fuchsin *n*, Anilinrot *n*, Azalein *n*
fuchsin sulfurous acid Fuchsinschweflige Säure *f*
fuchsite Fuchsit *m* (Min), Chromglimmer *m* (Min)
fuchsone Fuchson *n*
fuchsonimine Fuchsonimin *n*
fucitol Fucit *m*
fuconic acid Fuconsäure *f*
fucose Fucose *f*, Fukose *f*
fucosone Fucoson *n*
fucosterol Fucosterin *n*
fucoxanthin Fucoxanthin *n*
fuculose Fuculose *f*
fucusol Fucusol *n*
fucus vesiculosus Blasentang *m* (Bot)
fuel Brennstoff *m*, Brennmaterial *n*, Feuerungsmaterial *n*, Heizmaterial *n*, Heizmittel *n*, Heizstoff *m*, Kraftstoff *m* (Kfz), Treibstoff *m* (Kfz), **charge of** ~ Brennstoffgicht *f*, **consumption of** ~ Brennstoffverbrauch *m*, Feuerungsbedarf *m*, **heavy** ~ schwer flüchtiger Brennstoff, **high octane number** ~ klopffester Kraftstoff, **light [volatile]** ~ leicht flüchtiger Brennstoff, **low-grade** ~ minderwertiger Brennstoff, **non-volatile** ~ schwer flüchtiger Brennstoff, **solid** ~ Festkraftstoff *m*, **transport of** ~ Brennstoffzufuhr *f*

fuel-air mixture Kraftstoff-Luft-Gemisch *n*,
　Brennstoff-Luft-Gemisch *n*
fuel bed Brennstoffschicht *f*, **thickness of** ~
　Brennschichthöhe *f*
fuel briquet[te] Brennziegel *m*
fuel cartridge Brennstoffpatrone *f* (Atom)
fuel cell Brennstoffzelle *f*, Treibstoffraum *m*
fuel chamber Feuerungsraum *m*
fuel channel Brennstoffkanal *m*,
　Spaltstoffkanal *m* (Atom)
fuel charge tube Brennstoffladeröhre *f* (Atom)
fuel chemistry Brennstoffchemie *f*
fuel consumption Brennmaterialverbrauch *m*,
　Brennstoffverbrauch *m*,
　Kraftstoffverbrauch *m* (Kfz)
fuel cycle Brennstoffkreislauf *m* (Atom)
fuel economy Kraftstoffersparnis *f*
fuel efficiency Heizeffekt *m*
fuel element Brennelement *n*
fuel feed Benzinzuführung *f*,
　Brennstoffzuführung *f*
fuel feed pump Brennstofförderpumpe *f*
fuel filter Kraftstoffilter *n*, Kraftstoffreiniger *m*
fuel flow Brennstoffdurchfluß *m*
fuel gas Brenngas *n*, Heizgas *n*
fuel gauge indicator Benzinanzeiger *m* (Auto)
fuel heater Brennstoffvorwärmer *m*
fuel hopper Brennstofftrichter *m*
fuel injection Kraftstoffeinspritzung *f*
fuel-injection pump Brennstoffeinspritzpumpe *f*
fuel jet Kraftstoffdüse *f*
fuel leak Benzinleck *n*
fuel mixture Kraftstoffgemisch *n*
fuel oil Brennöl *n*, Heizöl *n*
fuel oil residue Brennölrückstand *m*
fuel pipe line Benzinleitung *f*
fuel pump Brennstoffpumpe *f*, Treibstoffpumpe *f*
fuel regeneration Brennstoffaufarbeitung *f*
fuel rod Brennelement *n* (Atom),
　Spaltstoffstab *m* (Atom)
fuels, additive for ~ Treibstoffzusatz *m*
fuel saving Kraftstoffersparnis *f*
fuel slug Spaltstoffblock *m*
fuel strainer Brennstoffilter *n*,
　Kraftstoffreiniger *m*
fuel supply Brennstoffzufuhr *f*,
　Kraftstoffzufuhr *f*
fuel tank Brennstoffbehälter *m*
fuel technology Brennstofftechnik *f*
fuel trap Kraftstoffabscheider *m*
fuel utilization Brennmaterialausnutzung *f*,
　Brennstoffausnutzung *f*
fuel value Brennwert *m*
fugacity Fugazität *f*
fugacity coefficient Fugazitätskoeffizient *m*
fulcrum Drehpunkt *m*, Hebepunkt *m*,
　Stützpunkt *m*
fulcrum pin Drehzapfen *m*

fulgenic acid Fulgensäure *f*, **anhydride of** ~
　Fulgid *n*
fulgide Fulgid *n*
fulgurite Fulgurit *m*, Blitzröhre *f*, Blitzsinter *m*
fuliginous rußig, rußartig
fuliginous fumes Rußdampf *m*
full voll
full-capping Runderneuerung *f* (Reifen)
full cure Ausvulkanisation *f*
fuller's earth Bleicherde *f*, fetter Ton *m*,
　Seifenerde *f*, Seifenton *m*
full gloss Hochglanz *m*
full heat Vollfeuer *n*
full jacket Vollmantel *m*
full load Vollast *f*
fullness Fülle *f*
full steam Volldampf *m*
full strength Vollton *n* (Pigment ohne
　Verschnittmittel)
full-view scale Vollsichtskala *f*
full-wave rectifier Doppelweggleichrichter *m*
　(Elektr), Netzgleichrichter *m* (Elektr)
fulminate Fulminat *n*, Knallpulver *n*
fulminate *a* knallsauer
fulminating mercury Knallquecksilber *n*
fulminating powder Knallpulver *n*
fulminating silver Knallsilber *n*
fulminic acid Knallsäure *f*
fulminuric acid Fulminursäure *f*
fulvalene Fulvalen *n*
fulvanol Fulvanol *n*
fulvene Fulven *n*
fulverin Fulverin *n*
fulvic acid Fulvinsäure *f*
fumaraldehyde Fumar[di]aldehyd *m*
fumaramide Fumaramid *n*
fumarase Fumarase *f* (Biochem)
fumarate Fumarat *n*, fumarsaures Salz *n*
fumarate *a* fumarsauer
fumardialdehyde Fumardialdehyd *m*
fumaric acid Fumarsäure *f*
fumarine Fumarin *n*, Protopin *n*
fumaruric acid Fumarursäure *f*
fumaryl chloride Fumarylchlorid *n*
fumble *v* [herum]tappen
fume Dampf *m*, Dunst *m*, Rauch *m*
fume *v* dampfen, rauchen, ~ **off** abrauchen
fume cupboard Abzug *m* (für Abgase),
　Laborabzug *m*
fume dispersion installation
　Entnebelungsanlage *f*
fume exhaust Rauchabzug *m*
fume hood Abzug *m* (für Abgase),
　Abzugshaube *f*, Laborabzug *m*
fumigant Desinfektionsmittel *n*, Räuchermittel *n*
fumigate *v* [aus]räuchern, ~ **with sulfur**
　ausschwefeln
fumigatin Fumigatin *n*
fumigating Räuchern *n*, Beschwefeln *n*

fumigating candle Räucherkerzchen *n*
fumigating essence Räucheressenz *f*
fumigating paper Räucherpapier *n*
fumigating powder Räucherpulver *n*
fumigation Ausräucherung *f*, Desinfektion *f*,
Räuchern *n*, ~ **with chlorine**
Chlorräucherung *f*
fuming Rauchen *n*
fuming *a* rauchend
fuming sulfuric acid rauchende Schwefelsäure *f*
fumivorous rauchverzehrend
function Funktion *f*, Wirkungsweise *f*, **algebraic**
~ algebraische Funktion *f*, **arbitrary** ~
willkürliche Funktion *f*, **continuous** ~ stetige
Funktion *f* (Math), **continuous integrable** ~
geschlossen integrierbare Funktion *f*,
discontinuous ~ unstetige Funktion *f* (Math),
elementary ~ elementare Funktion *f*, **elliptical**
~ elliptische Funktion *f*, elliptische
Funktion *f*, **even** ~ gerade Funktion *f*, **excess**
~ Überfunktion *f*, **explicit** ~ explizite
Funktion *f* (Math), **finite** ~ beschränkte
Funktion *f*, **implicit** ~ implizite Funktion *f*
(Math), **integrable** ~ integrierbare Funktion *f*
(Math), **integral rational** ~ ganze rationale
Funktion *f*, **inverse** ~ Umkehrfunktion *f*,
linear ~ lineare Funktion *f* (Math),
monotonic ascending ~ monoton steigende
Funktion *f*, **monotonic descending** ~ monoton
fallende Funktion *f*, **odd** ~ ungerade
Funktion *f*, **rational fractional** ~ gebrochen
rationale Funktion *f*, **transcendental** ~
transzendente Funktion *f*, ~ **of several**
variables Funktion *f* mehrerer Variabler
function *v* funktionieren, wirken
functional funktionell
functional capacity Leistungsvermögen *n*
functional center Funktionszentrum *n*
functional derivative Funktionalableitung *f*
functional determinant Funktionaldeterminante *f*
functional disorder Funktionsstörung *f*
functionality Funktionalität *f*
functional matrix Funktionalmatrix *f*
functional transformation
Funktionaltransformation *f*
functioning Funktionieren *n*
function plotter Funktionsschreiber *m*
fundamental fundamental, grundlegend,
grundsätzlich, zugrundeliegend
fundamental band Grundschwingung *f* (Spektr)
fundamental building block
Fundamentalbaustein *m*, Grundbaustein *m*
fundamental current Grundstrom *m* (Elektr)
fundamental equation Hauptgleichung *f*
fundamental experiment Grundversuch *m*
fundamental frequency Grundfrequenz *f* (Elektr)
fundamental lattice Grundgitter *n* (Krist)
fundamental law Grundgesetz *n*, Hauptsatz *m*
fundamental[ly] grundlegend

fundamentally different grundsätzlich
verschieden
fundamental oscillation Grundschwingung *f*
fundamental particle Elementarteilchen *n*
fundamental properties Grundeigenschaften *f pl*
fundamental research Grundlagenforschung *f*
fundamental rule Grundregel *f*
fundamental spectrum Grundspektrum *n*
fundamental structural unit Grundbaustein *m*
fundamental test Grundversuch *m*
fundamental theorem Fundamentalgesetz *n*
(Math)
fundamental unit Elementareinheit *f*,
Grundeinheit *f*
fundamental vibration Grundschwingung *f*
fundamental voltage Grundspannung *f* (Elektr)
fungicidal fungicid, pilztötend
fungicide Fungicid *n*, Antimykotikum *n*,
Fungizid *n*, Mittel *n* gegen Schimmelbildung,
pilztötendes Mittel *n*
fungiform pilzförmig
fungin Fungin *n*, Schwammstoff *m*
fungistatic fungistatisch
fungistatic substance Fungistatikum *n*
fungisterin Fungisterin *n*
fungisterol Fungisterin *n*
fungoid pilzartig, schwammig
fungoid growth Pilzbildung *f*, Pilzwucherung *f*
fungous fungös, pilzartig, schwammig
fungus (pl. fungi) Pilz *m*, Schwamm *m*
fungus cellulose Fungin *n*, Pilzcellulose *f*
funkite Funkit *m*
funnel Trichter *m*, Fülltrichter *m*, Gießloch,
Lunker *m* (Gieß), Rauchfang *m* (Techn),
neck of a ~ Trichterhals *m*, **to pour through a**
~ trichtern
funnel flask Trichterkolben *m*
funnel formation Trichterbildung *f*
funnel holder Trichterhalter *m* (Chem),
Filtergestell *n* (Chem), Filtrierstativ *n* (Chem)
funnel kiln Trichtertrockenofen *m*
funnel pipe Trichterrohr *n*
funnel-shaped trichterförmig
funnel stand Trichterhalter *m* (Chem),
Trichterstativ *n* (Chem)
funnel tube Trichterrohr *n*
fur Fell *n*, Pelz *m*, ~ **in boilers** Kesselstein *m*
furaldehyde Furaldehyd *m*
furan Furan *n*, Furfuran *n*
furan-2-carboxylic acid Brenzschleimsäure *f*
furancarbinol Furfurylalkohol *m*
furanidine Furanidin *n*
furanose ring Furanosering *m*
furanoside Furanosid *n*
furan plastics Kunststoffe *pl* aus Furanharzen
(pl)
furan ring Furanring *m* (Chem)
furazan Furazan *n*
furfuracrolein Furfuracrolein *n*

furfuracrylic acid Furfuracrylsäure *f*
furfural Furfural *n*, Furfuraldehyd *m*, Furfurol *n*
furfuralcohol Furfurylalkohol *m*
furfuraldazine Furfuraldazin *n*
furfuraldehyde Furaldehyd *m*, Furfural *n*,
  Furfurol *n*
furfuramide Furfuramid *n*
furfuran Furan *n*, Furfuran *n*
furfurol[e] Furfurol *n*, Furaldehyd *m*, Furfural *n*
furfurol production plant
  Furfurolgewinnungsanlage *f*
furfurol resin Furfurolharz *n*
furfuryl alcohol Furfurylalkohol *m*,
  Furanalkohol *m*
furfuryl aldehyde Furfurol *n*, Furfural *n*
furfuryl amide Furfuramid *n*
furfurylidene Furfuryliden *n*
furil Furil *n*
furilic acid Furilsäure *f*
furin[e] Furin *n*
furmethide Furmethid *n*
furnace Fabrikofen *m*, Feuerraum *m*, Ofen *m*,
  brick lining of a ~ Ofenausmauerung *f*,
  charging the ~ Ofenbeschickung *f*, **cross
  section of a** ~ Ofenquerschnitt *m*, **diameter of**
  ~ Ofendurchmesser *m*, **electric** ~ elektrischer
  Ofen, **interior of** ~ Ofeninnere *n*, **mouth of a**
  ~ Floßloch *n*, **rim of** ~ Ofenrand *m*, **roof of
  the** ~ Herdgewölbe *n*, **rotary** ~ Drehofen *m*,
  to charge the ~ den Ofen beschicken, **to shut
  down the** ~ den Ofen *m* abstellen, **to tap the**
  ~ den Ofen anbohren, **vertical heating** ~
  Herdtiefofen *m*, **working of a** ~ Ofengang *m*,
  ~ **for hardening** Härteofen *m*, ~ **for heating
  plates** Blechglühofen *m*, ~ **for melting native
  ore** Rohofen *m*, ~ **for refining iron**
  Anfrischfeuerung *f* (Metall), ~ **with cylinder
  grate** Walzenrostfeuerung *f*, ~ **with movable
  fire bars** Schüttelrostfeuerung *f*, ~ **with
  shaking grate** Schüttelrostfeuerung *f*, ~ **with
  two pits** Brillenofen *m*
furnace addition Schmelzzuschlag *m*
furnace arch Feuerungsgewölbe *n*
furnace black Ofenruß *m*, Flammruß *m*
furnace blast Hochofenwind *m*
furnace bottom Ofensohle *f*
furnace cadmia Gichtschwamm *m*,
  Ofengalmei *m*, Ofenschwamm *m*
furnace capacity Ofenfassung *f*
furnace chamber Ofenraum *m*
furnace charge Gicht *f*, Schmelzgut *n*
furnace crown Herdgewölbe *n*
furnace dust, reduction of ~
  Gichtstaubverminderung *f*
furnace efficiency Ofenleistung *f*,
  Ofenwirkungsgrad *m*
furnace ends Ofengekrätz *n*
furnace exhaust gas Ofenabgas *n*
furnace foundation Ofenfundament *n*

furnace gas Gichtgas *n*, Ofengas *n*, Rauchgas *n*
furnace grate Ofenrost *m*
furnace hand Hochofenarbeiter *m*
furnace house Ofenhalle *f*
furnace installation Feuerungsanlage *f*
furnace lead Herdblei *n*
furnace life Ofenhaltbarkeit *f*
furnace lining Ofenauskleidung *f*, Ofenfutter *n*,
  Schachtauskleidung *f*, **care of the** ~
  Schonung *f* des Ofenfutters
furnace load Herdbelastung *f*
furnace loss Abbrand *m*
furnaceman Heizer *m*, Hochofenarbeiter *m*,
  Schmelzer *m*
furnace operation Ofenbetrieb *m*
furnace output Ofenleistung *f*
furnace practice Ofenbetrieb *m*
furnace pressure Ofendruck *m*
furnace process, disturbance of the ~
  Ofengangstörung *f*
furnace product Ofengut *n*
furnace refining Raffination *f* im Schmelzfluß
furnace roof Ofenabdeckung *f*
furnace shell Ofenmantel *m*
furnace slag Ofenschlacke *f*
furnace soot Ofenruß *m*
furnace steel Schmelzstahl *m*
furnace top Ofenkopf *m*
furnace top bell Gichtglocke *f*
furnace top distributor Gichtverschluß *m*
furnace top ring Gichtring *m*
furnace transformer Ofenumformer *m*
furnace voltage Herdspannung *f*
furnish *v* liefern, ausstaffieren, ausstatten
furnishing fabric Dekorationsstoff *m*
furniture Einrichtung *f*, Möbel *pl*
furniture finish Möbellack *m*
furniture polish Möbelpolitur *f*
furniture varnish Möbellack *m*
furocoumarinic acid Furocumarinsäure *f*
furodiazole Furodiazol *n*
furoic acid Brenzschleimsäure *f*, Furoesäure *f*
furoin Furoin *n*
furol Furfurol *n*, Furfural *n*
furonic acid Furonsäure *f*
furostilbene Furostilben *n*
furoxane Furoxan *n*
furoyl, 2- ~ Pyromucyl *n*
furoylacetone Furoylaceton *n*
furoylbenzoylmethane Furoylbenzoylmethan *n*
furrow Nut *f*, Rinne *f*
furrowing Furchung *f*
furry flusig
furs Rauchwaren *pl*
further development Weiterentwicklung *f*
furtrethonium Furtrethonium *n*
furyl Furyl-
furyl carbinol Furfurylalkohol *m*
furylidene Furyliden *n*

fusain Fusain *n*
fusanol Fusanol *n*
fusarinine Fusarin[in] *n*
fuscin Fuscin *n*
fuscinic acid Fuscinsäure *f*
fuscite Fuscit *m*
fuse Zünder *m*, Sicherung *f* (Elektr),
Zündband *n*, **instantaneous** ~
Knallzündschnur *f*
fuse *v* [ein]schmelzen, sintern, verschmelzen, ~
off abschmelzen, ~ **on** aufschmelzen, ~
together verschmelzen, **difficult to** ~ schwer
schmelzbar
fuse box Sicherungsdose *f*, Sicherungskasten *m*
(Elektr)
fuse composition Zündersatz *m*
fused [ein]geschmolzen, ~ **on** aufgeschmolzen
fused catalyst Schmelzkontakt *m*
fused caustic soda Ätznatronschmelze *f*
fused joint Schmelzverbindung *f*
fused mass Schmelzfluß *m*
fused salt Salzschmelze *f*
fused salt electrolysis Schmelzelektrolyse *f*
fuse head Zündpille *f*
fuse housing Zündergehäuse *n*
fuse insert Sicherungseinsatz *m*
fuselage Rumpf *m* (Luftf)
fuse link Abschmelzstreifen *m*,
Sicherungsdraht *m*
fusel oil Fusel *m*, Fuselöl *n*, **containing** ~
fuselhaltig, **free from** ~ fuselfrei, **odor of** ~
Fuselgeruch *m*, **spirit containing** ~ Fusel *m*,
taste of ~ Fuselgeschmack *m*
fuse point Schmelzpunkt *m*
fuse strip Abschmelzstreifen *m*,
Schmelzstreifen *m*
fuse wire Schmelzdraht *m*, Abschmelzdraht *m*,
Schießkabel *n*
fusibility Schmelzbarkeit *f*, **easy** ~
Leichtschmelzbarkeit *f*
fusible schmelzbar, gießbar, **easily** ~ leicht
schmelzbar
fusible clay Schmelzerde *f*
fusible cone Brennkegel *m*, Schmelzkegel *m*,
Segerkegel *m*
fusible cutout Schmelzsicherung *f*
fusible glass Einschmelzglas *n*, Schmelzglas *n*
fusible metal Schnellot *n*
fusing Abbrand *m* (Elektr), Schmelzen *n*,
Sintern *n*, Verschweißen *n*,
Zusammenschmelzen *n*, **time of** ~
Abschmelzdauer *f*
fusing assistant Schmelzmittel *n*
fusing coefficient Abschmelzkonstante *f*
fusing constant Abschmelzkonstante *f*
fusing current Abschmelzstromstärke *f*
fusing fire Schmelzfeuer *n*
fusing-in Einschmelzen *n*

fusing point Schmelzpunkt *m*, Fusionspunkt *m*,
Schmelzgrad *m*
fusing temperature Erweichungstemperatur *f*
fusing voltage Schmelzspannung *f*
fusing zone Schmelzzone *f*
fusion Verschmelzung *f*, Abschmelzung *f*
(Techn), Aufschließen *n* (Anal), Fusion *f*,
Schmelzen *n* (Techn), Schmelzung *f* (Techn),
time of ~ Abschmelzzeit *f*, ~ **of metals**
Metallschmelze *f*, ~ **of slag**
Schlackenverschmelzung *f*
fusion curve Schmelzkurve *f*
fusion electrolysis Schmelzflußelektrolyse *f*
fusion energy Fusionsenergie *f* (Atom),
Verschmelzungsenergie *f*
fusion equilibrium Schmelzgleichgewicht *n*
fusion kettle Schmelzkessel *m*
fusion point Schmelzpunkt *m*
fusion pot Schmelzpfanne *f*
fusion process Schmelzgang *m*
fusion reaction Verschmelzungsreaktion *f*
fusion reactor Fusionsreaktor *m* (Atom)
fusion temperature Schmelzhitze *f*
fusion visco[si]meter Schmelzviskosimeter *n*
fusion welding Schmelzschweißung *f*
fustet Färberbaum *m* (Bot), Färbersumach *m*
(Bot)
fustic extract Gelbholzextrakt *m*
fustic wood Fustikholz *n*, Gelbholz *n*,
Kubaholz *n*
fustin Fustin *n*
futile nichtig, zwecklos
futility Nichtigkeit *f*
future possibility Zukunftsmöglichkeit *f*
fuzz *v* fasern
fuzziness Unreinheit *f*, Verschwommenheit *f*
fuzzy unscharf, verschwommen, undeutlich

G

gabardine Gabardine *f* (Text)
gabbro Gabbro *m* (Min), Schillerfels *m* (Min)
gaberdine Gabardine *f* (Text)
Gabriel's phthalimide synthesis Gabrielsche
   Phthalimidsynthese *f* (Chem)
gadget Vorrichtung *f*
gadoleic acid Gadoleinsäure *f*
gadolinia Gadolinerde *f*, Gadoliniumoxid *n*
gadolinite Gadolinit *m* (Min)
gadolinium (Symb. Gd) Gadolinium *n*
gadolinium bromide Gadoliniumbromid *n*
gadolinium chloride Gadoliniumchlorid *n*
gadolinium nitrate Gadoliniumnitrat *n*
gadolinium oxide Gadoliniumoxid *n*,
   Gadolinerde *f*
gadolinium sulfate Gadoliniumsulfat *n*
gafrinin Gafrinin *n*
gag Knebel *m*
gahnite Gahnit *m* (Min), Zinkspinell *m* (Min)
gaidic acid Gaidinsäure *f*
gain Gewinn *m*, Nutzen *m*, Verstärkung *f*
   (Elektr), Zunahme *f*
gain *v* erwerben, gewinnen
gain amplification Vorverstärkung *f*
gain control Verstärkerregelung *f* (Elektr),
   Amplitudenregelung *f* (Elektr)
gain equation Ausbeutegleichung *f* (Atom)
galactal Galaktal *n*
galactamine Galaktamin *n*
galactan Galaktan *n*
galactaric acid Galaktarsäure *f*
galactase Galaktase *f* (Biochem)
galactic system Milchstraßensystem *n* (Astr)
galactide Galaktid *n*
galactin Galaktin *n*
galactite Galaktit *m* (Min), Milchjaspis *m*
   (Min), Milchstein *m* (Min)
galactobiose Galaktobiose *f*
galactochloralic acid Galaktochloralsäure *f*
galactochloralose Galaktochloralose *f*
galactoflavine Galaktoflavin *n*
galactoheptulose Galaktoheptulose *f*
galactokinase Galaktokinase *f* (Biochem)
galactometasaccharine Galaktometasaccharin *n*
galactometer Galaktometer *n*, Milchmesser *m*
galactonic acid Galactonsäure *f*, Lactonsäure *f*
galactosamine Galactosamin *n*
galactosaminic acid Galaktosaminsäure *f*
galactosan Galaktosan *n*
galactose Galaktose *f*
galactose galacturonic acid
   Galaktosegalakturonsäure *f*
galactosidase Galaktosidase *f*
galactoside Galaktosid *n*, Cerebrosid *n*
galactosone Galaktoson *n*
galacturonic acid Galakturonsäure *f*

galaheptite Galaheptit *m*
galaheptonic acid Galaheptonsäure *f*
galaheptose Galaheptose *f*
galalith Kasein-Kunsthorn *n*
galanga oil Galgantöl *n*
galanga root Galgantwurzel *f*
galangin Galangin *n*
galanginidine Galanginidin *n*
galanginidinium hydroxide
   Galanginidiniumhydroxid *n*
galanthamine Galanthamin *n*
galanthaminic acid Galanthaminsäure *f*
galanthaminone Galanthaminon *n*
galanthidine Galanthidin *n*
galanthine Galanthin *n*
galaxy Milchstraße *f* (Astr)
galbacin Galbacin *n*
galban Galban *n*
galban resin Galbanharz *n*
galbanum Galbanharz *n*, Mutterharz *n*
galbanum oil Galbanumöl *n*
galbulin Galbulin *n*
galcatin Galcatin *n*
galegine Galegin *n*
galena Bleiglanz *m* (Min), Bleisulfid *n* (Min),
   Galenit *m* (Min), **fine ~** Hafnererz *n* (Min)
galenic bleiglanzhaltig
galenite Galenit *m* (Min), Bleiglanz *m* (Min),
   Bleisulfid *n* (Min)
galenobismuthite Galenobismutit *m* (Min)
galeopsis Hanfnesselkraut *n* (Bot)
galgravin Galgravin *n*
galicide Galicid *n*
Galilean number Galilei-Zahl *f*
galingale Galgant *m*, Galgantwurzel *f*
galingale oil Galgantöl *n*
galiosin Galiosin *n*
galipidine Galipidin *n*
galipine Galipin *n*
galipoidine Galipoidin *n*
galipot Galipot *m*, Gallipotharz *n*
galipot resin Gallipotharz *n*
galium Labkraut *n* (Bot)
gall Galle *f* (Med), **bitter as ~** gallenbitter
gall *v* gallieren
gallaceteine Gallacetein *n*
gallacetophenone Gallacetophenon *n*
gallal Gallal *n*
gallamide Gallamid *n*
gallamine blue Gallaminblau *n*
gallanilide Gallanilid *n*, Gallanol *n*
gallanol Gallanilid *n*, Gallanol *n*
gallate Gallat *n*, Salz *n* der Gallussäure (f)
gallate *a* gallussauer
gallbladder Gallenblase *f* (Med)
gallein Gallein *n*, Pyrogallolphthalein *n*

gallery Galerie *f*, Tribüne *f*
gallery tunnel Stollen *m* (Bergb)
galley [proof] Bürstenabzug *m*, Fahnenabzug *m* (Buchdr)
gallgreen gallengrün
gallic acid Gallussäure *f*
gallic chloride Gallium(III)-chlorid *n*, Gallichlorid *n*
gallic fermentation Gallussäuregärung *f*
gallic hydroxide Gallihydroxid *n*, Gallium(III)-hydroxid *n*
gallicin Gallicin *n*
gallic oxide Gallium(III)-oxid *n*, Gallioxid *n*
gallic salt Gallisalz *n*, Gallium(III)-Salz *n*
gallin Gallein *n*
galling Fressen *n* (Friktion)
gallinol Gallanilid *n*, Gallanol *n*
gallipharic acid Gallipharsäure *f*
gallipinic acid Gallipinsäure *f*
gallipot Galipot *m*
gallisine Gallisin *n*
gallium (Symb. Ga) Gallium *n*
gallium antimonide Galliumantimonid *n*
gallium arsenide Galliumarsenid *n*
gallium chloride Galliumchlorid *n*
gallium content Galliumgehalt *m*
gallium dichloride Gallochlorid *n*, Galliumchlorür *n* (obs), Gallium(II)-chlorid *n*
gallium(II) chloride Gallochlorid *n*, Galliumchlorür *n* (obs), Gallium(II)-chlorid *n*
gallium(III) chloride Gallium(III)-chlorid *n*, Gallichlorid *n*
gallium(III) hydroxide Gallihydroxid *n*, Gallium(III)-hydroxid *n*
gallium(III) oxide Gallium(III)-oxid *n*, Gallioxid *n*
gallium(III) salt Gallisalz *n*, Gallium(III)-Salz *n*
gallium(II) oxide Gallium(II)-oxid *n*, Galliummonoxid *n*, Gallooxid *n*
gallium ion Galliumion *n*
gallium monoxide Gallium(II)-oxid *n*, Galliummonoxid *n*, Gallooxid *n*
gallium sesquioxide Gallium(III)-oxid *n*, Gallioxid *n*
gallium sulfide Schwefelgallium *n*
gallium trichloride Gallium(III)-chlorid *n*, Gallichlorid *n*
gall-like gallig
gallnut Gallapfel *m* (Bot), Laubapfel *m*
gallnut extract Galläpfelextrakt *m*, **to treat with ~** gallieren
gallnut ink Gallustinte *f*
gallnuts, tincture of ~ Galläpfeltinktur *f*
gallocatechol Gallocatechin *n*
gallocyanin Gallocyanin *n*
galloflavin Galloflavin *n*
gallon Gallone *f* (Maß)
gallonitrile Gallonitril *n*
gallotannic acid Gallusgerbsäure *f*

gallotannin Gallusgerbsäure *f*
gallous chloride Gallium(II)-chlorid *n*, Gallochlorid *n*
gallous oxide Gallium(II)-oxid, Gallooxid *n*, Galliummonoxid *n*
galloyl Galloyl-
galloylbenzophenone Galloylbenzophenon *n*
gall stone Gallenstein *m* (Med)
galuteolin Galuteolin *n*
galvanic galvanisch
galvanic cell galvanisches Element *n*
galvanic coloring Galvanochromie *f*
galvanism Galvanismus *m*
galvanization Galvanisierung *f* (Med., Phys)
galvanize *v* verzinken (Techn), Metallüberzüge auf schmelzflüssigem Wege aufbringen (Techn)
galvanizing Verzinken *n* (Techn), **~ by dipping** heißes Verzinken *n*, Tauchgalvanisierung *f*
galvanizing bath Verzinkungswanne *f*, **hot ~** Metallsud *m*
galvanizing plant Verzinkerei *f*
galvanocautery Galvanokaustik *f*
galvanograph Galvano *n*
galvanographic galvanographisch
galvanography Galvanographie *f*
galvanomagnetic galvanomagnetisch
galvanomagnetism Galvanomagnetismus *m*
galvanometer Galvanometer *n*, **iron-clad ~** Panzergalvanometer *n*, **recording ~** Registriergalvanometer *n*, **reflecting ~** Spiegelgalvanometer *n*, **~ with ball-shaped shield** Kugelpanzergalvanometer *n*, **~ with hairpin shaped magnet** Haarnadelgalvanometer *n*
galvanometric galvanometrisch, galvanostatisch
galvanometry Galvanometrie *f*
galvanoplastic galvanoplastisch
galvanoplastic art Galvanoplastik *f*
galvanoplastic process Galvanoplastik *f*, **rapid ~** Schnellgalvanoplastik *f*
galvanoplastics Galvanoplastik *f*
galvanoscope Galvanoskop *n*
galvanostegy Galvanostegie *f*
galyl Galyl *n*
gamabufogenin Gamabufogenin *n*
gamatin Gamatin *n*
gambin Gambin *n*
gambir Gambir *n*
gambir catechol Gambircatechin *n*
gamboge Gummigutt *n*
gamboge butter Gambogebutter *f*
game Wildbret *n*
gamete Gamet *m*, Geschlechtszelle *f*, Keimzelle *f*
gametoblast Gametoblast *m* (Biol)
gametocyte Gametozyt *m*
gametocyte carrier Gametozytenträger *m*
gametophyte Gametophyt *m*
gamma acid Gammasäure *f*

gamma-active gamma-[radio]aktiv (Atom)
gamma-activity Gamma-Aktivität *f* (Atom)
gamma background Gammaspiegel *m*
gammacerane Gammaceran *n*
gamma counter tube Gammazählrohr *n*
gamma decay Gamma-Zerfall *m* (Atom)
gamma disintegration Gamma-Zerfall *m* (Atom)
gamma-globulin Gamma-Globulin *n*
gammagraphy Gammagraphie *f,*
   Gammaradiographie *f*
gamma iron Austenit *m*, Gammaeisen *n*
gamma irradiation Gammabestrahlung *f*
gamma particle Gammateilchen *n*
gamma position Gammastellung *f*
gamma quantum Gammaquant *n*
gamma radiation Gammastrahlung *f*
gamma-radioactive gamma-[radio]aktiv (Atom)
gamma radioactivity Gamma-Aktivität *f* (Atom)
gamma radiography Gammaradiographie *f*
gamma ray Gammastrahl *m*
gamma-ray activity Gammastrahlaktivität *f*
gamma-ray energy Gammaenergie *f* (Atom)
gamma-ray photon Gammastrahlphoton *n*
   (Atom)
gamma-ray quantum Gammaquant *n*
gamma-ray scattering Gammastrahlstreuung *f*
gamma-ray spectrometer Gammaspektrometer *n*
gamma-ray spectroscopy
   Gamma-Spektroskopie *f* (Atom)
gamma-ray spectrum Gammastrahlenspektrum *n*
gamma-ray therapy Gammastrahlentherapie *f*
gamma-resonance spectroscopy
   Gamma-Resonanzspektroskopie *f* (Spektr)
gamma-sensitive gammaempfindlich
gamma-sensitivity Gammaempfindlichkeit *f*
   (Atom)
gamma uranium Gamma-Uran *n* (Atom)
gammexane Gammexan *n*
   (Insektenvertilgungsmittel, HN)
gamont Gametozyt *m*
gamow factor Gamowfaktor *m*
gang Gang *m*, Gangart *f* (Web), Ganggestein *n*
   (Min)
gangliocyte Ganglienzelle *f* (Histol)
ganglion Ganglion *n*, Nervenknoten *m*
ganglion cell Ganglienzelle *f* (Histol)
ganglioside Gangliosid *n*
gangue Ganggestein *n* (Min), Gangmineral *n*
   (Min)
gangway Durchgang *m*
ganister Ganister *m* (Min)
ganoblast Adamantoblast *m* (Schmelzzelle)
ganomalite Ganomalith *m* (Min)
ganomatite Ganomatit *m* (Min),
   Gänsekotigerz *n* (Min)
ganophyllite Ganophyllit *m* (Min)
gantrisin Gantrisin *n*
gantry Rohrbrücke *f*
gantry crane Bockkran *m*

gap Abstand *m*, Lücke *f,* Öffnung *f,* Riß *m*,
   Spalt *m*, Spalte *f,* Zwischenraum *m*
gap conductivity Löcherleitung *f*
gap-filling fugenfüllend
gap filling adhesive Fugenkitt *m*
gap filling material Ausgleichsmasse *f*
gap length Lückenlänge *f*
gap length distribution Lückenlängenverteilung *f*
gap length measurement Lückenlängenmessung *f*
gaps, having ~ lückenhaft
garancin Krappfärbestoff *m*
garancine Garanzin *n*
garbage Abfall *m*, Müll *m*
garbage chute Müllschlucker *m*
garbage disposal Müllabfuhr *f*
garbage dump or pit Müllabladeplatz *m*
garbage removal Müllabfuhr *f*
garble *v* verstümmeln
gardenin Gardenin *n*
gardening Gartenbau *m*
garden mold Pflanzenerde *f*
gargle Mundwasser *n*, Gurgelwasser *n*
garlic Knoblauch *m* (Bot)
garlic-like knoblauchartig
garlic mustard oil Lauchhederichöl *n*
garlic oil Knoblauchöl *n*
garnet Granat *m* (Min), **artificial** ~
   Granatfluß *m*, **common** ~ gemeiner Granat *m*
   (Min), **green** ~ Kalkgranat *m* (Min), **mock** ~
   Glasgranat *m*, **white** ~ Leucit *m* (Min)
garnet carbuncle Karfunkelstein *m* (Min, obs)
garnetiferous granathaltig
garnet lac (Am. E.) Schellack *m*
garnet-like granatartig
garnierite Garnierit *m* (Min)
garnish *v* verzieren, garnieren
garnish plate Zierschild *n*
garryine Garryin *n*
gas (also see gases) Gas *n*, Benzin *n* (Am. E.),
   absorbed ~ absorbiertes Gas , **absorption of** ~
   Gasaufnahme *f,* **accumulation of** ~
   Gasansammlung *f,* **bottled** ~ Flaschengas *n*,
   burnt ~ verbranntes Gas *n*, **combustible** ~
   brennbares Gas *n*, Verbrennungsgas *n*,
   compressed ~ komprimiertes Gas *n*,
   Preßgas *n*, **current of** ~ Gasstrom *m*, **escape of**
   ~ Gasausströmung *f,* Gasentweichung *f,* **flask**
   for developing ~ Gasentwicklungsflasche *f,*
   foreign ~ Fremdgas *n*, **formation of** ~
   Gasentwicklung *f,* **generation of** ~
   Gasentwicklung *f,* **ideal** ~ ideales Gas *n*,
   ignition of ~ Gasentzündung *f,* **impermeable**
   to ~ gasundurchlässig, **inert** ~ indifferentes
   Gas *n*, inertes Gas *n*, **ionized** ~ ionisiertes
   Gas *n*, **lean** ~ Armgas *n*, geringwertiges
   Gas *n*, **lethal** ~ Giftgas *n*, **noxious** ~
   schädliches Gas *n*, **permanent** ~ permanentes
   Gas *n* (Phys), **permeability to** ~
   Gasdurchlässigkeit *f,* **permeable to** ~

gasdurchlässig, **poor** ~ geringwertiges Gas *n*,
producing ~ gasbildend, gaserzeugend,
production of ~ Gasentwicklung *f*,
Gaserzeugung *f*, **quality of the** ~
Gasbeschaffenheit *f*, **quantity of** ~
Gasmenge *f*, **reducing** ~ Reduktionsgas *n*,
residual ~ Gasrückstand *m*, **rich** ~
hochwertiges Gas *n*, **smell of** ~ Gasgeruch *m*,
to draw off ~ Gas absaugen, **to turn on the** ~
Gas aufdrehen, **toxic** ~ Giftgas *n*, **treated** ~
Reingas *n*, **type of** ~ Gasart *f*, **unburnt** ~
Frischgas *n*, **weight of** ~ Gasgewicht *n*, ~ **for**
heating Heizgas *n*, ~ **from low-temperature**
distillation Schwelgas *n*, ~ **from roasting**
Röstgas *n*, ~ **of combustion**
Verbrennungsgas *n*
gas *v* gasen
gas-absorbing gasabsorbierend
gas absorptiometer Gasabsorptionsmesser *m*
gas absorption Gasabsorption *f*, Gasaufnahme *f*
gas analyser Gasspurgerät *n*
gas analysis Gasanalyse *f*, Gasuntersuchung *f*
gas-analytical gasanalytisch
gas analyzing apparatus Gasanalysenapparat *m*
gas and water supply Gas- und
Wasserversorgung *f*
gas apparatus Gasapparat *m*
gas balance Gaswaage *f*
gas ballast pump Gasballastpumpe *f*
gas battery Gasbatterie *f* (Elektr), Gaselement *n*
(Elektr)
gas bell Gasglocke *f*, Gashalter *m*
gas black Gasruß *m*
gas bleaching Gasbleiche *f*
gas bombardment Gasbeschießung *f*
gas bottle Gasflasche *f*
gas bubble Gasbläschen *n*, Gasblase *f*
gas burette Gasbürette *f*
gas burner Gasbrenner *m*, Gaslampe *f*
gas calorimeter Gaskalorimeter *n*
gas candle Gaseinleitungskerze *f*
gas carbon Retortenkohle *f*
gas catcher Gasfang *m*
gas cell Gaselement *n* (Elektr), Gaskette *f*
(Elektr)
gas centrifuge Gaszentrifuge *f*
gas chromatogram Gaschromatogramm *n*
gas chromatography Gaschromatographie *f*
gas circulating pump Gasumlaufpumpe *f*
gas cleaner Gasreiniger *m*,
Gasreinigungsapparat *m*
gas cleaning Gasreinigung *f*
gas cleaning plant Gasreinigungsanlage *f*
gas coal Gaskohle *f*, Schmiedekohle *f*
gas coke Gaskoks *m*
gas collecting tube Gassammelröhre *f*
gas collector Gasfang *m*, Gassammler *m*
gas compressor Gasverdichter *m*

gas concentration tester
Gaskonzentrationsmesser *m*
gas condenser Gaskondensator *m*, Gaskühler *m*
gas conditioning Gasreinigung *f*
gas connection Gasanschluß *m*
gas constant Gaskonstante *f*
gas constituent Gasbestandteil *m*
gas consumption Gasverbrauch *m*
gas container Gasbehälter *m*
gas content Gasgehalt *m*
gas cooler Gaskühler *m*
gas current Gasstrom *m*
gas cutting Autogenschneiden *n*,
Brennschneiden *n*
gas cylinder Gasflasche *f*, Gaszylinder *m*
gas density Gasdichte *f*
gas density recorder Gasdichteschreiber *m*
gas detector Gasdetektor *m*
gas diffusion Gasdiffusion *f*
gas diffusion plant Gasdiffusionsanlage *f*
gas discharge Gasentladung *f* (Elektr)
gas discharge lamp Gasentladungslampe *f*
gas discharge tube Gasentladungsröhre *f*,
Kaltlichtröhre *f*
gas distribution Gasverteilung *f*
gas drying apparatus Gastrockner *m*
gas duct Gasweg *m*
gas dynamo Gasdynamo *m*
gaseity Gaszustand *m*
gas electrode Gaselektrode *f*
gas engine Gasmaschine *f*, Gasmotor *m*
gas engineering Gastechnik *f*
gas envelope Gashülle *f*, Schutzgas *n*
gaseous gasförmig, gasartig, gashaltig, gasig,
luftartig
gaseous atmosphere Gasatmosphäre *f*
gaseous condition Gaszustand *m*
gaseous diffusion Trennwanddiffusion *f*
gaseous fuel gasförmiger Brennstoff *m*,
Heizgas *n*
gaseous mixture Gasgemenge *n*
gaseousness Gaszustand *m*, Gasartigkeit *f*,
Gasförmigkeit *f*
gas equation, adiabatic ~ adiabatische
Gasgleichung *f*
gases, free of ~ gasfrei, **mixture of** ~
Gasgemisch *n*, **nitrous** ~ nitrose Gase *pl*, **real**
~ reale Gase *pl*, **theory of** ~ Gastheorie *f*
gas escape Gasausströmung *f*
gas exit pipe Gasableitungsrohr *n*, Gasfang *m*
gas exit tube Gasableitungsrohr *n*
gas expeller Entgaser *m*
gas explosion Gasexplosion *f*
gas-filled gasgefüllt
gas-filled relay Ionenschalter *m*
gas-fired gasbeheizt
gas-fired furnace Gasfeuerung *f*
gas firing Gasfeuerung *f*
gas fittings Gasanlage *f*, pl

gas flame Gasflamme *f*
gas flow control Gasströmungsregelung *f*
gas flue Gasabzug *m*
gas formation Gasbildung *f,* Gasentwicklung *f*
gas-forming gasbildend
gas furnace gasgefeuerter Ofen *m,* Gasofen *m*
gas gauge Gasdruckmesser *m,*
  Kraftstoffanzeiger *m* (Kfz)
gas generating apparatus
  Gasentwicklungsapparat *m*
gas generating bottle Gasentbindungsflasche *f*
gas generation Gaserzeugung *f*
gas generator Gasentwickler *m,*
  Gasentwicklungsapparat *m,* Gaserzeuger *m,*
  Gasgenerator *m,* Generator *m* zur
  Gaserzeugung
gas governor Gasregler *m*
gash klaffende Wunde *f*
gas-heated gasbeheizt, gasgeheizt
gas heater Gasofen *m*
gas heating Gasheizung *f,* Gasfeuerung *f*
gasholder Gasglocke *f,* Gassammler *m*
gas hold-up relativer Gasanteil *m* (Dest)
gasifiability Vergasbarkeit *f*
gasifiable vergasbar
gasification Vergasung *f,* Zerstäubung *f,* **fluidized**
  ~ Schwebevergasung *f,* ~ **in retorts**
  Retortenvergasung *f,* ~ **of coal**
  Kohlenvergasung *f*
gasification plant Vergasungsanlage *f*
gasifier Vergaser *m*
gasiform gasförmig, gasartig
gasify *v* vergasen, in Gas umwandeln
gasifying Vergasen *n*
gasifying heat Vergasungswärme *f*
gas ignition Gaszündung *f*
gas inlet Gaszuführung *f*
gas inlet capillary tube Gaszuführungskapillare *f*
gas inlet pipe Gaszuleitungsröhre *f*
gas inlet tube Gaszuleiter *m*
gas jet Gasdüse *f,* Gasstrahl *m*
gasket Dichtung *f,* Dichtungsmanschette *f,*
  Dichtungsring *m,* Flanschdichtung *f,*
  concentric ~ Manschettendichtung *f,*
  cylindrical ~ Lippendichtung *f,* **self-energized**
  ~ Formdichtung *f*
gasket ring Dichtungsring *m*
gasket sheet Dichtungsscheibe *f*
gasket swelling Quellen *n* der Dichtung
gaskinetic gaskinetisch
gas lamp Gaslampe *f*
gas law Gasgesetz *n*
gaslight Gasflamme *f,* Gaslampe *f,* Gaslicht *n,*
  incandescent ~ Gasglühlicht *n*
gas lighter Gasanzünder *m,* **automatic** ~
  Gasselbstzünder *m*
gas-lighting Gasbeleuchtung *f*
gas-like gasförmig
gas lime Gaskalk *m,* Grünkalk *m*

gas-liquefying plant Gasverflüssigungsanlage *f*
gas-liquid chromatography
  Gas-Flüssig-Chromatographie *f*
gas-liquid-extraction Gas-Flüssig-Extraktion *f*
gas liquor Gaswasser *n*
gas main pipe Gashauptleitung *f*
gas mantle Gasglühlichtkörper *m,*
  Gasglühlichtstrumpf *m,* Glühstrumpf *m*
gas mask Gasmaske *f,* Atemschutzgerät *n*
gas measurement Gasmessung *f*
gas-measuring tube Gasmeßrohr *n*
gas meter Gasuhr *f,* Gaszähler *m*
gas methanizing plant Gasmethanisieranlage *f*
gas mixing chamber Gasmischkammer *f*
gas mixture Gasgemisch *n,* Gasmischung *f,*
  non-ideal ~ nicht-ideale Gasmischung *f*
gas mixture controller Gasgemischregler *m*
gas molecule Gasmolekül *n*
gas motor Gas[kraft]maschine *f*
gas noise Ionenrauschen *n* (Elektr)
gas occlusion Gaseinschluß *m*
gas odorization Gasodorierung *f*
gas off-take Gasabzugsrohr *n,*
  Gichtgasabzugsrohr *n*
gas oil Gasöl *n*
gasoline Benzin *n,* Gasolin *n* (Chem),
  Kristallöl *n* (obs), **cracked** ~ Krackbenzin *n,*
  heavy ~ Schwerbenzin *n,* **medium heavy** ~
  Mittelbenzin *n,* ~ **from coal** Schwelbenzin *n,*
  ~ **used for cleaning purposes** Waschbenzin *n*
gasoline additives Kraftstoffzusätze *pl*
gasoline-air mixture Benzin-Luft-Gemisch *n*
gasoline consumption Benzinverbrauch *m*
gasoline container Benzinkanister *m*
gasoline engine Benzinmotor *m*
gasoline filter Benzinfilter *n,* Benzinreiniger *m*
gasoline float Benzinschwimmer *m*
gasoline gauge Benzinstandmesser *m,*
  Kraftstoffanzeiger *m*
gasoline level Benzinstand *m*
gasoline line Benzinleitung *f*
gasoline-oil mixture Benzin-Öl-Gemisch *n*
gasoline plant Benzin[gewinnungs]anlage *f*
gasoline pressure indicator Benzindruckmesser *m*
gasoline pump Benzinpumpe *f*
gasoline refining plant Benzinraffinerieanlage *f*
gasoline strainer Benzinfilter *n*
gasoline substitute Benzinersatz *m*
gasoline tank Benzintank *m*
gasoline trap Benzinabscheider *m*
gasoline vapor Benzindampf *m*
gasometer Gasometer *n,* Gasbehälter *m,*
  Gasglocke *f,* Gassammler *m*
gasometric gasometrisch, gasvolumetrisch
gasometry Gasmessung *f,* Gasometrie *f*
gas outlet Gasabzug *m*
gas outlet pipe Gasableitungsrohr *n*
gas oxygen lamp Leuchtgassauerstofflampe *f*
gas packing unter Gasschutz *m* verpacken

gas passage Gasdurchgang *m*, Gasweg *m*
gas permeability tester
Gasdurchlässigkeits[prüf]gerät *n*
gas phase Gasphase *f*, Gasraum *m*
gas phototube gasgefüllte Photozelle *f*
gas pipe Gasrohr *n*
gas pipeline Erdgasleitung *f*
gas pipet[te] Gaspipette *f*
gas piping Gasleitung *f*
gas pocket Gaseinschluß *m*
gas poisoning Gasvergiftung *f*
gas polarization Gaspolarisation *f*
gas pore Gaspore *f*
gas power engine Gaskraftmaschine *f*
gas pressure Gasdruck *m*, **positive ~**
Gasüberdruck *m*
gas pressure regulator Gasdruckregler *m*,
Gasdruckregulator *m*
gas producer Gaserzeuger *m*,
Gaserzeugungsapparat *m*, Gasgenerator *m*,
Generator *m* (Chem), Kraftgasanlage *f*
gas producing plant Gaserzeugungsanlage *f*
gas production Gasgewinnung *f*, Gasbereitung *f*
gasproof gasdicht, gassicher
gas puddling Gasfrischen *n*
gas purification Gasreinigung *f*
gas purifier Gasreiniger *m*
gas purifying apparatus Gasreiniger *m*
gas receiver Gasabsperrblase *f*
gas-recycle coking oven Spülgasschwelofen *m*
(Lurgi)
gas refrigeration cycle Gaskälteprozeß *m*,
isochoric ~ stilisierter Gaskälteprozeß *m*
gas refrigeration machine Kaltgasmaschine *f*
gas register Gasdruckregistrierapparat *m*
gas regulator Gasdruckregler *m*, Gaseinsteller *m*
gas reservoir Gasbehälter *m*
gas residue Gasrückstand *m*
gas retort Gasretorte *f*
gas retort carbon Gaskohle *f*
gas sample Gasprobe *f*
gas sampling valve Probengeber *m* für Gase
(Gaschromat)
gas scrubber Gaswäscher *m*
gas scrubbing Gaswäsche *f*
gas separation plant Gaszerlegungsanlage *f*
gas separator Gastrenner *m*
gassing Gasentwicklung *f*, Vergasen *n*,
Vergasung *f*
gas-solid chromatography
Gas-Festkörper-Chromatographie *f*
gas space Gasraum *m*, **volume of ~**
Gasraumvolumen *n*
gas stove Gasherd *m*, Gasofen *m*
gas suction plant Gasabsaugeanlage *f*
gassy gasartig, gashaltig
gastaldite Gastaldit *m* (Min)
gas tank Gasbehälter *m*, Gasometer *n*
gas tap Gashahn *m*

gas tar Gasteer *m*, Kohlenteer *m*
gas tee piece Gasteilungsröhre *f*
gas temperature Gastemperatur *f*
gas temperature regulator
Gastemperaturregler *m*
gas test Gasprobe *f*
gas tester Gasprüfer *m*, Gasprüfvorrichtung *f*,
Gasspürgerät *n*
gas testing apparatus Gasprüfvorrichtung *f*
gas thermometer Gasthermometer *n*
gas-tight gasdicht, gasundurchlässig
gas trace apparatus Gasspurenmesser *m*
gastric acid Magensäure *f*
gastric juice Magensaft *m*
gastric ulcer Magengeschwür *n* (Med)
gas tube Gasflasche *f*, Gasentladungsröhre *f*,
Ionenröhre *f*
gas turbine Gasturbine *f*
gas ultracentrifuge Ultra-Gaszentrifuge *f*
gas valve Gasventil *n*
gas volume Gasvolumen *n*
gas volumeter Gasvolumeter *n*
gas-volumetric gasvolumetrisch
gas washer Gaswäscher *m*, Skrubber *m*
gas water Gaswasser *n*
gas welding autogenes Schweißen *n*,
Gasschmelzschweißen *n*, Gasschweißen *n*
gas welding furnace Gasschweißofen *m*
gas works Gasanstalt *f*, Gaswerk *n*
gas yield Gasanfall *m*, Gasausbeute *f*
gate Eingußkanal *m*, Anguß *m*,
Eingußtrichter *m*, **direct ~** direkter Anguß *m*,
restricted ~ verjüngter Anguß *m*, **ring ~**
ringförmiger Anguß *m*
gate *v* einblenden (Elektron)
gate circuit Eingangskreis *m*
gate cutoff valve Angußabschneideventil *n*
gate cutter Angußabstanzer *m*
gate mark Angußstelle *f*
gate paddle agitator Rahmenrührer *m*
gate paddle mixer Gitterrührer *m*
gate pin Eingußmodell *n*
gate valve Absperrventil *n*, Absperrschieber *m*
gate valve housing Schiebergehäuse *n*
gather *v* [auf]sammeln, auffangen, lesen
(Früchte etc.), vereinigen
gathering anode Sammelanode *f*
gathering electrode Sammelelektrode *f*
gathering point Anfallstelle *f*
Gattermann aldehyde synthesis Gattermannsche
Aldehydsynthese *f*
gauge Meßgerät *n*, Anzeiger *m*, Eichmaß *n*,
Kaliber *n*, Maß *n*
gauge *v* eichen, [ab]messen; ausmessen,
kalibrieren
gauge block Meßklotz *m*
gauge cock Probierhahn *m*, Wasserstandshahn *m*
gauge group Eichgruppe *f*
gauge invariance Eichinvarianz *f*

gauge length Meßlänge *f*
gauge mark Eichstrich *m*, Eichnagel *m*
gauge pressure Manometerdruck *m*
gauger Eicher *m*, Eichmeister *m*, Meßgerät *n*
gauge reading Lehrenablesung *f*
gauge substance Eichsubstanz *f*
gauge tolerance Lehrentoleranz *f*
gauge transformation Eichtransformation *f*
gauge uniformity Maßgenauigkeit *f*
gauging Eichen *n*, Eichung *f*, Messung *f*,
    Passung *f*
gauging apparatus Kubizierapparat *m*
gauging chain Meßkette *f*
gaultheria oil Bergteeöl *n* (Bot), Gaultheriaöl *n*
    (Bot)
gaultherin Gaultherin *n*
gaultherolin Gaultherolin *n*
gauss Gauß *n* (Phys)
gauss algorithm Austauschalgorithmus *m*
Gauss eyepiece Gauß-Okular *n* (Opt)
Gaussian distribution Gaußsche Verteilung *f*
Gaussian image point Gaußscher Bildpunkt *m*
    (Opt)
gauze Gaze *f*, Flor *m*, Netz *n*
gauze filter Gazefilter *n*
gauze ribbon Gazeband *n*
Gay-Lussac law Gay-Lussacsches Gesetz *n*
Gay-Lussac tower Gay-Lussac-Turm *m* (Chem)
gaylussite Gaylussit *m* (Min), Natrocalzit *m*
    (Min)
GDP (Abk) Guanosindiphosphat *n*
geamine Geamin *n*
gear Getriebe *n*, Gang *m* (Auto),
    Getriebegang *m*, Getriebestufe *f* (Auto),
    Verzahnung *f*, disengaging ~ Ausrücker *m*,
    three-speed ~ dreistufiges Getriebe *n*
gear *v* übersetzen (Getriebe)
gear case Getriebegehäuse *n*, Kettenschutz *m*
    (Fahrrad)
gear chain Zahnkette *f*
gear changing Schalten *n*
gear coupling, dihedral ~ Bogenzahnkupplung *f*
gear cutting machine Zahnradfräsmaschine *f*,
    Verzahnmaschine *f* (Techn)
gear drive Zahnradantrieb *m*
gear[ed] clutch Getriebekupplung *f*
gear head motor Stirnradmotor *m*
gearing Getriebe *n*, Räderwerk *n*, Triebwerk *n*,
    Übersetzung *f*, Verzahnung *f*, intermediate ~
    Zwischengetriebe *n*
gearksutite Gearksutit *m* (Min)
gear lubricant Getriebeöl *n*
gear pump Zahnradpumpe *f*
gear rim Zahnkranz *m*
gear shaft Radwelle *f*
gearshift Schaltung *f*
gearshift bar Schaltstange *f*
gearshift[ing] Gangschaltung *f* (Auto)
gearshift knob Schaltknopf *m*

gear teeth, surface strength of ~
    Zahnflankfestigkeit *f*
gear transmission Übersetzungsgetriebe *n*
gear-type meter Ovalradzähler *m*
gear unit Getriebeeinheit *f*
gear wheel Zahnrad *n*, Getrieberad *n*
gear wheel molding machine
    Zahnräderformmaschine *f*
gedanite Mineralharz *n*
gedrite Gedrit *m* (Min)
gehlenite Gehlenit *m* (Min)
geic acid Ulminsäure *f*
geierite Geierit *m* (Min)
Geiger counter Geigersches Spitzenzählrohr *n*,
    Geigerzähler *m*
geigeric acid Geigersäure *f*
geigerin Geigerin *n*
Geiger-Mueller counter Geiger-Müller-Zähler *m*
Geiger threshold Geigerschwelle *f*
geijerin Geijerin *n*
geijerinic acid Geijerinsäure *f*
geikielite Geikielith *m* (Min), Ilmenit *m* (Min)
gein Gein *n*
geinic acid Geinsäure *n*
geissoschizine Geissoschizin *n*
geissoschizol Geissoschizol *n*
gel Gel *n*, Gallert *n*, Gallerte *f*, elastic ~
    elastisches Gel, to convert into a ~ gelieren,
    gelatinieren
gel *v* gelatinieren, gelieren
gelatification Gelierung *f*, Verwandlung *f* in
    Gallerte
gelatin Gallerte *f*, Gelatine *f*, bichromated ~
    Chromgelatine *f*, chromatized ~
    Chromgelatine *f*, dichromated ~
    Chromgelatine *f*, ~ containing silver bromide
    Bromsilbergelatine *f* (Phot), ~ of bones
    Knochenleim *m*
gelatin agar Gelatineagar *m*
gelatinase Gelatinase *f* (Biochem)
gelatinate *v* gelatinieren, gelieren, in Gallerte
    verwandeln
gelatination Gelierung *f*, Verwandlung *f* in
    Gallerte
gelatin bath Gelatinebad *n*
gelatin capsule Gelatinekapsel *f*
gelatin carbonite Gelatinecarbonit *m*
gelatin culture medium Gelatinenährboden *m*
gelatin dynamite Sprenggelatine *f*,
    Gelatinedynamit *n*
gelatine Gelatine *f*, Gallerte *f*, Sülze *f*
gelatin emulsion Gelatineemulsion *f*
gelatin filament Gelatinefaden *m*
gelatin foil Gelatinefolie *f*
gelatin [glue] Gelatineleim *m*
gelatinization Gelbildung *f*, Gelatinierung *f*
gelatinize *v* gelieren, gelatinieren, in Gallert
    verwandeln
gelatinizing agent Gelatinierungsmittel *n*

gelatinizing property Gelatinierungsvermögen *n*
gelatin-like substance Gallert *n*
gelatin mold Leimform *f*
gelatino-chloride paper Keltapapier *n* (Phot)
gelatinosulfurous bath Gelatineschwefelbad *n*
gelatinous gallertartig, gallertähnlich,
 gelatineartig, gelatinös, leimartig
gelatinousness gallertartige Beschaffenheit *f*
gelatinous substance Gallertmasse *f,*
 Gallertsubstanz *f*
gelatin process Gelatineprozeß *m*
gelatin sugar Leimsüß *n* (obs), Leimzucker *m*
 (obs)
gelation Gelbildung *f,* Gelierung *f,*
 Verwandlung *f* in Gallerte
gel chromatography Gelchromatographie *f*
gel electrophoresis Gelelektrophorese *f*
gel filtration Gelfiltration *f*
gel formation Gelbildung *f*
gelling Gelbildung *f,* Gelierung *f*
gelling agent Gelbildner *m,* Geliermittel *n*
gelling point Gelierpunkt *m,*
 Gel-Bildungstemperatur *f*
gelose Galaktan *n*
gel point Stockpunkt *m*
gel rubber Gelkautschuk *m*
gelsemic acid Gelseminsäure *f*
gelsemin[e] Gelsemin *n*
gelseminic acid Gelseminsäure *f*
gelseminine Gelseminin *n*
gelsemium Gelsemium *n* (Med)
gelsemium root Gelseminwurzel *f* (Bot)
gel time Gelierzeit *f*
gem Edelstein *m,* Perle *f*
gem-like edelsteinartig
gemmatin Gemmatin *n*
gemmation Knospenbildung *f* (Bot)
gene Erbeinheit *f* (Biol), Erbfaktor *m* (Biol),
 Gen *n* (Biol), **sex-linked** ~ Geschlechtsgen *n*
gene action Genwirkung *f*
gene activity Genaktivität *f*
genease Maltase *f* (Biochem)
gene change Genveränderung *f* (Biol)
gene map Genkarte *f*
gene mutation Genmutation *f* (Biol)
general allgemein
general action Allgemeinwirkung *f*
general effect Allgemeinwirkung *f*
general index Hauptregister *n*
generality Allgemeingültigkeit *f*
generalizable verallgemeinerungsfähig,
 allgemein anwendbar
generalization Verallgemeinerung *f*
generalize *v* verallgemeinern
general joinery Konstruktionsverleimung *f*
general knowledge Allgemeinwissen *n*
general purpose Allzweck-
general purpose computer
 Universal-Rechenmaschine *f*

general purpose electrode
 Allzweck[glas]elektrode *f*
general purpose probe Allzwecksonde *f*
general reaction Allgemeinreaktion *f*
general use, for ~ gemeinnützig
general validity Allgemeingültigkeit *f*
generate *v* abgeben (Chem), entwickeln (Chem),
 erzeugen (Chem), ~ **steam** Dampf erzeugen
generating flask Erzeugerkolben *m*
generating milling cutter Abwalzfräser *m*
generating station Elektrizitätswerk *n,*
 Primärstation *f*
generating vessel Entwicklungsgefäß *n*
generation Bildung *f* (Chem), Entwicklung *f*
 (Chem), Erzeugung *f* (Chem), Generation *f;,*
 heat of ~ Erzeugungswärme *f,* **spontaneous** ~
 Selbstzeugung *f,* ~ **of oxygen**
 Sauerstoffentwicklung *f,* ~ **of smoke**
 Rauchentwicklung *f*
generator Generator *m* (Elektr), Dynamo *m*
 (Elektr), Entwickler *m* (Chem), Erzeuger *m*
 (Chem) (gases from liquids), Austreiber *m,*
 continuous current ~ Gleichstromdynamo *m*
generator armature Dynamoanker *m*
generator bearing Dynamolager *n*
generator current Dynamostrom *m*
generator furnace Gasgenerator *m*
generator gas Generatorgas *n*
generator room Dynamoraum *m*
generator shaft Generatorwelle *f*
gene regulation Genregulation *f*
gene repression Genrepression *f*
generic name Gattungsname *m*
generic term Gattungsbegriff *m*
geneserethol Geneserethol *n*
geneserine Geneserin *f*
geneseroline Geneserolin *n*
geneseroline methine Geneserolinmethin *n*
genesis Entstehung *f,* Entwicklung *f,* Genese *f*
genetic genetisch
genetic factor Erbfaktor *m*
geneticist Genetiker *m*
genetics Genetik *f,* Abstammungslehre *f,*
 Erbforschung *f,* Vererbungslehre *f*
genetic trait Erbanlage *f*
genistein Genistein *n,* Prunetol *n*
genistin Genistin *n*
genital cycle Genitalzyklus *m,*
 Geschlechtszyklus *m*
genkwanin Genkwanin *n*
genome Chromosomensatz *m* (Biol)
genotype Erbmasse *f* (Biol), Erbtypus *m* (Biol),
 Genotypus *m* (Biol)
genotypical genotypisch (Biol)
genthite Genthit *m* (Min), Nickelgymnit *m*
 (Min)
gentiacauline Gentiacaulin *n*
gentiamarine Gentiamarin *n*
gentian Enzian *m* (Bot)

gentian bitter Enziantinktur *f*
gentian blue Anilinblau *n*, Enzianblau *n*
gentianic acid Gentisinsäure *f*
gentianin Gentianin *n*, Gentisin *n*
gentian oil Enzianöl *n*
gentianose Gentianose *f*
·gentian root Enzianwurzel *f* (Bot)
gentian spirit Enzianbranntwein *m*
gentian violet Gentianaviolett *n*
gentiobionic acid Gentiobionsäure *f*
gentiobiose Gentiobiose *f*
gentiogenin Gentiogenin *n*
gentiol Gentiol *n*
gentiopicrin Gentiopikrin *n*
gentioside Gentiosid *n*
gentisaldehyde Gentisinaldehyd *m*
gentisein Gentisein *n*
gentisic acid Gentisinsäure *f*
gentisin Gentisin *n*, Gentianin *n*
gentisinic acid Gentisinsäure *f*
gentisin quinone Gentisinchinon *n*
gentle gelinde, schonend
genuine echt, unverfälscht, gediegen, rein
genuineness Echtheit *f*, Unverfälschtheit *f*
geocentric geozentrisch
geocerinic acid Geocerinsäure *f*
geochemical geochemisch
geochemistry Geochemie *f*, Erdchemie *f*
geochronology Geochronologie *f*
geocronite Geokronit *m* (Min)
geode Geode *f* (Min)
geodesic parallel geodätisch parallel
geodesy Erdmessung *f*, Geodäsie *f*,
   Landvermessung *f*
geography Erdkunde *f*, Geographie *f*
geoid determination Geoidbestimmung *f*
geoid warping Geoidundulation *f*
geological erdgeschichtlich
geologist Geologe *m*
geology Erdgeschichte *f*, Geologie *f*
geomagnetic erdmagnetisch
geomagnetic field Erdmagnetfeld *n*
geomagnetism Erdmagnetismus *m*
geometrical isomerism Stereoisomerie *f*
geometrical mean geometrisches Mittel *n* (Math)
geometric isomer Konfigurationsisomer *n*
geometric mean geometrischer Mittelwert *m*
geometry Geometrie *f*, **analytical** ~ analytische
   Geometrie, **descriptive** ~ darstellende
   Geometrie, **solid** ~ Raumgeometrie *f*, ~ **of**
   coincidence Inzidenzgeometrie *f*, ~ **of general**
   similarity Ähnlichkeitsgeometrie *f*
geophysical geophysikalisch
geophysics Geophysik *f*
geosote Geosot *n*
geotectonic structure Großstruktur *f* (Geol)
geothermal gradient geothermische Tiefenstufe *f*
geothermic geothermisch
geothermometer Erdwärmemesser *m*

geranial Citral *n*, Geranial *n*
geranic acid Geraniumsäure *f*
geraniine Geraniin *n*
geraniol Geraniol *n*, Geranylalkohol *m*
geraniolene Geraniolen *n*
geranium essence Geraniumessenz *f*
geranium oil Geraniumöl *n*
geranyl Geranyl-
geranyl acetate Geranylacetat *n*
geranyl alcohol Geraniol *n*, Geranylalkohol *m*
geranyl amine Geranylamin *n*
geranyl methyl ether Geranylmethyläther *m*
gerhardtite Gerhardtit *m* (Min)
geriatrics Geriatrie *f* (Med)
germ Keim *m*, Ansteckungskeim *m*,
   Bakterium *n*, Keimling *m*
germacrane Germacran *n*
germander Gamander *m* (Bot)
germane Germaniumwasserstoff *m*
German gold Goldpulver *n*
germanic acid Germaniumsäure *f*
germanic chloride Germanium(IV)-chlorid *n*
germanicene Germanicen *n*
germanicol Germanicol *n*
germanicone Germanicon *n*
germanic oxide Germaniumdioxid *n*
germanic sulfide Germaniumsulfid *n*
germanin Germanin *n* (HN), Bayer 205 *n* (HN)
German industrial standards Deutsche
   Industrie-Norm *f* (DIN)
germanite Germanit *m* (Min)
germanium (Symb. Ge) Germanium *n*,
   Ekasilicium *n*
germanium chloride Germaniumchlorid *n*
germanium chloroform Germaniumchloroform *n*
germanium dioxide Germaniumdioxid *n*
germanium hydride Germaniumwasserstoff *m*
germanium monoxide Germanium(II)-oxid *n*,
   Germaniumoxydul *n* (obs)
germanium sulfide Germaniumsulfid *n*,
   Schwefelgermanium *n*
germanous oxide Germanium(II)-oxid *n*,
   Germaniumoxydul *n* (obs)
German Pharmacop(o)eia Deutsches
   Arzneibuch *n*
German saltpeter Knallsalpeter *m*
German silver Neusilber *n*, Argentan *n*,
   Packfong *n*, Perusilber *n*
German standards commission Deutscher
   Normenausschuß *m*
German standard specification Deutsche
   Industrie-Norm *f* (DIN)
German steel Schmelzstahl *m*
German tinder Zündschwamm *m*,
   Zunderschwamm *m*
germ carrier Bazillenträger *m*, Keimträger *m*
germ cell Keimzelle *f*
germ-free keimfrei
germ gland Geschlechtsdrüse *f*, Keimdrüse *f*

germicidal keimtötend
germicidal bath Desinfektionsbad *n*
germicide keimtötendes Mittel *n*,
　Fäulnismittel *n*, keimtötendes Mittel *n*
germicide *a* keimtötend
germiform keimförmig
germinant keimend
germinate *v* keimen, knospen; sich entwickeln
germinating Keimen *n*, **incapable of** ~
　keimunfähig
germinating apparatus Keimapparat *m*
germinating box Kastenkeimapparat *m*
germinating power Keimkraft *f*
germination Keimung *f* (Biol), Keimbildung *f*
　(Biol), **power of** ~ Keimfähigkeit *f*
germination inhibitor Keimhemmungsmittel *n*
germinative keimfähig
germine Germin *n*
germ killer Keimtöter *m*
germless keimfrei
germ [of a disease] Krankheitskeim *m* (Med)
germ plasm[a] Idioplasma *n*, Keimplasma *n*,
　change of the ~ Idiokinese *f*
germ-proof keimdicht
germ-proof filter Entkeimungsfilter *n*
germs, free of ~ keimfrei
geronic acid Geronsäure *f*
gersdorffite Gersdorffit *m* (Min),
　Arsennickelglanz *m* (Min),
　Nickelarsenikkies *m* (Min)
gesarol Gesarol *n* (HN)
gesnerin Gesnerin *n*
get *v* erhalten, bekommen
getaway power Anzugsvermögen *n* (Auto)
geyserite Geyserit *m* (Min)
gheddaic acid Gheddasäure *f*
ghetta acid Gheddasäure *f*
giant cell Riesenzelle *f*
giant chromosome Riesenchromosom *n*
giant erythrocyte Riesenerythrozyt *m*
giant growth Riesenwuchs *m*
giant molecule Riesenmolekül *n*
gib and cotter Doppelkeil *m*
gibberellenic acid Gibberellensäure *f*
gibberellic acid Gibberellsäure *f*
gibberene Gibberen *n*
gibberenone Gibberenon *n*
gibberic acid Gibbersäure *f*
Gibbs function Gibbssche Funktion *f*
Gibbs-Helmholtz equation Gibbs-Helmholtz'sche
　Gleichung *f*
gibbsite Gibbsit *m* (Min), Hydrargillit *m* (Min)
Gibbs' phase rule Gibbssche Phasenregel *f*
gibhead key Nasenkeil *m*
gieseckite Gieseckit *m* (Min)
gigantocyte Makrozyt *m*
gigantolite Gigantolith *m* (Min)
gilbertite Gilbertit *m* (Min)
gild *v* vergolden

gilded vergoldet
gilder's wax Glühwachs *n*, Vergoldungswachs *n*
gilding Vergolden *n*, Goldanstrich *m*,
　Goldauflage *f*, Vergoldung *f*, **cold** ~
　Kaltvergoldung *f*, **hot** ~ Sudvergoldung *f*, ~
　by contact Kontaktvergoldung *f*, ~ **on**
　water-size Leimvergoldung *f*, ~ **with gold leaf**
　Blattvergoldung *f*
gilding size Poliment *n*
gilding solution Vergoldungsflüssigkeit *f*
Gilead, balm of ~ Arabischbalsam *m*
gill Kieme *f* (Zool)
gill breathing Kiemenatmung *f* (Zool)
gillingite Gillingit *m*
gilsonite Gilsonit *m* (Min)
gilt edge Goldschnitt *m*
gilt paper Goldpapier *n*
gimlet Vorbohrer *m*
gin Wacholderschnaps *m*, Gin *m*, Hebewerk *n*
　(Techn)
ginger Ingwer *m* (Bot), Ingwergewürz *n*,
　pulverized ~ Ingwerpulver *n*, **tincture of** ~
　Ingwertinktur *f*
ginger brandy Ingwerlikör *m*
gingerol Gingerol *n*
ginger powder Ingwerpulver *n*
ginkgolic acid Ginkgolsäure *f*
ginner Baumwollentkörner *m*
ginnol Ginnol *n*
ginnone Ginnon *n*
giobertite Giobertit *m* (Min)
giorgiosite Giorgiosit *m* (Min)
giratory crusher Rundbrecher *m*
girder Träger *m*, Balken *m*, Durchzug *m*
girder bridge Trägerbrücke *f*
girder pass Trägerkaliber *n*
girdle Gürtel *m*
girgensonine Girgensonin *n*
girth ring Laufkranz *m*, Laufring *m*
gismondine Gismondin *m* (Min)
gismondite Gismondin *m* (Min)
gitalin Gitalin *n*
gitaloxigenin Gitaloxigenin *n*
gitigenine Gitigenin *n*
gitine Gitin *n*
gitogenic acid Gitogensäure *f*
gitogenine Gitogenin *n*
gitonine Gitonin *n*
gito[ro]side Gito[ro]sid *n*
gitoxigenin Gitoxigenin *n*
gitoxin Gitoxin *n*
gitoxoside Gitoxosid *n*
give *v* geben, ~ **notice** kündigen, ~ **off**
　abgeben, abwerfen, ~ **rise to** hervorrufen, ~
　warning kündigen
glacial eisartig, eisig, eiszeitlich
glacial acetic acid Eisessig *m*
glaciation Vereisung *f* (Geol)
glacier salt Gletschersalz *n*

gladiolic acid Gladiolsäure *f*
glance coal Glanzkohle *f,* Kohlenblende *f*
glance cobalt Glanzkobalt *m* (Min)
glancing angle Auffallwinkel *m*
gland Drüse *f* (Med), **endocrine** ~ endokrine
  Drüse, **exocrine** ~ exokrine Drüse, **lactiferous**
  ~ Milchdrüse *f,* **lymphatic** ~ Lymphdrüse *f,*
  mucous ~ Schleimdrüse *f,* **open** ~ exokrine
  Drüse, **sudoriparous** ~ schweißabsondernde
  Drüse
gland lobule Drüsenlappen *m* (Med)
glandular cell Drüsenzelle *f*
glandular cyst Drüsenzyste *f*
glandular disease Drüsenkrankheit *f*
glandular secretion Drüsenabsonderung *f,*
  Drüsensekret *n* (Med)
glare blendendes Licht *n*
glaring grell, blendend
glaserite Glaserit *m* (Min), Arcanit *m* (Min)
glass Glas *n,* **Bohemian** ~ böhmisches Glas,
  broken ~ Glasabfall *m,* Glasscherbe *f,* **bubble**
  in ~ Glasblase *f,* **cast** ~ gegossenes Glas,
  chemistry of ~ Glaschemie *f,* **colored** ~
  Farbglas *n,* **common** ~ Kalkglas *n,* **compressed**
  ~ komprimiertes Glas, **crushed** ~
  Flitterglas *n,* Glasmehl *n,* **formation of** ~
  Glasbildung *f,* **frosted** ~ Milchglas *n,*
  Mattglas *n,* Trübglas *n,* **ground** ~
  geschliffenes Glas, **hard as** ~ glashart,
  hardened ~ gehärtetes Glas, **hardly fusible** ~
  schwer schmelzbares Glas, **incrusted** ~
  Glasinkrustation *f,* **Jena** ~ Jenaer Glas,
  laminated ~ Mehrschichtenglas *n,*
  Verbundglas *n,* **molded** ~ Preßglas *n,* **pounded**
  ~ Flitterglas *n,* **powdered** ~ Glasstaub *m,*
  pressed ~ Preßglas *n,* **refined** ~ geläutertes
  Glas, **resembling** ~ glasartig, hyalin,
  reticulated ~ Fadenglas *n,* Filigranglas *n,*
  stained ~ Farbglas *n,* buntes Glas, **strainfree**
  ~ spannungsfreies Glas, **volcanic** ~
  Glasachat *m*
glass annealing Glasbrennen *n*
glass apparatus Glasapparat *m*
glass ball Glaskugel *f*
glass balloon Glasballon *m*
glass bead Glaskügelchen *n,* Glasperle *f*
glass bell Glasglocke *f*
glassblower Glasbläser *m*
glassblower's lamp Glasbläserlampe *f*
glassblowing Glasblasen *n*
glass box Glaskasten *m*
glass brick Glasstein *m*
glass bulb Glaskugel *f,* Glasballon *m,*
  Glasbirne *f*
glass capillary Glaskapillare *f*
glass carboy Glasballon *m*
glass case Glaskasten *m*
glass cement Glaskitt *m*
glass clear (PVC bottles) glasklar

glass cloth Glasgewebe *n,* Glasleinen *n*
glass cover Deckglas *n*
glass crucible Glastopf *m*
glass cutter Glasschneider *m*
glass cutting Glasschleifen *n,* Glasschneiden *n*
glass cylinder Glaszylinder *m*
glass devitrification Glasrekristallisation *f*
glass drawing Glasziehen *n*
glass dust Glasstaub *m*
glass electrode Glaselektrode *f*
glasses Brille *f*
glass etching Glasätzung *f*
glass fabric Glasseidengewebe *n*
glass factory Glasfabrik *f*
glass fiber Glasfaser *f,* Glasfaden *m*
glass fiber laminate Glasfaserschichtstoff *m*
glass fiber mat Glasseidenmatte *f*
glass fiber plastics Glasfaserkunststoffe *pl*
glass-fiber reinforced glasfaserverstärkt
glass fiber roving Glasseidenstrang *m*
glass filament Glasseidenfaden *m*
glass filter crucible Glasfiltertiegel *m*
glass filter disc Glasfilterplatte *f*
glass filter funnel Glasfilternutsche *f*
glass flashing Glasüberzug *m*
glass flask Glaskolben *m,* Glasflasche *f*
glass flux Glasfluß *m*
glass frit Glasfritte *f*
glass funnel Glastrichter *m*
glass furnace Glas[schmelz]ofen *m*
glass gall Glasgalle *f,* Glasschaum *m,*
  Glasschmutz *m*
glass gilding Glasvergoldung *f*
glass grinder Glasschleifer *m*
glass grinding Glasschleifen *n*
glass-guard Schutzglas *n*
glass-hard glashart
glass hardness Glashärte *f*
glass incrustation Glasinkrustation *f*
glass industry Glasindustrie *f*
glassine Pergamin *n*
glassine paper Pergaminpapier *n*
glassiness glasartige Beschaffenheit *f*
glass instrument Glasinstrument *n*
glass insulator Glasisolator *m*
glass jar Glasgefäß *n*
glass knife Glasmesser *n,* Glasschneider *m*
glass lens Glaslinse *f*
glass-like glasähnlich, glasartig, glasförmig,
  glasig
glass lubricator Ölerglas *n*
glassmaking Glasbereitung *f*
glass manufacture Glasfabrikation *f,*
  Glasherstellung *f*
glass marble Glasperle *f*
glass marking ink Glastinte *f*
glass melt Glasschmelze *f*
glass membrane Glasmembran *f*
glass [mineral] wool Glaswatte *f*

glass monofilament Glasseideneinzelfaden *m*
glass oil cup Ölerglas *n*
glass packing Glasemballage *f*
glass painting Glasmalerei *f*
glass pane Glasscheibe *f*
glass paper Glaspapier *n*
glass paste Glaspaste *f*
glass pin Glasstift *m*
glass plate Glasscheibe *f*, Glasplatte *f*, Glastafel *f*
glass point Glasspitze *f*
glass porcelain Milchglas *n*
glass pot Glasschmelzhafen *m*, Glastiegel *m*, Glastopf *m*
glass powder Glasmehl *n*, Glaspulver *n*
glass resonator Glasresonator *m*
glass retort Glasretorte *f*
glass rod Glasstab *m*, Rührstab *m*
glass screw Glasschraube *f*
glass silk Glasseide *f* (Text)
glass silvering Glasversilberung *f*
glass spar Glasspat *m* (Min)
glass splinter Glassplitter *m*
glass staining Glasfärben *n*
glass stem Glashalterung *f*
glass stopper Glasstöpsel *m*, Glasstopfen *m*
glass stress tester Glasspannungsprüfer *m*
glass substitute Glasersatz *m*
glass syringe Glasspritze *f*
glass thread Glasfaden *m*
glass tube Glasröhre *f*, Glasrohr *n*
glass tube holder Glasröhrenhalter *m*
glass tubing Glasrohr *n*
glass varnish Glaslack *m*
glass vessel Glasgefäß *n*
glassware Glasgeräte *pl*
glass waste Glasabfall *m*
glass wool Glaswolle *f*
glass working Glasbearbeitung *f*
glassworks Glasfabrik *f*, Glashütte *f*
glasswort Glaskraut *n*, Kalipflanze *f* (Bot)
glassy gläsern, glasähnlich, glasig, durchsichtig
glass yarn Glasseidengarn *n*
glassy feldspar Eisspat *m* (Min)
glauberite Glauberit *m* (Min)
Glauber's salt Glaubersalz *n*, Wundersalz *n* (obs)
glaucentrine Glaucentrin *n*
glaucine Glaucin *n*
glaucinic acid Glauciumsäure *f*
glaucium oil Glauciumöl *n*
glaucobilin Glaukobilin *n*
glaucochroite Glaukochroit *m* (Min)
glaucodot Glaukodot *m* (Min), Kobaltarsenkies *m* (Min)
glaucolite Glaukolith *m* (Min)
glauconic acid Glauconsäure *f*
glauconite Glaukonit *m* (Min)
glaucophane Glaukophan *m* (Min)
glaucophylline Glaukophyllin *n*

glaucoporphyrin Glaukoporphyrin *n*
glaucopyrite Glaukopyrit *m* (Min)
glaucosin Glaukosin *n*
glaucous gelblich grün, grünblau
glaze Glanz *m*, Glasur *f*, Lasur *f* (Malerei), **layer of** ~ Glasurschicht *f*, **precipitation** ~ Ausscheidungsglasur *f*
glaze *v* verglasen, glätten, glasieren, lasieren, mit einer Glasur überziehen, polieren, satinieren (Pap)
glaze ash Glasurasche *f*
glaze baking Einbrennen *n* der Glasur, Glasurbrand *m*
glaze color Glasurfarbe *f*
glazed glasiert, satiniert
glazed cardboard Glanzpappe *f*
glazed fabric Glanzstoff *m* (HN)
glazed gilding Glanzvergoldung *f*
glazed linen Glanzleinwand *f*
glazed paper Glanzpapier *n*, Atlaspapier *n*, Brillantpapier *n*, Glacépapier *n*, satiniertes Papier *n*, Satinpapier *n*
glaze factory Glasurenfabrik *f*
glaze frit Glasurfritte *f*
glaze sand Glasursand *m*
glazier Glaser *m*
glazier lead Scheibenblei *n*
glazier's diamond Glaserdiamant *m*
glazier's putty Glaserkitt *m*
glazier's scraper Ofenkrücke *f*
glazing Verglasung *f*, Glasieren *n*, Glasur *f*, Lasierung *f*, Lasur *f*, Politur *f*, Satinieren *n* (Pap)
glazing calender Satinierkalander *m* (Pap); Seidenglanzkalander *m* (Pap)
glazing composition Lasurfarbe *f*
glazing kiln Glasurofen *m*, Glattbrennofen *m*
glazing machine Glättmaschine *f*
glazing mass Glasurmasse *f*
glazing roller Glattwalze *f*, Satinierwalze *f*
glazing sheet Hochglanzfolie *f*
glazing varnish Glanzfirnis *m*, Glanzlack *m*
gleam Lichtschein *m*, Lichtschimmer *m*
gleam *v* schimmern, glitzern
gliadin Gliadin *n*
glide *v* gleiten
glide plane Gleitebene *f*, Gleitspiegelebene *f* (Krist)
glide reflection Gleitspiegelung *f*
glide relaxation Ergänzungsgleitung *f*
gliding plane Gleitfläche *f*
glimmer Lichtschimmer *m*, Glimmer *m* (Min), Mica *m* (Min), **yellow** ~ Katzengold *n* (Min)
glimmer *v* glimmen, schimmern
glimmering Glimmen *n*
glinkite Glinkit *m* (Min)
glisten *v* funkeln, flimmern, glänzen, gleißen, glitzern
glistening flimmernd, Geflimmer *n*

Glitsch valve Glitsch-Ventil *n* (Dest)
glitter Glanz *m*, Glitzern *n*
glitter *v* glitzern, flimmern, funkeln, glänzen,
  gleißen
glittering Geflimmer *n*
glittering ore Flittererz *n*
globe Kugel *f*, Ball *m*
globe condenser Kugelkühler *m*
globe-shaped kugelförmig, kugelrund
globe valve Kugelventil *n*
globin Globin *n*
globol Globol *n*
globose kugelförmig
globular rund, kugelartig, kugelförmig, kugelig
globular diorite Kugeldiorit *m* (Min)
globular molecule kugelförmiges Molekül *n*,
  Kugelmolekül *n*
globulin Globulin *n*
globulite Globulit *m*
globulol Globulol *n*
glockerite Glockerit *m* (Min)
glomelliferic acid Glomellifersäure *f*
glonoin Glonoin *n*, Glycerinnitrat *n*
gloss Glanz[effekt] *m*, Glasur *f*, Politur *f*, **high ~**
  Hochglanz *m*, **loss of ~** Blindwerden *n*,
  Glanzverlust *m*
gloss *v* firnissen, mit Firnis überziehen
gloss additive Glanzzusatz *m*
gloss color ink Glanzfarbe *f*
gloss improver Glanzverstärker *m*
gloss-improving glanzsteigernd
glossing Chevillieren *n*
glossing machine Chevilliermaschine *f*
gloss number Glanzzahl *f*
gloss oil Glanzöl *n*
gloss paint Glanzlack *m*
gloss-reducing agent Mattierungsmittel *n*
gloss retention Glanzerhaltung *f*
gloss starch Glanzstärke *f*
gloss tester Glanzmesser *m*
glossy [hoch]glänzend
glossy paste, extra ~ Hochglanzpaste *f*
glove box Handschuhschutzkammer *f* (Atom)
Glover acid Gloversäure *f*
Glover tower Gloverturm *m*
glow Glühen *n*, Glut´ *f*, Lichtschein *m*
glow *v* [aus]glühen, glimmen, leuchten
glow current Glimmstrom *m*
glow discharge Glimmentladung *f*
glow discharge light Glimmlicht *n*
glow electron Glühelektron *n*
glowing Glimmen *n*, Glühen *n*, Leuchten *n*
glowing *a* glühend, leuchtend
glowing ashes Glühasche *f*
glowing filament tube Glühdrahtröhre *f*
glowing heat Glühhitze *f*
glowing iron Glüheisen *n*
glowing red rotglühend
glowing substance Leuchtmasse *f*

glow lamp Glimmlampe *f*, Glühlampe *f*
glow lamp resistance Glühlampenwiderstand *m*
glow pipe Glührohr *n*
glow plug switch (for diesel)
  Glühanlaßschalter *m*
glow point Glimmspannung *f*
glow potential Glimmspannung *f*,
  Glimmentladungspotential *n* (Elektr)
glucagon Glucagon *n*
glucal Glucal *n*
glucamine Glucamin *n*
glucazidone Glucazidon *n*
glucic acid Glucinsäure *f*
glucinum Beryllium *n*, Glucinium *n* (obs)
glucinum bromide Berylliumbromid *n*
glucinum fluoride Berylliumfluorid *n*
glucinum nitrate Berylliumnitrat *n*
glucinum orthosilicate Berylliumorthosilikat *n*
glucinum oxide Berylliumoxid *n*, Glucinerde *f*
glucinum salt Berylliumsalz *n*
glucinum sulfate Berylliumsulfat *n*
glucitol Glucit *m*
glucoalyssin Glucoalyssin *n*
glucoarabin Glucoarabin *n*
glucoaubrietin Glucoaubrietin *n*
glucoberteroin Glucoberteroin *n*
glucocapparin Glucocapparin *n*
glucocheirolin Glucocheirolin *n*
glucocholic acid Glucocholsäure *f*
glucocochlearin Glucocochlearin *n*
glucoconringiin Glucoconringiin *n*
glucocorticosteroid Glucocorticosteroid *n*
glucocymarol Glucocymarol *n*
glucogallin Glucogallin *n*
glucogen Glykogen *n*
glucogenkwanin Glucogenkwanin *n*
glucoheptonic acid Glucoheptonsäure *f*
glucoheptulose Glucoheptulose *f*
glucoiberin Glucoiberin *n*
glucokinase Glucokinase *f* (Biochem)
glucokinin Insulin *n*
glucomannose Glucomannose *f*
glucometasaccharin Glucometasaccharin *n*
glucomethylose Glucomethylose *f*
gluconamide Gluconamid *n*
gluconeogenesis Gluconeogenese *f* (Biochem)
gluconic acid Gluconsäure *f*
gluconolactone Gluconolacton *n*
glucoproteid Glykoproteid *n*
glucoprotein Glykoproteid *n*
glucosaccharin Glucosaccharin *n*
glucosamine Glucosamin *n*, Chitosamin *n*
glucosaminic acid Glucosaminsäure *f*
glucosan Glucosan *n*
glucosazone Glykosazon *n*
glucose Glucose *f*, **~ from potato starch**
  Kartoffelzucker *m*
glucose-1-phosphate Glucose-1-phosphat *n*
glucose carrier Glucoseüberträger *m*

glucose cyanhydrin Glucosecyanhydrin *n*
glucose derivative Glucosederivat *n*
glucose-galactoside Lactobiose *f*, Lactose *f*
glucose metabolism Glucosestoffwechsel *m*
glucose oxidase Glucoseoxidase *f* (Biochem)
glucose oxime Glucoseoxim *n*
glucose phenylhydrazone
   Glucosephenylhydrazon *n*
glucose phosphate Glucosephosphat *n*
glucose phosphorylation
   Glucosephosphorylierung *f*
glucose transport Glucosetransport *m*
glucose uptake Glucoseaufnahme *f*
glucosidase Glucosidase *f* (Biochem)
glucoside Glykosid *n*
glucosometer Glucosometer *n*
glucosone Glucoson *n*
glucosulfamide Glucosulfamid *n*
glucosulfone Glucosulfon *n*
glucosuria Glycosurie *f*
glucovanillin Glucovanillin *n*, Vanillosid *n*
glucurolactone Glucurolacton *n*, Glucuron *n*
glucurone Glucurolacton *n*, Glucuron *n*
glucuronic acid Glucuronsäure *f*
glucuronidase Glucuronidase *f* (Biochem)
glue Leim *m*, Holzleim *m*, Kleber *m*,
   Klebstoff *m* (Leim), Kleister *m*, **coat of** ~
   Leimanstrich *m*, **cold** ~ Kaltleim *m*,
   hot-sealable ~ Schmelzkleber *m*, **liquid** ~
   flüssiger Kleber *m*, **mixed** ~ Klebmischung *f*,
   to cover with ~ leimen, ~ **from hides**
   Hautleim *m*
glue *v* [an]leimen, kleistern, [ver]kleben, ~ **on**
   aufleimen, festkleben, ~ **together**
   zusammenkleben, zusammenleimen, ~ **up**
   zuleimen
glued geleimt
glued joint Leimstelle *f*
glue film Klebefilm *m*, Klebefolie *f*
glue jelly Leimgallerte *f*
glue joint Leimfuge *f*
glue line Leimfuge *f*
glue machine Leimmaschine *f*
[glue] penetration Eindringen *n* (des Leims)
glue press Leimpresse *f*
glue priming Leimgrund *m*
glue solution Leimlösung *f*, Leimwasser *n*
glue spread Leimauftrag *m*, **double layer** ~
   beidseitiger Leimauftrag *m*, **single layer** ~
   einseitiger Leimauftrag *m*
glue spreader Leimauftragsmaschine *f*,
   Leimspachtel *f*
glue stock Leimgut *n*
glue water Leimwasser *n*, Planierwasser *n*
gluey klebrig, zähflüssig, leimartig
gluing Leimen *n*, Leimung *f*, Verleimen *n*
gluing machine Leimmaschine *f*,
   Anleimmaschine *f*
gluside Saccharin *n*

glutaconaldehyde Glutaconaldehyd *m*
glutaconic acid Glutaconsäure *f*
glutamal Glutamal *n*
glutamate dehydrogenase
   Glutamatdehydrogenase *f* (Biochem)
glutamate pyruvate transaminase
   Glutamat-Pyruvat-Transaminase *f* (Biochem)
glutamic acid Glutaminsäure *f*,
   Aminoglutarsäure *f*
glutaminase Glutaminase *f* (Biochem)
glutamine Glutamin *n*
glutaminic acid Glutaminsäure *f*
glutamylglutamic acid Glutamylglutaminsäure *f*
glutardialdehyde Glutardialdehyd *m*
glutaric acid Glutarsäure *f*
glutaric aldehyde Glutaraldehyd *m*
glutaroin Glutaroin *n*
glutathione Glutathion *n*
glutazine Glutazin *n*
glutelin Glutelin *n*
gluten Gluten *n* (Chem), Kleber *m*, klebrige
   Substanz *f*, ~ **contained in rye**
   Roggenkleber *m*, ~ **of oats** Haferkleber *m*
gluten bread Kleberbrot *n*
gluten protein Eiweißleim *m*
glutethimide Glutethimid *n*
glutiminic acid Glutiminsäure *f*
glutinene Glutinen *n*
glutinic acid Glutinsäure *f*
glutinol Glutinol *n*
glutinous klebrig, leimartig, leimig
glutokyrine Glutokyrin *n*
glutol Glutol *n*
glutose Glutose *f*
glybuthiazol Glybuthiazol *n*
glyc[a]emia Glykämie *f* (Med)
glycal hydrate Glykalhydrat *n*
glycamide Glycamid *n*
glycamine Glykamin *n*
glycans Glykane *pl*
glycarbylamide Glycarbylamid *n*
glyceraldehyde Glycerinaldehyd *m*
glyceraldehyde phosphate dehydrogenase
   Glycerinaldehydphosphatdehydrogenase *f*
   (Biochem)
glyceric acid Glycerinsäure *f*
glyceric aldehyde Glycerinaldehyd *m*
glyceride Glycerid *n*
glyceride isomerism Glyceridisomerie *f*
glycerin[e] Glycerin *n*
glycerin gelatine Glyceringelatine *f*
glycerin oil Glycerinöl *n*
glycerin ointment Glycerinsalbe *f*
glycerin soap Glycerinseife *f*
glycerol Glycerin *n*, Ölzucker *m* (obs)
glycerol caprate Caprin *n*
glycerol caproate Caproin *n*, Capronfett *n*
glycerol chlorohydrin Glycerinchlorhydrin *n*
glycerol diacetate Diacetin *n*

glycerol dibutyrate oleate Oleodibutyrin *n*
glycerol dibutyrate stearate Stearodibutyrin *n*
glycerol dichlorohydrin stearate
  Stearodichlorhydrin *n*
glycerol dilaurate palmitate Palmitodilaurin *n*
glycerol dinitrate Dinitroglycerin *n*
glycerol dinitrate explosive
  Dinitroglyzerinsprengstoff *m*
glycerol dioleate Diolein *n*
glycerol dioleate palmitate Dioleopalmitin *n*,
  Palmitodiolein *n*
glycerol dioleate stearate Dioleostearin *n*
glycerol dipalmitate Dipalmitin *n*
glycerol dipalmitate stearate Dipalmitostearin *n*,
  Stearodipalmitin *n*
glycerol diricinoleate Diricinolein *n*
glycerol distearate Distearin *n*
glycerol distearate myristate Myristodistearin *n*
glycerol distearate palmitate Distearopalmitin *n*,
  Palmitodistearin *n*
glycerol elaidate Elaidin *n*
glycerol eleostearate Eläostearin *n*
glycerol ester Glycerinester *m*
glycerol ethylidene ether Acetoglyceral *n*
glycerol formate Formin *n*
glycerol kinase Glycerinkinase *f* (Biochem)
glycerol monoacetate Mon[o]acetin *n*
glycerol monoethyl ether Monoäthylin *n*
glycerol monoformate Monoformin *n*
glycerol monooctadecenyl ether
  Selachylalkohol *m*
glycerol myristate Myristin *n*
glycerol palmitate Palmitin *n*
glycerol palmitate stearate oleate
  Oleopalmitostearin *n*
glycerol triacetate Triacetin *n*
glycerol tributyrate Tributyrin *n*
glycerol tricaprate Tricaprin *n*
glycerol tricaproate Tricaproin *n*
glycerol tricaprylate Tricaprylin *n*
glycerol trichlorohydrin Glycerintrichlorhydrin *n*
glycerol trilaurate Trilaurin *n*
glycerol trilinolate Trilinolein *n*, Trilinolin *n*
glycerol trimargarate Trimargarin *n*, Intarvin *n*
glycerol trimyristate Trimyristin *n*
glycerol trinitrate Glycerintrinitrat *n*, Trinitrin *n*
glycerol trioleate Triolein *n*, Ölsäureglycerid *n*,
  Olein *n*
glycerol tripalmitate Tripalmitin *n*
glycerol tripropionate Tripropionin *n*
glycerol tristearate Stearin *n*, Tristearin *n*
glycerol trivalerate Trivalerin *n*, Phocaenin *n*
glycerophosphatase Glycerinphosphatase *f*
  (Biochem)
glycerophosphoric acid Glycerinphosphorsäure *f*
glycerosazone Glycerosazon *n*
glycerose Glycerose *f*
glycerosone Glyceroson *n*
glycerosulfuric acid Glycerinschwefelsäure *f*

glycerotriphosphoric acid
  Glycerintriphosphorsäure *f*
glyceryl Glyceryl-
glyceryl acetate Triacetin *n*
glyceryl aldehyde Glycerinaldehyd *m*
glyceryl borate Boroglycerid *n*
glyceryl butyrate Tributyrin *n*
glyceryl caprate Tricaprin *n*
glyceryl caproate Tricaproin *n*
glyceryl caprylate Tricaprylin *n*
glyceryl dibutyrate Dibutyrin *n*
glyceryl dibutyrate oleate Oleodibutyrin *n*
glyceryl laurate Laurin *n*, Trilaurin *n*
glyceryl laurate dimyristate Laurodimyristin *n*
glyceryl laurate distearate Laurodistearin *n*
glyceryl linolate Trilinolein *n*
glyceryl linoleate distearate Linoleodistearin
glyceryl margarate Trimargarin *n*, Intarvin *n*
glyceryl monoacetate Acetin *n*
glyceryl myristate Trimyristin *n*
glyceryl nitrate Nitroglycerin *n*, Glonoin *n*,
  Glycerinnitrat *n*, Trinitrin *n*
glyceryl oleate Triolein *n*
glyceryl palmitate Tripalmitin *n*
glyceryl propionate Tripropionin *n*
glyceryl ricinoleate Triricinolein *n*
glyceryl stearate Tristearin *n*
glyceryl triacetate Glycerintriacetat *n*,
  Triacetin *n*
glyceryl tricaprate Caprin *n*
glyceryl tricaproate Caproin *n*
glyceryl [tri]oleate Ölsäureglycerid *n*, Olein *n*
glyceryl [tri]ricinoleate Ricinolein *n*
glyceryl valerate Trivalerin *n*
glycide Glycid[ol] *n*
glycidic acid Glycidsäure *f*
glycidol Glycid[ol] *n*, Epihydrinalkohol *m*
glycinaldehyde Glycinaldehyd *m*
glycinamide Glycinamid *n*
glycinate Glycinat *n*
glycine Glykokoll *n*, Aminoessigsäure *f*,
  Glycin *n*
glycine anhydride Glycinanhydrid *n*
glycine nitrile Glycinnitril *n*
glycine oxidase Glycinoxidase *f* (Biochem)
glycinic acid Glycinsäure *f*
glycocholeic acid Glykocholeinsäure *f*
glycocholic acid Glykocholsäure *f*
glycoclastic glykolytisch
glycocoll Glykokoll *n*, Aminoessigsäure *f*,
  Glycin *n*, Leimsüß *n* (obs), Leimzucker *m*
glycocyamidine Glykocyamidin *n*
glycocyamine Glykocyamin *n*
glycogallic acid Glykogallussäure *f*
glycogen Glykogen *n*, Leberstärke *f*
glycogenase Glykogenase *f* (Biochem)
glycogenesis Glykogenese *f* (Biochem),
  Zuckerbildung *f* (Biochem)
glycogen granule Glykogenkorn *n*

glycogenic acid Glykonsäure *f,* Gluconsäure *f*
glycogen metabolism Glykogenstoffwechsel *m*
glycogenolysis Glykogenolyse *f*
glycogenolytic glykogenspaltend
glycogenosis Glykogenspeicherkrankheit *f*
(Med)
glycogenous zuckererzeugend (Biol)
glycogen phosphorylase
Glykogenphosphorylase *f* (Biochem)
glycogen storage disease
Glykogenspeicherkrankheit *f* (Med)
glycogen synthetase Glykogensynthetase *f*
(Biochem)
glycol Glykol *n,* Äthandiol *n,* Äthylenalkohol *m,*
Diol *n*
glycol acetal Glykolacetal *n*
glycol aldehyde Glykolaldehyd *m*
glycol cephalin Glykolcephalin *n*
glycol chlorohydrin Glykolchlorhydrin *n*
glycolic acid Glykolsäure *f*
glycolic anhydride Glykolid *n*
glycolide Glykolid *n*
glycolipid Glykolipid *n*
glycolithocholic acid Glykolithocholsäure *f*
glycol[l]ic acid Glycolsäure *f*
glycol monoacetate Glykolmonoacetat *n*
glycol monosalicylate Spirosal *n*
glycolonitrile Glykolnitril *n*
glycolsulfuric acid Glykolschwefelsäure *f*
glycoluric acid Hydantoinsäure *f,*
Glykolursäure *f*
glycoluril Acetylendiurein *n,* Glykoluril *n*
glycolylurea Hydantoin *n*
glycolysis Glykolyse *f* (Biochem)
glycolytic glykolytisch
glycometabolism Zuckerstoffwechsel *m*
glyconic acid Glykonsäure *f*
glycopenia Zuckerarmut *f* (Blut)
glycoprotein Glykoprotein *n,* Glykoproteid *n,*
Mucoproteid *n*
glycoregulation
Kohlenhydratstoffwechselsteuerung *f*
glycosamine Glykosamin *n*
glycosazone Glykosazon *n*
glycosidase Glykosidase *f* (Biochem)
glycosidic linkage Glykosidbindung *f*
glycosine Glycosin *n*
glycosone Glykoson *n*
glycosphingoside Glykosphingosid *n*
glycosuria Glykosurie *f* (Med)
glycuronic acid Glykuronsäure *f*
glycyl Glycyl-
glycylcholesterol Glycylcholesterin *n*
glycylhistamine Glycylhistamin *n*
glycyrrhetinic acid Glycyrrhetinsäure *f*
glycyrrhizin Glycyrrhizin *n*
glyoxal Glyoxal *n,* Diformyl *n*
glyoxalase Glyoxalase *f* (Biochem)
glyoxalate cycle Glyoxalatzyklus *m*

glyoxalic acid Glyoxalsäure *f,* Glyoxylsäure *f*
glyoxaline Glyoxalin *n,* Imidazol *n*
glyoxalone Glyoxalon *n*
glyoxal sulfate Glyoxalsulfat *n*
glyoxime Glyoxim *n*
glyoxyl Glyoxyl-
glyoxylic acid Glyoxalsäure *f,* Glyoxylsäure *f*
glyphogene Glyphogen *n*
glyphographic glyphographisch
glyphography Glyphographie *f*
glyphylline Glyphyllin *n*
glyptal Glyptal *n* (HN)
glysal Spirosal *n*
gmelinite Gmelinit *m* (Min)
Gmelin's salt Gmelins Salz *n*
GMP (Abk) Guanosinmonophosphat *n*
gnaw off *v* abnagen
gneiss Gneis *m* (Geol), **primitive** ~ Urgneis *m*
gneissic gneisig
gneissoid gneisähnlich, gneisig
gnomonic gnomonisch
gnoscopine Gnoscopin *n*
goa powder Araroba *n,* Goapulver *n*
goatbush bark Bitterrinde *f*
goatsbeard Geißbartkraut *n* (Bot)
goat's rue Pestilenzkraut *n* (Bot), Peterskraut *n*
(Bot), Ziegenraute *f* (Bot)
goat tallow Ziegentalg *m*
goblet Becher *m*
goethite Goethit *m* (Min), Rubinglimmer *m*
(Min), Sammetblende *f* (Min)
goffer *v* gaufrieren, kräuseln
goitrin Goitrin *n*
goitrogen Goitrogen *n*
gold (Symb. Au) Gold *n,* Aurum *n* (Lat),
addition of ~ Goldzusatz *m,* **alloyed** ~
Karatgold *n,* **argentiferous** ~ Silbergold *n,*
bright ~ Glanzgold *n,* **brilliant** ~
Glanzgold *n,* **burnished** ~ Glanzgold *n,*
colloidal ~ Collaurin *n,* kolloidales Gold,
containing ~ goldhaltig, **fulminating** ~
Knallgold *n,* **native** ~ gediegenes Gold,
Jungferngold *n,* **removal of** ~ Entgoldung *f,*
shining like ~ goldglänzend, **tincture of** ~
Goldtinktur *f,* **to remove** ~ entgolden, ~
purified by parting Scheidegold *n*
gold alloy Goldlegierung *f*
gold amalgam Goldamalgam *n,* Quickgold *n*
gold anode, crude ~ Rohgoldanode *f*
gold assay Goldprobe *f*
gold assay heating apparatus
Goldprobenglühapparat *m*
gold assaying table Goldprobentafel *f*
gold balance Goldwaage *f*
gold bath, boiling ~ Goldsud *m*
gold blocking Blattvergoldung *f*
gold boiling flask Goldkochkölbchen *n*
gold bromide Goldbromid *n*
gold bronze Goldbronze *f*

gold bronze pigment Goldbronzepigment *n*
gold-brown goldbraun
gold calx Goldkalk *m*
gold carbide Goldcarbid *n*
gold cathode, refined ~ Feingoldkathode *f*
gold chips Goldabfall *m*
gold chloride Goldchlorid *n*
gold coin Goldmünze *f*
gold color Goldfarbe *f*
gold-colored goldfarben, goldgelb
gold compound Goldverbindung *f*
gold content Goldgehalt *m*
gold crystal Goldkristall *m*
gold cup Goldschale *f*
gold cyanide Goldcyanid *n*, Cyangold *n*
gold digging Goldgräberei *f*
gold dish Goldschale *f*
gold doping Goldzusatz *m*
gold dust Goldstaub *m*, Goldgrieß *m*
gold edge Goldschnitt *m*
golden golden, goldfarben
golden vein Goldader *f*
gold extraction Goldgewinnung *f*, Goldlaugerei *f*
goldfieldite Goldfieldit *m* (Min)
gold foil Goldfolie *f*, Blattgold *n*, Goldschlag *m*
gold fulminate Knallgold *n*
gold glimmer Katzengold *n* (Min)
gold grain Goldkorn *n*
gold hydroxide Goldhydroxid *n*
gold(I) compound Gold(I)-Verbindung *f*,
   Auroverbindung *f*, Goldoxydulverbindung *f*
   (obs)
gold(III) compound Auriverbindung *f*,
   Gold(III)-Verbindung *f*,
   Goldoxidverbindung *f* (obs)
gold(III) hydroxide Aurihydroxid *n*,
   Gold(III)-hydroxid *n*, Goldsäure *f*
gold(III) oxide Aurioxid *n*, Gold(III)-oxid *n*
gold(III) salt Aurisalz *n*, Gold(III)-Salz *n*,
   Goldoxidsalz *n* (obs)
gold ingot Goldbarren *m*
gold iodide Goldjodid *n*
gold(I) oxide Aurooxid *n*, Gold(I)-oxid *n*,
   Goldoxydul *n* (obs)
gold-iridium alloy Iridgold *n*
gold(I) salt Aurosalz *n*, Gold(I)-Salz *n*
gold leaching plant Goldlaugerei *f*
gold leaf Blattgold *n*, Goldblatt *n*, Goldschlag *m*
gold lettering Goldschrift *f*
goldlike goldähnlich, goldartig
gold-like brass Talmi *n*
gold luster Goldglanz *m*
gold mine Goldbergwerk *n*
gold monobromide Gold(I)-bromid *n*,
   Goldbromür *n* (obs)
gold monochloride Aurochlorid *n*,
   Goldchlorür *n* (obs), Gold(I)-chlorid *n*
gold monocyanide Gold(I)-cyanid *n*,
   Goldcyanür *n* (obs)

gold monoiodide Gold(I)-jodid *n*, Goldjodür *n*
   (obs)
gold monosulfide Gold(I)-sulfid *n*, Goldsulfür *n*
   (obs)
gold monoxide Gold(I)-oxid *n*, Goldoxydul *n*
   (obs)
gold number Goldzahl *f*
gold ore Golderz *n*, pounded ~ Goldschlich *m*
gold oxide Goldoxid *n*, Goldkalk *m*
gold paper Goldpapier *n*
gold parings Goldabfall *m*
gold parting Goldscheiden *n*, Goldscheidung *f*
gold phosphide Goldphosphid *n*
gold-plate *v* vergolden, goldplattieren
gold-plated vergoldet, goldplattiert
gold-plating Vergolden *n*, Goldplattierung *f*,
   Vergoldung *f*
gold potassium chloride Chlorgoldkalium *n*
gold powder Goldpulver *n*
gold precipitant Goldfällungsmittel *n*
gold precipitate Goldniederschlag *m*
gold priming varnish Goldgrundfirnis *m*
gold protoxide Gold(I)-oxid *n*, Goldoxydul *n*
   (obs)
gold purple Goldpurpur *m*
gold quartz Goldkies *m*
gold refiner Goldscheider *m*
gold refinery Goldscheideanstalt *f*
gold refining Goldscheidung *f*,
   Goldraffination *f*, Goldscheiden *n*, process of
   ~ Goldscheideverfahren *n*
gold selenide Goldselenid *n*
gold separation, assay flask for ~
   Goldkochkölbchen *n*
gold silver sulfide Goldsilbersulfid *n*
gold silver telluride Goldsilbertellurid *n*,
   Tellurgoldsilber *n*
gold size Goldgrundfirnis *m*, Goldleim *m*
gold slime Goldschlich *m*
goldsmith Goldschmied *m*
goldsmith's wash Goldkrätze *f*
gold sodium chloride Chlorgoldnatrium *n*,
   Natriumgoldchlorid *n*
gold solder Goldlot *n*
gold solution Goldlösung *f*
goldstone Glimmerquarz *m* (Min), Goldstein *m*
   (Min)
gold stripping Entgoldung *f*
gold sulfide Goldsulfid *n*, Goldschwefel *m*,
   Schwefelgold *n*, (Gold(III)-sulfid)
   Aurisulfid *n*
gold sweepings Goldkrätze *f*
gold tartrate Goldweinstein *m*
gold telluride Goldtellurid *n*, Tellurgold *n*
gold thread Fadengold *n*
gold tribromide Auribromid *n*,
   Gold(III)-bromid *n*
gold trichloride Aurichlorid *n*,
   Gold(III)-chlorid *n*

gold trioxide Aurioxid *n,* Gold(III)-oxid *n,* Goldtrioxid *n*
gold trisulfide Aurisulfid *n,* Gold(III)-sulfid *n*
gold varnish Goldfirnis *m,* Goldlack *m*
gold weight Goldgewicht *n*
gold wire Golddraht *m*
Golgi apparatus Golgi-Apparat *m* (Biol)
gommelin Dextrin *n*
gonad Geschlechtsdrüse *f,* Gonade *f,* Keimdrüse *f*
gonan Gonan *n*
goniometer Goniometer *n,* Winkelmesser *m* (für Flächenwinkel)
goniometric goniometrisch
goniometry Goniometrie *f,* Winkelmessung *f*
Gooch crucible Goochtiegel *m*
goods Ware[n pl] *f,* **fully fashioned** ~ Fertigware *f*
goods-in system Einlagersystem *n*
goods lift (Br. E.) Lastenaufzug *m*
goods-out system Auslagersystem *n*
goose-necked gekröpft (Techn)
gorceixite Gorceixit *m* (Min)
gorlic acid Gorlisäure *f*
goslarite Goslarit *m* (Min), weißer Vitriol *n*
gossypetin Gossypetin *n*
gossypetonic acid Gossypetonsäure *f*
gossypin Gossypin *n*
gossypitone Gossypiton *n*
gossypitrine Gossypitrin *n*
gossypitrone Gossypitron *n*
gossypol Gossypol *n*
gouge Hohleisen *n,* Drehdorneisen *n,* Haumeißel *m,* Hohlmeißel *m*
Goulard's extract Bleiessig *m*
Goulard's lotion Bleiwasser *n*
Goulard water Bleiwasser *n*
gout Gicht *f* (Med), **remedy for** ~ Gichtmittel *n* (Med)
gout paper Gichtpapier *n* (Med)
govern *v* regeln, lenken, regulieren
governing Regulierung *f*
governor Regler *m,* Regulator *m*
governor shaft Regulatorwelle *f*
governor sleeve Reglermuffe *f*
governor spring Regulatorfeder *f*
governor valve Reglerventil *n*
goyazite Goyazit *m* (Min)
grab Greifer *m,* Haltegriff *m*
grabbing crane Greiferkran *m*
grabbing [luffing] crane Greifer-Wippkran *m*
gradation Abstufung *f*
grade Grad *m;* Gefälle *n* (einer Straße), Klasse *f,* Note *f,* Qualität *f,* Rang *m;* Sorte *f;* Steigung *f* (Gefälle)
grade *v* abstufen; eichen, klassifizieren, ordnen; planieren (Techn), sortieren
grade efficiency Teilungszahl *f,* Trenngrad *m*
grade efficiency curve Trennkurve *f*

grader Planierer *m* (Techn), Scheider *m,* Sortiermaschine *f*
grade tunnel Freispiegeltunnel *n*
gradient Gradient *m;* Neigung *f,* Steigung *f,* **hydraulic** ~ Flüssigkeitsgradient *m* (Bodenkolonne)
gradient elution Gradientenelution *f*
gradient method Gradientenmethode *f,* ~ **for optimization of functions** Gradientenmethode *f* f. Funktionenoptimierung
grading by sifting Klassieren *n*
grading curve Korngrößenverteilung *f*
grading screen Klassiersieb *n*
gradually absatzweise, allmählich, schrittweise, stufenweise
graduate *v* abstufen, einteilen, graduieren, kalibrieren
graduated abgestuft, graduiert
graduated arc Gradbogen *m*
graduated buret[te] Meßbürette *f*
graduated circle Gradkreis *m,* Teilkreis *m*
graduated cylinder Meßzylinder *m*
graduated flask Meßkolben *m*
graduated jar Meßgefäß *n*
graduated pipet[te] Meßpipette *f,* Meßheber *m*
graduated plate Teilscheibe *f*
graduated tube Maßröhre *f*
graduation Abstufung *f,* Einteilung *f,* Gradeinteilung *f,* Gradierung *f* (Chem), Gradmessung *f* (Geol), Maßeinteilung *f,* Skala *f,* Stufenfolge *f*
graduation apparatus Gradierapparat *m,* Gradierwerk *n*
graduation furnace Gradierofen *m*
graduation house Gradierhaus *n*
graduation mark Ablesestrich *m,* Markierungsstrich *m*
graduation scale Gradteilung *f,* Kalibrierung *f*
graduation works Gradierwerk *n*
graduator Gradierapparat *m* (Chem), Gradmesser *m*
graft *v* [auf]pfropfen (Bot), Gewebe verpflanzen (Med), pfropfen (Polym)
graft copolymer Pfropfcopolymer *n*
grafting Pfropfung *f*
grafting wax Baumwachs *n,* Pfropfwachs *n*
graftonite Graftonit *m* (Min)
graft polymer Pfropfpolymerisat *n*
graft polymerization Pfropfpolymerisation *f*
grain Korn *n,* Ader *f* (Holz), Faser *f* (Holz), Faserrichtung *f* (Holz), Faserung *f* (Holz), Gefüge *n,* Getreide *n* (Bot), Getreideart *f* (Bot), Körnung *f,* Ledernarbung *f* (Leder), Mahlgut *n* (Techn), Narbe *f* (Leder), Narbung *f* (Leder), Pore *f* (Holz), Struktur *f,* Walzrichtung *f* (Techn), **character of** ~ Kornbeschaffenheit *f,* **coarse** ~ Grobkorn *n,* grobes Korn, **coarsening of** ~

Kornvergröberung *f,* **fine** ~ feines Korn,
fineness of ~ Kornfeinheit *f,* **size of** ~
Korngröße *f,* ~ **of a photographic layer**
Korn *n* einer photographischen Schicht, ~ **of**
emulsion Plattenkorn *n* (Phot), ~ **of sand**
Sandkorn *n*
grain *v* aussalzen (Seifenherstellung),
granulieren, körnen, masern, narben (Leder,
Papier)
grain alcohol Kornalkohol *m*
grain boundary Korngrenze *f* (Krist, Metall),
energy of ~ Korngrenzenenergie *f*
grain boundary diffusion Korngrenzendiffusion *f*
grain boundary migration
Korngrenzenwanderung *f*
grain boundary relaxation
Korngrenzenrelaxation *f* (Krist)
grain coarsening Kornvergröberung *f*
grain density Korndichte *f*
grain diameter Korndurchmesser *m*
grain dryer Getreidetrockner *m*
grained gekörnt, genarbt (Leder), körnig, narbig
(Leder), **medium** ~ mittelfeinkörnig
grained sugar Krümelzucker *m*
grain effects, printing ~ **on wood**
Maserdruckverfahren *n*
grain elevator Getreideheber *m*
grain growth Kornwachstum *n*
graining Maserung *f,* Grainieren *n,* Rauhen *n*
graining calender Grainierkalander *m,*
Narbkalander *m*
grain lead Kornblei *n*
grain leather Narbenleder *n*
grain number Kornzahl *f*
grain refinement Kornverfeinerung *f*
grains, in ~ körnig, ~ **of silver ashes**
Dunstsilber *n*
grain scales Kornwaage *f*
grain side Narbenseite *f* (Leder)
grain silo Getreidesilo *m*
grain size Korngröße *f*
grain spirits Kornalkohol *m,*
Kornbranntwein *m,* Kornschnaps *m*
grain structure Korngefüge *n,* Kornstruktur *f*
grain test Schnellgrießmethode *f*
grain tin Feinzinn *n,* Kristallzinn *n*
grain wood Aderholz *n*
grainy körnig, gekörnt; gemasert; grießig
gram (Am. E.) Gramm *n*
gram atom Grammatom *n*
gram-atomic weight Grammatomgewicht *n*
gram bottle Grammflasche *f*
gram calorie Grammkalorie *f*
gram equivalent Grammäquivalent *n,* Val *n*
gramicidin Gramicidin *n*
gramine Gramin *n*
gram ion Grammion *n*
grammatite Grammatit *m* (Min)
grammatitiferous grammatithaltig

gramme (Br. E.) Gramm *n*
grammite Grammit *m*
gram-molecular volume
Gramm-Molekülvolumen *n*
gram-molecular weight
Gramm-Molekülgewicht *n,* Molgewicht *n*
gram molecule Gramm-Molekül *n,*
Gramm-Mol *n,* Mol *n*
Gram-negative Gram-negativ (Bakt)
Gram-positive Gram-positive (Bakt)
Gram's stain Gramsche Färbung *f*
gram stain[ing] Gram-Färbung *f* (Biol)
gram weight Grammgewicht *n*
granary Getreidespeicher *m,* Kornspeicher *m*
granatal Granatal *n*
granatane Granatan *n*
granatanine Granatanin *n*
granatenine Granatenin *n*
granatine Granatin *n*
granatite Granatit *m* (Min)
granatoline Granatolin *n*
granatonine Granatonin *n*
Grancold pellet bonding Grancoldverfahren *n*
grandidierite Grandidierit *m* (Min)
granilite Granilit *m* (Min)
granite Granit *m* (Min), **giant** ~ Pegmatit *m*
(Min), **porphyroid** ~ Granitporphyr *m* (Min),
primitive ~ Urgranit *m,* **secondary** ~
Flözgranit *m*
granite-colored granitfarbig
granite formation Granitformation *f* (Geol)
granite-like granitartig, granitförmig
granite rock Granitfels *m*
granite sand Granitsand *m*
granitic granitartig, granitförmig
granitic insulator Granitisolator *m*
granitic quartz Granitquarz *m* (Min)
granitic rock Granitfels *m,* Granitgestein *n*
granitification Granitbildung *f* (Geol)
granitiform granitartig, granitförmig
granitite Granitit *m* (Min)
granitization Granitbildung *f* (Geol)
granitoid granitartig, granitähnlich
grant *v* bewilligen, gewähren, ~ **a concession or**
license konzessionieren
grantee Empfänger *m* einer Bewilligung,
Patentinhaber *m*
grantianic acid Grantianinsäure *f*
grantianine Grantianin *n*
granular gekörnt, granuliert; grießig, körnig
granular bed dust separator Drallschichtfilter *n*
granular bog iron ore Eisengraupen *f pl*
granular fracture iron Korneisen *n*
granularity Körnigkeit *f*
granular leucocyte Granulozyt *m*
granular limestone Erbsenstein *m* (Min)
granular ore Graupenerz *n*
granular size Korngröße *f*
granular sugar Krümelzucker *m*

granulate *v* körnen, granulieren, granieren, granulieren
granulated gekörnt, granuliert, körnig
granulated fracture körniger Bruch *m*
granulated metal Granalien *pl,* Metallkorn *n*
granulating Rauhen *n*
granulating crusher Granuliermühle *f*
granulating machine Granulierapparat *m*
granulating plant Granulationsanlage *f,* Granulieranlage *f*
granulation Körnen *n,* Granulation *f* (Med), Granulierung *f,* Körnung *f,* Kornbildung *f,* Kristallbildung *f* (Zuck)
granulator Granulator *m,* Granulierapparat *m*
granule Körnchen *n*
granules Granulat *n,* Grieß *m*
granulite Granulit *m* (Min), Weißstein *m* (Min)
granulitic granulitisch
granulocyte Granulozyt *m*
granulometer Körnigkeitsmesser *m*
granulometry Granulometrie *f*
granulose Granulose *f*
grape Weinbeere *f,* Weintraube *f*
grape juice Traubensaft *m*
grapelike traubenartig
grape marc brandy Tresterbranntwein *m*
grape marc wine Tresterwein *m*
grape mildew Rebenmeltau *m*
grape must Maische *f,* Most *m*
grapes, bunch of ~ Weintraube *f,* **marc of** ~ Weintreber *pl,* Weintrester *pl,* **pressing of** ~ Keltern *n,* **to press** ~ keltern
grape-seed oil Weinkernöl *n,* Önanthäther *m*
grape sugar Traubenzucker *m,* Dextrose *f,* Glucose *f*
graph Diagramm *n,* Kurve *f* (Math), Schaubild *n,* Zeichnung *f*
graph *v* graphisch darstellen
graphic[al] graphisch, anschaulich
graphic gold Schriftgold *n*
graphic granite Schriftgranit *m*
graphic tellurium Schrifterz *n*
graphite Graphit *m,* Bleischwärze *f,* Eisenschwärze *f,* Reißblei *n,* **coating with** ~ Graphitieren *n,* **deflocculated** ~ entflockter Graphit, **deposit of** ~ Graphitablagerung *f,* **flaky** ~ Schuppengraphit *m* (Min), **foliated** ~ blättriger Graphit, **ground** ~ Eisenschwärze *f,* Mineralschwarz *n,* **mixture of gold and** ~ Goldgraphit *m,* **poor in** ~ graphitarm, **powdered** ~ Graphitstaub *m,* **rich in** ~ graphitreich, **separation of** ~ Graphitausscheidung *f,* **synthetic** ~ Elektrographit *m*
graphite arc Graphitbogen *m*
graphite bearing Graphitlager *n*
graphite block Graphitblock *m*
graphite carbon Graphitkohle *f*
graphite crucible Graphit[schmelz]tiegel *m*

graphite electrode Graphitelektrode *f*
graphite eutectic Graphiteutektikum *n*
graphite formation Graphit[aus]bildung *f*
graphite layer Graphitschicht *f*
graphite-like graphitähnlich, graphitartig
graphite lubricant Graphitschmiere *f*
graphite lubrication Graphitschmierung *f*
graphite mill Graphitmühle *f*
graphite-moderated graphitmoderiert (Reaktor)
graphite[-moderated] reactor Graphitreaktor *m*
graphite moderator Graphitmoderator *m* (Atom), Graphitbremsmasse *f* (Atom)
graphite oil Graphitöl *n,* ~ **lubricating apparatus** Graphitölschmierapparat *m*
graphite pencil Graphitstift *m*
graphite powder Graphitpulver *n*
graphite pyrometer Graphitpyrometer *n,* Graphitthermometer *n*
graphite resistance furnace Graphitwiderstandsofen *m*
graphite tube Graphitrohr *n*
graphitic graphitisch, graphitartig, graphithaltig
graphitic acid Graphitsäure *f*
graphitic carbon Temperkohle *f*
graphitic clay Schieferkreide *f* (Min)
graphitic film Graphitüberzug *m*
graphitization Graphit[aus]bildung *f,* Graphitierung *f* (Metall), Temperkohleabscheidung *f*
graphitize *v* mit Graphit überziehen, graphitieren
graphitizing Graphitierung *f* (Metall)
graphitoid[al] graphitähnlich, graphitartig
graphometer Graphometer *n,* Winkelmesser *m*
grappling iron Finger *m* (Mech), Greifer *m*
grass bleaching natürliche Bleiche *f*
grass green grasgrün
grass oil Grasöl *n*
grate Feuerrost *m,* Gatter *n,* Gitter *n* (Rost), Rost *m,* **airspace between the** ~ Rostfuge *f,* **fixed** ~ fester Rost *m,* **flat** ~ Flachrost *m,* **horizontal** ~ Flachrost *m,* **revolving** ~ drehbarer Rost, **rotary** ~ drehbarer Rost, **sectional** ~ geteilter Rost *m,* **shaking** ~ beweglicher Rost *m,* **stationary** ~ fester Rost *m,* ~ **in one section** Ganzrost *m,* ~ **with external water cooling** Rieselrost *m*
grate *v* [zer]reiben
grate area Rostfläche *f*
grate bar Roststab *m,* Schienenrost *m*
grate cooling Rostkühlung *f*
grate fire Rostfeuer *n*
grate firing Rostfeuerung *f*
grate-kiln system Banddrehrohrofen *m*
grate load Rostbelastung *f*
grate opening Rostfuge *f,* Rostspalt *m*
grater Raspel *f,* Reibe *f,* Reibeisen *n*
grate surface Rostfläche *f*
graticule Fadenkreuz *n,* Strichgitter *n* (Opt)

grating Gitter *n* (Opt), **linear** ~ Strichgitter *n*
grating constant Gitterkonstante *f* (Phys)
grating diaphragm Gitterblende *f*
grating formula Gitterformel *f*
grating measurement Gittermessung *f*
grating spectrometer Gitterspektrometer *n*
gratiolin Gratiolin *n*
gratiolirhetine Gratiolirhetin *n*
gratiosolin Gratiosolin *n*
graulite Graulit *m* (Min)
gravel Kies *m*, Geröll *n*, Grieß *m* (Techn),
    Grobsand *m*, Schotter *m*, **deposition of** ~
    Kiesablagerung *f*, **fine** ~ Grand *m*
gravel-coated board bekieste Pappe *f*
gravel extraction plant Kieselgewinnungsanlage *f*
gravel-filled kiln Stückkiesofen *m*
gravel filter Kiesfilter *n*
gravelly kiesig, grießig, kieshaltig, sandreich
gravel pit Kiesgrube *f*
gravel screen Kiessieb *n*
gravel soil Kiesboden *m*
gravel stone Kieselstein *m*
graveoline Graveolin *n*
graver Stechmeißel *m*, Stichel *m*
Graves's (or Basedow's) disease Basedowsche
    Krankheit *f* (Med)
gravimeter Dichtemesser *m*, Gravimeter *n*,
    Schweremesser *m*, Schwerkraftmesser *m*
gravimeter batching Gewichtsdosierung *f*
gravimetric gewichtsanalytisch, gravimetrisch
gravimetric analysis Gewichtsanalyse *f*
gravimetry Gravimetrie *f*
graving tool Stichel *m*, Grabeisen *n*,
    Gradiereisen *n*
gravitate *v* gravitieren
gravitation Gravitation *f*, Gravitationskraft *f*,
    Schwerkraft *f*, **law of** ~ Gravitationsgesetz *n*
    (Phys), Fallgesetz *n* (Phys), **theory of** ~
    Gravitationstheorie *f*
gravitational acceleration
    Schwerebeschleunigung *f*,
    Erdbeschleunigung *f*, Fallbeschleunigung *f*
gravitational attraction Gravitationsanziehung *f*,
    Massenanziehung *f*, Schwereanziehung *f*
gravitational constant Fallbeschleunigung *f*,
    Gravitationskonstante *f*
gravitational excitation Gravitationserregung *f*
gravitational field Gravitationsfeld *n*,
    Schwerefeld *n*
gravitational flux Gravitationsfluß *m*
gravitational red shift
    Gravitationsverschiebung *f* nach Rot (Spektr)
gravitation potential Gravitationspotential *n*
gravitation torsion balance
    Gravitationsdrehwaage *f*
gravity Gravitation *f*, Schwere *f*, Schwerkraft *f*,
    acceleration of ~ Erdbeschleunigung *f*, **center
    of** ~ Schwerpunkt *m*, Baryzentrum *n* (Phys),
    Gleichgewichtspunkt *m*,

Schwerkraftzentrum *n*, **low center of** ~
    Haftung *f* am Erdboden, **specific** ~
    spezifisches Gewicht *n*
gravity anomaly Schwereanomalie *f*
gravity classifier Steigrohrsichter *m*
gravity closing Schließung *f* durch Eigengewicht
gravity feed Materialzuführung *f* durch Gefälle
gravity field Schwerefeld *n*, Schwerkraftfeld *n*
gravity formula Schwereformel *f*
gravity line Freispiegelleitung *f*,
    Schwerkraftbahn *f*
gravity mixer Schwerkraftmischer *m*
gravity network Schwerenetz *n*
gravity observation Schwerebeobachtung *f*
gravity roller conveyor Rollenbahn *f*
gravity settling chamber Staubkammer *f*
gravity settling process Absitzverfahren *n*
gravity survey Schwerevermessung *f*
gravity tank Fallbehälter *m*
gravity tube pump Fallrohrpumpe *f*
gravure Gravüre *f*, Klischeedruck *m*,
    Tiefdruck *m* (Buchdr)
gravure printing Kupfertiefdruck *m* (Buchdr)
gravure printing machine Tiefdruckmaschine *f*
    (Buchdr)
gray (see also grey) grau, **light** ~ hellgrau,
    weißgrau, **pale** ~ weißgrau, **yellowish** ~
    gelbgrau
gray-blue graublau
gray-brown graubraun
gray-colored graufarben, graufarbig
gray glass Rauchglas *n*
graying Vergrauung *f*
gray iron Gußeisen *n*
gray iron cast Grauguß *m*
gray iron fining Rohfrischen *n* (Metall)
gray iron foundry Graugießerei *f*
grayish fahlgrau
grayish black grauschwarz
gray manganese ore Graumanganerz *n* (Min)
gray metal gares Gußeisen *n*
gray pigment from clay slate Steingrau *n*
gray rot Graufäule *f*
gray silver Grausilber *n*
gray spiegel iron Grauspiegeleisen *n*
graystone Graustein *n* (Geol)
gray wacke Grauwacke *f* (Geol)
gray-white grauweiß
grease Fett *n*, Schmiere *f*, Schmiermittel *n*,
    extraction of ~ Entfettung *f*, **layer of** ~
    Fettschicht *f*, **odor of** ~ Fettgeruch *m*, **taste of**
    ~ Fettgeschmack *m*, **to purify** ~ Fett
    raffinieren
grease *v* [ein]schmieren, [an]fetten, einfetten,
    [ein]ölen
grease catcher Fettauffanggefäß *n*
grease cock Schmierhahn *m*
grease cup Staufferbüchse *f*, Schmierbüchse *f*,
    Schmiervase *f*

greased gefettet, [durch]geschmiert
grease extracting apparatus
   Entfettungsapparat *m*
grease extracting plant Entfettungsanlage *f*
grease extractor Fettabscheider *m*,
   Fettausscheider *m*
grease gun Fettschmierpresse *f*,
   Preßschmierpumpe *f*, Schmierpistole *f*
grease lubrication Fettschmierung *f*
grease luster, having a ~ fettglänzend
grease-packed bearing einmalig beim Einbau mit
   Fett geschmiertes Lager *n*
grease press Fettpresse *f*
greaseproof fettbeständig, fettdicht
greaser Schmierapparat *m*, Schmierer *m*
grease reservoir Schmierfettdepot *n*
grease separator Fettabscheider *m*,
   Fettausscheider *m*
grease solvent Fettlösungsmittel *n*
grease spot Fettfleck *m*
grease-spot photometer Fettfleckphotometer *n*
grease trap Fettabschneider *m*, Fettfang *m*
grease wool Fettwolle *f*
greasiness Fettigkeit *f*, Schmierigkeit *f*
greasing Schmieren *n*, Einfetten *n*
greasing apparatus Anfettapparat *m*
greasy fettartig, fett[ig], ölig, schmierig
greasy luster Fettglanz *m*
greasy quartz Fettquarz *m* (Min)
greasy rag Fettlappen *m*
greasy stain Fettfleck *m*
greasy wool Schweißwolle *f*
greasy yolk Fettschweiß *m*, Wollschweiß *m*
greater part Großteil *m*
green algae Grünalgen *pl* (Bot)
green bice Lasurgrün *n*
green blindness Grünblindheit *f*
green broom Färberginster *m* (Bot),
   Färberpfrieme *f* (Bot)
green [color] Grün *n*
green copperas Eisenoxydulsulfat *n*
green earth Grünerde *f* (Min), Veronesergrün *n*
   (Min)
green fodder dryer Grünfuttertrockner *m*
green garnet Grossular *m* (Min)
green gilding Grünvergoldung *f*
greenhouse Gewächshaus *n*, Treibhaus *n*
greenish grünlich, grünstichig
greenish black grünschwarz
greenish blue grünblau
greenish yellow grüngelb
green lead ore Buntbleierz *n* (Min)
green malt Grünmalz *n* (Brau), Malzkeime *pl*
greenockite Cadmiumblende *f* (Min)
greenovite Greenovit *m* (Min)
greensalt Grünsalz *n*
greensand Grünsand *m*
greensand casting Guß *m* in grünem Sand

greenstone Diorit *m* (Min), Grünstein *m* (Min),
   Jadeit *m* (Min)
greenstone slate Diabasschiefer *m* (Min)
green verditer Patina *f*, **coating of copper with ~**
   Patinierung *f*
green vitriol Eisenoxydulsulfat *n*, Eisenvitriol *n*
green weed Färbeginster *m* (Bot)
grenadine Grenadin *n*
grey (see also gray) grau
grey body emitter Graustrahler *m*
grey copper ore Fahlerz *n* (Min)
greying Vergrauung *f*
greyish fahlgrau
grey mold Grauschimmel *m*
grid Gitter *n* (Elektron, Techn), Netz *n* (Elektr),
   Rost *m* (Feuerrost), **coarse ~** weitmaschiges
   Gitter, **open ~** weitmaschiges Gitter,
   wide-meshed ~ weitmaschiges Gitter
grid accumulator Gitterakkumulator *m* (Elektr)
grid anode Netzanode *f*
grid anode capacity Gitteranodenkapazität *f*
grid bar bearer Rostschwelle *f*, Rostträger *m*
grid bias Gittervorspannung *f* (Elektr)
grid characteristic Gitterkennlinie *f*
grid circuit Gitter[strom]kreis *m* (Elektr)
grid condenser Gitterkondensator *m* (Elektr)
grid constant Gitterkonstante *f* (Radio)
grid-controlled gittergesteuert
grid current, reverse or negative ~ negativer
   Ionengitterstrom *m*
grid data Gitterwerte *m pl*
grid detector Gittergleichrichter *m* (Elektr)
grid drive Gittersteuerung *f*
grid electrode Gitterelektrode *f*
grid emission Gitteremission *f*
grid energy Gitterenergie *f* (Radio)
grid ionization chamber
   Gitterionisationskammer *f*
grid-iron valve Gitterventil *n* (Dest)
grid leak Gitterableitwiderstand *m* (Elektr)
grid modulation Gittermodulation *f* (Elektr)
grid of lines Liniennetz *n*, Raster *m*
grid plate Gitterboden *m* (Rektifikation)
grid point Gitterpunkt *m*
grid potential Gitterspannung *f* (Elektr)
grid ratio Rasterverhältnis *n*
grid rectifier Gittergleichrichter *m* (Elektr)
grid resistance Gitterwiderstand *m* (Elektr)
grid spinning Rostspinnverfahren *n*
grid voltage Gitterspannung *f* (Elektr)
Griesheim red Griesheimer Rot *n*
grifa Grifa *n*
griffithite Griffithit *m* (Min)
Griffith white Charltonweiß *n*
Grignard compound Grignard-Verbindung *f*
Grignard reaction Grignardsche Reaktion *f*
grill Bratrost *m*, Grill *m*
grill *v* braten, rösten, grillen
grilling Grillen *n*, Rösten *n*

grilon Grilon *n*
grind *v* [zer]reiben, mahlen, pulverisieren, schleifen (z. B. Messer, Glas), zermalmen, zerstoßen, ~ **coarsely** grobbrechen, ~ **in** einschleifen, einschaben, ~ **off** abschleifen, abschaben, ~ **on** aufschleifen, ~ **on the edge runner** kollern, ~ **out** ausschleifen, ~ **roughly** grobschleifen, [ab]schroten, ~ **to dust or powder** pulverisieren, einstäuben, ~ **with blast** absanden, ~ **with emery** abschmirgeln
grindability Vermahlungsfähigkeit *f*
grind and mix *v* kollern
grindelia Grindelienkraut *n* (Bot)
grindelol Grindelol *n*
grinder Brechwerk *n*, Schleifapparat *m*, Schleifmaschine *f*, Schleifstein *m*
grinding Reiben *n*, Mahlen *n*; Schleifen *n*, Zermahlen *n*, **coarse** ~ (or crushing) Grobzerkleinerung *f*, **degree of** ~ Mahlfeinheit *f*, Mahlgrad *m*, **fineness of** ~ Mahlfeinheit *f*, ~ **off** Abschleifung *f*, ~ **of lenses** Linsenschleifen *n*
grinding and polishing machine Schleif- und Poliermaschine *f*
grinding bench Schleifstuhl *m*
grinding board Schleifbrett *n*
grinding composition Schleifkomposition *f*
grinding disk Schleifscheibe *f*
grinding emery Schleifschmirgel *m*
grinding fineness, testing equipment for ~ Mahlfeinheitsprüfgerät *n*
grinding fixture, internal ~ Innenschleifvorrichtung *f*
grinding machine Schleifmaschine *f*, Zerkleinerungsmaschine *f*, **automatic** ~ Schleifautomat *m*, **belt for** ~ Schleifriemen *m*, **internal** ~ Innenschleifmaschine *f*
grinding material Schleifmittel *n*
grinding mill Brecher *m*, Mahlanlage *f*, Mahlwerk *n*
grinding motor Schleifmotor *m*
grinding operation Schleifarbeit *f*, Schleifvorgang *m*
grinding paste Einschleifpaste *f*
grinding path Mahlbahn *f*
grinding plant Mahlanlage *f*
grinding powder Schleifpulver *n*
grindings Abrieb *m*
grinding stock Mahlgut *n*
grinding surface Schleiffläche *f*
grinding tool Schleifwerkzeug *n*
grinding trestle Schleifbock *m*
grinding-type resin Pastenharz *n*
grinding unit Mahleinheit *f*
grinding wheel resin Schleifscheibenharz *n*
grinding width Schleifmaß *n*
grindstone Schleifstein *m*, Reibstein *m*
grip Griff *m*, Griffigkeit *f*, Heft *n* (Techn), Verbindungsstück *n* (Techn)

gripe *v* ergreifen, packen
grip for hand throttle Drehgriff *m* für Gaszug
griphite Greifstein *m* (Min)
gripping arrangement Einspannvorrichtung *f*, Spanneinrichtung *f*
gripping device Greifer *m*, Greifwerkzeug *n*
gripping power Greiffähigkeit *f*
gripping tongs Greifzange *f*
grip stopper Griffstopfen *m*
griqualandite Griqualandit *m* (Min)
grisan Grisan *n*
griseofulvic acid Griseofulvinsäure *f*
griseofulvin Griseofulvin *n*
grist Mahlgut *n*
gristle Knorpel *m* (Med)
gristles, to form ~ verknorpeln
gristly knorpelig
grit grober Sand *m*, Abrieb *m* (durch Abnutzung), Grieß *m* (Techn), Metallsand *m* (Techn)
grits Grieß *m* (Techn), Grütze *f*
gritty kiesig, grießig, körnig, sandig
groats Grütze *f*, Hafergrütze *f*
grochauite Grochauit *m* (Min)
groddeckite Groddeckit *m* (Min)
groin Gewölberippe *f*
grommet Durchführungshülse *f*, Durchführungsrohr *n* (Elektr), [Metall]Öse *f*, Metallring *m*
groove Rinne *f*, Auskehlung *f*, Auskerbung *f*, Aussparung *f*, Falz *m*, Furche *f*, Hohlkehle *f*, Kanal *m* (Abflußrinne), Nut *f*, Riffelung *f*, Rille *f*, Vertiefung *f*, **top width of** ~ obere Rillenbreite *f*
groove *v* auskehlen, einkerben, falzen, kannelieren, nuten, riefeln
groove angle Rillenwinkel *m* (Keilriemen)
groove cracking Rißbildung *f* im Profil, Profilriß *m*
grooved ausgekehlt, geriffelt, gerillt, rinnenförmig
grooved drum Rillentrommel *f*
grooved drum drier Rillenwalzentrockner *m*
grooved friction wheel Rillenreibrad *n*
grooved insulator Rillenisolator *m*
grooved rail Rillenschiene *f*
grooved roll Kaliberwalze *f*
grooved roller gear Kammwalzengetriebe *n*
grooved scraper Hohlschaber *m*
groover Einfeiler *m*, Nutwerkzeug *n*
grooves, to cut ~ Nuten *pl* stoßen, **to mill** ~ Nuten *pl* fräsen
groove spew (e. g. of a mo[u]ld) Austriebnut *f*
groove wheel Kehlrad *n*
grooving Furchung *f*, Kehlung *f*, Riffelung *f*, ~ **of rollers** Walzenriefelung *f*
grope *v* herumtappen
groppite Groppit *m* (Min)
groroilite Groroilith *m* (Min)

gross brutto, Gros *n*
gross molecular formula Bruttoformel *f*
gross price Bruttopreis *m*
gross reaction Bruttoreaktion *f*
gross tonnage Bruttotonnengehalt *m*
grossular Grossular *m*, grüner Granat *m*
grossularite Grossular *m* (Min),
   Kalk[ton]granat *m* (Min)
gross weight Bruttogewicht *n*, Grobgewicht *n*,
   Rohgewicht *n*
gross work Gesamtarbeit *f*
grothite Grothit *m* (Min)
ground Boden *m*, Erdschluß *m* (Elektr),
   Erdung *f* (Elektr), Grund *m*, Land *n*, **above** ~
   oberirdisch, **pervious** ~ durchlässiger
   Boden *m*, **wealth under** ~ Bodenschätze *pl*
   (Min)
ground *a* gemahlen, matt (Glas), mattgeschliffen
   (Glas), **coarsely** ~ grobgepulvert
ground *v* grundieren, erden (Elektr)
ground bacterium Bodenbakterie *f*
ground clamp Erdungsschelle *f*
ground coat Grundanstrich *m*, Unterschicht *f*
ground color Farbengrund *m*, Grundanstrich *m*,
   Grundierung *f*
ground connection Erdung *f*, Masseanschluß *m*
ground cover Schliffdeckel *m*
grounded geerdet (Elektr)
ground frame Grundgestell *n*
ground glass disk Mattscheibe *f* (Opt)
ground glass joint Glasschliff *m*
ground glass stopper Schliffstopfen *m*
ground humidity Bodenfeuchtigkeit *f*
ground ice Grundeis *n*
ground-in eingeschliffen
ground-in piston eingeschliffener Kolben *m*
ground ivy Gundelrebe *f* (Bot)
ground ivy oil Gundermannöl *n*
ground joint Schliff *m* (Glas), **male** ~
   Kernschliff *m* (Glas)
ground joint apparatus Schliffgerät *n* (Chem)
ground line Grundlinie *f* (Math)
ground meat Hackfleisch *n*
groundnut oil Arachisöl *n*, Erdnußöl *n*
ground oak bark Eichenmehl *n*
ground outline Grundriß *m*
ground phosphate Phosphatmehl *n*
ground plan Grundriß *m*
grounds Bodenhefe *f*, Satz *m*, Trester *pl*
ground state Grundzustand *m*
ground temperature Bodentemperatur *f*
ground water Grundwasser *n*, **level of** ~
   Grundwasserspiegel *m*
ground wave Bodenwelle *f*
ground wire Erdungsdraht *m*, Erd[ungs]leitung *f*
   (Elektr)

groundwork Fundament *n*, Grundmauern *pl*,
   Unterbau *m*
group Gruppe *f*, **chromophoric** ~ chromophore
   Gruppe, **functional** ~
   eigenschaftsbestimmende Gruppe,
   funktionelle Gruppe, **hydrophilic** ~
   hydrophile Gruppe *f*, **hydrophobic** ~
   hydrophobe Gruppe, **non-polar** ~ unpolare
   Gruppe, **of the same** ~ gruppengleich (Blut),
   polar ~ polare Gruppe, **prosthetic** ~
   prosthetische Gruppe, ~ **of lines**
   Liniengruppe *f*, ~ **of machines**
   Maschinensatz *m*
group activation Gruppenaktivierung *f*
group algebra Gruppenalgebra *f*
grouping Bündelung *f*, Zusammenstellung *f*, ~
   together Zusammenschluß *m*
group reagent Gruppenreagens *n*
group-transfer reaction
   Gruppentransferreaktion *f*
group velocity Gruppengeschwindigkeit *f*
grout Kalkguß *m*
grout *v* [**in**] einzementieren
grouting hole Eingußloch *n*
Grove cell Grove-Element *n* (Elektr)
grow *v* anwachsen, [aus]wachsen, gedeihen,
   züchten (Bakterien), zunehmen, ~
   exuberantly wuchern, ~ **old** altern, ~ **stale**
   abstehen (Metall); **stronger** anschwellen, ~
   together zusammenwachsen
grower Züchter *m*
grown, fully ~ ausgewachsen
growth Wachstum *n*, Vermehrung *f*, Wuchs *m*,
   Zunahme *f*, Zuwachs *m*, **center of** ~
   Wachstumszentrum *n*, **rank** ~ Wucherung *f*
   (Bot), **rate of** ~ Wachstumsgeschwindigkeit *f*,
   ~ **of crystals** Kristallwachstum *n*
growth curve Wachstumskurve *f*,
   Zunahmekurve *f*
growth factor Wachstumsfaktor *m* (Biol)
growth front Wachstumsfront *f* (Krist)
growth hormone Wachstumshormon *n*
growth-inhibiting wachstumshemmend,
   wachstumshindernd
growth promoter Wuchsstoff *m*
growth-promoting wachstumsfördernd (Biol)
growth-promoting substance Wuchsstoff *m*
growth spiral Wachstumsspirale *f*
grub Larve *f* (Zool), Made *f*, Raupe *f*
grub screw Kopfschraube *f* mit Schlitz,
   Madenschraube *f*, Stiftschraube *f*
grunauite Wismutnickelkobaltkies *m* (Min)
grunerite Grünerit *m* (Min)
grunlingite Grünlingit *m* (Min)
GTP (Abk) Guanosintriphosphat *n*
guacin Guacin *n*
guaco leaves Guacoblätter *n pl* (Bot)
guadalcazarite Guadalcazarit *m* (Min)
guaethol Guäthol *n*

guaiac Guajaköl *n* (Bot)
guaiacic acid Guajacinsäure *f,*
  Guajak[harz]säure *f*
guaiacin Guajacin *n,* Guajakessenz *f*
guaiacol Guajacol *n*
guaiacol carbonate Guajacolcarbonat *n,*
  Duotal *n* (HN)
guaiacol phthalein Guajacolphthalein *n*
guaiacol valerate Geosot *n*
guaiaconic acid Guajaconsäure *f*
guaiac resin Guajakharz *n,* **tincture of** ~
  Guajakharztinktur *f*
guaiac soap Guajakseife *f*
guaiac tincture Guajakharzlösung *f,*
  Guajaktinktur *f*
guaiacum Franzosenharz *n,* Franzosenholz *n,*
  Guajakholz *n*
guaiacum oil Guajaköl *n*
guaiacum wood Franzosenholz *n,* Guajakholz *n*
guaiacyl benzoate Guajacolbenzoat *n*
guaiacyl phosphate Guajacolphosphat *n*
guaiacyl salicylate Guajacolsalicylat *n*
guaiadol Guajadol *n*
guaiane Guajan *n*
guaiaretic acid Guajaretsäure *f,*
  Guajak[harz]säure *f*
guaiazulene Guajazulen *n*
guaiol Guajol *n*
guanamine Guanamin *n*
guanase Guanase *f* (Biochem)
guanazine Guanazin *n*
guanazole Guanazol *n*
guanidine Guanidin *n,* Imidoharnstoff *m*
guanidine base Guanidinbase *f*
guanidine carbonate Guanidincarbonat *n*
guanidine sulfate Guanidinsulfat *n*
guanidino Guanidino-
guanidylacetic acid Guanidylessigsäure *f*
guaniferous guanoführend
guanine Guanin *n,* Imidoxanthin *n*
guanine nucleotide Guaninnucleotid *n*
guanine pentoside Guaninpentosid *n*
guanite Guanit *m* (Min)
guano Guano *m*
guanomineral Guanomineral *n*
guanosin[e] Guanosin *n*
guanosine diphosphate Guanosindiphosphat *n*
guanosine monophosphate
  Guanosinmonophosphat *n,* **cyclic** ~
  zyklisches Guanosinmonophosphat *n*
guanosinephosphoric acid
  Guanosinphosphorsäure *f*
guanosine triphosphate Guanosintriphosphat *n*
guano superphosphate Guanosuperphosphat *n*
guanovulite Guanovulit *m* (Min)
guanyl Guanyl-
guanylglycine Guanylglycin *n*
guanylguanidine Diguanid *n,* Guanylguanidin *n*
guanylic acid Guanylsäure *f*

guanylthiourea Guanylthioharnstoff *m*
guanylurea Dicyandiamidin *n,*
  Guanylharnstoff *m*
guarana Guarana *f*
guaranine Guaranin *n*
guarantee Bürgschaft *f,* Garantie *f,*
  Gewährleistung *f*
guarantee *v* gewährleisten, haften, garantieren,
  gewährleisten
guard Abstreifmeißel *m* (Techn)
guard *v* behüten, ~ **against** schützen
guard chain Sicherheitskette *f*
guarded abgeschirmt
guard electrode Schutzelektrode *f*
guard plate Schutzblech *n*
guard rail Leitschiene *f,* Radlenker *m*
guard relay Sperrelais *n*
guard ring Hilfselektrode *f,* Schutzring *m*
  (Elektr)
guard wire Fangdraht *m*
guarinite Guarinit *m* (Min)
gudgeon Schildzapfen *m,* Zapfen *m*
gudgeon pin Kolbenbolzen *m*
guhr Gur *f*
guhr dynamite Gurdynamit *n*
guidance Führung *f,* Leitung *f*
guide Führung *f,* Führungselement *n,*
  Gleitschiene *f,* Gleitstange *f,* Richtschnur *f*
guide *v* [an]leiten, lenken
guide bar Führungsstange *f*
guide bearing Führungslager *n*
guide block Führungsbacken *m*
guide-line Richtlinie *f*
guide nozzle Leitdüse *f*
guide perforation Führungsloch *n*
guide pilot Führungszapfen *m*
guide pin Führungsstift *m,* Abhebestift *m*
guide pulley Führungsrolle *f,* Leitrolle *f*
guide pulley drive Winkeltrieb *m*
guide rail Führungsschiene *f,* Laufschiene *f,*
  Leitschiene *f*
guide rod Führungsstange *f*
guide roller Führungsrolle *f,* Führungswalze *f,*
  Leitrolle *f*
guide spindle Leitspindel *f*
guide sprocket Umlenkrad *n* (Stahlketten)
guide stone Leitstein *m*
guide value Orientierungswert *m*
guide vane Leitschaufel *f*
guide vanes Drallklappen *pl*
guide wheel Leitrad *n*
guiding field Führungsfeld *n*
Guignet's green Guignetgrün *n*
guillotine Guillotine *f,*
  Papierschneidemaschine *f,* Querschneider *m*
guillotine cutter Hackmaschine *f*
guillotine knife Hackmesser *n*
guinea green Guineagrün *n*
guinea pig (domestiziertes) Meerschweinchen *n*

guitermanite Guitermanit *m* (Min)
gulitol Gulit *m*
gully Abflußschacht *m*, Ablaufrinne *f*, Gully *m*
guloheptonic acid Guloheptonsäure *f*
gulonic acid Gulonsäure *f*
gulose Gulose *f*
gulosone Guloson *n*
guluronic acid Guluronsäure *f*
gum Gummi *n*, Gummiharz *n*, Kautschuk *m*, Klebemittel *n*, Kleister *m*, **containing** ~ gummihaltig, ~ **of silk** Bast *m*
gum *v* kleben; mit Gummilösung bestreichen, gummieren, leimen
gum acacia Arabingummi *n*
gum ammoniac Ammoniakgummi *n*, Ammoniakharz *n*
gum animé Flußharz *n*
gum arabic Gummiarabikum *n*, Akaziengummi *n*, Arabingummi *n*, Galamgummi *n*, Mimosengummi *n*
gum carane Mararaharz *n*
gum copal Flußharz *n*
gum dragon Tragantgummi *n*
gum elemi Elemiharz *n*
gum galbanum Galbanharz *n*
gum guaiac Franzosenharz *n*, Guajak *m*
gum juniper Wacholderharz *n*
gum kino Gabiragummi *m*
gum lac Lackharz *n*
gum-like gummiartig
[gum] mastic Mastix *m*
gummed gummiert (klebend)
gummed label Klebezettel *m*
gummed surface Gummierung *f* (klebende Fläche)
gumming Gummierung *f* (Klebstoff)
gummite Gummierz *n* (Min)
gummy gummiartig, gummihaltig, klebrig, zähflüssig
gummy material Leimstoff *m*
gumnite Gumnit *n*
gum resin Schleimharz *n*
gumresinous gummiharzig
gum rosin Naturharz *n*
gum[s] Zahnfleisch *n*
gum solution Gummilösung *f*
gum sugar Pektinose *f*
gum tragacanth Gummitragant *m*, Tragantgummi *n*
gum water Gummiwasser *n*, ~ **for marbling paper** Marmorierwasser *n* (Pap)
gumwax Gummiwachs *n*
gun Druckrohr *n* (Techn)
gun cotton Schießbaumwolle *f*, Collodiumwolle *f*
gunite Spritzbeton *m*
gunmetal Geschützbronze *f*, Geschützmetall *n*, Kanonenmetall *n*, Rotguß *m*
gunnera Färbernessel *f* (Bot), Gunnera *f* (Bot)
gunpowder Schießpulver *n*

gunpowder ore Pulvererz *n*
gunpowder press cake Pulverkuchen *m*
gunpowder smoke Pulverdampf *m*
gurjun [balsam] Gurjunbalsam *m*, Holzölfirnis *m*
gurjunene Gurjunen *n*
gurjunene alcohol Gurjunenalkohol *m*
gurjunene ketone Gurjunenketon *n*
gurjuresene Gurjuresen *n*
gurolite Gyrolith *m* (Min)
gushing Überschäumen *n* (b. Öffnen v. Flaschen etc.)
gusset Winkelstück *n*
gusseted sack Sack *m* in Kastenform
gustation Geschmackssinn *m*
gustatory cell Geschmackszelle *f*
gustatory nerve Geschmacksnerv *m*
gustatory organ Geschmacksorgan *n*
gut Darm *m* (Med), **artificial** ~ Kunstdarm *m*
guttapercha Guttapercha *f n*, Perchagummi *n*, **refined** ~ gereinigtes Guttapercha
guttapercha bottle Guttaperchaflasche *f*
guttapercha mastic Guttaperchakitt *m*
guttapercha mixture Guttaperchamischung *f*
guttapercha paper Guttaperchapapier *n*
guttapercha resin Guttaperchaharz *n*
guttapercha sheet Guttaperchapapier *n*
guttapercha solution Guttaperchalösung *f*
gutter Abflußrinne *f*, Abflußrohr *n*, Ablaufrinne *f*, Dachrinne *f*, Furche *f*, Gosse *f*, Gully *m*, Rinne *f*
gutter stone Hohlstein *m*
Gutzeit test Gutzeitsche Probe *f*
guvacine Guvacin *n*
guvacoline Guvacolin *n*
guy Führungskette *f*, Halteseil *n*, Leitstrick *m*
gymnemic acid Gymnemasäure *f*
gymnite Gymnit *m* (Min)
gymnogrammene Gymnogrammen *n*
gynaecology (Br. E.), gynecology (Am. E.) Gynäkologie *f* (Med)
gynaminic acid Gynaminsäure *f*
gynocardia oil Chaulmoograöl *n*, Gynocardiaöl *n*
gynocardic acid Gynocardsäure *f*
gynocardin Gynocardin *n*
gynocardinic acid Gynocardinsäure *f*
gynoval Gynoval *n*, Bornyval *n*
gypseous gipsartig
gypseous alabaster Alabastergips *m*
gypseous earth Gipserde *f*
gypseous marl Gipsmergel *m*
gypseous sand Gipssand *m*
gypseous stone Gipsstein *m* (Min)
gypsiferous gipshaltig
gypsite Gipsdruse *f* (Min)
gypsoplast Gipsabdruck *m*, Gipsguß *m*
gypsum Gips *m*, **compact** ~ körniger Gips, **containing** ~ gipshaltig, **crystallized** ~ Gipsdruse *f* (Min), Gipskristall *m*, **dead-burnt**

~ totgebrannter Gips, **earthy** ~ Gipserde *f*,
fibrous ~ Fasergips *m*, Atlasgips *m*,
Federgips *m*, Strahlgips *m*, **foliated** ~
Schiefergips *m* (Min), **foliated granular** ~
schuppig-körniger Gips, **granular** ~
Höhlengips *m*, **ground** ~ Lenzin *m*, **overburnt**
~ totgebrannter Gips, **powdered** ~
Gipsmehl *n*, **sparry** ~ Gipsblume *f*, **special** ~
Spezialgips *m*, **striate** ~ Federgips *m*
gypsum burning Gipsbrennen *n*
gypsum calcination Gipsbrennen *n*
gypsum cement Alabastergips *m*
gypsum industry Gipsindustrie *f*
gypsum model Gipsmodell *n*
gypsum paste Gipsbrei *m*
gypsum spar Gipsspat *m* (Min), Katzenglas *n*
 (Min)
gypsum wedge Gipskeil *m*
gyration Drehung *f*, Kreisbewegung *f*
gyrator Gyrator *m* (Elektr)
gyratory kreisend
gyratory crusher Kreiselbrecher *m*, Drehmühle *f*,
 Glockenmühle *f*, Rundbrecher *m*
gyratory screen Plansieb *n*, ~ **with supplementary**
 whipping action Taumelsieb *n*
gyrilone Gyrilon *n*
gyrocompass Kreiselkompaß *m*
gyrofrequency Kreiselfrequenz *f*
gyrograph Tourenzähler *m*
gyro-interaction Kreiselwechselwirkung *f*
gyrolite Gyrolith *m* (Min)
gyrolone Gyrolon *n*
gyromagnetic gyromagnetisch
gyrometer Gyrometer *n* (Phys)
gyro-mixer Kreiselmischer *m*
gyro-oscillation Kreiselschwingung *f*
gyrophoric acid Gyrophorsäure *f*
gyrophorine Gyrophorin *n*
gyroscope Gyroskop *n* (Phys), Kreisel *m* (Phys)
gyroscopic effect Kreiselwirkung *f*
gyroscopic motion Kreiselbewegung *f*
gyrostabilizer Kreisel *m*, Kreiselstabilisierung *f*
gyrostat Gyrostat *m*, Kreisel *m*

H

Haber-Bosch process Haber-Bosch-Verfahren *n*
habit Habitus *m* (Biol), Körperbeschaffenheit *f*,
Verhaltensweise *f*
habit effect Habituswirkung *f*
habituate *v* gewöhnen
habituation Gewöhnung *f*
hachure *v* schraffieren
H acid H-Säure *f*
haematic substance Blutbestandteil *m*
haemat- or haemato- see hemat-
haem- or haemo- see hem-
haemorrhage (Med) Blutsturz *m*, Blutung
hafnium (symb. Hf) Hafnium *n*
haft Griff *m*, Heft *n*
hagemannite Hagemannit *m* (Min)
Haidinger fringes Haidinger-Ringe *pl* (Opt)
haidingerite Haidingerit *m* (Min)
hail Hagel *m*
hainite Hainit *m* (Min)
hair Haar *n*, **color of the** ~ Haarfarbe *f*, **root of
the** ~ Haarwurzel *f*
hair cloth Haargewebe *n*
hair cord root Haarstrangwurzel *f* (Bot)
hair cracking Haarriß *m*
hairdressing Haarpflege *f*
hair fixature Haarfestiger *m*
hair hygrometer Fadenhygrometer *n*
hair keratin Haarkeratin *n*
hairline Anriß *m*, Haarstrich *m*
hairline crack Haarriß *m*
hair parasite Haarparasit *m* (Zool)
hairpin Haarnadel *f*
hair pyrites Haarkies *m* (Min)
hair setting lotion Haarfestiger *m*
hair tonic Haarpflegemittel *n* (Kosmetik),
Haarwasser *n* (Kosmetik)
hairy haarig
halation Lichtfleck *m* (Phot), Lichthof[bildung
f] *m* (Phot)
halazone Halazon *n* (HN)
half Hälfte *f*, Halbe *n*
half adder Halbaddierwerk *n*
half bleach Halbbleiche *f*
half-cell Halbelement *n* (Elektr), Halbzelle *f*
(Elektr)
half-chair form Halbsesselform *f* (Stereochem)
half-cooked halbgar
half-crush *v* halbmahlen
half cycle Halbperiode *f*
half decay period Halbwert[s]periode *f*
half dome Halbkuppel *f*
half-element Halbelement *n* (Elektr)
half-fused halbgesintert
half-grind *v* halbmahlen

half-life Halbwertszeit *f* (Atom), **biological** ~
biologische Halbwertszeit *f*, **effective** ~
effektive Halbwertszeit
half-life period Halbwertszeit *f* (Atom)
half-linen Halbleinen *n*
half-loop Halbschleife *f*
half measure Halbheit *f*
half-period Halbperiode *f*, Halbwertszeit *f*
(Atom)
half-power [band] width Halbwertbreite *f*
half-round halbrund
half-round file Halbrundfeile *f*
half-round iron Halbrundeisen *n* (Techn)
half-shade Halbschatten *m*
half-shade apparatus Halbschattenapparat *m*
(Opt)
half-shade plate Halbschattenplatte *f* (Opt)
half-shade polarimeter
Halbschattenpolarimeter *n*
half-shadow analyzer Halbschattenapparat *m*
half-shadow apparatus Halbschattenapparat *m*
half-shadow polarimeter
Halbschattenpolarisationsapparat *m*
half-shadow quartz plate
Halbschattenquarzplatte *f*
half-shadow system Halbschattensystem *n*
half-silvered halbdurchlässig verspiegelt (Opt)
half-sized halbgeleimt (Pap)
half-stuff Halbstoff *m* (Pap), Halbzellstoff *m*
half-stuff engine Halbzeugholländer *m* (Pap)
halftone Halbschatten *m*, Halbton *m*
halftone engraving Autotypie *f*
halftone process Halbtonverfahren *n*
half-track [drive] Halbkettenantrieb *m*
half-turn Halbdrehung *f*
half-value depth Halbwert[s]tiefe *f*
half-value layer Halbwert[s]schicht *f*
half-value period Halbwert[s]periode *f*
half-value time Halbwertszeit *f* (Atom)
halfwatt bulb Halbwattbirne *f* (Elektr)
halfwatt lamp Halbwattlampe *f* (Elektr)
half-wave Halbwelle *f*
half-wave potential Halbstufenpotential *n*
half-wave rectification Halbweggleichrichtung *f*
(Elektr)
half-wave rectifier Halbweggleichrichter *m*
(Elektr)
halfway halbwegs
half width Halbwertsbreite *f*, ~ **of interference
maxima** Halbwertsbreite *f* der Interferenz
(Opt)
halide Halogenid *n*, **metallic** ~ Metallhalogenid,
Halogenmetall *n*
halide crystal Halogenidkristall *m*
halidefree halogenidfrei

halide leak detector Halogenleckdetektor *m* (Chem)
halite Halit *m* (Min), Bergsalz *n* (Min)
hallerite Hallerit *m* (Min)
halliard block Fallblock *m*
halliard leading block Falleitblock *m*
hall mark Feingehaltsstempel *m* (der Londoner Goldschmiedeinnung)
halloysite Halloysit *m* (Min)
hallucinogen Halluzinogen *n* (Pharm)
Hallwachs effect lichtelektrische Elektronenemission *f*
halo Lichthof *m* (Astr), Strahlenring *m* (Astr)
halobenzoic acid Halogenbenzoesäure *f*
halochemical halochemisch
halochemistry Halochemie *f*
halochromism Halochromie *f* (Färb)
haloform reaction Haloformreaktion *f*
halogen Halogen *n*, Salzbildner *m*
halogen acid Halogen[wasserstoff]säure *f*
halogen alkane Halogenalkyl *n*
halogenatable halogenierbar
halogenate *v* halogenieren
halogenated, capable of being ~ halogenierbar
halogenation Halogenierung *f*
halogen atom Halogenatom *n*
halogen bulb Halogenlampe *f*
halogen carrier Halogenüberträger *m*
halogen compound Halogenverbindung *f*
halogen derivative Halogenderivat *n*
halogen[-quenched] counter Halogenzählrohr *n*
halogen quenching Halogenlöschung *f*
halogen-substituted indigo Halogenindigo *m*
halogen-substituted quinone Halogenchinon *n*
halogen-substituted xylene Haloxylol *n*
halogen substitution Halogensubstitution *f*
halography Halographie *f*
halohydrin Halohydrin *n*
haloid Halogensalz *n*
haloid acid Halogenwasserstoffsäure *f*
haloindigo Halogenindigo *m*
halometer Halometer *n*, Salzgehaltmesser *m*
halometric halometrisch
halometric test Salzgehaltmessung *f*
halometry Halometrie *f*, Salzgehaltmessung *f*
halonitrobenzene Nitrohalogenbenzol *n*
halophyte Halophyt *m* (Bot), Salzpflanze *f* (Bot)
halopyridine Halogenpyridin *n*
haloquinone Halogenchinon *n*
halostachine Halostachin *n*
halothane Halothan *n*
halotrichite Halotrichit *m* (Min), Faseralaun *m* (Min)
haloxylene Haloxylol *n*
haloxylin Haloxylin *n*
halozone Halozon *n*
halphen acid Halphensäure *f*
halt Halt *m*, Haltezeit *f*, Stillstand *m*
halve *v* halbieren

hamamelis, liquid extract of ~ Hamamelisblätterfluidextrakt *m* (Pharm)
hamamelis bark Hamamelisrinde *f*
hamamelis ointment Hamamelissalbe *f* (Pharm)
hamamelitannine Hamamelitannin *n*
hamamelonic acid Hamamelonsäure *f*
hamamelose Hamamelose *f*
hamartite Bastnäsit *m* (Min), Hamartit *m* (Min)
hambergite Hambergit *m* (Min)
Hamburg blue Hamburger Blau *n*
Hamburg white Hamburger Weiß *n*
Hamilton operator Hamilton-Operator *m*
hamlinite Hamlinit *m* (Min)
hammer Hammer *m*, **handle or shaft hole of a** ~ Hammerauge *n*, **striking face of a** ~ Hammerfinne *f*, ~ **operated by a crank** Kurbelhammer *m*
hammer *v* hämmern, schmieden, ~ **the bloom** zängen, ~ **well** durchhämmern
hammer crusher Hammerbrecher *m*
hammer cylinder Barzylinder *m*
hammer drill Bohrhammer *m*
hammer effect Hammer[schlag]effekt *m*
hammer face Hammerbahn *f*
hammer finish Hammerschlaglack *m*
hammer forging Freiformschmiedestück *n*
hammer handle Hammerstiel *m*
hammer-harden *v* kalthämmern, hartschlagen
hammer-hardening Kaltschmieden *n*
hammerhead Hammerbär *m*, Hammerkopf *m*
hammer inductor Hammerinduktor *m*
hammering spanner Schlagschlüssel *m*
hammering test Ausbreiteprobe *f*
hammer interrupter Hammerunterbrecher *m*
hammer mark Beule *f*
hammer metal finish Hammerschlaglack *m*
hammer mill Hammermühle *f* (Techn), Hammerwerk *n* (Techn), Schlagmühle *f*
hammer scale Hammerschlag *m*, Glühspan *m*, Zunder *m*
hammer shop Hammeranlage *f*, Hammerhütte *f*
hammersmith Hammerschmied *m*, Schmied *m*
hammer tup Hammerbär *m*, Hammerkopf *m*
hampdenite Hampdenit *m* (Min)
hamper Flaschenkorb *m*, Packkorb *m*, Tragkorb *m*
hamper *v* [be]hindern
hancockite Hancockit *m* (Min)
hand Zeiger *m* (Uhr) (Am. E.), Handlanger *m*, **supplied by** ~ handbeschickt, **worked by** ~ handbetätigt
hand adjuster Handsteller *m*
hand airpump Handluftpumpe *f*
hand apparatus Handapparat *m*
handbarrow Handkarren *m*, Trage *f*
hand bellows Handblasebalg *m*
hand blowpipe Handgebläse *n*
handbook Handbuch *n*
hand brake Handbremse *f*

hand brush Handfeger *m*
hand bucket Handeimer *m*
handcart Handkarren *m*
hand chisel Handmeißel *m*
hand cleanser Handreiniger *m*
hand control Handbedienung *f*
hand crane Armkran *m*
hand drilling machine Handbohrmaschine *f*
hand drive Handbetrieb *m*
handfed handbeschickt
handfired handbeschickt
hand fire extinguisher Handfeuerlöscher *m*
hand gear Handsteuerung *f*
handguard Handschutz *m*
handianolic acid Handianolsäure *f*
handicap *v* benachteiligen, [be]hindern
handicraft Handwerk *n*
handiwork Handarbeit *f*
hand ladle Handgießlöffel *m*, Handkelle *f*
handle Griff *m*, Haltegriff *m*, Heft *n*, Helm *m*,
Henkel *m*, Stiel *m*, **crooked** ~ Kurbel *f*, ~ **of**
peel Schwengel *m*
handle *v* anfassen, befühlen, behandeln,
bewältigen, gebrauchen, handhaben,
hantieren
handle bar Lenkstange *f*
handling Handhabung *f*, Behandlung *f*,
convenient ~ leichte Handhabung
hand lubrication Handschmierung *f*
hand-made handgefertigt, handgemacht, mit der
Hand angefertigt
handmade paper Büttenpapier *n*, Schöpfpapier *n*
hand mold Handform *f*
hand molding machine Handformmaschine *f*
hand-operated handbedient
hand operation Handbetrieb *m*
hand-picking Klaubarbeit *f* (Metall)
hand pneumatic rammer
Handpreßluftstampfer *m* (Techn)
hand-power machine Maschine *f* mit
Handbetrieb
handpress Handpresse *f*
handrail Geländer *n*, Handlauf *m*
hand rammer Handstampfer *m*
hand reamer Handreibahle *f*
hand regulation Handsteuerung *f*
hand safety guard Handschutzvorrichtung *f*
hand saver Schutzhandschuh *m*
hand saw Fuchsschwanz *m* (Techn)
hand scales Handwaage *f*
hand scoop Wurfschaufel *f*
hand screw press Handspindelpresse *f*
hand set Handapparat *m* (Elektr)
hand shears Handschere *f*
handspike Hebestange *f*, Handspake *f*,
Handspeiche *f*
hand switch Handschalter *m*
hand-tipping mechanism Handkippvorrichtung *f*
handwelding Handschweißen *n*

handwheel Handrad *n*
handwheel stamping press
Schlagradhandpresse *f*, Spindelpresse *f*
handwinch Handwinde *f*
handwork Handarbeit *f*
handy handlich, praktisch, geschickt, gewandt
hanger Hängeeisen *n*
hanging hängend
hanging bearing Hängelager *n*
hanging condenser Einhängekühler *m*
hanging filter Einhängefilter *n*
hanging of the charge Hängen *n* der Gicht
hanging position Schwebelage *f*
hanging stage Hängegerüst *n*
hank Docke *f* (Garn), Knäuel *n* (Garn),
Strang *m* (Garn)
hank dyeing apparatus Strangfärbeapparat *m*
hank dyeing holder Haspel *f*
hanksite Hanksit *m* (Min)
hannayite Hannayit *m* (Min)
hansa yellow Hansagelb *n*
Hanssen acid Hanssen-Säure *f*
haplo-bacterium Haplobakterium *n*
haploid haploid (Biol)
haploidy Haploidie *f* (Biol)
happen *v* passieren, sich abspielen, vorkommen,
sich ereignen
haptoglobin Haptoglobin *n*
hard hart, rauh, schwierig, **medium** ~ mittelhart
hard-baked hartgebrannt
hardboard Hartfaserplatte *f*
hard-burned hartgebrannt
hard-burned brick Hartbrandstein *m*
hard copper Hartkupfer *n*
hard crockery Halbporzellan *n*
hard-cutting alloy Hartschneidemetall *n*
hard-drying brilliant oil Harttrockenglanzöl *n*
hard-drying oil Harttrockenöl *n*
hard-drying time Durchtrocknungszeit *f*
harden *v* härten, abbinden (Zement), abbrennen
(Stahl), abschrecken (Met), auslagern
(Duralum), erhärten, erstarren, hart werden,
stählen, vergüten (Met), verhärten, ~ **by itself**
selbständig erhärten
hardenable härtbar
hardenable plastic Duroplast *n*
hardened gehärtet, ~ **by burning** hartgebrannt
hardened filter Hartfilter *n*
hardened glass Hartglas *n*
hardener Härtemittel *n*, Härter *m*, **cold** ~
Kalthärter *m*
hardening Härten *n*, Erhärten *n*, Erstarrung *f*,
Fixierung *f* (Phot), Härtung *f*, Hartwerden *n*,
Verhärten *n*, Verhärtung *f*, **degree of suitability**
for ~ Härteeignungseigenschaft *f*, **depth of** ~
Härtetiefe *f*, **interrupted** ~ unterbrochene
Härtung *f*, **process of** ~ Härtevorgang *m*, ~
and tempering Vergütung *f* (Stahl), ~ **by cold**
working Kaltverfestigung *f*, ~ **by sprinkling**

Spritzhärtung *f,* ~ **of liquid fats** Fetthärtung *f,*
~ **of steel** Härten *n* des Stahls, ~ **of the glaze**
Glattbrennen *n*
hardening agent Härtungsmittel *n*
hardening capacity Härtbarkeit *f,*
Härtungsfähigkeit *f,* Härtungsvermögen *n*
hardening constituent Härtebildner *m*
hardening crack Härteriß *m*
hardening furnace Härteofen *m,* Einsatzofen *m*
hardening liquid Ablöschflüssigkeit *f,*
Härteflüssigkeit *f,* Kühlflüssigkeit *f*
hardening mixture Härtemittel *n*
hardening operation Härtevorgang *m*
hardening plant Härteanlage *f*
hardening process Härtungsverfahren *n*
hardening room Erstarrungskammer *f,*
Erstarrungsraum *m*
hardening salt Härtesalz *n*
hardening strain Härtespannung *f,*
Härtungsspannung *f*
hardening stress, internal ~ Härtungsspannung *f*
hardening trough Einsatzkasten *m,* Härtetrog *m*
hard fiber Hartfaser *f*
hard filter Hartfilter *n*
hard-frozen hartgefroren
hard glass Hartglas *n*
hard glass beaker Hartglasbecher *m*
hard grain Hartkorn *n*
hard iron Harteisen *n*
hard lac resin Hartschellack *m*
hard metal Hartmetall *n*
hardness Festigkeit *f,* Härte *f,* **coefficient of** ~
Härtezahl *f,* **degree of** ~ Härtegrad *m,*
determination of ~ Härtebestimmung *f,*
Härtemessung *f,* **measurement of** ~
Härtemessung *f,* **non-carbonate** ~
Nichtcarbonathärte *f* (Wasser), **passive** ~
passive Härte, **permanent** ~ permanente
Härte, **temporary** ~ veränderliche Härte, **total**
~ Gesamthärte *f,* **unit of** ~ Härteeinheit *f,* ~
due to carbonate(s) Carbonathärte *f,* ~ **of**
water Wasserhärte *f*
hardness cooling rate curve Härtbarkeitskurve *f*
hardness drop tester Skleroskop *n*
hardness gauge Härtemesser *m*
hardness number Härtezahl *f*
hardness scale Härteskala *f*
hardness test Härteprobe *f*
hardness tester Härteprüfer *m,* Härteprüfgerät *n*
hardness test[ing] Härteprüfung *f*
hardness testing apparatus Härteprüfapparat *m*
hardness testing machine Härteprüfmaschine *f*
hardness testing rod Härteprüfstab *m*
hardness test procedure Härteprüfverfahren *n*
hard paper Hartpapier *n*
hard paper insulation Hartpapierisolation *f*
hard paraffin Hartparaffin *n*
hard pearlite Hartperlit *m* (Min)
hard porcelain Edelporzellan *n,* Hartporzellan *n*

hard-rolled hartgewalzt
hard rubber Hartgummi *n,* Vulkanit *n*
hard rubber goods Hartgummiwaren *pl*
hard rubber tube Hartgummirohr *n*
hard slag Härtling *m*
hard soap Kernseife *f,* Natronseife *f,* Stückseife *f*
hard solder Hartlot *n,* Messinglot *n,* Schlaglot *n*
hardsolder *v* hartlöten
hard-soldered hartgelötet
hard-soldering Hartlöten *n,* Hartlötung *f*
hard-soldering fluid Hartlötwasser *n*
hard solder wire Hartlötdraht *m*
hard steel Hartstahl *m*
hardware Bausteine *pl* (Comp), Eisengußware *f,*
Eisenwaren *pl*
hardware store Eisenwarenhandlung *f*
hard wax Hartwachs *n*
hard whiteware Hartsteingut *n*
hardystonite Hardystonit *m* (Min)
hard zinc Hartzink *n*
harm Beschädigung *f,* Schaden *m,* Verletzung *f*
harm *v* beschädigen, benachteiligen
harmalan Harmalan *n*
harmala red Harmalinrot *n*
harmaline Harmalin *n*
harman Harman *n,* Aribin *n*
harmful schädlich
harmine Harmin *n,* Yajein *n*
harmine dihydride Harmalin *n*
harminic acid Harminsäure *f*
harmless ungefährlich, unschädlich
harmol Harmol *n*
harmolic acid Harmolsäure *f*
harmonic harmonisch
harmonic analyzer Schwingungsanalysator *m*
harmonic dial Periodenuhr *f*
harmonic mean harmonischer Mittelwert *m*
harmonic response (Br. E.) Frequenzgang *m*
(Regeltechn)
harmonic series harmonische Reihe *f*
harmonious harmonisch
harmony Ebenmaß *n,* Einklang *m,* Harmonie *f*
harmotome Harmotom *m* (Min), Kreuzstein *m*
(Min)
harmyrine Harmyrin *n*
harness leather Blankleder *n*
harp screen Harfensieb *n*
harsh herb, scharf, rauh
harshness Härte *f*
hartshorn Hirschhorn *n*
hartshorn salt Hirschhornsalz *n*
harvest Ernte *f*
hashish Haschisch *n*
hasp Haspe *f,* Spule *f* (Garn)
hastanecine Hastanecin *n*
hasten *v* beschleunigen, [vor]eilen
hastingsite Hastingsit *m* (Min)
hasubanol Hasubanol *n*
hatch *v* schraffieren

hatchettenine Hatchettenin *n*
hatchettine Hatchettin *n*, Bergtalg *m*,
  Bergwachs *n*
hatchettite Bergwachs *n*, Mineraltalg *m*
hatchettolite Hatchettolith *m* (Min)
Hatchett's brown Hatchetts Braun *n*
hatching Schraffierung *f*
hauchecornite Hauchecornit *m* (Min)
hauerite Hauerit *m* (Min), Mangankies *m* (Min)
haughtonite Haughtonit *m* (Min)
haul *v* fördern (Bergb), ziehen, schleppen, ~ in
  einholen
haulage Förderung *f* (Bergb)
haulage rope Zugseil *n* (Seilbahn)
hauling bridge Förderbrücke *f*
hausmannite Glanzbraunstein *m* (Min),
  Hausmannit *m* (Min), Schwarzmanganerz *n*
  (Min)
hauyne Hauyn *m* (Min)
havana tobacco Kubatabak *m*
hawthorn Hagedorn *m* (Bot), Weißdorn *m*
  (Bot), liquid extract of ~
  Weißdornfluidextrakt *m*
hayesenite Hydroborocalcit *m* (Min)
hazard Gefährdung *f*, Gefahr *f*, Risiko *n*,
  biological ~ biologische Gefährdung *f*
hazardous gefährlich, riskant
haze Dunst *m*, leichter Nebel *m*, Schleier *m*
  (Phot), ~ on surfaces Hauch *m*
haze atmosphere Dunstatmosphäre *f*
hazel-colored haselnußbraun
hazelnut Haselnuß *f*, Hasel[nuß]staude *f* (Bot)
hazelnut oil Haselnußöl *n*
hazelnut size haselnußgroß
hazelwort Haselwurz *f* (Bot)
haziness Dunstigkeit *f*, Unschärfe *f*
hazing Anlaufen *n*, Schleierbildung *f*
hazy dunstig; verschwommen, unscharf
H-bomb Wasserstoffbombe *f*
head Kopf *m*, Ähre *f* (Getreide), Aufsatz *m*,
  Druck *m* (Techn), Knauf *m*, Kuppe *f*,
  Rotor *m* (Zentrifuge), ~ of a column Kopf *m*
  einer Kolonne
headache Kopfschmerzen *pl*
headache pencil Migränestift *m* (Pharm)
head bolting Deckelverschraubung *f*
head drum Abwurftrommel *f*, Antriebstrommel *f*,
  Kopftrommel *f*
heading Überschrift *f*
heading tool Nageleisen *n*
head lamp Scheinwerfer *m*
headless screw Gewindestift *m*, Madenschraube *f*
head loss Gefälleverlust *m*
head melter Schmelzmeister *m*
head-on radiation Direktstrahlung *f*
headpiece Aufsatz *m*, Oberteil *n*
headplate Kopfplatte *f*, Stirnplatte *f*
head pressure Extrusionsdruck *m*

head pulley Abwurftrommel *f*,
  Antriebstrommel *f*, Kopftrommel *f*
headquarters Hauptlager *n*, Hauptsitz *m*
headspace Kopfraum *m* (bei gefüllten
  Packungen)
head start Vorsprung *m*
head tin Feinzinn *n*
headwater Oberwasser *n*
headwaters Oberlauf *m*
head yeast Oberhefe *f* (Brau)
heal *v* heilen
healing ointment Heilsalbe *f* (Pharm)
healing plaster Heilpflaster *n* (Med)
healing serum Heilserum *n* (Med)
health Gesundheit *f*, for reasons of ~
  gesundheitshalber, injurious to ~
  gesundheitsgefährdend, gesundheitsschädlich,
  public ~ öffentliche Gesundheitspflege *f*
heap Haufen *m*
heap *v* aufstapeln, häufen, ~ up aufhäufen
heap charring Meilerverkohlung *f*, Verkohlung *f*
  in Meilern (pl)
heap coke Meilerkoks *m*
heap leaching Haufenlaugung *f*
heap roasting Haufenröstung *f*
heart Herz *n* (Med), Kern *m*, Mittelpunkt *m*
heart-burn Magenbrennen *n*
hearth Herd *m*, Arbeitsherd *m*, Esse *f*,
  Feuerstelle *f*, Schmelzraum *m*, bottom of ~
  Herduntergestell *n*, depth of ~ Herdtiefe *f*, ~
  heated with hot ashes Grudeherd *m*
hearth accretions Herdansätze *pl*
hearth arch Herdgewölbe *n*
hearth area efficiency Herdflächenleistung *f*
hearth ashes Herdasche *f*
hearth assay Herdprobe *f*
hearth block Bodenstein *m*
hearth casing Herdeinsatz *m*
hearth cinder Herdschlacke *f*
hearth electrode, consumption of the ~
  Aufzehrung *f* des Kohlebodens
hearth hole Brustöffnung *f* (Ofen)
hearth jacket Gestellmantel *m* (Ofen)
hearth ladle Herdlöffel *m*
hearth level Herdsohle *f*, Ofensohle *f*
hearth lining Herdfutter *n*, Auskleidung *f*
hearth mold Herdform *f*
hearth molding Herdformerei *f*
hearth opening Brustöffnung *f* (Ofen)
hearth plate Herdeisen *n*, Legeisen *n*
hearth process Herdfrischarbeit *f*
hearth refined iron Herdfrischeisen *n*
hearth refined steel Herdfrischstahl *m*
hearth refining Herdfrischen *n*
hearth ring machine Herdringmaschine *f*
hearth roaster Röstherd *m*
hearth surface Herdfläche *f*
heartrot Kernfäule *f*, Rotfäule *f*

heart transplant Herztransplantation *f* (Med),
  Herzverpflanzung *f* (Med)
heartwood Kernholz *n*
heat Hitze *f*, Glut *f*, Wärme *f*, **absorption of** ~
  Wärmeabsorption *f*, Wärmeaufnahme *f*,
  abstraction of ~ Wärmeentziehung *f*,
  accumulation of ~ Wärmeaufspeicherung *f*,
  Wärmestauung *f*, **action of** ~
  Hitzeeinwirkung *f*, Wärmeeinwirkung *f*, **added**
  ~ zugeführte Wärme, **amount of** ~
  Wärmemenge *f*, **convection of** ~
  Wärmeströmung *f*, **degree of** ~
  Feuerungsgrad *m*, Wärmegrad *m*, **development**
  of ~ Wärmeentwicklung *f*, **equivalent of** ~
  Hitzegegenwert *m*, **evolution of** ~
  Wärmeentwicklung *f*, **excess** ~ Überhitze *f*;
  Wärmeüberschuß *m*, **flow of** ~
  Wärmestrom *m*, **generation of** ~
  Wärmeerzeugung *f*, **increase of** ~
  Wärmeerhöhung *f*, **intensity of** ~
  Hitzegrad *m*, Wärmeintensität *f*, **interchange**
  of ~ Wärmeaustausch *m*, **latent** ~ latente
  Wärme, **loss of** ~ Wärmeverlust *m*,
  Wärmeabgabe *f*, **measurement of** ~
  Wärmemessung *f*, **mechanical equivalent of** ~
  Arbeitswert *m* der Wärme, **molar** ~ molare
  Wärme, **passage of** ~ Wärmeübergang *m*,
  potential ~ potentielle Wärme, **production of**
  ~ Wärmeerzeugung *f*, Thermogenese *f*,
  Wärmebildung *f*, **quantity of** ~
  Wärmemenge *f*, **radiant** ~ Wärmestrahlung *f*,
  radiation of ~ Wärme[aus]strahlung *f*,
  received ~ zugeführte Wärme, **resistance to** ~
  Wärmebeständigkeit *f*, **sensitive to** ~
  wärmeempfindlich, hitzeempfindlich, **source**
  of ~ Wärmequelle *f*, **specific** ~ spezifische
  Wärme, **stored** ~ aufgespeicherte Wärme,
  supply of ~ Wärmezufuhr *f*, **to radiate** ~
  Wärme *f* ausstrahlen, **to store** ~ Wärme *f*
  aufspeichern, **total** ~ Gesamtwärme *f*, **to**
  transmit ~ Wärme *f* durchlassen,
  transmission of ~ Wärmeübergang *m*,
  Wärmeübertragung *f*, **unit of** ~
  Wärmeeinheit *f*, **waste** ~ Überhitze *f*, **white** ~
  Weißglut *f*, ~ **conducted off** abgegebene
  Wärme, ~ **given off** abgegebene Wärme, ~ **of**
  absorption Absorptionswärme *f*, ~ **of**
  formation Bildungswärme *f*, ~ **of fusion**
  Schmelzwärme *f*, **latent** ~ **of fusion** latente
  Schmelzwärme (Therm), ~ **of reaction**
  Reaktionswärme *f*, Wärmetönung *f*, ~ **of**
  setting Abbindewärme *f* (Zement), ~ **of**
  solution Lösungswärme *f*, ~ **of sublimation**
  Sublimationswärme *f* (Therm), ~ **sensitivity**
  Hitzeempfindlichkeit *f*, ~ **supplied** zugeführte
  Wärme
heat *v* erhitzen, anfeuern (Kessel), aufheizen,
  befeuern, [be]heizen, erwärmen, ~ **gently**
  schwach erhitzen, ~ **red hot** ausglühen,

rotglühen, ~ **thoroughly** abglühen (Met),
  durchheizen, ~ **to red heat** rotglühen
heatable heizbar
heat absorption Wärmeabsorption *f*
heat absorption capacity
  Wärmeaufnahmefähigkeit *f*
heat abstraction Wärmeableitung *f*,
  Wärmeentzug *m*
heat accumulation Wärmestau *m*
heat accumulator Wärmesammler *m*
heat balance Wärmeausgleich *m*, Wärmebilanz *f*,
  Wärmehaushalt *m*
heat breakdown Wärmedurchschlag *m*
heat build-up Eigenerwärmung *f*
heat capacity Wärmeinhalt *m*, spezifische
  Wärme, Wärmekapazität *f*
heat change Wärmetönung *f* (Chem)
heat circulation Wärmeumlauf *m*,
  Wärmezirkulation *f*
heat coil Heizwendel *f*
heat compensation Wärmeausgleich *m*
heat conductance Wärmeleitung *f*,
  Wärmedurchlässigkeit *f*
heat-conducting wärmeleitend
heat conduction Wärmeleitung *f*
heat conductivity Wärmeleit[ungs]vermögen *n*
heat-conductivity curve Wärmeleitungskurve *f*
heat consumption Wärmeverbrauch *m*
heat convection Wärmekonvektion *f*
heat-deformable thermoplastisch
heat denaturation Wärmedenaturierung *f*
heat dilatation Wärmedehnung *f*
[heat] dilatometer Wärmedehnungsmesser *m*
heat dissipation Wärmezerstreuung *f*
heat distortion point Temperaturformstabilität *f*
heat distribution Wärmeverteilung *f*
heat drop Wärmeabfall *m*, Wärmegefälle *n*,
  adiabatic ~ adiabatische Wärmeabgabe, **total**
  ~ Gesamtwärmegefälle *n*
heat economy Wärmehaushalt *m*
heated erhitzt, **capable of being** ~ heizbar, ~
  thoroughly abgeglüht (Met)
heated manifold Heißkanalverteiler *m*
heat effect Wärmetönung *f* (Chem),
  Wärmewirkung *f* (Chem)
heat efficiency Wärmeleistung *f*
heat emission Wärmeabgabe *f*
heat emitting surface Wärmeabgabefläche *f*
heat energy Wärmeenergie *f*
heat engine Wärmekraftmaschine *f*
heat engineering Wärmetechnik *f*
heat equalization Wärmeausgleich *m*
heat equation Wärmegleichung *f*
heater Heizer *m*, Heizfaden *m* (Elektr),
  Heizkörper *m*, Heizvorrichtung *f*, ~ **in contact**
  with exhaust Mischungsvorwärmer *m*
heater band Bandheizkörper *m*; Heizband *n*
heater current Heiz[faden]strom *m* (Elektr)
heater drum Heiztrommel *f*

heater filament voltage Heizfadenspannung *f*
(Elektr)
heater head Heizkopf *m*
heater plug Glühkerze *f* (Techn)
heater shell Vorwärmemantel *m*
heater spiral Heizspirale *f* (Elektr)
heater trough Heizwanne *f*
heater tunnel Heizkanal *m*
heater wire Heizdraht *m* (Elektr)
heat exchange Wärmeaustausch *m*
heat exchange medium Heizflüssigkeit *f*
heat exchanger Wärmeaustauscher *m*, **cocurrent**
~ Gleichstrom-Wärmeaustauscher *m*,
countercurrent ~
Gegenstrom-Wärmeaustauscher *m*, **crossflow**
~ Kreuzstrom-Wärmeaustauscher *m*,
extended surface ~
Rippenrohr-Wärmeaustauscher *m*, **finned tube**
~ Rippenrohr-Wärmeaustauscher *m*, **plate** ~
Platten-Wärmeaustauscher *m*, **reversing** ~
Umschaltwärmeaustauscher *m*, **scraped**
surface ~ Kratz-Wärmeaustauscher *m*,
screw-flight ~
Schnecken-Wärmeaustauscher *m*, **segment** ~
Lamellen-Wärmeaustauscher *m*, **shell and**
tube ~
Glattrohrbündel-Wärmeaustauscher *m*,
Rohrbündelwärmeaustauscher *m*, **spiral** ~
Spiralwärmeaustauscher *m*, **split-tube** ~
Spaltrohrwärmeaustauscher *m*, **thin film** ~
Dünnschicht-Wärmeaustauscher *m*,
trickled-surface ~
Riesel-Wärmeaustauscher *m*, **tube bundle** ~
Rohrbündelwärmeaustauscher *m*, ~ **with tube**
bundles Rohrbündel-Wärmeaustauscher *m*
heat expansion Wärmeausdehnung *f*
heat factor Wärmefaktor *m*
heat flash Wärmeblitz *m*
heat flow Wärmeströmung *f*, **rate of** ~ **per unit**
area Wärmestromdichte *f*
heat flow layout Wärmeflußbild *n*
heat fluctuation Wärmeschwankung *f*
heat flux Wärmefluß *m*, Wärmeströmung *f*,
Wärmestrom *m*
heat focus Wärmebrennpunkt *m*
heat-generating station Wärmekraftwerk *n*
heat generator Wärmeerzeuger *m*
heat gradient Wärmegefälle *n*, Wärmeverlauf *m*
heat-impermeable wärmeundurchlässig
heat indicating paint Anlauffarbe *f*,
Heißlaufmeldungsfarbe *f*
heating Heizung *f*, Abglühen *n* (Met),
Befeuerung *f*, Beheizung *f*, Erhitzen *n*,
Erwärmen *n*, Erwärmung *f*, **dielectric** ~
dielektrische Erwärmung, **fault due to** ~
Erwärmungsfehler *m*, **internal** ~
Innenheizung *f*, **preliminary** ~ Vorglühen *n*,
rate of ~ Erwärmungsgeschwindigkeit *f*, **to**
start ~ anheizen, ~ **in a closed tube**

Glühröhrchenprobe *f*, ~ **in contact with**
exhaust Mischungsvorwärmung *f*, ~ **of the**
head Heizen *n* des Blockkopfes, ~ **under**
pressure Druckerhitzung *f*
heating apparatus Heizvorrichtung *f*,
Glühapparat *m*
heating appliance Heizapparat *m*
heating arrangement Anheizvorrichtung *f*
heating bath liquid Heizbadflüssigkeit *f*
heating body Heizkörper *m*
heating boiler Heizkessel *m*
heating capacity Heizfähigkeit *f*
heating chamber Heizkammer *f*, Herdraum *m*
heating channel Heizkanal *m*
heating coil Heizschlange *f*, Heizspirale *f*
heating coils, system of ~
Heizschlangensystem *n*
heating coke Füllkoks *m*
heating conductor Heizleiter *m*
heating current Heizstrom *m* (Elektr)
heating cylinder Heizzylinder *m*
heating effect Heizeffekt *m*, Heizwirkung *f*
heating element Heizkörper *m*
heating elements Heizregister *n*
heating filament Heizfaden *m* (Elektr),
oxide-coated ~ Heizfaden *m* mit Oxidschicht
heating flame Heizflamme *f*
heating flue Heizkanal *m*, Heizrohr *n*
heating furnace Glühofen *m*, Ausglühofen *m*
heating grid Heizgitter *n*
heating hearth Schmelzherd *m*
heating jacket Heizmantel *m*, Heizhaube *f*
heating lamp Heizlampe *f*
heating liquid Heizflüssigkeit *f*
heating mantle Heizpilz *m* (Labor)
heating material Heizmaterial *n*, Heizmittel *n*
heating medium Heizmedium *n*
heating method Beheizungsart *f*
heating oil Heizöl *n*
heating pad Heizkissen *n*
heating passage Heizkanal *m*
heating pattern Temperaturverlauf *m* (Therm)
heating period Erwärmungszeit *f*
heating pin Wärmeleitstift *m*
heating plant Heizanlage *f*
heating plate Heizplatte *f*
heating power Heizkraft *f*
heating rod Heizstab *m*
heating space Heizraum *m*
heating spiral Heizwendel *f*
heating station Heizwerk *n*
heating steam Heizdampf *m*
heating stove Glühofen *m*
heating surface Heizfläche *f*
heating tape Heizband *n*
heating tool welding Heizelementschweißen *n*
heating tube Heizgranate *f*, Siederohr *n*
heating tube bundle Heizrohrbündel *n*

heating-up of the bearing Warmlaufen *n* des Lagers
heating-up zone Vorwärmezone *f*
heating worm Heizschlange *f*
heating zone Heizzone *f*
heat input Wärmezufuhr *f*
heat-insulated wärmeisoliert
heat-insulating wärmeisolierend, wärmeundurchlässig
heat-insulating jacket Wärmeschutzhaube *f*
heat-insulating material Wärmeschutzmasse *f*
heat insulation Wärmeisolation *f,* Wärmeisolierung *f,* Wärmeschutz *m*
heat insulator Wärmeisolator *m,* Wärmeschutzmasse *f*
heat insulator casing Wärmeschutzmantel *m*
heat interchange Wärmeausgleich *m*
heat irradiation Wärmeeinstrahlung *f*
heat leak Wärmeundichtigkeit *f*
heat loss Wärmeverlust *m,* Verlustwärme *f,* Abzugswärme *f*
heat measurement Wärmemessung *f*
heat measuring device Wärmemeßeinrichtung *f*
heat motion Wärmebewegung *f*
heat of atomization Atomisierungswärme *f*
heat output Wärmeleistung *f*
heat pipe Wärmerohr *n*
heat pole Wärmepol *m*
heat-pressure gauge Thermomanometer *n*
heat producer Wärmebildner *m*
heat-producing wärmeerzeugend
heatproof wärmebeständig, feuerfest, hitzebeständig
heatproof quality Wärmebeständigkeit *f*
heat radiation Wärmestrahlung *f*
heat radiator Wärmestrahler *m*
heat radition sensing device Wärmestrahlungsfühler *m*
heat ray Wärmestrahl *m*
heat-reactive hitzereaktiv
heat reflection Wärmerückstrahlung *f*
heat regulation Thermoregulation *f,* Wärmeregulation *f*
heat regulator Wärmeregler *m,* **automatic ~** Thermostat *m*
heat release Wärmeabgabe *f,* Wärmefreimachung *f*
heat removal Wärmeabführung *f*
heat requirement Wärmebedarf *m*
heat reservoir Wärmebehälter *m,* Wärmespeicher *m*
heat resistance Wärmebeständigkeit *f,* Erhitzungswiderstand *m* (Elektr), Temperaturbeständigkeit *f*
heat-resistant wärmebeständig, hitzebeständig, thermoresistent, thermostabil
heat-resisting hitzebeständig, wärmebeständig
heat-resisting quality Hitzebeständigkeit *f,* Wärmebeständigkeit *f*

heat scale Wärmeskala *f*
heat seal Verschweißung *f*
heat-seal *v* [ver]schweißen (Kunststoff)
heat-sealing Heißverschweißen *n,* Heißsiegeln *n,* Schweißen *n* (Kunststoff), Verschweißen *n* (Kunststoff)
heat-sealing film Heißsiegelfolie *f*
heat-sealing lacquer Heißklebelack *m,* Heißsiegellack *m*
heat-sealing press Heißsiegelpresse *f*
heat sensitive temperaturempfindlich
heat-sensitive paint Anlauffarbe *f* (Met)
heat sensitizing agent Wärmesensibilisierungsmittel *n*
heat-setting Heißfixieren *n,* Thermofixieren *n,* Thermofixierung *f*
heat stability Wärmebeständigkeit *f,* Hitzebeständigkeit *f,* Wärmestabilität *f*
heat stable hitzebeständig
heat storage Wärme[auf]speicherung *f*
heat supply Wärmezufuhr *f*
heat test Warmprobe *f*
heat theorem, Nernst ~ Wärmetheorem *n* von Nernst
heat-tinting Anlaufen *n* bei erhöhter Temperatur, **resistance to ~** Anlaufbeständigkeit *f*
heat transfer Wärmeübertragung *f,* Wärmedurchgang *m,* Wärmeübergang *m,* **coefficient of ~** Wärmeaustauschkoeffizient *m,* **overall coefficient of ~** Wärmedurchgangskoeffizient *m,* Wärmedurchgangszahl *f,* **~ per unit surface area** Heizflächenbelastung *f*
heat transfer coefficient Wärmeübergangskoeffizient *m,* Wärmeübergangszahl *f*
heat transfer fluid Wärmeträger *m*
heat transfer medium Heizbadflüssigkeit *f,* Wärmeträger *m,* Wärmeübertragungsmittel *n*
heat transformation Wärmeumsatz *m*
heat transmission Wärmedurchgang *m,* **coefficient of ~** Wärmedurchgangszahl *f*
heat transport Wärmetransport *m*
heat-treat *v* warmbearbeiten, vergüten (Stahl)
heat-treating Vergüten *n* (Met)
heat treatment Temperaturbehandlung *f,* thermische Behandlung *f,* Vergütung *f* (Stahl), Wärmebehandlung *f,* Warmbehandlung *f*
heat tube Wärmerohr *n*
heat unit Wärmeeinheit *f*
heat value Heizwert *m,* Wärmewert *m*
heat variation Wärmeschwankung *f*
heat wave Hitzewelle *f*
heave *v* heben, hochwinden
heaviness Schwere *f*
Heaviside layer Heaviside-Schicht *f*

Heaviside region Heaviside-Schicht *f*
Heaviside's expansion rule Heaviside'sche
Regel *f*
heavy schwer, gewichtig, schwerflüssig
heavy chemical Schwerchemikalie *f*
heavy current Starkstrom *m* (Elektr)
heavy current engineering Starkstromtechnik *f*
heavy-duty dauerhaft, Hochleistungs-
heavy-duty circuit breaker Schalterrohr *n*
heavy-duty condenser leistungsstarker
Kondensator *m*
heavy-duty drive Hochleistungsantrieb *m*
heavy-duty homogenizer
Hochleistungshomogenisiermaschine *f*
heavy-duty oil Hochleistungsöl *n*
heavy-duty packings Hochleistungsfüllkörper *pl*
(Dest)
heavy-duty plug-and-socket connection
Kragensteckvorrichtung *f*
heavy-duty sack freitragender Sack *m*
heavy-duty surface milling machine
Hochleistungsflächenfräsmaschine *f*
heavy earth Schwerspat *m* (Min)
heavy electron schweres Elektron *n*
heavy ends Dicköl *n*
heavy goods Schwergut *n*
heavy hydrogen Deuterium *n*, schwerer
Wasserstoff *m*, Tritium *n*
heavy industry Schwerindustrie *f*
heavy liquid Schwereflüssigkeit *f*
heavy media separation Sinkscheidung *f*,
Sinkschwimmtrennung *f*
heavy metal Schwermetall *n*
heavy milker Hochleistungstier *n* (Rind)
heavy oil Dicköl *n*, Schweröl *n*
heavy-section dickwandig
heavy spar Baryt *m*, Bariumsulfat *n* (Min),
Schwerspat *m* (Min), **fibrous** ~ Faserbaryt *m*
(Min)
heavy-water pile Schwerwasserreaktor *m*
hebronite Hebronit *m* (Min)
hecogenin Hecogenin *n*
hecogenoic acid Hecogensäure *f*
hectare Hektar *n* (Maß)
hectine Hektin *n*
hectograph Hektograph *m*,
Vervielfältigungsmaschine *f*
hectograph carbon paper Hektokopierpapier *n*
hectography Hektographie *f*
hectowatt hour Hektowattstunde *f* (Elektr)
hedaquinium Hedaquinium *n*
hedenbergite Hedenbergit *m* (Min),
Kalkeisenaugit *m* (Min)
hederacoside Hederacosid *n*
hederagenin Hederagenin *n*
hederagonic acid Hederagonsäure *f*
hederic acid Hederinsäure *f*, Efeusäure *f*
hederine Hederin *n*, Efeubitter *n*
hederose Hederose *f*

hedge hyssop Gnadenkraut *n* (Bot),
Purgierkraut *n* (Bot)
hedge mustard oil Hederichöl *n*
hedyphane Hedyphan *m* (Min)
heel effect Anodenschatten *m* (Röntgenröhren)
Hefner candle Hefnerkerze *f* (Phys)
hegonon Hegonon *n*
height Höhe *f*, Größe *f*, ~ **in the clear** lichte
Höhe, ~ **of bed** Schichthöhe *f*, ~ **of fall**
Fallhöhe *f*, ~ **of packing** Schütthöhe *f*
(Füllkörperkolonne), ~ **of rise** Steighöhe *f*
height determination Höhenbestimmung *f*
heighten *v* erhöhen, steigern
height equivalent to theorical plate (HETP)
Übergangseinheit *f* (Füllkörperkolonne)
height equivalent to transfer unit (HTU)
Auswaschzahl *f* (Füllkörperkolonne)
height gain function Höhengewinnfunktion *f*
height indicator Altimeter *n*, Höhenmesser *m*
height of bed Schüttungshöhe *f*
height of fill Überdeckungshöhe *f* (Gaschromat)
height regulating device Absenkvorrichtung *f*
(Techn)
heintzite Heintzit *m* (Min)
Heisenberg's equation of motion Heisenbergsche
Bewegungsgleichung *f* (Quant)
Heisenberg's uncertainty principle
Heisenbergsche Unschärfe-Beziehung *f*
helcosol Helcosol *n*, Wismutpyrogallat *n*
helenalin Helenalin *n*
helenien Helenien *n*
helenin Helenin *n*, Alantcampher *m*, Alantol *n*
helenium Inula *f* (Bot)
heleurine Heleurin *n*
helianthic acid Helianthsäure *f*
helianthin Helianthin *n*
helianthrene Helianthren *n*
helianthrone Helianthron *n*
helical schraubenförmig, schneckenförmig,
spiralförmig
helical blower Propellergebläse *n*,
Schraubengebläse *n*
helical classifier Schraubensichter *m*
helical gear Schneckenrad *n*
helical piston pump Schraubenkolbenpumpe *f*
helical ribbon agitator Wendelrührer *m*
helical spring Schraubenfeder *f*
helical spur gear Schrägzahnstirnrad *n*
helical tooth Winkelzahn *m*
helicin Helicin *n*
helicoid Schraubenfläche *f*
helicoidal surface Schraubenfläche *f*
helicoidin Helicoidin *n*
helicopter Hubschrauber *m*
heliochromoscope Heliochromoskop *n*
heliodor Heliodor *m* (Min)
helioengraving Kupferlichtdruck *m*
helio fast red Helioechtrot *n*
helio fast yellow Helioechtgelb *n*

heliograph Heliograph *m*
heliographic printing Heliotypie *f,*
  Lichtpausverfahren *n*
heliography Heliographie *f*
heliogravure Kupferlichtdruck *m*
heliometer Heliometer *n* (Astr)
heliophyllite Heliophyllit *m* (Min)
helioscope Helioskop *n* (Astr)
heliotherapy Heliotherapie *f* (Med)
heliotridane Heliotridan *n*
heliotridine Heliotridin *n*
heliotrine Heliotrin *n*
heliotrope Heliotrop *m* (Min)
heliotropic acid Piperonylsäure *f*
heliotropin Heliotropin *n,* Piperonal *n*
heliotype color printing Lichtfarbendruck *m*
heliotyping Heliotypie *f* (Phot)
helium (Symb. He) Helium *n*
helium age Heliumalter *n*
helium atom Heliumatom *n*
helium bubble chamber Heliumblasenkammer *f*
helium film Heliumschicht *f*
helium gas Heliumgas *n*
helium ion collision Heliumionenstoß *m*
helium nucleus Heliumkern *m*
helium tube Heliumröhre *f*
helium wave function Heliumwellenfunktion *f*
helix Helix *f,* Schneckenlinie *f,* Spirale *f,*
  Wendel *f,* ~ **of tape wire** Bandschraube *f*
helix angle Steigungswinkel *m*
helix structure Helixstruktur *f* (Biochem)
hellandite Hellandit *m* (Min)
hellebore Nieswurz[el] *f* (Bot), **tincture of green**
  ~ Nieswurztinktur *f*
helleboreine Helleborein *n*
hellebore root Christwurzel *f* (Bot), **white** ~
  Germerwurzel *f* (Bot), Krätzwurzel *f* (Bot)
helleboretin Helleboretin *n*
helleborin Helleborin *n*
hellebrin Hellebrin *n*
helmet Kopfschutz *m,* Schutzhelm *m*
Helmholtz double layer Helmholtz-Schicht *f*
helmintholite Wurmstein *m* (Min)
helminthosporin Helminthosporin *n*
helmitol Helmitol *n*
helper Handlanger *m,* Helfer *m*
helpful behilflich
helve Griff *m,* Stiel *m*
helve hammer Aufwerfhammer *m,* schwerer
  Schmiedehammer *m*
helveticoside Helveticosid *n*
helvine Helvin *m* (Min)
hemafibrite Hämafibrit *m* (Min)
hemagate Blutachat *m* (Min)
hemagglutination Hämagglutination *f*
hemagglutinogen Hämoagglutinogen *n*
hemalbumin Hämalbumin *n*
hemanalysis Blutanalyse *f*
hemanthamine Hämanthamin *n*

hemanthidine Hämanthidin *n*
hemanthine Hämanthin *n*
hemataminic acid Hämataminsäure *f*
hematein Hämatein *n*
hematid rotes Blutkörperchen *n*
hematin Hämatin *n,* Blutrot *n*
hematin chloride Hämin *n*
hematine Hämatoxylin *n*
hematinic acid Hämatinsäure *f*
hematite Hämatit *m* (Min), Blutstein *m* (Min),
  Glanzeisenerz *n* (Min), Glaskopf *m* (Min),
  Roteisenerz *n* (Min), **brown** ~
  Eisensumpferz *n* (Min), **earthy** ~
  Roteisenocker *m* (Min), **green** ~ grüner
  Glaskopf *m,* **porous form of** ~
  Eisenschaum *m* (Min), **red** ~
  Roteisenstein *m* (Min)
hematite deposit Hämatitvorkommen *n*
hematite pig iron Hämatitroheisen *n* (Metall)
hematitic hamatitartig
hematocyte Hämatozyt *m*
hematocytolysis Hämatozytolyse *f*
hematogen[e] Blutbildner *m*
hematogenesis Hämogenese *f*
hematogenic blutbildend
hematogenic agent Blutbildungsmittel *n*
hematogenous blutbildend
hematoidin Bilirubin *n,* Hämatoidin *n*
hematolite Hämatolith *m* (Min)
hematommic acid Hämatommsäure *f*
hematopan Hämatopan *n*
hematoporphyrin Hämatoporphyrin *n*
hematostibiite Hämatostibiit *m* (Min)
hematoxylic acid Hämatoxylin *n*
hematoxylin Hämatoxylin *n*
hematoxylin solution Hämatoxylinlösung *f*
hematozoon Blutparasit *m,* Hämatozoon *n*
heme Häm *n*
hemellitenol Hemellitenol *n*
hemellitic acid Hemellithsäure *f*
hemellitol Hemellitol *n*
hemerythrin Hämerythrin *n*
hemiacetal Halbacetal *n*
hemialbumose Hemialbumose *f*
hemicellulose Halbzellulose *f,* Hemicellulose *f*
hemicolloid Semikolloid *n,* Übergangskolloid *n*
hemicycle Halbkreis *m*
hemicyclic hemizyklisch
hemiglobin Hämiglobin *n*
hemihedral hemiedrisch (Krist), halbflächig
  (Krist)
hemihedral form Hemiedrie *f* (Krist)
hemihedrism Hemiedrie *f* (Krist)
hemihedron Hemieder *n* (Krist),
  Halbflächner *m* (Krist)
hemihydrate Halbhydrat *n*
hemimellitene Hemimellitol *n*
hemimellitic acid Hemimellitsäure *f*
hemimellitol Hemimellitol *n*

hemimorphic hemimorph (Krist)
hemimorphism Hemimorphie *f* (Krist),
Hemimorphismus *m* (Krist)
hemimorphite Hemimorphit *m* (Min),
Kieselzinkerz *n* (Min)
hemimorphous hemimorph (Krist)
hemin Hämin *n*, Chlorhämatin *n*
hemipic acid Hemipinsäure *f*
hemipinic acid Hemipinsäure *f*
hemiprism Halbprisma *n* (Krist)
hemiprismatic halbprismatisch (Krist)
hemipyocyanin Hemipyocyanin *n*
hemisphere Halbkugel *f*
hemispherical halbkugelförmig, halbkugelig
hemispherical shape Halbkugelgestalt *f*
hemitoxiferine Hemitoxiferin *n*
hemitropal hemitropisch
hemitrope hemitropisch
hemitropy Hemitropie *f* (Krist)
hemlock Schierling *m* (Bot), Erdschierling *m*
(Bot)
hemlock poison Schierlingsgift *n*
hemlock resin Hemlockharz *n*
hemobilirubin Blutbilirubin *n*, Hämobilirubin *n*
hemochromogen[e] Hämochromogen *n*
hemocoagulative blutgerinnend
hemocuprein Hämocuprein *n*
hemocyanin Hämocyanin *n*
hemocyte Hämatozyt *m*
hemo[cyto]genesis Blutbildung *f*,
Hämozytogenese *f* (Blutkörperchenbildung)
hemocytology Hämatozytologie *f*
hemocytolysis Blutkörperchenauflösung *f*,
Hämozytolyse *f*
hemocytopoiesis Blutzellenbildung *f*
hemogenesis Hämogenese *f*
hemoglobin Hämoglobin *n*, Blutfarbstoff *m*
hemoglobin estimation
Hämoglobinbestimmung *f*
hemogram Blutbild *n*, Differentialblutbild *n*,
Hämogramm *n*
hemolysis Hämolyse *f*
hemoporphyrin Hämoporphyrin *n*
hemopyrrole Hämopyrrol *n*
hemopyrrolene phthalide
Hämopyrrolenphthalid *n*
hemopyrrolidine Hämopyrrolidin *n*
hemopyrroline Hämopyrrolin *n*
hemoquinic acid Hämochinsäure *f*
hemoquinine Hämochinin *n*
hemorrhage Bluterguß *m*, Blutung *f*
hemosiderin Hämosiderin *n*
hemostatic blutstillendes Mittel *n* (Pharm)
hemostatic *a* blutstillend
hemostyptic Hämostyptikum *n* (Pharm)
hemotoxin Blutgift *n*, Hämotoxin *n*
hemovanadium Hämovanadin *n*
hemp Hanf *m*
hemp belt Hanfriemen *m*

hemp brush Hanfbürste *f*
hemp cord Hanfseil *n*
hemp core Hanfeinlage *f*
Hempel burette Hempel-Bürette *f*
hemp fiber Hanffaser *f*
hemp layer Hanfeinlage *f*
hemp nettle Hanfnesselkraut *n* (Bot)
hemp oil Hanföl *n*
hemp packing Hanfdichtung *f*,
Hanfumwickelung *f*
hemp rope Hanfseil *n*, Hanftau *n*
hempseed Hanfkörner *n pl* (Bot), Hanfsamen *m*
(Bot)
hempseed oil Hanföl *n*
hemp strand Hanflitze *f*
henbane leaves Bilsenkrautblätter *n pl* (Bot)
henbane oil Bilsenkrautöl *n*
henbane seed Bilsenkrautsamen *m* (Bot)
hendecadiene Undecadien *n*
hendecadiyne Undecadiin *n*
hendecahedral elfflächig
hendecahedron Elfflächner *m*
hendecanal Undecylaldehyd *m*
hendecane Undecan *n*
hendecanoic acid Undecylsäure *f*
hendecanol Undecanol *n*
hendecene Undecylen *n*
hendecyl alcohol Undecanol *n*
heneicosane Heneikosan *n*
heneicosanoic acid Heneikosansäure *f*
heneicosylene Heneikosylen *n*
henna Henna *f* (Bot, Färb)
hentriacontane Hentriakontan *n*
hentriacontanone, 16- ~ Palmiton *n*
henwoodite Henwoodit *m* (Min)
hepar Leber *f*
hepar calcis Calciumsulfid *n*,
Kalkschwefelleber *f* (obs)
heparene Heparen *n*
heparin Heparin *n*
hepar sulfuris Kaliumsulfid n, Leberschwefel *m*
(obs)
hepar test Heparprobe *f* (Anal)
hepatectomy Hepatektomie *f* (Med)
hepatic Hepatikum *n* (Pharm)
hepatica Leberkraut *n* (Bot)
hepatic cell Leberzelle *f*
hepatic cinnabar Lebererz *n*
hepatic cobalt Leberkobalt *n* (Min)
hepatic ore Lebererz *n*
hepatic pyrites Leberkies *m* (Min)
hepatin Glykogen *n*, Leberstärke *f*
hepatite Hepatit *m*, Leberstein *m*
hepatitis Hepatitis *f* (Med)
hepatocyte Hepatozyt *m*
hepatolysis Hepatolyse *f* (Med),
Leberzellenzerfall *m* (Med)
heptacene Heptacen *n*
heptacontane Heptacontan *n*

heptacosane Heptakosan *n*
heptacosanoic acid Cerotinsäure *f*
heptacyclene Heptacyclen *n*
heptadecadiene Heptadecadien *n*
heptadecane Heptadecan *n*
heptadecanoic acid Margarinsäure *f,*
    Daturinsäure *f,* Heptadecansäure *f*
heptadecylamine Heptadecylamin *n*
heptadecylene Heptadecylen *n*
heptadiene Heptadien *n*
heptafulvalene Heptafulvalen *n*
heptagon Siebeneck *n*
heptagonal siebeneckig
heptahedral siebenflächig
heptahedron Siebenflächner *m*
heptahydrate Heptahydrat *n*
heptaldehyde Heptaldehyd *m,* Heptylaldehyd *m*
heptaldoxime Heptaldoxim *n*
heptalene Heptalen *n*
heptamethylene Heptamethylen *n*
heptaminol Heptaminol *n*
heptanal Heptaldehyd *m,* Heptylaldehyd *m,*
    Önanthaldehyd *m*
heptanaphthene Heptanaphthen *n*
heptane Heptan *n*
heptanedioic acid Pimelinsäure *f*
heptanoic acid Heptylsäure *f,* Önanthsäure *f*
heptanol Heptanol *n,* Heptylalkohol *m,*
    Önanthol *n*
heptanone, 4- ~ Dipropylketon *n*
heptanoylphenol Heptanoylphenol *n*
heptaric acid Heptarsäure *f*
heptatomic siebenatomig; siebenwertig
heptatriene Heptatrien *n*
heptavalence Siebenwertigkeit *f*
heptavalent siebenwertig
heptene Hepten *n,* Heptylen *n*
heptenylene glycol Heptenylenglykol *n*
heptindole Heptindol *n*
heptine Heptin *n,* Önanthin *n*
heptinic acid Heptinsäure *f*
heptode Heptode *f*
heptodecanone, 9- ~ Pelargon *n*
heptoic acid Heptylsäure *f*
heptoic aldehyde Heptylaldehyd *m*
heptose Heptose *f*
heptoxime Heptoxim *n*
heptyl Heptyl-
heptyl alcohol Heptanol *n,* Heptylalkohol *m*
heptyl aldehyde Heptaldehyd *m,*
    Heptylaldehyd *m,* Önanthaldehyd *m*
heptylamine Heptylamin *n*
heptyl bromide Heptylbromid *n*
heptylene Hepten *n,* Heptylen *n*
heptylic acid Heptylsäure *f*
heptylic aldehyde Heptylaldehyd *m*
heptyne Heptin *n*
heraclin Heraclin *n*
herapathite Herapathit *n*

herbacetin Herbacetin *n*
herbal kräuterartig
herbal ointment Kräutersalbe *f* (Pharm)
herbal remedy Kräutermittel *n* (Pharm)
herb extract Kräuterauszug *m*
herbicide Unkrautvertilgungsmittel *n,* Herbizid *n*
herb infusion Kräuteraufguß *m*
herbipoline Herbipolin *n*
herbivore Pflanzenfresser *m*
herbivorous pflanzenfressend, phytophag
herb liqueur Kräuterlikör *m*
herb salve Kräutersalbe *f* (Pharm)
herb spirits Kräuterschnaps *m*
herb tea Kräutertee *m*
herclavin Herclavin *n*
hercules rail Herkulesschiene *f*
herculin Herculin *n*
hercynite Hercynit *m* (Min)
herderite Herderit *m* (Min)
hereditary erblich
hereditary defect Erbfehler *m*
hereditary disease Erbkrankheit *f*
hereditary factor Erbfaktor *m*
heredity Vererbung *f* (Biol)
hermannite Hermannit *m* (Min)
hermetical hermetisch, luftdicht
hermetically sealed hermetisch verschlossen,
    vakuumdicht
hermetic seal Luftabschluß *m,* hermetischer
    Verschluß *m*
hermodactyl Hermesfinger *m* (Bot)
herniaria Harnkraut *n* (Bot)
herniarin Herniarin *n*
heroine Heroin *n,* Diacetylmorphin *n*
Héroult electric-arc furnace
    Héroult-Lichtbogenofen *m*
Héroult ore-smelting furnace
    Héroult-Schachtofen *m*
Héroult resistance furnace
    Héroult-Widerstandsofen *m*
herpes soap Flechtenseife *f*
herrengrundite Herrengrundit *m* (Min)
herrerite Herrerit *m* (Min)
herringbone gear Pfeilrad *n,* Pfeilradgetriebe *n*
herring oil Heringsöl *n*
herschelite Herschelit *m* (Min)
hertolan Hertolan *n* (Trübungsmittel)
hertz Hertz *n* (Elektr)
Hertz frequency unit Hertz-Frequenzeinheit *f*
Hertzian wave elektromagnetische Welle *f,*
    Hertzsche Welle *f*
herzynine Herzynin *n*
hesitation Bedenken *n,* Zögern *n,* **without ~**
    anstandslos
hesperetic acid Hesperetinsäure *f,*
    Hesperitinsäure *f*
hesperetin Hesperetin *n*
hesperidene Hesperiden *n*

hesperidin Hesperidin *n*, Pomeranzenbitter *n*
(obs)
hesperitinic acid Hesperetinsäure *f*,
Hesperitinsäure *f*, Isoferulasäure *f*
hesperonal calcium Hesperonalcalcium *n*
Hessian bordeaux Hessischbordeaux *n*
Hessian purple Hessischpurpur *n*
hessite Hessit *m* (Min)
Hess' law of constant summation of heat
Hess'scher Satz *m* der konstanten
Wärmemengen (pl)
hessonite Hessonit *m* (Min)
HET acid Het-Säure *f*
hetairite Hetairit *m* (Min)
hetero-atom Heteroatom *n*
heteroatomic heteroatomig
heteroauxin Heteroauxin *n*
heterocellular verschiedenzellig (Biol)
heterocharge Heteroladung *f* (Chem)
heterochromatic heterochromatisch,
verschiedenfarbig
heterochromous ungleichfarbig,
verschiedenfarbig, heterochrom
heterocodeine Heterokodein *n*
heterocyclic heterozyklisch
heterocyclic compounds Heterocyclen *pl*
heterodisperse heterodispers
heterodyne method Überlagerungsmethode *f*
heterogamete Heterogamet *n* (Biol)
heterogeneity Heterogenität *f*,
Ungleichartigkeit *f*, Ungleichförmigkeit *f*
heterogeneous heterogen, ungleichartig,
verschiedenartig
heterogeneousness Inhomogenität *f*
heterogeneous reactor Heterogenreaktor *m*
heterogenite Heterogenit *m* (Min)
heterologous heterolog, abweichend
heterolupane Heterolupan *n*
heterolysis Heterolyse *f*
heteromorphic heteromorph,
verschiedengestaltig
heteromorphite Federerz *n* (Min),
Heteromorphit *m* (Min)
heteromorphy Heteromorphie *f*
heterophylline Heterophyllin *n*
heteropolar heteropolar, wechselpolar
heteropolar bond Ionenbindung *f*
heteropoly acid Heteropolysäure *f*
heteropolycondensation
Heteropolykondensation *f*
heteropolymerization Heteropolymerisation *f*,
Mischpolymerisation *f*
heteropolysaccharide Heteropolysaccharid *n*
heteroquinine Heterochinin *n*
heterosis Heterosis *f* (Biol)
heterosite Heterosit *m* (Min)
heterostatic heterostatisch
heterotrophic heterotroph
heteroxanthine Heteroxanthin *n*

heterozygote Heterozygot *m* (Biol)
heterzygous heterozygot (Genpaar)
hetisine Hetisin *n*
hetocresol Hetokresol *n*
hetol Hetol *n*
heubachite Heubachit *m* (Min)
heulandite Heulandit *m* (Min), Blätterzeolith *m*
(Min)
hevea Heveabaum *m*, Kautschukbaum *m*
heveene Heveen *n*
hew *v* hauen, hacken, ~ **down** abhauen,
umfällen, ~ **off** abhauen
hewer Hauer *m*
hewettite Hewettit *m* (Min)
hexa-ammine cobalt(III) chloride
Luteokobaltchlorid *n*
hexa-atomic sechsatomig; sechswertig
hexabasic sechsbasisch
hexabromide number Hexabromidzahl *f*
hexabromide value Hexabromidzahl *f*
hexacene Hexacen *n*
hexachlorendomethylene tetrahydrophthalic acid
Het-Säure *f*
hexachlorobenzene Hexachlorbenzol *n*,
Perchlorbenzol *n*
hexachlorodisilane Hexachlordisilan *n*
hexachlorodisiloxane Hexachlorodisiloxan *n*
hexachloroethane Hexachloräthan *n*, Hexoran *n*,
Perchloräthan *n*
hexachlorophene Hexachlorophen *n*
hexachloro-platinic acid Hexachloroplatinsäure *f*
hexachronic acid Hexachronsäure *f*
hexacontane Hexakontan *n*
hexacosane Hexakosan *n*
hexacosanic acid Hexakosansäure *f*
hexacosanol Hexacosanol *n*, 1- ~
Cerylalkohol *m*
hexacyano-ferrate(II)-ion
Hexacyanoferrat(II)-ion *n*
hexacyclic hexacyclisch
hexadecadiene Hexadecadien *n*
hexadecadiine Hexadecadiin *n*
hexadecanal Palmitylaldehyd *m*
hexadecanate Palmitat *n*, Palmitinsäuresalz *n*
hexadecane Hexadecan *n*
hexadecanoic acid Palmitinsäure *f*
hexadecanol Hexadecanol *n*, 1- ~
Cetylalkohol *m*
hexadecene Hexadecylen *n*
hexadecenoic acid Zoomarinsäure *f*
hexadecine Hexadecin *n*
hexadecyl Cetyl-
hexadecyl alcohol Cetylalkohol *m*,
Hexadecylalkohol *m*
hexadecylene Ceten *n*, Cetylen *n*, Hexadecylen *n*
hexadecyne Hexadecin *n*
hexadecynoic acid, 7- ~ Palmitolsäure *f*
hexadiene Hexadien *n*
hexadienedioic acid, 2,4- ~ Muconsäure *f*

hexadienic acid Sorbinsäure *f*
hexadienoic acid, 2,4- ~ Sorbinsäure *f*
hexadiine Hexadiin *n*
hexadiyne Hexadiin *n*
hexaethylbenzene Hexaäthylbenzol *n*
hexaethyldisiloxane Hexaäthyldisiloxan *n*
hexafluoroacetylacetone
    Hexafluoracetylaceton *n*
hexafluorosilicic acid Hexafluorokieselsäure *f*
hexagon Hexagon *n*, Sechseck *n*, Sechskant *m*
hexagonal hexagonal, sechseckig, sechskantig
hexagonal holohedral hexagonalholoedrisch
    (Krist)
hexagonal nut Sechskantmutter *f*
hexagonal pyramidal hexagonalpyramidal
    (Krist)
hexagonal trapezohedral
    hexagonaltrapezoedrisch (Krist)
hexagon head with collar Sechskant *m* mit
    Ansatz
hexagon iron Sechskanteisen *n*
hexahedral hexaedrisch, sechsflächig
hexahedron Sechsflächner *m* (Krist),
    Hexaeder *n* (Krist), Kubus *m* (Krist),
    Würfel *m* (Krist)
hexahelicene Hexahelicen *n*
hexahydrate Hexahydrat *n*
hexahydrobenzene Cyclohexan *n*,
    Hexahydrobenzol *n*, Hexamethylen *n*
hexahydrofarnesol Hexahydrofarnesol *n*
hexahydrohumulene Hexahydrohumulen *n*
hexahydrophenol Cyclohexanol *n*, Hexalin *n*
hexahydropyrazine Piperazin *n*
hexahydropyridine Hexahydropyridin *n*,
    Piperidin *n*
hexahydrosalicylic acid Hexahydrosalicylsäure *f*
hexahydroxymethylanthraquinone
    Rhodocladonsäure *f*
hexaiododisilane Hexajoddisilan *n*
hexakisoctahedral hexakisoktaedrisch (Krist)
hexakisoctahedron Hexakisoktaeder *n* (Krist),
    Achtundvierzigflächner *m* (Krist)
hexakistetrahedron Hexakistetraeder *n* (Krist)
hexakontane Hexakontan *n*
hexal Hexal *n*
hexalin Hexalin *n*
hexalupine Hexalupin *n*
hexamecol Hexamekol *n*
hexamethonium bromide
    Hexamethoniumbromid *n*
hexamethoxydisiloxane Hexamethoxydisiloxan *n*
hexamethylbenzene Hexamethylbenzol *n*,
    Melliten *n*
hexamethylcyclotrisiloxane
    Hexamethylcyclotrisiloxan *n*
hexamethyldisilane Hexamethyldisilan *n*
hexamethyldisiloxane Hexamethyldisiloxan *n*
hexamethylene Hexamethylen *n*, Cyclohexan *n*,
    Hexahydrobenzol *n*

hexamethylene amine Hexamethylenamin *n*
hexamethylen[e]amine sulfosalicylate Hexal *n*
hexamethylene bromide Hexamethylenbromid *n*
hexamethylene diamine Hexamethylendiamin *n*
hexamethylene diisocyanate
    Hexamethylendiisocyanat *n*
hexamethylene glycol Hexamethylenglykol *n*
hexamethylenetetramine
    Hexamethylentetramin *n*, Aminoform *n*,
    Urotropin *n*
hexamethylenetetramine bromethylene
    Bromalin *n*
hexamethylenetetramine salicylate Saliformin *n*
hexamethylenetetramine tetraiodide Siomin *n*
hexamethylenimine Hexamethylenimin *n*
hexamethylolmelamine Hexamethylolmelamin *n*
hexametric hexametrisch
hexamine Hexamethylentetramin *n*
hexanal Capronaldehyd *m*
hexanaphthene Cyclohexan *n*,
    Hexahydrobenzol *n*
hexane Hexan *n*, **impure** ~ Kanadol *n*
hexanedioic acid Adipinsäure *f*
hexangular sechseckig, sechswinklig
hexanitrobiphenyl Hexanitrobiphenyl *n*,
    Hexanitrodiphenyl *n*
hexanitrodiphenylamine Dipikrylamin *n*
hexanitroethane Hexanitroäthan *n*
hexanoic acid Capronsäure *f*
hexanol Hexanol *n*
hexanone Hexanon *n*
hexaphene Hexaphen *n*
hexaphenyldisiloxane Hexaphenyldisiloxan *n*
hexaphenylethane Hexaphenyläthan *n*
hexapyrine Hexapyrin *n*
hexasymmetric hexasymmetrisch
hexatomic sechsatomig; sechswertig
hexatriene Hexatrien *n*
hexavalent sechswertig
hexavanadic acid Hexavanadinsäure *f*
hexene Hexen *n*, Hexylen *n*
hexetone Hexeton *n*
hexine Hexamethylentetramin *n*
hexinic acid Hexinsäure *f*
hexite Hexit *m*
hexobarbital Hexobarbital *n*
hexoctahedron Diamantoeder *n* (Krist),
    Hexoktaeder *n*
hexode Hexode *f* (Elektr)
hexogen Hexogen *n*
hexokinase Hexokinase *f*
hexone Hexon *n*
hexonic acid Hexonsäure *f*
hexophan Hexophan *n*
hexoran Hexachloräthan *n*, Hexoran *n*
hexose Hexose *f*
hexose diphosphate Hexosediphosphat *n*
hexose phosphate Hexosephosphat *n*
hexosidase Hexosidase *f* (Biochem)

hexuronic acid Hexuronsäure *f*
hexyl Hexyl-, Capryl-
hexylacetic acid Octylsäure *f*
hexyl alcohol Hexylalkohol *m*
hexylamine Hexylamin *n*
hexyl bromide Hexylbromid *n*
hexylcaine Hexylcain *n*
hexylene Hexen *n*, Hexylen *n*
hexylene glycol Hexylenglykol *n*
hexyl hydride Hexan *n*
hexylmethane Heptan *n*
hexylresorcinol Hexylresorcin *n*
hexyne Hexin *n*
heyderiol Heyderiol *n*
heyderioquinone Heyderiochinon *n*
hiascinic acid Hiascinsäure *f*
hibbenite Hibbenit *m* (Min)
hibernate *v* überwintern (Zool)
hibernation Überwinterung *f*, Winterschlaf *m*
hibiscus acid Hibiscussäure *f*
hibschite Hibschit *m* (Min)
hidden verborgen, verdeckt, versteckt
hiddenite Hiddenit *m* (Min)
hide Fell *n*, Tierhaut *f*, **raw ~** Rohhaut *f*
hide glue Hautleim *m*
hiding power Deckfähigkeit *f*, Deckkraft *f*,
  Deckvermögen *n*
hielmite Hjelmit *m* (Min)
hieratite Hieratit *m* (Min)
high hoch
high-alloy hochlegiert
high-angle conveyor belt Steilfördergurt *m*
high-bay warehouse Hochraumlager *n*
high-boiling hochsiedend
high clarity container Klarsichtbehälter *m*
high-contrast kontrastreich
high-current argon arc Argonhochstrombogen *m*
high current transformer Hochstromtrafo *m*
high density polyethylene
  Niederdruckpolyäthylen *n*
high-density storage Zeichendichte *f* (Computer)
high-density wood Preßholz *n*
high-duty hochbeansprucht
high-duty machine Hochleistungsmaschine *f*
high-energy energiereich, schnellfliegend
high energy nuclear physics Hochenergiephysik *f*
high energy working Hochleistungsverformung *f*
higher harmonic current Oberstrom *m*
higher harmonic voltage Oberspannung *f*
high explosive Brisanzstoff *m*, hochexplosiver
  Sprengstoff *m*
high-fidelity lautgetreu, Hi-Fi
high finish[ing] Hochveredelung *f*
high frame Hochspant *n*
high frame system Hochspantsystem *n*
high frequency Hochfrequenz *f*
high-frequency *a* hochfrequent
high-frequency bonding
  Hochfrequenzverleimung *f*

high-frequency cable Hochfrequenzkabel *n*
high-frequency chocke Hochfrequenzdrossel *f*
high-frequency current Hochfrequenzstrom *m*
high-frequency electrical engineering
  Hochfrequenztechnik *f*
high-frequency induction furnace
  Hochfrequenzinduktionsofen *m*
high-frequency iron core
  Hochfrequenzeisenkern *m*
high-frequency plasma torch
  Hochfrequenzplasmabrenner *m*
high-frequency range Hochfrequenzbereich *m*
high-frequency spectroscopy
  Hochfrequenzspektroskopie *f*
high-frequency titration Hochfrequenztitration *f*
high-frequency transformer
  Hochfrequenztransformator *m* (Elektr)
high-frequency welding
  Hochfrequenzschweißen *n*, dielektrisches
  Schweißen *n*
high-gain amplifier Hochleistungsverstärker *m*
high-gloss varnish Brillantlack *m*
high-grade hochwertig, hochgradig, reinst
high-grade fuel hochwertiger Brennstoff *m*
high-grade steel Edelstahl *m*, Qualitätsstahl *m*
high-impedance relay Drosselrelais *n*
high-intensity arc Hochstromkohlebogen *m*
high-intensity magnetic groove drum separator
  Starkfeld-Trommel-[Rillen-]Scheider *m*
high-lift platform truck Hochhubwagen *m*
high luster finished spiegelblank
highly compressed hochverdichtet
highly concentrated hochgradig
highly dangerous lebensgefährlich
highly explosive hochexplosiv, brisant
high-molecular hochmolekular
high performance apparatus
  Hochleistungsapparatus *f*
high polymer Hochpolymere *n*, Riesenmolekül *n*
high-polymer *a* hochpolymer
high potential Hochspannung *f*
high-power motor Hochleistungsmotor *m*
high-power pile Hochleistungsreaktor *m*
high-power reactor Hochleistungsreaktor *m*
high pressure Hochdruck *m*
high-pressure arc Hochdruckbogen *m* (Elektr)
high-pressure area Hochdruckgebiet *n* (Meteor)
high-pressure blade Hochdruckschaufel *f*
high-pressure blower Hochdruckgebläse *n*
high-pressure burner Hochdruckbrenner *m*
high-pressure casing Hochdruckmantel *m*
high-pressure cylinder Hochdruckzylinder *m*
high-pressure engine Hochdruckmaschine *f*
high-pressure fittings Hochdruckarmaturen *pl*
high-pressure glow discharge
  Hochdruckglimmentladung *f*
high-pressure hose Hochdruckschlauch *m*
high-pressure hydrogenation
  Hochdruckhydrierung *f*

high-pressure lamp Hochdrucklampe *f*
high-pressure main Hochdruckleitung *f*
high-pressure measuring equipment
  Hochdruckmeßgerät *n*
high-pressure mounting Hochdruckarmatur *f*
high-pressure piping Hochdruckleitung *f*
high-pressure plant Hochdruckanlage *f*
high-pressure polyethylene
  Hochdruckpolyäthylen *n*
high-pressure process Hochdruckverfahren *n*
high-pressure pump Hochdruckpumpe *f*
high-pressure reactor Hochdruckreaktor *m*
high-pressure rotary blower
  Hochdruckkapselgebläse *n*
high-pressure seal Druckdichtung *f,*
  Hochdruckdichtung *f*
high-pressure spiral blower
  Hochdruckschraubengebläse *n*
high-pressure stage Hochdruckstufe *f*
high-pressure steam Hochdruckdampf *m*
high-pressure [steam] boiler
  Hochdruckdampfkessel *m*
high-pressure synthesis Hochdrucksynthese *f*
high-pressure tire Hochdruckreifen *m*
high-pressure transport Hochdruckförderung *f*
high-pressure valve Hochdruckventil *n*
high-pressure water Preßwasser *n*
high-quality stainless steel korrosionsbeständiger
  Edelstahl *m*
high-quality steel Edelstahl *m*
high-quality tool steel Hochleistungsstahl *m*
high rack Hochregal *n*
high-resolution spectrometry hochauflösende
  Spektrometrie *f*
high-severity cracking
  Kurzzeit-Crackverfahren *n*
high-speed schnell, schnellaufend, schnellbewegt
high-speed event Vorgang *m* hoher
  Geschwindigkeit
high-speed machine Hochleistungsmaschine *f*
high-speed mixer Schnellrührer *m*
high-speed piston pump schnellaufende
  Kolbenpumpe *f*
high-speed rolling mill Schnellwalzwerk *n*
high-speed steel Schnell[dreh]stahl *m*
high-speed stirrer Schnellrührer *m*
high-sulfur coal Schwefelkohle *f*
high-temperature aging Wärmelagerung *f*
high-temperature alloy
  Hochtemperaturlegierung *f,* Superlegierung *f*
high-temperature chlorination
  Hochtemperaturchlorierung *f*
high-temperature grease Hochtemperaturfett *n*
high-temperature material
  Hochtemperaturwerkstoff *m*
high-temperature quenching Thermalhärtung *f*
high-temperature reactor
  Hochtemperaturreaktor *m*

high-temperature strength or stability
  Warmfestigkeit *f*
high-temperature yield strength
  Warmfließgrenze *f*
high tension Hochspannung *f*
high-tension *a* mit hoher Spannung
high tension arc Hochspannungsflamme *f*
high tension arc reaction
  Hochspannungsflammenreaktion *f*
high tension cable Hochspannungskabel *n*
high tension clamp Hochspannungsklemme *f*
high tension current Hochspannungsstrom *m*
high tension danger Hochspannungsgefahr *f*
high tension ignition Hochspannungszündung *f*
high tension insulator Hochspannungsisolator *m*
high tension line Hochspannungsleitung *f,*
  Starkstromleitung *f*
high tension power pack
  Hochspannungsnetzgerät *n*
high tension shell Hochspannungsgehäuse *n*
high tension terminal Hochspannungsklemme *f*
high tension transformer
  Hochspannungstransformator *m*
high tension unit
  Hochspannungserzeugungsanlage *f*
high vacuum Hochvakuum *n*
high-vacuum amplifier Hochvakuumverstärker *m*
high-vacuum engineering Hochvakuumtechnik *f*
high-vacuum grease Hochvakuumfett *n*
high-vacuum phototube Hochvakuumzelle *f*
high-vacuum pump Hochvakuumpumpe *f*
high-vacuum technique Vakuumtechnik *f*
high-vacuum technology Hochvakuumtechnik *f*
high-vacuum tube Hochvakuumröhre *f* (Elektr)
high voltage Hochspannung *f*
high-voltage *a* mit hoher Spannung (Elektr)
high-voltage battery Anodenbatterie *f*
high voltage current Hochspannungsstrom *m,*
  Starkstrom *m*
high voltage discharge tube
  Hochspannungsentladungsröhre *f*
high voltage plant Hochspannungsanlage *f*
high voltage power arc
  Hochspannungsstarkstrombogen *m*
hillaengsite Hillängsit *m* (Min)
hilt Griff *m,* Heft *n*
himbeline Himbelin *n*
hinder *v* [be]hindern, hemmen, verhindern
hindered-settling classifier
  Horizontalschlämmer *m*
hindrance Behinderung *f,* Hemmung *f,*
  Hindernis *n,* Hinderung *f,* **steric ~** sterische
  Hinderung *f*
hinge Angel *f,* Gelenk *n* (Masch); Scharnier *n*
hinged bolt Gelenkschraube *f,* Klappschraube *f*
hinged cover Klappdeckel *m*
hinged pipe Gelenkrohr *n*
hinged valve Drehklappe *f*
hinge fitting Gelenkanschluß *m*

hinge [joint] Scharnier *n*
hinge pad Scharnierunterlage *f*
hinge pin Gelenkbolzen *m*
hinokic acid Hinokisäure *f*
hinokiflavone Hinokiflavon *n*
hinokinin Hinokinin *n*
hinoki oil Hinokiöl *n*
hinokitol Hinokitol *n*
hinopurpurin Hinopurpurin *n*
hinsdalite Hinsdalit *m* (Min)
hint Hinweis *m*
hintzeite Hintzeit *m* (Min)
hiochic acid Hiochisäure *f*
hiortdahlite Hiortdahlit *m* (Min)
hip [berry] Hagebutte *f* (Bot)
hippeastrine Hippeastrin *n*
hipped roof Pultdach *n*
hippulin Hippulin *n*
hippurate Hippurat *n*, hippursaures Salz *n*
hippuric acid Hippursäure *f*, Benzoylglykokoll *n*,
  salt or ester of ~ Hippurat *n*
hippuricase Hippuricase *f* (Biochem)
hippuritic limestone Hippuritenkalk *m* (Min)
hippuryl Hippuryl-
hippuryllysine methyl ester
  Hippuryllysinmethylester *m*
hiptagenic acid Hiptagensäure *f*
hiptagin Hiptagin *n*
hircine Hircin *n*
H-iron Doppel-T-Eisen *n*
Hirsch funnel Hirsch-Filtertrichter *m*
hirsute haarig, zottig
hisingerite Hisingerit *m* (Min)
hispidin Hispidin *n*
hiss *v* zischen
hiss-free klingfrei (Rad)
hissing Zischen *n*
hissing mechanism Zischmechanismus *m*
histaminase Histaminase *f* (Biochem)
histamine Histamin *n*
histamine dihydrochloride
  Histamindihydrochlorid *n*
histamine oxidase Histaminoxydase *f*
histidase Histidase *f* (Biochem)
histidin[e] Histidin *n*
histidine dihydrochloride
  Histidindihydrochlorid *n*
histidinol Histidinol *n*
histochemistry Gewebechemie *f*, Histochemie *f*
histogenesis Gewebebildung *f* (Biol),
  Histogenese *f* (Biol), Histogenie *f* (Biol)
histogenic gewebebildend
histohaematin Histohämatin *n*
histologic[al] histologisch
histologist Histologe *m*
histology Gewebelehre *f*, Histologie *f*
histolysis Gewebezerstörung *f*, Gewebszerfall *m*,
  Histolyse *f*
histolytic histolytisch

histone Histon *n*
histoplasma Histoplasma *n*
histosite Gewebsparasit *m*, Histosit *m*
hit Schlag *m*, Stoß *m*, Treffer *m*
hit *v* schlagen, treffen
hitch Verbindungshaken *m*, Zugvorrichtung *f*
hitching Einhaken *n*
hit theory Treffertheorie *f*
hjelmite Hjelmit *m* (Min)
hoarfrost Rauhreif *m*
hob Prägestempel *m*, Stempel *m* eines
  Preßwerkzeuges
hobbing Prägen *n* (der Formteilkonturen im
  Werkzeug mit dem Prägestempel)
hobbing press Prägepresse *f*
hodograph Hodograf *m*
hoegbomite Högbomit *m* (Min)
Höppler hardness Kegelfließpunkt *m* nach
  Höppler
Hoffmann's drops Hoffmannstropfen *pl*
Hoffmann's tincture Hoffmannstropfen *pl*
Hofmann degradation Hofmannscher Abbau *m*
  (Chem)
Hofmann elimination Hofmann-Eliminierung *f*
  (Chem)
Hofmann rearrangement Hofmannsche
  Umlagerung *f*
hohmannite Hohmannit *m* (Min)
hoist Aufzug *m* (für Lasten), Elevator *m*,
  Hebevorrichtung *f*, Hebewerk *n*,
  Kranlaufwinde *f*
hoist *v* heben, hochwinden, hochziehen
hoist drive Hubwerk *n*
hoist frame Fördergerüst *n*, Förderturm *m*
hoisting cable Förderseil *n*
hoisting capacity Hubkraft *f* (Kran),
  Hubleistung *f* (Kran)
hoisting crane Hebekran *m*, Ladekran *m*
hoisting engine Hebemaschine *f*, Ladekran *m*
hoisting hook Förderhaken *m*
hoisting magnet Hebemagnet *m*
hoisting rope Aufzugsseil *n*, Förderseil *n*
hoisting unit Hubwerk *n*
hokutolite Hokutolit[h] *m* (Min)
holamine Holamin *n*
holaphyllamine Holaphyllamin *n*
hold Griff *m*, Halt *m*, Stütze *f*
hold *v* aufnehmen, fassen ([ent]halten), ~ **back**
  abhalten, ~ **down** unterdrücken
holdback carrier Rückhalteträger *m*
holdback tension Rückhaltespannung *f*
hold-down Gegenhalter *m* (Niederhalter)
hold-down groove Haltekerbe *f*, Haltenute *f*
hold-down plate Halteplatte *f*
holder Halter *m*, Einsatz *m*, Fassung *f* (Elektr),
  Ständer *m*, ~ **for screwing dies**
  Gewindekluppe *f*
holder block Formrahmen *m*

holding capacity Fassungsvermögen *n*
　(Lastwagen)
holding magnet Haltemagnet *m*
holding plate Griffplatte *f*
holding screw Halteschraube *f*
hold-out Porendichtigkeit *f*
hold-up Inhalt *m* (Füllkörperkolonne)
hole Loch *n*, Bohrung *f* (Techn), Leerstelle *f*
　(Chem), Öffnung *f,* **drilled** ~ Bohrloch *n,* **to**
　make a ~ lochen, ~ **in electron valence band**
　Defektelektron *n*
hole *v* durchlöchern
hole count check Lochzahlprüfer *m* (Comp)
hole density Fehlstellenkonzentration *f* in
　Flüssigkeiten
hole-forming pin Lochstift *m*
hole gauge Lochlehre *f*
holes, full of ~ löcherig
hole trap Defektelektronenhaftstelle *f*
Holland blue Holländerblau *n*
hollander Holländer *m* (Pap)
hollander *v* holländern
hollandite Hollandit *m* (Min)
hollow Blase *f* (Gieß), Höhlung *f,* Hohlraum *m,*
　Rinne *f,* Vertiefung *f*
hollow *a* hohl, dumpf (hohlklingend)
hollow *v* ausdrehen, aushöhlen, auskehlen
　(Techn), ~ **out** ausbauchen, aussparen
hollow adze Hohldechsel *f*
hollow body Hohlkörper *m*
hollow boring tool Hohlbohrstahl *m*
hollow cable Hohlseil *n*
hollow-cast hohlgegossen
hollow casting Hohlguß *m*
hollow chamfer Hohlkant *n*
hollow chisel Hohlmeissel *m,* Hohleisen *n*
hollow cylinder Hohlzylinder *m,* Hohlwalze *f*
hollow cylindrical cathode Hohlzylinderkathode *f*
hollow grate Hohlrost *m*
hollow gratebar Hohlroststab *m*
hollow groove Hohlkehle *f* (Masch)
hollow-ground hohlgeschliffen
hollow mill Außenfräser *m,* Stiftfräser *m*
hollow plane Hohlhobel *m*
hollow prism Hohlprisma *n*
hollow rod Hohl[pump]gestänge *n*
　(Ölgewinnung)
hollow roll Hohlwalze *f*
hollow rope Hohlseil *n*
hollow section Hohlprofil *n,* Hohlschnitt *m*
hollow shaft Hohlwelle *f*
hollow space Hohlraum *m*
hollow spar Hohlspat *m* (Min)
hollow sphere Hohlkugel *f*
hollow stone Hohlstein *m*
hollow stopper Hohlstopfen *m*
holly Stechpalme *f* (Bot)
holmite Holmit *m* (Min)
holmium (Symb. Ho) Holmium *n*

holmquistite Holmquistit *m* (Min)
holocaine Holocain *n*
holocrystalline holokristallin, vollkristallin
holo-enzyme Holoenzym *n* (Biochem)
holography Holographie *f*
holohedral ganzflächig (Krist), holoedrisch
　(Krist), vollflächig (Krist)
holohedral crystal Holoeder *n* (Krist)
holohedrism Holoedrie *f* (Krist)
holohedron Holoeder *n* (Krist), Vollflächner *m*
　(Krist)
holohedry Holoedrie *f* (Krist), Vollflächigkeit *f*
　(Krist)
holomorphic holomorph
holomycin Holomycin *n*
holothin Holothin *n*
holotrichite Federalaun *m* (Min)
holozymase Holozymase *f* (Biochem)
holt-melt adhesive Schmelzkleber *m*
homarine Homarin *n*
homatropine Homatropin *n,* Homotropin *n*
homilite Homilit *m* (Min)
homing Zielfindung *f* (Rdr)
homoallantoic acid Homoallantoinsäure *f*
homoallantoin Homoallantoin *n*
homoanisic acid Homoanissäure *f*
homoanthranilic acid Homoanthranilsäure *f*
homoanthroxanic acid Homoanthroxansäure *f*
homoapocamphoric acid
　Homoapocamphersäure *f*
homoarginine Homoarginin *n*
homoasparagine Homoasparagin *n*
homoatomic gleichatomig
homoberberine base Homoberberinbase *f*
homobetaine Homobetain *n*
homocaffeic acid Homokaffeesäure *f*
homocamphenilone Homocamphenilon *n*
homocamphor Homocampher *m*
homocamphoric acid Homocamphersäure *f*
homocaronic acid Homocaronsäure *f*
homochelidonine Homochelidonin *n*
homocholine Homocholin *n*
homochromatic homochrom, gleichfarbig,
　einfarbig
homochromous gleichfarbig, einfarbig,
　homochrom
homochrysanthemic acid
　Homochrysanthemsäure *f*
homoconiine Homoconiin *n*
homocumic acid Homocuminsäure *f*
homocuminic acid Homocuminsäure *f*
homocyclic homozyklisch
homocystamine Homocystamin *n*
homocysteine Homocystein *n*
homocystine Homocystin *n*
homodisperse homodispers
hom[o]eomorphism Gleichartigkeit *f* der
　Kristallform, Gleichförmigkeit *f,*
　Homöomorphie *f,* Isomorphie *f*

hom[o]eomorphous homöomorph (Krist)
hom[o]eopathic homöopathisch
homoeopathy Homöopathie *f*
homoeriodictyol Homoeriodictyol *n*
homofenchol Homofenchol *n*
homogallaldehyde Homogallusaldehyd *m*
homogeneity Homogenität *f,* Gleichartigkeit *f*
homogeneous homogen, artgleich, gleichartig, gleichmäßig, **to make** ~ homogenisieren
homogeneous iron Homogeneisen *n*
homogeneously mixed innig gemischt
homogeneousness Homogenität *f,* Gleichartigkeit *f*
homogeneous reactor Homogenreaktor *m*
homogeneous steel Homogenstahl *m*
homogenization Homogenisierung *f*
homogenize *v* homogenisieren
homogenizer Homogenisierapparat *m,* Emulgiermaschine *f,* Homogenisator *m*
homogenizing Homogenisieren *n*
homogenizing silo Homogenisiersilo *m*
homogentisic acid Homogentisinsäure *f*
homogeranic acid Homogeraniumsäure *f*
homogeraniol Homogeraniol *n*
homoguaiacol Homoguajacol *n*
homohedral homoedrisch (Krist)
homohedrism Homoedrie *f* (Krist)
homohedron Homoeder *n* (Krist)
homoheliotropin Homoheliotropin *n*
homohordenine Homohordenin *n*
homoleucine Homoleucin *n*
homolevulinic acid Homolävulinsäure *f*
homolinalool Homolinalool *n*
Homolka's base Homolkasche Base *f*
homologous homolog, entsprechend
homologue[s] Homologe[s] *n*
homology Homologie *f*
homolysine Homolysin *n*
homolysis Homolyse *f*
homomenthene Homomenthen *n*
homomenthone Homomenthon *n*
homomeroquinene Homomerochinen *n*
homomesitone Homomesiton *n*
homomesityl oxide Homomesityloxid *n*
homomorphic homomorph, gleichgestaltig, gleichförmig
homomorphism Homomorphismus *m* (Math)
homomorpholine Homomorpholin *n*
homomorphous gleichartig, gleichgestaltig
homomuscarine Homomuscarin *n*
homonataloin Homonataloin *n*
homoneurine Homoneurin *n*
homonicotinic acid Homonicotinsäure *f*
homonorcamphoric acid Homonorcamphersäure *f*
homonymic homonym
homonymous homonym, gleichnamig
homophorone Homophoron *n*
homophthalic acid Homophthalsäure *f*

homophthalimide Homophthalimid *n*
homopinene Homopinen *n*
homopinol Homopinol *n*
homopiperidine Homopiperidin *n*
homopiperonal Homopiperonal *n*
homopiperonyl alcohol Homopiperonylalkohol *m*
homopiperonyl amine Homopiperonylamin *n*
homopiperonylic acid Homopiperonylsäure *f*
homopolar gleichpolig, homöopolar, unpolar
homopolar generator Gleichpolgenerator *m* (Elektr)
homopolymerization Homopolymerisation *f*
homopolysaccharide Homopolysaccharid *n*
homopterocarpine Homopterocarpin *n*
homopyrocatechol Homobrenzcatechin *n*
homoquinoline Homochinolin *n*
homoquinolinic acid Homochinolinsäure *f*
homosalicylaldehyde Homosalicylaldehyd *m*
homosalicylic acid Homosalicylsäure *f*
homosaligenin Homosaligenin *n*
homosekikaic acid Homosekikasäure *f*
homoserine Homoserin *n*
homospecificity Homospezifität *f*
homoterephthalic acid Homoterephthalsäure *f*
homoterpenylic acid Homoterpenylsäure *f*
homoterpineol Homoterpineol *n*
homotetrophan Homotetrophan *n*
homotropine Homotropin *n*
homovanillic acid Homovanillinsäure *f*
homovanillin Homovanillin *n*
homoveratrole Homoveratrol *n*
homoveratryl alcohol Homoveratrylalkohol *m*
homoverbanene Homoverbanen *n*
homozygote Homozygot *m* (Biol)
homozygous homozygot, reinerbig
hone Wetzstein *m,* Abziehstein *m,* Schleifstein *m*
honey Honig *m,* **artificial** ~ Kunsthonig *m,* **clarified** ~ Honigseim *m,* **crystallized** ~ Zuckerhonig *m,* **liquid** ~ Honigseim *m*
honeycomb Wabe *f,* Bienenwabe *f,* Honigwabe *f*
honeycombed löcherig, lunkerig, wabenartig, blasig
honeycomb rectifier Zellengleichrichter *m*
honeycomb structure Wabenstruktur *f,* Bienenwabenstruktur *f*
honey-like honigähnlich
honeystone Honigstein *m* (Min), Honigtopas *m* (Min), Mellit *m* (Min)
honeywine Honigwein *m*
honing appliance Abziehvorrichtung *f*
honing machine Ziehschleifmaschine *f*
honor *v* [ver]ehren, beehren
hood Haube *f,* Abzug *m* (Techn), Abzugsschrank *m* (Techn), Deckel *m,* Dunstfang *m* (Techn), Kuppel *f,* Motorhaube *f,* **removable** ~ Kappe *f*
hook Haken *m*
hook *v* einhaken, zuhaken

hook belt fastener Riemenhaken *m,*
  Riemenklammer *f*
hooked krumm, hakenförmig
Hooke's law Hookesches Gesetz *n,*
  Elastizitätsgesetz *n*
hook fittings Hakengeschirr *n*
hook iron Hakeneisen *n*
hook-like hakenförmig
hook link chain Hakenkette *f*
hook lock Hakenschloß *n*
hook pliers Hakenzange *f*
hook spanner Hakenschlüssel *m*
hoop Reif *m,* Faßreifen *m,* Öse *f,* Reifen *m,*
  Ring *m* (barrel), Band *n*
hoop iron Bandeisen *n,* Eisenband *n,* Reifeisen *n*
hoop steel Bandstahl *m*
hop Hopfen *m* (Bot)
hop bitter Hopfenbitter *n,* Hopfenbitterstoff *m*
hop bitter acid Hopfen[bitter]säure *f*
hop cooling plant Hopfenkühlanlage *f*
hop cooling room Hopfenkühlraum *m*
hop dust Hopfenmehl *n*
hopeite Hopeit *m* (Min)
hopeless hoffnungslos, aussichtslos
hopene Hopen *m*
hopenol Hopenol *n*
hopenone Hopenon *n*
hop growing Hopfenbau *m*
hop kiln Hopfendarre *f*
hop oil Hopfenöl *n*
hopped beer Hopfenbier *n* (Brau)
hopper Fülltrichter *m,* Aufschüttrichter *m,*
  Trichteraufsatz *m*
hopper bracket Träger *m* des Trichters
hopper dryer Trichtertrockner *m*
hopper mill Trichtermühle *f*
hops Hopfen *m,* **aroma of** ~ Hopfenaroma *n,*
  infusion of ~ Hopfenaufguß *m*
hop storage Hopfenlagerung *f*
hop store refrigeration Hopfenlagerkühlung *f*
horbachite Horbachit *m* (Min)
hordeine Hordenin *n*
hordenine Hordenin *n*
horizon Horizont *m*
horizontal horizontal, waagrecht
horizontal commutator Plankollektor *m*
horizontal filter-bag centrifuge
  Stülpenbeutelzentrifuge *f*
horizontal grate Planrost *m*
horizontal [line] Horizontale *f,* Waag[e]rechte *f*
horizontal plane Horizontalebene *f*
horizontal rotary filter Schüsselfilter *n*
horizontal row (of a matrix) Horizontalreihe *f*
  (einer Matrix)
horizontal section Grundriß *m,*
  Horizontalschnitt *m*
horizontal-tube evaporator
  Horizontalrohrverdampfer *m*
hormonal hormonal

hormonal action Hormonwirkung *f*
hormonal induction hormonale Induktion *f*
hormonal treatment Hormonbehandlung *f*
  (Med)
hormone Hormon *n,* **administration of a** ~
  Hormongabe *f* (Med), **adrenal cortical** ~
  adrenokortikales Hormon, **adrenocorticotropic**
  ~ adrenocorticotropes Hormon, ACTH,
  antidiuretic ~ antidiuretisches Hormon,
  corpus luteum ~ gelbkörpererzeugendes
  Hormon, **diabetogenic** ~ diabetogenes
  Hormon, **follicle-stimulating** ~
  Follikelreifungshormon *n,* **follicular** ~
  Follikelhormon *n* (Biochem), **gonadotropic** ~
  gonadotropes Hormon, **luteinizing** ~
  Gelbkörperbildungshormon *n,* **parathyroid** ~
  parathyroides Hormon
hormone balance Hormongleichgewicht *n,*
  Hormonhaushalt *m*
hormone preparation Hormonpräparat *n*
hormone secretion Hormonabsonderung *f*
hormonic hormonal
hormonic activity Hormonwirkung *f*
horn Horn *n,* Hupe *f*
hornblende Hornblende *f* (Min), Schörlblende *f*
  (Min)
hornblende schist Hornblendeschiefer *m* (Min)
hornblende slate Hornblendeschiefer *m* (Min)
horn charcoal Hornkohle *f*
horn coal Hornkohle *f*
horn file Hornfeile *f*
hornlead Hornblei *n* (Min), Phosgenit *m* (Min)
hornlike hornähnlich, hornig
horn meal Hornmehl *n*
horn parchment Hornpergament *n*
horn quicksilver Quecksilberhornerz *n* (Min),
  Quecksilberspat *m* (Min)
horn rasp Hornfeile *f*
horn shavings Hornabfall *m,* Hornschabsel *pl,*
  Hornspäne *m pl*
horn sheet Hornblatt *n*
horn silver Hornsilber *n* (Min), Cerargyrit *m*
  (Min), Hornerz *n* (Min), Silberhornerz *n*
  (Min)
horn slate Hornschiefer *m* (Min)
hornstone Hornstein *m* (Min)
hornstone porphyry Horn[stein]porphyr *m* (Min)
horny hornartig, hornig
horny agate Hornachat *m* (Min)
horny skin Hornhaut *f*
horny tissue Horngewebe *n,* Keratingewebe *n*
horsebane Roßfenchel *m* (Bot)
horsebean Pferdebohne *f* (Bot)
horse chestnut oil Roßkastanienöl *n*
horse fat Pferdefett *n*
horse grease Pferdefett *n*
horsehair Roßhaar *n*
horsehair sieve Sieb *n* aus Roßhaar
horseheal Inula *f*

horsemint oil Monardaöl *n*
horsepower Pferdestärke *f*, PS *n* (Abk)
horseradish Meerrettich *m* (Bot)
horseradish oil Meerrettichöl *n*
horse serum Pferdeserum *n*
horseshoe Hufeisen *n*
horseshoe iron Hufstabeisen *n*
horseshoe magnet Hufeisenmagnet *m*
horsfordite Horsfordit *m* (Min)
hortiamine Hortiamin *n*
horticulture Gartenbau *m*
hortonolite Hortonolith *m* (Min)
hose Schlauch *m*, **acid-proof** ~ Säureschlauch *m*
hose clip Schlauchschelle *f*
hose coupling Schlauchanschluß *m*,
   Schlauchkupplung *f*
hosemiazide Hosemiazid *n*
hose wrapping machine
   Schlauchwickelmaschine *f*
host Wirt *m* (Biol), **intermediate** ~
   Zwischenwirt *m*, **to act as a** ~ als Wirt dienen
   (Biol)
host cell Wirtszelle *f* (Biol)
host change Wirtswechsel *m* (Biol)
host molecule Wirtsmolekül *n*
host organism Wirtsorganismus *m* (Biol)
hot heiß, hoch[radio]aktiv (Atom)
hot air Heißluft *f*
hot-air bath Heißluftbad *n*
hot-air cabinet Trockenschrank *m*
hot-air chamber Trockenkammer *f*
hot-air curing Heißluftvulkanisation *f*
hot-air drying Heißlufttrocknung *f*
hot-air engine Heißluftmaschine *f*,
   Luftexpansionsmaschine *f*,
   Luftkraftmaschine *f*
hot-air funnel Heißlufttrichter *m*
[hot-]air heating Luftheizung *f*
hot-air heating apparatus
   Luftheizungsapparat *m*, Luftheizvorrichtung *f*
hot-air main Heißwindleitung *f*
hot-air motor Heißluftmotor *m*
hot ashes Glutasche *f*
hotbed Frühbeet *n*, Mistbeet *n*
hot-bending test Warmbiegeprobe *f*
hot-blast Fön *m*, Heißluftgebläse *n*
hot-blast cupola Heißwindkupolofen *m*
hot-blast period Warmblaseperiode *f*
hot-blast stove Winderhitzer *m*
hot-blast valve Heißwindschieber *m*
hot-brittle heißbrüchig, warmbrüchig
hot bulb Glühkopf *m*, Glühhaube *f*
hot catchpot Heißabscheider *m*
hot cathode Glühelektrode *f*
hot cathode discharge Glühkathodenentladung *f*
hot charge warmer Einsatz *m*
hot chromatography Heißchromatographie *f*
hot coat Heißspritzlack *m*
hot-dipped tinplate feuerverzinntes Blech *n*

hot drawing Warmziehen *n*
hot-ductile warmdehnbar (Met)
hot ductility Warmdehnbarkeit *f* (Met)
hot-forming Warmformgebung *f*,
   Warmverformung *f*
hot-forming *a* heiß verformbar
hot-forming property Warmverformbarkeit *f*
hot-galvanize *v* feuerverzinken, tauchverzinken
hot galvanizing Ansieden *n*, Feuerverzinken *n*,
   Schmelztauchverfahren *n*
hot galvanizing method Ansiedeverfahren *n*
hot gas Heißgas *n*
hot gas sintering Heißgassinterverfahren *n*
hot gilding Warmvergoldung *f*
hothouse Glashaus *n*, Treibhaus *n*
hot lacquer Heißspritzlack *m*
hot lacquering process Heißspritzverfahren *n*
hotmelt adhesive Schmelzkleber *m*
hotmelt coating Aufschmelzüberzug *m*,
   Heißschmelzmasse *f*
hot-molding Warmpressen *n*
hot plate Heizplatte *f*, Wärmeplatte *f*,
   Warmhalteplatte *f*
hot-plate welding Heizplattenschweißen *n*
hot-press Heißpresse *f*
hot-press *v* dekatieren, warmpressen
hot-pressing Warmpressen *n*
hot-quenching method Thermalhärtung *f*
hot rinser Heißspüler *m*
hot-roll *v* warmwalzen
hot-rolled section Warmprofil *n*
hot-rolled steel strip Warmbandstahl *m*
hot-rolled wide strip Warmbreitband *n*
hot rolling Warmwalzen *n*
hot runner mold Spritzgußform *f*
hot saw Warmsäge *f*
hot scale Glührückstand *m*
hot-sealable heiß versiegelbar
hot-sealing Heißsiegelverfahren *n*
hot sealing adhesive Heißsiegelkleber *m*
hot-sealing apparatus Heißsiegelapparat *m*
hot-sealing wax Heißsiegelwachs *n*
hot-setting warmabbindend
hot-setting adhesive Warmklebstoff *m*
hot-shaping Warmformgebung *f*
hot-shoe welding Spiegelschweißen *n*
hot-short heißbrüchig, rotbrüchig, warmbrüchig
hot-shortness Warmbrüchigkeit *f*, Rotbruch *m*
hot silvering bath Ansiedesilberbad *n*
hot-solder *v* heißlöten
hot spraying Heißspritzen *n*,
   Heißsprühverfahren *n*
hot spring Thermalquelle *f*
hot stamping Heißprägen *n*
hot stamping lacquer Heißprägelack *m*
hot-steam valve Heißdampfventil *n*
hot-strain *v* warmrecken
hot straining Warmrecken *n*
hot strength Warmfestigkeit *f*

hot stretching Heißverstrecken *n*
hot-stuff *v* einbrennen (Leder)
hot test Warmprobe *f*
hot tinning Feuerverzinnen *n* (Metall),
Feuerverzinnung *f* (Metall)
hot transverse strength Heißbiegefestigkeit *f*
hot water apparatus Wasserwärmer *m*
hot water discharge from boiler
Heißwasservorlauf *m*
hot-water heating Warmwasserheizung *f*,
Zentralheizung *f*
hot-water pump Heißwasserpumpe *f*,
Warmwasserpumpe *f*
hot water return to boiler Heißwasserrücklauf *m*
hot-water supply Warmwasserversorung *f*
hot-water tank Warmwasserbehälter *m*,
Warmwasserspeicher *m*
hot-water tap Warmwasserhahn *m*
hot-wire thermogravitational column
Draht-Trennrohr *n*
hot-work *v* warm bearbeiten
hot workability Warmverarbeitungsfähigkeit *f*
hot-workable warmverformbar
hot-work hardening Heißverfestigung *f*
hot-working Warmbearbeitung *f*,
Warmformgebung *f*, Warmrecken *n*,
Warmverformung *f*
hot yield point Warmstreckgrenze *f*
hot-zinced feuerverzinkt
hour Stunde *f*
hourly rate Stundengeschwindigkeit *f*
housefront plaster Fassadenputz *m*
house paint, outside ~ Fassadenfarbe *f*
house vermin Hausungeziefer *n*
housing Gehäuse *n*, ~ **for two rolls**
Zweiwalzenständer *m*
hover *v* schweben
hover craft Luftkissenfahrzeug *n*,
Schwebefahrzeug *n*
howlite Howlith *m* (Min)
HP (Abk) Pferdestärke *f*
hub for magnetic tape Spulenkern *m* für
Magnetband
hub flange Nabenflansch *m*
hudsonite Hudsonit *m* (Min)
hue Farbton *m*, Farbenschattierung *f*, Nuance *f*,
Tönung *f*
Hübl number Jodzahl *f*
huebnerite Hübnerit *m* (Min)
hugelite Hügelit *m* (Min)
hull Hülse *f* (Bot), Samenhülse *f* (Bot)
hullite Hullit *m* (Min)
hull plating Rumpfbeplankung *f* (Schiff)
hulsite Hulsit *m* (Min)
hulupone Hulupon *n*
humate Humat *n*
humbertiol Humbertiol *n*
humboldtilite Humboldtilith *m* (Min)

humboldtine Eisenresinit *m* (Min),
Humboldtin *m*
humectant Anfeuchter *m*
humectation Anfeuchtung *f*
Hume-Rothery rule Hume-Rothery-Regel *f*
humic acid Huminsäure *f*, **salt or ester of** ~
Humat *n*
humic carbon Humuskohle *f*
humid naß, feucht
humidification Anfeuchtung *f*,
Feuchtegraderhöhung *f*, Feuchtwerden *n*, **heat
of** ~ Benetzungswärme *f*
humidify *v* anfeuchten, befeuchten
humidifying agent Befeuchtungsmittel *n*
humidity Feuchtigkeit *f*, Luftfeuchte *f*, Nässe *f*,
Naßgehalt *m*, Wassergehalt *m*, **absorption of**
~ Feuchtigkeitsaufnahme *f*, **degree of** ~
Feuchtigkeitsgrad *m*, **relative** ~ relative
Luftfeuchtigkeit *f*, **variation of** ~
Feuchtigkeitsschwankung *f*, ~ **of air**
Luftfeuchtigkeit *f*, **relative** ~ [of the air]
Luftfeuchtigkeitsgrad *m*
humidity chamber Feuchtkammer *f*
humidity condition Feuchtigkeitsbedingung *f*
humidity controller Feuchtigkeitsregler *m*
humidity feeler Feuchtigkeitsfühler *m*
humin Humin *n*
humite Humit *m* (Min), Chondrodit *m* (Min)
humor Feuchtigkeit *f*
humous humusartig
hump Höcker *m*
Humphrey separator Wendelscheider *m*
humulane Humulan *n*
humulene Humulen *n*
humulic acid Humulinsäure *f*
humulin Humulin *n*, Lupulin *n*
humulohydroquinone Humulohydrochinon *n*
humulon Humulon *n*
humuloquinone Humulochinon *n*
humus Humus *m*, Modererde *f*, **rich in** ~
humusreich
humus coal Humuskohle *f*
humus earth Humuserde *f*
humus layer Humusschicht *f*
hundredth Hundertstel *n*
hundred thousandth Hunderttausendstel *n*
hunnemanine Hunnemannin *n*
hurdle Faschine *f*, ~ **to dry malt** Darrhürde *f*
hureaulite Hureaulit *m* (Min)
hurl *v* schleudern, ~ **down** [ab]stürzen, zu
Boden werfen
huronite Huronit *m*
hurt *v* [be]schädigen, verletzen
husbandry Landwirtschaftskunde *f*
husk Gerüst *n* (Techn), Samenhülse *f*, Schale *f*,
Schote *f*
husk *v* schälen
husks Drusen *pl*, Trester *pl*
hussakite Hussakit *m* (Min)

hutchinsonite Hutchinsonit *m* (Min)
hyacinth Hyazinth *m* (Min), roter Zirkon *m*
 (Min)
hyacinthine crystal Hyazinthkristall *m*
hyacinth oil Hyazinthenöl *n*
hyalescence Glasartigkeit *f*
hyalescent durchsichtig *n* (wie Glas)
hyaline gläsern, glasig, hyalin
hyaline quartz Glasquarz *m* (Min)
hyalite Hyalit *m* (Min), Glasopal *m* (Min),
 Wasseropal *m* (Min)
hyalographic hyalographisch, glasätzend
hyalography Hyalographie *f*, Glasätzen *n*
hyaloid glasähnlich, glasig
hyaloidine Hyaloidin *n*
hyalophane Hyalophan *m* (Min),
 Bariumfeldspat *m* (Min)
hyalophotography Hyalophotographie *f*
hyalosiderite Hyalosiderit *m* (Min),
 Eisenchrysolith *m* (Min), Olivin *m* (Min)
hyalotechnical hyalotechnisch
hyalotechnics Hyalotechnik *f*
hyalotekite Hyalotekit *m* (Min)
hyalurgy Hyalurgie *f*
hyaluronic acid Hyaluronsäure *f*
hyaluronidase Hyaluronidase *f* (Biochem)
hybrid Hybrid *n*
hybrid binding Hybridbindung *f*
hybridization Hybridation *f* (Biol),
 Hybridisierung *f*, Kreuzung *f*
 (Bastardisierung)
hybridize hybridisieren (Biol)
hydantoic acid Hydantoinsäure *f*,
 Glykolursäure *f*
hydantoin Hydantoin *n*
hydnocarpic acid Hydnocarpussäure *f*
hydnocarpus oil Hydnocarpusöl *n*
hydnoresinotannol Hydnoresinotannol *n*
hydracetin Hydracetin *n*
hydracetylacetone Hydracetylaceton *n*
hydracid sauerstofffreie Säure,
 Wasserstoffsäure *f*
hydracrylic acid Hydracrylsäure *f*
hydralazine Hydralazin *n*
hydramine Hydramin *n*
hydrangenol Hydrangenol *n*
hydrangine Hydrangin *n*
hydrant Hydrant *m*
hydrargillite Hydrargillit *m* (Min)
hydrargyrism Quecksilbervergiftung *f* (Med)
hydrargyrol Hydrargyrol *n*
hydrargyrum (Lat., Symb. Hg) Quecksilber *n*
hydrastic acid Hydrastsäure *f*
hydrastine Hydrastin *n*
hydrastine sulfate Hydrastinsulfat *n*
hydrastinine Hydrastinin *n*
hydrastinine hydrochloride
 Hydrastininhydrochlorid *n*
hydrastis rhizome Hydrastisrhizom *n* (Bot)

hydratase Hydratase *f* (Biochem)
hydrate Hydrat *n*, ~ of baryta Barythydrat *n*
hydrate *v* hydratisieren, wässern
hydrate-containing hydrathaltig
hydrated hydrathaltig, hydratisch, hydratisiert,
 wasserhaltig
hydrated acid Hydratsäure *f*
hydrated barium oxide Ätzbaryt *m*
hydrated cellulose Hydratzellulose *f* (Chem)
hydrated ion hydratisiertes Ion *n*
hydrated iron(III) oxide Ferrihydroxid *n*
hydrated oxide Hydroxid *n*, Oxidhydrat *n*
hydrate isomerism Hydratisomerie *f*
hydrate water Hydratwasser *n*
hydration Hydratisierung *f*, Hydratation *f*,
 Hydratbildung *f*, Wasseranlagerung *f*, degree
 of ~ Hydratisierungsgrad *m*, heat of ~
 Hydratationswärme *f*
hydration water Hydratwasser *n*
hydratropic acid Hydratropasäure *f*
hydratropic alcohol Hydratropaalkohol *m*
hydraulic hydraulisch
hydraulic accumulator hydraulischer
 Akkumulator *m*
hydraulic brake hydraulische Bremse *f*,
 Wasserdruckbremse *f*
hydraulic cement Wasserzement *m*
hydraulic classifier Stromklassierer *m*
hydraulic cup Manschette *f*
hydraulic elevator hydraulischer Aufzug *m*
hydraulic excavator Hydraulikbagger *m*,
 Hydrobagger *m*
hydraulic extrusion Kolbenpressen *n*
hydraulic filter press Plattenpreßfilter *n*,
 Siebtrommel-Filter *n*
hydraulic fluid Hydraulikflüssigkeit *f*
hydraulic gradient hydraulischer Gradient *m*
hydraulic gypsum Estrichgips *m*
hydraulic lift hydraulischer Aufzug *m*
hydraulic lime Wasserkalk *m*
hydraulic medium Druckflüssigkeit *f*
hydraulic mortar Wassermörtel *m*
hydraulic plant, direct pumping ~
 Druckwasseranlage *f* mit Druckpumpen (pl),
 ~ with air bottle system Druckwasseranlage *f*
 mit Druckluftbelastung
hydraulic plunger Mönchskolben *m*
hydraulic power hydraulische Kraft *f*,
 Wasserkraft *f*
hydraulic press hydraulische Presse *f*,
 Wasserdruckpresse *f*
hydraulic pressure Flüssigkeitsdruck *m*,
 Wasser[säulen]druck *m*
hydraulic pressure gauge
 Wasserdruckmanometer *n*
hydraulic pressure head Wasserdruckhöhe *f*
hydraulic ram Druckwasserpumpe *f*,
 hydraulische Presse *f*, Mönchskolben *m*
hydraulic riveting Druckwassernietung *f*

hydraulics Hydraulik *f*, Mechanik *f* flüssiger Körper (pl)
hydraulic test Wasserdruckprobe *f*
hydraulic water Druckwasser *n*
hydraziacetic acid Hydraziessigsäure *f*
hydrazicarbonyl Hydrazicarbonyl *n*
hydrazide Hydrazid *n*
hydrazimethylene Hydrazimethylen *n*
hydrazine Hydrazin *n*
hydrazine bifluoride Hydrazinbifluorid *n*
hydrazine chloride Hydrazinchlorid *n*
hydrazine dihydrochloride Hydrazindihydrochlorid *n*
hydrazine dimalonic acid Hydrazindimalonsäure *f*
hydrazine fluoride, acid ~ Hydrazinbifluorid *n*
hydrazine hydrate Hydrazinhydrat *n*
hydrazine hydrogen fluoride Hydrazinbifluorid *n*
hydrazine sulfate Hydrazinsulfat *n*
hydrazine yellow Hydrazingelb *n*
hydrazoate Azid *n*
hydrazobenzene Hydrazobenzol *n*
hydrazo-compound Hydrazoverbindung *f*
hydrazodicarbonamide Hydrazodicarbonamid *n*, Hydrazoformamid *n*
hydrazodicarbonhydrazide Hydrazoformhydrazid *n*
hydrazoformamide Hydrazoformamid *n*, Hydrazodicarbonamid *n*
hydrazoformhydrazide Hydrazoformhydrazid *n*
hydrazoformic acid Hydrazoameisensäure *f*
hydrazoic acid Stickstoffwasserstoffsäure *f*
hydrazomethane Hydrazomethan *n*
hydrazone Hydrazon *n*
hydrazotoluene Hydrazotoluol *n*
hydride Hydrid *n*, Wasserstoffverbindung *f*
hydride transfer Hydrid-Übertragung *f*
hydrindacene Hydrindacen *n*
hydrindane Hydrindan *n*
hydrindene Hydrinden *n*, Indan *n*
hydrindic acid Hydrindinsäure *f*
hydrindole Hydrindol *n*
hydrindone Hydrindon *n*, Indanon *n*
hydrine Hydrin *n*
hydriodic acid Jodwasserstoffsäure *f*
hydroanisoin Hydroanisoin *n*
hydroanthracene Hydroanthracen *n*
hydroapatite Hydroapatit *m* (Min)
hydroaromatic hydroaromatisch
hydrobarometer Hydrobarometer *n*
hydrobenzamide Hydrobenzamid *n*
hydrobenzoin Hydrobenzoin *n*
hydroberberine Hydroberberin *n*
hydrobilirubin Hydrobilirubin *n*
hydrobiology Hydrobiologie *f*, Wasserbiologie *f*
hydrobixin Hydrobixin *n*
hydroboracite Hydroboracit *m* (Min)
hydrobornylene Hydrobornylen *n*
hydroborocalcite Hydroborocalcit *m* (Min)

hydroborofluoric acid Borfluorwasserstoffsäure *f*, Borflußsäure *f*
hydroboron Borhydrid *n*
hydrobromic acid Bromwasserstoffsäure *f*
hydrobromide Hydrobromid *n*
hydrocaffeic acid Hydrokaffeesäure *f*
hydrocamphene Hydrocamphen *n*
hydrocampnospermonol Hydrocampnospermonol *n*
hydrocaoutchouc Hydrokautschuk *m*
hydrocarbon Kohlenwasserstoff *m*, **halogenated** ~ Halogenkohlenwasserstoff *m*, **saturated** ~ Grenzkohlenwasserstoff *m*
hydrocarbonaceous kohlenwasserstoffartig, kohlenwasserstoffhaltig
hydrocarbonate Hydrogencarbonat *n*, saures Carbonat *n*
hydrocarbon gas Kohlenwasserstoffgas *n*
hydrocarbon lamp Kohlenwasserstofflampe *f*
hydrocarbons Kohlenwasserstoffe *pl*, **aliphatic** ~ aliphatische Kohlenwasserstoffe, **aromatic** ~ aromatische Kohlenwasserstoffe, **cyclic** ~ zyklische Kohlenwasserstoffe, **saturated** ~ gesättigte Kohlenwasserstoffe, **unsaturated** ~ ungesättigte Kohlenwasserstoffe
hydrocarbostyril Hydrocarbostyril *n*
hydrocardanol Hydrocardanol *n*
hydrocell Hydroelement *n*
hydrocellobial Hydrocellobial *n*
hydrocellulose Hydrocellulose *f*
hydrocellulose acetate Acetylhydrocellulose *f*, Hydrocelluloseacetat *n*
hydrocephalin Hydrocephaelin *n*
hydrocerite Hydrocerit *m*
hydrocerussite Hydrocerussit *m* (Min)
hydrochalcone Hydrochalkon *n*
hydrochelidonic acid Acetondiessigsäure *f*, Hydrochelidonsäure *f*
hydrochinidine Hydrocinchonidin *n*
hydrochloric salzsauer
hydrochloric acid Salzsäure *f*, Acidum hydrochloricum *n* (Lat), Chlorwasserstoff *m*
hydrochloric acid container Salzsäurebehälter *m*
hydrochloric acid mist Salzsäurenebel *m*
hydrochloric acid pickle Salzsäurebeize *f*
hydrochloric acid plant Chlorwasserstoffanlage *f*, Salzsäureanlage *f*
hydrochloride Hydrochlorid *n*, Chlorhydrat *n*
hydrochromone Hydrochromon *n*
hydrocinchonicine Hydrocinchonicin *n*, Hydrocinchotoxin *n*
hydrocinchonidine Hydrocinchonidin *n*
hydrocinchonine Hydrocinchonin *n*
hydrocinchotoxine Hydrocinchonicin *n*, Hydrocinchotoxin *n*
hydrocinnamaldehyde Hydrozimtaldehyd *m*
hydrocinnamic acid Hydrozimtsäure *f*
hydrocinnamic alcohol Hydrozimtalkohol *m*
hydrocinnamic aldehyde Hydrozimtaldehyd *m*

hydrocinnamide Hydrocinnamid *n*
hydrocinnamoin Hydrocinnamoin *n*
hydrocinnamyl alcohol Hydrozimtalkohol *m*
hydrocoffeic acid Hydrokaffeesäure *f*
hydrocollidine Hydrocollidin *n*
hydro-compound Hydroverbindung *f*
hydroconchinine Hydroconchinin *n*
hydrocortisone Hydrocortison *n*
hydrocotarnine Hydrocotarnin *n*
hydrocotoin Hydrocotoin *n*
hydrocoumaric acid Hydrocumarsäure *f*
hydrocoumarilic acid Hydrocumarilsäure *f*
hydrocoumarin Hydrocumarin *n*
hydrocracking Hydrocrackreaktion *f,*
Hydrospaltung *f*
hydrocresol Hydrocresol *n*
hydrocumarone Hydrocumaron *n*
hydrocupreane Hydrocuprean *n*
hydrocyanic acid Blausäure *f,*
Ameisensäurenitril *n,* Cyanwasserstoff *m,*
containing ~ blausäurehaltig, cyanidhaltig,
poisoning with ~ Blausäurevergiftung *f*
hydrocyanic acid poisoning Blausäurevergiftung *f*
hydrocyanide Hydrocyanid *n*
hydrocyclone Hydrozyklon *m,* **conical** ~
Spitzzyklon *m* (Bergb)
hydrodiffusion Hydrodiffusion *f*
hydrodolomite Hydrodolomit *m* (Min)
hydrodynamic[al] hydrodynamisch
hydrodynamics Hydrodynamik *f,*
Strömungslehre *f*
hydrodynamometer Hydrodynamometer *n*
hydroelectric hydroelektrisch
hydroelectric generator Wasserdynamo *m*
hydroelectricity Hydroelektrizität *f*
hydroelectric power plant Wasserkraftwerk *n*
hydroelectric power station Kraftspeicherwerk *n*
hydro-electrostatic machine
Dampfelektrisiermaschine *f*
hydroelement Hydroelement *n*
hydroextract *v* zentrifugieren, [aus]schleudern
hydroextraction Entwässern *n,* Entwässerung *f*
hydroextractor Zentrifuge *f,*
Schleuder[maschine] *f,*
Zentrifugaltrockenmaschine *f*
hydrofluoboric acid Borfluorwasserstoffsäure *f,*
Borflußsäure *f*
hydrofluocerite Hydrofluocerit *m* (Min)
hydrofluoric fluorwasserstoffsauer
hydrofluoric acid Fluorwasserstoffsäure *f,*
Flußsäure *f,* **salt of** ~ Fluorid *n*
hydrofluoric acid plant Flußsäureanlage *f*
hydrofluoride Hydrofluorid *n*
hydroforming Hydroformieren *n* (Chem)
hydroformylation Hydroformylierung *f*
hydrogalvanic hydrogalvanisch
hydrogel Hydrogel *n*
hydrogen (Symb. H) Wasserstoff *m,* **activated** ~
aktivierter Wasserstoff, **atomic** ~ atomarer

Wasserstoff, **charged with** ~
Wasserstoffbeladung *f* (Metall), **containing** ~
wasserstoffhaltig, **formation of** ~
Wasserstoffbildung *f,* **generation of** ~
Wasserstoffentwicklung *f,* **heavy** ~
Deuterium *n,* Tritium *n,* schwerer Wasserstoff,
nascent ~ naszierender Wasserstoff in statu
nascendi, **rich in** ~ wasserstoffreich
hydrogen absorption Wasserstoffaufnahme *f*
hydrogen acceptor Wasserstoffakzeptor *m*
hydrogen arc Wasserstofflichtbogen *m*
hydrogen arsenide Arsenwasserstoff *m*
hydrogenase Hydrogenase *f* (Biochem)
hydrogenate *v* hydrieren; (Fette) härten
hydrogenated hydriert
hydrogenation Hydrierung *f,*
Wasserstoffanlagerung *f,* **catalytic** ~
katalytische Hydrierung, **partial** ~ partielle
Hydrierung, **vapor-phase** ~
Gasphase-Hydrierung *f,* ~ **of coal**
Kohlehydrierung *f,* ~ **of oil** Ölhärtung *f*
(Chem)
hydrogenation apparatus Hydrierapparat *m*
hydrogenation cracking reaction
Hydrocrackreaktion *f*
hydrogenation of fats Fetthärtung *f,*
Fetthydrierung *f*
hydrogenation plant Hydrieranlage *f,*
Hydrierwerk *n*
hydrogenation process Hydrierungsverfahren *n*
(of fats), Härtungsverfahren *n* (für Fette)
hydrogenation process gasoline Hydrierbenzin *n*
hydrogenation under pressure Druckhydrierung *f*
hydrogen atmosphere Wasserstoffatmosphäre *f*
hydrogen atom Wasserstoffatom *n*
hydrogen bacteria Wasserstoffbakterien *pl*
hydrogen bomb Wasserstoffbombe *f,* H-Bombe *f*
hydrogen bond Wasserstoff[brücken]bindung *f,*
innermolecular ~ innermolekulare
Wasserstoffbrückenbindung, **intramolecular**
~ intramolekulare
Wasserstoffbrückenbindung
hydrogen bottle Wasserstoffflasche *f*
hydrogen bromide Bromwasserstoff *m*
hydrogen bubble chamber
Wasserstoffblasenkammer *f*
hydrogen chloride Chlorwasserstoff *m,*
generation of ~
Chlorwasserstoffentwicklung *f*
hydrogen compound Wasserstoffverbindung *f*
hydrogen container Wasserstoffbehälter *m*
hydrogen content Wasserstoffgehalt *m*
hydrogen continuum Wasserstoffkontinuum *n*
hydrogen cyanide Cyanwasserstoff *m,*
Ameisensäurenitril *n,* Blausäuregas *n*
hydrogen cylinder Wasserstoffflasche *f*
hydrogen dioxide Hydrogenperoxid *n,*
Wasserstoffperoxid *n*
hydrogen electrode Wasserstoffelektrode *f*

hydrogen embrittlement Beizsprödigkeit *Met,*
  Wasserstoffkrankheit *f* (Met)
hydrogen equivalent Wasserstoffäquivalent *n*
hydrogenerator Härtungsautoklav *m* (Fette)
hydrogen exchange reaction
  Wasserstoffaustauschreaktion *f*
hydrogen filling Wasserstoffüllung *f*
hydrogen flame Wasserstoffflamme *f*
hydrogen flow Wasserstoffstrom *m*
hydrogen fluoride Fluorwasserstoff *m*
hydrogen formation Wasserstoffbildung *f*
hydrogen generating plant
  Wasserstoffgewinnungsanlage *f*
hydrogen generator Wasserstoffentwickler *m,*
  Wasserstofferzeuger *m*
hydrogen halide Halogenwasserstoff *m*
hydrogenic wasserstoffähnlich
hydrogen inlet tube Wasserstoffzuleitungsröhre *f*
hydrogen iodide Jodwasserstoff *m*
hydrogen ion Wasserstoffion *n*
hydrogen ion activity Wasserstoffionenaktivität *f*
hydrogen ion concentration Wasserstoffzahl *f*
hydrogen ion exponent
  Wasserstoffionenexponent *m*
hydrogenize *v* hydrieren
hydrogenizing Hydrierung *f*
hydrogen lamp Wasserstofflampe *f*
hydrogen line Wasserstofflinie *f* (Spektr)
hydrogen molecule Wasserstoffmolekül *n*
hydrogen nucleus Wasserstoffkern *m*
hydrogenolysis Hydrogenolyse *f,*
  Hydrospaltung *f*
hydrogenous wasserstoffhaltig
hydrogen outlet tube
  Wasserstoffableitungsröhre *f*
hydrogen overvoltage Wasserstoffüberspannung *f*
hydrogen oxide Wasser *n*
hydrogen peroxide Hydrogenperoxid *n,*
  Wasserstoff[su]peroxid *n*
hydrogen peroxide solution Perhydrol *n*
hydrogen phosphide Phosphin *n,*
  Phosphorwasserstoff *m*
hydrogen plant Wasserstoffanlage *f*
hydrogen polysulfide Hydrogenpolysulfid *n,*
  Polyschwefelwasserstoff *m*
hydrogen producing plant
  Wasserstoffgewinnungsanlage *f*
hydrogen reduction Wasserstoffreduktion *f*
hydrogen selenide Selenwasserstoff *m*
hydrogen series Wasserstoffserie *f* (Spektr)
hydrogen silicide Siliciumwasserstoff *m*
hydrogen soldering Wasserstofflötung *f*
hydrogen spark gap Wasserstofffunkenstrecke *f*
hydrogen spectrum Wasserstoffspektrum *n*
hydrogen sulfide Schwefelwasserstoff *m,*
  compound of ~
  Schwefelwasserstoffverbindung *f*
hydrogen sulfide group
  Schwefelwasserstoffgruppe *f* (Anal)

hydrogen sulfide meter
  Schwefelwasserstoffmesser *m*
hydrogen sulfide precipitate
  Schwefelwasserstoffniederschlag *m*
hydrogen sulfocyanate Rhodanwasserstoff *m*
hydrogen superoxide Wasserstoffsuperoxid *n*
hydrogen swell Wasserstoffbombage *f*
[hydrogen] swelling Bombage *f* (Dose)
hydrogen tank Wasserstoffbehälter *m*
hydrogen tartrate Bitartrat *n*
hydrogen telluride Tellurwasserstoff *m*
hydrogen thiocyanate Rhodanwasserstoff *m*
hydrogen trinitride Stickstoffwasserstoff *m*
hydrogen uptake Wasserstoffaufnahme *f*
hydroginkgol Hydroginkgol *n*
hydrography Hydrographie *f*
hydrohalic acid Halogenwasserstoffsäure *f*
hydrohalogen compound
  Halogenwasserstoffverbindung *f*
hydrohematite Hydrohämatit *m* (Min)
hydrohydrastinine Hydrohydrastinin *n*
hydrojuglone Hydrojuglon *n*
hydrol Hydrol *n*
hydrolactal Hydrolactal *n*
hydrolapachol Hydrolapachol *n*
hydrolase Hydrolase *f* (Biochem)
hydrolith Hydrolith *n*
hydrologic hydrologisch
hydrology Gewässerkunde *f,* Hydrologie *f*
hydrolomatiol Hydrolomatiol *n*
hydrolysate Hydrolysat *n*
hydrolysis Hydrolyse *f,* Aufschluß *m,*
  hydrolytischer Abbau *m,* ~ **of wood**
  Holzverzuckerung *f*
hydrolysis precipitation Hydrolysenfällung *f*
hydrolysis rate Hydrolysegeschwindigkeit *f*
hydrolytic hydrolytisch
hydrolytic agent Aufschlußmittel *n*
hydrolyzable hydrolysierbar
hydrolyzation Hydrolysierung *f*
hydrolyzation process Aufschlußverfahren *n,*
  Hydrolyse *f*
hydrolyze *v* hydrolysieren, aufschließen
hydrolyzing *a* hydrolysierend
hydromagnesite Hydromagnesit *m* (Min)
hydromagnocalcite Hydromagnocalcit *m* (Min)
hydromechanical hydromechanisch
hydromechanics Hydromechanik *f*
hydromel Honigwasser *n,* **vinous** ~ Met *m*
hydromelanothallite Hydromelanothallit *m*
  (Min)
hydrometallurgy Hydrometallurgie *f,*
  Naßmetallurgie *f*
hydrometer Hydrometer *n,* Densimeter *n,*
  Dichtemesser *m* (für Flüssigkeiten),
  Flüssigkeitsspindel *f,* Flüssigkeitswaage *f,*
  Säuredichtemesser *m* (für Säuren),
  Säureprüfer *m* (für Säuren), Säurespindel *f*
  (für Säuren), Tauchwaage *f,* **graduated** ~

Skalenaräometer *n*, ~ **for determining the**
specific gravity of urine Urinometer *n* (Med)
hydrometric hydrometrisch
hydrometrograph Hydrometrograph *m*
hydrometry Hydrometrie *f*
hydronalium Hydronalium *n*
hydron blue Hydronblau *n*
hydronitric acid Stickstoffwasserstoffsäure *f*
hydronium ion Hydroniumion *n*
hydrooctinum Hydrooctinum *n*
hydroorotic acid Hydroorotsäure *f*
hydropathic hydropathisch (Med)
hydrophane Hydrophan *m* (Min), Edelopal *m*
(Min), Wasseropal *m* (Min)
hydrophanousness Durchsichtigkeit *f* im Wasser
hydrophilic hydrophil, wasserbindend
hydrophilizing Hydrophilierung *f* (Text)
hydrophobe hydrophob, wasserabstoßend
hydrophobia Rabies *f* (Tiermed)
hydrophobic hydrophob
hydrophobizing Hydrophobierung *f* (Text)
hydrophone Unterwasserhorchgerät *n*
hydrophthalic acid Hydrophthalsäure *f*
hydrophyte Hydrophyt *m*, Wasserpflanze *f*
hydropiperic acid Hydropiperinsäure *f*
hydropiperinic acid Hydropiperinsäure *f*
hydropiperoin Hydropiperoin *n*
hydropneumatic hydropneumatisch
hydro-pneumatic plant
Druckwasser-Druckluft-Anlage *f*
hydroprotopine Hydroprotopin *n*
hydropyrine Hydropyrin *n*, Grifa *n*
hydroquinane Hydrochinan *n*
hydroquinene Hydrochinen *n*
hydroquinidine Hydroconchinin *n*
hydroquinine Hydrochinin *n*
hydroquinol Hydrochinon *n*,
para-Dihydroxybenzol *n*
hydroquinol dimethyl ether
Hydrochinondimethyläther *m*
hydroquinol monomethyl ether
Hydrochinonmonomethyläther *m*
hydroquinone Hydrochinon *n*
hydroquinone developer
Hydrochinonentwickler *m*
hydroquinotoxine Hydrochinotoxin *n*
hydroquinoxaline Hydrochinoxalin *n*
hydro-refined product Hydrierraffinat *n*
hydroresorcinol Hydroresorcin *n*
hydrorhombinine Hydrorhombinin *n*
hydrorubber Hydrokautschuk *m*
hydroselenic acid Selenwasserstoffsäure *f*
hydrosilicate Hydrosilikat *n*
hydrosol Hydrosol *n*
hydrosorbic acid Hydrosorbinsäure *f*
hydrosphere Hydrosphäre *f*
hydrostatic hydrostatisch
hydrostatic pressure Druck *m* der ruhenden
Flüssigkeit

hydrostatics Hydrostatik *f*
hydrosulfate Hydrogensulfat *n*
hydrosulfide Hydrogensulfid *n*, Mercaptan *n*,
Thioalkohol *m*, Thiol *n*
hydrosulfite Hydrogensulfit *n*,
Natriumhydrosulfit *n*
hydrosulfuric acid Schwefelwasserstoff *m*
hydrotalcite Hydrotalkit *m* (Min)
hydrotechnical hydrotechnisch
hydrotechnics Hydrotechnik *f*
hydrotelluric acid Tellurwasserstoffsäure *f*
hydrotelluride Hydrotellurid *n*
hydrothermal hydrothermal
hydrotimeter Hydrotimeter *n*
hydrotoluoin Hydrotoluoin *n*
hydrotropy Hydrotropie *f*
hydroturbine Wasserturbine *f*
hydrouracil Hydrouracil *n*
hydrourushiol Hydrourushiol *n*
hydrous wäss[e]rig, wasserhaltig
hydrous ferrous sulfate Eisenvitriol *m*
hydrous wool fat Lanolin *n*
hydroxamic acid Hydroxamsäure *f*
hydroxanthommatin Hydroxanthommatin *n*
hydroxide Hydroxid *n*
hydroxide complex Hydroxokomplex *m*
hydroxocobalamin Aquocobalamin *n*
hydroxonic acid Hydroxonsäure *f*
hydroxy-9-octadecenoic acid, 12- ~
Ricinelaidinsäure *f*, Ricinolsäure *f*,
Ricinsäure *f*
hydroxyacanthine Oxyacanthin *n*
hydroxyacetic acid Glycolsäure *f*,
Oxyessigsäure *f*
hydroxyacetone Acetol *n*
hydroxy acid Hydroxylsäure *f*
hydroxyacridine Oxyacridin *n*
hydroxy aldehyde Oxyaldehyd *m*
hydroxy amino acid Hydroxyaminosäure *f*
hydroxyamphetamine Hydroxyamphetamin *n*
hydroxyanthracene Anthrol *n*
hydroxyanthranilic acid Hydroxyanthranilsäure *f*
hydroxyanthraquinone Hydroxyanthrachinon *n*,
Oxyanthrachinon *n*
hydroxyanthrone Oxanthranol *n*
hydroxyaphylline Hydroxyaphyllin *n*
hydroxyazobenzene Hydroxyazobenzol *n*
hydroxy azo compound Hydroxyazoverbindung *f*,
Oxyazoverbindung *f*
hydroxybarbituric acid, 5- ~ Dialursäure *f*
hydroxybenzaldehyde Hydroxybenzaldehyd *m*,
Oxybenzaldehyd *m*
hydroxybenzamide, 2- ~ Salicylamid *n*
hydroxybenzanthrone Hydroxybenzanthron *n*
hydroxybenzene Phenol *n*, Oxybenzol *n*
hydroxybenzoic acid Hydroxybenzoesäure *f*,
Oxybenzoesäure *f*, Phenolcarbonsäure *f*
hydroxybenzophenone Oxybenzophenon *n*
hydroxybenzyl alcohol Hydroxybenzylalkohol *m*

hydroxybenzyl cyanide Hydroxybenzylcyanid *n*
hydroxybioxindol Isatan *n*
hydroxybutanal Oxybutyraldehyd *m*
hydroxybutane diamide, 2- ~ Malamid *n*
hydroxybutanedioic acid Äpfelsäure *f*
hydroxybutanoic acid lactone, 4- ~
  Butyrolacton *n*
hydroxy butyraldehyde Oxybutyraldehyd *m*
hydroxybutyric acid Hydroxybuttersäure *f*,
  Oxybuttersäure *f*
hydroxycamphor Oxycampher *m*
hydroxycarboxylic acid Oxycarbonsäure *f*
hydroxychloroquine Hydroxychlorochin *n*
hydroxycinnamic acid Oxyzimtsäure *f*
hydroxycobalamine Hydroxykobalamin *n*
hydroxy compound Oxyverbindung *f*
hydroxycoumarin, 4- ~ Umbelliferon *n*
hydroxydecanoic acid Hydroxydecansäure *f*
hydroxydiphenyl Hydroxydiphenyl *n*
hydroxydiphenylamine Hydroxydiphenylamin *n*,
  Oxydiphenylamin *n*
hydroxydiphenylmethane Oxydiphenylmethan *n*,
  Benzylphenol *n*
hydroxydroserone Hydroxydroseron *n*
hydroxyethanoic acid Glycolsäure *f*
hydroxyethylamine Oxäthylamin *n*
hydroxyfatty acid Oxyfettsäure *f*
hydroxyhexadecanoic acid, 16- ~
  Juniperinsäure *f*
hydroxyhumulinic acid Hydroxyhumulinsäure *f*
hydroxyhydroquinone Hydroxyhydrochinon *n*
hydroxyketone Hydroxyketon *n*, Oxyketon *n*
hydroxyl Hydroxyl- *n*, **containing** ~
  hydroxylhaltig
hydroxylamine Hydroxylamin *n*,
  Oxyammoniak *n* (obs)
hydroxylamine hydrochloride
  Hydroxylaminchlorhydrat *n*
hydroxylammonium compound
  Hydroxylammoniumverbindung *f*,
  Oxyammoniumverbindung *f*
hydroxylapatite Hydroxylapatit *m*
hydroxylase Hydroxylase *f* (Biochem)
hydroxylate *v* hydroxylieren
hydroxylation Hydroxylierung *f*
hydroxyl group Hydroxylgruppe *f*,
  Hydroxylradikal *n*, **alcoholic** ~ alkoholische
  Hydroxylgruppe
hydroxyl ion Hydroxylion *n*
hydroxyl ion concentration
  Hydroxylionenkonzentration *f*
hydroxyl number Hydroxylzahl *f*
hydroxylpalmitone Hydroxylpalmiton *n*
hydroxyl radical Hydroxylradikal *n*
hydroxylysine Hydroxylysin *n*
hydroxymenthane Menthanol *n*, **3-** ~ Menthol *n*
hydroxymenthene Menthenol *n*
hydroxymenthylic acid Oxymenthylsäure *f*

hydroxymethoxy benzaldehyde
  Hydroxymethoxybenzaldehyd *m*
hydroxymethylbenzoic acid
  Oxymethylbenzoesäure *f*
hydroxymethyl-chrysazin, 3- ~ Rhabarberon *n*
hydroxynaphthalene Naphthol *n*
hydroxynaphthoic acid Oxynaphthoesäure *f*,
  Naphthol[carbon]säure *f*
hydroxynervonic acid Hydroxynervonsäure *f*
hydroxynitrobenzyl chloride
  Hydroxynitrobenzylchlorid *n*
hydroxyoctanthrene Octanthrenol *n*
hydroxyocthracene Octhracenol *n*
hydroxypalmitic acid, 16- ~ Juniperinsäure *f*
hydroxypentadecylic acid
  Hydroxypentadecylsäure *f*
hydroxyperezone Hydroxyperezon *n*
hydroxyphenanthrene Phenanthrol *n*
hydroxyphenylacetic acid
  Hydroxyphenylessigsäure *f*, Mandelsäure *f*
hydroxyphenyl naphthylamine
  Hydroxyphenylnaphthylamin *n*
hydroxypinic acid Hydroxypinsäure *f*
hydroxyproline Hydroxyprolin *n*, Oxyprolin *n*
hydroxypropanenitrile Milchsäurenitril *n*
hydroxypropanoic acid, 3- ~ Hydracrylsäure *f*
hydroxypropionic acid Hydroxypropionsäure *f*,
  Oxypropionsäure *f*
hydroxypropiophenone Hydroxypropiophenon *n*
hydroxy purine, 6- ~ Hypoxanthin *n*
hydroxypyrene Hydroxypyren *n*
hydroxypyridine Hydroxypyridin *n*
hydroxyquinaldic acid, 4- ~ Kynurensäure *f*
hydroxyquinoline Hydroxychinolin *n*,
  Oxychinolin *n*, **2-** ~ Carbostyril *n*, **4-** ~
  Kynurin *n*, **8-** ~ Oxin *n*
hydroxyquinone Oxychinon *n*
hydroxystearic acid Oxystearinsäure *f*
hydroxysuccinic acid Oxybernsteinsäure *f*
hydroxytoluene Oxytoluol *n*
hydroxytropan, 3- ~ Tropin *n*
hydroxytryptophane Hydroxytryptophan *n*
hydroxyvaline Hydroxyvalin *n*
hydrozincite Hydrozinkit *m* (Min), Zinkblüte *f*
  (Min)
hydurilic acid Hydurilsäure *f*
hyenanchin Hyenanchin *n*
hygiene Hygiene *f*
hygienic hygienisch
hygienic[al] gesundheitlich
hygienics Gesundheitslehre *f*, Hygiene *f*
hygrine Hygrin *n*, Cuscohygrin *n*
hygrinic acid Hygrinsäure *f*
hygrofix Hygrofix *n*
hygrograph Hygrograph *m*,
  Feuchtigkeitsschreiber *m*
hygrometer Feuchtigkeitsmeßinstrument *n*,
  Hygrometer *n*, Luftfeuchtigkeitsmesser *m*
hygrometric hygrometrisch

hygrometric condition Luftfeuchtigkeit *f*
hygrometry Hygrometrie *f*,
Feuchtigkeitsmessung *f*
hygromycin Hygromycin *n*
hygrophyllite Hygrophyllit *m* (Min)
hygroscope Hygroskop *n*,
Feuchtigkeitsanzeiger *m*
hygroscopic hygroskopisch, wasseranziehend
hygroscopic capacity Hygroskopizität *f*
hygroscopic coefficient Hygroskopizitätszahl *f*
hygroscopic instability Feuchtlabilität *f*
hygroscopicity Hygroskopizität *f*,
Wasseranziehungsvermögen *n*,
Wasseraufnahmefähigkeit *f*
hygroscopic quality
Wasseraufsaugungsvermögen *n*
hygroscopy Hygroskopie *f*
hygrostat Feuchtigkeitsregler *m*, Hygrostat *m*
hylotropic hylotrop
hyocholic acid Hyocholsäure *f*
hyoscine Hyoscin *n*
hyoscine hydrobromide Hyoscinbromhydrat *n*,
Hyoscinhydrobromid *n*
hyoscine sulfate Hyoscinsulfat *n*
hyoscyamine Hyoscyamin *n*, Daturin *n*
hyoscyamine hydrobromide
Hyoscyaminhydrobromid *n*
hyoscyamine hydrochloride
Hyoscyaminchlorhydrat *n*
hyoscyamine sulfate Hyoscyaminsulfat *n*
hypaphorine Hypaphorin *n*
hypargyrite Hypargyrit *m* (Min)
hyperacidity Hyperazidität *f* (Med),
Übersäuerung *f* (Med), **~ of the stomach**
Magensäureüberproduktion *f* (Med)
hyperbilirubin[a]emia erhöhter Bilirubingehalt im
Blut
hyperbilirubinuria erhöhter Bilirubingehalt im
Urin
hyperbola Hyperbel *f*, **rectangular ~**
gleichseitige Hyperbel
hyperbolic hyperbolisch
hyperbolic function Hyperbelfunktion *f* (Math),
inverse ~ Area-Funktion *f* (Math)
hyperbolic geometry Absolutgeometrie *f*
hyperbolic inverse Hyperbelinverse *f*
hyperbolic orbit Hyperbelbahn *f*
hyperboloid Hyperboloid *n*
hypercalcemia Hypercalcämie *f* (Med)
hyperchrome hyperchrom
hyperchromicity Hyperchromie *f* (Med)
hyperchromism Hyperchromie *f* (Med)
hypercomplex hyperkomplex
hyperconjugation Hyperkonjugation *f*
hyperemia Hyperämie *f* (Med)
hypereutectic hypereutektisch, übereutektisch
hypereutectoid Übereutektoid *n*
hyperfiltration Hyperfiltration *f*
hyperfine hyperfein

hyperfine spectrum Hyperfeinspektrum *n*
hyperfine structure Hyperfeinstruktur *f*,
Hyperfeinaufspaltung *f*
hyperfine structure coupling
Hyperfeinstrukturkopplung *f*
hyperfine structure multiplet
Hyperfeinstrukturmultiplett *n*
hyperfine structure operator
Hyperfeinstrukturoperator *m*
hyperfunction Überfunktion *f*
hypergeometric hypergeometrisch
hypergeometric distribution hypergeometrische
Verteilung *f*
hyperglyc[a]emia Blutzuckeranstieg *m*,
Hyperglykämie *f*,
Blutzuckererhöhung *f*
hypergol Hypergol *n*
hyperhypophysism Hypophysenüberfunktion *f*
(Med)
hypericin Hypericin *n*
hyperin Hyperin *n*
hyperinsulinism Hyperinsulinismus *m* (Med)
hypermetabolism erhöhter Stoffwechsel *m* (Med)
hypernucleus Hyperkern *m* (Atom)
hyperol Hyperol *n*
hyperon Hyperon *n* (Atom)
hyperpituitarism Hypophysenüberfunktion *f*
(Med)
hypersensitive überempfindlich
hypersensitiveness Überempfindlichkeit *f*
hypersensitization Hypersensibilisierung *f* (Phot)
hypersthene Hypersthen *m* (Min),
Labradorblende *f* (Min)
hypersthene rock Hypersthenfels *m* (Min)
hypersthenic hypersthenhaltig
hypersurface Hyperfläche *f* (Math)
hypersynchronous übersynchron
hypertension Hypertension *f* (Med),
Hypertonie *f* (Med)
hyperthyreosis Hyperthyreose *f* (Med)
hyperthyroidism Schilddrüsenüberfunktion *f*
(Med)
hypertonic hypertonisch (Pharm)
hypervitaminosis Hypervitaminose *f* (Med),
Vitaminüberschuß *m*
hyphomycetes Fadenpilze *pl*, Hyphomyzeten *pl*
hypinosis Fibrinmangel *m* (Med)
hypnal Hypnal *n*
hypnone Hypnon *n*
hypnotic Betäubungsmittel *n*, Hypnotikum *n*,
Schlafmittel *n* (Pharm)
hypnotic *a* hypnotisch
hypoacidity Säuremangel *m*
hypoactivity Unterfunktion *f* (Med)
hypobilirubin[a]emia Bilirubinmangel *m* (Med)
hypobromite Hypobromit *n*
hypobromous acid unterbromige Säure *f*, **salt of**
~ Hypobromit *n*
hypocalcia Kalkmangel *m* (Med)

hypochlorite Hypochlorit *n*
hypochlorite solution Hypochloritlösung *f*
hypochlorous unterchlorig
hypochlorous acid unterchlorige Säure *f*,
    Bleichsäure *f*, Hypochlorsäure *f*, salt of ~
    Hypochlorit *n*
hypochlorous anhydride Chlormonoxid *n*
hypochromicity Hypochromie *f* (Med)
hypodermic needle Injektionsnadel *f*
hypodermic tablets Tablettenimplantate *pl*
hypodermis Unterhautgewebe *n* (Med)
hypoeutectic Untereutektikum *n*
hypoeutectic *a* untereutektisch, untereutektoid
hypoferric an[a]emia Eisenmangelanämie *f*
hypofibrinogen[a]emia Fibrinogenmangel *m*
hypofunction Unterfunktion *f* (Med)
hypogaeic acid Hypogäasäure *f*, Physetolsäure *f*
hypogallic acid Hypogallussäure *f*
hypogene hypogen
hypogiyc[a]emia Hypoglykämie *f* (Med)
hypohormonal hormonarm
hypo-immunity Immunitätsschwäche *f*
hypoinsulinism Insulinmangel *m*,
    Insulinmangelkrankheit *f* (Med)
hypoiodite Hypojodit *n*
hypoiodous acid, salt of ~ Hypojodit *n*
hypoleucocytosis Leukozytenarmut *f* (Med)
hypoliposis Fettmangel *m*
hyponitrite Hyponitrit *n*
hyponitrous acid untersalpetrige Säure *f*, salt of
    ~ Hyponitrit *n*
hypophosphite Hypophosphit *n*
hypophosphoric acid Unterphosphorsäure *f*
hypophosphorous acid unterphosphorige Säure *f*,
    salt or ester of ~ Hypophosphit *n*
hypophysis Hypophyse *f*
hyposideremia Hyposiderämie *f* (Med)
hyposterol Hyposterol *n*
hyposulfite Hyposulfit *n*, Dithionit *n*
hyposulfurous acid Dithionigsäure *f*
hypotaurine Hypotaurin *n*
hypotension Hypotension *f* (Med)
hypotensive blutdrucksenkend
hypotensor blutdrucksenkendes Mittel *n*
hypotenuse Hypotenuse *f* (Math)
hypothalamus Hypothalamus *m* (Med)
hypothermia Hypothermie *f* (Biol)
hypothesis Hypothese *f*, Annahme *f*,
    Lehrmeinung *f*, ~ of compound nucleus
    Sandsackmodell *n* des Atomkerns
hypothetic[al] angenommen, hypothetisch
hypothyroidism Schilddrüsenunterfunktion *f*
hypothyrosis Schilddrüsenunterfunktion *f* (Med)
hypotrochoid Hypotrochoide *f* (Math)
hypovitaminosis Avitaminose *f* (Med),
    Hypovitaminose *f* (Med), Vitaminmangel *m*
    (Med)
hypoxanthine Hypoxanthin *n*, Sarkin *n*

hypoxanthine riboside Inosin *n*
hypsochromic hypsochrom
hypsometer Höhenmesser *m*, Hypsometer *n*
hyptolide Hyptolid *n*
hyraldite Hyraldit *n*
hyrgol Hyrgol *n*
hyssop Ysop *m* (Bot)
hyssop oil Ysopöl *n*
hystazarin Hystazarin *n*
hysteresis Hysterese *f*, Hysteresis *f*, coefficient of
    ~ Hysteresekoeffizient *m*, dielectric ~
    dielektrische Hysterese, dielektrische
    Nachwirkung, magnetic ~ magnetische
    Hysterese, ~ of a counter Alterung *f* eines
    Zählrohrs
hysteresis coefficient Hysteresekoeffizient *m*
hysteresis cycle Hystereseschleife *f*
hysteresis distortion durch den Hystereseeffekt
    verursachte Verzerrung (Magn)
hysteresis energy Hysteresearbeit *f* (Magn)
hysteresis loop Hystereseschleife *f*
hysteresis loss Hystereseverlust *m*
hythizine Hythizin *n*

I

iatrochemical iatrochemisch
iatrochemistry Iatrochemie *f*
iatrol Iatrol *n*
iba-acid Iba-Säure *f*
I-beam Doppel-T-Träger *m*
iberin Iberin *n*
iberite Iberit *m* (Min)
ibervirin Ibervirin *n*
ibit Ibit *n*
iboquine Ibochin *n*
ice Eis *n*, **artificial** ~ Kunsteis *n*, **manufactured**
  ~ Kunsteis *n*, **to cover with** ~ vereisen, mit
  Eis überdecken, **to turn into** ~ vereisen
ice *v* glasieren, vereisen
ice blocks Blockeis *m*
ice calorimeter Eiskalorimeter *n*
ice-cold eiskalt
ice-cooled eisgekühlt
ice cream Eiscreme *f*, Eiskreme *f*, Speiseeis *n*
ice crystal Eiskristall *m*
ice formation Eisbildung *f*, Vereisung *f* (Geol)
ice-free eisfrei, vereisungsfrei
ice funnel Eistrichter *m*
iceland moss Gallertmoos *n* (Bot), isländische
  Flechte *f* (Bot)
iceland spar [isländischer] Doppelspat *m* (Min),
  Kalkspat *m* (Min)
icelike eisähnlich, eisartig
ice-making Eisbereitung *f*
ice [-making] machine Eismaschine *f*
ice point Gefrierpunkt *m* (Wasser)
ice production Eiserzeugung *f*
ice target Auffänger *m* aus Eis
ice zone Eiszone *f*
ice zone melting Eiszonenschmelzen *n*
ichiba acid Ichibasäure *f*
ichnography Ichnographie *f*
ichthalbin Ichthalbin *n*, Ichthyoleiweiß *n*
ichthargan Ichthargan *n*
ichthulin Ichthulin *n*
ichthyocolla Fischleim *m*, Ichthyocoll *n*
ichthyoform Ichthyoform *n*
ichthyol Ichthyol *n*, Ammoniumsulfichthyolat *n*
ichthyol albuminate Ichthalbin *n*,
  Ichthyoleiweiß *n*
ichthyol formaldehyde Ichthyoform *n*
ichthyolite Ichthyolit *n*
ichthyol powder Ichthyolpuder *m*
ichthyol preparation Ichthyolpräparat *n*
ichthyol silver Ichthargan *n*
ichthyol soap Ichthyolseife *f*
ichthyol substitute Ichthyolersatz *m*
ichthyopterin Ichthyopterin *n*
icicane Icican *n*
icicle Eiszapfen *m*
icing Zuckerguß *m*

icing sugar Puderzucker *m*
iconometer Ikonometer *n*
iconoscopy Ikonoskopie *f*
icosahedral ikosaedrisch (Krist), zwanzigflächig
  (Krist)
icosahedron Ikosaeder *n* (Krist),
  Zwanzigflächner *m* (Krist)
icositetrahedral vierundzwanzigflächig
icositetrahedron Ikositetraeder *m* (Krist),
  Vierundzwanzigflächner *m* (Krist)
icthiamine Icthiamin *n*
icy eisig
idaein Idaein *n*
idaric acid Idozuckersäure *f*
iddingsite Iddingsit *m* (Min)
idea Einfall *m*, Idee *f*
ideal ideal
ideal crystal Idealkristall *m*
ideal state Idealzustand *m*
identic artgleich
identical identisch
identification Erkennung *f*, Identifizierung *f*,
  Kennzeichnung *f*
identification apparatus Kenngerät *n*
identification card Kennkarte *f*,
  Personalausweis *m*
identification limit Erfassungsgrenze *f*
identification number Kennummer *f*
identification tag Erkennungsmarke *f*
identify *v* identifizieren, erkennen
identity Gleichheit *f*, Identität *f*
identity operator Identitätsoperator *m*
identity period Identitätsabstand *m* (Krist)
idiochromasy Idiochromasie *f*
idiochromatic idiochromatisch
idioelectric idioelektrisch, selbstelektrisch
idiokinetic idiokinetisch
idiomere Idiomer *n*
idiomorphic idiomorph, eigengestaltig (Krist)
idiophanism Idiophanismus *m* (Krist)
idiophanous idiophan (Krist)
idioplasm Erbplasma *n*, Idioplasma *n*
idite Idit *m*
iditol Idit *m*
idle admittance Leerlaufadmittanz *f*
idle bush Leerlaufbuchse *f*
idle current Blindstrom *m*
idle friction Leerlaufreibung *f*
idle motion Leergang *m*, Leerlauf *m*
idle nozzle Leerlaufdüse *f*
idler Leerlaufrolle *f*, Spannrolle *f*, Stützrolle *f*,
  Tragrolle *f*, Tragscheibe *f*, Umlenktrommel *f*
idler arm Hilfslenker *m*
idle running Leergang *m*
idle stroke Leertakt *m* (Motor)
idle valve Leerlaufventil *n*

idle wheel Leerlaufrolle *f*
idling Leerlauf *m*
idling speed Leerlaufdrehzahl *f*
idocrase Idokras *m* (Min)
idonic acid Idonsäure *f*
idosaccharic acid Idozuckersäure *f*
idosaminic acid Idosaminsäure *f*
idose Idose *f*
IDP (Abk) Inosindiphosphat *n*
idrialite Idrialit *m* (Min)
iduronic acid Iduronsäure *f*
igasuric acid Igasursäure *f*
I-girder Doppel-T-Träger *m*
iglesiasite Iglesiasit *m*
igneous feurig, glühend
igneous dikes or veins pl (Geol)
igneous rocks Erstarrungsgestein *n* (Geol),
  Eruptivgestein *n* (Geol)
ignitability Entflammbarkeit *f*, Entzündbarkeit *f*
ignitable entflammbar, entzündbar,
  feuerfangend
ignite *v* [an]zünden, [sich] entzünden, zünden
igniter Zünder *m*, Anzünder *m*,
  Zündvorrichtung *f*
ignitibility Entzündbarkeit *f*, Zündfähigkeit *f*
ignitible entzündbar, zündfähig
igniting Anzünden *n*, Entzünden *n*, Zünden *n*
igniting agent Zündstoff *m*
igniting flame Zündflämmchen *n*, Zündflamme *f*
igniting power Zündkraft *f*
ignition Entzündung *f*, Zündmittel *n*, Zündung *f*
  (Auto, Elektr), Zündvorgang *m*, **adjustment of**
  ~ Zündungseinstellung *f*,
  Zündungsregulierung *f*, **advanced** ~
  Frühzündung *f*, **condition for** ~
  Zündbedingung *f*, **early** ~ Frühzündung *f*,
  late ~ Spätzündung *f*, verzögerte Zündung,
  low-tension ~ Niederspannungszündung *f*,
  retarded ~ Nachzündung *f*, Spätzündung *f*,
  spontaneous ~ Selbstentzündung *f*, **time of** ~
  Zündmoment *m*, Zündzeitpunkt *m*, **to adjust**
  the ~ die Zündung *f* einstellen, **type of** ~
  Zündungsart *f*, ~ **by compression**
  Kompressionszündung *f*, ~ **by low tension**
  fuses Glühzündung *f*
ignition accumulator Zündakkumulator *m*
  (Elektr)
ignition anode Zündanode *f*
ignition apparatus Zündapparat *m*
ignition battery Zündbatterie *f*
ignition cable Zündkabel *n*
ignition cable lacquer Zündkabellack *m*
ignition capsule Glühschälchen *n*
ignition circuit Zündungsstromkreis *m* (Elektr)
ignition coil Zündspule *f*
ignition compound Zündmasse *f*, Zündmittel *n*
ignition delay Zündverzug *m*
ignition device Zündeinrichtung *f*
ignition electrode Zündelektrode *f*

ignition flame passage Zündflammenleitung *f*
ignition heat Zündwärme *f*
ignition interference Zündstörung *f*
ignition key Zündschlüssel *m* (Kfz)
ignition lag Zündverzug *m*
ignition lead (Br. E.) Zündkabel *n* (Kfz)
ignition mixture Entzündungsgemisch *n*,
  Zündmasse *f*
ignition pipe Zündleitung *f*
ignition point Entzündungspunkt *m*,
  Zündtemperatur *f*
ignition point tester Brennpunktprüfer *m*
ignition quality Zündfähigkeit *f*
ignition residue Glührückstand *m*
ignition retardation Zündverzug *m*
ignition sequence Zündfolge *f*
ignition spark Zündfunken *m*
ignition switch Zündschalter *m*
ignition temperature Entflammungspunkt *m*,
  Entzündungspunkt *m*, Zündtemperatur *f*
ignition test Zündprobe *f*
ignition threshold Zündspannung *f*
ignition time Zündungszeit *f*
ignition velocity determination
  Zündgeschwindigkeitsbestimmung *f*
ignitron Ignitron *n*
ignorant unwissend
ignotine Ignotin *n*
ihleite Ihleit *m* (Min)
iletin (HN) Insulin *n*
ilexanthin Ilexanthin *n*
ilicin Ilicin *n*
ilicyl alcohol Ilicylalkohol *m*
ill krank, schlecht
illipe butter Illipebutter *f*
illipe fat Illipefett *n*
illipe oil Illipeöl *n*
illium Illium *n* (Leg)
ill-natured bösartig
illness Erkrankung *f*, Krankheit *f*
ill-smelling übelriechend
illuminant Beleuchtungsmittel *n*, Lichtquelle *f*
illuminate *v* beleuchten, anstrahlen, aufhellen,
  aufklären
illuminated manometer Leuchtmanometer *n*
illuminated period Leuchtdauer *f*
illuminated sign Lichtzeichen *n*
illuminating leuchtend, lichtgebend
illuminating alcohol Leuchtspiritus *m*
illuminating effect Leuchtwirkung *f*
illuminating equipment
  Beleuchtungseinrichtung *f*
illuminating gas Leuchtgas *n*, ~ **from coal**
  Steinkohlenleuchtgas *n*
illuminating lens Beleuchtungslinse *f*
illuminating power Leuchtkraft *f*
illuminating value Leuchtwert *m*

illumination Beleuchtung *f*, Aufhellung *f*, **unit of ~** Beleuchtungsintensität *f*, **vertical ~** Auflicht *n*
illuminator Leuchtkörper *m*, **monochromatic ~** monochromatische Lichtquelle *f* (Opt)
illustrate *v* illustrieren, veranschaulichen, abbilden
illustration Abbildung *f*, Erläuterung *f*, Illustration *f*, Veranschaulichung *f*
illustrative material Anschauungsmaterial *n*
ilmenite Ilmenit *m* (Min), Geikielith *m* (Min), Titaneisenerz *n* (Min)
ilmenium Ilmenium *n*
ilmenorutile schwarzes Titanoxid *n* (Min), Ilmenorutil *m* (Min)
ilsemannite Ilsemannit *m* (Min)
ilvaite Ilvait *m* (Min), Kieselkalkeisen *n* (Min), Yenit *m* (Min)
imabenzil Imabenzil *n*
image Bild *n*, Abbild *n*, **center of ~** Bildmittelpunkt *m*, **inverted ~** Umkehrbild *n*, **latent ~** latentes Bild *n*, **reflected ~** Spiegelbild *n*, **reversed ~** Umkehrbild *n*
image aberration Bildfehler *m*
image dissector Bildzerleger *m* (TV)
image distortion Bildverzerrung *f*
image equation Abbildungsgleichung *f*
image erection Bildaufrichtung *f*
image fault Abbildungsfehler *m* (Opt)
image field Bildfeld *n* (Opt)
image focus[s]ing Bildeinstellung *f*
image recording Bildaufzeichnung *f*
image sharpness Bildschärfe *f* (Phot)
imaginary imaginär, fiktiv
imaginary experiment Gedankenexperiment *n*
imaginary part (of a complex number) Imaginärteil *m* (einer komplexen Zahl)
imaginative phantasievoll, erfinderisch
imasatin Imasatin *n*
imasatinic acid Imasatinsäure *f*
imbalance Unausgeglichenheit *f*, Ungleichgewicht *n*
imbed *v* einbetten
imbibe *v* einsaugen
imbricaric acid Imbricarsäure *f*
imerinite Imerinit *m* (Min)
imesatin Imesatin *n*
Imhoff tank Emscherbrunnen *m*
imidazole Imidazol *n*
imidazole base Imidazolbase *f*
imidazoledione Hydantoin *n*
imidazole group Imidazolgruppe *f*
imidazole ring Imidazolring *m*
imidazoletrione Parabansäure *f*
imidazolidine Imidazolidin *n*
imidazoline Imidazolin *n*
imidazolone Imidazolon *n*
imid[e] Imid *n*
imide base Imidbase *f*

imidocarbamide Guanidin *n*, Imidocarbamid *n*
imidocarbonic acid Imidokohlensäure *f*
imido compound Imidoverbindung *f*
imido ester Imidsäureester *m*
imidogen Imidogruppe *f*
imido group Imidogruppe *f*
imidosulfonic acid Imidosulfonsäure *f*
imidoxanthin Imidoxanthin *n*
iminazolone Imidazolon *n*
imine Imin *n*
imine formation Iminbildung *f*
imino-base sekundäres Amin *n*
imino cyclooctane, 1,5- ~ Granatanin *n*
iminodiacetic acid Iminodiessigsäure *f*
iminodibenzyl Iminodibenzyl *n*
iminodiformic acid Iminodiameisensäure *f*
iminodipropionic acid Iminodipropionsäure *f*
iminoindigo Iminoindigo *m*
iminooxindole, 3- ~ Imesatin *n*
iminopyrine Iminopyrin *n*
imipramine Imipramin *n*
imitate *v* imitieren, nachahmen, imitieren, nachbilden
imitated imitiert, nachgeahmt, falsch, künstlich
imitation Imitation *f*, Nachahmung *f*, Nachbildung *f*
imitation gold Scheingold *n*, Halbgold *n*, Schaumgold *n*
imitation leather Lederimitation *f*
imitation morocco Saffianimitation *f*
imitation silver foil Metallsilber *n*
immaterial stofflos
immature unreif, unentwickelt
immeasurable unermeßlich, unmeßbar
immedial pure blue Immedialreinblau *n*
immedial yellow Immedialgelb *n*
immediate effect Sofortwirkung *f*
immense unermeßlich
immense number Unzahl *f*
immerse *v* eintauchen, einbetten
immersion Eintauchen *n*, Versenken *n*, **depth of ~** Eintauchtiefe *f*
immersion battery Tauchbatterie *f*
immersion chain Eintauchkette *f*
immersion condenser Einhängekühler *m*, Immersionskondensor *m* (Opt)
immersion filter Eintauchnutsche *f*
immersion fluid Immersionsflüssigkeit *f*
immersion heater Tauchsieder *m*
immersion measuring cell Eintauchmeßzelle *f*
immersion method Immersionsmethode *f*, Eintauchverfahren *n*
immersion oil Immersionsöl *n*
immersion pump Tauchpumpe *f*
immersion roll Tauchwalze *f*
immersion stem Tauchrohr *n*
immersion test Tauchversuch *m*
immersion time Eintauchzeit *f*

immiscibility Nichtmischbarkeit *f,*
Unvermischbarkeit *f*
immiscible nicht mischbar, unvermischbar
immobile unbeweglich
immovability Bewegungslosigkeit *f*
immovable unbeweglich, fest
immune immun, unempfänglich
immune body Immunkörper *m*
immune globuline Immunoglobulin *n*
immune protein Immunprotein *n*
immune response Immunoreaktion *f*
immunity Immunität *f,* Resistenz *f,* **acquired ~**
erworbene Immunität, **active ~** aktive
Immunität, **antibacterial ~** Immunität *f*
gegen Bakterien, **antimicrobic ~** Immunität *f*
gegen Bakterien, **antitoxic ~** Giftimmunität *f,*
congenital ~ angeborene Immunität, **inherent**
~ angeborene Immunität, **inherited ~** ererbte
Immunität, **permanent ~** bleibende
Immunität
immunization Schutzimpfung *f* (Med), **passive ~**
passive Immunisierung
immunization procedure
Immunisierungsverfahren *n*
immunize *v* immunisieren
immunizing unit Immun[isierungs]einheit *f*
immunobiochemical immunbiochemisch
immunobiological immunbiologisch
immunobiology Immunbiologie *f*
immunochemistry Immunchemie *f*
immuno-electrophoresis Immunelektrophorese *f*
immunoglobulin Immunoglobulin *n*
immunologic immunologisch, serologisch
immunological effect Immunisierungseffekt *m*
immunologist Immunologe *m*
immunology Immunitätsforschung *f,*
Immunitätslehre *f,* Immunologie *f*
immunosuppressives Immunsuppressiva *pl*
impact Aufprall *m,* Anprall *m,* Aufschlag *m,*
Prall *m,* Schlag *m,* Stoß *m,* Wucht *f,* **angle of**
~ Auftreffwinkel *m* (Mech), **energy of ~**
Auftreffenergie *f,* **resistance to ~**
Schlagwiderstand *m,* **velocity of ~**
Aufschlaggeschwindigkeit *f*
impact ball hardness Fallhärte *f*
impact ball hardness test Fallhärteprüfung *f*
impact ball hardness tester Fallhärteprüfer *m*
impact bending strength Schlagbiegefestigkeit *f*
impact bending test Schlagbiegeprobe *f,*
Schlagbiegeversuch *m*
impact break Stoßbruch *m*
impact broadening Stoßverbreiterung *f*
impact bruise Stoßverletzung *f*
impact cleaving Schlagspaltung *f*
impact compression test Schlagdruckversuch *m*
impact crusher Prallbrecher *m,* Schlagbrecher *m*
impact crushing Prallzerkleinerung *f*
impact detonator Aufschlagzünder *m*
impact elasticity Schlagelastizität *f*

impact energy Schlagarbeit *f*
impact experiment Stoßversuch *m*
impact failure Stoßdefekt *m*
impact fluorescence Stoßfluoreszenz *f* (Opt)
impact force Stoßkraft *f*
impact hardness Schlaghärte *f*
impact hardness test Schlaghärteprüfung *f*
impact hardness tester Schlaghärteprüfer *m*
impact ionization Stoßionisation *f*
impact loading Prallbeanspruchung *f*
impact mill Prallmühle *f*
impact mo[u]lding Schlagpreßverfahren *n*
impact noise insulation Trittschalldämmung *f*
impactor nozzle Pralldüse *f*
impact parameter Stoßparameter *m*
impact penetration test Durchstoßversuch *m*
impact polarization Stoßpolarisation *f*
impact pulling test Schlagzerreißversuch *m*
impact pulverizer Prallzerkleinerungsmaschine *f,*
Stoßmühle *f*
impact resilience Stoßelastizität *f*
impact resistance Schlagfestigkeit *f*
impact-resistant schlagfest
impact strength Schlagfestigkeit *f,*
Stoßfestigkeit *f*
impact stress Schlagbeanspruchung *f*
impact tearing test Schlagzerreißversuch *m*
impact tensile stress Schlagzugbeanspruchung *f*
impact tensile test Schlagzugversuch *m*
impact tension test Schlagzerreißversuch *m,*
Zugversuch *m* mit Schlagbeanspruchung
impact test Schlagprobe *f,* Fallversuch *m,*
continuous ~ Dauerschlagprobe *f*
impact tester Schlagprüfgerät *n*
impact value Kerbzähigkeit *f*
impact [wheel] mixer Turbomischer *m*
impair *v* beeinträchtigen, beschädigen,
verschlechtern
impaired function Unterfunktion *f* (Med)
impairment Benachteiligung *f,* Schädigung *f*
impalpable unfühlbar, ungreifbar
impart *v* verleihen, gewähren
impartial neutral, unparteiisch
impedance Impedanz *f* (Elektr),
Scheinwiderstand *m* (Elektr)
impedance coil Drosselspule *f*
impedance converter Impedanzwandler *m*
impedance-matching circuit link
Übertragungsglied *n*
impede *v* behindern
impediment Behinderung *f,* Abhaltung *f,*
Hindernis *n*
impeller Kreiselrührer *m,* Kreiselmischer *m,*
Kreiselrad *n,* Laufrad *n,* Schaufelrad *n,*
Zellenrad *n*
impeller blade Laufradflügel *m*
impeller breaker [mill] Pralltellermühle *f,*
Schneidmühle *f*

impeller mixer Impellermischer *m*,
Kreiselmischer *m*
impenetrability Undurchdringlichkeit *f*
impenetrable undurchdringlich, undurchlässig
imperatorin Imperatorin *n*, Peucedanin *n*
imperfect fehlerhaft, mangelhaft, unvollständig
imperfect crystal Realkristall *m*
imperfection Fehlerhaftigkeit *f*, Fehlstelle *f*
(Krist), Mangel *m*, ~ **of a crystal**
Kristallfehler *m*
imperial blue Imperialblau *n*
imperial green Kaisergrün *n*
imperial red Kaiserrot *n*
imperil *v* gefährden
impermeability Dichtigkeitseigenschaften *pl*,
Impermeabilität *f*, Undurchdringlichkeit *f*,
Undurchlässigkeit *f*, ~ **to air**
Luftundurchlässigkeit *f*
impermeability to gas Gasdichtigkeit *f*
impermeabilization Wasserdichtmachen *n*
impermeable undurchdringlich, undurchlässig
impervious undurchdringlich, undurchlässig, ~
to water wasserundurchlässig
imperviousness Undurchlässigkeit *f*, ~ **to water**
Wasserundurchlässigkeit *f*
impetuous heftig, tobend
impetus Anstoß *m*, Stoßkraft *f*
impinge *v* anprallen, zusammenstoßen, ~
[against] anstoßen
impingement Anprall *m*, Zusammenstoß *m*
implant Einlage *f* (Med), Implantat *n* (Med)
implausible unwahrscheinlich
implement Werkzeug *n*, Gerät *n*, Hilfsmittel *n*
implement finish Gerätelack *m*
implode *v* implodieren
implosion Implosion *f*
imply *v* bedeuten
imponderability Unwägbarkeit *f*
imponderable unwägbar, gewichtslos
imponderables Imponderabilien *f pl* (Phys)
importance Bedeutung *f*, Belang *m*, Wichtigkeit *f*
important wichtig, bedeutend, bedeutsam, **highly**
~ hochwichtig
imposing bedeutsam, eindrucksvoll
impoverish *v* verarmen
impoverishment Verarmung *f*, ~ **of acid**
Säureverarmung *f*
impracticable unausführbar, undurchführbar,
unbrauchbar
impractical unpraktisch, unhandlich
impregnate *v* imprägnieren, befruchten (Biol),
durchtränken, sättigen, tränken, ~ **with sulfur**
abschwefeln, ~ **wood with zinc chloride**
burnettisieren
impregnating Durchdringen *n*, Durchtränken *n*
impregnating agent Imprägnier[ungs]mittel *n*
impregnating bath Tränkbad *n*
impregnating calender Gummierkalander *m*

impregnating equipment Imprägnierapparate *m*
pl
impregnating fluid Imprägnierflüssigkeit *f*
impregnating liquid Imprägnierflüssigkeit *f*,
Tränkmasse *f*
impregnating machine Imprägniermaschine *f*
impregnating material Tränkmasse *f*
impregnating oil Imprägnieröl *n*
impregnating pan Imprägnierpfanne *f*
impregnating preparation Imprägniermittel *n*
impregnating substance Imprägnier[ungs]mittel *n*
impregnating tank Imprägnierkessel *m*
impregnating varnish Tränklack *m*
impregnating vat Imprägniertrog *m*,
Tränkbehälter *m*
impregnation Imprägnierung *f*, Durchdringung *f*,
Grundierung *f* (Malerei), Sättigung *f*,
Tränkung *f*, **flame-proof** ~ flammenfeste
Imprägnierung, **method of** ~
Imprägnierverfahren *n*, ~ **of wood**
Holzaufbereitung *f*, Holztränkung *f*
impregnation time Tränkungsdauer *f*
impregnation varnish Tränklack *m*
impress Abdruck *m*, Eindruck *m*
impress *v* aufdrucken, einprägen
impressed aufgedruckt
impression Eindruck *m*, Aufdruck *m*,
Matrizenhohlraum *m*, Schreibspur *f*,
Vertiefung *f*, **area of** ~ Eindruckfläche *f*,
depth of ~ Eindringungstiefe *f*
impression cylinder Druckzylinder *m* (Buchdr)
impression molding Kontaktpressen *n*
impression of a die Kontur *f* des Gesenks
imprint Aufdruck *m*, Eindruck *m*
imprint *v* aufdrucken, aufprägen, einprägen
improbability Unwahrscheinlichkeit *f*
improbable unwahrscheinlich
improper unecht (Math), ungeeignet
improper fraction unechter Bruch *m* (Math)
improve *v* verbessern, veredeln, verfeinern,
vergüten, vervollkommnen, vorraffinieren
(Blei), weiterverarbeiten (Text)
improved verbessert
improvement Verbesserung *f*, Veredelung *f*,
Verfeinerung *f*
improving Verbessern *n*, Veredeln *n*
improvise *v* extemporieren, improvisieren
improvised behelfsmäßig
impsonite Impsonit *m* (Min)
impulse Impuls *m*, Antrieb *m*, lineares
Moment *n* (Math), Reiz *m* (Med), Stoß *m*,
Stromstoß *m* (Elektr), **duration of** ~
Impulsdauer *f*, **setting in of** ~
Impulseinsatz *m*
impulse approximation Stoßapproximation *f*
impulse breakdown Stoßdurchschlag *m*
impulse centrifuge Freistrahlzentrifuge *f*
impulse delay Impulsverzögerung *f*
impulse discharge Stoßentladung *f*

impulse emission Impulsgabe *f*
impulse excitation Impulserregung *f,*
　　Stoßerregung *f*
impulse generator Impulsformer *m,*
　　Impulsgeber *m,* Wanderwellengenerator *m*
impulse response Impulsantwort *f* (Math)
impulse sequence Impulsfolge *f*
impulse turbine Aktionsturbine *f* (Mech),
　　Freistrahlturbine *f* (Mech),
　　Gleichdruckturbine *f* (Mech),
　　Pelton-Turbine *f* (Mech)
impulse wheel Aktionsturbine *f* (Mech),
　　Freistrahlturbine *f* (Mech), Gleichdruckrad *n*
　　(Mech), Gleichdruckturbine *f* (Mech),
　　Pelton-Turbine *f* (Mech)
impulse width or length Impulsbreite *f*
impulsing Impulsgabe *f*
impulsive force Triebkraft *f*
impure unrein (Chem)
impurities, interstitial ~
　　Einlagerungsfremdatome *n pl*
impurity Verunreinigung *f,* Begleitstoff *m,*
　　Fremdstoff *m*
impurity addition Fremdatomzusatz *m*
impurity band conduction Störbandleitung *f*
impurity effect Fremdstoffeinfluß *m*
impurity quenching Fremdausleuchtung *f*
imputrescibility Fäulnisbeständigkeit *f*
inability Unfähigkeit *f*
inaccuracy Fehler *m,* Ungenauigkeit *f,* **limit of** ~
　　Fehlergrenze *f*
inaccurate ungenau
inaction Bewegungslosigkeit *f*
inactivate *v* inaktivieren, passivieren
inactivation Inaktivierung *f*
inactive inaktiv, inert (Chem), reaktionsträge
　　(Chem), untätig, unwirksam, wirkungslos, **to**
　　render ~ inaktivieren, entaktivieren
inactivity Inaktivität *f,* Passivität *f,*
　　Reaktionsträgheit *f* (Chem), Trägheit *f*
　　(Chem), Unwirksamkeit *f,* Wirkungslosigkeit *f*
in advance im voraus
inanimate unbelebt, leblos
inappropriate unzweckmäßig, ungeeignet
inborn kongenital (Med)
inbreeding Inzucht *f*
incalculable unberechenbar, unbestimmbar
incandesce *v* glühen
incandescence Weißglühen *n,* Weißglut *f,* **zone of**
　　~ Brennzone *f*
incandescent weißglühend
incandescent body Glühkörper *m*
incandescent bulb Glühbirne *f*
incandescent burner Glühlichtbrenner *m*
incandescent cathode Glühkathode *f*
incandescent filament Glüh[lampen]faden *m*
incandescent lamp Glühlampe *f*
incandescent lamp photometer
　　Glühlampenphotometer *n*

incandescent light Glühlicht *n*
incandescent mantle Glühstrumpf *m*
incandescent valve Glühkathode *f*
incandescent welding Widerstandsschweißung *f*
incanine Incanin *n*
incaninic acid Incaninsäure *f*
incapacity Unfähigkeit *f*
incapsulation Abkapselung *f,* Einkapselung *f*
incarnatrin Incarnatrin *n*
incarnatyl alcohol Incarnatylalkohol *m*
incendiary bomb Brandbombe *f*
incendiary composition Brandsatz *m*
incense Weihrauch *m*
incense resin Weihrauchharz *n*
incentive Anreiz *m,* Ansporn *m*
inch Zoll *m* (Maß)
incidence Einfall *m* (Strahl), Eintritt *m* (Strahl),
　　angle of ~ Einfallswinkel *m,*
　　Auftreffwinkel *m,* Inzidenzwinkel *m,* **axis of**
　　~ Einfallslot *n*
incidence angle Einfallswinkel *m,*
　　Eintrittswinkel *m*
incidence plane Einfallsebene *f* (Opt)
incident einfallend
incident light Einfallicht *n* (Opt), Gegenlicht *n*
incinerate *v* einäschern, veraschen
incinerating Einäschern *n,* Veraschen *n*
incinerating capsule Glühschälchen *n*
incineration Einäscherung *f,* Veraschung *f,* ~ **of**
　　filter paper Einäschern *n* des Filters
incineration dish Veraschungsschale *f* (Chem)
incinerator Müllverbrennungsofen *m*
incipient anfänglich, beginnend
incipient crack Anbruch *m*
incipient fluidizing velocity (fluidized beds)
　　Lockerungsgeschwindigkeit *f*
incise *v* einschneiden
incision Einschnitt *m,* Schnitt *m*
incisive einschneidend
inclinable geneigt, schrägstellbar
inclination Neigung *f,* Inklination *f* (Magn),
　　Schräge *f,* Schrägstellung *f,* **angle of** ~
　　Inklinationswinkel *m,* Neigungswinkel *m,*
　　Steigungswinkel *m*
inclination balance Neigungswaage *f*
inclinatory needle Inklinationsnadel *f*
incline Neigung *f,* Neigungsebene *f*
incline *v* neigen, abschrägen, auslenken
inclined geneigt, abschüssig, schief, schräg
inclined arrangement Schräglage *f*
inclined belt Steigband *n*
inclined belt conveyor Schrägförderband *n*
inclined conveyer Schrägförderer *m*
inclined elevator Schrägaufzug *m*
inclined film Schrägfilm *m*
inclined hoist Schrägaufzug *m*
inclined lift Schrägaufzug *m*
inclined plane schiefe Ebene *f*
inclined slit Schrägspalt *m*

inclined track Schrägbahn *f*
inclined-tube evaporator
Schrägrohrverdampfer *m*
inclinometer Neigungsmesser *m*
include *v* beifügen, einschließen
included inbegriffen
inclusion Einschluß *m*, Inklusion *f*, ~ **of air**
Lufteinschluß *f*
inclusion complex Addukt *n*, Klathrat *n* (Chem)
inclusion compound Einschlußverbindung *f*
inclusion theorem Einschließungssatz *m*
inclusive inbegriffen
incoherence Inkohärenz *f*
incoherent inkohärent (Phys),
unzusammenhängend, zusammenhangslos
incombustibility Unverbrennbarkeit *f*
incombustible unverbrennbar, feuerfest
incomparable unvergleichbar
incompatibility Unverträglichkeit *f*,
Inkompatibilität *f*
incompatibility tensor Inkompatibilitätstensor *m*
incompatible gruppenfremd (Blut), typenfremd
(Blut), unverträglich
incomplete unvollständig, lückenhaft
incompletely burnt gas Schwelgas *n*
incompletely carbonized charcoal Rauchkohle *f*
incompleteness Halbheit *f*, Unvollständigkeit *f*
incompressibility Inkompressibilität *f*,
Raumbeständigkeit *f*,
Unzusammendrückbarkeit *f*
incompressible inkompressibel (Med),
unkomprimierbar, unverdichtbar
inconceivable undenkbar, unvorstellbar
incondensable unverdichtbar
incongruence Inkongruenz *f* (Stereochem),
Nichtübereinstimmung *f*
incongruent inkongruent (Math)
inconsiderable bedeutungslos, belanglos
inconsistency Inkonsequenz *f*, Sprunghaftigkeit *f*
inconsistent inkonsequent, widerspruchsvoll
inconstancy Unbeständigkeit *f*, Inkonstanz *f*,
Ungleichförmigkeit *f*, Ungleichmäßigkeit *f*
inconstant unbeständig, veränderlich,
inkonstant, wandelbar
inconvenient unbequem, ungünstig
incorporate *v* angliedern, einarbeiten, einbauen,
einverleiben, vereinigen, ~ **in plaster**
eingipsen
incorporation Angliederung *f* (Organisation),
Aufnahme *f*, Einverleibung *f*
incorrect fälschlich, falsch, unrichtig
incorrectness Fehlerhaftigkeit *f*, Unrichtigkeit *f*
incorrodible unkorrodierbar
incorruptible unverderblich
increase Anstieg *m*, Anwachsen *n*, Erhöhung *f*,
Steigerung *f*, Wachsen *n*, Wachstum *n*,
Zunahme *f*, Zuwachs *m*, ~ **in output or**
efficiency or performance
Leistungssteigerung *f*

increase *v* ansteigen, anschwellen, anwachsen,
heraufsetzen, steigen, wachsen, zunehmen
increased erhöht
increasing zunehmend
incredible undenkbar, unglaublich
increment Inkrement *n* (Math), Zuwachs *m*
incrust *v* inkrustieren (Min), mit einer Kruste
überziehen, verkrusten
incrustation Belag *m*, Inkrustation *f* (Med,
Geol), Inkrustierung *f*, Kesselstein *m* (Techn),
Kesselsteinablagerung *f* (Techn), Kruste *f*,
Krustenbildung *f*, Verkrustung *f*, ~ **near**
furnace top Gichtschwamm *m*
incrustation heat Sinterungshitze *f*
incubation Bebrütung *f*, Inkubation *f*
incubation time Inkubationszeit *f*
incubator Brutapparat *m*, Brutkasten *m*,
Brutschrank *m*, Wärmeschrank *m*
inculcate *v* einprägen
incurable unheilbar
indacene Indacen *n*
indamine Indamin *n*
indan Indan *n*, Hydrinden *n*
indanol Indanol *n*
indanone Indanon *n*, Hydrindon *n*
indanthrazine Indanthrazin *n*
indazine Indazin *n*
indazole Indazol *n*
indazole quinone Indazolchinon *n*
indecomposable unzerlegbar, unzersetzbar
indefinite unbegrenzt, unbeschränkt, unbestimmt
indefiniteness Unbestimmtheit *f*
indelible ink Kopiertinte *f*, Zeichentinte *f*
indene Inden *n*
indene resin Indenharz *n*
indent *v* einzahnen, [ein]kerben, einpressen,
einprägen
indentation Kerbe *f*, Auszackung *f*, Einbeulung *f*,
Eindruck *m*, Einkerbung *f*, **hardness by** ~
Eindruckhärte *f*
indentation cup Eindruckkalotte *f*
indentation hardness Eindruckhärte *f*,
Kugeldruckhärte *f*
indentation problem Kerbproblem *n*
indentation test Kerbschlagprobe *f*,
Kerbschlagversuch *m*
indentation value Kerbschlagfestigkeit *f*
indented gezahnt, ausgekerbt
indenting tool Eindruckstempel *m*
independence Unabhängigkeit *f*
independent selbständig, unabhängig
independent particle model Einteilchenmodell *n*
indestructibility Unzerstörbarkeit *f*
indestructible unzerstörbar, unzerreißbar
indeterminacy Unbestimmtheit *f*
indeterminacy principle
Unbestimmbarkeitsrelation *f* (Quant),
Unbestimmtheitsprinzip *n* (Quant)
indeterminate unbestimmt

index Index *m* (pl Indices), Kennziffer *f,*
  Register *n,* Tabelle *f*
index *v* mit einem Index versehen
index card Karteikarte *f*
index error Indexfehler *m*
indexing Indizierung *f*
indexing disk Schaltscheibe *f*
index value Gütezahl *f*
Indian berries Fischkörner *n pl*
Indian blue Indischblau *n*
Indian butter Gheebutter *f*
Indian corn Mais *m* (Bot)
Indian cress Kapuzinerkresse *f* (Bot)
Indian ink Ausziehtusche *f*
indianite Indianit *m* (Min)
Indian meal Maismehl *n*
Indian saffron Gelbwurzel *f* (Bot)
Indian stone Jaspis *m* (Min)
Indian wadding Kapok *m*
Indian yellow Indischgelb *n,* Azoflavin *n*
india rubber Gummi *m,* Kautschuk *m,*
  Radiergummi *m*
india rubber cement Gummikitt *m*
india rubber sheet Plattenkautschuk *m*
indican Indican *n*
indicate *v* anzeigen; andeuten, angeben,
  bedeuten, hinweisen
indicating agent Indikator *m*
indicating device Anzeigevorrichtung *f*
indicating instrument Anzeigegerät *n,*
  Anzeigeinstrument *n*
indicating pressure gauge Zeigermanometer *n*
indicating range Anzeigebereich *m*
indication Anzeichen *n,* Andeutung *f,* Angabe *f,*
  Anzeige *f,* Hinweis *m,* Indikation *f* (Med),
  error in ~ Falschweisung *f*
indication of direction Richtungsanzeige *f*
indication sensitivity Anzeigeempfindlichkeit *f*
indicator Anzeiger *m,* Indikator *m* (Chem),
  Indikatorinstrument *n,* Zeiger *m*
indicator cylinder Indikatorzylinder *m*
indicator diagram Indikatordiagramm *n*
indicator gauge Anzeigegerät *n*
indicator glow lamp Anzeigeglimmlampe *f*
indicator lag Anzeigeträgheit *f*
indicator paper Indikatorpapier *n* (Chem)
indicator pipe Indikatorrohr *n*
indicator piston Indikatorkolben *m*
indicator range Indikatorbereich *m*
indicator reading Zeigerablesung *f*
indicator signal Schauzeichen *n*
indicator spring Indikatorfeder *f*
indicatrix Indikatrix *f* (Math)
indices (pl. of index) Indizes *m pl*
indicolite Indikolit *m* (Min), blauer Turmalin *m*
  (Min)
indifferent gleichgültig, indifferent, neutral
indigestible unverdaulich, schwer verdaulich

indigo Indigo *m,* Indigoblau *n,* Indigotin *n,*
  halogenated ~ Halogenindigo *m,* **reduced** ~
  Indigoweiß *n,* **soluble** ~ Indigocarmin *n*
indigo bath Indigoküpe *f*
indigo blue Indigoblau *n,* Indigofarbstoff *m,*
  Indigotin *n,* Küpenblau *n*
indigo blue *a* indigoblau
indigo carmine Indigokarmin *n*
indigo copper Covellin *m* (Min),
  Kupferindigo *m* (Min)
indigo derivative Indigoderivat *n*
indigo disulfonic acid Indigodisulfonsäure *f*
indigo extract Indigoauszug *m,* Indigoextrakt *m*
indigo gluten Indigoleim *m*
indigoid Indigoid *n*
indigometer Indigogütemesser *m*
indigometry Indigomessung *f*
indigo plant Indigopflanze *f* (Bot)
indigo printing Indigodruck *m*
indigo purple Indigopurpur *m*
indigo red Indigorot *n,* Indirubin *n*
indigo salt Indigosalz *n*
indigo solution Indigolösung *f*
indigo substitute Indigoersatz *m*
indigosulfonic acid Indigosulfosäure *f,*
  Indigoschwefelsäure *f*
indigosulfuric acid Indigoschwefelsäure *f,*
  Indigosulfosäure *f*
indigo suspension Indigosuspension *f*
indigo synthesis Indigosynthese *f*
indigotate Indigosalz *n*
indigotin Indigotin *n,* Indigo *m,* Indigoblau *n*
indigo tincture Indigotinktur *f*
indigo white Indigweiß *n,* Leukoindigo *n*
indin Indin *n*
indirect indirekt
indirectly acting indirekt wirkend
indirectly fired indirekt beheizt
indirubin Indirubin *n,* Indigorot *n*
indispensability Unentbehrlichkeit *f*
indispensable unumgänglich
indistinct undeutlich, unscharf, **to become** ~
  verschwimmen (Opt)
indistinguishable gleichartig, nicht
  unterscheidbar
indium (Symb. In) Indium *n*
indium antimonide Indiumantimonid *n*
indium chelate Indiumchelat *n*
indium chloride Indiumchlorid *n*
indium compound Indiumverbindung *f*
indium content Indiumgehalt *m*
indium(II) oxide Indium(II)-oxid *n,*
  Indiummonoxid *n*
indium iodide Indiumjodid *n*
indium(I) oxide Indium(I)-oxid *n,*
  Indiumsuboxid *n* (obs)
indium monoxide Indium(II)-oxid *n,*
  Indiummonoxid *n*
indium selenide Indiumselenid *n*

indium sesquisulfide Indium(III)-sulfid *n*
indium suboxide Indium(I)-oxid *n,*
  Indiumsuboxid *n* (obs)
indium sulfate Indiumsulfat *n*
indium sulfide Indiumsulfid *n,* Schwefelindium *n*
indium telluride Indiumtellurid *n*
individual einzeln, individuell
individual absorption Eigenabsorption *f*
individual case Einzelfall *m*
individual efficiency Einzelwirkungsgrad *m*
individual excitation Eigenerregung *f*
individual frequency Eigenfrequenz *f*
individuality Eigenart *f,* Individualität *f*
individual[ly] individuell
individual motion Eigenbewegung *f*
individual stage Einzelschritt *m*
indivisibility Unteilbarkeit *f*
indivisible unteilbar, unzerlegbar
indoaniline Indoanilin *n*
indochromogen Indochromogen *n*
indoform Indoform *n*
indole Indol *n*
indole acetic acid Indolessigsäure *f,*
  Indolylessigsäure *f*
indolebutyric acid Indolbuttersäure *f*
indolenine Indolenin *n*
indolent faul, träge
indolepropionic acid Indolpropionsäure *f*
indoline Indolin *n*
indolinone Indolinon *n*
indolizidine Indolizidin *n*
indolizine Indolizin *n*
indolone Indolon *n*
indolyl Indolyl-
indolylalanine Tryptophan *n*
indone Hydrindon *n,* Indanon *n,* Indon *n*
indoor paint Innenanstrichfarbe *f*
indophenazine Indophenazin *n*
indophenine Indophenin *n*
indophenol Indophenol *n*
indophenol blue Indophenolblau *n*
indophenol oxidase Indophenoloxydase *f*
  (Biochem)
indoquinoline Indochinolin *n*
indoxyl Indoxyl *n*
indoxylic acid Indoxylsäure *f*
indoxylsulfuric acid Indoxylschwefelsäure *f*
induce *v* auslösen, erregen, induzieren
induced induziert, angefacht
induced coil Funkeninduktor *m*
induced current Induktionsstrom *m,*
  Sekundärstrom *m*
induced draft installation Saugzuganlage *f*
induced magnetization, coefficient of ~
  Suszeptibilität *f*
induced motor Asynchronmotor *m*
induced roll separator Walzenscheider *m*
induced voltage Sekundärspannung *f* (Elektr)
inducing Einleiten *n* (Med)

inducing current Primärstrom *m* (Elektr)
inductance Induktanz *f* (Magn), **mutual** ~
  Wechselinduktion *f*
induction Induktion *f,* Induzierung *f* (Elektr),
  coefficient of ~ Induktionskoeffizient *m,*
  heteropolar ~ Wechselpolinduktion *f,*
  homopolar ~ Gleichpolinduktion *f,* **law of** ~
  Induktionsgesetz *n,* **line of** ~
  Induktionslinie *f,* **magnetic** ~ magnetische
  Induktion, **mutual** ~ gegenseitige Induktion,
  Wechselinduktion *f,* **photochemical** ~
  photochemische Induktion, ~ **in air**
  Luftinduktion *f,* ~ **in iron** Eiseninduktion *f*
induction air Ansaugluft *f*
inductional induktiv (Elektr)
induction apparatus Induktionsapparat *m*
induction coil Funkeninduktor *m,*
  Induktionsrolle *f,* Induktionsspule *f*
induction current Induktionsstrom *m,*
  Erregerstrom *m,* Nebenstrom *m*
induction curve Induktionskurve *f*
induction electricity Induktionselektrizität *f*
induction flux Induktionsfluß *m*
induction-free induktionsfrei
induction furnace Induktionsofen *m*
induction hardening Hochfrequenzhärtung *f*
  (Techn)
induction heat Induktionswärme *f*
induction heating Erhitzung *f* durch Induktion,
  Induktionsheizung *f*
induction law Induktionsgesetz *n*
induction machine Induktionsmaschine *f*
induction motor Asynchronmotor *m,*
  Induktionsmotor *m*
induction period Induktionsperiode *f*
induction phenomenon Induktionserscheinung *f*
induction pressure Ladedruck *m* (Elektr)
inductive induktiv (Elektr)
inductive capacity Induktionskapazität *f,*
  Induktionsvermögen *n*
inductive circuit Induktionskreis *m* (Elektr)
inductively coupled induktiv gekoppelt
inductive resistance Impedanz *f*
inductivity Induktivität *f* (Elektr)
inductometer Induktometer *n*
inductophone Induktophon *n*
inductor Induktor *m,* Induktionsapparat *m,*
  Induktionsspule *f*
inductor wheel Induktorrad *n*
inductoscope Induktoskop *n*
induline Indulin *n*
induline scarlet Indulinscharlach *n*
indurate *v* [ver]härten; indurieren
industrial industriell, fabrikatorisch, gewerblich,
  technisch
industrial accident Betriebsunfall *m*
industrial chemicals Industriechemikalien *pl*
industrial chemist Betriebschemiker *m*
industrial disease Berufskrankheit *f*

industrial engineer Betriebstechniker *m*
industrial engineering Gewerbetechnik *f*
industrial federation Wirtschaftsverband *m*
industrial fittings Armaturen für Betriebe (pl)
industrial furnace Industrieofen *m*
industrial hygiene Gewerbehygiene *f*
industrialization Industrialisierung *f*
industrialize *v* industrialisieren
industrial monopoly Wirtschaftsmonopol *n*
industrial operation Industriebetrieb *m*
industrial plant Industrieanlage *f,*
   Industriebetrieb *m* (Fabrik)
industrial product Industrieerzeugnis *n*
industrial reactor Industriereaktor *m*
industrial research Betriebsforschung *f,*
   Industrieforschung *f*
industrial rubber Industriegummi *m*
industrial salt Gewerbesalz *n*
industrial scale großtechnischer Maßstab *m,* **on
   an** ~ großtechnisch
industrial specification sheet Normblatt *n*
industrial trucks Flurförderzeuge *pl*
industrial waste gas Industrieabgas *n*
industrial waste product Industrieabfallstoff *m*
industrial waste water Fabrikabwasser *n,*
   Industrieabwasser *n*
industrial worker Fabrikarbeiter *m*
industrious arbeitsam, betriebsam
industriousness Arbeitsamkeit *f,* Fleiß *m*
industry Industrie *f,* Gewerbe *n,* **branch of** ~
   Industriezweig *m,* **cosmetic** ~ kosmetische
   Industrie, **heavy** ~ Schwerindustrie *f,* **steel** ~
   Stahlindustrie *f*
industry standard Industrienorm *f*
inedible ungenießbar
ineffective unwirksam, wirkungslos
ineffectiveness Unwirksamkeit *f,*
   Wirkungslosigkeit *f*
inefficiency Wirkungslosigkeit *f*
inefficient unwirksam, wirkungslos
inelastic unelastisch
inequality Ungleichheit *f*
inequation Ungleichung *f* (Math)
inert edel (Gas), inaktiv (Chem), indifferent
   (Gas), inert (Gas), passiv (Chem),
   reaktionsträge (Chem), träge (Chem),
   wirkungslos
inertance Inertanz *f*
inert gas Schutzgas *n,* Edelgas *n,* Inertgas *n*
inert gas plant Schutzgasanlage *f*
inert gas shell Edelgasschale *f* (Chem)
inert gas structure Edelgasstruktur *f*
inertia Beharrungsvermögen *n,* Inertie *f,* träge
   Masse, Trägheit *f,* Trägheitsvermögen *n,* **axis
   of** ~ Trägheitsachse *f,* **center of** ~
   Trägheitsmittelpunkt *m,* **law of** ~
   Trägheitsgesetz *n,* **momentum of** ~
   Trägheitsmoment *n,* **state of** ~

   Beharrungszustand *m,* ~ **due to rotation**
   Drehungsträgheit *f*
inertia effect Beharrungswirkung *f*
inertiafree trägheitslos
inertialess trägheitslos
inertial force Trägheitskraft *f*
inertial resistance Trägheitswiderstand *m*
inertial wave Trägheitswelle *f*
inertia principle Beharrungsprinzip *n*
inert material Ballastmaterial *n*
inertness Passivität *f,* Reaktionsträgheit *f*
   (Chem), Trägheit *f* (Chem), **chemical** ~
   chemische Trägheit
inesite Inesit *m* (Min)
inexact ungenau, unscharf
inexhaustibility Unerschöpflichkeit *f*
inexhaustible unerschöpflich, unversiegbar
inexorable unerbittlich
inexpansible unausdehnbar
inexpedient ungeeignet, unzweckmäßig
inexpensive billig
inexperienced unerfahren
inexplicable unerklärbar, unerklärlich,
   unverständlich
inexplorable unerforschlich
inexplosive explosionssicher
inextensible unausdehnbar
inextinguishable unauslöschbar, unauslöschlich
infantile paralysis Kinderlähmung *f* (Med)
infect *v* anstecken, infizieren, infizieren,
   verseuchen
infection Ansteckung *f* (Med), Infektion *f*
   (Med), **center of** ~ Ansteckungsherd *m,*
   danger of ~ Ansteckungsgefahr *f,*
   Infektionsgefahr *f,* **focus of** ~
   Infektionsherd *m,* **reaction of** ~
   Infektreaktion *f,* **risk of** ~
   Ansteckungsgefahr *f* (Med),
   Infektionsgefahr *f,* **source of** ~
   Ansteckungsquelle *f,* Infektionsquelle *f*
infectious ansteckend (Med),
   krankheitsübertragend, übertragbar (Med)
infectious disease Infektionskrankheit *f*
infective ansteckend, infektiös
infective agent Infektionserreger *m*
inference Folgerung *f,* Rückschluß *m,* Schluß *m*
   (Math)
inferior minderwertig, ~ **in value** geringwertig,
   minderwertig
inferiority Inferiorität *f,* Minderwertigkeit *f*
inferior ore Pochgänge *m pl* (Bergb)
inferior quality Minderwertigkeit *f,* zweitklassige
   Qualität *f*
infestation Befall *m*
infilco process Infilco-Verfahren *n*
infiltrate *v* durchsetzen, [durch]tränken,
   eindringen, einsickern, infiltrieren
infiltrated air Falschluft *f*

infiltration Durchdringung *f,* Eindringen *n,*
Infiltration *f*
infinite grenzenlos, unbegrenzt, unendlich
infinitesimal unendlich klein
infinitesimal calculus Infinitesimalrechnung *f*
(Math)
infinity Unendlichkeit *f,* **to approach** ~
unendlich groß werden
infinity adjustment Unendlicheinstellung *f* (Opt)
inflame *v* entzünden, entflammen
inflammability Entzündlichkeit *f,* Brennbarkeit *f,*
Entzündbarkeit *f,* Feuergefährlichkeit *f*
inflammable entzündbar, brennbar,
entflammbar, feuergefährlich, [leicht]
entzündlich, **easily** ~ leicht entzündlich,
highly ~ leicht brennbar, leicht entzündlich,
spontaneously ~ selbstentzündlich
inflammable material Zündstoff *m*
inflammable ore Branderz *n*
inflammation Entflammung *f,* Entzündung *f*
(Med), **spontaneous** ~ Selbstentzündung *f*
inflammation tester Flammpunktprüfer *m*
inflammatory entzündlich (Med)
inflate *v* aufblähen, aufblasen, aufpumpen,
auftreiben
inflated aufgeblasen, luftgefüllt
inflation Aufblähung *f,* Aufblasen *n*
inflation sleeve Füllansatz *m*
inflation strength Schwellfestigkeit *f*
inflation tube Füllschlauch *m*
inflect *v* biegen, beugen
inflected gebogen, gebeugt
inflection Beugung *f,* Krümmung *f,* Wendung *f*
(Math), **point of** ~ Wendepunkt *m*
inflexibility Starre *f,* Unbiegsamkeit *f*
inflexible starr, steif, unbiegsam
inflexion Beugung *f,* Biegung *f,* Krümmung *f,*
Wendung *f* (Math)
inflow Zufluß *m*
inflow current Zulaufstrom *m*
inflow velocity Zuströmgeschwindigkeit *f*
influence Einfluß *m;* Beeinflussung *f,*
Einwirkung *f,* Influenz *f* (Elektr), **mutual** ~
gegenseitige Einwirkung, ~ **of air**
Lufteinwirkung *f,* ~ **of frost**
Frosteinwirkung *f*
influence *v* beeinflussen, einwirken
influenceable beeinflußbar
influence electricity Influenzelektrizität *f*
influent Zufluß *m*
influenza Grippe *f* (Med)
influenza virus Grippevirus *m*
influx Einfließen *n,* Einströmen *n,* Zustrom *m*
inform *v* benachrichtigen, informieren,
anzeigen, in Kenntnis setzen
information Information *f,* Angabe *f,* Auskunft *f,*
Bescheid *m,* Nachricht *f,* **exchange of** ~
Erfahrungsaustausch *m*

information center Beratungsstelle *f,*
Informationszentrum *n*
informative aufschlußreich, lehrreich
infra gravity wave Infraschwerewelle *f*
infrared infrarot, ultrarot, **opaque to** ~
ultrarotundurchlässig
infrared absorption Infrarotabsorption *f*
infrared absorption spectrum
Infrarotabsorptionsspektrum *n*
infrared analyzer Infrarotanalysator *m*
infrared block Ultrarotsperre *f*
infrared detector Infrarotdetektor *m*
infrared dryer Infrarottrockner *m*
infrared drying Infrarottrocknung *f*
infrared emission Ultrarotemission *f*
infrared filter Ultrarotfilter *m*
infrared heating Infrarotheizung *f*
infrared permeability Infrarotdurchlässigkeit *f*
infrared radiation Infrarotstrahlung *f*
infrared radiator Infrarotstrahler *m*
infrared removal Infrarotbeseitigung *f*
infrared spectroscopy Infrarotspektroskopie *f*
infrared spectrum Infrarotspektrum *n,*
IR-Spektrum *n*
infrared transmittance Infrarotdurchlässigkeit *f*
infrasonics Infraschall *m*
infringe *v* übertreten, verletzen
infringement Verletzung *f*
infundibular trichterförmig
infuse *v* aufgießen (z. B. Kräuter), einflößen,
eingießen, infundieren (z. B. Kräuter)
infusibility Unschmelzbarkeit *f*
infusible nicht schmelzbar, unschmelzbar
infusion Aufguß *m,* Infusion *f*
infusion bottle Infusionsflasche *f*
infusion of balm Melissentee *m*
infusion vessel Aufgußgefäß *n*
infusorial earth Kieselgur *f,* Berggur *f,*
Infusorienerde *f*
ingenious geistreich, geistvoll, genial, geschickt
ingenuity Scharfsinn *m*
ingestion Aufnahme *f* (Nahrung), Einnahme *f*
(Med)
ingestion [of food] Nahrungsaufnahme *f*
ingot Barren *m,* Block *m,* Eisenblock *m* (Techn),
Gußblock *m* (Techn), Ingot *m,* Kokille *f*
(Techn), Luppe *f,* **to clamp the** ~ den
Block *m* einspannen, **to grip the** ~ den
Block *m* einspannen
ingot casting Blockguß *m,* Kokillenguß *m*
(Stahlwerk)
ingot charging crane Blockeinsetzkran *m*
ingot copper Blockkupfer *n*
ingot crane Blockförderkran *m*
ingot gold Klumpengold *n*
ingot head heating device Blockkopfbeheizung *f*
ingot holding device Blockeinspannvorrichtung *f*
ingot iron Blockeisen *n,* Armco-Eisen *n* (HN)
ingot metal Flußeisen *n*

ingot mold Blockkokille *f,* Gußform *f,* Kokille *f,*
  head of ~ Muldenkopf *m*
ingot mold varnish Kokillenlack *m*
ingot pusher Block[aus]drücker *m*
ingot pushing device Blockdrücker *m*
ingot rolling mill Blockwalze *f*
ingot slicing machine Blockteilmaschine *f*
ingot steel Flußstahl *m,* Blockstahl *m* (Metall),
  Ingotstahl *m*
ingot steel wire Flußstahldraht *m*
ingot stripper Blockabstreifer *m*
ingot support Kokillentisch *m*
ingot tilter Blockhebetisch *m*
ingrained color Echtfarbe *f*
ingredient Bestandteil *m,* Ingrediens *n,* Zutat *f,*
  ~ of the mixture Mischungsbestandteil *m*
ingress pipe Einflußröhre *f*
inhale *v* einatmen, inhalieren
inharmonious unharmonisch
inherent innewohnend, angeboren
inherent filtration Eigenfilterung *f*
inherent value Eigenwert *m*
inherit *v* erben (Biol)
[in]heritable erblich
inheritance Erbgut *n,* Vererbung *f*
inhibit *v* [be]hindern, hemmen (Chem),
  inhibieren, verhindern, verzögern
inhibiting agent Inhibierungsmittel, Hemmittel *n*
inhibiting substance Hemmittel *n*
inhibition Hemmung *f* (Chem), Hinderung *f,*
  Inhibierung *f* (Chem), **allosteric ~**
  allosterische Hemmung *f,* **competitive ~**
  kompetitive Hemmung *f,* **irreversible ~**
  irreversible Hemmung *f,* **noncompetitive ~**
  nichtkompetitive Hemmung *f*
inhibitor Inhibitor *m,* Hemmer *m,* Hemmstoff *m,*
  Radikalfänger *m*
inhibitory center Inhibitionszentrum,
  Hemmungszentrum *n*
inhibitory effect Hemmwirkung *f*
inhibitory phase Inhibitionsphase *f*
inhomogeneity Inhomogenität *f,*
  Ungleichförmigkeit *f*
inhomogeneous inhomogen, nicht homogen,
  uneinheitlich, ungleichförmig
initial anfänglich, ursprünglich
initial amplification Anfangsverstärkung *f*
initial condition Anfangszustand *m*
initial cost Anschaffungskosten *pl*
initial detonating agent Initialsprengstoff *m*
initial direction Ausgangsrichtung *f*
initial dose Anfangsdosis *f* (Med)
initial equation Ausgangsgleichung *f* (Math)
initial explosive substance Initialzündmittel *n*
initial force Initialkraft *f*
initial igniting agent Initialzündmittel *n*
initial load Erstbelastung *f*
initial material Ausgangsprodukt *n,*
  Ausgangsstoff *m,* Urstoff *m*

initial member Anfangsglied *n*
initial part Anfangsstück *n*
initial phase Anfangsstadium *n*
initial point Anfangspunkt *m,* Nullpunkt *m*
  (Diagramm)
initial position Ausgangsposition *f,*
  Ausgangsstellung *f*
initial potential Ausgangspotential *n* (Elektr)
initial pressure Ausgangsdruck *m*
initial product Ausgangsprodukt *n*
initial solution Ausgangslösung *f*
initial speed Anfangs-,
  Ausgangsgeschwindigkeit *f*
initial state Anfangszustand *m,*
  Ausgangszustand *m,* Urzustand *m*
initial stress Vorspannung *f* (Mech)
initial temperature Anfangstemperatur *f,*
  Ausgangstemperatur *f*
initial tension Vorspannung *f*
initial value problem Anfangswertproblem *n*
  (Math)
initial voltage Anfangsspannung *f* (Elektr),
  Funkenpotential *n* (Elektr)
initial vortex Anfahrwirbel *m*
initiate *v* beginnen, einleiten
initiating power Initialkraft *f*
initiation Einführung *f,* Inangriffnahme *f*
initiator Initialzünder *m,* Initiator *m*
inject *v* einblasen (Dampf), einspritzen, impfen
  (Kristalle), injizieren (Med)
injectable solution Injektionslösung *f*
injected area Spritzlingsfläche *f*
injected article Spritzling *m*
injecting Einspritzen *n,* Einpressen *n*
injection Applikation *f* (Med), Einspritzung *f,*
  Injektion *f* (Med), Spritze *f* (Med),
  Spritzvorgang *m,* **~ of water**
  Wassereinspritzung *f*
injection block Probengebung *f* (Gaschromat)
injection capacity Spritzvolumen *n*
injection cock Einspritzhahn *m*
injection compound Spritzgußmasse *f*
injection cooler Einspritzkühler *m*
injection cooling Einspritzkühlung *f*
injection cycle Spritzzyklus *m*
  (Verarbeitungsgeschwindigkeit)
injection defect Spritzfehler *m*
injection die Spritzform *f*
injection [die] casting Spritzguß *m*
injection mold Spritzform *f* (Kunststoff),
  Spritzgußform *f,* **runnerless ~** angußlose
  Spritzgußform *f*
injection-mold *v* spritzgießen
injection-molded part Spritzgußteil *n*
injection molding Spritzgußverfahren *n*
  (Kunststoff), Spritzen *n* (Kunststoff),
  Spritzgießen *n*
injection molding cylinder Spritzgußzylinder *m*

injection molding machine Spritzgußmaschine *f*
injection [molding] nozzle Spritzdüse *f,*
Angußbuchse *f*
injection molding practice Spritzgußtechnik *f*
injection [molding] pressure Spritzdruck *m*
injection molding process Spritzgußverfahren *n*
injection molding technique Spritzgußtechnik *f*
injection needle Injektionsnadel *f* (Med)
injection nozzle Einspritzdüse *f*
injection plunger Spritzkolben *m*
injection port Eingabe *f* (Gaschromat),
Flüssigkeitseingabeort *m* (Gaschromat)
injection port heater Probeneinlaßheizung *f*
(Gaschromat)
injection preparation Injektionsmittel *n*
injection property Spritzfähigkeit *f*
[injection] syringe Injektionsspritze *f,* Spritze *f*
injection-type engine Einspritzmotor *m*
injector Injektor *m,* Dampfstrahlpumpe *f,*
Druckstrahlpumpe *f*
injector nozzle or cone Injektordüse *f*
injector valve Injektorventil *n*
injure *v* benachteiligen, schädigen, verletzen
injurious schädlich
injuriousness Schädlichkeit *f*
injury Schaden *m,* Beschädigung *f,* Verletzung *f*
(Med), Wunde *f,* **severe ~** schwere Wunde,
slight ~ leichte Wunde
ink Tinte *f,* Einschwärzfarbe *f,* **magnetic ~**
magnetische Tinte
ink *v* einfärben, einschwärzen
ink blue Tintenblau *n*
ink distribution Verreibung *f* der Farbe (Buchdr)
ink for corrugated board Slotterfarbe *f*
inking Einfärben *n*
inking brush Auftragspinsel *m*
ink pad Stempelkissen *n*
ink powder Tintenpulver *n*
ink receptivity Tintenaufnahme[fähigkeit] *f*
ink spot Tintenfleck *m*
ink stain Tintenfleck *m*
inkstone Atramentstein *m,* Tintenstein *m*
ink test Tintenprobe *f*
inlaying Einlegung *f,* Getäfel *n*
inlay work Einlegearbeit *f*
inlet Angußsteg *m* (Gieß), Eingang *m,*
Eingußkanal *m* (Gieß); Einlaß *m,*
Einlaßöffnung *f,* Eintritt *m,* Zugang *m,*
Zuleitung *f*
inlet case Einströmkasten *m*
inlet fitting Füllansatz *m*
inlet gas Frischgas *n*
inlet hole Eintrittsöffnung *f*
inlet pipe Einleitungsrohr *n,* Einfüllstutzen *m,*
Einlaßrohr *n*
inlet plate Einlaufboden *m* (Kolonne),
Einströmboden *m* (Kolonne)
inlet port Einlaßöffnung *f,* Eintrittsschlitz *m*
inlet sluice Einflußschleuse *f*

inlet strainer Eingußsieb *n,* Einlaufseiher *m*
inlet system Einlaßsystem *n* (Massenspektr)
inlet temperature Eintrittstemperatur *f*
inlet tube Zuleitungsrohr *n,* Einlaßstutzen *m*
inlet valve Einlaßventil *n,* Einströmventil *n*
innate angeboren, innewohnend
inner anode Innenanode *f*
inner bar Steg *m* (Techn)
inner dimension Innenmaß *n*
inner mandrel Innendorn *m*
inner metallization Innenmetallisierung *f*
inner orbital complex Innenorbitalkomplex *m*
inner point Innenpunkt *m*
inner pole casing Innenpolmantel *m*
inner pole frame Innenpolgestell *n*
inner tube Luftschlauch *m* (Reifen)
inner wall Innenwand *f*
inner width lichte Weite *f*
innovation Erneuerung *f,* Neuerung *f*
innoxious unschädlich
innumerable unzählbar
inoblast Fibroblast *m* (Histol)
inoculate *v* [ein]impfen, inokulieren (obs)
inoculation Impfen *n* (Krist), Impfung *f* (Med),
prophylactic ~ Schutzimpfung *f* (Med)
**inoculation method for the separation of
enantiomers** Impfmethode *f* zur
Racemat-Spaltung (Stereochem)
inoculum Impfstoff *m*
inocyte Bindegewebszelle *f* (Histol),
Fibroblast *m* (Histol)
inodorous geruchlos, duftlos, geruchfrei
inodorousness Geruchlosigkeit *f*
inolite Fadenstein *m* (Min)
inophyllic acid Inophyllsäure *f*
inorganic anorganisch
inorganic chemist Anorganiker *m*
inorganics anorganische Chemie *f*
inosine Inosin *n*
inosine diphosphate Inosindiphosphat *n*
inosine monophosphate Inosinmonophosphat *n*
inosine phosphorylase Inosinphosphorylase *f*
(Biochem)
inosine triphosphate Inosintriphosphat *n*
inosinic acid Inosinsäure *f*
inosite Inosit *m,* Fleischzucker *m* (obs)
inositol Inosit *m,* Fleischzucker *m* (obs),
Muskelzucker *m* (obs)
inositol monomethyl ether, 1- ~ Quebrachit *m*
inosose Inosose *f*
inoxidizable unoxidierbar
inoyite Inoyit *m*
in-pipe Einflußrohr *n*
in-plant coloring Selbst[ein]färbung *f*
input Eingabe *f,* Eingang *m* (Elektr), zugeführte
Menge *f*
input admittance Eingangsleitwert *m* (Elektr)
input amplification Vorverstärkung *f* (Elektr)
input amplifier Vorverstärker *m* (Elektr)

input capacitance Gitterkapazität *f* (Elektr)
input current Eingangsstrom *m* (Elektr)
input power Eingangsenergie *f* (Elektr),
  Eingangsleistung *f*
input resistance Eingangswiderstand *m*
input value Eingangswert *m*
input voltage Eingangsspannung *f*
inquire [into] *v* nachforschen
inquiry Anfrage *f,* Umfrage *f,* Untersuchung *f*
insalubrious gesundheitsschädlich
insanitary gesundheitsschädlich, unhygienisch,
  gesundheitsschädlich
inscribe *v* einschreiben, beschreiben
inscription Aufschrift *f,* Beschriftung *f*
insect Insekt *n,* Schädling *m,* Ungeziefer *n*
insect bite Insektenstich *m*
insect control Insektenbekämpfung *f,*
  Insektenvertilgung *f*
insect exterminator Insektizid *n*
insecticidal insektizid, insektentötend
insecticide Insektenvertilgungsmittel *n,*
  Insektizid *n,* Schädlingsbekämpfungsmittel *n,*
  gaseous ~ Räuchermittel *n* gegen Insekten
insecticide *a* insektizid
insectifuge Insektenvertreibungsmittel *n*
insect larva (pl larvae) Insektenlarve *f*
insectology Insektenkunde *f*
insect powder Insektenpulver *n*
insect wax Insektenwachs *n*
insecure unsicher
insecurity Unsicherheit *f*
insemination Besamung *f* (Biol), **artificial ~**
  künstliche Besamung
insensibility Unempfindlichkeit *f*
insensitive unempfindlich
insensitiveness Unempfindlichkeit *f*
inseparable untrennbar, unzertrennlich
in series hintereinander, in Serie
insert Einsatzstück *n,* Einlageteil *n,* Einsatz *m,*
  protruding type ~ Einpreßteil *n,* **through-type**
  ~ durchgehendes Einpreßteil
insert *v* einfügen, einschalten (Elektr),
  einschieben, einsetzen
insert adapter eingesetztes Paßstück *n*
inserted piece Einsatzstück *n*
insertion Einsatz *m,* Einlage *f,* Einschaltung *f*
  (Elektr), Einschiebung *f,* **point of ~**
  Einbaustelle *f*
insertion condenser Einsatzkühler *m*
insertion reaction Einschiebungsreaktion *f*
  (Chem)
insertion thermostat Eintauchthermostat *m*
insert mold Formeneinsatz *m,* Gesenkeinsatz *m*
insert pin Haltestift *m*
insert tube Ansatzrohr *n*
inside bark Bast *m*
inside calipers Hohlzirkel *m,* Lochtaster *m*
inside chaser Innenstrehler *m*
inside chasing tool Innenstrehler *m*

inside diameter Innendurchmesser *m,*
  Innenweite *f,* lichte Weite *f,* **~ of a pipe or**
  tube Rohrweite *f*
inside dimension Innenabmessung *f*
inside height lichte Höhe
insight Einsicht *f,* Erkenntnis *f*
insignificance Bedeutungslosigkeit *f*
insignificant bedeutungslos
insipid fade, schal
insipin Insipin *n*
insolubility Unauflöslichkeit *f,* Unlöslichkeit *f*
  (Chem)
insoluble unauflösbar, unauflöslich, unlöslich
  (Chem), **~ in alkali** laugenunlöslich, **~ in**
  water wasserunlöslich
insoluble matter Unlösliche[s] *n*
insomnia Schlaflosigkeit *f* (Med)
in space räumlich
inspect *v* beaufsichtigen, besichtigen, prüfen,
  untersuchen
inspection Aufsicht *f,* Durchsicht *f,*
  Inaugenscheinnahme *f,* Kontrolle *f,* Prüfung *f,*
  Untersuchung *f*
inspection hole Schauloch *n,* Schauöffnung *f*
inspection panel Schauöffnung *f*
inspection room Schaukammer *f*
inspection table Prüfstrecke *f*
inspection window Schauloch *n*
inspector Abnahmebeamter *m,* Prüfer *m*
inspiration Einatmung *f,* Inspiration *f*
inspire *v* begeistern
inspissate *v* eindampfen, eindicken
inspissated extract Dickauszug *m*
inspissating Eindicken *n*
inspissation Eindampfung *f;* Eindickung *f,* **~ of**
  oil Ölverdickung *f*
instability Instabilität *f,* Labilität *f,*
  Unbeständigkeit *f,* Unsicherheit *f,* **state of ~**
  Labilitätszustand *m,* **thermal ~** thermische
  Instabilität *f*
instable instabil, labil, unbeständig
install *v* installieren, anbringen, einbauen,
  einrichten
installation Anlage *f,* Aufstellung *f* (Maschine
  etc.), Betriebsanlage *f,* Einrichtung *f,*
  Installation *f,* Installierung *f*
installation cable, damp-proof ~
  Feuchtraumleitung *f*
installation dimension Einbaumaß *n*
instance Fall *m* (Beispiel)
instant Augenblick *m,* Moment *m*
instantaneous augenblicklich, momentan
instantaneous fuse empfindlicher Zünder *m*
instantaneous photograph Momentaufnahme *f*
  (Phot)
instantaneous release Momentauslöser *m* (Phot)
instantaneous value (Am. E.) Ist-Wert *m*
instep Rist *m* (Fuß), Spann *m*
instil[l] *v* einflößen, einträufeln, eintröpfeln

instilling Einflößen *n*, Eintröpfeln *n*
instinct Fingerspitzengefühl *n*, Instinkt *m*
institute of technology technische Hochschule *f*
institutional advertising Repräsentativwerbung *f*
instruct *v* anleiten, anweisen, belehren,
    unterrichten
instruction Anweisung *f*, Belehrung *f*,
    Vorschrift *f*
instruction book Bedienungsvorschrift *f*
instruction guidance Anleitung *f*
instruction manual Bedienungsanleitung *f*
instruction model Lehrmodell *n*
instructions technische Vorschriften *pl*, ~ for use
    Bedienungsvorschriften *pl*
instructions for use Benutzungsvorschriften *pl*
instructions to authors Autorenanweisungen *pl*
instrument Apparat *m*, Gerät *n*, Instrument *n*,
    Werkzeug *n*
instrumental analysis Instrumentalanalyse *f*
instrumental error Instrumentalfehler *m*
instrumentation Instrumentierung *f*
instrumentation engineer
    Betriebskontrollingenieur *m*
instrument board Schalttafel *f*,
    Instrumentenbrett *n*
instrument cover Instrumentenhaube *f*
instrument error Anzeigefehler *m*,
    Instrumentalfehler *m*
instrument housing Instrumentengehäuse *n*
instrument panel Armaturenbrett *n*
instrument reading, remote ~ Fernablesung *f*
instruments room Apparateraum *m*
insufficiency Unzulänglichkeit *f*
insufficient ungenügend, unzulänglich, ~
    temperature Untertemperatur *f*
insufflate *v* einblasen
insulanoline Insulanolin *n*
insularine Insularin *n*
insulate *v* isolieren
insulated isoliert, fully ~ vollisoliert, thermally
    ~ wärmeisoliert
insulating isolierend
insulating agent Absperrmittel *n*
insulating asphalt felt Asphaltisolierfilz *m*
insulating bell Isolierglocke *f*
insulating cardboard Isolierpappe *f*
insulating chamber Isolationskammer *f*
insulating cord Isolierschnur *f*
insulating covering Isolationshülle *f*,
    Isolierhülle *f*
insulating fabric Isoliergewebe *n*
insulating film or layer Schutzschicht *f*
insulating grid Isolierrost *m*
insulating handle Isoliergriff *m*
insulating hose Isolierschlauch *m*
insulating lacquer, electrical ~
    Elektroisolierlack *m*
insulating layer Isolierschicht *f*

insulating material Isolationsstoff *m*,
    Isoliermasse *f*, Isoliermaterial *n*,
    Isoliermittel *n*, electrical ~
    Elektroisoliermaterial *n*
insulating paint Isolationsanstrich *m*,
    Isolierlack *m*
insulating paper Isolierpapier *n*
insulating pitch Isolierpech *n*
insulating plate Isolierplatte *f*
insulating power Isolationsvermögen *n*,
    Isolierfähigkeit *f*
insulating property Isolierfähigkeit *f*
insulating quality Isolierfähigkeit *f*
insulating tape Isolationsband *n*, Isolierband *n*
insulating tube Isolierrohr *n*
insulating tube coiling machine
    Isolierrohrwickelmaschine *f*
insulating tubing Isolierschlauch *m*
insulating varnish Isolierfirnis *m*, Isolierlack *m*
insulation Isolation *f*, Isoliermaterial *n*,
    Isolierschutz *m*, Isolierung *f*, defect in ~
    Isolationsfehler *m*, fibrous ~
    Faserstoffisolation *f*, state of ~
    Isolationszustand *m*, ~ against radar detection
    Radarsicherung *f*
insulation board Faserpappe *f*
insulation coefficient Isolationsfaktor *m*
insulation detector Isolationsprüfer *m*
insulation material Isolierungsmaterial *n*
insulation resistance Isolationswiderstand *m*
insulation tester Isolationsprüfer *m*
insulation varnish Isolationslack *m*
insulator Isolator *m*, Isolationsmittel *n*,
    Isolationsstoff *m*, Isolierschicht *f*,
    Nichtleiter *m*, ~ for power current
    Starkstromisolator *m* (Elektr)
insulator surface Isolatoroberfläche *f*
insulin Insulin *n*, administration of ~
    Insulinzufuhr *f*, bound ~ gebundenes Insulin,
    test for ~ Insulinnachweis *m*
insulin deficiency Insulinmangel *m*
insulin deficiency diabetes
    Insulinmangeldiabetes *m*
insulinogenesis Insulinbildung *f*
insulin secretion Insulinausschüttung *f*
insulin shock [therapy] Insulinschock *m*
insulin unit Insulineinheit *f*
insurance Versicherung *f*
intaglio Intaglio *n* (Buchdr), Tiefrelief *n*
    (Buchdr)
intaglio *v* eingravieren, tiefätzen
intake Aufnahme *f*, Einflußröhre *f*, Einlauf *m*,
    ~ of food Nahrungsaufnahme *f*, ~ of water
    Wasseraufnahme *f*
intake channel Saugkanal *m*, Saugöffnung *f*
intake duct Saugkanal *m*
intake pipe Ansaugrohr *n*, Zulaufrohr *n*
intake port Saugstutzen *m*
intake valve Ansaugventil *n*, Einströmventil *n*

intake valve chamber Einlaßventilkammer *f*
in tandem hintereinander
intarvin Intarvin *n*
integer ganze Zahl *f* (Math)
integer *a* ganz, vollständig, ganzzahlig (Math)
integerrinecic acid Integerrinecinsäure *f*
integrable integrierbar
integral Integral *n*, **concept of the** ~
  Integralbegriff *m* (Math), **definite** ~
  bestimmtes Integral, **double** ~
  Doppelintegral *n*, **elliptical** ~ elliptisches
  Integral *n*, **indefinite** ~ unbestimmtes
  Integral, **line** ~ Linienintegral *n*, **multiple** ~
  mehrfaches Integral, **particular** ~ partikuläres
  Integral *n*, **singular** ~ singuläres Integral *n*,
  surface ~ Oberflächen-Integral *n*, **volume** ~
  Volumenintegral *n*, **~ over a plane area**
  Flächenintegral *n*
integral *a* ganz, ganzzahlig (Math), integrierend
integral action time (Br. E.) Nachstellzeit *f*
  (Regeltechn)
integral by-pass Innenumwegschaltung *f*
integral calculus Integralrechnung *f*
integral control Integralregelung *f*
integral curve Integralkurve *f*
integral domain Integritätsbereich *m*
integral equation Integralgleichung *f*
integral formula Integralformel *f*
integral invariant Integralinvariante *f*
integral kernel Integralkern *m*
integral reflection Gesamtreflexion *f*
integral relation Integralbeziehung *f*
integral representation Integraldarstellung *f*
integral theorem Integralsatz *m*
integral throttle expansion Drosseleffekt *m*
integrand Integrand *m*
integrant integrierend, wesentlich
integrate *v* ergänzen, integrieren (Math)
integrating integrierend
integrating circuit Integrierschaltung *f*
integrating gear Integrierwerk *n*
integration Integration *f*, **graphical** ~ graphische
  Integration, **intermittent** ~ absatzweise
  Summierung *f* (Comp), **numerical** ~
  numerische Integration *f*, **~ by parts** partielle
  Integration, **~ by substitution** Integration *f*
  durch Substitution, **~ of vector functions**
  Integration *f* v. Vektorfunktionen
integration constant Integrationskonstante *f*
integration device Integrationsgerät *n*
integration sign Integralzeichen *n*
integrator Integrator *m*, Integriergerät *n*,
  Planimeter *n*
integro-differential equation
  Integrodifferentialgleichung *f*
integument Haut *f*
intelligible verständlich
intend *v* beabsichtigen
[intended] use Verwendungszweck *m*

intense heftig, hochgradig, intensiv, angestrengt
intensification Steigerung *f*, Verstärkung *f*
intensifier Druckerhöher *m*, Verstärker *m* (Phot)
intensify *v* verstärken, steigern, intensivieren
intensifying screen Verstärkerschirm *m*
intensimeter Intensimeter *n*
intensity Intensität *f*, Heftigkeit *f*, Helligkeit *f*
  (Opt), Stärke *f*, Stromstärke *f* (Elektr),
  horizontal ~ Waagerechtintensität *f*, **level of**
  ~ Intensitätsniveau *n*, **original** ~
  Anfangsintensität *f*, **total** ~
  Gesamtintensität *f*
intensity anomaly Intensitätsanomalie *f*
intensity decrease Intensitätsabnahme *f*,
  Intensitätsschwächung *f*
intensity region Intensitätsbereich *m*
intensity sum rule Intensitätssummensatz *m*
intensive intensiv, durchgreifend, heftig, stark
interaccelerator Zwischenbeschleuniger *m*
interact *v* aufeinander wirken
interaction Wechselwirkung *f*, **mutual** ~
  gegenseitige Beeinflussung *f*
interaction force Wechselwirkungskraft *f*
interaction representation
  Wechselwirkungsdarstellung *f*
interatomic distance Atomabstand *m*,
  interatomarer Abstand *m*
intercalate *v* einlagern, einschieben, einschalten
intercalation Einlagerung *f*, Einschaltung *f*,
  Einschiebung *f*
intercalation compound
  Einlagerungsverbindung *f*
intercellular interzellulär
intercellular substance Interzellularsubstanz *f*
intercept Abschnitt (Geom)
intercepts, law of rational ~ Gesetz der
  rationalen Achsenabschnitte *m pl*
interchange Austausch *m*, Auswechslung *f*,
  Vertauschung *f*, **~ of sites** Platzwechsel *m*
interchange *v* austauschen, auswechseln,
  vertauschen
interchangeability Austauschbarkeit *f*,
  Vertauschbarkeit *f*
interchangeable austauschbar, auswechselbar,
  vertauschbar
interchangeable part Austauschteil *n*
interchange energy Austauschenergie *f*
interchanging Auswechslung *f*
intercommunicating vessel Verbindungsgefäß *n*
  (Med)
interconnection Zwischenschaltung *f*
intercrescence Verwachsung *f*,
  Zusammenwachsen *n*
intercrystalline interkristallin
intercrystalline crack interkristalliner Riß *m*
interdependent untereinander abhängig
interelectrode capacitance
  Zwischenelektrodenkapazität *f*
interest Anteil *m*, Beteiligung *f*, Interesse *n*

interface Begrenzungsfläche *f,*
Berührungsfläche *f,* Grenzfläche *f* (Krist),
Phasengrenzschicht *f,* **active at the ~**
grenzflächenaktiv
interface-active grenzflächenwirksam
interface potential Grenzflächenpotential *n*
interfacial angle Flächenschnittwinkel *m,*
Grenzflächenwinkel *m*
interfacial area Austauschfläche *f* (Dest)
interfacial diffusion Grenzflächendiffusion *f*
interfacial phenomenon
Grenzflächenerscheinung *f*
interfacial polarization
Grenzflächenpolarisation *f*
interfacial tension Grenzflächenspannung *f*
interfacial work Grenzflächen-Reibungsarbeit *f*
interfere *v* beeinträchtigen, intervenieren,
interferieren (Elektr), stören (Elektr)
interference Interferenz *f* (Phys), Rauschen *n*
(Radio), Störung *f,* **~ in double refraction**
Interferenzdoppelbrechung *f*
interference angle Interferenzwinkel *m*
interference current Störstrom *m* (Elektr)
interference drag
Wechselwirkungswiderstand *m* (Elektr)
interference elimination Entstörung *f,*
Störschutz *m*
interference factor Störgröße *f*
interference fading Interferenzschwund *m*
interference field Störfeld *n*
interference figure Interferenzbild *n,*
Interferenzfigur *f*
interference fringe Interferenzstreifen *m*
interference level Störpegel *m*
interference measuring apparatus Störmeßgerät *n*
interference pattern Interferenzbild *n* (Opt)
interference photograph Interferenzaufnahme *f*
interference radiation Störstrahlung *f*
interference spectrum Interferenzspektrum *n*
interferometer Interferometer *n* (Phys)
interferometry Interferometrie *f,*
Interferenzmeßverfahren *n*
interferon Interferon *n*
intergral, improper ~ uneigentliches Integral *n*
intergrowth Verwachsung *f* (Krist)
interhalide compound Interhalogenverbindung *f*
interionic distance Ionenabstand *m*
interior Innere[s] *n*
interior coating Innenanstrich *m*
interior view Innenansicht *f*
interlace *v* vernetzen, verflechten
interlaminar interlaminar
interlaminar bonding Schichtverband *m*
interlattice plane distance
Gitterebenenabstand *m* (Krist)
interlayer Zwischenschicht *f*
interlayer distance Schichtenabstand *m*
interleaf Durchschußblatt *n* (Buchdr),
Zwischenlage *f* (Buchdr)

interlining felt Einlagevlies *n*
interlink *v* verketten
interlinked verkettet
interlinked current Linienstrom *m*
interlinking Verkettung *f*
interlinking point Verkettungspunkt *m*
interlock *v* ineinandergreifen
interlocking *a* ineinandergreifend
intermediary dazwischenliegend, intermediär
intermediary body Zwischenkörper *m*
intermediary member Zwischenglied *n*
intermediary metabolism
Intermediärstoffwechsel *m*
intermediary stage Zwischenstufe *f*
intermediary step Zwischenschritt *m*
intermediate Zwischenglied *n,* Mittelstück *n,*
Zwischenstufe *f* (chem. Verbindung)
intermediate *a* dazwischenliegend, intermediär,
zwischenständig
intermediate air release Zwischenbelüftung *f*
intermediate color Mittelfarbe *f,* Zwischenfarbe *f*
intermediate [compound] Zwischenverbindung *f*
(Chem)
intermediate container Zwischenbehälter *m*
intermediate coolant Zwischenkühlmittel *n*
intermediate cooling Zwischenkühlung *f*
intermediate electrode Zwischenelektrode *f*
intermediate form Zwischenform *f*
intermediate layer Zwischenlage *f*
intermediate position Mittelstellung *f*
intermediate [product] Zwischenprodukt *n*
intermediate rolls Zwischenwalzwerk *n*
intermediate socket Mignonfassung *f* (Elektr)
intermediate space Zwischenraum *m*
intermediate state Übergangszustand *m*
intermediate tank Zwischenbehälter *m*
intermesh *v* vernetzen
intermeshed verkettet
intermeshing ineinandergreifend (Zahnräder)
intermetallic intermetallisch
interminable unendlich
intermingle *v* vermischen
intermission Pause *f,* Unterbrechung *f*
intermittence Aussetzen *n,* Diskontinuität *f,*
Unterbrechung *f*
intermittency effect Intermittenzeffekt *m* (Opt)
intermittent intermittierend, aussetzend,
diskontinuierlich, stoßweise, unterbrochen
intermittent motion ruckweise Bewegung *f,*
Wechselbewegung *f*
intermixing Durchmischung *f*
intermixture Beimischung *f,* Durchmischung *f*
intermolecular intermolekular,
zwischenmolekular
internal combustion engine
Verbrennungsmotor *m*
internal combustion turbine
Verbrennungsturbine *f*
internal condensation Inkohlung *f*

internal dust extraction Innenentstäubung *f*
internal expanding brake Innenbackenbremse *f*
internal fittings Einbauten *pl*
internal force innere Kraft *f*
internal friction Eigenreibung *f*
internal furnace Innenfeuerung *f*
internal gears Innenverzahnung *f*
internal heat of evaporation
    Disgregationswärme *f*
internal lining Innenauskleidung *f*
internal measure Innenmaß *n*
internal medicine innere Medizin *f*
internal mixer Innenmischer *m*
internal plant Innenanlage *f*
internal-pole dynamo Innenpoldynamo *m*
internal pressure Innendruck *m*
internal pressure test Innendruckversuch *m*
internal resistance Eigenwiderstand *m,*
    Innenwiderstand *m* (Elektr)
internal standard line Vergleichsspektrallinie *f*
    (Spektr)
internal strain innere Beanspruchung *f*
internal surface Innenfläche *f*
internal thread Innengewinde *n*
internal wiring Innenleitung *f*
international metric measures metrisches
    Maßsystem *n*
international unit internationale Einheit *f*
internuclear distance Kernabstand *m* (Atom)
interpack packing Interpack-Füllkörper *m*
interplanar interplanar, zwischen den Ebenen *pl*
interplanar spacing Ebenenabstand *m,*
    Netzebenenabstand *m*
interplay Wechselwirkung *f,* Zusammenspiel *n*
interpolar distance Polabstand *m,* Pollücke *f*
interpolate *v* interpolieren, einschalten
interpolation Interpolation *f*
interpole Zwischenpol *m*
interpose *v* dazwischenschalten (Techn),
    intervenieren
interpret *v* auswerten, interpretieren, auslegen
interpretation Auslegung *f*
interrelation Wechselbeziehung *f*
interrenin Interrenin *n*
    (Nebennierenrindenhormon)
interrupt *v* unterbrechen, abbrechen, aussetzen,
    stören
interrupted unterbrochen, diskontuinierlich,
    lückenhaft
interrupter Ausschalter *m,* Unterbrecher *m*
    (Elektr), **electrolytic** ~ elektrolytischer
    Unterbrecher *m*
interruption Unterbrechung *f,* Pause *f,*
    Stockung *f,* Störung *f,* **point of** ~
    Unterbrechungsstelle *f,* ~ **of operation**
    Betriebsunterbrechung *f*
interruption of work Arbeitsunterbrechung *f*
intersect *v* [durch]kreuzen, [durch]schneiden,
    sich schneiden

intersecting schneidend
intersecting plane Schnittebene *f*
intersecting point Schnittpunkt *m*
intersection Schnittpunkt *m,* Durchschnitt *m*
    (Math); Knotenpunkt *m,* Schnitt *m*
interspace Zwischenraum *m,* **atomic** ~ atomarer
    Zwischenraum *m*
intersperse *v* einstreuen
interspin crossing Spinumkehr *f*
interstage trap Zwischenabscheider *m*
interstice Zwischenraum *m,* Hohlraum *m,*
    Lücke *f,* Zwischengitterplatz *m* (Krist)
interstitial atom Zwischengitteratom *n* (Krist)
interstitial compound Einlagerungsverbindung *f,*
    Gitterhohlraumverbindung *f* (Krist)
interstitial ion Zwischengitterion *n*
interstitial mechanism
    Zwischengittermechanismus *m*
interstitial migration Zwischengitterwanderung *f*
interstitial pair Zwischengitterpaar *n*
interstitial position Zwischengitterplatz *m*
    (Krist)
interstitial structures Einlagerungsstrukturen *pl*
interstratified zwischengeschichtet
interstructure Zwischenwand *f*
intertwine *v* verschlingen, verflechten
intertwiningly verschlungen
interval Abstand *m,* Intervall *n,* Pause *f,*
    Zwischenpause *f,* Zwischenraum *m,* **angular**
    ~ Winkelintervall *n*
intervene *v* intervenieren
intervertebral disk Bandscheibe *f* (Med)
interweave *v* einflechten
intestinal flora Darmbakterien *pl* (Bakt)
intestine Darm *m* (Med)
intoxicant Rauschmittel *n*
intoxicate *v* berauschen, vergiften
intoxicating Vergiften *n*
intoxication Vergiftung *f,* Trunkenheit *f,*
    symptom of ~ Vergiftungserscheinung *f*
intra-atomic inneratomar, innerhalb eines Atoms
intracutaneous intrakutan
intramine Intramin *n*
intramolecular intramolekular; innerhalb eines
    Moleküls
intramuscular intramuskulär
intravenous intravenös
intravenous injection Injektion *f* i. d. Blutbahn
intricate kompliziert, schwierig
intrinsic wahr, eigen
intrinsic conduction Eigenleitung *f* (Phys)
intrinsic induction Magnetisierungsintensität *f*
intrinsic semiconductor Eigenhalbleiter *m*
introduce *v* einführen, einleiten, eintragen, ~ **an**
    alkyl alkylieren, ~ **benzoyl [into]** benzoylieren,
    ~ **two nitro groups** dinitrieren
introducing Einführen *n,* Einleiten *n,*
    Eintragen *n*

introduction Einführung _f_, Einleitung _f_,
  Eintragung _f_, Vorrede _f_
intumescence Intumeszenz _f_ (Med, Krist)
in tune in Resonanz
inula Inula _f_ (Bot)
inulase Inulase _f_ (Biochem)
inulin Inulin _n_, Alantin _n_, Alantstärke _f_
inulinase Inulase _f_ (Biochem), Inulinase _f_
  (Biochm)
invalid nichtig, ungültig
invalidity Arbeitsunfähigkeit _f_ (Med),
  Nichtigkeit _f_, Ungültigkeit _f_
invar Invar _n_ (Leg)
invariable invariabel, konstant, unveränderlich
invariant Invariante _f_ (Math)
invariant _a_ unveränderlich
Invar steel Invarstahl _m_
invent _v_ erfinden
invention Erfindung _f_, **according to the** ~
  erfindungsgemäß
inventors, protection of ~ Erfinderschutz _m_
inventory Bestandsverzeichnis _n_, Inventar _n_,
  Inventur _f_, Lageraufnahme _f_, Lagerbestand _m_
inverse entgegengesetzt, umgekehrt, verkehrt
inverse cosine Arkuskosinus _m_ (Math)
inverse cotangent Arkuskotangens _m_ (Math)
inverse function Umkehrfunktion _f_
inversely proportional umgekehrt proportional
inverse sine Arkussinus _m_ (Math)
inverse tangent Arkustangens _m_ (Math)
inverse voltage Sperrspannung _f_ (Elektr)
inversion Inversion _f_ (Math, Chem),
  Umkehrung _f_, **center of** ~
  Inversionszentrum _n_, **point of** ~
  Inversionspunkt _m_, Knickpunkt _m_, ~ **of**
  sugar Zuckerinversion _f_
inversion axis Inversionsachse _f_ (Krist)
inversion integral Umkehrintegral _n_ (Math)
inversion layer Inversionsschicht _f_
inversion point Inversionspunkt _m_
inversion temperature Inversionstemperatur _f_
inversion theorem Umkehrsatz _m_
invert _v_ umkehren, invertieren
invertase Invertase _f_ (Biochem), Saccharase _f_
  (Biochem), Sucrase _f_ (Biochem)
inverted umgekehrt
inverted image Umkehrbild, umgekehrtes Bild _n_
  (Opt)
invertible invertierbar, umkehrbar
invertin Invertin _n_, Invertase _f_
inverting mirror Umkehrspiegel _m_ (Opt)
inverting prism Umkehrprisma _n_
invert soap Invertseife _f_, kationenaktives
  Waschmittel _n_
invert sugar Invertzucker _m_
investigate _v_ untersuchen, erforschen,
  nachforschen, prüfen
investigation Erforschung _f_, Nachforschung _f_,
  Untersuchung _f_, **material for** ~

Untersuchungsmaterial _n_, **method of** ~
  Untersuchungsverfahren _n_, **range of** ~
  Untersuchungsbereich _m_
investigation board Untersuchungsausschuß _m_
investigator Forscher _m_, Nachforscher _m_
investment Anlage _f_, Beteiligung _f_, Investition _f_
investment foundry Ausschmelzgießerei _f_
in view of angesichts
invisibility Unsichtbarkeit _f_
invisible unsichtbar
invisible ink Geheimtinte _f_
involuntary unwillkürlich
involute Evolvente _f_ (Geom)
involution Involution _f_, rückläufige
  Entwicklung _f_, Schrumpfung _f_
inward bulge Einbuchtung _f_
inward curvature Einbuchtung _f_
iodanil Jodanil _n_
iodaniline Jodanilin _n_
iodanisole Jodoanisol _n_
iodargyrite Jodargyrit _m_ (Min)
iodate Jodat _n_, jodsaures Salz _n_
iodate _v_ jodieren
iodatometry Jodatometrie _f_ (Anal)
iodeosin Jodeosin _n_
iodic acid Jodsäure _f_, **salt of** ~ Jodat _n_
iodic anhydride Jodpentoxid _n_,
  Jodsäureanhydrid _n_
iodide Jodid _n_, Jodür _n_ (obs),
  Jodwasserstoffsalz _n_
iodiferous jodhaltig
iodiferous remedy Jodmittel _n_ (Pharm)
iodinate _v_ jodieren
iodination Jodierung _f_
iodine Jod _n_, **tincture of** ~ Jodtinktur _f_
iodine [absorption] value Jodzahl _f_
iodine addition product Jodadditionsprodukt _n_
iodine ampoule Jodampulle _f_
iodine azide Jodazid _n_
iodine balsam Jodbalsam _m_
iodine bromide Bromjod _n_, Jodbromid _n_
iodine chloride Chlorjod _n_, Jodchlorid _n_
iodine compound Jodverbindung _f_
iodine content Jodgehalt _m_
iodine cyanide Cyanjodid _n_, Jodcyan _n_
iodine dioxide Joddioxid _n_
iodine fluoride Fluorjod _n_
iodine intake Jodaufnahme _f_
iodine isotope Jodisotop _n_
iodine-like jodartig
iodine monobromide Jodmonobromid _n_,
  Jodbromür _n_ (obs)
iodine monochloride Jodmonochlorid _n_,
  Jodchlorür _n_ (obs)
iodine number Jodzahl _f_
iodine ointment Jodsalbe _f_
iodine pentafluoride Jodpentafluorid _n_
iodine pentoxide Jodpentoxid _n_,
  Jodsäureanhydrid _n_

iodine plant Jodanlage *f*
iodine poisoning Jodismus *m* (Med),
    Jodvergiftung *f*
iodine preparation Jodpräparat *n*
iodine soap Jodseife *f*
iodine solution Jodlösung *f*
iodine spring Jodquelle *f*
iodine test Jodprobe *f*
iodine tribromide Jodtribromid *n*
iodine trichloride Jodtrichlorid *n*,
    Dreifachchlorjod *n*
iodine value Jodzahl *f*, Verseifungszahl *f*
iodine vapor Joddampf *m*
iodine water Jodwasser *n*
iodinin Jodinin *n*
iodipin Jodipin *n*
iodism Jodismus *m*, Jodvergiftung *f*
iodite Jodit *n*
iodize *v* jodieren, mit Jod behandeln
iodized collodion Jodkollodium *n*
iodized paper Jodpapier *n*
iodized starch Jodstärke *f*
iodoacetate Jodacetat *n*
iodoalkane Jodalkyl *n*
iodoaminobenzene Jodanilin *n*
iodoaniline Jodanilin *n*
iodoaseptol Jodaseptol *n* (Med)
iodobenzene Jodbenzol *n*
iodobenzoic acid Jodbenzoesäure *f*
iodochlorohydroxyquinoline Vioform *n*
iodocinnamic acid Jodzimtsäure *f*
iodoethane Äthyljodid *n*, Jodäthyl *n*
iodoether Jodäther *m*
iodoform Jodoform *n*
iodoform gauze Jodoformgaze *f*
iodoformin Jodoformin *n*
iodoform test Jodoformprobe *f*
iodogorgoic acid Jodgorgosäure *f*
iodohydrin Jodhydrin *n*
iodohydroxynaphthoquinone
    Jodoxynaphthochinon *n*
iodol Jodol *n*, Tetrajodpyrrol *n*
iodol soap Jodolseife *f*
iodomethane Methyljodid *n*, Jodmethyl *n*
iodometric jodometrisch (Anal)
iodometry Jodometrie *f* (Anal), jodometrische
    Titration *f* (Anal)
iodonium Jodonium *n*
iodonium compound Jodoniumverbindung *f*
iodonium ion Jodonium-Ion *n*
iodophen Jodophen *n*, Nosophen *n*
iodophenol Jodphenol *n*
iodophenyldimethylpyrazolone
    Jodphenyldimethylpyrazolon *n*
iodophthalein Jodophthalein *n*
iodophthalic acid Jodphthalsäure *f*
iodopsin Jodopsin *n*
iodopyrine Jodopyrin *n*
iodoquinoline Jodchinolin *n*

iodosalicylic acid Jodsalicylsäure *f*
iodosobenzene Jodosobenzol *n*
iodosol Jodosol *n*, Jodthymol *n*
iodostannous acid Stannojodwasserstoffsäure *f*
iodo-starch paper Jodstärkepapier *n*
iodothiouracil Jodthiouracil *n*
iodothyrin Jodothyrin *n*
iodotoluene Jodtoluol *n*
iodous jodig
iodyrite Jodargyrit *m* (Min), Jodyrit *m* (Min)
iolite Iolith *m* (Min), Cordierit *m* (Min),
    Dichroit *m* (Min), Wassersaphir *m* (Min)
ion Ion *n*
ion activity Ionenaktivität *f*
ion activity coefficient
    Ionenaktivitätskoeffizient *m*
ion adsorption Ionenadsorption *f*
ionane Jonan *n*
ion binding Ionenbindung *f*
ion chain Ionenkette *f*
ion chamber Ionisationskammer *f*
ion charge number Ionenladungszahl *f*
ion cluster Ionenwolke *f*
ion color Ionenfarbe *f*
ion conductivity Ionenleitfähigkeit *f*
ion counter Ionisationsmesser *m*
ion current Ionenstrom *m*
ion-dipole complex Ionendipolkomplex *m*
ion drift velocity
    Ionenwanderungsgeschwindigkeit *f*
ion exchange Ionenaustausch *m*
ion exchange chromatography
    Ionenaustauschchromatographie *f*
ion exchange membrane
    Ionenaustauschermembran *f*
ion exchanger Ionenaustauscher *m*
ion exchange resin Austauscherharz *n*,
    ionenaustauschendes Harz *n*
ion flotation Ionenflotation *f*
ion gauge Ionisationsmanometer *n*
ionic ionisch
ionic atmosphere Ionenwolke *f*
ionic atomization pump Ionenzerstäuberpumpe *f*
ionic chain polymerization
    Ionenkettenpolymerisation *f*
ionic character Ionencharakter *m*
ionic charge Ionenladung *f*
ionic cleavage Ionisierung *f*
ionic concentration Ionenkonzentration *f*,
    Ionenstärke *f*, **total ~**
    Gesamtionenkonzentration *f*
ionic conductivity Ionenleitung *f*
ionic crystal Ionenkristall *m* (Krist)
ionic density Ionendichte *f*
ionic discharge Ionenentladung *f*
ionic energy Ionenenergie *f*
ionic equation Ionengleichung *f*
ionic equilibrium Ionengleichgewicht *n*
ionic form Ionenform *f*

ionic friction Ionenreibung *f*
ionic increment Ioneninkrement *n*
ionic micelles Ionenmizellen *pl*
ionic migration Ionenwanderung *f,*
Ionenbewegung *f*
ionic mobility Ionenbeweglichkeit *f*
ionic movement Ionenbewegung *f*
ionic product Ionenprodukt *n,*
Löslichkeitsprodukt *n*
ionic quantimeter Ionendosimeter *n*
ionic radius Ionenradius *m*
ionic ray Ionenstrahl *m*
ionic reaction Ionenreaktion *f*
ionic state Ionenzustand *m*
ionic strength Ionenstärke *f,*
Ionenkonzentration *f*
ionic theory Ionentheorie *f*
ionic valency Ionenwertigkeit *f*
ionic velocity Ionengeschwindigkeit *f*
ionidine Jonidin *n*
ion-ion recombination Ion-Ion-Rekombination *f*
ionisability Ionisierbarkeit *f*
ionisation limit Ionisierungsgrenze *f* (Phys)
ionium (Symb. Io) Ionium *n*
ionizable ionisierbar
ionization Ionisierung *f,* Ionisation *f,* **degree of**
~ Ionisationsgrad *m,* **heat of** ~
Ionisationswärme *f,* Ionisierungswärme *f,*
quarternary ~ Ionisation *f* vierter Ordnung,
~ **by collision** Stoßionisation *f,* ~ **by electrons**
Elektronenionisierung *f,* ~ **of air**
Luftionisation *f*
ionization balance Ionisationsgleichgewicht *n*
ionization chamber Ionisationskammer *f*
ionization constant Ionisationskonstante *f*
ionization current Ionisationsstrom *m*
ionization efficiency Ionisierungsausbeute *f*
ionization energy Ionisationsenergie *f,*
Ionisierungsenergie *f*
ionization gauge Ionisationsmanometer *n*
ionization potential Ionisationspotential *n,*
Ionisierungspotential *n*
ionization track Ionisierungsbahn *f*
ionization vacuum gauge
Ionisationsvakuummeter *n*
ionization voltage Ionisationsspannung *f,*
Ionisierungsspannung *f*
ionize *v* ionisieren, dissoziieren
ionized layer Ionisationsschicht *f*
ionizer Ionisator *m,* Ionisierungsmittel *n*
ionizing electrode Ionisierungselektrode *f*
ionizing power Ionisierungsvermögen *n*
ionizing wave Ionisierungswelle *f*
ion lattice Ionengitter *n* (Krist)
ion limit Ionengrenzwert *m*
ion loss Ionenabgabe *f*
ion migration Ionenwanderung *f*
ion movement Ionenbewegung *f*

ionogenic ionenbildend, ionenerzeugend,
ionogen
ionol Jonol *n*
ionone Jonon *n*
iononic acid Jononsäure *f*
ionophoresis Ionophorese *f*
ionosphere Ionosphäre *f*
ionosphere sound Ionosphärenschall *m*
ionospheric conductivity
Ionosphärenleitfähigkeit *f*
ionospheric soundings Ionosphärengeräusche *n*
pl
ion particle density Ionenteilchendichte *f*
ion path Ionenbahn *f*
ion pump Ionenpumpe *f*
ion retardation process
Ionenverzögerungsverfahren *n*
ion retarding field Ionenbremsfeld *n*
ions, pair of ~ Ionenpaar *n,* **production of** ~
Ionenbildung *f*
ion saturation current Ionensättigungsstrom *m*
ion transition Ionenübergang *m*
ion transport Ionentransport *m*
ion trap Ionenfalle *f* (Rdr)
iopanoic acid Jopansäure *f*
iophenoic acid Jophensäure *f*
iosene Josen *n*
iothion Jothion *n*
ipecac Brechwurz *f* (Bot, Med)
ipecacuanha root Ipecacuanhawurzel *f* (Bot),
Brechwurz *f* (Bot)
ipecacuanhic acid Ipecacuanhasäure *f*
ipomeanic acid Ipomeansäure *f*
ipomic acid Sebacinsäure *f,* Ipomsäure *f*
iproniazid Iproniazid *n*
ipuranol Ipuranol *n*
irene Iren *n*
iretol Iretol *n*
iridaceous plant Liliengewächs *n* (Bot)
iridesce *v* irisieren, schillern
iridescence Farbenschiller *m,* Irisieren *n,*
Schillern *n*
iridescense Farbenspiel *n*
iridescent schillernd, buntschillernd, irisierend
iridescent luster Schillerglanz *m*
iridescent substance Schillerstoff *m*
iridin Iridin *n*
iridium (Symb. Ir) Iridium *n,* **spongy** ~
Iridiumschwamm *m*
iridium chelate Iridiumchelat *n*
iridium chloride Iridiumchlorid *n*
iridium oxide Iridiumoxid *n*
iridium potassium chloride
Kaliumiridiumchlorid *n*
iridium sponge Iridiumschwamm *m*
iridium sulfide Iridiumsulfid *n,*
Schwefeliridium *n*
iridize irisieren
iridodicarboxylic acid Iridodicarbonsäure *f*

iridoskyrin Iridoskyrin *n*
iridosmine Osmiridium *n*
iriphan Iriphan *n*
iris Iris *f* (Auge), Regenbogenhaut *f* (Auge)
irisate irisieren
irisating salt Irisiersalz *n*
iris blue Irisblau *n*
Irish moss Karrag[h]eenmoos *n* (Bot)
irisin Irisin *n*
iris oil Irisöl *n*
iris shutter Iris-Blende *f* (Opt)
iris violet Irisviolett *n*
iron Eisen *n,* **acid** ~ saures Eisen, **added** ~
  Zusatzeisen *n,* **basic** ~ Thomaseisen *n*
  (Metall), **brittle** ~ sprödes Eisen, **cadmiated**
  ~ kadmiertes Eisen, **carburized** ~ gekohltes
  Eisen, **close-grained** ~ Feinkorneisen *n,*
  coarse-grained ~ Grobkorneisen *n,*
  commercial ~ technisches Eisen,
  constructional ~ Baueisen *n,* **decarburized** ~
  entkohltes Eisen, **fine-grained** ~
  Feinkorneisen *n,* **flat** ~ Flacheisen *n,* **forged**
  ~ geschmiedetes Eisen, **forging** ~
  schmiedbares Eisen, **galvanized** ~ verzinktes
  Eisen, **hammer-hardened** ~ hartgeschlagenes
  Eisen, **highly carbonated** ~ schaumiges
  Roheisen, **high-silicon** ~ hochsiliziertes Eisen,
  hot brittle ~ heißbrüchiges Eisen,
  rotbrüchiges Eisen, **mixed** ~ Mischeisen *n,*
  native ~ gediegenes Eisen *n,* **off-grade** ~
  Ausfalleisen *n,* **pure** ~ gediegenes Eisen,
  red-short ~ heißbrüchiges Eisen, rotbrüchiges
  Eisen, **refined** ~ Frisch[feuer]eisen *n,*
  Gareisen *n,* Raffinierstahl *m,* **rich in** ~
  eisenreich, **rusted** ~ verrostetes Eisen,
  separation of ~ Eisenausscheidung *f,* **silvery**
  ~ Glanzeisen *n,* **soft** ~ geschmeidiges Eisen,
  Weicheisen *n,* **tincture of** ~ Eisentinktur *f,*
  wedge-shaped ~ keilförmiges Eisen, **white**
  crystalline ~ kleinspiegeliges Weißeisen,
  wrought ~ geschmiedetes Eisen,
  Schmiedeeisen *n,* ~ **cast in a loam mold**
  Lehmguß *m*
iron *a* eisern
iron *v* abstrecken (Met)
iron acetate solution Eisenacetatlösung *f*
iron albuminate Eisenalbuminat *n*
iron alloy Eisenlegierung *f*
iron alum Eisenalaun *m*
iron ammonium chloride Eisensalmiak *m*
iron ammonium citrate Eisenammoncitrat *n*
iron ammonium sulfat Eisenammoniakalaun *m,*
  Eisenammoniumsulfat *n*
iron analysis Eisenanalyse *f,*
  Eisenuntersuchung *f*
iron arsenide Eisenarsenid *n,* Arseneisen *n*
iron asbestos Eisenasbest *m* (Min),
  Eisenamiant *m* (Min)
iron atom Eisenatom *n*

iron bacteria Eisenbakterien *pl*
[iron] ball Deul *m* (Metall), Eisenballen *m*
iron bars, large ~ Grobeisen *n*
iron bath Eisenbad *n*
iron-bearing eisenführend
iron black Eisenschwarz *n*
iron blast furnace Eisenhochofen *m*
iron block Eisenklumpen *m,* Eisensau *f*
iron boride Eisenborid *n,* Ferrobor *n*
iron bronze Eisenbronze *f*
iron building Eisenkonstruktion *f*
iron calcium phosphate, native ~ Richellit *m*
  (Min)
iron cap Eisenhaube *f*
iron carbide Eisencarbid *n*
iron carbonate Eisencarbonat *n*
iron carbonyl Eisencarbonyl *n,* Ferrocarbonyl *n*
iron caseïnate Caseïneisen *n*
iron casing Eisenkasten *m*
iron casting Eisenguß *m,* **malleable** ~
  Tempergußstück *n*
iron cell Eisenelement *n*
iron cement Eisenkitt *m,* Eisenzement *m*
iron chloride tincture Eisenchloridtinktur *f*
iron chrysolite Eisenchrysolith *m* (Min),
  Eisenolivin *m* (Min)
iron-clad eisenverkleidet, gepanzert
ironclad magnet Mantelmagnet *m*
iron clay Eisenton *m*
iron cobalt arsenide, native ~
  Arsenikkobaltkies *m* (Min), Arsenkobalt *n*
  (Min)
iron color Eisenfarbe *f*
iron-colored eisenfarbig
iron compound Eisenverbindung *f*
iron concrete Eisenbeton *m*
iron constantan element
  Eisenkonstantanelement *n*
iron content Eisengehalt *m*
iron core Eiseneinlage *f,* **(of a magnet)** ~
  Eisenkern *m* (eines Magneten)
iron core coil Eisenkernspule *f*
iron covering Blechmantel *m*
iron crystal Eisenkristall *m*
iron cyanide Eisencyanid *n,* Cyaneisen *n*
iron cyanogen compound Cyaneisenverbindung *f,*
  Eisencyanverbindung *f*
iron disulfide Eisendisulfid *n,*
  Doppeltschwefeleisen *n,* **native** ~ Markasit *m*
  (Min)
iron dross Eisensinter *m,* Hochofenschlacke *f*
iron drum Eisenfaß *n*
iron druse Eisendruse *f* (Min)
iron dust Eisen[feil]staub *m*
irone Iron *n*
iron earth Eisenerde *f*
iron electrode Eisenelektrode *f*
iron examination Eisenuntersuchung *f*
iron extract Eisenextrakt *m*

iron extraction Enteisenung *f*
iron filings Eisen[feil]späne *pl*, Eisenfeilicht *n*
iron flint Eisenkiesel *m* (Min), Eisenquarz *m* (Min)
iron flowers Eisenblumen *f pl*
iron fluoride Eisenfluorid *n*, basic ~ Eisenoxyfluorid *n*
iron founder Eisengießer *m*
iron foundry Eisengießerei *f*, Eisenschmelzhütte *f*
iron frame Eisengestell *n*
iron-free eisenfrei
iron gallate ink Eisengallustinte *f*
iron garnet Eisengranat *m* (Min)
iron geode Eisendruse *f* (Min)
iron glance Eisenglanz *m* (Min), Glanzeisenerz *n* (Min)
iron glazing Eisenglasur *f*
iron glue Eisenkitt *m*
iron glycerophosphate Eisenglycerophosphat *n*
iron grain Eisenkorn *n*
iron grating Eisengitter *n*
iron hammer scale Eisenhammerschlag *m*, Hammerschlag *m*
iron hematite, specular ~ Spiegeleisen *n* (Min)
iron hemisulfide Halbschwefeleisen *n*
iron hood Eisenmantel *m*
iron hoop Eisenband *n*
iron(II) acetate Ferroacetat *n*, Eisen(II)-acetat *n*
iron(II) arsenate Eisen(II)-arsenat *n*, Ferroarsenat *n*
iron(II) bromide Eisendibromid *n*, Eisenbromür *n* (obs), Eisen(II)-bromid, Ferrobromid *n*
iron(II) carbonate Eisen(II)-carbonat *n*, Ferrocarbonat *n*
iron(II) chloride Eisendichlorid *n*, Eisenchlorür *n* (obs), Eisen(II)-chlorid *n*, Ferrochlorid *n*
iron(II) compound Eisen(II)-Verbindung *f*, Eisenoxydulverbindung *f* (obs), Ferroverbindung *f*
iron(II) cyanide Ferrocyanid *n*, Eisencyanür *n* (obs), Eisen(II)-cyanid *n*
iron(II) fluoride Eisendifluorid *n*
iron(II) glycerophosphate Ferroglycerinphosphat *n*
iron(II) hydroxide Eisen(II)-hydroxid *n*, Eisenhydroxydul *n* (obs), Ferrohydroxid *n*
iron(III) acetate Eisen(III)-acetat *n*, Ferriacetat *n*
iron(III) arsenite Eisen(III)-arsenit *n*, Ferriarsenit *n*
iron(III) bromide Eisentribromid *n*, Ferribromid *n*
iron(III) chloride Eisen(III)-chlorid *n*, Eisensesquichlorid *n*, Eisentrichlorid *n*, Ferrichlorid *n*
iron(III) chromate Ferrichromat *n*
iron(III) citrate Eisen(III)-citrat *n*, Ferricitrat *n*

iron(III) compound Eisen(III)-Verbindung *f*, Ferriverbindung *f*
iron(III) cyanide Eisen(III)-cyanid *n*, Ferricyanid *n*
iron(III) fluoride Eisen(III)-fluorid *n*, Eisentrifluorid *n*
iron(III) hydroxide Ferrihydroxid *n*, Eisen(III)-hyroxid *n*, Eisenoxidhydrat *n*
iron(II)(III) iodide Ferriferrojodid *n*, Eisenjodürjodid *n* (obs)
iron(II)(III) oxide Ferriferrooxid *n*, Eisen(II)(III)-oxid *n*
iron(III) iodate Ferrijodat *n*
iron(III) iodide Eisen(III)-jodid *n*, Ferrijodid *n*
iron(III) lactate Ferrilactat *n*
iron(III) nitrate Ferrinitrat *n*
iron(II) iodide Ferrojodid *n*, Eisendijodid *n*, Eisen(II)-jodid *n*, Eisenjodür *n* (obs)
iron(II) ion Eisen(II)-Ion *n*, Ferroion *n*
iron(III) oxalate Ferrioxalat *n*
iron(III) oxide Eisen(III)-oxid *n*, Eisensesquioxid *n*, Ferrioxid *n*
iron(III) phosphate Ferriphosphat *n*
iron(III) rhodanide Ferrirhodanid *n*
iron(III) salt Eisen(III)-Salz *n*, Ferrisalz *n*
iron(III) sulfate Ferrisulfat *n*
iron(III) sulfide Ferrisulfid *n*, Eisen(III)-sulfid *n*, Eisensesquisulfid *n*
iron(III) tannate Eisen(III)-tannat *n*, Ferritannat *n*
iron(III) tartrate Eisen(III)-tartrat *n*, Ferritartrat *n*
iron(III) thiocyanate Eisen(III)-rhodanid *n*, Ferrirhodanid *n*
iron(III) valerate Ferrivalerianat *n*
iron(II) lactate Ferrolactat *n*
iron(II) nitrate Ferronitrat *n*
iron(II) oxalate Ferrooxalat *n*
iron(II) oxide Ferrooxid *n*, Eisen(II)-oxid *n*, Eisenmonoxid *n*, Eisenoxydul *n* (obs)
iron(II) phosphate Ferrophosphat *n*
iron(II) salt Eisen(II)-Salz *n*, Ferrosalz *n*
iron(II) sulfate Ferrosulfat *n*, Eisenoxydulsulfat *n* (obs)
iron(II) sulfide Ferrosulfid *n*, Eisen(II)-sulfid *n*, Eisenmonosulfid *n*, Eisensulfür *n* (obs)
iron(II) thiocyanate Eisen(II)-rhodanid *n*, Eisenrhodanür *n* (obs)
iron industry Eisenindustrie *f*
ironing Abstrecken *n* (Metall), Abstreckziehen *n*, Bügeln *n*
iron ink Eisentinte *f*
iron iodate Eisenjodat *n*
iron lacquer Eisenlack *m*
iron lead Bleieisenstein *m*
ironlike eisenähnlich, eisenartig
iron liquor Eisenbeize *f*
iron lode Eisengang *m* (Bergb)
iron loss Eisenabbrand *m*, Eisenverlust *m*

iron lozenges Eisentabletten *pl* (Pharm)
iron lung eiserne Lunge *f* (Med)
iron magnetic flux Eisenkraftfluß *m*
iron manganese silicide Ferromangansilicium *n*
iron manganese tungstate
  Eisenmanganwolframat *n*
iron mantle Eisenmantel *m*
iron meal Eisenmehl *n*
iron metallurgy Eisenhüttenkunde *f*
iron metasilicate Eisenmetasilicat *n*
iron mine Eisenbergwerk *n*
iron minium Eisenmennige *f*
iron modification Eisenmodifikation *f*
iron monosulfide Eisen(II)-sulfid *n*,
  Eisenmonosulfid *n*
iron monoxide Eisenmonoxid *n*, Ferrooxid *n*
iron mordant Eisenbeize *f*, Eisenbrühe *f*,
  Eisengrund *m*
iron muffle Muffel *f* aus Eisen
iron natrolite Eisennatrolit *m* (Min)
iron nickel accumulator
  Eisennickelakkumulator *m*
iron nitride Eisennitrid *n*
iron nitrite Eisennitrit *n*
iron ocher Eisenocker *m*, **red** ~
  Roteisenocker *m* (Min)
iron olivine Fayalit *m* (Min)
iron ore Eisenerz *n*, **argillaceous** ~ toniges
  Eisenerz, **botryoidal** ~ Schaleneisenstein *m*,
  columnar argillaceous red ~ Nagelerz *n*
  (Min), **compact** ~ derbes Eisenerz, **earthy** ~
  Eisenmulm *m* (Min), **fibrous** ~ faseriges
  Eisenerz, **friable** ~ mulmiges Eisenerz,
  granular argillaceous ~ Hirseerz *n* (Min),
  green ~ Grüneisenerde *f* (Min), **hepatic** ~
  Eisenlebererz *n* (Min), **lenticular** ~
  Pfennigstein *m* (Min), **manganiferous** ~
  manganhaltiges Eisenerz, Eisenmanganerz *n*,
  muddy ~ Eisenschlick *m*, **nickeliferous** ~
  nickelhaltiges Eisenerz, **porous** ~ poriges
  Eisenerz, **red** ~ Roteisenerz *n* (Min), **siliceous**
  ~ Kieseleisenstein *m* (Min), **specular** ~
  Glanzeisenerz *n* (Min), **yellow** ~
  Gelbeisenerz *n* (Min), **zinciferous** ~
  zinkhaltiges Eisenerz
iron ore cement Ferrarizement *m*
iron ore deposit Eisenerzlagerstätte *f* (Bergb),
  Eisenerzvorkommen *n*
iron ore mine Eisenerzgrube *f* (Bergb)
iron ore vein Eisengang *m* (Bergb)
iron oxalate Eisenoxalat *n*
iron oxide Eisenoxid *n*, **black** ~ Eisenmohr *m*
  (Min), **earthy** ~ Ocker *m*, **red** ~ Kaiserrot *n*,
  rich in ~ eisenoxidreich
iron oxide black Eisenoxidschwarz *n*
iron oxide layer Eisenoxidschicht *f*
iron oxide pigment Eisenoxidfarbe *f*
iron oxide red Eisenoxidrot *n*
iron oxyfluoride Eisenoxyfluorid *n*

iron paranucleinate Triferrin *n*
iron pastille Eisenpastille *f*
iron pentacarbonyl Eisenpentacarbonyl *n*
iron peptonate Eisenpeptonat *n*
iron phosphate Eisenphosphat *n*
iron phosphide Eisenphosphid *n*,
  Phosphoreisen *n*
iron pig Eisenklumpen *m*, Eisengans *f*
iron pit Eisenbergwerk *n*
iron plate Eisenplatte *f*, Blechtafel *f*,
  Eisenblech *n*
iron Portland cement Eisenportlandzement *m*
iron potassium alum Eisenkaliumalaun *m*
iron powder Eisenpulver *n*, Stahlpulver *n*
iron powder coil Eisenpulverkernspule *f*
iron precipitate Eisenniederschlag *m*
iron preparation Eisenpräparat *n*
iron-producing eisenschaffend
iron production Eisengewinnung *f* (Metall)
iron protecting paint Eisenschutzfarbe *f*
iron protosulfide Eisen(II)-sulfid *n*
iron protoxide Eisen(II)-oxid *n*, Eisenoxydul *n*
  (obs)
iron putty Eisenkitt *m*
iron pyrites Pyrit *m* (Min), Eisenkies *m* (Min),
  Schwefelkies *m* (Min), **cupriferous** ~
  kupferreicher Eisenkies, **white** ~
  Vitriolkies *m* (Min)
iron quartz Eisenquarz *m* (Min)
iron quinine citrate Eisenchininicitrat *n*
iron rake Kratze *f*, Rührhaken *m*
iron refinery Eisenfrischerei *f*
iron refinery slag Eisenfeinschlacke *f*
iron removal Enteisenung *f*
iron requirement Eisenbedarf *m*
iron rolling mill Eisenwalzwerk *n*
iron rubber Eisengummi *n*
iron runner Eisenabflußrinne *f*
iron rust Eisenrost *m*
iron saccharate Eisensaccharat *n*
iron salt Eisensalz *n*
iron sample Eisenprobe *f*
iron sand, titaniferous ~ Titaneisensand *m*
[iron] scale Hammerschlag *m*
iron scrap Eisenabfall *m*, Eisenabgang *m*
iron selenide Eisenselenid *n*, Seleneisen *n*
iron separation Eisenabscheidung *f*
iron separator Eisenabscheider *m*,
  Eisenseparator *m*
iron sesquichloride Eisensesquichlorid *n*
iron sesquioxide Ferrioxid *n*, Eisen(III)-oxid *n*,
  Eisensesquioxid *n*
iron sesquisulfide Eisen(III)-sulfid *n*,
  Eisensesquisulfid *n*
iron shavings Eisendrehspäne *m pl*
iron shell Blechmantel *m*
iron silicate Eisensilicat *n*
iron silicide Eisensilicid *n*, Siliciumeisen *n*
iron silver glance Sternbergit *m* (Min)

iron slab Bramme *f*
iron slag Eisenschlacke *f*
iron smelting Eisenverhüttung *f* (Metall)
iron sodium pyrophosphate
    Eisennatriumpyrophosphat *n*
iron sow Eisensau *f* (Metall)
iron spark Eisenfunke *m*
iron spinel Hercynit *m* (Min)
iron sponge Eisenschwamm *m*
iron spot Eisenfleck *m*
iron stain Eisenfleck *m*
iron stand Eisengestell *n*
ironstone Eisenerz *n* (Min), Eisenstein *m* (Min),
    calcareous ~ Kalkeisenstein *m* (Min), red ~
    Roteisenstein *m* (Min), yellow ~
    Gelbeisenstein *m* (Min)
iron sublimate Eisensublimat *n*
iron sulfate Eisensulfat *n*
iron sulfate plant Eisensulfatanlage *f*
iron sulfide Eisensulfid *n*, Eisenkies *m* (Min),
    Schwefeleisen *n*
iron syrup Eisensirup *m*
iron-tanned leather Eisenleder *n*
iron test Eisenprobe *f*
iron testing Eisenuntersuchung *f*
iron testing apparatus
    Eisenuntersuchungsapparat *m*
iron tonic Eisenmittel *n*
iron trade Eisenhandel *m*
iron tube Eisenrohr *n*
iron turnings Eisenspäne *m pl*
iron varnish Eisenlack *m*
iron vitriol Eisenvitriol *n*
iron vitriol solution Eisenvitriollösung *f*
ironware Eisenwaren *pl*, Metallwaren *pl*, small
    ~ Kleineisenzeug *n*
iron waste Eisenabbrand *m*, Eisenabgang *m*
iron water Eisenwasser *n*
iron wedge Eisenkeil *m*
iron weighting Eisenbeschwerung *f*
iron wire Eisendraht *m*
ironwood Eisenholz *n*
ironworks Eisenhütte *f*, Hammerwerk *n*
iron yellow Eisen[oxid]gelb *n*
irradiance Bestrahlungsstärke *f* (Opt)
irradiate *v* bestrahlen, anstrahlen, ausstrahlen;
    belichten (Phot), ~ acoustically beschallen
irradiated, lightly ~ schwach bestrahlt
irradiation Bestrahlung *f*, Ausstrahlung *f*,
    duration of ~ Bestrahlungsdauer *f* (Atom),
    intensity of ~ Belichtungsstärke *f* (Phot)
irradiation damage Bestrahlungsschaden *m*
irradiation equipment Bestrahlungsgeräte *n pl*
irradiation hole Bestrahlungskanal *m*
irradiation lamp Bestrahlungslampe *f*
irradiator Strahler *m*
irrational irrational
irrealizable undurchführbar
irreducible nicht reduzierbar

irregular unregelmäßig, abnorm, irregulär
    (Krist), regelwidrig, uneinheitlich,
    ungeordnet, ungleichförmig
irregularity Unregelmäßigkeit *f*, Abnormität *f*,
    Anomalie *f*, degree of ~
    Ungleichförmigkeitsgrad *m*
irrelevant belanglos
irreparable irreparabel, nicht wiederherstellbar
irrepressible unaufhaltsam
irrespective unbeschadet, ungeachtet
irresponsible unverantwortlich
irreversibility Irreversibilität *f*
irreversible irreversibel, nicht reversibel, nicht
    umkehrbar
irreversible cycle nicht umkehrbarer Vorgang *m*
irrevocable unwiderruflich
irrigate *v* bewässern, berieseln
irrigation Berieselung *f*, [künstliche]
    Bewässerung *f*
irrigation canal Bewässerungsgraben *m*
irrigation cooler Berieselungskühler *m*
irrigation plant Berieselungsanlage *f*
irrigation surface Berieselungsfläche *f*,
    Rieselfläche *f*
irrigation water Rieselwasser *n*
irritability Reizbarkeit *f*
irritant Reizmittel *n*
irritant *a* reizend
irritant action Reizwirkung *f*
irritate *v* irritieren, reizen, irritieren
irritation Reiz *m* (Med), Reizung *f* (Med)
irritative reizerregend (Med)
irrotational fluid motion wirbelfreie Strömung *f*
isaconitic acid Isaconitsäure *f*
isamic acid Isamsäure *f*
isamide Isamid *n*
isanic acid Isansäure *f*
isanolic acid Isanolsäure *f*
isaphenic acid Isaphensäure *f*
isatan Isatan *n*
isatanthrone Isatanthron *n*
isathophan Isathophan *n*
isatic acid Isatinsäure *f*
isatin Isatin *n*
isatin derivative Isatinderivat *n*
isatinecic acid Isatinecinsäure *f*
isatinic acid Isatinsäure *f*
isatinic acid lactam Isatin *n*
isatogen Isatogen *n*
isatogenic acid Isatogensäure *f*
isatoic acid Isatosäure *f*
isatoid Isatoid *n*
isatol Isatol *n*
isatoxime Isatoxim *n*
isatronic acid Isatronsäure *f*
isatropic acid Isatropasäure *f*
isatyde Isatyd *n*
isenthalpic expansion isenthalpischer
    Drosseleffekt *m*

isentrope Isentrope *f*
isentropic isentropisch
isentropic analysis Isentropenanalyse *f*
isentropic chance of state isentropische
 Zustandsänderung *f*
isethionic acid Isäthionsäure *f*
ishikawaite Ishikawait *m* (Min)
isinglass Fischleim *m*, Hausenblase *f*,
 Ichthyocoll *n*
island model Inselmodell *n*
islets of Langerhans Langerhanssche Inseln *pl*
 (Med)
isoaconitic acid Isaconitsäure *f*
isoadenine Isoadenin *n*
iso-agglutinin Isoagglutinin *n*
isoalloxazine Isoalloxazin *n*
isoamidone Isoamidon *n*
isoamyl Isoamyl- *n*
isoamyl acetate Isoamylacetat *n*
isoamylacetic acid Isoamylessigsäure *f*
isoamyl alcohol Isoamylalkohol *m*
isoamyl bromide Isoamylbromid *n*
isoamyl caprate Caprinsäureisoamylester *m*
isoamyl chloride Isoamylchlorid *n*
isoamylene Isoamylen *n*
isoandrosterone Isoandrosteron *n*
isoanethol Isoanethol *n*
iso-antibody Isoantikörper *m*
iso-antigen Isoantigen *n*
isoantipyrine Isoantipyrin *n*
isoascorbic acid Isoascorbinsäure *f*
isoasparagine Isoasparagin *n*
isobar Isobar *n* (Phys, Chem), Isobare *f*
 (Meteor), Linie *f* gleichen Drucks (Meteor)
isobarbituric acid Isobarbitursäure *f*
isobaric isobar[isch]
isobaric slope Druckgefälle *n* (Meteor)
isobars, group of ~ Isobarenschar *f*
isobebeerine Isobebeerin *n*
isobenzil Isobenzil *n*
isobianthrone Isobianthron *n*
isoborneol Isoborneol *n*
isobornyl acetate Isobornylacetat *n*
isobornyl isovalerate Gynoval *n*
isobutane Isobutan *n*
isobutanol Isobutylalkohol *m*
isobutene Isobutylen *n*
isobutyl Isobutyl-
isobutyl acetate Isobutylacetat *n*
isobutyl alcohol Isobutylalkohol *m*
isobutyl amine Isobutylamin *n*
isobutyl benzoate Isobutylbenzoat *n*
isobutyl bromide Isobutylbromid *n*
isobutyl chloride Isobutylchlorid *n*
isobutylene Isobutylen *n*
isobutylene bromide Isobutylenbromid *n*
isobutylidene Isobutyliden *n*
isobutyl iodide Isobutyljodid *n*
isobutyl mercaptan Isobutylmercaptan *n*

isobutyl oleate Tebelon *n*
isobutyl-p-aminobenzoate Cycloform *n*
isobutyl preparation Isobutylpräparat *n*
isobutyl salicylate Isobutylsalicylat *n*
isobutyl thiocyanate Isobutylrhodanid *n*
isobutyraldehyde Isobutyraldehyd *m*
isobutyric aldehyde Isobutyraldehyd *m*
isobutyric anhydride Isobuttersäureanhydrid *n*
isobutyric nitrile Isobuttersäurenitril *n*
isobutyronitrile Isobuttersäurenitril *n*
isobutyryl Isobutyryl-
isobutyryl chloride Isobutyrylchlorid *n*
isocadinene Isocadinen *n*
isocaine Isocain *n*
isocamphane Isocamphan *n*
isocamphoric acid Isocamphersäure *f*
isocamphoronic acid Isocamphoronsäure *f*
isocantharidine Isocantharidin *n*
isocaproic acid Isocapronsäure *f*
isocellular gleichzellig
isocenter Isozentrum *n*
isocetic acid Pentadecansäure *f*
isocetinic acid Isocetinsäure *f*
isochamic acid Isochamsäure *f*
isochavicinic acid Isochavicinsäure *f*
isochlorine Isochlorin *n*
isochlorogenic acid Isochlorogensäure *f*
isocholesterol Isocholesterin *n*
isochore Isochore *f*
isochoric isochor
isochroman Isochroman *n*
isochromat Isochromate *f*
isochromatic gleichfarbig, isochrom,
 isochromatisch
isochronous isochron, isochronisch
isochronous curve Isochrone *f*
isocinchomeronic acid Isocinchomeronsäure *f*
isocineole Isocineol *n*
isocinnamic acid Isozimtsäure *f*
isocitric acid Isocitronensäure *f*
isoclinal line Isokline *f*
isoclinal lines Linien *pl* gleicher Neigung
isocoffeine Isocoffein *n*
isocolchicine Isocolchicin *n*
isocoproporphyrin Isokoproporphyrin *n*
isocorybulbine Isocorybulbin *n*
isocorydine Isocorydin *n*
isocoumaran Isocumaran *n*
isocoumaranone Isocumaranon *n*
isocoumarin Isocumarin *n*
isocrotonic acid Isocrotonsäure *f*
isocryptoxanthin Isokryptoxanthin *n*
isocumene Isocumol *n*
isocyanate Isocyanat *n*
isocyanic acid Isocyansäure *f*
isocyanide Isonitril *n*
isocyanin Isocyanin *n*
isocyanuric acid Fulminursäure *f*,
 Isocyanursäure *f*

isocyclic isocyclisch
isocymene Metacymol *n*
isocytidine Isocytidin *n*
isocytosine Isocytosin *n*
isodehydroacetic acid Isodehydracetsäure *f*
isodesmic, simple ~ **crystal** einfach
 isodesmischer Kristall *m*
isodimorphism Isodimorphie *f* (Krist),
 Isodimorphismus *m* (Krist)
isodisperse isodispers
isodose Isodosis *f*
isodose recorder Isodosenschreiber *m*
isodrin Isodrin *n*
isodulcite Isodulcit *m*
isodural Isodural *n*
isodurene Isodurol *n*
isoduridine Isoduridin *n*
isoduryl Isoduryl-
isodynamic line Isodyname *f* (Phys)
isoelectric isoelektrisch
isoelectric point isoelektrischer Punkt *m*
isoelectronic isoelektronisch
isoemetine Isoemetin *n*
isoemodin Rhabarberon *n*
isoephedrine Isoephedrin *n*
isoerucic acid Brassidinsäure *f*, Isoerucasäure *f*
isoestrone Isoöstron *n*
isoeugenol Isoeugenol *n*
isoeugenyl acetate Isoeugenylacetat *n*
isofebrifugin Isofebrifugin *n*
isofenchoic acid Isofenchosäure *f*
isofenchol Isofenchol *n*
isoferulic acid Hesperitinsäure *f*, Isoferulasäure *f*
isofisetin Luteolin *n*
isoflavone Isoflavon *n*
isoform Isoform *n*
isofucosterol Isofucosterin *n*
isofulminate Isofulminat *n*
isogalbulin Isogalbulin *n*
isogalloflavin Isogalloflavin *n*
isogeny Stoffgleichheit *f*
isogeronic acid Isogeronsäure *f*
isogladiolic acid Isogladiolsäure *f*
isoglucal Isoglucal *n*
isoglutamine Isoglutamin *n*
isoglutathione Isoglutathion *n*
isogonal isogonal, winkeltreu
isogony Isogonie *f* (Krist)
isogram Isogramm *n*
isography Isographie *f*
isoguanine Isoguanin *n*
isohemanthamine Isohämanthamin *n*
isohemipic acid Isohemipinsäure *f*
isohexacosane Ceran *n*
isohexylnaphthazarin Isohexylnaphthazarin *n*
isohistidine Isohistidin *n*
isohomopyrocatechol Isohomobrenzcatechin *n*
isohumulinic acid Isohumulinsäure *f*

isohydric isohydrisch (Chem), korrespondierend
 (Chem)
isohydrobenzoin Isohydrobenzoin *n*
isohydry Isohydrie *f*
isoindigo Isoindigo *n*
isoindole Isoindol *n*
isoindoline Isoindolin *n*
isoionic gleichionisch
isokainic acid Isokainsäure *f*
isolate *v* isolieren, absondern, abtrennen, rein
 darstellen (Chem), scheiden (Chem)
isolated abgeschieden, abgesondert, abgesperrt,
 alleinstehend, isoliert (Chem), vereinzelt
isolating isolierend
isolation Isolation *f*, Isolierung *f*
isolation procedure Isolierungsmethode *f*
isolaureline Isolaurelin *n*
isoleucine Isoleucin *n*
isoleutherin Isoleutherin *n*
isolimonene Isolimonen *n*
isologous isolog
isolomatiol Isolomatiol *n*
isolysergic acid Isolysergsäure *f*
isolysergine Isolysergin *n*
isolysergol Isolysergol *n*
isolysine Isolysin *n*
isomalic acid Isoäpfelsäure *f*
isomaltitol Isomaltit *m*
isomaltose Isomaltose *f*
isomenthone Isomenthon *n*
isomer Isomer *n*, isomere Verbindung *f*,
 positional ~ Stellungsisomer *n* (Chem),
 trans-bridged ~ trans-Brückenisomer *n*
 (Stereochem)
isomerase Isomerase *f*
isomeric isomer
isomeric change Isomerisierung *f*
isomerism Isomerie *f*, **geometric[al]** ~
 geometrische Isomerie (Stereochem), **optical**
 ~ optische Isomerie (Krist), **possibility of** ~
 Isomeriemöglichkeit *f*
isomerization Isomerisierung *f*
isomerize *v* isomerisieren
isomerous isomer
isomers, mixture of ~ Isomerengemisch *n*
isometameric isometamer
isometric isometrisch, längentreu, maßgleich,
 tesseral (Krist)
isometric system Tesseralsystem *n* (Krist)
isomorphic isomorph
isomorphine Isomorphin *n*
isomorphism Formgleichheit *f*,
 Gleichgestaltigkeit *f*, Isomorphie *f*,
 Isomorphiebeziehung *f*, Isomorphismus *m*,
 double ~ Isodimorphie *f* (Krist)
isomorphous gleichgestaltig, isomorph
isomycomycin Isomycomycin *n*
isonicotinic acid Isonicotinsäure *f*
isonitrile Isonitril *n*

isonitrosobarbituric acid, 5- ~ Violursäure *f*
isonitroso compound Isonitrosoverbindung *f*
isooctane Isooctan *n*
isooctene Isoocten *n*
isooctylhydrocupreine Vuzin *n*
isoodyssic acid Isoodyssinsäure *f*
isooleic acid Isoölsäure *f*
isooxindigo Isooxindigo *n*
isopentane Isopentan *n*
isopentyl alcohol Isopentylalkohol *m*
isoperimetric isoperimetrisch, von gleichem
  Umfang
isoperistaltic isoperistaltisch
isophorol Isophorol *n*
isophthalic acid Isophthalsäure *f,*
  Metaphthalsäure *f*
isophthalimide Isophthalimid *n*
isopicric acid Isopikrinsäure *f*
isopiperic acid Isopiperinsäure *f*
isopleth Isoplethe *f* (Math)
isopoly acid Isopolysäure *f*
isoporic focus Isoporenfokus *m*
isopral Isopral *n*
isoprenaline Isoprenalin *n*
isoprene Isopren *n*
isoprene caoutchouc Isoprenkautschuk *m*
isoprene rule Isoprenregel *f*
isoprenoid Isoprenoid *n*
isopropanol Isopropylalkohol *m,*
  Dimethylcarbinol *n,* Isopropanol *n*
isopropenyl Isopropenyl-
isopropyl Isopropyl-
isopropylacetic acid Isopropylessigsäure *f*
isopropylacetylacetone Isopropylacetylaceton
isopropyl alcohol Isopropylalkohol *m,*
  Dimethylcarbinol *n,* Isopropanol *n*
isopropyl benzene Isopropylbenzol *n*
isopropylbenzoic acid Isopropylbenzoesäure *f*
isopropylbenzyl alcohol Cuminol *n*
isopropyl bromide Isopropylbromid *n*
isopropyl cyanide Isobuttersäurenitril *n*
isopropyl ether blend Isopropyläthergemisch *n*
isopropyl iodide Isopropyljodid *n*
isopropylmalonic acid Isopropylmalonsäure *f*
isopseudocholesterol Isopseudocholesterin *n*
isopulegol Isopulegol *n*
isopurpuric acid Isopurpursäure *f*
isopurpurogallone Isopurpurogallon *n*
isopyromucic acid Isobrenzschleimsäure *f*
isopyronene Isopyronen *n*
isoquinaldic acid Isochinaldinsäure *f*
isoquinamine Isochinamin *n*
isoquinoline Isochinolin *n*
isoreserpic acid Isoreserp[in]säure *f*
isoreserpone Isoreserpon *n*
isoretinene Isoretinen *n*
isorhamnose Isorhamnose *f*
isoriboflavine Isoriboflavin *n*
isorotation Isorotation *f* (Stereochem)

isosaccharin Isosaccharin *n*
isosafrole Isosafrol *n*
isosceles gleichschenklig (Dreieck)
isoserine Isoserin *n*
isosmotic isosmotisch
isostatic isostatisch
isostilbene Isostilben *n*
isostrophanthic acid Isostrophanthsäure *f*
isostructural isostrukturell
isostrychnic acid Isostrychninsäure *f*
isostrychnine Isostrychnin *n*
isosuccinic acid Isobernsteinsäure *f*
isotactic isotaktisch
isotestosterone Isotestosteron *n*
isotetralin Isotetralin *n*
isotherm Isotherme *f*
isothermal isotherm[isch]
isothermal chart, high altitude ~
  Höhenisothermenkarte *f*
isothermal curve Isotherme *f*
isothermal lines Linien *pl* gleicher Temperatur
isotherm of reaction Reaktionsisotherme *f*
isotherms Linien *pl* gleicher Temperatur
isothiazole Isothiazol *n*
isothiocyanate Isothiocyanat *n*
isothiocyanic acid, salt of ~ Isothiocyanat *n*
isotone Isoton *n*
isotonic isotonisch, von gleichem osmotischen
  Druck *m*
isotonicity Isotonie *f*
isotope Isotop *n,* **stable** ~ stabiles Isotop
isotope abundance Isotopenhäufigkeit *f*
isotope chemistry Isotopenchemie *f*
isotope dilution analysis
  Isotopenverdünnungsanalyse *f*
isotope dilution method
  Isotopen-Verdünnungsmethode *f*
isotope effect Isotopieeffekt *m*
isotope enrichment Isotopenanreicherung *f*
isotope fractionation Isotopentrennung *f*
isotope-labelled radiomarkiert
isotope production Isotopenherstellung *f*
isotope separation Isotopentrennung *f*
isotope separation plant
  Isotopentrennungsanlage *f*
isotope separation process
  Isotopentrennvorgang *m*
isotope weight Isotopengewicht *n*
isotopic abundance measurement
  Isotopenhäufigkeitsmessung *f*
isotopic composition Isotopenzusammensetzung *f*
isotopic compound Isotopenverbindung *f*
isotopic equilibrium Isotopengleichgewicht *n*
isotopic exchange reaction
  Isotopenaustauschreaktion *f*
isotopic indicator Leitisotop *n*
isotopic labeling Isotopenmarkierung *f*
isotopic mass Isotopengewicht *n*

isotopic mixture Isotopengemisch *n*
isotopic number Massenzahl *f*
isotopic ratio Isotopenhäufigkeitsverhältnis *n*,
  Isotopenverhältnis *n*
isotopic spin Isotopenspin *m*
[isotopic] tracer Leitisotop *n*
isotrehalose Isotrehalose *f*
isotron Isotron *n*
isotropic isotrop[isch]
isotropy Isotropie *f*
isotropy factor Isotropiefaktor *m*
isotryptamine Isotryptamin *n*
isotubaic acid Isotubasäure *f*, Rotensäure *f*
isotypical typengleich
isourea Pseudoharnstoff *m*
isovaleraldehyde Isovaleraldehyd *m*
isovalerate Isovaleriansäureester *m*, Salz *n* der
  Isovaleriansäure (f)
isovaleric acid Isovaleriansäure *f*,
  Isopropylessigsäure *f*
isovaleryl chloride Isovalerylchlorid *n*
isovaline Isovalin *n*
isovanillic acid Isovanillinsäure *f*
isoviolanthrone Isoviolanthron *n*
isoxazole Isoxazol *n*
isoxazoline Isoxazolin *n*
isozyme Isoenzym *n* (Biochem)
issue Ausgabe *f*, Ablauf *m*, Auslauf *m*,
  Streitobjekt *n*
issue *v* herauskommen, entströmen, ~ **from**
  herrühren (von); hervorgehen [aus], ~ **in**
  ergeben, resultieren
itabirite Itabirit *m* (Min),
  Eisenglimmerschiefer *m* (Min)
itacolumite Gelenkquarz *m* (Min), Itacolumit *m*
  (Min)
itaconic acid Itaconsäure *f*
Italian red Italienischrot *n*
itamalic acid Itamalsäure *f*
itch ointment Krätzsalbe *f* (Pharm)
item Posten *m*, Punkt *m*
itemize *v* [einzeln] aufzählen
iterate *v* iterieren
iteration method Iterationsverfahren *n*
iteration process Iterationsprozeß *m*
ITP (Abk) Inosintriphosphat *n*
itrol Itrol *n*
ittnerite Ittnerit *m* (Min)
I.U. (Abk) Internationale Einheit *f*,
  Immuneinheit *f*, Immunisierungseinheit *f*
iva oil Ivaöl *n*
ivory Elfenbein *n*
ivory *a* elfenbeinern
ivory black Elfenbeinschwarz *n*; gebranntes
  Elfenbein *n*, Knochenkohle *f*
ivory-colored elfenbeinfarbig
ivory nut Elfenbeinnuß *f* (Bot)
ivory paper Elfenbeinpapier *n*
ivory porcelain Elfenbeinporzellan *n*

ivory yellow Elfenbeingelb *n*
ivy Efeu *m* (Bot)
ixene Ixen *n*
ixolone Ixolon *n*
ixone Ixon *n*
Izod impact strength Kerbzähigkeit *f*

J

jaborandi oil Jaborandiöl *n* (Pharm)
jacinth Hyacinth *m* (Min)
jack Hebebock *m*, Hebevorrichtung *f*,
  Hebewinde *f*, Wagenheber *m*
jacket Mantel *m*, Auskleidung *f*, Hülle *f*,
  Isolierung *f*, Umhüllung *f*, Ummantelung *f*
jacket *v* verkleiden
jacket cooling Mantelkühlung *f*
jacket copper Mantelkupfer *n*
jacketed bottom Dampfboden *m*
jacketed mold Dampfkammerform *f*
jacketed pans Doppelkessel *m*
jacket piston meter Durchlaufkolbenmesser *m*
jacket sheet iron Mantelblech *n*
jacket tube Hüllrohr *n*, Mantelrohr *n*
jackscrew (Am. E.) Schraubheber *m*
jackshaft Vorgelegewelle *f*
jack switch Knebelschalter *m* (Elektr)
jack up *v* aufbocken, hochwinden
jacobine Jacobin *n*
jacobsite Jakobsit *m* (Min)
jacodine Jacodin *n*
„Jacquinot" advantage Apertur-Vorteil *m*
  (Spektr)
jade Jade *f* (Min), Nephrit *m* (Min)
jadeite Jadeit *m* (Min)
Jaffé's base Jaffé-Base *f*
jag Auszackung *f*, Zacken *m*, Zahn *m*
jag *v* auszacken
jagged ausgezackt, zackig
jail fever Fleckfieber *n* (Med), Typhus *m*
jaipurite Graukobalterz *n* (Min); Jaipurit *m*
  (Min)
jalap Jalape *f* (Bot)
jalapic acid Jalapinsäure *f*
jalapin Jalapin *n*
jalap [resin] Jalapenharz *n*
jalap root Jalapknolle *f* (Bot)
jalap soap Jalapseife *f*
jalpaite Jalpait *m* (Min)
jalpinolic acid Jalpinolsäure *f*
jam Konfitüre *f*, Marmelade *f*
jam *v* blockieren, drücken, einkeilen,
  einklemmen, einquetschen, festfressen,
  pressen, verklemmen, verstopfen
Jamaica pepper Jamaicapfeffer *m*
jamaicine Jamaicin *n*
jamba oil Jambaöl *n*
jambulol Jambulol *n*
jamesonite Jamesonit *m* (Min), Bergzundererz *n*
  (Min), Federerz *n* (Min), Querspießglanz *m*
  (Min)
jamming Festfressen *n*
Janus electrode Januselektrode *f*
Janus green Janusgrün *n* (Färb)
japaconine Japaconin *n*

japaconitine Japaconitin *n*
japan Japanlack *m*, Lack *m*
japan *v* mit Japanlack überziehen, schwarz
  streichen
Japan camphor Japancampher *m*
Japan earth Katechin *n* (Färb)
Japanese gelatin Agar-Agar *n*
japanic acid Japansäure *f*
Japan lacquer Japanlack *m*
japanner's gilding Lackvergoldung *f*
Japan tallow Japantalg *m*, Japanwachs *n*
Japan varnish Japanlack *m*
Japan wax Japanwachs *n*, Japantalg *m*
Japan wood Japanholz *n*
japonic acid Japonsäure *f*
jar Gefäß *n*, Krug *m*, Standgefäß *n*, weithalsige
  Flasche *f*
jar *v* knarren
jar diffusion Gefäßdiffusion *f*
jargon Jargon *m* (Min)
jarosite Jarosit *m* (Min)
jarring piston Rüttelkolben *m*
jasminaldehyde Jasminaldehyd *m*
jasmine oil Jasminöl *n*
jasmine root, yellow ~ Gelseminwurzel *f* (Bot)
jasmone Jasmon *n*
jasper Jaspis *m* (Min), blood-colored ~
  Röteljaspis *m* (Min)
jasper agate Jaspisachat *m* (Min)
jasperated china Jaspisporzellan *n*
jasper-colored jaspisfarben
jasper opal Jaspisopal *m* (Min)
jasper pottery Jaspissteingut *n*
jasper ware Jaspissteingut *n*
jaspery jaspisartig
jasponyx Jaspisonyx *m* (Min)
jatamansic acid Jatamansäure *f*
jatrorrhizine Jatrorrhizin *n*
jaulingite Jaulingit *m*
jaundice Gelbsucht *f* (Med)
Java cinnamon Javazimt *m*
javanicin Javanicin *n*
Java plum Jambulfrucht *f* (Bot)
Javelle's desinfecting liquor Javellesche Lauge *f*,
  Kaliumhypochloritlösung *f*
Javelle water Javellesche Lauge *f*, Eau de
  Javelle *n*, Kaliumhypochloritlösung *f*
jaw Backe *f*
jaw breaker Backenbrecher *m*,
  Zerkleinerungsmaschine *f*
jaw crusher Backenbrecher *m*
jaw crushing Vorzerkleinern *n*
jean Baumwollköper *m* (Text)
jecolein Jecolein *n*
jecorin Jecorin *n*

jefferisite Jefferisit *m* (Min)
jeffersonite Jeffersonit *m* (Min)
jellify *v* gallertartig werden
jelly Gallert *n*, Gallerte *f*, Gelee *n*, Sülze *f*, **to turn into** ~ gelieren
jelly *v* gelieren
jelly-fish Qualle *f* (Zool)
jellying Gelbildung *f*
jell[y]ing agent Gelierhilfe *f*
jelly-like gallertähnlich, gallertartig
jelly-like mass Gallertmasse *f*
jelly-like substance Gallertsubstanz *f*
Jena glass Jenaer Glas *n*
jenite Jenit *m*
jenkinsite Jenkinsit *m* (Min)
jequiritin Abrin *n*
jeremejewite Jeremejewit *m* (Min)
jerk Ruck *m*
jerkingly ruckweise
jerky ruckartig, ruckweise, stoßartig
jersey cloth Jersey *m* (Text)
jervaic acid Jervasäure *f*
jervane Jervan *n*
jervine Jervin *n*
jesaconitine Jesaconitin *n*
Jesuits' bark Chinarinde *f*, Jesuitenrinde *f*
Jesuits' powder Jesuitenpulver *n*
Jesuit tea Matéblätter *n pl*
jet Strahl *m*, Düse *f*, Gießkopf *m* (Gieß), Jett *m* *n* (Min), Verteilerdüse *f*, ~ **of liquid** Flüssigkeitsstrahl *m*
jet agitator Strahlmischer *m*
jet black jettschwarz
jet blender Strahlmischer *m*
jet burner Strahlbrenner *m*
jet carburetor Zerstäubungsvergaser *m*
jet condenser Einspritzkondensator *m*
jet cooler Einspritzkühler *m*
jet crystallizer Sprühkristallisator *m*
jet drier Düsentrockner *m*
jet dyeing Jet-Färberei *f*
jet engine Düsentriebwerk *n*, Düsenaggregat *n*
jet flame Stichflamme *f*
jet fuel Düsenkraftstoff *m*, Düsentreibstoff *m*
jet glass Jettglas *n*
jet mill Strahlmühle *f*
jet mixer Strahlmischer *m*
jet mold Spritzform *f*
jet molding Spritzverfahren *n*
jet molding nozzle Spritzdüse *f*
jet plane Düsenflugzeug *n*
jet power plant Strahltriebwerk *n*
jet-propelled plane Düsenflugzeug *n*
jet propulsion Düsenantrieb *m*, Strahlantrieb *m*
jet propulsion engine Rückstoßmotor *m*
jet propulsion principle Strahlrückstoßprinzip *n*
jet pump Ejektor *m*, Strahlpumpe *f*
jet regulator Strahlregler *m*
jet ring Düsenring *m*

jet scrubber Düsenwäscher *m*, Strahlwäscher *m*
jet solder method Spritzlötverfahren *n*
jet suction pump Saugstrahlpumpe *f*
jettisoning gear Schnellablaß *m*
jettison tank Benzintank *m* zum Abwerfen
jet tube reactor Strahldüsenreaktor *m*
jetty jettähnlich
jet unit Strahltriebwerk *n*
jewel Edelstein *m*, Juwel *m*
jewel[l]er's borax Juwelierborax *m*, Rindenborax *m*
jewel[l]er's red Polierrot *n*
jew's pitch Asphalt *m*
jib Ausleger *m*
jib crane Auslegerkran *m*
jib rotary crane Auslegerdrehkran *m*
jig Jigger *m* (Färb)
jig dyer Jiggerfärber *m*
jig table Schüttelrätter *m*, Schwingrätter *m*
jig welding Schweißen *n* (in einer Vorrichtung)
job Arbeit *f*, Beschäftigung *f*
jobbing ink Akzidenzfarbe *f*
jobbing varnish Universallack *m*
Job's method Jobsches Verfahren *n*
job worker Stückarbeiter *m*
jodival Jodival *n*
jodosterol Jodosterin *n*
jog formation Sprungbildung *f*
johannite Johannit *m* (Min), Uranvitriol *n* (Min)
johnstrupite Johnstrupit *m* (Min)
join *v* verbinden, aneinanderfügen, angliedern, beitreten (teilnehmen), fugen (Techn), ketten (Techn), koppeln, verknüpfen, zusammenfügen
joinability Verbindbarkeit *f*
joined verbunden, vereinigt, gepaart
joiner Schreiner *m*, Tischler *m*
joiner's bench Hobelbank *f*
joiner's glue Tischlerleim *m*
joiner's putty Holzkitt *m*
joining Verbinden *n*, Koppeln *n*, Verbindung *f*
joining element Verbindungselement *n*
joining flange Anschlußflansch *m*
joint Verbindung[sstelle] *f*, Bindeglied *n*, Fuge *f*; Gelenk *n*, Kupplung *f*, Verbindungsnaht *f*, Verbindungsstück *n*, **ball- and socket** ~ Kugelgelenk *n*, **permanent** ~ unlösbare Verbindung *f*, **universal** ~ Kardangelenk *n*, **watertight** ~ wasserdichte Verbindung *f*, **manufacturer's** ~ Herstellerverschluß *m* (z. B. Karton)
joint *a* gemeinsam
joint adapter Übergangsstück *n*
joint box Verbindungsmuffe *f* (Elektr)
joint connection Knotenverbindung *f*
joint glue Montageleim *m*
joint grease Dichtungsfett *n*
jointing Verbindung *f*

jointing mortar Fugmörtel *m*
jointing sleeve Verbindungsmuffe *f* (Elektr)
jointless fugenlos
jointless floor Spachtelfußboden *m*
joint packing Flanschendichtung *f*
joint pipe Gelenkrohr *n*
joint ring Dichtungsring *m*
jolt Ruck *m*, Stoß *m*
jolt *v* rütteln
jolting machine Rüttelmaschine *f*
jordanite Jordanit *m* (Min)
jordan mill Kegelmühle *f*
jordisite Jordisit *m* (Min)
joseite Joseit *m* (Min)
josephinite Josephinit *m* (Min)
joule Joule *n* (Maß)
Joulean heat Joulesche Wärme *f*
Joule's heat loss Joulescher Wärmeverlust *m*
Joule-Thomson effect Joule-Thomson-Effekt *m*
    (Techn)
journal Zapfen *m* (Techn), Zeitschrift *f,* **pointed**
    ~ Spitzzapfen *m,* ~ **on end of shaft**
    Stirnzapfen *m*
journal bearing Achslager *n*, Halslager *n*
journal box Achsbüchse *f*
journal friction Zapfenreibung *f*
juddite Juddit *m*
judg[e]ment Gutachten *n*, Urteil *n*
jug Krug *m*
juglone Juglon *n*, Nucin *n*
juglonic acid Juglonsäure *f*
juice Saft *m,* **to extract the** ~ **[from]** entsaften
juiciness Saftigkeit *f*
juicy saftig
julep Julep *m*
julienite Julienit *m* (Min)
juline Julin *n*
julocrotic acid Julocrotinsäure *f*
julolidine Julolidin *n*
juloline Julolin *n*
jump Satz *m*, Sprung *m*
jump *v* springen, ~ **over** überspringen
jump mechanism Sprungmechanismus *m*
junction Berührungspunkt *m* (Math),
    Knotenpunkt *m*, Lötstelle *f* (Techn),
    Übergang *m,* **cold** ~ kalte Lötstelle, ~ **point**
    Verbindung[sstelle] *f*
junction box Abzweigdose *f* (Elektr),
    Verbindungsmuffe *f* (Elektr)
junction clean-up Kurzschlußbeseitigung *f*
junction point Knotenpunkt *m*,
    Verkettungspunkt *m*
junipene Junipen *n*
juniper Wacholder *m* (Bot)
juniper camphor Wacholdercampher *m*
juniper gum Wacholderharz *n*
juniperic acid Juniperinsäure *f*
juniper oil Wacholderöl *n*
juniperol Juniperol *n*, Longiborneol *n*

juniper resin Wacholderharz *n*
juniper tar oil Cadeöl *n*
junk Altmaterial *n*
justification Begründung *f,* Berechtigung *f,*
    Rechtfertigung *f*
jute Jute[faser] *f,* **to batch** ~ batschen
jute ash Juteasche *f*
jute black Juteschwarz *n*
jute fiber Jutefaser *f*
jute yarn, heckled ~ Jutehechelgarn *n*
juxtaposition Aneinanderlagerung *f*

K

K-acid K-Säure *f*
kaemmererite Kämmererit *m* (Min)
kainic acid Kainsäure *f*
kainite Kainit *m* (Min)
kainosite Kainosit *m* (Min)
kairine Kairin *n*
kairoline Kairolin *n*
Kaiser oil Kaiseröl *n*
kakoxene Kakoxen *m* (Min)
kalignost (HN) Kalignost *n*,
    Natriumtetraphenyloborat *n*
kalinite Kalialaun *m* (Min), Kalinit *m* (Min)
kamacite Kamacit *m* (Min)
kamala Kamala *f* (Pharm)
kamarezite Kamarezit *m* (Min)
kamlolenic acid Kamlolensäure *f*
kammogenin Kammogenin *n*
kanamycin Kanamycin *n*
kanirin Kanirin *n*
kanosamine Kanosamin *n*
kanya butter Kanyabutter *f*
kanzuiol Kanzuiol *n*
kaolin Kaolin *n* (Min), Porzellanerde *f* (Min),
    Porzellanton *m* (Min), Weißtonerde *f* (Min)
kaolinite Kaolinit *m* (Min)
kaolinization Kaolinisierung *f*
kaolinize *v* kaolinisieren, in Kaolin verwandeln
Kaplan bucket Kaplanschaufel *f*
kapok oil Kapoköl *n*
karanjic acid Karanjisäure *f*
karanjol Karanjol *n*
karathane Karathan *n*
karelinite Karelinit *m* (Min)
karyinite Karyinit *m* (Min)
katuranin Katuranin *n*
kavain Methysticin *n*
kava root Kawawurzel *f* (Bot)
kawaic acid Kawasäure *f*
keel, hollow ~ Hohlkiel *m*
keel bolt Kielbolzen *m*
keeler Abwaschwanne *f*
keep *v* behalten, bewahren, ~ alive
    aufrechterhalten, am Leben erhalten, ~ away
    [from] abhalten, ~ back einbehalten, ~ going
    unterhalten (Reaktion, Verbrennung), ~ off
    abhalten, ~ up instandhalten
keeping properties Lagerbeständigkeit *f*
kefir Kefir *m*
keg Faß, Fäßchen *n*
keilhauite Keilhauit *m* (Min)
Kekulé formula Kekulé'sche Benzolformel *f*
kelene Kelen *n* (HN)
kelp Kelp *n*, Tang *m*, Varek *m*
kelp burner Salpetersieder *m*
kelp soda Vareksoda *f*
Kelvin effect Kelvin-Effekt *m* (Elektr)

Kelvin temperature Kelvintemperatur *f*
Kelvin [temperature] scale
    Kelvin[temperatur]skala *f*
kelyphite Kelyphit *m* (Min)
kentrolite Kentrolith *m* (Min)
kerasin Kerasin *n*
keratin Keratin *n*
keratin tissue Keratingewebe *n*
kerf Einschnitt *m*, Kerbe *f*
kermes Kermes[farbstoff] *m* (Färb)
kermes berry Kermesbeere *f* (Bot)
kermes grains Kermeskörner *n pl*
kermesite Kermesit *m* (Min), Antimonblende *f*
    (Min), Antimonzinnober *m* (Min),
    Rotspießglanz[erz] (Min), Spießblende *f*
    (Min)
kermes mineral Antimonkermes *m* (Min),
    Mineralkermes *m* (Min)
kermes red Kermesrot *n*
kermes scarlet Kermesscharlach *m*
kermisic acid Kermessäure *f*
kernel Kern *m* (Obst)
kernite Kernit *m* (Min)
kerosene Kerosin *n*, Leuchtöl *n*, fine ~
    Kaiseröl *n*
kerosene burner Petroleumbrenner *m*
kerosene lamp Petroleumlampe *f*
kerosine Kerosin *n*, Leuchtöl *n*
Kerr cell Kerrzelle *f*
kerstenite Kerstenit *m* (Min)
ketazine Ketazin *n*
ketene Keten *n*
ketimine Ketimin *n*
ketine Ketin *n*
keto acid Ketosäure *f*
keto-acidosis Ketoazidose *f* (Med)
ketoamine Ketoamin *n*
keto compound Ketoverbindung *f*
keto-enol equilibrium
    Keto-Enol-Gleichgewicht *n*
keto-enol tautomerism Keto-Enol-Tautomerie *f*
keto ester Ketoester *m*
keto form Ketoform *f*
ketogenesis Ketogenese *f*, Ketonbildung *f*
ketoglutarate dehydrogenase
    Ketoglutaratdehydrogenase *f* (Biochem)
ketoglutaric acid Ketoglutarsäure *f*
keto group Ketogruppe *f*
ketogulonic acid Ketogulonsäure *f*
ketohexokinase Ketohexokinase *f* (Biochem)
ketohexose Ketohexose *f*
ketoindoline, 2- ~ Oxindol *n*
ketol Ketol *n*, Ketonalkohol *m*
ketole Indol *n*
ketone Keton *n*, aromatic ~ aromatisches Keton
ketone alcohol Ketol *n*, Ketonalkohol *m*

ketone body Ketonkörper *m*
ketone imide Ketonimid *n*
ketone-like ketonartig
ketone oil Ketonöl *n*
ketone resin Ketonharz *n*
ketone splitting Ketonspaltung *f*
ketonic ketonartig
ketonic cleavage Ketonspaltung *f*
ketonuria Ketonurie *f* (Med),
    Ketonkörperausscheidung *f* (im Urin) (Med)
ketopantoic acid Ketopantosäure *f*
ketopentamethylene Ketopentamethylen *n*
ketopentose Ketopentose *f*
ketopinic acid Ketopinsäure *f*
ketose Ketose *f,* Ketozucker *m*
ketosis Ketose *f* (Med)
ketostearic acid Ketostearinsäure *f,* 6- ~
    Lactarinsäure *f*
ketosugar Ketose *f,* Ketozucker *m*
ketoxanthene, 9- ~ Xanthon *n*
ketoxime Ketoxim *n*
kettle Kessel *m*
kevatron Kevatron *n*
key Schlüssel *m,* Keil *m,* Taste *f,* hollow ~
    Hohlkeil *m*
key *v* [ver]keilen, ~ on aufkeilen
key arrangement Keilanordnung *f*
key coat Grundierung *f* für Weich-PVC
key component Schlüsselkomponente *f*
keyed festgekeilt
key face of nut Schlüsselfläche der Mutter
key-opening end Abrolldeckel *m*
key position Schlüsselstellung *f*
key seat Keilnut *f,* Keilschlitz *m*
keyseating machine Keilnutenziehmaschine *f*
key steel Keilstahl *m*
keystone Gewölbeschlußstein *m* (Arch)
keyway Keilnut *f,* Kerbnute *f*
keywaying machine Keilnutenstoßmaschine *f*
keyways, to cut ~ Nuten *pl* stoßen, to mill ~
    Nuten *pl* fräsen
keyword Stichwort *n*
khaki Khaki *n*
khaki color Khakifarbe *f*
khaki-colored khakifarben
khellactone Khellacton *n*
khellin Khellin *n*
khellinin Khellinin *n*
khellinone Khellinon *n*
khellinquinone Khellinchinon *n*
khellol Khellol *n*
kick out *v* herausschlagen
kid leather Glacéleder *n*
kidney Niere *f* (Med)
kidney ore Nierenerz *n* (Min)
kidney remedy Nierenmittel *n* (Pharm)
kidney-shaped nierenförmig
kidney stone Nephrit *m* (Min), Nierenstein *m*
    (Med)

kier Beuchkessel *m,* Bleichfaß *n*
kier bleaching Packbleiche *f*
kier boiling Beuchen *n*
kieselgu[h]r Kieselgur *f,* Berggur *f*
kill *v* [ab]töten, beruhigen (Metall)
killed beruhigt (Stahl)
killing Abtöten *n*
killinite Killinit *m* (Min)
kiln Brennofen *m,* Darre *f,* Darrofen *m,*
    Kühlofen *m* (Glas), Röstofen *m,* opening of a
    ~ Ofenöffnung *f,* ~ for roasting iron ore
    Eisensteinröstofen *m*
kiln brick feuerfester Ziegel *m*
kiln chamber Darraum *m*
kiln-dried malt Darrmalz *n*
kiln-dry *v* brennen, rösten, darren
kiln-drying Darrarbeit *f,* Darren *n*
kiln floor Darrboden *m*
kiln plant Darranlage *f*
kiloampere Kiloampere *n* (Elektr)
kilocalorie große Kalorie *f,* Kilo[gramm]kalorie *f*
kilodyne Kilodyn *n*
kilogauss Kilogauß *n* (Magn)
kilogram Kilogramm *n*
kilogramme (Br. E.) Kilogramm *n*
kilogram-meter Meterkilogramm *n*
kilogram weight Kilopond *n*
kilohertz Kilohertz *n* (Elektr)
kiloohm Kiloohm *n* (Elektr)
kilovolt Kilovolt *n* (Elektr)
kilowatt Kilowatt *n* (Elektr)
kilowatt hour Kilowattstunde *f* (Elektr)
kind Art *f,* Sorte *f*
kindle *v* entzünden, anfachen, anzünden,
    entflammen
kindling Entzünden *n,* Anzündmaterial *n*
kinematics Kinematik *f,* Bewegungslehre *f*
kinetic kinetisch
kinetic energy Bewegungsenergie *f,* kinetische
    Energie *f*
kinetics Kinetik *f,* steady-state ~ Kinetik *f* im
    stationären Zustand
kinetin Kinetin *n*
kingbolt Drehzapfen *m,* Hauptbolzen *m*
kingpin Hauptbolzen *m,* Spurzapfen *m*
king's blue Königsblau *n,* Leydenerblau *n*
king's yellow gelbe Arsenblende *f* (Min),
    Operment *n* (Min), Rauschgelb *n* (Min)
kink band Knickband *n*
kink band density Knickbanddichte *f*
kino Kinogummi *n,* Kinoharz *n*
kino gum Kinogummi *n,* Kinoharz *n*
kinoin Kinoin *n*
kinotannic acid Kinogerbsäure *f*
kinova Chinova *f*
Kipp's apparatus Kippscher Apparat *m* (Chem)
Kipp's generator Kippscher Apparat *m* (Chem)
kirsch Kirschwasser *n*

kish Eisenschaum *m* (Metall), Garschaum *m* (Metall), Graphit *m* (Min)
kit Werkzeugtasche *f*
kitchen utensil Küchengerät *n*
kitol Kitol *n*
Kjeldahl flask Kjeldahlkolben *m*
Kjeldahl nitrogen method Stickstoffbestimmung *f* nach Kjeldahl
klaproth[ol]ite Klaproth[ol]it *m* (Min), Kupferwismuterz *n* (Min)
klementite Klementit *m*
klipsteinite Klipsteinit *m*
klovosene Klovosen *n*
knack Kunstgriff *m*
knead *v* kneten
kneadability Knetbarkeit *f*
kneadable knetbar
kneader Kneter *m*, Knetmaschine *f*
kneader mixer Mischkneter *m*
kneader pump Knetpumpe *f*
kneading machine Kneter *m*, Knetmaschine *f*
kneading mill Knetapparat *m*
knebelite Knebelit *m* (Min), Eisenknebelit *m* (Min)
knee-toggle [clamping] Kniehebelschließsystem *n*
knife Messer *n*, **inserted** ~ eingesetztes Messer
knife block Messerwelle *f*
knife carrier Messerträger *m*
knife coating Rakelstreichverfahren *n*, Walzenstreichverfahren *n*
knife crusher Messerbrecher *m*
knife drum Messerwalze *f*
knife-edge Messerschneide *f*, Schneide *f*, Schneidmesser *n*
knife-edge support Schneidenlager *n*
knife-handle Messergriff *m*
knife holder Messerhalter *m*
knife roll Messerwalze *f*
knife scale Messerschale *f*
knife sledge Messerschlitten *m*
knife support Messerstütze *f*
knifing surfacer Ziehspachtel *f*
knit *v* stricken, wirken (Text)
knitted goods Wirkwaren *pl* (Text), Trikotagen *pl* (Text)
knitting, circular ~ Rundwirkerei *f* (Text)
knitting machine Strickmaschine *f*
knob Knauf *m*, Knopf *m*
knobby knotig, höckerig
knock Schlag *m*, Stoß *m*, **acceleration** ~ Beschleunigungs-Klopfen *n*, **high-speed** ~ Hochgeschwindigkeits-Kopfen *n*, **low-speed** ~ Beschleunigungs-Klopfen *n*
knock *v* schlagen, stoßen, ~ **off** abklatschen, abschlagen, abstoßen, ~ **over** umwerfen
knocker Klöppel *m*, Klopfer *m*
knocking of the motor Klopfen *n* des Motors
knocking-out Herausstoßen *n*
knockmeter Klopfdetektor *m*

knockout Ausdrückvorrichtung *f*, Ausstoßer *m*, Auswerfer *m*, **free water** ~ Vorabscheider *m* (Öl)
knock-out cam Auswerfernocke *f*
knockout die Stanzform *f*
knockout frame Ausdrückrahmen *m*
knockout pin Ausdrückstift *m*, Ausheberstift *m*, Auswerferstift *m*
knockout pin plate Ausdrückstiftplatte *f*, Auswerferplatte *f*
knock-out ring Auswerferring *m*
knock-over bar Abschlagschiene *f*
knock-resistant klopffest
knock-stable klopffest
knot Knoten *m*
knot *v* knüpfen, knoten
knot grass Knöterich *m*
knotting varnish Knotenlack *m*
knotty knotig, höckerig
knotty section Fladerschnitt *m*
knowledge Wissen *n*, Erkenntnis *f*, Kenntnis *f*
Knudsen flow Diffusion *f* freier Moleküle (pl), thermische Molekularströmung *f*
knurl nut Rändelmutter *f*
kobellite Kobellit *m* (Min)
Koch's bacillus Tuberkelbazillus *m*
koechlinite Koechlinit *m* (Min)
koenenite Könenit *m* (Min)
kogasin Kogasin *n*
kojic acid Kojisäure *f*
Ko-kneader Ko-Kneter *m*
kokum-butter Kokumbutter *f*
kokusagine Kokusagin *n*
kokusagininic acid Kokusagininsäure *f*
kola seed Cola[nuß] *f*
Kolle culture flask Kolle-Schale *f*
komarite Komarit *m* (Min)
konimeter Konimeter *n*
konimetric konimetrisch
koninckite Koninckit *m* (Min)
kopsane Kopsan *n*
kopsinane Kopsinan *n*
kopsine Kopsin *n*
kopsinilam Kopsinilam *n*
kopsinin Kopsinin *n*
Kordofan gum Gummiarabikum *n*, Kordofangummi *n*
korynite Korynit *m* (Min), Arsenantimonnickelglanz *m* (Min)
kosin Kussin *n*
Kossel-Sommerfeld displacement law Kossel-Sommerfeldsches Verschiebungsgesetz *n* (Spektr), Satz *m* der spektroskopischen Verschiebung (Spektr)
k[o]umiss Milchwein *m*
koussin Kussin *n*
kousso flowers Kussoblüten *f pl* (Bot)
kousso powder Kussopulver *n*
krablite Krablit *m*

kraft paper Kraft-Papier *n*, Packpapier *n*
krameria Ratanhiawurzel *f* (Bot)
krantzite Krantzit *m* (Min)
kraurite Kraurit *m* (Min)
Krebs cycle Citronensäurezyklus *m*,
  Tricarbonsäurezyklus *m*,
  Zitronensäurezyklus *m*
kreittonite Kreittonit *m* (Min)
kremersite Kremersit *m* (Min)
Kremnitz white Kremnitzerweiß *n*,
  Kremserweiß *n*
krennerite Krennerit *m* (Min)
kromycin Kromycin *n*
kromycol Kromycol *n*
kryogenin Kryogenin *n*
kryptocyanine Kryptocyanin *n*
krypton (Symb. Kr) Krypton *n*
K-shell K-Schale *f* (Atom)
kümmel Kümmel[branntwein] *m*
kunzite Kunzit *m* (Min)
Kupffer's cell Sternzelle *f* (Histol)
kurchamine Kurchamin *n*
kuromatsuene Kuromatsuen *n*
kuromatsuol Kuromatsuol *n*
kutkin Kutkin *n*
kwangoside Kwangosid *n*
kyanethine Kyanäthin *n*
kyanidine Kyanidin *n*
kyanite Kyanit *m* (Min), blauer Disthen *m*
  (Min)
kyanization Kyanisation *f*, Kyanisierung *f*
kyanize *v* kyanisieren
kyanizing Kyanisieren *n*
kyanmethin Kyanmethin *n*
kyaphenine Kyaphenin *n*
kymograph Kymograph *m*
kynurenic acid Kynurensäure *f*
kynureninase Kynureninase *f* (Biochem)
kynurenine Kynurenin *n*
kynurenine yellow Kynureningelb *n*
kynurin Kynurin *n*
kyrosite Kyrosit *m*

L

lab Labferment *n*, Labor *n*, Laboratorium *n*
lab coat Arbeitsanzug *m*, Laborkittel *m*
labdane Labdan *n*
labdanolic acid Labdanolsäure *f*
labdanum Ladanum *n*, Ladangummi *n*,
  Ladanharz *n*
label Aufkleber *m*, Aufklebezettel *m*,
  Aufschrift *f*, Beschriftung *f*,
  Bezeichnungsschild *n*, Etikett *n*
label *v* beschriften, auszeichnen, bekleben,
  bezeichnen, etikettieren, markieren
label lacquer Etikettenlack *m*
label[l]ing machine Etikettiermaschine *f*
labelling technique Markierungstechnik *f*
  (Isotopen)
label printing machine
  Etikettenbedruckmaschine *f*
lab ferment Labferment *n* (Biochem)
labiate Lippenblütler *m* (Bot)
labile labil, instabil, schwankend, unsicher,
  unbeständig (Chem), zersetzlich (Chem)
lability Labilität *f*, **state of** ~
  Labilitätszustand *m*
lab journal Laborjournal *n*, Versuchsprotokoll *n*
labor, division of ~ Arbeitsteilung *f*
laboratory Labor *n*, Laboratorium *n*,
  Versuchsraum *m*, **hot** ~ heißes Laboratorium
laboratory accident Laboratoriumsunfall *m*
laboratory assistant Laborant[in] *m f*
laboratory balance Laboratoriumswaage *f*
laboratory [ball] mill Labormühle *f*
laboratory bench Labor[atoriums]tisch *m*,
  Labortisch *m*
laboratory clamp Laborklemme *f*
laboratory column with rotating belt
  Drehbandkolonne *f*
laboratory course Praktikum *n*
laboratory crusher Labormühle *f*
laboratory drying oven Labortrockenschrank *m*
laboratory equipment Laboreinrichtung *f*
laboratory examination Laboruntersuchung *f*
laboratory experiment
  Laboratoriumsexperiment *n*
laboratory findings Laborbefund *m*
laboratory fittings Armaturen für Labors (pl)
laboratory furnace Laboratoriumsofen *m*
laboratory furniture Labormöbel *pl*
laboratory inspection Laboruntersuchung *f*
laboratory installation
  Laboratoriumseinrichtung *f*
laboratory investigation Laborprüfung *f*,
  Laboruntersuchung *f*
laboratory mixer Labormischer *m*
laboratory pump Laborpumpe *f*
laboratory size Laboratoriumsmaßstab *m*

laboratory stirrer Laborrührer *m*
laboratory study Laboruntersuchung *f*
laboratory superheater
  Laboratoriumsüberhitzer *m*
laboratory table Labortisch *m*, Arbeitstisch *m*
laboratory technician Chemotechniker[in] *m f*,
  Laborant[in] *m f*
laboratory test Labor[atoriums]versuch *m*,
  Laborprüfung *f*, Laboruntersuchung *f*
laboratory trial Labor[atoriums]versuch *m*
laboratory truck (Am. E.) Labor[kraft]wagen *m*
laboratory utensils Laboratoriumsgeräte *pl*
laboratory worker Laborhelfer *m*
laboratory yield Laborausbeute *f*
labor cost Lohnkosten *pl*
laborer Arbeiter *m*, Hilfsarbeiter *m*
labor-saving Arbeitseinsparung *f*, arbeitssparend
Labrador feldspar Labradorfeldspat *m* (Min),
  Labradorit *m* (Min)
labradorite Labradorit *m* (Min),
  Kalkoligoklas *m* (Min), Labradorfeldspat *m*
  (Min)
Labrador stone Labradorstein *m* (Min),
  Labradorfeldspat *m* (Min), Labradorit *m*
  (Min)
laburnine Laburnin *n*
labyrinth gap Labyrinthspalt *m*
labyrinth seal Labyrinthdichtung *f*
lac Lack[harz *n*] *m*, Imprägnierharz *n*
laccaic acid Laccainsäure *f*, Lacksäure *f*
laccaine Laccain *n*
laccainic acid Laccainsäure *f*, Lacksäure *f*
laccase Laccase *f* (Biochem)
lacceroic acid Laccersäure *f*
laccol Laccol *n*, Urushinsäure *f*
laccolite Lakkolith *m*
lac dye Färberlack *m*
lac ester Lackester *m*
lac extract Lackextrakt *m*
lachnophyllic acid Lachnophyllumsäure *f*
lachnophyllum ester Lachnophyllumester *m*
lachrymal gland Tränendrüse *f*
lack Mangel *m*, ~ **of air** Luftmangel *m*, ~ **of**
  luster Glanzlosigkeit *f*, ~ **of oxygen**
  Sauerstoffmangel *m*
lack *v* fehlen
lacking, to be ~ fehlen
lacmoid Lackmoid *n*, Lakmoid *n*,
  Resorcinblau *n*
lacmus Lackmus *m*
lac operone lac-Operon *n* (Biochem)
lacquer Lack *m*, Lackfarbe *f*, Lackfirnis *m*,
  brushing ~ Streichlack *m*, **coat of** ~
  Lacküberzug *m*, **protecting** ~ Schutzlack *m*,
  transparent ~ Lasurlack *m*
lacquer *v* lackieren

lacquer auxiliary Lackhilfsmittel *n*
lacquer base Lackbasis *f*
lacquer coat Lackschicht *f*
lacquer enamel Farblack *m*
lacquerer Lackierer *m*
lacquer film Lackhaut *f*
lacquer former Lackbildner *m*
lacquering by electrodeposition elektrostatische
  Lackierung *f*
lacquer layer, imperfections of the ~
  Lackfilmunterbrechungen *pl*
lacquer solvent Lackverdünner *m*
lac repressor lac-Repressor *m* (Biochem)
lacrimator Augenreizstoff *m*
lacroixite Lacroixit *m* (Min)
lac smearing machine Imprägniermaschine *f*
lactacidogen Lactacidogen *n*
lactalbumin Lactalbumin *n*
lactaldehyde Milchsäurealdehyd *m*
lactaldehyde reductase Laktaldehydreduktase *f*
  (Biochem)
lactam Lactam *n*, Laktam *n*
lactamide Lactamid *n*
lactaminic acid Lactaminsäure *f*
lactan Lactam *n*
lactarazulene Lactarazulen *n*
lactaric acid Lactarsäure *f*
lactarinic acid Lactarinsäure *f*
lactase Lactase *f* (Biochem)
lactate Lactat *n*, Laktat *n*, Milchsäureester *m*
lactate dehydrogenase Lactatdehydrogenase *f*
  (Biochem)
lactate racemase Laktatracemase *f* (Biochem)
lactation Laktation *f* (Med)
lacteal milchartig, milchig
lacthydroxamic acid Lacthydroxamsäure *f*
lactic acid Milchsäure *f*, **ester of** ~
  Milchsäureester *m*, **formation of** ~
  Milchsäurebildung *f*, **salt or ester of** ~
  Lactat *n*
lactic [acid] amide Lactamid *n*
lactic acid plant Milchsäureanlage *f*
lactic aldehyde Milchsäurealdehyd *m*
lactic fermentation Milchsäuregärung *f*
lactic nitrile Milchsäurenitril *n*
lactide Lactid *n*
lactiferous milchhaltig
lactim Lactim *n*
lactin Lactim *n*
lactitol Lactit *m*
lactobacillus (pl lactobacilli)
  Milchsäurebazillus *m* (Bakt)
lactobionic acid Lactobionsäure *f*
lactobiose Lactobiose *f*, Lactose *f*,
  Milchzucker *m*
lactobutyrometer Lactobutyrometer *n*,
  Milchfettmesser *m*
lactodensimeter Lactodensimeter *n*

lactoflavin Lactoflavin *n*, Riboflavin *n*, Vitamin
  B_2
lactogenic hormone Laktationshormon *n*
lactoglobulin Lactoglobulin *n*
lactolactic acid Dilactylsäure *f*
lactolide Lactolid *n*
lactometer Lactodensimeter *n*, Milchmesser *m*
lactonase Lactonase *f* (Biochem)
lactone Lacton *n*
lactonic acid Lactonsäure *f*
lactonic linkage Lactonbindung *f*
lactonitrile Lactonitril *n*, Milchsäurenitril *n*
lactophenine Lactophenin *n*
lactoprotein Laktoprotein *n*, Milcheiweiß *n*
lactosamine Lactosamin *n*
lactosazone Lactosazon *n*
lactoscope Lactoskop *n*
lactose Laktose *f*, Lactobiose *f*, Milchzucker *m*
lactosone Lactoson *n*
lactucarium Lactucarium *n*
lactulose Lactulose *f*
lactyl Lactyl-
lactyl guanidine Alakreatin *n*
lactylolactic acid Lactylmilchsäure *f*
lac varnish Firnislack *m*, Lackfirnis *m*
ladanum Ladanum *n*, Ladangummi *n*,
  Ladanharz *n*
ladder Leiter *f*, Laufmasche *f*
ladder chain Hakenkette *f*
ladderproof laufmaschenfest (Text), maschenfest
  (Text)
ladder-proofing agent Maschenfestmittel *n*
ladle Schöpflöffel *m*; Ausschöpfkelle *f*,
  Gußpfanne *f* (Gieß), Pfanne *f* (Gieß), **lip of**
  ~ Pfannenausguß *m*, **rim of** ~
  Pfannenrand *m*, ~ **for adding something**
  Nachsetzlöffel *m*, ~ **for ring carrier**
  Gabelpfanne *f*
ladle *v* schöpfen, ~ **out** ausschöpfen
ladle bail Pfannengehänge *n*
ladle capacity Pfanneninhalt *m*
ladle carriage Gießwagen *m*
ladle guide Pfannenführung *f*
ladle handle Pfannenbügel *m*
ladle sample Schöpfprobe *f*
ladle support Pfannenbügel *m*
ladle truck Gießwagen *m*
l[a]evorotation Linksdrehung *f*
l[a]evorotatory linksdrehend
lag Nacheilung *f* (Elektr), Verzögerung *f*, **angle
  of** ~ Nacheilwinkel *m* (Elektr), **thermal** ~
  thermische Verzögerung, **time** ~ zeitliche
  Nacheilung
lag [behind] *v* nacheilen (Elektr)
lagonite Lagonit *n*
lag phenomenon Verzögerungserscheinung *f*
Lagrangian function Lagrangesche Funktion *f*
  (Math)
lag screw Vier- oder Sechskantholzschraube *f*

laid-open specification Offenlegungsschrift *f*
(Pat)
laifan Laifan *n*
lake Pigmentfarbe *f*, Lack *m*, Lackfarbe *f*, round
~ Kugellack *m*
lake dye Färberlack *m*
lake dyestuff Lackfarbstoff *m*
lake former Lackbildner *m*
lake ore Eisensumpferz *n* (Min)
lake pigment Lackpigment *n*, Verschnittfarbe *f*
lambda Mikroliter *m*
lambert (unit of brigthness) Lambert *n* (Opt,
Maßeinheit der Helligkeit)
Lambert-Beer law Lambert-Beersches Gesetz *n*
lambertine Lambertin *n*
lambertite Lambertit *m* (Min)
lamella Blättchen *n*
lamella evaporator Lamellenverdampfer *m*
lamellar lamellar, blättrig, lamellenartig,
schichtartig
lamellar cleavage Blätterbruch *m*
lamellar exchanger
Lamellenwärmeaustauscher *m*
lamellar pyrites Blätterkies *m* (Min)
lamellar structure Lamellenstruktur *f*,
Schichtgefüge *n*
lamellar talc Talkglimmer *m* (Min)
lamellar tube Lamellenröhre *f*
lamellate[d] lamellar
lamellated peat Blättertorf *m*
lamelliform blättrig, plättchenartig
lamellose lamellar
lamina Blättchen *n*, Lamelle *f*, Plättchen *n*,
Schicht *f* (Holz)
laminar laminar (Strömung)
laminar flow Laminarströmung *f*,
Bandströmung *f*
laminaribiitol Laminaribiit *m*
laminate Kunststoff-Folie *f*,
Schicht[preß]stoff *m*, Verbundwerkstoff *m*
laminate *v* in dünne Blättchen *pl* aufspalten,
abschuppen, kaschieren (Pap), schichten,
walzen
laminate[d] *a* lamellar, blätterförmig, blättrig,
geschichtet, lamellenartig, plättchenartig,
schuppig, (Glas) zusammengesetzt
laminated bag Verbundfolienbeutel *m*
laminated brush Blätterbürste *f*
laminated core Blechkern *m*, Blechkernspule *f*
(Elektr)
laminated fabric Hartgewebe *n*
laminated film Kombinationsfolie *f*
laminated glass Schichtglas *n*
laminated iron core Eisenblätterkern *m*
laminated iron core coil Blätterkernspule *f*
laminated magnet Blättermagnet *m*,
Lamellenmagnet *m*
laminated material Schichtstoff *m*
laminated paper Hartpapier *n*

laminated plastic panel Schichtstoffplatte *f*
laminated sheet Schichtplatte *f*
laminated spring Blattfeder *f*
laminated structure Schichtgefüge *n*
laminated wood Lagenholz *n*, Schichtholz *n*
laminate skin Schichtstoffaußenhaut *f*
laminating press Presse *f* zum Kaschieren
laminating wax Kaschierwachs *n*
lamination Kaschierung *f*, Lamellenstruktur *f*,
Preßbahn *f*, Schichtenbildung *f*,
Schichtstofftechnik *f*, Schichtung *f*
lamination coating Kaschieren *n* (Pap)
laminography Schichtaufnahme *f* (Gewebe)
lamp Lampe *f*, gas-filled ~ gasgefüllte Lampe,
reflecting ~ Lampe *f* mit Reflektor
lampblack Lampenruß *m*, Büttenruß *m*,
Flammruß *m*, Kienruß *m*, Lampenschwarz *n*,
Ölruß *m*, Schwärze *f*, manufacture of ~
Rußbrennerei *f*
lampblack factory Rußhütte *f*
lamp furnace Lampenofen *m*
lamp glass Glaszylinder *m*
lamp holder Glühbirnenfassung *f*,
Lampenfassung *f*
lamp hour Lampenbrennstunde *f*
lamp house Lampengehäuse *n*
[lamp] kerosine Leuchtpetroleum *n*
lamp oil Lampenöl *n*
lamp plate Lampenteller *m*
lamp pole Beleuchtungsmast *m*
lamp socket Lampenfassung *f*
lanacyl violet Lanacylviolett *n*
lanane Lanan *n*
lanarkite Lanarkit *m* (Min)
lancelet Amphioxus *m* (Zool), Lanzettfischchen
n (Zool)
lanceol Lanceol *n*
lanceolatin Lanceolatin *n*
land Abquetschgrat *m*, Land *n* (of a mold),
Abquetschfläche *f* (eines Werkzeuges)
land *v* landen
land area Abquetschrand *m*
landing Landung *f*
landing flap Landeklappe *f* (Luftf)
landing speed Landegeschwindigkeit *f* (Luftf)
landing strip Landebahn *f* (Luftf)
landmark Grenzstein, Feldstein; Markstein *m*
langbanite Langbanit *m* (Min)
langbeinite Langbeinit *m* (Min)
Langmuir trough Langmuir Mulde *f*
lanital Lanital *n*
lanoceric acid Lanocerinsäure *f*
lanolin acid Lanolinsäure *f*
lanolin cream Lanolincreme *f*
lanolin[e] Lanolin *n*, Wollfett *n*, crude ~
Rohlanolin *n*
lanolin manufacture Lanolinerzeugung *f*
lanopalmic acid Lanopalminsäure *f*
lanostane Lanostan *n*

lanosterol Lanosterin *n*
lansfordite Lansfordit *m* (Min)
lantana oil Lantanaöl *n*
lantanuric acid Lantanursäure *f*
lanthanide Lanthanid *n*
lanthanide group Lanthanidenreihe *f*
lanthanum (Symb. La) Lanthan *n*
lanthanum ammonium nitrate
　　Lanthanammoniumnitrat *n*
lanthanum carbonate Lanthancarbonat *n*
lanthanum chloride Lanthanchlorid *n*
lanthanum nitrate Lanthannitrat *n*
lanthanum oxide Lanthanoxid *n*
lanthanum potassium sulfate
　　Lanthankaliumsulfat *n*
lanthanum sulfate Lanthansulfat *n*
lanthanum sulfide Lanthansulfid *n*
lanthionine Lanthionin *n*
lanthopine Lanthopin *n*
lantol Lanthopin *n*
lanum Lanolin *n*
lap Polierscheibe *f*, Überstand *m*
lap *v* abtuschieren, läppen, polieren, schleifen;
　　sich überlappen
lapachenole Lapachenol *n*
lapachoic acid Lapachol *n*, Tecomin *n*
lapachol Lapachol *n*, Lapachosäure *f*, Tecomin *n*
lapachone Lapachon *n*
lapathy root Grindwurzel *f* (Bot)
lapinone Lapinon *n*
lapis divinus Kupferalaun *m*
lapis lazuli Lapislazuli *m* (Min), Azurblau *n*
　　(Farbe), Azurit *m* (Min), Azurstein *m* (Min),
　　Lasurit *m* (Min)
Laplace's differential equation Laplacesche
　　Differentialgleichung *f*
Laplacian transformation
　　Laplacetransformation *f*
lappaconitine Lappaconitin *n*
lapped (cutting edges) geläppt (Schnittflächen)
lapping Läppen *n*
lapping abrasive Läppmittel *n*
lapping machine Läppmaschine *f*
lapping strip Deckband *n*
lapping wheel Läppscheibe *f*
lapse Ablauf *m*, Versäumnis *n*
lap seam Überlapptnaht *f* (Dose)
lap-seamed überlappt gelötet
lap time Anzugszeit *f*
lap-welded geschweißt
lap-welding Flachschweißung *f*,
　　Übereinanderschweißung *f*;
　　Überlappungsschweißung *f*
larch Lärche *f* (Bot)
larch agaric Lärchenschwamm *m*,
　　Löcherschwamm *m*
larch pitch Lärchenpech *n*
larch resin Lärchenbaumharz *n*
larch wood oil Lärchenholzöl *n*

lard Schweinefett *n*, Schweineschmalz *n*
lardaceous schmalzig, fettig
larderellite Larderellit *m* (Min)
lard lubrication Speckschmierung *f*
lard oil Specköl *n*
lardy schmalzig
large groß, ausgedehnt, umfassend
large angle grain boundary
　　Großwinkelkorngrenze *f*
large calorie große Kalorie *f*, Kilokalorie *f*
large-grained grobkörnig
large-leaved großblätterig
large-scale factory Produktionsfabrik *f*
large-scale manufacture Massenherstellung *f*
large-scale test Großversuch *m*
large-scale workings Großabbau *m*
laricinolic acid Laricinolsäure *f*
lariciresinol Lariciresinol *n*
larixic acid Larixinsäure *f*
larmor precession Larmorpräzession *f* (Atom)
larva (pl. larvae) Larve *f* (Biol)
larvicide Raupenvertilgungsmittel *n*
larynx Kehlkopf *m* (Anat)
laser, ruby ~ Rubin-Laser *m*, **solid ~**
　　Festkörper-Laser *m*
laser apparatus Lasergerät *n*
laser beam Laserstrahl *m*
laser crystal Laserkristall *m*
laser diode Laserdiode *f*
laser equipment Laseranlage *f*
laser impulse Laserimpuls *m*
laser [light beam] Laser *m*
laser method Lasertechnik *f*
laser multiplier Laserverstärker *m*
laserol Laserol *n*
laserone Laseron *n*
laserpitin Laserpitin *n*
laser source Laseranregung *f*
lash *v* [fest]binden
lashing Schnur *f*
lasiocarpic acid Lasiocarp[in]säure *f*
lasiocarpine Lasiocarpin *n*
lasionite Lasionit *m* (Min)
last *v* [an]dauern, fortbestehen
lastex Lastex *n* (HN)
lastex yarn Lastexfaden *m*
lasting beständig, dauerhaft, haltbar
lasting effect Dauerwirkung *f*
last pass Schlußkaliber *n*
last runnings Nachlauf *m*
latch Falleisen *n*, Schnappschloß *n*,
　　Sicherheitsschloß *n*
latch plate Druckstiftplatte *f*, Halteplatte *f*
late ignition Nachzündung *f*
latency Latenz *f*
latent latent, gebunden
lateral seitlich
lateral branch Nebenzweig *m*
lateral deflection Seitenverschiebung *f*

lateral expansion Querdehnung *f*
lateral face Seitenfläche *f*
lateral induction Seiteninduktion *f*
laterally seitwärts, seitlich
lateral oscillation Querschwingung *f*
lateral reinforcement Querbewehrung *f*
lateral stability Querstabilität *f*
lateral strength Querfestigkeit *f*
laterite Laterit *m* (Min)
latex Latex *m* (Bot), Milchsaft *m* (Bot)
latex mill Latexmühle *f*
latex paint Latexanstrich *m*
lathe Drehbank *f,* Abdrehmaschine *f,*
    Drehmaschine *f,* **high-speed** ~
    Schnelldrehbank *f*
lathe center Drehbankspitze *f*
lathe chuck Spannfutter *n*
lather Schaum *m* (Seife)
lathe spindle Leitspindel *f*
lathe tool Drehstahl *m*
lathosterol Lathosterin *n*
latitude Breite *f* (Geographie), **degree of** ~
    Breitengrad *m*
latrobite Latrobit *m*
lattice Gitter *n,* **atomic** ~ Atomgitter *n,*
    body-centered cubic ~
    kubisch-raumzentriertes Gitter (Krist), **closest**
    packed hexagonal ~ hexagonal dichteste
    Kugelpackung *f,* **cubic** ~ kubisches Gitter
    (Krist), **face-centered** ~ flächenzentriertes
    Gitter (Krist), **unit cell volume of a** ~
    Gitterzellenvolumen *n*
lattice absorption edge Gitterabsorptionskante *f*
lattice arrangement Gitterordnung *f*
lattice bond Gitterbindung *f*
lattice complex Gitterkomplex *m*
lattice conductivity Gitterleitfähigkeit *f*
lattice constant Gitterkonstante *f* (Phys)
lattice cooling stack Lattengradierwerk *n*
lattice defect Gitterfehlordnung *f* (Krist),
    Gitterfehlstelle *f* (Krist)
lattice defect production
    Gitterfehlstellenerzeugung *f*
lattice dimensions Gitterabmessungen *pl* (Krist)
lattice dislocation Gitterstörung *f*
lattice electrons Gitterelektronen *pl*
lattice energy Gitterenergie *f* (Phys)
lattice forces Gitterkräfte *pl*
lattice heat Gitterwärme *f*
lattice imperfection Gitterfehlstelle *f* (Krist),
    Gitterstörung *f* (Krist)
lattice jib Gitterausleger *m*
latticelike gitterförmig
lattice parameter Gitterparameter *m*
lattice plane Netzebene *f* (Krist)
lattice point Gitterbaustein *m* (Phys)
lattice position Gitterstelle *f*
lattice reactor Gitterreaktor *m*
lattice rearrangement Gitterverlagerung *f*

lattice relaxation Gitterrelaxation *f*
lattice scattering Gitterstreuung *f*
lattice site Gitterplatz *m*
lattice spacing Gitterabstand *m*
lattice spectrum Gitterspektrum *n*
lattice structure Gitterbau *m*
lattice unit Gittereinheit *f*
lattice vacancy Gitterfehlstelle *f,*
    Gitterleerstelle *f,* Gitterlücke *f*
lattice vibration Gitterschwingung *f*
lattice vibration superposition
    Gitterschwingungsüberlagerung *f*
lattice wave Gitterwelle *f*
latticework Fachwerk *n*
laubanite Laubanit *m*
laudanidine Laudanidin *n*
laudanine Laudanin *n*
laudanosine Laudanosin *n*
laudanosoline Laudanosolin *n*
laudanum Laudanum *n,* Opiumtinktur *f*
Laue diagram Lauediagramm *n*
Laue method Laue-Verfahren *n*
Laue spots Laue-Flecke *pl*
laughing gas Lachgas *n,* Distickstoffoxid *n,*
    Lustgas *n* (obs)
laumontite Laumontit *m* (Min), Schaumspat *m*
    (Min)
launch *v* abschießen (Rakete), vom Stapel lassen
    (Schiff)
launching cradle Ablaufapparat *m*
laundering, heavy ~ Grobwäsche *f*
laundry soap Kernseife *f*
lauraldehyde Laurinaldehyd *m,*
    Dodecylaldehyd *m*
lauralkonium Lauralkonium *n*
lauramide Lauramid *n*
laurane Lauran *n*
laurate Laurat *n,* Laurinsäureester *m*
laurel Lorbeer *m* (Bot), Lorbeeröl *n,* **volatile** ~
    Bayöl *n*
laurel berry Lorbeere *f* (Bot)
laurel camphor Japancampher *m*
laureline Laurelin *n*
laurel leaf Lorbeerblatt *n*
laurel oil Lorbeeröl *n*
laurel wax Myristin *n*
laurene Lauren *n*
laurenone Laurenon *n*
laurent acid Laurentsäure *f*
lauric acid Laurinsäure *f,* Dodecansäure *f,*
    Dodecylsäure *f*
lauric aldehyde Laurinaldehyd *m,*
    Dodecylaldehyd *m*
laurin Laurin *n*
laurionite Laurionit *m* (Min)
laurite Laurit *m* (Min)
laurodimyristine Laurodimyristin *n*
laurodistearin Laurodistearin *n*
lauroleic acid Lauroleinsäure *f*

laurolene Laurolen *n*
laurolene-3-carboxylic acid Lauronolsäure *f*
laurolitsine Laurolitsin *n*
laurone Lauron *n*
lauronitrile Lauronitril *n*
lauronolic acid Lauronolsäure *f*
laurotetanine Laurotetanin *n*
laurylalanylglycine Laurylalanylglycin *n*
lauryl alcohol Laurylalkohol *m*
laurylglycine Laurylglycin *n*
lautarite Lautarit *m* (Min)
Lauth's violet Thionin *n*
lautite Lautit *m*
lava Lava *f*, scoriaceous ~ Schlackenlava *f*,
  stream of ~ Lavastrom *m*, vitreous ~
  Lavaglas *n*, Glasachat *m*
lava flow Lavastrom *m*
lava-like lavaähnlich
lavandulic acid Lavandulylsäure *f*
lavandulol Lavandulol *n*
lava rocks Lavagestein *n*
lavender Lavendel *m* (Bot), Lavendelfarbe *f*,
  extract of ~ Lavendelgeist *m*
lavender blue Lavendelblau *n*
lavender-colored lavendelfarben
lavender flowers Lavendelblüten *pl* (Bot)
lavender oil Lavendelöl *n*
lavender water Lavendelwasser *n*
lavenite Lavenit *m* (Min)
law Gesetz *n*, Grundsatz *m*, Regel *f*
lawful rechtmäßig
law of averages Mittelwertsatz *m*
law of conservation of matter Gesetz *n* von der
  Erhaltung der Masse
law of constant summation of heat Satz *m* von
  den konstanten Wärmesummen (pl)
law of contact series Gesetz *n* der
  Spannungsreihe (Elektrochem)
law of least heating effect Gesetz *n* der kleinsten
  Stromwärme
law of mass action Massenwirkungsgesetz *n*,
  Gesetz *n* von Guldberg und Waage
law of mass attraction Massenanziehungsgesetz *n*
law of multiple proportions Gesetz *n* der
  multiplen Proportionen (Chem)
law of rational intercepts Gesetz *n* der rationalen
  Achsenabschnitte
law of spectroscopic displacement
  Kossel-Sommerfeldsches
  Verschiebungsgesetz *n* (Spektr), Satz *m* der
  spektroskopischen Verschiebung (Spektr)
lawrencite Lawrencit *m* (Min)
lawsone Lawson *n*
lawsonite Lawsonit *m* (Min)
law suit Prozeß *m* (Jur)
laxative Abführmittel *n* (Pharm), Laxativum *n*
  (Pharm)
lay *v* legen, ~ bare bloßlegen, ~ off ablegen, ~
  on auftragen

layer Schicht *f*, Blatt *n* (Anat), Flöz *n* (Bergb),
  Lage *f*, Lager *n* (Bergb), Lamelle *f*, absorbing
  ~ absorbierende Schicht, atmospheric ~
  atmosphärische Schicht, height of ~
  Schichthöhe *f*, intermediate ~
  Zwischenschicht *f*, monoatomic ~ einatomige
  Schicht, monoatomare Schicht, permeable ~
  durchlässige Schicht, photoelectric ~
  lichtelektrische Schicht, thickness of the ~
  Schichtdicke *f*, upper ~ Oberschicht *f*, ~ of
  ice Eisansatz *m*, ~ of metal Metallschicht *f*, ~
  of scale Kesselsteinkruste *f*
layer *v* [über]schichten
layer-corrosion Schichtkorrosion *f*
layer filtration Schichtenfiltration *f* (Zucker)
layer lattice Schichtengitter *n*, Schichtgitter *n*
  (Krist)
layer lattice structure Schichtgitterstruktur *f*
layer lines Schichtlinien *pl* (Krist)
layer molding Übereinanderpressen *n*
layer polarizer Einschichtpolarisator *m*
layers, in ~ schichtenweise, to arrange in ~
  [auf]schichten
layer separation Grenzschichtablösung *f*
layout Anordnungsschema *n*, Entwurf *m*,
  Grundriß *m*, Plan *m*
layout sketch Aufstellungsskizze *f*,
  Entwurfsskizze *f*
lazuline [blue] Lazulinblau *n*
lazulite Lazulit *m* (Min), Blauspat *m* (Min)
lazurite Lasurstein *m* (Min)
lazy faul, träge
lazy tongs Gelenkschere *f*, Nürnberger Schere *f*
leach Auslaugung *f*
leach *v* [aus]laugen, durchsickern
leachable laugungsfähig
leached out ausgelaugt
leacher Auslauger *m*
leaching Auslaugen *n*, Auslaugung *f*, Laugung *f*
  (caverns in rocks), Aussoltechnik *f*
leaching plant Lauge[n]anlage *f*
leaching pump Aussüßpumpe *f*
leaching tank Reinigungstank *m*
leaching vat Laugebehälter *m*, Laugenfaß *n*
leach residue Lauge[n]rückstand *m*
lead Ganghöhe *f* (Gewinde), Mine *f* (Bleistift),
  Vorsprung *m* (Symb. Pb), Blei *n* (Chem),
  angle of ~ Voreilwinkel *m*, argentiferous ~
  Reichblei *n*, commercial ~ Handelsblei *n;,*
  crude ~ Rohblei *n*, Werkblei *n*, crystallized
  ~ Bleifluß *m*, fulminating ~ Knallblei *n*, hard
  or antimonial ~ Hartblei *n*, lump of ~
  Bleibarren *m*, Bleibrocken *m*, native ~
  gediegenes Blei, precipitated ~
  Bleiniederschlag *m*, raw ~ Werkblei *n*, refined
  ~ Frischblei *n*, Weichblei *n*, rich ~
  Reichblei *n*, rolled ~ gewalztes Blei,
  Rollenblei *n*, Walzblei *n*, soft ~ Weichblei *n*,
  spongy ~ Bleischwamm *m*, to cover with ~

verbleien, **to line with** ~ ausbleien, **workable**
~ Werkblei *n*, **yellow** ~ Bleigelb *n*, ~ **free
from silver** Armblei *n*, ~ **obtained from dross**
Abzugsblei *n*
lead *v* [an]führen, leiten, mit Blei verkleiden
(Techn), voreilen (Elektr), ~ **back**
zurückführen, ~ **over** überleiten, ~ **through**
durchführen
lead accumulator Bleiakkumulator *m*, **charging
of the** ~ Aufladen *n* des Bleiakkumulators
lead acetate Bleiacetat *n*, Bleizucker *m*, **basic** ~
solution Bleiessig *m*
lead acetate ointment Bleizuckersalbe *f*
lead alloy Bleilegierung *f*
lead angle Voreilwinkel *m*
lead anode Bleianode *f*
lead antimonate Bleiantimonat *n*
lead antimony alloy Bleiantimonlegierung *f*
lead antimony glance Bleiantimonerz *n* (Min),
Bleiantimonglanz *m* (Min)
lead antimony sulfide, native ~ Jamesonit *m*
(Min), Meneghinit *m* (Min)
lead arsenate Bleiarsenat *n*, Flockenerz *n* (Min)
lead ash Bleiglätte *f*
lead basin Bleischale *f*
lead bath Bleibad *n*
lead-bearing bleiführend
lead bismuth alloy Bleiwismutlegierung *f*
lead body burden Bleibelastung *f* (Toxikologie)
lead borate Bleiborat *n*
lead borate glass Bleiboratglas *n*
lead bottle Bleiflasche *f*
lead brick Bleiblock *m*, Bleibrikett *n*
lead carbonate Bleikarbonat *n*, Bleispat *m*
(Min), **basic** ~ Bleiweiß *n*
lead carbonate ointment Bleiweißsalbe *f* (Pharm)
lead chamber Bleikammer *f*
lead chamber crystals Bleikammerkristalle *m pl*
lead chamber process Bleikammerverfahren *n*
lead chloride Bleichlorid *n*
lead chloride carbonate Bleihornerz *n* (Min),
Hornblei *n* (Min)
lead chromate Bleichromat *n*, Bleigelb *n*,
Chrompigment *n*, Leipziger Gelb *n*, Pariser
Gelb *n*, **basic** ~ Chromrot *n*
lead-coated verbleit
lead-coating Verbleien *n*, Verbleiung *f*
lead-coating plant Verbleiungsanlage *f*
lead colic Bleikolik *f* (Med)
lead color Bleifarbe *f*, Bleigrau *n*
lead-colored bleifarben, bleigrau
lead compound Bleiverbindung *f*
lead content Bleigehalt *m*
lead covering Bleimantel *m*, Bleiumhüllung *f*
lead crystal Bleiglas *n*
lead cyanamide Bleicyanamid *n*
lead deposit Bleischlamm *m* (Akku)
lead dichloride Bleidichlorid *n*

lead dioxide Bleidioxid *n*, Plumbioxid *n*
lead dioxide plate Bleisuperoxidplatte *f*
lead dish Bleischale *f*
lead dross Bleiabgang *m*, Bleikrätze *f*,
Bleischaum *m*
lead druse Bleidruse *f*
lead dryer Bleisikkativ *n*
lead dust Bleimehl *n*, Bleistaub *m*
lead dust accumulator Bleistaubakkumulator *m*
lead dust storage battery
Bleistaubakkumulator *m*
leaden bleiern
leader (recording tape) Vorspann *m*
leader pin Führungssäule *f*
lead fittings Bleiarmaturen *pl*
lead fluoride Bleifluorid *n*
lead fluosilicate Bleisilicofluorid *n*,
Kieselfluorblei *n*
lead foil Bleifolie *f*
lead fume Bleidampf *m*, Bleirauch *m*,
Flugstaub *m* [der Bleiöfen (pl)]
lead furnace Bleiofen *m*
lead fuse Bleisicherung *f*
lead gasket Bleidichtung *f*
lead gasoline Bleibenzin *n*
lead glance Bleiglanz *m* (Min), Bleisulfid *n*
(Min)
lead glass Bleiglas *n*, Kristallglas *n*
lead glass counter Bleiglaszähler *m*
lead glaze Bleiglasur *f*
lead gravel Bleigrieß *m*
lead-gray bleigrau
lead grit Bleigrieß *m*
lead hardening Bleihärtung *f*
leadhillite Leadhillit *m* (Min)
lead hood Bleihaube *f*
lead hydride Bleiwasserstoff *m*
lead hydroxide Bleihydroxid *n*
lead(II) compound Blei(II)-Verbindung *f*,
Plumboverbindung *f*
lead(II)(IV) oxide Mennige *f*
lead(II) nitrite Nitritglätte *f*
lead(II) salt Blei(II)-Salz *n*, Plumbosalz *n*
lead-in Zuführung *f* (Elektr), Zuleitung *f*
leading Blei[ein]fassung *f*
leading edge Steuerkante *f*, Vorderkante *f*
leading screw Leitspindel *f*
leading term Anfangsglied *n* (Math)
lead iodide Bleijodid *n*
lead ion Bleiion *n*
lead(IV) compound Blei(IV)-Verbindung *f*,
Plumbiverbindung *f*
lead(IV) oxide Plumbioxid *n*
lead(IV) salt Blei(IV)-Salz *n*, Plumbisalz *n*
lead jacket Bleimantel *m*
lead joint Bleidichtung *f*
leadlike bleiartig
lead line Lotleine *f*

lead lining Bleiauskleidung *f*, Bleieinlage *f*, Verbleiung *f*
lead lode Bleiader *f* (Bergb), Bleigang *m* (Bergb)
lead manganate Bleimanganat *n*
lead meal Bleimehl *n*
lead metaphosphate Bleimetaphosphat *n*
lead metasilicate Bleimetasilikat *n*
lead mold Bleiform *f*
lead molybdate, native ~ Gelbbleierz *n*, gelber Bleispat *m*, Wulfenit *m* (Min)
lead monoxide Bleimonoxid *n*, Bleiglätte *f*, **native** ~ Massicot *m* (Min)
lead mount Bleifassung *f* (Techn)
lead naphthenate Bleinaphthenat *n*
lead nitrate Bleinitrat *n*, Bleisalpeter *m*
lead ocher Bleiocker *m* (Min)
lead ointment Bleisalbe *f*
lead ore Bleierz *n*, **green** ~ Grünbleierz *n* (Min), **heavy** ~ Schwerbleierz *n* (Min), **red** ~ Rotbleierz *n* (Min), **white** ~ Weißbleierz *n* (Min), **yellow** ~ Gelbbleierz *n*, gelber Bleispat *m*
lead orthosilicate Bleiorthosilikat *n*
lead oxalate Bleioxalat *n*
lead oxide Bleioxid *n*, Bleiglätte *f*
lead oxychloride Bleioxychlorid *n*, Englischgelb *n*, Kasseler Gelb *n*
lead packing Bleidichtung *f*
lead paint Bleifarbe *f*
lead paper Bleipapier *n*
lead pencil Bleistift *m;* Graphitstift *m*
lead peroxide Bleidioxid *n*, Bleiperoxid *n*
lead peroxide plate Bleisuperoxidplatte *f*
lead persulfate Bleipersulfat *n*
lead pig Bleimulde *f*
lead pigment Bleifarbe *f*
lead pipe Bleiröhre *f*, Bleirohr *n*
lead plating Verbleiung *f*
lead poisoning Bleivergiftung *f* (Med)
lead powder Bleimehl *n*
lead primer Bleigrundierung *f*
lead protoxide Bleiglätte *f*, Bleimonoxid *n*
lead refinery Bleiraffinerie *f*
lead refining Bleiraffination *f*
lead refining plant Bleihütte *f*
lead regulus Bleikönig *m*, Bleiregulus *m*
lead resistance Zuleitungswiderstand *m* (Elektr)
lead-responsive bleiempfindlich
lead roasting process Bleiröstprozeß *m*
leads, to match ~ eine Leitung *f* mit ihrem Wellenwiderstand abschließen
lead salt Bleisalz *n*
lead scoria Bleiabgang *m*
lead screen Bleischirm *m*
lead screw spindle Schraubenspindel *f*
lead scum Abstrichblei *n* (Metall), Bleischaum *m* (Metall)
lead seal Bleiplombe *f*, Plombe *f*

lead-sealing Plombieren *n*
lead selenide Bleiselenid *n*, Selenblei *n*, **native** ~ Klausthalit *m* (Min)
lead sheathing Bleimantel *m*
lead shielding Bleiabschirmung *f*
lead shot Bleischrot *m*
lead siccative Bleisikkativ *n*
lead silicate Bleisilikat *n*
lead silver telluride Tellursilberblei *n* (Min)
lead skim Abstrichblei *n* (Metall)
lead slag Bleischlacke *f*
lead sleeve Bleimuffe *f*
lead slime Bleischlich *m*
lead smeltery Bleihütte *f*
lead smelting Bleiverhüttung *f*
lead smelting hearth Bleischmelzherd *m*
lead smoke Bleirauch *m*, Flugstaub *m* der Bleiöfen (pl)
lead solder Bleilot *n*, Lotblei *n*
lead soldering process Bleilötverfahren *n*
lead spar Bleispat *m* (Min), **red** ~ Rotbleierz *n*, **yellow** ~ gelber Bleispat, Gelbbleierz *n*
lead sponge Bleischwamm *m*
lead spring Bleibügel *m*
lead stearate Bleistearat *n*
lead storage battery Bleiakkumulator *m*
lead subacetate Bleisubacetat *n*
lead suboxide Bleioxydul *n* (obs)
lead sulfate Bleisulfat *n*, **iridescent** ~ Regenbogenerz *n* (Min)
lead sulfide Bleisulfid *n*, **native** ~ Bleiglanz *m* (Min)
lead tannate Bleitannat *n*
lead tannate ointment Tanninbleisalbe *f* (Pharm)
lead telluride Altait *m* (Min), Bleitellurid *n*, Tellurblei *n*
lead tempering bath Bleiofen *m*
lead tetraacetate Bleitetraacetat *n*
lead tetrachloride Bleitetrachlorid *n*
lead thiocyanate Bleirhodanid *n*
lead thiosulfate Bleithiosulfat *n*
lead-tin alloy Bleizinnlegierung *f*
lead titanate Bleititanat *n*
lead tree Bleibaum *m*
lead tube Bleiröhre *f*, Bleirohr *n*, Bleitube *f*
lead tungstate Bleiwolframat *n*, Scheelbleierz *n* (Min), Stolzit *m* (Min), Wolframbleierz *n* (Min)
lead vapor Bleidampf *m*
lead vein Bleiader *f* (Bergb), Bleigang *m* (Bergb)
lead vinegar Bleiessig *m*
lead vitriol Bleivitriol *n*, Bleisulfat *n*
lead waste Bleiabfälle *pl*, Bleikrätze *f*
lead water Bleiwasser *n*
lead weight Bleigewicht *n*
lead wire Bleidraht *m*, Zuführungsdraht *m* (Elektr), Zuleitung *f* (Elektr)
leady bleiartig, bleiern
lead-zinc accumulator Bleizinkakkumulator *m*

lead-zinc storage battery Bleizinkakkumulator *m*
leaf (pl. leaves) Blatt *n*, small ~ Blättchen *n*,
   thin ~ Blättchen *n*
leaf aluminum Blattaluminium *n*
leaf coloring matter Blattpigment *n*
leaf electrometer Blattelektrometer *n*
leaf fertilization Blattdüngung *f*
leaf filter Blattfilter *n*
leaf gold Blattgold *n*, imitation ~ Glanzgold *n*
leaflet Blättchen *n*, Prospekt *m*
leaf metal Blattmetall *n*, Folie *f*
leaf mold Lauberde *f*
leaf of holder Fassungslamelle *f*
leaf powder Blättchenpulver *n*
leaf silver Blattsilber *n*, false ~ Metallsilber *n*
leaf spring Blattfeder *f*
leafy belaubt, blätterig
leak Leck *n*, undichte Stelle *f*, Verluststrom *m*
   (Elektr)
leak *v* auslaufen (ausrinnen), durchsickern,
   entweichen (Gas), leck sein, rinnen, streuen
   (Elektr)
leakage Entweichen *n* (Gas), Schwund *m*,
   Verlust *m*, electromagnetic ~
   elektromagnetische Streuung, magnetic ~
   magnetische Streuung der Kraftlinien (pl),
   without ~ streuungslos, ~ of electricity
   Elektrizitätsverlust *m*
leakage coefficient Streufaktor *m*,
   Streukoeffizient *m*
leakage current Leckstrom *m* (Elektr),
   Ableitstrom *m*, Ableitungsstrom *m* (Elektr),
   Anzapfstrom *m*, Kriechstrom *m* (Elektr),
   Selbstentladestrom *m* (Elektr)
leakage path for electric current Kriechweg *m*
leakage resistance Ableitungswiderstand *m*,
   Streuwiderstand *m*
leakage steam Leckdampf *m*
leakage tester Lecksucher *m*
leakage voltage Kriechspannung *f*
leak detection Lecksuche *f*
leak detector Dichtigkeitsprüfgerät *n*,
   Lecksucher *m*, Lecksuchgerät *n*
leak detector head Lecksuchsonde *f*
leak detector tube Lecksuchröhre *f*
leakiness Undichtheit *f*
leaking leck, undicht
leakproof lecksicher, wasserdicht
leak resistance Abzweigwiderstand *m*
leak test Leckprüfung *f*
leaky leck, undicht
lean mager, arm (Gas), fettarm
lean [against] *v* anlehnen
lean concrete Magerbeton *m*
lean gas Schwachgas *n*
lean material Magerungsmittel *n*
leap Satz *m*, Sprung *m*
leap *v* springen, schnellen, ~ off abspringen
leap year Schaltjahr *n*

least-energy principle Prinzip *n* der kleinsten
   Wirkung (Mech)
least squares method Fehlerquadratmethode *f*
   (Math)
leather Leder *n*, dressed ~ Garleder *n*, high-class
   ~ Qualitätsleder *n*, oil-tanned ~
   Sämischleder *n*, untanned ~ Rohleder *n*
leatherboard Faserleder *n*, Lederpappe *f*
leather-coating color Lederdeckfarbe *f*
leather color Lederfarbe *f*
leather-colored lederfarben
leather cuttings Leimleder *n*
leather dressing Lederbereitung *f*,
   Lederpflegemittel *n*, Lederzurichtung *f*
leather [dressing] oil Lederöl *n*
leather driving belt Ledertreibriemen *m*
leather finish Lederlack *m*
leather finishing Lederzurichtung *f*
leather gasket Lederdichtung *f*
leather glue Lederleim *m*
leather hollow Lederhohlkehle *f*
leather industry Lederindustrie *f*
leather-like lederähnlich, lederartig
leather packing Lederdichtung *f*
leather pad Lederkissen *n*
leather polish Lederwichse *f*
leather ring Lederring *m*
leather sealing Lederdichtung *f*
leather sheets, artificial ~ Folienkunstleder *n*
leather strap Lederriemen *m*
leather substitute Lederersatz *m*,
   Lederimitation *f*
leather varnish Lederlack *m*
leathery lederähnlich, lederartig
leather yellow Phosphin *n*
leaven Gärmittel *n*, Gärstoff *m*, Hefe *f*,
   Sauerteig *m*
leaven *v* säuern (Teig)
leavening Gärungsstoff *m*, Säuerung *f* (Teig)
leaves Laub *n*
Lebanon cedar oil Libanonzedernöl *n*
Leblanc process Leblanc-Verfahren *n*
lecanoric acid Lecanorsäure *f*, Diorsellinsäure *f*
LeChatelier's principle (of least restraint)
   LeChateliersches Prinzip *n* (des kleinsten
   Zwanges)
lecithalbumin Lecithalbumin *n*
lecithin Lecithin *n*, pure ~ Reinlecithin *n*
lecithinase Lecithinase *f* (Biochem)
lecithin level Lecithinspiegel *m* (Med)
Leclanché cell Leclanché-Element *n*,
   Braunsteinelement *n*
lecontite Lecontit *m* (Min)
lecture Vorlesung *f*, Referat *n*, Vortrag
lecture demonstration Vorlesungsversuch *m*
lecture hall Hörsaal *m*
lecturer Dozent *m*, Lektor *m*
lectures, interdisciplinary course of ~
   Ringvorlesung *f*

ledeburite Ledeburit *m*
ledene Leden *n*
ledic acid Ledsäure *f*
ledienoside Ledienosid *n*
ledol Ledol *n*
ledum camphor Ledol *n*
leek Lauch *m* (Bot)
leek-green lauchgrün
lees Hefe *f*, Bodensatz *m*, Drusen *pl*, **potash from
  burnt ~ of wine** Drusenasche *f*
leeway Abtrift *f*
left-handed propeller Linksschraube *f*
left-handed quartz Linksquarz *m* (Min)
left-handed system Linkssystem *n*
left turn Linksdrehung *f* (Kfz)
leg Bein *n* (Gestell), Schenkel *m* (Winkel)
  (Zirkel)
legal gesetzlich
legal chemistry Gerichtschemie *f*
legalize *v* beurkunden, legalisieren
legible lesbar, leserlich
legitimate rechtmäßig
legs of spider (in blow-molding head) Steg *m*
  (Spritzguß)
legumin Legumin *n*, Avenin *n*, Pflanzencasein *n*
leguminous plants Hülsenfrüchte *pl*
lehrbachite Selenquecksilberblei *n* (Min)
Leidenfrost phenomenon Leidenfrostsches
  Phänomen *n*
leifite Leifit *m* (Min)
Leipzig yellow Leipzigergelb *n*
lemma Hilfssatz *m*
lemniscate Schleifenlinie *f* (Math)
lemon Zitrone *f*
lemon *a* zitronenfarben
lemonade Limonade *f*, Zitronenlimonade *f*
lemonade powder Limonadenpulver *n*
lemon balm Zitronenmelisse *f* (Bot)
lemon blossom Zitronenblüte *f* (Bot)
lemongrass oil Lemongrasöl *n*
lemon juice Zitronensaft *m*
lemon oil Citronenöl *n*, Zitronenöl *n*
lemonol Lemonol *n*
lemon peel Zitronenschale *f*, **candied ~**
  Zitronat *n*
lemon squeezer Zitronenpresse *f*
lemon thyme oil Zitrothymöl *n*
lemon tree Zitronenbaum *m*
lemon yellow Zitronengelb *n*
lemon yellow *a* zitronengelb
Lenard tube Lenardröhre *f*
lengenbachite Lengenbachit *m* (Min)
length Länge *f*, **overall ~** Gesamtlänge *f*, **unit of
  ~** Längeneinheit *f*, **~ between centers**
  Spitzenabstand *m*, **~ of grip** Klemmlänge *f*,
  ~ of stroke Hubbegrenzung *f*
length allowance Längentoleranz *f*
length dimension Längenmaß *n*
lengthen *v* verlängern, recken

lengthening Verlängerung *f*
lengthening lever Verlängerungshebel *m*
length gauge Längenmaß *n* (Meßgerät)
length preserving längentreu
lengthwise längs, der Länge nach
lengthwise direction Längsrichtung *f*
lenitive Abführmittel *n* (Pharm),
  Linderungsbalsam *m* (Pharm)
lenitive *a* mildernd (Med), schmerzlindernd
lenitive ointment Linderungssalbe *f* (Pharm)
lens Linse *f* (Opt), Objektiv *n* (Opt), **achromatic
  ~** achromatische Linse, **biconcave ~**
  bikonkave Linse, doppeltkonkave Linse,
  biconvex ~ bikonvexe Linse, doppeltkonkave
  Linse, **bifocal ~** bifokale Linse, **binary ~**
  Zwillingslinse *f* (Opt), **clouding of a ~**
  Linsentrübung *f*, **coated ~** beschichtete Linse,
  compound ~ zusammengesetzte Linse (Opt),
  concave ~ konkave Linse, **concavoconcave ~**
  doppeltkonkave Linse, **condensing ~** konvexe
  Linse, **convex ~** konvexe Linse,
  Sammellinse *f*, **convexoconcave ~**
  konvexkonkave Linse, **decentered ~**
  dezentrierte Linse, **diverging ~** konkave
  Linse, Zerstreuungslinse *f*, **double ~**
  Doppelobjektiv *n*, **electrostatic ~** elektrische
  Linse, **long-focus ~** langbrennweitige Linse,
  mounted ~ gefaßte Linse, **opacity of a ~**
  Linsentrübung *f*, **plano-concave ~**
  plankonkave Linse, **plano-convex ~**
  plankonvexe Linse, **positive ~** Sammellinse *f*
  (Opt), **short-focus ~** kurzbrennweitige Linse,
  supplementary ~ Zusatzlinse *f* (Opt),
  telephotographic ~ Teleobjektiv *n* (Phot),
  variable focus ~ Gummilinse *f* (Opt), Linse *f*
  mit veränderlichem Fokus, **wide-angle ~**
  Weitwinkelobjektiv *f* (Opt), **wide-aperture ~**
  Weitwinkelobjektiv *n* (Opt)
lens cap Objektivdeckel *m*
lens combination Linsensystem *n*
lens cooler Linsenkühler *m*
lens curvature Linsenkrümmung *f*
lens equation Linsengleichung *f*
lens error Linsenfehler *m*
lenses, system of ~ Linsensystem *n*
lens formula, Gaussian ~ Gaußsche
  Linsenformel *f* (Opt)
lens glass Linsenglas *n*
lens holder Objektivfassung *f*
lens opening Blendenöffnung *f*
lens power Stärke *f* einer Linse (Opt)
lens-shaped linsenartig, linsenförmig
lens tube Objektivfassung *f*
lenticular linsenartig, linsenförmig
lenticular astigmatism Linsenastigmatismus *m*
lentiform linsenförmig
lentil Linse *f* (Bot)
lentil-sized linsengroß
lentine Lentin *n*

leonecopalic acid Leonecopalsäure *f*
leonecopalinic acid Leonecopalinsäure *f*
leonhardite Leonhardit *m* (Min)
leonite Leonit *m* (Min)
leontin Leontin *n*
leopoldite Leopoldit *m* (Min)
lepargylic acid Lepargylsäure *f*
lepidene Lepiden *n*
lepidine Lepidin *n*
lepidinic acid Lepidinsäure *f*
lepidokrokite Lepidokrokit *m* (Min)
lepidolite Lepidolith *m* (Min),
   Lithiumglimmer *m* (Min), Schuppenstein *m*
   (Min)
lepidomelane Lepidomelan *m* (Min)
lepidone Lepidon *n*
lepranthine Lepranthin *n*
leprapinic acid Leprapinsäure *f*
leprosy Lepra *f* (Med)
leptaflorine Leptaflorin *n*
leptochlorite Leptochlorit *m* (Min)
leptocladine Leptocladin *n*
leptodactyline Leptodactylin *n*
leptogenin Leptogenin *n*
lepton Lepton *n*
leptoside Leptosid *n*
leptospermol Leptospermol *n*
leptospermone Leptospermon *n*
lerbachite Lerbachit *m*
lespedin Lespedin *n*
lessen *v* verringern, schwächen
lesson Lektion *f*
let *v* lassen, erlauben, ~ **down** herunterlassen,
   ~ **in** einlassen, ~ **off** ablassen (Gas),
   abschießen, ~ **off steam** Dampf ablassen, ~
   out entwischen lassen, ~ **stand** abstehen
   lassen, ~ **through** durchlassen (Licht)
let-down gas Entspannungsgas *n*
let-down vessel Entspannungsgefäß *n*
lethal letal, tödlich
lethal amount or dose Letaldosis *f*, tödliche
   Dosis *f*
lethal factor Letalfaktor *m* (Biol)
lethality Sterblichkeit *f*
lethargy Lethargie *f* (Med), Schlafsucht *f*
let-off stand Ablaufgestell *n*, Abwickelbock *m*
letter acids Buchstabensäuren *pl*
letterpress printing Buchdruck *m*
letterpress printing inks Buchdruckfarben *pl*
lettsomite Lettsomit *m* (Min)
lettuce, wild ~ Giftlattich *m* (Bot)
lettuce opium Lactucarium *n*, Lattichopium *n*
leucacene Leukacen *n*
leucaniline Leukanilin *n*
leucaurine Leukaurin *n*
leuchtenbergite Leuchtenbergit *m* (Min)
leucine Leucin *n*
leucine aminopeptidase Leucinaminopeptidase *f*
   (Biochem)

leucine aminotransferase
   Leucinaminotransferase *f* (Biochem)
leucinic acid Leucinsäure *f*
leucinol Leucinol *n*
leucite Leucit *m* (Min), Leuzit *m* (Min)
leucitoid Leucitoid *n*
leuco-agglutinin Leukoagglutinin *n*
leucoalizarin Leukoalizarin *n*
leucoatromentin Leukoatromentin *n*
leucoauramine Leukoauramin *n*
leuco base Leukobase *f*
leucochalcite Leukochalcit *m* (Min)
leuco-compound Leukoverbindung *f*
leucocyclite Leukocyklit *m*
leucocyte weißes Blutkörperchen *n*, Leukozyte *f*,
   weißes Blutkörperchen *n*
leucocyte count, increase of the ~
   Leukozytenanstieg *m*
leucocytes, breaking down of ~
   Leukozytenzerfall *m*, **formation of** ~
   Leukozytenbildung *f*
leucocytogenesis Leukozytogenese *f*
leucocytolysis Leukozytenzerfall *m* (Med),
   Leukozytolyse *f* (Med)
leucocytopenia Leukozytenverminderung *f*
   (Med)
leucocytosis Leukozytenanstieg *m* (Med),
   Leukozytenvermehrung *f* (Med),
   Leukozytose *f* (Med)
leucodrin Proteacin *n*
leucoellagic acid Leukoellagsäure *f*
leuco form Leukoform *f*
leucogallocyanin Leukogallocyanin *n*
leucoindigo Indigoweiß *n*, Leukoindigo *n*
leucoindophenol Leukoindophenol *n*
leucoline Leukolin *n*
leucol yellow Leukolgelb *n*
leucolysis Leukolyse *f* (Med),
   Leukozytenauflösung *f* (Med)
leucomaine Leukomain *n*
leucomalachite green Leukomalachitgrün *n*
leucomelone Leucomelon *n*
leucometer Leukometer *n*, Weißgradmesser *m*
leuconic acid Leukonsäure *f*
leuconin Leukonin *n*
leucopelargonidin Leukopelargonidin *n*
leucopenia Leukozytenarmut *f* (Med)
leucophane Leukophan *m* (Min)
leucopterin Leukopterin *n*
leucopyrite Leukopyrit *m* (Min)
leucoquinizarin Leukochinizarin *n*
leucorosaniline Leukorosanilin *n*
leucorosolic acid Leukorosolsäure *f*
leucosapphire weißer Saphir *m*
leucosphenite Leukosphenit *m* (Min)
leucothioindigo Leukothioindigo *n*
leucothionine Leukothionin *n*
leucotil Leukotil *m* (Min)

leucotrope Leukotrop *n*,
  Phenyldimethyl-benzylammoniumchlorid *n*
leuco vat dyestuff Leukoküpenfarbstoff *m*
leucoxene Leukoxen *m* (Min)
leucrose Leucrose *f*
leucylalanine Leucylalanin *n*
leucylalanylalanine Leucylalanylalanin *n*
leucylasparagine Leucylasparagin *n*
leucylaspartic acid Leucylasparaginsäure *f*
leucylcystine Leucylcystin *n*
leucylglycylalanine Leucylglycylalanin *n*
leucylglycylaspartic acid
  Leucylglycylasparaginsäure *f*
leucylleucine Leucylleucin *n*
leucylproline Leucylprolin *n*
leucyltryptophan[e] Leucyltryptophan *n*
leuk[a]emia Leukämie *f* (Med)
leuk[a]emic leukämisch
leuko- see leuco-
levan Lävan *n*
levarterenol Levarterenol *n*
level Ebene *f*, Höhe *f*, Horizontale *f*, Niveau *n*,
  Richtwaage *f* (Techn), Spiegel *m* (einer
  Flüssigkeit), Waagrechte *f*, Wasserwaage *f*
  (Techn), **difference of** ~
  Niveauunterschied *m*, **height of** ~
  Niveauhöhe *f*, **initial** ~ Anfangsniveau *n*, **to
  bring to a** ~ nivellieren
level *a* eben, horizontal, flach
level *v* ebnen, abtragen, egalisieren, nivellieren
  (Techn), planieren, ~ **off** abflachen,
  abschwächen, verflachen, ~ **up** nach oben
  ausgleichen, erhöhen
level broadening Niveauverbreiterung *f*
level density Termdichte *f*
level diagram Termschema *n*
level displacement Termbeeinflussung *f*
level dyeing auxiliary Egalisierhilfsmittel *n*
level dyeing property Egalfärbevermögen *n*
level indicator Flüssigkeitsstandanzeiger *m*,
  Füllstandanzeiger *m*
level[l]ing Ausgleichsvermögen *n* (Text),
  Egalisieren *n*, Einebnen *n*, Einebnung *f*,
  Einpegelung *f*, Nivellierung *f*, Planieren *n*,
  Verlauf *m* (Lack)
level[l]ing agent Egalisierer *m*, Egalisiermittel *n*,
  Verlaufmittel *n* (Lack)
level[l]ing blade Ausstreichmesser *n*
level[l]ing bottle Niveauflasche *f*
level[l]ing bulb Niveaukugel *f*, Nivelliergefäß *n*
level[l]ing dyestuff Egalisierfarbstoff *m*
level[l]ing power Egalisierungsvermögen *n*
level[l]ing property Verlauffähigkeit *f* (Lack)
level[l]ing rule Richtschiene *f*
level[l]ing screw Einstellschraube *f*,
  Nivellierschraube *f*, Stellschraube *f*
levelness Ebenheit *f*, Egalität *f*
level scheme Niveauschema *n*
level spacing Niveauabstand *m*

level structure Termstruktur *f*
level systematics Termsystematik *f*
level transportation Waag[e]rechtförderung *f*
  (Techn)
lever Hebel *m*, Brecheisen *n*, Hebearm *m*,
  Hebebaum *m*, Heber *m*, Hebevorrichtung *f*,
  arm of a ~ Hebelarm *m*, **law of the** ~
  Hebelgesetz *n* (Phys), **one-armed** ~
  einarmiger Hebel, **two-armed** ~ ·
  doppelarmiger (zweiarmiger) Hebel *m*
lever action Hebelwirkung *f*
leverage Hebekraft *f*, Hebelkraft *f*,
  Hebelübersetzung *f*, Hebelverhältnis *n*,
  Hebelwirkung *f*
lever apparatus Hebelvorrichtung *f*
lever arm Hebelarm *m*, Kurbelarm *m*
lever arrangement Hebelwerk *n*
lever balance Hebelwaage *f*
lever brake Hebelbremse *f*
lever closure Bügelverschluß *m*
lever commutator Hebelumschalter *m*
lever control Hebelsteuerung *f*
leverierite Leverierit *m* (Min)
lever lid tin Dose *f* mit Eindrückdeckel
lever press Hebelpresse *f*, Hebelpreßwerk *n*
lever release Hebelauslösung *f*
levers, arragement of ~ Hebelanordnung *f*,
  relation of ~ Hebelverhältnis *n*, **system of** ~
  Hebelwerk *n*
lever scale Hebelwaage *f*
lever shears Hebelschere *f*
lever shifter Hebelschalter *m*
lever starter Hebelanlasser *m*
lever switch Hebel[um]schalter *m*
lever transmission Hebelübersetzung *f*
lever-type Brinell machine Brinell-Presse *f* mit
  Hebelwaage
levigate *v* pulverisieren, schlämmen
levigation Pulverisierung *f*, Schlämmung *f*
levo-acid Linkssäure *f*
levocamphor Linkscampher *m*
levogyrate quartz Linksquarz *m* (Min)
levolactic acid Linksmilchsäure *f*
levopimaric acid Lävopimarsäure *f*
levopolarization Linkspolarisation *f*
levorotation Linksdrehung *f*
levorotatory linksdrehend
levorphanol Levorphan[ol] *n*
levotartaric acid Linksweinsäure *f*
levulic acid Lävulinsäure *f*
levulin Lävulin *n*
levulinaldehyde Lävulinaldehyd *m*
levulinic acid Lävulinsäure *f*
levulinic aldehyde Lävulinaldehyd *m*
levulochloralose Lävulochloralose *f*
levulosan Lävulosan *n*
levulose Fructose *f*, Lävulose *f*
levulosin Lävulosin *n*
levyne Levyn *m* (Min)

levynite Levyn *m* (Min)
Lewis acid Lewissäure *f*
Lewis base Lewisbase *f*
lewisite Lewisit *m* (Min), Lewisit *n*
　(Kampfstoff)
Leyden jar Leidener Flasche *f*
lherzolyte Lherzolyth *m*
liability Haftbarkeit *f,* Haftpflicht *f,* Haftung *f*
liable haftbar, to be ~ haften, ~ to lose color
　abfärbig
liberate *v* befreien, abgeben (Chem), entwickeln
　(Chem), in Freiheit setzen (Chem), frei
　machen
liberation Ausscheidung *f,* Freisetzung *f,*
　Freiwerden *n*
libethenite Libethenit *m* (Min)
libi-dibi Dividivi *pl*
libocedrene Libocedren *n*
licanic acid Licansäure *f*
licence (Br. E.) Lizenz *f,* Konzession *f*
licence fee (Br. E.) Lizenzgebühr *f*
licencer (Br. E.) Lizenzgeber *m*
license (Am. E.) Lizenz *f,* Bewilligung *f,*
　Konzession *f,* withdrawal of ~ (Am. E.)
　Lizenzentzug *m*
license *v* befugen, konzessionieren
license contact (Am. E.) Lizenzabkommen *n*
licensed lizenziert
licensee Lizenznehmer *m*
license fee (Am. E.) Lizenzgebühr *f* (Jur)
licenser Lizenzgeber *m*
licensing [Am. E.] Lizenzerteilung *f*
licensor Lizenzgeber *m*
lichen Flechte *f* (Bot, Med), [crustose] ~
　Borkenflechte *f*
lichen coloring matter Flechtenfarbstoff *m* (Bot)
lichenin Lichenin *n,* Flechtenstärkemehl *n*
lichen red Flechtenrot *n*
lichen starch Flechtenstärkemehl *n,* Lichenin *n*
lichesterinic acid Lichesterinsäure *f*
lick-roll process Pflatschen *n*
licorice Lakritze *f,* Süßholz *n*
licorice extract Süßholzextrakt *m*
licorice root Lakritzenholz *n*
licorice water Lakritzenwasser *n*
lid Deckel *m,* Klappe *f*
liebenerite Liebenerit *m* (Min)
Liebig condenser Liebigkühler *m*
liebigite Liebigit *m* (Min)
lievrite Lievrit *m* (Min)
life Leben *n,* Lebendigkeit *f,* Lebensdauer *f,*
　average ~ mittlere Lebensdauer, mean ~
　mittlere Lebensdauer
life cycle Lebenszyklus *m* (Biol)
life-destroying lebensvernichtend
life expectancy Lebenserwartung *f*
lifeless unbelebt, leblos
life of furnace Schmelzreise *f*
life-sustaining lebenserhaltend

lifetime Lebensdauer *f,* comparative ~
　vergleichbare Lebensdauer
lift Aufzug *m,* Fahrstuhl *m,* Hebewerk *n,* Lift *m,*
　equation of ~ Auftriebsformel *f,* height of ~
　Hubhöhe *f*
lift *v* [hoch]heben, anheben, ~ off abheben, ~
　up hochwinden, hochziehen
lift and force pump Saug- und Druckpumpe *f*
lift axis Auftriebsachse *f*
lift coefficient Auftriebkoeffizient *m*
lift conveyer Höhenförderer *m*
lifter Heber *m,* Hebedaumen *m,* Hebekopf *m,*
　Nocken *m*
lifter rod Hubstange *f*
lift force Auftriebskraft *f*
lift frame Fördergerüst *n,* Förderturm *m*
lift guide rail Aufzugsführungsschiene *f*
lift hammer Aufwerfhammer *m*
lifting Ausheben *n,* Heben *n,* Hochziehen *n,* ~
　out the block Ausheben *n* des Blocks
lifting apparatus Hebevorrichtung *f,*
　Hebegerät *n,* Hebezeug *n*
lifting appliance Hebezeug *n*
lifting arm Hebearm *m*
lifting arrangement Hebewerk *n*
lifting bar Hebezwinge *f*
lifting capacity Hebekraft *f,* Hubleistung *f*
lifting device Hebevorrichtung *f,*
　Abhebevorrichtung *f*
lifting force Hubkraft *f*
lifting frame Hebegerüst *n*
lifting magnet Hebemagnet *m,*
　Lasthebemagnet *m*
lifting motion Hubbewegung *f*
lifting power Hebekraft *f,* Tragfähigkeit *f*
lifting pump Hubpumpe *f*
lifting table Hebetisch *m,* Rollentisch *m*
lifting winch Hebewinde *f*
lift of sleeve Muffenhub *m*
lift pump Hebpumpe *f,* Saugpumpe *f*
lift rope Aufzugsseil *n*
ligand Ligand *m*
ligand-field theory Ligandenfeldtheorie *f*
ligand replacement Ligandenaustausch *m*
ligase Ligase *f* (Biochem)
ligature Binde *f,* Ligatur *f* (Buchdr)
light Licht *n,* absence of ~ Lichtausschluß *m,*
　absorption of ~ Lichtabsorption *f,* action of ~
　Lichteinwirkung *f,* artificial ~ künstliches
　Licht, circle of ~ Lichtkreis *m,* circularly
　polarized ~ zirkular polarisiertes Licht, cold
　~ kaltes Licht, diffraction of ~
　Lichtbeugung *f,* dimmed ~ Abblendlicht *n,*
　direct ~ Auflicht *n* (Elektronenmikroskop),
　direktes Licht, elliptically polarized ~
　elliptisch polarisiertes Licht, emergent ~
　austretendes Licht, exclusion of ~
　Lichtausschluß *m,* fast to ~ lichtecht,
　imperviousness to ~ Lichtundurchlässigkeit *f,*

incident ~ Auflicht *n*, einfallendes Licht, **intensity of** ~ Lichtintensität *f*, **in the absence of** ~ unter Lichtausschluß, **monochromatic** ~ monochromatisches Licht, einfarbiges Licht, **plane-polarized** ~ linear-polarisiertes Licht (Opt), **polarized** ~ polarisiertes Licht, **protection against** ~ Lichtschutz *m*, **reflected** ~ reflektiertes Licht *n*, **sensitive to** ~ lichtempfindlich, **stable to** ~ lichtbeständig, **to emit** ~ leuchten, **transmitted** ~ durchfallendes Licht, **ultraviolet** ~ ultraviolettes Licht, **unit of** ~ Lichteinheit *f*

light *a* leicht, dünnflüssig (Öl), hell, licht, schwach

light *v* anzünden, anfeuern, beleuchten, ~ **a fire** einheizen

light-absorbing lichtabsorbierend, lichtschluckend

light absorption Lichtabsorption *f*

light adaptation Lichtadaption *f* (Auge)

light admission port Lichteintrittsschacht *m*

light ashes Flugasche *f*

light barrier Lichtschranke *f*

light beam Lichtbündel *n*, Lichtstrahl *m*

light bulb Glühbirne *f*, Glühlampe *f*

light bundle Lichtbündel *n*

light-catalytically lichtkatalytisch

light ceiling Lichtdecke *f*

light-colored hellfarbig, lichtfarben

light corpuscle Lichtkorpuskel *f*

light density construction Leichtstoffbauweise *f*

light diffuser Lichtraster *m*, Raster *m*

light diffusive capacity Lichtzerstreuungsvermögen *n*

light diffusive power Lichtzerstreuungskraft *f*

light dispersive capacity Lichtzerstreuungsvermögen *n*

light dispersive power Lichtzerstreuungskraft *f*

light distillate fuel (LDF) Leichtbenzin *n*

lighted ausgeleuchtet

light efficiency Lichtausbeute *f*

lighten *v* erleichtern; sich aufhellen

light energy Lichtenergie *f*

lightening power Aufhellvermögen *n*

lighter Anzünder *m*, Feueranzünder *m*, Feuerzeug *n*

lighter fuel Feuerzeugbenzin *n*

lightfast lichtecht

light fastness Lichtechtheit *f*

lightfastness testing Lichtechtheitsprüfung *f*

light filter Lichtfilter *n*

light gap Lichtspalt *m*

light gasoline Leichtbenzin *n*

lighting Zünden *n*, Anzünden *n*, Beleuchtung *f*

lighting circuit Lichtleitung *f*

lighting device Anzünder *m*

lighting fittings Beleuchtungsarmaturen *f pl*

lighting gas Leuchtgas *n*

lighting hour Brennstunde *f* (Elektr)

lighting installation Beleuchtungsanlage *f*

lighting panel Beleuchtungstafel *f*

lighting set Lichtaggregat *n* (Elektr)

light-insensitive lichtunempfindlich

light intensity Beleuchtungsstärke *f*, Lichtintensität *f*, Lichtstärke *f*, **of high** ~ lichtstark, **of low** ~ lichtschwach

light intensity variation Lichtintensitätsschwankung *f*

light interference Lichtinterferenz *f*

light isomerization Lichtisomerisation *f*

light metal Leichtmetall *n*

light-metal casting Leichtmetallguß *m*

light-metal working Leichtmetallbearbeitung *f*

light meter Belichtungsmesser *m* (Phot), Helligkeitsmesser *m* (Phot), Lichtstärkemesser *m* (Phot), Luxmeter *n* (Phot)

lightning Blitz *m*

lightning arrester Blitzableiter *m*

lightning conductor Blitzableiter *m*

lightning flash Blitz[strahl] *m*

lightning rod Blitzableiter *m*

light oil Leichtöl *n*

light panel Leuchteinsatz *m*

light path Lichtweg *m*

light period Leuchtdauer *f*

light-permeable lichtdurchlässig

light petrol Leichtbenzin *n*

light point line recorder Lichtpunktlinienschreiber *m*

light probe technique Lichtsondentechnik *f*

lightproof lichtundurchlässig, lichtbeständig, lichtecht

light pulse Lichtimpuls *m*

light quantum Lichtquant *n*, Photon *n*

light radiation Lichtstrahlung *f*

light reaction Lichtreaktion *f* (Biochem)

light reflection Lichtreflex *m*

light resistance Lichtbeständigkeit *f*

light-resisting lichtecht

light respiration Lichtatmung *f* (Biochem)

light scattering Lichtstreuung *f*

light scattering method Lichtstreuungsmethode *f*

light section iron Kleineisen *n*

light-sensitive surface or layer lichtempfindliche Schicht *f*

light-sensitive tube Fotozelle *f*

light sensitivity Lichtempfindlichkeit *f*

light signal Lichtsignal *n*

light-sized kleinstückig

light slit Lichtspalt *m*

light source Lichtquelle *f*, **monochromatic** ~ monochromatische Lichtquelle (Opt), **standard** ~ Strahlungsnormale *f* (Opt)

light spectrum Lichtspektrum *n*

light spot Lichtfleck *m*

light spot galvanometer Lichtzeigergalvanometer *n*

light switch Lichtschalter *m*
light-tight lichtdicht, lichtundurchlässig
light transmission Lichtdurchlässigkeit *f*
light-transmitting lichtdurchlässig
light value Helligkeit *f*, Leuchtkraft *f*
light valve Lichtventil *n*
light velocity Lichtgeschwindigkeit *f*
light water reactor Leichtwasserreaktor *m*
light wave Lichtwelle *f*
lightweight aggregate concrete
    Leichtzuschlagbeton *m*
lightweight concrete Leichtbeton *m*
lightweight construction Leichtbau *m*
lightweight construction method Leichtbauweise *f*
light wheel set Leichtradsatz *m*
light year Lichtjahr *n*
light yield Lichtausbeute *f*
ligneous holzartig, holzig
ligneous asbestos Holzamiant *m*
ligneous fiber Holzfaser *f*
ligneous matter Holzteilchen *n*, Ligninkörper *m*
lignicidal holzzerstörend
lignification Verholzung *f*, **degree of** ~
    Verholzungsgrad *m*
ligniform amianthus Holzamiant *m*
lignify *v* verholzen
lignin Lignin *n*, Holz[faser]stoff *m*, **containing** ~
    holzstoffhaltig, **free from** ~ holzstofffrei
lignin plastic Ligninkunststoff *m*
lignite Braunkohle *f*, Lignit *m*, **crude** ~
    Rohbraunkohle *f*, **fibrous** ~
    Faserbraunkohle *f*, **tar from** ~
    Braunkohlenteer *m*
lignite bed Braunkohlenlager *n*
lignite breeze Braunkohlenklein *n*
lignite briquet[te] Braunkohlenbrikett *n*
lignite carbonization plant
    Braunkohlenschwelerei *f*
lignite coke Braunkohlenkoks *m*, Grudekoks *m*
lignite coking plant Braunkohlenschwelerei *f*
lignite distillation gas Braunkohlenschwelgas *n*
lignite dressing plant
    Braunkohlenaufbereitungsanlage *f*
lignite drying Braunkohlentrocknung *f*
lignite dust Braunkohlenlösche *f*
lignite-fired furnace Braunkohlenfeuerung *f*
lignite firing Braunkohlenfeuerung *f*
lignite furnace Braunkohlenfeuerung *f*
lignite low temperature coke
    Braunkohlenschwelkoks *m*
lignite mine Braunkohlenbergwerk *n*
lignite open cut Braunkohlentagebau *m*
lignite pitch Braunkohlen[teer]pech *n*
lignite shale Braunkohlenschiefer *m* (Min)
lignite tar Braunkohlenteer *m*
lignite wax Montanwachs *m*
lignitic braunkohlenhaltig
lignitic earth Erdkohle *f*

lignocellulose Lignocellulose *f*, Holzfaserstoff *m*,
    Holzzellstoff *m*
lignoceric acid Lignocerinsäure *f*
lignoceryl alcohol Lignocerylalkohol *m*
lignone Lignon *n*
lignone sulfonate Lignonsulfonat *n*
lignose Lignose *f*
lignosulfin Lignosulfit *n*
lignosulfite Lignosulfit *n*
lignosulfonic acid Lignosulfonsäure *f*
lignum vitae Franzosenholz *n*, Guajakholz *n*
ligroin[e] Ligroin *n*, Petroläther *m*
ligroin gas lamp Ligroingaslampe *f*
ligurite Ligurit *m* (Min)
ligustrin Syringin *n*, Lilacin *n*
ligustrum Liguster *m* (Bot)
like ähnlich, gleich
likelihood Wahrscheinlichkeit *f*
likely wahrscheinlich
likeness Ähnlichkeit *f*, Gleichheit *f*
lilac [color] Lila *n*
lilac-colored lila
lilacin Lilacin *n*, Syringin *n*
lilac oil Fliederöl *n*
lilagenin Lilagenin *n*
lillianite Lillianit *m* (Min)
lilolidine Lilolidin *n*
lily Lilie *f* (Bot)
lima wood Limaholz *n*
limb Glied *n* (Körper), Limbus *m* (Gradkreis)
limb darkening Randverdunkelung *f*
lime Kalk *m*, Äscherkalk *m*, Kalkerde *f*, **burnt** ~
    gebrannter Kalk *m*, Calciumoxid *n*, **carbolated**
    ~ Carbolkalk *m*, **carbonate of** ~
    Calciumcarbonat *n*, kohlensaurer Kalk *m*,
    caustic ~ gebrannter Kalk, **containing** ~
    kalkhaltig, **cream of** ~ Kalkbrei *m*,
    Kalkmilch *f*, **dead** ~ verwitterter
    (abgestorbener) Kalk, **deficient in** ~ kalkarm,
    fat ~ fetter Kalk, Weißkalk *m*, **free from** ~
    kalkfrei, **gray** ~ Graukalk *m*, **hydrated** ~
    gelöschter Kalk *m*, Calciumhydroxid *n*,
    micaceous ~ Glimmerkalk *m*, **milk of** ~
    Kalkbrühe *f*, **rich** ~ Fettkalk *m*, **rich in** ~
    kalkreich, **saponification by** ~
    Kalkverseifung *f*, **slaked** ~ Löschkalk *m*,
    Calciumhydroxid *n*, gelöschter Kalk *m*,
    solution of ~ Kalklösung *f*, **to extract** ~
    entkalken, **to mix with** ~ kalken, **to overburn**
    ~ Kalk totbrennen, **to soak in** ~ kalken,
    unslaked ~ ungelöschter Kalk, **white** ~
    Weißkalk *m*, ~ **for whitewashing**
    Tünchkalk *m*, ~ **slaked in the air** verwitterter
    (abgestorbener) Kalk
lime *v* kalken, äschern (Leder), einkalken,
    schwöden
lime basin Kalkloch *n*
lime bath Kalkbad *n*
lime bin Kalksilo *m*

lime boil Kalkbeuche *f*
lime bucking lye Kalkbeuche *f*
lime burner Kalkbrenner *m*
lime-burning kiln Kalkbrennofen *m*
lime cast Kalkverputz *m*
lime cement Kalkkitt *m*
lime cement mortar Kalkzementmörtel *m*
lime charging floor Kalkbeschickbühne *f*
lime charging gallery Kalkbeschickbühne *f*
lime chest Kalkkasten *m*
lime coating Kalkschleier *m*
lime concrete Kalkbeton *m*
lime cream Kalkbrei *m*, Schwödemasse *f*
lime crucible Kalktiegel *m*
lime crusher Kalkmühle *f*
lime feldspar Kalkfeldspat *m* (Min)
lime fertilizer Düngekalk *m*
limeflux Äschersatz *m*
lime [fruit] Limette *f* (Bot)
lime glass Kalkglas *n*
lime gravel Sandmergel *m*
lime grease Kalkfett *n*
lime harmotome Kalkharmotom *m* (Min)
lime hepar (obs) Kalkschwefelleber *f*
lime juice Limettenessenz *f*
lime kiln Kalkofen *m*, Kalkbrennerei *f*
lime kiln gas Kalkofengas *n*
limelight Kalklicht *n*
lime-like kalkartig
lime liquor Äscherbrühe *f*, Kalkbrühe *f*
lime marl kalkhaltiger Ton *m*, Kalkmergel *m*
lime mica Kalkglimmer *m* (Min)
lime milk Kalkmilch *f*
lime mill Kalkmühle *f*
lime mortar Kalkmörtel *m*, Luftmörtel *m*
lime mud Kalkschlamm *m*
limene Limen *n*
lime nitrogen Calciumcyanamid *n*,
  Kalkstickstoff *m*
lime oil Limettöl *n*
lime paint Kalkanstrich *m*
lime paste Kalkbrei *m*, Schwöde *f*
lime pit Äscher *m* (Gerb), Äschergrube *f*
  (Gerb), Kalkäscher *m* (Gerb), Kalkgrube *f*
  (Gerb), Schwödgrube *f* (Gerb)
lime powder Kalkpulver *n*
lime precipitate Kalkniederschlag *m*
lime process Äscherverfahren *n* (Gerb)
lime resistance Kalkbeständigkeit *f*
limes Grenzwert *m* (Math), Limes *m* (Math)
lime sand Kalksand *m*
lime sandstone Kalksandstein *m* (Min)
lime scum Kalkschaum *m*
lime sediment Kalkniederschlag *m*,
  Saturationsschlamm *m*
lime silo Kalksilo *m*
lime slag Kalkschlacke *f*
lime slaking Kalklöschen *n*
lime slaking apparatus Kalklöschapparat *m*

lime sludge Kalkschlamm *m*
lime soap Kalkseife *f*
lime-soda feldspar Schillerfels *m* (Min)
lime soil Kalkboden *m*
limestone Kalkstein *m* (Min), **argillaceous** ~
  Tonkalk *m* (Min), **bedded** ~ Flözkalk *m*,
  bituminous ~ Stinkkalk *m* (Min), **coarse** ~
  Grobkalk *m*, **fibrous** ~ Faserkalk *m* (Min),
  marly ~ Tonkalkstein *m*, **siliceous** ~
  Kieselkalkstein *m* (Min), **tufaceous** ~
  Kalktuff *m* (Min), Rindenstein *m* (Min)
limestone flux Kalk[stein]zuschlag *m*
limestone quarry Kalksteinbruch *m*
limestone vein Kalkader *f*
limestone whiteware Kalksteingut *n*
lime sulfur Schwefelkalk *m*
lime trowel Kalkkelle *f*
limette oil Limettöl *n*
limettin Limettin *n*
lime vat Schwödfaß *n* (Gerb)
limewash Kalktünche *f*, Kalkfarbe *f*,
  Kalkmilch *f*, Leimfarbe *f*
limewash *v* tünchen, weißen
lime-washed abgeläutert (Leder)
lime-washing Abläuterung *f*
limewash paint Kalkfarbe *f*, Tünchfarbe *f*
limewater Ätzkalklösung *f*, Kalkbrühe *f*,
  Kalkmilch *f*, Kalktünche *f*, **to cleanse with** ~
  anschwöden (Gerb)
limeworks Kalkbrennerei *f*, Kalkwerk *n*
liming Äschern *n*, Schwöden *n*
liming tub Äscherfaß *n* (Gerb)
limit Grenze *f*, Abgrenzung *f*, Begrenzung *f*,
  Limes *m* (Math), ~ **of resolution**
  Auflösungsgrenze *f*, Grenze *f* des
  Auflösungsvermögens (Opt), ~ **of**
  strain-hardening Kaltverfestigungsgrenze *f*, ~
  of tolerance Erträglichkeitsgrenze *f*
limit *v* begrenzen, einschränken
limitable abgrenzbar
limitation Grenze *f*, Begrenzung *f*,
  Beschränkung *f*, Einschränkung *f*
limit curve Grenzkurve *f*
limited begrenzt, beschränkt
limiter Begrenzer *m* (Elektr)
limiting begrenzend; höchstzulässig
limiting angle Grenzwinkel *m*
limiting area Grenzgebiet *n*
limiting case Grenzfall *m*
limiting concentration Grenzkonzentration *f*
limiting condition Grenzbedingung *f*,
  Grenzzustand *m*, Randbedingung *f*
limiting conductivity Grenzleitfähigkeit *f*
limiting current density Grenzstromdichte *f*
limiting diffusion current
  Diffusionsgrenzstrom *m*
limiting flow Grenzströmung *f*
limiting frequency Grenzfrequenz *f*
limiting line Grenzlinie *f*

limiting mobility Grenzbeweglichkeit *f*
limiting range of stress Dauerfestigkeit *f*
limiting ray Grenzstrahl *m*
limiting size Grenzmaß *n*
limiting state Grenzzustand *m*
limiting stress Grenzspannung *f*
limiting structures *pl* Grenzstrukturen *pl*
limiting temperature Grenztemperatur *f*
limiting value Grenzwert *m*
limiting values (Math) Limeswerte *pl*
limiting velocity Grenzgeschwindigkeit *f*
limit load Grenzbelastung *f*
limits, permissible ~ Toleranz *f*, ~ of
  integration Integrationsgrenzen *pl*
limit size Grenzmaß *n*
limit switch Endumschalter *m*
limit theorem Grenzwertsatz *m* (Math)
limit value Grenzwert *m*
limnite Limnit *m* (Min), Raseneisenerz *n* (Min)
limocitrin Limocitrin *n*
limocitrol Limocitrol *n*
limonene Limonen *n*
limonetrite Limonetrit *n*
limonite Limonit *m* (Min), Bergbraun *n* (Min),
  Brauneisenerz *n* (Min), Rasen[eisen]erz *n*
  (Min), Sumpferz *n* (Min), Wiesenerz *n* (Min),
  oolitic ~ Linsenerz *n* (Min)
limpid durchsichtig, [wasser]klar
limy kalkig
linaloa oil Linaloeöl *n*
linaloe oil Linaloeöl *n*
linalool Linalool *n*, Coriandrol *n*
linaloolene Linaloolen *n*
linalyl acetate Linalylacetat *n*
linalyl chloride Linalylchlorid *n*
linamarin Linamarin *n*
linarigenin Linarigenin *n*
linarine Linarin *n*
linarite Linarit *m* (Min), Kupferbleivitriol *n*
  (Min)
linchpin Achsennagel *m*
lindackerite Lindackerit *m* (Min)
lindelofidine Lindelofidin *n*
linden oil Lindenöl *n*
linderazulene Linderazulen *n*
linderic acid Lindersäure *f*
lindesite Lindesit *m*
lindsayite Lindsayit *m*
line Linie *f*, Reihe *f*, Strich *m*, Telephon- oder
  Telegraphenleitung *f* (Elektr), broken ~
  gestrichelte Linie, curved ~ krumme Linie,
  dash-dotted ~ strichpunktierte Linie, dashed
  ~ gestrichelte Linie, dotted ~ punktierte
  Linie, enhanced ~ verbreiterte Linie, full ~
  ausgezogene Linie, out of ~ schief, schräg,
  solid ~ ausgezogene Linie, underground ~
  unterirdische Stromleitung *f*, voltage of the ~
  Leitungsspannung *f*, ~ of curvature
  Krümmungslinie *f*, ~ of demarcation

Grenzlinie *f*, Scheidelinie *f*, ~ of magnetic
  force magnetische Kraftlinie *f*
line *v* [aus]füttern, auskleiden, verkleiden
linear linear, geradlinig
linear acceleration Linearbeschleunigung *f*
linear accelerator Linearbeschleuniger *m*
linear expansion Längenausdehnung *f*,
  coefficient of ~
  Längenausdehnungskoeffizient *m*
linear extension Längenausdehnung *f*
linear focus Strichfokus *m* (Opt)
linearity Linearität *f*
linearization Linearisierung *f*
linearized linearisiert
linearly dependent linear abhängig
linear measure Längenmaß *n*
linear measurement Längenmessung *f*
linear pinch lineare Einschnürung *f*
linear transformation lineare Transformation *f*
  (Math)
line block Strichklischee *n*
line broadening Linienverbreiterung *f* (Spektr)
lined ausgekleidet
line density Liniendichte *f* (Spektr)
line dipole Liniendipol *m*
line disturbance Netzstörung *f* (Elektr)
line index Zeilenindex *m*
line integral Linienintegral *n*
line interval Linienabstand *m* (Spektr)
linelaidic acid Linelaidinsäure *f*
linen Leinen *n* (Text), Leinwand *f* (Text),
  Linnen *n* (Text), Wäsche *f* (Text)
linen [rag] paper Leinenpapier *n*
line of action (Am. E.) Eingriffsstrecke *f*
  (Getriebe)
line pair Linienpaar *n* (Spektr)
line pressure Betriebsdruck *m* (Hydr),
  Leitungsdruck *m*
liner Ausfütterung *f*, Füllstück *n*, Futterrohr *n*
  (Reaktor), Innensack *m*, Innenseele *f*
  (Reifen), Laufbüchse *f*
line resistance Leitungswiderstand *m*
line reversal Linienumkehr *f*
lines, rich in ~ linienreich (Spektr), series of ~
  Linienreihe *f* (Spektr)
line screen Raster *m* (Spektr)
lineshaft Laufwelle *f*
lines of force, bundle of ~ Kraftlinienbündel *n*,
  deflection of ~ Kraftlinienablenkung *f*,
  direction of ~ Kraftlinienrichtung *f*
line spectrum Liniensprektrum *n*
line splitting Linienaufspaltung *f*
line voltage Leitungsspannung *f*
line width Linienbreite *f* (Spektr)
liniment Einreibemittel *n* (Pharm), Liniment *n*
  (Pharm), Salbe *f* (Pharm)
linin Linin *n*
lining Auskleidung *f*, Ausfütterung *f*,
  Beschlag *m* (Auskleidung), Futter *n*,

Futterstoff *m*, Innensack *m*, Verkleidung *f,*
acidproof ~ säurefeste Auskleidung, **film for**
~ Auskleidefolie *f,* **refractory** ~ feuerfeste
Auskleidung, **shaft** ~ Auskleidung *f* des
Schachtes
lining material for repairs Ausbesserungsmasse *f*
lining wax Kaschierwachs *n*
link Gelenk *n*, Glied *n*, Ring *m*, Verbindung *f,*
Verbindungsglied *n*, ~ **[of a chain]**
Kettenglied *n*
link *v* verbinden, verketten
linkage Verknüfung *f,* Angliederung *f,*
Kopplung *f* (Elektr), Verbindung *f,*
Verkettung *f*
linkage between nuclei Kernverknüpfung *f*
linkage electron Bindungselektron *n*
linkage force Bindungskraft *f*
linkage gear Koppelgetriebe *n*
link belt Gliederriemen *m*
link bolt Gelenkbolzen *m*
linking machine Kettelmaschine *f*
linking process Verbindungsprozeß *m*
link stone Schakenstein *m*
Linn[a]ean system Linnésches System *n*
linnaeite Linneit *m* (Min), Kobaltnickelkies *m*
(Min)
linocaffein Linocaffein *n*
linocinnamarin Linocinnamarin *n*
linoleate Linoleat *n*
linoleic acid Linolsäure *f,* Hanfsäure *f* (obs)
linoleine Linolein *n*
linolenic acid Linolensäure *f*
linolenyl alcohol Linolenylalkohol *m*
linoleodistearin Linoleodistearin
linoleone Linoleon *n*
linoleum Linoleum *n*
linoleum substitute Linoleumersatz *m*
linolic acid Linolsäure *f*
linosite Linosit *m*
linoxyn Linoxyn *n*
linseed Leinsamen *m* (Bot)
linseed decoction Leinsamenabkochung *f*
linseed oil Leinöl *n*, Dicköl *n*, **boiled** ~
Leinölfirnis *m*, Leinöllack *m*, **raw** ~
Rohleinöl *n*, **solid oxidized** ~ Linoxyn *n*,
vulcanized ~ Ölkautschuk *m*
linseed oil cake Leinkuchen *m*
linseed oil paint Leinölfarbe *f*
linseed oil varnish Leinölfirnis *m*, Leinöllack *m*
linseed standoil Leinöl-Standöl *n*
lint Fussel *f,* Scharpie *f*
linters Linters *pl*
linusinic acid Linusinsäure *f*
liothyronine Liothyronin *n*
liovil Liovil *n*
lip Ausguß *m*, Auskragung *f,* Wulstaustrieb *m*
lipaemia Lipämie *f* (Med)
liparite Liparit *m*
lipase Lipase *f* (Biochem)

lipid Lipid *n*, **nonsaponifiable** ~ nicht
verseifbares Lipid *n*, **saponifiable** ~
verseifbares Lipid *n*
lipid catabolism Lipidkatabolismus *m*
lipid[e] Lipid n, Lipoid *n*
lipid membrane Lipidmembran *f*
lipid metabolism Lipidstoffwechsel *m* (Physiol)
lipoamide Lipoamid *n*
lipoblast Fettzelle *f,* Lipoblast *m*
lipocatabolism Fettabbau *m* (Physiol)
lipochrome Lipochrom *n*
lipocyte Fettzelle *f,* Lipozyt *m*
lipogenesis Lipogenese *f* (Physiol)
lipogenous fettbildend
lipoic acid Liponsäure *f*
lipoid fettähnlich, fettartig, lipoid, Lipoid *n*
lipol Lipol *n*
lipolysis Fettspaltung *f,* Lipolyse *f*
lipolytic fettspaltend, lipolytisch
lipolytic agent Fettspalter *m*
lipometabolism Fetthaushalt *m*,
Fettstoffwechsel *m* (Biochem)
lipophilic lipophil, fettaffin
lipophobic lipophob
lipoprotein Lipoproteid *n*, Lipoprotein *n*
liposoluble fettlöslich
liposome Liposom *n*, Fetttröpfchen in Zellen
lipoxydase Lipoxidase *f* (Biochem)
lipped mortar Ausgußmörser *m*
lip relief angle (drilling) Freiwinkel *m*
lips (for sheeting or film dies) Kopfplatte *f*
liquate *v* [ab]darren, [ab]seigern, ausseigern
liquated copper Darrkupfer *n*
liquation Seigerung *f,* Ausseigern *n*,
Schmelzen *n*, Seigern *n* (copper), Darrarbeit *f,*
Darren *n*, **to separate by** ~ ausseigern
liquation disk Scheidekuchen *m*
liquation hearth Seigerherd *m*, Abdörrofen *m*
(Kupfer), Darrofen *m*
liquation lead Seigerblei *n*
liquation process Seigerprozeß *m*
liquation residue Seigerrückstand *m*
liquation slag Seigerschlacke *f*
liquefaction Verflüssigung *f,* Flüssigwerden *n*, ~
of gases Gasverflüssigung *f*
liquefied gas Flüssiggas *n*
liquefier Verflüssigungsapparat *m*
liquefy *v* verflüssigen
liquefying Verflüssigen *n*
liquefying agent Verflüssigungsmittel *n*
liquefying plant Verflüssigungsanlage *f*
liquefying point Tropfpunkt *m*
liqueur Likör *m*
liquid (also see liquids) Flüssigkeit *f,* flüssiger
Körper *m*, **heat of a**
~ Flüssigkeitswärme *f,* **ideal** ~ ideale
Flüssigkeit, **level of** ~ Flüssigkeitsstand *m*,
motion of a ~ Flüssigkeitsbewegung *f,*
Newtonian ~ Newtonsche Flüssigkeit,
optically void ~ optisch reine Flüssigkeit

(Opt), **quantity of** ~ Flüssigkeitsmenge *f,*
surface of a ~ Flüssigkeitsoberfläche *f,*
Flüssigkeitsspiegel *m,* **wetting** ~ benetzende
Flüssigkeit
liquid *a* flüssig, dünnflüssig (Öl)
liquid air flüssige Luft *f*
liquid circulation Flüssigkeitsumlauf *m*
liquid column Flüssigkeitssäule *f*
liquid cooler Flüssigkeitskühler *m*
liquid cooling Flüssigkeitskühlung *f*
liquid damping Flüssigkeitsdämpfung *f*
liquid density recorder
Flüssigkeitsdichteschreiber *m*
liquid distributor Flüssigkeitsverteiler *m* (Dest)
liquid-drop behavior Tröpfchenverhalten *n*
liquid-drop model Tröpfchenmodell *n*
liquid feed point Verteilungsstelle *f* (Dest)
liquid-filled thermometer
Flüssigkeitsthermometer *n*
liquid fuel flüssiger Brennstoff *m*
liquid gas plant Flüssiggasanlage *f*
liquid gradient Flüssigkeitsgradient *m*
liquidity Dünnflüssigkeit *f,* flüssiger Zustand *m*
liquid layer Flüssigkeitsschicht *f*
liquid level Flüssigkeitsspiegel *m,*
Flüssigkeitsstand *m*
[liquid] level indicator Flüssigkeitsanzeiger *m,*
Flüssigkeitsstandanzeiger *m*
liquid level measurement
Flüssigkeitsstandmessung *f*
liquid-liquid extraction
Flüssig-Flüssig-Extraktion *f*
liquid load factor F-Faktor *m* (Dest)
liquid magmatic liquidmagmatisch
liquid manure Düngejauche *f,* Gülle *f*
liquid measure Flüssigkeitsmaß *n*
liquid-metal cooling Flüssigmetallkühlung *f*
liquid meter Flüssigkeitsmesser *m*
liquid mirror reflection
Flüssigkeitsspiegelreflexion *f*
liquid mist removal Nebelabscheidung *f*
liquid mixing apparatus
Flüssigkeitsmischapparat *m*
liquid mixture Flüssigkeitsgemisch *n*
liquid petrol gas (PG or LP gas) Flüssiggas *n*
liquid phase Trennflüssigkeit *f* (Gaschromat)
liquid-phase cracking
Flüssigphase-Krackverfahren *n*
liquid-phase oxidation Flüssigphasenoxidation *f*
liquid-phase process Sumpfverfahren *n*
liquid-phase synthesis Flüssigphase-Synthese *f*
liquid polish Polierflüssigkeit *f*
liquid·ring [or seal] pump Flüssigkeitsringpumpe *f*
liquids, immiscible ~ nicht mischbare
Flüssigkeiten *pl,* **impermeable to** ~
flüssigkeitsundurchlässig, **miscible** ~
mischbare Flüssigkeiten *pl,* **pump for** ~
Flüssigkeitspumpe *f,* **raising of** ~

Flüssigkeitsförderung *f,* **sealed** ~
Flüssigkeitseinschlüsse *pl*
liquid scintillation counter
Flüssigkeits-Szintillationszähler *m*
liquid seal Flüssigkeitssperrung *f,*
Sperrflüssigkeit *f*
liquid seal pump Wasserringpumpe *f*
liquid-solid chromatography
Säulenchromatographie *f*
liquid solidification Erstarren *n* einer
Flüssigkeit, Einfrieren *n,* Festwerden *n* einer
Flüssigkeit, Gefrieren *n*
liquid surface Flüssigkeitsoberfläche *f*
liquidus curve Liquiduskurve *f*
liquid-vapor ratio
Flüssigkeits-Dampfverhältnis *n*
liquor Brühe *f,* Flotte *f* (Färb); Lauge *f,*
Laugenbad *n,* **crude** ~ Rohlauge *f,* **loss in** ~
Flottenverlust *m* (Färb)
liquor circulation Flottenzirkulation *f*
liquor circulation plant Laugenumwälzanlage *f*
liquor concentration Flottenkonzentration *f*
liquor distributor Laugenverteiler *m*
liquorice Lakritze *f*
liquor length ratio Flottenverhältnis *n* (Färb)
liquor pick-up Flottenaufnahme *f* (Färb)
liquor ratio Flottenverhältnis *n* (Färb)
liroconite Lirokonit *m* (Min), Linsenerz *n*
(Min), Linsenkupfer *n* (Min)
liskeardite Liskeardit *m* (Min)
Lissajous figures Lissajous-Figuren *pl*
lissamine dyestuff Lissaminfarbstoff *m*
lissolamine Lissolamin *n*
list Liste *f,* Tabelle *f,* Verzeichnis *n*
listoform soap Listoformseife *f*
list pot Abtropfpfanne *f*
liter (Am. E.) Liter *m n,* **weight per** ~
Litergewicht *n*
literature documentation
Literaturdokumentation *m,*
Literaturverfolgung *f*
literature search Literaturstudium *n*
liter flask Literkolben *m*
liters, capacity in ~ Literinhalt *m*
litharge Bleioxid *n,* Bleiglätte *f,* Bleiocker *m*
(Min), Bleischwamm *m,* **black impure** ~
Fußglätte *f,* **hard** ~ Frischglätte *f*
litharge plant Bleiglätteanlage *f*
lithia Lithiumoxid *n*
lithia feldspar Lithiumfeldspat *m* (Min)
lithia mica Lithiumglimmer *m* (Min)
lithionite Lithionit *m* (Min), Zinnwaldit *m*
(Min)
lithiophilite Lithiophilit *m* (Min)
lithiophorite Lithiophorit *m* (Min)
lithium (Symb. Li) Lithium *n*
lithium-12-hydroxy grease
Lithium-12-hydroxyfett *n*
lithium acetate Lithiumacetat *n*

lithium acetyl salicylate Apyron *n*, Grifa *n*, Lithiumacetylsalicylat *n*
lithium alkyl Lithiumalkyl *n*
lithium aluminum hydride Lithiumaluminiumhydrid *n*
lithium amide Lithiumamid *n*
lithium arsenate Lithiumarsenat *n*
lithium benzoate Lithiumbenzoat *n*
lithium bicarbonate Lithiumbicarbonat *n*
lithium borate glass Lithiumboratglas *n*
lithium borohydride Lithiumborhydrid *n*
lithium bromide Lithiumbromid *n*
lithium chelate Lithiumchelat *n*
lithium chloride Lithiumchlorid *n*
lithium compound Lithiumverbindung *f*
lithium content Lithiumgehalt *m*
lithium ethylanilide Lithiumäthylanilid *n*
lithium fluoride Lithiumfluorid *n*
lithium grease Lithiumfett *n*
lithium hydride Lithiumhydrid *n*
lithium hydrogen carbonate Lithiumbicarbonat *n*
lithium iodide Lithiumjodid *n*
lithium mica Lepidolith *m* (Min)
lithium nitride Lithiumnitrid *n*, Stickstofflithium *n*
lithium ore Lithiumerz *n*
lithium oxide Lithiumoxid *n*, Lithion *n*
lithium quinate Urosin *n*
lithium salicylate Lithiumsalicylat *n*
lithium silicate glass Lithiumsilikatglas *n*
lithium sulfate Lithiumsulfat *n*
lithium tantalate Lithiumtantalat *n*
lithium urate Lithiumurat *n*
lithocholic acid Lithocholsäure *f*
lithochromatics Chromolithographie *f*, Farbendruck *m*
lithocolla Steinkitt *m*, Steinleim *m*
lithograph Lithographie *f*, Steindruck *m*
lithograph *v* lithographieren
lithographic lithographisch
lithographic color Lithographiefarbe *f*, Plakatfarbe *f*
lithographic varnish Leinölfirnis *m*
lithography Lithographie *f*, Steindruck *m*
litholeine Litholein *n*
lithol fast yellow Litholechtgelb *n*
lithol scarlet Litholscharlach *n*
lithomarge Steinmark *n* (Min), Wundererde *f* (Min), ~ **containing iron** Eisensteinmark *n* (Min)
lithophosphor Leuchtstein *m*
lithopone Charltonweiß *n*, Lithopon *n*, Zinksulfidweiß *n*
lithosiderite Lithosiderit *m* (Min)
lithosphere Lithosphäre *f* (Geol), Gesteinsmantel *m* (Geol)
lithoxyl[e] versteinertes Holz *n*
litmopyrine Grifa *n*, Lithiumacetylsalicylat *n*
litmus Lackmus *m*

litmus blue Lackmusblau *n*
litmus liquor Lackmustinktur *f*
litmus paper Lackmuspapier *n*
litmus solution Lackmustinktur *f*
litre (Br. E.) Liter *m n*
litter Abfall *m*
live *a* lebend, lebendig, stromführend (Elektr)
liveingite Liveingit *m* (Min)
live ore Schütterz *n*
liver Leber *f*
liver-brown cinnabar Lebererz *n* (Min)
liver cell Leberzelle *f*
liver enzyme Leberenzym *n*
liver metabolism Leberstoffwechsel *m*
liver of sulfur (obs) Kalischwefelleber *f*
liver sugar Glykogen *n*
live steam direkter Dampf *m*, Frischdampf *m*
live steam line Frischdampfleitung *f*
live steam pipe Frischdampfrohr *n*
live steam piping Frischdampfleitung *f*
livestock Vieh *n*
live virus vaccine Lebendvirusvakzine *f*
living being Lebewesen *n*
living cells Frischzellen *pl*
living organism Lebewesen *n*, **in the** ~ in vivo (Lat)
livingstonite Livingstonit *m* (Min)
lixivial salt Laugensalz *n*
lixiviate *v* [aus]laugen, mit Lauge behandeln
lixiviated ausgelaugt
lixiviating tank Auslaugebehälter *m*
lixiviating vat Auslaugekasten *m*
lixiviation Auslaugung *f*, Laugung *f*
lixiviation residue Laugerückstand *m*
lixiviation vat Laugebottich *m*
lixivious laugenartig
lixivium Extrakt *m*, Lauge *f*, **crude** ~ Rohlauge *f*
lizard stone Serpentinmarmor *m* (Min)
load Last *f*, Beanspruchung *f*, Belastung *f*, Ladegewicht *n*, **additional** ~ Zusatzbelastung *f*, **axially symmetrical** ~ drehsymmetrische Belastung *f*, **dead** ~ Belastung *f* durch Eigengewicht, ruhende Last, **dummy** ~ künstliche Belastung (Elektr), **excess** ~ Überschußlast *f*, **half** ~ Halblast *f*, **increasing** ~ zunehmende Belastung, **inductive** ~ induktive Belastung, **intermittent shock** ~ stoßweise Belastung, **live** ~ wechselnde Belastung, bewegliche Last, **permanent** ~ ständige Belastung, **permissible** ~ zulässige Beanspruchung, **point of application of** ~ Lastangriffspunkt *m*, **rated** ~ zulässige Höchstbelastung, **ratio of** ~ Belastungsverhältnis *n*, **safe** ~ zulässige Belastung, **to run without** ~ leerlaufen, **uniformly distributed** ~ Gleichbelastung *f*
load *v* beladen, aufladen, belasten, bepacken, beschicken, beschweren, chargieren, speisen
load area Belastungsfläche *f*

load at break Bruchlast *f*
load capacity Belastbarkeit *f* (Elektr),
　Belastungsfähigkeit *f*, Tragfähigkeit *f*
load carrying equipment Förderanlage *f*
load change Laständerung *f*
load circuit Arbeitskreis *m*, Heizkreis *m*
load compensation Belastungsausgleich *m*
load cycles, number of ~ **to failure**
　Bruchlastspielzahl *f*
load diagram Belastungsdiagramm *n*
load distribution Lastverteilung
load duration Belastungsdauer *f*
loaded beladen, belastet, **heavily** ~ füllstoffreich,
　lightly ~ füllstoffarm
loaded valve Belastungsventil *n*
load equalization Belastungsausgleich *m*
loader Einfüller *m*, Ladegerät *n*, Verlader *m*,
　wheel ~ **or payloader** Radlader *m*
load-extension curve Lastdehnungskurve *f*
load factor Belastungsfaktor *m*, Belastung *f*,
　Belastungsgrad *m*, Belastungskoeffizient *m*
load fluctuation Belastungsschwankung *f*,
　Belastungswechsel *m*
load independence Lastunabhängigkeit *f*
loading Belastung *f*, Aufladung *f*, Beschicken *n*,
　Beschickung *f*, Ladung *f* (Last), Verladung *f*,
　initial ~ Vorbelastung *f*, **range of** ~
　Beanspruchungsbereich *m*, ~ **of the electrodes**
　Elektrodenbelastung *f*, ~ **with metallic salts**
　Metallbeschwerung *f*
loading capacity Belastungsvermögen *n*,
　Ladefähigkeit *f*
loading cavity Füllraum *m*
loading chamber Füllkammer *f*, Füllraum *m*
loading chute Laderutsche *f*
loading crane Ladekran *m*, Chargierkran *m*
loading curve Belastungskurve *f*
loading density Füllungsdichte *f*
loading device Ladevorrichtung *f*
loading diagram Beladeplan *m*
loading hopper Einfülltrichter *m*
loading nose Biegestempel *m* (Biegeprüfung)
　(bending test), Druckfinne *f*
loading per unit area Flächeneinheitslast *f*
loading plant Verladeanlage *f*
loading platform Laderampe *f*
loading point Aufgabestelle *f*, obere
　Belastungsgrenze *f*, Staugrenze *f* (Dest)
loading portal Ladeportal *n*
loading ramp Laderampe *f*, Verladerampe *f*
loading resistance Ballastwiderstand *m*
loading sequence Belastungsfolgen *pl*
loading space Füllraum *m*
loading station Beladestation *f*
loading table Beladeplan *m*
loading test Belastungsprobe *f*, Probebelastung *f*
loading time Belastungsdauer *f*
loading tray Füllblech *n*, Fülltablett *n*,
　Füllvorrichtung *f*

loading trough Einflußrinne *f*
load limit Belastungsgrenze *f*
load line Lademarke *f*
load phase Belastungsphase *f*
load resistance Arbeitswiderstand *m* (Elektr)
load-shift gearing Lastschaltgetriebe *n*
load spectrum Belastungsfolgen *pl*, ~ **under**
　service conditions
　Beanspruchungs-Charakteristik *f*
loadstone natürlicher Magnet *m*
load test Belastungsprobe *f*,
　Druckerweichungsprüfung *f* (Email), ~ **up to**
　breaking Belastungsprobe bis zum Bruch
load value Lastwert *m*
load variation Belastungsänderung *f* (Elektr),
　Belastungswechsel *m*
load voltage Belastungsspannung *f* (Elektr)
loaf Laib *m*
loaf sugar Hutzucker *m*
loam Lehm *m*
loam beater Lehmmesser *n*
loam cake, to prepare the ~ den Lehmkuchen *m*
　herrichten
loam casting Lehmguß *m*
loam coat Lehmschicht *f*
loam core Lehmkern *m*
loaming Lehmtünche *f*
loam lute Lehmkitt *m*
loam mold Lehmform *f*
loam molded casting Lehmformguß *m*
loam molding Lehmformerei *f*
loam mortar Lehmmörtel *m*
loam pit Lehmgrube *f*
loamy lehmartig, lehmhaltig, lehmig
loamy marl Lehmmergel *m*
loamy sand fetter Sand *m*
loangocopalic acid Loangocopalsäure *f*
loangocopalinic acid Loangocopalinsäure *f*
loangocopaloresene Loangocopaloresen *n*
lobaric acid Lobarsäure *f*
lobelane Lobelan *n*
lobelanidine Lobelanidin *n*
lobelanine Lobelanin *n*
lobelia Lobelie *f* (Bot), Lobelienkraut *n* (Bot)
lobelia tincture Lobelientinktur *f*
lobelidine Lobelidin *n*
lobelidiol Lobelidiol *n*
lobeline Lobelin *n* (Med, Chem)
lobelinic acid Lobelinsäure *f*
lobelionol Lobelionol *n*
lobelol Lobelol *n*
lobelone Lobelon *n*
lobe pump Wälzkolbenpumpe *f*
local örtlich
local action Lokalwirkung *f*
local an[a]esthesia Lokalanästhesie *f* (Med),
　örtliche Betäubung *f*
local an[a]esthetic Lokalanästhetikum *n*, örtliches
　Betäubungsmittel *n*

local cell Lokalelement *n*
localization Begrenzung *f,* Eingrenzung *f,*
  Lokalisation *f,* Lokalisierung *f*
localization theorem Lokalisierungssatz *m*
localize *v* lokalisieren, eingrenzen
localized corrosion Lokalkorrosion *f*
local value Stellenwert *m* (Math)
locaose Lokaose *f*
locate *v* eingrenzen, lokalisieren
locating pin Fixierstift *m*
location Platz *m,* Lage *f,* Ortung *f,* Standort *m*
location of mistakes Fehlereingrenzung *f*
lock Schloß *n,* Sperrvorrichtung *f,* Verschluß *m*
lock *v* zuschließen, arretieren; verriegeln,
  zusperren
lock bar Sperrschiene *f*
locked, capable of being ~ absperrbar
locker (abschließbarer) Schrank *m,* Schließfach *n*
lock[ing] Arretierung *f*
locking button Arretierungsknopf *m*
locking device Arretiervorrichtung *f,*
  Schließvorrichtung *f,* Sperrvorrichtung *f*
lock[ing] lever Feststellhebel *m*
locking pin Arretierstift *m*
locking pressure Schließdruck *m*
locking relay Sperrelais *n*
locking ring Schließring *m*
locking screw Arretierschraube *f,*
  Feststellschraube *f,* Sicherungsschraube *f,*
  Verschlußschraube *f*
lock knob Einrastknopf *m*
locknit maschenfest (Text)
locknut Gegenmutter *f,* Verschlußmutter *f*
lock ring Verschlußring *m*
lock washer Federring *m*
locomotive Lokomotive *f*
locus Ort *m,* **achromatic** ~ achromatisches
  Gebiet *n,* **[geometrical]** ~ geometrischer Ort *m*
locust bean Johannisbrot *n*
locuturine Locuturin *n*
lodal Lodal *n*
lode Ader *f* (Bergb), Flöz *n* (Bergb), Gang *m*
  (Bergb)
lode seam Erzader *f*
lodestone natürlicher Magnet *m*
lodged corn Lagergetreide *n*
lodging Lagern *n* (Getreide)
loellingite Arseneisen *n* (Min), Löllingite *m*
  (Min)
loess Löß *m* (Geol)
loeweite Löweit *m* (Min)
log Klotz *m,* Logarithmus *m* (Math)
loganin Loganin *n*
loganite Loganit *m*
logarithm Logarithmus *m* (Math), **Brigg's** ~
  Briggscher Logarithmus, dekadischer
  Logarithmus, **common** ~ Briggscher
  Logarithmus *m,* **decadic** ~ Briggscher
  Logarithmus, dekadischer Logarithmus,

decimal ~ Briggscher Logarithmus *m,*
five-figure ~ fünfstelliger Logarithmus,
inverse ~ Numerus *m,* **natural** ~ natürlicher
  Logarithmus, **to take the [of]** ~
  logarithmieren, ~ **to the base ten**
  Zehnerlogarithmus *m*
logarithmic[al] logarithmisch
logarithmic function Logarithmusfunktion *f*
log[arithmic] paper Logarithmenpapier *n*
logarithmic table Logarithmentafel *f* (Math)
logging Bohrlochmessungen *pl*
logical logisch, folgerichtig
log-log plot doppeltlogarithmisches Diagramm *n*
logwood Blauholz *n,* Blutholz *n,*
  Campecheholz *n*
logwood chips Blauholzspäne *pl*
logwood extract Blauholzextrakt *m*
logwood liquor Blauholztinktur *f*
loiponic acid Loiponsäure *f*
lomatiol Lomatiol *n*
lonchidite Lonchidit *m*
long-armed langarmig
long-chain langkettig
long-distance call Ferngespräch *n*
long-distance flight Langstreckenflug *m*
long-distance operation Fernantrieb *m*
long-duration test Dauerversuch *m*
longeron Rumpfholm *m* (Luft)
longevity lange Lebensdauer *f,* Langlebigkeit *f*
long-fibered langfaserig
long flame tube Langflammrohr *n*
long-handed tool Fernbedienungsgerät *n*
longibornane Longibornan *n*
longiborneol Longiborneol *n*
longidione Longidion *n*
longifdione Longifdion *n*
longifolic acid Longifolsäure *f*
longitude, degree of ~ Längengrad *m* (Geogr)
longitude determination Längenbestimmung *f*
longitudinal acceleration Längsbeschleunigung *f*
longitudinal axis Längsachse *f*
longitudinal blow-up ratio Längsblasverhältnis *n*
longitudinal covering machine
  Längsbedeckungsmaschine *f*
longitudinal crack Längsriß *m*
longitudinal cut Längsschnitt *m*
longitudinal deviation Längenabweichung *f*
longitudinal direction Längsrichtung *f*
longitudinal expansion Längs[aus]dehnung *f*
longitudinal extension Längsdehnung *f*
longitudinal feed Längsvorschub *m*
longitudinal frame Längsspant *n*
longitudinal girder Längsträger *m*
longitudinal groove Längsfurche *f,* Längsnut *f*
longitudinal induction Längsinduktion *f*
longitudinal knurls Längsrändel *pl*
longitudinal magnetization
  Längsmagnetisierung *f*

longitudinal oscillation Längsschwingung *f,*
Longitudinalschwingung *f*
longitudinal pitch Längsteilung *f*
longitudinal pressure Längsdruck *m*
longitudinal rib Längsrippe *f*
longitudinal slot Längsnut *f*
longitudinal tie Längsschwelle *f*
longitudinal vibration Longitudinalschwingung *f*
longitudinal view Längsansicht *f*
longitudinal wave Longitudinalwelle *f*
long-life preparation Dauerpräparat *n*
long-lived langlebig
long-oil langölig, ölreich
long-slot burner Langlochbrenner *m*
long-term langfristig, langwierig
long-time storage test Dauerlagerversuch *m*
long-tube evaporator Langrohr-Verdampfer *m*
long wave Langwelle *f*
long-wave range Langwellenbereich *m*
long-wave receiver Langwellenempfänger *m*
long-wave transmitter Langwellensender *m*
look Aussehen *n*
look *v,* ~ **for** suchen, ~ **forward to**
entgegensehen
look out! Achtung!
loom Webstuhl *m*
loom setting Webstuhleinstellung *f* (Web)
loop geschlossener Stromkreis *m* (Elektr), Öse *f,*
Schlaufe *f,* Schleife *f,* Schlinge *f,*
Schwingungsbauch *m* (Phys)
loop antenna Rahmenantenne *f*
loop dryer Girlandentrockner *m,*
Hängetrockner *m*
loop formation Schleifenbildung *f*
loop galvanometer Schleifengalvanometer *n*
loop oscillator Schleifenoszillograph *m*
loop strength Knotenfestigkeit *f,*
Schlingenfestigkeit *f*
loop system Ringschaltung *f*
loop tenacity Maschenfestigkeit *f*
loose lose, locker, schlaff, **to make** ~ lockern
loose connection Wackelkontakt *m* (Elektr)
looseleaf binder Loseblattsammelmappe *f*
loosen *v* [ab]lösen, locker werden, ~ **up**
auflockern
loosening Lockerung *f,* Auflockerung *f,*
Loslösung *f*
loose roller Losrolle *f*
loose weight Schüttgewicht *n*
lop *v* bekappen (Holz)
lophane Lophan *n*
lophenol Lophenol *n*
lophine Lophin *n*
lophocerine Lophocerin *n*
lophoite Lophoit *m*
lophophorine Lophophorin *n*
lorandite Lorandit *m* (Min)
loranskite Loranskit *m* (Min)
Lorentz transformation Lorentztransformation *f*
loretin Loretin *n*

lorry (Br. E.) Lastkraftwagen *m*
Loschmidt number Loschmidtsche Zahl *f*
lose *v* verlieren, ~ **black color** abschwärzen, ~
color abfärben
losophan Losophan *n*
loss Verlust *m,* Abnahme *f,* Ausfall *m,*
Schaden *m,* Schwinden *n,* Schwund *m,* **free of**
~ verlustfrei, **total** ~ Gesamtverlust *m,*
without ~ verlustlos, ~ **by evaporation**
Einzehrung *f,* ~ **from cooling**
Abkühlverlust *m,* ~ **of metal** Metallverlust *m*
loss angle Verlustwinkel *m* (Elektr)
loss damping Verlustdämpfung *f*
loss factor Verlustfaktor *m*
loss of material Substanzverlust *m*
lost verloren
lot Fabrikationspartie *f*
lotoflavin Lotoflavin *n*
loturine Harman *n*
lotusin Lotusin *n*
loud laut
loudness level Lautstärkepegel *m*
loudspeaker Lautsprecher *m*
louse (pl. lice) Laus *f*
louvre classifier Jalousiesichter *m*
lovage oil Liebstöckelöl *n*
lovage root Liebstöckelwurzel *f* (Bot)
low nieder, niedrig, tief
low-boiling niedrigsiedend, tiefsiedend
low-capacity hoist Kleinlastaufzug *m*
low-carbon kohlenstoffarm
low-contrast kontrastarm (Phot)
lower *v* [ab]senken, erniedrigen, herabsetzen,
herunterdrücken, verringern
lower box Unterkasten *m*
lowering Absenken *n,* Erniedrigung *f,* ~ **of vapor**
pressure Dampfdruckerniedrigung *f*
lowering device Senkvorrichtung *f*
lower layer Unterschicht *f*
lower part Unterstück *n*
low-fat milk Magermilch *f*
low fermentation yeast Unterhefe *f* (Brau)
lowflux reactor Niederflußreaktor *m*
lowflux reactor core Niederflußreaktorkern *m*
low frequency Niederfrequenz *f*
low-frequency *a* niederfrequent
low-frequency amplifier NF-Verstärker *m*
(Elektr)
low-frequency current Niederfrequenzstrom *m,*
Strom *m* geringer Frequenz
low-frequency voltage NF-Spannung *f* (Elektr)
low-grade arm (Erz), geringwertig, mager (Erz),
minderwertig
low-lift pumping station Vorpumpwerk *n*
low-lift truck Niederhubwagen *m*
low-melting leicht schmelzbar,
niedrigschmelzend, tiefschmelzend
low-molecular niedermolekular
low pass filter Drosselkette *f*

low-power reactor Niederleistungsreaktor *m*
low pressure Unterdruck *m*
low-pressure boiler Niederdruckkessel *m*
low-pressure column Niederdrucksäule *f*
low-pressure discharge Niederdruckentladung *f*
low-pressure fan Niederdrucklüfter *m*
low-pressure molding Niederdruckpressen *n*,
   Niederdruckpreßverfahren *n*
low-pressure plastic Niederdruckschichtstoff *m*
low-pressure plunger Niederdruckkolben *m*
low-pressure polyethylene
   Niederdruckpolyäthylen *n*
low-pressure steam Niederdruckdampf *m*
low-pressure steam engine
   Niederdruckdampfmaschine *f*
low-pressure tank Niederdruckkessel *m*
low-pressure tire Niederdruckreifen *m*
low-pressure vessel Niederdruckkessel *m*
low red heat Dunkelrotglut *f*
low-resistance niederohmig
low-speed niedertourig
low-temperature behavior Kälteverhalten *n*
low-temperature carbonization Schwelerei *f*
low-temperature carbonization coke
   Schwelkoks *m*
low-temperature carbonization of fines
   Staubschwelung *f*
low-temperature property Kältebeständigkeit *f*
low-temperature reactor
   Niedertemperaturreaktor *m*
low-temperature rectification
   Niedertemperaturfraktionierung *f*
low-temperature scrubbing Kaltwäsche *f*
low-temperature shift conversion
   Tieftemperaturkonvertierung *f*
low-temperature tar Schwelteer *m*, Tiefteer *m*,
   Tieftemperaturteer *m*
low-tension cable Niederspannungskabel *n*
low-tension current Niederspannungsstrom *m*
low-tension line Niederspannungsleitung *f*
low-tension plant Niederspannungsanlage *f*
low-tension shell Niederspannungsgehäuse *n*
low tide Ebbe *f*
low-viscosity dünnflüssig (Öl)
low-voltage niedervoltig
low-voltage arc Niedervoltbogen *m*
low-voltage bell Schwachstromglocke *f*
low-voltage current Niederspannungsstrom *m*,
   Schwachstrom *m*
low-voltage generator
   Niederspannungserzeuger *m*
low wine Lutter *m*
loxodromic line Loxodrome *f* (Math)
lozenge, [medicated] ~ Pastille *f*
L-shell L-Schale *f* (Atom)
luargol Luargol *n*
lubanol Lubanol *n*
lubricant Gleitmittel *n*, Schmiere *f*,
   Schmiermittel *n*, solid ~ Starrschmiere *f*

lubricant exudation Ausschwitzen *n* des
   Gleitmittels
lubricant testing equipment
   Schmiermittelprüfgerät *n*
lubricate *v* [ein]schmieren, [ein]fetten, einölen,
   ölen, ~ again nachschmieren
lubricating Einfetten *n*, Einschmieren *n*,
   Schmieren *n*
lubricating device Schmiervorrichtung *f*
lubricating effect Schmierwirkung *f*
lubricating grease Schmierfett *n*
lubricating liquid Schmierflüssigkeit *f*
lubricating material Schmiermaterial *n*
lubricating oil Schmieröl *n*, Maschinenöl *n*,
   Motorenöl *n*, film of ~ Schmierölschicht *f*,
   mineral ~ mineralisches Schmieröl
lubricating point Schmierstelle *f*
lubricating power Schmierfähigkeit *f*
lubricating property Schmiereigenschaft *f*
lubricating pump Schmier[öl]pumpe *f*
lubricating syringe Ölspritze *f*
lubricating value Schmierwert *m*
lubrication Schmierung *f*, Einfettung *f*, Fettung *f*,
   Ölung *f*, circulatory ~ Umlaufschmierung *f*
lubrication cock Schmierhahn *m*
lubrication connection Schmierstutzen *m*
lubrication nipple Schmiernippel *m*
lubricator Schmiervorrichtung *f*, Öler *m*,
   Schmierapparat *m*, Schmiermittel *n*, ~
   actuated by hand Handschmiervorrichtung *f*
lubricator glass Ölglas *n*
lubricity Schlüpfrigkeit *f*; Schmierfähigkeit *f*
lucanthone Lucanthon *n*
lucidol Lucidol *n*
luciferase Luciferase *f* (Biochem)
luciferin Luciferin *n*
lucigenin Lucigenin *n*
lucinite Lucinit *m* (Min)
lucite Lucit *n*
lucrative einträglich
lucullite Lukullit *m* (Min)
ludlamite Ludlamit *m* (Min)
ludwigite Ludwigit *m* (Min)
lueneburgite Lüneburgit *m* (Min)
luffing crane Wippkran *m*
luffing jib Nadelausleger *m*
luffing-jib crane Wippkran *m*
luffing-slewing crane Wippdrehkran *m*
lug Fahne *f* (Elektr), Knagge *f* (Techn), Öhr *n*
   (Techn), ~ (of a tire profile) Stollen *m* (eines
   Reifenprofils)
lugol's solution Lugolsche Lösung *f*
lugs (Am. E.) Obststeige *f*
lukewarm lauwarm, handwarm
lumazine Lumazin *n*
lumber Bauholz *n*, Nutzholz *n*
lumber preservation Holzimprägnierung *f*
lumen Lumen *n* (Einheit d. Lichtstroms)
lumichrome Lumichrom *n*

lumiestrone Lumiöstron *n*
lumiisolysergic acid Lumiisolysergsäure *f*
luminal Luminal *n*
luminesce *v* lumineszieren, leuchten
luminescence Lumineszenz *f*, Leuchten *n*
luminescence excitation Lumineszenzanregung *f*
luminescence microscope
  Lumineszenzmikroskop *n*
luminescent lumineszierend, leuchtend
luminescent coating Leuchtanstrich *m*
luminescent image Leuchtbild *n*
luminescent material Leuchtmaterial *n*
luminescent screen Leuchtschirm *m*
luminiferous lichterzeugend, lichtgebend
luminol Luminol *n*
luminophore Luminophor *n*
luminosity Hellbezugswert *m*, Helligkeit *f*,
  Leuchten *n*, Leuchtfähigkeit *f*, Leuchtkraft *f*,
  Lichtstärke *f*, **overall** ~ Gesamtleuchtkraft *f*
luminosity curve, absolute ~ absolute
  Helligkeitskurve *f*
luminous leuchtend, lichtgebend
luminous anode Leuchtanode *f*
luminous bacteria Leuchtbakterien *pl*
luminous bar Leuchtstab *m*
luminous body Lichtkörper *m*
luminous ceiling Leuchtdecke *f*
luminous color Leuchtfarbe *f*
luminous column Leuchtsäule *f*
luminous cone Lichtkegel *m*
luminous dial Leuchtzifferblatt *n*
luminous discharge current
  Glimmentladungsstrom *m*
luminous effect Leuchtwirkung *f*, Lichteffekt *m*
luminous energy Lichtenergie *f*,
  Strahlungsenergie *f*
luminous filament Leuchtfaden *m*
luminous intensity Leuchtkraft *f*, Helligkeit *f*,
  Lichtstärke *f*
luminous paint Leuchtfarbe *f*, Fluoreszenzfarbe *f*
luminous phenomenon Leuchterscheinung *f*
luminous pigment Leuchtpigment *n*
luminous quartz Leuchtquarz *m* (Min)
luminous spar Leuchtspat *m* (Min)
luminous spot Lichtpunkt *m*
luminous stimulus Lichtreiz *m*
luminous substance Leuchtstoff *m*
luminous transmittance Lichtdurchlässigkeit *f*
lumistanol Lumistanol *n*
lumisterol Lumisterin *n*, Lumisterol *n*
lump Brocken *m*, Klumpen *m*, Luppe *f* (Metall),
  Stück *n*, **little** ~ Klümpchen *n*
lump coal Stückkohle *f*, Würfelkohle *f*
lump ore Grubenklein *n*, Stückerz *n*, **proportion**
  of ~ Stückgehalt *m* (Erz)
lump quartz Stückquarz *m*
lumps, formation of ~ Klumpenbildung *f*
lump starch Bröckelstärke *f*

lump-sum Pauschale *f*, Pauschalgebühr *f*,
  Pauschalsumme *f*
lumpy klumpig
lunacridine Lunacridin *n*
lunamarine Lunamarin *n*
lunar caustic Höllenstein *m*, Ätzstein *m*
lunar rocks Mondgestein *n*
lunasine Lunasin *n*
lung cancer Lungenkrebs *m* (Med)
lungwort Lungenkraut *n* (Bot)
lunine Lunin *n*
lunnite Lunnit *m* (Min)
lupane Lupan *n*
lupanine Lupanin *n*
lupanoline Lupanolin *n*
lupene Lupen *n*
lupeol Lupeol *n*
lupeone Lupeon *n*
lupeose Lupeose *f*
lupetazin tartrate Lycetol *n*
lupetidine Lupetidin *n*
lupinane Lupinan *n*
lupinidine Lupinidin *n*, Spartein *n*
lupinine Lupinin *n*
lupininic acid Lupininsäure *f*
lupucarboxylic acid Lupucarbonsäure *f*
lupulic acid Lupulinsäure *f*
lupulin Lupulin *n*, Hopfenbitter *n*,
  Hopfenmehl *n*
lupuline hopfenähnlich
lupulinic acid Lupulinsäure *f*,
  Hopfenbittersäure *f*
lupulon Lupulon *n*
lupuloxinic acid Lupuloxinsäure *f*
lussatite Lussatit *m* (Min)
luster Glanz[effekt] *m*, Oberflächenglanz *m*,
  Schein *m*, Schimmer *m*, **unctuous** ~ fettiger
  Glanz
luster color Lüsterfarbe *f*
luster effect Glanzwirkung *f*
luster glaze Lüsterglasur *f*
lusterless glanzlos
luster pigments Glanzpigmente *pl*
lusterwash Metallglanzfarbe *f*
lustre, silky ~ Seidenglanz *m*
lustrous glänzend, metallglänzend, schimmernd,
  strahlend
lute Kitt *m*, Dichtungskitt *m*
lute *v* dichten, [ver]kitten
lutecium (Symb. Lu) Lutetium *n*, Cassiopeium *n*
lutein Lutein *n*
luteocobaltic chloride Luteokobaltchlorid *n*
luteo-compound Luteoverbindung *f*
luteol Luteol *n*
luteolin Luteolin *n*, Waugelb *n*
luteolinidine Luteolinidin *n*
luteotropin Luteotropin *n*
lutetium (Symb. Lu) Lutetium *n*
lutidine Lutidin *n*

lutidinic acid Lutidinsäure f
lutidone Lutidon n
luting Kitt m, Verkitten n
luting agent Dichtungskitt m
luting clay Kitterde f
luting wax Wachskitt m
lux Lux n (Einheit d. Beleuchtungsstärke)
Luxemburg pig iron Luxemburger Roheisen n
Lux gas balance Luxsche Gaswaage f
luxmeter Luxmeter n, Helligkeitsmesser m
luxullian Luxullian m (Min)
luzonite Luzonit m (Min)
lyase Lyase f (Biochem)
lycetol Lycetol n
lycoctonine Lycoctonin n
lycopene Lycopin n
lycoperdine Lucoperdin n
lycopersene Lycopersen n
lycophyll Lycophyll n
lycopin Lycopin n, Rubidin n
lycopodium Bärlapp m (Bot), Bärlappmehl n, Blitzpulver n (Techn), Hexenmehl n, Lykopodium n
lycopodium dust Lykopodiumsamen m
lycopodium powder Bärlappsamen m, Bärlappsporen pl
lycopodium seed Bärlappsamen m, Bärlappsporen pl
lycorene Lycoren n
lycorine Lycorin n, Narcissin n
lycoxanthin Lycoxanthin n
Lydian stone Lydit m (Min)
lydite Lydit m
lye Lauge f, Brühe f, **crude** ~ Rohlauge f, **fast to** ~ laugenecht, **feed of** ~ Laugenzuführung f, **resembling** ~ laugenartig
lye bath Laugenbad n
lye composition Laugenzusammensetzung f
lye container Laugenbehälter m
lye evaporator Laugenverdampfer m
lye hydrometer Laugenwaage f
lye processing Laugenverarbeitung f
lye recovery Laugenregenerierung f
lye-resisting laugenwiderstandsfähig
lye tank Laugenbehälter m
lye vat Laugenfaß n, Laugenkessel m
lygosin Lygosin n
Lyman series Lymanserie f (Spektr)
lymph Lymphe f (Med)
lymphatic gland Lymphdrüse f, Lymphknoten m
lymphatic vessel Lymphgefäß n (Med)
lymph node Lymphdrüse f, Lymphknoten m
lymphoblast Lymphoblast m
lymphocyte Lymphozyt m, Lymphzelle f
lymphocytopenia Lymphozytenarmut f (Med), Lymphozytenmangel m
lymphogenesis Lymphozytenbildung f
lymph vessel Lymphgefäß n
lyophilic lyophil

lyophilization Gefriertrocknung f, Lyophilisation f
lyophilize v gefriertrocknen, gefriertrocknen, lyophilisieren
lyophilized gefriergetrocknet
lyophilizer Gefriertrockenapparat m, Gefriertrocknungsanlage f
lyophobic lyophob
lyosol Lyosol n
lyotropic lyotrop (Salz)
lyrostibnite Kermesit m (Min)
lysalbinic acid Lysalbinsäure f
lysergene Lysergen n
lysergic acid Lysergsäure f
lysergide Lysergid n
lysergine Lysergin n
lysergol Lysergol n
lysidine Lysidin n
lysidine tartrate Lysidintartrat n
lysine Lysin n, Diaminocapronsäure f
lysine dihydrochloride Lysindihydrochlorid n
lysine racemase Lysinracemase f (Biochem)
lysinogen Lysinogen n
lysis Lyse f, Zerfall m, Zerstörung f
lysobacterium Lysobakterium n
lysocithine Lysozithin n
lysoform Lysoform n
lysolecithin Lysolecithin n
lysosome Lysosom-Körperchen n
lysozyme Lysozym n (Biochem)
lyxitol Lyxit m
lyxoflavine Lyxoflavin n
lyxomethylitol Lyxomethylit m
lyxonic acid Lyxonsäure f
lyxosamine Lyxosamin n
lyxose Lyxose f

M

maaliene Maalien *n*
macarite Macarit *n*
Macassar oil Makassaröl *n*
mace Muskatblüte *f* (Bot)
mace butter Muskatbutter *f,* Muskatfett *n*
mace oil Macisöl *n,* Muskatbalsam *m,*
   Muskat[blüten]öl *n*
macerals Macerale *pl*
macerate *v* einweichen, mazerieren
maceration Auslaugen *n* (Zucker),
   Einweichung *f,* Mazeration *f,* ~ of horn
   Hornbeize *f*
macerator Mazeriergefäß *n*
machaeric acid Machaersäure *f*
machaerinic acid Machaerinsäure *f*
Mache unit Mache-Einheit *f* (Radioaktivität)
machilene Machilen *n*
machilol Machilol *n*
machinability, free-cutting ~ spanabhebende
   Bearbeitbarkeit *f*
machine Maschine *f,* Maschinerie *f,*
   Mechanismus *m,* Vorrichtung *f,* **belt-driven** ~
   Maschine *f* für Riemenbetrieb, **by** ~
   maschinell, **driven by** ~ maschinell
   angetrieben, **farming or agricultural** ~
   landwirtschaftliche Maschine, **hydraulic** ~
   hydraulische Maschine
machine *v* maschinell bearbeiten
machine casting Maschinenguß *m*
machine construction Maschinenbau *m*
machine-cut peat Maschinentorf *m*
machine drive Maschinenantrieb *m*
machine factory Maschinenfabrik *f*
machine-made maschinell hergestellt
machine oil Maschinenöl *n*
machine paper Maschinenpapier *n*
machine puddling Maschinenpuddeln *n*
machinery Maschinenpark *m*
machine tool Werkzeugmaschine *f*
machine tool finish Werkzeugmaschinenlack *m*
machine unit Maschinenanlage *f*
machine work Maschinenarbeit *f*
machining Bearbeitung *f,* Verarbeitung *f*
machining allowance Verarbeitungstoleranz *f*
machining method Bearbeitungsmethode *f*
   (Techn)
machining step Bearbeitungsstufe *f*
machining time Durchlaufzeit *f*
machinist Maschinist *m*
Mach number Machzahl *f*
macilenic acid Macilensäure *f*
macilolic acid Macilolsäure *f*
Maclaurin['s] series Maclaurinsche Reihe *f*
MacLeod gauge MacLeod-Druckmesser *m*
macleyine Macleyin *n,* Protopin *n*
maclurin Maclurin *n,* Moringagerbsäure *f*

macroanalysis Makroanalyse *f*
macroapparatus Makroapparat *m*
macroaxis Makroachse *f* (Krist)
macrobacterium Makrobakterium *n*
macrobiosis Langlebigkeit *f,* Makrobiose *f*
macroblast Makroblast *m*
macrochemistry Makrochemie *f*
macrocosm Makrokosmos *m*
macrocrystalline makrokristallin
macrocyte Makrozyt *m*
macrogamete Makrogamet *m*
macroglobulin Makroglobulin *n*
macrogol Macrogol *n* (Kurzbezeichnung für
   Polyäthylenglykol)
macrokinetics Makrokinetik *f*
macrometeorology Makrometeorologie *f*
macro-method Makromethode *f*
macromolecular hochmolekular,
   makromolekular
macromolecule Makromolekül *n,* **cross-linked** ~
   vernetztes Makromolekül, **linear** ~
   Fadenmolekül *n,* lineares Makromolekül,
   rod-like ~ stäbchenförmiges Makromolekül,
   spherical ~ kugelförmiges Makromolekül
macrophage Makrophage *m*
macrophotography Makrophotographie *f*
macropone Macropon *n*
macrorheology Makrorheologie *f*
macroscopic makroskopisch
macroscopy Grobstrukturuntersuchung *f*
macrostructure Makrostruktur *f,* Grobgefüge *n,*
   Gußstruktur *f*
macrostructure analysis Grobstrukturanalyse *f*
macrotomic acid Makrotominsäure *f*
macrotomine Makrotomin *n*
macroturbulence Makroturbulenz *f*
maculate gefleckt (Bot)
maculine Maculin *n*
maculosidine Maculosidin *n*
madaralbane Madaralban *n*
madder Färberröte *f* (Bot), Färberrot *n,*
   Krapp *m* (Bot), **like** ~ krappartig, **root of** ~
   Färberröte *f* (Bot)
madder *v* mit Krapp färben
madder bleach Krappbleiche *f*
madder carmine Krappkarmin *n*
madder color Krappfarbe *f*
madder-colored krapprot
madder dye Krappfarbe *f*
madder-dyeing Krappfärben *n*
madder lake, red ~ Krappcarmin *n*
madder red Krapprot *n*
madder root Krappwurzel *f*
madder yellow Krappgelb *n*
Maddrell's salt Maddrellsches Salz *n*
madia oil Madiaöl *n*

madreporite Madreporit *m*
mafenide Mafenid *n*
mafura butter Mafuratalg *m*
mafura tallow Mafuratalg *m*
magdala red Magdalarot *n*, Naphthalinrot *n*
magenta Fuchsin *n*
magenta acid Magentasäure *f*
magenta red Fuchsin *n*, Magentarot *n*
magic magisch
magma Magma *n*
magmatic magmatisch
magnesia Bittererde *f* (Min), Magnesiumoxid *n*,
   bicarbonate of ~ saure Magnesia, **burnt** ~
   gebrannte Magnesia, **containing** ~
   magnesiahaltig, magnesiumoxidhaltig,
   hardness due to ~ Magnesiahärte *f*
magnesia bath Magnesiabad *n*
magnesia cement Magnesiazement *m*,
   Sorelzement *m*
magnesia hardness Magnesiahärte *f*
magnesia iron spinel Chlorospinell *m* (Min)
magnesia lime stone Bitterkalk *m* (Min)
magnesia mixture Magnesiamixtur *f*
magnesian magnesiahaltig, magnesiumhaltig
magnesian limestone Magnesiakalk *m* (Min)
magnesia red Magnesiarot *n* (Farbstoff)
magnesia soap Magnesiaseife *f*
magnesiochromite Magnesiochromit *m* (Min)
magn[esi]oferrite Magnesioferrit *m* (Min)
magnesioludwigite Magnesioludwigit *m* (Min)
magnesite Magnesit *m* (Min), Bitterspat *m*
   (Min), Magnesiumkarbonat *n*, **siliceous** ~
   Kieselmagnesit *m* (Min)
magnesite brick Magnesitstein *m*,
   Magnesitziegel *m*
magnesite mass Magnesitmasse *f*
magnesite spar Bitterspat *m* (Min)
magnesium (Symb. Mg) Magnesium *n*,
   containing ~ magnesiumhaltig
magnesium acetate Magnesiumacetat *n*
magnesium agricultural lime
   Magnesiummergel *m* (Agr)
magnesium alkyl Magnesiumalkyl *n*
magnesium alloy Magnesiumlegierung *f*
magnesium aluminate Magnesiumaluminat *n*
magnesium amalgam Magnesiumamalgam *n*
magnesium antimonide Magnesiumantimonid *n*
magnesium borate Magnesiumborat *n*,
   Antifungin *n* (Pharm)
magnesium bromide Magnesiumbromid *n*
magnesium bronze Magnesiumbronze *f*
magnesium carbonate Magnesiumcarbonat *n*,
   native ~ Magnesit *m* (Min)
magnesium chloride Magnesiumchlorid *n*
magnesium citrate Magnesiumcitrat *n*
magnesium compound Magnesiumverbindung *f*
magnesium content Magnesiumgehalt *m*
magnesium dioxide Magnesiumsuperoxid *n*
magnesium flashlight Magnesium[blitz]licht *n*

magnesium fluoride Magnesiumfluorid *n*
magnesium glycerophosphate
   Magnesiumglycerophosphat *n*
magnesium hardness Magnesiumhärte *f*
magnesium hydroxide Magnesiumhydroxid *n*,
   Ätzmagnesia *f*
magnesium hypochlorite solution
   Magnesiableichflüssigkeit *f*
magnesium iodide Magnesiumjodid *n*
magnesium iron carbonate, native ~
   Mesitinspat *m* (Min)
magnesium iron mica Magnesiaeisenglimmer *m*
   (Min)
magnesium lactate Magnesiumlactat *n*
magnesium lamp Magnesiumlampe *f*
magnesium light Magnesiumlicht *n*
magnesium lime, hydrated ~
   Magnesiumlöschkalk *m*, **mixed** ~
   Magnesium-Mischkalk *m*
magnesium marlstone Dolomitmergel *m* (Min)
magnesium mica Amberglimmer *m* (Min),
   Magnesiaglimmer *m* (Min)
magnesium monel Magnesium-Monel *n* (Leg)
magnesium nitrate Magnesiumnitrat *n*
magnesium nitride Stickstoffmagnesium *n*
magnesium oxide Magnesiumoxid *n*, Bittererde *f*
   (Min), Magnesia *f*
magnesium palmitate Magnesiumpalmitat *n*
magnesium pectolite Magnesiumpektolith *m*
   (Min)
magnesium peroxide Magnesiumsuperoxid *n*
magnesium phosphate Magnesiumphosphat *n*
magnesium quicklime Magnesiumbranntkalk *m*
magnesium ribbon Magnesiumband *n*
magnesium ricinate Magnesiumricinat *n*
magnesium salt Magnesiumsalz *n*
magnesium silicate Magnesiumsilicat *n*
magnesium silicide Siliciummagnesium *n*
magnesium silicofluoride
   Magnesiumsilicofluorid *n*
magnesium stearate Magnesiumstearat *n*
magnesium sulfate Magnesiumsulfat *n*,
   Bittersalz *n*, **native** ~ Kieserit *m* (Min)
magnesium sulfate calcining plant
   Magnesiumsulfatkalzinieranlage *f*
magnesium sulfide Magnesiumsulfid *n*
magnesium superoxide Magnesiumsuperoxid *n*
magnesium tape Magnesiumband *n*
magnesium tartrate Magnesiumtartrat *n*
magnesium wire Magnesiumdraht *m*
magneson Magneson *n*
magnet Magnet *m*, **artificial** ~ künstlicher
   Magnet, **bell-shaped** ~ Glockenmagnet *m*,
   induced ~ induzierter Magnet, **inducing** ~
   induzierender Magnet, **natural** ~ natürlicher
   Magnet, **nonpermanent** ~ fremderregter
   Magnet, **permanent** ~ permanenter Magnet,
   spherical ~ Kugelmagnet *m*, **straight** ~

Magnetstab m, **temporary** ~ temporärer
Magnet
magnet armature Magnetanker m
magnet carrier Magnetträger m
magnet core Magnetkern m, ~ **of an arc lamp**
Magnet m einer Bogenlampe
magnet crane Magnetkran m
magnetic magnetisch
magnetically controlled magnetgesteuert
magnetic anomaly magnetische
Unregelmäßigkeit f
magnetic axis Magnetachse f
magnetic bar Magnetstab m
magnetic brake Magnetbremse f
magnetic coil Magnetspule f
magnetic core memory Magnetkernspeicher m
magnetic cycle Magnetisierungszyklus m
magnetic density in space Raumdichte f des
Magnetismus
magnetic equator Indifferenzzone f (Magn)
magnetic field Magnetfeld n, **alternate** ~
magnetisches Wechselfeld, **intensity of** ~
magnetische Feldstärke, **non-uniform** ~
ungleichförmiges Magnetfeld n, **transverse** ~
transversales Magnetfeld, Quermagnetfeld n,
strength of ~ **at saturation**
Sättigungsfeldstärke f, ~ **in air**
Luftmagnetfeld n, ~ **of the earth**
Erdmagnetfeld n
magnetic flux Magnetfluß m, Kraftlinienfluß m,
Kraftlinienstrom m, magnetische Induktion f,
distribution of ~ Kraftflußverteilung f,
variation of ~ Kraftflußvariation f
magnetic idling roll Magnetrolle f
magnetic induction Magnetinduktion f,
magnetische Induktion f
magnetic intensity Magnetisierung f
magnetic iron [ore] Magnetit m (Min),
Magneteisenerz n (Min), Magneteisenstein m
(Min)
magnetic line of force magnetische Kraftlinie f,
Magnetkraftlinie f
magnetic moment magnetisches Moment n
magnetic needle Magnetnadel f
magnetic pole Magnetpol m
magnetic pyrite Magnetkies m (Min),
Magnetopyrit m (Min)
magnetic recorder, tape of ~
Magnetophonband n
magnetic resonance spectrum Spektrum n der
magnetischen Resonanz (Spektr)
magnetic reversal Ummagnetisierung f
magnetic rotation spectrum
Magnetorotationsspektrum n
magnetic separation Magnetscheidung f
magnetic separator Magnet[ab]scheider m,
Eisenscheider m
magnetic sound Magnetton m
magnetic stirrer Magnetrührer m

magnetic switch Magnetschalter m
magnetic system Magnetsystem n
magnetic tape Magnetophonband n,
Magnettonband n
magnetic transformation, point of ~
magnetischer Umwandlungspunkt m
magnetic unit magnetische Einheit f
magnetic valve Magnetventil n
magnet interrupter Magnetunterbrecher m
magnet iron Magneteisen n (Min)
magnetism Magnetismus m, **free** ~ freier
Magnetismus, **line of** ~
Magnetisierungslinie f, **remanent** ~
remanenter Magnetismus, **residual** ~
remanenter Magnetismus, **reversal of** ~
Ummagnetisierung f
magnetite Magnetit m (Min), Magneteisenerz n
(Min), Magneteisenstein m (Min), **earthy** ~
Eisenmohr m (Min)
magnetizability Magnetisierbarkeit f,
Magnetisierfähigkeit f
magnetizable magnetisierbar, magnetisch
erregbar
magnetization Magnetisierung f, **coefficient of** ~
Magnetisierungskoeffizient m, **cycle of** ~
Magnetisierungszyklus m, **direction of** ~
Magnetisierungsrichtung f, **intensity of** ~
Magnetisierungsstärke f, **method of** ~
Magnetisierungsart f, **residual** ~ remanente
Magnetisierung, **reversal of** ~
Ummagnetisierung f, **saturated** ~
Magnetisierung f bis zur Sättigung,
superposed ~ überlagerte Magnetisierung, **to
reverse the** ~ ummagnetisieren
magnetization curve Magnetisierungskurve f
magnetization line Magnetisierungslinie f
magnetize v magnetisieren, magnetisch machen
magnetizing Magnetisieren n, Magnetisierung f
magnetizing apparatus Magnetisiergerät n
magnetizing coil Magnetisierungsspirale f,
Magnetisierungsspule f
magnetocaloric magnetokalorisch
magneto-chemistry Magnetochemie f
magnetoelectric magnetoelektrisch
magnetoelectric generator Magnetinduktor m
magnetoelectricity Magnetelektrizität f
magnetoelectric machine
Magnetelektrisiermaschine f
magnetoelectric motor Magnetmotor m
magnetograph Magnetograph m
magnetohydrodynamics
Magnetohydrodynamik f (Abk. MHD)
magneto-ignition Magnetzündung f
magnetometer Magnetmesser m,
Magnetometer n
magnetometric magnetometrisch
magnetomotive magnetomotorisch
magneton Magneton n (Phys)
magnetooptical magnetooptisch

magnetophone Magnetophon *n*
magnetopyrite Magnetopyrit *m* (Min)
magnetoscope Magnetoskop *n*
magnetostatics Magnetostatik *f*
magnetostriction Magnetostriktion *f*
magnetostrictive magnetostriktiv
magnet pole Magnetpol *m*
magnetron Magnetron *n*,
    Hohlraumresonatorröhre *f,* Magnetfeldröhre *f*
magnet steel Magnetstahl *m*
magnet wheel Polrad *n*
magnet winding Magnetwicklung *f*
magnet yoke Magnetjoch *n*
magnification Vergrößerung *f,* **actual** ~
    Eigenvergrößerung *f,* **final** ~
    Endvergrößerung *f,* **longitudinal** ~
    Längsvergrößerung *f* (Opt), **real** ~
    Eigenvergrößerung *f*
magnification effectiveness
    Vergrößerungsnutzgrad *m* (Opt)
magnification factor Vergrößerungsfaktor *m,*
    Q-Faktor *m*
magnification scale Vergrößerungsmaßstab *m*
magnifier Lupe *f* (Opt), Vergrößerungsglas *n,*
    Verstärker *m* (Elektr)
magnify *v* vergrößern
magnifying apparatus Vergrößerungsgerät *n*
magnifying glass Vergrößerungsglas *n* (Opt),
    Brennglas *n* (Opt), Lupe *f* (Opt)
magnifying lens Lupe *f* (Opt),
    Vergrößerungslinse *f* (Opt)
magnifying mirror Vergrößerungsspiegel *m*
magnitude Ausdehnung *f,* Größe *f,* **order of** ~
    Größenordnung *f*
magnochromite Magnochromit *m* (Min)
magnocurarine Magnocurarin *n*
magnoferrite Magnoferrit *m* (Min)
magnolamine Magnolamin *n*
magnolaminic acid Magnolaminsäure *f*
magnolite Magnolith *m* (Min)
magnolol Magnolol *n*
magon Magon *n*
mahogany Mahagoni *n*
mahogany brown mahagonibraun
Maillard reaction Maillard-Reaktion *f*
maim *v* verstümmeln
main Hauptstromleitung *f* (Elektr), Leitung *f,*
    Netz *n* (Elektr), Rohrleitung *f*
main action Hauptwirkung *f*
main air duct Hauptwindleitung *f*
main axis Hauptachse *f*
main bearing Hauptlager *n,* Grundlager *n,*
    Wellenlager *n*
main busbar Hauptleitungsschiene *f*
main circuit Hauptstromkreis *m*
main column Hauptsäule *f*
main condenser Hauptkondensator *m*
main constituent Hauptbestandteil *m*
main depot Hauptniederlassung *f*

main dynamo Hauptlichtmaschine *f*
main flow Hauptströmung *f*
main fuse Hauptsicherung *f* (Elektr)
main gate valve Hauptabsperrarmatur *f*
main gear Hauptantrieb *m*
main girder Hauptträger *m*
main group Hauptgruppe *f* (Chem)
main hoist speed Haupthubgeschwindigkeit *f*
main index Hauptregister *n*
main jet Hauptdüse *f*
main line Hauptleitung *f*
main nozzle Hauptdüse *f*
main part Hauptbestandteil *m*
main pass Hauptzug *m*
main patent Hauptpatent *n*
main phase Hauptphase *f* (Elektr)
main piping Haupt[rohr]leitung *f*
main plane Haupttragfläche *f*
main [pressure] line Betriebsdruckleitung *f*
main product Hauptprodukt *n*
main ram Hauptstempel *m,* Preßkolben *m*
main reaction Hauptreaktion *f* (Chem)
main register Hauptregister *n*
main runner Graben *m*
mains Hauptleitung *f*
mains coil Netzspule *f*
mains connection Netzanschluß *m* (Elektr)
main series Hauptreihe *f* (Chem)
main shaft Hauptschacht *m,* Antriebswelle *f,*
    Hauptwelle *f*
main source Hauptquelle *f*
main spring Triebfeder *f*
main stream Hauptstrom *m*
main subject Hauptfach *n*
main switch Hauptschalter *m*
main switch desk Hauptschaltpult *n*
maintain *v* [aufrecht]erhalten, instandhalten,
    unterhalten (Reaktion, Verbrennung)
maintainable speed Dauergeschwindigkeit *f*
    (Kfz)
maintenance Aufrechterhaltung *f,* Erhaltung *f,*
    Instandhaltung *f,* Pflege *f,* Unterhaltung *f,*
    Wartung *f* (Techn), **cost of** ~
    Instandhaltungskosten *pl,* Wartungskosten *pl*
maintenance cost Wartungskosten *pl,*
    Instandhaltungskosten *pl*
maintenance routine work
    Instandhaltungsarbeiten *f pl*
main trunk Hauptstamm *m,* Hauptstrang *m*
    Förderband
main vaporizer Hauptverdampfer *m*
main voltage Hauptspannung *f* (Elektr),
    Netzspannung *f* (Elektr)
maisin Maisin *n*
maize Mais *m*
maize gluten Maiskleber *m*
maize oil Maisöl *n*
maize starch Maisstärke *f*
majolica Majolika *f*

majolica color Majolikafarbe *f*
major axis Hauptachse *f* (Math)
majority Großteil *m*, Mehrheit *f*
major loss prevention Anlagesicherung *f*
make *v* machen, darstellen (Chem), erzeugen,
    fabrizieren, anfertigen, ~ **acidic** ansäuern, ~
    a mold einen Abdruck herstellen, ~ **a tracing**
    kopieren, ~ **available** bereitstellen, ~ **easier**
    erleichtern, ~ **firm** [ver]festigen, ~ **heavy**
    erschweren, ~ **insensitive** desensibilisieren, ~
    out ausfertigen, ~ **tight** abdichten, ~ **up the**
    charge gattieren (Gieß)
make and break ignition Abreißzündung *f*
maker Hersteller *m*
makeshift Behelf *m*, Notbehelf *m*
makeshift *a* behelfsmäßig, provisorisch
makeshift construction Behelfskonstruktion *f*
make-up Aufmachung *f*, Schminke *f* (Kosm)
make-up gas Frischgas *n*
make-up piece Fassonrohr *n*, Formstück *n*
making useless Unbrauchbarmachen *n*
malabar tallow Malabartalg *m*
Malacca nut Malakkanuß *f*
malachite Malachit *m* (Min), Berggrün *n* (Min),
    Kupferspat *m* (Min), **calcareous** ~
    Kalkmalachit *m* (Min), **fibrous** ~
    Fasermalachit *m* (Min)
malachite green Malachitgrün *n*
malacolite Malakolith *m* (Min), Augit *m* (Min)
malakin Malakin *n*
malakon Malakon *n*
malamic acid Malamidsäure *f*
malamide Malamid *n*
malamidic acid Malamidsäure *f*
malanilic acid Malanilsäure *f*
malaria Malaria *f* (Med), **fight against** ~
    Malariabekämpfung *f*
malaria control Malariabekämpfung *f*
malarial fever Malariafieber *n* (Med)
malarial parasite Malariaerreger *m* (Zool)
malarin Malarin *n*
malate Malat *n*, Salz oder Ester der Äpfelsäure
malate dehydrogenase Malatdehydrogenase *f*
    (Biochem)
malate synthetase Malatsynthetase *f* (Biochem)
malcolmiin Malcolmiin *n*
maldistribution Fehlverteilung *f*
    (Füllkörperkolonne), Randgängigkeit *f*
maldonite Maldonit *m* (Min), Wismutgold *n*
    (Min)
male männlich
malealdehyde Maleinaldehyd *m*
maleate Maleat *n*, Salz oder Ester der
    Maleinsäure
maleate isomerase Maleatisomerase *f* (Biochem)
male die Gewindematrix *f*, Patrize *f*
maleic acid Maleinsäure *f*, **salt or ester of** ~
    Maleat *n*
maleic aldehyde Maleinaldehyd *m*

maleic anhydride Maleinsäureanhydrid *n*
maleic hydrazide Maleinsäurehydrazid *n*
maleic resin Maleinatharz *n*, Maleinsäureharz *n*
maleinimide Maleinimid *n*
maleinuric acid Maleinursäure *f*
male mold Patrize *f*
male mold method Positivverfahren *n*
male thread Außengewinde *n*
maletto bark Malettorinde *f*
maleylsulfathiazole Maleylsulfathiazol *n*
malformation Mißbildung *f*
malic acid Äpfelsäure *f*, Oxybernsteinsäure *f*,
    salt or ester of ~ Malat *n*
malic acid monoamide Malamidsäure *f*
malic amide Malamid *n*
malignancy Bösartigkeit *f* (Med), Malignität *f*
    (Med), Schädlichkeit *f*
malignant bösartig
malladrite Malladrit *m* (Min)
malleability Hämmerbarkeit *f*,
    Geschmeidigkeit *f* (Met), Schmiedbarkeit *f*,
    Verformbarkeit *f*
malleable [kalt] schmiedbar, geschmeidig,
    hämmerbar
malleable brass Neumessing *n*
malleable iron Schmiedeeisen *n*
malleabl[e]ize *v* glühfrischen (Metall)
mallein Mallein *n* (Impfstoff)
mallet Holzhammer *m*
mallow Malvenkraut *n* (Bot)
mallow leaves Malvenblätter *pl* (Bot)
malm künstlicher Mergel *m*, Malm *m* (Geol)
malnutrition Unterernährung *f*,
    Ernährungsfehler *m*, Ernährungsstörung *f*
malodor Gestank *m*
malodorous übelriechend
malol Malol *n*, Urson *n*
malonaldehyde Malonaldehyd *m*
malonaldehydic acid Malonaldehydsäure *f*
malonamic acid Malonamidsäure *f*
malonamide Malonamid *n*
malonamidic acid Malonamidsäure *f*
malonanilide Malonanilid *n*
malonate Malonat *n*, Malonester *m*, Salz oder
    Ester der Malonsäure
malonic acid Malonsäure *f*, **salt or ester of** ~
    Malonat *n*
malonic acid diethyl ester
    Malonsäurediethylester *m*
malonic acid ethyl ester Äthylmalonat *n*
malonic aldehyde Malonaldehyd *m*
malonic amide Malonamid *n*
malonic ester Malonsäurediäthylester *m*,
    Malonsäureester *m*
malonuric acid Malonursäure *f*
malonyl chloride Malonylchlorid *n*
malonyl-CoA Malonyl-Coenzym A *n* (Biochem)
malonyl transacetylase Malonyltransacetylase *f*
    (Biochem)

malonyl urea Malonylharnstoff *m;* Barbitursäure *f*
malt Malz *n,* **brittle** ~ Glasmalz *n,* **high-kilned** ~ Färbemalz *n,* **infusion of** ~ Malzaufguß *m,* **substitute for** ~ Malzsurrogat *n,* **to cure** ~ darren
malt *v* malzen
maltase Maltase *f* (Biochem)
malt barley Malzgerste *f*
malt beer Malzbier *n*
malt chamber Würzraum *m* (Brau)
malt coffee Malzkaffee *m*
malt crushing room Schroterei *f* (Brau)
malt diastase Malzdiastase *f* (Biochem), Maltin *n*
malt drying tray Malzdarrhorde *f*
maltdust Darrstaub *m*
malted bonbon Malzbonbon *m*
maltese cross tube Schattenkreuzröhre *f*
malt extract Malzauszug *m,* Malzextrakt *m*
malt floor Malztenne *f*
malt floor cooling Malztennenkühlung *f*
maltha Bergteer *m* (Min), Erdteer *m* (Min)
malt house Mälzerei *f*
malt husks Malztreber *pl*
malting Mälzerei *f,* Malzbereitung *f*
malting machine Malzentkeimungsmaschine *f*
maltitol Maltit *m*
malt kiln Malzdarre *f*
malt mill Malzmühle *f,* Schrotmühle *f*
maltobionic acid Maltobionsäure *f*
maltobiose Maltose *f,* Maltobiose *f,* Malzzucker *m*
maltodextrin Maltodextrin *n,* Amyloin *n*
maltonic acid Maltonsäure *f*
maltosazone Maltosazon *n*
maltose Maltose *f,* Maltobiose *f,* Malzzucker *m*
maltosone Maltoson *n*
malt-powder plant Malzpulveranlage *f*
malt residuum Malztreber *pl*
malt spirits Getreidebranntwein *m*
malt sprout Malzkeim *m*
malt starch Malzstärke *f*
malt sugar Malzzucker *m,* Maltose *f*
maltulose Maltulose *f*
malt vat Malzbottich *m*
malt vinegar Bieressig *m*
malvalic acid Malval[in]säure *f*
malvidin Malvidin *n*
malvin Malvin *n*
mammal Säugetier *n* (Zool)
mammary gland Brustdrüse *f,* Milchdrüse *f*
mammein Mammein *n*
manage *v* leiten, bewirtschaften, disponieren, handhaben, hantieren, verwalten
management Betriebsführung *f,* Geschäftsleitung *f,* Verwaltung *f*
manager Betriebsführer *m,* Geschäftsführer *m*
managing betriebsführend, leitend, verwaltend

managing director Betriebsdirektor *m*
Manchester brown Triaminoazobenzol *n,* Vesuvin *n* (Farbstoff)
Manchester yellow Naphthylamingelb *n*
mandarin[e] Mandarine *f* (Bot)
mandelate (salt or ester of mandelic acid) Mandelat *n*
mandelic acid Mandelsäure *f*
mandelic amide Mandelsäureamid *n*
mandelic nitrile Mandelsäurenitril *n*
mandelonitrilase Mandelnitrilase *f* (Biochem)
mandelonitrile Mandelsäurenitril *n,* Benzaldehydcyanhydrin *n*
mandragora Mandragora *f* (Bot), Alraunwurzel *f* (Bot)
mandrake Alraun *m* (Bot), Podophyllum *n*
mandrel Dorn *m,* Richtdorn *m,* Spritzdorn *m,* **hollow** ~ Hohldorn *m*
mandrel carrier Dornhalter *m,* Dornhalterung *f*
mandrel test Dornbiegeprobe *f*
maneuverability Lenkbarkeit *f,* Manövrierfähigkeit *f,* Wendigkeit *f*
maneuverable lenkbar, manövrierbar
mangabeira [rubber] Mangabeiragummi *m*
manganate Manganat *n*
manganese (Symb. Mn) Mangan *n,* **containing** ~ manganhaltig, **red** ~ Rhodochrosit *m* (Min); Rotstein *m* (Min), **siliceous** ~ Kieselmangan *n* (Min)
manganese acetate Manganacetat *n*
manganese alloy Manganlegierung *f*
manganese alum Manganalaun *m*
manganese ammonium phosphate Manganammoniumphosphat *n*
manganese bath method Manganbadmethode *f*
manganese binoxide Mangandioxid *n,* Mangan(IV)-oxid *n*
manganese bister Manganbraun *n*
manganese blue Manganblau *n* (Zementblau)
manganese borate Manganoborat *n*
manganese bronze Manganbronze *f*
manganese carbide Mangancarbid *n*
manganese carbonate, native ~ Rhodochrosit *m* (Min)
manganese chloride Manganchlorid *n*
manganese compound Manganverbindung *f*
manganese content Mangangehalt *m*
manganese dioxide Braunstein *m,* Mangandioxid *n,* Mangan(IV)-oxid *n*
manganese fluosilicate Mangansilicofluorid *n,* Kieselfluormangan *n*
manganese garnet Mangangranat *m* (Min)
manganese green Bariummanganat *n* (Min), Mangangrün *n* (Min)
manganese heptoxide Manganheptoxid *n,* Mangan(VII)-oxid *n*
manganese hydroxide Manganhydroxid *n*
manganese(II) Mangano-
manganese(II) acetate Manganoacetat *n*

manganese(II) carbonate Manganocarbonat *n*
manganese(II) compound
  Mangan(II)-Verbindung *f*,
  Manganoverbindung *f*,
  Manganoxydulverbindung *f* (obs)
manganese(III) Mangani-
manganese(III) compound
  Mangan(III)-Verbindung *f*,
  Manganiverbindung *f*
manganese(III) hydroxide Manganihydroxid *n*
manganese(II) iodide Mangan(II)-jodid *n*,
  Manganjodür *n* (obs)
manganese(II) ion Mangan(II)-ion,
  Manganoion *n*
manganese(III) oxide Mangansesquioxid *n*
manganese(III) phosphate Manganiphosphat *n*
manganese(III) salt Mangan(III)-Salz *n*,
  Manganisalz *n*
manganese(III) sulfate Manganisulfat *n*
manganese(II) phosphate Manganophosphat *n*
manganese(II) salt Mangan(II)-Salz *n*,
  Manganosalz *n*, Manganoxydulsalz *n* (obs)
manganese(II) sulfate Manganosulfat *n*
manganese(IV) chloride Mangantetrachlorid *n*
manganese(IV) oxide Braunstein *m*,
  Mangandioxid *n*
manganese lactate Manganlactat *n*
manganese monoxide Mangan(II)-oxid *n*,
  Manganoxydul *n* (obs)
manganese nitrate Mangannitrat *n*
manganese nodule Manganknolle *f* (Min)
manganese ore Manganerz *n* (Min), **black ~**
  Schwarzmanganerz *n* (Min), **grey ~**
  Manganit *m* (Min)
manganese oxalate Manganoxalat *n*
manganese preparation Manganpräparat *n*
manganese protoxide Mangan(II)-oxid *n*,
  Manganoxydul *n* (obs)
manganese pyrophosphate
  Manganpyrophosphat *n*
manganese selenide Manganselenid *n*
manganese sesquioxide Mangan(III)-oxid *n*,
  Mangansesquioxid *n*
manganese silicate, native ~ Rhodonit *m* (Min)
manganese silicide Silicomangan *n*
manganese spar Manganspat *m* (Min)
manganese spectrum Manganspektrum *n*
manganese steel Manganstahl *m*
manganese sulfide Alabandin *m* (Min),
  Mangansulfid *n*
manganese tetrachloride Mangantetrachlorid *n*
manganese trifluoride Mangantrifluorid *n*
manganese trioxide Mangantrioxid *n*,
  Mangansäureanhydrid *n*, Mangan(VI)-oxid *n*
manganese violet Manganviolett *n*
manganese(VI) oxide Mangansäureanhydrid *n*,
  Mangantrioxid *n*
manganese vitriol Manganvitriol *n*
[manganese] wad Manganschaum *m*

manganese white Manganweiß *n*
manganese zinc spar Manganzinkspat *m* (Min)
manganic Mangani-, Mangan(III)-
manganic acid Mangansäure *f*
manganic anhydride Mangansäureanhydrid *n*,
  Mangantrioxid *n*, Mangan(VI)-oxid *n*
manganic compound Mangan(III)-Verbindung *f*,
  Manganiverbindung *f*
manganic fluoride Mangantrifluorid *n*
manganic oxide Mangan(III)-oxid *n*,
  Mangansesquioxid *n*
manganic oxide hydrate, native ~ Manganit *m*
  (Min)
manganic phosphate Manganiphosphat *n*
manganic salt Mangan(III)-Salz *n*,
  Manganisalz *n*
manganic sulfide, native ~ Mangankies *m* (Min)
manganic(VI) acid Mangansäure *f*
manganicyanic acid
  Manganicyanwasserstoffsäure *f*
manganiferous manganhaltig
manganin Manganin *n*
manganite Manganit *m* (Min),
  Graumanganerz *n* (Min)
manganocalcite Manganocalcit *m* (Min)
manganocolumbite Manganocolumbit *m* (Min)
manganosite Manganosit *m* (Min),
  Mangan(II)-oxid *n* (Min)
manganospherite Manganosphärit *m* (Min)
manganostibiite Manganostibiit *m* (Min)
manganotantalite Manganotantalit *m* (Min)
manganous Mangano-, Mangan(II)-
manganous acetate Manganoacetat *n*
manganous borate Manganoborat *n*
manganous carbonate Manganocarbonat *n*,
  native ~ Manganspat *m* (Min)
manganous chloride Mangan(II)-chlorid *n*
manganous compound Mangan(II)-Verbindung *f*,
  Manganoverbindung *f*,
  Manganoxydulverbindung *f* (obs)
manganous hydroxide Manganohydroxid *n*,
  Manganhydroxydul *n* (obs),
  Manganoxydulhydrat *n* (obs)
manganous iodide Mangan(II)-jodid *n*
manganous ion Mangan(II)-ion *n*,
  Manganoion *n*
manganous oleate Manganooleat *n*
manganous oxide Mangan(II)-oxid *n*,
  Manganoxydul *n* (obs), **native ~**
  Manganosit *m* (Min)
manganous phosphate Mangan(II)-phosphat *n*,
  Manganophosphat *n*
manganous salt Mangan(II)-Salz *n*,
  Manganosalz *n*, Manganoxydulsalz *n* (obs)
manganous sulfate Manganosulfat *n*
manganous sulfide Mangan(II)-sulfid *n*, **native ~**
  Manganblende *f* (Min)
manganpectolite Manganpektolith *m* (Min)
mangiferin Mangiferin *n*

mangin Mangin *n*
mangle Mange[l] *f,* Rolle *f*
mangle *v* verstümmeln
Mangold's acid Mangoldsche Säure *f*
mangostin Mangostin *n*
mangrove bark Mangrovenrinde *f*
manhole Einsteigöffnung *f,* Einsteigschacht *m,*
   Kabelschacht *m,* Mannloch *n,*
   Reinigungsöffnung *f*
man-hour Arbeitsstunde *f*
manifest *v* äußern, verkünden
manifold Rohrverzweigung *f,* Sammelleitung *f,*
   Verteilerkanal *m,* Verteilungskopf *m*
manifold *a* mannigfach, mehrfach, vielfach
manifold *v* vervielfachen, vervielfältigen
manifold pressure control Ladedruckregler *m*
manifold pressure monitor or recorder
   Ladedruckschreiber *m*
manihot oil Manihotöl *n*
Manila hemp Manilahanf *m*
Manila paper Manilapapier *n,* Bastpapier *n*
manioc Kassavamehl *n,* Maniok *m*
manipulable bedienbar
manipulate *v* manipulieren, beeinflussen,
   handhaben, hantieren
manipulated variable Stellgröße *f*
manipulation Manipulation *f,* Bearbeitung *f,*
   Bedienung *f,* Betätigung *f,* Handhabung *f,*
   mechanical ~ mechanische Verarbeitung *f*
manipulator, remote ~ ferngesteuerter Greifer *m*
manna Manna *n*
mannan Mannan *n*
manneotetrose Manneotetrose *f*
manner Art *f,* Methode *f,* Weise *f*
Mannheim gold Kupfergold *n,* Mannheimer
   Gold *n,* Neugold *n*
Mannich reaction Mannich-Reaktion *f*
mannide Mannid *n*
manninotriose Manninotriose *f*
mannitan Mannitan *n*
mannite Mannit *m,* Mannazucker *m*
mannite ester Mannitester *m*
mannitol Mannit *m,* Mannazucker *m*
mannitol ester Mannitester *m*
mannitol nitrate Nitromannit *m*
mannochloralose Mannochloralose *f*
mannoheptonic acid Mannoheptonsäure *f*
mannoheptose Mannoheptose *f*
mannomustine Mannomustin *n*
mannonic acid Mannonsäure *f*
mannosaccharic acid Mannozuckersäure *f*
mannosamine Mannosamin *n*
mannose Mannose *f*
mannosidase Mannosidase *f*
mannurone Mannuron *n*
mannuronic acid Mannuronsäure *f*
manoeuvrability Manövrierfähigkeit *f,*
   Wendigkeit *f*

manometer Manometer *n,*
   Dampfdruckmesser *m,* Druckanzeiger *m,*
   Gasdruckmesser *m,* **differential** ~
   Feindruckmesser *m*
manometer scale Manometerskala *f*
manometric manometrisch
manometric observation
   Manometerbeobachtung *f*
manondonite Manondonit *m*
Mansfeld copper ore Mansfelder
   Kupferschiefer *m* (Min)
manthine Manthin *n*
mantissa Mantisse *f* (Math)
mantle Mantel *m,* Überform *f* (Gieß)
mantle bracket Eisenkranz *m*
mantle sheet steel Manteleisen *n*
manual Handbuch *n,* Bedienungsvorschrift *f,*
   Leitfaden *m*
manual control Handregelung *f* (Regeltechn)
manual labor Handarbeit *f*
manual loading station Handvorgaberegler *m*
manually manuell
manually controlled handbetrieben,
   handgesteuert
manually operated handbetrieben
manual manipulation Handbedienung *f*
manual operation Handbedienung *f*
manual welding Handschweißung *f*
manual work Handarbeit *f*
manufacture Fabrikat *n,* Anfertigung *f,*
   Fabrikation *f,* Fertigung *f,* Produktion *f,*
   serienmäßige Herstellung *f,* **course of** ~
   Herstellungsgang *m,* **process of** ~
   Darstellungsverfahren *n,* Herstellungsgang *m,*
   stage of ~ Verarbeitungsstufe *f,* ~ **of**
   man-made fibers Chemiefasererzeugung *f,* ~
   of oil Ölgewinnung *f,* ~ **of precision**
   instruments feinmechanische Industrie *f,* ~ **of**
   synthetic fibers Chemiefasererzeugung *f*
manufacture *v* anfertigen, fabrizieren, fertigen,
   herstellen, erzeugen, ~ **on an**
   industrial scale fabrikmäßig herstellen
manufactured hergestellt
manufactured article Fabrikware *f*
manufactured gas Industriegas *n*
manufacturer Fabrikant *m,* Hersteller *m,*
   Produzent *m*
manufacturer's joint Fabrikkante *f* (Karton)
manufacturing activities Produktionsprogramm *n*
manufacturing engineer Betriebsingenieur *m*
manufacturing engineering Fertigungstechnik *f*
manufacturing industry Fabrikwesen *n*
manufacturing machinery
   Herstellungsmaschinen *f pl*
manufacturing method Bearbeitungsmethode *f,*
   Herstellungsgang *m,* Herstellungsweise *f*
manufacturing process Fabrikationsprozeß *m,*
   Herstellungsverfahren *n*

manufacturing program
Fabrikationsprogramm *n*, Fabrikationsplan *m*
manufacturing scheme Fabrikationsplan *m*
manukene Manuken *n*
manure Dünger *m*, Dung *m*, Mist *m*, **artificial ~**
Kunstdünger *m*, **green ~** Gründünger *m*,
liquid ~ Jauche *f*
manure works Düngemittelfabrik *f*
manuring Düngen *n*, Düngung *f*, **green ~**
Gründüngen *n*
manuring salt Düngesalz *n*
many-colored farbenreich, bunt
many-electron spectrum
Mehrelektronenspektrum *n*
manysided vielseitig
manystage mehrstufig
map Landkarte *f*
maple honey Ahornhonig *m*
maple juice Ahornsaft *m*
maple molasses Ahornmelasse *f*
maple pulp Ahornzellstoff *m*
maple sugar Ahornzucker *m*
maple syrup Ahornsirup *m*
maple varnish Ahornlack *m*
maple wood Ahornholz *n*
mappine Mappin *n*
mar *v* beschädigen, entstellen
maranta Marantastärke *f*
marbelizing Marmorierung *f*
marble Marmor *m*, **figured ~** Bildmarmor *m*,
green ~ Kalkglimmerschiefer *m* (Min),
imitation ~ Gipsmarmor *m*, **ligneous ~**
Holzmarmor *m*, **pisolitic ~**
Erbsensteinmarmor *m* (Min), **shining ~**
Glanzmarmor *m*, **spotted ~** Augenmarmor *m*,
variegated ~ Bildmarmor *m*
marble cement Marmorzement *m*
marble-colored marmorfarbig
marbled marmoriert, gesprenkelt
marble glass Marmorglas *n*
marble gypsum Marmorgips *m* (Marmorzement)
marbleized marmoriert
marbleizing Marmorierung *f*
marble-like marmorähnlich, marmorartig
marble lime Marmorkalk *m*
marble plate Marmorplatte *f*
marble quarry Marmorbruch *m*
marble white Marmorweiß *n*
marbling Aderung *f* (Marmor)
marbling effect Marmoreffekt *m*
marcasite Markasit *m* (Min)
marc brandy Treberbranntwein *m*
maretin Maretin *n*
marfanil Marfanil *n*
margaric acid Margarinsäure *f*,
Heptadecansäure *f*
margarin[e] Margarine *f*
margarite Margarit *m* (Min), Bergglimmer *m*
(Min), Perlglimmer *m* (Min)

margarodite Margarodit *m* (Min)
margaron Margaron *n*
margarosanite Margarosanit *m* (Min)
margin Rand *m*, Spielraum *m*
marginal coefficient Randkoeffizient *m*
marginal note Randanmerkung *f*,
Randbemerkung *f*
marginal punched cards Randlochkarten *pl*
marginal sharpness Randschärfe *f* (Opt)
marginal vibration Grenzschwingung *f*
margin of manufacture Einpaßzugabe *f*
marignacite Marignacit *m* (Min)
marihuana Marihuana *n*
marijuana Marihuana *n*
marinate *v* marinieren, sauer einlegen
marine blue Marineblau *n*
marine finish Unterwasserlack *m*
marine glue Schiffsleim *m*
marine optics Meeresoptik *f*
marine varnish Schiffsfirnis *m*
maripa fat Maripafett *n*
marjoram oil Majoranöl *n*
mark Markierung *f*, Abdruck *m*, Kennzeichen *n*,
Kennzeichnung *f*, Merkmal *n*, Narbe *f*,
Zeichen *n*
mark *v* markieren, anzeichnen, kennzeichnen,
~ off abgrenzen, **~ out** abstecken, ausersehen
marked markiert, ausgeprägt
marker Anzeiger *m*, Signiereinrichtung *f*, **felt tip**
~ or felt point marker Filzschreiber *m*
marker color Kennfarbe *f*
market Markt *m*
marketable konkurrenzfähig, verkäuflich
marketing research Absatzforschung *f*
market research Marktforschung *f*
marking Markierung *f*, Bezeichnung *f*,
Kennzeichnung *f*, **~ off** (of prints)
Abklatschen *n*
marking gauge Parallelreißer *m*
marking ink Signiertinte *f*, Zeichentinte *f*
marking nut Malakkanuß *f* (Bot)
marking on products Produktkennzeichnung *f*
marking pin Signierstift *m*
markogenic acid Markogensäure *f*
Markovnikov rule Markownikoffsche Regel *f*
Markush structure Markush-Formel *f*
marl Mergel *m*, **calcareous ~** Kalkmergel *m*,
containing ~ mergelhaltig, **vitrifiable ~**
Glasmergel *m*
marling Mergeln *n*
marlite Marlit *m* (Min)
marlitic marlitartig
marl pit Mergelgrube *f*
marly mergelartig, mergelhaltig
marly clay Mergelton *m*
marly limestone Mergelkalk *m*
marly sandstone Mergelsandstein *m* (Min)
marmatite Marmatit *m* (Min),
Eisenzinkblende *f* (Min)

marmelosin Marmelosin *n*
marmesilic acid Marmesilsäure *f*
marmin Marmin *n*
marmolite Marmolith *m* (Min)
marmoraceous marmorähnlich, marmorartig
marmoration Aderung *f* (Marmor),
  Marmorierung *f*
marmoreal marmorähnlich, marmorartig
marproof kratzfest
mar-resistance Kratzfestigkeit *f*
marrianolic acid Marrianolsäure *f*
marrow Mark *n*
marrow cell Knochenmarkzelle *f*
marrubanol Marrubanol *n*
marrubenol Marrubenol *n*
marrubic acid Marrubinsäure *f*
Mars Mars *m* (Astr), **inhabitant of** ~
  Marsbewohner *m*
Marseilles soap Marseiller Seife *f*
marsh Sumpf *m*
marshalling section Stapelstrecke *f*
marsh gas Grubengas *n*, Sumpfgas *n*
marshite Marshit *m* (Min)
marshmallow flowers Eibischblüten *f pl* (Bot)
marshmallow leaves Eibischblätter *n pl* (Bot)
marshmallow root Altheewurzel *f* (Bot),
  Eibischwurzel *f* (Bot)
marshmallow syrup Altheesirup *m*
marsh ore Rasen[eisen]erz *n* (Min), Sumpferz *n*
  (Min)
marsh samphire Glaskraut *n* (Bot),
  Kalipflanze *f* (Bot)
Marsh['s] test Marshsche Arsenprobe *f*
marsh trefoil Fieberklee *m*
martensite Martensit *m*
martensite formation Martensitbildung *f*
martensite point Martensitpunkt *m*
Martian year Marsjahr *n* (Astr)
Martin blower Martingebläse *n*,
  Kapselhochdruckgebläse *n*
martinite Martinit *m* (Min)
Martin steel Martinstahl *m*
martite Martit *m* (Min)
Martius yellow Martiusgelb *n*
marvel Wunder *n*
marzipan Marzipan *n*
mascagnine Mascagnin *m* (Min)
mascagnite Mascagnin *m* (Min)
mascara Wimperntusche *f* (Kosm)
masculine männlich
maser Maser (Abk. für „Microwave
  Amplification by Stimulated Emission of
  Radiation)
mash Brei *m* (Pap), Maische *f*
mash *v* [ein]maischen, zermalmen, zerquetschen
mashing, second ~ Aufmaischen *n* (Zuck)
mashing temperature, final ~
  Abmaischtemperatur *f* (Brauw)
mash liquor Maischwasser *n*

mash tub thermometer Bottichthermometer *n*
  (Brau), Maischthermometer *n* (Brau)
mash yeast Maischhefe *f*
mask Maske *f*
mask *v* maskieren
maskelynite Maskelynit *m* (Min)
masking Maskierung *f*
masking coating Abdecklack *m*
masking material Abdeckmittel *n*
masking paper Abdeckpapier *n*
masking tape Kreppband *n*
mason Maurer *m*
mason *v* mauern
masonine Masonin *n*
Masonite process Masonite-Verfahren *n*
masopin Masopin *n*
mass Masse *f*, Menge *f*, **center of** ~
  Schwerpunkt *m*, **critical** ~ kritische Masse *f*,
  determination of ~ Massebestimmung *f*,
  distribution of ~ Massenverteilung *f*, **reduced**
  ~ reduzierte Masse (Mech), **sponge-like** ~
  schwammige Masse
mass absorption Massenabsorption *f*
mass acceleration Massenbeschleunigung *f*
mass action Massenwirkung *f*, **law of** ~
  Massenwirkungsgesetz *n*
mass application Massenanwendung *f*
mass attenuation coefficient
  Massenschwächungskoeffizient *m*
mass balance Mengenbilanz *f*
mass center Massemittelpunkt *m*
mass concentration Gewichtskonzentration *f*
mass correction Massenkorrektur *f*
mass density Massendichte *f*
mass difference Massenunterschied *m*
mass diffusion Fremdgasdiffusion *f* (Sieb,
  Lochblende, poröse Zwischenwand usw.)
mass disappearance Masseschwund *m*
mass-dyed spinngefärbt
mass-energy equation
  Masse-Energie-Gleichung *f*
mass-energy equivalence, equation of ~
  Masse-Energie-Äquivalentgleichung *f*
mass equation Massengleichung *f*
mass equivalent Masseäquivalent *n*
mass estimation Massenschätzung *f*
mass extinction coefficient
  Massenschwächungskoeffizient *m*
mass flow density Massenflußdichte *f*,
  Massenstromdichte *f*
massicot Massicot *m* (Min), Bleiocker *m* (Min)
massive massiv
massive ore Derberz *n*
mass loss Masseschwund *m*, Masseverlust *m*
mass nucleus Massenkern *m*
mass number Massenzahl *f* (Atom)
mass particle Masseteilchen *n*
mass poisoning Massenvergiftung *f*

mass polymerization Blockpolymerisation *f*,
Polymerisation in Masse
mass-produced goods Massenartikel *pl*
mass production Massenproduktion *f*,
Massenerzeugung *f*, Massenfabrikation *f*,
Massenfertigung *f*, serienmäßige Herstellung *f*
mass radiator Massenstrahler *m*
mass ratio Massenverhältnis *n*,
Mengenverhältnis *n*
mass scattering coefficient
Massenstreukoeffizient *m*
mass spectrograph Massenspektrograph *m*,
Massenspektrometer *n*, **velocity focusing** ~
Massenspektrograph *m* mit
Geschwindigkeitsfokussierung (Spektr)
mass spectrometer Massenspektrometer *n*
mass spectrometric massenspektrometrisch
mass spectroscopy Massenspektroskopie *f*
mass spectrum Massenspektrum *n* (Spektr)
mass stopping power Massenbremsvermögen *n*
mass synchrometer Massensynchrometer *n*
mass tone Vollton *m* (Pigment ohne
Verschnittmittel)
mass transfer Massetransport *m*,
Stoffaustausch *m*, Stoffaustauschprozeß *m*,
Stofftransport *m*, Stoffübergang *m*, **rate of** ~
Stoffübergangsgeschwindigkeit *f*, **resistance to**
~ Stoffdurchgangswiderstand *m*, ~ **from**
continuous phase to droplets Stoffübergang *m*
am umströmten Tropfen
mass transfer coefficient
Stoffübergangskoeffizient *m*,
Stoffübergangszahl *f*
mass transfer equipment
Austauscheinrichtungen *pl*
mass transfer resistance
Stoffaustauschwiderstand *m*
mass transfert accompanied by reaction
Stoffübergang *m* mit gleichzeitiger Reaktion
mass transportation Massentransport *m*
mass unit Masseneinheit *f*
mass vaccination Massenimpfung *f*
mast cell Mastzelle *f*
master Meister *m*
master *v* beherrschen, bewältigen
master arm Steuerarm *m*
master batch Vormischung *f*, Grundmischung *n*,
angereicherte Mischung
master controller Führungsregler *m* (Regeltechn)
master file Zentralkartei *f*
master gauge Vergleichslehre *f* (Techn)
mastering-pit Ledergrube *f*
master key Hauptschlüssel *m*
master mechanic Vorarbeiter *m*
masterpiece Meisterstück *n*
master record Stammsatz *m* (Informatik)
master steel pattern Tauchform *f*
master switch Hauptschalter *m*
masterwort oil Meisterwurzelöl *n*

mastic Kitt *m*, Spachtelmasse *f*, **to fasten with** ~
ankitten
masticadienonic acid Masticadienonsäure *f*
mastic asphalt Gußasphalt *m*
masticate *v* kneten (Gummi)
masticating Mastizieren *n* (Gummi)
mastication Mastikation *f*, Zerkleinerung *f*
mastic cement Mastixkitt *m*, Steinkitt *m*
masticin Masticin *n*
masticinic acid Masticinsäure *f*
mastic liquor Mastixbranntwein *m*
mastic resin Mastixharz *n*
mastic varnish Mastixfirnis *m*
mastocyte Mastzelle *f*
masurium (Symb. Tc) Technetium *n*,
Masurium *n* (obs)
masut Naphtharückstand *m*
mat blind (Metall), Matte *f*
mat *a* glanzlos, matt
matairesinol Matairesinol *n*
matairesinolic acid Matairesinolsäure *f*
matatabilactone Matatabilacton *n*
match Gegenstück *n*, Streichholz *n*, Zündholz *n*
match *v* angleichen, [an]passen, nachstellen
matchbox Streichholzschachtel *f*
matched angepaßt
matched metal die molding Heißpreßverfahren *n*
(Polyester)
match industry Zündholzindustrie *f*
matching Anpassung *f*, Nachstellung *f*
matching condenser Abgleichkondensator *m*
(Elektr)
matching transformer
Anpassungstransformator *m*
match plate Modellplatte *f*
maté leaves Matéblätter *pl*
materia Materie *f*, Sachgebiet *n*
material Material *n*, Bestandteil *m*, Stoff *m*,
Substanz *f*, Werkstoff *m*, **added** ~
Zusatzsubstanz *f*, **consumption of** ~
Werkstoffbedarf *m*, **depleted** ~ erschöpftes
Material (Atom), **fatigue of** ~
Werkstoffermüdung *f*, **feed of** ~
Werkstoffzuführung *f*, **fibrous** ~
Faserwerkstoff *m*, **fire-proof** ~ feuerfester
Werkstoff, **laminated** ~ geschichteter
Werkstoff, **non-fissionable** ~ nicht spaltbares
Material *n* (Atom), **polymeric** ~ polymerer
Werkstoff, **property of** ~
Materialeigenschaft *f*, **strain or tension of** ~
Materialspannung *f*, **strength of** ~
Werkstoffestigkeit *f*, **tannic acid-resisting** ~
gerbsäurebeständiger Werkstoff, **to remove** ~
aussparen, ~ **to be bleached** Bleichgut *n*, ~ **to**
be dried Trockengut *n*, ~ **to be ground**
Mahlgut *n*, ~ **to be separated** Scheidegut *n*
material *a* materiell, stofflich
material accumulation Werkstoffanhäufung *f*
material balance Stoffbilanz *f*

material consumption Materialverbrauch *m*
material defect Materialfehler *m*
material economy Materialausnutzung *f*
material proof Sachbeweis *m*
material requirements Materialbedarf *m*
material roll Aufwickelwalze *f*
materials handling Fördertechnik *f*
materials technology Werkstoffkunde *f*
material stress Materialbeanspruchung *f*
material testing Materialprüfung *f,*
    Werkstoffprüfung *f*
material testing apparatus Materialprüfgerät *n,*
    Werkstoffprüfapparat *m*
mat etching agent Mattätzmittel *n*
matezitol Matezit *m*
mat glaze Mattglasur *f*
mathematical mathematisch
mathematician Mathematiker *m*
mathematics Mathematik *f,* **applied** ~
    angewandte Mathematik, **pure** ~ reine
    Mathematik
maticin Maticin *n*
matico leaves Matikoblätter *n pl*
matico oil Matikoöl *n*
matildite Matildit *m* (Min)
mating surface Abquetschfläche *f*
matlockite Matlockit *m* (Min)
mat molding Mattenpreßverfahren *n*
matricaria camphor Matricariacampher *m*
matricaria ester Matricariaester *m*
matricarianol Matricarianol *n*
matricaric acid Matricariasäure *f*
matricarin Matricarin *n*
matridine Matridin *n*
matrine Matrin *n*
matrinic acid Matrinsäure *f*
matrinidine Matrinidin *n*
matrix Matrize *f,* Formeneinsatz *m,* Gangart *f*
    (Min), Gießform *f,* Mater *f* (Papier), Matrix *f*
    (Math), Prägeform *f,* Preßform *f,* Stanzform *f,*
    order of a ~ Rang einer Matrize *f,* **regular** ~
    reguläre Matrize *f,* **skew symmetrical** ~
    schiefsymmetrische Matrize *f*
matrix holder Matrizenhalter *m*
matte, concentrated ~ Spurstein *m*
matted mattiert
matter Materie *f,* Gegenstand *m,* Stoff *m,*
    Substanz *f,* **structure of** ~ Aufbau der
    Materie, **volatile** ~ flüchtiger Bestandteil
matteucinol Matteucinol *n*
matting Mattierung *f*
matting sand Mattiersand *m*
matt salt Ammoniumbifluorid *n*
maturation division Eireifungsteilung *f* (Biol)
mature reif, fällig
mature *v* reifen, ablagern, altern (Magnet),
    reifen
mature[d] abgelagert, reif
maturing time Reifezeit *f*

maturity Reife *f*
maucherite Maucherit *m* (Min)
mauleonite Mauleonit *m*
mauvaniline Mauvanilin *n*
mauve Malvenfarbe *f*
mauvein Mauvein *n*
maximal maximal, höchst
maximal pressure Maximaldruck *m*
maximal value Höchstwert *m*
maximum Höchstmaß *n,* Maximum *n*
maximum *a* höchst, maximal
maximum amplification Verstärkbarkeitsgrenze *f*
maximum amplitude Maximalamplitude *f*
maximum boost Höchstladedruck *m*
maximum brittleness Alterungsgrenze *f*
maximum capacity Maximalleistung *f*
maximum content Höchstgehalt *m*
maximum current Maximalstrom *m* (Elektr)
maximum deviation Maximalabweichung *f*
maximum discharge Höchstentladung *f*
maximum dose Höchstdosis *f* (Pharm),
    Maximaldosis *f* (Pharm)
maximum duty Höchstbelastung *f*
maximum efficiency Höchstleistung *f*
maximum frequency Höchstfrequenz *f*
maximum intensity Intensitätsmaximum *n*
maximum load Höchstbelastung *f,*
    Bruchbelastung *f,* Höchstlast *f,* Knicklast *f,*
    point of ~ Bruchgrenze *f*
maximum output Höchstleistung *f,*
    Maximalausbeute *f,* Maximalleistung *f,*
    Spitzenleistung *f*
maximum permissible höchstzulässig
maximum pressure Höchstdruck *m*
maximum range Maximalreichweite *f*
maximum safe value höchstzulässiger Wert *m*
maximum speed Maximalgeschwindigkeit *f,*
    Höchstdrehzahl *f*
maximum stress Höchstbeanspruchung *f*
maximum temperature Höchsttemperatur *f,*
    Temperaturmaximum *n*
maximum valence Höchstwertigkeit *f* (Element)
maximum value Maximalwert *m;* Scheitelwert *m,*
    Spitzenwert *m*
maximum voltage Höchstspannung *f* (Elektr),
    Maximalspannung *f* (Elektr)
maximum weight Höchstgewicht *n,*
    Maximalgewicht *n*
maximum weight of load Höchstladegewicht *n*
Maxwell bridge Maxwellbrücke *f*
Maxwell[ian] distribution Maxwellsche
    Verteilung *f*
mayonnaise Mayonnaise *f*
mazapilite Mazapilit *m*
mazarine blue Mazarinblau *n*
mazut Masut *n,* Naphtharückstand *m*
m/c direction Längsrichtung *f*
m-cresol Metakresol *n*
mead Met *m,* Honigwein *m*

meadow saffron Herbstzeitlose *f* (Bot)
meager mager
meal Mehl *n*, Schrotmehl *n*, **coarse ~**
 Grandmehl *n*
meal cooling Schrotkühlung *f*
meal cooling unit Schrotkühlanlage *f*
mealiness Mehlartigkeit *f*
mealy mehlartig, mehlig
mean Durchschnitt *m*, Mittel *n*
mean *a* durchschnittlich
mean *v* bedeuten
mean annual temperature
 Temperaturjahresmittel *n*
mean error Streuung *f* (Math)
meaning Bedeutung *f*, Sinn *m*
meaningless bedeutungslos
mean value Durchschnittswert *m*, Mittelwert *m*
mean value theorem Mittelwertsatz *m*
mean velocity Durchschnittsgeschwindigkeit *f*
measles Masern *pl* (Med), **German ~** Röteln *pl*
measles vaccine Masernimpfstoff *m*
measurable meßbar
measure Maß *n*, Ausmaß *n*, Format *n*,
 Maßeinheit *f*, **~ of capacity** Hohlmaß *n*,
 Raummaß *n*
measure *v* [ab]messen, ausmessen, bemessen,
 eichen, vermessen, **~ again** nachmessen, **~
 the depth of water** loten
measured length Meßstrecke *f*
measured value Meßwert *m*
measurement Messung *f*, Abmessung *f*,
 Vermessung *f*, **actual ~** Ist-Maß *n*, **method of
 ~** Meßverfahren *n*, **range of ~**
 Meßbereich *m*, **result of ~** Meßergebnis *n*, **to
 take a ~** eine Messung machen, **~ of length**
 Längenmessung *f*
measurement and control Regeltechnik *f*
measuring Messen *n*, [Ver]Messung *f*, **accurate ~**
 Präzisionsmessung *f*
measuring apparatus Meßapparat *m*, Meßgerät *n*
measuring arrangement Meßanordnung *f*
measuring beaker Meßbecher *m*
measuring bridge Meßbrücke *f* (Elektr)
measuring cell Meßzelle *f*
measuring cell capacitance Meßzellenkapazität *f*
measuring chain Meßkette *f*
measuring circuit Meßkreis *m*
measuring cup Meßbecher *m*
measuring cylinder Meßzylinder *m*
measuring desk Meßpult *n*
measuring device Meßvorrichtung *f*
measuring element Meßwerk *n*
measuring error Meßfehler *m*
measuring fault Meßfehler *m*
measuring flask Meßkolben *m*
measuring fluid Meßflüssigkeit *f*
measuring gauge Meßlehre *f*
measuring glass Mensurglas *n*

measuring instrument Meßinstrument *n*
measuring location Meßort *m*
measuring means or unit Meßglied *n*
 (Regeltechn)
measuring method Meßmethode *f*
measuring microscope Meßmikroskop *n*
measuring pipet[te] Meßpipette *f*, Meßheber *m*
measuring point Meßort *m*, Meßpunkt *m*,
 Meßstelle *f*
measuring point selector Meßstellenumschalter *m*
measuring quantity Meßgröße *f*
measuring range Meßbereich *m*
measuring resistance Meßwiderstand *m*
measuring result Meßergebnis *n*
measuring rod Maßstab *m*
measuring rule Maßstab *m*
measuring section Meßstrecke *f*
measuring stick Maßstab *m*
measuring system Maßsystem *n*, Meßmethode *f*
measuring tape Maßband *n*, Bandmaß *n*
measuring tool Meßwerkzeug *n*
measuring transformer Meßtransformator *m*
measuring transmitter or transducer
 Meßumformer *m* (Regeltechn)
measuring tube Meßröhre *f*, Meßrohr *n*
measuring vessel Meßgefäß *n*
measuring wire Gefälldraht *m*
meat Fleisch *n*, **canned ~** Dosenfleisch *n*,
 Fleischkonserve *f*, **pickled ~** Pökelfleisch *n*,
 preserved ~ Fleischkonserve *f*, **smoked ~**
 Rauchfleisch *n*, **tinned ~** Büchsenfleisch *n*,
 Fleischkonserve *f*
meat extract Fleischextrakt *m*
meat juice Fleischsaft *m*
meat marking inks Fleischstempelfarben *pl*
meat meal mill Fleischmehlmühle *f*
meat poisoning Fleischvergiftung *f*
meat shredding machine
 Fleischzerfaserungsmaschine *f*
mecamylamine Mecamylamin *n*
Mecca balsam Mekkabalsam *m*
mechanic Maschinist *m*, Mechaniker *m*,
 Schlosser *m*
mechanical mechanisch, automatisch,
 maschinell, maschinenmäßig
mechanical drawing Düsenziehverfahren *n*
mechanical dressing Aufbereitung *f* durch
 Maschinen (pl)
mechanical dressing process
 Trockenaufbereitung *f*
mechanical engineer Maschinenbauer *m*,
 Maschineningenieur *m*
mechanical engineering Maschinenbau *m*
mechanical equivalent Arbeitsäquivalent *n*
mechanically controlled mechanisch betätigt
mechanical mop Schwabbelmaschine *f*
mechanical puddling Maschinenpuddeln *n*
mechanical pusher Ausdrückmaschine *f*
 (Kokerei)

mechanical shaker Schüttelmaschine *f*
mechanical wood pulp paper Holzschliffpapier *n*
mechanician Mechaniker *m*, Monteur *m* (Elektr)
mechanics Mechanik *f*, Bewegungslehre *f*, ~ **of rigid bodies** Mechanik *f* fester Körper (pl)
mechanism Mechanismus *m*, mechanische Vorrichtung *f*
mechanist Maschinist *m*, Mechaniker *m*
mechanization Mechanisierung *f*
mechanize *v* mechanisieren
mechanochemistry Mechanochemie *f*
meclozine Meclozin *n*
mecocyanine Mekocyanin *n*
meconic acid Mekonsäure *f*, Hydroxychelidonsäure *f*
meconidin Mekonidin *n*
meconin Mekonin *n*
meconinic acid Mekoninsäure *f*
median perpendicular Mittelsenkrechte *f*
medical medizinisch, ärztlich
medical check-up Untersuchung *f* (Med)
medical formula Arzneiformel *f*
medical man Arzt *m*
medical plant Arzneipflanze *f*, Heilpflanze *f*
medical profession Ärzteschaft *f*
medical science Heilkunde *f*, Medizin *f*
medicament Arznei *f*, Heilmittel *n* (Med), Medikament *n* (Med), Medizin *f* (Med)
medicated heilkräftig, Arzneistoffe enthaltend
medicinal medizinisch, medizinal, heilsam
medicinal herbs Heilkräuter *pl*
medicinal plant Arzneipflanze *f*
medicinal substance Arzneistoff *m*
medicinal well Heilquelle *f*
medicine Medizin *f*, Heilkunde *f*, Heilmittel *n* (Med), Heilstoff *m*, Medikament *n*, **forensic** ~ forensische Medizin, gerichtliche Medizin, **physical** ~ physikalische Medizin
medicine bottle Arzneiflasche *f*
medicine capsule Arzneikapsel *f* (Pharm)
medicine glass Mediziinglas *n*
medicobotanical medizinisch-botanisch
medicochemical medizinisch-chemisch
medicolegal gerichtsmedizinisch
medinal Diäthylbarbitursäure *f*, Medinal *n* (HN)
mediocre mittelmäßig
mediocrity Mittelmäßigkeit *f*
medium Medium *n* (Phys, Chem), Mitte *f*, Mittel *n*, **complete** ~ Vollkultur *f* (Biol)
medium drawing Mittelzug *m*
medium fit Bewegungssitz *m* (Techn)
medium frequency Mittelfrequenz *f*
medium-heavy oil Mittelöl *n*
medium plate Mittelblech *n*
medium setting cement Mittelbinder *m*
medium size Mittelgröße *f*
medium wave range Mittelwellenbereich *m*
medium waves Mittelwellen *pl*

medrylamine Medrylamin *n*
medulla Mark *n*
medullary markartig
medullary substance Marksubstanz *f*
medullary tube Markröhrchen *n*
medullary tumor Knochenmarktumor *m* (Med)
meerschaum Meerschaum *m*, Sepiolith *m*
meet *v* [zusammen]treffen
meeting Tagung *f*
megabar Megabar *n*
megabromite Megabromit *m*
megacarpine Megacarpin *n*
megacycle Megahertz *n* (Elektr)
megaerg Megaerg *n* (Phys)
megafarad Megafarad *n* (Elektr)
megagamete Makrogamet *m* (Biol)
megaloblast Makroblast *m*
megalocyte Makrozyt *m*, Riesenblutkörperchen *n*
megalopolis Ballungszentrum *n*
megaphen Megaphen *n*
megaphone Lautsprecher *m*, Megaphon *n*
megavolt Megavolt *n* (Elektr)
megohm Megohm *n* (Elektr)
megohmmeter Megohmmeter *n*
meiosis Meiose *f* (Biol), Reduktionsteilung *f* (Biol)
melacacidin Melacacidin *n*
melaconite Melakonit *m* (Min), Schwarzkupfererz *n* (Min)
melam Melam *n*
melamazine Melamazin *n*
melamine Melamin *n*, Cyanurtriamid *n*, Tricyansäuretriamid *n*
melamine formaldehyde Melaminformaldehyd *m*
melamine [formaldehyde] resin Melamin[formaldehyd]harz *n*
melampyrine Dulcit *m*, Melampyrin *n*, Melampyrit *m*
melange Melange *f*, Mischung *f*
melange effect Melangeeffekt *m*
melange printing Vigoureuxdruck *m*
melaniline Melanilin *n*
melanine Melanin *n*
melanite Melanit *m* (Min), Eisengranat *m* (Min)
melanocarcinoma Melanokarzinom *n* (Med)
melanocerite Melanocerit *m* (Min)
melanochroite Melanochroit *m* (Min)
melanocytes Melanocyten *pl* (Pigmentzellen)
melanogallic acid Melangallussäure *f*
melanoidin Melanoidin *n*
melanolite Melanolith *m* (Min)
melanophlogite Melanophlogit *m* (Min)
melanothallite Melanothallit *m* (Min)
melanotropin Melanotropin *n* (Hormon)
melanterite Melanterit *m* (Min), Atramentstein *m* (Min)
melanureic acid Melanurensäure *f*

melaphyre Melaphyr *m* (Min)
melatonin Melatonin *n*
Meldola blue Meldolablau *n*
melem Melem *n*
melene Melen *n*
meletin Quercetin *n*
melezitose Melicitose *f*
melibiase Melibiase *f* (Biochem)
melibiitol Melibiit *m*
melibionic acid Melibionsäure *f*
melibiosazone Melibiosazon *n*
melibiose Melibiose *f*
melibiosone Melibioson *n*
melibiulose Melibiulose *f*
melicitose Melicitose *f*
melicopidine Melicopidin *n*
melicopine Melicopin *n*
melidoacetic acid Melidoessigsäure *f*
melilite Melilith *m* (Min)
melilot Steinklee *m* (Bot)
melilotic acid Melilotsäure *f*
melilotin Melilotin *n*
melilotol Melilotol *n*
melinite Melinit *m* (Min)
melinonine Melinonin *n*
melisimplexin Melisimplexin *n*
melisimplin Melisimplin *n*
melissa leaves Melissenblätter *pl*
melissa oil Melissenöl *n*
melissic acid Melissinsäure *f*
melissic alcohol Myricylalkohol *m*
melissone Melisson *n*
meliternin Meliternin *n*
melitose Melitose *f*, Raffinose *f*
melitriose Melitriose *f*, Raffinose *f*
melittin Melittin *n* (Peptid)
melizitose Melicitose *f*
mellassic acid Melassensäure *f*
mellein Mellein *n*
melleous honigähnlich
mellibiase Mellibiase *f*
mellimide Mellimid *n*
mellite Mellit *m* (Min), Honigstein *m* (Min),
  Honigtopas *m* (Min)
mellitene Melliten *n*
mellitic acid Benzolhexacarbonsäure *f*,
  Honigsäure *f*, Mellitsäure *f*
mellitol Mellitol *n*
mellon[e] Mellon *n*
mellophanic acid Mellophansäure *f*
mellow mürbe, reif, abgelagert, mild, weich
melonite Melonit *m* (Min), Tellurnickel *n* (Min)
melon oil Melonenöl *n*
melon seed oil Melonenöl *n*
Melotte fusible alloy Melotte-Metall *n*
melphalan Melphalan *n*
melt Schmelze *f*, Schmelzfluß *m*, crude ~
  Rohschmelze *f*, difficult to ~ schwer
  schmelzbar

melt *v* schmelzen, sich auflösen, verschmelzen,
  zerfließen, ~ completety garschmelzen (Met),
  ~ down einschmelzen, weichfeuern (Metall),
  zusammenschmelzen, ~ off abschmelzen, ~
  out ausschmelzen, ~ through durchschmelzen,
  ~ together zusammenschmelzen
meltability Schmelzbarkeit *f*
melt-back method Rückschmelzverfahren *n*
meltdown slag Einschmelzschlacke *f*
melted, to be ~ down aufschmelzen (Chem)
melted fat Schmalz *n*
melted snow or ice Schmelzwasser *n*
melt extrusion Schmelzspinnen *n*
melt-grown crystals Schmelzlinge *pl*
melt index Schmelzindex *m*
melting Schmelze *f*, Schmelzen *n*,
  Verschmelzung *f*, high ~ hochschmelzend,
  process of ~ Schmelzgang *m*, ~ down
  Einschmelzen *n*, ~ of slags
  Schlackenverschmelzung *f*, ~ under a white
  slag Schmelzen *n* ohne Oxydation
melting bath Schmelzbad *n*
melting centrifuge Schmelzschleuder *f*
melting charge Schmelzgut *n*
melting cone Brennkegel *m*, Schmelzkegel *m*,
  Segerkegel *m*
melting crucible Schmelztiegel *m*
melting down process
  Zusammenschmelzverfahren *n*
melting enthalpy Schmelzenthalpie *f*
melting flask Einschmelzkolben *m*
melting furnace Schmelzofen *m*
melting heat Schmelzwärme *f*
melting litharge Schichtglätte *f*
melting loss Abbrand *m*
melting operation Schmelzgang *m*, conduct of the
  ~ Schmelzführung *f*
melting out process Ausschmelzverfahren *n*
melting pan Klärpfanne *f*
melting plant Schmelzanlage *f* (Met)
melting point Schmelzpunkt *m*, Fusionspunkt *m*,
  Verflüssigungspunkt *m*, depression of ~
  Schmelzpunkterniedrigung *f*, having a high ~
  hochschmelzend, incongruent ~
  inkongruenter Schmelzpunkt, lowering of the
  ~ Schmelzpunkterniedrigung *f*, mixed ~
  Mischschmelzpunkt *m*
melting point apparatus
  Schmelzpunktbestimmungsapparat *m*
melting point capillary Schmelzpunktröhrchen *n*
melting point diagram Schmelzdiagramm *n*
melting point pressure curve
  Schmelzdruckkurve *f*
melting point tube Schmelzpunktröhrchen *n*
melting pot Schmelztiegel *m*, Gießtiegel *m*,
  Schmelzkessel *m*
melting pressure Schmelzdruck *m*
melting process Schmelzverfahren *n* (Met)

melting range Schmelzbereich *m*,
  Schmelzintervall *n*
melting rate Schmelzgeschwindigkeit *f*
melting stock Beschickungsgut *n*,
  Beschickungsmaterial *n*
melting stove Schmelzherd *m*
melting substance Schmelzkörper *m*
melting table Abschmelztisch *m*
melting temperature Schmelzhitze *f*,
  Schmelztemperatur *f*
melting zone Schmelzzone *f*
melt spinning Schmelzspinnen *n*
member Element n, Glied *n*, Mitglied *n*, Teil *m*
members, structural ~ strukturelle
  Bestandteile *pl*
membership Mitgliedschaft *f*
membrane Membran *f*, Häutchen *n*, **compound** ~
  mehrschichtige Membran, **elastic** ~ elastische
  Membran, **fibrous** ~ Faserhaut *f*, **intercellular**
  ~ interzelluläre Membran, **semipermeable** ~
  halbdurchlässige Membran, semipermeable
  Membran
membrane composition
  Membranzusammensetzung *f*
membrane diffusion Osmose *f*
membrane electrode Membranelektrode *f*
membrane lipid Membranlipid *n*
membrane permeability
  Membrandurchlässigkeit *f*
membrane potential Membranpotential *n*
membrane protein Membranprotein *n*
membranous filter Membranfilter *n*
memo Aktennotiz *f*
memorandum Aktennotiz *f*, Aktenvermerk *m*
memory drum Speichertrommel *f* (Elektron)
memory effect Speichereffekt *m* (extrusion),
  Strangaufweitung *f*
memory location Gedächtnisspeicher *m*
  (Rechenmaschine)
memory or spider line Markierungsstreifen *m*
menac[c]anite Menaccanit *m* (Min)
menachanite Menaccanit *m* (Min)
menadiol Menadiol *n*
menadione Menadion *n*, Vitamin K₃
menakanite Menaccanit *m* (Min)
menaphthylamine Menaphthylamin *n*
menaphthylbromide Menaphthylbromid *n*
mendelevium (Symb. Md o. Mv) Mendelevium *n*
Mendelian law Vererbungsgesetz, Mendelsches
  Gesetz *n*
Mendel's law Mendelsches Gesetz *n*,
  Vererbungsgesetz *n*
mendipite Mendipit *m* (Min)
mendozite Mendozit *m* (Min)
meneghinite Meneghinit *m* (Min)
mengite Mengit *m*
menhaden oil Menhadenöl *n*

menilite Menilit *m* (Min), Knollenopal *m*
  (Min), Leberopal *m* (Min)
meninges Hirnhaut *f* (Med)
meningitis Hirnhautentzündung *f* (Med)
meniscus Meniskus *m*
menisperine Menisperin *n*
menispermine Menispermin *n*
menstruum Arzneimittelträger *m*
mental geistig
menthacamphor Menthol *n*
menthane Menthan *n*
menthanol Menthanol *n*
menthanone Menthanon *n*
menthene Menthen *n*
menthenol Menthenol *n*
menthenone Menthenon *n*
menthofuran Menthofuran *n*
menthol Menthol *n*
mentholide Mentholid *n*
menthol salicylate Salimenthol *n*
menthol valerate Valeriansäurementholester *m*,
  Validol *n*
menthone Menthon *n*
menthonol Menthonol *n*
menthospirine Menthospirin *n*
menthyl Menthyl-
menthyl acetate Menthylacetat *n*
menthylamine Menthylamin *n*
menthylamine hydrochloride
  Menthylaminhydrochlorid *n*
menthyl ester Menthylester *m*
menthyl lactate Menthyllactat *n*
menthyl nitrobenzoate Menthylnitrobenzoat *n*
menthyl valerate Menthylvalerat *n*
mention *v* anführen, erwähnen
menyanthin Menyanthin *n*
menyanthol Menyanthol *n*
meperidine Meperidin *n*
mephenesin Mephenesin *n*
mephentermine Mephentermin *n*
mephenytoin Mephenytoin *n*
meprobamate Meprobamat *n*
meprylcaine Meprylcain *n*
mepyramine Mepyramin *n*
meralluride Merallurid *n*
meranzin Meranzin *n*
meratin Meratin *n*
merbromin Mercurochrom *n*
mercaptamine Mercaptamin *n*
mercaptan Mercaptan *n*, Thioalkohol *m*, Thiol *n*
mercaptide Mercaptid *n*
mercapto Sulfhydryl-
mercaptobenzimidazole Mercaptobenzimidazol *n*
mercaptobenzothiazole Mercaptobenzothiazol *n*
mercaptoethanolamine Mercaptoäthanolamin *n*
mercaptol Mercaptol *n*
mercaptomerin Mercaptomerin *n*
mercaptomethane Methylsulfhydrat *n*
mercaptopurine Mercaptopurin *n*

mercaptoquinoline Mercaptochinolin *n*
mercapturic acid Mercaptursäure *f*
mercerization Mercerisation *f*
(Baumwollveredelungsverf)
mercerization auxiliary Mercerisierhilfsmittel *n*
mercerization [process] Merzerisation *f,*
Merzerisierung *f*
mercerize *v* merzerisieren
mercerizer Merzerisieranlage *f*
mercerizing Merzerisieren *n*
mercerizing process Merzerisierverfahren *n*
merchandise Ware *f*
merchant bar Raffinierstahl *m*, Stabeisen *n*
merchant bar iron Handelseisen *n*
mercuderamide Mercuderamid *n*
mercurial Quecksilbermittel *n* (Pharm)
mercurial *a* merkurial (Med), quecksilberartig,
quecksilberhaltig
mercurial balsam Quecksilberbalsam *m*
mercurial gilding Quecksilbervergoldung *f*
mercurialism Quecksilbervergiftung *f*
mercurialization Quecksilberbehandlung *f,*
Quecksilbersättigung *f*
mercurial level Quecksilberstand *m*
mercurial nickel plating
Quecksilbervernickelung *f*
mercurial ointment Quecksilbersalbe *f* (Pharm)
mercurial plaster Quecksilberpflaster *n*
mercurial poisoning Quecksilbervergiftung *f*
(Med)
mercurial preparation Quecksilbermittel *n*
(Pharm); Quecksilberpräparat *n* (Pharm)
mercurial pressure Quecksilberdruck *m*
mercurial remedy Quecksilbermittel *n* (Pharm)
mercurial soap Quecksilberseife *f*
mercurial solution Quecksilberwasser *n*
mercurial soot Quecksilberruß *m*
mercurial thermometer
Quecksilberthermometer *n*
mercuric ammonium chloride
Quecksilberchloridamid *n*
mercuric benzoate Quecksilberbenzoat *n,*
Mercuribenzoat *n*
mercuric carbolate Quecksilberphenolat *n*
mercuric chloride Quecksilber(II)-chlorid *n,*
Mercurichlorid *n*, Sublimat *n*
mercuric chloride solution Sublimatlösung *f*
mercuric chloroiodide Quecksilberchlorojodid *n*
mercuric compound
Quecksilber(II)-Verbindung *f,*
Mercuriverbindung *f*
mercuric cyanide Mercuricyanid *n*
mercuric fulminate Knallquecksilber *n,*
Platzquecksilber *n* (obs),
Quecksilberfulminat *n*
mercuric iodide Quecksilber(II)-jodid *n,*
Mercurijodid *n*
mercuric lactate Mercurilactat *n*
mercuric nitrate Mercurinitrat *n*

mercuric oxide Quecksilber(II)-oxid *n,*
Mercurioxid *n*
mercuric oxycyanide Quecksilberoxycyanid *n,*
Mercurioxycyanid *n*
mercuric phenate Mercuriphenolat *n*
mercuric phenolate Mercuriphenolat *n*
mercuric phenoxide Mercuriphenolat *n*
mercuric phosphate Mercuriphosphat *n,*
Quecksilber(II)-phosphat *n*
mercuric porphin chelate
Quecksilberporphinchelat *n*
mercuric potassium cyanide
Kaliumquecksilber(II)-cyanid *n*
mercuric potassium iodide
Kaliumquecksilber(II)-jodid *n*
mercuric rhodanide Mercurirhodanid *n*
mercuric salicylate Mercurisalicylat *n,*
Quecksilbersalicylat *n*
mercuric succinimide Quecksilbersuccinimid *n*
mercuric sulfate Mercurisulfat *n,*
Quecksilber(II)-sulfat *n*
mercuric sulfate ethylene diamine Sublamin *n*
mercuric sulfide Quecksilber(II)-sulfid *n,*
Mercurisulfid *n*, **black** ~ Quecksilbermohr *m*
(Min), **red** ~ Zinnober *m* (Min)
mercuric thiocyanate Mercurirhodanid *n*
mercuricyanic acid
Mercuricyanwasserstoffsäure *f*
mercuriferous quecksilberführend,
quecksilberhaltig
mercurisaligenin Mercurisaligenin *n*
mercurize *v* merkurieren
mercurobutol Mercurobutol *n*
mercurochrome Mercurochrom *n*
mercurophen Mercurophen *n*
mercurous acetate Mercuroacetat *n,*
Quecksilber(I)-acetat *n*
mercurous chloride Quecksilber(I)-chlorid *n,*
Kalomel *n*, Mercurochlorid *n*, **native** ~
Hornquecksilber *n* (Min), Quecksilberspat *m*
(Min)
mercurous chromate Chromzinnober *m*
mercurous compound
Quecksilber(I)-Verbindung *f,*
Mercuroverbindung *f*
mercurous cyanide Quecksilber(I)-cyanid *n,*
Quecksilbercyanür *n* (obs)
mercurous iodide Quecksilber(I)-jodid *n,*
Mercurojodid *n*
mercurous iodobenzene-p-sulfonate Anogen *n*
mercurous nitrate Mercuronitrat *n,*
Quecksilber(I)-nitrat *n*
mercurous oxide Quecksilber(I)-oxid *n,*
Quecksilberoxydul *n* (obs)
mercurous phosphate Mercurophosphat *n*
mercurous potassium cyanide
Kalium-quecksilber(I)-cyanid *n*
mercurous salt Quecksilber(I)-Salz *n*
mercurous sulfate Mercurosulfat *n*

mercurous sulfide Quecksilber(I)-sulfid *n*
mercurous thiocyanate Quecksilber(I)-rhodanid *n*
mercury Quecksilber *n*, Merkur *m* (Astr),
   coating of ~ Quecksilberauflage *f,* colloidal ~
   kolloidales Quecksilber, containing ~
   quecksilberhaltig, fulminating ~
   Knallquecksilber *n*, Quecksilberfulminat *n*,
   native ~ Jungfernquecksilber *n*,
   Quecksilbergur *f,* sliming of ~ Verbutterung *f*
   des Quecksilbers
mercury alkyl Quecksilberalkyl *n*
mercury alloy Amalgam *n*,
   Quecksilberlegierung *f*
mercury amide chloride
   Quecksilberchloridamid *n*
mercury antimony sulfide
   Quecksilberantimonsulfid *n*
mercury arc Quecksilberlichtbogen *m*
mercury barometer Quecksilberbarometer *n*
mercury bath Quecksilberbad *n*
mercury bichloride Mercurichlorid *n*,
   Quecksilber(II)-chlorid *n*
mercury biiodide Mercurijodid *n*,
   Quecksilber(II)-jodid *n*
mercury bucket Schwarzbottich *m*
mercury bulb Quecksilberkapsel *f*
mercury circuit breaker
   Quecksilberkippschalter *m* (Elektr)
mercury column Quecksilberfaden *m*,
   Quecksilbersäule *f*
mercury compound Quecksilberverbindung *f*
mercury contact Quecksilberkontakt *m*
mercury content Quecksilbergehalt *m*
mercury diethyl Quecksilberdiäthyl *n*
mercury diiodide Quecksilber(II)-jodid *n*,
   Mercurijodid *n*
mercury dimethyl Quecksilberdimethyl *n*
mercury discharge tube, high-pressure ~
   Quecksilberhochdruckentladungsröhre *f*
mercury ejector pump
   Quecksilberdampfstrahl-Pumpe *f*
mercury electrode Quecksilber-Elektrode *f,*
   dropping ~ Quecksilbertropfelektrode *f*
mercury filling Quecksilberfüllung *f*
mercury fulminate knallsaures Quecksilber *n*,
   Quecksilberfulminat *n*
mercury furnace Quecksilberofen *m*
mercury glidin Quecksilberglidin *n*
mercury halide Halogenquecksilber *n*
mercury(I) chloride Quecksilber(I)-chlorid *n*,
   Kalomel *n*, Mercurochlorid *n*
mercury(I) compound
   Quecksilber(I)-Verbindung *f,*
   Mercuroverbindung *f*
mercury(II) chloride Quecksilber(II)-chlorid *n*,
   Mercurichlorid *n*, Sublimat *n*
mercury(II) compound Mercuriverbindung *f,*
   Quecksilber(II)-Verbindung *f*
mercury intensifier Quecksilberverstärker *m*

mercury lamp Quecksilberlampe *f*
mercury manometer Quecksilbermanometer *n*
mercury monochloride Quecksilber(I)-chlorid *n*,
   Kalomel *n*, Mercurochlorid *n*
mercury naphtholate Naphtholquecksilber *n*
mercury nitrate Quecksilbernitrat *n*
mercury ointment Quecksilbersalbe *f* (Pharm)
mercury ore Quecksilbererz *n* (Min)
mercury peptonate Quecksilberpeptonat *n*
mercury phenolate Phenolquecksilber *n*
mercury phenoxide Phenolquecksilber *n*
mercury phosphate Quecksilberphosphat *n*
mercury point lamp Quecksilberpunktlampe *f*
mercury poisoning Quecksilbervergiftung *f*
mercury p-phenyl thionate Hydrargyrol *n*
mercury precipitate Quecksilberniederschlag *m*
mercury pressure Quecksilberdruck *m*,
   Quecksilbersäule *f*
mercury pressure gauge Quecksilbermanometer *n*
mercury process Quecksilberverfahren *n*
mercury protochloride Mercurochlorid *n*
mercury pump Quecksilberpumpe *f*
mercury safety valve
   Quecksilber-Rückschlagventil *n*
mercury salicylarsenite Enesol *n*
mercury salicylate Quecksilbersalicylat *n*
mercury salt Quecksilbersalz *n*
mercury seal Quecksilberverschluß *m*
mercury selenide Selenquecksilber *n*
mercury solution Quecksilberlösung *f*
mercury spectrum, continuous ~
   Quecksilberkontinuum *n*
mercury succinimide Quecksilbersuccinimid *n*
mercury sulfate Quecksilbersulfat *n*
mercury sulfide Quecksilbersulfid *n*, black ~
   Mineralmohr *m* (Min), Mohr *m* (Min), red ~
   Zinnober *m* (Min)
mercury sulfide selenide, native ~ Onofrit *m*
   (Min)
mercury switch Quecksilber[kipp]schalter *m*
mercury tannate Quecksilbertannat *n*
mercury thiocyanate Quecksilberrhodanid *n*
mercury thread Quecksilberfaden *m*,
   Thermometerfaden *m*
mercury tongs Quecksilberzange *f*
mercury trap Quecksilberfalle *f*
mercury trough Quecksilberwanne *f*
mercury vapor Quecksilberdampf *m*
mercury vapor lamp Quecksilberdampflampe *f*,
   high pressure ~
   Quecksilberhochdrucklampe *f*
mercury vapor pump
   Quecksilberhochvakuumpumpe *f*
mercury [vapor] rectifier
   Quecksilberdampfgleichrichter *m*,
   Quecksilbergleichrichter *m* (Elektr)
mercury vapor stream Quecksilberdampfstrom *m*
mercury vessel Quecksilbergefäß *n*
mercury vitriol Quecksilbervitriol *n*

mergal Mergal *n*
merger Fusionierung *f*
merging point Durchdringungsstelle *f*
meridional flux Meridionaltransport *m*
merimine Merimin *n*
meriquinone Merichinon *n*
merocyanine Merocyanin *n*
merolignin Merolignin *n*
meromorphic meromorph (Math)
meroquinene Merochinen *n*
merosinigrin Merosinigrin *n*
merotropism Merotropie *f*
merotropy Merotropie *f*
Merrifield technique Merrifield-Technik *f*
(Festphasen-Peptidsynthese)
mersalyl Mersalyl *n*
mersol Mersol *n*
mersolate Mersolat *n*
merthiolate Merthiolat *n*
mesaconic acid Mesaconsäure *f*
mesantenic acid Mesantensäure *f*
mescaline Mescalin *n*
mescaline sulfate Mescalinsulfat *n*
mesembran Mesembran *n*
mesembrene Mesembren *n*
mesembrine Mesembrin *n*
mesembrol Mesembrol *n*
mesh Gewebe *n* (Web), Masche *f*
meshing Drahtgewebe *n*, Eingriffsverhältnisse *pl*
(Getriebe)
mesh screen Maschensieb *n*
mesh sieve Maschensieb *n*
mesh size Maschenweite *f*
mesidine Mesidin *n*
mesitene lactone Mesitenlacton *n*
mesitine spar Mesitin[spat] *m* (Min)
mesitite Mesitin[spat] *m* (Min)
mesitoic acid Mesitoesäure *f*
mesitol Mesitol *n*
mesitonic acid Mesitonsäure *f*
mesityl Mesityl-
mesitylaldehyde Mesit[yl]aldehyd *m*
mesitylene Mesitylen *n*
mesitylene alcohol Mesitol *n*
mesitylenic acid Mesitylensäure *f*
mesitylic acid Mesitylsäure *f*
mesitylinic acid Mesitylensäure *f*
mesitylol Mesitylen *n*
mesityl oxide Mesityloxid *n*
meso atom Mesoatom *n*
mesobilirubin Mesobilirubin *n*
mesochlorine Mesochlorin *n*
mesocolloid Mesokolloid *n* (Phys)
mesocorydaline Mesocorydalin *n*
mesoerythritol Mesoerythrit *m*
meso form Mesoform *f*
mesohem Mesohäm *n*
mesoinositol Mesoinosit *m*
mesoionic mesoionisch

mesolite Mesolith *m* (Min),
Kalknatronzeolith *m* (Min)
mesomeric mesomer
mesomeric effect Mesomerieeffekt *m*
mesomerism Mesomerie *f*
mesomorphous mesomorph
meson Meson *n*, Mesotron *n*
meson capture Mesoneneinfang *m*
meson decay Mesonenzerfall *m*
meson detection Mesonennachweis *m*
meson field Mesonenfeld *n*
mesonic atom Mesonenatom *n*
mesonic level Mesonterm *m*
mesonic term Mesonterm *m*
meson physics Mesonenphysik *f*
meson production Mesonenerzeugung *f*
meson-proton scattering Mesonprotonstreuung *f*
meson theory Mesonentheorie *f* (Atom)
meson track Mesonenbahn *f*
meson yield Mesonenausbeute *f*
mesophase Mesophase *f*
mesophilic mesophil
mesophylline Mesophyllin *n*
mesoporphyrine Mesoporphyrin *n*
mesoporphyrinogene Mesoporphyrinogen *n*
mesopyropheophorbide Mesopyrophäophorbid *n*
mesopyrrochlorine Mesopyrrochlorin *n*
mesorcin Mesorcin *n*
mesorcinol Mesorcin *n*
mesotan Mesotan *n*
mesotartaric acid Mesoweinsäure *f*
mesotartrate Mesotartrat *n*
mesothorium Mesothorium *n*
mesotron Meson *n*, Mesotron *n*
mesotron shower Mesotronenschauer *m*
mesotropism Mesotropie *f*
mesotype Mesotyp *m* (Min)
mesoxalaldehydic acid Mesoxalaldehydsäure *f*
mesoxalic acid Mesoxalsäure *f*
mesoxalyl Mesoxalyl-
mesoxalylurea Mesoxalylharnstoff *m*
mesoxophenin Mesoxophenin *n*
mesoyohimbine Mesoyohimbin *n*
Mesozoic Mesozoikum *n*
message Nachricht *f*
messelite Messelith *m* (Min)
messenger ribonucleic acid
Messenger-Ribonucleinsäure *f*
mesulfen Mesulfen *n*
mesylchloride Mesylchlorid *n*
meta acetaldehyde Metaldehyd *m*
meta acid Metasäure *f*
metabiosis Metabiose *f* (Biol)
metabisulfite Metabisulfit *n*
metabolic metabolisch
metabolic abnormality Stoffwechselanomalie *f*
metabolic disease Stoffwechselerkrankung *f*
metabolic inhibition Stoffwechselblockierung *f*

metabolic intermediate
Stoffwechselzwischenprodukt *n*
metabolic pathway Stoffwechselweg *m*
metabolic process Stoffwechselablauf *m*
metabolic rate Stoffwechselgröße *f*
metabolic regulation Stoffwechselregulation *f*
metabolism Metabolismus *m*, Stoffwechsel *m*,
   disordered ~ gestörter Stoffwechsel, **general**
   ~ Gesamtstoffwechsel *m*, **intermediary ~**
   intermediärer Stoffwechsel,
   Zwischenstoffwechsel *m*, **regenerative ~**
   Regenerationsstoffwechsel *m*, **reversible ~**
   umkehrbarer Stoffwechsel *m*, **total ~**
   Gesamtstoffwechsel *m*, **~ of minerals**
   Mineralstoffwechsel *m*, **~ of the tissues**
   Gewebestoffwechsel *m*
metabolite Metabolit *m*, Stoffwechselprodukt *n*
metaborate Metaborat *n*
metaboric acid Metaborsäure *f*
metabrushite Metabrushit *m* (Min)
metacarbonic acid Metakohlensäure *f*
metacenter Metazentrum *n*
metacentric metazentrisch
metacetone Metaceton *n*
metachloral Metachloral *n*
metachlorobenzoic acid Metachlorbenzoesäure *f*
metachromatism Metachromasie *f*
metachrome dye Metachromfarbe *f*
metachrome mordant Metachrombeize *f*
metachrome yellow Metachromgelb *n*
metacinnabarite Metacinnabarit *m* (Min),
   Metazinnober *m* (Min)
meta compound Metaverbindung *f*
metacrolein Metakrolein *n*
metacrystalline metakristallin
metacycline Metacyclin *n*
metacyclophane Metacyclophan *n*
meta-directing metadirigierend (Chem)
metaferric oxide Metaeisenoxid *n*
metaformaldehyde Metaformaldehyd *m*
metagallic acid Metagallussäure *f*
metagenesis Metagenese *f* (Biol)
metagenin Metagenin *n*
metahemipic acid Metahemipinsäure *f*
metal Metall *n*, **added ~** Zusatzmetall *n*, **base ~**
   unedles Metall, **crude ~** Rohmetall *n*,
   deposition of ~ Metallabscheidung *f*, **foreign**
   ~ Fremdmetall *n*, **ladling out the ~**
   Ausschöpfen *n* des Metalls, **light [weight] ~**
   Leichtmetall *n*, **like ~** metallähnlich,
   metallartig, **molten ~** Metallschmelze *f*, **native**
   ~ Jungfernmetall *n*, **noble ~** edles Metall,
   poor in ~ metallarm, **powder ~**
   Sintermetall *n*, **precious ~** edles Metall, **sheet**
   of ~ Metallblatt *n*, **soft ~** Weichmetall *n*, **to**
   scour ~ Metall abbeizen, **white ~**
   Weißmetall *n*
metal acetylide Metallacetylid *n*, Metallcarbid *n*
metal alkyl Metallalkyl *n*

metal alloy Metallegierung *f*
metal amide Metallamid *n*
metal ammonia compound
   Metallammoniakverbindung *f*
metalation Metallierung *f* (Chem)
metal azide Metallazid *n*
metal bar Metallstange *f*
metal bonding Metallkleben *n*
metal-braided metallumsponnen
metal buffer Metallpuffer *m*
metal button Metallkönig *m*
metal carbide Metallkarbid *n*
metal carbonyl Metallcarbonyl *n*
metal carboxide Metallcarbonyl *n*
metal casing Blechmantel *m*, Metallgehäuse *n*
metal casting Gußstück *n*
metal chelate Metallchelat *n*
metal chips Metallabfall *m*, Metallspäne *pl*
metal clamp Metallklemme *f*
metal-coat *v* metallisieren
metal coating Metallbelag *m*, Metallisieren *n*,
   Metallüberzug *m*
metal-coating of plastics
   Kunststoffmetallisierung *f*
metal-colored metallfarbig
metal coloring Metallochromie *f*
metal composition Mischmetall *n*
metal conditioner Reaktionsprimer *m*
metal content[s] Metallgehalt *m*
metal cutting spanabhebende
   Metallbearbeitung *f*
metal-cutting *a* spanabhebend
metal-cutting material Schneidmetall *n*
metal-cutting pliers Metallschere *f*
metal-cutting saw Metallsäge *f*
metal degassing Metallentgasung *f*
metal degreasing Metallentfettung *f*
metaldehyde Metaldehyd *m*
metal deposit Metallniederschlag *m*
metal dust Metallstaub *m*
metal effect threads Metalleffektfäden *pl*
metal electrode Metallelektrode *f*
metal exchange rate
   Metallaustauschgeschwindigkeit *f*
metal fatigue testing
   Dauerschwingfestigkeitsversuch *m*
metal filament Metallfaden *m*
metal filament lamp Metallfadenlampe *f*
metal filings Metallspäne *pl*, Metallkrätze *f*
metal film Metallschicht *f*
metal finishing lathe Metalldrehbank *f*
metal flange Metallflansch *m*
metal foil Blattmetall *n*, Metallfolie *f*, **coating of**
   ~ Metallfolienüberzug *m*
metal foundry Metallgießerei *f*
metal fragment Metallsplitter *m*, Metallstück *n*
metal grain Metallkorn *n*
metal guide Leitblech *n* (Techn)
metal halide Halogenmetall *n*

metal hydroxide Metallhydroxid *n*
metal impurity Begleitmetall *n*
metal ion Metallion *n*
metal lacquer Metallack *m*
metallammine Metallammoniakverbindung *f*
metallic metallisch, metallähnlich, metallartig,
  metallen
metallic antimony Antimonmetall *n*, metallisches
  Antimon *n*
metallic arsenic Arsenmetall *n*, metallisches
  Arsen *n*
metallic ashes Metallasche *f*
metallic azide Metallazid *n*
metallic bath Metallbad *n*
metallic bismuth Wismutmetall *n*
metallic brush Metallpinsel *m*
metallic calx Metallkalk *m*
metallic carbide Metallkarbid *n*
metal[lic] color Metallfarbe *f*
metallic coloring Metallfärbung *f*
metallic composition Metallmischung *f*
metal[lic] compound Metallverbindung *f* (Chem)
metallic content[s] Metallgehalt *m*
metal[lic] cyanide Metallcyanid *n*
metallic dust Flugstaub *m*
metallic foil Metallfolie *f*
metallic hydride Metallhydrid *n*
metallicity metallische Eigenschaft *f*,
  Metallartigkeit *f*
metallic luster Metallglanz *m*
metallic mirror Metallspiegel *m*
metallic mixture Metallgemisch *n*
metallic molybdenum Molybdänmetall *n*
metallic ore Metallerz *n*
metallic oxide Metalloxid *n*
metallic oxide cathode Metalloxidkathode *f*
metallic packing Metalldichtung *f*
metallic paper Metallpapier *n*
metallic pigment Metallfarbkörper *m*,
  Metallpigment *n*
metallic point Metallspitze *f*
metallic precipitate Metallniederschlag *m*
metallic printing Bronzedruck *m* (Buchdr)
metallic property Metalleigenschaft *f*
metallic regulus Metallkönig *m*
metallic sand Metallsand *m*
metallic screen Metallschirm *m*
metallic soap Metallseife *f*
metallic sodium metallisches Natrium *n*,
  Natriummetall *n*
metallic solution Metallbad *n*
metallic sounding metallisch klingend
metallic sulfide Metallsulfid *n*
metal[lic] thermometer Metallthermometer *n*
metallic thread Metallfaser *f*
metallic titanium Titanmetall *n*
metallic uranium Uranmetall *n*
metalliferous erzhaltig, metallführend,
  metallhaltig

metalliform metallähnlich, metallförmig
metallin Metallin *n*
metalline metallähnlich, metallartig, metallen
metal[l]ization Metallimprägnierung *f*,
  Metallisierung *f*, Metallspritzverfahren *n*,
  outer ~ Außenmetallisierung *f*, **~ by burning
  in** Einbrennmetallisierung *f*
metal[l]ize *v* metallisieren; mit Metallsalzen
  imprägnieren
metal[l]izing Metallisierung *f*,
  Metallbedampfung *f*
metallochemistry Metallchemie *f*
metallochromy Metallfärbung *f*,
  Metallochromie *f*
metallographic metallographisch
metallographic examination
  Metalluntersuchung *f*
metallography Metallographie *f*, Metallkunde *f*
metalloid Metalloid *n*, Nichtmetall *n*
metalloid *a* metallartig, metalloid
metallo-organic metallorganisch
metallo-organic pigment halbmineralisches
  Pigment *n*
metalloplastic galvanoplastisch
metalloporphine Metallporphin *n*
metalloprotein Metallprotein *n*
metallurgic[al] hüttenmännisch, metallurgisch
metallurgical coke Hüttenkoks *m*
metallurgical dust Metallhüttenstaub *m*
metallurgical engineering Hüttentechnik *f*
metallurgical operations Verhüttung *f* (Metall)
metallurgical plant Hüttenwerk *n*
metallurgical smoke Hüttenrauch *m*
metallurgist Metallurg[e] *m*
metallurgy Metallurgie *f*, Hüttenkunde *f*,
  Hüttenwesen *n*, Metallgewinnung *f*,
  Metallkunde *f* (Erzscheidekunde)
metal matrix Metallmatrize *f*
metal mercaptan Mercaptid *n*
metal mirror Metallspiegel *m*
metal mist Metallnebel *m*
metal mixture Metallmischung *f*
metal ore Metallerz *n*
metal oxide Metalloxid *n*
metal pick-up Metallaufnahme *f* (v.
  Lebensmitteln)
metal pipe Metallrohr *n*
metal-plate *v* metallisieren
metal-plating Metallauflage *f*, Metallüberzug *m*
metal primer Metallgrundierung *f*
metal printing Argentindruck *m*
metal processing Metallbearbeitung *f*
metal production Metallgewinnung *f*
metal pyrometer Metallpyrometer *n*
metal reinforcement Metallarmierung *f*
metal residue Metallrückstand *m*
metal ring Metallring *m*
metal saw Eisensäge *f*
metal scrap Bruchmetall *n*, Schrott *m*

metal single crystal Metalleinkristall *m*
metal slag Metallschlacke *f*
metal splinter Metallsplitter *m*
metal spraying Metallspritzverfahren *n*,
Spritzmetallisieren *n*
metal spray method Metallspritzmethode *f*
metal-spun metallumsponnen
metal strapping Bandeisen *n* (Verpackung)
metal-substrate complex
Metall-Substrat-Komplex *m* (Biochem)
metal sulfide Metallsulfid *n*
metal-to-glass seal Metall-Glas-Kitt *m*
metal vapor Metalldampf *m*
metal vapor lamp Metalldampflampe *f*
metal waste Metallkrätze *f*
metal wire Metalldraht *m*
metal worker Metallarbeiter *m*
metal-working eisenverarbeitend, [spanlose]
Metallbearbeitung *f*
metal-working *a* metallverarbeitend
metameric metamer
metamerism Metamerie *f*
metamery Metamerie *f*
metamorphic metamorph[isch]
metamorphism Metamorphismus *m*,
Metamorphose *f*
metamorphose *v* metamorphosieren (Biol), sich
verwandeln (Biol)
metamorphosis Metamorphose *f* (Biol),
Gestaltsveränderung *f* (Biol), Verwandlung *f*
(Biol)
metamyelocyte Metamyelozyt *m*
metanethol Metanethol *n*
metanicotine Metanicotin *n*
metanilic acid Metanilsäure *f*
metaniline yellow Metanilgelb *n*
metanil yellow Metanilgelb *n*
metapeptone Metapepton *n*
metaphase Metaphase *f* (Biol)
metaphen Metaphen *n*
metaphenylene blue Metaphenylenblau *n*
metaphosphate Metaphosphat *n*
metaphosphinic acid Metaphosphinsäure *f*
metaphosphoric acid Metaphosphorsäure *f*
metapilocarpine Metapilocarpin *n*
metapiptone Metapipton *n*
metaplasm Metaplasma *n* (Biol)
meta position Metastellung *f* (Chem)
metapyroracemic acid Metabrenztraubensäure *f*
metaraminol Metaraminol *n*
metasaccharic acid Metazuckersäure *f*
metasaccharin Metasaccharin *n*
metasaccharonic acid Metasaccharonsäure *f*
metasaccharopentose Metasaccharopentose *f*
metasantonin Metasantonin *n*
metasilicate Metasilikat *n*
metasilicic acid Metakieselsäure *f*
metasolvan Metasolvan *n*
metasomatic metasomatisch

metastability Metastabilität *f*
metastable metastabil, halbbeständig
metastannic acid Metazinnsäure *f*
metastasis Metastase *f* (Med)
metastasis of cancer Krebsmetastase *f* (Med)
metastyrene Metastyrol *n*
metastyrolene Metastyrol *n*
metatartaric acid Metaweinsäure *f*
metatitanic acid Metatitansäure *f*
metatorbernite Metatorbernit *m* (Min)
metatungstic acid Metawolframsäure *f*
metavanadate Metavanadat *n*
metavanadic acid Metavanadinsäure *f*, **salt of** ~
Metavanadat *n*
metaxite Metaxit *m* (Min)
metaxylene Metaxylol *n*
metaxylidine Metaxylidin *n*
metazeunerite Metazeunerit *m* (Min)
metazirconic acid Metazirkonsäure *f*
meteloidine Meteloidin *n*
meteor Meteor *m* (Astr), Sternschnuppe *f* (Astr)
meteoric meteorisch (Astr)
meteoric iron Meteoreisen *n*
meteoric light Meteorlicht *n*
meteoric stone Meteorstein *m*
meteorite Meteorit *m*
meteor-like meteorähnlich
meteorograph Meteorograph *m*
meteorographic meteorographisch
meteorography Meteorographie *f*
meteorologic meteorologisch
meteorological map Wetterkarte *f*
meteorological observation Wetterbeobachtung *f*
meteorology Meteorologie *f*, Wetterkunde *f*,
Witterungskunde *f*
meteorometer Meteorometer *n*
meteor trail Meteorschweif *m* (Astr)
meter Meßinstrument *n*, Meßuhr *f*, Zähler *m*,
Zählwerk *n* (Am. E.), Meter *m n*
meter board Zählertafel *f* (Elektr)
meter-candle Lux *n* (Maß)
metering Messen *n*
metering clockwork Meßuhr *f*
metering device Dosiereinrichtung *f*
metering equipment Meßausrüstung *f*
metering pump Dosierpumpe *f*
metering tank Dosiertank *m*
meter reading Zählerablesung *f*
methacetin Methacetin *n*, Acetanisidin *n*
methacholine Methacholin *n*
methacrolein Methacrolein *n*
methacrylate Methacrylat *n*,
Methacrylsäureester *m*
methacrylic acid Methacrylsäure *f*
methacrylic ester Methacrylsäureester *m*
methacrylic methylester
Methacrylsäuremethylester *m*
methadone Methadon *n*
methamphetamine Methamphetamin *n*

methamphetamine hydrochloride
  Methamphetaminhydrochlorid *n*
methanal Formaldehyd *m*
methanamide Formamid *n*
methanation Methanisierung *f*
methane Methan *n*, Grubengas *n*, Sumpfgas *n*
methane dicarboxylic acid Malonsäure *f*
methane phosphonic acid Methylphosphonsäure *f*
methane siliconic acid Methansiliconsäure *f*
methane sulfonic acid Methansulfonsäure *f*
methane thiol Methylmerkaptan *n*
methanoic acid (obs) Ameisensäure *f*
methanol Methanol *n*, Holzgeist *m*,
  Methylalkohol *m*
methanolic methylalkoholisch, methanolisch
methanol plant Methanolanlage *f*
methanometer Grubengasanzeiger *m*,
  Grubengasmeßgerät *n*, Methanometer *n*
met[h]azonic acid Methazonsäure *f*
methemoglobin Methämoglobin *n*
methene Methylen *n*, Methen *n*
methene disulfonic acid Methionsäure *f*
methenyl Methin *n*
methine Methin *n*
methine group Methingruppe *f*
methiodal Methiodal *n*
methionic acid Methionsäure *f*
methionine Methionin *n*
methionol Methionol *n*
method Methode *f*, System *n*, Verfahren *n*,
  Verfahrensweise *f*, **indirect ~** indirektes
  Verfahren, **~ of least squares** Methode *f* der
  kleinsten Quadrate (pl), **~ of operation**
  Betriebsweise *f*, **~ of preparation**
  Herstellungsverfahren *n*
methodical methodisch, planmäßig, systematisch
method of counting Zählverfahren *n*
method of least squares Methode der kleinsten
  Fehlerquadrate *pl*
methodology Methodik *f*
methone Methon *n*
methoxide Methylat *n*
methoxyacetic acid Methoxyessigsäure *f*
methoxybenzene Anisol *n*
methoxy coumarin, 7- ~ Herniarin *n*
methoxy derivative Methoxyderivat *n*
methoxylation Methoxylierung *f*
methoxy[l] determination Methoxylbestimmung *f*
methoxy[l] group Methoxy[l]gruppe *f*
methoxylmethyl salicylate Mesotan *n*
methoxymethane Dimethyläther *m*,
  Methyläther *m*
methoxyphedrine Methoxyphedrin *n*
methoxyphenamine Methoxyphenamin *n*
methoxypropionic acid Methoxypropionsäure *f*
methronic acid Methronsäure *f*
methronol Methronol *n*
methyl Methyl-
methylacetanilide Methylacetanilid *n*, Exalgin *n*

methyl acetate Methylacetat *n*
methyl acrylate Methylacrylat *n*,
  Acrylmethylester *m*
methyladnamine Methyladnamin *n*
methylal Methylal *n*
methyl alcohol Methylalkohol *m*, Holzgeist *m*,
  Methanol *n*
methyl alcoholic methylalkoholisch,
  methanolisch
methylamine Methylamin *n*
methylamine hydrochloride
  Methylaminhydrochlorid *n*
methylaminoacetic acid Sarkosin *n*
methylaniline Methylanilin *n*
methyl anthranilate Methylanthranilat *n*
methylanthraquinone Methylanthrachinon *n*
methylarsonic acid Methylarsonsäure *f*
methylate Methylat *n*
methylate *v* methylieren
methylated methyliert, **capable of being ~**
  methylierbar
methylated spirit denaturierter Methylalkohol *m*
methylation Methylierung *f*
methylatropine nitrate Eumydrin *n* (HN)
methyl benzanthracene Methylbenzanthracen *n*
methylbenzene Toluol *n*
methyl benzoate Methylbenzoat *n*,
  Benzoesäuremethylester *m*
methylbenzoic acid Toluylsäure *f*
methyl benzophenanthrene
  Methylbenzphenanthren *n*
methyl benzoylacetone Methylbenzoylaceton *n*
methyl benzyl chloride Methylbenzylchlorid *n*
methyl benzyl glyoxime Methylbenzylglyoxim *n*
methyl benzyl ketone Methylbenzylketon *n*
methyl butanal Methylbutanal *n*
methyl butane Methylbutan *n*
methyl butanol Methylbutanol *n*
methyl butene dioic acid Citraconsäure *f*
methyl butene dioic anhydride
  Citraconsäureanhydrid *n*
methyl butenolide Methylbutenolid *n*
methyl butylacetic acid Methylbutylessigsäure *f*
methyl butyl ether Methylbutyläther *m*
methyl butyl ketone Methylbutylketon *n*
methyl camphorate Camphersäuremethylester *m*
methyl caoutchouc Methylkautschuk *m*
methyl caprate Caprinsäuremethylester *m*
methyl caproate Capronsäuremethylester *m*
methyl carbamate Methylurethan *n*, Urethylan *n*
methyl carbazole Methylcarbazol *n*
methyl carbitol Methylcarbitol *n*
methyl carbylamine Methylcarbylamin *n*
methyl cellosolve Methylcellosolve *f*
methyl cellulose Methylcellulose *f*
methyl chloride Methylchlorid *n*
methyl chloroacetate Methylchloracetat *n*
methyl chlorosilane Methylchlorsilan *n*
methyl cholanthrene Methylcholanthren *n*

methyl cinnamate Methylcinnamat *n*
methyl compound Methylverbindung *f*
methyl coniine Methylconiin *n*
methyl cresylate Kresylmethyläther *m*
methyl crotonic acid Methylcrotonsäure *f*
methyl cyanide Acetonitril *n*, Methylcyanid *n*
methyl cyanoacetate Methylcyanacetat *n*
methyl cyclohexane Methylcyclohexan *n*
methyl cyclohexanol Methylcyclohexanol *n*,
   Methylhexalin *n*
methyl-cyclopentadecanone, 3- ~ Muskon *n*
methyl cytosine Methylcytosin *n*
methyl decalone Methyldecalon *n*
methyl diethylmethane Methyldiäthylmethan *n*
methyl diiodosalicylate Sanoform *n*
methylene Methylen *n*, Methen *n*
methylene aminoacetonitrile
   Methylenaminoacetonitril *n*
methylene blue Methylenblau *n*
methylene butanedioic acid Itaconsäure *f*
methylene chloride Methylenchlorid *n*,
   Dichlormethan *n*
methylene digallic acid Methylendigallussäure *f*
methylene diiodide Methylenjodid *n*
methylene disulfonic acid Methionsäure *f*
methylene ditannin Methylenditannin *n*,
   Tannoform *n*
methylene glycol Methylenglykol *n*
methylene group Methylengruppe *f*
methylene guaiacol Methylenguajakol *n*
methylene imine Methylenimin *n*
methylene iodide Methylenjodid *n*
methylene oxide Formaldehyd *m*
methylene protocatechuic acid Piperonylsäure *f*
methylene saccharic acid Methylenzuckersäure *f*
methylene succinic acid Itaconsäure *f*
methylene sulfate Methylensulfat *n*
methylene violet Methylenviolett *n*
methylenimine Methylenimin *n*
methyl eosin Methyleosin *n*
methyl ester Methylester *m*
methyl ether Dimethyläther *m*, Holzäther *m*
   (obs), Methyläther *m*
methylethylacetylene Valerylen *n*
methyl ethyl ether Methyläthyläther *m*
methyl ethyl ketone Methyläthylketon *n*
methyl ethyl pyrrole Methyläthylpyrrol *n*
methyl formanilide Methylformanilid *n*
methyl fructopyranoside
   Methylfructopyranosid *n*
methylfumaric acid Mesaconsäure *f*
methylfuran, 2- ~ Silvan *n*
methyl gallate Gallicin *n*,
   Gallussäuremethylester *m*
methyl gentiobiose Methylgentiobiose *f*
methyl glucamine Methylglucamin *n*
methyl glucoside Methylglucosid *n*
methyl glycine Sarkosin *n*
methyl glycol Methylglykol *n*

methyl glycoside Methylglykosid *n*
methyl glyoxal Methylglyoxal *n*
methyl group Methylgruppe *f*
methyl group acceptor
   Methylgruppenakzeptor *m*
methyl group donor Methylgruppendonor *m*
methyl group transfer
   Methylgruppenübertragung *f*
methyl guanidine Methylguanidin *n*
methylguanidoacetic acid Kreatin *n*
methyl halide Methylhalogenid *n*
methyl hemin Methylhämin *n*
methyl heptyl ketone Methylheptylketon *n*
methyl hexalin Methylhexalin *n*
methyl hexane Heptan *n*
methyl hydride Methan *n*
methyl hydrocupreine Hydrochinin *n*
methyl hydroquinone Methylhydrochinon *n*
methyl hydroxyacetophenone
   Methylhydroxyacetophenon *n*
methylindole Skatol *n*
methyl inositol Mytilit *m*
methyl iodide Methyljodid *n*
methyl isobutyl ether Methylisobutyläther *m*
methyl isobutyl ketone Methylisobutylketon *n*
methylisophthalic acid, 4- ~ Xylidinsäure *f*
methyl isopropylbenzene
   Methylisopropylbenzol *n*
methyl isopropyl ketone Methylisopropylketon *n*
methyl isopropylnaphthalene
   Methylisopropylnaphthalin *n*
methyl isopropylphenanthrene
   Methylisopropylphenanthren *n*
methyl isopropylphenol Methylisopropylphenol *n*
methyl isoquinoline Methylisochinolin *n*
methyl isothiocyanate Methylsenföl *n*
methyljuglone Plumbagin *n*
methylketol Methylketol *n*
methylmaleic acid Citraconsäure *f*
methylmaleic anhydride
   Citraconsäureanhydrid *n*
methyl malonate Methylmalonat *n*
methylmalonic acid Methylmalonsäure *f*
methyl mannopyranoside
   Methylmannopyranosid *n*
methyl methacrylate
   Methacrylsäuremethylester *m*,
   Methylmethacrylat *n*
methyl methoxy[l]-benzoquinone
   Methylmethoxybenzochinon *n*
methyl morphimethine Methylmorphimethin *n*
methylmorphine Codein *n*
methyl mustard oil Methylsenföl *n*
methyl naphthalene Methylnaphthalin *n*
methyl naphthalene sulfonic acid
   Methylnaphthalinsulfonsäure *f*
methyl naphthoquinone Methylnaphthochinon *n*,
   2- ~ Menadion (Vitamin K₃)
methyl naphthyl ketone Methylnaphthylketon *n*

methyl nitrobenzoate Methylnitrobenzoat *n*
methyl nonyl ketone Methylnonylketon *n*
methyl number Methylzahl *f*
methylol melamine Methylolmelamin *n*
methylol methylene urea
Methylolmethylenharnstoff *m*
methyl orange Methylorange *n*
methyl oxalate Methyloxalat *n*
methyl oxine Methyloxin *n*
methyloxytetrahydroquinoline Kairin *n*
methyl pentanal Methylpentanal *n*
methyl pentane Methylpentan *n*
methyl pentanol Methylpentanol *n*
methyl pentene Methylpenten *n*
methyl pentynol Methylpentynol *n*
methyl phenanthrene Methylphenanthren *n*
methyl phenanthrene quinone
Methylphenanthrenchinon *n*
methyl phenethylamine sulfate
Methylphenäthylaminsulfat *n*
methylphenobarbital Methylphenobarbital *n*
methyl phenol Kresol *n*
methyl phenyl ether Anisol *n*
methyl phenyl hydrazine
Methylphenylhydrazin *n*
methyl phenyl ketone Methylphenylketon *n*
methyl phenyl nitrosamine
Methylphenylnitrosamin *n*
methyl phenyl silicone Methylphenylsilicon *n*
methyl phosphine Methylphosphin *n*
methyl phytylnaphthoquinone
Methylphytylnaphthochinon *n*
methyl picramide Methylpikramid *n*
methylpiperidine Pipecolin *n*
methyl propyl ketone Methylpropylketon *n*
methylpsychotrine Methylpsychotrin *n*
methylpurpuroxanthin, 4- ~ Rubiadin *n*
methyl pyrazolecarboxylic acid
Methylpyrazolcarbonsäure *f*
methylpyridine Picolin *n*, Methylpyridin *n*
methyl pyridinium hydroxide
Methylpyridiniumhydroxid *n*
methyl pyridone Methylpyridon *n*
methyl pyrrole Methylpyrrol *n*
methyl quinizarin Methylchinizarin *n*
methyl quinoline Methylchinolin *n*, 4- ~
Lepidin *n*
methyl radical Methylradikal *n*
methyl red Methylrot *n*
methylreductinic acid Methylreductinsäure *f*
methylreductone Methylredukton *n*
methylresorcinol, 4- ~ Kresorcin *n*, 5- ~
Orcin *n*
methylrosaniline Methylrosanilin *n*
methyl salicylate Gaultherolin *n*,
Methylsalicylat *n*
methylsalicylic acid Kresotinsäure *f*,
Methylsalicylsäure *f*
methyl silane Methylsilan *n*

methyl silicate Methylsilicat *n*
methyl silicone Methylsilicon *n*
methyl silicone oil Methylsiliconöl *n*
methyl silicone resin Methylsiliconharz *n*
methyl silicone rubber Methylsiliconkautschuk *m*
methylstearic acid Methylstearinsäure *f*
methylsuccinic acid Brenzweinsäure *f*
methyl sulfate Dimethylsulfat *n*
methyl sulfide Dimethylsulfid *n*
methyl sulfoxide Dimethylsulfoxid *n*
methyl tannin Methylotannin *n*
methyl tartrate Methyltartrat *n*
methyl tetralone Methyltetralon *n*
methyl thiazole carboxylic acid
Methylthiazolcarbonsäure *f*
methyl thienyl ketone Methylthienylketon *n*
methyl thiocyanate Methylrhodanid *n*
methylthiouracil Methylthiouracil *n*
methylthymol blue Methylthymolblau *n*
methyl toluene sulfonate Methyltoluolsulfonat *n*
methyl trichlorosilane Methyltrichlorsilan *n*
methyl trihydroxyanthraquinone
Methyltrihydroxyanthrachinon *n*
methyl trinitrophenylnitramine
Methyltrinitrophenylnitramin *n*
methyl umbelliferone Herniarin *n*,
Methylumbelliferon *n*
methyl undecyl ketone Methylundecylketon *n*
methyl uracil Methyluracil *n*, 5- ~ Thymin *n*
methyl urethane Methylurethan *n*, Urethylan *n*
methyl vinyl ketone Methylvinylketon *n*
methyl violet Methylviolett *n*
methylxanthogenic acid Methylxanthogensäure *f*
methyl yohimbate Yohimbin *n*
methymycin Methymycin *n*
methysticine Methysticin *n*
methysticol Methysticol *n*
meticillin Meticillin *n* (Antibiotikum)
metiram Metiram *n* (Fungizid)
metisazone Metisazon *n*
metixene Metixen *n* (Pharm)
metobromurone Metobromuron *n* (Herbizid)
metofenazate Metofenazat *n* (Med)
metol Metol *n* (HN)
metoleic acid Metoleinsäure *f*
metonal Metonal *n*
metoquinone Metochinon *n*
metoxazine Metoxazin *n*
metozine Metozin *n*
metre (Br. E.) Meter *m n*
metric metrisch
metric system Dezimalsystem *n*
metric thread metrisches Gewinde *n*
metriol Metriol *n*
mevaldic acid Mevaldsäure *f*
mevalonic acid Mevalonsäure *f*
mevalonic lactone Mevalonolacton *n*
meyerhofferite Meyerhofferit *m* (Min)
meymacite Meymacit *m* (Min)

mezcaline Mezcalin *n*
mezereon bark Seidelbastrinde *f* (Bot)
miargyrite Miargyrit *m* (Min),
Silberantimonglanz *m* (Min)
miascite Miascit *m*
miasm[a] Miasma *n* (Med)
miazine Miazin *n*
mica Glimmer *m* (Min), Fraueneis *n* (Min),
Jungfernglas *n,* Marienglas *n* (Min), Mika *f*
m (Min), **black** ~ Magnesiaglimmer *m,*
schwarzer Glimmer *m,* **colorless** ~
Katzensilber *n* (Min), **common** ~
Silberglimmer *m* (Min), **containing** ~
glimmerhaltig, **exfoliated** ~ Blähglimmer *m*
(Min), **potash** ~ weißer Glimmer (Min),
pressed ~ Preßglimmer *m,* **striated** ~
Strahlenglimmer *m* (Min), **wound on** ~ auf
Glimmer aufgewickelt, **yellow** ~
Goldglimmer *m* (Min), Katzenglas *n* (Min),
Katzengold *n* (Min)
micaceous glimmerartig, glimmerhaltig
micaceous cerussite Bleiglimmer *m* (Min)
micaceous clay Glimmerton *m* (Min)
micaceous copper Kupferglimmer *m* (Min)
micaceous iron ore Eisenglimmer *m* (Min)
micaceous lamina Glimmerblättchen *n*
mica condenser Glimmerkondensator *m*
mica gluing machine Glimmerklebemaschine *f*
micanite Mikanit *n*
micanite linen Mikanitleinwand *f*
micanite paper Mikanitpapier *n*
mica ozone tube Ozonglimmerröhre *f*
mica plate Glimmerplatte *f*
mica schist Glimmerschiefer *m* (Min)
mica sheet Glimmerscheibe *f*
mica slate Glimmerschiefer *m* (Min), Phyllit *m*
(Min)
mica spectacles Glimmerbrille *f*
micell[e] Micelle *f,* Mizelle *f*
micelle formation Micellenbildung *f*
michaelsonite Michaelsonit *m* (Min)
Michael's reaction Michaelreaktion *f*
Michler's base Michlers Base *f,* Tetrabase *f*
Michler's ketone Michlers Keton *n*
micranthine Micranthin *n*
micro-amperemeter Mikroamperemeter *n*
microanalysis Mikroanalyse *f* (Anal)
microanalysis balance Mikroanalysenwaage *f*
microapparatus Mikroapparat *m*
microautoradiography Mikroautoradiographie *f*
microbalance Mikrowaage *f*
microbar Mikrobar *n*
microbarograph Mikrobarograph *m*
microbe Kleinlebewesen *n,* Mikrobe *f*
microbeam of X-rays Mikroröntgenstrahl *m*
microbeam technique Feinstrahlmethode *f*
microbe culture Mikrobenzüchtung *f*
microbial mikrobisch
microbic durch Mikroben verursacht

microbicidal antibiotisch
microbicide Konservierungsmittel *n*
microbin Mikrobin *n*
microbiological mikrobiologisch
microbiologist Mikrobiologe *m*
microbiology Mikrobiologie *f*
microbioscope Bakterienmikroskop *n*
microburet[te] Mikrobürette *f*
microburner Mikrobrenner *m*
microchemical mikrochemisch
microchemistry Mikrochemie *f*
microchronometer Mikrochronometer *n*
microcidin Mikrocidin *n*
microcline Mikroklin *m* (Min), gemeiner
Feldspat (Min)
microcosm Mikrokosmos *m*
microcosmic mikrokosmisch
microcosmic salt Natriumammoniumphosphat *n*
microcosmic salt bead Phosphorsalzperle *f*
microcrack Mikroriß *m*
microcrystal Mikrokristall *m*
microcrystalline mikrokristallin
microcrystalline wax Mikrowachs *n*
microcurie Mikrocurie *n*
microdensitometer Mikrodensitometer *n*
microencapsulation Mikroverkapselung *f*
microextractor Mikroextraktor *m*
microfarad Mikrofarad *n* (Elektr)
microfilm Mikrofilm *m* (Phot)
microfilter Mikrofilter *n*
microfinish Feinstbearbeitung *f*
microflaw Haarriß *m*
microgamete Mikrogamet *m*
microgametocyte Mikrogametozyt *m*
microgram (Am. E.) Mikrogramm *n*
microgramme (Br. E.) Mikrogramm *n*
micrography Mikrographie *f*
micro high-pressure equipment
Mikrohochdruckgerät *n*
microhm Mikrohm *n* (Elektr)
microkinetics Mikrokinetik *f*
microlite Mikrolith *m* (Min)
microliter Mikroliter *m*
micromanometer Feindruckmanometer *n*
micromerograph Mikromerograph *m*
micrometer Mikrometer *n* (ein Millionstel
Meter), Feinmesser *m,* Mikrometerschraube *f*
micrometer adjustment Mikrometereinstellung *f*
micrometer cal[l]iper Feinmeßlehre *f,*
Feinstellschraube *f,* Schraublehre *f*
micrometer eye-piece Fadenmikrometer *n* (Opt)
micrometer screw Meßschraube *f,*
Mikrometerschraube *f*
micro-method Mikromethode *f*
micrometric mikrometrisch
micrometric eyepiece Meßokular *n*
micrometry Mikrometrie *f*
micro-micro farad Picofarad *n*

micromicron Mikromikron *n* (ein Millionstel Mikron)
micromillimeter Millimikron *n*
micromole Mikromol *n*
micron Mikrometer *n*, Mikron *n*
micronizer Feinstmahlvorrichtung *f*
micronizing Mikronisieren *n*
micronucleus Mikronukleus *m*, Mikrokern *m*
micronutrient Mikronährstoff *m*
microorganism Mikroorganismus *m*, Kleinlebewesen *n*, Mikrobe *f* (Biol)
microparasite Mikroparasit *m*
microperthite Mikroperthit *m* (Min)
microphage Mikrophage *m*
microphone Mikrophon *n*
microphotograph Mikrobild *n*, Mikrophotographie *f*
microphotographic mikrophotographisch
microphotography Mikrophotographie *f*
microphotometer Mikrophotometer *n*
microphyllic acid Microphyllinsäure *f*
micropipet|te| Mikropipette *f*
micropolishing Mikropolieren *n*
micropore Mikropore *f*
microradiograph Mikroröntgenbild *n*
microrheology Mikrorheologie *f*
microscale Mikroskala *f*
microscale test Kleinversuch *m*
microscope Mikroskop *n*, **binocular** ~ binokulares Mikroskop, **compound** ~ mehrlinsiges Mikroskop, zusammengesetztes Mikroskop *n*, **reflecting** ~ Spiegelmikroskop *n*
microscope *v* mikroskopieren, mikroskopisch untersuchen
microscope image Mikroskopbild *n*
microscope slide box Objektträgerkasten *m*
microscopic mikroskopisch
microscopical stain Mikrofarbstoff *m*
microscopical testing Mikroskopieren *n*
microscopic eyepiece Okularglas *n*
microscopy Mikroskopie *f*
microsecond Mikrosekunde *f*
microsieving Mikrosiebung *f*
microslip Feingleitung *f*
microsome Mikrosom *n*
microsommite Mikrosommit *m* (Min)
microspectroscope Mikrospektroskop *n*
microstand Mikrostativ *n*
microstructure Mikrostruktur *f*, Feingefüge *n*, mikroskopische Struktur *f*
microswitch Kontaktschalter *m* (Elektr), Mikroschalter *m*
microtesting equipment Mikroprüfgeräte *n pl*
microtome Mikrotom *n*
microtome blade Mikrotommesser *n*
microtomy Dünnschnittverfahren *n* (Mikroskopie), Mikrotomie *f* (Mikroskopie)
microtron Mikrotron *n* (Atom)

microtube Mikroröhre *f*
microturbulence Mikroturbulenz *f*
microvolt Mikrovolt *n* (Elektr)
microvolumetric mikrovolumetrisch
microwave Mikrowelle *f*, Ultrakurzwelle *f*
microwave breakdown Mikrowellendurchschlag *m*
microwave dielectrometer Mikrowellendielektrometer *n*
microwave discharge Mikrowellenentladung *f*
microwave interferometry Mikrowelleninterferometrie *f*
microwave measurement Mikrowellenmessung *f*
microwave propagation Mikrowellenausbreitung *f*
microwave region Mikrowellenbereich *m*
microwave resonance Mikrowellenresonanz *f*
microwave spectroscopy Mikrowellenspektroskopie *f*
microwave spectrum Mikrowellenspektrum *n*
microwave technique Mikrowellentechnik *f*
microwax Mikrowachs *n*
microzoon Kleinlebewesen *n*
middle Mitte *f*
middle box Mittelkasten *m*
middle conductor Mittelleiter *m*
middle fraction Mittellauf *m* (Dest)
middle oil Mittelöl *n*
middle piece Mittelstück *n*
middle ring Zwischenring *m*
middle roll Mittelwalze *f*
middle runnings Mittellauf *m* (Dest)
middlings Nachmehl *n*
midget step switch Zwergstufenschalter *m*
mid point Mittelpunkt *m*
mid vertical Mittelsenkrechte *f*
miedziankite Miedziankit *n*
miersite Miersit *m* (Min)
mignonette Reseda *f* (Bot)
mignonette green Resedagrün *n*
mignonette oil Resedaöl *n*
migrate *v* wandern
migration Wanderung *f*, Ortswechsel *m*, Platzwechsel *m*, Verschiebung *f* (Chem), **direction of** ~ Wanderungssinn *m* (Ionen), **sense of** ~ Wanderungssinn *m* (Ionen)
migration loss Wanderungsverlust *m*
migration velocity Wanderungsgeschwindigkeit *f*
migratory cell Wanderzelle *f*
MIG welding MIG-Schweißen *n* (Abk. für Metall-Inertgas-Schweißen)
mikado yellow Mikadogelb *n*
milarite Milarit *m* (Min)
mild mild, sanft, schonend
mildew Mehltau *m*, Meltau *m*, Moder *m*, Schimmel *m*
mildew *v* modern
mildewed schimmelfleckig, moderig
mildew-resistant schimmelfest

mild steel Weichstahl *m*
mild steel wire Flußstahldraht *m*
mileage Kilometerleistung *f*
milfoil Schafgarbe *f* (Bot)
milk Milch *f,* **canned** ~ Dosenmilch *f,* **condensed**
  ~ Kondensmilch *f,* **curdled** ~ Sauermilch *f,*
  dry ~ Trockenmilch *f,* **evaporated** ~
  Dosenmilch *f,* evaporierte Milch, **fat** ~
  Vollmilch *f,* **pasteurized** ~ pasteurisierte
  Milch, **skimmed** ~ Magermilch *f,* abgerahmte
  Milch, **sterilized** ~ sterilisierte Milch, **tinned**
  ~ Dosenmilch *f,* **unskimmed** ~ Vollmilch *f,*
  vegetable ~ pflanzliche Milch, ~ **of lime**
  Kalkwasser *n,* Naßkalk *m*
milk agaric Milchpilz *m*
milk albumin Lactalbumin *n*
milk condensing plant Milchkondensieranlage *f*
milk cooler Milchkühler *m*
milk cooling plant Milchkühlanlage *f*
milk fermentation Milchgärung *f*
milk glass Milchglas *n*
milk indicator Milchindikator *m*
milkiness Trübung *f*
milking grease Melkfett *n*
milk of magnesia Magnesiamilch *f* (Pharm)
milk powder Milchpulver *n,* Trockenmilch *f*
milk protein Milcheiweiß *n*
milk serum Milchserum *n*
milk sugar Milchzucker *m,* Laktobiose *f,*
  Laktose *f*
milk sugar plant Milchzuckeranlage *f*
milk white Milchweiß *n*
milk-white agate Milchachat *m* (Min)
milk-white opal Milchopal *m* (Min)
milkwort Polygala *f* (Bot)
milky milchig, milchartig
milky quartz Milchquarz *m* (Min)
mill Mühle *f,* Fabrik *f,* Hammerwerk *n* (Techn),
  Walzwerk *n* (Techn), Werk *n,* **contrarotation
  ball-race-type pulverizing** ~
  Federkraft-Kugelmühle *f,* **fluid energy or jet**
  ~ Strahlmühle *f;* **sand or attrition** ~
  Attritor *m* (Sandmühle), ~ **for wet grinding**
  Naßmühle *f*
mill *v* mahlen, fräsen (Techn), walken (Leder,
  Text.), walzen (Techn), zerreiben, ~ **off**
  abfräsen, ~ **out** ausfräsen, ~ **threads**
  Gewinde fräsen
mill base Mahlgut *n* (Farben)
millboard dichte Pappe *f*
mill dust Mehlstaub *m*
milled head Rändelkopf *m*
millerite Millerit *m* (Min), Haarkies *m* (Min),
  Nickelkies *m* (Min)
mill grinder Fräswerkzeug *n,* Schlitzfräser *m*
mill housing Walzenständer *m*
milliampere Milliampere *n* (Elektr)
milliard (Br. E.) Milliarde *f*
millibar Millibar *n*

millicurie Millicurie *n*
milligram (Am. E.) Milligramm *n*
milligramme (Br. E.) Milligramm *n*
milliliter (Am. E.) Milliliter *m n*
millilitre (Br. E.) Milliliter *m n*
millimeter Millimeter *m n*
millimeter graph paper Millimeterpapier *n*
millimicron Millimikron *n*
millimole Millimol *n* (Chem)
milling Mahlen *n,* Fräsen *n* (Techn),
  Mahlarbeit *f,* Rändeln *n* (Techn), **fast to** ~
  walkecht, **high-speed** ~
  Hochgeschwindigkeitsfräsen *n*
milling cutter Fräser *m,* Fräswerkzeug *n*
milling cycle Mahlgang *m*
milling fixture Fräsvorrichtung *f*
milling machine Fräsmaschine *f,*
  Piliermaschine *f,* **automatic** ~ Fräsautomat *m*
  (Techn)
milling ore Pocherz *n*
milling plant Mahlanlage *f*
milling process Mahlverfahren *n*
milling section Mahlraum *m*
milling work Fräsarbeit *f*
millionth [part] Millionstel *n*
milliroentgen Milliröntgen *n*
mill iron Puddelroheisen *n*
millisecond Millisekunde *f*
millivolt Millivolt *n* (Elektr)
Millon's reagent Millons-Reagens *n*
mill scale Glühspan *m,* Hammerschlag *m*
mill setting Walzenspalteinstellung *f*
millstone Mühlstein *m*
mill train Walzstraße *f,* Walzstrecke *f*
milori blue Miloriblau *n*
milt Milz *f* (Anat)
mimeograph Mimeograph *m,*
  Vervielfältigungsmaschine *f*
mimetesite Mimetesit *m* (Min), Arsenbleierz *n*
  (Min), Grünbleierz *n* (Min), Mimetit *m*
  (Min)
mimetite Mimetit *m* (Min), Arsenbleierz *n*
  (Min), Bleiblüte *f* (Min), Mimetesit *m* (Min),
  Traubenblei *n* (Min)
mimosine Mimosin *n*
minaline Minalin *n*
minasragrite Minasragrit *m* (Min)
minced meat Hackfleisch *n*
mincer Hackmaschine *f*
mind, to bear in ~ berücksichtigen
mine Bergwerk *n* (Bergb), Mine *f* (Bergb),
  Zeche *f* (Bergb)
mine coal Förderkohle *f,* Zechenkohle *f*
mine damp Grubenwetter *n*
mine dust Grubenstaub *m*
mine explosion Grubenexplosion *f* (Bergb)
mine gas Grubengas *n*
miner Bergarbeiter *m,* Bergmann *m*
mineral Mineral *n*

mineral *a* mineralisch
mineral analysis Mineralanalyse *f*
mineral black Grubenschwarz *n*
mineral blue Mineralblau *n*, Kupferlasur *f*,
  Wolframblau *n*
mineral caoutchouc Elaterit *m*, Erdharz *n*
mineral carbon Graphit *m*
mineral charcoal Fusain *n*
mineral chemistry Mineralchemie *f*
mineral coal Steinkohle *f*
mineral color Erdfarbe *f*, Mineralfarbe *f*
mineral coloring matter Mineralfarbstoff *m*
mineral content[s] Mineralstoffgehalt *m*
mineral dust Gesteinstaub *m*
mineral fiber Mineralfaser *f*
mineral filler Gesteinsmehl *n*
mineral flax Faserasbest *m* (Min)
mineral green Erdgrün *n*
mineralizable vererzbar
mineralization Mineralisation *f*, Vererzung *f*
mineralize *v* vererzen
mineralizer Vererzungsmittel *n*
mineral lubricating oil Mineralschmieröl *n*
mineral matter Mineralmasse *f* (Straßenbau)
mineral mordant Mineralbeize *f*
mineral naphtha Bergnaphtha *f n*, Bergöl *n*
mineralogical mineralogisch
mineralogy Mineralogie *f*, Gesteinskunde *f*
mineral oil Erdöl *n*, Mineralöl *n*, Petroleum *n*,
  refined ~ Mineralölraffinat *n*
mineral oil reservoir Erdöllager *n*
mineral pigment Farberde *f*, Mineralfarbe *f*
mineral pitch Asphalt *m*, Bergharz *n*, Erdpech *n*
mineral purple Ocker *m*
minerals, collection of ~ Mineraliensammlung *f*
mineral spring Heilquelle *f*, Mineralquelle *f*
mineral tallow Erdtalg *m*, Erdwachs *n*
mineral tanning Mineralgerbung *f*
mineral tar Erdteer *m*
mineral vein Erzader *f*, Mineralgang *m* (Bergb)
mineral water Mineralwasser *n*, **acidulous ~**
  Säuerling *m*
mineral wax Bergwachs *n* (Min), Ceresin *n*
  (Min), Erdwachs *n* (Min), Ozokerit *m* (Min)
mineral white Mineralweiß *n*
mineral wool Bergwolle *f*, Schlackenwolle *f*
mineral yellow Kaisergelb *n*, Patentgelb *n*
miner's lamp Grubenlampe *f*
miner's lung Anthrakose *f* (Med)
minette Minette *f* (Min)
mine ventilation Grubenbewetterung *f* (Bergb)
mingle *v* sich vermischen, vermengen,
  verschmelzen
mingler Maischtrog *m*
mingling Vermischen *n*
miniature camera Kleinbildkamera *f* (Phot)
mini-computer Klein-Computer *m*
minimal polynomial Minimalpolynom *n*

minimax concept (in factorial design)
  Minimaxkonzept *n* (b. Versuchsplanung)
minimize *v* auf das Kleinstmaß zurückführen,
  verringern
minimum Minimum *n*, Kleinstwert *m*
minimum beam, definition of ~
  Minimumstrahldefinition *f*, **marking of ~**
  Minimumstrahlkennzeichnung *f*
minimum content Minimalgehalt *m*
minimum current Minimalstrom *m*
minimum deflection Mindestdurchbiegung *f*
minimum efficiency Mindestleistung *f*
minimum output Mindestleistung *f*
minimum overvoltage Mindestüberspannung *f*
  (Elektr)
minimum pressure Mindestdruck *m*
minimum requirement Mindestanforderung *f*
minimum speed Mindestgeschwindigkeit *f*
minimum thermometer
  Minimum-Thermometer *n*
minimum value Minimalwert *m*, Kleinstwert *m*,
  Tiefstwert *m*
minimum voltage Minimalspannung *f* (Elektr)
minimum work Mindestarbeit *f*
mining Bergbau *m*, Grubenbau *m*, **open-cast ~**
  Tagebau *m*
mining explosive Bergbausprengmittel *n*
mining industry Montanindustrie *f*
mining method Abbaumethode *f* (Bergb)
mining rail Grubenschiene *f*
mining vehicle Grubenfahrzeug *n*
minioluteic acid Minioluteinsäure *f*
Ministry of Agriculture
  Landwirtschaftsministerium *n*
minium Mennige *f*, Bleioxyduloxid *n*, Minium *n*
minium-colored mennigefarben
mink Nerz *m*
mink fat Nerzöl *n*
minor determinant Unterdeterminante *f* (Math)
minority Minderheit *f*
mint Minze *f* (Bot)
mint camphor Menthol *n*
minus minus; negativ
minus sign Minuszeichen *n* (Math)
minute Minute *f*
minute hand Minutenzeiger *m*
minutes, for several ~ minutenlang
miosis Meiose *f* (Biol), Reduktionsteilung *f*
  (Biol)
miotic Miotikum *n* (Pharm)
mirabilite Mirabilit *m* (Min)
miracil Miracil *n*
miracle Wunder *n*, Wunderwerk *n*
mirage Fata Morgana *f* (Opt), Luftspiegelung *f*
  (Opt)
mirbane essence Mirbanessenz *f*, Mirbanöl *n*
mirbane oil Mirbanöl *n*, Mirbanessenz *f*,
  Nitrobenzol *n*
mire Schlamm *m*

mirene Miren *n*
mirror Spiegel *m*, **concave** ~ Hohlspiegel *m*
mirror *v* [ab]spiegeln
mirror bronze Spiegelbronze *f*
mirror finish Hochglanz *m*
mirror finish *v* hochglanzpolieren
mirror foil Folie *f* eines Spiegels
mirror galvanometer Spiegelgalvanometer *n*
mirror glass Spiegelglas *n*
mirror image Spiegelbild *n*
mirror image function Spiegelbildfunktion *f*
mirror-inverted spiegelbildlich
mirror-panelled spiegelverkleidet
mirror plane Spiegelebene *f*
mirror scale Spiegelskala *f*
mirror symmetry Spiegelsymmetrie *f*
miry schlammig
misadjustment Fehleinstellung *f*
misalignment Fluchtungsfehler *m*
miscalculation Rechenfehler *m*
miscella Miscella *f*
miscellaneous verschiedenartig
misch metal Mischmetall *n* (Cer-Mischmetall)
miscibility Mischbarkeit *f*
miscibility gap Mischungslücke *f* (in flüssig-flüssig-Systemen)
miscible mischbar, **completely** ~ vollkommen mischbar (Chem)
misdirect *v* fehlleiten
mishap leichter Unfall *m*
misinterpretation Mißdeutung *f*
mislead *v* fehlleiten
mispickel Mispickel *m* (Min), Arsenkies *m* (Min), Arsenopyrit *m* (Min)
misplaced material Fehlaustrag *m* (Siebtechn)
misplaced size Fehlkorn *n* (Siebtechn)
miss *v* verfehlen, versäumen
missile Fernlenkgeschoß *n*, Geschoß *n*, **air-to-underwater** ~ Luft-Unterwasser-Rakete *f*, **surface-to-air** ~ Boden-Luft-Rakete *f*, **surface-to-surface** ~ Boden-Boden-Rakete *f*
missing abwesend, fehlend
mist Nebel *m*, Dunst *m*, **formation of** ~ Nebelbildung *f*, **to eliminate the** ~ entnebeln
mistake Fehler *m*, Irrtum *m*
mist eliminating Entnebeln *n*
mist elimination Entnebelung *f*
mist eliminator Entnebelungsanlage *f*
mist flow Nebelströmung *f*
mistletoe Mistel *f* (Bot)
mist-like nebelartig
misusage Mißbrauch *m*
misuse Mißbrauch *m*
misuse *v* mißbrauchen
mite Milbe *f* (Zool)
mite infection Milbeninfektion *f*
mite killer Akarizid *n*, milbentötendes Mittel *n*

miter Fuge *f* (Techn), Gefüge *n* (Techn), Gehre *f* (Techn), Gehrung *f* (Techn)
miter gear Winkelrad *n*
miter joint Gehrungsfuge *f*
miter plane Gehrungshobel *m*
miter valve Kegelventil *n*
miter weld Gehrungsschweißung *f*
mithridate Mithridat *n*
miticide Mitizid *n*
mitigant mildernd
mitigate *v* lindern, mildern
mitigation Linderung *f*, Milderung *f*
mitis green Mitisgrün *n*
mitochondria Mitochondrien *pl*
mitochondria distribution Mitochondrienverteilung *f*
mitochondria membrane Mitochondrienmembran *f*
mitochondria permeability Mitochondriendurchlässigkeit *f*
mitomycin Mitomycin *n*
mitosis indirekte Zellkernteilung, Mitose *f*
mitotic poison Mitosegift *n*
mitragyne Mitragynin *n*
mitragynine Mitragynin *n*
mitraphyllol Mitraphyllol *n*
mitraversine Mitraversin *n*
mix Gemisch *n*
mix *v* [ver]mischen, kreuzen (Biol), legieren (Met), melieren (Web), vermengen, zusammenschütten, ~ **in** einrühren, ~ **with** beimengen, zumischen
mixable mischbar
mixed gemischt, vermischt
mixed adhesive Zweikomponentenkleber *m*
mixed-bed exchanger Mischbettaustauscher *m*
mixed bleaching Mischbleichen *n*
mixed color Mischfarbe *f*
mixed-crystal alloy Mischkristallegierung *f*
mixed-crystal formation Mischkristallbildung *f*
mixed crystals, series of ~ Mischkristallreihe *f*
mixed culture Mischkultur *f*
mixed feed Mengfutter *n*
mixed-flow impeller Mischflußschnellrührer *m*
mixed gas Halbwassergas *n*, Mischgas *n*
mixed metal Mischmetall *n*
mixed phase cracking Gemischtphase-Krackverfahren *n*
mixed process Halbtrockenverfahren *n* (Zement)
mixed salt Mischsalz *n*
mixer Mischapparat *m*, Mischer *m*, Rührer *m*, **continuous** ~ Fließmischer *m*, **double motion** ~ gegenläufiges Doppelrührwerk *n*, **fluidized bed** ~ Fließbettmischer *m*, **high-speed** ~ Schnellmischer *m*, **impeller or gyro** ~ Kreiselmischer *m*, **intensive** ~ Innenmischer *m*, **pressurized** ~ Druckmischer *m*, **simple** ~ einfacher Mischer *m*, **spiral or helical blade** ~

Wendelrührer *m*, **tilted** ~ schrägstehender
Mischer *m*, **dry-blend** ~ **or dry-coloring mixer**
Taumelmischer *m*, **intensifier-type** ~ **or**
intensive mixer Intensivmischer *m*
mixer platform Mischerbühne *f*
mixer-settler Mischer-Abscheider *m*
mixer-settler tower Turmextraktor *m*
mixing Mischen *n*, Vermengung *f*, Vermischen *n*,
heat of ~ Mischungswärme *f*, **incapable of** ~
nicht mischbar, ~ **by liquid jets**
Strahlmischen *n*, ~ **gases and liquids**
Begasen *n* (v. Flüssigkeiten), ~ **of the raw**
materials Gattieren *n* der Rohmaterialien (n
pl) (Met)
mixing and granulating machine Misch- und
Granuliermaschine *f*
mixing apparatus Mischapparat *m*
mixing beater Mischholländer *m*
mixing bell Mischkugel *f*
mixing chamber Mischraum *m*
mixing cone Mischtrichter *m*
mixing control Mischregler *m*
mixing dump Mischhalde *f*
mixing installation Mischanlage *f*
mixing jet Mischdüse *f*
mixing kettle Rührwerkskessel *m*
mixing machine Mischmaschine *f*
mixing plate Mischplatte *f*
mixing power Mischvermögen *n*
mixing process Mischungsvorgang *m*
mixing proportion Mischungsverhältnis *n*
mixing screw Mischschnecke *f*
mixing stockpiles Haldenvermischung *f*
mixing trough Mischbottich *m*
mixing tube Mischrohr *n*
mixing unit for bathwater Mischbatterie *f* für
Bad
mixing valve Mischventil *n*, Regler *m*
mixing vat Anmachbottich *m*
mixing vessel Mischgefäß *n*, Mischkessel *m*
mixing water Anmachwasser *n*
mixite Mixit *m* (Min)
mixtion Mischung *f* (Pap)
mixture Mischung *f*, Gemenge *n*, Gemisch *n*,
Mixtur *f* (Pharm), Vermengung *f*,
Vermischung *f*, **combustible** ~ brennbares
Gemisch, **composition of a** ~
Mischungsverhältnis *n*, **constituent part or**
ingredient of a ~ Gemengeanteil *m*, **eutectic**
~ eutektisches Gemisch, **homogeneous** ~
einheitliches Gemisch, **inflammable** ~
brennbares Gemisch, **ratio of the components**
of a ~ Mischungsverhältnis *n*, **sample of** ~
Mischungsprobe *f*, ~ **of clay und culm**
Gestübe *n*, ~ **of hydrochloric und nitric acid**
Aqua regia, ~ **of lime and coal**
Kalkkohlegemisch *n*, ~ **of ores** Gattierung *f*,
~ **of ores and fluxes** Möller *m*, ~ **of oxalic**

and malic acid Kichererbsensäure *f* (obs), ~
[of substances] Stoffgemisch *n*
mixtures, rule of ~ Mischungsregel *f*
mixture shutter Gemischklappe *f*
mizzonite Mizzonit *m* (Min)
MKS system MKS-System *n* (Abk. für
Meter-Kilogramm-Sekunde)
mobile fahrbar, freizügig
mobile film Mischfilm *m*
mobile film column Mischfilmkolonne *f*
mobile film drum Mischfilmtrommel *f*
mobile phase Fließmittel *n* (Gaschromat)
mobility Beweglichkeit *f*, Leichtflüssigkeit *f*,
limitation of ~ Beweglichkeitsgrenze *f* (Krist)
mobility equation Beweglichkeitsgleichung *f*
mobility tensor Beweglichkeitstensor *m*
mobilization Mobilisierung *f*
mocha stone Moosachat *m* (Min)
mochras Malabargummi *n*
mochyl alcohol Mochylalkohol *m*
mock-up Attrappe *f*
modacrylic fibers Modacrylfasern *pl*
modal fibers Modalfasern *pl*
model Vorbild *n*, Modell *n*, Muster *n*,
Schablone *f*
model *v* modellieren, [nach]bilden
model[l]ing material or clay Knetmasse *f*
model theory Modelltheorie *f*
mode of action Wirkungsweise *f*
mode of formation Bildungsweise *f*
moderate gemäßigt, mäßig
moderate *v* mildern, abbremsen (Atom),
beruhigen, lindern, mäßigen, moderieren
(Atom), verlangsamen (Atom)
moderating material Bremssubstanz *f* (im
Reaktor)
moderation Abbremsung *f* (Atom), Mäßigung *f*,
Milderung *f*, Moderierung *f* (Atom),
Verlangsamung *f* (Atom), ~ **of neutrons**
Bremsung *f* von Neutronen (n pl)
moderator Bremssubstanz *f*
modern modern, zeitgemäß
modernize *v* modernisieren
modest anspruchslos, bescheiden, maßvoll
modifiable abwandelbar, modifizierbar
modification Modifikation *f*, Abänderung *f*,
Abwandlung *f*, Modifizierung *f*
modifications Umbauten *pl*
modify *v* abändern, abwandeln, modifizieren
modular equation Modulargleichung *f*
modular function Modulfunktion *f*
modular system Bausteinsystem *n*
modulate *v* abstimmen, modulieren
modulation Abstimmung *f*, Aussteuerung *f*,
Modulation *f*
modulation control Modulationsregler *m*
modulation frequency Modulationsfrequenz *f*
modulator crystal Modulatorkristall *m*
modulator stage Modulationsstufe *f*

module Modul *m*
modulus Modul *m*, ~ **of elasticity**
  Elastizitätsmodul *m* (Phys), E-Modul *m*
  (Phys)
modulus in flexure, Young's ~
  Elastizitätsmodul *m* berechnet aus dem
  Biegeversuch
Mössbauer effect Mößbauer Effekt *m*
mohair Mohair *n*
Mohr's balance Mohrsche Waage *f*
  (Mohr-Westphalsche Waage)
Mohr's salt Mohrsches Salz *n*,
  Ammoniumferrosulfat *n*
Mohs' hardness Mohssche Härte *f*
Mohs' scale of hardness Mohssche Härteskala *f*
moiety Hälfte *f*
moissanite Moissanit *m* (Min)
moist feucht, naß
moisten *v* anfeuchten, befeuchten, benetzen,
  besprengen
moistenable benetzbar
moistener Anfeuchter *m*
moistening Anfeuchten *n*, Befeuchten *n*,
  Benetzen *n*
moistening agent Befeuchtungsmittel *n*
moistening chamber Befeuchtungskammer *f*,
  Feuchtkammer *f*
moistening plant Befeuchtungsanlage *f*
moistening power Benetzungsfähigkeit *f*
moisture Feuchtigkeit *f*, **amount of** ~
  Feuchtigkeitsgehalt *m*, **bound** ~ gebundene
  Feuchtigkeit, **deposit of** ~
  Feuchtigkeitsniederschlag *m*, **free** ~ freier
  Wassergehalt, **percentage of** ~
  Feuchtigkeitsgehalt *m*, Wassergehalt *m*
moisture-attracting wasseranziehend
moisture content Feuchtigkeitsgehalt *m*,
  Wassergehalt *m*, **critical** ~ kritische
  Gutfeuchte *f*, **normal** ~
  Normalwassergehalt *m*
moisture contents Naßgehalt *m*
moisture control Feuchtigkeitsregelung *f*
moisture determination apparatus
  Feuchtigkeitsmesser *m*
moisture expansion Feuchtigkeitsausdehnung *f*,
  Quellen *n*
moisture pick-up Feuchtigkeitsaufnahme *f*
moisture-repellent hydrophob
moisture-resistant feuchtigkeitsbeständig
moisture test Naßprobe *f*
moisture tester Feuchtigkeitsprüfer *m*
moisture vapor transmission rate
  Wasserdampfdurchlässigkeit *f*
molality Molalität *f*
molar molar
molar concentration Molkonzentration *f*
molar fraction Molenbruch *m*

molar heat Molwärme *f*
molarity Molarität *f*
molar quantum Molquant *n*
molar ratio Molverhältnis *n*
molar refractivity Molrefraktion *f*
molar solution Mollösung *f*
molar volume Molvolumen *n*
molar weight Mol[ar]gewicht *n*
molasses Melasse *f*, Zuckerdicksaft *m*, **extraction**
  of sugar from ~ Melasseentzuckerung *f*
molasses mash or wash Melassemaische *f*
molasses pulp Melasseschnitzel *pl*
molasses pump Melassepumpe *f*
mold Abguß *m* (Gieß), Düngeerde *f* (Agr),
  Gießform *f* (Gieß), Gußform *f* (Gieß),
  Matrize *f* (Techn), Modell *n* (Techn),
  Modererde *f* (Agr), Preßform *f* (Techn),
  Schimmelpilz *m* (Bot), **cast iron** ~ gußeiserne
  Form, **doming** ~ gewölbte Form, **double force**
  ~ Form *f* mit Doppelstempel, **layer of** ~
  Humusschicht *f*, **open sand** ~ Herdform *f*,
  permanent ~ bleibende Form, **positive** ~
  Füllform *f*, **split[-cavity]** ~ Backenform *f*,
  mehrteilige Form, **stripper plate** ~ Form *f* mit
  Abstreiferplatte **to ram up a** ~ eine Gußform
  aufstampfen
mold *v* formen, bilden, gestalten, gießen (Gieß),
  kneten (Teig), modellieren, pressen,
  schimmeln, ~ **in** einpressen, **to begin to** ~
  anschimmeln
moldability Formbarkeit *f*
moldable [ver]formbar, **easily** ~ knetbar
mold base Formrahmen *m*
mold block Formblock *m*
mold carriage Formwagen *m*
mold casting Formguß *m*
mold [cavity] Matrize *f*
mold charge Füllmaterial *n*
mold closing pressure Formenschließdruck *m*
mold closure Schließen *n* der Form
mold composition Formstoff *m*
mold construction Formenbau *m*
mold curing Formheizung *f*
mold designer Formenkonstrukteur *m*
mold dope Formeneinstreichmittel *n*
mold dryer Formentrockner *m*
molded article Formkörper *m*, Formling *m*,
  Formteil *n*, Preßteil *n*
molded cylinders of [activated char]coal
  Formkohle *f*
molded fiber board Hartfaserplatte *f*
molded-in stress innere Spannung *f*
molded insulating material Isolierpreßstoff *m*
molded part Formteil *n*
molded peat Streichtorf *m*
molded piece Formstück *n*, Preßling *m*,
  Preßteil *n*, Spritzpreßling *m*
molder Former *m*, Gießer *m*
molder *v* (Am. E.) [ver]modern

molder's black Gießereischwärze *f*
molder's hammer Modellhammer *m*
molder spade Formerspaten *m*
molder's tool Formerwerkzeug *n*
mold formation Schimmelbildung *f*
mold frame Formrahmen *m*
mold [fungus] Schimmelpilz *m*
mold growth Schimmelbildung *f*
molding Formerei *f*, Formgebung *f*, Formung *f*,
  Schimmelbildung *f*, **high-frequency** ~
  Pressen *n* mit Hochfrequenzvorwärmung,
  high-pressure ~ Pressen *n* unter hohem
  Druck, **laminated** ~ Formstück *n* aus
  Schichtstoff, **open sand** ~ Herdformerei *f*,
  short ~ nicht ausgeformtes Formteil *n*; ~ **in**
  boxes Kastenformerei *f*, **low-temperature**
  matched die ~ **or cold press molding**
  Kaltpreßverfahren *n* (Polyester)
molding board Formplatte *f*, Knetbrett *n*
molding box Formkasten *m*
molding clay Formerde *f*, Formerton *m*,
  Formton *m*
molding composition Preßmasse *f*,
  Preßmischung *f*
molding compound Abgußmasse *f*, Preßmasse *f*,
  Preßmischung *f*, **fabric-filled** ~
  Cordpreßmasse *f*, **fast-curing** ~
  Schnellpreßmasse *f*
molding cycle Preßvorgang *m*
molding fault Preßfehler *m*
molding flask Gußflasche *f*
molding hall Formerei *f*
molding hole Dammgrube *f*
molding loam Formlehm *m*
molding machine Formmaschine *f*, **upright** ~
  Bockformmaschine *f*, ~ **with rotary table**
  Drehtischformmaschine *f*
molding machinery Formmaschinenanlage *f*
molding material Formmasse *f*, Preßmasse *f*
molding pin Formerstift *m*
molding plane Gesimshobel *m*, Kehlhobel *m*
molding plug Preßstempel *m*
molding powder Preßpulver *n*
molding press Formpresse *f*, Kunststoffpresse *f*
mold[ing] pressure Preßdruck *m*
molding resin Preßharz *n*
molding's sand Formsand *m*, Gieß[erei]sand *m*,
  Klebsand *m*; Modellsand *m*, **dressing of the** ~
  Aufbereitung *f* des Formsandes
molding shrinkage Formenschwindmaß *n* (Plast)
molding technique Preßtechnik *f*,
  Preßverfahren *n*
molding tool Preßwerkzeug *n*
molding wax Bossierwachs *n*
mold insert Formeinsatz *m*, Gesenkeinsatz *m*,
  Matrizeneinsatz *m*
mold-like schimmelartig
mold locking force Werkzeugzuhaltekraft *f*

mold lubricant Formeneinstreichmittel *n*,
  Formtrennmittel *n*
mold maker Werkzeugmacher *m*
mold mark durch das Werkzeug verursachte
  Fehlstelle *f*, Markierung *f* des Werkzeuges
mold parting agent Formtrennmittel *n*
mold parting line Trennfuge *f* der Form
mold press Formpresse *f*
mold release Formentrennmittel *n*
mold release device Entformungsvorrichtung *f*
mold ring Fußheizring *m*
mold shrinkage Formenschwindmaß *n* (Plast)
moldy schimmelig, moderig, **[slightly]** ~
  angeschimmelt, **to become** ~ schimmeln
moldy peat Modertorf *m*
moldy smell Modergeruch *m*
mole Mol *n*
molecular molekular
molecular adhesion Moleküladhäsion *f*
molecular aggregate Molekülaggregat *n*
molecular arrangement Molekülanordnung *f*
molecular asymmetry Molekülasymmetrie *f*
molecular attraction Molekularanziehung *f*
molecular beam Molekularstrahl *m*
molecular beam experiment
  Molekularstrahlexperiment *n*
molecular beam method
  Molekularstrahlmethode *f*
molecular biology Molekularbiologie *f*
molecular chain, linear ~ Fadenmolekül *n*
molecular cluster Molekülaggregat *n*,
  Molekülkomplex *m*
molecular complex Molekülverbindung *f*
molecular conductivity Molekularleitfähigkeit *f*
molecular diameter Moleküldurchmesser *m*
molecular dispersion Molekulardispersion *f*
molecular distillation Kurzwegdestillation *f*,
  Molekulardestillation *f*
molecular effect Molekularwirkung *f*
molecular energy Molekularenergie *f*
molecular evaporator Kurzwegverdampfer *m*
molecular fission Molekularzertrümmerung *f*
molecular force Molekularkraft *f*
molecular formula, empirical ~ Bruttoformel *f*
molecular friction Molekularreibung *f*
molecular heat Molekularwärme *f*
molecularity Molekularität *f*,
  Molekularzustand *m*
molecular lattice Molekülgitter *n*
molecular layer Molekülschicht *f*
molecular magnet Elementarmagnet *m*
molecular magnetism Molekularmagnetismus *m*
molecular model Molekülmodell *n*
molecular movement Molekularbewegung *f*
molecular orbital Molekülorbital *n*
molecular pressure Molekulardruck *m*
molecular ray Molekularstrahl *m*
molecular recoil Molekülrückstoß *m*

molecular refraction
  Molekularbrechungsvermögen *n*,
  Molekularrefraktion *f*
molecular repulsion Molekularabstoßung *f*
molecular retardation Molekularverzögerung *f*
molecular rotation Moleküldrehung *f*,
  Molekülrotation *f*, Molrotation *f*
molecular rotation spectrum
  Molekülrotationsspektrum *n*
molecular scattering Molekularstreuung *f*
molecular sieve Molekularsieb *n*
molecular spectrum Molekülspektrum *n*
molecular state Molekularzustand *m*
molecular structure Molekülaufbau *m*,
  Molekülverband *m*
molecular transformation Molekülumlagerung *f*
molecular velocity Molekülgeschwindigkeit *f*
molecular volume Molvolumen *n*
molecular weight Molekulargewicht *n*, **of high** ~
  hochmolekular
molecular weight determination
  Molekulargewichtsbestimmung *f*
molecular weight distribution
  Molekulargewichtsverteilung *f*
molecule Molekül *n*, Molekel *f*, **activated** ~
  aktiviertes Molekül, **core of a** ~
  Molekülrumpf *m*, **vibration of a** ~
  Molekülschwingung *f*
molecule formation Molekülbildung *f*
mole percent Molprozent *n*
molest *v* belästigen
mollification Erweichung *f*
mollify *v* erweichen
mollisin Mollisin *n*
molluscicides, molluscacides Molluskizide *pl*
molluscide Molluscid *n*
Molotov cocktail Molotowcocktail *m*
molten geschmolzen, schmelzflüssig
molten charge warmer Einsatz *m*
molten electrolyte Badschmelze *f*
molybdate molybdänsaures Salz *n*, Molybdat *n*
molybdate fiery red Mineralfeuerrot *n*
molybdenite Molybdänit *m* (Min),
  Molybdänglanz *m* (Min)
molybdenous compound
  Molybdän(II)-Verbindung *f*
molybdenous iodide Molybdändijodid *n*
molybdenous salt Molybdän(II)-Salz *n*
molybdenum (Symb. Mo) Molybdän *n*
molybdenum alloy Molybdänlegierung *f*
molybdenum blue Molybdänblau *n*
molybdenum chloride Molybdänchlorid *n*
molybdenum content Molybdängehalt *m*
molybdenum diiodide Molybdändijodid *n*
molybdenum dioxide Molybdändioxid *n*
molybdenum filament Molybdänfaden *m*
molybdenum glance Molybdänglanz *m* (Min)
molybdenum glass Molybdänglas *n*
molybdenum metal Molybdänmetall *n*

molybdenum oxide Molybdänoxid *n*
molybdenum silver Molybdänsilber *n*
molybdenum steel Molybdänstahl *m*
molybdenum sulfide Molybdänsulfid *n*
molybdenum trioxide Molybdäntrioxid *n*,
  Molybdänsäureanhydrid *n*
molybdic acid Molybdänsäure *f*, **salt of** ~
  Molybdat *n*
molybdic anhydride Molybdänsäureanhydrid *n*,
  Molybdäntrioxid *n*
molybdic ocher Molybdänocker *m* (Min),
  Molybdit *m* (Min)
molybdic oxide Molybdäntrioxid *n*,
  Molybdänsäureanhydrid *n*
molybdite Molybdit *m* (Min),
  Molybdänocker *m* (Min)
molybdomenite Molybdomenit *m* (Min)
molybdophyllite Molybdophyllit *m* (Min)
molybdosodalite Molybdosodalith *m* (Min)
molybdous compound
  Molybdän(II)-Verbindung *f*
molybdous iodide Molybdändijodid *n*
molybdous salt Molybdän(II)-Salz *n*
molysite Molysit *m* (Min)
moment Augenblick *m*, Moment *m*, Moment *n*
  (Phys), ~ **of force** Kraftmoment *n*, ~ **of**
  inertia Trägheitsmoment *n*, ~ **of rotation**
  Drehmoment *n*
momentary momentan, kurzzeitig
moment-resisting biegungsfest
momentum Bewegungsgröße *f*, Impuls *m*,
  Moment *n* (Phys), Stoßkraft *f*, Wucht *f*, **linear**
  ~ lineares Moment, **magnetic** ~
  magnetisches Moment, **static** ~ statisches
  Moment
momentum and energy exchange Impuls- und
  Energieaustausch *m*
momentum change Impulsänderung *f*
momentum determination Impulsbestimmung *f*
momentum distribution Impulsverteilung *f*
momentum measurement Impulsmessung *f*
momentum principle Impulssatz *m*
momentum space integral Impulsraumintegral *n*
momentum space representation
  Impulsraumdarstellung *f*
momentum transfer Impulsübertragung *f*
momentum transfer collision Stoß *m* mit
  Impulsübertragung
monacetin Acetin *n*, Monoacetin *n*
monad Monade *f* (Zool)
monardin Monardin *n*
monarsone Monarson *n*
monastral pigment Monastralpigment *n*
monatomic monoatomar, einatomig
monaxial einachsig
monazite Monazit *m* (Min)
monazite sand Monazitsand *m* (Min)
Mond gas Mondgas *n*
Mond gas producer Mondgasgenerator *m*

Monel [metal] Monelmetall *n*
monesia bark Monesiarinde *f*
monesia extract Monesiaextrakt *m*
monesin Monesin *n*
monetite Monetit *m* (Min)
monimolite Monimolit[h] *m* (Min)
monite Monit *m*
monitor Überwachungsgerät *n*, Kontrollgerät *n*,
　Mithörer *m*, Monitor *m*, Warngerät *n*
monitor *v* abhören, mithören
monitoring Kontrolle *f*, Überwachung *f*
monitoring equipment Überwachungsgeräte *n pl*
monitron Monitron *n*
monkey Affe *m*
monkey wrench Universalschraubenschlüssel *m*,
　Engländer *m* (Techn)
monkshood Akonit *m* (Bot, Pharm), Eisenhut *m*
　(Bot, Pharm)
monninine Monninin *n*
monoacetate Monoacetat *n*
monoacetin Monoacetin *n*
monoacid einsäurig
monoamide Monoamid *n*
monoamine Monoamin *n*
monoamine oxidase Monoaminoxidase *f*
　(Biochem)
monoammonium phosphate primäres
　Ammoniumphosphat *n*
monoatomic einatomig
monoatomicity Einatomigkeit *f*
monobarium silicate Monobariumsilikat *n*
monobasic einbasig, einbasisch
monobenzone Monobenzon *n*
monoblast Monoblast *m* (junger Monozyt)
monobromated camphor Bromcampher *m*
monobromoisovaleryl urea
　Monobromisovalerylharnstoff *m*
monobromopropionic acid
　Monobrompropionsäure *f*
monocalcium phosphate
　Monocalciumphosphat *n*
monocaproin Monocaproin *n*
monocarboxylic acid Monocarbonsäure *f*
monocellular organism Einzeller *m*
monochlorhydrate Monochlorhydrat *n*
monochlor[o]acetic acid Monochloressigsäure *f*
monochloroamine Monochloramin *n*
monochlor[o]hydrin Monochlorhydrin *n*
monochlorosilane Monochlorsilan *n*
monochord Monochord *n*
monochroic monochromatisch
monochromatic monochromatisch, einfarbig
monochromatic filter Monochromator *m*
monochromatization Monochromatisierung *f*
monochromator Monochromator *m*
monochrome monochrom, einfarbig
monochrome yellow Monochromgelb *n*
monochromic monochrom, einfarbig
monoclinic monoklin

monocoque structure Schalenbauteil *n*
monocrotalic acid Monocrotal[in]säure *f*
monocrotaline Monocrotalin *n*
monocrotic acid Monocrot[in]säure *f*
monocrystal Einkristall *m*, Monokristall *m*
monocrystal point Einkristallspitze *f*
monocyclic monozyklisch
monocylindrical einzylindrig
monocyte Monozyt *m*
monocyte-lymphocyte rate
　Monozyten-Lymphozyten-Verhältnis *n*
monodenteriobenzene Monodenteriobenzol *n*
monoenergetic monoenergetisch
monoethylin Monoäthylin *n*
monofilament Einzelfaden *m*
monoformin Monoformin *n*
monogenesis Monogenese *f*, Urzellentheorie *f*
monogenic monogen, gemeinsamen Ursprungs
monoglyceride Monoglycerid *n*
monograph Monographie *f*
monoheteroatomic monoheteroatomig
monohydrate Monohydrat *n*
monohydric alcohol einwertiger Alkohol *m*
monohydrochloride Monochlorhydrat *n*
monohydrogen phosphate sekundäres Phosphat *n*
　(Monohydrogenphosphat)
monoisotopic monoisotop
monolaurin Monolaurin *n*
monolayer monomolekulare Schicht *f*
monolupine Monolupin *n*
monomer Monomere *n*
monomeric monomer
monomerism Monomerie *f*
monomethylamine Monomethylamin *n*
monomethylol urea Monomethylolharnstoff *m*
monomolecular einmolekular, monomolekular
monomorphic monomorph, gleichgestaltig
monomyristin Monomyristin *n*
mononitrate Mononitrat *n*
mononitrophenol Mononitrophenol *n*
mononuclear einkernig
mononucleotide Mononucleotid *n*
monoolein Monoolein *n*
monopalmitin Monopalmitin *n*
monoperphthalic acid Monoperphthalsäure *f*,
　Phthalmonopersäure *f*
monophase einphasig
monophase system Einphasensystem *n*
monophenetidine citrate Monophenetidincitrat *n*
monophosphothiamine Monophosphothiamin *n*
monopoly Monopol *n*, Alleinverkauf *m*
monopotassium phosphate
　Monokaliumphosphat *n*
monorhein Monorhein *n*
monorheinanthrone Monorheinanthron *n*
monosaccharide Monosaccharid *n*
monosilane Monosilan *n*
monosilicate Monosilicat *n*
monosodium carbonate Mononatriumcarbonat *n*

monosodium citrate Mononatriumcitrat *n*
monosodium phosphate Mononatriumphosphat *n*
monosodium salt Mononatriumsalz *n*
monosodium sulfate Mononatriumsulfat *n*
monosodium sulfite Mononatriumsulfit *n*
monostearin Monostearin *n*
mono-substituted product
  Monosubstitutionsprodukt *n*
monosulfonic acid Monosulfonsäure *f*
monoterpene Monoterpen *n*
monotonous monoton, einförmig, eintönig
monotony Monotonie *f*
monotrichic eingeißlig (Zool)
monotrichous eingeißlig (Zool)
monotropic monotrop
monotropy Monotropie *f*
monotungsten carbide Monowolframcarbid *n*
monovalence Einwertigkeit *f* (Chem)
monovalent einwertig (Chem), monovalent
monoxide Monoxid *n*
monozygote Monozygot *m*
monrolite Monrolith *m* (Min)
montanic acid Montansäure *f*
montanine Montanin *n*
montanite Montanit *m* (Min)
montanol Montanol *n*
montanone Montanon *n*
montan pitch Montanpech *n*
montan resin Montanharz *n*
montan wax Montanwachs *n*
montanyl alcohol Montanylalkohol *m*
montebrasite Montebrasit *m* (Min)
monte-jus Montejus *m*, Saftheber *m*
montmorillonite Montmorillonit *m* (Min)
montroydite Montroydit *m* (Min)
monuron Monuron *n*
moonstone Mondstein *m* (Min)
moor coal Moor[braun]kohle *f*
moor peat Moortorf *m*
moraine Moräne *f* (Geol)
morbid krankhaft
morbidity Kränklichkeit *f*, Morbidität *f*
mordant Beize *f*, Farbbeize *f*, Kaustikum *n*
  (Chem), **dyeing on a** ~ beizenfärbend,
  preliminary ~ Vorbeize *f*, **weak** ~ Vorbeize *f*
mordant *a* beißend, kaustisch, scharf
mordant *v* beizen, fixieren (Farbe)
mordant action Beizkraft *f*, Beizwirkung *f*
mordant auxiliary Beizhilfsmittel *n*
mordant color Beizfarbe *f*
mordant dye Beizenfarbstoff *m*
mordant dyeing Beizenfärben *n*
mordant dyestuff Beizenfarbstoff *m*
mordant printing Beizendruck *m*
mordant yellow Beizengelb *n*
mordenite Mordenit *m* (Min)
morenosite Morenosit *m* (Min)
morganite Morganit *m* (Min)
morin Morin *n*

morinda Morinde *f* (Bot)
morindone Morindon *n*
moringatannic acid Moringagerbsäure *f*
moringine Moringin *n*
morinite Morinit *m* (Min)
morion Morion *m* (Min)
morocco (leather) Saffian[leder] *n*
morolic acid Morolsäure *f*
moronal Moronal *n*
moroxite Moroxit *m* (Min)
morphan Morphan *n*
morphanthridine Morphanthridin *n*
morphenol Morphenol *n*
morphia Morphin *n*, Morphium *n*
morphic acid Morphinsäure *f*
morphigenine Morphigenin *n*
morphimethine Morphimethin *n*
morphine Morphin *n*, Morphium *n*
morphine acetate Morphinacetat *n*
morphine hydrochloride Morphinhydrochlorid *n*
morphine meconate Morphinmekonat *n*
morphine methylbromide Morphosan *n*
morphine methyl ether Kodein *n*
morphine poisoning Morphinismus *m* (Med),
  Morphiumvergiftung *f* (Med)
morphine sulfate Morphinsulfat *n*
morphine valerate Morphinvalerianat *n*
morphinism Morphinismus *m* (Med),
  Morphiumvergiftung *f* (Med)
morphinone Morphinon *n*
morphinum Morphin *n*
morphium Morphium *n*
morphogenesis Morphogenese *f*
morphol Morphol *n*
morpholine Morpholin *n*
morphologic morphologisch
morphological change Formveränderung *f*
morphology Formenlehre *f*, Morphologie *f*
morpholone Morpholon *n*
morpholquinone Morpholchinon *n*
morphoran Morphoran *n*
morphosane Morphosan *n*
morphothebaine Morphothebain *n*
Morse apparatus Morseapparat *m*
Morse code Morsealphabet *n*
mortal tödlich, sterblich
mortality Mortalität *f*, Sterblichkeit[sziffer] *f*
mortar Mörtel *m*, Mörser *m*, Reibschale *f*,
  acid-proof ~ Säuremörtel *m*, **hard** ~
  Steinmörtel *m*, **non-hydraulic** ~ Luftmörtel *m*,
  quickly hardening ~ schnellbindender
  Mörtel, **water** ~ hydraulischer Mörtel
mortar sand Mörtelsand *m*
mortise Nut *f*, Keilnut *f*
mortise *v* auslochen, einzapfen, verzapfen
mortise chisel Stemmeisen *n*
morula cell Blastomere *n* (Biol)
morvenite Morvenit *m* (Min)
mosaic Mosaik *n*

mosaic gold Muschelgold *n*, Musivgold *n*,
   Zinndisulfid *n*
mosaic silver Musivsilber *n*
mosaic structure Mosaikstruktur *f*
mosandrite Mosandrit *m* (Min)
moschus Moschus *m*
Moseley's law Moseleysches Gesetz *n*
mosesite Mosesit *m* (Min)
moslene Moslen *n*
mosquito Moskito *m* (Zool), Stechmücke *f*
mosquito repellent Mückenvertreibungsmittel *n*
moss Moos *n* (Bot)
moss agate Moosachat *m* (Min)
moss green Moosgrün *n*
mossite Mossit *m* (Min)
moss starch Flechtenstärkemehl *n*, Lichenin *n*
moth Nachtfalter *m* (Zool)
mother cell Mutterzelle *f*, Stammzelle *f*
mother liquor Mutterlauge *f*
mother-of-pearl Perlmutter *f*
mother tincture Urtinktur *f*
mother vat Mutterfaß *n*
mother yeast Mutterhefe *f*
moth powder Mottenpulver *n*
mothproof mottensicher
mothproofing agent Mottenschutzmittel *n*
motion Bewegung *f*, Gang *m* (einer Maschine),
   alternate back and forward ~ hin- und
   hergehende Bewegung, anticlockwise ~
   Gegenlauf *m*, center of ~ Drehpunkt *m*,
   characteristic ~ Eigenbewegung *f*, constrained
   ~ erzwungene Bewegung (Mech), continuous
   ~ fortschreitende Bewegung, double ~
   gegenläufige Bewegung, law of ~
   Bewegungsgesetz *n*, rate of ~
   Bewegungsgeschwindigkeit *f*, reciprocating ~
   hin- und hergehende Bewegung, setting in ~
   Ingangsetzung *f*, state of ~
   Bewegungszustand *m*, thermic ~ thermische
   Bewegung, to set in ~ in Gang setzen, up and
   down ~ auf- und niedergehende Bewegung
motion equation Bewegungsgleichung *f*
motionless bewegungslos, regungslos, still,
   unbeweglich
motion picture Film *m*
motive Anlaß *m*, Beweggrund *m*, Motiv *n*,
   without any ~ ohne jeden Anlaß
motive power Antriebskraft *f*, bewegende Kraft *f*,
   Triebkraft *f*
motley clay Buntton *m*
motor Motor *m*, Antriebsmaschine *f*,
   direct-current ~ Gleichstromelektromotor *m*,
   geared ~ Getriebemotor *m*, polyphase ~
   Mehrphasenmotor *m*, supercharged ~
   Kompressormotor *m*, ~ of external rotor type
   Außenläufermotor *m*, the ~ runs idle der
   Motor läuft leer, ~ with short-circuit rotor
   Kurzschlußmotor *m*
motor armature Motoranker *m*

motor convertor Motorumformer *m*
motor drive Motorantrieb *m*
motor fuel synthesis Treibstoffsynthese *f*
motor gasoline Kraftstoff *m*
motor hood Motorhaube *f*
motorization Motorisierung *f*
motorize *v* motorisieren
motor oil Motorenöl *n*
motor power Motorleistung *f*
motor shaft Motorwelle *f*
motor speed Motordrehzahl *f*
motor spirit (Br. E.) Kraftstoff *m*, knock-resistant
   ~ klopffester Kraftstoff, premium grade ~
   Super-Kraftstoff, regular grade ~
   Normal-Kraftstoff
motor starter Anlasser *m*
motor vehicle Kraftfahrzeug *n*
motor works Motorenfabrik *f*
mottled geflammt (Keram, Holz), gefleckt
   (Keram), gesprenkelt, marmoriert, maseriert
mottles Marmorierung *f*
mottling Aderung *f*, Marmorierung *f*,
   Maserung *f*, Sprenkelung *f*
mottramite Mottramit *m* (Min)
mould (Br. E.) see mold
moulder *v* (Br. E.) see molder
mo[u]lding machine, reciprocating screw injection
   ~ Schneckenkolbenspritzgußmaschine *f*
mo[u]lding plaster Modellgips *m*
moulting hormone Häutungshormon *n*
mount Fassung *f*, Gestell *n*, Spannleiste *f*,
   Träger *m*
mount *v* [an]steigen, aufsetzen, montieren;
   aufstellen, fassen (Edelstein)
mountain ash Eberesche *f* (Bot)
mountain balm Bergbalsam *m* (Bot)
mountain blue Azurit *m* (Min), Bergblau *n*;
   Chessylit[h] *m* (Min); Kupferlasur *f* (Min)
mountain chalk Bergkreide *f*
mountain cork Bergkork *m*, Korkasbest *m*
mountain flax Amiant *m* (Min), Bergflachs *m*
   (Min), ligneous ~ holziger Bergflachs
mountain flour Berggur *f*, Bergmehl *n*
mountain green Berggrün *n*, Kupfergrün *n*
mountain leather Bergleder *n* (Min)
mountain peat Bergtorf *m*
mountain soap Erdseife *f* (Min), Saponit *m*
   (Min)
mountain tallow Bergtalg *m*, Bergwachs *n*
mountain tinder Bergzunder *m*
mountain yellow Berggelb *n*
mounting Gestell *n*, Armatur *f*, Befestigung *f*,
   Fassung *f*, Fixierung *f* (Präparat), Montage *f*,
   Montierung *f*, Rahmen *m*, ~ of optical
   instruments Fassung *f* optischer Instrumente
   (pl)
mounting frame Montagerahmen *m*
mounting panel Grundbrett *n*

mounting plate Aufspannplatte *f,*
  Befestigungsplatte *f*
mountings Beschläge *m pl*
mounting stud Befestigungsbolzen *m*
mouth Öffnung *f,* Gichtöffnung *f* (Hochofen),
  Mündung *f* (Röhre, Fluß), Mundstück *n*
mouthpiece Mundstück *n*
mouthwash Mundwasser *n*
movability Beweglichkeit *f,* Verschiebbarkeit *f*
movable beweglich, fahrbar, verstellbar
move *v* bewegen, verlagern, verschieben,
  versetzen, ~ forward vorrücken, ~ to and
  fro hin- und herbewegen
moveable beweglich
movement Bewegung *f,* Lauf *m* (einer
  Maschine), direction of ~
  Bewegungsrichtung *f,* freedom of ~
  Bewegungsfreiheit *f,* ~ of the rider
  Reiterverschiebung *f*
moving bed Wanderbett *n* (technische
  Reaktionsführung)
moving belt production Bandarbeit *f*
moving coil Drehspule *f* (Elektr)
moving-coil ammeter Drehspulenstrommesser *m*
moving-coil galvanometer
  Drehspulgalvanometer *n*
moving-coil instrument Drehspulinstrument *n*
moving-coil mirror galvanometer
  Drehspulspiegelgalvanometer *n*
moving-magnet galvanometer
  Nadelgalvanometer *n* (Elektr)
movrine Movrin *n*
mucamide Mucamid *n*
mucedine Mucedin *n*
mucic acid Schleimsäure *f*
muciferous schleimbildend
mucigen Mucigen *n*
mucigenous schleimbildend
mucilage Gummilösung *f,* Leim *m,*
  Pflanzenschleim *m,* Schleim *m*
mucilaginous schleimig, seimig
mucin Mucin *n*
mucinogen Mucinogen *n*
mucobromic acid Mucobromsäure *f*
mucochloric acid Mucochlorsäure *f*
mucoid Mucoid *n*
mucoidin Mucoidin *n*
mucoinositol Mucoinosit *m*
mucoitin sulfate Mucoitinsulfat *n*
mucoitinsulfuric acid Mucoitinschwefelsäure *f*
mucolytic schleimlösend (Med)
muconic acid Muconsäure *f*
mucopeptide Mucopeptid *n*
mucopolysaccharidase Mucopolysaccharidase *f*
  (Biochem)
mucopolysaccharide Mucopolysaccharid *n*
mucoprotein Mucoproteid *n*
mucosa (pl mucosae) Schleimhaut *f* (Med)
mucosin Mucosin *n*

mucous schleimig
[mucous] bursa Schleimbeutel *m* (Med)
mucous membrane Schleimhaut *f*
mucous secretion Schleimabsonderung *f*
mucous tissue Schleimgewebe *n*
mucus Schleim *m*
mucusane Mucusan *n*
mud Kot *m,* Schlamm *m,* Schlick *m,* Schmutz *m,*
  accumulation of ~ Verschlammung *f,* deposit
  of ~ Schlammablagerung *f*
mudaric acid Mudarsäure *f*
mudarin Mudarin *n*
mud box Schlammkasten *m,* Reinigungskasten *m*
mud cock Schlammablaßhahn *m*
muddy lehmig, schlammhaltig, schlammig
muddy ground Schlammboden *m*
mud filter Schlammfilter *n*
mudguard Kotflügel *m* (Kfz), Schutzblech *n*
mud gun Stichlochstopfmaschine *f*
muffle Farbenschmelzofen *m,* Muffel *f* (Metall),
  cast iron ~ Muffel *f* aus Eisen, fire clay ~
  Muffel *f* aus Schamotte
muffle *v* [ab]dämpfen (Akust)
muffled dumpf, abgedämpft
muffle furnace Muffelofen *m,* Glühretorte *f,*
  double ~ Doppelmuffelofen *m,* gas-fired ~
  Gasmuffelofen *m*
muffler Schalldämpfer *m* (Kfz)
muffle roaster Muffelofen *m*
muffling Abdämpfung *f* (Akust), Dämpfung *f*
  (Akust)
mug Krug *m*
mugwort Beifuß *m* (Bot)
mugwort oil Beifußöl *n*
muhuhu oil Muhuhu-Öl *n*
mulberry tree Maulbeerbaum *m*
mulch *v* mulchen
muller Tellerreiber *m*
muller mixer Kollergang *m,* Rollquetscher *m*
mulling time Kollerzeit *f*
mulse Honigwein *m*
mult[i]angular vieleckig
multicavity magnetron Vielschlitzmagnetron *n*
multicellular mehrzellig, vielzellig
multicellular voltmeter Multizellularvoltmeter *n*
multicentered bonding Mehrzentrenbindung *f*
multicentered reaction Mehrzentrenreaktion *f*
multi-chamber centrifuge Kammerzentrifuge *f*
multicircuit switch Serienschalter *m* (Elektr)
multicolored mehrfarbig
multicolor finish Mehrfarbenlack *m*
multicolor printing Mehrfarbendruck *m*
multi-compartment drum filter
  Trommelsaugfilter *n,* Trommelzellenfilter *n*
multicomponent distillation
  Mehrkomponentendestillation *f*
multicomponent equilibrium
  Mehrstoffgleichgewicht *n*

multicomponent mixture Mehrstoffgemisch *n*,
Vielstoffgemisch *n*
multicomponent system Mehrstoffgemisch *n*,
Vielstoffgemisch *n*
multiconductor Vielfachleiter *m*
multicore vieladrig
multicore cable Mehrleiterkabel *n*
multicoupling Mehrfachschaltung *f*
multicurrent dynamo Mehrfachstromerzeuger *m*
multicyclone Multizyklon *m*
multicylinder mehrzylindrig
multicylinder dryer Mehrzylindertrockner *m*
multi-daylight press Stapelpresse *f*
multidimensional mehrdimensional
multi-electrode system Mehrelektrodensystem *n*
multi-electrode tube Mehrelektrodenröhre *f*
multiengine[d] mehrmotorig
multienzyme complex Multienzymkomplex *m*
multienzyme system Multienzymsystem *n*
multiflame burner Mehrfachbrenner *m*
multiform vielgestaltig, vielförmig
multigroup theory Multigruppentheorie *f*
multi-impression mold Mehrfachform *f*
multilayer Mehrfachschicht *f*
multilayered mehrschichtig, vielschichtig
multinutrient fertilizer Mehrnährstoffdünger *m*
multipack Sammelpackung *f*
multipart mehrteilig
multiphase mehrphasig
multiphase converter Mehrphasenumformer *m*
(Elektr)
multiphase current Mehrphasenstrom *m* (Elektr)
multiphase current generator
Mehrphasenstromerzeuger *m* (Elektr)
multiphase current plant
Mehrphasenstromanlage *f*
multiphase system Mehrphasensystem *n*
(Elektr), **balanced** ~ angeglichenes
Mehrphasensystem
multiple Mehrfache[s] *n* (Math), Vielfache[s] *n*
(Math), **integral** ~ ganzes Vielfaches, **least
common** ~ das kleinste gemeinsame
Vielfache, **odd** ~ ungerades Vielfaches, ~ **of
a number** Vielfache[s] *n* einer Zahl (Math)
multiple *a* mehrfach, vielfach, vielfältig
multiple bond Mehrfachbindung *f*
multiple cavity mold Mehrfachform *f*
multiple coil condenser Intensivkühler *m*
multiple collision Vielfachstoß *m*
multiple contact Vielfachkontakt *m*
multiple crucible method Mehrtiegelverfahren *n*
multiple cut lathe Vielschnittdrehbank *f*
multiple decay Aufzweigung *f* (Atom)
multiple delay discriminator
Vielfachverzögerungsdiskriminator *m*
multiple-disc brake Lamellenbremse *f*
multiple grid tube Mehrgitterröhre *f*
multiple hearth roaster Etagenofen *m*,
Etagenröstofen *m*

multiple-intertube burner Stahlbrenner *m*
multiple opening press Etagenpresse *f*
multiple paddle agitator
Mehrfachbalkenrührer *m*
multiple pendulum Mehrfachpendel *n*
multiple piston arrangement
Mehrkolbenanordnung *f*
multiple-plate brake Lamellenbremse *f*
multiple purpose Mehrzweck *m*
multiple scattering Vielfachstreuung *f*
multiple slip Mehrfachgleitung *f*
multiple stage boiler Staffelkessel *m*
multiple-stage demagnetization
Mehrstufenentmagnetisierung *f*
multiple system Parallelschaltungssystem *n*
multiplet Multiplett *n* (Spektr)
multiplet component Multiplettanteil *m*
multiplet intensity rule Intensitätsregel *f* für ein
Multiplett (Spektr)
multiplet splitting Multiplettaufspaltung *f*
multiplet structure Multiplettstruktur *f*
multiple-tube cooler Rohrbündelkühler *m*
multiple-way stopcock Vielwegehahn *m* (Chem)
multiple wire system Mehrleitersystem *n*
multiplex mehrfach, vielfach
multiplicand Multiplikand *m* (Math)
multiplication Multiplikation *f* (Math), **sign of**
~ Malzeichen *n* (Math),
Multiplikationszeichen *n* (Math), ~ **of power**
Kraftvervielfachung *f*
multiplication constant
Multiplikationskonstante *f*
multiplication factor Multiplikationsfaktor *m*
multiplication theorem Multiplikationstheorem *n*
multiplicative vervielfältigend
multiplicity Vielfältigkeit *f*, Vielfalt *f*,
Vielwertigkeit *f* (Math), Vielzahl *f*
multiplier Multiplikator *m* (Math),
Verstärker *m* (Phys), Vervielfacher *m* (Phys)
multiplier photoelectric cell
Vervielfacherphotozelle *f*
multi-ply mehrschichtig
multiply *v* multiplizieren (Math), vermehren,
vervielfachen, vervielfältigen
multi-point recorder Punktdrucker *m*
multipolar mehrpolig, vielpolig
multipolar armature Vielpolanker *m*
multipolar dynamo Vielpoldynamo *m*
multipolarity Multipolordnung *f*
multipole field Multipolfeld *n*
multipole potential Multipolpotential *n*
multipole radiation Multipolstrahlung *f*
multipole transition Multipolübergang *m*
multipurpose Mehrzweck-
multipurpose *a* universell
multipurpose computer Universalrechner *m*
multipurpose furnace Mehrzweckofen *m*
multipurpose machine Universalmaschine *f*
multirotation Multirotation *f*

multiscaler Vielfachzähler *m*
multiscrew extruder Mehrschneckenpresse *f*
multistage mehrstufig, vielstufig
multistage compression mehrstufige
  Verdichtung *f*
multistage grate Staffelrost *m*
multi-stage impulse ribbon blender
  MIG-Rührer *m*
multistage rocket [Mehr]Stufenrakete *f*
multistep reduction gear Stufengetriebe *n*
multivalence Mehrwertigkeit *f* (Chem),
  Vielwertigkeit *f* (Chem)
multivalent höherwertig (Chem), mehrwertig
  (Chem), vielwertig (Chem)
multivector Multivektor *m*
multivitamin preparation Multivitaminpräparat *n*
multiwire counter tube Vieldrahtzählrohr *n*
mu meson My-Meson *n*, Myon *n*
munduloxic acid Munduloxsäure *f*
muntz metal Muntzmetall *n*
muon My-Meson *n*, Myon *n*
muon capture Myoneneinfang *m*
muon pair production Myonenpaarerzeugung *f*
muramic acid Muraminsäure *f*, Muramsäure *f*
murami[ni]dase Murami[ni]dase *f* (Biochem)
murchisonite Murchisonit *m* (Min)
murexan Uramil *n*
murexide Murexid *n*, Ammonpurpurat *n*,
  Purpurcarmin *n*
murexide assay Murexidprobe *f* (Anal)
murexine Murexin *n*
muriatic acid (commercial grades) Salzsäure *f*
murky dunkel, düster
muromontite Muromontit *m* (Min)
muropeptide Muropeptid *n*
murrayin Murrayin *n*
muscarine Muscarin *n*
muscarone Muscaron *n*
muscarufin Muscarufin *n*
muscle Muskel *m*, resting ~ ruhender Muskel *m*
muscle adenylic acid Muskeladenylsäure *f*
muscle cell Muskelzelle *f*
muscle contraction Muskelkontraktion *f*
muscle excitation Muskelanregung *f*
muscle extract Muskelextrakt *m*
muscle fiber Muskelfaser *f*
muscle fibrin Syntonin *n*
muscle phosphorylase Muskelphosphorylase *f*
  (Biochem)
muscle pigment Muskelfarbstoff *m*
muscle relaxation Muskelentspannung *f*
muscle sugar Inosit *m*
muscle tissue Muskelgewebe *n*
muscone Muskon *n*
muscovado Muskovade *f*
muscovite Muskovit *m* (Min), Kaliglimmer *m*
  (Min), Phengit *m* (Min), Silberglimmer *m*
  (Min)
muscular fiber Muskelfaser *f*

muscular performance Muskelleistung *f*
muscular system Muskulatur *f*
musculine Muskulin *n*
mushroom Pilz *m*
mushroom insulator Pilzisolator *m*
mushroom mixer Pilzmischer *m*
mushroom poisoning Pilzvergiftung *f*
mushroom stone Schwammstein *m*
mushroom valve Muschelschieber *m*
musk Moschus *m*
muskine Muskon *n*
musk ketone Moschusketon *n*
musk root Sumbulwurzel *f*
musk seed Moschuskörner *pl*
musk seed oil Moschuskörneröl *n*
muslin Musselin *m* (Text)
muslin polishing wheel Nesselpolierscheibe *f*
must Most *m*
mustard Senf *m*, Mostrich *m*, ground ~
  Senfpulver *n*
mustard gas Senfgas *n*, Yperit *n*
mustard gas sulfone Senfgassulfon *n*
mustard oil Senföl *n*, cyanated ~ Cyansenföl *n*
musty moderig, schimmelartig, schimmelig,
  vermodert
mutagen Mutagen *n*
mutant Mutante *f*, durch Mutation entstandene
  Variante
mutarotation Mutarotation *f*
mutase Mutase *f* (Biochem)
mutation Genänderung *f* (Biol),
  Genveränderung *f* (Biol), Mutation *f* (Biol)
mutation rate Mutationsgeschwindigkeit *f*
mutatochrome Mutatochrom *n*
mutatoxanthin Mutatoxanthin *n*
muthmannite Muthmannit *m* (Min)
Muthmann's liquid Acetylentetrabromid *n*
mutilate *v* verstümmeln
mutton Hammelfleisch *n*
mutton tallow Hammeltalg *m*
mutual gegenseitig, wechselseitig
mutual action Synergismus *m*
mutual inductance Gegeninduktion *f*
mutual inductivity Gegeninduktion *f*
mutual potential energy
  Wechselwirkungsenergie *f*
myanesin Myanesin *n*
mycaminitol Mycaminit *m*
mycaminose Mycaminose *f*
mycarose Mycarose *f*
mycelium Myzel[ium] *n*, Pilzgeflecht *n*
mycetism Pilzvergiftung *f*
mycobacterium Mycobakterium *n*
mycocerosic acid Mycocerosinsäure *f*
mycoctonine Mycoctonin *n*
mycodextrane Mykodextran *n*
mycogalactan Mykogalaktan *n*
mycoine Mycoin *n*
mycol Mykol *n*

mycolipenic acid Mycolipensäure *f*
mycology Mykologie *f*, Pilzkunde *f*
mycomycin Mycomycin *n*
mycophenolic acid Mycophenolsäure *f*
mycoprotein Mycoprotein *n*
mycosamine Mycosamin *n*
mycose Mykose *f*
mycosis Pilzkrankheit *f* (Med)
mycosterol Mykosterin *n*
mydriasine Mydriasin *n*
mydriatine Mydriatin *n*
myelin Myelin *n*
myelitis Rückenmarkentzündung *f* (Med)
myeloblast Knochenmarkleukozyt *m*
myeloid markartig
myeloid cell Knochenmarkzelle *f*
myeloid tissue Markgewebe *n* (Med)
myelolymphocyte Knochenmarklymphozyt *m*
myelomonocyte Knochenmarkmonozyt *m*
myeloperoxidase Myeloperoxidase *f* (Biochem)
myocyte Muskelzelle *f*
myofibril Myofibrille *f*
myoglobin Myoglobin *n*, Muskelfarbstoff *m*
myoh[a]ematin Myohämatin *n*
myoinosamine Myoinosamin *n*
myokinase Myokinase *f* (Biochem)
myosin Myosin *n*, Muskeleiweiß *n*
myosmine Myosmin *n*
myrcene Myrcen *n*
myrcenol Myrcenol *n*
myrcia wax Myrtenwachs *n*
myrental Myrental *n*
myricetin Myricetin *n*
myricin Myricin *n*
myricyl Myricyl-
myricyl alcohol Myricylalkohol *m*
myricyl palmitate Myricin *n*
myristic acid Myristinsäure *f*
myristica oil Muskatbalsam *m*
myristicin Myristicin *n*
myristicin aldehyde Myristicinaldehyd *m*
myristicin glycol Myristicinglykol *n*
myristicinic acid Myristicinsäure *f*
myristic ketone Myriston *n*
myristicol Myristicol *n*
myristin Myristin *n*
myristoleic acid Myristoleinsäure *f*
myristolic acid Myristolsäure *f*
myristone Myriston *n*
myristyl alcohol Myristylalkohol *m*
myronic acid Myronsäure *f*
myrosase Myrosin *n*
myrosin Myrosin *n*
myrosinase Myrosin *n*
myroxin Myroxin *n*
myrrh Myrrhe *f* (Bot)
myrrh balm Myrrhenbalsam *m*
myrrh oil Myrrhenöl *n*
myrrh tincture Myrrhentinktur *f*

myrtanol Myrtanol *n*
myrtenal Myrtenal *n*
myrtenic acid Myrtensäure *f*
myrtenic aldehyde Myrtenal *n*
myrtenol Myrtenol *n*
myrticolorin Myrticolorin *n*
myrtillidin Myrtillidin *n*
myrtillin Myrtillin *n*
myrtillogenic acid Myrtillogensäure *f*
myrtle berries, essence of ~ Myrtenessenz *f*
myrtle green Myrtengrün *n*
myrtle oil Myrtenöl *n*
myrtle wax Myristin *n*, Myrtenwachs *n*
myrtol Myrtol *n*
mytilite Mytilit *m*
mytilitol Mytilit *m*
mytilotoxin Mytilotoxin *n*
myxobacteria Myxobakterien *pl* (Biol)
myxophyceae Spaltalgen *pl*
myxoxanthin Myxoxanthin *n*

N

nacarat Nakarat[farbe f] *n* (Färb)
nacre Perlmutt *n*, Perlmutter *f*
nacreous perlmutterartig
nacreous effect Perlmutteffekt *m*
nacreous luster Perlmuttglanz *m*
nacreous particles Fischsilberpigmente *pl*,
  Perlmutter-Pigmentteilchen *pl*
nacreous pigment Fischsilberpigment *n*
NAD (Abk) Diphosphopyridinnucleotid *n*
nadorite Nadorit *m* (Min)
NADP (Abk) Triphosphopyridinnucleotid *n*
  (Biochem)
naegite Naegit *m* (Min)
nagyagite Nagyagit *m* (Min), Blättertellur *n*
  (Min), Graugolderz *n* (Min), Tellurglanz *m*
  (Min)
nail Nagel *m*
nail *v* [fest]nageln
nail brush Nagelbürste *f*
nail-holding property Nagelbarkeit *f*
nail iron Nageleisen *n*
nail pass Nagelkaliber *n*
nail polish Nagellack *m* (Kosm)
nail polish remover Nagellackentferner *m*
  (Kosm)
nails, to fasten by ~ festnageln
naked unbewaffnet (Auge)
nakrite Nakrit *m* (Min)
nalorphine Nalorphin *n*
namakochrome Namakochrom *n*
name Benennung *f* (Chem)
name *v* benennen
name plate Schild *n*
nandinine Nandinin *n*
nantokite Nantokit *m* (Min)
napelline Napellin *n*
naphtha Erdöl *n*, Leuchtöl *n*, Naphtha *n*, **crude**
  ~ Rohnaphtha *n*, Rohpetroleum *n*,
  distillation product of ~ Erdöldestillat *n*,
  native ~ Rohnaphtha *n*, Rohpetroleum *n*
naphthacene Naphthacen *n*
naphthacene quinone Naphthacenchinon *n*
naphthacridine Naphthacridin *n*,
  Dibenzacridin *n*
naphthacridinedione Anthrachinonacridin *n*
naphthacyl black Naphthacylschwarz *n*
naphthalane Naphthalan *n*
naphthaldazine Naphthaldazin *n*
naphthaldehyde Naphthaldehyd *m*,
  Naphthoealdehyd *m*
naphthaldehydic acid Naphthaldehydsäure *f*
naphthalene Naphthalin *n*, **crude** ~
  Rohnaphthalin *n*
naphthalene blue Naphthalinblau *n*
naphthalene camphor Naphthalincampher *m*
naphthalene carboxylic acid Naphthoesäure *f*

naphthalene derivative Naphthalinderivat *n*
naphthalenediol, 1,2- ~
  Naphthobrenzcatechin *n*, 1,3- ~
  Naphthoresorcin *n*
naphthalene dyestuff Naphthalinfarbstoff *m*
naphthalene indigo Naphthalinindigo *m*
naphthalene red Naphthalinrot *n*, Magdalarot *n*
naphthalene separator Naphthalinabscheider *m*
naphthalenesulfonic acid
  Naphthalinsulfonsäure *f*
naphthalenic acid Naphthalinsäure *f*
naphthalic acid Naphthalsäure *f*
naphthalide Naphthalid *n*
naphthalidine Naphthylamin *n*
naphthalimide Naphthalimid *n*
naphthalin Naphthalin *n*
naphthalol Naphthalol *n*
naphthamine Naphthamin *n*
naphthane Naphthan *n*
naphthanene Naphthanen *n*
naphthanisol Naphthanisol *n*
naphthanol Naphthanol *n*
naphthanthracene Naphthanthracen *n*
naphthanthracridine Naphthanthracridin *n*
naphthanthraquinone Naphthanthrachinon *n*
naphthanthroxanic acid Naphthanthroxansäure *f*
naphtha residue Erdölrückstand *m*
naphtharson Naphtharson *n*
naphthazarine Naphthazarin *n*
naphthazine Naphthazin *n*
naphthenate Naphthenat *n*
naphthene Naphthen *n*
naphthenic acid Naphthensäure *f*
naphthhydrindene Naphthhydrinden *n*
naphthidine Naphthidin *n*
naphthimidazole Naphthimidazol *n*
naphthindazole Naphthindazol *n*
naphthindene Naphthinden *n*
naphthindigo Naphthindigo *n*
naphthindole Naphthindol *n*
naphthindoline Naphthindolin *n*
naphthindone Naphthindon *n*
naphthindoxyl Naphthindoxyl-
naphthionate Naphthionat *n*
naphthionic acid Naphthionsäure *f*
naphthisatin Naphthisatin *n*
naphthisatin chloride Naphthisatinchlorid *n*
naphthoacridine Naphthoacridin *n*
naphthocarbostyril Naphthocarbostyril *n*
naphthochromanone Naphthochromanon *n*
naphthofluorene Naphthofluoren *n*
naphthofuchsone Naphthofuchson *n*
naphthofuran Naphthofuran *n*
naphthohydroquinone Naphthohydrochinon *n*
naphthoic acid Naphthoesäure *f*
naphthoic aldehyde Naphthoealdehyd *m*

naphthol Naphthol *n*, Oxynaphthalin *n*
naphthol-4-sulfonic acid, 1- ~ Nevillesäure *f*
naphthol-8-sulfonic acid, 2- ~ Bayersche Säure, Bayer-Säure *f* (HN), Croceinsäure *f*
naphthol allyl ether Naphtholallyläther *m*
naphthol benzoate Naphtholbenzoat *n*
naphthol black Naphtholschwarz *n*
naphthol blue Naphtholblau *n*
naphtholcarboxylic acid Oxynaphthoesäure *f*, Naphtholcarbonsäure *f*
naphthol dye[stuff] Naphtholfarbstoff *m*
naphthol orange Naphtholorange *n*
naphthol phthalein Naphtholphthalein *n*
naphthol salicylate Naphtholsalicylat *n*
naphtholsalol Naphtholsalol *n*
naphthol solution Naphthollösung *f*
naphthol sulfonate Naphtholsulfonat *n*
naphtholsulfonic acid Naphtholsulfonsäure *f*
naphthol yellow Naphtholgelb *n*
naphthonitrile Naphthonitril *n*
naphthophenazine Naphthophenazine *n*, Phenonaphthazin *n*
naphthophenofluorindine Naphthophenofluorindin *n*
naphthopicric acid Naphthopikrinsäure *f*
naphthopiperazine Naphthpiperazin *n*
naphthopurpurin Naphthopurpurin *n*
naphthopyrane Naphthopyran *n*
naphthopyrazine Naphthopyrazin *n*
naphthopyrocatechol Naphthobrenzcatechin *n*
naphthopyrone Naphthopyron *n*
naphthoquinaldine Naphthochinaldin *n*
naphthoquinhydrone Naphthochinhydron *n*
naphthoquinoline Naphthochinolin *n*
naphthoquinone Naphthochinon *n*
naphthoresorcinol Naphthoresorcin *n*
naphthostyril Naphthostyril *n*
naphthotetrazole Naphthotetrazol *n*
naphthothianthrene Naphthothianthren *n*
naphthothiazine Naphthothiazin *n*
naphthothiazole Naphthothiazol *n*
naphthothioflavone Naphthothioflavon *n*
naphthothioindigo Naphthothioindigo *n*
naphthothioxole Naphthothioxol *n*
naphthotriazine Naphthotriazin *n*
naphthotriazole Naphthotriazol *n*
naphthoxazine Naphthoxazin *n*, Phenoxazin *n*
naphthoxazole Naphthoxazol *n*
naphthoylbenzoic acid Naphthoylbenzoesäure *f*
naphthoyl chloride Naphthoylchlorid *n*
naphthoylpropionic acid Naphthoylpropionsäure *f*
naphthoyltrifluoroacetone Naphthoyltrifluoraceton *n*
naphthuric acid Naphthursäure *f*
naphthyl Naphthyl-
naphthylamine Naphthylamin *n*, Aminonaphthalin *n*

naphthylamine-4-sulfonic acid, 1- ~ Naphthionsäure *f*
naphthylamine-6-sulfonic acid, 1- ~ Clevesäure *f*
naphthylamine hydrochloride Naphthylaminchlorhydrat *n*
naphthylamine red Naphthylaminrot *n*
naphthylamine sulfonic acid Naphthylaminsulfonsäure *f*
naphthylamine yellow Naphthylamingelb *n*
naphthyl benzoate Naphtholbenzoat *n*
naphthyl blue Naphthylblau *n*
naphthyl chloride Chlornaphthalin *n*, Naphthylchlorid *n*
naphthyl cyanide Naphthonitril *n*
naphthylene diamine Naphthylendiamin *n*
naphthylene sulfonylide Naphthylensulfonylid *n*
naphthyl mercaptan Thionaphthol *n*
naphthyl methyl bromide Menaphthylbromid *n*
naphthyl methyl ketone Acetonaphthon *n*
naphthyl phenylamine Naphthylphenylamin *n*
naphthyl salicylate Naphthalol *n*, Naphthylsalicylat *n*, 2- ~ Betol *n*
naphthyl thiourea Naphthylthioharnstoff *m*
naphthyl trichlorosilane Naphthyltrichlorsilan *n*
naphthyridine Naphthyridin *n*
napkin paper Serviettenpapier *n*
Naples red Neapelrot *n*
Naples yellow Neapelgelb *n*
napoleonite Napoleonit *m* (Min), Corsit *m* (Min)
nappa [leather] Nappaleder *n*
napping mill Rauhmaschine *f*
narcein[e] Narcein *n*
narceine hydrochloride Narceinhydrochlorid *n*
narceine sodium salicylate Antispasmin *n*
narceonic acid Narceonsäure *f*
narcine Narcein *n*
narcissamine Narcissamin *n*
narcissidine Narcissidin *n*
narcissine Narcissin *n*, Lycorin *n*
narcophine Narkophin *n*
narcotic Betäubungsmittel *n*, Narkotikum *n*, Rauschgift *n*
narcotic *a* narkotisch, betäubend
narcotine Narkotin *n*
narcotine hydrochloride Narkotinhydrochlorid *n*
narcotization Narkotisierung *f*
narcotize *v* narkotisieren
narcotizing narkotisierend
narcotoline Narcotolin *n*
nargol Nargol *n*
naringenin Naringenin *n*
naringeninic acid Naringeninsäure *f*
naringetol Naringetol *n*
naringin Naringin *n*
narrow schmal, eng, knapp
narrow *v* sich verengen, ~ down einengen
narrow film Schmalfilm *m* (Phot)

narrowing Einengung *f*, Verengung *f*,
Verjüngung *f*
narrow-meshed engmaschig
narrow-neck[ed] bottle Enghalsflasche *f*
narrow V-belt Schmalkeilriemen *m*
nartazine Nartazin *n*
narwedine Narwedin *n*
nascent entstehend, naszierend (Chem)
nascent state Entstehungszustand *m*
nasonite Nasonit *m* (Min)
nasturtium Kapuzinerkresse *f* (Bot)
natalensine Natalensin *n*
nataloin Nataloin *n*
native gediegen (Min), nativ, natürlich
native copper Bergkupfer *n*
native substance Naturstoff *m*
natroborocalcite Natroborocalcit *m* (Min)
natrocalcite Natrocalcit *m* (Min)
natrolite Natrolith *m* (Min), Faserzeolith *m*
(Min), Nadelzeolith *m* (Min)
natrophilite Natrophilit *m* (Min)
natural natürlich, nativ, roh (Wolle), wirklich,
naturgemäß, eigen
natural-circulation boiler Durchlaufkessel *m*,
Naturumlaufkessel *m*
natural color Eigenfarbe *f*, Naturfarbe *f*
natural colored naturfarben
natural convection Eigenkonvektion *f* (Therm)
natural drying Lufttrocknung *f*
natural dyestuff Naturfarbstoff *m*
natural exposure test Freilagerversuch *m*
natural fat Naturfett *m*
natural fiber Naturfaser *f*
natural frequency Eigenfrequenz *f*, ~ **in bending**
Biegeeigenschwingung *f*
natural gas Erdgas *n*, **source of** ~
Erdgasvorkommen *n*
natural gas plant Erdgasanlage *f*
naturalist Naturforscher *m*, Naturkundiger *m*
natural law Naturgesetz *n*
naturally occurring in der Natur vorkommend
natural number natürliche Zahl *f*
natural phenomenon Naturerscheinung *f*
natural power Naturkraft *f*
natural process Naturerscheinung *f*
natural product Naturprodukt *n*
natural radiation Eigenstrahlung *f*
natural resin Naturharz *n*
natural science Naturwissenschaft *f*,
Naturforschung *f*, Naturkunde *f*
natural selection natürliche Zuchtwahl *f* (Biol)
natural silk Naturseide *f*, **waste from** ~
Naturseidenabfall *m*
natural state Naturzustand *m*
natural steel Wolfsstahl *m*, Schmelzstahl *m*
natural stone Naturstein *m*
natural substance Naturstoff *m*
natural tannin Naturgerbstoff *m*

natural uranium graphite reactor
Natururan-Graphit-Reaktor *m*
natural vibration Eigenschwingung *f*
natural water Rohwasser *n*
natural weathering test Freilagerversuch *m*
nature Natur *f*, Beschaffenheit *f*, Eigenart *f*,
Wesen *n*, **law of** ~ Naturgesetz *n*, **true to** ~
naturgetreu, ~ **of a charge** Ladungssinn *m*
nature of surface Oberflächenbeschaffenheit *f*
naught Null *f*
naumannite Naumannit *m* (Min), Selensilber *n*
(Min)
nausea Brechreiz *m*, Übelkeit *f*
nauseating ekelerregend
navigation Navigation *f*, Schiffahrt *f*
navy blue Marineblau *n*
n-butane Normalbutan *n*
near nah[e]
near point Nahpunkt *m* (Opt)
neatness Sauberkeit *f*
neat's foot oil Klauenöl *n*, Huffett *n*,
Knochenöl *n*
nebula Nebel *m* (Astr)
nebularine Nebularin *n*
nebular line Nebellinie *f*
nebular shell Nebelhülle *f*
necessary notwendig
necessary article Bedarfsgegenstand *m*
necessity Notwendigkeit *f*, Bedürfnis *n*
neck Hals *m*, ~ **of a converter** Hals *m* der
Bessemerbirne, ~ **of a shaft** Hals *m* einer
Welle
neck down *v* verstrecken (Kunststoffäden)
neck graduation Halsteilung *f*
neck groove Halskerbe *f*
neck-in Randeinzug *m* (Extrusion)
necking [down] Verstreckung *f*
necrobiosis Nekrobiose *f*, Zellenabsterben *n*
necrosamine Necrosamin *n*
need Bedarf *m*, Bedürfnis *n*
need *v* benötigen, bedürfen
needle Nadel *f*, Zeiger *m* (Kompaß), **point of a**
~ Nadelspitze *f*
needle bearing Nadellager *n*
needle biopsy Nadelbiopsie *f* (Med)
needle bottom Nadelboden *m*
needle deflection Zeigerausschlag *m*
needle electrode Spitzenelektrode *f*
needle galvanometer Nadelgalvanometer *n*
(Elektr), Zeigergalvanometer *n* (Elektr)
needle iron ore Nadeleisenerz *n* (Min)
needle lubricator Nadelschmierer *m*,
Nadelschmiergefäß *n*
needle mold Nadelform *f*
needle ore Nadelerz *n* (Min)
needle probe Nadelsonde *f*
needle shape Nadelform *f*
needle-shaped nadelförmig
needle spar Aragonit *m* (Min)

needle stone Nadelstein *m* (Min), Natrolith *m* (Min)
needle tin Nadelzinnerz *n* (Min)
needle valve Nadelventil *n*, **miniature ~** Kleinstnadelventil *n*
needle zeolite Nadelzeolith *m* (Min), Nadelstein *m* (Min)
neem oil Margosöl *n*
negative Negativ *n* (Phot)
negative *a* negativ, verneinend
negative bias negative Vorspannung *f*
negative electrode Kathode *f*, negative Elektrode *f*
negative lens Konkavlinse *f* (Opt)
negatively charged negativ geladen
negative paper Negativpapier *n* (Phot)
negative plate Minuselektrode *f*, Minusplatte *f* (Akku)
negative pole Minuspol *m*
negative ray Kathodenstrahl *m*
negative sign Minuszeichen *n* (Math)
negative valence negative Wertigkeit *f*
negativity Negativität *f* (Elektr)
negatron Negatron *n* (Phys)
neglect Vernachlässigung *f*
neglect *v* vernachlässigen
neglected unberücksichtigt
negligence Fahrlässigkeit *f*, Nachlässigkeit *f*
negligent achtlos, gleichgültig, fahrlässig
negligible belanglos, vernachlässigbar
neighborhood Nachbarschaft *f*, Umgebung *f*
neighboring (Br. E. = neighbouring) benachbart, vizinal
neighboring atom Nachbaratom *n*
neighboring group participation Nachbargruppeneffekt *m*
neighboring position Nachbarstellung *f* (Atom)
nemalite Nemalith *m* (Min)
nemaphyllite Nemaphyllit *m* (Min)
nematic nematisch (Krist)
nematocide Nematocid *n*
nemotin Nemotin *n*
nemotinic acid Nemotinsäure *f*
neoabietic acid Neoabietinsäure *f*
neoagarobiose Neoagarobiose *f*
neoagarobitol Neoagarobit *m*
neoamyl alcohol Neoamylalkohol *m*
neoarsphenamine Neoarsphenamin *n*
neoaspartic acid Neoasparaginsäure *f*
neobotogenin Neobotogenin *n*
neocarthamin Neocarthamin *n*
neocerotic acid Neocerotinsäure *f*
neochanin Neochanin *n*
neochlorogenic acid Neochlorogensäure *f*
neocinchophen Neocinchophen *n*
neococcin Neucoccin *n*
neocolemannite Neocolemannit *m*
neocuproine Neocuproin *n*
neocyanine Neocyanin *n*, Allocyanin *n*

neodiarsenol Neodiarsenol *n*
neodymia Neodymoxid *n*
neodymium (Symb. Nd) Neodym *n*
neodymium carbide Neodymcarbid *n*
neodymium chloride Neodymchlorid *n*
neodymium citrate Neodymcitrat *n*
neodymium content Neodymgehalt *m*
neodymium oxide Neodymoxid *n*
neodymium sulfate Neodymsulfat *n*
neodymium sulfide Neodymsulfid *n*
neoergosterol Neoergosterin *n*
neogen Neogen *n*
neoglycerol Neoglycerin *n*
neohecogenin Neohecogenin *n*
neoherculin Neoherculin *n*
neohexane Neohexan *n*
neoinosamine Neoinosamin *n*
neoinositol Neoinosit *m*
neoisoverbanol Neoisoverbanol *n*
neolactose Neolactose *f*
neolan blue Neolanblau *n*
neoline Neolin *n*
neolite Neolith *m*
neolithic neolithisch, jungsteinzeitlich
Neolithic Period Jungsteinzeit *f*, Neolithikum *n*
neomenthol Neomenthol *n*
neomycin Neomycin *n*
neon (Symb. Ne) Neon *n*
neon content Neongehalt *m*
neonicotine Anabasin *n*
neon indicator Neonindikator *m*
neon lamp Neonlampe *f*
neon light Neonlicht *n*
neon light[ing] Neonbeleuchtung *f*
neon liquefier Neonverflüssigungsmaschine *f*
neon tube Neonröhre *f*, Leuchtröhre *f*
neopelline Neopellin *n*
neopentane Neopentan *n*
neopentyl alcohol Neopentylalkohol *m*
neopentyl bromide Neopentylbromid *n*
neopentyl glycol Neopentylglykol *n*
neopentyl rearrangement Neopentyl-Umlagerung *f* (Chem)
neophyl chloride Neophylchlorid *n*
neophytadiene Neophytadien *n*
neopine Neopin *n*
neoplasm Geschwulstbildung *f* (Med), Neoplasma *n* (Med)
neoprene Neopren *n*
neoreserpic acid Neoreserpsäure *f*
neoretinene Neoretinen *n*
neosaman Neosaman *n*
neosine Neosin *n*
neostigmine Neostigmin *n*
neostrychnine Neostrychnin *n*
neotannyl Neotannyl-
neotantalite Neotantalit *m* (Min)
neoteben Neoteben *n*
neotrehalose Neotrehalose *f*

neotyrosine Neotyrosin *n*
neovitamin Neovitamin *n*
Neozoic Neuzeit *f* (Geol), Neozoikum *n* (Geol), neozoisch (Geol)
nepetalic acid Nepetalsäure *f*
nepetalinic acid Nepetalinsäure *f*
nepetic acid Nepetsäure *f*
nepetolic acid Nepetolsäure *f*
nepetonic acid Nepetonsäure *f*
nepheline Nephelin *m* (Min), Fettstein *m* (Min)
nephelinic nephelinartig
nephelinite Nephelinit *m* (Min), Nephelinbasalt *m* (Min)
nephelite Nephelin *m* (Min), Eläolith *m* (Min), Fettstein *m* (Min)
nephelometer Nephelometer *n*, Trübungsmesser *m*
nephelometric nephelometrisch
nephelometric analysis Nephelometrie *f*
nephrine Nephrin *n*
nephrite Nephrit *m* (Min), Beilstein *m* (Min), Jade *f* (Min)
nephritic Nierenmittel *n* (Pharm)
nephritic wood Grießholz *n*
nephritoid Nephritoid *m*
nephrosteranic acid Nephrosteransäure *f*
nephrosterinic acid Nephrosterinsäure *f*
nepodin Nepodin *n*
nepouite Nepouit *m* (Min)
Neptune Neptun *m* (Astr)
neptune blue Neptunblau *n*
neptunite Neptunit *m* (Min)
neptunium (Symb. Np) Neptunium *n*
neral Neral *n*
neriantin Neriantin *n*
Nernst burner Nernstbrenner *m*
Nernst effect Nernsteffekt *m*
Nernst glower Nernststift *m*
Nernst heat theorem Nernstsches Wärmetheorem *n*
Nernst lamp Nernstlampe *f*
Nernst rod Nernststift *m*
Nernst theory Nernstsches Verteilungsgesetz *n*
nero-antico Nero antico *m* (Min), schwarzer Porphyr *m* (Min)
nerol Nerol *n*
neroli camphor Nerolicampher *m*
nerolidol Nerolidol *n*, Peruviol *n*
nerolin Nerolin *n*
neroli oil Neroliöl *n*, Pomeranzenblütenöl *n*
nerve Nerv *m* (Med), **autonomic** ~ Nerv *m* des autonomen Systems, **sympathetic** ~ Sympathikus[nerv] *m*
nerve cell Nervenzelle *f*
nerve center Nervenzentrum *n*
nerve fiber Nervenfaser *f*
nerve gas Nervengas *n*
nerve tissue Nervengewebe *n*
nervon Nervon *n*

nervonic acid Nervonsäure *f*
nervous depressant Nervenberuhigungsmittel *n* (Pharm)
nervous disease Nervenleiden *n* (Med)
nervous irritation Nervosität *f*
nervousness Nervosität *f*
nervous system, autonomic ~ autonomes Nervensystem *n*, vegetatives Nervensystem *n*, **central** ~ Zentralnervensystem *n*, **parasympathetic** ~ parasympathisches Nervensystem *n*, **peripheral** ~ peripheres Nervensystem *n*, **sympathetic** ~ sympathisches Nervensystem *n*
neryl alcohol Nerylalkohol *m*
nesquehonite Nesquehonit *m* (Min)
Nessler's reagent Nesslers Reagens *n*
net Netz *n*, Gewebe *n*, Gitter *n* (Drahtnetz)
net efficiency Nutzleistung *f*, Nutzeffekt *m*
net energy Nutzenergie *f*
net gain Reingewinn *m*
net load Nettolast *f*
net profit Reingewinn *m*
net-shaped netzförmig
net-shaped electrode Netzelektrode *f*
nettle Nessel *f* (Bot)
nettle poison Nesselgift *n*
nettle-rash Nesselfieber *n*
nettle silk Nesselseide *f* (Text)
net transport Nettotransport *m*
net weight Füllgewicht *n*, Nettogewicht *n*, Reingewicht *n*
network Netz[werk] *n*, Stromnetz *n* (Elektr), Vernetzung *f*
network of mains Leitungsnetz *n*
network polymer Netzpolymer *n*
net yield Reingewinn *m*
Neuberg blue Neubergblau *n*
Neuburg siliceous chalk Neuburger Kieselkreide *f* (Min)
Neuburg siliceous whiting Neuburger Kieselweiß *n*
neuralgia Neuralgie *f* (Med)
neuraltheine Neuralthein *n*
neuraminic acid Neuraminsäure *f*
neuraminidase Neuraminidase *f* (Biochem)
neuridin Neuridin *n*
neurine Neurin *n*
neurobiology Neurobiologie *f*
neurochemistry Neurochemie *f*
neurocyte Nervenzelle *f*, Neuron *n*
neurodine Neurodin *n*
neurofebrin Neurofebrin *n*
neurofibril Nervenfibrille *f*
neurohormone Neurohormon *n*
neuroleptic Neuroleptikum *n* (Pharm)
neurology Nervenlehre *f* (Med), Neurologie *f* (Med)
neuron Neuron *n*, Nervenzelle *f*
neuronal Neuronal *n*

neuroparalysis Nervenlähmung *f* (Med)
neuropathy Nervenleiden *n* (Med)
neurophysiology Neurophysiologie *f*
neuroplasm Neuroplasma *n*
neurosporene Neurosporin *n*
neurotoxin Nervengift *n*
neurotropine Neurotropin *n*
neurovirus Neurovirus *m*
neutral neutral, inaktiv (Chem), ungeladen (Chem)
neutral axis Nullinie *f*, Schwerelinie *f*
neutral blue Neutralblau *n*
neutrality Neutralität *f*
neutralization Neutralisation *f*, Absättigung *f*, Entkopplung *f* (Elektr), Neutralisierung *f*, **heat of** ~ Neutralisationswärme *f* (Therm), Neutralisierungswärme *f*
neutralization enthalpy Neutralisationsenthalpie *f*
neutralization number Neutralisationszahl *f*
neutralization plant Neutralisieranlage *f*
neutralize *v* neutralisieren, abstumpfen, entkoppeln (Elektr)
neutralizing neutralisierend
neutralizing titration Neutralisationstitration *f*
neutral meson Neutretto *n*, Ny-Meson *n*
neutralon Neutralon *n*
neutral point Neutralpunkt *m*, Indifferenzpunkt *m*
neutral position Nullstellung *f*, Ruhelage *f*, Ruhestellung *f*
neutral red Neutralrot *n*
neutral salt Neutralsalz *n*
neutral tint Neutralfarbe *f*
neutral wire Nulleiter *m* (Elektr), **earthed** ~ geerdeter Nulleiter
neutral zone Indifferenzzone *f* (Elektr)
neutretto neutrales Meson *n* (Phys), Neutretto *n* (Phys)
neutrino Neutrino *n* (Atom)
neutron Neutron *n* (Atom), **delayed [fission]** ~ verzögertes Neutron (Atom), **epithermal** ~ epithermisches Neutron, **excess** ~ Überschußneutron *n*, **high-energy** ~ Neutron *n* hoher Energie, **intermediate** ~ mittelschnelles Neutron, **secondary** ~ Sekundärneutron *n*
neutron absorber Neutronenfänger *m*, Neutronenabsorber *m*
neutron-absorbing neutronenabsorbierend
neutron absorption Neutronenabsorption *f*
neutron activation analysis Neutronenaktivierungsanalyse *f*
neutron age Neutronenalter *n* (Atom)
neutron-alpha reaction Neutronen-Alpha-Reaktion *f*
neutron attenuation Neutronenabschwächung *f*
neutron beam Neutronenbündel *n*, Neutronenstrahl *m*

neutron binding energy Neutronenbindungsenergie *f* (Atom)
neutron bombardment Neutronenbeschießung *f*, Neutronenbeschuß *m*
neutron breeder reactor, fast ~ Schnellneutronen-Brutreaktor *m* (Atom)
neutron bullet Neutronengeschoß *n*
neutron capture Neutroneneinfang *m*, **parasitic** ~ parasitärer Neutroneneinfang *m*
neutron counter Neutronenzähler *m*
neutron cross section Neutronenquerschnitt *m*
neutron current Neutronenstrom *m* (Atom)
neutron decay Neutronenzerfall *m*
neutron density Neutronendichte *f* (Atom)
neutron density distribution Neutronendichteverteilung *f*
neutron detection Neutronennachweis *m*
neutron detection technique Neutronennachweistechnik *f*
neutron diffraction Neutronenbeugung *f*
neutron diffraction pattern Neutronenbeugungsaufnahme *f*
neutron diffusion Neutronendiffusion *f*
neutron dosage measurement Neutronendosismessung *f*
neutron efficiency Neutronenausbeute *f*
neutron emission, delayed ~ verzögerte Neutronenemission *f* (Atom)
neutron energy Neutronenenergie *f*
neutron flux Neutronenfluß *m*, **fast** ~ Schnellneutronenfluß *m*
neutron flux measurement Neutronenflußmessung *f*
neutron-gamma reaction Neutronen-Gamma-Reaktion *f*
neutron generator Neutronengenerator *m*
neutron irradiation Neutronenbestrahlung *f*
neutron monitor Neutronenüberwachungsgerät *n*
neutron optics Neutronenoptik *f*
neutron physics Neutronenphysik *f*
neutron polarization Neutronenpolarisation *f*
neutron-proton reaction Neutron-Proton-Reaktion *f*
neutron radiation Neutronenstrahlung *f*
neutron radiography Neutronenradiographie *f*
neutron reflector Neutronenreflektor *m*
neutron resonance Neutronenresonanz *f*
neutron resonance line Neutronenresonanzlinie *f*
neutron robber Neutronenfänger *m*
neutrons, high-speed ~ schnelle Neutronen *pl*, **loss of** ~ Neutronenverlust *m*, **multiplicity of** ~ Neutronenvervielfachung *f*, **shooting out** ~ neutronenspeiend, **slow** ~ langsame Neutronen *pl*, thermische Neutronen *pl*, **source of** ~ Neutronenquelle *f*, **stray** ~ vagabundierende Neutronen *pl*, **thermal** ~ langsame Neutronen *pl*, thermische Neutronen *pl*
neutron scattering Neutronenstreuung *f*

neutron shield Neutronenschutz *m*
neutron shielding Neutronenabschirmung *f*
neutron spectrometer Neutronenspektrometer *n*
neutron spectroscopy Neutronenspektroskopie *f*
neutron spectrum Neutronenspektrum *n*
neutron-transparent neutronendurchlässig
neutron treatment Neutronenbehandlung *f*
neutron velocity Neutronengeschwindigkeit *f*
(Atom)
neutron wave length Neutronenwellenlänge *f*
neutron yield Neutronenausbeute *f*
Neville acid Nevillesäure *f*
new blue Neublau *n*
new building Neubau *m*
new construction Neukonstruktion *f*
new definition Neudefinition *f*
new formation Neubildung *f*
new fuchsin Neufuchsin *n*
new red Neurot *n*
newton Newton *n* (Maß)
Newtonian flow Newtonsche Strömung *f*
Newtonian fluid Newtonsche Flüssigkeit *f*
Newton's alloy Newton-Legierung *f*
Newton['s] rings Newtonsche Ringe *pl* (Opt)
new yellow Neugelb *n*
Ngai-camphor Ngai-Campher *m*
niacin Niacin *n*, Nicotinsäure *f*
nibbling saw Dekupiersäge *f*
niccolite Rotnickelkies *m* (Min),
Arsennickelkies *m* (Min), Kupfernickel *n*
(Min), Niccolit *m* (Min)
nichine Nichin *n*
nick bend test Kerbbiegeprobe *f*
nickel (Symb. Ni) Nickel *n*, **addition of** ~
Nickelzusatz *m*, **colloidal** ~ kolloidales
Nickel, **containing** ~ nickelhaltig, **finely
divided** ~ feinverteiltes Nickel, **pure** ~
Reinnickel *n*
nickel *v* vernickeln
nickel ammonium compound
Nickelammoniumverbindung *f*
nickel arsenide Arsennickel *n* (Min)
nickel bath Vernickelungsbad *n*, **boiling** ~
Nickelsud *m*
nickel bloom Nickelblüte *f* (Min), Annabergit *m*
(Min)
nickel bromide Nickelbromid *n*
nickel bronze Nickelbronze *f*
nickel carbonyl Nickelcarbonyl *n*
nickel chelate Nickelchelat *n*
nickel chloride Nickelchlorid *n*
nickel coating, black ~ Schwarznickelüberzug *m*
nickel cobalt alloy Nickelkobaltlegierung *f*
nickel content Nickelgehalt *m*
nickel copper Nickelkupfer *n*
nickel crucible Nickeltiegel *m*
nickel cyanide complex Nickelcyanidkomplex *m*
nickel filings Nickelfeilspäne *m pl*
nickel formate Nickelformiat *n*

nickel glance Gersdorffit *m* (Min),
Nickelarsenkies *m* (Min), Nickelglanz *m*
(Min)
nickel green Nickelgrün *n*
nickel gymnite Nickelgymnit *m* (Min),
Genthit *m* (Min)
nickel hardening method
Nickelhärtungsverfahren *n*
nickelic compound Nickel(III)-Verbindung *f*
nickelic hydroxide Nickel(III)-hydroxid *n*
nickelic oxide Nickel(III)-oxid *n*
nickelic salt Nickel(III)-Salz *n*
nickelicyanic acid Nickelicyanwasserstoffsäure *f*
nickeliferous nickelhaltig
nickeliferous steel Nickelstahl *m*
nickeline Kupfernickel *n* (Min)
nickel iodate Nickeljodat *n*
nickel iron Nickeleisen *n*
nickel-iron battery Akkumulator *m* mit
Laugenfüllung, Nickeleisenbatterie *f*
nickel-like nickelartig
nickelling, hot contact ~
Kontaktansiedevernicklung *f*, **light** ~
Leichtvernicklung *f*, **rapid** ~
Schnellvernicklung *f*
nickel lining Nickelauskleidung *f*
nickel manganese Nickelmangan *n*
nickel monoxide Nickel(II)-oxid *n*,
Nickelmonoxid *n*
nickel mordant Nickelbeize *f*
nickel mounting Nickelbeschlag *m*
nickel nitrate Nickelnitrat *n*
nickel ocher Annabergit *m* (Min)
nickelocyanic acid
Nickelocyanwasserstoffsäure *f*
nickel ore Nickelerz *n* (Min)
nickelous carbonate Nickel(II)-carbonat *n*
nickelous chloride Nickel(II)-chlorid *n*
nickelous compound Nickel(II)-Verbindung *f*
nickelous cyanide Nickel(II)-cyanid *n*
nickelous hydroxide Nickel(II)-hydroxid *n*
nickelous nickelic oxide Nickel(II)(III)-oxid *n*,
Nickelonickelioxid *n*
nickelous nitrate Nickel(II)-nitrat *n*
nickelous oxide Nickel(II)-oxid *n*,
Nickelmonoxid *n*
nickelous salt Nickel(II)-Salz *n*,
Nickeloxydulsalz *n* (obs)
nickelous sulfate Nickel(II)-sulfat *n*,
Nickelvitriol *n*
nickelous thioglycolate Nickelthioglykolat *n*
nickel oxalate Nickeloxalat *n*
nickel-plate *v* vernickeln
nickel-plated vernickelt
nickel-plating Vernick[e]lung *f*,
Nickelüberzug *m*, Vernickeln *n*, **hard** ~
Hartvernickelung *f*, **hot** ~
Sudvernick[e]lung *f*, **solid** ~
Solidvernicklung *f*, ~ **of sheet zinc**

Zinkblechvernick[e]lung *f*, ~ **of zinc**
Zinkvernickelung *f*
nickel production Nickelgewinnung *f*
nickel pyrites Nickelkies *m* (Min), Millerit *m*
(Min)
nickel quinaldinate Nickelchinaldinat *n*
nickel sheet Nickelblech *n*
nickel silicate, native hydrated ~ Röttisit *m*
(Min)
nickel silver Neusilber *n*
nickel speiss Nickelspeise *f* (Min)
nickel steel Nickelstahl *m*, Nickelflußeisen *n*
nickel steel casting Nickelstahlguß *m*
nickel sulfantimonide, native ~
Nickelantimonglanz *m* (Min)
nickel sulfate Nickelsulfat *n*
nickel sulfide Nickelsulfid *n*
nickel telluride Tellurnickel *n* (Min)
nickel tetracarbonyl Nickeltetracarbonyl *n*
nickel vessel Nickelgefäß *n*
nickel vitriol Nickelvitriol *n*, Morenosit *m*
(Min), Nickelsulfat *n*
Nicol's prism Nicolsches Prisma *n* (Opt)
nicomorphine Nicomorphin *n*
nicopholine Nicopholin *n*
nicotein[e] Nicotein *n*
nicotelline Nicotellin *n*
nicotianine Nicotianin *n*
nicotimine Nicotimin *n*
nicotinamide Nicotinamid *n*, Nikotinamid *n*
nicotinamide adenine dinucleotide
Nikotinamidadenindinukleotid *n*
nicotinamide mononucleotide
Nikotinamidmononukleotid *n*
nicotin[e] Nikotin *n*, **free from** ~ nikotinfrei,
sensitive to ~ nikotinempfindlich, **with a low**
~ **content** nikotinarm
nicotineamide adenine dinucleotide phosphate
Triphosphopyridinnucleotid *n* (NADP, TPN)
nicotine content Nikotingehalt *m*
nicotine poisoning Nikotinvergiftung *f* (Med)
nicotine salt Nikotinsalz *n*
nicotinic acid Nikotinsäure *f*
nicotinic methylbetaine Trigonellin *n*
nicotinism Nikotinvergiftung *f* (Med)
nicotinuric acid Nikotinursäure *f*
nicotinyl alcohol Nikotinylalkohol *m*
nicotyrine Nicotyrin *n*
Niehaus' therapy Frischzellentherapie *f*
niello Schwarzschmelz *m*
nigella seeds Schwarzkümmel *m*
nigerine Nigerin *n*
nigerose Nigerose *f*
night-blind nachtblind
nightblindness Nachtblindheit *f*
night blue Nachtblau *n*
night current Nachtstrom *m*
nightshade, deadly ~ Tollkirsche *f* (Bot)
night shift Nachtschicht *f*

nigraniline Nigranilin *n*, Nigrosin *n*
nigrine Nigrin *n*
nigrometer Nigrometer *n*
nigrosine Nigrosin *n*, Anilinschwarz *n*,
Nigranilin *n*
nikethamide Coramin *n*, Nikethamid *n*
nile blue Nilblau *n*
nihydrin Ninhydrin *n*
ninhydrine reaction Ninhydrinreaktion *f*
niobate Niobat *n*
niobic acid Niobsäure *f*
niobic anhydride Niobpentoxid *n*
niobic compound Niob(V)-Verbindung *f*
niobic fluoride Niobpentafluorid *n*
niobic oxide Niobpentoxid *n*, Niob(V)-Oxid *n*
niobite Niobit *m* (Min), Columbit *m* (Min)
niobium (Symb. Nb) Niob[ium] *n*, Columbium *n*
niobium hydride Niobwasserstoff *m*
niobium pentafluoride Niobpentafluorid *n*
niobium pentoxide Niobpentoxid *n*
niobous compound Niob(III)-Verbindung *f*
niobous oxide Niob(III)-Oxid *n*
nioxime Nioxim *n*
nipecotic acid Nipecotinsäure *f*
Niphos process Niphos-Verfahren *n*
(Korrosionsschutz)
nip [off] *v* abzwicken
nip of the rolls Walzenmund *m*, Walzenspalt *m*
nippers Beißzange *f*, Kneifzange *f*
nipping press Abpreßmaschine *f* (Buchdr)
nipple Nippel *m*, Schmiernippel *m*
nip pressure Druck *m* im Walzenspalt
nip roll Abzugswalze *f*, Haltewalze *f*
nip rolls Abzug *m* (Extrusion)
nip roll stand Klemmwalzenbock *m*
nirvanine Nirvanin *n*
nirvanol Nirvanol *n*
niter Salpeter *m*, Kaliumnitrat *n*
niter efflorescence Salpeterblumen *pl*
niter paper Salpeterpapier *n*
niter pot Salpeterhafen *m*
niter production Salpetererzeugung *f*
niter works Salpeterhütte *f*
niton Niton *n*, Radiumemanation *f*, Radon *n*
nitragin Nitragin *n*
nitramide Nitramid *n*
nitramine Nitramin *n*
nitranilic acid Nitranilsäure *f*
nitraniline Nitranilin *n*
nitraniline hydrochloride Nitranilinchlorhydrat *n*
nitraniline orange Nitranilinorange *n*
nitraniline red Nitranilinrot *n*
nitranilinesulfonic acid Nitranilinsulfonsäure *f*
nitranisol Nitranisol *n*
nitrate Nitrat *n*, salpetersaures Salz *n*, ~ **of**
potash Kalisalpeter *m*
nitrate *v* nitrieren, mit Salpetersäure behandeln
nitrate film Nitratfilm *m* (Phot)
nitrate mordant Nitratbeize *f*

nitrate plant Nitratanlage *f*
nitrate reductase Nitratreduktase *f* (Biochem)
nitratine Nitratin *n*
nitrating nitrierend
nitrating acid Nitriersäure *f*
nitrating apparatus Nitrierapparat *m*
nitrating centrifuge Nitrierzentrifuge *f*
nitrating mixture Nitriergemisch *n*
nitrating plant Nitrieranlage *f*
nitrating process Nitrierungsvorgang *m*, **duration of the** ~ Nitrierdauer *f*
nitrating temperature Nitriertemperatur *f*
nitrating vessel Nitriertopf *m*
nitration Nitrierung *f* (Chem), **degree of** ~ Nitrierungsgrad *m*, **depth of** ~ Nitriertiefe *f*, **stage of** ~ Nitrierungsstufe *f*
nitration furnace Nitrierofen *m*
nitration paper Nitrierpapier *n*
nitration product Nitrierungsprodukt *n*
nitrazine yellow Nitrazingelb *n*
nitrene Nitren *n*
nitrenium ion Nitrenium-Ion *n*
nitric salpetersauer
nitric acid Salpetersäure *f*, Aqua fortis (Lat), Scheidewasser *n* (obs), **containing** ~ salpetersäurehaltig, **ester of** ~ Salpetersäureester *m*, **fuming** ~ rauchende Salpetersäure, **salt of** ~ Nitrat *n*, salpetersaures Salz *n*
nitric acid bath Salpetersäurebad *n*
nitric acid plant Salpetersäureanlage *f*
nitric acid vapor Salpetersäuredampf *m*
nitric anhydride Distickstoffpentoxid *n*, Salpetersäureanhydrid *n*
nitric oxide Stick[stoff]oxid *n*
nitric oxide equilibrium Stickoxidgleichgewicht *n*
nitridation Nitrierung *f* (Met)
nitridation hardness Nitrierungshärte *f*
nitride Nitrid *n*
nitride *v* nitrieren (Met), nitrierhärten
nitriding Nitrierhärten *n* (Met), Nitridieren *n*, Nitrieren *n* (Met)
nitriding action Nitrierwirkung *f* (Met)
nitriding box Nitrierkasten *m*
nitriding equipment Nitrieranlage *f* (Metall)
nitriding furnace Nitrierofen *m*
nitriding process Nitrierhärteverfahren *n* (Met), Nitrierhärtung *f* (Met)
nitrifiable nitrierbar
nitrification Nitrierung *f*, Nitrifizierung *f* (Bakt)
nitrify *v* nitrieren, nitrifizieren (Bakt)
nitrifying nitrierend, nitrifizierend (Bakt)
nitrifying bacteria nitrifizierende Bakterien *pl*, Stickstoffbakterien *pl*
nitrile Nitril *n*
nitrile base Nitrilbase *f*, tertiäres Amin *n*
nitrile rubber Nitrilkautschuk *m*
nitrilotriacetic acid Nitrilotriessigsäure *f*
nitrine Nitrin *n*

nitrite Nitrit *n*, salpetrigsaures Salz *n*
nitrite-free nitritfrei
nitrite mother liquor Nitritmutterlauge *f*
nitrite pickling salt Nitritpökelsalz *n*
nitrite reductase Nitritreduktase *f* (Biochem)
nitrite[s], free from ~ nitritfrei
nitrito-group Nitritogruppe *f*
nitroacetic acid Nitroessigsäure *f*
nitroacetophenone Nitroacetophenon *n*
nitroalkane Nitroalkan *n*
nitroaniline Nitroanilin *n*
nitroanisole Nitroanisol *n*
nitroanthraquinone Nitroanthrachinon *n*
nitrobacteria Nitrobakterien *pl*
nitrobarbituric acid, 5- ~ Dilitursäure *f*
nitrobarite Barytsalpeter *m*
nitrobenzaldehyde Nitrobenzaldehyd *m*
nitrobenzene Nitrobenzol *n*, Mirbanöl *n*
nitrobenzene diazonium chloride Nitrobenzoldiazoniumchlorid *n*
nitrobenzoic acid Nitrobenzoesäure *f*
nitrobenzyl chloride Nitrobenzylchlorid *n*
nitrobromoform Brompikrin *n*
nitrobutane Nitrobutan *n*
nitrocellulose Nitrozellulose *f*, Collodiumwolle *f*, Kollodiumwolle *f*, Schießbaumwolle *f*
nitrocellulose coating color Nitrozellulosedeckfarbe *f*
nitro[cellulose] lacquer Nitrolack *m*
nitrocellulose rayon Nitroseide *f*
nitrochlorobenzene Nitrochlorbenzol *n*
nitrochloroform Chlorpikrin *n*
nitrochlorotoluene Nitrochlortoluol *n*
nitrocinnamic acid Nitrozimtsäure *f*
nitrococussic acid Nitrococussinsäure *f*
nitro-compound Nitroverbindung *f*, Nitrokörper *m*
nitro-copper Nitrokupfer *n*
nitrocymene Nitrocymol *n*
nitro-derivative Nitroderivat *n*
nitrodimethyl aniline Nitrodimethylanilin *n*
nitrodiphenyl Nitrodiphenyl *n*
nitroethane Nitroäthan *n*
nitro-explosive Nitrosprengstoff *m*
nitro-fatty acid Nitrofettsäure *f*
nitroform Nitroform *n*
nitrofural Nitrofural *n*
nitrogelatin Nitrogelatine *f*
nitrogen (Symb. N) Stickstoff *m*, **containing** ~ stickstoffhaltig, **free from** ~ stickstofffrei, **oxidation of** ~ Stickstoffoxidation *f*, **saturation with** ~ Stickstoffsättigung *f*, **total** ~ Gesamtstickstoff *m*
nitrogen absorption Stickstoffaufnahme *f*
nitrogen bridge Stickstoffbrücke *f*
nitrogen bromide Bromstickstoff *m*
nitrogen chloride Chlorstickstoff *m*
nitrogen compressor Stickstoffkompressor *m*
nitrogen-containing stickstoffhaltig

nitrogen-containing acid Stickstoffsäure *f*
nitrogen content Stickstoffgehalt *m*
nitrogen cycle Stickstoffkreislauf *m*
nitrogen determination Stickstoffbestimmung *f*
nitrogen dioxide Distickstoffdioxid *n*, Stickstoffdioxid *n*
nitrogen excretion Stickstoffausscheidung *f*
nitrogen fixation Stickstoffixierung *f* (Biol), enzymatic ~ enzymatische Stickstoffixierung *f*, **symbiontic** ~ symbiontische Stickstoffixierung *f*
nitrogen-fixing plant Stickstoffsammler *m* (Bot)
nitrogen flow Stickstoffstrom *m*
nitrogen halide Stickstoffhalogenverbindung *f*
nitrogen-hardening Nitrierhärten *n* (Met), Nitrierhärtung *f* (Met)
nitrogen-hardening process Nitrierhärteverfahren *n*
nitrogen hydride Stickstoffwasserstoff[säure f] *m*
nitrogen iodide Jodstickstoff *m*
nitrogen metabolism Stickstoffstoffwechsel *m*
nitrogen monoxide Distickstoffoxid *n*, Lachgas *n*, Stickoxydul *n* (obs)
nitrogenous stickstoffhaltig
nitrogen[ous] fertilizer Stickstoffdünger *m*
nitrogen pentoxide Distickstoffpentoxid *n*, Salpetersäureanhydrid *n*
nitrogen peroxide Stickstoffdioxid *n*, Stickstoffperoxid *n*
nitrogen sesquioxide Distickstofftrioxid *n*, Salpetrigsäureanhydrid *n*
nitrogen silicide Stickstoffsilicid *n*
nitrogen sulfide Schwefelstickstoff *m*
nitrogen supply Stickstoffgabe *f* (Agr)
nitrogen tetroxide Distickstofftetroxid *n*, Stickstofftetroxid *n*
nitrogen trioxide Stickstofftrioxid *n*
nitroglycerin Nitroglycerin *n*, Glycerintrinitrat *n*, Trinitrin *n*
nitroglycerol Nitroglycerin *n*, Glycerintrinitrat *n*
nitro-group Nitrogruppe *f*
nitroguanidine Nitroguanidin *n*
nitro-hydrochloric acid Aqua regia (Lat), Königswasser *n*
nitroleum Glycerinnitrat *n*
nitrolic acid Nitrolsäure *f*
nitrolim[e] Calciumcyanamid *n*, Kalkstickstoff *m*
nitromagnesite Magnesiasalpeter *m* (Min), Nitromagnesit *m* (Min)
nitromannite Nitromannit *m*
nitrometal Nitrometall *n*
nitrometer Nitrometer *n*, Salpeterwaage *f*
nitromethane Nitromethan *n*
nitromethyl aniline Nitromethylanilin *n*
nitromethyl anthraquinone Nitromethylanthrachinon *n*
nitromethyl naphthalene Nitromethylnaphthalin *n*
nitromethyl propane Nitromethylpropan *n*

nitron Nitron *n*
nitronaphthalene Nitronaphthalin *n*
nitronaphthalene sulfonic acid Nitronaphthalinsulfonsäure *f*
nitronaphthalene trisulfonic acid Nitronaphthalintrisulfonsäure *f*
nitronaphthoquinone Nitronaphthochinon *n*
nitronic acid Nitronsäure *f*
nitronitrosobenzene Nitronitrosobenzol *n*
nitronitroso dye Nitronitrosofarbstoff *m*
nitronium ion Nitroniumion *n*
nitroparaffin Nitroparaffin *n*
nitrophenol Nitrophenol *n*
nitrophenylacetic acid Nitrophenylessigsäure *f*
nitrophenylacetylene Nitrophenylacetylen *n*
nitrophenylcarbamyl chloride Nitrophenylcarbamylchlorid *n*
nitrophenylhydrazine Nitrophenylhydrazin *n*
nitrophenylpropiolic acid Nitrophenylpropiolsäure *f*
nitrophilous nitrophil
nitrophthalic acid Nitrophthalsäure *f*
nitro powder Nitropulver *n*
nitropropane Nitropropan *n*
nitropropene Nitropropen *n*
nitroprussiate Nitroprussidverbindung *f*
nitroprusside Nitroprussidverbindung *f*
nitropyrene Nitropyren *n*
nitropyridine Nitropyridin *n*
nitroquinoline Nitrochinolin *n*
nitroquinone Nitrochinon *n*
nitrosaccharose Knallzucker *m*
nitrosalicylic acid Nitrosalicylsäure *f*
nitrosamine Nitrosamin *n*
nitrosamine red Nitrosaminrot *n*
nitrosation Nitrosierung *f*
nitrose Nitrose *f*, nitrose Säure *f*
nitrosilk Nitroseide *f*
nitrosite Nitrosit *n*
nitrosobase Nitrosobase *f*
nitrosobenzene Nitrosobenzol *n*
nitrosobutane Nitrosobutan *n*
nitroso compound Nitrosoverbindung *f*
nitrosodimethylamine Nitrosodimethylamin *n*
nitrosodiphenylamine Diphenylnitrosamin *n*
nitroso dyestuff Nitrosofarbstoff *m*
nitroso group Nitrosogruppe *f*
nitrosohydroxylamine oxide Nitrosohydroxylaminoxid *n*
nitrosoindoxyl Isatoxim *n*
nitrosomethyl aniline Nitrosomethylanilin *n*
nitrosomethyl toluidine Nitrosomethyltoluidin *n*
nitrosomethyl urethane Nitrosomethylurethan *n*
nitrosonaphthol Nitrosonaphthol *n*
nitrosophenol Nitrosophenol *n*
nitrosophenyl hydrazine Nitrosophenylhydrazin *n*
nitrosophenyl hydroxylamine Kupferron *n*, Nitrosophenylhydroxylamin *n*

nitroso radical Nitrosoradikal *n*
nitrosoresorcinol Nitrosoresorcin *n*
nitrososulfonic acid Nitrososulfonsäure *f*
nitrosourethane Nitrosourethan *n*
nitrostarch Nitrostärke *f*
nitrosteel Nitrierstahl *m*
nitrosulfamide Nitrosulfamid *n*
nitrosulfathiazole Nitrosulfathiazol *n*
nitrosulfonic acid Nitrosulfonsäure *f*,
    Nitrosylschwefelsäure *f*
nitrosulfuric acid Nitrosylschwefelsäure *f*,
    Mischsäure *f* (obs), Nitrose *f* (Techn)
nitrosyl chloride Nitrosylchlorid *n*
nitrosylsulfuric acid Nitrosylschwefelsäure *f*
nitrosynthetic lacquer Kombinationslackfarbe *f*
nitrotartaric acid Nitroweinsäure *f*
nitrotoluene Nitrotoluol *n*
nitrotoluene sulfonic acid
    Nitrotoluolsulfonsäure *f*
nitrotoluidine Nitrotoluidin *n*
nitrourea Nitroharnstoff *m*
nitrourethane Nitrourethan *n*
nitrous salpeterartig, salpeterhaltig, salpetrig
    (Chem), salpetrigsauer (Chem)
nitrous acid salpetrige Säure *f*
nitrous anhydride Distickstofftrioxid *n*,
    Salpetrigsäureanhydrid *n*,
    Stickstoffsesquioxid *n*
nitrous earth Salpetererde *f*
nitrous fumes nitrose Gase *pl*
nitrous oxide Distickstoffoxid *n*, Lachgas *n*,
    Lustgas *n*, Stickoxydul *n* (obs)
nitr[ox]yl chloride Nitrylchlorid *n*
nitroxylene Nitroxylol *n*
nitroxyl radical Nitroxylgruppe *f*
nivalic acid Nivalsäure *f*
niveau Niveau *n*
nivenite Nivenit *m* (Min)
nivitin Nivitin *n*
nixie tube Nixierohr *n*
NMR-spectrum Kernresonanzspektrum *n*
nobelium (Symb. No) Nobelium *n*
noble edel (Met)
noble gas Edelgas *n*
noble gas configuration Edelgaskonfiguration *f*
noble metal Edelmetall *n*
nocardamin Nocardamin *n*
noctal Noctal *n*
nodal line Knotenlinie *f* (Akust)
nodal point Knotenpunkt *m* (Phys),
    Schwingungsknoten *m* (Phys)
node Wurzelknoten *m* (Bot)
nodular knotig
nodular ore Nierenerz *n* (Min)
nodule Blase *f* (Glas), Klümpchen *n* (Med),
    Knolle *f* (Bot), Knoten *m* (Bot), Niere *f*
    (Geol)
no-fines concrete haufwerkporiger Beton *m*

noise Lärm *m*, Rauschen *n*, Störgeräusch *n*,
    combating of ~ Lärmbekämpfung *f*, **injury**
    caused by ~ Lärmschädigung *f*, **sensitive to** ~
    lärmempfindlich, **source of** ~ Lärmquelle *f*
noise ban Lärmverbot *n*
noise disturbance Lärmbelästigung *f*
noise eliminator Störschutz *m* (Radio)
noise factor Rauschfaktor *m*
noise figure Rauschzahl *f*
noise intensity [or level] Lärmintensität *f*
noiseless geräuschlos, ruhig
noise level Geräuschpegel *m*
noise prevention Lärmschutz *m*
noise suppression Geräuschdämpfung *f*
noise voltage Rauschspannung *f* (Elektron)
noisy laut, lärmend
no-load Leerlauf *m*
no-load current Leerlaufstrom *m*
no-load friction Leergangreibung *f*
no-load speed Leerlaufdrehzahl *f*
no-load test Leerlaufversuch *m*
no-load work Eigenarbeit *f*
nomenclature Nomenklatur *f*, **chemical** ~
    chemische Nomenklatur, **Geneva** ~ Genfer
    Nomenklatur (Chem), **Geneva system of** ~
    Genfer Nomenklatursystem *n* (Chem), ~ **of**
    organic compounds Nomenklatur *f*
    organischer Verbindungen (pl)
nominal current Nennstrom *m* (Elektr)
nominal load Nennlast *f*
nominal[ly] nominell
nominal pressure Nenndruck *m*
nominal size Nennmaß *n*, Nennweite *f*,
    Sollmaß *n*
nominal value Nennwert *m*, Nominalwert *m*
nominal width Nennweite *f*
nomogram Funktionsleiter *f*, Nomogramm *n*
    (Math), **right angle** ~
    Rechtwinkelnomogramm *n*
nomograph Nomogramm *n* (Math)
nomography Nomographie *f*
nonabrasiveness Abriebfestigkeit *f*
nonabrasive quality Abriebfestigkeit *f*
nonacosane Nonakosan *n*
non-additivity Nichtadditivität *f*
nonadecane Nonadekan *n*
nonadiene Nonadien *n*
nonadilactone Nonadilacton *n*
nonagon Neuneck *n* (Math)
non-air-inhibited [polyester] finish glänzend
    auftrocknender Lack *m* (auf Polyesterbasis)
non-alkali alkalifrei
nonamethylene glycol Nonamethylenglykol *n*
nonanal Pelargonaldehyd *m*
nonane Nonan *n*
nonanoic acid Pelargonsäure *f*
nonanoyl Pelargonyl-
nonappearance Nichterscheinen *n*
nonaqueous nichtwäss[e]rig

nonatriacontane Nonatriakontan *n*
nonbalanced unausgeglichen
non-bleeding überspritzecht
nonbreakable unzerbrechlich
noncalendered matt (Pap)
noncapacitive kapazitätsfrei
noncarbonaceous kohlefrei
nonchalking abkreidefest, nicht kreidend
noncircular unrund
nonclinkering schlackenfrei (Kohle)
non-clogging pump Freistrompumpe *f*
noncombustible unverbrennbar
noncompensated unausgeglichen
noncompetitive konkurrenzlos, nicht kompetitiv
  (Kinetik)
noncondensing engine Auspuffmaschine *f*
nonconducting nichtleitend
nonconducting composition Isolationsstoff *m*,
  Isoliermasse *f*
nonconducting material Isoliermittel *n*,
  Wärmeschutzmasse *f*
nonconductor Nichtleiter *m*, Isolator *m*
noncorrodibility Rostbeständigkeit *f*
noncorrodible korrosionsbeständig,
  rostbeständig
noncorroding korrosionsfrei
noncorrosive korrosionsfrei
noncreasing knitterfest
noncrystalline nichtkristallin
noncutting shaping spanlose Formgebung *f*
nondecylic acid Nondecylsäure *f*
nondelay fuse empfindlicher Zünder *m*
nondeposit bottle Einwegflasche *f*
nondestructive zerstörungsfrei
nondiaphanous lichtundurchlässig
nondimensional dimensionslos
nondimensional quantity absoluter Zahlenwert *m*
nondimensional ratio absoluter Zahlenwert *m*
nondistorting verzerrungsfrei
nondrinkable water Nutzwasser *n*
nondrying nicht trocknend
nonelastic unelastisch
non-electrolyte Anelektrolyt *m*,
  Nichtelektrolyt *m*
nonene Nonylen *n*
non-equilibrium state
  Ungleichgewichtszustand *m*
noneuclidian nicht euklidisch
nonexplosive nicht explosiv
nonfelting nicht [ver]filzend
nonferrous alloy NE-Legierung *f*,
  Nichteisenlegierung *f*
nonferrous metal Nichteisenmetall *n*,
  Buntmetall *n*, NE-Metall *n*
nonfibrous nicht faserig
nonfilterable unfiltrierbar
nonflammable nicht brennbar
nonfoaming nicht schäumend
nonfraying nicht ausfasernd

nonfusible nicht schmelzbar
nongaseous nicht gasförmig
nongasifiable nicht vergasbar
nonhomogeneous nicht homogen, durchwachsen
nonhygroscopic hydrophob
nonicing vereisungsfrei
nonideal nicht ideal
nonignition, zone of ~ Zündlücke *f*
nonindicating nicht anzeigend
noninductive [selbst]induktionsfrei
nonine Nonin *n*
noninflammability Unbrennbarkeit *f*
noninflammable nicht brennbar, nicht
  entflammbar, nicht feuergefährlich,
  unbrennbar (nicht entzündbar)
nonius Nonius *m*
nonlinear nichtlinear
nonlinearity Nichtlinearität *f*
nonluminous nichtleuchtend
non-magnetic antimagnetisch, unmagnetisch
nonmarring kratzfest
nonmeltable unschmelzbar
nonmetal Nichtmetall *n*
nonmetallic nichtmetallisch
nonmetallic element Metalloid *n*
nonmicrophonic klingfrei (Rad)
nonmiscibility Nichtmischbarkeit *f*
nonmiscible unmischbar
non-Newtonian medium Nicht-Newtonsches
  Medium *n*
non-nitrogenous stickstofffrei
nonoscillating schwingungsfrei
nonoxidizable nicht oxidierbar
nonperiodic nicht periodisch, unperiodisch
nonpoisonous ungiftig, giftfrei
nonpolar unpolar
nonpolarizable unpolarisierbar
nonpolarized unpolarisiert, neutral
nonporous blasenfrei
non-pressure pipe druckloses Rohr *n*
nonradiative strahlungslos
nonreactive unempfindlich
nonreactivity Unempfindlichkeit *f*
nonrecurrent action einmaliger Vorgang *m*
nonresinous harzfrei
nonresponsiveness Unempfindlichkeit *f*
nonreturnable bottle Einwegflasche *f*
non-return-flow wind tunnel Windkanal *m*
  offener Bauart
nonreturn valve Rückschlagventil *n*,
  Rückströmsperre *f*
nonreversible nicht umkehrbar
nonrigid unstarr, weich
nonrusting nicht rostend, rostbeständig
nonscaling zunderbeständig, zunderfest
nonscaling property Zunderbeständigkeit *f*
nonseparable nicht separierbar
nonsequential sprunghaft
nonshattering splitterfrei, unzerbrechlich

nonshrinkable nicht schrumpfend
nonshrinking krumpffrei
nonskid rutschfest, rutschsicher, trittfest
nonskid depth Gravierungstiefe *f,* Profiltiefe *f*
nonskid design Gleitschutzprofil *n*
nonskid flooring rutschfester Bodenbelag *m*
nonslip trittfest, schiebefest
nonsplintering splitterfrei
nonstaining nicht verfärbend
nonswelling quellfest
nonterminating unendlich
nontoxic ungiftig, atoxisch, giftfrei
nontransparency Undurchsichtigkeit *f*
nontransparent undurchsichtig
nontronite Nontronit *m* (Min)
nonuniform ungleichförmig, ungleichmäßig
nonuniformity Ungleichförmigkeit *f,*
 Ungleichmäßigkeit *f*
nonuniform strain ungleichmäßige
 Deformation *f* (Med)
nonvolatile nicht flüchtig, nicht vergasbar
nonvortical wirbelfrei
non-workable unverformbar
nonwoven fabric Faservlies *n*
nonyl Nonyl-
nonyl alcohol Nonylalkohol *m*
nonyl aldehyde Nonylaldehyd *m*
nonylene Nonylen *n*
nonylic acid Nonylsäure *f*
nonylphenol Nonylphenol *n*
nonyne Nonin *n*
noose Schlinge *f,* [zuziehbare] Schleife *f*
nootkatone Nootkaton *n*
nopadiene Nopadien *n,* Nopen *n*
nopaline Nopalin *n*
nopene Nopadien *n,* Nopen *n*
nopinane Nopinan *n*
nopinene Nopinen *n*
nopinic acid Nopinsäure *f*
nopinol Nopinol *n*
nopinone Nopinon *n*
nopol Nopol *n*
noradrenaline Noradrenalin *n*
norapomorphine Norapomorphin *n*
noraporphine Noraporphin *n*
noratropine Noratropin *n*
norbixin Norbixin *n*
norbornane Norbornan *n*
norbornene Norbornen *n*
norborneol Norborneol *n*
norcamphane Norcamphan *n*
norcamphene Norcamphen *n*
norcamphidine Norcamphidin *n*
norcamphor Norcampher *m*
norcamphoric acid Norcamphersäure *f*
norcarane Norcaran *n*
norcaryophyllenic acid Norcaryophyllensäure *f*
norcodeine Norkodein *n*
norcorydaline Norcorydalin *n*

nordenskioldine Nordenskiöldin *m* (Min)
nordesoxyephedrine Nordesoxyephedrin *n*
Nordhausen acid Oleum
nordmarkite Nordmarkit *m*
norecgonine Norecgonin *n*
norgeranic acid Norgeraniumsäure *f*
Norge saltpeter Norgesalpeter *m*
norhydrastinine Norhydrastinin *n*
norhyoscyamine Norhyoscyamin *n*
norite Norit *m* (Geol)
norleucine Norleucin *n*
norlupinane Norlupinan *n*
norlupinone Norlupinon *n*
norm Norm *f,* Regel *f*
normal normal, normgerecht, ordentlich,
 regelrecht (to a curve), Normale *f*
normal acceleration Normalbeschleunigung *f*
normal acid Normalsäure *f*
normal butane Normalbutan *n*
normal calomel electrode
 Normalkalomelelektrode *f*
normal candle Normalkerze *f*
normal concentration Normalkonzentration *f*
normal density Normaldichte *f*
normal distribution Normalverteilung *f*
normal dodecane Normaldodecan *n*
normal eicosane Normaleikosan *n*
normal electrode potential of a metal
 Normalpotential *n* eines Metalls
normal element Normalelement *n* (Elektr)
normal heptane Normalheptan *n*
normal heptylic acid Normalheptylsäure *f*
normal hydrogen electrode
 Normalwasserstoffelektrode *f*
normality Normalität *f*
normalization Normalisierung *f,*
 Normenaufstellung *f,* Normung *f*
normalize *v* normalisieren, normen,
 spannungsfrei glühen (Techn)
normalizing Normalglühen *n* (Met)
normal [line] Normale *f* (Geom)
normally normalerweise
normal mode or vibration Normalschwingung *f*
normal output Nennleistung *f,* Normalleistung *f*
normal pentane Normalpentan *n*
normal plane (to a curve) Normalebene *f*
normal pressure Normaldruck *m*
normal solution Normallösung *f* (Chem)
normal spectrum Beugungsspektrum *n* (Spektr),
 Normalspektrum *n* (Spektr)
normal speed Normalgeschwindigkeit *f,*
 Betriebsdrehzahl *f*
normal state Normalzustand *m*
normal strength Normalstärke *f*
normal stress Normalbeanspruchung *f,*
 Normalspannung *f*
normal temperature Normaltemperatur *f*
normal tension Normalspannung *f*
normal torque Normaldrehmoment *n*

normal voltage Normalspannung *f*
norm gear Schraubenradgetriebe *n*
normoblast Normoblast *m* (Med)
normocyte Normozyt *m* (Med)
normorphine Normorphin *n*
normuscarine Normuscarin *n*
normuscarone Normuscaron *n*
normuscone Normuscon *n*
nornarceine Nornarcein *n*
nornicotine Nornicotin *n*
noropianic acid Noropiansäure *f*
norpenaldic acid Norpenaldsäure *f*
norphedrin Norphedrin *n*
norpinic acid Norpinsäure *f*
norprogesterone Norprogesteron *n*
norpseudotropine Pseudonortropin *n*
norsolanellic acid Norsolanellsäure *f*
northebaine Northebain *n*
northern lights Nordlicht *n* (Astr)
north light Nordlicht *n* (Astr)
North Star Nordstern *m* (Astr), Polarstern *m*
northupite Northupit *m* (Min)
nortricyclene Nortricyclen *n*
nortropane Nortropan *n*
nortropine Nortropin *n*
nortropinone Nortropinon *n*
norvaline Norvalin *n*
norvaline amide Norvalinamid *n*
nose Nase *f*, Schnabel *m* (Techn), Vorsprung *m* (Techn)
nosean Nosean *m* (Min)
nose-heaviness Kopflastigkeit *f*
nose-heavy kopflastig
nose piece Revolver *m* (Mikroskop)
nose radius Spitzenradius *m*
nosophen Nosophen *n*, Jodophen *n*
notarization notarielle Beurkundung *f*
notation Eintragung *f*, Bezeichnung *f*, Definitionsweise *f*, **binary** ~ binäre Zahlendarstellung *f*, **decimal** ~ dezimale Zahlendarstellung *f*
notation system Bezeichnungssystem *n*
notch Kerbe *f*, Auskerbung *f*, Aussparung *f*, Einfeilung *f*, Einkerbung *f*, Einschnitt *m*, Falz *m*, Kimme *f*, Scharte *f*
notch *v* [ein]kerben, auszacken, einfeilen, einschneiden, falzen, kimmen
notched [ein]gekerbt, zackig
notched-bar bending test Einkerbbiegeversuch
notched-bar bend test Kerbfaltversuch *m*
notched-bar impact bending test Kerbschlagbiegeversuch *m*
notched-bar [impact] strength Kerbschlagfestigkeit *f*
notched-bar impact test Kerbschlagprobe *f*, Kerbzähigkeitsprüfung *f*
notched bar tensile test Kerbzugprobe *f*, Kerbzugversuch *m*
notched-bar toughness Kerbzähigkeit *f*

notched-tube tensile test Rohrkerbzugversuch *m*
notch factor Kerbeinflußzahl *f*
notch impact strength Kerbschlagzähigkeit *f*
notch impact test Kerbschlagprobe *f*
notching tool Kröse *f*
notch sensitivity Kerbempfindlichkeit *f*
notch toughness Kerbfestigkeit *f*
note Bemerkung *f*, Notiz *f*, Vermerk *m*
note *v* bemerken, notieren
notebook Notizbuch *n*
noteworthy beachtenswert, bemerkenswert
notice Bemerkung *f*, Notiz *f*, Vermerk *m*
notice *v* bemerken, wahrnehmen
noticeable bemerkenswert, wahrnehmbar
notifiable meldepflichtig, anzeigepflichtig
notify *v* benachrichtigen
notion Begriff *m*
nought Null *f*
nourish *v* [er]nähren
nourishing nahrhaft
nourishment Ernährung *f*, Nährmittel *n*, Nahrung *f*
nova Nova *f* (Astr)
novacine Novacin *n*
novaculite Wetzschiefer *m* (Min)
novaine Carnitin *n*, Novain *n*, Vitamin B *n*
novel neu[artig]
novelty Neuheit *f*
noviose Noviose *f*
novobiocin Novobiocin *n*
novocaine Novocain *n*
novo-iodine Novojodin *n*
novolak Novolak *n*
noxious schädlich
nozzle Düse *f*, Ansatzrohr *n*, Ausflußöffnung *f*, Ausgußschnauze *f*, Mundstück *n*, Schlauchtülle *f*, Schnauze *f*, Strahlrohr *n*, Stutzen *m*, **calibrated** ~ Meßdüse *f*, ~ **of a blowpipe** Lötrohrspitze *f*
nozzle adapter Düsenpaßstück *n*
nozzle block Düsenblock *m*
nozzle box Düsenkammer *f*
nozzle discharge centrifuge Düsen-Tellerzentrifuge *f*, Teller-Düsenzentrifuge *f*
nozzle dryer Düsentrockner *m*
nozzle meter Düsenmesser *m*
nozzle pipe Düsenstock *m*
nozzle plate Düsenplatte *f*
nozzle pulverizer Strahlprallmühle *f*
nozzle register Angußbuchse *f*
nozzle regulator Düsenregler *m*
nozzle section Düsenquerschnitt *m*
nozzle shut-off device Düsenverschluß *m* (Spritzguß)
nozzle valve Düsenventil *n*
n-pentane Normalpentan *n*
NPK fertilizer NPK-Dünger *m* (Abk. für Stickstoff, Phosphor, Kalium)

NQR spectroscopy NQR-Spektroskopie *f*
  (Kernquadrupolresonanz-Spektroskopie)
N-terminal amino acid N-terminale
  Aminosäure *f*
nuance Nuance *f*
nucin Nucin *n*, Juglon *n*
nuclear nuklear
nuclear age Atomzeitalter *n*
nuclear alignment Kernausrichtung *f* (Atom)
nuclear attraction Kernanziehung *f* (Atom)
nuclear axis Kernachse *f* (Atom)
nuclear bombardment Kernbeschießung *f*
  (Atom), Kernbeschuß *m* (Atom)
nuclear capture Kerneinfang *m*
[nuclear] chain reaction Kernkettenreaktion *f*
nuclear chain reactor Kernreaktor *m*
nuclear charge Kernladung *f*
nuclear charge distribution
  Kernladungsverteilung *f*
nuclear charge number Kernladungszahl *f*
  (Atom)
nuclear collision Kernstoß *m*
nuclear component Kernbestandteil *m* (Atom)
nuclear decay Kernzerfall *m* (Atom)
nuclear demagnetization
  Kernentmagnetisierung *f* (Atom)
nuclear density Atomkerndichte *f*
nuclear disintegration Kernzerfall *m* (Atom)
nuclear distortion Kernverformung *f*
nuclear division Kernteilung *f*
nuclear energy Atomenergie *f*, Kernenergie *f*
nuclear energy level Kernenergieniveau *n*
nuclear energy level diagram
  Kernenergieniveaudiagramm *n*
nuclear energy rocket Kernrakete *f*
nuclear engineering Kerntechnik *f*
nuclear evaporation Kernverdampfung *f* (Atom)
nuclear excitation Kernanregung *f* (Atom)
nuclear explosion Kernexplosion *f*
nuclear fast red Kernechtrot *n*
nuclear ferromagnetism
  Kernferromagnetismus *m*
nuclear field Kernfeld *n* (Atom), **repulsing** ~
  abstoßendes Kernfeld *n*
nuclear fission Kernspaltung *f* (Atom),
  Spaltung *f* des Atomkerns (Atom)
nuclear fission energy Kernspaltungsenergie *f*
  (Atom)
nuclear force Kernkraft *f* (Atom)
nuclear formula Kernregel *f*
nuclear fragment Kernbruchstück *n*,
  Kernsplitter *m*
nuclear fuel Kernbrennstoff *m* (Atom),
  denatured ~ denaturierter Kernbrennstoff *m*
  (Atom)
nuclear fusion Kernfusion *f* (Atom),
  Kernverschmelzung *f* (Atom)
nuclear fusion reactor Kernfusionsreaktor *m*
  (Atom)

nuclear induction spectrograph
  Kerninduktionsspektrograph *m*
nuclear industry Kernindustrie *f* (Atom)
nuclear isobar Kernisobar *n*
nuclear isomer Kernisomer *n*
nuclear isomerism Kernisomerie *f* (Atom)
nuclear level Kernniveau *n*
nuclear magnetic kernmagnetisch
nuclear magnetic moment magnetisches
  Kernmoment *n*
nuclear magnetic resonance kernmagnetische
  Resonanz *f*, magnetische Kernresonanz *f*
nuclear magnetic resonance spectrograph
  Kernresonanzspektrograph *m*,
  Kernresonanzspektrum *n*
nuclear magnetic resonance [spetroscopy]
  Kernresonanzspektrometrie *f*
nuclear magneton Kernmagneton *n* (Atom)
nuclear matter Kernmaterie *f*
nuclear medicine Nuklearmedizin *f*
nuclear membrane Kernmembran *f* (Biol)
nuclear metallurgy Kernmetallurgie *f*
nuclear model Einteilchenmodell *n*
nuclear moment Kernmoment *n*
nuclear particle Kernteilchen *n*
nuclear photodisintegration Kernphotozerfall *m*
  (Atom)
nuclear photoelectric effect Kernphotoeffekt *m*
  (Atom)
nuclear physicist Atomphysiker *m*,
  Kernphysiker *m*
nuclear physics Atomphysik *f*, Kernphysik *f*
nuclear pile Kernreaktor *m*, Atommeiler *m*
nuclear polymerism Kernpolymerie *f*
nuclear potential Kernpotential *n*
nuclear power Atomkraft *f*
nuclear power plant Kernkraftwerk *n*
nuclear power station Atomkraftwerk *n*,
  Kernkraftwerk *n*
nuclear precession Präzession *f* des Atomkerns
nuclear property Kerneigenschaft *f*
nuclear protein Zellkern-Protein *n*
nuclear proton Kernproton *n* (Atom)
nuclear purity Nuklearreinheit *f*
nuclear quadrupole coupling
  Kernquadrupolkopplung *f*
nuclear radiation Kernstrahlung *f*
nuclear radius Kernhalbmesser *m*
nuclear reaction Kernreaktion *f*
nuclear reactor Atomreaktor *m*
nuclear rearrangement Kernumgruppierung *f*
nuclear region Kernnähe *f*
nuclear research Kernforschung *f*,
  Nuklearforschung *f*
nuclear resonance energy Kernresonanzenergie *f*
nuclear resonance fluorescence
  Kernresonanzfluoreszenz *f*
nuclear rocket propulsion Kernraketenantrieb *m*
nuclear scattering Kernstreuung *f*

nuclear shell Kernschale *f*
nuclear shell model Kernschalenmodell *n*
nuclear size Atomkerngröße *f*
nuclear spectroscopy Kernspektroskopie *f*
   (Atom)
nuclear spectrum Kernspektrum *n*
nuclear spin Kernspin *m*, Kerndrall *m*,
   Kerndrehung *f*
nuclear spin coupling Kernspinkopplung *f*
nuclear spin quantum number
   Kernspinquantenzahl *f*
nuclear stain Kernfarbstoff *m* (Hist)
nuclear state Kernzustand *m*
nuclear structure Kernaufbau *m* (Atom),
   Kernstruktur *f*
nuclear susceptibility Kernsuszeptibilität *f*
nuclear symmetry Kernsymmetrie *f*
nuclear synthesis Kernsynthese *f*
nuclear technology Atomtechnik *f*, Kerntechnik *f*
nuclear test Atomversuch *m*
nuclear testing ground Atomversuchsgelände *n*
nuclear theory Kerntheorie *f*
nuclear transformation Kernumwandlung *f*,
   artificial ~ künstliche Kernumwandlung *f*,
   enforced ~ erzwungene Kernumwandlung *f*
nuclear transition Kernübergang *m*
nuclear transmutation Kernumwandlung *f*
nuclear war Atomkrieg *m*
nuclear weapon Atomwaffe *f*, Kernwaffe *f*
nuclease Nuclease *f* (Biochem)
nucleate boiling Blasensieden *n*,
   Blasenverdampfung *f*
nucleating agent Keimbildner *m*
nucleation Kernbildung *f*, Bildung *f* von
   Kristallisationskeimen (pl)
nuclei, condensed ~ kondensierte Kerne *pl*
   (Chem)
nucleic acid Nukleinsäure *f*
nuclein Nuklein *n*
nucleinic acid (obs) Nukleinsäure *f*
nucleogenesis Nukleogenese *f*
nucleolus Nukleolus *m*
nucleon Nukleon *n* (Atom), Kernteilchen *n*
   (Atom), **evaporation** ~ Ausdampfnukleon *n*
nucleonics Kerntechnologie *f*, Nukleonik *f*
nucleon number Nukleonenzahl *f* (Atom)
nucleon width Nukleonenbreite *f*
nucleophilic nukleophil
nucleophilic addition nucleophile Addition *f*
nucleoplasm Kernplasma *n*
nucleoprotein Kernprotein *n*, Nukleoproteid *n*,
   Nukleoprotein *n*
nucleosidase Nukleosidase *f* (Biochem)
nucleoside Nukleosid *n*
nucleoside kinase Nukleosidkinase *f* (Biochem)
nucleoside monophosphate
   Nukleosidmonophosphat *n*
nucleoside transferase Nukleosidtransferase *f*
   (Biochem)

nucleosin Nukleosin *n*
nucleotidase Nukleotidase *f* (Biochem)
nucleotide Nukleotid *n*
nucleotide unit Nukleotid-Einheit *f*
nucleus (pl. nuclei) Kern *m*, Atomkern *m*
   (Chem), Zellkern *m* (Biol), **artificial
   radioactive** ~ künstlich radioaktiver Kern *m*
   (Atom), **even-even** ~ Kern *m* mit Protonen
   und Neutronen von gerader Zahl, **excited** ~
   angeregter Kern, **haploid** ~ Kern *m* mit
   halber Chromosomenzahl (Biol), **intermediate**
   ~ Zwischenkern *m*, **odd-even** ~ Kern *m* mit
   Protonen von ungerader und Neutronen von
   gerader Zahl (Atom), **odd-odd** ~ Kern *m* mit
   Protonen und Neutronen von ungerader Zahl
   (Atom), **size of** ~ Kerngröße *f*, ~ **of a crystal**
   Kristallkern *m*
nucleus formation Kernbildung *f*
nucleus model Kernmodell *n* (Atom)
nuclide Nuklid *n* (Phys)
nudic acid Nudinsäure *f*
nugget of gold Goldklumpen *m*
nuisance level Immisionen *pl*
null nichtig, ungültig
null element Nullelement *n*
null indicator Nullindikator *m*
nullity Nichtigkeit *f*, Ungültigkeit *f*
null operator Nulloperator *m*
null or zero matrix Nullmatrize *f*
number Zahl *f*, Anzahl *f*, Nummer *f*, **complex** ~
   komplexe Zahl *f*, **conjugated complex** ~
   konjugiert komplexe Zahl *f*, **even** ~ gerade
   Zahl, **fractional** ~ gebrochene Zahl,
   imaginary ~ imaginäre Zahl, **irrational** ~
   irrationale Zahl, **maximum or highest** ~
   Höchstzahl *f*, **odd** ~ ungerade Zahl, **rational**
   ~ rationale Zahl, **real** ~ reelle Zahl,
   three-figure ~ dreistellige Zahl, **total** ~
   Gesamtzahl *f*, **uneven** ~ ungerade Zahl *f*, ~ **in
   brackets** eingeklammerte Zahl, ~ **of blows**
   Schlagzahl *f*, ~ **of cycles** Periodenzahl *f*, ~ **of
   ends per inch** Fadendichte *f* (Reifen), ~ **of
   faces** Flächenzahl *f* (Krist), ~ **of meshes**
   Maschenzahl *f*, ~ **of oscillations** Frequenz *f*,
   ~ **of periods** Periodenzahl *f*, ~ **of revolutions**
   Drehzahl *f*, ~ **of strokes** Hubzahl *f*
number *v* zählen, beziffern, numerieren
number average Zahlenmittel *n*
numbering Bezifferung *f*, Numerierung *f*
number system Zahlensystem *n*, **binary** ~ binäres
   Zahlensystem, **decimal** ~ dezimales
   Zahlensystem
numeral Ziffer *f*
numeral ratio Zahlenverhältnis *n*
numeration Bezifferung *f*, Numerierung *f*
numerator (of a fraction) Zähler *m* (eines
   Bruches)
numerical numerisch, zahlenmäßig
numerical control Datensteuerung *f*

numerical equation numerische Gleichung *f,*
  Zahlengleichung *f* (Math)
numerical example Zahlenbeispiel *n*
numerical index Zahlenindex *m*
numerical value Zahlenwert *m*
numerical verifier Lochprüfer *m* (Comp)
numerometry Numerometrie *f*
  (Titrationsmethode)
nupharine Nupharin *n*
Nuremberg violet Manganviolett *n,* Nürnberger
  Violett *n*
nut Nuß *f,* Mutter *f* (Techn), Schraubenmutter *f*
  (Techn), **clamping or retaining** ~
  Haltemutter *f*
nut coal Nußkohle *f*
nut iron Gewindeeisen *n,* Muttereisen *n*
nutmeal Nußkernmehl *n*
nutmeg Muskatnuß *f*
nutmeg butter Muskatbutter *f,* Muskatbalsam *m,*
  Muskatfett *n*
nutmeg oil Muskat[nuß]öl *n*
nut mill Nußmühle *f*
nut oil Nußöl *n*
nutrient Nährstoff *m,* Nährmittel *n*
nutrient *a* nährend, nahrhaft
nutrient content[s] Nährstoffgehalt *m*
nutrient liquid Nährflüssigkeit *f*
nutrient medium Nährboden *m* (Bakt)
nutrient yeast Nährhefe *f*
nutriment Nahrung *f,* Nährmittel *n,*
  Nahrungsmittel *n*
nutrition Ernährung *f,* Nahrung *f,* **science of** ~
  Nahrungsmittelkunde *f*
nutritional contents Nährgehalt *m*
nutritional deficiency Ernährungsmangel *m*
nutritional requirements Nahrungsbedarf *m*
nutritious nahrhaft
nutritiousness Nahrhaftigkeit *f*
nutritive nahrhaft
nutritive liquid Nährflüssigkeit *f*
nutritive preparation Nährpräparat *n*
nutritive salt Nährsalz *n*
nutritive solution Nährlösung *f*
nutritive substance Nährstoff *m,*
  Nahrungsmittel *n*
nutritive value Nährwert *m*
nutritive yeast Nährhefe *f*
nutrose Nutrose *f*
nutsch and pan filter Nutschen-Filter *n*
nux vomica Brechnuß *f,* Krähenauge *n,* **extract of**
  ~ Brechnußextrakt *m*
nyctal Nyctal *n*
nydrazid Nydrazid *n*
nylon Nylon *n*
nylon fabric Nylongewebe *n* (Text)
nylon flake Nylonschnitzel *n m*
nylon plastic Nylonkunststoff *m,*
  Polyamidkunststoff *m*
nylon-reinforced nylonverstärkt

O

oak Eiche *f,* Eichenbaum *m*
oak apple Gallapfel *m,* Laubapfel *m* (Bot)
oak bark Eichenrinde *f,* ~ for tanning
Lohrinde *f*
oak dust Eichenholzmehl *n*
oak extract Eichenholzextrakt *m*
oakum Werg *n*
o-aminobenzoic acid ortho-Aminobenzoesäure *f,*
Anthranilsäure *f*
oat crusher Haferquetsche *f*
oat flakes Haferflocken *pl*
oatmeal Hafermehl *n*
oats Hafer *m*
oat starch Haferstärke *f*
obese adipös, dick, beleibt
obesity Fettsucht *f* (Med)
object Gegenstand *m,* Objekt *n,* ~ to be
examined Untersuchungsobjekt *n*
object *v* Einspruch erheben, beanstanden,
reklamieren
object carrier drum Objektträgertrommel *f*
object clamp Objektklammer *f*
object finder Objektsucher *m*
object glass Objektiv *n,* two-lens ~ zweiteiliges
Objektiv
objection Beanstandung *f,* Einwendung *f,*
Reklamation *f*
objectionable nicht einwandfrei
objective Objektiv *n* (Opt), focus of the ~
Objektivbrennpunkt *m,* ~ of good light
transmitting capacity lichtstarkes Objektiv, ~
of short focal length kurzbrennweitiges
Objektiv
objective *a* objektiv, sachlich
object-preserving objekttreu
oblate abgeplattet
oblateness Abplattung *f*
oblige *v* verpflichten
oblique schief, schräg, schrägwinklig
oblique-angled schiefwinklig
oblique angular schiefwinklig
oblique draft gauge Schrägrohrzugmesser *m*
oblique extruder head Schrägspritzkopf *m*
oblique set valve Schrägsitzventil *n*
oblique tube Schrägrohr *n*
obliquity Schräge *f*
obliterate *v* auslöschen, austilgen, ausradieren
oblitine Oblitin *n*
oblong Rechteck *n*
oblong *a* länglich, rechteckig (Math)
oblong size Langformat *n*
obscurine Obscurin *n*
observable wahrnehmbar
observation Beobachtung *f,* Wahrnehmung *f,*
error of ~ Beobachtungsfehler *m,* field of ~
Beobachtungsfeld *n*

observational method Beobachtungsmethode *f*
observation data Beobachtungsergebnis *n*
observation hole Schauloch *n*
observation point Beobachtungsstelle *f,*
Meßstelle *f*
observation tube Beobachtungsrohr *n*
observatory Observatorium *n*
observe *v* beobachten, bemerken, wahrnehmen
observed value Beobachtungswert *m*
observer Beobachter *m,* point ~ punktförmiger
Beobachter
obsidian Obsidian *m* (Min), Glasachat *m* (Min),
Glaslava *f* (Min)
obstacle Hindernis *n*
obstetrician Geburtshelfer *m*
obstetrics Geburtshilfe *f*
obtain *v* erhalten, bekommen, erlangen,
erwerben, gewinnen
obtrusive aufdringlich
obtusatic acid Obtusatsäure *f*
obtuse stumpf (Winkel)
obtuse-angled stumpfwinklig
obtusifolin Obtusifolin *n*
obtusilic acid Obtusilsäure *f*
obversely umgekehrt
obvious deutlich, offensichtlich,
selbstverständlich
occasion Anlaß *m,* Gelegenheit *f*
occasional gelegentlich
occlude *v* okkludieren, absorbieren,
einschließen
occlusion Abschließung *f,* Absorption *f* (Chem),
Einschließung *f,* Okklusion *f* (Chem)
occlusion capacity Okklusionsvermögen *n*
occlusion of gases Blasenbildung (Chem)
occupation Beruf *m,* Gewerbe *n*
occupation rule Besetzungsvorschrift *f*
occupy *v* bekleiden (Amt), beschäftigen,
besitzen, einnehmen (Platz, Raum)
occur *v* vorkommen, eintreten, sich abspielen,
sich ereignen
occurrence Vorkommen *n,* Auftreten *n,*
Begebenheit *f,* Ereignis *n*
occurring, naturally ~ natürlich vorkommend
oceanic ozeanisch
oceanium (obs) Hafnium *n*
oceanography Meereskunde *f,* Ozeanographie *f*
ocean water Meerwasser *n*
ocher (Am. E.) Ocker *m* (Min), calcareous ~
Kalkocker *m* (Min), containing ~
ockerhaltig, yellow ~ Berggelb *n* (Min),
Gelberde *f* (Min)
ocherous ockerfarben, ockerhaltig
ocher-yellow ockergelb
o-chlorobenzaldehyde
ortho-Chlorbenzaldehyd *m*

o-chlorobenzoic acid ortho-Chlorbenzoesäure *f*
o-chlorotoluene ortho-Chlortoluol *n*
ochracin Ochracin *n*
ochre (Br. E.) Ocker *m*
ocimene Ocimen *n*
o-compound Orthoverbindung *f*
ocreine Ocrein *n*
octaamylose Octaamylose *f*
octabasic achtbasisch
octacosane Oktakosan *n*
octacosanol Octacosanol *n*
octadecane Octadecan *n*
octadecanoic acid Octadecansäure *f*,
　Stearinsäure *f*
octadecanol Octadecylalkohol *m*
octadecatrienoic acid Octadecatriensäure *f*
octadecen-1-ol, 9- ~ Oleinalkohol *m*
octadecenoic acid, 9- ~ Ölsäure *f*, Oleinsäure *f*
octadecine Octadecin *n*
octadecyl alcohol Octadecylalkohol *m*
octadecyl bromide Octadecylbromid *n*
octadecylene Anthemen *n*
octadecyne Octadecin *n*
octadecynoic acid, 5- ~ Taririnsäure *f*, **9-** ~
　Stearolsäure *f*
octadiene, 1,4- ~ Conylen *n*
octagon Achteck *n*
octagonal achteckig, achtkantig
octagonal column Achtecksäule *f*
octagonal spanner Achtkantschlüssel *m*
octagon bar Achtkantstab *m*
octahedral achtflächig (Krist), oktaedrisch
　(Krist)
octahedral symmetry Oktaedersymmetrie *f*
　(Krist)
octahedrite Anatas *m* (Min), Oktaedrit *m* (Min)
octahedron Oktaeder *n* (Krist), Achtflächner *m*
　(Krist)
octahedron structure Oktaederstruktur *f*
octahydroanthracene Okthracen *n*
octahydronaphthalene Oktalin *n*
octahydrophenanthrene Oktanthren *n*
octalupine Octalupin *n*
octamethyl lactose Octamethyllactose *f*
octamethyl sucrose Octamethylsaccharose *f*
octamylamine Octamylamin *n*
octanal Octylaldehyd *m*
octane Octan *n*, Oktan *n*
octanedioic acid Korksäure *f*, Suberinsäure *f*
octane number Oktanzahl *f*, Klopfwert *m*
octane rating Oktanwert *m*
octanoic acid Octylsäure *f*
octanol Octanol *n*, Octylalkohol *m*
octant diagram Oktanten-Diagramm *n*
octanthrene Oktanthren *n*
octanthrenol Oktanthrenol *n*
octanthrenone Oktanthrenon *n*
octant rule Octantenregel *f*, Oktanten-Regel *f*
octastearyl sucrose Octastearylsaccharose *f*

octatrienol Octatrienol *n*
octavalence Achtwertigkeit *f*
octavalent achtwertig, achtbindig
octene Caprylen *n*, Octen *n*, Octylen *n*
octet rule Oktettregel *f*
octet shell Achterschale *f* (Chem)
octhracene Okthracen *n*
octhracenol Okthracenol *n*
octhracenone Okthracenon *n*
octodecanoic acid Stearinsäure *f*
octodecanoyl Stearyl-
octopamine Octopamin *n*
octopine Octopin *n*
octopole excitation Oktopolanregung *f*
octopole transition Oktopolübergang *m*
octose Octose *f*, Oktose *f*
octupole excitation Oktopolanregung *f*
octupole transition Oktopolübergang *m*
octyl Octyl-
octyl alcohol Octylalkohol *m*
octylaldehyde Octylaldehyd *m*
octylamine Octylamin *n*
octylene Octylen *n*
octylic acid Octylsäure *f*
octyl nitrite Octylnitrit *n*
octylphenol Octylphenol *n*
octyne, 1- ~ Capryliden *n*
ocular Okular *n*
odd unpaarig, ungerade
odd-even ungerade-gerade (Atom)
odd-numbered ungeradzahlig
odd-odd ungerade-ungerade (Atom)
odontolite Zahntürkis *m* (Min)
odontolith Zahntürkis *m* (Min)
odor Duft *m* (Am. E.), Geruch *m*
odoriferous substance Duftstoff *m*
odorless duftlos, geruchfrei, geruchlos
odorous duftig, wohlriechend
odorous substance Duftstoff *m*
odour (Br. E.) Geruch *m*
odyssic acid Odyssinsäure *f*
odyssin Odyssin *n*
oedema Ödem *n*
oellacherite Öllacherit *m* (Min)
oenanthine Önanthin *n*
oenometer Önometer *n*
[o]esophagus Speiseröhre *f* (Anat)
oestradiol Östradiol *n*
oestriol Östriol *n*
oestrone Östron *n*
o-ethylphenol Phlorol *n*
offal Abfall *m*
offal timber Abfallholz *n*
off-center unmittig
off-center position Unmittigkeit *f*
offensive offensiv
offer Angebot *n*
offer *v* [an]bieten
off-flavor Beigeschmack *m*

office computers Anlagen der mittleren
  Datentechnik *f*
official offiziell, amtlich, behördlich
officinal weight Arzneigewicht *n*
off-position Nullstellung, Abschaltstellung *f*
offset gekröpft (Techn), Offsetdruck *m*
offset ink Offsetfarbe *f*
offset printing Offsetdruck *m*
offset process Offsetverfahren *n* (Buchdr)
offset yield strength Dehngrenze *f*
off-sites Nebenanlagen *pl*
ohm Ohm *n* (Elektr)
ohmic drop of voltage Ohmscher
  Spannungsabfall *m*
ohmic loss Ohmscher Verlust *m*,
  Stromwärmeverlust *m*
ohmic resistance Ohmscher Widerstand *m*
ohmmeter Ohmmeter *n* (Elektr),
  Widerstandsmesser *m* (Elektr)
Ohm's law Ohmsches Gesetz *n* (Elektr)
o-hydroxyanisole Guajacol *n*
o-hydroxybenzoic acid Salicylsäure *f*
oiazine Oiazin *f*
oil Öl *n*, Erdöl *n*, **aromatic ~** Duftöl *n*, **bodied ~**
  Dicköl *n*, **boiled ~** Ölfirnis *m*, **cold test ~**
  kältebeständiges Öl, **containing ~** ölhaltig,
  drying ~ trocknendes Öl, **essential ~**
  ätherisches Öl, **fixed ~** gehärtetes Öl,
  frost-resisting ~ kältebeständiges Öl, **heavy ~**
  dickes Öl, **hydrogenated ~** gehärtetes Öl, **light**
  ~ dünnflüssiges Öl, **sulfated ~** sulfoniertes
  Öl, **sulfurized ~** geschwefeltes Öl, **to remove**
  ~ entölen, **volatile ~** ätherisches Öl, leicht
  flüchtiges Öl, **~ for molds** Formenöl *n*, **~**
  under pressure Drucköl *n*
oil *v* [ein]ölen, einfetten, schmieren, **~**
  thoroughly durchölen
oil absorption Ölabsorption *f*, Ölzahl *f*
oil absorption capacity Ölaufnahmefähigkeit *f*
oil acidimeter Ölacidimeter *n*
oil atomizer Ölzerstäuber *m*
oil base paint Ölanstrichfarbe *f*
oil basin Ölbehälter *m*
oil bath Ölbad *n*
oil-bearing erdölhaltig
oil-binding property Ölbindevermögen *n*
oil black Ölruß *m*, Ölschwarz *n*
oil bleaching Ölbleiche *f*
oil burner Ölbrenner *m*
oil burning Ölfeuerung *f*
oil-burning *a* ölgefeuert
oil-burning furnace Ölflammofen *m*
oil cake Ölkuchen *m*
oil can Ölkanne *f*, Schmierkanne *f*
oil catcher Ölfänger *m*
oil catchpan Ölauffangwanne *f*
oil chalk Ölkreide *f*
oil chamber Ölkammer *f*
oil change Ölwechsel *m*

oil chemistry Erdölchemie *f*
oil circulating tank Ölkreislaufbehälter *m*
oil circulation Ölumlauf *m*
oil circulation lubrication Ölumlaufschmierung *f*
oil cloth Wachstuch *n*
oil cloth varnish Wachstuchlack *m*
oil color Ölfarbe *f*
oil color pencil Ölfarbstift *m*
oil compatibility Ölverträglichkeit *f*
oil consumption Ölverbrauch *m*
oil container Ölkanister *m*
oil[-cooled] transformer Öltransformator *m*
oil crayon Ölkreidestift *m*
oil crust Ölkruste *f*
oil cup Ölgefäß *n*, Schmiervase *f*
oil damping Öldämpfung *f*
oil deodorization Öldesodorisierung *f*
oil dipstick Öl[stands]meßstab *m*
oil dish Ölfänger *m*
oil distilling process Öldestillationsverfahren *n*
oil drain channel Ölkanal *m*
oil drain[er] Ölablaß *m*
oil dregs Ölrückstand *m*
oil dressing Fettappretur *f* (Leder)
oil drill Ölbohrer *m*
oil drop Öltröpfchen *n*
oiled geölt
oil ejector Öldampfstrahlpumpe *f*
oil enamel Öllack *m*
oiler (Am. E.) Ölkanne *f*, Öltanker *m*
oil extraction Ölextraktion *f*, Ölgewinnung *f*
oil extraction plant Ölextraktionsanlage *f*
oil extractor Ölextraktor *m*, Ölschleuder *f*
oil factory Ölwerk *n*
oil feed Ölzufluß *m*
oil filling Ölfüllung *f*
oil film Ölfilm *m*, Ölhaut *f*, Schmierfilm *m*
oil filter Ölfilter *m n*, Ölreinigungsvorrichtung *f*
oil-fired ölgefeuert
oil-fired furnace Petroleumfeuerung *f*
oil firing Ölfeuerung *f*
oil flotation process Ölschwemmverfahren *n*
oilfoot Ölbodensatz *m*
oil-forming ölbildend
oil fume Öldunst *m*
oil furnace Ölfeuerung *f*
oil gas Ölgas *n*
oil gasification, gas from ~ Ölspaltgas *n*
oil gauge Ölstandsmesser *m*
oil gilding Ölvergoldung *f*
oil groove Ölnute *f*, Schmiernut *f*
oil ground Ölgrund *m*
oil hardening Ölhärtung *f*
oil heating Ölheizung *f*, Ölfeuerung *f*
oil heating plant Ölheizanlage *f*
oil hole Schmierloch *n*
oil hydrometer Ölwaage *f*
oil-immersion objective Immersionsobjektiv *n*
  (Opt)

oil-impregnated ölimprägniert
oil industry Erdölindustrie *f*
oiliness Fettigkeit *f*, Schmierfähigkeit *f*
oiling Ölen *n*, Einfetten *n*, Schmieren *n*
oil injection Öleinspritzung *f*
oil insulation Ölisolation *f*, Ölisolierung *f*
oil insulator Ölisolator *m*
oil-in-water emulsion Öl-in-Wasser-Emulsion *f*
oil lamp Petroleumlampe *f*
oil leather Ölleder *n*
oil level Ölstand *m*
oil level check Ölstandskontrolle *f*
oil level gauge Ölmeßstab *m*, Ölstandsanzeiger *m*
oil lubrication Ölschmierung *f*
oil mill Ölmühle *f*, Ölwerk *n*
oil mordant Netzbeize *f* (Färb), Ölbeize *f*
oil mud Ölschlamm *m*
oil of pennyroyal Poleyöl *n*
oil outlet Ölablaß *m*
oil overflow pipe Ölüberlaufrohr *n*
oil paint Ölfarbe *f*, coat of ~
    Ölfarbenanstrich *m*, full-gloss ~ glänzende
    Ölfarbe, heat-resisting ~ hitzebeständige
    Ölfarbe, undercoat ~ Ölvorstreichfarbe *f*
oil paper Ölpapier *n*
oil paste Ölpaste *f*
oil pipe Ölleitung *f*
oil piping Erdölleitung *f*
oil pollution Ölpest *f*
oil press Ölpresse *f*
oil pressing Ölschlagen *n*
oil pressure Öldruck *m*
oil-pressure gauge Öldruckanzeiger *m*,
    Ölmanometer *n*
oil pressure pipe Öldruckleitung *f*
oil primer Ölgrundierung *f*
oil priming Ölgrundierung *f*
oil processing Ölverarbeitung *f*
oil producer Erdölerzeuger *m*
oil-producing countries Öllieferländer *pl*
oil product Erdölprodukt *n*
oil production Erdölförderung *f*
oilproof ölbeständig, ölundurchlässig
oil pump Ölpumpe *f*, high vacuum ~
    Hochvakuumölpumpe *f*
oil purification Ölreinigung *f*
oil purifier Ölreiniger *m*
oil recovery factor Entölungsgrad *m*
oil recovery plant Ölrückgewinnungsanlage *f*
oil rectifier Ölrückgewinnungsanlage *f*
oil refinery Erdölraffinerie *f*
oil refining Erdölverarbeitung *f*, Ölraffination *f*,
    Ölraffinierung *f*
oil relief valve Ölüberdruckventil *n*
oil reservoir Ölbehälter *m*, Ölkammer *f*
oil residue Ölrückstand *m*, Ölschlamm *m*
oil resin Ölharz *n*
oil-resistant ölbeständig
oil retainer ring Öldichtungsring *m*

oil return pipe Ölrücklaufrohr *n*
oil ring Ölring *m*, Schmierölring *m*
oil seal Ölabdichtung *f*
oil sediment Ölsatz *m*, Ölschlamm *m*
oil separation Ölabscheidung *f*
oil separator Entöler *m*, Ölabscheider *m*,
    Ölausscheider *m*, ~ for steam
    Dampfentöler *m*
oil shale Ölschiefer *m* (Min)
oil shock absorber Ölstoßdämpfer *m*
oil siccative Öltrockner *m*
oil skin Ölhaut *f*
oil sludge Ölschlamm *m*
oil smoke Öldampf *m*
oil soap Ölseife *f*
oil-soluble öllöslich
oil-soluble dye Oleosolfarbe *f*
[oil] source rock Muttergestein *n* (Erdöl)
oil spot Ölfleck *m*
oil spray Ölstaub *m*
oil stain Ölfleck *m*
oilstone Ölstein *m*, Abziehstein *m*
oil substitute Ölersatz *m*
oil supply Ölzufluß *m*
oil tank Ölbehälter *m*
oil tanker Öltanker *m*, Öltankschiff *n*
oil-tanned fettgar
oil tanning Fettgerbung *f*
oil tester Ölprüfer *m*
oil thrower Ölschleuderring *m*
oil-tight öldicht
oil transfer pump Ölförderpumpe *f*
oil trap Ölabscheider *m*
oil trough Ölsammler *m*
oil tube Ölleitung *f*
oil vapor Öldampf *m*, Öldunst *m*, Ölstaub *m*
oil varnish Ölfirnis *m*, Öllack *m*
oil way Schmiernut *f*
oil well Ölquelle *f*
oil white Ölweiß *n*
oil whizzer Ölschleuder *f*
oil wiper Ölabstreicher *m*
oily ölig, ölartig, schmierig
oily scum Ölschaum *m*
oily taste Ölgeschmack *m*
ointment Salbe *f* (Med)
ointment base Salbengrundlage *f*
ointment for burns Brandbalsam *m*
ointment for wounds Wundsalbe *f*
okanin Okanin *n*
old gold Altgold *n*
oldhamite Oldhamit *m* (Min)
old metal Altmetall *n*
oleaginous ölartig, ölig
oleander leaves Oleanderblätter *n pl* (Bot)
oleandrin Oleandrin *n*
oleandronic acid Oleandronsäure *f*
oleandrose Oleandrose *f*
oleanol Oleanol *n*

oleanolic acid Caryophyllin *n*, Oleanolsäure *f*
oleanone Oleanon *n*
oleate ölsaures Salz *n*, Oleat *n*
olefin[e] Olefin *n*
olefine alcohol Olefinalkohol *m*
olefine halide Olefinhalogenid *n*
olefine metal complex Olefinmetallkomplex *m*
olefinic olefinisch
oleic acid Ölsäure *f*, containing ~ ölsäurehaltig
oleic acid soap Ölseife *f*
oleiferous ölhaltig
olein Olein *n*, Ölsäureglycerid *n*, Triolein *n*
oleinic acid Ölsäure *f*, Oleinsäure *f*
oleodipalmitin Oleodipalmitin *n*
oleodistearin Oleodistearin *n*
oleograph Öldruck *m* (Bild); Ölfarbendruck *m*
   (Bild)
oleography Öl[farben]druck *m*
   (Bilddruckverfahren)
oleomargarine Margarine *f*, Oleomargarinen *n*
oleometer Ölmesser *m*, Ölwaage *f*
oleone Oleon *n*
oleopalmitostearin Oleopalmitostearin *n*
oleophobic oleophob
oleophobizing Oleophobierung *f*
oleoresin Fettharz *n*, Ölharz *n*, Oleoresin *n*,
   Weichharz *n*
oleostearic acid Oleostearinsäure *f*
oleum rauchende Schwefelsäure *f*, Nordhauser
   Schwefelsäure *f*, Oleum *n*
oleyl alcohol Oleylalkohol *m*
oleyl ethanesulfonic acid Oleyläthansulfonsäure *f*
oleyl sodium sulfate Oleylnatriumsulfat *n*
olibanum Weihrauch *m*
olibanum oil Olibanöl *n*
oligoclase Oligoklas *m* (Min)
oligocyth[a]emia Erythrozytenverminderung *f*
   (Med), Oligozythämie *f* (Med)
oligomeric oligomer
oligomerization Oligomerisation *f*
oligomycin Oligomycin *n*
oligonspar Oligonspat *m* (Min)
oligonucleotide Oligonukleotid *n*
oligopeptide Oligopeptid *n*
oligosaccharide Oligosaccharid *n*
oligotrophic nährstoffarm
olivacine Olivacin *n*
olive Olive *f* (Bot)
olive-green olivgrün
olivenite Olivenit *m* (Min), Olivenerz *n* (Min)
olive oil Olivenöl *n*, lowest grade of ~
   Höllenöl *n*
olive ore Olivenerz *n* (Min), Olivenit *m* (Min)
olivetol Olivetol *n*
olivetoric acid Olivetorsäure *f*
olive tree Ölbaum *m* (Bot)
olivil Olivil *n*
olivine Olivin *m* (Min), Chrysolith *m* (Min)
ombuin Ombuin *n*

omega-meson Omega-Meson *n*
omegatron Omegatron *n* (Phys)
omission Unterlassung *f*, Versäumnis *n*
omit *v* auslassen, versäumen, weglassen
ommochrome Ommochrom *n*
omphacite Omphazit *m* (Min)
once-through boiler Zwangsdurchlaufkessel *m*
oncogenesis Tumorbildung *f* (Med)
oncogenic tumorbildend
oncogenous tumorbildend
oncolytic tumorzellenzerstörend
oncotic onkotisch (kolloid-osmotisch)
ondograph Wellenschreiber *m*
one-bath tanning Einbadgerbung *f* (Gerb)
one-carbon fragment
   Ein-Kohlenstoff-Fragment *n*
one-component adhesive
   Einkomponentenkleber *m*
one-component system
   Einkomponentensystem *n*, Einstoffsystem *n*
one-digit einstellig (Math)
one-dimensional eindimensional
one-electron state Einelektronenzustand *m*
one-factor-at-a-time method
   Ein-Faktor-Methode *f* (bei Optimierung)
one-figure einstellig (Math)
one-fire color Monobrandfarbe *f*
one-flow cascade cycle OFC-Verfahren *n*
one-flue boiler Einflammrohrkessel *m*
onegite Onegit *m* (Min)
one-jig process Durchfahrtechnik *f* (Galv)
one-layer filter Einfachfilter *n*
one-phase system Einphasensystem *n*
one-shot lubrication Zentralschmierung *f*
one-sided einseitig
one-stage einstufig
one-way stopcock Einweghahn *m*
onion Zwiebel *f* (Bot)
onium structure Oniumstruktur *f* (Chem)
on-line digital process control computer
   Computer *m* für die direkte digitale
   Verfahrensregelung
onocerin Onocerin *n*, Onocol *n*
onocol Onocerin *n*, Onocol *n*
on-off control Zweipunktregelung *f*
on-off controller (Am. E.) Zweipunktregler *m*
on-off switch Ein- und Ausschalter *m*
onofrite Onofrit *m* (Min),
   Selenschwefelquecksilber *n* (Min)
ononetin Ononetin *n*
ononin Ononin *n*
ononitol Ononit *n*
onset Beginn *m*, Einsetzen *n*
on-stream analysis Prozeßanalysentechnik *f*
onyx Onyx *m* (Min)
onyx-agate Onyxachat *m* (Min)
onyx-marble Onyxmarmor *m* (Min)
oolite Oolith *m* (Geol), Rogenstein *m* (Geol)
oolite formation Oolithformation *f* (Geol)

oolitic hematite Hirseerz *n* (Min)
oolitic iron stone Eisenoolith *m* (Min)
oolitic ore Linsenerz *n* (Min)
oolitiferous oolithhaltig
ooporphyrine Ooporphyrin *n*
oosporein Oosporein *n*
ooze Beize *f* (Gerb), Schlick *m*
ooze *v* sickern, ~ **out** durchsickern,
   herausströmen, langsam auslaufen
ooze leather Lohbrühleder *n*
oozer (Am. E.) undichte Dose *f*
oozy schlammig, schlammhaltig
opacifier Trübungsmittel *n* (Email)
opacifying power Deckvermögen *n*
opacity Deckfähigkeit *f* (Techn), Deckkraft *f*
   (Techn), Lasurfähigkeit *f* (Techn),
   Lichtundurchlässigkeit *f* (Phys), Opazität *f*,
   Undurchsichtigkeit *f*
opal Opal *m* (Min), **brown opaque** ~
   Leberopal *m* (Min), **ligneous** ~ Holzopal *m*
   (Min), **transparent** ~ Hydrophan *m* (Min)
opal agate Opalachat *m* (Min)
opal allophane Opalallophan *m* (Min)
opal blue Opalblau *n*
opalesce *v* opaleszieren, opalisieren, bunt
   schillern
opalescence Opaleszenz *f*, Farbenspiel *n*,
   Opalglanz *m*, Opalisieren *n*, Schillern *n*
opalescent opalisierend, schillernd, **to render** ~
   opalisieren
opal glass Milchglas *n*, Opalglas *n*
opaline Opalin *n*
opaline *a* opalartig, bunt schillernd
opaline luster Opalglanz *m*
opalize *v* opalisieren
opal jasper Opaljaspis *m* (Min)
opal matrix Opalmutter *f*
opal varnish Opalfirnis *m*
opaque undurchsichtig, deckfähig (Techn),
   getrübt (Glas), lichtundurchlässig (Phys), opak
opaque green Deckgrün *n*
opaqueness Undurchsichtigkeit *f*,
   Lichtundurchlässigkeit *f*
opaque white Deckweiß *n*
open *v* öffnen, aufschließen, aufsperren,
   durchbrechen, ~ **into** münden
open-air bleaching Luftbleiche *f*
open area Öffnungsverhältnis *n* (Dest)
open assembly time (adhesives) offene Zeit *f*
open band semiconductor Offenbandhalbleiter *m*
opencast Tagebau *m* (Bergb)
open-cast working Tagebaugewinnung *f*
open-hearth furnace Siemens-Martin-Ofen *m*
open-hearth iron Martin[fluß]eisen *n*
open-hearth pig iron Siemens-Martin-Roheisen *n*
open-hearth practice
   Siemens-Martin-Verfahren *n*
open-hearth process Siemens-Martin-Prozeß *m*,
   Herdfrischen *n*

open-hearth refining Herdfrischen *n*
open-hearth steel Martinstahl *m*,
   Herdfrischstahl *m*
opening Öffnung *f*, Apertur *f* (Opt), Einleiten *n*,
   Loch *n*, Lücke *f*, Mündung *f* (Röhre), Spalt *m*,
   ~ **of a double bond** Aufrichtung *f* einer
   Doppelbindung (Chem), ~ **the tap hole**
   Aufbringen *n* des Abstichloches
opening ram Rückdruckkolben *m*
opening stroke Werkzeugöffnungshub *m*
open loop Regelstrecke *f*
open-pit ore mining Erztagebau *m*
open planning Großraumbau *m*
open-plan or open-style office Großraumbüro *n*
open sore offene Wunde *f*
open steam cure Freidampfheizung *f* (Gummi)
open steam curl Freidampfvulkanisation *f*
open-surface cooler Rieselkühler *m*
open top can Falzdeckeldose *f*, Konservendose *f*
open-work durchbrochen
open workings Tagebau *m* (Bergb)
operable betriebsfähig, durchführbar
operate *v* arbeiten, bedienen, betätigen
   (maschinell), funktionieren, handhaben,
   operieren (Med)
operating condition Arbeitsbedingung *f*
operating conditions Betriebsdaten *pl*
operating current Betriebsstrom *m*
operating data Betriebsdaten *pl*
operating engineer Betriebsingenieur *m*
operating expenses Betriebsunkosten *pl*
operating figures Betriebskennzahlen *f pl*
operating gear Betätigungsvorrichtung *f*
operating installation Betriebseinrichtung *f*
operating instructions Bedienungsanleitung *f*,
   Betriebsvorschriften *pl*
operating lever Bedienungshebel *m*,
   Betätigungshebel *m*
operating line Arbeitskurve *f* (Atom),
   Austauschgerade *f* (Rektifikation),
   Bilanzlinie *f* (McCabe-Thiele-Diagramm)
operating load Betriebsbeanspruchung *f*
operating mechanism Bewegungsvorrichtung *f*
operating panel Schalttafel *f*
operating platform Bedienungsbühne *f*
operating pressure Betriebsdruck *m*
operating pulpit Steuerbühne *f*
operating range (of the final control element)
   Stellbereich *m* (Regeltechn)
operating time Laufzeit *f*
operating trouble Betriebsstörung *f*
operating valve Steuerschieber *m*
operating voltage Betriebsspannung *f*
operation Arbeits[vor]gang *m*, Arbeitsweise *f*,
   Bedienung *f*, Betätigung *f*, Betrieb *m*,
   Eingriff *m* (Med), Handhabung *f*, Prozeß *m*,
   Verfahren *n*, Vorgang *m*, **continuous** ~
   Dauerbetrieb *m*, **cyclic** ~ periodischer
   Vorgang, **mathematical** ~ mathematische

Operation *f*, **method of** ~ Wirkungsweise *f*,
mode of ~ Arbeitsweise *f*, **periodic** ~
periodischer Vorgang, **range of** ~
Arbeitsbereich *m*, **surgical** ~ chirurgische
Operation *f* (Med), **to be in** ~ in Betrieb sein
operational betrieblich, betriebsbedingt
operational altitude Betriebshöhe *f*
operational analysis Verfahrensanalyse *f*
operational procedure Betriebsablauf *m*
operation research Verfahrensuntersuchung *f*
operations, sequence of ~ Arbeitsablauf *m*
operative betrieblich
operative attenuation Betriebsdämpfung *f*
operator Bedienungsmann *m*,
Betriebstechniker *m*
operator's stand Bedienungsstand *m*
operon Operon *n* (Biochem)
ophelic acid Opheliasäure *f*
ophicalcite Ophicalcit *m* (Min)
ophiotoxin Ophiotoxin *n*
ophioxylin Ophioxylin *n*
ophite Ophit *m* (Min), Schlangenstein *m* (Min)
ophthalmology Augenheilkunde *f*
opiane Opian *n*
opianic acid Opiansäure *f*
opianine Narcotin *n*
opianyl Mekonin *n*
opiate Morphiumpräparat *n*, Opiat *n*
opiated syrup Opiumsirup *m*
opiazone Opiazon *n*
opinion Meinung *f*, Ansicht *f*
opium Opium *n*, **addiction to** ~ Opiumsucht *f*,
extract of ~ Opiumextrakt *m*, **tincture of** ~
Opiumtinktur *f*
opium addict Opiumsüchtiger *m*
opium alkaloid Opiumalkaloid *n*
opiumism Opiumsucht *f*
opium plaster Opiumpflaster *n*
opium poisoning Opiumvergiftung *f* (Med)
opium smell Opiumgeruch *m*
opium water Opiumwasser *n*
opopanax Panaxgummi *n*, Opopanax *m*
o-position Orthostellung *f* (Chem)
oppose *v* gegenüberstellen, bekämpfen,
entgegenrichten, entgegensetzen
opposed entgegengesetzt
opposing gegensinnig
opposing force Gegenkraft *f*
opposite entgegengesetzt, ungleichnamig (Elektr)
opposite connection Gegenschaltung *f*
oppositely charged entgegengesetzt geladen
oppositely directed gegenläufig
opposite sides gegenüberliegende Seiten *pl*
opsopyrrole Opsopyrrol *n*
optic[al] optisch
optical aberration Abbildungsfehler *m* (Opt)
optical activity optische Aktivität *f*
optical bleaching agent optischer Aufheller *m*
optical brightener optischer Aufheller *m*

optical coincidence cards Sichtlochkarten *pl*
optical glass Linsenglas *n*
optical image optisches Bild *n*
optical isomers optische Isomere *pl*
optically absorptive lichtschluckend
optically active optisch aktiv
optically inactive optisch inaktiv
optically negative optisch negativ
optically positive optisch positiv
optically sensitive lichtempfindlich
optical microscope Lichtmikroskop *n*
optical microscopy Lichtmikroskopie *f*
optical path Lichtweg *m*
optical phenomenon Lichterscheinung *f*
optical reader printing inks Belegleserfarben *pl*
[optical] refraction Lichtbrechung *f*
optical refractive power
Lichtbrechungsvermögen *n*
optical resolution Antipodentrennung *f*
optical stimulus Sehreiz *m*
optician Optiker *m*
optic nerve Sehnerv *m* (Anat)
optics Lehre *f* vom Licht, Optik *f*
optimal optimal
optimality principle Optimalitätsprinzip *n*
optimeter Optimeter *n*
optimization Optimierung *f*
optimization of functions
Funktionenoptimierung *f*
optimization of parameters
Parameteroptimierung *f*
optimum Bestwert *m*, Optimum *n*
optimum *a* optimal
option Optionsrecht *n*
optional wahlweise
optoelectronics Optoelektronik *f* (Elektrooptik)
optoquinic acid Optochinsäure *f*
oral oral
orange Orange *f* (Bot)
orange bitter Pomeranzenlikör *m*
orange-colored orange[farben], rotgelb
orange flower oil Neroliöl *n*,
Pomeranzenblütenöl *n*
orange juice Apfelsinensaft *m*, Orangensaft *m*
orange lead Orangemennige *f*, Bleisafran *m*
orange minium Orangemennige *f*
orange ocher Orangeocker *m*
orange peel Orangenschale *f*
orange peel oil Apfelsinenschalenöl *n*,
Orangenschalenöl *n*
orange seed oil Orangensamenöl *n*
orange yellow Orangegelb *n*
orangite Orangit *m* (Min)
orbit Elektronenbahn *f* (Atom), Kreisbahn *f*,
photocentric ~ Bahn *f* des Lichtzentrums, ~
(of the electron) Umlaufbahn *f* (des
Elektrons)
orbital Orbital *n* (Atom), **anti-bonding** ~
antibindendes Orbital

orbital angular momentum Bahndrehimpuls *m* (Atom)
orbital diameter Bahndurchmesser *m* (Atom)
orbital electron Bahnelektron *n*, Hüllenelektron *n*
orbital ellipse Bahnellipse *f* (Astr)
orbitally degenerate bahnentartet (Atom)
orbital moment Bahnmoment *n*
orbital moment components Bahnmomentanteile *pl*
orbital momentum Bahnimpuls *m*
orbital motion Bahnbewegung *f* (Atom), Umlaufbewegung *f* (Atom)
orbital plane Bahnebene *f*
orbital radius Bahnradius *m* (Elektron)
orbital revolution Bahnumdrehung *f*
orbitals, overlapping ~ überlappende Orbitale *pl*
orbital theory Orbitaltheorie *f*
orbital transition (of an electron) Bahnübergang *m* (eines Elektrons)
orbital velocity Bahngeschwindigkeit *f* (Atom)
orbital wave function Wellenbahnfunktion *f* (Atom)
orbit correction Bahnkorrektur *f*
orbit shift coil Ablenkspule *f*
orcein Orcein *n*, Flechtenrot *n*
orchil Orseille *f*
orchil extract Orseilleextrakt *m*
orcin Orcin *n*
orcinol Orcin *n*
orcirufamin Orcirufamin *n*
orcyl aldehyde Orcylaldehyd *m*
order Bestellung *f*, Auftrag *m*, Befehl *m*, Ordnung *f*, Rang *m*, Reihe *f*, Reihenfolge *f*, ~ **of interference** Beugungszahl *f* (Opt); ~ **of reaction** Reaktionsordnung *f*
order *v* befehlen, bestellen, ordnen
order-disorder transformation Ordnungsunordnungsumwandlung *f*
order form Bestellschein *m*
ordering transition Ordnungsumwandlung *f*
orderly ordentlich
order parameter Ordnungsparameter *m*
order-pick container Kommissionierbehälter *m*
order picker Kommissioniergerät *n*
order-picking bays Kommissionierstollen *pl*
order-picking section Kommissionierbereich *m*
order-picking warehouse Kommissionierlager *n*
order processing Auftragsbearbeitung *f*
order stacker Kommissionierstapler *m*
ordinal number Ordinalzahl *f*, Ordnungzahl *f*
ordinance Verordnung *f*
ordinary gewöhnlich, gemein
ordinary glass Normalglas *n*
ordinary soap Kernseife *f*
ordinate Ordinate *f* (Math)
ordinate axis Ordinatenachse *f* (Math), Y-Achse *f* (Math)

ore Erz *n*, **addition of** ~ Erzzusatz *m*, **calcined** ~ geröstetes Erz, **dressed** ~ Erzschlich *m*, **easily fusible** ~ leicht schmelzbares Erz, **easily reducible** ~ leicht reduzierbares Erz, **enrichment of** ~ Erzanreicherung *f*, **graded** ~ Stufenerz *n*, **high-grade** ~ reichhaltiges Erz, **incompletely roasted** ~ unvollkommen geröstetes Erz, **low-grade** ~ Armerz *n*, minderhaltiges Erz, **magnetic** ~ magnetisches Erz, **mixed** ~ Mischerz *n*, **preparation of** ~ Erzaufbereitung *f*, **pure** ~ Stufenerz *n* (Min), **raw** ~ ungeröstetes Erz, Frischerz *n*, **rich in** ~ erzreich, **roasted** ~ geröstetes Erz, Garerz *n*, **self-fluxing** ~ selbstgehendes Erz, **siliceous** ~ kieselreiches Erz, **type of** ~ Erzart *f*, ~ **in beds** Flözerz *n*, ~ **in veins** Flözerz *n*, ~ **to be calcined** Röstgut *n*, ~ **to be roasted** Röstgut *n*, ~ **to be smelted** Hüttenspeise *f*
ore-bearing erzführend, erzhaltig
ore body Erzstock *m*
ore breaker Erzbrecher *m*
ore-breaking plant Erzzerkleinerungsanlage *f*
ore briquette Erzpreßstein *m*, Erzziegel *m*
ore briquetting Erzbrikettierung *f*
ore burdening Erzmöllerung *f*
ore charge Erzgicht *f*
ore crusher Erzbrecher *m*, Erzquetsche *f*
ore crusher plant Erzzerkleinerungsanlage *f*
ore-crushing Erzzerkleinerung *f*
ore deposit Erzlagerstätte *f*
ore distributer Erzverteiler *m*
ore dressing Erzaufbereitung *f*, Anreicherung *f* der Erze (pl), Aufbereitungsverfahren *n* (Met)
ore-dressing plant Erzaufbereitungsanlage *f*
ore dust Erzstaub *m*, Mulm *m*
ore formation Erzformation *f*
ore grinder Erzzerreiber *m*
ore leaching pit Erzlaugegrube *f*
ore leaching plant Erzlaugerei *f*
ore loading device Erzladevorrichtung *f*
ore lode Erzader *f*
ore mine Erzbergwerk *n*
ore mining Erzförderung *f*, Erzgewinnung *f*
ore process Erzfrischverfahren *n*
ore puddling Erzpuddeln *n*
ore pulp Erztrübe *f*
ore roasting Erzröstung *f*
ore screener Erzscheider *m*
ore separator Erzscheider *m*
ore slag Erzschlacke *f*
ore slime Erzschlamm *m*, Pochschlich *m*
ore sludge Erzschlamm *m*
ore smelting Erzschmelzen *n*
ore-smelting furnace Erzschmelzofen *m*
ore storing place Erzlagerplatz *m*
ore vein Erzader *f*
ore wagon Erzförderwagen *m*
ore washing Erzwäsche *f*, Waschen *n* der Erze (pl)

orexin Orexin *n*
orexin hydrochloride Orexinchlorhydrat *n*
orexin tannate Orexintannat *n*
organelle Organelle *f* (Biol)
organic organisch
organic-chemical organisch-chemisch
organic chemist Organiker *m*
organic chemistry organische Chemie *f*
organic compound organische Verbindung *f*
organic pigment Pigmentfarbstoff *m*
organism Lebewesen *n* (Biol), Organismus *m*
organization Organisation *f*
organize *v* organisieren
organogel Organogel *n*
organographic organographisch (Med)
organography Organbeschreibung *f* (Med),
  Organographie *f* (Med)
organoleptic[ally] organoleptisch
organoleptic test Sinnesprüfung *f*
organoleptic testing Geschmacksprüfung *f*
organologic organologisch (Med)
organology Lehre *f* von den Organen,
  Organologie *f* (Med)
organo-magnesium compound
  Organo-magnesium-Verbindung
  Grignard-Verbindung *f*
organometal Organometall *n*
organometallic metallorganisch
organoplastic organoplastisch
organosilanediol Organosilandiol *n*
organosilicon compound
  Organosiliciumverbindung *f*
organosiloxane Organosiloxan *n*
organosol Organosol *n*
orientation Orientierung *f*, Ortung *f*, **of same ~**
  gleichgerichtet, **preferred** ~ bevorzugte
  Richtung *f*
orientation dependence
  Orientierungsabhängigkeit *f*
orientation relation Orientierungsbeziehung *f*
orientation-responsive orientierungsabhängig
orientation-sensitive orientierungsabhängig
orientation triangle Orientierungsdreieck *n*
oriented orientiert
orifice Öffnung *f*, Austrittsöffnung *f*, Blende *f*
  (Phot), Düsenaustritt *m*, Mündung *f*
orifice relief Düsenaustritt *m*, Erweiterung *f* des
  Düsenkanals
origan Dostkraut *n* (Bot)
origanum Dostkraut *n* (Bot)
origanum oil Origanumöl *n*
origin Ursprung *m*, Abkunft *f*, Abstammung *f*,
  Entstehung *f*, Herkunft *f*,
  Koordinatennullpunkt *m* (Math), Quelle *f*,
  mode of ~ Entstehungsart *f*
original original, anfänglich, ursprünglich
original constituent Ausgangsprodukt *n*
original equipment Erstausrüstung *f*,
  Grundausrüstung *f*

original force Urkraft *f*
original material Ausgangsmaterial *n*,
  Ausgangsstoff *m*, Ausgangssubstanz *f*,
  Ursubstanz *f*
original packing Originalpackung *f*
original product Ausgangsprodukt *n*
original state Urzustand *m*
original substance Ausgangssubstanz *f*,
  Ursubstanz *f*
original temperature Ausgangstemperatur *f*
original titer Urtiter *m* (Anal)
original volume Anfangsvolumen *n*
originate *v* entstehen, hervorgehen [aus]
originating firm Herstellerfirma *f*
origin distortion Nullpunktanomalie *f*
origin of species, theory of the ~
  Abstammungslehre *f*
orixine Orixin *n*
orizabin Jalapin *n*
orlean Orlean *n*
orlon Orlon *n*
ormolu Malergoldfarbe *f*, Muschelgold *n*
ormolu varnish Goldlackimitation *f*
ormosine Ormosin *n*
ormosinine Ormosinin *n*
ornament Ornament *n*, Verzierung *f*
ornament *v* verzieren, schmücken, bemustern
ornamental glass Kunstglas *n*
ornithine Ornithin *n*
ornithine cycle Ornithinzyklus *m*
ornithine transaminase Ornithintransaminase *f*
  (Biochem)
ornithine transcarbamylase
  Ornithintranscarbamylase *f* (Biochem)
ornithology Ornithologie *f*, Vogelkunde *f*
ornithuric acid Ornithursäure *f*
orobol Orobol *n*
orogenesis Gebirgsbildung *f* (Geol),
  Orogenese *f* (Geol)
orotic acid Orotsäure *f*
orotidine phosphate Orotidinphosphat *n*
orotidylic acid Orotidylsäure *f*
oroxylin Oroxylin *n*
orphenadrine Orphenadrin *n*
orpiment Auripigment *n* (Min), Operment *n*
  (Min), Rauschgelb *n* (Min), Reißgelb *n*
  (Min), **red** ~ Rauschrot *n* (Min), Realgar *m*
  (Min)
orris oil Irisöl *n*
Orr white Charltonweiß *n*
Orsat apparatus Orsat'scher Apparat *m*
orseille Orseille *f*
orseille weed Färberflechte *f* (Bot)
orse[i]llin Orseillin *n*, Roccellin *n*
orselle Orseille *f*
orsellinate Orsellinat *n*
orsellinic acid Orseillinsäure *f* (Orsellinsäure),
  salt or ester of ~ Orsellinat *n*
orsudan Orsudan *n*

orthanilic acid Orthanilsäure *f*
orthite Orthit *m* (Min), Allanit *m* (Min)
orthoacetic acid Orthoessigsäure *f*
ortho acid Orthosäure *f*
orthoantimonic acid Orthoantimonsäure *f*
orthoarsenic acid Orthoarsensäure *f*
orthoaxis Orthoachse *f* (Krist),
   Orthodiagonale *f* (Krist)
orthoboric acid Orthoborsäure *f*
orthocarbonic acid Orthokohlensäure *f*
orthocarbonic ester Orthokohlensäureester *m*
orthochromatic orthochromatisch, farbrichtig
orthochromatic plate Orthochromplatte *f*
orthocinnamic acid Orthozimtsäure *f*
orthoclase Orthoklas *m* (Min), [monokliner]
   Kalifeldspat *m* (Min)
orthoclase porphyry Orthoklasporphyr *m* (Min)
ortho compound Orthoverbindung *f*
orthodiagonal orthodiagonal (Krist)
orthodiagonal axis Orthoachse *f* (Krist),
   Orthodiagonale *f* (Krist)
ortho-directing in ortho-Stellung dirigierend,
   orthodirigierend (Chem)
orthoform Orthoform *n*
orthoformate Orthoameisensäureester *m*
orthoformic acid Orthoameisensäure *f*
orthoformic ester Orthoameisensäureester *m*
orthogonal orthogonal, rechtwinklig, rechteckig
orthogonality Orthogonalität *f*, Rechteckigkeit *f*,
   Rechtwinkligkeit *f*
orthogonality property
   Orthogonalitätseigenschaft *f*
orthogonality relation Orthogonalitätsrelation *f*
orthogonalization process
   Orthogonalisierungsprozeß *m*
orthogonal matrix orthogonale Matrize *f*
orthogonal system Orthogonalsystem *n*
ortho-hydrogen Orthowasserstoff *m*
orthonitric acid Orthosalpetersäure *f*
orthop[a]edic orthopädisch
ortho-para conversion
   Ortho-Para-Umwandlung *f* (Wasserstoff)
orthophonic lautgetreu
orthophosphoric acid Orthophosphorsäure *f*
ortho position Orthostellung *f* (Chem)
orthoprism Orthoprisma *n* (Krist)
orthorhombic orthorhombisch (Krist)
orthoscope Orthoskop *n* (Med)
orthosilicate Orthosilicat *n*
orthosilicic acid Orthokieselsäure *f*, **salt or ester
   of ~** Orthosilicat *n*
ortho silicoformic acid
   Orthosiliciumameisensäure *f*
orthotropy Orthotropie *f*
ortizon Ortizon *n*, Hyperol *n*
ortol Ortol *n*
osajin Osajin *n*
osazone Osazon *n*

oscillate *v* hochfrequente Schwingungen
   erzeugen (Elektr), oszillieren, pendeln,
   schwingen, vibrieren
oscillating oszillierend
oscillating axle Schwingachse *f*
oscillating circuit Schwingungskreis *m*
oscillating crystal method
   Schwingkristallmethode *f* (Krist)
oscillating disk viscosimeter
   Schwingviskosimeter *n*
oscillating grate Schüttelrost *m*
oscillating grate oven Hubofen *m*
oscillating screen centrifuge
   Schwingsiebzentrifuge *f*
oscillating system, degenerate ~ entartetes
   Schwingungssystem *n*
oscillating twisting machine
   Torsionsschwinggerät *n*
oscillation Oszillation *f*, Ladungswechsel *m*
   (Elektr), Schwanken *n*, Schwingung *f*,
   Vibration *f*, **angle of ~** Ausschubwinkel *m*,
   aperiodic ~ aperiodische Schwingung, **center
   of ~** Schwingungszentrum *n*, **continuous ~**
   gleichförmige Schwingung, ungedämpfte
   Schwingung, **damped ~** gedämpfte
   Schwingung, **duration of ~**
   Schwingungsdauer *f*, **dying ~** abklingende
   Schwingung, **forced ~** erzwungene
   Schwingung, **free ~** Eigenschwingung *f*, freie
   Schwingung, **harmonic ~** harmonische
   Schwingung, **partial ~** Teilschwingung *f*,
   period of ~ Schwingungsperiode *f*, **plane of ~**
   Schwingungsebene *f*, **sustained ~**
   ungedämpfte Schwingung, **time of ~**
   Periodendauer *f*, **undamped ~** ungedämpfte
   Schwingung, **~ around a center**
   Drehschwingung *f*
oscillation amplitude Schwingungsamplitude *f*
oscillation amplitude limit Grenzamplitude *f*
oscillation discharge oszillierende Entladung
oscillation energy Schwingungsenergie *f*
oscillation frequency Schwingungsfrequenz *f*,
   Schwingungszahl *f*
oscillation mechanism
   Schwingungsmechanismus *m*
oscillation phase Schwingungsphase *f*
oscillation time Schwingungsperiode *f*
oscillation torsion test Drehschwingversuch *m*
oscillator Oszillator *m*,
   Hochfrequenzgenerator *m*, **piezoelectric ~**
   piezoelektrischer Oszillator
oscillator frequency Oszillatorfrequenz *f*
oscillator tube Oszillatorröhre *f*
oscillatory oszillierend, schwingungsfähig
oscillatory circuit Schwingkreis *m* (Elektr)
oscillatory discharge oszillierende Entladung *f*
oscillatory instability Schwingungsinstabilität *f*
oscillatory transformer Oszillationsumformer *m*
oscillogram Oszillogramm *n*, Wellenbild *n*

oscillograph Oszillograph *m*, **cathode ray ~**
Kathodenstrahloszillograph *m*, **double-wire
loop ~** Zweischleifenoszillograph *m*,
electrostatic ~ elektrostatischer Oszillograph
oscillograph amplifier
Oszillographenverstärker *m*
oscillograph curve Oszillogramm *n*
oscillographic technique Oszillographentechnik *f*
oscilloscope Oszilloskop *n*,
[Niederspannungs]Oszillograph *m*
oscilloscope method, triggered ~
Synchroskopmethode *f*
oscine Oscin *n*
osculating plane Schmiegungsebene *f*
osculation Oskulation *f* (Math)
osmane Osman *n*
osmate Osmat *n*
osmic acid Osmiumsäure *f*, **salt or ester of ~**
Osmat *n*
osmic compound Osmium(IV)-Verbindung *f*
osmiridium Osmiridium *n*
osmium (Symb. Os) Osmium *n*, **containing ~**
osmiumhaltig
osmium alloy Osmiumlegierung *f*
osmium content Osmiumgehalt *m*
osmium dioxide Osmiumdioxid *n*
osmium filament Osmiumfaden *m*
osmium lamp Osmiumlampe *f*
osmium monoxide Osmium(II)-oxid *n*,
Osmiummonoxid *n*
osmium peroxide Osmiumperoxid *n*,
Osmiumtetr[a]oxid *n*
osmium protoxide Osmium(II)-oxid *n*,
Osmiumoxydul *n* (obs)
osmium tetroxide Osmiumperoxid *n*,
Osmiumtetr[a]oxid *n*, Osmium(VIII)-oxid *n*
osmolarity Osmolarität *f*
osmometer Osmometer *n*, Osmosemesser *m*
osmometry Osmometrie *f*
osmond iron Osmundeisen *n*
osmondite Osmondit *m*
osmose Osmose *f*
osmosis Osmose *f*
osmosis apparatus Osmoseapparat *m*
osmotic osmotisch
osmotic equilibrium osmotisches
Druckgleichgewicht *n*
osmous compound Osmium(II)-Verbindung *f*
osmund furnace Osmundofen *m*
osmund iron Osmundeisen *n*
osone Oson *n*
osotriazole Osotriazol *n*
ossein Ossein *n*, Knochengallerte *f*
ossification Ossifikation *f* (Med),
Verknöcherung *f* (Med)
osteocarcinoma Knochenkarzinom *n* (Med),
Knochenkrebs *m* (Med), Osteokarzinom *n*
(Med)
osteochemistry Osteochemie *f*

osteocyte Knochenzelle *f*
osteogenesis Knochenbildung *f*, Osteogenese *f*
osteogenous knochenbildend
osteogeny Knochenbildung *f*, Osteogenese *f*
osteolite Osteolith *m* (Min)
osteology Knochenlehre *f* (Med)
osteolysis Knochenauflösung *f*, Osteolyse *f*
osthenol Osthenol *n*
osthol Osthol *n*
ostholic acid Ostholsäure *f*
ostracite Muschelversteinerung *f*
ostranite Ostranit *m* (Min)
ostreasterol Ostreasterin *n*
ostruthin Ostruthin *n*
ostruthol Ostruthol *n*
Ostwald's dilution law Ostwaldsches
Verdünnungsgesetz *n*
Ostwald's theory of indicators Ostwaldsche
Indikatortheorie *f*
osyritrin Osyritrin *n*
otavite Otavit *m* (Min)
otobite Otobit *n*
Otto engine Ottomotor *m*
ottrelite Ottrelith *m* (Min)
ouabain Ouabain *n*
ounce Unze *f*
outbalance *v* übertreffen
outburst Ausbruch *m*
outdoor außenseitig
outdoor exposure Außenbewitterung *f*
outdoor paint Außen[anstrich]farbe *f*
outer casing Außenmantel *m*
outer circuit Außenleitung *f*
outer container Außenbehälter *m*
outer crucible Außentiegel *m*
outermost shell Außenschale *f* (Atom)
outer orbit Außenbahn *f* (Elektron)
outerpole armature Außenpolanker *m*
outer tube Außenrohr *n*
outer wall Außenwand *f*
outfit Ausrüstung *f*, Ausstattung *f*, Einrichtung *f*
outflow Absonderung *f* (Med), Ausfluß *m*,
Auslauf *m*, Ausströmung *f*
outflowing ausfließend, ausströmend
outflow time Auslaufzeit *f*
outflow velocity Ausflußgeschwindigkeit *f*,
Ausströmgeschwindigkeit *f*
outgassing Entgasen *n*, Entgasung *f*
outgoing air Abluft *f*
outgrowth Auswuchs *m* (Med)
outlay Ausgaben *pl*, Betriebskosten *pl*
outlet Auslauf *m*, Abflußmöglichkeit *f*,
Abflußrinne *f*, Ablaßöffnung *f*, Ablauf *m*,
Abzug *m*, Abzugsöffnung *f*, Abzugsrohr *n*,
Anschluß *m* (Elektr), Austritt *m*,
Austrittsöffnung *f*
outlet box Anschlußkasten *m* (Elektr)
outlet cock Ausflußhahn *m*
outlet hopper Ablauftrichter *m*

outlet piece Vorstoß *m*
outlet pipe Ausflußrohr *n*, Entnahmerohr *n*
outlet plate Abnahmeboden *m* (Kolonne)
outlet port Austrittsschlitz *m*
outlet temperature Ablauftemperatur *f*,
    Austrittstemperatur *f*
outlet tray Abnahmeboden *m* (Kolonne)
outlet trough Abflußrinne *f*
outlet tube Abfluß *m*, Ablaßrohr *n*,
    Auslaufrohr *n*
outlet valve Abflußventil *n*, Abzugsklappe *f*,
    Abzugsventil *n*, Austrittsventil *n*
outlier Ausreisser *m* (Statist)
outline Umriß *m*, Abriß *m*, Entwurf *m*, Kontur *f*,
    Skizze *f*, **to make a rough** ~ flüchtig entwerfen
outline *v* entwerfen, skizzieren
outmatch *v* übertreffen
out of action, to put ~ außer Betrieb setzen
out of order gestört
out-of-phase phasenverschoben
output Arbeitsleistung *f*, Ausstoß *m*,
    Ausstoßleistung *f*, Ertrag *m*,
    Erzeugungsmenge *f*, Förderquantum *n*
    (Bergb), Förderung *f* (Bergb), Leistung *f*,
    Nutzleistung *f*, Produktion *f*, **continuous** ~
    Dauerleistung *f*, **control of** ~
    Leistungsregelung *f*, **large** ~
    Massenerzeugung *f*, ~ **per hour**
    Stundenleistung *f*
output boost Leistungssteigerung *f*
output channel Ausflußkanal *m*
output electrode Abnahmeelektrode *f*
output maximum Leistungsmaximum *n*
output meter Leistungsmesser *m*
output pentode Fünfpolendröhre *f*
output terminal Entnahmebüchse *f*
outset Anfang *m*, Beginn *m*
outside-air temperature Außenlufttemperatur *f*
outside calipers Außentaster *m*
outside diameter Außendurchmesser *m*
outside dimension Außenabmessung *f*,
    Außenmaß *n*
outside pipeline Außen[rohr]leitung *f*
outside pressure Außendruck *m*
outside sprinkling Außenberieselung *f*
outside thread Außengewinde *n*
outside world Außenwelt *f*
outstanding hervorragend
outward electrode Außenelektrode *f*
ouvarovite Chromgranat *m* (Min)
oval oval, eiförmig
ovalbumin Ovalbumin *n*, Eialbumin *n*
ovalene Ovalen *n*
oval gear meter Ovalradzähler *m*
oval pass Ovalkaliber *n*
oval [-shaped] länglich rund
ovanene Ovanen *n*
oven Ofen *m*, Backofen *m*
oven charge Ofenfüllung *f*

oven drying Ofentrocknung *f*
oven floor Ofensohle *f*
oven soot Ofenruß *m*
overabundant übermäßig, überschüssig,
    überzählig
overacidification Übersäuerung *f*
overacidify *v* übersäuern
overactivity Überfunktion *f*
overall Arbeitsanzug *m*
overall *a* gesamt, brutto
overall conversion Gesamtumsatz *m*
overall effect Gesamtwirkung *f*
overall efficiency Totalnutzeffekt *m*
overall length Gesamtlänge *f*
overall mass transfer coefficient
    Stoffdurchgangszahl *f*
over-all reaction Bruttoreaktion *f*
overburden *v* überladen, überlasten
overburn *v* (lime) (Kalk) totbrennen
overburning Totbrennen *n*
overcharge Überbelastung *f*
overcharge *v* überbelasten
overcolor *v* überfärben
overcome *v* beheben, bewältigen
overcompression ratio
    Überkompressionsverhältnis *n*
overcure Nachvulkanisation *f*
overcurrent Überstrom *m* (Elektr)
overcurrent relay Überlastungsschutz *m* (Elektr)
overdrive Schnellgang *m*
overdry *v* übertrocknen
overdye *v* überfärben, zu stark färben
overexpose *v* überbelichten (Phot)
overexposure Überbelichtung *f* (Phot)
overfeeding Überdosierung *f*, Überfütterung *f*
overfeed tenter Schrumpfrahmen *m*
overferment *v* übergären
overflow Überlauf *m*, Formenaustrieb *m*
    (Reifen) (injection molding), Schwimmhäute *f*
overflow *v* überlaufen, überquellen,
    überschwemmen
overflow alarm Überlaufwasserstandmelder *m*
overflow centrifuge Überlaufzentrifuge *f*
overflowing Überlaufen *n*
overflow line Überlaufleitung *f*
overflow pipe Überlaufrohr *n*
overflow record Folgesatz *m* (Informatik)
overflow relay Abschaltrelais *n*
overflow trap Überlaufventil *n*
overflow tube Überlaufrohr *n*
overflow valve Überströmventil *n*
overflow vessel Überlaufgefäß *n*
overglaze Aufglasur *f*, Überglasur *f*, zweite
    Glasur *f*
overglaze color Aufglasurfarbe *f*
overgrind *v* totmahlen
overground totgemahlen; über dem Erdboden
overgrowth Überwachsung *f* (Krist)
overhaul *v* überholen, überprüfen, reparieren

overhauled instandgesetzt
overhauling Überholung *f*, gründliche
Überprüfung *f*, **in need of** ~
überholungsbedürftig
overhead Dampfstrom *m* (Atom), oben,
oberirdisch
overhead contact system Oberleitung *f*
overhead conveyer trolley Hängebahnlaufkatze *f*
(Techn)
overhead crane Hängekran *m*
overhead crane and track
Hängekran-Hängebahn-Anlage *f*
overhead handling bodenfreie Beförderung
overhead line Freileitung *f*
overhead pipe junction Stichrohrbrücke *f*
overhead pipelines Rohrbrücke *f*
overhead reservoir Hochbehälter *m*
overhead tank Hochbehälter *m*
overhead travelling crane Brückenkran *m*
overhead wire Hochleitung *f* (Elektr)
overhead wiring Oberleitung *f*
overheat *v* überhitzen (Dampf)
overheating of material Materialüberhitzung *f*
overinflation Überschreitung *f* des Innendrucks
(Reifen)
overlap Überlagerung *f*, Überlappung *f*,
Überschneidung *f*
overlap *v* sich überschneiden, überlappen
overlap model Überlappungsmodell *n*
overlapping Überdecken *n*, Überlagerung *f*,
Überlappung *f*, ~ **of orbitals** Überlappung *f*
von Orbitalen (pl)
overlapping *a* überlappend
overlay mat Oberflächenvlies *n* (glass fiber)
overload Arbeitsüberlastung *f* (Techn),
Überbeanspruchung *f*, Überbelastung *f*
overload *v* über[be]lasten, überladen
overload capacity Überlastbarkeit *f*
overload current Überlaststrom *m*
overloading Über[be]lastung *f*
overload protection Überlastungsschutz *m*
overload release Überlastauslöser *m*
overload test Forcierungsversuch *m*
overlying darüberliegend
overmeasure Übermaß *n*
overoxidation Überoxidierung *f*
overpotential Überpotential *n*
overpressure Überdruck *m*
overpressure valve Sicherheitsventil *n*,
Überdruckventil *n*
overprint *v* überdrucken
overprinting Bedrucken *n*
overprint varnish Überdrucklack *m*,
Überzugslack *m*
overproduction Überproduktion *f*
overrefined übergar (Metall)
overrelaxation, method of ~
Überrelaxationsmethode *f*
override *v* überschreiten

overroasting Totrösten *n*
oversaturate *v* übersättigen
oversaturation Übersättigung *f*
oversecretion Überproduktion *f* (Drüse)
oversize product Rückstand *m* (Sieb)
oversteering Übersteuerung *f*
overstrain *v* überbeanspruchen
overstrain[ing] Überbeanspruchung *f*
overstress *v* überbeanspruchen
overstress[ing] Überbeanspruchung *f*
overtanned totgegerbt
overtime Überstunden *pl*
overtime work Mehrarbeit *f*
overtop *v* hervorragen
overturn *v* umkippen, umstoßen
overvoltage Überspannung *f*, Überpotential *n*
overvoltage protection Überspannungsschutz *m*
(Elektr)
overweight Übergewicht *n*
overwork Arbeitsüberlastung *f*, Überstunden *pl*
overwork *v* sich abarbeiten, überanstrengen
ovicide Ovizid *n*
oviduct Eileiter *m* (Med)
ovocyte Eizelle *f*
ovoid eiförmig
ovolarvicide Ovolarvizid *n*
ovoplasm Eiplasma *n*
ovoverdin Ovoverdin *n*
ovulation Ovulation *f*, Eiausstoßung *f*
ovulation inhibitor Ovulationshemmer *m*
ovum Eizelle *f*, **fertilized** ~ befruchtete Eizelle
oxacid Oxysäure *f*
oxadiazole Oxadiazol *n*, Oxydiazol *n*
oxalacetic ester Oxalessigsäureester *m*
oxalaldehyde Glyoxal *n*
oxalaldehydic acid Glyoxylsäure *f*
oxalate Oxalat *n*, Oxalsäureester *m*,
Oxalsäuresalz *n*
oxalate developer Oxalatentwickler *m*
oxalate exchange Oxalataustausch *m*
oxalato-chelate Oxalatochelat *n*
oxalcitraconic acid Oxalcitraconsäure *f*
oxalene Oxalen *n*
oxalene diuramidoxime Oxalendiuramidoxim *n*
oxalic acid Oxalsäure *f*, Kleesäure *f* (obs), **salt or**
ester of ~ Oxalat *n*
oxalic acid monoamide Oxamidsäure *f*
oxalic acid monoanilide Oxanilsäure *f*
oxalic acid plant Oxalsäureanlage *f*
oxalic ester Oxalsäureester *m*
oxalic monoureide Oxalursäure *f*
oxalimide Oximid *n*
oxalite Oxalit *m* (Min)
oxaloacetic acid Oxalessigsäure *f*
oxaloacetic transaminase
Oxalessigsäuretransaminase *f* (Biochem)
oxalonitrile Dicyan *n*, Oxalnitril *n*
oxal[o]succinic acid Oxalbernsteinsäure *f*
oxaluria Oxalurie *f* (Med)

oxaluric acid Oxalursäure *f*
oxalyl Oxalyl-
oxalyl chloride Oxalylchlorid *n*
oxalylurea Oxalylharnstoff *m*, Parabansäure *f*
oxamethane Oxamethan *n*
oxamic acid Oxamidsäure *f*
oxamide Oxamid *n*
oxamine blue Oxaminblau *n*
oxamine dye Oxaminfarbstoff *m*
oxaminic acid Oxamidsäure *f*
oxamycin Oxamycin *n*
oxanamide Oxanamid *n*
oxanilic acid Oxanilsäure *f*
oxanilide Oxanilid *n*
oxanthranol Oxanthranol *n*
oxanthrone Oxanthron *n*
oxaphor Oxaphor *n*
oxapropanium Oxapropanium *n*
oxazine Oxazin *n*
oxazine dye Oxazinfarbstoff *m*
oxazole Oxazol *n*
oxazolidine Oxazolidin *n*
oxazolidone Oxazolidon *n*
oxazoline Oxazolin *n*
oxazolone Oxazolon *n*
oxdiazine Oxdiazin *n*
oxdiazole Oxdiazol *n*, Furazan *n*, Furodiazol *n*
oxepane Oxepan *n*
oxepin Oxepin *n*
oxetane Oxetan *n*
oxetone Oxeton *n*
oxidant Oxidationsmittel *n*, Oxidans *n* (pl
   Oxidantien), oxidierendes Agens *n*
oxidase Oxidase, Oxydase *f (Biochem)*
oxidation Oxidation *f*, Oxydation, Frischen *n*
   (Metall), **anodic** ~ anodische Oxidation,
   biological ~ biologische Oxidation, **capable o**
   ~ oxidationsfähig, oxidierbar, **catalytic** ~
   katalytische Oxidation, **degree of** ~
   Oxidationsgrad *m*, Oxidationsstufe *f*, **heat of**
   ~ Oxidationswärme *f*, **low temperature** ~
   Tieftemperaturoxidation *f*, **product of** ~
   Oxidationsprodukt *n*, **selective** ~ selektive
   Oxidation, **stage of** ~ Oxidationsstufe *f*
oxidation chamber Oxidationsraum *m*
oxidation inhibitor Oxidationsinhibitor *m*
oxidation number Oxidationszahl *f*
oxidation potential Oxidationspotential *n*
oxidation process Oxidationsprozeß *m*,
   Oxidationsvorgang *m*, Oxidieren *n*
oxidation product Oxidationsprodukt *n*
oxidation-reduction Oxidoreduktion *f*
oxidation-reduction cell
   Oxidations-Reduktions-Kette *f*
oxidation-reduction electrode
   Oxidations-Reduktions-Elektrode *f*,
   Redoxelektrode *f*
oxidation-reduction pair
   Oxidations-Reduktions-Paar *n*

oxidation-reduction potential
   Oxidations-Reduktions-Potential *n*,
   Redoxpotential *n*
oxidation-reduction reaction
   Oxidations-Reduktions-Reaktion *f*
oxidation resistance Oxidationsbeständigkeit *f*
oxidation stability Oxidationsbeständigkeit *f*
oxidation state Oxidationszustand *m*
oxidation step Oxidationsstufe *f*
oxidation zone Verbrennungszone *f*
oxide Oxid *n*, Oxyd *n*, **acidic** ~ saures Oxid,
   amphoteric ~ amphoteres Oxid, **basic** ~
   basisches Oxid, **film of** ~ Oxidhaut *f*, **higher**
   ~ höherwertiges Oxid, **layer of** ~
   Oxidschicht *f*, **lower** ~ niedrigeres Oxid,
   Oxydul *n* (obs)
oxide brown Oxidbraun *n*
oxide cathode Oxidkathode *f*
oxide-coated mit einem Oxid überzogen
oxide coating Oxidüberzug *m*, Oxidbeschlag *m*
oxide color Oxidfarbe *f*
oxide electrode Oxidelektrode *f*
oxide film Oxidbelag *m*, Oxidschicht *f*
oxide-forming oxidbildend
oxide inclusion Oxideinschluß *m*
oxide[s], free from ~ oxidfrei
oxide scale Glühspanschicht *f*
oxide wax Oxidwachs *n*
oxide yellow Oxidgelb *n*
oxidizability Oxidationsfähigkeit *f*,
   Oxidierbarkeit *f*
oxidizable oxidationsfähig, oxidierbar, **easily** ~
   leicht oxidierbar
oxidize *v* oxidieren, frischen, (Metall) in ein Oxid
   verwandeln, ~ **electrolytically** eloxieren
oxidized oil Dicköl *n*
oxidized surface Oxidhaut *f*
oxidizer Oxidationsmittel *n*
oxidizing Oxidieren *n*
oxidizing *a* oxidierend, **strongly** ~ stark
   oxidierend
oxidizing action Oxidationswirkung *f*
oxidizing agent Oxidationsmittel *n*
oxidizing catalyst Oxidationskatalysator *m*
oxidizing effect Oxidationswirkung *f*
oxidizing flame Oxidationsflamme *f*
oxidizing furnace Oxidationsofen *m*
oxidizing oven Oxidationsofen *m*
oxidizing process Frischverfahren *n* (Metall)
oxidizing slag Oxidationsschlacke *f*
oxidizing substance Oxidationsmittel *n*
oxido-reductase Oxidoreduktase *f* (Biochem)
oxime Oxim *n*
oxamide Oximid *n*
oxindigo Oxindigo *n*
oxindole Oxindol *n*
oxine Oxin *n*
oxirane Oxiran *n*
oxoadrenochrome Oxoadrenochrom *n*

oxoaristic acid Oxoaristsäure *f*
oxobutanedioic acid Oxalessigsäure *f*
oxochroman, 2- ~ Melilotol *n*
oxoctenol Oxoctenol *n*
oxoethanoic acid Glyoxylsäure *f*
oxohemanthidine Oxohämanthidin *n*
oxoheptanedioic acid, 4- ~
   Hydrochelidonsäure *f*
oxohydrindene Indanon *n*
oxomalonic acid Mesoxalsäure *f*
oxomenthane Menthanon *n*, 3- ~ Menthon *n*
oxomenthene Menthenon *n*
oxone Oxon *n*
oxonic acid Oxonsäure *f*
oxonite Oxonit *n*
oxonium salt Oxoniumsalz *n*
oxooctanthrene Oktanthrenon *n*
oxoocthracene Okthracenon *n*
oxopentanal, 4- ~ Lävulinaldehyd *m*
oxopentanoic acid, 4- ~ Lävulinsäure *f*
oxophenarsine Oxophenarsin *n*
oxopropanedioic acid Mesoxalsäure *f*
oxosynthesis Oxosynthese *f*
oxozone Oxozon *n*
oxthiane Oxthian *n*
oxthiazole Oxthiazol *n*
oxthine Oxthin *n*
oxthiole Oxthiol *n*
oxyacetylene blowpipe
   Acetylen-Sauerstoff-Gebläse *n*
oxyacetylene cutter autogener Schneidapparat *m*,
   Brennschneider *m*
oxyacetylene cutting autogenes Schneiden *n*
oxyacetylene cutting device
   Acetylen-Schnittbrenner *m*
oxyacetylene torch Acetylengebläse *n*
oxyacetylene welding
   Acetylen-Sauerstoff-Schweißung *f*
oxyacetylene welding torch
   Acetylenschweißbrenner *m*
oxyacid Oxysäure *f*, Sauerstoffsäure *f*
oxyammonia Hydroxylamin *n*
oxybiotin Oxybiotin *n*
oxybutyric aldehyde Aldol *n*
oxycamphor Oxycampher *m*
oxycellulose Oxycellulose *f*, Oxyzellulose *f*
oxychloride Oxychlorid *n*
oxychlorination Oxychlorierung *f*
oxycinchophen Oxycinchophen *n*
oxycodone Oxycodon *n* (Analgetikum)
oxy compound Oxyverbindung *f*
oxydiethanoic acid Diglykolsäure *f*
oxyethylsulfonic acid Isäthionsäure *f*
oxygen (Symb. O) Sauerstoff *m*, absence of ~
   Sauerstoffausschluß *m*, circulating ~
   Blutsauerstoff *m* (Med), containing ~
   sauerstoffhaltig, enriching with ~
   Sauerstoffanreicherung *f*, generation of ~
   Sauerstoffentwicklung *f*, Sauerstoffabgabe *f*,

lacking in ~ sauerstoffarm, lean in ~
sauerstoffarm, migration of ~
Sauerstofftransport *m*, production of ~
Sauerstoffgewinnung *f*, release of ~
Sauerstoffabgabe *f*, removal of ~
 Desoxidation *f*, Desoxidierung *f*, ~ in the
form of ozone Ozonsauerstoff *m*
oxygen absorbent Sauerstoffänger *m*
oxygen absorption Sauerstoffaufnahme *f*
oxygen-acetylene cutting autogenes Schneiden *n*
oxygen-acetylene welding autogenes Schweißen *n*
oxygen acid Sauerstoffsäure *f*
oxygen analyzer Sauerstoffmesser *m*,
   Sauerstoffprüfer *m*
oxygenase Oxygenase *f* (Biochem)
oxygenate *v* mit Sauerstoff anreichern
oxygenated sauerstoffbeladen
oxygenation Oxygenierung *f*,
   Sauerstoffanreicherung *f*
oxygen bridge Sauerstoffbrücke *f*
oxygen carrier Sauerstoff[über]träger *m*
oxygen compound Sauerstoffverbindung *f*
oxygen compression Sauerstoffkompression *f*
oxygen concentration Sauerstoffkonzentration *f*
oxygen content Sauerstoffgehalt *m*
oxygen converter steel
   Sauerstoff-Konverterstahl *m*
oxygen cycle Sauerstoffkreislauf *m*
oxygen cylinder Sauerstoffbombe *f*,
   Sauerstoffflasche *f*
oxygen deficiency Sauerstoffmangel *m*
oxygen-deficient sauerstoffarm
oxygen demand Sauerstoffverbrauch *m*
oxygen displacement Sauerstoffverschiebung *f*
oxygen electrode Sauerstoffelektrode *f*
oxygen-enriched mit Sauerstoff angereichert,
   sauerstoffbeladen
oxygen formation Sauerstoffbildung *f*
oxygen-free sauerstofffrei
oxygen generator Sauerstofferzeuger *m*,
   Sauerstoffgewinnungsapparat *m*
oxygen-hydrogen welding Knallgasschweißen *n*
oxygenic sauerstoffhaltig, oxydisch
oxygen inhaling apparatus
   Sauerstoff[atmungs]gerät *n*
oxygen inlet tube Sauerstoffzuleitungsröhre *f*
oxygen jet Sauerstoffstrahl *m*
oxygen mask Sauerstoffmaske *f*
oxygen meter Sauerstoffmesser *m*
oxygen outlet tube Sauerstoffableitungsröhre *f*
oxygen plant Sauerstoff[gewinnungs]anlage *f*
oxygen pole Sauerstoffpol *m*, Anode *f*
oxygen pressure Sauerstoffdruck *m*
oxygen producer Sauerstofferzeuger *m*
oxygen recorder Sauerstoffschreiber *m*
oxygen removal Sauerstoffentzug *m*
oxygen requirement Sauerstoffverbrauch *m*
oxygen steelmaking process
   Sauerstoff-Konverterverfahren *n*

oxyhemoglobin Oxyhämoglobin *n*
oxyhydrogen blowpipe Knallgasgebläse *n*
oxyhydrogen cell Knallgaskette *f*
oxyhydrogen flame Knallgasflamme *f*
oxyhydrogen gas Knallgas *n*, **formation of** ~
  Knallgasbildung *f*
oxyhydrogen light Knallgaslicht *n*
oxyhydrogen welding Knallgasschweißung *f*,
  Sauerstoffschweißung *f*,
  Wasserstoff-Sauerstoff-Schweißung *f*
oxyluciferin Oxyluciferin *n* (Biochem)
oxyluminescence Chemolumineszenz *f*
oxymalonic acid Oxymalonsäure *f*,
  Tartronsäure *f*
oxymel Essighonig *m*, Oxymel *n*, Sauerhonig *m*
oxymorphone Oxymorphon *n*
oxynarcotine Oxynarcotin *n*
oxynicotine Oxynicotin *n*
oxynitroso radical Nitritogruppe *f*
oxypolygelatin Oxypolygelatine *f*
oxyquercetin Myricetin *n*
oxysalt Oxysalz *n*
oxysparteine Oxyspartein *n*
oxysulfuric acid Oxyschwefelsäure *f*
oxythiamine Oxythiamin *n*
oxytocinase Oxytocinase *f* (Biochem)
ozobenzene Ozobenzol *n*
ozocerite Erdwachs *n*, Ceresin *n*,
  Mineralwachs *n*, Ozokerit *m*
ozokerite Bergwachs *n*, Ceresin *n*
ozonator Ozonerzeuger *m*
ozone Ozon *n*, **containing** ~ ozonhaltig,
  generating ~ ozonerzeugend, **method of**
  producing ~ Ozondarstellungsmethode *f*,
  producing ~ ozonerzeugend
ozone annihilation Ozonzerstörung *f*
ozone apparatus Ozonapparat *m*
ozone bleach Ozonbleiche *f*
ozone[-containing] water Ozonwasser *n*
ozone content Ozongehalt *m*
ozone formation Ozonbildung *f*
ozone generation Ozonentwicklung *f*,
  Ozonerzeugung *f*
ozone generator Ozonentwickler *m*,
  Ozonerzeuger *m*
ozone layer Ozonschicht *f* (Meteor)
ozone paper Ozonpapier *n*,
  Jodkalistärkepapier *n*
ozone plant Ozonanlage *f*
ozone production Ozonentwicklung *f*,
  Ozonerzeugung *f*
ozone [test] paper Jodkalistärkepapier *n*,
  Ozonreagenspapier *n*
ozone tube Ozonröhre *f*
ozonide Ozonid *n*
ozoniferous ozonerzeugend, ozonhaltig
ozonification Ozonisierung *f*, Verwandlung *f* in
  Ozon

ozonization Ozonisation *f*, Behandlung *f* mit
  Ozon, Ozonisierung *f*, ~ **of air**
  Luftozonisierung *f*, **plant for** ~ **of air**
  Luftozonisierungsanlage *f*
ozonize *v* mit Ozon behandeln, ozonisieren
ozonized oxygen Ozonsauerstoff *m*
ozonizer Ozonerzeuger *m*, Ozonisator *m*
ozonolysis Ozonolyse *f*, Ozonspaltung *f*
ozonometer Ozonometer *n*, Ozonmesser *m*
ozonometric ozonometrisch
ozonometry Ozonmessung *f*, Ozonometrie *f*
ozonoscope Ozonoskop *n*

P

pace Tempo *n*, Geschwindigkeit *f*
pachnolite Pachnolith *m* (Min)
pachymic acid Pachymsäure *f*
p-acid para-Säure *f*
pack Ballen *m*, Bündel *n*, Satz *m* (Papier), ~
  subdivided into compartments
  Stegverpackung *f*
pack *v* [ab]dichten (Techn), abpacken,
  einpacken, emballieren, verpacken,
  zusammenpressen
package Paket *n*, Verpackung *f*, deceptive ~ or
  „bluff" package Mogelpackung *f*
package *v* abpacken
packaging Abpackung *f*, Verpacken *n*,
  commercial ~ handelsübliche Verpackung *f*
packaging bag Abpackbeutel *m*
packaging department
  Konfektionierungsabteilung *f*
packaging machine Abpackmaschine *f*
packaging material Verpackungsmaterial *n*
packaging plant Abfüllbetrieb *m*
packed bubble column Sumpfreaktor *m*
packed column Füllkörperkolonne *f*,
  Füllkörpersäule *f*
packed tower Füllkörperturm *m*
packed tube Füllkörperrohr *n*
packfong Packfong *m*
pack hardening Einsatzhärten *n*
pack hardening process
  Kasten-Einsatzhärteverfahren *n*
packing Verpacken *n*, Abdichtung *f* (Techn),
  Abpackung *f*, Ausfüllmasse *f* (Techn),
  Dichtung *f* (Techn), Füllkörper *pl* (Techn),
  Packung *f* (Techn); Verpackung *f*, depth of ~
  Dichtungstiefe *f*, height of ~ Schichthöhe *f*,
  ~ for fractionating columns
  Kolonnenfüllmaterial *n*
packing cardboard Dichtungspappe *f*
packing case Versandkiste *f*, Versandschachtel *f*
packing component Packungsanteil *m* (Phys)
packing density Fülldichte *f*, Packungsdichte *f*
packing drum Dichtungswalze *f*
packing effect Packungseffekt *m* (Phys),
  Massendefekt *m* (Phys)
packing hemp Dichtwerg *n*
packing joint Dichtungsfuge *f*,
  Flanschendichtung *f*
packing machine Abpackmaschine *f*
packing material Dichtungsmaterial *n*,
  Ausfüllstoff *m*, Verpackungsmaterial *n*
packing paper Packpapier *n*
packing piece Distanzring *m*
packing press Packpresse *f*, Bündelpresse *f*
packing ring Dichtungsring *m*, Buchsring *m*,
  Dichtungsscheibe *f*
packing sheet Packtuch *n*

packing surface Dichtungsfläche *f*
packing unit Verpackungseinheit *f*
packing valve Dichtungsklappe *f*
packing washer Abdichtungsring *m*
pad Bausch *m*, Kissen *n*, Polster *n*, Tampon *m*
pad *v* polstern, wattieren
pad bath Klotzbad *n* (Färb)
padding Polsterung *f*, Klotzen *n* (Färb),
  Klotzverfahren *n* (Färb)
padding machine Klotzmaschine *f*
padding process Klotzverfahren *n* (Färberei)
paddle Paddel *n*, Schaufel *f*
paddle agitator Paddelrührer *m*
paddle dryer Schaufeltrockner *m*
paddle mixer Paddelrührer *m*, Schaufelrührer *m*
paddle pit Haspelgrube *f*
paddle stirrer Schaufelrührer *m*
paddle wheel Schaufelrad *n*
pad saw Fuchsschwanz *m* (Techn)
pad steam technique Klotzdämpfverfahren *n*
page Seite *f* (Buch)
pagoda stone Pagodenstein *m* (Min),
  Bildstein *m* (Min)
pagodite Bildstein *m* (Min)
pail Kübel *m*, Eimer *m*
pain Schmerz *m*, relief of ~ Schmerzlinderung *f*
painful schmerzhaft
pain-killer Schmerzbetäubungsmittel *n* (Pharm),
  Schmerzmittel *n* (Pharm)
painless schmerzlos
paint Farbe *f*, Anstrich *m*, Anstrichfarbe *f*, coat
  of ~ Anstrich *m*, finish ~ Deckanstrich *m*,
  fire-proof ~ feuerfeste Farbe, heavily loaded
  ~ hochpigmentierter Lack *m*, tropical ~
  Tropenfarbe *f*, ~ for outside use
  Außen[anstrich]farbe *f*
paint *v* [an]malen, [an]streichen, lackieren
paint coat [Farb]Anstrich *m*
painter Maler *m*
painter's color Malerfarbe *f*
painter's glazing Malerglasur *f*
painter's gold Malergold *n*, Malergoldfarbe *f*
painter's priming Malergrundierung *f*
painter's tool Anstrichgerät *n*
painter's varnish Malerfirnis *m*
paint factory Farbenfabrik *f*
paint film Anstrichfilm *m*
painting Bild *n*, Anstrich *m*, Malerarbeiten *pl*,
  Streichen *n*, waterproof ~ wasserfester
  Anstrich, weatherproof ~ wetterfester
  Anstrich
painting system Anstrichsystem *n*
painting tool Anstreichgerät *n*
paint mixer Farbenmischzylinder *m*
paint remover Abbeizmittel *n*,
  Farbenabbeizmittel *n*

paint removing agent Farbenabbeizmittel *n*
paint resin Lackharz *n*
paint roller Malerwalze *f*
paint shop Lackiererei *f*
paint slips Malschlicker *m*
paint-spraying plant Farbspritzanlage *f*
paint-spraying system Lackierspritzverfahren *n*
paint stripper Abbeizmittel *n*
paint testing equipment Anstrichprüfgeräte *pl*
paint, varnish and lacquer industry
   Lackindustrie *f*
paint work Malerarbeit *f*
pair Paar *n*, ~ **of diaphragms** Blendenpaar *n*, ~
   of planes Ebenenpaar *n*
pair *v* paaren, paarweise anordnen
pair annihilation Paarvernichtung *f* (Atom)
pair conversion Paarumwandlung *f*
pair creation Paarbildung *f*, Paarerzeugung *f*
paired gepaart
pair formation Paarbildung *f*
pairing Paarung *f*, ~ **of chromosomes**
   Chromosomenpaarung *f*, ~ **of nuclei**
   Kernpaarung *f*
pairing energy Paarungsenergie *f*
[pair of] scissors Schere *f*
pairs, in ~ paarweise
paktong Packfong *n*
palagonite Palagonit *m*
palatability Genießbarkeit *f*
palatable genießbar, schmackhaft,
   wohlschmeckend
palatine red Palatinrot *n*
palatinite Palatinit *m*
palatinose Palatinose *f*
pale blaß, bleich, weißlich
pale blue fahlblau
pale fracture, of ~ weißbrüchig
paleness Fahlheit *f*, Farblosigkeit *f*
paleolithic period Paläolithikum *n* (Geol)
paleomagnetism Paläomagnetismus *m*
paleozoic Paläozoikum *n*
palette Palette *f*, Farbenteller *m*
pale yellow blaßgelb, fahlgelb, mattgelb
palingenesis Palingenese *f* (Entwicklung nach
   dem biogenetischen Grundgesetz)
palisander Palisanderholz *n*
palisander wood oil Palisanderholzöl *n*
palite Palit *n*, Chlormethylchloroformiat *n*
palladic acid Palladiumsäure *f*
palladic chloride Palladium(IV)-chlorid *n*
palladic compound Palladium(IV)-Verbindung *f*
palladic oxide Palladium(IV)-Oxid *n*
pallad[i]ous compound
   Palladium(II)-Verbindung *f*
pallad[i]ous oxide Palladium(II)-oxid *n*
palladium (Symb. Pd) Palladium *n*, **colloidal** ~
   kolloides Palladium
palladium alloy Palladiumlegierung *f*
palladium amalgam Palladiumamalgam *n*

palladium asbestos Palladiumasbest *m*
palladium black Palladiummohr *m*,
   Palladiumschwarz *n*
palladium content Palladiumgehalt *m*
palladium dioxide Palladiumdioxid *n*,
   Palladium(IV)-oxid *n*
palladium gold Palladgold *n* (Min),
   Palladiumgold *n*
palladium hydride Palladiumwasserstoff *m*
palladium monoxide Palladium(II)-oxid *n*,
   Palladiumoxydul *n* (obs)
palladium nitrate Palladiumnitrat *n*
palladium ore Palladiumerz *n* (Min)
palladium silicide Palladiumsilicid *n*
palladium sponge Palladiumschwamm *m*
palladous bromide Palladium(II)-bromid *n*,
   Palladiumbromür *n* (obs)
palladous chloride Palladium(II)-chlorid *n*,
   Palladochlorid *n*
palladous hydroxide Palladium(II)-hydroxid *n*,
   Palladohydroxid *n*
palladous iodide Palladium(II)-jodid *n*,
   Palladiumjodür *n* (obs)
palladous oxide Palladium(II)-oxid *n*,
   Palladiumoxydul *n* (obs)
palladous salt Palladium(II)-Salz *n*,
   Palladiumoxydulsalz *n* (obs)
pallasite Pallasit *m* (Min)
pallet saver Palettensparer *m*
pallet truck Gabelhubwagen *m*
pallet turning clamp Palettenwendeklammer *f*
palliative Linderungsmittel *n* (Pharm)
palliative balsam Linderungsbalsam *m* (Pharm)
Pall rings Pallringe *pl*
palm Palme *f* (Bot)
palmarosa oil Palmarosaöl *n*
palmate rubine Palmatrubin *n*
palmatine Palmatin *n*
palmellin Palmellin *n*
palmerite Palmerit *m* (Min)
palmic acid Palmitinsäure *f*
palmitate Palmitat *n*, Palmitinsäureester *m*,
   palmitrinsaures Salz *n*
palmitenone Palmitenon *n*
palmitic acid Palmitinsäure *f*, Cetylsäure *f*
palmitic aldehyde Palmitylaldehyd *m*
palmitin Palmitin *n*
palmitin candle Palmitinkerze *f*
palmitin soap Palmitinseife *f*
palmitodichlorohydrin Palmitodichlorhydrin *n*
palmitodilaurin Palmitodilaurin *n*
palmitodiolein Palmitodiolein *n*
palmitodistearin Palmitodistearin *n*
palmitoleic acid Palmitoleinsäure *f*
palmitolic acid Palmitolsäure *f*
palmitone Palmiton *n*
palmitostearoolein Oleopalmitostearin *n*
palmitoyl coenzyme A Palmitoylcoenzym A *n*
palmityl Palmityl-

palmitylalanine Palmitylalanin *n*
palm [kernel] oil Palmkernöl *n*
palm nut meal Palmkernmehl *n*
palm nut oil Palmkernöl *n*
palm oil Palmbutter *f,* Palmöl *n*
palm oil soap Palmölseife *f*
palm toddy Palmbranntwein *m*
palpate *v* befühlen, abtasten
palpation Betasten *n*
paltreubine Paltreubin *n*
pamaquin Plasmochin *n*
pamoic acid Pamosäure *f*
pan Pfanne *f,* Auffanggefäß *n,* Schale *f,*
Schüssel *f,* ~ for weights Waagschale *f*
pan acid Pfannensäure *f*
panacon Panacon *n*
Panama bark Panamarinde *f*
pan arrest (of a balance) Schalenarretierung *f*
(Waage)
panary fermentation Brotgärung *f*
pan boiling control apparatus
Kochkontrollapparat *m*
panchromatic panchromatisch (Phot),
empfindlich für alle Farben (Phot)
panchromatism Panchromasie *f* (Phot)
panclastite Panklastit *n*
pancratine Pancratin *n*
pancreas Bauchspeicheldrüse *f* (Med),
Pankreas *n* (Med)
pancreatic amylase Pankreasamylase *f*
(Biochem), Amylopsin *n* (Biochem)
pancreatic diastase Pankreasdiastase *f*
pancreatic enzyme Pankreasenzym *n*
pancreatic function Pankreasfunktion *f* (Med)
pancreatic juice Bauchspeichel *m,*
Pankreassaft *m*
pancreatic lipase Pankreaslipase *f* (Biochem)
pancreatin Pankreatin *n*
pancreatolipase Pankreaslipase *f* (Biochem)
pancreozymin Pankreozymin *n*
(Gewebshormon)
pandan straw Pandanstroh *n*
pandermite Pandermit *m* (Min)
pane Platte *f,* Scheibe *f* (Glas), Tafel *f*
panel Füllung *f* (z. B. Tür), Schaltkasten *m*
(Elektr), Täfelung *f*
panel board Füllbrett *n,* Schaltbrett *n* (Elektr)
panel[l]ed verkleidet, getäfelt
panel[l]ing Fachwerk *n,* Täfelung *f,* Verkleidung *f*
panel mounting Einschubrahmen *m*
pan filter Nutschenfilter *n*
panflavine Panflavin *n*
pan grinder Kollergang *m,* Kollermühle *f*
pan holder Schalenbügel *m,* height of ~
Schalenbügelhöhe *f*
paniculatine Paniculatin *n*
panitol Panit *m*
pan mill Kollergang *m*

pan mixer Mörtelmischmaschine *f* (Bauw),
Pfannenmischer *m,* Tellermischer *m*
pannaric acid Pannarsäure *f*
panose Panose *f*
pan scale Pfannenstein *m*
pansy Stiefmütterchen *n* (Bot)
pantetheine Pantethein *n*
pantethine Pantethin *n*
panthenol Panthenol *n*
pantoate Pantoat *n*
pantocain Pantokain *n* (Pharm)
pantograph Pantograph *m* (Zeichengerät),
Storchschnabel *m* (Zeichengerät)
pantograph arm Storchschnabelarm *m*
pantoic acid Pantosäure *f*
pantometer Pantometer *n*
pantonine Pantonin *n*
pantopon Pantopon *n* (Pharm), addicted to ~
pantoponsüchtig
pantoscope Pantoskop *n,* Weitwinkellinse *f*
pantoscopic pantoskopisch
pantotheine Panthothein *n*
pantothenic acid Pantothensäure *f*
pantothine Pantothin *n*
papain Papain *n* (Biochem)
papaver Molin *m* (Bot, Pharm)
papaverine Papaverin *n*
papaveroline Papaverolin *n*
papaya Papayafrucht *f* (Bot)
papayotin Papain *n* (Biochem)
paper Papier *n,* 50% rag-content ~
Halbhadernpapier *n,* absorbent ~
Fließpapier *n,* airproof ~ luftdichtes Papier,
antitarnish ~ rostschützendes Papier *n,* bits of
~ Papierschnitzel *n pl,* black-out ~
lichtdichtes Papier, blotting ~ Fließpapier *n,*
Löschpapier *n,* corrugated ~ Wellpappe *f,*
double-albuminized ~
Brillantalbuminpapier *n,* enameled ~
Kreidepapier *n,* endless ~ Papierbahn *f,*
fire-resisting ~ feuerfestes Papier, flame-
resisting ~ flammfestes Papier *n,* flame
retardant ~ schwer entflammbares Papier *n,*
glazed ~ Firnispapier *n,* gestrichenes Papier,
grease-proof ~ fettdichtes Papier, gummed ~
gummiertes Papier, handmade ~
Büttenpapier *n,* handgeschöpftes Papier, hard
~ Pappe *f,* impregnated ~ imprägniertes
Papier, insulating ~ Isolierpapier *n,*
metallized ~ metallisiertes Papier, oiled ~
Ölpapier *n,* geöltes Papier *n,* one-side-coated
~ einseitig gestrichenes Papier, photocopying
~ Photokopierpapier *n,* photographic ~
lichtempfindliches Papier *n,* photographisches
Papier, processed ~ veredeltes Papier,
sensitized ~ lichtempfindliches Papier *n,* sized
~ geleimtes Papier, smoothing ~
Schmirgelpapier *n,* tracing ~ Pauspapier *n,*
unsized ~ ungeleimtes Papier, waterproof ~

wasserfestes Papier, **wet-strength** ~ naßfestes
Papier, **wood-free** ~ holzfreies Papier, ~
without wood-pulp holzfreies Papier
paper bag Tüte *f,* Papierbeutel *m*
paper board Pappe *f*
paper calender Papierkalander *m*
paper carrier Papierunterlage *f*
paper chart feed Papiervorschub *m*
paper chromatogram Papierchromatogramm *n*
paper chromatography Papierchromatographie *f,*
  ascending ~ aufsteigende
  Papierchromatographie, **descending** ~
  absteigende Papierchromatographie
paper cloth Papiergewebe *n*
paper coal Papierkohle *f,* Blätterkohle *f*
paper coating Oberflächenleimung *f* (Papier),
  Schutzdecke *f* (Papier)
paper cover Umschlag *m* (Papier)
paper cup Papierbecher *m*
paper disk Papierscheibe *f*
paper electrophoresis Papierelektrophorese *f*
paper filler Papierfüllstoff *m*
paper filter Papierfilter *n*
paper finishing Papierveredelung *f*
paper industry Papierindustrie *f*
paper knife Papiermesser *n*
paper lacquer Papierlack *m*
paper lining Papierauskleidung *f*
papermaking Papierfabrikation *f*
papermaking auxiliary Papierhilfsmittel *n*
papermaking machine Papiermaschine *f*
paper manufacture Papierfabrikation *f*
paper mill Papierfabrik *f*
paper processing Papierveredelung *f*
[paper] pulp Papierbrei *m,* Papiermasse *f,*
  mechanical ~ mechanische Papiermasse,
  semi-chemical ~ halbchemische Papiermasse
[paper] rags Hadern *m pl*
paper roll Papierrolle *f*
paper sack Papiersack *m*
paper shredding machine Papierwolf *m*
paper sizing Papierleimung *f*
paper strip Papierstreifen *m*
paper tab Papierzunge *f*
paper tape Lochstreifen *m* (Comp),
  Papierband *n*
paper tape reader Lochstreifenleser *m* (Comp)
paper tape to magnetic tape converter
  Lochstreifen-Magnetbandumwandler *m*
  (Comp)
paper tape unit Lochstreifeneinheit *f* (Comp)
paper testing Papierprüfung *f*
paper testing equipment Papierprüfgeräte *pl*
paper thread Papierfaden *m*
paper web Papierbahn *f*
paperweight Papierbeschwerer *m,*
  Briefbeschwerer *m*
paper-working papierverarbeitend
paper yarn Papiergarn *n*

papilionaceous flower Schmetterlingsblütler *m*
  (Bot)
Papin's digestor Papinscher Topf *m*
paprica Paprika *m* (Bot)
paprica oil Paprikaöl *n*
paprika Paprika *m* (Bot)
papyrine Papyrin *n*
paraacetaldehyde Paraldehyd *m*
para acid para-Säure *f*
parabanic acid Oxalylharnstoff *m,*
  Parabansäure *f*
para blue Parablau *n*
parabola Parabel *f* (Math)
parabolic mirror Parabolspiegel *m*
paraboloidal reflector Parabolspiegel *m*
paraboloid of revolution Rotationsparaboloid *n*
para butter Parabutter *f*
parabuxin Parabuxin *n*
paracasein Paracasein *n*
paracentric parazentrisch
paracetamol Paracetamol *n*
parachute Fallschirm *m*
para compound para-Verbindung *f*
paraconic acid Paraconsäure *f*
paracotoin Paracotoin *n*
Para cress Parakresse *f* (Bot)
paracrystal Parakristall *m*
paracyanogen Paracyan *n*
paracyclophane Paracyclophan *n*
para-dihydroxybenzene Hydrochinon *n*
para-directing paradirigierend (Chem)
paraffin Paraffin *n,* Grenzkohlenwasserstoff *m*
  (Chem), **crude** ~ Rohparaffin *n,* **liquid** ~
  Paraffinöl *n,* **native** ~ Erdwachs *n,* **soft** ~
  Weichparaffin *n,* **solid** ~ Erdwachs *n,* **to coat
  with** ~ paraffinieren, mit Paraffin überziehen,
  to wax with ~ paraffinieren
paraffin *v* paraffinieren, mit Paraffin behandeln
paraffin bath Paraffinbad *n*
paraffin candle Paraffinkerze *f*
paraffinic acid Paraffinsäure *f*
paraffin impregnation Paraffintränkung *f*
paraffin lubricating oil Paraffinschmieröl *n*
paraffin oil Paraffinöl *n,* Leuchtpetroleum *n*
paraffin ointment Paraffinsalbe *f*
paraffin paper Paraffinpapier *n*
paraffin residue Paraffinrest *m* (Chem)
paraffin series Paraffinreihe *f*
paraffin varnish Paraffinlack *m*
paraffin wax Paraffin *n,* festes Paraffin, **plug of**
  ~ Paraffinpfropfen *m*
[paraffin-]wax sizing Wachsleimung *f* (Pap)
paraform Paraform *n,* Paraformaldehyd *m*
paraformaldehyde Paraformaldehyd *m,*
  Paraform *n*
parafuchsin Parafuchsin *n*
paragenesis Paragenese *f*
paragenetic paragenetisch
paraglobulin Paraglobulin *n*

paragonite Paragonit *m* (Min),
  Natronglimmer *m* (Min)
Paraguay tea Matéblätter *n pl*
parahopeite Parahopeit *m* (Min)
para hydrogen Parawasserstoff *m*
paralactic acid Paramilchsäure *f*
paralaurionite Paralaurionit *m* (Min)
paralbumin Paralbumin *n*
paraldehyde Paraldehyd *m*
paraldol Paraldol *n*
parallactic parallaktisch
parallax Parallaxe *f,* **free from** ~ parallaxenfrei
parallel Parallele *f,* **to connect in** ~ parallel
  schalten (Elektr), **to run** ~ parallel laufen
parallel *a* parallel, gleichlaufend
parallel circuit Parallelkreis *m* (Elektr)
parallel-connected nebeneinander geschaltet
  (Elektr), parallel geschaltet (Elektr)
parallel connection Nebeneinanderschaltung *f*
  (Elektr), Parallelschaltung *f* (Elektr)
parallel coupling Parallelschaltung *f*
parallel displacement Parallelverschiebung *f*
parallel experiment Parallelversuch *m*
parallel flow gas burner
  Parallelstrom-Gasbrenner *m*
parallel flow heat exchanger
  Gleichstromwärmeaustauscher *m*
parallel flow principle Gleichstromprinzip *n*
parallel guide Parallelführung *f*
parallelism Parallelität *f,* Gleichlauf *m*
parallel key Flachkeil *m* (Techn)
parallel [line] Parallele *f*
parallelogram Parallelogramm *n,* ~ **of forces**
  Kräfteparallelogramm *n*
parallel reaction Parallelreaktion *f*
parallel resistance Parallelwiderstand *m*
parallel resonance Parallelresonanz *f*
parallel row Parallelreihe *f*
parallel ruler Parallellineal *n*
parallel shift Parallelverschiebung *f*
parallel system Parallelschaltungssystem *n*
  (Elektr)
parallex compensation Parallaxenausgleich *m*
paraluminite Paraluminit *m* (Min)
paralysis Lähmung *f* (Med), Paralyse *f* (Med)
paralyzation Lähmung *f* (Med)
paralyze *v* lähmen, lahmlegen
param Dicyandiamid *n*
paramagnetic paramagnetisch
paramagnetism Paramagnetismus *m*
paramandelic acid Paramandelsäure *f*
paramecium Pantoffeltierchen *n* (Zool)
parameter Parameter *m*
parameter integral Parameterintegral *n*
parametric equation Parametergleichung *f*
  (Math)
parametric representation
  Parameterdarstellung *f,* ~ **of a function**
  Parameterdarstellung einer Funktion *f*

parametrization technique
  Parametrierungstechnik *f*
paramide Paramid *n*
paramine brown Paraminbraun *n*
param[o]ecium Pantoffeltierchen *n* (Zool)
paramorphic paramorph (Krist)
paramorphine Paramorphin *n*
paramorphism Paramorphismus *m*
paramorphous paramorph (Krist)
paramucosin Paramucosin *n*
para-naphthalene (obs) Anthracen *n*
paranil Paranil *n*
parapectic acid Parapektinsäure *f*
parapeptone Parapepton *n,* Syntonin *n*
para-position Parastellung *f* (Chem)
pararosaniline Pararosanilin *n*
pararosolic acid Pararosolsäure *f*
para rubber Paragummi *n,* Parakautschuk *m*
para rubber oil Paragummiöl *n*
parasite Parasit *m* (Biol), Schmarotzer *m* (Biol),
  specific ~ an einen spezifischen Wirt
  gebundener Parasit, ~ **of the intestine**
  Enterozoon *n* (pl. Enterozoen)
parasite current Störstrom *m* (Elektr)
parasitic parasitisch, parasitenartig,
  schmarotzend
parasitic disease Parasitenbefall *m*
parasiticide Antiparasitikum *n,*
  parasitentötendes Mittel *n*
parasitize *v* parasitieren, schmarotzen
parasitology Parasitologie *f,* Parasitenkunde *f*
parasorbic acid Parasorbinsäure *f*
para state Parazustand *m*
parastilbite Parastilbit *m*
paratacamite Paratakamit *m* (Min)
paratartaric acid Paraweinsäure *f*
parathiazine Parathiazin *n*
parathion Parathion *n,* E 605 (HN)
parathyroid hormone
  Nebenschilddrüsenhormon *n,* Parathormon *n*
paratitol Paratit *m*
paratype Paratyp *m* (Biol)
paratyphoid Parathyphus *m* (Med)
paratyphoid vaccine Paratyphusvakzine *f*
paravivianite Paravivianit *m* (Min)
paraxanthine Paraxanthin *n,* Eblanin *n*
paraxial orbit Paraxialbahn *f*
parboil *v* abbrühen, ankochen, halb kochen
parboiling Halbkochen *n*
parcel Ballen *m,* Paket *n*
parcel *v* [in Pakete] packen
parch *v* ausdörren, rösten, dörren
parchment Pergament *n,* **imitation** ~ Pergamin *n*
parchment color Pergamentfarbe *f*
parchment glue Pergamentleim *m*
parchmentize *v* pergamentieren
parchmentizing machine Pergamentiermaschine *f*
parchment making Pergamentbereitung *f*

parchment paper Pergamentpapier *n,*
Kunstpergament *n*
parchment pipe Pergamentschlauch *m*
pare *v* [ab]schälen, abrinden
pareira Pareirawurzel *f* (Bot)
parent acid Stammsäure *f*
parent atom Ausgangsatom *n*
parent cell Mutterzelle *f,* Stammzelle *f*
parent compound Stammsubstanz *f,*
Stammverbindung *f*
parentheses runde Klammern *pl*
parent isotope Ausgangsisotop *n,* Mutterisotop *n*
parent patent Hauptpatent *n*
parent state Ausgangszustand *m*
parent substance Stammsubstanz *f,*
Muttersubstanz *f,* Stammkörper *m*
parent yeast Mutterhefe *f*
pargasite Pargasit *m* (Min)
parget Kalkanwurf *m,* Putz *m*
paricine Paricin *n*
paridin Paridin *n*
parietal cell Wandzelle *f*
parietin Parietin *n,* Physciasäure *f,* Physcion *n*
pariglin Smilacin *n*
parillic acid Parillin *n*
parillin Parillin *n*
parinaric acid Parinarsäure *f*
parings Schabsel *n pl,* Schnitzel *pl,* Späne *pl,* ~
of skin Leimleder *n*
Paris black Pariser Schwarz *n*
Paris blue Berliner Blau *n,* Pariser Blau *n*
Paris green Pariser Grün *n,* Kaisergrün *n,*
Königsgrün *n,* Kupferarsenacetat *n,*
Schweinfurter Grün *n*
parisite Parisit *m* (Min)
Paris lake Pariser Lack *m*
Paris yellow Kaisergelb *n,* Pariser Gelb *n,*
Bleichromat *n*
parity Parität *f*
parity change Paritätsänderung *f*
parity conservation Paritätserhaltung *f*
parity selection rule Paritätsauswahlregel *f*
parkerize *v* parke[ri]sieren
Parkes process Parkesieren *n*
parkia oil Parkiaöl *n*
parmetol Parmetol *n*
parodontosis Parodontose *f* (früher Paradentose)
parodyne Parodyn *n*
paromamine Paromamin *n*
paromose Paromose *f*
paroxazine Paroxazin *n*
paroxypropione Paroxypropion *n*
parquetry sealing Parkettbodenversiegelung *f*
parquine Parquin *n*
parrot green Papageigrün *n*
part Teil *m n,* Abschnitt *m*
part *v* scheiden, trennen
partial partiell, teilweise, unvollkommen,
unvollständig

partial boiling Halbkochen *n,* Souplieren *n*
(Seide)
partial carbonizer Schwelzylinder *m*
partial cargo Teilladung *f*
partial charring Ankohlen *n*
partial cross section Partialquerschnitt *m*
partial fraction Partialbruch *m* (Math)
partial fraction rule Partialbruchregel *f* (Math)
partial fractions, decomposition into ~
Partialbruchzerlegung *f* (Math)
partial load Teilladung *f*
partial pressure Partialdruck *m,* **ultimate** ~
Endpartialdruck *m*
partial reduction Halbreduktion *f* (Chem,
Metall)
partial shadow Halbschatten *m*
partial valence Partialvalenz *f*
partial view Teilansicht *f*
participate *v* sich beteiligen, teilnehmen
participation Beteiligung *f,* Mitwirkung *f,*
Teilnahme *f*
particle Teilchen *n,* Baustein *m,* Korpuskel *n,*
Partikel *f,* Stoffteilchen *n,* **accelerated** ~
beschleunigtes Teilchen, **ionizing** ~
ionisierendes Teilchen, **scattered** ~
Streuteilchen *n,* **subatomic** ~ subatomares
Teilchen
particle accelerator Teilchenbeschleuniger *m*
particle board Holzwerkstoffe *pl*
particle bombardment Teilchenbeschuß *m*
particle charge Teilchenladung *f*
particle density Teilchenzahldichte *f*
particle detection Teilchennachweis *m*
particle diameter, mean ~ mittlerer
Teilchendurchmesser *m*
particle energy Teilchenenergie *f*
particle image Teilchenbild *n*
particle impact Teilchenstoß *m*
particle momentum Teilchenimpuls *m*
particle motion Teilchenbewegung *f*
particles, compression of ~
Teilchenzusammenpressung *f,* **displacement of**
~ Teilchenverschiebung *f,* **number of** ~
Teilchenzahl *f*
particle shape Teilchengestalt *f*
particle shift Teilchenverschiebung *f*
particle size Teilchengröße *f,* Korngröße *f,*
Partikelgröße *f*
particle size analysis Feinheitsanalyse *f*
particle size determination
Korngrößenbestimmung *f*
particle size distribution Korngrößenverteilung *f*
particle structure Teilchenstruktur *f*
particle velocity Teilchengeschwindigkeit *f*
particular ausgeprägt, besonders, partikulär
particular case Einzelfall *m*
particularity Besonderheit *f,* Eigentümlichkeit *f*
particularize *v* partikularisieren, spezifizieren,
ausführlich angeben

parting Scheidung *f* (Chem), Trennung *f* (Chem)
parting agent Trennmittel *n*
parting compound Trennmittel *n*, Gleitmittel *n*
parting compound line Trennfuge *f*
parting device Scheidevorrichtung *f*
parting furnace Scheideofen *m*
parting line Teilungslinie *f*, Trennlinie *f*
parting process Scheideverfahren *n* (Met)
parting silver Scheidesilber *n*
partition Fach *n*, Scheidewand *f*, Trenneinsatz *m*, Zwischenboden *m*, Zwischenwand *f*, **semipermeable ~** halbdurchlässige Scheidewand *f*
partition coefficient Verteilungskoeffizient *m* (Chromatogr)
partition line Trennfuge *f*
parts, consisting of several ~ mehrteilig
parvoline Parvolin *n*
Pascal law Pascalsches Gesetz *n* (Mech), Druckfortpflanzungsgesetz *n* (Mech)
Paschen series Paschenserie *f* (Spektr)
pascoite Pascoit *m* (Min)
pass Durchgang *m*, Walzenfurche *f* (Techn)
pass *v*, **~ by** verstreichen (Zeit), **~ over** übergehen (Dest), überleiten, **~ through** durchleiten, durchströmen, hindurchtreten
passage Durchgang *m*, Durchfluß *m*, Durchlaß *m*, Durchtritt *m*, Durchzug *m*, Übergang *m*, **circular ~** Rundkanal *m*, **~ to the limit** Grenzübergang *m* (Math)
passage opening Durchgangsöffnung *f*
passing station or post Durchlaufstation *f* (Förderer)
passivate *v* passivieren
passivating agent Passivierungsmittel *n*
passivation Passivierung *f*
passivation potential Passivierungspotential *n*
passive passiv
passive hardness Abnutzungshärte *f*
passiveness Passivität *f*
passivity Passivität *f*
pass template drawing Kalibrierungszeichnung *f*
paste Brei *m* (Pap), breiige Masse *f*, Klebematerial *n*, Kleister *m*, Paste *f*, Teig *m*, **to make into a ~** aufschlämmen
paste *v* anschlämmen, anteigen, [be]kleben; kleistern, **~ on** aufkleben, aufkleistern, **~ together** verkitten
pasteboard Karton *m*, Pappe *f*
pasteboard box Karton *m*, Pappschachtel *f*
paste grinding machine Pastenanreibemaschine *f*
pastel [color] Pastellfarbe *f*
pastel shade Pastellton *m*
paste resin Harzpaste *f*, Pastenharz *n*
Pasteur effect Pasteureffekt *m* (Biochem)
pasteurization Entkeimung *f* (Milch), Pasteurisierung *f*
pasteurize *v* pasteurisieren, sterilisieren

pasteurizer Pasteurisierapparat *m*
pasteurizing Pasteurisieren *n*
pasteurizing plant Pasteurisieranlage *f*
pastille Pastille *f* (Med)
pasting auxiliary Anteigemittel *n*
pasting press Klebepresse *f*
pasting process Pasten *n*, Pasting-Verfahren *n*
pasty breiartig, breiig, klebrig, kleisterig, pappig, teigig
patch Pflaster *n*
patch *v* flicken
patching compound Reparaturmasse *f* (Reifen)
patchoulene Patschulen *n*
patchouli Patschuli[parfüm] *n*, **essence of ~** Patschuliessenz *f*, **smell of ~** Patschuliduft *m*
patchouli alcohol Patschulialkohol *m*
patchouli oil Patschuliöl *n*
patent Patent *n*, **expired ~** abgelaufenes Patent, **filed ~** angemeldetes Patent, **foreign ~** Auslandspatent *n*, **~ of addition** Zusatzpatent *n*, **protection of or by a ~** Patentschutz *m*, **subject matter of a ~** Patentgegenstand *m*, **to apply for a ~** ein Patent anmelden, ein Patent *n* anmelden, **to infringe on a ~** ein Patent verletzen, **to secure a ~** [for] ein Patent erwirken [für], **to take out a ~** patentieren, **~ applied for** angemeldetes Patent, **a ~ has been granted** ein Patent ist erteilt worden
patent *v* patentieren, gesetzlich schützen
patentability Patentfähigkeit *f*
patentable patentfähig, patentierbar
patent act Patentgesetz *n* (Jur)
patent annuity Patentgebühr *f* (Jahresgebühr)
patent applicant Patentanmelder *m*, Patentsucher *m*
patent application Patentanmeldung *f*, Patentgesuch *n*
patent application published for opposition Auslegeschrift *f*
patent attorney (Am. E.) Patentanwalt *m*
patent blue Patentblau *n*
patent claim Patentanspruch *m*
patent core nail Patentkernnagel *m*
patent drawing Patentzeichnung *f*
patented gesetzlich geschützt (durch ein Patent), patentiert, gesetzlich geschützt, **~ at home and abroad** patentiert im In- und Ausland
patentee Patentinhaber *m*, Patentträger *m*
patent exploitation Patentverwertung *f*
patent fastener Druckknopf *m*
patent fee Patentgebühr *f*
patent grant Patenterteilung *f*
patent infringement Patentverletzung *f*
patent-infringing patentverletzend
patent in suit Klagepatent *n* (Jur)
patent law Patentgesetz *n* (Jur)
patent leather Lackleder *n*, **chrome-tanned ~** Chromlackleder *n*
patent leather enamel Glanzlederlack *m*

patent legislation Patentgesetzgebung *f*
patent nickel alloy Patentnickel *n*
patent office Patentamt *n*
patent opposition Patenteinspruch *m*
patentor Patentgeber *m*
patent plaster Edelputz *m*
patent procedure Patentverfahren *n*
patent register Patentregister *n*
patent right Patentrecht *n* (Jur)
patent rolls Patentregister *n*
patent specification Patentbeschreibung *f*
patent suit Patentklage *f* (Jur)
patent yellow Patentgelb *n*, Englischgelb *n*
paternoster Paternoster *m* (Techn)
path Bahn *f*, Laufbahn *f*, Weg *m*, Wegstrecke *f*,
 average free ~ mittlere freie Weglänge *f*, **mean
 free** ~ mittlere freie Weglänge *f*
path [length] Weglänge *f*
path length distribution Weglängenverteilung *f*
path of contact (Br. E.) Eingriffsstrecke *f*
 (Getriebe)
pathogenic krankheitserregend, pathogen
pathological krankhaft
pathologist Pathologe *m* (Med)
pathology Pathologie *f* (Med),
 Krankheitslehre *f* (Med), **comparative** ~
 vergleichende Pathologie, **general** ~
 allgemeine Pathologie, **physiological** ~
 physiologische Pathologie
pathometabolism Pathometabolismus *m*
pathophysiology pathologische Physiologie *f*
path reversal Bahnumkehr *f*
path spin Bahndrehimpuls *m* (Atom)
pathway Weg *m* (Metabolismus)
patience Geduld *f*
patina Edelrost *m*, Patina *f*, **to cover with** ~
 patinieren
patrinite Patrinit *m* (Min)
patrix (pl. patrices) Patrize *f*
patronite Patronit *m* (Min)
pattern Muster *n*, Dessin *n* (Stoff), Lehre *f*
 (Meßgerät), Modell *n*, Probe *f*, Profil *n* (der
 Lauffläche) (Reifen), Schablone *f* (Techn),
 Schnittmuster *n* (Schneiderei), Vorbild *n*,
 outline of ~ Modellumriß *m*
pattern half Modellhälfte *f*
patterning wheel Musterrad *n*
pattern maker Kernmacher *m* (Gieß)
pattern plate Modellplatte *f*
pattern selector lever Musterstopperhebel *m*
pattern varnish Modellack *m*
Pattinson's process Pattinsonieren *n*
patulin Patulin *n*
paucine Paucin *n*
Pauli [exclusion] principle Pauli-Prinzip *n*,
 Pauliverbot *n* (Phys)
pause Pause *f*, Anhalt *m*, Haltezeit *f*,
 Unterbrechung *f*

pause *v* innehalten, pausieren
pave *v* [be]pflastern
pavement Pflaster *n* (Straßenbau)
pavine Pavin *n*
pawl Klinke *f*, Sperrklinke *f*, ~ **with roller**
 Klinke *f* mit Rolle
pawl coupling Klinkenkupplung *f*
pawl release Klinkenauslösung *f*
payment Bezahlung *f*, Begleichung *f*
pay ore abbauwürdiges Erz *n*
payroll accounting Lohn- und
 Gehaltsabrechnung *f*
paytine Paytin *n*
p-compound para-Verbindung *f*
p-conductor p-Leiter *m*
p-diaminobenzophenone Michlers Keton *n*
pea Erbse *f* (Bot)
peach Pfirsich *m* (Bot)
peach blossom color Pfirsichblütenfarbe *f*
peach-colored pfirsichfarben
pea coal Grießkohle *f*, Perlkohle *f*
peacock ore Bornit *m* (Min), Buntkupfererz *n*
 (Min)
peak Gipfel *m*, Höhepunkt *m*, Maximum *n*,
 Spitze *f*
peak current Spitzenstrom *m*
peaker strip Differenzierspule *f*
peak factor Scheitelfaktor *m*
peak load Belastungsspitze *f*, Höchstbelastung *f*,
 Spitzenlast *f*
peak output Höchstleistung *f*, Spitzenleistung *f*
peak [value] Scheitelwert *m*
peak velocity Spitzengeschwindigkeit *f*
peak voltage Spannungshöchstwert *m* (Elektr),
 Spitzenspannung *f* (Elektr)
pealike erbsenförmig
peanut oil Erdnußöl *n*, Arachisöl *n*
pea ore Bohnerz *n*, Erbsenerz *n* (Min)
pearceite Pearceit *m* (Min)
pearl Perle *f*
pearl ash Perlasche *f*, Holzkohlenasche *f*
pearl-colored perlfarben
pearl effect Fischsilbereffekt *m*
pearl essence pigment Fischsilberpigment *n*
pearl form Perlform *f*
pearling Perlen *n*
pearlite Perlit *m* (Min)
pearl-like perlartig
pearl mica Perlglimmer *m* (Min)
pearl moss Perlmoos *n*
pearl powder Perlweiß *n*
pearl shell Perlmutter *f*
pearl sinter Perlsinter *m*
pearl spar Perlspat *m* (Min), Bitterkalk *m* (Min)
pearl white Perlweiß *n*
pearly perlenartig, gekörnt
pearly luster Perlenglanz *m*, Perlmutterglanz *m*
pear-shaped birnenförmig
pear vinegar Birnenessig *m*

pea-shaped erbsenförmig
pea-sized erbsengroß
peastone Erbsenstein *m* (Min), Oolith *m* (Min)
peat Torf *m*, Erdtorf *m*, **containing** ~ torfhaltig, **fibrous** ~ Fasertorf *m*, Wurzeltorf *m*, **hand-cut** ~ Handtorf *m*, **powdered** ~ Torfmehl *n*, **pressed** ~ Preßtorf *m*
peat ashes Torfasche *f*
peat brick Torfstein *m*
peat briquet[te] Torfbrikett *n*
peat coal Lignit *m*
peat digging Torfgewinnung *f*
peat dust Torfmull *m*, Torfstaub *m*
peat gas Torfgas *n*
peat gas producer Torfgasgenerator *m*
peat mold Torferde *f*
peat slab Torfplatte *f*
peat water Moorwasser *n*
peaty torfartig
peaty odor Torfgeruch *m*
peaty soil Moorerde *f*
pebble Kiesel[stein] *m*, grober Kies *m*
pebble-bed sintering Wirbelsintern *n*
pebble filter Kiesfilter *n*
pebble heater Kieselmassewärmer *m*
pebble mill Kugelmühle *f*
pebble powder Kieselpulver *n*
pebbles Geröll *n*, **pounded** ~ Kieselmehl *n*
pebbly ground Kieselgrund *m*
peckhamite Peckhamit *m* (Min)
pectase Pektase *f* (Biochem)
pectic pektinsauer
pectic acid Pektinsäure *f*
pectin Pektin *n*, Pflanzengallerte *f*
pectinase Pektinase *f*
pectin depolymerase Pektindepolymerase *f* (Biochem)
pectinite Pektinit *m*
pectin sugar Pektinose *f*
pectolite Pektolith *m* (Min)
pectolytic pektinspaltend
pectoral tea Brusttee *m* (Pharm)
pectose Pektose *f*
pectosinic acid Pektosinsäure *f*
peculiar besonders, eigenartig
peculiarity Besonderheit *f*, Eigenart *f*, Eigentümlichkeit *f*
pedal pad Pedalgummi *m*
pedestal Sockel *m*, Ständer *m*
pedestal bearing Stehlager *n*
pediatrics Pädiatrie *f* (Med), Kinderheilkunde *f* (Med)
peel Schale *f* (Früchte etc.), Pelle *f* (Früchte etc.), **balanced** ~ Beschickschwengel *m*, Chargierschwengel *m*, ~ **[of fruit]** Fruchtschale *f*
peel *v* [ab]schälen, [ab]blättern, [ab]schilfern, abschuppen, entrinden
peeled abgeschält (Früchte etc.)

peeling Schälen *n*, **[lacquer]** ~ Abschälen *n* (des Lackes)
peel-off coating Abdecklack *m*
peel strength (electroplated articles) Haftfestigkeit *f*
peen (of a hammer) Pinne *f* (eines Hammers) (Techn)
peening ridge Stemmkante *f*
peep hole Schauloch *n*, Schauöffnung *f*
peg Dübel *m*, Holznagel *m*, Splint *m*, Stift *m*, ~ **for drawing boards** Heftzwecke *f*
pegan Pegan *n*
peganite Peganit *m* (Min), Variscit *m* (Min)
pegene Pegen *n*
pegenone Pegenon *n*
pegging rammer Dämmholz *n*, Spitzstampfer *m*
peg impactor mill Stiftmühle *f*
pegmatite Pegmatit *m*, Ganggranit *m*
pegmatite anhydrite Pegmatitanhydrit *m* (Min)
pegmatitic pegmatitartig
pegmatoid pegmatitartig
pegmatolite Pegmatolith *m*
pegnin Pegnin *n*
pelargonaldehyde Pelargonaldehyd *m*, Nonylaldehyd *m*
pelargonate Pelargonat *n*, Salz oder Ester der Pelargonsäure *f*
pelargone Pelargon *n*
pelargonic pelargonsauer
pelargonic acid Pelargonsäure *f*, **salt or ester of** ~ Pelargonat *n*
pelargonic ether Pelargonäther *m*
pelargonidin glucoside Pelargonin *n*
pelargonin Pelargonin *n*
pelargonyl Pelargonyl-
pelentanoic acid Pelentansäure *f*
Peligot tube Peligotröhre *f*
pellagra Pellagra *n* (Med)
pellet Kügelchen *n*, Pille *f*, Tablette *f*
pellet *v* tablettieren
pellet drier Granulatschleuder *f*
pelletierine Pelletierin *n*, Punicin *n*
pelleting Pelletieren *n*
pelleting press Tablettenpresse *f*, Tablettiermaschine *f*
pelleting property Tablettierfähigkeit *f*
pelletizer Granuliermaschine *f*, Krümelmaschine *f*
pellets Granulat *n*
pellet size Tablettengröße *f*
pellicle Häutchen *n* (Techn), Überzug *m*, ~ **of oxygen** Oxidhäutchen *n*
pellitorine Pellitorin *n*
pellotine Pellotin *n*
peloconite Pelokonit *m*
pelt Fell *n*, Tierpelz *m*
Pelton turbine Pelton-Turbine *f*, Strahlturbine *f*

Pelton wheel Pelton-Turbine *f* (Mech),
  Aktionsturbine *f* (Mech), Freistrahlturbine *f*
  (Mech), Gleichdruckturbine *f* (Mech)
peltry Rauchwaren *pl*, Pelzwerk *n*
penaldic acid Penald[in]säure *f*
penam Penam *n*
pencatite Pencatit *m*
pencil Stift *m*, feiner Pinsel *m*, Schreibstift *m*, ~
  for marking glass Fettstift *m*
pencil carbon Handdurchschreibepapier *n*
pencil slate Griffelschiefer *m*
pendent hängend
pending schwebend
pending patent angemeldetes Patent *n*
pendulate *v* pendeln
pendulum Perpendikel *n*, Pendel *n*, **ballistic** ~
  ballistisches Pendel, **conical** ~ kegelförmiges
  Pendel (Mech), **mathematic or simple** ~
  mathematisches oder einfaches Pendel,
  physical or compound ~ physikalisches oder
  zusammengesetztes Pendel *n*, **amplitude of** ~
  swing Pendelausschlag *m*
pendulum clock Pendeluhr *f*
pendulum deflection Pendelausschlag *m*
pendulum governor Pendelregler *m*
pendulum hammer Pendelhammer *m*
pendulum motion Pendelbewegung *f*
pendulum suspension Pendelaufhängung *f*
penetrability Durchdringbarkeit *f*
penetrable durchdringbar, durchgängig,
  durchlässig
penetrableness Durchdringbarkeit *f*
penetrate *v* durchdringen, eindringen
penetrating durchdringend, stechend (Geruch)
penetrating power Durchdringungsvermögen *n*
  (Röntgenstrahlen)
penetration Aufziehen *n* (Flüssigkeit),
  Durchdringen *n*, Durchdringung *f*,
  Durchfärbung *f* (Text), Durchschlag *m*
  (Geschoß), Eindringen *n*,
  Eindringungsvermögen *n*, Penetration *f*,
  Tiefenwirkung *f* (eines Abbeizmittels), **depth
  of** ~ Eindringtiefe *f*, **point of** ~
  Durchschlagspunkt *m*, ~ **of dampness**
  Durchfeuchtung *f*, ~ **of moisture**
  Feuchtigkeitsdurchschlag *m*
penetration capacity Durchdringungsfähigkeit *f*
penetration coefficient Durchgriff *m*
penetration complex Durchdringungskomplex *m*
  (Chem)
penetration depth Eindringtiefe *f*
penetration dyeing Durchfärbung *f*
penetration energy, total ~ **[at failure]**
  Durchstoßfestigkeit *f*
penetration index Penetrometerzahl *f*
penetration model Penetrationsmodell *n*
penetration notch Einbrandkerbe *f*
penetration point Durchdringungsstelle *f*

penetration twin Durchkreuzungszwilling *m*
  (Krist)
penetration twins Durchdringungszwillinge *pl*
  (Krist)
penetrative durchdringend, durchgreifend
penetrative dyeing agent Durchfärbemittel *n*
penetrative effect Tiefenwirkung *f*
penetrative power Durchschlagskraft *f*
penetrometer Penetrationsmesser *m*,
  Härtemesser *m*, Penetrometer *n*
penetron Meson *n*
penfieldite Penfieldit *m* (Min)
penicillamine Penicillamin *n*
penicillanic acid Penicillansäure *f*
penicillenic acid Penicillensäure *f*
penicillin Penicillin *n*, **insensitive to** ~
  penicillinunempfindlich, **resistant to** ~
  penicillinresistent, **sensitive to** ~
  penicillinempfindlich
penicillinase Penicillinase *f* (Biochem)
penicillin flask Penicillinkolben *m*
penicillin level Penicillinspiegel *m*
penicillin-resistant penicillinresistent
penicilliopsin Penicilliopsin *n*
penicilloic acid Penicillosäure *f*
penillic acid Penillsäure *f*
penilloaldehyde Penilloaldehyd *m*
penilloic acid Penillasäure *f*
penillonic acid Penillonsäure *f*
pennine Pennin *m* (Min)
penninite Pennin *m* (Min)
pennite Pennit *m*
pennol Pennol *n*
pennone Pennon *n*
pennyhead stopper Lappengriffstopfen *m*
pennyroyal oil Poleiöl *n*
pen point Federspitze *f*
pentabasic fünfbasisch
pentacarbocyclic pentakarbozyklisch
pentacene Pentacen *n*
pentacetylglucose Pentacetylglucose *f*
pentachloroethane Pentachloräthan *n*, Pentalin *n*
pentachlorothiophenol Pentachlorthiophenol *n*
pentacontane Pentacontan *n*
pentacosane Pentacosan *n*
pentacyanic acid Pentacyansäure *f*
pentadecane Pentadecan *n*
pentadecanoic acid Pentadecansäure *f*
pentadiene Pentadien *n*, Piperylen *n*
pentaerythritol Pentaerythrit *m*
pentaerythritol tetranitrate (PETN)
  Pentaerythrittetranitrat *n*
pentaglycerol Pentaglycerin *n*
pentagon Fünfeck *n*, Pentagon *n*
pentagonal fünfeckig, fünfkantig, pentagonal
pentagonal hemihedral pentagonalhemiedrisch
pentagon dodecahedron Pentagondodekaeder *n*
pentagon icositetrahedron
  Pentagonikositetraeder *n*

pentagrid valve or tube Pentagridröhre *f*
pentahedral fünfflächig, pentaedrisch
pentahedron Fünfflächner *m*, Pentaeder *n*
pentahomoserine Pentahomoserin *n*
pental Pental *n*
pentalene Pentalen *n*
pentaline Pentalin *n*
pentamethonium Pentamethonium *n*
pentamethylbenzene Pentamethylbenzol *n*
pentamethylene Pentamethylen *n*
pentamethylene bromide Pentamethylenbromid *n*
pentamethylene diamine Pentamethylendiamin *n*,
    Kadaverin *n*
pentamethylene sulfide Pentamethylensulfid *n*
pentamethylglucose Pentamethylglucose *f*
pentanal Pentanal *n*, Valeraldehyd *m*
pentane Pentan *n*
pentanedioic acid Glutarsäure *f*
pentanediol Pentandiol *n*
pentane normal thermometer
    Pentannormalthermometer *n*
pentanethiol Pentanthiol *n*
pentanoic acid Valeriansäure *f*
pentanol Pentanol *n*
pentanone Pentanon *n*
pentanoyl Valeryl-
pentanoyl bromide Valerylbromid *n*
pentanoyl chloride Valerylchlorid *n*
pentanthrimide Pentanthrimid *n*
penta peptide Pentapeptid *n*
pentaphene Pentaphen *n*
pentaphenylphosphorus Pentaphenylphosphor *m*
pentasulfide Pentasulfid *n*
pentathionate Pentathionat *n*
pentathionic acid Pentathionsäure *f,* salt or ester
    of ~ Pentathionat *n*
pentatomic fünfatomig
pentatriacontane Pentatriakontan *n*
pentatriacontanone, 18- ~ Stearon *n*
pentavalence Fünfwertigkeit *f*
pentavalent fünfwertig
pentazine Pentazin *n*
pentene Penten *n*
pentenedioic acid Glutaconsäure *f*
pentetrazole Pentetrazol *n*, Corazol *n*
penthiazolidine Penthiazolidin *n*
penthiazoline Penthiazolin *n*
penthiofurane Penthiophen *n*
penthiophene Penthiophen *n*
pentine Pentin *n*
pentite Pentit *m*
pentitol Pentit *m*
pentlandite Pentlandit *m* (Min),
    Eisennickelkies *m* (Min)
pentode Pentode *f* (Elektr), Fünfpolröhre *f*
    (Elektr)
pentonic acid Pentonsäure *f*
pentosan Pentosan *n*
pentose Pentose *f*

pentose cycle Pentosezyklus *m* (Biochem)
pentose isomerase Pentoseisomerase *f* (Biochem)
pentose phosphate Pentosephosphat *n*
pentoxazoline Pentoxazolin *n*
pentoxide Pentoxid *n*
pentulose Pentulose *f*
pentyl- Amyl-, Pentyl-
pentyl alcohol Pentylalkohol *m*
pentylamin Amylamin *n*
pentyl nitrite Amylnitrit *n*
pentyl oxide Amyläther *m*
pentyne Pentin *n*
peonine Päonin *n*
peonol Päonol *n*
peony [flower] Päonie *f* (Bot), Pfingstrose *f* (Bot)
pepper Pfeffer *m,* to season with ~ [ein]pfeffern
pepper alkaloid Pfefferalkaloid *n*
peppermint Pfefferminze *n* (Bot)
peppermint brandy Pfefferminzschnaps *m*
peppermint scent Pfefferminzgeruch *m*
peppermint smell Pfefferminzgeruch *m*
peppery pfefferartig, scharf, beißend
pepsin Pepsin *n* (Biochem), containing ~
    pepsinhaltig
pepsinogen Pepsinogen *n* (Biochem)
peptidase Peptidase *f* (Biochem)
peptide Peptid *n,* synthetic ~ synthetisches
    Peptid
peptide analysis Peptidanalyse *f*
peptide bond Peptidbindung *f*
peptide formation Peptidbildung *f*
peptide hydrolysis Peptidhydrolyse *f*
peptide separation Peptidtrennung *f*
peptide synthesis Peptidsynthese *f*
peptidoglycan Peptidoglykan *n*
peptidyl puromycin Peptidylpuromycin *n*
peptidyl transferase Peptidyltransferase *f*
    (Biochem)
peptization Peptisation *f*
peptize *v* peptisieren
peptone Pepton *n*
peptonic peptonhaltig
peptonization Peptonisierung *f*
peptonize *v* peptisieren, peptonisieren, in
    Pepton verwandeln
peptonoid peptonartig
peptotoxine Peptotoxin *n*
peracetic acid Peressigsäure *f*
peracid Persäure *f*
perautan Perautan *n*
perbenzoic acid Perbenzoesäure *f*
perborate Perborat *n*
perboric acid Perborsäure *f,* salt or ester of ~
    Perborat *n*
perbromic acid Perbromsäure *f*
perbutyric acid Perbuttersäure *f,* Butterpersäure *f*
percaine Percain *n*
percarbonate Perkarbonat *n*
percarbonic acid Perkohlensäure *f*

perceivable merklich, wahrnehmbar
perceive *v* wahrnehmen, merken
percent Prozent *n*, Hundertstel *n*, ~ **by volume**
Volum[en]prozent *n*, ~ **by weight**
gewichtsprozentig
percent *a* prozentig
percentage Prozentgehalt *m*, Hundertsatz *m*,
Prozentsatz *m*
percentage increase prozentuale Zunahme *f*
percentage purity Reinheitsgrad *m*
percent by volume volum[en]prozentig
perceptibility Wahrnehmbarkeit *f*
perceptible wahrnehmbar
perception Wahrnehmung *f*, Auffassung *f*,
Empfindung *f*, Empfindungsvermögen *n*,
Wahrnehmungsvermögen *n*, **limit of** ~
Wahrnehmungsgrenze *f*
perchlorate Perchlorat *n*, überchlorsaures Salz *n*
perchloric acid Perchlorsäure *f*, **monohydrated** ~
Perchlorsäuremonohydrat *n*, **salt or ester of** ~
Perchlorat *n*
perchloric anhydride Chlorheptoxid *n*,
Perchlorsäureanhydrid *n*
perchlorobenzene Perchlorbenzol *n*
perchlorobenzophenone Perchlorbenzophenon *n*
perchloroethane Perchloräthan *n*
perchloroethylene Perchloräthylen *n*
perchloromethyl mercaptan
Perchlormethylmercaptan *n*
perchromate Perchromat *n*
perchromic acid Perchromsäure *f*, **salt of** ~
Perchromat *n*
percolate *v* filtern, durchseihen, durchsickern,
filtrieren, kolieren, läutern
percolating Filtern *n*, Läutern *n*, Seihen *n*
percolating *a* läuternd
percolating water Sickerwasser *n*
percolation Filtern *n*, Durchfluß *m* (Geol),
Durchseihen *n*, Durchsickern *n*, Filtration *f*,
Perkolation *f*
percolation range Sickerungsstrecke *f*
percolator Perkolator *m*, Filtrierbeutel *m*,
Filtriersack *m*, Filtriertuch *n*, Seiher *m*
percussion Schlag *m*, Erschütterung *f*,
Perkussion *f*, Stoß *m*, **sensitivity to** ~
Schlagempfindlichkeit *f*
percussion borer Schlagbohrer *m*, Stoßbohrer *m*
percussion cap Sprengkapsel *f*, Zündhütchen *n*
percussion drill Stoßbohrer *m* (Bergb)
percussion figure Perkussionsfigur *f*,
Schlagfigur *f*
percussion force Durchschlagskraft *f*
percussion fuse Aufschlagzünder *m*
percussion powder Knallpulver *n*
percussion primer Perkussionszünder *m*
percussion priming Perkussionszündung *f*
percussion wave Stoßwelle *f* (Explosion)
percussive force Perkussionskraft *f*, Stoßkraft *f*
percutaneous perkutan

percylite Percylit *m* (Min)
pereirine Pereirin *f*
perezon Perezon *n*, Pipitzahoinsäure *f*
perfect einwandfrei, fehlerfrei, vollendet,
vollkommen
perfect *v* vervollkommnen, vollenden
perfect condition Idealzustand *m*
perfection Perfektion *f*, Vervollkommnung *f*,
Vollkommenheit *f*
perforate *v* perforieren, durchbohren,
durchlöchern, lochen
perforated löcherig, perforiert, durchbrochen
perforated basket centrifuge
Siebtrommelzentrifuge *f*
perforated belt Lochband *n*
perforated bottom Durchschlagboden *m*,
Nadelboden *m*
perforated brick Lochziegel *m*
perforated card system Hollerithsystem *n*
perforated cup Ambergsieb *n* (Histol)
perforated glass-fiber felts
Glasvlieslochbahnen *pl*
perforated metal ladle durchlöcherter
Metallöffel *m*
perforated plate Lochplatte *f*
perforated screen Lochblende *f*
perforated siren [disk] Lochsirene *f*
perforated tray Schlitzboden *m*
perforating machine Perforiermaschine *f*,
Lochmaschine *f*, Lochstanze *f*
perforation Durchlöcherung *f*, Perforation *f*
perform *v* verrichten, ausführen,
bewerkstelligen, leisten
performance Ausführung *f*, Erfüllung *f*,
Leistung *f* (Techn), Leistungsfähigkeit *f*
(Techn), Nutzeffekt *m* (Techn), Verhalten *n*,
Verrichtung *f*, Wirkungsweise *f*, **top** ~
Spitzenleistung *f*
performance characteristics
Gebrauchseigenschaften *f pl*
performance coefficient Leistungsfaktor *m*
performance level Leistungsbereich *m*
performance limit Leistungsgrenze *f*
performance margin Leistungsspielraum *m*
performance testing Leistungsfähigkeitsprobe *f*
performic acid Perameisensäure *f*,
Ameisenpersäure *f*
performic acid oxidation
Perameisensäureoxidation *f*
perfume Parfüm *n*, Duftstoff *m*, Wohlgeruch *m*
perfume *v* parfümieren, duften
perfumed wohlriechend, duftreich
perfuming Geruchsbehandlung *f*, Parfümieren *n*
perfuming pan Rauchpfanne *f*
perfusion Perfusion *f*
perfusion system Perfusionssystem *n*
pergenol Pergenol *n*
perhydrocarotene Perhydrocarotin *n*
perhydrogenate *v* perhydrieren

perhydrogenize v perhydrieren
perhydroindole Perhydroindol n
perhydrol Perhydrol n
perhydrolycopene Perhydrolycopin n
perhydronaphthalene Perhydronaphthalin n
perhydroretene Retenperhydrid n
periacenaphthindane Periacenaphthindan n
pericardium Herzbeutel m
periclase Periklas m (Min)
periclasite Periklas m (Min)
pericline Periklin m (Min)
pericondensed perikondensiert
pericyclocamphane Pericyclocamphan n
peridot Peridot m (Min), Chrysolith m (Min),
  Olivin m (Min)
perihelion Perihelium n (Astron)
perillaldehyde Perillaaldehyd m
perilla oil Perillaöl n
perillene Perillen n
perillic acid Perillasäure f
perillic alcohol Perillaalkohol m
perimeter Umfang m, Umkreis m
perimidine Perimidin n
perimidone Perimidon n
perinaphthane Perinaphthan n
perinaphthanone Perinaphthanon n
perinaphthoxazine Perinaphthoxazin n
period Periode f, Zeitraum m, **half a** ~ halbe
  Periode, **long** ~ lange Periode, **medium** ~
  mittlere Periode, **short** ~ kleine Periode, ~ **of**
  oscillation Schwingungsdauer f, ~ **of slow**
  combustion Schwelperiode f, ~ **of unloading**
  Löschfrist f, ~ **of working charge**
  Durchsetzzeit f
periodate Perjodat n
periodate degradation Perjodatabbau m
periodic acid Perjodsäure f, **salt of** ~ Perjodat n
periodical abwechselnd, periodisch
periodic arrangement Periodensystem n (Chem)
periodic boundary condition Randbedingung f
  der Periodizität
periodicity Frequenz f (Elektr), Periodizität f,
  degree of ~ Periodizitätsgrad m
periodicity condition Periodizitätsbedingung f
periodic parallelogram
  Periodenparallelogramm n
periodic system Periodensystem n (Chem),
  periodisches System n
periodic table Periodensystem n (Chem)
periodic volume Periodizitätsvolumen n
period indicator Periodenanzeiger m
period of service Betriebsperiode f
peripheral kernfern (Atom), peripher[isch]
peripheral velocity Umkreisgeschwindigkeit f
periphery Peripherie f, Umfang m, Umkreis m
periplocin Periplocin n
periplogenin Periplogenin n
perishable leicht verderblich, verderblich
perished abgestanden (Stahl)

peristalsis Peristaltik f (Med)
peristaltic pump Rollkolbenpumpe f,
  Schlauchpumpe f
peristaltin Peristaltin n
peristerite Peristerit m (Min)
peritectic peritektisch
peritectic period Peritektikum n
peritoneum Bauchfell n (Med)
periwinkle Immergrün n (Bot)
Perkin's reaction Perkinsche Reaktion f
perlite Perlit m (Min)
permanence Permanenz f, Beharrungszustand m,
  Dauerhaftigkeit f
permanent permanent, beständig, bleibend,
  dauerhaft, fortdauernd, haltbar, ständig
permanent finishing Hochveredlung f,
  Hochveredlungsmittel n (Text)
permanent green Permanentgrün n
permanent hardness Resthärte f (Wasser)
permanent load Dauerbeanspruchung f
permanently finished hochveredelt
permanent magnet Dauermagnet m
permanent magnetism permanenter
  Magnetismus m
permanent-magnet steel Dauermagnetstahl m
permanent mold Dauerform f
permanent properties Dauereigenschaften pl
permanent red Permanentrot n
permanent set bleibende Durchbiegung f,
  bleibende Verformung f
permanent set apparatus Dauerzugapparat m
permanent stress Dauerbeanspruchung f
permanent wave preparation
  Dauerwellenpräparat n
permanent white Permanentweiß n,
  Bariumsulfat n, Barytweiß n
permanganate Permanganat n
permanganic acid Permangansäure f,
  Übermangansäure f
permeability Permeabilität f,
  Durchdringbarkeit f, Durchlässigkeit f,
  Magnetisierungszahl f (Magn), **change of** ~
  Permeabilitätsänderung f, **reversible** ~
  reversible Permeabilität, umkehrbare
  Permeabilität, ~ **at low magnetizing forces**
  Permeabilität f bei kleinen Feldstärken (pl),
  ~ **for water** Wasserdruchlässigkeit f, ~ **of the**
  capillary walls Permeabilität f der
  Kapillarwände (f pl)
permeability coefficient
  Durchlässigkeitskoeffizient m
permeability for water Wasserdurchlässigkeit f
permeability test Durchlässigkeitsversuch m
permeable durchdringbar, durchlässig,
  permeabel, porös, ~ **to light rays**
  lichtdurchlässig
permeable to air luftdurchlässig
permeameter Durchlässigkeitsmesser m,
  Diffusionsgerät n

permeance magnetische Leitfähigkeit *f*,
Magnetisierungszahl *f*
permease Permease *f* (Biochem)
permeate *v* durchdringen, durchsetzen
permissible erlaubt, zulässig
permission Bewilligung *f*, Konzession *f*
permit Bewilligung *f*, Genehmigung *f*
permit *v*, ~ to make or sell under license
lizenzieren
permitted erlaubt
permittivity Dielektrizitätskonstante *f*
permselectivity Permselektivität *f*
permutation Permutation *f* (Math),
Vertauschung *f*
permutations and combinations
Anzahlfunktionen *pl*
permutator Kommutatorgleichrichter *m*
permute *v* permutieren (Math), vertauschen
(Math)
permutite process Permutit-Verfahren *n*
Pernambuco rubber Pernambukokautschuk *m*
Pernambuco wood Pernambukholz *n*
pernigraniline Pernigranilin *n*
pernitrosocamphor Pernitrosecampher *m*
peronine Peronin *n*
peropyrene Peropyren *n*
perosmic oxide Osmium(VIII)-oxid *n*
peroxidase Peroxidase *f* (Biochem),
Peroxydase *f* (Biochem)
peroxidation Oxidation *f* zum Peroxid,
Peroxidierung *f*
peroxide Peroxid *n*, Superoxid *n*, formation of ~
Peroxidbildung *f*
peroxide bleaching Oxidationsbleiche *f*, fastness
to ~ Peroxidbleichechtheit *f*
peroxide catalysis Peroxidkatalyse *f*
peroxide rearrangement Peroxid-Umlagerung *f*
peroxide sediment Superoxidschlamm *m*
peroxocobaltic complex Peroxokobaltkomplex *m*
peroxo-group Peroxogruppe *f*
peroxyacetic acid Peressigsäure *f*
peroxyacid Persäure *f*, Peroxysäure *f*
peroxybenzoic acid Perbenzoesäure *f*
peroxyborate Perborat *n*
peroxyboric acid Perborsäure *f*
peroxybutyric acid Butterpersäure *f*,
Perbuttersäure *f*
peroxycarbonic acid Perkohlensäure *f*
peroxychromic acid Perchromsäure *f*
peroxy derivative Peroxyderivat *n*
peroxydiphosphoric acid
Peroxydiphosphorsäure *f*
peroxydisulfuric acid Peroxydischwefelsäure *f*
peroxydol Peroxydol *n*
peroxyformic acid Perameisensäure *f*,
Ameisenpersäure *f*
peroxygenation Peroxygenierung *f*
peroxymonosulfuric acid
Peroxymonoschwefelsäure *f*
peroxyphosphate Peroxyphosphat *n*

peroxypropionic acid Perpropionsäure *f*
peroxy salt Persalz *n*
peroxysulfate Persulfat *n*
peroxysulfuric acid Carosche Säure *f*,
Perschwefelsäure *f*, salt or ester of ~
Persulfat *n*
perpendicular *a* senkrecht, lotrecht,
perpendikulär
perpendicular [line] Senkrechte *f*, Vertikale *f*
perphthalic acid Perphthalsäure *f*
perpropionic acid Perpropionsäure *f*
perrhenate Perrhenat *n*
perruthenate Perrutheniat *n*
perry Birnenwein *m*
persalt Persalz *n* (Chem)
perseulose Perseulose *f*
perseverance Ausdauer *f*, Beharrlichkeit *f*
persevere *v* beharren
Persian berry Kreuzdornbeere *f* (Bot)
Persian brown Persischbraun *n*
persico[t] Persiko *m* (Likör)
persist *v* beharren, fortbestehen
persistence Beharrlichkeit *f*, Beharrung *f*
persistence characteristic Abklingcharakteristik *f*
persistency Beharrlichkeit *f*
persistent ausdauernd, beharrlich
persistent spectrum Grundspektrum *n*
personal hygiene Körperpflege *f*
personify *v* verkörpern
personnel Belegschaft *f*, Personal *n*
perspective Perspektive *f*
perspective *a* perspektivisch
perspiration Ausdünstung *f* (Schweiß),
Schweiß *m*, Schweißbildung *f*, Schwitzen *n*
perspiration fastness Schweißechtheit *f*
perspiration-resistant schweißecht
perspire *v* ausdünsten, schwitzen
persulfate Persulfat *n*
persulfocyanic acid Persulfocyansäure *f*
persulfomolybdic acid Persulfomolybdänsäure *f*
persulfuric acid Perschwefelsäure *f*, Carosche
Säure *f*
perthiocarbonic acid Perthiokohlensäure *f*
perthite Perthit *m* (Min)
perthitic perthitähnlich
per thousand Promille *n*
pertinax Pertinax *n* (HN)
pertinent angemessen, sachgemäß
pertonal Pertonal *n*
perturbation Schwankung *f*, Störung *f*, magnetic
~ magnetische Störung
perturbation calculation Störungsrechnung *f*
perturbation energy Störungsenergie *f*
perturbation-insensitive störungsunempfindlich
perturbation method Störungsmethode *f*
perturbation-sensitive störungsempfindlich
perturbation theory Störungstheorie *f*
perturbed gestört
Peru balsam Perubalsam *m*

peruol Peruol *n*
Peruvian bark Chinarinde *f,* Fieberrinde *f,*
Perurinde *f,* **extract of** ~
Chinarindenextrakt *m*
peruvin Peruvin *n*
peruviol Peruviol *n,* Nerolidol *n*
pervade *v* durchdringen
pervious durchlässig, durchdringbar, undicht
(Techn), ~ **to water** wasserdurchlässig
perviousness Durchlässigkeit *f*
pervitin Pervitin *n*
perylene Perylen *n*
perylene quinone Perylenchinon *n*
pest Schädling *m* (Biol), Insekt *n* (Biol)
pest control Schädlingsbekämpfung *f*
pest control product
Schädlingsbekämpfungsmittel *n*
pesticide Schädlingsbekämpfungsmittel *n,*
Pflanzenschutzmittel *n*
pestilence Pestilenz *f,* Pest *f,* Seuche *f*
pestilential verpestend, ansteckend
pestle Mörserkeule *f,* Pistill *n* (Chem), Stößel *m*
pest[s] Ungeziefer *n*
petalite Petalit *m* (Min)
Petri dish Petrischale *f*
petrifaction Petrefakt *n,* Versteinerung *f*
petrifiable versteinerungsfähig
petrified versteinert
petrify *v* versteinern
petrifying Versteinern *n*
petrochemical industry Petrochemie *f*
petrochemistry Petrochemie *f*
petrol (Br. E.) Benzin *n,* Treibstoff *m,* **heavy** ~
Schwerbenzin *n,* **medium heavy** ~
Mittelbenzin *n*
petrolatum Petrolatum *n,* Vaselin *n,* Vaseline *f*
petrol engine Benzinmotor *m*
petroleum Petroleum *n,* Erdöl *n,* Mineralöl *n,*
containing ~ erdölhaltig, petroleumhaltig,
crude ~ Rohpetroleum *n,* **heating with** ~
Petroleumheizung *f*
petroleum chemistry Erdölchemie *f,*
Petrolchemie *f*
petroleum coke Ölkoks *m*
petroleum cracking process
Petroleumkrackverfahren *n*
petroleum distillate Erdöldestillat *n*
petroleum distilling apparatus
Petroleumdestillierapparat *m*
petroleum engine Petrolmotor *m*
petroleum ether Petroläther *m*
petroleum gas Petroleumgas *n*
petroleum hydrocarbons
Benzinkohlenwasserstoffe *pl*
petroleum injector Petroleumeinspritzer *m*
petroleum jelly Vaselin *n,* Vaseline *f*
petroleum pitch Petroleumasphalt *m*
petroleum plant Erdölanlage *f*
petroleum product Erdölerzeugnis *n*

petroleum refining plant Petroleumraffinerie *f*
petroleum spray Petroleumnebel *m*
petroleum spring Petroleumquelle *f*
petroleum stove Petroleumkochapparat *m,*
Petroleumkocher *m*
petroleum tester Petroleumprüfer *m*
petroleum vapor Petroleumdampf *m*
petrol gauge Benzinstandmesser *m*
petrolic acid Petrolsäure *f*
petroline Petrolin *n*
petrolize *v* petrolisieren, mit Petroleum
behandeln
petrol meter Kraftstoffmesser *m*
petrology Gesteinskunde *f*
petrol plant Benzinanlage *f*
petrol pump Benzinpumpe *f*
petrol substitute Benzinersatz *m*
petrol vapor Benzindampf *m*
petrophysical gesteinsphysikalisch
petroselaidic acid Petroselaidinsäure *f*
petroselinic acid Petroselinsäure *f*
petzite Petzit *m* (Min)
peucedanin Peucedanin *n,* Imperatorin *n*
pewter Zinn *n* (für Hausgeräte), Zinngerät *n,*
hard ~ Hartzinn *n,* ~ **for soldering** Lötzinn *n*
pewter *a* zinnern
phacolite Phakolit *m* (Min), Chabasit *m* (Min)
phages Phagen *pl*
phagocyte Phagozyt *m*
phagocytolysis Phagozytenzerfall *m*
phagocytosis Phagozytose *f*
phanodorm Phanodorm *n*
pharmaceutical Arznei *f,* Arzneimittel *n*
pharmaceutical *a* pharmazeutisch
pharmaceutical chemistry Pharmakochemie *f,*
Heilmittelchemie *f*
pharmaceutical industry Arzneimittelindustrie *f*
pharmaceutical legislation Arzneimittelgesetz *n*
pharmaceutical product Medikament *n* (Pharm)
pharmaceuticals Apothekerwaren *f pl*
pharmaceutical science Apothekerkunst *f*
pharmaceutics Arzneimittelkunde *f,*
Pharmazeutik *f*
pharmaceutist Apotheker *m,* Pharmazeut *m*
pharmacist Apotheker *m,* Pharmazeut *m*
pharmacochemistry Pharmakochemie *f*
pharmacokinetics Pharmakokinetik *f*
pharmacolite Pharmakolith *m* (Min),
Arsenikblüte *f* (Min)
pharmacological pharmakologisch
pharmacological research
Arzneimittelforschung *f*
pharmacologist Pharmakologe *m*
pharmacology Pharmakologie *f,*
Arzneimittelkunde *f,* Arzneimittellehre *f*
pharmacomania Arzneimittelsucht *f*
pharmacop[o]eia Pharmakopöe *f,* amtl.
Arzneibuch *n,*

pharmacosiderite Pharmakosiderit *m* (Min),
Würfelerz *n* (Min)
pharmacy Apotheke *f,* Arzneimittelkunde *f,*
Pharmazie *f*
pharynx Rachenhöhle *f*
phase Periode *f* (Elektr), Phase *f,* **artificial** ~
Hilfsphase *f,* **dispersed** ~ Dispersionsphase *f,*
disperse Phase, **in** ~ phasengleich, **initial** ~
Anfangsphase *f,* **intermediate** ~
Zwischenphase *f,* **opposite** ~ entgegengesetzte
Phase, **out of** ~ außer Phase, phasenungleich,
stationary ~ stationäre Phase
phase advance Phasenvoreilung *f*
phase angle Phasenwinkel *m*
phase balance Phasengleichheit *f*
phase behavio[u]r Phasenverhalten *n*
phase boundary Phasengrenze *f*
phase boundary potential Phasengrenzpotential *n*
phase change Phasenänderung *f,*
Phasenumwandlung *f*
phase changer Phasenschieber *m*
phase coincidence Phasengleichheit *f,*
Phasenübereinstimmung *f*
phase compensation Phasenausgleich *m*
phase-contrast image Phasenkontrastbild *n*
phase-contrast method
Phasenkontrastverfahren *n* (Mikroskopie)
phase-contrast microscope
Phasenkontrastmikroskop *n*
phase correction Phasenausgleich *m*
phase current Phasenstrom *m*
phased abgestuft
phase-dependent phasenabhängig
phase diagram Phasendiagramm *n,*
Zustandsdiagramm *n*
phase difference Phasendifferenz *f,*
Phasenunterschied *m,* Phasenverschiebung *f,*
angle of ~ Phasenverschiebungswinkel *m*
phase-difference microscope
Phasenkontrastmikroskop *n*
phase displacement Phasenverschiebung *f,*
Phasenunterschied *m,* **angle of** ~
Phasenverschiebungswinkel *m*
phase distortion Phasenverzerrung *f*
phase distribution Phasenverteilung *f*
phase equilibrium Phasengleichgewicht *n*
phase fading Phasenschwund *m*
phase grating Phasengitter *n*
phase indicator Phasenanzeiger *m* (Elektr)
phase inversion Phasenumkehr *f*
phase inverter Phasenumkehrer *m*
phase inverter [tube] Phasenumkehrröhre *f*
(Elektron)
phase jump Phasensprung *m*
phase lag Phasennacheilung *f,*
Phasenverzögerung *f,* **angle of** ~
Phasenverzögerungswinkel *m*
phase lead Phasenvoreilung *f,* **angle of** ~
Phasenvoreilungswinkel *m*

phasemeter Phasenmesser *m*
phaseolunatin Phaseolunatin *n*
phase opposition Gegenphasigkeit *f*
phase pattern Phasendiagramm *n*
phase regulation Phasenregulierung *f*
phase regulator Phasenregler *m*
phase resistance Phasenwiderstand *m*
phase resonance Phasenresonanz *f*
phase retardation Phasenverzögerung *f*
phase rule [Gibbssche] Phasenregel *f*
phases, difference of ~ Phasenunterschied *m,*
interconnection of ~ Phasenverkettung *f,*
interlinking of ~ Phasenverkettung *f,* **number**
of ~ Phasenzahl *f*
phase-sensitive phasenempfindlich
phase separation Phasenentmischung *f,*
Phasentrennung *f*
phase shift analysis Phasenwinkelanalyse *f*
phase shift control Phasenanschnittsteuerung *f*
phase shifter Phasenschieber *m*
phase shift[ing] Phasenverschiebung *f*
phase-shifting transformer
Synchron-Phasenschieber *m*
phase space Phasenraum *m*
phase titration Phasentitration *f*
phase transformation Phasentransformation *f*
phase transition Phasenübergang *m,*
Phasenumwandlung *f,* **first order** ~
Phasenumwandlung *f* erster Ordnung
phase transition point Phasenwechselpunkt *m*
phase unbalance Phasenungleichheit *f*
phase velocity Phasengeschwindigkeit *f*
phase voltage Phasenspannung *f*
phasotron Phasotron *n*
pheanthine Phäanthin *n*
pH electrode pH-Elektrode *f*
phellandral Phellandral *n*
phellandrene Phellandren *n*
phellandric acid Phellandrinsäure *f*
phellogenic acid Phellogensäure *f*
phellonic acid Phellonsäure *f*
phenacaine Holocain *n*
phenacemide Phenacemid *n*
phenacetein Phenacetolin *n*
phenacetin Phenacetin *n*
phenacetolin Phenacetolin *n*
phenacetornithuric acid Phenacetornithursäure *f*
phenaceturic acid Phenacetursäure *f*
phenacite Phenakit *m* (Min)
phenacyl Phenacyl-
phenacylamine Phenacylamin *n*
phenacyl bromide Phenacylbromid *n*
phenalan Phenalan *n*
phenalene Phenalen *n*
phenamine Phenamin *n,* Phenokoll *n*
phenamine blue Phenaminblau *n*
phenanthrene Phenanthren *n*
phenanthrene dibromide Phenanthrendibromid *n*
phenanthrene quinone Phenanthrenchinon *n*

phenanthrenesulfonic acid
 Phenanthrensulfonsäure *f*
phenanthridine Phenanthridin *n*
phenanthridone Phenanthridon *n*
phenanthrol Phenanthrol *n*
phenanthroline Phenanthrolin *n*
phenanthrone Phenanthron *n*
phenanthrophenazine Phenophenanthrazin *n*
phenanthroylpropionic acid
 Phenanthroylpropionsäure *f*
phenanthrylamine Phenanthrylamin *n*
phenarsazine Phenarsazin *n*
phenate Phenolat *n*
phenate of lime Carbolkalk *m*
phenazine Phenazin *n*
phenazine methosulfate Phenazinmethosulfat *n*
phenazone Phenazon *n*
phenazonium hydroxide Phenazoniumhydroxid *n*
phendioxin Phendioxin *n*
pheneserine Pheneserin *n*
phenethyl Phenäthyl-
phenetidine Phenetidin *n*
phenetole Phenyläthyläther *m*,
 Äthylphenyläther *m*, Phenetol *n*
phengite Phengit *m* (Min), Fensterglimmer *m*
 (Min), Muskovit *m* (Min)
phenhomazine Phenhomazin *n*
phenic acid Phenol *n*, Carbolsäure *f*
phenicarbazide Phenicarbazid *n*
phenicin Phönizin *n*
phenicite Phönizit *m*
pheniodol Pheniodol *n*
phenmorpholine Phenmorpholin *n*
phenobarbital Phenobarbital *n*, Luminal *n* (HN)
phenobutiodil Phenobutiodil *n*
phenocoll Phenokoll *n*, Phenamin *n*
phenocoll chloride Phenokollchlorid *n*
phenocoll salicylate Phenokollsalicylat *n*
phenocryst Einsprengling *m* (Krist)
phenocyanin Phenocyanin *n*
phenol Phenol *n*, Carbol *n*, Carbolsäure *f*,
 Oxybenzol *n*, **aqueous solution of** ~
 Carbolwasser *n*
phenol aceteine Phenolacetein *n*
phenol apparatus Phenolapparat *m*
phenolarsonic acid Phenolarsonsäure *f*
phenolate Phenolat *n*
phenol content Phenolgehalt *m*
phenoldisulfonic acid Phenoldisulfonsäure *f*
phenol dye Phenolfarbstoff *m*
phenol formaldehyde resin
 Phenolformaldehydharz *n*
phenol glycoside Phenolglykosid *n*
phenol homolog[ue] Phenolhomologe[s] *n*
phenol hydroxyl Phenolhydroxyl *n*
phenolic phenolisch
phenolic adhesive Phenolharzklebstoff *m*,
 Phenolharzleim *m*
phenolic cement Phenolharzkleber *m*

phenolic foam Phenolharzschaum *m*
phenolic laminated sheet
 Phenolharzschichtstoff *m*
phenolic molding compound or composition
 Phenolharzpreßmischung *f*
phenolic plastic Phenolharzkunststoff *m*
phenolic resin Phenolharz *n*, Phenoplast *m*
phenolic varnish Phenolharzlack *m*
phenol ointment Carbolsalbe *f* (Pharm)
phenol oxidase Phenoloxidase *f* (Biochem)
phenolphthalein Phenolphthalein *n*
phenolphthalein paper Phenolphthaleinpapier *n*
phenol red Phenolrot *n*
phenolsulfonate Phenolsulfonat *n*
phenolsulfonic acid Phenolsulfonsäure *f*
phenoluria Phenolausscheidung *f* (im Harn)
 (Med)
phenomenological[ly] phänomenologisch
phenomenon Phänomen *n*, Erscheinung *f*,
 accompanying ~ Begleiterscheinung *f*,
 aperiodic ~ aperiodischer Vorgang, **attendant**
 ~ Begleiterscheinung *f*, Nebenerscheinung *f*,
 reverse ~ Umkehrerscheinung *f* (Phot),
 secondary ~ Begleiterscheinung *f*,
 Nebenerscheinung *f*, ~ **caused by light**
 Lichterscheinung *f*, ~ **of diffraction**
 Beugungserscheinung *f*
phenonaphthacridine Phenonaphthacridin *n*
phenonaphthazine Phenonaphthazin *n*,
 Naphthophenazine *n*
phenonium ion Phenonium-Ion *n*
phenophenanthrazine Phenophenanthrazin *n*
phenophosphazinic acid Phenophosphazinsäure *f*
phenoplast Phenoplast *n*, Phenolharz *n*
phenopyrine Phenopyrin *n*
phenoquinone Phenochinon *n*
phenosafranine Phenosafranin *n*, Safranin *n*
phenose Phenose *f*
phenostal Phenostal *n*
phenosuccin Phenosuccin *n*, Pyrantin *n*
phenothiazine Phenothiazin *n*, Phenthiazin *n*,
 Thiodiphenylamin *n*
phenothioxin Phenoxthin *n*
phenotype Phänotypus *m*, Erscheinungsform *f*
phenoval Phenoval *n*
phenoxarsine Phenoxarsin *n*
phenoxathiin Phenox[a]thiin *n*
phenoxazine Phenoxazin *n*, Naphthoxazin *n*
phenoxide Phenolat *n*
phenoxin Kohlenstofftetrachlorid *n*
phenoxthine Phenoxthin *n*
phenoxyacetic acid Phenoxyessigsäure *f*
phenoxypropyl bromide Phenoxypropylbromid *n*
phenpiazine Chinoxalin *n*
phenselenazine Phenselenazin *n*
phenselenazone Phenselenazon *n*
phenthiazine Phenthiazin *n*
phenurone Phenuron *n*
phenyl Phenyl-

phenylacetaldehyde Phenylacetaldehyd *m*
phenylacetamide Phenylacetamid *n*
phenylacetic acid Phenylessigsäure *f*
phenylacetone Phenylaceton *n*
phenylacetonitrile Benzylcyanid *n*,
 Phenylessigsäurenitril *n*
phenylaceturic acid Phenylacetursäure *f*
phenylacetyl chloride Phenylessigsäurechlorid *n*
phenylacridine Phenylacridin *n*
phenylacrolein Phenylakrolein *n*
phenylalanine Phenylalanin *n*
phenylamine Anilin *n*, Phenylamin *n*
phenylaminoazobenzene
 Phenylamidoazobenzol *n*
phenylaniline Diphenylamin *n*
phenylarsine Phenylarsin *n*
phenylate Phenylat *n*, Phenolat *n*
phenylate *v* phenylieren
phenyl azide Phenylazid *n*, Triazobenzol *n*
phenylbenzoate Phenylbenzoat *n*
phenylbenzoquinone Phenylbenzochinon *n*
phenylboric acid Benzolboronsäure *f*
phenyl bromide Brombenzol *n*
phenyl brown Phenylbraun *n*
phenylbutadiene Phenylbutadien *n*
phenylbutene Phenylbuten *n*
phenylbutyric acid Phenylbuttersäure *f*
phenylcacodyl Phenylkakodyl *n*
phenylcaproic acid Phenylcapronsäure *f*
phenylcarbamic acid Carbanilsäure *f*
phenylcarbinol Benzylalkohol *m*
phenyl carbonate Diphenylcarbonat *n*
phenyl chloride Chlorbenzol *n*, Phenylchlorid *n*
phenylchloroform Benzotrichlorid *n*,
 Phenylchloroform *n*
phenylchlorosilane Phenylchlorsilan *n*
phenylcyanate Phenylcyanat *n*
phenyldiamine Phenyldiamin *n*
phenyldithiocarbamate
 Phenyldithiocarbaminat *n*
phenyldithiocarbamic acid
 Phenyldithiocarbaminsäure *f*
phenylene Phenylen-
phenylene blue Phenylenblau *n*, Indamin *n*
phenylene brown Phenylenbraun *n*
phenylenediamine Diaminobenzol *n*,
 Phenylendiamin *n*
phenylene diazosulfide Phenylendiazosulfid *n*
phenylene disulfide Phenylendisulfid *n*
phenylene residue Phenylenrest *m*
phenylene sulfonylide Phenylensulfonylid *n*
phenylene urea Phenylenharnstoff *m*
phenyl ester Phenolester *m*
phenylethanol Phenyläthylalkohol *m*
phenylether Phenoläther *m*
phenylethyl alcohol Phenyläthylalkohol *m*,
 Benzylcarbinol *n*
phenylethyl amine Phenyläthylamin *n*
phenylethylene Styrol *n*, Cinnamen *n*

phenylethyl ether Phenyläthyläther *m*, Phenetol *n*
phenylethyl formamide Phenyläthylformamid *n*
phenyl fatty acid Phenylfettsäure *f*
phenylfluorone Phenylfluoron *n*
phenylformamide Formanilid *n*
phenylglycine Phenylglycin *n*
phenylglycocoll Phenylglycin *n*
phenylglycolic acid Phenylglykolsäure *f*
phenylglycollic acid Mandelsäure *f*
phenylheptatrienal Phenylheptatrienal *n*
phenylhydrazine Phenylhydrazin *n*
phenylhydrazinesulfonic acid
 Phenylhydrazinsulfosäure *f*
phenylhydrazone Phenylhydrazon *n*
phenylhydroxylamine Phenylhydroxylamin *n*
phenyl iodochloride Phenyljodidchlorid *n*
phenylisopropylamine Phenylisopropylamin *n*
phenylisopropylamine sulfate
 Phenylisopropylaminsulfat *n*
phenylisothiocyanate Phenylisothiocyanat *n*,
 Phenylsenföl *n*
phenylketonuria Phenylketonurie *f* (Med)
phenyl lithium Phenyllithium *n*
phenyl magnesium bromide
 Phenylmagnesiumbromid *n*
phenylmalonate Phenylmalonat *n*
phenylmethylketone Acetophenon *n*,
 Phenylmethylketon *n*
phenylmethylpyrazolone
 Phenylmethylpyrazolon *n*
phenyl mustard oil Phenylsenföl *n*
phenylnaphthylamine Phenylnaphthylamin *n*
phenylnitroethane Phenylnitroäthan *n*
phenylnitromethane Phenylnitromethan *n*
phenyl-n-propyl ketone Butyrophenon *n*
phenylone Phenylon *n*
phenyl orange Phenylorange *n*
phenylosazone Phenylosazon *n*
phenyloxamic acid Oxanilsäure *f*
phenylparaconic acid Phenylparaconsäure *f*
phenylpentadienal Phenylpentadienal *n*
phenylphenazonium dye
 Phenylphenazoniumfarbstoff *m*
phenylpiperidine Phenylpiperidin *n*
phenylpolyenal Phenylpolyenal *n*
phenylpropanolamine Phenylpropanolamin *n*
phenylpropenal, 3- ~ Cinnamylaldehyd *m*
phenylpropiolic acid Phenylpropiolsäure *f*
phenylpropionate Phenylpropionat *n*
phenylpropionic acid Phenylpropionsäure *f*
phenylpropylmalonic acid
 Phenylpropylmalonsäure *f*
phenylpyridine Phenylpyridin *n*
phenyl pyruvate Phenylpyruvat *n*
phenylsalicylate Phenylsalicylat *n*, Salol *n*
phenylserine Phenylserin *n*
phenyl silicate Phenylsilikat *n*
phenylsilicon chloride Phenylsiliciumchlorid *n*
phenyl silicone Phenylsilicon *n*

phenylsulfamic acid Phenylsulfaminsäure *f*
phenylsulfuric acid Phenylschwefelsäure *f*
phenylthiohydantoic acid
Phenylthiohydantoinsäure *f*
phenylthiohydantoin Phenylthiohydantoin *n*
phenyltoluenesulfonate Phenyltoluolsulfonat *n*
phenyltolylketone Tolylphenylketon *n*
phenyltrichlorosilane Phenyltrichlorsilan *n*
phenylurea Phenylharnstoff *m*
phenylurethane Phenylurethan *n*, Euphorin *n*
phenylvaleric acid Phenylvaleriansäure *f*
pheochlorophyll Phäochlorophyll *n*
pheophorbide Phäophorbid *n*
pheophorbine Phäophorbin *n*
pheophytine Phäophytin *n*
pheoporphyrin Phäoporphyrin *n*
pheromone Pheromon *n*
pH gradient pH-gradient *m*
phial Flakon *m n*, kleine Flasche *f*,
Medizinglas *n*, Phiole *f*
philadelphite Philadelphit *m* (Min)
phillipsite Phillipsit *m* (Min), Kalkharmotom *m*
(Min)
phillygenine Phillygenin *n*
philosopher's wool (obs) Philosophenwolle *f*,
Zinkoxid *n*
phlegmatize *v* (to make e.g. organic peroxides
less liable to decomposition by means of inert
compounds) phlegmatisieren
phlobaphene Phlobaphen *n*
phlogiston theory Phlogistontheorie *f*
phlogopite Amberglimmer *m* (Min),
Phlogopit *m* (Min), Rhombenglimmer *m*
(Min)
phloionic acid Phloionsäure *f*
phloionolic acid Phloionolsäure *f*
phloracetophenone Phloracetophenon *n*
phloretic acid Phloretinsäure *f*
phloretin Phloretin *n*
phlorizin Phlorizin *n*
phloroacetophenone Phloroacetophenon *n*
phlorol Phlorol *n*
phloroquinyl Phlorchinyl-
phlorrhizin Phlorrhizin *n*
phloxin Phloxin *n*
pH-measurement pH-Messung *f*
pH-meter pH-Meßgerät *n*, pH-Meter *n*,
Potentiometer *n*
phocenin Phocaenin *n*
pholedrine Pholedrin *n*
pholerite Pholerit *m* (Min)
phon Phon *n*
phonolite Phonolith *m* (Min), Klingstein *m*
(Min)
phonometer Schallmeßgerät *n*
phonometry Schallmessung *f*
phonon Phonon *n*, Schallquantum *n*
phonon entropy Schallquantenentropie *f*
pH optimum pH-Optimum *n*

phorone Phoron *n*
phoronic acid Phoronsäure *f*
phosgene Phosgen *n*, Carbonylchlorid *n*,
Kohlensäurechlorid *n*
phosgene decomposition Phosgenzerfall *m*
phosgene formation Phosgenbildung *f*
phosgenite Phosgenit *m* (Min), Bleihornerz *n*
(Min), Hornbleierz *n* (Min)
phosocresol Phosokresol *n*
phosphagen Phosphagen *n*
phosphanthrene Phosphanthren *n*
phosphatase Phosphatase *f* (Biochem), **alkaline**
~ alkalische Phosphatase
phosphate Phosphat *n*, phosphorsaures Salz *n*,
high-temperature ~ Glühphosphat *n*, **primary**
~ primäres Phosphat *n*,
Dihydrogenphosphat *n*, **secondary** ~
sekundäres Phosphat,
Monohydrogenphosphat *n*, **tertiary** ~
tertiäres Phosphat
phosphate *v* phosphatieren
phosphate acidity Phosphatacidität *f*
phosphate bath Phosphatbad *n*
phosphate bearing phosphatführend
phosphate bond, energy-rich ~ energiereiche
Phosphatbindung *f*
phosphate buffer Phosphatpuffer *m*
phosphate chalk Phosphatkreide *f*
phosphate coating Phosphatüberzug *m*
phosphate fertilizer Phosphatdüngemittel *n*,
Phosphatdünger *m*
phosphate mill Phosphatmühle *f*
phosphate plant Phosphatanlage *f*
phosphate powder Phosphatmehl *n*
phosphate slag Thomas-Schlacke *f*
phosphate transfer Phosphatübertragung *f*
phosphatic phosphathaltig, phosphatisch
phosphatide Phosphatid *n*, Phospholipoid *n*
phosphatidyl choline Phosphatidylcholin *n*
phosphatidyl ethanolamine
Phosphatidyläthanolamin *n*
phosphatidyl serine Phosphatidylserin *n*
phosphatiferous phosphatführend
phosphating Phosphatierung *f*
phosphatization Phosphatierung *f*
phosphatize *v* phosphatieren (Techn)
phosphazobenzene Phosphazobenzol *n*
phosphenic acid Phosphensäure *f*
phosphenyl Phosphenyl-
phosphenyl chloride Phosphenylchlorid *n*
phosphenylic acid Phosphenylsäure *f*
phosphide Phosphid *n*
phosphine Phosphin *n*, Phosphorwasserstoff *m*
phosphite Phosphit *n*, phosphorigsaures Salz *n*
phosphoarginine Phosphoarginin *n*
phosphobacterium Leuchtbakterium *n*
phosphocerite Phosphocerit *m* (Min),
Rhabdophan *m* (Min)

phosphocholine Phosphocholin *n*
phosphocreatine Phosphokreatin *n*
phosphodiesterase Phosphodiesterase *f*
(Biochem)
phosphoenolpyruvic acid
Phosphoenolbrenztraubensäure *f*
phosphofructokinase Phosphofructokinase *f*
(Biochem)
phosphoglucoisomerase
Phosphoglucoisomerase *f* (Biochem)
phosphoglucomutase Phosphoglucomutase *f*
(Biochem)
phosphogluconic acid Phosphogluconsäure *f*
phosphoglyceraldehyde
Phosphoglycerinaldehyd *m*
phosphoglycerate kinase
Phosphoglyceratkinase *f* (Biochem)
phosphoglyceric acid Phosphoglycerinsäure *f*
phosphoglyceromutase Phosphoglyceratmutase *f*
(Biochem)
phosphohomoserine Phosphohomoserin *n*
phosphokinase Phosphokinase *f* (Biochem)
phospholipase Phospholipase *f* (Biochem)
phospholipid[e] Phospholipoid *n*, Phosphatid *n*
phosphomolybdic acid Phosphormolybdänsäure *f*
phosphonic acid Phosphonsäure *f*
phosphonium base Phosphoniumbase *f*
phosphonium chloride Phosphoniumchlorid *n*
phosphonium iodide Phosphoniumjodid *n*,
Jodphosphonium *n*
phosphoprotein Phosphoprotein *n*,
Phosphorproteid *n*
phosphorate *v* mit Phosphor verbinden *m*,
phosphorisieren
phosphorated oil Phosphoröl *n* (Pharm)
phosphor bronze Phosphorbronze
phosphorchalcite Phosphorchalcit *m* (Min)
phosphoresce *v* phosphoreszieren, nachleuchten
phosphorescence Chemolumineszenz *f* (Chem),
Phosphoreszenz *f*
phosphorescence spectrum
Phosphoreszenzspektrum *n*
phosphorescent phosphoreszierend
phosphorescent stone Leuchtstein *m*
phosphorescent substance Leuchtmaterial *n*
phosphoretted hydrogen Phosphorwasserstoff *m*
phosphoribitol Phosphoribit *m*
phosphoribomutase Phosphoribomutase *f*
phosphoribose Phosphoribose *f*
phosphoric phosphorhaltig
phosphoric acid Phosphorsäure *f*, salt or ester of
~ Phosphat *n*, syrupy ~ sirupöse
Phosphorsäure
phosphoric acid plant Phosphorsäureanlage *f*
phosphoric anhydride Phosphorpentoxid *n*,
Phosphorsäureanhydrid *n*
phosphoric bromide Phosphorpentabromid *n*,
Phosphor(V)-bromid *n*

phosphoric chloride Phosphorpentachlorid *n*,
Phosphor(V)-chlorid *n*
phosphoric ester Phosphorsäureester *m*
phosphoric iodide Phosphor(V)-jodid *n*
phosphoric oxide Phosphor(V)-oxid *n*,
Phosphorpentoxid *n*,
Phosphorsäureanhydrid *n*
phosphoric pig iron Phosphorroheisen *n*
phosphoric sulfide Phosphorpentasulfid *n*,
Phosphor(V)-sulfid *n*
phosphorite Phosphorit *m* (Min), Apatit *m*
(Min)
phosphorize *v* mit Phosphor verbinden *m*,
phosphorisieren
phosphorogenic Phosphoreszenz erzeugend
phosphorography Phosphorographie *f*
phosphorolysis Phosphorolyse *f*
phosphoronitridic acid Nitrilophosphorsäure *f*
phosphorous phosphorig[sauer]
phosphorous acid phosphorige Säure *f*
phosphorous anhydride
Phosphorigsäureanhydrid *n*,
Phosphortrioxid *n*
phosphorous bromide Phosphor(III)-bromid *n*,
Phosphortribromid *n*
phosphorous chloride Phosphor(III)-chlorid *n*,
Phosphortrichlorid *n*
phosphorous iodide Phosphor(III)-jodid *n*
phosphorous oxide Phosphortrioxid *n*
phosphorous sulfide Phosphortrisulfid *n*,
Phosphorsesquisulfid *n*
phosphorous trihalide Phosphortrihalogen *n*
phosphoruranylite Phosphoruranylit *n*
phosphorus (Symb. P) Phosphor *m*, containing ~
phosphorhaltig, determination of ~
Phosphorbestimmung *f*, elimination of ~
Phosphorabscheidung *f*, free from ~
phosphorfrei, red ~ roter Phosphor,
separation of ~ Phosphorabscheidung *f*, violet
~ violetter Phosphor, white or yellow ~
farbloser Phosphor
phosphorus content Phosphorgehalt *m*
phosphorus crystal Phosphorkristall *m*
phosphorus hydride Phosphorin *n*,
Phosphorwasserstoff *m*
phosphorus inflammation Phosphorentzündung *f*
phosphorus intake Phosphoraufnahme *f*
phosphorus match Phosphorzündholz *n*
phosphorus metabolism Phosphorstoffwechsel *m*
(Physiol)
phosphorus nitride Phosphornitrid *n*
phosphorus oxychloride Phosphoroxychlorid *n*
phosphorus paste Phosphormasse *f*,
Phosphorpaste *f*
phosphorus pentabromide
Phosphorpentabromid *n*,
Pentabromphosphor *m*,
Phosphor(V)-bromid *n*

phosphorus pentachloride
  Phosphorpentachlorid *n*,
  Phosphor(V)-chlorid *n*
phosphorus pentahalide
  Phosphorpentahalogenid *n*
phosphorus pentaiodide Phosphor(V)-jodid *n*
phosphorus pentasulfide Phosphorpentasulfid *n*,
  Phosphor(V)-sulfid *n*
phosphorus pentoxide Phosphorpentoxid *n*,
  Phosphorsäureanhydrid *n*,
  Phosphor(V)-oxid *n*
phosphorus poisoning Phosphorvergiftung *f*
  (Med)
phosphorus powder Phosphorpulver *n*
phosphorus sesquisulfide Phosphorsesquisulfid *n*,
  Phosphortrisulfid *n*
phosphorus spoon Phosphorlöffel *m*
phosphorus spy Phosphorspion *m*
phosphorus steel Phosphorstahl *m*
phosphorus sulfobromide
  Phosphorsulfobromid *n*
phosphorus sulfochloride Phosphorsulfochlorid *n*
phosphorus tribromide Phosphor(III)-bromid *n*,
  Phosphortribromid *n*, Tribromphosphor *m*
phosphorus trichloride Phosphor(III)-chlorid *n*,
  Phosphortrichlorid *n*
phosphorus triiodide Phosphor(III)-jodid *n*
phosphorus trioxide Phosphortrioxid *n*,
  Phosphorigsäureanhydrid *n*,
  Phosphor(III)-oxid *n*
phosphorus trisulfide Phosphortrisulfid *n*,
  Phosphorsesquisulfid *n*
phosphorus vapor Phosphordampf *m*
phosphoryl Phosphoryl-
phosphorylase Phosphorylase *f* (Biochem)
phosphorylase kinase Phosphorylasekinase *f*
  (Biochem)
phosphorylase phosphatase
  Phosphorylasephosphatase *f* (Biochem)
phosphorylation Phosphorylierung *f*, **oxidative ~**
  oxidative Phosphorylierung
phosphoryl chloride Phosphorylchlorid *n*,
  Phosphoroxychlorid *n*
phosphoserine Phosphoserin *n*
phosphosiderite Phosphosiderit *m* (Min)
phosphosphingoside Phosphosphingosid *n*
phosphotartaric acid Phosphorweinsäure *f*
phosphotransacetylase Phosphotransacetylase *f*
  (Biochem)
phosphotransferase Phosphotransferase *f*
  (Biochem)
phosphotungstate Phosphorwolframat *n*
phosphotungstic acid Phosphorwolframsäure *f*
phostamic acid Phostamsäure *f*
phostonic acid Phostonsäure *f*
phot Phot *n* (Opt)
photene Photen *n*
photo Photographie *f*, Lichtbild *n*, Photo *n*,
  Photoabzug *m*

photoabsorption band Photoabsorptionsband *n*
photo-active photosensitiv
photobacteria Leuchtbakterien *pl*
photobiolgy Photobiologie *f*
photocatalysis Photokatalyse *f*
photocathode Kathode *f* einer Photozelle,
  Photokathode *f*, Rasterkathode *f* (Telev)
photocell Photozelle *f* (Elektr), **gas-filled ~**
  gasgefüllte Photozelle
photocenter Lichtzentrum *n*
photochemical photochemisch, lichtchemisch
photochemical isomerization
  Photo-Isomerisierung *f*
photochemistry Photochemie *f*
photochlorination Lichtchlorierung *f*,
  Photochlorierung *f*
photochromatic photochromatisch
photoconduction Photoleitung *f*, photoelektrische
  Leitung *f*
photoconductive effect innerer Photoeffekt *m*
photoconductive sensitivity
  Photoleitempfindlichkeit *f*
photoconductivity lichtelektrische Leitfähigkeit *f*,
  Photoleitfähigkeit *f*
photoconductivity cell Photoleitfähigkeitszelle *f*
photoconductor Photoleiter *m*
photocopier Fotokopiergerät *n*
photocopy Fotokopie *f*, Lichtpause *f*
photocopying apparatus Fotokopiergerät *n*
photocopy[ing] paper Lichtpauspapier *n*
photocurrent Photoemissionsstrom *m*
photocurrent decay Photostromabfall *m*
photocurrent stimulation Photostromanregung *f*
photodeuteron Photodeuteron *n*
photodielectric lichtdielektrisch,
  photodielektrisch
photodisintegration Kernphotoeffekt *m*
photodissociation Photolyse *f*, Photozersetzung *f*
photodynamic photodynamisch
photoelastic spannungsoptisch
photoelasticity Spannungsoptik *f*
photoelectric lichtelektrisch, photoelektrisch
photoelectrical effect, internal ~ innerer
  Photoeffekt *m*
photo[electric] cell Photozelle *f*, lichtelektrische
  Zelle *f*
photoelectric cell amplifier
  Photozellenverstärker *m*
photo[electric] current Photostrom *m*
photo[electric] effect, external ~ äußerer
  lichtelektrischer Effekt *m*, äußerer
  Photoeffekt *m*, **inner** ~ innerer Photoeffekt *m*
photoelectric emission äußerer Photoeffekt *m*,
  lichtelektrische Elektronemission *f*
photoelectricity Photoelektrizität *f*,
  Lichtelektrizität *f*
photoelectric limiting potential lichtelektrisches
  Grenzpotential *n*

photoelectric threshold lichtelektrischer Grenzwert *m*, lichtelektrischer Schwellenwert *m*
photo[electric] tube lichtelektrische Zelle *f*, Photozelle *f*
photoelectric yield photoelektrisches Emissionsvermögen *n*
photoelectromagnetic photoelektromagnetisch
photoelectron Photoelektron *n*, Leuchtelektron *n*
photoelectron multiplier Photoelektronenvervielfacher *m*
photoemission Photoemission *f*
photoemission current Photoemissionsstrom *m*
photoemissive cell Photozelle *f* mit äußerem lichtelektrischen Effekt
photoemissive effect äußerer Photoeffekt *m*
photoemissive gas-filled cell Glimmzelle *f*
photoengraving Chemigraphie *f*
photofission Photospaltung *f*
photofluorography Photofluorographie *f*
photogalvanography Photogalvanographie *f*
photogen Photogen *n*
photogenic lichterzeugend, photogen
photogen lamp Photogenlampe *f*
photoglow tube Glimmzelle *f*
photogrammetry Photogrammetrie *f*, Meßbildverfahren *n*, Phototopographie *f*
photograph Photographie *f*, Lichtbild *n*, **wide-angle** ~ Weitwinkelaufnahme *f* (Phot)
photograph *v* photographieren
photographer Photograph *m*
photographic photographisch
photographic film speed Empfindlichkeit *f* des photographischen Films
photographic paper Photopapier *n*
photographic picture Lichtbild *n*, Photographie *f*
photographic plate Photoplatte *f*
photographic plate photometer Schwärzungsphotometer *n*
photographic printing Lichtdruck *m*
photographic tracing Lichtpause *f*
photography Photographie *f*, Lichtbildkunst *f*
photogravure Photogravüre *f*, Lichtkupferdruck *m*
photogravure printing machine Tiefdruckmaschine *f*
photohalide Photohalogen *n*
photoinitiator Photoinitiator *m*
photoionization Photoionisierung *f*
photoionization efficiency Photoionisierungsausbeute *f*
photolithographic photolithographisch
photolithography Photolithographie *f*, Lichtsteindruck *m*
photoluminescence Photolumineszenz *f*
photolysis Photolyse *f*
photolytic cell Elektrolytzelle *f*
photomagnetic photomagnetisch

photomagnetism Photomagnetismus *m*, Lichtmagnetismus *m*
photomechanical color printing Lichtfarbendruck *m*
photomechanical printing Lichtdruck *m*
photomeson Photomeson *n*
photometer Photometer *n*, Belichtungsmesser *m*, Licht[stärke]messer *m*, Luxmeter *n*
photometer lamp Photometerlampe *f*
photometric photometrisch
photometrically, to measure ~ photometrieren
photometry Photometrie *f*, Lichtstärkemessung *f*
photomicrograph Mikroaufnahme *f*, Mikrophoto[graphie f] *n*
photomicrography Photomikrographie *f*, Mikrophotographie *f*
photomultiplier Photoverstärker *m*
photon Photon *n*, Lichtquant *n*
photon absorption Photonenabsorption *f*
photoneutron Photoneutron *n*
photonitrosation Photonitrosierung *f*
photonuclear effect Kernphotoeffekt *m* (Atom)
[photo]nuclear interaction Kernwechselwirkung *f*
photonuclear reaction Kernphotoreaktion *f*
photonucleon Photonukleon *n*
photooxidation Photooxidation *f*
photooximation Photooximierung *f*
photooxygenation Photooxygenierung *f*
photophilous lichtliebend (Bakt)
photophoresis Photophorese *f*
photophosphorylation Photophosphorylierung *f*
photopolymerization Photopolymerisation *f*, lichtinduzierte Polymerisation *f*
photoprint photographischer Abzug *m*, Photokopie *f*
photoproton Photoproton *n*
photoreaction Photoreaktion *f*
photoreactor Photoreaktor *m*
photoreduction Photoreduktion *f*
photoresistance lichtelektrischer Widerstand *m*; Photowiderstand *m*
photoresistive cell Widerstandsphotozelle *f*
photorespiration Photoatmung *f*
photosantonic acid Photosantonsäure *f*
photosantonin Photosantonin *n*, Chromosantonin *n*
photoscope Photoskop *n*
photosensitive lichtempfindlich, photosensitiv
photosensitive surface Photoschicht *f*
photosensitivity Lichtempfindlichkeit *f*
photosensitize *v* lichtempfindlich machen
photosensitizer Photosensibilisator *m*, Sensibilisator *m*
photosphere Photosphäre *f*
photostable lichtbeständig, lichtunempfindlich
photostat Lichtpause *f*
photo supplies photographische Bedarfsartikel *pl*
photosynthesis Photosynthese *f*
photosynthetic photosynthetisch

phototaxis Phototaxis *f*
phototherapy Lichttherapie *f*
phototimer Schaltuhr *f*
phototropic phototrop[isch]
phototropism Phototropie *f*
phototropy Phototropie *f*
phototube Photoelektronenröhre *f*
phototypography Phototypographie *f*
phototypy Phototypie *f*, Lichtdruck *m*
photovoltaic cell Elektrolytzelle *f*,
  Sperrschichtzelle *f*
photozincographic photozinkographisch
photozincography Photozinkographie *f*,
  Zinkätzung *f*
phrenosin Phrenosin *n*
phrenosinic acid Phrenosinsäure *f*
pH scale pH-Skala *f*
pH-shift pH-Verschiebung *f*
phthalacene Phthalacen *n*
phthalaldehyde Phthalaldehyd *m*
phthalaldehydic acid Phthalaldehydsäure *f*
phthalamic acid Phthalamidsäure *f*
phthalamide Phthalamid *n*
phthalan Phthalan *n*
phthalandione Phthalsäureanhydrid *n*
phthalanil Phthalanil *n*
phthalanilic acid Phthalanilsäure *f*
phthalanilide Phthalanilid *n*
phthalate Phthalat *n*, Salz oder Ester der
  Phthalsäure *m*
phthalazine Phthalazin *n*
phthalazone Phthalazon *n*
phthalein Phthalein *n*
phthalic phthalsauer
phthalic acid Phthalsäure *f*, Alizarinsäure *f*
  (obs), **commercial** ~ Handelsphthalsäure *f*
phthalic aldehyde Phthalaldehyd *m*
phthalic anhydride Phthalsäureanhydrid *n*
phthalic ester Phthalsäureester *m*,
  Phthalsäuredimethylester *m*
phthalic resin Phthalsäureharz *n*
phthalide Phthalid *n*
phthalidene acetic acid Phthalidenessigsäure *f*
phthalimide Phthalimid *n*
phthalimidine Phthalimidin *n*
phthalimidoacetic acid Phthalimidoessigsäure *f*
phthaline Phthalin *n*
phthalocyanine Phthalocyanin *n*
phthalocyanine blue Phthalocyaninblau *n*
phthalocyanine green Phthalocyaningrün *n*
phthalocynine dye Phthalocyaninfarbstoff *m*
phthalonic acid Phthalonsäure *f*
phthalonimide Phthalonimid *n*
phthalonitrile Phthalonitril *n*
phthalophenone Phthalophenon *n*
phthaloxime Phthaloxim *n*
phthalyl Phthalyl-
phthalyl chloride Phthalylchlorid *n*
phthalylglycine Phthalylglycin *n*

phthalyl hydrazide Phthalylhydrazid *n*
phthalyl sulfathiazole Phthalylsulfathiazol *n*
phthalyl synthesis Phthalylsynthese *f*
phthiocol Phthiokol *n*
phthioic acid Phthionsäure *f*
phthoric acid (obs) Fluorwasserstoff *m*
phulaxite Phulaxit *n*
pH-value pH-Wert *m*, **determination of the** ~
  pH-Messung *f*
phycitol Phycit *m*
phycobilin Phykobilin *n*
phycochrome Phykochrom *n*
phycocyanin Phykocyanin *n*
phycocyanogen Phykocyan *n*
phycoerythrin Phykoerythrin *n*
phycomycetes Phykomyceten *pl*, Algenpilze *pl*
phycophein Phykophäin *n*
p-hydroxycoumaric acid Umbellsäure *f*
p-hydroxyphenylalanine Tyrosin *n*
phygon Phygon *n*
phyllite Phyllit *m* (Min), Urtonschiefer *m* (Min)
phyllochlorine Phyllochlorin *n*
phyllocyanin Blattblau *n*
phylloerythrin Phylloerythrin *n*
phyllohemin Phyllohämin *n*
phylloporphyrin Phylloporphyrin *n*
phyllopyrrole Phyllopyrrol *n*
phylloquinone Phyllochinon *n* (Vitamin K)
phyllorin Phyllorin *n*
phyostigmine Phyostigmin *n*
physalin Physalin *n*
physcic acid Physciasäure *f*, Physcion *n*
physcion Physciasäure *f*, Physcion *n*
physeteric acid Physeter[in]säure *f*
physetoleic acid Physetolsäure *f*
physical körperlich, physikalisch
physical appearance Habitus *m* (Biol)
physical characteristic Stoffwert *m*
physical-chemical physikalisch-chemisch
physical chemist Physikochemiker *m*
physical chemistry physikalische Chemie *f*
physical condition Aggregatzustand *m*
physical form Aggregatform *f* (Phys)
physical state Aggregatzustand *m*
physician Arzt *m*
physicist Physiker *m*
physicochemical physikalisch-chemisch
physics Physik *f*, **chemical** ~ Chemiephysik *f*,
  solid-state ~ Festkörperphysik *f*
physiography Physiographie *f*
physiological physiologisch
physiology Physiologie *f*, **applied** ~ angewandte
  Physiologie, **comparative** ~ vergleichende
  Physiologie, **experimental** ~ experimentelle
  Physiologie, **pathological** ~ pathologische
  Physiologie
physodalic acid Physodalsäure *f*
physodalin Physodalin *n*
physol Physol *n*

physostigmine Physostigmin *n*, Eserin *n*
physostigmol Physostigmol *n*
phytadiene Phytadien *n*
phytane Phytan *n*
phytanol Phytanol *n*
phytase Phytase *f* (Biochem)
phytene Phyten *n*
phytenic acid Phytensäure *f*
phytic acid Phytinsäure *f*
phytin Phytin *n*
phytochemical phytochemisch
phytochemistry Pflanzenchemie *f*, Phytochemie *f*
phytoene Phytoen *n*
phytohormone Pflanzenhormon *n*,
    Phytohormon *n*
phytol Phytol *n*
phytolacca toxin Phytolaccatoxin *n*
phytology Pflanzenkunde *f*
phytone Phyton *n*
phytoparasite Pflanzenparasit *m*, Phytoparasit *m*
phytophagous phytophag, pflanzenfressend
phytoplasm Phytoplasma *n*
phytoprotein Pflanzeneiweiß *n*
phytosterin Phytosterin *n*
phytosterol Phytosterin *n*
phytosterolin Phytosterolin *n*
phytotoxic phytotoxisch
phytotoxin Pflanzengift *n*
piaselenole Piaselenol *n*
piauzite Piauzit *m*
piazine Piazin *n*
piazothiole Piazothiol *n*
picamar Pikamar *n*, Teerbitter *n*
picein Picein *n*
picene Picen *n*
picene quinone Picenchinon *n*
pichurim bean Pichurimbohne *f* (Bot)
pichurim oil Pichurimtalgsäure *f*
picite Picit *m*
pick Pickel *m* (Techn), Schuß *m* (Web),
    Spitzhacke *f*
pick *v* ablesen (Früchte etc.), klauben (Erze),
    lesen (Erze), scheiden (Erze), ~ out
    ausklauben (Erze), herausgreifen
picked ore Scheideerz *n*
pickel bath Pickelbrühe *f*
pickeringite Pickeringit *m* (Min)
picking belt Klaubband *n* (Erze), Leseband *n*
    (Erze)
picking ore Klauberz *n*
picking or sorting table Lesetisch *m* (Erze)
picking plant Sortieranlage *f*
picking table Klaubetisch *m*
pickle Beize *f* (Met), Beizflüssigkeit *f* (Met),
    Essigkonserve *f*, Gelbbrenne *f* (Met),
    Salzlake *f*, Salzlauge *f*, ~ for rendering a dull
    surface Mattbeize *f*

pickle *v* abbeizen (Met), dekapieren (Met),
    einsalzen, entzundern (Met), marinieren,
    pökeln
pickled gepökelt, eingesalzen
pickled sheet Mattblech *n*
pickle lag Anlaufzeit *f* der
    Wasserstoffentwicklung (Metall)
pickler Gelbbrenner *m* (Met)
pickling Abbeizen *n* (Met), Beizbehandlung *f*
    (Met), Dekapieren *n* (Met), Gelbbrennen *n*
    (Met), Pickeln *n* (Häute), Pökeln *n*, ~ the
    metal Blankmachen *n* des Metalls
pickling acid Brennsäure *f*, removing the ~
    Poltern *n* (Metall)
pickling agent Abbeizmittel *n*, Beizmittel *n*
pickling appliances Beizgeräte *n pl*
pickling basket Beizkorb *m*
pickling bath Beizbad *n* (Met), Dekapierbad *n*
    (Met), Vorbrenne *f* (Met), cuprous ~
    Cuprodekapierbad *n*, ~ containing an
    inhibitor Sparbeize *f*
pickling bath heater Beizbaderwärmer *m*
pickling brine Pökel *m*
pickling cellar Pökelkeller *m*
pickling compound Beizzusatz *m*
pickling house Beizerei *f*
pickling inhibitor Beizenfetter *m*,
    Sparbeizzusatz *m*
pickling liquor Beizlauge *f*
pickling odor Pökelgeruch *m*
pickling plant Abbrenneinrichtung *f* (Met),
    Beizanlage *f* (Met), wide strip ~
    Breitbandbeizanlage *f*
pickling solution Dekapierflüssigkeit *f*
pickling tank Beiztank *m*
pickling time Pökeldauer *f*
pickling tub Pökelfaß *n*
pickling vat Abbrennkessel *m*, Beizkasten *m*,
    Pökelkufe *f*
pickling water, waste ~ Beizabwasser *n*
pickling wheel Beizrad *n*
pickup Anzugsvermögen *n* (Auto) (of a
    phonograph), Tonabnehmer *m*
pick up *v* aufheben
pick-up carrier Abfanggraben *m* (Wasser)
pickup groove Haltekerbe *f*, Haltenute *f*
pickup roll Aufnahmewalze *f*, Auftragwalze *f*
pick-up velocity Abtastgeschwindigkeit *f*
pico-farad Picofarad *n*
picoline Pikolin *n*
picoline ferroprotoporphyrin
    Picolineisenprotoporphyrin *n*
picolinic acid Picolinsäure *f*
picolyl- Picolyl-
picotite Picotit *m* (Min)
picramic acid Pikraminsäure *f*
picramide Pikramid *n*
picranilide Pikranilid *n*
picranisic acid Pikrinsäure *f*

picrate Pikrat *n*
picric acid Pikrinsäure *f,* Bittersäure *f* (obs),
Trinitrophenol *n,* **crude** ~ Rohpikrinsäure *f*
picric acid complex Pikrinsäurekomplex *m*
picrin Pikrin *n*
picrite Pikrit *m*
picrocine Pikrocin *n*
picrocinic acid Pikrocinsäure *f*
picrocininic acid Pikrocininsäure *f*
picrocrocin Picrocrocin *n*
picroic acid Pikrosäure *f*
picroilmenite Pikroilmenit *m*
picrol Pikrol *n*
picrolichenin Flechtenbitter *n*
picrolite Pikrolith *m* (Min), Bitterstein *m* (Min)
picrolonic acid Pikrolonsäure *f*
picromerite Pikromerit *m* (Min)
picromycin Pikromycin *n*
picronitric acid Pikrinsäure *f*
picropharmacolite Pikropharmakolith *m* (Min)
picropodophyllin Pikropodophyllin *n*
picroroccellin Pikroroccellin *n*
picrosmine steatite Pikrosminsteatit *m* (Min)
picrotin Pikrotin *n*
picrotinic acid Pikrotinsäure *f*
picrotitanite Pikrotitanit *m* (Min)
picrotoxin Pikrotoxin *n*
picryl Pikryl-
picrylacetate Pikrylacetat *n*
picrylamine Pikramid *n*
picryl chloride Pikrylchlorid *n*
pictol Metol *n*
picture Bild *n,* Abbildung *f,* Aufnahme *f* (Phot),
distorted ~ Zerrbild *n*
picture disturbance Bildstörung *f* (Telev)
pictures, to take ~ photographieren
picture screen Bildfeld *n* (Telev), Bildschirm *m*
picture tube Bildröhre *f* (Telev), Fernsehröhre *f*
picylene Picylen *n*
piece Stück *n,* **curved** ~ Bogenstück *n,* **formed** ~
Formling *m,* ~ **for use** Gebrauchsstück *n*
piece *v* [an]stücken, ausbessern
piece-dyed stückgefärbt, im Stück gefärbt
piece goods Stückware *f,* **plain shade** ~
Unistückware *f*
piece number Stückzahl *f*
piece rate wages Akkordlohn *m,* Stücklohn *m*
pieces, consisting of two ~ zweiteilig, **to break up
into** ~ grobbrechen, **to fall into** ~
auseinanderfallen, **to take to** ~
auseinandernehmen
piece work Akkordarbeit *f*
piece worker Akkordarbeiter *m,* Stückarbeiter *m*
piedmontite Piemontit *m* (Min),
Manganepidot *m* (Min)
pierce *v* durchbohren, [durch]lochen,
durchlöchern, durchstechen, durchstoßen,
perforieren, ~ **open** aufstechen
piercer Bohrer *m,* Lochdorn *m,* Locheisen *n*

piercing stechend, scharf, schrill
piercing mandrel Lochdorn *m*
piezoelectric piezoelektrisch, druckelektrisch,
kristallelektrisch
piezoelectric crystal Piezokristall *m* (Krist),
Schwingquarz *m* (Elektr)
piezoelectric effect Piezoeffekt *m*
piezoelectricity Piezoelektrizität *f,*
Druckelektrizität *f,* Kristallelektrizität *f*
piezometer Piezometer *n,* Druckmesser *m*
piezometry Piezometrie *f*
piezotropy Piezotropie *f*
pig Block *m* (Techn), Massel *f,* Roheisen *n*
pig-and-ore process Roheisen-Erz-Verfahren *n*
pig-and-scrap process
Schrott-Roheisen-Verfahren *n,*
Siemens-Martin-Prozeß *m*
pig bed Gießbett *n,* Masselbett *n*
pig boiling Puddeln *n* (Metall),
Schlackenfrischen *n* (Metall)
pig breaker Masselbrecher *m,* Fallwerk *n*
pig disk Scheibeneisen *n*
pigeon-hole Fach *n* (Schrank)
pig iron Masseleisen *n,* Gußeisen *n,* Roheisen *n,*
basic ~ Thomasroheisen *n,* **close-grained** ~
feinkörniges Roheisen, **coarsely crystalline** ~
Grobspiegeleisen *n,* **forge** ~
Frischereiroheisen *n,* **graphitic** ~ graues
Roheisen, **gray** ~ Graueisen *n,* graues
Roheisen, **high silicon** ~ Glanzeisen *n,*
Schwarzeisen *n,* **kishy** ~ schwarzes Roheisen,
malleable ~ schmiedbares Gußeisen,
Temperroheisen *n,* Weichguß *m,* **mottled white**
~ Forelleneisen *n,* Kerneisen *n,* **open hearth**
~ Stahlroheisen *n,* **phosphorous** ~
phosphorsaures Roheisen, **porous** ~ lückiges
Floß *n,* **pure** ~ geläutertes Roheisen, **white** ~
gemeines Eisen, Matteisen *n,* Treibeisen *n,*
Weißeisen *n,* ~ **for castings**
Gießereiroheisen *n,* ~ **for puddling**
Puddelroheisen *n,* ~ **for refining**
Frischfeuerroheisen *n,* Herdfrischroheisen *n,*
Holzkohlenroheisen *n,* ~ **rich in silicon**
Glanzeisen *n,* Schwarzeisen *n,* ~ **saturated
with carbon** schaumiges Roheisen
pig iron bath Roheisenbad *n*
pig iron casting Grauguß *m*
pig iron charge Roheisencharge *f,*
Roheisengichtsatz *m*
pig iron mixer Roheisenmischer *m*
pig iron-ore process Roheisen-Erz-Verfahren *n*
pig iron production Roheisenerzeugung *f*
pig iron-scrap process
Roheisen-Schrott-Verfahren *n*
pig lead Blockblei *n,* Bleigans *f,* Muldenblei *n,*
Ofenblei *n*
pigment Pigment *n,* Farbe *f,* Farbkörper *m,*
Farbstoff *m*
pigment *v* pigmentieren, [sich] färben

pigmentation Pigmentierung *f*
pigment bacterium Pigmentbakterie *f*
pigment binder Pigmentbindemittel *n*
pigment brown Pigmentbraun *n*
pigment chrome yellow Pigmentchromgelb *n*
pigment disperser Pigmentverteiler *m*
pigment[ed] cell Pigmentzelle *f*
pigment fast red Pigmentechtrot *n*
pigmenting bath Pigmentierbad *n*
pigment loading, paint with high ~
     hochpigmentierter Lack *m*
pigment paste Pigmentteig *m*
pigment printing Pigmentdruck *m*
pigment scarlet Pigmentscharlach *n*
[pigment] toner Lackfarbstoff *m*,
     Pigmentfarbstoff *m*
pig metal Floß *n* (Metall)
pig mold Floßenbett *n*
pig nickel Blocknickel *n*
pig [of iron] Massel *f*
pilarite Pilarit *m*
pile Haufen *m*, Meiler *m*, Säule *f* (Elektr),
     Schweißpaket *n* (Metall), Stapel *m*, Stoß *m*,
     bare ~ Reaktor *m* ohne Reflektor
pile *v* aufschichten, [auf]stapeln, ~ up
     akkumulieren, aufhäufen, aufschichten,
     aufstapeln, sich häufen
pile and weld *v* gerben (Stahl)
pile coking Meilerverkokung *f*
pile driver Fallwerk *n*
pile fabric Florgewebe *n*
pilé liquor Pilékläre *f*
pilé sugar Pilézucker *m*
pilé sugar crusher Pilébrechwerk *n*
pilferproof cap Schraubverschluß *m* mit
     Originalitätssicherung
piling Stapelung *f*, Bohlenzaun *m* (Techn),
     Paketieren *n*
pilinite Pilinit *m*
pilite Pilit *m*
pill Pille *f*, Tablette *f*
pillar Pfeiler *m*, Säule *f*, Stütze, Träger *m*
pillar bracket Säulenarm *m*
pill box Pillenschachtel *f*
pillow Kissen *n*
pillow block Lagerblock *m* (Am. E.), Stehlager *n*
pillow sack Flachsack *m*
pilocarpic acid Pilocarpinsäure *f*
pilocarpidine Pilocarpidin *n*
pilocarpine Pilocarpin *n*
pilocarpine hydrochloride
     Pilocarpinhydrochlorid *n*
pilocarpine nitrate Pilocarpinnitrat *n*
pilopic acid Pilopsäure *f*
pilosine Pilosin *n*
pilosinine Pilosinin *n*
pilot beam Leitstrahl *m*
pilot burner Sparbrenner *m*

pilot flame Sparflämmchen *n*, Sparflamme *f*,
     Zündflamme *f*
pilot frequency Steuerfrequenz *f*
pilot lamp Kontrollampe *f*, Pilotlicht *n*
     (Mikroskop)
pilot light Zündflamme *f*
pilot model Versuchsmodell *n*
pilot plant Versuchsanlage *f*, Modellanlage *f*,
     Musterbetrieb *m*, Technikum *n*
pilot plant scale Großversuchsmaßstab *m*
pilot pressure Vorfülldruck *m*, Vorsteuerdruck *m*
pilot relay Steuerrelais *n* (Elektr)
pilot valve Schaltventil *n*, Vorfüllventil *n*
pimanthrene Pimanthren *n*
pimarabietic acid Pimarabietinsäure *f*
pimaric acid Pimarsäure *f*
pimelic acid Pimelinsäure *f*
pimelic aldehyde Pimelinaldehyd *m*
pimelinketone Cyclohexanon *n*, Pimelinketon *n*
pimelite Pimelit[h] *m* (Min)
piment Piment *m*
pimento Allerleigewürz *n* (Bot),
     Nelkenpfeffer *m* (Bot)
pi-meson Pi-Meson *n*, Pion *n*
pimpernel root Bibernellwurzel *f* (Bot)
pimpinella saponin Pimpinellasaponin *n*
pimpinellin Pimpinellin *n*
pimple Finne *f* (Med), Pickel *m* (Med)
pin Nadel *f*, Bolzen *m* (Techn), Drehachse *f*,
     Dübel *m* (Techn), Stift *m* (Techn),
     Tragzapfen *m* (Techn)
pinabietic acid Pinabietinsäure *f*
pinabietin Pinabietin *n*
pinaciolite Pinakiolith *m* (Min)
pinacoid Pinakoid *n* (Krist)
pinacol Pinakol *n*
pinacolin Pinakolin *n*
pinacolin rearrangement Pinakolin-Umlagerung *f*
pinacolone rearrangement
     Pinakolin-Umlagerung *f*
pinacolyl chloride Pinakolinchlorid *n*
pinacyanol Pinacyanol *n*
pinane Pinan *n*
pinanol Pinanol *n*
pinastric acid Pinastrinsäure *f*
pinaverdol Pinaverdol *n*
pinboard programming
     Stecknadelprogrammierung *f* (Comp)
pincers Zange *f*, Federzange *f*, Kneifzange *f*,
     Pinzette *f*
pinch Kniff *m*
pinch *v*, ~ off abzwicken, abklemmen
pinch bar Hebeeisen *n*, Wagonrücker *m*
pinchbeck Goldkupfer *n*, Talmi *n*
pinch clamp Quetschhahn *m*
pinchcock Quetschhahn *m*
pinch effect Pinch-Effekt *m*,
     Einschnürungseffekt *m*
pinchers Blockzange *f*, Kneifzange *f*

pinch valve Quetschventil *n*
pinch zone Zone *f* konstanter Zusammensetzung
pin drill Zapfenbohrer *m*
pine Föhre *f* (Bot)
pineal gland Zirbeldrüse *f* (Med)
pineapple essence Ananasäther *m*,
  Ananasessenz *f*
pineapple fiber Ananasfaser *f*
pine bark Fichtenrinde *f*
pin electrode Stiftelektrode *f*
pinene Pinen *n*, Australen *n*, Lauren *n*
pine needle extract Fichtennadelextrakt *m*,
  Kiefernnadelextrakt *m*,
  Tannennadelextrakt *m*
pine needle oil Fichtennadelöl *n*
pinene hydrochloride Pinenchlorhydrat *n*
pine oil Fichtenöl *n*, Kienöl *n*
pine pitch Kienteerpech *n*
pine resin Fichtenharz *n*, Kiefernharz *n*,
  Kienharz *n*, Kolophonium *n*
pine soot Fichtenruß *m*, Kienruß *m*
pine tar Fichtenteer *m*, Kienteer *m*
pinewood cellulose Fichtenholzzellstoff *m*
pinewood oil Kienholzöl *n*
pinewood reaction Fichtenspanreaktion *f* (Chem)
pin gear Schaftverzahnung *f*
pinguite Pinguit *m* (Min)
pinhole Fadenlunker *m*, Krater *m* (Farbe),
  Nadelöhr *n*
pinhole camera Lochkamera *f*
pinhole detector Funkinduktor *m*,
  Lochsuchgerät *n*
pinhole diaphragm Lochblende *f*
pinholes Lochfraß *m*
pinholing Kraterbildung *f*, Lochfraßkorrosion *f*,
  Porenbildung *f*
pinicolic acid Pinicolsäure *f*
pinidine Pinidin *n*
pinifolic acid Pinifolsäure *f*
pinion Kammwalze *f*, Ritzel *n*
pinion housing Kammwalzgerüst *n*
pinion neck Kammwalzenzapfen *m*
pinion shaft Ritzelwelle *f*
pinion steel Triebstahl *m*
pinite Pinit *m*; Fichtenzucker *m*
pinitol Pinit *m*
pink rosa[farben]
pink gilding Rosavergoldung *f*
pinking salt Rosiersalz *n*
pinkroot Wurmkraut *n* (Bot)
pinksalt Pinksalz *n*, Ammoniumzinnchlorid *n*,
  Zinnammoniumchlorid *n*
pinnoite Pinnoit *m* (Min)
pinocamphane Pinocamphan *n*
pinocampholenic acid Pinocampholensäure *f*
pinocamphone Pinocamphon *n*
pinocamphoric acid Pinocamphersäure *f*
pinocarveol Pinocarveol *n*
pinocarvone Pinocarvon *n*

pinocytosis Flüssigkeitsaufnahme *f* (Zelle),
  Pinozytose *f*
pinol Pinol *n*
pinolene Pinolen *n*
pinol hydrate Sobrerol *n*
pinolite Pinolit *m*
pinolol Pinolol *n*
pinolone Pinolon *n*
pinonaldehyde Pinonaldehyd *m*
pinonic acid Pinonsäure *f*
pinononic acid Pinononsäure *f*
pinophanic acid Pinophansäure *f*
pin-point burner Nadelbrenner *m*
pinpoint corrosion Lochfraß *m*
pin-point gate Nadelpunktanguß *m* (Gieß),
  Punktanguß *m*
pin riveting Stiftnietung *f*
pinselic acid Pinselinsäure *f*
pinselin Pinselin *n*
pin stenter Nadelspannrahmen *m*
pintadoite Pintadoit *m* (Min)
pin weir Nadelwehr *n*
pinworm Madenwurm *m* (Zool)
piny tallow Pinientalg *m*
pion Pi-Meson *n*, Pion *n*
pioneer Bahnbrecher *m*, Pionier *m*
pioneering experiment Fundamentalversuch *m*
pip Kern *m* (Obst)
pipe Rohr *n*, Leitung *f*, Leitungsrohr *n*,
  Lunker *m* (Gieß), Röhre *f*, **armed** ~
  Panzerrohr *n*, **length of** ~ Rohrlänge *f*,
  pneumatic ~ pneumatische Pfeife *f*, **seamless**
  ~ nahtlose Röhre, ~ **with swivel elbows**
  Gelenkrohr *n*
pipe *v* Lunker bilden (Gieß)
pipe arrangement Rohranordnung *f*
pipe attachment thermometer
  Rohranliegethermometer *n*
pipe bend Rohrknie *n*
pipe burst Rohrbruch *m*
pipe choking Rohrverstopfung *f*
pipe clay bildsamer Ton *m* (Min), Kaolin *n*
  (Min), Pfeifenton *m* (Min)
pipe clip Rohrschelle *f*
[pipe] coil Rohrschlange *f*
pipecoleine Pipecolein *n*
pipecoline Pipecolin *n*
pipecolinic acid Pipecolinsäure *f*
pipecolyl Pipecolyl-
pipe compensator Rohrausgleicher *m*
pipe condenser Röhrenkondensator *m*
pipe connection Rohransatz *m*, Rohranschluß *m*
pipe connector Rohrverbindung *f*
pipe coupling Rohrkupplung *f*
pipe cross Kreuzstück *n*
pipe diameter Rohrdurchmesser *m*
pipe discharge Rohrentladung *f*
pipe ejector mixer Rohrrührer *m*
pipe elbow Rohrkrümmer *m*

pipe elimination Lunkerverhütung *f* (Gieß)
pipe eliminator Lunkerverhütungsmittel *n*
  (Gieß)
pipe eradicator Lunkerverhütungsmittel *n*
  (Gieß)
pipe expansion joint Rohrausdehnungsstück *n*
pipe fastening Rohrbefestigung *f*
pipe filter Röhrenfilter *m*
pipe fitter Rohrleger *m*
pipe fittings Rohrarmaturen *pl*
pipe flange Rohrflansch *m*
pipe flow Rohrströmung *f*
pipe friction Rohrreibung *f*
pipe glaze Pfeifenfirnis *m*
pipe hanger Rohrhalter *m*
pipe joint Rohrverbindung *f*,
  Rohrverschraubung *f*
pipe junction Rohransatz *m*, Rohrweiche *f*
pipeless lunkerfrei, lunkerlos
pipeline Rohrleitung *f*, **large diameter** ~
  Großrohrleitung *f*
pipeline fittings Rohrleitungsarmaturen *f pl*
pipeline mixer Durchflußmischer *m*
pipeline system Leitungsnetz *n*
pipe lining Rohrauskleidung *f*
pipe-pushing hydraulisches Vorpreßverfahren *n*
piperazidine Piperazin *n*
piperazine Piperazin *n*, Diäthylendiamin *n*
piperettic acid Piperettinsäure *f*
piperettine Piperettin *n*
piperic acid Piperinsäure *f*
piperidazine Piperidazin *n*
piperidine Piperidin *n*, Hexahydropyridin *n*
piperidine-3,4-dicarboxylic acid Loiponsäure *f*
piperidine blue Piperidinblau *n*
piperidine-N-carboxylic acid Pipecolinsäure *f*
piperidinium hydroxide Piperidiniumhydroxid *n*
piperidone Piperidon *n*
piperidyl hydrazine Piperidylhydrazin *n*
piperidyl urethane Piperidylurethan *n*
piperil Piperil *n*
piperilic acid Piperilsäure *f*
piperimidine Piperimidin *n*
piperine Piperin *n*
piperinic acid Piperinsäure *f*
piperitone Piperiton *n*
piperolidine Piperolidin *n*
piperonal Piperonal *n*, Heliotropin *n*
piperonyl alcohol Piperonylalkohol *m*
piperonyl aldehyde Piperonal *n*
piperonyl chloride Piperonylchlorid *n*
piperonylic acid Piperonylsäure *f*
piperoxan Piperoxan *n*
piperyl Piperyl-
piperylene Piperylen *n*
piperyl-piperidine Piperin *n*
piperylurethane Piperylurethan *n*
pipes, formation of ~ Trichterbildung *f*, **to draw**
  ~ Rohre *pl* ziehen

pipe socket Röhrenmuffe *f*
pipe solder Röhrenlot *n*
pipestone Pfeifenstein *m* (Min)
pipe support Rohrbefestigung *f*, Rohrhaken *m*,
  Rohrhalter *m*
pipe system Rohrnetz *n*
pipe thread Rohrgewinde *n*
pipe tongs Rohrzange *f*
pipet[te] Pipette *f*, Stechheber *m*, **auxiliary** ~
  Hilfspipette *f*, **delivery** ~ Auslaufpipette *f*, **to**
  measure with a ~ pipettieren, **to remove with a**
  ~ herauspipettieren, **to transfer with a** ~
  pipettieren, ~ **with globe stopcock**
  Kugelhahnpipette *f*, **weighing** ~ **with stopcock**
  Kugelhahnpipette *f*, Kugelhahnstechheber *m*
pipet[te] *v* pipettieren, ~ **out** herauspipettieren
pipette holder Pipettenhalter *m*,
  Pipettenständer *m*
pipette stand Pipettenständer *m*
pipetting apparatus Pipettiergerät *n*
pipe union Muffenverbindung *f*
pipe wrench Rohrzange *f*
piping Rohrleitung *f*, Rohrnetz *n*,
  Leitungssystem *n*, Lunkerbildung *f* (Gieß),
  open-air ~ Freileitung *f*
piping layout Rohrleitungsanlage *f*
pipitzahoic acid Pipitzahoinsäure *f*
pipitzol Pipitzol *n*
pipradrol Pipradrol *n*
piprinhydrinate Piprinhydrinat *n*
pirn Schußspule *f* (Web)
pirn *v* schußspulen (Web)
pirn winder Schußspulmaschine *f* (Web)
pirssonite Pirssonit *m* (Min)
pisang wax Pisangwachs *n*, Bananenwachs *n*
pisanite Pisanit *m* (Min)
piscidic acid Piscid[in]säure *f*
piscidin Piscidin *n*
pisiform erbsenförmig
pisiform iron ore Erbsenerz *n* (Min)
pisolite Pisolith *m* (Min), Erbsenstein *m* (Min)
pisolitic pisolithartig, erbsensteinhaltig
pisolitiferous pisolithhaltig
p-isopropylbenzaldehyde Cuminaldehyd *m*
p-isopropylbenzoic acid Cuminsäure *f*
pistachio green Pistaziengrün *n*
pistachio nuts Pistazien *pl*
pistachio oil Pistazienöl *n*
pistachios Pistazien *pl*
pistacite Pistazit *m* (Min), Epidot *m* (Min)
pistil Griffel *m* (Bot)
pistol pipe Pistolenröhre *f*
piston Kolben *m* (Mech), **auxiliary** ~
  Hilfskolben *m*, **clearance of** ~ Kolbenspiel *n*,
  downstroke of a ~ Kolbenniedergang *m*,
  stroke of ~ Kolbenhub *m*, Kolbenweg *m*
piston area Kolbenfläche *f*
piston ascent Kolbenaufgang *m*
piston blade Kolbenflügel *m*

piston diaphragm Kolbenmembran *f*
piston displacement Hubraum *m*,
  Zylindervolumen *n*
piston extension Kolbenansatz *m*
piston pin Kolbenbolzen *m*
piston pin bushing Kolbenbolzenlager *n*
piston pressure Kolbendruck *m*
piston pressure diagram Kolbenkraftdiagramm *n*
piston pump Kolbenpumpe *f*
piston ring Kolbenring *m* (Mech)
piston rod Kolbenstange *f*, Pleuelstange *f*
piston steam engine Kolbendampfmaschine *f*
piston stroke Kolbenhub *m*, Kolbenbewegung *f*
piston stroke volume Kolbenhubraum *m*
piston travel Kolbenweg *m*
piston valve Kolbenventil *n*
piston vane Kolbenflügel *m*
piston wrench Kolbenschlüssel *m*
pit Bergwerk *n* (Bergb), Grube *f*, Kern *m*
  (Obst), Preßfehler *m*, Rostgrübchen *n*,
  Schacht *m* (Bergb)
pita fiber Pitafaser *f*
pita hemp Pitahanf *m*
pit burning Grubenverkohlung *f*
pit cable Abteufkabel *n* (Bergb)
pitch Ganghöhe *f* (Gewinde), Pech *n* (Min),
  Tonhöhe *f* (Akust), **common black** ~
  Schiffspech *n*, **decreasing** ~ abnehmende
  Ganghöhe, **hard** ~ Glaspech *n*, Steinpech *n*,
  liquid ~ Harzpech *n*, **progressive** ~
  zunehmende Ganghöhe, ~ **in casks**
  Faßpech *n*, ~ **of a screw** Steigung *f* des
  Schraubengewindes, ~ **of a tone** Höhe *f* eines
  Tones
pitch *v* pechen, [ver]pichen
pitch ball Pechkugel *f*
pitch black pechschwarz
pitchblende Pechblende *f* (Min), Uraninit *m*
  (Min), Uranpecherz *n* (Min)
pitch cake Pechkuchen *m*
pitch circle Lochkreis *m*, Teilkreis *m* (Zahnrad),
  Wälzkreis *m*
pitch coal Pechkohle *f*
pitchcoat *v* verpichen
pitch cone Grundkegel *m*
pitch diameter Gewindedurchmesser *m*,
  Teilkreisdurchmesser *m*
pitched thread Pechdraht *m*
pitcher Krug *m*
pitch error Steigungsfehler *m*
pitch garnet Pechgranat *m* (Min)
pitching temperature Anstelltemperatur *f*
pitching tool Bergeisen *n*,
  Geradhängevorrichtung *f*
pitching vessel Anstellbottich *m* (Brau)
pitching yeast Anstellhefe *f*
pitch kiln Pechofen *m*
pitch-like pechartig
pitchline *v* verpichen

pitch line velocity Umfangsgeschwindigkeit *f* im
  Wälzkreis
pitch oil Pechöl *n*
pitch oven Pechofen *m*
pitch peat Pechtorf *m*, Specktorf *m*
pitch pine Harzkiefer *f* (Bot)
pitchstone Pechstein *m* (Min), Resinit *m* (Min)
pitchy pechartig, pechig
pitchy iron ore Kolophoneisenerz *n* (Min)
pit coal Steinkohle *f*, Grubenkohle *f*,
  Schwarzkohle *f*
pit fire Grubenbrand *m* (Bergb)
pit furnace Schachtofen *m*
pit gravel Grubenkies *m*
pith Mark *n* (Bot)
pit helmet Grubenhelm *m*
pith-like markartig (Bot)
pit-lime Grubenkalk *m*
pitman Bergmann *m*
Pitot nozzle Staudüse *f*
Pitot tube Pitotsche Röhre *f*, Stauröhre *f*
pit sand Grubensand *m*
pittacol Pittacol *n*
pit tannage Grubengerbung *f*
pitticite Eisenpecherz *n* (Min),
  Kolophoneisenerz *n* (Min), Pittizit *m* (Min),
  Skorodit *m* (Min)
pitting Ausbröckeln *n*, grübchenartige
  Rostanfressung *f*, Grübchenbildung *f* (Korr),
  Kraterbildung *f*, Lochfraß *m* (Korr),
  Porenbildung *f*
pitting corrosion Lochfraßkorrosion *f*
pituitary gland Hirnanhangdrüse *f*, Hypophyse *f*
  (Med)
pituitary hormone Hypophysenhormon *n*
pituitary lobe, posterior ~
  Hypophysenhinterlappen *m* (Med)
pit water Grubenwasser *n*
pivalaldehyde Pivalinaldehyd *m*
pivalic acid Pivalinsäure *f*
pivaloin Pivaloin *n*
pivalone Pivalon *n*
pivalophenone Pivalophenon *n*
PIV drive Regelgetriebe *n*
pivot Angelpunkt *m*, Drehpunkt *m*,
  Drehzapfen *m*, Spurzapfen *m*, Türangel *f*
pivot bearing Zapfenlager *n*
pivot element Pivotelement *n* (Math)
pivoting Zapfenlagerung *f*
pivoting clamp Pendelhalter *m*
pivoting slider Kippreiter *m*
pivot joint Drehgelenk *n*
pivot [journal] Spurzapfen *m*
pivot nozzle Zapfendüse *f*
pivot pin Zapfennadel *f*
pivot thrust Zapfendruck *m*
pizein Pizein *n*
placards, color for printing ~ Plakatfarbe *f*
place Ort *m*, Platz *m*, Stelle *f*

place _v_ legen, setzen, stellen
placebo Placebo _n_ (Pharm), Suggestionsmittel _n_
place isomerism Stellungsisomerie _f_ (Chem)
placer gold Alluvialgold _n_, Waschgold _n_
place value Stellenwert _m_ (Math)
placing Anbringung _f_, ~ **the crucible**
  Einsetzen _n_ des Tiegels
plagihedral schiefflächig (Krist)
plagiocitrite Plagiocitrit _m_
plagioclase Plagioklas _m_ (Min)
plagioclastic plagioklastisch
plagionite Plagionit _m_ (Min)
plague Pest _f_ (Med), Seuche _f_ (Med)
plain Ebene (Geol), Flachland _n_ (Geol)
plain _a_ unlegiert (Met)
plain bearing Gleitlager _n_
plain grinding Rundschleifen _n_
plain milling cutter Walzenfräser _m_ (Techn)
plain pipet[te] Vollpipette _f_ (Anal)
plain roll Glattwalze _f_
plain weave Leinenbindung _f_ (Text)
plait _v_ flechten
plaiting Falten _n_, Flechten _n_
plait point Falt[ungs]punkt _m_
plait point curve Faltpunktkurve _f_
plan Plan _m_, Entwurf _m_, Grundriß _m_
  (Zeichnung), Programm _n_, Riß _m_
  (Zeichnung), Schema _n_
plan _v_ beabsichtigen, planen
plancheite Plancheit _m_ (Min)
Planck's constant Plancksche Konstante _f_,
  Plancksches Wirkungsquantum _n_
Planck's [elementary] quantum of action
  Plancksches Wirkungsquantum _n_
Planck's law of radiation Plancksches
  Strahlungsgesetz _n_
plane Ebene _f_, Hobel _m_ (Techn),
  Tragfläche _f_ (Luftf), **horizontal** ~ waagrechte
  Fläche, **principal** ~ Hauptebene _f_ (Opt),
  tangential ~ Tangentialebene _f_; ~ **of polarized**
  light Ebene _f_ des polarisierten Lichtes, ~ **of**
  projection Projektionsebene _f_, Rißebene _f_, ~
  of rotation Drehungsebene _f_, ~ **of symmetry**
  Symmetrieebene _f_, ~ **o. planar arrangement**
  ebene Anordnung _f_
plane _a_ eben, flach
plane _v_ [ein]ebnen, abrichten (Bleche),
  abschlichten (Techn), [aus]hobeln (Techn),
  glätten, planieren, schlichten
plane bit Hobeleisen _n_
plane flange Planflansch _m_
plane geometry Planimetrie _f_ (Math)
plane grating Plangitter _n_
plane ground joint Planschliff _m_
plane iron Hobeleisen _n_
plane mirror Planspiegel _m_
plane-parallel planparallel
plane-polarized geradlinig polarisiert, linear
  polarisiert

planer Hobel _m_
planer block Hobelwelle _f_
planerite Planerit _m_ (Min)
plane symmetry Flächensymmetrie _f_,
  Plansymmetrie _f_
planet Planet _m_ (Astr)
planetarium Planetarium _n_
planetary planetarisch (Astr)
planetary agitator Planetenrührer _m_
planetary gear Planetengetriebe _n_,
  Umlaufgetriebe _n_
planetary mill Planetenmühle _f_
planetary mixer Planetenmischer _m_
planetary orbit Planetenbahn _f_ (Astr)
planetary [paddle] mixer Planetenrührwerk _n_
planetary stirrer Planetenrührwerk _n_
planetary system Planetensystem _n_
planimeter Flächenmesser _m_, Planimeter _n_
planimeter _v_ planimetrieren
planimetry Planimetrie _f_ (Math)
planing Abhobeln _n_, Hobeln _n_
planing bench Hobelbank _f_
planing cut Planschnitt _m_
planing fixture Hobelvorrichtung _f_
planing machine Hobelmaschine _f_
planish _v_ [hochglanz]polieren, planieren,
  schlichten
planishing hammer Glanzhammer _m_,
  Schlichthammer _m_
planishing knife Geradeisen _n_
planispiral flachgewunden
plank Bohle _f_, [Schiffs]Planke _f_
planking Beplankung _f_, Verschalung _f_
plankton Plankton _n_ (Zool)
planning Planen _n_, Planung _f_
plano-concave plankonkav
plano-convex plankonvex
planoferrite Planoferrit _m_ (Min)
plan of layout Lageplan _m_
plansifter Plansichter _m_
plant Fabrik _f_, Betriebsanlage _f_,
  Fertigungsanlage _f_, Gewächs _n_, Pflanze _f_
  (Bot), Werk _n_, **cultivated** ~ Kulturpflanze _f_
  (Bot), **large-scale** ~ Fabrikationsanlage _f_
plantain Wegerich _m_ (Bot)
plant boiler Betriebskessel _m_
plant chemistry Pflanzenchemie _f_
plant conditions Betriebsverhältnisse _n pl_
plant disease Pflanzenkrankheit _f_
plant engineer Betriebsingenieur _m_
planteobiose Planteobiose _f_
planteose Planteose _f_
plant extract Pflanzenextrakt _m_
plant gelatin Pflanzengallerte _f_
plant growth retarder Hemmstoff _m_ (Bot)
plant growth substance Pflanzenwuchsstoff _m_
plant hormone Pflanzenhormon _n_
plant juice Pflanzensaft _m_
plant location study Standortfrage _f_

plant manager Betriebsleiter *m*
plant oil Pflanzenöl *n*
plant pest Pflanzenschädling *m*
plant physiology Pflanzenphysiologie *f*
plant preservation chemical
   Pflanzenschutzmittel *n*
plant product pflanzliches Produkt *n*
plant protection Pflanzenschutz *m*
plant protective [product] Pflanzenschutzmittel *n*
plant scale Betriebsmaßstab *m*
plant unit Großanlage *f*
plan view Aufriß *m*
plaque Plaque *f* (Med)
plasm Plasma *n* (Biol)
plasma Plasma *n* (Biol)
plasma arc cutting Plasmaschneiden *n*
plasma arc torch Lichtbogenplasmabrenner *m*
plasma burner Plasmabrenner *m*
plasma cascade torch Kaskadenbrenner *m*,
   Plasmakaskadenbrenner *m*
plasma cell Plasmazelle *f*
plasma chemistry Plasmachemie *f*
plasmacyte Plasmazelle *f*
plasma interaction Plasmawechselwirkung *f*
plasma jet spraying Plasmaspritzverfahren *n*
plasma layer Plasmaschicht *f*
plasma membrane Plasmamembran *n*,
   Ektoplasma *n*
plasma oscillation Plasmaschwingung *f*
plasma physics Plasmaphysik *f*
plasma propulsion Plasmaantrieb *m*
plasma protein Plasmaprotein *n*
plasma torch Plasmabrenner *m*
plasma welding Plasmaschweißen *n*
plasmin Fibrinolysin *n*
plasmochin Plasmochin *n*
plasmodium Malariaerreger *m*, Plasmodium *n*
plasmogamy Zellverschmelzung *f*
plasmolysis Plasmolyse *f*, Zellschrumpfung *f*
plasmolytic plasmolytisch
plasmoquine Plasmochin *n*
plaster Estrich *m*, Gipsmörtel *m* (Bauw),
   Kalkanwurf *m* (Bauw), Pflaster *n* (Med),
   Putz *m* (Bauw), Wundpflaster *n* (Med),
   hardening of ~ Gipserhärtung *f*, **light-weight**
   ~ Porengips *m*, **orthopaedic** ~
   orthopädischer Gips, **to fill with** ~ ausgipsen,
   unsifted ~ grober Gips, **marble chip filled** ~
   or aggregate plaster Waschputz *m*
plaster *v* vergipsen, in Gips legen, verputzen
plaster cast Gipsabdruck *m*, Gipsabguß *m*
plaster casting Gipsguß *m*
plaster coat Maurerschutzlack *m*
plaster electrode Gipselektrode *f*
plaster fixation Eingipsung *f*
plastering Stuckarbeit *f*, Verputz *m*
plaster kiln Gipsbrennofen *m*
plaster model Gipsmodell *n*
plaster mold Gipsform *f*

plaster of Paris [gebrannter] Gips,
   Alabastergips *m*, Stuck *m*
plaster stone Gips[stein] *m* (Min)
plaster wall Kalkwand *f*
plastic Kunststoff *m*, Plastik *n*, Preßstoff *m*,
   foamed ~ Poroplast *n*, Porenstoff *m*,
   laminated ~ Schicht[preß]stoff *m*
plastic *a* [ver]formbar, knetbar; modellierbar,
   plastisch, **fully** ~ vollplastisch
plastic alloy Knetlegierung *f*
plastic case Kunststoffbehälter *m*
plastic coated kunststoffbeschichtet
plastic coating Kunststoffbeschichtung *f*
plastic cold working Kaltverformung *f*
plastic container Kunststoffbehälter *m*
plastic dispersion Kunststoffdispersion *f*
plastic foam Schaumstoffkörper *m*
plastic foil Plastikfolie *f*
plasticine Knetgummi *m n*
plasticity Plastizität *f*,
   Formänderungsvermögen *n*, Formbarkeit *f*,
   Verformbarkeit *f*
plasticize *v* plastifizieren, plastizieren,
   weichmachen (Kunststoff), ~ **on the surface**
   angelatinieren, vorplastifizieren
plasticized PVC Weich-PVC *n*
plasticizer Fließmittel *n* (Gummi),
   Weichmacher *m* (Kunststoff), ~ **with solvent**
   properties gelatinierender Weichmacher *m*
plasticizer blend Weichmachermischung *f*
   (Kunststoff)
plasticizing Plastifizierung *f*, Weichmachen *n*
plasticizing capacity Verflüssigungsleistung *f*
plasticizing rate Weichmachungsgrad *m*
plastic limit Plastizitätsgrenze *f*
plastic lining Kunststoffauskleidung *f*
plastic masses, liquid ~ Fluidoplaste *pl*
plastic material Kunststoff *m*
plastic metal Plastikmetall *n*
plastic molding compound Kunststoffpreßmasse *f*
plastic pipe Kunststoffrohr *n*
plastic-proofed kunststoffimprägniert
plastic refractory clay Klebsand *m*
plastics Kunststoffe *pl*, **made of** ~ aus
   Kunststoff hergestellt, **pressure setting** ~
   druckhärtbare Plaste *pl*, **thermosetting** ~
   hitzehärtbare Plaste, ~ **capable of being**
   hardened härtbare Plaste
plastic sheet Kunststoffolie *f*, Kunststoffplatte *f*
plastic stopper Kunststoffstopfen *m*
plastic tank Kunststofftank *m*
plastic workability
   Formveränderungsvermögen *n*
plastic working plastische Verformung *f*
plastify *v* plastifizieren, erweichen
plastigel Plastigel *n*
plastilina Plastilin *n*
plastiline Plastilin *n*
plastisol Plastisol *n*

plastisol molding Verarbeitung *f* von Plastisol
plastite Plastit *n*
plastocyanin Plastocyanin *n*
plastograph Plastograph *m*
plastomer Kunststoff *m*
plastometer Plastizitätsmesser *m*, Plastometer *n*
plastoquinone Plastochinon *n*
plate Platte *f*, Blech *n* (Techn), Blechtafel *f*
  (Techn), Boden *m* (Rektifikation), Schild *n*,
  Tafel *f*, Teller *m*, **double-reduced** ~ doppelt
  reduziertes Blech *n*, **dual coated** ~
  differenzverzinntes Blech *n*, **heavy** ~
  Grobblech *n*, **perforated** ~ gelochtes Blech,
  thick ~ Grobblech *n*, **worked** ~ getriebenes
  Blech, ~ **with patterns** Formplatte *f*,
  Modellplatte *f*
plate *v* plattieren, überziehen
plate-and-frame press Rahmenfilterpresse *f*
plateau slope Plateauneigung *f*
plate bar Platine *f*
plate battery Anodenbatterie *f*
plate bending machine Blechbiegemaschine *f*
plate capacitance Anodenkapazität *f*
plate cell Tellerzelle *f*
plate clutch Plattenkupplung *f*
plate column Bodenkolonne *f*
plate condensor Plattenkondensator *m*
plate conveyer Plattenband *n*
plate culture Plattenkultur *f*
plate cutting machine Blechschneidemaschine *f*
plated plattiert, metallüberzogen
plated box Panzerkarton *m*
plated finish Bügelzurichtung *f* (Leder)
plate dissipation Anodenverlustleistung *f*
plate dryer Tellertrockner *m*
plate electrode Plattenelektrode *f*
plate evaporator Plattenverdampfer *m*
plate exchanger Plattenaustauscher *m*
plate freezer Plattengefrierapparat *m*
plate gauge Blechlehre *f*, Groblehre *f*
plate girder construction Vollwandbauweise *f*
plate glass Flachglas *n*, Spiegelglas *n*
plate heating furnace Blechglühofen *m*
plate holder Kassette *f* (Phot), Plattenhalter *m*
  (Phot), Plattenkassette *f* (Phot)
platelet Plättchen *n* (Blut)
plate mark Feingehaltsstempel *m*
plate mill Blechstraße *f*, Blechstrecke *f*
plate molding Plattenformerei *f*
platen Platte *f*
platen filter Plattenfilter *n*
platen mark durch die Preßplatte verursachte
  Fehlstelle *f*, Markierung *f* der Preßplatte
platen press Plattenpresse *f*, Tiegeldruckpresse *f*,
  Walze *f*
plate peak voltage Anodenspitzenspannung *f*
plate regenerator Scheibenregenerator *m*
plate resistance Anodenwiderstand *m*
plate resistor Anodenwiderstand *m*

plate roll Blechwalze *f*
plate rolling Blechwalzen *n*
plate [rolling] mill Blechwalzwerk *n* (Grobblech)
plates, number of ~ Bodenzahl *f* (Rektifikation)
plate shears Blechschere *f*
plate size Plattengröße *f*
plate spring Blattfeder *f*
plate supply Anodenspeisung *f*
plate tinning furnace Blechverzinnungsofen *m*
plate tower Bodenkolonne *f*
plate-type electrostatic machine
  Plattenelektrisiermaschine *f*
plate-type electrostatic separator
  Freifallscheider *m*
plate-type heat exchanger
  Plattenwärmeaustauscher *m*
plate valve Plattenventil *n*
plate welding Blechschweißung *f*
platform Plattform *f*, Laderampe *f*, Laufbühne *f*,
  Tribüne *f*, **finger or derrickman's** ~
  Aushängebühne *f*
platform car Plattformwagen *m*
platform conveyer Wandertisch *m*
platinammonium chloride
  Platinammoniumchlorid *n*
platinate Platinat *n*
plating Belag *m*, Beplattung *f*, Metallauflage *f*,
  Metallisieren *n*, Plattieren *n*, ~ **with gold leaf**
  Blattvergoldung *f*
plating bath Plattierungsbad *n*,
  Galvanoplastikbad *n*
plating plant Plattierungsanlage *f*
platinic acid Platinsäure *f*, **salt or ester of** ~
  Platinat *n*
platinic chloride Platin(IV)-chlorid *n*
platinic compound Platiniverbindung *f*,
  Platin(IV)-Verbindung *f*
platinic hydroxide Platin(IV)-hydroxid *n*
platinic oxide Platin(IV)-oxid *n*
platinic salt Platinisalz *n*, Platin(IV)-salz *n*
platinicyanic acid Platinicyanwasserstoffsäure *f*
platiniferous platinhaltig
platinization Platinierung *f*
platinize *v* platinieren, mit Platin überziehen
platinized platiniert
platinized asbestos Platinasbest *m*
platinized charcoal Platinkohle *f*
platinizing Platinieren *n*
platinocyanic acid Platincyanwasserstoff *m*,
  Platinocyanwasserstoffsäure *f*
platinoid Platinoid *n*
platinoid *a* platinartig
platinotype paper Platinpapier *n*
platinotype process Platindruck *m*
platinous chloride Platin(II)-chlorid *n*,
  Platinochlorid *n*
platinous compound Platin(II)-Verbindung *f*,
  Platinoverbindung *f*
platinous cyanide Platin(II)-cyanid *n*

platinous hydroxide Platin(II)-hydroxid *n*
platinous oxide Platin(II)-oxid *n*, Platinoxydul *n* (obs)
platinum (Symb. Pt) Platin *n*, bright ~ Glanzplatin *n*, brilliant ~ Glanzplatin *n*, burnished ~ Glanzplatin *n*, containing ~ platinhaltig, crude ~ Rohplatin *n*, spongy ~ Platinmohr *m*, Platinschwamm *m*, to coat with ~ platinieren
platinum alloy Platinlegierung *f*
platinum basin Platinschale *f*
platinum black Platinmohr *m*, Platinschwarz *n*
platinum boat Platinschiffchen *n*
platinum boiler Platinkessel *m*
platinum bronze Platinbronze *f*
platinum coating Platinüberzug *m*
platinum coil Platinspirale *f*
platinum compound Platinverbindung *f*
platinum cone Platinkegel *m*, Platinkonus *m*
platinum contact Platinkontakt *m*
platinum content Platingehalt *m*
platinum crucible Platintiegel *m*
platinum dichloride Platin(II)-chlorid *n*, Platinochlorid *n*
platinum dicyanide Platin(II)-cyanid *n*
platinum dish Platinschale *f*
platinum electrode Platinelektrode *f*
platinum foil Platinblech *n*
platinum group Platinreihe *f*
platinum hydrosol Platinhydrosol *n*
platinum ingot Platinbarren *m*
platinum iridium Platiniridium *n*
platinum knife Platinmesser *n*
platinum monoxide Platin(II)-oxid *n*, Platinoxydul *n* (obs)
platinum muffle Platinmuffel *f*
platinum nickel cell Platinnickelelement *n*
platinum ore Platinerz *n*
platinum pin Platinnadel *f*
platinum-plate *v* platinieren
platinum-plated platiniert
platinum-plating Platinierung *f*
platinum point Platinkontakt *m*, Platinspitze *f*
platinum potassium salt Platinkaliumsalz *n*
platinum refining Platinscheidung *f*
platinum residue Platinrückstand *m*
platinum retort Platinretorte *f*
platinum rhodium couple Platinrhodiumelement *n*
platinum rhodium wire Platinrhodiumdraht *m*
platinum sheet Platinblech *n*
platinum silicide Platinsilicid *n*
platinum solution Platinlösung *f*
platinum spatula Platinspatel *m*
platinum spiral Platinspirale *f*
platinum sponge Platinschwamm *m*
platinum spoon Platinlöffel *m*
platinum still Platinretorte *f*
platinum substitute Platinersatz *m*

platinum sulfide Platinsulfid *n*
platinum tetrachloride Platin(IV)-chlorid *n*, Platintetrachlorid *n*
platinum vessel Platingefäß *n*
platinum ware Platingerät *n*
platinum wire Platindraht *m*
platinum wire loop Platindrahtöse *f*
platinum zinc element Platinzinkelement *n*
plattnerite Plattnerit *m* (Min), Braunbleioxid *n* (Min), Schwerbleierz *n* (Min)
platynecic acid Platynecinsäure *f*
platynecine Platynecin *n*
platynite Platynit *m* (Min)
play Spielraum *m* (Bewegungsfreiheit)
pleat Falte *f*
pleat *v* falten, plissieren
pleated sheet structure Faltblattstruktur *f*
pleating Falten *n*, Plissieren *n*
pleiadiene Pleiadien *n*
Pleistocene [period] Eiszeitalter *n*, Pleistozän *n*
plenargyrite Plenargyrit *m* (Min)
plenum chamber Luftkammer *f* (Phys)
pleochroic pleochroitisch
pleochroism Pleochroismus *m* (Krist)
pleochromatic pleochroitisch
pleochromatism Pleochroismus *m* (Krist)
pleonast Pleonast *m* (Min)
pleonectite Pleonektit *m*
plessite Plessit *m* (Min), Gersdorffit *m* (Min), Nickelarsenglanz *m* (Min)
plexigel paste Plexigelpaste *f*
plexiglass (HN) Plexiglas *n*, Akrylglas *n*
pliability Biegsamkeit *f*, Geschmeidigkeit *f*
pliable biegsam, geschmeidig
pliancy Biegsamkeit *f*
pliant biegsam, geschmeidig
pliers Zange *f*, Drahtzange *f*, Falzzange *f*, Kneifzange *f*
plioform wax Plioform *n*
plodder Auspreßkneter *m*, Strangpresse *f*
plot Diagramm *n*, graphische Darstellung *f*, Plan *m*
plot *v* auftragen (Kurven etc.), aufzeichnen, graphisch darstellen, ~ a value against another einen Wert in Abhängigkeit von einem anderen darstellen
plot layout model Entwurfsmodell *n*
plotter Zeichenautomat *m*, Zeichengerät *n*
plotting, logarithmic ~ logarithmische Darstellung *f*
plotting paper Millimeterpapier *n*
plough (Br. E.) Pflug *m*
plough *v* pflügen
ploughblade mixer Druvatherm-Reaktor *m*
plow (Am. E.) Pflug *m*
plow *v* pflügen
plug Ablaßstopfen *m*, Bolzen *m*, Dübel *m*, Matrizenteil *n* (Techn), Patrize *f* (Techn), Pfropfen *m* (Med), Stecker *m* (Elektr),

Stöpsel *m* (Elektr), Tampon *m* (Med),
Verschlußstück *n*, **fusible** ~ Bleisicherung *f*,
~ **of a cock** Küken *n* (Chem), ~ **of three-way
cock** Dreiwegehahnküken *n*
plug *v* plombieren (Zahn), ~ **in** einstecken
(Elektr), ~ [up] verstopfen
plug box Steckdose *f* (Elektr), Steckkontakt *m*
(Elektr)
plug cylinder Kolbenzylinder *m*
plug flow Kolbenblasenströmung *f*,
Kolbenströmung *f*, Pfropfenströmung *f*
plug-flow reactor Kolbenstromreaktor *m*, ideales
Strömungsrohr *n*
plug fuse Stöpselsicherung *f*
plugging (Am. E.) Gegenstrombremsung *f*
(Mech)
plug head Stöpselkopf *m*
plug-in socket Steckfassung *f*
plug ramming machine Bodenstampfmaschine *f*
plug screw Absperrschraube *f*
plug valve Hahn *m* (Rohrleitung)
plug welding Lochschweißung *f*
plum Pflaume *f*
plumb Lot *n*, Senkblei *n*
plumb *v* loten, verlöten (mit Blei)
plumbagin Plumbagin *n*
plumbago Graphit *m* (Min), Reißblei *n* (Min)
plumbate Plumbat *n*
plumb bob Lot *n*, Richtblei *n*, Richtlot *n*,
Senkblei *n*, Senklot *n*
plumbeous bleiartig, bleiern
plumber Installateur *m*, Klempner *m*,
Spengler *m*
plumber's solder Lötzinn *n*
plumbic bleihaltig
plumbic compound Blei(IV)-Verbindung *f*
plumbic oxide Blei(IV)-oxid *n*, Plumbioxid *n*
plumbic salt Blei(IV)-Salz *n*, Plumbisalz *n*
plumbiferous bleiführend, bleihaltig
plumbing Rohrleitung *f*, Installation *f*,
Klempnerarbeiten *f pl*
plumbite Plumbit *n*
plumb line Lot *n*, Lotschnur *f*, Richtschnur *f*;
Senkblei *n*
plumb line deflection Lotabweichung *f*
plumbocalcite Plumbocalcit *m* (Min)
plumboferrite Plumboferrit *m* (Min)
plumbogummite Plumbogummit *m* (Min),
Gummi[blei]spat *m* (Min)
plumboniobite Plumboniobit *n* (Min)
plumbous bleihaltig
plumbous compound Blei(II)-Verbindung *f*,
Plumboverbindung *f*
plumbous-plumbic oxide Mennige *f*
plumbous salt Blei(II)-Salz *n*, Plumbosalz *n*
plumb rule Senkwaage *f*
plumieria bark Plumierarinde *f*
plumieric acid Plumierasäure *f*
plumieride Plumierid *n*

plummet Bleilot *n*, Lot *n*, Senkblei *n*
plumosite Plumosit *m* (Min)
plunge *v* [ein]tauchen
plunge battery Tauchbatterie *f*
plunge cell Tauchelement *n*
plunge cylinder Tauchzylinder *m*
plunger Plungerkolben *m*, Tauchkolben *m*,
Tauchspule *f* (Elektr)
plunger magnet Saugmagnet *m*
plunger mold Preßspritzform *f*
plunger molding Preßspritzen *n*
plunger [or screw] forward time Nachdruckzeit *f*
plunger piston Tauchkolben *m*
plunger piston pump Tauchkolbenpumpe *f*
plunger pump Plungerpumpe *f*,
Tauchkolbenpumpe *f*
plunger retainer Stemmplatte *f*
plunger test Dornprüfung *f*
plunging Eintauchen *n*
plurality Mehrzahl *f*, Vielzahl *f*
plural scattering Mehrfachstreuung *f*
plush Plüsch *m* (Text)
plush copper Chalkotrichit *m* (Min)
plus minus scale Plus-Minus-Skala *f*
plus minus tolerance Plusminustoleranz *f*
plus sign Additionszeichen *n*, Pluszeichen *n*
Pluto Pluto *m* (Astr)
plutonium (Symb. Pu) Plutonium *n*
plutonium breeder Plutoniumbrutreaktor *m*
plutonium pile Plutoniumreaktor *m*
plutonyl ion Plutonylion *n*
ply Lage *f*, Einlage *f* (Reifen), Gewebelage *f*,
Schicht *f*
ply adhesion Lagenhaftung *f*
ply down stitcher Anrollvorrichtung *f*
(Reifenwickelei)
ply glue Sperrholzkleber *m*, Sperrholzleim *m*
ply separation Lagentrennung *f*
plywood Sperrholz *n*, Furnierholz *n*,
Schichtholz *n*, **molded** ~ Formpreßholz *n*
plywood clamping ring Sperrholzspannring *m*
pneumatic Reifen *m* (Kfz)
pneumatic *a* pneumatisch, druckluftbetätigt,
luftgefüllt
pneumatic brake Druckluftbremse *f*
pneumatic[-conveyor] drier Stromtrockner *m*
pneumatic drill Preßluftbohrer *m*
pneumatic hammer Preßlufthammer *m*,
Drucklufthammer *m*
pneumatic lift pneumatischer Aufzug *m*
pneumatic pressure Luftdruck *m*
pneumatic pump Luftpumpe *f*
pneumatic rammer Preßluftstampfmaschine *f*
pneumatics Pneumatik *f* (Aeromechanik)
pneumatic tire Luftreifen *m*, Pneumatik *m*
pneumatic tube conveyor Rohrpostanlage *f*,
shuttle for ~ Rohrpostbehälter *m*
pneumatic tube system Rohrpost *f*

pneumatolysis Pneumatolyse *f* (Geol)
pneumococcosis Pneumokokkeninfektion *f*
pneumonia Lungenentzündung *f* (Med)
p-n-transition p-n-Übergang *m*
poacher Bleichholländer *m* (Pap)
poaching engine Bleichholländer *m* (Pap)
pocket Tasche *f*, Hohlraum *m*
pocket ammeter Taschenamperemeter *n*
pocket belt conveyor Gurttaschenförderer *m*
pocket book Taschenbuch *n*
pocket collecting electrode
    Fangraumniederschlagselektrode *f*
pocket edition Taschenausgabe *f*
pocket size Taschenformat *n*
pocket voltmeter Taschenvoltmeter *n*
pock wood Guajacholz *n*, Pockholz *n*
pod Samenhülse *f*, Schote *f*
podocarpic acid Podocarpinsäure *f*
podolite Podolit *m* (Min)
podophyllic acid Podophyll[in]säure *f*
podophyllin [resin] Podophyllin[harz] *n*
podophyllotoxin Podophyllotoxin *n*
podophyllum Podophyllum *n* (Pharm)
poecilitic buntsandsteinartig
point Punkt *m*, Spitze *f* (Nadel), ~ of
    attachment Befestigungsstelle *f*, ~ of attack
    Angriffspunkt *m*, ~ of ignition
    Flammpunkt *m*, ~ of introduction
    Einführungsstelle *f*, ~ of view Anschauung *f*,
    Gesichtspunkt *m*, steel ~ or hardened point
    verstählte Spitze *f*
point *v* [an]spitzen, weisen
point charge Punktladung *f*
point contact Punktberührung *f*,
    Spitzenkontakt *m*
point defect Punktfehlordnung *f*
point dipole Punktdipol *m*
point discharge Spitzenentladung *f*
pointed spitz[ig]
pointed rabble Brechstange *f*
point effect Spitzenwirkung *f*
point efficiency Punktwirkungsgrad *m*
    (Rektifikation)
point electrode Spitzenelektrode *f*
pointer Zeiger *m*
pointer galvanometer Drehspulgalvanometer *n*
pointer reading Zeigerablesung *f*
pointer setting Zeigerstellung *f*
point gauge Spitzenlehre *f*
point image punktförmige Abbildung *f*
point lattice Punktgitter *n*, ~ in space räumliches
    Punktgitter, ~ in the plane ebenes Punktgitter
point material Spitzenmaterial *n*
point of contact Auflagepunkt *m*
point of intersection Kreuzpunkt *m*, motion of ~
    Knotenpunktbewegung *f*, velocity of ~
    Knotenpunktgeschwindigkeit *f*
point of support Auflagepunkt *m*
point of view Standpunkt *m*

point position Punktlage *f*
point-shaped punktförmig
point slope method Polygonzugverfahren *n*
point source of light punktförmige Lichtquelle *f*
point support Spitzenlagerung *f*
point surface transformation
    Punktflächentransformation *f*
point suspension Spitzenlagerung *f*
point transformation Punkttransformation *f*
poise Poise *n* (Viskositätseinheit)
Poiseuille's formula Poiseuillesches Gesetz *n*
poison Gift *n*, Giftstoff *m*, extraction of ~
    Entgiftung *f*, paralysing ~ lähmendes Gift,
    slow ~ schleichendes Gift
poison *v* vergiften
poison bait Ködergift *n*
poisoned wheat Giftweizen *m* (gegen Ratten)
poison gas Gaskampfstoff *m*, Giftgas *n*,
    Kampfgas *n*
poison gland Giftdrüse *f*
poison grain Giftkorn *n*
poisoning Vergiften *n*, Vergiftung *f*, symptom of
    ~ Vergiftungserscheinung *f* (Med), ~ by gas
    Gasvergiftung *f*, ~ of the catalyst
    Vergiftung *f* des Katalysators
poison oak, extract of ~ Giftsumachextrakt *m*
poisonous giftig, toxisch
poisonous action Giftwirkung *f*
poisonous effect Giftwirkung *f*
poisonous matter Giftstoff *m*
Poisson's equation Poissonsche Gleichung *f*
Poisson's ratio Poissonsche Konstante *f*
poke *v* stochern, stoßen
poker Schürstange *f*, Herdeisen *n*, Rühreisen *n*
polar polar
polar angle Polarwinkel *m*
polar attraction Polanziehung *f*
polar axis Polarachse *f* (Geom)
polar blackout Polverdunklung *f*
polar circle Polarkreis *m*
polar coordinate Polarkoordinate *f* (Math)
polarimeter Polarimeter *n*
polarimetric polarimetrisch
polarimetry Polarimetrie *f*
polariscope Polariskop *n*
polarity Polarität *f*, reversal of ~ Polumkehr *f*, to
    reverse the ~ umpolarisieren
polarity indicator Stromrichtungsanzeiger *m*
polarity paper Polreagenspapier *n*
polarizability Polarisierbarkeit *f*
polarizability tensor Polarisierbarkeitstensor *m*
polarizable polarisierbar
polarization Polarisation *f*, Polarisierung *f*, angle
    of ~ Polarisationswinkel *m*, circular ~
    zirkulare Polarisation, elliptical ~ elliptische
    Polarisation, left-handed ~
    Linkspolarisation *f*, phenomenon of ~
    Polarisationserscheinung *f*, plane of ~
    Polarisationsebene *f*, plane or linear ~ lineare

Polarisation, **reversal of** ~ Umpolarisation *f,* **right-handed** ~ Rechtspolarisation *f,* **state of** ~ Polarisationszustand *m*
polarization apparatus Polarisationsapparat *m*
polarization capacity Polarisationskapazität *f*
polarization color Polarisationsfarbe *f*
polarization constant Polarisationskonstante *f*
polarization current Polarisationsstrom *m*
polarization curve Polarisationskurve *f*
polarization detection Polarisationsnachweis *m*
polarization electrode Polarisationselektrode *f*
polarization energy Polarisationsenergie *f*
polarization fading Polarisationsschwund *m*
polarization resistance Polarisationswiderstand *m*
polarization state Polarisationszustand *m*
polarization unit vector Einheitsvektor *m* für Polarisation (Elektr)
polarization voltage Polarisationsspannung *f*
polarize *v* polarisieren
polarized polarisiert, **elliptically** ~ elliptisch polarisiert
polarizer Polarisator *m,* Polarisationsprisma *n*
polarizing polarisierend
polarizing grating Polarisationsgitter *n*
polarizing instrument Polarisationsinstrument *n*
polarizing microscope Polarisationsmikroskop *n*
polarizing prism Polarisationsprisma *n*
polarogram Polarogramm *n*
polarograph Polarograph *m*
polarographic polarographisch
polarography Polarographie *f,* polarographische Bestimmung *f*
polar orbit Polarbahn *f*
polar planimeter Polarplanimeter *n*
polar region Polarkreis *m*
polar tension Polspannung *f*
pole Pfosten *m,* Pol *m* (Elektr, Phys), Stab *m,* Stange *f,* **consequent** ~ Folgepol *m,* **inducing** ~ induzierender Pol, **length of** ~ Pollänge *f,* **negative** ~ Minuspol *m,* negativer Pol, **positive** ~ Pluspol *m,* positiver Pol
pole *v* polen
pole arc, length of ~ Polbogenlänge *f*
pole armature Polanker *m*
polecat fat Iltisfett *n*
pole center Polmitte *f*
pole changer Polwechsler *m*
pole changing [control] Polumschaltung *f*
pole-changing switch Polwechsler *m*
pole core Polkern *m*
pole dislocation Polversetzung *f*
pole edge Polkante *f*
pole end Polende *n*
pole face Polfläche *f*
pole finder Polsucher *m*
pole finding paper Polsuchpapier *n*
pole formation Polbildung *f*
pole induction Polinduktion *f*

pole paper Pol[reagens]papier *n*
pole reversal Polumkehr *f*
pole ring Polbüchse *f,* Polring *m*
poles, change of ~ Polwechsel *m,* **distance between** ~ Polabstand *m,* Polentfernung *f,* **like** ~ gleichnamige Pole *pl,* **opposite** ~ ungleichnamige Pole *pl,* **pair of** ~ Polpaar *n,* **production of** ~ Polerzeugung *f,* **reversion of the** ~ Umpolung *f,* **similar** ~ gleichnamige Pole *pl,* **to reverse the** ~ umpolen
pole saturation Polsättigung *f*
pole shoe Polschraube *f*
pole shoe bore Polschuhbohrung *f*
pole shoe tip Polschuhspitze *f*
pole strength Polstärke *f*
pole terminal Polklemme *f* (Elektr)
polianite Polianit *m* (Min), Pyrolusit *m* (Min)
poling (of lead) Polen *n* (des Bleis)
polio inoculation Polioimpfung *f* (Med)
polio[myelitis] Kinderlähmung *f* (Med), Poliomyelitis *f* (Med)
poliomyelitis vaccine Polioimpfstoff *m* (Med)
polio serum Polioserum *n*
polio vaccine Polioimpfstoff *m,* Poliovakzine *f* (Med)
polish Politur *f,* Glätte *f,* Glanz *m,* Glanzmittel *n,* Poliermittel *n,* Poliersand *m* (Techn), **brilliant** ~ Hochglanzpolitur *f,* **high luster** ~ Hochglanzpolitur *f,* **resistance to** ~ Schleifhärte *f* (Techn), **to give the first** ~ vorpolieren, **to rough** ~ feuerschleifen (Techn)
polish *v* polieren, abschleifen (Techn), abschmirgeln (Techn), blank reiben, gerben (Met), glänzend machen, glätten, putzen (Schuhe), ~ **with acid** ätzpolieren, ~ **with emery** abschmirgeln
polishable polierbar
polished glatt, poliert
polished section Schliffbild *n*
polisher Polierer *m,* Glätter *m,* Polierscheibe *f,* Polierstahl *m*
polishing Polieren *n,* Chevillieren *n* (Kunstseide), Glanzschleifen *n,* Hochglanzgebung *f;* Schmirgeln *n*
polishing cask Pulverglättfaß *n*
polishing chalk Polierkalk *m*
polishing cloth Polierfilz *m,* Poliertuch *n*
polishing composition Poliermasse *f*
polishing disc Polierscheibe *f*
polishing drum Poliertrommel *f*
polishing dust Polierstaub *m*
polishing liquid Polierflüssigkeit *f*
polishing machine Poliermaschine *f*
polishing material Poliermittel *n,* Schleifmittel *n*
polishing oil Polituröl *n*
polishing paste Polierpaste *f*
polishing pickle Glanzbrenne *f*
polishing plate Polierblech *n*

polishing powder Polierpulver *n*
polishing room Polierraum *m*
polishing slate Polierschiefer *m* (Min)
polishing spirit Politurspiritus *m*
polishing stone Polierstein *m*
polishing tool Glättwerkzeug *n*
polishing wax Bohnerwachs *n*, Polierwachs *n*
polishing wheel Polierscheibe *f*
polishing wool Putzwolle *f*
pollen Blütenstaub *m* (Bot), Pollen *m* (Bot)
pollucite Pollucit *m* (Min), Pollux *m* (Min)
pollute *v* verunreinigen, beschmutzen
polluted verunreinigt
pollution Verschmutzung *f*, Verunreinigung *f*,
  atmospheric ~ Verunreinigung *f* der Luft,
  source of ~ Verunreinigungsherd *m*
pollux Pollux *m* (Min), Pollucit *m* (Min)
polonium (Symb. Po) Polonium *n*
polyacenaphthylene Polyacenaphthylen *n*
polyacid Polysäure *f*
polyacrylamide Polyacrylamid *n*
polyacrylamide electrophoresis
  Polyacrylamidelektrophorese *f*
polyacrylate Polyacrylat *n*,
  Polyacrylsäureester *m*
polyacryl fiber Polyacrylfaser *f*
polyacrylic acid Polyacrylsäure *f*
polyacrylic resin Polyacrylharz *n*
polyacrylonitrile Polyacrylnitril *n*
polyacrylonitrile fiber Polyacrylnitrilfaser *f*,
  PAN-Faser *f*
polyaddition Polyaddition *f*
polyadelphite Polyadelphit *m* (Min)
polyaffinity theory Polyaffinitätstheorie *f*
polyalanine Polyalanin *n*
polyalcohol Polyalkohol *m*
polyalkyl methacrylate
  Polymethacrylsäurealkylester *m*
polyalthic acid Polyalthsäure *f*
polyamide Polyamid *n*
polyamide fiber Polyamidfaser *f*
polyamine Polyamin *n*
polyargyrite Polyargyrit *m* (Min)
polyarsenite Polyarsenit *m* (Min)
polyatomic mehratomig, vielatomig
polyazo dyestuff Polyazofarbstoff *m*
polybasic mehrbasisch, vielbasisch
polybasicity Mehrbasigkeit *f*
polybasite Polybasit *m* (Min), Eugenglanz *m*
  (Min)
polybutene Polybuten *n*, Polybutylen *n*
polybutylene Polybuten *n*, Polybutylen *n*
polychloroprene Polychloropren *n*
polychlorotrifluoroethylene
  Polychlortrifluoräthylen *n*
polychroism Polychroismus *m* (Opt)
polychroite Polychroit *n*
polychromatic polychrom, mehrfarbig, vielfarbig
polychromatic process Kohledruck *m*

polychrome polychrom, vielfarbig
polychromic acid Polychromsäure *f*
polychromy Polychromie *f*
polycinnamic acid Polyzimtsäure *f*
polycondensation Polykondensation *f*
polycondensation product Polykondensat *n*,
  Polykondensationsprodukt *n*
polycrase Polykras *m* (Min)
polycrasite Polykrasit *m* (Min)
polycrystal Vielkristall *m*
polycrystalline polykristallin
polycyclic polyzyklisch
polycylindrical mehrzylindrig
polydimensional allseitig
polydisperse polydispers
polydispersity Polydispersität *f*
polydymite Polydymit *m* (Min)
polyelectrolyte Polyelektrolyt *m*
polyenergetic polyenergetisch
polyester Polyester *m*
polyester fabric Polyestergewebe *n*
polyester fiber Polyesterfaser *f*
polyester putty Glasfaserkitt *m*
polyester resin Polyesterharz *n*
polyester urethane Polyesterurethan *n*
polyether Polyäther *m*
polyethylene Polyäthylen *n*
polyethylene film Polyäthylenfilm *m*
polyethylene glycol Polyäthylenglykol *n*
polyethylene polyamine Polyäthylenpolyamin *n*
polyfoil vielblätterig
polyfunctional polyfunktionell
polygala Kreuzblume *f* (Bot), Polygala *f* (Bot)
polygalic acid Polygalin *n*
polygalin Polygalin *n*, Polygalasäure *f*
polyglucuronic acid Polyglucuronsäure *f*
polyglycerol Polyglycerin *n*
polyglycol Polyglykol *n*
polyglycolide Polyglykolid *n*
polygon Polygon *n* (Math), Vieleck *n* (Math)
polygonal vieleckig, polygonal
polygonization Polygonisierung *f*
polyhalide Polyhalogenid *n*
polyhalite Polyhalit *m* (Min)
polyhedral polyedrisch, vielflächig
polyhedron Polyeder *n*, Vielflächner *m*
polyheteroatomic polyheteroatomig
polyhydric mehrere Hydroxylgruppen
  enthaltend (Alkohol)
polyhydric alcohol mehrwertiger Alkohol,
  Polyalkohol *m*, Polyol *n*
polyhydroxy alcohol Polyalkohol *m*
polyhydroxy aldehyde Polyhydroxy[l]aldehyd *m*
polyhydroxy ketone Polyhydroxyketon *n*
polyisobutylene Polyisobutylen *n*
polymer Polymer[e] *n*, Polymerisat *n*,
  Polymerisationsprodukt *n*, **low temperature** ~
  Tieftemperaturpolymerisat *n*, **mixed** ~
  Mischpolymer *n*

polymerase Polymerase *f* (Biochem)
polymer dispersion Kunststoffdispersion *f*
polymer gasoline Polymerbenzin *n*
polymeric polymer
polymeride Polymer[e] *n*, Polymerisat *n*
polymerism Polymerie *f*
polymerizable polymerisierbar
polymerizate Polymerisat *n*,
    Polymerisationsprodukt *n*
polymerization Polymerisation *f*, **bead** ~
    Perlpolymerisation *f*, **degree of** ~
    Polymerisationsgrad *m*, **grain** ~
    Perlpolymerisation *f*, ~ **in solution**
    Lösungspolymerisation *f*
polymerization catalyst
    Polymerisationskatalysator *m*
polymerization emulsion
    Polymerisationsemulsion *f*
polymerization inhibitor
    Polymerisationsinhibitor *m*
polymerization reactor Polymerisationsanlage *f*
polymerize *v* polymerisieren
polymerized, highly ~ hochpolymer
polymers, linear ~ lineare Polymere *pl*
polymethacrylate Polymethakrylsäureester *m*
polymethylene Polymethylen *n*
polymethylene diamine Polymethylendiamin *n*
polymethylene halide Polymethylenhalogenid *n*
polymignite Polymignit *m* (Min)
polymnite Polymnit *m*
polymolybdic acid Polymolybdänsäure *f*
polymorphic polymorph, heteromorph
polymorphism Polymorphie *f*,
    Polymorphismus *m*, Vielgestaltigkeit *f*
polymorphous polymorph, vielgestaltig
polymorphy Polymorphismus *m*
polymyxin Polymyxin *n*
polyneuridinic acid Polyneuridinsäure *f*
polynitro complex Polynitrokomplex *m*
polynomial Polynom *n* (Math)
polynomial *a* vielgliedrig (Math)
polynuclear mehrkernig, vielkernig
polynucleotide Polynucleotid *n*
polyol Polyol *n*
polyoma virus Polyomavirus *n*
polyoxyethylene Polyoxyäthylen *n*
polyoxymethylene Polyoxymethylen *n*
polypeptidase Polypeptidase *f* (Biochem)
polypeptide Polypeptid *n*
polypeptide chain Polypeptidkette *f*
polyphase mehrphasig
polyphase armature Mehrphasenanker *m*
polyphase current Mehrphasen[wechsel]strom *m*,
    Drehstrom *m*
polyphase furnace Mehrphasenofen *m*
polyphase generator Mehrphasengenerator *m*
polyphenoloxidase Polyphenoloxydase *f*
    (Biochem)
polyphosphatase Polyphosphatase *f* (Biochem)

polyphosphate Polyphosphat *n*
polyphosphoric acid Polyphosphorsäure *f*
polyporenic acid Polyporensäure *f*
polyporic acid Polyporsäure *f*
polypropylene Polypropylen *n*
polypropylene glycol Polypropylenglykol *n*
polyribose nucleotide Polyribonukleotid *n*
polyribosom Polyribosom *n*
polysaccharase Polysaccharase *f* (Biochem)
polysaccharide Polysaccharid *n*, **sulfated** ~
    sulfatisiertes Polysaccharid
polysaccharide synthesis Polysaccharidsynthese *f*
poly[sacchar]ose Polyose *f*, Polysaccharose *f*
polysalicylide Polysalicylid *n*
polysilicic acid Polykieselsäure *f*
polysiloxan Polysiloxan *n*
polysome Polyribosom *n*, Polysom *n*
polystyrene Polystyrol *n*
polystyrene resin Polystrolharz *n*
polysulfide Polysulfid *n*
polysulfide rubber Polysulfidkautschuk *m*
polysymmetry Polysymmetrie *f*
polytechnical school Polytechnikum *n*
polytechnics Polytechnik *f*
polytelite Polytelit *m*
polyterpene Polyterpen *n*
polytetrafluoroethylene Polytetrafluoräthylen *n*,
    Teflon *n* (HN)
polythionate Polythionat *n*
polythionic acid Polythionsäure *f*
polytrifluorochloroethylene
    Polytrifluorchloräthylen *n*
polytropic polytrop
polyvalence Mehrwertigkeit *f* (Chem),
    Vielwertigkeit *f*
polyvalent mehrwertig (Chem), vielwertig
polyvinyl acetal Polyvinylacetal *n*
polyvinyl acetate Polyvinylacetat *n*
polyvinyl alcohol Polyvinylalkohol *m*
polyvinyl alkyl ether Polyvinylalkyläther *m*
polyvinyl carbazole Polyvinylcarbazol *n*
polyvinyl chloride Polyvinylchlorid *n*, PVC *n*,
    postchlorinated ~ nachchloriertes
    Polyvinylchlorid
polyvinyl cyanide Polyvinylcyanid *n*
polyvinyl methyl ether Polyvinylmethyläther *m*
polyvinyl pyrrolidone Polyvinylpyrrolidon *n*
pomade Pomade *f*, Fettsalbe *f*
pomatum Pomade *f*
Pompey red Pompejanischrot *n*
Ponceau red Ponceaurot *n*
ponder *v* abwägen, bedenken
ponderability Wägbarkeit *f* (Phys)
ponderable wägbar
ponderous schwer, gewichtig
pool Becken *n*
pool boiling Sieden *n* bei freier Konvektion
poor arm, mager (Bergb), mangelhaft, schlecht,
    ungenügend, **to become** ~ verarmen

poor gas Schwachgas *n*
poor lime Magerkalk *m*
poplar Pappel *f* (Bot)
poplar oil Pappelöl *n*
poplin Popelin *m* (Text)
poppy Mohn *m* (Bot) (Bot, Pharm)
poppy oil Mohnöl *n*
poppy seed Mohnsamen *m* (Bot)
population genetics Bevölkerungsgenetik *f*
populene Populen *n*
populin Populin *n*, Benzoylsalicin *n*
populinate *v* mit Populin behandeln
porcelain Porzellan *n*, fusible ~
Kryolithporzellan *n*, hard ~
Feldspatporzellan *n*, Steinporzellan *n*,
manufacture of ~ Porzellanmanufaktur *f*,
opaque ~ feines Steingut *n*, soft ~
Weichporzellan *n*, Fritt[en]porzellan *n*, ~ for
chemical-technical purposes
chemisch-technisches Porzellan, ~ for
punching Stanzporzellan *n*
porcelain beaker Porzellanbecher *m*
porcelain boat Porzellanschiffchen *n*
porcelain cement Porzellankitt *m*
porcelain clay Porzellanerde *f*, Kaolin *n*,
Porzellanton *m*
porcelain color Porzellan-Schmelzfarbe *f*
porcelain container Porzellanbehälter *m*
porcelain crucible Porzellantiegel *m*
porcelain cup Porzellanschale *f*
porcelain enamel Weißemail *n*
porcelain evaporating basin
Porzellanabdampfschale *f*
porcelain funnel Porzellantrichter *m*,
Saugfilter *m*
porcelain insulator Porzellanisolator *m*
porcelain jasper Porzellanjaspis *m* (Min),
Porzellanit *m* (Min)
porcelain kiln Porzellanbrennofen *m*
porcelain manufacture Porzellanfabrikation *f*
porcelain pan Porzellankasserolle *f*
porcelain paste Porzellanmasse *f*
porcelain ring Porzellanring *m*
porcelain spar Porzellanspat *m* (Min)
porcelain tank Porzellanbehälter *m*
porcelain trough Porzellanwanne *f*
porcelain tube Porzellanröhre *f*, Porzellanrohr *n*
porcelain ware Porzellangerät *n*,
Porzellanwaren *pl*
porcelaneous porzellanartig
porcellanite Porzellanit *m* (Min),
Porzellanjaspis *m* (Min)
pore Pore *f*, size of ~ Porengröße *f*
pore conductivity Porenleitfähigkeit *f*
pore fluid Porenflüssigkeit *f*
pore fluid inelasticity
Porenflüssigkeitsinelastizität *f*
pore size Porengröße *f*, Porenweite *f*
pore space Porenraum *m*, Hohlraum *m*

pore testing agent Porenprüfmittel *n*
poriferous porös, porig
pork Schweinefleisch *n*
poroidine Poroidin *n*
poromeric material Fließleder *n*
porosimetry Porosimetrie *f*
porosity Porosität *f*, Durchlässigkeit *f*,
Lunkerung *f* (Techn), Porenweite *f*,
Undichtigkeit *f*, degree of ~ Porositätsgrad *m*,
Undichtigkeitsgrad *m*, of fine ~ kleinluckig
(Techn)
porosity determination equipment
Porenprüfgerät *n*
porous porös, durchlässig, löcherig, porig,
schaumig (Gieß), schwammig, coarsely ~
grobporig
porous concrete Schaumbeton *m*
porous earthenware Tongut *n*
porous iron Eisenschwamm *m*
porousness Porosität *f*, Undichtigkeit *f*
porous point Lunkerstelle *f*
porous pot Tonzylinder *m*
porous soap Schwammseife *f*
porous spot Lunkerstelle *f*
porpezite Porpezit *m* (Min), Palladiumgold *n*
(Min)
porphin[e] Porphin *n*
porphin ring Porphingerüst *n*, Porphinring *m*
porphin structure Porphingerüst *n*
porphobilinogen Porphobilinogen *n*
porphyrazine Porphyrazin *n*
porphyrexide Porphyrexid *n*
porphyric acid Euxanthon *n*
porphyric basalt Basaltporphyr *m* (Min)
porphyrilic acid Porphyrilsäure *f*
porphyrilin Porphyrilin *n*
porphyrin Porphyrin *n*
porphyrin biosynthesis Porphyrinbiosynthese *f*
porphyrine ring Porphyrinring *m*
porphyrin metabolism Porphyrinstoffwechsel *m*
porphyrite Porphyrit *m* (Min)
porphyritic porphyrartig
porphyritic schist Porphyrschiefer *m* (Min)
porphyritic trap Trapporphyr *m* (Min)
porphyroid Porphyroid *n*
porphyropsin Porphyropsin *n*
porphyroxine Porphyroxin *n*
porphyry Porphyr *m* (Min), argillaceous ~
Tonporphyr *m* (Min), black ~ schwarzer
Porphyr, secondary ~ Flözporphyr *m* (Min),
tufaceous ~ Fleckenporphyr *m* (Min)
port Öffnung *f*
portable tragbar, beweglich
portable box Transportkasten *m*
portable mixer Anklemmrührer *m*
portal jib crane Portalkran *m*
portative force Tragkraft *f*
portion Teil *m*, Anteil *m*, Portion *f*, Quantum *n*
portions, in ~ portionsweise

portland blast furnace cement
Portlandhochofenzement *m*
portland cement Portlandzement *m*
portland clinker Portlandklinker *m*
portland limestone Portlandkalk *m*
position Lage *f,* Ort *m,* Stand *m,* Stellung *f*
(Chem), **change of** ~ Lageänderung *f,*
Ortswechsel *m,* **disengaged** ~
Ausrückstellung *f,* **horizontal** ~ liegende
Stellung, **inclined** ~ geneigte Lage, **initial** ~
Anfangslage *f,* **normal** ~ Normalstellung *f,*
off ~ Ausschaltstellung *f,* **original** ~
Ausgangslage *f,* **upright** ~
Hochkantstellung *f,* **vertical** ~
Senkrechtstellung *f,* **to be in a** ~ [to] imstande
sein
positional lagemäßig
positional accuracy Positionsgenauigkeit *f*
positioner Stellmotor *m,* Stellorgan *n*
(Regeltechnik), Stellwerk *n*
positioner booster Anstellverstärker *m*
position fixing Ortung *f*
positioning bolt Anstellschraube *f*
position isomerism Stellungsisomerie *f*
position measurement Ortsmessung *f*
position vector Ortsvektor *m*
positive Positiv *n* (Phot)
positive *a* positiv
positive displacement meter Volumenzähler *m*
positive electrode Anode *f*
positive lens Konvexlinse *f* (Opt)
positive mold Füllraum-Werkzeug *n,*
Positivform *f*
positron Positron *n,* positives Elektron *n*
positron decay Positronenzerfall *m* (Atom)
positron disintegration Positronenzerfall *m*
(Atom)
positron emission Positronenemission *f* (Atom),
Positronenstrahlung *f*
positron emitter Positronstrahler *m*
positronium Positronium *n*
positron track Positronenbahn *f*
poskine Poskin *n*
possibility Möglichkeit *f*
post Ständer *m,* Pfosten *m*
post-acceleration Nachbeschleunigung *f*
postage Porto *n*
post-chlorinate *v* nachchlorieren
post-chlorinated nachchloriert
poster Plakat *n*
poster pillar Plakatsäule *f*
post-exposure Nachbelichtung *f* (Phot)
post-fermentation Nachgärung *f*
postformed molding nachgeformtes Formteil *n*
postforming Nachformen *n,* nachträgliche
Formung *f*
postforming sheet formbare Folie *f*
posthypophysis Hypophysenhinterlappen *m*
(Anat)

post-mortem Obduktion *f*
postpone *v* verschieben, aufschieben
postponed aufgeschoben
post-treat *v* nachbehandeln
postulate Postulat *n,* Voraussetzung *f*
postulate *v* postulieren, voraussetzen, fordern
posture Haltung *f,* Stellung *f*
pot Kochtopf *m*
potable trinkbar
potable water Trinkwasser *n*
pot annealing furnace Topfglühofen *m*
potash Pottasche *f,* Kaliumcarbonat *n,* **caustic** ~
Kaliumhydroxid *n,* Ätzkali *n,* **pure** ~
Perlasche *f*
potash alum Aluminiumkaliumsulfat *n,*
Kali[um]alaun *m*
potash bulb Kaliapparat *m,* Kalikugel *f*
potash calcining Pottaschebrennen *n*
potash factory Pottaschesiederei *f*
potash feldspar gemeiner Feldspat *m* (Min),
Kalifeldspat *m* (Min), Orthoklas *m* (Min)
potash fertilizer Kalidünger *m,* Kalidüngesalz *n*
potash-free kalifrei
potash fusion Kalischmelze *f*
potash glass Kaliglas *n*
potash iron alum Kalieisenalaun *m*
potash lead glass Kalibleiglas *n*
potash melt Kalischmelze *f*
potash mica Kaliglimmer *m* (Min)
potash mine Kalibergwerk *n*
potash niter Kalisalpeter *m*
potash salt Kalisalz *n,* **bed of** ~ Kalisalzlager *n*
potash sifter Laugenaschensieber *m*
potash soap Kaliseife *f,* Schmierseife *f*
potash soft soap Kalischmierseife *f*
potash solution Pottaschelösung *f*
potash vat Pottascheküpe *f*
potash water Pottaschewasser *n*
potash waterglass Kali[um]wasserglas *n*
potash works Kaliwerk *n*
potassic kaliumhaltig
potassiferous kaliumhaltig, kalihaltig
potassium (Symb. K) Kalium *n*
potassium acetate Kaliumacetat *n*
potassium alum Aluminiumkaliumsulfat *n,*
Kali[um]alaun *m*
potassium aluminate Kalitonerde *f,*
Kaliumaluminat *n*
potassium aluminum silicate, native ~
Kalifeldspat *m* (Min)
potassium aluminum sulfate Kalialaun *m,*
Kaliumaluminiumsulfat *n,* **native** ~
Kalinit *m* (Min)
potassium ammonium antimonic bitartrate
Antiluetin *n*
potassium ammonium nitrate
Kaliammonsalpeter *m*
potassium ammonium tartrate
Kaliumammoniumtartrat *n*

potassium antimonate Kaliumantimonat *n*
potassium antimonyl tartrate
Antimonkaliumtartrat *n*, Brechweinstein *m*
potassium arsenate Kaliumarsenat *n*
potassium arsenite Kaliumarsenit *n*
potassium aurichloride
Kaliumgold(III)-chlorid *n*
potassium aurobromide Aurokaliumbromid *n*,
Kaliumgold(I)-bromid *n*
potassium aurocyanide Aurokaliumcyanid *n*,
Goldkaliumcyanür *n* (obs)
potassium bicarbonate Kaliumbikarbonat *n*,
Kaliumhydrogenkarbonat *n*
potassium bichromate Kaliumbichromat *n*
potassium binoxalate Kaliumbioxalat *n*
potassium bioxalate Kaliumbioxalat *n*,
Sauerkleesalz *n* (obs)
potassium bisulfate Kaliumbisulfat *n*
potassium bisulfite Kaliumbisulfit *n*
potassium bitartrate Kaliumbitartrat *n*, Cremor
Tartari *m*
potassium borate glass Kaliumboratglas *n*
potassium borofluoride Kaliumborfluorid *n*
potassium borotartrate Boraxweinstein *m*
potassium bromate Kaliumbromat *n*
potassium bromide Kaliumbromid *n*
potassium carbonate Kaliumkarbonat *n*,
Pottasche *f*, **acid ~** Kaliumbikarbonat *n*
potassium chlorate Kaliumchlorat *n*
potassium chloride Kaliumchlorid *n*
potassium chlorite Kaliumchlorit *n*
potassium chloroplatinate
Kaliumchloroplatinat *n*
potassium chromate Kaliumchromat *n*, **red ~**
Kaliumdichromat *n*
potassium chromium sulfate
Kaliumchromalaun *m*
potassium citrate Kaliumcitrat *n*
potassium compound Kaliumverbindung *f*
potassium cuprocyanide
Kaliumkupfer(I)-cyanid *n*
potassium cyanate Kaliumcyanat *n*
potassium cyanide Cyankalium *n*,
Kaliumcyanid *n*
potassium cyanide liquor Cyankaliumlauge *f*
potassium cyanocobaltate(III)
Kobalticyankalium *n*
potassium cyanocuprate(I)
Kaliumkupfer(I)-cyanid *n*
potassium cyanoferrate(III) Ferricyankalium *n*
potassium cyanoplatinite
Kaliumplatin(II)-cyanid *n*
potassium dichromate Kaliumbichromat *n*,
Kaliumdichromat *n*
potassium dihydrogen phosphate
Kaliumdihydrogenphosphat *n*
potassium double salt Kaliumdoppelsalz *n*
potassium ferrate Kaliumferrat *n*
potassium ferric oxalate Kaliumferrioxalat *n*

potassium ferric sulfate Eisenalaun *m*,
Kaliumferrisulfat *n*
potassium ferricyanide Kaliumferricyanid *n*,
Ferricyankalium *n*,
Kaliumeisen(III)-cyanid *n*, rotes
Blutlaugensalz *n*
potassium ferrocyanide Kaliumferrocyanid *n*,
Ferrocyankalium *n*, gelbes Blutlaugensalz *n*,
Kaliumeisen(II)-cyanid *n*
potassium fluoborate Kaliumborfluorid *n*
potassium fluoride Kaliumfluorid *n*
potassium fluosilicate Kaliumsilicofluorid *n*,
Kieselfluorkalium *n*
potassium glycerophosphate
Kaliumglycerophosphat *n*
potassium guaiacol sulfonate
Kaliumsulfoguajacol *n*, Thiocol *n* (HN)
potassium halide Kaliumhalogenid *n*
potassium hydride Kaliumwasserstoff *m*
potassium hydrogen carbonate
Kaliumbicarbonat *n*
potassium hydrogen oxalate Kaliumbioxalat *n*
potassium hydrogen sulfite Kaliumbisulfit *n*
potassium hydrogen tartrate Kaliumbitartrat *n*
potassium hydroxide Kaliumhydroxid *n*,
Ätzkali *n*
potassium hydroxide solution Kalilauge *f*
potassium hypochlorite solution Eau de Javelle *n*,
Kaliumhypochloritlösung *f*
potassium iodide Kaliumjodid *n*
potassium iodide ointment Kaliumjodidsalbe *f*
(Pharm)
potassium iodide starch Kaliumjodidstärke *f*
potassium iodide starch paper
Jod[kali]stärkepapier *n*
potassium iridichloride Kaliumiridichlorid *n*
potassium iron(III) sulfate Ferrikaliumsulfat *n*
potassium iron(II) sulfate Ferrokaliumsulfat *n*
potassium magnesium sulfate
Kaliummagnesiumsulfat *n*
potassium manganate Kaliummanganat *n*
potassium manganic sulfate Manganalaun *m*
potassium mercuricyanide
Kaliumquecksilber(II)-cyanid *n*
potassium mercurocyanide
Kaliumquecksilber(I)-cyanid *n*
potassium mercury iodide
Kaliumquecksilberjodid *n*
potassium metal Kaliummetall *n*
potassium monosulfide Kaliummonosulfid *n*
potassium monoxide Kaliummonoxid *n*
potassium myronate Sinigrin *n*
potassium niobate Kaliumniobat *n*
potassium nitrate Salpeter *m*, Kalisalpeter *m*,
Kaliumnitrat *n*
potassium nitrate superphosphate
Kalisalpetersuperphosphat *n*
potassium oxalate Kaliumoxalat *n*, **acid ~**
Kaliumbioxalat *n*

potassium oxide Kaliumoxid *n*
potassium perchlorate Kaliumperchlorat *n*
potassium periodate Kaliumperjodat *n*
potassium permanganate Kaliumpermanganat *n*
potassium persulfate Kaliumpersulfat *n*
potassium phenolate Phenolkalium *n*
potassium phenolsulfonate
    Kaliumphenolsulfonat *n*
potassium phosphate Kaliumphosphat *n*
potassium platinichloride
    Kaliumplatin(IV)-chlorid *n*
potassium platinochloride
    Kaliumplatin(II)-chlorid *n*
potassium platinocyanide
    Kaliumplatin(II)-cyanid *n*
potassium polysulfide Kaliumpolysulfid *n*
potassium porphin chelate
    Kaliumporphinchelat *n*
potassium prussiate, red ~ Ferricyankalium *n*,
    Kaliumeisen(III)-cyanid *n*,
    Kaliumferricyanid *n*, **yellow** ~
    Ferrocyankalium *n*, Kaliumeisen(II)-cyanid *n*,
    Kaliumferrocyanid *n*
potassium pyrosulfite Kaliumpyrosulfit *n*
potassium regulus Kaliumregulus *m*
potassium salt Kaliumsalz *n*, Kalisalz *n*, **deposit
    of** ~ Kalisalzlager *n*, ~ **for fertilizing**
    Kalidüngesalz *n*
potassium selenate Kaliumselenat *n*
potassium silicate Kaliumsilicat *n*,
    Kaliumwasserglas *n*
potassium silicofluoride Kaliumsilicofluorid *n*,
    Kieselfluorkalium *n*
potassium soap Kaliumseife *f*
potassium sodium carbonate
    Kaliumnatriumcarbonat *n*
potassium sodium feldspar Kalinatronfeldspat *m*
    (Min)
potassium sodium tartrate
    Kaliumnatriumtartrat *n*, Rochellesalz *n*
potassium sulfate Kaliumsulfat *n*, **acid** ~
    Kaliumbisulfat *n*
potassium sulfide Kaliumsulfid *n*,
    Schwefelkalium *n*
potassium sulfite Kaliumsulfit *n*, **acid** ~
    Kaliumbisulfit *n*
potassium tantalate Kaliumtantalat *n*
potassium tartrate Kaliumtartrat *n*,
    Kremortartari *m*, **acid** ~ Kaliumbitartrat *n*
potassium tetroxalate Kaliumtetraoxalat *n*
potassium thiocyanate Kaliumrhodanid *n*
potassium thiosulfate Kaliumthiosulfat *n*
potassium trisulfide Kaliumtrisulfid *n*
potassium tungstate Kaliumwolframat *n*
potassium vapor Kaliumdampf *m*
potassium xanthogenate Kaliumxanthogenat *n*
potato Kartoffel *f*
potato brandy Kartoffelbranntwein *m*

potato flour Kartoffelmehl *n*, **dextrinized** ~
    Kartoffelwalzmehl *n*
potato peel Kartoffelschale *f*
potato spirit Kartoffelbranntwein *m*
potato starch Kartoffelstärke *f*
potato sugar Kartoffelzucker *m*
potency Potenz *f*, Wirksamkeit *f*
potent kräftig, stark
potential Potential *n* (Elektr), **additional** ~
    Zusatzpotential *n*, **change in** ~
    Potentialsprung *m*, **chemical** ~ chemisches
    Potential, **compensation of** ~
    Potentialausgleich *m*, **disruptive** ~
    Funkenpotential *n*, **drop of** ~
    Potentialabfall *m*, **excess** ~ Überpotential *n*,
    fall of ~ Potentialabfall *m*, **increase in** ~
    Potentialanstieg *m*
potential *a* potentiell
potential barrier Energiewall *m*, Potentialberg *m*,
    Potentialrand *m* (Elektron),
    Potentialschwelle *f*
potential difference Potentialdifferenz *f*,
    Potentialunterschied *m*, Spannungsabfall *m*,
    Spannungsunterschied *m*
potential distribution Potentialverteilung *f*
potential drop Spannungsgefälle *n*
potential energy potentielle Energie *f*,
    Lageenergie *f* (Mech)
potential equation Potentialgleichung *f*
potential field Potentialfeld *n*
potential-forming potentialbildend
potential gradient Potentialgefälle *n*,
    Potentialgradient *m*
potential hole Potentialmulde *f* (Elektron)
potentiality Möglichkeit *f*, Potentialität *f*
potential jump Potentialsprung *m*
potential position Potentiallage *f*
potential rise Potentialanstieg *m*
potential surface Potentialfläche *f*
potential threshold Potentialschwelle *f*
potential value Potentialwert *m*
potential well Potentialmulde *f*, Potentialtopf *m*
potential well depth Potentialtopftiefe *f*
potentiometer Kompensator *m* (Elektr),
    Potentiometer *n*
potentiometric potentiometrisch
potentiometric controller
    Kompensationsregler *m*
potentiometry Potentiometrie *f*
pot furnace Tiegel[schmelz]ofen *m*
potlife Topfzeit *f*
pot metal Schmelzfarbglas *n*
pot press Kachelpresse *f*, Ringpresse *f*,
    Schachtelpresse *f*, Trogpresse *f*
potsherd Scherbe *f*
pot steel process Tiegelstahldarstellung *f*
potstone Lavezstein *m* (Min)
potter Töpfer *m*, Hafner *m*
potter's clay Töpferton *m*, Töpfererde *f*

potter's ore Glasurerz *n*, Töpfererz *n*
potter's wheel Töpferscheibe *f*
pottery Tongut *n*, Keramik *f*, Keramikwaren *pl*,
Steingut *n*, Steinzeug *n*, Töpferei *f*,
Tonwaren *f pl*
pot time Topfzeit
potting Einbetten *n*
potting medium Einbettmasse *f*
pot-type molding Preßspritzen *n*
pouch Beutel *m*
poudrette Fäkaldünger *m*, Mistpulver *n*
pound (lb.) Pfund *n* (Gewichtseinheit)
pound *v* zerkleinern, zerstoßen
pounded glass Glasglanz *m*
pounder Stößel *m*, Pistill *n*, Stampfer *m*
pounding tool Ausschlageisen *n*
pour *v* gießen, strömen, ~ **in** eingießen,
einschütten, einströmen, ~ **into another**
container umschütten, ~ **off** abfüllen (Brau);
abgießen, ~ **on** aufschütten, ~ **out** ausgießen,
ausschütten, herausströmen, ~ **together**
zusammengießen, zusammenschütten
pourability Gießbarkeit *f* (Metall),
Gießfähigkeit *f* (Metall)
pourable gießbar
poured, capable of being ~ gießfähig
pouring Gießen *n*, ~ **in** Eingießen *n*, Einguß *m*,
~ **off** Abguß *m*, Abgießen *n*, ~ **on** Anguß *m*,
Aufschüttung *f*
pouring compound Gußmasse *f*
pouring device Gießvorrichtung *f*
pouring head with sunk basin Einguß *m* mit
Vormulde
pouring-in hole Eingußöffnung *f*
pouring ladle Gießpfanne *f*
pouring level Abstichsohle *f*
pouring practice Eingußtechnik *f*
pouring spout Ausgußrinne *f*
pouring spout or lip Ausgußschnauze *f*
pouring temperature Vergießtemperatur *f*
pouring test Auslaufprobe *f*
pour point Stockpunkt *m*
pour sintering Schüttsintern *n*
powder Pulver *n*, Mehl *n*, Puder *m*, Staub *m*,
coarse ~ Grieß *m*, **coarsely ground** ~ grob
gemahlenes Pulver, **explosive** ~ brisantes
Pulver, **fine-grained** ~ feinkörniges Pulver,
finely ground ~ fein gemahlenes Pulver,
small-grained ~ feinkörniges Pulver,
smokeless ~ rauchloses Pulver, ~ **for burns**
Brandpuder *m* (Pharm)
powder *v* pulverisieren, mahlen, [ein]pudern,
einstäuben
powder anode Pulverglühanode *f*
powder barrel Pulverfaß *n*
powder blender Pulvermischer *m*
powder blue Kobaltfarbe *f*
powder charge Pulverladung *f*
powder coating Pulverbeschichtung *f*

powdered pulverisiert
powdered alum Alaunmehl *n*
powdered coal Pulverkohle *f*
powdered fiber Faserpulver *n*
powdered lead Bleigrieß *m*
powdered manure Düngepulver *n*
powdered milk Milchpulver *n*
powdered stone Gesteinsmehl *n*
powdered sugar Puderzucker *m*
powder explosion Pulverexplosion *f*
powder fire extinguisher Pulverlöscher *m*
(Feuerlöscher)
powdering Bestreuen *n*, Feinmahlen *n*,
Kreiden *n*, Pulverisieren *n*
powder keg Pulverfaß *n*
powder metallurgy Pulvermetallurgie *f*,
Sintermetallurgie *f*
powder mortar Pulvermörser *m*
powder pattern Pulverdiagramm *n* (Krist)
powder press cake Pulverpreßkuchen *m* (Gieß)
powder process Einstaubverfahren *n*
powder sintering Pulversintern *n*
powder test Pulverprobe *f*
powdery pulverisiert, pulverartig, pulverförmig,
pulverig
powellite Powellit *m* (Min)
power Kraft *f*, Antriebskraft *f* (Techn),
Befugnis *f*, Brennstärke *f* (Linse), Energie *f*
(Elektr), Leistungsvermögen *n* (Phys),
Potenz *f* (Math), Stärke *f*,
Vergrößerungskraft *f* (Opt), Vermögen *n*,
apparent ~ scheinbare Leistung, **loss of** ~
Energieverlust *m* (Elektr), Leistungsabfall *m*
(Elektr), **polarizing** ~ polarisierende
Wirkung, **to raise to a higher** ~ potenzieren
(Math), **to raise to the second** ~ zur zweiten
Potenz erheben, quadrieren, **to raise to the -th**
~ in die -nte Potenz erheben, ~ **of**
comprehension Fassungsvermögen *n*, ~ **of ten**
Zehnerpotenz *f* (Math)
power amplification Leistungsverstärkung *f*
power amplifier Leistungsverstärker *m*,
Kraftverstärker *m*
power-and-free conveyor Schleppkreisförderer *m*
power-and-free overhead conveyor Stauförderer *m*
power balance Energiebilanz *f*
power brake Leistungsbremse *f*, Servobremse *f*
power cable Starkstromkabel *n* (Elektr)
power circuit Starkstromleitung *f*
power coefficient Leistungszahl *f*
power component Wattkomponente *f* (Elektr)
power connection Netzanschluß *m* (Elektr)
power constant Leistungskonstante *f*
power consumer Stromverbraucher *m*
power consumption Energieverbrauch *m*,
Stromverbrauch *m*, **minimum** ~ **for a given**
mixing time Optimierung *f* nach der kleinsten
Mischarbeit
power control Drosselventil *n*, Leistungsregler *m*

power current Starkstrom *m*, Kraftstrom *m*
power curve Leistungskurve *f*
power cutback Leistungsverminderung *f*
power demand Leistungsbedarf *m* (Elektr)
power diagram Leistungsdreieck *n*
power dissipation Energiezerstreuung *f*
power distribution Kraftverteilung *f*
power distribution curve Testgütekurve *f* (Statist)
power drive Kraftantrieb *m*
power-driven maschinell angetrieben,
  mechanisch betätigt
power drop Leistungsabfall *m*
power engineering Starkstromtechnik *f*
power factor Leistungsfaktor *m*
power-factor indicator Phasenmesser *m*
power feed cable Stromzuführungskabel *n*
power fuel Treibstoff *m*
powerful kräftig, stark
power function (in testing hypothesis)
  Gütefunktion *f* (in Statistik)
power gas Kraftgas *n*, Treibgas *n*
power gas [generating] plant
  Kraftgas[erzeugungs]anlage *f*
power generation Krafterzeugung *f*
power generator Generator *m* für Kraftstrom
power input Antriebsleistung *f*, aufgenommene
  Leistung *f*, Leistungsaufnahme *f*,
  Leistungsaufwand *m*
powerized storage line Stauförderer *m*
power jet Treibstrahl *m*
power line Stromleitung *f* (Elektr)
power load Strombelastung *f*
power loss Energieverlust *m*, Leistungsverlust *m*
power mains Leitungsnetz *n* (Elektr)
power measurement Leistungsmessung *f*
power number (of a mixer)
  Widerstandskoeffizient *m* (eines Rührers)
power output Energieausbeute *f*
power output stability Intensitätskonstanz *f*
  (Elektr)
power pack Netz[anschluß]gerät *n*
power pile Leistungsreaktor *m*
power piping Hochdruckleitung *f*
power plant Kraftwerk *n*
power rating Krafteinschätzung *f*
power reactor Kernkraftwerk *n*,
  Leistungsreaktor *m*
power rectifying valve
  Hochleistungsgleichrichterröhre *f*
power requirements Leistungsbedarf *m*
power saving Kraftersparnis *f*
power-saving *a* kraftsparend
power sensitivity Leistungsempfindlichkeit *f*
power series Potenzreihe *f* (Math)
power series expansion
  Potenzreihenentwicklung *f*
power source Stromquelle *f*
power station Elektrizitätswerk *n*, Kraftwerk *n*

power supply Stromversorgung *f*,
  Energieversorgung *f*, **stabilized** ~
  Spannungskonstanthalter *m*
power supply unit Netzgerät *n* (Elektr),
  Netzteil *m* (Elektr)
power system Starkstromnetz *n* (Elektr)
power take-off Aufsteckgetriebe *n*
power transformer Leistungstransformator *m*
power transmission Energieübertragung *f*
power transmission plant Transmissionsanlage *f*
power unit motor operator (Am. E.) Stellmotor *m*
pozzuolana Pozz[u]olanerde *f*
pozzuolana mortar Porzellanmörtel *m*
pozzuolanic pozzuolanartig
p-phenylenediamine p-Phenylendiamin *n*,
  Ursol *n*
p-position para-Stellung *f*
p-propenylphenol Anol *n*
practicability Anwendbarkeit *f*,
  Durchführbarkeit *f*
practicable ausführbar, durchführbar
practical praktisch, angewandt, anwendbar
practical course Praktikum *n*
practical experience Betriebserfahrung *f*
practice Praxis *f* (Med, pl Praxen), **relating to** ~
  praxisbezogen
practice *v* praktizieren (Med)
practise *v* ausüben, praktizieren (Med)
pramocaine Pramocain *n*
praseodymium (Symb. Pr) Praseodym *n*
praseodymium carbide Praseodymcarbid *n*
praseodymium chloride Praseodymchlorid *n*
praseodymium content Praseodymgehalt *m*
praseodymium dioxide Praseodymdioxid *n*
praseodymium oxide Praseodymoxid *n*
praseodymium selenate Praseodymselenat *n*
praseodymium sulfate Praseodymsulfat *n*
praseodymium sulfide Praseodymsulfid *n*
praseolite Praseolith *m* (Min)
preamplifier Vorverstärker *m*
precalculate *v* vorausberechnen, vorkalkulieren
precalculated vorausberechnet
precalculation Vorausberechnung *f*,
  Vorkalkulation *f*
precast concrete Fertigbeton *m*
precast concrete unit Betonfertigteil *n*
precaution Sicherheitsmaßnahme *f*, Vorsicht *f*,
  Vorsichtsmaßnahme *f*
precautionary measure Vorsichtsmaßnahme *f*
precession Präzession *f*
precious edel, kostbar, wertvoll
precious metal Edelmetall *n*
precious metal smelting Reichschmelzen *n*
precious opal Edelopal *m* (Min)
precious steel Edelstahl *m*
precious stone Edelstein *m*
precious wood Edelholz *n*
precipitability Ausfällbarkeit *f*, Fällbarkeit *f*
  (Chem), Niederschlagbarkeit *f* (Chem)

precipitable [aus]fällbar, niederschlagbar
precipitant Fällungsmittel *n*
precipitate Niederschlag *m*, Ausfällung *f*,
Bodenkörper *m*, Bodensatz *m*,
Fällungsprodukt *n*, Präzipitat *n*, **active** ~
aktiver Niederschlag, **curdy** ~ käsiger
Niederschlag, **fine-granular** ~ feinkörniger
Niederschlag, **flocculent** ~ flockiger
Niederschlag, **poor in** ~ niederschlagsarm,
red ~ rotes Präzipitat *n* (Chem), **white fusible**
~ weißes schmelzbares Präzipitat *n* (Chem),
white infusible ~ weißes unschmelzbares
Präzipitat *n* (Chem), ~ **of metallic sulfide**
Metallsulfidniederschlag *m*
precipitate *v* [aus]fällen, abscheiden,
ausscheiden, niederschlagen, präzipitieren, ~
by addition of acid aussäuern
precipitated chalk Schlämmkreide *f*
precipitating Ausfällen *n*, Fällen *n*,
Präzipitieren *n*
precipitating agent Fällungsmittel *n*
precipitating bath Fällbad *n*
precipitating liquid Fällflüssigkeit *f* (Chem)
precipitating process Niederschlagsverfahren *n*
precipitating vat Fällbottich *m*,
Präzipitierbottich *m*
precipitating vessel Fällkessel *m*,
Präzipitiergefäß *n*
precipitation Fällung *f*, Ausfällen *n*,
Ausfällung *f*, Ausscheiden *n*, Niederschlag *m*
(Regen), Präzipitation *f*, **atmospheric** ~
atmosphärischer Niederschlag, **fractionated** ~
fraktionierte Fällung, **heat of** ~
Präzipitationswärme *f*, **law of** ~
Fällungsregel *f*, ~ **by electrolysis**
elektrolytische Fällung, ~ **of excess of copper**
Kupferaushärtung *f*, ~ **of impurities**
Ausfällung *f* von Verunreinigungen (pl)
precipitation apparatus Fällapparat *m*, Fäller *m*
precipitation cask Präzipitierfaß *n*
precipitation electrode Niederschlagselektrode *f*
precipitation hardening Aushärtung *f* (Metall),
Ausscheidungshärtung *f*, Seigerungshärtung *f*
precipitation method Fällmethode *f*
precipitation polymerization
Fällungspolymerisation *f*
precipitation process Niederschlagsarbeit *f*
precipitation reaction Fällungsreaktion *f*
precipitation tank Niederschlagstrog *m*
precipitation vessel Niederschlagsgefäß *n*
precipitator Niederschlagsapparat *m*,
Ausfällapparat *m*, Fäller *m*
precipitin Präzipitin *n*
precipitous abschüssig
precipitron Precipitron *n*
precise präzise, genau
precise adjustment Feinregulierung *f*
precision Genauigkeit *f*, Präzision *f*
precision adjusting valve Feinstellventil *n*

precision adjustment Feinstellung *f*
precision balance Feinwaage *f*, Präzisionswaage *f*
precision blower Präzisionsgebläse *n*
precision bore burette KPG-Feinbürette *f*
precision buret[te] Feinbürette *f* (Anal)
precision casting Präzisionsguß *m*
precision finished casting Präzisionsfertigguß *m*
precision glass stirrer shaft KPG-Rührwelle *f*
precision glass tube Präzisionsglasrohr *n*
precision instrument Präzisionsinstrument *n*,
feinmechanisches Instrument *n*,
Feinmeßgerät *n*
precision lathe Feindrehbank *f*,
Präzisionsdrehbank *f*
precision measurement Feinmessung *f*, **technique
of** ~ Feinmeßtechnik *f*
precision measuring Präzisionsmessung *f*
precision mechanics Feinmechanik *f*,
Präzisionsmechanik *f*
precision needle valve Nadelfeinregulierventil *n*
precision tool Präzisionswerkzeug *n*
precision value Präzisionswert *m*
precision work Feinarbeit *f* (Techn),
Präzisionsarbeit *f*
preclassifying Vorsortieren *n*
precoat Anschwemmschicht *f*
precoat filter Anschwemmfilter *m*
precombustion Vorverbrennung *f*
precombustion chamber Vorkammer *f* der
Verbrennung (Motor)
precompress *v* vorverdichten
precondensate Vorkondensat *n*
precondensation Vorkondensation *f*
precondition Vorbedingung *f*
precontrol Vorregelung *f*
precool *v* vorkühlen
precooler Vorkühler *m*
precooling Vorkühlung *f*
precure Anvulkanisation *f*
precure *v* anvulkanisieren
precuring Anvulkanisation *f*
precursor Vorläufer *m*
predetermine *v* vorausberechnen,
vorausbestimmen
predict *v* voraussagen
prediction Voraussage *f*, Vorhersage *f*
predischarge Vorentladung *f*
predissociation Prädissoziation *f*
predissociation spectrum
Prädissoziationsspektrum *n*
prednisolone Prednisolon *n*
prednisone Prednison *n*
predominant überwiegend
predry *v* vortrocknen
predryer Vortrockner *m*
predrying Vortrocknung *f*
pre-emergence herbicide Vorauflaufherbizid *n*
pre-evaporation Vorverdampfung *f*
pre-evaporator Vorverdampfer *m*

prefabricate v vorfertigen
prefabricated unit Fertigteil n
prefabrication Vorfertigung f, Montagebau m
preferable vorzüglich
preference Bevorzugung f, Vorzug m
preferential bevorzugt
prefiring Vorfeuerung f
preflame ignition Vorreaktion f im
  Brenngemisch
preform Vorformling m
preform v vorbilden, vorformen
preforming tool Tablettierwerkzeug n
preform process Vorformverfahren n
pregerminate v vorkeimen (Biol)
pregnancy test Schwangerschaftstest m
pregnane Pregnan n
pregnanediol Pregnandiol n
pregnanedione Pregnandion n
pregnenolone Pregnenolon n
pregrattite Pregrattit m (Min)
preheat v vorheizen, vorwärmen
preheated vorgewärmt, angewärmt
preheater Vorwärmer m, Anwärmer m
preheating Vorerhitzung f, Vorwärmen n,
  Vorwärmung f
preheating cabinet Vorwärmschrank m
preheating chamber Vorwärmkammer f
preheating furnace Vorwärmeofen m
preheating oven Vorwärmeofen m
preheating zone Vorwärmezone f
prehemataminic acid Prähämataminsäure f
prehnite Prehnit m (Min)
prehnitenol Prehnitenol n
prehnitic acid Prehnitsäure f
prehnitol Prehnitol n
pre-ignition Frühzündung f, vorzeitige
  Zündung f
prejudiced befangen, voreingenommen
preliminary vorhergehend, vorläufig
preliminary alarm Voralarm m
preliminary cleaning Vorreinigung f
preliminary condenser Vorkondensator m
preliminary desiccation Vorentwässerung f
preliminary design Vorentwurf m
preliminary drying Vortrocknung f
preliminary engineering Projektierung f
preliminary experiment Vorversuch m
preliminary fermentation Angärung f (Chem)
preliminary filtering Vorfiltern n
preliminary heating Vorwärmung f
preliminary heating zone Vorwärmezone f
preliminary investigation Voruntersuchung f
preliminary leaching Vorlaugung f
preliminary measures vorbereitende
  Maßnahmen pl
preliminary pickle Vorbrenne f
preliminary pressure Vordruck m
preliminary purification Vorreinigung f
preliminary purifier Vorreiniger m

preliminary reaction Vorreaktion f
preliminary reduction Vorreduktion f
preliminary tannage Vorgerbung f
preliminary tanning Vorgerbung f
preliminary test Vorprobe f, Vorprüfung f,
  Vorversuch m
preliminary washing drum Vorwaschtrommel f,
  Rauhwaschtrommel f
preliminary work Vorarbeit f
premagnetization Vormagnetisierung f
premature vorzeitig, verfrüht, voreilig
premium Prämie f
premium gasoline Superbenzin n
premix v vormischen
premix burner Injektorbrenner m,
  Vormischbrenner m
premixed vorgemischt
prenol Prenol n
prenyl chloride Prenylchlorid n
preparation Vorbereitung f, Zubereitung f,
  Ansatz m (Chem), Bereitung f, Darstellung f
  (Chem), Herstellung f, Präparat n (Chem),
  manner of ~ Aufbereitungsart f, **method of** ~
  Herstellungsmethode f, **process of** ~
  Darstellungsverfahren n, ~ **in wax**
  Wachspräparat n; ~ **of alum**
  Alaunsieden n, ~
  of wood Holzaufbereitung f
preparation plant Aufbereitungsanlage f
preparation room Vorbereitungsraum m
preparation[s] for work Arbeitsvorbereitung f
preparation tube Präparateröhrchen n
preparative chemical Vorbehandlungsmittel n
preparatory work Vorarbeit f
prepare v [vor]bereiten, anfertigen, anrichten,
  aufbereiten (Erz), bereitstellen, darstellen
  (Chem), herrichten, herstellen, präparieren,
  zubereiten
prepared einsatzbereit, hergestellt, zubereitet
preparedness Bereitschaft f
preparing salt Grundiersalz n, Präpariersalz n
preparing vessel Ansatzbottich m
prephenic acid Prephensäure f
preplasticizing method Vorplastifiziermethode f
prepolarization Vorpolarisierung f
prepolish v vorpolieren
preponderant überwiegend, vorwiegend
preprint Vorabdruck m (Buchdr)
prepurified vorgereinigt
prepurify v vorreinigen
prereaction Vorreaktion f
prereduce v vorreduzieren
prereduction Vorreduktion f
prerequisite Voraussetzung f, Vorbedingung f
preresistor Vorwiderstand m (Elektr)
preroast v vorrösten
preroasting Vorrösten n (Metall)
preroasting furnace Vorröstofen m
prescription Rezept n (Pharm), **by** ~ **only**
  rezeptpflichtig

preselect *v* vorwählen
presence Anwesenheit *f,* Beisein *n,* Gegenwart *f,* Vorhandensein *n*
present anwesend
present *v* einreichen
present value Gegenwartswert *m*
preservation Aufbewahrung *f,* Erhaltung *f,* Frischhaltung *f,* Haltbarmachung *f,* Konservierung *f,* ~ **of food** Lebensmittelkonservierung *f*
preservation of the species Arterhaltung *f* (Biol), Arterhaltungstrieb *m* (Biol)
preservative Konservierungsmittel *n,* Fäulnismittel *n,* keimtötendes Mittel *n,* Schutzmittel *n* (Med)
preservative *a* fäulnisverhütend, konservierend
preservative method Konservierungsmethode *f*
preserve *v* [auf]bewahren, aufheben, einmachen (Obst etc.), erhalten, haltbar machen, konservieren, ~ **food** Nahrungsmittel haltbar machen
preserved food, fully ~ Vollkonserve *f*
preserved fruit Obstkonserven *f pl*
preserved in sugar eingezuckert
preserve[s] Eingemachtes *n,* Konserve[n pl] *f,* Präserve *f*
preserving Konservieren *n,* Haltbarmachen *n*
preserving agent Konservierungsmittel *n*
preserving jar Einweckglas *n*
preserving package Frischhaltepackung *f*
preserving process Konservierungsverfahren *n*
preset *v* vorher festlegen
preshape *v* vorbilden, vorformen
pre-shave lotion Rasierwasser *n* (Kosmetik)
presinter *v* vorsintern
presizing Vorleimen *n* (Papier)
presorting Vorsortieren *n*
press Presse *f,* Quetsche *f*
press *v* pressen, bügeln (Stoff), drücken, einen Druck ausüben, ~ **down a lever** einen Hebel niederdrücken, ~ **in** einpressen, ~ **into** hineinpressen, ~ **on** andrücken, ~ **out** auspressen, ~ **through** durchdrücken, durchpressen, ~ **together** zusammendrücken
press board Preßspan *m*
press button Druckknopf *m*
press cake Preßkuchen *m*
press cooler Druckkühler *m*
press cork Preßkorken *m*
press cross beam Preßholm *m*
press cylinder Preßzylinder *m*
pressed article Preßling *m*
pressed brick Preßziegel *m*
pressed charcoal Preßkohle *f*
pressed glass Preßglas *n*
pressed hard glass Preßhartglas *n*
pressed juice Preßsaft *m*
pressed part Preßteil *n*
pressed pulp Preßrückstand *m*

presser plate Gegendruckplatte *f*
pressing Pressen *n,* Satinieren *n* (Pap), Stanzen *n* (Techn), **cold** ~ Kaltpressen *n*
pressing time, additional ~ Durchwärmzeit *f* (Holz)
pressman Drucker *m*
press mold Preßform *f*
press molding machine Preßformmaschine *f*
press-on cap Aufdrückdeckel *m* (Dose)
press plate Preßplatte *f*
press plunger Preßkolben *m*
press polish Hochglanz *m*
press ram Preßkolben *m*
press section Preßpartie *f*
press separator Scheidepresse *f*
press sintering furnace Drucksinterofen *m*
press sweating process Preßschwitzverfahren *n*
press time (hot bonding) Preßzeit *f*
press tool Preßwerkzeug *n*
press tub Kelterbütte *f*
pressure Druck *m,* Druckkraft *f,* Spannung *f* (Elektr), **axial** ~ Axialdruck *m,* **balance of** ~ Druckausgleich *m,* **center of** ~ Druckmittelpunkt *m,* **change of** ~ Druckänderung *f,* **constant** ~ gleichbleibender Druck, **continuous** ~ Dauerdruck *m,* **decrease of** ~ Druckerniedrigung *f,* **differential** ~ Wirkdruck *m,* **equal** ~ gleicher Druck, **equalization of** ~ Druckausgleich *m,* **excess hydrostatic** ~ hydrostatischer Überdruck, **generation of** ~ Druckentwicklung *f,* Druckerzeugung *f,* **high** ~ Hochdruck *m,* **hydrostatic** ~ hydrostatischer Druck, Flüssigkeitsdruck *m,* **increase in** ~ Druckanstieg *m,* **initial** ~ Anfangsdruck *m,* **internal** ~ innerer Druck (Mech), **lateral** ~ Seitendruck *m,* **line of constant** ~ Isobare *f,* **low** ~ Niederdruck *m,* **lowest** ~ Minimaldruck *m,* **osmotic** ~ osmotischer Druck, **partial** ~ Partialdruck *m,* Teildruck *m,* **reduction of** ~ Druckabnahme *f,* **release of** ~ Druckentlastung *f,* **resistant to** ~ druckfest, **sensitive to** ~ druckempfindlich, **total** ~ Gesamtdruck *m,* **transmission of** ~ Druckausbreitung *f,* **variation in** ~ Druckänderung *f,* ~ **above the atmospheric** Überdruck *m,* ~ **in terms of millimeters of mercury** Quecksilbersäulendruck *m* in Millimetern (pl), ~ **on lever or handle** Kurbeldruck *m*
pressure altitude Barometerhöhe *f*
pressure angle of tooth profile Eingriffswinkel *m* der Zahnflanke
pressure bag molding Niederdruckverfahren *n* (Polyester)
pressure bell Druckglocke *f*

pressure boiler, superhigh ~
Höchstdruckkessel *m*
pressure bottle Druckflasche *f*
pressure broadening Druckverbreiterung *f*
(Spektr)
press[ure] casting Preßguß *m*
pressure chamber Druckkammer *f*,
Druckraum *m*
pressure code for pipelines Druckstufen *pl* für
Rohrleitungen
pressure coefficient Druckkoeffizient *m*
pressure-compensating druckausgleichend
pressure connection Druckstutzen *m*
pressure controller Druckregler *m*
pressure converter Druckwandler *m*
pressure cooker Schnellkochtopf *m*
pressure curve Druckkurve *f*, Drucklinie *f*,
Spannungslinie *f*
pressure cylinder Druckwalze *f*, Druckzylinder *m*
pressure decrease Druckverminderung *f*
pressure-density relation
Druck-Dichte-Beziehung *f*
pressure-dependent druckabhängig
pressure die casting Druckguß *m*
pressure difference Druckunterschied *m*
pressure difference meter
Druckunterschiedsmesser *m*
pressure diffusion Druckdiffusionsverfahren *n*
pressure digester Druckkocher *m*
pressure dispenser Druckzerstäuber *m*
pressure distillation Druckdestillation *f*
pressure drop Druckabfall *m*, Druckverlust *m*,
Druckverminderung *f*
pressure drum filter Drucktrommelfilter *n*
pressure duct Druckkanal *m*
pressure effect Preßeffekt *m*
pressure element Druckdose *f*, Druckmeßzelle *f*
pressure equalization Druckausgleich *m*
pressure equalizer Druckvermittler *m*
pressure equilibrium Druckgleichgewicht *n*
pressure figure Druckfigur *f* (Krist)
pressure filling Druckfüllung *f* (Aerosoldose)
pressure filter Druckfilter *m*, Drucknutsche *f*
pressure filtration Druckfiltration *f*
pressure flask Druckflasche *f*, Druckkolben *m*
pressure fluid Druckflüssigkeit *f*,
Preßflüssigkeit *f*
pressure forming Formstanzen *n*
pressure gas Druckgas *n*
pressure gauge Druckmesser *m*,
Druckanzeiger *m*, Druckaufnehmer *m*,
Manometer *n*
pressure gauge pipe Manometerrohr *n*
pressure gauge position Manometerstand *m*
pressure gauge spring Manometerfeder *f*
pressure gradient Druckgradient *m*,
Druckgefälle *n*
pressure increase Druckzunahme *f*,
Druckentwicklung *f*, Drucksteigerung *f*

pressure indicator Druckanzeiger *m*
pressure joint Druckstutzen *m*
pressure leaf filter Druckblattfilter *n*,
Zentrifugalreinigungsfilter *n*
pressure line Druckleitung *f*
pressure loss Druckverlust *m*,
Spannungsverlust *m*
pressure lubrication Druckschmierung *f*,
Preßschmierung *f*
pressure-marking resistance Schreibfestigkeit *f*
(Lacke)
pressure measurement Druckmessung *f*
pressure oil wash Drucköllwäsche *f*
pressure pack Aerosolverpackung *f*,
Druckbehälter *m*
pressure pad Druckleiste *f*, Druckscheibe *f*,
Gummikissen *n*, Preßkissen *n*
pressure passage Druckkanal *m*
pressure pipe Druckrohr *n*
pressure piping Druckleitung *f*
pressure plate Andruckplatte *f* (Phot)
pressure polymerization Druckpolymerisation *f*
pressure pot Druckgefäß *n*
pressure-proof druckfest
pressure pulverizer Gebläseschlägermühle *f*
pressure pump Druckpumpe *f*
pressure ram Druckkolben *m*
pressure range Druckbereich *m*
pressure ratio Druckverhältnis *n*
pressure reducing valve Entlastungsventil *n*,
Entspannungsventil *n*
pressure reduction valve
Druckminderungsventil *n*,
Druckreduzierventil *n*
pressure regulator Druckregler *m*,
Druckregulator *m*
pressure release Entspannung *f*
pressure relief Druckausgleich *m*
[pressure-]relief valve Sicherheitsventil *n*,
Überdruckventil *n*
pressure reservoir Druckkessel *m*
pressure resistance Druckfestigkeit *f*
pressure-responsive druckabhängig
pressure rise Druckanstieg *m*, Drucksteigerung *f*,
Druckzunahme *f*
pressure roll Druckwalze *f*, Andruckwalze *f*,
Preßwalze *f*
pressure roller Druckwalze *f*, Andruckwalze *f*
pressure sensitive adhesive Haftkleber *m*,
Selbstkleber *m*
pressure sensitive tape Haftklebeband *n*
pressure setting (of plastics) Druckhärtung *f*
(von Kunststoffen)
pressure shift (spectral lines)
Druckverschiebung *f* (Spektrallinien)
pressure side Druckseite *f*
pressure spray process Druckspritzverfahren *n*
pressure spread Druckausbreitung *f*
pressure spring Druckfeder *f*

pressure strain Druckbeanspruchung *f*
pressure swing process Druckwechselverfahren *n*
   (Füllkörperkolonne)
pressure tank Druckkessel *m*, Autoklav *m*
pressure test Druckprobe *f*, Härteprobe *f*
pressure transmitter Druckübersetzer *m*
pressure tubing Vakuumschlauch *m*
pressure-type capacitor
   Hochdruckkondensator *m*
pressure unit Druckeinheit *f*
pressure valve Druckventil *n*
pressure vessel Druckbehälter *m*, Druckgefäß *n*
pressure volume diagram
   Druckvolumendiagramm *n*
pressure water Druckwasser *n*
pressure wave Druckwelle *f*
pressure-welded druckverschweißt
pressure welding Druckschweißung *f*
pressurized mixer Druckmischer *m*
pressurized pack Aerosolverpackung *f*,
   Druckpackung *f*
pressurized water reactor Druckwasserreaktor *m*
press vat Kelterbütte *f*
press welding Preßschweißen *n*
pre-tan *v* vorgerben
pretreat *v* vorbehandeln
pretreatment Vorbehandlung *f*
prevail *v* [vor]herrschen
prevailing vorherrschend
prevalent vorherrschend
prevent *v* verhindern, verhüten, vorbeugen
prevention Prophylaxe *f* (Med), Verhinderung *f*,
   ~ of backfiring Rückschlaghemmung *f*
prevention of accidents, regulations for ~
   Unfallverhütungsvorschriften *pl*
preventive Abwehrmittel *n* (Med),
   Prophylaktikum *n* (Med), Verhütungsmittel *n*
preventive measure Vorbeugungsmaßnahme *f*
previous examination Vorprüfung *f*
prevulcanization Anvulkanisation *f*
prevulcanize *v* anvulkanisieren
price Preis *m*, **~ ex works** Fabrikpreis *m*
price list Preisliste *f*
price tag Preisschild *n*
prick punch Vorstecher *m* (Am. E.), Körner *m*
primaquine Primachin *n*
primary primär
primary air Primärluft *f*
primary alcohol primärer Alkohol *m*
primary cancer Primärkrebs *m* (Med)
primary cell Primärelement *n*
primary circuit Primär[strom]kreis *m*
primary cleaning Grobreinigung *f*
primary coil Primärspule *f*
primary color Primärfarbe *f*
primary cooler Vorkühler *m*
primary crusher Grobbrecher *m*
primary crystallization Primärkristallisation *f*
primary current Primärstrom *m* (Elektr)

primary electron Primärelektron *n*
primary etching Primärätzung *f*
primary excitation Primäranregung *f*
primary filter Vorfilter *n*
primary furnace Vorfrischofen *m*
primary industry Grundstoffindustrie *f*
primary material Ausgangsmaterial *n*,
   Ausgangsstoff *m*
primary matter Urstoff *m*, Ursubstanz *f*
primary neutron Primärneutron *n*
primary or detecting element Meßfühler *m*
   (Regeltechn)
primary particle Primärteilchen *n*
primary period Primärformation *f* (Geol)
primary potential Vorspannung *f* (Elektr)
primary product Ausgangserzeugnis *n*
primary pulse Primärimpuls *m*
primary purification Vorfrischen *n* (Metall)
primary radiation Primärstrahlung *f*
primary reaction Primärreaktion *f*
primary rocks Urgestein *n*
primary spectrum Primärspektrum *n* (Spektr)
primary station Hauptwerk *n*
primary structure Primärgefüge *n*
primary treatment Vorbehandlung *f*
primary voltage Primärspannung *f*
prime *v* grundieren, Farbengrund geben
prime-coated grundiert
prime coating Vorstreichen *n*
prime cost Anschaffungskosten *pl*,
   Selbstkostenpreis *m*
prime matter Urmaterie *f*
prime mover Antriebsmaschine *f*
prime number Primzahl *f* (Math)
primer Zündhütchen *n*, Anlaßkraftstoff *m*
   (Techn), Grundanstrich *m*, Grundierfarbe *f*
   (Malerei), Grundierlack *m*, Grundierung *f*
   (Malerei), Initialexplosivstoff *m*,
   Initialzündmittel *n*, Vorlack *m* (Malerei),
   Zündpatrone *f*
primer sealer Grundstrich *m* mit porenfüllenden
   Eigenschaften
primer tubing Anlaßrohrleitung *f*
primer valve Anlaßventil *n*
primeval ursprünglich
priming Einspritzen *n* (Mot), erster Anstrich *m*,
   Grundierung *f*, Zündung *f* (Bergb)
priming apparatus Zündapparat *m*
priming cartridge Zündpatrone *f*
priming coat Grundanstrich *m*;
   Grundieranstrich *m*, Grundierung *f*
priming color Grundierfarbe *f*
priming composition Initialzündmittel *n*,
   Zündsatz *m*
priming explosive Initialexplosivstoff *m*,
   Initialsprengstoff *m*
priming oil paint Ölgrundierung *f*
priming paint Grundierfarbe *f*, Vorstreichfarbe *f*
priming pan Anfeuerungstopf *m*

priming plug Füllschraube *f*
priming pump Ansaugpumpe *f*
priming reaction Beladungsreaktion *f*
priming valve Anstechventil *n*,
   Sicherheitsventil *n*
priming varnish Grundierfirnis *m*
primitive primitiv, ursprünglich
primitive form Kernform *f* (Gieß), Urform *f*
primitive man Urmensch *m*
primrose Primel *f* (Bot)
primulin Primulin *n*, Aureolin *n*
primulin base Primulinbase *f*
primulindisulfonic acid Primulindisulfonsäure *f*
primulin dye Primulinfarbstoff *m*
primulin red Primulinrot *n*
primverin Primverin *n*
principal agent Hauptagens *n*
principal azimuth Hauptazimut *m*
principal binding medium Hauptbindemittel *n*
principal circuit Hauptstromkreis *m*
principal constituent Hauptbestandteil *m*
principal focal distance Hauptbrennweite *f*
principal incidence, angle of ~
   Haupteinfallswinkel *m* (Opt)
principal normal Hauptnormale *f* (einer Kurve)
principal planes, distance between ~
   Hauptebenenabstand *m*
principal point Hauptpunkt *m*
principal point refraction
   Hauptpunktbrechwert *m*
principal point triangulation
   Hauptpunkttriangulation *f*
principal spectrum Hauptspektrum *n*
principal stress Hauptspannung *f*
principal valence Hauptvalenz *f* (Chem)
principal valency Hauptvalenz *f* (Chem)
principle Prinzip *n*, Gesetz *n*,
   Grundbestandteil *m* (Chem), Grundregel *f*,
   Grundsatz *m*, Regel *f*, **~ of invariance**
   Invarianzprinzip *n* (Math), **~ of least action**
   Prinzip *n* des geringsten Zwanges, **~ of**
   relativity Relativitätsprinzip *n*, **~ of**
   reversibility Reversibilitätsprinzip *n* (Opt)
print [Ab]Druck *m*, Abzug *m* (Phot),
   Druckkattun *m* (Text), Pause *f* (Zeichnung),
   Photokopie *f* (Phot), Positiv *n* (Phot), **white**
   ~ Weißpause *f*
print *v* [ab]drucken, abziehen (Phot), bedrucken,
   pausen, **~ chiné** chinieren, **~ [off]**
   abklatschen (Buchdr), **~ over** überdrucken, **~**
   upon bedrucken
printed gemustert (Text)
printed circuit board Leiterplatte *f*
printer Drucker *m*
printer carriage tape Lochband *n* (Comp)
printer's ink Buchdruckerschwärze *f*,
   Druckerschwärze *f*
printer's varnish Buchdruckerfirnis *m*

printing Aufdruck *m*, Bedrucken *n*,
   Buchdruck *m*, **continuous stationary form** ~
   Endlosdruck *m*, **to finish** ~ ausdrucken
printing chiné Chinierung *f*
printing color Druckfarbe *f*
printing in black Schwarzdruck *m* (Buchdr)
printing ink Buchdruckerfarbe *f*,
   Druckerschwärze *f*, **glossy** ~
   Glanzdruckfarbe *f*
printing ink drying Druckfarbentrocknung *f*
printing machine Druckmaschine *f* (Buchdr)
printing method Druckverfahren *n*
printing oil Drucköl *n* (Buchdr)
printing-out paper Kopierpapier *n*,
   Zelloidinpapier *n* (Phot)
printing paper Kopierpapier *n*
printing paste Druckteig *m* (Buchdr)
printing plate, intagliated ~ Tiefdruckplatte *f*
printing press Druckerpresse *f*
printing process Druckverfahren *n*
printing roller Druckwalze *f* (Buchdr)
printing style Schreibweise *f*
print-out process Auskopierprozeß *m* (Phot)
Prinz acid Prinz-Säure *f*
priority Vorrang *m*
priority claim Prioritätsbeanspruchung *f*
   (Patent)
priority right Prioritätsrecht *n* (Patent)
prism Prisma *n* (Opt), **direct-vision** ~
   geradsichtiges Prisma, **image-inverting** ~
   bildumkehrendes Prisma, **reversing** ~
   Wendeprisma *n* (Opt), **right-angled** ~
   Umkehrprisma *n*
prismatic prismatisch
prismatic spectrum Prismenspektrum *n*,
   Brechungsspektrum *n*
prismatine Prismatin *m* (Min)
prism[at]oid Prismatoid *n* (Geom)
prisms, set of three ~ Dreiprismensatz *m*
prism surface Prismenfläche *f*
private communication Privatmitteilung *f*
privilege Privileg *n*, Vergünstigung *f*, Vorrecht *n*
prize Prämie *f*, [Sieger]Preis *m*, **to give a** ~
   prämieren
probability Wahrscheinlichkeit *f*, **calculus of** ~
   Wahrscheinlichkeitsrechnung *f*, **conditional** ~
   bedingte Wahrscheinlichkeit *f*, **~ of error**
   Fehlerwahrscheinlichkeit *f*, **~ of hits**
   Trefferwahrscheinlichkeit *f*
probability curve Häufigkeitskurve *f* (Statist),
   Wahrscheinlichkeitskurve *f* (Statist)
probability density Wahrscheinlichkeitsdichte *f*
probability distribution
   Wahrscheinlichkeitsverteilung *f*
probability factor Wahrscheinlichkeitsfaktor *m*
probability integral
   Wahrscheinlichkeitsintegral *n*
probability law Wahrscheinlichkeitsgesetz *n*
probability wave Wahrscheinlichkeitswelle *f*

probable wahrscheinlich
probably wahrscheinlich
probarbital Probarbital *n*
probe Meßkopf *m,* Sonde *f* (Med)
probe *v* sondieren (Med)
probe area Sondenausdehnung *f*
probe characteristic Sondencharakteristik *f*
probe technique Sondentechnik *f*
probing head Tastkopf *m*
problematic problematisch, zweifelhaft, ungewiß
problematic situation Problematik *f*
procaine Procain *n,* Syncain *n*
procarboxypeptidase Procarboxypeptidase *f*
  (Biochem)
procedure Verfahren *n,* Arbeitsgang *m,*
  Behandlungsverfahren *n,* Prozedur *f,*
  Versuchsanordnung *f*
proceed *v* gedeihen
proceeding Verfahren *n*
procellose Procellose *f*
process Verfahren *n,* Prozeß *m,* Reaktionsfolge *f*
  (Chem), Vorgang *m,* **adiabatic** ~
  adiabatischer Prozeß, **biological** ~
  biologischer Prozeß, **chemical** ~ chemischer
  Prozeß, chemischer Vorgang, **continuous** ~
  Fließbetrieb *m* (Reaktionstechnik),
  kontinuierlicher Prozeß, **controlled** ~
  gesteuerter Prozeß, **irreversible** ~ irreversibler
  Vorgang *n,* **metallurgical** ~ metallurgisches
  Verfahren, **principle of** ~
  Verfahrensprinzip *n,* **reversible** ~
  umkehrbarer Vorgang, **wet** ~ nasses
  Verfahren, ~ **of development**
  Entwicklungsverfahren *n* (Phot), ~ **of**
  formation Bildungsvorgang *m*
process *v* verarbeiten, bearbeiten, [chemisch]
  behandeln, haltbar machen (Lebensmittel),
  herstellen, ~ **further** weiterverarbeiten
process analysis Prozeßanalyse *f*
process control Betriebskontrolle *f,*
  Verfahrensregelung *f*
process control computer Prozeßrechner *m*
process data Betriebskennzahlen *f pl*
process design Verfahrensplanung *f*
process engineering Prozeßentwicklung *f*
process engraving Autotypie *f*
process flow Fertigungsfluß *m*
[process] flow sheet Verfahrensschema *n*
processibility Verarbeitbarkeit *f*
processing Verarbeitung *f,* fabrikationsmäßige
  Herstellung *f,* Veredelung *f* (Techn), **further**
  ~ Weiterverarbeitung *f,* **method of** ~
  Verarbeitungsmethode *f*
processing industry Veredelungsindustrie *f*
processing instruction Verarbeitungsrichtlinie *f*
processing plant Aufbereitungsanlage *f,*
  Verarbeitungsanlage *f*
processing temperature
  Verarbeitungstemperatur *f*

process optimization Prozeßoptimierung *f*
process printing Mehrfarbendruck *m*
process steam Betriebsdampf *m*
process technology Verfahrenstechnik *f*
process water Brauchwasser *n*
prochlorite Fächerstein *m* (Min), Prochlorit *m*
  (Min)
procure *v* beschaffen, verschaffen
produce *v* hervorbringen, entwickeln, erzeugen,
  herstellen, hervorrufen, produzieren,
  verursachen, ~ **in quantity** serienmäßig
  herstellen, ~ **steam** Dampf *m* erzeugen
producer Erzeuger *m,* Generator *m* (Techn),
  Hersteller *m,* Produzent *m*
producer gas Generatorgas *n,* Kraftgas *n*
producer gas plant Generatorgasanlage *f*
producer gas process Generatorgasprozeß *m*
producer operation Generatorbetrieb *m*
producer temperature Generatortemperatur *f*
product Produkt *n,* Erzeugnis *n,* Fabrikat *n,*
  crude ~ Rohprodukt *n,* **final** ~
  Fertigerzeugnis *n,* **finished** ~
  Fertigerzeugnis *n,* Fertigprodukt *n,*
  precipitated ~ Ausscheidungsprodukt *n,* **raw**
  ~ Rohprodukt *n,* **semi-finished** ~
  Halbfertigprodukt *n,* **semi-manufactured** ~
  Halbfertigerzeugnis *n,* **separated** ~
  Ausscheidungsprodukt *n,* ~ **of disintegration**
  Verwitterungsgebilde *n,* ~ **of reaction**
  Reaktionsprodukt *n*
product capacity Mengenleistung *f*
product gas (cracking process) Spaltgas *n*
production Produktion *f,* Anfertigung *f,*
  Ausstoß *m* (Techn), Erzeugung *f,*
  Fabrikation *f,* Fertigung *f,* Herstellung *f,*
  continuous ~ Fließarbeit *f,* **large-scale** ~
  Großproduktion *f,* serienmäßige Herstellung,
  method of ~ Herstellungsweise *f,* **process of** ~
  Darstellungsverfahren *n,* **total** ~
  Gesamterzeugung *f,* ~ **of fiber**
  Fasergewinnung *f*
production capacity Produktionskapazität *f*
production control Betriebskontrolle *f,*
  Betriebsüberwachung *f,*
  Fabrikationskontrolle *f*
production costs Gestehungskosten *pl*
production engineer Betriebsingenieur *m*
production engineering Fertigungsplanung *f*
production line Förderband *n,* Transportband *n*
production lot Fabrikationsnummer *f*
production manager Fabrikationsleiter *m*
production permit Produktionserlaubnis *f*
production plan Fabrikationsprogramm *n*
production plant Gewinnungsanlage *f*
production potential Produktionsmöglichkeit *f*
production program Fertigungsprogramm *n*
production reactor Produktionsreaktor *m*
production research Betriebsforschung *f*
production scheme Fabrikationsprogramm *n*

production supervision Betriebsüberwachung *f*
production-type lathe Produktionsdrehbank *f*
production unit Produktionsanlage *f*
productive ergiebig, ertragreich, produktiv
productiveness Ergiebigkeit *f*, Fruchtbarkeit *f*,
  Leistungsfähigkeit *f*
productivity Leistungsfähigkeit *f* (Betrieb),
  Produktivität *f*
proenzyme Proenzym *n* (Biochem)
profession Beruf *m*, Gewerbe *n*
professional berufsmäßig, fachmännisch
professional education Fachausbildung *f*
professional training Berufsausbildung *f*
profile Profil *n*, Durchschnitt *m*, Querschnitt *m*,
  ~ **of a line** Linienprofil *n*
profile *v* fassonieren, profilieren
profile album Formenbuch *n*, Profilalbum *n*
profiled joint Dichtungsprofil *n*
profile[d] tube Profilrohr *n*
profile fiber Profilfaser *f*
profile iron Fassoneisen *n*
profile mill cutter Profilwalze *f*
profile miller Formfräsmaschine *f*
profile milling Formfräsen *n* (Techn)
profile milling cutter Formfräser *m*
profile shift Profilverschiebung *f*
profiling Formgebung *f*
profiling calender Profilkalander *m*
profiling machine Kopierfräsmaschine *f*
profiling [work] Fassonarbeit *f* (Techn)
profit Gewinn *m*, Nutzen *m*, **margin of** ~
  Verdienstspanne *f*
profitable einträglich, gewinnbringend,
  nutzbringend
proflavine Proflavin *n* (HN)
profound[ly] gründlich
progesterone Progesteron *n*,
  Corpus-luteum-Hormon *n*,
  Gelbkörperhormon *n*, Progestin *n*
progestin Progesteron *n*, Progestin *n*
program Programm *n*
program *v* programmieren
program control Programmregelung *f*,
  Zeitplanregelung (Regeltechn)
program-controlled programmgesteuert
program[m]ed programmiert
programmer Programmierer *m*
programming Programmieren *n*,
  Programmierung *f*, **linear** ~ lineare
  Programmierung
program[m]ing device Programmiereinrichtung *f*
program[m]ing language Programmiersprache *f*
program[m]ing system Programmiersystem *n*
program register Steuerbefehlspeicher *m*
  (Comp)
progress Fortschritt *m*

progression Reihe *f*, Progression *f* (Math),
  Reihenentwicklung *f* (Math), **arithmetic** ~
  arithmetische Reihe
progressive fortschrittlich
progressive cutting machine
  Stufenschneidemaschine *f*
progressively schrittweise
progressive wave Wanderwelle *f*
proinsulin Proinsulin *n*
project Projekt *n*, Planung *f*
project *v* projektieren, hervorragen, planen,
  projizieren, vorspringen
project execution Projektabwicklung *f*
projectile Geschoß *n*, Wurfgeschoß *n*
projectiles, speed of ~
  Geschoßgeschwindigkeit *f*
projecting Projektierung *f*
projecting *a* vorragend, vorstehend
projecting apparatus Projektionsapparat *m*
projection Bildwiedergabe *f* (Phot), Projektion *f*
  (Phot), Vorsprung *m*, Wurf *m* (of a pipe etc.),
  Halsansatz *m*, **plane of** ~ Bildebene *f*
projectionist Vorführer *m* (Film, Dias)
projection plane Projektionsebene *f*
projection reading scale Projektionsablesung *f*
projection screen Bildschirm *m;*
  Projektionsschirm *m*
projection welding Buckelschweißung *f*
projective projektiv
projector Projektionsapparat *m*, Projektor *m*
prolactin Laktationshormon *n*, Prolaktin *n*
prolamine Prolamin *n*
prolectite Prolektit *m*
proliferate *v* wuchern
proliferation Proliferation *f*, [üppiges]
  Wachstum *n*, Wucherung *f*
proliferation phase Proliferationsphase *f*
  (Physiol)
prolinase Prolinase *f* (Biochem)
prolin[e] Prolin *n*
proline dehydrogenase Prolindehydrogenase *f*
  (Biochem)
prolinol Prolinol *n*
prolong *v* verlängern, ausdehnen
prolongation Verlängerung *f*, [Aus]Dehnung *f*
promazine Promazin *n*
promethazine Promethazin *n*
promethium (Symb. Pm) Promethium *n*
promising aussichtsreich, **very** ~
  vielversprechend
promote *v* [be]fördern, begünstigen
promoter chemischer Beschleuniger *m*
promotion Beförderung *f*, Beschleunigung *m*
  (Reaktionskinetik), Förderung *f*
promotive förderlich, fördernd
prone anfällig
proneness Anfälligkeit *f*
prong adjuster (forklift) Zinkenverstellgerät *n*
pronounced ausgesprochen

prontosil Prontosil *n*
proof Beweis *m*, Beweisstück *n*, Nachweis *m*,
  Probe *f*, Probeabzug *m* (Phot, Buchdr),
  Prüfung *f*, **method of** ~ Beweisführung *f*
proof *a* beständig, undurchlässig
proof press Andruckmaschine *f* (Buchdr)
proof puller Fahnenabzieher *m*
proof stress Prüfspannung *f* (Mech)
prop Stütze *f*, Strebe *f*, Stützbalken *m*
prop *v* [ab]stützen, absteifen, versteifen
propadiene Allen *n*
propaesin Propäsin *n*
propaganda gift Werbegeschenk *n*
propagate *v* ausbreiten, fortpflanzen, verbreiten
propagate field Ausbreitungsfeld *n* (Comp),
  Ausdehnungsfeld *n* (Comp)
propagation Ausbreitung *f*, Fortpflanzung *f*
  (Phys, Biol), Verbreitung *f*, Vermehrung *f*,
  direction of ~ Ausbreitungsrichtung *f*,
  Fortpflanzungsrichtung *f*, **velocity of** ~
  Ausbreitungsgeschwindigkeit *f*, ~ **of electrical**
  waves Fortpflanzung *f* elektrischer Wellen
  (pl), ~ **of pressure** Druckausbreitung *f*, ~ **of**
  the species Arterhaltung *f* (Biol)
propagation constant Fortpflanzungskonstante *f*
propagation field Fortbewegungsfeld *n*
propagation of error Fehlerfortpflanzung *f*, **law**
  of ~ Fehlerfortpflanzungsgesetz *n*
propagation parameter Ausbreitungsparameter *m*
propagation problem Ausbreitungsproblem *n*
propagation velocity
  Fortpflanzungsgeschwindigkeit *f* (Phys)
propanal Propanal *n*, Propionaldehyd *m*
propanamide Propionamid *n*
propane Propan *n*
propanedioic acid Malonsäure *f*
propanediol Propylenglykol *n*
propane gas Propangas *n*
propane nitrile Propionitril *n*
propane tricarboxylic acid Tricarballylsäure *f*
propanetriol Glycerin *n*
propanoic acid Propionsäure *f*
propanol Propanol *n*
propargyl Propargyl-
propargyl alcohol Propargylalkohol *m*
propargyl aldehyde Propargylaldehyd *m*
propargylic acid Propargylsäure *f*, Propiolsäure *f*
propellant Treibgas *n*, Treibmittel *n*,
  Treibstoff *m*
propeller Propeller *m*, Luftschraube *f*,
  Schiffsschraube *f*, **axis of a** ~ Propellerachse *f*
propeller mixer Propellermischer *m*,
  Propellerrührer *m*, Schraubenrührer *m*
propeller pump Propellerpumpe *f*,
  Rotationspumpe *f*
propeller thrust Schraubenschub *m*
propeller-type pump Zentrifugalpumpe *f*,
  Rotationspumpe *f*
propelling force Triebkraft *f*

propellor mixer Propellerrührer *m*
propenal Propenal *n*, Acrylaldehyd *m*,
  Akrolein *n*
propene Propen *n*, Propylen *n*
propene-1,2,3-tricarboxylic acid Aconitsäure *f*
propene oxide Propylenoxid *n*
propene sulfide Propylensulfid *n*
propenoic acid Acrylsäure *f*
propenol Propenol *n*
propenyl Propenyl-
propenyl-2,4,5-trimethoxybenzene Asaron *n*
propenyl alcohol Propenylalkohol *m*
propenylidene Propenyliden *n*
propeptone Hemialbumose *f*
proper eigen, richtig, passend
proper fraction echter Bruch *m* (Math)
proper mass Eigenmasse *f*
properties, change of ~
  Eigenschaftsveränderungen *f pl*
property Eigenschaft *f*, Eigentum *n*, **additive** ~
  additive Eigenschaft (Phys), **chemical** ~
  chemische Eigenschaft,
  concentration-dependent ~
  konzentrationsabhängige Eigenschaft,
  metalloid ~ halbmetallische Eigenschaft
prophetin Prophetin *n*
prophylactic Abwehrmittel *n* (Med);
  Prophylaktikum *n* (Med)
prophylactic *a* vorbeugend (Med)
prophylactic dose Schutzdosis *f* (Med)
prophylaxis Prophylaxe *f* (Med)
propine Propin *n*
propinol Propargylalkohol *m*
propiolactone Propiolacton *n*
propiolaldehyde Propiolaldehyd *m*
propiolate propiolsauer
propiolic acid Propargylsäure *f*, Propiolsäure *f*
propion Propion *n*
propionaldazine Propionaldazin *n*
propionaldehyde Propionaldehyd *m*
propionamide Propionamid *n*
propionamidine Propionamidin *n*
propionate Propionat *n*, Propionsäureester *m*,
  propionsaures Salz *n*
propionic acid Propionsäure *f*
propionic anhydride Propionsäureanhydrid *n*
propionitrile Propionitril *n*, Äthylcyanid *n*
propionyl Propionyl-
propionyl chloride Propionylchlorid *n*
propionyl-CoA Propionyl-CoA *n*
propionyl peroxide Propionylperoxid *n*
propiophenone Propiophenon *n*
propolis Bienenharz *n*, Klebwachs *n*,
  Pichwachs *n*, Propolis *f*
propolis resin Propolisharz *n*
proponal Proponal *n*
proportion Verhältnis *n*, Anteil *m*, Ebenmaß *n*,
  Mengenverhältnis *n*, Proportion *f*, Umfang *m*,
  ~ **of ingredients** Mengungsverhältnis *n*, ~ **of**

sizes Größenverhältnis *n*, ~ of voids
Porenanteil *m*
proportional proportional, verhältnismäßig,
directly ~ direkt proportional, **inversely** ~
umgekehrt proportional
proportional band Proportionalbereich *m*
(Regeltechn)
proportional control factor (Am. E.)
Übertragungsfaktor *m* (Regeltechnik)
proportional counter Proportionalzählrohr *n*
proportionality Proportionalität *f,* **assumption of**
~ Proportionalitätsannahme *f*
proportionality factor Proportionalitätsfaktor *m*,
Verhältniszahl *f*
proportionality range Proportionalbereich *m*
proportional offset P-Abweichung *f*
(Regeltechn)
proportional power proportional geregelte
Stromversorgung *f*
proportionate angemessen, anteilig,
proportional, verhältnismäßig
proportioning Proportionieren *n*
proportioning pump Dosierpumpe *f,*
Mischpumpe *f*
proposition Vorschlag *m*, Lehrsatz *m*,
Vorhaben *n*
proprietor Eigentümer *m*, Inhaber *m*
propulsive jet Antriebsstrahl *m*
propyl Propyl-
propylacetate Propylacetat *n*
propylacetylene Propylacetylen *n*
propylal Propylal *n*
propyl alcohol Propanol *n*, Propylalkohol *m*
propylamine Propylamin *n*
propylbenzene Propylbenzol *n*
propyl bromide Propylbromid *n*
propyl chloride Propylchlorid *n*
propyl cyanide Butyronitril *n*
propylene Propylen *n*
propylene chlorohydrin Propylenchlorhydrin *n*
propylene diamine Propylendiamin *n*
propylene glycol Propylenglykol *n*
propylene oxide Propylenoxid *n*
propylene sulfide Propylensulfid *n*
propylenimine Propylenimin *n*
propyl ether Propyläther *m*
propyl formate Propylformiat *n*
propylidene Propyliden *n*
propyl iodide Propyljodid *n*
propyl mercaptan Propylmercaptan *n*
propyl nitrobenzoate
Nitrobenzoesäurepropylester *m*
propyloxine Propyloxin *n*
propylpiperidine, 2- ~ Koniin *n*
propylpyrogallol dimethyl ether Pikamar *n*,
Teerbitter *n*
propylsilicone Propylsilicon *n*
propylthiouracil Propylthiouracil *n*
propynal Propiolaldehyd *m*

propyne Allylen *n*
propynoic acid Propargylsäure *f,* Propiolsäure *f*
prorennin Prorennin *n*
prosapogenin Prosapogenin *n*
prosecution Prüfverfahren *n* (Jur)
prospect Aussicht *f*
prospect *v* schürfen (Bergb)
prospective zukünftig, voraussichtlich
prospectus Prospekt *m*, Werbe[druck]schrift *f*
prostagladins Prostagladine *pl*
prostata gland Prostatadrüse *f*
prosthetic prosthetisch
protactinium (Symb. Pa) Protactinium *n*
protagon Protagon *n*
protaminase Protaminase *f*
protamine Protamin *n*
protargol Protargol *n*
proteacin Proteacin *n*
protease Protease *f* (Biochem)
protect *v* abschirmen, [be]schützen, bewahren
protected geschützt
protected by law gesetzlich geschützt
protecting cap Schutzhaube *f,* Schutzkappe *f*
protecting cover Schutzhaube *f*
protecting glass Schutzglas *n*
protecting glasses Schutzbrille *f*
protecting goggles Schutzbrille *f*
protecting mask Schutzmaske *f*
protecting plate Schutzplatte *f*
protecting spectacles Schutzbrille *f*
protecting varnish Deckfirnis *m*, Schutzfirnis *m*
protection Abschirmung *f* (Techn), Schutz *m*, ~
of designs or copyrights Musterschutz *m*, ~ **of**
iron Eisenschutz *m*
protective schützend
protective casing Schutzhülse *f*
protective clothing Schutz[be]kleidung *f*
protective coat Schutzbelag *m*
protective coating Schutzanstrich *m*,
Schutzüberzug *m*
protective colloid Schutzkolloid *n*
protective contact Schutzkontakt *m* (Elektr)
protective contact socket Schukosteckdose *f*
(Elektr, HN)
protective covering Schutzhülle *f*
protective device Schutzvorrichtung *f*
protective effect Schutzwirkung *f*
protective funnel Schutztrichter *m*
protective gas Schutzgas *n*
protective grating Schutzgitter *n*
protective isolation Schutzisolierung *f*
protective jacket Schutzmantel *m*
protective layer Schutzschicht *f*
protective rampart Schutzwall *m*
protective strip Stoßleiste *f*
protective switch Schutzschalter *m*
protective tissue Deckgeflecht *n*, Schutzgewebe *n*
protective tube Schutzrohr *n*, Verkleidungsrohr *n*
protective tubing Isolierschlauch *m*

protective wall Schutzwand *f*
proteid Proteid *n*
protein Protein *n*, Eiweiß *n*, Eiweißkörper *m*,
  conjugated ~ konjugiertes Protein *n*,
  constitutive ~ konstitutives Protein,
  containing ~ eiweißhaltig, proteinhaltig,
  crystalline ~ kristallines Protein *n*, **denatured**
  ~ denaturiertes Protein, **fibrous** ~
  Faserprotein *n*, fibrilläres Protein, **foreign** ~
  Fremdeiweiß *n*, körperfremdes Protein *n*,
  globular ~ globuläres Protein, **native** ~
  natives Protein, **oligomeric** ~ oligomeres
  Protein *n*, **polymeric** ~ polymeres Protein *n*,
  poor in ~ eiweißarm, **rich in** ~ eiweißreich,
  variety of ~ Eiweißart *f*
protein analysis Proteinanalyse *f*
proteinase Proteinase *f* (Biochem)
protein back-bone Proteingerüst *n*
protein coagulation Eiweißgerinnung *f*
protein complex Proteinkomplex *m*
protein composition Proteinzusammensetzung *f*
protein compound Proteinverbindung *f*
protein conformation Proteinkonformation *f*
protein content Eiweißgehalt *m*
protein deficiency Eiweißmangel *m*
protein degradation Eiweißabbau *m*,
  Eiweißspaltung *f*, Proteinabbau *m*
protein degradation product
  Eiweißabbauprodukt *n*
protein denaturation Proteindenaturierung *f*
protein diet Eiweißdiät *f*
protein fiber Eiweißfaser *f* (Text)
protein fractionation Proteinfraktionierung *f*
protein hydrolysis Proteinhydrolyse *f*
protein level Eiweißspiegel *m*, Proteinspiegel *m*
protein-like eiweißartig
protein material Eiweißkörper *m*
protein metabolism Eiweißhaushalt *m*,
  Eiweißstoffwechsel *m*
protein minimum Eiweißminimum *n*
protein molecule Eiweißmolekül *n*
protein nitrogen Eiweißstickstoff *m*
protein precipitation Proteinfällung *f*
protein purification Proteinreinigung *f*
protein renaturation Proteinrenaturierung *f*
protein requirement Eiweißbedarf *m*
protein separation Proteintrennung *f*
protein structure Proteinstruktur *f*
protein subunit Proteinuntereinheit *f*
protein synthesis Proteinsynthese *f*
protein turbidity Eiweißtrübung *f*
protein turnover Proteinumsatz *m*
proteinuria Eiweißausscheidung *f* (im Urin)
proteohormone Proteohormon *n*
proteolysis Proteolyse *f*, Abbau von Eiweiß,
  Eiweißabbau *m*, **product of** ~
  Eiweißspaltprodukt *n*
proteolyte Proteolyt *m*
proteolytic proteolytisch, eiweißspaltend

protest [against] *v* beanstanden
prothrombin Prothrombin *n* (Biochem),
  Thrombozym *n* (Biochem)
prothrombokinase Prothrombokinase *f*
  (Biochem)
protium Protium *n*
protoanemonin Protoanemonin *n*
protobastite Protobastit *m* (Min)
protoberberine Protoberberin *n*
protocatechualdehyde Protocatechualdehyd *m*
protocatechuic acid Protocatechusäure *f*
protocetraric acid Protocetrarsäure *f*
protoclase Protoklas *m*
protocrocin Protocrocin *n*
protoechinulinic acid Protoechinulinsäure *f*
protoglucal Protoglucal *n*
protohem Protohäm *n*
proton Proton *n*, Wasserstoffkern *m*, **high-speed**
  ~ schnelles Proton, **thermal** ~ thermisches
  Proton
proton acceleration Protonenbeschleunigung *f*
proton accelerator Protonenbeschleuniger *m*
proton acceptor Protonenakzeptor *m*
proton affinity Protonenaffinität *f*
protonation Protonierung *f*
proton bombardment Protonenbeschießung *f*
proton bullet Protonengeschoß *n*
proton donor Protonendonator *m*
proton excess Protonenüberschuß *m*
proton exchange Protonenaustausch *m*
proton jump Protonensprung *m*
proton mass Protonenmasse *f*
proton number Protonenzahl *f*
proton path Protonenbahn *f*
proton precession frequency
  Protonenpräzessionsfrequenz *f*
proton range Protonenreichweite *f*
proton repulsion Protonenabstoß *m*
protons, number of ~ Protonenzahl *f* (Atom)
proton spectrum Protonenspektrum *n*
proton spin Protonenspin *m*
proton synchrotron Protonensynchrotron *n*
proton track Protonenbahn *f*, Protonenspur *f*
proton transfer Protonenübertragung *f*
protopapaverine Protopapaverin *n*
protophilic protophil (Lösungsmittel)
protopine Protopin *n*, Fumarin *n*, Macleyin *n*
protoplasm[a] Protoplasma *n* (Biol)
protoplast Protoplast *m* (Biol)
protoporphyrin Protoporphyrin *n*
protoquercitol Protoquercit *m*
prototropy Prototropie *f*
prototype Prototyp *m*, Urform *f*, Urtyp *m*
protoveratrine Protoveratrin *n*
protovermiculite Protovermiculit *m* (Min)
protoxide Protoxid *n*
protozoon (pl. protozoa) Protozoon *n* (pl
  Protozoen), Einzeller *m*
protract *v* hinausziehen

protractor Schmiege *f*, Transporteur *m* (Geom), Winkelmesser *m* (Geom)
protrude *v* hervorragen, hervorstehen
protrusion Ausbuchtung *f*
protuberance Ausstülpung *f*, Hervortreten *n*
proustite Proustit *m* (Min), Arsenrotgültigerz *n* (Min), Arsensilberblende *f* (Min)
prove *v* beweisen, begründen, belegen, erweisen, nachweisen, sich bewähren, sich herausstellen
provide *v* ausstatten, beschaffen, besorgen, versorgen, ~ **with springs** abfedern
provided that falls
provided with ausgerüstet mit
proving ground Versuchsfeld *n*
provision Vorkehrung *f*, Maßnahme *f*, Vorsorge *f*
provisional vorläufig
provisional estimate Voranschlag *m*
provisions Lebensmittelvorräte *pl*
provitamin Provitamin *n*, Vitaminvorläufer *m*
provocation Reiz *m*, Reizung *f*
proximate benachbart
proximity effect Eigenkapazitätseffekt *m*
proximity rule Proximitätsregel *f*
proximity switch Näherungsschalter *m*
prulaurasin Prulaurasin *n*
prunasin Prunasin *n*
prunella salt Prunellensalz *n*
prunetol Prunetol *n*, Genistein *n*
prunin Prunin *n*
prunol Malol *n*, Urson *n*
Prussian blue Berlinerblau *n*, Ferriferrocyanid *n*
Prussian brown Preußischbraun *n*
Prussian white Berlinerweiß *n*
prussiate Prussiat *n*
prussiate of potash, red ~ Kaliumeisen(III)-cyanid *n*, rotes Blutlaugensalz *n*, **yellow** ~ gelbes Blutlaugensalz *n*, Kaliumeisen(II)-cyanid *n*
prussic acid Blausäure *f*, Cyanwasserstoffsäure *f*
psammite Psammit *m* (Geol)
pseudacetic acid Propionsäure *f*
pseudaconitine Pseudoaconitin *n*
pseudoacid Pseudosäure *f*
pseudoadenosine Pseudoadenosin *n*
pseudo alkali Pseudoalkali *n*
pseudoamethyst Pseudoamethyst *m* (Min), violetter Flußspat *m* (Min)
pseudoapatite Pseudoapatit *m* (Min)
pseudoaspidin Pseudoaspidin *n*
pseudoasymmetric pseudo-asymmetrisch
pseudoaxial quasi-axial
pseudo-base Pseudobase *f*
pseudobrookite Pseudobrookit *m* (Min)
pseudobutyl alcohol Pseudobutylalkohol *m*
pseudobutyl chloride Pseudobutylchlorid *n*
pseudocartilage Stützgewebe *n* (Histol)
pseudocatalysis Pseudokatalyse *f*
pseudocatalytic pseudokatalytisch
pseudocatalyzer Pseudokatalysator *m*

pseudocholesterol Pseudocholesterin *n*
pseudochrysolite Pseudochrysolith *m* (Min)
pseudoconhydrine Pseudoconhydrin *n*
pseudocrystal Pseudokristall *m* (Krist)
pseudocrystalline pseudokristallin
pseudocrystallite Pseudokristall *m* (Krist)
pseudocumene Pseudocumol *n*
pseudocumenesulfonic acid Pseudocumolsulfonsäure *f*
pseudocumenol Pseudocumenol *n*
pseudocumidine Pseudocumidin *n*
pseudocumyl Pseudocumyl *n*
pseudocyanine Pseudocyanin *n*
pseudodiazoacetic acid Pseudodiazoessigsäure *f*
pseudodiosgenin Pseudodiosgenin *n*
pseudoemerald Pseudosmaragd *m* (Min)
pseudogalena falscher Bleiglanz *m* (Min)
pseudogarnet Pseudogranat *m* (Min)
pseudogaylussite Pseudogaylussit *m* (Min)
pseudoglucal Pseudoglucal *n*
pseudohalogen Pseudohalogen *n*
pseudohexagonal pseudohexagonal (Krist)
pseudoionone Pseudojonon *n*
pseudoleucine Pseudoleucin *n*
pseudolibethenite Pseudolibethenit *m* (Min)
pseudolimonene Pseudolimonen *n*
pseudomalachite Pseudomalachit *m* (Min), Phosphorchalcit *m* (Min)
pseudometallic luster unvollkommener Metallglanz *m*
pseudomonas Pseudomonasbakterien *pl*
pseudomonotropic pseudomonotrop
pseudomonotropy Pseudomonotropie *f*
pseudomorphosis Pseudomorphose *f*
pseudomorphous pseudomorph
pseudomorphous crystal Afterkristall *m*
pseudopelletierine Pseudopelletierin *n*
pseudophite Pseudophit *n*
pseudopinene Pseudopinen *n*
pseudoplastic pseudoplastisch, quasiplastisch
pseudoplastic flow Fließerweichung *f*
pseudopurpurin Pseudopurpurin *n*
pseudoracemic pseudoracemisch
pseudosaccharin Pseudosaccharin *n*
pseudosphere Pseudosphäre *f*
pseudostrychnine Pseudostrychnin *n*
pseudosymmetry Pseudosymmetrie *f*
pseudotannin Pseudogerbstoff *m*
pseudotensor Pseudotensor *m*
pseudoternary pseudoternär
pseudotetragonal pseudotetragonal (Krist)
pseudotropine Pseudotropin *n*
pseudourea Pseudoharnstoff *m*
pseudouric acid Pseudoharnsäure *f*
pseudoxanthine Pseudoxanthin *n*
psicaine Psicain *n*
psicose Psicose *f*

psilomelane Psilomelan *m* (Min), schwarzer
Glaskopf *m* (Min), Schwarzmanganerz *n*
(Min)
psittacinite Psittazinit *m* (Min)
psoralen Psoralen *n*
psoralic acid Psoralsäure *f*
psoromic acid Psoromsäure *f*
p-state p-Zustand *m*
psychopharmacologic agents Psychopharmaka *pl*
psychotrine Psychotrin *n*
psychrometer Psychrometer *n*,
Luftfeuchtigkeitsmesser *m*
psychrometric psychrometrisch
psychrometry Psychrometrie *f*,
Feuchtigkeitsmessung *f*
psylla acid Psyllasäure *f*
psylla alcohol Psyllaalkohol *m*
pteridine Pteridin *n*
pterin Pterin *n*
pterocarpine Pterocarpin *n*
pteroic acid Pteroinsäure *f*
pterolite Pterolith *m* (Min)
pteropterin Pteropterin *n*
pterostilbene Pterostilben *n*
pteroylglutamic acid Pteroylglutaminsäure *f*
ptilolite Ptilolith *m* (Min)
ptomain[e] Ptomain *n*, Fäulnisalkaloid *n*,
Leichenalkaloid *n*, Leichengift *n*
ptyalase Ptyalin *n* (Biochem)
ptyalin Ptyalin *n* (Biochem)
ptychotis oil Ptychotisöl *n*
p-type semiconductor p-Halbleiter *m*
puberulic acid Puberulsäure *f*
puberulonic acid Puberulonsäure *f*
public address system Lautsprecheranlage *f*
public announcement Bekanntgabe *f*
publication Bekanntmachung *f*, Publikation *f*,
Veröffentlichung *f*, **monthly** ~
Monatsschrift *f*, **technical** ~ Fachblatt *n*
publicity department Werbeabteilung *f*
publish *v* bekanntgeben, veröffentlichen
pucherite Pucherit *m* (Min)
pucker *v* fälteln, kräuseln
puckered faltig
puckering Faltenwerfen *n*
pudding stone Puddingstein *m* (Min)
puddle *v* im Flammofen frischen (Met),
puddeln (Met), ~ **by hand** handpuddeln
puddle ball Puddelluppe *f*, Rohluppe *f*
puddle cinder Puddelschlacke *f*
puddle[d] iron Luppeneisen *n*, Puddeleisen *n*
puddler Puddler *m*
puddling Puddeln *n*, Flammofenfrischen *n*, ~
for crystalline iron Kornpuddeln *n*
puddling bar Brechstange *f*
puddling basin Puddelherd *m*
puddling door Kratzentür *f*
puddling furnace Puddelofen *m*,
Eisenfrischflammofen *m*, **double** ~

Doppelpuddelofen *m*, **rotary** ~
Puddeldrehofen *m*
puddling hearth Puddelherd *m*
puddling process Puddelverfahren *n*,
Frischarbeit *f*
puddling slag Puddelschlacke *f*
puddling tools Gezähe *n*
puddling works Puddelhütte *f*, Puddelwerk *n*
puering Beizen *n* (Gerb), Beizung *f* (Gerb)
puering bath Beizbad *n* (Gerb)
puering vat Beizkufe *f* (Gerb)
puffball Bovist *m* (Bot)
puff-drying Explosionstrocknung *f*
puffed up aufgeblasen, aufgedunsen
puff off *v* verpuffen
pufierite Pufierit *m*
pug *v* mischen und kneten
pugmill Kollergang *m*, Lehmknetmaschine *f*,
Mischmühle *f*
pukateine Pukatein *n*
pulegane Pulegan *n*
pulegone Pulegon *n*
pull Zug *m*
pull *v* zerren; ziehen, ~ **back** zurückziehen, ~
off abziehen, ~ **out** ausreißen, ~ **the trigger**
abdrücken (Gewehr)
pull-back ram Rückdruckkolben *m*,
Rückzugkolben *m*
pull cord Zugleine *f*
pull electrode Zugelektrode *f*
puller screw Abziehschraube *f*
pulley Flaschenzug *m*, Rolle *f*, **fixed** ~ feste
Rolle, **loose** ~ lose Riemenscheibe *f*, **movable**
~ bewegliche Rolle, **split** ~ geteilte
Riemenscheibe *f*, **tight** ~ feste
Riemenscheibe *f*
pulley block Blockscheibe *f*, Rollenzug *m*
pulley molding machine
Riemenscheibenformmaschine *f*
pulley patterns, set of ~
Riemenscheibenmodellsatz *m*
pulleys, set of ~ Flaschenzug *m*
pulley wheel Rolle *f*
pulling device Abzugsvorrichtung *f*
pulling down Abbruch *m*
pulling force Zugkraft *f*
pulling rope Zugseil *n*
pull rod Spannwalze *f*
pull roll Abzugswalze *f*
pull test Zugprobe *f*, Reckprobe *f*, Zugprüfung *f*
pulmonary circulation kleiner Kreislauf *m* (Med)
pulp Fleisch *n* (Früchte), Ganzstoff *m*,
Ganzzeug *n* (Pap), gewaschenes
Erz *n* (Bergb), Mark *n* (Bot), Papierbrei *m*
(Pap), Pülpe *f*, Pulpe *f* (Pap), Schlich *m*
(Bergb), **bleached** ~ gebleichter Zellstoff,
brown mechanical ~ Braunschliff *m*, **free from**
~ holzschlifffrei, **high boiled** ~ weicher
Zellstoff, **high-grade** ~ hochwertiger Zellstoff,

low-boiled ~ harter Zellstoff, **low-grade** ~ minderwertiger Zellstoff, **processed** ~ veredelter Zellstoff, **refined** ~ veredelter Zellstoff, **semichemical** ~ halbchemischer Zellstoff, ~ **from rags** Lumpenbrei *m* (Papier)

pulp *v*, ~ **in a hollander** holländern, ~ **rags in a hollander** holländern

pulpability Aufschließbarkeit *f* (Pap)
pulp bale Zellstoffballen *m*
pulp beating Zellstoffmahlung *f*
pulp bleaching Zellstoffbleiche *f*
pulp board Zellstoffkarton *m*, Zellstoffpappe *f*
pulp catcher Faserfänger *m* (Zuck), Papiermassefänger *m* (Pap), Stoffänger *m* (Pap)
pulp chest Rührbütte *f*
pulp coloring Büttenfärbung *f*
pulp conveyor Stofförderanlage *f* (Pap)
pulp digester Zellstoffkocher *m*
pulp drying Zellstofftrocknung *f*
pulp drying machine Zellstofftrockenmaschine *f*
pulp engine Holländer *m* (Pap)
pulper Einstampfmaschine *f* (Papier)
pulp felt Zellstoffilz *m*
pulp-handling centrifugal pump Stoffkreiselpumpe *f*
pulping Aufschluß *m* (Papier), Zellstoffaufschluß *m* (Pap), **preparation of waste for** ~ Abfallstoffaufbereitung *f* (Pap)
pulping machine Einstampfmaschine *f* (Papier)
pulping process Papieraufbereitungsprozeß *m*
pulp meter Zeugregler *m* (Pap)
pulp press water Preßwasser *n* (Zuck)
pulp processing Zellstoffveredelung *f* (Pap)
pulp production Zellstoffherstellung *f*
pulp sizing Büttenleimung *f*
pulp slime Zellstoffschleim *m*
pulp slurry Faserbrei *m*
pulp vat Lumpenbütte *f* (Pap)
pulp wadding Zellstoffwatte *f*
pulp yarn Papierstoffgarn *n*
pulque Agavenbranntwein *m*
pulsate *v* pulsieren
pulsating pulsierend, stoßweise
pulsating-bed drier Wirbelstoßtrockner *m*
pulsating column Pulsierkolonne *f*
pulsating current Wellenstrom *m* (Elektr)
pulsation Pulsieren *n*
pulsator Pulsator *m*
pulse Impuls *m* (Elektr), kurzer Stromstoß *m* (Elektr)
pulse amplifier Impulsverstärker *m*
pulse clipper Impulsbegrenzer *m*
pulse corona Koronadurchbruch *m*
pulsed bed Pulsationsbett *n*
pulsed discharge Stoßentladung *f*
pulse duration Impulsbreite *f*
pulse forming stage Impulsbildungsstufe *f*

pulse height spectrum Impulsgrößenspektrum *n*
pulse-lengthening impulsverlängernd
pulse operation Impulsfolge *f*
pulse propagation Stoßfortpflanzung *f*
pulse size Stoßgröße *f*
pulse size distribution Stoßgrößenverteilung *f*
pulse stripper Impulsabtrenner *m*, Impulsbegrenzer *m*
pulse traveling Impulsfortpflanzung *f*
pulse triggering Impulsauslösung *f*
pulse voltage, peak of ~ Impulsamplitude *f* (Telev)
pulsing Impulsgabe *f*
pulsometer Pulsometer *n*
pulverable pulverisierbar
pulverizable pulverisierbar
pulverization Pulverisierung *f*, Feinstmahlung *f*, Zerstäubung *f* (Flüssigkeiten)
pulverize *v* pulverisieren, pulvern, zermahlen, zerreiben, zerstäuben (Flüssigkeiten), zerstoßen
pulverized pulverisiert, staubförmig, pulverig
pulverized coal burner Staubbrenner *m*
pulverized fuel Brennstaub *m*
pulverized ore Pochmehl *n*
pulverizer Zerkleinerungsmaschine *f*, Zerstäuber *m*
pulverizing Pulverisieren *n*, Feinmahlen *n*
pulverizing equipment Pulverisieranlage *f*, Mahlanlage *f*
pulverizing machine Pulverisiermaschine *f*
pulverizing plant Zerkleinerungsanlage *f*
pulverulent pulverig, feinpulverig, pulverförmig, staubartig
pulvic acid Pulvinsäure *f*
pumice Bimsstein *m*
pumice *v* mit Bimstein abreiben
pumice cloth Bimssteintuch *n*
pumice concrete Bimsbeton *m*
pumice gravel Bimskies *m*
pumiceous bimssteinähnlich
pumice powder Bimssteinmehl *n*
pumice sand Bimssand *m*
pumice soap Bimssteinseife *f*
pumice stone Bimsstein *m*, **to rub with** ~ mit Bimsstein abreiben
pumice stone paper Bimssteinpapier *n*
pump Pumpe *f*, **involute** ~ Evolventenpumpe *f*, **positive displacement** ~ Verdrängerpumpe *f*, **type of** ~ Pumpenbauart *f*, ~ **for extreme pressures** Höchstdruckpumpe *f*, **vapor** ~ **or jet pump** Treibmittelpumpe *f*, **rotating piston** ~ **or rotary pump** Kreiskolbenpumpe *f*, **liquid-seal** ~ **or water-ring pump** Flüssigkeitsring-Pumpe *f*
pump *v* pumpen, ~ **in** einpumpen, ~ **out** abpumpen, auspumpen, ~ **up** aufpumpen
pump chamber Pumpenzylinder *m*

pump circulation Umpumpbetrieb *m*
pump coupling Pumpengelenk *n*
pump fluid Treibmittel *n* (Pumpe)
pump fluid filling Treibmittelfüllung *f* (Pumpe)
pump fluid return pipe Treibmittelrücklauf *m*
  (Pumpe)
pump house Pumpenhaus *n*
pump impeller Pumpenflügel *m*
pumping plant Pumpenanlage *f*
pumping set Pumpanlage *f*
pumping speed Pumpgeschwindigkeit *f*
pumping station Pumpanlage *f*
pumpkin seed oil Kürbiskernöl *n*
pump piston Pumpenkolben *m*
pump plunger Pumpenkolben *m*
pump rods Pumpengestänge *n*
pump valve Pumpenventil *n*
pump wheel Pumprad *n*
punch Locheisen *n*, Ausschlageisen *n*,
  Durchschlag *m*, Körner *m*, Lochstempel *m*,
  Prägestempel *m*, Stanzstempel *m*, **floating** ~
  beweglicher Stempel, pendelnder, **loose** ~
  loser Stempel *m*
punch *v* [aus]stanzen, [an]körnen, durchbohren,
  [durch]lochen, durchlöchern, durchschlagen,
  lochstanzen, ~ **holes** löchern
punch bar bearing Stanzerlager *n*
punch card Lochkarte *f*
punch card column Lochkartenspalte *f* (Comp)
punched out [aus]gestanzt
punched tape Lochstreifen *m*
punching Stanzen *n*
punching die Lochstanze *f*
punching machine Stanzmaschine *f*, Lochstanze *f*,
  Lochstanzmaschine *f*
punching quality Stanzfähigkeit *f*
punchings Stanzabfälle *pl*
punching support Stanzplatte *f*
punching test Durchstoß-Test *m*
punch line Stanzlinie *f*
punch mark Ankörnung *f*, Körner *m*
punch-marked angekörnt
punch press Lochstanze *f*, Stanzpresse *f*
punctiform punktförmig
punctual pünktlich; punktförmig
punctuation mark Interpunktionszeichen *n*
puncture Durchschlag *m* (Elektr),
  Durchstechen *n*, Punktion *f* (Med)
puncture *v* durchbohren, durchlöchern,
  durchstechen, punktieren (Med)
puncture cutout Durchschlagssicherung *f*
  (Elektr)
puncture gauging chain Einstichmeßkette *f*
puncture method Anstichmethode *f*
puncture-proof durchschlagsicher (Elektr),
  pannensicher (Reifen)
puncture resistance Durchstoßfestigkeit *f*,
  Sticheinreißfestigkeit *f*

puncture-sealing safety tube pannensicherer
  Schlauch
pungent beißend (Geruch), scharf (Geruch),
  stechend (Geruch)
punicine Punicin *n*, Pelletierin *n*
punish *v* [be]strafen
punk Zündschwamm *m*, Zunderholz *n*
punnets (Br. E.) Obststeige *f*
purchase Einkauf *m*, **terms of** ~
  Bezugsbedingungen *pl*
purchase *v* [ein]kaufen
pure rein, echt (Chem), gediegen (Met), lauter,
  sauber
pure blue Reinblau *n*
pure carbon Reinkohle *f*
pure element (without isotopes) Reinelement *n*
  (ohne Isotope)
pure gas Reingas *n*
pure gold Feingold *n*
pure product column Reinkolonne *f*
purest reinst
pure yeasting machine Hefereinzuchtapparat *m*
purgative Abführmittel *n* (Pharm),
  Laxiermittel *n* (Pharm)
purgative herb Purgierkraut *n*
purgative salt Purgiersalz *n*
purge *v* reinigen
purging agaric Lärchenschwamm *m* (Bot)
purging air Spülluft *f*
purging buckthorn Färberbeere *f* (Bot)
purification Reinigung *f*, Klärung *f*, Läuterung *f*,
  Purifikation *f*, Reindarstellung *f* (Chem), ~ **by**
  carbonation Saturationsscheidung *f* (Zucker),
  method of ~ Reinigungsmethode *f*, ~ **of**
  enzymes Reinigung *f* von Enzymen (pl), ~ **of**
  organic compounds Reinigung *f* organischer
  Verbindungen (pl), ~ **of sewage disposal**
  Abwasserreinigung *f*, ~ **of the liquor**
  Laugenreinigung *f*, **process of** ~
  Reinigungsprozeß *m*
purification plant Reinigungsanlage *f*
purification process Reinigungsprozeß *m*
purified rein, gereinigt, **in a preliminary way** ~
  vorgereinigt
purifier Reinigungsmittel *n*, Reiniger *m*,
  Reinigungsapparat *m*
purify *v* reinigen, frischen (Metall), läutern,
  raffinieren, säubern, ~ **by smelting**
  ausschmelzen, ~ **in a preliminary way**
  vorreinigen
purifying Reinigen *n*
purifying agent Reinigungsmittel *n*,
  Läuterungsmittel *n*
purifying apparatus Reinigungsapparat *m*
purifying mass Reinigungsmasse *f*
purifying material Reinigermasse *f*
purifying plant Reinigungsanlage *f*
purifying process Windfrischverfahren *n*
  (Metall)

purifying salve Reinigungssalbe *f*
purifying tank Reinigungsbottich *m*
purifying vessel Läuterungsgefäß *n*,
Läuterungskessel *m*
purin[e] Purin *n*
purine base Purinbase *f*
purine biosynthesis Purinbiosynthese *f*
purine derivative Purinderivat *n*
purine phosphoribosyl transferase
Purinphosphoribosyltransferase *f* (Biochem)
purin nucleotide Purinnukleotid *n*
purinoic acid Purinoesäure *f*
purity Reinheit *f*, Echtheit *f*, Feinheit *f* (Metall),
spectral ~ spektrale Reinheit *f*, test for ~
Reinheitsprobe *f*
puromycin Puromycin *n*
purple blaurot, Purpur *m*, Purpurfarbe *f*
purple *a* purpurn, purpurfarben
purple carmine Purpurcarmin *n*
purple-colored purpurfarben, purpurrot
purple coloring Purpurfärbung *f*
purple lake Purpurlack *m*
purple of cassius Goldpurpur *m* (Chem)
purple ore Kiesabbrand *m*, Purpurerz *n*
purple red Purpurrot *n*
purple violet Purpurviolett *n*
purple wood Amarantholz *n*, Purpurholz *n*
purplish schist Purpurschiefer *m* (Min)
purpose Absicht *f*, Verwendungszweck *m*,
Zweck *m*
purposeful zweckdienlich, zweckmäßig
purpurate Purpurat *n*
purpurate indicator Purpuratindikator *m*
purpurea glycoside Purpureaglykosid *n*
purpuric acid Purpursäure *f*
purpurine Purpurin *n*
purpurite Purpurit *m* (Min)
purpurogallin Purpurogallin *n*
purpurogenone Purpurogenon *n*
purpuroxanthene Purpuroxanthen *n*
purse silk Kordonettseide *f* (Text)
pus Eiter *n* (Med)
pus cell Eiterzelle *f*
puschkinite Puschkinit *m*
push Stoß *m*
push *v* stoßen, drängen, drücken, schieben,
treiben, ~ back zurückstoßen, ~ [down] (a
button) (einen Knopf) niederdrücken, ~
forward vorwärtsstoßen, ~ through
durchstoßen
push button Druckknopf *m*, Drucktaste *f*
push-button control Druckknopfsteuerung *f*,
Drucktastenbedienung *f*
push-button switch Druckknopfschalter *m*
pushing trough Schub[förder]rinne *f*
push-off Abstoß *m*, Abstoßen *n*
push-out package Durchdrückpackung *f*
push-pull Gegentakt-, Zug-Druck-
push-pull mechanism Stoß-Zug-Mechanismus *m*
push-pull oscillator Gegentaktsender *m*

push rod Stösselstange *f*
pushrod pin bearing Schieberlager *n*
push screw Abdrückschraube *f*
pustule Bläschen *n*
putrefaction Fäulnis *f*, Faulen *n*, Verfall *m*,
Verwesung *f*, product of ~ Fäulnisprodukt *n*
putrefaction germ Fäulniskeim *m*
putrefaction process Fäulnisprozeß *m*
putrefactive fäulniserregend
putrefactive agent Fäulniserreger *m*
putrefactive alkaloid Fäulnisalkaloid *n*
putrefactive fermentation Fäulnisgärung *f*
putrefactive odor Fäulnisgeruch *m*
putrefy *v* [ver]faulen, verwesen
putrefying Faulen *n*, Verwesen *n*
putrescence Fäulnis *f*, Faulen *n*, Verwesen *n*
putrescent faulend
putrescibility Verfaulbarkeit *f*
putrescible verfaulbar
putrescine Putrescin *n*
putrid faul, moderig, to become ~ faulen
putridity Fäulnis *f*, Moder *m*
putridness Fäulnis *f*
putrid odor Fäulnisgeruch *m*
putrid spot Faulfleck *m*
putrid spots, having ~ faulfleckig
putty Fensterkitt *m*, Kitt *m*, Klebstoff *m* (Kitt),
Spachtelkitt *m*
putty *v* [ein]kitten, verkitten
putty chaser Kollergang *m*
putty industry Kittindustrie *f*
putty knife Kittmesser *n*, Spachtel *m f*
puttylike kittartig
Puzzolan earth Puzzolanerde *f*
pycnite Pyknit *m* (Min), Schörlit *m* (Min)
pycnochlorite Pyknochlorit *m* (Min)
pycnometer Pyknometer *n*
pycnometric pyknometrisch
pycnotrop Pyknotrop *m* (Min)
pyoctanine Pyoktanin *n*
pyocyanin Pyocyanin *n*
pyocyte Eiterzelle *f*
pyolipic acid Pyolipinsäure *f*
pyracene Pyracen *n*
pyraconitine Pyraconitin *n*
pyracridone Pyracridon *n*
pyracyclene Pyracyclen *n*
pyramid Pyramide *f*, truncated ~ abgestumpfte
Pyramide
pyramidal pyramidenförmig
pyramidal carbonate of lime Pyramidenspat *m*
(Min)
pyramidal octahedron Pyramidenoktaeder *n*
pyramidon Pyramidon *n*
pyran Pyran *n*
pyranene Pyranen *n*
pyranose Pyranose *f*
pyranoside Pyranosid *n*
pyranthrene Pyranthren *n*

pyranthridine Pyranthridin *n*
pyranthridone Pyranthridon *n*
pyranthrone Pyranthron *n*
pyrantin Pyrantin *n*, Phenosuccin *n*
pyranyl Pyranyl-
pyrargyrite Pyrargyrit *m* (Min),
Antimonsilberblende *f* (Min), dunkles
Rotgültigerz *n* (Min), Rubinblende *f* (Min)
pyrazine Pyrazin *n*
pyrazinoic acid Pyrazinsäure *f*
pyrazole Pyrazol *n*
pyrazole blue Pyrazolblau *n*
pyrazolic acid Pyrazolsäure *f*
pyrazolidine Pyrazolidin *n*
pyrazolidone Pyrazolidon *n*
pyrazoline Pyrazolin *n*
pyrazolone Pyrazolon *n*
pyrazolone dye Pyrazolonfarbstoff *m*
pyrelene Pyrelen *n*
pyrene Pyren *n*
pyrenequinone Pyrenchinon *n*
pyrethric acid Pyrethrinsäure *f*
pyrethrin Pyrethrin *n*
pyrethrol Pyrethrol *n*
pyrex [glass] Pyrexglas *n* (HN)
pyrheliometer Pyrheliometer *n*
pyridazine Pyridazin *n*
pyridazone Pyridazon *n*
pyridil Pyridil *n*
pyridilic acid Pyridilsäure *f*
pyridin[e] Pyridin *n*
pyridine carboxylic acid Picolinsäure *f*
pyridine derivative Pyridinderivat *n*
pyridine mesoporphyrin Pyridinmesoporphyrin *n*
pyridine nucleus Pyridinkern *m*
pyridine oxide Pyridinoxid *n*
pyridine red Pyridinrot *n*
pyridine ring Pyridinkern *m*
pyridine sulfonic acid Pyridinsulfonsäure *f*
pyridine tricarboxylic acid, 2,4,5- ~
Berberonsäure *f*
pyridinium hydroxide Pyridiniumhydroxid *n*
pyridofluorene Pyridofluoren *n*
pyridone Pyridon *n*
pyridophthalane Pyridophthalan *n*
pyridophthalide Pyridophthalid *n*
pyridostilbene Pyridostilben *n*
pyridoxal Pyridoxal *n*
pyridoxal phosphate Pyridoxalphosphat *n*
pyridoxin[e] Pyridoxin *n*, Adermin *n*, Vitamin
B₆ *n*
pyridyl Pyridyl-
pyridylazonaphthol Pyridylazonaphthol *n*
pyriform birnenförmig
pyrilium salt Pyriliumsalz *n*
pyrimidine Pyrimidin *n*
pyrimidine base Pyrimidinbase *f*
pyrimidine ring Pyrimidinring *m*
pyrimidone Pyrimidon *n*

pyrimidyl Pyrimidyl-
pyrindan Pyrindan *n*
pyrindol Pyrindol *n*
pyrite Pyrit *m* (Min), Eisenkies *m* (Min),
Katzengold *n* (Min), radiated ~ Markasit *m*
(Min), white iron ~ Graueisenerz *n* (Min)
pyrite kiln Kiesbrenner *m*, Kiesofen *m*
pyrites Pyrit *m* (Min), calcined ~
Kiesabbrand *m*, capillary ~ Haarkies *m*
(Min), roasted ~ Kiesabbrand *m*, ~
containing silver Gelf *m* (Min)
pyrite[s] burner Kiesbrenner *m*, Pyritofen *m*
pyrite[s] cinder Pyritabbrand *m*
pyrite[s] furnace Pyritofen *m*
pyrithiamine Pyrithiamin *n*
pyritic pyritartig, kiesartig, kiesig
pyritic process, partial ~ Halbpyritschmelzen *n*
(Metall)
pyritiferous kieshaltig, pyrithaltig
pyroabietic acid Pyroabietinsäure *f*
pyroacid Brenzsäure *f*, Pyrosäure *f*
pyroantimonate Pyroantimonat *n*
pyroantimonic acid Pyroantimonsäure *f*
pyroarsenic acid Pyroarsensäure *f*
pyroaurite Pyroaurit *m* (Min)
pyrobelonite Pyrobelonit *m* (Min)
pyrocatechin Brenzcatechin *n*, Brenzkatechin *n*,
Katechol *n*, Pyrocatechin *n*, Pyrokatechin *n*
pyrocatechol Brenzkatechin *n*, Pyrokatechin *n*
pyrocatechol dimethyl ether
Brenzcatechindimethyläther *m*
pyrocatechol phthalein Brenzcatechinphthalein *n*
pyrochlore Pyrochlor *m* (Min)
pyrochroite Pyrochroit *m* (Min)
pyrochromate Dichromat *n*, Pyrochromat *n*
pyrocinchonic acid Pyrocinchonsäure *f*
pyrocoll Pyrokoll *n*
pyrocomane Pyrokoman *n*
pyrodin Pyrodin *n*, Hydracetin *n*
pyroelectric pyroelektrisch
pyroelectricity Pyroelektrizität *f*
pyrogallic pyrogallussauer
pyrogallic acid Pyrogallol *n*, Pyrogallussäure *f*
pyrogallol Pyrogallol *n*, Pyrogallussäure *f*
pyrogallol monoacetate Eugallol *n*
pyrogallol phthalein Gallein *n*,
Pyrogallolphthalein *n*
pyrogallol triacetate Pyrogalloltriacetat *n*
pyrogen Pyrogen *n*
pyrogen dye Pyrogenfarbstoff *m*
pyroglutamic acid Pyroglutaminsäure *f*
pyrography Pyrographie *f*, Brandmalerei *f*
pyrogravure Brandmalerei *f*
pyroligneous spirit Holzspiritus *m*
pyrolignite Pyrolignit *n*
pyrolusite Pyrolusit *m* (Min), Braunstein *m*
(Min), Mangandioxid *n* (Min), Polianit *m*
(Min), Weichmanganerz *n* (Min)
pyrolysis Brenzreaktion *f*, Pyrolyse *f*

pyrolyze *v* pyrolysieren
pyromagnetic pyromagnetisch
pyromeconic acid Pyromekonsäure *f*
pyromellitic acid Pyromellitsäure *f*
pyrometallurgy Pyrometallurgie *f*
pyrometer Pyrometer *n*, Glutmesser *m*,
    Hitze[grad]messer *m*
pyrometer compensating circuit
    Pyrometerausgleichsleitung *f*
pyrometric cone Brennkegel *m*
pyrometry Pyrometrie *f*, Hitzemessung *f*
pyromorphite Blaubleierz *n* (Min),
    Braunbleierz *n* (Min), Pyromorphit *m* (Min),
    green ~ Grünblei[erz] *n* (Min), variegated ~
    Buntbleierz *n* (Min)
pyromorphous pyromorph
pyromucate Pyromucat *n*
pyromucic pyroschleimsauer
pyromucic acid Brenzschleimsäure *f*,
    Pyromuconsäure *f*, Pyroschleimsäure *f*, salt or
    ester of ~ Pyromucat *n*
pyronaphtha Pyronaphtha *n*
pyrone Pyron *n*, 1,4- ~ Pyrokoman *n*
pyronine Pyronin *n*
pyronine dye Pyroninfarbstoff *m*
pyronone Pyronon *n*
pyrope Pyrop *m* (Min)
pyrophanite Pyrophanit *m* (Min)
pyrophanousness Durchsichtigkeit *f* im Feuer
pyropheophorbide Pyrophäophorbid *n*
pyrophoric pyrophor, luftentzündlich
pyrophorus Pyrophor *n*, Luftzünder *m*
pyrophosphatase Pyrophosphatase *f*
pyrophosphate Pyrophosphat *n*
pyrophosphoric acid Pyrophosphorsäure *f*,
    Diphosphorsäure *f*
pyrophosphorous acid pyrophosphorige Säure *f*
pyrophotography Pyrophotographie *f*
pyrophyllite Pyrophyllit *m* (Min)
pyrophysalite Pyrophysalit *m* (Min)
pyropissite Pyropissit *m*
pyroquinine Pyrochinin *n*
pyroracemic acid Brenztraubensäure *f*
pyroracemic aldehyde Methylglyoxal *n*
pyrorthite Pyrorthit *m*
pyrosal Salipyrin *n*
pyrosclerite Pyrosklerit *m*
pyroscope Pyroskop *n*
pyrosmalite Pyrosmalit *m* (Min)
pyrostain Gelbschleier *m*
pyrostilbite Pyrostilbit *m* (Min)
pyrostilpnite Pyrostilpnit *m* (Min)
pyrosulfate Pyrosulfat *n*, Disulfat *n*
pyrosulfite Pyrosulfit *n*
pyrosulfuric acid Dischwefelsäure *f*,
    Pyroschwefelsäure *f*
pyrosulfurous pyroschwefelig
pyrosulfuryl chloride Pyrosulfurylchlorid *n*

pyrotartaric acid Brenzweinsäure *f*,
    Pyroweinsäure *f*
pyrotartrate Pyrotartrat *n*
pyrotechnical pyrotechnisch
pyrotechnics Pyrotechnik *f*
pyroterebic acid Brenzterebinsäure *f*
pyrotritaric acid Pyrotritarsäure *f*
pyrotritartaric acid Uvinsäure *f*
pyrovinic acid Brenzweinsäure *f*
pyroxene Pyroxen *m* (Min)
pyroxenic pyroxenhaltig
pyroxeniferous pyroxenhaltig
pyroxonium salt Pyroxoniumsalz *n*
pyroxylin Pyroxylin *n*, Kollodiumwolle *f*
pyroxylin filament Pyroxylinfaden *m*
pyroxylin varnish Pyroxylinlack *m*
pyrrhoarsenite Pyrrhoarsenit *m* (Min)
pyrrhosiderite Pyrrhosiderit *m* (Min)
pyrrhotite Magnetkies *m* (Min),
    Magnetopyrit *m* (Min)
pyrril Pyrril *n*
pyrrocoline Pyrrocolin *n*
pyrrodiazole Pyrrodiazol *n*
pyrrol[e] Pyrrol *n*
pyrrol[e] blue Pyrrolblau *n*
pyrrolidine Pyrrolidin *n*
pyrrolidine carboxylic acid Prolin *n*
pyrrolidone Pyrrolidon *n*
pyrroline carboxylic acid Pyrrolincarbonsäure *f*
pyrrolizidine Pyrrolizidin *n*
pyrrolone Pyrrolon *n*
pyrromethene Pyrromethen *n*
pyrroporphyrin Pyrroporphyrin *n*
pyrroyl Pyrroyl-
pyrryl Pyrryl-
pyrthiophanthrone Pyrthiophanthron *n*
pyruvaldehyde Methylglyoxal *n*
pyruvate Pyruvat *n*
pyruvate decarboxylase Pyruvatdecarboxylase *f*
    (Biochem)
pyruvate dehydrogenase Pyruvatdehydrogenase *f*
    (Biochem)
pyruvate dehydrogenase complex
    Pyruvatdehydrogenasekomplex *m* (Biochem)
pyruvate kinase Pyruvatkinase *f* (Biochem)
pyruvate oxidase Pyruvatoxydase *f* (Biochem)
pyruvic acid Brenztraubensäure *f*
pyruvic aldehyde Methylglyoxal *n*
pyruvyl Pyruvyl-
pyruvylalanine Pyruvylalanin *n*
pyruvylglycine Pyruvylglycin *n*
pyrylene Pyrylen *n*
Pythagorean pythagareisch
Pythagorean theorem Pythagoreischer
    Lehrsatz *m* (Math), Satz *m* von Pythagoras
    (Math)
pyvuril Pyvuril *n*

Q

quadrangle Viereck *n*
quadrangular viereckig, vierseitig
quadrant Quadrant *m*
quadrant balance Flächengewichtswaage *f*
quadrant electrometer Quadrantelektrometer *n*
quadrant iron Quadranteisen *n*, Säuleneisen *n*
quadratic quadratisch
quadratically, to decrease ~ quadratisch
abnehmen
quadrature Quadratur *f*
quadribasic vierbasig
quadricyclene Quadricyclen *n*
quadrilateral vierseitig
quadrillion (Am. E.) Billiarde *f*
quadrimolecular viermolekular
quadrivalence Vierwertigkeit *f*
quadrivalent vierwertig
quadruple vierfach, vierzählig
quadruple point Quadrupelpunkt *m*
quadruple twin Achter *m*
quadruplex system Vierfachbetrieb *m*
quadrupole force Quadrupolkraft *f*
quake *v* beben
qualification Bewertung *f*, Qualifikation *f*
qualified qualifiziert, tauglich
qualify *v* befähigen, qualifizieren, sich eignen
qualimeter Qualimeter *n*
qualitative qualitativ
qualitative analysis qualitative Analyse *f*
quality Qualität *f*, Beschaffenheit *f*, Eigenart *f*,
Eigenschaft *f*, Fähigkeit *f*, Güte *f*,
Güteeigenschaft *f*, **coefficient of** ~
Qualitätskoeffizient *m*, **degree of** ~
Gütegrad *m*, **deviation in** ~
Qualitätsabweichung *f*, **difference in** ~
Qualitätsunterschied *m*, **improvement of** ~
Gütesteigerung *f*, Qualitätsverbesserung *f*, **of**
first class ~ hochwertig, **property influencing**
~ Güteeigenschaft *f*, **requirement as to** ~
Güteanforderung *f*, **rust-preventing** ~
rostschützende Wirkung *f*
quality constant Qualitätsmaß *n*
quality control Güteüberwachung *f*,
Qualitätskontrolle *f*, Qualitätsüberwachung *f*
quality factor Gütezahl *f*
quality index Güteziffer *f*
quality level, acceptable ~ annehmbare
Qualitätsgrenze *f*
quality preservation Qualitätserhaltung *f*
quality retention Qualitätserhaltung *f*
quality seal Gütezeichen *n*
quality standard Gütenorm *f*
quality test Gütetest *m*, Qualitätsprüfung *f*
quanta Quanten *pl* (Phys)
quantic quantenhaft
quantifiable quantitativ bestimmbar

quantification Quantitätsbestimmung *f*
quantimeter Quantimeter *n*
quantitative quantitativ, mengenmäßig
quantitative analysis quantitative Analyse *f*
quantitative composition Mengenverhältnis *n*
quantitative determination Mengenbestimmung *f*,
Quantitätsbestimmung *f*
quantitative ratio Mengenverhältnis *n*
quantity Quantität *f*, Anzahl *f*, Menge *f*,
Quantum *n*, **actual** ~ Istmenge *f*, **auxiliary** ~
Hilfsgröße *f*, **directed** ~ gerichtete Größe *f*,
negligible ~ zu vernachlässigende Größe *f*,
oriented ~ gerichtete Größe *f*, **supplied** ~
Liefermenge *f*, **theoretical** ~ Sollmenge *f*,
thermodynamic ~ thermodynamische
Größe *f*, **total** ~ Gesamtmenge *f*, **unknown** ~
unbekannte Größe *f*, ~ **of air** Luftmenge *f*, ~
of deposit Ablagerungsmenge *f*, ~ **to be**
measured Meßgröße *f*
quantity galvanometer Quantitätsgalvanometer *n*
quantity of light Lichtmenge *f*
quantity production Massenproduktion *f*,
Massenfertigung *f*, serienmäßige Herstellung *f*
quantity stop Anschlagraste *f*
quantization Quantelung *f*, Quantisierung *f*
quantization of momentum Impulsquantelung *f*
quantize *v* quanteln
quantized gequantelt
quantosome Quantosom *n* (Biol)
quantum Quant *n* (Phys), Quantum *n*
quantum biochemistry Quantenbiochemie *f*
quantum biology Quantenbiologie *f*
quantum chemistry Quantenchemie *f*
quantum condition Quantenbedingung *f*
quantum effect Quanteneffekt *m*
quantum electrodynamics
Quantenelektrodynamik *f*
quantum hypothesis Quantenhypothese *f*
quantum jump Quantensprung *m*
quantum leakage Quantenverlust *m*
quantum-mechanical quantenmechanisch
quantum mechanics Quantenmechanik *f*
quantum number Quantenzahl *f*, **azimuthal** ~
azimutale Quantenzahl, **inner** ~ innere
Quantenzahl, **principal** ~
Hauptquantenzahl *f*, **total** ~
Gesamtquantenzahl *f*
quantum orbit Quantenbahn *f*
quantum path Quantenbahn *f*
quantum physics Quantenphysik *f*
quantum resonance Quantenresonanz *f*
quantum scattering Quantenstreuung *f*
quantum state Quantenzustand *m*
quantum statistics Quantenstatistik *f* (Quant)
quantum theory Quantentheorie *f*

quantum transition Quantensprung *m*,
   Quantenübergang *m*
quantum weight Quantengewicht *n*
quantum yield Quantenausbeute *f*
quarry Steinbruch *m*
quarry stone Bruchstein *m*
quartation Quartation *f*
quarter Viertel *n*, ~ **of a circle** Quadrant *m*
quarter bend [pipe] Krümmer *m*
quarter-period Viertelperiode *f*
quarterturn belt Halbkreuzriemen *m*
quartet[te] Quartett *n*
quartz Quarz *m* (Min), **auriferous** ~
   Goldquarz *m* (Min), **bituminous** ~
   Stinkquarz *m* (Min), **blue** ~ Saphirquarz *m*
   (Min), **fetid** ~ Stinkquarz *m* (Min), **fibrous** ~
   Faserquarz *m* (Min), Faserstein *m* (Min),
   Strahlenquarz *m* (Min), **fused** ~
   geschmolzener Quarz, **piezoelectric** ~
   piezoelektrischer Quarz, **racemic** ~ optisch
   inaktiver Quarz, ~ **in lumps** Stückquarz *m*
quartz carbon mixture Quarzkohlegemisch *n*
quartz clock Quarzuhr *f*
quartz crystal Kristallquarz *m* (Min),
   Quarzkristall *m* (Min)
quartz crystal clock Quarzuhr *f*
quartz cuvette Quarzküvette *f*
quartz-fiber manometer
   Quarzfadenmanometer *n*,
   Quarzfadendruckmesser *m*
quartz filament Quarzfaden *m*
quartz glass Quarzglas *n*
quartz grains Quarzkörner *n pl*
quartz gravel Quarzkiesel *m*
quartziferous quarzhaltig
quartzine Quarzin *n*
quartzite Quarzit *m* (Min),
   Grauwackenquarz *m* (Min), Körnerquarz *m*
   (Min)
quartz lamp Analysenlampe *f*, Quarzlampe *f*
quartz lens Quarzlinse *f*
quartz mercury vapor lamp
   Quecksilber-Quarzlampe *f*
quartz oscillator Quarzoszillator *m*
quartzose quarzartig, quarzhaltig, quarzig
quartzose sand Quarzsand *m* (Min)
quartzose schist Quarzschiefer *m* (Min)
quartzous quarzartig, quarzhaltig, quarzig
quartz pisolite Quarzpisolith *m* (Min)
quartz porphyry Quarzporphyr *m* (Min)
quartz powder Quarzmehl *n*
quartz resonator Schwingquarz *m* (Elektr)
quartz rock Kieselgestein *n*
quartz sand Quarzsand *m* (Min)
quartz sinter Kieselsinter *m*, Quarzsinter *m*
quartz slate Quarzschiefer *m* (Min)
quartz thread Quarzfaden *m*
quartz vein Quarzgang *m*
quartz ware Quarzgut *n*

quartz wedge Quarzkeil *m*
quartz wool Quarzwatte *f*, Quarzwolle *f*
quartzy quarzhaltig, quarzig
quartzy agate Quarzachat *m* (Min)
quartzy sandstone Quarzsandstein *m*
quarz-like quarzartig
quasi-atom Quasiatom *n* (Phys)
quasi-chemical quasichemisch
quasielastic quasielastisch
quasi equilibrium Quasigleichgewicht *n*
quasilinear quasilinear
quasilinearization Quasilinearisieren *n*
quasi-molecule Quasimolekül *n*
quasi momentum Quasiimpuls *m*
quasi neutrality Quasineutralität *f*
quasistable quasistabil
quasistatic quasistatisch
quasistationary quasistationär
quasiviscous quasiviskos
quasiviscous flow Fließerweichung *f*
quassia Bitteresche *f* (Bot)
quassia wood Bitterholz *n*, Quassiaholz *n*, **extract**
   of ~ Quassiaextrakt *m*
quassin Quassin *n*, Quassiabitter *n*
quaternary quartär, quaternär
quaternary mixture Vierstoffgemisch *n*
quaternary steel Quaternärstahl *m*
quaternary structure Quartärstruktur *f*
quaternary system Vierstoffsystem *n*
quaterrylene Quaterrylen *n*
quebrachamine Quebrachamin *n*
quebrachine Quebrachin *n*
quebrachite Quebrachit *m*
quebrachitol Quebrachit *m*
quebracho Quebracho *m* (Bot)
quebracho bark Quebrachorinde *f*
quebracho extract Quebrachoextrakt *m*
quebracho-tannic acid Quebrachogerbsäure *f*
quebracho wood Quebrachoholz *n*
quench *v* [aus]löschen, abkühlen (Met),
   abschrecken (Met), härten (Met)
quenchable [aus]löschbar
quench-age *v* aushärten
quench aging Abschreckalterung *f*
quenched gelöscht, abgeschreckt
quenched charcoal Löschkohle *f*
quenched spark Löschfunken *m*
quench hardening Abschreckhärtung *f*
quenching Löschen *n*, Abschrecken *n*,
   Abschreckung *f*, Auslöschung *f*, Dämpfung *f*,
   ~ **and tempering** Vergütung *f* (Stahl)
quenching agent Abschreckmittel *n*
quenching bath Abschreckbad *n*, Härtebad *n*
quench[ing] circuit Löschkreis *m*,
   Löschschaltung *f*
quenching method Abschreckmethode *f*
quenching oil Härteöl *n*
quenching properties Löscheigenschaften *f pl*
quenching stress Härtungsspannung *f*

quenching tank Abschreckbottich *m*
quenching temperature Abschrecktemperatur *f*
quenching time Abschreckzeit *f*
quenching tower Löschturm *m*
quench pulse Löschimpuls *m*
quench tank Abschreckbehälter *m*
quench tube reactor Hordenofen *m*
quenstedtite Quenstedtit *m* (Min)
quercetagetin Quercetagetin *n*
quercetin Quercetin *n*
quercetinic acid Quercetinsäure *f*
quercetone Querceton *n*
quercimeritrine Quercimeritrin *n*
quercin Quercin *n*
quercitannic acid Eichengerbsäure *f*
quercite Quercit *m*, Eichelzucker *m*
quercitol Quercit *m*
quercitrin Quercitrin *n*
quercitron Färbereiche *f* (Bot), Quercitron *n*
quercitron bark Färberrinde *f*, Quercitronrinde *f*
quercitron lake Quercitronlack *m*
question *v* [aus]fragen
questionable bedenklich, fraglich
quetenite Quetenit *m* (Min)
quick rasch, schnell
quick-acting schnellwirkend
quick-action stop valve Schnellschlußventil *n*
quick clamp Schnellklemme *f*
quick-curing schnellhärtend (Kunstharz)
quicken *v* beleben, beschleunigen
quick freezing Schnellgefrieren *n*
quicklime gebrannter Kalk *m*, Löschkalk *m*,
    ungelöschter Kalk *m*
quick match Zündband *n*
quick-opening schnellöffnend
quick procedure Schnellverfahren *n*
quick-run filter Schnellfilter *n*
quicksand Schwimmsand *m*, **layer of** ~
    Schwimmsandschicht *f*
quick-setting schnell[ab]bindend
quicksilver Quecksilber *n*, **native** ~
    Quecksilberstein *m* (Min)
quicksilver ore Quecksilbererz *n* (Min)
quick test Vorprobe *f*
quick water Quickwasser *n*
quiescent period Ausgärzeit *f* (Stahl)
quiet ruhig, still
quiet *v* beruhigen
quietol Quietol *n*
quillai[a] bark Quillajarinde *f*
quillaic acid Quillajasäure *f*
quina Fieberrinde *f* (Bot)
quinacetophenone Chinacetophenon *n*
quinacridine Chinacridin *n*
quinacridone Chinacridon *n*
quinaldic acid Chinaldinsäure *f*
quinaldinate Chinaldinat *n*
quinaldine Chinaldin *n*
quinaldine blue Chinaldinblau *n*

quinalizarin Chinalizarin *n*
quinamicine Chinamicin *n*
quinamine Chinamin *n*
quinane Chinan *n*
quinanisol Chinanisol *n*
quinanthridine Chinanthridin *n*
quinazine Chinoxalin *n*
quinazoline Chinazolin *n*
quinazolone Chinazolon *n*
quince Quitte *f* (Bot)
quince juice Quittensaft *m*
quince mucilage Quittenschleim *m*
quince oil Quittenöl *n*
quindoline Chindolin *n*
quinene Chinen *n*
quinethyline Chinäthylin *n*
quinetum Chinetum *n*
quinhydrone Chinhydron *n*
quinhydrone electrode Chinhydronelektrode *f*
quinic acid Chinasäure *f*
quinicine Chinicin *n*
quinidamine Chinidamin *n*
quinidane Chinidan *n*
quinidine Chinidin *n*, Conchinin *n*
quinidine sulfate Chinidinsulfat *n*
quinindene Chininden *n*
quinindole Chinindol *n*
quinindoline Chinindolin *n*
quinine Chinin *n*
quinine acetate Chininacetat *n*
quinine alkaloid Chininalkaloid *n*
quinine anisate Chininanisat *n*
quinine antimonate Chininantimonat *n*
quinine aspirin Chininaspirin *n*
quinine base Chinabase *f*
quinine bisulfate Chininbisulfat *n*
quinine chromate Chininchromat *n*
quinine citrate Chinincitrat *n*
quinine ethyl carbonate Chininäthylcarbonat *n*,
    Euchinin *n*
quinine ferricyanide Chininferricyanid *n*
quinine hydrobromide Chininbromhydrat *n*,
    Chininhydrobromid *n*
quinine hydrochloride Chininhydrochlorid *n*
quinine hydrogen sulfate Chininbisulfat *n*
quinine hydroiodide Chininhydrojodid *n*
quinine lactate Chininlactat *n*
quinine nitrate Chininnitrat *n*
quinine phosphate Chininphosphat *n*
quinine salicylate Chininsalicylat *n*, Salochinin *n*
quinine succinate Chininsuccinat *n*
quinine sulfate Chininsulfat *n*, **acid** ~
    Chininbisulfat *n*
quinine sulfate periodide Herapathit *n*
quinine valerate Chininvalerianat *n*
quinine wine Chinawein *m*
quininone Chininon *n*
quinisatin Chinisatin *n*
quinisatinic acid Chinisatinsäure *f*

quinism Chininvergiftung *f*
quinisocaine Chinisocain *n*
quinite Chinit *m*
quinitol Chinit *m*
quinizarin Chinizarin *n*
quinizarin green Chinizaringrün *n*
quinodimethane Chinodimethan *n*
quinoidine Chinoidin *n*
quinolin[e] Chinolin *n*
quinoline aldehyde Chinolinaldehyd *m*
quinoline alkaloid Chinolinalkaloid *n*
quinoline base Chinolinbase *f*
quinoline blue Chinolinblau *n*, Cyanin *n*,
   Cyaninfarbstoff *m*
quinoline carboxylic acid Chinolincarbonsäure *f*
quinoline derivative Chinolinderivat *n*
quinoline dicarboxylic acid, 2,3- ~
   Akridinsäure *f*
quinoline dye Chinolinfarbstoff *m*
quinoline hydrochloride Chinolinhydrochlorid *n*
quinoline quinone Chinolinchinon *n*
quinoline red Chinolinrot *n*
quinoline salicylate Chinolinsalicylat *n*
quinoline strychenone Chinolinstrychenon *n*
quinoline sulfate Chinolinsulfat *n*
quinoline yellow Chinolingelb *n*
quinolinic acid Chinolinsäure *f*
quinolinol Chinolinol *n*, Oxychinolin *n*
quinolizidine Chinolizidin *n*
quinolizidone Chinolizidon *n*
quinolizine Chinolizin *n*
quinolizone Chinolizon *n*
quinologist Chinologe *m*
quinology Chinologie *f*
quinolone Chinolon *n*
quinolyl Chino[l]yl-
quinomethane Chinomethan *n*
quinone Chinon *n*, Benzochinon *n*, **halogenated**
   ~ Halogenchinon *n*
quinone diimine Chinondiimin *n*
quinone dioxime Chinondioxim *n*
quinone monoxime Chinonmonoxim *n*
quinonimine Chinonimin *n*
quinophthalone Chinophthalon *n*
quinopyrine Chinopyrin *n*
quinosol Chinosol *n*
quinotannic acid Chinagerbsäure *f*
quinoticine Chinoticin *n*
quinotidine Chinotidin *n*
quinotine Chinotin *n*
quinotinone Chinotinon *n*
quinotoxine Chinotoxin *n*
quinotropine Chinotropin *n*
quinova bitter Chinovabitter *n*
quinovene Chinoven *n*
quinovic acid Chinovasäure *f*
quinovin Chinovin *n*, Chinovabitter *n*
quinovose Chinovose *f*
quinoxaline Chinoxalin *n*

quinoxanthene Chinoxanthen *n*
quinoyl- Chino[l]yl-
quinquevalence Fünfwertigkeit *f*
quinquiphenyl Quinquiphenyl *n*
quinrhodin Chinrhodin *n*
quintessence Auszug *m*, Quintessenz *f*
quintuple fünffach
quintuple point Quintupelpunkt *m*
quinuclidine Chinuclidin *n*
quisqueite Quisqueit *m* (Min)
quiver *v* zittern
quota Anteil *m*
quotation Angebot *n*
quotation mark Anführungszeichen *n*
quote *v* anführen, zitieren
quotient Quotient *m* (Math)

R

rabbet Fuge *f*, Nut *f*
rabbet *v* falzen
rabbet plane Hohlkehlhobel *m*
rabble Kratze *f*, Rührhaken *m*, Schürstange *f*
rabelaisin Rabelaisin *n*
rabies Rabies *f* (Tiermed), Tollwut *f*
rabies vaccine Tollwutvakzine *f*
race Rasse *f*
racemate Racemat *n* (Stereochem)
racemethorphan Racemethorphan *n*
racemic racemisch
racemic compound Racemat *n* (Stereochem)
racemic modification Racemform *f*
racemism Racemie *f*, racemischer Zustand *m*
racemization Racemisierung *f*, thermal ~
  thermische Racemisierung, ~ by Walden
  inversion Racemisierung *f* durch Waldensche
  Umkehr
racemize *v* racemisieren
racemoramide Racemoramid *n*
racemorphan Racemorphan *n*
rachitis Rachitis *f* (Med), englische Krankheit *f*
  (Med)
racing of the machine Durchgehen *n* der
  Maschine
racing tire Rennreifen *m* (Kfz)
rack Gestell *n*, Regal *n*, Zahnstange *f* (Techn),
  hook-in ~ einhängbares Regal *n*, shelves in a
  ~ Regalfächer *pl*
[rack] aisle Regalgang *m*, Regalgasse *f*
rack and pinion Stangentriebwerk *n*,
  Zahnstangengetriebe *n*
rack brace Bohrknarre *f*
rack gear Stangentriebwerk *n*
racking cellar Abziehkeller *m*
racking plant Abfüllanlage *f* (f. Fässer)
racking pump Abfüllpumpe *f* (Brauw)
racking square Abfüllbütte *f*
rack-running order-picking system
  Kommissionierfördersystem *n*
rack stacker Regalförderzeug *n*
rack tool Kammstahl *m*
radar Funkmeßgerät *n*, Funkmeßtechnik *f*,
  Radar *n*
radar antenna Radarantenne *f*
radar [apparatus] Radargerät *n*
radar control Radarkontrolle *f*
radar-controlled radargesteuert
radar engineering Radartechnik *f*
radar installation Radaranlage *f*
radar screen Radarbildschirm *m*, Radarschirm *m*
radar surface range
  Radaroberflächenreichweite *f*
radar unit Radaranlage *f*
radar warning system Radarwarnsystem *n*
raddeanine Raddeanin *n*

radial radial
radial acceleration Radialbeschleunigung *f*
radial bag filter Rundfilter *m*
radial bearing Radiallager *n*, Querlager *n*
radial clearance Schneckenspiel *n*
radial coil armature Innenpolanker *m*
radial crack Radialriß *m*
radial drill Radialbohrer *m*
radial drilling machine Radialbohrmaschine *f*
radial eigenfunction Radialeigenfunktion *f*
radial expansion Radialausdehnung *f*
radial flow Radialstrom *m*
radial-flow compressor Radialverdichter *m*
radial-flow impeller Turbinenmischer *m*
radial-flow scrubber Radialstromwäscher *m*
radial force Radialkraft *f*
radial integral Radialintegral *n*
radial load Radialbelastung *f*
radial method Radialmethode *f*
radial mode Radialschwingung *f*
radial motion Radialbewegung *f*
radial motion chamber Radialkammer *f*
radial propeller Radialpropeller *m*
radial runout Radialschlag *m* (Kfz)
radial velocity Radialgeschwindigkeit *f*
radiance Lichtglanz *m*, Lichtschein *m*,
  Strahlung *f*
radiant strahlend
radiant flux density Bestrahlungsstärke *f* (Opt)
radiant heat Strahlungswärme *f*
radiant heating system
  Wärmestrahlungsheizung *f*
radiant intensity Strahlungsintensität *f*
radiant-type boiler Flammrohrkessel *m*
radiate *v* [aus]strahlen, durchstrahlen, Strahlen
  aussenden
radiate[d] strahlenförmig, strahlig
radiated barite Strahlbaryt *m* (Min)
radiated pyrite Markasit *m* (Min)
radiated spar Federspat *m* (Min)
radiating strahlend
radiating efficiency Strahlungsausbeute *f*
radiating power Strahlungsvermögen *n*
radiating surface Ausstrahlungsfläche *f*
radiation Strahlung *f*, Ausstrahlung *f*,
  Bestrahlung *f*, Strahlenemission *f*,
  characteristic ~ Eigenstrahlung *f*, coherent ~
  kohärente Strahlung, cosmic ~ kosmische
  Strahlung, damaged by ~ strahlengeschädigt,
  danger of ~ Strahlungsgefährdung *f*,
  fluorescent ~ Fluoreszenzstrahlung *f*, free of
  ~ strahlungsfrei, hardness of ~
  Strahlenhärte *f*, heat of ~ Strahlungswärme *f*,
  heterogeneous ~ heterogene Strahlung,
  homogeneous ~ homogene Strahlung,
  incoming ~ Einstrahlung *f*, intensity of ~

Strahlungsintensität *f,* **ionizing** ~ ionisierende
Strahlung, **long-distance** ~ Fernbestrahlung *f,*
loss by ~ Strahlungsverlust *m,* **magnetic [field]**
~ magnetische Strahlung, **opacity for** ~
Strahlenundurchlässigkeit *f,* **primary** ~
Primärstrahlung *f,* **radioactive** ~ radioaktive
Strahlung, **range of** ~ Strahlungsbereich *m,*
secondary ~ Sekundärstrahlung *f,* **source of** ~
Strahlungsquelle *f,* **standard of** ~
Strahlungsnorm *f,* ~ **of heat**
Wärmestrahlung *f,* ~ **of light** Lichtstrahlung *f*
radiation absorption Strahlungsabsorption *f*
radiation activity Strahlenaktivität *f*
radiation analyzer Strahlenanalysator *m*
radiation bombardment Strahlungseinwirkung *f*
radiation capacity Ausstrahlungsvermögen *n*
radiation catalysis Strahlungskatalyse *f*
radiation characteristic
Strahlungscharakteristik *f*
radiation chart Strahlungsdiagramm *n*
radiation coefficient Strahlungskoeffizient *m*
radiation counter tube Strahlungszählrohr *n*
radiation damage Schädigung *f* durch
Bestrahlung
radiation damping Strahlungsdämpfung *f*
radiation density Strahlungsdichte *f*
radiation detection instrument
Strahlennachweisgerät *n*
radiation diagram Strahlungsdiagramm *n*
radiation dosage Strahlendosis *f,*
Strahlendosierung *f*
radiation dose Strahlendosis *f*
radiation dryer Strahlungstrockner *m*
radiation effect Strahlungseffekt *m*
radiation efficiency Strahlungsleistung *f*
radiation elimination Strahlenauslöschung *f*
radiation energy Strahlenenergie *f;*
Strahlungsenergie *f*
radiation fluxmeter, total ~
Gesamtstrahlungsdosimeter *n*
radiation furnace Strahlungsofen *m*
radiation hazard Strahlengefahr *f*
radiation hazards, exposed to ~
strahlengefährdet
radiation heating Strahlungsheizung *f*
radiation-induced strahlungsinduziert
radiation-initiated strahlungsinduziert
radiation injury Verletzung *f* durch Bestrahlung
radiation intensitometer Strahlenstärkemesser *m*
radiation law Strahlungsgesetz *n*
radiation length Strahlungsreichweite *f* (Atom)
radiationless strahlungsfrei, strahlungslos
radiation loss Ausstrahlverlust *m*
radiation measuring equipment
Strahlungsmeßgerät *n*
radiation [measuring] instrument
Strahlungsmeßinstrument *n*
radiation meter Strahlungsmeßgerät *n*

radiation monitor Strahlenüberwachungsgerät *n,*
Strahlenwarngerät *n*
radiation monitoring device Strahlenwarngerät *n*
radiation mutation Bestrahlungsmutation *f*
radiation of light Lichtausstrahlung *f*
radiation potential Strahlungspotential *n*
radiation preservation Strahlenkonservierung *f*
radiation product Bestrahlungsprodukt *n*
radiation protection Strahlenschutz *m,*
Strahlungsabschirmung *f*
radiation pyrometer Strahlungspyrometer *n*
radiation receiver Empfängerkammer *f*
radiation resistance Strahlungswiderstand *m*
radiation-resistant strahlenresistent
radiation screen Strahlungsabschirmung *f*
radiation shield Strahlungsabschirmung *f*
radiation shielding Strahlenabschirmung *f,*
strahlungsabschirmend
radiation standard Strahlungsnormal *n*
radiation sterilization
Bestrahlungssterilisierung *f*
radiation surplus Strahlungsüberschuß *m*
radiation survey Strahlungsüberwachung *f*
radiation theory Strahlungstheorie *f*
radiation-therapy Bestrahlungstherapie *f* (Med)
radiation transition Strahlungsübergang *m*
radiation transmission Strahlungsdurchgang *m*
radiation treatment Strahlenbehandlung *f,*
Strahlentherapie *f* (Med)
radiation width Strahlungsbreite *f*
radiative capture Einfangprozeß *m* (Atom)
radiative transition Strahlungsübergang *m*
radiator Heizkörper *m,* Radiator *m,* Strahler *m,*
cellular-type ~ Lamellenkühler *m* (Kfz)
radiator cover Kühlerhaube *f*
radiator grill Kühlergrill *m,* Kühlerrost *m*
radiator hose Kühlerschlauch *m* (Kfz)
radiator paint Heizkörperfarbe *f*
radiator shutter Kühlerjalousie *f*
radical Radikal *n* (Chem), Rest *m* (Chem),
aliphatic ~ aliphatisches Radikal, **aromatic** ~
aromatisches Radikal, **free** ~ freies Radikal,
long-lived ~ langlebiges Radikal, **short-lived**
~ kurzlebiges Radikal
radical *a* radikal, durchgreifend
radical chain reaction Radikalkettenreaktion *f*
radical expression Wurzelausdruck *m* (Math)
radical index Wurzelexponent *m* (Math)
radical polymerization Radikalpolymerisation *f*
radicals, migration of ~ Radikalwanderung *f*
radical sign Wurzelzeichen *n*
radical substitution Substitution *f* von
Radikalen (pl)
radicinin Radicinin *n*
radio Radio[apparat] *m,* Funk *m,* Rundfunk *m*
radio *v* funken
radioactinium (Symb. RdAc) Radioaktinium *n*
radioactive radioaktiv, **highly** ~ hochradioaktiv,
to render ~ radioaktiv machen

radioactive decontamination radioaktive Entseuchung *f*
radioactive effluents radioaktive Abwässer *pl*
radioactive element Radioelement *n*, radioaktives Element *n*
radioactive equilibrium radioaktives Gleichgewicht *n* (Atom)
radioactive fall-out Atomregen *m*
radioactive isotope Radioisotop *n*
radioactively contaminated atomverseucht
radioactive material radioaktiver Stoff *m*
radioactive nuclide Radionuklid *n*
radioactive ore detector Prospektionszähler *m*
radioactive phosphorus Radiophosphor *m*
radioactive pollution of the air Luftverpestung *f* durch Radioaktivität
radioactive substance radioaktiver Stoff *m*
radioactive tracer Radioindikator *m*
radioactive waste radioaktive Abfallstoffe *pl*, Atommüll *m*
radioactivity Radioaktivität *f*, **artificial** ~ künstliche Radioaktivität, **induced** ~ induzierte Radioaktivität, **~ of the air** Luftradioaktivität *f*
radioactivity detection Radioaktivitätsnachweis *m*
radioautography Autoradiographie *f*, Strahlungsphotographie *f*
radiobiological radiobiologisch
radiobiology Radiobiologie *f*
radio cabinet Radiogehäuse *n*
radiocarbon Radiokohlenstoff *m*, radioaktiver Kohlenstoff *m*
radiocarbon-dating Radiokohlenstoff-Datierung *f*
radiochemist Radiochemiker *m*
radiochemistry Radiochemie *f*
radiochrometer Radiochrometer *n*
radio communication Funkverbindung *f*
radio control drahtlose Fernsteuerung *f*, Funk[fern]steuerung *f*
radio-control *v* fernsteuern
radioelement radioaktives Element *n*
radiofrequency Hochfrequenz *f*, Radiofrequenz *f*
radiofrequency spectrometer Radiofrequenzspektrometer *n*
radiogram Funktelegramm *n*, Radiogramm *n* (Phys), Röntgenaufnahme *f* (Phys)
radiograph Radiogramm *n* (Phys), Röntgenaufnahme *f* (Phys), Röntgenbild *n* (Phys)
radiographic röntgenographisch
radiography Radiographie *f*; Röntgenphotographie *f*
radio interferometer Radiointerferometer *n*
radioisotope Radioisotop *n*, radioaktives Isotop *n*
radioisotope technique Leitisotopenmethode *f*
radiolite Radiolith *m* (Min)

radiologic[al] radiologisch
radiological result[s] Röntgenbefund *m*
radiologist Radiologe *m*, Röntgenologe *m*
radiology Radiologie *f*, Röntgenkunde *f*, Strahlenlehre *f*
radiolucency Röntgenstrahlendurchlässigkeit *f*
radiolucent röntgenstrahlendurchlässig
radioluminescence Radiolumineszenz *f*
radiolysis Radiolyse *f*
radiometer Radiometer *n*, Strahlenmesser *m*
radiometric radiometrisch
radionuclide Radionuklid *n*
radiopaque undurchlässig für Röntgenstrahlen (pl)
radiophotograph[y] Radiophotographie *f*
radio receiver Rundfunkempfänger *m*, **heterodyne** ~ Überlagerungsempfänger *m*
radioresistance Strahlenresistenz *f*
radio resistor Radiowiderstand *m*
radioscope Strahlungssucher *m*
radiosensitive strahlenempfindlich
radiosensitivity Strahlungsempfindlichkeit *f*
radio [set] Radio *m*, Radioapparat *m*
radio station Funkanlage *f*
radiostrontium Strontium 90 *n*
radiotelegraphy drahtlose Telegraphie *f*
radiotelephone Funksprechgerät *n*
radiotellurium Radiotellur *n*
radiotherapist Radiologe *m*, Röntgenologe *m*
radiotherapy Radiumbehandlung *f* (Med), Röntgenbestrahlung *f*, Röntgentherapie *f*
radiothorium (Symb. RdTh) Radiothorium *n*
radiothorium content Radiothoriumgehalt *m*
radiotolerance Strahlungsfestigkeit *f*
radiotracer Radioindikator *m*
radium (Symb. Ra) Radium *n*, **containing** ~ radiumhaltig, **sample of** ~ Radiumprobe *f*, **[spontaneous] disintegration of** ~ Radiumzerfall *m*, **supply of** ~ Radiumquelle *f*
radium atom Radiumatom *n*
radium bromide Radiumbromid *n*
radium-contaminated radiumverseucht
radium emanation Radiumemanation *f*, Radon *n*
radium iodide Radiumjodid *n*
radium radiation Radiumstrahlung *f*
radium rays Radiumstrahlen *m pl*
radium refinery Radiumgewinnungsanlage *f*
radium-resistant radiumresistent
radium source Radiumquelle *f*
radium specimen Radiumpräparat *n*
radium therapy Radiumtherapie *f* (Med)
radius Radius *m*, Halbmesser *m*, **effective** ~ Wirkungshalbmesser *m*, **rolling** ~ wirksamer dynamischer Halbmesser, **~ of bend** Krümmungshalbmesser *m*, **~ of curvature** Rundungshalbmesser *m*, **~ of operation** Reichweite *f*
radius bar Leitstange *f*
radius gauge Halbmesserlehre *f*

radius rod Leitstange *f*
radius vector Radiusvektor *m*
radix (pl = radices or radixes) Wurzel *f* (Bot)
radon (Symb. Rn) Radon *n*, Radiumemanation *f*
radon seed Radonhohlnadel *f*
rafaelite Rafaelit *m*
raffia Raffiabast *m*
raffinate Raffinat *n* (Öl)
raffinose Raffinose *f*
rafter Sparren *m* (Techn)
rag Lumpen *m*
rag bleaching Lumpenbleiche *f*
rag boiler Hadernkocher *m* (Pap)
rag bolt Hackbolzen *m*, Steinschraube *f*
rag cutter Hadernschneider *m* (Pap)
rag cutting machine Hadernschneider *m* (Pap)
rag engine Holländer *m* (Pap)
rag paper Lumpenpapier *n*
rag pulling oil Reißöl *n*
rags calender Fetzenkalander *m* (Gummi)
rag sorter Hadernsortierer *m* (Pap)
ragweed oil Ambrosiaöl *n*
rail Schiene *f*, Führungsschiene *f*, Gleitbahn *f*
(für Werkzeugmaschinen und Förderanlagen),
heavy ~ Vollbahnschiene *f*, ~ **for trunk lines**
Vollbahnschiene *f*
railing Geländer *n*, Gatter *n*
rail installation Gleisanlage *f*
rail joint baseplates
Schienenstoßzwischenlagen *pl*
rail pass Schienenkaliber *n*
rail spike Schienennagel *m*
rail tank car Kesselwagen *m*
raimondite Raimondit *m* (Min)
rainbow agate Regenbogenachat *m* (Min)
rainbow quartz Regenbogenquarz *m* (Min)
rain drop Regentropfen *m*
rainfall Niederschlag *m* (Regen)
rain gauge Regenmesser *m*
rain water Regenwasser *n*
raise *v* [hoch]heben, anbauen (Pflanzen),
aufrichten, aufstellen, erhöhen (Preis),
hochwinden, züchten (Tiere)
raised erhöht
raisin Rosine *f*
raising of water Wasserförderung *f*
rake Rechen *m*, Feuerkrücke *f*, Rühreisen *n*,
Scharre *f*
rake angle Spanwinkel *m*
raker blade centrifuge Raumschaufelzentrifuge *f*
rake the fire *v* das Feuer schüren
raking arm Krählarm *m* (Eindicker)
ralstonite Ralstonit *m* (Min)
ram Kolben *m* (Techn), Rammblock *m*
ram *v* einen Druck ausüben;, rammen,
feststampfen, ~ **down** einstampfen
ramalic acid Ramalsäure *f*
ramalinolic acid Ramalinolsäure *f*
Raman effect Ramaneffekt *m* (Spektr)

Raman line Ramanlinie *f* (Spektr)
Raman shift Ramanverschiebung *f* (Spektr)
Raman spectrum Ramanspektrum *n* (Spektr)
ram extruder Kolbenstrangpresse *f*
ramie Ramie *f* (Bot), Chinagras *n* (Bot)
ramie [fiber] Ramiefaser *f*
ramie yarn Ramiegarn *n*
ramification Verzweigung *f*
ramified verzweigt
ramify *v* verzweigen
ramirite Ramirit *m*
rammed concrete gestampfter Beton *m*
rammelsbergite Rammelsbergit *m* (Min)
rammer Rammer *m*, Stampfer *m*, **flat** ~
Flachstampfer *m*
ramming volume Stampfvolumen *n*
ramp Rampe *f*, geneigte Fläche *f*
ram pressure Stempeldruck *m*
ramson Bärenlauch *m* (Bot)
rancid ranzig, **tendency to become** ~
Ranzigwerden *n*
rancidity Ranzigkeit *f*, Ranzigwerden *n*
rancidness Ranzigkeit *f*
randanite Randanit *m*
randite Randit *m*
random ungeordnet, willkürlich
random cause Zufall *m*
random noise Rauschen *n*
random orientation Zufallsorientierung *f*
random sample Zufallsstichprobe *f*
random series Zufallsfolge *f*
random test Stichprobe *f*
random variable Zufallsgröße *f*
range Bereich *m*, Kette *f*, Reichweite *f*,
Spielraum *m*, **critical** ~ kritischer Bereich,
maximum ~ maximale Reichweite (Atom)
range effect Reichweiteneffekt *m*
range energy Reichweitenenergie *f*
range finder Entfernungsmesser *m*; Telemeter *n*
range measurement Reichweitenmessung *f*
range of convergence (of power series)
Konvergenzintervall *m* (v. Potenzreihen)
range reduction Reichweitenverkürzung *f*
ranite Ranit *m*
rank Rang *m*
rank *a* üppig; ranzig
rank *v* einreihen
rankness Üppigkeit *f*, Ranzigkeit *f*
Ranque effect Ranque-Effekt *m*
ranunculin Ranunculin *n*
raolin Raolin *n*
rapanone Rapanon *n*
rape Raps *m* (Bot)
rape oil Rapsöl *n*
rapeseed cake Rübölkuchen *m*
rapeseed oil Rapsöl *n*
raphanin Raphanin *n*
raphilite Raphilit *m*
rapid rasch, schnell

rapid-acting press Schnellpresse *f*
rapid analysis Schnellanalyse *f*
rapid charge Schnellaufladung *f*
rapid circulation evaporator
Schnellumlaufverdampfer *m*
rapid cooling system Schnellkühlung *f*
rapid coupling Schnellschlußkupplung *f*
rapid dipping liquid Schnellbeize *f*
rapid discharge Schnellentladung *f*
rapid discharger Schnellentlader *m*
rapid fast dyestuff Rapidechtfarbe *f*
rapid freezing room Schnellgefrierraum *m*
rapid ice Rapid-Eis *n*
rapidity Eile *f*, Schnelligkeit *f*
rapid method Schnellverfahren *n*
rapid (or accelerated) method
Abkürzungsverfahren *n*
rapid shut-down Sicherheitsabschaltung *f*
rapid visco[si]meter Schnellviskosimeter *n*
rapinic acid Rapinsäure *f*
rare edel (Chem), selten
rare earth metal Seltenerdmetall *n*
rarefaction Verdünnung *f* (Gas), ~ **of air**
Luftverdünnung *f*
rarefiability Verdünnbarkeit *f* (Gas)
rarefiable verdünnbar (Gas)
rarefied air verdünnte Luft *f*
rarefy *v* verdünnen (Gas)
rare gas Edelgas *n*
rare gas rectifier Edelgasgleichrichter *m*
Raschig rings Raschigringe *pl*
rasp Grobfeile *f*, Raspel *f*, Rauhkratze *f*,
Reibeisen *n*
rasp *v* raspeln, schaben, ~ **off** abraspeln
raspberry essence Himbeeressenz *f*
raspberry syrup Himbeersirup *m*
raspings Raspelspäne *pl*
raspite Raspit *m* (Min)
raster [scan] microscope Rastermikroskop *n*
rastolyte Rastolyt *m*
ratchet Sperrhaken *m*, Sperrklinke *f*
ratchet drill Bohrknarre *f*
ratchet tooth Sperrzahn *m*
ratchet wheel Sperrad *n*
rate Geschwindigkeit *f*, Feinheit *f* (Gewinde),
Tarif *m*
rate *v* beurteilen, bewerten
rate action differenzierend wirkender Einfluß *m*
(Regeltechn) (Am. E.), Vorhalt *m*
(Regeltechn)
rate change Geschwindigkeitswechsel *m*
rate constant Geschwindigkeitskonstante *f*,
Reaktionsgeschwindigkeitskonstante *f*
rate-determining geschwindigkeitsbestimmend
rated frequency Nennfrequenz *f* (Elektr)
rated load Nennbelastung *f*
rated speed Nenndrehzahl *f*
rate law Zeitgesetz *n* (einer Reaktion)

rate of reaction Umsatzgeschwindigkeit *f*,
equivalent ~
Äquivalentreaktionsgeschwindigkeit *f*
rate time (Am. E.) Vorhaltzeit *f*
rathite Rathit *m* (Min)
ratify *v* bestätigen
rating Beurteilung *f*, Bewertung *f*, Leistung *f*
(Masch), Schätzung *f*
ratio Verhältnis *n*, **inverse** ~ umgekehrtes
Verhältnis, ~ **of tension to thrust**
Zugdruckverhältnis *n*
rational rational, rationell
rationality Rationalität *f*
rationality law (of crystal parameters)
Rationalitätsgesetz *n* (der Kristallparameter)
rationalization Rationalisierung *f*
rat poison Rattengift *n*
ratsbane Rattengift *n*
rattle *v* klappern; rattern
rattle snake Klapperschlange *f*
raubasine Raubasin *n*
raugustine Raugustin *n*
rauhimbine Rauhimbin *n*
raumitorine Raumitorin *n*
raunescic acid Raunescinsäure *f*
raunescine Raunescin *n*
raunormic acid Raunormsäure *f*
raunormine Raunormin *n*
raupine Raupin *n*
rauwolscane Rauwolscan *n*
rauwolscine Rauwolscin *n*
rauwolscone Rauwolscon *n*
rauwolsinic acid Rauwolsinsäure *f*
ravel *v* entwirren, auftrennen, ~ **out** ausfasern
ravenilin Ravenilin *n*
raw roh, unbearbeitet, ungefärbt (Tuch),
ungekocht
raw cotton Rohbaumwolle *f*
raw fabric Rohgewebe *n* (Text)
raw gas Rohgas *n*
raw glass Rohglas *n*
rawhide pinion Rohhautritzel *n*
raw material Rohmaterial *n*,
Ausgangswerkstoff *m*, Grundstoff *m*,
Rohstoff *m*
raw oil Rohöl *n*
raw ore Roherz *n*
raw potash Schweißasche *f*
raw product Roherzeugnis *n*
raw sheet iron Rohblech *n*
raw silk Ekrüseide *f*, Rohseide *f*
raw silk thread Rohseidenfaden *m* (Text)
raw slag Rohschlacke *f*
raw smelting Roharbeit *f* (Metall),
Rohschmelzen *n* (Metall)
raw sugar Rohzucker *m*, **second** ~
Rohzuckernachprodukt *n*
raw wool Rohwolle *f*

ray (see also rays) Strahl *m*, Lichtstrahl *m*,
  complex ~ zusammengesetzter Strahl (Phys),
  extraordinary ~ außerordentlicher Strahl,
  incident ~ einfallender Strahl, ordinary ~
  ordentlicher Strahl, reflected ~ reflektierter
  Strahl, ~ of light Lichtstrahl *m*
ray deflector Strahlablenker *m*
ray disintegration Strahlenzerfall *m*
ray filter Strahlenfilter *m*
ray hardness Strahlenhärte *f*
ray interferometer Strahleninterferometer *n*
Rayleigh scattering Rayleighsche
  Streustrahlung *f*
rayon Kunstseide *f* (Text), Reyon *m n* (Text)
rayon filament Kunstseidenfaden *m*
ray quantum Strahlenquant *n*
ray retardation Strahlenbremsung *f*
rays, beam of ~ Strahlenbündel *n*, bundle of ~
  Strahlenbündel *n*, cosmic ~ kosmische
  Strahlen *pl*, course of ~ Strahlengang *m*
  (Opt), infrared ~ infrarote Strahlen *pl*, path of
  ~ Strahlengang *m* (Opt), positive ~
  Anodenstrahlung *f*, ultraviolet ~ ultraviolette
  Strahlen *pl*
ray therapy Strahlenbehandlung *f* (Med),
  Strahlentherapie *f* (Med)
razor blade Rasierklinge *f*
reabsorb *v* resorbieren
reabsorbable resorbierbar
reabsorption Resorption *f*
reach Ausladung *f* (Kran), Reichweite *f*
reach truck Schiebemaststapler *m*
react, to allow to ~ einwirken lassen, to cause to
  ~ umsetzen (Chem)
react *v* einwirken (Chem), entgegenwirken,
  reagieren
reactance Blindwiderstand *m* (Elektr),
  Reaktanz *f* (Elektr), positive ~ induktive
  Reaktanz, ~ due to capacity
  Kapazitätsreaktanz *f*
reactance coefficient Reaktanzfaktor *m*
reactance current Blindstrom *m*
reactance matrix Blindwiderstandsmatrix *f*
reactant Reaktionsmittel *n*, Reaktionspartner *m*,
  Reaktionsteilnehmer *m*
reacting reagierend, capable of ~ reaktionsfähig
reacting capacity Reaktionsfähigkeit *f*
reaction Reaktion *f*, Einwirkung *f* (Chem),
  Gegenwirkung *f*, Rückwirkung *f*,
  Umsetzung *f* (Chem), Umwandlung *f*
  (Chem), acid ~ saure Reaktion, allergic ~
  allergische Reaktion (Med), anaplerotic ~
  anaplerotische Reaktion *f*, balanced ~
  unvollständige Reaktion, basic ~ basische
  Reaktion, bimolecular ~ bimolekulare
  Reaktion, coupled ~ gekoppelte Reaktion,
  zusammengesetzte Reaktion, course of ~
  Reaktionsablauf *m*, delayed ~ verzögerte
  Reaktion *f*, electrochemical ~

elektrochemische Reaktion, endergonic ~
endergonische Reaktion *f*, exergonic ~
exergonische Reaktion *f*, exothermic ~
exotherme Reaktion, first-order ~ Reaktion *f*
erster Ordnung, gaseous ~ Gasreaktion *f*,
heterogeneous ~ heterogene Reaktion,
homogeneous ~ homogene Reaktion,
incomplete ~ unvollständige Reaktion,
initiating ~ Startreaktion *f*, intermediate ~
Zwischenreaktion *f* (Chem), irreversible ~
irreversible Reaktion, monomolecular ~
monomolekulare Reaktion, neutral ~
neutrale Reaktion, order of ~
Reaktionsordnung *f*, overall ~
Gesamtreaktion *f*, partial ~ Teilreaktion *f*
(Chem), photochemical ~ photochemische
Reaktion, rapid ~ schnellverlaufende
Reaktion, rate-limiting ~
geschwindigkeitsbestimmende Reaktion *f*,
reversible ~ reversible Reaktion, umkehrbare
Reaktion, secondary ~ Sekundärreaktion *f*,
stereokinetically controlled ~ stereokinetisch
gesteuerte Reaktion, successive ~
Stufenreaktion *f*, total ~ Gesamtreaktion *f*,
trimolecular ~ trimolekulare Reaktion,
vigorous ~ stürmische Reaktion
reaction avalanche Reaktionslawine *f*
reaction chain Reaktionskette *f*
reaction chamber Reaktionskammer *f*
reaction cluster Reaktionsknäuel *n*
reaction control Reaktionslenkung *f*
reaction coupling Rückkopplung *f*
reaction effect Reaktionswirkung *f*
reaction energy Reaktionsenergie *f*
reaction entropy Reaktionsentropie *f*
reaction equation Reaktionsgleichung *f*
reaction flask Reaktionskolben m,
  Zersetzungskolben *m*
reaction gas Reaktionsgas *n*
reaction ground coat process
  Reaktionsgrundverfahren *n* (Polyester)
reaction hearth Arbeitsherd *m*
reaction inhibition Reaktionshemmung *f*
reaction isotherm Reaktionsisotherme *f*
reaction kinetics Reaktionskinetik *f*
reaction liquid Reaktionsflüssigkeit *f*
reaction mechanism Reaktionsmechanismus *m*
reaction mixture Reaktionsgemisch *n*
reaction motor Reaktionsmotor *m*
reaction pressure Reaktionsdruck *m*
reaction product Reaktionsprodukt *n*
reaction-propulsion jet Rückstoßdüse *f*
reaction rate Reaktionsgeschwindigkeit *f*
reaction region Reaktionsbereich *m*
reaction scheme Reaktionsschema *n*
reaction space Reaktionsraum *m*
reaction technology Reaktionstechnik *f*
reaction time Einwirkungszeit *f*,
  Reaktionsdauer *f*, Reaktionszeit *f*

reaction tower Reaktionsturm *m*
reaction turbine Überdruckturbine *f*
reaction type Reaktionstyp *m*
reaction value Anlaufwert *m* (Regeltechn)
reaction vessel Reaktionsgefäß *n*
reaction volume Reaktionsvolumen *n*
reaction zone Wirkungszone *f*
reactivate *v* reaktivieren
reactivation Reaktivierung *f*, Wiederbelebung *f*
reactive reaktionsfähig (Chem), reaktiv, gegenwirkend, rückwirkend
reactive current Blindstrom *m* (Elektr)
reactive ground coat technique Kontaktgrundverfahren *n*
reactive power Blindleistung *f* (Elektr)
reactive voltage Blindspannung *f*
reactivity Reaktivität *f*, Reaktionsfähigkeit *f*, Reaktionsvermögen *n*
reactor Reaktor *m* (Phys), **adiabatic** ~ adiabatischer Reaktor *m*, **beryllium-moderated** ~ berylliummoderierter Reaktor, **boiling water** ~ Siedewasserreaktor *m*, **carbon dioxide-cooled** ~ CO-gekühlter Reaktor, **cold** ~ kalter Reaktor, **enriched fuel** ~ angereicherter Reaktor, **epithermal** ~ epithermischer Reaktor, **fast-breeding** ~ Schnellbrutreaktor *m*, **fluidized fuel** ~ Flüssigspaltstoffreaktor *m*, **gas-cooled** ~ gasgekühlter Reaktor, **graphite-moderated** ~ graphitmoderierter Reaktor, **homogeneous** ~ homogener Reaktor, **ideal mixed** ~ ideal gemischter Reaktor *m*, **intermediate** ~ mittelschneller Reaktor, **liquid-fuel** ~ Reaktor *m* mit flüssigem Brennstoff, **liquid metal fuelled** ~ Flüssigmetallreaktor *m*, **mobile** ~ beweglicher Reaktor, **primary** ~ primärer Reaktor, **secondary** ~ sekundärer Reaktor, Sekundärreaktor *m*, **wetted wall column or thin-film** ~ Dünnschichtreaktor *m*, **recycle** ~ **or loop reactor** Schlaufenreaktor *m*, **tubular** ~ **or tube reactor** Rohrreaktor *m*
reactor blanket Reaktorbrutmantel *m*
reactor-clarifier Flockungsklärbecken *n*
reactor-clarifier for water treatment Schlammkontaktanlage *f*
reactor control Reaktorsteuerung *f*
reactor control room Reaktorschaltwarte *f*
reactor core Reaktorkern *m*
reactor engineering Reaktortechnologie *f*
reactor feed Spaltstoff *m* (Atom)
reactor fuel Kernbrennstoff *m* (Atom), Spaltstoff *m* (Atom)
reactor loading Reaktorbeschickung *f*
reactor physics Reaktorphysik *f*
reactor poison Reaktorgift *n*
reactor power Reaktorleistung *f*
reactor shell Reaktormantel *m*
reactor shielding Reaktorabschirmung *f*

reactor technology Reaktortechnik *f*, Reaktortechnologie *f*
reactor waste Reaktorabfälle *pl*, Spaltstoffabfälle *m pl*
read *v* lesen, ~ **off** ablesen, ~ **out** ablesen
readability Ablesbarkeit *f*
readable ablesbar, lesbar
reader Datensichtgerät *n*
readiness Bereitschaft *f*, Bereitwilligkeit *f*
reading Ablesung *f*, Anzeige *f* (Gerät), Meßwert *m*, ~ **by mirror** Spiegelablesung *f*
reading device Ablesevorrichtung *f*
reading error Ablesefehler *m*
reading glass Lupe *f*
reading lens Lupe *f*
reading microscope Meßmikroskop *n*
readjust *v* nachregeln
readjusting device Nachstellvorrichtung *f*
readjustment Nachstellung *f*, Neueinstellung *f*
readout timer Schaltuhr *f* mit Zeitdrucker
ready bereit[willig], ~ **for use** gebrauchsfertig
ready for action einsatzbereit
ready-made gebrauchsfertig
ready-mixed mortar Fertigmörtel *m*
ready-to-eat dish Fertiggericht *n*
reagent Reagens *n*, Reagenz *n*, Reaktionsmittel *n*
reagent bottle Reagentienflasche *f*
reagent feeding Reagenzdosierung *f*
reagent paper Reagenspapier *n*
reagent room Reagentienraum *m*
reagent solution Reagenslösung *f*
real faktisch, real, tatsächlich
realgar Realgar *m* (Min), Arsendisulfid *n* (Min), Arsenrubin *m* (Min), Rauschrot *n* (Min), rote Arsenblende *f* (Min)
realign *v* wieder ausrichten, wiederausrichten
realignment Wiederausrichtung *f*
real image reelles Bild *n* (Opt)
reality Realität *f*, Wirklichkeit *f*
realizable realisierbar
realization Verwirklichung *f*
realize *v* verwirklichen
ream Ries *n* (Papiermaß)
ream *v* ausbohren, ausfräsen, erweitern (Bohrloch)
reamer Reibahle *f*, Aufdornwerkzeug *n*
reaming fixture Reibvorrichtung *f*
reanneal *v* nachglühen (Met)
reannealing Nachglühen *n* (Met)
rear-mounted ripper Heckausreißer *m*
rearrange *v* umgruppieren, umlagern, umorganisieren
rearrangement Neuordnung *f*, Umgruppierung *f*, Umlagerung *f* (Chem), **atomic** ~ atomare Umlagerung, **intramolecular** ~ intramolekulare Umlagerung, **nuclear** ~ Umgruppierung *f* im Kern, ~ **via a**

carbonium ion Umlagerung *f* über ein Carbonium-Ion, ~ **with participation of neighboring groups** Umlagerung *f* unter Nachbargruppenbeteiligung
rear wall Rückwand *f*
reason Grund *m*, Anlaß *m*, Ursache *f*
reasonable vernünftig, angemessen
reassembly Wiederzusammenfügen *n*
reassure *v* beruhigen
Réaumur scale Réaumurskala *f*
Réaumur thermometer Réaumurthermometer *n*
rebate Rabatt *m*
reboiler Aufwärmer *m* (Atom)
rebound Zurückprallen *n*, Abprall *m*, Rückprall *m*, Rückstoß *m*
rebound *v* zurückprallen, abprallen, abspringen
rebound elasticity Rückprallelastizität *f*
rebound electrons Prallelektronen *pl*
rebound hardness Skleroskophärte *f*, Sprunghärte *f*
rebuild *v* umbauen, umkonstruieren
rebuilding Umbau *m*, Wiederaufbau *m*
recalculation Umrechnung *f*
recalibrate *v* nacheichen
recalibration Nacheichung *f*, Neueichung *f*
recanescic acid Recanescinsäure *f*
recanescic alcohol Recanescinalkohol *m*
recanescine Recanescin *n*
recap *v* runderneuern (Reifen)
recapping Besohlen *n* (Reifen), Runderneuerung *f* (Reifen)
recarburize *v* aufkohlen (Metall)
recarburizer kohlender Zusatz *m*
recast *v* umgießen, umschmelzen
recasting Wiedereinschmelzen *n*
receipt Empfangsbestätigung *f*, Quittung *f*
receive *v* bekommen, erhalten
receiver Empfänger *m*, Auffangbehälter *m* (Techn), Behälter *m* (Techn), Empfangsgerät *n* (Radio), Rezipient *m* (Chem), Sammelgefäß *n* (Techn), Vorlage *f* (Chem)
receiver adapter Destilliervorstoß *m*, ~ **with vacuum connection** Vakuumvorstoß *m*
receiver bottle Auffangflasche *f*
receiving aerial Empfangsantenne *f*
receiving circuit Empfängerstromkreis *m*
receiving flask Auffangkolben *m*
receiving set Empfänger *m*
receiving tube Auffangröhre *f*
receptacle Behälter *m*, Gefäß *n*, Sammelgefäß *n*
reception Aufnahme *f*, Empfang *m*
receptivity Aufnahmefähigkeit *f*, Aufnahmevermögen *n*, Rezeptivität *f*, ~ **for dyes** Färbefähigkeit *f*
receptor Rezeptor *m* (Biochem)
recess Vertiefung *f*, Ausbuchtung *f*, Auskehlung *f*, Aussparung *f*, Kerbe *f*, **depth of** ~ Gesenktiefe *f*

recess *v* ausbuchten, aussparen
recessed plate press Kammerfilterpresse *f*
recessed square Innenvierkant *n*
recess for radiators Heiznische *f*
recessive rezessiv (Biol)
recharge *v* wieder aufladen (Elektr), wieder beschicken
recipe Rezept *n*
recipient Rezipient *m* (Chem)
recipient vessel Vorlage *f*
reciprocal reziproker Wert *m* (Math)
reciprocal *a* gegenseitig, reziprok, umgekehrt, wechselseitig
reciprocal action Wechselwirkung *f*
reciprocal conversion Wechselumsetzung *f*
reciprocal integration Kehrwertintegration *f* (Math)
reciprocal pusher centrifuge Schubzentrifuge *f*
reciprocal [value] Kehrwert *m*
reciprocate *v* hin- und herbewegen (Kolben)
reciprocating compressor Hubkolbenverdichter *m*, Kolbenkompressor *m*, Kolbenverdichter *m*
reciprocating drier Schubwendetrockner *m*
reciprocating engine Kolbendampfmaschine *f*
reciprocating motion Hin- und Herbewegung *f*, Hubbewegung *f*, Wechselbewegung *f*
reciprocating plate extraction column Siebbodenkolonne *f* für flüssig-flüssig Extraktion
reciprocating proportioning pump Dosierkolbenpumpe *f*
reciprocating pump Kolbenpumpe *f*, Kurbelpumpe *f*
reciprocating-screw injection molding machine Schubschneckenspritzgußmaschine *f*
reciprocity Reziprozität *f*, Wechselseitigkeit *f*
reciprocity theorem Reziprozitätstheorem *n*
recirculation Rückleitung *f*
recirculation pump Umwälzpumpe *f*
reclaim Regenerat *n* (Gummi)
reclaim *v* regenerieren (Gummi), rückgewinnen, zurückerhalten, zurückgewinnen
reclaimed ground neugewonnenes Land *n*, Schwemmland *n*
reclaimed rubber Regeneratgummi *m*
reclaimed wool Reißwolle *f*
reclaiming Regeneration *f* (Gummi), Rückgewinnung *f*, Wiedergewinnung *f*
reclaiming agent Regeneriermittel *n* (Gummi)
reclamation Landgewinnung *f*, Rückgewinnung *f*
recoating Neubelegen *n* (Walze)
recognition Erkennung *f*
recognize *v* anerkennen, identifizieren
recoil Rückprall *m*, Prellklotz *m* (Techn), Rückstoß *m* (Mot), **power of** ~ Rückstoßkraft *f*
recoil *v* abprallen
recoil atom Rückstoßatom *n*

recoil buffer (wedge) Hemmkeil *m*
recoil counter Rückstoßzähler *m*
recoil detection Rückstoßnachweis *m*
recoil electron Rückstoßelektron *n*
recoil energy Rückstoßenergie *f*
recoil fragment Rückstoßbruchstück *n*
recoil ionization Rückstoßionisierung *f*
recoil liquid Bremsflüssigkeit *f*
recoil method Rückstoßmethode *f*
recoil nucleus Rückstoßkern *m* (Atom)
recoil proton Rückstoßproton *n*
recoil spectrum Rückstoßspektrum *n* (Atom)
recoil spindle Prellbolzen *m*
recoil spring Rückstoßfeder *f*, Vorholfeder *f*
recombination Rekombination *f*,
   Wiedervereinigung *f*, **preferential** ~
   bevorzugte Rekombination, **radiative** ~
   Strahlungsrekombination *f*
recombination center Rekombinationszentrum *n*
recombination coefficient
   Rekombinationskoeffizient *m*
recombination continuum
   Rekombinationskontinuum *n*
recombination law Wiedervereinigungsgesetz *n*
recombination velocity
   Rekombinationsgeschwindigkeit *f* (Elektr)
recombine *v* rekombinieren, wieder vereinigen
recommend *v* befürworten, empfehlen
recommendation Referenz *f*
recomputation Umrechnung *f*
reconcentration Wiederanreicherung *f*
recondition *v* wieder instandsetzen
reconditioning Wiederinstandsetzung *f*
reconstruct *v* umbauen; rekonstruieren
reconstruction Rekonstruktion *f*, Umbau *m*,
   Wiederaufbau *m*
record Aufzeichnung *f*, Bericht *m*, Bestleistung *f*,
   Dokument *n*, Register *n*, Schallplatte *f*
record *v* registrieren, aufschreiben, aufzeichnen
recorder Registrierapparat *m*; registrierendes
   Instrument *n*; Schreiber *m*; Schreibgerät *n*,
   integrating ~ Schreiber *m* mit Integrator
recorder controller Schreiber *m* mit Regelsatz
recorder panel Registriertafel *f*
recording Aufzeichnung *f*, Registrierung *f*
recording apparatus Schreiber *m*,
   Registriervorrichtung *f*
recording barometer Barograph *m*
recording field Aufsprechfeld *n*
recording head Aufsprechkopf *m*
recording head gap Aufsprechkopfspalt *m*
recording hygrometer Hygrograph *m*
recording instrument registrierendes
   Instrument *n*, Registriergerät *n*
recording manometer Registriermanometer *n*
recording mechanism Registriervorrichtung *f*,
   Zählwerk *n*
recording paper Registrierpapier *n*
recording process Anzeigeverfahren *n*

recording scales Registrierwaage *f*
recording tachometer Drehzahlenschreiber *m*
recording tape Tonband *n*
recording thermometer Registrierthermometer *n*
records Archiv *n*, Protokoll *n*
recover *v* wiedergewinnen, zurückerhalten,
   zurückgewinnen
recoverable zurückgewinnbar
recovered acid Abfallsäure *f*
recovery Erholung *f*, Rückgewinnung *f*,
   Wiedergewinnung *f*, Zurückgewinnung *f*, ~
   of solvent[s] Lösungsmittelrückgewinnung *f*
recovery effect Erholungseffekt *m*
recovery plant Rückgewinnungsanlage *f*
recovery power Regenerationsvermögen *n*
recovery process Erholungserscheinung *f*
recovery rate Erholungsgeschwindigkeit *f*
recovery time Regelzeit *f* (Regeltechn) (of a
   Geiger counter), Erholungszeit *f* (eines
   Geiger-Zählrohrs)
recrystallization Rekristallisation *f*,
   Umkristallisierung *f*
recrystallization twins
   Rekristallisationszwillinge *pl* (Krist)
recrystallize *v* rekristallisieren, umkristallisieren
recrystallizing Umkristallisieren *n*
rectangle Rechteck *n*
rectangular rechteckig, orthogonal
rectangular can Rechteckdose *f*, Vierkantdose *f*
rectifiable rektifizierbar (Techn)
rectification Berichtigung *f*, Gleichrichtung *f*
   (Elektr), Läuterung *f* (Chem), Rektifikation *f*
   (Chem), Richtigstellung *f*, **azeotropic** ~
   azeotrope Rektifikation, **vacuum** ~
   Vakuumrektifikation *f*, ~ **at normal pressures**
   Normaldruckrektifikation *f*
rectification plant Rektifizieranlage *f*
rectification section Rektifikationszone *f*
rectified gleichgerichtet (Elektr), rektifiziert
   (Chem)
rectifier Destillationsapparat *m* (Chem),
   Gleichrichter *m* (Elektr),
   Rektifizierapparat *m* (Chem), **full-wave** ~
   Vollweggleichrichter *m* (Elektr), **gas-filled** ~
   gasgefüllter Gleichrichter, **half-wave** ~
   Halbwellengleichrichter *m* (Elektr),
   semi-conductor ~ Halbleitergleichrichter *m*
rectifier cell Sperrschichtzelle *f*
rectifier characteristic Gleichrichterkennlinie *f*
rectifier photocell Halbleiterphotozelle *f*
rectifier plant Gleichrichteranlage *f*
rectifier valve Gleichrichterröhre *f*
rectify *v* korrigieren, berichtigen, destillieren
   (Chem), entfuseln (Chem), gleichrichten
   (Elektr), rektifizieren (Chem)
rectifying Rektifizieren *n*
rectifying apparatus Rektifizierapparat *m*
rectifying coil Ausgleichsspule *f*

rectifying column Fraktionierturm *m,*
   Rektifikationskolonne *f,* Rektifiziersäule *f*
rectifying crystal Gleichrichterkristall *m*
rectifying section Verstärkerteil *m*
   (Rektifikationskolonne)
rectilinear geradlinig
rectilinear manipulator
   Koordinatenmanipulator *m*
recuperate *v* wiederherstellen
recuperation Rückgewinnung *f* (Elektr,
   Energie), Wiederherstellung *f* (from
   mechanical damage), Ausheilen *n*
recuperator Rekuperator *m*
recur *v* wiederkehren
recurrence Wiederholung *f*
recurrent periodisch auftretend
recursion Rekursion *f*
recursion formula Rekursionsformel *f*
recycle *v* rückführen
recycle gas Umwälzgas *n*
recycle reactor Kreislaufreaktor *m,* Schlaufen-
   Reaktor *m*
recycling Abfälleverwertung *f,*
   Abfallverwertung *f,* Altmaterialverwertung *f*
red rot, **bright** ~ hellrot, **deep** ~ hochrot, **fiery**
   ~ feuerrot, **glowing** ~ glutrot, **pale** ~ fahlrot
red alga Rotalge *f* (Bot)
red antimony Antimonblende *f* (Min)
red arsenic Realgar *m* (Min), rote Arsenblende *f*
   (Min)
red bark rote Chinarinde *f*
red blood cell Erythrozyt *m*
red blood corpuscle rotes Blutkörperchen *n*
red brass Rotguß *m*
red bronze Rotguß *m*
red cell count Erythrozytenzählung *f*
red chalk Rötel *m* (Min), **color made of** ~
   Rötelfarbe *f*
red charcoal Rotkohle *f*
red [color] Rot *n*
red coloration Rotfärbung *f*
red-colored rotfarbig
red copper Rotkupfer *n* (Min)
redden *v* röten
reddingite Reddingit *m* (Min)
reddish rötlich
reddish brown rötlich braun, rotbraun
red earth Roterde *f*
redecorate *v* neu streichen, renovieren
redefinition Neudefinition *f*
red enamel Rotlack *m*
redesign *v* umkonstruieren
red filter Rotfilter *n*
red-free light Rotfreilicht *n*
red gilding Rotvergoldung *f*
red gold Rotgold *n*
red hardness Rotglühhärte *f,* Rotgluthärte *f*
red heat Rotglut *f,* **loss at** ~ Glühverlust *m,*
   stable at ~ glühbeständig, **to heat to** ~

rotglühen, **auf Rotglut erhitzen, to work at** ~
   warm bearbeiten
red heat test Rotbruchprobe *f*
red-hot rotglühend
redhot iron Glüheisen *n*
red iron ore Hämatit *m* (Min)
redissolve *v* wieder auflösen
redistil[l] *v* umdestillieren
redistillation Redestillation *f*
redistribution experiments
   Kommutierungsversuche *pl*
redistributor Verteiler *m,* Wiederverteiler *m*
red lead Mennige *f,* Bleioxyduloxid *n,* **coat of** ~
   Mennigeanstrich *m*
red lead cement Mennigekitt *m*
red lead oxide Bleioxyduloxid *n*
red lead paint Mennigefarbe *f*
red lead putty Mennigekitt *m*
red light Rotlicht *n*
red liquor Rotbeize *f* (Färb), Tonbeize *f*
red litharge Goldglätte *f*
red lye Rotlauge *f*
red mine stone Hämatit *m* (Min)
red mordant Rotbeize *f* (Färb)
redness Röte *f*
red ocher Rötelerde *f* (Min)
red orpiment Realgar *m* (Min)
redox electrode Redoxelektrode *f*
redox equilibrium Redoxgleichgewicht *n*
redox indicator Redoxindikator *m*
redox measurement Redoxmessung *f*
redox polymerization Redoxpolymerisation *f*
redox potential
   Oxidations-Reduktions-Potential *n,*
   Redoxpotential *n*
redox reaction
   Oxidations-Reduktions-Reaktion *f,*
   Redoxreaktion *f*
redox system Redox-System *n*
red pepper Schotenpfeffer *m*
red precipitate Quecksilber(II)-oxid *n*
red print Rotpause *f*
red purple Rotviolett *n*
redress *v* gleichrichten (Elektr)
red rot Rotfäule *f*
redrying machine Fermentationsmaschine *f*
   (Tabak)
red shift Rotverschiebung *f* (Spektr)
red-short rotbrüchig, warmbrüchig
red-shortness Rotbruch *m,* Rotbrüchigkeit *f*
red silver ore Rotsilbererz *n* (Min)
red sludge Rotschlamm *m*
red-spotted rotfleckig
red-stained rotfleckig
red toner Rottöner *m*
reduce *v* abnehmen, abschwächen (Phot),
   anfrischen (Met), ermäßigen, erniedrigen,
   frischen (Blei), herabsetzen, reduzieren

(Chem), verkleinern, verringern, ~ **in value**
entwerten, ~ **to small pieces** zerkleinern
reduced reduziert (Chem), vermindert
reduced pigment Verschnittfarbe *f*
reduced pressure verminderter Druck *m*
reducer Abschwächer *m* (Phot),
  Reduktionsmittel *n* (Chem), Reduzierstück *n*
  (Techn), Verjüngungsrohrstutzen *m* (Techn)
reducibility Reduzierbarkeit *f*
reducible reduzierbar, **easily** ~ leicht
  reduzierbar
reducing reduzierend
reducing agent Reduktionsmittel *n*,
  reduzierendes Agens *n*
reducing carbon Reduktionskohle *f*
reducing coupling Reduzierkupplung *f,*
  Reduktionsverbindungsstück *n*
reducing fitting Reduzierstück *n*
reducing flange Reduzierflansch *m,*
  Stauflansch *m*
reducing medium Abschwächer *m* (Phot)
reducing piece Reduzierstück *n*
reducing pipe Reduzierstück *n*, Übergangsrohr *n*
reducing power Reduktionsvermögen *n*,
  Reduzierfähigkeit *f*
reducing pressure valve Druckreduzierventil *n*
reducing process Frischarbeit *f* (Blei),
  Reduktionsvorgang *m*
reducing slag Reduktionsschlacke *f*
reducing tee[-piece] Reduzier-T-Stück *n*
reducing valve Reduzierventil *n*
reductant reduzierendes Agens *n*,
  Reduktionsmittel *n*
reductase Reduktase *f* (Biochem)
reductic acid Reduktinsäure *f*
reduction Reduktion *f* (Chem), Abschwächung *f*
  (Phot), Anfrischung *f* (Met), Aufhellung *f*
  (Lack), Erniedrigung *f*, Herabsetzung *f*,
  Reduzierung *f*, Verkleinerung *f*,
  Verringerung *f*, **electrolytic** ~ elektrolytische
  Reduktion, **heat of** ~ Reduktionswärme *f*,
  velocity of ~ Reduktionsgeschwindigkeit *f*
reduction division Meiose *f* (Biol),
  Reduktionsteilung *f* (Biol)
reduction equivalent Reduktionsäquivalent *n*
reduction flame Reduktionsflamme *f*
reduction furnace Reduktionsofen *m* (Metall),
  Reduzierofen *m*
reduction gas Reduktionsgas *n*
reduction gear Reduktionsgetriebe *n*,
  Untersetzungsgetriebe *n*
reduction graduation Reduktionsteilung *f* (Phys)
reduction-oxidation cell
  Reduktions-Oxidations-Kette *f*
reduction potential Reduktionspotential *n*
reduction process Niederschlagsarbeit *f*,
  Reduktionsarbeit *f*
reduction ratio Abbaugrad *m* (b. Zerkleinerung)
reduction scale Verkleinerungsmaßstab *m*

reduction scheme Reduktionsschema *n*
reduction slag Reduktionsschlacke *f*
reduction smelting Reduktionsschmelzen *n*
reduction stage Reduktionsstufe *f*
reduction temperature Reduktionstemperatur *f*
reduction valve Reduzierventil *n*
reduction zone Reduktionszone *f*
reductive reduzierend (Chem)
reductive agent Reduktionsmittel *n*
reductodehydrocholic acid
  Reduktodehydrocholsäure *f*
reductone Redukton *n*
red violet Kardinalrot *n*
redwood Rotholz *n*
redye *v* auffärben, nachbeizen, nachfärben,
  umfärben
reed Schilfrohr *n* (Bot)
reel Haspel *f,* Rolle *f,* Spule *f,* Winde *f*
reel *v* aufspulen, aufrollen, aufspulen, haspeln,
  ~ **off** abhaspeln, abspulen, abwickeln, ~ **up**
  aufwickeln
re-emission Sekundäremission *f*
re-establishment Wiederherstellung *f*
re-examination Nachprüfung *f*
re-examine *v* nachprüfen
refer *v* hinweisen, verweisen
reference Bezug *m*, Bezugnahme *f*, Hinweis *m*,
  Literaturangabe *f*, Quellenangabe *f*,
  Referenz *f*, Verweis *m*, **standard of** ~
  Vergleichsmaßstab *m*
reference body Vergleichskörper *m*
reference circle Bezugskreis *m*
reference electrode Bezugselektrode *f,*
  Vergleichs-Elektrode *f*, Referenzelektrode *f*
reference element Bezugselement *n*
reference ellipsoid Referenzellipsoid *n*
reference fluid Vergleichsflüssigkeit *f*
reference frequency Bezugsfrequenz *f* (Elektr)
reference fuel Bezugskraftstoff *m*
reference gas Vergleichsgas *n*
reference gauge Vergleichslehre *f* (Techn)
reference input Sollwert *m*
reference level Vergleichspegel *m*
reference library Handbücherei *f*
reference line Bezugslinie *f*, Referenzlinie *f*,
  Vergleichslinie *f*
reference magnitude Bezugsgröße *f*
reference point Bezugspunkt *m*
reference sample Vergleichsprobe *f*
reference solution Bezugslösung *f*
reference standard Bezugsnormal *n*
reference stars Bezugssterne *m pl*
reference stimulus Eichreiz *m*
reference system Bezugssystem *n*
reference table Nachschlagetabelle *f*
reference temperature Bezugstemperatur *f*
reference type strength Richttyptiefe *f*
reference value Bezugswert *m*

reference work Nachschlagewerk *n*
refill Nachfüllpackung *f*, Nachfüllung *f*
refill *v* wieder [auf]füllen
refillable valve Nachfüllventil *n*
refilling Nachfüllung *f*
refine *v* verfeinern, aufbereiten (Met), frischen (Eisen), läutern (Techn), raffinieren (Techn), reinigen (Met), scheiden (Met), seigern (Met), veredeln, ~ **again** nachseigern (Met), ~ **thoroughly** garfrischen (Metall)
refined raffiniert, gefrischt (Eisen), geläutert
refined copper Feinkupfer *n*, Garkupfer *n*
refined gold Brandgold *n*
refined iron Qualitätseisen *n*, Feineisen *n*, Holzkohleneisen *n*
refined lead Armblei *n*
refined or toughpitch copper hammergares Kupfer *n*
refined platinum Feinplatin *n*
refined product Raffinat *n*
refined salt Edelsalz *n*
refined silver Brandsilber *n*
refined steel Edelstahl *m*, Herdfrischstahl *m*
refined sugar Feinzucker *m*, Zuckerraffinade *f*
refined zinc Feinzink *n*
refinement Feinheit *f*, Läuterung *f*, Raffination *f*, Raffinationsbehandlung *f*, Raffinierung *f*, Vered[e]lung *f*, Verfeinerung *f*
refiner Feinbrenner *m*; Frischer *m* (Eisen), Raffineur *m*, Stoffmühle *f* (Pap)
refinery Raffinerie *f*, Frischerei *f* (Metall), Raffinationsanlage *f*, Scheideanstalt *f* (Metall)
refinery pig iron Herdfrischroheisen *n*
refinery process Herdfrischarbeit *f*, Herdfrischprozeß *m*
refinery slag Feinschlacke *f*, Gargekrätz *n*, Garkrätze *f*
refinery waste Raffinationsabfall *m*
refining Raffinierung *f*, Feinen *n* (Met), Frischen *n* (Eisen), Läuterung *f* (Chem) (Glas), Raffinieren *n* (Zucker), Reinigen *n* (Met), Seigern *n* (Met), Veredeln *n*, **electrolytical** ~ elektrolytische Reinigung *f*, **first** ~ Rohfrischen *n* (Metall), ~ **in the open hearth** Herdfrischen *n*, ~ **[of steel] by top-blowing** Oberwindfrischen *n*, ~ **of waste metal** Krätzfrischen *n*, ~ **with iron ore** Erzglühfrischen *n*
refining assay Garprobe *f*
refining bath Fällbad *n*
refining boiler Läuterungsgefäß *n*, Läuterungskessel *m*
refining cinders Eisenfeinschlacke *f*
refining cupel Abtreibkapelle *f* (Metall)
refining fire Läuterfeuer *n*, Reinigungsfeuer *n*
refining foam Garschaum *m*
refining forge Feinofen *m*
refining forge slag Frischfeuerschlacke *f*

refining furnace Anfrischofen *m*, Feinofen *m*, Frischofen *m*, Garofen *m*, Raffinierofen *m*, Treibofen *m*
refining hearth Feinherd *m*, Garherd *m*, Kapelle *f*
refining kiln Läuterofen *m*
refining melt Abtreibschmelze *f* (Metall)
refining method Läutermethode *f*
refining plant Fällanlage *f*, Raffinationsanlage *f*, Raffinieranlage *f*
refining process Raffinationsverfahren *n*, Feinprozeß *m* (Met), Läuterungsprozeß *m*, Scheideverfahren *n* (Met), Veredelungsverfahren *n*, Vered[e]lungsvorgang *m*
refining puddling Feinpuddeln *n* (Metall)
refining slag Garschlacke *f*
refining smelting Raffinationsschmelzen *n*
refining test Garprobe *f*
refining treatment Raffinationsbehandlung *f*
refining trough Anfrischtrog *m*
refining vessel Anfrischgefäß *n*
refining works Frischhütte *f*
refit *v* wiederherstellen
reflect *v* reflektieren, widerspiegeln, zurückstrahlen
reflectance Reflexionsstärke *f*, Weißgehalt *m* (Email)
reflected reflektiert, zurückgestrahlt
reflecting field (vacuum tube) Bremsfeld *n*
reflecting film Reflexfolie *f*
reflecting galvanometer Spiegelgalvanometer *n* (Elektr)
reflecting goniometer Reflexionsgoniometer *n*
reflecting luster Spiegelglanz *m*
reflecting microscope Reflexionsmikroskop *n*
reflecting power Reflexionsvermögen *n*
reflecting prism Reflexionsprisma *n*
reflecting projector Episkop *n*
reflecting surface Reflexionsebene *f*
reflection Reflexion *f*, Abstrahlung *f*, Reflex *m* (Physiol), Spiegelbild *n*, Spiegelung *f*, Widerschein *m*, Zurückstrahlung *f*, **angle of** ~ Reflexionswinkel *m*, ~ **of light** Reflexion *f* des Lichtes
reflection abundance Reflexreichtum *m*
reflection coefficient Reflexionskoeffizient *m*
reflection densitometer Aufsichtschwärzungsmesser *m*
reflection factor Reflexionsvermögen *n*
reflection grating Reflexionsgitter *n* (Opt)
reflection invariance Spiegelungsinvarianz *f*
reflection plane Spiegelebene *f*
reflection polarizer Reflexionspolarisator *m*
reflectivity Reflexionsdämpfung *f*, Reflexionsvermögen *n*
reflectometer Reflexionsmesser *m*
reflector Reflektor *m*, Rückstrahler *m*, Spiegel *m*
reflector aperture Spiegelöffnung *f*

reflex Reflex *m*
reflex galvanometer Reflexionsgalvanometer *n*,
  Spiegelgalvanometer *n*
reflexion Reflex *m*, Reflexion *f*, Rückstrahlung *f*
reflex klystron Reflexklystron *n* (Elektr)
reflex [motion] Reflexbewegung *f*
reflux Rückfluß m, Rücklauf *m* (Rektifikation),
  Zurückfließen *n*
reflux condenser Rückflußkühler *m;*
  Rücklaufkühler *m*
reflux ratio Rücklaufverhältnis *n*
  (Rektifikation)
reformation Rückbildung *f,* Umgestaltung *f*
reformed gasoline Reformierbenzin *n*
refract *v* brechen (Opt)
refracted ray Brechungsstrahl *m*
refracting [licht]brechend
refracting angle Brechungswinkel *m*
refracting layer Brechungsschicht *f*
refraction Brechung *f* (Opt),
  Brechungsvermögen *n* (Opt), Refraktion *f*
  (Opt), **angle of** ~ Brechungswinkel *m*,
  Refraktionswinkel *m*, **axis of** ~
  Brechungsachse *f,* **coefficient of** ~
  Refraktionskoeffizient *m*, **diffuse** ~ diffuse
  Brechung (Opt), **dispersed** ~ Streubrechung *f,*
  equivalent of ~ Refraktionsäquivalent *n*,
  index of ~ Brechungsindex *m*,
  Refraktionsindex *m*, **law of** ~
  Brechungsgesetz *n,* **modified index of** ~
  modifizierter Brechungsindex (Opt), **molar** ~
  molare Brechung, **ordinary** ~ ordentliche
  Brechung, **plane of** ~ Brechungsebene *f*
  (Opt), **power of** ~ Brechungsvermögen *n*
  (Opt), **specific** ~ spezifische Refraktion, ~ **of**
  light Lichtbrechung *f*
refraction index Brechungsexponent *m*,
  Brechungsindex *m*
refraction power, molar ~ molares
  Brechungsvermögen *n*
refractive [licht]brechend
refractive aberration Brechungsabweichung *f*
refractive index Brechungsexponent *m*,
  Brechungsindex *m*, Refraktionsindex *m*, **leap**
  of ~ Brechungsindexsprung *m*, **principal** ~
  Hauptbrechungszahl *f* (Opt), **real part of the**
  ~ Brechungsindexrealteil *m*
refractive medium brechendes Medium *n*
refractive power Brechungsvermögen *n*
refractivity Brechungsvermögen *n*, **molar** ~
  molares Brechungsvermögen
refractometer Brechungsmesser *m* (Opt),
  Refraktometer *n* (Opt), **dipping or immersion**
  ~ Eintauchrefraktometer *n*
refractometric refraktometrisch
refractometry Brechungsindexbestimmung *f,*
  Refraktometrie *f*
refractor Refraktor *m* (Opt)

refractoriness Feuerbeständigkeit *f,*
  Feuerfestigkeit *f*
refractory feuerbeständig, feuerfest,
  hitzebeständig, schwer schmelzbar, **highly** ~
  hochfeuerfest
refractory brick Schamottestein *m*
refractory material feuerfester Stoff *m*
refractory materials Feuerfestmaterialien *n pl*
refractory porcelain Feuerporzellan *n*
refractory sand Glühsand *m*
refrain [from] *v* Abstand nehmen [von]
refrangibility Brechbarkeit *f*
refresh *v* auffrischen, erfrischen
refreshing and cleaning tissues
  Frischhaltetücher *pl*
refrigerant Kühlmittel *n*, Abkühlungsmittel *n*,
  Kältemittel *n*
refrigerant liquid controls Drosselvorrichtung *f*
  (f. Kältetechn)
refrigerate *v* kühlen, tiefkühlen
refrigerating Kühlen *n*
refrigerating agent Kältemittel *n*
refrigerating aggregate Kühlaggregat *n*
refrigerating capacity Kälteleistung *f*
refrigerating machine Kältemaschine *f*
refrigerating pipe Kühlrohr *n*
refrigerating salt Kühlsalz *n*
refrigeration Kühlung *f*, Kälteerzeugung *f*
refrigeration distribution network
  Kältefernversorgung *f*
refrigeration equipment Kältemaschine *f*
refrigeration plant Kühlanlage *f*
refrigeration surface Abkühlungsfläche *f*
refrigerative kälteerzeugend
refrigerator Kühlschrank *m*, Abkühlkessel *m*
  (Techn), Eisschrank *m*, Kältemaschine *f*
refrigerator truck Kühlwagen *m*
refuel *v* [auf]tanken
refueling Brennstofferneuerung *f,* Tanken *n*
refusal Ablehnung *f,* Abschlag *m*,
  Verweigerung *f*
refuse Abfall *m*, Abraum *m*, Gekrätz *n* (Met),
  Müll *m*
refuse *v* ablehnen, abweisen, ausschlagen,
  verweigern
refuse collecting plant Abfallsammelanlage *f*
refuse destruction plant
  Abfallvernichtungsanlage *f*
refuse destructor Abfallvernichtungsanlage *f*
refuse disposal or destruction Abfallvernichtung *f*
refuse disposal plant Abfallvernichtungsanlage *f*
refuse fat Fettabfälle *m pl*
refuse fuel Abfallbrennstoff *m*
refuse tank Abfallbehälter *m*
regaining Rückgewinnung *f*
regard *v* betrachten, berücksichtigen
regelation Regelation *f,* Wiedergefrieren *n*
regenerate *v* fortpflanzen (Biol), regenerieren,
  wiedergewinnen

regenerated material Regenerat *n*
regenerated rubber Regeneratgummi *m*
regeneration Regeneration *f,* Fortpflanzung *f*
(Biol), Regenerierung *f* (Techn),
Rückgewinnung *f* (Techn), Wiederbelebung *f,*
Wiedergewinnung *f* (Techn), **process of** ~
Regenerationsprozeß *m,* **product of** ~
Regenerationsprodukt *n,* **velocity of** ~
Neubildungsgeschwindigkeit *f,* ~ **of the bath
liquor** Auffrischen *n* des Bades
regeneration plant Regenerieranlage *f*
regeneration velocity
Neubildungsgeschwindigkeit *f*
regenerative cell Regenerativelement *n*
regenerative chamber Regenerativkammer *f*
regenerative circuit Rückkopplungskreis *m*
regenerative furnace Regenerativofen *m,*
Speicherofen *m*
regenerative power Regenerationsfähigkeit *f*
regenerative principle Regenerativprinzip *n*
regenerative process Regenerationsvorgang *m*
regenerator Regenerativfeuerung *f,*
Regenerator *m,* Wärmespeicher *m*
region Bereich *m,* Gebiet *n,* **epithermal** ~
epithermisches Gebiet, **visible** ~ sichtbares
Gebiet (Opt)
register Register *n,* Tabelle *f,* Verzeichnis *n,*
Zählwerk *n,* **mechanical** ~ mechanisches
Zählwerk
register *v* registrieren, aufzeichnen, eintragen,
protokollieren (Jur)
registered gesetzlich geschützt (z. B.
Warenzeichen)
registering instrument Registrierinstrument *n*
registration Registrierung *f,* Eintragung *f*
registry Registrierung *f*
regression Zurückgehen *n,* **edge of** ~ Gratlinie *f*
(Geom), **point of** ~ Umkehrpunkt *m*
regression line Regressionslinie *f*
regression model Regressionsmodell *n*
regroup *v* umgruppieren, umlagern
regrouping Umgruppierung *f*
regulable regulierbar
regulable gear unit Regelgetriebe *n*
regular regulär, gleichmäßig, normal,
regelmäßig, stetig
regularity Gesetzmäßigkeit *f,* Gleichförmigkeit *f,*
Regelmäßigkeit *f*
regularity tester Gleichheitsprüfer *m*
regular transformation reguläre
Transformation *f* (Math)
regulate *v* regeln, lenken, regulieren, steuern
regulating, automatically ~ selbstregelnd
regulating apparatus Reguliervorrichtung *f*
regulating cock Regulierhahn *m*
regulating device Reguliervorrichtung *f*
regulating flow fittings Regelarmaturen *f pl*
regulating range Verstellbereich *m*
regulating resistance Regelwiderstand *m*

regulating rod Regulierstange *f*
regulating screw Regulierschraube *f,*
Stellschraube *f*
regulating slide Regulierschieber *m*
regulating step Regelstufe *f*
regulating substance Reglersubstanz *f*
regulating switch Regelschalter *m* (Elektr)
regulating unit Stellorgan *n* (Regeltechnik)
regulating valve Regler *m,* Reglerventil *n*
regulating voltage Regulierspannung *f*
regulation Vorschrift *f,* Anordnung *f,*
Einstellung *f,* Regelung *f,* Regulation *f* (Biol),
Regulierung *f,* Steuerung *f,* **automatic** ~
Selbststeuerung *f,* **supplementary** ~
Zusatzbestimmung *f;* **system with inherent** ~
(Br. E.) Regelstrecke mit Ausgleich *m*
regulation desk Steuerpult *n*
regulation process Regelungsvorgang *m*
regulations, according to ~ vorschriftsmäßig
regulation with fixed set Festwertregelung *f*
(Regeltechn)
regulator Regulator *m,* Regler *m* (Elektr),
Reglersubstanz *f,* Reguliervorrichtung *f*
regulatory enzyme Regulatorenzym *n*
regulatory gene Regulatorgen *n*
reguline regulinisch
regulus Regulus *m*
reheat *v* nachlassen (Met), tempern (Met),
wieder erhitzen
reheat change Nachschwindung *f*
reheating Wiedererwärmen *n*
reheating furnace Glühofen *m,*
Nachwärmofen *m,* **crane for the** ~
Tiefofenkran *m*
reheating furnace slag Schweißofenschlacke *f*
rehmannic acid Rehmannsäure *f*
reichardtite Reichardtit *m* (Min)
reignition Nachzündung *f*
reinfection Reinfektion *f* (Med)
reinforce *v* verstärken, absteifen, abstützen,
bewehren, kräftigen
reinforced verstärkt
reinforced concrete Eisenbeton *m*
reinforcement Armierung *f,* Verstärkung *f,*
Versteifung *f*
reinforcement iron Moniereisen *n*
reinforcing filler Verstärkerfüllstoff *m*
reinforcing iron (for concrete) Betoneisen *n*
reinforcing material Verstärkungsmaterial *n*
reinforcing mesh Metallnetzeinlage *f*
reinite Reinit *m* (Min)
reissue patent Abänderungspatent *n* (Jur)
reject *v* ablehnen, abstoßen, abweisen,
verwerfen, zurückweisen
rejection Ablehnung *f,* Ausschuß *m,*
Verwerfung *f*
rejector circuit Sperrstromkreis *m*
rejects Abfallware *f,* Ausschuß *m*
reject tray Auswerfertisch *m*

rejigging (electroplating) Umsteckverfahren *n;*
  method without ~ Durchfahrtechnik *f* (Galv)
relapse Rückfall *m*
related [to] bezogen auf, verwandt
relation Beziehung *f,* Bezug *m,* Relation *f*
  (Math), **mutual** ~ Wechselbeziehung *f,* ~ **of**
  solubility Löslichkeitsverhältnis *n*
relationship Verwandtschaft *f,*
  Wechselbeziehung *f*
relative relativ, verhältnismäßig, ~ **to** bezogen
  auf
relative accuracy Relativgenauigkeit *f*
relative aperture Öffnungsverhältnis *n*
relative displacement Relativverschiebung *f*
relative motion Relativbewegung *f*
relativistic relativistisch
relativity Relativität *f,* **law of** ~
  Relativitätsgesetz *n*
relativity principle Relativitätsprinzip *n*
relax *v* entspannen, ausspannen, lockern
relaxant Relaxans *n* (Pharm)
relaxation Relaxation *f,* Entspannung *f,* **method**
  of ~ Relaxationsverfahren *n* (Iterationsverf)
relaxation behavior Relaxationsverhalten *n*
relaxation distance Abbremsungslänge *f* (Atom)
relaxation frequency Kippfrequenz *f*
relaxation length Abbremsungslänge *f* (Atom)
relaxation oscillation Kippschwingung *f*
relaxation oscillator Kippgenerator *m,*
  Kippgerät *n*
relaxation time Relaxationszeit *f*
relay Relais *n* (Elektr), **auxiliary** ~ Hilfsrelais *n,*
  instantaneous-release ~ Relais *n* mit
  Schnellauslösung, **quick-operating** ~ Relais *n*
  mit Schnellauslösung, **slow-release** ~ Relais *n*
  mit verzögerter Auslösung, **time-delay** ~
  Relais *n* mit verzögerter Auslösung, **ultrafine**
  ~ Ultrafeinrelais *n*
relay station Umlenkstation *f*
relay tube Schaltröhre *f*
release Abgabe *f* (Energie), Auslöser *m* (Techn),
  Auslösung *f,* Entbindung *f,* **manual** ~
  Handauslösung *f*
release *v* freigeben, abgeben (Energie), auslösen,
  entbinden, entfesseln, entlassen, freisetzen, ~
  the mold entformen, ~ **the tension** entspannen
release agent Trennmittel *n,* Entschalungsöl *n,*
  Gleitmittel *n*
release button Auslöseknopf *m*
release current Auslösestrom *m*
release lever Auslösehebel *m*
release mechanism Auslösevorrichtung *f*
release note Lieferfreigabeschein *m*
release paper Trennpapier *n* (Klebebänder)
  (with pressure-sensitive adhesives),
  Schutzpapier *n*
release sleeve Auslösemuffe *f*
releasing mechanism Auslösemechanismus *m*

releasing of electrons Elektronenaustritt *m,*
  Elektronenfreigabe *f*
reliability Aussagefähigkeit *f,* Zuverlässigkeit *f,*
  ~ **in service** Betriebszuverlässigkeit *f*
reliable zuverlässig, ~ **in operation**
  betriebssicher
relief Relief *n,* Aussparung *f*
relief engraving Hochätzung *f*
relief [or clearance] angle Freiwinkel *m*
relief printing Reliefdruck *m,* Hochdruck *m*
relief printing plate Hochdruckmutterplatte *f,*
  Hochplatte *f*
relief valve Überdruckventil *n,*
  Entlüftungsventil *n,* Stoßventil *n*
relieve *v* erleichtern, lindern
relining Neubelegen *n* (Walze)
reload *v* wieder beschicken
reloading Wiederbelastung *f*
relocate *v* versetzen, neu anordnen
relocation Verlagerung *f,* Versetzung *f*
relugite Relugit *n*
remain *v* [ver]bleiben, verharren, verweilen
remainder Rest *m,* Rückstand *m,* Überrest *m,* ~
  of an infinite series Restglied *n* einer
  unendlichen Reihe (Math)
remaining remanent (Magn), restlich,
  verbleibend
remanence Remanenz *f* (Magn)
remanent remanent (Magn)
remark Bemerkung *f*
remark *v* bemerken
remarkable auffallend, beachtenswert,
  bemerkenswert
remedy Arznei *f* (Pharm), Gegenmittel *n*
  (Pharm), Heilmittel *n* (Pharm),
  Medikament *n* (Pharm), Medizin *f* (Pharm),
  ~ **for burns** Brandmittel *n* (Pharm), ~ **for**
  external application äußerlich anzuwendendes
  Mittel *n* (Pharm)
remedy *v* abhelfen
remelt *v* umschmelzen, wieder einschmelzen
remelting Wiedereinschmelzen *n*
remelting furnace Umschmelzofen *m*
remelting process Umschmelzverfahren *n*
remelt metal Umschmelzmetall *n*
remerine Roemerin *n*
remnant Rest *m*
remodel *v* umgestalten, umarbeiten
remolinite Atakamit *m* (Min)
remote entfernt
remote action Fernwirkung *f*
remote area monitoring system
  Fernüberwachungssystem *n*
remote control Fernbedienung *f,*
  Ferneinstellung *f,* Fernlenkung *f,*
  Fernsteuerung *f,* **mechanical** ~ mechanische
  Fernsteuerung *f*
remote control equipment Fernsteueranlage *f,*
  Fernsteuereinrichtung *f* (Regeltechn)

remote-controlled fernbedient, ferngesteuert
remote control measuring installation
   Fernmeßeinrichtung *f*
remote control panel Fernsteuertafel *f*
remote effect Fernwirkung *f*
remote handling Fernbedienung *f*
remote handling equipment
   Fernbedienungsgerät *n*
remote indicator Fernanzeiger *m*
remote manipulating equipment
   Fernbedienungsgerät *n*
remote operation equipment
   Fernbedienungsgerät *n*
remote transmission Fernübertragung *f*
remote viewing Fernbeobachtung *f*
removable ablösbar, abnehmbar, auswechselbar,
   herausnehmbar, transportabel, **easily** ~ leicht
   entfernbar
removal Beseitigung *f*, Abführung *f*,
   Entfernung *f*, Verdrängung *f*, Wegnahme *f*, ~
   of clinker Rostschlagen *n*, ~ **of electrons**
   Entzug *m* von Elektronen (pl), ~ **of oxygen**
   Entziehung *f* von Sauerstoff, ~ **of paint**
   Abbeizung *f*, ~ **of water** Wasserentziehung *f*
remove *v* beseitigen, abführen, abmontieren,
   abnehmen, abräumen, beheben, entfernen,
   wegnehmen, wegräumen, wegschaffen, ~
   grease degrassieren, ~ **lead** entbleien, ~ **mud**
   abschlämmen, ~ **silver [from]** entsilbern, ~
   slag entschlacken, ~ **slime** entschleimen, ~
   sludge entschlammen, ~ **the parentheses**
   (brackets, braces) die Klammern auflösen, ~
   tin entzinnen
remove lime [from] *v* abkalken
remover Entferner *m*
removing the metal Ausbringen *n* des Metalls
renardite Renardit *m* (Min)
renaturation Renaturierung *f*
rend *v* [ein]reißen
render *v* machen, abstatten, ~ **alkaline** alkalisch
   machen, ~ **impure** verunreinigen, ~
   ineffective lahmlegen, ~ **passive** passivieren, ~
   ~ **pliable** assouplieren, ~ **safe** entschärfen
rendering insoluble Unlöslichmachen *n*
rendering operative Inbetriebsetzung *f*
renew *v* erneuern, auffrischen, neu beginnen,
   renovieren
renewal Erneuerung *f*, ~ **of acid**
   Säureauffrischung *f*, ~ **of air** Luftwechsel *m*
renghol Renghol *n*
reniform nierenförmig
rennase Labferment *n* (Biochem), Rennin *n*
   (Biochem)
rennet Lab *n*
rennet bag Labmagen *m*
rennet casein Labkasein *n*
rennin Rennin *n*, Labferment *n*
renormalization of mass Renormierung *f* der
   Masse (Quant)

renovasculine Renovasculin *n*
renovate *v* renovieren, erneuern
renovation Renovierung *f*, Erneuerung *f*
renoxydine Renoxydin *n*
rensselaerite Rensselärit *m* (Min)
rent Riß *m*, Spalt *m*, Sprung *m*
reorganization Neugestaltung *f*
reorganize *v* umdisponieren, umorganisieren
reorientation Umorientierung *f*
repair Reparatur *f*, Instandsetzung *f*, **in need of**
   ~ reparaturbedürftig
repair *v* reparieren, ausbessern, instandsetzen,
   wiederherstellen
repairs Instandsetzungsarbeiten *f pl*, **in want of**
   ~ ausbesserungsbedürftig, reparaturbedürftig
repair shop Reparaturwerkstatt *f*
repair work Instandsetzung *f*,
   Instandsetzungsarbeiten *pl*
repandine Repandin *n*
repandulinic acid Repandulinsäure *f*
reparable reparabel, instandsetzungsfähig
repayment Tilgung *f*, Rückzahlung *f*
repeat *v* wiederholen
repeated wiederholt
repeated direct stress test Dauerzugversuch *m*
repeated flexural stress
   Dauerbiegebeanspruchung *f*
repeated flexural test Dauerknickversuch *m*
repeated impact energy Dauerschlagarbeit *f*
repeated impact tension test
   Dauerschlagzugversuch *m*
repeated impact testing machine
   Dauerschlagwerk *n*
repeated tension test Dauerzugversuch *m*
repel *v* abstoßen, abweisen, entgegenwirken,
   like poles ~ **each other** gleichnamige Pole
   stoßen einander ab
repellent abstoßend, Phobiermittel *n*
repelling action of electron charge
   Eigenabstoßung *f* der Elektronen (pl)
repelling power Abstoßungskraft *f*
repercussion Rückprall *m*, Rückstoß *m*
repetition Wiederholung *f*
replace *v* ersetzen, austauschen, auswechseln,
   substituieren (Chem), vertreten
replaceable ersetzbar, substituierbar (Chem),
   vertretbar
replacement Austausch *m* (Ersatz), Ersatz *m*,
   Umtausch *m*
replacement part Ersatzteil *n*
replenish *v* ergänzen, nachfüllen, wieder
   auffüllen
replenishing cup Einfülltopf *m*
replica Abguß *m* (Gieß)
replicate Wiederholungsversuch *m*
replica technique Folienabdruckverfahren *n*
   (Mikroskopie)
replication Replikation *f*, **conservative** ~
   konservative Replikation *f*, **disperse** ~

disperse Replikation *f*, **semi-conservative** ~
semikonservative Replikation *f*
replication rate Replikationsgeschwindigkeit *f*
reply Antwort *f*
reply *v* antworten, entgegnen
report Referat *n*, Bericht *m*, Knall *m*, **analytical**
~ Analysenbericht *m*, **to go off with a** ~
knallen
Reppe synthesis Reppe-Synthese *f* (Chem)
reprecipitation Umfällung *f*, Wiederausfällung *f*,
Wiederausflockung *f*
represent *v* darstellen, ~ **graphically** graphisch
darstellen
representation Darstellung *f*, Abbildung *f*
representative Repräsentant *m*, Vertreter *m*
repression Repression *f* (Biochem),
Zurückdrängung *f*, **coordinative** ~
koordinative Repression *f*, **genetic** ~
genetische Repression *f*, **catabolic** ~ **or**
catabolite repression katabolische Repression *f*
repressor Repressor *m*
repressor-inducer complex
Repressor-Induktor-Komplex *m*
reprint Nachdruck *m*, Sonderdruck *m*
reprinting ink Umdruckfarbe *f*
reprocessing Wiederaufbereitung *f*
reproduce *v* abbilden, abklatschen (Buchdr),
fortpflanzen (Biol), hervorbringen (Biol),
nachbilden, reproduzieren, vervielfältigen
reproducibility Reproduzierbarkeit *f*
reproducible reproduzierbar
reproducing method Reproduktionsverfahren *n*
reproduction Reproduktion *f*, Fortpflanzung *f*
(Biol), Kopie *f*, Vermehrung *f* (Biol),
Vervielfältigung *f*, Wiedergabe *f* (Elektr),
process of ~ Reproduktionsverfahren *n*
reproductive factor Multiplikationsfaktor *m*
repulse *v* abstoßen, abweisen
repulsion Abstoßung *f*, Rückstoß *m*, **effect of** ~
Abstoßungseffekt *m*, **electrostatic** ~
elektrostatische Abstoßung, **mutual** ~
gegenseitige Abstoßung
repulsion power Rückstoßkraft *f*
repulsive energy Abstoßungsenergie *f*
repulsive force Abstoßungskraft *f*
repulsive potential Abstoßungspotential *n*
repurifier Nachreiniger *m*
reputation Ansehen *n*
request Bitte *f*
request stop willkürlicher Stopp *m* (Comp)
require *v* beanspruchen, bedürfen, erfordern
requirement Anforderung *f*, Bedarf *m*,
Bedürfnis *n*, Erfordernis *n*, **minimum** ~
geringster Bedarf
requisite Bedarfsartikel *m*, Bedarfsgegenstand *m*
re-reel *v* umspulen
rerun routine Wiederholprogramm *n* (Comp)
resacetophenone Resacetophenon *n*
resaldol Resaldol *n*

resale Wiederverkauf *m*
resaurin Resaurin *n*
resazurin Resazurin *n*
rescinnamine Rescinnamin *n*
rescue *v* retten, befreien
research Forschung *f*, Forschungsarbeit *f*,
Nachforschung *f*, Untersuchung *f*, **basic** ~
Grundlagenforschung *f*, **line of** ~
Forschungsrichtung *f*, **result of** ~
Forschungsergebnis *n*
research center Forschungszentrum *n*
research contract Forschungsauftrag *m*
research department Forschungsabteilung *f*
researcher Forscher *m*
research institute Forschungsanstalt *f*,
Forschungsinstitut *n*
research laboratory Forschungslaboratorium *n*,
Untersuchungslabor *n*
research method Forschungsmethode *f*
research pile Forschungsreaktor *m*
research program Forschungsprogramm *n*
research project Forschungsprojekt *n*
research purpose Forschungszweck *m*
research reactor Forschungsreaktor *m*,
Reaktor *m* für Forschungszwecke
research [work] Forschungsarbeit *f*
research worker Forscher *m*
reseda Reseda *f* (Bot)
reseda oil Resedaöl *n*
resemblance Ähnlichkeit *f*
resemble *v* ähneln, gleichen
resembling ähnlich
resembling gelatin gallertartig
reserpan Reserpan *n*
reserpic acid Reserp[in]säure *f*
reserpic alcohol Reserpinalkohol *m*
reserpiline Reserpilin *n*
reserpine Reserpin *n*
reserpinediol Reserpindiol *n*
reserpone Reserpon *n*
reserpoxydine Reserpoxidin *n*
reserve Reserve *f*, Rücklage *f*, Vorrat *m*
reserve *v* zurückbehalten
reserve capacity Kapazitätsreserve *f*
reserve carbohydrate Reservekohlenhydrat *n*
reserve fund Sicherungsrücklage *f*
reserve piece Ersatzstück *n*
reserve tank Reservetank *m*
reservoir Reservoir *n*, Behälter *m*,
Sammelbecken *n*, Sammelgefäß *n*,
Speicher *m*, Tank *m*, Vorratsgefäß *n*,
Wasserbehälter *m*, **elevated** ~
Hochbehälter *m*
reservoir engineering Lagerstättenphysik *f*
reservoir pressure Lagerstättendruck *m*
reset Nachstellung *f*
reset *v* nachstellen
reset action Rückführwirkung *f*
reset amount Rückführgröße *f*

reset rate Rückführgeschwindigkeit *f*
reset time (Am. E.) Nachstellzeit *f* (Regeltechn)
residence time Aufenthaltszeit *f*, Haltezeit *f*,
　Verweilzeit *f*
residence time distribution
　Verweilzeitverteilung *f*
residual remanent (Phys), restlich, übrig
residual acid Säurerückstand *m*
residual activity Restaktivität *f*
residual bar forces Reststabkräfte *f pl*
residual charge Restladung *f*
residual current Reststrom *m*
residual deviation Restablenkung *f*
residual gas Restgas *n*
residual hardness Resthärte *f* (Wasser)
residual image Nachleuchtbild *n*
residual intensity Restintensität *f*
residual liquid Restlösung *f*
residual liquor or lye Restlauge *f*
residual magnetism Remanenz *f*,
　Restmagnetismus *m*
residual melt Restschmelze *f*
residual moisture Restfeuchtigkeit *f*
residual neutron Restneutron *n*
residual [nuclear] radiation Reststrahlung *f*
　(Atom)
residual oil Restöl *n*, Rückstandsöl *n*
residual pressure Restdruck *m*
residual resistance Restwiderstand *m*
residual stress Restspannung *f*
residual valence Restvalenz *f*
residual voltage Remanenzspannung *f*
residuary rückständig
residuary product Abfallprodukt *n*
residue Rest *m*, Abraumstoff *m* (Techn),
　Blasenrückstand *m* (Dest), Bodensatz *m*,
　Rückstand *m*, Trester *pl* (Techn),
　Überbleibsel *n*, **aliphatic** ~ aliphatischer Rest,
　free from ~ rückstandsfrei, **high-boiling** ~
　hochsiedender Rückstand, **insoluble** ~
　unlöslicher Rückstand, **total** ~
　Gesamtrückstand *m*, **without** ~ restlos, ~
　from glowing Glührückstand *m*, ~ **of the**
　atomic nucleus Atomrumpf *m*
residue after evaporation Eindampfrückstand *m*
residue analysis Rückstandsanalyse *f*
residues Abbrand *m*
residuum Rückstand *m*, Bodensatz *m*
resilience Elastizität *f*, Abprallung *f*,
　Federkraft *f*, Rückfederung *f*, Rückprall *m*,
　Zurückschnellen *f*
resilience tester Rückprallprüfgerät *n*
resiliency Elastizität *f*
resilient elastisch
resin Harz *n*, **artificial** ~ künstliches Harz,
　Kunstharz *n*, **chelating** ~ komplexierendes
　Harz, **crude** ~ Halbharz *n*, **disc of** ~
　Harzkuchen *m*, **extraction of** ~ Entharzung *f*,
　flow of ~ Harzfluß *m*, **formation of** ~

Harzbildung *f*, **free from** ~ harzfrei, **hard** ~
Hartharz *n*, **rich in** ~ harzreich, **soft** ~
Weichharz *n*, **solid** ~ Hartharz *n*, **synthetic** ~
künstliches Harz, Kunstharz *n*, **tapping of** ~
Harzgewinnung *f*, **thermosetting** ~
hitzehärtbares Harz, **to deprive of** ~
entharzen, **to remove** ~ entharzen
resin *v* harzen, mit Harz behandeln (Techn)
resin acid Harzsäure *f*
resin adhesive Harzkleber *m*, Klebharz *n*,
　synthetic ~ Kunstharzklebstoff *m*
resin alcohol Harzalkohol *m*
resinalite Harzstein *m*
resinate Resinat *n*, Harzester *m*
resinate *v* harzen, mit Harz imprägnieren
　(Techn), ~ **wine** Wein *m* mit Harz würzen
resin binder Harzträger *m*
resin-bonded harzverleimt
resin cement Harzkitt *m*
resin cerate Basilicumsalbe *f*
resin channel Harzgang *m*, Harzkanal *m*
resin color Harzfarbe *f*
resin content Harzgehalt *m*
resin deposit (in wood) Harzgalle *f*
resin distilling plant Harzdestillationsanlage *f*
resineon Resineon *n*
resin essence Harzessenz *f*
resin finish Kunstharzausrüstung *f*,
　Kunstharzüberzugslack *m*
resin gall Harzgalle *f*
resin gas Harzgas *n*
resin glue Harzleim *m*
resinic acid Harzsäure *f*
resinic body Harzkörper *m*
resiniferous harzhaltig
resinification Harzbildung *f*, Verharzung *f*
resinified, to become ~ harzig werden
resiniform harzförmig
resinify *v* verharzen, harzig machen
resin-injection mo[u]lding Einspritzverfahren *n*
　(Polyester)
resin-like harzähnlich, harzartig
resin melting Harzschmelzen *n*
resin melting pan Harzkessel *m*
resin melting plant Harzschmelzanlage *f*
resin milk Harzlösung *f*
resinoelectric negativ elektrisch
resinoid Resinoid *n*
resinoid *a* harzähnlich, harzartig
resin oil Harzöl *n*, Harznaphtha *n*, Retinol *n*
resin ointment Königssalbe *f*
resinol Resinol *n*
resinotannol Resinotannol *n*
resinous harzartig, harzhaltig, harzig, **[highly]** ~
　harzreich
resinous cement Harzkitt *m*
resinous coal Harzkohle *f*
resinous compound Harzmasse *f*
resinous electricity Harzelektrizität *f*

resinous juice Harzsaft *m*
resinous luster Harzglanz *m*
resinous pine Kien *m*
resinous substance Harzkörper *m*, Harzstoff *m*
resinous tanning agent Harzgerbstoff *m*
resinous tar Harzteer *m*
resin pocket Harznest *n*, Harztasche *f*
resin product Harzprodukt *n*
resin rod Harzstange *f*
resin rubber Harzgummi *n*
resin size Harzleim *m*
resin smearing machine Beharzungsmaschine *f*
resin soap Harzseife *f*
resin solvent Harzlösungsmittel *n*
resin torch Harzfackel *f*
resin varnish Harzfirnis *m*, Harzlack *m*,
    Tränkharz *n*
resiny harzig
resist Ätzpapp *m*
resist *v* widerstehen
resistance Beständigkeit *f* (Techn), Festigkeit *f*
    (Techn), Resistenz *f*, Widerstand *m* (Elektr),
    apparent ~ scheinbarer Widerstand,
    coefficient of ~ Festigkeitskoeffizient *m*,
    combined ~ Kombinationswiderstand *m*,
    dielectric ~ dielektrischer Widerstand,
    differential ~ innerer Widerstand, effective ~
    effektiver Widerstand, Wirkwiderstand *m*,
    increase of ~ Widerstandserhöhung *f*,
    inductive ~ Induktionswiderstand *m*,
    induktiver Widerstand, inherent ~ natürliche
    Widerstandsfähigkeit *f*, internal ~ innerer
    Widerstand, limit of ~ Festigkeitsgrenze *f*,
    magnetic ~ magnetischer Widerstand,
    modulus of ~ Festigkeitszahl *f*, moment of ~
    Widerstandsmoment *n*, negative ~ negativer
    Widerstand, non-inductive ~ induktionsfreier
    Widerstand, Ohmic ~ Ohmscher Widerstand,
    real ~ Wirkwiderstand *m*, specific ~
    spezifischer Widerstand, thermal ~
    thermischer Widerstand, to short-circuit the ~
    den Widerstand kurzschließen, total ~
    Gesamtwiderstand *m*, unit of ~
    Widerstandseinheit *f* (Elektr), ~ due to sizing
    Leimfestigkeit *f*, ~ during charge
    Ladewiderstand *m*, ~ of the material
    Festigkeit *f* des Werkstoffes, ~ to abrasion
    Abreibefestigkeit *f*, ~ to acid[s]
    Säureechtheit *f*, ~ to air Luftbeständigkeit *f*,
    ~ to bending Biegefestigkeit *f*,
    Biegungswiderstand *m*, ~ to climatic
    conditions Klimabeständigkeit *f*, ~ to flow
    Fließwiderstand *m*, ~ to frost
    Frostbeständigkeit *f*, ~ to infection
    Infektionsresistenz *f*, ~ to reaction
    Reaktionswiderstand *m*, ~ to scrubbing
    Walkwiderstand *m*, ~ to separation
    Trennfestigkeit *f*, ~ to shearing
    Scherfestigkeit *f*, ~ to spreading

Ausbreitungswiderstand *m*, ~ to tarnish
    Anlaufbeständigkeit *f*, ~ to weathering
    Widerstandsfähigkeit *f* gegen atmosphärische
    Korrosion
resistance alloy Widerstandslegierung *f*
resistance amplifier Widerstandsverstärker *m*
resistance anomaly Widerstandsanomalie *f*
resistance body Widerstandskörper *m*
resistance box Stöpselkasten *m* (Elektr),
    Widerstandskasten *m* (Elektr)
resistance bridge Meßbrücke *f* (Elektr),
    Widerstandsbrücke *f* (Elektr)
resistance capacity Widerstandskapazität *f*
resistance coefficient Widerstandszahl *f* (Dest)
resistance coil Widerstandsspule *f*
resistance component Widerstandskomponente *f*
resistance furnace Widerstandsofen *m*
resistance grid Widerstandsgitter *n*
resistance-heated widerstandsbeheizt
resistance heating Widerstandsheizung *f*
resistance increase Widerstandszunahme *f*
resistance loss Ohmscher Verlust *m*, Joulescher
    Wärmeverlust *m*, Stromwärmeverlust *m*
resistance measurement Widerstandsmessung *f*
resistance plate Widerstandsplatte *f*
resistance reduction furnace
    Widerstands-Reduktionsofen *m*
resistance support Widerstandsträger *m*
resistance thermometer
    Widerstandsthermometer *n*
resistance to chipping Kantenfestigkeit *f*
    (Tabletten)
resistance to tracking Kriechfestigkeit *f*
resistance to wear Abnutzungsbeständigkeit *f*
resistance welding Widerstandsschweißung *f*
resistance wire Heizdraht *m*,
    Widerstandsdraht *m* (Elektr)
resistant beständig, widerstandsfähig, ~ to
    chemicals chemikalienbeständig, ~ to fracture
    bruchfest, ~ to frost frostsicher, ~ to water
    wasserfest
resistant to bending biegefest
resisting force Widerstandskraft *f*
resistive circuit Widerstandsschaltung *f*
resistivity Widerstandsfähigkeit *f*
resistor Widerstand *m* (Elektr), exponential ~
    spannungsabhängiger Widerstand
resocyclopharol Resocyclopharol *n*
resodiacetophenone Resodiacetophenon *n*
resoflavin Resoflavin *n*
resogalangin Resogalangin *n*
resol Resol *n*
resolin yellow Resolingelb *n*
resol resin Resolharz *n*
resolution Auflösung *f* (Opt), Bildauflösung *f*,
    horizontal ~ Horizontalauflösung *f*, input ~
    or resolution sensitivity (Am. E.) or
    sensitiveness (Br. E.) Ansprechwert *m*
    (Regeltechn)

resolution of optical forms Antipodentrennung *f*
resolvability Auflösbarkeit *f*
resolvable auflösbar
resolve *v* auflösen, trennen, zerlegen
resolving power Auflösungsvermögen *n* (Opt),
Bildauflösungsvermögen *n*, Trennvermögen *n*
(Gaschromat)
resolving time Auflösezeit *f*
resonance Resonanz *f*, Nachhall *m*, Widerhall *m*,
steric inhibition of ~ sterische
Resonanzhinderung *f*
resonance amplifier Resonanzverstärker *m*
resonance band Resonanzband *n*
resonance capture Resonanzeinfang *m*
resonance circuit Resonanzstromkreis *m*
resonance condition Resonanzbedingung *f*
resonance effect Resonanzwirkung *f*
resonance energy Delokalisationsenergie *f*,
Mesomerieenergie *f* (Chem),
Resonanzenergie *f*
resonance fluorescence Resonanzfluoreszenz *f*
resonance frequency Resonanzfrequenz *f*
resonance hybrid Resonanzhybrid *n*
resonance level Resonanzniveau *n*
resonance line Resonanzlinie *f* (Spektr)
resonance neutron Resonanzneutron *n* (Atom)
resonance overlap Resonanzüberlagerung *f*
resonance photon Resonanzphoton *n*
resonance radiation Resonanzstrahlung *f*
resonance range Resonanzbereich *m*
resonance resistance Resonanzwiderstand *m*
resonance scatterer Resonanzstreuer *m*
resonance scattering Resonanzstreuung *f*
resonance screen Resonanzsieb *n*
resonance spectrum Resonanzspektrum *n*
(Spektr)
resonance-stabilized resonanzstabilisiert
resonance state Resonanzzustand *m*
resonance term Resonanzglied *n*
resonance transfer Resonanzübertragung *f*
resonance vibration Resonanzschwingung *f*
resonant in Resonanz
resonant cavity tube Hohlraumresonatorröhre *f*
resonant frequency variation
Resonanzfrequenzänderung *f*
resonating structure mesomere Grenzstruktur *f*
(Chem)
resonator Resonator *m*
resoorobol Resoorobol *n*
resorb *v* resorbieren
resorbin Resorbin *n*
resorcin Resorcin *n*
resorcin blue Resorcinblau *n*
resorcinism Resorcinvergiftung *f* (Med)
resorcinol Resorcin *n*, Metadioxybenzol *n*
resorcinol monoacetate Euresol *n* (HN)
resorcinol phthalein Resorcinphthalein *n*
resorcitol Resorcit *m*
resorcyl aldehyde Resorcylaldehyd *m*

resorcylic acid Resorcylsäure *f*
resorption Resorption *f*
resorufin Resorufin *n*
resosantal Resosantal *n*
resound *v* widerhallen
resource Hilfsquelle *f*
respect Achtung *f*, Beziehung *f*, Rücksicht *f*
respect *v* achten
respectful respektvoll
respective betreffend, jeweilig
respectively beziehungsweise
respiration Atmen *n*, Atmung *f*, **cellular** ~
Zellatmung *f* (Biol), **inhibition of** ~
Atmungshemmung *f*
respiration filter Atemfilter *m*
respiration rate Atmungsgeschwindigkeit *f*
respirator Atemschutzgerät *n* (Techn),
Gasmaske *f*, Staubmaske *f*
respiratory respiratorisch
respiratory center Atmungszentrum *n*
respiratory chain Atmungskette *f* (Biochem)
respiratory enzyme Atmungsferment *n*
(Biochem)
respiratory exchange Luftaustausch *m* (Lunge)
respiratory frequency Atemfrequenz *f*
respiratory poison Atemgift *n*
respiratory tract Atemweg *m*
resplendent strahlend, schimmernd
respond *v* antworten, entgegnen
response Antwort *f*, Ansprechvermögen *n*,
Reaktion *f*
response time Anlaufzeit *f*, Reaktionszeit *f*
response value Zielgröße *f*
responsibility Haftbarkeit *f*, Verantwortung *f*
responsible haftbar, verantwortlich, **to be** ~
haften
rest Rest *m*, Ruhe *f*, **at** ~ ruhend
rest *v* ruhen
restain *v* aufbeizen (Möbel)
rest energy Ruheenergie *f*
restharrow root Hauhechelwurzel *f* (Bot)
restlessness Unruhe *f*
rest mass Ruhemasse *f*
restoration Restaurierung *f*, Wiederherstellung *f*,
Wiederinstandsetzung *f*
restorative Belebungsmittel *n* (Med)
restore *v* erneuern, restaurieren,
wiederherstellen
rest position Ruhelage *f*
rest potential Ruhepotential *n*
restrain *v* einschränken, zurückhalten
restrainer Sparbeize *f*
restraining coil Haltespule *f*
restraint Beschränkung *f*, Behinderung *f*
restrict *v* beschränken
restricted gate Abreißkegel *m* (verjüngter
Angußkanal)
restriction Beschränkung *f*, Einschränkung *f*
restrictive measure Einschränkungsmaßnahme *f*

restrictor Durchflußregler *m* (Gaschromat),
Nadelventil *n* (Gaschromat)
restrictor bar Fließbalken *m*
restrictor valve Drosselventil *n*
resublimate *v* resublimieren
result Ergebnis *n*, Befund *m*, Fazit *n* (Math),
Nachwirkung *f*, Resultat *n*, **initial** ~
Anfangsergebnis *n*, ~ **of experiment**
Prüfungsergebnis *n*
result *v* resultieren, ausfallen, ausgehen, sich
ergeben
resultant Resultante *f*, Resultierende *f*
resume *v* wieder aufnehmen
resumption Wiederaufnahme *f*
resveratrol Resveratrol *n*
ret *v* rösten; verrotten
retail price Ladenpreis *m*
retain *v* behalten, beibehalten, einbehalten,
sperren (Techn)
retainer Haltebügel *m*, Halteplatte *f*
retainer plate Matrizenplatte *f*
retainer ring Haltering *m*, Spannmutter *f*,
Spannring *m*
retaining core Stützkern *m*
retaining nut Befestigungsmutter *f*
retaining plate Flacheisenverankerung *f*
retaining spring Haltefeder *f*
retard *v* verlangsamen, abbremsen, hemmen,
verzögern
retardant vehicle Depotmittel *n*
retardation Verzögerung *f*, Abbremsung *f*
(Verzögerung), Bremsung *f*, Verlangsamung *f*,
period of ~ Verzögerungszeit *f*, ~ **of electrons**
Elektronenbremsung *f*, ~ **of stimulation**
Reizhemmung *f*
retardation angle Verzögerungswinkel *m*
retardation factor Retardierungsfaktor *m*
retardation potential Verzögerungspotential *n*
retarder Retarder *m*, Verzögerer *m*
retarding-field electrode Bremsfeldelektrode *f*
retarding power Bremsvermögen *n*
retene Reten *n*
retene perhydride Retenperhydrid *n*
retene quinone Retenchinon *n*
retention Retention *f*, Zurückbehalten *n*,
intensity of ~ Haftstärke *f*, ~ **of configuration**
Retention *f* der Konfiguration (Stereochem)
retention cup Haftnäpfchen *n*
retention index Retentionsindex *m*
(Gaschromat)
retention mixer Speichertankmischer *m*
retentive festhaltend
retentivity Koerzitivkraft *f*
retentivity of hardness Härtebeständigkeit *f*
rethrin Rethrin *n*
reticular netzförmig
reticular structure Netzstruktur *f*
reticulate[d] netzförmig
reticulin Reticulin *n*

reticulocytes Reticulozyten *pl*
retina Netzhaut *f* (Med)
retinal Retinal *n*
retinasphalt Retinasphalt *m*
retinene Retinen *n*
retort Retorte *f*, Destillierkolben, **bulb of a** ~
Retortenbauch *m*, **mouth of a** ~
Retortenmündung *f*, ~ **for dry distillation**
Schwelretorte *f*
retort carbon Retortengraphit *m*, Retortenkohle *f*
retort clamp Retortenklemme *f*
retort coal Retortenkohle *f*
retort coking Retortenverkokung *f*
retort contact process
Retortenkontaktverfahren *n*
retort furnace Retortenofen *m*
retort graphite Retortengraphit *m*
retort process Muffelverfahren *n*
retort stand Bunsenstativ *n*, Retortengestell *n*,
Retortenhalter *m*
retouch *v* retuschieren
retouch[ing] Retusche *f*
retouching varnish Retuschierlack *m*
retract *v* zurückziehen
retractable einziehbar
retractable-fork reach truck Schubgabelstapler *m*
retractile einziehbar
retracting mechanism Einziehvorrichtung *f*
retransformation Rückverwandlung *f*
retread *v* runderneuern (Reifen)
retreading Runderneuerung *f* (Reifen)
retroaction Rückwirkung *f*
retroactive rückwirkend
retrograde motion Rückwärtsbewegung *f*
retronecanol Retronecanol *n*
retronecanone Retronecanon *n*
retronecic acid Retronecinsäure *f*
retronecine Retronecin *n*
retting Rösten *n* (Flachs etc.), Verrotten *n*
retting pit Rottegrube *f*, Einweichgrube *f*
return *v* wiederkehren, zurückkehren
returnable container Mehrwegbehälter *m*
return current Rückstrom *m*
return flow Rückfluß *m*
return-flow atomization Rücklaufzerstäubung *f*
return-flow wind tunnel Windkanal *m*
geschlossener Bauart
return line Rückführleitung *f*
return motion Rückbewegung *f*, Rückgang *m*
return passage Überströmkanal *m*
return pin Rückstoßstift *m*
return sludge Rücklaufschlamm *m*
return spring Ausdrückbolzenfeder *f*
(Rückzugfeder), Rückdruckfeder *f*,
Rückstellfeder *f*, Rückzugfeder *f*
return station Umkehrstation *f*
return stroke Rückbewegung *f* (Kolben),
Rücklauf *m* (Kolben)
reunion Wiedervereinigung *f*

reusable wiederverwertbar
reuse *v* wieder verwenden
reused cotton [from rags] Reißbaumwolle *f*
reveal *v* enthüllen, bloßlegen
reverberate *v* reverberieren (Techn), widerhallen (Akust), zurückstrahlen
reverberation Nachklang *m* (Akust), Reverberieren *n* (Techn), Widerhall *m* (Akust), Zurückwerfen *n*
reverberatory flame Stürzflamme *f*
reverberatory furnace Flammofen *m*, Schmelzflammofen *m*, gas-fired ~ Gasflammofen *m*
reverberatory smelting Flammofenarbeit *f*
reversal Umkehr[ung] *f*, Wendung *f*
reversal [of direction] Richtungsumkehr *f*
reversal spectrum Inversionsspektrum *n* (Spektr)
reverse Gegenteil *n*
reverse *a* umgekehrt
reverse *v* umkehren, ändern, umpolen (Elektr), umsteuern, wenden
reverse bending test Hin- und Herbiegeversuch *m*
reverse current Rückstrom *m*, Gegenstrom *m*
reverse curve Gegenkrümmung *f*
reversed umgekehrt
reversed bending Wechselbiegung *f*
reversed current Gegenstrom *m*
reverse double gear wheel Rücklaufdoppelrad *n*
reverse gear Wendegetriebe *n*
reverse motion Gegenlauf *m*
reverse osmosis Hyperfiltration *f*, umgekehrte Osmose *f*
reverser Wendeschalter *m* (Elektr)
reverse reaction Gegenreaktion *f*, Rückreaktion *f*
reverse reactive ground coat technique Umkehrkontaktgrundverfahren *n*
reverse roll coater Spachtelwalze *f*
reverse roll coating Umkehrwalzenbeschichtung *f*
reverse scavenging Umkehrspülung *f* (Mot)
reverse shaft Kehrwelle *f*
reverse side Kehrseite *f*, Rückseite *f* (des Gewebes) (Text)
reversibility Reversibilität *f*, Umkehrbarkeit *f*, ~ of poles Polumkehrbarkeit *f*
reversible umkehrbar, reversibel, umschaltbar
reversible pendulum Reversionspendel *n*
reversing contactor Umschaltschütz *m* (Elektr)
reversing exchanger Rekuperator *m* (Metall)
reversing gear Umkehrgetriebe *n*, Umsteuerungsvorrichtung *f*, Wendegetriebe *n*
reversing handle Umstellhebel *m*
reversing lever Umstellhebel *m*
reversing mill Umkehrwalzwerk *n*
reversing mirror Umkehrspiegel *m*
reversing nozzle Umlenkungsdüse *f*
reversing plate mill Reversierblechwalzwerk *n*
reversing pole Wendepol *m*
reversing prism Umkehrprisma *n* (Opt)

reversing rolling mill Kehrwalzwerk *n*, Reversierwalzwerk *n*
reversing shaft Umsteuerwelle *f*
reversing switch Umkehrschalter *m*, Wendeschalter *m*
reversing valve Umschaltventil *n*, Umstellklappe *f*
reversion Umkehrung *f*, Umpolung *f* (Elektr)
reversion procedure Umkehrverfahren *n*
review Rezension *f*
review *v* rezensieren
reviewer Rezensent *m*
revise *v* revidieren
revised design Neukonstruktion *f*
revision Überprüfung *f*
revive *v* wiederbeleben, auffrischen, avivieren (Chem), wiederherstellen, ~ the litharge Bleiglätte *f* frischen
revivification Regeneration *f* (Katalysator), Wiederbelebung *f*, Wiederherstellung *f*
reviving Avivage *f* (Chem), Wiederbeleben *n*
revolution Umdrehung *f*, Drehung *f*, Kreisbewegung *f*, Kreislauf *m*, Rotation *f*, rotierende Bewegung *f*, period of ~ Umlaufzeit *f*, surface of ~ Rotationsfläche *f*
revolutionary umstürzend, umwälzend
revolution counter Drehzahlmesser *m*, Tourenzähler *m*, Umdrehungszähler *m*
revolution indicator Umdrehungsanzeiger *m*
revolutions, number of ~ Umdrehungszahl *f*, Drehzahl *f*, Tourenzahl *f*, with a low number of ~ niedertourig, ~ per minute Umdrehungen *pl* in der Minute
revolution telltale Umdrehungsanzeiger *m*
revolve *v* rotieren, kreisen, sich drehen
revolver Revolver *m*
revolver lathe Revolverdrehbank *f*
revolving kreisend, rotierend
revolving bath Karussellbad *n* (Galv)
revolving coil Drehspule *f*
revolving crystal Drehkristall *m*
revolving cylindrical furnace Drehrohrofen *m*
revolving-drum dense-medium vessel (dense-medium separation) Trommel[sink]scheider *m*
revolving filter Trommelfilter *n*, Zellenfilter *n*
revolving furnace Drehofen *m*
revolving grate Drehrost *m*
revolving plane Drehebene *f*
revolving puddling Drehpuddeln *n*
revolving puddling furnace Drehpuddelofen *m*
revolving reverberatory furnace Drehflammofen *m*
revolving screen Walzsieb *n*
rewash *v* nachwaschen
rewind *v* umspulen, umwickeln
rewinding machine Aufwickelmaschine *f*, Umrollmaschine *f*
rework *v* umarbeiten

Reynold's number Reynoldsche Zahl *f*
rezbanyite Rezbanyit *m* (Min)
rhabarberone Rhabarberon *n*
rhabdite Rhabdit *m* (Min)
rhabdolith Rhabdolith *m* (Min)
rhabdophane Rhabdophan *m* (Min)
rhaeticite Rhätizit *m* (Min)
rhagite Rhagit *m* (Min)
rhamnal Rhamnal *n*
rhamnazin Rhamnazin *n*
rhamnegin Rhamnegin *n*
rhamnetin Rhamnetin *n*
rhamnicoside Rhamnicosid *n*
rhamnin Rhamnin *n*
rhamnite Rhamnit *m*
rhamnitol Rhamnit *m*
rhamnocitrin Rhamnocitrin *n*
rhamnofluorin Rhamnofluorin *n*
rhamnoheptose Rhamnoheptose *f*
rhamnohexose Rhamnohexose *f*
rhamnol Rhamnol *n*
rhamnonic acid Rhamnonsäure *f*
rhamnose Rhamnose *f*, Isodulcit *m*
rhamnoside Rhamnosid *n*
rhamnosterin Rhamnosterin *n*
rhamnoxanthin Rhamnoxanthin *n*
rhamnulose Rhamnulose *f*
rhaponticin Rhaponticin *n*
rhapontigenin Rhapontigenin *n*
rhapontin Rhapontin *n*
rhatany Ratanhia *f* (Bot)
rhatany root Ratanhiawurzel *f* (Bot)
rhatany tincture Ratanhiatinktur *f*
rhein Rhein *n*
rhein amide Rheinamid *n*
rhein chloride Rheinchlorid *n*
rhenic [acid] anhydride Rhenium(VI)-oxid *n*
rhenium (Symb. Re) Rhenium *n*
rhenium heptoxide Rheniumheptoxid *n*
rhenium trioxide Rheniumtrioxid *n*,
  Rhenium(VI)-oxid *n*
rheochrysin Rheochrysin *n*
rheologic[al] rheologisch
rheological property Fließeigenschaft *f*
rheology Rheologie *f*, Fließlehre *f*,
  Strömungslehre *f*
rheometer Rheometer *n* (Med)
rheopectic rheopektisch
rheopexy Rheopexie *f* (Kolloidchem)
rheoscope Elektroskop *n*, Stromanzeiger *m*
rheoscopic elektroskopisch
rheostat Rheostat *m*, Regelwiderstand *m*,
  Regulierwiderstand *m*, Schiebewiderstand *m*,
  Vorschaltwiderstand *m*
rheostatic braking Widerstandsbremsung *f*
rheotron Elektronenbeschleuniger *m*,
  Rheotron *n*
rhesus factor Rhesusfaktor *m*, Rh-Faktor *m*
rhesus monkey Rhesusaffe *m*

rhetsine Rhetsin *n*
rhetsinine Rhetsinin *n*
rheumatic rheumatisch (Med)
rheumatism Rheumatismus *m* (Med)
rheumemodin Rheumemodin *n*
Rh-factor Rhesusfaktor *m*
rhigolene Rhigolen *n*
rhinanthin Rhinanthin *n*
rhinology Rhinologie *f* (Med)
rhizocarpic acid Rhizocarpsäure *f*
rhizoid Rhizoid *n* (Bot)
rhizome Rhizom *n* (Bot)
rhizonic acid Rhizonsäure *f*
rhizoninic acid Rhizoninsäure *f*
rhizopterine Rhizopterin *n*
rhodacene Rhodacen *n*
rhodaform Rhodaform *n*
rhodamine Rhodamin *n*
rhodanate Rhodanid *n*
rhodanic acid Rhodanin *n*, Rhodanwasserstoff *m*
rhodanide Rhodanid *n*
rhodanine Rhodanin *n*
rhodanizing Rhodiumplattierung *f*
rhodeasapogenin Rhodeasapogenin *n*
rhodeite Rhodeit *m*
rhodeohexonic acid Rhodeohexonsäure *f*
rhodeol Rhodeit *m*
rhodeonic acid Rhodeonsäure *f*
rhodeorhetin Convolvulin *n*
rhodeose Rhodeose *f*
rhodexin Rhodexin *n*
rhodinal Rhodinal *n*
rhodinic acid Rhodinsäure *f*
rhodinol Rhodinol *n*
rhodite Rhodiumgold *n* (Min)
rhodium (Symb. Rh) Rhodium *n*, **colloidal** ~
  kolloidales Rhodium, **metallic** ~
  Rhodiummetall *n*
rhodium bath Rhodiumbad *n*
rhodium chloride Rhodiumchlorid *n*
rhodium compound Rhodiumverbindung *f*
rhodium content Rhodiumgehalt *m*
rhodium oxide Rhodiumoxid *n*
rhodium-plated rhodiniert
rhodium salt Rhodiumsalz *n*
rhodizite Rhodizit *m* (Min)
rhodizonic acid Rhodizonsäure *f*
rhodochlorine Rhodochlorin *n*
rhodochrosite Rhodochrosit *m* (Min),
  Himbeerspat *m* (Min), Manganspat *m* (Min),
  Rosenspat *m* (Min)
rhodocladonic acid Rhodocladonsäure *f*
rhododendrol Rhododendrol *n*
rhodol Metol *n*, Rhodol *n*
rhodolite Rhodolith *m* (Min)
rhodonite Rhodonit *m* (Min), Mangankiesel *m*
  (Min), Rotspat *m* (Min)
rhodophite Rhodophit *m*
rhodophyllite Rhodophyllit *m*

rhodopin Rhodopin *n*
rhodoporphyrine Rhodoporphyrin *n*
rhodopsin Rhodopsin *n*, Sehpurpur *m*
rhodosamine Rhodosamin *n*
rhodoxanthin Rhodoxanthin *n*
rhoduline Rhodulin *n*
rhodusite Rhodusit *m*
rhoeadine Rhöadin *n*
rhoeagenine Rhöagenin *n*
rhoenite Rhönit *m* (Min)
rhoifolin Rhoifolin *n*
rhombic rhombisch, rautenförmig, rhombenförmig
rhombic dodecahedron Rhombendodekaeder *n* (Krist)
rhombic mica Rhombenglimmer *m* (Min)
rhombifoline Rhombifolin *n*
rhombinine Rhombinin *n*
rhomboclase Rhomboklas *m* (Min)
rhombohedral rhomboedrisch (Krist)
rhombohedron Rhomboeder *n* (Krist), Rautenflächner *m* (Krist)
rhombohemihedral rhombischhemiedrisch
rhombohemimorphous rhombischhemimorph
rhomboholohedral rhombischholoedrisch
rhomboid Rhomboid *n*
rhomboid *a* rhomboidisch
rhomboidal rautenförmig, rhomboidisch
rhomboid filling stones Rhomboederfüllsteine *m pl*
rhombopyramidal rhombischpyramidal
rhomb spar Rautenspat *m* (Min)
rhombus Rhombus *m*, Raute *f*
Rh-positive rhesus[faktor]-positiv
Rh-testing Rhesusfaktorbestimmung *f*
rhubarb Rhabarber *m* (Bot)
rhubarb syrup Rhabarbersirup *m*
rhubarb tincture Rhabarbertinktur *f*
rhus varnish Rhuslack *m*
rhyacolite Rhyakolith *m* (Min)
rhyolite Rhyolith *m*
rhythm Rhythmus *m*
rhythmic rhythmisch
rib Rippe *f*, **decorative** ~ Rippe *f* im Seitendekor (Reifen)
ribamine Ribamin *n*
ribazole Ribazol *n*
ribbed geriffelt, gerippt
ribbed cooler Rippenkühler *m*
ribbed funnel Riffeltrichter *m*, Rippentrichter *m*
ribbed glass Riffelglas *n*
ribbon Band *n*
ribbon agate Bandachat *m* (Min)
ribbon blender Zwangsmischer *m*
ribbon coil Bandspule *f*
ribbon cutting Bandschnitt *m*
ribbon jasper Bandjaspis *m* (Min)
ribbon lacquer Bandlack *m*
ribbon mixer Bandmischer *m*

ribitol Ribit *m*
ribodesose Ribodesose *f*
riboflavin Riboflavin *n*, Lactoflavin *n*, Vitamin B₂ *n*
riboflavin phosphate Riboflavinphosphat *n*
riboketose Riboketose *f*
ribonic acid Ribonsäure *f*
ribonuclease Ribonuclease *f* (Biochem)
ribonucleic, ribosomal ~ acid ribosomale Ribonucleinsäure *f*, **viral ~ acid** virale Ribonucleinsäure *f*
ribonucleic acid Ribonucleinsäure *f*
ribosamine Ribosamin *n*
ribose Ribose *f*
ribosome Ribosom *n*
ribosome dissociation Ribosomendissoziation *f*
ribosome distribution Ribosomenverteilung *f*
ribosome function Ribosomenfunktion *f*
ribosome initiation complex Ribosomenausgangskomplex *m*
ribosome reconstitution Ribosomenrekonstitution *f*
ribulose Ribulose *f*
ribulose diphosphate Ribulosediphosphat *n*
rice Reis *m*
rice oil Reisöl *n*
rice spirit Reisbranntwein *m*
rice starch Reisstärke *f*
rich reich, satt (Farbe)
rich coal Fettkohle *f*
richellite Richellit *m* (Min)
rich in color farbenreich
rich ore Edelerz *n*
rich slag Frischschlacke *f*, Reichschlacke *f*
richterite Richterit *m* (Min)
ricin Ricin *n*
ricinate Ricinoleat *n*
ricinelaidic acid Ricinelaidinsäure *f*, Ricinuselaidinsäure *f*
ricinenic acid Ricinensäure *f*
ricinic acid Ricinolsäure *f*, Ricinsäure *f*
ricinine Ricinin *n*
ricinoleate Ricinoleat *n*
ricinoleic acid Ricinolsäure *f*, Ricinusölsäure *f*
ricinolein Ricinolein *n*
ricinostearolic acid Ricinstearolsäure *f*
rickardite Rickardit *m* (Min)
rickets Rachitis *f* (Med), Englische Krankheit *f* (Med)
rickety rachitisch
rid *v* befreien
riddellic acid Riddellinsäure *f*
riddle grobes Sieb *n*, Durchwurf *m*, Rätter *m*, Schüttelsieb *n*
riddle *v* rättern (Bergb)
rider Reiter *m*
rider adjustment Reiterversetzung *f*
ridge Grat *m*, Kante *f*
ridge line analysis Kammlinienanalyse *f* (Statist)

riebeckite Riebeckit *m* (Min)
Riemann's criterion Riemannsche
Integrierbarkeit *f*
rig Prüfstand *m*
right Recht *n*, Befugnis *f*, **to be** ~ stimmen,
richtig sein, **to set** ~ berichtigen
right *a* richtig
right angle bend Rohrbogen *m*
right-angled rechtwinklig
right-angle valve Eckventil *n*
right-handed rechtsgängig
right-handed rotation Rechtsdrehung *f*
right-hand rule Rechtehandregel *f* (Phys)
right-hand thread Rechtsgewinde *n*
right proportion Gleichmaß *n*
rigid steif, fest, hart, stabil, starr, unbiegsam
rigid body displacement
Starrkörperverschiebung *f*
rigidity Unbeweglichkeit *f*, Biegfestigkeit *f*,
Festigkeit *f*, Härte *f*, Starre *f*, Starrheit *f*,
Steifheit *f*
rigidity modulus Schubmodul *m*
rigid sheet Hartfolie *f*
rigor Härte *f*
rigorous rigoros, energisch
rim Rand *m*, Felge *f* (Techn), Kante *f*, Kranz *m*
(Techn), **flat base** ~ Flachbettfelge *f*, **straight
side** ~ Flachbettfelge *f*, **three-piece** ~
dreiteilige Felge, **upper** ~ Oberkante *f*
rim *v* einfassen
rim angle (of surface tension) Randwinkel *m*
(der Oberflächenspannung)
rim base Felgenboden *m*
rim contour Felgenprofil *n*
rim decarbonization Randentkohlung *f*
rim fermentation Randgärung *f*
rim flange Felgenhorn *n*, **removable** ~
Felgenhornring *m*
rim ledge Felgenschulter *f*
rim pattern Kranzmodell *n*
rim width between the flanges Felgenmaulweite *f*
ring Ring *m*, Hof *m* (Opt), Öse *f*, Reif *m*,
Reifen *m*, **aromatic** ~ aromatischer Ring,
five-membered ~ Fünfring *m* (Chem),
strainless ~ spannungsfreier Ring (Chem), ~
for preserving jars Einkochring *m*
ring acylation Ringacylierung *f*
ring agate Ringachat *m* (Min)
ring balance Ringwaage *f*
ring burner Heizkranz *m*, Kronenbrenner *m*,
Ringbrenner *m*, Rundbrenner *m*
ring-chain tautomerism
Ring-Ketten-Tautomerie *f* (Chem)
ring channel Ringkanal *m*
ring cleavage Ring[auf]spaltung *f*
ring closure Ringschluß *m*, Ringbildung *f*
ring-cluster structure Ring-Ketten-Struktur *f*
ring compound Ringverbindung *f* (Chem),
zyklische Verbindung *f* (Chem)

ring conduit Ringleitung *f*
ring covering machine
Ringummantelungsmaschine *f*
Ringer's solution Ringersche Lösung *f*
ring expanding or drift test
Ringaufdornversuch *m*
ring extension Ringerweiterung *f*
ring folding test Ringfaltversuch *m*
ring formation Ringbildung *f* (Chem),
Zyklisierung *f* (Chem)
ring gauge Kaliberring *m*
ring gauges Lehrringe *pl*
ring grating Ringgitter *n*
ringing inductor Lautinduktor *m*
ring magnet Ringmagnet *m*
ring mechanism Ringmechanismus *m*
ring opening Ringöffnung *f*
ring packing Füllung *f* mit Raschigringen (pl)
(Dest)
ring-roller mill Dreiwalzenringmühle *f*,
Federrollenmühle *f*, Ringmühle *f*
ring roll mill Ringwalzenmühle *f*
ring roll press Ringwalzenpresse *f*
ring rupture Ringaufspaltung *f*
rings Ringfüllkörper *pl* (Chem)
ring shape Ringform *f*
ring-shaped ringförmig
ring structure Ringstruktur *f*
ring tensile test Ringzugversuch *m*
ring weight Belastungsring *m*
rinkite Rinkit *m* (Min)
Rinmann's green Rinmanns Grün *n*,
Kobaltgrün *n*
rinneite Rinneit *m* (Min)
rinse *v* [ab]spülen, abwaschen, ausspülen,
entseifen, nachspülen
rinsing Abspülen *n*, Spülen *n*, Spülung *f*
rinsing agent Spülmittel *n*
rinsing liquid Spülflüssigkeit *f*
rinsing vat Auswaschbottich *m*
rinsing water Spülwasser *n*
rionite Rionit *m*
rip Riß *m*
rip *v* aufschlitzen, ~ **open** aufreißen
ripe reif, mürbe, **to become** ~ reifen
ripen *v* altern (Wein etc.), reifen
ripeness Reife *f*, **degree of** ~ Reifegrad *m*
ripening, time for ~ Reifungsdauer *f*
ripidolite Ripidolith *m* (Min), Fächerstein *m*
(Min), Klinochlor *m* (Min)
ripper Ausreisser *m* (Bauw)
ripple tray Wellsiebboden *m* (Kolonne)
ripple varnish Kräusellack *m*
ripple voltage Brummspannung *f*, pulsierende
Spannung *f*
rise Ansteigen *n*, Anstieg *m*, Anwachsen *n*,
Aufstieg *m*, Steigerung *f*, **capillary** ~ kapillare
Steighöhe

rise *v* [an]steigen, aufsteigen, sich erheben, zunehmen
riser Steiger *m* (Gieß), Steigrohr *n* (Techn)
rise time Durchschaltzeit *f* (Elektr)
risic acid Risinsäure *f*
rising [auf]steigend
rising film evaporator Kletterverdampfer *m*
rising pipe Steigrohr *n*
rising velocity Steiggeschwindigkeit *f* (v. Gasblasen)
risk Risiko *n*
risky gewagt
rissic acid Risinsäure *f*
ristin Ristin *n*
rittingerite Rittingerit *m* (Min)
rival *v* rivalisieren, konkurrieren
rivalry Rivalität *f*, Konkurrenz *f*
rivanol Rivanol *n*
rivelling Kräuseln *n* (Farbe)
river Fluß *m*, Strom *m*
river gravel Flußkies *m*
river sand Flußsand *m*
river silt Flußschlamm *m*
rivet Niet *m*, Niete *f*
rivet *v* [ver]nieten, annieten, festmachen
riveted [an]genietet
riveted chain Stiftkette *f*
rivet[ed] joint Nietverbindung *f*
rivet head Nietkopf *m*, countersunk ~ Nietsenkkopf *m*
riveting Nieten *n*, Nietnaht *f*, Nietung *f*
riveting set Nietstempel *m*
riveting tongs Nietzange *f*
riveting tool Nietwerkzeug *n*
rivet iron Nieteisen *n*
rivet joint Nietung *f*
rivets, spacing of ~ Nietteilung *f*
rivet shank Nietschaft *m*
rivotite Rivotit *m* (Min)
RNA biosynthesis RNS-Biosynthese *f*
RNA polymerase RNS-Polymerase *f* (Biochem)
road Straße *f*
road construction Straßenbau *m*
road marker Straßenmarkierung *f*
road paint Straßenmarkierungsfarbe *f*
road surface Straßenbelag *m*
road test Straßenerprobung *f*
roar Gebrüll *n*, Lärm *m*
roast, final ~ Fertigröstung *f*
roast *v* abschwelen (Techn), ausglühen (Techn), braten, rösten (Techn), ~ again nachrösten, ~ slightly anrösten, ~ [thoroughly] abrösten
roasted geröstet, abgeschwelt
roasted malt Röstmalz *n*
roasted ore Abbrand *m*
roaster Röster *m*, Röstofen *m*
roaster air fan Rostluftgebläse *n*
roaster gas Röstgas *n*

roasting Rösten *n*, Röstung *f*, chlorinating ~ chlorierendes Rösten, flux for ~ Röstzuschlag *m*, loss due to ~ Röstverlust *m*, oxidation ~ oxidierendes Rösten, partial ~ Teilröstung *f*, period of ~ Brennzeit *f*, product of ~ Röstprodukt *n*, residue from ~ Röstrückstand *m*, ~ in heaps Haufenröstung *f*
roasting apparatus Röstapparat *m*
roasting bed Röstbett *n*
roasting chamber Röstkammer *f*
roasting charge Röstgut *n*
roasting dish Glühschale *f*, Röstscherben *m*
roasting furnace Brennofen *m*, Röstofen *m*
roasting gas Röstgas *n*
roasting hearth Röstherd *m*
roasting installation Röstanlage *f*
roasting kiln Röstofen *m*
roasting pan Brennpfanne *f*
roasting plant Röstanlage *f*
roasting practice Röstbetrieb *m*
roasting process Röstprozeß *m*, Röstarbeit *f*, Röstverfahren *n*
roasting product Rösterzeugnis *n*
roasting residue Röstrückstand *m*
roasting temperature Rösttemperatur *f*
roast-reaction process Röstreaktionsarbeit *f*
roast-reduction process Röstreduktionsarbeit *f*
robinin Robinin *n*
robinobiose Robinobiose *f*
robot Roboter *m*
roburite Roburit *m* (Sprengstoff)
robust robust, kräftig
roccellic acid Roccellsäure *f*
roccelline Roccellin *n*, Orseillerot *n*
Rochelle salt Rochellesalz *n*, Kaliumnatriumtartrat *n*
rock Fels *m*, Gestein *n*, hard as ~ felsenhart, kind of ~ Felsart *f*, ore-bearing ~ erzführendes Gestein
rock *v* hin- und herbewegen
rock agate Felsenachat *m* (Min)
rock alum Bergalaun *m* (Min)
rock butter Steinbutter *f* (Min)
rock cork Bergkork *m* (Min)
rock crystal Bergkristall *m* (Min), Quarzkristall *m* (Min)
rock drill Gesteinsbohrer *m*
rock drilling Gesteinsbohren *n*
rock drilling machine Gesteinsbohrmaschine *f*
rocker Kurvenscheibe *f*, Wippe *f* (Techn)
rocker arm Kipphebel *m*, Schwenkarm *m*
rocker arm shaft Kipphebelwelle *f*
rocker arm wear Kipphebelverschleiß *m*
rocket Rakete *f*, dry-fuelled ~ Rakete *f* mit festem Brennstoff, liquid-fuelled ~ Rakete *f* mit flüssigem Brennstoff, liquid-propellant ~ Rakete *f* mit flüssigem Brennstoff, single-stage ~ Einstufenrakete *f*,

solid-propellant ~ Rakete *f* mit festem
Brennstoff
rocket drive Raketenantrieb *m*
rocketery Raketenforschung *f*
rocket head Raketenkopf *m*
rocket launching site Raketenabschußbasis *f*
rocket pile Raketenreaktor *m*
rocket propellant Raketentreibstoff *m*
rocket-propelled spaceship Raketenraumschiff *n*
rocket propulsion Raketenantrieb *m*
rocket research Raketenforschung *f*
rock flint Bergkiesel *m* (Min), Feuerstein *m*
(Min), Flint *m* (Min)
rocking appliance Schüttelapparat *m*
rocking lever Schwinghebel *m*
rocking trough Schwingrinne *f*
rock-like felsartig
rock lime Bergkalk *m* (Min), Bergkreide *f* (Min)
rock oil Erdöl *n*, Naphtha *f n*, Petroleum *n*,
Steinöl *n*
rock salt Steinsalz *n*
rock salt mine Salzbergwerk *n*
rock soap Bergseife *f* (Min)
Rockwell hardness Rockwell-Härte *f*
Rockwell hardness number Rockwell-Härtezahl *f*
Rockwell hardness test Rockwell-Härteprüfung *f*
rockwood Bergholz *n* (Min), Holzasbest *m*
(Min)
rock wool Mineralwolle *f*, Schlackenwolle *f*,
Steinwolle *f*
rocky felsig
rod Stab *m*, Stäbchen *n*, Stange *f*, **drawing of** ~
Stabziehen *n*
rod bending test Dornbiegeprobe *f*
rod cell Stäbchenzelle *f* (Histol)
rod connecting clamp
Stangenverbindungsklemme *f*
rod control Stangensteuerung *f*
rod electrode Stabelektrode *f*
rodent Nagetier *n* (Zool)
rodinal Rodinal *n*
rod insulator Stabisolator *m*
rod magnet Zylindermagnet *m*
rod mill Rohrmühle *f*, Stabmühle *f*
rod-shaped stäbchenförmig
rod stock Stabmaterial *n*
rod thermostat Stabthermostat *m*
roe Rogen *m*
roentgen apparatus Röntgenapparat *m*
roentgen film Röntgenfilm *m*
roentgenogram Röntgenaufnahme *f*,
Röntgenbild *n*
roentgenography Röntgenographie *f*,
Röntgenphotographie *f*
roentgenology Röntgenologie *f*
roentgenometer Röntgenstrahlenmesser *m*
roentgenoscope Röntgenoskop *n*,
Röntgenapparat *m*

roentgenoscopy Röntgenoskopie *f*,
Röntgenuntersuchung *f*
roentgenotherapy Röntgenbestrahlung *f*
roentgen spectrum Röntgenspektrum *n*
roentgen [unit] Röntgen *n*, Röntgeneinheit *f*
roepperite Röpperit *m* (Min)
roestone Rogenstein *m* (Min)
roll Rolle *f* (Papier etc.), Stange *f* (Schwefel),
Walze *f* (Techn), **chilled** ~ Hartwalze *f*,
direct-coupled ~ direkt angetriebene Walze,
offset ~ vorgelagerte Walze, ~ **of fabric**
Materialrolle *f*, ~ **of uncured rubber** Puppe *f*
(Gummi)
roll *v* kalandern (Pap), rollen, strecken (Techn),
wälzen, walzen (Techn), ~ **in** einwalzen, ~
off abrollen, wegrollen, abwalzen, ~ **on return
pass** zurückwalzen, ~ **out** auswalzen
rollable walzbar (Techn)
rollback routine Wiederholprogramm *n* (Comp)
roll bending Bombieren *n* (Techn),
Walzendurchbiegung *f*
roll blade Walzenmesser *n*
roll bleaching Aufdockbleiche *f*
roll brush Walzenbürste *f*
roll clearance or gap Walzenspalt *m*
roll coater Aufwalzvorrichtung *f*
roll coating Walzenauftragverfahren *n*
roll coating varnish Walzenlack *m*
roll crossing Walzenschrägverstellung *f*
roll drafting Kalibrieren *n* (der Walzen pl)
rolled gewalzt
rolled alloy Walzlegierung *f*
rolled copper plate Walzkupferplatte *f*
rolled gold Dublee *n*, Dubleegold *n*
rolled iron Walzeisen *n*
rolled iron beam Walzeisenträger *m*
rolled-on cap aufgerollter Deckel *m* (Dose)
rolled shape Walzprofil *n*
rolled sheet Walzfolie *f*
rolled sheet metal gewalztes Blech *n*
rolled tube gewickeltes Rohr *n*
roller Rolle *f*, Gleitrolle *f*, Trommel *f*, Walze *f*,
Zylinder *m*, **cover for** ~ Walzenbezug *m*
roller application Aufwalzen *n* (Folien)
roller bearing Rollenlager *n*, Walzenlager *n*,
parallel ~ zylindrisches Rollenlager, **taper** ~
konisches Rollenlager
roller calender Rollkalander *m*
roller carriage arm Walzenstuhl *m*
roller chain Rollenkette *f*
roller-coated walzlackiert
roller coater Walzlackiermaschine *f*
roller coating Walzverfahren *n*
roller composition Walzenmasse *f* (Buchdr)
roller conveyor turntable Drehrollerbahn *f*
roller covering Walzenbezug *m*
roller drier Walzentrockner *m*
roller feed Walzenvorschub *m*
roller-flight conveyor Rollenkettenförderer *m*

roller frame Walzenstuhl *m* (Buchdr)
roller guide Rollschlitten *m*
roller mill Walzenmühle *f,* Walzenstuhl *m*
roller press Walzenpresse *f*
roller printing paper Walzendruckpapier *n*
roller process Walzverfahren *n*
rollers Walzwerk *n*
roller squeegee Gummiquetschwalze *f*
roller table Walztisch *m*
roller vat Rollenkasten *m,* Rollenkufe *f*
roller weir Walzenwehr *n*
roll face wirksame Walzenbreite *f*
rolling Walzen *n* (Techn), **path of ~**
   Walzbahn *f,* **to make smooth by ~** abwalzen
rolling ability Walzbarkeit *f*
rolling circle Walzkreis *m*
rolling crusher Walzenbrecher *m*
rolling dryer Walzentrockner *m*
rolling friction Rollreibung *f* (Mech)
rolling ingot Walzbarren *m*
rolling mill Walzwerk *n*
rolling mill engine Walzenzugmaschine *f*
rolling mill operation Walzbetrieb *m*
rolling mill practice Walzbetrieb *m*
rolling operation Walzvorgang *m*
rolling pin Rollholz *n* (Techn)
rolling plate Walkplatte *f*
rolling press Rollenpresse *f,* Walzenpresse *f*
rolling pressure Walzdruck *m*
rolling resistance rollende Reibung *f* (Mech),
   Rollwiderstand *m*
rolling ring crusher Mantelbrecher *m*
rolling sheet Walzfell *n,* Walzhaut *f,* Walztafel *f*
rolling stand Walzgerüst *n*
rolling surface Walzfläche *f*
roll jaw crusher Backen-Kreiselbrecher *m*
roll lathe Walzendrehbank *f* (Techn)
roll mandrel Walzdorn *m*
roll neck Walzenzapfen *m*
roll-nip adjustment Walzenspalteinstellung *f*
roll opening Kaliber *n*
roll pass Kaliber *n*
roll pot Walzkessel *m*
rolls, pressure of the ~ Walzendruck *m,* **rotation
   of** ~ Walzenbewegung *f,* **set of** ~
   Walzstrecke *f,* **train of** ~ Walzstraße *f,*
   Walzstrecke *f,* **wear of** ~ Walzenverschleiß *m*
roll scale Walzzunder *m* (Met)
roll-shaped rollenförmig
roll sheet iron Rollenblech *n*
roll sulfur Stangenschwefel *m*
roll torque Walzendrehmoment *n*
roll turning Eindrehen *n* der Kaliber (n pl)
roll vat Walzkessel *m*
Roman candle Römerkerze *f*
Roman cement Patentzement *m,*
   Romanzement *m,* Wassermörtel *m*
Roman numeral römische Ziffer
romeite Romeit *m* (Min)

romerite Römerit *m* (Min)
rongalite Rongalit *n*
roof boarding Holzschalung *f* (Dach)
roofing material Verdeckstoff *m*
roofing paper Dachpappe *f*
roofing slate Tafelschiefer *m*
[roofing] tile Dachziegel *m*
room Raum *m,* **~ for experiments**
   Versuchsraum *m*
room temperature Raumtemperatur *f,*
   Zimmertemperatur *f*
room thermo-regulator Raumtemperaturregler *m*
root Wurzel *f,* **cube ~** dritte Wurzel, **extraction
   of a ~** Radizieren *n* (Math),
   Wurzel[aus]ziehen *n* (Math), **index of a ~**
   Wurzelexponent *m* (Math), **square ~**
   Quadratwurzel *f,* **to extract a ~** eine Wurzel
   ziehen, radizieren (Math)
root branching Wurzelverzweigung *f*
root filling Wurzelfüllung *f*
root mean square quadratischer Mittelwert *m*
root rubber Wurzelkautschuk *m*
Roots pump Rootspumpe *f*
rootstock Rhizom *n* (Bot)
root tannin Wurzelgerbstoff *m*
rope Seil *n,* Strick *m,* Tau *n,* **coil of ~**
   Seilwindung *f,* **flat ~** Flachseil *n,* **flat balance
   ~** Flachunterseil *n,* **round ~** Rundseil *n,* **solid
   ~** Vollseil *n,* **to unwind a ~** ein Seil
   abwickeln, **to wind up a ~** ein Seil aufwickeln,
   turn of ~ Seilwindung *f*
rope belt Kordelriemen *m*
rope drive Seilantrieb *m*
rope drop hammer Seilfallhammer *m*
rope drum Seiltrommel *f*
rope loop Seilschlinge *f*
rope lubricator Seilschmierer *m*
rope pulley Seilrolle *f,* Seiltrommel *f*
rope railway Drahtseilbahn *f*
rope sheave Seilrolle *f*
rope splicing Seilspleißung *f*
rope winch Seilwinde *f,* Kabelwinde *f*
ropy klebrig, fadenziehend, kahmig, **to be ~**
   Fäden ziehen
rosamine Rosamin *n*
rosaniline Rosanilin *n*
rosaniline hydrochloride Fuchsin *n*
rosasite Rosasit *m* (Min)
roscherite Roscherit *m* (Min)
roscoelite Roscoelit *m* (Min)
rose rosafarben
rose copper Rosenkupfer *n,* Scheibenkupfer *n*
Rose crucible Rosetiegel *m*
rose honey Rosenhonig *m*
rosein Rosein *n,* Fuchsin *n*
roselin Anilinrot *n*
roselite Roselith *m* (Min)
rosellane Rosellan *m* (Min), Rosit *m* (Min)

rosemary Rosmarin *m* (Bot), **oil of ~**
Rosmarinöl *n*
rosemary leaves Rosmarinblätter *n pl*
rosemary ointment Rosmarinsalbe *f*
rosenolic acid Rosenolsäure *f*
rose oil Rosenöl *n*
rose petals Rosenblätter *n pl*
rose quartz Rosenquarz *m* (Min)
roses, fragrance of ~ Rosenduft *m*
Rose's metal Rosesches Metall *n*
rose spar Rosenspat *m* (Min)
rose steel Rosenstahl *m*
rosette Rosette *f*
rosette copper Rosettenkupfer *n*,
Scheibenkupfer *n*
rose vinegar Rosenessig *m* (Pharm)
rose vitriol Kobaltvitriol *m*
rosewood Rosenholz *n*
rosewood oil Rosenholzöl *n*
rosin Harz *n*, Fichtenharz *n*, Geigenharz *n*,
Kiefernharz *n*, Kolophonium *n*
rosin cerate Harzcerat *n*, Harzsalbe *f*,
Königssalbe *f*
rosindole Rosindol *n*
rosindone Rosindon *n*
rosinduline Rosindulin *n*
rosindulone Rosindon *n*
rosin ester Harzester *m*
rosin oil Harzöl *n*, Terpentinharzöl *n*
rosin soap Harzseife *f*
rosin spirit Harzessenz *f*, Harzspiritus *m*
rosiny harzig
rosite Rosit *m* (Min), Rosellan *m* (Min)
rosmarinic acid Rosmarinsäure *f*
rosoic acid Rosoesäure *f*
rosolic acid Rosolsäure *f*
rosslerite Rößlerit *m* (Min)
rosthornite Rosthornit *m*
rosy rosafarben
rot Anbrüchigkeit *f*, Fäule *f*, Fäulnis *f*, **~ of
walls** Mauerfraß *m*
rot *v* [ver]faulen, modern, verwesen, verwittern,
~ through durchfaulen
rotalix tube Rotalixdrehanodenröhre *f*
rotary drehbar, kreisend
rotary alternating axis Drehspiegelachse *f*
rotary bin Drehbunker *m*
rotary blower Umlaufgebläse *n*
rotary casting Schleuderguß *m*
rotary cement kiln Zementdrehofen *m*
rotary column Rotationskolonne *f*
rotary compressor Drehkolbenverdichter *m*
rotary condenser Drehkondensator *m*
rotary converter Einankerumformer *m*,
Drehumformer *m*
rotary crane Drehkran *m*
rotary crusher Glockenmühle *f*
rotary cutter Schälmaschine *f*
rotary cutters Schneidmühle *f*

rotary dispersion Rotationsdispersion *f*,
Rotationsstreuung *f*
rotary drum filter Trommeldrehfilter *n*
rotary dryer Trockentrommel *f*, Drehtrockner *m*
rotary evaporator Rotations-Verdampfer *m*
rotary extractor Karussellextrakteur *m*
rotary film evaporator Rota-Filmverdampfer *m*
rotary filter Drehfilter *m*, Trommelfilter *n*
rotary furnace Drehofen *m*, Trommelofen *m*
rotary head generator Drehrostgenerator *m*
rotary hearth furnace Tellerofen *m*
rotary hose Spülbohrschlauch *m*
rotary impulse Drehimpuls *m*
rotary inversion axis Drehinversionsachse *f*
rotary kiln Drehofen *m*
rotary knife cutter Schneidmühle *f*
rotary luffing crane Drehwippkran *m*,
Wippdrehkran *m*
rotary magnet Drehmagnet *m*
rotary mixer Mischtrommel *f*, Rollfaß *n*,
Trommelmischer *m*
rotary momentum Drehimpuls *m*
rotary motion Drehbewegung *f*, rotierende
Bewegung *f*, Umlaufbewegung *f*
rotary piston Drehkolben *m*
rotary piston engine Kreiskolbenmotor *m*
(Wankel)
rotary piston manometer
Druckkolbenmanometer *n*
rotary piston meter Ringkolbenzähler *m*
rotary piston pump Drehkolbenpumpe *f*
rotary polarization Drehpolarisation *f*
rotary potentiometer Drehpotentiometer *n*
rotary pressure filter Druckdrehfilter *n*
rotary puddling furnace Wendeofen *m*
rotary pump Drehkolbenpumpe *f*,
Kreiskolbenpumpe *f*, Umlaufpumpe *f*
rotary rim Drehkranz *m*
rotary roaster Trommelofen *m*
rotary shelf drier Tellertrommeltrockner *m*
rotary slide valve Drehschieber *m*
rotary squeezer Luppenmühle *f*
rotary switch Drehschalter *m*
rotary table Drehtisch *m*
rotary table furnace Drehtischofen *m*
rotary tabletting press Tischpresse *f*
rotary thermostat Umlaufthermostat *m*
rotary tower crane Turmdrehkran *m*
rotary tube furnace Drehrohrofen *m*
rotary-vane feeder Zellenradschleuse *f*
rotary vane pump Drehschieberpumpe *f*
rotary velocity Umlaufgeschwindigkeit *f*
rotatable clamp Drehmuffe *f*
rotate *v* rotieren, kreisen, sich drehen,
umlaufen, **~ [around]** umkreisen
rotating kreisend, rotierend
rotating bucket elevator Wurfbecherwerk *n*
rotating calcining oven Drehkalzinierofen *m*
rotating carrier Drehscheibe *f*

rotating-crystal method Drehkristallmethode *f*
(Krist)
rotating cylinder Wendezylinder *m*
rotating-disc monochromator
Drehscheibenmonochromator *m*
rotating field Drehfeld *n*, **direction of** ~
Drehfeldrichtung *f*, **speed of the** ~
Drehfeldgeschwindigkeit *f*
rotating joint Drehgelenk *n*
rotating motion Drehbewegung *f*
rotating piston pump Kreiskolbenpumpe *f*
rotating plate condenser Drehkondensator *m*
rotating punch Revolverstanze *f*
rotating scratch brush Umlaufkratzbürste *f*
rotating spindle Drehachse *f*
rotating valve Umlaufventil *n*
rotating vane Blendenrad *n*
rotating wetted wall evaporator
Rotationsdünnschichtverdampfer *m*
rotation Rotation *f*, Drehbarkeit *f*,
Drehbewegung *f*, Drehung *f*, Kreislauf *m*,
Rotationsbewegung *f*, rotierende Bewegung *f*,
Umdrehung *f*, Umlauf *m*, **angle of** ~
Rotationswinkel *m*, **axis of** ~ Drehachse *f*,
Rotationsachse *f*, **characteristic** ~
Eigendrehbewegung *f*, **direction of** ~
Rotationsrichtung *f*, Drehsinn *m*, **energy of** ~
Rotationsenergie *f*, **free** ~ freie Rotation *f*,
individual ~ Eigendrehung *f*, **molecular** ~
molekulares Drehungsvermögen, **moment of**
~ Drehmoment *n*, **optical** ~ optische
Drehung *f*, optisches Drehungsvermögen *n*,
radius of ~ Rotationsradius *m*, **reverse** ~
entgegengesetzte Drehung *f*, **sense of** ~
Rotationsrichtung *f*, **specific** ~ spezifische
Drehung *f*, **speed of** ~
Rotationsgeschwindigkeit *f*,
Umdrehungsgeschwindigkeit *f*,
Umlaufgeschwindigkeit *f*, **symmetry of** ~
Rotationssymmetrie *f*, **velocity of** ~
Rotationsgeschwindigkeit *f*, ~ **of axes**
Drehung *f* des Achsenkreuzes, ~ **of the**
vibration plane Rotation *f* der
Schwingungsfläche (Krist)
rotational axis Rotationsachse *f*
rotational band spectrum
Rotationsbandenspektrum *n*
rotational broadening Rotationsverbreiterung *f*
rotational constant Rotationskonstante *f*
(Spektr)
rotational distortion Rotationsverzerrung *f*
rotational flattening Rotationsabplattung *f*
rotational freedom Rotationsfreiheit *f*
rotational frequency Umlauffrequenz *f*
rotational isomers Rotationsisomere *pl*
rotational level Rotationsniveau *n*
rotational molding of plastics
Schleuderverfahren *n* (Kunststoff)

rotational momentum, characteristic ~
Eigendrehimpuls *m*
rotational quantum number
Rotationsquantenzahl *f*
rotational spectrum Rotationsspektrum *n*
rotational speed Drehgeschwindigkeit *f*,
Umlaufgeschwindigkeit *f*
rotational sterilisation Rotationssterilisation *f*
rotational symmetry Drehsymmetrie *f*
rotational viscosimeter Rotationsviskosimeter *n*
rotational wave Wirbelwelle *f*
rotation cooling Umlaufkühlung *f*
rotation ellipsoid Umdrehungsellipsoid *n*
rotation evaporator Rotationsverdampfer *m*
rotation hyperboloid Umdrehungshyperboloid *n*
rotation-inversion axis
Rotations-Inversionsachse *f* (Krist)
rotation molding Rotationspressen *n*
rotation of crops Fruchtfolge *f*
rotation paraboloid Umdrehungsparaboloid *n*
rotation polarization Rotationspolarisation *f*
rotation vibrational spectrum
Rotationsschwingungsspektrum *n*
rotator rotierender Apparat *m*, Wender *m*
rota[to]ry rotierend
rotatory dispersion curve
Rotationsdispersionskurve *f*
rotatory force Drehungskraft *f*
rotatory motion Rotationsbewegung *f*
rotatory power Drehungsvermögen *n*
rota[to]ry pump Rotationspumpe *f*,
Kreiselpumpe *f*
rotatory rocket Rotationsrakete *f*
rotenic acid Rotensäure *f*
rotenolol Rotenolol *n*
rotenolone Rotenolon *n*
rotenone Rotenon *n*
rotenonone Rotenonon *n*
roteol Roteol *n*
Rotex [eccentric roll] crusher Rotex-Brecher *m*
rothoffite Rothoffit *m* (Min)
rotogravure Rotationsdruck *m*
Roto-Louvre drier Roto-Louvre-Trockner *m*
rotor Rotor *m*, Laufrad *n*, **short circuit** ~
Käfiganker *m*, **squirrel cage** ~ Käfiganker *m*
rotor cage impeller Trommelkreiselrührer *m*
rotor current Rotorstrom *m*
rotor end plate Verschleißscheibe *f*
(Innenmischer)
rotor shaft Schaufelwelle *f*
rot-preventing fäulnisverhindernd
rotproofness Fäulnisbeständigkeit *f*
rotten anbrüchig, faul, morsch, schwammig
(Holz), verdorben, **to get** ~ verderben
rottenness Fäule *f*, Fäulnis *f*, Moder *m*,
Morschheit *f*
rotten odor Fäulnisgeruch *m*
rottenstone Bimsstein *m*, Tripel *m*
rotting faulend, Verrottung *f*

rotting vat Faulbütte *f*
rottisite Röttisit *m* (Min)
rottlerin Rottlerin *n*
rotundine Rotundin *n*
rough rauh, derb, grob, narbig, roh (Stein, Diamant), uneben, ungeschliffen (Stein, Diamant)
roughcast Berapp *m* (Bauw), Rauhputz *m* (Bauw), Rohguß *m* (Gieß)
roughcast *v* berappen
roughcast glass Rohglas *n*
rough drawing Rohentwurf *m*
roughen *v* anrauhen, aufrauhen
roughening Anrauhen *n*, Aufrauhen *n*
rough grinding mill Schrotmühle *f*
roughing Vordrehen *n*
roughing cutter Vorschneidfräser *m*
roughing filter Vorfilter *n*
roughing flotation Grobflotation *f*
roughing mill Vorwalzwerk *n*, Blockwalzwerk *n*, Vorstraße *f*
roughing pump line Grobpumpleitung *f*
roughing rolls Luppenwalzwerk *n*
roughing strand of rolls Vorwalzstrecke *f*
roughing tool Schruppstahl *m*
roughing train Luppenwalzwerk *n*
roughness Derbheit *f*, Rauheit *f*, Rauhtiefe *f*, Unebenheit *f*
roughness-height Rauhtiefe *f*
roughness number, maximum ~ Rauhtiefe *f*
roughness spectrum Rauhigkeitsspektrum *n*
rough ore from the mine Grubenerz *n*
rough plaster Berapp *m* (Bauw)
rough polish *v* vorschleifen
rough roll Stachelwalze *f*, Zackenwalze *f*
rough spelter Rohzink *n*
rough steel Schmelzstahl *m*
rough wood Rohling *m* (Holz)
round rund, kreisförmig
round *v*, ~ **off** abkanten, abrunden (Zahl), unrund werden, ~ **out** ausbeulen, ~ **up** aufrunden (Math)
round belt Rundriemen *m*
round-belt pulley Schnürrad *n*
rounded [off] abgerundet
round file Rundfeile *f*, Hohlfeile *f*
round filter Rundfilter *m*
round hole sieve or screen Rundlochsieb *n*
rounding Rundung *f*
rounding-off error Abrundungsfehler *m* (Math)
rounding tool Abrundwerkzeug *n*
round iron Rundeisen *n*, Stabeisen *n*, Stangeneisen *n*
round key Rundkeil *m*
round spirit level Dosenlibelle *f*
route coded zielgesteuert
routed zielgesteuert
routine allgemein gebräuchlich, routinemäßig
routine testing Reihenuntersuchung *f*

routing Ausfräsen *n*
routing machine Fräsmaschine *f*
roving Vorgarn *n*
row Reihe *f*
row matrix Zeilenmatrix *f*
row of tuyères Düsenreihe *f*
royal blue Königsblau *n*, Englischblau *n*
royalty Lizenz *f*, Lizenzgebühr *f*, **free of** ~ lizenzfrei
royalty basis, on a ~ lizenzweise
rub *v* reiben, scheuern, wischen, ~ **down** abschleifen, ~ **in** einreiben, ~ **off** abreiben, abfärben (Farbe), abscheuern (durch Abnutzung), ~ **out** auswischen, radieren (mit Gummi), ~ **with ointment** einsalben
ruban Ruban *n*
rubatoxanone Rubatoxanon *n*
rubazonic acid Rubazonsäure *f*
rubber Gummiring *m*, Gummi *n m*, Gummiband *n*, Kautschuk *m*, **cellular** ~ Moosgummi *m*, **chlorinated** ~ Chlorkautschuk *m*, **cold** ~ Tieftemperaturkautschuk *m*, **containing** ~ gummihaltig, **crude** ~ Rohgummi *m*, Rohkautschuk *m*, **expanded** ~ Moosgummi *m*, **hard** ~ Hartkautschuk *m*, **natural** ~ Naturkautschuk *m*, **raw** ~ Rohgummi *m*, Rohkautschuk *m*, **reclaimed** ~ Regenerat *n* (Gummi), **regenerated** ~ regeneriertes Gummi, **soft** ~ Weichgummi *m*, **synthetic** ~ Kunstgummi *m*, **unvulcanized** ~ Rohgummi *m*, Rohkautschuk *m*, **vulcanized** ~ vulkanisiertes Gummi, **weak** ~ Kautschuk *m* ohne Nerv
rubber adapter Gummiverbindung *f*
rubber adhesive Gummikleber *m*
rubber apron Gummischürze *f*
rubber bag molding Gummisackverfahren *n*
rubber ball Gummiball *m*
rubber band Gummiband *n*
rubber belt Gummiriemen *m*, Gurtband *n*
rubber blanket Gummituch *n*
rubber cake Kautschukkuchen *m*
rubber cap Gummikappe *f*, Gummipfropfen *m*
rubber cement Kautschukkitt *m*
rubber-coated gummiert (mit Gummi versehen)
rubber-coating Gummierung *f*
rubber colors Gummifarben *pl*
rubber cork Gummistopfen *m*
rubber crumb Altgummimehl *n*
rubber cushion Gummikissen *n*
rubber-cushioned gummigefedert
rubber diaphragm Gummimembran *f*
rubber diaphragm chamber Gummimembrankammer *f*
rubber expansion bag Orsat-Blase *f*
rubber extraction Kautschukgewinnung *f*
rubber fabric Kautschukgewebe *n*

rubber gasket Gummidichtung *f,*
Gummidichtungsring *m,* Gummimanschette *f*
rubber glove Gummihandschuh *m*
rubber hose Gummischlauch *m*
rubber insulated gummiisoliert
rubber insulation Gummiisolierung *f*
rubberize *v* gummieren
rubberizing machine Gummierkalander *m*
rubber joint Gummidichtung *f*
rubber kneader Gummikneter *m*
rubber latex Kautschuklatex *m,*
Rohgummimilch *f*
rubberlike gummiartig, kautschukähnlich
rubber-lined gummiert (mit Gummi versehen)
rubber mass Kautschukteig *m*
rubber-metal bond Gummimetallverbindung *f*
rubber milk Kautschukmilchsaft *m*
rubber oil Kautschuköl *n*
rubber packing Gummidichtung *f*
rubber pad Gummikissen *n,* Preßkissen *n,*
Reibkissen *n*
rubber paint Kautschukfarbe *f*
rubber part, molded ~ Gummiformteil *n*
rubber paving Gummipflaster *n*
rubber plate Gummiplatte *f*
rubber plug Gummipfropfen *m*
rubber poison Gummigift *n,* Kautschukgift *n*
rubber policeman Gummiwischer *m*
rubber resin Gummiharz *n*
rubber ring Gummimanschette *f*
rubber scrap Gummiabfall *m*
rubber sleeve Gummimanschette *f,*
Gummimuffe *f*
rubber solution Gummilösung *f,*
Kautschuklösung *f*
rubber spreader Streichmaschine *f* (Gummi)
rubber stopper Gummistöpsel *m,*
Gummistopfen *m*
rubber substitute Faktis *m* (HN),
Gummiersatz *m*
rubber suction ball Gummisauger *m*
rubber testing equipment Gummiprüfgerät *n*
rubber thread Gummifaden *m*
rubber tree Kautschukbaum *m,* Heveabaum *m*
rubber tube Gummirohr *n*
rubber tube connection Schlauchansatz *m,*
Schlauchverbindung *f*
rubber tubing Gummischlauch *m*
rubber varnish Gummifirnis *m,* Gummilack *m,*
Kautschuklack *m*
rubber washer Gummischeibe *f*
rubber waste Kautschukabfall *m*
rubbery gummiartig
rubbing Reibung *f,* **fast to** ~ reibecht, ~ **against**
each other Aneinanderreiben *n,* ~ **in**
Einreibung *f,* ~ **off** Abreibung *f*
rubbing action Reibungswirkung *f*
rubbing fastness Reibechtheit *f*
rubbing primer Schleifgrund *m*

rubbing surface Reibfläche *f*
rubbish Abfall *m,* Abraum *m,* Müll *m,* Schutt *m*
rubbish dump Abfallhaufen *m,* Schutthaufen *m*
rubble Geröll *n*
rubble pavement Kopfsteinpflaster *n*
rubeane Rubeanwasserstoff[säure]
rubeanic acid Rubeansäure *f,*
Rubeanwasserstoff[säure]
rubellite Rubellit *m* (Min)
ruberite Cuprit *m* (Min)
ruberythric acid Ruberythrinsäure *f*
ruberythrinic acid Ruberythrinsäure *f*
rub-fast wischfest
rubiadin Rubiadin *n*
rubianic acid Ruberythrinsäure *f*
rubicelle Rubicell *m* (Min)
rubicene Rubicen *n*
rubichloric acid Rubichlorsäure *f*
rubichrome Rubichrom *n*
rubidine Rubidin *n*
rubidium (Symb. Rb) Rubidium *n,* **containing** ~
rubidiumhaltig
rubidium alum Rubidiumalaun *m*
rubidium aluminum sulfate Rubidiumalaun *m*
rubidium bromide Rubidiumbromid *n*
rubidium chloroplatinate
Rubidiumplatinchlorid *n*
rubidium compound Rubidiumverbindung *f*
rubidium dioxide Rubidiumsuperoxid *n*
rubidium hydride Rubidiumhydrid *n*
rubidium iodate Rubidiumjodat *n*
rubidium iodide Rubidiumjodid *n*
rubidium peroxide Rubidiumsuperoxid *n*
rubidium strontium age
Rubidiumstrontiumalter *n*
rubidium sulfate Rubidiumsulfat *n*
rubiginic acid Rubiginsäure *f*
rubiginol Rubiginol *n*
rubixanthin Rubixanthin *n*
rubremetine Rubremetin *n*
rubrene Rubren *n*
rubreserine Rubreserin *n*
rubroskyrin Rubroskyrin *n*
ruby Rubin *m* (Min)
ruby arsenic Realgar *m* (Min)
ruby blende Rubinblende *f* (Min)
ruby[-colored] rubinfarben, rubinrot
ruby copper Cuprit *m* (Min), Rotkupfererz *n*
(Min)
ruby glass Rubinglas *n*
ruby mica Rubinglimmer *m* (Min)
ruby silver [dunkles] Rotgültigerz *n* (Min)
ruby spinel Rubinspinell *m* (Min)
ruby sulfur Realgar *m* (Min), rote
Arsenblende *f* (Min), Rubinschwefel *m*
(Min)
rudder Ruder *n,* Steuer *n*
rudder frame Ruderrahmen *m*
ruddle Rötel *m* (Min), Rötelerde *f* (Min)

rude rauh, schroff
rudeness Derbheit *f*
rue Raute *f* (Bot)
rue oil Rautenöl *n*
ruficoccin Ruficoccin *n*
rufigallic acid Rufigallussäure *f*
rufin Rufin *n*
rufiopin Rufiopin *n*
rufol Rufol *n*
rug underlay Teppichunterlage *f*
ruin Verderb *m*, Zerfall *m*
ruin *v* zerstören, verderben, zugrunde richten
rule Maßstab *m*, Lineal *n*, Norm *f*, Regel *f*,
  Richtscheit *n* (Techn), Verordnung *f*,
  Vorschrift *f*, **empirical** ~ empirische Regel,
  general ~ allgemeine Regel, ~ **of mutual**
  exclusion Regel *f* der gegenseitigen
  Ausschließung (Spektr), ~ **of three**
  Dreisatz *m* (Math)
rule *v* beherrschen, **according** ~ gesetzmäßig
rule of signs Zeichenregel *f* (Math)
ruler Abrichtlineal *n*, Richtscheit *n*
rum Rum *m*
rum distillery Rumbrennerei *f*
rumen Pansen *m*
Rumford's photometer Rumford-Photometer *n*,
  Schattenphotometer *n*
rumicin Rumicin *n*
ruminant Wiederkäuer *m* (Zool)
ruminate *v* wiederkäuen
rumpfite Rumpfit *m* (Min)
rumpled knitterig
run Förderstrecke *f*, Lauf *m*, ~ **of a pipe**
  Verlauf *m* einer Rohrleitung
run *v* laufen, fließen, ~ **a blind** (or blank test)
  einen Blindversuch durchführen, ~ **counter to**
  entgegenlaufen, ~ **down** ablaufen
  (Mechanismus), einrennen, ~ **hot** warmlaufen
  (Mot), ~ **idle** leer laufen, ~ **in** einlaufen,
  einströmen, ~ **off** abfließen, ablaufen, ~ **out**
  ablaufen (zeitlich), ausfließen, auslaufen, ~
  through durchlaufen
runaway Durchgehen *n* (Reaktor)
run-in Einlaufspur *f*
runner Abstichrinne *f* (Metall),
  Angußverteiler *m*, Einspritzkanal *m*,
  Gießkopf *m* (Gieß), Laufmasche *f*, Laufrad *n*
  (Techn), Laufrolle *f* (Techn), Steg *m*
  (Spritzguß), Zuflußkanal *m* (Gieß)
runnerless molding
  Vorkammerdurchspritzverfahren *n*
running Destillat *n* (Dest), Fließen *n*, Gang *m*
  (einer Maschine), ~ **off** Abrinnen *n*, ~ **of the**
  dye Verlaufen *n* der Farbe (Färb), ~ **without**
  load Leerlauf *m*
running board Trittbrett *n*
running glaze Laufglasur *f*
running properties Laufeigenschaften *pl*
running wave Wanderwelle *f*

run-of-mine ore Fördererz *n* (Min)
run oil Ablauföl *n*
runout, lateral ~ Axialschlag *m* (eines Rades),
  ~ **of a gear** Abnutzung *f* eines Getriebes
run-proof [lauf]maschenfest (Text)
run steel Flußstahl *m*
runway Landebahn *f* (Luftf), Startbahn *f* (Luftf)
rupture Bruch *m*, Riß *m*, Zerplatzen *n*,
  elongation at ~ Bruchdehnbarkeit *f*,
  Bruchdehnung *f*, **lasting** ~ Dauerbruch *m*,
  modulus of ~ Bruchmodul *m*,
  Zerreißmodul *m*, **moment of** ~
  Bruchmoment *n*
rupture *v* brechen, zerreißen
rupture disc Berstscheibe *f*, Brechplatte *f*,
  Reißscheibe *f*
rupture limit Bruchgrenze *f*
rupture strength Reißfestigkeit *f*
rupture stress Zerreißspannung *f*
rush Andrang *m*, ~ **of current** Stromstoß *m*
rush *v* hetzen
russet rötlichbraun
Russian leather Juchtenleder *n*
rust Rost *m*, **deposit of** ~ Rostansatz *m*,
  formation of ~ Rostansatz *m*, **free from** ~
  rostfrei, **layer of** ~ Rostschicht *f*, **tendency to**
  ~ Korrosionsneigung *f*, **to begin to** ~
  anrosten, ~ **from external sources**
  Fremdrost *m*
rust *v* [ver]rosten, einrosten, ~ **off** abrosten, ~
  through durchrosten
rust cement Eisenkitt *m*
rust-colored rostfarben
rust creep Unterrostung *f*
rusted eingerostet
rust formation Rostbildung *f*
rust-free rostfrei
rusting Verrosten *n*, Rosten *n*
rust inhibition Rostverhütung *f*
rust inhibitor Rostschutzmittel *n*
rustle *v* knistern (z. B. Seide), rascheln
rustless korrosionsfrei, rostfrei
rustless iron rostfreies Eisen *n*
rustling Knistern *n*, Rascheln *n*
rust mark Rostfleck *m*
rust-preventing rostverhütend
rust prevention Rostverhütung *f*
rust preventive Rostschutzmittel *n*
rust preventive paint Rostschutzanstrich *m*,
  Rostschutzfarbe *f*
rustproof nichtrostend, rostbeständig, rostfrei,
  rostsicher
rust-proofing Rostschutz *m*
rust-proofing grease Rostschutzfett *n*
rust-proofing process Rostschutzverfahren *n*
rustproof property Korrosionsbeständigkeit *f*
rust protection Korrosionsschutz *m*,
  Rostschutz *m*
rust removal Entrostung *f*, Rostentfernung *f*

rust remover Entrostungsmittel *n*,
  Rostentferner *m*
rust-removing agent Entrostungsmittel *n*,
  Rostentfernungsmittel *n*
rust-resistant rostbeständig
rust-resisting property Korrosionsbeständigkeit *f*
rust stain Rostfleck *m*
rusty rostig, verrostet, **to get** ~ rosten
rusty brown rostbraun
rutecarpine Rutaecarpin *n*
ruthenate Ruthenat *n*
ruthenic acid Rutheniumsäure *f,* **salt or ester of**
  ~ Ruthenat *n*
ruthenic oxide Ruthenium(IV)-oxid *n*
ruthenium (Symb. Ru) Ruthenium *n*
ruthenium compound Rutheniumverbindung *f*
ruthenium content Rutheniumgehalt *m*
ruthenium sulfide, native ~ Laurit *m* (Min)
rutherford (unit of radioactivity) Rutherford *n*
  (Radioaktivitätseinheit)
rutherfordine Rutherfordin *m* (Min)
rutherfordite Rutherfordit *m* (Min)
rutile Rutil *m* (Min)
rutin Rutin *n*
rutinic acid Rutinsäure *f*
rutinose Rutinose *f*
rutoside Rutosid *n*
rutylene Rutylen *n*
RVT rule Reaktionsgeschwindigkeit-Tempera-
  tur-Regel *f* (reaction velocity-time rule),
  RGT-Regel *f*
Rx only (Abk) rezeptpflichtig
Rydberg constant Rydbergkonstante *f*
rye Roggen *m* (Bot)
rye bread Roggenbrot *n*
rye flour Roggenmehl *n,* Schwarzmehl *n*
rye starch Roggenstärke *f*

S

sabadilla seed Sabadillsamen *m* (Bot)
sabadine Sabadin *n*
sabadinine Sabadinin *n*
saber flask Säbelkolben *m* (Chem)
sabinaketone Sabinaketon *n*
sabinane Sabinan *n*
sabinene Sabinen *n*
sabinenic acid Sabinensäure *f*
sabinic acid Sabininsäure *f*
sabinol Sabinol *n*
sabromin Sabromin *n*
sac Sack *m*, Beutel *m* (Bot, Zool)
saccharase Saccharase *f*, Invertase *f*, Invertin *n*
saccharate Saccharat *n*
saccharetin Saccharetin *n*
saccharic acid Zuckersäure *f*
saccharide Saccharid *n*
sacchariferous zuckerhaltig
saccharification Verzuckerung *f*, Zuckerbildung *f*
saccharify *v* verzuckern, zu Zucker werden
saccharimeter Saccharimeter *n*,
    Zucker[gehalts]messer *m*
saccharimetry Saccharimetrie *f*, Zuckermessung *f*
saccharin Saccharin *n*, o-Benzoesäuresulfimid *n*,
    Süßstoff *m*, **containing** ~ saccharinhaltig,
    sodium salt of ~ Kristallose *f*
saccharine zuckerartig, zuckerhaltig, zuckerig
saccharine fermentation Zuckergärung *f*
saccharine juice Zuckersaft *m*
saccharine tincture Zuckertinktur *f*
saccharinic acid Zuckersäure *f*
saccharinity Zuckerartigkeit *f*, Zuckerhaltigkeit *f*
saccharinol Saccharin *n*
saccharite Saccharit *m*
saccharize *v* in Zucker verwandeln
saccharobiose Sucrose *f*, Saccharobiose *f*
saccharoidal zuckerartig
saccharometabolism Zuckerstoffwechsel *m*
saccharometer Saccharometer *n*,
    Zuckermesser *m*
saccharomycete Hefepilz *m*, Saccharomyzet *m*
saccharonic acid Saccharonsäure *f*
saccharose Sucrose *f*, Rohrzucker *m*,
    Saccharobiose *f*, Saccharose *f*
saccharose phosphorylase
    Saccharosephosphorylase *f* (Biochem)
saccharum Zucker *m*
S-acid S-Säure *f* (Färb)
sack Sack *m*, **to put into a** ~ einsacken
sack *v* absacken, in Säcke füllen, einsacken
sack beating machine Sackausklopfmaschine *f*
sack conveyor Sackförderer *m*
sack filler Einsackmaschine *f*
sack filling machine Sackfüllmaschine *f*
sack-filling plant Absackanlage *f*
sacking Sackleinen *n*, Sackleinwand *f*

sacking machine Absackmaschine *f*
sacred bark Sagradarinde *f*
sacrificial anode Opferanode *f*
sadden *v* dunkel färben (Färb), nachdunkeln
saddle Auflageschuh *m* (Seilbahn), Sattel *m*
saddle bearing Schwengellager *n*
saddle scale Sattelwaage *f*
saddle-shaped coil Sattelschlange *f*
saddle spring wire Sattelfederdraht *m*
safe sicher, zulässig, ~ **to operate** betriebssicher
safeguard Schutz *m*, Sicherung *f*
safeguard *v* sichern
safelight Dunkelleuchte *f*
safety Sicherheit *f*, **factor of** ~
    Sicherheitsmaßnahme *f*, ~ **against explosion**
    accidents Explosionssicherheit *f*
safety bolt Sicherheitsriegel *m*
safety bottle Sicherheitsflasche *f*
safety cap Sicherheitskappe *f*
safety catch Fangvorrichtung *f*
safety circuit Sicherheitskreis *m*
safety cutout Schmelzsicherung *f*
safety device Schutzvorrichtung *f*,
    Sicherheitsvorrichtung *f*
safety diaphragm Sicherheitsmembran *f*,
    Berstscheibe *f*, Brechplatte *f*, Reißscheibe *f*
safety explosive Sicherheitssprengstoff *m*
safety factor Sicherheitsfaktor *m*
safety funnel Schutztrichter *m*
safety fuse Abschmelzsicherung *f*,
    Patentzündschnur *f*, Sicherheitszünder *m*
safety gas Schutzgas *n*
safety glass Sicherheitsglas *n*, Verbundglas *n*
safety hook Sicherheitshaken *m*
safety interlock Sicherheitsblockierung *f*
safety ladder Sicherheitsleiter *f*
safety lamp Wetterlampe *f* (Bergb);
    Schlagwetteranzeiger *m* (Bergb),
    Sicherheitslampe *f*
safety lock Sicherheitsverschluß *m*
safety measure Schutzmaßnahme *f*
safety net Schutznetz *n*
safety pipe Sicherheitsrohr *n*
safety provisions Arbeitsschutz *m*
safety regulations Schutzvorschriften *f pl*
safety regulator Sicherheitsregler *m*
safety rule Sicherheitsvorschrift *f*
safety screw Sicherungsschraube *f*
safety shut-down Sicherheitsabschaltung *f*
safety slotting Feinprofilierung *f* (Reifen)
safety spring Federsicherung *f*
safety switch Sicherheitsschalter *m*,
    Kontrollschalter *m*, Schutzschalter *m*
safety tube Sicherheitsröhre *f*, Sicherheitsrohr *n*
safety valve Sicherheitsventil *n*,
    Rückschlagventil *n*, Sicherheitsklappe *f*,

Überdruckventil *n*, **high lift** ~
Hochhubsicherheitsventil *n*
safety valve weight Sicherheitsventilbelastung *f*
safety zone Sicherheitsabstand *m*,
Sicherheitszone *f*
saffian Saffian *n*
safflorite Safflorit *m* (Min), Arsenkobalt *n*
(Min)
safflor red Carthamin *n*
safflower Saflor *m* (Bot), Färberdistel *f* (Bot)
safflower oil Safloröl *n*
safflower red Saflorrot *n* (Färb)
saffron Safran *m* (Bot), **containing** ~
safranhaltig
saffron color Safrangelb *n*
saffron-colored safranfarben, safrangelb
saffron-like safranähnlich
saffron oil Safranöl *n*
saffron substitute Safransurrogat *n*
saffrony safrangelb; safranähnlich
safranal Safranal *n*
safranine Safranin *n*
safraninol Safraninol *n*
safranol Safranol *n*
safrole Safrol *n*
safrosin Safrosin *n*
sag *v* absacken, durchbiegen, durchhängen
sagapenum Sagapengummi *n*
sage Salbei *m* (Bot)
sagenite Sagenit *m* (Min)
sage oil Salbeiöl *n*
sagger Brennkapsel *f* (Keram)
sagger clay Kapselton *m* (Keram)
sago Sago *m;* Palmmehl *n*
saiodin Sajodin *n*
sakebiose Sakebiose *f*
salability Verkäuflichkeit *f*
salable verkäuflich
salacetamide Salacetamid *n*
salacetol Salacetol *n*
salad oil Salatöl *n*, Tafelöl *n*
salamander Eisensau *f* (Metall)
salamide Salicylamid *n*
sal ammoniac Salmiak *m*, Ammoniumchlorid *n*,
flowers of ~ Salmiakblumen *pl*, **solution of** ~
Ammoniumchloridlösung *f*, ~ **for soldering**
Lötasche *f*
sal ammoniac cell Salmiakelement *n*
salantol Salacetol *n*
salary Gehalt *n*
salazinic acid Salazinsäure *f*
salazosulfamide Salazosulfamid *n*
salcomine Salcomin *n*
sale Verkauf *m*
salep Salep *m*
sales department Verkaufsabteilung *f*
salesman Verkäufer *m*, Reisender *m*
sales number Verkaufsnummer *f*
salicil Salicil *n*

salicin Salicin *n*, Saligenin *n*
salicyl Salicyl-
salicylacetic acid Salicylessigsäure *f*
salicylal Salicylaldehyd *m*
salicylalcohol Salicylalkohol *m*, Saligenin *n*
salicylaldehyde Salicylaldehyd *m*
salicylamide Salicylamid *n*
salicylanilide Salicylanilid *n*
salicylate Salicylat *n*, Salz oder Ester der
Salicylsäure *f*
salicylate *v* mit Salicylsäure behandeln
salicylated talc Salicylstreupulver *n*
salicylated tallow Salicyltalg *m*
salicylic acid Salicylsäure *f*, **salt or ester of** ~
Salicylat *n*
salicylic aldehyde Salicylaldehyd *m*
salicylic cotton Salicylwatte *f*
salicylic preparation Salicylpräparat *n*
salicylic soap Salicylsäureseife *f*
salicylide Salicylid *n*
salicylosalicylic acid Diplosal *n*, Disalicylsäure *f*
salicylquinine Salochinin *n*
salicyl yellow Salicylgelb *n*
salifebrin Salicylanilid *n*
saliferous salzbildend; salzhaltig
saliferous clay Salzton *m*
salifiable salzbildend
salification Salzbildung *f*
saliformine Saliformin *n*
saligenin Salicin *n*, Saligenin *n*
salimenthol Salimenthol *n*, Samol *n*
salimeter Salzgehaltmesser *m*, Salzmesser *m*,
Solwaage *f*
salinaphthol Salinaphthol *n*, Naphthalol *n*
salinazid Salinazid *n*
saline Saline *f*
saline *a* salzhaltig, salzig
saline deposits Abraumsalze *pl*
saline flux Salzfluß *m*
saline infusion Kochsalzinfusion *f* (Med)
saline manure Düngesalz *n*
salineness Salzhaltigkeit *f*, Salzigkeit *f*
saline particle Salzteilchen *n*
saline soil Salzboden *m*
saline solution Salzlösung *f*
saline water Salinenwasser *n*
saliniform salzartig
salinigrin Salinigrin *n*
salinity Salzgehalt *m*, Salzhaltigkeit *f*,
Salzigkeit *f*
salinity diagram Salzgehaltdiagramm *n*
salinometer Lakeprober *m*, Salinometer *n*,
Salz[gehalt]messer *m*, Salzwaage *f*, Solwaage *f*
salinometry Salzmessung *f*
salipurpol Salipurpol *n*
salipyrine Salipyrin *n*
saliretin Saliretin *n*
saliva Speichel *m*
salivary gland Speicheldrüse *f* (Med)

salivin Ptyalin *n* (Biochem)
sallow fahl[gelb]
salmiac Ammoniumchlorid *n*, Salmiak *m n*
salmine Salmin *n*
salmite Salmit *m*
salmon Lachs *m* (Zool)
salmon-colored lachsfarben
salmonella bacilli Salmonellen *pl* (Bakt)
salmonellae Salmonellen *pl* (Bakt)
salmonellas Salmonellen *pl* (Bakt)
salmonellosis Salmonellenerkrankung *f* (Med),
    Salmonellose *f* (Med)
salmon oil Lachsöl *n*
salocoll Phenokollsalicylat *n*
salophen Salophen *n*
saloquinine Salochinin *n*
salosalicylide Disalicylid *n*, Salosalicylid *n*
sal prunella Prunellensalz *n*
salpyrin Salipyrin *n*
salseparin Smilacin *n*
salsolidine Salsolidin *n*
salsoline Salsolin *n*
salt Salz *n*, alkaline ~ alkalisches Salz,
    crystalline ~ kristallisiertes Salz, inner ~
    inneres Salz, layer of ~ Salzschicht *f*, rich in
    ~ salzreich, to add more ~ nachsalzen,
    undissociated ~ undissoziiertes Salz, ~ of a
    heavy metal Schwermetallsalz *n*, ~ of carbonic
    acid Karbonat *n*, ~ of sulfuric acid Sulfat *n*,
    ~ of tartaric acid Tartrat *n*
salt *v* salzen, einsalzen, ~ lightly ansalzen, ~
    out aussalzen
salt addition Salzzusatz *m*
salt ash Salzasche *f*
salt bath Salzbad *n*
salt bath casehardening Salzbadeinsatzhärtung *f*
salt bath pot Salzbadtiegel *m*
salt-bearing salzhaltig
salt bed Salzlager *n*
salt boiler Salzsieder *m*
salt bridge Salzbrücke *f*
salt brine Salzlake *f*, Salzlauge *f*
salt bucket Salzkübel *m*
salt cake technisches Natriumsulfat *n*
salt clay Salzton *m*
salt contents Salzgehalt *m*
salt crust Salzhaut *f*, Salzkruste *f*, Salzrinde *f*
salt deposit Salzniederschlag *m*,
    Salzablagerung *f*, Salzlager *n*
salt dressing Salzbeize *f*
salt drying oven Salztrockenofen *m*
salt earth Salzerde *f*
salted eingepökelt, gesalzen
salt effect Salzeffekt *m*
salt equilibrium Salzhaushalt *m* (Physiol)
saltern Saline *f*
salt film Salzhaut *f*
salt flux Salzfluß *m* (Gieß)

salt formation Salzbildung *f*, internal ~ innere
    Salzbildung
salt former Salzbildner *m*
salt-forming salzbildend
salt-forming substance Salzbildner *m*
salt-free salzfrei
salt glaze Salzglasur *f*
saltiness Salzigkeit *f*
salting Einsalzen *n*, Salzzuschlag *m*, ~ out
    Aussalzen *n*
salting out Aussalzen *n* (Seifenherstellg)
salting out effect Aussalzeffekt *m*
salt layer Salzlager *n*, Salzschicht *f*
salt-like salzähnlich, salzartig
salt making Salzbereitung *f*
salt meat Pökelfleisch *n*
salt melt Salzschmelze *f*
salt mine Salzbergwerk *n*
salt pair Salzpaar *n*
salt pairs, reciprocal ~ reziproke Salzpaare *pl*
salt pan scale Salzpfannenstein *m*
saltpeter Salpeter *m*, Kalisalpeter *m*,
    Kaliumnitrat *n*, containing ~ salpeterhaltig,
    crude ~ Rohsalpeter *m*, cubic ~
    Würfelsalpeter *m*
saltpeter boiler Salpeterkessel *m*
saltpeter earth Salpetererde *f*, native ~
    Gayerde *f* (Min)
saltpeter flowers Salpeterblumen *pl*
saltpeter solution Salpeterlösung *f*
saltpeter sweepings Fegesalpeter *m*,
    Kehrsalpeter *m*
saltpeter test Salpeterprobe *f*
saltpeter works Salpetersiederei *f*
saltpetre (Br. E.) Salpeter *m*, Chilesalpeter *m*
salt pit Salzgrube *f*
salt production Salzgewinnung *f*
salt refuse Abfallsalz *n*
salt requirement Salzbedarf *m* (Physiol)
salt residue Salzrückstand *m*
salt sample Salzprobe *f*
salt solution Salzlösung *f*, physiological ~
    physiologische Kochsalzlösung *f*
salt-spray test Salz-Sprüh-Test *m* (Korrosion)
salt spring Salzquelle *f*, Sole *f*
salt sweepings Kehrsalz *n*
salt tank Salzbehälter *m*
salt test Salzprobe *f*
salt vein Salzader *f*
salt water Salzwasser *n*, Sole *f*
saltwater bath Salzwasserbad *n*
saltwater elevator Solheber *m*
saltwater-proof seewasserbeständig
saltwater refrigeration Salzwasserkühlung *f*
salt well Salzquelle *f*
salt works Saline *f*, Salzsiederei *f*
salty salzig
salty taste Salzgeschmack *m*
salumin Salumin *n*

salvage Rettung *f*, Bergung *f*,
 Wiedergewinnung *f* (Techn)
salvage *v* verwerten, rückgewinnen
salvarsan Arsphenamin *n*, Salvarsan *n* (HN)
salve Salbe *f*
salve for chilblains Frostsalbe *f* (Pharm)
salve spatula Salbenspatel *m*
salvia Salbei *m* (Bot)
salvy salbenartig
samandarine Samandarin *n*
samaric chloride Samarium(III)-chlorid *n*
samaric compound Samarium(III)-Verbindung *f*
samarium (Symb. Sm) Samarium *n*
samarium carbide Samariumcarbid *n*
samarium chloride Samariumchlorid *n*
samarium oxide Samariumoxid *n*
samarium sulfate Samariumsulfat *n*
samarium trichloride Samarium(III)-chlorid *n*
samarous compound Samarium(II)-Verbindung *f*
samarskite Samarskit *m* (Min)
samatose Samatose *f*
Sambesi black Sambesischwarz *n*
sambunigrin Sambunigrin *n*
samin Samin *n*
samiresite Samiresit *m* (Min)
samol Salimenthol *n*, Samol *n*
sample Muster *n*, Probe *f*, Probestück *n*,
 Prüfling *m* (Objekt), ~ at random
 Stichprobe *f*
sample *v* eine Probe nehmen, prüfen
sample application Anwendungsbeispiel *n*
sample bottle Probeflasche *f*
sample box Probeschachtel *f*
sample [material] Probegut *n*
sample preparation Probenvorbereitung *f*
sampler Auffangbehälter *m*, Probenehmer *m*
samples, to take ~ Proben pl entnehmen
sample splitting Probenteilung *f*
sampling Probeentnahme *f*, Probeneingabe *f*
 (Gaschromat), Probeziehen *n*
sampling equipment Probenehmer *m*
sampling length Bezugslänge *f*
sampling material Probegut *n*
sampling point Entnahmestelle *f*
sampling probe Entnahmesonde *f*
samsonite Samsonit *m* (Min)
sand Sand *m*, argillaceous ~ Tonsand *m*,
 auriferous ~ Goldsand *m*, coarse ~ Grand *m*,
 fat ~ fetter Formsand *m*, gravelly ~
 Kiessand *m*, layer of ~ Sandschicht *f*, loamy
 ~ fetter Formsand *m*, micaceous ~
 Glimmersand *m*, ~ containing little clay
 magerer Sand
sand *v* mit Sand scheuern, polieren oder
 schleifen, schleifen, schmirgeln
sandalwood Sandelholz *n*
sandalwood oil Sandelholzöl *n*
sandarac[h] resin Sandarakharz *n*
sand bath Sandbad *n*

sand-bearing sandführend
sandblast Sandstrahl *m*, Sandstrahlgebläse *n*
sandblast *v* sandstrahlen
sandblast apparatus Sandstrahlgebläse *n*
sandblasting Sandstrahlen *n*
sandblast nozzle Sandstrahldüse *f*
sandblast plant Sandstrahlanlage *f*
sandblast protective equipment
 Sandstrahlschutzgerät *n*
sandblast unit Sandstrahlgebläse *n*
sand casting Sandguß *m*
sand cement Sandzement *m*
sand cleaning Altsandaufbereitung *f* (Gießerei)
sand-colored sandfarben
sand discharge pipe Sandauslaufrohr *n*
sand filter Sandfilter *n*
sand filtration Sandfiltration *f*
sandglass Sanduhr *f*
sand grinding stone Sandschleifstein *m*
sanding Sandstrahlen *n*
sanding sealer Autospachtel *f*, Schleifgrund *m*
sandiver Glasgalle *f*, Glasschaum *m*
sandlike sandartig
sand-lime brick Kalksandstein *m* (Bauw)
Sandmeyer['s diazo-] reaction
 Sandmeyer-Reaktion *f* (Chem)
sand mold Sandform *f*
sand molder Former *m*
sand nozzle Sanddüse *f*
sand packing Sandverdichtung *f*
sandpaper Sandpapier *n*, Schmirgelpapier *n*
sandpaper *v* abschmirgeln, abschleifen
sandstone Sandstein *m*, argillaceous ~
 Tonsandstein *m*, kaoliniferous ~
 Kaolinsandstein *m*, resembling ~
 sandsteinartig, siliceous ~ Kieselsandstein *m*,
 variegated ~ bunter Sandstein
sandstone [in blocks] Quadersandstein *m*
sandstone stratum Sandsteinlager *n* (Geol)
sand stratum Sandschicht *f*
sandwich construction Verbundbauweise *f*
sandwich panel Verbundplatte *f*
sandy sandartig, sandig
sandy marl Sandmergel *m*
sandy shale Sandschiefer *m* (Min)
sandy soil Sandboden *m*
sanfor finish Sanforausrüstung *f* (Text)
sanforizing [process] Sanforisieren *n* (Text)
sanguinarine Sanguinarin *n*
sanicle [leaves] Sanikel *m* (Bot)
sanidine Sanidin *m* (Min), Eisspat *m* (Min)
sanitary gesundheitlich, hygienisch, sanitär
sanitary engineering Gesundheitstechnik *f*
sanitary science Hygiene *f*
sanitation Gesundheitspflege *f*
sanitizer Desinfektionsmittel *n*
sanoform Sanoform *n*
sanshoamide Sanshoamid *n*
sanshool Sanshool *n*

santalal Santalal *n*
santalbic acid Santalbinsäure *f*
santalene Santalen *n*
santalic acid Santalsäure *f*
santalin Santalin *n*, Sandelrot *n*
santal oil Sandelholzöl *n*
santalol Santalol *n*
santene Santen *n*
santene hydrate Santenhydrat *n*
santenic acid Santensäure *f*
santenol Santenol *n*
santenonalcohol Santenonalkohol *m*
santonica Wurmkraut *n* (Bot)
santonic acid Santonsäure *f*
santonin Santonin *n*
santoninic acid Santoninsäure *f*
santorin earth Santorinerde *f*
santowax Santowax *n*
sap Saft *m* (Pflanzen), rich in ~ saftreich
sapanwood Sap[p]anholz *n*
sap content Saftgehalt *m*
sapimine Sapimin *n*
sapogenine Sapogenin *n*
saponaceous seifenhaltig, seifig
saponaceous clay Seifenton *m*
saponarin Saponarin *n*
saponifiability Verseifbarkeit *f*
saponifiable verseifbar
saponification Verseifung *f* (Chem), ~ of esters
   Esterverseifung *f*, ~ of fat[s] Fettverseifung *f*
saponification flask Verseifungskolben *m*
   (Chem)
saponification number Verseifungszahl *f* (Chem)
saponification value Verseifungszahl *f* (Chem)
saponify *v* verseifen (Chem)
saponin Saponin *n*, Seifenstoff *m*
saponite Saponit *m* (Min)
saporval Saporval *n*
saposalicylic ointment Saposalicylsalbe *f*
sapotalene Sapotalin *n*
sapovaseline Sapovaseline *f*
sapphire Saphir *m*, artificial green ~ Amaryl *m*
   (Min), white cloudy ~ Milchsaphir *m* (Min)
sapphire quartz Sapphirquarz *m* (Min)
sapphire whiskers
   Aluminiumoxid-Einkristallfasern *pl*
sapphirine Saphirin *m* (Min)
sapphirine *a* saphirähnlich, saphirblau
sappiness Saftigkeit *f*
sappy saftig
sapropel Faulschlamm *m* (Geol), Sapropel *n*
sapropelic coal Sapropelkohle *f*
saprophytes Saprophyten *pl*
saprozoic organisms pl
sap wood Splintholz *n*
saran Saran *n*
sarcine Sarkin *n*
sarcocol Sarkokol *n*
sarcocollin Sarkokollin *n*

sarcolactic acid Fleischmilchsäure *f* (Biochem)
sarcolite Sarkolith *m* (Min)
sarcolysine Sarkolysin *n*
sarcoma (pl sarcomas or sarcomata) Sarkom *n*
   (Med)
sarcoside Sarkosid *n*
sarcosine Sarkosin *n*
sarcosine acid Sarkosinsäure *f*
sarcosine anhydride Sarkosinanhydrid *n*
sarcosine nitrile Sarkosinnitril *n*
sarcosporidia Sarkosporidien *pl*
sardonyx Sardonyx *m* (Min)
sargasterol Sargasterin *n*
sarin Sarin *n*
sarkine Hypoxanthin *n*, Sarkin *n*
sarkinite Sarkinit *m* (Min)
sarkomycin Sarkomycin *n*
sarmentogenin Sarmentogenin *n*
sarmentose Sarmentose *f*
sarracinic acid Sarracinsäure *f*
sarsaparilla Sarsaparille *f* (Bot),
   Sarsaparillwurzel *f* (Pharm)
sarsasaponin Sarsasaponin *n*
sartorite Sartorit *m* (Min)
sash Schieberahmen *m*
sash window Schiebefenster *n*
sassoline Sassolin *m* (Min)
sassolite Sassolin *m* (Min)
satellite Satellit *m* (Astr), Trabant *m* (Astr)
satellite line Begleitlinie *f*
satellite pulse Nebenimpuls *m*
satin Satin *m* (Text)
satin paper Atlaspapier *n*
satin spar Atlasspat *m* (Min), Atlasstein *m*
   (Min)
satin weave Satingewebe *n* (Text)
satiny atlasglänzend, seidig
satisfaction Befriedigung *f*
satisfactory befriedigend, ausreichend
satisfy *v* abhelfen, befriedigen, ~ an equation
   einer Gleichung entsprechen
sativic acid Sativinsäure *f*
saturability Sättigungsvermögen *n*
saturant Imprägniermittel *n*, Sättigungsmittel *n*
saturate *v* sättigen (Chem), durchsetzen,
   [durch]tränken, saturieren (Chem)
saturated gesättigt (Chem), highly ~
   hochgesättigt (Chem), ~ in the cold state kalt
   gesättigt
saturated activity Sättigungsaktivität *f* (Atom)
saturated aliphatic acid Paraffinsäure *f*
saturated steam Sattdampf *m*
saturated vapor Sattdampf *m*
saturated with water wassergesättigt
saturating agent Sättigungsmittel *n*
saturating apparatus Sättigungsapparat *m*
saturating forces Sättigungskräfte *pl*
saturation Sättigung *f*, Durchdringung *f*,
   Durchsetzung *f*, Durchtränkung *f*,

Karbonisation *f* (Zucker), Saturation *f* (Zucker), **degree of** ∼ Sättigungsgrad *m*, **limit of** ∼ Sättigungsgrenze *f*, **state of** ∼ Sättigungszustand *m*, ∼ **of air** Luftsättigung *f*
saturation capacity Sättigungskapazität *f*
saturation current Sättigungsstromstärke *f*
saturation curve Sättigungskurve *f*
saturation factor Sättigungskoeffizient *m*
saturation isomerism Sättigungsisomerie *f*
saturation limit Sättigungsgrenze *f*
saturation magnetization Sättigungsmagnetisierung *f*
saturation point Sättigungsgrenze *f*, Sättigungspunkt *m*
saturation pressure Sättigungsdruck *m*
saturation tank Saturateur *m* (Zucker)
saturation temperature Sättigungstemperatur *f*
saturation value Sättigungswert *m*
saturation vessel Saturationsgefäß *n*
saturator Sättigungsapparat *m*, Saturateur *m* (Zucker), Saturationsgefäß *n* (Zucker)
sauce tobacco *v* Tabak beizen
saucing Beizen *n* (Tabak)
sausage poisoning Wurstvergiftung *f*
sausage skin Wursthaut *f*
saussurite Saussurit *m* (Min)
save *v* sparen, retten
save-all Rückgewinnungsanlage *f*, Stoffänger *m* (Pap), **stuff from the** ∼ Fangstoff *m*
save-oil Ölsammler *m*
savin Sadebaum *m* (Bot)
saving Einsparung *f*
saving in material Materialersparnis *f*
savinin Savinin *n*
savin oil Sadebaumöl *n*
savory Bohnenkraut *n* (Bot)
savory *a* wohlschmeckend
saw Säge *f*, **cross-cut circular** ∼ Querkreissäge *f*, **dimensional** ∼ Formatkreissäge *f*, **multiple circular** ∼ Mehrfachkreissäge *f*
saw blade Sägeblatt *n*
saw bow Sägebogen *m*
sawdust Sägemehl *n*, Sägespäne *pl*
saw frame Sägebogen *m*, Sägegatter *n*
sawhorse representation Sägebock-Formel *f* (Stereochem)
saw veneer Sägefurnier *n*
sawwort Färberscharte *f* (Bot)
Saxon blue Neublau *n*, Sächsischblau *n*
Saybolt second Sayboltsekunde *f*
scacchite Scacchit *m* (Min)
scaffold Gerüst *n*, Hängegerüst *n*
scaffolding of the charge Hängen *n* der Gicht
scaffold pole Rüstbaum *m*, Rüststange *f*, Rundholz *n*
scalar field Skalarfeld *n*
scalar product Skalarprodukt *n* (zweier Vektoren)
scald Brand *m*, Verbrühung *f*

scald *v* verbrühen, [aus]brühen, verbrennen
scalding kettle Abbrühkessel *m*
scale Waagschale *f*, Gradeinteilung *f*, Kesselstein *m* (Techn), Kruste *f* (Techn); Maßstab *m*, Schuppe *f* (Biol), Skala *f*, Zunder *m* (Techn), **at an enlarged** ∼ in vergrößertem Maßstab, **at a reduced** ∼ in verkleinertem Maßstab, **deposit of** ∼ Kesselsteinablagerung *f*, **double graduated** ∼ doppelseitig kalibrierte Skala, **graduation of** ∼ Skaleneinteilung *f*, **in** ∼ maßstabgerecht, **layer of** ∼ Kesselsteinschicht *f*, **rough** ∼ Grobteilung *f*, **true to** ∼ maßstabgerecht, ∼ **of hardness** Härteskala *f*, ∼ **of reduction** Verkleinerungsmaßstab *m*
scale *v* schuppen, ∼ **off** abblättern (Med), abbröckeln, abklopfen (z. B. Kesselstein), ablösen, abschälen, abschuppen, abzundern
scale division Skalenintervall *m*, Skalenteilung *f*
scale formation Krustenbildung *f*, Zunderbildung *f* (Met), **prevention of** ∼ Kesselsteinverhütung *f*
scale housing Waagengehäuse *n*
scale interval Skalenintervall *m*
scale microscope Skalenmikroskop *n*
scalenohedral skalenoedrisch (Krist)
scalenohedron Skalenoeder *n* (Krist)
scalepan Waagschale *f*
scaler Zähler *m* (Elektron)
scale range Meßbereich *m*, Regelbereich *m*
scale reading Skalenablesung *f*
scale remover Kesselsteinbeseitigungsmittel *n*
scales Waage *f*, **adjusting** ∼ Ausgleichswaage *f*
scale solvent Kesselsteinlösungsmittel *n*
scale-up Maßstabsvergrößerung *f*, Vergrößerung *f*
scaling Abblättern *n*, Schuppenbildung *f*, **high temperature** ∼ Verzunderung *f*, **resistance to** ∼ Zunderbeständigkeit *f*
scaling furnace Beizofen *m*
scaling loss Abbrand *m*
scaling up Modellübertragung *f*
scalpel Skalpell *n*
scalping (screening to remove foreign matter) Skalping *n*
scaly blätterig, schuppig
scaly hematite Eisenmann *m* (Min)
scammonin Skammonin *n*
scammony Skammonienharz *n*
scan *v* abtasten
scan area Abtastfläche *f*
scandium (Symb. Sc) Scandium *n*
scandium oxide Scandiumoxid *n*
scandium sulfate Scandiumsulfat *n*
scanner Abtaster *m*, Abtastgerät *n*, Abtastvorrichtung *f*, Detektor *m*
scanning Abtasten *n*, Abtastung *f*
scanning aperture Abtastöffnung *f* (Telev)
scanning beam Abtaststrahl *m*, Taststrahl *m*

scanning belt Abtastband *n*
scanning device Abtastgerät *n*, Bildabtaster *m* (Telev)
scanning disk Abtastscheibe *f* (Telev)
scanning electron microscope Rasterelektronenmikroskop *n*
scanning field Abtastfeld *n*
scanning frequency Abtastfrequenz *f*
scanning generator Kippgenerator *m*
scanning light spot abtastender Lichtfleck *m*
scanning line Abtastzeile *f*
scanning linearity Abtastlinearität *f* (Telev)
scanning microscope Rastermikroskop *n*, Abtastmikroskop *n*, Meßmikroskop *n*
scanning screen Rasterschirm *m*
scanning speed Abtastgeschwindigkeit *f*
scanning voltage Abtastspannung *f* (Telev)
scantling Sparren *m*
scanty knapp, dürftig
scapolite Skapolith *m* (Min), Wernerit *m* (Min)
scar Narbe *f*, Schramme *f*
scar *v* ritzen, schrammen
scarce knapp, kärglich
scarcity Knappheit *f*
scarf *v* ausschärfen (Techn)
scarf joint Falzverbindung *f*, Schaftung *f*
scarf milling machine Ausschärffräsmaschine *f*
scarlet Scharlach *m*, Scharlachrot *n*, **deep ~** Kardinalrot *n*
scarlet *a* scharlachfarben
scarlet dyeing Scharlachfärben *n*
scarlet fever Scharlach *m* (Med)
scarred narbig
scarring Narbenbildung *f*
scatol Skatol *n*
scatter *v* [zer]streuen, ausbreiten, verstreuen
scattered, back ~ zurückgestreut
scattered radiation Streustrahlung *f*, **protection against** ~ Streustrahlenschutz *m*
scattered rays Streustrahlen *pl*
scatter electron Streuelektron *n*
scatter fading Streuschwund *m*
scattering Streuung *f*, Zerstreuung *f*, **acoustic ~** akustische Streuung (Akust), **effect of ~** Streueffekt *m*, **inelastic ~** unelastische Streuung, **large-angle ~** Streuung *f* im weiten Winkel
scattering amplitude Streuamplitude *f*
scattering angle Streuwinkel *m*
scattering center Streuzentrum *n*
scattering coefficient Streukoeffizient *m*
scattering continuum Streukontinuum *n*
scattering cross section Streuungsquerschnitt *m*
scattering factor Streufaktor *m*
scattering matrix Streumatrix *f*
scattering power Streuvermögen *n*, Zerstreuungsvermögen *n*
scatter range Streubereich *m*
scavenge *v* spülen

scavenging Spülung *f*
scavenging valve Spülventil *n*
sceleranecic acid Sceleranecinsäure *f*
sceleratine Sceleratin *n*
sceletal muscle Skelettmuskel *m*
scent Duft *m*, Geruch *m*
scent *v* aromatisieren, parfümieren
scented wohlriechend, duftend
scent gland Duftdrüse *f* (Zool)
scenting Aromatisierung *f*
scentless duftlos, geruchlos
schapbachite Schapbachit *m* (Min)
schappe silk Schappeseide *f*
schappe spinning Schappespinnerei *f*
Schardinger enzyme Xanthinoxydase *f* (Biochem)
schedule Aufstellung *f*, Liste *f*, Plan *m*, Tabelle *f*, Verzeichnis *n*
Scheele's green Scheelesches Grün *n*
scheelite Scheelit *m* (Min), Scheelerz *n* (Min), Scheelspat *m* (Min), Schwerstein *m* (Min)
schefferite Schefferit *m* (Min)
schematic schematisch
scheme Plan *m*, Programm *n*, Projekt *n*, Schema *n*, System *n*
scheme *v* entwerfen, projektieren
Schiff's base Schiffsche Base *f*
Schiff's solution fuchsinschwefelige Säure *f*
schiller spar Schillerspat *m* (Min)
schist Schiefer *m* (Geol), **argillaceous ~** Tonschiefer *m* (Min)
schistosome Bilharzia *f* (Zool, Med)
schistosomiasis Bilharziose *f* (Med)
schistous schieferartig, schieferhaltig, schiefrig
schistous amphibolite Hornblendeschiefer *m* (Min)
schistous basalt Basaltschiefer *m*
schistous diorite Grünsteinschiefer *m* (Min)
schistous sandstone Sandschiefer *m* (Min)
schizomycetes Schizomyzeten *pl*, Spaltpilze *pl*
schizomycetic fermentation Spaltpilzgärung *f*
schizophyceae Spaltalgen *pl*
schlich Schlamm *m*, Schlick *m*
schlieren Schlieren *pl*
schlieren diaphragm Schlierenblende *f*
schlieren method Schlierenverfahren *n*
schlieren optics Schlierenoptik *f* (Opt)
schlieren photograph Schlierenaufnahme *f* (Phot)
schlieren photography Schlierenphotographie *f*
Schlippe's salt Schlippesches Salz *n*
schneebergite Schneebergit *m* (Min)
Schoellkopf's acid Schöllkopfsche Säure *f*
schoenite Schönit *m* (Min)
Schoop's metal spraying process Schoopsches Metallspritzverfahren *n*
schorl Schörl *m* (Min)
schorlaceous schörlähnlich
schorlite Schörlit *m* (Min)

schorlomite Schorlomit *m* (Min)
Schotten-Baumann reaction
Schotten-Baumannsche Reaktion *f*
Schottky effect Schottky-Effekt *m*
schreibersite Schreibersit *m*,
Phosphornickeleisen *n* (Min)
schroeckingerite Schröckingerit *m* (Min)
Schrödinger oscillation equation Schrödingersche
Schwingungsgleichung *f*
Schumann region Schumanngebiet *n* (Spektr)
schungite Schungit *m* (Min)
schwazite Schwazit *m* (Min)
Schweinfurt green Schweinfurter Grün *n*
science Wissenschaft *f*, ~ **of mineral deposits**
Lagerstättenkunde *f*, ~ **of nutrition**
Ernährungskunde *f*
Science Research Council (Br. E.)
Forschungsrat *m*
scientific [natur]wissenschaftlich
scientific glassware Laborgeräte *n pl* aus Glas
scientific literature Fachliteratur *f*
scientific research Naturforschung *f*
scientist Wissenschaftler *m*, Naturforscher *m*
scillabiose Scillabiose *f*
scillain Scillain *n*
scillaren Scillaren *n*
scillarenin Scillarenin *n*
scillaridin Scillaridin *n*
scillaridinic acid Scillaridinsäure *f*
scilliglaucoside Scilliglaucosid *n*
scillin Scillin *n*
scillipicrin Scillipikrin *n*
scilliroside Scillirosid *n*
scillitin Scillitin *n* (Pharm)
scillitoxin Scillitoxin *n*
scintigram Szintigramm *n*
scintillant Szintillationssubstanz *f*
scintillate *v* szintillieren, Funken sprühen
scintillating funkelnd, funkensprühend,
schillernd
scintillation Funkeln *n*, Szintillation *f* (Phys)
scintillation counter Szintillationszähler *m*,
Strahlungsintensitätsmesser *m*, Szintillator *m*
scintillation liquid Szintillationsflüssigkeit *f*
scintillation plastic Kunststoffszintillator *m*
scintillation response
Szintillationsansprechvermögen *n*
scintillation spectrometer
Szintillationsspektrometer *n*
scintillation spectrometry
Szintillationsspektrometrie *f*
scintillator Szintillator *m*
scintilloscope Szintilloskop *n*
scission Spaltung *f*
scissors Handschere *f*
sclaren Sclaren *n*
scleretinite Skleretinit *m* (Min)
sclerometer Sklerometer *n*, Härtemesser *m*,
Ritzhärtemesser *m*, Ritzhärteprüfer *m*

sclerometer test Ritzhärteprüfung *f*
sclerometric hardness Ritzhärte *f*
scleroprotein Skleroprotein *n*
scleroscope Skleroskop *n*
scleroscope hardness Kugelfallhärte *f*,
Rückprallhärte *f*, Skleroskophärte *f*
scleroscope hardness test Kugelfallprobe *f*
scleroscope test Fallhärteprüfung *f*
sclerosis Sklerose *f* (Med), Verhärtung *f*
(Zellgewebe)
sclerotic acid Sklerotinsäure *f*
sclerotinic acid Sklerotinsäure *f*
sclerotinol Sclerotinol *n*
scolecite Skolezit *m* (Min)
scolopsite Skolopsit *m* (Min)
scombrin Scombrin *n*
scoop Ausschöpfkelle *f*, Schaufel *f*, Schöpfer *m*,
Schöpfgefäß *n*, Schöpflöffel *m*
scoop *v*, ~ **off** abschöpfen, ~ **up** aufhäufen
scooping thermometer Schöpfthermometer *n*
scoop wheel Schöpfrad *n*, Wurfrad *n*
scoparin Scoparin *n*
scoparone Scoparon *n*
scope Bereich *m*, Reichweite *f*, Umfang *m*
scopine Scopin *n*
scopolamine Skopolamin *n*, Hyoscin *n*
scopolamine hydrobromide
Scopolaminhydrobromid *n*
scopolamine methylbromide
Scopolaminmethylbromid *n*
scopolamine methylnitrate
Scopolaminmethylnitrat *n*
scopoleine Scopolein *n*
scopoletin Scopoletin *n*, Gelseminsäure *f*
scopoline Scopolin *n*
scorbutic skorbutisch (Med)
scorbutus Skorbut *m* (Med)
scorch *v* versengen, ausdörren, durchschmoren
(Kabel), verbrennen
scorching Anvulkanisation *f*, Versengen *n*
score, depth of ~ Rauhtiefe *f*
score *v* einritzen, kerben
scoria Schlacke *f*, Erzschaum *m*,
Metallschlacke *f*
scoriaceous schlackenartig, schlackenreich,
schlackig
scorification Läuterverfahren *n* (Gold, Silber),
Schlackenbildung *f*, Verschlackung *f*
scorification test Ansiedeprobe *f*
scorifier Ansiedescherben *m*, Röstscherben *m*
scoriform schlackenförmig
scorify *v* [ver]schlacken
scoring Einschnitt *m*, Kerbbildung *f*,
Riefenbildung *f*, Ritzen *n*
scoring machine Ritzmaschine *f*
scorodite Skorodit *m* (Min), Knoblaucherz *n*
(Min), Pittizit *m* (Min)
Scotch tape Tesafilm *m* (HN)

scour *v* scheuern, [ab]reiben, abschruppen,
　blank reiben, entfetten (Wolle), reinigen,
　säubern, ~ **off** abscheuern
scoured entfettet (Wolle)
scourer Scheuergerät *n*
scourge Geißel *f*
scouring Scheuern *n*, Entbasten *n*, Entfetten *n*,
　~ **of wool** Reinigen *n* der Wolle
scouring agent Entbastungsmittel *n*
scouring cloth Putzlappen *m*
scouring liquor Waschlauge *f*
scouring material Putzmittel *n*
scouring powder Fleckenreinigungspulver *n*,
　Scheuerpulver *n*
scouring sand Fegesand *m*
scouring stick Fleckstift *m*
scouring stone Fleckstein *m*
scrap Abbruch *m* (Met), Abfall *m*, Ausschuß *m*,
　Fetzen *m*, Rest *m*, Schnitzel *m n*, **additon of** ~
　Schrottzugabe *f*, **malleable** ~
　Temperschrott *m*, **utilization of** ~
　Altmaterialverwertung *f*
scrap *v* verschrotten, abwracken
scrap coke Koksabfall *m*
scrap copper Altkupfer *n*, Bruchkupfer *n*
scrape *v* abziehen (Gerb), kratzen, schaben,
　scharren, ~ **off** abkratzen, abschaben,
　abspachteln, ~ **out** auskratzen, ausschaben
scraped off abgestrichen (Metall)
scraped shell cooler Kratzkühler *m*
scraped-surface heat exchanger
　Kratzwärmeaustauscher *m*
scraper Kratzer *m*, Abschab[e]eisen *n*,
　Abstreicheisen *n*, Kratzbürste *f*, Kratzeisen *n*,
　Schaber *m*, Schabmesser *n*, Scharre *f*, **flat** ~
　Flachschaber *m*
scraper conveyor Kratzerförderer *m*
scraper knife Schabemesser *n*
scraper loader Kratzlader *m*
scraper rope Schrapperseil *n*
scraping Kratzen *n*, Schaben *n*, Scharren *n*
scraping device Schabevorrichtung *f*
scraping iron Abschab[e]eisen *n*, Kratzeisen *n*
scraping knife Abstreifmesser *n*, Schabemesser *n*
scraping-off knife Abstreichmesser *n*
scrapings Geschabsel *n*, Schabsel *n pl*, Späne *pl*,
　~ **of liquation** Seigerkrätze *f*
scrap iron Schrott *m*, Abfalleisen *n*, Alteisen *n*
scrap lead Bruchblei *n*
scrap leather Lederabschabsel *pl*
scrap melting Schrottverhüttung *f*
scrap metal Schrott *m*, Altmetall *n*,
　Bruchmetall *n*, Metallabfall *m*,
　Umschmelzmetall *n*, **smelting of** ~
　Altmetallverhüttung *f*
scrapped, due to be ~ abbruchreif (Masch)
scrapping Schrottzugabe *f*, Verschrottung *f*
scrap preparation plant
　Schrottaufbereitungsanlage *f*

scrap recovery Schrottverwertung *f*
scrap rubber Abfallgummi *m*, Altgummi *m*
scraps Abfall *m*
scrap silver Bruchsilber *n*
scratch Kratzer *m*, Abschürfung *f*; Schramme *f*
scratch *v* [ein]ritzen, kratzen, schrammen, ~ **off**
　abkratzen, ~ **out** auskratzen
scratch brush Kratzbürste *f*
scratch brushing Kratzen *n*
scratch brushing machine Kratzmaschine *f*
scratch coat Unterputz *m*
scratcher Kratzeisen *n*
scratch hardness Ritzhärte *f*
scratch-hardness number Ritzhärtezahl *f*
scratching Kratzen *n*, **susceptibility to** ~
　Ritzbarkeit *f*, **susceptible to** ~ ritzbar
scratching machine Ritzmaschine *f*
scratch method Ritzverfahren *n*
scratch resistance Ritzhärte *f*, Kratzfestigkeit *f*
scratch-resistant kratzfest
scratch test Ritzversuch *m*
scratch tester Ritzhärteprüfer *m*
screen Schirm *m*, Abschirmung *f*, Drahtnetz *n*
　(Techn), Durchwurf *m* (Techn), Feuerrost *m*
　(Techn), Filter *n* (Opt), [großes] Sieb *n*
　(Techn), Raster *m* (Phot), Schutzblech *n*,
　Wandschirm *m*, **rotary** ~ Rundsieb *n*
screen *v* abschirmen, [durch]sieben (Techn),
　rättern (Techn), ~ **out** aussieben
screen beater mill Siebschlagmühle *f*
screen belt dryer Laufbandtrockner *m*,
　Siebbandtrockner *m*
screen bottom Siebbelag *m*
screen centrifuge decanter Siebdekanter *m*
screen classifier Kanalradsichter *m*,
　Siebplansichter *m*
screen cloth Siebgewebe *n*
screen dryer Siebtrockner *m*
screened abgeschirmt
screened ore Scheideerz *n*
screen efficiency Siebgütegrad *m*
screener Knotenfänger *m* (Pap),
　Siebvorrichtung *f*
screen grid Schirmgitter *n*, Schutzgitter *n*
screening Sieben *n*, **loss due to** ~
　Aufbereitungsverlust *m*
screening action Schirmwirkung *f*
screening agent Lichtschutzmittel *n*
screening analysis Siebanalyse *f*
screening [ball] mill Siebkugelmühle *f*
screening constant Abschirmkonstante *f*,
　Abschirmungskonstante *f*
screening correction Abschirmkorrektur *f*
screening device Scheideanlage *f*,
　Scheidevorrichtung *f*
screening effect Abschirmeffekt *m*,
　Abschirmungseffekt *m*, Abschirmwirkung *f*
screening fraction Siebfraktion *f*
screening magnet Scheidemagnet *m*

screening material Siebgut *n*
screening number Abschirmungszahl *f*
screening plant Siebanlage *f*
screenings Ausgesiebte[s] *n*, Gesiebte[s] *n*,
Siebrückstand *m*
screening wire Siebdraht *m*
screen lining Siebbelag *m*
screen microscope Rastermikroskop *n*
screen opening Blendenöffnung *f*
screen sizing Kornklassierung *f*
screen system Rastersystem *n*
screen varnish Schablonenlack *m*
screw Schraube *f*; Schnecke *f*, **adjustable** ~
Stellschraube *f*, **button-head** ~
Halbrundkopfschraube *f*, **double-flighted** ~
doppelgängige Schnecke, **endless** ~
Schraube *f* ohne Ende, **left-hand** ~
linksgängige Schraube, **round-headed** ~
Halbrundschraube *f*, **sectional** ~
Gliederschnecke *f*, **single-flighted** ~
eingängige Schnecke, **square-thread** ~
flachgängige Schraube, **tightening** ~
Spannschraube *f*, **two-thread** ~
doppelgängige Schraube, **variable root** ~
kernprogressive Schnecke *f*, ~ **with head**
Kopfschraube *f*
screw *v* [ver]schrauben, ~ **in** einschrauben, ~
off abschrauben, ~ **on** anschrauben,
aufschrauben, ~ **together**
zusammenschrauben, verschrauben,
zusammenschrauben, ~ **up** zuschrauben
screw axis Schraubenachse *f* (Krist)
screw cap Schraubdeckel *m*, Überwurfmutter *f*
screw clamp Klemmschraube *f*,
Schraubenklemme *f*, Schraubzwinge *f*
screw closure Schraubverschluß *m*
screw compressor Schraubenverdichter *m*
screw conveyor Förderschnecke *f*
screw-conveyor drier Spiralbandtrockner *m*
screw-conveyor drum drier
Schneckentrommeltrockner *m*
screw-conveyor extractor Schneckenextrakteur *m*
screw cutting machine
Gewindeschneidmaschine *f*
screw diameter Schneckendurchmesser *m*
screw dies Gewindeschneidbacken *pl*,
Kluppenbacken *pl*
screw dislocation Schraubenversetzung *f*
screw displacement pump
Schraubenspindelpumpe *f*
screw-down mechanism Anstellmechanismus *m*
screwdriver Schraubenzieher *m*
screw dryer Schneckentrockner *m*
screwed connection Gewindeanschluß *m*,
Rohrverschraubung *f*
screwed nipple Gewindenippel *m*
screwed pin Gewindestift *m*
screwed pipe joint Schraubverbindung *f*
screwed socket Gewindemuffe *f*

screw extruder Schnecken[strang]presse *f*
screw extrusion Schneckenpressen *n*
screw feeder Schneckenschleuse *f*
screw fitting Verschraubung *f*
screw flight Schneckengang *m*
screw-flights evaporator Schneckenverdampfer *m*
screw-flights heat exchanger
Schneckenwärmeaustauscher *m*
screw head Schraubenkopf *m*, **[half-]round** ~
halbrunder Schraubenkopf
screw hook Hakenschraube *f*,
Schraub[en]haken *m*
screwing die stock Gewindekluppe *f*
screw-in jacket Einschraubhülse *f*
screw jack (Br. E.) Schraubheber *m*
screw machine Fassondrehbank *f*
screw mixer Schneckenmischer *m*
screw-on socket Überschraubmuffe *f*
screw pinchcock Schraubenquetschhahn *m*
screw pipe joint Rohrverschraubung *f*
screw pitch Gewindesteigung *f*
screw plate Gewindeeisen *n*
screw plug Gewindestopfen *m*,
Verschlußschraube *f*
screw press Spindelpresse *f*
screw pump Schneckenpumpe *f*,
Schraubenpumpe *f*
screw reactor Schneckenreaktor *m*
screw spanner Schraubenschlüssel *m*
screw speed Schneckendrehzahl *f*
screw stock Gewindekluppe *f*
screw tap Gewindebohrer *m*
screw taps, steel for ~ Gewindebohrstahl *m*
screw thread Schraubengewinde *n*,
Schneckengewinde *n*, **backlash of** ~
Flankenspiel *n*, **depth of a** ~ Gangtiefe *f*, ~
basing on the inch system Zollgewinde *n*
screw tool Gewindestahl *m*
screw type extrusion machine
Schneckenstrangpresse *f*
screw type injection molding machine
Schneckenspritzgußmaschine *f*
screw with half-turn compression (i. e. screw that
gives a sudden increase in compression at the
beginning of the metering section by a
reduction in the depth of flight)
Kurzkompressionsschnecke *f*
scribing tool Reißnadel *f*
scroll conveyor centrifuge
Schnecken-Siebzentrifuge *f*
scroll saw Dekupiersäge *f*
scrooping Avivage *f* (Seide)
scrub *v* scheuern, reinigen
scrubbable scheuerfest
scrubber Skrubber *m*, Sprühapparat *m* (für
Absorption), Turmwäscher *m*, Wäscher *m* für
Gas

scrubbing Scheuern *n*, **resistant to ~**
  abwaschbar, **~ with oil under pressure**
  Druckölwäsche *f*
scrubbing process Waschvorgang *m*
scrubbing tower Berieselungsturm *m*,
  Waschturm *m*
scrub resistance Scheuerfestigkeit *f*
scuffing Fressen *n* (Techn)
scuff resistance Abriebfestigkeit *f*,
  Scheuerfestigkeit *f*, Schreibfestigkeit *f*
scuff rib Scherleiste *f*
scullery Spültisch *m*
sculptor's plaster Stuckgips *m*
scum Schaum *m*, Abschaum *m*, Abstrich *m*,
  Kahm *m*, Schlamm *m* (Zucker)
scum *v* abschäumen, abschlacken, abschöpfen,
  schäumen
scum basket Schaumkorb *m*
scum cock Schaumhahn *m*
scummer Abstreifer *m* (f. Schlacke)
scumming Abschäumen *n*, Schäumen *n*
scum pan Schlammpfännchen *n*
scum pipe Schaumröhre *f*
scum riser Schaumkopf *m*, Schaumtrichter *m*
scurvy Skorbut *m* (Med)
scutamol Scutamol *n*
scutellarin Scutellarin *n*
scyllitol Scyllit *m*
scylloquercitol Scylloquercit *m*
sea foam Meerschaum *m*
sea gauge Tiefenmesser *m* (Nav)
seal Robbe *f* (Zool), Schutzanstrich *m*, Siegel *n*,
  Verschluß *m*, **air-tight ~** luftdichter
  Verschluß, **controlled gap ~** Spaltdichtung *f*,
  hermetic ~ luftdichter Verschluß, **labyrinth ~**
  Labyrinthdichtung *f*, **slip ring ~**
  Gleitringdichtung *f*, **threaded shaft or**
  hydrodynamic ~ Gewindewellendichtung *f*
seal *v* versiegeln; abdichten, abschmelzen,
  einschmelzen, plombieren, **~ hermetically**
  hermetisch verschließen, luftdicht
  verschließen, **~ off** abkapseln
sealed versiegelt, zugeschmolzen
sealed tube Einschlußrohr *n*
sealed tube furnace Kanonenofen *m*
sealer Versiegelungsmittel *n*, Einlaßgrund *m*,
  Porenfüller *m*, Porenschließer *m*
sea level Meeresspiegel *m*
sealing Abdichten *n*, Verkitten *n*, Verschließen *n*
sealing alloy Einschmelzlegierung *f*
sealing chamber Vergußkammer *f*
sealing composition Füllmasse *f*
sealing compound Dichtungsmasse *f*,
  Dichtungsmaterial *n*, Verbindungskitt *m*,
  Vergußmasse *f*
sealing flange Flanschendichtung *f*,
  Verschlußflansch *m*
sealing joint Dichtungsfuge *f*
sealing lacquer Dichtungslack *m*

sealing ledge Dichtungsleiste *f*
sealing liquid Sperrflüssigkeit *f*
sealing machine Verschließmaschine *f*
sealing primer Einlaßgrund *m*
sealing profile Dichtungsprofil *n*
sealing ring Dichtungsring *m*
sealing run Kappnaht *f*
sealing screw Dichtungsschraube *f*
sealing sleeve Dichtungsmuffe *f*
sealing system, automatic multiple-station ~ (for
  plastics) Etagen-Schiebetischanlage *f*
  (Kunststoffe)
sealing tape Dichtungsstreifen *m*,
  Klebestreifen *m*
sealing tube Einschmelzrohr *n*
sealing water Sperrwasser *n*
sealing wax Siegellack *m*, Klebwachs *n*, **stick of**
  ~ Siegellackstange *f*
seal oil Robbentran *m*
seam Saum *m*, Falz *m* (Dose), Flöz *n* (Geol),
  Fuge *f*, Lager *n* (Geol), Naht *f*, Narbe *f*
seam *v* bördeln, abkanten, falzen
seamed-on aufgefalzt
seaming Bördeln *n*
seaming machine Blechfalzmaschine *f*
seamless nahtlos
seamless weld fugenlose Verschweißung *f*
seam [of ores] Lagerstätte *f* (Bergb)
seam weld Schweißnaht *f*
seam welder Nahtschweißmaschine *f*
seam welding Nahtschweißen *n*
sea ooze Meerschlamm *m*
sea peat Meertorf *m*
search Suche *f*, Untersuchung *f*
search *v* [ab]suchen, forschen [nach], **~ out**
  ausforschen, ausfindig machen
search light Scheinwerfer *m*
search method Suchmethode *f*
search radar Suchradar *n*
sea salt Meersalz *n*
sea salt refinery plant Meersalzraffinieranlage *f*
sea sand Meersand *m*, Seesand *m*
sea sedge Sandsegge *f* (Bot)
season *v* ablagern (Holz), abschmecken
  (würzen), austrocknen (Holz), auswittern
  (Holz), würzen, **~ thoroughly** durchwürzen,
  ~ timber Holz lagern
seasoned abgelagert (Holz)
seasoning Austrocknen *n*, Auswitterung *f*,
  Gewürz *n*, Würze *f*
seasoning kiln Trockenofen *m* (Holz)
season shade Modeton *m*, Saisonfarbe *f*
seat Sitz *m*, Auflagefläche *f*, Teller *m* (Ventil)
sea tang Seetang *m* (Bot)
seating surface Auflagefläche *f*
sea water Meerwasser *n*, Salzwasser *n*
sea water desalination Seewasserentsalzung *f*
sea water distillation apparatus
  Seewasserdestillierapparat *m*

seaweed Meeresalge *f,* Seetang *m,* Tang *m*
seaweed fiber Algenfaser *f*
sebacamide Sebazinsäureamid *n*
sebaceous talgartig, talghaltig
sebaceous gland Talgdrüse *f* (Med)
sebacic acid Sebacinsäure *f*
sebacil Sebacil *n*
sebacoin Sebacoin *n*
sebaconitrile Sebaconitril *n*
secaclavine Secaclavin *n*
secaline Secalin *n*
secant Sekante *f*
seclude *v* ausschließen
seclusion Isolierung *f*
secobarbital Secobarbital *n*
second Sekunde *f*
secondary sekundär, zweitrangig
secondary accelerator Zweitbeschleuniger *m*
secondary action Nebenwirkung *f*
secondary air Sekundärluft *f,* Beiluft *f*
secondary alcohol sekundärer Alkohol *m*
secondary alloy Altmetallegierung *f*
secondary battery Akkumulator *m*
secondary bond Nebenbindung *f*
secondary burst Sekundärdurchbruch *m*
secondary cell Sekundärelement *n*
secondary circuit Sekundärkreis *m* (Elektr)
secondary color Nebenfarbe *f,* zusammengesetzte
    Farbe *f*
secondary component Nebenbestandteil *m*
secondary constituent Nebenbestandteil *m*
secondary discharge Nebenentladung *f*
secondary effect Nebeneffekt *m,*
    Nebenwirkung *f*
secondary emission cathode
    Sekundäremissionskathode *f*
secondary emission current
    Sekundäremissionsstrom *m*
secondary emission factor
    Sekundäremissionsfaktor *m*
secondary fermentation Nachgärung *f*
secondary flow Sekundäremission *f,*
    Sekundärströmung *f*
secondary function Nebenfunktion *f*
secondary gluing Montageleimung *f*
secondary infection Sekundärinfektion *f*
secondary particle Sekundärteilchen *n*
secondary pipe (in an ingot) sekundärer
    Lunker *m*
secondary quality Nebeneigenschaft *f*
secondary quantum number Nebenquantenzahl *f,*
    Drehimpulsquantenzahl *f*
secondary reaction Nebenreaktion *f* (Chem)
secondary stress Sekundärspannung *f*
secondary structure Sekundärstruktur *f* (Chem)
secondary treatment Nachbehandlung *f*
secondary valence Nebenvalenz *f*
secondary voltage Sekundärspannung *f* (Elektr)
secondary winding Sekundärwicklung *f*

second baking Nachbrand *m*
second boiling Nachsud *m*
second charge Nachladung *f* (Akku)
second power zweite Potenz *f*
second product Nachprodukt *n*
second runnings Nachlauf *m*
seconds pendulum Sekundenpendel *n*
second wort Nachwürze *f*
secrete *v* abscheiden, absondern, ausscheiden,
    sezernieren
secretin Sekretin *n*
secretion Sekretion *f,* Absonderung *f,* Sekret *n*
secretor Sekretionsorgan *n*
section Abschnitt *m,* Teil *m,* **horizontal** ~
    Flachschnitt *m,* Horizontalschnitt *m*
sectional drawing Querschnittszeichnung *f*
sectional model Schnittmodell *n*
sectional plane Schnittebene *f*
sectional steel Profilstahl *m*
section diameter Reifenbreite *f* (Kfz)
section height Reifenhöhe *f* (Kfz)
section iron Fassoneisen *n,* Formeisen *n*
section-lined schraffiert
section rolling Profilwalzen *n*
sections, by ~ streckenweise
sections of boshes Rastformen *f pl* (Hochofen)
section steel Formstahl *m*
section through rank Inkohlungsprofil *n*
sector Kreisausschnitt *m,* Sektor *m,* ~ **of a circle**
    Kreissektor *m*
sector field Sektorfeld *n*
secular hundertjährig
secular determinant Säkulardeterminante *f*
secular equation Säkulargleichung *f*
secular equilibrium Dauergleichgewicht *n*
    (Atom)
secular retardation Säkularverzögerung *f*
secure *v* sichern, befestigen
securing screw Sicherungsschraube *f*
security Sicherheit *f,* Garantie *f,* Gewähr *f,*
    measure of ~ Sicherheitsmaßnahme *f,* **to give**
    [for] ~ bürgen
sedamine Sedamin *n*
Sedan black Sedanschwarz *n*
sedanolic acid Sedanolsäure *f*
sedanolide Sedanolid *n*
sedatin Valeridin *n*
sedative Beruhigungsmittel *n* (Pharm),
    Nervenberuhigungsmittel *n* (Pharm),
    Sedativum *n* (Pharm), Tranquil[l]izer *m*
    (Pharm)
sedative *a* sedativ (Med)
sedative agent Beruhigungsmittel *n* (Pharm)
sedative poison lähmendes Gift *n*
sedative salt Sedativsalz *n*
sediment Sediment *n,* Ablagerung *f,*
    Bodenkörper *m,* Bodensatz *m,*
    Niederschlag *m,* Satz *m,* **without** ~

rückstandsfrei, ~ **from carbonation**
Saturationsschlamm *m*
sediment *v* sedimentieren, sich niederschlagen
sedimentary sedimentär
sedimentary rock Ablagerung *f* (Geol)
sedimentary rocks Sedimentgestein *n*
sedimentation Sedimentation *f,*
Sedimentbildung *f,* Bodensatzbildung *f,*
Schlämmverfahren *n* (Techn)
sedimentation aid Flockungsmittel *n,*
Sedimentierhilfsmittel *n*
sedimentation coefficient
Sedimentationskoeffizient *m*
sedimentation column Sedimentationssäule *f*
sedimentation equilibrium
Sedimentationsgleichgewicht *n*
sedimentation inhibitor
Absetzverhinderungsmittel *n*
sedimentation test Sedimentationsversuch *m*
sedimentation velocity
Sedimentationsgeschwindigkeit *f,*
Sinkgeschwindigkeit *f*
sedimentology Ablagerungskunde *f*
sedinine Sedinin *n*
sedoheptitol Sedoheptit *m*
sedoheptose Sedoheptose *f*
sedoheptulose Sedoheptulose *f*
sedoheptulosone Sedoheptuloson *n*
sedridine Sedridin *n*
seebachite Seebachit *m* (Min)
Seebeck effect Seebeckeffekt *m,*
thermoelektrischer Effekt *m*
Seebeck voltage Thermospannung *f*
seed Saat *f* (Bot), Samen *m* (Bot)
seed[-corn] Saatgut *n*
seed crystal Impfkristall *m,* Kristallkeim *m*
seed disinfection Samenbeizung *f*
seed dressing Beizmittel *n* (Saatgut),
Saatbeizmittel *n,* Saatgutbeize *f*
seed fish Laichfisch *m* (Zool)
seed growing Saatzucht *f*
seed husk Samenschale *f*
seeding Impfung *f* (Krist)
seeding suspension Kristallkeimsuspension *f*
seed leaf Keimblatt *n* (Bot)
seedling Keimling *m*
seeing capacity Sehschärfe *f*
seek *v* suchen
seesaw pan Kipppfanne *f*
seethe *v* sieden, [ab]brühen, kochen
Seger cone Segerkegel *m*
segment Abschnitt *m,* Kreisabschnitt *m,*
Segment *n,* ~ **of a circle** Kreissegment *n,* ~ **of
a sphere** Kugelabschnitt *m*
segmental scale Hilfsskala *f*
segmentation Furchung *f,* Segmentation *f*
segment heat exchanger
Lamellenwärmeaustauscher *m*
segregate entmischen (Metall)

segregate *v* absondern, seigern (Metall)
segregating Seigern *n*
segregation Abtrennung *f,* Aussonderung *f,* **line
of** ~ (in an ingot) Seigerungszone *f,* ~ **of the
electrolyte** Entmischung *f* des Elektrolyten
segregation factor Segregationsgrad *m*
Seignette salt Kaliumnatriumtartrat *n,*
Seignettesalz *n*
seismic ray Erdbebenstrahl *m*
seismic wave Erdbebenwelle *f*
seismograph Seismograph *m,*
Erdbebenmesser *m,*
Erdbebenregistrierinstrument *n*
seismological observatory Erdbebenwarte *f*
seismology Seismologie *f,* Erdbebenkunde *f*
seismometer Erdbebenmesser *m,* Seismometer *n*
seismometric[al] seismometrisch
seismometry Seismometrie *f*
seismoscope Seismoskop *n,* Erdbebenanzeiger *m*
seize *v* ergreifen, abfangen, beschlagnahmen,
fassen, festfressen (Lager), ~ **the mordant** die
Beize annehmen
seizing (of a mold) Festfressen *n*
seizure Ergreifung *f* (stoppage of a machine as a
result of excessive friction), Verschweißen *n*
(Met)
sekikaic acid Sekikasäure *f*
selacholeic acid Selacholeinsäure *f,*
Nervonsäure *f*
selachyl alcohol Selachylalkohol *m*
selagine Selagin *n*
selaginol Selaginol *n*
selbite Selbit *m*
seldom selten
select *v* [aus]wählen
selected area diffraction Feinbereichsbeugung *f*
selecting magnet Wahlmagnet *m*
selection Auswahl *f,* Auslese *f,* Wahl *f,* **theory of
natural** ~ Selektionstheorie *f*
selection principle Auswahlprinzip *n*
selection rule Auswahlregel *f*
selective selektiv, wahlweise
selective corrosion Lokalkorrosion *f*
selective emitter Selektivstrahler *m*
selective hardening Teilhärtung *f*
selective procedure Auswahlverfahren *n*
selectivity Selektivität *f,* Trennschärfe *f*
selector Wähler *m* (Techn)
selector disk Wählerscheibe *f*
selenate Selenat *n*
selenazine Selenazin *n*
selenazole Selenazol *n*
selenic acid Selensäure *f,* **hydrated** ~
Selensäurehydrat *n*
selenic acid hydrate Selensäurehydrat *n*
selenic anhydride Selentrioxid *n*
selenic compass Selenkompaß *m*
selenide Selenid *n*
seleniferous selenhaltig

selenin acid Seleninsäure *f*
selenindigo Selenindigo *n*
selenious selenig
selenious acid, salt or ester of ~ Selenit *n*
selenious anhydride Selendioxid *n*,
 Selenigsäureanhydrid *n*
selenite Selenit *m* (Min), Blättergips *m* (Min),
 Gipsspat *m* (Min), Marienglas *n* (Min),
 Selenit *n* (Chem)
selenitiferous selenithaltig
selenium (Symb. Se) Selen *n*
selenium barrier cell Selensperrschichtzelle *f*
selenium bismuth glance Selenwismutglanz *m*
 (Min)
selenium cell Selenzelle *f*
selenium chloride Selenchlorid *n*
selenium coating Selenüberzug *m*
selenium compound Selenverbindung *f*
selenium content Selengehalt *m*
selenium cyanide Selencyanid *n*
selenium dioxide Selendioxid *n*,
 Selenigsäureanhydrid *n*
selenium film Selenschicht *f*
selenium glass Selenfilter *n* (Phot), Selenglas *n*
selenium hexafluoride Selenhexafluorid *n*
selenium lattice Selengitter *n*
selenium layer Selenschicht *f*
selenium ore Selenerz *n* (Min)
selenium oxide Selenoxid *n*
selenium receiver Selenempfänger *m*
selenium rectifier Selengleichrichter *m*
selenium slime Selenschlamm *m*
selenium sol Selensol *n*
selenium sulfide Schwefelselen *n*
selenium tetrachloride Selentetrachlorid *n*
selenium trioxide Selentrioxid *n*
selenobenzoic acid Selenobenzoesäure *f*
selenocyanic acid Selencyansäure *f*
selenocyanogen Selenocyan *n*
selenofuran Selenofuran *n*, Selenophen *n*
selenoid switch Magnetschalter *m*
selenonaphthene Selenonaphthen *n*
selenophene Selenofuran *n*, Selenophen *n*
selenophenol Selenophenol *n*
selenopyronine Selenopyronin *n*,
 Selenoxanthen *n*
selenosalicylic acid Selenosalicylsäure *f*
selenoxanthene Selenopyronin *n*,
 Selenoxanthen *n*
selensulfur Selenschwefel *m* (Min)
selenurea Selenharnstoff *m*
self-acting automatisch, selbsttätig
self-adherent selbstklebend
self-adhering selbstklebend
self-adhesive selbsthaftend (Klebband)
self-adhesive tape Klebestreifen *m*
self-backing electrode Dauerelektrode *f*
self-balancing sich selbst abgleichend
self-cleaning sich selbst reinigend

self-colored eigenfarbig, maßeingefärbt,
 naturfarben
self-computing chart Nomogramm *n*
self-consuming sich selbst verzehrend
self-contained in sich geschlossen
self-contained drive Einzelantrieb *m*
self-contained paper Einschichtpapier *n*
 (Buchdr)
self-contained press Presse *f* mit Einzelantrieb
self-curing selbstvulkanisierend, selbsthärtend
self-damping Eigendämpfung *f* (Mech)
self-digestion Autodigestion *f*
self-discharge Selbstentladung *f*
self-discharge current Selbstentladestrom *m*
self-discharger Selbstentlader *m*
self-evident selbstverständlich
self-exchange Selbstaustausch *m*
self-excitation Eigenerregung *f*, Selbsterregung *f*
self-excited eigenerregt
self-feeding furnace Schüttfeuerung *f*
self-fluxing selbstgängig
self-heating Selbsterhitzung *f*
self-heating oxide cathode
 Aufheizpastenkathode *f*
self-igniter Selbstzünder *m*
self-igniting selbstzündend
self-ignition Selbstzündung *f*
self-inductance Selbstinduktivität *f*
self-inductance coil Selbstinduktionsspule *f*
self-induction Eigeninduktion *f* (Elektr),
 Selbstinduktion *f* (Eletr), **coefficient of** ~
 Selbstinduktionskoeffizient *m*, **law of** ~
 Selbstinduktionsgesetz *n*
self-induction voltage
 Selbstinduktionsspannung *f*
self-inductive selbstinduktiv
self-intoxication Autointoxikation *f*
self-inversion Selbstumkehr *f*
self-locking selbstsperrend
self-lubricating selbstschmierend
self-lubrication Selbstschmierung *f*
self-lubricator Selbstöler *m*
self-magnetic eigenmagnetisch
self-modulation Eigendämpfung *f*,
 Eigenmodulation *f*
self-oxidation Autoxidation *f*, Selbstoxidation *f*
self-polymerization Autopolymerisation *f*
self-preservation, instinct of ~ Erhaltungstrieb *m*
self-propelled loader Ladegerät *n* mit
 Eigenfahrantrieb
self-quenching selbstlöschend
self-reactance Eigenreaktanz *f*
self-recording automatisch registrierend
self-registering thermometer
 Registrierthermometer *n*
self-regulating selbstregelnd
self-regulating process (Am. E.) Regelstrecke mit
 Ausgleich *m*

self-regulation Selbstregelung *f* (Reaktor), Selbststeuerung *f* (Biol)
self-repulsion Selbstabstoßung *f*
self-reversal Selbstumkehr *f,* ~ **(of spectral lines)** Selbstumkehrung *f* (von Spektrallinien)
selfsealing Kaltsiegelung *f,* selbst[ab]dichtend, Selbstklebung *f*
self-shielding selbstabschirmend
self-stabilization Eigenstabilisierung *f*
self-superheating Selbstüberhitzung *f*
self-supporting freitragend
self-sustaining selbsterhaltend
self-tannage Alleingerbung *f*
self-tannin Alleingerbstoff *m*
selinane Selinan *n*
selinene Selinen *n*
sell *v* verkaufen
sellaite Sellait *m* (Min)
Seltzer [water] Selterswasser *n*
selvage Kante *f* (Web), Webkante *f* (Web)
selvage tension Kantenspannung *f*
selvedge Webkante *f* (Web)
semiactive halbaktiv
semialbumose Hemialbumose *f*
semi-amplitude Halbamplitude *f*
semianthracite Halbanthrazit *m*
semiapertural angle Halböffnungswinkel *m*
semiautomatic halbautomatisch, halbmechanisch
semiaxis Halbachse *f* (Geom)
semibatch halbkontinuierlich
semi-beam ausladender oder freitragender Balken
semibenzene Semibenzol *n*
semibituminous halbfett (Kohle)
semibreadth Halbwert[s]breite *f*
semicarbazide Semicarbazid *n*
semicarbazide hydrochloride Semicarbazidhydrochlorid *n*
semicarbazidobenzamide, 3- ~ Kryogenin *n*
semicarbazone Semicarbazon *n*
semi-chemical pulp Halbzellstoff *m*
semi-chrome tanning Semichromgerbung *f*
semicircle Halbkreis *m,* Winkelmesser *m*
semicircular halbkreisförmig
semicoke Grudekoks *m,* Halbkoks *m,* Schwelkoks *m*
semicoking Halbkokung *f*
semicolloid Halbkolloid *n,* Semikolloid *n*
semiconducting halbleitend
semiconduction property Halbleitereigenschaft *f*
semiconductor Halbleiter *m* (Elektr)
semiconductor cell Halbleiterzelle *f*
semiconductor layer Halbleiterschicht *f*
semiconductor memory Halbleiterspeicher *m*
semiconductor surface Halbleiteroberfläche *f*
semicontinuous halbkontinuierlich
semiconvergent Halbkonvergente *f*
semicrystal Halbkristall *m* (Krist)
semicrystalline halbkristallin

semicured halb vulkanisiert
semicycle Halbperiode *f*
semicylindrical halbzylindrisch
semidine Semidin *n*
semidine rearrangement Semidin-Umlagerung *f*
semidry halb trocken
semidull halbmatt
semiempirical halbempirisch
semifinished goods Halbfertigwaren *f pl,* Halbzeug *n*
semifinished material Halbzeug *n*
semifinished products Halbzeug *n*
semifluid breiartig, zähflüssig
semifused halb geschmolzen
semigas Halbgas *n*
semigas furnace Halbgasfeuerung *f*
semigloss Halbglanz *m*
semigloss paint Halbmattlack *m*
semihydrate Semihydrat *n*
semi-industrial halbtechnisch
semiliquidity Halbflüssigkeit *f*
semilog einfachlogarithmisch
semi-luxuries Genußmittel *n pl*
semimanufactured goods Halbzeug *n*
semimechanical halbmechanisch, halbautomatisch
semimetal Halbmetall *n*
semimetallic halbmetallisch
semimicro analysis Halbmikroanalyse *f*
semimicro apparatus Halbmikroapparat *m*
seminitrile Halbnitril *n*
seminormal halbnormal
seminose Seminose *f*
semiopal Halbopal *m* (Min), Holzopal *m* (Min)
semioscillation Halbperiode *f*
semioxamazide Semioxamazid *n*
semiparabolic superstructure Halbparabelfachwerkträger *m*
semiperiod Halbperiode *f*
semipermeable halbdurchlässig, semipermeabel
semipermeable wall halbdurchlässige Wand *f*
semipolar halbpolar, semipolar
semiporcelain Halbporzellan *n*
semiprecious stone Halbedelstein *m*
[semi-]preserve Halbkonserve *f*
semiproduction basis Halbfließbandfertigung *f*
semiproducts Halbzeug *n*
semiracemic halbracemisch
semireduction Halbreduktion *f* (Chem, Metall)
semirigid halbstarr
semirigid container halbstarrer Behälter *m*
semisintered halb gesintert
semiskilled angelernt
semislicing halbschnittig (Zuck)
semisolid halbfest
semispan Halbspannweite *f*
semispherical halbkugelig
semisteel Halbstahl *m,* Gußeisen *n* mit Stahlschrottzusatz

semisynthetic halbsynthetisch
semitransparency Halbdurchsichtigkeit *f*
semi-transparent halbdurchscheinend,
  halbdurchsichtig
semivulcanized halb vulkanisiert
semiwater gas Halbwassergas *n*, Mischgas *n*
semiwet halbnaß
semolina Grieß *m*
sempervirine Sempervirin *n*
semseyite Semseyit *m* (Min)
senaite Senait *m* (Min)
senarmontite Senarmontit *m* (Min)
send *v* [ver]senden, ~ **out** aussenden, ~ **through**
  durchschicken
sender Adressant *m*, Absender *m*, Sender *m*
  (Elektr)
senecic acid Senecinsäure *f*
senecifolidine Senecifolidin *n*
senecifoline Senecifolin *n*
senecine Senecin *n*
senecio alkaloid Senecioalkaloid *n*
senecioic acid Seneciosäure *n*
senecionine Senecionin *n*
seneciphyllic acid Seneciphyll[in]säure *f*
seneciphylline Seneciphyllin *n*
senega, liquid extract of ~ Senegafluidextrakt *m*
senega extract Senegaextrakt *m*
senega root Senegawurzel *f* (Bot)
senega syrup Senegasirup *m*
senegin Senegin *n*
senna leaves Sennesblätter *pl* (Bot)
sennidine Sennidin *n*
sennitol Sennit *m*
sennoside Sennosid *n*
sensation Empfindung *f*
sense Sinn *m*
sensing element Fühlglied *n*, Meßkopf *m*
sensitive empfindlich, sensitiv, **extremely** ~
  höchstempfindlich, **highly** ~
  hochempfindlich, ~ **to acids**
  säureempfindlich, ~ **to air** luftempfindlich, ~
  to frost frostempfindlich, ~ **to water**
  wasserempfindlich
sensitive element Fühlglied *n*
sensitive green Sensitivgrün *n*
sensitiveness Sensibilität *f*, ~ **to acids**
  Säureempfindlichkeit *f*
sensitive time Ansprechzeit *f* (Atom)
sensitivity Empfindlichkeit *f*, Sensibilität *f*,
  Ansprechempfindlichkeit *f*, Empfindlichkeit *f*,
  Empfindungsvermögen *n*, **color** ~ spektrale
  Empfindlichkeit, **degree of** ~
  Empfindlichkeitsgrad *m*, **monochromatic** ~
  monochromatische Empfindlichkeit, **range of**
  ~ Empfindlichkeitsbereich *m*, ~ **to noise**
  Lärmempfindlichkeit *f*, ~ **to water**
  Wasserempfindlichkeit *f*
sensitivity center Empfindlichkeitszentrum *n*
  (Phot)

sensitivity limit Empfindlichkeitsgrenze *f*
sensitivity measure Empfindlichkeitsmaß *n*
sensitization Sensibilisierung *f*
sensitize *v* [licht]empfindlich machen,
  sensibilisieren (Phot)
sensitized lichtempfindlich (Phot)
sensitizer Sensibilisator *m*
sensitizing dye Sensibilisierungsfarbstoff *m*
sensitizing maximum
  Sensibilisierungsmaximum *n*
sensitocolorimeter Sensitokolorimeter *n*
sensitometer Empfindlichkeitsmesser *m* (Opt)
sensitometry Empfindlichkeitsmessung *f* (Opt),
  Sensitometrie *f* (Opt)
sensory sensorisch
sensory testing Sinnesprüfung *f*
separability Scheidbarkeit *f*, Trennbarkeit *f*
separable [ab]trennbar, scheidbar, separabel,
  spaltbar, zerlegbar
separate *v* trennen, [ab]lösen, [ab]scheiden,
  absondern, [ab]spalten, ausscheiden, loslösen,
  separieren, ~ **by adding salt** aussalzen, ~ **by**
  filtering abfiltern, ~ **into components**
  entmischen
separate commutation Fremdführung *f*
separated product Abscheidungsprodukt *n*
  (Chem)
separate excitation Sondererregung *f*
separate lubricator Einzelöler *m*
separate oiling Einzelschmierung *f*
separate pot mold Mehrfachwerkzeug *n* mit
  getrennten Füllräumen
separating Abscheiden *n*, Absondern *n*,
  Abspalten *n*, Trennen *n*, **rough** ~ (of ores)
  rohes Sortieren [der Erze pl], ~ **into**
  components Entmischen *n*
separating agent Ausscheidungsmittel *n*,
  Scheidemittel *n*, Schleppmittel *n*
  (Rektifikation)
separating and shaking apparatus Scheide- und
  Schüttelvorrichtung *f*
separating apparatus Scheidevorrichtung *f*
separating buret[te] Scheidebürette *f*
separating capacity Trennschärfe *f*
separating column Trennsäule *f*
separating cut Trennschnitt *m* (Schweiß)
separating flask Scheidekolben *m*
separating funnel Scheidetrichter *m*
separating layer Trennungsschicht *f*
separating line Trennlinie *f*
separating liquid Scheideflüssigkeit *f*
separating plant Sortieranlage *f*,
  Sichtungsanlage *f*
separating process Scheideverfahren *n* (Met),
  Scheidevorgang *m*
separating sieve Scheidesieb *n*
separating valve Trennventil *n*
separating vessel Abscheidegefäß *n*,
  Scheidegefäß *n*

separating wall Scheidewand *f*
separation Trennung *f*, Abscheidung *f*,
  Absonderung *f*, Abspaltung *f*, Abtrennung *f*,
  Ausscheiden *n*, Entmischung *f*, Loslösung *f*,
  Scheiden *n*, **dense-media** ~
  Schwertrübetrennung *f*, **heat of** ~
  Trennungswärme *f*, **heavy-medium** ~
  Schwertrübetrennung *f*, **means of** ~
  Trennungsmittel *n*, **method of** ~
  Trennungsmethode *f*, **process of** ~
  Trennungsvorgang *m*, **wet** ~ nasse
  Scheidung *f*, **work of** ~ Abtrennungsarbeit *f*,
  ~ **by absorption chromatography**
  absorptionschromatographische Trennung, ~
  by addition of salt Aussalzen *n*, ~ **by hand**
  Handscheidung *f*, ~ **in flakes** Ausflockung *f*,
  ~ **of flow** Ablösung *f* (b. Strömungen), ~ **of**
  gases Gastrennung *f*, ~ **of oil from water**
  Wasserentölung *f*
separation column Trennsäule *f*
separation factor Trennfaktor *m*, **ideal** ~
  theoretischer Trennfaktor *m* (Atom)
separation liquid Trennungsflüssigkeit *f*
separation method Trennungsverfahren *n*
separation nozzle process Trenndüsenverfahren *n*
separation plant, chemical ~ chemische
  Trennanlage *f*
separation point Scheidepunkt *m*
separation potential Trennungspotential *n*
separation process Entmischungsvorgang *m*,
  Trennungsgang *m*, Trennungsverfahren *n*
separation product Ausscheidungsprodukt *n*
separation sharpness Trennschärfe *f*
separator Abscheidegefäß *n*, Abscheider *m*,
  Scheider *m*, Scheidevorrichtung *f*,
  Trennvorrichtung *f*, **submerged belt** ~
  Tauchbandscheider *m*
separator head Verteilerkopf *m*
separator trap Scheidegefäß *n*
separator tube Trennrohr *n* (Atom)
separatory funnel Scheidetrichter *m*
sepia Sepia *f* (Zool)
sepia brown Sepiabraun *n*
sepiolite Sepiolith *m* (Min)
sepsine Sepsin *n*
sepsis Sepsis *f* (Med), Blutvergiftung *f* (Med)
septic Fäulniserreger *m*
septic *a* septisch, fäulniserregend, faulend
septic disease Sepsis *f* (Med)
septic tank Faulgrube *f*, Kleinklärwerk *n*
septivalent siebenwertig
septum Scheidewand *f* (Bot, Zool)
sequel Folge[erscheinung] *f*
sequence Reihenfolge *f*, Abfolge *f*,
  Aufeinanderfolge *f*, Folge *f*, ~ **of charging**
  Chargierfolge *f*, ~ **of numbers** Zahlenfolge *f*
sequence analysis Sequenzanalyse *f*
sequence homology Sequenzhomologie *f*
sequence of operations Funktionsablauf *m*

sequence switch cam Steuerungsdaumen *m*
sequence timer Folgezeitschalter *m*
sequential test Folgeprüfung *f*
sequestering Maskierung *f*
sequoyitol Sequoyit *m*
seredine Seredin *n*
serendibite Serendibit *m* (Min)
serially hintereinander
serially connected in Reihe geschaltet
serial machine Serienmaschine *f*
serial number Werksnummer *f*, Seriennummer *f*
serial sectioning microtome
  Serienschnittmikrotom *n*
serial weighings Wägeserie *f*
sericin Sericin *n*
sericin coating Sericinschicht *f*
sericin layer Sericinschicht *f*
sericite Sericit *m* (Min)
sericite schist Sericitschiefer *m* (Min)
series Serie *f*, Folge *f* (Math), Reihe *f* (Math),
  Reihenfolge *f*, Zahlenreihe *f* (Math),
  arithmetic ~ arithmetische Reihe, **homologous**
  ~ homologe Reihe (Chem), **in** ~
  serienmäßig, der Reihe nach, **infinite** ~
  unendliche Reihe (Math), **to connect in** ~
  hintereinander schalten, ~ **of elements** (in
  accordance with electrochemical potentials)
  elektrochemische Spannungsreihe *f*, ~ **of**
  experiments Versuchsreihe *f*, ~ **of pulses**
  Impulsfolge *f*, ~ **of tests** Versuchsreihe *f*
series arc lamp Hauptschlußbogenlampe *f*
series block furnace Reihenblockofen *m*
series characteristic Seriencharakteristik *f*
series connection Reihenschaltung *f*,
  Serienschaltung *f*
series dynamo Hauptstromdynamo *m* (Elektr)
series excitation Hauptschlußerregung *f*,
  Hauptstromerregung *f*, Serienerregung *f*
series furnace Serienofen *m*
series limit Seriengrenze *f* (Spektrum)
series motor Hauptschlußmotor *m*,
  Reihenschlußmotor *m*
series of measurement Meßreihe *f*
series production Reihenfertigung *f*,
  Serienproduktion *f*
series resistance vorgeschalteter Widerstand *m*
series rheostat Serienwiderstand *m*
series spectrum Serienspektrum *n*
series switch Serienschalter *m*
series wound motor Hauptschlußmotor *m*
serine Serin *n*
serine deaminase Serindesaminase *f* (Biochem)
serine dehydratase Serindehydratase *f* (Biochem)
serinol Serinol *n*
serious bedenklich, ernsthaft
serodiagnosis Serodiagnose *f* (Med)
serolactaminic acid Serolactaminsäure *f*
serologic serologisch
serology Serologie *f*

seromycin Seromycin *n*
seronegative seronegativ (Med)
seropositive seropositiv (Med)
serotonin Serotonin *n*
serpentine Serpentin *m* (Min), **fibrous** ~
Serpentinasbest *m* (Min)
serpentine asbestos Serpentinasbest *m* (Min)
serpentinic acid Serpentinsäure *f*
serpentinite Serpentinit *m* (Min)
serratamic acid Serratam[in]säure *f*
serrate *v* auszacken
serrate[d] sägeartig, ausgekerbt, geriffelt, gezackt
serration Auszackung *f*, Zacken *m*
serum Serum *n*, **anticytotoxic** ~
anti[retikulär-]zytotoxisches Serum, **immune**
~ Immunserum *n*, **artificial** ~ **preparation**
Blutersatz *m*
serum albumin Serumalbumin *n*
serum diagnosis Serodiagnose *f* (Med)
serum globulin Serumglobulin *n*
serum lipoprotein Serumlipoproteid *n*
serum protein Serumeiweiß *n*
serve *v* [be]dienen
service Bedienung *f*, Kundendienst *m*, Wartung *f*
service *v* warten (Techn)
serviceable gebrauchsfähig
serviceable life Nutzungsdauer *f*
service conditions Gebrauchsbedingungen *f pl*
service durability Gebrauchsfestigkeit *f*
service evaluation test Gebrauchswertprüfung *f*
service life Dauerhaltbarkeit *f*; Haltbarkeit *f*;
Lebensdauer *f*
service pipe Zuleitungsrohr *n*
service pressure Betriebsdruck *m*
service requirement Betriebserfordernis *f*
service routine Wartungsprogramm *n*
service saddle Anbohrschelle *f*
service station Tankstelle *f*
service stress Betriebsbeanspruchung *f*
service water Gebrauchswasser *n*
servicing Instandhaltung *f*
servicing time Wartungszeit *f*
serving tray Serviertablett *n*
servoamplifier Servoverstärker *m*,
Hilfsverstärker *m*
servocomponent Servokomponente *f*
servo-control Folgeregelung *f* (Regeltechn),
Servosteuerung *f*
servodrive Servoantrieb *m*, Hilfsantrieb *m*
servo-follower Folgeregler *m* (Regeltechn)
servomechanism Servomechanismus *m*
servomotor Stellmotor *m*, Servomotor *m*
servopotentiometer Servopotentiometer *n*
servosystem Servoregelsystem *n*
sesame oil Sesamöl *n*, **brominated** ~ Bromipin *n*
sesame oil soap Sesamölseife *f*
sesamin Sesamin *n*
sesamol Sesamol *n*
sesamolin Sesamolin *n*

seselin Seselin *n*
sesqui anderthalb, eineinhalb (Chem)
sesquibasic anderthalbbasisch
sesquibenihene Sesquibenihen *n*
sesquicarbonate Sesquicarbonat *n*
sesquichloride Sesquichlorid *n*
sesquioxide Sesquioxid *n*
sesquisalt Sesquisalz *n*
sesquiterpene Sesquiterpen *n*
set Satz *m*, Garnitur *f*, **permanent** ~ bleibende
Formänderung *f*, permanente Veränderung *f*,
~ **of boilers** Einsatzkessel *pl*
set *v* setzen, abbinden (Zement), ansetzen
(Termin), binden (Mörtel), einfassen
(Edelstein), erhärten (Mörtel), ~ **down**
abstellen, ~ **free** entfesseln, freigeben,
freimachen, ~ **going** betätigen (von Hand), ~
in einsetzen, beginnen, eintreten, ~ **in plaster**
in Gips legen, ~ **on fire** anzünden, ~ **out**
ausrecken (Häute), ~ **to work** in Betrieb
setzen
setacyl blue Setacylblau *n*
setback Rückschlag *m*
set hammer Setzhammer *m*
set head Setzkopf *m*
set of functions Funktionensystem *n*, **complete** ~
vollständiges Funktionensystem *n*, **orthogonal**
~ orthogonales Funktionensystem *n*,
orthonormal ~ orthonormiertes
Funktionensystem *n*
set pin Paßstift *m*
set point Fixpunkt *m*, Sollwert *m*
set point adjustment Sollwerteinstellung *f*
setscrew Befestigungsschraube *f*, Druckschraube,
Einstellschraube *f*, Feststellschraube *f*,
Grenzschraube *f*, Stellschraube *f*
setting Abbindung *f* (Leim); Einstellung *f*,
Erhärten *n* (Techn), Erstarren *n* (Techn),
Fassung *f* (Edelstein), **cold** ~ Kalthärtung *f*,
~ **in motion** Inbetriebsetzung *f*
setting agent Härter *m* (Kunststoffe)
setting angle Anstellwinkel *m*
setting cure Vorvulkanisation *f* (Gummi)
setting device Einstellvorrichtung *f*
setting period Härtungszeit *f*
setting point Erstarrungspunkt *m*
setting quality Abbindefähigkeit *f*
setting retarder Abbindeverzögerer *m*
setting temperature Härtetemperatur *f*
setting time Erstarrungszeit *f*, Gelierzeit *f* (of
cement), Abbindezeit *f* (Zement)
setting up Eindicken *n* (Farbe), Montage *f*
settle *v* absitzen (Niederschlag), ausgleichen,
auslagern (Bier), begleichen, beilegen (Streit),
entscheiden, klären (Flüssigkeit),
sedimentieren (Niederschlag), sich setzen
(Niederschlag)
settled state Beharrungszustand *m*

settlement Abkommen *n*, Abmachung *f*,
Klärung *f* (Flüssigkeit)
settler Abscheider *m* (Dispersion),
Absetzbehälter *m*, Klärfaß *n*
settling Absetzen *n*, Klären *n*, Trennverfahren *n*
(für Dispersionen)
settling apparatus Klärapparat *m*
settling bassin Absatzbassin *n*
settling box Reinigungskasten *m*
settling cask Klärfaß *n*
settling cone Stoffänger *m* (Pap)
settling filter Anschwemmfilter *m*
settling pit Absitzgrube *f*
settlings Satz *m* (Niederschlag)
settling sump Klärsumpf *m*
settling tank Absatzbecken *n*, Absetzbassin *n*,
Absetzbehälter *m*, Klärbecken *n*,
Klärbottich *m*, Klärteich *m*, Setzkasten *m*,
secondary (or final) ~ Nachklärbecken *n*
settling time Absetzzeit *f*
settling vat Absetzbottich *m*, Klärbottich *m*,
Setzbottich *m*
settling velocity Sinkgeschwindigkeit *f*
settling vessel Absetzbehälter *m*, Absetzgefäß *n*,
Nachgärungskasten *m*
settling volume Absitzvolumen *n*
setup Aufbau *m*, Aufstellung *f*, ~ **for a
calculating machine** Rechenschaltung *f*
set up *v* aufstellen, aufsetzen, errichten,
montieren
set-up costs Rüstkosten *pl*
sevenfold siebenfach
seven-membered siebengliedrig
seven-membered ring Siebenring *m* (Chem)
sever *v* [ab]scheiden, absondern, [ab]trennen,
lösen
sew *v* heften (Bücher), nähen
sewage Abwasser *n*, Sickerwasser *n*, **salvage of** ~
Abwasserverwertung *f*, **supervision of** ~
Abwasserkontrolle *f*
sewage clarification Abwasserklärung *f*
sewage disposal plant Abwasseranlage *f*
sewage flow Abwasserzufluß *m*
sewage pit Sickergrube *f*
sewage plant Kläranlage *f*
sewage sludge Klärschlamm *m*
sewage water Abwasser *n*,
Kanalisationsabwasser *n*, Kloakenwasser *n*
sewage water pump Abwasserpumpe *f*
sewer Abwasserkanal *m*, Graben *m*, Kloake *f*
sewerage system Kanalisierung *f*
sewer gas Faulschlammgas *n*, Klärgas *n*,
Kloakengas *n*
sewer gas plant Klärgasanlage *f*
sewing Nähen *n*
sexangular sechseckig
sex chromosome Geschlechtschromosom *n*
sex hormone Geschlechtshormon *n*,
Sexualhormon *n*, **female** ~ weibliches

Geschlechtshormon, **male** ~ männliches
Geschlechtshormon
sextol Sextol *n*
sextol phthalate Sextolphthalat *n*
sextol stearate Sextolstearat *n*
seybertite Seybertit *m* (Min)
shackle Schäkel *m*
shackle hook Wirbelhaken *m*
shade Schatten *m*, Farbtönung *f* (Farbe),
Farbton *m* (Farbe), Lichtschutz *m*, Nuance *f*
(Farbe), Schattierung *f* (Farbe), **change in** ~
Farbtonänderung *f*, Farbtonverschiebung *f*,
depth of ~ Farbtiefe *f*
shade *v* ablichten; abstufen (Farbe), nuancieren
(Farbe), schönen, tönen (Farbe), verdunkeln
shade card Spinnfarbenkarte *f*
shading Farbabstufung *f*, Farbenschattierung *f*,
Farbtönung *f*
shading paint Blendschutzanstrich *m*,
Sonnenschutzfarbe *f*
shadow Schatten *m*
shadow cloth Schattiergewebe *n*
shadow cone Begrenzungskegel *m*
shadow image Schattenbild *n*
shadow picture Schattenbild *n*
shadow projection microscopy
Schattenmikroskopie *f*
shaft Achse *f* (Techn), Schacht *m* (Bergb),
Schaft *m*, Stiel *m* (Werkzeug), Welle *f*
(Techn), **articulated** ~ Gliederwelle *f*, **flexible**
~ Federwelle *f*, **free** ~ glatte Welle, **to sink
the** ~ den Schacht abteufen, ~ **of furnace**
Ofenschacht *m*, ~ **with key way** genutete
Welle
shaft bearing Wellenlager *n*
shaft bender Gabelbieger *m*
shaft brick Schachtstein *m*
shaft building, section for ~
Grubenausbauprofil *n*
shaft coupling Wellenkupp[e]lung *f*
[shaft] crucible furnace Tiegelschachtofen *m*
shaft diameter Schachtdurchmesser *m*
shaft furnace Schachtofen *m*, ~ **for hardening**
Schachthärteofen *m*, ~ **for reheating**
Schachtglühofen *m*
shaft governor Achsenregler *m*
shaft hammer Stielhammer *m*
shafting Wellenleitung *f*
shaft installation Schachtanlage *f*
shaft journal Wellenzapfen *m*
shaft kiln Schachtofen *m*
shaft leather Fahlleder *n*
shaft lining Schachtauskleidung *f*,
Schachtausmauerung *f*
shaft plant Schachtanlage *f*
shaft pump Schachtpumpe *f*
shaft top Schachtaufsatz *m*
shaggy struppig
shagreen genarbtes (körniges) Leder *n*

shake Riß *m* (Holz)
shake *v* [auf]schütteln, beben, rütteln, ~ out ausschütteln, ~ up aufrütteln
shaken rissig (Holz)
shaker Schüttelvorrichtung *f*
shaker conveyor Schüttelrutsche *f*
shaker screen Schüttelsieb *n*
shaker sifter Schüttelsieb *n*
shaking Durchschütteln *n*, Schütteln *n*
shaking apparatus Schüttelapparat *m*, Schüttelvorrichtung *f*
shaking autoclave Schüttelautoklav *m*
shaking effect Schüttelwirkung *f*
shaking grate Schüttelrost *m*
shaking machine Schüttelmaschine *f*
shaking motion Rüttelbewegung *f*, Schüttelbewegung *f*
shaking movement Schüttelbewegung *f*
shaking out Ausschütteln *n*
shaking screen Wurfsieb *n*
shaking shoot Schüttelrutsche *f*
shaking sieve Rüttelsieb *n*, Schüttelsieb *n*, Schwingsieb *n*
shaking speed Schüttelfrequenz *f*
shaking table Rätter *m* (Bergb), Rütteltisch *m* (Bergb)
shaking trough Schüttelrinne *f*
shale Schiefer *m* (Min)
shale oil Schiefer[schwel]öl *n*
shale tar Schieferteer *m*
shallow flach, seicht
shallowness Untiefe *f*
shallow water wave Seichtwasserwelle *f*
shammy leather Chamois *n* (Leder), Sämischleder *n*
shampoo Haarwaschmittel *n*
shank Schaft *m*, Stiel *m* (Bot), Unterschenkel *m* (Anat)
shank cutter Schaftfräser *m*
shank diameter Schaftdurchmesser *m*
shank length Schenkellänge *f*
Shannon's sampling theory Abtasttheorem *n* (nach Shannon)
shantung Schantung *m* (Text)
shapability Formbarkeit *f*
shape Fasson *f*, Form *f*, Gestalt *f*, change of ~ Formänderung *f*, stability of ~ Formbeständigkeit *f*
shape *v* formen, bilden, fassonieren (Techn), gestalten, modellieren (Techn)
shapeable formbar
shape factor (of particles) Formfaktor *m*
shapeless formlos, gestaltlos
shapelessness Formlosigkeit *f*
shape-retaining formstabil
shaper plate Formplatte *f*
shaping Formgebung *f*, Formung *f*, Verformung *f*
shaping method Formungsmethode *f*

shaping pass Formkaliber *n*
shaping ring Vorformring *m*
shaping steam Bombierdampf *m*
shaping [work] Fassonarbeit *f* (Techn)
share Teil *m*, Anteil *m*, Beitrag *m*
sharp scharf, beißend (z. B. Geruch), schneidend (z. B. Wind), schrill (Ton), spitz[ig]
sharp bend Knick *m*
sharp-edged scharf begrenzt, scharfkantig
sharpen *v* schärfen, [an]spitzen, schleifen, zuspitzen
sharpener Bleistiftspitzer *m*
sharp fire paint Scharffeuerfarbe *f*
sharply defined scharf begrenzt
sharpness Schärfe *f*
sharp V-thread Spitzgewinde *n*
shatter *v* zerschlagen, zersplittern, zertrümmern
shattering power Brisanz *f*
shatter oscillation Zerreißfrequenz *f*
shatterproof splitterfrei (Glas), unzerbrechlich (Glas)
shatterproofness Unzerbrechlichkeit *f*
shatter test (coke) Stürzprobe *f*
shattuckite Shattuckit *m* (Min)
shaving lather Rasierschaum *m* (Kosmetik)
shaving lotion Rasierwasser *n* (Kosmetik)
shaving machine Abziehmaschine *f*
shavings Späne *m pl*, Drehspäne *pl*, Hobelspäne *pl*, Schabsel *n pl*
shea butter Schibutter *f*
sheaf Garbe *f*, Haspel *f*
shear Abscherung *f*, Scherbeanspruchung *f*, Scherung *f*, Schub *m*, elasticity in ~ Schubelastizität *f*, resistance to ~ Schubwiderstand *m*
shear *v* [ab]schneiden, scheren
shear action Scherwirkung *f*
shear area Abscherfläche *f*
shear blade Scherblatt *n*
shear coefficient Schubzahl *f*
shear edge Abquetschfläche *f*, Abquetschrand *m*
shear fracture Scherungsbruch *m*
shearing Abscherung *f*, Scheren *n*, modulus of ~ Schermodul *m*, Schubelastizitätsmodul *m*
shearing disc mixer Scherscheibenmischer *m*
shearing force Scherbeanspruchung *f*, Scherkraft *f*, Schubkraft *f*
shearing instability Scherungsinstabilität *f*
shearing limit Schergrenze *f*
[shearing] modulus of elasticity Schubelastizitätsmodul *m*
shearing resilience Scherspannung *f*
shearing strain Scherbeanspruchung *f*
shear[ing] strength Scherfestigkeit *f*, Schubfestigkeit *f*
shearing stress Scherbeanspruchung *f*, Scherkraft *f*, Schubbeanspruchung *f*
shearing stress line Schublinie *f*
shearing test Abscherversuch *m*

shear modulus Schubmodul *m*, Torsionsmodul *m*
shear [off] *v* abscheren
shear pin Scherbolzen *m*
shears Schere *f*, Blechschere *f*
shear steel Gerbstahl *m*
shear stress Schubspannung *f*
sheath Scheide *f*, Hülse *f*, Mantel *m*,
Schutzhülle *f*
sheathe *v* umhüllen, ummanteln
sheathing Ummantelung *f*, Verkleidung *f*,
Verschalung *f*, ~ **of cables**
Kabelummantelung *f*
sheathing compound Kabelmasse *f*
sheave Rolle *f* (Flaschenzug), Scheibe *f*
(Flaschenzug)
shed *v* vergießen, verschütten, abwerfen
sheen Glanz *m*, Politur *f*, **high** ~ Hochglanz *m*
sheet Blatt *n* (Papier), Blech *n* (Techn),
Bogen *m* (Papier), [dünne] Platte *f* (Techn),
Folie *f* (Techn)
sheet aluminum Aluminiumblech *n*
sheet calender Bogenkalander *m*
sheet construction Blechkonstruktion *f*
sheet copper Kupferblech *n*, Blattkupfer *n*,
Kupferblatt *n*
sheet cork Korkscheibe *f*
sheet filter Schichtenfilter *n*, **drum type** ~
Trommelschichtenfilter *n*
sheet furnace Blechwärmofen *m*
sheet gauge Blechlehre *f*
sheet gelatin Gelatinefolie *f*
sheeting Holzverschalung *f*, Folienmaterial *n*
sheeting calender Folienkalander *m*
sheeting die Breitschlitzdüse *f*
sheet ingot Walzbarren *m*
sheeting out roller Ausziehwalze *f*
sheet iron Eisenblech *n*, Blechtafel *f*, **profiled** ~
Formblech *n*, **thick** ~ Grobblech *n*, **tinned** ~
Weißblech *n*
sheet iron anode Eisenblechanode *f*
sheet iron belt Eisenblechring *m*
sheet iron ring Eisenblechring *m*
sheet iron scraps Ausschußblech *n*
sheet iron shell Blechmantel *m*
sheet lead Bleiblech *n*, Walzblei *n*
sheet lining machine Bogenkaschiermaschine *f*
sheet metal Blech *n*, **rolled** ~ Walzblech *n*
sheet metal container Blechbehälter *m*
sheet metal edge Blechkante *f*
sheet metal lining Verkleidung *f* aus Blech
sheet metal printing Blechdruck *m*
sheet metal scrap Blechabfälle *m pl*
sheet metal strip Blechstreifen *m*
sheet metal tube Blechrohr *n*
sheet metal varnish Blechlack *m*
sheet metal working Blechbearbeitung *f*,
Blechverarbeitung *f*
sheet mica Glimmerfolie *f*, Plattenglimmer *m*
sheet rolling Blechwalzen *n*

sheet [rolling] mill Blechwalzwerk *n* (Feinblech),
Feinstraße *f* (Metall)
sheet rubber Blattgummi *m*, Plattenkautschuk *m*
sheet steel Stahlblech *n*, Blattstahl *m*
sheet tin Zinnblech *n*
sheet width Tafelbreite *f*
sheet zinc Zinkblech *n*
shelf Bord *n*, Fach *n*, Regal *n*
shelf dryer Hordentrockner *m*
shelf life Haltbarkeit *f*, Lagerbeständigkeit *f*,
Lagerfähigkeit *f*, Umschlagzeit *f*
shelf-type reactor Hordenofen *m*
shell Hülle *f*, Außenmantel *m*
(Wärmeaustauscher), Gehäuse *n*, Hülse *f*
(Bot), Schale *f* (Elektronen) (Früchte etc.),
Schote *f* (Bot), Umhüllung *f*, Ummantelung *f*,
closed ~ vollbesetzte Schale (Atom), **inner** ~
innere Schale (Atom), **outer** ~ äußere Schale
(Atom), ~ **of bosh** Rastmantel *m*
shell *v* [ab]schälen, enthülsen, ~ **off** abblättern
(Med)
shellac *v* mit Schellack streichen
shellac[k] Schellack *m*, **bleached** ~ gebleichter
Schellack, **white** ~ gebleichter Schellack
shellac substitute Achatschellack *m*
shellac varnish Schellackfirnis *m*
shell-and-tube condenser
Rohrbündelkondensator *m*
shell and tube heat exchanger
Glattrohrbündel-Wärmeaustauscher *m*
shellane Schellan *n*
shell distribution Schalenverteilung *f*
sheller Enthülser *m*
shell form Hohlform *f*
shell gold Malergold *n*
shell iron Gehäuseeisen *n*
shell limestone Muschelkalk[stein] *m* (Min)
shell marble Muschelmarmor *m* (Min),
Drusenmarmor *m* (Min)
shell marl Muschelerde *f*, Muschelmergel *m*
shell model Schalenmodell *n* (Atom)
shellolic acid Schellolsäure *f*
shell reamer Hohlreibahle *f*
shell transformer Manteltransformator *m*
shell-type magnet Mantelmagnet *m*
shelly muschelig
shelter Bunker *m*, Schutz *m*
shelter *v* schützen
shelves Gestell *n* (Regal)
sherardize *v* sherardisieren, trocken verzinken
sherardizing process Sherardisier-Verfahren *n*
sheridanite Sheridanit *m* (Min)
shibuol Shibuol *n*
shield *v* [ab]schirmen, ausblenden (Elektr)
shielded abgeschirmt
shielded enclosure Schutzkammer *f*
shielding Abschirmung *f*, Mantel *m*,
Schutzbedeckung *f*, Schutzhülle *f*
shielding chamber Abschirmkammer *f*

shielding constant Abschirmkonstante *f*
shield[ing] diaphragm Schirm *m*
shift Schicht *f* (Arbeitsschicht), Verschiebung *f,*
  Versetzung *f,* Wechsel *m,* **horizontal** ~
  Horizontalverschiebung *f,* **output per** ~
  Schichtleistung *f,* ~ **of alkyl groups**
  Umlagerung *f* von Alylgruppen (pl), ~ **of**
  spectral lines Spektrallinienverschiebung *f*
shift *v* die Lage ändern, schalten (Getriebe), sich
  verlagern, umschalten (Techn), verschieben,
  wandern, wechseln
shifter Ausrücker *m* (Techn)
shifting device Ausrückvorrichtung *f*
shift lever Schalthebel *m,* Umstellhebel *m*
shift plate Schaltblech *n*
shift register Schieberegister *n* (Comp)
shikimic acid Shikimisäure *f*
shikimol Safrol *n*
shikonin Shikonin *n*
shim Beilegscheibe *f,* Paßring *m*
shimmer *v* schimmern, flimmern
shim rod Anpassungsstab *m,* Trimmstab *m*
shine Schein *m,* Lichtschein *m*
shine *v* scheinen, glänzen, leuchten, ~ **through**
  durchscheinen, durchstrahlen
shingle Schindel *f,* Meerkies *m*
shingle *v* zängen (Techn)
shingling Zängearbeit *f*
shining flimmernd, leuchtend
shining soot Glanzruß *m*
shiny glänzend, ~ **with oil** ölglänzend
ship *v* befördern; verladen, verschiffen,
  versenden
shipbuilding Schiffbau *m*
shipbuilding material Schiffbaumaterial *n*
shiploader Schiffsbelader *m*
shipment Lieferung *f,* Transport *m,*
  Verschiffung *f,* Versendung *f*
shipper (Am. E.) Spediteur *m*
shipping cask Versandfaß *n*
ship's bottom paint Unterwasserfarbe *f*
shipwreck Schiffbruch *m*
shirlan Shirlan *n*
shisonin Shisonin *n*
shock Stoß *m,* Erschütterung *f,* Schlag *m*
  (Elektr), Schock *m,* Zusammenprall *m,*
  resistance to ~ Schlagfestigkeit *f,*
  Stoßfestigkeit *f,* **susceptibility to** ~
  Stoßempfindlichkeit *f*
shock-absorbent erschütterungsfrei
shock absorber Schwingmetall *n,*
  Stoßdämpfer *m* (Auto)
shock absorber leg Federbein *n*
shock-absorbing stoßdämpfend
shock absorption Stoßdämpfung *f*
shock bending test Schlagbiegeprobe *f*
shock crushing test Stauchprobe *f*
shock elasticity Stoßelastizität *f*
shock excitation Stoßanregung *f,* Stoßerregung *f*

shock load Stoßbelastung *f*
shock motion ruckweise Bewegung *f* (Med)
shock pendulum Schlagpendel *n*
shock period Stoßdauer *f*
shock-proof erschütterungsfest, stoßfest
shock-proofness Stoßfestigkeit *f*
shock seed Anregekristalle *m pl*
shock setting Schockfixierung *f*
shock strength Stoßstärke *f*
shock stress Beanspruchung *f* auf
  Schlagfestigkeit, Schlagbeanspruchung *f*
shock test Schlagprobe *f,* Schlagversuch *m,* ~ **on**
  notched bar Kerbschlagprobe *f*
shock tube Druckwellenreaktor *m*
  (Gastechnologie), Schock[wellen]reaktor *m*
  (Gastechnologie), Stoßreaktor *m*
  (Gastechnologie)
shock wave Druckwelle *f,* Stoßwelle *f*
shock wave luminescence
  Verdichtungsstoßleuchten *n*
shock wave reactor Druckwellenreaktor *m*
  Schock[wellen]reaktor *m,* Stoßreaktor *m*
  (Gastechnologie)
shoddy Reißwolle *f,* Shoddy *n*
shoe blacking Schuhschwärze *f,* Schuhcreme *f*
shoe brake Backenbremse *f*
shoemakers thread Pechdraht *m*
shoe polish Schuhkreme *f,* Schuhwichse *f*
shogaol Shogaol *n*
shonanic acid Shonansäure *f*
shoot Keim *m* (Bot), Rutsche *f* (Techn),
  Sproß *m* (Bot)
shoot *v* schießen, ~ **out** herausschleudern
shop Laden *m,* Werkstatt *f* (Techn)
shop drawing Werkstattzeichnung *f*
shop test Werkstattprüfung *f*
Shore durometer Shore-Härteprüfer *m*
Shore hardness Shore-Härte *f*
Shore's dynamic indentation test
  Shore-Fallprobe *f*
short kurz, brüchig (Met), klein, knapp, mürbe
  (Gebäck etc.), spröde (Met)
shortage Mangel *m,* Fehlbestand *m,*
  Knappheit *f,* Verknappung *f*
short-armed kurzarmig
short-bed reactor Kurzschichtreaktor *m*
short circuit Kurzschluß *m,* **dead** ~ vollständiger
  Kurzschluß, **duration of** ~ Kurzschlußzeit *f*
short-circuit *v* kurzschließen
short-circuit brake Kurzschlußbremse *f*
short-circuit brush Kurzschlußbürste *f*
short-circuit carbon Kurzschlußkohle *f*
short-circuit contact Kurzschlußkontakt *m*
short-circuit current Kurzschlußstrom *m*
short-circuit disk Kurzschlußscheibe *f*
short-circuited kurzgeschlossen
short-circuiter Kurzschließer *m*
short-circuit excitation Kurzschlußerregung *f*

short-circuit furnace Kurzschlußofen *m*
short-circuit loop Kurzschlußschleife *f*,
  Kurzschlußwindung *f*
short-circuit loss Kurzschlußverlust *m*
short-circuit potential Kurzschlußpotential *n*
short-circuit resistance Kurzschlußwiderstand *m*
short-circuit ring Kurzschlußring *m*
short-circuit rotor Kurzschlußanker *m*
short-circuit spark Kurzschlußfunke *m*
short-circuit voltage Kurzschlußspannung *f*
shortcoming Unzulänglichkeit *f*, Mangel *m*
short-crested kurzkammig
shorten *v* [ab]kürzen, verkürzen, ~ by forging
  stauchen
shortened verkürzt
shortening Schrumpfung *f* (Techn), Stauchung *f*
  (Techn), Verkürzung *f*
short evaporator coil Steilrohrverdampfer *m*
short-fibered kurzfaserig
short-hair *v* putzen (Häute)
short-lived kurzlebig
short-necked kurzhalsig
short-necked round-bottom flask
  Kurzhalsrundkolben *m* (Chem)
shortness Brüchigkeit *f* (Met), Knappheit *f*,
  Sprödigkeit *f* (Met)
short paragraph kleine Textanzeige *f* (Werbung)
short-range disorder Nahunordnung *f*
short-range focusing Naheinstellung *f* (Opt)
short-range limit Kurzreichweitengrenze *f*
short-range order Nahordnung *f*
short-range order theory Nahordnungstheorie *f*
short rotary furnace Kurztrommelofen *m*
shorts Nebenprodukte *pl* (Techn)
short supply ungünstige Lieferlage *f*
short-term kurzzeitig
short thread Kurzgewinde *n*
short-time test Kurzversuch *m*
shortwave Kurzwelle *f*
shortwave *a* kurzwellig
shortwave radiation Kurzwellenstrahlung *f*
shortwave set Kurzwellenapparat *m*
shortway distiller Kurzwegdestillator *m*
shot Aufnahme *f* (Phot), Injektion *f* (Med),
  Schuß *m*
shotblast *v* sandstrahlen
shot-lubrication system Eindruckschmierung *f*
shot mold Kapsel *f* zum Stückkugelguß
shot molding Schrotpressen *n*
shot noise Schroteffekt *m* (Elektr)
shot weight Füllgewicht *n*, Schußgewicht *n*
shoulder Schulter *f*; Achsel *f*
shoulder area Schulterpartie *f*
shoulder strap Achselstück *n*
shove *v* schieben, stoßen, ~ in einschieben
shovel Schaufel *f*, Schippe *f*, to stir with a ~
  umschaufeln
shovel excavator Löffelbagger *m*

show *v* [an]zeigen, anweisen, aufweisen, ~
  through durchschlagen (Farbe)
shower Schauer *m*, Dusche *f*, ~ of electrons
  Elektronenschwarm *m*
show room Ausstellungsraum *m*
shred Bruchstück *n*
shred *v* zerfasern, zerkleinern, zerschnitzeln
shredder Zerfaserer *m* (Pap)
shrimp Krabbe *f* (Zool)
shrink *v* [ein]schrumpfen, eingehen, einlaufen,
  krimpen, krumpfen (Text), schwinden, sich
  zusammenziehen, zusammenschrumpfen, ~
  on aufschrumpfen
shrinkage Einschrumpfen *n*, Abnahme *f*
  (Schrumpfung), Einlaufen *n* (Text),
  Schrumpfung *f*, Schwinden *n*, Schwund *m*,
  Zusammenschrumpfen *n*, amount of ~
  Schwindmaß *n*, resistance to ~
  Einlauffestigkeit *f* (Text), ~ of cloth
  Eingehen *n* des Gewebes (Text), ~ of the
  steel Werfen *n* des Stahls
shrinkage block Schrumpfstück *n*,
  Schrumpfvorrichtung *f*
shrinkage block or jig Richtvorrichtung *f*
shrinkage cavity Lunkerhohlraum *m*,
  Schwindlunker *m*
shrinkage gauge Schwindmaß *n*
shrinkage spot Schwundstelle *f*
shrink film Schrumpffolie *f*
shrink form Richtform *f* (beim Akkukasten)
shrinkhole Lunker *m*; Lunkerhohlraum *m*,
  Lunkerstelle *f*
shrinkholes, free of ~ lunkerfrei, lunkerlos
shrinking Lunkerbildung *f* (Techn),
  Schrumpfung *f*
shrinking lacquer Schrumpflack *m*
shrinking thread Schrumpfgewinde *n*
shrink packaging Schrumpfverpackung *f*
shrink proofing Krumpffestmachen *n* (Text)
shrink resistance Krumpffestigkeit *f*
shrink-resistant einlaufecht, krumpfecht
shrink wrap film Folienhaube *f*
shrivel *v* [ein]schrumpfen, runzeln,
  zusammenschrumpfen
shrivel varnish Kräusellack *m*
shrunk-on pipe joint Einsteckklebeverbindung *f*
shunt Nebenschluß *m*, Shunt *m*, inductive ~
  induktiver Nebenschluß, magnetischer
  Nebenschluß, magnetic ~ magnetischer
  Nebenschluß, to put in ~ shunten
shunt *v* einen Nebenschluß bilden, shunten, ~
  off abzweigen (Elektr)
shunt box Vorschaltkasten *m* (Elektr)
shunt circuit Nebenschlußstromkreis *m*
shunt connection Nebenschlußschaltung *f*
shunt current Nebenschlußstrom *m*
shunted parallelgeschaltet (Elektr)
shunted current Abzweigstrom *m*
shunt excitation Nebenschlußerregung *f*

shunt magnet Nebenschlußmagnet *m*
shunt ratio Nebenschlußverhältnis *n*
shunt regulation Nebenschlußregelung *f*
shunt resistance Abzweigwiderstand *m*,
  Nebenwiderstand *m*
shunt wire Abzweigdraht *m*
shunt-wound arc lamp Nebenschlußbogenlampe *f*
shunt-wound motor Nebenschlußmotor *m*
shut *v* [ver]schließen, zumachen, [zu]sperren, ~
  off absperren, abdrosseln, abschalten,
  abschließen, abstellen (z. B. Maschine)
shutdown Außerbetriebsetzung *f*,
  Betriebsstillegung *f*, Betriebsstörung *f*,
  Betriebsunterbrechung *f*
shutoff Abstellvorrichtung *f*
shutoff cock Absperrhahn *m*
shutoff device Absperrvorrichtung *f*
shut-off nozzle Absperrdüse *f*, Verschlußdüse *f*
shutoff time Stillstandszeit *f*
shutoff valve Absperrventil *n*
shutter Spund *m* (Gieß), Verschluß *m* (Phot),
  Verschlußblende *f* (Phot)
shutter flap Verschlußklappe *f* (Phot)
shutter lever Blendenhebel *m* (Opt)
shutter mask Abdeckblende *f* (Opt)
shutter release Verschlußauslösung *f* (Phot)
shutting down Außerbetriebsetzung *f*
shutting off the blast Abstellen *n* des Gebläses
shuttle Schiffchen *n*, Weberschiffchen *n*,
  Webschütze *m*
shuttle *v* hin- und herbewegen
shuttle platform Verschiebebühne *f*
sialic acid Sialsäure *f*
sialyl transferase Sialyltransferase *f* (Biochem)
Siamese siamesisch
siaresinolic acid Siaresinolsäure *f*
siberite Siberit *m* (Min)
siccative Sikkativ *n*, Trockenmittel *n*,
  Trockenpulver *n*
siccative oil Trockenöl *n*
siccative varnish Trockenfirnis *m*
sick krank
sickle cell Sichelzelle *f*
sickle-cell hemoglobin Sichelzellenhämoglobin *n*
sickle form Sichelform *f*
sicklerite Sicklerit *m* (Min)
sickness Erkrankung *f*, Krankheit *f*, Übelkeit *f*
side Seite *f*, Neben-
side airlift agitator Seitenmischluftrührer *m*
side bars or stanchions Holme *pl*
side bearing Längslager *n*
side bracket bearing Flanschlager *n*
side by side nebeneinander
side chain Seitenkette *f* (Chem)
side chamber Seitenkammer *f*
side-channel compressor Ringgebläse *n*
side-channel pump Seitenkanalpumpe *f*
side cutting tool Seitendrehmeißel *m*
side drain Nebenkanal *m*

side effect Nebenwirkung *f*
side flue Seitenzug *m*
side frequency Nebenfrequenz *f*,
  Nachbarfrequenz *f*
side-inverted seitenverkehrt
side issue Randerscheinung *f*
side jib Seitenausleger *m*
side leakage Flankenstreuung *f*
side line Nebenbetrieb *m*
side loader Seitengabelstapler *m*
side-loading forklift Quergabelstapler *m*,
  Seitengabelstapler *m*
side milling cutter Walzenfräser *m*
side neck Seitenhals *m*
side-outlet [pipe] clamp Anbohrschelle *f*
side projection Profil *n*
side reaction Nebenreaktion *f*
sidereal year Sternjahr *n* (Astr)
side rest Abstellplatte *f*
side ring Hornring *m* (der Felge)
siderin yellow Sideringelb *n*
siderite Meteorgestein *n* (Min), Eisenspat *m*
  (Min), Siderit *m* (Min), decomposed ~
  Blauerz *n* (Min)
siderocalcite eisenhaltiger Dolomit *m* (Min)
side roll Futterwalze *f*
sideroscope Sideroskop *n*
siderosis Metallstaublunge *f* (Med)
siderurgical cement Eisenportlandzement *m*
sideseam, interlocked ~ gefalzte Längsnaht *f*
  (Dose)
side surface Seitenfläche *f*
side tool Seitenstahl *m*
side view Seitenansicht *f*
side wall Seitenwand *f*
sideways seitwärts, seitlich
siding (Am. E.) Fassade *f*
sidio quartz glass utensil Sidioquarzglasgerät *n*
Sidot's blende Sidot-Blende *f*
siegenite Siegenit *m* (Min)
Siemens butterfly valve Siemenssche
  Wechselklappe *f*
Siemens-Martin furnace Siemens-Martin-Ofen *m*
Siemens-Martin process
  Siemens-Martin-Prozeß *m*,
  Herdfrischprozeß *m*
Siemens-Martin steel works Martinstahlhütte *f*
Siemens regenerative open-hearth furnace
  Siemens-Regenerativfeuerung *f*
sieve Sieb *n*, Durchwurf *m*, Rätter *m*, coarse ~
  Rohsieb *n*, cylindrical ~ Trommelsieb *n*,
  rotary ~ Siebtrommel *f*, wide meshed ~
  Grobsieb *n*, grobes Sieb *n*
sieve *v* [durch]sieben
sieve analysis Siebanalyse *f*
sieve bend Bogensieb *n*
sieve dryer Siebtrockner *m*
sieve grate Siebrost *m*
sieve plate Siebboden *m*, Siebplatte *f*

sieve plate column Siebplattenkolonne *f*,
　pulse-type ~ pulsierende Siebbodenkolonne *f*
sieve residue Siebrückstand *m*
sieve residues Grieben *pl*
sieves, set of ~ Siebsatz *m*
sieve tray Siebboden *m*
sieving Sichtung *f*
sieving filter Siebfilter *m*
sieving machine Siebmaschine *f*
sift *v* [durch]sieben, ~ **[off]** absieben, ~ **out**
　aussieben
sifter Sieb *n*
sifting Durchsieben *n*, Sieben *n*
sifting device Sichter *m* (Bergb),
　Siebvorrichtung *f*
sight Sehkraft *f*
sight [at] *v* anvisieren
sight-feed lubricator Öltropfgefäß *n*, Tropföler *m*
sight glass Schauglas *n*
sight hole Schauöffnung *f*
sighting apparatus Visierapparat *m*
sighting mark Index *m* (bei Maßstäben)
sighting mark error Indexfehler *m*
sigma bond Sigma-Bindung *f*
sigma complex Sigma-Komplex *m*
sigma pile Sigmareaktor *m*
sign Zeichen *n*, Anzeichen *n*, Kennzeichen *n*,
　Merkmal *n*, Symbol *n*, Vorzeichen *n* (Math),
　change of ~ Vorzeichenwechsel *m* (Math),
　determination of ~ Vorzeichenbestimmung *f*
　(Math), **negative** ~ negatives Vorzeichen,
　positive ~ positives Vorzeichen
sign *v* unterschreiben, unterzeichnen, signieren
signal Signal *n*, Zeichen *n*, **acoustic** ~
　akustisches Signal
signal alarm Signaleinrichtung *f*
signal current amplifier Meßstromverstärker *m*
signal fire Blinkfeuer *n*
signal flare Lichtsignal *n*
signal installation Signalanlage *f*
signal lamp Kontrollampe *f*, Signallampe *f*
signal[l]ing Signalgabe *f*, Signalisieren *n*
signalling instrument Signalkontaktgeber *m*
signal paint Markierungsfarbe *f*
signal strip Signalstreifen *m*
signal transmission Signalgabe *f*
signature Unterschrift *f*
sign board Schild *n*, Firmenschild *n*
sign convention Vorzeichenfestsetzung *f* (Math)
sign determination Vorzeichenbestimmung *f*
　(Math)
significance Bedeutung *f*, **level of** ~
　Signifikanzzahl *f* (Statistik)
significant bedeutend, bedeutsam
signification Bedeutung *f*
signify *v* bedeuten, bezeichnen
sign of charge determination
　Ladungsvorzeichenbestimmung *f*

signs, explanation of ~ Zeichenerklärung *f*,
　reversal of ~ Vorzeichenumkehrung *f*, **rule of**
　~ Vorzeichenregel *f*
signum function Signumfunktion *f* (Math)
silage [fodder] Silofutter *n*
silane Silan *n*
silanetriol Silantriol *n*
silencer Schalldämpfer *m*
silencing Schalldämpfung *f*
silent geräuschlos, ruhig, still, **to become** ~
　verstummen
silent discharge Glimmentladung *f*
silex Kiesel *m*, hitzebeständiges Glas *n*
silex white Kieselweiß *n*
silfbergite Silfbergit *m* (Min)
silhouette procedure Schattenbildverfahren *n*
silica Kieselerde *f*, Siliciumdioxid *n*, **containing**
　~ kieselhaltig, **fused** ~ Quarzglas *n*
silica content of limestone Kieselsäuregehalt *m*
　des Kalksteins
silica crystal Kieselkristall *m*
silica flask Quarzkolben *m* (Chem)
silica gel Silikagel *n*
silicane Silican *n*
silica refractory Quarzschamottestein *m*
silicate Silikat *n*, ~ **with chain structure**
　Kettensilikat *n*, ~ **with framework structure**
　Gerüstsilikat *n*
silicate paint Wasserglasfarbe *f*
silicate slag Silikatschlacke *f*
silication Silikatbildung *f*, **degree of** ~
　Silizierungsgrad *m*
silica triangle Quarzdreieck *n*
silica tubing Quarzrohr *n*
siliceous kiesel[erde]haltig, kieselsäurehaltig
siliceous blende Kieselblende *f* (Min)
siliceous chalk Siliciumkreide *f*
siliceous earth Kieselerde *f* (Min)
siliceous feldspar Kieselspat *m* (Min)
siliceous flux Kieselfluß *m*
siliceous rock Kieselgestein *n*
siliceous schist Kieselschiefer *m* (Min)
siliceous sinter Kieselsinter *m*, Quarzsinter *m*,
　Sinterquarz *m*
silicic acid Kieselsäure *f*, **salt of** ~ Silikat *n*
silicic anhydride Kieselsäureanhydrid *n*,
　Siliciumoxid *n*, Silika *f*
silicicolin Silicicolin *n*
silicide Silicid *n*
siliciferous kieselerdehaltig, siliciumhaltig
silicification Silizierung *f*, Verkieselung *f*, **degree**
　of ~ Silizierungsgrad *m*
silicify *v* verkieseln
siliciolite Siliciolith *m* (Min)
siliciophite Siliciophit *m* (Min)
silicious see siliceous
silicium (Symb. Si) Silicium *n*
silicoacetic acid Silicoessigsäure *f*
silicocalcareous kieselkalkhaltig

silicochloroform Siliciumchloroform *n*
silicoethane Siliciumäthan *n*, Silicoäthan *n*
silicofluoric acid Fluorkieselsäure *f,*
    Siliciumfluorwasserstoffsäure *f,* **salt of** ~
    Silicofluorid *n*
silicofluoride Silicofluorid *n*,
    Fluorsiliciumverbindung *f*
silicoformic acid Silicoameisensäure *f*
silicol Silikol *n*
silicomanganese Silicomangan *n*
silico-manganese steel Mangansiliciumstahl *m*
silicomethane Siliciummethan *n*, Silican *n*
silicon (Symb. Si) Silicium *n*, Silizium *n*
silicon boride Siliciumborid *n*
silicon bromide Siliciumbromid *n*
silicon bronze Siliciumbronze *f*
silicon bronze wire Siliciumbronzedraht *m*
silicon carbide Karborund *n*,
    Kohlenstoffsilicium *n*, Siliziumcarbid *n*
silicon chloride Siliciumchlorid *n*
silicon chloride bromide Siliciumchlorbromid *n*
silicon compound Siliciumverbindung *f*
silicon content Siliciumgehalt *m*
silicon dioxide Siliciumdioxid *n*,
    Kieselsäureanhydrid *n*, Silika *f*
silicon dressing plant Silikaaufbereitungsanlage *f*
silicone Silicon *n*
silicone grease Siliconschmierfett *n*
silicone oil Siliconöl *n*
silicone resin Siliconharz *n*
silicone rubber Siliconkautschuk *m*
silicon fluoride Siliciumfluorid *n*
silicon fluorine compound
    Siliciumfluorverbindung *f*
silicon hydride Silican *n*, Siliciumwasserstoff *m*
siliconic acid Siliconsäure *f*
silicon iodide Siliciumjodid *n*
silicon iodoform Siliciumjodoform *n*
silicon iron Ferrosilicium *n*
siliconize *v* silizieren
siliconizing Aufsilizieren *n*
silicon manganese steel Siliciummanganstahl *m*
silicon monoxide Siliciummonoxid *n*
silicon nitride Stickstoffsilicid *n*
silicon steel Siliciumstahl *m*
silicon tetrabromide Siliciumtetrabromid *n*
silicon tetrachloride Tetrachlorsilicium *n*
silicon tetrafluoride Siliciumtetrafluorid *n*
silicon tetrahydride Siliciummethan *n*
silicon tetraiodide Siliciumtetrajodid *n*
silicon tetraphenyl Tetraphenylsilicium *n*
silicooxalic acid Siliciumoxalsäure *f,*
    Silicooxalsäure *f*
silicopropane Silicopropan *n*
silicosis Silikose *f* (Med)
silicotungstic acid Kieselwolframsäure *f*
silite Silit *n* (HN)
silk Seide *f* (Text), **artificial** ~ Kunstseide *f*
    (Text), Reyon *m*, **degummed** ~ Cuiteseide *f,*

half boiled ~ Soupleseide *f,* **natural** ~ echte
    Seide, **oiled** ~ Ölseide *f,* **pure** ~ Reinseide *f,*
    raw ~ Rohseide *f* (Text), **weighted** ~
    beschwerte Seide
silk gelatin Sericin *n*
silk glue Sericin *n*
silkworm Seidenraupe *f* (Zool)
silky asbestos Amiant *m* (Min)
sillimanite Sillimanit *m* (Min), Faserstein *m*
    (Min), Fibrolith *m* (Min)
silo Silo *m* (Agr)
silo, silage pits Gärsilo *m*
siloxane Siloxan *n*
siloxen Siloxen *n*
siloxicon Siloxicon *n*, Carbosiliciumoxid *n*
silt Schlamm *m*, Kesselschlamm *m*
silt *v* verschlammen; aufschwemmen
silting Verschlammung *f*
silty schlammig
silundum Silundum *n*
silver (Symb. Ag) Silber *n*, **bright** ~
    Glanzsilber *n*, **capillary** ~ Haarsilber *n*
    (Min), **coined** ~ gemünztes Silber, **colloidal** ~
    kolloidales Silber, Credesches Silber *n*, **crude**
    ~ Rohsilber *n*, **deposited** ~ Fällsilber *n*,
    deposit of ~ Silberniederschlag *m*, **grain of** ~
    Silberkorn *n*, **native** ~ gediegenes Silber,
    powdered ~ Silberpulver *n*, **precipitated** ~
    Fällsilber *f,* **refined** ~ Raffinatsilber *n*,
    separation of ~ Silberscheidung *f,* **solid** ~
    massives Silber, **test for** ~ Silberprobe *f,*
    unrefined ~ Rohsilber *n*, **vitreous** ~
    Glanzsilber *n*, ~ **containing dross**
    Aftersilber *n*, ~ **from lead ore** Werksilber *n*
silver *a* silbern
silver *v* versilbern
silver acetate Silberacetat *n*
silver acetylide Silberacetylid *n*
silver albuminate Silberalbuminat *n*
silver alloy Silberlegierung *f*
silver amalgam Silberamalgam *n*,
    Amalgamsilber *n*
silver antimonide Silberantimonid *n*
silver antimony glance Silberantimonglanz *m*
    (Min), Miargyrit *m* (Min)
silver antimony sulfide Silberantimonsulfid *n*
silver arsenate Silberarsenat *n*
silver asbestos Silberasbest *m* (Min)
silver assay Silberprobe *f*
silver azide Silberazid *n*
silver balance Silberwaage *f*
silver bichromate Silberbichromat *n*
silver bismuth sulfide, native ~ Matildit *m* (Min)
silver brick Silbersau *f*
silver bromide Silberbromid *n*
silver bromide collodion Bromsilberkollodium *n*
silver bromide gelatin Bromsilbergelatine *f*
    (Phot)

silver bromide gelatin paper
Bromsilbergelatinepapier *n*
silver bromide paper Bromsilberpapier *n* (Phot)
silver carbide Silbercarbid *n*
silver caseinate Argonin *n*
silver chlorate Silberchlorat *n*
silver chloride Silberchlorid *n*, **incrustation of** ~
Chlorsilberkruste *f*, **native** ~ Hornerz *n*
(Min), Hornsilber *n* (Min)
silver chloride bath Chlorsilberbad *n*
silver chloride battery Chlorsilberelement *n*
silver chloride gelatin Chlorsilbergelatine *f*
silver chloride gelatin paper
Chlorsilbergelatinepapier *n*
silver cladding Silberplattierung *f*
silver-coated mirror Silberspiegel *m*
silver coin Silbermünze *f*
silver-colored silberfarben
silver content Silbergehalt *m*
silver copper glance Silberkupferglanz *m* (Min)
silver crucible Silbertiegel *m*
silver cyanide Silbercyanid *n*
silver cyanide bath Silbercyanidbad *n*
silver dichromate Silberbichromat *n*
silver dithionate Silberdithionat *n*
silver enamel Silberlack *m*
silver enanthate Silberönanthat *n*
silver filings Silberfeilspäne *pl*
silver fir Edeltanne *f*, Weißtanne *f*
silver fir oil Weißtannenöl *n*
silver fluoride Silberfluorid *n*, Fluorsilber *n*
silver foil Silberfolie *f*, Blattsilber *n*, **imitation** ~
Rauschsilber *n*
silver formate Silberformiat *n*
silver foundry Silberhütte *f*
silver fulminate knallsaures Silber *n*,
Knallsilber *n*
silver glance Silberglanz *m* (Min), Argentit *m*
(Min), **black** ~ Melanglanz *m* (Min), **brittle**
~ Sprödglaserz *n* (Min), **earthy** ~
Silberschwärze *f* (Min)
silver grey silbergrau
silver halide Silberhalogenid *n*, Halogensilber *n*
silver hydroxide Silberhydroxid *n*
silvering Versilbern *n*, Versilberung *f*, ~ **by**
contact Kontaktversilberung *f*, ~ **by rubbing**
on Anreibeversilberung *f*
silvering bath, hot ~ Silbersud *m*, **hot light** ~
Silberweißbad *n*, Silberweißsud *m*
silver ingot Barrensilber *n*, Silbersau *f*
silver iodate Silberjodat *n*
silver iodide Silberjodid *n*, **native** ~ Jodyrit *m*
(Min)
silver lactate Silberlactat *n*, Actol *n*
silver leaf Blattsilber *n*
silver malate Silbermalat *n*
silver mine Silbermine *f* (Bergb)
silver nitrate Silbernitrat *n*, Ätzsilber *n*,
Höllenstein *m*

silver nitrate bath Höllensteinbad *n*
silver nitrate solution Silbernitratlösung *f*,
Höllensteinlösung *f*
silver nitride Stickstoffsilber *n*
silver ore Silbererz *n* (Min), **black** ~
Schwarzerz *n* (Min), **brittle** ~ Melanglanz *m*
(Min), Sprödglaserz *n* (Min), **poor** ~
Armstein *m* (Min), **red** ~ Rotgüldigerz *n*, **rich**
~ Formerz *n* (Min), **white** ~
Weißgüldigerz *n* (Min)
silver orthophosphate Silberorthophosphat *n*
silver oxalate Silberoxalat *n*
silver oxide Silberoxid *n*
silver paper Silberpapier *n*
silver pecipitate Silberniederschlag *m*
silver peroxide Silberperoxid *n*
silver phosphate Silberphosphat *n*
silver-plate *v* versilbern
silver-plated versilbert
silver-plating Versilbern *n*, Versilberung *f*, **bright**
~ Glanzversilberung *f*, **hard** ~
Hartversilberung *f*, **heavy** ~
Starkversilberung *f*, **solid** ~
Solidversilberung *f*
silver powder Silberpulver *n*, weiße Bronze *f*
silver precipitant Silberfällungsmittel *n*
silver propionate Silberpropionat *n*
silver pyrophosphate Silberpyrophosphat *n*
silver refinery Silberscheideanstalt *f*
silver refining Silberraffination *f*,
Silberscheidung *f*, **apparatus for** ~
Silberraffinationsvorrichtung *f*
silver refining hearth Silberbrennherd *m*
silver sand Silbersand *m*
silver selenide Selensilber *n*, Silberselenid *n*
silver separation Silberscheidung *f*
silver single crystal Silbereinkristall *m*
silver soap Metallseife *f*
silver solder Silberlot *n*
silver steel Silberstahl *m*
silver sulfide Silbersulfid *n*, Schwefelsilber *n*,
Weicherz *n* (Min)
silver tailings Silberschlamm *m*
silver telluride Tellursilber *n*
silver thread Fadensilber *n*
silver tinsel Lametta *f n*, Rauschsilber *n*
silver wire Silberdraht *m*
silver work Silberarbeit *f*
silver works Silberhütte *f*
silvery silberartig, silberglänzend, silberig,
silbern
silvestrene Silvestren *n*
silylamine Silylamin *n*
similar ähnlich, gleichartig, gleichförmig,
gleichnamig
similarity Ähnlichkeit *f*, Gleichartigkeit *f*, **law of**
~ Ähnlichkeitsgesetz *n*, **principle of** ~
Ähnlichkeitstheorie *f*
similarity principle Ähnlichkeitsmechanik *f*

similarity rule Ähnlichkeitsregel *f*
similarity theorem Ähnlichkeitssatz *m*
similar transformation [of matrices]
Ähnlichkeitstransformation *f* (Math)
similitude Ähnlichkeit *f*, **principle of** ~
Ähnlichkeitsprinzip *n*, **ratio of** ~
Ähnlichkeitsverhältnis *n*
similor Halbgold *n*, Similor *n*
simmer *v* leicht kochen, sieden
simmering [gelinde] Sieden *n*
simonellite Simonellit *n*
simple glide Einfachgleitung *f*
simplification Vereinfachung *f*,
Rationalisierung *f*
simplify *v* vereinfachen
simulate *v* nachahmen, vortäuschen
simulated vorgetäuscht
simulation Nachahmung *f*, Simulation *f*
simulator Simulator *m*
simultaneous gleichzeitig; simultan
sinalbin Sinalbin *n*
sinamine Sinamin *n*
sinapic acid Sinapinsäure *f*
sinapine Sinapin *n*
sinapisine Sinapisin *n*
sinapolin Sinapolin *n*
sinapyl alcohol Sinapinalkohol *m*
sine Sinus *m* (Math)
sine condition Sinusbedingung *f*
sine current Sinusstrom *m*
sine curve Sinuskurve *f*, Sinuslinie *f*
sine function Sinusfunktion *f* (Math)
sine galvanometer Sinusbussole *f*
sine series Sinusreihe *f*
sine-shaped sinusförmig
sine transformation Sinustransformation *f*
sinew Sehne *f* (Anat)
singe *v* versengen
singeing on hot plates Plattensengen *n*
single einzeln, einfach, einmalig
single *v* **out** heraussuchen
single-acting einfachwirkend
single-bath process Einbadverfahren *n*
single bead tire Einkernreifen *m*
single belt Einzelriemen *m*
single-blade einschauflig
single boiler Einfachkessel *m*
single bond Einfachbindung *f* (Chem)
single-cavity mold Einfachform *f*
single-celled einzellig
single collision Einzelstoß *m*
single column Einzelkolonnenanordnung *f*
(Gaschromat)
single-column press Einständerpresse *f*
single control shower mixer Mischbatterie *f*
single-core einadrig
single crystal Einkristall *m*
single-crystal surface Einkristalloberfläche *f*
single-cylinder einzylindrig

single dislocation Einzelversetzung *f*
single dividing method Einzelteilverfahren *n*
single drive Einzelantrieb *m*
single electrode potential
Einzelelektrodenpotential *n*
single-electrode system Halbelement *n*
single-filament lamp Einfadenlampe *f* (Elektr)
single-flamed einflammig
single-flow turbine Einstromturbine *f*
single-girder crane Einträgerkran *m*
single-grid tube (Br. E. valve) Eingitterröhre *f*
single-handed eigenhändig; einhändig
single heater Einzelheizer *m*
single hole nozzle Einlochdüse *f*
single-impression mold Einfachform *f*
single impression mo[u]ld Einfachwerkzeug *n*
single-layered einschichtig
single-leaf electrometer Einblattelektrometer *n*
single-lens einlinsig (Opt)
single-level formula Einniveauformel *f*
single lifting table Einzelaushubtisch *m*
single line process Einstrangverfahren *n*
single-lobe pump Sperrschieberpumpe *f*,
Trennflügelpumpe *f*
single manifold die einfache Breitschlitzdüse *f*
single nozzle Einzeldüse *f*
single opening Einzelmündung *f*
single part Einzelteil *n*
single particle level Einteilchenniveau *n*
single particle model Einteilchenmodell *n*
single particle shell model
Einteilchenschalenmodell *n*
single particle transition Einteilchenübergang *m*
single particle wave function
Einteilchenwellenfunktion *f*
single-phase einphasig, monophasisch
single-phase alternator Einphasengenerator *m*
single-phase armature Einphasenanker *m*
single-phase commutator motor
Einphasenkollektormotor *m*
single-phase current Einphasenstrom *m*
single-phase current plant
Einphasenstromanlage *f*
single-phase furnace Einphasenofen *m*
single-phase generator Einphasendynamo *m*
single-phase induction motor
Einphaseninduktionsmotor *m*
single-phase series motor
Einphasenserienmotor *m*
single-phase shunt motor
Einphasennebenschlußmotor *m*
single-phase system Einphasensystem *n*,
Einstoffsystem *n*
single-phase wiring Einphasenleitung *f*
single-piece work Einzelanfertigung *f*
single-point threading tool
Gewindeformdrehstahl *m*
single potential Einzelpotential *n*,
Einzelspannung *f*

single process Einzelvorgang *m*
single-purpose machine Einzweckmaschine *f*
(Techn)
single-refracting einfachbrechend
single refraction Einfachbrechung *f* (Opt)
single regulation Einzelregelung *f*
single-rod glass electrode Einstabglaselektrode *f*
single roller ring mill Einrollenmühle *f*
single-rotor screw pump Spindelpumpe *f*
single row einreihig
single-stage einstufig
single-stage recycle Einstufenrückführung *f*
single-step method Einzelschrittverfahren *n*
singlet Singulett *n*
single-threaded eingängig
singlet linkage Einelektronenbindung *f*
single-toggle crusher Einschwingenbrecher *m*
single transformer Einzeltransformator *m*
(Elektr)
singlet state Singulettzustand *m*
singlet system Singulettsystem *n*
single-valued einwertig (Math)
single valve Einzelschieber *m*
single-zone reactor Einzonenreaktor *m*
singlings Nachlauf *m*, Lutter *m* (Dest)
singular matrix singuläre Matrize *f*
sinigrin Sinigrin *n*
sinistrin Sinistrin *n*
sink Ausguß *m*, Abfluß *m*, Abflußrohr *n*,
Gießrinne *f* (Techn), Spültisch *m*
sink *v* [ver]sinken, abteufen (Techn), versenken,
~ [away] absinken, ~ down absacken
sinkaline Cholin *n*
sink and float process Sinkscheideverfahren *n*
sink and float separation
Schwimm-Sink-Aufbereitung *f*
sink basin Ausgußbecken *n*
sinker Senkkörper *m*
sinking Sinken *n*, Versinken *n*
sinking trestle Abteufgerüst *n*
sinking work Abteufarbeit *f*
sink mark Einfallstelle *f*, Einsackstelle *f*, Mulde *f*
sink plug Ausgußstopfen *m*
sink spot Einfallstelle *f*
sinomenine Sinomenin *n*
sinoside Sinosid *n*
sinter Sinter *m*
sinter *v* fritten, sintern
sinter cooler Sinterkühler *m*
sintered cake Sinterkuchen *m*
sintered disc filter funnel Glasfilternutsche *f*
sintered material Sinterwerkstoff *m*
sintered metal Sintermetall *n*
sintered powder metal Sintereisen *n*
sintering Sintern *n*, Fritten *n*
sintering coal Sinterkohle *f*
sintering furnace Sinterofen *m*
sintering heat Sinterungshitze *f*
sintering plant Sinteranlage *f*

sintering point Erweichungspunkt *m*
sintering process Sintervorgang *m*,
Sinterröstverfahren *n*
sinter roasting Sinterröstung *f*
sinusoid Sinuskurve *f* (Math)
sinusoidal sinusförmig
siomine Siomin *n*
siphon Siphon *m*, Flüssigkeitsheber *m*,
Heberrohr *n*, Saugheber *m*
siphon *v* ausheben
siphonage Abhebern *n*, Aushebern *n*
siphon barometer Heberbarometer *n*
siphon lubricator Dochtschmierer *m*
siphon piping Heberleitung *f*
siphon spillway Siphonüberlauf *m*
siphon tube Stechheber *m*
siphon vessel Hebergefäß *n*
siren Sirene *f*
siren sound Sirenenton *m*
siren valve Sirenenventil *n*
sirup (see also syrup) Sirup *m*
sisal Agavenfaser *f*, Sisal[hanf] *m*
sisalagenin Sisalagenin *n*
sisal hemp Sisalhanf *m*
sismondine Sismondin *m* (Min)
site Platz *m*, Lage *f*, Stelle *f*
site plan Lageplan *m*
sitology Diätkunde *f*; Ernährungswissenschaft *f*
sitostane Sitostan *n*
sitostanol Sitostanol *n*
sitostanone Sitostanon *n*
sitostene Sitosten *n*
sitosterol Sitosterin *n*
situated befindlich, gelegen
situation Zustand *m*, Lage *f*, Situation *f*
six-membered sechsgliedrig
six-membered ring Sechsring *m* (Chem)
six-phase star connection
Sechsphasensternschaltung *f*
six-phase system Sechsphasensystem *n*
six-sided sechsseitig
six-way cable Sechsfachkabel *n*
size Größe *f*, Maß *n*, Nummer *f*, Abmessung *f*,
Appretur *f*, Format *n*, Kleister *m* (Techn),
Leim *m* (Techn), Leimbrühe *f*,
Schlichtleim *m* (Text), increase in ~
Größenzunahme *f*, intermediate ~
Zwischengröße *f*
size *v* appretieren, kalibrieren, leimen (mit
Kleister), schlichten (Web), sieben, sortieren
sized geleimt
size distribution Größenverteilung *f*
size enlargement Stückigmachen *n*
size factor Größenfaktor *m*
size of print Bildformat *n*
size reduction by cutters Schneidzerkleinerung *f*,
Zerfasern *n*
sizing Größenbestimmung *f*, Dimensionierung *f*,
Korngrößentrennung *f* (Techn), Leimen *n*

(Pap), Leimung *f* (Pap), Sortierung *f* nach der
Größe (Techn), **fast to** ~ schlichtecht (Web),
~ **by sifting** Klassieren *n*
sizing agent Schlichtmittel *n* (Web)
sizing bath Schlichtebad *n*
sizing calender Gummierkalander *m*
sizing machine Schlichtmaschine *f* (Text)
sizing material Appreturmasse *f,*
    Schlichtmittel *n* (Web)
sizing press Leimwerk *n*
sizing roll Kalibrierwalze *f*
sizing test Leimgradprüfung *f*
sizing trough Anfeuchtgrube *f*
sizing vat Anfeuchtgrube *f*
sizzle *v* zischen
sizzling zischend
skatole Skatol *n*
skatoxyl Skatoxyl *n*
skein Knäuel *m* (Wolle), Strang *m* (Wolle)
skeleton Skelett *n*, Gerippe *n*, Knochengerüst *n*
skeleton diagram Prinzipschaltung *f*
sketch Skizze *f,* Entwurf *m*, Rohentwurf *m*,
    Umriß *m*, Zeichnung *f*
sketch *v* skizzieren, entwerfen, [flüchtig]
    entwerfen, skizzieren, zeichnen
skew schräg, [wind]schief
skew rays Schrägstrahlen *pl* (Opt)
skew rolling mill Schrägwalzwerk *n*
skew-symmetric[al] schiefsymmetrisch
skid Ausgleiten *n*, Seitwärtsgleiten *n*
skilled gewandt, erfahren
skilled worker gelernter Arbeiter *m*,
    Facharbeiter *m*
skillet Gußtiegel *m*, Schmelztiegel *m*
skim *v* abschöpfen, abstreichen (Met),
    entrahmen (Milch), skimmen (Met), ~ **off**
    entrahmen, abrahmen, abschäumen,
    abschöpfen, ~ **the milk** die Milch entrahmen
skim coat Skimmschicht *f*
skimmed abgestrichen, ~ **off** abgerahmt (Milch)
skimmed milk Magermilch *f*
skimmer Abstreicheisen *n* (Techn),
    Abstreichlöffel *m;* Schaumhaken *m* (Techn),
    Schaumkelle *f,* Schaumlöffel *m,*
    Schlackenüberlauf *m* (Techn)
skimmetine Skimmetin *n*
skimmianine Skimmianin *n*
skimmianinic acid Skimmianinsäure *f*
skimmin Skimmin *n*
skimming Abstrich *m*, Schaum *m*, Skimmen *n*
skimming agent Abschäummittel *n*
skimming device Abschäumer *m*
skimming ladle Abschäumlöffel *m*,
    Abstreichlöffel *m*, Schaumkelle *f*
skimmings Abgeschäumte[s] *n*, Abschaum *m*,
    Abstreifergut *n;* Armoxide *pl* (Blei)
skimming sieve Abschäumsieb *n*
skimming spoon Schaumlöffel *m*

skin Haut *f,* Balg *m* (Tier), Fell *n* (Tier), Pelle *f*
    (Bot), **outside of the** ~ Blumenseite *f* (Gerb)
skin *v* abhäuten, enthäuten, schälen
skin carcinoma Hautkrebs *m* (Med)
skin cream Hautcreme *f*
skin decontamination ointment
    Hautentgiftungssalbe *f*
skin disease Hautkrankheit *f*
skin effect Hauteffekt *m* (Elektr),
    Hautwirkung *f* (Elektr)
skin friction coefficient Druckverlustziffer *f* (v.
    Rohren), Rohrströmungziffer *f*
skin fungus Hautpilz *m*
skin glue Lederleim *m*
skin grafting Hauttransplantation *f* (Med)
skinning Abfeimen *n* (Glas), Hautbildung *f*
skin nutrient Hautnährcreme *f* (Kosm)
skin oil Hautöl *n* (Kosm)
skin pack Konturpackung *f,* Sichtpackung *f*
skin parasite Hautparasit *m*
skin parings Fellabfälle *m pl*, Hautabfälle *m pl*
skin plate PVC-beschichtetes Blech *n*
skin poison Hautgift *n*
skin-protecting agent Hautschutzmittel *n*
skin-protecting preparation
    Hautschutzpräparat *n*
skin specialist Dermatologe *m* (Med)
skin thermometer Hautthermometer *n*
skip *v* auslassen
skip hoist with tipping bucket Schrägaufzug *m*
    mit Kippgefäß
skip phenomenon Sprungerscheinung *f*
skirted cone Kegelkern *m* mit Verlängerung
skirting board Wandleiste *f,* Fußleiste *f*
skive *v* [auf]spalten (Leder, Fell)
skull Bär *m* (Metall), Eisensau *f* (Metall)
skutterudite Skutterudit *m* (Min),
    Arsenkobalt *n* (Min), Speiskobalt *m* (Min)
sky blue Himmelblau *n*
skylight Oberlicht *n*
skyrin Skyrin *n*
sky wave Raumwelle *f*
skywriter Himmelsschreiber *m* (Luftf)
slab Platte *f,* Tafel *f*
slab bloom Bramme *f*
slab extruder Fellspritzmaschine *f*
slab ingot Bramme *f*
slab milling Flachnutenfräsen *n*
slack Kohlengrus *m*, Afterkohle *f,*
    Kohlenklein *n*, Kohlenmulm *m*
slack *a* schlaff
slack *v* [Kalk] löschen
slack coal Kohlengrus *m*, Afterkohle *f*
slacken *v* lockern, entspannen, locker werden,
    schlaff werden
slacking Abschlacken *n*
slack silk Plattseide *f*
slag Schlacke *f,* Abzug *m* (Metall), Gekrätz *n*,
    acid ~ saure Schlacke, **basic** ~

Thomasschlacke *f*, **capacity for forming** ~
Verschlackungsfähigkeit *f*, **change of** ~
Schlackenwechsel *m*, **coarse metal** ~
Rohsteinschlacke *f*, **containing** ~
schlackenhaltig, **dry** ~ zähflüssige Schlacke,
flow of ~ Schlackentrift *f*, **fluid** ~
dünnflüssige Schlacke, **formation of** ~
Schlackenbildung *f*, **free from** ~
schlackenfrei, **fusible** ~ schmelzbare
Schlacke, **ground** ~ Schlackenmehl *n*, [ground]
basic ~ Thomasmehl *n*, **layer of** ~
Schlackenschicht *f*, **oxidizing** ~
Frischschlacke *f*, **poor** ~ Rohschlacke *f*,
removal of ~ Abschlacken *n*, **rich** ~
Garschlacke *f*, **rich in** ~ schlackenreich,
semipasty ~ zähflüssige Schlacke, **sticky** ~
zähflüssige Schlacke, **to form** ~ schlacken,
Schlacken bilden, **to rake out** ~ abschlacken,
Schlacke *f* abziehen, **to remove** ~
abschlacken, Schlacke *f* abziehen, **to skim off**
~ Schlacke *f* abziehen, **to tap off the** ~
abschlacken, **viscous** ~ zähflüssige Schlacke,
wet ~ dünnflüssige Schlacke, **white** ~
Reduktionsschlacke *f*, ~ **from foundry**
Hüttenafter *n*, ~ **from roasting**
Röstschlacke *f*, ~ **of liquation**
Krätzschlacke *f*, ~ **poor in iron** eisenarme
Schlacke
slag *v* verschlacken, Schlacken bilden, ~ **out**
abschlacken
slag addition Schlackenzuschlag *m*
slag ball Schlackenkugel *f*
slag bed Garschlackenboden *m*, Schwalboden *m*
slag block Schlackenblock *m*
slag bottom Garschlackenboden *m*,
Schwalboden *m*
slag brick Schlackenstein *m*
slag bucket Schlackenkübel *m*
slag cake Schlackenkuchen *m*
slag cement Hüttenzement *m*,
Schlackenzement *m*
slag chamber Schlackenfang *m*,
Schlackenkammer *f*
slag concrete Schlackenbeton *m*
slag crust Schlackenkruste *f*
slag discharge Schlackenabfluß *m*
slag dump Schlackenhalde *f*
slag-forming schlackenbildend
slag-forming constituent Schlackenbildner *m*
slag-forming period Schlackenbildungsperiode *f*,
Feinperiode *f*
slag furnace Schlackenherd *m*
slagged out abgeschlackt
slagging Verschlackung *f*, **danger of** ~
Verschlackungsgefahr *f*
slagging spout Abschlackrinne *f*
slag granulation Schlackengranulation *f*
slaggy schlackenhaltig, schlackenreich, schlackig
slag hearth Schlackenherd *m*

slag hole Schlackenauge *n*,
Schlacken[stich]loch *n*
slag inclusion Schlackeneinschluß *m*
slag iron Schlackeneisen *n*
slag ladle Schlackenpfanne *f*
slag lead Krätzblei *n*, Schlackenblei *n*
slag notch Abschlacköffnung *f*
slag pit Schlackengrube *f*
slag plate Schlackenblech *n*
slag pocket Schlackenfang *m*,
Schlackenkammer *f*
slag Portland cement Eisenportlandzement *m*
slag press Schlackenpresse *f*
slag process Sinterprozeß *m*, Sintervorgang *m*
slag puddling Schlackenpuddeln *n*,
Fettpuddeln *n*
slag puddling furnace Fettpuddelofen *m*
slag ratio Schlackenzahl *f* (CaO/SiO₂)
slag removal plant Entschlackungsanlage *f*
slag sand Schlackensand *m*
slag separation Schlackenabsonderung *f*
slag skimmer Schlackenabscheider *m*
slag smelting Schlackenarbeit *f*
slag spout Schlackenabstichrinne *f*
slag stone Schlackenstein *m*
slag streak Schlackenstreifen *m*
slag thread Schlackenfaden *m*
slag trap Schlackenfänger *m*
slag wag[g]on Schlackenwagen *m*
slag washing [process] Schwalarbeit *f*
slag wool Hüttenwolle *f*, Mineralwolle *f*,
Schlackenwolle *f*
slakable [ab]löschbar (Kalk)
slake *v* auslöschen (Feuer), löschen, ~ **lime**
Kalk löschen
slaked gelöscht (Kalk)
slaked lime Äscher *m* (Gerb), gelöschter
Kalk *m*, Löschkalk *m*
slaking Löschen *n* (Kalk)
slant Schräge *f*, Neigung *f*
slant *v* abschrägen
slant[ing] schief, schräg
slanting position Schräglage *f*
slash Schlitz *m*
slash *v* [auf]schlitzen
slat Latte *f*, Leiste *f*
slate Schiefer *m* (Geol), **argillaceous** ~
Tonschiefer *m* (Geol), **calcareous** ~
Kalkschiefer *m* (Geol), **green** ~
Grünschiefer *m*, **hard calcareous** ~
Schiefermarmor *m* (Min), **pale gray** ~
Fahlstein *m* (Min), **powdered** ~
Schiefermehl *n*, **primitive** ~ Urschiefer *m*
slate clay Schieferton *m* (Min)
slate coal Blätterkohle *f*, Blattkohle *f*
slate-like schieferähnlich, schieferartig, schiefrig
slate slab Schieferplatte *f*
slate spar Schieferspat *m* (Min)
slate vat Schieferbottich *m*

slaty schieferartig, schieferhaltig, schiefrig
slaty clay Blätterton *m*
slaty coal Schieferkohle *f*
slaty lead Schieferblei *n* (Min)
slaty marl Mergelschiefer *m* (Min)
slaty talc Talkschiefer *m* (Min)
sled Schlitten *m*
sledge Schlitten *m*
sledge hammer Zuschlaghammer *m*,
  Vorschlaghammer *m*
sleeker Dämmbrett *n*, Glättmaschine *f*
sleeper pass Schwellenkaliber *n*
sleeping draught Schlafmittel *n* (Pharm)
sleeping sickness Schlafkrankheit *f* (Med)
sleeve Gleitmuffe *f* (Techn), Hülse *f* (Techn)
sleeve coupling Muffenkupplung *f*
sleeve dipole Manschettendipol *m*
sleeve joint Muffenverbindung *f*
sleeve nut Schraubmuffe *f*
sleeve rod Ventilstange *f*
slewing crane Schwenkkran *m*
slice Scheibe *f*, Schnitte *f*
slice *v* [in Scheiben] schneiden
sliced film geschälte Folie *f*, Schälfolie *f*
slicer Schneidmaschine *f*
slick of waste metal Krätzschlich *m* (Gieß)
slide Diapositiv *n* (Phot), Gleitbahn *f* (Techn),
  Gleitelement *n*, Schieber *m* (Techn)
slide *v* gleiten, rutschen, schieben, ~ **off**
  abrutschen
slide-back voltmeter vergleichendes Voltmeter *n*
slide bar Gleitschiene *f*, Gleitspur *f*
slide bearing Gleitlager *n*
slide block Gleitklotz *m*, **adjusting** ~
  Stellschlitten *m*
slide box Gleitbüchse *f* (Elektr)
slide carriage Schieber *m*
slide gauge Schieblehre *f*
slide holder Objektträger *m*
slider Schiebekontakt *m*, Schieber *m*
slide rail Gleitschiene *f*
slide resistance Schiebewiderstand *m*
slide rod Schieberstange *f*
slide rule Rechenschieber *m* (Math)
slide valve Absperrschieber *m*, Gleitventil *n*,
  Schieber *m*, **balanced** ~
  Entlastungsschieber *m*, **flat** ~
  Flachschieber *m*
slide valve diagram Schieberdiagramm *n*
slide valve gear Schiebersteuerung *f*
slide valve spindle Schieberstange *f*
slide-wire potentiometer
  Schleifdrahtkompensator *m*
sliding rutschig, verschiebbar
sliding belt Gleitriemen *m*
sliding block Gleitblock *m*
sliding box Schiebekassette *f*
sliding contact Gleitkontakt *m*, Schleifkontakt *m*
sliding door Schiebetür *f*

sliding friction Gleitreibung *f*, **coefficient of** ~
  Gleitreibungskoeffizient *m* (Mech), **wear**
  caused by ~ Gleitverschleiß *m*
sliding grate Schieberost *m*
sliding grating Schiebegitter *n*
sliding magnet Gleitmagnet *m*
sliding microtome Schlittenmikrotom *n*
sliding nut Gleitmutter *f*
sliding plate Gleitplatte *f*
sliding pressure Gleitspannung *f*
sliding punch Schiebestempel *m*,
  Verschiebestempel *m*
sliding rail Laufschiene *f*
sliding resistance Gleitwiderstand *m*
sliding roof Schiebedach *n*
sliding sleeve Ausrückmuffe *f*
sliding switch Schaltschieber *m*
sliding-vane pump Flügelzellenpumpe *f*
sliding-vane rotary compressor
  Vielzellenverdichter *m*
sliding weight Schiebgewicht *n*, Laufgewicht *n*,
  balance with ~ Laufgewichtswaage *f*
sliding weight balance Laufgewichtswaage *f*
sliding window Schiebefenster *n*
slight leicht, schwach
slime Schlamm *m*, Pochtrübe *f* (Techn),
  Schleim *m* (Bot, Zool), **to coat with** ~ mit
  Schleim oder Schlamm überziehen
slime *v* mit Schleim oder Schlamm überziehen,
  abschleimen
slime-forming schleimbildend (Bakt)
slime layer Schlammschicht *f*
slime pit Schlammgrube *f*, Schlammherd *m*
slime plate Schlammteller *m*
slimes Laugerückstand *m*
slime separator Schlammabscheider *m*
slime table Schlammherd *m*
slimy schlammig, schleimartig, schleimig
sling Befestigungsbügel *m*, Schleuder *f*,
  Schlinge *f*
slip Ausgleiten *n*, Gleiten *n*, **angle of** ~
  Gleitwinkel *m* (Mech), **side** ~ seitliches
  Rutschen *n*
slip *v* [aus]gleiten, rutschen, ~ **off** abgleiten,
  abrutschen
slip angle Schräglaufwinkel *m* (Reifen)
slip band Gleitband *n*
slip band distribution Gleitbandverteilung *f*
slip band spacing Gleitbandabstand *m*
slip cap Schiebedeckel *m*
slip cover Überstülphaube *f*
slip fittings Gleitarmaturen *pl*,
  Gleitausrüstungsteile *pl*
slip joint Gleitfuge *f*, Gleitverbindung *f*
slip line field Gleitlinienfeld *n*
slip line length Gleitlinienlänge *f*
slip line pattern Gleitlinienbild *n*
slip lines Gleitlinien *pl*, **net of** ~
  Gleitliniennetz *n*

slip motion Gleitbewegung *f*
slip-on collar Aufschweißbund *m*
slip-on flange Aufschweißflansch *m*
slip-on ring Aufsteckring *m*
slip-on travelling gear Aufsteckfahrantrieb *m* (Kran)
slippage along the wall Wandgleitung *f*
slipper animalcule Pantoffeltierchen *n*
slipper block Blattzapfen *m*
slipper shoe Blattzapfen *m*
slippery glatt, glitschig, schlüpfrig
slipping of the belt Gleiten *n* des Riemens, Schlüpfen *n* des Riemens
slipping surface Gleitfläche *f*
slip plane Gleitebene *f,* Gleitfläche *f*
slip plane blocking Gleitebenenblockierung *f*
slip resistance Gleitsicherheit *f,* Rutschsicherheit *f*
slip ring Gleitring *m,* Schleifring *m*
slip ring armature Schleifringanker *m*
slip ring rotor Schleifringläufer *m*
slip speed Schlüpfgeschwindigkeit *f*
slip-type expansion joint Dehnungsstopfbuchse *f*
slip zone Gleitzone *f*
slit Schlitz *m,* Öffnung *f,* Spalt *m,* width of ~ Spaltbreite *f* (Opt)
slit *v* [auf]schlitzen, aufschneiden, ~ slightly anritzen
slit and tongue joint Kerbenfügung *f*
slit height Spalthöhe *f* (Opt)
slit illumination Spaltausleuchtung *f* (Opt)
slit image Spaltbild *n* (Opt)
slit orifice Schlitzdüse *f*
slit sieve Schlitzsieb *n*
slitter Längsschneider *m*
slit width Spaltbreite *f* (Opt)
sloe Schlehe *f* (Bot)
sloe tree Schlehdorn *m* (Bot)
slogan Schlagwort *n,* Werbeformel *f*
slope Abhang *m,* Böschung *f,* Gefälle *n,* Neigung *f,* Neigungsebene *f,* Schräge *f*
slope *v* abfallen (Straße etc.), abschrägen, böschen
sloping schräg
sloping grate schräger Rost *m,* Schrägrost *m*
sloping position Schräglage *f*
slot Spalte *f,* Keilnut *f* (Techn), Kerbe *f* (Techn), Nut *f* (Techn), Schlitz *m,* elongated ~ Längsschlitz *m*
slot *v* nuten
slot atomizer Schlitzzerstäuber *m*
slot burner Schnittbrenner *m*
slot cutter Nutenfräser *m*
slot depth Schlitztiefe *f*
slot die Schlitzwerkzeug *n,* Breitschlitzdüse *f,* Schlitzform *f*
slot guide Schlitzführung *f*
slot machine Automat *m*
slot milling machine Langlochfräsmaschine *f*

slot sprayer Schlitzzerstäuber *m*
slotted cards Schlitzlochkarten *pl*
slotted shutter Schlitzverschluß *m*
slotted [wall] grab Schlitzwandgreifer *m*
slotting machine Nutenfräser *m,* Nutenstoßmaschine *f*
slotting saw Stoßsäge *f*
slot wedge Nutverschlußkeil *m*
slot welding Dübelschweißung *f*
slot width Schlitzbreite *f*
slow langsam
slow decomposition Verwesung *f*
slow down *v* verlangsamen, abbremsen
slow-dyeing langsam ziehend (Färb)
slowing down Verlangsamung *f,* Abbremsung *f* (Atom), Moderierung *f* (Atom)
slowing-down area Bremsfläche *f*
slowing-down density Bremsdichte *f* (Atom)
slowing-down process Abbremsprozeß *m* (Atom)
slowing-down time Bremszeit *f* (Reaktor)
slow match Lunte *f*
slow-motion picture Zeitlupenaufnahme *f* (Phot)
slow neutron fission Spaltung *f* mit langsamen Neutronen (pl)
slow-setting cement Langsambinder *m*
slow-speed niedertourig
slubs Flusen *f pl* (Web)
sludge Schlamm *m,* Aufschlämmung *f,* Bodensatz *m,* Faulschlamm *m* (Techn), Matsch *m,* preparation of the ~ Schlammaufbereitung *f*
sludge bag Schlammbeutel *m*
sludge concentration meter Schlammkonzentrationsmesser *m*
sludge draining Schlammentwässerung *f*
sludge pump Schlammpumpe *f*
sludge removal plant Entschlammungsanlage *f*
sludge valve Schlammventil *n*
sludgy schlammig
slug Rohling *m* (zum Verpressen), Rohmetall *n*
slug flow Brecherströmung *f*
sluggish [reaktions]träge
sluice Absperrglied *n*
sluice valve Absperrschieber *m,* Schieber *m,* parallel faced ~ Parallelabsperrschieber *m*
slurry Schlamm *m,* Aufschlämmung *f,* Brei *m*
slurrying Aufschlämmen *n*
slurry reactor Schlammreaktor *m,* Suspensionsreaktor *m*
slurry substance breiige Masse *f*
slurry thickener Schlammwassereindicker *m*
slush Schlamm *m,* Schmiere *f* (Techn)
small klein
small ampoule Amphiole *f* (Pharm)
small-angle grain boundary Kleinwinkelkorngrenze *f*
small-angle scattering Kleinwinkelstreuung *f*
small coal Feinkohle *f,* Grießkohle *f*
small combustion pipe Glühröhrchen *n*

small-grained kleinkörnig
small intestine Dünndarm *m*
small-notched kleinzackig
small pieces feinstückig
smallpox Pocken *pl* (Med), Blattern *pl* (Med)
smallpox virus Pockenvirus *m*, Pockenerreger *m*
smalls Grießkohle *f*, Grubenklein *n*
small section rolling mill Feineisenstraße *f*
small-sized kleinstückig
small-toothed kleinzackig
smalogenin Smalogenin *n*
smalt Smalte *f*, Kaiserblau *n*, Kobaltblau *n*,
      Schmaltblau *n*, Schmalte *f* (Kobaltschmelze),
      Schmelzblau *n*, **coarsest** ~ Blausand *m*
smalt green Kobaltgrün *n*
smaltine Smaltin *m* (Min), Smaltit *m* (Min),
      weißer Speiskobalt *m* (Min)
smaltite Smaltit *m* (Min), Smaltin *m* (Min),
      weißer Speiskobalt *m* (Min)
smaragdine smaragdfarben
smaragdite Smaragdit *m* (Min)
smash *v* zertrümmern
smashing Zertrümmerung *f*
smashing probability
      Zertrümmerungswahrscheinlichkeit *f*
smear Abstrich *m* (Med)
smear *v* [ein]schmieren, anschmieren,
      beschmieren, einreiben, überstreichen,
      verschmieren
smear head Schmierkopf *m*
smearing Schmierung *f*, Einreibung *f*
smectite Fetton *m* (Min), Seifenerde *f* (Min),
      Seifenton *m* (Min)
smegma Smegma *n* (Med)
smell Geruch *m*, Geruchssinn *m*, **absence of the**
      sense of ~ Anosmie *f* (Med), **bad** ~
      Gestank *m*, **free from** ~ geruchfrei, geruchlos,
      pungent ~ stechender Geruch, **sense of** ~
      Geruchssinn *m*
smell *v* riechen
smelt *v* abschmelzen (Met), [ein]schmelzen
      (Metall), verhütten (Metall)
smelter coke Schmelzkoks *m*
smelter gas Hüttengas *n*
smelter smoke Hüttenrauch *m*
smelting Verhüttung *f* (Metall),
      Ausschmelzung *f* (Metall), Hüttenarbeit *f*
      (Metall), Schmelzen *n* (Metall), **coarse metal**
      ~ Rohsteinschmelzen *n*, **method of** ~
      Schmelzverfahren *n* (Met), **process of** ~
      Schmelzverlauf *m*, **product ready for** ~
      verhüttbares Gut *n*, **ready for** ~ verhüttbar,
      state of ~ Schmelzverlauf *m*, ~ **of crude**
      metals Rohschmelzen *n*
smelting charge Gicht *f* (Metall) (Schmelzgut)
smelting coke Hüttenkoks *m*, Schmelzkoks *m*
smelting finery Frischhütte *f*
smelting flux Schmelzfluß *m*

smelting flux electrolysis
      Schmelzflußelektrolyse *f*
smelting furnace Schmelzofen *m* (Met)
smelting house Schmelzerei *f*
smelting plant Schmelzanlage *f*
smelting practice or schedule Schmelzführung *f*
smelting process Schmelzarbeit *f*,
      Verhüttungsvorgang *m*
smelting works Schmelzhütte *f*, Hüttenwerk *n*
smilacin Smilacin *n*
smilagenin Smilagenin *n*
smirnovine Smirnovin *n*
smithite Smithit *m* (Min)
smith's coal Schmiedekohle *f*, Gaskohle *f*
smith's hearth Schmiedeherd *m*
smithsonite Smithsonit *m* (Min), Zinkspat *m*
      (Min), **siliceous** ~ Kieselgalmei *m* (Min)
smog Nebel *m* (mit Rauch), **layer of** ~
      Dunstglocke *f*
smoke Rauch *m*, Dampf *m*, **absence of** ~
      Rauchlosigkeit *f*, **development of** ~
      Rauchbildung *f*, **diminution of** ~
      Rauchverdünnung *f*, **formation of** ~
      Rauchbildung *f*, **generation of** ~
      Rauchbildung *f*, **giving little** ~ rauchschwach,
      thick ~ Qualm *m*, **vapors of** ~
      Rauchschwaden *pl*
smoke *v* rauchen, ausräuchern (Raum), blaken,
      dampfen, räuchern, rußen, schwärzen, ~ **out**
      ausräuchern
smoke abatement Rauchbekämpfung *f*
smoke alarm Rauchmelder *m*
smoke black Schwärze *f*
smoke box Rauchkammer *f*
smoke box gas Rauchkammergas *n*
smoke chamber Rauchkammer *f*
smoke-colored rauchfarben
smoke combustion Rauchverzehrung *f*
smoke condenser Flugstaubkondensator *m*,
      Rußkammer *f*
smoke consumer Rauchverzehrer *m*
smoke-consuming rauchverzehrend
smoke-consuming device
      Rauchverbrennungseinrichtung *f*
smoke consumption Rauchverzehrung *f*
smoke-cured rauchgar
smoke density Rauchdichte *f*, Rauchstärke *f*
smoke density meter Rauchdichtemesser *m*
smoke disperser Qualmabzugsrohr *n*
smoke-dried geräuchert, rauchgar
smoked salmon substitute Lachsersatz *m*
smoke gauge Kapnoskop *n*
smoke generator Raucherzeuger *m*
smokeless rauchfrei
smoke monitor Rauchmeldegerät *n*
smoke nuisance Rauchbelästigung *f*
smoke point Rauchpunkt *m*, **determination of** ~
      Rauchpunktbestimmung *f*
smoke prevention Rauchverhütung *f*

smokeproof rauchdicht
smoker's cancer Raucherkrebs *m*
smoker's cough Raucherhusten *m* (Med)
smoke scale Rauchskala *f*
smoke-screen acid Nebelsäure *f*
smoke stack Schornstein *m*, Schlot *m*
smokestone Rauchquarz *m* (Min)
smoke-tight rauchdicht
smoke tube boiler Rauchrohrkessel *m*
smoking Rauchen *n*
smoking *a* rauchend
smoky rauchig, qualmig
smoky black rauchschwarz
smoky quartz Rauchquarz *m* (Min)
smoky topaz Rauchtopas *m* (Min)
smolder *v* schwelen, glimmen (unter der Asche), qualmen
smoldering Schwelen *n*, Glimmen *n*, **resistance to** ~ Glutsicherheit *f*, **to continue** ~ nachschwelen
smoldering fire Schwelbrand *m*, Qualmfeuer *n*
smoldering zone Schwelzone *f*
smooth eben, glatt, gleichmäßig
smooth *v* glätten, hobeln, planieren, schleifen, schlichten, ~ **down** sich beruhigen
smoother Glätter *m*, Glättpresse *f* (Pap)
smoothing Glätten *n*, Glatthobeln *n*, Schlichten *n*
smoothing choke Abflachungsdrossel *f*
smoothing equipment Glättvorrichtung *f*
smoothing file Glättfeile *f*
smoothing iron Glätteisen *n*
smoothing plane Abrichthobel *m*
smoothing press Glättpresse *f*
smoothing roll Glättwalze *f*
smoothing roller Glattwalze *f*
smoothing tool Schlichtstahl *m*
smoothness Glätte *f*, Gleichmäßigkeit *f*
smother Rauch *m*
smut Getreidebrand *m* (Bot), Kornbrand *m* (Bot), Kornfäule *f* (Bot)
smut *v* beschmutzen, berußen
snail Schnecke *f* (Zool)
snail's pace Schneckentempo *n*
snake Schlange *f* (Zool)
snake poison Schlangengift *n*
snakeroot Schlangenwurzel *f* (Bot)
snakestone Schlangenstein *m* (Min)
snap flask Abschlagformkasten *m*
snap freezing schnelles Gefrieren *n* durch Besprühen mit flüssigem Stickstoff
snap hammer Schellhammer *m*
snap hook Karabinerhaken *m*
snap off *v* abspringen
snap-on end can Lötdeckeldose *f*
snap-ring needle bearing Wälzlager-Schnappkäfig *m*
snapshot Momentaufnahme *f* (Phot), Schnappschuß *m*
snap test Kurzprüfung *f*

snarl *v* verwirren (Garn)
sneeze gas Reizgas *n*
sneezewort Nieskraut *n* (Bot)
sniff *v* schnüffeln, schnuppern
snout Schnauze *f*
snow Schnee *m*, **melted** ~ Schneewasser *n*
snow water Schneewasser *n*
snow-white schneeweiß
soak *v* einweichen, durchnässen, sich voll Wasser saugen, wässern, ~ **in** einsickern, einziehen (Flüssigkeit), ~ **in lime water** einkalken, ~ **up** einsaugen
soakaway Sickergrube *f*
soaking Einweichen *n*
soaking agent Einweichmittel *n*
soaking pit Einweichgrube *f*
soaking tub Einweichbottich *m*
soap Seife *f*, **hard** ~ Kernseife *f*, **potash** ~ Schmierseife *f*, **sodium** ~ Kernseife *f*, **soft** ~ Schmierseife *f*
soap *v* abseifen, einseifen
soap bark Seifenrinde *f*
soap boiler Seifensieder *m*
soap boiler's lye Seifensiederlauge *f*
soap bubble Seifenblase *f*
soap charge Seifensud *m*
soap consumption Seifenverbrauch *m*
soap content Seifengehalt *m*
soap emulsion Seifenemulsion *f*
soap flake Seifenflocke *f*
soap improver Seifenveredelungsmittel *n*
soap lather Seifenschaum *m*
soap mill Piliermaschine *f*
soap powder Seifenpulver *n*
soap rock Wascherde *f*
soap solution Seifenlauge *f*, Seifenlösung *f*
soapstone Seifenstein *m* (Min), Speckstein *m* (Min), **powdered** ~ Talkumpuder *m*, **pulverized** ~ Talkum *n*
soapstone machine Talkumiermaschine *f*
soap substitute Seifenersatz *m*
soap suds Seifenlauge *f*, Seifenwasser *n*
soap water Seifenwasser *n*
soap weed Seifenkraut *n* (Bot)
soap works Seifensiederei *f*, Seifenfabrik *f*
soapy seifig, seifenartig
soar *v* [auf]steigen
sobralite Sobralit *m*
sobrerol Sobrerol *n*
socket Fassung *f* (Elektr), Muffe *f*, Rohransatz *m*, Steckdose *f* (Elektr), Steckhülse *f*, Verbindungsmuffe *f*, Zapfenlager *n*
socket[ed] pipe Muffenrohr *n*
socket end of a pipe Hals *m* einer Röhre
socket key Aufsteckschlüssel *m*
socket wrench Steckschlüssel *m*
soda Soda *f*, Natriumcarbonat *n*, Natrium carbonicum *n* (Lat), **containing** ~ sodahaltig,

crude ~ Rohsoda *f,* **crystallized** ~
Kristallsoda *f,* **finely pulverized** ~ Feinsoda *f,*
preparation of ~ Sodadarstellung *f*
soda alum Natriumalaun *m,* Natronalaun *m*
soda ash Rohsoda *f,* Sodaasche *f*
soda bath Sodabad *n*
soda bleaching lye Natronbleichlauge *f*
soda cellulose Natroncellulose *f,*
Natronzellstoff *m*
soda content Sodagehalt *m*
soda crystals Kristallsoda *f,* Waschkristalle *pl*
soda extract[ion] Sodaauszug *m*
soda factory Sodafabrik *f*
soda feldspar Natronfeldspat *m* (Min)
soda lake Natronsee *m*
soda lime Natronkalk *m*
soda-lime glass Natronkalkglas *n,* Normalglas *n*
sodalite Sodalith *m* (Min)
soda lye Sodalauge *f*
sodamide Natriumamid *n*
soda niter Natronsalpeter *m*
soda powder Sodapulver *n*
soda pulp Natroncellulose *f,* Natronzellstoff *m*
soda recovery Sodarückgewinnung *f* (Papier)
soda residue Sodarückstand *m*
soda soap Natronseife *f*
soda solution Sodalösung *f*
soda vat Sodaküpe *f*
soda water Sodawasser *n,* Mineralwasser *n,*
Sprudel *m*
soda waterglass Natronwasserglas *n*
soda works Sodafabrik *f*
soddyite Soddyit *m* (Min)
sodium (Symb. Na) Natrium *n,* **containing** ~
natriumhaltig
sodium acetate Natriumacetat *n,* Natrium
aceticum *n* (Lat)
sodium acetylide Natriumacetylid *n*
sodium alcoholate Natriumalkoholat *n*
sodium alkylate Natriumalkylat *n*
sodium alum Natriumalaun *m,* Natronalaun *m*
sodium aluminate Natriumaluminat *n*
sodium aluminum fluoride
Aluminiumnatriumfluorid *n,* **native** ~
Kryolith *m* (Min)
sodium amalgam Natriumamalgam *n*
sodium amide Natriumamid *n*
sodium ammonium nitrate
Natronammonsalpeter *m*
sodium ammonium phosphate
Natriumammoniumphosphat *n*
sodium ammonium sulfate
Natriumammoniumsulfat *n*
sodium amytal Natriumamytal *n*
sodium arsenate Natriumarsenat *n*
sodium arsenite Natriumarsenit *n*
sodium ascorbate Natriumascorbat *n*
sodium azide Natriumazid *n*

sodium benzene sulfonate
Natriumbenzolsulfonat *n*
sodium benzoate Natriumbenzoat *n,* Natrium
benzoicum *n* (Lat)
sodium bicarbonate Natriumbicarbonat *n,*
Natriumhydrogencarbonat *n*
sodium bichromate Natriumdichromat *n*
sodium bisulfate Natriumbisulfat *n,*
Mononatriumsulfat *n*
sodium bisulfite Natriumbisulfit *n,*
Mononatriumsulfit *n*
sodium bitartrate Natriumbitartrat *n*
sodium borate Natriumborat *n*
sodium borohydride Natriumborhydrid *n*
sodium bromide Natriumbromid *n*
sodium carbolate Natriumphenolat *n*
sodium carbonate Natriumcarbonat *n,*
kohlensaures Natrium *n,* Natrium
carbonicum *n* (Lat), Soda *f,* **acid** ~
Natriumbicarbonat *n,*
Mononatriumcarbonat *n,*
Natriumhydrogencarbonat *n,* **crude** ~
Rohsoda *f,* **native monohydrated** ~
Thermonatrit *m* (Min), **powdered** ~
Sodapulver *n*
sodium caseinate Natriumcaseinat *n*
sodium cellulose Natroncellulose *f,*
Natronzellstoff *m*
sodium chlorate Natriumchlorat *n*
sodium chloride Natriumchlorid *n,* Kochsalz *n,*
containing ~ kochsalzhaltig
sodium chloride content Kochsalzgehalt *m*
sodium chloride lattice Steinsalzgitter *n* (Krist)
sodium chloride solution Kochsalzlösung *f,*
Natriumchloridlösung *f*
sodium cholate Natriumcholat *n*
sodium choleinate Natriumcholeinat *n*
sodium cinnamate Natriumcinnamat *n,* Hetol *n*
sodium compound Natriumverbindung *f*
sodium cresylate Kresolnatron *n*
sodium cyanamide Natriumcyanamid *n*
sodium cyanide Natriumcyanid *n*
sodium cyanoferrate(III) Ferricyannatrium *n*
sodium cyclamate Natriumcyclamat *n*
sodium dehydrocholate Natriumdehydrocholat *n*
sodium dichromate Natriumdichromat *n*
sodium diethylbarbiturate
Natriumdiäthylbarbiturat *n*
sodium diiodosalicylate Natriumdijodsalicylat *n*
sodium dioxide Natriumperoxid *n*
sodium diuranate Natriumdiuranat *n*
sodium electrolysis Natriumelektrolyse *f*
sodium ethoxide Natriumäthylat *n*
sodium ethylate Natriumäthylat *n*
sodium ethyl sulfate Natriumäthylsulfat *n*
sodium ferric saccharate Natriumferrisaccharat *n*
sodium ferricyanide Ferricyannatrium *n*

sodium ferrocyanide Ferrocyannatrium *n*
sodium flame Natriumflamme *f*
sodium fluoaluminate
   Aluminiumnatriumfluorid *n*
sodium fluorescein Uranin *n*
sodium fluoride Natriumfluorid *n*
sodium fluosilicate Natriumsilicofluorid *n*
sodium formate Natriumformiat *n*
sodium formate plant Natriumformiatanlage *f*
sodium glycerophosphate
   Natriumglycerophosphat *n*
sodium-graphite reactor
   Natrium-Graphit-Reaktor *m*
sodium hexametaphosphate
   Natriumhexametaphosphat *n*
sodium hippurate Natriumhippurat *n*
sodium hydrazide Natriumhydrazid *n*
sodium hydride Natriumhydrid *n*
sodium hydrogen carbonate
   Natriumbicarbonat *n*,
   Mononatriumcarbonat *n*,
   Natriumhydrogencarbonat *n*
sodium hydrogen sulfate Natriumbisulfat *n*,
   Mononatriumsulfat *n*,
   Natriumhydrogensulfat *n*
sodium hydrogen sulfide
   Natriumhydrogensulfid *n*
sodium hydrogen sulfite Natriumbisulfit *n*,
   Mononatriumsulfit *n*,
   Natriumhydro[gen]sulfit *n*
sodium hydrogen tartrate Natriumbitartrat *n*
sodium hydrosulfite Natriumhydrosulfit *n*
sodium hydroxide Natriumhydroxid *n*,
   Ätznatron *n*
sodium hydroxide solution Natronlauge *f*,
   Ätznatronlösung *f*
sodium hypochlorite Natriumhypochlorit *n*,
   solution of ~ Bleichwasser *n*
sodium hypochlorite solution Eau de
   Labarraque *n*
sodium hypophosphite Natriumhypophosphit *n*
sodium hyposulfite Natriumhyposulfit *n*
sodium iodate Natriumjodat *n*
sodium iodide Natriumjodid *n*
sodium ion Natriumion *n*
sodium iron alum Natroneisenalaun *m*
sodium iron sulfate Natroneisenalaun *m*
sodium leading-off tube Natriumableitungsrohr *n*
sodium line Natriumlinie *f* (Spektr)
sodium lygosinate Lygosinnatrium *n*
sodium malonate Natriummalonat *n*
sodium metaarsenite Natriummetaarsenit *n*
sodium metabolism Natriumstoffwechsel *m*
   (Physiol)
sodium metal metallisches Natrium *n*,
   Natriummetall *n*
sodium metanilate Natriummetanilat *n*
sodium metaphosphate Natriummetaphosphat *n*
sodium methyl sulfate Natriummethylsulfat *n*

sodium microcline Natronmikroklin *m* (Min)
sodium naphtholate Naphtholnatrium *n*
sodium niobate Natriumniobat *n*
sodium nitrate Natriumnitrat *n*, Chilesalpeter *m*,
   Natronsalpeter *m*, salpetersaures Natrium *n*
sodium nitrite Natriumnitrit *n*
sodium nitrobenzene sulfonate
   Natriumnitrobenzolsulfonat *n*
sodium nitrobenzoate Natriumnitrobenzoat *n*
sodium nitroferricyanide Natriumnitroprussiat *n*,
   Nitroprussidnatrium *n*
sodium nitroprussiate Natriumnitroprussiat *n*,
   Nitroprussidnatrium *n*
sodium nitroprusside Natriumnitroprussiat *n*,
   Nitroprussidnatrium *n*
sodium nosophen Nosophennatrium *n*
sodium nucleinate Natriumnukleinat *n*
sodium oleate Natriumoleat *n*
sodium orthoclase Natronorthoklas *m* (Min)
sodium pantothenate Natriumpantothenat *n*
sodium perborate Perborax *m*, Peroxydol *n*
sodium peroxide Natriumperoxid *n*,
   Natriumhyperoxid *n*, Natriumsuperoxid *n*
sodium peroxyborate Perborax *m*
sodium phenate Natriumphenolat *n*
sodium phenolate Natriumphenolat *n*,
   Phenolnatrium *n*
sodium phenol sulfonate
   Natriumphenolsulfonat *n*
sodium phenoxide Phenolnatrium *n*
sodium phenyl ethyl barbiturate
   Natriumphenyläthylbarbiturat *n*
sodium phosphate Natriumphosphat *n*,
   phosphorsaures Natrium *n*
sodium phosphide Natriumphosphid *n*,
   Phosphornatrium *n*
sodium phosphite Natriumphosphit *n*
sodium platinum chloride
   Natriumplatinchlorid *n*
sodium pliers Natriumzange *f*
sodium plumbate Natriumplumbat *n*
sodium polyphosphate Natriumpolyphosphat *n*
sodium polysulfide Natriumpolysulfid *n*
sodium porphine chelate Natriumporphinchelat *n*
sodium potassium tartrate
   Natriumkaliumtartrat *n*, Seignettesalz *n*
sodium press Natriumpresse *f*
sodium pump Natriumpumpe *f* (Biochem)
sodium pyrophosphate Natriumpyrophosphat *n*
sodium pyruvate Natriumpyruvat *n*
sodium saccharinate Natriumsaccharinat *n*
sodium salicylate Natriumsalicylat *n*,
   Salicylnatron *n*
sodium salt Natriumsalz *n*
sodium santoninate Santoninnatrium *n*
sodium silicate Natriumsilikat *n*, Wasserglas *n*
sodium silicofluoride Natriumsilicofluorid *n*
sodium stannate Natriumstannat *n*,
   Grundiersalz *n*, Präpariersalz *n*

sodium succinate Natriumsuccinat *n*
sodium sulfantimonate Schlippesches Salz *n*
sodium sulfate Natriumsulfat *n*, **acid** ~
Natriumbisulfat *n*; Mononatriumsulfat *n*,
crystalline ~ Glaubersalz *n*
sodium sulfate decahydrate Glaubersalz *n*
sodium sulfide Natriumsulfid *n*
sodium sulfite Natriumsulfit *n*, **acid** ~
Natriumbisulfit *n*, Mononatriumsulfit *n*
sodium tartrate Natriumtartrat *n*, **acid** ~
Natriumbitartrat *n*
sodium tetraborate Borax *m*
sodium tetrametaphosphate
Natriumtetrametaphosphat *n*
sodium tetraphenylporphine
Natriumtetraphenylporphin *n*
sodium thiocyanate Natriumrhodanid *n*
sodium thiosulfate Fixiernatron *n*,
Natriumthiosulfat *n*
sodium toluene sulfonate
Natriumtoluolsulfonat *n*
sodium tongs Natriumzange *f*
sodium tungstate Natriumwolframat *n*
sodium uranate Natriumuranat *n*
sodium valerate Natriumvalerianat *n*
sodium vapor Natriumdampf *m*
sodium vapor high-pressure lamp
Natriumdampf-Hochdrucklampe *f*
sodium-vapor lamp Natrium[dampf]lampe *f*
sodium wire Natriumdraht *m*
sodium wolframate Natriumwolframat *n*
sod oil Degras *n*
soft weich, locker, sanft, **half** ~ halbweich,
medium ~ mittelweich
soft annealing Weichglühen *n*
soft center[ed] steel Weichkernstahl *m*
soft coal Braunkohle *f*, Weichkohle *f*
soft drink alkoholfreies Getränk *n*
soften *v* weich machen, abschwächen,
adoucieren (Farben), aufweichen, enthärten,
erweichen, mildern, weich oder biegsam
werden
softener Aufweichmittel *n*, Enthärtungsmittel *n*,
Weichmacher *m* (Kunststoff)
softening Erweichen *n*, Weichmachen *n*,
Weichwerden *n*, **intermediate** ~
Zwischenglühung *f*, ~ **by steeping**
Einweichen *n*
softening agent Weichmacher *m*
softening furnace Vorraffinierofen *m*
softening of water, chemical ~ chemische
Wasserenthärtung *f*, **complete** ~
Vollentsalzung *f* von Wasser
softening point Erweichungspunkt *m*,
Fließschmelztemperatur *f*
softening process Chevillieren *n*
softening range Erweichungsbereich *m*,
Erweichungszone *f*
soft iron core Weicheisenkern *m*

soft lead impress Weichbleiabdruck *m*
softness Weichheit *f*
soft rot Moderfäulepilz *m*
soft soap Schmierseife *f*, Kaliumseife *f*
soft solder Schnellot *n*, Weichlot *n*
soft-solder *v* mit Weichlot löten
soft steel Flußstahl *m*
sogdianose Sogdianose *f*
soil Acker *m* (Agr), Boden *m* (Agr), Erde *f*
(Agr), Land *n* (Erdboden), Schmutz *m*, **chalky**
~ kalkiger Boden, **limy** ~ kalkiger Boden,
nature of the ~ Bodenbeschaffenheit *f*, **type of**
~ Bodenart *f*
soil *v* beschmutzen, beflecken, besudeln,
verunreinigen
soil analysis Bodenanalyse *f*
soil cement stabilization Bodenvermörtelung *f*
soil colloid Bodenkolloid *n*
soil corrosion Bodenkorrosion *f*
soil examination Bodenuntersuchung *f*
soil investigation Bodenuntersuchung *f*
soil nutrient Bodennährstoff *m*
soil reaction Bodenreaktion *f*
soil stabilization Bodenvermörtelung *f*
soil technology Bodentechnologie *f*
soil thermometer Erdbodenthermometer *n*
soil utilization Bodennutzung *f*
soja [bean] Sojabohne *f* (Bot)
soja [bean] oil Sojabohnenöl *n*
sojourn probability
Aufenthaltswahrscheinlichkeit *f*
sol Sol *n*, **flocculated** ~ ausgeflocktes Sol,
thixotropic ~ thixotropes Sol
solabiose Solabiose *f*
soladulcidine Soladulcidin *n*
solandrine Solandrin *n*
solanellic acid Solanellsäure *f*
solanesol Solanesol *n*
solanidine Solanidin *n*
solanin Solanin *n*
solanocapsine Solanocapsin *n*
solanorubin Solanorubin *n*
solar chromosphere Sonnenchromosphäre *f*
solar eclipse Sonnenfinsternis *f* (Astr)
solar energy Sonnenenergie *f*
solar heat Sonnenwärme *f*
solarization Solarisation *f* (Phot)
solar photosphere Sonnenphotosphäre *f*
solar radiation Sonnenbestrahlung *f*,
Sonnenstrahlung *f*
solar spectrum Sonnenspektrum *n*
solarstearin Solarstearin *n*
solar system Sonnensystem *n* (Astr)
solasulfone Solasulfon *n*
solatriose Solatriose *f*
sold verkauft
solder Lötmetall *n*, Lot *n*, **low tin** ~ Lot *n* (mit
niedrigem Zinngehalt), **quick** ~ Schnellot *n*,
soft ~ Schnellot *n*, Weichlot *n*

solder v löten, ~ **on** anlöten, auflöten
solderable lötbar
soldered [an]gelötet, **capable of being** ~ lötbar
soldered joint Lötfuge f, Lötstelle f, Lötverbindung f
soldered junction Lötstelle f, Lötverbindung f
soldering Löten n, Lötstelle f, Lötung f, **autogenous** ~ autogene Lötung
soldering acid Lötsäure f
soldering block Lötblock m
soldering board Lötbrett n
soldering compound Lötpulver n
soldering fluid Lötwasser n
soldering frame Lötgestell n
soldering furnace Lötofen m
soldering iron Lötkolben m
soldering lamp Lötlampe f
soldering material Lot n
soldering metal Lötmetall n
soldering pan Lötpfanne f
soldering salt Lötsalz n
soldering seam Lötfuge f, Lötnaht f, Lötstelle f
soldering tweezers Lötzange f
soldering zinc Lötzink n
solder splashes Lotspritzer pl
sole accelerator Alleinbeschleuniger m
sole distributor Alleinvertreter m
sole leather Pfundleder n
solenoid Solenoid n (Elektr), Drahtspirale f (Elektr), Induktionsspule f (Elektr)
sole plate Fußplatte f, Sohlplatte f
sole producer Alleinhersteller m
solfatarite Solfatarit m (Min)
solid fester Körper m, Festkörper m
solid a fest, derb, gediegen, haltbar, hart, kompakt, massiv, starr
solid angle Raumwinkel m
solid bed Festbett n (Katalysator)
solid body Festkörper m
solid-bowl or imperforate-basket centrifuge Vollwandzentrifuge f
solid carbon Homogenkohle f
solid content[s] Festgehalt m
solid diffusion Festkörperdiffusion f
solidification Erstarren n, Erstarrung f, Festwerden n, Verfestigung f, ~ **of the bath** Einfrieren n des Bades, ~ **of the block** Blockeinfrierung f
solidification curve Erstarrungskurve f
solidification point Erstarrungspunkt m
solidification range Erstarrungsbereich m
solidification structure Erstarrungsgefüge n
solidification temperature Erstarrungstemperatur f
solidified erstarrt
solidified lava erstarrte Lava f
solidify v erstarren, verfestigen, fest werden
solidifying Härten n, Hartwerden n
solidifying point Stockpunkt m

solidity Dichte f, Festigkeit f, Haltbarkeit f, Stabilität f
solid jacket Vollmantel m
solid-liquid rocket Feststoff-Flüssigkeits-Rakete f
solid lubricants Festschmierstoffe m pl
solid matter Feststoff m, Festsubstanz f
solid measure Raummaß n
solid ore deposit Erzstock m
solid phase Bodenkörper m (Chem), Festphase f
solid phase technique Festphasentechnik f
solid piston Scheibenkolben m
solid-piston pump Scheibenkolbenpumpe f
solid propellant rocket Feststoffrakete f
solid rubber Festkautschuk m
solids, extraction of ~ Festkörperextraktion f, **total** ~ Gesamtgehalt m an Feststoffen (pl)
solid sampler Feststoff-Probengabe f
solids content Feststoffgehalt m
solid-state board Einschubrahmen m (als Träger gedruckter Schaltungen)
solid-state chemistry Festkörperchemie f
solid-state reaction Festkörperreaktion f
solid-state terminology Festkörperterminologie f
solid-stem thermometer Stabthermometer n
solid support Trägermaterial n (Gaschromat)
solidus curve Soliduskurve f
soliquid Dispersion f von festen Teilchen (pl) (Phys)
solitary vereinzelt
solketal Solketal n
solochrome dyestuff Solochromfarbstoff m
soloric acid Solorsäure f
solstice Sonnenwende f (Astr)
solubility Löslichkeit f, Lösbarkeit f (Math), **determination of** ~ Löslichkeitsbestimmung f, **influence on** ~ Löslichkeitsbeeinflussung f, **limited** ~ beschränkte Löslichkeit, **limit of** ~ Löslichkeitsgrenze f, **low** ~ Schwerlöslichkeit f, **molar** ~ molare Löslichkeit, **partial** ~ beschränkte Löslichkeit
solubility pressure Löslichkeitsdruck m
solubility product Löslichkeitsprodukt n
solubilization Aufschluß m (Chem)
solubilization process Lösungsprozeß m
solubilize v anlösen, aufschließen, löslich machen
solubilizer Aufschlußmittel n, Hilfslösungsmittel n, Lösungsvermittler m
soluble löslich, [auf]lösbar, **easily** ~ leichtlöslich (Chem), **readily** ~ leichtlöslich (Chem)
soluble gun cotton Kollodiumwolle f, Pyroxylin n
soluble matter, total ~ Gesamtlösliche[s] n
solurol Solurol n
solute gelöster Stoff m, Gelöste[s] n
solution Auflösung f (Chem), Lösung f (Chem, Math), **alcoholic** ~ alkoholische Lösung, **aqueous** ~ wässrige Lösung, **centinormal** ~ 1/100 normale Lösung, **concentrated** ~

659

konzentrierte Lösung, **decinormal** ~ 1/10
normale Lösung, **dilute** ~ verdünnte Lösung,
equimolecular ~ äquimolekulare Lösung,
ethereal ~ ätherische Lösung, **heat of** ~
Auflösungswärme *f,* **hundredth normal** ~
1/100 normale Lösung, **ideal** ~ ideale
Lösung, **initial** ~ Anfangslösung *f,* **molar** ~
molare Lösung, **non-ideal** ~ nichtideale
Lösung, **normal** ~ normale Lösung, **one tenth
normal** ~ 1/10 normale Lösung, **original** ~
Anfangslösung *f,* **saturated** ~ gesättigte
Lösung, **solid** ~ feste Lösung (Krist),
supersaturated ~ übersättigte Lösung,
unsaturated ~ ungesättigte Lösung
solution adhesive Lösungsmittelkleber *m*
solution annealing Lösungsglühen *n* (Metall)
solution enthalpy Lösungsenthalpie *f*
solution equilibrium Lösungsgleichgewicht *n*
solution-grown crystal Lösungskristall *m*
solution heat treatment Vergütungsglühen *n*
(Leichtmetall)
solution polymerization
Lösungsmittelpolymerisation *f,*
Lösungspolymerisation *f*
solution potential Lösungspotential *n*
solution pressure Lösungsdruck *m*
solutions, isobaric ~ Lösungen *pl* mit gleichem
Dampfdruck (Phys)
solution theory Lösungstheorie *f*
solvable auflösbar (Math)
solvate *v* anlösen, solvatisieren
solvated solvatisiert
solvating envelope Solvathülle *f*
solvation Solvatation *f,* Solvatisierung *f,* **energy
of** ~ Solvatationsenergie *f*
Solvay process Ammoniaksodaprozeß *m,*
Solvay-Verfahren *n*
solve *v* lösen (Math)
solvency Lösevermögen *n*
solvent Lösungsmittel *n,* Solvens *n,* **dissociating**
~ dissoziierendes Lösungsmittel, **non-aqueous**
~ nichtwässriges Lösungsmittel, **nonpolar** ~
unpolares Lösungsmittel, **polar** ~ polares
Lösungsmittel, **residual** ~
Lösungsmittelrückstand *m,* **selective** ~
differenzierendes Lösungsmittel, selektives
Lösungsmittel, **volatile** ~ flüchtiges
Lösungsmittel, **with low** ~ **content**
lösungsmittelarm
solvent *a* lösend
solvent extraction Lösungsmittelextraktion *f*
solvent-free lösungsmittelfrei
solvent mixture Lösungsmittelgemisch *n*
solvent molding Tauchformen *n*
solvent naphtha Lösungsbenzol *n,*
Solventnaphtha *n*
solvent power Auflösungsvermögen *n* (Chem),
Lösungskraft *f*

solvent recovery Lösungsmittelrückgewinnung *f,*
Rückgewinnung *f* des Lösungsmittels
solvents, fastness to ~ Lösungsmittelechtheit *f,*
fast to ~ lösungsmittelbeständig
solvent surplus Lösungsmittelüberschuß *m*
solvent vapor Lösungsmitteldampf *m*
solvent water Lösungswasser *n*
solvolysis Solvolyse *f*
somalin Somalin *n*
somatotrophin Somatotrophin *n*
sombrerite Sombrerit *m* (Min)
somnirol Somnirol *n*
somnitol Somnitol *n*
somnoform Somnoform *n*
sonicate *v* beschallen
sonication Beschallung *f*
sonic atomizer Schallzerstäuber *m*
sonic screen Schallsieb *n*
soot Ruß *m,* **covering with** ~ Berußen *n,* **deposit
of** ~ Rußansatz *m,* **formation of** ~
Rußbildung *f,* **lustrous form of** ~ Glanzruß *m,*
producing ~ rußend, **to coat with** ~ anrußen,
to produce ~ rußen
soot *v* rußen, verrußen
soot collector Rußvorlage *f*
soot flake Rußflocke *f*
soothing remedy Linderungsmittel *n* (Pharm)
sooting Verrußen *n*
sootless nichtrußend
soot receiver Rußvorlage *f*
sooty rußig, berußt, rauchschwarz
sooty coal Rußkohle *f*
sooty color, of ~ rußfarben
sophocarpidine Sophocarpidin *n*
sophoranol Sophoranol *n*
sophoricol Sophoricol *n*
sophorine Sophorin *n*
sophoritol Sophorit *m*
sophorose Sophorose *f*
soporific Schlafmittel *n* (Pharm)
soporific *a* einschläfernd, narkotisch
soranjidiol Soranjidiol *n*
sorbate Sorbat *n,* Salz oder Ester der
Sorbinsäure *f*
sorbic acid Sorbinsäure *f,* **salt or ester of** ~
Sorbat *n*
sorbicillin Sorbicillin *n*
sorbieritol Sorbierit *m*
sorbin Sorbin *n,* Sorbinose *f*
sorbinose Sorbin *n,* Sorbinose *f*
sorbite Sorbit *m* (Chem)
sorbitol Sorbit *m* (Chem)
sorbosazone Sorbosazon *n*
sorbose Sorbose *f,* Sorbin *n,* Sorbinose *f*
sorburonic acid Sorburonsäure *f*
sorbyl alcohol Sorbylalkohol *m*
Sorel cement Sorelzement *m*
sorghum oil Sorghumöl *n*
sorigenin Sorigenin *n*

sorinin Sorinin *n*
sorption Sorption *f*
sorrel salt Kaliumbioxalat *n*
sort Sorte *f*, Art *f*
sort *v* sortieren, auslesen, klauben (Bergb), ~
   out aussortieren
sorter Ausleser *m*, Sortierer *m*
sorting Sortieren *n*, Auslesen *n*, **mechanical** ~
   mechanische Sortierung *f*, ~ **of ores**
   Erzscheidung *f*
sorting band Leseband *n* (Bergb)
sorting device Scheidevorrichtung *f*
sorting machine Sortiermaschine *f*
sorting magnet Scheidemagnet *m*
sorting table Sortiertisch *m*, Klaubetisch *m*
sosoloid Dispersion *f* von festen Teilchen (pl)
   (Phys)
sound Schall *m*, Klang *m*, Laut *m*, Ton *m*,
   intensity of ~ Schallstärke *f*, **level of** ~
   Lautstärkepegel *m*, **propagation of** ~
   Schallfortpflanzung *f*, **refraction of** ~
   Schallbrechung *f* (Akust), **source of** ~
   Schallquelle *f*, **velocity of** ~
   Schallgeschwindigkeit *f*, **absorption or**
   attenuation of ~ **in solids**
   Körperschalldämmung *f*
sound *a* gesund
sound *v* loten, auspeilen, sondieren (Med),
   tönen (Akust)
sound absorber Schalldämpfer *m*
sound-absorbing schalldämpfend,
   schallschluckend
sound-absorbing paint Antidröhnlack *m*,
   schalldämpfende Anstrichfarbe *f*,
   Schallschlucklack *m*
sound absorption Schallabsorption *f*
sound absorption coefficient
   Schallabsorptionsgrad *m;*
   Schallschluckgrad *m*
sound amplifier Lautverstärker *m* (Akust)
sound analysis Schallanalyse *f*
sound attenuation Schalldämpfung *f*
sound damper Schalldämpfer *m*
sound-damping schalldämpfend
sound-deadening ceiling Schallschluckdecke *f*
sound-deadening composition Antidröhnlack *m*
sound-deadening paint schalldämpfende
   Anstrichfarbe *f*, Schallschlucklack *m*
sounded abgelotet
sound frequency Tonfrequenz *f*
sounding lead Lot[blei] *n*, Senkblei *n*
sounding rod Visitiereisen *n*
sound-insulating schallisolierend
sound insulation Schallisolation *f*,
   Schallisolierung *f*
sound insultation Schallschutz *m*
sound intensity Lautstärke *f*, Schallintensität *f*
sound intensity meter Lautstärkemesser *m*
sound level Schallpegel *m*

sound locator Horchgerät *n*
sound measurement Lautstärkemessung *f*
soundness Gesundheit *f*
sound perception Lautempfindung *f*,
   Schallwahrnehmung *f*
sound permeability Schalldurchlässigkeit *f*
sound pressure Schalldruck *m*
sound producer Tonerreger *m*
soundproof schalldicht, geräuschsicher
sound-proofing schalldämpfend
sound propagation Schallausbreitung *f*,
   Schallfortpflanzung *f*
sound recording Schallaufzeichnung *f*,
   Tonaufnahme *f*
sound reduction index Schalldämmaß *n*
sound reproduction Tonwiedergabe *f* (Akust)
sound reproduction technique
   Tonwiedergabetechnik *f* (Akust)
sound spectrum Schallspektrum *n* (Akust),
   Tonspektrum *n* (Akust)
sound stimulus Schallreiz *m*
sound track Tonrille *f* (Akust), Tonspur *f*
   (Akust)
sound transmission Schalldurchlässigkeit *f*
sound transmission loss Schalldämmaß *n*
sound volume Lautstärke *f*
sound wave Schallwelle *f*
souple silk Soupleseide *f*
sour *a* sauer
sour *v* ansäuern, sauer werden
sour bath Sauerbad *n*
sour bleaching Sauerbleiche *f*, Naßbleiche *f*
source Quelle *f*, ~ **of current** Stromquelle *f*, ~ **of**
   light Lichtquelle *f*, ~ **of fire** Feuerherd *m*
   (Brandherd)
source range Quellenbereich *m*
sour dough Sauerteig *m*
souring Ansäuern *n*, Aussäuerung *f* (Chem),
   Sauerwerden *n*
sourish säuerlich
sourish sweet süßsauer
sourness Säure *f*
sour wine Säuerling *m*
sow Eisenklumpen *m* (Metall)
sow *v* säen (Agr)
sowing Saatbestellung *f* (Agr)
Soxhlet apparatus Soxhletapparat *m*
Soxhlet extractor Soxhlet-Extraktor *m*
soyanal Soyanal *n*
soyasapogenol Sojasapogenol *n*
soybean Sojabohne *f* (Bot)
soybean oil meal Sojaextraktionsschrot *n*
soz[o]iodol Sozojodol *n*, Sozojodolsäure *f*
sozoiodolic acid Sozojodol *n*
sozolic acid Sozolsäure *f*
space Abstand *m*, Platz *m*, Raum *m*, **economy in**
   ~ Raumersparnis *f*, **lack of** ~ Platzmangel *m*,
   restricted ~ Platzmangel *m*, **saving in** ~

Raumersparnis *f,* **unit of** ~ Raumeinheit *f,* ~ **filled with rarefied air** luftverdünnter Raum *m*
space-centered raumzentriert (Krist)
space charge Raumladung *f*
space-charge accumulation Raumladungsanhäufung *f*
space-charge capacity Raumladungskapazität *f*
space-charge cloud Raumladungswolke *f*
space-charge density Raumladungsdichte *f*
space-charge detector Raumladungsdetektor *m*
space-charge distortion Raumladungsverzerrung *f*
space-charge effect Raumladungseffekt *m*
space-charge grid Raumladungsgitter *n*
space charges, concentration or accumulation of ~ Raumladungswolke *f*
space-charge zone (solid-state phys) Raumladungszone *f*
space cooling Raumkühlung *f*
space coordinate Raumkoordinate *f*
spacecraft Raumfahrzeug *n*
space curvature Raumkrümmung *f*
space curve Raumkurve *f*
space diagonal Raumdiagonale *f*
space diagram Raumbild *n*, Raumgebilde *n*
space dimension Raumdimension *f*
space filling Raumerfüllung *f* (Atom)
space filling atom model Atommodell *n*, Kalottenmodell *n*
space focusing Raumfokussierung *f*
space formula Raumformel *f*
space grid charge Gitterraumladung *f*
space group Raumgruppe *f* (Krist)
space groups, table of ~ Raumgruppentabelle *f*
space group symbols Raumgruppensymbolik *f*
space lattice Raumgitter *n* (Krist), **cubic** ~ kubisches Raumgitter (Krist)
space lattice plane Raumgitterebene *f*
space lattice structure Raumgitterstruktur *f* (Krist)
space navigation Weltraumschiffahrt *f*
spacer Abstandhalter *m*, Abstandsplatte *f,* Distanzring *m*, Distanzrolle *f,* Distanzscheibe *f,* Zwischentaste *f* (Schreibmaschine)
spacer block Distanzblock *m*
spacer bushing Distanzbuchse *f*
space reflection Raumspiegelung *f*
space requirement Platzbedarf *m*
space research Raumforschung *f*
space rocket Weltraumrakete *f*
spacer ring Zwischenring *m*
spaceship Raumfahrzeug *n*, Raumschiff *n*, Weltraumschiff *n*
space shuttle wiederverwendbare Raumfähre *f*
space station Weltraumstation *f*
space-time structure Raumzeitstruktur *f*
space-time yield Raum-Zeit-Ausbeute *f*
space transformation Raumtransformation *f*

space velocity Katalysatorbelastung *f*
space wave Raumwelle *f*
spacing Abstand *m*, ~ **of holes** Lochabstand *m*
spacing cam Vorschubdaumen *m*
spacing collar Beilegering *m*, Spindelring *m*
spacing contact Ruhekontakt *m*
spacing piece Distanzstück *n*
spacing position Trennstellung *f*
spacing ring Distanzring *m*
spallation Mehrfachzertrümmerung *f* (Phys) (of neutrons), Absplitterung *f* (von Neutronen)
spalling Schälen *n* (Met) (wear particularly in roller bearings when flakes of metal are removed at the surface), Abblättern *n* (Met)
span Spanne *f,* Spannweite *f,* Zeitspanne *f*
span adjustment Meßschritteinstellung *f*
spangle Flitter *m*
spangolite Spangolith *m* (Min)
span head Kegelkopf *m*
spanner (Br. E.) Schraubenschlüssel *m*
spanner bolt Schlüsselschraube *f*
spar Rundholz *n*, Spat *m* (Min), **gypseous** ~ blättriger Gips *m*, Gipsblume *f*
sparassol Sparassol *n*
spare battery Reservebatterie *f*
spare boiler Ersatzkessel *m*
spare bolt Ersatzbolzen *m*
spare engine Hilfsmaschine *f*
spare heater plate Ersatzheizplatte *f*
spare part Ersatzteil *n*, Reserveteil *n*
spare piece Ersatzstück *n*
spare roll Ersatzrolle *f*
spare roller Ersatzrolle *f*
spare tube Austauschröhre *f*
sparger Aufgußapparat *m* (Brau)
sparing out Aussparung *f*
spark Funke[n] *m*, Feuerfunke *m*, ~ **on closing** Schließungsfunke *m*, ~ **on opening** Öffnungsfunke *m*
spark *v* durchschlagen (Elektr), Funken sprühen
spark advance Frühzündung *f*
spark arrestor Funkenlöscher *m*, Funkenlöschscheibe *f*
spark chamber Funkenkammer *f*
spark coil Funkeninduktor *m*
spark condenser Störschutzkondensator *m* (Elektr)
spark conductor Funkenleiter *m*
spark current Durchschlagsstrom *m*
spark discharge Funkenentladung *f*
spark discharge continuum Funkenkontinuum *n*
spark extinguisher Funkenlöscher *m*, Funkenlöschscheibe *f*
spark gap Funkenstrecke *f,* Elektrodenabstand *m*
spark ignition Funkenzündung *f*
sparking Funkenbildung *f,* Durchschlagen *n*, Funkensprühen *n*, **moment of** ~ Zündmoment *m*

sparking plug Funkengeber *m*, Zündkerze *f*
sparking potential Funkenpotential *n* (Elektr)
sparking voltage Funkenspannung *f*
spark killer Funkenlöscher *m*
sparkle *v* funkeln, flimmern, Funken sprühen,
  glitzern, moussieren, schäumen
spark length Funkenlänge *f*
sparkless funkenlos
sparkless commutation, limit of ~
  Funkengrenze *f*
sparklet Fünkchen *n*
spark-like funkenähnlich
sparkling funkelnd, flimmernd; moussierend,
  schäumend
sparkling wine Schaumwein *m*
sparkover Überschlag *m* (Elektr)
sparkover voltage Überschlagsspannung *f*
spark path Funkenstrecke *f*
spark plug Zündkerze *f* (Kfz)
spark plug ignition Kerzenzündung *f*
spark plug socket Kerzenstecker *m*
spark plug spanner Kerzenschlüssel *m*
spark plug wire (Am. E.) Zündkabel *n* (Kfz)
spark plug wrench Kerzenschlüssel *m*
spark potential Funkenpotential *n*
spark quenching apparatus
  Funkenlöschvorrichtung *f*
sparks, emitting ~ funkensprühend, **emitting of**
  ~ Funkensprühen *n*, **formation of** ~
  Funkenbildung *f*
spark spectrum Funkenspektrum *n*
spark-suppressing funkenverhindernd
spark tracking Kriechwegbildung *f* (durch
  Gleitfunken)
spark transmitter Funkengeber *m*
spar milling machine Holmfräsmaschine *f*
sparry spatig, spatartig
sparry gypsum Gipsspat *m* (Min), Fraueneis *n*
  (Min), Katzenglas *n* (Min)
sparry limestone Schieferkalkstein *m* (Min)
sparteine Spartein *n*
spartyrine Spartyrin *n*
spar varnish Bootslack *m*, Decklack *m*
spasm Krampf *m* (Med), Spasmus *m* (Med)
spasmodic krampfartig
spasmolytic krampflösend
spathic spatig
spathic chlorite Chloritspat *m* (Min)
spathic iron Siderit *m* (Min)
spathic iron ore Siderit *m* (Min), Eisenspat *m*
  (Min)
spathic ore Spaterz *n* (Min)
spathulatine Spathulatin *n*
spatial räumlich
spatial charge Raumladung *f*
spatial expansion Raumausdehnung *f*
spatial formula Raumformel *f*
spatial isomerism Raumisomerie *f*
spatial requirement Raumbedarf *m*

spatter *v* [be]spritzen, verspritzen
spattering Spritzen *n*
spatula Spachtel *m f*, Spatel *m f*, **grooved** ~
  Furchenspatel *f*
spatula-shaped spatelförmig
spawn Laich *m* (Zool)
spawning time Laichzeit *f* (Zool)
speaker Lautsprecher *m* (Elektr),
  Vortragende(r) *f*
speaker cone Lautsprechermembran *f*
spear pyrites Speerkies *m* (Min)
spear shape Speerform *f*
spear-shaped speerförmig
special besonders, speziell
special alloy Speziallegierung *f*
special branch Fachgebiet *n*, ~ **of science**
  Fachwissenschaft *f*
special brochure Sonderprospekt *m*
special construction Sonderanfertigung *f*,
  Sonderkonstruktion *f*
special design Sonderausführung *f*,
  Spezialausführung *f*
special equipment Sonderausrüstung *f*,
  Spezialausrüstung *f*
special iron Qualitätseisen *n*
specialist Spezialist *m*, Fach-, Fachmann *m*,
  Sachkundiger *m*, Sachverständiger *m*
specialize *v* spezialisieren
special knowledge Fachkenntnis *f*
special leaflet Merkblatt *n*
special product Sonderprodukt *n*
special production Sonderanfertigung *f*
special purpose fixture
  Sonderzweckvorrichtung *f*
special reagents Spezialreagentien *pl*
special rules Sondervorschriften *pl*
special size Sondergröße *f*
special spanner Fassonschlüssel *m*
special steel Qualitätsstahl *m*, Spezialstahl *m*
special subject Fachgebiet *m*
special type Spezialausführung *f*
species Art *f*, Spezies *f*, **of the same** ~ artgleich
species difference Artunterschied *m*
species-specific artspezifisch
species specificity Artspezifität *f*
specific spezifisch
specification Spezifikation *f*, Beschreibung *f* (z.
  B. Patent), Bezeichnung *f*, Einzeldarstellung *f*,
  Mindestanforderung *f*
specification factor Bestimmungsgröße *f* (Techn)
specifications Betriebsvorschriften *f pl*,
  Lieferbedingungen *f pl*, Normen *f pl*
specific conductivity spezifische Leitfähigkeit *f*,
  spezifisches Leitvermögen *n*
specific energy Eigenenergie *f*
specific gravity spezifisches Gewicht *n* (Phys),
  Wichte *f* (Phys)
specific gravity bottle Pyknometer *n*,
  Dichtemesser *m*; Grammflasche *f*

specific gravity spindle Senkspindel *f*
specific heat spezifische Wärme *f* (Phys)
specific inductive capacity
  Dielektrizitätskonstante *f*
specificity Spezifität *f*
specific pressure drop spezifischer
  Druckverlust *m* (Dest)
specific volume spezifisches Volumen *n* (Phys)
specific weight spezifisches Gewicht *n* (Phys)
specify *v* spezifizieren, angeben
specimen Muster *n*, Exemplar *n*, Probe *f*,
  Probestück *n*
specimen bottle Probeflasche *f*
specimen holder Objektträger *m*
specimen page Musterseite *f* (Buchdr),
  Probeseite *f* (Buchdr)
specimen print Probeabzug *m* (Buchdr)
specimen region Objektbereich *m*
specimen room Präparateraum *m*
speck Fleck *m*, Fleck[en] *m*
speckle Fleck[en] *m*
speckled scheckig; gesprenkelt
speckled wood Maserholz *n*
spectacle furnace Brillenofen *m*
spectacles Brille *f*
spectral spektral
spectral analysis Spektralanalyse *f*
spectral characteristic Spektralcharakteristik *f*
  (Spektr)
spectral color Spektralfarbe *f*
spectral decomposition spektrale Zerlegung *f*
spectral distribution Spektralverteilung *f*
spectral line Spektrallinie *f*
spectral lines, folding of ~
  Spektrallinienfaltung *f*, **shift of** ~
  Spektrallinienverschiebung *f* (Spektr)
spectral purity spektrale Reinheit *f* (Opt)
spectral range Spektralbereich *m*
spectral region Spektralbereich *m*
spectral sensitivity Spektralempfindlichkeit *f*
  (Phot)
spectral series Spektrallinienfolge *f*
spectral shift Spektralverschiebung *f*
spectroanalytic spektralanalytisch
spectrogram Spektrogramm *n*
spectrograph Spektrograph *m*
spectroheliogram Spektroheliogramm *n*
spectrohelioscope Spektrohelioskop *n*
spectrometer Spektralapparat *m*, Spektrometer *n*
spectrometric spektrometrisch,
  spektralanalytisch
spectrometry Spektrometrie *f*
spectrophotometer Spektralphotometer *n*
spectroscope Spektroskop *n*
spectroscopic spektroskopisch,
  spektralanalytisch
spectroscopy Spektroskopie *f*
spectrum (pl. spectra) Spektrum *n* (pl.
  Spektren), **chromatic** ~ Farbenspektrum *n*,

continuous ~ kontinuierliches Spektrum,
discontinuous ~ diskontinuierliches
  Spektrum, **secondary** ~ Sekundärspektrum *n*
spectrum analysis Spektralanalyse *f*
specular [cast] iron Spiegeleisen *n*
specular hematite Spekularit *m* (Min)
specular iron Spekularit *m* (Min), Eisenglanz *m*
  (Min), Roteisenstein *m* (Min)
specular iron ore Eisenglanz *m* (Min),
  Spiegelerz *n* (Min)
specularite Spekularit *m* (Min)
specular reflectance Glanzwinkelreflexion *f*
  (Opt)
specular stone Marienglas *n* (Min)
speculum metal Spiegelmetall *n*, Glanzmetall *n*
speed Geschwindigkeit *f*, Schnelligkeit *f*,
  Tempo *n*, **at high** ~ mit hoher
  Geschwindigkeit, **constant** ~ gleichförmige
  Geschwindigkeit, **critical** ~ kritische
  Geschwindigkeit, **even** ~ Gleichgang *m*, **limit
  of** ~ Geschwindigkeitsgrenze *f*, **loss of** ~
  Geschwindigkeitsabfall *m*, **mean or average** ~
  durchschnittliche Geschwindigkeit, **proper** ~
  Eigengeschwindigkeit *f*, **top or maximum** ~
  Höchstgeschwindigkeit *f*, **uniform** ~
  gleichförmige Geschwindigkeit, **unit of** ~
  Geschwindigkeitseinheit *f*, ~ **of a lens**
  Helligkeit *f* einer Linse (Opt), ~ **of formation**
  Bildungsgeschwindigkeit *f*, ~ **of light**
  Lichtgeschwindigkeit *f*
speed-controlled geschwindigkeitsgesteuert
speed counter Tourenzähler *m*
speed drive, adjustable ~ regelbarer Antrieb *m*
speed governor Geschwindigkeitsregler *m*
speed indicator Tourenzähler *m*, Tachometer *n m*
speeding up Beschleunigung *f*
speed limit, exceeding the ~
  Geschwindigkeitsüberschreitung *f*
speedometer Geschwindigkeitsmesser *m*,
  Tachometer *n m*, Tourenzähler *m*
speed ratio Geschwindigkeitsverhältnis *n*
speed reducer Reduktionsgetriebe *n*
speed regulating device Drehzahlregler *m*
speed regulation Drehzahlregelung *f*
speed regulator Beschleunigungsregler *m*
speed-setting device Regler *m* (Drehzahl)
speed transmitter Geschwindigkeitsgeber *m*
speed up *v* beschleunigen
spelter Rohzink *n*, **refined** ~ Raffinatzink *n*
spencerite Spencerit *m* (Min)
spend *v* aufwenden, verbrauchen
spent acid Abfallsäure *f*, Abgangssäure *f*
spent grains Malztreber *pl*
spent lye Abfallauge *f*
spent malt Malztreber *pl*
spent yeast Brannthefe *f*
sperm Samen *m*, Samenflüssigkeit *f*
spermaceti Spermacet *n*, Walrat *m*
spermatic cell Samenzelle *f*

spermatic nucleus Samenkern *m* (Zool)
spermatocidal samentötend
spermatogenesis Spermatogenese *f,*
  Samenbildung *f*
spermatophyta Spermatophyten *pl* (Bot)
spermatozoon (pl spermatozoa) Spermatozoon *n*
  (pl. Spermatozoen)
sperm cell Samenzelle *f*
spermicidal Spermizid (Pharm)
spermicidal *a* samentötend
spermidine Spermidin *n*
spermine Spermin *n*
sperm oil Spermöl *n,* Walratöl *n*
spermostrychnine Spermostrychnin *n*
sperrylite Sperrylith *m* (Min)
spessartite Mangangranat *m* (Min),
  Spessartin *m* (Min), Spessartit *m* (Geol)
spew Austrieb *m*
spew channel Auslaufrille *f*
spew groove Austriebrille *f*
sphaerosiderite Sphärosiderit *m* (Min)
sphalerite Sphalerit *m* (Min), falscher
  Bleiglanz *m* (Min), Zinkblende *f* (Min),
  fibrous ~ Schalenblende *f* (Min)
sphene Sphen *m* (Min)
sphenoid Sphenoid *n* (Krist)
sphenoid *a* keilähnlich
sphere Kugel *f,* Ball *m,* Sphäre *f,* **diameter of a** ~
  Kugeldurchmesser *m,* **solid** ~ Vollkugel *f,*
  surface of a ~ Kugeloberfläche *f,* ~ **of action**
  Wirkungssphäre *f,* ~ **of reflection**
  Reflexionskugel *f*
sphere packing Kugelpackung *f* (Krist), **closest**
  ~ dichteste Kugelpackung
spherical kugelförmig; kugelig, sphärisch (Math)
spherical bacteria Kugelbakterien *pl*
spherical cap Kalotte *f* (Chem), Kugelkalotte *f*
  (Chem)
spherical condenser Kugelkühler *m*
spherical coupling Kugelkupplung *f*
spherical function Kugelfunktion *f*
spherical harmonic analysis
  Kugelfunktionsentwicklung *f*
spherical indentation Kugeleindruck *m*
spherical joint Kugelkupplung *f,*
  Kugelschliffverbindung *f*
spherical mirror Kugelspiegel *m*
spherical mold Kugelform *f*
spherical molecule Kugelmolekül *n*
spherical protein Kugelprotein *n*
spherical radiator Kugelstrahler *m*
spherical receiver Kugelvorlage *f*
spherical section Kugelschnitt *m*
spherical segment Kugelsegment *n*
spherical shape Kugelgestalt *f*
spherical shell Kugelschale *f*
spherical surface Kugeloberfläche *f*
spherical symmetry Kugelsymmetrie *f*
spherical triangle Kugeldreieck *n*

sphericity Kugelgestalt *f,* Sphärizität *f*
spherocarpine Sphaerocarpin *n*
spherocobaltite Kobaltspat *m* (Min)
spherocyte Kugelzelle *f*
spheroid Rotationsellipsoid *n,* Sphäroid *n*
spheroidal kugelig, kugelähnlich, sphäroidisch
spheroidizing Weichglühen *n* (Techn)
spherolite Sphärolith *m* (Geol)
spherometer Sphärometer *n*
spherophorin Sphaerophorin *n*
spherophorol Sphärophorol *n*
spherophysine Sphärophysin *n*
spherosiderite Sphärosiderit *m* (Min)
spherulite Nierenstein *m* (Min)
sphingoin Sphingoin *n*
sphingolipid Sphingolipid *n*
sphingomyelin Sphingomyelin *n*
sphingosine Sphingosin *n*
sphondin Sphondin *n*
sphragide Sphragid *m* (Min)
spica Ähre *f* (Bot)
spice Gewürz *n,* Gewürzstoff *m,* Würze *f*
spice *v* würzen
spiced wine Gewürzwein *m*
spices, odor of ~ Würzgeruch *m*
spice wood oil Benzoelorbeeröl *n*
spiculisporic acid Spiculisporsäure *f*
spicy würzig
spicy flavor, to deprive of ~ entwürzen
spicy odor Würzgeruch *m*
spider Armkreuz *n* (Techn), Drehkreuz *n*
  (Techn), Spinne *f* (Zool), Tragkreuz *n*
  (Techn)
spider cell Sternzelle *f* (Histol)
spider leg Dornhaltersteg *m*
spiegel Spiegeleisen *n* (Techn)
spigot Zapfen *m*
spigot joint Zapfenverbindung *f,*
  Muffen[rohr]verbindung *f*
spike Bolzen *m,* Hakennagel *m*
spike oil Spiköl *n*
spilanthol Spilanthol *n*
spill *v* vergießen, verschütten
spillway Überlauf *m*
spilosite Fleckschiefer *m* (Geol)
spin Spin *m* (Phys), Eigendrehbewegung *f*
  (Phys), **dependence on** ~ Spinabhängigkeit *f*
spin *v* rotieren, spinnen (z. B. Wolle)
spinacene Squalen *n*
spinacine Spinacin *n*
spinal cord Rückenmark *n* (Anat)
spin alignment Spinausrichtung *f* (Atom)
spin angular moment Spindrehimpuls *m* (Atom)
spinasterol Spinasterin *n*
spinazarin Spinazarin *n*
spin bath Spinnbad *n*
spin conservation Spinerhaltung *f*
spin coupling Spinkopplung *f*
spin density Spindichte *f*

spin direction Spinrichtung *f* (Atom)
spin distribution Spinverteilung *f*
spindle Achse *f* (Techn), Spindel *f,* Welle *f*
(Techn)
spindle bearing Spindellagerung *f,* Zapfenlager *n*
spindle box Kammwalzenständer *m*
spindle cell Spindelzelle *f*
spindle flange Spindelflansch *m*
spindle housing Kammwalzenständer *m*
spindle oil Leichtöl *n*
spindle press Spindelpresse *f*
spindle socket Spindelfuß *m*
spindle tree Spindelbaum *m* (Bot)
spindle valve Spindelventil *n*, T-Ventil *n*
spin doublet Spindublett *n* (Atom)
spin doubling Spinverdopplung *f*
spine Wirbelsäule *f* (Med)
spin-echo-method Spin-Echo-Verfahren *n*
(Atom)
spin effect Spineinfluß *m*
spinel Spinell *m* (Min)
spinel ruby Saphirrubin *m* (Min)
spinel twin Spinellzwilling *m* (Krist)
spin energy Spinenergie *f*
spin exchange interaction
Spinaustauschwechselwirkung *f*
spin flip Spinumklappung *f,* Umklappen *n* des
Spins
spin flip scattering Spinstreuung *f*
spin function Spinfunktion *f*
spin lattice relaxation Spingitterrelaxation *f*
spin magnetism Spinmagnetismus *m*
spin matrix Spinmatrize *f*
spin momentum Spinmoment *n*
spin momentum density Spinmomentdichte *f*
spin multiplet Spinmultiplett *n* (Atom)
spin multiplicity Spin-Multiplizität *f*
spinneret Spinndüse *f* (Kunstfaserherstellung)
spinning Spinnen *n*, direct ~
Direktspinnverfahren *n*
spinning band column Drehbandkolonne *f*
spinning blend Spinnmischung *f*
spinning bobbin Spinnspule *f*
spinning dope Spinnflüssigkeit *f*
spinning electron Drehelektron *n*
spinning head Spinnkopf *m*
spinning machine Spinnmaschine *f*
spinning mill Spinnerei *f*
spinning nozzle Spinndüse *f*
spinning paste Spinnpaste *f*
spinning ring Spinnring *m*
spinning roller Spinnzylinder *m*
spinning solution Spinnlösung *f*
spinning system Spinnverfahren *n*
spin number Spinzahl *f* (Atom)
spinochrome Spinochrom *n*
spin operator Spinoperator *m*
spin orbit Spinbahn *f*
spin [orbit] coupling Spinbahnkopplung *f*

spin orbit interaction Spinbahnwechselwirkung *f*
spin orbit splitting Spinbahnaufspaltung *f*
spinor calculus Spinorkalkül *n*
spinor field Spinorfeld *n*
spin polarization Spinpolarisation *f*
spin population Spinpopulation *f*
spin quantum number Spinquantenzahl *f* (Atom)
spin-spin interaction
Spin-Spin-Wechselwirkung *f* (Atom)
spinthariscope Spinthariskop *n* (Phys)
spinulosin Spinulosin *n*
spin variation Drehimpulsänderung *f*
spin wave method Spinwellenmethode *f*
spin welding Reibungsschweißen *n*,
Rotationsschweißen *n*
spiraein Spiräin *n*
spiraeoside Spiraeosid *n*
spiral Spirale *f*
spiral *a* gewunden, spiralförmig
spiral agitator Bandrührer *m*
spiral applicator Spiralauftragwalze *f,*
Spiralrakel *f*
spiral classifier Spiralsichter *m*,
Spiralwindsichter *m*
spiral condenser Schlangenkühler *m*
spiral conductor Wendelleiter *m* (Elektr)
spiral gears Schnecke *f* (Techn)
spirally fluted spiralverzahnt
spiral mixer Wendelrührer *m*
spiral nebula Spiralnebel *m* (Astr)
spiral orbit spectrometer
Spiralbahnspektrometer *n*
spiral spring Spiralfeder *f,* Schraubenfeder *f,*
Wickelfeder *f*
spiral stirrer Rührschnecke *f*
spiral stopper Spiralstopfen *m*
spiral tube Schlangenrohr *n*, Spiralrohr *n*
spiral-tube heat exchanger
Spiralrohrwärmeaustauscher *m*
spiral tube pneumatic drier
Drallrohr-Trockner *m*
spiral wheel Schneckenrad *n*
Spirillum (pl Spirilla) Schraubenbakterium *n*
spirit Branntwein *m*, Spiritus *m*, **crude** ~
Rohspiritus *m*, **mixed** ~ Mischspiritus *m*, **raw**
~ Rohspiritus *m*, **sugared** ~
Zuckerbranntwein *m*, ~ **of balm**
Melissengeist *m*
spirit blue alkoholische Lösung von Anilinblau *n*
spirit lacquer Spritlack *m*
spirit lamp Spirituslampe *f*
spirit level Libelle *f*, Nivellierwaage *f*,
Wasserwaage *f*
spirit of the age Zeitgeist *m*
spirit of wine Weingeist *m*
spirits alkoholisches Getränk *n*, Spirituosen *pl*
spirit-soluble spritlöslich
spirituous alkoholisch, geistig, sprithaltig
spirit varnish Spirituslack *m*, Spritlack *m*

spirocide Spirocid *n*
spiro-compound Spiro-Verbindung *f*
spirocyclan Spirocyclan *n*
spirocyclic spirocyclisch
spiroheptane Spiroheptan *n*
spirohexane Spirohexan *n*
spiropentane Spiropentan *n*
spirosal Spirosal *n*
spit *v* spucken, spritzen; spratzen (Techn)
spitting Spratzen *n* (Techn)
spittle Speichel *m*
splash *v* [ver]spritzen, besprenkeln
splash head, double bulb sloping ~
  Doppeltropfenfänger *m*
splash head adapter Tropfenfängeraufsatz *m*
splashing Herumspritzen *n*
splash lubrication Tauchschmierung *f,*
  Spritzschmierung *f*
splayed circular pass Spitzbogenkaliber *n*
spleen Milz *f* (Anat)
splendid glanzvoll, hervorragend
splice Falzung *f*
splice *v* spleißen, falzen, splissen
splice point Falzstelle *f*
splicer adhesive Fugenleim *m*
splicing, tapeless ~ of veneers
  Furnierfugenverleimung *f*
spline Feder *f* für Keilnut, Keil *m,* Splint *m*
spline grinder Keilwellenschleifmaschine *f*
splines Keilverzahnung *f*
spline shaft Keilwelle *f,* Nutenwelle *f*
splint Schiene *f* (Med), Span *m* (Techn),
  Versteifung *f* (Techn)
splint coal Schieferkohle *f* (Min)
splinter Splitter *m,* Bruchstück *n,* Span *m*
splinter *v* zersplittern (Holz)
splintering off Absplittern *n*
splinter-proof nichtsplitternd, splittersicher
split Spalt *m,* Riß *m*
split *v* [auf]spalten, auseinanderreißen, keilen
  (Techn), schlitzen, spleißen (Techn), teilen, ~
  off abspalten, ~ up aufspalten, zerlegen
split cavity blocks Gesenkteile *pl*
split coupling Schalenkupplung *f*
split log Halbholz *n*
split mandrel Spaltdorn *m*
split mold mehrteilige Preßform *f,*
  zusammengesetztes Werkzeug *n*
split-phase motor Spaltpolmotor *m*
split pin Splint *m*
split pole motor Spaltpolmotor *m*
split product Spaltprodukt *n,* Spaltstück *n*
splits Gesenkteile *pl*
splitting Aufspaltung *f,* Aufsplitterung *f,*
  Spaltung *f,* Teilung *f,* Zerfall *m,* ~ off
  Abspalten *n,* Abspaltung *f,* ~ of fatty acids
  Spaltung *f* von Fettsäuren
splitting *a* [ab]spaltend
split wall tile Spaltplatte *f*

splutter *v* spritzen, sprühen
spodumene Spodumen *m* (Min), Triphan *m*
  (Min)
spoil Aushub *m* (Baugruben)
spoil *v* beschädigen, verderben, ruinieren,
  faulen
spoilage Verderb *m*
spoiled casting Ausschuß *m* (Gieß)
spoke wire Speichendraht *m*
sponge Schwamm *m*
sponge charcoal Schwammkohle *f*
sponge filter Schwammfilter *m*
sponge rubber Schaumgummi *m,*
  Schwammgummi *m*
spongesterol Spongesterin *n*
spongin Spongin *n*
sponginess Schwammigkeit *f,* Porosität *f,* ~ of
  cast iron Schwammigkeit *f* des Gußeisens
spongosine Spongosin *n*
spongothymidine Spongothymidin *n*
spongouridine Spongouridin *n*
spongy schwammig, blasig, locker, porös,
  schwammartig
spongy copper Schwammkupfer *n*
spongy iron Eisenschwamm *m*
spongy nickel Nickelschwamm *m*
spongy peat Schwammtorf *m*
spongy wood Schwammholz *n*
spontaneous spontan, freiwillig
spontaneous discharge Selbstentladung *f* (Akku)
spontaneous evolution of heat Selbsterwärmung *f*
spontaneous generation Abiogenese *f* (Biol),
  Urzeugung *f* (Biol)
spontaneous heating Selbsterwärmung *f*
spontaneous ignition temperature
  Selbstentzündungstemperatur *f*
spool Haspel *f,* Spule *f*
spool *v* spulen, aufwickeln
spooling oil Spulöl *n*
spoon Löffel *m*
spoon agitator Löffelrührer *m*
sporadic vereinzelt, sporadisch
spore Spore *f* (Biol)
spore formation Sporenbildung *f*
spore of a bacillus Bazillenspore *f*
sporicidal sporentötend
spot Fleck[en] *m,* Ort *m,* Punkt *m,* Stelle *f,* dull
  ~ matte Stelle
spot *v* beflecken, betupfen, tüpfeln
spot analysis Tüpfelanalyse *f* (Anal)
spot check Einzelprobe *f,* Stichprobe *f*
spot distortion Fleckverformung *f*
spot glueing Punktleimverfahren *n*
spotlight Scheinwerfer *m*
spot method Tüpfelmethode *f* (Anal)
spot plate Tüpfelplatte *f* (Anal)
spot remover Fleckenentfernungsmittel *n*
spot repair Laufflächenreparatur *f*

spotted fleckig, geflammt (Keram, Holz), gefleckt, scheckig
spot test Tüpfelprobe *f,* Tüpfelreaktion *f*
spot-weld *v* punktschweißen
spot-welded punktgeschweißt
spot welder Punktschweißmaschine *f*
spot welding Punktschweißen *n,* Punktschweißung *f,* Punktschweißverfahren *n*
spot welding machine Punktschweißmaschine *f*
spout Ausguß *m,* Füllrohr *n,* Schnauze *f*
sprat oil Sprottenöl *n*
spray Sprühdose *f,* Flugwasser *n,* Schaum *m,* Sprühwasser *n,* Zerstäubungsmittel *n*
spray *v* zerstäuben, aufdüsen, aufsprühen, aufstäuben, berieseln; besprengen, [be]spritzen, [be]sprühen, verstäuben, ~ **on** aufspritzen
spray booth Spritzkabine *f*
spray can Sprühdose *f*
spray carburetor Zerstäubungsvergaser *m*
spray closure Spritzverschluß *m*
spray column Sprühkolonne *f*
spray cooler Berieselungskühler *m*
spray discharge Sprühentladung *f*
spray dried rubber Sprühkautschuk *m*
spray drier Rieseltrockner *m,* Sprühtrockner *m,* Zerstäubungstrockner *m*
spray-drying Sprühtrocknung *f,* Trocknen *n* durch Zerstäuben, ~ **with hot air** Heißsprühverfahren *n*
sprayer Zerstäuber *m,* Berieselungsapparat *m,* Sprühapparat *m,* **tubular** ~ Rohrzerstäuber *m*
spray gun Spritzpistole *f,* Sprühpistole *f,* Zerstäuberpistole *f*
spraying Versprühen *n,* Aufspritzen *n,* Berieselung *f,* Zerstäuben *n*
spraying cleaner Zerstäubungsreiniger *m*
spraying device Berieselungsvorrichtung *f,* Spritzapparat *m*
spray[ing] lacquer Spritzlack *m*
spraying mixture Spritzmasse *f*
spraying pressure Spritzdruck *m*
spraying process Sprühverfahren *n* (Färberei)
spraying technique Spritzverfahren *n*
spray-lacquer *v* spritzlackieren
spray lacquering Spritzlackierung *f*
spray nozzle Spritzdüse *f,* **[hollow & full] cone** ~ Drallkammerdüse *f* (Hohl- und Vollkegel)
spray-on finish Spritzlack *m*
spray tower Berieselungsturm *m,* Sprühkolonne *f,* Sprühturm *m* (Verfahrenstech) (for crystallization), Spritzturm *m*
spray-up method (polyesters) Faserspritzverfahren *n*
spray washer Sprühwäscher *m*
spray water Spritzwasser *n*

spread Ausbreitung *f,* Verbreitung *f,* ~ **of a light source** Streuung *f* einer Lichtquelle
spread *v* ausstrecken, auftragen (z. B. Farbe), entfalten, spreizen, verbreiten (Krankheit), verschleppen (Krankheit), verstreichen (z. B. Farbe), verteilen, ~ **apart** auseinanderspreizen, ~ **on** aufstreichen, ~ **out** ausbreiten
spreadability Verlauffähigkeit *f* (Lack)
spread coater Streichmaschine *f*
spread coating Streichverfahren *n*
spreader Verteiler *m,* Spritzdüse *f,* Zerstäuber *m*
spreader roll Auftragwalze *f*
spread footing liegender Rost *m*
spreading coefficient Spreitungskoeffizient *m* (Therm)
spreading knife Streichmesser *n*
spreading machine Streichmaschine *f*
spreading range Streubereich *m*
spreading velocity Streichgeschwindigkeit *f*
spread of flame Glutfestigkeit *f*
spring Feder *f* (Techn), Quelle *f* (Gewässer), **compensating** ~ Ausgleichfeder *f,* **triangular** ~ Dreieckfeder *f*
spring *v* springen, ~ **back** zurückfedern, ~ **off** abspringen
spring action Federwirkung *f*
spring balance Federwaage *f*
spring beam Prellklotz *m*
spring bolt Federriegel *m*
[spring] bow compasses Nullenzirkel *m*
spring bows Federzirkel *m,* Nullenzirkel *m*
spring brine Quellsole *f*
spring catch Schnappverschluß *m*
spring clamp Quetschhahn *m*
spring clip Quetschhahn *m*
spring constant Federkonstante *f*
spring contact federnder Kontakt *m*
spring couple pin Federverschlußbolzen *m*
spring dividers Federzirkel *m*
spring dynamometer Federzugmesser *m*
spring ejector Ausdrückstift *m* (mit Federkraft betätigt)
springer Scheinbombage *f* (Dose)
spring flex rule Rollbandmaß *n*
spring galvanometer Federgalvanometer *n*
spring guide block Federgleitblock *m*
spring hammer Federhammer *m*
spring hardness Federhärte *f*
spring hinge Federangel *f*
springiness Federelastizität *f,* Schnellkraft *f,* Sprungkraft *f*
spring latch Federklinke *f*
spring lever Federheber *m*
spring load Federbelastung *f*
spring-loaded bolt Federschraube *f*
spring-loaded governor Federregler *m*
spring-loaded shuttle valve Einfachwechselventil *n*

spring pendulum Federpendel *n*
spring pressure Federdruck *m*
spring pressure gauge Federmanometer *n*,
Plattenfedermanometer *n*
spring regulator Federregulator *m*
spring roller bearing Federrollenlager *n*
spring safety hook Karabinerhaken *m*
spring salt Quellsalz *n*, Solsalz *n*
spring shackle Federbund *m*
spring suspension Federaufhängung *f*, Federung *f*
spring tension Federspannung *f*
spring valve Federventil *n*
spring washer Federdichtung *f*, Federring *m*,
Federteller *m*
spring water Quellwasser *n*, Brunnenwasser *n*,
acidulous ~ Sauerbrunnen *m*
springy federnd, elastisch
sprinkle *v* benetzen, [be]sprengen, bespritzen,
bestreuen, sprenkeln, sprühen
sprinkler Berieselungsapparat *m*,
Spritzapparat *m*
sprinkler can Streudose *f*
sprinkler plant Sprinkleranlage *f*
sprinkler system Sprinkleranlage *f*
sprinkling Benetzen *n*, Beregnung *f*,
Berieselung *f*, Besprühen *n*
sprinkling apparatus Benetzvorrichtung *f*
sprinkling can Gießkanne *f*
sprocket gear Kettenrad *n*
sprocket wheel Kettenrad *n*, Kettenzahnrad *n*
sprout Keim *m* (Bot), Sproß *m* (Bot)
sprout *v* keimen (Bot), sprießen (Bot)
spruce charcoal Fichtenholzkohle *f*
spruce resin Fichtenharz *n*, Fichtenpech *n*
sprue Anguß[kanal] *m*, Eingußtrichter *m*,
Gießloch *n*
sprue bush Angußbuchse *f*
sprue bushing Angußbuchse *f*, Spritzmulde *f*
sprue ejector Angußdrückstift *m*
sprue lock pin Angußauswerferstift *m*,
Angußdruckstift *m*
sprue pouring head Eingußkanal *m*
sprue puller Angußabreißer *m*
sprue slug Angußkegel *m*, Angußstück *n*
spun glass gesponnenes Glas *n*, Glaswolle *f*
spun gold Goldgespinst *n*
spun rayon Kunstseide *f*
spur gear Stirnrad *n* (Masch)
spur gear[ing] Stirnradgetriebe *n*
spurger Gasverteiler *m*
spurious falsch, unecht
spurious coupling Streukoppelung *f*
spurious pulses Fehlstöße *m pl*
spurrite Spurrit *m* (Min)
spurt *v* ausspritzen (Flüssigkeit)
spur wheel Stirnrad *n* (Masch)
spur wheel drive Stirnradantrieb *m* (Masch)
spy atom Spionenatom *n*
spy sample Spionenprobe *f*

squalane Squalan *n*
squalene Squalen *n*
squamatic acid Squamatsäure *f*
squamous schuppig
square Quadrat *n* (Math), Vierkant *m* (Techn)
square *a* quadratisch, viereckig
square *v* zur zweiten Potenz erheben (Math),
quadrieren (Math), ~ **out** ausquadrieren
(Math)
square adjustment Quadratausgleich *m*
square bar Vierkanteisen *n* (Techn)
[square] brackets eckige Klammern *pl*
square cascade Stufenkaskade *f*
square centimeter Quadratzentimeter *n*
square-cut adhesion test Gitterschnittprüfung *f*
square decimeter Quadratdezimeter *n*
square file Vierkantfeile *f*
square fluctuation Schwankungsquadrat *n*
square head Vierkantkopf *m*
square-headed bolt Vierkantschraube *f*
square iron Vierkanteisen *n* (Techn)
square jaw clutch Zapfenkupplung *f*
square kilometer Quadratkilometer *n*
square matrix quadratische Matrize *f*
square measure Flächenmaß *n*
square meter Quadratmeter *n*
square millimeter Quadratmillimeter *n*
square nut Vierkantmutter *f*
square root Quadratwurzel *f* (Math)
squares, method of least ~ Methode *f* der
kleinsten Quadrate (pl), **sum of** ~
Quadratsumme *f*
square signal Rechteckimpuls *m*
square spanner (Br. E.) Vierkantschlüssel *m*
(Techn)
square thread Flachgewinde *n*
square-threaded flachgängig
square-topped pulse Rechteckimpuls *m*
square washer Vierkantscheibe *f*
square wave voltage Rechteckspannung *f*
square wire Vierkantdraht *m*
square wrench (Am. E.) Vierkantschlüssel *m*
(Techn)
squaric acid Quadratsäure *f*
squash *v* zerdrücken (zu Brei), zerquetschen
squawroot Krebswurzel *f* (Bot)
squeak *v* quietschen, knarren
squeegee (Am. E.) Quetschwalze *f*
squeeze *v* quetschen, abklatschen (Techn),
drücken, einquetschen, pressen, ~ **[off]**
abquetschen, ~ **out** ausdrücken, auspressen,
ausquetschen, ~ **through** durchquetschen
squeeze bottle Spritzflasche *f*
squeezer Preßmaschine *f*, Quetsche *f*,
Quetschmaschine *f*
squeeze roll[er] Quetschwalze *f*

squeezing effect Abquetscheffekt *m*,
Quetschwirkung *f*
squeezing factor Auspreßmaschine *f*
squeezing machine Quetschmaschine *f*
squib Feuersprüher *m*
squill Meerzwiebel *f* (Bot)
squirrel cage rotor Kurzschlußanker *m*
squirt *v* spritzen, ~ **in** einspritzen
squirting Spritzen *n*
squirting pipe Abspritzrohr *n*
stab *v* [durch]stechen
stabbing thermometer Einsteckthermometer *n*
stab holes pl
stabilite Stabilit *n*
stability Stabilität *f*, Beständigkeit *f*, Bestand *m*,
Dauerhaftigkeit *f*, Festigkeit *f*, Haltbarkeit *f*,
Kippsicherheit *f*, **degree of** ~
Stabilitätsgrad *m*, **dimensional** ~
Formbeständigkeit *f*, **limit of** ~
Stabilitätsgrenze *f*, Festigkeitsgrenze *f*, ~ **in
air** Luftbeständigkeit *f*, ~ **in the position of
rest** Kippsicherheit *f* der Ruhelage
stability check Stabilitätsprüfung *f*
stability condition Stabilitätsbedingung *f*
stability constant Stabilitätskonstante *f*
stability ratio Festigkeitszahl *f*
stability rule Stabilitätsregel *f*
stabilization Stabilisierung *f*, Konstanthaltung *f*
stabilization resistance
Stabilisierungswiderstand *m*
stabilize *v* stabilisieren, festigen, haltbar
machen, konstant halten
stabilizer Stabilisator *m*, Kippsicherung *f*
stabilizer tube Stabilisationsröhre *f*
stabilizing Haltbarmachen *n*, Konstanthaltung *f*
stabilizing bath Stabilisierungsbad *n*
stable stabil, beständig, fest, haltbar; standfest,
~ **in air** luftbeständig, ~ **in water**
wasserbeständig, ~ **toward acids**
säurebeständig
stachydrine Stachydrin *n*
stachyose Stachyose *f*
stack Stapel *m*
stack *v* [auf]stapeln, aufeinanderschichten,
aufschichten, häufeln
stacking Stapelung *f*
stacking fault energy Stapelfehlerenergie *f*
stacking fault formation Stapelfehlerbildung *f*
stacking lift Stapellift *m*
stacking operator Stapeloperator *m*
stacking rack Stapelgestell *n*
stack molding Übereinanderpressen *n*
staff Belegschaft *f*
staff cell Stabzelle *f* (Histol), Stäbchenzelle *f*
(Histol)
staffelite Staffelit *m* (Min)
stage Gerüst *m* (Techn), Stadium *n*, Stufe *f*,
lowering ~ Senkbühne *f*, ~ **of decomposition**
Abbaustufe *f* (Chem), ~ **of development**

Entwicklungsstadium *n*, ~ **of the slide**
Objekttisch *m*
stage compressor Stufenkompressor *m*
stage micrometer Kreuztischmikrometer *n*
stages, in ~ stufenweise
stagewise absatzweise
stagewise operation Satzbetrieb *m*
stagger *v* staffeln
staggered gestaffelt [angeordnet], versetzt
[angeordnet]
staggered riveting Versatznietung *f*
stagnant stagnierend, stehend, stillstehend
stagnant volume Totvolumen *n*
stagnate *v* faulen (Wasser), stagnieren
stagnation Stagnation *f*, Stockung *f*
stagnation point Staupunkt *m*
stain Fleck[en] *m*, Beizfarbe *f*, Beizflüssigkeit *f*,
Beizmittel *n*
stain *v* fleckig werden, beflecken, beizen,
färben, grundieren, kontrastieren (Biol); ~ **off**
abfärben, ~ **slightly** antönen, anfärben, ~
superficially anfärben
stainability Färbbarkeit *f*
stainable färbbar
stained farbig, fleckig
staining Beizen *n*, Färben *n*, Färbung *f*,
Fleckigwerden *n*, **fast to** ~ abklatschecht
staining block Färbenapf *m*
staining test Farbprüfung *f*
stainless fleckenfrei; korrosionsbeständig
(Stahl), korrosionsfrei (Stahl), nichtrostend
(Stahl), rostbeständig (Stahl); rostfrei (Stahl)
stainless property Korrosionsbeständigkeit *f*,
Rostbeständigkeit *f*
stainless steel Edelstahl *m*
stain removal Detachur *f*, **clay for** ~
Fleckstein *m*
stain remover Fleckenentferner *m*,
Fleckenentfernungsmittel *n*
stair Stufe *f*, Treppe *f*
stair covering Treppenbelag *m*
stake Formerstift *m*
stake off *v* abstecken
stalactite hängender Tropfstein *m*, Stalaktit *m*
stalactitic stalaktitisch, tropfsteinartig
stalactitic formation Tropfsteinbildung *f*
stalagmite nach oben wachsender Tropfstein *m*,
Stalagmit *m*
stalagmitical stalagmitartig
stalagmometer Stalagmometer *n*
stalagmometry Stalagmometrie *f*
stale schal, abgestanden, fade
stall roasting Stadelröstung *f*
stamen Staubgefäß *n* (Bot)
stamp Stempel *m*, Lochstempel *m* (Techn),
Pochstempel *m* (Techn), Prägestempel *m*
(Techn), Stanze *f* (Techn)
stamp *v* stempeln, aufdrucken, [ein]stampfen
(Techn), feststampfen (Techn), markieren,

pochen (Erz), prägen (Techn) (z. B. Blech),
pressen (Techn), stanzen (Techn), ~ **out**
ausstanzen

stamped part Stanzteil *n*

stamper Stampfer *m*, Pochstempel *m*, Stössel *m*

stamping Ausstanzen *n*, Stanzen *n*, **waste from** ~
Stanzabfälle *pl*

stamping calender Prägekalander *m*

stamping device Stanzvorrichtung *f*

stamping die Prägestanze *f*

stamping enamel Stanzlack *m*

stamping form Stampfform *f*

stamping ink Stempelfarbe *f*

stamping iron Pochstempel *m*

stamping lacquer Prägelack *m*

stamping machine Lochmaschine *f*,
Stanzmaschine *f*

stamping ore Pocherz *n*

stamping-out machine Stanzmaschine *f*

stamping press Prägepresse *f*, Prägewerk *n*

stampings Stanzabfälle *pl*

stamping tool Stanzwerkzeug *n*

stamp mill Pochmühle *f*, Pochwerk *n*

stamp milling Verpochen *n*

stamp mortar Pochtrog *m*

stamp pulp Pochtrübe *f*

stamp rock Pochgestein *n*

stanchion Pfosten *m*, Stütze *f*

stand Ständer *m*, Ablage *f* (Ablagetisch),
Abstelltisch *m*, Gestell *n*, Objektträger *m*
(Mikroskop), Standort *m*, Stativ *n*, Stütze *f*,
Tribüne *f*, **to allow to** ~ stehen lassen, **to let** ~
stehen lassen

stand *v* stehen, ~ **apart or aside** abstehen, ~ **by**
beistehen, ~ **opposite** gegenüberstehen, ~ **out**
hervorragen, ~ **still** stillstehen, stocken

standard Standard *m*, Feingehalt *m*
(Edelmetall), Norm *f*, ~ **of alloy**
Münzgehalt *m*

standard *a* normal, normgerecht, serienmäßig

standard acid Titriersäure *f*

standard adjustment Standardeinstellung *f*

standard atmosphere Normalatmosphäre *f*

standard bar Kontrollstab *m*

standard candle power Normalkerze *f*

standard capacity Normalleistung *f*,
Standardkapazität *f*

standard cell Normalelement *n* (Elektr),
Normalzelle *f* (Elektr), Standardelement *n*
(Elektr)

standard condition Normalzustand *m*

standard conditions Normalbedingungen *pl*

standard couple Fixierungspaar *n*

standard cross section Normalprofil *n*

standard design Normalausführung *f*,
Standardausführung *f*

standard deviation Standardabweichung *f*
(Statistik)

standard dimension Normalformat *n*

standard electrode Normalelektrode *f*,
Standardelektrode *f*

standard enthalpy Standardenthalpie *f*

standard entropy Standardentropie *f*

standard equipment Normalausrüstung *f*

standard errors, calculation of ~
Standardfehlerberechnung *f*

standard filter Normalfilter *n*

standard frequency Normalfrequenz *f*

standard gas Normalgas *n*

standard gauge Prüfmaß *n*

standard glass Normalglas *n*

standard gold Probiergold *n*

standard ground joint Normschliff *m*

standard hydrogen electrode
Normalwasserstoffelektrode *f*

standardizable eichfähig

standardization Normung *f*, Eichung *f*,
Normalisierung *f*, Normenaufstellung *f*,
Standardisierung *f*, Vereinheitlichung *f*, ~ **of**
the circuit Meßkreiseichung *f*

standardize *v* standardisieren, eichen,
kalibrieren, normen, normieren,
vereinheitlichen

standardized genormt

standard light source Standardlichtquelle *f* (Opt)

standard liquid Vergleichsflüssigkeit *f*

standard load Normalbelastung *f*

standard measure Eichmaß *n*, Einheitsmaß *n*
(Techn), Normalmaß *n*, Urmaß *n*

standard method Einheitsmethode *f*

standard moisture Normalfeuchtigkeit *f*

standard normal distribution
Standardnormalverteilung *f* (Statistik)

standard of measure Richtmaß *n*

standard [or prototype] meter Urmeter *n*

standard packing Einheitspackung *f*,
Originalpackung *f*

standard pitch Normstimmton *m* (Akust)

standard plane Kristallfläche *f* mit
Miller-Indices (pl) (Krist)

standard potential Normalpotential *n*

standard pressure Normaldruck *m*

standard property Standardeigenschaft *f*

standard radiator Normalstrahler *m*,
Vergleichsstrahler *m*

standard resistance Normalwiderstand *m*,
Vergleichswiderstand *m*

standards Normen *f pl*, **conforming to** ~
normgerecht

standard setting Standardeinstellung *f*

standard silver Probesilber *n*

standard size Normalformat *n*, Normalgröße *f*,
Normmaß *n*

standard solution Titrierlösung *f*,
Normallösung *f*, Vergleichslösung *f*

standard specification Norm *f*, Normvorschrift *f*,
according to ~ normgerecht

standard strength Normalstärke *f,* **~ of a solution** Titer *m* einer Lösung
standard system Einheitssystem *n*
standard table Normaltabelle *f*
standard temperature Normaltemperatur *f*
standard thread Einheitsgewinde *n,* Normalgewinde *n*
standard thread gauge Normalgewindelehre *f*
standard tin Probezinn *n*
standard titrimetric substance Urtitersubstanz *f* (Anal)
standard type Normalausführung *f,* Standardmodell *n*
standard type construction Einheitsbauart *f* (Arch)
standard weight Eichgewicht *n,* Normalgewicht *n,* Probegewicht *n*
standard Weston cadmium cell Cadmium-Normalelement *n*
stand base Stativfuß *m*
stand-by battery Reservebatterie *f*
stand-by boiler Ersatzkessel *m*
standby channels Leer-Rohrnetz *n*
stand-by engine Hilfsmaschine *f,* Reservemaschine *f*
stand-by machine Austauschmaschine *f*
standby store Reservespeicher *m* (Comp)
stand-by system Notaggregat *n*
standing stehend
stand oil Standöl *n,* Dicköl *n*
stand oil boiling plant Standölkochanlage *f*
standpipe Standrohr *n,* Steigrohr *n*
standpoint Anschauungsweise *f,* Standpunkt *m*
stand rod Stativstange *f*
standstill Stillstand *m*
stannane Zinnwasserstoff *m*
stannary Zinnbergwerk *n*
stannate Stannat *n,* Salz *n* der Zinnsäure
stannic acid Zinnsäure *f,* **salt of ~** Stannat *n*
stannic anhydride Zinnsäureanhydrid *n,* Zinndioxid *n*
stannic bromide Zinntetrabromid *n,* Stannibromid *n,* Zinn(IV)-bromid *n*
stannic chloride Zinntetrachlorid *n,* Stannichlorid *n,* Zinnbutter *f,* Zinn(IV)-chlorid *n*
stannic compound Zinn(IV)-Verbindung *f*
stannic fluoride Zinntetrafluorid *n*
stannic hydroxide Zinn(IV)-hydroxid *n,* Stannihydroxid *n*
stannic iodide Zinntetrajodid *n,* Stannijodid *n,* Zinn(IV)-jodid *n*
stannic oxide Zinn(IV)-oxid *n,* Stannioxid *n,* Zinndioxid *n,* Zinnsäureanhydrid *n*
stannic salt Stannisalz *n,* Zinn(IV)-salz *n*
stannic sulfide Zinndisulfid *n,* Stannisulfid *n,* Zinn(IV)-sulfid *n*
stannic thiocyanate Rhodanzinn *n*
stanniferous zinnführend, zinnhaltig

stannite Zinnkies *m* (Min), Stannin *n* (Min), Stannit *m* (Min)
stannous acetate Stannoacetat *n,* Zinn(II)-acetat *n*
stannous bromide Zinndibromid *n,* Zinn(II)-bromid *n*
stannous chloride Zinndichlorid *n,* Zinn(II)-chlorid *n*
stannous chromate Stannochromat *n*
stannous compound Zinn(II)-Verbindung *f*
stannous hydroxide Zinn(II)-hydroxid *n,* Stannohydroxid *n*
stannous iodide Stannojodid *n,* Zinn(II)-jodid *n*
stannous oxalate Stannooxalat *n*
stannous oxide Zinn(II)-oxid *n,* Stannooxid *n,* Zinnmonoxid *n,* Zinnprotoxid *n*
stannous salt Stannosalz *n,* Zinn(II)-Salz *n*
stannous sulfate Stannosulfat *n*
stannous sulfide Zinn(II)-sulfid *n,* Stannosulfid *n*
stannous tartrate Stannotartrat *n*
staphisagrine Staphisagrin *n*
staphisagroine Staphisagroin *n*
staphisaine Staphisagrin *n*
staphylococcal infection Staphylokokkeninfektion *f* (Med)
staphylococcal strain Staphylokokkenstamm *m*
staple Heftklammer *f,* Klammer *f* (Heftklammer), Rohstoff *m,* Rohwolle *f*
staple *v* sortieren
staple conveyer Stapelförderer *m*
stapled closure Heftverschluß *m*
staple fastening Heften *n* mit Klammern (pl), Verklammerung *f*
staple fiber Spinnfaser *f*
staple food Grundnahrungsmittel *n,* Hauptnahrungsmittel *n*
staple length Stapellänge *f*
stapler Heftmaschine *f*
stapling room Heftraum *m*
star cell Sternzelle *f* (Histol)
starch Stärke *f,* Stärkemehl *n,* **containing ~** stärke[mehl]haltig, **excess of ~** Stärkeüberschuß *m,* **powdered ~** Stärkemehl *n,* Stärkepulver *n,* **variety of ~** Stärkeart *f*
starch *v* stärken (mit Stärke)
starch blue Stärkeblau *n*
starch content Stärkegehalt *m*
starch degradation Stärkeabbau *m*
starch equivalent Stärkeeinheit *f*
starch flour Stärkemehl *n,* Satzmehl *n*
starch gel Stärkegel *n*
starch gel electrophoresis Stärkegel-Elektrophorese *f*
starch gloss Stärkeglanz *m*
starch granule Stärkekorn *n*
starch gum Dextrin *n*
starch iodide Jodstärke *f*
starch iodide paper Jodstärkepapier *n*

starch iodide test Jodstärkeprobe *f*
starch paper Stärkepapier *n*
starch paste Stärkekleister *m*
starch powder Stärkemehl *n*, Stärkepulver *n*
starch solution Stärkelösung *f*
starch sugar Stärkezucker *m*, Stärkesirup *m*
starch syrup Stärkesirup *m*, Stärkezucker *m*
starch test Stärkereaktion *f*
starch water Stärkewasser *n*
starchy stärke[mehl]haltig
star circuit Sternschaltung *f* (Elektr)
star cluster Sternhaufen *m* (Astr)
star-connected sterngeschaltet (Elektr)
star connection Sternschaltung *f* (Elektr),
    Y-Schaltung *f* (Elektr)
star-delta switch Sterndreieckschalter *m* (Elektr)
star feeder Zellenradschleuse *f*
star polyhedron Sternflächner *m*
star sapphire Sternsaphir *m* (Min)
start Beginn *m*, Start *m*
start *v* andrehen (Techn), anfangen, anlassen
    (Motor), anstellen, aufnehmen (beginnen),
    beginnen, einleiten, ingangsetzen
start and stop button Betätigungsknopf *m*
starter Anlasser *m* (Kfz), Anlaßmotor *m*
starter nozzle Anlaßdüse *f*
starter resistance Anlaßwiderstand *m*
starting Anlassen *n* (Motor), Inangriffnahme *f*,
    Inbetriebsetzung *f*, Ingangsetzung *f*, ~
    without shock stoßfreies Ingangsetzen *n*
starting bath Ansatzbad *n*
starting compound Ausgangsverbindung *f*
starting crank Andrehkurbel *f*
starting current Anfahrstrom *m*
starting cut Anschnitt *m* (Techn)
starting dynamo Anlaßgenerator *m*
starting energy Anfahrenergie *f*, Anfahrkraft *f*
starting equation Ausgangsgleichung *f*
starting lever Anfahrhebel *m*, Anlaßhebel *m*,
    Einschalthebel *m*
starting machine Anlaßmaschine *f*
starting material Ausgangsmaterial *n*,
    Ausgangsstoff *m*, Rohstoff *m*
starting period Anlaufzeit *f*
starting point Anfangspunkt *m*, Ansatzpunkt *m*,
    Ausgangsbasis *f*, Ausgangspunkt *m*
starting position Startstellung *f*
starting product Ausgangserzeugnis *n*,
    Ausgangsprodukt *n*, Vorprodukt *n*
starting sheet Mutterblech *n*
starting switch Anlasser *m*, Anlaßschalter *m*
starting time Anlaufzeit *f*
starting transformer Anlaßtransformator *m*
    (Elektr)
starting valve Anlaßventil *n*, Startventil *n*
starting voltage Einsatzspannung *f*
start-up Inbetriebnahme *f*, Inbetriebsetzung *f*,
    Ingangsetzung *f*
start-up procedure Anfahrvorgang *m*

starvation flotation Hungerflotation *f*
star voltage Sternspannung *f* (Elektr)
stassfurtite Staßfurtit *m* (Min)
statcoulomb Statcoulomb *n*
state Zustand *m*, Beschaffenheit *f*, Stadium *n*,
    active ~ aktiver Zustand, **amorphous** ~
    amorpher Zustand, **crystalline** ~ kristalliner
    Zustand, **degenerate** ~ entarteter Zustand,
    density of ~ Zustandsdichte *f*, **diagram of** ~
    Zustandsdiagramm *n*, **disordered** ~
    ungeordneter Zustand, **equation of** ~
    Zustandsgleichung *f*, **excited** ~ angeregter
    Zustand (Atom), **free** ~ freier Zustand,
    gaseous ~ Gaszustand *m*, gasförmiger
    Zustand, **highly excited** ~ hochangeregter
    Zustand, **intermediate** ~ Zwischenzustand *m*,
    liquid ~ flüssiger Zustand, **metastable** ~
    metastabiler Zustand, **quasi-stationary** ~
    quasi-stationärer Zustand, **solid** ~ fester
    Zustand, **steady** ~ stationärer Zustand,
    uncombined ~ freier Zustand, ~ **of affairs**
    Sachlage *f*, Sachverhalt *m*, ~ **of aggregation**
    Aggregatzustand *m*
state *v* angeben, behaupten, konstatieren
statement Angabe *f*, Aussage *f*, Behauptung *f*
statement of application of funds
    Bewegungsbilanz *f*
state of coordination Koordinationsart *f*
static statisch, ruhend
statically determinable statisch bestimmbar
static [ball] indentation test
    Kugeldruckhärteuntersuchung *f*
static energy Ruheenergie *f*
static friction Haftreibung *f*, **coefficient of** ~
    Haftreibungskoeffizient *m* (Mech)
static hardness test Kugeldruckprobe *f*
static load Ruhebelastung *f*
static mixer statischer Mischer *m*
static position Ruhestellung *f*
statics Statik *f*
static test Belastungsprobe *f*,
    Belastungsversuch *m*
stationary stationär, beständig, fest[stehend],
    gleichbleibend, ruhend, standortgebunden,
    stillstehend, unbeweglich
stationary phase Trennfüllung *f* (Gaschromat)
statistical statistisch
statistician Statistiker *m*
statistics Statistik *f*
stator Stator *m* (Elektr)
stator current Ständerstrom *m*, Statorstrom *m*
status Zustand *m*, Status *m*
status nascendi Entstehungszustand *m*
statvolt Statvolt *n*
Staudinger function Viskositätszahl *f*
Staudinger index Grenzviskositätszahl *f*
staurolite Staurolith *m* (Min), Kreuzstein *m*
    (Min)

stay Strebe *f* (Techn), Stütze *f* (Techn)
stay *v* verweilen, abstützen, verspannen, versteifen, verstreben
stay bolt Stehbolzen *m*
stay plate Unterlagsplatte *f*
steadiness Gleichmäßigkeit *f*, Stetigkeit *f*
steady gleichmäßig; andauernd, fest, gleichbleibend, gleichförmig, laminar (Strömung), stabil, ständig
steady position Ruhelage *f*
steady-state characteristic Kennlinie *f* (Regeltechn)
steady-state condition Beharrungszustand *m* (Regeltechn)
steady-state gain (Br. E.) Übertragungsfaktor *m* (Regeltechnik)
steady-state operation stationärer Betriebszustand *m*
steady voltage Gleichspannung *f* (Elektr)
steam Dampf *m*, Dunst *m*, Wasserdampf *m*, **conveyance of** ~ Dampfüberführung *f*, Dampfüberleitung *f*, **dry** ~ Trockendampf *m*, **energy of** ~ Dampfenergie *f*, **formation of** ~ Dampfbildung *f*, **production of** ~ Dampfproduktion *f*, **saturated** ~ gesättigter Dampf, **separation of oil from** ~ Dampfentölung *f*, **state of** ~ Dampfzustand *m*, **superheated** ~ überhitzter Dampf, Trockendampf *m*, **temperature of** ~ Dampftemperatur *f*, **to inject** ~ Wasserdampf *m* einblasen, **utilization of** ~ Dampfausnützung *f*, **volume of** ~ Dampfvolumen *n*, **weight of** ~ Dampfgewicht *n*, **wetness of** ~ Dampffeuchtigkeit *f*, ~ **for heating purposes** Heizdampf *m*
steam *v* dämpfen, dekatieren, dünsten, [ver]dampfen, ~ **down** eindampfen, ~ **off** abdampfen, ~ **out** ausdampfen
steam accumulator Dampfspeicher *m*
steam admission valve Dampfeinlaßventil *n*
steam and air mixture Luftdampfgemisch *n*
steam apparatus Dampfapparat *m*
steam atomizer Öldruck-Dampfdruckzerstäuber *m*
steam autoclave, heating in the ~ Dampflagerungsverfahren *n*
steam bath Dampfbad *n*
steam blower Dampfgebläse *n*
steam boiler Dampfkessel *m*, Wasserdampfentwickler *m*, Wasserdampferzeuger *m*
steam boiler feed water Dampfkesselspeisewasser *n*
steam bubble Dampfblase *f*
steam chamber Dampfbehälter *m*, Dampfreservoir *n*
steam channel Heizkanal *m*
steam chest Dampfkasten *m*

steam circulation system Dampfumwälzeinrichtung *f*
steam coal Flammkohle *f*, Dampfkesselkohle *f*
steam cock Dampfhahn *m*
steam coil Dampf[heiz]schlange *f*, Heizschlange *f*
steam collector Dampfsammler *m*
steam color Dampffarbe *f* (Färb)
steam compression Dampfdruck *m*
steam condenser pipe Dampfniederschlagsrohr *n*
steam consumption Dampfbedarf *m*
steam converter Dampfumformer *m*
steam cooker Dämpfer *m*
steam cooking apparatus Dampfkochapparat *m*
steam cooler Dampfkühler *m*
steam cracker Röhrenspaltofen *m*
steam cracking process Röhrenspaltverfahren *n*
steam crane Dampfkran *m*
steam cut-off Dampfabsperrung *f*
steam cut-off valve Dampfabsperrschieber *m*
steam cylinder Dampfzylinder *m*
steam density Dampfdichte *f*
steam developer Dampfentwickler *m*
steam diaphragm pump Dampfpumpe *f*
steam discharge Dampfabführung *f*, Dampfableitung *f*
steam distillation Destillation *f* mit Wasserdampf, Wasserdampfdestillation *f*
steam distributor Dampfverteiler *m*
steam-dried dampfgetrocknet
steam drive Dampfbetrieb *m*
steam-driven generator Dampfdynamo *m*
steam drum Oberkessel *m*
steam dryer Dampftrockenapparat *m*, Dampftrockner *m*
steam-drying device Dampfentwässerungsapparat *m*
steam dynamo Dampfdynamo *m*
steamed mechanical wood pulp Dampfholzschliff *m*
steam ejector unit Dampfstrahlluftsaugeaggregat *n*
steam engine Dampfmaschine *f*, **uniflow** ~ Gleichstromdampfmaschine *f*
steam escape pipe Dampfabblaserohr *n*
steam exhaust Dampfauspuff *m*
steam exhaustor Dampfsauger *m*
steam exhaust port Dampfaustrittsöffnung *f*
steam expansion Dampfausdehnung *f*
steam filter Dampffilter *m*
steam fittings Dampfarmaturen *pl*
steam flow Dampfdurchtritt *m*, Dampfstrom *m*
steam flowmeter Dampfstrommesser *m*
steam flow opening Dampfdurchtritt *m*
steam friction Dampfreibung *f*
steam funnel Dampftrichter *m*
steam gate valve Dampfabsperrschieber *m*
steam gauge Dampfdruckmesser *m*, Dampfuhr *f*
steam-generating dampferzeugend
steam-generating heat Dampfbildungswärme *f*

steam generation Dampfbildung *f,*
  Dampfentwicklung *f,* Dampferzeugung *f*
steam generator Dampfentwickler *m,*
  Dampfentwicklungsapparat *m,*
  Dampferzeuger *m,* Dampfkessel *m*
steam governor Dampfgeschwindigkeitsmesser *m*
steam hammer Dampfhammer *m*
steam heating Dampfheizung *f,* Zentralheizung *f*
steam heating apparatus
  Dampfheizungsvorrichtung *f*
steam heating pipe Dampfheizrohr *n*
steam hose Dampfschlauch *m*
steaming Dampfen *n,* Eindampfen *n*
steam injection Dampfstimulation *f*
steam injection process Dampfstoßverfahren *n*
steam injector Dampfstrahlapparat *m*
steam inlet Dampfeintritt *m*
steam inlet pipe Dampfzuleitungsrohr *n*
steam inlet port Dampfeinströmungskanal *m*
steam jacket Dampfhülle *f,* Dampfmantel *m*
steam jacket heating Dampfmantelheizung *f*
steam jet Dampfstrahl *m*
steam jet aspirator Dampfstrahlsauger *m*
steam jet atomizer Dampfstrahlzerstäuber *m*
steam jet blower Dampfstrahlgebläse *n*
steam-jet ejector Wasserdampfstrahl-Pumpe *f*
steam-jet refrigeration machine
  Dampfstrahlkältemaschine *f*
steam jet sand blast Dampfsandstrahlgebläse *n*
steam jet sprayer Dampfstrahlzerstäuber *m*
steam kiln Dampfdarre *f*
steam line Dampfleitungsrohr *n*
steam lubrication Dampfschmierung *f*
steam measuring Dampfmessung *f*
steam meter Dampfmesser *m,* Dampfuhr *f*
steam nozzle Dampfdüse *f*
steam outlet Dampfausströmung *f,*
  Dampfentnahme *f*
steam pipe Dampf[leitungs]rohr *n*
steam piping Dampfleitung *f*
steam piston Dampfkolben *m*
steam plant Dampf[kraft]anlage *f,*
  Dampfmaschinenanlage *f*
steam plate Heizplatte *f*
steam point Verdampfungspunkt *m*
steam port Dampfweg *m,* Dampfkanal *m*
steam pot Dampftopf *m*
steam power Dampfkraft *f*
steam power forging press
  Dampfdruckschmiedepresse *f*
steam power station Dampfkraftwerk *n*
steam pressure Dampfdruck *m,* **reduction of** ~
  Dampfdruckverminderung *f*
steam pressure pump Dampfdruckpumpe *f*
steam pressure register Registriermanometer *n*
steam printing Dampfdruckerei *f*
steamproof dampfdicht
steam pump Dampfpumpe *f*
steam purple Dampfpurpur *m*

steam radiator Dampfheizkörper *m*
steam recorder Dampfschreiber *m*
steam reducing valve Dampfreduzierventil *n*
steam regulating valve Dampfzulaßventil *n*
steam requirement Dampfbedarf *m*
steam roasting furnace Dampfröstofen *m*
steam safety valve Dampfsicherheitsventil *n*
steam separator Dampf[wasser]abscheider *m*
steam slide valve Dampf[einlaß]schieber *m*
steam sterilizer Dampftopf *m*
steam stop valve Dampfabsperrventil *n*
steam superheater Dampfüberhitzer *m*
steam superheating Dampfüberhitzung *f*
steam supply Dampfzuführung *f,*
  Dampfzuleitung *f*
steam supply pipe Dampfzuführungsrohr *n,*
  Dampfzuleitungsrohr *n*
steam table Dampftabelle *f* (Techn)
steam-tight dampfdicht
steam tin Waschzinn *n*
steam trap Dampfentwässerer *m,*
  Kondensationswasserabscheider *m,*
  Kondenstopf *m,*
  Niederschlagswasserabscheider *m*
steam tube Dampfrohr *n*
steam turbine Dampfturbine *f*
steam valve Dampfventil *n*
steam winch Dampfwinde *f*
steamy dampfig, dunstig
stearaldehyde Stearaldehyd *m*
stearamide Stearamid *n*
stearate Stearat *n,* Salz oder Ester der
  Stearinsäure *f*
stearic acid Stearinsäure *f,* Octadecansäure *f,*
  Talgsäure *f* (obs), **salt or ester of** ~ Stearat *n*
stearin Stearin *n,* Glycerintristearat *n*
stearin cake Stearinkuchen *m*
stearin candle Stearinkerze *f*
stearin mold Stearinform *f*
stearin pitch Stearinpech *n*
stearin soap Stearinseife *f*
stearodibutyrin Stearodibutyrin *n*
stearodichlorohydrin Stearodichlorhydrin *n*
stearodipalmitin Stearodipalmitin *n*
stearolic acid Stearolsäure *f*
stearone Stearon *n*
stearoptene Stearopten *n*
stearoxylic acid Stearoxylsäure *f*
stearyl Stearyl-
stearylalanine Stearylalanin *n*
stearylglycine Stearylglycin *n*
steatite Steatit *m* (Min), Seifenstein *m* (Min),
  Speckstein *m* (Min)
steatitic specksteinartig
steel Stahl *m,* **annealed** ~ vergüteter Stahl, **basic**
  ~ Thomasstahl *m,* **burnt** ~ übergarer Stahl,
  casehardened soft ~ Compoundstahl *m,*
  cementation of ~ Zementstahlbereitung *f,*
  conversion into ~ Stählung *f,* **dead soft** ~

niedriggekohlter Stahl, **effervescent** ~
unruhiger Stahl, **granulated** ~ Kornstahl *m*,
hard ~ Hartstahl *m*, **hard as** ~ stahlhart,
hardened ~ gehärteter Stahl, **hardening of** ~
Stahlhärtung *f*, **hardness of** ~ Stahlhärte *f*,
heat-resistant ~ hitzebeständiger Stahl,
heat-treated ~ gehärteter Stahl, vergüteter
Stahl, **high** ~ Hartstahl *m*, **hypereutectoid** ~
übereutektoider Stahl, **killed** ~ beruhigter
Stahl, **low-alloy** ~ niedriglegierter Stahl,
low-carbon ~ Schmiedeeisen *n*,
niedriggekohlter Stahl, **low-carbon free-cutting**
~ kohlenstoffarmer Schnellautomatenstahl,
mild ~ Weichstahl *m*, Thomasstahl *m*, **natural**
~ Rohstahl *m*, **noncorroding** ~
rostbeständiger Stahl, **overblown** ~ übergarer
Stahl, **overheated** ~ überhitzter Stahl, **plain** ~
unlegierter Stahl, **puddling of** ~
Stahlpuddeln *n*, **quenched and drawn** ~
abgeschreckter und angelassener Stahl, **raw** ~
Rohstahl *m*, **refined** ~ Raffinierstahl *m*,
running ~ unruhiger Stahl, **rustless** ~
rostbeständiger Stahl, **rustproof** ~
nichtrostender Stahl, **sectional [or structural]**
~ Profilstahl *m*, **semi-killed** ~ halbberuhigter
Stahl, **simple** ~ unlegierter Stahl, **soft** ~
Weichstahl *m*, **soft free-machining** ~
Schnellautomatenweichstahl *m*, **stainless** ~
korrosionsbeständiger Stahl, rostbeständiger
Stahl, rostfreier Stahl, **structural** ~
Baustahl *m*, **structural alloy** ~ legierter
Baustahl, **superfine** ~ Rosenstahl *m*,
tempering of ~ Stahlhärtung *f*, **to convert into**
~ verstählen, **to edge with** ~ verstählen, **to**
point with ~ anstählen, verstählen, **to soften**
~ Stahl dekarbonisieren, **to temper** ~ Stahl
brennen, **weldable** ~ schweißbarer Stahl,
manufacture of ~ **by fusion**
Flußeisenerzeugung *f*, ~ **for dies and stamps**
Döpperstahl *m*, ~ **for high-temperature service**
warmfester Stahl, ~ **for punches**
Lochstempelstahl *m*, ~ **ready for casting**
unmittelbar vergießbarer Stahl, ~ **with a high**
creep limit warmfester Stahl, ~ **with good**
high-temperature characteristics warmfester
Stahl
steel *v* [an]stählen
steel alloy Stahllegierung *f*
steel anode Stahlanode *f*
steel ball Stahlkugel *f*
steel beam Stahlträger *m*
steel blue Stahlblau *n*
steel bottle Stahlflasche *f*
steel bronze Stahlbronze *f*
steel casting Stahl[form]guß *m*
steel cementing furnace Zementstahlofen *m*
steel concrete Stahlbeton *m*
steel converting furnace Zementierofen *m*

steel cylinder Bombe *f* (Chem), Stahlflasche *f*,
Stahlzylinder *m*
steel disk Stahlscheibe *f*
steel dowel Stahldübel *m*
steel edge Stahlschneide *f*
steel-face *v* anstählen
steel foundry Stahlgießerei *f*
steel frame Stahlrahmen *m*, **tubular** ~
Stahlrohrgestell *n*
steel framework Einschaleisen *n*
steel furnace Stahlofen *m*
steel gray stahlgrau
steel hoop Stahlreif[en] *m*
steeling Anstählen *n*
steel ingot Stahlblock *m*
steel-like stahlähnlich
steel magnet Stahlmagnet *m*
steelmaking Stahlbereitung *f* (Metall)
steel manufacture Stahlerzeugung *f*
steel mortar Stahlmörser *m*
steel ore Stahlerz *n*
steel pig Stahleisen *n*
steel plate Stahlblech *n*
steel-plating Verstählen *n*, Verstählung *f*
steel process Stahlbereitungsprozeß *m*
steel production Stahlerzeugung *f*,
Stahlgewinnung *f*
steel puddling Stahlpuddeln *n*
steel rail Stahlschiene *f*
steel rule Stahlschablone *f*
steel rule die Bandstahlschnitt *m*
steel sample Stahlprobe *f*
steel scrap Stahlschrott *m*
steel sheet Stahlblech *n*
steel stanchion Stahlträger *m*
steel strip Bandstahl *m*, Stahlstreifen *m*
steel superstructures Stahlhochbauten *m pl*
steel tool Drehstahl *m*
steel turnings Stahldrehspäne *m pl*
steel wire Stahldraht *m*
steel wire brush Stahlkratzbürste *f*
steel wool Stahlwolle *f*
steel works, basic ~ Thomasstahlwerk *n*
steelyard Laufgewichtswaage *f*, Schnellwaage *f*
steep Bad *n*, Beuchwasser *n*, Lauge *f*
steep *a* steil, abschüssig
steep *v* eintauchen, anbrühen, durchnässen,
durchtränken, einweichen, ~ **in alum**
alaunen, ~ **in lye** [ein]laugen
steeped in lye eingelaugt
steeping Einweichen *n*, ~ **in lye** Einlaugen *n*, ~
in water Einwässerung *f*
steeping bath Beizbad *n*
steeping liquor Beizflüssigkeit *f*
steeping method Rottmethode *f*
steeping trough Einquellbottich *m*,
Vormaischbottich *m*
steeping tub Quellbottich *m*
steeping vat Quellbottich *m*

steepness Steilheit *f*
steering arm Steuerhebel *m*
[steering axle] tie rod Spurstange *f*
steering column Lenksäule *f* (Kfz)
steering device Lenkvorrichtung *f*
steering gear Lenkvorrichtung *f*
steering wheel Lenkrad *n*, Steuer[rad] *n*
steering wheel axle Steuerradachse *f*
steinmannite Steinmannit *m* (Min)
stellar atmosphere Sternatmosphäre *f*
stellar cluster Sternhaufen *m* (Astr)
stellar spectrum Sternspektrum *n*
stellerite Stellerit *m* (Min)
stelznerite Stelznerit *m* (Min)
stem Stamm *m*, Stiel *m*
stem correction Fadenkorrektur *f*
   (Thermometer)
stem guide Spindelführung *f*
stench Gestank *m*
stencilling Schablonendruck *m*
step Schritt *m*, Stufe *f*, in ~ in Phase [mit]
step *v* treten
step bearing Spurlager *n*
step by step schrittweise, stufenweise
step control Stufenregelung *f*
step controller Stufenregler *m*
step-down spannungserniedrigend
step down [the voltage] *v* hinuntertransformieren
   (Spannung)
step-down transformer Abwärtstransformator *m*
   (Elektr)
step function Stufenfunktion *f*
step function response (Brit. E.)
   Übergangsfunktion *f* (Regeltechn)
step groove Stufennut *f*
stephanine Stephanin *n*
stephanite Stephanit *m* (Min),
   Antimonsilberglanz *m* (Min), Sprödglaserz *n*
   (Min)
step height Stufenhöhe *f*
step input Sprungeingang *m*
step length Stufenlänge *f*
step lens Stufenlinse *f* (Opt)
stepless stufenlos
stepped gear Stufenzahnrad *n*
stepped roll Staffelwalze *f*, Stufenwalze *f*
step process Stufenprozeß *m*
step pulley Stufenscheibe *f*
step-shaped terrassenartig
step size Schrittgröße *f*, Schrittweite *f*
step switch Stufenschalter *m*
step-up spannungserhöhend (Elektr)
step-up cure Stufenheizung *f* (Gummi)
step-up motor Stufenmotor *m*
stepwise stufenweise, schrittweise
stepwise elution Stufeneluierung *f*
stepwise mechanism Stufenmechanismus *m*
sterane Steran *n*
steranthrene Steranthren *n*

stercobilin Stercobilin *n*
stercorite Stercorit *m* (Min)
sterculic acid Sterculsäure *f*
stereochemical stereochemisch
stereochemistry Stereochemie *f*
stereogram raumbildliche Darstellung *f*
stereogram description
   Stereogrammbeschreibung *f*
stereographic stereographisch
stereoisomer Stereoisomer *n*, Raumisomer *n*
stereoisomerase Stereoisomerase *f* (Biochem)
stereo-isomeric stereoisomer, raumisomer
stereoisomerism Stereoisomerie *f*,
   Raumisomerie *f*
stereometer Stereometer *n*
stereometric stereometrisch
stereometry Stereometrie *f*, Raumgeometrie *f*
   (Math)
stereoregulated polymerization stereoregulierte
   Polymerisation *f*
stereoscope Stereoskop *n*
stereoscopic stereoskopisch
stereoscopic picture Raumbild *n*
stereoselectivity Stereoselektivität *f*
stereospecificity Stereospezifität *f*
stereospectrogram Stereospektrogramm *n*
   (Spektr)
steric sterisch; räumlich
steric factor sterischer Faktor *m*
sterid Steroid *n*
sterile steril, keimfrei, keimunfähig, taub
   (Bergb), tot (Bergb)
sterile filtration EK-Filtration *f*
   (Entkeimungsfiltration), Sterilfiltration *f*
sterility Unfruchtbarkeit *f* (Biol), Sterilität *f*
sterilization Sterilisation *f*, Entkeimung *f*,
   Keimtötung *f*
sterilization plant Entkeimungsanlage *f*
sterilize *v* sterilisieren, desinfizieren, entkeimen
sterilized steril
sterilizer Einweckapparat *m*, Sterilisator *m*,
   Sterilisierapparat *m*
sterilizing keimtötend
sterilizing agent Sterilisationsmittel *n*
sterilizing apparatus Einweckapparat *m*
sternbergite Sternbergit *m* (Min),
   Eisensilberglanz *m* (Min), Silberkies *m* (Min)
sternite Sternit *m*
Stern multiplier Sternscher Vervielfacher *m*
steroid Steroid *n*
steroid hormone Steroidhormon *n*
steroid skeleton Steroidgerüst *n*
sterol Sterin *n*
sterosan Sterosan *n*
sterro metal Eichmetall *n*, Sterrometall *n*
stethoscope Stethoskop *n* (Med)
stevic acid Stevinsäure *f*
stew *v* schmoren
stewartite Stewartit *m* (Min)

stewing, to extract by ~ ausschmoren
stibamine glucoside Stibaminglucosid *n*
stib[i]ate Antimonat *n*, Stibiat *n* (obs)
stibiconite Stibiconit *m* (Min), Antimonocker *m* (Min)
stibine Antimonwasserstoff *m*, Stibin *n*
stibinoaniline Stibinoanilin *n*
stibinobenzene Stibinobenzol *n*
stibiotantalite Stibiotantalit *m* (Min)
stibium (Symb. Sb) Antimon *n*
stibnite Stibnit *m* (Min), Grauspießglanz *m* (Min)
stibonic acid Stibonsäure *f*
stibosamine Stibosamin *n*
stichtite Stichtit *m* (Min)
stich welding Heftschweißen *n*
stick Stange *f* (Kosm), Stock *m*
stick *v* [an]haften, kleben, ~ **in** einstecken, ~ **on** ankleben, anstecken, befestigen, ~ **together** aneinanderhaften, zusammenkleben
stick cement Stangenkitt *m*
sticker Aufkleber *m*, Klebezettel *m*
stick gauge Fühlstab *m*
stickiness Klebrigkeit *f*
sticking [an]haftend, klebend
sticking charge Haftladung *f*
sticking plaster Heftpflaster *n* (Med), Klebpflaster *n*
sticking potential Haftpotential *n* (Elektr)
sticking thermometer Haftthermometer *n*
sticking wax Klebewachs *n*
stick of corrosion inhibitor Korrosionsschutzstab *m*
stick potash Stangenkali *n*
stick sulfur Stangenschwefel *m*
sticky klebrig, haftfähig, leimig, pappig
stiff steif, ~ **against torsion** drehsteif
stiff brush Kratzbürste *f*
stiffen *v* [ver]steifen, absteifen, verfestigen, verstärken
stiffener Versteifungsmittel *n*, Versteifungsstreifen *m*
stiffening Versteifung *f*, Bewehrung *f*, Verstärkung *f*
stiffening ring Aussteifungsring *m*
stiffness Steifheit *f*, Festigkeit *f*, Starre *f*, Starrheit *f*, Steifigkeit *f*, **degree of** ~ Steifheitsgrad *m*, ~ **in flexure** Biegesteifheit *f*
stiffness modulus Steifigkeitsmodul *m*
stigmastane Stigmastan *n*
stigmasterol Stigmasterin *n*
stigmatic stigmatisch (Opt)
stigmator Stigmator *m* (Elektronenmikroskop)
stilbazole Stilbazol *n*
stilbazoline Stilbazolin *n*
stilbene Stilben *n*
stilbene yellow Stilbengelb *n*
stilbestrol Stilböstrol *n*
stilbite Stilbit *m* (Min), Strahlzeolith *m* (Min)

still Retorte *f*, Destillationsgefäß *n*, Destillierapparat *m*, Destillierkolben *m*
still *a* regungslos, still
still head Aufsatz *m* (Dest), Brennhelm *m* (Dest)
stillhouse Schnapsbrennerei *f*
stillingic acid Stillingsäure *f*
stillopsidin Stillopsidin *n*
still temperature Kopftemperatur *f* (Dest)
stilpnomelane Stilpnomelan *m* (Min), Melanglimmer *m* (Min)
stilpnosiderite Stilpnosiderit *m* (Min)
stimulant Stimulans *n*, Anregungsmittel *n*, Belebungsmittel *n*, Reizmittel *n*
stimulate *v* anregen, anreizen, stimulieren
stimulating anregend, belebend, stimulierend
stimulating effect Reizwirkung *f*
stimulating substance Reizstoff *m*
stimulation Anregung *f*, Reizung *f*
stimulus Reiz *m*, Stimulans *n*
stimulus conduction Erregungsleitung *f* (Nerv), Reizfortpflanzung *f* (Nerv)
sting *v* stechen (Insekt)
stinkstone Stinkkalk *m* (Min), Stinkstein *m* (Min)
stipitatic acid Stipitatsäure *f*
stipple paint plastische Farbe *f*
stippling paint Tupffarbe *f*
stipulate *v* ausbedingen, festsetzen
stipulation Abmachung *f*, Bedingung *f*
stir *v* rühren, ~ **the fire** [das Feuer] schüren, ~ **together** zusammenrühren
stirred ball mill Rührwerks-Mühle *f*
stirred pressure vessel Rührdruckgefäß *n*
stirred vessel cascades Rührkesselkaskaden *pl*
stirrer Rührer *m*, Rührapparat *m*, Rühreisen *n*, Rührstab *m*, Rührvorrichtung *f*, Rührwerk *n*, **electromagnetic** ~ elektromagnetischer Rührer, **mechanical** ~ mechanischer Rührer
stirrer drive Rührwerksantrieb *m*
stirrer stand Rührstativ *n*
stirring Rühren *n*
stirring apparatus Rührapparat *m*, Rührwerk *n*
stirring arm Rührarm *m*
stirring device Rührvorrichtung *f*, Rührwerk *n*
stirring hole Schüröffnung *f*
stirring machine Rührmaschine *f*
stirring motor Rühr[werk]motor *m*
stirring rate Rührgeschwindigkeit *f*
stirring rod Rührstab *m*
stirring tub Rührbütte *f*
stirring unit Rührwerk *n*
stitch Stich *m*, Masche *f*, Nadelstich *m*
stitch *v* nähen; heften
stitch holding property Nadelausreißfestigkeit *f*
stitching Nähen *n*
stitch tear resistance Stichausreißfestigkeit *f*
stizolobic acid Stizolobsäure *f*

stock Bestand *m*, Lagerbestand *m*,
Materialvorrat *m*, Rasse *f* (Biol), Vorrat *m*,
fully compounded ~ Fertigmischung *f*
stock *v* aufstapeln (Vorräte)
stock bottle Vorratsflasche *f*
stock crane Lagerkran *m*
stock distributing gear Schüttvorrichtung *f*
stock farming Viehzucht *f*
stock guide Abstreifbacke *f*
stocking cutter Vorschneidfräser *m*
stock line gauge Gichtanzeiger *m*
stock liquor Stammflotte *f*
stockpile Lagerplatz *m*
stockpile *v* aufspeichern, aufstapeln
stock pyrometer Einstichpyrometer *n*
(Extrusion)
stock rail Anschlagschiene *f*
stock receipt Wareneingang *m*
stock solution Stammlösung *f* (Chem),
Vorratslösung *f* (Chem)
stocktaking Inventur *f*
stock vat Stammküpe *f*
stock vessel Standgefäß *n*
stoichiometric[al] stöchiometrisch (Chem)
stoichiometry Stöchiometrie *f* (Chem)
stoke *v* schüren, beschicken, ~ **mechanically**
[den Rost] mechanisch beschicken
stoke hole Schürloch *n*, Heiztür *f*
stoker Heizer *m*, Ofenkrücke *f*
stoker firing Mühlenfeuerung *f*, selbsttätige
Rostfeuerung *f*
stokesite Stokesit *m* (Min)
Stokes' rule Stockessche Regel *f*
Stokes' theorem Stokesscher Satz *m*
stoking Aufschüttung *f*, Rostbeschickung *f*,
Rostfeuerung *f*, **technique of** ~
Feuerungstechnik *f*
stoking grate Schürrost *m*
stoking platform Heizerstand *m*
stolzite Stolzit *m* (Min), Wolframbleierz *n*
(Min)
stomach Kern *m* (Obst), **acidity of the** ~
Magensäuregehalt *m*
stomach acidity Magensäure *f*
stomachic bitter Magenbitter *m*
stomachic drops Magentropfen *m pl* (Pharm)
stone Kern *m* (Obst), Stein *m*, **hard as** ~
steinhart, **porous** ~ Filterstein *m*
stone ashes Steinasche *f*
stone bolt Steinschraube *f*
stone breaker Erzquetsche *f*, Steinbrecher *m*
stone burnisher Achatglättmaschine *f*
stone china feines Steingut *n*
stone flax Asbest *m*
stone-like steinartig
stone marl Steinmergel *m*
stone pillar Steinpfeiler *m*
stone pitch Steinpech *n*
stone powder Gesteinsmehl *n*

stone ring Steinring *m*
stones Gestein *n*, **brittle** ~ faules Gestein
stone spindle Mühleisen *n*
stone wall Mauersockel *m*, Steinsockel *m*
stoneware Steingut *n*, **feldspathic** ~
Halbporzellan *n*
stoneware vessel Steinzeuggefäß *n*
stony steinig; steinartig
stool Stuhl *m*, Hocker *m*
stop Halten *n*, Anschlag[punkt] *m*, Ende *n*,
Pause *f*, Stillstand *m*, Unterbrechung *f*
stop *v* [an]halten, abstellen, arretieren,
aufhalten, außer Betrieb setzen, eine Tätigkeit
einstellen, innehalten, stillstehen, stocken, ~
running auslaufen
stop bath Unterbrechungsbad *n* (Phot)
stop button Anschlagbutzen *m*
stopcock Hahn *m*, Absperrhahn *m*, **ground-in** ~
eingeschliffener Hahn, **regulation by** ~
Hahnsteuerung *f*
stopcock control Hahnregulierung *f*
stopcock grease Hahnfett *n*
stopcock manifold Hahnbrücke *f*
stop collar Anschlag *m*, ~ **of cap**
Kappenanschlag *m*
stop dog Anschlagbolzen *m*
stop lever Abstellhebel *m*, Rasthebel *m*
stop light Stopplicht *n*
stop mechanism Abstellvorrichtung *f*
stop nut Nietmutter *f*
stop-over Aufenthalt *m*
stoppage Blockierung *f*
stoppage of the flow Abstoppen *n* der Strömung
stopper Stöpsel *m*, Flaschenverschluß *m*,
Pfropfen *m*, Stopfen *m*, Verschlußstopfen *m*,
solid ~ Vollstopfen *m*
stopper *v* verschließen, zukorken
stopper lifting device Stopfenhebevorrichtung *f*
stopper nozzle Stopfenausguß *m*
stopper rod Ventilstange *f*
stopper screw Absperrschraube *f*
stop pin Anschlagbolzen *m*, Anschlagstift *m*,
Vorsteckstift *m*
stopping Halten *n*, ~ **the blast** Abstellen *n* des
Gebläses
stopping device Anhaltevorrichtung *f*,
Arretierung *f*, Arretiervorrichtung *f*
stopping electron Bremselektron *n*
stopping equivalent Bremsäquivalent *n*
stopping formula Bremsformel *f*
stopping number Bremszahl *f*
stopping point Haltepunkt *m*
stopping potential Anhaltpotential *n*,
Bremspotential *n*
stopping power Bremsvermögen *n*
stop screw Anschlagschraube *f*, Grenzschraube *f*
stop spindle Anschlagspindel *f*
stop switch Abstellschalter *m*

stop valve Absperrventil *n*, Sperrventil *n*, Stauventil *n*
stop watch Stoppuhr *f*
storable lagerbeständig
storage Lagerung *f*, Aufbewahrung *f*, Speicherung *f*, **cold ~** Kaltlagerung *f*, **duration of ~** Aufbewahrungsdauer *f*, **time of ~** Aufbewahrungsdauer *f*, **~ of electrical energy** Elektrizitätsansammlung *f*
storage basement Lagerkeller *m*
storage basin Reservoir *n*, Tank *m*
storage battery Akkumulator *m*, Sammelbatterie *f*, Speicherbatterie *f*
storage battery cell Akkumulatorenelement *n*, Akkumulatorzelle *f*
storage battery voltage Akkumulatorspannung *f*
storage bottle Aufbewahrungsflasche *f*
storage capacity Speichervermögen *n*
storage cell Sekundärelement *n*
storage circuit Umlauflager *n*
storage flask Vorratsgefäß *n*
storage in blocks Blocklagerung *f*
storage life Lagerbeständigkeit *f*, Lagerfähigkeit *f*
storage place Ablage *f*
storage plant Speicheranlage *f*
storage rope Magazinseil *n*
storage shelf Abstellbrett *n*
storage stability Lagerfähigkeit *f*
storage tank Lagertank *m*, Sammelbehälter *m*
storage test Lagerversuch *m*, **accelerated ~** beschleunigter Lagerversuch *m*
storage tube Speicherröhre *f*
storage unit Gedächtnisspeicher *m* (Rechenmaschine)
storage vessel Aufbewahrungsbehälter *m*, Lagergefäß *n*, Vorratsbehälter *m*
storax Styrax *m* (Balsam), Storax *m* (Balsam)
storax oil Storaxöl *n*
storax ointment Storaxsalbe *f*
storax saponin Storaxsaponin *n*
store Laden *m*, Lager[haus] *n*
store *v* [ein]lagern, ablagern, aufbewahren, speichern, **~ up** aufspeichern, aufstapeln
store cellar Lagerkeller *m*
stored abgelagert, aufgespeichert
stored malt Altmalz *n*
storehouse Lager[haus] *n*, Vorratshaus *n*
storeroom Vorratsraum *m*
storey (Br. E.) Stockwerk *n*
storing Lagerung *f*; Speichern *n*
storing property Lagerfähigkeit *f*
stout handfest; massiv
stout beer Starkbier *n* (dunkles)
stout-walled starkwandig
stovaine Stovain *n*
stovarsol Stovarsol *n*
stove Ofen *m*, Herd *m*, **hot-air ~** Cowper *m*
stove enamel Ofenemaillelack *m*

stove polish Ofenglanz *m*, Herdputzmittel *n*
stove varnish Einbrennfirnis *m*
stoving tempern (Plast)
stoving enamel Einbrennemaille *f*
stoving lacquer Einbrennlack *m*
stoving temperature Einbrenntemperatur *f*
stow *v* [ver]stauen, **~ away** wegräumen
stowage Stauen *n*, Stauung *f*
stowing Stauen *n*
straddle carrier Portalstapler *m*
straight gerade, aufrecht
straight-chain geradkettig
straightedge Lineal *n*, Abrichtlineal *n*, Abstreichholz *n*, Richtscheit *n*
straighten *v* ausrichten, geradebiegen, geraderichten
straightening hammer Schlagwerkzeug *n*
straightening machine Richtmaschine *f*
straightening of the sheet Ebnen *n* der Blechtafel
straightening plate Richtplatte *f*
straightening press Richtpresse *f*
straightening roll Richtwalze *f*
straight-grained geradfaserig
straight line Gerade *f* (Math)
straight-lined geradlinig
straight pin Zylinderstift *m*
straight pipe Normalrohr *n*
straight-through process Durchfahrtechnik *f* (Galv)
straight-through reactor Durchlaufreaktor *m*
straight-through valve Durchgangsventil *n*
straight-way cock Durchgangshahn *m*
straight wrap Längsummantelung *f* (Drahtkern)
strain Beanspruchung *f*, Belastung *f*, Deformation *f*, Stamm *m* (Bakt), Verzerrung *f*, **bending or flexural ~** Biegeformung *f*, **elastic ~** elastische Beanspruchung *f*, **homogeneous ~** homogene Formänderung *f*, **non-uniform ~** ungleichmäßige Verformung *f* (Mech), **severe ~** hohe Beanspruchung *f*, **steric ~** sterische Spannung *f*, **~ of discharge** Entladebeanspruchung *f*
strain *v* [ab]seihen (Flüssigkeit), anspannen, anstrengen, beanspruchen, deformieren, [durch]seihen (Flüssigkeit), filtern, filtrieren (Flüssigkeit), spannen, [ver]zerren, verformen
strain aging Reckalterung *f*
strain bolt Sperrbolzen *m*
strain coefficient Schubzahl *f* (in tension), Dehnzahl *f*
strain distribution Deformationsverteilung *f*
strain double refraction Spannungsdoppelbrechung *f*
strained abgeseiht, filtriert, gefiltert, verzerrt
strained honey Schleuderhonig *m*
strain energy Verzerrungsenergie *f*
strainer Filter *n*, Knotenfänger *m* (Pap), Seiher *m*, Sieb *n*, Siebfilter *n*,

Siebvorrichtung *f;* Spannvorrichtung *f*
(Techn)
strainer press Seiherpresse *f*
strain factor Spannungsfaktor *m*
strain figure Fließfigur *f*
strain gauge Dehnungsmeßstreifen *m,*
Spannungsprüfer *m*
strain hardening Streckhärtung *f,* Verfestigung *f*
durch Recken
strain indicator Spannungsanzeiger *m*
straining Durchseihen *n,* Filtrieren *n,* Seihen *n*
straining bag Filtriersack *m*
straining chamber Filterkammer *f,* Siebkammer *f*
straining cloth Filtriertuch *n*
straining dish Siebschale *f*
straining filtration Siebfiltration *f*
straining frame experiment Zugversuch *m*
straining vat Läuterbottich *m*
strainless spannungsfrei
strain photograph Spannungsphoto *n*
strain rate Verformungsgeschwindigkeit *f*
strain rod Pressenholm *m*
strain-stress curve Zug- und Druckkurve *f*
strain theory Spannungstheorie *f*
stramonium, extract of ~ Stechapfelextrakt *m*
strand Faser *f,* Litze *f* (Techn), Spinnfaden *m,*
Strähne *f* (Haar)
strap Riemen *m,* Gurt *m*
strap brake Bandbremse *f*
strap connection Laschenverbindung *f*
strap iron Bandeisen *n*
strapping Anschnallen *n,* Verpackungsband *n*
strass Straß *m*
stratification Aufschichtung *f,*
Schichtenbildung *f,* Schichtung *f*
stratified schichtenweise, überschichtet
stratified flow Schichtströmung *f*
stratified sampling Gruppenauswahl *f* (Statist)
stratify *v* [auf]schichten, Schichten bilden
stratigraphy Schichtenfolge *f* (Geol)
stratosphere Stratosphäre *f*
stratum Schicht *f,* Flöz *n* (Bergb), Lage *f,* **upper**
~ Oberschicht *f* (Geol)
stratus cloud Schichtwolke *f*
straw Strohhalm *m*
strawberry aldehyde Erdbeeraldehyd *m*
strawboard Strohpappe *f*
straw cellulose Strohzellulose *f*
straw-colored strohfarbig
straw fiber Strohfaser *f*
straw plait Strohgeflecht *n*
straw pulp Strohzellstoff *m* (Pap)
straw rope spinning machine
Strohseilspinnmaschine *f*
stray vagabundierend
stray *v* streuen (Elektr)
stray coupling Streukoppelung *f*
stray current Fremdstrom *m,* Störstrom *m*
(Elektr), Streustrom *m* (Elektr)

stray electron Fremdelektron *n*
stray field Streufeld *n*
stray field screening Streufeldabschirmung *f*
stray flux Streufluß *m*
stray induction Streuinduktion *f*
stray light Falschlicht *n* (Opt), Streulicht *n*
(Opt)
stray neutron Vagabundierneutron *n* (Atom)
stray radiation Streustrahlung *f*
stray rays Streustrahlen *pl*
stray reactance Streureaktanz *f*
stray voltage Streuspannung *f*
streak Abstrich *m* (Bakt), Ader *f* (Holz),
Holzmaserung *f* (Holz), Maser *f* (Holz),
Schliere *f* (Chem), Streifen *m*
streaked streifig, aderig, gemasert (Holz)
streak formation Streifenbildung *f*
streakiness Gestreiftheit *f*
streak lightning Linienblitz *m*
streaky gestreift
stream Strom *m,* Fluß *m,* Strahl *m*
stream *v* fließen, strömen, ~ **in** einströmen
stream gold Flußgold *n*
stream hardening Strahlhärtung *f*
streamline Stromlinie *f,* Stromlinienform *f*
streamlined stromlinienförmig, windschnittig
streamline flow Bandströmung *f*
streamline form Stromlinienform *f*
streamline shape Stromlinienform *f*
stream tin Seifenzinn *n,* Zinnseife *f*
street sweepings Straßenkehricht *m*
strengite Strengit *m* (Min)
strength Kraft *f,* Bruchfestigkeit *f* (Techn),
Festigkeit *f* (Techn), Stärke *f,*
Wirkungskraft *f* (Phys), **degree of** ~
Stärkegrad *m,* **loss of** ~ Festigkeitsverlust *m,*
~ **at low temperature** Kältefestigkeit *f,* ~ **of**
material Materialfestigkeit *f,* ~ **of solution**
Konzentration *f* der Lösung
strength coefficient Festigkeitszahl *f*
strengthen *v* [ver]stärken, aussteifen (Techn),
bewehren (Techn), kräftigen, stark machen,
verfestigen, verstrammen (Techn)
strengthened verstärkt
strengthener Verstärkungsrippe *f* (Techn)
strengthening Stärken *n,* Verstrammung *f*
(Techn)
strength limit Bruchgrenze *f*
strength test Festigkeitsprüfung *f*
strength tester Festigkeitsprüfer *m*
strepsilin Strepsilin *n*
streptamine Streptamin *n*
streptidine Streptidin *n*
streptobiosamine Streptobiosamin *n*
streptococcus (pl streptococci) Streptokokkus *m*
(pl Streptokokken)
streptokinase Streptokinase *f* (Biochem)
streptolysin Streptolysin *n*
streptomycin Streptomycin *n*

streptose Streptose *f*
streptovitacin Streptovitacin *n*
stress Beanspruchung *f,* Belastung *f,* Kraft *f,*
Spannung *f,* **highest** ~
Höchstbeanspruchung *f,* **intermittent** ~
stoßweise Beanspruchung, **internal** ~
Eigenspannung *f,* innere Spannung *f,* **kind of**
~ Beanspruchungsart *f,* **limiting range of** ~
Beanspruchungsgrenze *f,* **limit of** ~
Beanspruchungsgrenze *f,* **pulsating** ~
stoßweise Beanspruchung, **removal of** ~
Entspannen *n,* **repetition of** ~
Dauerbeanspruchung *f,* **safe working** ~
zulässige Beanspruchung, **tangential** ~
Tangentialbeanspruchung *f*
stress *v* beanspruchen, überlasten
stress-anneal *v* spannungsfrei machen (durch
Ausglühen)
stress birefringence Spannungsdoppelbrechung *f*
stress calculation Festigkeitsberechnung *f*
stress coating Dehnlinienverfahren *n*
stress concentration Druckansammlung *f*
stress corrosion cracking
Spannungsrißkorrosion *f*
stress crack Spannungsriß *m*
stress cracking Spannungskorrosion *f*
(Polyäthylen), Spannungsrißbildung *f*
stress crazing Spannungsriß *m*
stress field Spannungsfeld *n*
stress-free spannungsfrei
stress function Spannungsfunktion *f*
stress graph Spannungsdiagramm *n*
stress intensity factor
Spannungsintensitätsfaktor *m*
stress property Festigkeitseigenschaft *f*
stress ratio Mittelspannungsanteil *m*
stress relaxation Spannungsrelaxation *f*
stress relaxation rate
Entspannungsgeschwindigkeit *f*
stress relaxation resistance
Entspannungswiderstand *m*
stress relaxation time Entspannungszeit *f*
stress relaxation yield limit
Entspannungskriechgrenze *f*
stress-relieving anneal Entspannungsglühen *n*
stress reversal Wechselbeanspruchung *f*
stress-strain diagram
Spannungsdehnungsdiagramm *n,*
Zerreißdiagramm *n*
stress-strain relation
Spannungsdehnungsbeziehung *f*
stress tensor Spannungstensor *m*
stress uniformity, designed for ~ auf
gleichmäßige Beanspruchung hin konstruiert
stretch Dehnen *n,* Strecken *n,* **modulus of** ~
Reckmodul *m*
stretch *v* [aus]strecken, [aus]dehnen, [aus]weiten,
recken, spannen, verstrecken
(Kunststoffäden)

stretchability Dehnbarkeit *f,* Reckfähigkeit *f,*
Verstreckbarkeit *f*
stretchable ausweitbar, dehnbar, spannbar
stretch die Streck[zieh]form *f*
stretched ausgezogen, gespannt
stretcher Spannrahmen *m* (Techn)
stretch forming Streckformen *n*
stretching Dehnung *f,* Reckung *f,* Spannen *n,*
Streckung *f,* Streckziehen *n,* Verstrecken *n,*
resistance to ~ Streckfestigkeit *f*
stretching bolt Spannschloß *n*
stretching device, "across' ~
Breitreckvorrichtung *f* (Film)
stretching force Streckkraft *f*
stretching machine Reckmaschine *f*
stretching property Dehnbarkeit *f*
stretching screw Spannschloß *n*
stretching strain Zugbeanspruchung *f,*
Zugverformung *f*
stretch shrinkage curve
Streckungs-Schrumpfungs-Kurve *f*
stretch spinning Streckspinnen *n*
stretch test Streckprobe *f,* Reckprobe *f*
strew *v* [be]streuen
strewing Bestreuen *n*
stria Streifen *m*
striated gestreift
striation Schlierenbildung *f* (Opt)
strickle board Kernschablone *f*
strigovite Strigovit *m* (Min)
strike Schlag *m,* Abstreichholz *n* (Metall),
Arbeitseinstellung *f*
strike *v* schlagen, anprallen, treffen, ~ **against**
anstoßen, auftreffen, ~ **off** abklatschen,
abschlagen
strike lever Klöppelhebel *m*
striker Hammerbär *m,* Schlagfinne *f,*
Schlagpendel *n*
striker bar Anschlagschiene *f*
striking auffallend
striking edge Schlagfinne *f*ᶜ
striking energy Auftreffenergie *f,* Schlagarbeit *f*
striking pendulum Schlagpendel *n*
striking surface Schlagfläche *f*
striking velocity Aufschlaggeschwindigkeit *f*
string Schnur *f;* Bindfaden *m,* Saite *f*
stringer Längsträger *m,* Stützbalken *m*
string galvanometer Fadengalvanometer *n;*
Saitengalvanometer *n*
string test Fadenprobe *f* (Zuck)
stringy Fäden ziehend, faserig, klebrig,
zäh[flüssig]
strip Streifen *m,* Band *n,* Leiste *f*
strip *v* abstreifen, abtakeln (Schiff), abziehen,
entrinden, bloßlegen, ~ **off** abstreifen
strip chart Diagrammstreifen *m*
strip coating abziehbarer Lack *m*
stripe Streifen *m*
striped gestreift

strip form Streifenform *f*
strip fuse Streifensicherung *f*
strip heater Heizband *n*
strip heating Heizung mit Bandheizkörper *m*,
  Bandheizung *f*
stripiness Gestreiftheit *f*
strip iron Reifeisen *n*
strip magnet Bandmagnet *m*
strip material Bandmaterial *n*
strip mill Bandagewalzwerk *n*,
  Reifenwalzwerk *n*
strippable coating Folienlack *m*
stripped atom Atomkern *m* ohne Elektronen,
  hochionisiertes Atom *n*
stripper Abscheider *m* (Atom), Abstreifer *m*,
  Schälvorrichtung *f*
stripper plate Abstreifplatte *f*
stripper plate mold Abstreiferform *f*
stripping Abstreifen *n* (vom Stempel),
  Stripping *n* (Atom) (of a mold),
  Abspannen *n* (eines Werkzeuges), ~ **of the**
  mold Verreißen *n* der Form
stripping agent Abziehmittel *n*
stripping auxiliary Abbeizzusatz *m*
stripping bath Abziehbad *n*
stripping capacity Leistung *f* (Bagger)
stripping column Abtreibkolonne *f*,
  Abtriebsäule *f*
stripping effect Abzieheffekt *m*
stripping of tin Entzinnung *f*
stripping paper Abziehpapier *n*
stripping plate Abstreifplatte *f*
stripping plate molding machine
  Durchziehformmaschine *f*,
  Durchzugformmaschine *f*
stripping zone Abtriebsteil *n*
strip steel Bandstahl *m*
strip winder Strip-Förderer *m*
strip winding machine Bandwickelmaschine *f*
strive *v* sich bemühen, streben
strobe light Blitzgerät *n*
stroboside Strobosid *n*
stroke Schlag *m*, Hieb *m*, Hub *m* (Techn),
  Schlaganfall *m* (Med), Takt *m* (Motor),
  down[ward] ~ Niedergang *m* (Kolben), **effect**
  of ~ Schlagwirkung *f*, **forward** ~
  Kolbenhingang *m*, **height of** ~ Hubhöhe *f*,
  length of ~ Hubhöhe *f*, Hublänge *f*, **reversal**
  of ~ Hubwechsel *m*, **up and down** ~
  doppelter Hub, ~ **with hammer**
  Hammerschlag *m*
stroke compensator Stoßausgleicher *m*
stroke counter Hubzähler *m*
stroke lift Hub *m*
stroke ratio Hubverhältnis *n*
stroke volume Hubraum *m*
stroma Grundgewebe *n* (Histol), Stützgewebe *n*
  (Histol)

stromeyerite Stromeyerit *m* (Min),
  Kupfersilberglanz *m* (Min)
strong stark, fest (haltbar, stark), kräftig
strong current Starkstrom *m*
strontia Strontian *m* (Min), Strontianerde *f*
  (Min), Strontiumoxid *n* (Min)
strontianite Strontianit *m* (Min)
strontium (Symb. Sr) Strontium *n*
strontium acetate Strontiumacetat *n*
strontium age Strontiumalter *n*
strontium carbide Strontiumcarbid *n*
strontium carbonate Strontiumcarbonat *n*, **native**
  ~ Strontianit *m* (Min)
strontium chloride Strontiumchlorid *n*
strontium content Strontiumgehalt *m*
strontium hydride Strontiumwasserstoff *m*
strontium hydroxide Strontiumhydroxid *n*
strontium iodide Strontiumjodid *n*
strontium isotope Strontiumisotop *n*
strontium nitrate Strontiumnitrat *n*
strontium oxide Strontiumoxid *n*, Strontian *m*
  (Min), Strontianerde *f* (Min)
strontium perborate Strontiumperborat *n*
strontium sulfate Strontiumsulfat *n*, Zölestin *m*
  (Min)
strontium tungstate Strontiumwolframat *n*
strontium white Strontiumweiß *n*
strophanthic acid Strophanthsäure *f*
strophanthidin Strophanthidin *n*
strophanthidinic acid Strophanthidinsäure *f*
strophanthin Strophanthin *n*
strophanthinic acid Strophanthinsäure *f*
strophanthobiose Strophanthobiose *f*
strophanthoside Strophanthosid *n*
strophanthotriose Strophanthotriose *f*
strophanthum Strophanthin *n*
strophanthus seed oil Strophanthusöl *n*
structural strukturell, baulich
structural adhesive Montageleim *m*
structural analysis Strukturanalyse *f* (Chem)
structural casting Bauguß *m*
structural change Gefüge[ver]änderung *f*,
  Strukturänderung *f* (Chem)
structural composition Gefügeaufbau *m*
structural constitution Gefügeaufbau *m*
structural examination Gefügeuntersuchung *f*
structural fatigue Strukturermüdung *f*
structural fiber Faserpappe *f*
structural formula Strukturformel *f* (Chem)
structural iron Baueisen *n*
structural ironwork Eisenkonstruktion *f*
structural isomerism Strukturisomerie *f*
structural member Konstruktionsteil *n*
structural protein Strukturprotein *n*
structural rail Konstruktionsschiene *f*
structural steel Baustahl *m*, Formstahl *m*
structural strength Baufestigkeit *f*
structural transformation Gefügeneubildung *f*,
  Gefügeumwandlung *f*

structural type Strukturtyp *m* (Krist)
structural unit Struktureinheit *f*
structure Aufbau *m* (Chem), Bau *m*, Bauart *f*,
  Bauwerk *n*, Gefüge *n*, Gliederung *f*,
  Struktur *f* (Chem), **angular** ~ gewinkelte
  Struktur, **aromatic** ~ aromatische Struktur,
  cellular ~ zellenartige Struktur, **change in** ~
  Strukturwandel *m*,
  Strukturveränderung *f* (Chem), **close-grained**
  ~ dichtes Gefüge, **coarse or gross** ~
  Grobgefüge *n*, **determination of** ~
  Strukturbestimmung *f*, **fibrous** ~ faserige
  Struktur, **fine** ~ Kleingefüge *n*, **fine-scaled** ~
  feinschuppige Struktur, **foliated** ~ blättrige
  Struktur, **granular** ~ körnige Struktur,
  honeycombed ~ zellenartige Struktur, **lamellar**
  ~ Lamellenstruktur, **primary** ~ primäre
  Struktur, **proof of** ~ Strukturbeweis *m*,
  secondary ~ sekundäre Struktur, **slaty** ~
  schieferartige Struktur, **tertiary** ~
  Tertiärstruktur *f*, **type of** ~ Gefügeart *f*, ~ **of**
  matter Aufbau *m* der Materie, Stoffaufbau *m*
structure-dependent strukturabhängig
structured film Mehrlagenfolie *f*,
  Mehrschichtfolie *f*
structure factor Strukturfaktor *m*
structureless gefügelos, strukturlos
structure-sensitive strukturempfindlich
struggle *v* kämpfen
strut Strebe *f* (Techn)
strut *v* abstützen, versteifen, verstreben
strut[ting] Verstrebung *f*
struverite Strüverit *m* (Min)
struvite Struvit *m* (Min)
strychnic acid Strychninsäure *f*
strychnidine Strychnidin *n*
strychnine Strychnin *n*
strychnine acetate Strychninacetat *n*
strychnine hydrochloride
  Strychninhydrochlorid *n*
strychnine nitrate Strychninnitrat *n*
strychnospermine Strychnospermin *n*
Stuart model Atomkalotte *f*
stucco Gipsmörtel *m*, Putzkalk *m*, Stuck *m*,
  Stuck[gips] *m*
stud Knopf *m*
stud [bolt] Stiftschraube *f*
studlink chain Stegkette *f*
study Studie *f*, Beobachtung *f*, Untersuchung *f*
stuff Stoff *m*, Feinzeug *n* (Pap), Material *n*,
  Rohstoff *m*
stuff catcher Stoffänger *m* (Pap), Zeugfänger *m*
  (Pap)
stuffing Füllung *f*, Fettschmiere *f* (Techn),
  Füllmasse *f*, Füllmaterial *n*, Füllsel *n*
stuffing box Stopfbüchse *f* (Masch)
stuffing box compensator
  Stopfbüchsenausgleicher *m*

stuffing box stud Stopfbüchsenschraube *f*
  (Masch)
stuffing bush Stopfbüchse *f* (Masch)
stuffing machine Füllmaschine *f*,
  Stopfmaschine *f*
stump Strunk *m*, Stumpf *m*
stunt *v* verkümmern
stuppeic acid Stuppeasäure *f*
sturdiness Festigkeit *f*, Robustheit *f*
sturdy robust, stabil, fest
sturin Sturin *n*
stutzite Tellursilberblende *f* (Min)
style griffelartiges Instrument *n*, Fasson *f*,
  Griffel *m* (Bot), Stichel *m*, Stil *m*, Stilart *f*
styling Formgebung *f*
stylopine Stylopin *n*
stylotype Stylotyp *m* (Min)
stylus Griffel *m*, Nadel *f* (Tonabnehmer)
styphnic acid Styphninsäure *f*
styptic blutstillendes Mittel *n* (Pharm),
  Styptikum *n*
styptic *a* blutstillend
stypticine Stypticin *n*, Kotarninhydrochlorid *n*
stypticite Stypticit *m*
styptic pencil Alaunstift *m*
styptic wool Eisenchloridwatte *f* (Pharm)
styptol Styptol *n*
styracine Styracin *n*
styracite Styracit *m*
styracitol Styracit *m*
styracol Styracol *n*
styrax Storax *m*, Styrax *m*
styrenated styrolisiert
styrene Styrol *n*, **crude** ~ Rohstyrol *n*
styrene plastic Styrolkunststoff *m*
styrene resin Styrolharz *n*
styrol Styrol *n*
styrolene Styrol *n*
styrolene alcohol Cinnamylalkohol *m*
styron Styron *n*
styryl Styryl-
styryl alcohol Cinnamylalkohol *m*
styryl ketone Styrylketon *n*
subacid säuerlich
subacidity Säuerlichkeit *f*
subaphylline Subaphyllin *n*
subathizone Subathizon *n*
subatomic subatomar
subatomic particle Nuklearteilchen *n*
subchloride Subchlorid *n*
sub-commitee Unterausschuß *m*
subcooling Unterkühlung *f*
subcritical unterkritisch
subculture Tochterkultur *f* (Bakt)
subcutaneous cell tissue Unterzellgewebe *n*
  (Med)
subcutaneous[ly] subkutan
sub-determinant Subdeterminante *f* (Math)
subdivided unterteilt

subdivision Aufgliederung *f,* Unterteilung *f*
subdue *v* [ab]dämpfen (Licht)
subduing Abdämpfung *f* (Licht)
suberane Suberan *n,* Cycloheptan *n,*
　Heptamethylen *n*
suberate Suberat *n*
suberene Cyclohepten *n*
suberic acid Korksäure *f,* Suberinsäure *f*
suberin Korkstoff *m,* Suberin *n*
suberoin Suberoin *n*
suberol Cycloheptanol *n,* Suberol *n*
suberone Cycloheptanon *n,* Suberon *n*
suberonic acid Suberonsäure *f*
suberose korkähnlich, korkig
suberous korkähnlich
suberyl alcohol Suberylalkohol *m*
suberyl arginine Suberylarginin *n*
subfreezing temperature Minustemperatur *f*
subgroup Nebengruppe *f,* Untergruppe *f* (Blut)
subjacent darunterliegend
subject Gegenstand *m,* Thema *n*
subject index Sachregister *n*
subjective[ly] subjektiv
subjects taught Lehrstoff *m*
subject [to] *v* aussetzen, unterwerfen
sublamine Sublamin *n*
sublattice Teilgitter *n,* Untergitter *n*
sublevel Unterstufe *f*
sublimability Sublimierbarkeit *f*
sublimable sublimierbar
sublimate Sublimat *n,* corrosive ~
　Quecksilber(II)-chlorid *n*
sublimate *v* sublimieren
sublimate bath Sublimatbad *n*
sublimating vessel Sublimiergefäß *n*
sublimation Sublimation *f* (Chem),
　Verflüchtigung *f* (fester Stoffe) (Chem),
　entrainer or carrier ~ Trägergassublimation *f,*
　heat of ~ Sublimationswärme *f*
sublimation apparatus Sublimationsapparat *m*
sublimation coefficient
　Sublimationskoeffizient *m*
sublimation cooling Sublimationskühlung *f*
sublimation curve Sublimationskurve *f*
sublimation enthalpy Sublimationsenthalpie *f*
sublimation nucleus Sublimationskern *m*
sublimation point Sublimationspunkt *m*
sublimation pressure Sublimationsdruck *m*
sublimation retort Sublimierretorte *f*
sublimation temperature
　Sublimationstemperatur *f*
sublimator Sublimierofen *m*
sublime *v* sublimieren (Chem)
sublimer Sublimator *m*
subliming, fastness to ~ Sublimierechtheit *f*
subliming furnace Sublimierofen *m*
subliming pot Sublimiertopf *m*
submaster controller Folgeregler *m*
submerge *v* untertauchen, versenken

submerged arc welding process
　Unterpulverschweißverfahren *n*
submerged bearing Unterwasserlager *n*
submerged channel induction furnace
　Rinnenofen *m*
submerged coil condenser Tauchkondensator *m*
submerged evaporator Tauchverdampfer *m*
submerged-flame process
　Tauchflammverfahren *n*
submerged-piston pump Tauchkolben-Pumpe *f*
submerged pump Unterwasserpumpe *f*
submerged-tube evaporator
　Tauchrohrverdampfer *m*
submetallic luster Halbmetallglanz *m*
submicron Submikron *n*
submicroscopic submikroskopisch
submicrostructure Submikrostruktur *f*
submit *v* unterwerfen, unterziehen
subnitrate Subnitrat *n,* basisches Nitrat *n*
subnormal Subnormale *f* (Math)
subnormal *a* unternormal
subordinate untergeordnet
subordinate quantum number
　Nebenquantenzahl *f*
suboxide Suboxid *n*
subpressure Unterdruck *m*
subpressure zone Unterdruckgebiet *n*
subresin Halbharz *n*
subroutine Unterprogramm *n* (Comp)
subsalt basisches Salz *n*
subscribe *v* abonnieren
subscriber Abonnent *m*
subscript tiefgestellter Index *m*
subscription Abonnement *n*
subscription fee Beitrag *m* (Zahlung)
subsequence Teilfolge *f* (Math)
subsequent anschließend, nachfolgend
subsequent charging Nachbeschickung *f*
subsequent ripening Nachreifen *n*
subshell Unterschale *f* (Atom)
subside *v* absenken
subsidence Absenken *n*
subsidiary battery Verstärkungsbatterie *f*
　(Elektr)
subsidiary company Tochtergesellschaft *f*
subsidiary flow Nebenströmung *f*
subsidize *v* bezuschussen
subsist *v* existieren
subsoil Untergrund *m* (Erde)
subspace Unterraum *m*
subspecies Untergruppe *f* (Zool, Bot)
substance Substanz *f,* Masse *f,* Materie *f,*
　Quadratmetergewicht *n* (Papier), Stoff *m,*
　added ~ Zusatzmittel *n,* Zusatzstoff *m,*
　amount of ~ Substanzmenge *f,* **anionic** ~
　anionenaktive Substanz, **cationic** ~
　kationenaktive Substanz, **change of** ~
　Substanzveränderung *f,* **chemically pure** ~
　chemisch reine Substanz, **colorless** ~ farblose

Substanz, **pure** ~ Reinsubstanz *f,* ~ **to be analyzed** Analysensubstanz *f*
substantial beträchtlich, wesentlich
substantiate *v* begründen
substantive dye Direktfarbstoff *m*
substantivity Substantivität *f*
substituent Substituent *m* (Chem)
substitute Ersatz *m,* Austauschstoff *m,* Surrogat *n*
substitute *v* einsetzen, ersetzen, substituieren (Chem), vertreten
substitute material Ersatzstoff *m,* Austauschwerkstoff *m*
substitute product Austauschprodukt *n*
substitute tool Austauschwerkzeug *n*
substituting substituierend
substitution Substitution *f,* Austausch *m* (Ersatz), Substituierung *f,* Verdrängung *f,* **electrophilic** ~ elektrophile Substitution, **nucleophilic** ~ nucleophile Substitution, **to react by** ~ substituierend wirken
substitution product Substitutionsprodukt *n*
substitution reaction Substitutionsreaktion *f,* Verdrängungsreaktion *f*
substrate Fällungsunterlage *f,* Farbträger *m,* Schichtträger *m,* Substrat *n* (Chem)
substrate inhibition Substrathemmung *f* (Biochem)
substrate saturation Substratsättigung *f* (Biochem)
substratum Substrat *n* (Biol), Untergrund *m*
substructure Unterbau *m,* Unterstruktur *f*
subtangent Subtangente *f* (Math)
subterranean unterirdisch
subterranean cable Erdkabel *n*
subterraneous fire Erdbrand *m*
subtilin Subtilin *n*
subtilisin Subtilisin *n* (Biochem)
subtitle Untertitel *m*
subtle zart, dünn
subtract *v* abziehen (Math), subtrahieren (Math)
subtraction Subtraktion *f* (Math)
subtraction color Subtraktionsfarbe *f*
subtractive color Subtraktionsfarbe *f*
subtractive process Subtraktivverfahren *n* (Phot)
subtractor Subtraktivfilter *m* (Phot)
subtrahend Subtrahend *m* (Math)
subunit Untereinheit *f*
success, partial ~ Teilerfolg *m*
successful erfolgreich, bewährt
succession Aufeinanderfolge *f,* Folge *f,* Reihe *f,* Reihenfolge *f,* Sukzession *f*
successive aufeinanderfolgend
successively schrittweise, hintereinander
succinamic acid Succinamidsäure *f*
succinamide Succinamid *n*
succinanilide Succinanilid *n*
succinate Succinat *n*

succinate dehydrogenase Succinatdehydrogenase *f* (Biochem)
succinate thiokinase Succinatthiokinase *f* (Biochem)
succindialdehyde Succindialdehyd *m*
succinhydrazide Succinhydrazid *n*
succinic acid Bernsteinsäure *f*
succinic acid dibenzyl ester Benzylsuccinat *n*
succinic anhydride Bernsteinsäureanhydrid *n*
succinimide Succinimid *n*
succinimidine Succinimidin *n*
succinite Bernstein *m* (Min), Goldgranat *m* (Min), Succinit *m* (Min)
succinoabietic acid Succinoabietinsäure *f*
succinoabietinolic acid Succinoabietinolsäure *f*
succinoabietol Succinoabietol *n*
succinodehydrogenase Succinodehydrogenase *f* (Biochem)
succinonitrile Bernsteinsäurenitril *n*
succinonitrilic acid Bernsteinnitrilsäure *f*
succinoresene Succinoresen *n*
succinoresinol Succinoresinol *n*
succinoylsulfathiazole Succinoylsulfathiazol *n*
succinyl Succinyl-, Bernsteinsäure-
succinyl chloride Bernsteinsäurechlorid *n,* Succinylchlorid *n*
succinylcholine Succinylcholin *n*
succinyl CoA Succinyl-CoA *n* (Biochem)
succinyl coenzyme A Succinyl-Coenzym A, Succinyl-CoA *n* (Biochem)
succinylfluorescein Succinylfluorescein *n*
succinylosuccinic acid Succinylbernsteinsäure *f*
succulent saftig, saftreich
suck *v* saugen, ~ **in** ansaugen, einsaugen, ~ **off** abnutschen (Chem), absaugen, ~ **out** aussaugen, ~ **through** durchsaugen, ~ **up** aufsaugen, absorbieren
sucked off abgenutscht (Chem)
sucker Sauger *m*
sucker rod Pumpenstange *f*
sucking Saugen *n,* Ansaugen *n*
sucking action Saugwirkung *f*
sucking and forcing pump Saug- und Druckpumpe *f*
sucking-off Absaugen *n*
sucrase Invertase *f* (Biochem), Invertin *n* (HN), Saccharase *f* (Biochem), Sucrase *f* (Biochem)
sucrate Saccharat *n,* Sucrat *n*
sucrol Sucrol *n*
sucrose Sucrose *f,* Rohrzucker *m,* Runkelrübenzucker *m,* Saccharose *f*
sucrose [density] gradient Sucrosedichtegradient *m* (Biochem)
sucrose gradient centrifugation Sucrosegradientenzentrifugation *f*
suction Ansaugen *n,* Saugen *n,* Sog *m,* **resistance to** ~ Saugwiderstand *m*
suction air Saugluft *f,* Saugwind *m*
suction air chamber Saugwindkessel *m*

suction air conveyor Saugluftförderanlage *f*
suction air plant Saugluftanlage *f*
suction air vessel Saugwindkessel *m*
suction apparatus Ansaugvorrichtung *f,*
Saugapparat *m,* Sauger *m*
suction basket Saugkorb *m*
suction bend Saugkrümmer *m*
suction bottle Saugflasche *f*
suction box Saugkasten *m*
suction brush Saugbürste *f*
suction capacity Ansaugleistung *f*
suction cell filter Saugzellenfilter *n*
suction chamber Saugkammer *f*
suction crucible Saugtiegel *m*
suction cupola Saugkupolofen *m,*
Saugkuppelofen *m*
suction device Saugvorrichtung *f*
suction effect Saugwirkung *f*
suction fan Absaugventilator *m,*
Ansauggebläse *n*
suction filter Nutsche *f,* Saugfilter *m*
suction filter apparatus Nutschapparat *m*
suction-filtered abgenutscht
suction filter press Saugfilterpresse *f*
suction filtration Saugfiltration *f*
suction flask Saugkolben *m* (Chem)
suction force Saugkraft *f*
suction funnel Nutsch[en]trichter *m*
suction gas Sauggas *n*
suction gas motor Sauggasmotor *m*
suction height Saughöhe *f*
suction line Saugleitung *f*
suction nozzle Saugdüse *f*
suction orifice Ansaugöffnung *f*
suction passage Saugkanal *m*
suction pipe Saugrohr *n,* Ansaugrohr *n*
suction pipe connection Saugrohranschluß *m*
suction pipe[te] Saugpipette *f*
suction plant Absaug[e]anlage *f*
suction port Saugkanal *m*
suction power Saugleistung *f*
suction pressure Saugdruck *m*
suction process Saugvorgang *m*
suction pump Saugpumpe *f*
suction side Saugseite *f*
suction strainer Nutsche *f,* Saugfilter *m*
suction stroke Ansaugehub *m,* Saughub *m*
suction tube Ansaugrohr *n,* Saugrüssel *m*
suction valve Ansaugventil *n,* Saugventil *n*
sudan Sudan *n*
sudden[ly] plötzlich, schlagartig
sudorific schweißtreibend
suds Schaum *m* (Seife), Seifenlauge *f*
suede Plüschleder *n*
sueded gerauht (Leder)
sue [for] *v* verklagen [wegen]
suet Talg *m,* Schmer *m n,* Unschlitt *n*
suffer *v* [er]leiden
sufficient ausreichend, genügend, hinreichend

suffocate *v* ersticken
suffocating erstickend
suffocation Erstickung *f,* **symptom of** ~
Erstickungserscheinung *f*
sugar Zucker *m,* **burnt** ~ verbrannter Zucker,
containing ~ zuckerhaltig, **crystallized** ~
Kristallzucker *m,* **dextrorotatory** ~
rechtsdrehender Zucker, **extraction of** ~
Entzuckerung *f,* **granulated** ~
Kristallzucker *m,* **levorotatory** ~
linksdrehender Zucker, **production of** ~
Zuckererzeugung *f,* **reducing** ~ reduzierender
Zucker (Chem), **refined** ~ Raffinade *f,* **to**
carbonate ~ Zucker saturieren (Zucker), **to**
preserve in ~ einzuckern, **unrefined** ~
Rohzucker *m,* ~ **in urine** Harnzucker *m,* ~ **of**
lead Bleizucker *m* (Chem), ~ **refined with**
steam Dampfzucker *m*
sugar *v* süßen, zuckern
sugar acid Zuckersäure *f*
sugar analysis Zuckeranalyse *f*
sugar beet Zuckerrübe *f*
sugar beet chip Zuckerrübenschnitzel *n*
sugar breaker Knotenbrecher *m* (Zuck)
sugar candy Kandiszucker *m*
sugar cane Zuckerrohr *n,* **crushed** ~ Bagasse *f*
sugar coal Zuckerkohle *f*
sugar-coat *v* überzuckern
sugar-coated tablet Dragée *n*
sugar compound Zuckerverbindung *f*
sugar content Zuckergehalt *m,* ~ **of the blood**
Blutzuckergehalt *m*
sugar crusher Zuckerquetsche *f*
sugar factory Zuckerfabrik *f*
sugar honey Zuckerhonig *m*
sugar industry Zuckerindustrie *f*
sugar-like zuckerähnlich, zuckerartig
sugar manufacture Zuckerfabrikation *f*
sugar mill Zuckerrohrmühle *f*
sugar production Zuckerherstellung *f*
sugar refinery Zuckerraffinerie *f*
sugar refining thermometer
Zuckerkochthermometer *n*
sugar residue Zuckertrester *pl*
sugar solution Zuckerlösung *f*
sugar substitute Zuckerersatz *m*
sugar syrup decolorizing plant
Zuckersaftentfärbungsanlage *f*
sugary zuckerartig, zuckerhaltig, zuckerig
suggest *v* anregen, vorschlagen
suggested formulation Richtrezeptur *f*
suint Wollfett *n,* Wollschmiere *f,*
Wollschweiß *m,* **potash from** ~
Wollschweißasche *f*
suint ash Schweißasche *f*
suint content Schweißgehalt *m* (Wolle)
suit *v* anpassen, eignen
suitable passend, geeignet, angebracht,
angemessen, zutreffend, zweckdienlich, **most**

~ bestgeeignet, ~ **for exploiting or working**
abbauwürdig (Bergb)
sulfacetamide Sulfacetamid *n*
sulfachrysoidine Sulfachrysoidin *n*
sulfadiazine Sulfadiazin *n*
sulfadicramide Sulfadicramid *n*
sulfaethidole Sulfaethidol *n*
sulfaguanidine Sulfaguanidin *n*
sulfamethoxypyridazine
Sulfamethoxypyridazin *n*
sulfamic acid Amidoschwefelsäure *f,*
Amidosulfosäure *f,* Sulfamidsäure *f*
sulfamide Sulfamid *n*
sulfanilamide Sulfanilamid *n*
sulfanilic acid Sulfanilsäure *f*
sulfanilide Sulfanilid *n*
sulfanthrol Sulfanthrol *n*
sulfapyridine Sulfapyridin *n*
sulfapyrimidine Sulfapyrimidin *n*
sulfatase Sulfatase *f* (Biochem)
sulfate Sulfat *n,* **containing** ~ sulfathaltig
sulfate *v* sulfatieren, mit Schwefelsäure
behandeln
sulfate cellulose Sulfatzellstoff *m* (Pap)
sulfate content Sulfatgehalt *m*
sulfate ester Schwefelsäureester *m*
sulfate ion Sulfation *n*
sulfate kraft paper Natronkraftpapier *n*
sulfate pulp Sulfatcellulose *f* (Pap),
Sulfatzellstoff *m* (Pap)
sulfate water Sulfatwasser *n*
sulfathiazole Sulfathiazol *n*
sulfating Sulfatieren *n*
sulfating agent Sulfatisierungsmittel *n*
sulfation Sulfatierung *f*
sulfatize *v* sulfatieren (durch Rösten etc.)
sulfatizing Sulfatieren *n* (durch Rösten etc.)
sulfhydrate Hydrosulfid *n,* Sulfohydrat *n*
sulfhydryl Sulfhydryl-
sulfidal Sulfidal *n*
sulfide Sulfid *n,* **containing** ~ sulfidisch
sulfide inclusion Sulfideinschluß *m*
sulfide ore sulfidisches Erz *n*
sulfide stress cracking Spannungsrißkorrosion *f*
(Stahl)
sulfidic sulfidisch
sulfimide Sulfimid *n*
sulfindigotic acid Indigoschwefelsäure *f*
sulfine Sulfin *n*
sulfinic acid Sulfinsäure *f*
sulfite Sulfit *n*
sulfite cellulose Sulfitcellulose *f* (Pap),
Sulfitzellstoff *m* (Pap)
sulfite digestion Sulfitaufschluß *m*
sulfite digestor Sulfitkocher *m* (Pap)
sulfite liquor Sulfitlauge *f*
sulfite liquor plant Sulfitlaugeanlage *f* (Pap)
sulfite lye Sulfitlauge *f*
sulfite oxidase Sulfitoxidase *f* (Biochem)

sulfite process Sulfitprozeß *m* (Pap),
Sulfitverfahren *n* (Pap)
sulfite pulp Sulfitcellulose *f* (Pap),
Sulfitzellstoff *m* (Pap)
sulfite reductase Sulfitreduktase *f* (Biochem)
sulfite waste liquor Sulfitablauge *f*
sulfoacetic acid Sulfoessigsäure *f*
sulfoantimonic acid Schwefelantimonsäure *f*
sulfobenzide Sulfobenzid *n*
sulfobenzimide Sulfobenzimid *n*
sulfobenzoic acid Sulfobenzoesäure *f*
sulfobenzoic acid imide Saccharin *n*
sulfoborite Sulfoborit *m* (Min)
sulfocarbanilide Diphenylthioharnstoff *m,*
Sulfocarbanilid *n*
sulfocarbimide Isothiocyansäure *f*
sulfocarbolic acid Phenolsulfonsäure *f*
sulfochloride Sulfochlorid *n,*
Sulfonsäurechlorid *n*
sulfocinnamic acid Sulfozimtsäure *f*
sulfocyanate Rhodanid *n*
sulfocyanic acid Rhodanwasserstoff *m,*
Sulfocyansäure *f*
sulfocyanide Rhodanid *n*
sulfoform Sulfoform *n*
sulfo-group Sulfogruppe *f*
sulfoichtyolic acid Sulfoichthyolsäure *f*
sulfolane Sulfolan *n*
sulfolysis Sulfolyse *f*
sulfonal Sulfonal *n*
sulfonamide Sulfonamid *n,* Sulfonsäureamid *n*
sulfonatable sulfonierbar
sulfonate Sulfonat *n*
sulfonate *v* sulfonieren, sulfurieren
sulfonation Sulfonierung *f,* Sulfurierung *f*
sulfonation plant Sulfonieranlage *f*
sulfone Sulfon *n*
sulfone carboxylic acid Sulfoncarbonsäure *f*
sulfonic acid Sulfonsäure *f*
sulfonic group Sulfogruppe *f*
sulfonium compound Sulfoniumverbindung *f*
sulfonium radical Sulfoniumradikal *n*
sulfonmethane Sulfonal *n*
sulfonyldiacetic acid Sulfonyldiessigsäure *f*
sulfophenic acid Phenolsulfonsäure *f*
sulfophenol Sulfocarbolsäure *f*
sulfopurpuric acid Purpurschwefelsäure *f*
sulforaphen Sulforaphen *n*
sulfosalicylaldehyde Sulfosalicylaldehyd *m*
sulfosalicylic acid Sulfosalicylsäure *f*
sulfosol Sulfosol *n*
sulfourea Thioharnstoff *m*
sulfoxide radical Sulfoxidradikal *n*
sulfoxylic acid Sulfoxylsäure *f*
sulfur (Symb. S) Schwefel *m* (Am. E.), **capillary**
~ Haarschwefel *m* (Min), **containing** ~
schwefelhaltig, **crude** ~ Rohschwefel *m,*
fibrous native ~ Haarschwefel *m* (Min),
flowers of ~ Schwefelblüte *f,* **free from** ~

schwefelfrei, **liver of** ~ Schwefelleber *f,*
monoclinic ~ monokliner Schwefel, **native** ~
Bergschwefel *m* (Min), Jungfernschwefel *m*
(Min), **plastic** ~ plastischer Schwefel, **poor in**
~ schwefelarm, **powdered** ~
Schwefelpulver *n,* **rhombic** ~ rhombischer
Schwefel, **roll of** ~ Schwefelstange *f,*
Stangenschwefel *m,* **separation of** ~
Schwefelabscheidung *f,*
Schwefelausscheidung *f,* **sublimated** ~
Schwefelblüte *f,* **to fumigate with** ~
ausschwefeln, **total** ~ Gesamtschwefel *m,* **to**
treat with ~ schwefeln, **wettable** ~
Netzschwefel *m*
sulfur *v* schwefeln, ausschwefeln, einschwefeln
sulfurate *v* mit Schwefel behandeln, schwefeln
sulfurating Schwefeln *n*
sulfuration Schwefeln *n,* Schwefelung *f*
sulfur bacteria Schwefelbakterien *pl*
sulfur-bearing rock Schwefelgestein *n*
sulfur bichloride Schwefeldichlorid *n*
sulfur bridge Schwefelbrücke *f* (Chem)
sulfur bromide Bromschwefel *m,*
Schwefelbromid *n*
sulfur chamber Schwefelkammer *f*
sulfur chloride Chlorschwefel *m,*
Schwefelchlorid *n*
sulfur color Schwefelfarbe *f*
sulfur-colored schwefelfarben
sulfur combustion furnace
Schwefelverbrennungsofen *m*
sulfur compound Schwefelverbindung *f*
sulfur content Schwefelgehalt *m*
sulfur determination Schwefelbestimmung *f*
(Anal)
sulfur dichloride Schwefeldichlorid *n*
sulfur dioxide Schwefeldioxid *n,*
Schwefligsäureanhydrid *n*
sulfur dioxide production plant
Schwefeldioxidgewinnungsanlage *f*
sulfur dross Schwefelschlacke *f*
sulfur dye Schwefelfarbstoff *m*
sulfur earth Schwefelerde *f*
sulfur[e]ous (Am. E.) schwef[e]lig
sulfureousness Schwefelhaltigkeit *f*
sulfur[eous] water Schwefelwasser *n*
sulfuretin Sulfuretin *n*
sulfuret[t]ed geschwefelt
sulfuretted hydrogen Schwefelwasserstoff *m*
sulfur extraction plant
Schwefelextraktionsanlage *f*
sulfur fluoride Fluorschwefel *m*
sulfur fume Schwefelrauch *m*
sulfur fumigation Ausschwefeln *n*
sulfur halide Halogenschwefel *m,*
Schwefelhalogen *n*
sulfur heptoxide Schwefelheptoxid *n*
sulfuric schwefelsauer

sulfuric acid Schwefelsäure *f,* **free from** ~
schwefelsäurefrei, **fumes of** ~
Schwefelsäuredämpfe *pl,* **fuming** ~ rauchende
Schwefelsäure *f,* Oleum *n,* **manufacture of** ~
Schwefelsäurefabrikation *f,* **salt or ester of** ~
Sulfat *n*
sulfuric acid carboy Schwefelsäureballon *m*
sulfuric acid demijohn Schwefelsäureballon *m*
sulfuric acid plant Schwefelsäureanlage *f*
sulfuric anhydride Schwefelsäureanhydrid *n,*
Schwefeltrioxid *n*
sulfuric ether Diäthyläther *m*
sulfur impression Schwefelabdruck *m*
sulfuring Schwefeln *n,* Schwefelung *f*
sulfuring room Schwefelkammer *f*
sulfur iodide Jodschwefel *m*
sulfurization Schwefelung *f,* Ausschwefelung *f*
(Chem), Schwefeln *n,* Vulkanisation *f*
sulfurize *v* [ein]schwefeln, mit Schwefel
behandeln, vulkanisieren
sulfurized geschwefelt
sulfurized oil Faktoröl *n*
sulfurizing Schwefeln *n*
sulfur loss Schwefelverlust *m*
sulfur melting pan Schwefelpfanne *f*
sulfur mine Schwefelgrube *f*
sulfur monochloride Schwefelmonochlorid *n*
sulfur odor Schwefelgeruch *m*
sulfur ointment Schwefelsalbe *f* (Pharm)
sulfurol black Sulfurolschwarz *n*
sulfur ore Schwefelerz *n* (Min)
sulfurous schwefelig
sulfurous acid schwefelige Säure *f,* **salt of** ~
Sulfit *n*
sulfurous anhydride Schwefeldioxid *n,*
Schwefligsäureanhydrid *n*
sulfur pit Schwefelgrube *f*
sulfur pockmarks Schwefelpocken *pl*
sulfur precipitate Schwefelniederschlag *m*
sulfur print Schwefelabdruck *m*
sulfur production plant
Schwefelgewinnungsanlage *f*
sulfur purification Schwefelreinigung *f*
sulfur refinery Schwefelhütte *f,* Schwefelwerk *n*
sulfur refining furnace Schwefelläuterofen *m*
sulfur removal Entschwefelung *f*
sulfur remover Entschwefelungsmittel *n*
sulfur salt Schwefelsalz *n*
sulfur sesquioxide Dischwefeltrioxid *n,*
Schwefelsesquioxid *n*
sulfur spring Schwefelquelle *f*
sulfur staining Marmorierung *f* (Dosen),
Schwefelverfärbung *f* (Dosen)
sulfur test Schwefelprobe *f,* **Baumann** ~
Baumannsche Schwefelprobe *f*
sulfur tetrachloride Schwefeltetrachlorid *n*
sulfur tetroxide Schwefeltetroxid *n*
sulfur threads Fadenschwefel *m*
sulfur treatment Schwefelung *f*

sulfur trioxide Schwefelsäureanhydrid *n*,
  Schwefeltrioxid *n*
sulfur vapor Schwefeldampf *m*
sulfur yellow Schwefelgelb *n*
sulfuryl Sulfuryl-
sulfuryl chloride Sulfurylchlorid *n*
sullage Asche *f* (Techn), Schlacke *f*
sulochrin Sulochrin *n*
sulphate Sulfat *n*
sulphite schwefligsauer (Salz) (Br. E.), Sulfit *n*
sulphite *v* sulfitieren
sulphur (Br. E.) Schwefel *m*
sulphur[e]ous (Br. E.) schwef[e]lig
sulvanite Sulvanit *m* (Min)
sum Summe *f*, Betrag *m*, Fazit *n*, **algebraic[al]** ~
  algebraische Summe, ~ **of the digits**
  Quersumme *f*
sum *v* summieren, addieren, ~ **up** addieren
sumac[h] Sumach *m* (Bot, Gerb)
sumac[h] liquor Sumachtrübe *f*
sumac[h] tanning Sumachgerbung *f*
sumaresinol Sumaresinol *n*
sumaresinolic acid Sumaresinol *n*
sumatrol Sumatrol *n*
sumbul oil Moschuswurzelöl *n*
sumbul root Moschuswurzel *f* (Bot, Pharm),
  Sumbulwurzel *f*
summarize *v* zusammenstellen, zusammenfassen
summarizing, selective ~ selektive Summierung
  (Comp)
summary Abriß *m*, Übersicht *f*,
  Zusammenfassung *f*, Zusammenstellung *f*
summated current Integralstromstärke *f*
summation Summierung *f*
summation band Summierungsband *n* (Spektr)
summation equation Summengleichung *f*
summation formula Summenformel *f*
summation meter Summenzähler *m* (Comp)
summation of heat, law of constant ~ **of Hess**
  Satz der konstanten Wärmesummen von Hess
summation sign Summationszeichen *n* (Math)
summit Gipfel[punkt] *m*, Höhepunkt *m*,
  Kuppe *f*, Spitze *f*
sump Sumpf *m*, Kolonnensumpf *m*
sump hole Sickeranlage *f*
sum rule Summensatz *m*
sum total Fazit *n* (Math)
sun continuum Sonnenkontinuum *n*
sun cure Heliotherapie *f*
sundial Sonnenuhr *f*
sundries Verschiedenes *n*
sunflower oil Sonnenblumenöl *n*
sunk [screw] head versenkter Schraubenkopf *m*
sunk spot Einsackstelle *f*
sunlight Sonnenlicht *n*
sunproof wax Wachs *n* zur Verhinderung von
  Lichtrissen (pl)
sunscreen chemical Lichtschutzmittel *n*
sunshine recorder Heliothermometer *n*

sunshine roof Schiebedach *n*
sun spot Sonnenfleck *m* (Astr)
sunstone Sonnenstein *m* (Min), Orthoklas *m*
  (Min)
suntan Sonnenbräune *f*
superacidity Übersäuerung *f*, Hyperacidität *f*,
  Peracidität *f*
superacidulate *v* übersäuern
superacidulation Übersäuerung *f*
superalloy hochwarmfeste Legierung *f*
superannuation Verjährung *f*
supercalender Hochkalander *m*,
  Satinierkalander *m* (Pap)
supercalendering Superkalandrieren *n* (Pap)
supercentrifuge Superzentrifuge *f*
supercharge *v* überladen, vorverdichten (Mot)
supercharged engine Kompressionsmotor *m*
supercharger Vorverdichter *m*
supercharging equipment Aufladeeinrichtung *f*
superconduction electron Supraleitelektron *n*
superconductive supraleitend
superconductivity Supraleitfähigkeit *f*
superconductor Supraleiter *m*, **hard** ~
  nicht-idealer Supraleiter (Elektr), **non-ideal** ~
  nicht-idealer Supraleiter (Elektr)
supercooled unterkühlt
supercooling Unterkühlung *f*
supercritical überkritisch
superelastic superelastisch
superficial oberflächlich
superficial current Oberflächenstrom *m*
superficial hardening Oberflächenhärtung *f*
superficiality Oberflächlichkeit *f*
superficial oxydation Oberflächenoxidation *f*
superficial structure Oberflächenstruktur *f*
superficial tanning Angerbung *f*
superfinishing Feinstbearbeitung *f*
superfluid supraflüssig
superfluous überflüssig
superfractionation Feinfraktionierung *f*
superheat *v* überhitzen
superheated steam überhitzter Dampf *m*
superheated steam cooler Heißdampfkühler *m*
superheated steam fittings
  Heißdampfarmaturen *pl*
superheater Überhitzer *m* (Dampfmasch)
superheater coil Dampfheizschlange *f*,
  Überhitzerschlange *f*
superheater surface Überhitzerheizfläche *f*
superheating Überheizen *n*, Überhitzung *f*,
  degree of ~ Überhitzungsgrad *m*, **intermediate**
  ~ Zwischenüberhitzung *f*, **temperature of** ~
  Überhitzungstemperatur *f*
superheating heat Überhitzungswärme *f*
superheating plant Überhitzeranlage *f*
superimposable, to be ~ sich zur Deckung
  bringen lassen (Stereochem)
superimpose *v* überlagern
superimposed übereinanderliegend, überlagert

superinfection Reinfektion *f* (Med)
superintend *v* beaufsichtigen
superintendent Abteilungsleiter *m*,
  Betriebsleiter *m*, Betriebsoberingenieur *m*
superior höher[wertig], vorzüglich
superior alloy steel Edelstahl *m*
superiority Überlegenheit *f*, Vorzüglichkeit *f*
supermultiplet Supermultiplett *n*
supernatant Überstand *m* (Chem)
supernatant *a* auf der Oberfläche schwimmend
supernormal übernormal
superoxide Peroxid *n*, Superoxid *n*
superpalite Superpalit *n*
superphosphate Superphosphat *n*
superpolyamide Superpolyamid *n*
superpolyester Superpolyester *m*
superpose *v* übereinanderlagern, überlagern
superposed turbine Vorschaltturbine *f*
superposition Superposition *f*, Überlagerung *f*,
  principle of optical ~
  Superpositionsprinzip *n* der Optik (Opt), ~
  of vibrations Schwingungsüberlagerung *f*
superposition principle Superpositionsprinzip *n*
superpressure zone Überdruckgebiet *n*
superproton Hyperon *n* (Atom)
super-regenerative receiver
  Geradeausempfänger *m*
supersaturate *v* übersättigen
supersaturated übersättigt
supersaturating Übersättigen *n*
supersaturation Übersättigung *f*
superscript hochgestellter Index *m*
supersensitive überempfindlich, hochempfindlich
supersensitivity Überempfindlichkeit *f*
supersession Verdrängung *f*, Abschaffung *f*
supersonic equipment Ultraschallgeräte *n pl*
supersonic flaw detector Ultraschallprüfgerät *n*
supersonic frequency Ultraschallfrequenz *f*
supersonic sounding Ultraschallecholotung *f*
supersonic speed Ultraschallgeschwindigkeit *f*
supersonic wave Ultraschallwelle *f*
superspeed Übergeschwindigkeit *f*
supervise *v* beaufsichtigen, überwachen
supervision Aufsicht *f*, Beaufsichtigung *f*,
  Kontrolle *f*, Überwachung *f*
supervisor Aufseher *m*
supervisory personnel Überwachungspersonal *n*
superweapon Superwaffe *f*
supinidine Supinidin *n*
supinine Supinin *n*
supplant *v* verdrängen
supplement Anhang *m*, Beilage *f*, Ergänzung *f*,
  Nachtrag *m*, Zusatz *m*
supplement *v* ergänzen
supplemental amount Zusatzmenge *f*
supplementary ergänzend, zusätzlich
supplementary agent Zusatzmittel *n*
supplementary feed cock Nebenspeisehahn *m*
supplementary rule Zusatzbestimmung *f*

suppleness Geschmeidigkeit *f*
supplier Lieferant *m*, Lieferer *m*, Lieferfirma *f*
supply Beschaffung *f*, Lieferung *f*, Versorgung *f*,
  Vorrat *m*, Zufluß *m*, Zufuhr *f*, **area of** ~
  Konsumgebiet *n*, **condition of** ~
  Lieferbedingung *f*, **date of** ~ Lieferfrist *f*,
  place of ~ Konsumstelle *f*, **source of** ~
  Bezugsquelle *f*, **terms of** ~ Lieferbedingung *f*
supply *v* versorgen, beschaffen, liefern, speisen,
  zuführen
supply circuit Speisekreis *m* (Elektr),
  Zuleitungsstromkreis *m* (Elektr)
supply duct Zuleitungsrohr *n*
supply main Zuführungsleitung *f*
supply network Leitungsnetz *n*
supply pipe Zuführungsrohr *n*
supply program Lieferplan *m*
supply schedule Lieferplan *m*
support Abstützung *f* (Techn), Auflage *f*
  (Techn), Beistand *m*, Halt *m*, Haltegerüst *n*
  (Techn), Halter *m* (Techn), Sprosse *f*
  (Techn), Ständer *m* (Techn), Stativ *n* (Techn),
  Träger *m* (Techn), Unterstützung *f*, **condition**
  of ~ Auflagebedingung *f*, **point of** ~
  Stützpunkt *m*
support *v* [unter]stützen, abstützen, befürworten,
  Halt geben, halten, tragen
supporting arm Tragarm *m*
supporting base Träger *m*
supporting block Auflagebock *m*
supporting bracket Tragpratze *f*
supporting coil Tragspindel *f*
supporting electrode Trägerelektrode *f*
supporting electrolyte Trägerelektrolyt *m*
supporting facility Halterungsvorrichtung *f*
supporting flange Auflageflansch *m*
supporting frame Tragrahmen *m*
supporting gas dryer Schwebegastrockner *m*
supporting girder Tragbalken *m*
supporting girders Trägernetz *n*
supporting journal Tragzapfen *m*
supporting power Tragkraft *f*
supporting ring Unterlagsring *m*
supporting rod Tragstange *f*
supporting roll Stützwalze *f*
supporting table Auflagetisch *m*
supporting tissue Stützgewebe *n* (Histol)
support plate Zwischenplatte *f*
support post Stütze *f*
supports Grubenausbau *m*, **distance between** ~
  Auflageentfernung *f*
suppose *v* annehmen
supposing angenommen
suppository Zäpfchen *n* (Pharm)
suppress *v* dämpfen (Schwingungen)
suppression Dämpfung *f*, Unterdrückung *f*
suppression filter Sperrfilter *n* (Opt)
suppressor Suppressor *m* (Biochem)
suppressor mutation Suppressormutation *f*

supraconduction Supraleitung *f*
supraconductivity Supraleitfähigkeit *f*
supraconductor Supraleiter *m*
supramolecular supramolekular
suprarenal cortex Nebennierenrinde *f* (Anat)
suprarenin Adrenalin *n*
suprasterol Suprasterin *n*
suramin Suramin *n*
surcharge Überlast *f*
surface Oberfläche *f*, **angle of** ~
Flächenwinkel *m*, **center of gravity of a** ~
Flächenschwerpunkt *m*, **character of the** ~
Oberflächenbeschaffenheit *f*, **charging of the**
~ Oberflächenbeladung *f*, **cleanliness of** ~
Oberflächenreinheit *f*, **corrugated** ~ wellige
Oberfläche, **curvature of a** ~
Flächenkrümmung *f*, **equipotential** ~
Fläche *f* gleichen Potentials, **even** ~ glatte
Oberfläche, **normal of** ~ Flächennormale *f*,
pitted ~ löcherige Oberfläche, **porous** ~
poröse Oberfläche, **rough** ~ rauhe
Oberfläche, **sensitization of the** ~
Oberflächenbeladung *f*, **state of** ~
Oberflächenzustand *m*, **symmetry of** ~
Flächensymmetrie *f*, ~ **of a body**
Körperoberfläche *f*
surface *v* flachdrehen (Techn)
surface action Oberflächenwirkung *f*
surface-active oberflächenaktiv
surface-active agent oberflächenaktives Mittel *n*
surface activity Oberflächenaktivität *f*
surface affinity Oberflächenaffinität *f*
surface an[a]esthetic Oberflächenanästhetikum *n*
[surface] area Grundfläche *f*
surface blowhole Außenlunker *m*
surface boiling örtliches Sieden *n*
surface carburettor Oberflächenvergaser *m*
surface cell Deckzelle *f*
surface charge Oberflächenladung *f*
surface-charge density
Oberflächenladungsdichte *f* (Phys)
surface chemistry Oberflächenchemie *f*
surface coating resin Deckanstrichharz *n*
surface combustion Oberflächenverbrennung *f*
surface complex Oberflächenkomplex *m*
surface compound Oberflächenverbindung *f*
surface concentration
Oberflächenkonzentration *f*
surface condensation Oberflächenkondensation *f*
surface condenser Oberflächenkondensator *m*,
Röhrenkondensator *m*
surface condition Oberflächenbeschaffenheit *f*
surface conductance Oberflächenleitwert *m*
surface conduction Oberflächenleitung *f*
surface conductivity Oberflächenleitfähigkeit *f*
surface contamination Oberflächenverseuchung *f*
surface cooler Oberflächenkühler *m*
surface cooling Oberflächenberieselung *f*,
Oberflächenkühlung *f*

surface crack Oberflächenriß *m*
surface culture Oberflächenkultur *f*
surface decarburization Oberflächenentkohlung *f*
surface defect Oberflächenfehler *m*,
Außenlunker *m*
surface density Flächendichte *f*,
Oberflächenladungsdichte *f* (Elektr)
surface density of charge Ladungsflächendichte *f*
surface development Oberflächenentwicklung *f*
surface discharge Oberflächenentladung *f*
surface distribution Oberflächenverteilung *f*
surface effect Oberflächeneffekt *m*,
Oberflächeneinfluß *m*,
Oberflächenerscheinung *f*
surface element Flächenelement *n*,
Flächenteilchen *n*
surface energy Oberflächenenergie *f*
surface enrichment Oberflächenveredelung *f*
surface excavator Flachbagger *m*
surface fermentation Obergärung *f* (Brau)
surface film Oberflächenfilm *m*
surface finish Oberflächenausführung *f*,
Oberflächenbeschaffenheit *f*
surface floor-based handling equipment
Flurfördermittel *n* (Flurförderer)
surface friction Oberflächenreibung *f*
surface gauge Flächenlehre *f*, Parallelreißer *m*
surface glow Glimmhaut *f*
surface grinder Flächenschleifmaschine *f*,
Planschleifmaschine *f*
surface grinding Planschleifen *n*
surface-hardened oberflächengehärtet
surface hardening Oberflächenhärtung *f*
surface hardness Oberflächenhärte *f*
surface imperfection Oberflächenfehler *m*
surface influence Oberflächeneinfluß *m*
surface integral Flächenintegral *n* (Math),
Oberflächenintegral *n*
surface interaction Oberflächenwechselwirkung *f*
surface ionization Oberflächenionisierung *f*
surface irregularity Oberflächenfehler *m*
surface lattice Oberflächengitter *n*
surface layer Deckschicht *f*
surface magnetization
Oberflächenmagnetisierung *f*
surface measurement Flächenmessung *f*
surface migration Oberflächenwanderung *f*
surface of impingement Prallfläche *f*
surface orientation Oberflächenorientierung *f*
surface oxidation Oberflächenoxidation *f*
surface pattern Oberflächenstruktur *f*
surface potential [of a solution]
Oberflächenpotential *n* [einer Lösung]
surface preparation Oberflächenvorbehandlung *f*
surface preservation Oberflächenkonservierung *f*
surface pressure Oberflächendruck *m*
surface properties Oberflächenbeschaffenheit *f*
surface quality Oberflächengüte *f*

surfacer Hobelmaschine *f,* **reinforced** ~ Armierungsspachtel *f*
surface-reactive oberflächenaktiv
surface refinement Oberflächenveredelung *f*
surface renewal Oberflächenerneuerung *f*
surface replica Oberflächenabdruck *m*
surface resistance Oberflächenwiderstand *m*
surface roughness Oberflächenrauhheit *f*
surface-scratching test Ritzhärteprobe *f*
surface sheet Deckblatt *n,* Deckbogen *m*
surface size Oberflächenleim *m* (Pap)
surface sizing Oberflächenleimung *f* (Pap)
surface stability Oberflächenfestigkeit *f*
surface state Oberflächenzustand *m*
surface strength Oberflächenfestigkeit *f*
surface structure Oberflächenbau *m*
surface temperature Außentemperatur *f,* Oberflächentemperatur *f*
surface tension Oberflächenspannung *f,* **lowering of** ~ Erniedrigung der Oberflächenspannung
surface tension meter Oberflächenspannungsmesser *m*
surface test Oberflächenmethode *f*
surface texturing Oberflächengestaltung *f*
surface treatment Oberflächenbearbeitung *f,* Oberflächenbehandlung *f,* Oberflächenveredelung *f*
surface water Oberflächenwasser *n* (Geol), Tagwasser *n* (Bergb)
surface waviness Schlieren *pl* auf der Oberfläche
surface weathering Verwitterung *f* der Oberfläche
surface yeast Oberhefe *f* (Brau)
surfacing mat Oberflächenvlies *n* (glass fiber)
surfactant oberflächenaktives Mittel *n,* Tensid *n*
surge chamber Mischkammer *f* (Gas)
surge generator Wanderwellengenerator *m*
surgery Chirurgie *f*
surge tank Zwischenbehälter *m*
surge voltage Stoßspannung *f*
surgical chirurgisch
surgical spirits Franzbranntwein *m* (Pharm)
surinamine Surinamin *n*
surpalite Diphosgen *n,* Surpalit *n*
surplus Überschuß *m,* Übermaß *n*
surplus *a* überschüssig
surrender Abtretung *f*
surrogate Ersatzpräparat *n* (Pharm)
surround *v* umgeben
survey Überblick *m,* Übersicht *f*
survey *v* beaufsichtigen, vermessen (Land etc.)
survey[ing] Vermessung *f* (Land etc.), Vermessungskunde *f*
survey instrument Überwachungsgerät *n*
survivor Überlebende[r] *m f*
susannite Susannit *m* (Min)
susceptance Blindleitwert *m* (Elektr)
susceptance matrix Blindleitwertsmatrix *f*

susceptibility Anfälligkeit *f,* Empfänglichkeit *f,* Magnetisierbarkeit *f* (Phys), Suszeptibilität *f* (Phys)
susceptibility maximum Suszeptibilitätsmaximum *n*
susceptible anfällig, angreifbar, beeinflußbar
susceptible to stimuli reizempfindlich
suspend *v* aufhängen, aufschlämmen (Techn), aussetzen (zeitlich), suspendieren (Chem)
suspended fein verteilt, [frei]hängend, schwebend, ~ **in a rotating position** drehbar aufgehängt
suspended body Schwebekörper *m*
suspended centrifuge Hängekorbzentrifuge *f*
suspended knife sledge Hängeschlitten *m*
suspended particle Schwebeteilchen *n*
suspended particle dryer Zerstäubungstrockner *m*
suspended platform Schwebebühne *f*
suspended position Schwebelage *f*
suspending agent Schwebemittel *n*
suspension Aufhängung *f* (Techn), Aufschlämmung *f* (Chem), Aufschwemmung *f* (Chem), Federung *f* (Techn), Gehänge *n* (Techn), Suspension *f* (Chem), **aqueous** ~ wässrige Suspension, **bifilar** ~ Zweidrahtaufhängung *f,* **coarsely dispersed** ~ grob disperse Suspension, **in** ~ schwebend
suspension agent Antiabsetzmittel *n,* Stellmittel *n* (Email)
suspension band system Spannbandsystem *n*
suspension device Aufhängevorrichtung *f*
suspension filter Einhängefilter *m*
suspension furnace Schwebeschmelzofen *m*
suspension girder Hängebalken *m*
suspension hook Aufhängebügel *m,* Einhängehaken *m,* Endhaken *m*
suspension lugs Aufhängenasen *pl*
suspension machine Suspendiermaschine *f*
suspension method Schwebemethode *f* (zur Dichtebestimmung)
suspension point Aufhängepunkt *m* (Waagschale)
suspension polymerization Suspensionspolymerisation *f*
suspension power Schwebefähigkeit *f*
suspension ring Hängering *m*
suspension track Hängebahn *f*
suspension wire Aufhängedraht *m*
suspensoid Suspensoid *n,* Suspensionskolloid *n*
suspicious clue Verdachtsmoment *n*
suspicious fact Verdachtsmoment *n*
sustain *v* halten, unterstützen, unterhalten (Reaktion, Verbrennung)
sustained loading Langzeitbeanspruchung *f*
sustained release action Depoteffekt *m* (Pharm)
suxamethonium Suxamethonium *n*
suxethonium Suxethonium *n*
svanbergite Svanbergit *m* (Min)

Svedberg unit Svedberg-Einheit *f*
  (Ultrazentrifugation)
swaged forging Gesenkschmiedestück *n*
swaging Gesenkschmieden *n*
swallow *v* [ver]schlucken
swamp Sumpf *m*
swamp ore Sumpferz *n*, Rasen[eisen]erz *n* (Min)
swartziol Swartziol *n*
swash plate mixer Wankscheiben-Mischer *m*
sweat Schweiß *m*
sweat *v* [aus]schwitzen
sweat gland schweißabsondernde Drüse *f*,
  Schweißdrüse *f*
sweating Ausschwitzen *n*, Schwitzen *n*
Swedish bar iron Osmundeisen *n*
sweep band Ablenkungsband *n* (Elektr)
sweeper bristles Besenborsten *pl*
sweep frequency Ablenkfrequenz *f* (Telev)
sweep generator Kippgenerator *m*
sweeping electrode Reinigungselektrode *f*,
  Ziehelektrode *f*
sweepings Kehricht *m*
sweep speed Laufgeschwindigkeit *f* (Elektr)
sweet chestnut Edelkastanie *f*
sweet cider Süßmost *m*
sweeten *v* [ab]süßen, **~ slightly** ansüßen
sweetening agent Süßstoff *m*
sweetish süßlich
sweetleaf Färbersüßblatt *n* (Bot)
[sweet] marjoram Majoran *m*
sweet oil Speiseöl *n*
sweet potato Batate *f*
sweets Konfekt *n*
sweet water spindle Absüßspindel *f*
sweetwood bark Cascarilla *f* (Bot)
swell *v* [an]schwellen, aufblähen, ausbauchen,
  quellen, **~ up** aufblähen, aufquellen
swelling Schwellen *n*, Anschwellen *n*,
  Aufblähung *f*, Aufschwellen *n*, Bombage *f*
  (Dose), Geschwulst *n*, Quellen *n*, **resistance to**
  ~ Quellbeständigkeit *f*
swelling auxiliary Quellhilfsmittel *n*
swelling behavior Quellverhalten *n*
swelling capacity Quellfähigkeit *f*,
  Quellvermögen *n*, Schwellkraft *f*
swelling colloid Quellungskolloid *n*
swelling heat Quellungswärme *f*
swelling liquor Schwellbeize *f*
swelling of a mold Atmen *n* eines Werkzeuges,
  Aufgehen *n* des Werkzeuges
swelling test Anschwellprobe *f*
swell-resistant quellfest
swell-starch flour Quellmehl *n*
swertinin Swertinin *n*
swietenose Swietenose *f*
swim *v* schwimmen
swimming Schwimmen *n*
swing Ausschlag *m* (Pendel)

swing *v* schwenken, schwingen, **~ open**
  aufklappen
swing bolt Gelenkschraube *f*, Klappschraube *f*
swing bucket Hängegefäß *n*
swing crane Drehkran *m*
swing diffuser Pendelbelüfter *m*
swinging pendelnd, schwingend, **to cease ~**
  ausschwingen, im Schwung nachlassen
swinging arm Schwenkarm *m*
swinging driller Schwenkbohrmaschine *f*
swinging key Schwenktaster *m*
swinging sieve Schüttelsieb *n*, Schwingsieb *n*
swing mirror Drehspiegel *m*
swing-out ausschwenkbar
swing out *v* ausschwingen, voll schwingen
swingpan Kipppfanne *f*
swing pipe Schwenkrohr *n*
swing screw Gelenkschraube *f*
swing [sledge] mill Schwingmühle *f*
swirling motion Durchwirbelung *f*,
  Wirbelbewegung *f*
swirl nozzle Wirbeldüse *f*
switch Schalter *m* (Elektr), Wechsel *m*, Weiche *f*
  (Eisenbahn), **auxiliary ~** Hilfsschalter *m*,
  cutout ~ Ausschalter *m*, **intermediate ~**
  Zwischenschalter *m*, **quick-action ~**
  Schnappschalter *m*, **remote ~**
  Fernschaltapparat *m*, **reversed ~** Schalter *m*
  mit Umkehrwirkung, **single-pole ~** einpoliger
  Schalter, **triple-pole ~** dreipoliger Schalter
switch *v* [um]schalten, **~ off** abschalten,
  abstellen, ausschalten (Licht), **~ on** andrehen
  (Elektr), anschalten, einschalten (Licht)
switch arc Schaltbogen *m*
switch base Schaltersockel *m*
switchboard Schaltbrett *n*, Schalttafel *f*
switchboard diagram Schaltbild *n*,
  Schaltschema *n*
switch box Schaltdose *f*, Schaltkasten *m*
switch cabinet Schaltschrank *m*
switch clock Schaltuhr *f*
switch covering Schalterkappe *f*
switch desk, central ~ Hauptschaltpult *n*
switch device Schaltvorrichtung *f*
switching Schalten *n*, Schaltung *f*, Umkehrung *f*,
  ~ off Abschaltung *f*, **~ on** Einschalten *n*,
  Einschaltung *f*
switching action Schalttätigkeit *f*
switch knob Drehknopf *m*
switch lever Stellvorrichtung *f*
switch mechanism Schaltmechanismus *m*
switch panel Schalttafel *f*
switch post installation Stellwerksanlage *f*
swivel Wirbel *m*, Drehzapfen *m*
swivel *v* schwenken
swivel closure Drehverschluß *m*
swivel hook Wirbelhaken *m*
swivel[l]ing ausschwenkbar
swivel-mounted schwenkbar

swivel pin Spurzapfen *m*
swollen [an]geschwollen, aufgedunsen
sycoceryl alcohol Sycocerylalkohol *m*
sydnone Sydnon *n*
syenite Syenit *m* (Geol)
syenitic syenithaltig
sylvanite Sylvanit *m* (Min), Schrifterz *n* (Min),
  Schrifttellur *n* (Min), Tellurgoldsilber *n*
sylvic acid Sylvinsäure *f*
sylvine Sylvin *m* (Min)
symbiont Symbiont *m* (Biol)
symbiosis Symbiose *f* (Biol)
symbiotic symbio[n]tisch
symbol Symbol *n*, Formelzeichen *n*, Sinnbild *n*,
  Zeichen *n*, **chemical** ~ chemisches Zeichen
symbols, explanation of ~ Zeichenerklärung *f*,
  key to ~ Zeichenerklärung *f*
symmetric[al] symmetrisch
symmetry Symmetrie *f*, Ebenmaß *n*,
  Gleichmaß *n*, **axis of** ~ Drehspiegelachse *f*,
  Symmetrieachse *f*, **center of** ~
  Symmetriezentrum *n*, **characteristic** ~
  Eigensymmetrie *f*, **degree of** ~
  Symmetriegrad *m*, **element of** ~
  Symmetrieelement *n*, **external** ~ äußere
  Symmetrie, **highest** ~ höchste Symmetrie,
  internal ~ innere Symmetrie, **property of** ~
  Symmetrieeigenschaft *f*, **statistic** ~ statistische
  Symmetrie, **x-fold** ~ x-zählige Symmetrie
symmetry factor Symmetriezahl *f*
symmetry operator Symmetrieoperator *m*
symmetry symbols Symmetriesymbolik *f*
Symons standard cone crusher
  Symons-Brecher *m*
sympatholytin Sympatholytin *n*
sympatol Sympatol *n*
symptom Symptom *n*, Erscheinung *f*,
  Krankheitssymptom *n*
synaldoxime Synaldoxim *n*
synanthrose Lävulin *n*, Synanthrose *f*
synaptase Emulsin *n*
syncaine Syncain *n*
synchro-balance printing Abgleichschreibfolge *f*
synchronism Synchronismus *m*, Gleichlauf *m*,
  Gleichzeitigkeit *f*
synchronization Synchronisation *f*,
  Synchronisierung *f*
synchronize *v* synchronisieren, abstimmen,
  gleichschalten
synchronizing coil Abstimmspule *f*
synchronous synchron, gleichlaufend,
  gleichzeitig, **to run** ~ synchron laufen
synchronous clock Synchronuhr *f*
synchronous computer Synchronrechner *m*
synchronous motor Synchronmotor *m*
synchronous muscle synchroner Muskel *m*
synchronous speed Synchrongeschwindigkeit *f*
synchro transmitter Synchronübertrager *m*
synchrotron Synchrotron *n* (Phys)

syn-configuration syn-Konfiguration *f*
  (Stereochem)
syndiazotate Syndiazotat *n*
syndicate Syndikat *n*
syndrome Syndrom *n* (Med)
synephrine Synephrin *n*
synergism Synergie *n* (Med)
syngenite Syngenit *m* (Min)
syn-isomer Syn-Form *f*
synonymous gleichbedeutend
synopsis Übersicht *f*
synthalin Synthalin *n*
synthesis Synthese *f* (Chem), Aufbau *m* (Chem),
  Darstellungsverfahren *n* (Chem), **asymmetric**
  ~ asymmetrische Synthese, **stereoselective** ~
  stereoselektive Synthese, **total** ~
  Totalsynthese *f*
synthesis gas Synthesegas *n*
synthesis gas plant Synthesegasanlage *f*
synthesize *v* synthetisieren
synthesizing Synthetisierung *f*
synthetase Synthease *f* (Biochem)
synthetic synthetisch, **fully** ~ vollsynthetisch
synthetically, to prepare ~ synthetisch herstellen
synthetic fiber Chemiefaden *m*, Synthesefaser *f*
synthetic gut künstlicher Darm *m*
synthetic leather Kunstleder *n*
synthetic process synthetisches Verfahren
synthetic resin Edelkunstharz *n*
synthetic resin cement Kunstharzleim *m*
synthetic rubber Kunstkautschuk *m*
synthetize *v* synthetisieren
syntonin Syntonin *n*
syoyu-aldehyde Syoyu-Aldehyd *m*
syphon Abfüllheber, Siphon *m*
syphon pipe Siphonrohr *n*
syphon pump Heberpumpe *f*
syringe Spritze *f*, Injektionsspritze *f*
syringenin Syringenin *n*
syringic acid Syringasäure *f*
syringin Lilacin *n*, Syringin *n*
syringyl alcohol Syringaalkohol *m*
syrup Sirup *m*, Dicksaft *m*
syrup resin Harzsirup *m*
syrupy dickflüssig
syrupy consistency Dickflüssigkeit *f*
system System *n*, Schema *n*, **antisymmetric** ~
  antisymmetrisches System, **binary** ~ binäres
  System, **catoptric** ~ katoptrisches System
  (Opt), **closed** ~ abgeschlossenes System,
  dioptric ~ dioptrisches System (Opt),
  heterogeneous ~ heterogenes System,
  homogeneous ~ homogenes System, **metric** ~
  metrisches System
system analysis Systemanalyse *f*
systematic[al] systematisch, planmäßig
systematization Systematisierung *f*
system of units, absolute ~ absolutes
  Maßsystem *n*

system residence Systemspeicher m (Comp)
Szilard-Chalmers method
   Szilard-Chalmers-Verfahren n

T

tab Aufreißband *n*, Aufreißstreifen *m*
tabacoresene Tabacoresen *n*
tabacoresinol Tabacoresinol *n*
tabasheer Tabaschir *m* (Bot)
tabashir Tabaschir *m* (Bot)
tab gate Bandanguß *m*, Dachanguß *m*
table Liste *f,* Tabelle *f,* Tisch *m,* ~ **of contents**
  Inhaltsangabe *f,* Sachregister *n,* ~ **of factors**
  Faktorentabelle *f* (Math), ~ **of logarithms**
  Logarithmentafel *f* (Math)
table area Tischfläche *f*
table covering Tischbelag *m*
table diamond Tafelstein *m*
table filter Planfilter *n*, Tellerfilter *n*
table lamp Tischleuchte *f*
table model Tischgerät *n*
table press Preßtisch *m*
table salt Speisesalz *n*
tablet Pastille *f* (Med), Tablette *f* (Med), Tafel *f*
tablet compressing machine Tablettenpresse *f,*
  Tablettiermaschine *f*
tablet disintegration tester
  Tablettenzerfallbarkeitsprüfer *m*
tableting machine Tablettiermaschine *f*
table top Tischplatte *f*
tablet press Pastillenpresse *f*
tablet production Tablettenherstellung *f*
table travel Tischhub *m*
tablets for sustained release action
  Depot-Tabletten *pl*
tabtoxinine Tabtoxinin *n*
tabular tabellarisch, tafelförmig
tabularize *v* tabellarisieren
tabular spar Tafelspat *m* (Min)
tabulate *v* tabellarisch zusammenstellen,
  tabellarisieren
tabulation Tabellarisierung *f*
tabun Tabun *n*
tacamahac Tacamahak *n* (Harz)
tachhydrite Tachhydrit *m* (Min)
tachiol Tachiol *n*
tachograph Tachograph *m*
tachometer Geschwindigkeitsmesser *m,*
  Tachometer *n,* **distant reading** ~
  Ferndrehzahlmesser *m*
tachysterol Tachysterol *n*
tackifier Mittel zur Erhöhung der Klebrigkeit *f,*
  Klebrigmacher *m*
tackiness Klebrigkeit *f,* Haftfähigkeit *f*
tacking rivet Heftniet *m*
tack welding Heftschweißen *n*
tacky klebrig
tacky dry trockenklebrig
taconite Takonit *m* (Min)
tacrine Tacrin *n*
tactoid Taktoid *n* (Kolloidchem)

tag Etikett *n*, Preisschild *n*, Preiszettel *m*
tagatose Tagatose *f*
tagging compound Markierungsverbindung *f*
tagging experiment Markierungsexperiment *n*
tagilite Tagilit *m* (Min)
tail Ausläufer *m*, Ende *n*, Endstück *n*,
  Schwanz *m*
tail drum Umlenktrommel *f*
tail flap Schmutzfänger *m*
tail gas Endgas *n*
tail hammer Schwanzhammer *m*
tailings Rückstände *m pl*, Abfall *m*, Erzabfälle *m*
  pl, Flotationsabgänge *pl*
tail pulley Umlenktrommel *f*
tailshaft bracket Wellenbock *m*
tainiolite Tainiolith *m* (Min)
takatonine Takatonin *n*
Taka[w]o base Taka[w]obase *f*
take *v* nehmen, fassen, ergreifen, ~ **apart**
  abmontieren, zerlegen, ~ **away** entziehen,
  wegnehmen, ~ **off** abnehmen, entfernen, ~
  out herausnehmen, ~ **part** teilnehmen, ~
  place sich abspielen, stattfinden, ~ **up** in
  Angriff nehmen
take care! Achtung!
take-off Abflug *m* (Luftf)
take-off device Abzugsvorrichtung *f*
take-off roll Abrollwalze *f,* Abwickelwalze *f,*
  Abzugswalze *f*
take-off speed Abfluggeschwindigkeit *f* (Luftf)
take-up Nachspannung *f* (Keilriemen),
  Spannvorrichtung *f*
take-up mechanism Abziehvorrichtung *f*
take-up roll Abnahmewalze *f,* Aufwickelwalze *f*
Talbot bands Talbot-Bänder *pl* (Spektr)
talc Talk *m* (Min), **yellow** ~ Goldtalk *m*
talcose talkartig
talcose quartz Talkquarz *m* (Min)
talcous talkartig
talcous clay Talkton *m*
talc schist Talkschiefer *m* (Min)
talcum Talk *m* (Min)
talcum powder Körperpuder *m*, Talkpulver *n*
talent Begabung *f,* Veranlagung *f*
talite Talit *m*
talitol Talit *m*
tall hoch, groß
tallianine Tallianin *n*
tallness Größe *f,* Höhe *f*
tall oil Tallöl *n*
tall oil varnish Tallöllack *m*
talloleic acid Tallölsäure *f*
tallow Talg *m*, Unschlitt *n*, **containing** ~
  talghaltig, **refined** ~ Feintalg *m*, ~ **in casks**
  Faßtalg *m*
tallow *v* [be]talgen

tallow-like talgartig
tallow oil Talgöl *n*, Härteöl *n*
tallow soap Talgseife *f*, Unschlittseife *f*
tallowy talgartig
talmi [gold] Talmi *n*
talolactone Talolacton *n*
talomethylose Talomethylose *f*
talomucic acid Taloschleimsäure *f*
talonic acid Talonsäure *f*
talosamine Talosamin *n*
talose Talose *f*
tamanite Tamanit *m* (Min)
tamarind Tamarinde *f* (Bot)
tamarugite Tamarugit *m* (Min)
tamp *v* verstopfen (Bohrloch)
tamped asphalt Stampfasphalt *m*
tamped volume measuring appliance
  Stampfvolumenmesser *m*
tamped weight Stampfgewicht *n*
tamper Stampfer *m*
tamperproof closure Verschluß *m* mit
  Originalitätssicherung
tampicin Tampicin *n*
tampicolic acid Tampicolsäure *f*
tamping compound Stampfmasse *f*
tampon Tampon *m* (Med)
tan Gerbmittel *n*, Eichenlohe *f*, **to steep in** ~
  lohen
tan *v* gerben, beizen, tannieren, ~ **a second time**
  nachgerben
tanacetene Tanaceten *n*
tanacetone Tanaceton *n*
tanacetophorone Tanacetophoron *n*
tanbark Lohe *f*, Gerberlohe *f*, Lohrinde *f*
tan color Lohfarbe *f*
tandem arrangement Reihenanordnung *f*
tandem connection Hintereinanderschaltung *f*
tandem engine Tandemmaschine *f*
tandem pump Tandempumpe *f*
tan earth Loherde *f*
tang Tang *m* (Bot)
tangent Tangente *f* (Math), Tangens *m* (Math),
  inflectional ~ Wendetangente *f*
tangent *a* sich berührend, tangential
tangential tangential
tangential acceleration
  Tangentialbeschleunigung *f*
tangential casting Tangentialguß *m* (Gieß)
tangential cut Tangentialschnitt *m*
tangential force Tangentialkraft *f* (Mech)
tangential plane Berührungsebene *f* (Math)
tangential stress Scherbeanspruchung *f*;
  Schubspannung *f*
tangential velocity Tangentialgeschwindigkeit *f*
tangent law Tangentensatz *m* (Math)
tangent line load Drehmoment *n* eines
  Zahnrades
tangent point Berührungspunkt *m* (Math)

tangent[s], point of intersection of ~
  Tangentenschnittpunkt *m*
tangent theorem Tangentensatz *m* (Math)
tanghinine Tanghinin *n*
tank Tank *m*, Behälter *m*, Bottich *m*,
  Reservoir *n*, Wanne *f*, **high level** ~
  Hochbehälter *m*, **high-pressure** ~
  Hochdruckkessel *m*
tank coating Tankanstrich *m*
tank cupola Kupolofen *m* mit erweitertem Herd
tank furnace Wannenofen *m*
tank sheet iron Behälterblech *n*
tank wagon Tankwagen *m*, Behälterwagen *m*,
  Kesselwagen *m*
tan liquor Gerb[e]brühe *f*, Gerblohe *f*
tannable gerbbar
tannage Gerbung *f*, **supersonic** ~
  Ultraschallgerbung *f*
tannal Tannal *n*
tannalbin Tannalbin *n*, Tanninalbuminat *n*
tannase Tannase *f* (Biochem)
tannate Tannat *n*, gerbsaures Salz *n*
tanned lohgar (Gerb)
tanner Gerber *m*
tanner's bark Eichenlohe *f*, Gerberlohe *f*
tanner's liquor Gerb[e]brühe *f*
tanner's pit Äscher *m* (Gerb), Äschergrube *f*
  (Gerb), Kalkäscher *m*
tanner's tallow Gerbertalg *m*
tanner's waste Gerbereiabfälle *n pl*
tannery Gerbanlage *f*, Gerberei *f*, Lohgerberei *f*
tannic gerbstoffartig
tannic acid Gerbsäure *f*, Gallusgerbsäure *f*,
  Tannin *n*, **salt or ester of** ~ Tannat *n*, **to
  mordant with** ~ tannieren
tanniferous gerbsäurehaltig
tannigen Tannigen *n*, Diacetyltannin *n*,
  Tannogen *n*
tannin Tannin *n*, Gallusgerbsäure *f*, Gerbsäure *f*,
  integral ~ Austauschgerbstoff *m*, **replacement**
  ~ Austauschgerbstoff *m*
tannin albumate Tannalbin *n*,
  Tanninalbuminat *n*
tanning Gerben *n*, Gerbung *f*, **rapid** ~
  Schnellgerbung *f*, **to finish** ~ ausgerben
tanning agent Gerbmittel *n*, Gerbstoff *m*
tanning auxiliary Gerbereihilfsprodukte *pl*,
  Gerbhilfsmittel *n*
tanning drum Gerbfaß *n*
tanning extract Gerb[stoff]auszug *m*,
  Gerb[stoff]extrakt *m*
tanning liquor Gerb[e]brühe *f*
tanning matter, determination of ~
  Gerbstoffbestimmung *f*
tanning method, quick ~ Schnellgerbmethode *f*
tanning pit Gerbgrube *f*
tanning plant Gerbanlage *f*
tanning process Gerbprozeß *m*
tanning vat Gerbkufe *f*

tannin mordant Gerbstoffbeize *f*
tannin ointment Tanninsalbe *f* (Pharm)
tannoform Methylenditannin *n*, Tannoform *n*
tannogen Tannogen *n*
tannometer Gerbsäuremesser *m*
tannon Tannon *n*, Tannopin *n*
tannopin Tannon *n*, Tannopin *n*
tannyl Tannyl-
tan ooze Gerb[e]brühe *f*
tan pickle Lohbrühe *f*
tan powder Lohpulver *n*
tansy Rainfarnkraut *n* (Bot), Wurmkraut *n* (Bot)
tansy flower Rainfarnblüte *f* (Bot)
tantalate Tantalat *n*
tantalic acid Tantalsäure *f*, **salt of** ~ Tantalat *n*
tantalic anhydride Tantalpentoxid *n*, Tantal(V)-oxid *n*
tantalic chloride Tantalpentachlorid *n*
tantalic compound Tantal(V)-Verbindung *f*
tantalic oxide Tantalpentoxid *n*, Tantal(V)-oxid *n*
tantalite Tantalit *m* (Min), Columbeisen *n* (Min), Harttantalerz *n* (Min)
tantalous compound Tantal(III)-Verbindung *f*
tantalum (Symb. Ta) Tantal *n*
tantalum dioxide Tantaldioxid *n*
tantalum fluoride Tantalfluorid *n*
tantalum metal Tantalmetall *n*
tantalum ore Tantalerz *n* (Min)
tantalum pentachloride Tantalpentachlorid *n*
tantalum pentoxide Tantalpentoxid *n*, Tantal(V)-oxid *n*
tantalum tetroxide Tantaltetroxid *n*
tan vat Gerbtrog *m*, Lohfaß *n*, Lohgrube *f*
tanyard Lohgerberei *f*
tap Hahn *m*, Anstich *m* (Brau), Auslaufventil *n*, Gewindebohrer, Wasserhahn *m*, Zapfhahn *m*
tap *v* abstechen (Metall), abzapfen (Flüssigkeit), anzapfen (Brau)
tap borer Zapfenbohrer *m*
tap cinder Schlacke *f*, Garschlacke *f*, Puddelschlacke *f*, Rohschlacke *f*
tape channel Lochbandkanal *m* (Comp)
tape-controlled carriage Lochbandvorschub *m* (Comp)
taped closure Klebestreifenverschluß *m*
tape measure Bandmaß *n*, Meßband *n*, Metermaß *n*
taper Verjüngung *f* (Techn), Kegel *m* (verjüngter Teil), Kerze *f*
taper *v* sich verjüngen, spitz zulaufen, zuspitzen
taper cone Verjüngung *f*
tapered konisch, spitz zulaufend
tapered overlap Schaftung *f*
tapered thread Trapezgewinde *n*
tapering Abschrägung *f*, Konizität *f*, Zuspitzung *f*
taper pin Kegelstift *m*

taper roller bearing Kegelrollenlager *n*
tape sealing Bandsiegeln *n*
tapeworm Bandwurm *m* (Med)
taphole Abstichloch *n* (Metall), Abstichöffnung *f* (Metall), Stichloch *n* (Metall), **freezing of** ~ Verstopfung *f* des Stichloches, **stopping up of** ~ Verstopfung *f* des Stichloches
taphole displacement Stichlochversetzung *f*
taphole gun Stichlochstopfmaschine *f*
tapioca Cassawastärke *f*, Tapioka *f*
tapiolite Tapiolith *m* (Min)
tapped blind hole Sackloch *n* mit Gewinde
tapped coil Abzweigspule *f*
tapper Abstecher *m*
tappet Daumen *m*, Hebedaumen *m*, Hebekopf *m*
tappet drum Daumentrommel *f*
tappet roller Anschlagrolle *f*, Schlagrolle *f*
tapping Anzapfung *f*, Abstich *m*, Abzweigung *f*, ~ **of iron** Eisenabstich *m*, ~ **of the slag** Ablassen *n* der Schlacke (Met) Schlackenabstich *m*, ~ **with a crowbar** Stangenabstich *m*
tapping bar Sticheisen *n*
tapping hole Abstichloch *n* (Metall), Gußloch *n*, Stichöffnung *f* (Metall)
tapping interval Abstichpause *f*
tapping iron Rengel *m*
tapping knife Zapfmesser *n*
tapping machine Gewindeschneidmaschine *f*
tapping point Anzapfstelle *f*
tapping probe Entnahmesonde *f*
tappings Puddelschlacke *f*
tapping sample Abstichprobe *f*
tapping screw Schneidschraube *f*
tapping shovel Abstichschaufel *f*
tapping slag Abstichschlacke *f*
tapping spout Abstichrinne *f* (Metall)
tapping switch Abzweigschalter *m*
tapping turbine Anzapfturbine *f*
tap plug Abzweigstecker *m*
tap water Leitungswasser *n*
tar Teer *m*, **coat of** ~ Teeranstrich *m*, **low temperature** ~ Urteer *m*, **original** ~ Urteer *m*, ~ **extracted by low-temperature process** Schwelteer *m*
tar *v* teeren
tarapacaite Tarapacait *m* (Min)
tar asphalt Teerasphalt *m*
taraxacerin Taraxacerin *n*
taraxacin Taraxacin *n*, Löwenzahnbitter *n*
taraxacum extract Löwenzahnextrakt *m*
taraxanthin Taraxanthin *n*
tar boiling plant Teerkocherei *f*
tarbuttite Tarbuttit *m* (Min)
tar coating Teeranstrich *m*, Teerüberzug *m*
tar collector Teersammler *m*
tar constituent Teerbestandteil *m*
tar content Teergehalt *m*

tar dregs Teersatz *m*
tare Tara *f*, Leergewicht *n*
tare *v* ausgleichen, [aus]tarieren
tare balance Tarierwaage *f*
tare cup Tarierbecher *m*
tar enamel coating Goudronüberzug *m*,
   Teerschutzüberzug *m*
tare shot Tarierschrot *m*
tare weight Verpackungsgewicht *n*
tar extractor Teerabscheider *m*
tar formation Teerbildung *f*
tar fume Teerdampf *m*
tar furnace Teerfeuerung *f*
tar gas Teergas *n*
target Zielscheibe *f*, Antikathode *f*
   (Röntgenröhre), Auffangschirm *m*, Target *n*
   (Atomphys), Ziel *n*, **fixed** ~ unbewegliches
   Ziel
target area Aufprallfläche *f* (Elektronen)
target atom Zielatom *n*
target current Targetstrom *m*
target speed Zielgeschwindigkeit *f*
target theory Treffertheorie *f*
targusic acid Lapachol *n*
tar incrustation Teerkruste *f*
taring device Tariervorrichtung *f*
tariric acid Taririnsäure *f*
tar mist Teernebel *m*
tarnish Anlaufen *n*, auf Metall
tarnish *v* anlaufen (Met), machen, mattieren
   (Techn), sich beschlagen, trüben
tarnishable mattierbar
tarnished angelaufen (Met), blind (Met), matt
   (Met)
tarnishing Anlaufen *n*, Blindwerden *n*,
   Glanzverlust *m*, Mattbrennen *n*, Mattierung *f*
tarnish proofness Anlaufbeständigkeit *f*
tar oil Teeröl *n*
tar paper Teerpapier *n*
tar pit Teergrube *f*
tar pitch Teerpech *n*
tarragon Estragon *m* (Bot)
tarragon oil Estragonöl *n*
tarragon vinegar Estragonessig *m*
tarred board Teerpappe *f*
tarred oakum Teerwerg *n*
tarred tape Teerband *n*
tarring Teeren *n*, Einteeren *n*, Teerung *f*
tarry teerartig
tarry residue Teerrückstand *m*
tar sediment Teersatz *m*
tar separator Teerabscheider *m*
tart herb, sauer, scharf
tartar Weinstein *m*, Zahnstein *m*, **cream of** ~
   Cremortartari *m*, Kaliumbitartrat *n*, **formation**
   of ~ Weinsteinbildung *f*
tartar content Weinsteingehalt *m*
tartar emetic Brechweinstein *m*,
   Kaliumantimonyltartrat *n*

tartar emetic substitute Brechweinsteinersatz *m*
tartareous weinsteinartig
tartaric acid Weinsäure *f*, Dioxybernsteinsäure *f*,
   levorotatory ~ linksdrehende Weinsäure
tartarize *v* tartarisieren, mit Weinstein
   behandeln
tartness Säure *f*, Herbheit *f*, Schärfe *f*
tartrate Tartrat *n*, weinsaures Salz *n*, **acid**
   potassium ~ Cremor Tartari *m*, **potassium**
   hydrogen ~ Cremor Tartari *m*
tartrazine Tartrazin *n*
tartronic acid Tartronsäure *f*, Oxymalonsäure *f*
tartronuric acid Tartronursäure *f*
tartronylurea Tartronylharnstoff *m*
tar vapor Teerdampf *m*, Teernebel *m*
tar water Teerwasser *n*
task Aufgabe *f*, Pensum *n*
tasmanite Tasmanit *m*
tasmanone Tasmanon *n*
taspininic acid Taspininsäure *f*
taste Geschmack *m*, **aromatic** ~
   Würzgeschmack *m*, **crude** ~
   Rohgeschmack *m*, **free from** ~ geschmackfrei,
   sense of ~ Geschmack[s]sinn *m*, **spicy** ~
   Würzgeschmack *m*, **with respect to** ~
   geschmacklich, ~ **of the cask**
   Faßgeschmack *m*
taste *v* abschmecken, schmecken, probieren,
   kosten, ~ **rotten** faul schmecken
taste cell Geschmackskörperchen *n*
taste improvement Geschmacksverbesserung *f*
tasteless geschmackfrei, geschmacklos
tastelessness Geschmacklosigkeit *f*
tasting, to test by ~ abschmecken
tasty schmackhaft
taurine Taurin *n*
taurocholate Taurocholat *n*
taurocholeic acid Taurocholeinsäure *f*
taurocholic acid Taurocholsäure *f*, **salt or ester of**
   ~ Taurocholat *n*
taurocyamine Taurocyamin *n*
taurolithocholic acid Taurolithocholsäure *f*
taurylic acid Taurylsäure *f*
taut straff, stramm
tautomeric tautomer
tautomeric derivative Tautomer[e] *n*
tautomerism Tautomerie *f*
taut suspension system Spannbandsystem *n*
tawed alaungar (Gerb)
tawer Alaungerber *m*, Weißgerber *m*
tawery Alaungerberei *f*, Weißgerberei *f*
tawny gelbbraun
taxicatigenin Taxicatigenin *n*
taxicatin Taxicatin *n*
taxine Taxin *n*
taxis Taxis *f* (Biol)
Taylor's series Taylorsche Reihe *f* (Math)
T bar T-Eisen *n*
T beam T-Träger *m*

T die T-Düse *f*
tea Tee *m*, **diuretic** ~ harntreibender Tee
 (Pharm), **infusion of** ~ Teeaufguß *m*
teach *v* lehren, unterrichten
teak [wood] Teakholz *n*
teallite Teallit *m* (Min)
tear Riß *m*, **path of** ~ Reißweg *m*
tear *v* [zer]reißen, ~ **asunder** durchreißen,
 auseinanderreißen, ~ **down** einreißen,
 niederreißen, ~ **loose** losreißen (Elektronen),
 ~ **off** abreißen, ~ **open** aufreißen, ~ **out**
 ausreißen
tear gas Tränengas *n*, Reizgas *n*
tear[ing] limit Zerreißgrenze *f*
tearing strength Einreißfestigkeit *f*
tearing strength test Reißfestigkeitsprobe *f*
tearing test Reißprobe *f*, Reißversuch *m*
tear-off cap Abreißkapsel *f*
tear-off package Abreißverpackung *f*
tear-proof zerreißfest
tear resistance Reißfestigkeit *f*,
 Ausreißfestigkeit *f*, Einreißfestigkeit *f*,
 Zerreißfestigkeit *f*
tear-shaped tropfenförmig
tear speed Zerreißgeschwindigkeit *f*,
 Bruchgeschwindigkeit *f*
tear strip Aufreißstreifen *m*
tear-strip can Reißbanddose *f*
tear tab Abreißlasche *f*
tease *v* [auf]rauhen (Tuch), fasern
tebelon Tebelon *n*
technetium (Sym. Tc) Technetium *n*
technical technisch, Fach-, fachmännisch
technical adviser Fachberater *m*
technical analysis technische Analyse *f*
technical college Fachschule *f*
technical education Fachausbildung *f*
technical expression Fachausdruck *m*
technical highschool Polytechnikum *n*
technical journal Fachzeitschrift *f*
technical literature Fachliteratur *f*
technical school Technikum *n*
technical science Technik *f*
technical studies Fachstudium *n*
technical term Fachausdruck *m*, Terminus
 technicus (Lat)
technical terminology Fachsprache *f*
technical terms Fachsprache *f*
technical worker Facharbeiter *m*
technician Techniker *m*
technicolor process Technicolorverfahren *n*
 (HN)
technique Methode *f*, Technik *f*, Verfahren *n*
technochemical technisch-chemisch
technological technologisch
technology Technologie *f*
teclu burner Teclubrenner *m* (Chem)
tecomin Tecomin *n*
tectonic tektonisch (Geol)

tectonic movement Gebirgsbewegung *f*
tectonics Tektonik *f* (Geol)
tectoquinone Tectochinon *n*
tedious langwierig, zeitraubend
tee iron T-Eisen *n*
teeming nozzle Ausflußöffnung *f*
tee square Reißschiene *f*
teething troubles Kinderkrankheiten *pl* (Med)
teflon Teflon *n* (HN)
Teichmann's crystals Teichmannsche
 Häminkristalle *pl*
teichoic acid Teichonsäure *f*
telecommunication[s] engineering
 Fernmeldetechnik *f*
telecontrol Fernsteuerung *f*
telecounter Fernzählwerk *n*
telefocus Fernfokus *m*
telefocus cathode Fernfokuskathode *f*
telegram Telegramm *n*
telegraph *v* telegraphieren
telegraphy Telegraphie *f*, **wireless** ~ drahtlose
 Telegraphie
telemeter Entfernungsmesser *m*, Telemeter *n*
telemotor Telemotor *m*
telepathine Telepathin *n*
telephone Telefon *n*
telephone and telegraph engineering
 Fernmeldetechnik *f*
telephone circuit Telefonleitung *f*
telephotography Fernphotographie *f*,
 Telephotographie *f*
telephotometry Telephotometrie *f*
teleprinter Fernschreiber *m*
telescope Teleskop *n*, Fernrohr *n*
telescope *v* ineinanderschieben
telescope carriage Teleskopwagen *m*
telescoped ineinandergeschachtelt
telescopic teleskopisch, ausziehbar,
 zusammenlegbar
telescopic excavator Teleskopbagger *m*
telescopic tube Ausziehröhre *f*, Ausziehrohr *n*
telescoping ineinandergreifend
teletherapy Ferntherapie *f*, Teletherapie *f*
telethermometer Fernthermometer *n*
television Fernsehen *n*
television set Fernsehapparat *m*; Fernsehgerät *n*
television tube Bildröhre *f* (Telev),
 Fernsehröhre *f*
telex Fernschreibnetz *n*
telex *v* kabeln
telfairic acid Telfairiasäure *f*
telltale gauge Axiometer *n*
telltale light Kontrollampe *f*
tellurate Tellurat *n*, tellursaures Salz *n*
tellurhydrate Hydrotellurid *n*
telluric acid Tellursäure *f*, **salt of** ~ Tellurat *n*
telluric acid hydrate Tellursäurehydrat *n*
telluric anhydride Tellurtrioxid *n*
telluric compound Tellur(VI)-Verbindung *f*

telluride Tellurid *n*
telluriferous tellurführend, tellurhaltig
tellurite Tellurit *m* (Min), Salz *n* der tellurigen Säure (Chem), Tellurit *n* (Chem)
tellurium (Symb. Te) Tellur *n*
tellurium compound Tellurverbindung *f*
tellurium content Tellurgehalt *m*
tellurium dioxide Tellurdioxid *n*, Tellurigsäureanhydrid *n*
tellurium glance Nagyagit *m* (Min), Tellurglanz *m* (Min)
tellurium monoxide Tellurmonoxid *n*
tellurium ore Tellurerz *n* (Min)
tellurium salt Tellursalz *n*
tellurium trioxide Tellurtrioxid *n*
tellurocyanic acid Tellurcyansäure *f*
tellurophenol Tellurophenol *n*
tellurous tellurig (Chem)
tellurous acid tellurige Säure *f*, **salt of** ~ Tellurit *n*
tellurous acid anhydride Tellurdioxid *n*
tellurous compound Tellur(IV)-Verbindung *f*
teloidine Teloidin *n*
teloidinone Teloidinon *n*
telomer Telomer[e] *n*
telomerization Telomerisation *f*
telomerize *v* telomerisieren
teloschistin Teloschistin *n*
temiscamite Temiskamit *m* (Min)
temper Härte *f* (Met), Härtegrad *m* (Met), **degree of** ~ Härtegrad *m*, ~ **of steel** Härte *f* des Stahles
temper *v* tempern (Metall), abschrecken (Metall), adoucieren (Gußeisen), anlassen (Metall), ausglühen (Metall), härten (Metall), löschen (Metall), rasch abkühlen (Glas)
temperable härtbar
temperate gemäßigt, mäßig
temperature Temperatur *f*, **absolute** ~ absolute Temperatur, **ambient** ~ Umgebungstemperatur *f*, **average** ~ mittlere Temperatur, **change of** ~ Temperaturänderung *f*, **characteristic** ~ Eigentemperatur *f*, **constancy of** ~ Temperaturkonstanz *f*, **critical** ~ kritische Temperatur, **drop in** ~ Temperaturabfall *m*, **elevated** ~ erhöhte Temperatur, **equalization of** ~ Temperaturausgleich *m*, **excess** ~ Übertemperatur *f*, **fluctuation in** ~ Temperaturschwankung *f*, **influence of** ~ Temperatureinfluß *m*, **initial** ~ Anfangstemperatur *f*, **internal** ~ Innentemperatur *f*, **line of constant** ~ Isotherme *f*, **low** ~ Tieftemperatur *f*, **lowering of** ~ Temperatursenkung *f*, **lowest** ~ Tiefsttemperatur *f*, **maximum** ~ maximale Temperatur, **mean** ~ Durchschnittstemperatur *f*, **mittlere** Temperatur, **minimum** ~ niedrigste

Temperatur, Temperaturminimum *n*, **outdoor** ~ Außentemperatur *f*, **outside** ~ Außentemperatur *f*, **rated** ~ Nenntemperatur *f*, **rise of** ~ Temperaturanstieg *m*, **variation of** ~ Temperaturänderung *f*, ~ **of reaction** Reaktionstemperatur *f*
temperature adjustment Temperatureinstellung *f*
temperature alarm device Temperaturalarmvorrichtung *f*
temperature bulb Wärmefühler *m*
temperature coefficient Temperaturkoeffizient *m*, Wärmekoeffizient *m*
temperature color scale Farbtemperaturskala *f* (Opt)
temperature compensation Temperaturausgleich *m*
temperature control Temperaturregler *m*, Temperaturwächter *m*
temperature decrease Temperaturabnahme *f*
temperature dependence Temperaturabhängigkeit *f*
temperature dependent temperaturabhängig
temperature drop Temperaturabnahme *f*, Temperaturerniedrigung *f*, Temperaturgefälle *n*
temperature effect Temperatureinfluß *m*, Temperaturbeeinflussung *f*, Temperatureffekt *m*
temperature gradient Temperaturgradient *m*, Wärmegefälle *n*
temperature increase Temperaturerhöhung *f*, Temperaturzunahme *f*
temperature-independent temperaturunabhängig
temperature indicating paint Anlauffarbe *f* (Met)
temperature influence Temperaturbeeinflussung *f*
temperature interval Temperaturintervall *n*
temperature ionization Wärmeionisation *f*
temperature jump Temperatursprung *m*
temperature limit Temperaturgrenze *f*
temperature measurement Temperaturmessung *f*
temperature observation Temperaturbeobachtung *f*
temperature range Temperaturbereich *m*
temperature recorder Registrierthermometer *n*, Temperaturschreiber *m*
temperature reduction Temperatursenkung *f*
temperature registration device Temperaturregistrierapparat *m*
temperature regulator Temperaturregler *m*
temperatures, resistant to high ~ hochtemperaturbeständig
temperature scale Temperaturskala *f*
temperature setting Temperatureinstellung *f*
temperature tracer Temperaturfühler *m*
temperature variation Temperaturschwankung *f*
temper carbon Härtungskohle *f*, Temperkohle *f*
temper color Anlaßfarbe *f*, Anlauffarbe *f*, Glühfarbe *f*

tempered gehärtet (Met)
temper etching Anlaßätzung *f*
temper hardening Anlaßhärtung *f*
tempering Tempern *n* (Met), Anlassen *n* (Met),
  Glühen *n* (Met), Härten *n* (Met),
  Temperaturbehandlung *f* (Met), **temperature
  of** ~ Anlaßtemperatur *f*
tempering agent Härtemittel *n*, Tempermittel *n*
tempering bath Härtebad *n*
tempering brittleness Blaubrüchigkeit *f* (Metall)
tempering color Anlaßfarbe *f*
tempering flame furnace Härteflammofen *m*
tempering forge Härteschmiedefeuer *n*
tempering furnace Härteofen *m*,
  Adoucierofen *m*, Anlaßofen *m*
tempering hardness Anlaßhärte *f*
tempering liquid Ablöschflüssigkeit *f*,
  Härteflüssigkeit *f*
tempering plant Härteanlage *f*
tempering powder Härtepulver *n*
tempering quality Vergütbarkeit *f*
tempering salt Anlaßsalz *n*
tempering stove Härteofen *m*
tempering temperature Anlauftemperatur *f*,
  Nachglühtemperatur *f*
tempering water Löschwasser *n* (Met)
temper jacket Temperiermantel *m*
template Matrize *f* (Biochem), Schablone *f*,
  dividing ~ Teilschablone *f*
template pipe Paßrohr *n*
template reaction Matrizenreaktion *f*
Temple's estimation of eigenvalues
  Einschließungssatz *m* (v. Temple)
templet Schablone *f*, Lehre *f*, Leitblech *n*
  (Techn)
tempo Tempo *n*, Geschwindigkeit *f*
temporary kurzzeitig, provisorisch, temporär,
  vorübergehend
tenacious bruchfest (Phys), festhaltend, zäh
tenacity Zähigkeit *f*, Bruchfestigkeit *f* (Phys),
  Klebrigkeit *f*, Reißfestigkeit *f* (Phys),
  Zähfestigkeit *f* (Phys)
tendency Tendenz *f*, Neigung *f*, **downward** ~
  fallende Tendenz, ~ **to crystallize**
  Kristallisationsneigung *f*
tender Lieferangebot *n*, **invitation for** ~
  Anfrage *f*
tender *a* weich, zart
tendon Sehne *f* (Anat)
tennantite Tennantit *m* (Min), Arsenfahlerz *n*
  (Min), Graukupfererz *n* (Min)
tenon Zapfen *m* (Zimmerei)
tenoning machine Zapfenschneidmaschine *f*
tenorite Tenorit *m* (Min)
tenosine Tenosin *n*
tense gespannt
tensibility Spannbarkeit *f*
tensible dehnbar
tensile dehnbar

tensile and compressive stress Zug- und
  Druckspannung *f*
tensile bar Zugstab *m*
tensile creep test Zeitstand-Zugversuch *m*
tensile fatigue test Zugermüdungsversuch *m*
tensile impact strength Zugschlagzähigkeit *f*
tensile ring Zerreißring *m*
tensile shear stress Scherzugversuch *m*
tensile shock test Schlagzerreißversuch *m*
tensile strain Beanspruchung *f* auf Zug
tensile strength Dehnfestigkeit *f*, Reißfestigkeit *f*,
  Zugfestigkeit *f*, ~ **at low temperature**
  Kältebruchfestigkeit *f*
tensile strength test Zugfestigkeitsuntersuchung *f*
tensile stress Beanspruchung *f* auf Zug,
  Reckspannung *f*, Zugbeanspruchung *f*,
  resistance to ~ Zerreißfestigkeit *f*, **ultimate** ~
  Zerreißbelastung *f*, ~ **at yield**
  Streckspannung *f*
tensile test Zerreißprobe *f*, Reckprobe *f*,
  Zugprobe *f*, **hot** ~ Warmzerreißprobe *f*
tensile yield Streckgrenze *f*
tensiometer Spannungsprüfeinrichtung *f*
  (Techn)
tension Spannung *f*, Zugkraft *f*, Zugspannung *f*,
  degree of ~ Spannungsgrad *m*, **free of** ~
  spannungsfrei, **without** ~ spannungslos, ~
  due to pressure Druckspannung *f*
tension and compression test Zugdruckversuch *m*
tension brittleness Spannsprödigkeit *f*
tension cleaving Zugspaltung *f*
tension-compression fatigue testing machine
  Zugdruckdauerprüfmaschine *f*
tension difference Spannungsunterschied *m*
  (Techn)
tension indicator Spannungsmesser *m*
tensioning roll Zugwalze *f*
tension resonance Spannungsresonanz *f*
tension ring Stellring *m*
tension screw Spannbügel *m*
tension spring Spannungsfeder *f*, Zugfeder *f*
tension test Zerreißprobe *f*, **static** ~
  Zugversuch *m* mit ruhender Last mit
  statischer Beanspruchung
tension tester Zerreißmaschine *f*
tension testing Zerreißprüfung *f*
tension weight Spanngewicht *n*
tensor Tensor *m*
tensor algebra Tensoralgebra *f*
tensor condition Tensorbedingung *f*
tensor density Tensordichte *f*
tensor force Tensorkraft *f*
tentative probeweise, versuchsweise
tentative method Versuchsmethode *f*
tentative standard Normenentwurf *m*
tenter Spannrahmen *m*
tenter *v* spannen (Appretur)
tenter dryer Bahnentrockner *m*,
  Spannrahmentrockner *m*

tentering Spannen *n*
tentering frame Spannrahmen *m*
tenth [part] Zehntel *n*
tenuazonic acid Tenuazonsäure *f*
tenuiorin Tenuiorin *n*
tephroite Tephroit *m* (Min)
tephrosic acid Tephrosinsäure *f*
tephrosin Tephrosin *n*
tepid lauwarm
teraconic acid Teraconsäure *f*
terapinic acid Terapinsäure *f*
teratolite Teratolith *m* (Min)
terbium (Symb. Tb) Terbium *n*
terbutol resin Terbutolharz *n*
terebene Tereben *n*
terebic acid Terpentinsäure *f*
terebinic acid Terpentinsäure *f*
terecamphene Terecamphen *n*
terephthal aldehyde Terephthalaldehyd *m*
terephthal green Terephthalgrün *n*
terephthalic acid Terephthalsäure *f*
terephthalonic acid Terephthalonsäure *f*
terephthalophenone Terephthalophenon *n*
terephthalopinacone Terephthalopinakon *n*
terephthalyl alcohol Terephthalalkohol *m*
teresantalic acid Teresantalsäure *f*
teresantalol Teresantalol *n*
terlinguaite Terlinguait *m* (Min)
term Fachausdruck *m*, Ausdruck *m* (Math),
  Bedingung *f*, Glied *n* (Math), Niveaustufe *f*
  (Atom), **anomalous ~** anomaler Term
  (Spektr), **broader ~** Sammelbegriff *m*,
  collective ~ Sammelbegriff *m*, **displaced ~**
  anomaler Term (Spektr), **metastable ~**
  metastabiler Term
term diagram Termschema *n*
term energy Termenergie *f*
termierite Termierit *m* (Min)
terminal Abzweigklemme *f* (Elektr),
  Anschlußklemme *f* (Elektr), Klemme *f*
  (Elektr)
terminal *a* endständig (Chem)
terminal board Klemmbrett *n*
terminal clamp Anschlußklemme *f*
terminal compound Grenzverbindung *f*
terminal group Endgruppe *f*
terminal group analysis
  Endgruppenbestimmung *f*
terminal head Kabelendverschluß *m* (Elektr)
terminal potential difference Klemmspannung *f*
terminal screw Klemmschraube *f* (Elektr)
terminal strip Lüsterklemme *f*
terminal velocity Endgeschwindigkeit *f*,
  Schwebegeschwindigkeit *f*
terminal voltage Klemmspannung *f*
terminate *v* ablaufen (zeitlich), begrenzen
  (räumlich)
terminating reaction Abbruchreaktion *f*
term influence Termbeeinflussung *f*

terminolic acid Terminolsäure *f*
terminology Terminologie *f*, Fachausdrücke *m pl*
terminus Endpunkt *m*
terms of delivery or supply Lieferbedingungen *f*
  pl
term splitting Termaufspaltung *f*
term value Termlage *f*, Termwert *m*
ternary ternär, dreistoffig (Chem), dreizählig
ternary alloy Dreistofflegierung *f*
ternary mixture Dreistoffgemisch *n*
ternary system Dreistoffsystem *n*,
  Dreistoffgemisch *n*
terpacid Terpacid *n*
terpane Terpan *n*
terpene Terpen *n*
terpene chemistry Terpenchemie *f*
terpene-free terpenfrei
terpene group Terpengruppe *f*
terpeneless terpenfrei
terpene-like terpenähnlich
terpenic acid Terpensäure *f*
terpenolic acid Terpenolsäure *f*
terpenylic acid Terpenylsäure *f*
terphenyl Terphenyl *n*
terpilene Terpilen *n*
terpilene hydride Terpilenhydrid *n*
terpin[e] Terpin *n*
terpine hydrate Terpinhydrat *n*
terpinene Terpinen *n*
terpineol Terpineol *n*
terpinol Terpinol *n*
terpinolene Terpinolen *n*
terpinylene Terpilen *n*
terra cotta Terrakotta *f* (Keram)
terra japonica gelbes Katechin *n* (Färb)
terrazzo Terrazzo *m*, Zementmosaik *n*
terreic acid Terreinsäure *f*
terrein Terrein *n*
terrestrial irdisch
terrestrial gamma radiation
  Erdgammastrahlung *f*
terrestrial magnetism Erdmagnetismus *m*
terrestric acid Terrestrinsäure *f*
terry cloth Frottee *n* (Text)
terrylene Terrylen *n* (HN)
tertiary tertiär
tesseral tesseral, isometrisch (Krist)
test Test *m*, Untersuchung *f*, Experiment *n*,
  Nachweis *m*, Probe *f*, Prüfung *f*, Versuch *m*,
  duration of ~ Versuchsdauer *f*, **small scale ~**
  Kleinversuch *m*
test *v* [aus]probieren, erproben, testen,
  untersuchen
testable prüfbar
test acid Titriersäure *f*
testane Testan *n*
test ashes Kapellenasche *f*, Klärstaub *m*
test bar Probestab *m*, Versuchsstab *m*
test bench Rollenprüfstand *m*

test certificate Prüfschein *m*, Prüfungszeugnis *n*
test cock Probierhahn *m*
test condition Versuchsbedingung *f*
test conditions Abnahmebedingungen *pl*
test data Prüfdaten *pl*
test distribution curve Testverteilung *f* (Statist)
tested bewährt, erprobt
tester Prüfer *m*, Eichmeister *m*, Prüfapparat *m*,
　Prüfgerät *n*
test evaluation Versuchsauswertung *f*
test gas Meßgas *n*
test glass Reagensglas *n*
test head Meßkopf *m*
testify *v* beglaubigen
testimony Zeugnis *n*
testing Probenahme *f*, Prüfwesen *n;* Revision *f*,
　object submitted for ~ Prüfstück *n*
testing apparatus Prüfapparat *m*,
　Bestimmungsapparat *m*
testing arrangement Prüfvorrichtung *f*
test[ing] bench Prüfstand *m*
testing equipment Prüfgerät *n*,
　Untersuchungsgerät *n*
testing laboratory Prüfanstalt *f*,
　Prüflaboratorium *n*
test[ing] method Prüfverfahren *n*, Prüfungsart *f*
testing of material Materialprüfung *f*,
　nondestructive ~ zerstörungsfreie
　Werkstoffprüfung *f*
testing pendulum Prüfpendel *n*
testing plant Versuchsanlage *f*
testing position Prüfstellung *f*
testing pressure Prüfdruck *m*
testing process Prüfverfahren *n*
testing rules Prüfungsvorschriften *pl*
test[ing] stand Prüfstand *m*
testing time Probezeit *f*
testing transformer Prüftransformator *m*
testing voltage Prüfspannung *f* (Elektr)
test load Probebelastung *f*, Versuchslast *f*
test material Versuchsmaterial *n*
test model Versuchsmodell *n*
test mouse Versuchsmaus *f*
testosterone Testosteron *n*
test paper Probepapier *n*, Reagenspapier *n*,
　Reagenzpapier *n*
test piece Probestück *n*, Prüfling *m* (Objekt),
　Versuchsstück *n*
test pieces, selection of ~ Probeauswahl *f*
test pressure Probedruck *m*
test procedure Meßverfahren *n*
test protocol Versuchsprotokoll *n*
test reactor Versuchsreaktor *m*
test report Prüfbericht *m*, Prüfungsbericht *m*,
　Versuchsbericht *m*
test result Prüfergebnis *n*, Prüfungsergebnis *n*,
　Versuchsergebnis *n*
test rod Probestab *m*, Versuchsstab *m*
test run Probelauf *m*

tests, series of ~ Untersuchungsreihe *f*
test sample Probe *f*
test specification Prüfbestimmung *f*
test strip Probestreifen *m*
test strip apparatus Teststreifenapparatur *f*
test time Laufzeit *f*, Versuchsdauer *f*
test tube Reagenzglas *n*
test-tube brush Reagenzglasbürste *f*
test-tube holder Reagenzglashalter *m*
test-tube rack Reagenzglasgestell *n*
test tube research Reagenzglasversuch *m*
test tube stand Reagenzglasgestell *n*
test value Meßwert *m*
test weight Probegewicht *n*
test weld Probeschweißung *f*
test wire Prüfdraht *m*
tetanine Tetanin *n*
tetanthrene Tetanthren *n*
tetanthrenone Tetanthrenon *n*
tetanus Tetanus *m* (Med), Starrkrampf *m*
　(Med), Wundstarrkrampf *m* (Med)
tetanus antitoxin Tetanusantitoxin *n*
tetanus bacillus Tetanusbazillus *m*
tetanus inoculation Tetanusschutzimpfung *f*
　(Med)
tetanus toxin Tetanustoxin *n*
tetany Tetanie *f* (Med)
tetartohedron Tetartoeder *n* (Krist)
tetraacetyldextrose Tetraacetyldextrose *f*
tetrabarbital Tetrabarbital *n*
tetrabase Tetrabase *f*
tetrabasic vierbasig, vierbasisch
tetraborane Borbutan *n*
tetraboric acid Tetraborsäure *f*
tetrabromoethane Acetylentetrabromid *n*,
　Tetrabromäthan *n*
tetrabromostearic acid Tetrabromstearinsäure *f*
tetracaine Tetracain *n*
tetracarbinol Tetracarbinol *n*
tetracarboxybutane Butantetracarbonsäure *f*
tetracene Tetracen *n*
tetrachlorobenzene Tetrachlorbenzol *n*
tetrachlor[o]ethane Acetylentetrachlorid *n*,
　Tetrachloräthan *n*
tetrachloromethane Chlorkohlenstoff *m*,
　Kohlenstofftetrachlorid *n*
tetrachloro-p-benzoquinone Chloranil *n*
tetrachloroquinol Chloranol *n*
tetrachloroquinone Tetrachlorchinon *n*
tetrachlorosilane Tetrachlorsilicium *n*
tetracontane Tetrakontan *n*
tetracosane Tetrakosan *n*
tetracovalent tetrakovalent
tetracyclone Tetracyclon *n*
tetradecane Tetradecan *n*
tetradecanoic acid Myristinsäure *f*
tetradecanol, 1- ~ Myristylalkohol *m*
tetradentate vierzahnig

tetradymite Tellurwismut *n* (Min),
Tetradymit *m* (Min)
tetraethylene pentamine Tetraäthylenpentamin *n*
tetraethyl lead Bleitetraäthyl *n*
tetrafluoromethane Fluorkohlenstoff *m*
tetrafunctional tetrafunktionell
tetragonal tetragonal
tetrahedral tetraedrisch, vierflächig
tetrahedral symmetry Tetraedersymmetrie *f*
tetrahedrite Tetraedrit *m* (Min), Fahlerz *n*
(Min), Graukupfererz *n* (Min),
Kupferfahlerz *n* (Min), ~ **containing mercury**
Quecksilberfahlerz *n*
tetrahedron Tetraeder *n* (Krist), Vierflächner *m*
(Krist)
tetrahexacontane Tetrahexakontan *n*
tetrahydrate Tetrahydrat *n*
tetrahydroabietic acid Tetrahydroabietinsäure *f*
tetrahydrobiopterin Tetrahydrobiopterin *n*
tetrahydrofolic acid Tetrahydrofolsäure *f*
tetrahydroform Trimethylenimin *n*
tetrahydrofuran Tetrahydrofuran *n*
tetrahydrofurfuryl alcohol
Tetrahydrofurfurylalkohol *m*
tetrahydrolilole Lilolidin *n*
tetrahydronaphthalene Tetralin *n* (HN)
tetrahydropyrrol[e] Pyrrolidin *n*,
Tetrahydropyrrol *n*
tetrahydroquinoline Tetrahydrochinolin *n*
tetrahydroserpentine Tetrahydroserpentin *n*
tetrahydrothiophene Butylensulfid *n*
tetrahydroxyanthraquinone, 1,2,5,6- ~
Rufiopin *n*
tetrahydroxyflavone Luteolin *n*
tetraiodoethylene Dijodoform *n*
tetraiodofluorescein Jodeosin *n*
tetraiodophenolphthalein Jodophen *n*,
Nosophen *n*
tetraiodopyrrole Jodol *n*
tetrakishexahedron Tetrakishexaeder *n* (Krist)
tetralin[e] Tetralin *n* (HN)
tetralite Tetryl *n*
tetralol Tetralol *n*
tetralone Tetralon *n*
tetralupine Tetralupin *n*
tetrameric tetramer
tetramethyl-3-hydroxybenzene, 1,2,4,5- ~
Durenol *n*
tetramethyl alloxantine Amalinsäure *f*
tetramethylammonium hydroxide
Tetramethylammoniumhydroxid *n*
tetramethylammonium iodide
Tetramethylammoniumjodid *n*
tetramethylated base Tetramethylbase *f*
tetramethylbenzene, 1,2,4,5- ~ Durol *n*
tetramethylene Tetramethylen *n*
tetramethylene bromide Tetramethylenbromid *n*
tetramethylene diamine Tetramethylendiamin *n*,
Putrescin *n*

tetramethylene glycol Tetramethylenglykol *n*
tetramethylethylene Tetramethyläthylen *n*
tetramethylethyleneglycol Pinakol *n*
tetramethylglucose Tetramethylglucose *f*
tetramethylmethane Tetramethylmethan *n*
tetramic acid Tetramsäure *f*
tetramine Tetramin *n*
tetrammine copper(II) sulfate
Kupfertetramminsulfat *n*
tetranitrochrysazin Tetranitrochrysazin *n*
tetranitromethane Tetranitromethan *n*
tetranitromethylaniline Tetranitromethylanilin *n*
tetranitrophenol Tetranitrophenol *n*
tetranthrimide Tetranthrimid *n*
tetraphene Tetraphen *n*
tetraphenoxysilane Tetraphenoxysilan *n*
tetraphenylene Tetraphenylen *n*
tetraphenylethylene Tetraphenyläthylen *n*
tetraphenylfurfurane Lepiden *n*
tetraphenylhydrazine Tetraphenylhydrazin *n*
tetraphenylmethane Tetraphenylmethan *n*
tetraphenyl-p-pyrimidine Amaron *n*
tetraphenyl pyrazine Amaron *n*
tetrasodium pyrophosphate
Tetranatriumpyrophosphat *n*
tetrasymmetric tetrasymmetrisch
tetrathionate Tetrathionat *n*
tetrathionic acid Tetrathionsäure *f*, **salt of** ~
Tetrathionat *n*
tetratomic vieratomig
tetratriacontane Tetratriakontan *n*
tetravalence Vierwertigkeit *f*
tetravalent vierwertig
tetrazene Tetrazen *n*
tetrazine Tetrazin *n*
tetrazole Tetrazol *n*
tetrazone Tetrazon *n*
tetrazotic acid Tetrazotsäure *f*
tetrinic acid Tetrinsäure *f*
tetrode Tetrode *f*
tetrol Furan *n*
tetrolic acid Tetrolsäure *f*
tetronal Tetronal *n*
tetronic acid Tetronsäure *f*
tetrose Tetrose *f*
tetroxane Tetroxan *n*
tetroxide Tetroxid *n*
tetryl Tetryl *n*
tetrylammonium Tetrylammonium *n*
tetryzoline Tetryzolin *n*
tetuin Tetuin *n*
teucrin Teucrin *n*
texasite Texasit *m* (Min)
textile Gewebe *n*
textile auxiliary Textilhilfsmittel *n*
textile coating Stofflack *m* (Text)
textile dressing Textilvered[e]lung *f*
textile fiber Textilfaser *f*
textile finishing Textilvered[e]lung *f*

textile finishing agent Textilvered[e]lungsmittel *n*
textile industry Textilindustrie *f*
textile printer Tuchdrucker *m* (Text)
textile printing Zeugdruck *m*
textile printing plant Zeugdruckerei *f* (Text)
textiles Textilien *pl*
textile thread Textilfaser *f*
textilite yarn Textilit *n*
texture Gewebe *n*, Faserung *f*, Gefüge *n*,
    Struktur *f*, Textur *f*, change of ~
    Strukturveränderung *f*, fine ~ Kleingefüge *n*,
    homogeneous ~ gleichmäßiges Gefüge,
    portion of the ~ Gefügebestandteil *m*
textured plaster Reibeputz *m*
thalamus Fruchtboden *m* (Bot)
thalenite Thalenit *m* (Min)
thalidomide Thalidomid *n*, Contergan *n*
thalleioquinoline Thalleiochinolin *n*
thallic chloride Thallium(III)-chlorid *n*
thallic compound Thallium(III)-Verbindung *f*
thallic iodide Thallium(III)-jodid *n*
thallic ion Thallium(III)-ion *n*
thallic oxide Thallium(III)-oxid *n*, Thallioxid *n*
thalline Thallin *n*
thalline sulfate Thallinsulfat *n*
thallium (Symb. Tl) Thallium *n*
thallium alum Thallium[aluminium]alaun *m*
thallium chloride Thalliumchlorid *n*
thallium content Thalliumgehalt *m*
thallium formate Thalliumformiat *n*
thallium glass Thalliumglas *n*
thallium iodide Thalliumjodid *n*
thallium malonate Thalliummalonat *n*
thallium monobromide Thallium(I)-bromid *n*
thallium monochloride Thallium(I)-chlorid *n*,
    Thallochlorid *n*
thallium monoiodide Thallium(I)-jodid *n*
thallium oxide Thalliumoxid *n*
thallium selenate Thalliumselenat *n*
thallium sesquioxide Thallium(III)-oxid *n*
thallium sulfide Thalliumsulfid *n*
thallium trichloride Thallium(III)-chlorid *n*
thallium triiodide Thallium(III)-jodid *n*
thallous aluminum sulfate
    Thallium[aluminium]alaun *m*
thallous bromide Thallium(I)-bromid *n*
thallous carbonate Thallium(I)-carbonat *n*
thallous chlorate Thallochlorat *n*,
    Thallium(I)-chlorat *n*
thallous chloride Thallium(I)-chlorid *n*,
    Thallochlorid *n*
thallous compound Thalloverbindung *f*,
    Thallium(I)-Verbindung *f*
thallous hydroxide Thallium(I)-hydroxid *n*
thallous iodate Thallojodat *n*
thallous iodide Thallium(I)-jodid *n*
thallous ion Thalloion *n*, Thallium(I)-ion *n*
thallous oxide Thallium(I)-oxid *n*, Thallooxid *n*
thallous sulfide Thallium(I)-sulfid *n*

thalmine Thalmin *n*
thamnolic acid Thamnolsäure *f*
thanatol Guäthol *n*
thapsic acid Thapsiasäure *f*
thaumasite Thaumasit *m* (Min)
thaw *v* [auf]tauen, ~ off abtauen
thawing Tauen *n*
thawing agent Auftaumittel *n*
thaw point Taupunkt *m*
theanine Theanin *n*
thebaine Paramorphin *n*, Thebain *n*
thebaine hydrochloride Thebainhydrochlorid *n*
thebenine Thebenin *n*
theine Koffein *n*, Thein *n*
thelephoric acid Thelephorsäure *f*
thenaldehyde Thenaldehyd *m*
thenalidine Thenalidin *n*
thenardite Thenardit *m* (Min)
Thenard's blue Kobaltblau *n*, Thenards Blau *n*
thenoic acid Thenoesäure *f*
thenoylacetone Thenoylaceton *n*
thenylamine Thenylamin *n*
theobromine Theobromin *n*
theobromine sodium formate Theophorin *n*
theobromine sodium iodide Eustenin *n*
theobromine sodium salicylate Diuretin *n*
theocine Theocin *n*
theodolite Theodolit *m*
theolactine Theolactin *n*
theophorine Theophorin *n*
theophylline Theophyllin *n*, Theocin *n* (HN)
theorem Lehrsatz *m*, Satz *m*, ~ of impulse
    Impulssatz *m* (Math)
theoretic[al] theoretisch
theory Lehre *f*, Theorie *f*, ~ of combinations
    Kombinatorik *f*, ~ of relativity
    Relativitätstheorie *f*
theory of errors Fehlerrechnung *f*
theory of fractures Bruchvorgang *m*
theory of gases, kinetic ~ kinetische Gastheorie *f*
theralite Theralith *m* (Geol)
therapeutic Therapeutikum *n*
therapeutic *a* therapeutisch
therapeutic agent Heilmittel *n* (Med)
therapeutic[al] heilkräftig
therapeutic effect Heilwirkung *f*
therapeutic purpose Heilzweck *m*
therapeutics Therapeutik *f* (Med), Therapie *f*
    (Med)
therapeutic value Heilkraft *f*
therapist Therapeut *m*
therapy Therapie *f* (Med)
there and back hin und zurück
thermal thermisch
thermal absorption Wärmeabsorption *f*
thermal agitation Wärmebewegung *f*
thermal agitation noise Wärmerauschen *n*
thermal analysis Thermoanalyse *f*
thermal balance Wärmebilanz *f*

thermal breakdown Wärmedurchschlag *m*
thermal circuit breaker Thermoschalter *m*
thermal conduction Wärmeleitung *f*
thermal conductivity Wärmeleitfähigkeit *f,*
Wärmeleitfähigkeitskoeffizient *m,*
Wärmeleitvermögen *n,* **coefficient of** ~
Wärmeleitzahl *f,* spezifisches
Wärmeleitvermögen *n* (Therm)
thermal-conductivity cell
Wärmeleitfähigkeitsdetektor *m*
thermal-conductivity gauge
Wärmeleitungsmanometer *n*
thermal cycle Wärmekreislauf *m*
thermal cycling test Temperaturwechseltest *m*
thermal degradation or decomposition
thermischer Abbau *m*
thermal diffusion Thermodiffusion *f,*
Wärmeausbreitung *f*
thermal diffusion ratio
Thermodiffusionsverhältnis *n*
thermal diffusivity thermisches
Diffusionsvermögen *n* (Therm),
Wärmeausbreitungsvermögen *n*
thermal efficiency Wärmewirkungsgrad *m,*
Wärmeausbeute *f*
thermal energy Wärmeenergie *f*
thermal [energy] yield Wärmeausbeute *f*
thermal equilibrium Wärmegleichgewicht *n*
thermal excitation thermische Anregung *f* (Phys)
thermal expansion Wärmeausdehnung *f,*
coefficient of ~
Wärmeausdehnungskoeffizient *m*
thermal flow Wärmefluß *m*
thermal force Thermokraft *f*
thermal impulse welding
Wärmeimpulsschweißen *n*
thermal insulation Wärmeisolierung *f*
thermal insulation board Wärmeschutzplatte *f*
thermal ionization Wärmeionisation *f*
thermal phenomenon Wärmephänomen *n*
thermal power Heizwert *m*
thermal power station Wärmekraftanlage *f*
thermal process Wärmevorgang *m*
thermal property Temperaturverhalten *n*
thermal radiation Wärmestrahlung *f*
thermal refinement Vergütung *f* (Stahl)
thermal relay Thermorelais *n*
thermal resistance Wärmedämmwert *m,*
Wärmedurchlaßwiderstand *m,*
Wärme[leit]widerstand *m*
thermal shield Wärmeschutzvorrichtung *f*
thermal shock test Hitzeschockprobe *f*
thermal spalling resistance
Temperaturwechselbeständigkeit *f*
thermal spectrum Wärmespektrum *n*
thermal spring Thermalquelle *f,* Therme *f*
thermal stability Wärmebeständigkeit *f,*
Wärmestabilität *f*
thermal stimulus Wärmereiz *m*

thermal stratification Temperaturschichtung *f*
thermal stress Wärmespannung *f*
thermal sulfur bath Thermalschwefelbad *n*
thermal switch Wärmeschalter *m*
thermal transmission Hitzedurchlässigkeit *f*
thermal transmittance
Wärmedurchgangskoeffizient *m,*
Wärmedurchgangszahl *f*
thermal treatment Wärmebehandlung *f*
thermal unit Wärmeeinheit *f,* Kalorie *f*
thermal water Thermalwasser *n*
thermic thermisch, kalorisch
thermic protection Wärmeschutz *m*
thermion Thermion *n,* Wärmeion *n*
thermionic thermionisch, glühelektrisch
thermionic amplifier Glühkathodenverstärker *m*
thermionic constant Glühemissionskonstante *f*
thermionic current Thermionenstrom *m*
thermionic discharge Glühelektronenentladung *f*
thermionic electron source
Glühelektronenquelle *f*
thermionic emission thermische Emission *f,*
Glühelektronenemission *f,*
Thermionenaussendung *f*
thermionic emission property
Glühemissionseigenschaft *f*
thermionic tube Glühkathodenröhre *f*
thermistor Thermistor *m* (Elektr)
thermit[e] Thermit *n*
thermite process Thermitverfahren *n*
thermite welding aluminothermische
Schweißung *f,* Thermitschweißung *f*
thermit welding of aluminum
Alutherm-Verfahren *n*
thermo-ammeter Thermoamperemeter *n*
thermobarometer Thermobarometer *n*
thermobattery Thermobatterie *f*
thermocautery Thermokaustik *f* (Med)
thermochemical thermochemisch
thermochemistry Thermochemie *f*
thermochrome crayon Thermochromstift *m*
thermochrome paint Thermochromfarbe *f*
thermocolor Thermofarbe *f*
thermocompensator Thermokompensator *m*
thermocompression Thermokompression *f*
thermocouple Thermoelement *n,*
Temperaturfühler *m,* Thermokette *f*
thermocurrent thermoelektrischer Strom *m*
thermocycling test Temperaturwechseltest *m*
thermodiffusion Thermodiffusion *f,*
Wärmediffusion *f*
thermodin Thermodin *n*
thermodynamic thermodynamisch
thermodynamic potential thermodynamisches
Potential *n*
thermodynamics Thermodynamik *f,*
Wärmelehre *f*
thermoelastic thermoelastisch

thermoelastic coefficient
Thermoelastizitätskoeffizient *m* (Therm)
thermoelasticity Thermoelastizität *f*
thermoelectric thermoelektrisch, wärmeelektrisch
thermoelectric couple Thermoelement *n*
thermoelectric current Thermostrom *m*
thermoelectric effect Thermoeffekt *m*,
thermoelektrischer Effekt *m*
thermoelectric force Thermokraft *f*
thermoelectric galvanometer
Thermogalvanometer *n*
thermoelectricity Thermoelektrizität *f*,
Elektrothermie *f*
thermoelectric motor Thermomotor *m*
thermoelectric power Thermokraft *f*
thermoelement Thermoelement *n*,
Temperaturfühler *m*, Thermokette *f*
thermoexcitory temperaturerhöhend,
wärmeerzeugend
thermofixation Thermofixieren *n*
thermoforming Thermoverformung *f*
thermoforming machine Tiefziehmaschine *f*
thermogenerator Wärmeerzeuger *m*
thermogenesis Thermogenese *f*
thermograph Thermograph *m*,
Registrierthermometer *n*
thermohardening hitzehärtend
thermolabile thermolabil, wärmeunbeständig
thermolamp Wärmelampe *f*
thermoluminescence Thermolumineszenz *f*
thermoluminescent thermolumineszent
thermolysis Thermolyse *f* (Chem)
thermolytic thermolytisch
thermomagnetic thermomagnetisch
thermomagnetic motor Thermomotor *m*
thermomagnetism Wärmemagnetismus *m*
thermometer Thermometer *n*, **clinical** ~
Fieberthermometer *n*, **elbow box** ~
Kastenwinkelthermometer *n*, **general purpose**
~ Allgebrauchsthermometer *n*, **recording** ~
aufzeichnendes Thermometer, **remote control**
~ Fernthermometer *n*, **straight** ~
Stockthermometer *n*, **straight box** ~
Kastenstockthermometer *n*, **unsheathed** ~ **on**
scale plate Plattenthermometer *n*
thermometer bulb Termometerkugel *f*
thermometer column Thermometersäule *f*
thermometer mounting Thermometerfassung *f*
thermometer pocket Thermometertasche *f*
thermometer reading Thermometerstand *m*
thermometer scale Thermometerskala *f*
thermometer sheath Thermometerschutzrohr *n*
thermometer tube Thermometerrohr *n*
thermometric thermometrisch
thermometric column Thermometerfaden *m*
thermometric reading Thermometerstand *m*
thermometry Temperaturmessung *f*,
Thermometrie *f*
thermonatrite Thermonatrit *m* (Min)

thermonuclear thermonuklear
thermophil[ic] thermophil (Bakt)
thermophil[ic] bacteria thermophile Bakterien *pl*
thermophone Thermophon *n*
thermophyllite Thermophyllit *m* (Min)
thermophysics Thermophysik *f*
thermopile Thermobatterie *f*
thermoplastic Thermoplast *m*
thermoplastic *a* thermoplastisch, warm
verformbar
thermoplastic adhesive Schmelzklebstoff *m*,
Thermokleber *m*
thermoplasticity Warmverformbarkeit *f*
thermoplastics thermoplastische Werkstoffe *pl*
thermoprene Thermopren *n*
thermopsine Thermopsin *n*
thermoreceptor Thermorezeptor *m*
thermo-recording paint Anlauffarbe *f* (Met)
thermoregulation Temperaturregelung *f*
thermoregulator Thermoregler *m*,
Wärmeregler *m*, Wärmeregulator *m*
thermorelay Thermorelais *n*
thermoresistant thermostabil
thermos [bottle] Thermosflasche *f*
thermoscope Thermoskop *n*
thermoset duroplastisch
thermosetting Hitzehärten *n*, Thermofixieren *n*
thermosetting *a* durch Wärme härtbar,
hitzehärtbar
thermosetting adhesive Schmelzklebstoff *m*
thermosetting plastic Duroplast *n*
thermosiphon principle Thermosiphonprinzip *n*
thermosiphon reboiler Thermosiphonheizung *f*
(Kolonne)
thermostable hitzebeständig, thermoresistent,
thermostabil
thermostat Regler *m* (f. Wärme), Thermostat *m*,
Wärmeregler *m*, **round** ~ Rundthermostat *m*
thermostatic thermostatisch
thermostatic control Thermoregler *m*
thermoswitch Bimetallschalter *m*
thesine Thesin *n*
thesinic acid Thesinsäure *f*
thesis Dissertation *f*, **[doctoral]** ~ Doktorarbeit *f*
theta function Thetafunktion *f*
theta meson Theta-Meson, Theton *n*
theta polarization Theta-Polarisation *f* (Elektr)
theveresin Theveresin *n*
thevetin Thevetin *n*
thevetose Thevetose *f*
thiacetarsamide Thiacetarsamid *n*
thiacetic acid Thiacetsäure *f*
thiachromone Thiachromon *n*
thialdine Thialdin *n*
thiamide Thiamid *n*
thiaminase Thiaminase *f* (Biochem)
thiamin[e] Thiamin *n*, Aneurin *n*, Vitamin B_1 *n*
thiamine hydrochloride Thiaminhydrochlorid *n*
thiamine mononitrate Thiaminmononitrat *n*

thiamine pyrophosphate Thiaminpyrophosphat *n*
thiamorpholine Thiamorpholin *n*
thianaphthene Thianaphthen *n*
thianthrene Thianthren *n*
thiapyran Thiapyran *n*
thiaxanthene Thiaxanthen *n*
thiazane Thiazan *n*
thiazine Thiazin *n*
thiazole Thiazol *n*
thiazolidine Thiazolidin *n*
thiazoline Thiazolin *n*
thiazolyl Thiazolyl-
thiazosulfone Thiazosulfon *n*
thick dick
thicken *v* dick werden, eindampfen, eindicken
thickener Eindicker *m*, Verdickungsmittel *n*
thickening Dickwerden *n*, Eindicken *n*, ~ **by boiling** Einkochen *n*, ~ **by evaporation** Eindampfen *n*, ~ **of oil** Ölverdickung *f*, ~ **of the melt** Eindicken *n* der Badschmelze
thickening agent Verdickungsmittel *n*
thickening machine Eindicker *m*
thickening substance Eindickungsmittel *n*
thick juice Dicksaft *m*
thick mash Dickmaische *f*
thickness Dicke *f*, Stärke *f*, **difference in** ~ Dickenabweichung *f*
thick oil Dicköl *n*
thick-walled dickwandig, starkwandig
thick-walled rubber tubing Druckschlauch *m*
thienyl Thienyl-
thienyl chloride Thienylchlorid *n*
thienylmethylamine Thienylmethylamin *n*
thietane Thietan *n*
thiete Thiet *n*
thiirane Thiiran *n*
thimble Fingerhut *m*
thin dünn
thin *v* verdünnen
thin bed process Dünnbettverfahren *n*
thin-bodied dünnflüssig (Öl)
thin-film evaporator Dünnschichtverdampfer *m*
thin-film heat exchanger Dünnschichtwärmeaustauscher *m*
thin-film oven test Dünnschicht-Ofenversuch *m*
thin-film recitifier Dünnschichtrektifikator *m* (Dest)
thin-film sublimer Dünnfilmsublimator *m*
thin juice Dünnsaft *m* (Zuck)
thin-layer cell Dünnschichtzelle *f*
thin-layer chromatography Dünnschichtchromatographie *f*, **two-dimensional** ~ zweidimensionale Dünnschichtchromatographie *f*
thin-layer electrophoresis Dünnschichtelektrophorese *f*
thin layer screening Dünnschichtsiebung *f*
thinner Verdünner *m*, Verdünnungsmittel *n*
thinness Dünne *f*

thinning Ausfall *m* (Haar), Verdünnen *n*, Verdünnung *f*
thinolite Thinolith *m* (Min)
thin plate Feinblech *n*
thin section Dünnschnitt *m*
thin sheet Feinblech *n*
thin-sheeted dünntafelig
thin-walled dünnwandig
thioacetic acid Thioessigsäure *f*
thioacetone Thioaceton *n*
thioacid Thiosäure *f*, **anhydride of a** ~ Thioanhydrid *n*
thioalcohol Thiol *n*, Schwefelalkohol *m*, Thioalkohol *m*
thioanhydride Thioanhydrid *n*
thioaniline Thioanilin *n*
thioanisol Thioanisol *n*
thioantimonate Thioantimonat *n*
thioantimonic acid Schwefelantimonsäure *f*, Thioantimonsäure *f*
thioarsenate Thioarsenat *n*
thioarsenic acid Thioarsensäure *f*, **salt of** ~ Thioarsenat *n*
thioarsenite Thioarsenit *n*
thioarsenous acid thioarsenige Säure *f*, **salt of** ~ Thioarsenit *n*
thiobacteria Thiobakterien *pl*
thiobarbituric acid Thiobarbitursäure *f*
thiobenzanilide Thiobenzanilid *n*
thiobenzoic acid Thiobenzoesäure *f*
thiocacodylic acid Thiokakodylsäure *f*
thiocarbamide Thiocarbamid *n*, Thioharnstoff *m*
thiocarbanilide Thiocarbanilid *n*
thiocarbazide Thiocarbazid *n*
thiocarbin Thiocarbin *n*
thiocarbinol Thiocarbinol *n*
thiocarbonic acid Thiokohlensäure *f*
thiocarbonyl Thiocarbonyl-
thiocarbonyl chloride Thiophosgen *n*
thiochroman Thiochroman *n*
thiochromene Thiochromen *n*
thiochromone Thiochromon *n*
thiocol Thiocol *n* (HN)
thiocresol Thiokresol *n*
thioctic acid Thioctinsäure *f*
thiocyanate Rhodanid *n*
thiocyanatoauric(I) acid Aurorhodanwasserstoffsäure *f*
thiocyanatoauric(III) acid Aurirhodanwasserstoffsäure *f*, Gold(III)-rhodanwasserstoffsäure *f*
thiocyanic acid Rhodanwasserstoff[säure *f*] *m*, Thiocyansäure *f*, **salt of** ~ Rhodanid *n*
thiocyanin Thiocyanin *n*
thiocyanogen Rhodan *n*
thiocyanogen compound Rhodanverbindung *f*, Thiocyanverbindung *f*
thiodiacetic acid Thiodiessigsäure *f*
thiodiazine Thiodiazin *n*

thiodiazole Thiodiazol *n*
thiodiglycol Thiodiglykol *n*
thiodiglycolic acid Thiodiglykolsäure *f*
thiodiphenylamine Thiodiphenylamin *n*
thiodipropionic acid Thiodipropionsäure *f*
thioether Thioäther *m*
thioflavone Thioflavon *n*
thioform Thioform *n*
thioformanilide Thioformanilid *n*
thioformic acid Thioameisensäure *f*
thiogen black Thiogenschwarz *n*
thiogen dye Thiogenfarbstoff *m*
thiogermanic acid Thiogermaniumsäure *f*
thioglucose Thioglucose *f*
thioglycol Thioglykol *n*
thioglycolic acid Thioglykolsäure *f*
thiohydantoin Thiohydantoin *n*
thioindigo Thioindigo *n*
thioindigo red Thioindigorot *n*
thioindole Thioindol *n*
thioindoxyl Thioindoxyl *n*
thioketone Thioketon *n*
thiokinase Thiokinase *f* (Biochem)
thiokol Thiokol *n* (HN)
thiol Thiol *n*, Mercaptan *n*, Thioalkohol *m*
thiolactic acid Thiomilchsäure *f*
thiolane Thiolan *n*
thiolase Thiolase *f* (Biochem)
thiolic acid Thiolsäure *f*
thiolvaline Thiolvalin *n*
thiolysis Thiolyse *f*
thiolytic thiolytisch
thiomalic acid Thioäpfelsäure *f*
thiomalonic acid Thiomalonsäure *f*
thionalid Thionalid *n*
thionaphthene Benzothiophen *n*, Thionaphthen *n*
thionaphthol Thionaphthol *n*
thionessal Thionessal *n*
thionic acid Thionsäure *f*
thionine Thionin *n*
thionine dye Thioninfarbstoff *m*
thionocarbamic acid Thioncarbaminsäure *f*
thionocarbonic acid Thionkohlensäure *f*
thionuric acid Thionursäure *f*
thionyl Thionyl-
thionyl bromide Thionylbromid *n*
thionyl chloride Thionylchlorid *n*
thiooxindole Thiooxindol *n*
thiooxine Thiooxin *n*
thiopegan Thiopegan *n*
thiophane Thiophan *n*
thiophanthrene Thiophanthren *n*
thiophanthrone Thiophanthron *n*
thiophene Thiophen *n*
thiophenine Thiophenin *n*
thiophenol Thiophenol *n*
thiophosgene Thiophosgen *n*
thiophosphate Thiophosphat *n*
thiophosphoric acid Thiophosphorsäure *f*

thiophthalide Thiophthalid *n*
thiophthene Thiophthen *n*
thiopyran Thiopyran *n*
thioquinanthrene Thiochinanthren *n*
thiosalicylic acid Thiosalicylsäure *f*
thio salt Thiosalz *n*
thiosemicarbazide Thiosemicarbazid *n*
thioserine Thioserin *n*
thiosinamine Thiosinamin *n*
thiostannate Thiostannat *n*
thiostannic acid Thiozinnsäure *f*, **salt of** ~ Thiostannat *n*
thiosulfate Thiosulfat *n*
thiosulfuric acid Thioschwefelsäure *f*
thiotenol Thiotenol *n*
thiotetrabarbital Thiotetrabarbital *n*
thiothiamine Thiothiamin *n*
thiouracil Thiouracil *n*
thiourazole Thiourazol *n*
thiourea Thioharnstoff *m*
thiourea resin Thioharnstoffharz *n*
thiourethane Thiourethan *n*
thiouridine Thiouridin *n*
thioxane Thioxan *n*
thioxanthene Thioxanthen *n*
thioxanthone Thioxanthon *n*
thioxene Thioxen *n*
thioxine Thioxin *n*
thioxine black Thioxinschwarz *n*
thioxole Thioxol *n*
third power dritte Potenz *f* (Math), **to raise to the** ~ kubieren
third wire Nulleiter *m* (Elektr)
thiuram Thiuram *n*
thiuram disulfide Thiuramdisulfid *n*
thixotropic thixotrop
thixotropy Thixotropie *f*
tholloside Thollosid *n*
thollosidic acid Thollosidsäure *f*
Thomas converter Thomasbirne *f* (Metall), Thomaskonverter *m* (Metall)
Thomas low-carbon steel Thomasflußeisen *n* (Metall)
Thomas meal Thomasmehl *n*, Schlackenmehl *n*
Thomas pig iron Thomasroheisen *n*
Thomas process Thomas-Verfahren *n*
Thomas slag Thomasschlacke *f*
Thomas steel Thomasstahl *m*
thomsenolite Thomsenolith *m* (Min)
Thomson effect Thomson-Effekt *m*
thomsonite Thomsonit *m* (Min)
thong Lederriemen *m*
thoria Thorerde *f* (Min), Thoriumoxid *n* (Min)
thorianite Thorianit *m* (Min)
thoriated filament Thoriumfaden *m*
thorite Thorit *m* (Min)
thorium (Symb. Th) Thorium *n*
thorium breeder Thorium-Brutreaktor *m*
thorium compound Thoriumverbindung *f*

thorium content Thoriumgehalt *m*
thorium dioxide Thorerde *f* (Min),
   Thoriumdioxid *n*
thorium emanation Thoriumemanation *f*
thorium nitrate Thoriumnitrat *n*
thorium oxide Thoriumoxid *n*, Thorerde *f* (Min)
thorium precipitate Thoriumniederschlag *m*
thorium radioactive series
   Thorium-Zerfallsreihe *f*
thorium silicate, native hydrated ~ Orangit *m*
   (Min)
thorium sulfate Thoriumsulfat *n*
thorium sulfide Thoriumsulfid *n*
thorn Dorn *m*, Stachel *m*
thorn apple Stechapfel *m* (Bot)
thoron Thoriumemanation *f* (Symb. Tn),
   Thoron *n*
thoroughbred reinrassig (Tier)
thorough[ly] gründlich, gründlich, eingehend
thortveitite Thortveitit *m* (Min)
thousandth [part] Tausendstel *n*
thread Faden *m*, Faser *f*, Gang *m* (Schraube),
   Gewinde *n* (Techn), **cut** ~ geschnittenes
   Gewinde, **depth of** ~ Gewindetiefe *f*, **double**
   ~ doppelgängiges Gewinde, zweigängiges
   Gewinde, **end of the** ~ Gewindeauslauf *m*,
   form of ~ Gewindeprofil *n*, **inside** ~
   Innengewinde *n*, **left-handed** ~ linksgängiges
   Gewinde, **length of** ~ Fadenlänge *f*,
   Gewindelänge *f*, **metric** ~ metrisches
   Gewinde, **milled** ~ gefrästes Gewinde,
   non-reversibility of the ~ Selbsthemmung *f*
   des Gewindes, **pitch of** ~ Gewindesteigung *f*,
   rolled ~ gewalztes Gewinde, **round** ~
   Rundgewinde *n*, **single** ~ einfaches Gewinde
   (eingängig), **triple** ~ dreigängiges Gewinde,
   waxed ~ Wachsfaden *m*
threadbare fadenscheinig
thread base Gewindefuß *m*
thread count Fadenzahl *f*
thread counter Fadenzähler *m* (Text)
thread cutting Gewindeschneiden *n*
thread-cutting lathe Gewindedrehbank *f*
thread depth Gangtiefe *f*
thread diameter Gewindedurchmesser *m*
threaded bush Gewindebuchse *f*
threaded coupling Gewindestück *n*
threaded shaft seal Gewindewellendichtung *f*
thread[ed] sleeve Schraubmuffe *f*
thread electrometer Fadenelektrometer *n*
thread fungus Fadenpilz *m*
thread galvanometer Fadengalvanometer *n*
thread gauge Gewindelehre *f*
thread guide Fadenführer *m*
threading machine Gewindeschneidmaschine *f*
thread-like fadenförmig
thread locking compound
   Schraubensicherungslack *m*
thread mill Gewindefräsmaschine *f*

thread milling cutter Gewindefräser *m*
thread plug Gewindestöpsel *m*
thread suspension Fadenaufhängung *f*
threadworm Fadenwurm *m* (Zool)
three-beam problem Dreistrahlproblem *n*
three-body collision Dreierstoß *m*
three center bond Dreizentrenbindung *f*
three-centered orbital Dreizentren-Orbital *n*
three-colored dreifarbig
three-color photography
   Dreifarbenphotographie *f*
three-color print Dreifarbendruck *m*
three-color screen Dreifarbenraster *m*
three-component alloy Dreistofflegierung *f*
three-component system Dreistoffsystem *n*
three-core cable Dreileiterkabel *n*
three-cornered dreikantig, dreiwinkelig
three-cylinder engine Drillingmaschine *f*
three-cylinder motor Dreizylindermotor *m*
three-dimensional dreidimensional, räumlich
three-dimensional curve Raumkurve *f*
three-dimensional lattice Raumgitter *n* (Krist)
three-dimensional quality Dreidimensionalität *f*,
   Plastizität *f*
three-dimensional structure Raumgebilde *n*
three-electrode tube Triode *f*
three-faced dreiflächig
three-filament lamp Dreifadenlampe *f* (Elektr)
three-finger rule Dreifingerregel *f* (Elektr)
three-flame dreiflammig
threefold dreifach, dreizählig
threefold collision Dreierstoß *m*
three-footed dreifüßig
three-gas-pass boiler Dreizugkessel *m*
three-high blooming mill Trioblockwalzwerk *n*
three-high mill Triowalzwerk *n*
three-legged dreischenkelig
three-membered dreigliedrig
three-membered ring Dreiring *m* (Chem)
three-necked flask Dreihalskolben *m* (Chem)
three-part dreiteilig
three-phase dreiphasig (Elektr)
three-phase alternator Drehstromgenerator *m*
three-phase armature Drehstromanker *m*
three-phase asynchronous motor
   Dreiphaseninduktionsmotor *m*
three-phase circuit Dreiphasenschaltung *f*
three-phase current Drehstrom *m*,
   Dreiphasenstrom *m*
three-phase current generator
   Drehstromgenerator *m*
three-phase current output Drehstromleistung *f*
three-phase current plant Drehstromanlage *f*
three-phase field Dreiphasenfeld *n*
three-phase four-wire system
   Dreiphasenvierleitersystem *n*
three-phase generator Drehstromdynamo *m*,
   Dreiphasenstromerzeuger *m*
three-phase motor Drehstrommotor *m*

three-phase plant Drehstromanlage *f*
three-phase six wire system
Dreiphasensechsleitersystem *n*
three-phase star connected system
Dreiphasendreileitersternschaltung *f*
three-phase system Drehstromsystem *n*,
Dreiphasensystem *n*
three-phase wire Drehstromleitung *f*
three-photon annihilation
Dreiquantenvernichtung *f*
three-pin plug Dreistiftstecker *m*
three-ply dreischichtig
three-point contact Dreipunkt-Kontakt *m*
three-point mounting Dreipunktaufstellung *f*
three-pole construction Dreipolanordnung *f*
three-quarter preserve Dreiviertelkonserve *f*
three-roll mill Dreiwalzenstuhl *m*
three-sided dreiseitig
three-speed motor Dreistufenmotor *m*
three-square file Dreikantfeile *f*
three-square scraper Dreikantschaber *m*
three-storied furnace Dreietagenofen *m*
three-valued dreiwertig (Math)
three-way cock Dreiwegehahn *m*
three-way connection Dreiwegverbindung *f*
three-way diverting valve Dreiwegverteilerventil *n*
three-way stop-cock Dreiwegehahn *m*
three-way valve Dreiwegventil *n*
three-wire system Dreileitersystem *n*
threitol Threit *m*
threo form threo-Form *f* (Stereochem)
threomine deaminase Threomindesaminase *f*
(Biochem)
threo modification threo-Form *f* (Stereochem)
threonic acid Threonsäure *f*
threonine Threonin *n*
threonine aldolase Threoninaldolase *f*
(Biochem)
threoninol Threoninol *n*
threooxazoline Treooxazolin *n*
threopentulose Threopentulose *f*
threosamine Threosamin *n*
threose Threose *f*
threpsology Ernährungskunde *f*
threshold Schwelle *f*
threshold current Schwellenstrom *m* (Elektr)
threshold detector Schwellendetektor *m*
threshold dose Toleranzdosis *f*, kritische Dosis *f*
threshold energy Einsatzenergie *f* (Phys),
Schwellenenergie *f* (Atom)
threshold field curve Schwellenwertkurve *f*
threshold measurement Schwellenwertmessung *f*
threshold of hearing Hörschwelle *f*
threshold potential Schwellenpotential *n*,
Schwellenspannung *f*
threshold sensitivity Schwellenempfindlichkeit *f*
threshold value Schwellenwert *m*
threshold wave Grenzwelle *f*

throat Gicht *f* (Metall), Hals *m*, Rachen *m*, ~ of
the converter Mündung *f* der Birne
throat bar Schneidbalken *m*
throat flame Gichtflamme *f*
throat opening Gichtöffnung *f*
throat stopper Gichtverschluß *m*
throat swab Rachenabstrich *m* (Med)
throat temperature Gichttemperatur *f*
thrombase Thrombin *n*
thrombin Thrombin *n*
thrombocyte Thrombozyt *m* (Med)
thrombocyte value Thrombozytenwert *m*
thrombocytosis Thrombozytenanstieg *m* (Med)
thrombokinase Thrombokinase *f* (Biochem)
thromboplastin Thromboplastin *n* (Biochem)
thrombosis Thrombose *f* (Med)
thrombus Blutgerinnsel *n* (Med), Blutpfropf *m*
(Med)
throttle Drossel *f*, Stauflansch *m*
throttle *v* [ab]drosseln, einschnüren, ~ down
abdrosseln (teilweise)
throttle diagram Drosseldiagramm *n*
throttle diaphragm Drosselscheibe *f*, Staurand *m*
throttle flange Drosselflansch *m*
throttle governor Drosselregulator *m*
throttle plate Luftklappe *f* (des Vergasers)
throttle slide Drosselschieber *m*
throttle valve Ausströmungsregulator *m*,
Drosselklappe *f*, Drosselventil *n*, Luftklappe *f*
(des Vergasers)
throttling Abdrosseln *n*, Drosselung *f*,
Drosselverfahren *n*
throttling range Proportionalbereich *m*
throttling valve Sparventil *n*
through bolt Durchgangsschraube *f*,
Durchsteckschraube *f*
through hole Durchgangsloch *n*
throughput Durchsatz *m*, Durchsatzleistung *f*
„throughput" advantage Apertur-Vorteil *m*
(Spektr)
through-way valve Durchgangsventil *n*
throw Wurf *m*
throw *v* werfen, ~ away verwerfen
(Niederschlag), ~ into gear einkuppeln, ~ off
abschleudern, abwerfen, ~ on anwerfen, ~
out ausstoßen, auswerfen, herausschleudern
(of gear), ausrücken, ~ through durchwerfen
throw off Abwurf *m*
throw-out lever Ausrückerhebel *m*
throw-over switch Umschalter *m*
thrust Stoß *m*, axialer Druck *m*, Schub *m*, center
of ~ Antriebsmittelpunkt *m*, upward ~
Auftrieb *m*
thrust *v* stoßen, ~ out ausstoßen
thrust ball bearing Druckkugellager *n*,
Kugeldrucklager *n*
thrust bearing Drucklager *n*, Spurlager *n*
thrust block Drucklager *n*
thrust collar Druckring *m*

thrust force Schubkraft *f*
thrust generator Druckerzeuger *m*
thrust loading Schubbelastung *f*
thrust performance Schubleistung *f*
thrust plane Schürffläche *f*
thrust ring Stützring *m*
thrust section Antriebsteil *n*
thrust shaft Druckwelle *f*
thrust washer Druckscheibe *f*
thujaketone Thujaketon *n*
thujaketonic acid Thujaketonsäure *f*
thujamenthol Thujamenthol *n*
thujane Thujan *n*
thuja oil Thujaöl *n*
thujene Thujen *n*
thujetic acid Thujetinsäure *f*
thujetin Thujetin *n*
thujigenin Thujigenin *n*
thujin Thujin *n*
thujol Thujol *n*
thujone Thujon *n*
thujorhodin Rhodoxanthin *n*
thujyl Thujyl-
thujyl alcohol Thujylalkohol *m*
thujyl amine Thujylamin *n*
thulite Thulit *m* (Min), Zoisit *m* (Min)
thulium (Symb. Tm o. Tu) Thulium *n*
thumb nut Flügelmutter *f*
thumb rule Faustregel *f*
thumb screw Daumenschraube *f*,
    Flügelschraube *f*, Knebelschraube *f*
thumbtack Heftzwecke *f*, Reißnagel *m* (Am. E.),
    Reißnagel *m*
thundertube Blitzröhre *f*
thuringite Thuringit *m* (Min)
thylakoid Thylakoid *n*
thylakoid membrane Thylakoidmembran *f*
thyme, liquid extract of ~
    Thymianfluidextrakt *m*, **mother of** ~
    Feldthymiankraut *n* (Bot)
thyme camphor Thymiancampher *m*, Thymol *n*
thyme-like thymianartig
thymene Thymen *n*
thyme oil Quendelöl *n*, **wild** ~ Feldthymianöl *n*
thymianic acid Thymiansäure *f*
thymidine Thymidin *n*
thymidine diphosphate Thymidindiphosphat *n*
thymidine monophosphate
    Thymidinmonophosphat *n*
thymidine triphosphate Thymidintriphosphat *n*
thymidylic acid Thymidylsäure *f*
thymine Thymin *n*
thyminic acid Solurol *n*
thyminose Thyminose *f*
thymoform Thymoform *n*
thymol Thymol *n*, Thymiancampher *m*
thymol blue Thymolblau *n*,
    Thymolsulfonphthalein *n*
thymol indophenol Thymolindophenol *n*

thymol iodide Jodthymol *n*, Aristol *n*, Jodosol *n*
thymol phthalein Thymolphthalein *n*
thymol sulfonphthalein Thymolblau *n*,
    Thymolsulfonphthalein *n*
thymomenthol Thymomenthol *n*
thymonucleic acid Thymonucleinsäure *f*
thymoquinone Thymochinon *n*
thymotal Thymotal *n*
thymotic aldehyde Thymotinaldehyd *m*
thymotol Thymotol *n*
thymus Thymus *m* (Med)
thymus gland Thymusdrüse *f* (Med)
thynnin Thynnin *n*
thyratron Thyratron *n* (Elektr)
thyresol Thyresol *n*
thyroacetic acid Thyroessigsäure *f*
thyrobutyric acid Thyrobuttersäure *f*
thyroformic acid Thyroameisensäure *f*
thyroglobulin Thyroglobulin *n*
thyroid activity Schilddrüsenfunktion *f* (Med)
thyroid cancer Schilddrüsenkrebs *m* (Med)
thyroid function Schilddrüsenfunktion *f*
thyroid gland Schilddrüse *f* (Med)
thyroidine Thyroidin *n*
thyroiodine Thyroxin *n*
thyronamine Thyronamin *n*
thyronine Thyronin *n*
thyronucleoprotein Thyronucleoproteid *n*
thyropenia Schilddrüsenunterfunktion *f* (Med)
thyropropionic acid Thyropropionsäure *f*
thyroxine Thyroxin *n*, Schilddrüsenhormon *n*
ticonium Tikonium *n*
tide Gezeit *f*, **turn of the** ~ Flutwechsel *m*
tide computing machine
    Gezeitenrechenmaschine *f*
tide gauge Flutmesser *m*
tides Gezeiten *pl*
tidy rein, sauber
tie Bindeglied *n*
tie *v* [an]binden, festmachen, knüpfen, koppeln,
    ~ **off** abschnüren, ~ **[together]** verknüpfen,
    ~ **up** abbinden (abschnüren), lahmlegen
tie bolt Verbindungsbolzen *m*
tie-in (tire) Einlagenumschlag *m* (Reifen)
tie line (in phase diagram) Verbindungslinie *f*
tiemannite Tiemannit *m* (Min),
    Selenquecksilber *n* (Min)
Tiemann-Reimer reaction Tiemann-Reimersche
    Reaktion *f*
tie rod Zugstange *f*, Haltestrebe *f*
tie rod head assembly Lenkspurstangenkopf *m*
tight undurchlässig, dicht, fest, gespannt, knapp,
    straff
tight cure Ausvulkanisation *f*
tighten *v* [ab]dichten, andrehen (Schraube),
    anziehen, spannen, straffen
tight fit Feinpassung *f*
tightness Dichte *f*, Festigkeit *f*, Straffheit *f*
tight pulley Festscheibe *f*

tiglaldehyde Guajol *n*, Tiglinaldehyd *m*
tiglic acid Tiglinsäure *f*, **salt or ester of** ~
  Tiglinat *n*
tiglic alcohol Tiglinalkohol *m*
tiglic aldehyde Guajol *n*, Tiglinaldehyd *m*
tigloidine Tigloidin *n*
tigogenin Tigogenin *n*
tilasite Tilasit *m* (Min)
tile Fliese *f*, Kachel *f*, Ziegel *m*, **unburnt** ~
  Rohziegel *m*
tile clay Ziegelton *m*
tile ore Ziegelerz *n* (Min), Kupferocker *m*
  (Min), Rotkupfererz *n* (Min)
tiliacorine Tiliacorin *n*
Tillman's reagent Tillmans-Reagens *n*
tilt *v* gerben (Stahl), kanten, kippen, neigen,
  schrägstellen, verkanten, ~ **over** umkippen
tiltable kippbar, schrägstellbar
tilt brace Hammerband *n*
tilted geneigt, schräggestellt
tilted [cylinder] mixer Schrägtrommelmischer *m*
tilted drum mixer Schrägtrommelmischer *m*
tilted steel Gerbstahl *m*
tilter Kantvorrichtung *f*, Kippvorrichtung *f*
tilt hammer Aufwerfhammer *m*
tilting apparatus Kippvorrichtung *f*
tilting device Kippvorrichtung *f*
tilting force (bulldozer) Kippkraft *f*
tilting head press Kippkopfpresse *f*,
  Scharnierpresse *f*
tilting open hearth furnace Kippofen *m*
tilting pan Kippkessel *m*
tilting pan filter Karussellnutschenfilter *n*,
  Kippwannenfilter *n*
tilting stage Drehschwingungsrahmen *m*
tiltmeter Neigungsmesser *m*
timber Balken *m*, Bauholz *n*, Nutzholz *n*
timber drying plant Holztrockenanlage *f*
timbre Klangfarbe *f* (Akust)
time Zeit *f*, **at the same** ~ gleichzeitig,
  determination of ~ Zeitbestimmung *f*,
  equation of ~ Zeitgleichung *f*, **for a short** ~
  kurzfristig, kurzzeitig, **period of** ~ Zeitdauer *f*,
  Zeitraum *m*, **to stop the** ~ die Zeit abstoppen,
  variable with ~ zeitlich veränderlich, ~ **of**
  direct contact Verweilzeit *f*, ~ **of flight**
  Laufzeit *f*, ~ **of operation** Betriebsdauer *f*, ~
  of passage Laufzeit *f*
time *v* die Zeit abstoppen
time-activity curve Zeit-Umsatz-Kurve *f* (Kin)
time average Zeitmittel *n*
time channel analyzer Zeitkanalanalysator *m*
time constant Zeitkonstante *f*
time-consuming zeitraubend
time-cycle operation Zeitverhalten *n*
time-dependent zeitabhängig
time exposure Zeitaufnahme *f* (Phot)
time-for-fracture curve Zeitbruchlinie *f*
time function element Zeitglied *n* (Regeltechn)

time fuse Zeitzünder *m*
time interval Zeitabstand *m*, Zeitintervall *m*,
  Zeitspanne *f*
time lag Zeitdifferenz *f*, zeitliche Verzögerung *f*
time-lag relay Verzögerungsrelais *n*, Zeitrelais *n*
time line Zeitlinie *f*
timely zeitgemäß
time-of-flight mass spectrograph
  Laufzeitmassenspektrograph *m*
time-of-flight measurement Flugzeitmessung *f*,
  Laufzeitmessung *f*
time-of-flight method Flugzeitmethode *f*
time-of-flight spectrometer
  Flugzeitspektrometer *n*,
  Laufzeitspektrometer *n*
time-of-flight technique Laufzeitmethode *f*
time pattern control (Am. E.) Zeitplanregelung
  (Regeltechn)
time program control Zeitplanregelung
  (Regeltechn)
time quenching Zeithärtung *f*
timer Stoppuhr *f*, Zeitmesser *m*
time recorder Zeitschreiber *m*
time release Zeitauslösung *f*
time resolution Zeitauflösungsvermögen *n*
time response Übergangsfunktion *f* (Regeltechn)
time reversal operator Zeitumkehroperator *m*
time scale Zeitskala *f*
time schedule Terminplan *m*
time sequence Zeitfolge *f*
time shift Zeitverschiebung *f*
time stamping machine Zeitmarkengeber *m*
time switch Schaltuhr *f*, Zeitschalter *m*
time-table Stundenplan *m*
time value, mean ~ Zeitmittelwert *m*
time vector Zeitvektor *m*
time yield limit Zeitdehnspannung *f*,
  Zeitkriechgrenze *f*, Zeitstandkriechgrenze *f*
timing Synchronisierung *f*, zeitliche Steuerung *f*
timing device Zeitmesser *m*
timing shaft Steuerwalze *f*
tin (Symb. Sn) Zinn *n*, Blechbüchse *f*, **containing**
  ~ zinnführend, zinnhaltig, **crackling of** ~
  Zinnknirschen *n*, **fine** ~ Zinnschlich *m*, **finest**
  English ~ Rosenzinn *n*, **flowers of** ~
  Zinnblumen *f pl*, **granulated** ~ Tropfzinn *n*,
  Zinngranalien *f pl*, **laminated** ~ Walzzinn *n*,
  raw ~ Werkzinn *n*, **refuse of** ~ Zinnabfall *m*,
  rich in ~ zinnreich, **spongy** ~
  Zinnschwamm *m*, **standard** ~ Kronzinn *n*, **to**
  line with ~ verzinnen, **containing** ~ **oxide**
  zinnoxidhaltig, ~ **with high lead content**
  Halbzinn *n*
tin *a* zinnern
tin *v* verzinnen
tin alloy Zinnlegierung *f*
tin ammonium chloride
  Ammoniumzinnchlorid *n*, Pinksalz *n*,
  Zinnammoniumchlorid *n*, Zinnsalmiak *m*

tin ashes Zinnasche *f,* Zinnkrätze *f*
tin-bearing zinnführend, zinnhaltig
tin bottle Blechflasche *f*
tin box Blechkasten *m,* Blechschachtel *f*
tin bromide Zinnbromid *n*
tincal Tinkal *m* (Min)
tin calx Zinnkalk *m*
tin can Dose *f,* Konservendose *f*
tin case Blechkasten *m*
tin cast Zinnguß *m*
tin chloride Zinnchlorid *n*
tin chloride bath Chlorzinnbad *n*
tin-coated verzinnt
tin-coating Verzinnen *n*
tin content Zinngehalt *m*
tin cry Zinngeschrei *n* (Chem), Zinnknirschen *n* (Chem)
tin crystals Zinngraupen *f pl*
tinction Färbung *f* (Färb)
tinctorial power Färbekraft *f*
tinctorial strength Färbekraft *f,* Farbstärke *f*
tinctorial value Farbkraft *f*
tincture alkoholischer Auszug *m* (Pharm), Tinktur *f* (Pharm), ~ **of benzoin** Benzoetinktur *f,* ~ **of nux vomica** Brechnußtinktur *f*
tinder Zunder *m,* Feuerschwamm *m,* Holzzunder *m,* **ashes of** ~ Zunderasche *f*
tin dibromide Zinndibromid *n,* Zinn(II)-bromid *n*
tin dichloride Zinndichlorid *n,* Zinn(II)-chlorid *n*
tin diiodide Zinn(II)-jodid *n,* Stannojodid *n*
tin dioxide Stannioxid *n,* Zinndioxid *n,* Zinn(IV)-oxid *n,* Zinnsäureanhydrid *n,* **native** ~ Bergzinnerz *n* (Min)
tin disulfide Zinndisulfid *n,* Stannisulfid *n,* Zinn(IV)-sulfid *n*
tin dross Zinngekrätz *n*
tin dust Zinnstaub *m*
tin filings Zinnfeilicht *n,* Zinnfeilspäne *m pl*
tin foil Blattzinn *n,* Stanniol[papier] *n,* Zinnfolie *f*
tinfoil coat[ing] Zinnverspiegelung *f*
tinfoil lacquer Stanniollack *m*
tin founder Zinngießer *m*
tin funnel Blechtrichter *m*
tinge Farbtönung *f,* Schattierung *f*
tinge *v* anfärben
tin glaze Zinnglasur *f*
tin grate Zinngitter *n*
tin green Zinngrün *n*
tin hydride Stannan *n,* Zinnwasserstoff *m*
tin hydroxide Zinnhydroxid *n*
tin iodide Zinnjodid *n*
tin lacquer Silberlack *m*
tin-like zinnartig
tin lode Zinnader *f*
tin metal, mossy ~ Zinnwolle *f*
tin monosulfide Stannosulfid *n,* Zinn(II)-sulfid *n*

tin monoxide Zinnmonoxid *n,* Stannooxid *n,* Zinn(II)-oxid *n,* Zinnprotoxid *n*
tin mordant Zinnbeize *f*
tinned verzinnt
tinning Reiblöten *n,* Verzinnen *n,* Verzinnung *f*
tinning bath Zinnbad *n*
tinning vat Weißblechkessel *m*
tinny zinnartig
tin ore Zinnerz *n* (Min), **alluvial** ~ Waschzinn *n,* **fibrous** ~ Holzzinnerz *n* (Min)
tin ore refuse Zinnafter *m*
tin oxide Zinnoxid *n*
tin pest Zinnpest *f*
tin phosphate, weighting with ~ Zinnphosphatbeschwerung *f*
tin pipe Zinnrohr *n*
tin plague Zinnpest *f*
tin plate verzinntes Eisenblech, Weißblech *n,* Zinnblech *n,* **crystallized** ~ Perlmutterblech *n*
tin-plate *v* verzinnen
tinplate enamel Blechlack *m*
tinplate lacquer Blechlack *m*
tinplate scrap Weißblechabfall *m*
tinplate varnish Blechlack *m*
tinplate works Verzinnungsanstalt *f*
tin-plating Verzinnen *n,* Verzinnung *f,* ~ **in hot bath** Weißsudverzinnung *f*
tin protoxide Zinn(II)-oxid *n,* Zinnprotoxid *n*
tin pyrites Zinnkies *m* (Min), Stannin *m* (Min), Stannit *m* (Min)
tin refuse Zinngekrätz *n,* Zinnkrätze *f*
tin salt Zinnsalz *n*
tin sample Zinnprobe *f*
tin scum Zinnabstrich *m*
tinsel Flittersilber *n*
tinsel wire Flitterdraht *m*
tin slimes Zinnschlich *m*
tin solder Lötzinn *n;* Weichlot *n,* Weißlot *n,* Zinnlot *n*
tinstone Bergzinnerz *n* (Min), Kassiterit *m* (Min), Zinnstein *m* (Min)
tin sulfide Zinnsulfid *n*
tint Farbnuance *f,* Farbton *m,* Rasterton *m* (Buchdr), Schattierung *f,* Tönung *f*
tint *v* abtönen, nuancieren, tönen
tinted glass Rauchglas *n*
tin test Zinnprobe *f* (Anal)
tin tetrabromide Stannibromid *n,* Zinn(IV)-bromid *n,* Zinntetrabromid *n*
tin tetrachloride Zinntetrachlorid *n,* Stannichlorid *n,* Tetrachlorzinn *n,* Zinn(IV)-chlorid *n*
tin tetrafluoride Zinntetrafluorid *n*
tin tetraiodide Zinntetrajodid *n,* Stannijodid *n,* Zinn(IV)-jodid *n*
tin thiocyanate Rhodanzinn *n*
tinting Abtönen *n* (Färb)
tinting color Abtönfarbe *f* (Färb)
tinting plant Färbeanlage *f*

tinting power Abtönvermögen *n*, Farbstärke *f*
tinting strength Farbstärke *f*
tinting value Aufhellungswert *m*, Farbzahl *f*
tintometer Farbenmesser *m*
tin tree Zinnbaum *m*
tint tone Aufhellung *f*
tin vein Zinnader *f*
tin vessel Zinngerät *n*
tin waste Zinnabfall *m*
tin wire Zinndraht *m*
tip Spitze *f*, **tendency to** ~ Kippbestreben *n*
tip *v*, ~ **over** umfallen, umkippen
[tip] clearance (of gear) Kopfspiel *n*
tipped bestückt (Werkzeug)
tipping the box Entleerung *f* des Kastens, ~ **the molds** Kippen *n* der Formen (pl)
tipping bridge Wippbrücke *f*
tipping bucket Kippbecher *m*
tipping device Kipper *m*, Kippvorrichtung *f*
tipping furnace Kippofen *m*
tipping motion Kippbewegung *f*
tipping platform Plattformkipper *m*
tipping stage Kippbühne *f*, Stürzbühne *f*
tire Reifen *m* (Kfz), **body of the** ~ Gewebeunterbau *m* (Reifen), **canvas** ~ Vollgewebereifen *m*, **green** ~ Reifenrohling *m*, **low-pressure** ~ Ballonreifen *m*, **mud and snow** ~ Matsch- und Schneereifen *m*, **non-skid** ~ profilierter Reifen, **oversize** ~ Reifen *m* in Übergröße, **solid** ~ Blockreifen *m*, **solid rubber** ~ Vollgummireifen *m*, **tubeless** ~ schlauchloser Reifen
tire *v*, ~ **out** sich abarbeiten
tire buffing machine Rauhmaschine *f* (Reifen)
tire capping materials Reifenrunderneuerungsmaterialien *n pl*
tire casing Lauffläche *f* (Reifen), Reifendecke *f*
tire mill Bandagewalzwerk *n*, Reifenwalzwerk *n*
tire mold lubricant Einstreichmittel *n* für Reifenformen (pl)
tire reclaim, whole ~ Ganzreifenregenerat *n*
tiring Altern *n*, Ermüdung *f*
T iron T-Eisen *n*, T-Träger *m*
tissue Gewebe *n* (Biol), **breaking down of** ~ Histolyse *f*, Gewebszerfall *m*, **intermuscular** ~ intermuskuläres Gewebe (Med), **subcutaneous** ~ Unterhautgewebe *n* (Med)
tissue culture Gewebekultur *f* (Biol)
tissue extract Gewebeextrakt *m*
tissue fluid Gewebeflüssigkeit *f*
tissue injury Gewebeschaden *m* (Biol), **latent** ~ Gewebespätschaden *m* (Biol)
tissue metabolism Gewebestoffwechsel *m*
tissue paper Seidenpapier *n*
tissue parasite Gewebsparasit *m*
tissue protein Gewebeeiweiß *n*
tissue regeneration Gewebeneubildung *f*

tissue respiration Gewebeatmung *f* (Biol)
tissue slice Gewebeschnitt *m* (Biol)
titanate Titanat *n*
titania Titandioxid *n* (Min)
titanic acid Titansäure *f*
titanic anhydride Titandioxid *n*, Titansäureanhydrid *n*
titanic iron ore Titaneisenerz *n* (Min), Titaneisenstein *m* (Min)
titanic oxide Titandioxid *n*, Titan(IV)-oxid *n*
titanic salt Titan(IV)-salz *n*, Titanisalz *n*
titaniferous titanführend, titanhaltig
titaniferous ferromanganese Eisenmangantitan *n*
titaniferous iron Titaneisen *n*
titaniferous magnetic iron oxide Menaccanit *m* (Min)
titaniferous ore Titanerz *n* (Min)
titanite Titanit *m* (Min), Sphen *m* (Min)
titanium (Symb. Ti) Titan *n*, **containing** ~ titanhaltig
titanium ammonium formate Titanammoniumformiat *n*
titanium ammonium oxalate Titanammonoxalat *n*
titanium chloride Titanchlorid *n*
titanium content Titangehalt *m*
titanium dioxide Titandioxid *n*, Titan(IV)-oxid *n*, Titansäureanhydrid *n*, **native** ~ Oktaedrit *m* (Min), Rutil *m* (Min)
titanium dioxide filling material Titandioxidfüllstoff *n*
titanium formate Titanformiat *n*
titanium metal Titanmetall *n*
titanium ore Titanerz *n* (Min)
titanium oxide Titanoxid *n*
titanium pigment Titanpigment *n*
titanium potassium oxalate Titankaliumoxalat *n*
titanium salt Titansalz *n*
titanium steel Titanstahl *m*
titanium thermit Titanthermit *n*
titanium vapor pump Titanverdampferpumpe *f*
titanium white Titanweiß *n*
titanolite Titanolith *m* (Min)
titanolivine Titanolivin *m* (Min)
titanomorphite Titanomorphit *m* (Min)
titanous hydroxide Titan(III)-hydroxid *n*
titanous salt Titan(III)-salz *n*
titer Titer *m*
title Titel *m*, Benennung *f*
title page Titelblatt *n*
titrate *v* titrieren, ~ **back** zurücktitrieren
titrating Titrieren *n*
titrating acid Titriersäure *f*
titrating analysis Maßanalyse *f*
titrating apparatus Titrierapparat *m*, Titriervorrichtung *f*
titrating solution Titrierflüssigkeit *f*
titration Maßanalyse *f*, Titration *f*, Titrierung *f*, **analysis by** ~ Titrieranalyse *f*, **back** ~

Rücktitration *f* (Chem), **electrometric** ~
elektrometrische (potentiometrische)
Titration, **final point of a** ~
Umschlagspunkt *m,* **iodometric** ~
jodometrische Titration, **potentiometric** ~
potentiometrische Titration
titration apparatus Titrierapparat *m*
titration cell Titrierzelle *f*
titration curve Titrationskurve *f*
titration method Titrationsverfahren *n,*
Titriermethode *f*
titration solution Titerflüssigkeit *f*
titration standard Titer *m* einer Lösung
titration value Titrierzahl *f*
titrimeter Potentiometer *n* für Titration (HN)
titrimetric maßanalytisch, titrimetrisch
titrimetric standard Urtiter *m* (Anal)
titrimetry Maßanalyse *f,* Titrimetrie *f*
titriplex (commercial name for ethylene diamine
tetraacetic acid) Titriplex *n* (HN)
toad flax Leinkraut *n* (Bot)
toad venom Krötengift *n*
toast *v* rösten (Brot), toasten
tobacco mosaic virus Tabakmosaikvirus *m*
tobacco seed oil Tabaksamenöl *n*
tobacco tar Tabakteer *m*
Tobias acid Tobiassäure *f*
tocamphyl Tocamphyl *n*
tocol Tocol *n*
tocopherol Tocopherol *n*
tocopheroxide Tocopheroxid *n*
tocopherylquinone Tocopherylchinon *n*
tocopurple Tocopurpur *n*
toe-in Vorspur *f* (Kfz)
toggle Knebel *m*
toggle crusher Kniehebelbrecher *m*
toggle joint Kniegelenk *n*
toggle [lever] press Kniehebelpresse *f,*
Kniegelenkpresse *f*
toilet article Toilettenartikel *m*
toilet soap Feinseife *f*
tokorogenic acid Tokorogensäure *f*
tokorogenin Tokorogenin *n*
tolane Tolan *n*
tolane red Tolanrot *n*
tolazoline Tolazolin *n*
tolbutamide Tolbutamid *n*
toledo blue Toledoblau *n*
tolerance Toleranz *f,* Abmaß *n,* Fehlergrenze *f,*
Maßabweichung *f,* Spielraum *m,* [zulässige]
Abweichung *f,* **extent of** ~ Toleranzfeld *n,*
permissible ~ zulässige Abweichung, **zone of**
~ Toleranzfeld *n*
tolerance dose Toleranzdosis *f*
tolerance limit Verträglichkeitsgrenze *f* (Med)
tolerance test Belastungsprobe *f* (Med),
Verträglichkeitsprobe *f* (Med)
tolidine Tolidin *n*
tolil Tolil *n*

tolite Tolit *n*
tolonium Tolonium *n*
tolpronine Tolpronin *n*
tolualdehyde Tolualdehyd *m*
toluamide Toluamid *n*
toluene Toluol *n*
toluenediamine Toluoldiamin *n*
toluenediazonium hydroxide
Toluoldiazoniumhydroxid *n*
toluene musk Toluolmoschus *m*
toluenesulfinic acid Toluolsulfinsäure *f*
toluenesulfochloride Toluolsulfochlorid *n*
toluenesulfonamide Toluolsulfonamid *n*
toluenesulfonanilide Toluolsulfonanilid *n*
toluenesulfonic acid Toluolsulfonsäure *f*
toluenesulfonyl chloride Toluolsulfonylchlorid *n*
toluic acid Toluylsäure *f*
toluidine Toluidin *n,* Aminotoluol *n*
toluidine blue Toluidinblau *n*
toluidine orange Toluidinorange *n*
toluidinesulfonic acid Toluidinsulfonsäure *f*
tolunitrile Tolunitril *n*
toluphenone Toluphenon *n*
toluquinaldine Toluchinaldin *n*
toluquinhydrone Toluchinhydron *n*
toluquinol Toluchinol *n*
toluquinoline Toluchinolin *n*
toluquinone Toluchinon *n*
toluresitannol Toluresitannol *n*
tolusafranine Tolusafranin *n*
tolu tincture Tolutinktur *f*
toluylene Toluylen *n*
toluylene blue Toluylenblau *n*
toluylene diamine Toluylendiamin *n*
toluylene red Toluylenrot *n*
toluylene red base Toluylenrotbase *f*
tolyl Tolyl-
tolyl acetate Kresylacetat *n*
tolyl black Tolylschwarz *n*
tolyl carbinol Tolylcarbinol *n*
tolyl carbonate Kresylcarbonat *n*
tolylhydrazine Tolylhydrazin *n*
tolylmethyl ketone Tolylmethylketon *n*
tolylphenyl ketone Tolylphenylketon *n*
tolyl phosphate Kresylphosphat *n*
tolylsulfuric acid Kresylschwefelsäure *f*
tolyl toluenesulfonate Kresyltoluolsulfonat *n*
tolypyrine Tolypyrin *n*
tomatidine Tomatidin *n*
tomatine Tomatin *n*
tombac Tombak *m* (Legierung)
tombac powder gelbe Bronze *f*
tomentous filzig (Bot)
tomentum Filz *m* (Bot)
tommy Stellstift *m* (Mech)
tommy screw Knebelschraube *f*
tone Farbton *m* (Färb), Laut *m* (Akust),
Nuance *f* (Färb), Ton *m* (Akust)
tone *v* nuancieren, schönen, tonen (Phot)

tone controller Tonregler *m* (Akust)
toner Toner *m* (Phot)
tongs Zange *f*
toning Tönen *n* (Färb), Tonen *n* (Phot)
toning and fixing bath Tonfixierbad *n* (Phot)
toning and fixing salt Tonfixiersalz *n* (Phot)
tonka bean Tonkabohne *f* (Bot)
tonnage Tonnengehalt *m*
tonsil Mandel *f* (Med)
tonsilitis Mandelentzündung *f* (Med)
tool Werkzeug *n*, Gerät *n*, **cutting ~**
spanabhebendes Werkzeug, **plastic ~**
Werkzeug *n* aus Kunststoff, **~ for noncutting**
shaping Werkzeug *n* für spanlose Formung
toolbox Werkzeugkiste *f*
tool [disc] grinder Werkzeugschleifmaschine *f*
tool gauge Schneidstahllehre *f*
tool holder Werkzeughalter *m*
tool hook Bohrhaken *m*
tooling fixture Bearbeitungsvorrichtung *f*
tool kit Werkzeugkasten *m*
tool lathe Werkzeugmacherdrehbank *f*
toolroom Werkzeugmacherei *f*
tool set Werkzeugsatz *m*
tool steel Werkzeugstahl *m*, Diamantstahl *m*,
alloy ~ legierter Werkzeugstahl, **carbon ~**
unlegierter Werkzeugstahl, **high-speed ~**
Schnelldrehstahl *m*, **special-alloy ~**
hochlegierter Werkzeugstahl, **unalloyed ~**
unlegierter Werkzeugstahl
tooth Zahn *m*, Zacken *m*, **breadth of ~**
Zahnbreite *f*, **depth of ~** Zahntiefe *f*, **molar ~**
Backenzahn *m*, **root of the ~** Zahnwurzel *f*,
space of the ~ Zahnlücke *f*
tooth-belt gear Zahnriemenrad *n*
tooth block holder Modellträger *m*,
Schablonenhalter *m*
toothed gezähnt, zackig
toothed belt Zahnriemen *m*
toothed disk Zahnscheibe *f*
toothed gearing Zahnradgetriebe *n*
toothed rim Zahnkranz *m*
toothed roll Stachelwalze *f*
toothed segment Zahnsegment *n*
toothed spring disc federnde Zahnscheibe
tooth face Zahnkopf *m*
tooth filling Zahnfüllung *f*, Zahnplombe *f*
toothing Zahnung *f*
toothpaste Zahnpasta *f*
tooth powder Zahnpulver *n*
tooth profile Zahnprofil *n*, Zahnform *f*
tooth space Zahnlücke *f*
tooth thickness Zahndicke *f*
top Deckel *m*, Gipfel *m*, Höhepunkt *m*, Kopf *m*,
Oberseite *f*, Spitze *f*, **~ of crucible**
Obergestell *n* (Ofen), **~ of hearth**
Obergestell *n* (Ofen), **~ removable** Kappe *f*
top *v* bekappen (Holz), überfärben

topaz Topas *m* (Min), **false ~** Gelbquarz *m*
(Min), **Oriental ~** orientalischer Topas *m*
(Min), **smoky ~** Rauchtopas *m* (Min)
topazolite Topazolith *m* (Min), gelber Granat *m*,
Goldgranat *m* (Min)
top blowing tuyère Oberwinddüse *f*
top-blown converter Aufblaskonverter *m*
top capping Besohlen *n* (Reifen)
top circle Kopfkreis *m*
topcoat Deckanstrich *m*, Deckfarbe *f*,
Deckschicht *f*, letzter Anstrich *m*,
Oberschicht *f*, Schlußstrich *m*,
Überzugslack *m*
top edge Oberkante *f*
top ejection Abdrücken *n* von oben (Abstreifen
vom Stempel)
top feed filter Innenzellen-Filter *n*,
Oben-Aufgabe-Filter *n*
top fermentation Obergärung *f* (Brau)
top fermentation yeast Oberhefe *f* (Brau)
top-fermenting obergärig
top fertilizer Kopfdünger *m*
top finish Deckappretur *f* (Leder)
top flame Gichtflamme *f*
top gas Gichtgas *n*
top-heaviness Kopflastigkeit *f*
top-heavy kopflastig, oberlastig
topical aktuell
top layer Oberschicht *f*
top layer formation Deckschichtenbildung *f*
top light Oberlicht *n*
topochemistry Topochemie *f*
topological topologisch (Geom)
top part Oberteil *n*
top plate Deckplatte *f*, Kopfplatte *f*
top product Kopfprodukt *n*
top ram press Oberdruckpresse *f*,
Oberkolbenpresse *f*
top roll Oberwalze *f*
top speed Spitzengeschwindigkeit *f*
top surface Stirnfläche *f*
top treading Besohlen *n* (Reifen)
top view Ansicht *f* von oben
top weight Obergewicht *n*, Oberlast *f*
toramin Toramin *n*
torbanite Torbanit *m*
torbernite Torbernit *m* (Min), Uranglimmer *m*
(Min)
torch Brenner *m* (Schweiß), Fackel *f*,
low-pressure ~ Injektorbrenner *m* (Schweiß)
torch hardening Oberflächenhärtung *f*
tormentil extract Tormentillextrakt *m*
tormentil tannic acid Tormentillgerbsäure *f*
torn rissig
toroidal condenser Toroidkondensator *m*
torque Drehmoment *n*, Drehbeanspruchung *f*,
Drehfestigkeit *f*, Drehkraft *f*, **angle of ~**
Torsionswinkel *m*
torque amplifier Drehmomentverstärker *m*

torque compensation Drehmomentausgleich *m*
torque elasticity Torsionselastizität *f*
torque resistance Verdrehfestigkeit *f*
torrefaction Darren *n*, Dörren *n*, Rösten *n*
torrefy *v* rösten, dörren
torreol Torreol *n*
torsibility Drehfestigkeit *f*
torsion Drehkraft *f* (Phys), Drehung *f*, Torsion *f*
(Phys), Windung *f*, **angle of** ~
Torsionswinkel *m*, **modulus of** ~
Verdrehungsmodul *n*, **moment of** ~
Drehmoment *n*, **resistance to** ~
Torsionswiderstand *m*
torsional angle Verdrehungswinkel *m*
torsional elasticity Verdrehungselastizität *f*
torsional force Torsionskraft *f*,
Verdrehungskraft *f*
torsional frequency
Torsionsschwingungsfrequenz *f*
torsional mode Drehschwingung *f*
torsional moment Drehmoment *n*,
Torsionsmoment *n*, Verdrehungsmoment *n*
torsional oscillation Drehschwingung *f*
torsional oscillation apparatus
Torsionsschwingungsapparatur *f*
torsional pendulum test
Torsionsschwingungsversuch *m*
torsional rigidity Torsionsmodul *m*
torsional shear, yield point of ~
Verdrehungsgrenze *f*
torsional strain Verdrehung *f*
torsional strength Drehfestigkeit *f*,
Torsionsfestigkeit *f*
torsional stress Beanspruchung *f* auf Verdrehen,
Drehbeanspruchung *f*, Drehspannung *f*,
Torsionsbeanspruchung *f*
torsional tension Drehspannung *f*
torsional tester Torsionsprüfmaschine *f*
torsional vibrator Torsionsschwingzentrifuge *f*
torsional wave Torsionsschwingung *f*
torsion angle Drehungswinkel *m*,
Torsionswinkel *m*
torsion balance Drehwaage *f*, Torsionswaage *f*
torsion bar Drehstabfeder *f*
torsion constant Torsionskonstante *f*
torsion electrometer Torsionselektrometer *n*
torsion failure Verdrehungsbruch *m*
torsionfree torsionsfrei
torsion meter Torsionsmesser *m*
torsion pendulum Drehpendel *n*,
Torsionspendel *n*
torsion spring Torsionsfeder *f*
torsion suspension Torsionsaufhängung *f*,
Drehgestellfederung *f*
torsion test Torsionsversuch *m*,
Verdrehungsprobe *f*
tortoise shell Schildpatt *n*
torus Wulst *m* (Arch)
toss energy (in impact testing) Schleuderenergie *f*

tossing Rührverfahren *n* (beim Schlämmen)
tosyl·blocking group Tosylschutzgruppe *f*
tosyl chloride Tosylchlorid *n*
total restlos, total
total *v* addieren, summieren
total acid Gesamtsäure *f*
total immersion test Dauertauchversuch *m*
(Korr)
total-immersion thermometer
Eintauchthermometer *n*
totality Gesamtheit *f*
total liabilities Fremdkapital *n*
total reflection Totalreflexion *f*,
Totalspiegelung *f* (Opt), **angle of** ~
Totalrückstrahlwinkel *m* (Opt)
total reflectometer Totalreflektometer *n*
total step method Gesamtschrittverfahren *n*
total transfer Summenübertragung *f* (Comp)
touch *v* befühlen, berühren
touchwood Holzzunder *m*
tough bruchfest (Techn), hammergar (Techn),
zäh
toughen *v* abhärten, zäh machen, zäh werden, ~
by poling dichtpolen, ~ **copper** Kupfer polen
toughness Zähigkeit *f*, Bruchfestigkeit *f* (Techn),
Festigkeit *f*
toughness test Zähigkeitsprobe *f*
tourmalin[e] Turmalin *m* (Min), **blue** ~
Indigolith *m* (Min), Indigostein *m* (Min)
tourmaline tongs Turmalinzange *f*
tow Schleppseil *n*, Werg *n*
tow *v*, ~ **away** abschleppen
towel Handtuch *n*
tower Kolonne *f* (Dest), Säule *f* (Dest), Turm *m*
tower liquor Turmlauge *f* (Pap)
towing equipment Abschleppgerät *n*
towing load Taulast *f*
town gas Stadtgas *n*
tox[a]emia Toxämie *f* (Med)
toxalbumen Toxalbumin *n*
toxalbumin Toxalbumin *n*
toxic giftig, toxisch
toxic agent Giftstoff *m*
toxicogenic Gift erzeugend
toxicologic[al] toxikologisch
toxicology Giftkunde *f*, Toxikologie *f*
toxiferine Toxiferin *n*
toxigene Toxigen *n*
toxin Gift *n*, Toxin *n*
toxoflavin Toxoflavin *n*
toxoid Toxoid *n*, Immunstoff *m*
toxopyrimidine Toxopyrimidin *n*
T-piece Dreiwegstück *n*, T-Stück *n*
T-pipe T-Rohr *n*
trace Spur *f*, Fährte *f*, Spur *f* (kleine Menge), ~
of acid Säurespur *f*
trace *v* abstecken; aufspüren, [auf]zeichnen,
pausen, skizzieren, ~ **back** *v* ableiten,
zurückführen

trace detection Spurennachweis *m*
tracer Taster *m*, Fühler *m*, Isotopenindikator *m* (Chem); Stahlstift *m* (Techn); Tracer *m* (Chem), **radioactive** ~ radioaktiver Indikator *m*
tracer atom Indikatoratom *n*
tracer bullet Rauchspurmunition *f*
tracer chemistry Spurenchemie *f*
tracer element Indikatorelement *n*, Spurenelement *n*
tracer head Fühlkopf *m*
tracer isotope Isotopenindikator *m*
tracer method Indikatorenmethode *f*, Indikatorverfahren *n*
tracer technique Leitisotopenmethode *f*, Tracertechnik *f*
traces, in ~ spurenweise
trachea Luftröhre *f* (Med)
tracheid Tracheide *f* (Bot)
trachyte Trachyt *m* (Geol)
trachyte porphyry Trachytporphyr *m* (Geol)
tracing Pause *f*, Pauszeichnung *f*, Zeichnung *f*
tracing cloth Pausleinen *n*, Pausleinwand *f*
tracing film Pausfilm *m*
tracing paper Pauspapier *n*
tracing recorder Nachlaufschreiber *m*
tracing routine Überwachungsprogramm *n* (Comp)
track Fährte *f*, Gleis *n* (Eisenbahn), Schiene *f* (Eisenbahn), Spur *f* (Bahn)
track curvature Bahnkrümmung *f*
tracking Kriechwegbildung *f*
tracking error Spurverzerrung *f*
tracking resistance Kriechstromfestigkeit *f*
track joint Schienenlasche *f*
track photometry Bahnspurenphotometrie *f*
track property Spureigenschaft *f*
track rod Spurstange *f*
track wheel Laufrad *n*, Spurrolle *f*
tractile streckbar
tractility Streckbarkeit *f*
traction Zug *m*, Zugkraft *f*
traction lever Zughebel *m*
traction rope Zugseil *n* (Seilbahn)
tractive force Zugkraft *f*, Anfahrkraft *f*, Anzugskraft *f*
tractive power Zugleistung *f*
tractor Schlepper *m*, Traktor *m*, Zugmaschine *f*
tractrix Traktrix *f* (Math)
trade Gewerbe *n*, Handel *m*
trade designation Handelsbezeichnung *f*
trademark Fabrikmarke *f*, Fabrikzeichen *n*, Handelszeichen *n*, Schutzmarke *f*, Warenzeichen *n*, **registered** ~ eingetragene Schutzmarke *f*
trademarks, protection of registered ~ Markenschutz *m*

trade name Handelsbezeichnung *f*, Handelsname *m*, **copyrighted** ~ gesetzlich geschützter Handelsname
traffic Verkehr *m*
traffic accident Verkehrsunfall *m*
traffic paint Straßenmarkierungsfarbe *f*
tragacanth Tragant *m* (Bot)
tragacanth mucilage Tragantschleim *m*
tragantine Tragantin *n*
trail Fährte *f*, Spur *f*, T-Schiene *f*
trailer Anhänger *m* (Auto)
trailing edge Hinterkante *f*
train abrichten, schulen
trainee Praktikant *m*
training Übung *f*, **further** ~ Fortbildung *f*
train of machines Maschinenpark *m*
train of thought Gedankengang *m*
train oil Tran *m*, Fischtran *m*, **containing** ~ tranhaltig, tranig, **smell of** ~ Trangeruch *m*
train oil soap Transeife *f*
trajectory Bahnkurve *f*, Fallkurve *f*, Flugbahn *f*, Fluglinie *f*, Wurfbahn *f*
trajectory parabola Fallparabel *f* (Phys)
trammel Ellipsenzirkel *m* (Math)
tranquilizer Beruhigungsmittel *n* (Pharm)
tranquillizer Psychosedativum *n* (Pharm), Tranquil[l]izer *m* (Pharm)
transacetylase Transacetylase *f* (Biochem)
transacetylation Transacetylierung *f* (Biochem)
transaction Abwicklung *f*
transaldolase Transaldolase *f* (Biochem)
transaminase Transaminase *f* (Biochem)
transaminate *v* transaminieren
transamination Transaminierung *f* (Biochem)
trans arrangement Transanordnung *f*
trans-butenedioic acid Fumarsäure *f*
transcalent wärmedurchlässig
transceiver Sendeempfangsapparat *m*, Sende- und Empfangsgerät *n*
transcendental transzendent (Math)
transconductance Übergangsleitwert *m*, Durchgriff *m*
transcriber Übertragungsgerät *n*
transcription Transkription *f* (Biochem), **inhibition of** ~ Transkriptionshemmung *f* (Biochem)
transducer Umwandler *m* (Elektr)
transductor magnetischer Verstärker *m*
transesterification Umesterung *f*
transfer Übergang *m*, Überleitung *f*, Übertragung *f*, **principle of** ~ Übertragungsprinzip *n*
transfer *v* überführen, übertragen, ~ **by pouring** umschütten
transferase Transferase *f* (Biochem)
transfer characteristic Übergangsleitwert *m*
transfer circuit Verbindungsleitung *f*
transfer[ence] Übertragung *f*

transference number Ionenüberführungszahl *f*
(Elektrochem)
transference of heat Wärmefortpflanzung *f*
transfer function (Am. E.)
Übertragungsfunktion *f* (Regeltechn)
transfer gear Übergabekopf *m*
transfer mold Preßspritzform *f*
transfer molding Spritzpressen *n*,
Transferpressen *n*
transfer of profits Ertragsaustausch *m*
transfer paper Umdruckpapier *n*
transfer picture Abziehbild *n*
transfer pipet[te] Vollpipette *f* (Anal)
transfer plunger Preßspritzkolben *m*,
Transferkolben *m*
transfer pot Transfer-Druckkammer *f*
transfer potential Übertragungspotential *n*
transfer press Schließpresse *f*
transfer pump Umwälzpumpe *f*
transfer rate Übertragungsrate *f* (Comput),
Übertragungsverhältnis *n*
transfer ribonucleic acid
Transfer-Ribonucleinsäure *f*
transfer unit Übergangseinheit *f*
(Füllkörperkolonne), **height of** ~
Bodenäquivalent *n*, **overall gas-phase** ~
gasseitige Übergangseinheit *f* (Dest)
transform Bildfunktion *f* (Math), trans-Form *f*
(Stereochem)
transform *v* transformieren (Elektr), überführen,
umbilden, umformen, umspannen (Elektr),
verwandeln
transformable umwandelbar
transformation Transformation *f* (Elektr, Med),
Transposition *f* (Math), Umänderung *f*,
Umformung *f*, Umsetzung *f* (Chem),
Umwandlung *f*, Verwandlung *f*, **degree of** ~
Umwandlungsverhältnis *n*, **heat of** ~
Umwandlungswärme *f*, **radioactive** ~
radioaktive Umwandlung, **rate of** ~
Umwandlungsgeschwindigkeit *f*, ~ **of**
elements Elementumwandlung *f*, ~ **of energy**
Energieumwandlung *f*
transformation diagram
Umwandlungsdiagramm *n*
transformation matrix Transformationsmatrix *f*
transformation point Umwandlungspunkt *m*,
Übergangspunkt *m*
transformation process Umwandlungsprozeß *m*
transformation product Umwandlungsprodukt *n*
transformation range Umwandlungsbereich *m*
transformation ratio Übersetzungsverhältnis *n*
transformer Transformator *m* (Elektr),
Umformer *m* (Elektr), Umspanner *m*
(Elektr), **reducing** ~
Reduziertransformator *m*, **regulating** ~
Reguliertransformator *m*, **step-down** ~
Reduziertransformator *m*, **step-up** ~
Aufwärtstransformator *m*

transformer capacity Umspannleistung *f*
transformer coil Transformatorspule *f*
transformer current Transformatorenstrom *m*
transformer station Transformatorenstation *f*
transgress *v* überschreiten
transient vorübergehend
transient condition Übergangszustand *m*
transient phenomenon Einschwingvorgang *m*
(Elektr)
transient response Ansprechverzögerungszeit *f*,
Übergangsverhalten *n*,
Übertragungsverhalten *n* (Am. E.),
Übergangsfunktion *f* (Regeltechn)
transient state Übergangszustand *m*,
vorübergehender Zustand *m*
transient time Einschwingzeit *f* (Elektr)
transient voltage Ausgleichspannung *f*
transillumination Durchleuchtung *f*
trans-isomer Transform *f* (Stereochem)
transistor Transistor *m* (Elektr), **point contact** ~
Spitzentransistor *m*
transistor apparatus Transistorgerät *n*
transit Durchgang *m*
transition Übergang *m* (Chem, Phys),
Zustandsveränderung *f* (Chem, Phys),
forbidden ~ verbotener Übergang,
Übergangsverbot *n*, **forced or nonspontaneous**
~ erzwungener Übergang, **heat of** ~
Überführungswärme *f*, Übergangswärme *f*,
intrashell ~ Übergang *m* innerhalb einer
Schale (Atom), **permitted** ~ erlaubter
Übergang, ~ **without radiation** strahlenloser
Übergang
transitional cell Übergangszelle *f* (Biol)
transitional period Übergangszeit *f*
transitional stage Übergangsstadium *n*
transitional state Übergangszustand *m*
transition color Übergangsfarbe *f*
transition element Übergangselement *n* (Chem)
transition frequency Sprungfrequenz *f*
transition half life Übergangshalbwertzeit *f*
transition limestone Grauwackenkalkstein *m*
transition metal Übergangsmetall *n*
transition phenomenon Übergangserscheinung *f*
transition point Übergangspunkt *m*,
Umwandlungspunkt *m*
transition probability
Übergangswahrscheinlichkeit *f*
transition range Umwandlungsbereich *m*
transition region Übergangsbereich *m*,
Überlappungsbereich *m*
transition resistance Übergangswiderstand *m*
transition state Übergangszustand *m*
transition temperature Übergangstemperatur *f*,
Umwandlungstemperatur *f*
transition zone Übergangszone *f*,
Umwandlungszone *f*
transit phase angle Laufwinkel *m*
(Laufzeitwinkel)

transit time Durchgangszeit f
transketolase Transketolase f (Biochem)
translation Translation f, Übertragung f,
  regulation of ~ Translationsregulation f
  (Biochem)
translational energy Translationsenergie f
translational motion Translationsbewegung f
translation group Translationsgruppe f (Krist)
translation plane Translationsebene f (Krist),
  Translationsfläche f (Krist)
translatory motion Parallelverschiebung f
translatory velocity
  Translationsgeschwindigkeit f
translucency Durchscheinen n,
  Durchsichtigkeit f
translucent durchscheinend, lichtdurchlässig,
  opak, transluzent
translucent agate Eisachat m (Min)
transmarine cable Überseekabel n
transmethylation Transmethylierung f
transmissibility Durchlässigkeit f (Phys)
transmissible übertragbar (Med)
transmission Übertragung f, Fortpflanzung f
  (Phys), Lichtdurchlässigkeit f (Phys),
  Transmission f (Techn), Übersetzung f
  (Techn), Weiterleitung f, **direct** ~ direkte
  Übertragung, **ratio of** ~
  Übersetzungsverhältnis n, ~ **of power**
  Kraftübertragung f
transmission belt Treibriemen m
transmission characteristic (Brit. E.)
  Übertragungsfunktion f (Regeltechn)
transmission cross section
  Transmissionsquerschnitt m
transmission drive Transmissionsantrieb m
transmission factor
  Durchlässigkeitskoeffizient m
transmission gear Übersetzungsgetriebe n,
  Zahnradübersetzung f
transmission grate Durchlaßgitter n
transmission line, high-voltage overhead ~
  Hochspannungsfreileitung f
transmission line method Meßleitungsverfahren n
transmission oil Getriebeöl n
transmission shaft Getriebewelle f
transmit v übertragen, durchlassen (Licht),
  funken, übermitteln, übersenden, vererben
  (Biol)
transmittance Fortpflanzung f (Phys)
transmitted light Durchlicht n, **[viewed] by** ~ in
  der Durchsicht f
transmitter Sendeapparat m, Sender m,
  Übertrager m
transmitter valve or tube Senderöhre f
transmitting aerial Sendeantenne f
transmitting and receiving set Sende- und
  Empfangsgerät n
transmitting station Sendestation f

transmutation Transmutation f (Biol),
  Umwandlung f (Atom), **radioactive** ~
  radioaktive Umwandlung, ~ **of atoms**
  Umwandlung f der Atome (pl)
transmute v umwandeln; verwandeln
transparence Durchsichtigkeit f
transparency Durchsichtigkeit f,
  Lichtdurchlässigkeit f, Transparenz f
transparent durchsichtig, diaphan,
  durchscheinend, lasierend, lichtdurchlässig,
  transparent
transparent pack Klarsichtpackung f
transparent paint Lacklasurfarbe f
transparent paper Ölpapier n
transparent sheet Klarsichtfolie f
transparent varnish color Lacklasurfarbe f
transpeptidase Transpeptidase f (Biochem)
transphosphatase Transphosphatase f (Biochem)
transphosphorylase Transphosphorylase f
  (Biochem)
transphosphorylation Transphosphorylierung f
transpire v ausdünsten, schwitzen
transplant v verpflanzen
transplantation of tissue Gewebeverpflanzung f
transport Beförderung f, Transport m, **active** ~
  aktiver Transport m, **coupled** ~ gekoppelter
  Transport m, **faciliated** ~ erleichterter
  Transport m, ~ **of charge** Gichtbeförderung f,
  ~ **of masses** Massenbeförderung f
transport v [weiter]befördern, transportieren
transportable transportierbar, förderbar,
  versandfähig
transportation Beförderung f, **means of** ~
  Transportmittel n, Verkehrsmittel n
transport box Transportkasten m
transport case Transportkasten m
transport catalysis Übertragungskatalyse f
transporter Beförderer m, Verlader m
transporter bridge Verladebrücke f
transporting installation
  Beförderungsvorrichtung f
transport number of ions
  Ionenüberführungszahl f (Elektrochem)
transport protein Transportprotein n
transport vessel Transportgefäß n
transpose v umlagern
trans position Trans-Stellung f (Stereochem),
  Umlagerung f (Chem)
trans-sulfurase Transsulfurase f (Biochem)
transuranic element Transuran n
transversal Transversale f (Math)
transversal a quer, transversal
transversal axis Querachse f
transversal effect Quereffekt m (Krist)
transversality condition
  Transversalitätsbedingung f (Math)
transversal strength Querfestigkeit f
transverse quer[laufend], schräg, transversal
transverse axis Querachse f

transverse bending test Querbiegeversuch *m*
transverse blow-up ratio Querblasverhältnis *n*
transverse borehole Querbohrung *f*
transverse contraction Querzusammenziehung *f*
transverse direction Querrichtung *f*
transverse extension Querdehnung *f*
transverse finning Querberippung *f*
transverse fold Querfalte *f*
transverse oscillation Transversalschwingung *f*
transverse section Querschnitt *m*
transverse spacing Querteilung *f*
transverse strain Querbeanspruchung *f*
transverse strength Biegefestigkeit *f*,
  Bruchfestigkeit *f*
transverse test Biegeprüfung *f* (Gußeisen)
transverse vibration Querschwingung *f*
transverse wave Transversalwelle *f*
transverter Umrichter *m*
trap Abscheider *m*, Falle *f*, Haftstelle *f* (Phys),
  Scheidevorrichtung *f*
trap *v* [ab]fangen, einfangen
trap door Klappe *f*
trapeziform trapezförmig
trapezohedron Trapezoeder *n* (Krist)
trapezoid Trapez *n*
trapezoidal trapezförmig
trapezoidal rule Trapezregel *f*
trapezoidal washer Trapezring *m*
trapped eingefangen
trapped slag Schlackeneinschluß *m*
trapping of air Lufteinschluß *m*
trap tuff Trapptuff *m* (Geol)
trash Auswurf *m*, Abfall *m*
trass Duckstein *m* (Geol), Traß *m* (Geol)
trass concrete Traßbeton *m*
traumaticin[e] Traumaticin *n*
travel Hub *m*, Radlauf *m*
travel *v* wandern, ~ **through** durchsetzen
traveling crab Laufkatze *f*
traveling crane Laufkran *m*
traveling grate Wanderrost *m*
traveling of ions Ionenwanderung *f*
traveling platform Schiebebühne *f*
traveling stoker feed Wanderrostbeschickung *f*
travelling gear Fahrantrieb *m*
travelling-grate stoker Wanderrost *m*
travelling pan filter Bandzellenfilter *n*,
  Kapillarband-Filter *n*
travel time Laufzeit *f*
traverse *v* durchkreuzen
traverse frame Rahmenquerspant *m*
traverser Schiebebühne *f*
travertine Travertin *m* (Min)
tray Austauschboden *m* (Rektifikation),
  Boden *m* (Rektifikation), Tablett *n*, **wire mesh**
  ~ Rostboden *m* (Dest)
tray column Bodenkolonne *f*
tray converter Hordenkontaktkessel *m*

tray [or plate] efficiency
  Bodenverstärkungsverhältnis *n*
trays, equivalent theoretical number of ~
  äquivalente theoretische Bodenzahl *f*, **number**
  of actual ~ Zahl der einzubauenden Böden,
  number of theoretical ~ Trennstufenzahl *f*
  (Dest)
[tray] spacing Bodenabstand *m* (Dest)
tray sublimer Hordensublimator *m*
treacle Sirup *m*, Zuckerdicksaft *m*,
  Zuckerhonig *m*, **containing** ~ siruphaltig
treacly sirupartig
tread Lauffläche *f* (Reifen), Reifenprofil *n*
  (Reifen)
tread *v* treten
tread cut Schnitt *m* in der Lauffläche
tread design Lauffflächenprofil *n*
tread gum Lauffflächenplatte *f*
treadle bar Sperrschiene *f*
tread radius Lauffflächenkrümmung *f*
tread skirt Auslauf *m* des rohen Laufstreifens
tread skiver Ablangmaschine *f* für Laufstreifen
  (pl)
treasure Kleinod *n*, Schatz *m*
treat *v* bearbeiten, behandeln, ~ **subsequently**
  nachbehandeln, ~ **with lime** abkalken,
  einkalken
treatise [wissenschaftliche] Abhandlung *f*
treatment Aufbereitung *f*, Bearbeitung *f*,
  Behandlung *f*, Verarbeitung *f*, **further** ~
  Weiterbehandlung *f*, **method of** ~
  Behandlungsverfahren *n*, Behandlungsweise *f*,
  preliminary ~ Vorbehandlung *f*, **subsequent**
  ~ Weiterbehandlung *f*, Weiterverarbeitung *f*
treatment and settling vessel Aufbereitungs- und
  Absetzbehälter *m*
treble dreifach
tree Baum *m*
treelike baumähnlich
tree resin Baumharz *n*
trehalase Trehalase *f* (Biochem)
trehalose Trehalose *f*
trellis [work] Gitterwerk *n*
tremble *v* beben, zittern
tremolite Tremolit *m* (Min), Weißhornblende *f*
  (Min)
tremuloidin Tremuloidin *n*
trench Graben *m*, tiefe Rinne *f*
trend Neigung *f*, Tendenz *f*
trespass *v* übertreten
tretamine Tretamin *n*
triacanthine Triacanthin *n*
triacetate Triacetat *n*
triacetin Glycerintriacetat *n*, Triacetin *n*
triacetone diamine Triacetondiamin *n*
triacontane Triakontan *n*
triacontanol Triacontanol *n*
triacontylene Triakontylen *n*
tri[acyl]glyceride Triacylglycerid *n*

triakisoctahedron Triakisoktaeder *n* (Krist)
trial Probe *f,* Versuch *m*
trial and error method empirisches
　Ermittlungsverfahren *n,*
　Näherungsverfahren *n*
trial calculation Proberechnung *f*
trialkylsilane Trialkylsilan *n*
trial run Probelauf *m*
triaminoazobenzene Anilinbraun *n,*
　Bismarckbraun *n,* Vesuvin *n*
triaminobenzoic acid Triaminobenzoesäure *f*
triaminotriphenylmethane Leukanilin *n*
triangle Dreieck *n,* Winkel *m* (Techn),
　equilateral ~ gleichseitiges Dreieck, **isosceles**
　~ gleichschenkliges Dreieck, **rectangular** ~
　rechtwinkliges Dreieck, **scalene** ~
　ungleichseitiges Dreieck, **spherical** ~
　sphärisches Dreieck
triangle connection Dreieckschaltung *f*
triangle of forces Kräftedreieck *n*
triangular dreieckig, dreiseitig, dreiwinkelig
triangular cross section Dreiecksquerschnitt *m*
triangular diagram Dreiecksdiagramm *n*
triangular electrode Dreieckselektrode *f*
triangular file Dreikantfeile *f*
triangular groove Dreikantnut *f*
triangular inequality Dreiecksungleichung *f*
triangular matrix Dreiecksmatrize *f* (Math)
triangular mixer Dreikantrührer *m*
trianthrimide Trianthrimid *n*
triarylmethyl carbanion
　Triarylmethyl-Carbanion *n*
triarylmethyl carbonium ion
　Triarylmethyl-Carboniumion *n*
triarylmethyl radical Triarylmethyl-Radikal *n*
triatomic dreiatomig
triaxial dreiachsig
triazine Triazin *n*
triazo- Azido-, Triazo-
triazobenzene Triazobenzol *n*
triazole Triazol *n,* **1,2,3-** ~ Osotriazol *n*
triazolidine Triazolidin *n*
triazoline Triazolin *n*
triazolone Triazolon *n*
triazolyl Triazolyl-
tribasic dreibasig, dreibasisch
tribenzylamine Tribenzylamin *n*
tribromoacetaldehyde Bromal *n*
tribromoacetic acid Tribromessigsäure *f*
tribromoaniline Tribromanilin *n*
tribromoethanal, 2,2,2- ~ Bromal *n*
tribromohydrin Tribromhydrin *n*
tribromomethane Bromoform *n*
tribromonitromethane Bromopikrin *n*
tribromophenol Tribromphenol *n*
tributary function Nebenfunktion *f*
tributyrin Tributyrin *n,* Butyrin *n*
tricalcium phosphate Tricalciumphosphat *n*
tricaprin Tricaprin *n*

tricaproin Tricaproin *n*
tricaprylin Tricaprylin *n*
tricarballylic acid Tricarballylsäure *f*
tricarboxylic acid cycle Tricarbonsäurezyklus *m,*
　Zitronensäurezyklus *m*
tricatin Tricatin *n*
trichalcite Trichalcit *m* (Min)
trichina (pl. trichinae) Trichine *f*
trichinoscope lamp Trichinoskoplampe *f*
trichinosis Trichinose *f* (Med)
trichinous trichinös (Med)
trichite Trichit *m*
trichloroacetaldehyde Chloral *n*
trichloroacetic acid Trichloressigsäure *f*
trichlorobutylidene Trichlorbutyliden *n*
trichloroethanal Chloral *n*
trichloroethane Trichloräthan *n*
trichloroethylene Trichloräthylen *n*
trichloroethylurethan Voluntal *n*
trichlorohydrin Trichlorhydrin *n*
trichloroisopropyl alcohol Isopral *n*
trichloromethane Chloroform *n*
trichloromethyl chloroformate Diphosgen *n*
trichloronitromethane Chlorpikrin *n*
trichlorosilane Siliciumchloroform *n,*
　Trichlorsilan *n*
trichobacterium Fadenbakterium *n* (Bakt),
　Trichobakterium *n* (Bakt)
trichodesmic acid Trichodesminsäure *f*
trichosanic acid Trichosansäure *f*
trichotomy Dreiteiligkeit *f*
trichroism Dreifarbigkeit *f* (Opt),
　Trichroismus *m* (Opt)
trichromatic dreifarbig
trichromatic colorimeter
　Dreifarbenkolorimeter *n*
trichromatic filter Dreifarbenfilter *n*
trichromatism Dreifarbigkeit *f*
tricin Tricin *n*
trick Trick *m*
trickle *v* tröpfeln, rieseln, rinnen, sickern, ~
　through durchsickern
trickle-bed reactor Rieselreaktor *m*
trickle charger Kleinlader *m* (Elektr)
trickled-surface heat exchanger
　Rieselwärmeaustauscher *m*
trickle flow Rieselströmung *f*
trickling Berieselung *f,* Rieseln *n*
trickling water Rieselwasser *n*
triclinic triklin (Krist)
tricosane Trikosan *n*
tricosanol Trikosylalkohol *m*
tricosanone, 12- ~ Lauron *m*
tricosyl alcohol Trikosylalkohol *m*
tricot Trikot *m* (Text)
tricresol Trikresol *n*
tricresyl phosphate Trikresylphosphat *n*
tricyanic acid Cyanursäure *f,* Tricyansäure *f*
tricyanogen chloride Tricyanchlorid *n*

tricyanomethane Cyanoform *n*
tricyclamol Tricyclamol *n*
tricyclazine Tricyclazin *n*
tricyclene Tricyclen *n*
tricyclenic acid Tricyclensäure *f*
tricymylamine Tricymylamin *n*
tridecanal Tridecylaldehyd *m*
tridecane Tridecan *n*
tridecanone, 7- ~ Önanthon *n*
tridecylaldehyde Tridecylaldehyd *m*
tridecylene Tridecylen *n*
tridecylic acid Tridecylsäure *f*
tridymite Tridymit *m* (Min)
tried bewährt, erprobt
trielaidin Trielaidin *n*
triethanolamine Triäthanolamin *n*
triethylamine Triäthylamin *n*
triethylbenzene Triäthylbenzol *n*
triethylene diamine Triäthylendiamin *n*
triethylene glycol Triäthylenglykol *n*
triethylene rhodamine Triäthylenrhodamin *n*
triethylphosphine Triäthylphosphin *n*
triethylsilane Triäthylsilan *n*
triferrin Triferrin *n*
trifle Bagatelle *f,* Kleinigkeit *f*
trifluoromethane Fluoroform *n*
triformol Triformol *n*
trigger Abzughahn *m* (einer Spritzpistole),
   Auslösehebel *m,* Auslöser *m* (Phot),
   Startreagens *n* (Kettenreaktion)
trigger circuit Kippschaltung *f* (Elektr),
   Triggerschaltung *f* (Elektr)
trigger current Auslösestrom *m*
trigger electrode Steuerelektrode *f*
triggering impulse Auslöseimpuls *m*
trigger lag Auslöseverzug *m*
trigger magnet Auslösemagnet *m*
trigger mechanism Auslösevorrichtung *f*
trigger [off] *v* auslösen
triglyceride Triglycerid *n*
triglycol Triglykol *n*
trigonal dreieckig, trigonal (Krist)
trigonalbipyramidal trigonalbipyramidal (Krist)
trigondodecahedron Trigondodekaeder *n* (Krist)
trigonelline Trigonellin *n*
trigonometer Trigonometer *n*
trigonometric trigonometrisch
trigonometrical function, inverse ~
   Arc-Funktion *f* (Math)
trigonometric function Winkelfunktion *f*
trigonometry Trigonometrie *f* (Math)
trihedral dreiflächig, dreikant
trihedron Dreiflächner *m* (Krist)
trihexosan Trihexosan *n*
trihydroxyanthraquinone, 1,2,6- ~
   Flavopurpurin *n,* **1,2,7-** ~ Anthrapurpurin *n*
trihydroxybenzene Trihydroxybenzol *n,* **1,2,3-** ~
   Pyrogallol *n,* Pyrogallussäure *f*
trihydroxybenzoic acid, 3,4,5- ~ Gallussäure *f*

trihydroxycyanidine Tricyansäure *f*
trihydroxypurine Trihydroxypurin *n*
trihydroxytriphenylmethane Leukaurin *n*
trihyroxyflavanone, 4,5,7- ~ Naringenin *n*
triiodomethane Jodoform *n*
triisopropylphenol Triisopropylphenol *n*
trilactin Trilactin *n*
trilateral dreiseitig
trilaurin Trilaurin *n*
trilex rim Trilexfelge *f*
trilinear coordinate Dreieckskoordinate *f*
trilinolein Trilinolein *n*
trilite Trilit *n,* Trinitrotoluol *n*
trilling Drillingskristall *n* (Krist)
trillion (Am. E.) Billion *f,* (Br. E.) Trillion *f*
trilon Trilon *n*
trim (original size minus size after trimming)
   Beschnitt *m* (Buchdr)
trim *v* abkanten (Techn), garnieren, ordnen,
   zurechtschneiden (Techn), zurichten (Techn)
trimargarin Trimargarin *n*
trimellitic acid Trimellitsäure *f*
trimer Trimer *n*
trimeric trimer
trimerite Trimerit *m* (Min)
trimesic acid Trimesinsäure *f*
trimesitinic acid Trimesinsäure *f*
trimetaphan Trimetaphan *n*
trimetaphosphoric acid Trimetaphosphorsäure *f*
trimethadione Trimethadion *n*
trimethindinium Trimethindinium *n*
trimethoxybenzoic acid
   Trimethoxybenzoesäure *f,* **2,4,5-** ~
   Asaronsäure *f*
trimethyl-1-cyclopentene, 1,2,3- ~ Laurolen *n*
trimethyl-2,6-dihydroxypurine, 1,3,7- ~ Thein *n*
trimethylacetaldehyde Pivalinaldehyd *m*
trimethylacetic acid Pivalinsäure *f*
trimethylacetophenone Trimethylacetophenon *n*
trimethylallantoin, 1,3,6- ~ Kaffolin *n*
trimethylamine Trimethylamin *n,* Secalin *n*
trimethylamine oxidase Trimethylaminoxidase *f*
   (Biochem)
trimethylamine oxide Kanirin *n*
trimethylammonium bromide
   Trimethylammoniumbromid *n*
trimethylaniline, 2,4,5- ~ Pseudocumidin *n,*
   2,4,6- ~ Mesidin *n*
trimethylbenzene Trimethylbenzol *n,* **1,2,3-** ~
   Hemellitol *n,* **1,2,4-** ~ Pseudocumol *n,* **1,3,5-**
   ~ Mesitylen *n*
trimethylbenzoic acid, 2,4,5- ~ Durylsäure *f*
trimethylboron Bormethyl *n,* Trimethylbor *n*
trimethylbutane Trimethylbutan *n*
trimethylcellulose Trimethylcellulose *f*
trimethylchlorosilane Trimethylchlorsilan *n*
trimethylcyclohexane Trimethylcyclohexan *n*
trimethylene Trimethylen *n*

trimethylene bromohydrin
   Trimethylenbromhydrin *n*
trimethylene cyanide Trimethylencyanid *n*
trimethylene diamine Trimethylendiamin *n*
trimethylene glycol Trimethylenglykol *n*
trimethylene imine Trimethylenimin *n*
trimethylene oxide Trimethylenoxid *n*
trimethylene trinitramine Hexogen *n*
trimethyl ester Trimethylester *m*
trimethylethylene Pental *n*
trimethylglucose Trimethylglucose *f*
trimethylglycocoll Betain *n*,
   Trimethylglykokoll *n*
trimethylhydroquinone Trimethylhydrochinon *n*
trimethylmethane Trimethylmethan *n*
trimethylnaphthalene Trimethylnaphthalin *n*
trimethylpentane Trimethylpentan *n*
trimethylphenol, 2,4,6- ~ Mesitol *n*
trimethylsilane Trimethylsilan *n*
trimethylstibine Trimethylstibin *n*
trimethylsuccinic acid Trimethylbernsteinsäure *f*
trimethylxanthine Trimethylxanthin *n*, **1,3,7-** ~
   Coffein *n*, Thein *n*
trim line Trimmkante *f*
trimming Besatz *m*, Borte *f*, Einfassung *f*,
   Entgraten *n*, Zurichten *n* (blockmaking),
   Ausschnitt *m*
trimming condenser Abgleichkondensator *m*
trimming cutter Beschneidmaschine *f*
trimming machine Zuschneidemaschine *f*
trimming press Abkantpresse *f*
trimmings Abfälle *pl*, Schneidabfall *m*
trimming tool Abgratwerkzeug *n*
trimolecular trimolekular
trimorphine Trimorphin *n*
trimorphism Dreigestaltigkeit *f*, Trimorphie *f*
   (Krist)
trimorphous dreigestaltig
trim size Schneidmaß *n*
trimyristine Trimyristin *n*
trinitrin Nitroglycerin *n*, Trinitrin *n*
trinitroanthraquinone Trinitroanthrachinon *n*
trinitrobenzene Trinitrobenzol *n*
trinitrobenzoic acid Trinitrobenzoesäure *f*
trinitrobutyltoluene Trinitrobutyltoluol *n*
trinitrochlorobenzene Trinitrochlorbenzol *n*
trinitrofluorenone Trinitrofluorenon *n*
trinitroglycerine Nitroglycerin *n*
trinitromethane Nitroform *n*
trinitrophenol Trinitrophenol *n*
trinitrophenylacridine Trinitrophenylacridin *n*
trinitroresorcinol Trinitroresorcin *n*
trinitrostilbene Trinitrostilben *n*
trinitrotoluene Trinitrotoluol *n*
trinitroxylene Trinitroxylol *n*
trinol Trinitrotoluol *n*, Trinol *n*
trinomial dreigliedrig (Math)
trinucleotide Trinucleotid *n*
trioctyl phosphate Trioctylphosphat *n*

triode Dreielektrodenröhre *f* (Elektr), Triode *f*
   (Elektr)
triolein Triolein *n*, Olein *n*
trional Trional *n*
triose Triose *f*
triose isomerase Trioseisomerase *f* (Biochem)
triose mutase Triosemutase *f* (Biochem)
triose phosphate Triosephosphat *n*
triose phosphate dehydrogenase
   Triosephosphatdehydrogenase *f* (Biochem)
triosephosphate isomerase
   Triosephosphatisomerase *f* (Biochem)
triose reductone Trioseredukton *n*
trioxan[e] Trioxan *n*
trioxide Trioxid *n*
trioxymethylene Trioxymethylen *n*
tripalmitin Tripalmitin *n*
tripedal dreifüßig
tripelargonin Tripelargonin *n*
tripelennamine Tripelennamin *n*
trip-free mechanism Fortschaltungsmechanik *f*
triphane Triphan *m* (Min), Spodumen *m* (Min)
triphase current Dreiphasenstrom *m*
triphase system Dreiphasensystem *n*
triphenin Triphenin *n*
triphenylamine Triphenylamin *n*
triphenylbenzene Triphenylbenzol *n*
triphenylbromosilane Triphenylbromsilan *n*
triphenylcarbinol Triphenylcarbinol *n*
triphenylchloromethane Tritylchlorid *n*
triphenylene Triphenylen *n*
triphenylimidazole, 2,4,5- ~ Lophin *n*
triphenylmethane Triphenylmethan *n*
triphenylmethane dye
   Triphenylmethanfarbstoff *m*
triphenylmethyl Triphenylmethyl-, Trityl-
triphenylmethyl radical
   Triphenylmethylradikal *n*
triphenylphosphate Triphenylphosphat *n*
triphenylphosphine Triphenylphosphin *n*
triphenylphosphite Triphenylphosphit *n*
triphenylphosphonium bromide
   Triphenylphosphoniumbromid *n*
triphenylguanidine Triphenylguanidin *n*
triphenylrosaniline Triphenylrosanilin *n*
triphenylsilane Triphenylsilan *n*
triphenylsilanol Triphenylsilanol *n*
triphenylsilicon chloride
   Triphenylsiliciumchlorid *n*
triphenylstibine Triphenylstibin *n*
triphenylstibine sulfide Sulfoform *n*
triphosphopyridine nucleotide
   Triphosphopyridinnucleotid *n* (TPN)
triple dreifach, dreizählig
triple bond Dreifachbindung *f* (Chem),
   Acetylenbindung *f* (Chem)
triple coincidence Dreifachkoinzidenz *f*
triple collision Dreierstoß *m*
triple condenser Dreifachkühler *m*

triple effect Drillingswirkung *f*
triple-flame dreiflammig
triple helix Tripelhelix *f*
triple linkage Dreifachbindung *f*
triple melting-down process Dreimalschmelzerei *f*
triple point Tripelpunkt *m* (Phys)
triple point pressure Tripelpunktsdruck *m*
triple recorder Dreifachschreiber *m*
triple roller grinding mill
   Dreiwalzenreibemaschine *f*
triple roller mill Dreiwalzwerk *n*
triple roller mill stand Dreiwalzenständer *m*
triple salt Tripelsalz *n*
triple-salt pigment Dreisalzpigment *n*
triple split Dreifachaufspaltung *f*
triplet Triplett *n*
triplet series Triplettserie *f*
triplet spectrum Triplettspektrum *n*
triplet state Triplettzustand *m*
triplet system Triplettsystem *n*
triple tube Dreifachröhre *f*
triplex glass Sicherheitsglas *n*
triplite Triplit *m* (Min), Eisenpecherz *n*
triploidite Triploidit *m* (Min)
tripod Dreibein *n*, Dreifuß *m*
tripod stand Dreifußstativ *n*
tripod starter Dreifußanlasser *m*
tripoli Tripel *m*, Polierschiefer *m*, Tripelerde *f*
tripoli slate Tripelschiefer *m* (Min)
tripolyphosphoric acid Tripolyphosphorsäure *f*
tripper Abwurfvorrichtung *f*, Abwurfrolle *f*,
   Abwurftrommel *f*
trippkeite Trippkeit *m* (Min)
trip relay Auslöserelais *n*
triprolidine Triprolidin *n*
tripropionin Tripropionin *n*
tripropylamine Tripropylamin *n*
tripropylsilane Tripropylsilan *n*
trip switch Auslöseschalter *m*
triptane Triptan *n*
triptycene Triptycen *n*
tripyrrole Tripyrrol *n*
triricinolein Triricinolein *n*, Ricinolein *n*
trisaccharide Trisaccharid *n*
trisazo-dye Trisazofarbstoff *m*
trisect *v* dreiteilen
trisilane Trisilan *n*
trisilylamine Trisilylamin *n*
trisodium citrate Trinatriumcitrat *n*
trisodium phosphate Trinatriumphosphat *n*
tristearin Tristearin *n*, Stearin *n*
tristimulus method Dreibereichsverfahren *n*
   (Färb)
tristimulus value Normfarbwert *m*
trisulfane Trisulfan *n*
tritane Tritan *n*
trite abgedroschen
triterpene Triterpen *n*
triterpenoid Triterpenoid *n*

trithiane Trithian *n*
trithioacetophenone Trithioacetophenon *n*
trithiobenzaldehyde Trithiobenzaldehyd *m*
trithiocarbonic acid Trithiokohlensäure *f*
trithioformaldehyde Trithioformaldehyd *m*
trithiol Trithiol *n*
trithiolane Trithiolan *n*
trithione Trithion *n*
trithionic acid Trithionsäure *f*
tritiated tritiummarkiert, mit Tritium behandelt
tritisporin Tritisporin *n*
tritium (Symb. T) Tritium *n*
tritium beta rays Tritiumbetastrahlen *m pl*
tritol Tritol *n*
triton Triton *n*
tritopine Tritopin *n*
triturate *v* pulverisieren, verreiben, zermahlen,
   zermalmen, zerreiben
triturating apparatus Reibapparat *m*
trituration Pulverisierung *f*, Zerreibung *f*
trityl Triphenylmethyl-, Trityl-
trityl alcohol Tritylalkohol *m*
trityl chloride Tritylchlorid *n*
triundecylin Triundecylin *n*
trivalence Dreiwertigkeit *f* (Chem)
trivalent dreiwertig (Chem), dreibindig (Chem)
trivalerin Trivalerin *n*, Phocaenin *n*
trivial geringfügig, trivial
trivial name Trivialname *m*
troche Pastille *f*
trochoid Trochoide *f* (Math)
trochoidal analyzer Trochoidalanalysator *m*
trochoidal pump Trochoidenpumpe *f*
trochotron Trochotron *n* (Phys)
troegerite Trögerit *m* (Min)
Troeger's base Trögersche Base *f*
troilite Troilit *m* (Min)
trolite Trolit *m* (Kunststoff)
trolleite Trolleit *m* (Min)
trolley Kontaktrolle *f* (Techn), Laufrolle *f*
   (Stromabnehmerrolle), Rollwagen *m*
trolley jib Katzenausleger *m* (Kran),
   Laufkatzenausleger *m*
trolley track Laufkatzenfahrbahn *f*
trolley wheel Kontaktrolle *f*
trollixanthin Trollixanthin *n*
trona [salt] Tronasalz *n*
troostite Troostit *m* (Min), Hartperlit *m* (Min)
tropacocaine Tropacocain *n*
tropaic acid Tropasäure *f*
tropaic aldehyde Tropaaldehyd *m*
tropan Tropan *n*
tropanol Tropanol *n*, 3- ~ Tropin *n*
tropanone Tropanon *n*
tropein[e] Tropein *n*
tropeolin Tropäolin *n*
tropic acid Tropasäure *f*
tropical climate Tropenklima *n*

tropical conditions, resistant to ~ tropenfest,
 stability to ~ Tropenfestigkeit *f*
tropic aldehyde Tropaaldehyd *m*
tropicalized tropengeschützt
tropics Tropen *pl*, **insulation for the** ~
 Tropenisolierung *f*
tropidine Tropidin *n*
tropigenine Tropigenin *n*
tropigline Tropiglin *n*
tropilidene Tropiliden *n*
tropilidin Tropilidin *n*
tropine Tropin *n*
tropine carboxylic acid Ecgonin *n*
tropinic acid Tropinsäure *f*
tropinone Tropinon *n*
tropocollagen Tropokollagen *n*
tropolone Tropolon *n*
tropomyosin Tropomyosin *n*
tropone Tropon *n*
troponin Troponin *n*
troposphere Troposphäre *f*
tropylium ion Tropylium-Ion *n*
trotyl Trinitrotoluol *n*, Trotyl *n*
trouble Störung *f*
trouble *v* belästigen, stören
trouble-location problem Störsuchaufgabe *f*
troubleproof störungsfrei
trouble shooting Fehlersuche *f*,
 Störungsbeseitigung *f*, Störungssuche *f*
trouble spot Störungszentrum *n*
trough Mulde *f*, Tal *n* (einer Kurve), Trog *m*,
 Walkbassin *n*, Wanne *f*, **pneumatic** ~
 pneumatische Wanne, ~ **(of a wave)**
 Wellental *n*
trough charging crane Muldenchargierkran *m*,
 Muldensetzkran *m*
trough conveyor Trogförderband *n*
trough dryer Muldentrockner *m*
troughed-belt conveyor Kurvenbandförderer *m*
troughed conveyor belt Muldengurtförderer *m*
troughing idler Spurrolle *f*
trough kneader Trogkneter *m*
trough lining Trogfutter *n*, Badfutter *n*
trough mixer Trogmischer *m*
trowel Hohlspatel *m*, Maurerkelle *f*
trowel *v* aufspachteln
troy weight Goldgewicht *n*, Karatgewicht *n*
truck Handwagen *m*, Laufrad *n* (Techn),
 Rollwagen *m* (Am. E.), Lastkraftwagen *m*
truck balance Wagonwaage *f*
truck crane Autokran *m*
truck dryer Hordentrockner *m*
truck load Wagenladung *f*
truck rim Lastwagenfelge *f*
true wahr, richtig, tatsächlich, **out of** ~ unrund
true [up] *v* ausrichten
truffle Trüffel *f* (Bot)
trumpet Eingußtrichter *m*
truncate abstumpfen (Math)

truncated gestutzt
truncation Abstumpfung *f* (Math)
trunk Rumpf *m*, Säulenschaft *m* (Arch),
 Schacht *m* (Techn), Stamm *m* (Baum)
trunk cable Fernleitungskabel *n*
trunk call Ferngespräch *n*
trunk line Fernleitung *f*, Hauptleitung *f* (Teleph)
trunnion Zapfen *m*, Drehzapfen *m*
truss Gestell *n*, Hängewerk *n*, Konsole *f*,
 Tragbalken *m*
trustee Bevollmächtigter *m*
truxelline Truxellin *n*
truxene Truxen *n*
truxillic acid Truxillsäure *f*
truxilline Truxillin *n*
truxinic acid Truxinsäure *f*
try *v* versuchen, [aus]probieren, ~ **out** erproben
trypaflavin[e] Trypaflavin *n*
trypan blue Trypanblau *n*, Kongoblau *n*
trypan red Trypanrot *n*
trypan violet Trypanviolett *n*
tryparsamide Tryparsamid *n*
trypsin Trypsin *n* (Biochem)
trypsinogen Trypsinogen *n* (Biochem)
tryptamine Tryptamin *n*
tryptathionine Tryptathionin *n*
tryptazan Tryptazan *n*
tryptic activity tryptische Aktivität *f*
tryptone Trypton *n*
tryptophanase Tryptophanase *f* (Biochem)
tryptophan decarboxylase
 Tryptophandecarboxylase *f* (Biochem)
tryptophan[e] Tryptophan *n*
tryptophanol Tryptophanol *n*
tryptophan oxidase Tryptophanoxidase *f*
 (Biochem)
tryptophan synthetase Tryptophansynthease *f*
 (Biochem)
tryptophol Tryptophol *n*
tschermigite Tschermigit *m* (Min)
Tschugajew's reagent Tschugajeffs Reagens *n*
tsetse fly Tsetsefliege *f* (Zool)
T-shaped T-förmig
T square Reißschiene *f*
tsuzuic acid Tsuzusäure *f*
T tube T-Rohr *n*
tuaminoheptane Tuaminoheptan *n*
tub Wanne *f*, Bottich *m*, Bütte *f*, Kübel *m*,
 Zuber *m*
tubaic acid Tubasäure *f*
tubanol Tubanol *n*
tubatoxin Rotenon *n*
tubazid Tubazid *n*
tube Rohr *n*, Röhre *f* (Elektr), Schlauch *m*,
 acid-proof ~ Säureschlauch *m*, **air-cooled** ~
 luftgekühlte Röhre, **ascending** ~ Steigrohr *n*,
 cemented ~ Rohr *n* mit Klebnaht, **gas-filled**
 ~ gasgefüllte Röhre, **hollow** ~ Hohlröhre *f*,
 machine laminated ~ bearbeitetes Rohr *n* aus

Schichtpreßstoff, **molded laminated** ~
formgepreßtes Rohr *n* aus Schichtpreßstoff,
packed ~ Rohr *n* mit Füllkörpern, **sealed-off**
~ abgeschmolzene Röhre, **water-cooled** ~
wassergekühlte Röhre
tube amplifier Röhrenverstärker *m*
tube brush Röhrenwischer *m*
tube bundle Rohrbündel *n*
tube bundle evaporator
Rohrbündelverdampfer *m*
tube casting Röhrenguß *m*
tube clamp Schlauchklemme *f*
tube converter Röhrenkontaktkessel *m*
tube cutter Rohrabschneider *m*
tube drawing mill Rohrziehstraße *f*
tube evaporator Röhrenverdampfer *m*
tube extruding press Rohrpresse *f*,
Schlauchmaschine *f*
tube extrusion machine Rohrpresse *f*,
Schlauchmaschine *f*
tube filling machine Tubenfüllmaschine *f*
tube fittings Rohrarmaturen *pl*
tube furnace Röhrenofen *m*, Rohrofen *m*
tube generator Röhrengenerator *m*
tube ignition Glührohrzündung *f*
tube joint Rohrverbindung *f*
tube level Röhrenlibelle *f*
tube-like röhrenartig
tube mill Rohrmühle *f*, Trommelmühle *f*
tube ozonizer Röhrenozonisator *m*
tube plug Rohrstöpsel *m*
tubercle Tuberkel *m* (Med)
tubercle bacillus Tuberkelbazillus *m*
tubercular tuberkulös (Med)
tuberculin Tuberkulin *n*
tuberculosis Tuberkulose *f* (Med)
tuberculostearic acid Tuberkulostearinsäure *f*
tuberculous tuberkulös (Med)
tuberculum Tuberkel *m* (Med)
tube reactor Röhrenofen *m*
tube rectifier Röhrengleichrichter *m*
tube rolling Röhrenwalzen *n*
tubes, forwarding [or transmission] ~
Fahrrohre *pl*
tube support Röhrenträger *m*
tube support plate Rohrboden *m*
tube valve Röhrenventil *n*
tube vice Rohrschraubstock *m*
tube winding machine Schlauchwickelmaschine *f*
tub fermentation Bottichgärung *f*
tubing Rohrleitung *f*, Rohrmaterial *n*, **acid-proof**
~ Säureschlauch *m*
tubing clip Schlauchklemme *f*
tubing machine Schlauchmaschine *f*
tubular röhrenartig, röhrenförmig
tubular bag Schlauchbeutel *m*
tubular boiler Röhrenkessel *m*, Heizrohrkessel *m*
tubular bowl centrifuge Superzentrifuge *f*
tubular bowl separator Röhrenseparator *m*

tubular box spanner Rohrsteckschlüssel *m*
tubular condenser Röhrenkondensator *m*
tubular cooler Röhrenkühler *m*
tubular core Rohrkern *m*
tubular furnace Röhrenofen *m*
tubular fuse Röhrensicherung *f*
tubular insulation Rohrisolierung *f*
tubular lamp Röhrenlampe *f*
tubular magnet Röhrenmagnet *m*
tubular pressure filter Rohrdruckfilter *n*
tubular reactor Strömungsrohr *n*
tubular vaporizer Röhrenverdampfer *m*
tubulated flask mit Ansatzrohr
tudor accumulator Tudorakkumulator *m*
tufa Tuff *m* (Geol), **calcareous** ~ Kalktuff *m*
(Geol)
tufaceous tuffartig
tufaceous earth Tufferde *f*
tufaceous limestone Tuffkalk *m*
tuff Tuff *m* (Geol)
tug *v* zerren, schleppen
tugboat Schleppboot *n*, Schlepper *m*
tula metal Tulametall *n*
tumble *v* stürzen, stolpern
tumbler Bleichtrommel *f* (Techn), Wasserglas *n*
tumbler centrifuge Taumelzentrifuge *f*
tumbler switch Kippschalter *m* (Elektr)
tumbler tin-plating Rollfaßverzinnung *f*
tumbling barrel Rollfaß *n*, Scheuertrommel *f*
tumbling mill Trommelmühle *f*, Walzmühle *f*,
autogenous ~ Autogenmühle *f*
tumbling mixer Mischtrommel *f*,
Trommelmischer *m*
tumeric paper Curkumapapier *n*
tumeric test Curkumaprobe *f*
tumor Geschwulst *n*, **experimental** ~
experimenteller Tumor, **growth of the** ~
Tumorwachstum *n*, **malignant** ~ bösartiger
Tumor (Med)
tumor cell Tumorzelle *f*
tumor cells, dislocation of ~
Tumorzellenverschleppung *f* (Med)
tumo[u]r Tumor *m* (Med), Geschwulst *n* (Med),
Wucherung *f* (Med)
tumulosic acid Tumulosinsäure *f*
tune *v* stimmen (Akust), ~ **down** abschwächen
(Akust)
tuned circuit Abstimmkreis *m*
tung oil Holzöl *n*, Tungöl *n*
tung standoil Holzstandöl *n*
tungstate Wolframat *n*
tungsten (Symb. W) Wolfram *n*, **metallic** ~
Wolframmetall *n*, **single crystal of** ~
Wolframeinkristall *m*
tungsten arc lamp Wolframbogenlampe *f*
tungsten bronze Safranbronze *f*,
Wolframbronze *f*
tungsten carbide Wolframcarbid *n*
tungsten content Wolframgehalt *m*

tungsten electrode Wolframelektrode f
tungsten filament Wolframfaden m
tungsten lamp Wolframlampe f
tungsten metal Wolframmetall n
tungsten ore Wolframerz n (Min)
tungsten rectifier Wolframgleichrichter m (Elektr)
tungsten steel Wolframstahl m
tungsten tip Wolframspitze f
tungsten trioxide Wolframtrioxid n
tungsten white Wolframweiß n
tungsten wire Wolframfaden m
tungstic acid Wolframsäure f, salt of ~ Wolframat n
tungstic anhydride Wolframsäureanhydrid n, Wolframtrioxid n
tungstic ocher Wolframocker m (Min)
tungstite Wolframocker m (Min)
tungstoboric acid Borwolframsäure f
tuning Abstimmen n, clearness or selectivity or sharpness of ~ Abstimmschärfe f
tuning apparatus Abstimmvorrichtung f
tuning capacitor Abstimmkondensator m
tuning circuit Abstimmkreis m
tuning coil Abstimmspule f
tuning dial Abstimmskala f
tuning error Abgleichfehler m (Elektr)
tuning fork Stimmgabel f
tuning indication tube Abstimmanzeigerohr f
tuning knob Abstimmknopf m
tuning scale Abstimmskala f
tunnel Stollen m (Bergb), Tunnel m
tunnel-cap tray Tunnelboden m
tunnel drier Tunneltrockner m, Durchlauftrockner m, Kanaltrockner m, Schachttrockner m
tunnel freezer Gefriertunnel m
tunnel furnace Durchsatzofen m
tunneling effect Tunneleffekt m (Atom)
tunnel kiln Tunnelofen m
turanite Turanit m (Min)
turanose Turanose f
turbid trüb[e], getrübt, to become ~ sich trüben
turbidimeter Trübungsmesser m, Turbidimeter n
turbidimetric analysis Nephelometrie f
turbidity Trübung f, Schleierbildung f
turbidity factor Trübungsfaktor m
turbidity point Trübungstemperatur f
turbine Turbine f, axial-flow ~ Axial[strom]turbine f
turbine blade Turbinenschaufel f
turbine efficiency Turbinenwirkungsgrad m
turbine flowmeter, turbometer or ~ Turbinen-Durchflußmesser m
turbine mixer Schaufelradmischer m, Turbinenmischer m, Turbinenrührer m, Turbomischer m
turbine nozzle Turbinendüse f
turbine output Turbinenleistung f

turbine plant Turbinenanlage f
turbine rotor Turbinentrommel f
turbine shaft Turbinenwelle f
turbine sublimer Turbinensublimator m
turbine water meter Einflügelradzähler m
turboblower Turbogebläse n, Kreiselgebläse n
turboburner Turbobrenner m
turbocompressor Turbokompressor m, Kreiselverdichter m
turbodrier Ringetagentrockner m, Turbinentrockner m, Turbotrockner m
turbodynamo Turbodynamo m
turbogenerator Turbogenerator m
turbomixer Turbomischer m
turbomolecular pump Turbomolekularpumpe f
turboviscosimeter Turboviskosimeter n
turbulence Turbulenz f, Durchwirbelung f (Phys), Unruhe f, Wirbelbewegung f (Phys), degree of ~ Turbulenzgrad m
turbulence chamber Wirbelkammer f
turbulence effect Turbulenzeinfluß m
turbulent turbulent
turbulent layer (dynamics) Wirbelschicht f
turicine Turicin n
Turkey brown Türkischbraun n
Turkey red Türkischrot n, Alizarinaltrot n
turmeric Curcuma[wurzel] f (Bot), Gelbwurzel f (Bot)
turmeric paper Curcumapapier n
turmeric test Curcumaprobe f
turmerone Turmeron n
turn Ausschlag m (Waage), Drehung f, Reihenfolge f, Turnus m, Wendung f, sudden ~ Ruck m, ~ (of a screw) Gang m (einer Schraube)
turn v drehen, wenden, ~ aside [sich] abwenden, ~ away [sich] abwenden, ~ off abschalten, abstellen, zudrehen (Hahn), ~ on andrehen, aufdrehen, ~ over umdrehen, wälzen
turnable drehbar
turnbuckle Spannschloß n
Turnbull's blue Turnbulls Blau n
turn-down ratio Arbeitsbereich m (Dest)
turned off abgestellt
turner Dreher m (Techn)
turnerite Turnerit m (Min)
turning Drehen n, Drehung f
turning a kreisend
turning basin Wendebecken n
turning blade Wendeschaufel f
turning chisel Drehmeißel m
turning crane Drehkran m
turning force Drehkraft f
turning gear Schwenkgetriebe n
turning lathe Drehbank f
turning machine Drehwerk n
turning moment Drehmoment n

turning motion Drehbewegung *f*
turning point Drehpunkt *m,* Umkehrpunkt *m,*
Wendepunkt *m* (Math)
turnings pl
turning tool Drehstahl *m,* Abdrehstahl *m,*
Drehdorneisen *n,* Drehwerkzeug *n*
turnover Umsatz *m*
turnover number of an enzyme Wechselzahl eines
Enzyms
turnover pattern plate Wendeplatte *f*
turnover top table Wendeplatte *f*
turns, number of ~ Windungszahl *f*
turnsole Lackmus *m* (Chem)
turnsole blue Lackmusblau *n*
turntable Drehscheibe *f,* Drehtisch *m*
turpentine Terpentin *n,* **mordant based on** ~
Terpentinbeize *f,* **oil of** ~ Terpentinöl *n,*
spirits of ~ Terpentinalkohol *m,*
Terpentinöl *n*
turpentine camphor Terpentincampher *m*
turpentine ointment Terpentinsalbe *f* (Pharm)
turpentine pitch Terpentinpech *n*
turpentine resin Terpentinharz *n*
turpentine substitute Terpentinersatz *m*
turpentine varnish Terpentinfirnis *m,*
Terpentinlack *m*
turquoise Türkis *m* (Min)
turquoise blue Türkisblau *n*
turquoise green Turkisgrün *n*
turret dryer (leather) Trockenturm *m*
turret lathe Revolverdrehbank *f*
tussalvine Tussalvin *n*
tutocaine Tutocain *n*
tutty Gichtschwamm *m,* Ofenbruch *m,*
Ofengalmei *m,* Ofenschwamm *m,*
Zinkschwamm *m*
tuyère Formnase *f* (Gieß), Luftdüse *f,*
Winddüse *f,* **conduit pipe of the** ~
Nasengasse *f*
tuyère gate Düsenabsperrschieber *m*
tuyère hole Windöffnung *f* (Metall)
tuyère nose Formrüssel *m*
tuyère nozzle der Form
tuyère opening Formöffnung *f*
tuyère plate Formzacken *m* (Gieß)
tuyère saddle Formsattel *m*
tuyère slag Formschlacke *f*
tweezers Federzange *f,* Haarzange *f,* Pinzette *f*
twelve dozen Gros *n*
twice doppelt, zweimal
twilight Zwielicht *n*
twill Köper *m* (Text), Drillich *m* (Text),
Köperbindung *f* (Text)
twin Zwilling *m*
twin air supply Zweitluftzufuhr *f*
twin block Zwillingsblock *m*
twin cable Zweileiterkabel *n*
twin centrifugal pump Doppelschleuderpumpe *f*

twin chamber steam trap
Zweikammerkondenstopf *m*
twin clamp Doppelschelle *f*
twin conductor Doppelleiter *m*
twin crystal Doppelkristall *m* (Krist),
Zwillingskristall *m* (Krist)
twin cylinder mixer Zwillingstrommelmischer *m*
twine Schnur *f*
twin-feed spray equipment
Zweikomponentenspritzvorrichtung *f*
twin formation Zwillingsbildung *f*
twin loading portal Doppelladeportal *n*
twinning Zwillingsbildung *f*
twinning axis Zwillingsachse *f* (Krist)
twin pump Zwillingspumpe *f*
twin roller mill Zweiwalzenmühle *f*
twin rollers Duowalze *f*
twin rolling mill Zweiwalzwerk *n*
twin-rotor mixer Zweiwellenrührer *m,*
Zweiwellenrührgerät *n*
twin screw Doppelschnecke *f*
twin-screw pressure mixer
Doppelschrauben-Druckmischer *m*
twin-shell or Vee mixer Hosenmischer *m*
twin socket Doppelsteckdose *f* (Elektr)
twin-threaded doppelstrangig
twin-threaded helix structure doppelstrangige
Helixstruktur *f*
twin worm Doppelschnecke *f*
twin-worm mixer Doppelschneckenmischer *m*
twirl *v* quirlen, wirbeln
twist Drehung *f,* Torsion *f,* Verdrehung *f,*
Windung *f,* **absence of** ~ Drallfreiheit *f,* **angle**
of ~ Torsionswinkel *m,*
Verdrehungswinkel *m*
twist *v* [ver]drehen
twistable [ver]drehbar
twist drill Drehbohrer *m,* Spiralbohrer *m*
twisted verdreht
twisted tooth spur gear Schrägzahnstirnrad *n*
twist factor Drehungsfaktor *m*
twisting Drehen *n,* Verdrehen *n*
twisting force Drehkraft *f,* Torsionskraft *f*
twisting resistance Torsionswiderstand *m*
twisting test Torsionsversuch *m*
twist setting Zwirneinstellung *f* (Web)
Twitchell's reagent Twitchells Reagens *n*
two-bath process Zweibadverfahren *n*
two-beam problem Zweistrahlproblem *n*
two-body collison Zweierstoß *m*
two-body formalism Zweikörperformalismus *m*
two-body problem Zweikörperproblem *n*
two-cell accumulator Doppelakkumulator *m*
two-center bond Zweizentrenbindung *f* (Atom)
two-center problem Zweizentrenproblem *n*
two-colored zweifarbig
two-color printing Zweifarbendruck *m* (Buchdr)
two-compartment mill Zweikammermühle *f*

two-component balance
Zweikomponentenwaage *f*
two-component finish Reaktionslack *m*
two-component mixture Zweistoffgemisch *n*
two-cycle engine Zweitaktmaschine *f*
two-cylinder engine Zweizylindermotor *m*
two-dimensional zweidimensional
twofold doppelt, zweifach
two-girder crane Zweiträgerkran *m*
two lipped end mill
Zweischneider-Langlochfräser *m*
two-meson system Zweimesonensystem *n*
two-necked zweihalsig
two-phase zweiphasig
two-phase [alternating] current
Zweiphasenstrom *m*
two-phase armature Zweiphasenanker *m*
two-phase equilibrium
Zweiphasengleichgewicht *n*
two-phase flow Zweiphasenströmung *f*
(Hydrodynamik)
two-phase four wire system
Zweiphasenvierleitersystem *n*
two-phase generator Zweiphasendynamo *m*,
Zweiphasenstromerzeuger *m*
two-phase motor Zweiphasenmotor *m*
two-phase printing process
Zweiphasendruckverfahren *n*
two-phase system Zweiphasensystem *n*
two-phase three wire system
Zweiphasendreileitersystem *n*
two-phase transformer
Zweiphasentransformator *m*
two-phase wiring Zweiphasenleitung *f*
two-plate capacitor Zweischichtenkondensator *m*
two-polar zweipolig
two-position controller (Am. E.)
Zweipunktregler *m*
two-roll attrition mill
Zweirollen-Scheibenmühle *f*
two-roller mill Duowalzwerk *n*
two-roll mill Zweiwalzenstuhl *m*
two-sided zweiseitig
two-stage zweistufig
two-stage governor Zweistufenregler *m*
two-stage process Zweistufenprozeß *m*
two-step control (Br. E.) Zweipunktregelung *f*
two-step controller Zweipunktregler *m*
two-step reduction Zweistufenreduktion *f*
two-thread worm Zweigangschnecke *f*
two-tone color Doppeltonfarbe *f*
two-way cock Zweiwegehahn *m*
two-way pump Zweiwegepumpe *f*
two-way radio Sende- und Empfangsgerät *n*
two-way switch Wechselschalter *m*
two-way thread Zweiganggewinde *n*
two-way valve Zweiwegeventil *n*, Doppelventil *n*,
Wechselventil *n*
two-wire system Zweileitersystem *n*

two-zone model Zweizonenmodell *n*
two-zone reactor Zweizonenreaktor *m*
tychite Tychit *m* (Min)
tylophorine Tylophorin *n*
tylose Methylcellulose *f*
tymp plate Tümpeleisen *f*
tymp sheet steel Tümpelblech *n*
tymp stone Tümpelstein *m*
Tyndall cone Tyndallkegel *m*
Tyndall effect Tyndalleffekt *m*
type Typ *m*, Art *f*, Sorte *f*, Type *f* (Buchstabe,
Letter), **spaced** ~ gesperrte Schrift *f* (Buchdr)
type metal Letternmetall *n*, Schriftmetall *n*
type plate Typenschild *n*
typesetting Setzen *n* (Buchdr)
type size Kegelhöhe *f* (Buchdr)
type specimen Musterexemplar *n*
typewriter Schreibmaschine *f*
typewriter ribbon Farbband *n*
typhoid bacilli Typhusbakterien *pl*
typhoid fever Typhus *m* (Med)
typhoid vaccine Typhusvakzine *f*
typhoon Taifun *m*, Wirbelsturm *m*
typhotoxin Typhotoxin *n*
typhus Fleckfieber *n* (Med)
tyramine Tyramin *n*
tyre (Br. E.) Reifen *m* (Kfz)
tyrite Tyrit *m* (Min)
tyrocidine Tyrocidin *n*
tyrolite Tyrolit *m* (Min), Kupferschwamm *m*
(Min)
tyrosal Salipyrin *n*
tyrosinase Tyrosinase *f* (Biochem)
tyrosin[e] Tyrosin *n*
tyrosine decarboxylase Tyrosindecarboxylase *f*
(Biochem)
tyrosine transaminase Tyrosintransaminase *f*
(Biochem)
tyrosinol Tyrosinol *n*
tyrosinosis Tyrosinosis *f* (Med)
tyrosol Tyrosol *n*
tyrotoxicon Tyrotoxicon *n*, Tyrotoxin *n*
tyrotoxin Tyrotoxicon *n*, Tyrotoxin *n*
tysonite Tysonit *m* (Min)
tyvelose Tyvelose *f*

U

uabain Ouabain *n*
ubiquinone Ubichinon *n*
uhligite Uhligit *m* (Min)
U-iron U-Eisen *n*
Ukita's acid Ukita-Säure *f*
ulcer Geschwür *n*
uleine Ulein *n*
ulexine Ulexin *n*
ulexite Ulexit *m* (Min)
ullmannite Ullmannit *m* (Min),
　　Antimonnickelkies *m* (Min),
　　Nickelantimonglanz *m* (Min),
　　Nickelspießglanz *m* (Min)
ulmic acid Ulminsäure *f*
ulmin Ulmin *n*
ulmin brown Ulminbraun *n*
ulrichite Ulrichit *m* (Min)
ultimate bearing stress Höchstbelastung *f*
ultimate elongation Dehnungsgrenze *f*
ultimate flexural stress Grenzbiegespannung *f*
ultimate pressure Enddruck *m*
ultimate strength Dehnungsgrenze *f*,
　　Endstabilität *f*
ultimate stress Bruchbelastung *f*
ultimate stress limit Bruchgrenze *f*
ultimate total pressure Endtotaldruck *m*
ultracentrifuge Ultrazentrifuge *f*, **analytical** ~
　　analytische Ultrazentrifuge, **preparative** ~
　　präparative Ultrazentrifuge
ultracondenser Ultrakondensor *m* (Opt)
ultrafilter Ultrafilter *m*
ultrafiltration Ultrafiltration *f*
ultra gravity wave Ultraschwerewelle *f*
ultrahigh frequency Ultrahochfrequenz *f*
ultrahigh vacuum Höchstvakuum *n*
ultraionization Ultraionisierung *f*
ultramarine Azurblau *n*, Lasurblau *n*,
　　Ultramarin *n* (Chem), **native** ~ Lasurstein *m*
　　(Min)
ultramarine *a* ultramarin[blau]
ultramarine blue Ultramarin[e]blau *n*
ultramicrochemical ultramikrochemisch
ultramicroscope Ultramikroskop *n*
ultramicroscopic ultramikroskopisch
ultramicroscopy Ultramikroskopie *f*
ultrapore Feinpore *f*, Ultrapore *f*
ultraquinine Ultrachinin *n*
ultrasensitive überempfindlich
ultrasensitivity Überempfindlichkeit *f*
ultrashort wave Ultrakurzwelle *f*
ultrashort wave transmitter
　　Ultrakurzwellensender *m*
ultrasonic apparatus Ultraschallapparat *m*,
　　Ultraschallgerät *n*
ultrasonic attenuation
　　Ultraschallabschwächung *f*

ultrasonic biometry Ultraschallbiometrie *f*
ultrasonic generator Ultraschallerzeuger *m*
ultrasonics Lehre vom Ultraschall *f*
ultrasonic velocity Ultraschallgeschwindigkeit *f*
ultrasonic wave Überschallwelle *f*
ultrasound generator Ultraschallerzeuger *m*
ultraviolet Ultraviolett *n*
ultraviolet *a* ultraviolett
ultraviolet inhibitor Ultraviolettinhibitor *m*
ultraviolet lamp Ultraviolettlampe *f*,
　　Höhensonne *f*
ultraviolet light Ultraviolettlicht *n*
ultraviolet light *a*, **transparent to** ~
　　ultraviolettdurchlässig
ultraviolet radiation Ultraviolettbestrahlung *f*
ultraviolet ray emitter Ultraviolettstrahler *m*
ultraviolet spectrum UV-Spektrum *n*
umbellatine Umbellatin *n*
umbellic acid Umbellsäure *f*
umbelliferae Doldenblütler *pl*
umbelliferone Umbelliferon *n*
umbelliferose Umbelliferose *f*
umbelliprenin Umbelliprenin *n*
umbellularic acid Umbellularsäure *f*
umbellulone Umbellulon *n*
umber Umber *m* (Min), Umbererde *f* (Min)
umbilicaric acid Umbilicarsäure *f*
umbilicin Umbilicin *n*
umbrella-type bubble cap Regenschirmglocke *f*
　　(Dest)
umbrella-type bubble-cap tray
　　Regenschirmglockenboden *m* (Dest)
unable außerstande, unfähig
unadulterated unverfälscht
unaffected unangegriffen
unalloyed unlegiert
unalterable unveränderlich
unaltered unverändert
unambigous unzweideutig
unambiguity Eindeutigkeit *f*
unambiguous eindeutig, unzweideutig
unanimous einhellig, einstimmig
unannealed ungetempert (Met)
unattacked unangegriffen
unbalance Gleichgewichtsstörung *f*, Unwucht *f*
unbalanced nicht im Gleichgewicht,
　　unausgeglichen
unbalance eccentricity, degree of ~
　　Rundlauffehler *m*
unbearable unerträglich
unbelievable undenkbar
unbind *v* entfesseln
unbleached ungebleicht
unbranched unverzweigt
unbreakable unzerbrechlich, bruchsicher
unburnt ungebrannt (Ziegel)

uncertain ungewiß, unsicher
uncertainty Ungenauigkeit *f,* Ungewißheit *f,*
   Unsicherheit *f*
uncertainty condition Unschärfebedingung *f*
uncertainty factor Unsicherheitsfaktor *m*
uncertainty principle Unsicherheitsprinzip *n*
uncertainty relation Unschärferelation *f* (Phys),
   Unsicherheitsrelation *f* (Phys)
unchangeable unveränderlich
unchanged unverändert
unclean unrein, unsauber
uncleanness Unreinheit *f*
uncoil *v* abrollen, abspulen, abwickeln
uncolored ungefärbt
uncombined ungebunden
uncompleted unvollkommen
unconditional bedingungslos, unbedingt
uncontrolled ungezügelt
uncouple *v* loslösen, auskuppeln, loskuppeln
uncoupler Entkoppler *m* (Biochem)
uncoupling device Abschlagevorrichtung *f*
uncover *v* abdecken, aufdecken, entblößen
uncrystallized unkristallisiert
unctuous ölig, fettig; salbenartig
undamaged unbeschädigt; unversehrt
undamped ungedämpft
undecadiene Undecadien *n*
undecadiine Undecadiin *n*
undecane Undecan *n*
undecanoic acid Undecansäure *f*
undecanol Undecanol *n*
undecene Undecylen *n*
undecine Undecin *n*
undecyl Undecyl-
undecyl alcohol Undecanol *n*
undecyl bromide Undecylbromid *n*
undecylene Undecylen *n*
undecylenic acid Undecylensäure *f*
undecylic acid Undecylsäure *f*
undecylic aldehyde Undecylaldehyd *m*
undefinable unbestimmbar, undefinierbar
undefined undefiniert
underburned charcoal Blindkohle *f*
undercarriage Fahrgestell *n*
undercoat Deckgrund *m,* Grundanstrich *m,*
   Grundierung *f*
undercool *v* unterkühlen
undercooling Unterkühlung *f*
underdone halbgar
underdriven centrifuge Hubbodenzentrifuge *f*
underexcitation Untererregung *f*
underexcited untererregt
underexpose *v* unterbelichten (Phot)
underexposed unterbelichtet (Phot)
underexposure Unterbelichtung *f* (Phot)
underfeeding Unterdosierung *f* (Techn),
   Unterernährung *f*
underglaze color Unterglasurfarbe *f*
undergo *v* erleiden

undergrate firing Unterfeuerung *f*
underground erdverlegt, unterirdisch
underground cable Erdkabel *n*
underground gasification (coal) Flözvergasung *f*
underground mining Untertag[e]bau *m* (Bergb)
underground water Grundwasser *n*
underground working Untertag[e]bau *m* (Berg),
   Untertagebetrieb *m* (Bergb)
underinflation des Innendrucks (Reifen)
underlying darunterliegend
undermine *v* unterhöhlen
underoxidized unteroxydiert
underpass Unterführung *f*
underside Unterseite *f*
understand *v* begreifen, verstehen
understanding Verständnis *n*
undertake *v* unternehmen
undertaking Unternehmen *n*
undertone Unterton *m* (Farbe)
undertread des Laufstreifens
underwater coating Unterwasseranstrich *m*
underwater paint Unterwasserfarbe *f*
underweight Mindergewicht *n,* Untergewicht *n*
undestroyed unzerstört
undeterminable unbestimmbar
undiffracted ungebeugt (Opt)
undiluted unverdünnt
undissociated undissoziiert
undissolved ungelöst (Chem)
undistorted unverzerrt
undressed casting Rohguß *m*
undulate *v* schwingen, wogen, wallen
undulated wellenförmig
undulation Schwingung *f,* Vibration *f,*
   Wellenbewegung *f*
undulation theory Undulationstheorie *f*
undulatory wellenartig
undulatory current Undulationsstrom *m*
undulatory line Wellenlinie *f*
undulatory motion Wellenbewegung *f*
undyed ungefärbt (Tuch)
unecconomic unwirtschaftlich
unequal ungleich[förmig]
unequal tee Verengerungs-T-Stück *n*
unessential unwesentlich
unesterified unverestert
uneven uneben
uneven dyeing Farbunruhe *f*
unevenness Unebenheit *f,* Ungleichheit *f*
unexpected unerwartet
unexplainable unerklärbar, unerklärlich
unexposed unbelichtet
unfading farbecht
unfermentable unvergärbar
unfermented unfermentiert
unfiltered ungefiltert
unfold *v* entfalten
unfounded grundlos
ungear *v* auskuppeln

unglazed unglasiert
ungrease v entfetten
ungulinic acid Ungulinsäure f
unhandy unhandlich
unhardened ungehärtet
unharmonious unharmonisch
unhesitating anstandslos
uniaxial einachsig (Krist, Techn)
uniaxiality Einachsigkeit f (Krist)
unicellular einzellig
unidimensional eindimensional
unidirected gleichgerichtet
unification Vereinheitlichung f
unified einheitlich, vereinheitlicht
unified thread Einheitsgewinde n
unifilar suspension Eindrahtaufhängung f,
   Einfadenaufhängung f (Elektr)
uniflow turbine Einstromturbine f
uniform einförmig, einheitlich, gleichbleibend,
   gleichförmig, gleichmäßig, unveränderlich
uniformalization Vereinheitlichung f
uniformalize v vereinheitlichen
uniform attack ebenmäßiger Angriff m
uniformity Einheitlichkeit f, Gleichförmigkeit f,
   Gleichmäßigkeit f, Stetigkeit f
uniform system Einheitssystem n
unilateral einseitig
unimaginable undenkbar, unvorstellbar
unimolecular einmolekular, monomolekular
unimportant unwesentlich, belanglos
uninterrupted lückenlos, ununterbrochen
uninuclear einkernig
union Verbindung f, Vereinigung f, Verknüfung f
union fabric Mischgewebe n (Text)
uniplanar ring ebener Ring m (Chem)
unipolar einpolig f
unipolar dynamo Einpoldynamo m,
   Unipolardynamo m
unipolar induction Unipolarinduktion f
unipolarity Einpoligkeit f, Unipolarität f
unique einmalig, einzigartig
uniqueness Eindeutigkeit f (Math)
unit Einheit f, Aggregat n (Techn), Anlage f
   (Techn), Bezugseinheit f, Einer m (Math),
   Maßeinheit f (Phys), **absolute** ~ absolute
   Einheit f, ~ **of conductivity** Einheit f des
   Leitvermögens, ~ **of measure** Maßeinheit f, ~
   of quantity Mengeneinheit f, ~ **of time**
   Zeiteinheit f, ~ **of volume** Raumeinheit f
unit arc Einheitsbogen m
unitary einheitlich
unit cell Elementarzelle f, Gittereinheit f
unit charge Einheitsladung f
unit construction Normalausführung f
unit construction system Baukastensystem n
unit crystal Einheitskristall m
unit diagram Einheitsdiagramm n
unite v vereinigen, koppeln, verbinden (Chem)
unit flange Einheitsflansch m

unit function Einheitsfunktion f
unit magnetic pole magnetischer Einheitspol m
unit operation Grundoperation f
unit place Einerstelle f
unit plane mit Miller-Indices pl (Krist)
unit pressure Einheitsdruck m
unit price Einheitspreis m
unity Einheit f
univalence Einwertigkeit f (Chem)
univalent einwertig (Chem), einbindig (Chem),
   monovalent
universal allseitig, universal
universal galvanometer Universalgalvanometer n
universal gauge Universallehre f
universal grinding machine
   Universalschleifmaschine f
universal indicator paper
   Universalindikatorpapier n
universal joint Universalgelenk n,
   Gelenkverbindung f (Techn),
   Kardangelenk n, Kreuzgelenkkupplung f
universal manometer Universalmanometer n
universal microscope Universalmikroskop n
universal remedy Universalarznei f (Pharm)
universal rolling mill Universalwalzwerk n
universal screw spanner
   Universalschraubenschlüssel m
universal shaft Gelenkwelle f (Techn)
universal testing machine
   Universalprüfmaschine f
universal three-axis stage Universaldrehtisch m
universal varnish Universallack m
universe Universum n, Weltall n
university, technical ~ technische Hochschule f
[university] enrol[l]ment Immatrikulation f
unknown Unbekannte f n
unknown a unbekannt
unlatch v ausklinken
unlike ungleich[namig], verschieden[artig]
unlime v abkalken, entkalken
unliming Entkalken n, Entkalkung f
unlimited grenzenlos, unbegrenzt, unbeschränkt,
   unendlich
unload v abladen, ausladen, entladen, entlasten
unloading Ausladen n, Entladen n
unloading control Überlaufkontrolle f
unloading valve Entlastungsventil n
unlock v aufschließen, aufsperren, öffnen
unmalleable spröde (Met)
unmanned unbemannt (Satellit)
unmarked nicht markiert
unmeltable unschmelzbar
unmixed unvermischt
unmixing Entmischen n
unmodified unverändert
unoccupied unbesetzt
unoriented nicht orientiert
unoxidizable nicht oxydierbar
unpack v auspacken

unpaired ungepaart
unplasticized weichmacherfrei
unpolarized unpolarisiert
unpolished rauh
unprecedented beispiellos
unprejudiced unvoreingenommen
unpretending anspruchslos
unprime v entschärfen
unprovable unbeweisbar
unpublished unveröffentlicht
unquantized ungequantelt
unquenchable unlöschbar
unravel v ausfasern, zerfasern (Pap)
unravelling machine Zerfaserer m (Pap)
unreal unwirklich
unrefined roh (z. B. Öl)
unreliable unzuverlässig
unresolved unaufgelöst
unroasted ungeröstet
unroll v abrollen, abwickeln
unsafe unsicher, gefährlich
unsafety Unsicherheit f
unsalted ungesalzen
unsaturated ungesättigt
unsavoriness Geschmacklosigkeit f
unscarred narbenlos
unscientific unwissenschaftlich
unscorified unverschlackt
unscrew v abschrauben, abdrehen,
   aufschrauben, herausschrauben, losschrauben
unseasoned ungewürzt
unshrinkable einlaufecht, nicht einlaufend
unsized ungeleimt (Papier)
unskilled worker Hilfsarbeiter m
unslakable unlöschbar (Kalk)
unslaked ungelöscht (Kalk)
unsmeltable unverhüttbar (Metall)
unsolder v ablöten, auflöten, loslöten
unsoldered ungelötet
unsolvable unlösbar
unsolved ungelöst (z. B. Problem)
unstable instabil (Chem), labil, unbeständig
   (Chem), unsicher, zersetzlich (Chem)
unstained ungefärbt
unsteadiness Unruhe f, Unstetigkeit f
unsteady instationär, schwankend, unsicher,
   unbeständig, ungleichmäßig
unstressing Entspannen n
unsuitable unbrauchbar, ungeeignet,
   unzweckmäßig
unsupported selbsttragend, trägerlos, ungestützt
unsupported film selbsttragende Folie f
unsurpassed unübertrefflich
unsusceptibility Unempfindlichkeit f
unsymmetrical asymmetrisch, unsymmetrisch
untanned ungegerbt
untearable unzerreißbar
untie v lösen, losbinden
untight undicht

untimber v abholzen
untin v entzinnen
untrimmed unbeschnitten
untrue unrund
untwist v aufdrehen, lösen
unusual ungewöhnlich
unweldable unschweißbar
unwind v abrollen, abspulen, abwickeln
unwind stand Abwickelbock m
unworkable unverhüttbar (Metall)
unwrap v auspacken
unwrought roh (Stein, Diamant), unbearbeitet
upas Upas n
upcurrent aufsteigender Luftstrom m,
   Aufwind m
updraft Aufstrom m, Aufwind m
up-draught sintering machine
   Bandsintermaschine f
upfold v auffalten
upgrade v veredeln, verfeinern
upgrading Verbesserung f, Verfeinerung f
upholstery Polstermöbel pl
upkeep Instandhaltung f
up-milling Gegenlauffräsen n
upper buddle Oberfaß n (Metall)
upper connection Kopfkontakt m
upper heat Oberfeuer n
upper layer Deckschicht f
upper part Oberteil n
upper side Oberseite f
upperside run conveyor Obertrumförderer m
upper stratum Deckschicht f
upright aufrecht, senkrecht
upset Stauchungsübergang m
upsetting Stauchung f
upsetting effect Stauchwirkung f
upsetting machine Stauchmaschine f
upsetting test Stauchprobe f, Stauchversuch m
upstream water Oberwasser n
upstroke Aufwärtshub m
upstroke press Unterkolbenpresse f
uptake Aufnahme f, Steigkanal m (Techn),
   Zugkanal m (Techn)
up-to-date zeitgemäß, modern
upward aufwärts
upward motion Aufwärtsbewegung f
upward stream Aufstrom m
uracil Uracil n
uraconite Uranocker m (Min)
uracylic acid Uracylsäure f
uralin Chloralurethan n
uralite Uralit m (Min)
uralorthite Uralorthit m (Min)
uramil Uramil n, Murexan n
uramildiacetic acid Uramildiessigsäure f
uramine Guanidin n
uranate Uranat n
uranediol Urandiol n
urania blue Uraniablau n

uranic acid Uransäure *f,* **salt of** ~ Uranat *n*
uranic anhydride Uransäureanhydrid *n*
uranic compound Uran(VI)-Verbindung *f*
uranic fluoride Uranhexafluorid *n*
uranic ocher Uranocker *m* (Min)
uranic oxide Uransäureanhydrid *n,*
Uran(VI)-oxid *n*
uranide Uranid *n*
uraniferous uranhaltig
uranin Uranin *n*
uraninite Uraninit *m* (Min), Uranpechblende *f*
(Min), Uranpecherz *n* (Min)
uranite Uranit *m* (Min), Schweruranerz *n* (Min)
uranitiferous uranithaltig
uranium (Symb. U) Uran *n,* **containing** ~
uranhaltig, **enriched** ~ angereichertes Uran,
lumped ~ in Blöcken pl, **enriched** ~ **reactor**
mit angereichertem Uran
uranium acetate Uranacetat *n*
uranium acetylide Urancarbid *n*
uranium carbide Urancarbid *n*
uranium cartridge Uranstabhülse *f*
uranium compound Uranverbindung *f*
uranium content Urangehalt *m*
uranium decay series Uranzerfallsreihe *f*
uranium deposit Uranvorkommen *n*
uranium dioxide Urandioxid *n,* Uran(IV)-oxid *n*
uranium fission Uranspaltung *f* (Atom)
uranium glass Uranglas *n*
uranium glimmer Uranglimmer *m* (Min)
uranium graphite reactor Urangraphitreaktor *m*
uranium hexafluoride Uranhexafluorid *n*
uranium intensifier Uranverstärker *m*
uranium isotope Uranisotop *n* (Atom)
uranium lead Uranblei *n* (Min)
uranium metal Uranmetall *n*
uranium mica Uranglimmer *m* (Min)
uranium nucleus Urankern *m*
uranium ore Uranerz *n* (Min)
uranium peroxide Uranperoxid *n*
uranium phosphate Uranphosphat *n*
uranium pile Uranmeiler *m,* Uranbrenner *m,*
Uranreaktor *m*
uranium salt Uransalz *n*
uranium series Uranreihe *f*
uranium tetrafluoride Urantetrafluorid *n*
uranium trioxide Uransäureanhydrid *n,*
Uran(VI)-oxid *n*
uranium yellow Urangelb *n*
uranolepidite Uranolepidit *m* (Min)
uranoniobite Uranniobit *m* (Min)
uranophane Uranophan *m* (Min)
uranopilite Uranopilit *m* (Min)
uranospherite Uranosphärit *m* (Min)
uranospinite Uranospinit *m* (Min)
uranothallite Uranothallit *m* (Min)
uranothorite Uranothorit *m* (Min)
uranotil Uranotil *m* (Min)

uranous compound Uran(IV)-Verbindung *f,*
Uranoverbindung *f*
uranous hydroxide Uran(IV)-hydroxid *n,*
Uranohydroxid *n*
uranous oxide Urandioxid *n,* Uran(IV)-oxid *n*
uranous salt Uran(IV)-Salz *n,* Uranosalz *n*
uranyl Uranyl-
uranyl acetate Uranylacetat *n*
uranyl hydrogen phosphate
Uranylhydrogenphosphat *n*
uranyl nitrate Uranylnitrat *n*
uranyl phosphate, acid ~
Uranylhydrogenphosphat *n*
uranyl radical Uranylradikal *n*
uranyl spectrum Uranylspektrum *n*
uranyl sulfate Uranylsulfat *n*
urase Urease *f* (Biochem)
urate harnsaures Salz *n,* Urat *n,* **acid** ~ Biurat *n,*
hydrogen ~ Biurat *n*
urazine Urazin *n*
urazole Urazol *n*
urbanite Urbanit *m* (Min)
urea Harnstoff *m*
urea cycle Harnstoffzyklus *m*
urea formaldehyde resin
Harnstoff-Formaldehyd-Kunstharz *n*
urea quinate Urol *n*
urea resin Harnstoff-Formaldehyd-Kunstharz *n*
urea synthesis Harnstoffsynthese *f*
urechochrome Urechochrom *n*
ureide Ureid *n*
urethane Urethan *n*
urethane resin Urethanharz *n*
urethylan Urethylan *n,* Methylurethan *n*
uretidine Uretidin *n*
uretine Uretin *n*
urezin Urezin *n*
urgency Dringlichkeit *f*
urgent dringend, dringlich
uric acid Harnsäure *f*
uricase Uricase *f* (Biochem)
uridine Uridin *n*
uridine diphosphate Uridindiphosphat *n*
uridine diphosphate galactose
Uridindiphosphatgalactose *f* (Biochem)
uridine diphosphate glucose
Uridindiphosphatglucose *f*
uridine monophosphate Uridinmonophosphat *n*
uridine phosphoric acid Uridinphosphorsäure *f*
uridine triphosphate Uridintriphosphat *n*
uridylic acid Uridylsäure *f*
uridyl transferase Uridyltransferase *f* (Biochem)
urinalysis Harnanalyse *f,* Harnuntersuchung *f*
urinary calculus Harnstein *m*
urinary constituent Harnbestandteil *m*
urine Harn *m,* Urin *m,* **testing of** ~
Harnuntersuchung *f*
urine analysis Harnanalyse *f,*
Harnuntersuchung *f*

urine test Harnuntersuchung *f*
urinose harnartig
urinous harnartig, urinartig
urobenzoic acid Hippursäure *f*
urobilin Hydrobilirubin *n*, Urobilin *n*
urobilinogen Urobilinogen *n*
urocanase Urocanase *f* (Biochem)
urocanin Urocanin *n*
urocaninic acid Urocaninsäure *f*
urochrome Urochrom *n*
uroerythrin Uroerythrin *n*
urol Urol *n*
uron Uron *n*
uronic acid Uronsäure *f*
uroporphyrin Uroporphyrin *n*
urorosein Urorosein *n*
urosine Urosin *n*
urothion Urothion *n*
urotropine Urotropin *n*,
   Hexamethylentetramin *n*
urotropine quinate Chinotropin *n*
urotropine salicylate Saliformin *n*
urotropine tannin Tannopin *n*
uroxanic acid Uroxansäure *f*
ursane Ursan *n*
ursol Ursol *n*
ursolic acid Ursolsäure *f*, Malol *n*, Urson *n*
urson Urson *n*, Malol *n*
urusene Urusen *n*
urushenol Urushenol *n*
urushic acid Urushinsäure *f*
urushiol Urushiol *n*
urusite Urusit *m*
usability Verwendbarkeit *f*,
   Verwendungsfähigkeit *f*
usable gebrauchsfähig, verwendbar
use Gebrauch *m*, Anwendung *f*, Ausnutzung *f*,
   Benutzung *f*, Gebrauch *m*, Nutzanwendung *f*,
   Nutzung *f*, Verwendung *f*, continuous ~
   Dauereinsatz *m*, repeated ~
   Wiederverwendung *f*
use *v* anwenden, gebrauchen, benutzen,
   verwenden, ~ up aufarbeiten, aufbrauchen,
   verbrauchen, verzehren
used air Abluft *f*
used material Altmaterial *n*
used oil Ablauföl *n*
used rubber Altgummi *m*
useful brauchbar, nützlich, nutzbringend,
   zweckdienlich, zweckmäßig, to be ~ taugen
useful capacity Nutzkapazität *f*
useful current Wirkstrom *m*
useful effect Nutzeffekt *m*
usefulness Brauchbarkeit *f*, Tauglichkeit *f*,
   Verwendbarkeit *f*, Verwendungsfähigkeit *f*
useful work Nutzarbeit *f*
useless unbrauchbar, aussichtslos, nutzlos
user Benutzer *m*

uses, range of ~ Anwendungsbereich *m*,
   Anwendungsgebiet *n*
U-shaped U-förmig
usnetic acid Usnetsäure *f*
usnetinic acid Usnetinsäure *f*
usnic acid Usninsäure *f*
usnolic acid Usnolsäure *f*
ustilic acid Ustilsäure *f*
usual gewöhnlich, normal
utensil Gerät *n*, Werkzeug *n*
utensils Geschirr *n*
utilities Energieversorgung *f*
utility Brauchbarkeit *f*
utility model Gebrauchsmuster *n*
utilizable auswertbar, nutzbar, verwendbar
utilization Ausnutzung *f*, Nutzbarmachung *f*,
   Nutzung *f*, Verwertung *f*, ~ of heat
   Wärme[aus]nutzung *f*, ~ of waste
   Abfallverwertung *f*, ~ of waste heat
   Abhitzeverwertung *f*
utilization coefficient Ausnutzungskoeffizient *m*
utilization factor Ausnutzungsfaktor *m*
utilize *v* ausnutzen, verwerten
utter *v* äußern
U-tube U-Rohr *n*, U-Bogen *m*
U-tube manometer U-Rohr-Manometer *n*
uvanite Uvanit *m* (Min)
uvaol Uvaol *n*
uvarovite Uwarowit *m* (Min), Chromgranat *m*
   (Min)
uvic acid Uvinsäure *f*
uvinic acid Uvinsäure *f*
uviol glass Uviolglas *n* (Opt)
uvitic acid Uvitinsäure *f*
uvitonic acid Uvitonsäure *f*
uwarowite Uwarowit *m* (Min), Chromgranat *m*
   (Min)
uzarin Uzarin *n*

V

vacancy Fehlstelle *f,* Leerstelle *f,* Lücke *f,*
  unbesetzter Platz *m*
vacancy diffusion Leerstellendiffusion *f*
vacancy formation Leerstellenbildung *f*
vacancy migration Leerstellenwanderung *f*
vacant unbesetzt, leer
vaccenic acid Vaccensäure *f*
vaccinate *v* impfen
vaccinated [schutz]geimpft
vaccination Impfung *f* (Med), Schutzimpfung *f,*
  Vakzination *f,* antityphoid ~
  Typhusschutzimpfung *f* (Med)
vaccine Impfserum *n,* Impfstoff *m,* Vakzine *f,*
  mixed ~ Mischvakzine *f*
vaccine production Impfstoffgewinnung *f,*
  Impfstoffherstellung *f*
vacciniin Vacciniin *n*
vacuole Vakuole *f* (Biol)
vacuometer Vakuummesser *m*
vacuum Vakuum *n,* Luftleere *f,* luftleerer
  Raum *m,* Luftunterdruck *m,* coarse ~
  Grobvakuum *n,* high ~ Hochvakuum *n,*
  medium high ~ Feinvakuum *n,* partial ~
  Unterdruck *m*
vacuum arc furnace Vakuum-Lichtbogenofen *m*
vacuum attachment Vakuumanschluß *m*
vacuum bag molding
  Vakuumgummisackverfahren *n,*
  Vakuumverfahren *n* (Polyester)
vacuum brake Unterdruckbremse *f*
vacuum breaker Rückschlagventil *n*
vacuum cabinet Vakuumschrank *m*
vacuum cell Vakuumzelle *f*
vacuum chamber Unterdruckkammer *f,*
  Vakuumkammer *f*
vacuum cleaner Staubsauger *m*
vacuum coating im Vakuum
vacuum coating machine Metallisieranlage *f* (für
  Kunststoffe)
vacuum concrete Vakuumbeton *m*
vacuum connection Sauganschluß *m,* zum
  Vakuum
vacuum conveyor tube Saugluftförderer *m*
vacuum correction Vakuumkorrektur *f*
vacuum desiccator Vakuumexsikkator *m*
vacuum distillation Vakuumdestillation *f*
vacuum drum dryer Vakuumwalzentrockner *m*
vacuum drum filter Vakuumtrommelfilter *n*
vacuum dryer Vakuumtrockner *m*
vacuum drying oven Vakuumtrockenofen *m*
vacuum electric arc LBV-Verfahren *n*
  (Schmelzverf)
vacuum evaporation Vakuumverdampfung *f*
vacuum evaporator Vakuumverdampfer *m,*
  Unterdruckverdampfer *m*
vacuum fermentation Unterdruckgärung *f* (Brau)

vacuum filter Vakuumfilter *n*
vacuum filtration Vakuumfiltration *f*
vacuum flask Vakuumkolben *m*
vacuum forming Tiefziehen *n,*
  Vakuum[ver]formung *f*
vacuum freeze dryer Gefriertrockner *m,*
  Vakuumtiefkühltrockner *m*
vacuum furnace Vakuumofen *m*
vacuum gauge Druckmeßgerät *n* für
  Vakuumtechnik, Unterdruckmesser *m,*
  Vakumeter *n,* Vakuummeßinstrument *n*
vacuum grating spectrograph
  Vakuumgitterspektrograph *m*
vacuum grease Vakuumfett *n*
vacuum jacket Vakuummantel *m*
vacuum leaf filter Tauch-Filter *n*
vacuum manometer Unterdruckmanometer *n*
vacuum measuring instrument
  Vakuummeßinstrument *n*
vacuum melting plant Vakuumschmelzanlage *f*
vacuum oil Vakuumöl *n*
vacuum package Vakuumverpackung *f*
vacuum paddle dryer Vakuumschaufeltrockner *m*
vacuum physics Vakuumphysik *f*
vacuum plant Saugluftanlage *f,* Vakuumanlage *f*
vacuum polarization diagram
  Vakuumpolarisationsdiagramm *n*
vacuum process Vakuumverfahren *n*
vacuum pump Vakuumpumpe *f*
vacuum receiver Vakuumvorlage *f*
vacuum rectifier Vakuumgleichrichter *m* (Elektr)
vacuum regulator Vakuumregler *m*
vacuum rotary dryer Vakuumtrommeltrockner *m*
vacuum seal Vakuumverschluß *m*
vacuum shelf dryer Vakuumtrockenschrank *m*
vacuum tank Vakuumkessel *m*
vacuum tap Vakuumhahn *m*
vacuum tester Vakuumprüfer *m*
vacuum-tight vakuumdicht
vacuum tube Vakuumröhre *f*
vacuum wash Saugwäsche *f*
vagabond current Streustrom *m*
vague undeutlich, vage
vakerin Vakerin *n*
valancy band Valenzband *n*
valence Valenz *f,* Wertigkeit *f,* concept of ~
  Valenzbegriff *m,* directed ~ Valenzrichtung *f,*
  of higher ~ höherwertig, stoichiometric ~
  stöchiometrische Wertigkeit, total ~
  Gesamtwertigkeit *f*
valence angle Valenzwinkel *m*
valence band Valenzband *n*
valence bond Valenzbindung *f*
valence dash Valenzstrich *m* (Chem)
valence dash formula Valenzstrichformel *f*
  (Chem)

valence electron kernfernes Elektron *n*,
  Valenzelektron *n*
valence orbital Valenzorbital *n*
valence shell Valenzschale *f* (Atom),
  Außenschale *f* (Atom)
valence stage Wertigkeitsstufe *f*
valencianite Valencianit *m* (Min) ·
valency Valenz *f*, Wertigkeit *f*, coordinative ~
  koordinative Valenz, localized ~ lokalisierte
  Valenz, saturated ~ gesättigte Valenz,
  unsaturated ~ nicht abgesättigte Valenz
valency-controlled valenzgesteuert
valency electron Valenzelektron *n*
-valent -wertig
valentinite Valentinit *m* (Min),
  Weißspießglanzerz *n* (Min)
valeral Valeraldehyd *m*
valeraldazine Valeraldazin *n*
valeraldehyde Valeraldehyd *m*, Amylaldehyd *m*
valeraldoxime Valeraldoxim *n*
valeramide Valeramid *n*
valeramidine Valeramidin *n*
valeranilide Valeranilid *n*
valerate Valerianat *n*
valerene Valeren *n*
valerian Baldrian *m* (Bot, Med), extract of ~
  Baldrianextrakt *m*
valerianaceous baldrianartig
valerianate Valerianat *n*
valerian extract Baldrianextrakt *m*
valerianic acid Valeriansäure *f*
valerian oil Baldrianöl *n*
valerian rhizome Baldrianwurzel *f* (Bot)
valerian root Baldrianwurzel *f* (Bot)
valerian tincture Baldriantinktur *f*
valeric acid Valeriansäure *f*, salt or ester of ~
  Valerianat *n*
valeric anhydride Valeriansäureanhydrid *n*
valeridin Valeridin *n*
valeroidine Valeroidin *n*
valerol Valerol *n*
valerolactone Valerolakton *n*
valerone Valeron *n*
valeronitrile Valeronitril *n*
valerophenone Valerophenon *n*
valeryl Valeryl-
valeryl bromide Valerylbromid *n*
valeryl chloride Valerylchlorid *n*
valerylene Valerylen *n*
valeryl nitrile Valerylnitril *n*
valerylphenetidine Valeridin *n*
valid gültig
validity Gültigkeit *f*, Gültigkeit *f*,
  Rechtsgültigkeit *f*, limit of ~
  Gültigkeitsgrenze *f*, range of ~
  Geltungsbereich *m*, region of ~
  Gültigkeitsbereich *m*
validol Validol *n*, Valeriansäurementholester *m*
valine Valin *n*

valinol Valinol *n*
valinomycin Valinomycin *n*
vallesine Vallesin *n*
valley Tal *n*
valonia Walone *f* (Gerb)
valtropine Valtropin *n*
valuable wertvoll, gehaltreich
valuation Bewertung *f*, Schätzung *f*,
  Wertbestimmung *f*
valuation adjustment Bewertungsausgleich *m*
value Wert *m*, Größe *f*, absolute ~
  Absolutwert *m*, accidental ~ Zufallswert *m*,
  actual ~ Istwert *m*, admissible ~ zulässiger
  Wert, approximate ~ Annäherungswert *m*,
  angenäherter Wert, basic or fundamental ~
  Grundwert *m*, constant ~ Festwert *m*,
  konstanter oder unveränderlicher Wert,
  critical ~ kritischer Wert, determination of ~
  Wertbestimmung *f*, experimental ~
  experimenteller Wert, intermediate ~
  Zwischenwert *m*, most probable ~
  wahrscheinlichster Wert, numerical ~
  numerischer Wert (Math), optimum ~
  optimaler Wert, peak or maximum ~
  Höchstwert *m*, reciprocal ~ reziproker Wert,
  relative ~ Bezugswert *m*, to be of ~ taugen,
  true ~ Istwert *m*
value *v* [ab]schätzen, beurteilen
valve Ventil *n*, Absperrglied *n*,
  Absperrvorrichtung *f*, Klappe *f*, auxiliary ~
  Hilfsventil *n*, counter-pressure ~
  Gegendruckventil *n*, double-seat ~
  doppelseitiges Ventil, mechanically operated
  ~ mechanisch gesteuertes Ventil, ~ with
  inclined seat Schrägsitzventil *n*, ~ with two
  bends Schlangenventil *n*
valve adjustment Ventileinstellung *f*
valve arrangement, controllable ~
  Extruderventile *pl*
valve block Ventilblock *m*
valve bonnet Ventilhaube *f*, Ventiloberteil *n*
valve box Ventilgehäuse *n*
valve box cover Ventil[gehäuse]deckel *m*
valve cage Ventilkäfig *m*
valve chamber Ventilgehäuse *n*
valve clearance Ventilspiel *n*
valve cone Ventilkegel *m*
valve core Ventileinsatz *m*
valve disc Ventilteller *m*
valve effect Sperrwirkung *f*
valve flap Ventilklappe *f*
valve gear Ventilsteuerung *f*, Kolbensteuerung *f*
valve governor Regler *m* (Reglerventil)
valve grinding Ventileinschleifen *n*
valve guard Fängerglocke *f*
valve guide Ventilführung *f*
valve housing Ventilgehäuse *n*, Ventilkörper *m*
valve leather Ventilleder *n*

valve lever Ventilhebel *m*
valve lift Ventilhub *m*, Ventilstoß *m*
valve liner Ventilführungsrohr *n*
valve oil can Ventilölkanne *f*
valve outlet Ventilauslaß *m*
valve piston Saugkolben *m*
valve regulation Ventilsteuerung *f*
valve relief Ventilentlastung *f*
valve seal Ventilfett *n*, Ventilverschluß *m*
valve spring Ventilfeder *f*
valve stem Ventilspindel *f*, Ventilstange *f*
valve train wear Kipphebelverschleiß *m*
valve travel Ventilspiel *n*
valve tray Ventilboden *m* (Rektifikation)
valve voltmeter Röhrenvoltmeter *n*
valyl Valeryldiäthylamid *n*, Valyl *n*
vanadate Vanadat *n*
vanadic vanadiumhaltig
vanadic acid Vanadinsäure *f*, salt of ∼
   Vanadat *n*
vanadic anhydride Vanadinsäureanhydrid *n*,
   Vanadiumpentoxid *n*
vanadic chloride Vanadiumchlorid *n*
vanadiferous vanadiumhaltig
vanadinite Vanadinit *m* (Min),
   Vanadinbleierz *n* (Min), Vanadinbleispat *m*
   (Min)
vanadium (Symb. V) Vanadium *n*, Vanadin *n*,
   containing ∼ vanadiumhaltig
vanadium chloride Vanadiumchlorid *n*
vanadium content Vanadingehalt *m*
vanadium dichloride Vanadiumdichlorid *n*,
   Vanadium(II)-chlorid *n*
vanadium nitride Vanadinstickstoff *m*
vanadium pentoxide Vanadiumpentoxid *n*,
   Vanadinsäureanhydrid *n*
vanadium sesquioxide Vanadinsesquioxid *n*
vanadium steel Vanadinstahl *m*,
   Vanadiumstahl *m*
vanadometric vanadometrisch
vanadyl salt Vanadylsalz *n*
vanadyl sulfate Vanadylsulfat *n*
vanderoside Vanderosid *n*
van der Waals adsorption Van der Waalssche
   Adsorption *f*
van der Waals forces Van der Waalssche
   Kräfte *pl*
Van der Waals interactions Van der Waalssche
   Wechselwirkungen *pl*
van der Waals radius Van der Waals-Radius *m*
vane Flügel *m*, Turbinenschaufel *f*
vane setting Flügelstellung *f*
vane water meter Einflügelradzähler *m*
vane wheel Schaufelrad *n*
vanilla Vanille *f* (Bot)
vanilla flavor Vanillearoma *n*
vanilla pod Vanillenschote *f*
vanilla substitute Vanilleersatz *m*
vanillic acid Vanillinsäure *f*

vanillin Vanillin *n*
vanillin sugar Vanillinzucker *m*
vanilloside Vanillosid *n*, Glucovanillin *n*
vanillyl alcohol Vanillylalkohol *m*
vanish *v* verschwinden, ausbleiben
vanishing impulse transmitter
   Verschwindimpulsgeber *m*
vanishing line Fluchtlinie *f*
vanishing point Fluchtpunkt *m* (Opt)
vanthoffite Vanthoffit *m* (Min)
Van't Hoff's isochore Van't Hoffsche Isochore *f*
Van't Hoff's principle of mobile equilibrium
   Prinzip *n* von Van't Hoff
vanyldisulfamide Vanyldisulfamid *n*
vapor Dampf *m*, Dunst *m*, **cold** ∼ Kaltdampf *m*,
   moisture of ∼ Dampffeuchtigkeit *f*, **to clear
   from** ∼ entdampfen, **to treat with** ∼
   bedampfen
vapor adsorption Dampfadsorption *f*
vapor barrier Dampfsperre *f*
vapor bath Dunstbad *n*
vapor capacity factor Dampfbelastungsfaktor *m*
vapor-compression refrigeration machine
   Kaltdampfmaschine *f*
vapor cure Dunstvulkanisation *f*
vapor density Dampfdichte *f*
vapor diffusion pump Dampfstrahler *m*
vapor hood Dunsthaube *f*
vaporimeter Vaporimeter *n*
vaporizability Verdampfbarkeit *f*,
   Vergasbarkeit *f*
vaporizable verdampfbar, verdunstbar, vergasbar
vaporization Dampfbildung *f*, Verdampfung *f*,
   Verdunstung *f*, Vergasung *f*, **heat of** ∼
   Verdampfungswärme *f*,
   Verdunstungswärme *f*, **rate of** ∼
   Verdampfungsgeschwindigkeit *f*
vaporization plant Bedampfungsanlage *f*
vaporize *v* verdampfen, eindampfen,
   verdunsten, vergasen
vaporizer Vergaser *m*, Dampfentwickler *m*,
   Eindampfkessel *m*, Verdampfer *m*,
   Verdampfungsapparat *m*, Zerstäuber *m*
vaporizing Verdampfen *n*, Abdunsten *n*,
   Vergasen *n*
vapor jet diffusion pump
   Dampfstrahldiffusionspumpe *f*
vapor lock (in fuel systems)
   Dampfblasenbildung *f* (in
   Kraftstoffsystemen)
vapor nozzle Dampfventil *n*
vaporous dampfig, dunstig, dampfförmig
vaporous envelope Dunsthülle *f*
vaporousness Dunstigkeit *f*
vapor outlet Dampfausgang *m*
vapor phase Gasphase *f*
vapor-phase corrosion Dampfphasenkorrosion *f*
vapor-phase inhibitor paper
   Korrosionsschutzpapier *n*

vapor pressure Dampfdruck *m*
vapor pressure recorder Dampfdruckschreiber *m*
vapor pressure thermometer
  Dampfdruckthermometer *n*
vapor pump Treibmittelpumpe *f*
vapor riser Dampfdurchtrittsöffnung *f* (Dest),
  Dampfkamin *m* (Dest), Gasdurchtrittshals *m*
  (Dest)
vapor supply pipe Dampfleitung *f*
vapor tension Dampfspannung *f*
vapor throughput Dampfdurchsatz *m* (Dest)
vapor-tight dampfdicht
vapor velocity Dampfgeschwindigkeit *f*
vapory dampfig, dunstig
variability Variabilität *f*, Veränderlichkeit *f*
variable Veränderliche *f* (Math), Unbestimmte *f*
  (Math), Variable *f* (Math), **dependent** ~
  abhängige Veränderliche (Math), **controlled**
  ~ (Am. E.), **controlled condition** (Brit. E.)
  bleibende Regelgröße *f*
variable *a* veränderlich, abwandelbar,
  abwechselnd, regelbar (Techn), schwankend,
  unbeständig, variabel, wandelbar, **infinitely** ~
  regelbar
variable [gain] amplifier Regelverstärker *m*
variable of state Zustandsvariable *f*
variable resistance Rheostat *m*
variable speed gear[ing] Regelgetriebe *n*
variable speed motor Regelmotor *m*
variable transformer Regeltransformator *m*
  (Elektr)
variance Abweichung *f*, **analysis of** ~
  Varianzanalyse *f*
variant Abart *f* (Biol), Variante *f*
variation Veränderung *f*, Abänderung *f*, Abart *f*
  (Biol), Abwandlung *f*, Variation *f* (Biol),
  Wechsel *m*, **allowable** ~ Toleranz *f*, ~ **of load**
  Belastungsschwankung *f*
variation[al] principle Variationsprinzip *n*
variations, range of ~ Schwankungsbereich *m*
varicolored verschiedenfarbig, vielfarbig
varicose vein Krampfader *f* (Med)
variegated sandstone Buntsandstein *m* (Min)
variety Abart *f*, Abwechslung *f*,
  Mannigfaltigkeit, Spielart *f*, Variante *f*,
  Verschiedenheit *f*, Vielseitigkeit *f*
variety reduction Typenbeschränkung *f*
variola Pocken *pl* (Med)
variolaric acid Variolarsäure *f*
variolation Pockenimpfung *f*
variolite Variolit *m* (Min), Blatterstein *m*,
  Pockenstein *m* (Min)
variometer Drehdrossel *f* (Elektr)
variscite Variszit *m* (Min)
varnish Firnis *m*, Anstrichmittel *n*,
  Dekorationslack *m*, Glasur *f* (Töpferei),
  Lack *m*, **acid-proof** ~ säurefester Lack,
  brushing ~ Streichlack *m*, **clear** ~
  Lasurlack *m*, Transparentlack *m*, **coat of** ~

Lackanstrich *m*, **dull clear** ~ Mattklarlack *m*;,
  fireproof ~ Feuerlack *m*, **interior** ~
  Innenlack *m*, **layer of** ~ Lackschicht *f*, **white**
  ~ Weißlack *m*, ~ **for incandescent lamps**
  Glühlampenlack *m*, ~ **for iron** Eisenfirnis *m*,
  ~ **with oil length of 1:1** einfettiger Lack *m*
varnish *v* firnissen, glasieren, lackieren
varnish brush Lackierpinsel *m*
varnish coating Deckfirnis *m*, Lackierung *f*
varnish color Firnisfarbe *f*, Lackfarbe *f*
varnish crusher Harzmühle *f*
varnished fabric Lackgewebe *n*
varnished paper Lackpapier *n*
varnisher Lackierer *m*
varnish ester Lackester *m*
varnishing machine Lackiermaschine *f*
varnishing resin Lackharz *n*
varnish-like firnisartig, lackartig
varnish mill Harzmühle *f*
varnish paint Lackfarbe *f*
varnish remover Lackabbeizmittel *n*,
  Lackentferner *m*
varnish sediment Lacksatz *m*
varnish stain Lackbeize *f*
varnish substitute Firnisersatz *m*
varnish thinner Lackverdünner *m*
varnish works Lackfabrik *f*
vary *v* abändern, [ab]wechseln, abweichen,
  schwanken, variieren, verändern
varying abwechselnd, abweichend, verschieden,
  wechselnd
vascular calcification Gefäßverkalkung *f* (Med)
vascular system Gefäßsystem *n*
vasculose Vasculose *f*
vaseline-filled cable gestopftes Kabel *n*
vaseline [oil] Vaselin *n*
vasenol Vasenol *n*
vasicine Vasicin *n*
vasicinone Vasicinon *n*
vasoconstriction Gefäßverengung *f* (Med)
vasoconstrictor Gefäßverengungsmittel *n*
  (Pharm)
vasodilatation Gefäßerweiterung *f* (Med)
vasodilator Gefäßerweiterungsmittel *n* (Pharm)
vasoliment Vasoliment *n*
vasopressin Vasopressin *n*
vasopressor Kreislaufmittel *n* (Pharm)
vasotonine Vasotonin *n*
vat Bottich *m*, Bütte *f*, Holländerkasten *m*,
  Kübel *m*, Küpe *f* (Färb), Trog *m*, Wanne *f*
vat *v* [ver]küpen (Färb)
vat blue Küpenblau *n*
vat coloring Büttenfärbung *f*
vat dye Küpenfarbstoff *m*
vat dyeing Küpenfärben *n*, Küpenfärberei *f*
vat dyestuff Küpenfarbstoff *m*
vat paper Büttenpapier *n*
vat press Büttenpresse *f*
vat print Küpendruck *m*

vat retardant Küpenverzögerer *m*
vat sizing Büttenleimung *f*
vault Gewölbe *n* (Arch)
vault *v* [sich] wölben
vauqueline Strychnin *n*
vauquelinite Vauquelinit *m* (Min)
V-belt Keilriemen *m*, V-Riemen, **endless** ~
  endloser Keilriemen, **open-end** ~ endlicher
  Keilriemen
V-belt drive Keilriemenantrieb *m*
V-belt pulley Keilriemenscheibe *f*
veal Kalbfleisch *n*
veatchine Veatchin *n*
vector Vektor *m*, **characteristic** ~ Eigenvektor *m*,
  flux of ~ Vektorfluß *m*, **unit cell** ~
  Basisvektor *m*
vector addition Vektoraddition *f*
vector analysis Vektoranalyse *f*
vector diagram Vektordiagramm *n*,
  Zeigerdiagramm *n*
vector equation Vektorgleichung *f*
vectorial vektoriell
vector[ial] field Vektorfeld *n*
vector model (of the atom) Vektormodell *n* (des
  Atoms)
vector quantity Vektorgröße *f*
vector representation vektorielle Darstellung *f*
vector set Vektorensatz *m*
vector space Vektorraum *m*
vector sum of currents Integralstromstärke *f*
vee belt V-Riemen
Vee mixer Hosenmischer *m*
vee notch V-Kerbe *f*
vegetable Gemüse *n*
vegetable *a* pflanzlich, vegetabilisch
vegetable albumin Pflanzenalbumin *n*
vegetable black Rebenschwarz *n*
vegetable butter Pflanzenfett *n*
vegetable casein Pflanzencasein *n*
vegetable coloring matter Pflanzenfarbstoff *m*
vegetable down Kapok *m*
vegetable dye Pflanzenfarbe *f*
vegetable earth Düngeerde *f*
vegetable extract Pflanzenauszug *m*
vegetable fat Pflanzenfett *n*, Pflanzenöl *n*
vegetable fiber Pflanzenfaser *f*
vegetable ivory Elfenbeinnuß *f*
vegetable jelly Pektin *n*
vegetable manure Gründünger *m*
vegetable mold Pflanzenerde *f*
vegetable oil Pflanzenfett *n*, pflanzliches Öl *n*
vegetable parchment Echtpergamentpapier *n*
vegetable poison Pflanzengift *n*
vegetable protein Pflanzeneiweiß *n*
vegetable resin Naturharz *n*
vegetables, canned ~ Dosengemüse *n*, **tinned** ~
  Dosengemüse *n*
vegetable salt Kräutersalz *n*
vegetable silk Pflanzenseide *f*

vegetable soil Humus *m*, Humuserde *f*
vegetable sulfur Bärlappmehl *n*,
  Bärlappsamen *m*
vegetable tallow Pflanzentalg *m*
vegetable wax Pflanzenwachs *n*
vegetaline Vegetalin *n*
vegetation Vegetation *f*, Pflanzenwachstum *n*,
  Pflanzenwelt *f*, **zone of** ~ Vegetationsgürtel *m*
vegetation period Vegetationsperiode *f*
vegetative vegetativ
vegetative period Vegetationszeit *f*
vehemence Heftigkeit *f*
vehement heftig
vehicle Beförderungsmittel *n*, Fahrzeug *n*
vehicle capacity, exploitation of ~
  Fahrzeugvolumen-Ausnutzung *f*
vehicles, fleet of ~ Fahr[zeug]park
veil *v* verhüllen
vein Ader *f* (Med), Blutader *f* (Med), Flöz *n*
  (Bergb), Gang *m* (Bergb), Holzmaserung *f*
  (Holz), Maser *f* (Holz), Vene *f* (Med)
vein *v* masern
veined aderig, gemasert, marmoriert
veined wood Maserholz *n*
vein ore Gangerz *n*
vellein Vellein *n*
velocitron Laufzeitspektrograph *m*
velocity Geschwindigkeit *f*, Schnelligkeit *f*,
  average ~ mittlere Geschwindigkeit, **decrease
  of** ~ Geschwindigkeitsabnahme *f*, **excess** ~
  Übergeschwindigkeit *f*, **final** ~
  Endgeschwindigkeit *f*, **initial** ~
  Anfangsgeschwindigkeit *f*, ~ **of a falling body**
  Fallgeschwindigkeit *f* (Phys), ~ **of discharge**
  Entladungsgeschwindigkeit *f*, ~ **of ionization**
  Ionenbildungsgeschwindigkeit *f*, ~ **of
  reaction** Reaktionsgeschwindigkeit *f*, ~ **of
  rotation** Drehgeschwindigkeit *f*
velocity component
  Geschwindigkeitskomponente *f*
velocity constant Geschwindigkeitskonstante *f*
velocity curve Geschwindigkeitskurve *f*
velocity dependence
  Geschwindigkeitsabhängigkeit *f*
velocity distribution
  Geschwindigkeitsverteilung *f*
velocity gradient Geschwindigkeitsgefälle *n*,
  Geschwindigkeitsgradient *m*
velocity of underground flow
  Fortschrittsgeschwindigkeit *f*
velocity recorder Geschwindigkeitsmesser *m*
velocity resolution Geschwindigkeitsauflösung *f*
velocity selector Geschwindigkeitsselektor *m*
velocity stage Geschwindigkeitsstufe *f*
velocity triangle Geschwindigkeitsdreieck *n*
velvet Samt *m* (Text)
velvet black samtschwarz
velvet copper ore Kupfersamterz *n* (Min)
velvet finish Samtappretur *f* (Text)

velvet leather Plüschleder *n*
velvet-like samtglänzend
velvety samtartig, samtglänzend
vena Vene *f* (Med)
vending machine Automat *m* (Verkaufsautomat)
veneer Furnier *n*
veneer *v* furnieren
veneer glue Furnierleim *m*, Preßholzkleber *m*,
  Sperrholzkleber *m*
veneering ein Furnier *n* aufbringen
veneering adhesive Furnierleim *m*
veneering press Furnierpresse *f*
veneer strips Möbelfolien *pl*
veneer wood Furnierholz *n*
Venetian mosaic Terrazzo *m*, Zementmosaik *n*
venetian red Venezianischrot *n*
venetian turpentine Lärchenharzöl *n*
Venice sumach Färberbaum *m*, Färbersumach *m*
venom Schlangengift *n*, Tiergift *n*
venomous giftig
vent Abzugsöffnung *f*, Ablaßöffnung *f*, Abzug *m*
  (für Abgase), Entlüftung *f*, Lüftung *f*
vent *v* entlüften, ventilieren
vent hole Luftloch *n*
vent hole can Füllochdose *f*
ventilate *v* [be]lüften, durchlüften, entlüften,
  ventilieren
ventilated roof Kaltdach *n*
ventilating duct Luftkanal *m*
ventilating shaft Lüftungsschacht *m*
ventilating system Lüftungsanlage *f*
ventilation Ventilation *f*, Belüftung *f*,
  Bewetterung *f* (Bergb), Entlüftung *f*,
  Lüftung *f*, Luftzufuhr *f*, Rauchabzug *m*,
  external ~ Fremdbelüftung *f*
ventilation equipment Entlüftungsanlage *f*
ventilation hood Belüftungshaube *f*
ventilation plant Belüftungsanlage *f*
ventilation system Belüftungsanlage *f*
ventilator Ventilator *m*, Entlüftungsschlitz *m*,
  Gebläse *n*, Lüftungsanlage *f*;
  Wetterschacht *m* (Bergb)
vent (of a mo[u]ld) Entlüfungsloch *n*
vent pipe Abzugsrohr *n*, Entlüftungsrohr *n*
vents (in mo[u]ld) Entlüfungseinsätze *pl*
venture Risiko *n*, Wagnis *n*
venturi burner Venturibrenner *m*
venturi fluidized bed Venturi-Wirbelschichtbett *f*
venturine Glimmerquarz *m* (Min)
venturi scrubber Venturiwäscher *m*
venturi tube Venturidüse *f*
veracevine Veracevin *n*
veranthridine Veranthridin *n*
veratral Veratral *n*
veratramine Veratramin *n*
veratric acid Veratrinsäure *f*
veratril Veratril *n*
veratrine Veratrin *n*
veratrine resin Veratrinharz *n*

veratrine sulfate Veratrinsulfat *n*
veratroidine Veratroidin *n*
veratrole Veratrol *n*,
  Brenzcatechindimethyläther *m*
veratrosine Veratrosin *n*
veratroyl chloride Veratroylchlorid *n*
veratrum alkaloid Veratrumalkaloid *n*
veratryl alcohol Veratrylalkohol *m*
veratryl chloride Veratrylchlorid *n*
verbanol Verbanol *n*
verbanone Verbanon *n*
verbascose Verbascose *f*
verbazide Verbazid *n*
verbena Verbene *f* (Bot), Eisenkraut *n* (Bot)
verbenalin Verbenalin *n*, Verbenalosid *n*
verbenalinic acid Verbenalinsäure *f*
verbenaloside Verbenalosid *n*, Verbenalin *n*
verbena oil Verbenaöl *n*
verbene Verbene *f* (Bot)
verbenol Verbenol *n*
verdigris Grünspan *m*, crystallized ~
  Grünspanblumen *pl*, crystals of ~
  Grünspanblumen *pl*, neutral ~
  Kupferacetat *n*
verditer, green ~ Erdgrün *n*
verification Bestätigung *f*, Beurkundung *f*,
  Echtheitsprüfung *f*
verify *v* beweisen, kontrollieren, prüfen,
  bestätigen
vermicidal wurmtötend
vermicide Wurmmittel *n* (Pharm)
vermiculate wurmstichig
vermiculite Vermiculit *m* (Min)
vermifuge Wurmmittel *n* (Pharm),
  Anthelminthikum *n* (Pharm)
vermilion Zinnoberfarbe *f*, Zinnoberrot *n*
vermin Schädling *m*, Ungeziefer *n*
vermin destruction Ungezieferbekämpfung *f*
vermin extirpation Ungezieferbekämpfung *f*
vermuth Wermut *m*
vernier Nonius *m*, Feinmesser *m*, Vernier *m*
vernier calliper[s] Schieblehre *f*
vernier reading Feinablesung *f*, Noniusablesung *f*
vernier scale Noniuseinteilung *f*, Noniusskala *f*
vernine Vernin *n*
vernolic acid Vernolsäure *f*
Verona green Veronesergrün *n*
veronal Veronal *n* (HN)
verosterine Verosterin *n*
versatile vielseitig
versatility Vielseitigkeit *f*, ~ of service vielseitige
  Verwendbarkeit *f*
versenol Versenol *n*
vertebra Wirbelknochen *m* (pl vertebrae),
  Wirbel *m* (Anat)
vertebral column Wirbelsäule *f* (Med)
vertebrate Wirbeltier *n* (Zool)
vertex Scheitel[punkt] *m*, Spitze *f*

vertical Senkrechte *f*, **deflection of the** ~ Lotschwankung *f*
vertical *a* lotrecht, senkrecht, vertikal
vertical angle Scheitelwinkel *m*
vertical axis Vertikalachse *f*
vertical bearing Stehlager *n*
vertical conveyor Senkrechtförderer *m*
vertical diffusion Vertikalausbreitung *f*
vertical distribution Höhenverteilung *f*
vertical drive Senkrechtantrieb *m*
vertical line Senkrechte *f*
vertical magnet Hebemagnet *m*
vertical pipe Fallrohr *n*, Standrohr *n*
vertical plunger Vertikalkolben *m*
vertical polarization Vertikalpolarisation *f*
vertical recovery bend Destillierbrücke *f*
vertical resolution Senkrechtauflösung *f* (Telev)
vertical scale Höhenmaßstab *m*
vertical screw mixer Umlaufschneckenmischer *m* (for silos), Schneckenmischer *m*
vertical section Aufriß *m*
vertical shaper Senkrechthobel *m*
vertical shell-and-tube condenser Turmverflüssiger *m*
vertical shift Vertikalverschiebung *f*
vertical span Stützhöhe *f*
vertical upward motion Hebbewegung *f*
vertigo Gleichgewichtsstörung *f* (Med)
vervain Eisenkraut *n* (Bot)
vervain oil Verbenaöl *n*
vesicant blasenziehend (Med)
vesicatory blasenziehend
vesicle Bläschen *n*, Blase *f*, Wasserblase *f*
vesicular bläschenförmig, blasig
vesiculation Bläschenbildung *f* (Med)
vessel Becken *n*, Behälter *m*, Gefäß *n*, Tank *m*, ~ **under pressure** Druckgefäß *n*
vessel closure Gefäßverschluß *m*
vestrylamine Vestrylamin *n*
vesuvian Vesuvian *m* (Min), Idokras *m* (Min)
vesuvianite Vesuvian *m* (Min), Idokras *m* (Min)
vesuvin Vesuvin *n* (Farbstoff)
veterinarian Tierarzt *m*, Veterinär *m*
veterinary Tierarzt *m*, Veterinär *m*
veterinary hospital Tierklinik *f*
veterinary medicine Veterinärmedizin *f*
veterinary science Veterinärmedizin *f*
veterinary surgeon Tierarzt *m*, Veterinär *m*
vetivalene Vetivalen *n*
vetivazulene Vetivazulen *n*
vetiverone Vetiveron *n*
vetivone Vetivon *n*
viability Ertragskraft *f*
viaduct Überführung *f*
vial Glasfläschchen *n*, Ampulle *f*, Arzneifläschchen *n*
vial mouth Bördelrand *m*
viboquercitol Viboquercit *m*

vibrate *v* vibrieren, beben, oszillieren, pendeln, schwingen
vibrating Vibrieren *n*
vibrating ball mill Schwingmühle *f*
vibrating conveyor Schwingförderer *m*
vibrating grid electrode Gittervibratorelektrode *f*
vibrating screen Schüttelsieb *n*, Schwingsieb *n*, Vibrationssieb *n*
vibrating sieve Rüttelsieb *n*
vibrating spiral elevator Wendelwuchtförderer *m*
vibrating table Rütteltisch *m*, Schüttelherd *m*, Schütteltisch *m*, Schwingherd *m*, Schwingrätter *m*
vibrating trough Schwingrinne *f*
vibration Vibration *f*, Oszillation *f*, Schwingung *f*, **forced** ~ erzwungene Schwingung, **free** ~ freie Schwingung, **individual** ~ Eigenschwingung *f*, **natural** ~ Eigenschwingung *f*, **partial** ~ Teilschwingung *f*, **sympathetic** ~ Resonanzschwingung *f*
vibration absorption Vibrationsdämpfung *f*
vibrational energy Vibrationsenergie *f*
vibrational quantum number Oszillationsquantenzahl *f*, Schwingungsquantenzahl *f*
vibrational spectrum Schwingungsspektrum *n*
vibrational state Schwingungszustand *m*
vibration damping Schwingungsdämpfung *f*
vibration direction Schwingungsrichtung *f*, **principal** ~ Hauptschwingungsrichtung *f*
vibration electrometer Vibrationselektrometer *n*
vibration excitation Schwingungserregung *f*
vibration exciter Schwingungserreger *m*
vibration frequency Schwingungszahl *f*
vibration galvanometer Vibrationsgalvanometer *n* (Elektr)
vibration indicator Schwingungsmesser *m*
vibration insulation Schwingungsisolierung *f*
vibrationless erschütterungsfrei
vibration measuring apparatus Schwingungsmesser *m*
vibration mill Schwingmühle *f*
vibration mixer Vibrationsrührer *m*
vibration node Schwingungsknoten *m*
vibration phase Schwingungsphase *f*
vibration quantum Schwingungsquant *n*
vibration rotation spectrum Rotationsschwingungsspektrum *n*
vibration superposition Schwingungsüberlagerung *f*
vibration test, free ~ Ausschwingungsversuch *m*
vibrator Vibrator *m*, Rüttelapparat *m*, Rüttler *m*
vibrator electrode Vibratorelektrode *f*
vibratory oszillierend, schwingungsfähig
vibratory motion Zitterbewegung *f*
vibrograph Schwingungsaufzeichner *m*, Vibrograph *m*
viburnitol Viburnit[ol] *m*

viburnum bark Viburnumrinde *f* (Bot)
vicianine Vicianin *n*
vicianose Vicianose *f*
vicilin Vicilin *n*
vicinal benachbart, vizinal
vicinal effect Vizinaleffekt *m*
vicinal face Vizinalfläche *f* (Krist)
vicinal function Vizinalfunktion *f*
vicinal surface Vizinalfläche *f*
vicine Vicin *n*
Vicker's hardness Vickers-Härte *f*
Victoria blue Viktoriablau *n*, **new** ~
   Neuviktoriablau *n*
Victoria green Malachitgrün *n*
vidal black Vidalschwarz *n*
Vienna green Wienergrün *n*
view Anschauung *f*, Ansicht *f*, Aussicht *f*, **full** ~
   Gesamtansicht *f*, **in** ~ **[of]** in Anbetracht
view *v* besichtigen
viewer Sucher *m* (Opt)
view finder Bildsucher *m* (Phot), Sucher *m*
   (Phot)
view finder lens Sucherobjektiv *n*
viewing microscope Beobachtungsmikroskop *n*
viewing screen Negativschaukasten *m*
vignette *v* vignettieren (Opt)
vignetting effect Vignettierungseffekt *m* (Opt)
vigorous heftig, kräftig, **to become** ~ werden
vigoureux printing Vigoureuxdruck *m*
villarsite Villarsit *m* (Min)
villiaumite Villiaumit *m* (Min)
vinamar Vinamar *n*
vinasse Schlempe *f* (Techn)
vinbarbital Vinbarbital *n*
vincaine Vincain *n*
vine black Frankfurter Schwarz *n*,
   Rebenschwarz *n*
vinegar Essig *m*, **aromatic** ~ Kräuteressig *m*,
   Räucheressig *m*, **flower of** ~ Essigschaum *m*,
   mother of ~ Mutteressig *m*
vinegar essence Essigessenz *f*
vinegarlike essigartig
vinegar making Essigbereitung *f*
vinegar mother Essigpilz *m*
vinegar plant Essiganlage *f*
vinegar vapors pl
vinegar water Essigwasser *n*
vine-grower Winzer *m*
vinetine Oxyacanthin *n*
vineyard Weinberg *m*
vinhaticoic acid Vinhaticosäure *f*
vinol Vinylalkohol *m*
vinous fermentation Weingärung *f*
vintage Weinernte *f*
vintager Winzer *m*
vinyl Vinyl-
vinyl acetal Vinylacetal *n*
vinyl acetate Vinylacetat *n*
vinylacetic acid Vinylessigsäure *f*

vinyl acetylene Vinylacetylen *n*
vinyl alcohol Vinylalkohol *m*
vinylation Vinylierung *f*
vinyl benzene Styrol *n*
vinyl benzoate Vinylbenzoat *n*
vinyl bromide Vinylbromid *n*
vinyl butyrate Vinylbutyrat *n*
vinyl carbazol Vinylcarbazol *n*
vinyl chloride Vinylchlorid *n*, Chloräthylen *n*
vinyl compound Vinylverbindung *f*
vinyl crotonate Vinylcrotonat *n*
vinyl cyanide Acrylnitril *n*
vinylene Vinylen-
vinyl ester Vinylester *m*
vinyl formate Vinylformiat *n*
vinylidene Vinyliden-
vinylidene chloride Vinylidenchlorid *n*
vinylidene plastic Vinylidenkunststoff *m*
vinyl iodide Vinyljodid *n*
vinyl plastics Vinoplaste *pl*, Vinylkunststoffe *pl*
vinyl polymers Vinylpolymere *pl*
vinyl propionate Vinylpropionat *n*
vinyl pyridine Vinylpyridin *n*
vinyl pyrrolidone Vinylpyrrolidon *n*
vinyl resin Vinylharz *n*
vinyon Vinyon *n*
vioform Vioform *n*
violanthrene Violanthren *n*
violanthrone Violanthron *n*, Dibenzanthron *n*
violaquercitrin Violaquercitrin *n*
violaxanthin Violaxanthin *n*
violence Heftigkeit *f*
violent heftig
violet blaurot, Veilchen *n* (Bot), Violett *n*
violet *a* violett
violine Violin *n* (Chem)
violuric acid Violursäure *f*
violutin Violutin *n*
violxanthin Violxanthin *n*
viral strain Virusstamm *m*
virensic acid Virenssäure *f*
virescent grünlich
virgin gediegen (Met)
virgin copper Kupferröte *f*
virgin lead Jungfernblei *n*
virial Virial *n* (Phys)
virial coefficient Virialkoeffizient *m*
virial equation Virialgleichung *f*
virial theorem Virialsatz *m*
viridian Guignetgrün *n*
viridian green Veronesergrün *n*
viridifloric acid Viridiflorinsäure *f*
viridiflorine Viridiflorin *n*
viridine Viridin *n*
virologist Virologe *m*
virology Virologie *f*, Viruskunde *f*
virosis Viruskrankheit *f* (Med)
virotoxic virotoxisch
virtual image virtuelles Bild *n* (Opt)

virtual value Effektivwert *m*
virulence Virulenz *f* (Med)
virus (pl viruses) Virus *m* (pl Viren)
virus culture Virenkultur *f*
virus disease Viruskrankheit *f* (Med), Virose *f* (Med)
virus protein Virusproteid *n* (Virusprotein)
virus strain Virusstamm *m*
virus vaccine Virusimpfstoff *m*
visammin Visammin *n*
visamminol Visamminol *n*
viscid klebrig, schleimig, viskos, zähflüssig, **very ~** hochviskos
viscidity Dickflüssigkeit *f*
viscin Viscin *n*
visco-elastic viskoelastisch
visco-elasticity Viskoelastizität *f*
viscometer Viskosimeter *n*, Viskositätsmesser *m;*, **efflux ~** Auslaufviskosimeter *n*, **forced ball ~** Kugelfallviskosimeter *n*, **suspended-level Ubbelohde ~** Ubbelohde-Viscosimeter *n* mit hängendem Kugelniveau
viscometry Viskosimetrie *f*
visco-plastic viskoplastisch
viscose Viskose *f*, **raw ~** Rohviskose *f*
viscose *a* , **structurally ~** strukturviskos
viscose filament Viskosefaden *m*
viscose pump Viskosepumpe *f*
viscose rayon Viskoseseide *f*, Chardonnetseide *f*, Viskosefaser *f*
viscose sponge Viskoseschwamm *m*
viscosimeter Viskosimeter *n*, Viskositätsapparat *m*, Viskositätsmesser *m*, **efflux ~** Auslaufviskosimeter *n*
viscosimetric viskosimetrisch
viscosimetry Viskosimetrie *f*
viscosity Viskosität *f*, innere Reibung *f*, Zähflüssigkeit *f*, **absolute ~** absolute Viskosität, **anomalous ~** anomale Viskosität, **apparent ~** scheinbare Viskosität, **correction for ~** Viskositätskorrektur *f*, **degree of ~** Viskositätsgrad *m*, **dynamic ~** dynamische Viskosität, **high ~** Schwerflüssigkeit *f*, **internal ~** Eigenviskosität *f*, **of low ~** niedrigviskos, **plastic ~** plastische Viskosität, **structural ~** Strukturviskosität *f*
viscosity coefficient Viskositätskonstante *f*
viscosity depressant Viskositätsverminderer *m*
viscosity effect Viskositätseffekt *m*
viscosity index Viskositätsindex *m*
viscosity relationship Viskositätsbeziehung *f*
viscous viskos, dickflüssig, schwerflüssig, sirupartig, zähflüssig, **highly ~** hochviskos
viscous material evaporator Dickstoffverdampfer *m*
viscousness Zähflüssigkeit *f*
vise Aufspannblock *m*, Schraubstock *m*, Zwinge *f*
visibility Sichtbarkeit *f*

visibility meter Sichtmesser *m*
visible sichtbar, wahrnehmbar (mit dem Auge), **to render ~** machen, **~ to the naked eye** mit dem bloßen Auge wahrnehmbar
vision, **defect of ~** Sehfehler *m*, **field of ~** Sehfeld *n*, **power of ~** Sehschärfe *f*, **range of ~** Sichtweite *f*
visnagin Visnagin *n*
visual visuell, optisch
visual colorimeter Dreifarbenmeßgerät *n*
visual distance Sehweite *f*
visual faculty Sehkraft *f*
visualize *v* machen
visual point Blickpunkt *m* (Opt)
visual threshold Sehschwelle *f*
vitachrome Vitachrom *n*
vital lebensnotwendig, lebenswichtig; vital, wesentlich
vitality Keimkraft *f* (Biol)
vitamin, **antiscorbutic ~** antiskorbutisches Vitamin, **fat-soluble ~** fettlösliches Vitamin *n*, **water-soluble ~** wasserlösliches Vitamin
vitamin B₆ Pyridoxin *n*
vitamin D₂ Calciferol *n*
vitamin deficiency [disease] Avitaminose *f*, Hypovitaminose *f*
vitamin E Tocopherol *n*
vitamine A₁ Axerophthol *n*
vitamin enrichment Vitaminierung *f*
vitaminize *v* vitaminisieren
vitamin K₂ Farnochinon *n*
vitaminology Vitaminkunde *f*
vitamin precursor Provitamin *n*
vitamins, **rich in ~** vitaminreich
vitamin unit Vitamineinheit *f*
vitellin Vitellin *n*
vitelline membrane Dotterhaut *f*
vitexine Vitexin *n*
vitiate *v* verunreinigen
vitiatine Vitiatin *n*
viticulture Weinbau *m*
vitreoelectric positiv elektrisch, glaselektrisch
vitreosity Glasartigkeit *f*
vitreous gläsern, glasartig, glasig
vitreous electricity Glaselektrizität *f*
vitreous luster Glasglanz *m*
vitreousness glasartige Beschaffenheit *f*, Glasartigkeit *f*
vitreous porcelain Glasporzellan *n*
vitreous sand Glassand *m*
vitreous silver Silberglaserz *n* (Min)
vitreous state Glaszustand *m*
vitrifiable color Schmelzfarbe *f*
vitrifiable earth Glaserde *f*
vitrification Sintern *n*, Sinterung *f*, Verglasung *f*, Vitrifikation *f*
vitrified clay pipe Steinzeugrohr *n*
vitriform glasartig

vitrify v dicht brennen (Keramik), glasieren, sintern, verglasen, vitrifizieren

vitrinite Vitrinit m

vitriol Vitriol n, **blue** ~ blauer Vitriol n, **containing** ~ vitriolhaltig, **formation of** ~ Vitriolbildung f, **green** ~ grüner Vitriol n, **solution of** ~ Vitriollösung f, **white** ~ weißer Vitriol n, Zinksulfat n, Zinkvitriol n, ~ **of copper** blauer Vitriol n

vitriolate v vitriolisieren

vitriolation Vitriolbildung f

vitriolic vitriolartig, vitriolhaltig

vitriolic lye Vitriollauge f

vitriolic ore Vitriolerz n (Min)

vitriolize v vitriolisieren

vitriol works Vitriolhütte f, Vitriolsiederei f

vivianite Vivianit m (Min), Blaueisenerz n (Min), Eisenblauerz n (Min)

vivification Belebung f

vivify v beleben

voacangic acid Voacanginsäure f

vocal cord Stimmband n (Anat)

voglite Voglit m (Min)

void Hohlraum m, Leere f, Lücke f

void a ungültig, leer, nichtig

void v leeren

void content Luftporenanteil m

volatile flüchtig, ätherisch, verdampfbar, **highly** ~ hochflüchtig, **not** ~ schwerflüchtig, **readily** ~ leichtflüchtig

volatile in steam dampfflüchtig

volatileness Flüchtigkeit f (Chem)

volatility Flüchtigkeit f, Verdampfbarkeit f, **low** ~ Schwerflüchtigkeit f, ~ **in steam** Wasserdampfflüchtigkeit f

volatility test apparatus Verdampfungsprüfer m

volatilization Abdampfen n, Verdampfung f, Verdunsten n, Verflüchtigung f

volatilization roasting Verdampfungsröstung f

volatilization temperature Verdampfungstemperatur f

volatilize v sich verflüchtigen

volborthite Volborthit m (Min)

volcanic vulkanisch

volcanic ashes Lavaguß m

volcanic eruption Vulkanausbruch m

volcanic glass Glaslava f

volcanic glass rock Obsidian m (Min)

volcanic rocks Auswurfgestein n

volcanic schorl Vulkanschörl m

volcanic tuff Traß m

volcano Vulkan m

volcanology Vulkanologie f

volemite Volemit m

volemitol Volemit m

volt Volt n (Elektr)

voltage Spannung f (Elektr), **active** ~ Wirkspannung f (Elektr), **additional** ~ Zusatzspannung f, **continuous current** ~ Gleichspannung f (Elektr), **depending on the** ~ spannungsabhängig, **excessive** ~ Überspannung f (Elektr), **excess of** ~ Spannungsüberschuß m (Elektr), **external** ~ Fremdspannung f, **fall of** ~ Spannungsrückgang m, **fluctuation of** ~ Spannungsschwankung f, **induced** ~ induzierte Spannung, **limit of** ~ Spannungsgrenze f, **low** ~ Niederspannung f (Elektr), **lowered value of** ~ der Spannung, **nominal** ~ Nennspannung f, **puncture or rupturing** ~ Durchschlagspannung f, ~ **in open circuit** Ruhespannung f

voltage characteristic Spannungscharakteristik f

voltage collapse Spannungszusammenbruch m

voltage compensation Spannungsausgleich m (Elektr)

voltage component Spannungskomponente f

voltage curve Spannungskurve f

voltage decrease Spannungserniedrigung f

voltage difference Spannungsdifferenz f (Elektr), **sudden** ~ Spannungssprung m

voltage drop Spannungsabfall m, Spannungsgefälle n, Spannungsverlust m

voltage equation Spannungsgleichung f

voltage fluctuation Spannungsänderung f (Elektr)

voltage gradient Spannungsgefälle n

voltage increase Spannungserhöhung f

voltage indicator Spannungsprüfer m (Elektr)

voltage nodal point Spannungsknotenpunkt m

voltage reducing transformer Abwärtstransformator m (Elektr)

voltage regulation Spannungsregulierung f

voltage regulator Potentialregler m (Elektr), Spannungsregler m

voltage stabilizer Spannungsstabilisator m

voltage unit Spannungseinheit f (Elektr)

voltage variation Spannungsänderung f (Elektr)

voltage vector Spannungsvektor m

voltaic voltaisch, galvanisch

voltaic cell Voltaelement n (Elektr)

voltaic element Voltaelement n (Elektr)

volta induction coil Voltainduktor m

voltaism Galvanismus m

voltaite Voltait m (Min)

voltameter Voltameter n (Elektr)

voltametric voltametrisch

voltammeter Voltamperemeter n

volt-ampere Voltampere n (Elektr)

volt-ampere characteristic Stromspannungskennlinie f

voltmeter Spannungsmesser m (Elektr), Voltmeter n (Elektr), **electrostatic** ~ elektrostatisches Voltmeter, **thermionic** ~ elektronisches Voltmeter

volt second Voltsekunde f (Elektr)

voltzine Voltzin m (Min)

volume Band *m*, Fassungsraum *m*, Inhalt *m*,
Rauminhalt *m*, Volumen *n* (pl Volumina),
change of ~ Volumenänderung *f*, **critical** ~
kritisches Volumen, **decrease in** ~
Volumenverminderung *f*, **excess** ~
Überschußvolumen *n*, **incompressible** ~
inkompressibles Volumen, **increase in** ~
Volumenvermehrung *f*; Volumenzunahme *f*,
loss in ~ Volumenverlust *m*, **partial specific**
~ partielles spezifisches Volumen, **specific** ~
spezifisches Volumen, **unit of** ~
Volumeneinheit *f*
volume calculation Raumberechnung *f*
volume contraction Volumenkontraktion *f*,
Volumenverminderung *f*
volume control Mengenregler *m*
volume density Volumendichte *f*
volume effect Volumeneffekt *m*
volume energy Volumenenergie *f*
volume expansion Volumenausdehnung *f*
volume indicator Lautstärkemesser *m*
volume integral Volumenintegral *n* (Math)
volume ionization Volumenionisierung *f*
volumenometer Volumenmesser *m*,
Volumenometer *n*
volume percentage scale Volum[en]prozentskala *f*
volume percent content
Volum[en]prozentgehalt *m*
volume regulator Lautstärkeregler *m*
volumescope Volumenanzeiger *m*
volume-space velocity
Raum-Volumen-Geschwindigkeit *f*
volumeter Volumeter *n*, Durchflußmesser *n*,
Volumenmesser *m*
volumetric volumetrisch, maßanalytisch,
titrimetrisch
volumetric analysis Maßanalyse *f*, volumetrische
Analyse *f*
volumetric apparatus Titrierapparat *m*
volumetric capacity Fassungsvermögen *n*
volumetric flask Meßkolben *m*
volumetric percentage Volum[en]prozentgehalt *m*
volumetric pipet[te] Vollpipette *f* (Anal)
volumetric precipitation analysis
Fällungsanalyse *f*
volumetric proportions in mixtures
Mischungsvolumen *n*
volumetric weight Raumgewicht *n*
voluminous umfangreich, voluminös
voluntal Voluntal *n*
voluntary freiwillig
volute compasses Spiralenzirkel *m*
volute spring Wickelfeder *f*
vortex burner Wirbelbrenner *m*
vortex classifier Drehströmungsentstauber *m*
vortex crystallizer Wirbelkristallisator *m*
vortex tube Wirbelrohr *f*
vortex type separator Wirbelsichter *m*
vorticellae Glockentierchen *pl*

vorticity Wirbelströmung *f*
vorticity potential Wirbelwert *m*
vorticity tensor Drehgeschwindigkeitstensor *m*
vouch [for] *v* bürgen [für]
V-shaped V-förmig
vulcanite Ebonit *n*, Hartgummi *n*, Vulkanit *n*
vulcanizable vulkanisierbar
vulcanizate, cold ~ Kaltvulkanisat *n*
vulcanization Vulkanisation *f*, Vulkanisierung *f*
vulcanization works Vulkanisieranstalt *f*
vulcanize *v* vulkanisieren, ~ **on**
aufvulkanisieren
vulcanized vulkanisiert
vulcanized fiber Vulkanfiber *f* (HN)
vulcanized product, cold ~ Kaltvulkanisat *n*
vulcanized rubber Hartgummi *n*
vulcanizer Vulkanisator *m*, Vulkanisierapparat *m*
vulcanizing Vulkanisieren *n*
vulcanizing agent Vulkanisationsbeschleuniger *m*
vulcanizing kettle Vulkanisierkessel *m*
vulcanizing press Vulkanisierpresse *f*
vulgar gemein
vulpic acid Vulpinsäure *f*, Chrysopikrin *n*
vulpinite Vulpinit *m* (Min)
vuzine Vuzin *n*

W

wacke Wacke *f* (Geol)
Wackenroder's solution Wackenrodersche
  Flüssigkeit *f*
wad Bausch *m*, Pfropfen *m*, Wad *n* (Min),
  Wattebausch *m*
wadding Schutzpolster *n*, Stoßschutz *m*, Watte *f*,
  Wattierung *f*, **high-bulk [or high-loft]** ~
  Füllvlies *n*
wad ore Waderz *n* (Min)
wafers Oblaten *pl*
wage earner Lohnempfänger *m*
wage per piece Stücklohn *m*
wages Lohn *m*
wagnerite Wagnerit *m* (Min)
Wagner-Meerwein rearrangement
  Wagner-Meerwein-Umlagerung *f* (Chem)
wainscot Getäfel *n*, Täfelung *f*, Wandtäfelung *f*
waiting time Wartezeit *f*
wake Kielwasser *n*
Walden inversion Waldensche Umkehrung *f*
  (Chem)
wall Mauer *f*, Wand *f*, **light admitting** ~
  lichtdurchlässige Wand, **thickness of the** ~
  Wandstärke *f*
wall bracket Wandarm *m*, Wandhalter *m*,
  Wandstütze *f*, Winkelarm *m*
wall catalysis Wandkatalyse *f*
wall charge Wandladung *f*
wall charge density Wandladungsdichte *f*
wall console Wandkonsole *f*
wall corner fillet Mauereckleiste *f*
wall covering Wandbekleidung *f*
wall crane Konsolkran *m*
wall effect Wandeffekt *m* (Rheologie),
  Wandeinfluß *m* (Rheologie)
wall energy Wandenergie *f* (Krist)
wall fan Wandlüfter *m*
wall friction Wandreibung *f*
wall-ironing Tiefziehen *n*
wall paint Wandfarbe *f*
wall panelling Wandbekleidung *f*,
  Wandtäfelung *f*
wall paper Tapete *f*
wallpaper underlay Tapetenunterlage *f*
wall pellitory Mauerkraut *n* (Bot)
wall plug Steckdose *f* (Elektr), Wandstecker *m*
  (Elektr)
wall primer Wandgrundierung *f*
wall recombination Wandrekombination *f*
  (Chem)
wall rim Mauerkranz *m*
wall saltpeter Mauersalpeter *m*
wall slab Wandplatte *f*
wall switch board Wandschalttafel *f*
wall thickness Wanddicke *f*, Wandstärke *f*
wall tile Wandkachel *f*

wall trimming Wanddurchbruch *m*
wall tube Durchführungsrohr *n*
walnut Walnuß *f*, Nußbaumholz *n*
walnut oil Walnußöl *n*
walpurgit Walpurgit *m* (Min)
waltherite Waltherit *m* (Min)
want Bedürfnis *n*
wapplerite Wapplerit *m* (Min)
wardite Wardit *m* (Min)
ware Ware *f*, Geschirr *n*, **brown** ~ braunes
  Steingut *n*
warehouse Lagerhaus *n*, **high-rack** ~ **or high-bay**
  warehouse Hochregallager *n*, **flow** ~ **or transit**
  warehouse Durchlauflager *n*
warehouse *v* lagern
war gas Gaskampfstoff *m*, Kampfgas *n*
war industry Kriegsindustrie *f*
warm *a* warm
warm *v* [er]wärmen, ~ **thoroughly**
  durchwärmen, ~ **up** aufwärmen
warm air Warmluft *f*
war material industry Rüstungsindustrie *f*
warm bleach Warmbleiche *f*
warming up Anwärmen *n*
warming up period Anheizungsdauer *f*
warmth Wärme *f*
warm-up mill Vorwärmewerk *n*
warning! Achtung!
warning device Warngerät *n*
warning filter Warnfilter *m*
warning indicator Warngerät *n*
warning light Warnlicht *n*
warning signal Voralarm *m*
warp Kette *f* (Web)
warp *v* krümmen, scheren (Web), verziehen,
  zetteln (Web)
warp beam Kettbaum *m*
warp dyeing machine Kettenfärbeapparat *m*
warp effect Ketteneffekt *m* (Web)
warping Verwerfen *n*
warping machine Zettelmaschine *f*
warp knitting Kettstuhlwirkerei *f*
warp stripe Kettstreifen *m* (Web)
warp stripiness Kettstreifigkeit *f* (Web)
warrant Befugnis *f*
warranty Garantie *f*, Gewährleistung *f*
warrenite Warrenit *m* (Min)
warwickite Warwickit *m* (Min)
wash Alluvialschutt *m* (Geol)
wash *v* waschen, [ab]spülen, reinigen,
  schlämmen (Erze), schwemmen, ~ **away**
  fortspülen, wegspülen, wegwaschen, ~ **off**
  abwaschen, ~ **out** auswaschen, ausschlämmen,
  ausspülen
washability Abwaschbarkeit *f*, Waschbarkeit *f*
washable [ab]waschbar, waschecht
wash bottle Reinigungsflasche *f*, Waschflasche *f*

washer Waschvorrichtung *f,* Abdichtung *f*
(Techn), Beilegscheibe *f,* Berieselungsturm *m*
(Chem), Dichtung *f* (Techn),
Dichtungsring *m* (Techn), Halbholländer *m*
(Pap), Unterlegscheibe *f* (Techn)

washing Wäsche *f,* Läuterung *f* (Chem),
Naßaufbereitung *f* (Techn); Naßreinigung *f*
(Techn), Reinigen *n,* Reinigung *f,*
Verwaschung *f* (Geol), Waschen *n,*
Waschung *f,* **fastness to** ~ Waschfestigkeit *f,*
preliminary ~ Vorwäsche *f,* Vorwaschen *n,*
removable by ~ auswaschbar, ~ **of ores**
Erzschlämmen *n*

washing acid Waschsäure *f*

washing apparatus Schlämmvorrichtung *f*

wash[ing] bottle Spritzflasche *f,* Waschflasche *f,*
spiral ~ Spiralwaschflasche *f*

washing column Berieselungsturm *m,*
Waschturm *m*

washing crystals Waschkristalle *pl*

washing device Waschvorrichtung *f*

washing engine Halbholländer *m* (Pap),
Waschholländer *m* (Pap)

washing liquid Waschflotte *f*

washing liquor Waschlauge *f*

washing machine Waschmaschine *f,* **open width**
~ Breitwaschmaschine *f*

washing plant Spülanlage *f*

washing powder Waschpulver *n*

washing process Waschverfahren *n,*
Schlämmverfahren *n,* Waschvorgang *m*

washing room Spülraum *m*

washings Waschwasser *n* (Chem)

washing slag Darrgekrätz *n*

washing soap Waschseife *f*

washing table Abwaschtisch *m*

washing tank Waschkasten *m*

washingtonite Washingtonit *m* (Min)

washing tower Waschturm *m*

washing tub Waschbottich *m,*
Schlämmbottich *m,* Schlämmfaß *n*

washing tube Aussüßrohr *n*

washing zone Waschzone *f*

wash leather Fensterleder *n,* Waschleder *n*

wash ore Wascherz *n*

wash out Ausspülung *f*

wash polish Wischwachsemulsion *f*

wash primer Beizgrundierung *f,*
Grundiermittel *n,* Haftgrund *m,*
Haftgrund[ier]mittel *n*

washproof waschecht

wash solution Waschlösung *f*

wash tower Berieselungskühler *m*
(Benzolgewinnung)

wash water Waschwasser *n,* Spülwasser *n*

waste Abfall *m,* Abfallgut *n,* Abbrand *m*
(Metall), Abraum *m* (Bergb),
Ausscheidungsprodukt *n* (Biol), Ausschuß *m,*

Fabrikationsabfall *m,* Gekrätz *n* (Metall),
Müll *m,* Schutt *m,* Verlust *m,* Verschnitt *m*

waste *v* verschwenden

waste acid Abfallsäure *f,* Abgangssäure *f*

waste air Abluft *f*

waste channel Abflußkanal *m*

waste coal Abfallkohle *f*

waste copper Kupferabfälle *m pl*

waste cotton Putzbaumwolle *f*

waste disposal Abfallbeseitigung *f*

waste dump Halde *f* (Bergb)

waste fat Abfallfett *n*

waste fuel Abfallbrennstoff *m*

waste gas Ab[zugs]gas *n,* Feuergas *n,*
Hochofengas *n*

waste gas analysis Abgasanalyse *f*

waste gas feed heater Abgasvorwärmer *m,*
Economiser *m*

waste gas furnace mit Hochofengas

waste gas heat Gichtgaswärme *f*

waste gas superheater Abgasüberhitzer *m*

waste gas utilization Abgasverwertung *f*

waste grease Abfallfett *n*

waste heap ore Haldenerz *n*

waste heat Abhitze *f,* Abwärme *f*

waste heat boiler Abhitzekessel *m*

waste heat engine Abwärmekraftmaschine *f*

waste heat flue Abhitzekanal *m*

waste heat kiln Abwärmeofen *m*

waste heat loss Abwärmeverlust *m*

waste heat recovery plant
Abwärmeverwertungsanlage *f*

waste heat utilization Abwärmeverwertung *f*

waste iron Abfalleisen *n*

waste land recovery project Ödlandbegrünung *f*

waste liquor Ablauge *f*

waste lye Abfallauge *f,* Ablauge *f*

waste material Abfallstoff *m,* Abfallware *f*

waste metal Metallabfall *m,* Metallgekrätz *n*

waste oil Abfallöl *n,* Aböl *n,* Altöl *n*

waste ore Erztrübe *f*

waste paper Abfallpapier *n,* Altpapier *n,*
Ausschußpapier *n,* Einstampfpapier *n,*
Makulatur *f,* Papierausschuß *m*

waste pipe Abwasserrohr *n,* Abzugsrohr *n,*
Ausgußrohr *n*

waste product Abfallerzeugnis *n,*
Abfallprodukt *n,* Ausstoßprodukt *n*

waste quota Abfallquote *f*

waste rubber Abfallgummi *m,* Altgummi *m*

waste sheet Makulaturbogen *m*

waste slag Haldenschlacke *f*

waste steam Abdampf *m*

waste steam heating Abdampfheizung *f*

waste steam pipe Dampfableitungsrohr *n*

waste steam utilization Abdampfverwertung *f,*
Brüdenverwertung *f*

waste utilization Abfälleverwertung *f*

waste utilization plant Abfallverwertungsanlage *f*

waste valve Sumpfablaßventil *n*
waste washings Abfallgut *n* (Min)
waste water Abwasser *n*, **crude** ~
  Rohabwasser *n*, **utilization of** ~
  Abwasserverwertung *f*, ~ **from factories**
  Industrieabwasser *n*
waste water clarification Abwasserklärung *f*
waste water clarifying plant
  Abwasserkläranlage *f*
waste water container Schmutzwasserbehälter *m*
waste water disposal Abwasserbeseitigung *f*
waste water organisms pl
waste water purification Abwasserreinigung *f*
waste water purifying plant
  Abwasserreinigungsanlage *f*
waste wood Abfallholz *n*
waste wool Abfallwolle *f*, Putzwolle *f*, Shoddy *n*
watch Uhr *f*
watch *v* aufpassen
watch glass Uhrglas *n*
watch out! Achtung!
water Wasser *n*, **aerated** ~ gashaltiges Wasser,
  amount of ~ Wassermenge *f*, **bound** ~
  gebundenes Wasser, **carbonic** ~
  kohlensäurehaltiges Wasser, **chalybeate** ~
  eisenhaltiges Wasser, **chlorinated** ~
  chloriertes Wasser, **containing** ~ wasserhaltig,
  discharge of ~ Wasserabfluß *m*, **displacement
  of** ~ Wasserverdrängung *f*, **distilled** ~
  destilliertes Wasser, **extracellular** ~
  extrazelluläres Wasser, **filtered** ~ gefiltertes
  Wasser, **free from** ~ wasserfrei, **hard** ~ hartes
  Wasser, **hard calcareous** ~ hartes, kalkhaltiges
  Wasser, **heavy** ~ schweres Wasser, **impervious
  to** ~ wasserundurchlässig, **intracellular** ~
  intrazelluläres Wasser, **lack of** ~
  Wassermangel *m*, **resistant to** ~
  wasserbeständig, **running or flowing** ~
  fließendes Wasser, **soft** ~ weiches Wasser,
  soluble in ~ wasserlöslich, **stagnant** ~
  stehendes Wasser, **to remove** ~ entziehen,
  untreated ~ Rohwasser *n*, ~ **for general use**
  Gebrauchswasser *n*, ~ **for industrial purposes**
  Nutzwasser *n*, ~ **of crystallization**
  Kristallwasser *n*
water *v* wässern, befeuchten, [be]sprengen,
  bewässern
water-absorbing capacity
  Wasserabsorptionsvermögen *n*,
  Wasseraufnahmevermögen *n*
water absorption Wasseraufnahme *f*
water-absorptive capacity
  Wasseraufnahmevermögen *n*
water analysis Wasseranalyse *f*
water-attracting wasseranziehend
water balance Wasserhaushalt *m* (Physiol)
water bath Wasserbad *n* (Chem)
water bath ring Wasserbadring *m*
water bidistiller Wasserbidestillator *m*

water blister Wasserblase *f*
water-blue wasserblau
water bottle Wasserflasche *f*
water bubble Wasserblase *f*
water burst Wasserdurchbruch *m*
water calender Wasserkalander *m*
water calorimeter Wasserkalorimeter *n*
water channel Wasserkanal *m*
water clarifier Wasserkläranlage *f*
water color Aquarellfarbe *f*, Wasserfarbe *f*
water column Wassersäule *f*
water column pressure Wassersäulendruck *m*
water conduit Wasserleitung *f*
water consumption Wasserverbrauch *m*
water container Wasserbehälter *m*,
  Wassergefäß *n*
water contamination Wasserverseuchung *f*
water content Wassergehalt *m*
water content determination
  Wassergehaltbestimmung *f*
water-cooled wassergekühlt
water cooler Wasserkühler *m*
water cooling Wasserkühlung *f*
water cooling jacket Wasserkühlmantel *m*
water cooling plant Wasserkühlanlage *f*
water cushion Wasserpolster *n*
water cutlet [pipe] Wasserabflußrohr *n*
water deficiency Wassermangel *m* (Physiol)
water degassing Wasserentgasung *f*
water demand Wasserbedarf *m*
water determination apparatus
  Wasserbestimmungsapparat *m*
water distilling apparatus
  Wasserdestillierapparat *m*
water droplet Wassertröpfchen *n*
water elimination Wasseraustritt *m*
water equilibrium Wasserhaushalt *m* (Physiol)
water exhaust pump Wasserabsaugpumpe *f*
water fennel Wasserfenchel *m* (Bot),
  Roßfenchel *m* (Bot)
water filter Wasserfilter *m*
water gas Wassergas *n*, **enriched** ~
  angereichertes Wassergas, **production or
  generation of** ~ Wassergaserzeugung *f*
water gas outlet Wassergasabzug *m*
water gas plant Wassergasanlage *f*
water gas producer Wassergaserzeuger *m*
water gas tar Wassergasteer *m*
water gauge Pegel *m*, Wasserstandsanzeiger *m*,
  Wasserstandsmesser *m*, Wasseruhr *f*
water glass Wasserglas *n* (Chem) (Gefäß)
water glass cement Wasserglaskitt *m*
water glass paint Wasserglasanstrichfarbe *f*
water glass soap Wasserglasseife *f*
water glass solution Wasserglaslösung *f*
water green Wassergrün *n*
water hardening Wasserhärtung *f*
water hose Wasserschlauch *m*
wateriness Wässerigkeit *f*

watering Berieselung *f*
water inlet Wassereinlaß *m*, Wasserzufluß *m*
water inlet pipe Wasserzuflußrohr *n*
water-in-oil emulsion Wasser-in-Öl-Emulsion *f*
water insolubility Wasserunlöslichkeit *f*
water irrigation Wasserberieselung *f*
water jacket Wasserkühlmantel *m*,
  Wasserkühlung *f*
water-jacketed furnace Wassermantelofen *m*
water jet Wasserstrahl *m*
water jet nozzle Wasserstrahldüse *f*
water jet pump Wasserstrahlpumpe *f*,
  Wasserstrahlsauger *m*
water level Nivellierwaage *f* (Techn),
  Wasserspiegel *m*, Wasserstand *m*,
  Wasserwaage *f* (Techn)
water level indicator Wasserstandsanzeiger *m*,
  Wasserstandsmesser *m*
water level measurement Wasserstandsmessung *f*
water level regulator Wasserstandsregler *m*
water-like wasserähnlich
water lily Wasserschwertlilie *f* (Bot)
water lime Wasserkalk *m*
water main Wasserleitung *f*
watermark Wasserzeichen *n* (Pap), impressed ~
  geprägtes Wasserzeichen (Pap)
watermark apparatus Wasserzeichenapparat *m*
  (Pap)
water meter Wasseruhr *f*, Wassermesser *m*,
  Wasserzähler *m*
[water] mixer Mischbatterie *f*
water molecule Wassermolekül *n*
water opal Wasseropal *m* (Min)
water paint Wasser[emulsions]farbe *f*,
  non-washable ~ Leimfarbe *f*, oilbound ~
  Wasser-Öl-Farbe *f*
water pipe Wasserrohr *n*
water piping Wasserleitung *f*
water pollution Gewässerverschmutzung *f*,
  Wasserverseuchung *f*
water power Wasserkraft *f*
water power engine Wasserkraftmaschine *f*
water power plant Wasserkraftanlage *f*
water preheater Wasservorwärmer *m*
water pressure Wasserdruck *m*, regulation of ~
  Wasserdruckregelung *f*
waterproof wasserundurchlässig,
  wasserbeständig, wasserdicht, wasserfest
waterproof *v* abdichten, wasserdicht machen,
  imprägnieren (wasserfest machen)
waterproofing Imprägnierung *f*
waterproofness Wasserbeständigkeit *f*,
  Wasserfestigkeit *f*
water pump Wasserpumpe *f*
water purification Wasserreinigung *f*
water purification plant
  Wasserreinigungsanlage *f*
water receiver Wasservorlage *f*
water recooling system Wasserrückkühlung *f*

water removal Entwässern *n*
water-repellent coating wasserabweisende
  Beschichtung *f*
water-repellent wasserabstoßend
water-repellent effect Abperleffekt *m*
water requirement Wasserbedarf *m*
water reserve Wasservorrat *m*
water residue Wasserrest *m*, Wasserrückstand *m*
water resistance Wasserbeständigkeit *f*,
  Feuchtigkeitsbeständigkeit *f*
water-resistant wasserbeständig, naßecht
water-resisting property Wasserbeständigkeit *f*
water sapphire Wassersaphir *m* (Min)
water-screening plant
  Wassergroßreinigungsanlage *f*
water screw Wasserschnecke *f*
water seal Wasserverschluß *m*
water separation Wasserabscheidung *f*
water separator Entwässerungsvorrichtung *f*,
  Wasserabscheider *m*
water skin Wasserhaut *f*
water slurry wässrige Suspension *f*
water softener Wasserenthärter *m*
water softening Wasserenthärtung *f*
water softening plant Wasserenthärtungsanlage *f*
water solubility Wasserlöslichkeit *f*
water-soluble wasserlöslich
water spot Wasserfleck *m* (Pap)
water spray Wassersprühregen *m*
water spraying Wasserberieselung *f*
water-stabilized wasserstabilisiert
water stain Wasserbeize *f*
water still Wasserdestillierapparat *m*
water storage tank or basin Wasserspeicher *m*
water supply Wasserversorgung *f*,
  Wasservorrat *m*
water supply pipe Wasserzuleitungsrohr *n*
water surface Wasseroberfläche *f*,
  Wasserspiegel *m*
water syringe Wasserspritze *f*
water tank Wasserbecken *n*, Wasserbehälter *m*,
  Wasserkessel *m*
water testing Wasserprüfung *f*
water thermometer Wasserthermometer *n*
watertight wasserdicht, wasserfest,
  wasserundurchlässig
water tower Wasserturm *m*
water trap Wasserabscheider *m*
water treatment Wasseraufbereitung *f*,
  Wasserbehandlung *f*
water treatment plant Kläranlage *f*
water trough Wassertrog *m*, Wasserwanne *f*
water-tube boiler Großwasserraum-Kessel *m*,
  Wasserrohrkessel *m*
water tubing Wasserschlauch *m*
water vapor Wasserdampf *m*, Wasserdunst *m*,
  impermeable to ~ wasserdampfdicht
water vapor addition Wasserdampfzusatz *m*

water vapor permeability
Wasserdampfdurchlässigkeit *f*
water vessel Wassergefäß *n*
water wheel Wasserrad *n*
watery dünnflüssig (Öl), wässerig
watt Watt *n* (Elektr)
wattage Wattleistung *f,* Wattverbrauch *m*
watt component Wattkomponente *f*
watt consumption Wattverbrauch *m*
watt current Wirkstrom *m*
watt density (extruder heaters) Heizleistung *f*
watt-hour Wattstunde *f*
wattless blind, wattlos (Elektr)
wattless current Blindstrom *m*
wattmeter Leistungsmesser *m* (Elektr),
Wattmeter *n*
watt-second Wattsekunde *f* (Elektr)
wave Welle *f,* **electromagnetic ~**
elektromagnetische Welle, **elliptically
polarized ~** elliptisch polarisierte Welle, **peak
of ~** Wellenberg *m,* **progressive ~** wandernde
Welle, **stationary ~** stehende Welle
wave band Wellenband *n*
wave character Welleneigenschaft *f*
wave character[istic] Wellennatur *f*
wave constant Wellenkonstante *f*
wave crest Wellenberg *m,* Wellenkamm *m*
wave disturbance Wellenstörung *f*
wave drag Wellenwiderstand *m*
wave equation Schwingungsgleichung *f,*
Wellengleichung *f*
wave frequency Wellenfrequenz *f*
wave front Wellenfront *f*
wave function Wellenfunktion *f*
wave function formalism
Wellenfunktionsformalismus *m*
wave guide Hohlleiter *m* (Elektr)
wave length Wellenlänge *f,* **dependence on ~**
Wellenlängenabhängigkeit *f,* **of long ~**
langwellig, **region of ~**
Wellenlängenbereich *m*
wavelike wellenförmig
wave line Wellenlinie *f*
wavellite Wavellit *m* (Min)
wave loop Wellenbauch *m*
wavemaker Wellenerzeuger-Einrichtung *f*
wave measurement Wellenmessung *f*
wave-mechanical wellenmechanisch
wave mechanics Wellenmechanik *f*
wavemeter Frequenzmesser *m,* Wellenmesser *m*
wave motion Wellenbewegung *f*
wave packet Wellenpaket *n* (Phys)
wave path Wellenbahn *f* (Atom)
wave propagation Wellenausbreitung *f,*
Wellenfortpflanzung *f*
wave range Wellenbereich *m*
wave rectifier Wellengleichrichter *m* (Elektr)

waves, absorption of ~ Wellenabsorption *f,*
generation of ~ Wellenanregung *f,*
propagation of ~ Wellenfortpflanzung *f*
wave steepness Wellensteilheit *f*
wave theory Wellenlehre *f,* **~ of light**
Wellentheorie *f* des Lichtes (Opt)
wave vector Wellenvektor *m*
wave velocity Wellengeschwindigkeit *f*
wavy wellig
wax Wachs *n,* **bleached ~** gebleichtes Wachs,
soft ~ Weichparaffin *n*
wax *v* [ein]wachsen, bohnern
wax backing Paraffinkaschierung *f,*
Wachskaschierung *f*
wax cement Klebewachs *n,* Wachskitt *m*
wax cloth Wachstuch *n*
wax color Wachsfarbe *f*
wax-colored wachsfarben
wax contents Wachsgehalt *m*
wax crayon Fettstift *m*
wax emulsion Wachsemulsion *f*
waxing Einwachsen *n,* Wachsen *n*
wax insulated wire Wachsdraht *m*
wax light Wachskerze *f*
wax-like wachsähnlich, wachsartig
wax matrix Wachsmatrize *f*
wax melting house Wachsschmelzerei *f*
wax melting pot Wachsschmelzkessel *m*
wax model Wachsmodell *n*
wax mold Wachsform *f*
wax ointment Wachssalbe *f*
wax opal Wachsopal *m* (Min)
wax painting Wachsmalerei *f*
wax paper Wachspapier *n,* Butterbrotpapier *n*
wax pencil Fettstift *m*
wax polish Wachspoliermittel *n*
wax pouring table Wachsgießtisch *m*
wax size Wachsleim *m* (Pap)
wax stain Wachsbeize *f*
wax varnish Wachsfirnis *m*
waxy wachsähnlich, wachshaltig
wax-yellow wachsgelb
waxy yolk Wachsschweiß *m*
way Weg *m,* Art *f,* Bahn *f,* Führungsbahn *f*
weak schwach, verdünnt (z. B. Säure)
weak current line Schwachstromleitung *f*
weaken *v* [ab]schwächen, verdünnen
weakening Abschwächung *f,* Schwächung *f*
weak function Unterfunktion *f* (Med)
weak gas Schwachgas *n*
weak lye Laugenwasser *n*
wealthy reich, wohlhabend
wean *v* entwöhnen
wear Abnutzung *f,* Abrieb *m,* Verschleiß *m,* **sign
of ~** Abnutzungserscheinung *f,* **~ by friction**
Abnutzung *f* durch Reibung, **~ of chain**
Kettenverschleiß *m,* **~ on the tools**
Abnutzung *f* der Werkzeuge

wear *v* tragen (Kleidung), ~ **off** sich
 verbrauchen, ~ **out** abnutzen, erschöpfen,
 verschleißen
wearability Abnutzbarkeit *f,*
 Abnutzungsbeständigkeit *f*
wear [and tear] Verschleiß *m,* **resistance against**
 ~ Verschleißhärte *f*
wear hardness Abnutzungshärte *f*
weariness Ermüdung *f*
wearing capacity Abnutzbarkeit *f*
wearing plate Verschleißplatte *f*
wearing property Trageigenschaft *f* (Text)
wearing quality Abnutzbarkeit *f,*
 Verschleißverhalten *n*
wearing surface Abnutzungsfläche *f*
wearing test Abnutzungsprüfung *f*
wearisome langwierig
wear resistance Scheuerfestigkeit *f,*
 Verschleißfestigkeit *f*
wear-resistant abriebbeständig, verschleißfest,
 highly ~ hochverschleißfest
wear strip Verschleißband *n*
wear test Abnutzungsprobe *f,*
 Verschleißprüfung *f*
weather Wetter *n,* **period of poor** ~
 Schlechtwetterperiode *f*
weather *v* auswittern (Chem), verwittern
weather-ageing Alterung *f* durch Verwittern
weathered verwittert
weathered layer Verwitterungsschicht *f*
weather factor Witterungseinfluß *m*
weathering Verwitterung *f,* Auswitterung *f,*
 Bewitterung *f,* **accelerated** ~
 Kurzbewitterung *f*
weathering test, natural ~
 Bewitterungsprüfung *f*
weather map Wetterkarte *f*
weather observation Wetterbeobachtung *f*
weatherproof wetterbeständig
weather-proofness Wetterfestigkeit *f*
weather-protective coating
 Wetterschutzanstrich *m*
weather resistance Wetterbeständigkeit *f;*
 Wetterfestigkeit *f*
weather-resistant wetterbeständig
weather-resistant coating Wetterschutzanstrich *m*
weather-resisting wetterbeständig,
 witterungsbeständig
weather strip Fensterdichtung *f*
weave Gewebe *n* (Web), **tight** ~ feines Gewebe
 (Web)
weave *v* weben, wirken
weaver's knot Weberknoten *m*
weaving Weben *n*
webbing Gurtband *n*
web dryer Bahnentrockner *m*
weber Weber *n* (Elektr)
web guide Bahnenführung *f* (Web),
 Gewebeführung *f* (Web)

webnerite Webnerit *m* (Min)
websterite Websterit *m* (Min)
wedelic acid Wedelsäure *f*
wedelolactone Wedelolacton *n*
wedge Hemmkeil *m,* Keil *m,* **neutral** ~
 Graukeil *m* (Opt)
wedge *v* [fest]keilen, ~ **in** einkeilen
wedge angle Keilwinkel *m*
wedge bolt Stellkeil *m*
wedged in eingekeilt
wedge grip Spannbacke *f*
wedge inlet Keileinguß *m*
wedgelike keilförmig
wedge photometer Keilphotometer *n*
wedges, to fasten by ~ festkeilen
wedge-shaped keilförmig
wedge type V-belt Schmalkeilriemen *m*
wedgwood ware Basaltsteingut *n*
weed Unkraut *n*
weed killer Unkrautvertilgungsmittel *n*
weft Einschlagfaden *m* (Web)
weft bobbin Schußspule *f* (Web)
Wehnelt control grid Wehneltblende *f*
wehrlite Wehrlit *m* (Min), Spiegelglanz *m* (Min)
weibullite Weibullit *m* (Min)
weigh *v* wägen, wiegen, ~ **in** einwiegen
weighability Wägbarkeit *f*
weighable wägbar
weighed sample Einwaage *f*
weighing Wägen *n,* Wägung *f,* **rapid** ~
 Schnellwägen *n*
weighing accuracy Wägegenauigkeit *f*
weighing and filling machine Abfüllwaage *f*
weighing and measuring machine Dosimeter *n*
weighing appliance Wägevorrichtung *f*
weighing boat, alumin[i]um ~ aus Aluminium
weighing bottle Wägefläschchen *n,*
 Wägegläschen *n*
weighing device Wägevorrichtung *f*
weighing dish Waagschale *f*
weighing-in spoon Einwägelöffel *m*
weighing machine Waage *f*
weighing pipet[te] Wägepipette *f* (Anal)
weighing room Wägezimmer *n*
weighing table Wägetisch *m*
weighing tube Wägeröhrchen *n*
weigh-out store Anbruchslager *n*
weight Gewicht *n,* Gewichtsstück *n,* Last *f,*
 Schwere *f,* **analysis by** ~ Gewichtsanalyse *f,*
 change in ~ Gewichtsänderung *f,* **constant** ~
 Gewichtskonstanz *f,* **decrease in** ~
 Gewichtsabnahme *f,* Gewichtsverminderung *f,*
 determination of ~ Gewichtsbestimmung *f,*
 excess ~ Übergewicht *n,* **increase in** ~
 Gewichtszunahme *f,* **initial** ~ Einwaage *f,* **loss**
 in ~ Gewichtsabnahme *f,* Gewichtseinbuße *f,*
 Gewichtsverlust *m,* **part by** ~ Gewichtsteil *m,*
 percent by ~ Gewichtsprozent *n,* **proportion**
 by ~ Gewichtsverhältnis *n,* **specific** ~

spezifisches Gewicht, **unit of** ~
Gewichtseinheit *f*, **variation of** ~
Gewichtsänderung *f*, ~ **of a single piece**
Stückgewicht *n*, ~ **of charge** Einsatzgewicht *n*,
Schmelzgewicht *n*, ~ **of unit volume**
Volumengewicht *n*
weight *v* belasten, beschweren
weight alcoholometer Gewichtsalkoholometer *n*
weight arm Lastarm *m*
weight batching Gewichtsdosierung *f*
weight belt feeder Bandwaage *f*
weight buret[te] Wägebürette *f* (Anal)
weight control Gewichtskontrolle *f*
weight feeding Gewichtsdosierung *f*
weight gain Gewichtszunahme *f*
weight increase Gewichtserhöhung *f*
weight indicator Gewichtsanzeige *f*
weighting Beschweren *n*, Beschwerung *f*, ~ **with**
metallic salts Metallbeschwerung *f*
weighting agent Beschwerungsmittel *n*
weighting material Beschwerungsmittel *n*
weighting method Beschwerungsverfahren *n*
weightless leicht; schwerelos
weight loading Gewichtsbelastung *f*
weight percentage Gewichtsprozent *n*
weight ratio Gewichtsverhältnis *n*
weight ring Belastungsring *m*
weights, set of ~ Gewichtssatz *m*, Wägesatz *m*
weight saving Gewichtsersparnis *f*
weight variation Gewichtsschwankung *f*
weighty gewichtig, schwer[wiegend]
weir Wehr *n*
weissite Weißit *m* (Min)
weld Schweißung *f*, **location of** ~ Schweißstelle *f*
weld *v* schweißen, ~ **together**
zusammenschweißen
weldability Schweißbarkeit *f*
weldable schweißbar
weldable steel Schweißstahl *m*
weld cracking Schweißrissigkeit *f*
weld decay nach dem Schweißen
weld-deposited coating Schweißplattierung *f*
welded joint Schweißnaht *f*,
Schweißverbindung *f*
welder Schweißer *m*
weld-harden *v* schweißhärten
welding Schweißen *n*, Schweißarbeit *f*,
autogenous ~ autogene Schweißung *f*, **defect**
in ~ Schweißfehler *m*, **electric** ~ elektrische
Schweißung, **heated tool** ~
Heizkeilschweißen *n*, **heated wedge** ~
Heizkeilschweißen *n*, **linear** ~ geradliniges
Schweißen *n*, **molten metal** ~
Gießschweißen *n*, ~ **of plastics**
Kunststoffschweißen *n*
welding compound Schweißmittel *n*
welding elbow Einschweißkrümmer *m*
welding electrode Schweißelektrode *f*
welding fire Schweißfeuer *n*

welding flux Schweißmittel *n*
welding furnace Schweißofen *m*
welding gun Schweißbrenner *m*, Schweißgerät *n*
welding heat Schweißhitze *f*,
Schweißtemperatur *f*
welding machine Schweißmaschine *f*
welding metal Schweißmetall *n*
welding-neck collar Vorschweißbund *m*
welding-neck flange Vorschweißflansch *m*
welding plant Schweißanlage *f*
welding process Schweißverfahren *n*
welding rod Schweißdraht *m*, Schweißstab *m*
welding sand Schweißsand *m*
welding seam Schweißnaht *f*
welding set Schweißsatz *m*
weld[ing] steel Schweißstahl *m*
welding test Schweißprobe *f*
welding tip Schweißdüse *f*
welding torch Schweißbrenner *m*, Schweißgerät *n*
welding unit Schweißaggregat *n*
welding wire Schweißdraht *m*
weld iron Schweißeisen *n*
weld joining Verbindungsschweißen *n*
weldless nahtlos
weld line Bindenaht *f*, Schweißlinie *f*
weld[mark] Schweißstelle *f*, fehlerhafte
weld-seam tester Schweißnahtprüfgerät *n*
well Brunnen *m*, Quelle *f* (Öl), Schacht *m*
(Bergb), Zisterne *f*
well base rim Tiefbettfelge *f*
well brine Quellsole *f*
well defined eindeutig
well drilling Bohren *n* (Öl)
well-established feststehend
well-formed wohlausgebildet
well head Verflanschung *f*
well-ordered übersichtlich
well salt Quellsalz *n*
well-seasoned trocken (Holz)
well-shaped wohlausgebildet
wellsite Wellsit *m* (Min)
well water Brunnenwasser *n*
welt Rahmen *m* (Schuh)
welt[ing] Wulstrahmen *m*
wernerite Wernerit *m* (Min), Skapolith *m* (Min)
Wessely's anhydride Wesselys Anhydrid *n*
Weston cell Weston-Element *n*
Westphal balance Westphalsche Waage *f*
wet feucht, naß, ~ **and cold** naßkalt
wet *v* anfeuchten, befeuchten, benetzen
wet abrasion resistance Naßscheuerfestigkeit *f*,
Waschfestigkeit *f* (Lacke)
wet air pump Naßluftpumpe *f*
wet analysis Naßanalyse *f*
wet application Naßauftragung *f*
wet battery Naßbatterie *f* (Elektr)
wet bleach Naßbleiche *f*
wet bonding Naßverklebung *f*
wet catalysis process Naßkatalyseverfahren *n*

wet cleaning Naßaufbereitung *f,*
  Naßbehandlung *f*
wet compression machine
  Naßkompressionsmaschine *f*
wet contact process Naßkatalyseverfahren *n*
wet crushing Naßzerkleinerung *f*
wet-dial water meter Naßläufer *m*
wet dressing Naßbeize *f*
wet edge time (paints) offene Zeit *f*
wet fastness property Naßechtheit *f*
wet feed inlet Naßgutzuführung *f*
wet filter Naßfilter *n*
wet finishing Naßappretur *f*
wet fog Sprühnebel *m*
wet grinding Feuchtmahlung *f,* Naßmahlen *n,*
  Naßmahlung *f*
wet heat shrinking Naßhitzeausschrumpfung *f*
wet litharge Fließglätte *f*
wetness Nässe *f,* degree of ~
  Feuchtigkeitsgehalt *m*
wet-out (of glass fiber mats by polyester)
  Imprägnierung *f*
wet oxidation Naßoxidation *f*
wet picking Naßsortierung *f*
wet pressing Naßpressen *n*
wet processing Naßbehandlung *f*
wet puddling Naßpuddeln *n*
wet purification Naßreinigung *f*
wet rot Kellerschwamm *m*
wet rub fastness Naßwischfestigkeit *f*
wet scuffing resistance Naßabriebfestigkeit *f*
wet separation Naßscheidung *f*
wet-spinning Naßspinnen *n*
wet stamp mill Naßpochwerk *n*
wet steam Naßdampf *m*
wet strength Naßfestigkeit *f*
wettability Benetzbarkeit *f*
wettable benetzbar
wetted-wall column Dünnschichtkolonne *f,*
  Fallfilmkolonne *f,* Naßwandkolonne *f,*
  Rieselfilmkolonne *f*
wetted-wall tower Rieselfilm-Destillationsturm *m*
wet tensile strength Naßreißfestigkeit *f*
wetting Anfeuchten *n,* Benetzen *n,* Benetzung *f,*
  fastness to ~ Naßechtheit *f,* heat of ~
  Benetzungswärme *f*
wetting agent Benetzungsmittel *n,* ~ for
  mercerizing Merzerisiernetzmittel *n*
wetting drum Befeuchtungstrommel *f*
wetting out Durchtränken *n*
wetting powder Benetzungspulver *n*
wetting power Benetzungskraft *f*
wetting properties pl
wet treatment, fastness to ~
  Naßbehandlungsechtheit *f,* fast to ~ naßecht
whale-oil Walfischtran *m*
wharangin Wharangin *n*
wheat Weizen *m*
wheat bran Weizenkleie *f,* Weizennachmehl *n*

wheat flour Weizenmehl *n*
wheat germ oil Weizenkeimöl *n*
wheat malt Weizenmalz *n*
wheat starch Weizenstärke *f*
Wheatstone['s] bridge Wheatstonesche Brücke *f*
  (Elektr)
wheat straw Weizenstroh *n*
wheat [straw] pulp Weizenstrohzellstoff *m*
wheel Rad *n,* Scheibe *f,* pitch of the ~
  Radteilung *f,* solid ~ Vollrad *n,* toothed ~
  Zahnrad *n*
wheel alignment tester Spurprüfgerät *n*
wheel and tire assembly Aggregatradreifen *m*
wheel barrow Schiebkarren *m*
wheel base Achsabstand *m,* Radabstand *m,*
  Radstand *m*
wheel bore Radbohrmaschine *f*
wheel disc Radscheibe *f*
wheeldozer Raddozer *m*
wheel drive Radantrieb *m*
wheel flap Axialschlag *m*
wheel grease Wagenschmiere *f*
wheel intersection diameter
  Schnittpunktdurchmesser *m* (Reifen)
wheel lathe Räderdrehbank *f*
wheel ore Rädelerz *n* (Min)
wheel rim Radkranz *m*
wheel rolling mill Räderwalzwerk *n*
wheel set Radsatz *m*
wheel spider Radstern *m*
wheel turning lathe Raddrehbank *f*
wheelwork Räderwerk *n*
whet *v* wetzen, schleifen
whetstone Schleifstein *m,* Wetzstein *m*
whewellite Whewellit *m* (Min)
whey Käsewasser *n,* Molke *f*
whirl Wirbel *m*
whirl *v* wirbeln
whirling hygrometer Schleuderpsychrometer *n*
whirl sintering Wirbelsintern *n*
whirl sorting plant Wirbelsichter *m*
whirl stabilization Wirbelstabilisierung *f*
whiskers Einkristallfäden *pl*
whisky Whisky *m*
whistle Pfeife *f* (Akust)
whistle *v* pfeifen
whistle valve Pfeifventil *n*
white arsenic Arsentrioxid *n,* Schabenpulver *n*
white blood corpuscle weißes Blutkörperchen *n*
white break Weißbruch *m* (PVC)
white copper Neusilber *n*
white discharge Ätzweiß *n*
white finish Weißlack *m*
white frost Rauhreif *m*
white glass Kreideglas *n*
white gold Weißgold *n*
white guarea Bosse *f* (Holz)
white heat Weißglühhitze *f,* Weißglut *f*
white-heat *v* weißglühen

white-hot weißglühend
white lead Bleiweiß *n*
white lead paint Bleiweißfarbe *f*
white lead putty Bleiweißkitt *m*
white lead substitute Bleiweißersatz *m*
white level indicator Weißgehaltmesser *m*
white line [between passes] Leerzeile *f* (Buchdr)
white liquor Läuterbeize *f*
white metal Lagermetall *n*, Weißeisen *n*
whiten *v* bleichen, weißen, weiß machen
whiteness Weißgrad *m* (Pap)
whiteness retention Weißtonerhaltung *f*
whitening, siliceous ~ Kieselweiß *n*
white oil Paraffinöl *n*
white petrolatum Alvolen *n*
white precipitate ointment
  Quecksilberpräzipitatsalbe *f*
white [rubber] substitute weißer Faktis *m*
white spirit Lackbenzin *n*, Leichtbenzin *n*,
  Testbenzin *n*
whitestone Granulit *m* (Min)
white vitriol Zinkvitriol *n*
whitewash geschlämmte Kreide *f*,
  Kalkanstrich *m*, Kalkmilch *f*, Tünche *f*
whitewash *v* kalken, [über]tünchen, weißen
whitewash paint Kalkfarbe *f*
whitewasher Tüncher *m*
whiting Schlämmkreide *f*, Kreidegrund *m*
whitish weißlich
whitneyite Whitneyit *m* (Min)
whiz *v* zischen
whizzer Schleudertrockner *m*,
  Streuwindsichter *m*, Turbosichter *m*,
  Umluftsichter *m*
whole Gesamtheit *f*
whole *a* ganz
wholesale Großhandel *m*
wholesale manufacturing Massenherstellung *f*
wholesale trade Großhandel *m*
whole tire reclaim Ganzreifenregenerat *n*
whooping cough Keuchhusten *m* (Med)
wick Docht *m*
wick lubricator Dochtschmierer *m*
wick-oiling Dochtschmierung *f*
widdrenal Widdrenal *n*
widdrene Widdren *n*
widdrenic acid Widdrensäure *f*
wide breit, weit
wide band amplifier Breitbandverstärker *m*
wide base rim Breitfelge *f*
wide-flanged breitflanschig
wide-meshed weitmaschig
wide-mouthed bottle Pulverglas *n*, weithalsige
  Flasche *f*
widen *v* ausweiten, erweitern
wide-necked weithalsig
widening Verbreiterung *f*
widespread weitverbreitet

width Breite *f*, Weite *f*, **inner** ~ lichte Weite,
  overall ~ Gesamtbreite *f*, ~ **in the clear** lichte
  Weite, ~ **of field** Bildfeldgröße *f*, ~ **of hole**
  Lochweite *f*
Wieland-Gumlich aldehyde
  Wieland-Gumlich-Aldehyd *m*
Wien constant Wiensche Konstante *f*
Wien shift Wiensche Verschiebung *f* (Phys)
Wien's radiation law Wiensches
  Strahlungsgesetz *n*
Wijs number Jodzahl *f*
wild indigo Färberwaid *m* (Bot)
wild oats Flughafer *m* (Bot)
wild thyme Quendel *m* (Bot)
wild woad Färbergras *n* (Bot)
willagenin Willagenin *n*
willardiine Willardiin *n*
willemite Willemit *m* (Min)
willing bereitwillig
willingness Bereitwilligkeit *f*
willow Haderndrescher *m* (Pap)
willow bark Weidenrinde *f*
willyamite Willyamit *m* (Min)
Wilson cloud chamber Wilsonkammer *f*
Wilson cloud track method
  Wilsonnebelspurmethode *f*
wiltshireite Wiltshireit *m* (Min)
winch Winde *f*, Förderhaspel *f* (Bergb),
  Haspel *f*, Hebehaspel *f*, Kurbel *f*
winch beck Haspelkufe *f*
winch rope Haspelseil *n*
wind Wind *m*
wind *v* aufrollen, [auf]wickeln, haspeln, spulen,
  winden, ~ **around** umwickeln, ~ **off**
  abwickeln, abhaspeln, ~ **on** aufwickeln, ~ **up**
  aufwickeln, aufhaspeln, aufspulen,
  hochziehen
wind box Windkasten *m*
wind chest Windlade *f*
wind-drying Windtrocknung *f*
wind erosion Winderosion *f*
wind furnace Blasofen *m*, Windofen *m*,
  Zugofen *m*
wind gauge Anemometer *n*,
  Wind[geschwindigkeits]messer *m*
wind indicator Windanzeiger *m*
winding Wick[e]lung *f* (Elektr), **bifilar** ~ bifilare
  Wicklung, **cross-sectional area of** ~
  Wicklungsquerschnitt *m*, **direction of** ~
  Wicklungssinn *m*, **dual-strand** ~ bifilare
  Wicklung, **high-resistance** ~ hochohmige
  Wicklung
winding drum Seiltrommel *f*, Wickelrolle *f*,
  Wickeltrommel *f*
winding engine Förderkran *m*
winding machine Aufwickelmaschine *f*
windings, number of ~ Windungszahl *f*
winding-up mechanism Aufziehwerk *n*
windlass Winde *f*, Drehhaspel *f*, Hebewinde *f*

windlass rope Haspelseil *n*
windmill Windmühle *f*
wind-open end Abrolldeckel *m*
window glass Fensterglas *n*
windpipe Luftröhre *f* (Med)
wind pressure Luftdruck *m*
windproof windundurchlässig
windscreen Windschutzscheibe *f*
windscreen wash assembly
  Scheibenwaschanlage *f*
windscreen wiper Scheibenwischer *m*
windshield Windschutzscheibe *f*
wind-sifted windgesichtet
wind tunnel Windkanal *m*
wind tunnel [balance] aerodynamische Waage *f*
wind-up roll Aufwickelwalze *f*
wind-up stand Aufwickelbock *m*
wind velocity Windgeschwindigkeit *f*,
  Windstärke *f*
wind velocity indicator
  Windgeschwindigkeitsmesser *m*
wine Wein *m*, **adulteration of** ~
  Weinfälschung *f*, **camphorated** ~
  Campherwein *m*, **containing** ~ weinhaltig,
  medicated ~ Kräuterwein *m*, Würzwein *m*,
  sour ~ Sauerwein *m*, **spiced** ~ Würzwein *m*,
  white ~ Weißwein *m*, ~ **spiced with cloves**
  Nägeleinwein *m*
wine cask Weinfaß *n*
wine cellar Weinkeller *m*
wine content Weingehalt *m*
wine fusel oil Weinfuselöl *n*
wine lees Weinhefe *f*, **calcined** ~ Drusenasche *f*
wine press Kelter *f*, Kelterpresse *f*, Korbpresse *f*,
  Traubenpresse *f*
wine storage Weinlagerung *f*
wine vinegar Weinessig *m*
wine yeast Weinhefe *f*
wing Flügel *m*
wing burner Flachbrenner *m*
wing callipers Bogenzirkel *m*
winged head Flügelkopf *m*
wing[ed] nut Flügelmutter *f*
wing screw Flügelschraube *f*
wing tip Flügelspitze *f*
wing tip float Flügelspitzenschwimmer *m*
Winkler buret[te] Winklerbürette *f* (Anal)
winning Abbau *m* (Bergbau)
winnow *v* ausstäuben
winnower Wurfsichter *m*
wintergreen oil Gaultheriaöl *n*
wipe *v* wischen, ~ **off** abwischen, abputzen,
  abstreichen, ~ **out** auslöschen, austilgen,
  auswischen
wiper Abstreifer *m*, Abstreifring *m*
wiping pad Abwischbausch *m*
wiping paint Einlaßemaille *f*
wiping ring Abstreifring *m*
wiping solvent Abwischlösemittel *n*

wire Draht *m*, Kabel *n* (Elektr),
  Leitungsdraht *m* (Elektr), **bare** ~ blanker
  Draht, **diameter of** ~ Drahtstärke *f*, **enamelled**
  ~ Lackdraht *m*, **flat** ~ Flachdraht *m*, **gauge**
  of ~ Drahtnummer *f*, **helix of thin** ~
  Drahtschraube *f*, **insulated** ~ isolierter Draht,
  quality of ~ Drahtsorte *f*, **resistance of a** ~
  Drahtwiderstand *n*, **small section** ~
  Formdraht *m*, **tinned** ~ verzinnter Draht,
  waxed ~ Wachsdraht *m*
wire basket Drahtkorb *m*
wire bead Drahtkern *m*
wire brush Drahtbürste *f*, Kratzbürste *f*
wire cable Drahtseil *n*
wire cloth Drahtgewebe *n*
wire coating Kabelüberzug *m*
wire cord tire Stahlcordreifen *m*
wire core Drahteinlage *f*
wire covering Drahtummantelung *f*
wire covering compound Kabelmasse *f*
wire cutters Beißzange *f* (f. Draht),
  Drahtschneider *m*, Drahtzange *f*
wire drawing Drahtziehen *n*, Schliffriefen *pl*
wire drawing machine bench Drahtziehbank *f*
wire enamel Drahtemaille *f*, Drahtlack *m*
wire gauge Drahtlehre *f*
wire gauze Drahtgaze *f*, Drahtgeflecht *n*,
  Drahtgewebe *n*, Drahtnetz *n*
wire gauze cathode Drahtnetzkathode *f*
wire gauze diaphragm Drahtnetzdiaphragma *n*
wire gauze electrode Drahtnetzelektrode *f*
wire glass Drahtglas *n*
wire grating Drahtgitter *n*
wire gun Spritzpistole *f*
wire insulating ribbon Kabelband *n*
wire insulating tape Kabelband *n*
wire lacquer Drahtlack *m*
wire lattice Drahtgitter *n*
wireless (Br. E.) Radio *m*, Radioapparat *m*,
  Rundfunk *m*
wireless *a* drahtlos
wireless *v* (Br. E.) funken
wireless receiver (Br. E.) Rundfunkempfänger *m*
wire loop Drahtschleife *f*
wire mesh Drahtnetzgewebe *n*
wire mesh screen Drahtsieb *n*
wire-mesh tray Netzboden *m* (Dest)
wire mill Drahtstrecke *f*
wire nail Drahtstift *m*
wire netting Drahtgeflecht *n*, Drahtgewebe *n*,
  Drahtnetz *n*
wire network Drahtnetz *n*
wire pliers Drahtzange *f*
wire-reinforced belt Stahlseilgurtband *n*
wire reinforcement Drahteinlage *f*
wire release Drahtauslöser *m*
wire rod Drahteisen *n*
wire rolling Drahtwalzen *n*
wire rope Drahtseil *n*

wire shears Drahtschere *f*
wire sheathing Drahtbewehrung *f*
wire sieve Drahtsieb *n*
wire spiral Drahtspirale *f*
wire-stitched drahtgeheftet
wire stretcher Drahtspanner *m*
wire testing Drahtprüfung *f*
wire tinning plant Drahtverzinnerei *f*
wire triangle Drahtdreieck *n*
wire winding machine Drahtwickelmaschine *f*
wire work Drahtgeflecht *n*
wire working Drahtbearbeitung *f*
wire worm Drahtwurm *m* (Zool)
wiring Verdrahtung *f*
wiring diagram Schaltbild *n*, Schaltschema *n*
wiserite Wiserit *m* (Min)
withamite Withamit *m* (Min)
withdraw *v* entnehmen, entziehen, zurückziehen
withdrawal Entnahme *f*, ~ **of steam**
　Dampfentnahme *f*
withdrawal capacity, total fluid ~
　Naßölförderkapazität *f* (Ölgewinnung)
withdrawal sleeve Abziehhülse *f*
wither *v* welken, absterben, dörren
witherite Witherit *m* (Min)
withhold *v* einbehalten, vorenthalten
withstand *v* widerstehen
wittichenite Wittichenit *m* (Min),
　Wismutkupfererz *n* (Min)
woad Waid *n* (Färb)
woad blue Waidblau *n*
wobble *v* flattern
wobbler Kupplungszapfen *m*
woehlerite Wöhlerit *m* (Min)
Woehler test Dauerschwingversuch *m*
wogonine Wogonin *n*
wolchite Antimonkupferglanz *m* (Min)
wolfachite Wolfachit *m* (Min)
wolfram (Symb. W) Wolfram *n*, **metallic** ~
　Wolframmetall *n*
wolfram blue Wolframblau *n*
wolfram bronze Safranbronze *f*
wolframite Wolframit *m* (Min)
wolfram metal Wolframmetall *n*
wolfram ore Wolframerz *n*
wolfsbergite Wolfsbergit *m* (Min)
wollastonite Wollastonit *m* (Min), Tafelspat *m*
　(Min)
wonder Wunder *n*
wood Holz *n*, **branchless** ~ astfreies Holz,
　branchy ~ astreiches Holz, **chips of** ~
　Holzabfälle *pl*, **compregnated** ~ Kunstholz *n*,
　compressed ~ Hartholz *n*, Kunstholz *n*,
　decayed ~ faules Holz, Schwammholz *n*,
　densified ~ Hartholz *n*, Kunstholz *n*,
　fine-grained ~ feinfaseriges Holz, **green** ~
　frisches Holz, grünes Holz, **hard** ~
　Hartholz *n*, **impregnated** ~ getränktes Holz,
　improved ~ nachbehandeltes Holz, **live** ~

frisches Holz, grünes Holz, **petrified** ~
versteinertes Holz, **rotten** ~ morsches Holz,
soft ~ Weichholz *n*, **stained** ~ gebeiztes Holz,
streaky or curled ~ Maserholz *n*
wood agate Holzachat *m* (Min)
wood alcohol Holzgeist *m*, Methanol *n*,
　Methylalkohol *m*
wood ashes Holzasche *f*
wood ash lye Holzaschenlauge *f*
wood board Holzpappe *f*
wood-boring tool Holzbohrer *m*
wood cellulose Holzzellstoff *m*, Holzzeug *n*
wood cement Holzkitt *m*
wood charcoal Holzkohle *f*
wood chip Holzspan *m*
wood construction Holzbauweise *f*
wood copper Olivenerz *n* (Min), Olivenit *m*
　(Min)
wood culture Waldwirtschaft *f*
wood distillation Holzdestillation *f*
wood dust Holzmehl *n*
wooden air conduit Holzlutte *f*
wooden air duct Holzlutte *f*
wooden box hölzerner Formkasten *m*,
　Holzkasten *m*, Holzkiste *f*, **removable** ~
　Holzeinsatzkasten *m*
wooden chest Holzkasten *m*
wooden deck Holzschalung *f* (Dach)
wooden disc Holzscheibe *f*
wooden form Holzform *f*
wooden frame hölzerner Formkasten *m*
wooden handle Holzstiel *m*
wooden jaw Holzbacke *f*
woodenness Holzigkeit *f*
wooden paddle Rührholz *n*
wooden peg Holzdübel *m*, Holzstift *m*
wooden rule Holzmaßstab *m*
wooden shaft Holzwelle *f*
wooden stirrer Rührholz *n*
wood failure Holzbruch *m*
wood fiber Holzfaser *f*
wood-fiber board Holzfaserplatte *f*
wood fiber yarn Holzfasergarn *n*
wood finish Holzpolierlack *m*
wood fire Holzfeuer *n*
wood-fired furnace Holzfeuerung *f*
woodfree holzfrei
wood gas Holzgas *n*
wood gas generator Holzgasgenerator *m*
wood impregnation Holzimprägnierung *f*
wood impregnation plant Holzimprägnieranlage *f*
wood lacquer Holzlack *m*
wood meal Holzmehl *n*
wood meal mill Holzmehlmühle *f*
wood-milling cutter Holzfräser *m* (Techn)
wood mordant Holzbeize *f*
wood oil Holzöl *n*
wood oil sediment Holzölsatz *m*
wood oil varnish Holzölfirnis *m*

wood opal Holzopal *m* (Min)
wood paper Holzpapier *n*
wood paste Holzmasse *f*
wood peat Holztorf *m*
wood plastic composite
   Holz-Kunststoff-Kombination *f*
wood powder Holzmehl *n*, Holzpulver *n*
wood preservation Holzkonservierung *f*
wood preservative Holzschutzmittel *n*
wood primer Holzgrundierung *f*
wood processing Holzveredlung *f*
wood pulp Holzstoff *m*, Holzzeug *n*, Pülpe *f*,
   Pulpe *f*, **brown** ~ Braunschliff *m*, **free from** ~
   holzstofffrei
wood pulp filter Zellstoffilter *m*
wood pulp paper Holzfaserpapier *n*
wood pulp shredding machine
   Holzstoffaserungsmaschine *f*
wood resin Baumharz *n*, Holzharz *n*
wood rock holziger Bergflachs *m* (Min)
woodruff Waldmeister *m* (Bot)
wood saccharification Holzverzuckerung *f*
Wood's alloy Woodsches Metall *n*
wood screw Holzschraube *f*
wood shavings Hobelspäne *n pl*
wood-slat conveyor Holzplattenband *n*
wood soot Holzruß *m*
wood sorrel Sauerklee *m* (Bot)
wood spirit Holzalkohol *m*, Holzgeist *m*
wood stain Holzbeize *f*
wood staining Holzbeizen *n*
woodstone versteinertes Holz *n*, Xylolith *m*
   (Min)
wood sugar Holzzucker *m*, Xylose *f*
wood tin Holzzinn[erz] *n* (Min)
wood trimmer Bestoßmaschine *f*
wood varnish Holzfirnis *m*
wood veneer Holzfurnier *n*
wood vinegar Holzessig *m*
woodwardite Woodwardit *m* (Min)
wood waste Holzabfälle *pl*
wood wool Holzwolle *f*
wood wool filter Holzwollfilter *n*
wood working Holzbearbeitung *f*,
   Holzverarbeitung *f*
wood-working *a* holzverarbeitend,
   holzbearbeitend
wood [-working] industry Holzindustrie *f*,
   holzverarbeitende Industrie *f*
wood-working machine
   Holzbearbeitungsmaschine *f*
woody holzartig, holzig
woof Durchschuß *m* (Web), Eintrag *m* (Web)
wool Wolle *f*, **artificial** ~ Kunstwolle *f*,
   Zellwolle *f*, **resembling** ~ wollähnlich, **spun** ~
   Wollgarn *n* (Web), **synthetic** ~ Zellwolle *f*, **to**
   grease the ~ Wolle *f* schmälzen
wool ashes Wollasche *f*
wool fat Wollfett *n*

wool felt Wollfilz *m*
wool fiber Wollfaser *f*
wool grease Wollfett *n*
wool grease emulsion Wollfettemulsion *f*
wool grease recovery Wollfettgewinnung *f*
wool grease scum Wollfettschaum *m*
wool-like wollähnlich
woolly wollartig, wollig
wool oil Schmalzöl *n*, Spicköl *n*, Wollöl *n*
wool oiling der Wolle, der Wolle
wool scouring Wollwäsche *f*
wool substitute Woll-Ersatzstoff *m*
wool washing Wollwäsche *f*
wool wax Wollwachs *n*
wool yarn Wollgarn *n* (Web)
wool yolk Wollschmiere *f*, Wollschweiß *m*
woorara Curare *n*
Wootz steel Wootzstahl *m*
wording Ausdruck *m*, Fassung *f* (Lit)
work Arbeit *f*, Tätigkeit *f*, Werk *n*, **expenditure of**
   ~ Arbeitsaufwand *m*, **loss of** ~
   Arbeitsverlust *m*, **performance of** ~
   Arbeitsleistung *f*, Arbeitsverrichtung *f*, **piece**
   of ~ Werkstück *n*, **to be at** ~ in Betrieb sein,
   to set to ~ in Betrieb setzen, **total** ~
   Gesamtarbeit *f*, **unit of** ~ Arbeitseinheit *f*; ~ **of**
   deformation Verformungsarbeit *f*
work *v* arbeiten, abbauen (Bergb), bearbeiten,
   betätigen (von Hand), funktionieren,
   hantieren, wirken, ~ **off** abarbeiten, ~ **out**
   ausarbeiten, ~ **together** zusammenarbeiten,
   zusammenwirken, ~ **up** aufarbeiten, ~ **well**
   funktionieren
workability Bearbeitbarkeit *f*, Verarbeitbarkeit *f*
workable abbauwürdig (Bergb), bauwürdig,
   bearbeitbar, betriebsfähig, brauchbar,
   verformbar, ~ **in cold state** verformbar
workbench Arbeitstisch *m* (Techn), Werkbank *f*
work capacity Arbeitskapazität *f*
work dress Arbeitsanzug *m*
worker Arbeiter *m*, Holbzeugholländer *m* (Pap),
   qualified ~ qualifizierter Arbeiter,
   semi-skilled ~ angelernter Arbeiter, **skilled** ~
   gelernter Arbeiter, **unskilled** ~ ungelernter
   Arbeiter
work function Austrittsarbeit *f*
   (Festkörperphysik)
work function of electrons
   Elektronenaustrittsarbeit *f*
work hardening Verfestigung *f*, **limit of** ~
   Kaltverfestigungsgrenze *f*
work hardening curve Verfestigungskurve *f*
work holding device Spannwerkzeug *n*
working Verarbeitung *f*, Abbau *m* (Bergb),
   Bearbeitung *f*, Funktionieren *n*, **capable of** ~
   arbeitsfähig, **method of** ~ Arbeitsweise *f*
working *a* betrieblich, brauchbar, in Betrieb
working boiler Betriebskessel *m*
working capacity Arbeitsleistung *f* (Techn)

working committee Arbeitsausschuß *m*
working condition Arbeitsbedingung *f*
working conditions pl
working cylinder Arbeitszylinder *m*
working diagram Arbeitsdiagramm *n*
working drawing Werkstattzeichnung *f*
working electrode Arbeitselektrode *f,*
strombelastete Elektrode *f*
working expenses pl
working gear Bewegungsvorrichtung *f*
working group Arbeitskreis *m*
working hours Betriebszeit *f*
working hypothesis Arbeitshypothese *f*
working instruction Arbeitsanweisung *f*
working instructions Bedienungsanleitung *f*
working intensity Arbeitsintensität *f*
working load Betriebsbelastung *f* (Elektr),
Drucklast *f*
working method Arbeitsverfahren *n,*
Bearbeitungsverfahren *n*
working part Werkstück *n*
working period Betriebszeit *f*
working piston Arbeitskolben *m*
working pit Förderschacht *m* (Bergb)
working platform Arbeitsbühne *f,* Kanzel *f*
working power Arbeitskraft *f*
working pressure Arbeitsbelastung *f*
working process Arbeitsgang *m,*
Arbeitsverfahren *n*
working rate, high ~ starke Beanspruchung *f*
working regulations Arbeitsordnung *f*
working resistance Arbeitswiderstand *m*
working safety Betriebssicherheit *f*
working speed Arbeitsgeschwindigkeit *f*
working stress Nutzbeanspruchung *f,* **permissible**
~ Belastbarkeit *f* (Mech)
working surface Arbeitsfläche *f,* Angriffsfläche *f*
working temperature Betriebstemperatur *f*
working up Aufarbeitung *f*
working valve Bedienungshahn *m*
working velocity Arbeitsgeschwindigkeit *f*
working voltage Arbeitsspannung *f,*
Betriebsspannung *f*
workman Arbeiter *m*
workmanlike fachgemäß, meisterhaft
work piece Werkstück *n*
work room Arbeitsraum *m*
works Betriebsanlage *f,* Fabrik *f,* Werk *n*
workshop Arbeitsraum *m* (Werkstatt),
Werkstatt *f*
workshop drawing Werkstattzeichnung *f*
works management Betriebsorganisation *f*
work softening Verformungsentfestigung *f*
works standards Fabriknormen *f pl*
work studies Arbeitswissenschaft *f*
worktable Arbeitstisch *m*
work tolerance Werkstoleranz *f*
world Welt *f*
world production Welterzeugung *f*

world's supply Weltvorrat *m*
world-wide weltumspannend
worm Schraubengewinde *n,* Antriebsschnecke *f,*
Rohrschlange *f,* Wurm *m* (Zool), ~ **(of a**
screw) Gewinde *n*
worm bark Wurmrinde *f* (Bot)
worm conveyor Förderschnecke *f*
worm-destroying wurmvertilgend
worm drier Schneckentrockner *m*
worm-eaten wurmstichig
worm extruder Schneckenpresse *f*
worm gear drive Schneckenantrieb *m*
worm gear[ing] Schneckengetriebe *n*
worm gear[wheel] Schneckenrad *n*
worm moss Wurmmoos *n* (Bot)
worm powder Wurmpulver *n*
wormseed Wurmsamen *m* (Bot)
worm's eye view Ansicht von unten
worm shaft Schneckenachse *f,* Schneckenwelle *f*
worm type dryer Trockenschnecke *f*
worm wheel Schneckenrad *n*
wormwood, common ~ Wermutkraut *n* (Bot)
wormwood oil Wermutöl *n*
worn abgenutzt, verbraucht, abgegriffen, ~ **out**
verschlissen
worsted Kammgarn *n* (Text)
worsted spinning Kammgarnspinnerei *f*
wort Würze *f* (Brau), **original** ~ Stammwürze *f*
(Brau)
wort cooler Würz[e]kühler *m*
wort vat Würzkufe *f* (Brau)
wound Wunde *f*
wound fever Wundfieber *n* (Med)
woven glass fiber cloth Glasseidengewebe *n*
wrap Deckband *n*
wrap *v* einwickeln, einhüllen, umwickeln,
verpacken
wrapped cable besponnenes Kabel *n*
wrapper Deckblatt *n* (Zigarre), Verpackung *f*
wrapping Umkarton *m,* Umverpackung *f,*
Verpackung *f*
wrapping machine Einwickelmaschine *f*
wrapping paper Einpackpapier *n,*
Einschlagpapier *n,* Einwickelpapier *n,*
Packpapier *n,* **grey** ~ Traßpapier *n*
wreck *v* abwracken, zertrümmern
wrecking bar Brecheisen *n*
wrench (Am. E.) Schraubenschlüssel *m,* **cock or**
tap ~ Hahnschlüssel *m,* **hooked** ~
Hakenschlüssel *m*
wrench bolt Schlüsselschraube *f*
wrench hammer Schlagschlüssel *m*
wring *v* auswringen, ~ **out** auswinden
wringer Quetsche *f*
wringing fit Edelpassung *f,* Feinpassung *f*
wrinkle Falte *f,* Knitter *m*
wrinkle *v* runzeln, zusammenschrumpfen
wrinkled faltenreich, gekräuselt
wrinkle finish Kräusellack *m,* Runzellack *m*

wrinkle forming Faltenbildung *f,* Kräuselung *f*
wrinkle-free faltenlos
wrinkle resistance Knitterfestigkeit *f*
wrinkle-resistant knitterfest
wrinkles, formation of ~ Runzelbildung *f*
wrinkle varnish Kräusellack *m*
wrinkling Faltenbildung *f,* Hochziehen *n,*
   Runzelbildung *f*
write *v* schreiben; ~ **down** aufschreiben
writer Verfasser *m*
writing ink Schreibtinte *f*
writing paper Schreibpapier *n*
wrong[ly] irrtümlich
wrought alloy Schmiedelegierung *f*
wrought iron Schmiedeeisen *n,* Schweißeisen *n*
wrought-iron *a* schmiedeeisern
wrought iron head Hammerkopf *m*
wrought iron plate Schweißeisenblech *n*
wrought iron ring Schmiedeeisenring *m*
wrought iron scrap Schmiedeeisenabfälle *m pl*
wrought steel Schmiedestahl *m,* Schweißstahl *m*
wulfenite Wulfenit *m* (Min), Gelbbleierz *n*
   (Min), Molybdänbleispat *m* (Min)
Wulff-Bock crystallizer Kristallisierwiege *f*
Wurster salt Wurstersches Salz *n*
wurtzilite Wurtzilit *m* (Min)
wurtzite Wurtzit *m* (Min), Strahlenblende *f*
   (Min)
wurtzite lattice Wurtzitgitter *n* (Krist)

X

xanthaline Xanthalin *n*
xanthamylic acid Xanthamylsäure *f*
xanthan hydride Xanthanwasserstoff *m*
xanthate Xanthat *n*, Xanthogenat *n*
xanthazol Xanthazol *n*
xanthein Xanthein *n*
xanthene Xanthen *n*
xanthene dyestuffs Xanthenfarben *pl*
xanthenol Xanthenol *n*
xanthenone Xanthenon *n*, Xanthon *n*
xanthic acid Xanthogensäure *f*, salt or ester of ~
  Xanthogenat *n*
xanthin[e] Xanthin *n*, Krappgelb *n* (Techn)
xanthine oxidase Xanthinoxidase *f* (Biochem)
xanthione Xanthion *n*
xanthite Xanthit *m*
xanthoapocyanine Xanthoapocyanin *n*
xanthoarsenite Xanthoarsenit *m* (Min)
xanthocillin Xanthocillin *n*
xanthogen amide Xanthogenamid *n*
xanthogenate Xanthogenat *n*, Xanthat *n*
xanthogene Xanthogen *n*
xanthogenic acid Xanthogensäure *f*
xanthohumol Xanthohumol *n*
xanthoma Xanthom *n* (Med)
xanthomatosis Xanthomatose *f* (Med)
xanthone Xanthon *n*, Dibenzopyron *n*,
  Xanthenon *n*
xanthonic acid Xanthogensäure *f*
xanthoperol Xanthoperol *n*
xanthophyll Blattgelb *n* (obs), Xanthophyll *n*
xanthophyllite Xanthophyllit *m* (Min)
xanthopicrin Xanthopikrin *n*
xanthopicrite Xanthopikrit *n*
xanthoprotein Xanthoprotein *n*
xanthoprotein reaction Xanthoproteinreaktion *f*
xanthopterine Xanthopterin *n*
xanthopuccine Xanthopuccin *n*
xanthopurpurin Purpuroxanthen *n*,
  Xanthopurpurin *n*
xanthoraphine Xanthoraphin *n*
xanthorhamnin Xanthorhamnin *n*
xanthosiderite Xanthosiderit *m* (Min)
xanthosin Xanthosin *n*
xanthotoxin Xanthotoxin *n*
xanthoxylene Xanthoxylen *n*, Xanthoxylin *n*
xanthoxylin Xanthoxylen *n*, Xanthoxylin *n*
xanthurenic acid Xanthurensäure *f*
xanthydrol Xanthydrol *n*, Xanthenol *n*
xanthyl Xanthyl-
xanthylic acid Xanthylsäure *f*
x-axis X-Achse *f*
xenene Biphenyl *n*, Diphenyl *n*
xenomorphic xenomorph (Krist)
xenon (Symb. X or Xe) Xenon *n*
xenon arc Xenonbogen *m*

xenon bubble chamber Xenonblasenkammer *f*
xenon high pressure lamp
  Xenonhochdrucklampe *f*
xenon lamp Xenonlampe *f*
xenon tube Xenonröhre *f*
xenotime Xenotim *m* (Min), Ytterspat *m* (Min)
xenylamine Xenylamin *n*
xeroform Xeroform *n*
xerography Xerographie *f*
xeromorphic xeromorph
xerophilic xerophil
xerophyte Xerophyt *m* (Bot)
ximenic acid Ximensäure *f*
ximenynic acid Ximeninsäure *f*
xi-minus antiparticle Xi-minus-Antiteilchen *n*
xi-particle Xi-Teilchen *n*
xonotlite Xonotlit *m* (Min)
X-radiation Röntgenstrahlung *f*
X ray Röntgenstrahl *m*
x-ray *v* röntgen, durchleuchten
X-ray absorption spectrum
  Röntgenabsorptionsspektrum *n*
X-ray analysis Röntgenanalyse *f*, direct ~
  Röntgenstrukturanalyse *f*
X-ray apparatus Durchleuchtungsapparat *m*,
  Röntgenapparat *m*
X-ray beam Röntgenstrahlenbündel *n*
X-ray diagnostics Röntgendiagnostik *f*
X-ray diagram Röntgendiagram *n*
X-ray diffraction method
  Röntgenbeugungsmethode *f*
X-ray diffraction pattern
  Röntgenbeugungsaufnahme *f*,
  Röntgendiagramm *n*
X-ray emission spectrum
  Röntgen-Emissionsspektrum *n* (Spektr)
X-ray equipment Röntgenanlage *f*,
  Röntgeneinrichtung *f*
X-ray examination Röntgenuntersuchung *f*
X-ray film Röntgenaufnahme *f*, Röntgenfilm *m*
X-ray flash Röntgenblitz *m*
X-ray flash interference
  Röntgenblitzinterferenz *f*
x-raying Durchleuchtung *f* mit Röntgenstrahlen
X-ray laboratory Röntgenlaboratorium *n*
X-ray level Röntgenterm *m*
X-ray line Röntgenlinie *f*
X-ray negative Röntgennegativ *n*
X-ray photon Röntgenstrahlphoton *n*
X-ray [picture] Röntgenaufnahme *f*,
  Röntgenbild *n*, Röntgenogramm *n*
X-ray positive Röntgenpositiv *n*
X-ray radiometer Röntgenstrahlenmesser *m*
X-rays, hard ~ harte Röntgenstrahlen *pl*, soft ~
  weiche Röntgenstrahlen *pl*
X-ray spectrograph Röntgenspektrograph *m*

X-ray spectrum Röntgenspektrum *n*
X-ray structural analysis
Röntgenstrukturanalyse *f*
X-ray structure Röntgenfeinstruktur *f* (Krist)
X-ray treatment Röntgenbehandlung *f* (Med)
X-ray tube Röntgenröhre *f*
xylal Xylal *n*
xylan Xylan *n*, Holzgummi *m*
xylaric acid Xylarsäure *f*
xylem Holzgewebe *n* (Bot), Xylem *n* (Bot)
xylene Xylol *n*
xylenesulfonic acid Xylolsulfonsäure *f*
xylenol Xylenol *n*
xylenolphthalein Xylenolphthalein *n*
xylenol resin Xylenolharz *n*
xylic acid Xylylsäure *f*, **2,3-** ~ Hemellithsäure *f*
xylidic acid Xylidinsäure *f*
xylidide Xylidid *n*
xylidine Xylidin *n*
xylidine red Xylidinrot *n*
xylidinic acid Xylidinsäure *f*
xylitan Xylitan *n*
xylitol Xylit *m*
xylobiose Xylobiose *f*
xylocaine Xylocain *n*
xylochloral Xylochloral *n*
xylochloralose Xylochloralose *f*
xylocolla Holzleim *m*
xyloflavine Xyloflavin *n*
xylography Holzdruck *m*
xylohydroquinol Xylohydrochinon *n*
xyloid holzartig
xyloidine Nitrostärke *f*, Xyloidin *n*
xyloketose Xyloketose *f*
xylolith Xylolith *m* (Min)
xylonic acid Xylonsäure *f*
xylonite Celluloid *n* (HN)
xylopine Xylopin *n*
xylopinine Xylopinin *n*
xyloquinol Xylochinol *n*
xyloquinone Xylochinon *n*
xylorcinol Xylorcin *n*
xylose Holzzucker *m*, Xylose *f*
xylosone Xyloson *n*
xylotile Bergholz *n* (Min), Xylotil *m* (Min)
xylulose Xylulose *f*
xylulose phosphate Xylulosephosphat *n*
xylyl Xylyl-
xylyl chloride Xylylchlorid *n*
xylylene Xylylen-
xylylenediamine Xylylendiamin *n*
xylylene dichloride Xylylendichlorid *n*
xylylene urea Xylylenharnstoff *m*

Y

yajeine Harmin *n*, Yajein *n*
yangolactone Jangolacton *n*
yangonin Jangonin *n*
yard Lagerplatz *m*, Yard *n* (Maß)
yardstick Maßstab *m*
yarn Garn *n*, **blended** ~ Mischgarn *n*, **count of**
~ Garnnummer *f*, **grey** ~ rohes Garn
yarn break Fadenbruch *m*
yarn count Garnnummer *f*
yarn-dyed im Garn gefärbt
yarn dyeworks Garnfärberei *f*
yatren Loretin *n*
yaw *v* gieren
y-axis Ordinate *f*, Y-Achse *f*
y-branch Rohrabzweigstück *n*
Y-connection Sternschaltung *f* (Elektr),
Y-Schaltung *f* (Elektr)
Y-current Sternstrom *m*
yearly capacity Jahreserzeugung *f*
yearly output Jahreserzeugung *f*
yeast Bierhefe *f* (Brau), Faßbärme *f* (Brau),
Hefe *f*, **bottom** ~ untergärige Hefe, **crop of** ~
Hefeernte *f*, **pressed** ~ Preßhefe *f*, **top** ~
obergärige Hefe, **type of** ~ Hefeart *f*
yeast adenylic acid Hefeadenylsäure *f*
yeast bitter Hefenbitter *n*
yeast cell Hefezelle *f*
yeast culture Hefekultur *f*, Hefezucht *f*, **pure** ~
Hefereinzucht *f*
yeast extract Hefeextrakt *m*
yeast fungus Hefepilz *m*
yeast germ Hefenkeim *m*
yeast growth Wachstum *n* der Hefe
yeast-like hefeähnlich, hefig
yeast plant Hefeanlage *f*
yeast propagator Hefeapparat *m*
yeast room Heferaum *m*
yeast scum Hefeschaum *m*
yeast stopper Hefepfropfen *m*
yeast syrup Hefensirup *m*
yeasty hefeartig, hefig
yellow gelb[farbig], **light** ~ hellgelb, **pale** ~
weißgelb, **to turn** ~ vergilben
yellow *v* färben, vergilben
yellow balsam Erdschellack *m*
yellow brass Gelbkupfer *n*
yellow camomile Färberkamille *f* (Bot),
Färberblume *f* (Bot)
yellow catechu Gambir *n*
yellow coloring Gelbfärbung *f*
yellow copper Gelbkupfer *n*
yellow earth Ocker *m* (Min)
yellow fever Gelbfieber *n* (Med)
yellow hue Gelbstich *m*
yellow incandescence Gelbglut *f*

yellowing Gelbbeizen *n* (Metall), Gelbfärben *n*
(Färb), Vergilben *n*, Vergilbung *f*
yellow [iron] ocher Gelberde *f* (Min)
yellowish gelblich
yellow metal Gelbmetall *n*
yellow mica Goldglimmer *m* (Min)
yellowness Gelbheit *f*
yellow ocher Gelberde *f* (Min)
yellow resin Gelbharz *n*
yellow stain Gelbschleier *m*
yellowwood extract Gelbholzextrakt *m*
yenite Ilvait *m* (Min), Jenit *m* (Min)
yield Ausbeute *f* (Chem), Ertrag *m*,
Nutzleistung *f*, Produktion *f*, **increase in** ~
Ausbeuteerhöhung *f*, **photoelectric** ~
photoelektrische Ausbeute, **total** ~
Gesamtausbeute *f*
yield *v* abgeben (Phys), abwerfen, einbringen,
ergeben, hervorbringen, nachgeben
yield condition Fließbedingung *f*
yield curve Ausbeutekurve *f*
yield force Fließkraft *f*
yield phenomenon Streckgrenzeneffekt *m*,
Streckgrenzenerscheinung *f*
yield point Fließgrenze *f*, Fließpunkt *m*,
Streckgrenze *f*, ~ **in compression**
Quetschgrenze *f*; ~ **in torsion** Verdrehgrenze *f*,
~ **of torsional shear** Torsionsgrenze *f*
yield resistance Fließwiderstand *m*
yield restriction Fließeinschränkung *f*
yield strength Streckgrenze *f*
yield stress Elastizitätsgrenze *f*
yield stress ratio Fließverhältnis *n*
yield value Fließwert *m*
ylang-ylang oil Ylang-Ylang-Öl *n*
ylem Urmaterie *f*, Urplasma *n*, Urstoff *m*
yoghurt Joghurt *n*
yohimbine Yohimbin *n*
yohimb[oa]ic acid Yohimbinsäure *f*
yohimbone Yohimbon *n*
yoke block Rahmenverstärkung *f*
yolk Eigelb *n*, Wollschweiß *m*
yolk membrane Dotterhaut *f*
yolk wax Schweißwachs *n* (Wolle)
yonogenic acid Yonogen[in]säure *f*
yperite Senfgas *n*, Yperit *n*
ytterbia Ytterbin *n*, Ytterbinerde *f*
ytterbite Ytterbit *m*
ytterbium (Symb. Yb) Ytterbium *n*
ytterbium content Ytterbiumgehalt *m*
ytterbium oxide Ytterbin *n*, Ytterbinerde *f*
yttrialite Yttrialith *m* (Min)
yttriferous ytterhaltig
yttrium (Symb. Y) Yttrium *n*, **containing** ~
ytterhaltig
yttrium chloride Yttriumchlorid *n*

yttrium content Yttriumgehalt *m*
yttrium earth Yttererde *f*
yttrium nitrate Yttriumnitrat *n*
yttrium oxide Yttriumoxid *n,* Yttererde *f*
yttrium salt Yttersalz *n*
yttrocerite Ytterflußspat *m* (Min), Yttrocerit *m*
  (Min)
yttrocolumbite Yttrocolumbit *m* (Min)
yttrocrasite Yttrokrasit *m* (Min)
yttrofluorite Yttrofluorit *m* (Min)
yttroilmenite Yttroilmenit *m* (Min)
yttrotantalite Yttrotantalit *m* (Min)
yttrotitanite Yttrotitanit *m* (Min), Keilhauit *m*
  (Min)
Y tube Y-Rohr *n*
yucca fiber Yuccafaser *f*
yuccasaponin Yuccasaponin *n*
Y-voltage Y-Spannung *f* (Elektr)

Z

zaffer Glasurblau *n*, Zaffer *m*
zapon lacquer Zaponlack *m*
zapon varnish Zaponlack *m*
zaratite Zaratit *m* (Min), Nickelsmaragd *m* (Min)
zeaxanthin Zeaxanthin *n*
zebromal Zebromal *n*
zedoary Zitwer *m* (Bot)
zedoary seed Zitwersamen *m*
Zeeman effect Zeemaneffekt *m*
zein Zein *n*, Maiskleber *m*
Zeisel's method Zeiselsche Methode *f*
zenith Zenit *m*, Scheitelpunkt *m*
zenith distance Scheitelabstand *m*
zeolite Zeolith *m* (Min), crystallized ~ Keimspat *m* (Min), fibrous ~ Faserzeolith *m* (Min), Haarzeolith *m* (Min), foliated ~ Strahlzeolith *m* (Min), blättriger Zeolith, mealy ~ Zeolitherde *f*, vitreous ~ Glaszeolith *m* (Min)
zeolitic zeolithhaltig
zeophyllite Zeophyllit *m* (Min)
zephiran Zephiran *n*
zero Null *f*, Anfangspunkt *m*, absolute ~ absoluter Nullpunkt, to adjust to ~ auf Null einstellen
zero *v* auf Null einstellen
zero adjustment Null[punkt]einstellung *f*
zero adjustment control Nullpunktkontrolle *f*
zero axis Nullinie *f*
zero charge, potential of ~ Nullpotential *n*
zero current der Stromstärke
zero deviation Nullpunktabweichung *f*
zero drift Nullpunktverschiebung *f*
zero error Nullpunktabweichung *f*
zero instrument Nullinstrument *n*
zero-level sensitivity Nullpunktempfindlichkeit *f*
zero line Nullinie *f*
zero method Nullpunktmethode *f*
zero moment Nullmoment *n*
zero point Nullpunkt *m*, Neutralpunkt *m*
zero point energy Nullpunktenergie *f*
zero point motion Nullpunktunruhe *f*
zero point volume Nullpunktsvolumen *n*
zero position Nullage *f*, Nullstellung *f*
zero potential Nullpotential *n*
zero range Nullreichweite *f*
zero reading Nullablesung *f*
zero series Nullfolge *f* (Math)
zero setting Nullpunktschieber *m*
zerovalent nullwertig
zero value Nullwert *m*, ~ of the current der Stromstärke
zero voltage Nullspannung *f* (Elektr)
zero wire Nulleiter *m* (Elektr)
zeta potential Zetapotential *n*

zethrene Zethren *n*
zeunerite Zeunerit *m* (Min)
Ziegler's catalyst Ziegler-Katalysator *m*
Ziehl's stain Carbolfuchsin *n*
zierin Zierin *n*
zierone Zieron *n*
zigzag configuration Zick-Zack-Konfiguration *f* (Stereochem)
zinc (Symb. Zn) Zink *n*, butter of ~ Zinkbutter *f*, containing ~ zinkhaltig, extraction of ~ Zinkgewinnung *f*, flowers of ~ Zinkblumen *f pl*, Zinkwolle *f*, granulated ~ pl, raw ~ Werkzink *n*, spongy ~ Zinkschwamm *m*, strip of ~ Zinkstreifen *m*, to coat with ~ verzinken
zinc *v* verzinken
zinc accumulator Zinkakkumulator *m*
zinc acetate Zinkacetat *n*
zinc alkyl Zinkalkyl *n*
zinc alloy Zinklegierung *f*
zinc alum Zinkalaun *n*
zinc amalgam Zinkamalgam *n*
zincate Zinkat *n*
zinc bath Zinkbad *n*
zinc blende Zinkblende *f* (Min), Faserblende *f* (Min), Sphalerit *m* (Min)
zinc blende lattice Zinkblendegitter *n* (Krist)
zinc bloom Zinkblüte *f* (Min), Hydrozinkit *m* (Min)
zinc calx Zinkkalk *m*
zinc carbamate Zinkcarbamat *n*
zinc carbonate Zinkcarbonat *n*
zinc cement Zinkkitt *m*
zinc chloride Zinkchlorid *n*, Zinkbutter *f*
zinc chromate Zinkchromat *n*, Zinkgelb *n*
zinc citrate Zinkcitrat *n*
zinc-coating Verzinken *n*, Verzinkung *f*
zinc compound Zinkverbindung *f*
zinc contact Zinkkontakt *m*
zinc content Zinkgehalt *m*
zinc crust Reichschaum *m*
zinc cyanide Zinkcyanid *n*
zinc cyanoferrate(II) Zinkferrocyanid *n*
zinc deposit Gichtschwamm *m*, Ofenbrand *m*
zinc desilverization Zinkentsilberung *f*
zinc die-casting Zinkspritzguß *m*
zinc diethyl Zinkäthyl *n*
zinc dimethyl Zinkmethyl *n*
zinc disk Zinkscheibe *f*
zinc dross Zinkasche *f*, Zinkgekrätz *n*, Zinkkalk *m*
zinc dust Zinkstaub *m*
zinc electrode Zinkelektrode *f*
zinc electrolysis Zinkelektrolyse *f*
zinc etching Zinkätzung *f*, Photochemigraphie *f*
zinc ethide Zinkäthyl *n*

zinc ethyl Zinkäthyl *n*
zinc ethyl sulfate Zinkäthylsulfat *n*
zinc ferrocyanide Zinkferrocyanid *n*
zinc ferrous cyanide Zinkferrocyanid *n*
zinc filings pl
zinc fluoride Zinkfluorid *n*
zinc fluosilicate Zinksilicofluorid *n*,
　Kieselfluorzink *n*
zinc foil Zinkfolie *f*
zinc formate Zinkformiat *n*
zinc fume Zinkrauch *m*
zinc furnace Zinkofen *m*
zinc green Zinkgrün *n* (Malerei)
zinciferous zinkhaltig
zinc iodate Zinkjodat *n*
zinc iodide Zinkjodid *n*
zinc iodide starch paper Jodzinkpapier *n*,
　Zinkjodidstärkepapier *n*
zinc iodide starch solution Jodzinkstärkelösung *f*
zinc iron cell Zinkeisenelement *n*
zincite Rotzinkerz *n* (Min)
zin[c]kenite Zinckenit *m* (Min),
　Bleiantimonglanz *m* (Min)
zincky zinkartig
zinc lactate Zinklaktat *n*
zinc metasilicate Zinkmetasilikat *n*
zinc methyl Zinkmethyl *n*
zinc nitrate Zinknitrat *n*
zincograph *v* auf Zink ätzen
zincography Zinkographie *f*, Zink[hoch]ätzung *f*
zincoid zinkartig
zinc ointment Zinksalbe *f*
zinc oleate Zinkoleat *n*
zincon Zincon *n*
zinc ore Zinkerz *n* (Min), red ~ Rotzinkerz *n*,
　siliceous ~ Kieselzinkerz *n*
zinc ore roasting plant Zinkerzrösthütte *f*
zinc-organic zinkorganisch
zinc orthosilicate Zinkorthosilikat *n*
zincosite Zinkosit *m* (Min)
zinc oxide Zinkoxid *n*, impure ~ Hüttennichts *n*
　(Min), native ~ Spartalit *m* (Min)
zinc oxide plant Zinkoxidanlage *f*
zinc paint Zinkfarbe *f*
zinc perhydrol Zinkperhydrol *n*
zinc peroxide Zinkperhydrol *n*, Zinksuperoxid *n*
zinc phosphide Phosphorzink *n*
zinc phosphite Zinkphosphit *n*
zinc plate Zinkblech *n*
zinc-plate *v* verzinken
zinc-plating Verzinken *n*, Verzinkung *f*
zinc pole Zinkpol *m* (Anode)
zinc powder Zinkmehl *n*
zinc production Zinkgewinnung *f*
zinc refining Zinkraffination *f*
zinc-rich paint Zinkstaubfarbe *f*
zinc rod Zinkstab *m*
zinc salicylate Zinksalicylat *n*
zinc salt Zinksalz *n*

zinc scum Zinkschaum *m*
zinc selenide Zinkselenid *n*
zinc silicate Zinksilikat *n*, hydrous ~
　Kieselgalmei *m* (Min)
zinc silicofluoride Zinksilicofluorid *n*
zinc smeltery Zinkhütte *f*
zinc spar Zinkspat *m* (Min)
zinc spinel Zinkspinell *m* (Min), Gahnit *m*
　(Min)
zinc sponge Zinkschwamm *m*
zinc sulfate Zinksulfat *n*, Zinkvitriol *n*
zinc sulfide Zinksulfid *n*
zinc sulfide powder Zinksulfidpulver *n*
zinc sulfide screen Zinksulfidschirm *m*
zinc tetraphenylporphine
　Zinktetraphenylporphin *n*
zinc tin amalgam Zinkzinnamalgam *n*
zinc triple salt Zinktripelsalz *n*
zinc voltameter Zinkvoltameter *n*
zinc white Deckweiß *n*, Zinkweiß *n*
zinc works Zinkhütte *f*
zinc yellow Zinkgelb *n*
zingerone Zingeron *n*
zingiberene Zingiberen *n*
zingiberol Zingiberol *n*
zinkenite Zinkenit *m* (Min), Bleiantimonerz *n*
　(Min)
zinkite Zinkit *m* (Min)
zinnwaldite Zinnwaldit *m* (Min)
zippeite Zippeit *m* (Min)
zipper Reißverschluß *m*
zircon Zirkon *m* (Min), Jargon *m* (Min),
　Ostranit *m* (Min), transparent red ~
　Hyacinth *m* (Min)
zirconate Zirkonat *n*
zirconia Zirkonerde *f* (Min), Zirkonoxid *n*
　(Min)
zirconic acid Zirkonsäure *f*
zirconic anhydride Zirkon[ium]oxid *n*
zirconite Zirkonit *m* (Min)
zirconium (Symb. Zr) Zirkonium *n*, metallic ~
　Zirkon[ium]metall *n*
zirconium chloride Zirkonchlorid *n*
zirconium compound Zirkon[ium]verbindung *f*
zirconium dioxide Zirkon[ium]dioxid *n*,
　Zirkonerde *f* (Min)
zirconium fluoride Zirkontetrafluorid *n*
zirconium glass Zirkonglas *n*
zirconium hydride Zirkoniumhydrid *n*
zirconium hydroxide Zirkon[ium]hydroxid *n*
zirconium lamp Zirkonlampe *f*
zirconium metal Zirkon[ium]metall *n*
zirconium nitrate Zirkon[ium]nitrat *n*
zirconium oxide Zirkon[ium]oxid *n*
zirconium oxychloride Zirkon[ium]oxychlorid *n*
zirconium phosphate Zirkonphosphat *n*
zirconium phosphide Zirkonphosphid *n*
zirconium silicide Zirkonsilicid *n*
zirconium sulfate Zirkonsulfat *n*

zirconium tetrafluoride Zirkontetrafluorid *n*
zirconium white Zirkonweiß *n*
zircon light Zirkonlicht *n*
zirconyl Zirkonyl-
zirconyl chloride Zirkonylchlorid *n*,
 Zirkon[ium]oxychlorid *n*
zirkelite Zirkelit *m* (Min)
Z-iron Z-Eisen *n*
zodiac Tierkreis *m* (Astr)
zodiacal light Zodialkallicht *n* (Astr)
zoisite Zoisit *m* (Min)
zonal aberration Zonenfehler *m*
zonally averaged zonal gemittelt
zone Zone *f*, Bereich *m*, Gürtel *m*, Streifen *m*, ~
 of incandescence Glühzone *f*
zone axis Zonenachse *f*
zone control Stufenregelung *f*
zone electrophoresis Zonenelektrophorese *f*
zone heating Zonenheizung *f*
zone index Zonenindex *m*
zone marking paint Straßenmarkierungsfarbe *f*
zone melting Zonenschmelzen *n*
zone melting process Zonenschmelzverfahren *n*
zone [or zonal] centrifugation
 Zonenzentrifugation *f*
zone plane Zonenebene *f* (Krist)
zone refining Zonenschmelzverfahren *n*
zone refrigeration process
 Zonengefrierverfahren *n*
zones, rule of ~ Zonenregel *f*, **sequence of** ~
 Zonenfolge *f* (Halbleiter)
zonochlorite Zonochlorit *m* (Min)
zoochemistry Zoochemie *f*
zoolite Zoolith *m* (Geol)
zoolithic zoolithisch
zoological zoologisch
zoology Tierkunde *f*, Zoologie *f*
zoomaric acid Zoomarinsäure *f*
zooparasite Tierparasit *m*
zorgite Zorgit *m*
zunyite Zunyit *m* (Min)
zurlite Zurlit *m*
zwischenferment (glucose 6-phosphate
 dehydrogenase) Zwischenferment *n*
 (Biochem)
zwitter ion Zwitterion *n*
zygadenine Zygadenin *n*
zygadite Zygadit *m*
zygote befruchtete Eizelle *f*
zymase Zymase *f* (Biochem)
zymochemistry Gärungschemie *f*
zymogen Zymogen *n* (Biochem)
zymogenic gärungsfördernd, zymogen
zymogenous gärungsfördernd
zymohexase Zymohexase *f* (Biochem)
zymology Zymologie *f*, Gärungschemie *f*,
 Gärungslehre *f*
zymometer Gärungsmesser *m*
zymonic acid Zymonsäure *f*

zymosterol Zymosterin *n*
zymotechnical gärungstechnisch
zymotechnics Gärungstechnik *f*
zymotechnology Gärungstechnik *f*
zymurgy Gärungschemie *f*

Appendix
Anhang

Abbreviations used in this Dictionary
Verzeichnis der in diesem Wörterbuch verwandten Abkürzungen

Abk	Abkürzung	abbreviation
a, adj.	Adjektiv	adjective
adv	Adverb	adverb
Aerodyn	Aerodynamik	aerodynamics
Agr	Landwirtschaft	agriculture
Akust	Akustik	acoustics
Alch	Alchemie	alchemy
allg	allgemein	generally
Am. E.	amerikanisches Englisch	American English
Anat	Anatomie	anatomy
Arch	Architektur	architecture
Astr	Astronomie	astronomy
Atom	Atomterminologie	atomic terminology
Bakt	Bakteriologie	bacteriology
Bauw	Bauwesen	civil engineering
Bergb	Bergbau	mining
Biochem	Biochemie	biochemistry
Biol	Biologie	biology
Bot	Botanik	botany
Brau	Brauereiwesen	brewing
Br. E.	britisches Englisch	British English
Buchdr	Buchdruck	typography
Chem	Chemie, chemisch	chemistry, chemical
Comp	Computertechnologie	computer technology
Cyb	Kybernetik	cybernetics
Dest	Destillation	distillation
Elektr	Elektrizität	electricity
Elektrochem	Elektrochemie	electrochemistry
Elektron	Elektronik	electronics
f	weiblich	feminine
Färb	Färberei	dyeing
Galv	Galvanometrie	galvanometry
Gen	Genetik	genetics
Geogr	Geographie	geography
Geol	Geologie	geology
Geom	Geometrie	geometry
Gerb	Gerberei	tanning
Gieß	Gießerei	foundry
Gummi	Gummiindustrie	rubber industry
Histol	Histologie	histology
HN	Handelsname (ohne Garantie)	tradename (without guarantee)
Hydrodyn	Hydrodynamik	hydrodynamics
Jur	Rechtswissenschaft	jurisprudence
Keram	Keramik	ceramics
Kfz	Kraftfahrzeug	motor vehicle
Kin	Kinetik	kinetics
Kosm	Kosmetik	cosmetics
Krist	Kristallographie	crystallography
Lat	Latein	Latin
Leg	Legierung	alloy
Luftf	Luftfahrt	aeronautics
m	maskulin, männlich	masculine
Mag(n)	Magnetismus	magnetism
Masch	Maschinenwesen	mechanical engineering
Maß	Maßeinheit	measuring unit

Math	Mathematik	mathematics
Mech	Mechanik	mechanics
Med	Medizin	medicine
Met(all)	Metall, Metallurgie	metal, metallurgy
Meteor	Meteorologie	meteorology
Min	Mineralogie	mineralogy
Mol. Biol	Molekularbiologie	molecular biology
Mot	Motor	motor
n	neutrum, sächlich	neuter
o	oder	or
obs	veraltet	obsolete
Opt	Optik	optics
Pap(ier)	Papierindustrie	paper industry
Pathol	Pathologie	pathology
Pharm	Pharmakologie	pharmacology
Phot	Photographie	photography
Phys	Physik	physics
Physiol	Physiologie	physiology
Plast	Kunststoffe	plastics
pl	plural	plural
Quant	Quantentheorie	quantum theory
Rad	Radio	radio
Rdr	Radar	radar
Regeltechn	Regelungstechnik	control engineering
Rönt	Röntgenspektroskopie	X-ray spectroscopy
Schiff	Schiffahrt	maritime terminology
Schweiß	Schweißtechnik	welding
Spektr	Spektroskopie	spectroscopy
Statist	Statistik	statistics
Stereochem	Stereochemie	stereochemistry
Symb	Symbol	symbol
Techn	Technik	technology
Telef	Telefon	telephone
Telev	Fernsehen	television
Text	Textilindustrie	textile industry
Therm	Thermodynamik	thermodynamics
Vet	Tiermedizin	veterinary medicine
vgl	vergleiche	confer
v	Verb	verb
Web	Weberei	weaving
Zahnmed	Zahnmedizin	dentistry
z. B.	zum Beispiel	for instance
Zool	Zoologie	zoology
Zuck	Zuckerindustrie	sugar industry

Chemical Elements
Chemische Elemente

English Englisch	German Deutsch	Symbol Symbol	Atomic Number Ordnungszahl
actinium	Actinium	Ac	89
alumin(i)um	Aluminium	Al	13
americium	Americium	Am	95
antimony	Antimon	Sb	51
argon	Argon	Ar(A)	18
arsenic	Arsen	As	33
astatine	Astat	At	85
barium	Barium	Ba	56
berkelium	Berkelium	Bk	97
beryllium	Beryllium	Be	4
bismuth	Bismuth	Bi	83
boron	Bor	B	5
bromine	Brom	Br	35
cadmium	Cadmium	Cd	48
calcium	Calcium	Ca	20
californium	Californium	Cf	98
carbon	Kohlenstoff	C	6
cerium	Cer	Ce	58
cesium	Caesium	Cs	55
chlorine	Chlor	Cl	17
chromium	Chrom	Cr	24
cobalt	Cobalt	Co	27
copper	Kupfer	Cu	29
curium	Curium	Cm	96
dysprosium	Dysprosium	Dy	66
einsteinium	Einsteinium	Es(E)	99
erbium	Erbium	Er	68
europium	Europium	Eu	63
fermium	Fermium	Fm	100
fluorine	Fluor	F	9
francium	Francium	Fr	87
gadolinium	Gadolinium	Gd	64
gallium	Gallium	Ga	31
germanium	Germanium	Ge	32
gold	Gold	Au	79
hafnium	Hafnium	Hf	72
helium	Helium	He	2
holmium	Holmium	Ho	67
hydrogen	Wasserstoff	H	1
indium	Indium	In	49
iodine	Iod	I	53
iridium	Iridium	Ir	77
iron	Eisen	Fe	26
krypton	Krypton	Kr	36
lanthanum	Lanthan	La	57
lawrencium	Lawrencium	Lr	103
lead	Blei	Pb	82
lithium	Lithium	Li	3
lutetium	Lutetium	Lu	71

English / Englisch	German / Deutsch	Symbol / Symbol	Atomic Number / Ordnungszahl
magnesium	Magnesium	Mg	12
manganese	Mangan	Mn	25
mendelevium	Mendelevium	Md(Mv)	101
mercury	Quecksilber	Hg	80
molybdenum	Molybdän	Mo	42
neodymium	Neodym	Nd	60
neon	Neon	Ne	10
neptunium	Neptunium	Np	93
nickel	Nickel	Ni	28
niobium (columbium)	Niob	Nb	41
nitrogen	Stickstoff	N	7
nobelium	Nobelium	No	102
osmium	Osmium	Os	76
oxygen	Sauerstoff	O	8
palladium	Palladium	Pd	46
phosphorus	Phosphor	P	15
platinum	Platin	Pt	78
plutonium	Plutonium	Pu	94
polonium	Polonium	Po	84
potassium	Kalium	K	19
praseodymium	Praseodym	Pr	59
promethium	Promethium	Pm	61
protactinium	Protactinium	Pa	91
radium	Radium	Ra	88
radon	Radon	Rn	86
rhenium	Rhenium	Re	75
rhodium	Rhodium	Rh	45
rubidium	Rubidium	Rb	37
ruthenium	Ruthenium	Ru	44
samarium	Samarium	Sm	62
scandium	Scandium	Sc	21
selenium	Selen	Se	34
silver	Silber	Ag	47
silicon	Silicium	Si	14
sulfur	Schwefel	S	16
strontium	Strontium	Sr	38
tantalum	Tantal	Ta	73
technetium	Technetium	Tc	43
tellurium	Tellur	Te	52
terbium	Terbium	Tb	65
thallium	Thallium	Tl	81
thorium	Thorium	Th	90
thulium	Thulium	Tm	69
tin	Zinn	Sn	50
titanium	Titan	Ti	22
tungsten	Wolfram	W	74
uranium	Uran	U	92
vanadium	Vanadium	V	23
xenon	Xenon	Xe	54
ytterbium	Ytterbium	Yb	70
yttrium	Yttrium	Y	39
zinc	Zink	Zn	30
zirconium	Zirkonium	Zr	40

British and American Measures and Weights
Britische und amerikanische Maße und Gewichte

1. Linear Measure – Längenmaße

1 inch			=	2.54 cm
1 foot	=	12 inches	=	30.48 cm
1 yard	=	3 feet or 36 inches	=	0.914 m
1 mile	=	1760 yards	=	1.609 km

2. Square Measure – Flächenmaße

1 square inch			=	6.45 cm²
1 square foot	=	144 square inches	=	929.02 cm²
1 square yard	=	1296 square inches	=	8361.2 cm²
1 acre	=	4840 square yards	=	4046.8 m²
1 square mile	=	640 acres	=	2.59 km²

3. Cubic Measure – Raummaße

1 cubic inch			=	16.38 cm³
1 cubic foot	=	1728 cubic inches	=	0.0283 m³
1 cubic yard	=	46656 cubic inches	=	0.764 m³

4. Measure of Capacity – Hohlmaße

		liquid measure Flüssigkeitsmaß	dry measure Trockenmaß

British – britisch

		liquid measure	dry measure
1 fluid ounce			= 0.028 l
1 pint			= 0.56 l
1 (imperial) quart		= 2 pints	= 1.13 l
1 (imperial) gallon		= 4 quarts	= 4.54 l
1 bushel			= 36.36 l

American – amerikanisch

		liquid measure	dry measure
1 fluid ounce		= 29.57 ml	
1 pint		= 0.473 l	= 0.550 l
1 quart	= 2 pints	= 0.946 l	= 1.101 l
1 gallon	= 4 quarts	= 3.785 l	= 4.405 l
1 bushel			= 35.23 l
1 barrel petroleum	= 42 gallons	= 158,97 l	

5. Avoirdupois – Handelsgewichte

1 grain			=	0.064 g
1 ounce			=	28.35 g
1 pound	=	16 ounces	=	453.59 g
1 (short) hundredweight (Am)	=	100 pounds	=	45.35 kg
1 (long) hundredweight (Br)	=	112 pounds	=	50.80 kg
1 (short) ton (Am)	=	2000 pounds	=	907.18 kg
1 (long) ton (Br)	=	2240 pounds	=	1016.05 kg

6. Apothecaries' Weight – Apothekergewichte

1 grain			=	0.064 g
1 dram	=	60 grains	=	3.88 g
1 ounce	=	8 drams	=	31.10 g
1 pound	=	12 ounces	=	373.24 g

Conversion Table of Temperatures (from −20 to 100° Celsius)
Temperaturumrechnungstabelle (von −20 bis 100° Celsius)

Fahrenheit °F	Celsius °C	Fahrenheit °F	Celsius °C	Fahrenheit °F	Celsius °C
212	100	136.4	58	64.4	18
208.4	98	132.8	56	60.8	16
204.8	96	129.2	54	57.2	14
201.2	94	125.6	52	53.6	12
197.6	92	122	50	50	10
194	90				
		118.4	48	46.4	8
190.4	88	114.8	46	42.8	6
186.8	86	111.2	44	39.2	4
183.2	84	107.6	42	35.6	2
179.6	82	104	40	32	0
176	80				
		100.4	38	28.4	−2
172,4	78	96.8	36	24.8	−4
168.8	76	93.2	34	21.2	−6
165.2	74	89.6	32	17.6	−8
161.6	72	86	30	14	−10
158	70				
		82.4	28	10.4	−12
154.4	68	78.8	26	6.8	−14
150.8	66	75.2	24	3.2	−16
147.2	64	71.6	22	−0.4	−18
143.6	62	68	20	−4	−20
140	60				

Rules for Converting Temperatures
Temperaturumrechnungsregeln

1. Celsius into Fahrenheit
   Celsius in Fahrenheit

$$x \,°C = (32 + \frac{9}{5}x) \,°F$$

2. Fahrenheit into Celsius
   Fahrenheit in Celsius

$$x \,°F = (x - 32) \frac{5}{9} \,°C$$

Periodic Table of the Elements
Periodensystem der Elemente

1a	2a	3b	4b	5b	6b	7b	8	8	8	1b	2b	3a	4a	5a	6a	7a	0
1 **H** 1.0079																	2 **He** 4.00260
3 **Li** 6.94	4 **Be** 9.01218											5 **B** 10.81	6 **C** 12.011	7 **N** 14.0067	8 **O** 15.9994	9 **F** 18.998403	10 **Ne** 20.17
11 **Na** 22.98977	12 **Mg** 24.305											13 **Al** 26.98154	14 **Si** 28.0855	15 **P** 30.97376	16 **S** 32.06	17 **Cl** 35.453	18 **Ar** 39.948
19 **K** 39.0983	20 **Ca** 40.08	21 **Sc** 44.9559	22 **Ti** 47.90	23 **V** 50.941	24 **Cr** 51.996	25 **Mn** 54.9380	26 **Fe** 55.847	27 **Co** 58.9332	28 **Ni** 58.71	29 **Cu** 63.546	30 **Zn** 65.38	31 **Ga** 69.72	32 **Ge** 72.59	33 **As** 74.9216	34 **Se** 78.96	35 **Br** 79.904	36 **Kr** 83.80
37 **Rb** 85.467	38 **Sr** 87.62	39 **Y** 88.9059	40 **Zr** 91.22	41 **Nb** 92.9064	42 **Mo** 95.94	43 **Tc** 98.9062	44 **Ru** 101.07	45 **Rh** 102.9055	46 **Pd** 106.4	47 **Ag** 107.868	48 **Cd** 112.41	49 **In** 114.82	50 **Sn** 118.69	51 **Sb** 121.75	52 **Te** 127.60	53 **I** 126.9045	54 **Xe** 131.30
55 **Cs** 132.9054	56 **Ba** 137.33	57* **La** 138.9055	72 **Hf** 178.49	73 **Ta** 180.947	74 **W** 183.85	75 **Re** 186.2	76 **Os** 190.2	77 **Ir** 192.22	78 **Pt** 195.09	79 **Au** 196.9665	80 **Hg** 200.59	81 **Tl** 204.37	82 **Pb** 207.2	83 **Bi** 208.9808	84 **Po** (209)	85 **At** (210)	86 **Rn** (222)
87 **Fr** (223)	88 **Ra** 226.0254	89** **Ac** (227)	104 — (261)	105 — (262)													

*Lanthanoids *Lanthanoide	58 **Ce** 140.12	59 **Pr** 140.9077	60 **Nd** 144.24	61 **Pm** (145)	62 **Sm** 150.4	63 **Eu** 151.96	64 **Gd** 157.25	65 **Tb** 158.9254	66 **Dy** 162.50	67 **Ho** 164.9304	68 **Er** 167.26	69 **Tm** 168.9342	70 **Yb** 173.04
													71 **Lu** 174.97
**Actinoids **Actinoide	90 **Th** 232.0381	91 **Pa** 231.0359	92 **U** 238.029	93 **Np** 237.0482	94 **Pu** (244)	95 **Am** (243)	96 **Cm** (247)	97 **Bk** (247)	98 **Cf** (251)	99 **Es** (254)	100 **Fm** (257)	101 **Md** (258)	102 **No** (259)
													103 **Lr** (260)

Numbers in parentheses are mass numbers of most stable isotope of that element.
Ziffern in Klammern sind die Massenzahlen des stabilsten Isotops des betreffenden Elements.